***hy-*** U-shaped: hyoid bone
***hyal-*** glass: hyaline cartilage
***hyper-*** above, beyond: hypertonic
***hypo-*** under, below: hypotonic
***im-*** negative prefix: imbalance
***immun-*** free, exempt: immunity
***in-*** *see* im-
***inflamm-*** to set on fire: inflammation
***infra-*** beneath: infraorbital nerve
***inhal-*** to breathe in: inhalation
***insul-*** island: insula
***inter-*** among, between: interphase
***intra-*** inside: intracellular
***irid-*** rainbow, colored circle: iris
***is-*** equal: isotope
***jugul-*** neck, throat: jugular vein
***junct-*** yoke, join: junctional fiber
***juxta-*** near, beside: juxtaglomerular nephron
***kerat-*** horn: keratin
***kyph-*** hump, humpbacked: kyphosis
***labi-*** lip: labia
***labr-*** lip: glenoidal labrum
***labyrinth-*** maze: bony labyrinth
***lacrima-*** tears: lacrimal gland
***lact-*** milk: prolactin
***lacun-*** pool: lacuna
***lamell-*** thin plate: lamella
***lanug-*** down: lanugo
***laten-*** hidden: latent
***-lemm*** rind, peel: neurolemma
***leuk-*** white: leukocyte
***lingu-*** tongue: lingual tonsil
***lip-*** fat: lipids
***-logy*** study of: physiology
***lord-*** bent back: lordosis
***lun-*** moon: semilunar valve
***lute-*** yellow: macula lutea
***ly-*** loose, dissolve: lysosome
***-lyt*** dissolvable: electrolyte
***macr-*** large: macrophage
***macula*** spot, stain: macula lutea
***mal-*** bad, abnormal: malnutrition
***malle-*** hammer: malleus
***mamm-*** breast: mammary gland
***man-*** hand: manubrium
***mandib-*** jaw: mandible
***mast-*** breast: mastitis
***maxill*** jawbone: maxillary bone
***meat-*** a passage: auditory meatus
***medi-*** middle: adrenal medulla
***melan-*** black: melanin
***men-*** month: menstrual
***mening-*** membrane: meninges
***ment-*** mind: mental foramen
***mes-*** middle: mesoderm
***meta-*** after, beyond, accompanying: metabolism
***meter*** measure: calorimeter
***micr-*** small: microfilament
***mict-*** to pass urine: micturition
***mid-*** middle: midbrain
***milli-*** one-thousandth: millimeter
***mit-*** thread: mitosis
***mon-*** one: monozygotic
***mons-*** an eminence: mons pubis
***morul-*** mulberry: morula
***mot-*** move: motor neuron
***multi-*** many: multipolar neuron
***mut-*** change: mutation
***my-*** muscle: myofibril
***nas-*** nose: nasal
***nat-*** to be born: prenatal
***ne-*** new, young: neonatal period
***nephr-*** kidney: nephron
***neur-*** nerve: neuron
***neuter*** neither one nor the other: neutral
***nod-*** knot: nodule
***nucle-*** kernel: nucleus
***nutri-*** nourish: nutrient
***obes-*** fat: obesity
***occipit-*** back of the head: occipital lobe
***oculi-*** eye: orbicularis oculi
***odont-*** tooth: odontoid process
***-oid*** form: odontoid process
***olfact-*** to smell: olfactory
***olig-*** few, small: oligodendrocyte
***oo-*** egg: oogenesis
***or-*** mouth: oral cavity
***orb-*** circle: orbital
***-osis*** abnormal condition: leukocytosis
***oss-*** bone: osseous tissue
***-ous*** having qualities of: cancerous
***ov-*** egg: synovial fluid
***palpebra*** eyelid: levator palpebrae superioris
***papill-*** nipple: papillary muscle
***para-*** beside: parathyroid glands
***pariet-*** wall: parietal membrane
***path-*** disease, sickness: pathogen
***pell-*** skin: pellagra
***pelvi-*** basin: pelvic cavity
***peri-*** around: pericardial membrane
***pept-*** digest: peptic ulcer
***phag-*** eat: phagocytosis
***phen-*** show, be seen: phenotype
***phleb-*** vein: phlebitis
***phot-*** light: photoreceptor
***phren-*** mind, midriff: phrenic nerve
***pia-*** pious, gentle: pia mater
***pino-*** drink: pinocytosis
***pleur-*** rib, side: pleural membrane
***plex-*** strike: choroid plexus
***plic-*** fold: plicae circularis
***poie-*** make, produce: hematopoiesis
***poly-*** many: polyunsaturated
***popl-*** ham (knee): popliteal artery
***por-*** passage, channel: pore
***post-*** after: postnatal period
***pre-*** before: prepatellar bursa
***prime*** first: primordial follicle
***pro-*** before: prophase
***prox-*** nearest: proximal
***pseud-*** false: pseudostratified epithelium
***pter-*** wing: pterygoid process
***puber-*** adult: puberty
***pulmo-*** lung: pulmonary artery
***pyl-*** door, orifice: pyloric sphincter
***quadr-*** four: quadriceps femoris
***ramus*** branch: gray ramus
***rect-*** straight: rectum
***ren-*** kidney: renal cortex
***ret-*** net: rete testis
***reticul-*** network: sarcoplasmic reticulum
***rhin-*** nose: rhinitis
***sacchar-*** sugar: monosaccharide
***saltator*** dancer: saltatory conduction
***sarc-*** flesh: sarcoplasm
***scler-*** hard: sclera
***seb-*** grease: sebaceous gland
***sect-*** cut: section
***sella-*** saddle-shaped: sella turcica
***semi-*** half: semitendinosus
***sen-*** old: senescence
***sens-*** feeling: sensory neuron
***sin-*** hollow: sinus
***-some*** body: ribosome
***sorpt-*** to soak up: absorption
***squam-*** scale: squamous epithelium
***sta-*** halt, make stand: hemostasis
***strat-*** spread out: substrate
***stria-*** groove: striated muscle
***sub-*** under: substrate
***sulc-*** furrow: sulcus
***super-*** above: superior
***supra-*** above: supraspinatus muscle
***sutur-*** sewing: suture
***syn-*** together: synthesis
***syndesm-*** binding together: syndesmosis
***systol-*** contraction: systole
***tachy-*** rapid: tachycardia
***tal-*** ankle: talus
***tetan-*** stiff: tetanic
***thalam-*** chamber: thalamus
***theo-*** sheath: theca externa
***therm-*** heat: thermoreceptor
***thromb-*** clot: thrombocyte
***thyr-*** shield: thyroid gland
***toc-*** childbirth: oxytocin
***tom-*** cut: anatomy
***ton-*** stretch: isotonic
***trans-*** across: transverse
***tri-*** three: tricuspid valve
***trigon-*** triangle: trigone
***trop-*** turn, react: adrenocorticotropic
***troph-*** nurture: hypertrophy
***tuber-*** swelling: tuberculosis
***tympan-*** drum: tympanic membrane
***umbil-*** navel: umbilical cord
***un-*** one: unipolar neuron
***ur-*** urine: ketonuria
***uter-*** womb: uterine tube
***vag-*** to wander: vagus nerve
***valent*** having power: electrovalent
***vas-*** vessel: vasa recta
***vesic-*** bladder: vesicle
***vill-*** hair: villus
***visc-*** internal organs: viscera
***vitre-*** glass: vitreous humor
***voluntar-*** of free will: voluntary muscle
***xiph-*** sword: xiphoid process
***zon-*** belt: zona pellucida
***zy***

HOLE'S

# HUMAN ANATOMY & PHYSIOLOGY


DAVID SHIER
*Washtenaw Community College*

JACKIE BUTLER
*Grayson County Community College*

RICKI LEWIS
*The University at Albany*

Boston, Massachusetts Burr Ridge, Illinios Dubuque, Iowa
Madison, Wisconsin New York, New York San Francisco, California St. Louis, Missouri


SEVENTH EDITION

*WCB/McGraw-Hill*

*A Division of The* **McGraw-Hill** *Companies*

HUMAN ANATOMY & PHYSIOLOGY



4 5 6 7 8 9 0 QPD/QPD 9 0 9 8 7

ISBN 0-697-20959-8

Editor *Colin H. Wheatley*
Developmental Editor *Kristine Noel*
Production Editor *Marla K. Irion*
Designer *Christopher E. Reese*
Photo Editor *Janice Hancock*
Permissions Coordinator *Vicki Krug*
Art Processor *Brenda A. Ernzen*



Library of Congress Catalog Card Number: 94-72824

http://www.mhhe.com

# Brief Contents

# Contents



## Unit One Levels of Organization

### Chapter 1 Introduction to Human Anatomy and Physiology



### Chapter 2 Chemical Basis of Life

## Chapter 3 Cells



## Chapter 4 Cellular Metabolism



## Chapter 5 Tissues

# UNIT TWO SUPPORT AND MOVEMENT

## CHAPTER 6 SKIN AND THE INTEGUMENTARY SYSTEM



## CHAPTER 7 SKELETAL SYSTEM

## CHAPTER 12
## SOMATIC AND SPECIAL SENSES



## CHAPTER 13
## ENDOCRINE SYSTEM

## UNIT FOUR TRANSPORT

### CHAPTER 14 BLOOD



### CHAPTER 15 CARDIOVASCULAR SYSTEM

## CHAPTER 16 LYMPHATIC SYSTEM AND IMMUNITY

# UNIT FIVE ABSORPTION AND EXCRETION

## CHAPTER 17 DIGESTIVE SYSTEM

## Chapter 18 Nutrition and Metabolism



## Chapter 19 Respiratory System

## Chapter 20

## Urinary System



## Chapter 21

## Water, Electrolyte, and Acid-Base Balance

# UNIT SIX THE HUMAN LIFE CYCLE

## CHAPTER 22 REPRODUCTIVE SYSTEM

## Chapter 23 Human Growth and Development



## Chapter 24 Genetics



## Reference Plates

# CLINICAL APPLICAT

# PREFACE

Advertisers trying to describe a "new-and-improved" product face a dilemma: how to convince consumers that the item is better than it was, without disparaging its past value. The advertiser's dilemma aptly describes the challenge we confronted in revising John Hole's *Human Anatomy and Physiology*. How would we ever improve upon such a classic?

## A New Team

We began where most revisions begin—addressing reviewers' concerns. We made corrections, updated terminology, and adjusted the writing style, while carefully retaining the flavor and breadth of coverage of the original. We added exciting opening vignettes to the chapters, sprinkled in many brand-new boxes, and weaved throughout the text cellular and molecular explanations of macroscopic structures and functions.

The three of us—a physiologist, a microbiologist, and a geneticist—were strangers when brought together just two summers ago. Immersed in this project, we became great friends, and a true collaboration was born. Our hope is that the result not only retains the comprehensiveness of John Hole's work, but also embraces our diverse training, our many teaching experiences, and our shared enthusiasm.

## What's New

Hole's *Human Anatomy and Physiology* is still *Hole's Human Anatomy and Physiology*—but with a sharper focus and appearance. Here is what you will find.

**Introductory Vignettes** Readers are drawn immediately into the chapter's topic. Students will meet a fascinating cast of characters, from Judith R., a car accident victim in the emergency room of the hospital that opens chapter 1, to the final chapter's Orwell, a newborn in the year 2010 who undergoes a series of genetic tests—all of which are in experimental stages today. Along the way, readers will meet a carbohydrate-loading runner at the Boston marathon; Donald C., a survivor of a terrible fire; the 2 million-year-old skeleton of the Taung child; a famous violinist with incredibly flexible joints; a former president with Alzheimer's disease; Phineas Gage and Karen Ann Quinlan, whose tragic accidents taught us much about brain function; Christy Henrich, a gymnast who died of anorexia; and the unforgettable story and photo of the first child to receive insulin to treat his diabetes.

**Clinical Applications** Many of these intriguing essays not only introduce the latest entrants in medical technology, but also dip back into history to reexamine the medicinal leech; analyze the dangers from weapons from past wars; and use DNA fingerprinting to identify royal remains in Russia. Clinical Applications discussing the new and sometimes controversial include tissue engineering; the polymerase chain reaction; breaching the blood-brain barrier; preimplantation diagnosis; joint replacements; growth hormone treatment; assisted reproductive technologies; and gene mapping.

Other Clinical Applications continue the theme of molecular and cellular explanations for physiological events. These include "Disease at the Organelle Level" and "Molecular Underpinnings of Cardiovascular Disease." Coverage of pathology highlights well-known disorders that have been in the news recently, such as migraines, chronic and cancer pain, respiratory illness from smoking, prostate enlargement, and breast cancer, as well as the unusual, such as smell and taste disorders, rapid aging conditions, and prophyria and King George III.

CLINICAL APPLICATION 22.4

## Breast Cancer Update

### FINDING THE LUMP

Few discoveries are as terrifying to a woman as finding a lump in her breast. The woman may first notice the irregularity as a small area of thickening, or, she may detect a dimple, a change in contour, or a nipple that is flatter than usual, points in an unusual direction, or produces a discharge.

The next step is having a thorough physical exam. The doctor palpates the breast and performs a mammogram, which is an X-ray scan that pinpoints the location and approximate extent of abnormal tissue (fig. 22C). An ultrasound scan might be done to distinguish between a cyst and a tumor. If an area might be cancerous, the next step is a biopsy, in which a very thin needle samples the affected tissue.

Eighty percent of the time, a breast lump is a sign of fibrocystic breast disease, which is benign. It may be a fluid-filled sac of glandular tissue (a cyst) or a solid, fibrous mass of connective tissue (a fibroadenoma). Treatment includes taking vitamin E or synthetic androgens under a doctor's care, or lowering caffeine intake and examining unusual lumps with mammograms or biopsies, because women with fibrocystic breasts are 1.6 times more likely to develop breast cancer than are other women.

### TREATMENT

If biopsied breast cells are cancerous, treatment is usually surgical. A *lumpectomy* removes a small tumor and some surrounding tissue; a *modified mastectomy* removes an entire breast; and a *radical mastectomy* removes the breast and surrounding lymph and muscle tissue. Follow-up treatment varies depending upon the extent of the tumor. If abnormal tissue extends to nearby lymph nodes, chemotherapy and possibly radiation therapy will follow surgery. Chemotherapy is also indicated if a bone scan or other test reveals that the cancer has already spread. This happens in 7% of newly diagnosed cases of breast cancer.

The types of estrogen and progesterone receptors in the cancer cells determine, which drugs are used. Some drugs block estrogen or progesterone receptors, which blocks signals that tell the cells to divide. Tumor cells that lack hormone receptors are associated with a poor prognosis. Women with these tumors receive some type of chemotherapy even if the cancer has not spread to adjacent lymph nodes.

### CAUSES OF BREAST CANCER

A woman who has a close relative who has had breast cancer is at a higher risk of developing the condition than a woman without a family history of breast cancer. Experimental genetic tests can tell whether a woman with a strong family history will develop the inherited form of the illness. This was the case for a young woman whose mother and sister had died of breast cancer. Another sister had recently been diagnosed. Just before she was to undergo breast removal to avoid her feared fate, a genetic test showed that she had not inherited the responsible gene. The woman's cousin, who thought she could not inherit breast cancer because her father was related to the affected family, found that she had indeed inherited the gene. A mammogram revealed a tiny tumor, and surgery saved her life.

Ten percent of all breast cancers are inherited, with the two major responsible genes—called BRCA 1 and BRCA 2—discovered in 1993 and 1994. Researchers are still searching for the environmental triggers of the remaining 90% of cases. One candidate is prolonged exposure to estrogen. The estrogen link was first recognized in the 1970s, when researchers realized that young women who had had their ovaries removed very rarely developed breast cancer. Did their lack of estrogen protect against breast cancer? It appears so.

Prolonged exposure to estrogen can occur in a variety of ways:

- Early menarche (first period) and late menopause (last period).
- Pesticide residues and other pollutants. These contaminants stimulate cell division like estrogens do. Wildlife populations exposed to environmental estrogens demonstrate reproductive problems, such as infertility, undersized genitalia, and thin egg shells.
- Having no children, or having a first child after the age of 30.
- Not breastfeeding. Women who do not breastfeed have a higher breast cancer risk than women who do.

### BREAST CANCER STATISTICS

Nearly 3 million women in the United States currently have breast cancer. People frequently misunderstand the oft-quoted figure that a woman's lifetime risk of developing breast cancer is 1 in 8, believing that at any given time 1 in 8 U.S. women have breast cancer. This figure, however, refers to the lifetime risk for women who live to be 95 (chart 22D).

Statistics indicate that breast cancer is on the rise. The lifetime risk of 1 in 8 was only 1 in 16 in the 1940s. It has increased by 1% a year since then, and by 4% a year since 1987. At least some of that rise is due to more widespread and earlier diagnosis, a result of public health programs to educate women about breast self-exam and mammography. Thanks to these efforts, most women will recover from breast cancer. However, the illness is the second most common cancer in women after lung cancer and causes 46,000 deaths in the United States each year.

### PREVENTION

Health agencies advise mammograms once every two years for women aged 40 to 49, and yearly tests after that. A single "baseline" mammogram should be taken between age 35 and 40, as a basis of comparison. For a woman with a family history of breast cancer, this timetable is moved up. A mammogram can spot a tumor two years before a woman can feel it.

Thanks to the U.S. government's Women's Health Initiative, begun in 1993, we may soon learn enough about breast cancer to more effectively prevent and treat it. Some 70,000 women are participating in one or both of two investigations: The first study will test the ability of a diet low in fat and high in fruits and vegetables to prevent breast cancer. In the second study, healthy women with family histories of breast cancer are taking daily a drug called tamoxifen. This drug plugs up estrogen receptors in breast cells, preventing cell division.

Another candidate for breast cancer prevention is RU486, the drug used as an abortion pill because it prevents implantation by binding progesterone. This action may also prevent breast cancer. Tamoxifen and RU486 are currently being tested as breast cancer treatments too. Tamoxifen, for example, decreases risk of cancer recurrence by 30%.

In addition, at least two dozen substances are being tested for their ability to direct the immune system to fight breast cancer cells. And the Women's Health Initiative is attempting to sort through the various purported environmental causes of breast cancer, including alcohol, pesticides, birth control pills, and electromagnetic fields.

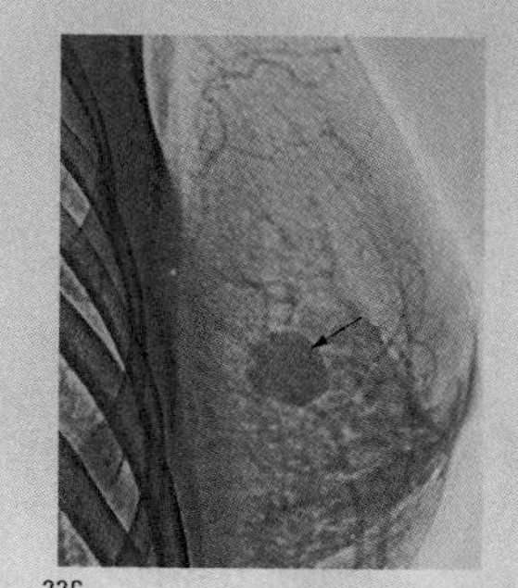

FIGURE 22C
Mammogram (X-ray film) of a breast with a tumor (arrow).

CHART 22D Breast Cancer Risk

| By Age | Odds | By Age | Odds |
|---|---|---|---|
| 25 | 1 in 19,608 | 60 | 1 in 24 |
| 30 | 1 in 2,525 | 65 | 1 in 17 |
| 35 | 1 in 622 | 70 | 1 in 14 |
| 40 | 1 in 217 | 75 | 1 in 11 |
| 45 | 1 in 93 | 80 | 1 in 10 |
| 50 | 1 in 50 | 85 | 1 in 9 |
| 55 | 1 in 33 | 95 or older | 1 in 8 |

**InnerConnections** These graphics appear at the end of selected chapters. They review the chapter material at a glance. More importantly, as their name states, they conceptually link the highlighted system to every other, reinforcing the dynamic interplays between groups of organs. InnerConnections can be used as springboards for class discussion, ideas for further study or term papers, review of chapter concepts, and reinforcement of the "big picture" in learning and applying the study of anatomy and physiology.

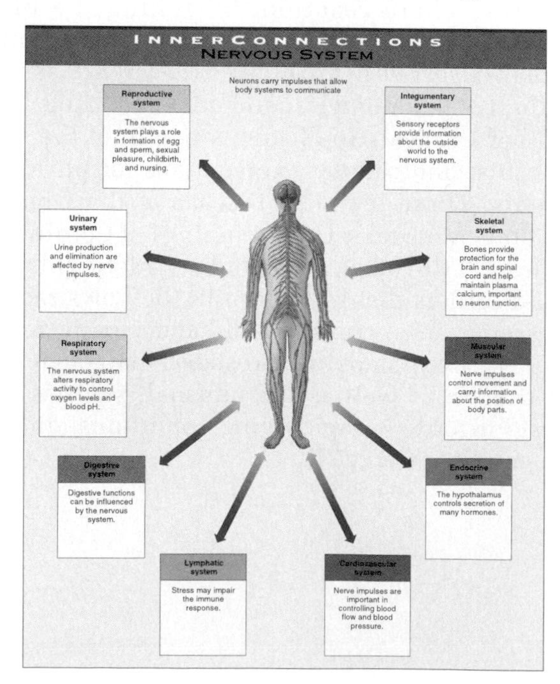


**Chapter Order** Following many reviewers' comments and our own instincts, we altered the chapter sequence to make more functional sense. The chapters are grouped as follows:

**Levels of Organization**

1. Introduction to Human Anatomy and Physiology
2. Chemical Basis of Life
3. Cells
4. Cellular Metabolism
5. Tissues

**Support and Movement**

6. Skin and the Integumentary System
7. Skeletal System
8. Joints and the Skeletal System
9. Muscular System

**Integration and Coordination**

10. Nervous System I: Basic Structure and Function
11. Nervous System II: Divisions of the Nervous System
12. Somatic and Special Senses
13. Endocrine System

**Transport**

14. Blood
15. Cardiovascular System
16. Lymphatic System and Immunity

**Absorption and Excretion**

17. Digestive System
18. Nutrition and Metabolism
19. Respiratory System
20. Urinary System
21. Water, Electrolyte, and Acid-Base Balance

**The Human Life Cycle**

22. Reproductive Systems
23. Human Growth and Development
24. Genetics

## Content and Emphasis Changes

- The concept and theme of homeostasis are established firmly in chapter 1 and continue throughout the text in special figures featuring a normal range icon for quick identification.
- The presentation of chemistry and cells (chapters 2 and 3) is extensively updated, including greater detail in some areas, clearer discussion of transport across membranes, and several new figures.
- Cellular energetics and genetic information, two of the most difficult topics (chapter 4), have been redone, with new and clearer art. Detail has been clarified without compromising readability.
- Chapter 5, Tissues, has new micrographs and corresponding artwork, with fresh examples and added icons to help the student locate tissues in the human body.
- Chapter 6 presents an updated discussion of temperature regulation and fever in terms of the internal environment and hypothalamic setpoint, a theme also touched on in later chapters.
- Muscle, nerve, and endocrine functions are explained more clearly, focusing on mechanisms.
- The many new figures in chapter 15, Cardiovascular System, enable the student to integrate the various structures and functions of this complex organ system.
- Updated discussion of the relationships between T cells, B cells, and macrophages highlights chapter 16.
- Chapter 19 offers improved discussion of the role of intrapleural pressure in ventilation. Reworked figures present $CO_2$ transport and the chloride shift and various hemoglobin dissociation curves.
- Rewritten sections on the countercurrent multiplier system and control of glomerular filtration rate, along with new figures, make chapter 20 easier to comprehend.
- The chapters on reproduction, growth, development, and genetics (chapters 22–24) are substantially updated, including much material on current and future medical technology. The final chapter, Genetics, is completely new, emphasizing health care and bioethical challenges.

**Illustration Program** Perhaps the most obvious change in this edition is the revamped art program. Detail, clarity, and consistency have been added, with frequent use of icons to orient the reader and to establish a sense of scale. Color is consistent from chapter to chapter. If a lymphatic pathway is green in one chapter, it is green elsewhere, also. Careful selection of new micrographs and rendering of new art better correlate the two. The presentation of mitosis is a good example of the improved correspondence between micrographs and art.

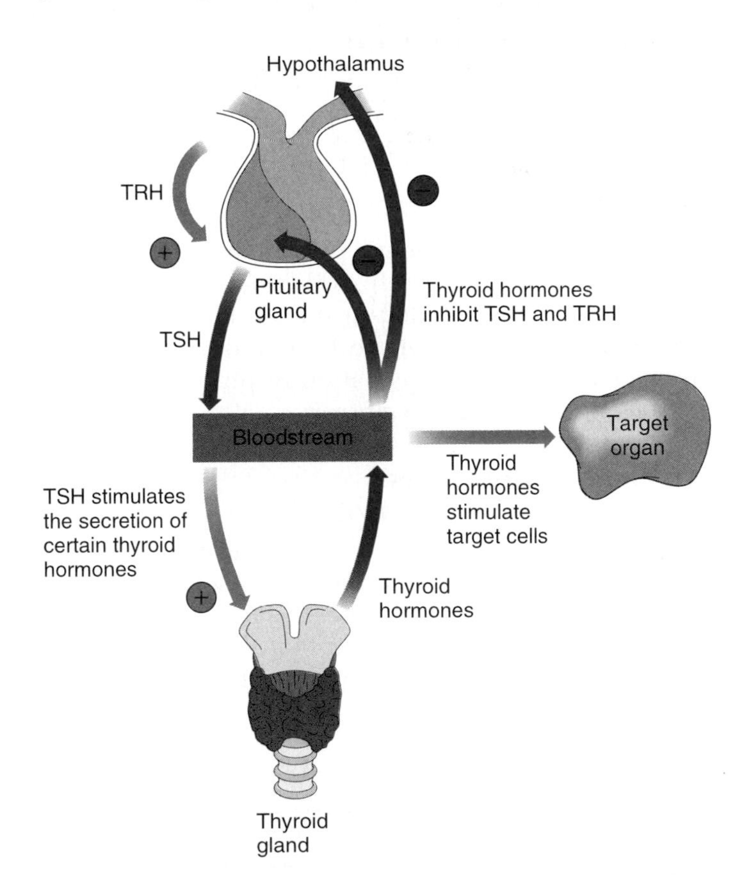

## Learning Aids for the Student

This textbook includes a variety of aids to the reader that should make your study of human anatomy and physiology more effective and enjoyable. These aids are included to help you master the basic concepts of human anatomy and physiology that are needed before progressing to more difficult material.

### Aids to Understanding Words

Aids to understanding words, found on the end sheets, and in Appendix D, helps build your vocabulary. This section includes root words, stems, prefixes, and suffixes that help you discover word meanings. Each root and an example word that uses the root are defined. Knowing the roots from these lists will help you discover and remember scientific word meanings.

### Chapter Objectives

Before you begin to study a chapter, carefully read the chapter objectives. These indicate what you should be able to do after mastering the information within the narrative. The review exercises at the end of each chapter are phrased as detailed objectives, and it is helpful to read them also before beginning your study. Both the chapter objectives and the review exercises are guides that indicate important sections of the narrative.

### Introductory Vignettes

At the beginning of every chapter are introductory vignettes that draw you immediately into the topic. These vignettes are all interesting real-life stories.

### Key Terms

Key terms and their phonetic pronunciations at the beginning of each chapter help build your science vocabulary. The words included in the list are used within the chapter and are likely to be found in subsequent chapters as well. An explanation of phonetic pronunciation is provided in the glossary.

### Review Questions Within the Narrative

Review questions occur at the end of major sections within each chapter. When you reach such questions, try to answer them. If you succeed, then you probably understand the previous discussion and are ready to proceed. If you have difficulty, reread that section before proceeding.

### Illustrations and Charts

Numerous illustrations and charts occur in each chapter and are placed near their related textual discussion. They are designed to help you visualize structures and processes, to clarify complex ideas, to summarize sections of the narrative, or to present pertinent data.

There are also sets of special reference plates to which you may want to refer. One set is designed to illustrate the structure and location of the major internal organs of the body. Another set depicts the structural detail of the human skull, and others help you locate major features of the body surface and visualize organs exposed by the dissection of a cadaver.

### Boxed Information

Short paragraphs in colored boxes occur throughout each chapter. Some of these paragraphs contain information that will help you apply the ideas presented in the narrative to clinical situations. Others provide information about changes that occur in the body's structure and function as a person passes through the various phases of the human life cycle. These will help you understand how certain body conditions change as a person grows older.

### Clinical Applications

Throughout the chapters are longer boxed sections entitled "Clinical Applications." These intriguing essays not only introduce the latest entrants in medical terminology, but also dip back into history.

### InnerConnections

"InnerConnections," found at the end of selected chapters, conceptually link the highlighted body system to every other system and reinforce the dynamic interplays between groups of organs. These graphic representations stress the "big picture" in learning and applying the concepts and facts of anatomy and physiology.

### Clinical Terms

At the end of many chapters are lists of related terms often used in clinical situations. These terms, along with their phonetic pronunciations and brief definitions, will be useful additions to your understanding of medical terminology.

### Critical Thinking Questions

The questions at the end of each chapter will help you gain experience in applying information to clinical situations.

### Review Exercises

The review exercises at the end of each chapter will check your understanding of the major ideas presented in the narrative. After studying the chapter,

## CHART LIFE SCIENCES ANIMATIONS (LSA) VIDEOTAPES

| Figure Number | LSA Number | Title |
|---|---|---|
| 2.3 | 1 | Formation of an Ionic Bond |
| 3.3 | 2 | Journey into a Cell |
| 3.12 | 4 | Cellular Secretion |
| 3.32 | 3 | Endocytosis |
| 3.35 | 12 | Mitosis |
| 4.8 | 11 | ATP as an Energy Carrier |
| 4.10 | 5 | Glycolysis |
| 4.11 | 6 | Oxidative Respiration |
| 4.12 | 7 | The Electron Transport Chain |
| 4.26 | 16 | Transcription of a Gene |
| 4.27 | 17 | Protein Synthesis |
| 4.28 | 15 | DNA Replication |
| 9.2 | 29 | Levels of Muscle Structure |
| 9.9 | 31 | Regulation of Muscle Contraction |
| 9.10 | 30 | Sliding Filament Model of Muscle Contraction |
| 10.4 | 22 | Formation of Myelin Sheath |
| 10.15 | 23 | Saltatory Nerve Conduction |
| 10.19 | 24 | Signal Integration |
| 10.21 | 25 | Reflex Arcs |
| 12.15 | 27 | The Organ of Corti |
| 12.20 | 26 | Organ of Static Equilibrium |
| 13.6 | 28 | Peptide Hormone Action |
| 14.22 | 40 | A, B, O Blood Types |
| 15.1 | 37 | Blood Circulation |
| 15.17 | 38 | Production of Electrocardiogram |
| 16.18 | 42 | Structure and Function of Antibodies |
| 16.21 | 41 | B Cell Immune Response |
| 16.22 | 42 | Structure and Function of Antibodies |
| 17.4 | 33 | Peristalsis |
| 17.41 | 34 | Digestion of Carbohydrates |
| 17.42 | 35 | Digestion of Proteins |
| 17.43 | 36 | Digestion of Lipids |
| 22.5 | 19 | Spermatogenesis |
| 22.20 | 20 | Oogenesis |
| 23.1 | 21 | Human Embryonic Development |
| 24.4 | 13 | Meiosis |
| 24.5 | 14 | Crossing Over |
| 24.6 | 13 | Meiosis |

reread the review exercises. If you can perform the tasks suggested, you have accomplished the goals of the chapter.

### Multimedia Tie-ins

This seventh edition introduces two new and exciting learning tools.

At the end of most chapters, a CD-ROM icon appears along with a statement listing a module of the *Explorations in Human Anatomy and Physiology* software that is appropriate for the chapter. *Explorations in Human Anatomy and Physiology* CD-ROM can be purchased by you for additional learning. The CD-ROM is also available to qualified adopters.

A videotape icon appears in many of the figure legends throughout the chapters. This icon alerts you that an animation related to that figure is available from the *WCB Life Science Animations* videotape series. Students can purchase the videotapes, and the tapes are available to qualified adopters.

### Appendixes, Glossary, and Index

The appendixes contain a variety of useful information. They include the following:

**A.** Periodic table of elements.

**B.** Lists of various units of measurements, their equivalents, and a description of how to convert one unit into another.

**C.** Lists of clinical laboratory tests commonly performed on human blood and urine. These lists include test names, normal adult values, and brief descriptions of their clinical significance.

**D.** Aids to understanding words.

The glossary defines the more important textual terms and provides their phonetic pronunciations. It also contains an explanation of phonetic pronunciation. The index is complete and comprehensive.

## SUPPLEMENTARY MATERIALS

The following supplementary materials are designed to help the instructor plan class work and presentations and to aid students in their learning activities.

1. *Laboratory Manual for Hole's Human Anatomy and Physiology* by Terry R. Martin is designed to accompany the seventh edition of *Hole's Human Anatomy and Physiology.* The lab manual has been thoroughly revised. This edition incorporates bar codes corresponding to images from the Slice of Life videodisc.
2. *Instructor's Manual and Test Item File* by Jeffrey and Karianne Prince contains chapter overviews, instructional techniques, suggested schedules, discussions of chapter elements, lists of related films, and directories of suppliers of audiovisual and laboratory materials. It also contains test

items for each chapter of the text that are designed to evaluate student understanding of the subject matter presented.

3. *Student Study Guide* by Nancy A. Sickles Corbett contains chapter overviews, chapter objectives, focus questions, mastery tests, study activities, and answer keys corresponding to the chapters of the text.
4. *Microtest* is a computerized testing service offered free upon request to qualified adopters of this textbook. A complete test item file is available on computer diskette for use with DOS, Windows, and Macintosh computers.
5. *Transparencies* include a set of 300 acetate sheets that complement classroom lectures or can be used for short quizzes.
6. *Student Study Art Notebook* contains the 300 illustrations from the transparency set. With this notebook students no longer have to worry about whether they will be able to see leader lines and labels in a large lecture hall.
7. *Visuals Textbank* is a set of 100 transparency masters. These transparencies feature line art from the text with labels deleted for student quizzing or practice.
8. *Extended Lecture Outline Software* consists of detailed outlines of each chapter on disk. Instructors can add their own lecture notes for convenience in lecture preparation. It is available for use with IBM or Macintosh computers.
9. *Instructor's Manual for the Laboratory Manual* by Terry R. Martin provides a chart to correlate the laboratory manual with chapters in the textbook. For each exercise there is a list of required materials, the approximate time for completion, topics for discussion, and answers to the questions in the lab report.
10. *QuickStudy* by Louis Giacinti is a computerized study guide containing test questions on a diskette in the form of true-false, multiple-choice and fill-in-the-blank. This is available for use with IBM and Macintosh computers.
11. *Histology Color Slides* include a set of seventy micrographs of tissues, organs, and other body features described in the textbook.
12. *Electronic Image Bank* is a CD-ROM containing virtually all of the art from the textbook. The CD-ROM contains an easy-to-use program that enables you to move quickly among the images, show or hide labels, and create your own multimedia presentation.
13. *Intelitool Supplement to accompany the Laboratory Manual* by Terry R. Martin contains four Intelitool laboratory exercises on muscle physiology, reflex physiology, electrocardiography, and spirometry.
14. *Answer Key for Chapter Review Exercises* by Connie Vinton-Schoepske and Kermit D. Harless provides answers to the review exercises at the end of each chapter.

Other learning aids available from Wm. C. Brown Publishers include:

15. *WCB Life Science Animations Videotape Series* is a series of five videotapes containing fifty-three animations that cover many of the key processes discussed in a physiology course.
16. *Explorations in Human Anatomy and Physiology CD-ROM* consists of fifteen interactive modules that stress human physiology. Students can actively investigate vital processes as they explore each module, which is filled with color, sound, and movement.
17. *The Dynamic Human CD-ROM* illustrates the important relationships between anatomical structures and their functions in the human body. Realistic computer visualization and three-dimensional visualizations are the premier features of this learning tool.
18. *The Dynamic Human Videodisc* contains over twenty-five animations, 130 histological micrographs, clinical footage, and line art from *Hole's Human Anatomy and Physiology.* A bar code directory is also available.
19. *Anatomy and Physiology Videodisc* is a four-sided videodisc containing more than thirty animations of physiological processes, as well as line art and micrographs. A bar code directory is also available.
20. *Study Cards for Anatomy and Physiology* by Kent M. Van De Graaff et al. is a boxed set of (300) 3-by-5-inch cards. It serves as a well-organized and illustrated synopsis of the structure and function of the human body. The Study Cards offer a quick and effective way for students to review human anatomy and physiology.
21. *StudyFlash: Anatomy and Physiology Interactive* by Christopher V. Cooper is a computer-based software tool to aid in the learning of anatomy and physiology. The program contains line art figures, along with tutorial aids, as well as a glossary. The computer diskettes are for use with IBM compatible computers.

22. *Knowledge Map of Human Anatomy Systems* by Craig Gundy is a computer tutorial of body systems and is available for the Macintosh. Each disk can be purchased separately or as a complete set. At the end of most chapters, a screened box appears informing the reader what Gundy software applies to that chapter.

23. *WCB Anatomy and Physiology Video Series* consists of the following:
    a. Human Skeletal Musculature System;
    b. Introduction to the Human Cadaver and Prosection;
    c. Introduction to Cat Dissection: Musculature;
    d. Blood Cell Counting, Identification, & Grouping;
    e. Internal Organs and the Circulatory System of the Cat; and
    f. Review of the Human Skeletal System

24. *Survey of Infectious and Parasitic Diseases* by Kent M. Van De Graaff is a black-and-white booklet that presents the essential information on one hundred of the most common and clinically significant diseases. A one-page presentation that includes pronunciation, derivation, definition, life cycle, description, signs and symptoms, laboratory diagnoses, and prevention/treatment is devoted to each of these diseases.

25. *Coloring Guide to Anatomy and Physiology* by Robert and Judith Stone emphasizes learning through the process of color association. The Coloring Guide provides a thorough review of anatomical and physiological concepts.

26. *Atlas of the Skeletal Muscles* by Robert and Judith Stone is a guide to the structure and function of human skeletal muscles. The illustrations help students locate muscles and understand their actions.

27. *Case Histories in Human Physiology* by Donna Van Wynsberghe and Gregory Cooley affords students opportunities for integrating their thinking and for problem solving. An answer key is available to instructors.

28. *Basic Health Science Chemistry: A Review and Workbook* by Joan Creager provides a thorough review of basic chemistry relevant to the life sciences, with concepts presented in the context of human physiology. It will serve as a study aid for students in undergraduate curricula in the health sciences and as a review tool for those entering graduate or professional schools.


## Acknowledgments

Any textbook is the result of hard work by a large team. Although we directed the revision, many "behind-the-scenes" people at Wm. C. Brown were indispensable to the project. We would like to thank our editorial team of Kevin Kane, Colin Wheatley, Craig Marty, and Kris Noel; our production team, which included Marla Irion, Donna Slade, Kennie Harris, Chris Reese, Brenda Ernzen, Janice Hancock, and Shirley Lanners; the artists of Illustrious, Inc.; the copyeditor Julie Bach; and most of all, John Hole, for giving us the opportunity and freedom to continue his classic work. We also thank our wonderfully patient families for their support.

*David Shier*
*Jackie Butler*
*Ricki Lewis*


## Reviewers

We would like to acknowledge the valuable contributions of the reviewers for the seventh edition who read either portions or all of the manuscript as it was being prepared, and who provided detailed criticisms and ideas for improving the narrative and the illustrations. They include the following:

*Susan M. Behling*
Concordia University Wisconsin

*Charles H. Bennett*
Kentucky State University

*Barbara A. Bernardi*
Springfield College in Illinois

*Moges Bizuneh*
IVY Tech State College

*Brenda C. Blackwelder*
Central Piedmont Community College

*Stanton Braude*
Washington University
University of Missouri at St. Louis

*Wanda L. Buckland*
Dabney S. Lancaster Community College

*Judith Carpenter*
Columbus State Community College

*Melvin C. Chambliss*
Michigan State University

*F. Jeffrey Chyatte*
University of Maryland

*Karen M. Cianci*
Houghton College

*Rosanne M. Ciccia*
D'Youville College
*Nancy A. Sickles Corbett*
Rutgers The State University of New Jersey
*James E. Cordes*
Louisiana State University at Eunice
*Michael Corral*
Darrow School
*Jean Cremins*
Massachusetts Bay Community College
*Opal H. Dakin*
Hinds Community College
*Patricia R. Daron*
Northern Virginia Community College
*Winifred B. Dickinson*
Franciscan University of Steubenville
*Michael A. Dorset*
Cleveland State Community College
*Victor P. Eroschenko*
University of Idaho
*L. Fleming Fallon*
Jameson Hospital
Columbia University School of Public Health
*Bruce A. Fisher*
Roane State Community College
*Kate Fleury*
Lake Washington Technical College
*Pamela B. Fouché*
Walters State Community College
*Ralph F. Fregosi*
The University of Arizona
*William S. Garlick*
Arizona State University
*Phyllis Gee*
University of Manitoba
*Mike Gehner*
Xavier University
*H. R. Giesman*
North Iowa Area Community College
*Sister Terence Glum*
University of Mary
*Keith R. Graham*
Lutheran College of Health Professions
*Darryl V. Grennell*
Alcorn State University
*Kevin Jon Gyolai*
North Dakota State College of Science
*Ruth L. Hays*
Clemson University
*Jimmie F. Hughey*
St. John's University
*Robert L. Jochen*
Blue Ridge Community College
*Jerry M. Johnson*
Western Baptist College
*Ronald L. Johnson*
Arkansas State University
*Drusilla B. Jolly*
Forsyth Technical Community College
*Joan H. Jones*
Naugatuck Valley Community-Technical College
*Brian E. Jordan*
Lansing Community College
*Kamal I. Kamal*
Valencia Community College–West Campus
*Dwight Kamback*
Northampton Community College
*Judith Kasperek*
Pitt Community College
*Gary Kennedy*
Lethbridge Community College
*Frank G. Kitakis*
Wayne County Community College
*John E. Kovaleski*
Indiana State University
*Jeffrey R. LaDuca*
Canisius College
*Billie S. Lane*
Chattanooga State Technical Community College
*Gina Langley*
Eastern New Mexico University-Ruidoso
*Mary T. Leonard*
University of Dayton
*Mary Katherine Lockwood*
University of New Hampshire
*D. M. Logan*
York University
*Bonita L. Longo*
Community Hospital School of Nursing
*Charmayne Mack*
Rosary College
*Dennis Malek*
Triton College
*Terry R. Martin*
Kishwaukee College
*William J. Mathena*
Kaskaskia College
*Pamela S. McLaughlin*
Madisonville Community College
*Michael C. Meyers*
Montana State University

*Robert D. Muckel*
Doane College
*Shirley Mulcahy*
San Diego Mesa College
*Tara Narayansingh*
University of Manitoba
*J. Felix Palmer*
Tulane University
*Brian K. Paulson*
California University of Pennsylvania
*Carlos F. A. Pinkham*
Norwich University
*Pam Rhyne*
Kennesaw State College
*Kristi Sather-Smith*
Hinds Community College
*Robert A. Sharp*
Aquinas College
*Clyde F. Smith*
Odessa College
*Jean E. Smith*
Carroll College
*Shirley N. Smith*
Lansing Community College
*Paulette R. Snyder*
Erie Community College, North
*Janet E. Steele*
University of Nebraska at Kearney
*Stuart S. Sumida*
California State University–San Bernardino
*Donald L. Terpening*
Ulster County Community College
*William R. Tobin, Jr.*
Erie Community College South Campus
*Robin Vance*
Union College
*Dianne L. Vermillion*
University of Rochester
*Margaret G. Wade*
Midland College
*Robert C. Wall*
Lake-Sumter Community College
*Garry M. Wallace*
Northwest College
*Leslie Jayne Wallace*
Baker College of Owosso
*Alan R. Wasmoen*
Iowa Central Community College
*Carl F. Wellstead*
West Virginia Institute of Technology
*Philip C. Whitford*
Capital University
*Barbara Wineinger*
Vincennes University Jasper
*Clarence C. Wolfe*
Northern Virginia Community College–Annandale Campus
*Ricky K. Wong*
Los Angeles Trade-Technical College
*Diana L. Wyman*
New Hampshire Technical College

## Special Contributors

*Louis A. Giacinti*
Milwaukee Area Technical College
*Charles J. Grossman*
Xavier University
Research Service, Veterans Affairs Medical Center
*Virginia Rivers*
Truckee Meadows Community College
*Kenneth S. Saladin*
Georgia College
*D. M. Van Wynsberghe*
University of Wisconsin-Milwaukee
*Leslie J. Wiemerslage*
Belleville Area College
*Eric A. Wise*
Santa Barbara City College

CHAPTER 1

# Introduction to Human Anatomy and Physiology

## Chapter Objectives

AFTER YOU HAVE STUDIED THIS CHAPTER, YOU SHOULD BE ABLE TO:

1. Define *anatomy* and *physiology,* and explain how they are related.
2. List and describe the major characteristics of life.
3. List and describe the major requirements of organisms.
4. Define *homeostasis* and explain its importance to survival.
5. Describe a homeostatic mechanism.
6. Explain what is meant by levels of organization.
7. Describe the locations of the major body cavities.
8. List the organs located in each major body cavity.
9. Name the membranes associated with the thoracic and abdominopelvic cavities.
10. Name the major organ systems and list the organs associated with each.
11. Describe the general functions of each organ system.
12. Properly use the terms that describe relative positions, body sections, and body regions.

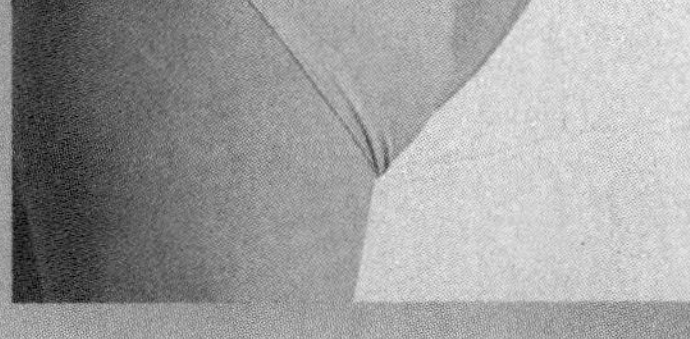

THE HEALTHY HUMAN BODY IS A WONDROUS BIOLOGICAL MACHINE.

## Key Terms

THE ACCENT MARKS USED IN THE PRONUNCIATION GUIDES ARE DERIVED FROM A SIMPLIFIED SYSTEM OF PHONETICS STANDARD IN MEDICAL USAGE. THE SINGLE ACCENT (′) DENOTES THE MAJOR STRESS, MEANING THAT EMPHASIS IS PLACED ON THE MOST HEAVILY PRONOUNCED SYLLABLE IN THE WORD. THE DOUBLE ACCENT (″) INDICATES SECONDARY STRESS. A SYLLABLE MARKED WITH A DOUBLE ACCENT RECEIVES LESS EMPHASIS THAN THE SYLLABLE THAT CARRIES THE MAIN STRESS, BUT MORE EMPHASIS THAN NEIGHBORING UNSTRESSED SYLLABLES.

**anatomy** (ah-nat′o-me)
**appendicular** (ap″en-dik′u-lar)
**axial** (ak′se-al)
**cardiovascular** (kahr″de-o-vas′ku-lur)
**digestion** (di-jest′yun)
**excretion** (ek-skre′shun)
**homeostasis** (ho″me-ō-sta′sis)
**metabolism** (mĕ-tab′o-lizm)
**negative feedback** (neg′ah-tiv fēd′bak)
**organelle** (or″gan-el′)
**organism** (or′gah-nizm)
**pericardial** (per″ĭ-kar′de-al)
**peritoneal** (per″ĭ-to-ne′al)
**physiology** (fiz″e-ol′o-je)
**pleural** (ploo′ral)
**reproduction** (re″pro-duk′shun)
**respiration** (res″pĭ-ra′shun)
**thoracic** (tho-ras′ik)
**visceral** (vis′er-al)

UNIT ONE

Judith R. had not been wearing a seat belt when the accident occurred because she had to drive only a short distance. She hadn't anticipated the intoxicated driver in the oncoming lane who swerved right in front of her. Thrown several feet, she now lay near her wrecked car as emergency medical technicians immobilized her neck and spine. Terrified, Judith tried to assess her condition. She didn't think she was bleeding, and nothing hurt terribly, but she felt a dull ache in the upper right part of her abdomen.

Minutes later, in the emergency room, a nurse gave Judith a quick exam, checking her blood pressure, pulse and breathing rate, and other vital signs and asking questions. These vital signs reflect underlying metabolic activities necessary for life, and are important in any medical decision. Because Judith's vital signs were stable, and she was alert, knew who and where she was, and didn't seem to have any obvious life-threatening injuries, transfer to a trauma center was not necessary. However, Judith continued to report abdominal pain. The attending physician ordered abdominal X rays, knowing that about a third of patients with abdominal injuries show no outward sign of a problem. As part of standard procedure, Judith received oxygen and intravenous fluids, and a technician took several tubes of blood for testing.

A young physician approached and smiled at Judith as assistants snipped off her clothing. The doctor carefully looked and listened, and gently poked and probed. She was looking for cuts, red areas called hematomas where blood vessels had broken, even treadmarks on the skin. Had Judith been wearing her seat belt, the doctor would have checked for characteristic "seat belt contusions," crushed bones or burst hollow organs caused by the twisting constrictions that can occur at the moment of impact when a person wears a seat belt. Finally, the doctor measured the girth of Judith's abdomen. If her abdomen swelled later on, this could indicate a complication, such as infection or internal bleeding.

On the basis of a hematoma in Judith's upper right abdomen and the continued pain coming from this area, the emergency room physician ordered a computerized tomography (CT) scan. The scan revealed a lacerated liver. Judith underwent emergency surgery to remove the small portion of this vital organ that included the tear.

When Judith awoke from surgery, a different physician was scanning her chart, looking up frequently. The doctor was studying her medical history for any notation of a disorder that might impede healing. Judith's history of slow blood clotting, he noted, might slow her recovery from surgery. Next, the physician looked and listened. A bluish discoloration of Judith's side might indicate bleeding from her pancreas, kidney, small intestine, or aorta (the artery leading from the heart). A bluish hue near the navel would also be a bad sign indicating bleeding from the liver or spleen. Her umbilical area was somewhat discolored.

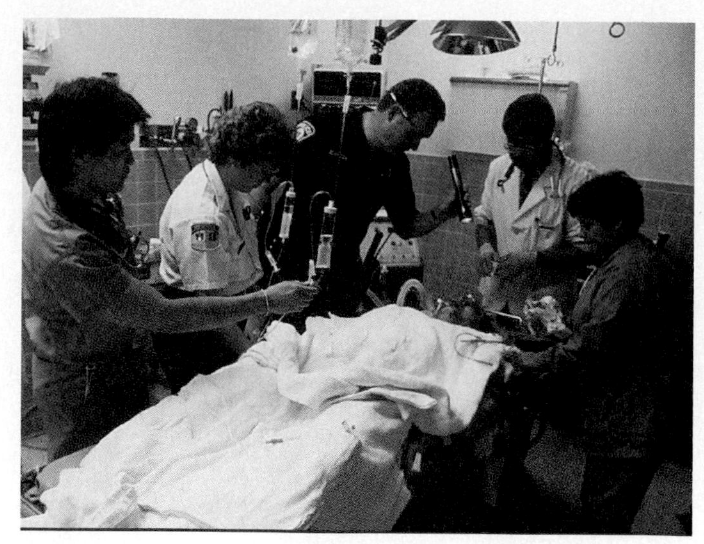

The difference between life and death may depend on a health care professional's understanding of the human body.

The doctor gently tapped Judith's abdomen and listened carefully to sounds from her digestive tract. A drumlike resonance could mean that a hollow organ had burst, whereas a dull sound might indicate internal bleeding. Judith's abdomen produced dull sounds throughout. Plus, her abdomen had swollen, the pain intensifying when the doctor gently pushed on the area. With Judith's heart rate increasing and blood pressure falling, bleeding from the damaged liver was a definite possibility.

Blood tests confirmed the doctor's suspicions. Because the blood is a complex mixture of biochemicals, it serves as a barometer of health. Injury or illness disrupts the body's maintenance of specific levels of various biochemicals. This maintenance is called homeostasis. Judith's blood tests revealed that her body had not yet recovered from the accident. Levels of clotting factors produced by her liver were falling, and blood was oozing from her incision, a sign of impaired clotting. Judith's blood glucose remained elevated, as it had been in the emergency room. Her body was still reacting to the injury.

Based on Judith's blood tests, heart rate, blood pressure, reports of pain, and the physical exam, the doctor sent her back to the operating room. Sure enough, the part of her liver where the injured portion had been removed was still bleeding. When the doctors placed packing material at the wound site, the oozing gradually stopped. Judith returned to the recovery room and, as her condition stabilized, to her room. This time, all went well, and a few days later she was able to go home. The next time she drove, Judith wore her seat belt!

Imagine yourself as one of the health care professionals who helped identify Judith R.'s injury and get her on the road back to health. How would you know what to look, listen, and feel for? How would you place the signs and symptoms into a bigger picture that would suggest the appropriate diagnosis? Nurses, doctors, technicians, and other integral members of health care teams must have a working knowledge of the many intricacies of the human body. How can they begin to understand the astounding complexity of the human body? The study of human anatomy and physiology is a daunting but fascinating and ultimately life-saving challenge.

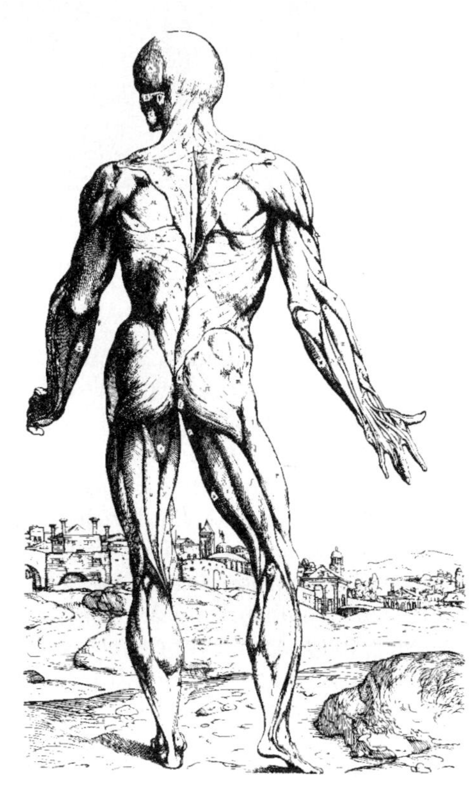

**FIGURE 1.1**
The study of the human body has a long history, as this illustration from the second book of *De Humani Corporis Fabrica* by Andreas Vesalius, issued in 1543, indicates. Note the similarity to the anatomical position (described later in this chapter).

Our understanding of the human body has a long and interesting history (fig. 1.1). It began with our earliest ancestors, who must have been as curious about how their bodies worked as we are today. At first their interests most likely concerned injuries and illnesses, because healthy bodies demand little attention from their owners. Although they did not have emergency rooms to turn to, primitive people certainly suffered from occasional aches and pains, injured themselves, bled, broke bones, and developed diseases. At first, healers relied heavily on superstitions and notions about magic. However, as they tried to help the sick, these early medical workers began to discover useful ways of examining and treating the human body. They observed the effects of injuries, noticed how wounds healed, and attempted to determine the causes of deaths by examining dead bodies. They also found that certain herbs and potions could sometimes be used to treat coughs, headaches, and other common problems. These long-ago physicians began to wonder how these substances, the forerunners of modern drugs, affected body functions in general.

As people began asking more questions and seeking answers the stage was set for the development of modern medical science. Techniques for making accurate observations and performing careful experiments evolved and knowledge of the human body expanded rapidly.

This new knowledge required a new, specialized language. Early medical providers devised many new terms to name body parts, describe their locations, and explain their functions. These terms, most of which originated from Greek and Latin, formed the basis for the language of anatomy and physiology. (A list of some of the modern medical and applied sciences appears on pages 23–25.)

1. What factors probably stimulated an early interest in the human body?
2. What kinds of activities helped promote the development of modern medical science?

## ANATOMY AND PHYSIOLOGY

As you read this book you will begin to learn how the human body maintains life by studying two major areas of medical science, **anatomy** and **physiology**. Anatomy deals with the **structure** (morphology) of body parts—what are their forms and how are they arranged? Physiology considers the **functions** of these body parts—what do they do and how do they do it? Although anatomists tend to rely more on examination of the body and physiologists more on experimentation, together their efforts have provided us with a solid foundation upon which to build an understanding of how our bodies work as living organisms.

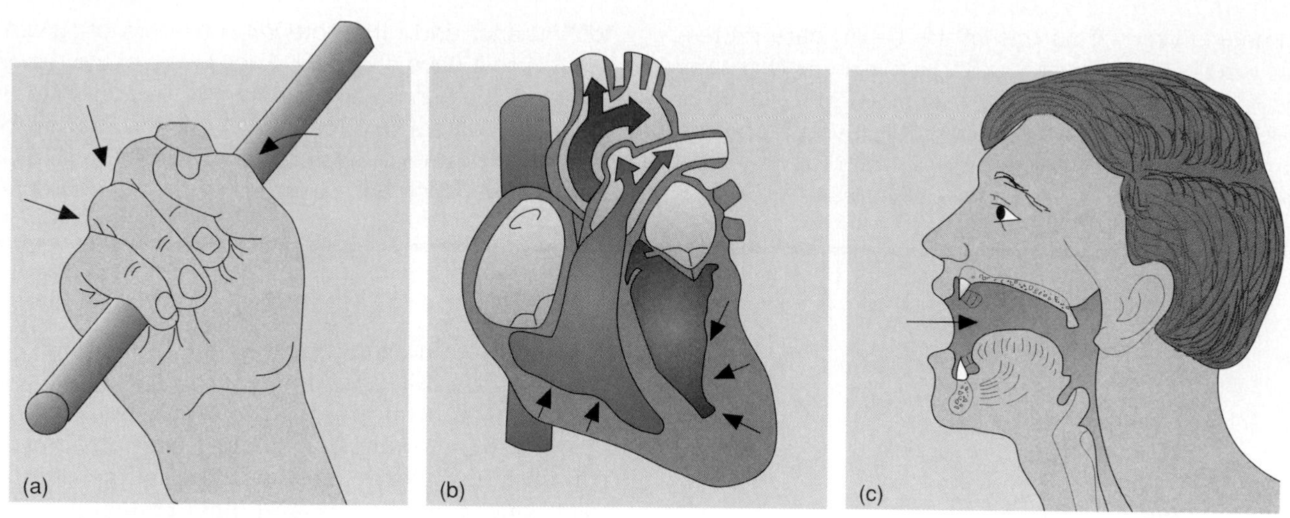


**FIGURE 1.2**

The structures of body parts are closely related to their functions: (*a*) the hand is adapted for grasping; (*b*) the heart for pumping blood; and (*c*) the mouth for receiving food.

Actually, it is difficult to separate the topics of anatomy and physiology because anatomical structures are so closely associated with their functions. These parts are arranged to form a well-organized unit—the **human organism**—and each part plays a role in the operation of the unit as a whole. This functional role depends upon the way the part is constructed. For example, the arrangement of parts in the human hand with its long, jointed fingers makes possible the ability to grasp. The heart's powerful muscular walls are structured to contract and propel blood out of the chambers and into blood vessels, and the valves associated with the vessels and chambers ensure that the blood will move in the proper direction. The shape of the mouth enables it to receive food; teeth are shaped so that they break solid foods into smaller pieces; and the muscular tongue and cheeks are constructed to help mix food particles with saliva and prepare them for swallowing (fig. 1.2).

1. What are the differences between an anatomist and a physiologist?
2. Why is it difficult to separate the topics of anatomy and physiology?
3. List several examples to illustrate the idea that the structure of a body part is closely related to its function.

## CHARACTERISTICS OF LIFE

A scene such as Judith R.'s accident and injury underscores the delicate balance that must be maintained in order to sustain life. In those seconds at the limits of life—the birth of a baby, a trauma scene, or the precise instant of death following a long illness—we often think about just what combination of qualities constitutes this state that we call life. Indeed, although this text addresses the human body, the most fundamental characteristics of life are shared by all **organisms.** As living organisms we can move and respond to our surroundings. We start out as small individuals and then grow, eventually to possibly reproduce. We gain energy by taking in or ingesting food, by breaking it down or digesting it, and by absorbing and assimilating it. We can then use it to grow and repair, or extract energy from it by the process of respiration. Substances circulate throughout the internal environment of our bodies. Finally, we send wastes from the body by excretion. Taken together, these physical and chemical events or reactions are referred to as **metabolism.** Chart 1.1 summarizes these characteristics of life.

At the accident scene and throughout Judith R.'s hospitalization following her head-on automobile collision, health care workers repeatedly monitored her **vital signs**—observable body functions that reflect metabolic activities essential for life. Vital signs indicate that a person is alive. Assessment of vital signs includes measuring body temperature and blood pressure and monitoring rates and types of pulse and breathing movements. Absence of vital signs signifies death. A person who has died displays no spontaneous muscular movements (including those of the breathing muscles and beating heart), does not respond to stimuli (even the most painful that can be ethically applied), exhibits no reflexes (such as the knee-jerk reflex and pupillary reflexes of the eye), and generates no brain waves (demonstrated by a flat encephalogram, which reflects a lack of metabolic activity in the brain).

## CHART 1.1 CHARACTERISTICS OF ANIMAL LIFE

| Process | Examples | Process | Examples |
|---|---|---|---|
| Movement | Change in position of the body or of a body part; motion of an internal organ | Digestion | Breakdown of food substances into simpler forms |
| Responsiveness | Reaction to a change taking place inside or outside the body | Absorption | Passage of substances through membranes and into body fluids |
| Growth | Increase in body size without change in shape | Circulation | Movement of substances from place to place in body fluids |
| Reproduction | Production of new organisms and new cells | Assimilation | Changing of absorbed substances into chemically different forms |
| Respiration | Obtaining oxygen, using oxygen in releasing energy from foods, and removing carbon dioxide | Excretion | Removal of wastes produced by metabolic reactions |

1. What are the characteristics of life?
2. How are the characteristics of life related to metabolism?

## MAINTENANCE OF LIFE

With the exception of an organism's reproductive structures, which function to perpetuate the species, all body structures and functions are directed toward achieving one goal—maintaining the life of the organism.

### Requirements of Organisms

Life depends upon certain environmental factors summarized in the following list:

1. **Water** is the most abundant substance in the body. It is required for a variety of metabolic processes, and it provides the environment in which most of them take place. Water also transports substances within organisms and is important in regulating body temperature.
2. **Food** refers to substances that provide organisms with necessary chemicals (nutrients) in addition to water. Nutrients supply energy and raw materials for building new living matter.
3. **Oxygen** is a gas that makes up about one-fifth of the air. It is used in the process of releasing energy from nutrients. The energy, in turn, is used to drive metabolic processes.
4. **Heat** is a form of energy. It is a product of metabolic reactions, and the rate at which these reactions occur is partly governed by the amount of heat present. Generally, the greater the amount of heat, the more rapidly chemical reactions take place. *Temperature* is a measurement of the amount of heat present.
5. **Pressure** is an application of force on an object or substance. For example, the force acting on the outside of a land organism due to the weight of air above it is called *atmospheric pressure.* In humans, this pressure plays an important role in breathing. Similarly, organisms living under water are subjected to *hydrostatic pressure*—a pressure exerted by a liquid—due to the weight of water above them. In complex organisms, such as humans, heart action produces blood pressure (another form of hydrostatic pressure), which forces blood through blood vessels.

Although organisms need water, food, oxygen, heat, and pressure, these factors alone are not enough to ensure survival. Both the quantities and the qualities of such factors are also important. Chart 1.2 summarizes the major requirements of organisms.

### Homeostasis

Some organisms exist as single **cells,** the smallest living units. Consider the amoeba, the simple, one-celled organism found in lakes and ponds (fig. 1.3). Despite its simple structure compared to a human, an amoeba has very specific requirements that must be met if it is to survive. As long as the outside world, its **environment,** supports its requirements, it flourishes. As environmental factors such as temperature, water composition, and food availability become unsatisfactory, its survival may be threatened. Although the amoeba has a limited ability to move from one place to another, environmental changes are likely to affect the whole pond, and with no place else to go, the amoeba dies.

In contrast to the amoeba, we humans are comprised of about 70 trillion cells that form their own environment inside our bodies. This **internal environment** protects our cells (and us!) from changes in the outside world that would kill isolated cells such as the amoeba. Our cells interact to keep this internal environment

## CHART 1.2 REQUIREMENTS OF ORGANISMS

| Factor | Characteristic | Use |
|---|---|---|
| Water | A chemical substance | For metabolic processes, as a medium for metabolic reactions, to transport substances, and to regulate temperature |
| Food | Various chemical substances | To supply energy and raw materials for the production of necessary substances and for the regulation of vital reactions |
| Oxygen | A chemical substance | To release energy from food substances |
| Heat | A form of energy | To help regulate the rates of metabolic reactions |
| Pressure | A force | Atmospheric pressure for breathing; hydrostatic pressure to help move blood |

relatively constant, despite an ever-changing outside environment. This maintenance of a stable internal environment is called **homeostasis,** and it is so important that most of our metabolic energy is spent on it. Many of the tests performed on Judith R. during her hospitalization assessed her body's return to homeostasis.

To better understand this idea of maintaining a stable internal environment, imagine a room equipped with a furnace and an air conditioner. Suppose the room temperature is to remain near 20° C (68° F), so the thermostat is adjusted to a **set point** of 20° C. Because a thermostat is sensitive to temperature changes, it will signal the furnace to start and the air conditioner to stop whenever the room temperature drops below the set point. If the temperature rises above the set point, the thermostat will cause the furnace to stop and the air conditioner to start. As a result, a relatively constant temperature will be maintained in the room (fig. 1.4).

A similar **homeostatic mechanism** regulates body temperature in humans (fig. 1.5). The "thermostat" is a temperature-sensitive region in a temperature control center of the brain called the hypothalamus. In healthy persons, the set point of the brain's thermostat is at or near 37° C (98.6° F).

If a person is exposed to a cold environment and the body temperature begins to drop, the temperature control center of the brain senses this change and triggers heat-generating and heat-conserving activities. For example, small groups of muscle cells may be stimulated to contract involuntarily, an action called shivering. Such muscular contractions produce heat, which helps warm the body. At the same time, blood vessels in the skin constrict so that blood flow there is reduced and heat is retained in deeper tissues.

If a person becomes overheated, the temperature control center of the brain triggers a series of changes that promote loss of body heat. For example, it stimulates sweat glands in the skin to secrete watery perspiration. As the water evaporates from the surface, heat is carried away and the skin is cooled. At the same time, the brain center causes blood vessels in the skin to dilate. This allows the blood that carries heat from deeper tissues to reach the surface where more heat is lost to the outside. Also, the brain center stimulates increased heart rate, causing a greater volume of blood to move into the surface vessels, and it stimulates an increase in breathing rate, allowing more heat-carrying air to be expelled from the lungs. Body temperature regulation is discussed in more detail in chapter 6.

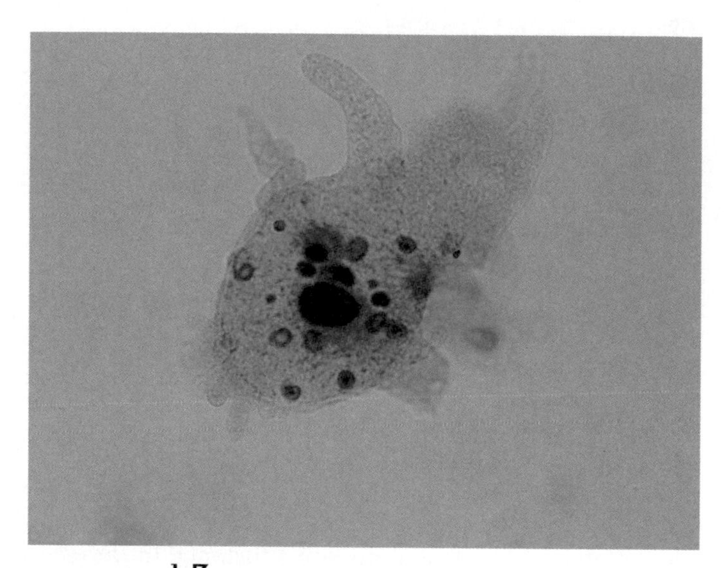

**FIGURE 1.3**
The amoeba is an organism consisting of a single cell (50×).

Another homeostatic mechanism regulates the blood pressure in the blood vessels (arteries) leading away from the heart. In this instance, pressure-sensitive areas (sensors) within the walls of these vessels sense changes in blood pressure and signal a

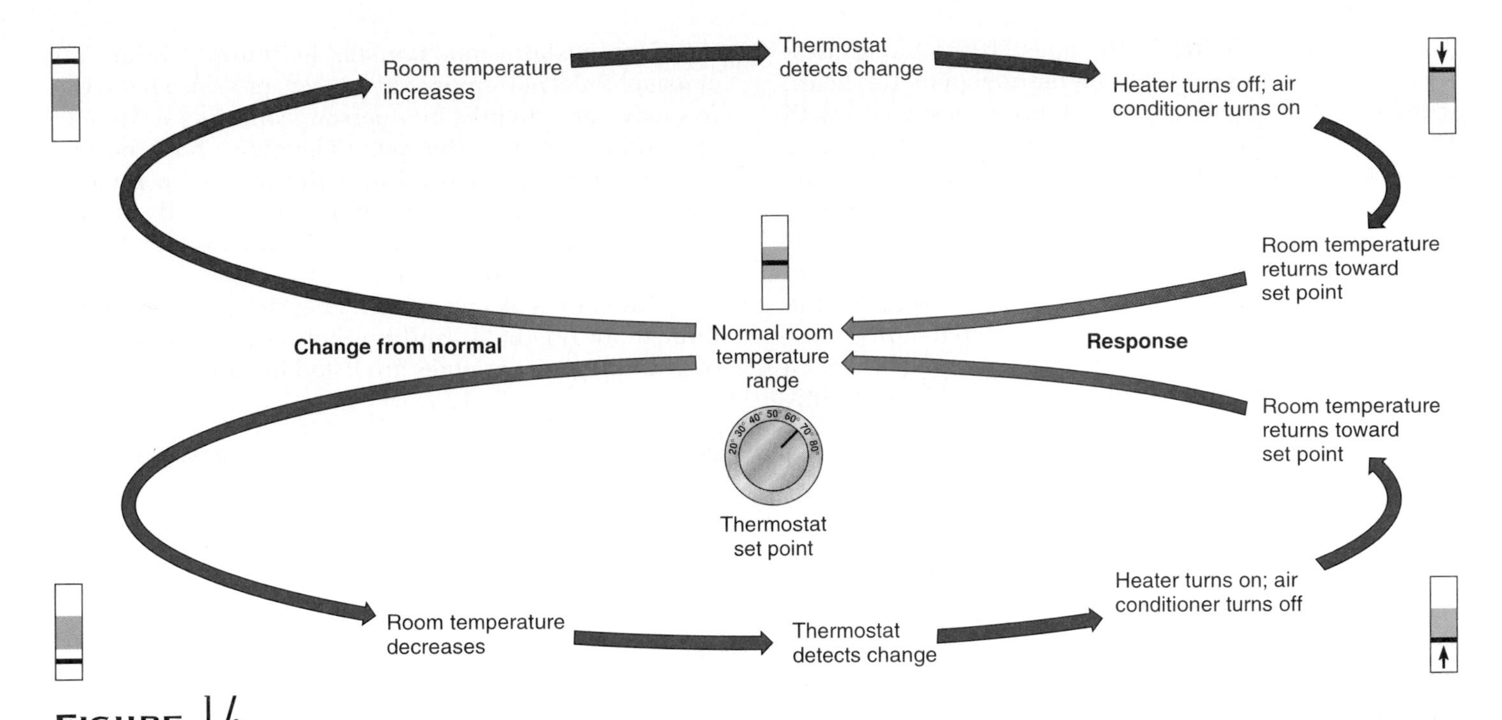


**FIGURE 1.4**

A thermostat that can signal an air conditioner and a furnace to turn on or off maintains a relatively stable room temperature. This system is an example of a homeostatic mechanism.

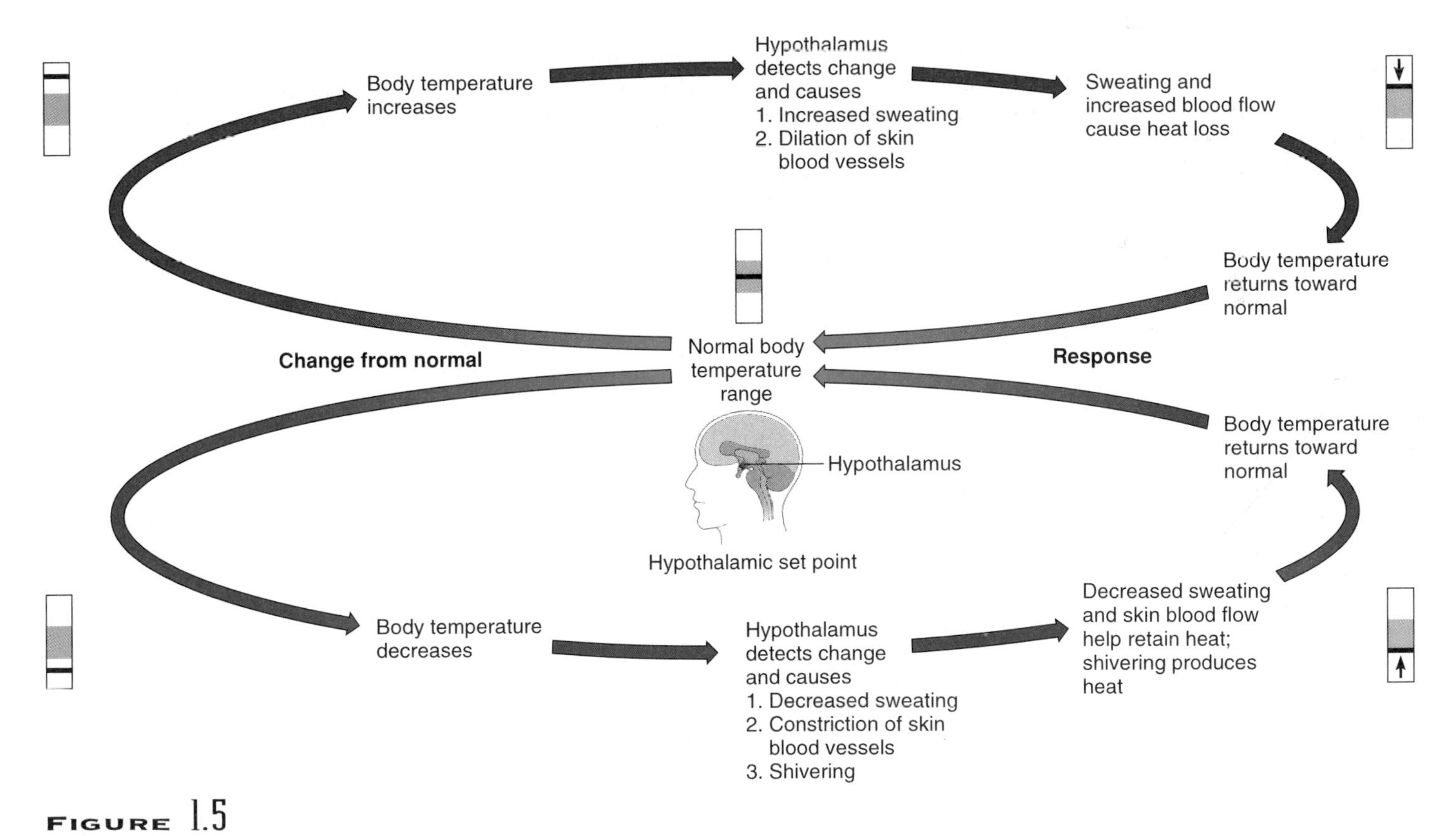


**FIGURE 1.5**

The homeostatic mechanism that regulates body temperature is an example of negative feedback.

pressure control center in the brain. If the blood pressure is above the set point, the brain signals the heart, causing its chambers to contract more slowly and with less force. Because of decreased heart action, less blood enters the blood vessels, and the pressure inside the vessels decreases. If the blood pressure is dropping below the set point, the brain center signals the heart to contract more rapidly and with greater force so that the pressure in the vessels increases. Chapter 15 discusses blood pressure regulation in more detail.

Similarly, a homeostatic mechanism regulates the concentration of the sugar glucose in the blood. In this case, cells within an organ called the pancreas determine the set point. If, for example, the amount of blood sugar increases following a meal, the pancreas detects this change and releases a chemical (insulin) into the blood. This substance causes sugar to move from the blood into various body cells and to be stored in the liver and muscles. As this occurs, the concentration of blood sugar decreases, and when it reaches the normal set point, the pancreas ceases its release of insulin. If, on the other hand, the blood sugar concentration becomes abnormally low, the pancreas detects this change and releases a different chemical (glucagon) that causes sugar to be released from storage into the blood. The regulation of the blood sugar concentration is discussed in more detail in chapter 13.

In each of these examples, homeostasis is maintained by a self-regulating control mechanism that can receive signals (or feedback) about changes away from the normal set point and can cause reactions that tend to return conditions to normal. Since the changes away from the normal state stimulate responses in the opposite direction, the responses are called *negative,* and the homeostatic control mechanism is said to act by a process of **negative feedback.** Negative feedback is discussed further in chapter 4.

Sometimes changes occur that stimulate still other similar changes. Such a process that causes movement away from the normal state is called a *positive feedback mechanism.*

Although most feedback mechanisms in the body are negative, a positive system operates for a short time when a blood clot forms, because the chemicals present in a clot promote still more clotting (see chapter 14). Another illustration of positive feedback is milk production. If a baby suckles with greater force or duration, the mother's mammary glands respond by making more and more milk.

Because positive feedback mechanisms usually produce unstable conditions, they are most commonly associated with diseases and may lead to death.

Homeostatic mechanisms maintain a relatively constant internal environment, yet physiological values may vary slightly in a person from time to time or from one person to the next. Therefore, both normal values for an individual and the idea of a **normal range** for the general population are clinically important. The normal range icons in figures 1.4 and 1.5 are intended to reinforce this concept.

Numerous examples of homeostasis are presented throughout this book, and normal ranges for a number of physiological variables are listed in Appendix C.

1. What requirements of organisms are provided from the external environment?
2. What is the relationship between oxygen use and heat production?
3. Why is homeostasis so important to survival?
4. Describe three homeostatic mechanisms.

## Levels of Organization

Early investigators, limited in their ability to observe small parts, focused their attention on the larger body structures. Studies of small parts had to await invention of magnifying lenses and microscopes, which came into use about 400 years ago. These tools revealed that larger body structures were made up of smaller parts, which, in turn, were composed of even smaller ones.

Today, scientists recognize that all materials, including those that comprise the human body, are composed of chemicals. These substances are made up of tiny, invisible particles called **atoms,** which are commonly bound together to form larger particles called **molecules;** small molecules may combine in complex ways to form larger molecules called **macromolecules.**

Within the human organism, the basic unit of structure and function is a microscopic cell. Although individual cells vary in size and shape, all have certain traits in common. All human cells contain structures called **organelles** that carry on specific activities. These organelles are composed of aggregates of large molecules, including proteins, carbohydrates, lipids, and nucleic acids. In fact, all cells contain a complete set of genetic instructions but use only a subset of them, providing cells with specialized functions.

Cells are organized into layers or masses that have common functions. Such a group of cells forms a **tissue.** Groups of different tissues form **organs**—complex structures with specialized functions—and groups of organs that function closely together comprise **organ systems.** Organ systems make up an **organism.** A body part can

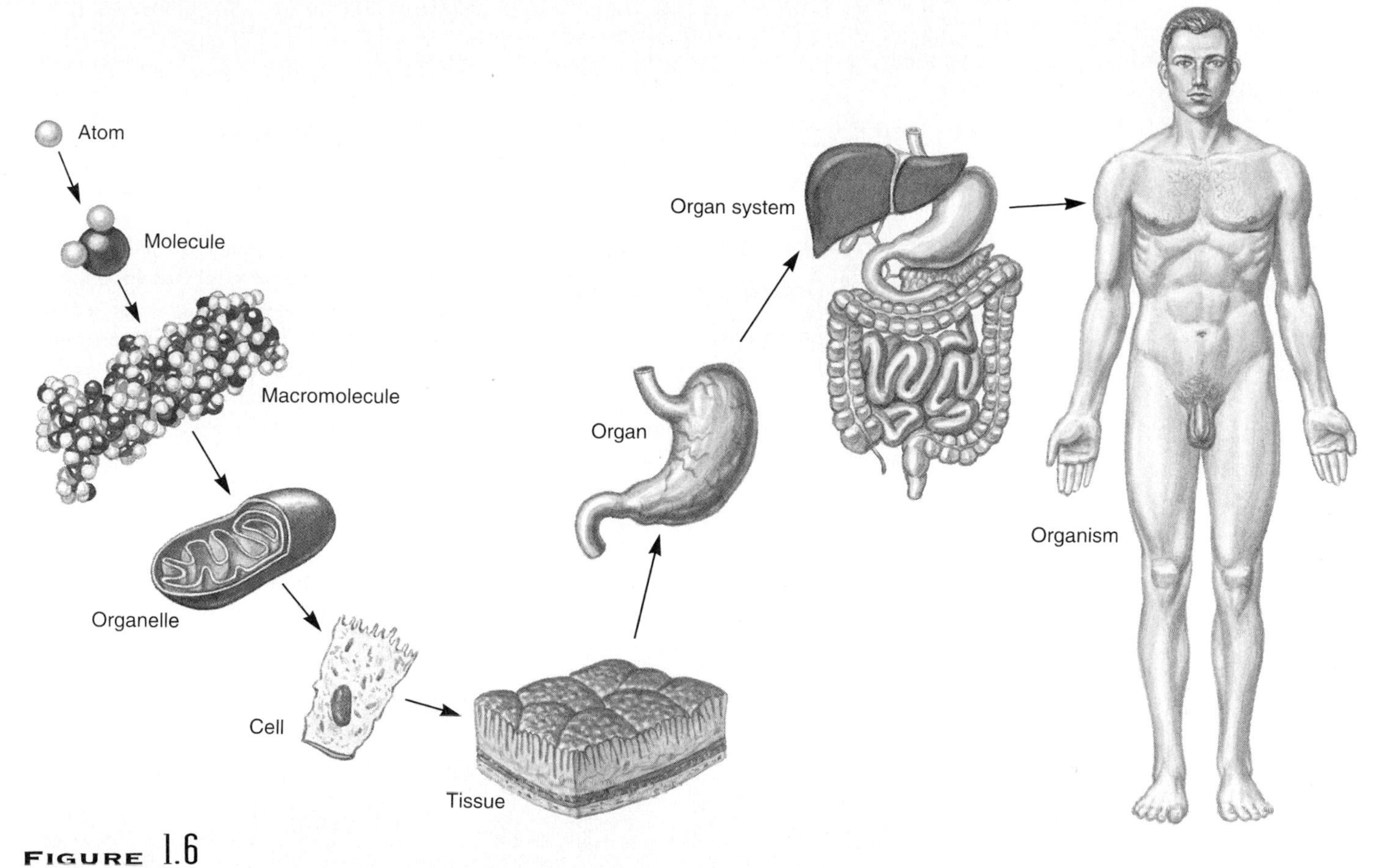


**FIGURE 1.6**
A human body is composed of parts within parts, which increase in complexity from the level of the atom to the whole organism.

be described at different levels. The heart, for example, contains muscle, fat, and nervous tissue. These tissues, in turn, are constructed of cells, which contain organelles. All of the structures of life are, ultimately, composed of chemicals (fig. 1.6). Clinical Application 1.1 describes two technologies used to visualize differences between tissues.

Chapters 2–6 discuss these levels of organization in more detail. Chapter 2 describes the atomic and molecular levels; chapter 3 deals with organelles and cellular structures and functions; chapter 4 explores cellular metabolism; chapter 5 describes tissues; and chapter 6 presents membranes as examples of organs, and the skin and its accessory organs as an example of an organ system. Beginning with chapter 7, the structures and functions of each of the organ systems are described in detail. Chart 1.3 lists the levels of organization and corresponding illustrations in this textbook.

1. How does the human body illustrate levels of organization?
2. What is an organism?
3. How do body parts that occupy different levels of organization vary in complexity?

## ORGANIZATION OF THE HUMAN BODY

The human organism is a complex structure composed of many parts. The major features of the human body include various cavities, a set of membranes, and a group of organ systems.

### Body Cavities

The human organism can be divided into an **axial portion,** which includes the head, neck, and trunk, and an **appendicular portion,** which includes the upper and lower limbs. Within the axial portion are two major cavities—a **dorsal cavity** and a larger **ventral cavity.** The organs contained within such a cavity are called **viscera.** The dorsal cavity can be subdivided into two parts—the **cranial cavity,** which houses the brain, and the **vertebral canal** (spinal cavity), which contains the

## Clinical Application 1.1

# Ultrasonography and Magnetic Resonance Imaging: A Tale of Two Patients

The two patients enter the hospital medical scanning unit hoping for opposite outcomes. Vanessa Q., who has suffered several early pregnancy losses, hopes that an ultrasound exam will reveal a living embryo in her still-flat abdomen. Michael P., a 16-year-old who has been suffering from terrible headaches, is to undergo a magnetic resonance imaging (MRI) scan to assure his physician (and himself!) that the cause of the headache is not a brain tumor.

Both ultrasound and magnetic resonance imaging scans are noninvasive procedures that provide images of soft internal structures. Ultrasonography uses high-frequency sound waves that are beyond the range of human hearing. A technician gently presses a device called a transducer that emits sound waves against the skin and moves it slowly over the surface of the area being examined, in this case, Vanessa's abdomen (fig. 1A).

The sound waves travel into the body, and when they reach a border between structures of slightly different densities, some of the waves reflect back to the transducer. Prior to the exam, Vanessa drank several glasses of water. Her filled bladder will intensify the contrast between her uterus (and its contents) and nearby organs. Other sound waves continue into deeper tissues, and some of them are reflected back by still other interfaces. As the reflected sound waves reach the transducer, they are converted into electrical impulses that are amplified and used to create a sectional image of the body's internal structure on a viewing screen. This is known as a sonogram (fig. 1B).

Glancing at the screen, Vanessa yelps in joy. The image looks only like a fuzzy lima bean with a pulsating blip in the middle, but she knows it is the image of an embryo—and its heart is beating!

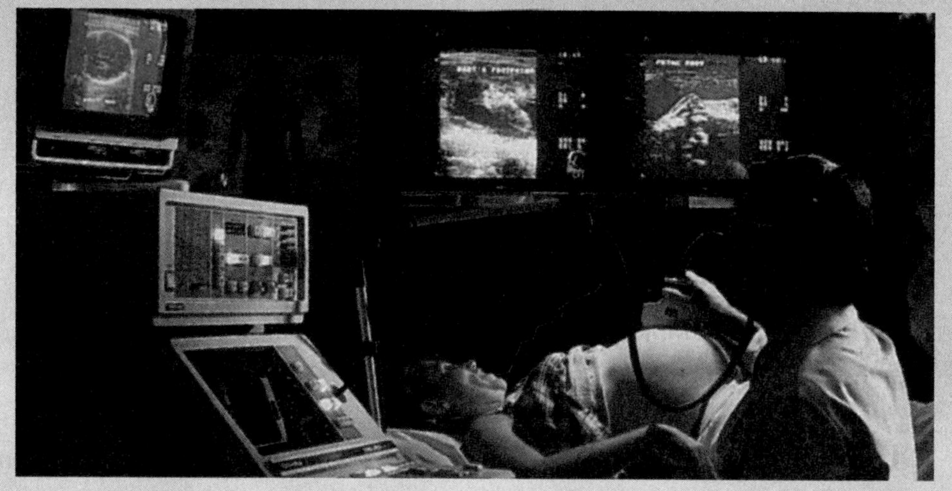

**Figure 1A**
Ultrasonography uses reflected sound waves to visualize internal body structures.

**Chart 1.3 Levels of Organization**

| Level | Example | Illustration |
|---|---|---|
| Atom | Hydrogen atom, lithium atom | Figure 2.1 |
| Molecule | Water molecule, glucose molecule | Figure 2.9 |
| Macromolecule | Protein molecule, DNA molecule | Figure 2.18 |
| Organelle | Mitochondrion, Golgi apparatus, nucleus | Figure 3.12 |
| Cell | Muscle cell, nerve cell | Figure 3.2 |
| Tissue | Simple squamous epithelium, loose connective tissue | Figure 5.1 |
| Organ | Skin, femur, heart, kidney | Figure 7.2 |
| Organ system | Integumentary system, skeletal system, digestive system | Figure 7.17 |
| Organism | Human | Figure 23.25 |

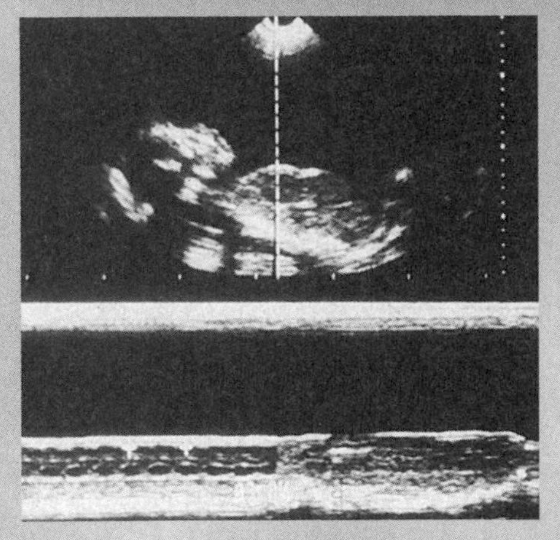

**FIGURE 1B**
This image resulting from an ultrasonographic procedure reveals the presence of a fetus in the uterus.

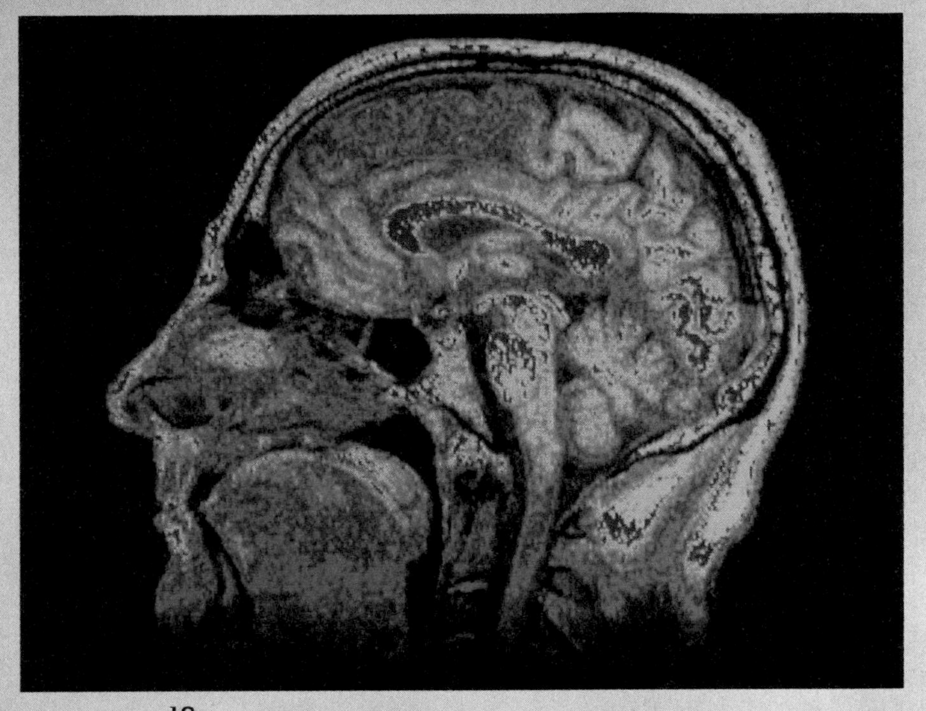

**FIGURE 1C**
Falsely colored MRI image of a human head and brain (sagittal section).

Vanessa's ultrasound exam takes only a few minutes, whereas Michael's MRI scan takes an hour. First he receives an injection of a dye that provides contrast so that a radiologist examining the scan can distinguish certain brain structures. Then, a nurse wheels the narrow bed on which Michael lies into a chamber surrounded by a powerful magnet and a special radio antenna. The chamber, which looks like a metal doughnut, is the MRI instrument. As Michael settles back and closes his eyes, a technician activates the device.

The magnet generates a magnetic field that alters the alignment and spin of certain types of atoms within Michael's brain. At the same time, a second rotating magnetic field causes particular types of atoms (such as the hydrogen atoms in body fluids and organic compounds) to release weak radio waves with characteristic frequencies. The nearby antenna receives and amplifies the radio waves, which are then processed by a computer. Within a few minutes, the computer generates a sectional image based on the locations and concentrations of the atoms being studied (fig. 1C). The device continues to produce data, painting portraits of Michael's brain in the transverse, coronal, and sagittal sections.

Michael and his parents nervously wait two days for the expert eyes of a radiologist to interpret the MRI scan. Happily, the scan shows normal brain structure. Whatever is causing Michael's headaches, it is not a possibly life-threatening brain tumor.

spinal cord and is surrounded by sections of the backbone (vertebrae). The ventral cavity consists of a **thoracic cavity** and an **abdominopelvic cavity.** Figure 1.7 shows these major body cavities.

The thoracic cavity is separated from the lower abdominopelvic cavity by a broad, thin muscle called the **diaphragm.** When it is at rest, this muscle curves upward into the thorax like a dome. When it contracts during inhalation, it presses down upon the abdominal viscera. The wall of the thoracic cavity is composed of skin, skeletal muscles, and various bones. Within it are the lungs and a region between the lungs, called the **mediastinum.** The mediastinum separates the thorax into two compartments that contain the right and left lungs. The remaining thoracic viscera—heart, esophagus, trachea, and thymus gland—are located within the mediastinum.

The abdominopelvic cavity, which includes an upper abdominal portion and a lower pelvic portion, extends from the diaphragm to the floor of the pelvis.

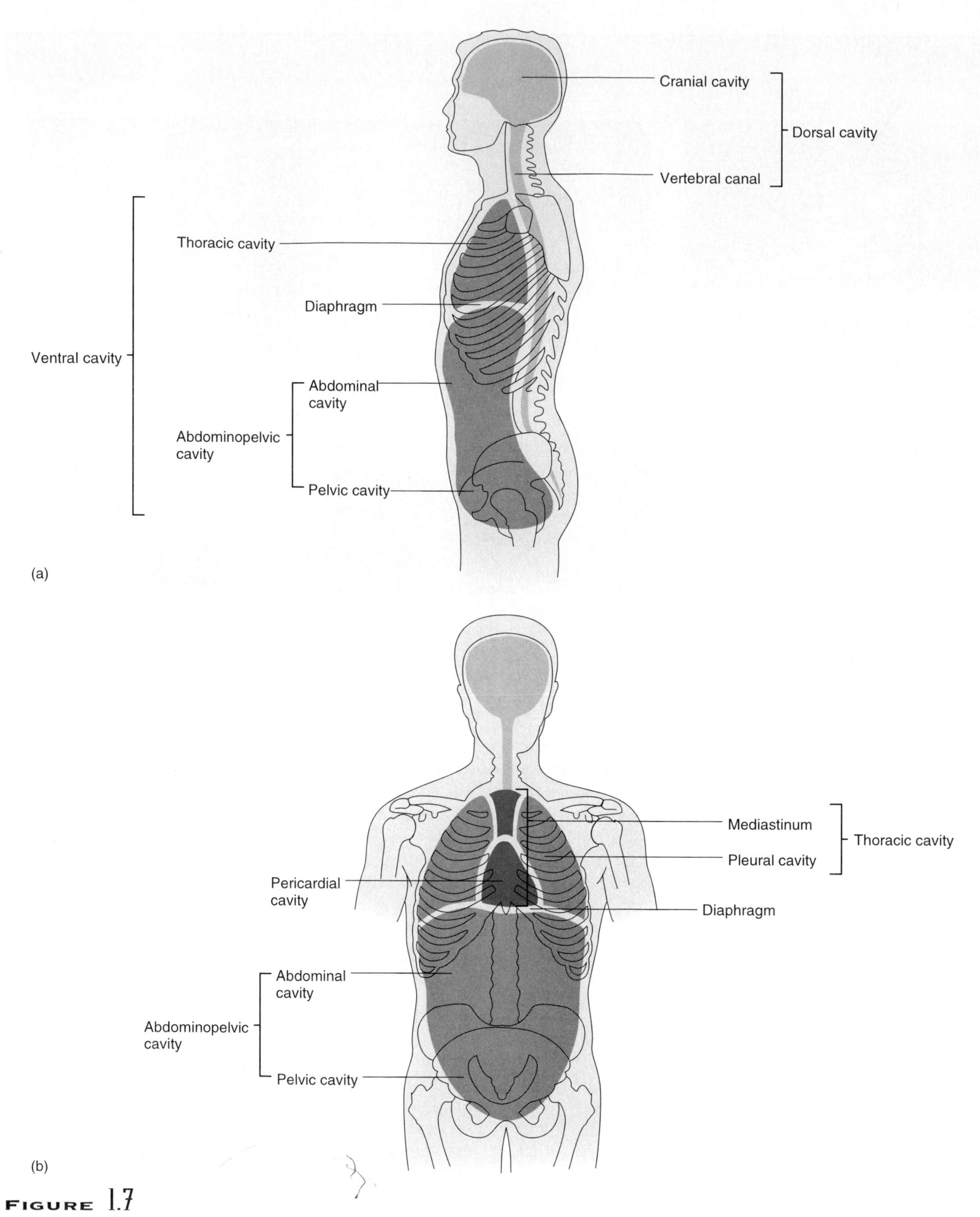


**FIGURE 1.7**
Major body cavities. (*a*) Lateral view. (*b*) Coronal view.

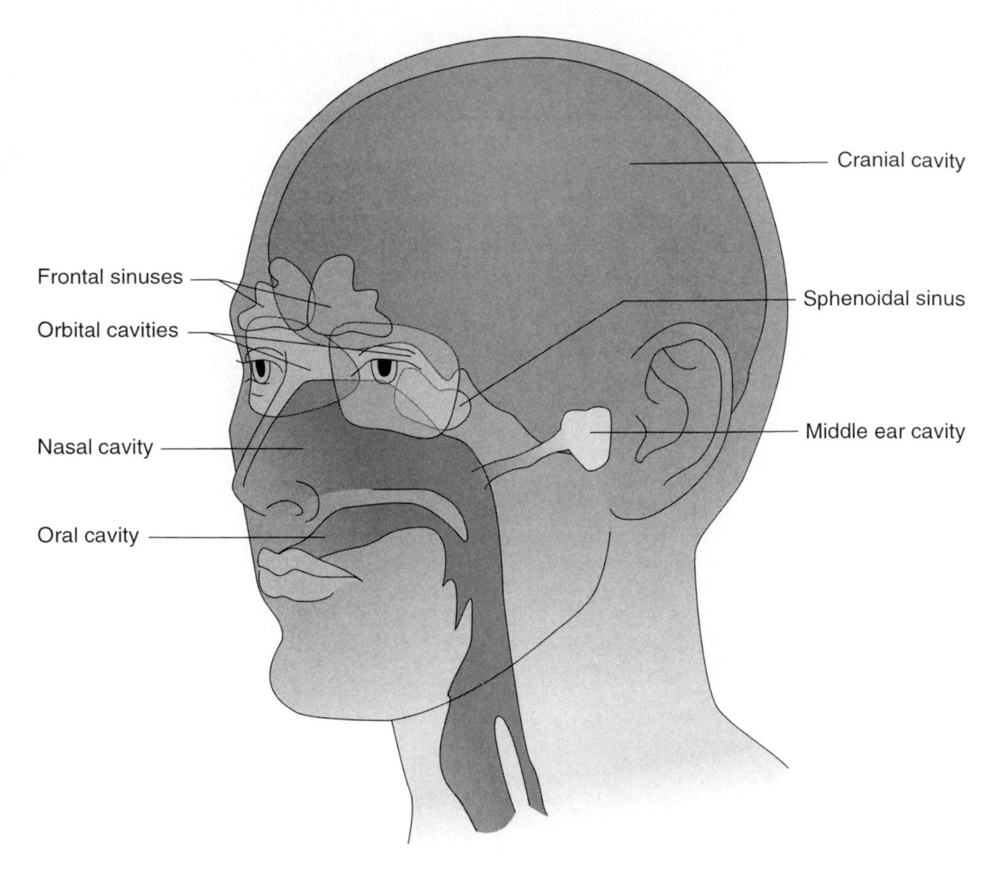


**FIGURE 1.8**
The cavities within the head include the cranial, oral, nasal, orbital, and middle ear cavities, as well as several sinuses.

Its wall consists primarily of skin, skeletal muscles, and bones. The viscera within the **abdominal cavity** include the stomach, liver, spleen, gallbladder, and the small and large intestines.

The **pelvic cavity** is the portion of the abdominopelvic cavity enclosed by the pelvic bones. It contains the terminal end of the large intestine, the urinary bladder, and the internal reproductive organs.

Smaller cavities within the head include the following (fig. 1.8):

1. *Oral cavity,* containing the teeth and tongue.
2. *Nasal cavity,* located within the nose and divided into right and left portions by a nasal septum. Several air-filled sinuses are connected to the nasal cavity. These include the sphenoidal and frontal sinuses (see chapter 7).
3. *Orbital cavities,* containing the eyes and associated skeletal muscles and nerves.
4. *Middle ear cavities,* containing the middle ear bones.

## Thoracic and Abdominopelvic Membranes

The walls of the right and left thoracic compartments, which contain the lungs, are lined with a membrane called the *parietal pleura.* The lungs themselves are covered by a similar membrane called the *visceral pleura.* (Note: *Parietal* refers to the membrane attached to the wall of a cavity, and *visceral* refers to one that is associated with an internal organ, such as a lung.)

The parietal and visceral **pleural membranes** are separated by a thin film of watery fluid (serous fluid) that they secrete. Although there is normally no actual space between these membranes, the potential space between them is called the *pleural cavity.*

The heart, which is located in the broadest portion of the mediastinum, is surrounded by **pericardial membranes.** A thin *visceral pericardium* (epicardium) covers the heart's surface and is separated from a much thicker, fibrous *parietal pericardium* by a small amount of serous fluid. The potential space

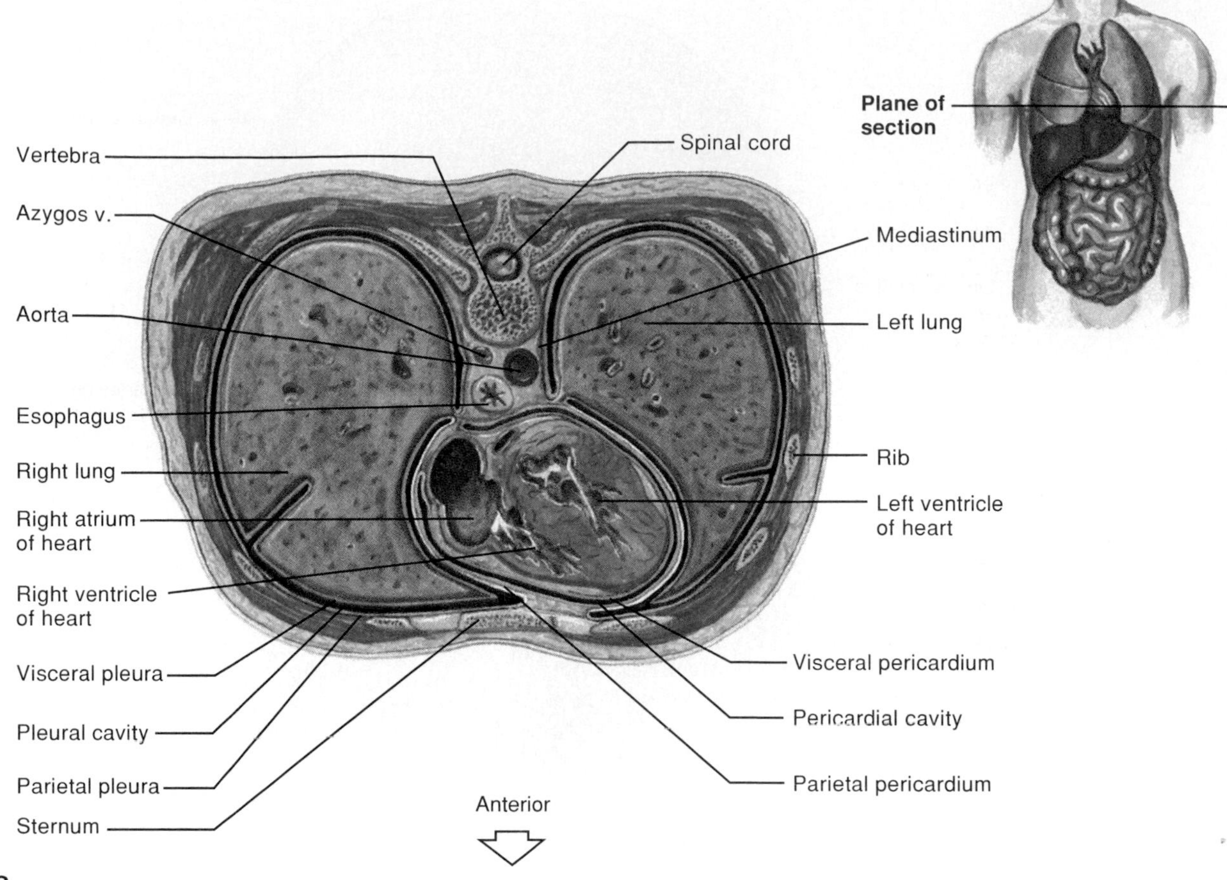


**FIGURE 1.9**
A transverse section through the thorax reveals the serous membranes associated with the heart and lungs (superior view).

between these membranes is called the *pericardial cavity*. Figure 1.9 shows the membranes associated with the heart and lungs.

In the abdominopelvic cavity, the lining membranes are called **peritoneal membranes.** A *parietal peritoneum* lines the wall, and a *visceral peritoneum* covers each organ in the abdominal cavity. The potential space between these membranes is called the *peritoneal cavity* (fig. 1.10).

1. What does the term *visceral* mean?
2. What organs occupy the dorsal cavity? The ventral cavity?
3. Name the cavities of the head.
4. Describe the membranes associated with the thoracic and abdominopelvic cavities.
5. Distinguish between the parietal and visceral peritoneum.

## Organ Systems

The human organism consists of several organ systems. Each system includes a set of interrelated organs that work together to provide specialized functions. The maintenance of homeostasis depends on these organ systems working together in a coordinated manner. A figure called **"InnerConnections"** featured at the ends of certain chapters ties together the ways in which organ systems interact. As you read about each organ system, you may want to consult the illustrations of the human torso provided in reference plates 1–7 and locate some of the features listed in the descriptions.

### Body Covering

The organs of the **integumentary system** (fig. 1.11) include the skin and accessory organs such as the hair, nails, sweat glands, and sebaceous glands. These parts protect underlying tissues, help regulate body temperature, house a variety of sensory receptors, and synthesize certain products. The integumentary system is discussed in chapter 6.

### Support and Movement

The organs of the skeletal and muscular systems support and move body parts.

The **skeletal system** (fig. 1.12) consists of the bones as well as the ligaments and cartilages that bind the bones together at joints. These parts provide frameworks and protective shields for softer tissues, serve as

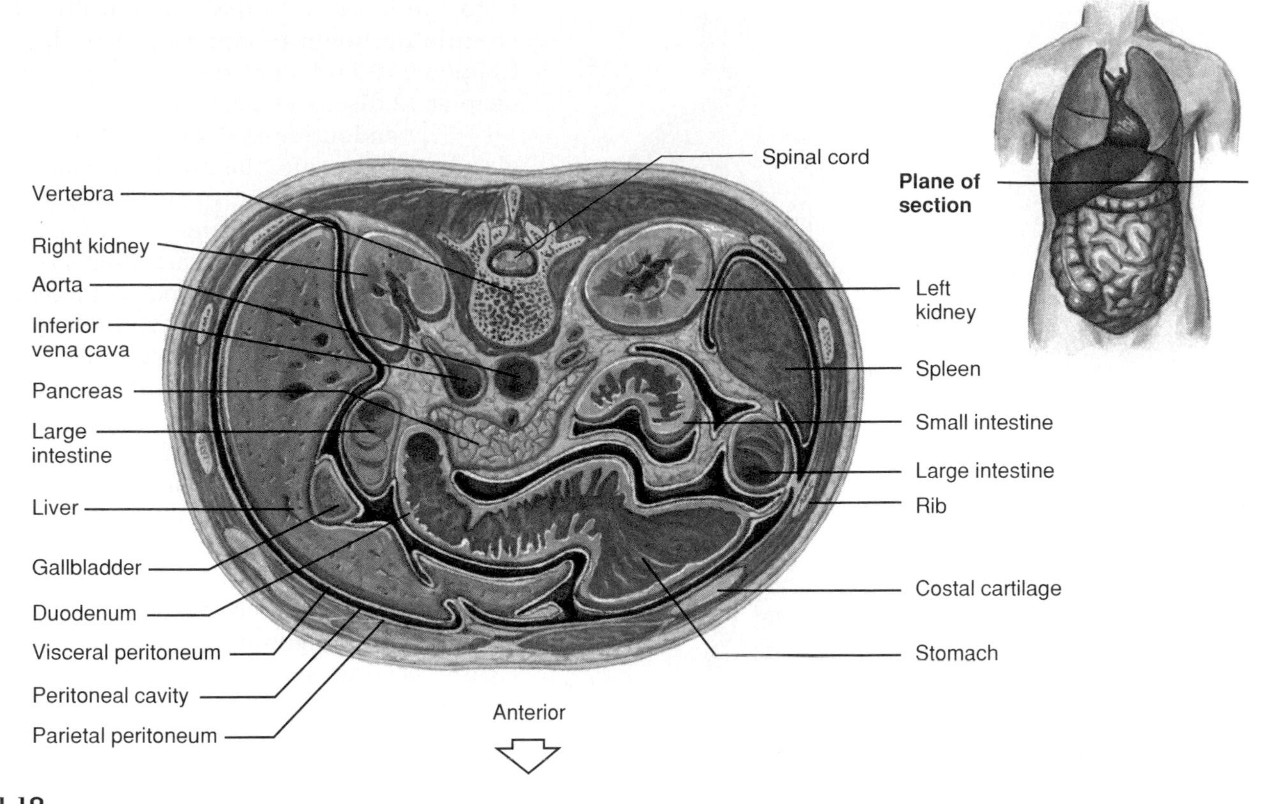


**FIGURE 1.10**
A transverse section through the abdomen (superior view). Note that the large intestine appears twice.

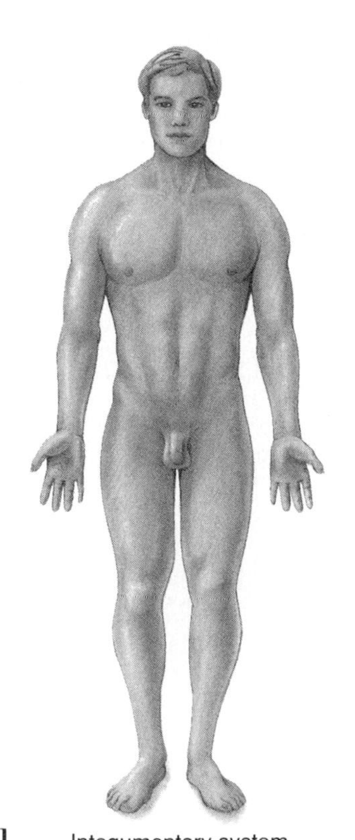


**FIGURE 1.11**
The integumentary system forms the body covering.

attachments for muscles, and act together with muscles when body parts move. Tissues within bones also produce blood cells and store inorganic salts.

The muscles are the organs of the **muscular system** (fig. 1.12). By contracting and pulling their ends closer together, they provide the forces that cause body movements. They also help maintain posture and are the main source of body heat.

The skeletal and muscular systems are discussed in chapters 7, 8, and 9.

## Integration and Coordination

For the body to act as a unit, its parts must be integrated and coordinated. The nervous and endocrine systems control and adjust various organ functions from time to time to maintain homeostasis.

The **nervous system** (fig. 1.13) consists of the brain, spinal cord, nerves, and sense organs. Nerve cells within these organs use electrochemical signals called *nerve impulses* (action potentials) to communicate with one another and with muscles and glands. Each impulse produces a relatively short-term effect on its target. Some nerve cells act as specialized sensory receptors that can detect changes occurring inside and outside the body. Others receive the impulses transmitted from these sensory units and interpret and act on the information received. Still others carry impulses

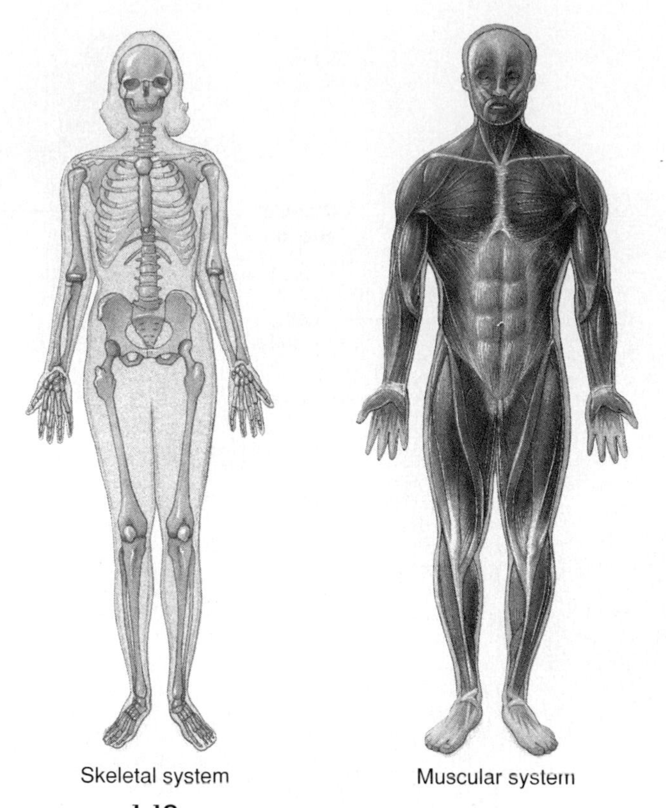


**FIGURE 1.12**
Organ systems associated with support and movement.

from the brain or spinal cord to muscles or glands, stimulating them to contract or to secrete products. Chapters 10 and 11 discuss the nervous system, and chapter 12 discusses sense organs.

The **endocrine system** (fig. 1.13) includes all the glands that secrete chemical messengers called *hormones.* The hormones, in turn, travel away from the glands in body fluids such as blood or tissue fluid. Usually a particular hormone affects only a particular group of cells, which is called its *target tissue.* The effect of a hormone is to alter the metabolism of the target tissue. Compared to nerve impulses, hormonal effects occur over a relatively long time period.

Organs of the endocrine system include the pituitary, thyroid, parathyroid, and adrenal glands, as well as the pancreas, ovaries, testes, pineal gland, and thymus gland. These are discussed further in chapter 13.

### Transport

Two organ systems are responsible for transporting substances throughout the internal environment. The **cardiovascular system** (fig. 1.14) includes the heart, arteries, veins, capillaries, and blood. The heart is a muscular pump that helps force blood through the blood vessels. The blood transports gases, nutrients, hormones, and wastes. It carries oxygen from the lungs and nutrients from the digestive organs to all

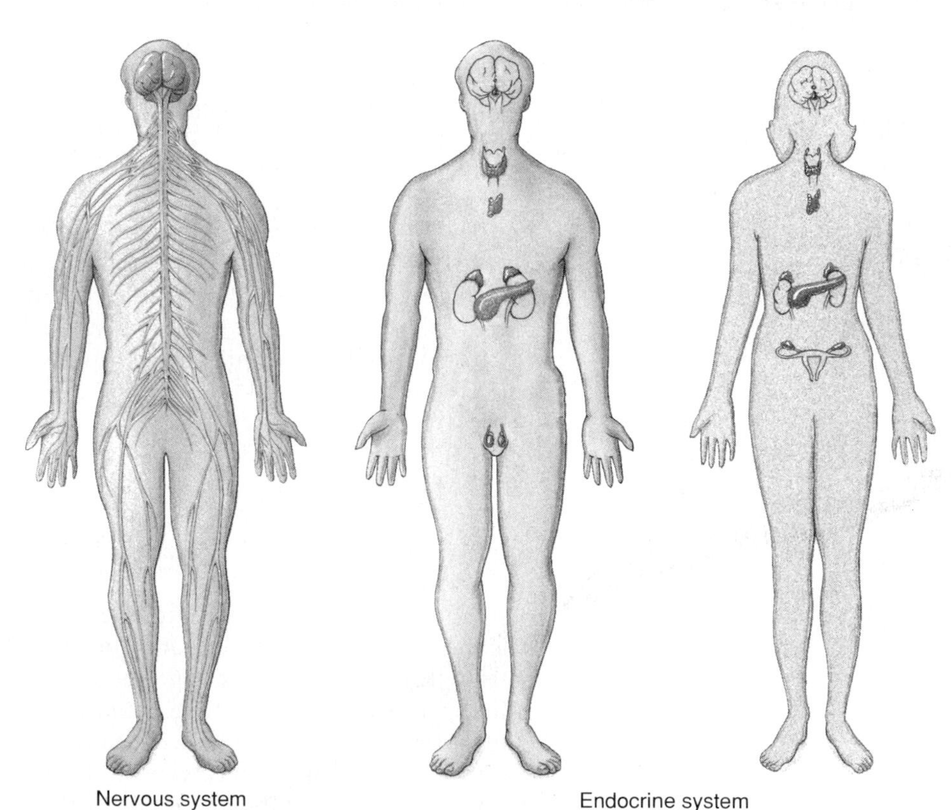


**FIGURE 1.13**
Organ systems associated with integration and coordination.

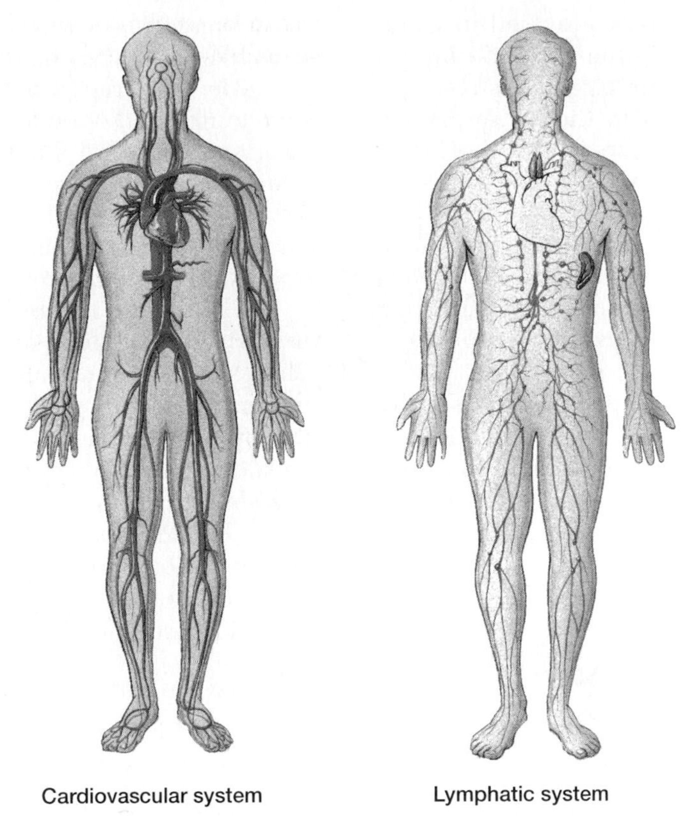


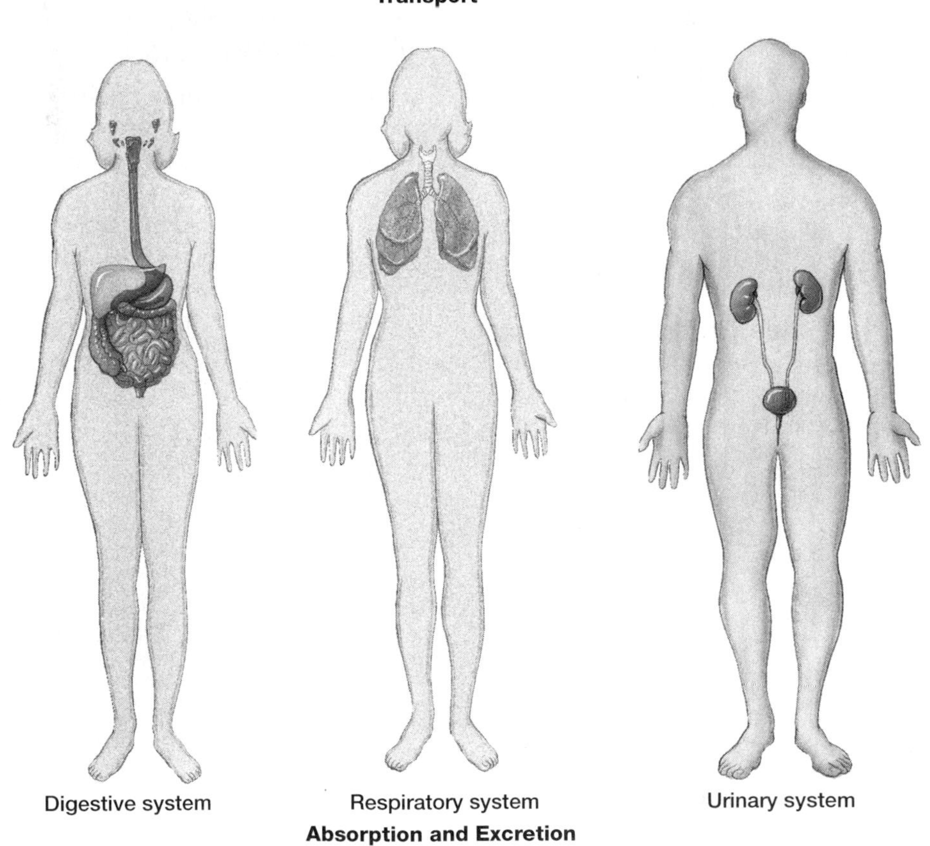


**FIGURE 1.14**
Organ systems associated with transport and with absorption and excretion.

body cells, where these substances are used in metabolic processes. The blood also transports hormones from endocrine glands to their target tissues, and carries wastes from body cells to the excretory organs, where the wastes are removed from the blood and released to the outside. Blood and the cardiovascular system are discussed in chapters 14 and 15.

The **lymphatic system** (fig. 1.14) is also involved with transport. It is composed of the lymphatic vessels, lymph fluid, lymph nodes, thymus gland, and spleen. This system transports some of the fluid from the spaces within tissues (tissue fluid) back to the bloodstream and carries certain fatty substances away from the digestive organs. Cells of the lymphatic system are called lymphocytes, and they defend the body against infections by removing disease-causing microorganisms and viruses from the tissue fluid. The lymphatic system is discussed in chapter 16.

### Absorption and Excretion

Organs in several systems absorb nutrients and oxygen and excrete various wastes. The organs of the **digestive system** (fig. 1.14), for example, receive foods from the outside. Then they break down food molecules into simpler forms that can pass through cell membranes and thus be absorbed. Materials that are not absorbed are eliminated by being transported back to the outside. Certain digestive organs (chapter 17) also produce hormones and thus function as parts of the endocrine system.

The digestive system includes the mouth, tongue, teeth, salivary glands, pharynx, esophagus, stomach, liver, gallbladder, pancreas, small intestine, and large intestine. Chapter 18 discusses nutrition.

The organs of the **respiratory system** (fig. 1.14) take air in and out and exchange gases between the blood and the air. More specifically, oxygen passes from air within the lungs into the blood, and carbon dioxide leaves the blood and enters the air. The nasal cavity, pharynx, larynx, trachea, bronchi, and lungs are parts of this system, which is discussed in chapter 19.

The **urinary system** (fig. 1.14) consists of the kidneys, ureters, urinary bladder, and urethra. The kidneys remove wastes from the blood and assist in maintaining the body's water and electrolyte balance. The product of these activities is urine. Other portions of the urinary system store urine and transport it outside the body. Chapter 20 discusses the urinary system.

Sometimes the urinary system is called the *excretory system*. However, excretion, or waste removal, is also a function of the respiratory, digestive, and integumentary systems.

### Reproduction

Reproduction is the process of producing offspring (progeny). Cells reproduce when they divide and give rise to new cells. The **reproductive system** (fig. 1.15) of an organism, however, produces whole new organisms like itself (see chapter 22).

The male reproductive system includes the scrotum, testes, epididymides, vasa deferentia, seminal vesicles, prostate gland, bulbourethral glands, urethra, and penis. These structures produce and maintain the male sex cells, or sperm cells (spermatozoa). The male reproductive system also transfers these cells from their site of origin into the female reproductive tract.

The female reproductive system consists of the ovaries, uterine tubes, uterus, vagina, clitoris, and vulva. These organs produce and maintain the female sex cells, or eggs cells (ova), receive the male cells, and transport the male and female cells within the female system. The female reproductive system also supports development of embryos and functions in the birth process.

Chart 1.4 summarizes the organ systems, the major organs that comprise them, and their major functions in the order you will read about them in this book. Figure 1.16 illustrates the organ systems in humans.

1. Name the major organ systems and list the organs of each system.
2. Describe the general functions of each organ system.

## ANATOMICAL TERMINOLOGY

To communicate effectively with one another, investigators over the ages have developed a set of terms with precise meanings. Some of these terms concern the relative positions of body parts, others refer to imaginary planes along which cuts may be made, and still others describe various body regions.

When such terms are used, it is assumed that the body is in the **anatomical position;** that is, it is standing erect, the face is forward, and the arms are at the sides, with the palms forward.

### Relative Position

Terms of relative position are used to describe the location of one body part with respect to another. They include the following:

1. **Superior** means a part is above another part, or closer to the head. (The thoracic cavity is superior to the abdominopelvic cavity.)
2. **Inferior** means situated below another part, or toward the feet. (The neck is inferior to the head.)
3. **Anterior** (or ventral) means toward the front. (The eyes are anterior to the brain.)
4. **Posterior** (or dorsal) is the opposite of anterior; it means toward the back. (The pharynx is posterior to the oral cavity.)

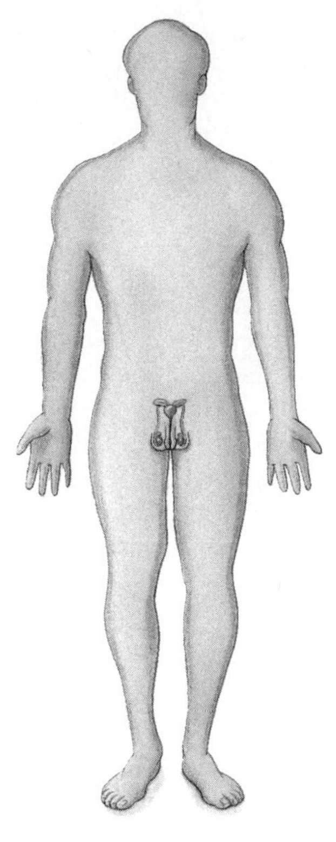
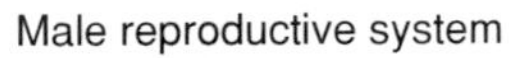
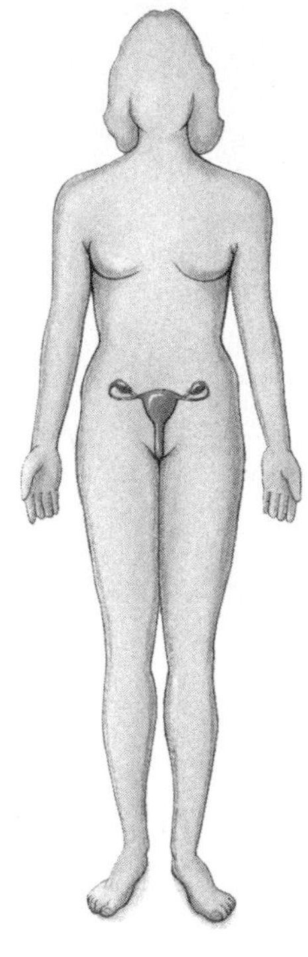


**FIGURE 1.15**
Organ systems associated with reproduction.

5. **Medial** relates to an imaginary midline dividing the body into equal right and left halves. A part is medial if it is closer to this line than another part. (The nose is medial to the eyes.)
6. **Lateral** means toward the side with respect to the imaginary midline. (The ears are lateral to the eyes.) **Ipsilateral** pertains to the same side (the spleen and the descending colon are ipsilateral), while **contralateral** refers to the opposite side (the spleen and the gallbladder are contralateral).
7. **Proximal** is used to describe a part that is closer to the trunk of the body or closer to another specified point of reference than another part. (The elbow is proximal to the wrist.)
8. **Distal** is the opposite of proximal. It means a particular body part is farther from the trunk or farther from another specified point of reference than another part. (The fingers are distal to the wrist.)
9. **Superficial** means situated near the surface. (The epidermis is the superficial layer of the skin.) **Peripheral** also means outward or near the surface. It is used to describe the location

**CHART 1.4 ORGAN SYSTEMS**

| Organ System | Major Organs | Major Functions |
|---|---|---|
| Integumentary | Skin, hair, nails, sweat glands, sebaceous glands | Protect tissues, regulate body temperature, support sensory receptors |
| Skeletal | Bones, ligaments, cartilages | Provide framework, protect soft tissues, provide attachments for muscles, produce blood cells, store inorganic salts |
| Muscular | Muscles | Cause movements, maintain posture, produce body heat |
| Nervous | Brain, spinal cord, nerves, sense organs | Detect changes, receive and interpret sensory information, stimulate muscles and glands |
| Endocrine | Glands that secrete hormones (pituitary gland, thyroid gland, parathyroid glands, adrenal glands, pancreas, ovaries, testes, pineal gland, and thymus gland) | Control metabolic activities of body structures |
| Digestive | Mouth, tongue, teeth, salivary glands, pharynx, esophagus, stomach, liver, gallbladder, pancreas, small and large intestines | Receive, break down, and absorb food; eliminate unabsorbed material |
| Respiratory | Nasal cavity, pharynx. larynx, trachea, bronchi, lungs | Intake and output of air, exchange of gases between air and blood |
| Cardiovascular | Heart, arteries, veins, capillaries | Move blood through blood vessels and transport substances throughout body |
| Lymphatic | Lymphatic vessels, lymph nodes, thymus, spleen | Return tissue fluid to the blood, carry certain absorbed food molecules, defend the body against infection |
| Urinary | Kidneys, ureters, urinary bladder, urethra | Remove wastes from blood, maintain water and electrolyte balance, store and transport urine |
| Reproductive | Male: scrotum, testes, epididymides, vasa deferentia, seminal vesicles, prostate gland, bulbourethral glands, urethra, penis | Produce and maintain sperm cells, transfer sperm cells into female reproductive tract |
| | Female: ovaries, uterine tubes, uterus, vagina, clitoris, vulva | Produce and maintain egg cells; receive and transport sperm cells; support development of an embryo and function in birth process |

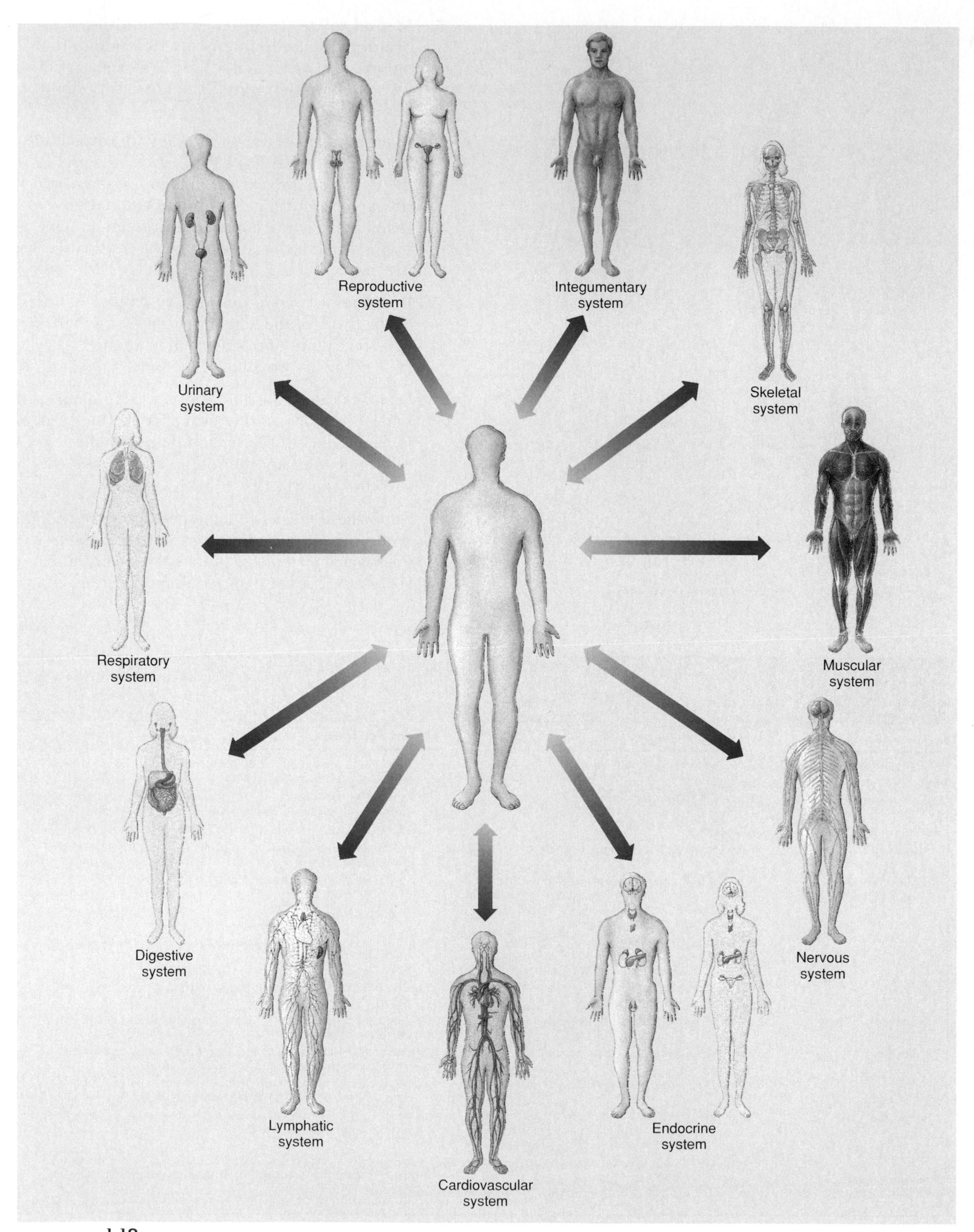


**FIGURE 1.16**
The organ systems in humans.

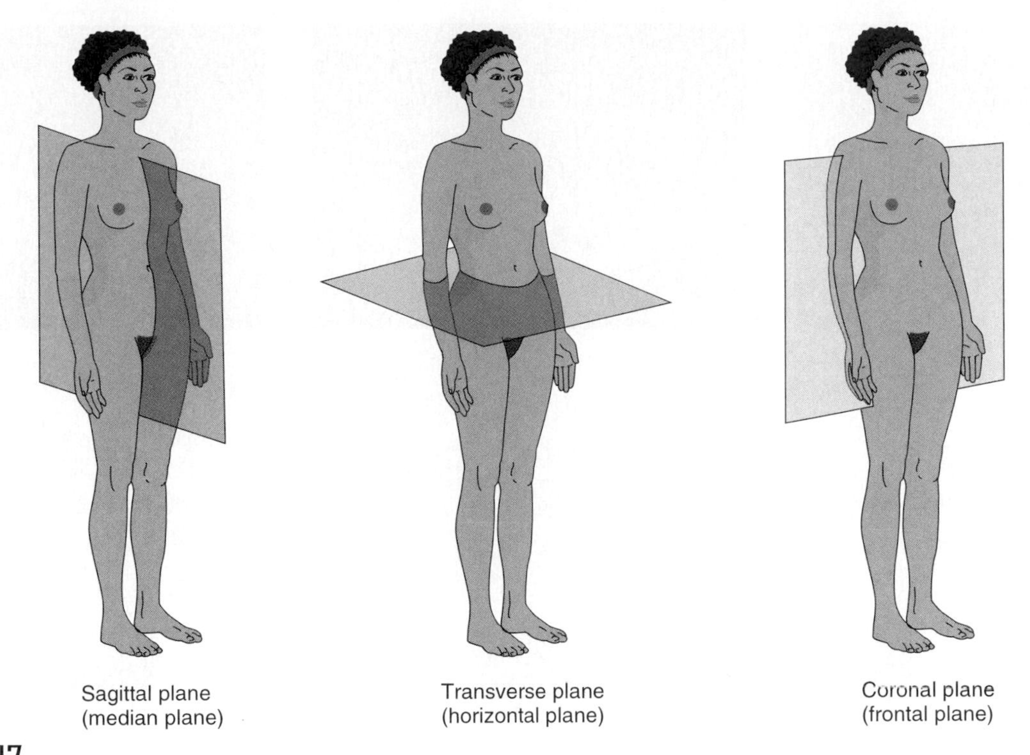


**FIGURE 1.17**
To observe internal parts, the body may be sectioned along various planes.

of certain blood vessels and nerves. (The nerves that branch from the brain and spinal cord are peripheral nerves.)

10. **Deep** is used to describe parts that are more internal. (The dermis is the deep layer of the skin.)

## Body Sections

To observe the relative locations and arrangements of internal parts, it is necessary to cut or section the body along various planes (figs. 1.17 and 1.18). The following terms are used to describe such planes and sections:

1. **Sagittal** refers to a lengthwise cut that divides the body into right and left portions. If a sagittal section passes along the midline and divides the body into equal parts, it is called median (midsagittal).
2. **Transverse** (or horizontal section) refers to a cut that divides the body into superior and inferior portions.
3. **Coronal** (or frontal) refers to a section that divides the body into anterior and posterior portions.

Sometimes a cylindrical organ such as a blood vessel is sectioned. In this case, a cut across the structure is called a *cross section,* an angular cut is called an *oblique section,* and a lengthwise cut is called a *longitudinal section* (fig. 1.19).

## Body Regions

A number of terms designate body regions. The abdominal area, for example, is subdivided into the following regions, as shown in figure 1.20:

1. **Epigastric region** The upper middle portion.
2. **Left** and **right hypochondriac regions** On each side of the epigastric region.
3. **Umbilical region** The central portion.
4. **Left** and **right lumbar regions** On each side of the umbilical region.
5. **Hypogastric region** The lower middle portion.
6. **Left** and **right iliac (or inguinal) regions** On each side of the hypogastric region.

The abdominal area also may be subdivided into the following four quadrants, as illustrated in figure 1.21:

1. **Right upper quadrant** (RUQ).
2. **Right lower quadrant** (RLQ).

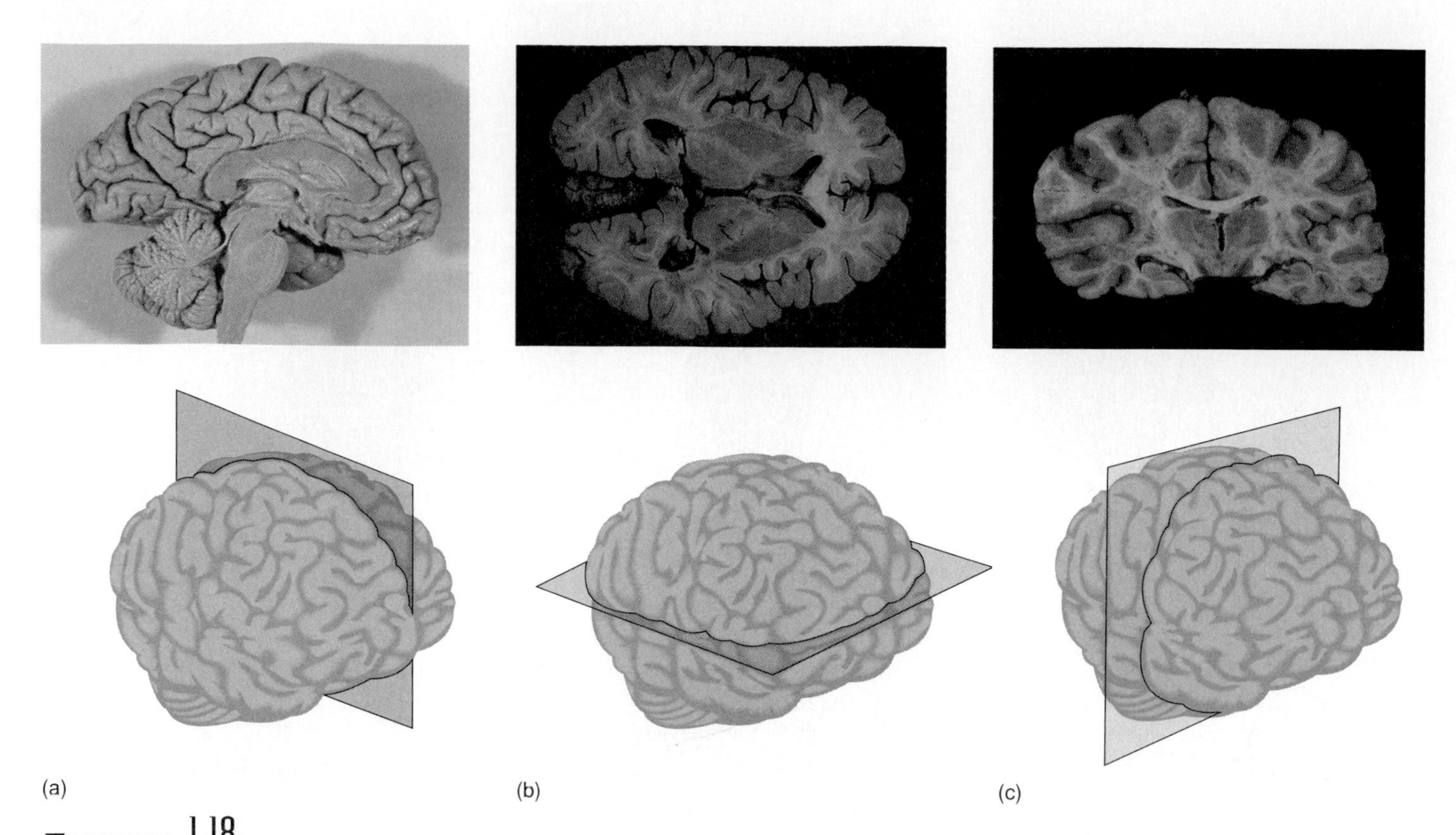


**FIGURE 1.18**
A human brain sectioned along (*a*) the sagittal plane, (*b*) the transverse plane, and (*c*) the coronal plane.

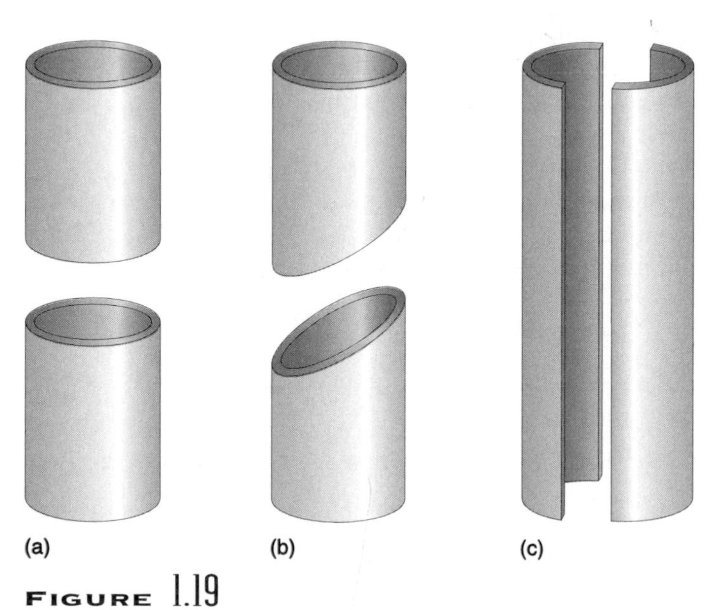


**FIGURE 1.19**
Cylindrical parts may be cut in (*a*) cross section, (*b*) oblique section, or (*c*) longitudinal section.

3. **Left upper quadrant** (LUQ).
4. **Left lower quadrant** (LLQ).

The following terms are commonly used when referring to various body regions. Figure 1.22 illustrates some of these regions.

**abdominal** (ab-dom′ĭ-nal) the region between the thorax and pelvis.
**acromial** (ah-kro′me-al) the point of the shoulder.
**antebrachial** (an″te-bra′ke-al) the forearm.
**antecubital** (an″te-ku′bĭ-tal) the space in front of the elbow.
**axillary** (ak′sĭ-ler″e) the armpit.
**brachial** (bra′ke-al) the arm.
**buccal** (buk′al) the cheek.
**carpal** (kar′pal) the wrist.
**celiac** (se′le-ak) the abdomen.
**cephalic** (sĕ-fal′ik) the head.
**cervical** (ser′vĭ-kal) the neck.
**costal** (kos′tal) the ribs.
**coxal** (kok′sal) the hip.
**crural** (kro͞o͞r′al) the leg.
**cubital** (ku′bĭ-tal) the elbow.
**digital** (dij′ĭ-tal) the finger.
**dorsum** (dor′sum) the back.
**femoral** (fem′or-al) the thigh.
**frontal** (frun′tal) the forehead.
**genital** (jen′i-tal) the reproductive organs.
**gluteal** (gloo′te-al) the buttocks.
**inguinal** (ing′gwĭ-nal) the depressed area of the abdominal wall near the thigh (groin).
**lumbar** (lum′bar) the region of the lower back between the ribs and the pelvis (loin).
**mammary** (mam′er-e) the breast.

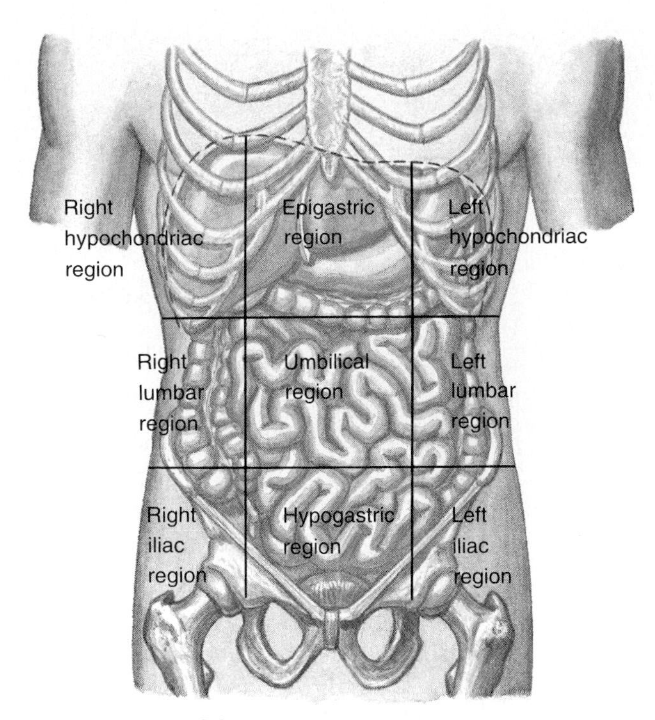


**FIGURE 1.20**
The abdominal area is subdivided into nine regions.

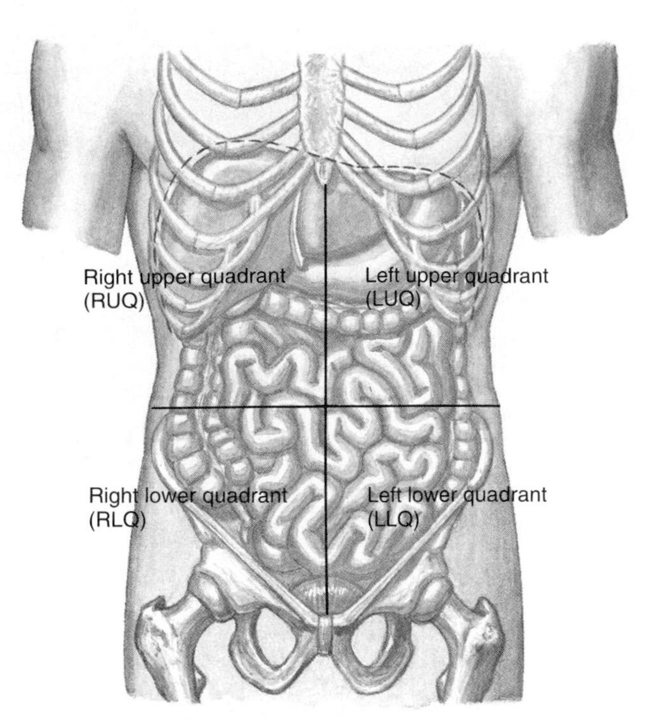


**FIGURE 1.21**
The abdominal area may be subdivided into four quadrants.

**mental** (men′tal) the chin.
**nasal** (na′zal) the nose.
**occipital** (ok-sip′ĭ-tal) the lower posterior region of the head.
**oral** (o′ral) the mouth.
**orbital** (or′bi-tal) the eye cavity.
**otic** (o′tik) the ear.
**palmar** (pahl′mar) the palm of the hand.
**patellar** (pah-tel′ar) the front of the knee.
**pectoral** (pek′tor-al) the chest.
**pedal** (ped′al) the foot.
**pelvic** (pel′vik) the pelvis.
**perineal** (per″ĭ-ne′al) the region between the anus and the external reproductive organs (perineum).
**plantar** (plan′tar) the sole of the foot.
**popliteal** (pop″lĭ-te′al) the area behind the knee.
**sacral** (sa′kral) the posterior region between the hipbones.
**sternal** (ster′nal) the middle of the thorax, anteriorly.
**tarsal** (tahr′sal) the instep of the foot.
**umbilical** (um-bil′ĭ-kal) the navel.
**vertebral** (ver′te-bral) the spinal column.

1. Describe the anatomical position.
2. Using the appropriate terms, describe the relative positions of several body parts.
3. Describe three types of sections.
4. Describe the nine regions of the abdomen.
5. Explain how the names of the abdominal regions describe their locations.

## Some Medical and Applied Sciences

**cardiology** (kar″de-ol′o-je) Branch of medical science dealing with the heart and heart diseases.
**cytology** (si-tol′o-je) Study of the structure, function, and diseases of cells.
**dermatology** (der″mah-tol′o-je) Study of skin and its diseases.
**endocrinology** (en″do-krĭ-nol′o-je) Study of hormones, hormone-secreting glands, and the diseases involving them.
**epidemiology** (ep″ĭ-de″me-ol′o-je) Study of the factors determining the distribution and frequency of the occurrence of health-related conditions within a defined human population.
**gastroenterology** (gas″tro-en″ter-ol′o-je) Study of the stomach and intestines, as well as the diseases involving them.
**geriatrics** (jer″e-at′riks) Branch of medicine dealing with elderly persons and their medical problems.
**gerontology** (jer″on-tol′o-je) Study of the process of aging and the various problems of elderly persons.
**gynecology** (gi″nĕ-kol′o-je) Study of the female reproductive system and its diseases.
**hematology** (hem″ah-tol′o-je) Study of blood and blood diseases.
**histology** (his-tol′o-je) Study of the structure and function of tissues.
**immunology** (im″u-nol′o-je) Study of the body's resistance to disease.

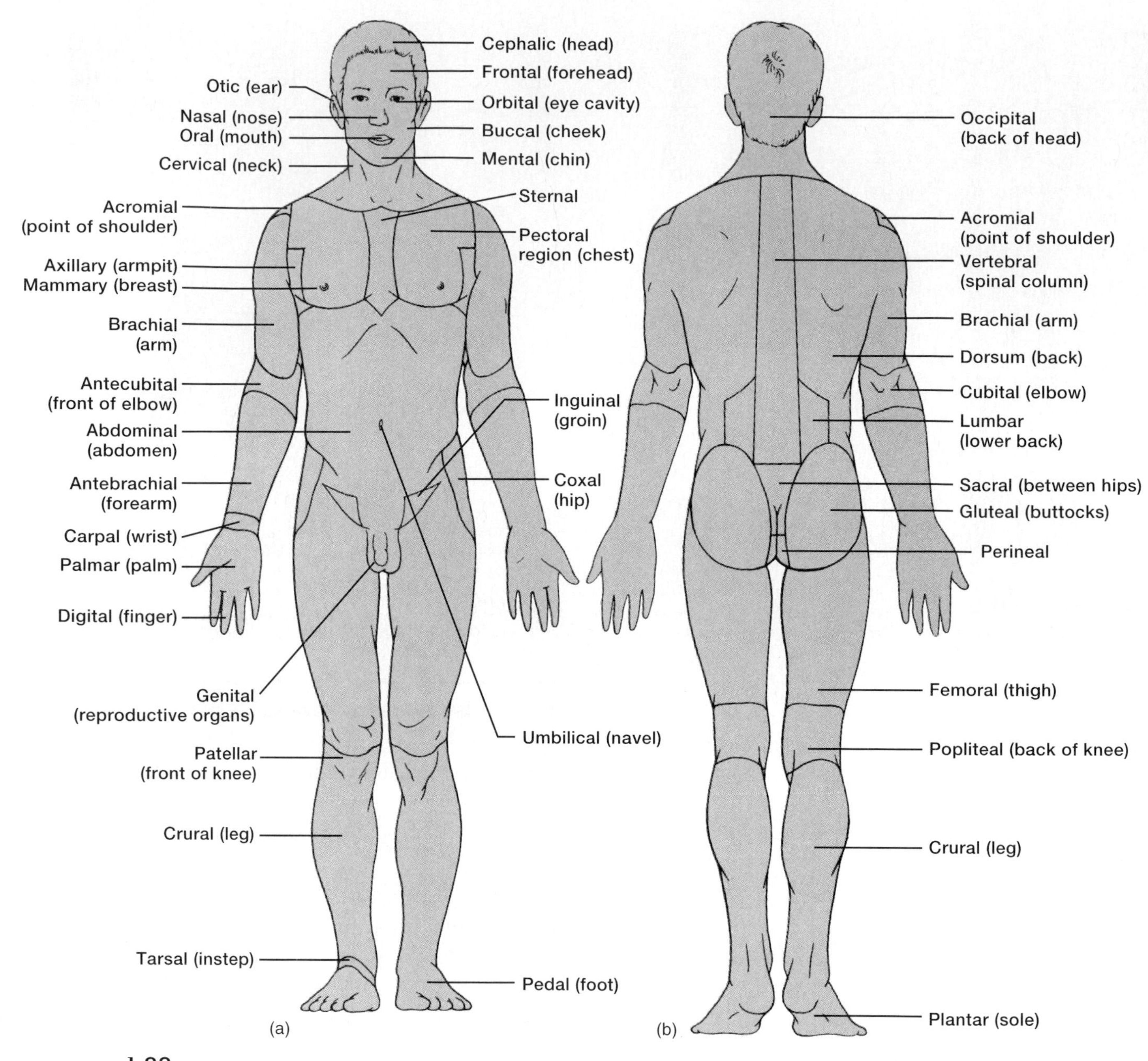


**FIGURE 1.22**
Some terms used to describe body regions. (*a*) Anterior regions; (*b*) posterior regions.

**neonatology** (ne″o-na-tol′o-je) Study of newborn infants and the treatment of their disorders.
**nephrology** (nĕ-frol′o-je) Study of the structure, function, and diseases of the kidneys.
**neurology** (nu-rol′o-je) Study of the nervous system in health and disease.
**obstetrics** (ob-stet′riks) Branch of medicine dealing with pregnancy and childbirth.
**oncology** (ong-kol′o-je) Study of cancers.
**ophthalmology** (of″thal-mol′o-je) Study of the eye and eye diseases.
**orthopedics** (or″tho-pe′diks) Branch of medicine dealing with the muscular and skeletal systems, and their problems.
**otolaryngology** (o″to-lar″in-gol′o-je) Study of the ear, throat, larynx, and diseases of these parts.
**pathology** (pah-thol′o-je) Study of structural and functional changes within the body produced by diseases.
**pediatrics** (pe″de-at′riks) Branch of medicine dealing with children and their diseases.
**pharmacology** (fahr″mah-kol′o-je) Study of drugs and their uses in the treatment of diseases.

**podiatry** (po-di′ah-tre) Study of the care and treatment of the feet.

**psychiatry** (si-ki′ah-tre) Branch of medicine dealing with the mind and its disorders.

**radiology** (ra″de-ol′o-je) Study of X rays and radioactive substances, as well as their uses in diagnosing and treating diseases.

**toxicology** (tok″sĭ-kol′o-je) Study of poisonous substances and their effects upon body parts.

**urology** (u-rol′o-je) Branch of medicine dealing with the urinary and male reproductive systems, and their diseases.

## Chapter Summary

### Introduction (page 2)

1. When a patient arrives at a hospital with an unknown injury, medical staff must rapidly apply their knowledge of human anatomy and physiology to correctly diagnose the problem.
2. Early interest in the human body probably developed as people became concerned about injuries and illnesses.
3. Early doctors began to learn how certain herbs and potions affected body functions.
4. In ancient times, it was believed that natural processes were caused by spirits and supernatural forces.
5. A set of terms originating from Greek and Latin formed the basis for the language of anatomy and physiology.

### Anatomy and Physiology (page 3)

1. Anatomy deals with the form and arrangement of body parts.
2. Physiology deals with the functions of these parts.
3. The function of a part depends upon the way it is constructed.

### Characteristics of Life (page 4)

Characteristics of life are traits shared by all organisms.

1. These characteristics include:
   a. Movement—changing body position or moving internal parts.
   b. Responsiveness—sensing and reacting to internal or external changes.
   c. Growth—increasing in size without changing in shape.
   d. Reproduction—producing offspring.
   e. Respiration—obtaining oxygen, using oxygen to release energy from foods, and removing gaseous wastes.
   f. Digestion—breaking down food substances into forms that can be absorbed.
   g. Absorption—moving substances through membranes and into body fluids.
   h. Circulation—moving substances through the body in body fluids.
   i. Assimilation—changing substances into chemically different forms.
   j. Excretion—removing body wastes.
2. Together these activities constitute metabolism.

### Maintenance of Life (page 5)

The structures and functions of body parts maintain the life of the organism.

1. Requirements of organisms
   a. Water is used in a variety of metabolic processes, provides the environment for metabolic reactions, and transports substances.
   b. Food supplies energy, raw materials for building substances, and chemicals necessary in vital reactions.
   c. Oxygen is used in releasing energy from food materials; this energy drives metabolic reactions.
   d. Heat is a product of metabolic reactions and helps govern the rates of these reactions.
   e. Pressure is an application of force; in humans, atmospheric and hydrostatic pressures help breathing and blood movements, respectively.
   f. Survival of an organism depends upon the quantities and qualities of these factors.
2. Homeostasis
   a. If an organism is to survive, the conditions within its body fluids must remain relatively stable.
   b. The tendency to maintain a stable internal environment is called homeostasis.
   c. Homeostatic mechanisms regulate body temperature, blood pressure, and blood sugar concentration.
   d. Homeostatic mechanisms employ negative feedback.

### Levels of Organization (page 8)

The body is composed of parts that can be considered at different levels of organization.

1. Material substances are composed of atoms.
2. Atoms join together to form molecules.
3. Organelles consist of aggregates of large molecules.
4. Cells, which are composed of organelles, are the basic units of structure and function of the body.
5. Cells are organized into layers or masses called tissues.
6. Tissues are organized into organs.
7. Organs form organ systems.
8. Organ systems constitute the organism.
9. These parts vary in complexity progressively from one level to the next.

### Organization of the Human Body (page 9)

1. Body cavities
   a. The axial portion of the body contains the dorsal and ventral cavities.
      (1) The dorsal cavity includes the cranial cavity and vertebral canal.
      (2) The ventral cavity includes the thoracic and abdominopelvic cavities, which are separated by the diaphragm.
   b. The organs within a body cavity are called viscera.
   c. Other body cavities include the oral, nasal, orbital, and middle ear cavities.

2. Thoracic and abdominopelvic membranes
   Parietal membranes line cavities; visceral membranes cover organs.
   a. Thoracic membranes
      (1) Pleural membranes line the thoracic cavity and cover the lungs.
      (2) Pericardial membranes surround the heart and cover its surface.
      (3) The pleural and pericardial cavities are potential spaces between these membranes.
   b. Abdominopelvic membranes
      (1) Peritoneal membranes line the abdominopelvic cavity and cover the organs inside.
      (2) The peritoneal cavity is a potential space between these membranes.
3. Organ systems
   The human organism consists of several organ systems. Each system includes a set of interrelated organs.
   a. Integumentary system
      (1) The integumentary system provides the body covering.
      (2) It includes the skin, hair, nails, sweat glands, and sebaceous glands.
      (3) It functions to protect underlying tissues, regulate body temperature, house sensory receptors, and synthesize various substances.
   b. Skeletal system
      (1) The skeletal system is composed of bones and the ligaments and cartilages that bind bones together.
      (2) It provides framework, protective shields, and attachments for muscles; it also produces blood cells and stores inorganic salts.
   c. Muscular system
      (1) The muscular system includes the muscles of the body.
      (2) It is responsible for body movements, maintenance of posture, and production of body heat.
   d. Nervous system
      (1) The nervous system consists of the brain, spinal cord, nerves, and sense organs.
      (2) It functions to receive impulses from sensory parts, interpret these impulses, and act on them by causing muscles or glands to respond.
   e. Endocrine system
      (1) The endocrine system consists of glands that secrete hormones.
      (2) Hormones help regulate metabolism by stimulating target tissues.
      (3) It includes the pituitary, thyroid, parathyroid, and adrenal glands; the pancreas, ovaries, testes, pineal gland, and thymus gland.
   f. Digestive system
      (1) The digestive system receives foods, breaks down nutrients into forms that can pass through cell membranes, and eliminates the materials that are not absorbed.
      (2) Some digestive organs produce hormones.
      (3) It includes the mouth, tongue, teeth, salivary glands, pharynx, esophagus, stomach, liver, gallbladder, pancreas, small intestine, and large intestine.
   g. Respiratory system
      (1) The respiratory system provides for the intake and output of air, and for the exchange of gases between the blood and the air.
      (2) It includes the nasal cavity, pharynx, larynx, trachea, bronchi, and lungs.
   h. Cardiovascular system
      (1) The cardiovascular system includes the heart, which pumps blood, and the blood vessels, which carry blood to and from body parts.
      (2) Blood transports oxygen, nutrients, hormones, and wastes.
   i. Lymphatic system
      (1) The lymphatic system is composed of lymphatic vessels, lymph nodes, thymus, and spleen.
      (2) It transports lymph from tissue spaces to the bloodstream and carries certain fatty substances away from the digestive organs. Lymphocytes defend the body against disease-causing agents.
   j. Urinary system
      (1) The urinary system includes the kidneys, ureters, urinary bladder, and urethra.
      (2) It filters wastes from the blood and helps maintain fluid and electrolyte balance.
   k. Reproductive systems
      (1) The reproductive system enables an organism to produce progeny.
      (2) The male reproductive system includes the scrotum, testes, epididymides, vasa deferentia, seminal vesicles, prostate gland, bulbourethral glands, urethra, and penis, which produce, maintain, and transport male sex cells.
      (3) The female reproductive system includes the ovaries, uterine tubes, uterus, vagina, clitoris, and vulva, which produce, maintain, and transport female sex cells.

## Anatomical Terminology (page 18)

Terms with precise meanings are used to help investigators communicate effectively with one another.

1. Relative position
   These terms are used to describe the location of one part with respect to another part.
2. Body sections
   Body sections are planes along which the body may be cut to observe the relative locations and arrangements of internal parts.
3. Body regions
   Various body regions are designated by special terms.

## Critical Thinking Questions

1. In many states, death is defined as "irreversible cessation of total brain function." How is death defined in your state? How is this definition related to the characteristics of life?
2. In health, body parts interact to maintain homeostasis. Illness may threaten homeostasis, requiring treatments. What treatments might be used to help control a patient's (a) body temperature, (b) blood oxygen concentration, and (c) water content?
3. Suppose two individuals have benign (noncancerous) tumors that produce symptoms because they occupy space and crowd adjacent organs. If one of these persons has a tumor in her ventral cavity and the other has a tumor in his dorsal cavity, which would be likely to develop symptoms first? Why?
4. If a patient complained of a "stomachache" and pointed to the umbilical region as the site of the discomfort, what organs located in this region might be the source of the pain?
5. How could the basic needs of a human be provided for a patient who is unconscious?
6. Assuming that the same information could be obtained by either method, what would be the advantage of using ultrasonography rather than X rays to visualize a fetus in the uterus?

## Review Exercises

### Part A

1. Briefly describe the early development of knowledge about the human body.
2. Distinguish between the activities of anatomists and physiologists.
3. How does a biological structure's form determine its function? Give an example.
4. List and describe ten characteristics of life.
5. Define *metabolism*.
6. List and describe five requirements of organisms.
7. Explain how the idea of homeostasis is related to the five requirements you listed in item 6.
8. Distinguish between heat and temperature.
9. What are two types of pressures that may act upon organisms?
10. How are body temperature, blood pressure, and blood sugar concentration controlled?
11. In what ways do homeostatic mechanisms act by negative feedback?
12. How does the human body illustrate the levels of anatomical organization?
13. Distinguish between the axial and appendicular portions of the body.
14. Distinguish between the dorsal and ventral body cavities, and name the smaller cavities that occur within each.
15. What are the viscera?
16. Where is the mediastinum?
17. Describe the locations of the oral, nasal, orbital, and middle ear cavities.
18. How does a parietal membrane differ from a visceral membrane?
19. Name the major organ systems, and describe the general functions of each.
20. List the major organs that comprise each organ system.
21. In what body region did Judith R.'s injury occur?

### Part B

1. Name the body cavity housing each of the following organs:
   a. stomach
   b. heart
   c. brain
   d. liver
   e. trachea
   f. rectum
   g. spinal cord
   h. esophagus
   i. spleen
   j. urinary bladder
2. Write complete sentences using each of the following terms correctly:
   a. superior
   b. inferior
   c. anterior
   d. posterior
   e. medial
   f. lateral
   g. ipsilateral
   h. contralateral
   i. proximal
   j. distal
   k. superficial
   l. peripheral
   m. deep
3. Prepare a sketch of a human body, and use lines to indicate each of the following sections:
   a. sagittal
   b. transverse
   c. coronal
4. Prepare a sketch of the abdominal area and indicate the location of each of the following regions:
   a. epigastric
   b. umbilical
   c. hypogastric
   d. hypochondriac
   e. lumbar
   f. iliac
5. Prepare a sketch of the abdominal area and indicate the location of each of the following regions:
   a. right upper quadrant
   b. right lower quadrant
   c. left upper quadrant
   d. left lower quadrant
6. Provide the common name for the region described by the following terms:
   a. acromial
   b. antebrachial
   c. axillary
   d. buccal
   e. celiac
   f. coxal
   g. crural
   h. femoral
   i. genital
   j. gluteal
   k. inguinal
   l. mental
   m. occipital
   n. orbital
   o. otic
   p. palmar
   q. pectoral
   r. pedal
   s. perineal
   t. plantar
   u. popliteal
   v. sacral
   w. sternal
   x. tarsal
   y. umbilical
   z. vertebral

# THE HUMAN ORGANISM

■ *The following series of illustrations shows the major organs of the human torso. The first plate illustrates the anterior surface and reveals the superficial muscles on one side. Each subsequent plate exposes deeper organs, including those in the thoracic, abdominal, and pelvic cavities.*

*Chapters 6–22 of this textbook describe the organ systems of the human organism in detail. As you read them, you may want to refer to these plates to help visualize the locations of organs and the three-dimensional relationships among them. You may also want to study the photographs of human cadavers in the reference plates that follow chapter 24. These photographs illustrate many of the larger organs of the human body.*

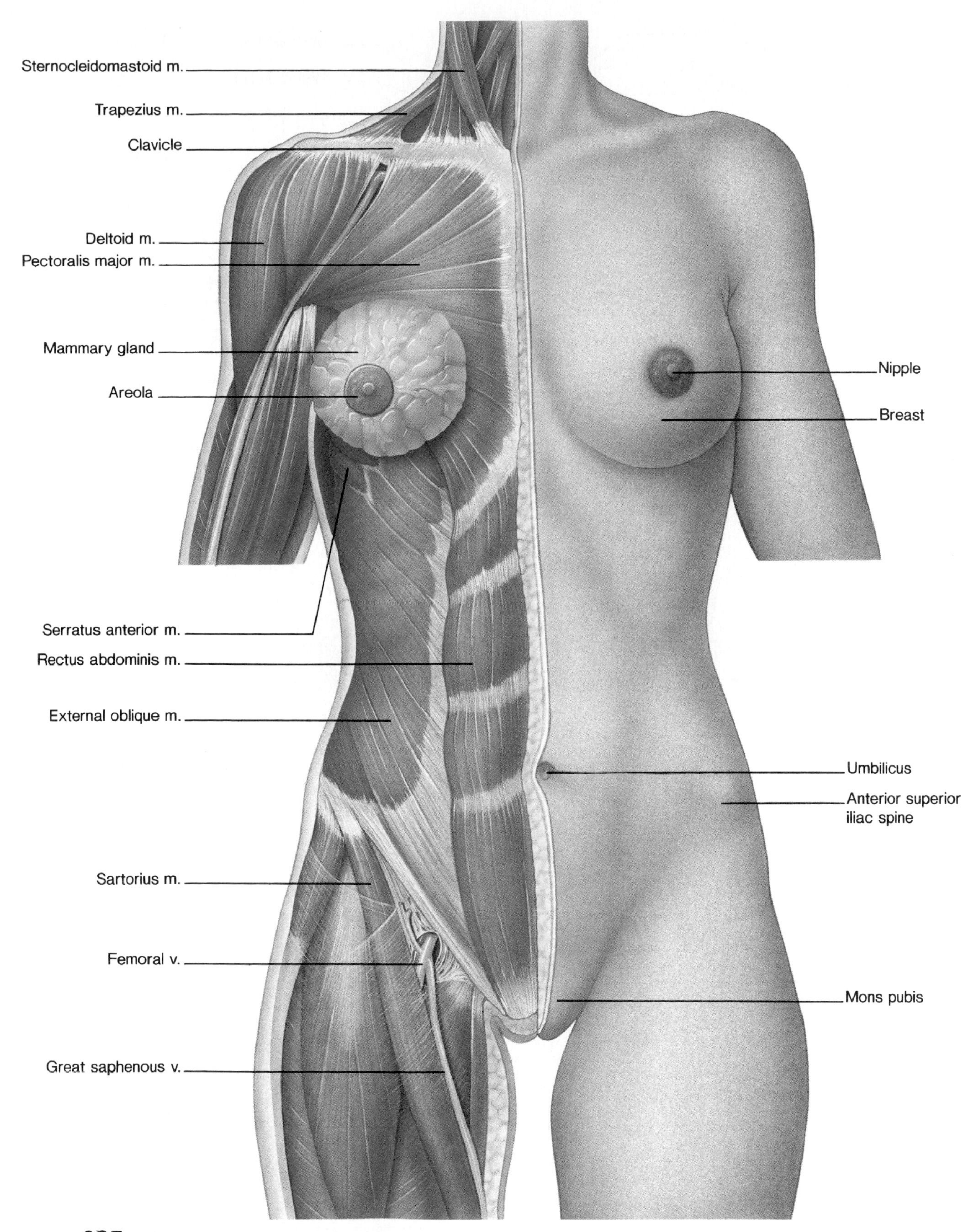


**PLATE ONE**

Human female torso, showing the anterior surface on one side and the superficial muscles exposed on the other side. (*m.* stands for *muscle; v.* stands for *vein*)

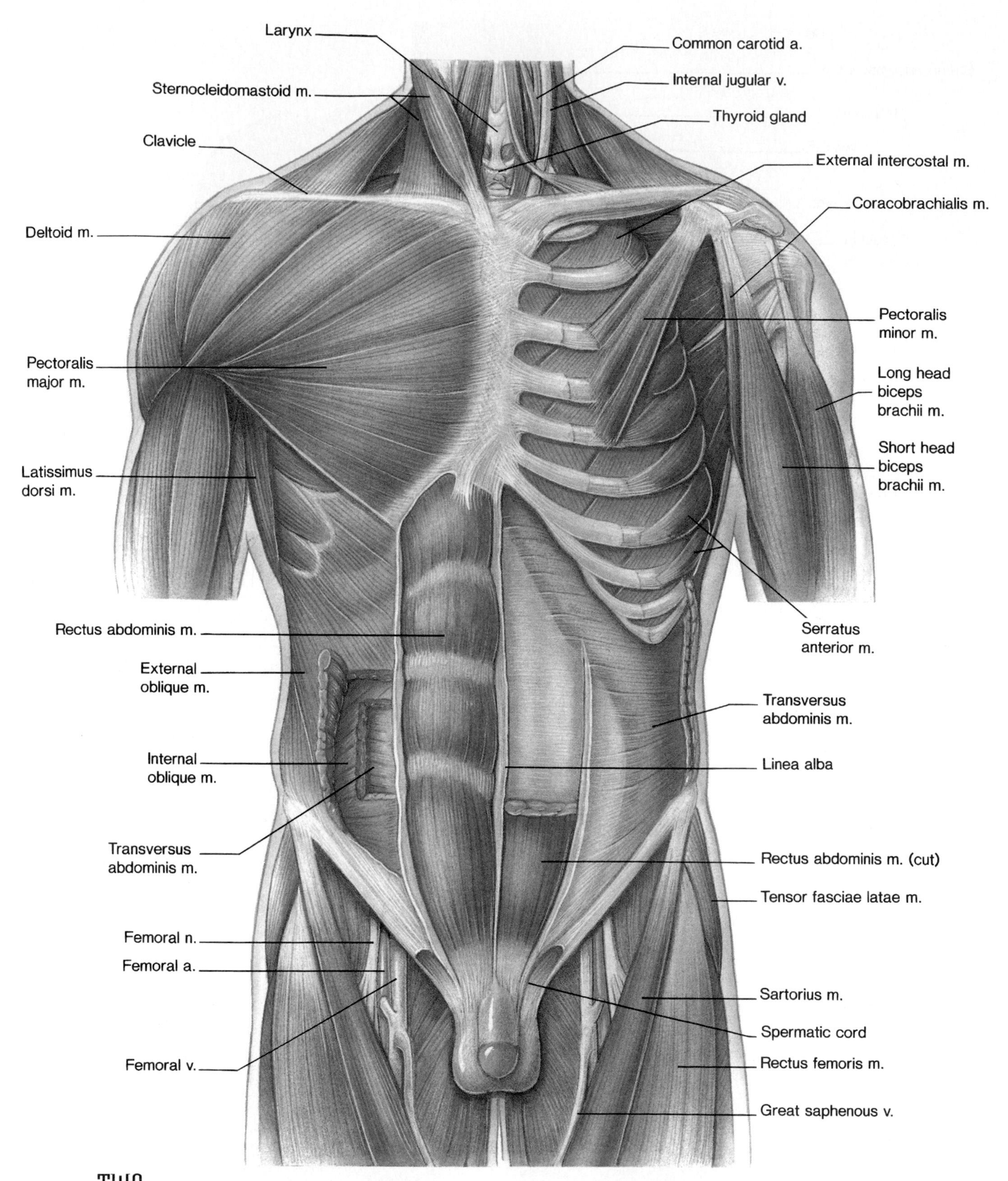


**PLATE TWO**

Human male torso, with the deeper muscle layers exposed. (*n.* stands for *nerve; a.* stands for artery)

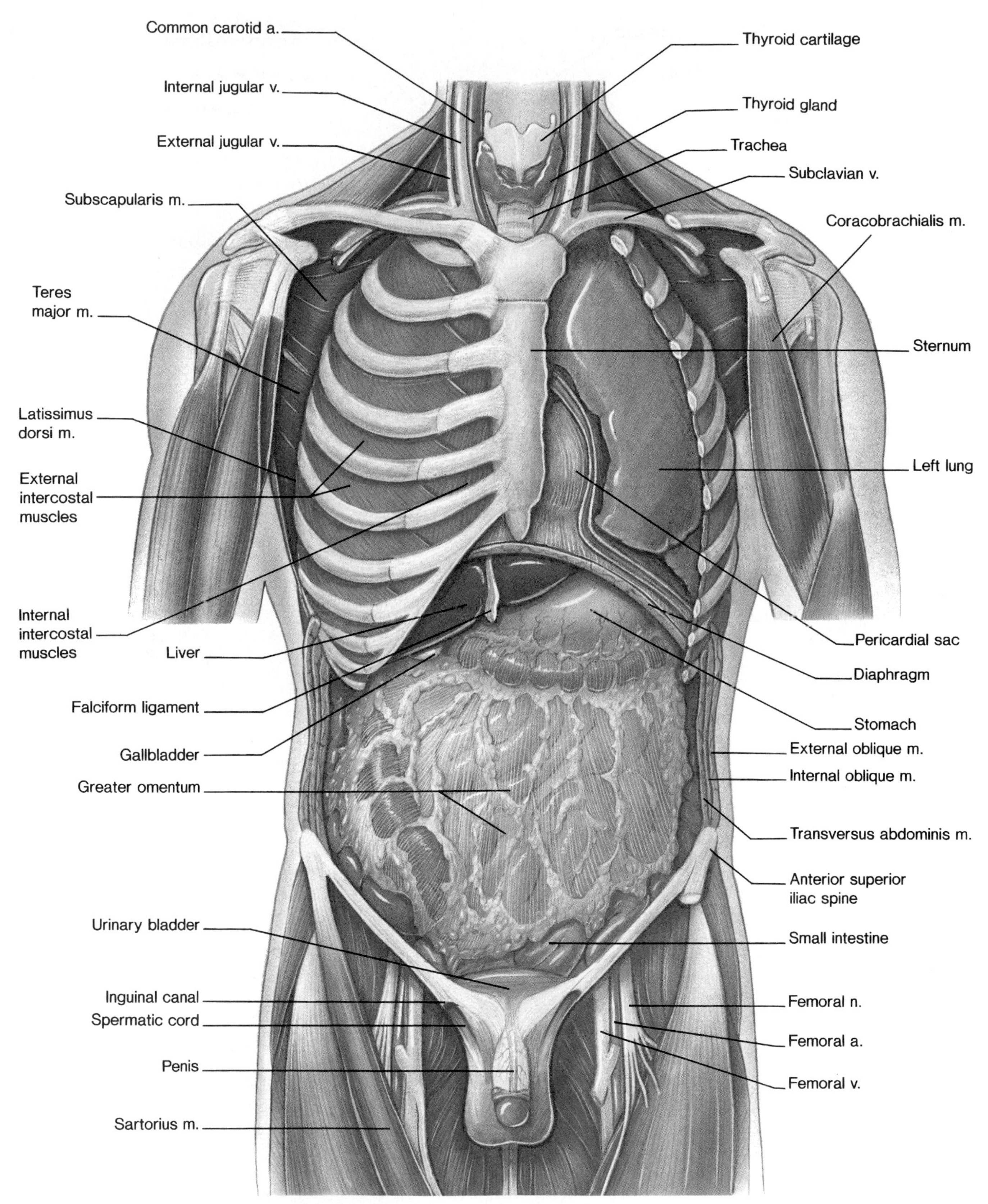


**PLATE THREE**

Human male torso, with the deep muscles removed and the abdominal viscera exposed.

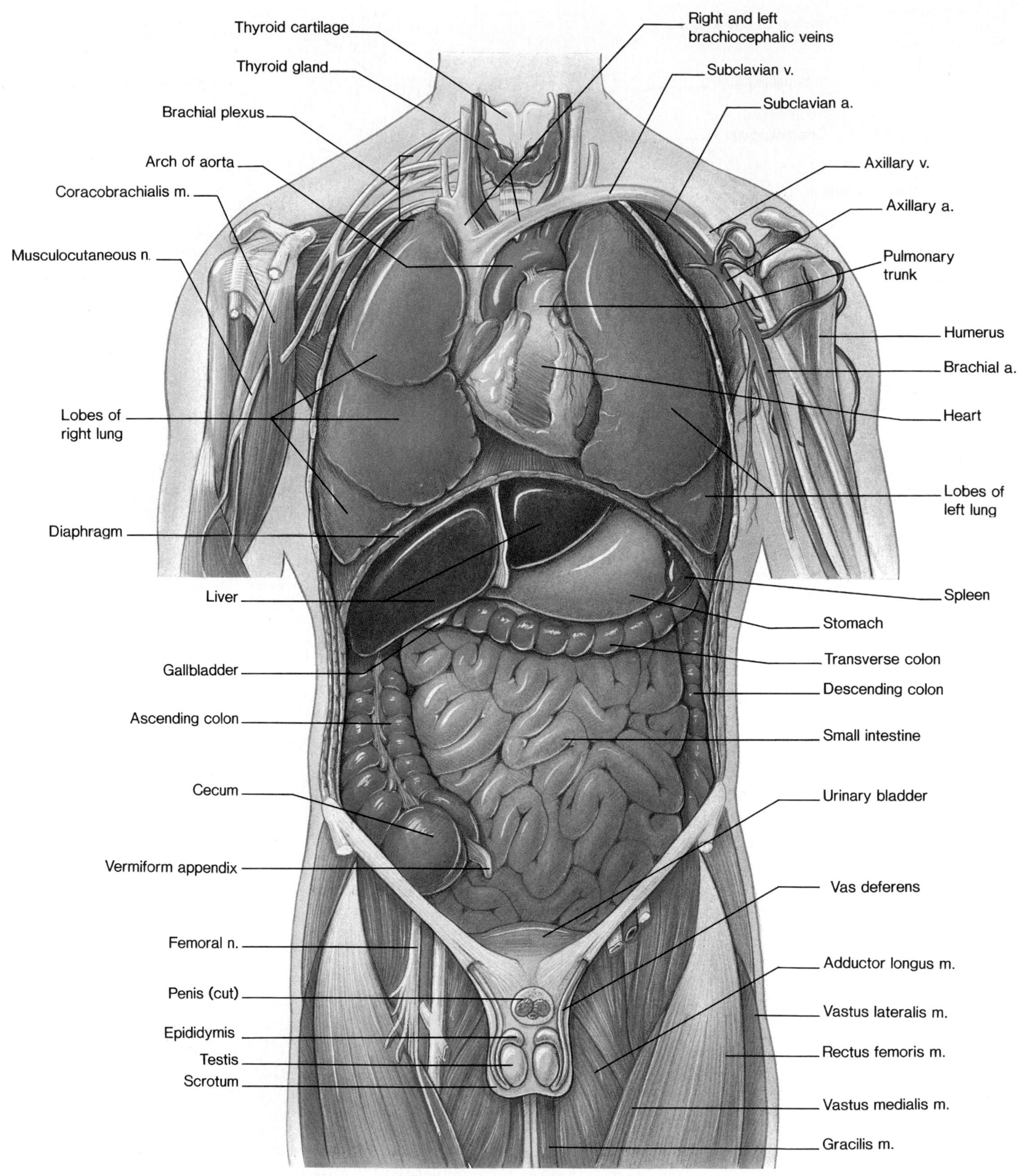


**PLATE FOUR**

Human male torso, with the thoracic and abdominal viscera exposed.

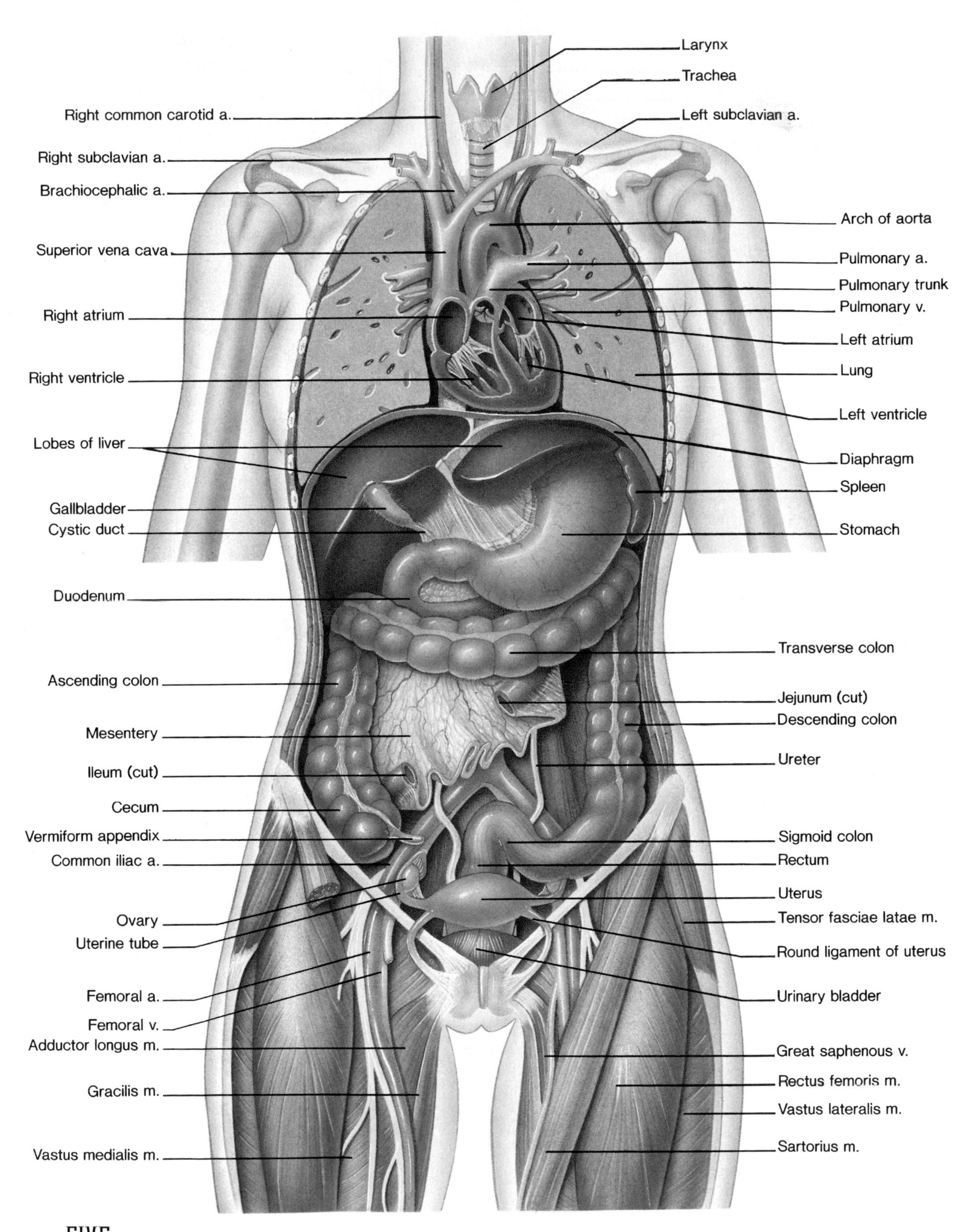


**PLATE FIVE**

Human female torso, with the lungs, heart, and small intestine sectioned and the liver reflected (lifted back).

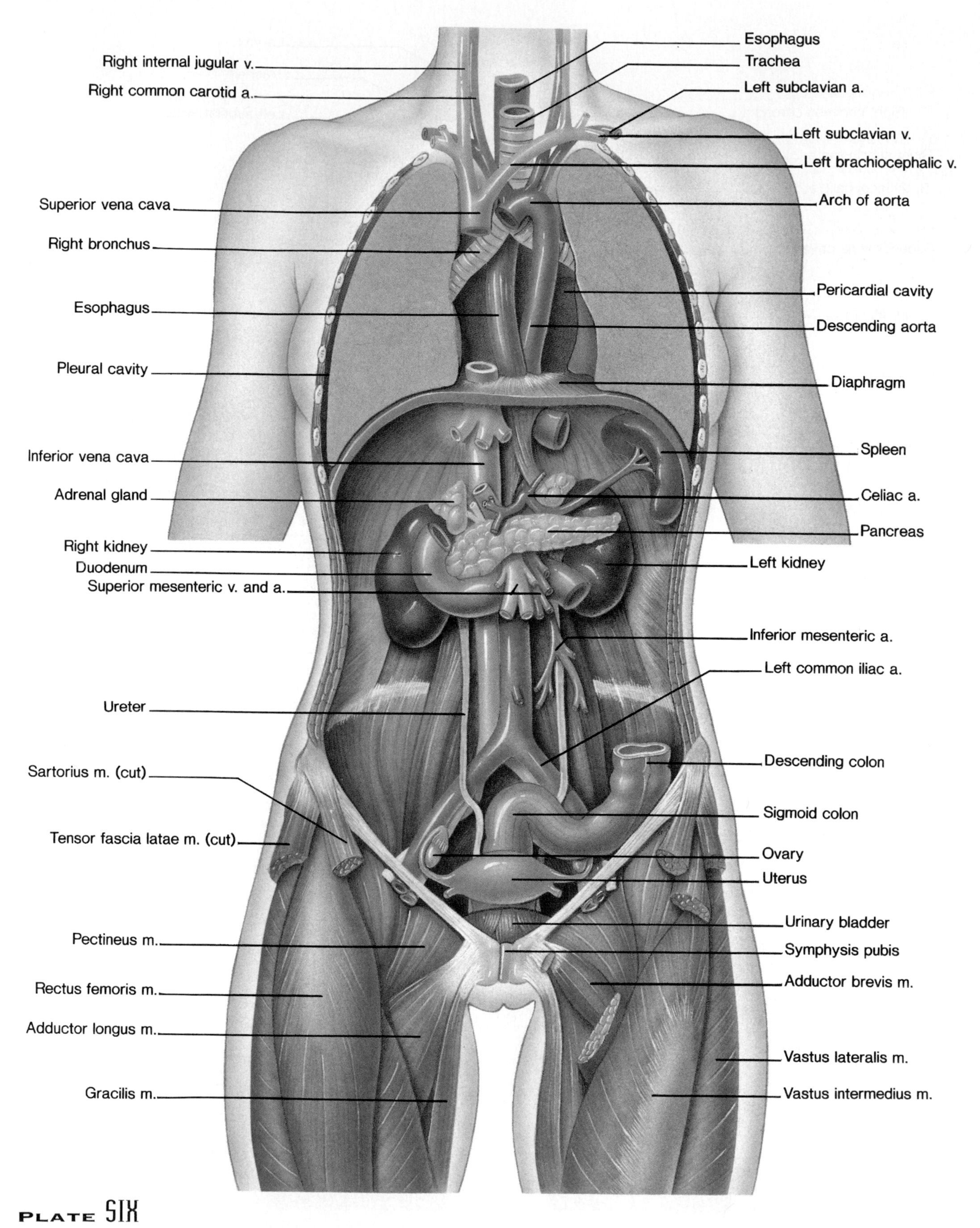


**PLATE SIX**

Human female torso, with the heart, stomach, liver, and parts of the intestine and lungs removed.

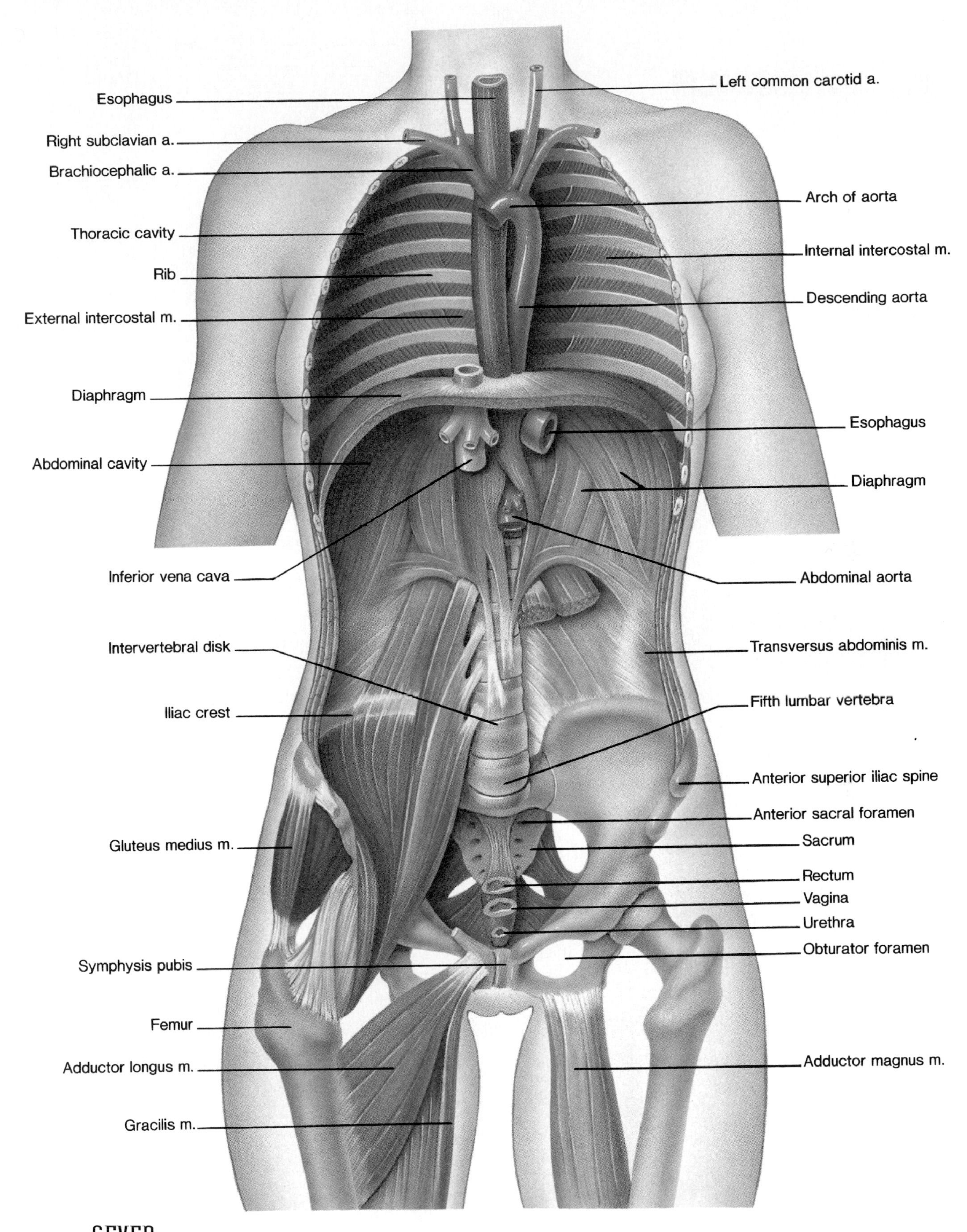


**PLATE SEVEN**
Human female torso, with the thoracic, abdominal, and pelvic viscera removed.

2

CHAPTER

# CHEMICAL BASIS OF LIFE

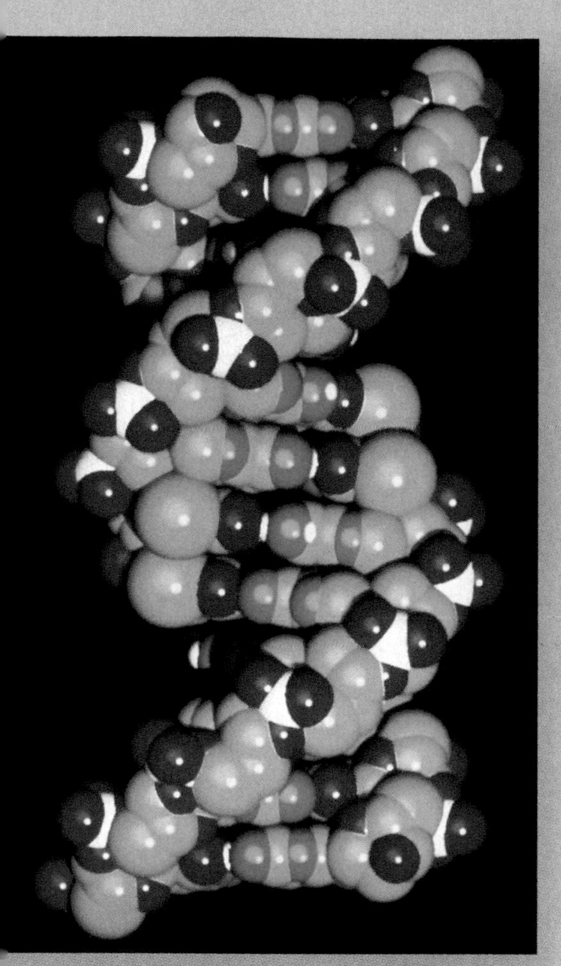

RECIPE FOR A HUMAN BEING: ABOUT 99% CARBON, HYDROGEN, NITROGEN, OXYGEN, PHOSPHORUS, AND SULFUR. THE REST INCLUDES SMALL AMOUNTS OF CALCIUM, POTASSIUM, SODIUM, CHLORIDE, AND MAGNESIUM, AND EVEN SMALLER AMOUNTS OF ZINC, IRON, MANGANESE, COPPER, IODINE, COBALT, FLUORIDE, CHROMIUM, AND SELENIUM. OF COURSE, THE ANATOMY AND PHYSIOLOGY OF A HUMAN IS NOT QUITE AS SIMPLE AS A LIST OF INGREDIENTS, BUT IF WE WERE SOMEHOW, TO BE REDUCED TO OUR CHEMICAL CONSTITUENTS, THIS IS WHAT THEY WOULD INCLUDE.

## CHAPTER OBJECTIVES

AFTER YOU HAVE STUDIED THIS CHAPTER, YOU SHOULD BE ABLE TO:

1. Explain how the study of living material is dependent on the study of chemistry.
2. Describe the relationships between matter, atoms, and molecules.
3. Discuss how atomic structure is related to the ways in which atoms interact.
4. Explain how molecular and structural formulas are used to symbolize the composition of compounds.
5. Describe three types of chemical reactions.
6. Discuss the concept of pH.
7. List the major groups of inorganic substances that are common in cells.
8. Describe the general roles played in cells by various types of organic substances.

## KEY TERMS

**atom** (at′om)
**carbohydrate** (kar″bo-hi′drāt)
**compound** (kom′-pownd)
**decomposition** (de″kom-po-zish′un)
**electrolyte** (e-lek′tro-līt)
**inorganic** (in″or-gan′ik)
**ion** (i′on)
**isotope** (i′so-top)
**lipid** (lip′id)
**molecule** (mol′ĕ-kūl)
**nucleic acid** (nu-kle′ik as′id)
**organic** (or-gan′ik)
**protein** (pro′te-in)
**synthesis** (sin′thĕ-sis)

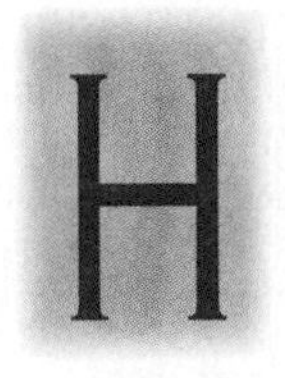

Human bodies are composed of chemicals. Much of our physiology is devoted to maintaining certain levels of particular chemicals in various parts of our anatomy. Consider the consequences of too much or too little of just one type of chemical the body needs in very small amounts—copper.

Ingrid W. was a happy and healthy high school senior when symptoms of Wilson disease, an inborn tendency to absorb too much copper from food, began. It started with stomachaches and headaches. Physicians noting her inflamed liver suspected hepatitis, but then more peculiar symptoms arose. Ingrid became unsteady, her handwriting changed, and her voice took on a slurred, low-pitched, gravelly quality. The young woman desperately sought help, only to face a slew of incorrect diagnoses—multiple sclerosis, Parkinson's disease, schizophrenia. Finally, a psychiatrist noticed greenish rings around her irises, a telltale sign that copper metabolism was awry.

Fortunately, Ingrid could be treated. A drug, penicillamine, enabled her body to rid itself of the poisonous buildup of copper. The drug's action was vividly obvious when Ingrid's urine turned the color of bright new pennies! Unfortunately, penicillamine could only halt further symptoms from developing; much damage had already been done.

Today, the once vivacious student lives in the geriatric ward of a state mental hospital, unable to talk or walk. Although her lopsided grin and drooling make Ingrid seem mentally deficient, she is alert, and communicates using a computer. Without penicillamine, she would not be alive.

Lack of copper began to visibly affect Wayne D. far earlier than the copper excess caused Ingrid's first symptoms. Wayne appeared normal at birth, but within the first few weeks of life he began to display seemingly unrelated symptoms. He had yellowish skin, persistently low body temperature, and extreme lethargy. Most striking was his hair, which rapidly changed from normal baby fuzz to an odd white kinky stubble. When he was a month old, he had the first of many seizures.

Wayne had Menkes disease, also known as "kinky-hair syndrome." Although his small intestine cells could absorb copper from food, they couldn't release it into his bloodstream. Despite a healthy diet, Wayne was literally starving for copper.

Lack of copper explained the boy's symptoms. Without copper to align the protein rods in hair, the hair falls apart. Wayne's lethargy, seizures, and eventual mental retardation were caused by impaired functioning of messenger chemicals (neurotransmitters) in his brain, which require copper. As Wayne grew, his bones bent inward as if he had scurvy, a vitamin C deficiency disease, because an enzyme needed to utilize vitamin C also depends on copper. The lack of copper continued to slow his growth and impair various functions until he died before the age of 10 years.

Biochemical imbalances lie behind many medical conditions. Too little iron causes anemia; too much iron causes hemochromatosis, which can impair organ function. Too little of a blood clotting factor results in a bleeding disorder; excess can dangerously impede circulation. In multiple sclerosis, nerves are stripped of their fatty coats, disrupting transmission of messages to muscle; in Tay-Sachs disease, nerves are obliterated in excessive fatty insulation, effectively shutting down the nervous system. In the hundreds of disorders called "inborn errors of metabolism," a single type of missing or abnormal enzyme (a type of protein) can devastate the human body.

Because chemicals constitute our bodies and direct virtually all the processes and events that make us alive, the study of anatomy and physiology must begin with chemistry.

---

Chemistry considers the composition of substances and how they change. Although it is possible to study anatomy without much reference to chemistry, it is essential for understanding physiology, because body functions depend on cellular functions that result from chemical changes.

As interest in the chemistry of living organisms grew, and knowledge of the subject expanded, a subdivision of science called biological chemistry, or **biochemistry,** emerged. Biochemistry has been important not only in helping explain physiological processes but also in developing many new drugs and methods for treating diseases.

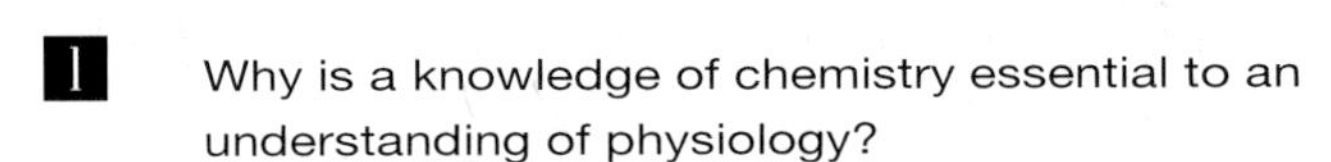

**1** Why is a knowledge of chemistry essential to an understanding of physiology?

**2** What is biochemistry?

## Structure of Matter

**Matter** is anything that has weight and takes up space. This includes all the solids, liquids, and gases in our surroundings as well as in our bodies. All matter is in the form of particles that are organized in specific ways. Chart 2.1 lists some particles of matter and their characteristics.

**CHART 2.1 Some Particles of Matter**

| Name | Characteristic | Name | Characteristic |
|---|---|---|---|
| Atom | Smallest particle of an element that has the properties of that element | Neutron ($n^0$) | Particle with about the same weight as a proton; uncharged and thus electrically neutral; found within a nucleus |
| Electron ($e^-$) | Extremely small particle with almost no weight; carries a negative electrical charge and is in constant motion around an atomic nucleus | Ion | Particle that is electrically charged because it has gained or lost one or more electrons |
| Proton ($p^+$) | Relatively large atomic particle; carries a positive electrical charge and is found within a nucleus | Molecule | Particle formed by the chemical union of two or more atoms |

## Elements and Atoms

All matter is composed of fundamental substances called **elements.** As of early 1995, 111 such elements are known, although naturally occurring matter on earth includes only 92 of them. Among these elements are such common materials as iron, copper, silver, gold, aluminum, carbon, hydrogen, and oxygen. Although some elements exist in a pure form, they occur more frequently in chemical combinations called compounds.

Elements needed in large amounts—such as carbon, hydrogen, oxygen, nitrogen, sulfur, and phosphorus—are termed **bulk elements.** These elements make up more than 95% (by weight) of the human body (chart 2.2). Those required in small amounts are called **trace elements.** Many trace elements are important parts of enzymes, usually proteins, that regulate the rates of chemical reactions in living things. Some elements that are toxic in large amounts, such as arsenic, may actually be vital in very small amounts, and these are called **ultratrace elements.** Note that copper, which causes such extreme symptoms when in excess in Wilson disease and when deficient in Menkes disease, is a trace element.

Elements are composed of tiny particles called **atoms,** which are the smallest complete units of the elements. Although the atoms that make up each element are chemically identical to one another, they differ from the atoms that make up other elements. Atoms vary in size, weight, and the way they interact with one another. Some, for instance, can combine either with atoms like themselves or with other kinds of atoms; other types of atoms lack this ability.

## Atomic Structure

An atom consists of a central portion called the **nucleus** and one or more **electrons** that are in constant motion around the nucleus. The nucleus contains one or more relatively large particles called **protons** and **neutrons,** whose weights are about equal, but which are otherwise quite different (fig. 2.1).

Electrons, which are extremely small and have almost no weight, carry a single, negative electrical charge ($e^-$). Protons each carry a single, positive electrical charge ($p^+$). Neutrons are uncharged and thus are electrically neutral ($n^0$).

Because the nucleus contains the protons, this part of an atom is always positively charged. However, the number of electrons outside the nucleus equals the number of protons, so a complete atom is electrically uncharged or *neutral.*

The atoms of different elements contain different numbers of protons. The number of protons in the atoms of a particular element is called its **atomic number.** Hydrogen, for example, whose atoms contain one proton, has the atomic number 1; carbon, whose atoms have six protons, has the atomic number 6.

The weight of an atom of an element is due primarily to the protons and neutrons in its nucleus, because the electrons have so little weight. For this reason, an atom of carbon with six protons and six neutrons weighs about twelve times as much as an atom of hydrogen, which has only one proton and no neutrons.

The number of protons plus the number of neutrons in each of its atoms is essentially equal to the **atomic weight** of an element. Thus, the atomic weight of hydrogen is approximately 1, and the atomic weight of carbon is approximately 12 (chart 2.3).

## Isotopes

All the atoms of a particular element have the same atomic number because they have the same number of protons and electrons. However, the atoms of an element vary in the number of neutrons in their nuclei; thus, they vary in atomic weight. For example, all oxygen atoms have eight protons in their nuclei. Some, however, have eight neutrons (atomic weight 16), others have nine neutrons (atomic weight 17), and still others have ten neutrons (atomic weight 18). Atoms that have the same atomic numbers but different atomic weights

## CHART 2.2 Major Elements in the Human Body (by Weight)

| Major Elements | Symbol | Approximate Percentage of the Human Body |
|---|---|---|
| Oxygen | O | 65.0 |
| Carbon | C | 18.5 |
| Hydrogen | H | 9.5 |
| Nitrogen | N | 3.2 |
| Calcium | Ca | 1.5 |
| Phosphorus | P | 1.0 |
| Potassium | K | 0.4 |
| Sulfur | S | 0.3 |
| Chlorine | Cl | 0.2 |
| Sodium | Na | 0.2 |
| Magnesium | Mg | 0.1 |
| | Total | 99.9% |
| **Trace Elements** | | |
| Cobalt | Co | |
| Copper | Cu | |
| Fluorine | F | Together |
| Iodine | I | less than 0.1% |
| Iron | Fe | |
| Manganese | Mn | |
| Zinc | Zn | |

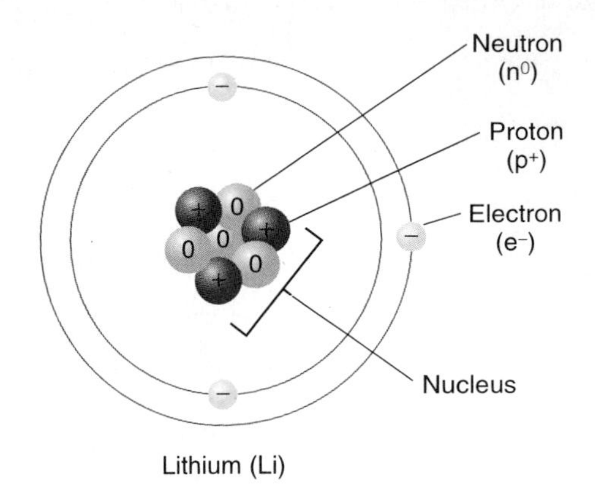


**FIGURE 2.1**
This simplified representation of an atom of lithium includes three electrons in motion around a nucleus that contains three protons and four neutrons. Circles depict electron shells.

## CHART 2.3 Atomic Structure of Elements 1 Through 12 (See Also Appendix A)

| Element | Symbol | Atomic Number | Atomic Weight | Protons | Neutrons | Electrons in Shells: First | Second | Third |
|---|---|---|---|---|---|---|---|---|
| Hydrogen | H | 1 | 1 | 1 | 0 | 1 | | |
| Helium | He | 2 | 4 | 2 | 2 | 2 | (inert) | |
| Lithium | Li | 3 | 7 | 3 | 4 | 2 | 1 | |
| Beryllium | Be | 4 | 9 | 4 | 5 | 2 | 2 | |
| Boron | B | 5 | 11 | 5 | 6 | 2 | 3 | |
| Carbon | C | 6 | 12 | 6 | 6 | 2 | 4 | |
| Nitrogen | N | 7 | 14 | 7 | 7 | 2 | 5 | |
| Oxygen | O | 8 | 16 | 8 | 8 | 2 | 6 | |
| Fluorine | F | 9 | 19 | 9 | 10 | 2 | 7 | |
| Neon | Ne | 10 | 20 | 10 | 10 | 2 | 8 | (inert) |
| Sodium | Na | 11 | 23 | 11 | 12 | 2 | 8 | 1 |
| Magnesium | Mg | 12 | 24 | 12 | 12 | 2 | 8 | 2 |

are called **isotopes** of an element. Because a sample of an element is likely to include more than one isotope, the atomic weight of the element is often presented as the average weight of the isotopes present.

The ways atoms interact with one another are due largely to the number of electrons they possess. Because the number of electrons in an atom is equal to its number of protons, all the isotopes of a particular element have the same number of electrons and react chemically in the same manner. Therefore, for example, any of the isotopes of oxygen can play the same role in the metabolic reactions of an organism.

Although some of the isotopes of an element may be stable, others may have unstable atomic nuclei that decompose, releasing energy or pieces of themselves. Such unstable isotopes are called *radioactive,* and the energy or atomic fragments they give off are called *atomic radiation.* Examples of elements that have radioactive isotopes include oxygen, iodine, iron, phosphorus, and cobalt. Some radioactive isotopes are used to detect and treat disease (Clinical Application 2.1).

Atomic radiation includes three common forms called alpha ($\alpha$), beta ($\beta$), and gamma ($\gamma$). Alpha radiation consists of particles from atomic nuclei, each of

**Clinical Application**

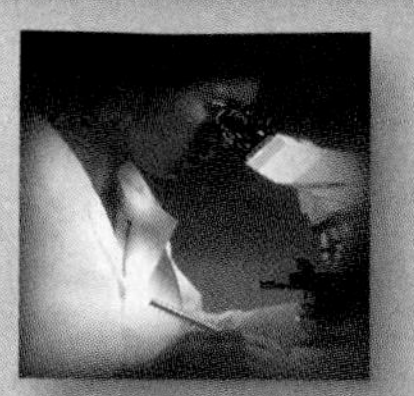

## Radioactive Isotopes Provide Windows on Physiology

Vicki L. arrived early at the nuclear medicine department of the health center. As she sat in an isolated cubicle, a doctor in full sterile dress approached with a small metal canister marked with numerous warnings. The doctor carefully unscrewed the top, inserted a straw, and watched as the young woman sipped the fluid within. It tasted like stale water, but was actually a solution containing a radioactive isotope, iodine-131.

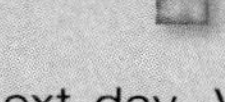

Vicki's thyroid gland had been removed three months earlier, and this test was to determine whether any active thyroid tissue remained. The thyroid is the only part of the body to metabolize iodine, so if Vicki's body retained any of the radioactive drink, it would mean that some of her cancerous thyroid gland remained. By using a radioactive isotope, her physicians could detect iodine uptake using a scanning device called a scintillation counter (fig. 2A). Figure 2B illustrates iodine-131 uptake in a complete thyroid gland.

The next day, Vicki returned for the scan, which showed that a small amount of thyroid tissue was indeed left and was functioning. This meant another treatment would be necessary. Vicki would drink more of the radioactive iodine, enough to destroy the remaining tissue. This time Vicki had her stale drink in an isolation room, which was lined with paper to keep her from contaminating the floor, walls, and furniture. The same physician administered the radioactive iodine. Vicki's physician had this job

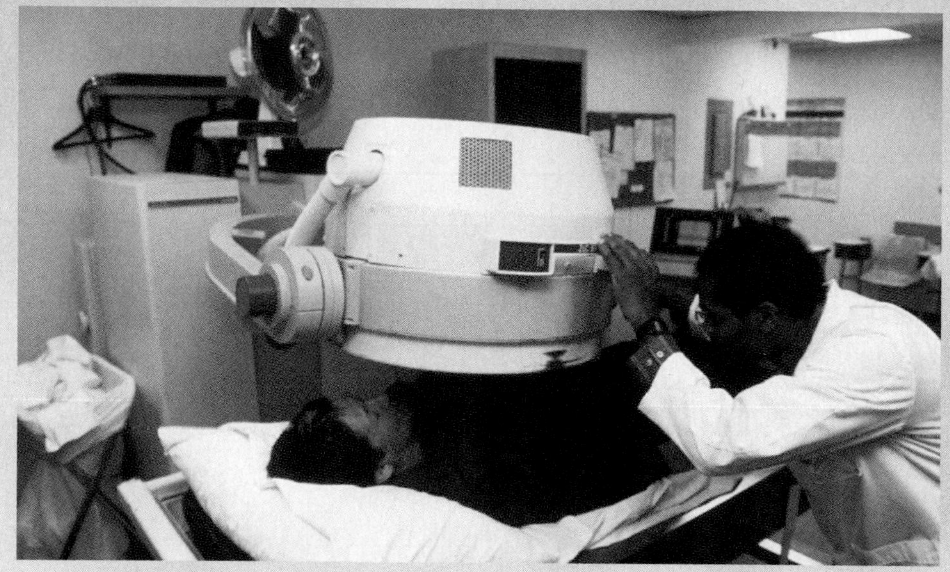

**Figure 2A**
Physicians use scintillation counters such as this to detect radioactive isotopes.

which includes two protons and two neutrons, that travel relatively slowly and have weak ability to penetrate matter. Beta radiation consists of much smaller particles (electrons) that travel more rapidly and penetrate matter more deeply. Gamma radiation is similar to X-ray radiation and is the most penetrating of these forms.

Each kind of radioactive isotope produces one or more of these forms of radiation with particular energy levels.

1. What is the relationship between matter and elements?
2. What elements are most common in the human body?
3. How are electrons, protons, and neutrons positioned within an atom?
4. What is an isotope?
5. What is atomic radiation?

### Bonding of Atoms

Atoms may combine with other atoms by forming **bonds.** When atoms form such bonds, they gain or lose electrons, or share electrons with other atoms.

because his own thyroid had been removed many years earlier, and therefore the radiation couldn't harm him.

After two days in isolation, Vicki was sent home with a list of rather odd directions. She was to stay away from her children and pets, wash her clothing separately, use disposable utensils and plates, and flush the toilet three times each time she used it. These precautions would minimize her contaminating her family—mom was radioactive!

Iodine-131 is a medically useful radioactive isotope because it has a short *half-life,* a measurement of the time it takes for half of an amount of an isotope to decay to a nonradioactive form. The half-life of iodine-131 is 8.1 days. With the amount of radiation in Vicki's body dissipating by half every 8.1 days, after three months there would be hardly any left. Doctors hoped that along with the radioactive iodine would go the remaining unhealthy thyroid cells.

Isotopes of other elements have different half-lives. The half-life of iron-59 is 45.1 days; that of phosphorus-32 is 14.3 days; that of cobalt-60 is 5.26 years; and that of radium-226 is 1,620 years.

A form of thallium-201 with a half-life of 73.5 hours is commonly used to detect disorders in the blood vessels supplying the heart muscle or to locate regions of damaged heart tissue after a heart attack. Gallium-67, with a half-life of 78 hours, is used to detect and monitor the progress of certain cancers and inflammatory illnesses. These medical procedures involve injecting the isotope into the blood and following its path using detectors that record images on paper or film.

Radioactive isotopes are also used to assess kidney function, estimate the concentrations of hormones in body fluids, measure blood volume and red blood cell mass, and study changes in bone density. Cobalt-60 is a radioactive isotope used to treat some cancers. The cobalt emits radiation that damages cancer cells more readily than it does healthy cells.

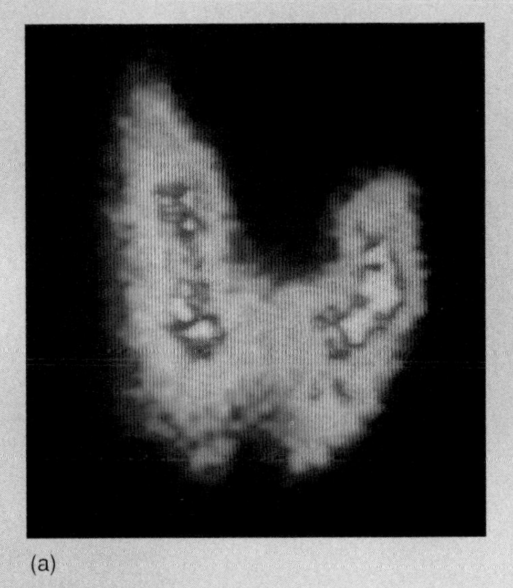
(a)

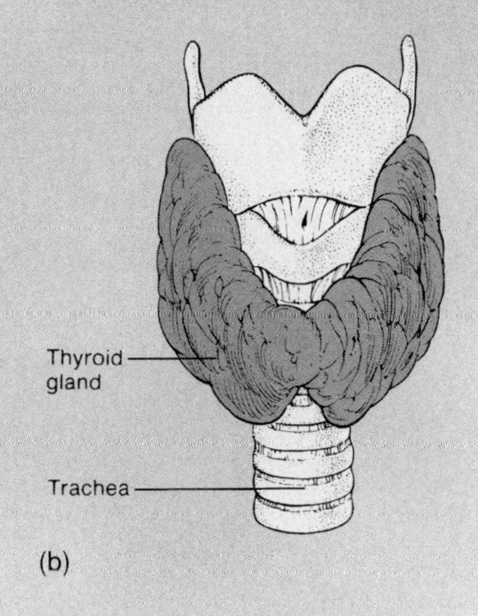

(b)

**FIGURE 2B**

(*a*) A scan of the thyroid gland twenty-four hours after the patient receives radioactive iodine. Note how closely the scan in (*a*) resembles the shape of the thyroid gland as depicted in (*b*).

The electrons of an atom are arranged in one or more *shells* around the nucleus. The maximum number of electrons that each of the first three inner shells can hold for elements of atomic number of 18 and under is as follows:

| | |
|---|---|
| First shell (closest to the nucleus) | 2 electrons |
| Second shell | 8 electrons |
| Third shell | 8 electrons |

More complex atoms may have as many as eighteen electrons in the third shell.

Simplified diagrams such as those in figure 2.2 are used to show electron configuration in atoms. Notice that the single electron of a hydrogen atom is located in the first shell, the two electrons of a helium atom fill its first shell, and the three electrons of a lithium atom are arranged with two in the first shell and one in the second shell.

Atoms such as helium, whose outermost electron shells are filled, have stable structures and are chemically inactive or inert (they cannot form chemical bonds). Atoms with incompletely filled outer shells, such as those of hydrogen or lithium, tend to gain, lose, or share electrons in ways that empty or fill their outer shells. In this way they achieve stable structures.

An atom of sodium, for example, has eleven electrons: two in the first shell, eight in the second shell,

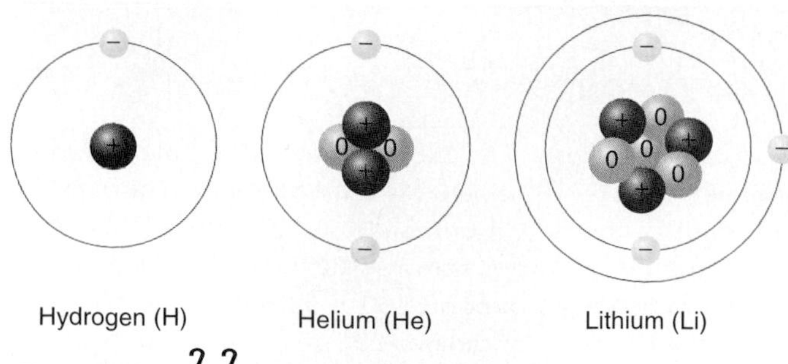


**FIGURE 2.2**

The single electron of a hydrogen atom is located in its first shell. The two electrons of a helium atom fill its first shell. The three electrons of a lithium atom are arranged with two in the first shell and one in the second shell.

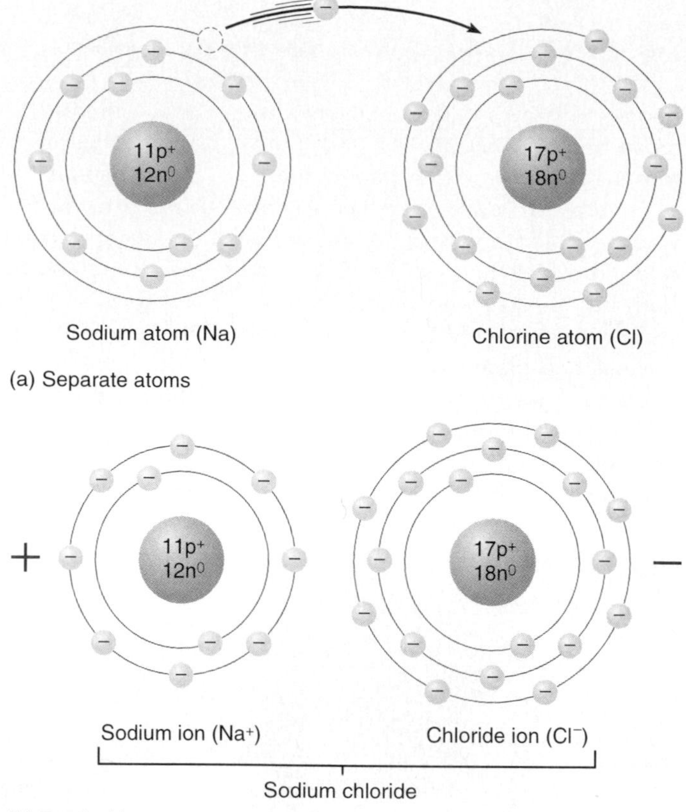


**FIGURE 2.3**

(*a*) If a sodium atom loses an electron to a chlorine atom, the sodium atom becomes a sodium ion and the chlorine atom becomes a chloride ion. (*b*) These oppositely charged particles attract electrically and join by an ionic bond.

and one in the third shell. This atom tends to lose the single electron from its outer shell, which leaves the second shell filled and the form stable (fig. 2.3*a*).

A chlorine atom has seventeen electrons arranged with two in the first shell, eight in the second shell, and seven in the third shell. An atom of this type will tend to accept a single electron, thus filling its outer shell and achieving a stable form.

Because each sodium atom tends to lose a single electron and each chlorine atom tends to accept a single electron, sodium and chlorine atoms will react together. During this reaction, a sodium atom loses an electron and is left with eleven protons ($11^+$) in its nucleus and only ten electrons ($10^-$). As a result, the atom develops a net electrical charge of $1^+$ and is symbolized $Na^+$. At the same time, a chlorine atom gains an electron, which leaves it with seventeen protons ($17^+$) in its nucleus and eighteen electrons ($18^-$). Thus, it develops a net electrical charge of $1^-$ and is symbolized $Cl^-$.

Atoms that have become electrically charged by gaining or losing electrons are called **ions,** and ions with opposite electrical charges attract one another. When this happens, a chemical bond called an **ionic bond** (electrovalent bond) forms between them. Sodium ions ($Na^+$) and *chloride* ions ($Cl^-$) uniting in this manner form the compound sodium chloride (NaCl), or table salt (fig. 2.3*b*).

Similarly, a hydrogen atom may lose its single electron and become a hydrogen ion ($H^+$). Such an ion can bond with a chloride ion ($Cl^-$) to form hydrogen chloride (HCl, hydrochloric acid).

Atoms may also bond together by sharing electrons rather than by gaining or losing them. A hydrogen atom, for example, has one electron in its first shell, but needs two to achieve a stable structure. It may fill this shell by combining with another hydrogen atom in such a way that the two atoms share a pair of electrons. As figure 2.4 shows, the two electrons then encircle the nuclei of both atoms, and each atom achieves a stable form. In this case the chemical bond between the atoms is called a **covalent bond.** One pair of electrons shared is a *single covalent bond;* two pairs of electrons shared is a *double covalent bond.*

Another type of chemical bond, called a *hydrogen bond,* is described later in this chapter. Clinical Application 2.2 examines how radiation that moves electrons can affect human health.

## Molecules and Compounds

When two or more atoms bond together, they form a new kind of particle called a **molecule.** The numbers and kinds of atoms in a molecule can be represented by a **molecular formula.** Such a formula consists of the symbols of the elements in the molecule together with numbers to indicate how many atoms of each element are present. For example, the molecular formula for water is $H_2O$, which means there are two

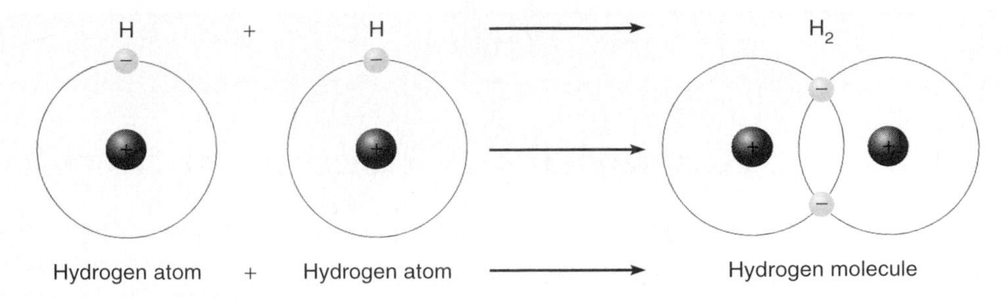


**FIGURE 2.4**
A hydrogen molecule forms when two hydrogen atoms share a pair of electrons and join by a covalent bond.

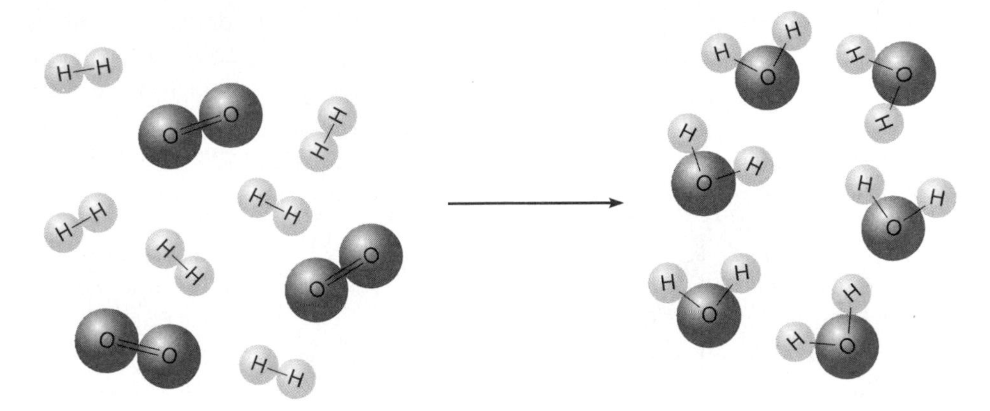

**FIGURE 2.5**
Under certain conditions, hydrogen molecules can combine with oxygen molecules to form water molecules.

atoms of hydrogen and one atom of oxygen in each molecule. The molecular formula for the sugar glucose is $C_6H_{12}O_6$, which means there are six atoms of carbon, twelve atoms of hydrogen, and six atoms of oxygen in a glucose molecule.

If atoms of the same element combine, they produce molecules of that element. Gases of hydrogen ($H_2$), oxygen ($O_2$), and nitrogen ($N_2$) consist of such molecules. If atoms of different elements combine, molecules of substances called **compounds** form. Two atoms of hydrogen, for example, can combine with one atom of oxygen to produce a molecule of the compound water ($H_2O$), as shown in figure 2.5. Table sugar, baking soda, natural gas, beverage alcohol, and most medical drugs are examples of compounds.

A molecule of a compound always contains definite kinds and numbers of atoms. A molecule of water ($H_2O$), for instance, always contains two hydrogen atoms and one oxygen atom. If two hydrogen atoms combine with two oxygen atoms, the compound formed is not water, but hydrogen peroxide ($H_2O_2$).

1. What is an ion?
2. Describe two ways that atoms may combine with other atoms.
3. Distinguish between a molecule and a compound.

Usually the atoms of each element form a specific number of chemical bonds. Hydrogen atoms form single bonds, oxygen atoms form two bonds, nitrogen atoms form three bonds, and carbon atoms form four bonds. The bonding capacities of these atoms can be represented by using symbols and lines as follows:

**—H** **—O—**

**—N—** (with a line below) **—C—** (with lines above and below)

These representations can be used to show how atoms bond and arrange in various molecules. Single lines represent single bonds and double lines represent double bonds. Illustrations of this type are called **structural formulas** (fig. 2.6).

## Chemical Reactions

When chemical reactions occur, bonds between atoms, ions, or molecules form or break, producing new combinations. For example, when two or more atoms (reactants) bond to form a more complex structure (end product), as when hydrogen and oxygen atoms bond to form molecules of water, the reaction is called **synthesis.** Such a reaction can be symbolized this way:

$$A + B \rightarrow AB$$

**CLINICAL APPLICATION**

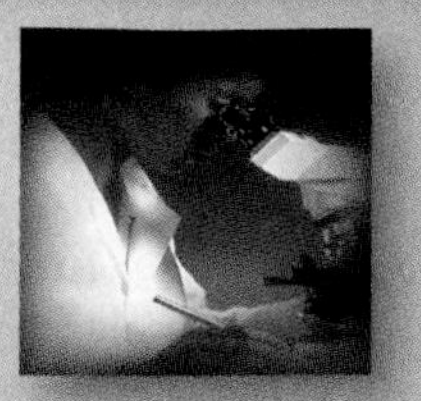

# Ionizing Radiation: A Legacy of the Cold War

Alpha, beta, and gamma radiation are called ionizing radiation because their energy adds or removes electrons from atoms (fig. 2C). Electrons dislodged by ionizing radiation can affect nearby atoms, disrupting physiology at the chemical level in a variety of ways—causing cancer, clouding the eye lens, and interfering with normal growth and development.

In the United States, most people are exposed to very low levels of ionizing radiation, mostly from background radiation, which originates from natural environmental sources (see chart 2A). This is not true, however, for people unfortunate enough to live near the sites of atomic weapons manufacture. Epidemiologists are now studying recently uncovered medical records that document illnesses linked to long-term exposure to ionizing radiation in a 1,200 square kilometer area in former East Germany. It is a frightening tale.

From a distance, the lake near Oberrothenback, Germany, appears inviting, but looks are deceiving. The lake contains enough toxins to kill thousands of people, its water polluted with heavy metals, low-level radioactive chemical waste, and 22,500 tons of arsenic. Radon, a radioactive by-product of uranium, permeates the soil. High death rates among farm animals and pets have been traced to drinking from the polluted lake. Cancer rates and respiratory disorders among the human residents nearby are far above normal. This isn't surprising, given the region's rather toxic history.

The deadly lake in Oberrothenback once served as a dump for a factory that produced "yellow cake," a term for processed uranium ore, which was used to build atomic bombs for the former Soviet Union. In the early 1950s, nearly half a million workers labored here and in surrounding areas in factories and mines. Records released in 1989, after the reunification of Germany, reveal that workers were given perks, such as alcoholic beverages and better wages, to work in the more dangerous areas. They paid a heavy price: tens of thousands of them died of lung ailments.

Today, the meticulously kept health records may help answer a long-standing question: What are the effects of exposure to long-term, low-level ionizing radiation? Until now, the risks of such exposure have been extrapolated from health statistics amassed for the victims, survivors, and descendants of the atomic blasts in Japan in the Second World War. But a single exposure, such as a bomb blast, may not have the same effect on the human body as extended exposure, such as the uranium workers experienced. The cold war may be over, but a lethal legacy of its weapons remains.

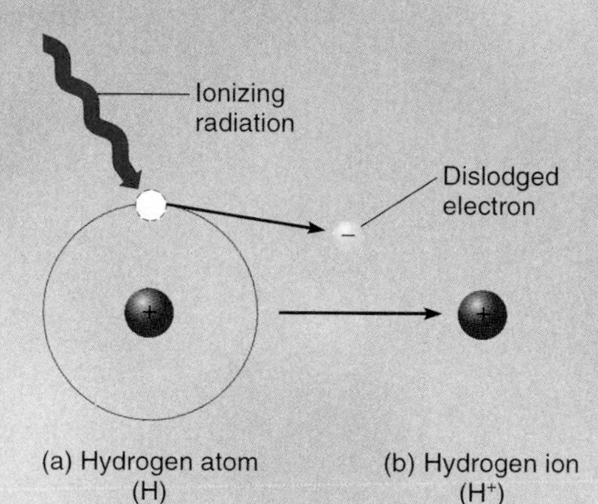


**FIGURE 2C**
(*a*) Ionizing radiation may dislodge an electron from an electrically neutral hydrogen atom; (*b*) without its electron, the hydrogen atom becomes a positively charged hydrogen ion ($H^+$).

**CHART 2A SOURCES OF IONIZING RADIATION**

| | |
|---|---|
| Background (Natural Environmental) | Cosmic rays from space<br>Radioactive elements in earth's crust<br>Rocks and clay in building materials<br>Radioactive elements naturally in the body (phosphorus-40, carbon-14) |
| Medical and Dental | X rays<br>Radioactive substances |
| Other | Atomic and nuclear weapons<br>Mining and processing radioactive minerals<br>Radioactive fuels in nuclear power plants<br>Radioactive elements in consumer products (luminescent dials, smoke detectors, color TV components) |

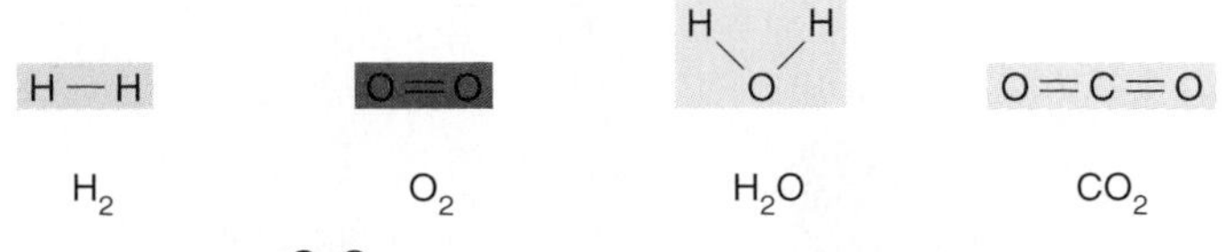


**FIGURE 2.6**
Structural formulas of molecules of hydrogen, oxygen, water, and carbon dioxide. Note the double covalent bonds.

If the bonds within a reactant molecule break so that simpler molecules, atoms, or ions form, the reaction is called **decomposition.** Thus, molecules of water can decompose to yield the products hydrogen and oxygen. Decomposition is symbolized as follows:

$$AB \rightarrow A + B$$

Synthetic reactions are particularly important in growth of body parts and repair of worn or damaged tissues, which involve the buildup of larger molecules from smaller ones. On the other hand, decomposition reactions occur when food substances are digested and energy is released from them.

A third type of chemical reaction is an **exchange reaction.** In this reaction, parts of two different kinds of molecules trade positions. The reaction is symbolized as follows:

$$AB + CD \rightarrow AD + CB$$

An example of an exchange reaction is when an acid reacts with a base, producing water and a salt. This type of reaction is discussed in the following section.

Many chemical reactions are reversible. This means the end product (or products) of the reaction can change back to the reactant (or reactants) that originally underwent the reaction. A **reversible reaction** can be symbolized using a double arrow, as follows:

$$A + B \rightleftarrows AB$$

Whether a reversible reaction proceeds in one direction or another depends on such factors as the relative proportions of reactant (or reactants) and end product (or end products) as well as the amount of energy available for the reaction. How quickly such a reaction proceeds may be affected by the presence or absence of **catalysts.** A catalyst is a particular atom or molecule that can change the rate of a reaction without itself being consumed by the reaction.

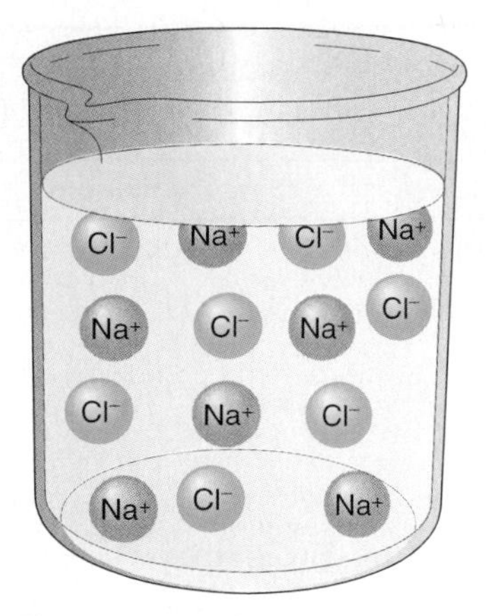


**FIGURE 2.7**
Crystals of table salt (NaCl) dissolve in water, releasing sodium ions ($Na^+$) and chloride ions ($Cl^-$).

## Acids, Bases, and Salts

Some compounds release ions (ionize) when they dissolve in water or react with water molecules. Sodium chloride (NaCl), for example, ionizes into sodium ions ($Na^+$) and chloride ions ($Cl^-$) when it dissolves (fig. 2.7). This reaction is represented by the following:

$$NaCl \rightarrow Na^+ + Cl^-$$

Because the resulting solution contains electrically charged particles (ions), it will conduct an electric current. Substances that release ions in water are, therefore, known as **electrolytes.** Electrolytes that release hydrogen ions ($H^+$) in water are called **acids.** For example, in water the compound hydrochloric acid (HCl) releases hydrogen ions ($H^+$) and chloride ions ($Cl^-$):

$$HCl \rightarrow H^+ + Cl^-$$

Electrolytes that release ions that combine with hydrogen ions are called **bases.** The compound sodium hydroxide (NaOH) in water releases hydroxyl ions ($OH^-$). The hydroxyl ions, in turn, can combine with hydrogen ions to form water. Thus, sodium hydroxide is a base:

$$NaOH \rightarrow Na^+ + OH^-$$

(Note: Some ions, such as $OH^-$, contain two or more atoms. However, such a group usually behaves like a single atom and remains unchanged during a chemical reaction.)

## CHART 2.4 TYPES OF ELECTROLYTES

| Characteristic | | Examples |
|---|---|---|
| Acid | Ionizes to release hydrogen ions ($H^+$). | Carbonic acid, hydrochloric acid, acetic acid, phosphoric acid |
| Base | Ionizes to release ions that can combine with hydrogen ions. | Sodium hydroxide, potassium hydroxide, magnesium hydroxide, aluminum hydroxide |
| Salt | Substance formed by the reaction between an acid and a base. | Sodium chloride, aluminum chloride, magnesium sulfate |

Acids and bases can react to form water and electrolytes called **salts.** For example, hydrochloric acid and sodium hydroxide react to form water and sodium chloride:

$$HCl + NaOH \rightarrow H_2O + NaCl$$

Chart 2.4 summarizes the three types of electrolytes.

## Acid and Base Concentrations

Concentrations of acids and bases affect the chemical reactions that constitute many life processes, such as those controlling breathing rate. Thus, the concentrations of these substances in body fluids are of special importance.

Hydrogen ion concentration can be measured in grams of ions per liter of solution. However, because hydrogen ion concentration can cover such a wide range (gastric juice is .01 grams $H^+$/liter; household ammonia has .00000000001 grams $H^+$/liter) a shorthand system called the **pH scale** has been developed. This system keeps track of the number of decimal places in a hydrogen ion concentration without having to write them out. Thus, for example, a solution with a hydrogen ion concentration of 0.1 grams per liter has a pH value of 1.0; a concentration of 0.01 g $H^+$/l has pH 2.0; 0.001 g $H^+$/l has pH 3.0, and so forth. Note that between each whole number on the pH scale, which extends from pH 0 to pH 14.0, there is a tenfold difference in hydrogen ion concentration. Also note that as the hydrogen ion concentration increases, the pH value decreases.

In pure water, which ionizes only slightly, the hydrogen ion concentration is 0.0000001 g/l, and the pH is 7.0. Because water ionizes to release equal numbers of acidic hydrogen ions and basic hydroxyl ions, it is said to be *neutral.*

$$H_2O \rightarrow H^+ + OH^-$$

Solutions with more hydrogen ions than hydroxyl ions are said to be *acidic,* that is, they have pH values of less than 7.0 (fig. 2.8). Solutions with fewer hydrogen ions than hydroxyl ions are said to be *basic* (alkaline); that is, they have pH values of more than 7.0.

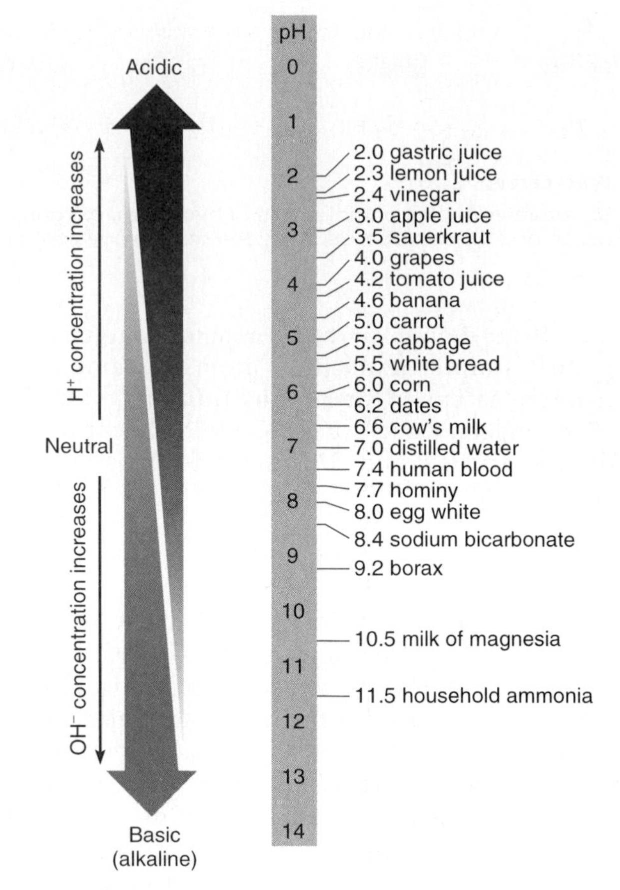


FIGURE 2.8

As the concentration of hydrogen ions ($H^+$) increases, a solution becomes more acidic, and the pH value decreases. As the concentration of hydrogen ion acceptors (such as hydroxyl or bicarbonate ions) increases, a solution becomes more basic, and the pH value increases. Note the pH of some common substances.

Chart 2.5 summarizes the relationship between hydrogen ion concentration and pH. The regulation of hydrogen ion concentrations in the internal environment is discussed in chapter 21.

Many fluids in the human body function within a narrow pH range. Illness results when pH changes. The normal pH of blood, for example, is 7.35 to 7.45. Blood pH of 7.5 to 7.8, called **alkalosis,** makes one feel agitated and dizzy. This can be caused by breathing rapidly at high altitudes, taking too many antacids, high fever, or anxiety. **Acidosis,** in which blood pH falls to 7.0 to 7.3, makes one feel disoriented and fatigued, and breathing may become difficult. This condition can result from severe vomiting, diabetes, brain damage, impaired breathing, and lung and kidney disease.

1. What is a molecular formula? A structural formula?
2. Describe three kinds of chemical reactions.
3. Compare the characteristics of an acid with those of a base.
4. What is pH?

## Chemical Constituents of Cells

The chemicals that enter into metabolic reactions or are produced by them can be divided into two large groups. Those that contain both carbon and hydrogen atoms are called **organic;** the rest are called **inorganic.**

Generally, inorganic substances dissolve in water or react with water to release ions; thus they are *electrolytes.* Some organic compounds dissolve in water also, but as a group they are more likely to dissolve in organic liquids like ether or alcohol. Organic compounds that dissolve in water usually do not release ions, and are therefore called *nonelectrolytes.*

### Inorganic Substances

Among the inorganic substances common in cells are water, oxygen, carbon dioxide, and a group called inorganic salts.

#### Water

Water ($H_2O$) is the most abundant compound in living material and is responsible for about two-thirds of the weight of an adult human. It is the major component of blood and other body fluids, including those within cells.

When substances dissolve in water, relatively large pieces of the substances break into smaller ones, and eventually, molecular-sized particles result. These tiny particles, which may be ions, are much more likely to react with one another than were the original large pieces. Consequently, most metabolic reactions occur in water.

Water also plays an important role in transportation of chemicals within the body. Blood, which is more than 90% water, carries many vital substances, such as oxygen, sugars, salts, and vitamins, from the organs of the digestive and respiratory systems to the cells. Water also carries waste materials, such as carbon dioxide and urea, from these cells to the lungs and kidneys, respectively, which remove them from the blood and release them to the outside of the body.

In addition, water can absorb and transport heat. Thus, blood carries the heat released from muscle cells during exercise from deeper parts of the body to the surface. At the same time, water released by skin cells in the form of perspiration can carry heat away by evaporation.

**Chart 2.5 Hydrogen Ion Concentrations and pH**

| Grams of $H^+$ per Liter | pH | |
|---|---|---|
| 1.0 | 0 | ↑ |
| 0.1 | 1 | |
| 0.01 | 2 | |
| 0.001 | 3 | Increasingly acidic |
| 0.0001 | 4 | |
| 0.00001 | 5 | |
| 0.000001 | 6 | |
| 0.0000001 | 7 | Neutral—neither acidic nor basic |
| 0.00000001 | 8 | |
| 0.000000001 | 9 | |
| 0.0000000001 | 10 | |
| 0.00000000001 | 11 | Increasingly basic |
| 0.000000000001 | 12 | |
| 0.0000000000001 | 13 | |
| 0.00000000000001 | 14 | ↓ |

#### Oxygen

Molecules of oxygen gas ($O_2$) enter the internal environment through the respiratory organs and are transported throughout the body by the blood, especially by red blood cells. Within cells, organelles use oxygen to release energy from nutrient molecules. The released energy is used to drive the cell's metabolic activities. A continuing supply of oxygen is necessary for cell survival and, ultimately, for the survival of the organism.

NO (nitric oxide) and CO (carbon monoxide) are two small chemicals with bad reputations. NO is found in smog, cigarettes, and acid rain. CO is a colorless, odorless, lethal gas that is notorious for causing death when it leaks from home heating systems or exhaust pipes in closed garages. But researchers have recently found that both NO and CO are important in physiology as biological messenger molecules. NO is involved in digestion, memory, immunity, respiration, and circulation. CO functions in the spleen, which recycles old red blood cells, and in the parts of the brain that control memory, smell, and vital functions. Neuroscientists now have two new messenger molecules to investigate, two tiny chemicals that were right under their noses!

#### Carbon Dioxide

Carbon dioxide ($CO_2$) is a simple, carbon-containing inorganic compound. It is produced as a waste product when energy is released during certain metabolic processes. As it moves from the cells into the surrounding body fluids and blood, most of the carbon dioxide reacts with water to form a weak acid (carbonic acid, $H_2CO_3$). This acid ionizes, releasing hydrogen ions ($H^+$) and bicarbonate ions ($HCO_3^-$), which the blood carries to the respiratory organs. There the chemical reactions reverse, and carbon dioxide gas is produced, eventually to be exhaled.

### Inorganic Salts

Inorganic salts are abundant in body fluids. They are the sources of many necessary ions, including ions of sodium ($Na^+$), chloride ($Cl^-$), potassium ($K^+$), calcium ($Ca^{+2}$), magnesium ($Mg^{+2}$), phosphate ($PO_4^{-3}$), carbonate ($CO_3^{-2}$), bicarbonate ($HCO_3^-$), and sulfate ($SO_4^{-2}$). These ions play important roles in metabolic processes, including those involved in maintenance of proper water concentrations in body fluids, pH, blood clotting, bone development, energy transfer within cells, and muscle and nerve functions.

These electrolytes must be present in certain concentrations, both inside and outside cells, to maintain homeostasis. Such a condition is called **electrolyte balance.** Disrupted electrolyte balance occurs in certain diseases, and modern medical treatment places considerable emphasis on restoring it.

Chart 2.6 summarizes the functions of some of the inorganic substances that commonly occur in cells.

1. What is the difference between an organic molecule and an inorganic molecule?
2. What is the difference between an electrolyte and a nonelectrolyte?
3. Define electrolyte balance.

## Organic Substances

Important groups of organic substances found in cells include carbohydrates, lipids, proteins, and nucleic acids.

### Carbohydrates

**Carbohydrates** provide much of the energy that cells require. They also supply materials required to build certain cell structures, and they often are stored as reserve energy supplies.

Carbohydrates are water-soluble molecules that contain atoms of carbon, hydrogen, and oxygen. These molecules usually have twice as many hydrogen as oxygen atoms, the same ratio of hydrogen to oxygen as in water molecules ($H_2O$). This ratio is easy to see in the molecular formulas of the carbohydrates glucose ($C_6H_{12}O_6$) and sucrose ($C_{12}H_{22}O_{11}$).

Carbohydrates are classified by size. Simple carbohydrates, or **sugars,** include the **monosaccharides** (single sugars) and **disaccharides** (double sugars). A monosaccharide may include from three to seven carbon atoms, and may be arranged in a straight chain or a ring (fig. 2.9). Monosaccharides include glucose (dextrose), fructose, and galactose. Disaccharides consist of two six-carbon units. Sucrose (table sugar) and lactose (milk sugar) are disaccharides.

Complex carbohydrates, also called **polysaccharides,** are built of simple carbohydrates (fig. 2.10). Plant starch is an example. Starch molecules consist of highly branched chains, each containing from twenty-four to thirty glucose units.

Animals, including humans, synthesize a polysaccharide similar to starch called *glycogen.* Its molecules also consist of branched chains of sugar units, but each chain contains only about a dozen glucose units.

### Lipids

**Lipids** are a group of organic chemicals that are insoluble in water but soluble in organic solvents, such as ether and chloroform. Lipids include a number of compounds, such as fats, phospholipids, and steroids, that have vital functions in cells and are important constituents of cell membranes (see chapter 3). The most common lipids, however, are the *fats.* They are used primarily to supply energy for cellular activities. In fact, fat molecules can supply more energy gram for gram than can carbohydrate molecules. This is why eating a fatty diet leads to weight gain.

Like carbohydrates, fat molecules are composed of carbon, hydrogen, and oxygen atoms. However, they contain a much smaller proportion of oxygen than do carbohydrates. This is illustrated by the formula for the fat known as *tristearin,* $C_{57}H_{110}O_6$.

The building blocks of fat molecules are **fatty acids** and **glycerol.** These smaller molecules are united so that each glycerol molecule combines with three fatty acid molecules. The result is a single fat molecule or *triglyceride* (fig. 2.11).

The glycerol portion of every fat molecule is the same, yet there are many kinds of fatty acids and, therefore, many kinds of fats. Although all fatty acid molecules contain a carboxyl group (COOH) at the end of a chain of carbon atoms, these molecules differ in the lengths of their carbon atom chains (although such chains usually contain an even number of carbon atoms). The chains also may vary in the ways the carbon atoms are combined. In some cases, the carbon atoms are all joined by single carbon-carbon bonds. This type of fatty acid is said to be *saturated;* that is, each carbon atom is bound to as many hydrogen atoms as possible and is thus saturated with hydrogen atoms. Other fatty acid chains have not bound their maximum number of hydrogen atoms. Therefore, they have one or more double bonds between carbon atoms. Those fatty acids with one double bond are called *monounsaturated fatty acids,* and those with two or more double bonds are said to be *polyunsaturated.* Similarly, fat molecules

## CHART 2.6 Inorganic Substances Common in Cells

| Substance | Symbol or Formula | Functions |
|---|---|---|
| ***I. Inorganic Molecules*** | | |
| Water | $H_2O$ | Major component of body fluids (chapter 21); medium in which most biochemical reactions occur; transports various chemical substances (chapter 14); helps regulate body temperature (chapter 6) |
| Oxygen | $O_2$ | Used in the release of energy from glucose molecules (chapter 4) |
| Carbon dioxide | $CO_2$ | Waste product that results from metabolism (chapter 4); reacts with water to form carbonic acid (chapter 19) |
| ***II. Inorganic Ions*** | | |
| Bicarbonate ions | $HCO_3^-$ | Helps maintain acid-base balance (chapter 21) |
| Calcium ions | $Ca^{+2}$ | Necessary for bone development (chapter 7); needed for muscle contraction (chapter 9) and blood clotting (chapter 14) |
| Carbonate ions | $CO_3^{-2}$ | Component of bone tissue (chapter 7) |
| Chloride ions | $Cl^-$ | Helps maintain water balance (chapter 21) |
| Hydrogen ions | $H^+$ | Determines blood pH (chapter 14) |
| Magnesium ions | $Mg^{+2}$ | Component of bone tissue (chapter 7); needed for certain metabolic processes (chapter 18) |
| Phosphate ions | $PO_4^{-3}$ | Needed for synthesis of ATP, nucleic acids, and other vital substances (chapter 4); component of bone tissue (chapter 7); helps maintain polarization of cell membranes (chapter 10) |
| Potassium ions | $K^+$ | Needed for polarization of cell membranes (chapter 10) |
| Sodium ions | $Na^+$ | Needed for polarization of cell membranes (chapter 10); helps maintain water balance (chapter 21) |
| Sulfate ions | $SO_4^{-2}$ | Helps maintain polarization of cell membranes (chapter 10); helps maintain acid-base balance (chapter 21) |

(a) (b) (c)

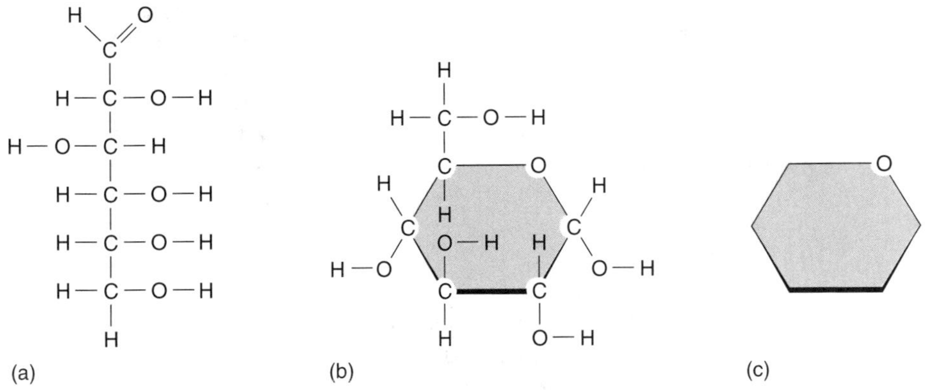

**FIGURE 2.9**
(*a*) Molecules of the monosaccharide glucose ($C_6H_{12}O_6$) may have a straight chain of carbon atoms. (*b*) More commonly, glucose molecules form a ring structure. (*c*) This shape symbolizes the ring structure of a glucose molecule.

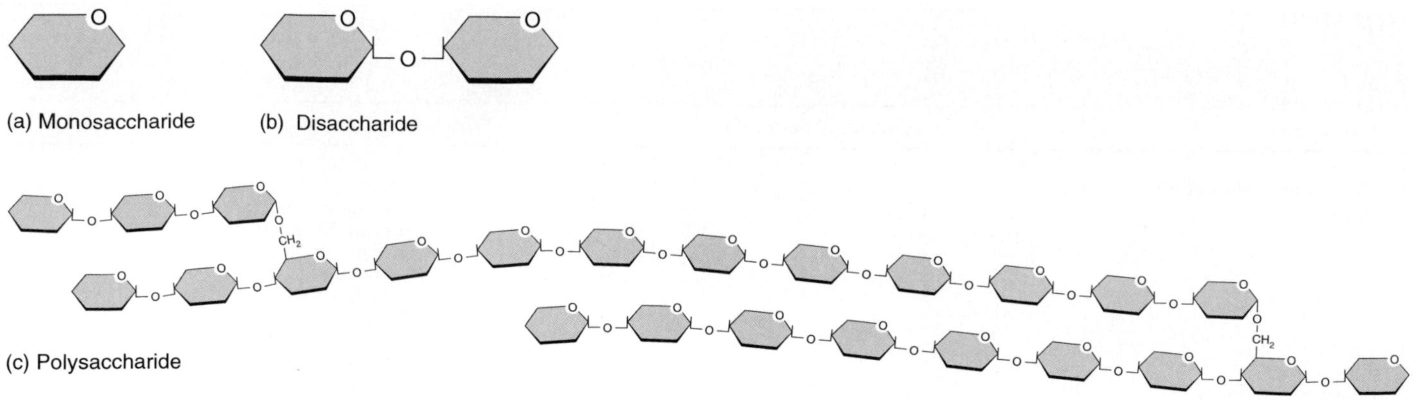

(a) Monosaccharide (b) Disaccharide

(c) Polysaccharide

**FIGURE 2.10**

(*a*) A monosaccharide molecule consisting of one 6-carbon atom building block; (*b*) a disaccharide molecule consisting of two of these building blocks; (*c*) a polysaccharide molecule consisting of many building blocks, which may form branches.

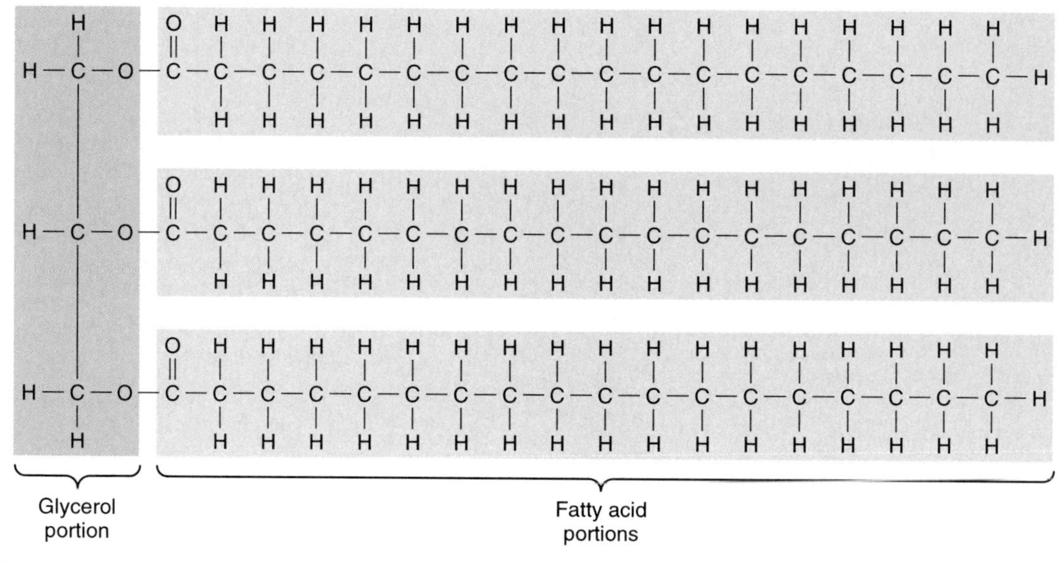


**FIGURE 2.11**

A triglyceride molecule consists of a glycerol and three fatty acid "tails."

that contain only saturated fatty acids are called **saturated fats,** and those that include unsaturated fatty acids are called **unsaturated fats** (fig. 2.12). Each kind of fat molecule has its own special properties.

A *phospholipid* molecule is similar to a fat molecule in that it contains a glycerol portion and fatty acid chains. The phospholipid, however, has only two fatty acid chains, and in place of the third one is a portion containing a phosphate group. This phosphate portion is soluble in water (hydrophilic) and forms the "head" of the molecule, while the fatty acid portion is insoluble in water (hydrophobic) and forms a "tail." Figure 2.13 illustrates the molecular structure of cephalin, a phospholipid in blood.

*Steroid* molecules are complex structures that include interconnected rings of carbon atoms (fig. 2.14). Among the more important steroids are cholesterol,

A diet that contains a high proportion of saturated fat seems to increase a person's chance of developing atherosclerosis, a serious disease that obstructs certain blood vessels. For this reason, many nutritionists recommend that wherever possible we substitute polyunsaturated fats for dietary saturated fats.

As a rule, saturated fats are more abundant in fatty foods that are solids at room temperature, such as butter, lard, and most other animal fats. Unsaturated fats, on the other hand, are likely to be plentiful in fatty foods that are liquids at room temperature, such as soft margarine and various seed oils, including corn oil and soybean oil. There are exceptions, however, since coconut and palm oils are relatively high in saturated fat.

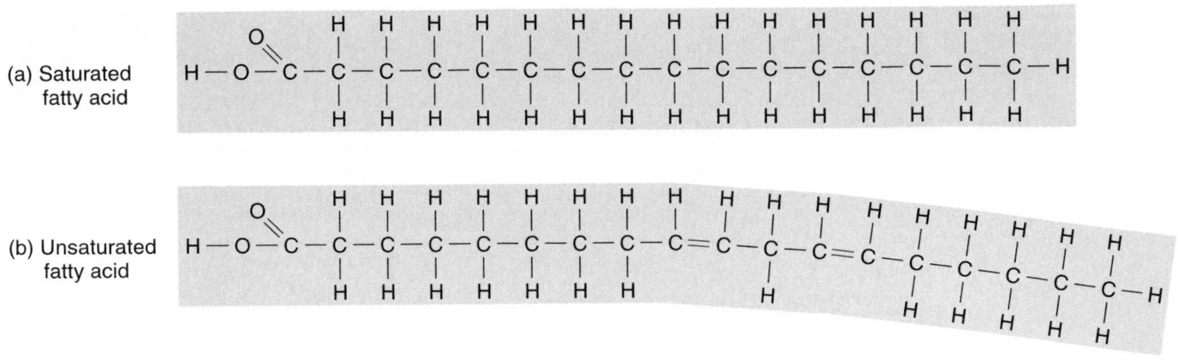

(a) Saturated fatty acid

(b) Unsaturated fatty acid

**FIGURE 2.12**
(*a*) A molecule of saturated fatty acid and (*b*) a molecule of unsaturated fatty acid. Double bonds are shown in red. Note that they cause a "kink" in the shape of the molecule.

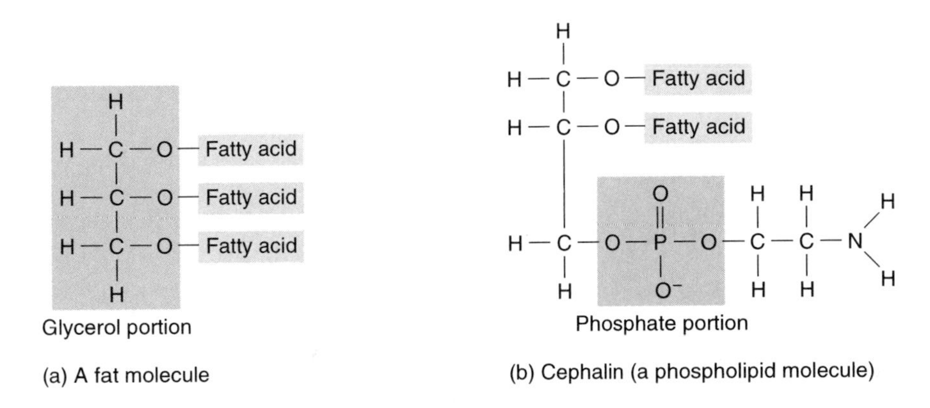


**FIGURE 2.13**
(*a*) A fat molecule (triglyceride) contains a glycerol and three fatty acids. (*b*) In a phospholipid molecule, a phosphate containing group replaces one fatty acid.

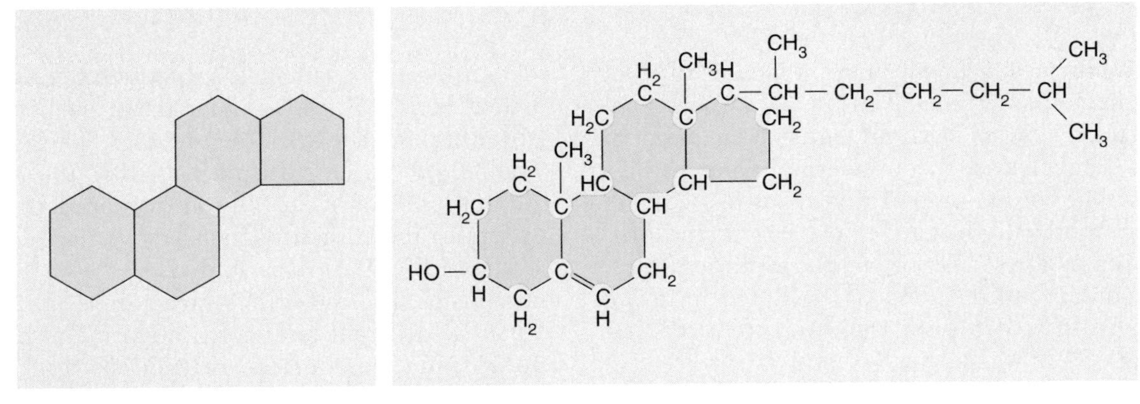

(a) Structure of a steroid

(b) Cholesterol

**FIGURE 2.14**
(*a*) The general structure of a steroid. (*b*) The structural formula for cholesterol, a steroid widely distributed in the body.

which is in all body cells and is used to synthesize other steroids; sex hormones, such as estrogen, progesterone, and testosterone; and several hormones from the adrenal glands. These steroids are discussed in chapters 13, 14, and 22.

Chart 2.7 summarizes the molecular structures and characteristics of lipids.

## Proteins

Some **proteins** serve as structural materials, energy sources, and chemical messengers (hormones). Others function as receptors on cell surfaces that bond to particular kinds of molecules, or act as weapons (antibodies) against substances that are foreign to the body. Still other proteins play vital roles in metabolic

## CHART 2.7 IMPORTANT GROUPS OF LIPIDS

| Group | Basic Molecular Structure | Characteristics |
|---|---|---|
| Triglycerides | Three fatty acid molecules bound to a glycerol molecule | Most common lipid in the body; stored in fat tissue as an energy supply. Fat tissue also provides insulation beneath the skin. |
| Phospholipids | Two fatty acid molecules and a phosphate group bound to a glycerol molecule (may also include a nitrogen-containing molecule attached to the phosphate group) | Used as structural components in cell membranes; large amounts are in the liver and parts of the nervous system. |
| Steroids | Four interconnected rings of carbon atoms | Widely distributed in the body with a variety of functions; includes cholesterol, sex hormones, and certain hormones of the adrenal glands. |

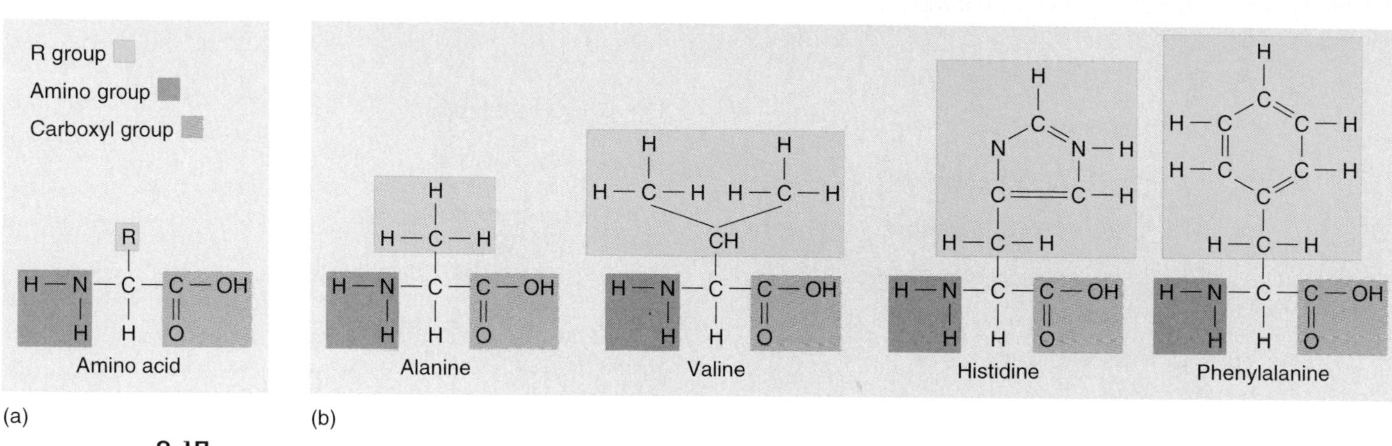


**FIGURE 2.15**

(*a*) The general structure of an amino acid. The portion in red and blue are common to all amino acids. (*b*) Some representative amino acids and their structural formulas. Each amino acid molecule has a particular shape due to the arrangement of its parts. Note the amino groups and carboxyl groups that are common to all amino acid molecules. Also note the different R groups.

processes as **enzymes.** Enzymes are molecules that act as catalysts in living systems. That is, they speed specific chemical reactions without being consumed in the process. (Enzymes are discussed in chapter 4.)

Like carbohydrates and lipids, proteins consist of atoms of carbon, hydrogen, and oxygen. In addition, proteins always contain nitrogen atoms, and sometimes contain sulfur atoms as well. The building blocks of proteins are smaller molecules called **amino acids.**

Each amino acid molecule has an amino group ($NH_2$) at one end and a carboxyl group (COOH) at the other end. Between these end groups is a single carbon atom (alpha carbon), and it, in turn, is bound to a hydrogen atom and to another group of atoms called a *side chain* or "R" group. The composition of the side chain distinguishes one kind of amino acid from another (fig. 2.15). About twenty different kinds of amino acids occur in proteins found in living things.

Proteins have complex shapes, yet the way they are put together is surprisingly simple. Each type of protein molecule contains specific numbers and kinds of amino acids connected by peptide bonds (special covalent bonds) in a particular linear sequence (primary structure), as shown in figure 2.16. These amino acid chains vary in length from less than 100 to more than 50,000 amino acids. The amino acid chain usually twists to form a coil (secondary structure) (fig. 2.17), and it, in turn, may be folded up into a unique three-dimensional form (tertiary structure) (fig. 2.18). Consequently, different kinds of protein molecules have different shapes that are related to their particular functions. Some protein molecules are long and fibrous, such as the keratin protein that forms hair, and the threads of fibrin protein that knit a blood clot. Many proteins are globular. Myoglobin and hemoglobin, which transport oxygen in muscle and blood, respectively, are globular, as are many enzymes.

Weak attractions called **hydrogen bonds** are important in maintaining the shape, or **conformation,** of a protein. These bonds form when electrons are not shared evenly in covalent bonds. As the result of this unequal sharing, oxygen tends to become very slightly negative and hydrogen very slightly positive.

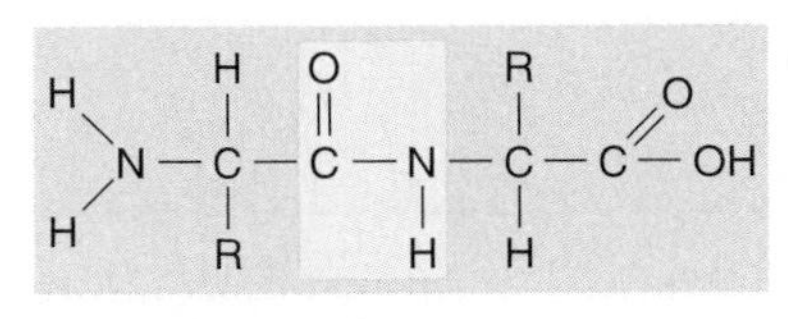


(a)

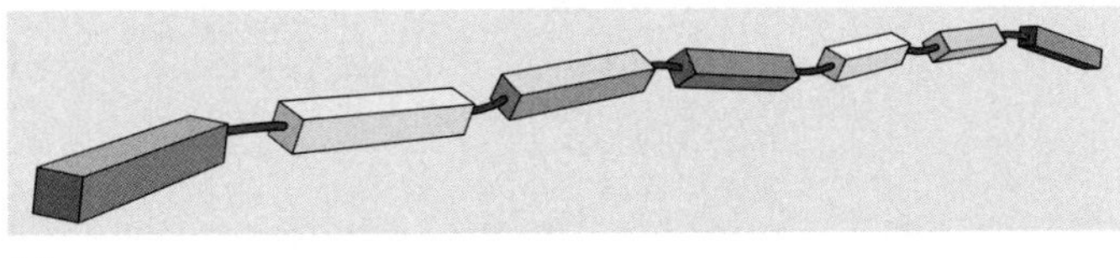

(b)

**FIGURE 2.16**

Primary structure of a protein. (*a*) A peptide bond between two amino acids. (*b*) A portion of the primary structure or amino acid sequence of a protein. Each different color represents a different amino acid.

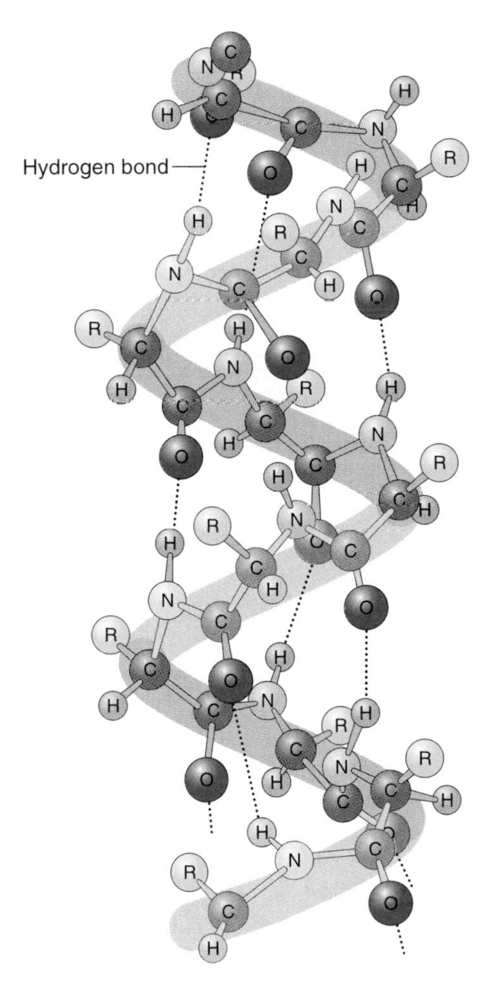


**FIGURE 2.17**

Secondary structure of a protein. The amino acid chain of a protein molecule sometimes twists, forming a coil held together by hydrogen bonds.

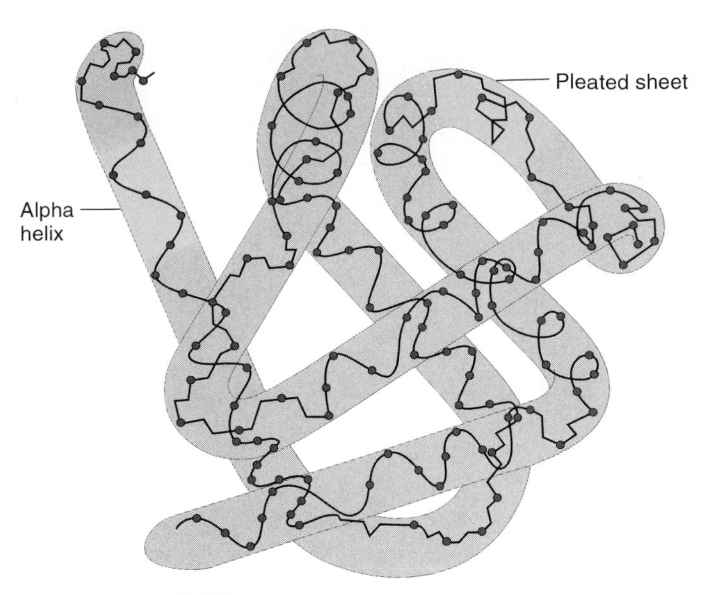


**FIGURE 2.18**

Tertiary structure of a protein. The coiled amino acid chain of a protein molecule folds into a unique three-dimensional structure.

Hydrogen atoms in one part of a protein may then attract oxygen atoms elsewhere in the molecule (see fig. 2.17). These forces contribute to the molecule's overall shape.

Various treatments can cause the secondary and tertiary structures of a protein's conformation to fall apart, or *denature*. Because the primary structure (amino acid sequence) remains, sometimes the protein can regain its shape when normal conditions return. High temperature, radiation, pH changes, and certain chemicals (such as urea) can denature proteins.

A familiar example of irreversible protein denaturation is the response of the protein albumin to heat, for example, cooking an egg white. A permanent wave that curls hair also results from protein denaturation. Chemicals first break apart the tertiary structure formed when sulfur-containing amino acids attract each other within keratin molecules. This relaxes the hair. When the chemicals are washed out and the hair set, the sulfur bonds reform, but in different places, changing the appearance of the hair.

## Nucleic Acids

**Nucleic acids** constitute genes, the instructions that control a cell's activities. These molecules are very large and complex. They contain atoms of carbon, hydrogen, oxygen, nitrogen, and phosphorus, which form building blocks called **nucleotides.** Each nucleotide consists of a 5-carbon sugar (ribose or deoxyribose), a

## CHART 2.8 ORGANIC COMPOUNDS IN CELLS

| Compound | Elements Present | Building Blocks | Functions | Examples |
|---|---|---|---|---|
| Carbohydrates | C,H,O | Simple sugar | Provide energy, cell structure | Glucose, starch |
| Lipids | C,H,O (often P) | Glycerol, fatty acids, phosphate groups | Provide energy, cell structure | Triglycerides, phospholipids, steroids |
| Proteins | C,H,O,N (often S) | Amino acids | Provide cell structure, enzymes, energy | Albumins, hemoglobin |
| Nucleic acids | C,H,O,N,P | Nucleotides | Store information for the synthesis of proteins, control cell activities | RNA, DNA |

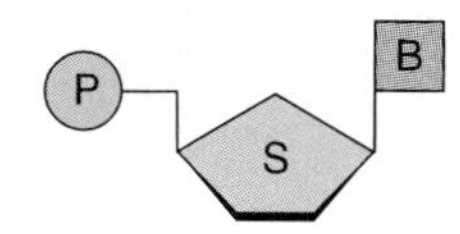


**FIGURE 2.19**
A nucleotide consists of a 5-carbon sugar (S), a phosphate group (P), and an organic base (B).

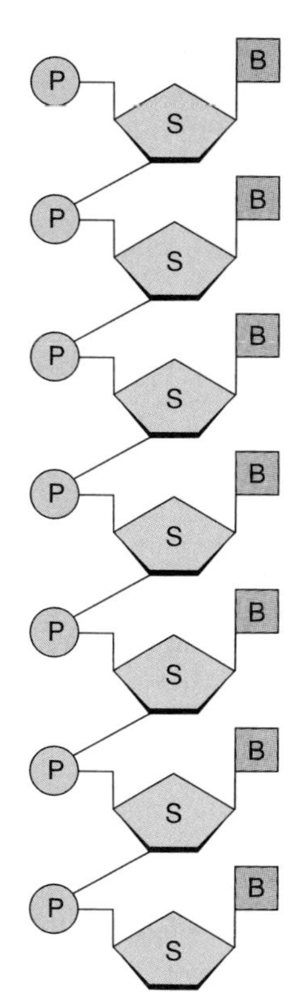


**FIGURE 2.20**
A schematic representation of a polynucleotide chain. A nucleic acid molecule consists of one (RNA) or two (DNA) such chains of nucleotides.

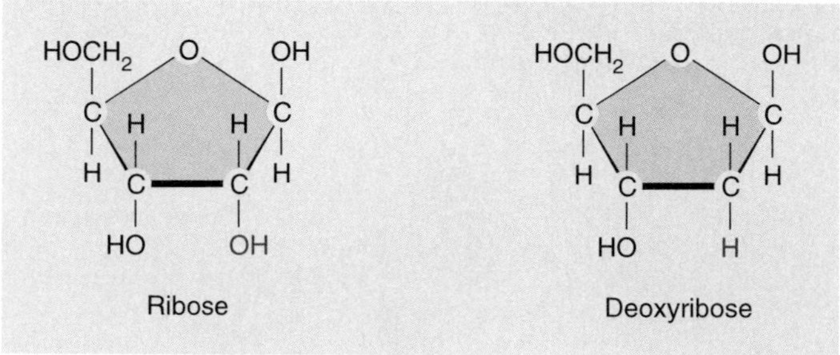

phosphate group, and one of several organic bases (fig. 2.19). Such nucleotides, linked in a chain, form a **polynucleotide** (fig. 2.20).

Ribose

Deoxyribose

**FIGURE 2.21**
The molecules of ribose and deoxyribose differ by a single oxygen atom.

There are two major types of nucleic acids. One type is composed of molecules whose nucleotides contain ribose sugar; it is called **RNA** (ribonucleic acid) and it exists as a single polynucleotide chain. The nucleotides of the second type contain deoxyribose sugar; nucleic acid of this type is called **DNA** (deoxyribonucleic acid) and it forms a double polynucleotide chain. Figure 2.21 compares the structure of ribose and deoxyribose, which differ by one oxygen atom. DNA and RNA also differ in the types of bases they contain.

DNA molecules store information in a type of molecular code. Cells use this information to construct specific protein molecules, which have a wide variety of functions. RNA molecules help to synthesize proteins.

DNA molecules have a unique ability to make copies of, or replicate, themselves. They replicate prior to cell reproduction, and each newly formed cell receives an exact copy of its parent's DNA molecules. Chapter 4 discusses the storage of information in nucleic acid molecules, its use in the manufacture of protein molecules, and the function of these proteins in controlling metabolic reactions.

Chart 2.8 summarizes the four groups of organic compounds. Figure 2.22 shows three-dimensional (space-filling) models of some important molecules, illustrating their shapes.

(a)

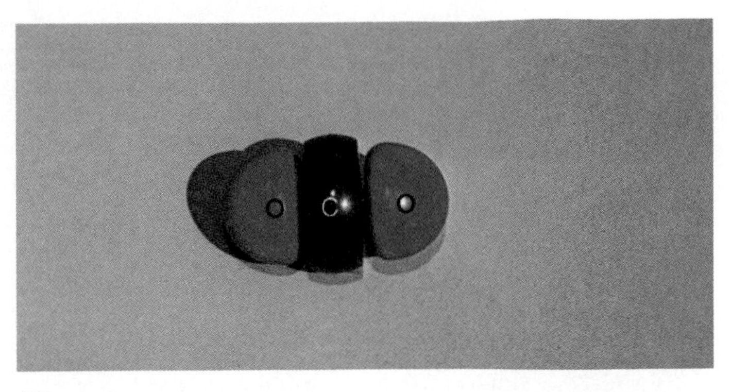


(b)

(c)

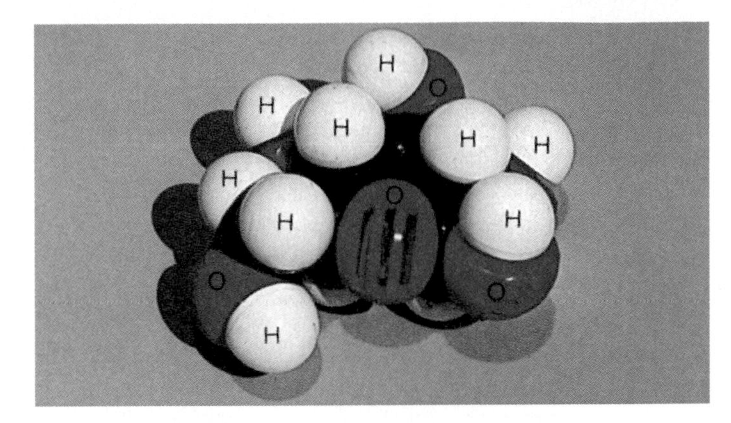


(d)

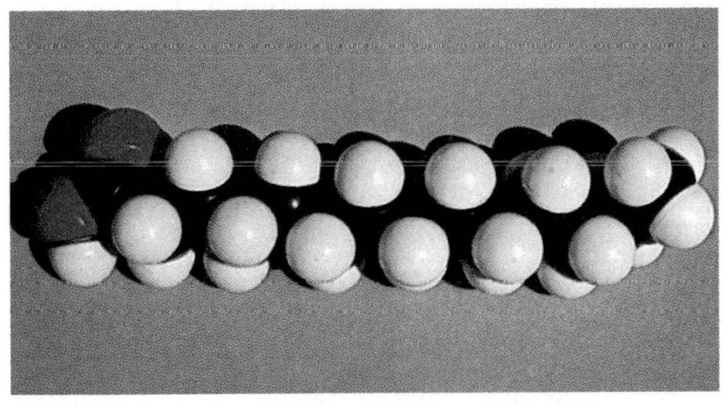

(e)

**FIGURE 2.22**

Three-dimensional (space-filling) models of several important molecules: (*a*) water, (*b*) carbon dioxide, (*c*) glycine (an amino acid), (*d*) glucose (a monosaccharide), (*e*) a fatty acid, (*f*) collagen (a protein). The models are not shown in relative scale. White = hydrogen, red = oxygen, blue = nitrogen, black = carbon.

(f)

Clinical Application 2.3 describes two techniques used to view human anatomy and physiology.

**1** Compare the chemical composition of carbohydrates, lipids, proteins, and nucleic acids.

**2** How does an enzyme affect a chemical reaction?

**3** What is likely to happen to a protein molecule that is exposed to excessive heat or radiation?

**4** What are the functions of DNA and RNA?

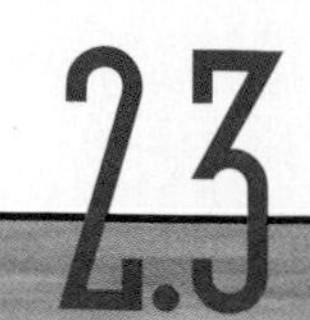

CLINICAL APPLICATION 2.3

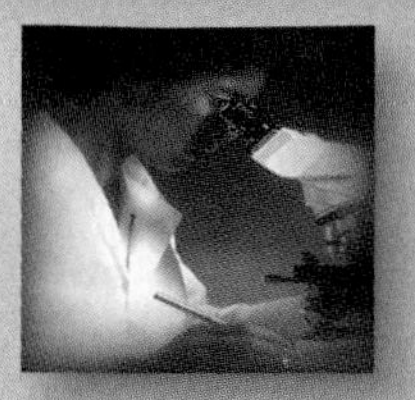

## CT Scanning and PET Imaging

Physicians use two techniques—computerized tomography (CT) scanning and positron emission tomography (PET imaging)—to paint portraits of anatomy and physiology, respectively.

In CT scanning, an X-ray emitting device is positioned around the region of the body being examined. At the same time, an X-ray detector is moved in the opposite direction on the other side of the body. As these parts move, an X-ray beam passes through the body from hundreds of different angles. Because tissues and organs of varying composition absorb X rays differently, the intensity of X rays reaching the detector varies from position to position. A computer records the measurements made by the X-ray detector and combines them mathematically. This creates on a viewing screen a sectional image of the internal body parts.

While ordinary X-ray techniques produce two-dimensional images, a CT scan provides three-dimensional information. The CT scan can also differentiate clearly between soft tissues of slightly different densities, such as the liver and kidneys, which cannot be seen in a conventional X-ray image. Thus, a CT scan can often spot abnormal tissue, such as a tumor. For example, a CT scan can tell whether a sinus headache that does not respond to antibiotic therapy is caused by a drug resistant infection or a tumor.

PET imaging uses radioactive isotopes that naturally emit positrons, which are atypical positively charged electrons, to detect biochemical activity in a specific body part (fig. 2D). Useful isotopes in PET imaging include carbon-11, nitrogen-13, oxygen-15, and fluorine-18. When one of these isotopes releases a positron, it interacts with a nearby negatively charged electron. The two particles destroy each other, an event called annihilation. At the moment of destruction, two gamma rays appear and move away from each other in opposite directions. Special equipment detects the gamma radiation.

To produce a PET image of biochemically active tissue, a person is injected with a metabolically active compound that includes a bound positron-emitting isotope. To study the brain, for example, a person is injected with glucose containing fluorine-18. After the brain takes up the isotope-tagged compound, the person rests her head within a circular array of radiation detectors. A device records each time two gamma rays are emitted simultaneously and travel in opposite directions (the result of annihilation). A computer collects and combines the data and generates a cross-sectional image. The image indicates the location and relative concentration of the radioactive isotope in different regions of the brain, and can be used to study those parts metabolizing glucose.

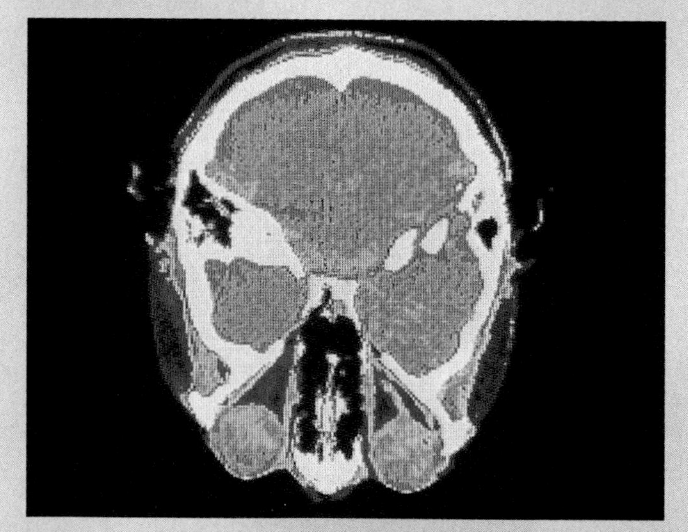

(a)

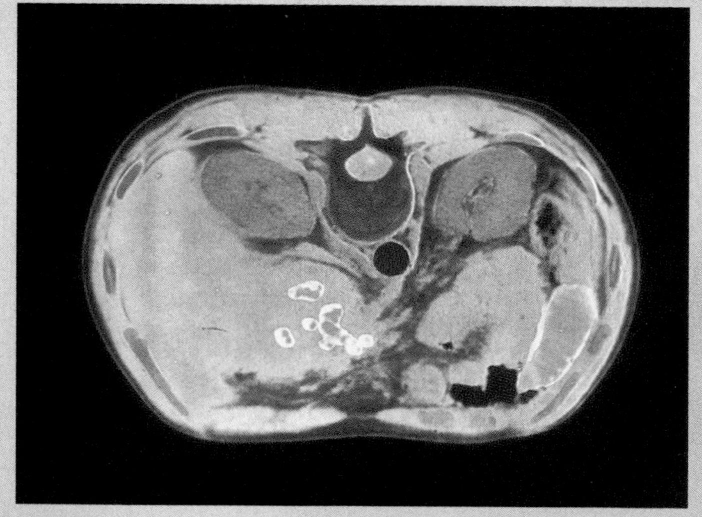

(b)

**FIGURE 2D**
CT scans of (*a*) the head and (*b*) the abdomen.

PET images reveal the parts of the brain that are affected in such disorders as Huntington disease, Parkinson's disease, epilepsy, and Alzheimer's disease, and they are used to study blood flow in vessels supplying the brain and heart. The technology is invaluable for detecting the physiological bases of poorly understood behavioral disorders, such as obsessive-compulsive disorder. In this odd condition, a person repeatedly performs a certain behavior, such as washing hands, showering, locking doors, or checking to see that the stove is turned off. PET images of people with this disorder reveal intense activity in two parts of the brain that are quiet in the brains of unaffected individuals. Knowing the site of altered brain activity can help researchers develop more directed drug therapy.

In addition to highlighting biochemical activities behind illness, PET scans allow biologists to track normal brain physiology. Figure 2E shows that different patterns of brain activity are associated with learning and with reviewing something already learned.

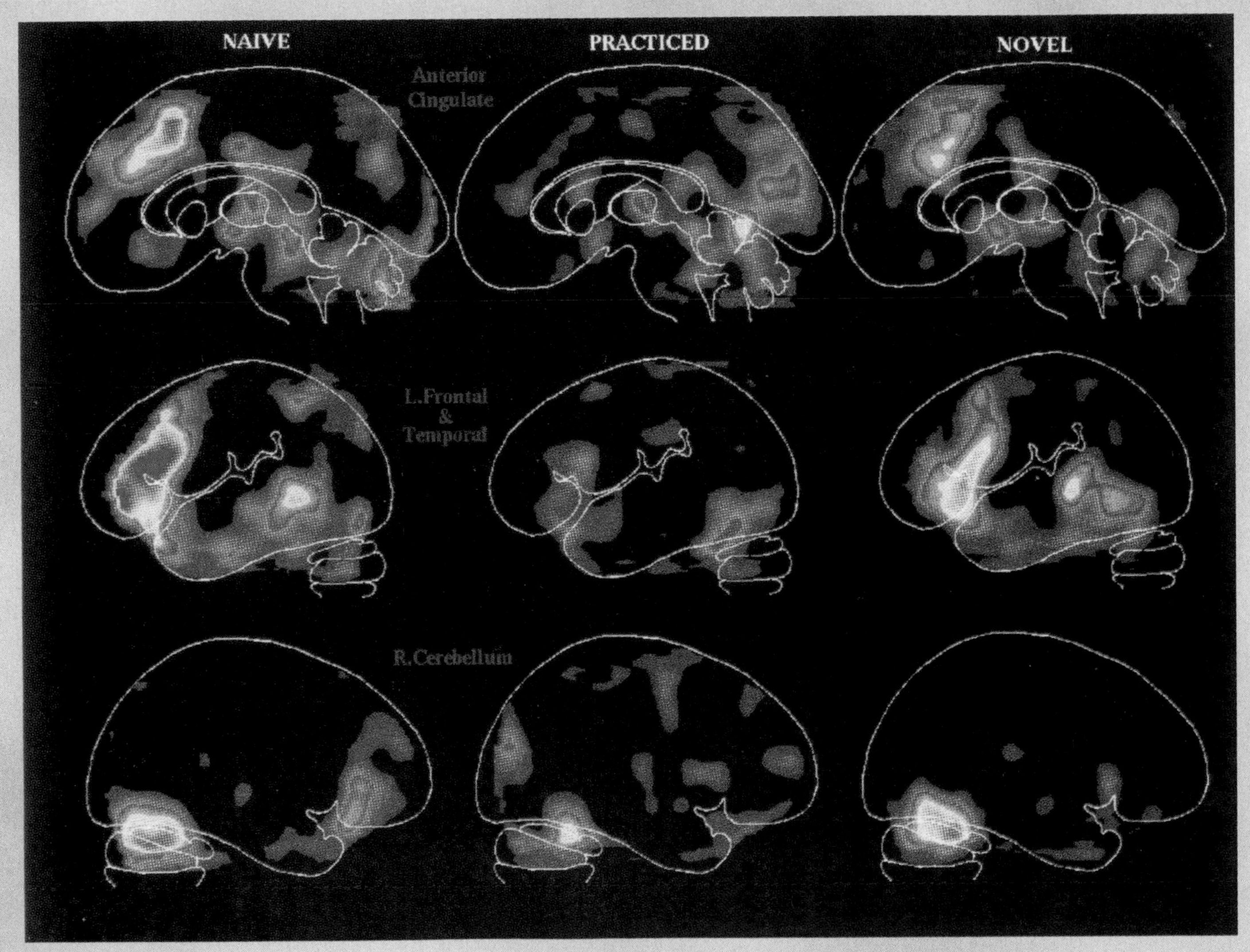


**FIGURE 2E**

These PET images demonstrate brain changes that accompany learning. The top and bottom views show different parts of the same brain. The "naive" brain on the left has been given a list of nouns and asked to visualize each word. In the middle column, the person has practiced the task, so he can picture the nouns with less brain activity. In the third column, the person receives a new list of nouns. Learning centers in the brain show increased activity.

## CHAPTER SUMMARY

### Introduction (page 37)

Chemistry deals with the composition of substances and changes in their composition. Biochemistry is the chemistry of living things.

### Structure of Matter (page 37)

Matter is anything that has weight and takes up space.

1. Elements and atoms
   a. Naturally occurring matter on earth is composed of ninety-two elements.
   b. Elements occur most frequently in chemical combinations called compounds.
   c. Elements are composed of atoms.
   d. Atoms of different elements vary in size, weight, and ways of interacting.
2. Atomic structure
   a. An atom consists of electrons surrounding a nucleus, which contains protons and neutrons. The exception is hydrogen, which contains only a proton in its nucleus.
   b. Electrons are negatively charged, protons positively charged, and neutrons uncharged.
   c. A complete atom is electrically neutral.
   d. The atomic number of an element is equal to the number of protons in each atom; the atomic weight is equal to the number of protons plus the number of neutrons in each atom.
3. Isotopes
   a. Isotopes are atoms with the same atomic number but different atomic weights (due to differing numbers of neutrons).
   b. All the isotopes of an element react chemically in the same manner.
   c. Some isotopes are radioactive and release atomic radiation.
4. Bonding of atoms
   a. When atoms combine, they gain, lose, or share electrons.
   b. Electrons are arranged in shells around a nucleus.
   c. Atoms with completely filled outer shells are inactive, whereas atoms with incompletely filled outer shells tend to gain, lose, or share electrons and thus achieve stable structures.
   d. Atoms that lose electrons become positively charged; atoms that gain electrons become negatively charged.
   e. Ions with opposite charges attract and join by ionic bonds; atoms that share electrons join by covalent bonds.
5. Molecules and compounds
   a. Two or more atoms of the same element joining form a molecule of that element. Atoms of different elements uniting form a molecule of a compound.
   b. Molecules contain definite kinds and numbers of atoms.
   c. A molecular formula represents the numbers and kinds of atoms in a molecule.
   d. A structural formula represents the arrangement of atoms within a molecule.
6. Chemical reactions
   a. In a chemical reaction, bonds between atoms, ions, or molecules break or form.
   b. Three kinds of chemical reactions are: synthesis, in which larger molecules form from smaller particles; decomposition, in which smaller particles form from larger molecules; and exchange reactions, in which parts of two different molecules trade positions.
   c. Many reactions are reversible. The direction of a reaction depends upon the proportion of reactants and end products, the energy available, and the presence or absence of catalysts.
7. Acids, bases, and salts
   a. Compounds that ionize when they dissolve in water are electrolytes.
   b. Electrolytes that release hydrogen ions are acids, and those that release hydroxyl or other ions that react with hydrogen ions are bases.
   c. Acids and bases react together to form water and electrolytes called salts.
8. Acid and base concentrations
   a. The concentration of hydrogen ions ($H^+$) and hydroxyl ions ($OH^-$) in a solution can be represented by pH.
   b. A solution with equal numbers of $H^+$ and $OH^-$ is neutral and has a pH of 7.0; a solution with more $H^+$ than $OH^-$ is acidic (pH less than 7.0); a solution with fewer $H^+$ than $OH^-$ is basic (pH more than 7.0).
   c. There is a tenfold difference in hydrogen ion concentration between each whole number in the pH scale.

### Chemical Constituents of Cells (page 47)

Molecules containing carbon and hydrogen atoms are organic and are usually nonelectrolytes; other molecules are inorganic and are usually electrolytes.

1. Inorganic substances
   a. Water is the most abundant compound in cells. Many chemical reactions take place in water. Water transports chemicals and heat, and helps release excess body heat.
   b. Oxygen releases energy needed for metabolic activities from glucose and other molecules.
   c. Carbon dioxide is produced when energy is released during metabolic processes.
   d. Inorganic salts provide ions needed in a variety of metabolic processes.
   e. Electrolytes must be present in certain concentrations inside and outside of cells.
2. Organic substances
   a. Carbohydrates provide much of the energy required by cells; their basic building blocks are simple sugar molecules.
   b. Lipids, such as fats, phospholipids, and steroids, supply energy and are used to build cell parts; their basic building blocks are molecules of glycerol and fatty acids.
   c. Proteins serve as structural materials, energy sources, hormones, cell surface receptors, antibodies, and enzymes.

(1) Enzymes initiate or speed chemical reactions without being consumed themselves.
(2) The building blocks of proteins are amino acids.
(3) Proteins vary in the numbers and kinds of amino acids they contain, the sequences in which these amino acids are arranged, and their three-dimensional structures.
(4) Protein molecules can be denatured by exposure to excessive heat, radiation, electricity, or certain chemicals.

d. Nucleic acids form genes, and thus control cell activities.
(1) The two major kinds are RNA and DNA.
(2) Nucleic acid molecules are composed of building blocks called nucleotides.
(3) DNA molecules store information that is used by cell parts to construct specific kinds of protein molecules.
(4) RNA molecules help synthesize proteins.
(5) DNA molecules are replicated and passed from parent to offspring cells during cell reproduction.

**Explorations in Human Anatomy and Physiology CD-ROM**
The module accompanying Chapter Two is #2 Active Transport.

## Critical Thinking Questions

1. What acidic and alkaline substances do you encounter in your everyday life? What foods do you eat regularly that are acidic? What alkaline foods do you eat?
2. Using the information on page 50 to distinguish between saturated and unsaturated fats, try to list all of the sources of saturated and unsaturated fats you have eaten during the past twenty-four hours.
3. How would you reassure a patient who is about to undergo CT scanning for evaluation of a tumor, and who fears becoming a radiation hazard to family members?
4. Various forms of ionizing radiation, such as that released from X-ray tubes and radioactive substances, are commonly used in the treatment of cancer, yet such exposure can cause adverse effects, including the development of cancers. How would you explain the value of radiation therapy to a cancer patient in light of this seeming contradiction?
5. How would you explain the importance of amino acids and proteins in a diet to a person who is following a diet composed primarily of carbohydrates?
6. What clinical laboratory tests with which you are acquainted involve a knowledge of chemistry?

## Review Exercises

1. Distinguish between chemistry and biochemistry.
2. Define *matter.*
3. Explain the relationship between elements and atoms.
4. Define *compound.*
5. List the four most abundant elements in the human body.
6. Describe the major parts of an atom.
7. Distinguish between protons and neutrons.
8. Explain why a complete atom is electrically neutral.
9. Distinguish between atomic number and atomic weight.
10. Define *isotope.*
11. Explain what is meant by *atomic radiation.*
12. Describe how electrons are arranged within atoms.
13. Explain why some atoms are chemically inert.
14. Distinguish between an ionic bond and a covalent bond.
15. Distinguish between a single covalent bond and a double covalent bond.
16. Explain the relationship between molecules and compounds.
17. Distinguish between a molecular formula and a structural formula.
18. Describe three major types of chemical reactions.
19. Explain what is meant by *reversible reaction.*
20. Define *catalyst.*
21. Define *acid, base, salt,* and *electrolyte.*
22. Explain what is meant by *pH,* and describe the pH scale.
23. Distinguish between organic and inorganic substances.
24. Describe the roles played by water and by oxygen in the human body.
25. List several ions cells require, and describe their general functions.
26. Define *electrolyte balance.*
27. Describe the general characteristics of carbohydrates.
28. Distinguish between simple and complex carbohydrates.
29. Describe the general characteristics of lipids.
30. Distinguish between saturated and unsaturated fats.
31. Describe the general characteristics of proteins.
32. Describe the function of an *enzyme.*
33. Explain how protein molecules may become denatured.
34. Describe the general characteristics of nucleic acids.
35. Explain the general functions of nucleic acids.

3

CHAPTER

# CELLS

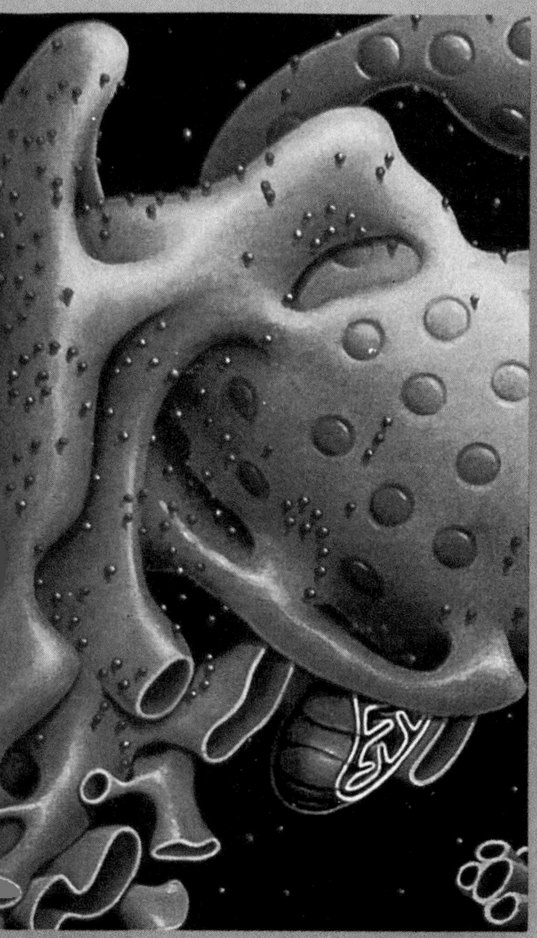

AN UNUSUAL, EXPLODED VIEW OF A CELL'S INTERIOR.

## CHAPTER OBJECTIVES

AFTER YOU HAVE STUDIED THIS CHAPTER, YOU SHOULD BE ABLE TO:

1. Explain how cells differ from one another.
2. Describe the general characteristics of a composite cell.
3. Explain how the structure of a cell membrane is related to its function.
4. Describe each kind of cytoplasmic organelle and explain its function.
5. Describe the cell nucleus and its parts.
6. Explain how substances move into and out of cells.
7. Describe the cell cycle.
8. Explain how a cell reproduces.
9. Describe several controls of cell reproduction.

## KEY TERMS

**active transport** (ak′tiv trans′port)

**centrosome** (sen′tro-sōm)

**chromosome** (kro′mo-sōm)

**cytoplasm** (si′to-plazm)

**cytoskeleton** (si′to-skel-i-tun)

**differentiation** (dif″er-en″she-a′shun)

**diffusion** (dĭ-fu′zhun)

**endocytosis** (en″do-si-to′sis)

**endoplasmic reticulum** (en′do-plaz′mik re-tik′u-lum)

**equilibrium** (e″kwi-lib′re-um)

**exocytosis** (ex-o-si-to′sis)

**extracellular** (eks″trah-sel′u-lar)

**facilitated diffusion** (fah-sil″i-tat′ed dĭ-fu′zhun)

**filtration** (fil-tra′shun)

**Golgi apparatus** (gol′je ap″ah-ra′tus)

**intracellular** (in″trah-sel′u-lar)

**lysosome** (li′so-sōm)

**micrometer** (mi′kro-me″ter)

**mitochondrion** (mi″to-kon′dre-on); plural, **mitochondria** (mi″to-kon′dre-ah)

**mitosis** (mi-to′sis)

**nucleolus** (nu-kle′o-lus)

**nucleus** (nu′kle-us)

**osmosis** (oz-mo′sis)

**permeable** (per′me-ah-bl)

**phagocytosis** (fag″o-si-to′sis)

**pinocytosis** (pi″no-si-to′sis)

**ribosome** (ri′bo-sōm)

**vesicle** (ves′i-k′l)

Something is definitely wrong with the cell deep within Eliot J.'s kidney. Unlike other cells that form the neat linings of the microscopic kidney tubules, this errant cell is rounder, oilier, somehow less distinctive in appearance. It simply doesn't seem to belong.

Although the man whose kidney harbors this aberrant cell is still quite healthy, a profound change has already occurred at the molecular level within that one peculiar cell. A tiny but powerful part of that cell's genetic instructions—a segment of DNA constituting a tumor suppressor gene—has been deactivated, and as a result, other genetic controls are going haywire. Genes that endow a kidney tubule cell with its special characteristics are silenced. Other genes that revert the cell to an embryonic form are activated. As a result of these changes, the cell divides more often than surrounding kidney cells. Soon, the one odd cell becomes two, the two become four, the four become eight, and on it goes. Then this enlarging tumor secretes biochemicals that establish a new blood supply just to nurture its renegade growth.

Eventually, when Eliot experiences kidney failure, physicians will scan the kidneys and find the tumor. It may already consist of millions of cells, but surgery may be able to remove all of it. A follow-up course of treatment with an immune system biochemical called interleukin-2 will give Eliot J. an even chance of surviving.

Cancer, like this abnormal kidney growth, is a body-wide illness, its treatment a race against time and cell division to remove or contain its growth. But cancer begins at the molecular and cellular levels. Understanding how cells function, how they specialize, and how they "know" when to divide lies at the root of many areas of medical science. Cell biology can explain both anatomy and physiology by breaking these biological sciences down to their component parts.

---

An adult human body consists of about seventy-five trillion **cells.** These cells have much in common, yet those in different tissues vary in a number of ways.

Cells vary considerably in size. We measure cell sizes in units called *micrometers.* A micrometer equals one thousandth of a millimeter and is symbolized μm. A human egg cell is about 140 μm in diameter and is just barely visible to an unaided eye. This is large when compared to a red blood cell, which is about 7.5 μm in diameter, or the most common white blood cells, which vary from 10 to 12 μm in diameter. On the other hand, smooth muscle cells can be between 20 and 500 μm long (fig. 3.1).

Cells also vary in shape, and typically their shapes are closely related to their functions (fig. 3.2). For instance, nerve cells often have long, threadlike extensions that transmit nerve impulses from one part of the body to another. Epithelial cells that line the inside of the mouth are thin, flattened, and tightly packed, somewhat like floor tiles. They are protective cells that shield underlying cells. Muscle cells, which contract to pull structures closer together, are slender and rodlike, with their ends attached to the parts they move. They are filled with contractile proteins. An adipose cell is little more than a blob of fat. A B lymphocyte is an antibody factory.

## A Composite Cell

It is not possible to describe a typical cell, because cells vary so greatly in size, shape, content, and function. We can, however, consider a hypothetical composite cell that includes many known cell structures (fig. 3.3).

A cell consists of two major parts—the nucleus and the cytoplasm. The nucleus is innermost and is enclosed by a thin membrane called the nuclear envelope. The cytoplasm is a mass of fluid that surrounds the nucleus and is itself encircled by an even thinner cell membrane (also called plasma membrane). Within the cytoplasm are specialized structures called cytoplasmic organelles that perform specific metabolic functions. The nucleus, on the other hand, directs the overall activities of the cell. The nucleus is also considered to be an organelle.

> A cell with a nucleus, such as those of the human body, is termed *eukaryotic,* meaning "true nucleus." In contrast are the *prokaryotic* ("before nucleus") cells of bacteria. These streamlined bacterial cells lack a nucleus and other membrane-bound organelles. Simpler than eukaryotic cells, the bacteria are nevertheless quite a successful life form—as anyone who has ever had a strep throat or sinus infection can attest!

1. Give two examples to illustrate that the shape of a cell is related to its function.
2. Name the two major parts of a cell.
3. What are the general functions of these two parts?

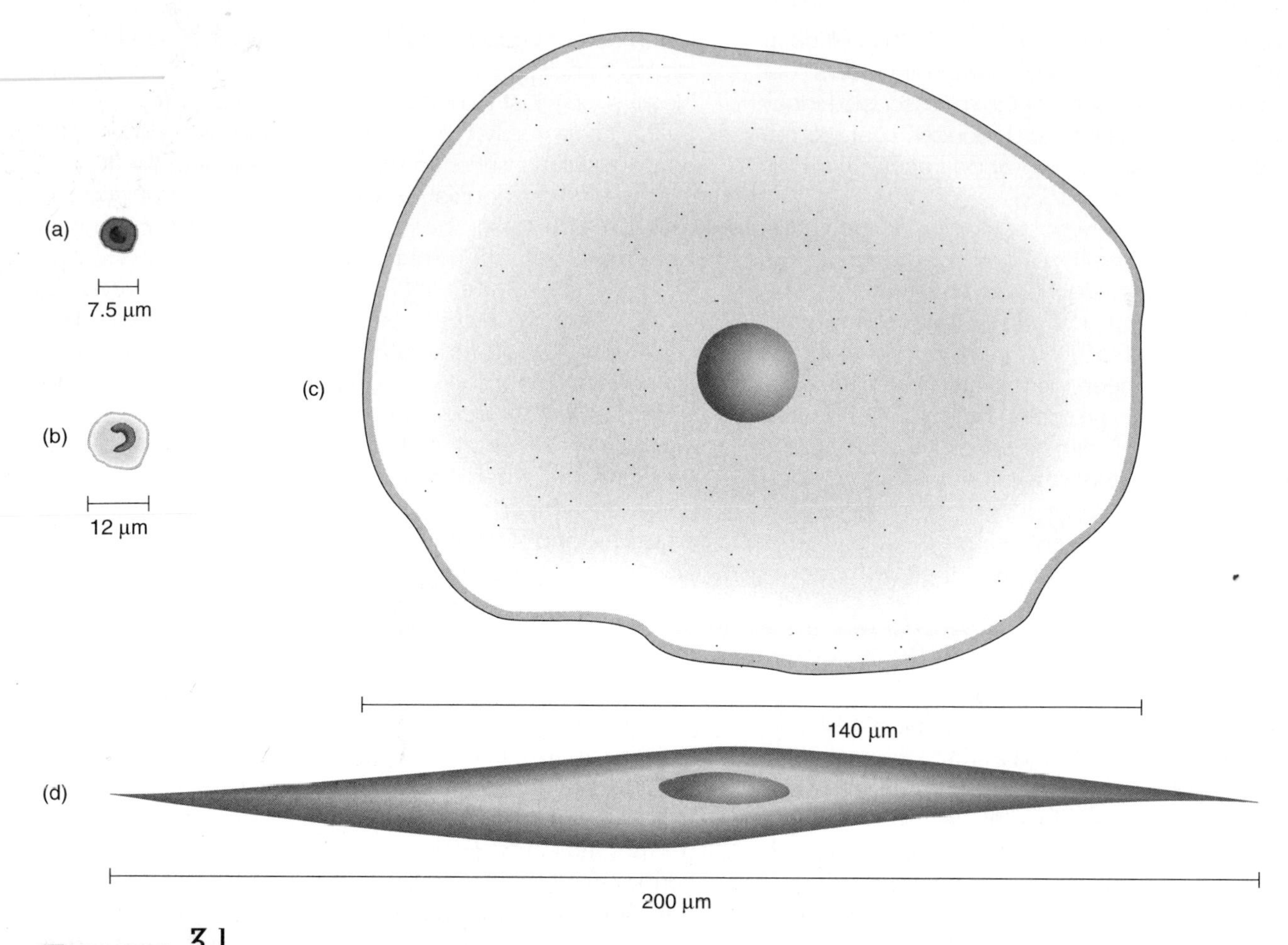


**FIGURE 3.1**

Cells vary considerably in size. This illustration shows the relative sizes of four types of cells. (*a*) Red blood cell, 7.5 μm in diameter; (*b*) white blood cell, 10–12 μm in diameter; (*c*) human egg cell, 140 μm in diameter; (*d*) smooth muscle cell, 20–500 μm in length.

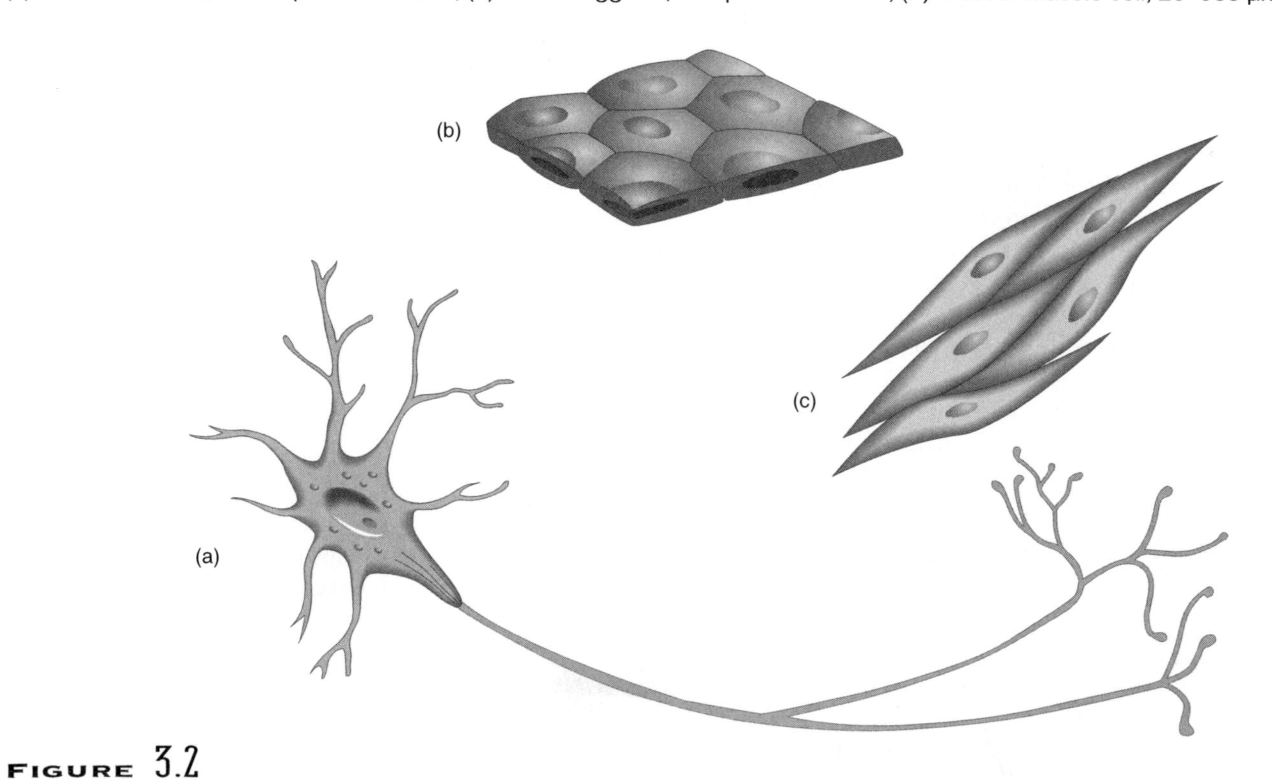


**FIGURE 3.2**

Cells vary in shape and function. (*a*) A nerve cell transmits impulses from one body part to another. (*b*) Epithelial cells protect underlying cells. (*c*) Muscle cells pull structures closer.

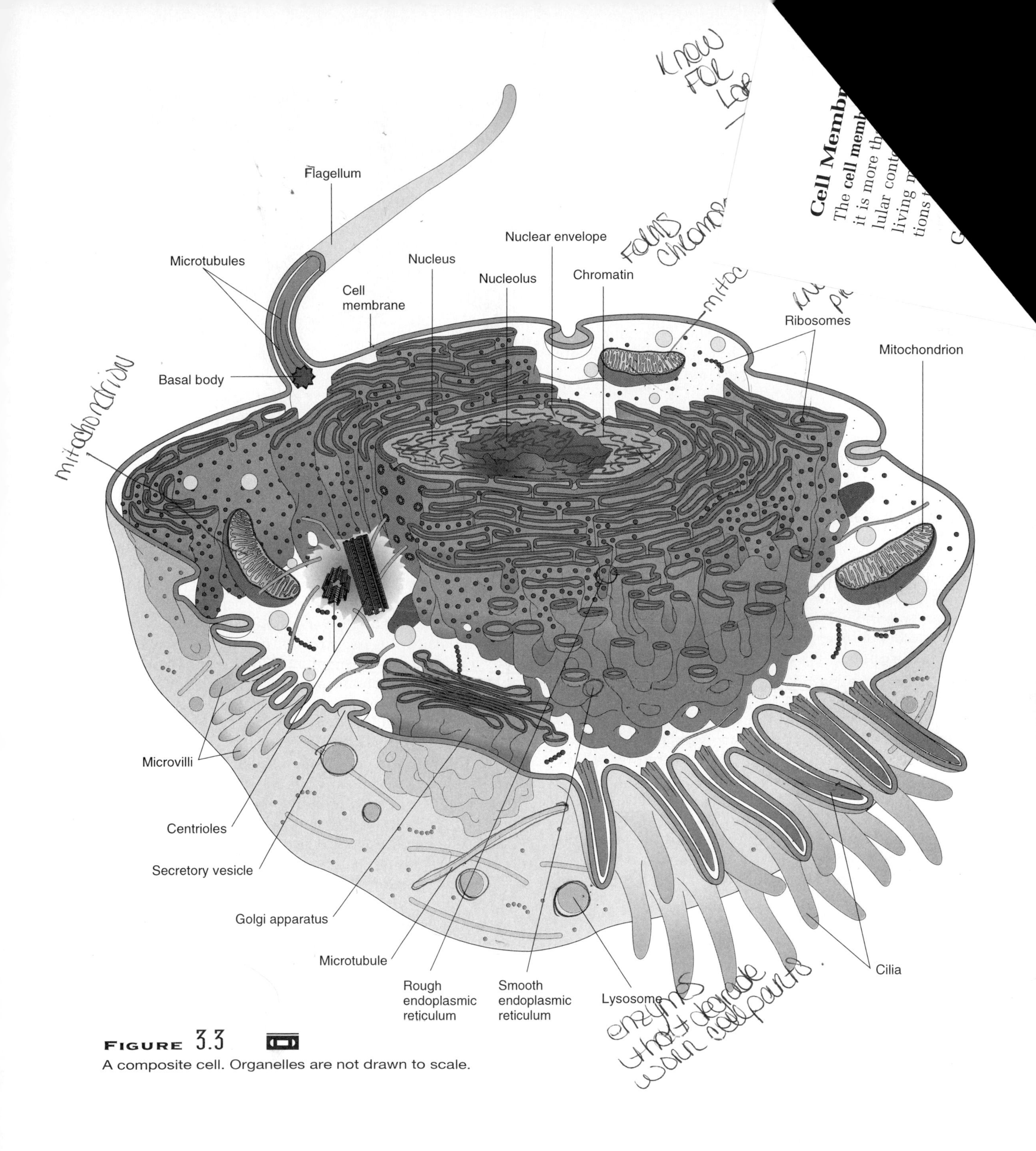


**Figure** 3.3

A composite cell. Organelles are not drawn to scale.

## ...ane

...rane is the outermost limit of a cell, but ...n a simple boundary surrounding the cel... ...nts. It is an actively functioning part of the ...aterial, and many important metabolic reac... ...ake place on its surfaces.

### ...eneral Characteristics

The cell membrane is extremely thin—visible only with the aid of an electron microscope (fig. 3.4)—but it is flexible and somewhat elastic. It typically has complex surface features with many outpouchings and infoldings that increase surface area. The membrane quickly seals minute breaks, but if it is extensively damaged, the cell contents escape, and the cell dies.

> The maximum effective magnification possible using a light microscope is about 1,200×. A transmission electron microscope (TEM) provides an effective magnification of nearly 1,000,000×, while a scanning electron microscope (SEM), can provide about 50,000×.
>
> Photographs of microscopic objects (micrographs) produced using the light microscope and the transmission electron microscope are typically two-dimensional, but those obtained with the scanning electron microscope have a three-dimensional quality (fig. 3.5).

In addition to maintaining the integrity of the cell, the membrane controls the entrance and exit of substances, allowing some in while excluding others. A membrane that functions in this manner is *selectively permeable.* The cell membrane is crucial because it is a conduit between the cell and the extracellular fluids in the body's internal environment. It allows the cell to receive and respond to incoming messages, a process called **signal transduction.**

### Membrane Structure

Chemically, the cell membrane is composed mainly of lipids and proteins, although it also contains a small quantity of carbohydrate. Its basic framework consists of a double layer (bilayer) of phospholipid molecules (see chapter 2 and fig. 2.13*b*). These molecules are arranged so that their water-soluble (hydrophilic) "heads," containing phosphate groups, form the surfaces of the membrane, and their water-insoluble (hydrophobic) "tails," consisting of fatty acid chains, make up the interior of the membrane (fig. 3.6). The lipid molecules are relatively free to move sideways within the plane of the membrane, and together they form a thin, but stable fluid film.

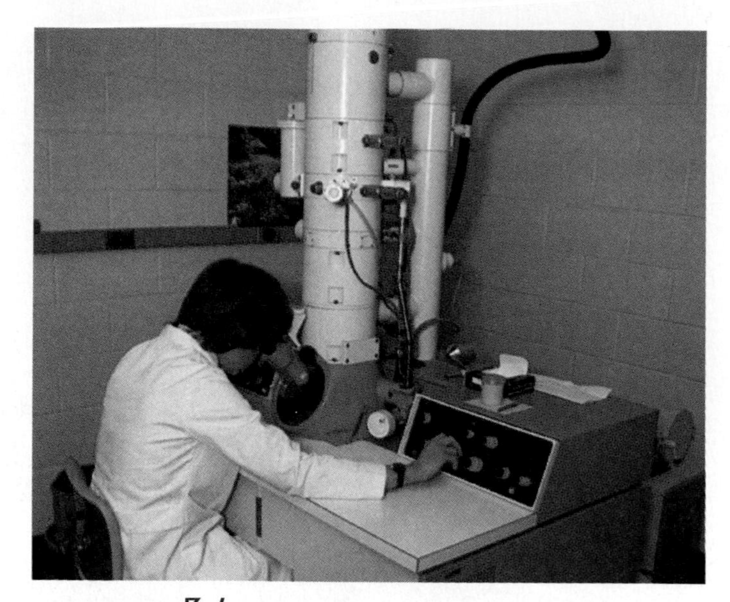

**FIGURE 3.4**
A transmission electron microscope.

Because the interior of the membrane consists largely of the fatty acid portions of the phospholipid molecules, it is oily. Molecules that are soluble in lipids, such as oxygen, carbon dioxide, and steroid hormones, can pass through this layer easily; however, the layer is impermeable to water soluble molecules, such as amino acids, sugars, proteins, nucleic acids, and various ions. Many cholesterol molecules embedded in the interior of the membrane also help make the membrane impermeable to water soluble substances. In addition, the relatively rigid structure of the cholesterol molecules helps stabilize the membrane.

Although the cell membrane includes only a few types of lipid molecules, it contains many kinds of proteins (fig. 3.7). These proteins are responsible for the special functions of the membrane, and they can be classified according to their shapes. One group of proteins, for example, consists of tightly coiled, rodlike molecules embedded in the bilayer of phospholipids. These fibrous proteins may completely span the membrane; that is, they may extend outward from its surface on one side, while their opposite ends communicate with the cell's interior. Such proteins often function as *receptors* that are specialized to combine with specific kinds of molecules, such as hormones (see chapter 13).

Another group of proteins are more compact and globular. Some of these proteins, called *integral proteins,* are embedded in the interior of the phospholipid bilayer. Typically, they span the membrane and form narrow passageways, or *channels,* through which small molecules and ions can cross the otherwise impermeable phospholipid bilayer. For example, some of these integral proteins form "pores" in the membrane that allow water molecules to pass through. Others are

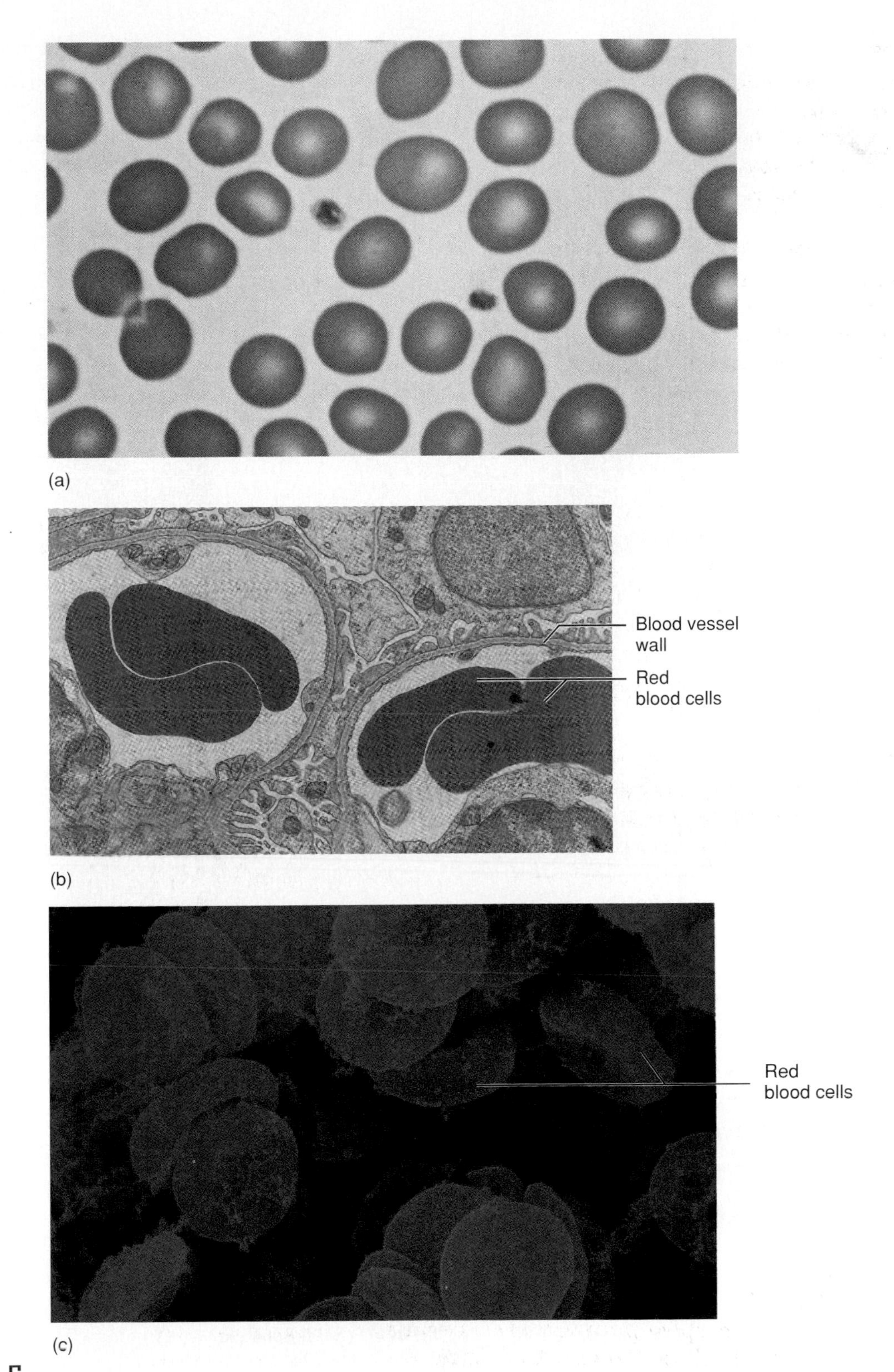


**FIGURE 3.5**

Human red blood cells as viewed using (*a*) a light microscope (1,400×), (*b*) a transmission electron microscope (3,800×), and (*c*) a scanning electron microscope (3,200×). The cells in (*a*) are stained to make them easier to see.

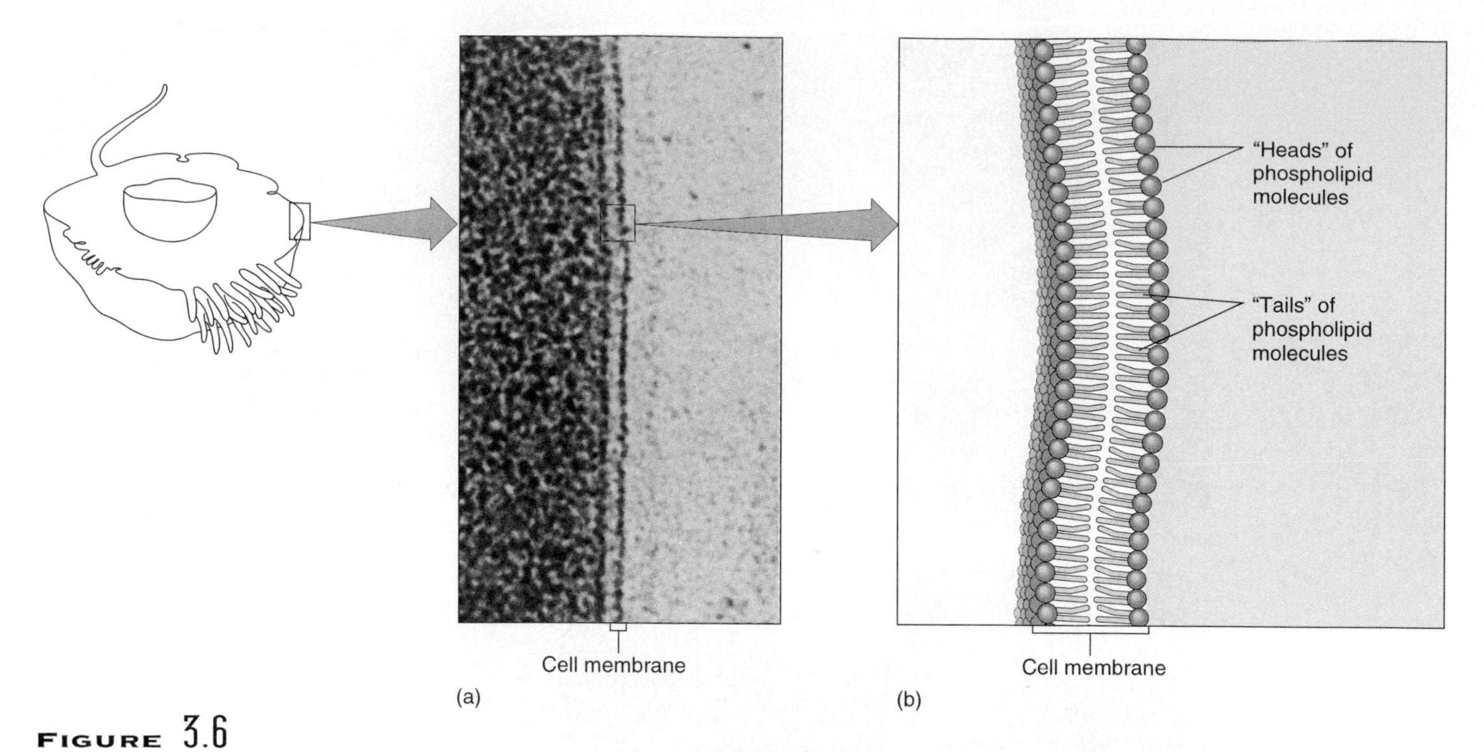


**FIGURE 3.6**
(*a*) A transmission electron micrograph of a cell membrane (250,000×); (*b*) the framework of the membrane consists of a double layer of phospholipid molecules.

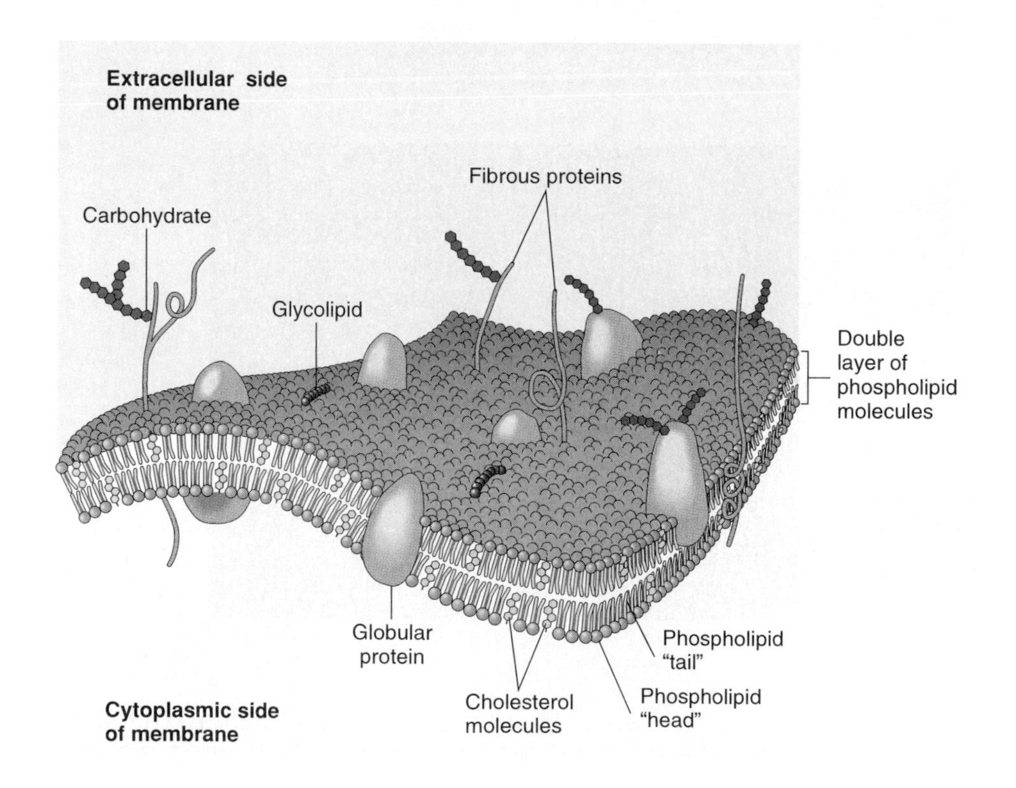


**FIGURE 3.7**
The cell membrane is composed primarily of phospholipids, with proteins scattered throughout the lipid bilayer and associated with its surfaces.

## Clinical Application 3.1

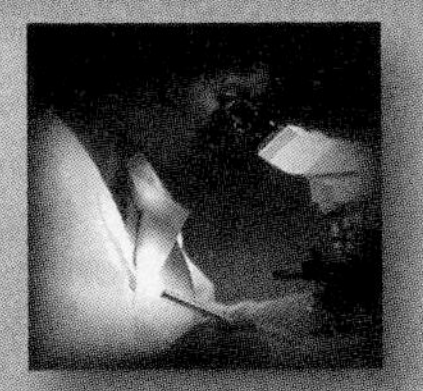

### The Blood-Brain Barrier

Perhaps nowhere else in the body are cells attached as firmly and closely as they are in the 400-mile network of capillaries in the brain. The walls of these microscopic blood vessels are but a single cell thick. They form sheets that fold into minute tubules. Studies in 1969 using the electron microscope revealed that the cell membranes overlap to form a barrier of tight junctions. Unlike the cells forming capillary walls elsewhere in the body, which are pocked with vesicles and window-like portals called clefts, the cells comprising this blood-brain barrier have few vesicles, and no clefts.

Why do the capillaries in the brain form such an impenetrable barrier? This anatomical arrangement shields delicate brain tissue from toxins in the bloodstream and from biochemical fluctuations that could be overwhelming if the brain had to continually respond to them. But all this protection has a downside—the brain is strictly off-limits to many therapeutic drugs. By studying the types of molecules embedded in the membranes of the cells forming the barrier, researchers are developing clever ways to sneak drugs into the brain. They can tag drugs to substances that can cross the barrier, design drugs to fit natural receptors in the barrier, or inject substances that temporarily relax the tight junctions forming the barrier.

The need to control the blood-brain barrier for drug delivery is compelling, for there are several increasing sources of brain disease in our population:

- Neurodegenerative diseases such as stroke, Alzheimer's disease, and Parkinson's disease increase in prevalence as the population ages.
- Brain tumors are more common as people survive other cancers because of improved treatments and earlier detection. Each year 15,000 people develop cancer in the brain as the primary site, but 150,000 develop brain tumors due to spread of a cancer that began elsewhere.
- AIDS has introduced several brain infections, such as Cryptococcal meningitis, caused by a fungus, and toxoplasmosis encephalitis, caused by a protozoan.
- Genetic research has identified the genes behind several disorders affecting the brain, leading the way to more targeted therapies.

more selective, forming channels that allow only particular substances to enter. In muscle and nerve cells, for example, selective channels control the movements of sodium and potassium ions, which are important in muscle contraction and nerve impulse conduction (see chapters 9 and 10).

Other globular proteins, called *peripheral proteins,* are associated with the surface of the cell membrane. These proteins function as enzymes (see chapter 4), and many are part of signal transduction.

### Intercellular Junctions

Many cells, such as blood cells, are separated from each other in fluid-filled spaces (intercellular spaces). In other cases, however, cells are tightly packed, their cell membranes connected by **intercellular junctions.**

In one type of intercellular junction, called a *tight junction,* the membranes of adjacent cells converge and fuse. The area of fusion surrounds the cell like a belt, and the junction closes the intercellular space between the cells. Cells that form sheetlike layers, such as those that line the inside of the digestive tract, often are joined by tight junctions. The linings of tiny blood vessels in the brain are extremely tight (Clinical Application 3.1).

Another type of intercellular junction, called a *desmosome,* rivets or "spot welds" adjacent skin cells, so they form a reinforced structural unit. The membranes of certain other cells, such as those in heart muscle and muscle of the digestive tract, are interconnected by tubular channels called *gap junctions.* These channels link the cytoplasm of adjacent cells and allow ions, nutrients (such as sugars, amino acids, and nucleotides), and other small molecules to move between them (fig. 3.8). Chart 3.1 summarizes these intercellular junctions.

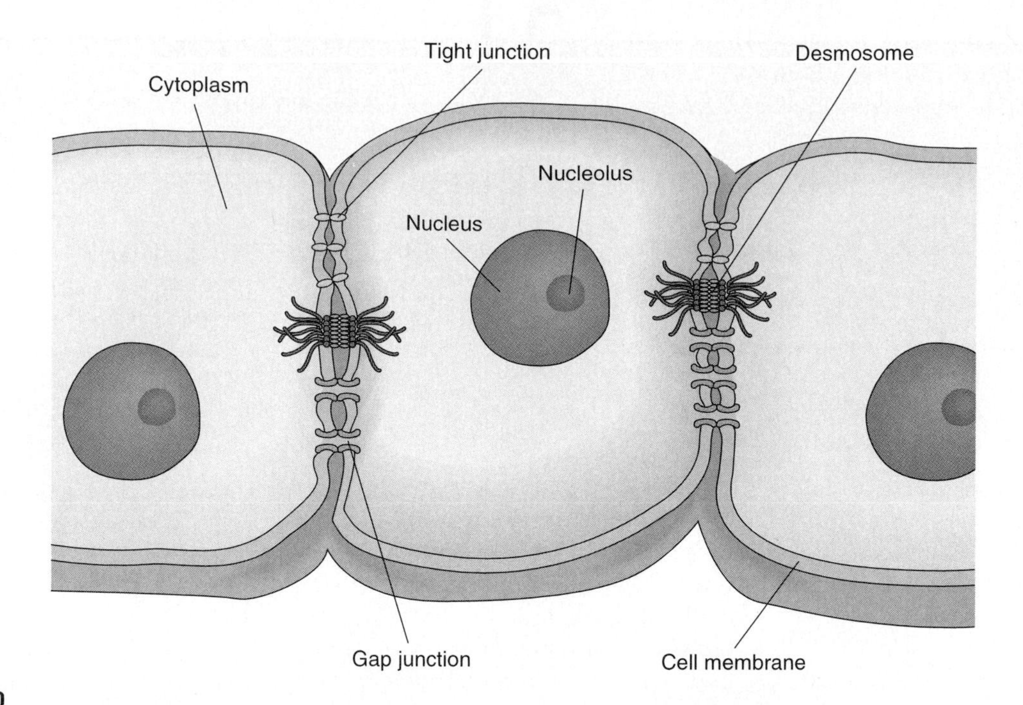


**FIGURE 3.8**
Some cells are joined by intercellular junctions, such as tight junctions that fuse neighboring membranes together, desmosomes that serve as "spot welds," or gap junctions that allow small molecules to move between the cytoplasm of adjacent cells.

### Cell Adhesion Molecules

Often cells must interact dynamically and transiently, rather than form permanent attachments. Proteins called **cell adhesion molecules,** or **CAMs** for short, guide cells on the move. Consider what happens when a white blood cell must travel in the bloodstream to the site of an injury, where it is needed to fight infection. Imagine that such a cell must reach a woody splinter embedded in a person's palm (fig. 3.9).

Once near the splinter, the white blood cell must slow down in the turbulence of the bloodstream. A CAM called a *selectin* does this by coating the white blood cell and providing traction. The white blood cell slows to a roll, and binds to carbohydrates on the inner capillary surface. Clotting blood, bacteria, and decaying tissue at the injury site release biochemicals (chemoattractants) that beckon the white blood cell. Finally, a CAM called an *integrin* contacts an adhesion receptor protein protruding into the capillary space near the splinter and pushes up through the capillary cell membrane, grabbing the passing slowed white blood cell and pulling it between adjacent tile-like cells of the capillary wall. White blood cells collecting at an injury site produce inflammation and, with the dying bacteria, constitute pus.

## Cytoplasm

When viewed through a light microscope, **cytoplasm** usually appears as clear jelly with specks scattered throughout. However, a transmission electron microscope (see fig. 3.4), which produces much greater magnification and ability to distinguish fine detail (resolution), reveals that cytoplasm contains networks of membranes and organelles suspended in a clear liquid (cytosol). It also contains abundant protein rods and tubules that constitute a supportive network called the **cytoskeleton.**

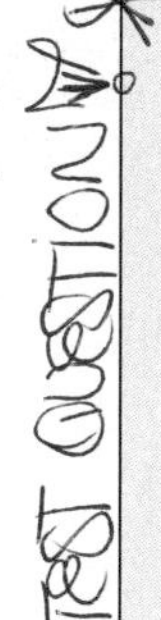


Dallas teenager Brooke Blanton was born lacking the CAMs that enable white blood cells to adhere to blood vessel walls. As a result, her sores do not heal, never forming pus because white blood cells never arrive at injury sites. Brooke's earliest symptoms were teething sores that did not heal. Today Brooke must be very careful to avoid injury or infection, for her white blood cells, although plentiful and healthy, simply zip past her wounds.

The activities of a cell occur largely in its cytoplasm, where nutrient molecules are received, processed, and used. In other words, cytoplasm is a site of metabolic reactions, in which the following organelles play specific roles:

1. **Endoplasmic reticulum.** The endoplasmic reticulum (ER) is a complex organelle composed of membrane-bound flattened sacs, elongated canals, and fluid-filled vesicles. These membranous parts are interconnected, and they

## CHART 3.1 Types of Intercellular Junctions

| Type | Function | Location |
|---|---|---|
| Tight junctions | Close intercellular space between cells by fusing cell membranes together | Cells that line inside of the small intestine |
| Desmosomes | Bind cells together by forming "spot welds" between adjacent membranes | Cells of the outer skin layer |
| Gap junctions | Form tubular channels between cells that allow substances to be exchanged | Muscle cells of the heart and digestive tract |

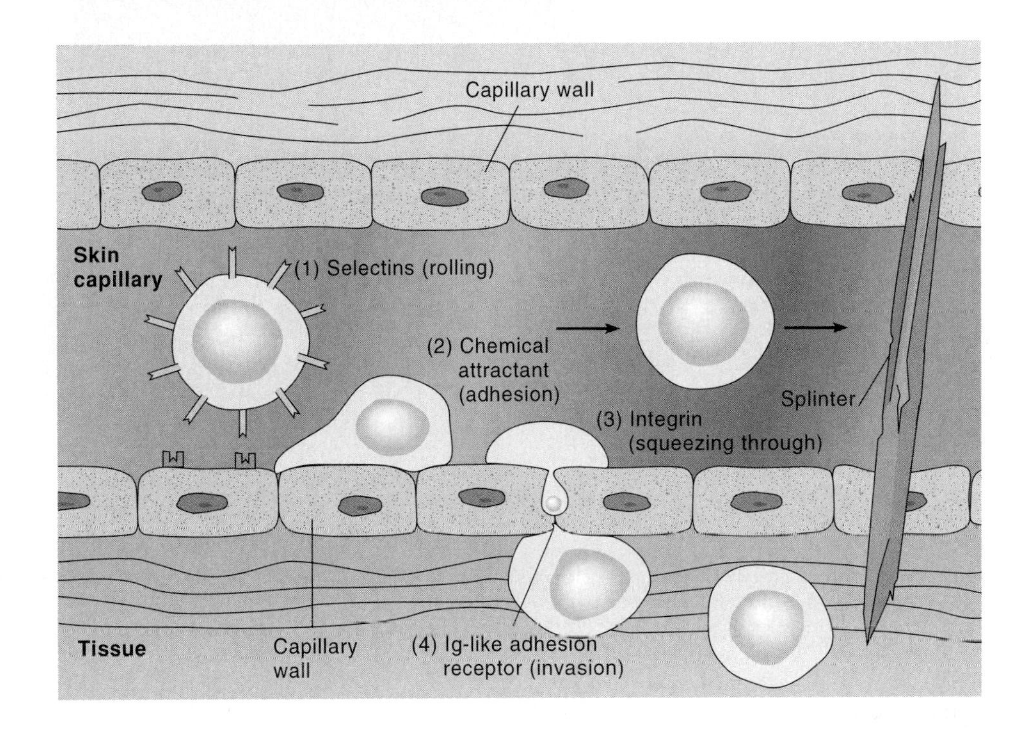


**FIGURE 3.9**
Cell adhesion molecules (CAMs) assist some cells in moving. When a white blood cell must leave the circulation to help fight an infection, it is first slowed from barreling through a capillary at 2,500 micrometers per second to rolling slowly at 50 micrometers per second by binding proteins called selectins (1). Next, specific chemical attractants cause the white blood cell to adhere at the site of an injury (2). An integrin protein (3) then squeezes through the capillary to grab the white blood cell, and with an adhesion receptor protein helps the cell squeeze through the tight junctions of the capillary lining to invade the tissue (4).

communicate with the cell membrane, the nuclear envelope, and certain cytoplasmic organelles. ER is widely distributed through the cytoplasm, providing a tubular transport system for molecules throughout the cell.

The endoplasmic reticulum also plays a role in the synthesis of protein and lipid molecules, some of which may be assembled into new membranes. Commonly, the outer membranous surface of the ER is studded with many tiny, spherical organelles called *ribosomes* that give the ER a textured appearance when viewed with an electron microscope. Such endoplasmic reticulum is termed *rough* ER. Endoplasmic reticulum that lacks ribosomes is called *smooth* ER (fig. 3.10).

The ribosomes of rough ER are sites of protein synthesis. The proteins may then move through the canals of the endoplasmic reticulum to the Golgi apparatus for further processing. Smooth ER, on the other hand, contains enzymes important in lipid synthesis.

2. **Ribosomes.** Besides being found on the endoplasmic reticulum, some ribosomes are scattered freely throughout the cytoplasm. All ribosomes are composed of protein and RNA, and they function in the synthesis of proteins (see chapter 4).
3. **Golgi apparatus.** The Golgi apparatus is composed of a stack of half a dozen or so flattened, membranous sacs called *cisternae.* This organelle refines, packages, and delivers proteins synthesized by the ribosomes associated with the ER (fig. 3.11).

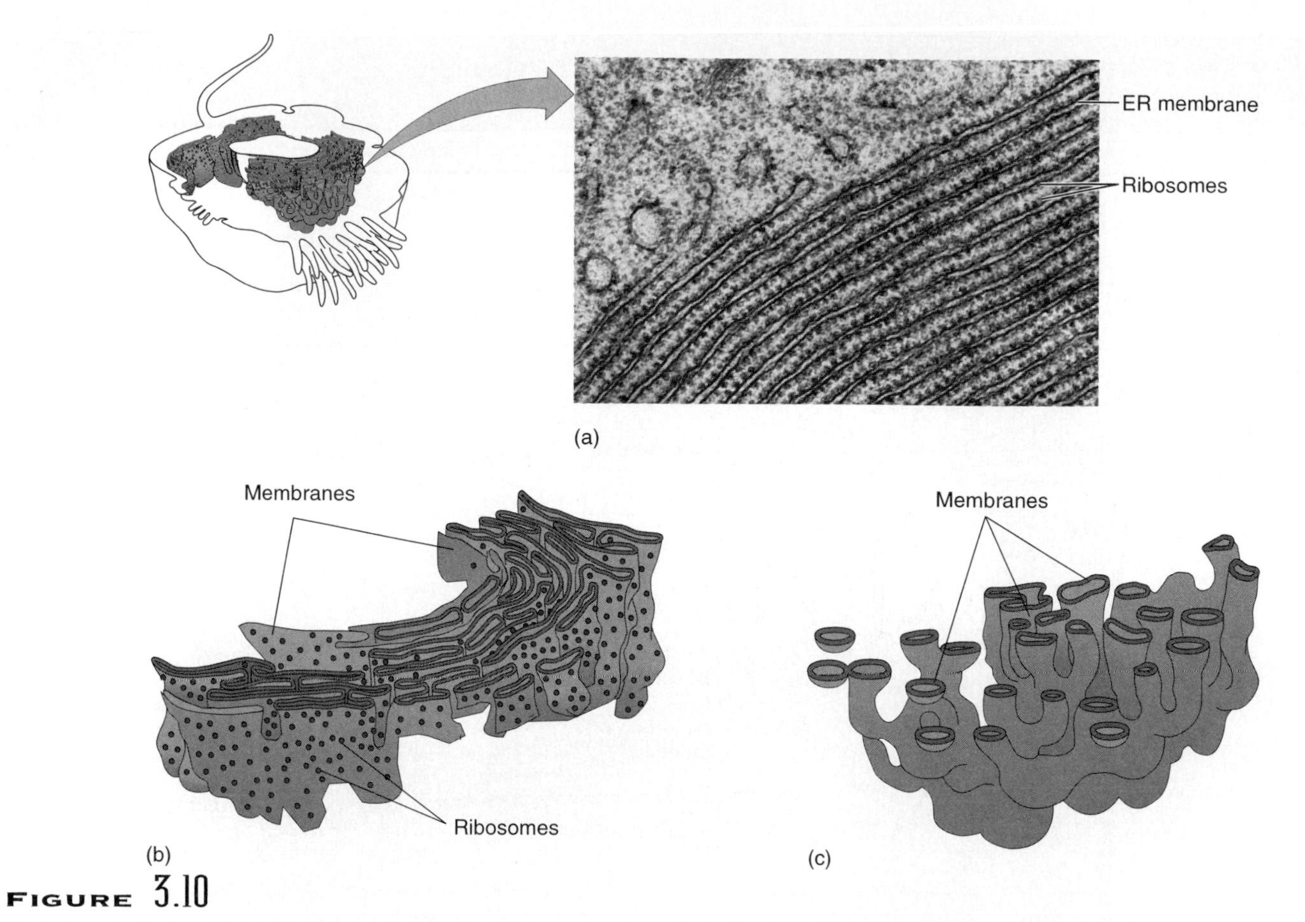


**FIGURE 3.10**

(*a*) A transmission electron micrograph of rough endoplasmic reticulum (ER) (100,000×). (*b*) Rough ER is dotted with ribosomes, while (*c*) smooth ER lacks ribosomes.

Proteins arrive at the Golgi apparatus enclosed in tiny vesicles composed of membrane from the endoplasmic reticulum. These sacs fuse to the membrane at the beginning or innermost end of the Golgi apparatus, which is specialized to receive proteins. Previously, these protein molecules were combined with sugar molecules, and thus they are *glycoproteins.*

As the glycoproteins pass from layer to layer through the Golgi stacks, they are modified chemically. For example, some sugar molecules may be added or removed from them. When they reach the outermost layer, the altered glycoproteins are packaged in bits of Golgi apparatus membrane that bud off and form transport vesicles. Such a vesicle may then move to the cell membrane, where it fuses with the membrane and releases its contents to the outside of the cell as a secretion. Other vesicles may transport glycoproteins to organelles within the cell (fig. 3.12).

In some cells, including certain liver cells and white blood cells (lymphocytes), glycoprotein molecules are secreted as rapidly as they are synthesized. However, in other cells, such as those that manufacture protein hormones, the vesicles containing the newly synthesized molecules remain in the cytoplasm and release the stored substances only when the cells are stimulated. (Hormone secretion is discussed in chapter 13.)

Note that new membrane originates in the endoplasmic reticulum. Portions of this membrane may then be transferred to the Golgi apparatus. From there, some of the membrane may move outward as part of a vesicle and join the cell membrane.

1. What is a selectively permeable membrane?
2. Describe the chemical structure of a cell membrane.
3. What are the different types of intercellular junctions?
4. What are some of the events of cell adhesion?
5. What are the functions of the endoplasmic reticulum?
6. Describe how the Golgi apparatus functions.

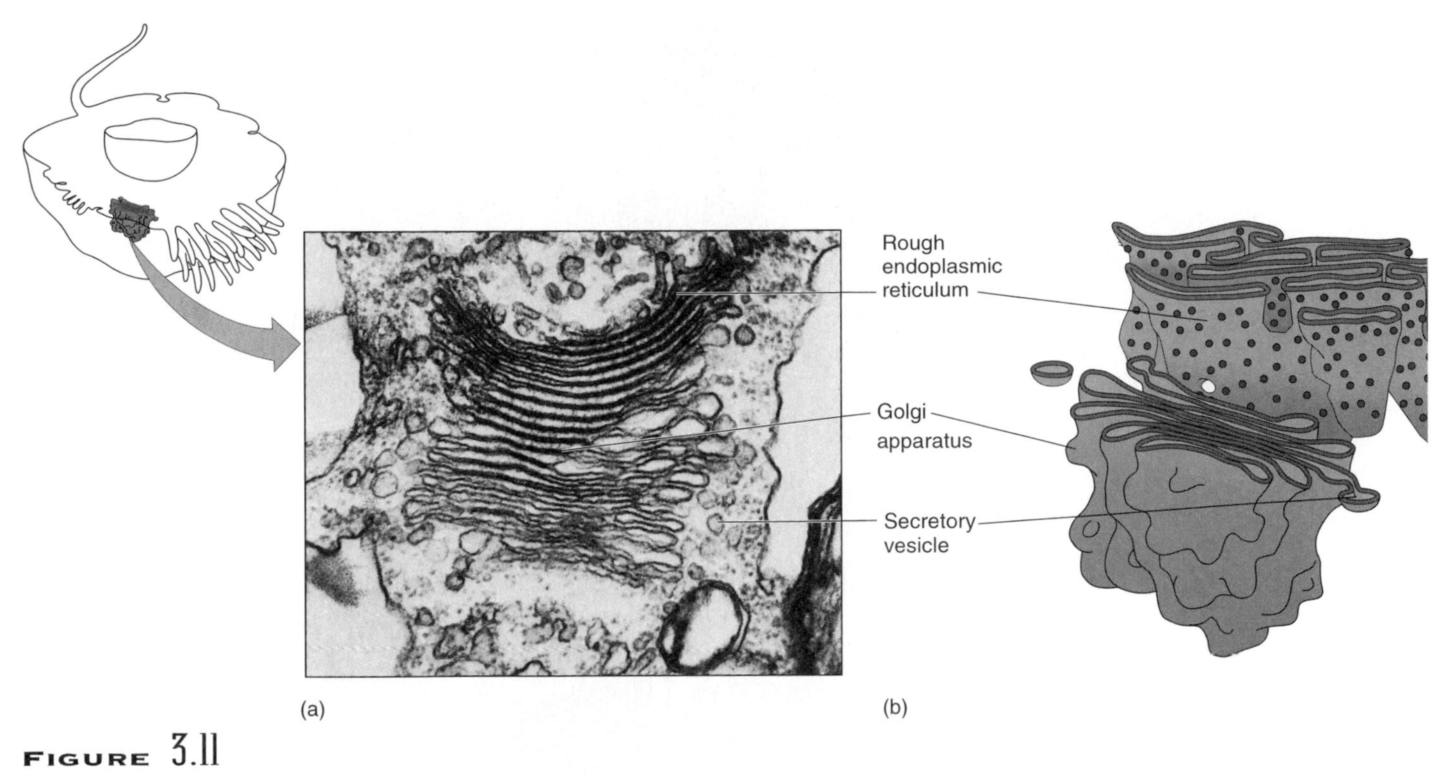


**FIGURE 3.11**

(*a*) A transmission electron micrograph of a Golgi apparatus (36,000×). (*b*) The Golgi apparatus consists of membranous sacs that continually receive vesicles from the endoplasmic reticulum and produce vesicles for secretion.

4. **Mitochondria.** Mitochondria are elongated, fluid-filled sacs 2–5 μm long. They often move about slowly in the cytoplasm and can reproduce by dividing. A mitochondrion also contains a small amount of DNA that encodes information for making a few kinds of proteins and specialized RNA. However, most mitochondrial proteins are encoded in the DNA of the nucleus. These proteins are synthesized elsewhere in the cell and later enter the mitochondria.

   The membrane of a mitochondrion has two layers—an outer membrane and an inner membrane. The inner membrane is folded extensively to form shelflike partitions called *cristae.* Small, stalked particles that contain enzymes are connected to the cristae. These enzymes and others dissolved in the fluid within the mitochondrion control many of the chemical reactions that release energy from glucose and other organic molecules. The mitochondria transform this newly released energy into a chemical form, the molecule adenosine triphosphate (ATP), that organelles can use (see fig. 3.13 and chapter 4). For this reason mitochondria are sometimes called the "powerhouses" of cells.

A typical cell has about 1,700 mitochondria, but cells with very high energy requirements, such as muscle, have many thousands of mitochondria. This is why a common symptom of illnesses affecting mitochondria is muscle weakness. Symptoms of these "mitochondrial myopathies" include exercise intolerance and weak and flaccid muscles.

Mitochondria are particularly fascinating to biologists because they provide glimpses into the past. Mitochondria are passed to offspring from mothers only, because these organelles are excluded from the part of a sperm that enters an egg cell. Evolutionary biologists study sequences of genes in mitochondria to trace human origins, back to a long ago group of ancestors metaphorically called "mitochondrial Eve."

Mitochondria may provide clues to a past far more remote than the beginnings of humankind. According to the widely accepted endosymbiont theory, mitochondria are the remnants of once free-living bacteria-like cells that were swallowed up by primitive eukaryotic cells. These bacterial passengers remain in our cells today, where they participate in energy reactions.

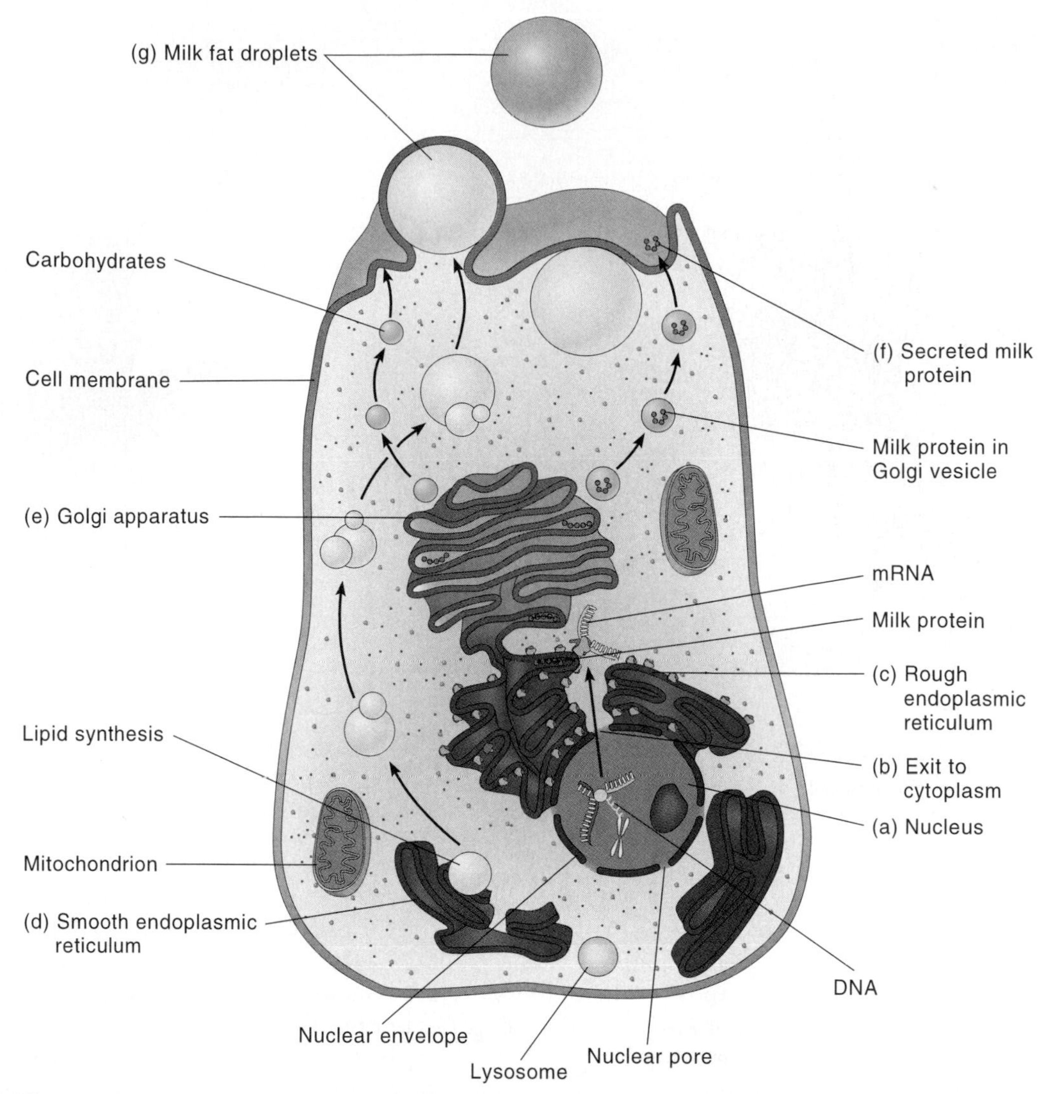


**FIGURE 3.12**

Milk secretion illustrates how organelles interact to synthesize, transport, store, and export biochemicals. Secretion begins in the nucleus (*a*), where messenger RNA molecules bearing genetic instructions for production of milk proteins exit through nuclear pores to the cytoplasm (*b*). Most proteins are synthesized on membranes of the rough endoplasmic reticulum (ER) (*c*), using amino acids in the cytoplasm. Lipids are synthesized in the smooth ER (*d*), and sugars are synthesized, assembled, and stored in the Golgi apparatus (*e*). An active mammary gland cell releases milk proteins from vesicles that bud off of the Golgi apparatus (*f*). Fat droplets pick up a layer of lipid from the cell membrane as they exit the cell (*g*). When the baby suckles, he or she receives a chemically complex secretion—milk.

5. **Lysosomes.** Lysosomes are the "garbage disposals" of the cell, whose function is to dismantle debris. They are sometimes difficult to identify because their shapes vary so greatly. However, they commonly appear as tiny, membranous sacs (fig. 3.14). These sacs contain powerful enzymes that break down protein, carbohydrate, and nucleic acids as well as foreign particles composed of these substances. Certain white blood cells, for example, engulf bacteria that are then digested by the lysosomal enzymes. This is one way that white blood cells help prevent bacterial infections.

   Lysosomes also destroy worn cellular parts. In fact, lysosomes of some scavenger cells may engulf and digest entire body cells that have been injured. How the lysosomal membrane is able to withstand being digested itself is not well understood, but this organelle sequesters enzymes that can function only under very acidic conditions, preventing them from destroying the cellular contents around them. Human lysosomes contain forty or so different types of enzymes. An abnormality in just one type of lysosomal enzyme can be devastating to health (see Clinical Application 3.2).

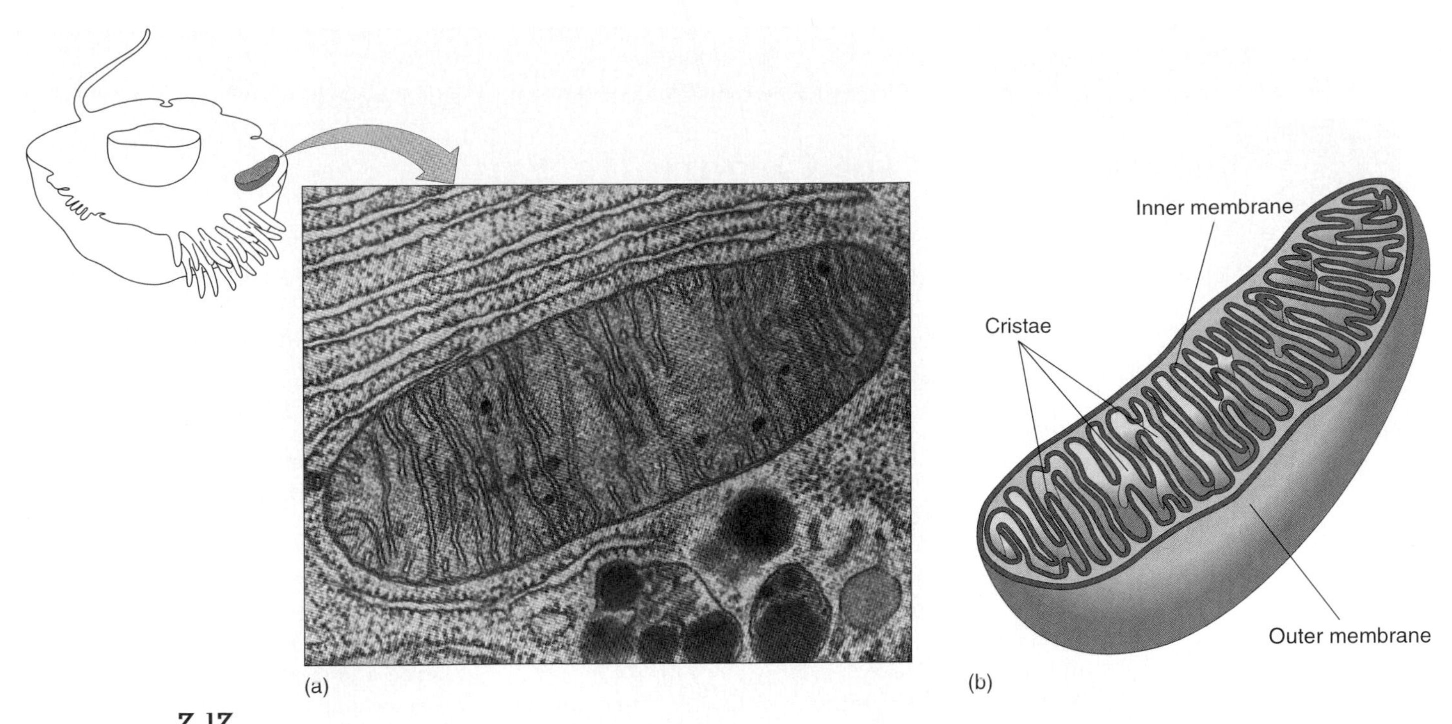


**FIGURE 3.13**
(*a*) A transmission electron micrograph of a mitochondrion (79,000×). (*b*) Cristae form partitions within this saclike organelle.

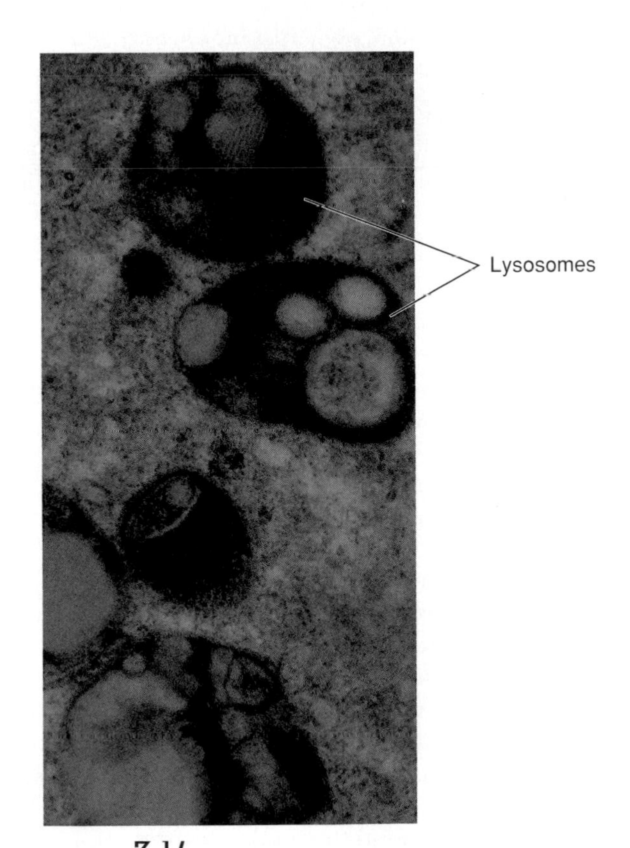


**FIGURE 3.14**
In this falsely colored transmission electron micrograph, lysosomes appear as membranous sacs (14,100×).

> Lysosomal digestive activity seems to be responsible for decreasing the size of body tissues at certain times. Such regression in size occurs in the maternal uterus following the birth of an infant, in the maternal breasts after the weaning of an infant, and in skeletal muscles during periods of prolonged inactivity.

6. **Peroxisomes.** Peroxisomes are membranous sacs that resemble lysosomes in size and shape. Although found in all human cells, peroxisomes are most abundant in cells of the liver and kidneys. Peroxisomes contain enzymes, called peroxidases, that catalyze metabolic reactions that release hydrogen peroxide ($H_2O_2$) as a by-product. Peroxisomes also contain an enzyme called catalase, which decomposes hydrogen peroxide, which is toxic to cells.

   The outer membrane of a peroxisome contains some forty types of enzymes, which catalyze a variety of biochemical reactions, including:

   - synthesis of bile acids, which are used in fat digestion
   - breakdown of lipids called very long chain fatty acids
   - degradation of rare biochemicals
   - detoxification of alcohol

## Clinical Application 3.2

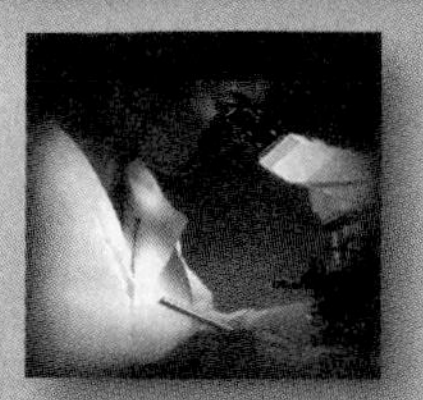

# Disease at the Organelle Level

German physiologist Rudolph Virchow first hypothesized cellular pathology—disease at the cellular level—in the 1850s. Today, new treatments for many disorders are a direct result of understanding a disease process at the cellular level. Here, we take a look at how three abnormalities—in cell membranes, in peroxisomes, and in lysosomes—cause whole-body symptoms.

### Cystic Fibrosis and the Cell Membrane

Cystic fibrosis (CF) was first described in medical journals in 1938 as a defect in the channels leading from certain glands, resulting in a variety of problems—chokingly thick mucus in the lungs and frequent infection there; a clogged pancreas, preventing digestive juices from reaching the intestines; and salty sweat. A child with CF is often small and sickly, and until the recent availability of biochemical tests, was often initially diagnosed simply as having "failure to thrive."

Earlier descriptions of CF mentioned the characteristic of salty sweat. A seventeenth century English saying states, "A child that is salty to taste will die shortly after birth." But this symptom would be a telling clue in explaining how the genetic abnormality causes symptoms felt at a whole-body level.

Researchers identified the cellular defect behind cystic fibrosis in 1989 as abnormal channels in lung and pancreas cells that trap salt within cells. The salty cellular interiors draw moisture in from surrounding tissue, drying out the mucus until it is so sticky that it clogs organs. Several new treatments, including a healthy gene introduced into the lungs in a nasal spray, target the illness at its cellular source.

### Adrenoleukodystrophy (ALD) and Peroxisomes

For young Lorenzo Odone, the first sign of adrenoleukodystrophy was disruptive behavior in school. When he became lethargic, weak, and dizzy, his teachers and parents realized that his problem was not just temper tantrums. His skin darkened, blood sugar levels plummeted, heart rhythm altered, and the levels of electrolytes in his body fluids became imbalanced. He lost control over his limbs as his nervous system continued to deteriorate. Lorenzo's parents took him to many doctors. Finally one of them tested for an enzyme normally manufactured in peroxisomes.

Abnormal peroxisomal enzymes can drastically affect health.

7. **Centrosome.** A centrosome (central body) is a structure located in the cytoplasm near the Golgi apparatus and nucleus. It is nonmembranous and consists of two hollow cylinders called *centrioles* built of tubelike proteins called microtubules. The centrioles usually lie at right angles to each other and function in cell reproduction. During this process, the centrioles move away from one another and take positions on either side of the nucleus. There they aid in distributing *chromosomes,* which carry DNA information, to the newly forming cells (fig. 3.15).

   Centrioles also form parts of the hairlike cellular projections called *cilia* and *flagella.*

8. **Cilia** and **flagella.** Cilia and flagella are motile processes that extend outward from the surfaces of certain cells. They are structurally similar and differ mainly in their length and the number present. Both contain a constant number of microtubules arranged in a distinct cylindrical pattern.

   Cilia occur in large numbers on the free surfaces of some epithelial cells. Each cilium is a tiny, hairlike structure about 10 μm long, which is attached just beneath the cell membrane to a modified centriole called a *basal body.*

   Cilia are arranged in precise patterns. They have a "to-and-fro" type of movement that is coordinated so that rows of cilia beat one after the other, producing a wave that sweeps across the ciliated surface. This action propels fluids, such as mucus, over the surface of tissues that form the lining of the respiratory tract (fig. 3.16). Chemicals in cigarette smoke destroy cilia, causing respiratory ailments.

   A cell usually has only one flagellum, which is much longer than a cilium. A flagellum begins its characteristic undulating, wavelike motion

Lorenzo's peroxisomes lacked the second most abundant protein in the outer membrane of this organelle. Normally, the missing protein transports an enzyme into the peroxisome. The enzyme controls breakdown of a type of very long chain fatty acid. Without the enzyme, the fatty acid builds up in cells in the brain and spinal cord, eventually stripping these cells of their fatty sheaths, made of a substance called myelin, that are vital for nerve transmission. Death comes in a few years.

For Lorenzo and many other sufferers of ALD, eating a type of triglyceride from rapeseed oil slows buildup of the very long chain fatty acids for a few years, stalling symptoms. But the treatment eventually impairs blood clotting and other vital functions, and fails to halt the progression of the illness.

### Tay-Sachs Disease and Lysosomes

Michael was a pleasant, happy infant who seemed to be developing normally until about six months of age. Able to roll over and sit for a few seconds, suddenly he seemed to lose those abilities. Soon, he no longer turned and smiled at his mother's voice, as he had before, and he did not seem as interested in his mobile as he once was. Concerned about Michael's reversals in development, his anxious parents took him to the doctor. It took exams by several specialists to diagnose Michael's Tay-Sachs disease, because, thanks to screening programs in the population groups known to have this inherited illness, fewer than ten new cases appear each year. Michael's parents were not among those ethnic groups and previously had no idea that they both were carriers of the gene that causes this very rare illness.

A neurologist clinched her suspicion of Tay-Sachs by looking into Michael's eyes, where she saw the telltale "cherry red spot" indicating the illness. A look at his cells provided further clues—the lysosomes, tiny enzyme-filled sacs, were swollen to huge proportions. Michael's lysosomes lacked one of the forty types of lysosomal enzymes, resulting in a "lysosomal storage disease" that built up fatty material on his nerve cells. His nervous system would continue to fail, and he would be paralyzed and unable to see or hear by the time he died, before the age of four.

The cellular and molecular signs of Tay-Sachs disease—the swollen lysosomes and missing enzyme—had been present long before Michael began to lag developmentally. The next time his parents expected a child, they had her tested before birth for the enzyme deficiency. They learned, happily, that she would be a carrier like themselves, but not ill.

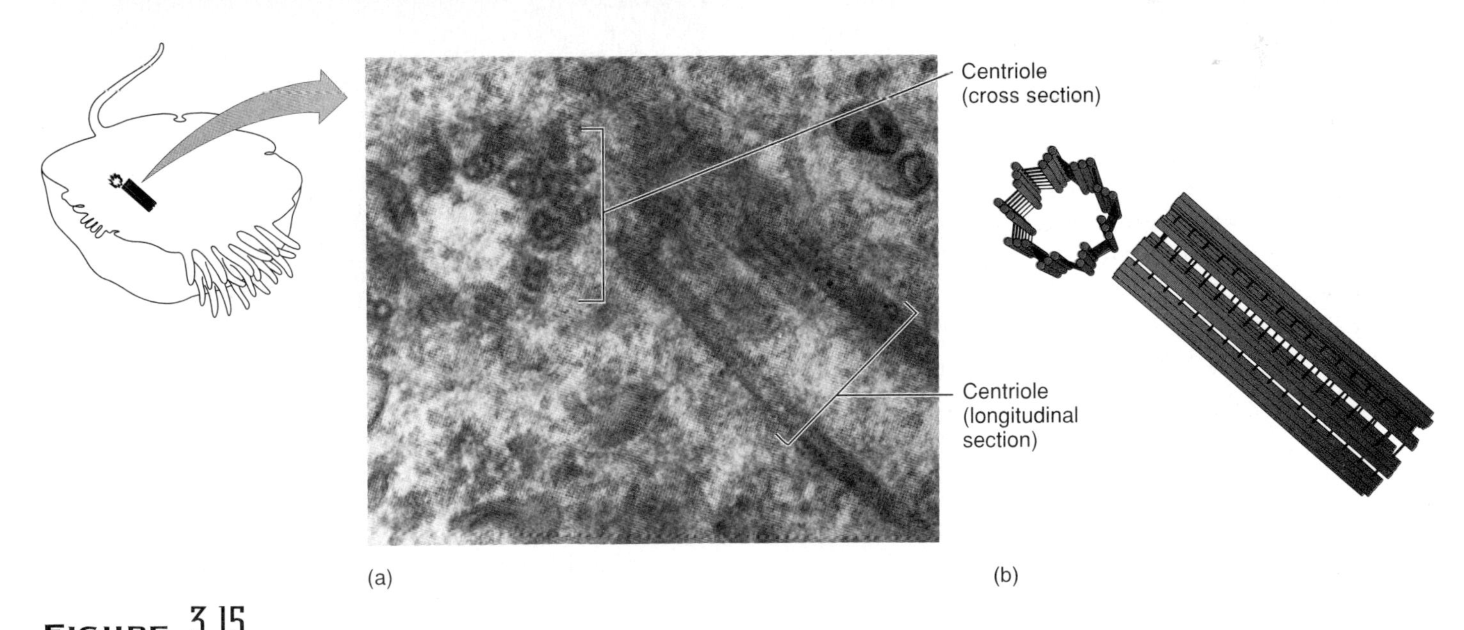


**Figure 3.15**
(*a*) A transmission electron micrograph of the two centrioles in a centrosome (142,000×). (*b*) Note that the centrioles lie at right angles to one another.

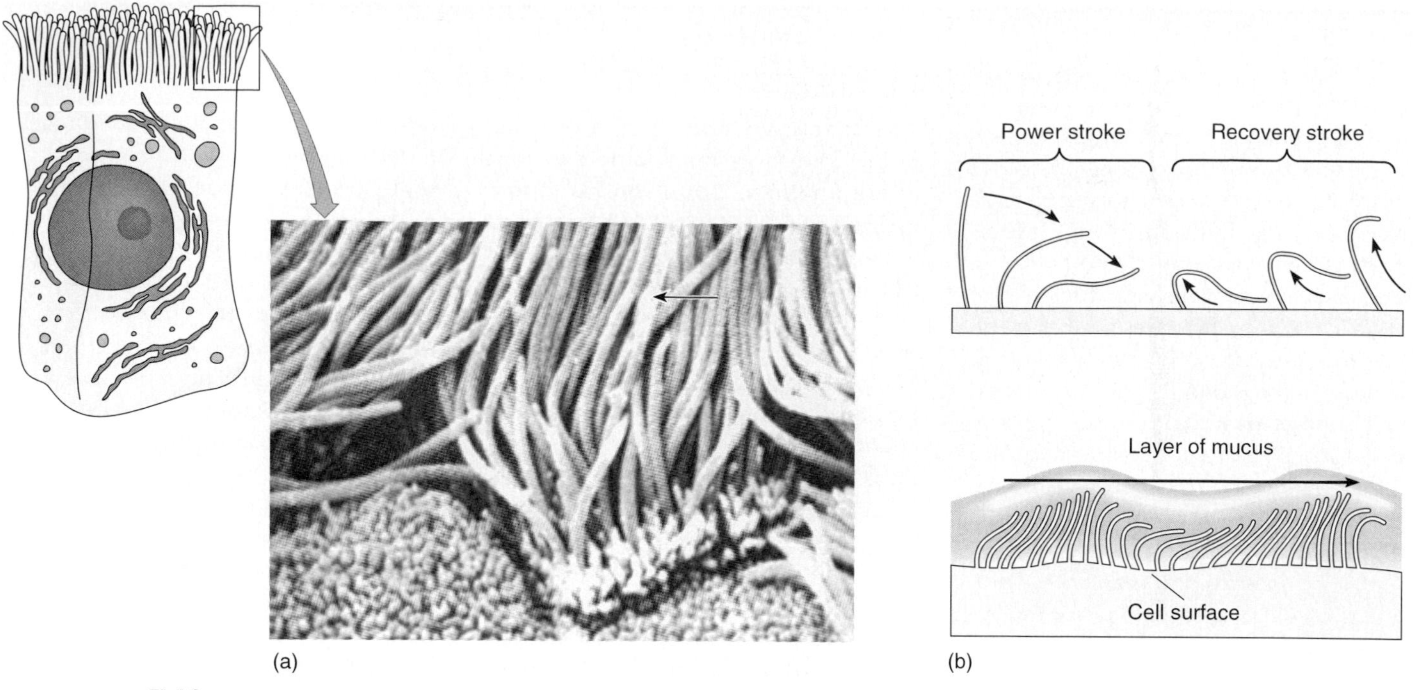


**FIGURE 3.16**
(*a*) Cilia, such as these (arrow), are common on the surfaces of certain cells that form the inner lining of the respiratory tract (10,000×). (*b*) Cilia have a power stroke and a recovery stroke creating a "to-and-fro" movement that sweeps fluids across the tissue surface.

at its base. The tail of a sperm cell, for example, is a flagellum that causes the sperm's swimming movements (see fig. 3.17 and chapter 22).

9. **Vesicles.** Vesicles (vacuoles) are membranous sacs that vary in size and contents. They may be formed when a portion of the cell membrane folds inward and pinches off. As a result, a tiny, bubblelike vesicle, containing some liquid or solid material that was formerly outside the cell, appears in the cytoplasm. The Golgi apparatus and ER also form vesicles.

    The processes of endocytosis and exocytosis, which involve vesicle formation, are discussed in a subsequent section of this chapter.

10. **Microfilaments and microtubules.** Two types of thin, threadlike structures found within the cytoplasm are microfilaments and microtubules.

    Microfilaments are tiny rods of the protein actin, arranged in meshworks or bundles. They cause various kinds of cellular movements. In muscle cells, for example, they constitute *myofibrils,* which help these cells shorten or contract. In other cells, microfilaments associate with the inner surface of the cell membrane and aid in cell motility (fig. 3.18).

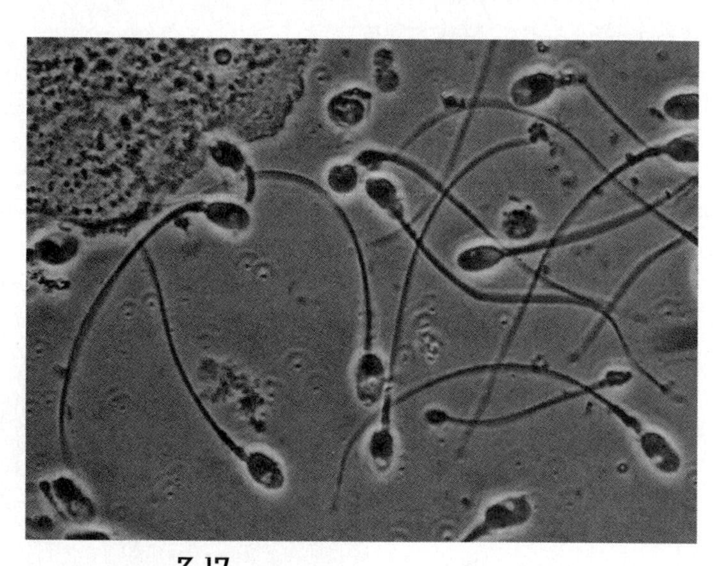

**FIGURE 3.17**
Light micrograph of human sperm cells (1,200×). Flagella form the tails of these cells.

Microtubules are long, slender tubes with diameters two or three times greater than those of microfilaments, and are composed of the globular protein tubulin. They are usually stiff, forming the cytoskeleton, which helps maintain the shape

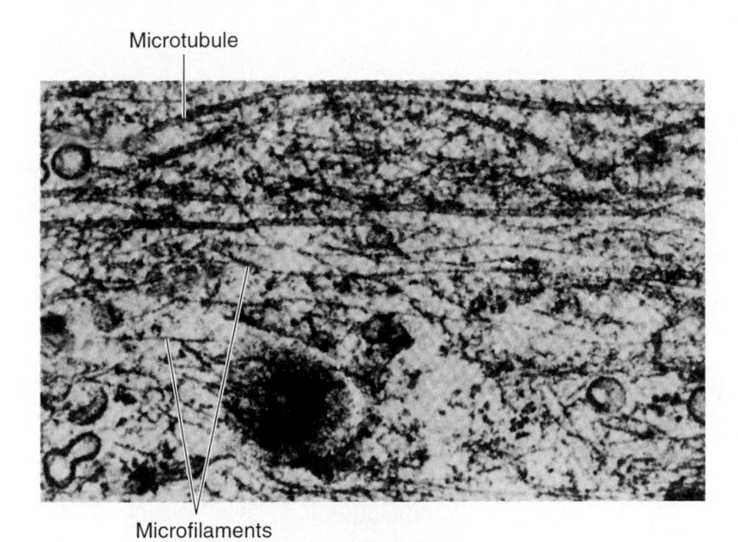


**FIGURE 3.18**

A transmission electron micrograph of microfilaments and microtubules within the cytoplasm.

of the cell (fig. 3.19). Microtubules interact with each other to provide movement, such as in cilia and flagella (see figs. 3.16 and 3.17).

Microtubules also move organelles within the cell. For instance, they are assembled in the cytoplasm during cellular reproduction and distribute chromosomes to the newly forming cells, a process described in more detail later in this chapter.

11. **Other structures.** In addition to these functional organelles, cytoplasm may contain masses of lifeless chemicals called *inclusions.* These usually remain in a cell temporarily. Inclusions include stored nutrients such as glycogen and lipids and pigments such as melanin in the skin.

1. Why are mitochondria sometimes called the "powerhouses" of cells?
2. How do lysosomes function?
3. Describe the functions of microfilaments and microtubules.
4. Distinguish between organelles and inclusions.

## Cell Nucleus

A **nucleus** is a relatively large, usually spherical structure that directs the activities of the cell. It is enclosed in a double-layered **nuclear envelope,** which consists of an inner and an outer lipid bilayer membrane. These

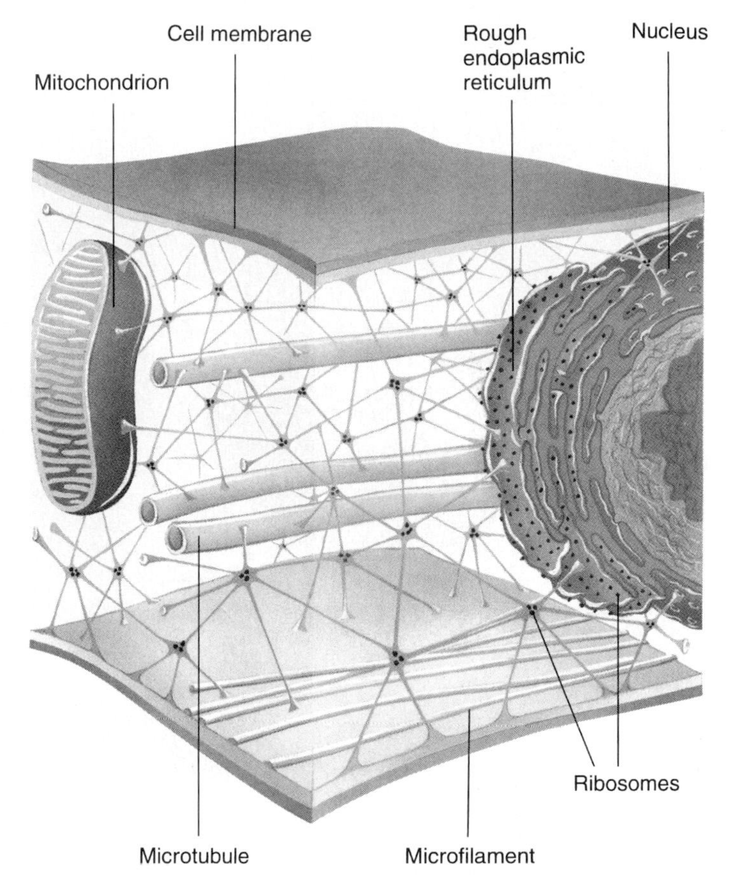


(a)

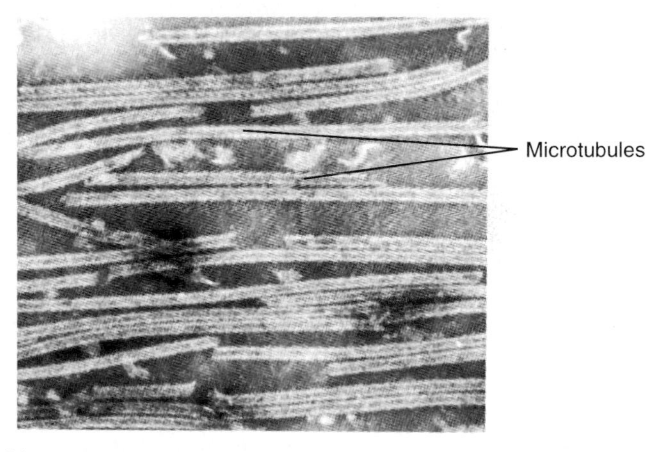


(b)

**FIGURE 3.19**

(*a*) Microtubules help maintain the shape of a cell by forming an internal scaffolding, or cytoskeleton, within the cytoplasm; (*b*) a transmission electron micrograph of microtubules (45,000×).

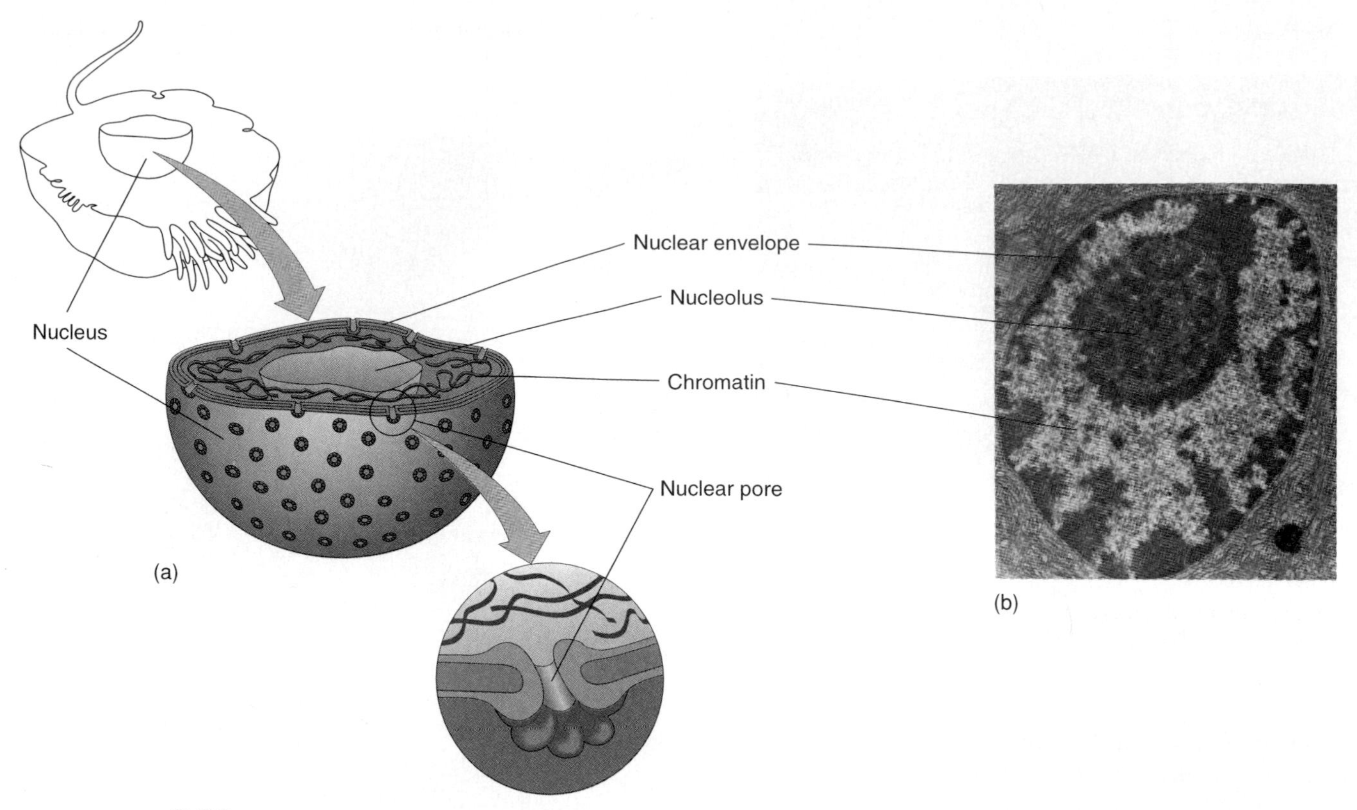


**FIGURE 3.20**
(*a*) The pores in the nuclear envelope allow certain substances to pass between the nucleus and the cytoplasm. (*b*) A transmission electron micrograph of a cell nucleus (8,000×). It contains a nucleolus and masses of chromatin.

two membranes have a narrow space between them, but are joined at places that surround relatively large openings called **nuclear pores.** These pores are not mere perforations, but channels consisting of more than 100 different types of proteins. Nuclear pores allow certain dissolved substances to move between the nucleus and the cytoplasm (fig. 3.20), most notably molecules of messenger RNA that carry genetic information.

The nucleus contains a fluid (nucleoplasm) in which other structures float. These structures include the following:

1. **Nucleolus.** A nucleolus ("little nucleus") is a small, dense body composed largely of RNA and protein. It has no surrounding membrane and is formed in specialized regions of certain chromosomes. It is the site of ribosome production. Once ribosomes form, they migrate through the nuclear pores to the cytoplasm. A cell may have more than one nucleolus. The nuclei of cells that synthesize large amounts of protein, such as those of glands, may contain especially large nucleoli.
2. **Chromatin.** Chromatin consists of loosely coiled fibers in the nuclear fluid. When cell reproduction begins, these fibers become more tightly coiled to form rodlike chromosomes. Chromatin fibers are composed of continuous DNA molecules wrapped around clusters of eight molecules of proteins called histones, giving the appearance of beads on a string. The DNA molecules contain the information for synthesis of proteins that promote cellular life processes.

Chart 3.2 summarizes the structures and functions of organelles.

> Cells die in different ways. *Apoptosis* is one form of cell death, in which the cell manufactures an enzyme that cuts up DNA not protected by histones. This is an active process in that a new substance is made. Apoptosis, a form of programmed cell death, is important in shaping the embryo, in maintaining organ form during growth, and in the development of the immune system and the brain.
>
> *Necrosis* is a type of cell death that is a passive response to severe injury. Typically proteins lose their characteristic shapes, and the cell membrane deteriorates as the cell swells and bursts.
>
> Unlike apoptosis, necrosis causes great inflammation.

## CHART 3.2 Structures and Functions of Organelles

| Organelle | Structure | Function |
|---|---|---|
| Cell membrane | Membrane composed mainly of protein and lipid molecules | Maintains integrity of the cell, controls the passage of materials into and out of the cell, and provides for signal transduction |
| Endoplasmic reticulum | Complex of interconnected membrane-bound sacs, canals, and vesicles | Transports materials within the cell, provides attachment for ribosomes, and synthesizes lipids |
| Ribosomes | Particles composed of protein and RNA molecules | Synthesize proteins |
| Golgi apparatus | Group of flattened, membranous sacs | Packages and modifies protein molecules for transport and secretion |
| Mitochondria | Membranous sacs with inner partitions | Release energy from food molecules and transform energy into usable form |
| Lysosomes | Membranous sacs | Contain enzymes capable of digesting worn cellular parts or substances that enter cells |
| Peroxisomes | Membranous vesicles | Contain enzymes called peroxidases, important in the breakdown of many organic molecules |
| Centrosome | Nonmembranous structure composed of two rodlike centrioles | Helps distribute chromosomes to new cells during cell reproduction and initiates formation of cilia |
| Cilia | Motile projections attached to basal bodies beneath the cell membrane | Propel fluids over cellular surface |
| Flagella | Motile projections attached to basal bodies beneath the cell membrane | Enable sperm cells to move |
| Vesicles | Membranous sacs | Contain various substances that recently entered the cell and store and transport newly synthesized molecules |
| Microfilaments and microtubules | Thin rods and tubules | Support cytoplasm and help move substances and organelles within the cytoplasm |
| Nuclear envelope | Porous double membrane that separates the nuclear contents from the cytoplasm | Maintains the integrity of the nucleus and controls the passage of materials between the nucleus and cytoplasm |
| Nucleolus | Dense, nonmembranous body composed of protein and RNA molecules | Site of ribosome formation |
| Chromatin | Fibers composed of protein and DNA molecules | Contains cellular information for synthesizing proteins needed in carrying on life processes |

1. How are the nuclear contents separated from the cytoplasm of a cell?
2. What is the function of the nucleolus?
3. What is chromatin?

## Movements into and out of the Cell

The cell membrane is a barrier that controls which substances enter and leave the cell. Oxygen and nutrient molecules enter through this membrane, whereas carbon dioxide and other wastes leave through it. These movements involve *physical* (or nonliving) processes such as diffusion, facilitated diffusion, osmosis, and filtration, and *physiological* (or living) mechanisms such as active transport, endocytosis, and exocytosis. The mechanisms of cell membrane transport are important for understanding many aspects of physiology.

### Diffusion

**Diffusion** is the process by which molecules or ions spontaneously move from regions where they are in higher concentrations toward regions where they are in lower concentrations.

Under natural conditions, molecules and ions constantly move at high speeds. Each particle travels in a separate path along a straight line until it collides and bounces off some other particle. Then it moves in another direction, only to collide again and change direction once more. Such random motion mixes molecules.

For example, when sugar (a solute) is put into a glass of water (a solvent), as illustrated in figure 3.21, the sugar seems to remain in high concentration at the bottom of the glass. Then it slowly disperses into solution. As this is happening, the moving water and sugar molecules are either colliding with one another or missing one another, and in time the sugar and water molecules will be evenly mixed. This mixing process, by which the sugar molecules spread from the region where they are in higher concentration

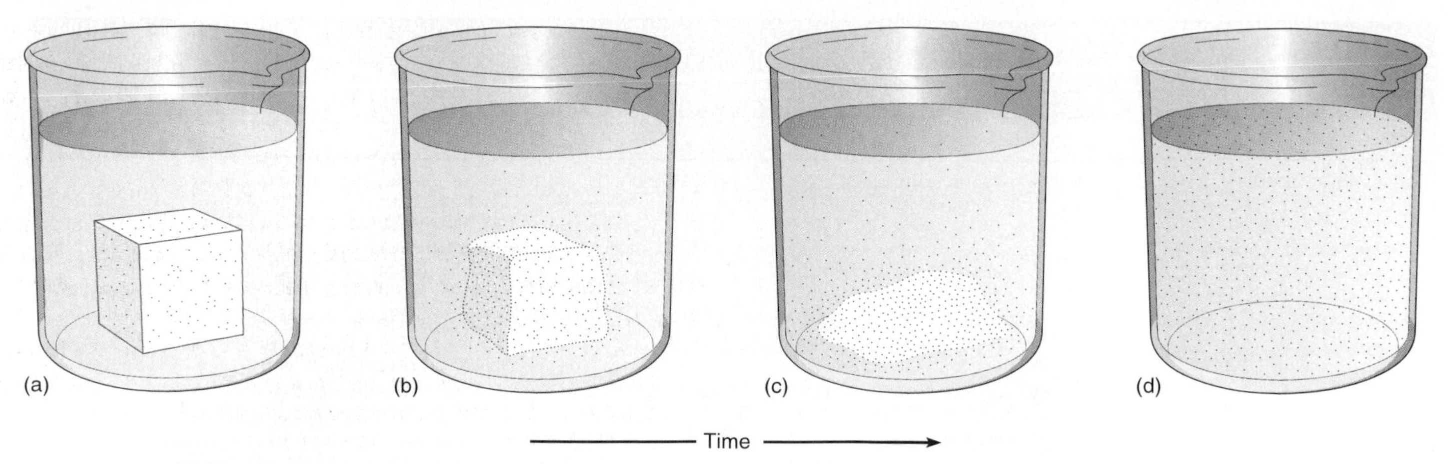


**FIGURE 3.21**

An example of diffusion (*a, b,* and *c*). A sugar cube placed in water slowly disperses, the sugar molecules diffusing from regions where they are more concentrated toward regions where they are less concentrated. (*d*) Eventually, the sugar molecules distribute evenly throughout the water.

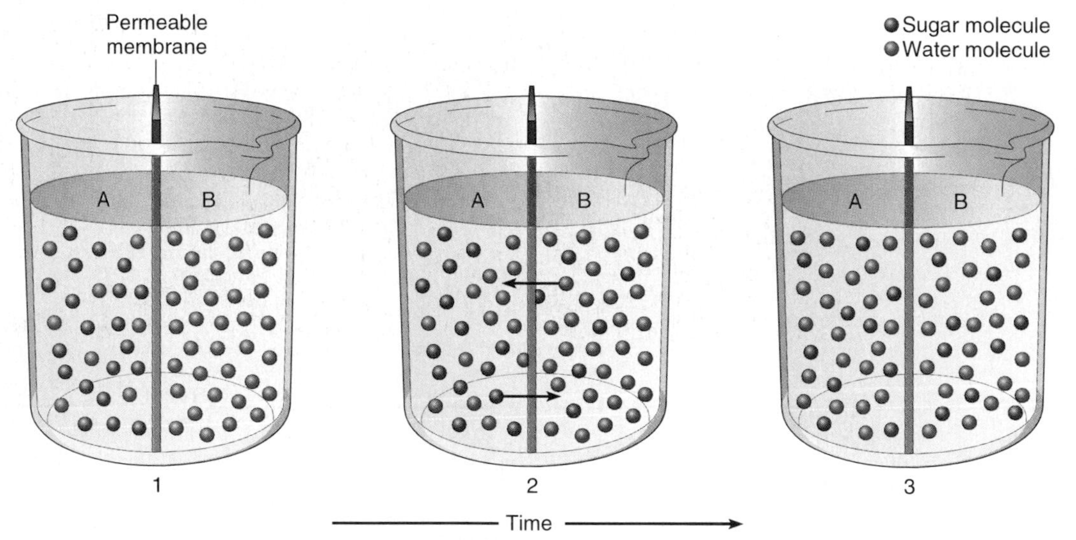


**FIGURE 3.22**

(1) The container is separated into two compartments by a membrane that is permeable to water and sugar molecules. Compartment A contains both types of molecules, while compartment B contains only water molecules. (2) As a result of molecular motions, sugar molecules tend to diffuse from compartment A into compartment B. Water molecules tend to diffuse from compartment B into compartment A. (3) Eventually equilibrium is achieved.

toward the regions where they are less concentrated, is diffusion. As a result of diffusion, the sugar eventually uniformly distributes in the water. This state of uniform distribution of molecules is called *equilibrium.* Although the molecules continue to move after equilibrium is achieved, their concentrations no longer change.

To better understand how diffusion accounts for the movement of various molecules through a cell membrane, imagine a container of water that is separated into two compartments by a membrane that has numerous pores large enough for water and sugar molecules to pass through (fig. 3.22). The sugar molecules are placed in one compartment (A) but not in the other (B). Because the sugar molecules move randomly in all directions, more diffuse from compartment A through the pores in the membrane and into compartment B than move in the other direction. Thus the sugar molecules diffuse from an area of higher concentration to an area of lower concentration. At the same time, water molecules tend to diffuse from compartment B (where they are in greater concentration) through the pores into compartment A (where they are in lesser concentration). Eventually, equilibrium will be achieved, with equal concentrations of water and of sugar molecules in each compartment.

Similarly, it is by diffusion that oxygen and carbon dioxide molecules are exchanged between the air

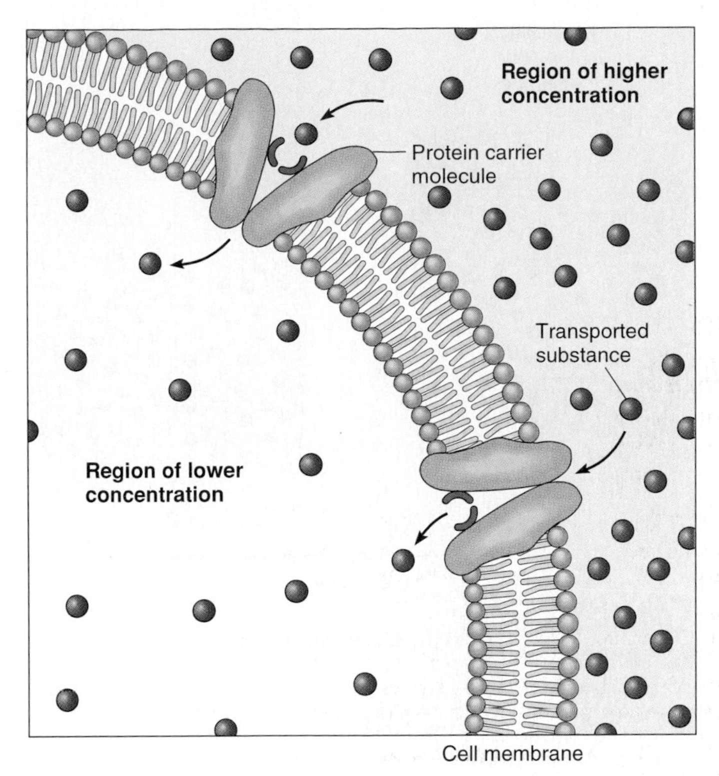


**FIGURE 3.23**
Some substances move into or out of cells by facilitated diffusion, transported by carrier molecules from a region of higher concentration to one of lower concentration.

and the blood in the lungs, and between the blood and the cells of various tissues. Oxygen molecules tend to diffuse through cell membranes and enter cells because normally these molecules are more highly concentrated on the outside than on the inside of the membranes. Carbon dioxide molecules tend to diffuse through cell membranes and leave cells because they are normally more concentrated on the inside than on the outside.

Factors that influence diffusion rate include the distance over which the diffusion will occur and the concentrations, weights, and temperatures of the diffusing molecules. Generally, diffusion occurs more rapidly over a shorter distance, when the concentration of the diffusing substance is greater, when the molecular weight is lower, and when the temperature is higher. A medical procedure to treat kidney failure, called hemodialysis, is based on the principle of diffusion. It is discussed in chapter 20.

## Facilitated Diffusion

Most sugars and amino acids are insoluble in lipids, and they are too large to pass through membrane pores. However, these molecules may still enter through the membrane of some cells by a process called **facilitated diffusion.** In the facilitated diffusion of glucose, for example, glucose combines with a protein carrier molecule at the surface of the membrane. This union of glucose and carrier molecule results in a change in the shape of the carrier that moves glucose to the other side of the membrane. The glucose portion is released, and the carrier molecule can return to its original shape to pick up another glucose molecule. The hormone *insulin,* discussed in chapter 13, promotes facilitated diffusion of glucose through the membranes of certain cells.

Facilitated diffusion is similar to simple diffusion in that it can only move molecules from regions of higher concentration toward regions of lower concentration. However, unlike simple diffusion the rate at which facilitated diffusion can occur is limited by the number of carrier molecules in the cell membrane (fig. 3.23).

## Osmosis

**Osmosis** is a special case of diffusion. It occurs whenever water molecules diffuse from a region of higher water concentration to a region of lower water concentration across a selectively permeable membrane, such as a cell membrane. In the following example, assume that the selectively permeable membrane is permeable to water molecules (the solvent), but impermeable to glucose molecules (the solute).

In solutions, a higher concentration of solute (glucose in this case) means a lower concentration of water; a lower concentration of solute means a higher concentration of water. This is because the solute molecules take up space that water molecules would otherwise occupy.

Just like molecules of other substances, molecules of water will diffuse from areas of higher concentration to areas of lower concentration. In figure 3.24, the presence of glucose in compartment A means that the water concentration there is less than the concentration of pure water in compartment B. Therefore, water diffuses from compartment B across the selectively permeable membrane and into compartment A. In other words, water moves from compartment B into compartment A by osmosis. Glucose, on the other hand, cannot diffuse out of compartment A because the selectively permeable membrane is impermeable to it.

Note in figure 3.24 that as osmosis occurs, the level of water on side A rises. This ability of osmosis to generate enough pressure to lift a volume of water is called *osmotic pressure.*

The greater the concentration of nonpermeable solute particles (glucose in this case) in a solution, the *lower* the water concentration of that solution and the *greater* the osmotic pressure. Water always tends to diffuse toward solutions of greater osmotic pressure.

Since cell membranes are generally permeable to water, water has equilibrated by osmosis throughout the body, and the concentration of water and solutes everywhere in the intracellular and extracellular fluids is essentially the same. Therefore, the osmotic pressure of the intracellular and extracellular fluids is the same. Any solution that has the same osmotic pressures as body fluids is called **isotonic.**

Solutions that have a higher osmotic pressure than the body fluids are called **hypertonic.** If cells are put into a hypertonic solution, there will be a net movement of water by osmosis out of the cells into the surrounding solution, and the cells shrink. Conversely, cells put into a **hypotonic** solution, which has a lower osmotic pressure than the body fluids, tend to gain water by osmosis and swell. Although cell membranes are somewhat elastic, the cells may swell so much that they burst. Figure 3.25 illustrates the effects of the three types of solutions on red blood cells.

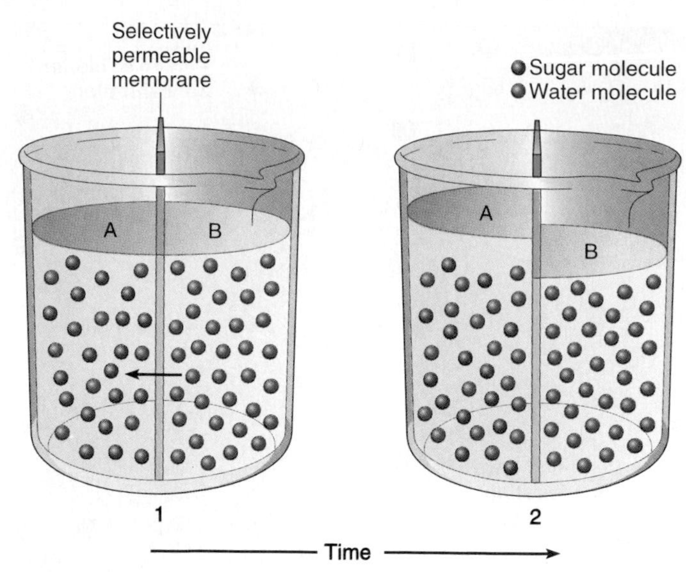


**FIGURE 3.24**

Osmosis. (1) A selectively permeable membrane separates the container into two compartments. At first, compartment A contains water and sugar molecules, while compartment B contains only water. As a result of molecular motions, water diffuses by osmosis from compartment B into compartment A. Sugar molecules remain in compartment A because they are too large to pass through the pores of the membrane. (2) Also, because more water is entering compartment A than is leaving it, water accumulates in this compartment. The level of liquid rises on this side.

> It is important to control the concentration of solute in solutions that are infused into body tissues or blood. Otherwise, osmosis may cause cells to swell or shrink, damaging them. For instance, if red blood cells are placed in distilled water (which is hypotonic to them), water will diffuse into the cells and they will burst (hemolyze). On the other hand, if red blood cells are exposed to 0.9% NaCl solution (normal saline), the cells will remain unchanged because this solution is isotonic to human cells. Similarly, a 5% solution of glucose is isotonic to human cells. (The lower percentage is needed with NaCl to produce an isotonic solution, in part because NaCl ionizes in solution more completely and produces a greater number of solute particles than does glucose.)

## Filtration

Molecules move through membranes by diffusion or osmosis because of their random movements. In other instances, molecules are forced through membranes by hydrostatic pressure called *blood pressure,* which is greater on one side of the membrane than on the other. This process that forces molecules through membranes is called **filtration.**

Filtration is commonly used to separate solids from water. One method is to pour a mixture of solids and water onto filter paper in a funnel (fig. 3.26). The paper serves as a porous membrane through which the small water molecules can pass, leaving the larger solid particles behind. Hydrostatic pressure, which is created by the weight of water due to gravity, forces the water molecules through to the other side.

Similarly, in the body, tissue fluid forms when water and dissolved substances are forced out through the thin, porous walls of blood capillaries, but larger particles such as blood protein molecules are left inside (fig. 3.27). The force for this movement comes from blood pressure, created largely by heart action, which is greater within the vessel than outside it. Filtration is discussed further in chapters 15 and 20.

1. What kinds of substances most readily diffuse through a cell membrane?
2. Explain the differences between diffusion, facilitated diffusion, and osmosis.
3. Distinguish among hypertonic, hypotonic, and isotonic solutions.
4. Explain how filtration occurs within the body.

## Active Transport

When molecules or ions pass through cell membranes by diffusion, facilitated diffusion, or osmosis, their net movement is from regions of higher concentration to regions of lower concentration. Sometimes, however, the net movement of particles passing through membranes is in the opposite direction; that is, the particles move from a region of lower concentration to one of higher concentration.

Sodium ions, for example, can diffuse slowly through cell membranes. Yet, the concentration of

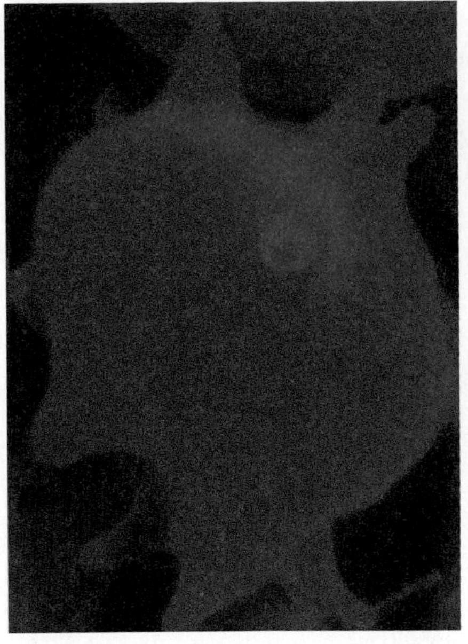

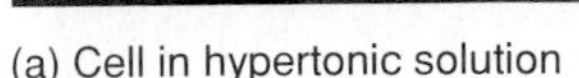

(a) Cell in hypertonic solution

(b) Cell in hypotonic solution

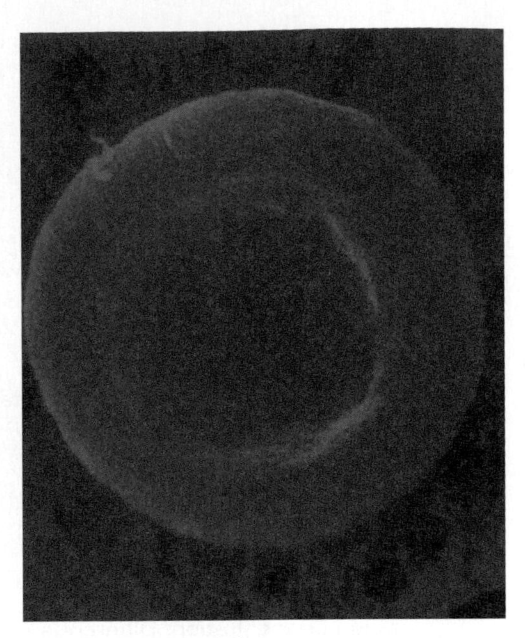

(c) Cell in isotonic solution

**FIGURE 3.25**

(*a*) If red blood cells are placed in a hypertonic solution, more water leaves than enters, and the cells shrink. (*b*) In a hypotonic solution, more water enters than leaves, and the cells swell and may burst. (*c*) In an isotonic solution, water enters and leaves the cells in equal amounts, and their sizes remain unchanged (figure *a* 18,000×; figure *b* 18,000×; figure *c* 18,000×).

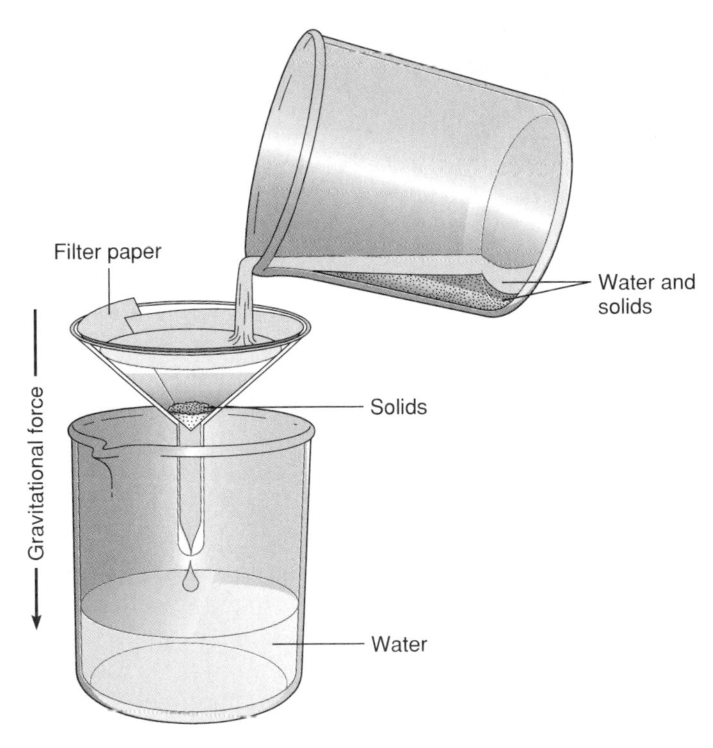


**FIGURE 3.26**

In this example of filtration, gravity provides the force that pulls water through filter paper, while tiny openings in the paper retain the solids. It is a little like the drip method of preparing coffee.

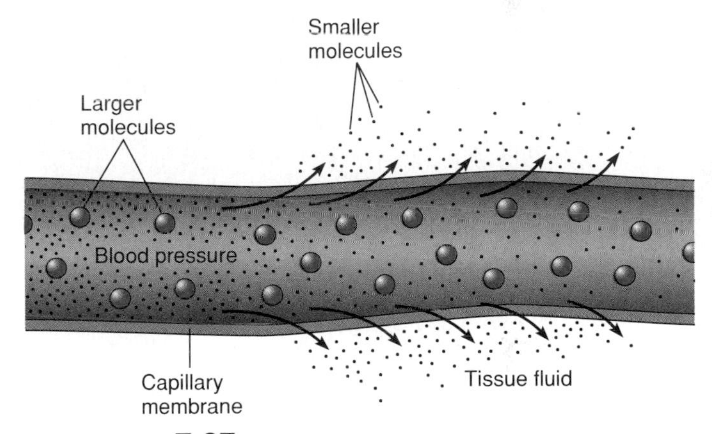


**FIGURE 3.27**

In this example of filtration, blood pressure forces smaller molecules through tiny openings in the capillary wall. The larger molecules remain inside.

these ions typically remains many times greater on the outside of cells (in the extracellular fluid) than on the inside of cells (in the intracellular fluid). This is because sodium ions are continually moved through the cell membrane from the regions of lower concentration (inside) to the regions of higher concentration (outside). Movement of this type is called **active transport.** It requires energy that derives from cellular metabolism. In fact, up to 40% of a cell's energy supply may be used for active transport of particles through its membranes.

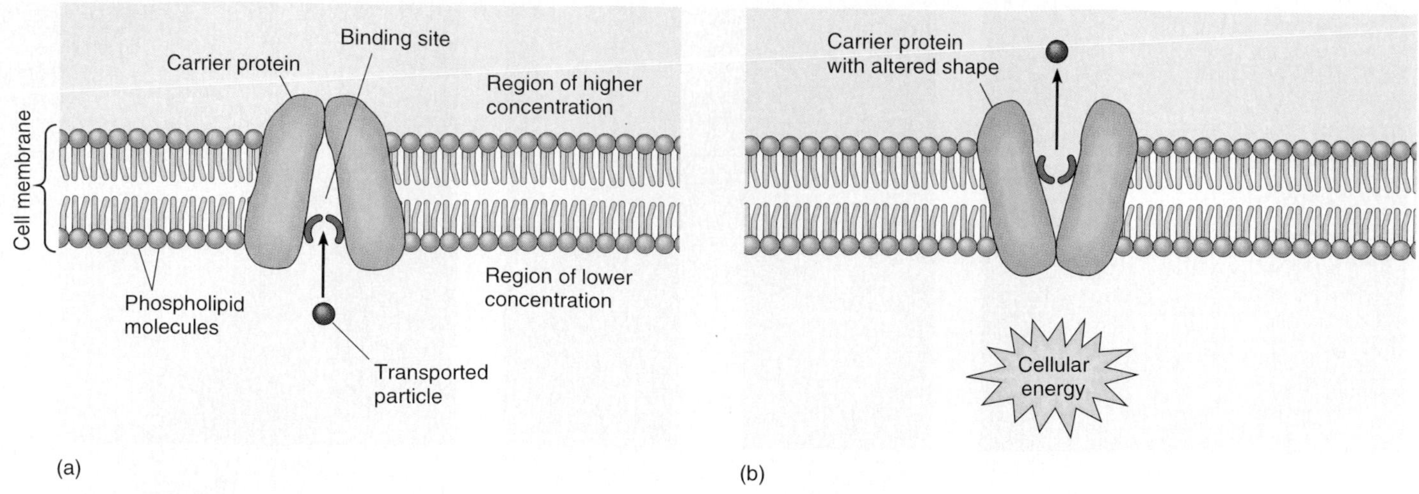


**FIGURE 3.28**
(*a*) During active transport, a molecule or ion combines with a carrier protein, whose shape is altered as a result. (*b*) This process, which requires energy, transports the particle through the cell membrane from an area of low concentration to an area of high concentration. Different substances move into or out of cells by this process.

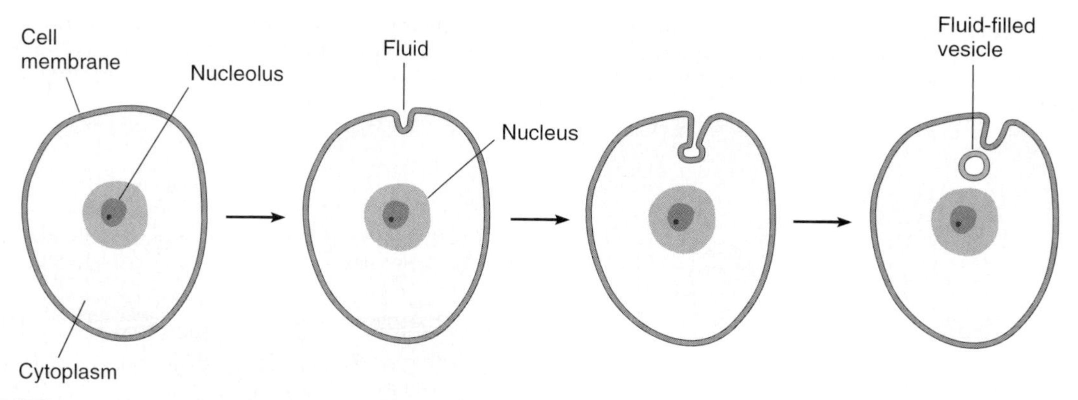


**FIGURE 3.29**
A cell may take in a tiny droplet of fluid from its surroundings by pinocytosis.

Active transport is similar to facilitated diffusion in that it uses carrier molecules found within cell membranes. As figure 3.28 shows, these carrier molecules are proteins that have binding sites that combine with the specific particles being transported. Such a union triggers the release of cellular energy, and this energy alters the shape of the carrier protein. As a result, the "passenger" molecules move through the membrane. Once on the other side, the transported particles are released, and the carrier molecules can accept other passenger molecules at their binding sites. Because they transport substances from regions of low concentration to regions of higher concentration, these carrier proteins are sometimes called "pumps." A sodium pump, for example, transports sodium ions out of cells.

Particles that are moved across cell membranes by active transport include sugars, amino acids, and sodium, potassium, calcium, and hydrogen ions. Some of these substances are actively transported into cells, and others are transported out. Movements of this type are important to cell survival, particularly maintenance of homeostasis. Some of these movements are described in subsequent chapters.

## Endocytosis

Two processes use cellular energy to move substances into (endocytosis) or out of (exocytosis) a cell without actually crossing the cell membrane. In **endocytosis,** molecules or other particles that are too large to enter a cell by diffusion or active transport may be conveyed within a vesicle formed from a section of the cell membrane. In **exocytosis,** the reverse process allows a substance stored in a vesicle to be secreted from the cells.

The three forms of endocytosis are pinocytosis, phagocytosis, and receptor-mediated endocytosis. In **pinocytosis,** cells take in tiny droplets of liquid from their surroundings (fig. 3.29). When this happens, a small portion of cell membrane indents (invaginates).

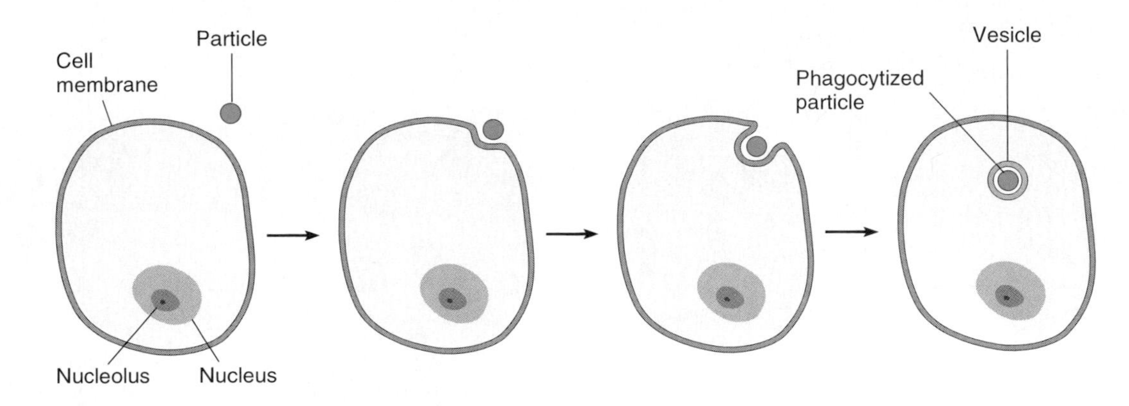


**FIGURE 3.30**
A cell may take in a solid particle from its surroundings by phagocytosis.

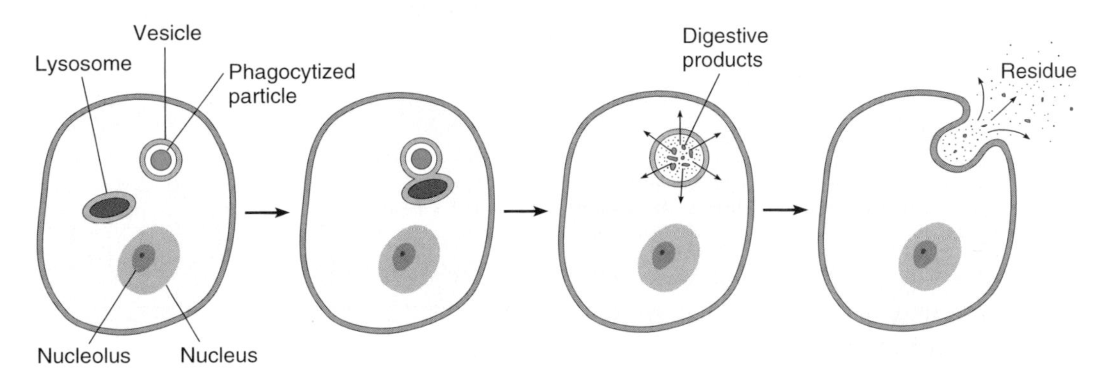


**FIGURE 3.31**
When a lysosome combines with a vesicle that contains a phagocytized particle, its digestive enzymes may destroy the particle. The products of this intracellular digestion diffuse into the cytoplasm. Any residue may be expelled from the cell by exocytosis.

The open end of the tubelike part thus formed seals off and produces a small vesicle about 0.1 μm in diameter. This tiny sac detaches from the surface and moves into the cytoplasm.

For a time, the vesicular membrane, which was part of the cell membrane, separates its contents from the rest of the cell; but eventually, the membrane breaks down and the liquid inside becomes part of the cytoplasm. In this way, a cell is able to take in water and the particles dissolved in it, such as proteins, that otherwise might be unable to enter because of their relatively large size.

**Phagocytosis** is similar to pinocytosis, but the cell takes in solids rather than liquid. Certain kinds of cells, including some white blood cells, are called *phagocytes* because they can take in solid particles such as bacterial cells and cellular debris. When a phagocyte first encounters such a particle, the particle becomes attached to the phagocyte's cell membrane. This stimulates a portion of the membrane to project outward, surround the particle, and slowly draw it inside. The part of the membrane surrounding the solid detaches from the cell's surface, and a vesicle containing the particle forms (fig. 3.30). Such a vesicle may be several micrometers in diameter.

Usually, a lysosome soon combines with such a newly formed vesicle, and lysosomal digestive enzymes decompose the vesicular contents (fig. 3.31). The products of this decomposition may then diffuse out of the lysosome and into the cytoplasm, where they may be used as raw materials in metabolic processes. Any remaining residue may be expelled from the cell by exocytosis. In this way, phagocytic cells can dispose of foreign objects like dust particles, remove damaged cells or cell parts that are no longer functional, or destroy bacteria that might otherwise cause infections.

Although pinocytosis is thought to provide only a minor route for substances to enter cells, phagocytosis is an important line of defense against invasion by disease-causing microorganisms.

Pinocytosis and phagocytosis engulf anything. In contrast is the more discriminating **receptor-mediated endocytosis,** which moves very specific kinds of particles into the cell. In receptor-mediated endocytosis, protein molecules extend through the cell membrane and are exposed on its outer surface. These proteins are receptors to which specific substances (ligands) from the fluid surroundings of the cell can bind. Thus, when molecules that are capable of binding to the

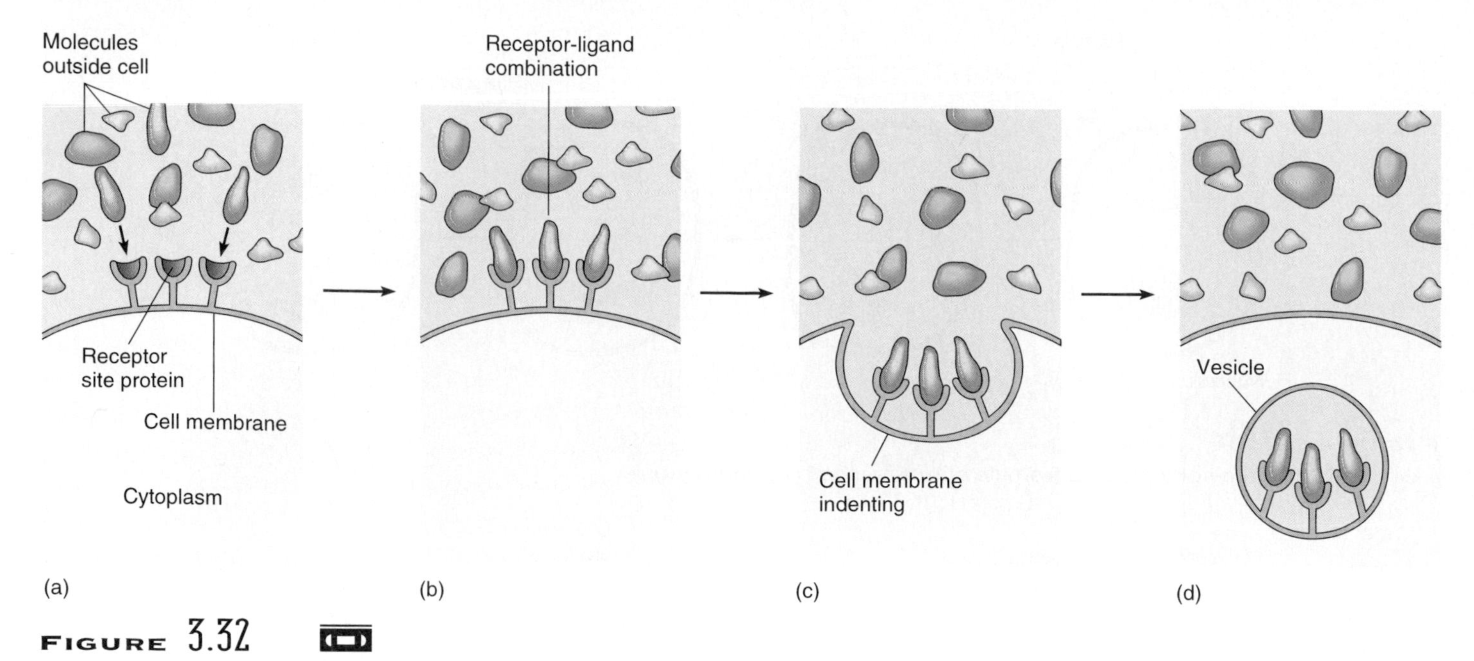


**FIGURE 3.32**

Receptor-mediated endocytosis. (*a*) A specific substance binds to a receptor site protein; (*b* and *c*) the combination of the substance with the receptor site protein stimulates the cell membrane to indent; (*d*) the resulting vesicle transports the substance into the cytoplasm.

receptor sites are present, they are the ones that are selected to enter the cell, and other kinds of molecules are left outside (fig. 3.32).

For example, cholesterol molecules that have been synthesized in liver cells are packaged in relatively large spherical particles called *low-density lipoproteins* (LDL). An LDL particle is surrounded by a coating that contains a binding protein called *apoprotein-B*. The membranes of various body cells possess receptor sites for apoprotein-B. When LDL particles are released by the liver into the blood, the body cells with apoprotein-B receptor sites can recognize the LDL particles and bind to them.

The formation of such a receptor-ligand combination somehow stimulates the cell membrane to indent and form a vesicle that contains the LDL particle. The vesicle carries the LDL particle to a lysosome, where enzymes digest it and release the cholesterol molecules for cellular uses.

Receptor-mediated endocytosis is particularly important because it allows a cell to remove and process specific kinds of substances from its surroundings, even when these substances are present in very low concentrations.

As a toddler, Stormie Jones already had a blood serum cholesterol level six times normal. Before she died at age 10, she had suffered several heart attacks and had undergone two cardiac bypass surgeries, several heart valve replacements, and finally a heart-liver transplant. The transplant lowered her blood cholesterol to a near-normal level, but she died from the multiple traumas suffered over her short lifetime.

Stormie had the severe form of familial hypercholesterolemia (FH), meaning simply too much cholesterol in the blood. Her liver cells lacked LDL receptors. Blocked from entering cells, cholesterol accumulated in her bloodstream, forming the plaques that caused her heart disease.

Stormie Jones was one in a million. Far more common are the one in 500 people who have the milder form of FH, in which liver cells have half the normal number of LDL receptors. These individuals are prone to suffer heart attacks in early adulthood. However, they can delay symptom onset by taking precautions to avoid cholesterol buildup, such as exercising, following a low-fat diet, and not smoking. (These precautions may also be beneficial to individuals not suffering from FH.)

## CHART 3.3 MOVEMENTS INTO AND OUT OF THE CELL

| Process | Characteristics | Source of Energy | Example |
|---|---|---|---|
| *I. Physical Processes* | | | |
| A. Diffusion | Molecules or ions move from regions of higher concentration toward regions of lower concentration. | Molecular motion | Exchange of oxygen and carbon dioxide in the lungs |
| B. Facilitated diffusion | Molecules are moved through a membrane by carrier molecules from a region of higher concentration to one of lower concentration. | Molecular motion | Movement of glucose through a cell membrane |
| C. Osmosis | Water molecules move from regions of higher concentration toward regions of lower concentration through a selectively permeable membrane. | Molecular motion | Distilled water entering a cell |
| D. Filtration | Smaller molecules are forced through porous membranes from regions of higher pressure to regions of lower pressure. | Hydrostatic pressure | Molecules leaving blood capillaries |
| *II. Physiological Processes* | | | |
| A. Active transport | Molecules or ions are carried through membranes by carrier molecules from regions of lower concentration toward regions of higher concentration. | Cellular energy | Movement of various ions and amino acids through membranes |
| B. Endocytosis | | | |
| 1. Pinocytosis | Membrane engulfs minute droplets of liquid from surroundings. | Cellular energy | Membrane-forming vesicles containing particles of relatively large molecular size dissolved in water |
| 2. Phagocytosis | Membrane engulfs solid particles from surroundings. | Cellular energy | White blood cell membrane engulfing bacterial cell |
| 3. Receptor-mediated endocytosis | Membrane engulfs selected molecules combined with receptor proteins. | Cellular energy | Cell removing cholesterol-containing LDL particles from its surroundings |
| C. Exocytosis | Vesicles fuse with membrane and release contents outside of the cell. | Cellular energy | Protein secretion, neurotransmitter release |

### Exocytosis

Exocytosis is essentially the reverse of endocytosis. In exocytosis, substances made within the cell are packaged into a vesicle, which then fuses with the cell membrane, thereby releasing its contents outside the cell. Cells secrete some proteins by this process. Nerve cells use exocytosis to release the neurotransmitter chemicals that signal other nerve cells, muscle cells, or glands.

Chart 3.3 summarizes the types of movement into and out of the cell.

1. What type of mechanism is responsible for maintaining unequal concentrations of ions on opposite sides of a cell membrane?
2. How are the processes of facilitated diffusion and active transport similar? How are they different?
3. What is the difference between pinocytosis and phagocytosis?
4. Describe receptor-mediated endocytosis.

## THE CELL CYCLE

The series of changes that a cell undergoes from the time it forms until it reproduces is called the **cell cycle** (fig. 3.33). Superficially, this cycle seems rather simple—a newly formed cell grows for a time and then divides in half to form two new cells, which in turn may grow and divide. Yet the details of the cycle are quite complex, involving interphase, mitosis, cytoplasmic division, and differentiation.

### Interphase

Before a cell actively divides, it must amass important biochemicals and duplicate much of its contents, so that two cells can form from one. This period of preparedness is called **interphase.** Once thought to be a time of rest, interphase is actually a time of great synthetic activity. During interphase, the cell maintains its routine "housekeeping" functions yet also takes on the tremendous task of replicating its genetic material. It also duplicates membranes, ribosomes, lysosomes, peroxisomes, and mitochondria. Interphase is divided into three phases. The "S" phase, when DNA replicates, is bracketed by two gap or growth periods, called $G_1$ and $G_2$, during which other structures are duplicated (fig. 3.34).

$G_2$ phase (gap)
Prophase
Metaphase
Anaphase
Telophase
Mitosis
$G_1$ phase (gap)
S phase (DNA synthesis)
Interphase

**FIGURE 3.33**
The cell cycle.

### Mitosis

In cell reproduction, a cell divides to form two new, complete cells. This process, called **mitosis,** includes two separate steps: division of the nuclear contents, which is called karyokinesis, and division of the cytoplasm, or cytokinesis. (Chapters 22 and 24 describe another type of cell division called *meiosis,* which produces sex cells.)

Division of the nucleus by mitosis is, of necessity, very precise, because the nucleus contains information (in the form of DNA molecules) that "tells" the cell how to carry on life processes. Each new cell resulting from mitosis must have a complete copy of this information in order to survive. Although the chromosomes have already been copied in interphase, it is in mitosis that the copies are evenly divided between the two new cells.

Mitosis is often described in stages, but the process is actually continuous without marked changes between one step and the next (fig. 3.35).

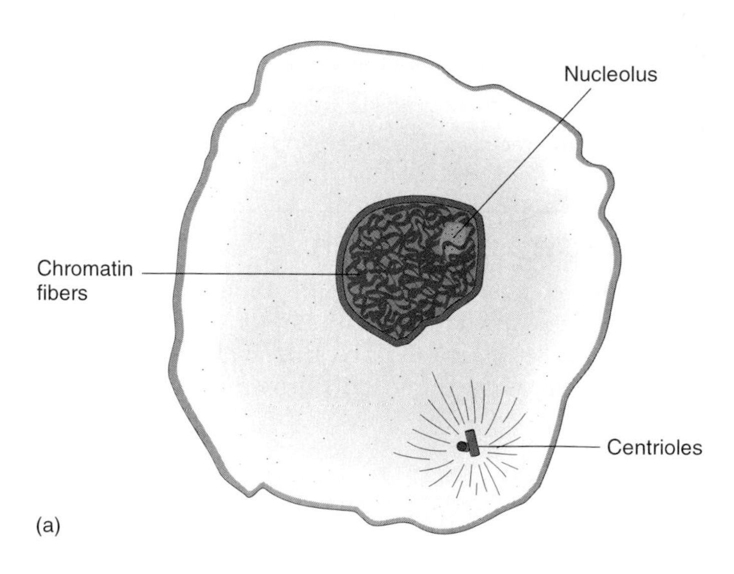


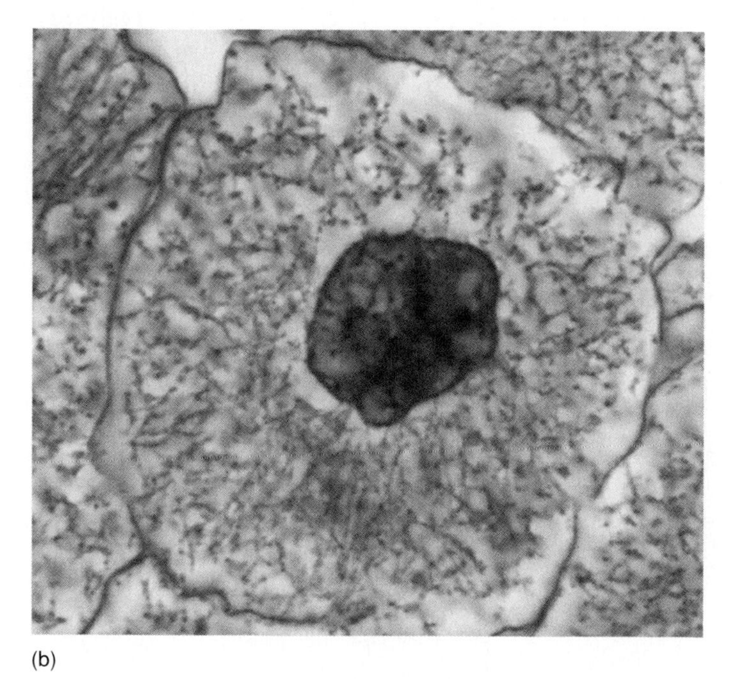

(b)

**FIGURE 3.34**
(*a*) Interphase lasts until a cell begins to undergo mitosis. (*b*) A micrograph of a cell in interphase (250×). Centrioles and chromatin fibers are not clearly visible at this magnification.

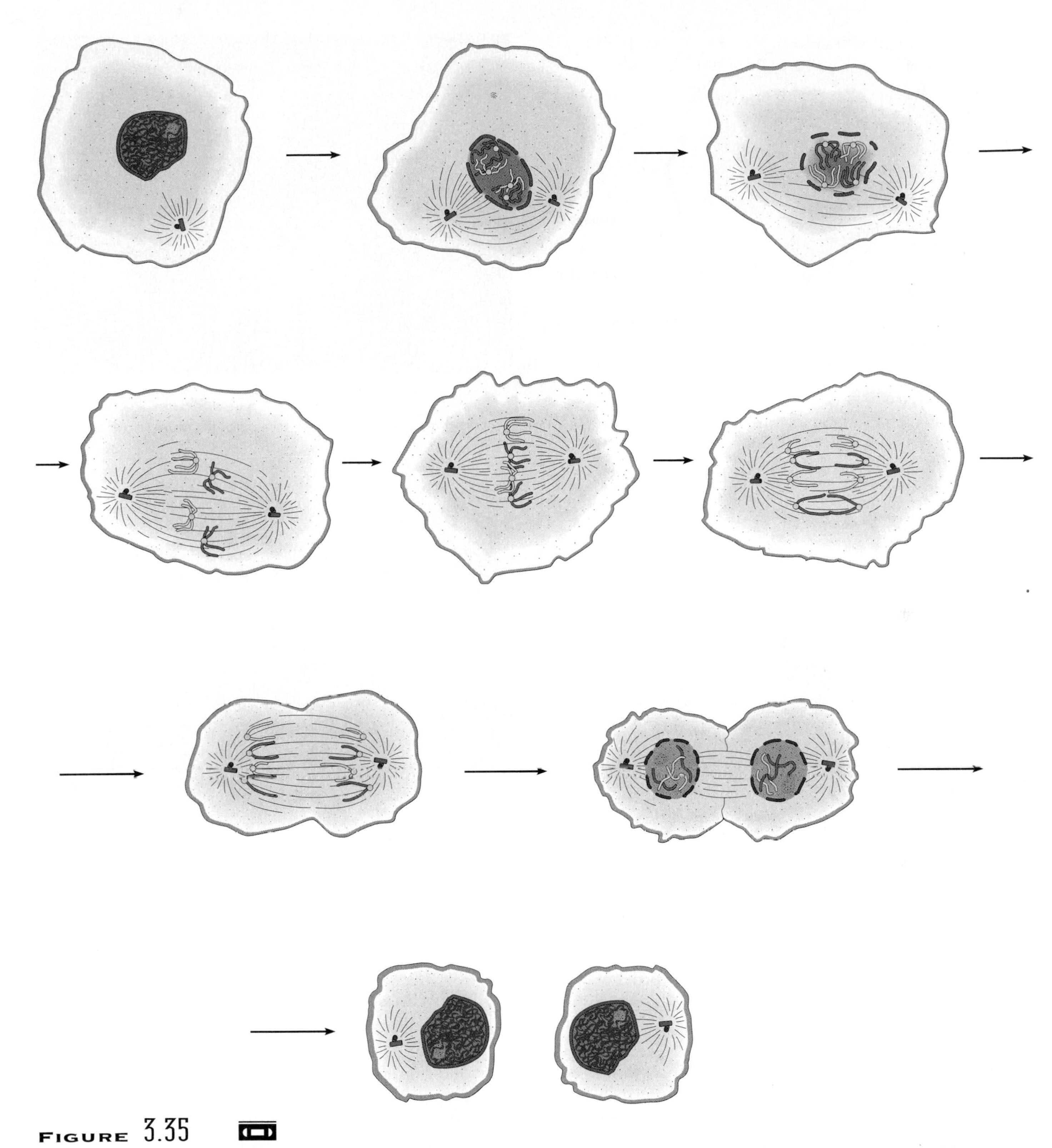

**FIGURE 3.35**

Mitosis is a continuous process during which the replicated genetic material is divided into two equal portions. After reading about mitosis, identify the phases of the process and the cell parts shown in this diagram.

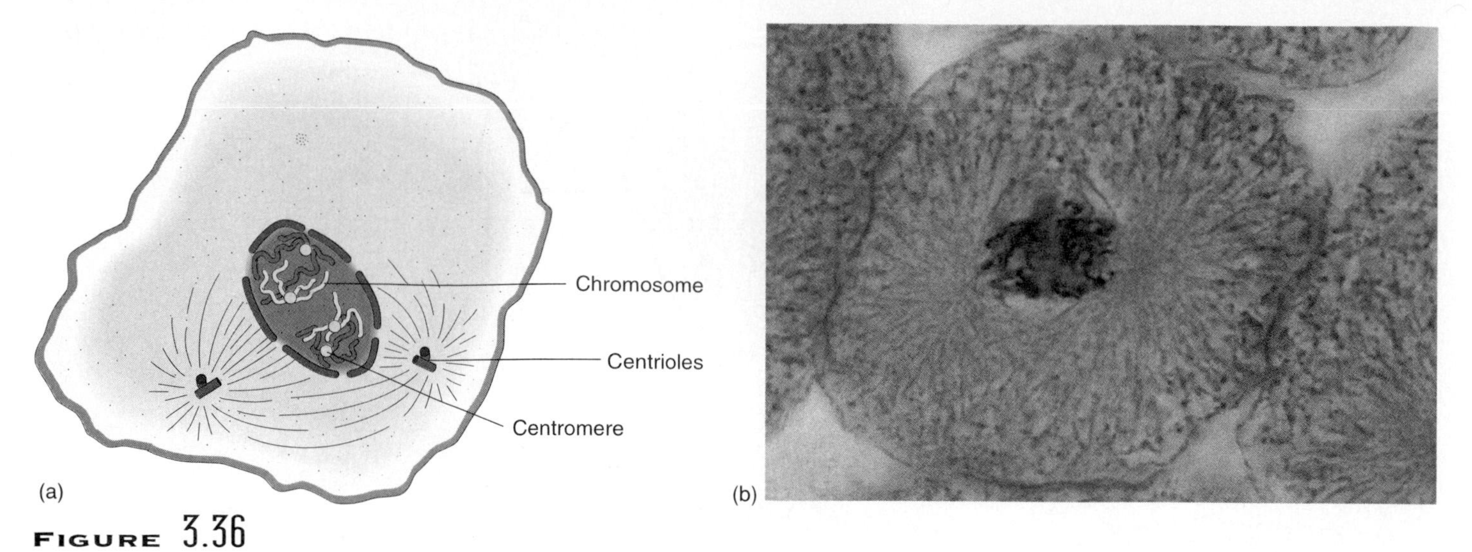


**FIGURE 3.36**

(*a*) In prophase, chromosomes form from chromatin in the nucleus, and the centrioles move to opposite sides of the cell. (*b*) A micrograph of a cell in prophase (250×).

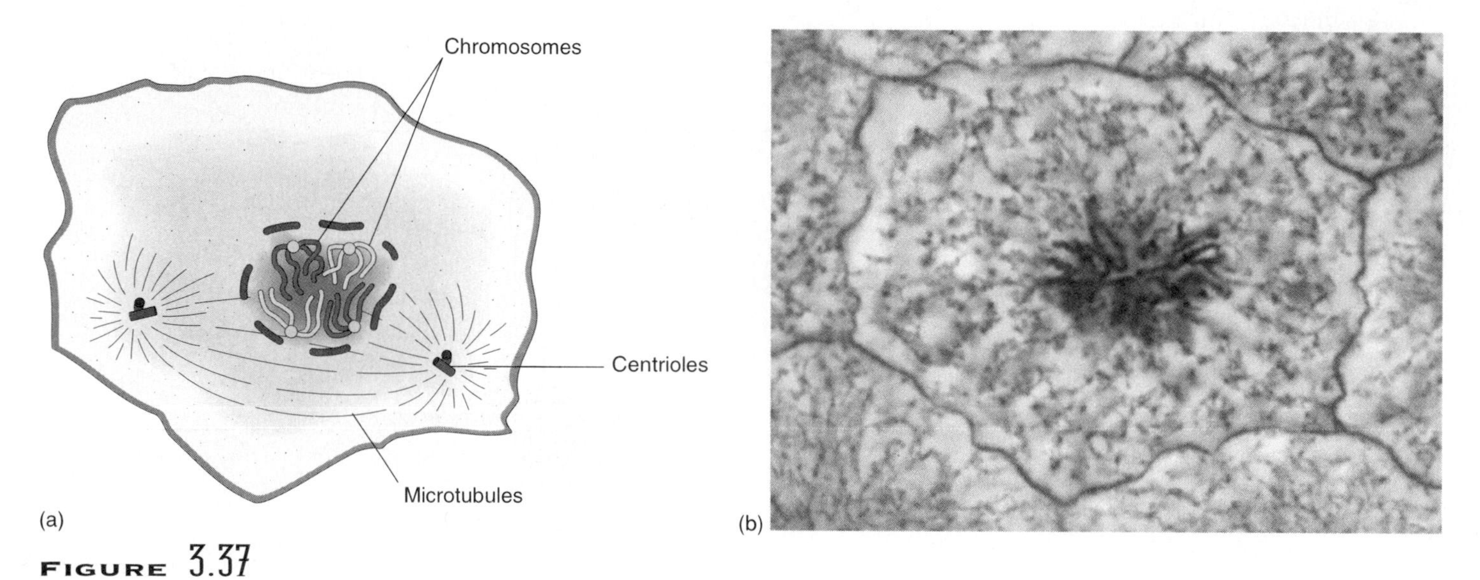


**FIGURE 3.37**

(*a*) Later in prophase, the nuclear envelope and nucleolus disappear. (*b*) A micrograph of a cell in late prophase (polar view) (280×).

The idea of stages is useful, however, to indicate the sequence in which major events occur. The stages of mitosis are:

1. **Prophase.** One of the first indications that a cell is going to reproduce is the appearance of *chromosomes.* These structures form as fibers of chromatin condense into tightly coiled rods. During interphase, the DNA molecules replicate (duplicate), so that each chromosome is composed of two identical structures, called chromatids, that are temporarily attached by a region on each called a *centromere.*

   The centrioles of the centrosome replicate just before the onset of mitosis, and during prophase, the two newly formed pairs of centrioles move to opposite sides of the cell. Soon the nuclear envelope and the nucleolus disperse and are no longer visible. Microtubules are assembled from tubulin proteins in the cytoplasm, and these structures associate with the centrioles and chromosomes (figs. 3.36 and 3.37). A spindle-shaped array of microtubules (spindle fibers) forms between the centrioles as they move apart.

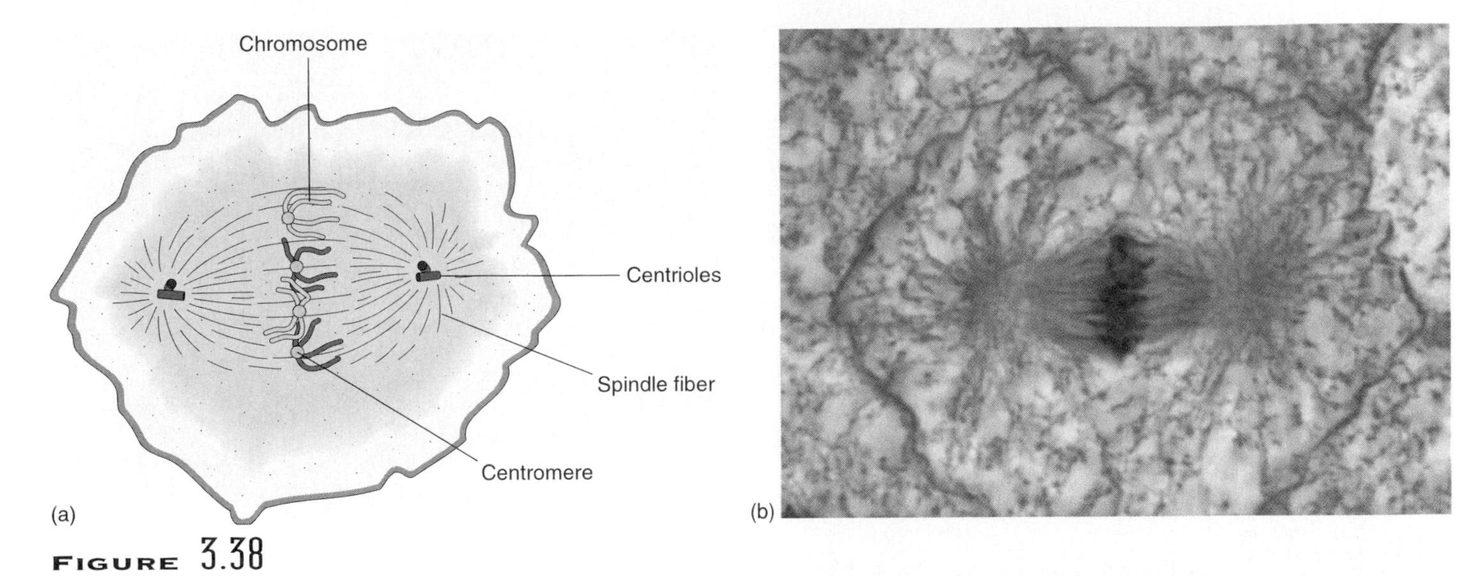


**FIGURE 3.38**

(*a*) In metaphase, chromosomes line up midway between the centrioles. (*b*) A micrograph of a cell in metaphase (280×).

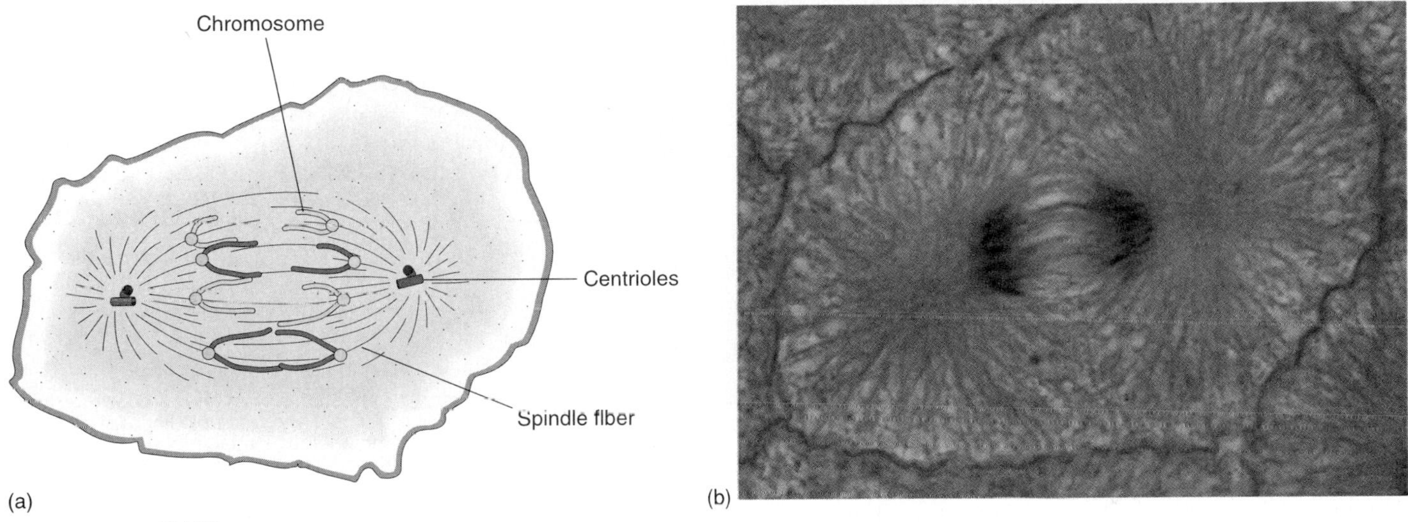


**FIGURE 3.39**

(*a*) In anaphase, centromeres divide, and the spindle fibers that have become attached to them pull the chromatids, now called chromosomes, toward the centrioles. (*b*) A micrograph of a cell in anaphase (280×).

2. **Metaphase.** The chromosomes align about midway between the centrioles, as a result of microtubule activity. Spindle fibers attach to the centromeres of the chromosomes so that a fiber accompanying one chromatid attaches to one side of a centromere, and a fiber accompanying the other chromatid attaches to the other side (fig. 3.38).
3. **Anaphase.** Soon the centromeres of the chromatids separate, and these identical chromatids become individual chromosomes. The separated chromosomes now move in opposite directions, and once again the movement results from microtubule activity. The spindle fibers shorten and pull their attached chromosomes toward the centrioles at opposite sides of the cell (fig. 3.39).
4. **Telophase.** The final stage of mitosis begins when the chromosomes complete their migration toward the centrioles. It is much like prophase, but in reverse. As the chromosomes approach the centrioles, they begin to elongate and unwind

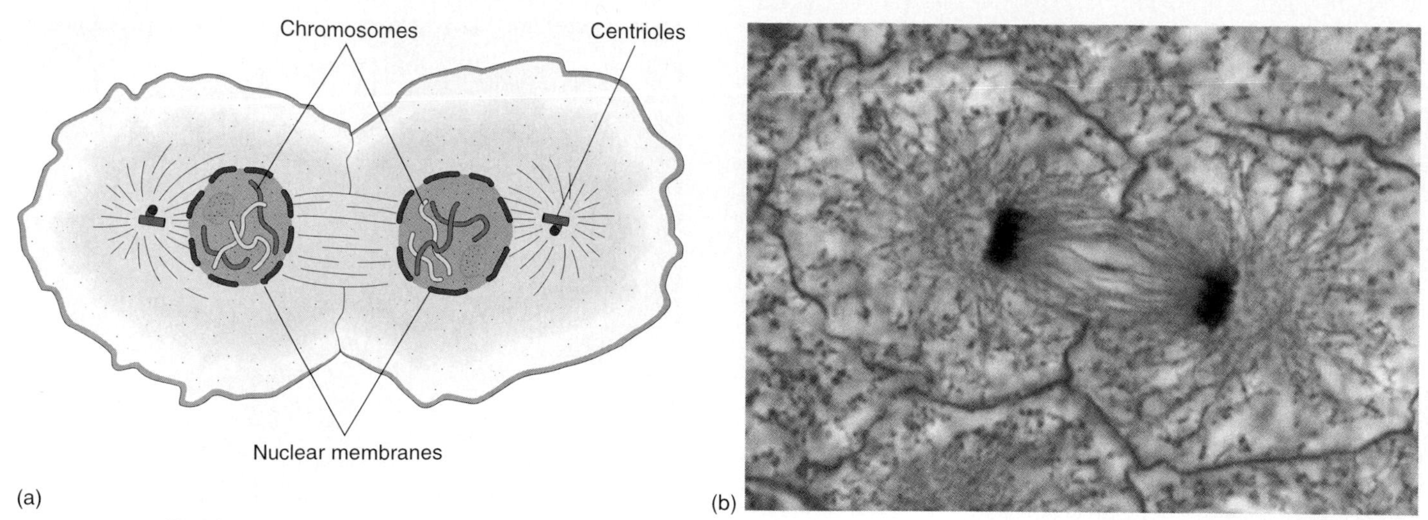


**FIGURE 3.40**

(*a*) In telophase, chromosomes elongate to become chromatin threads, and the cytoplasm begins to distribute between the two forming progeny cells. The replicate chromatids have separated to form the chromosomes of the newly formed cells. (*b*) A micrograph of a cell in telophase (280×).

from rodlike structures to threadlike structures. A nuclear envelope forms around each chromosome set, and nucleoli appear within the newly formed nuclei. Finally, the microtubules disassemble into free tubulin molecules (fig. 3.40).

Chart 3.4 summarizes the phases of mitosis.

**CHART 3.4 MAJOR EVENTS IN MITOSIS**

| Stage | Major Events |
|---|---|
| Prophase | Chromatin condenses into chromosomes; centrioles move to opposite sides of cytoplasm; nuclear membrane and nucleolus disperse; microtubules appear and associate with centrioles and duplicate chromatids. |
| Metaphase | Chromosomes align midway between the centrioles; spindle fibers from the centrioles attach to the centromeres of each chromosome. |
| Anaphase | Centromeres separate, and duplicate chromatids of the chromosomes separate; spindle fibers shorten and pull the individual chromosomes toward centrioles. |
| Telophase | Chromosomes elongate and form chromatin threads; nuclear membranes appear around each chromosome set; nucleoli appear; microtubules disappear. |

## Cytoplasmic Division

Cytoplasmic division (cytokinesis) begins during anaphase when the cell membrane starts to constrict around the middle. It continues to do so through telophase. The musclelike contraction of a ring of actin microfilaments pinches off two cells from one. The microfilaments assemble in the cytoplasm and attach to the inner surface of the cell membrane. The contractile ring is positioned at right angles to the microtubules that pulled the chromosomes to opposite ends of the cell during mitosis. As the ring pinches inward, it separates the two newly formed nuclei and divides about half of the organelles into each of the new cells.

Although the newly formed progeny cells may differ slightly in size and number of organelles and inclusions, they have identical chromosomes and thus contain identical DNA information. Except for size, they are copies of the parent cell (fig. 3.41).

## Cell Differentiation

Because all body cells form by mitosis and contain the same DNA information, they might be expected to look and act alike; obviously, they do not.

A human begins life as a single cell—a fertilized egg cell. This cell reproduces by mitosis to form two new cells; they, in turn, divide into four cells; the four become eight; and so forth. Then, in the third to eighth weeks, the cells specialize, developing distinctive structures and beginning to function in different ways. Some become skin cells, others become bone cells, and still others become nerve cells (fig. 3.42). By the time of birth, a human has more than 200 types of cells. The process by which cells develop different structures and specialized functions is called **differentiation.**

Cell specialization reflects genetic control. Special proteins activate some genes and repress others,

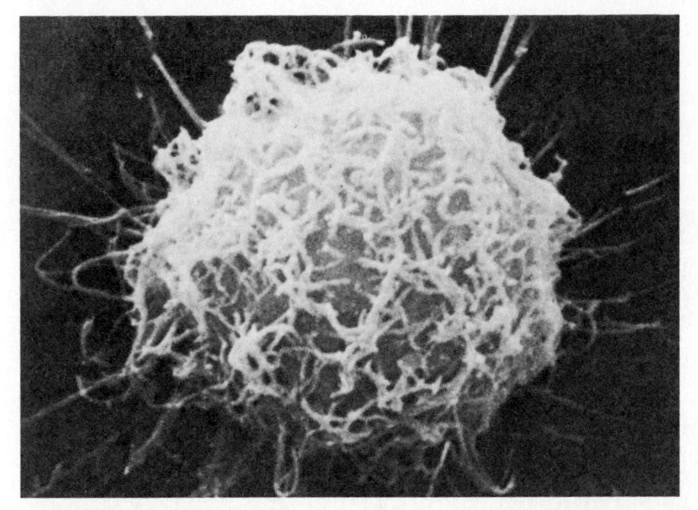

(a)

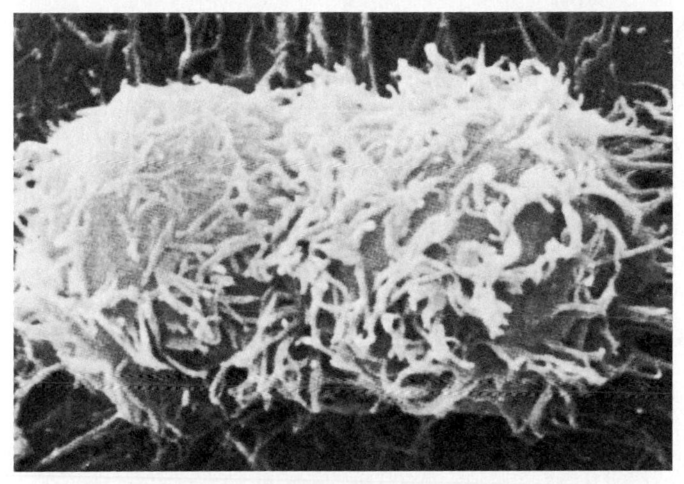

(b)

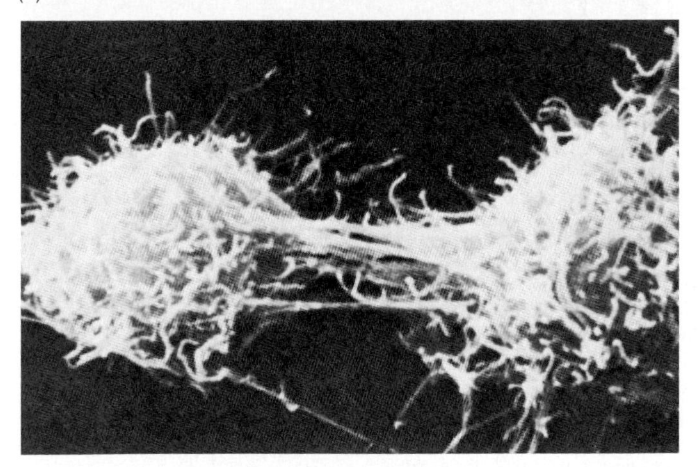

(c)

**FIGURE 3.41**
Following mitosis, the cytoplasm of the parent cell divides in two, as seen in these scanning electron micrographs (figure *a* 3,750×; figure *b* 3,750×; figure *c* 3,190×).

(From *Scanning Electron Microscopy in Biology*, by R. G. Kessel and C. Y. Shih. © 1976 Springer-Verlag.)

controlling the repertoire of biochemicals in the cell, and therefore its characteristics. In a nerve cell, the genes controlling neurotransmitter synthesis would be activated; in a bone cell, these genes would be silenced because it does not use neurotransmitters, but genes encoding the protein collagen, a major component of bone, would be very active. A differentiated cell can be compared to a library. It contains a complete collection of information, but only some of that information is accessed. Different cell types interact in ways that aid survival of the organism.

1. Why is precise division of nuclear materials during mitosis important?
2. Describe the events that occur during mitosis.
3. Name the process by which some cells become muscle cells and others become nerve cells.
4. How does DNA control differentiation?

## Control of Cell Reproduction

How often a cell reproduces is strictly controlled and varies with cell type.

Skin cells, blood-forming cells, and cells that line the intestine, for example, reproduce often. In contrast, cells of the liver reproduce a particular number of times, and then cease reproducing. If the number of liver cells is reduced by injury or surgery, the remaining cells are stimulated to reproduce again. Still other cells, such as nerve cells, apparently lose their ability to reproduce as they differentiate; therefore, damage to nerve cells usually permanently impairs nerve function.

Most types of human cells can divide up to about 50 times. Adherence to this limit can be startling. A connective tissue cell from a human fetus divides 35 to 63 times, the average being about 50 times. However, a similar cell from an adult divides only 14 to 29 times, as if the cells "know" how long they have existed, and how much longer they can exist.

A physical basis for this mitotic clock is the DNA at the tips of chromosomes, where the same six-nucleotide sequence repeats many hundreds of times. With each mitosis, up to 200 repeats are lost. When the chromosome tips wear down to a certain point, this somehow signals the cell to cease reproducing.

Other external and internal factors influence when a cell divides. Within cells, waxing and waning levels of

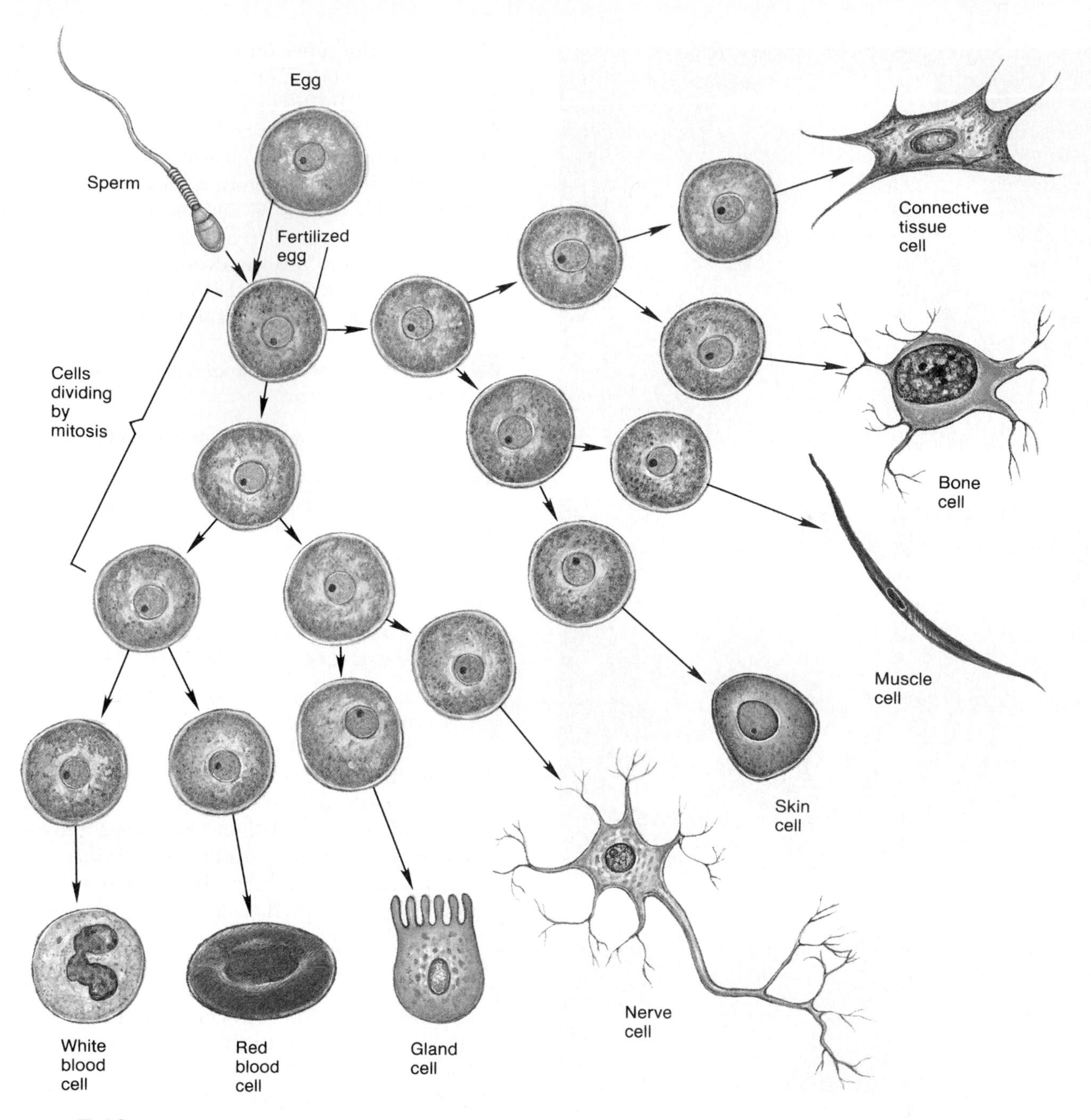


**FIGURE 3.42**
The trillions of cells in an adult human derive from the original fertilized egg cell by mitosis. As different genes turn on or off in different cells, the characteristics of specific cell types emerge. (Relative cell sizes are not to scale.)

proteins called kinases and cyclins control cell division. Another internal influence is simply cell size, specifically the ratio between the surface area provided by the cell membrane and the cell volume. The larger the cell, the more nutrients it needs to maintain the activities of life. However, a cell's surface area limits the amount of nutrients that can enter. Because volume increases faster than does surface area, a cell can grow too large to efficiently obtain nutrients. A cell can solve this growth problem by dividing. The resulting cells are smaller than the parent cell, and thus have a more favorable surface area-to-volume relationship.

External controls of cell division include hormones and growth factors. Hormones are biochemicals manufactured in a gland and transported in the bloodstream to a site where they exert an effect. Hormones signal mitosis in the lining of a woman's uterus each month, building up the tissue to nurture a possible pregnancy. Similarly, hormones stimulate cell division in her breasts when their function as milk-producing glands will soon be needed.

Growth factors are biochemicals that are like hormones in function but act closer to their sites of synthesis. Epidermal growth factor, for example, is responsible for growth of new skin beneath the scab on a skinned knee. This growth factor is also produced in salivary glands, explaining why an animal's licking a wound may speed healing.

Growth factors synthesized in the laboratory using genetic engineering techniques are used as drugs. Epidermal growth factor (EGF), for example, can speed healing of a wounded or transplanted cornea, a one-cell-thick layer covering the eye. Normally these cells do not divide. However, a damaged cornea treated with EGF divides to restore a complete cell layer. EGF is also used to help the body accept skin grafts and to stimulate healing of skin ulcers that occur as a complication of diabetes.

Space availability is another external factor that influences the timing and rate of cell division. Healthy cells do not divide if they are surrounded by other cells, a phenomenon called density dependent inhibition.

**1** How do cells vary in their rates of reproduction?

**2** What factors seem to control the number of times and the rate at which cells reproduce?

## Health Consequences of Loss of Cell Division Control

Control of cell division is absolutely crucial to health. With too infrequent mitosis, an embryo could not develop, a child could not grow, and wounds would not heal. Too frequent cell division produces an abnormal growth, or neoplasm, which may form a disorganized mass called a **tumor.**

Tumors are of two types. A *benign* tumor remains in place like a lump, eventually interfering with the function of healthy tissue. A *malignant,* or cancerous tumor looks quite different—it extends into surrounding tissue, resembling a crab with outreaching claws, which is where the name cancer comes from. Cancer cells, if not stopped, eventually reach the circulation and spread, or metastasize, to other sites. Chart 3.5 lists characteristics of cancer cells, and figure 3.43 illustrates how cancer cells infiltrate healthy tissue.

Cancer is actually a collection of disorders, distinguished by their site of origin and the affected cell type. Many cancers are treatable with surgery, radiation, or chemicals or immune system substances used as drugs. Experimental gene therapy is being used to fight cancer by giving tumor cells surface molecules that attract an attack by the body's immune system.

**CHART 3.5 Characteristics of Cancer Cells**

Loss of cell cycle control
Heritability (a cancer cell reproduces to form more cancer cells)
Transplantability (a cancer cell implanted into another individual will cause cancer to develop)
Dedifferentiation (loss of specialized characteristics)
Loss of density-dependent inhibition
Ability to induce local blood vessel formation (angiogenesis)
Invasiveness
Ability to spread (metastasis)

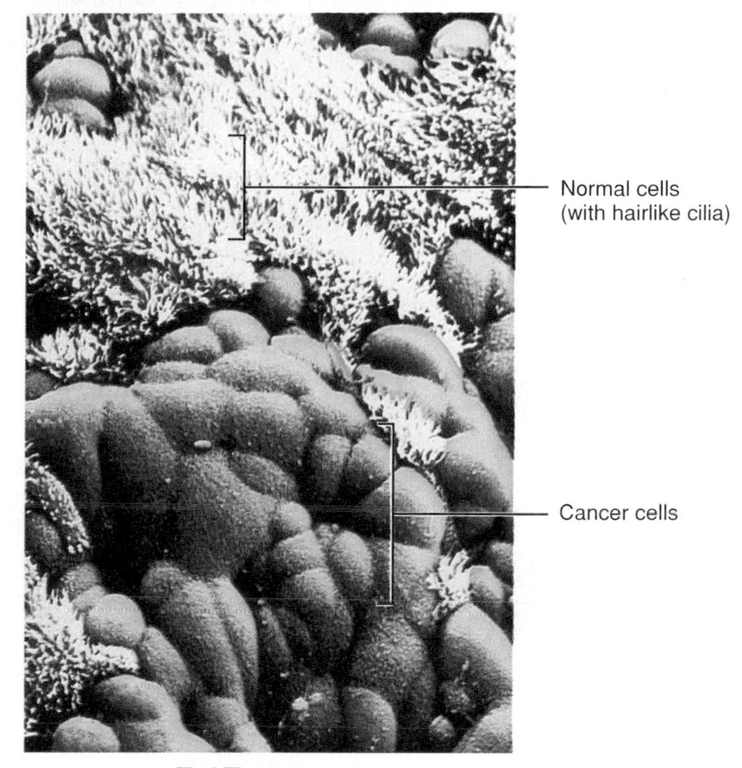


**FIGURE 3.43**
A cancer cell is rounder and less specialized than surrounding healthy cells. It secretes biochemicals that cut through nearby tissue, and others that form blood vessels that nurture the tumor's growth (58,800×).

At least two types of genes cause cancer. **Oncogenes** activate other genes that increase cell division rate. **Tumor suppressor genes** normally hold mitosis in check. When tumor suppressor genes are removed or otherwise inactivated, this lifts control of the cell cycle, and uncontrolled growth leading to cancer results (fig. 3.44). Environmental factors, such as exposure to toxic chemicals or radiation, may induce cancer by altering (mutating) oncogenes and tumor suppressor genes. Cancer may also be the consequence of a failure of normal cell death, resulting in overgrowth.

Normal anatomy and physiology—in other words, health—ultimately depend upon both the quality and quantity of the cells that comprise the human body.

1. How can too infrequent cell division affect health?
2. How can too frequent cell division affect health?
3. What is the difference between a benign and a cancerous tumor?
4. How do genes cause cancer?
5. How can factors in the environment cause cancer?

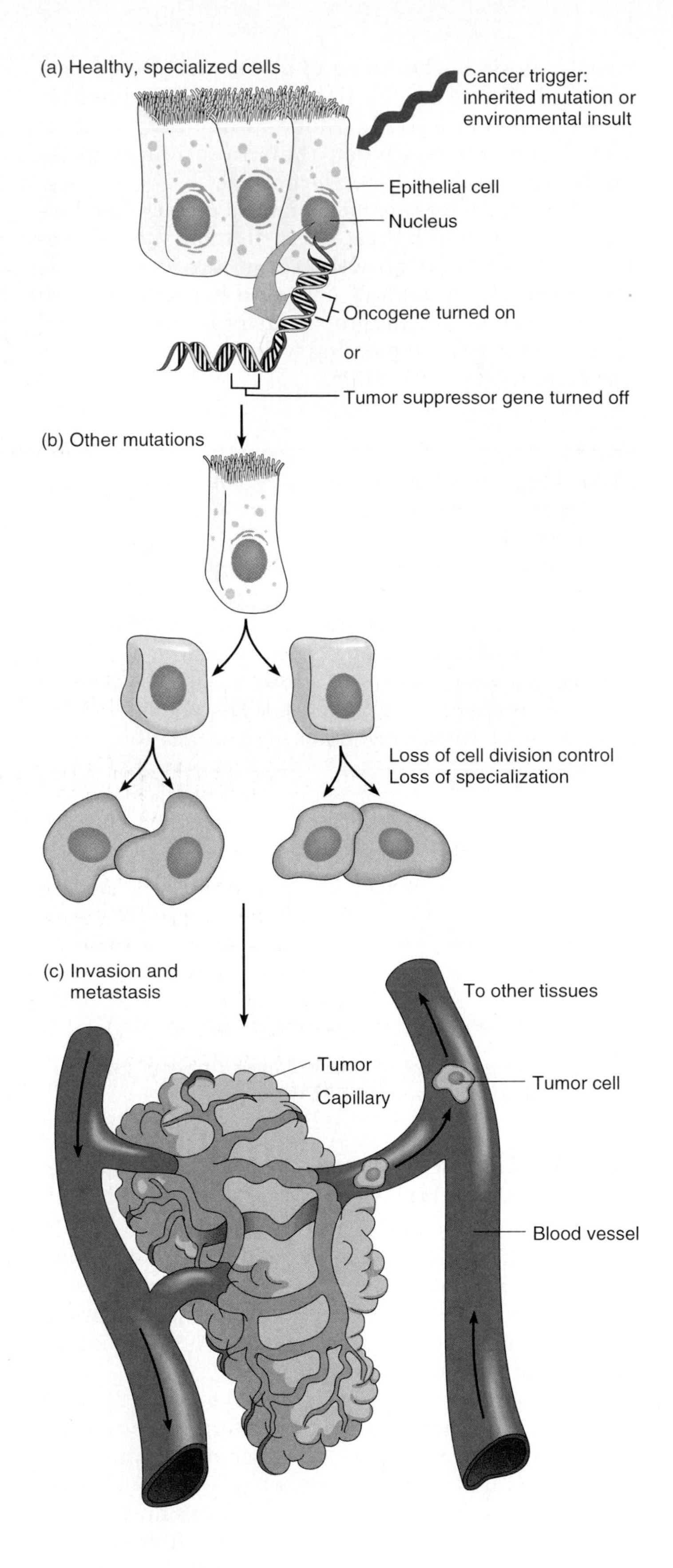


**FIGURE 3.44**

(*a*) In a healthy cell, oncogenes are not expressed and tumor suppressor genes are expressed. As a result, cell division rate is under control. Cancer begins in a single cell when an oncogene is turned on or a tumor suppressor gene is turned off. This initial step may result from an inherited mutation or from exposure to radiation, viruses, or chemicals that cause cancer. (*b*) Malignancy often results from a series of genetic alterations (mutations). An affected cell divides more often than the cell type it descends from and eventually loses its specialized characteristics. (*c*) Cancers grow and spread by inducing formation of blood vessels to nourish them, and then breaking away from their original location. The renegade cells often alter their genetic makeup and surface characteristics as they travel. This changeable nature is why many treatments eventually cease to work, or a supposedly vanquished cancer shows up someplace in the body other than where it originated.

## CHAPTER SUMMARY

### Introduction (page 61)

Cells vary considerably in size, shape, and function. The shapes of cells are closely related to their functions.

### A Composite Cell (page 61)

1. A cell includes a nucleus, cytoplasm, and a cell membrane.
2. Cytoplasmic organelles perform specific vital functions, but the nucleus controls the overall activities of the cell.
3. Cell membrane
   a. The cell membrane forms the outermost limit of the living material.
   b. It acts as a selectively permeable passageway that controls the movements of substances between the cell and its surroundings, and thus is the site of signal transduction.
   c. It includes protein, lipid, and carbohydrate molecules.
   d. The cell membrane framework consists mainly of a double layer of phospholipid molecules.
   e. Molecules that are soluble in lipids pass through the membrane easily, but water-soluble molecules do not.
   f. Cholesterol molecules help stabilize the membrane.
   g. Proteins are responsible for the special functions of the membrane.
      (1) Rodlike proteins form receptors on cell surfaces.
      (2) Globular proteins form channels for passage of various ions and molecules.
   h. Some cells are connected by means of specialized intercellular junctions called tight junctions, desmosomes, and gap junctions.
   i. Cell adhesion molecules oversee some cell interactions and movements.
4. Cytoplasm
   a. Cytoplasm contains networks of membranes and organelles suspended in fluid.
   b. Endoplasmic reticulum is composed of interconnected membranous sacs, canals, and vesicles that provide a tubular communication system and an attachment for ribosomes; it also functions in the synthesis of proteins, lipids, and hormones.
   c. Ribosomes are particles of protein and RNA that function in protein synthesis.
   d. The Golgi apparatus is composed of a stack of flattened, membranous sacs that package glycoproteins for secretion.
   e. Mitochondria are membranous sacs that contain enzymes involved with releasing energy from nutrient molecules and transforming energy into a usable form.
   f. Lysosomes are membranous sacs containing digestive enzymes that can destroy certain substances that enter cells.
   g. Peroxisomes are membranous, enzyme-containing vesicles.
   h. Cilia and flagella are motile processes that extend outward from some cell surfaces.
      (1) Cilia are numerous tiny, hairlike structures that wave, moving fluids across cell surfaces.
      (2) Flagella are longer processes; the tail of a sperm cell is a flagellum that enables the cell to swim.
   i. The centrosome is a nonmembranous structure consisting of two centrioles that aids in the distribution of chromosomes during cell reproduction.
   j. Vesicles are membranous sacs containing substances that recently entered the cell.
   k. Microfilaments and microtubules are threadlike processes that aid cellular movements and provide support and stability to cytoplasm.
   l. Cytoplasm may contain nonliving cellular products, such as nutrients and pigments, called inclusions.
5. Cell nucleus
   a. The nucleus is enclosed in a double-layered nuclear envelope including nuclear pores that controls the movement of substances between the nucleus and cytoplasm.
   b. A nucleolus is a dense body of protein and RNA that functions in the production of ribosomes.
   c. Chromatin is composed of loosely coiled fibers of protein and DNA that condense to become chromosomes during cell reproduction.

### Movements Into and Out of the Cell (page 79)

Movement of substances into and out of the cell may involve physical or physiological processes.

1. Diffusion
   a. Diffusion is movement of molecules or ions from regions of higher concentration toward regions of lower concentration.
   b. It is responsible for exchanges of oxygen and carbon dioxide within the body.
   c. Factors that increase the rate of diffusion include short distance, high concentration and low molecular weight of diffusing molecules, and high temperature.
2. Facilitated diffusion
   a. Facilitated diffusion uses carrier molecules in the cell membrane.
   b. This process moves substances only from regions of higher concentration to regions of lower concentration.
3. Osmosis
   a. Osmosis is a special case of diffusion in which water molecules diffuse from regions of higher water concentration to lower water concentration through a selectively permeable membrane.
   b. Osmotic pressure increases as the number of particles dissolved in a solution increases.

c. Cells lose water when placed in hypertonic solutions and gain water when placed in hypotonic solutions.
d. A solution is isotonic when it contains the same concentration of dissolved particles as the cell contents.

4. Filtration
   a. In filtration, molecules move from regions of higher hydrostatic pressure toward regions of lower hydrostatic pressure.
   b. Blood pressure filters water and dissolved substances through porous capillary walls.
5. Active transport
   a. Active transport moves molecules or ions from regions of lower concentration to regions of higher concentration.
   b. It requires cellular energy and carrier molecules in the cell membrane.
6. Endocytosis
   a. In pinocytosis, a cell membrane engulfs tiny droplets of liquid.
   b. In phagocytosis, a cell membrane engulfs solid particles.
   c. In receptor-mediated endocytosis, receptor proteins combine with specific molecules in the cell surroundings. The membrane engulfs the combinations.
7. Exocytosis
   a. Exocytosis is the reverse of endocytosis.
   b. In exocytosis, vesicles containing secretions fuse with the cell membrane, releasing the substances to the outside.

## The Cell Cycle (page 87)

1. The cell cycle includes interphase, mitosis, cytoplasmic division, and differentiation.
2. Interphase
   a. Interphase is the stage when a cell grows, cellular DNA replicates, and the cell forms new organelles.
   b. It terminates when the cell begins to undergo mitosis.
3. Mitosis
   a. Mitosis is the division and distribution of DNA to new cells during cell reproduction.
   b. The stages of mitosis include prophase, metaphase, anaphase, and telophase.
4. The cytoplasm divides into two portions following mitosis.
5. Cell differentiation is the specialization of structures and functions.

## Control of Cell Reproduction (page 93)

1. Cellular reproductive capacities vary greatly.
2. Chromosome tips that shorten with each mitosis provide a mitotic clock, usually limiting the number of divisions to 50.
3. Cell division is limited, and controlled by both internal and external factors.
4. As a cell grows, its surface area increases to a lesser degree than its volume, and eventually the area becomes inadequate for the needs of the living material within the cell. When a cell divides, the new cells have more favorable surface area-volume relationships.
5. Cancer is the consequence of a loss of cell cycle control.

## Critical Thinking Questions

1. Which process—diffusion, osmosis, or filtration—is most closely related to each of the following situations?
   a. The injection of a drug that is hypertonic to the tissues stimulates pain.
   b. A person with extremely low blood pressure stops producing urine.
   c. The concentration of urea in the dialyzing fluid of an artificial kidney is kept low.
2. What characteristic of cell membranes may account for the observation that fat-soluble substances like chloroform and ether cause rapid effects on cells?
3. A person exposed to excessive X rays may develop a decrease in white blood cell number and an increase in susceptibility to infections. In what way are these effects related?
4. Exposure to tobacco smoke causes cilia to become immobile and perhaps to degenerate. Why might this explain why tobacco smokers have an increased incidence of respiratory infections?
5. How would you explain the function of phagocytic cells to a patient with a bacterial infection?
6. How is knowledge of cellular reproduction important to an understanding of:
   a. growth
   b. wound healing
   c. cancer
7. Why are enlarged lysosomes a sign of a serious illness?

## Review Exercises

1. Use specific examples to illustrate how cells vary in size.
2. Describe how the shapes of nerve, epithelial, and muscle cells are related to their functions.
3. Name the major components of a cell, and describe their relationships to one another.
4. Discuss the structure and functions of a cell membrane.
5. How do cilia, flagella, and cell adhesion molecules provide cellular movement?
6. Distinguish between organelles and inclusions.
7. Define *selectively permeable.*
8. Describe the chemical structure of a membrane.
9. Explain how the structure of a cell membrane is related to its permeability.
10. Explain the function of membrane proteins.
11. Describe three kinds of intercellular junctions.
12. Describe the structures and functions of each of the following:
    a. endoplasmic reticulum
    b. ribosome
    c. Golgi apparatus
    d. mitochondrion
    e. lysosome
    f. peroxisome
    g. cilium
    h. flagellum
    i. centrosome
    j. vesicle
    k. microfilament
    l. microtubule
13. Describe the structure of the nucleus and the functions of its contents.
14. Distinguish between diffusion and facilitated diffusion.
15. Name four factors that increase the rate of diffusion.
16. Explain how diffusion aids in the exchange of gases within the body.
17. Define *osmosis.*
18. Explain what is meant by *osmotic pressure.*
19. Explain how the number of solute particles in a solution affects its osmotic pressure.
20. Distinguish between solutions that are hypertonic, hypotonic, and isotonic.
21. Define *filtration.*
22. Explain how filtration moves substances through capillary walls.
23. Explain why active transport is called a physiological process, while diffusion is called a physical process.
24. Explain the function of carrier molecules in active transport.
25. Distinguish between pinocytosis and phagocytosis.
26. Describe *receptor-mediated endocytosis.* How might it be used to deliver drugs across the blood-brain barrier?
27. List the phases in the cell cycle. Why is interphase *not* a time of cellular rest?
28. Name the two processes included in cell reproduction.
29. Describe the major events of mitosis.
30. Explain how the cytoplasm divides during cellular reproduction.
31. Explain what happens during *interphase.*
32. Define *differentiation.*
33. Explain how differentiation may involve the repression of DNA information.
34. How does genetic control underlie cancer causation?

4

CHAPTER

# Cellular Metabolism

Energy use at the subcellular level powers the muscles of the athlete—as well as the vital functions of life.

## Chapter Objectives

After you have studied this chapter, you should be able to:

1. Distinguish between anabolic and catabolic metabolism.
2. Explain how enzymes control metabolic processes.
3. Explain how cellular respiration releases chemical energy.
4. Describe how energy is made available for cellular activities.
5. Describe the general metabolic pathways of carbohydrates, lipids, and proteins.
6. Explain how metabolic pathways are regulated.
7. Describe how genetic information is stored within DNA molecules.
8. Explain how genetic information is used in the synthesis of proteins.
9. Describe how DNA molecules are replicated.
10. Explain how genetic information can be altered and how such a change may affect an organism.

## Key Terms

**aerobic respiration** (a-er-ō′bik res″pi-ra′shun)
**anabolism** (an″ah-bol′lizm)
**anaerobic respiration** (an″a-er-ō′bik res″pi-ra′shun)
**catabolism** (kat″ah-bol′liz-m)
**coenzyme** (ko-en′zīm)
**deamination** (de-am″ĭ-na′shun)
**dehydration synthesis** (de″hi-dra′shun sin′the-sis)
**DNA**
**energy** (en′er-je)
**enzyme** (en′zīm)
**gene** (jēn)
**genetic code** (je-net′ik kōd)
**glycolysis** (gli-kol′ĭ-sis)
**hydrolysis** (hi-drol′ĭ-sis)
**metabolism** (me-tab′o-liz-m)
**mutation** (mu-ta′shun)
**oxidation** (ok″si-da′shun)
**reduction** (re-duk′shun)
**replication** (re″pli-ka′shun)
**RNA**
**substrate** (sub′strāt)
**transcription** (tranz-krip′-shun)
**translation** (tranz-lay′shun)

It is a Sunday night in mid-April in Boston, and the restaurant is packed with skinny, raggedy-dressed people. The food at the pasta bar disappears almost as soon as the servers can put it down, as the patrons gobble up mountains of spaghetti and fistfuls of crusty bread and wash it down with soft drinks. The crowd is back the next morning, to consume stacks of pancakes.

Carbohydrate loading, as these healthy athletes are doing, is a rite of passage of sorts for the annual Boston marathon. By following this dietary regimen, the runners pack their muscles with glycogen, so that by the time they hit Heartbreak Hill—a grueling four-hill section at mile 20 of the 26.2 mile race—they have just enough energy to finish. Runners whose muscles are not bolstered with pre-race pasta feasts, particularly males whose fat supplies are generally less than women's, are more likely to "hit the wall" at the 20 mile mark. Their muscles literally run out of glycogen, and only the strongest of wills enables them to finish the race.

Glycogen consists of long chains of the simple sugar glucose. Biochemical reactions in body cells, and particularly in muscle cells, gradually break down glucose molecules, releasing the energy that holds its chemical bonds together. It is this energy that the cell uses to power the great variety of activities that constitute the living state, even when that includes very strenuous exercise. Although all of our cells break down glucose all of the time, we are rarely as aware of the process as is a marathon runner when rubbery limbs and sheer exhaustion indicate that he or she has hit the wall.

A marathon runner provides a vivid illustration of biological energy usage, both on the cellular and whole-body levels. Yet in every human cell, even in the most sedentary individual, thousands of chemical reactions essential to life occur every second. A special type of protein called an **enzyme** controls the pace of each reaction. The sum total of chemical reactions within the cell constitutes **metabolism.**

Many metabolic reactions occur one after the other, with the products of one reaction serving as starting materials of another, forming intricate pathways and cycles that may intersect by sharing intermediate compounds. As a result, metabolism in its entirety may seem enormously complex. However, individual pathways of metabolism are fascinating to study because they reveal how cells function—in essence, how chemistry becomes biology. This chapter explores how metabolic pathways supply a cell with energy, and how other biochemical processes enable a cell to produce proteins—including the enzymes that make all of metabolism possible.

## Metabolic Processes

Metabolic reactions and pathways are of two types. In **anabolism,** larger molecules are constructed from smaller ones, requiring input of energy. In **catabolism,** larger molecules are broken down into smaller ones, releasing energy.

### Anabolism

Anabolic metabolism provides all substances required for cellular growth and repair. For example, cells often join many simple sugar molecules (monosaccharides) to form larger molecules of glycogen by an anabolic process called *dehydration synthesis.* In this process, the larger carbohydrate molecule is built by bonding monosaccharide molecules together into a chain. This is what happens in the runner's muscle cells, using monosaccharides derived from his or her carbo-rich meal of the evening before the race. When adjacent monosaccharide units are joined, an –OH (hydroxyl group) from one monosaccharide molecule and an –H (hydrogen atom) from an –OH group of another are removed. As the –H and –OH react to produce a water molecule, the monosaccharides unite by a shared oxygen atom, as shown in figure 4.1 (read from left to right). As the process repeats, the molecular chain extends.

Similarly, glycerol and fatty acid molecules join by dehydration synthesis in fat (adipose) tissue cells to form fat molecules. In this case, three hydrogen atoms are removed from a glycerol molecule, and an –OH group is removed from each of three fatty acid molecules, as shown in figure 4.2 (read from left to right). The result is three water molecules and a single fat molecule, whose glycerol and fatty acid portions are bound by shared oxygen atoms.

Cells also build protein molecules by joining amino acid molecules by dehydration synthesis. When two amino acid molecules are united, an –OH from one and an –H from the $-NH_2$ group of another are removed. A water molecule forms, and the amino acid molecules join by a bond between a carbon atom and a nitrogen atom (fig. 4.3; read from left to right). This type of bond, which is called a *peptide bond,* holds the amino acids together. Two amino acids bound together form a *dipeptide,* and many joined in a chain form a *polypeptide.* Generally, a polypeptide consisting of 100

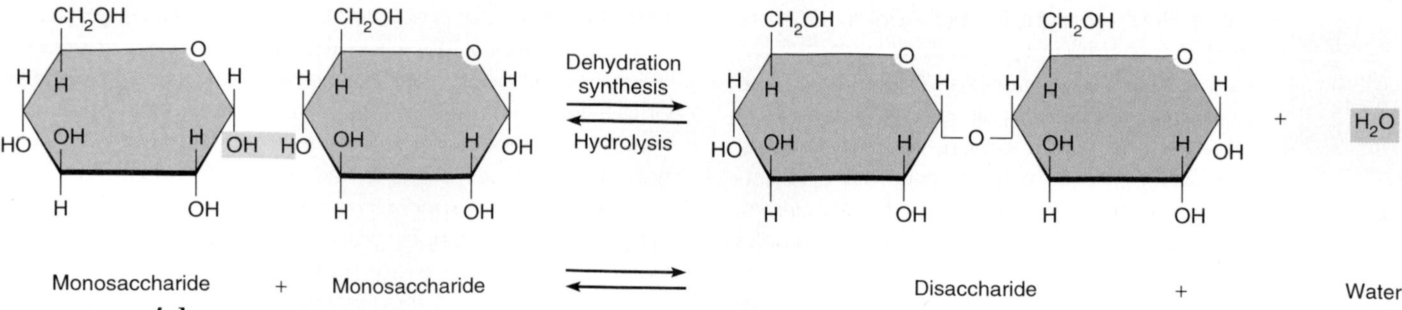


**FIGURE 4.1**

Two monosaccharides may join by dehydration synthesis to form a disaccharide. In the reverse reaction, the disaccharide is hydrolyzed to two monosaccharides.

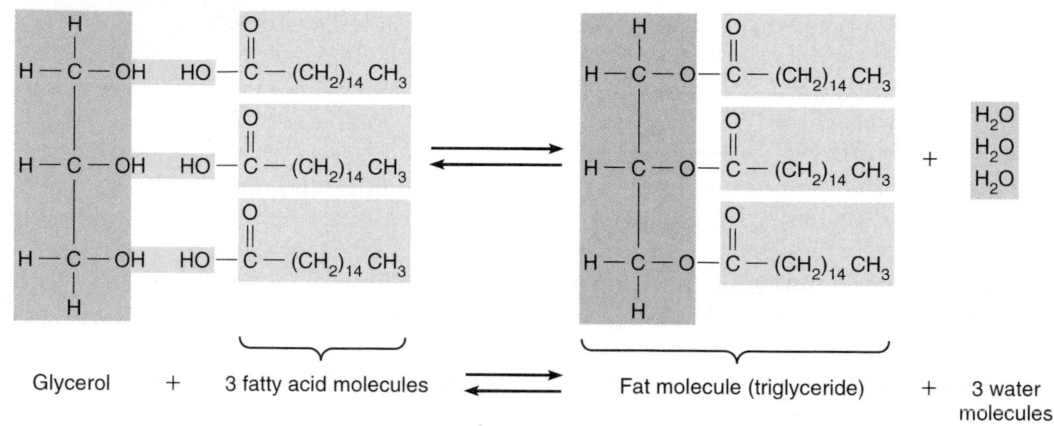


**FIGURE 4.2**

A glycerol molecule and three fatty acid molecules may join by dehydration synthesis to form a fat molecule (triglyceride). In the reverse reaction, fat is hydrolyzed to three fatty acids and glycerol.

or more amino acid molecules is called a *protein,* although the boundary distinguishing between polypeptides and proteins is not defined precisely.

## Catabolism

Physiological processes that break down larger molecules into smaller ones constitute catabolic metabolism. An example is **hydrolysis,** which can decompose carbohydrates, lipids, and proteins. A water molecule splits these substances into two simpler parts.

The hydrolysis of a disaccharide, for instance, results in two monosaccharide molecules (fig. 4.1; read from right to left). In this case, the bond between the simple sugars breaks, and the water molecule supplies a hydrogen atom to one sugar molecule and a hydroxyl group to the other. Thus, hydrolysis is the reverse of dehydration synthesis:

**Dehydration synthesis**
**Monosaccharide + Monosaccharide ⇌ Disaccharide + Water**
**Hydrolysis**

Hydrolysis breaks down carbohydrates into monosaccharides; fats into glycerol and fatty acids (fig. 4.2; read from right to left); proteins into amino acids (fig. 4.3; read from right to left); and nucleic acids into nucleotides.

Although water molecules provide the –H and –OH necessary for hydrolysis, it does not happen automatically. For example, water-soluble substances like the disaccharide sucrose (table sugar) *dissolve* in a glass of water, but do not undergo hydrolysis. Hydrolysis (like dehydration synthesis) requires the help of enzymes, which are discussed in the next section.

Both catabolism and anabolism must be carefully controlled so that the breakdown or energy-releasing reactions occur at rates that are adjusted to the requirements of the building up or energy-utilizing reactions. Any disturbance in this balance is likely to damage or kill cells.

1. What general functions do anabolism and catabolism serve?
2. What substance does the anabolic metabolism of monosaccharides form? Of amino acids? Of glycerol and fatty acids?
3. Distinguish between dehydration synthesis and hydrolysis.

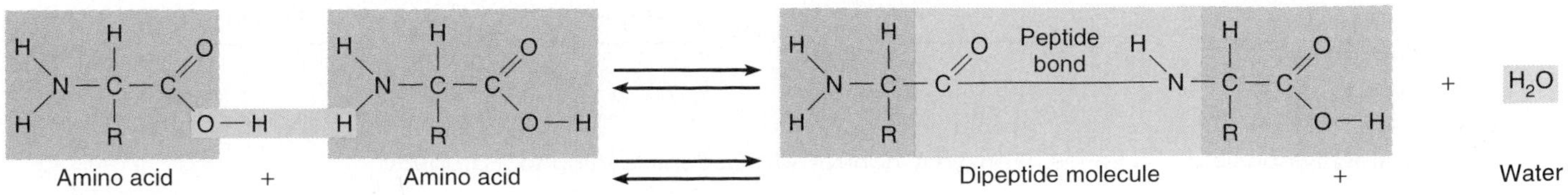

Amino acid + Amino acid ⇌ Dipeptide molecule + Water ($H_2O$)

Peptide bond

**FIGURE 4.3**

When two amino acid molecules unite by dehydration synthesis, a peptide bond forms between a carbon atom and a nitrogen atom. In the reverse reaction, a dipeptide is hydrolyzed to two amino acids.

## Control of Metabolic Reactions

Although different kinds of cells may conduct specialized metabolic processes, all cells perform certain basic reactions, such as the buildup and breakdown of carbohydrates, lipids, proteins, and nucleic acids. These reactions include hundreds of very specific chemical changes that must occur in a particular sequence. Enzymes control the rates of these metabolic reactions.

### Enzyme Action

Like other chemical reactions, metabolic reactions require energy (activation energy) before they proceed. This is why heat is used to increase the rates of chemical reactions in laboratories. Heat energy increases the rate at which molecules move and the frequency of molecular collisions. These collisions increase the likelihood of interactions among the electrons of the molecules that can form new chemical bonds. The temperature conditions in cells are usually too mild to adequately promote the reactions of life. Enzymes make these reactions possible.

Enzymes are almost invariably globular proteins (see chapter 2) that promote specific chemical reactions within cells by lowering the activation energy needed to start these reactions. Enzymes can speed, or catalyze, metabolic reactions by a factor of a million or more.

Enzymes are required in very small quantities because as they work, they are not consumed and can, therefore, function repeatedly. Also, each enzyme has *specificity,* acting only on a particular kind of substance, which is called its **substrate.** For example, the substrate of an enzyme called catalase (found in the peroxisomes of liver and kidney cells) is hydrogen peroxide, a toxic by-product of certain metabolic reactions. This enzyme's only function is to decompose hydrogen peroxide into water and oxygen, helping prevent accumulation of hydrogen peroxide that might damage cells.

Cellular metabolism includes hundreds of different chemical reactions, each controlled by a specific kind of enzyme. Often sequences of enzyme-controlled reactions, called **metabolic pathways,** lead to synthesis or breakdown of particular biochemicals (fig. 4.4). Thus, hundreds of different kinds of enzymes are present in every cell, and each enzyme must be able to "recognize" its specific substrate. This ability to identify a substrate depends upon the shape of an enzyme molecule. That is, each enzyme's polypeptide chain is twisted and coiled into a unique three-dimensional form, or **conformation,** that fits the special shape of its substrate molecule.

> You may have noticed the action of the enzyme catalase if you have ever used hydrogen peroxide to cleanse a wound. Injured cells release catalase, and when hydrogen peroxide contacts them, bubbles of oxygen are set free. The resulting foam removes debris from inaccessible parts of the wound.

During an enzyme-catalyzed reaction, regions of the enzyme molecule called **active sites** temporarily combine with portions of the substrate, forming an enzyme-substrate complex. This interaction strains chemical bonds in the substrate in a way that makes a particular chemical reaction more likely to occur. When it does, the enzyme is released in its original form, able to bind another substrate molecule (fig. 4.5).

Enzyme catalysis can be summarized as follows:

**Substrate molecule** + **Enzyme molecule** → **Enzyme-substrate complex** → **Product (changed substrate)** + **Enzyme molecule (unchanged)**

The speed of an enzyme-catalyzed reaction depends on the number of enzyme and substrate molecules in the cell. The reaction occurs more rapidly if the concentration of the enzyme or the concentration of the substrate increases. Also, the efficiency of different kinds of enzymes varies greatly. Thus, some enzymes can process only a few substrate molecules per second, while others can handle thousands or nearly a million substrate molecules per second.

Enzyme names are often derived from the names of their substrates, with the suffix *-ase* added. For example, a lipid-splitting enzyme is called *lipase,* a

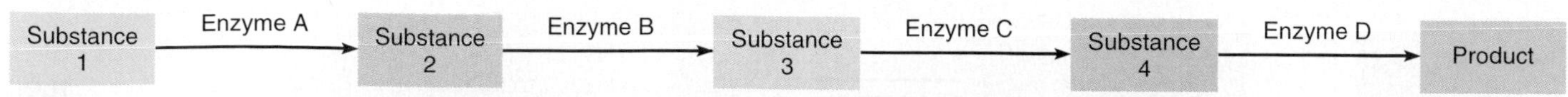


**FIGURE 4.4**
A metabolic pathway consists of a series of enzyme-controlled reactions leading to a product.

protein-splitting enzyme is *protease,* and a starch- (amylum) splitting enzyme is *amylase.* Similarly, *sucrase* is an enzyme that splits the sugar sucrose, *maltase* splits the sugar maltose, and *lactase* splits the sugar lactose.

## Cofactors and Coenzymes

Often an enzyme is inactive until it combines with a nonprotein component that either helps the active site attain its appropriate shape or helps bind the enzyme to its substrate. Such a substance, called a **cofactor,** may be an ion of an element, such as copper, iron, or zinc, or it may be a small organic molecule, called a **coenzyme.** Coenzymes are often composed of vitamin molecules or incorporate vitamin molecules into their structures.

**Vitamins** are essential organic substances that cannot be synthesized (or cannot be synthesized in sufficient quantities) by human cells and therefore must come from the diet. Since vitamins provide coenzymes that can, like enzymes, function again and again, cells need vitamins in very small quantities. Vitamins are discussed in chapter 18.

## Factors that Alter Enzymes

Almost all enzymes are proteins, and like other proteins, they can be denatured by exposure to excessive heat, radiation, electricity, certain chemicals, or fluids with extreme pH values. For example, many enzymes become inactive at 45° C, and nearly all of them are denatured at 55° C. Some poisons are chemicals that denature enzymes. Cyanides, for instance, can interfere with respiratory enzymes and damage cells by halting their energy-releasing processes.

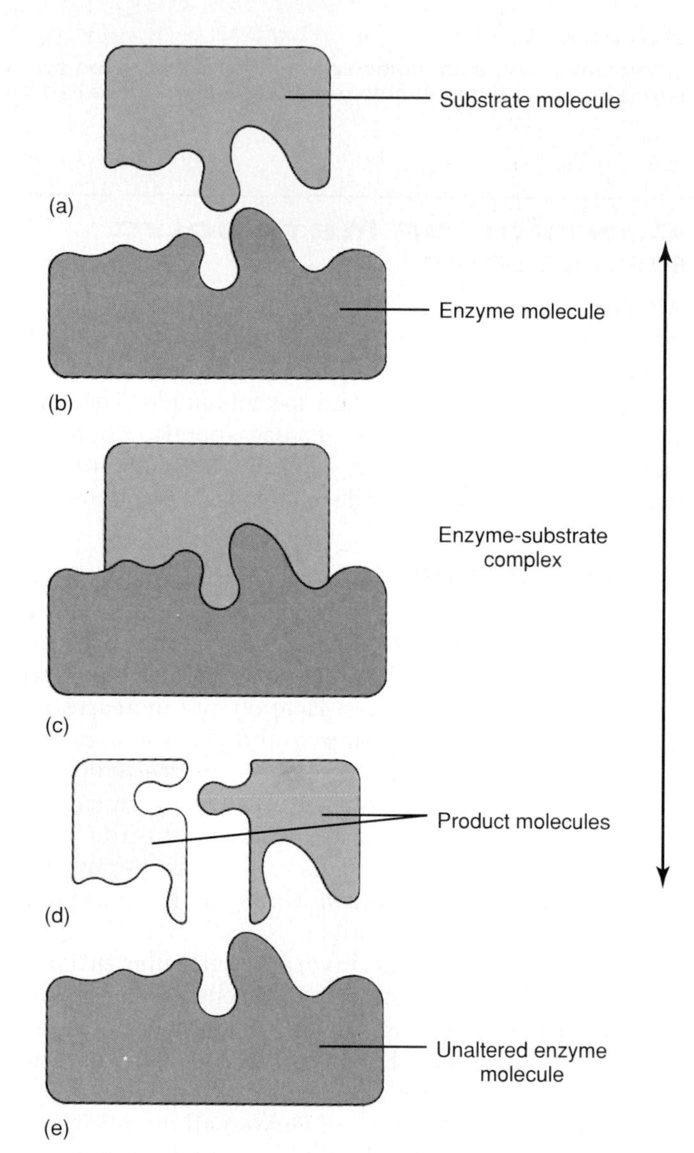


**FIGURE 4.5**
(*a*) The shape of a substrate molecule fits (*b*) the shape of the enzyme's active site. (*c*) When the substrate molecule combines temporarily with the enzyme, a chemical reaction occurs. The result is (*d*) product molecules and (*e*) an unaltered enzyme. Many enzymatic reactions are reversible.

> The antibiotic drug penicillin interferes with enzymes that enable certain bacteria to construct cell walls. As a result, the bacteria die. In this manner, penicillin protects against certain bacterial infections.

**1** What is an enzyme?

**2** How can an enzyme control the rate of a metabolic reaction?

**3** How does an enzyme "recognize" its substrate?

**4** What is the role of a cofactor?

**5** What factors can denature enzymes?

## ENERGY FOR METABOLIC REACTIONS

**Energy** is the capacity to change matter or move something; that is, it is the ability to do work. Therefore, we recognize energy by what it can do. Common forms of energy include heat, light, sound, electrical energy, mechanical energy, and chemical energy.

Although energy cannot be created or destroyed, it can be converted from one form to another. An ordinary incandescent light bulb changes electrical energy to heat and light, and an automobile engine converts the chemical energy in gasoline to heat and mechanical energy.

Whenever changes occur in the human body—which is a characteristic of life—energy is being transferred. Thus, all metabolic reactions involve energy in some form.

### Release of Chemical Energy

Most metabolic processes depend on chemical energy. This form of energy is held in the bonds between the atoms of molecules and is released when these bonds break. Burning releases the chemical energy of many substances. Such a reaction is usually started by applying sufficient heat. As the substance burns, molecular bonds break, and energy escapes as heat and light.

Similarly, glucose molecules are "burned" in cells. This process is called **oxidation.** The energy released by oxidation of glucose is used to promote cellular metabolism. There are, however, some important differences between the oxidation of substances inside cells and the burning of substances outside them.

Burning usually requires a relatively large amount of energy to begin, and most of the energy released escapes as heat or light. In cells, enzymes initiate oxidation by decreasing the amount of energy (activation energy) needed to start the process. Also, by transferring energy to special energy-carrying molecules, cells are able to capture in the form of chemical energy about half of the energy released. The rest escapes as heat, which helps maintain body temperature.

The process by which energy is released from molecules such as glucose and is transferred to other molecules is called **cellular respiration.** It involves chemical reactions that must occur in a particular sequence, each one controlled by a different enzyme. Some of these enzymes are in the cell's cytosol, while others are in the mitochondria.

**1** What is energy?

**2** How does cellular oxidation differ from burning?

**3** Define cellular respiration.

## AN OVERVIEW OF CELLULAR RESPIRATION

### Anaerobic Respiration

When a 6-carbon glucose molecule breaks down during the first part of cellular respiration, enzymes control a series of ten reactions, collectively called **glycolysis,** that breaks glucose into two 3-carbon pyruvic acid molecules. This phase of the process occurs in the cytosol, and because it does not require the presence of oxygen, it is called **anaerobic respiration.**

Although some energy is needed to activate the reactions of anaerobic respiration, more energy is released than is used. The excess is used to synthesize an energy-carrying molecule, called **ATP** (adenosine triphosphate) (fig. 4.6).

### Aerobic Respiration

Following the anaerobic phase of cellular respiration, oxygen must be available in order for the process of energy extraction to continue. For this reason, the second phase is called **aerobic respiration.** It takes place within the mitochondria, and transfers considerably more energy to ATP molecules.

After glucose breaks down, carbon dioxide molecules and hydrogen atoms remain. The carbon dioxide diffuses out of the cell as a waste, and the hydrogen atoms combine with oxygen to form water molecules. Thus, the final products of glucose oxidation are carbon dioxide, water, and energy.

### ATP Molecules

For each glucose molecule that is decomposed completely, up to thirty-eight molecules of ATP can be produced. Two of these are the result of anaerobic respiration, but the rest are formed during the aerobic phase.

Each ATP molecule consists of three main parts—an adenine, a ribose, and three phosphates in a chain (fig. 4.7). (Note the similarity between the general structure of an ATP molecule and the structure of a nucleotide within a nucleic acid molecule described in chapter 2; see fig. 2.19.)

As energy is released during cellular respiration, some of it is captured in the bond of the end phosphate (high-energy phosphate bond) of an ATP molecule. This stored chemical energy may be quickly transferred to another molecule involved in some metabolic process. When such an energy transfer occurs, the terminal, high-energy phosphate bond of the ATP molecule breaks, releasing the energy of this bond. Energy stored in ATP molecules is used whenever cellular

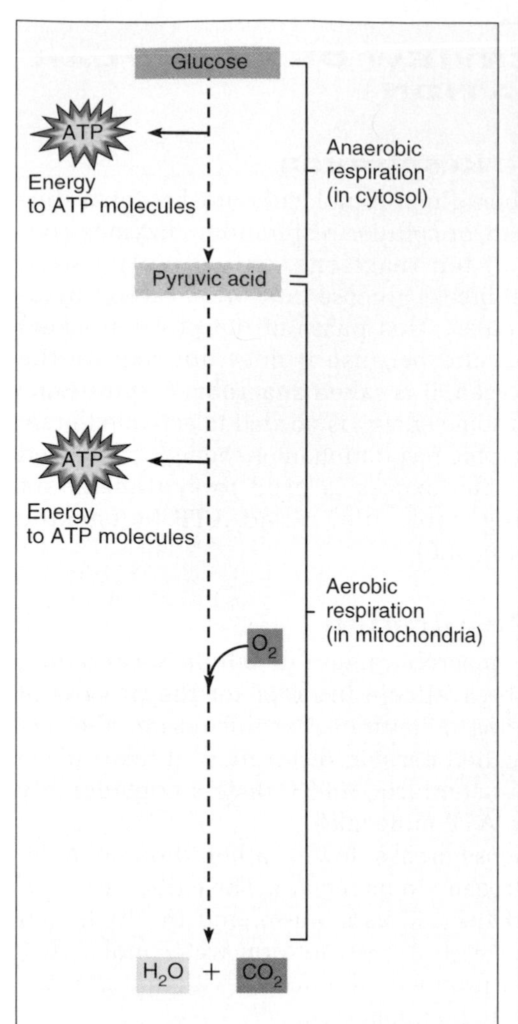


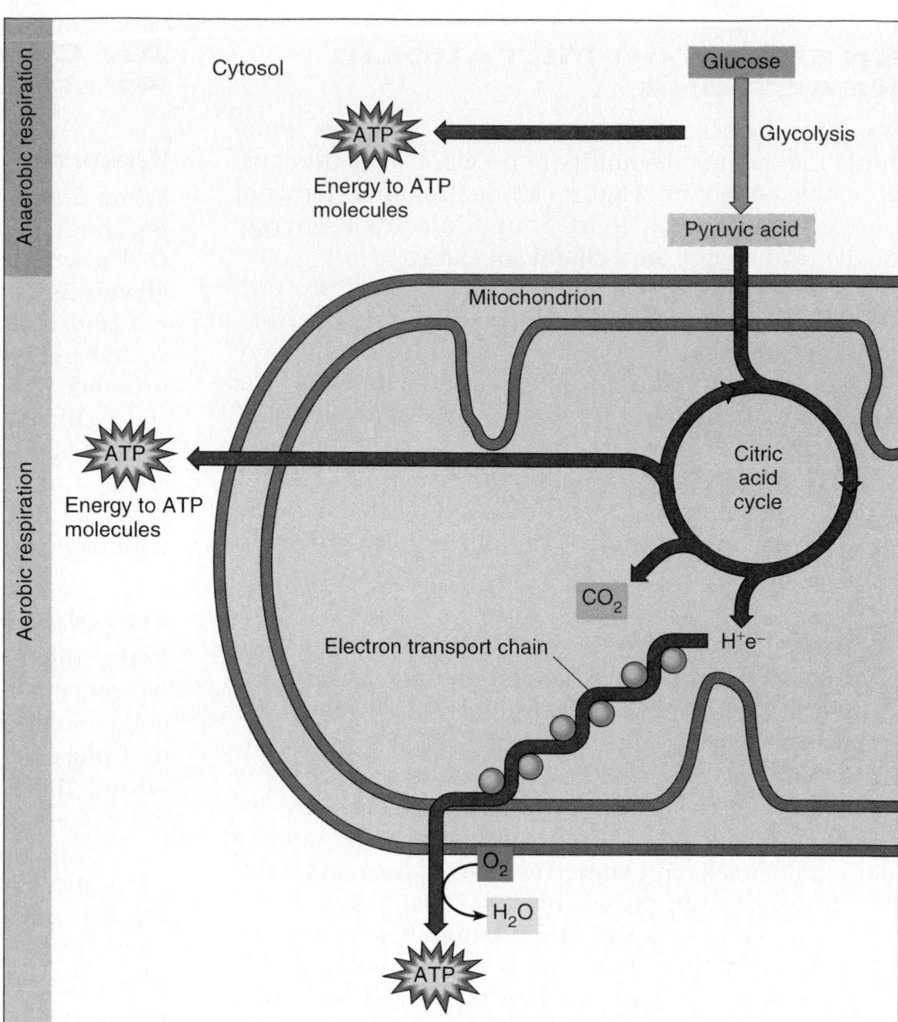


**FIGURE 4.6**

Anaerobic respiration occurs in the cytosol and does not require the presence of oxygen, whereas aerobic respiration occurs in the mitochondria only in the presence of oxygen.

work is performed, such as when muscle cells contract, when membranes carry on active transport, or when cells synthesize and release secretions.

An ATP molecule that has lost its terminal phosphate becomes an **ADP** (adenosine diphosphate) molecule. However, the ADP molecule can convert back into an ATP by capturing some energy and a phosphate. Thus, as figure 4.8 shows, ATP and ADP molecules shuttle back and forth between the energy-releasing reactions of cellular respiration and the energy-utilizing reactions of the cell.

ATP is not the only kind of energy-carrying molecule within a cell, but it is the primary one. Without ATP, cells die quickly.

1. What is anaerobic respiration? Aerobic respiration?
2. What happens to the energy released by cellular respiration?
3. What are the final products of cellular respiration?
4. What is the function of ATP molecules?

## A Closer Look at Cellular Respiration

Anabolic and catabolic pathways in cells usually consist of a number of different steps that must occur in a particular sequence, such as the reactions of aerobic respiration. The enzymes controlling these reactions must act in a specific order. Such precision of activity suggests that the enzymes are physically positioned in the exact sequence as that of the reactions they control. Indeed, the enzymes responsible for aerobic respiration are located in tiny, stalked particles on the membranes (cristae) within the mitochondria (see chapter 3).

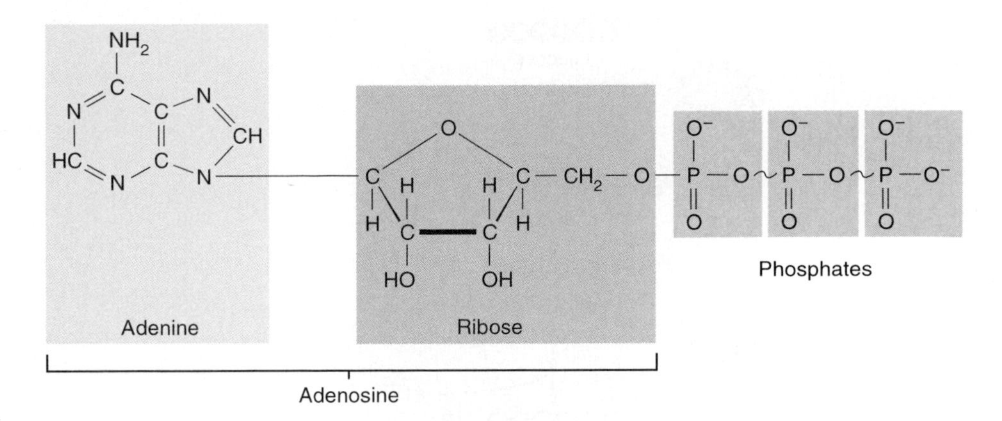


**FIGURE 4.7**

An ATP molecule consists of an adenine, a ribose, and three phosphates. The wavy lines connecting the last two phosphates represent high-energy chemical bonds. When broken, these bonds release energy quickly.

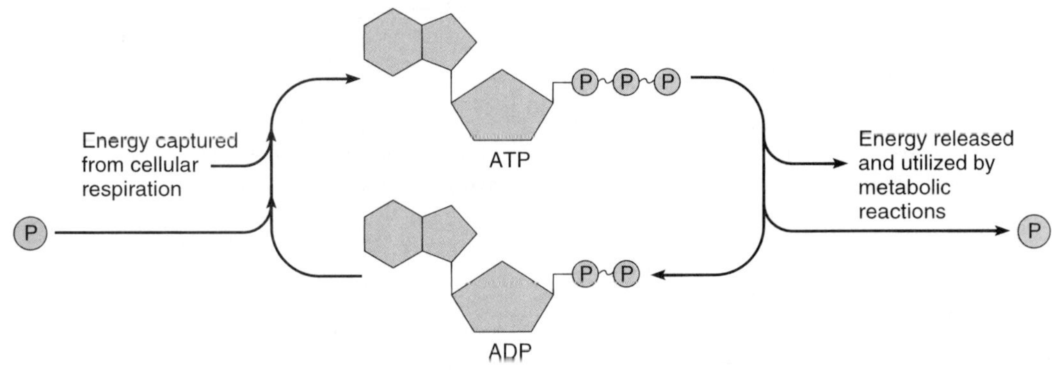


**FIGURE 4.8**

ATP provides energy for anabolic reactions and is regenerated by catabolic reactions.

Metabolic pathways are usually interconnected so that certain molecules may enter more than one pathway. For example, carbohydrate molecules from foods may enter catabolic pathways and be used to supply energy, or they may enter anabolic pathways and be stored as glycogen, fat, or other molecules (fig. 4.9).

## Carbohydrate Pathways

The average human diet consists largely of carbohydrates, which are digested into monosaccharides such as glucose. These substances are used primarily as cellular energy sources, which means they usually enter the catabolic pathways of cellular respiration. First, glucose is broken down to pyruvic acid in the cytoplasm, as discussed previously. Next, the pyruvic acid is broken down to yield a 2-carbon acetyl group that combines with a molecule of coenzyme A to form a substance called *acetyl coenzyme A.* It, in turn, is transported into a mitochondrion to form several intermediate products by a complex series of chemical reactions known as the **citric acid cycle** (Krebs cycle). During

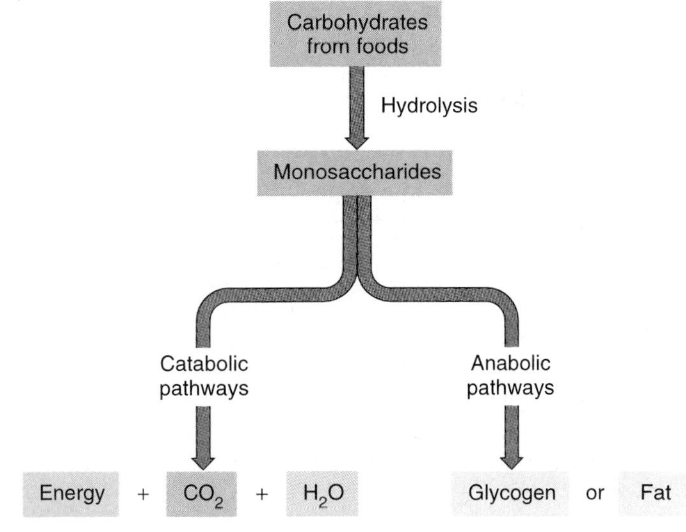


**FIGURE 4.9**

Hydrolysis breaks down carbohydrates from foods into monosaccharides. The resulting molecules may enter catabolic pathways and be used as energy sources, or they may enter anabolic pathways and be converted to glycogen or fat.

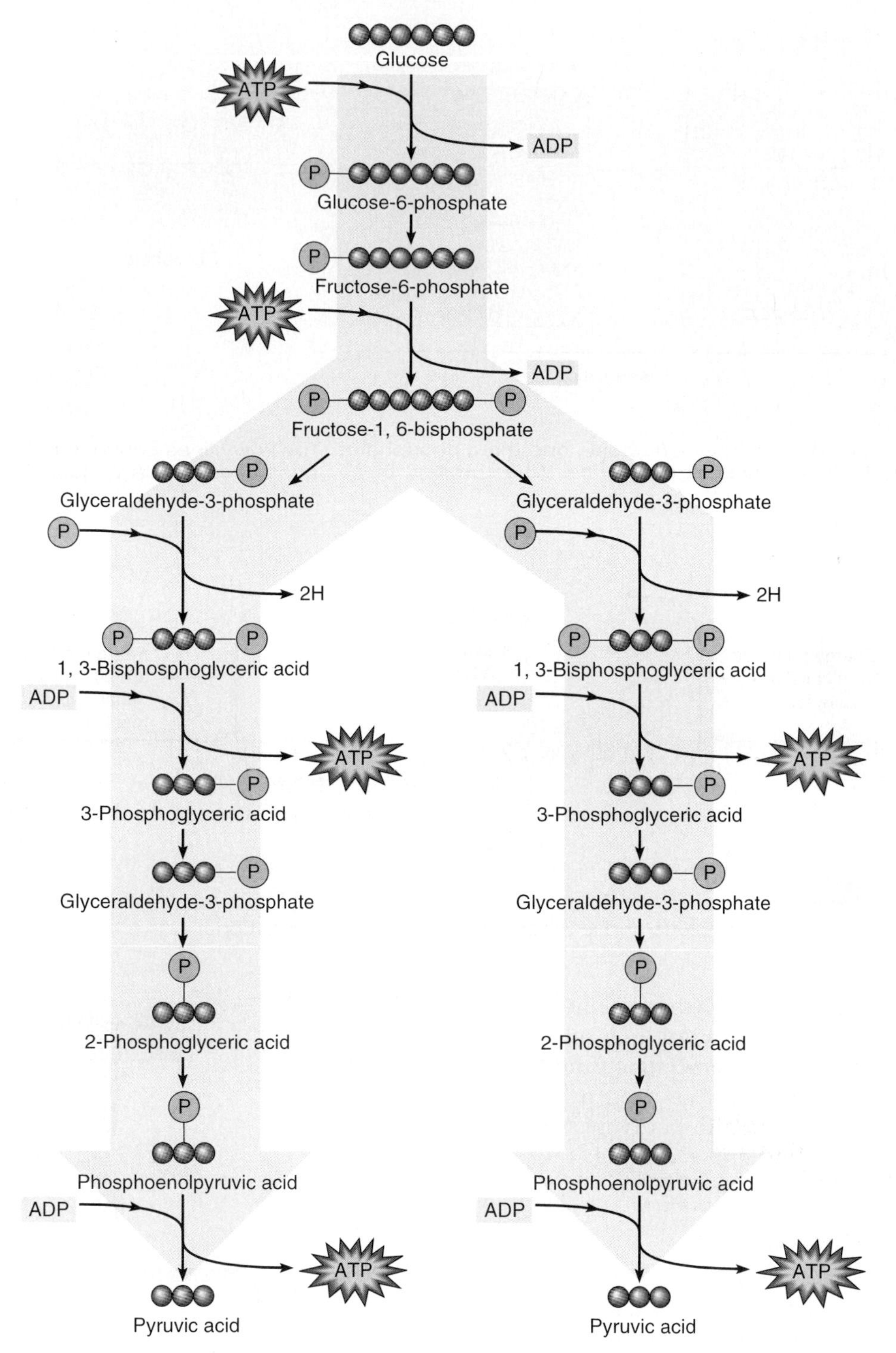


**FIGURE 4.10**

Chemical reactions of glycolysis. There is a net production of 2 ATP molecules from each glucose molecule. The four hydrogen atoms released provide electrons that may be used to generate ATP in the electron transport system, described later. (Note: By convention, carbon atoms are sometimes numbered from right to left.)

these reactions, energy is released and some of it is transferred to molecules of ATP, but the rest is lost as heat. The end products of this oxidation process are carbon dioxide and water. The next sections take a closer look at the breakdown of glucose, which provides much of the energy in the human body.

## Glycolysis

Figure 4.10 illustrates the chemical reactions of glycolysis. In the early steps of this metabolic pathway, the original glucose molecule is altered by the addition of phosphate groups (a process called *phosphorylation*) and by the rearrangement of its atoms. ATP supplies the phosphate groups and the energy needed for these reactions. The result is a molecule of fructose combined with two phosphate groups (fructose-1,6-bisphosphate). It is changed into two 3-carbon molecules (glyceraldehyde-3-phosphate), each of which in turn is converted to pyruvic acid, as follows:

**a.** An inorganic phosphate group is added to glyceraldehyde-3-phosphate to form 1,3-bisphosphoglyceric acid, releasing two hydrogen atoms.

**b.** 1,3-bisphosphoglyceric acid is changed to 3-phosphoglyceric acid. As this occurs, some energy in the form of a high-energy phosphate is transferred from the 1,3-bisphosphoglyceric acid to an ADP molecule, converting the ADP to ATP.

**c.** A slight alteration of 3-phosphoglyceric acid occurs to form 2-phosphoglyceric acid.

**d.** A change in 2-phosphoglyceric acid converts it into phosphoenolpyruvic acid.

**e.** Finally, a high-energy phosphate is transferred from the phosphoenolpyruvic acid to an ADP molecule, converting it to ATP. A molecule of pyruvic acid remains.

Thus, one 6-carbon molecule of glucose is ultimately broken down to two molecules of pyruvic acid. Also, a total of four hydrogen atoms are released (step *a*), and four ATP molecules are formed (two in step *b* and two in step *e*). However, because two molecules of ATP are used early in glycolysis, there is a net gain of only two ATP molecules during this phase of cellular respiration.

In the absence of oxygen, the resulting pyruvic acid molecules may be converted into **lactic acid.** Lactic acid is produced in human muscle cells that are working so strenuously that their production of pyruvic acid exceeds the oxygen supply. In this condition of "oxygen debt," the muscle cells are forced to depend on the less efficient anaerobic pathway, which provides fewer ATPs per glucose molecule than does aerobic respiration. The accumulation of lactic acid contributes to the feeling of muscle fatigue and cramps. Walking after cramping at the end of a race can make a runner feel better by hastening the depletion of lactic acid. When oxygen is again available in sufficient quantity, lactic acid is converted back to pyruvic acid in the liver. From this pyruvic acid, the body can extract more energy.

In the presence of oxygen, each pyruvic acid molecule is oxidized to an acetyl group, which then combines with a molecule of coenzyme A (obtained from the vitamin pantothenic acid) to form acetyl coenzyme A. As this occurs, two more hydrogen atoms are released for each molecule of acetyl coenzyme A formed. The acetyl coenzyme A is then broken down by means of the citric acid cycle, which is illustrated in figure 4.11.

Lactic acid formation occurs in an interesting variety of circumstances. Coaches measure lactic acid levels in swimmers' and sprinters' blood to assess their physical condition. Lactic acid accumulates to triple normal levels in the bloodstreams of children who cry vigorously when they are being prepared for surgery, but not in children who are calm and not crying. This suggests that lactic acid formation accompanies stress. Cancer cells deep within a tumor, far from oxygen-delivering blood, also produce lactic acid by metabolizing glucose. The lactic acid forms a "halo" around the cancer cells, which inactivates immune system biochemicals that would otherwise fight the cancer.

Because obtaining energy for cellular metabolism is vital, disruptions in glycolysis or the reactions that follow it can devastate health. Clinical Application 4.1 tells how medical sleuths traced a boy's unusual combination of symptoms to a block in glycolysis.

## Citric Acid Cycle

An acetyl coenzyme A molecule enters the citric acid cycle (Krebs cycle) by combining with a molecule of oxaloacetic acid to form citric acid. As citric acid is produced, coenzyme A is released and thus can be used again and again in the formation of acetyl coenzyme A from pyruvic acid molecules. The citric acid is then changed by a series of reactions back into oxaloacetic acid, and the cycle may be repeated.

Various steps in the citric acid cycle release carbon dioxide and hydrogen atoms. More specifically, for each glucose molecule metabolized in the presence of oxygen, two molecules of acetyl coenzyme A enter the citric acid cycle. As a result of the cycle, four carbon dioxide molecules and sixteen hydrogen atoms are released. At the same time, two more molecules of ATP form.

The released carbon dioxide dissolves in the cytoplasm and leaves the cell, eventually entering the bloodstream. Most of the hydrogen atoms released from the citric acid cycle, and those released during glycolysis and during the formation of acetyl coenzyme A, supply electrons used to produce ATP.

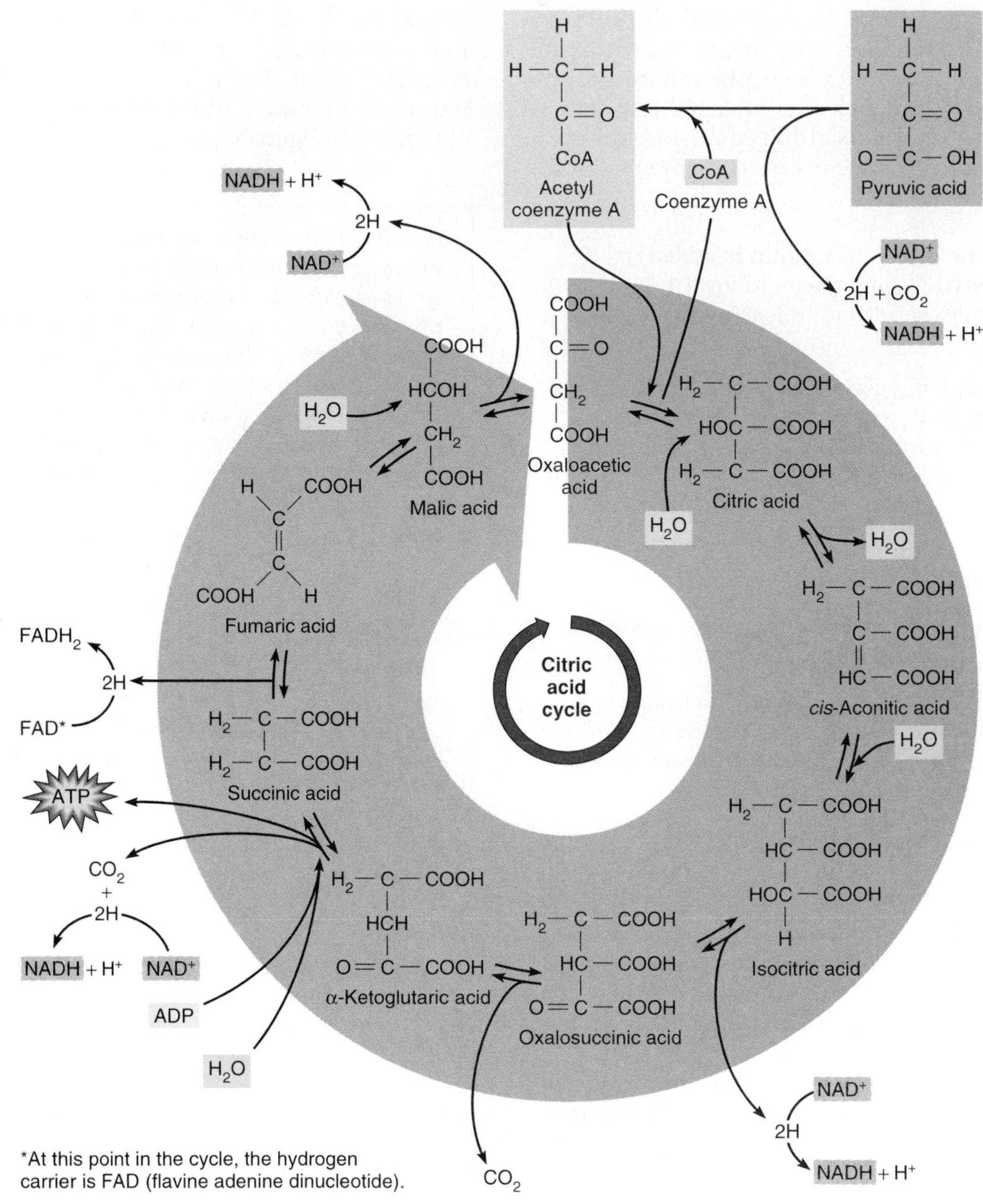


**FIGURE 4.11**

Chemical reactions of the citric acid cycle. NADH molecules carrying hydrogens are highlighted.

## ATP Synthesis

Note in figure 4.11 that the hydrogen atoms released from various metabolic reactions are passed in pairs to hydrogen carriers. One of these carriers is $NAD^+$ (nicotinamide adenine dinucleotide). When $NAD^+$ accepts a pair of hydrogen atoms, one of the atoms becomes a hydrogen ion and the other bonds to $NAD^+$ to form NADH, as follows:

$$NAD^+ + 2H \rightarrow NADH + H^+$$

$NAD^+$ is a coenzyme obtained from a vitamin (niacin), and when it combines with hydrogen, it is said to be *reduced*. (**Reduction** results from the addition of hydrogen or the gain of electrons; it is the opposite of oxidation.) Another hydrogen acceptor, FAD (flavine adenine dinucleotide), acts in a similar manner, combining with hydrogen to form $FADH_2$ (fig. 4.11).

In their reduced states, the hydrogen carriers NADH and $FADH_2$ hold most of the energy once held by the original glucose molecule.

## Clinical Application 4.1

# Overriding a Block in Glycolysis

Michael P. was noticeably weak from his birth. He didn't move much, had poor muscle tone and difficulty breathing, and grew exhausted merely from the effort of feeding. At the age of two and a half months, he suffered his first seizure, staring and jerking his limbs for several frightening minutes. Despite medication, his seizures continued, occurring more frequently.

The doctors were puzzled because the results of most of Michael's many medical tests were normal—with one notable exception. His cerebrospinal fluid (the fluid that bathes the brain and spinal cord) was unusually low in glucose and lactic acid. These deficiencies told the physicians that Michael's cells were not performing glycolysis or anaerobic respiration. Hypothesizing that a profound lack of ATP was causing the boy's symptoms, medical researchers decided to intervene beyond the block in the boy's metabolic pathway, taking a detour to energy production. When Michael was seven and a half months old, he began a diet rich in certain fatty acids. Within four days, he appeared to be healthy for the very first time! The diet had resumed aerobic respiration at the point of acetyl coenzyme A formation by supplying an alternative to glucose. Other children with similar symptoms have since enjoyed spectacular recoveries similar to Michael's thanks to the dietary intervention, but doctors do not yet know the long-term effects of the therapy. This medical success story, however, illustrates the importance of the energy pathways—and how valuable our understanding of them can be.

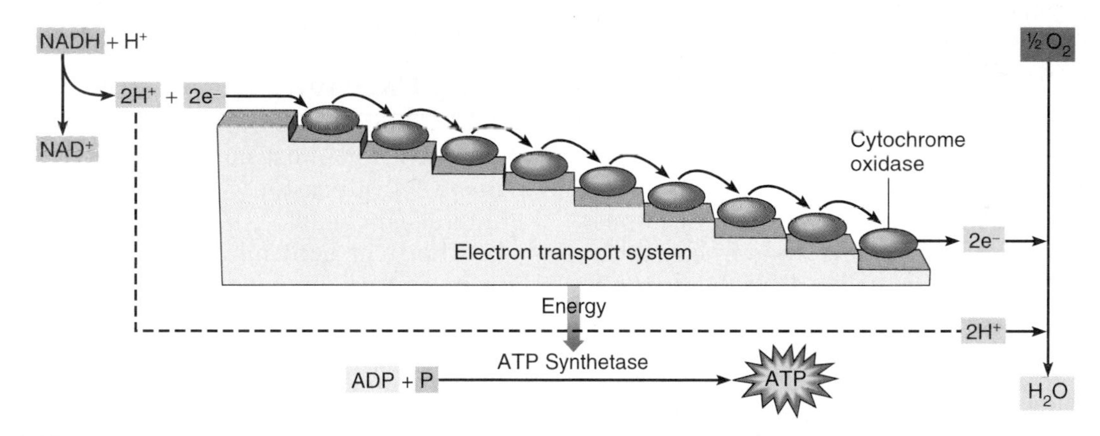


**FIGURE 4.12**
A summary of ATP synthesis by oxidative phosphorylation.

Figure 4.12 shows that when hydrogen is released from NADH, $NAD^+$ reappears and can once again accept hydrogens. Since this reaction removes hydrogen, the $NAD^+$ is said to be *oxidized.* (**Oxidation** results from the removal of hydrogen or the loss of electrons; it is the opposite of reduction.) At the same time, two electrons from the original pair of hydrogen atoms are passed to a sequence of electron carriers.

The molecules that act as electron carriers comprise an **electron transport system.** As electrons are passed from one carrier to another, the carriers are alternately reduced and oxidized as they accept or release electrons. The transported electrons gradually lose energy as they proceed down the chain.

Among the members of the electron transport system are several proteins, including a set of iron-containing molecules called **cytochromes.** The cytochromes are located in the inner membranes of the mitochondria (see chapter 3). The folds of the inner mitochondrial membrane provide surface area on which the energy reactions take place. In an active muscle cell, the inner mitochondrial membrane, if stretched out, would be about forty-five times as long as the cell membrane!

The final cytochrome of the electron transport chain (cytochrome oxidase) gives up a pair of electrons and causes two hydrogen ions (formed at the beginning of the sequence) to combine with an atom of oxygen. This process produces a water molecule:

$$2e^- + 2H^+ + \frac{1}{2} O_2 \rightarrow H_2O$$

Thus, oxygen is the final electron acceptor.

Cyanide ($CN^-$) kills by binding to the electron acceptor in the electron transport system that passes electrons to oxygen. Although there are antidotes, death is usually nearly instantaneous. Cyanide is used as a pesticide, for extracting silver and gold from ores, in electroplating, and in the textile and rubber industries.

Another respiratory poison, 2,4-dinitrophenol (DNP), "uncouples" passage of electrons down the electron transport system from ATP synthesis. As a result, electrons move, but no ATP forms. When a person breathes DNP, the metabolic rate races as electrons are passed with no energy payoff. On a whole-body level, temperature soars, sweat pours, nausea and vomiting are severe, and the person collapses and dies. DNP is used in dye manufacture, wood preservation, and as an insecticide.

Note in figure 4.12 that at the same time electrons pass through the electron transport system, energy is released. Some of this energy is used by the enzyme *ATP synthetase* to combine phosphate and ADP by a high-energy bond (phosphorylation), forming ATP.

Also note in figures 4.10 and 4.11 that twelve pairs of hydrogen atoms are released during the complete breakdown of one glucose molecule—two pairs from glycolysis, two pairs from the conversion of pyruvic acid to acetyl coenzyme A (one pair from each of two pyruvic acid molecules), and eight pairs from the citric acid cycle (four pairs for each of two acetyl coenzyme A molecules).

Oxidation (loss of electrons) of ten pairs of these hydrogen atoms produces thirty ATP molecules, while metabolism of the other two pairs forms four ATP molecules. Also, there is a net gain of two ATP molecules during glycolysis, and two ATP molecules form when the two acetyl coenzyme A molecules enter the citric acid cycle. Thus, a maximum of thirty-eight ATP molecules form for each glucose molecule metabolized.

Because this process of forming ATP involves both the oxidation of hydrogen atoms and the bonding of phosphate to ADP, it is called **oxidative phosphorylation.**

## Carbohydrate Storage

Whenever excess glucose is present, it may enter anabolic carbohydrate pathways and be linked into storage forms such as glycogen.

Most cells can produce glycogen. However, liver and muscle cells store the greatest amounts. Following a meal, when blood glucose concentration is relatively high, liver cells obtain glucose from the blood and convert it to glycogen. Between meals, when blood glucose concentration is lower, the reaction is reversed, and glucose is released into the blood. This mechanism ensures that cells will have a continual supply of glucose to support respiration.

Glucose can also be converted into fat molecules, which are later deposited in fat tissues. This happens when a person takes in more carbohydrates than can be stored as glycogen or are needed for normal activities. The body has an almost unlimited capacity to perform this type of anabolism, so overeating can result in becoming obese (overweight).

### Lipid Pathways

Foods contain lipids in the form of phospholipids, cholesterol, or, most commonly, fats called **triglycerides.** A triglyceride consists of a glycerol portion and three fatty acids.

The liver controls lipid metabolism, removing lipids from the circulation and altering their molecular structures. For example, the liver can shorten or lengthen the carbon chains of fatty acid molecules or introduce double bonds into these chains, thus converting fatty acids from one form to another.

Lipids provide a variety of physiological functions; however, fats are used mainly to supply energy. Gram for gram, fats contain more than twice as much chemical energy as carbohydrates or proteins. This is why people trying to lose weight are advised to minimize fats in their diets.

Before energy can be released from a triglyceride molecule, the molecule must undergo hydrolysis. As shown in figure 4.13, some of the resulting fatty acid portions can then be converted into molecules of acetyl coenzyme A by a series of reactions called **beta oxidation.**

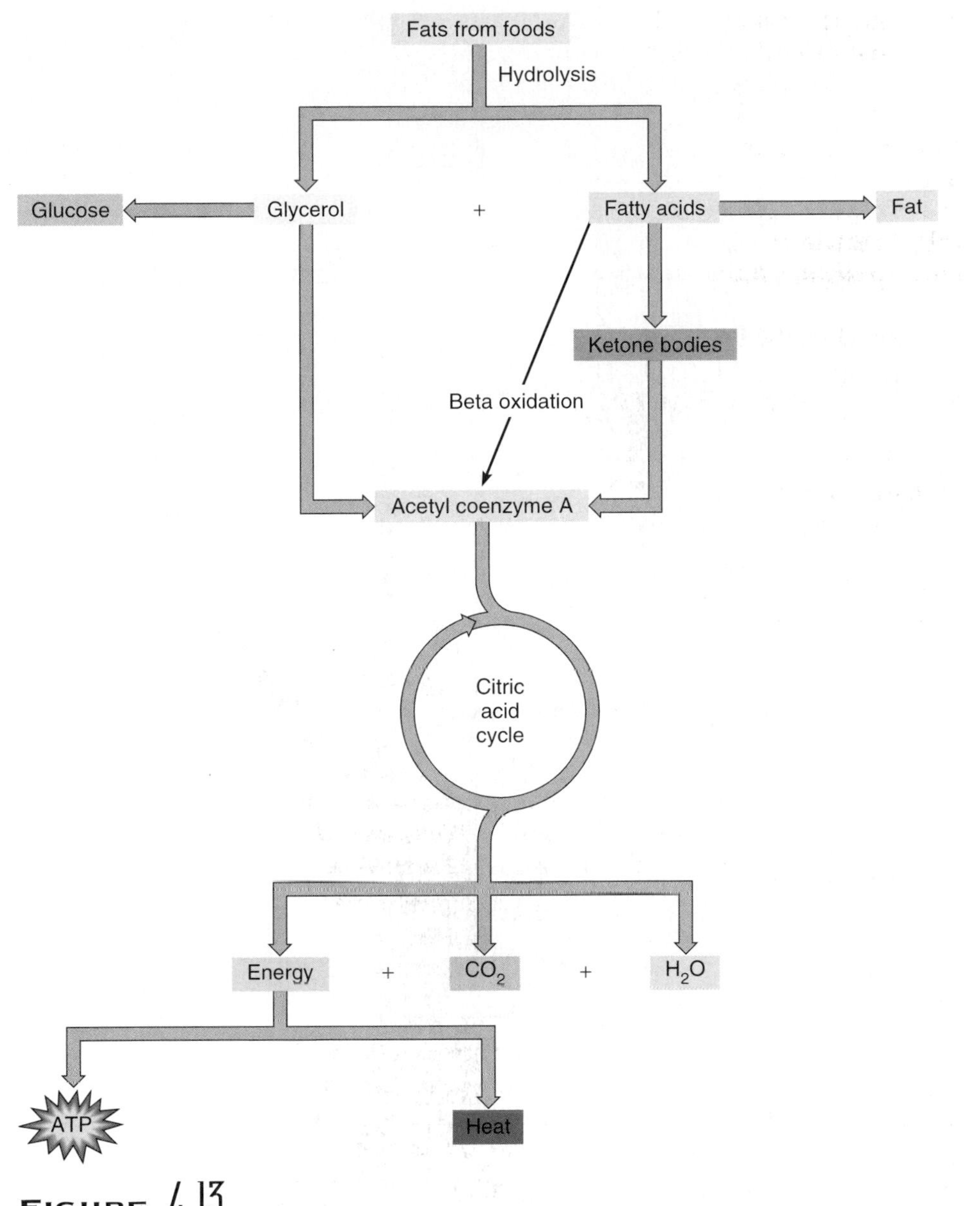


**FIGURE 4.13**
We digest fats from foods into glycerol and fatty acids, which may enter catabolic pathways and be used as energy sources.

In the first phase of beta oxidation, the fatty acids are converted into activated forms. This change requires energy from ATP and a special group of enzymes called thiokinases. Each of the enzymes in this group can act upon a fatty acid with a particular carbon chain length.

Once fatty acid molecules have been activated, other enzymes called **fatty acid oxidases** that are located within mitochondria break them down. This phase of the reactions removes segments of fatty acid chains (containing two carbon atoms each). Some of these segments are converted into acetyl coenzyme A molecules. Other segments are converted into compounds called **ketone bodies,** such as acetone, which later may be changed to acetyl coenzyme A. In either case, the resulting acetyl coenzyme A can be oxidized by the citric acid cycle. The glycerol portions of the triglyceride molecules can also enter metabolic pathways leading to the citric acid cycle, or they can be used to synthesize glucose.

It is also possible for glycerol and fatty acid molecules resulting from the hydrolysis of fats to be changed back into fat molecules by anabolic processes and to be stored in fat tissue. Additional fat molecules can be synthesized from excess molecules of glucose or amino acids.

When ketone bodies form faster than they can be decomposed, some of them are eliminated through the lungs and kidneys. Consequently, the breath and urine may develop a fruity odor due to the presence of the ketone acetone. This sometimes happens when a person fasts, forcing body cells to metabolize fat, in order to lose weight. Persons suffering from diabetes mellitus also are likely to metabolize excessive amounts of fats, and they too may have acetone in the breath and urine. At the same time, they may develop a serious imbalance in pH called acidosis, which is due to an overaccumulation of still other acidic ketone bodies.

1. What is a metabolic pathway?
2. How do cells use carbohydrates?
3. What must happen to fat molecules before they can supply energy?

## Protein Pathways

When dietary proteins are digested, the resulting amino acids are absorbed and transported by the blood to cells. Many of these amino acids are used to form new protein molecules, as specified by DNA.

Protein molecules may also supply energy. To do this, they must first be broken down into amino acids. The amino acids then undergo **deamination,** a process that occurs in the liver that removes the nitrogen-containing portions ($-NH_2$ groups) from the amino acids. These $-NH_2$ groups are converted later into a waste called **urea.**

> The liver produces urea from amino groups formed by deamination of amino acids. The blood carries urea to the kidneys, where it is excreted in urine. Certain kidney disorders impair the ability to remove urea from the blood, raising the blood urea concentration. A blood test called blood urea nitrogen (BUN) determines the blood urea concentration, and is often used to evaluate kidney function.

The remaining deaminated portions of the amino acids break down by one of several pathways. Some of these pathways lead to formation of acetyl coenzyme A, and others lead more directly to steps of the citric acid cycle. As energy is released from the cycle, some of it is captured in molecules of ATP (fig. 4.14). If energy is not needed immediately, the deaminated portions of the amino acids may be changed into glucose or fat molecules in still other metabolic pathways.

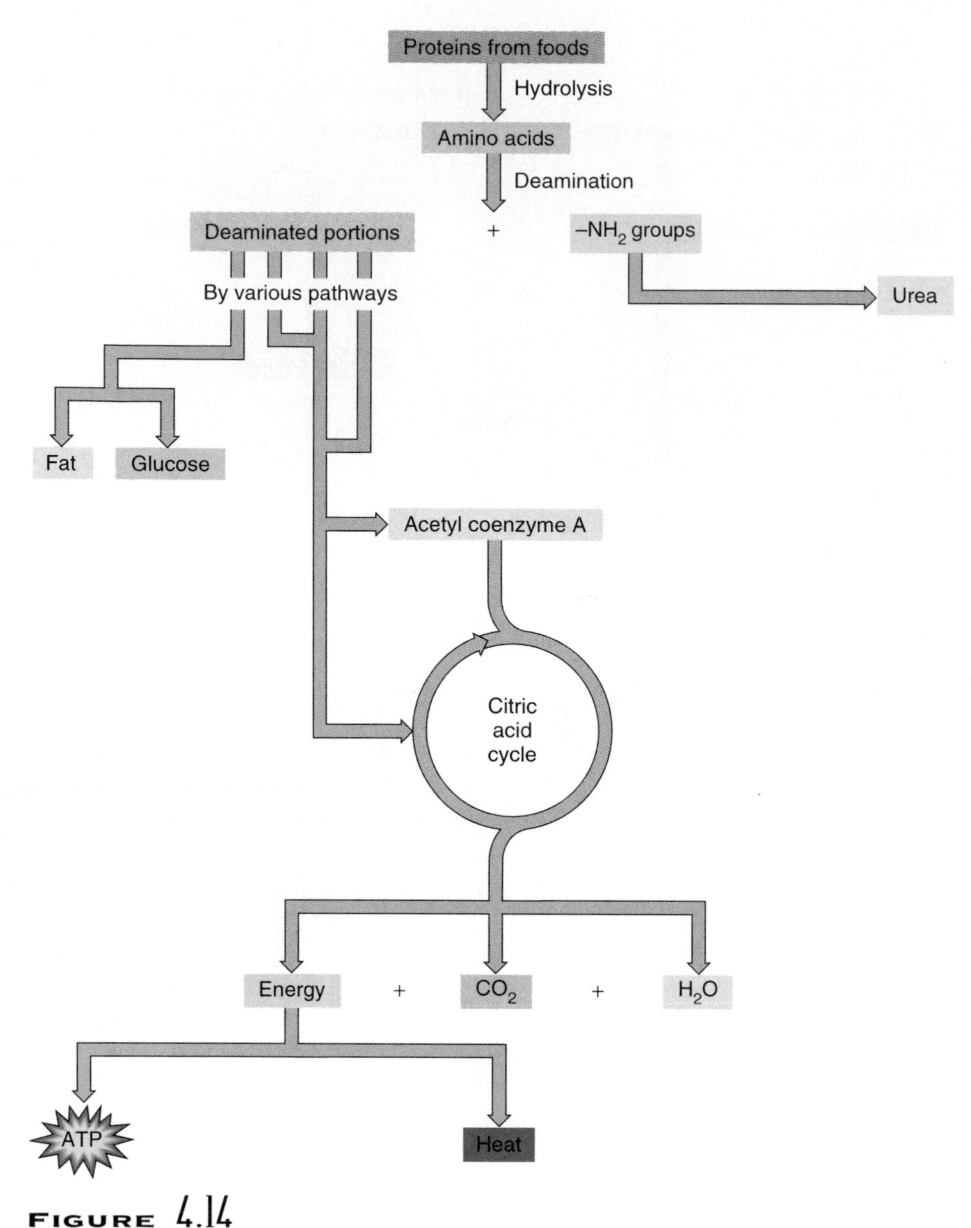


**FIGURE 4.14**
We digest proteins from foods into amino acids, but before these smaller molecules can be used as energy sources, they must be deaminated.

Glucose can be changed back into some amino acids if certain nitrogen-containing molecules are available. However, about eight necessary amino acids cannot be synthesized in adequate amounts in human cells and so must be provided in the diet. For this reason, these are called **essential amino acids.** They are discussed in chapter 18.

## Regulation of Metabolic Pathways

The rate at which a metabolic pathway functions is often determined by a regulatory enzyme responsible for one of its steps. This regulatory enzyme is present in limited quantity. Consequently, it can become saturated with substrate molecules whenever the substrate concentration increases above a certain amount. Once the enzyme is saturated, increasing the amount of substrate will no longer affect the reaction rate. In this way, a single enzyme in a pathway can control the whole pathway.

As a rule, such a **rate-limiting enzyme** is the first enzyme in a series. This position is important because some intermediate substance of the pathway might accumulate if an enzyme occupying another location in the sequence were rate limiting.

Often the final product of a metabolic pathway inhibits the rate-limiting regulatory enzyme. This type of control is called negative feedback. Accumulating product inhibits the pathway, and synthesis of the product

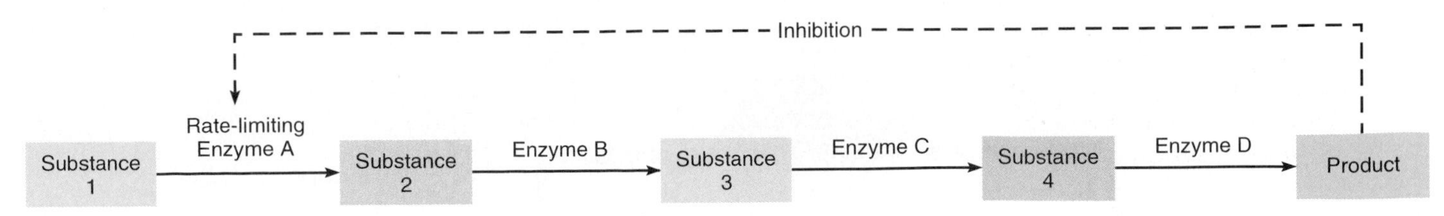


**FIGURE 4.15**
A negative feedback mechanism may control a rate-limiting enzyme in a metabolic pathway. The product of the pathway inhibits the enzyme.

falls. When the amount of product decreases, the inhibition lifts and more product is synthesized. This keeps the rate of production relatively stable (fig. 4.15).

Figure 4.16 summarizes catabolism of proteins, carbohydrates, and fats.

1. How do cells utilize proteins?
2. What is a rate-limiting enzyme?
3. How can negative feedback control a metabolic pathway?

## Nucleic Acids and Protein Synthesis

Because enzymes control the metabolic pathways that enable cells to survive, cells must possess information for producing these specialized proteins. Many other proteins are important in physiology as well, such as blood proteins, the proteins that form muscle and connective tissue, and the antibodies that protect against infection. The information that instructs a cell to synthesize a particular protein is held in the sequence of building blocks of **deoxyribonucleic acid** (DNA), the genetic material. As we will see later in this chapter, the correspondence between a unit of DNA information and a particular amino acid constitutes the genetic code.

### Genetic Information

Children resemble their parents because of inherited traits, but what actually passes from parents to a child is *genetic information,* in the form of DNA molecules from the parents' sex cells. As an offspring develops, the information is copied from cell to cell by mitosis. Genetic information "tells" cells how to construct specific protein molecules, which, in turn, function as structural materials, enzymes, or other vital substances.

The portion of a DNA molecule that contains the genetic information for making a particular protein is called a **gene.** Thus, inherited traits are determined by the genes contained in the parents' sex cells that fuse to form the first cell of an offspring's body.

Recall from chapter 2 that nucleotides are the building blocks of nucleic acids. A nucleotide consists of a 5-carbon sugar (ribose or deoxyribose), a phosphate group, and one of several organic, nitrogen-containing (nitrogenous) bases (fig. 4.17). In a DNA nucleotide, the base may be one of four types: adenine, thymine, cytosine, or guanine (fig. 4.18). DNA nucleotides form long strands, or **polynucleotide chains,** by alternately joining their sugar and phosphate portions, which provides a "backbone" structure (fig. 4.19).

A DNA molecule consists of two polynucleotide chains. The nitrogenous bases project from the sugar-phosphate backbone of one strand and bind by hydrogen bonds to nitrogenous bases of the second strand (fig. 4.20). The resulting structure is somewhat like a ladder, in which the uprights represent the sugar and phosphate backbones of the two strands and the rungs represent the nitrogenous bases. Notice that the sugars forming the two backbones point in opposite directions. For this reason, the two strands are called *antiparallel.*

A DNA molecule is sleek and symmetrical because the bases pair in only two combinations, such that the width of the overall structure is constant. Adenine (A), a two-ring structure, binds to thymine (T), a single-ring structure. Guanine (G), a two-ring structure, binds to cytosine (C), a single-ring structure (fig. 4.21). These pairs—A with T and G with C—are called **complementary base pairs.** Because of this phenomenon, the sequence of one DNA strand can always be derived from the other, by following the "base pairing rules." For example, if the sequence of one strand of the DNA molecule is G, A, C, T, then the complementary strand's sequence is C, T, G, A.

The sequence of bases in only one of the DNA strands encodes the information for a protein. The double-stranded DNA molecule twists to form a double helix, resembling a spiral staircase (fig. 4.22). An individual DNA molecule may be several million base pairs long. Investigators can use DNA to identify individuals (Clinical Application 4.2, see page 120).

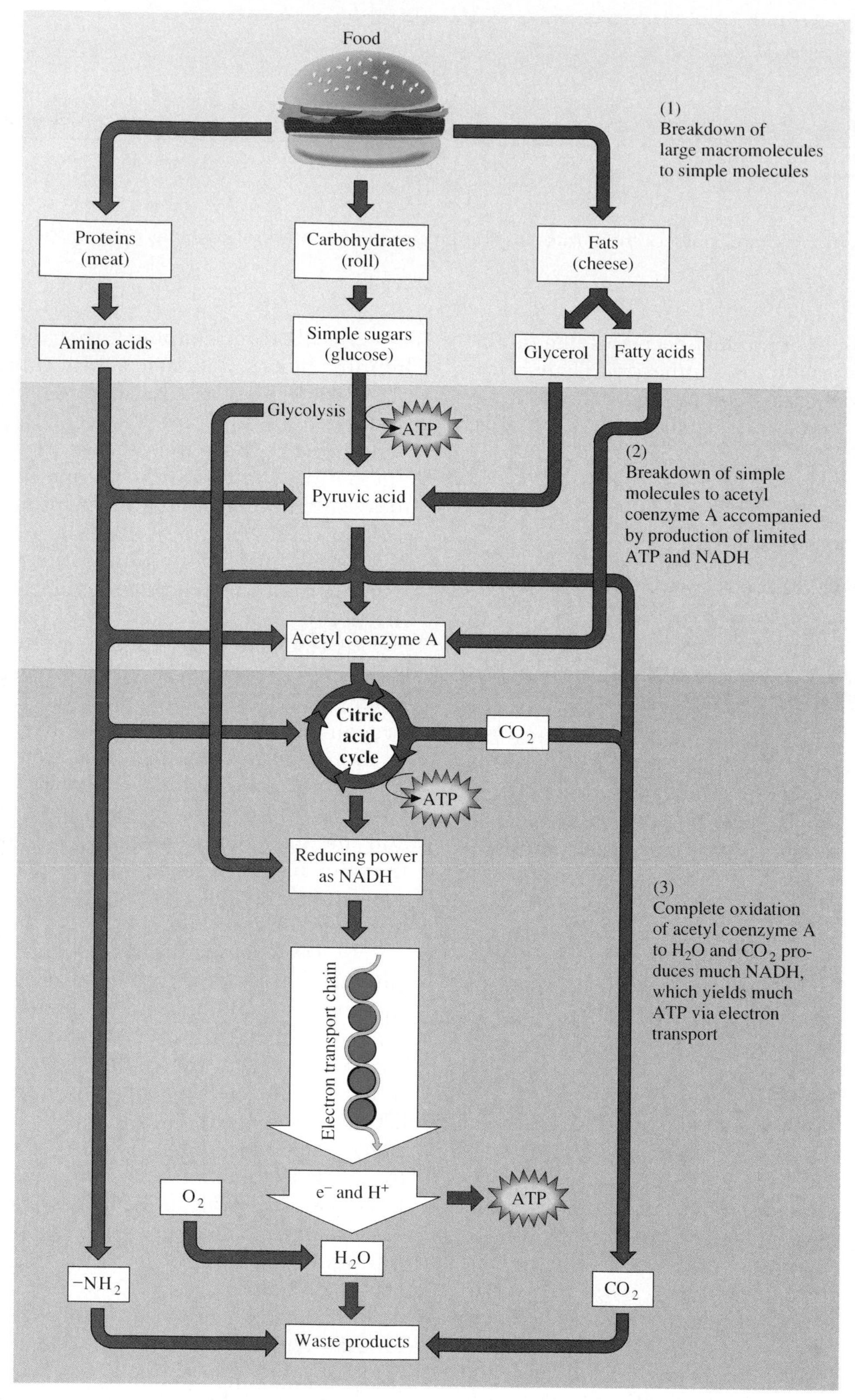


**FIGURE 4.16**
A summary of the breakdown of proteins, carbohydrates, and fats.

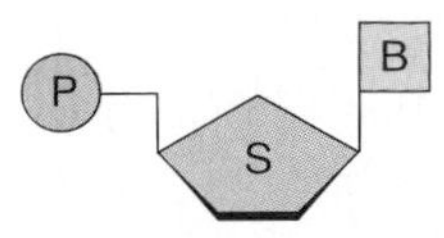


**FIGURE 4.17**
Each nucleotide of a nucleic acid consists of a 5-carbon sugar (S), a phosphate group (P), and an organic, nitrogenous base (B).

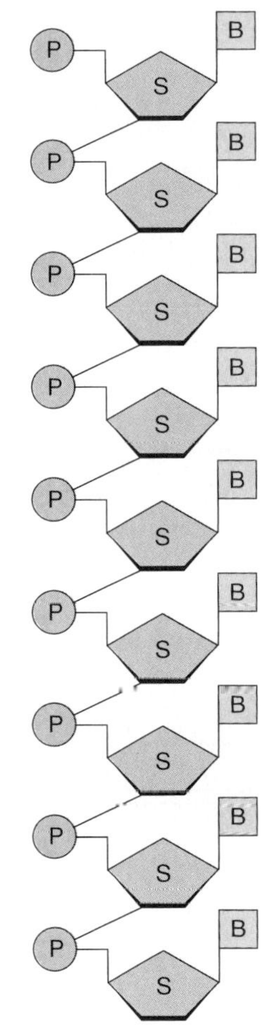


**FIGURE 4.19**
A single strand of DNA consists of a chain of nucleotides connected by a sugar-phosphate backbone.

## Genetic Code

Genetic information specifies how to position the amino acids correctly in a polypeptide chain. Each of the twenty different amino acids is represented in a DNA molecule by a triplet code, a particular sequence of three nucleotides. That is, the sequence C, G, T in a DNA strand represents one kind of amino acid; the sequence G, C, A represents another kind; and T, T, A

Adenine Thymine

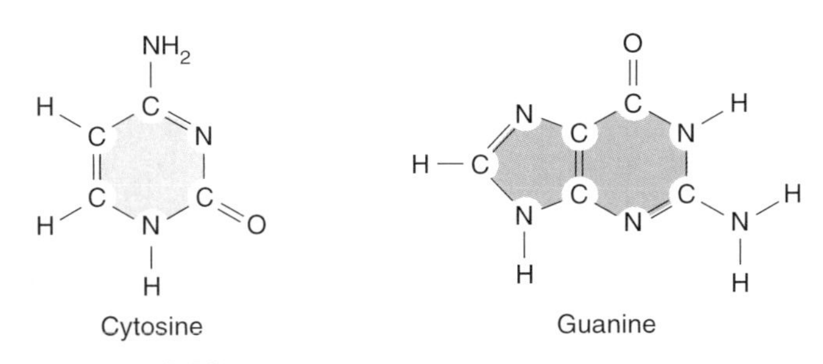


**FIGURE 4.18**
Each nucleotide of a DNA molecule contains one of the four nitrogenous bases adenine, thymine, cytosine, or guanine.

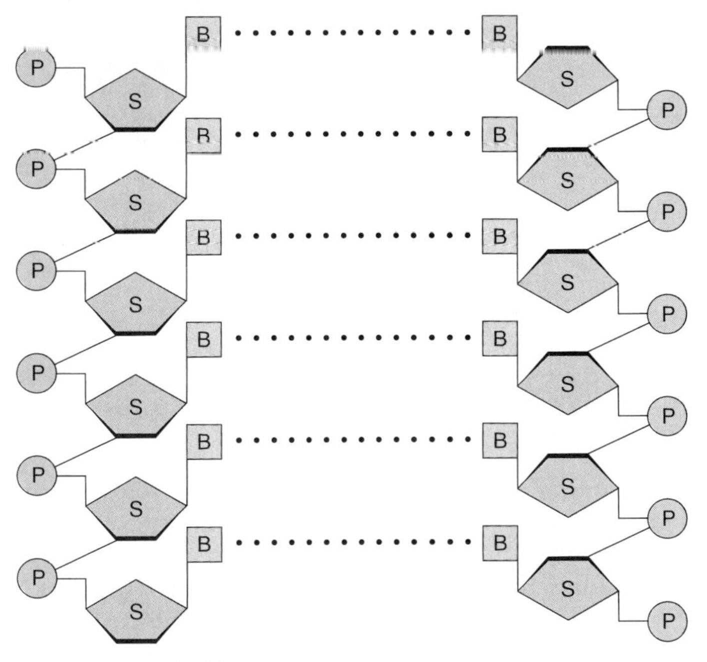


**FIGURE 4.20**
DNA consists of two polynucleotide chains. The nitrogenous bases of one strand are weakly held to the nitrogenous bases of the second strand by hydrogen bonds (dotted lines). Note that the sugars point in opposite directions—that is, the strands are antiparallel.

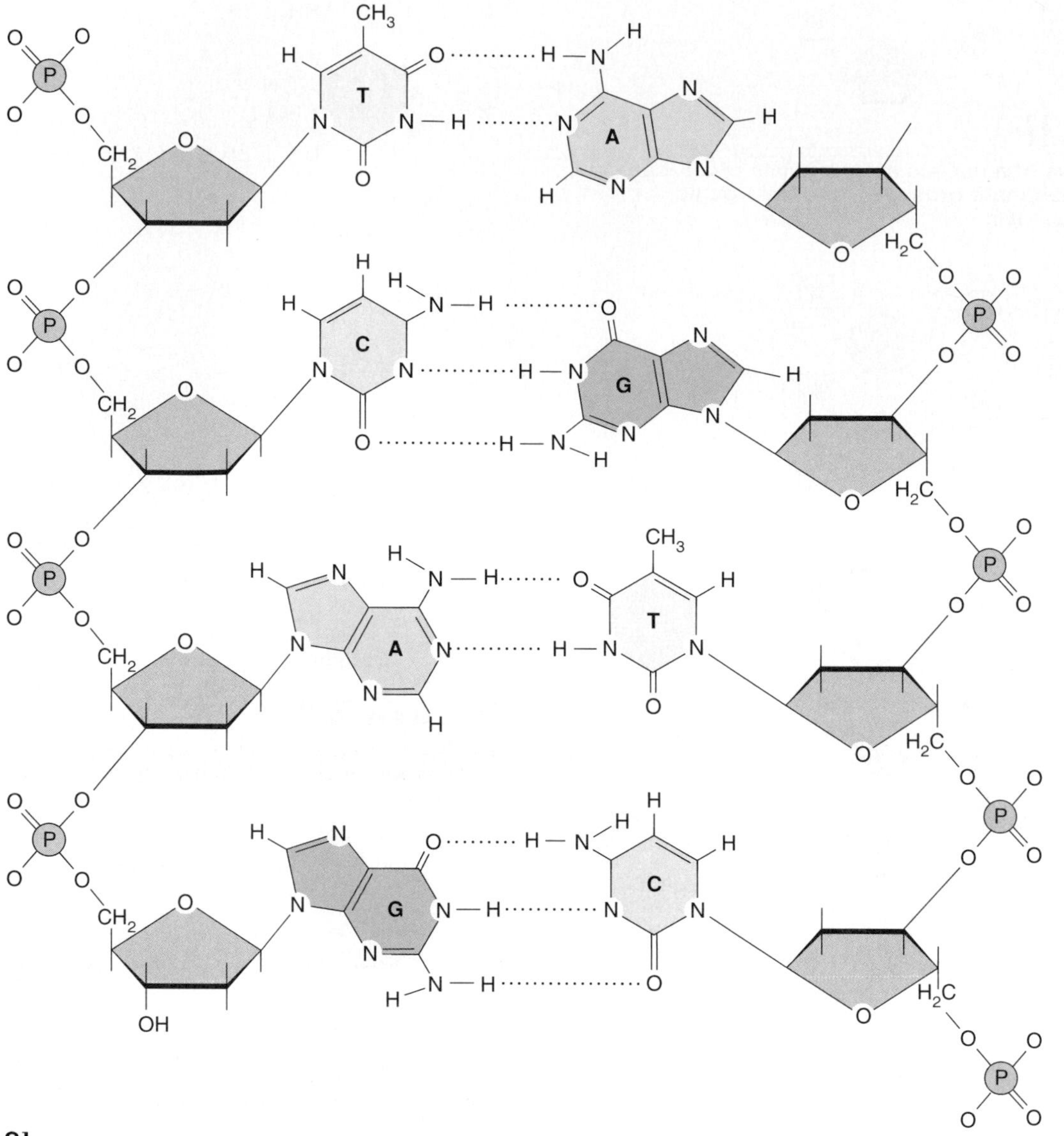

**FIGURE 4.21**

The nucleotides of a double-stranded DNA molecule pair so that an adenine (A) of one strand hydrogen bonds to a thymine (T) of the other strand, and a guanine (G) of one strand hydrogen bonds to a cytosine (C) of the other. The dotted lines represent hydrogen bonds.

still another kind (chart 4.1). Other sequences represent instructions for beginning or ending the synthesis of a protein molecule.

Thus, the sequence of nucleotides in a DNA molecule denotes the arrangement of amino acids constituting a protein molecule, as well as indicating how to start or stop the protein's synthesis. This method of storing information for protein synthesis is the **genetic code.** Because DNA molecules are located in the nucleus and protein synthesis occurs in the cytoplasm, the genetic information must somehow get from the nucleus into the cytoplasm. RNA molecules accomplish this.

> The genetic code is said to be universal, because all species on earth use the same DNA base triplets to specify the same amino acids. Researchers deciphered the code in the 1960s. When the media mentions an individual's genetic code, or that scientists are currently breaking the code, what they really are referring to is the sequence of DNA bases comprising a certain gene—not the genetic code (the correspondence between DNA triplet and amino acid) itself.

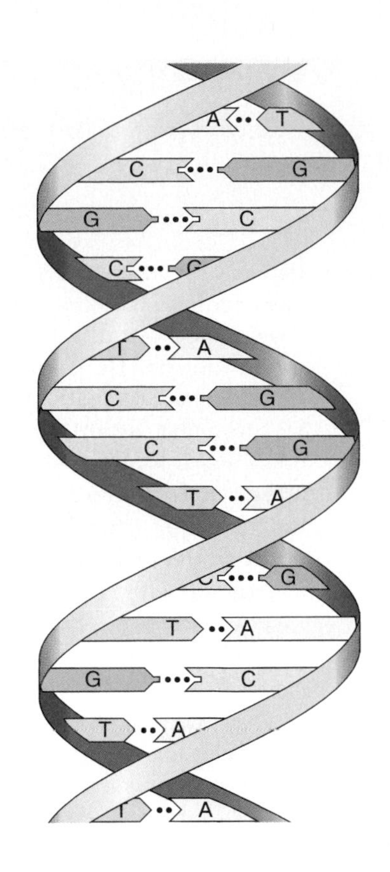


(a)

(b)

**FIGURE 4.22**

(*a*) The molecular ladder of a double-stranded DNA molecule twists into a double helix. (*b*) A model of a portion of a DNA molecule.

## CHART 4.1 Some Nucleotide Sequences of the Genetic Code

| Amino Acids | DNA Sequence | Complementary RNA Sequence |
|---|---|---|
| Alanine | CGT | GCA |
| Arginine | GCA | CGU |
| Asparagine | TTA | AAU |
| Aspartic acid | CTA | GAU |
| Cysteine | ACA | UGU |
| Glutamic acid | CTT | GAA |
| Glutamine | GTT | CAA |
| Glycine | CCG | GGC |
| Histidine | GTA | CAU |
| Isoleucine | TAG | AUC |
| Leucine | GAA | CUU |
| Lysine | TTT | AAA |
| Methionine | TAC | AUG |
| Phenylalanine | AAA | UUU |
| Proline | GGA | CCU |
| Serine | AGG | UCC |
| Threonine | TGC | ACG |
| Tryptophan | ACC | UGG |
| Tyrosine | ATA | UAU |
| Valine | CAA | GUU |
| **Instructions** | | |
| Start protein synthesis | TAC | AUG |
| Stop protein synthesis | ATT | UAA |

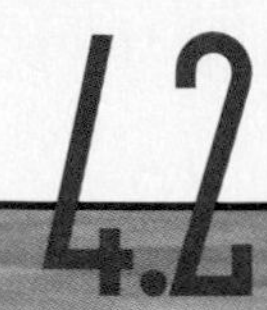

CLINICAL APPLICATION 4.2

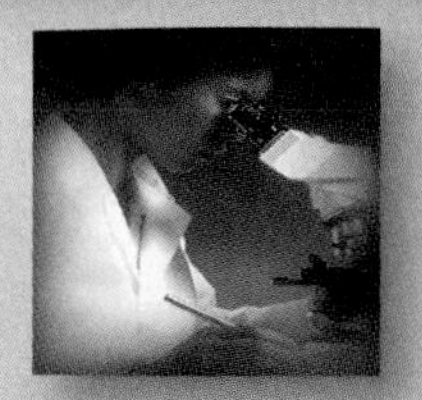

## DNA Makes History

In July 1918, the last tsar of Russia, Nicholas II, and his family, the Romanovs, met gruesome deaths at the hands of Bolsheviks in a town in the Ural Mountains of central Russia. Captors led the tsar, tsarina, four daughters and one son, plus the family physician and three servants, to a cellar and shot them, bayoneting those who did not die quickly enough. The executioners stripped the bodies and loaded them onto a truck, which would take them to a mine shaft where they would be disposed of. But the truck broke down, and the bodies were instead placed in a shallow grave, then damaged with sulfuric acid so that they could not be identified.

In July 1991, two Russian amateur historians found the grave, and based on its location, alerted the government that the long-sought bodies of the Romanov family might have been found. An official forensic examination soon determined that the skeletons were from nine individuals. The sizes of the skeletons indicated that three were children. The porcelain, platinum, and gold in the teeth of some of the skeletons suggested that they were royalty. The facial bones were so decomposed from the acid that conventional forensic tests were not possible. But one very valuable type of evidence remained—DNA. Forensic scientists extracted DNA from bone cells and mass-produced it for study using a technique called the polymerase chain reaction (PCR) described in Clinical Application 4.3.

By identifying DNA sequences specific to the Y chromosome, which is found only in males, the DNA detectives could tell which of the skeletons were from males. Then they delved into the DNA found in mitochondria. Because these organelles are passed from mother to offspring, identifying a mitochondrial DNA pattern in a woman and children would establish her as their mother. This was indeed so for one of the women (with impressive dental work) and the children.

But a mother, her children, and some companions does not a royal family make. The researchers had to find a connection between the skeletons and the royal family. Again they turned to DNA. Genetic material from one of the male skeletons shared certain rare DNA sequences with DNA from living descendants of the Romanovs. This man also had aristocratic dental work and shared DNA sequences with the children! The mystery of the fate of the Romanovs was apparently solved, thanks to the help of a most fascinating molecule—deoxyribonucleic acid.

### RNA Molecules

**RNA** (ribonucleic acid) molecules differ from DNA molecules in several ways. RNA molecules are usually single-stranded, and their nucleotides contain ribose rather than deoxyribose sugar. Like DNA, RNA nucleotides each contain one of four organic bases, but while adenine, cytosine, and guanine nucleotides occur in both DNA and RNA, thymine nucleotides are found only in DNA. In place of thymine nucleotides, RNA molecules contain uracil (U) nucleotides (fig. 4.23).

The first step in the delivery of information from the nucleus to the cytoplasm is the synthesis of a type of RNA called **messenger RNA** (mRNA). In messenger RNA synthesis, RNA nucleotides form complementary base pairs with a strand of DNA. However, just as the words in a sentence must be read in the correct order to make sense, the base sequence of a strand of DNA must be "read" in the correct direction. Furthermore, only one of the two antiparallel strands of DNA contains the genetic message. An enzyme called RNA polymerase determines the correct DNA strand and the right direction for RNA synthesis.

In mRNA synthesis, RNA polymerase binds to a promoter, a DNA base sequence that begins a gene. As a result of RNA polymerase binding, a section of the double-stranded DNA molecule unwinds and pulls apart, exposing a portion of the gene. RNA polymerase then moves along the strand, exposing other portions of the gene. At the same time, a molecule of mRNA forms as RNA nucleotides complementary to those arranged along the DNA strand are strung together. For example, if the sequence of DNA bases is A, C, A, A, T, G, C, G, T, A, the complementary bases in the developing mRNA molecule will be U, G, U, U, A, C, G, C, A, U, as shown in figure 4.24. (The other strand of DNA is not used in this process, but it is important in DNA replication, discussed later in the chapter.)

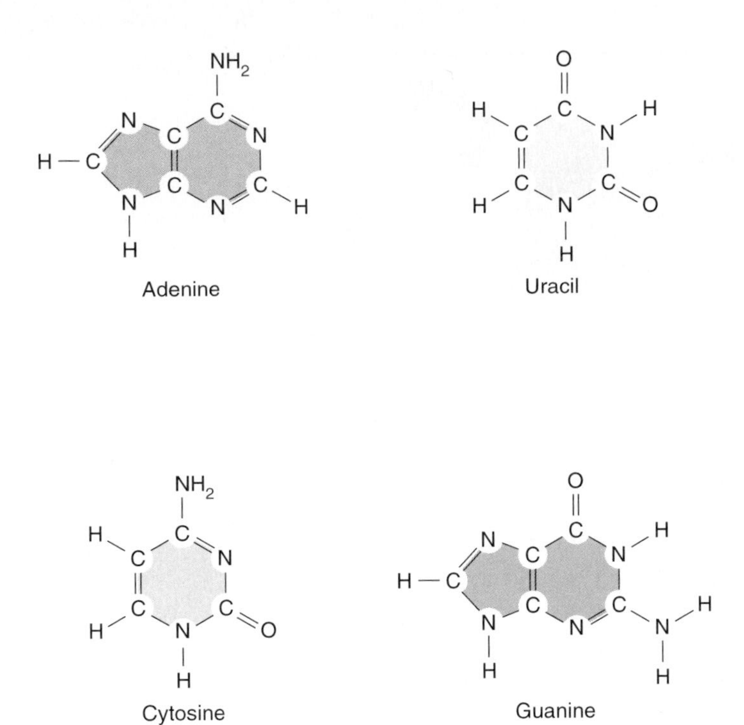


**FIGURE 4.23**
Each nucleotide of an RNA molecule contains one of four nitrogenous bases: adenine, uracil, cytosine, or guanine.

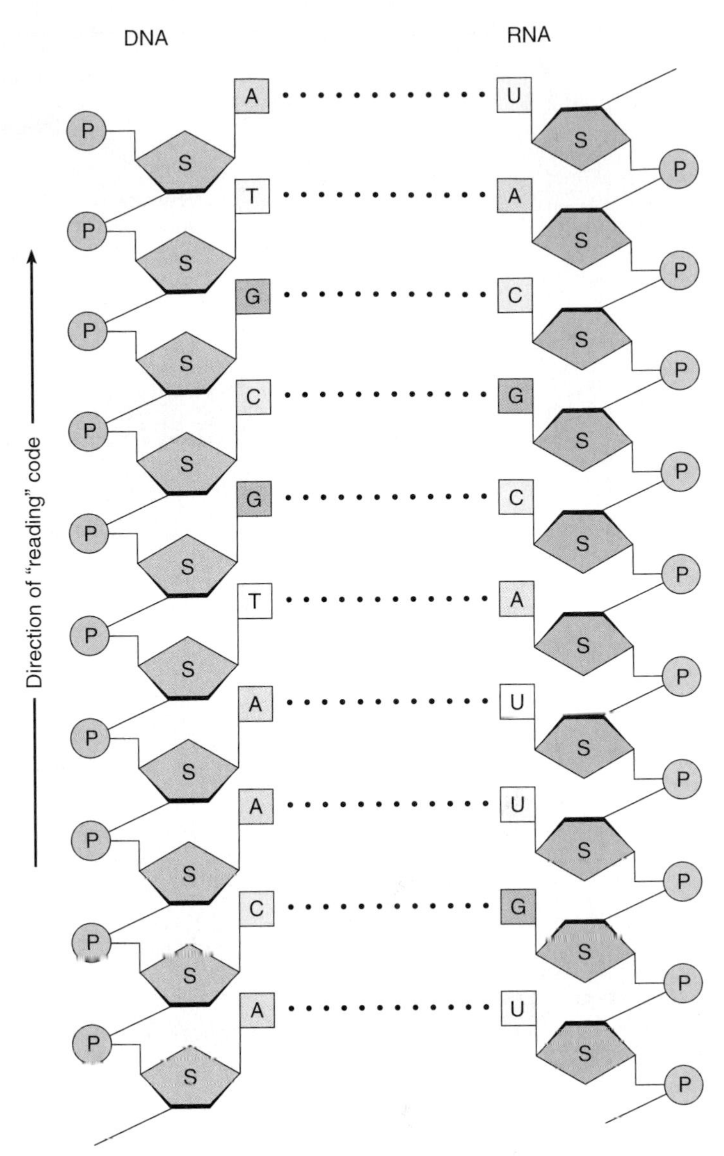


**FIGURE 4.24**
When an RNA molecule is transcribed from a strand of DNA, complementary nucleotides bond as in a double-stranded molecule of DNA, with one exception: RNA contains uracil nucleotides (U) in place of thymine nucleotides (T).

RNA polymerase continues to move along the DNA strand, exposing portions of the gene, until it reaches a special DNA base sequence (termination signal) that signals the end of the gene. At this point, the RNA polymerase releases the newly formed mRNA molecule and leaves the DNA. The DNA then rewinds and assumes its previous double-helix structure. This process of copying DNA information into the structure of an mRNA molecule is called **transcription.**

Messenger RNA molecules can be hundreds or even thousands of nucleotides long. They exit the nucleus through the nuclear pores, and enter the cytoplasm (fig. 4.25). There they associate with ribosomes and act as patterns, or templates, for synthesizing proteins. Protein synthesis is called **translation** (fig. 4.26).

Because an amino acid corresponds to a sequence of three nucleotides in a DNA molecule, the same amino acid is represented in the transcribed messenger RNA by the complementary set of three nucleotides. Such a triplet of nucleotides in a messenger RNA molecule is called a **codon** (see chart 4.1). Note that there are sixty-four possible triplets consisting of DNA bases, to encode twenty different amino acids. This means that more than one codon can specify the same amino acid, a point we will return to soon.

Chart 4.2 compares DNA and RNA molecules.

Until the early 1980s, all enzymes were thought to be proteins. Then, researchers found that a bit of RNA that they thought was contaminating a reaction in which RNA molecules are shortened actually contributed the enzymatic activity. The RNA enzymes were named "ribozymes." Because certain RNA molecules can carry information as well as function as enzymes—two biologically important properties—they may have been a bridge between the nonliving and the living on earth long ago.

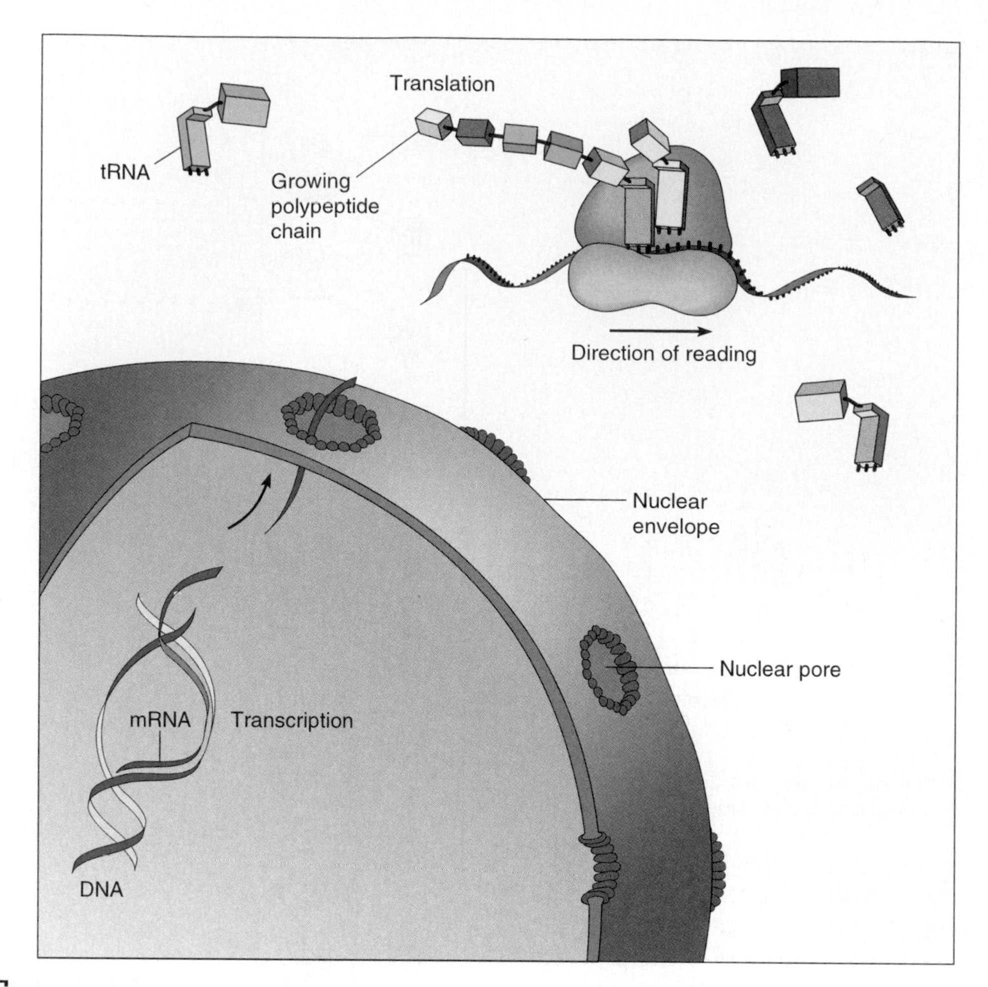


**FIGURE 4.25**
An mRNA molecule transcribed from a section of DNA exits the nucleus and enters the cytoplasm, where it associates with a ribosome to begin translation.

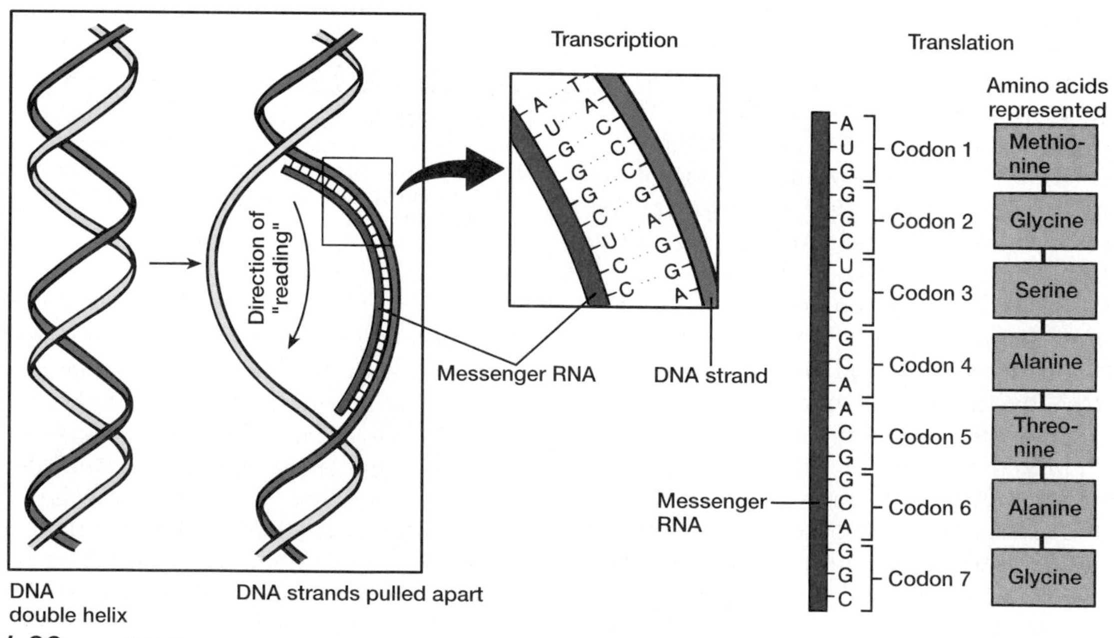


**FIGURE 4.26**
DNA information is transcribed into mRNA, which in turn is translated into a sequence of amino acids.

## Chart 4.2 A Comparison of DNA and RNA Molecules

| | DNA | RNA |
|---|---|---|
| *Main location* | Part of chromosomes, in nucleus | Cytoplasm |
| *5-carbon sugar* | Deoxyribose | Ribose |
| *Basic molecular structure* | Double-stranded | Single-stranded |
| *Organic bases included* | Adenine, thymine, cytosine, guanine | Adenine, uracil, cytosine, guanine |
| *Major functions* | Contains genetic code for protein synthesis, replicates itself prior to cell division | Messenger RNA carries transcribed DNA information to cytoplasm and acts as template for synthesis of protein molecules; transfer RNA carries amino acids to messenger RNA; ribosomal RNA provides structure for ribosomes |

### Protein Synthesis

Before a protein molecule can be synthesized, the correct amino acids must be present in the cytoplasm to serve as building blocks. Furthermore, these amino acids must be positioned in the proper locations along a strand of messenger RNA. A second kind of RNA molecule, synthesized in the nucleus and called **transfer RNA** (tRNA), aligns amino acids in the sequence specified by the DNA. A transfer RNA molecule consists of only seventy to eighty nucleotides and has a complex three-dimensional shape.

Since twenty different kinds of amino acids make up proteins in living things, there must be at least twenty different kinds of transfer RNA molecules to serve as guides. Each type of transfer RNA molecule is associated with one and only one type of amino acid. However, before a transfer RNA molecule can pick up its particular kind of amino acid, the amino acid must be activated. This step is catalyzed by special enzymes. ATP provides the energy to form a bond between the amino acid and its transfer RNA molecule.

Each type of transfer RNA includes a region at one end that contains three nucleotides in a particular sequence. These nucleotides bond a specific complementary nucleotide sequence on mRNA, a codon. Thus, the set of three nucleotides in the transfer RNA molecule is called an **anticodon.** In this way, the transfer RNA carries its amino acid to a correct position on a messenger RNA strand. This action occurs in close association with a ribosome.

A ribosome is a tiny particle of two unequal-sized subunits composed of **ribosomal RNA** (rRNA) and protein. The smaller subunit of a ribosome binds to a molecule of messenger RNA near the codon at the beginning of the messenger strand. This action allows a transfer RNA molecule with the complementary anticodon to bring the amino acid it carries into position and temporarily join to the ribosome. A second transfer RNA molecule (with its activated amino acid) complementary to the second codon on mRNA then binds to an adjacent site on the ribosome. A peptide bond forms between the two amino acids. Then the first transfer RNA molecule releases its amino acid and returns to the cytoplasm (fig. 4.27). This process repeats again and again as the ribosome moves along the messenger RNA, knitting together a chain of amino acids.

By this process, amino acids are added one at a time to the developing protein molecule. The enzymes necessary for the bonding of these amino acids are found in the larger subunit of the ribosome. This subunit also holds the growing chain of amino acids as the protein molecule forms.

A molecule of messenger RNA is usually associated with several ribosomes at the same time. Thus, several protein molecules, each in a different stage of development, may be present at any given moment.

As the protein molecule develops, proteins called *chaperones* fold it into its unique shape, and when the process is completed, it is released as a separate functional molecule. The transfer RNA molecules, ribosomes, mRNA, and of course the enzymes can function repeatedly in protein synthesis.

ATP molecules provide the energy for protein synthesis. Because a protein may consist of many hundreds of amino acids, and the energy from three ATP molecules is required to link each amino acid to the growing chain, a large proportion of a cell's energy supply supports protein synthesis.

The quantity of a particular protein that a cell synthesizes is generally proportional to the quantity of the corresponding messenger RNA present. Control of such protein production, therefore, involves the rate at which messenger RNA is synthesized in the nucleus (transcription) and the rate at which the messenger RNA is destroyed by enzymes (ribonucleases) in the cytoplasm. Proteins called *transcription factors* activate certain genes, thereby controlling which proteins a cell produces, and how much. Transcription factors

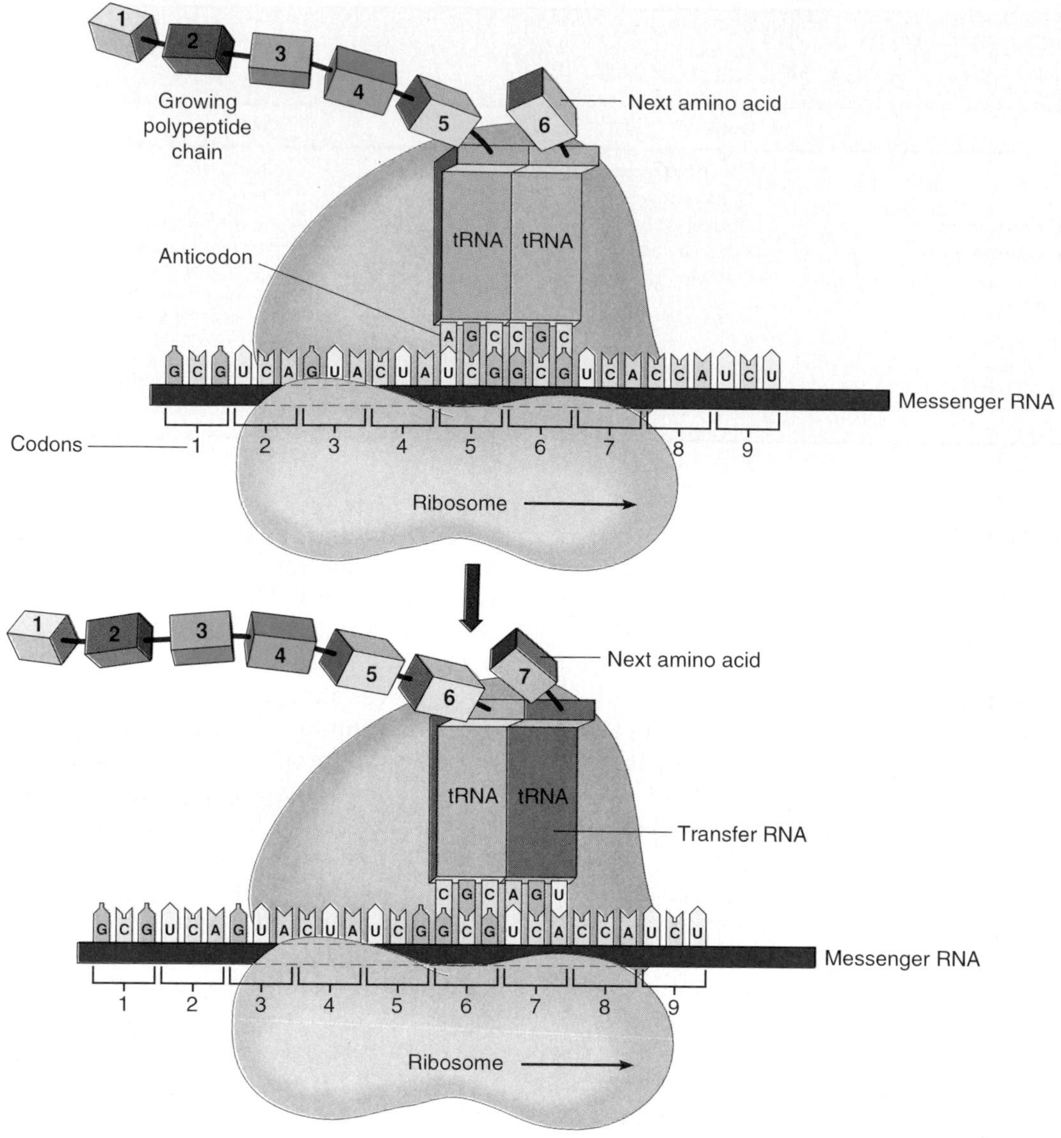


**FIGURE 4.27**

Molecules of transfer RNA (tRNA) attach to and carry specific amino acids, aligning them in the sequence determined by the codons of mRNA. These amino acids, connected by peptide bonds, form the polypeptide chain of a protein molecule.

are activated by extracellular signals such as hormones and growth factors. They are key components of the process of signal transduction.

Chart 4.3 summarizes protein synthesis.

1. What is the function of DNA?
2. How is information carried from the nucleus to the cytoplasm?
3. How are protein molecules synthesized?

> Some antibiotic drugs fight infection by interfering with bacterial protein synthesis, RNA transcription, or DNA replication. Rifampin is a drug that blocks bacterial transcription by binding to RNA polymerase, preventing the gene's message from being transmitted. Streptomycin is an antibiotic that binds a bacterium's ribosomal subunits, braking protein synthesis to a halt. Quinolone blocks an enzyme that unwinds bacterial DNA, preventing both transcription and DNA replication. Humans have different ribosomal subunits and transcription and replication enzymes than bacteria, so the drugs do not affect these processes in us.

## Chart 4.3 Protein Synthesis

**Transcription (Within the Nucleus)**

1. RNA polymerase binds to the base sequence of a gene.
2. This enzyme unwinds a portion of the DNA molecule, exposing part of the gene.
3. RNA polymerase moves along one strand of the exposed gene and catalyzes synthesis of an mRNA molecule, whose nucleotides are complementary to those of the strand of the gene.
4. When RNA polymerase reaches the end of the gene, the newly formed mRNA molecule is released.
5. The DNA molecule rewinds and closes the double helix.
6. The mRNA molecule passes through a pore in the nuclear envelope and enters the cytoplasm.

**Translation (Within the Cytoplasm)**

1. A ribosome binds to the mRNA molecule near the codon at the beginning of the messenger strand.
2. A tRNA molecule that has the complementary anticodon brings its amino acid to the ribosome.
3. A second tRNA brings the next amino acid to the ribosome.
4. A peptide bond forms between the two amino acids, and the first tRNA is released.
5. This process is repeated for each codon in the mRNA sequence as the ribosome moves along its length, forming a chain of amino acids.
6. As the chain of amino acids grows, it folds, with the help of chaperone proteins, into the unique shape of a functional protein molecule.
7. The completed protein molecule is released. The mRNA molecule, ribosome, and tRNA molecules are recycled.

### DNA Replication

When a cell reproduces, each newly formed cell needs a copy of the parent cell's genetic information (DNA) so it will be able to synthesize the proteins necessary to build cellular parts and carry on metabolism. DNA **replication** occurs during interphase of the cell cycle.

As replication begins, hydrogen bonds break between the complementary base pairs of the double strands comprising the DNA molecule. Then the double-stranded structure pulls apart and unwinds, exposing nucleotide bases. New nucleotides pair with the exposed bases, forming hydrogen bonds. Enzymes then knit together the sugar-phosphate backbone. In this way, a new strand of complementary nucleotides extends along each of the old (original) strands. Two complete DNA molecules result, each with one new and one original strand (fig. 4.28). During mitosis, the two DNA molecules that are held within the chromatids of each of the chromosomes are separated so that one of these DNA molecules passes to each of the new cells.

An enzyme called DNA polymerase carries out DNA replication. Clinical Application 4.3 discusses the polymerase chain reaction, a method for mass-producing, or amplifying, genes that has revolutionized biomedical science.

## Changes in Genetic Information

The amount of genetic information held within a set of human chromosomes is enormous, equal to twenty sets of *Encyclopaedia Britannica*.

Since each of the trillions of cells in an adult body results from mitosis, this large amount of information had to be replicated many times and with a high degree of accuracy. DNA can peruse itself for errors and correct them, but occasionally a mistake occurs or the DNA structure is damaged, and the genetic information is altered. Such a change is called a **mutation.**

### Nature of Mutations

Mutations can originate in a number of ways. For example, during DNA replication, a base may pair incorrectly within the newly forming strand, or some extra bases may be built into its structure. Or, sections of DNA strands may be deleted, moved to other regions of the molecule, or even attached to other chromosomes. In any case, the consequences are similar—genetic information is changed. If a protein is constructed from this information, its molecular structure is likely to be faulty and the protein nonfunctional. Figure 4.29 shows how the change of one base may cause sickle cell disease.

Fortunately, cells can detect damage in their DNA molecules, and use enzymes to repair damage occurring in a single strand of DNA. These **repair enzymes** can clip out defective nucleotides and fill the resulting gap with nucleotides complementary to those on the remaining strand of DNA, restoring the original structure of the double-stranded DNA molecule.

On the other hand, both strands of the DNA molecule may be damaged in the same region. This type of change is unlikely to be repaired, and if the cell survives and reproduces, the newly formed cells are likely to receive copies of the mutation.

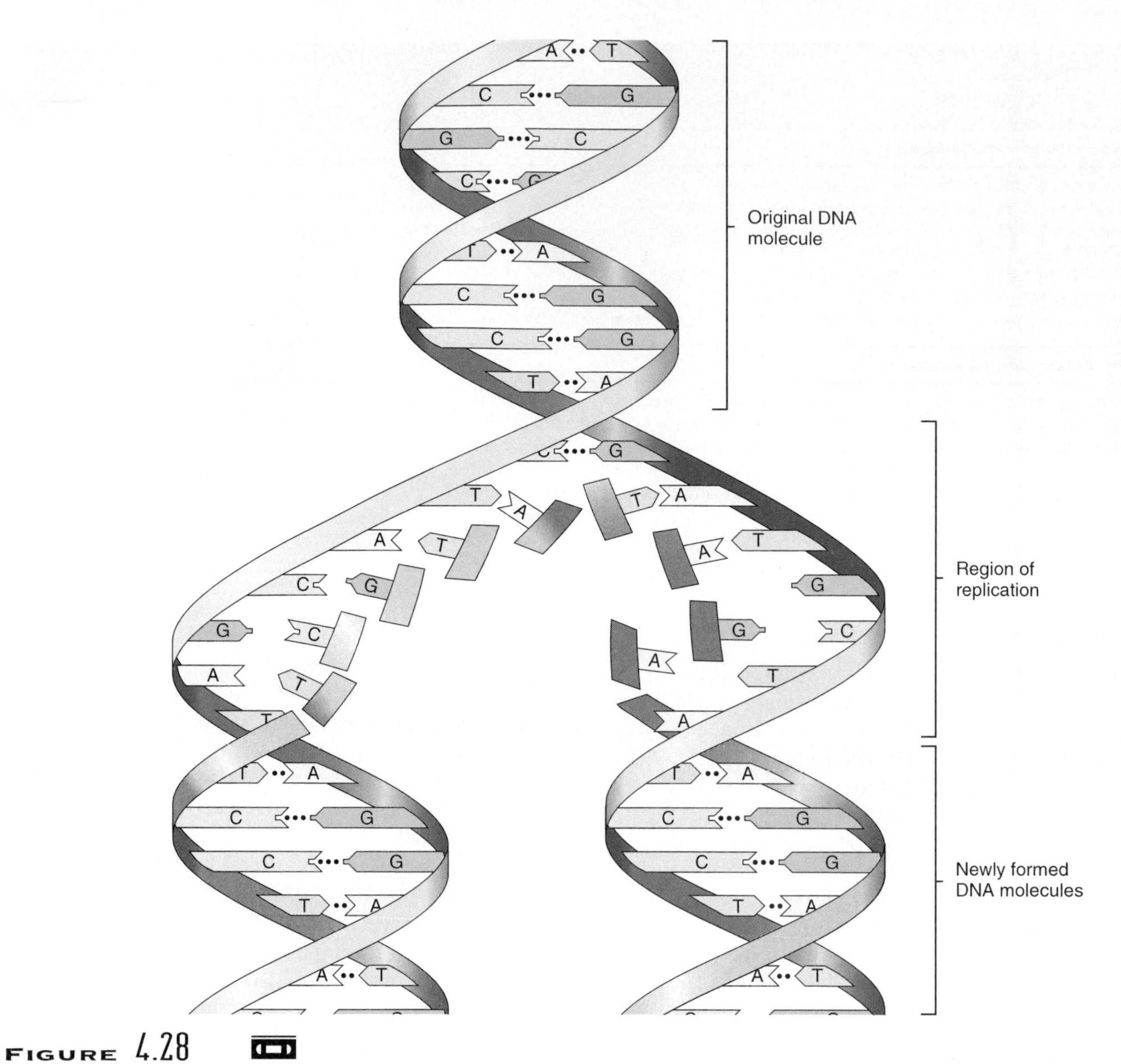


**FIGURE 4.28**

When a DNA molecule replicates, its original strands separate locally. A new strand of complementary nucleotides forms along each original strand.

## Effects of Mutations

The nature of the genetic code provides some protection against mutation. Sixty-one codons specify the twenty types of amino acids, and therefore some amino acids correspond to more than one codon type. Usually, two or three codons specifying the same amino acid differ only in the third base of the codon. A mutation changing that base encodes the same amino acid. For example, the DNA triplet codes GGA and GGG specify the amino acid proline. If a mutation changes the third position of GGA to a G, the amino acid for that position in the encoded protein does not change—it is still proline.

If a mutation alters a base in the second position, very often the substituted amino acid is very similar in overall shape to the normal one, and the protein is not changed significantly enough to affect its function. This mutation, too, would go unnoticed. (An exception is the mutation shown in fig. 4.29.) Yet another protection against mutation is that a person has two copies of each chromosome, and therefore of each gene. If one copy is mutated, the other may provide enough of the gene's normal function to maintain health. Finally, it also makes a difference whether a mutation occurs in a body cell of an adult or in a cell that is part of a developing embryo. In an adult, a mutant cell might not be noticed

## Clinical Application 4.3

# Gene Amplification Revolutionizes Biomedicine

The polymerase chain reaction (PCR) is a procedure that borrows a cell's mechanism for DNA replication, allowing researchers to make many copies of a gene of interest. Starting materials are

- two types of short DNA pieces known to bracket the gene of interest, called primers
- a hefty supply of DNA bases
- the enzymes that replicate DNA

A simple test procedure rapidly builds up copies of the gene. Here's how it works.

In the first step of PCR, heat is used to separate the two strands of the target DNA. Next, the temperature is lowered and the two short DNA primers are added. The primers bind by complementary base pairing to the separated target strands. In the third step, DNA polymerase and bases are added.

The enzyme adds bases to the primers and builds a sequence complementary to the target sequence. The newly synthesized strands then act as templates in the next round of replication, which is initiated immediately by raising the temperature. All of this is done in an automated device called a thermal cycler that controls the key temperature changes.

The pieces of DNA accumulate geometrically. The number of amplified pieces of DNA equals $2^n$, where n equals the number of temperature cycles. After just twenty cycles, one million copies of the original sequence float in the test tube. Some of the diverse applications of PCR are listed in chart 4A.

PCR's greatest strength is that it works on crude samples of rare and minute sequences, such as a bit of brain tissue on the bumper of a car, which in one criminal case lead to identification of a missing person. PCR's greatest weakness, ironically, is its exquisite sensitivity. A blood sample submitted for diagnosis of an infection contaminated by leftover DNA from a previous run, or a stray eyelash dropped from the person running the reaction, can yield a false positive result. The technique is also limited in that a user must know the sequence to be amplified.

## Chart 4A PCR Applications

**PCR has been used to amplify:**

- Genetic material from HIV in a human blood sample when infection has been so recent that antibodies are not yet detectable.
- A bit of DNA in a preserved quagga (a relative of the zebra) and a marsupial wolf, which are recently extinct animals.
- DNA in sperm cells found in the body of a rape victim so that specific sequences could be compared to those of a suspect in the crime.
- Genes from microorganisms that cannot be grown or maintained in culture for study.
- Mitochondrial DNA from various modern human populations. Comparisons of mitochondrial DNA sequences indicate that *Homo sapiens* originated in Africa, supporting fossil evidence.
- DNA from the brain of a 7,000-year-old human mummy, which indicated that native Americans were not the only people to dwell in North America long ago.
- DNA sequences unique to moose in hamburger meat, proving that illegal moose poaching had occurred.
- DNA sequences in maggots in a decomposing human corpse, enabling forensic scientists to determine the time of death.
- DNA in deteriorated road kills and carcasses washed ashore, to identify locally threatened species.
- DNA in products illegally made from endangered species, such as powdered rhinoceros horn, sold as an aphrodisiac.
- DNA sequences in animals that are unique to the bacteria that cause Lyme disease, providing clues to how the disease is transmitted.
- DNA from genetically altered microbes that are released in field tests, to follow their dispersion.
- DNA from a cell of an 8-celled human preembryo, to diagnose cystic fibrosis.
- Y chromosome specific DNA from a human egg fertilized in the laboratory to determine the sex.
- A papillomavirus DNA sequence present in, and possibly causing, an eye cancer.

because many normally functioning cells surround it. In the embryo, however, the abnormal cell might become the ancestor of great numbers of cells that are forming the body of a child. In fact, all the cells of the child's body could be defective if the mutation were present in the fertilized egg.

Mutations may occur spontaneously, if a chemical quirk causes a base in an original DNA strand to be in an unstable form just as replication occurs there. Certain chemical substances, called **mutagens,** cause such mutations. Chart 4.4 lists some mutagens.

A type of disorder called an "inborn error of metabolism" results from inheriting mutations that alter an enzyme. Such an enzyme block in a biochemical pathway has two general effects: the substance that the enzyme normally acts on builds up, and the substance resulting from the enzyme's normal action becomes scarce. It is a little like blocking a garden hose: water pressure builds up behind the block, but no water comes out after it.

The biochemical ups and downs triggered by an inborn error of metabolism can drastically affect health. The specific symptoms depend upon which pathways and substances are affected. Figure 4.30 shows how blocks of different enzymes in one biochemical pathway lead to different sets of symptoms. Clinical Application 4.4 describes phenylketonuria (PKU), one of these inborn errors about which we know a great deal.

DNA replication has built-in reactions that detect errors in the base sequence of the new strand and repair them. If DNA is not repaired, illness may result. A class of disorders affects DNA repair. One such condition is xeroderma pigmentosum (XP).

When other youngsters burst out of their homes on a sunny day to frolic outdoors, a child who has XP must cover up as completely as possible, wearing pants and long sleeves even in midsummer, and must apply sunscreen on every bit of exposed skin. Moderate sun exposure leads easily to skin sores or cancer. Even with all the precautions, the child's skin is a sea of freckles.

1. How are DNA molecules replicated?
2. What is a mutation?
3. How do mutations occur?
4. What kinds of mutations are of greatest concern?

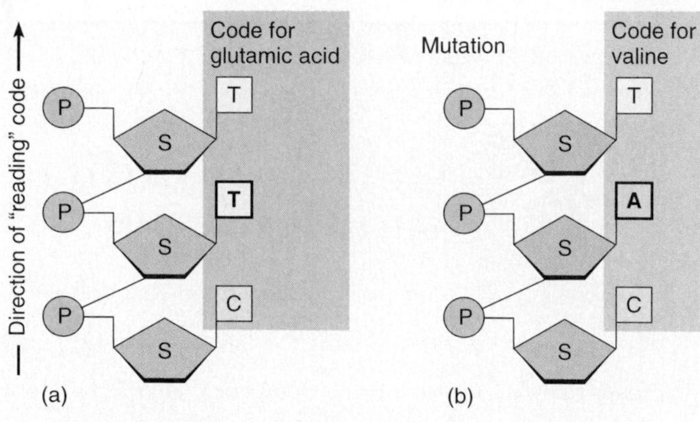


**FIGURE 4.29**

(*a*) The DNA code for the amino acid glutamic acid is CTT. (*b*) If something happens to change the first thymine in this section of the molecule to adenine, the DNA code changes to CAT, which specifies the amino acid valine. The resulting mutation, occurring in the genetic code for the protein hemoglobin, leads to sickle cell disease.

**CHART 4.4 COMMONLY ENCOUNTERED MUTAGENS**

| Mutagen | Source |
|---|---|
| Aflatoxin B | Fungi growing on peanuts and other foods |
| 2-amino 5-nitrophenol | Hair dye components |
| 2,4-diaminoanisole | |
| 2,5-diaminoanisole | |
| 2,4-diaminotoluene | |
| p-phenylenediamine | |
| Caffeine | Cola, tea, coffee |
| Furylfuramide | Food additive |
| Nitrosamines | Pesticides, herbicides, cigarette smoke |
| Proflavine | Antiseptic in veterinary medicine |
| Sodium nitrite | Smoked meats |
| Tris (2,3-dibromopropyl phosphate) | Flame retardant in children's sleepwear |

## THE HUMAN GENOME PROJECT

Researchers throughout the world are currently systematically locating, identifying, and deciphering the functions of each of the 70,000 genes that encode the biochemical blueprints of the human body. With each discovery, biologists can begin to explain how symptoms on a whole-body level result from abnormalities at the cellular and molecular levels. Although this genetic approach to understanding disease may apply directly to rare, inherited illnesses, the underlying

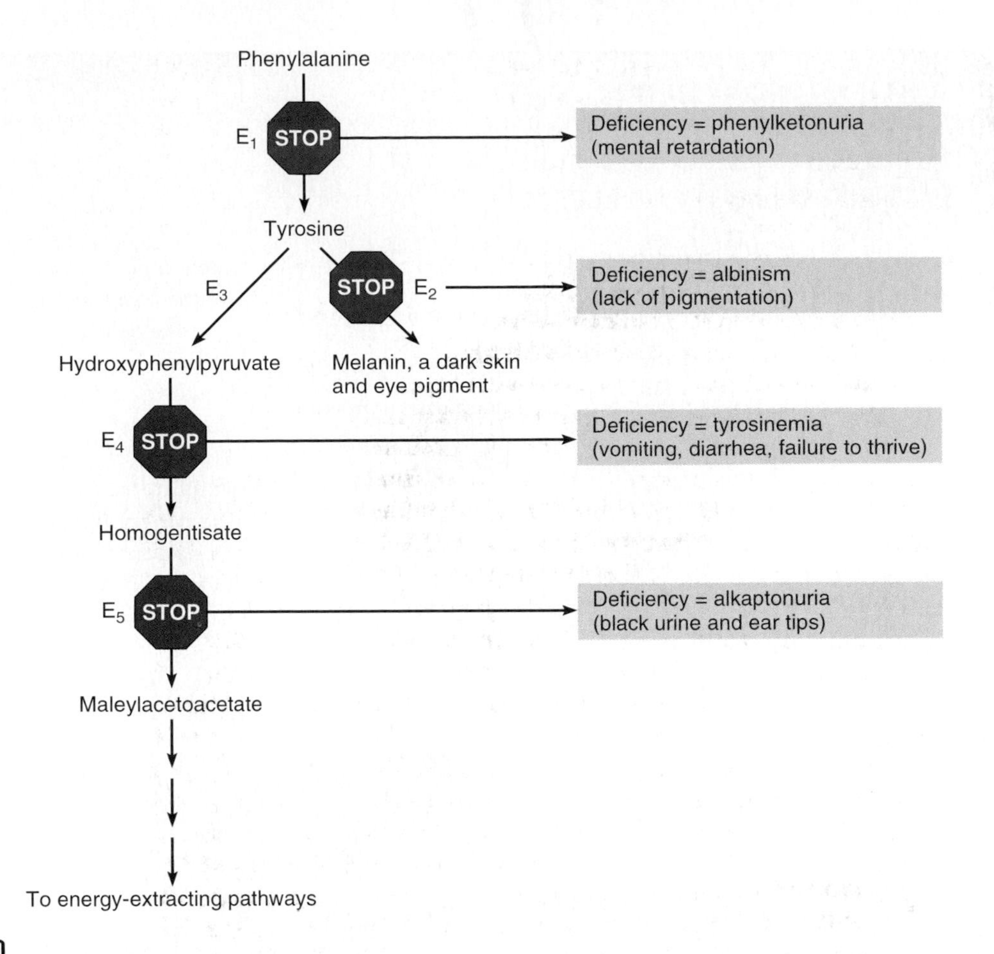


**FIGURE 4.30**

Four inborn errors of metabolism result from blocks affecting four enzymes in the pathway for the breakdown of phenylalanine, an amino acid. PKU results from a block of the first enzyme of the pathway ($E_1$). A block at $E_2$ leads to buildup of the amino acid tyrosine, and lack of its breakdown product, the pigment melanin, causes the pink eyes, white hair, and white skin of albinism. A block at $E_4$ can be deadly in infancy, and at $F_6$ leads to alkaptonuria, which causes severe joint pain and blackish deposits in the palate, ears, and eyes.

mechanisms revealed will also explain normal physiology. This was the case for Duchenne muscular dystrophy, and what it told us about normal muscle function.

Duchenne muscular dystrophy is an inherited illness of boys that greatly impairs their muscular strength. Most sufferers are wheelchair bound by adolescence and die from failure of the respiratory muscles before the age of 20. In 1986, discovery of the gene causing the disorder led to a protein—but it wasn't the actin and myosin proteins that physiologists had known for decades make up most of a muscle. The newly identified protein, named dystrophin, is present in normal muscle in vanishingly small amounts, which is why it had eluded detection for so long. But dystrophin has a critical role in muscle function—it supports muscle cell membranes from within. Without it, muscle structure falls apart, leading to muscle failure.

Unraveling the molecular underpinnings of disease lies at the forefront of biomedical research and is something that health care professionals of the future will become increasingly aware of. It is also changing the face of the traditional studies of anatomy and physiology. Wherever possible, this book provides molecular explanations of how the human body works.

1. What is the human genome project?
2. What will be the practical results of the human genome project?

## Clinical Application 4.4

# Phenylketonuria

In Oslo, Norway, in 1934, an observant mother of two mentally retarded children noticed that their soiled diapers had an odd, musty odor. She mentioned this to Ivar Folling, a relative who was a physician and biochemist. Folling was intrigued. Analyzing the children's urine, he found large amounts of the amino acid phenylalanine, which is usually present only in trace amounts because normally an enzyme hastens a chemical reaction that breaks it down. The children lacked this enzyme.

Because researchers knew that the mental retardation of PKU is caused by a buildup of phenylalanine, which is a normal component of protein, they wondered whether a diet very low in phenylalanine might prevent the mental retardation. The diet would include the other nineteen types of amino acids so that normal growth, which requires protein, could occur. The diet would theoretically alter the body's chemistry in a way that would counteract the overabundance of phenylalanine caused by the two faulty genes.

In 1963, theory became reality when a dietary treatment for this otherwise devastating inherited illness was devised (fig. 4A). The diet is very restrictive and difficult to follow, but it does prevent the otherwise inevitable mental retardation from developing. However, treated children may still have learning disabilities. We still do not know how long people with PKU should adhere to the diet, but it may be for their entire lives.

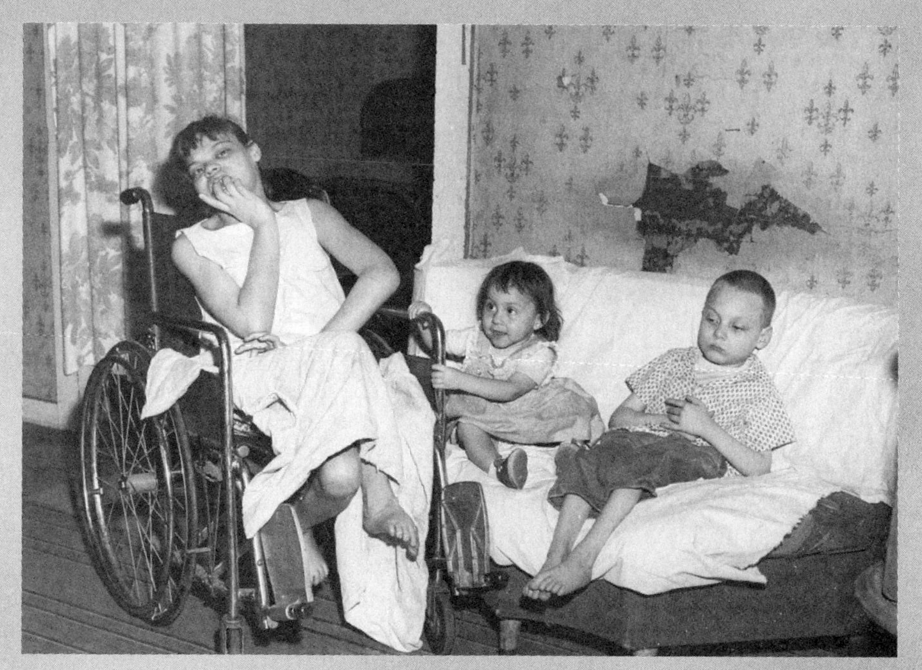

**Figure 4A**
These three siblings have each inherited PKU. The older two siblings—the girl in the wheelchair and the boy on the right—are mentally retarded, because they were born before a diet that prevents symptoms became available. The child in the middle, although she also has inherited PKU, is of normal intelligence because she was lucky enough to be born after the diet was invented.

## Chapter Summary

### Introduction (page 101)

A cell continuously carries on metabolic processes.

### Metabolic Processes (page 101)

1. Anabolism
   a. Anabolism consists of constructive processes in which smaller molecules are used to build larger ones.
   b. In dehydration synthesis, hydrogen atoms and hydroxyl groups are removed, water is formed, and smaller molecules bind by sharing atoms.
   c. Complex carbohydrates are synthesized from monosaccharides, fats are synthesized from glycerol and fatty acids, and proteins are synthesized from amino acids.

2. Catabolism
   a. Catabolism consists of decomposition processes in which larger molecules are broken down into smaller ones.
   b. In hydrolysis, a water molecule supplies a hydrogen atom to one portion of a molecule and a hydroxyl group to a second portion; the bond between these two portions breaks.
   c. Complex carbohydrates decompose into monosaccharides, fats decompose into glycerol and fatty acids, and proteins decompose into amino acids.

## Control of Metabolic Reactions (page 103)

Enzymes control metabolic reactions.

1. Enzyme action
   a. Metabolic reactions require energy to start.
   b. Enzymes are proteins that promote metabolic reactions.
   c. An enzyme acts upon a molecule by temporarily combining with it and distorting its chemical structure.
   d. The shape of an enzyme molecule fits the shape of its substrate molecule.
   e. When an enzyme combines with its substrate, the substrate changes, resulting in a product. The enzyme is released in its original form.
   f. The rate of enzyme-controlled reactions depends upon the number of enzyme and substrate molecules present and the efficiency of the enzyme.
   g. Enzymes are usually named according to their substrates, with *-ase* added.
2. Cofactors and coenzymes
   a. Cofactors are necessary parts of some enzyme molecules.
   b. A cofactor may be an ion or a small organic molecule called a coenzyme.
   c. Vitamins, which provide coenzymes, usually cannot be synthesized by human cells in adequate amounts.
3. Factors that alter enzymes
   a. Enzymes are proteins and can be denatured.
   b. Factors that may denature enzymes include excessive heat, radiation, electricity, certain chemicals, and extreme pH values.

## Energy for Metabolic Reactions (page 105)

Energy is a capacity to produce change or to do work. Common forms of energy include heat, light, sound, electrical energy, mechanical energy, and chemical energy. Whenever changes take place, energy is being transferred from one part to another.

1. Release of chemical energy
   a. Most metabolic processes utilize chemical energy that is released when molecular bonds are broken.
   b. The energy released from glucose during cellular respiration is used to promote metabolism.
   c. Enzymes in the cytoplasm and mitochondria control cellular respiration.

## An Overview of Cellular Respiration (page 105)

1. Anaerobic respiration
   a. The first phase of glucose decomposition occurs in the cytoplasm and does not require oxygen.
   b. Some of the energy released is transferred to molecules of ATP.
2. Aerobic respiration
   a. The second phase of glucose decomposition occurs within the mitochondria and requires oxygen.
   b. Considerably more energy is transferred to ATP molecules during this phase than during the anaerobic phase.
   c. The final products of glucose decomposition are carbon dioxide, water, and energy.
3. ATP molecules
   a. Thirty-eight molecules of ATP can be produced for each glucose molecule that is decomposed.
   b. Energy is captured in the bond of the terminal phosphate of each ATP molecule.
   c. Captured energy is released when the terminal phosphate bond of an ATP molecule is broken.
   d. An ATP molecule that loses its terminal phosphate becomes an ADP molecule.
   e. An ADP molecule can be converted to an ATP molecule by capturing some energy and a phosphate.

## A Closer Look at Cellular Respiration (page 106)

Metabolic processes usually have a number of steps that must occur in a specific sequence. A sequence of enzyme-controlled reactions is called a metabolic pathway. Typically, metabolic pathways are interconnected.

1. Carbohydrate pathways
   a. Carbohydrates may enter catabolic pathways and be used as energy sources.
   b. Glycolysis converts a 6-carbon glucose molecule into two 3-carbon pyruvic acid molecules, resulting in a small net gain of ATP.
   c. The citric acid cycle is a complex series of reactions decomposing molecules, releasing carbon dioxide and hydrogen atoms, and forming ATP molecules.
   d. Hydrogen atoms from the citric acid cycle are converted to hydrogen ions, which, in turn, combine with oxygen to form water molecules.
   e. Electrons from hydrogen atoms enter an electron transport system. Energy released from the system is used to form ATP.
   f. For each glucose molecule metabolized, a maximum of thirty-eight ATP molecules are formed.
   g. When present in excess, carbohydrates may enter anabolic pathways and be converted into glycogen or fat.

## Carbohydrate Storage

1. Lipid pathways
   a. Most dietary fats are triglycerides.
   b. Before fats can be used as an energy source they must be converted into glycerol and fatty acids.
   c. Beta oxidation decomposes fatty acids.
      (1) Beta oxidation activates fatty acids and breaks them down into segments containing two carbon atoms each.
      (2) Fatty acid segments are converted into acetyl coenzyme A, which can be oxidized by means of the citric acid cycle.
   d. Fats can be synthesized from glycerol and fatty acids and from excess glucose or amino acids.

2. Protein pathways
    a. Proteins are used as building materials for cellular components, as enzymes, and as energy sources.
    b. Before proteins can be used as energy sources, they must be broken down into amino acids, and the amino acids deaminated.
    c. The deaminated portions of amino acids can be broken down into carbon dioxide and water, or converted into glucose or fat.
    d. Human cells cannot synthesize eight essential amino acids in adequate amounts. These amino acids must come from foods.
3. Regulation of metabolic pathways
    a. A rate-limiting enzyme may regulate a metabolic pathway.
    b. A negative feedback mechanism in which the product of a pathway inhibits the regulatory enzyme may control the regulatory enzyme.
    c. The rate of product formation usually remains stable.

## Nucleic Acids and Protein Synthesis (page 115)

DNA molecules contain information that tells a cell how to synthesize proteins, including enzymes.

1. Genetic information
    a. DNA information specifies inherited traits that are passed from parents to offspring.
    b. A gene is a portion of a DNA molecule that contains the genetic information for making one kind of protein.
    c. The nucleotides of a DNA strand are arranged in a particular sequence.
    d. The nucleotides are paired with those of the second strand in a complementary fashion.
2. Genetic code
    a. The sequence of nucleotides in a DNA molecule represents the sequence of amino acids in a protein molecule.
    b. RNA molecules transfer genetic information from the nucleus to the cytoplasm.
3. RNA molecules
    a. RNA molecules are usually single-stranded, contain ribose instead of deoxyribose, and contain uracil nucleotides in place of thymine nucleotides.
    b. Messenger RNA molecules, which are synthesized in the nucleus, contain a nucleotide sequence that is complementary to that of an exposed strand of DNA.
    c. Messenger RNA molecules move into the cytoplasm, become associated with ribosomes, and act as patterns for the synthesis of protein molecules.
4. Protein synthesis
    a. Molecules of transfer RNA position amino acids along a strand of messenger RNA.
    b. A ribosome binds to a messenger RNA molecule and allows a transfer RNA molecule to recognize its correct position on the messenger RNA.
    c. The ribosome contains enzymes needed for the synthesis of the developing protein and holds the protein until it is completed.
    d. As the protein develops, it folds into a unique shape.
    e. ATP provides the energy for these molecular changes.
5. DNA replication
    a. Each new cell requires a copy of the parent cell's genetic information.
    b. DNA molecules are replicated during interphase of the cell cycle.
    c. Each new DNA molecule contains one old strand and one new strand.

## Changes in Genetic Information (page 125)

The amount of information within a DNA molecule is very large. Occasionally a change called a mutation occurs in the genetic information.

1. Nature of mutations
    a. Mutations include a variety of kinds of changes within DNA molecules.
    b. A protein synthesized from damaged DNA information may be nonfunctional.
    c. Repair enzymes can correct some forms of DNA damage.
2. Effects of mutations
    a. There are some built-in protections against the effects of mutations.
    b. A mutation in a sex cell or fertilized egg or early embryo may have a more severe effect than a mutation in an adult because a greater proportion of the individual's cells are affected.

## The Human Genome Project (page 128)

Researchers from many nations are cooperating to identify and study all human genes.

1. Understanding the function of individual genes will help us fight inherited disorders.
2. Understanding gene function can explain normal anatomy and physiology.

## Critical Thinking Questions

1. Because enzymes are proteins, they can denature. Relate this to the fact that changes in the pH value of body fluids during an illness may be life-threatening.
2. Some weight-reducing diets drastically limit the dieter's intake of carbohydrates but allow liberal use of fat and protein foods. What changes would such a diet cause in the cellular metabolism of the dieter? What changes could be noted in the urine of such a person?
3. How would you explain the fact that vitamins that function as coenzymes in cells are needed in extremely low concentrations?
4. What changes in concentrations of oxygen and carbon dioxide would you expect to find in the blood of a person who is forced to exercise on a treadmill beyond his or her normal capacity? How might these changes affect the pH of the person's blood?

5. How do the antibiotic actions of penicillin and streptomycin differ?
6. A student is accustomed to running 3 miles each afternoon at a slow, leisurely pace. One day, she runs a mile as fast as she can. Afterwards she is winded, with pains in her chest and leg muscles. She thought she was in great shape! What has she experienced, in terms of energy metabolism?
7. Fructose intolerance is an inherited disorder in which a missing enzyme makes a person unable to utilize fructose, a simple sugar abundant in fruit. Infants with the condition have very low mental and motor function. Older children are very lethargic and mildly mentally retarded. By adulthood, the nervous system deteriorates, eventually causing mental illness and death.

   Molecules that are derived from fructose are intermediates in the first few reactions of glycolysis. The enzyme missing in people with fructose intolerance would normally catalyze these reactions. Considering this information about the whole-body and biochemical effects of fructose intolerance, suggest what might be happening on a cellular level to these people.
8. Write the sequence of the complementary strand of DNA to the sequence A, G, C, G, A, T, T, G, C, A, T, G, C. What is the sequence of mRNA that would be transcribed from the given sequence?
9. Gout is an illness involving joint pain that was once attributed to being a lazy glutton. We now know that the pain is caused by accumulated crystals of a substance called uric acid in the joints, and the buildup results from a malfunctioning enzyme. What type of illness is gout?

## REVIEW EXERCISES

1. Define *anabolism* and *catabolism.*
2. Distinguish between dehydration synthesis and hydrolysis.
3. What is a *peptide bond?*
4. Define *enzyme.*
5. How does an enzyme interact with its substrate?
6. List three factors that increase the rates of enzyme-controlled reactions.
7. How are enzymes named?
8. Define *cofactor.*
9. Explain why humans need vitamins in their diets.
10. Explain how an enzyme may be denatured.
11. Define *energy.*
12. Explain how the oxidation of molecules inside cells differs from the burning of substances outside cells.
13. Define *cellular respiration.*
14. Distinguish between anaerobic and aerobic respiration.
15. Explain the importance of ATP to cellular processes.
16. Describe the relationship between ATP and ADP molecules.
17. What is a *metabolic pathway?*
18. Describe the process of *glycolysis.*
19. Review the major events of the citric acid cycle.
20. How are carbohydrates stored?
21. Define *beta oxidation.*
22. Explain how fats may serve as energy sources.
23. Define *deamination* and explain its importance.
24. Explain why some amino acids are called essential.
25. Explain how one enzyme can regulate a metabolic pathway.
26. Describe how a negative feedback mechanism can help control a metabolic pathway.
27. Explain what is meant by *genetic information.*
28. Describe the relationship between a DNA molecule and a gene.
29. Describe the general structure of a DNA molecule.
30. Explain how a DNA molecule stores genetic information.
31. Distinguish between messenger RNA and transfer RNA.
32. Distinguish between transcription and translation.
33. Explain the function of a ribosome in protein synthesis.
34. Distinguish between a codon and an anticodon.
35. Explain how a DNA molecule is replicated.
36. Define *mutation,* and explain how mutations may originate.
37. Define *repair enzyme.*
38. Explain how a mutation may affect an organism's cells.

5

CHAPTER

# TISSUES

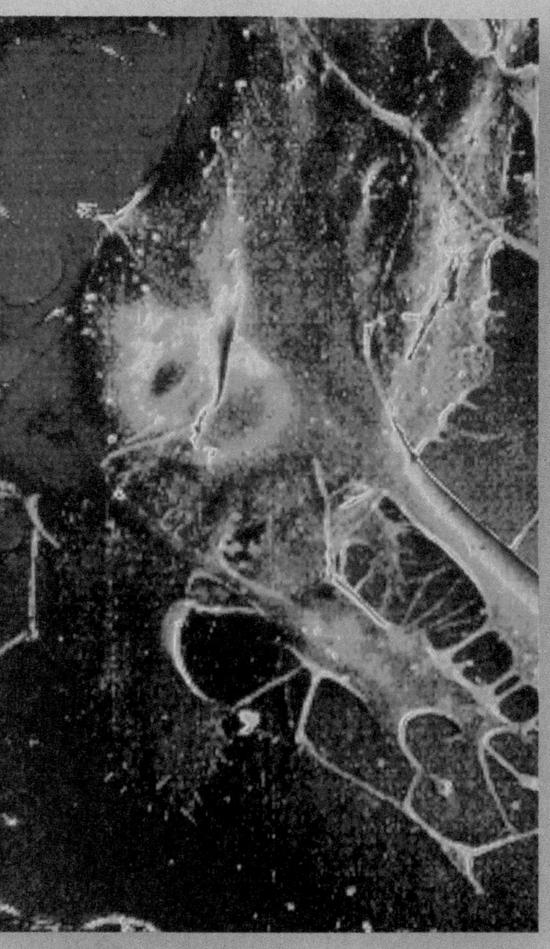

FIBROBLASTS ARE A MAJOR COMPONENT OF CONNECTIVE TISSUE (1,000×).

## CHAPTER OBJECTIVES

AFTER YOU HAVE STUDIED THIS CHAPTER, YOU SHOULD BE ABLE TO:

1. Describe the general characteristics and functions of epithelial tissue.
2. Name the types of epithelium and identify an organ in which each is found.
3. Explain how glands are classified.
4. Describe the general characteristics of connective tissue.
5. Describe the major cell types and fibers of connective tissue.
6. List the types of connective tissue within the body.
7. Describe the major functions of each type of connective tissue.
8. Distinguish among the three types of muscle tissue.
9. Describe the general characteristics and functions of nervous tissue.

## KEY TERMS

**adipose tissue** (ad′ĭ-pōs tish′u)
**cartilage** (kar′ti-lij)
**chondrocyte** (kon′dro-sīt)
**connective tissue** (kŏ-nek′tiv tish′u)
**epithelial tissue** (ep″ĭ-the′le-al tish′u)
**fibroblast** (fi′bro-blast)
**fibrous tissue** (fi′brus tish′u)
**macrophage** (mak′ro-fāj)
**muscle tissue** (mus′el tish′u)
**nervous tissue** (ner′vus tish′u)
**neuroglia** (nu-rog′le-ah)
**neuron** (nu′ron)
**osteocyte** (os′te-o-sīt)
**osteon** (os′te-on)
**reticuloendothelial tissue** (rĕ-tik″u-lo-en″do-the′le-al tish′u)

The decision to perform a Caesarean section comes so quickly that Natalie R. hardly has time to realize that she will be undergoing major surgery. The umbilical cord has wrapped around the fetus' neck, and he must be delivered as soon as possible. Fortunately, Natalie already has a catheter in place in her back, so it takes only seconds to prepare her for surgery. As she asks a nurse when the operation will begin, the obstetrician has already cut through her skin in a smile-shaped line low in the pubic area. He deftly cuts through and moves aside layers of fat, then muscle. Finally, the swollen, muscular uterus appears, jabs from the fetal limbs within rippling its surface. The physician makes another neat incision and lifts out the squalling infant—carefully untwisting the cord from around his neck.

The physician encountered several different tissues in delivering the baby. The epithelial and connective tissues of the skin look quite different from adipose (fat) tissue, which is quite different from muscles, nerves, or blood. The baby himself is a marvel of cellular specialization—nine months ago he was a single cell. Today, his trillion-plus cells are of the more than 200 types that make up a human body.

**CHART 5.1 TISSUES**

| Type | Function | Location | Distinguishing Characteristics |
|---|---|---|---|
| Epithelial | Protection, secretion, absorption, excretion | Cover body surfaces (inside and out), compose glands | Lack blood vessels, reproduce readily, cells are tightly packed |
| Connective | Bind, support, protect, fill spaces, store fat, produce blood cells | Widely distributed throughout the body | Mostly have good blood supply, cells are spaced farther apart with matrix in between |
| Muscle | Movement | Attached to bones, in the walls of hollow internal organs, heart | Contractile |
| Nervous | Transmit impulses for coordination, regulation, integration, and sensory reception | Brain, spinal cord, nerves | Cells connect to each other and other body parts |

In all complex organisms, cells are organized into layers or groups called **tissues.** Although the cells of different tissues vary in size, shape, arrangement, and function, those within a tissue are quite similar.

Usually tissue cells are separated by nonliving, intercellular materials that the cells secrete. These intercellular materials vary in composition from one tissue to another and may be solid, semisolid, or liquid. For example, a solid separates bone tissue cells, whereas a liquid separates blood tissue cells.

The tissues of the human body include four major types: *epithelial tissues, connective tissues, muscle tissues,* and *nervous tissues.* These tissues are organized into organs that have specialized functions. Chart 5.1 compares the four major tissue types.

This chapter examines epithelial and connective tissues. Other kinds of tissues will be discussed in subsequent chapters.

1 What is a tissue?

2 List the four major types of tissue.

## Epithelial Tissues

### General Characteristics

**Epithelial tissues** are widespread throughout the body. They cover all body surfaces—inside and out—and are the major tissues of glands.

Since epithelium covers organs, forms the inner lining of body cavities, and lines hollow organs, it always has a free surface—one that is exposed to the outside or to an open space internally. The underside of this tissue is anchored to connective tissue by a thin, nonliving layer called the **basement membrane** (basal lamina).

One of the ways that cancer cells spread is by secreting a substance that dissolves basement membranes. This enables cancer cells to invade adjacent tissue layers.

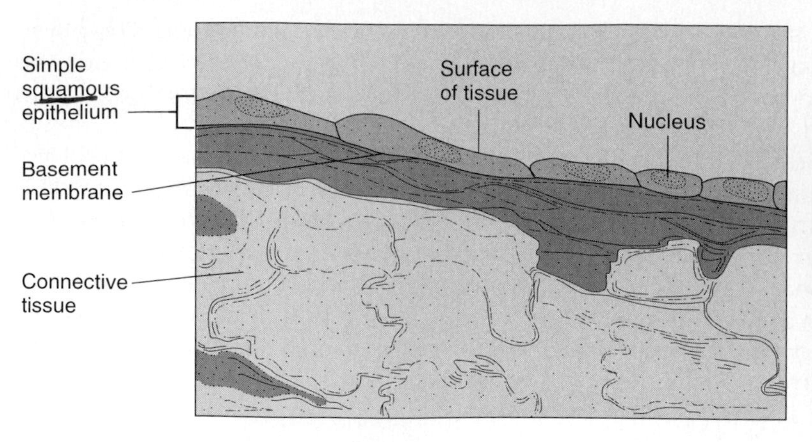


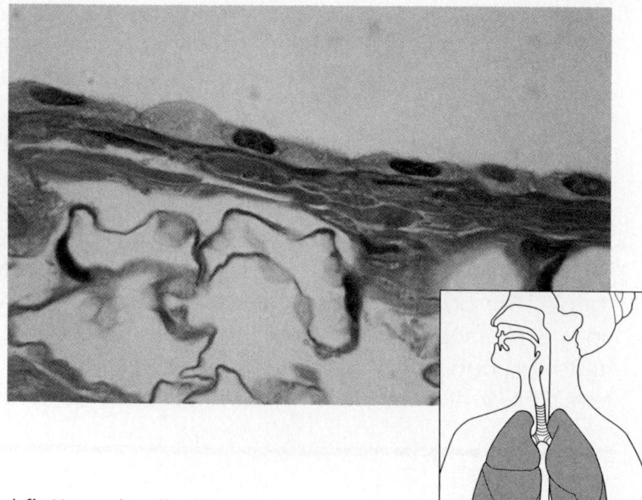

**FIGURE 5.1**
Simple squamous epithelium consists of a single layer of tightly packed flattened cells (250×).

As a rule, epithelial tissues lack blood vessels. However, they are nourished by substances that diffuse from underlying connective tissues, which are well supplied with blood vessels.

Epithelial cells reproduce readily. As a result, injuries to epithelium are likely to heal rapidly as new cells replace lost or damaged ones. Skin cells and the cells that line the stomach and intestines are continually being damaged and replaced.

Epithelial cells are tightly packed, with little intercellular material between them. In many places, desmosomes attach one to another (see chapter 3). Consequently, these cells provide effective protective barriers in such structures as the outer layer of the skin and the inner lining of the mouth. Other epithelial functions include secretion, absorption, and excretion.

Epithelial tissues are classified according to the specialized shapes, arrangements, and functions of their cells. Epithelial tissues that are composed of single layers of cells are *simple;* those with many layers of cells are *stratified;* those with thin, flattened cells are *squamous;* those with cubelike cells are *cuboidal;* and those with elongated cells are *columnar.* In the following descriptions, note that the free surfaces of epithelial cells are modified to reflect their specialized functions.

## Simple Squamous Epithelium

**Simple squamous epithelium** consists of a single layer of thin, flattened cells. These cells fit tightly together, somewhat like floor tiles, and their nuclei are usually broad and thin (fig. 5.1).

Substances pass rather easily through simple squamous epithelium, which occurs commonly at sites of diffusion and filtration. For instance, simple squamous epithelium lines the air sacs of the lungs where oxygen and carbon dioxide are exchanged. It also forms the walls of capillaries, lines the insides of blood and lymph vessels, and covers the membranes that line body cavities. However, because it is so thin and delicate, simple squamous epithelium can be damaged relatively easily.

## Simple Cuboidal Epithelium

**Simple cuboidal epithelium** consists of a single layer of cube-shaped cells. These cells usually have centrally located, spherical nuclei (fig. 5.2).

Simple cuboidal epithelium covers the ovaries and lines the kidney tubules and ducts of certain glands, such as the salivary glands, pancreas, and liver. In the kidneys, it functions in secretion and absorption; in glands, it participates in secreting glandular products.

## Simple Columnar Epithelium

The cells of **simple columnar epithelium** are elongated; that is, they are longer than they are wide. This tissue is composed of a single layer of cells whose nuclei are usually located at about the same level, near the basement membrane (fig. 5.3).

Simple columnar epithelium occurs in the linings of the uterus and organs of the digestive tract, including the stomach and intestines. Because its cells are elongated, this tissue is relatively thick, protecting underlying tissues. It also secretes digestive fluids and absorbs nutrients from digested foods.

Simple columnar cells, specialized for absorption, often have many minute, cylindrical processes extending

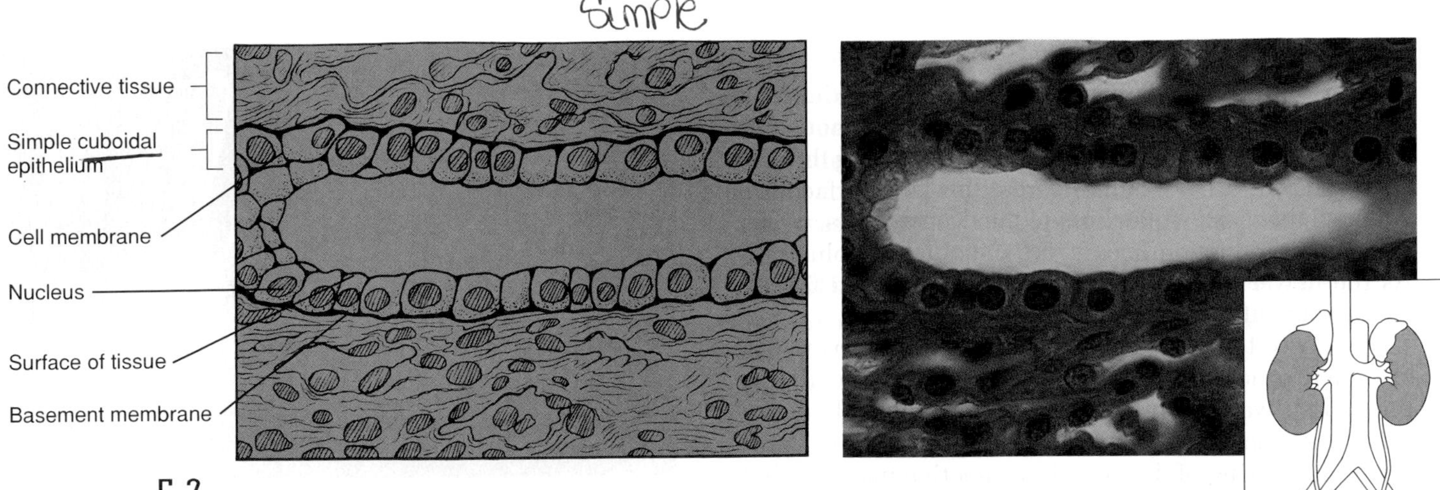


**Figure 5.2**
Simple cuboidal epithelium consists of a single layer of tightly packed cube-shaped cells (250×).

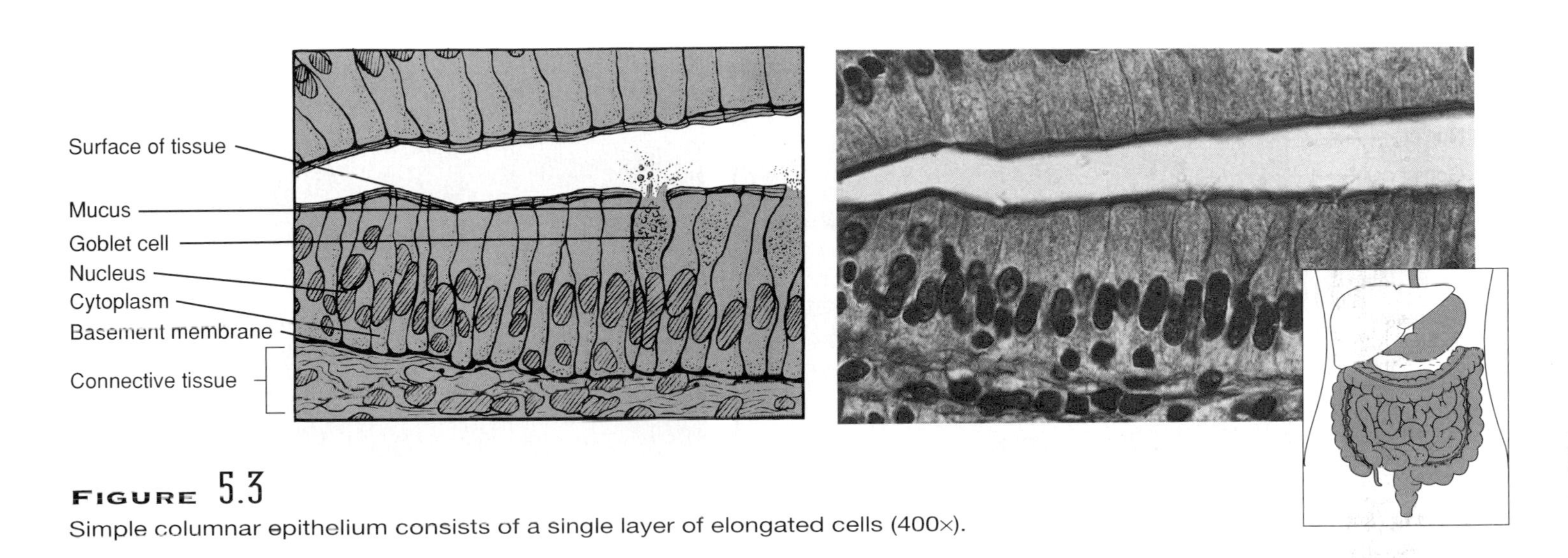


**Figure 5.3**
Simple columnar epithelium consists of a single layer of elongated cells (400×).

outward from their cell surfaces. These processes, called *microvilli,* are from 0.5 to 1.0 µm in length. They increase the surface of the cell membrane where it is exposed to the substances being absorbed (fig. 5.4).

Typically, specialized, flask-shaped glandular cells are scattered among the columnar cells of this tissue. These cells, called *goblet cells,* secrete a protective fluid called *mucus* onto the free surface of the tissue (see fig. 5.3).

## Pseudostratified Columnar Epithelium

The cells of **pseudostratified columnar epithelium** appear stratified or layered, but they are not. The layered effect occurs because their nuclei are located at two or more levels within the cells of the tissue. However, the cells, which vary in shape, all reach the basement membrane, even though some of them may not contact the free surface.

These cells commonly possess *cilia,* which are 7–10 µm in length. The cilia extend from the free surfaces of the cells, and they move constantly. Goblet cells are scattered throughout this tissue, and cilia sweep away the mucus they secrete (fig. 5.5).

Pseudostratified columnar epithelium lines the passages of the respiratory and reproductive systems. In the respiratory passages, the mucus-covered linings are sticky and trap particles of dust and microorganisms that enter with the air. The cilia move the mucus and its captured particles upward and out of the airways. In the reproductive tubes, cilia aid in moving sex cells from one region to another.

## CHART 5.3 Types of Exocrine Glands

| Type | Characteristics | Example |
|---|---|---|
| Unicellular glands | A single secretory cell | Mucus-secreting goblet cell (see fig. 5.3) |
| Multicellular glands | Glands that consist of many cells | |
| Simple glands | Glands that communicate with surface by means of unbranched ducts | |
| **1.** Simple tubular gland | Straight tubelike gland that opens directly onto surface | Intestinal glands of small intestine (see fig. 17.3) |
| **2.** Simple coiled tubular gland | Long, coiled, tubelike gland; long duct | Eccrine (sweat) glands of skin (see fig. 6.11) |
| **3.** Simple branched tubular gland | Branched, tubelike gland; duct short or absent | Mucous glands in small intestine (see fig. 17.3) |
| **4.** Simple branched alveolar gland | Secretory portions of gland expand into saclike compartments arranged along duct | Sebaceous gland of skin (see fig. 6.9) |
| Compound glands | Glands that communicate with surface by means of branched ducts | |
| **1.** Compound tubular gland | Secretory portions are coiled tubules, usually branched | Bulbourethral glands of male (see fig. 22.1) |
| **2.** Compound alveolar gland | Secretory portions are irregularly branched tubules with numerous saclike outgrowths | Salivary glands (see fig. 17.12) |

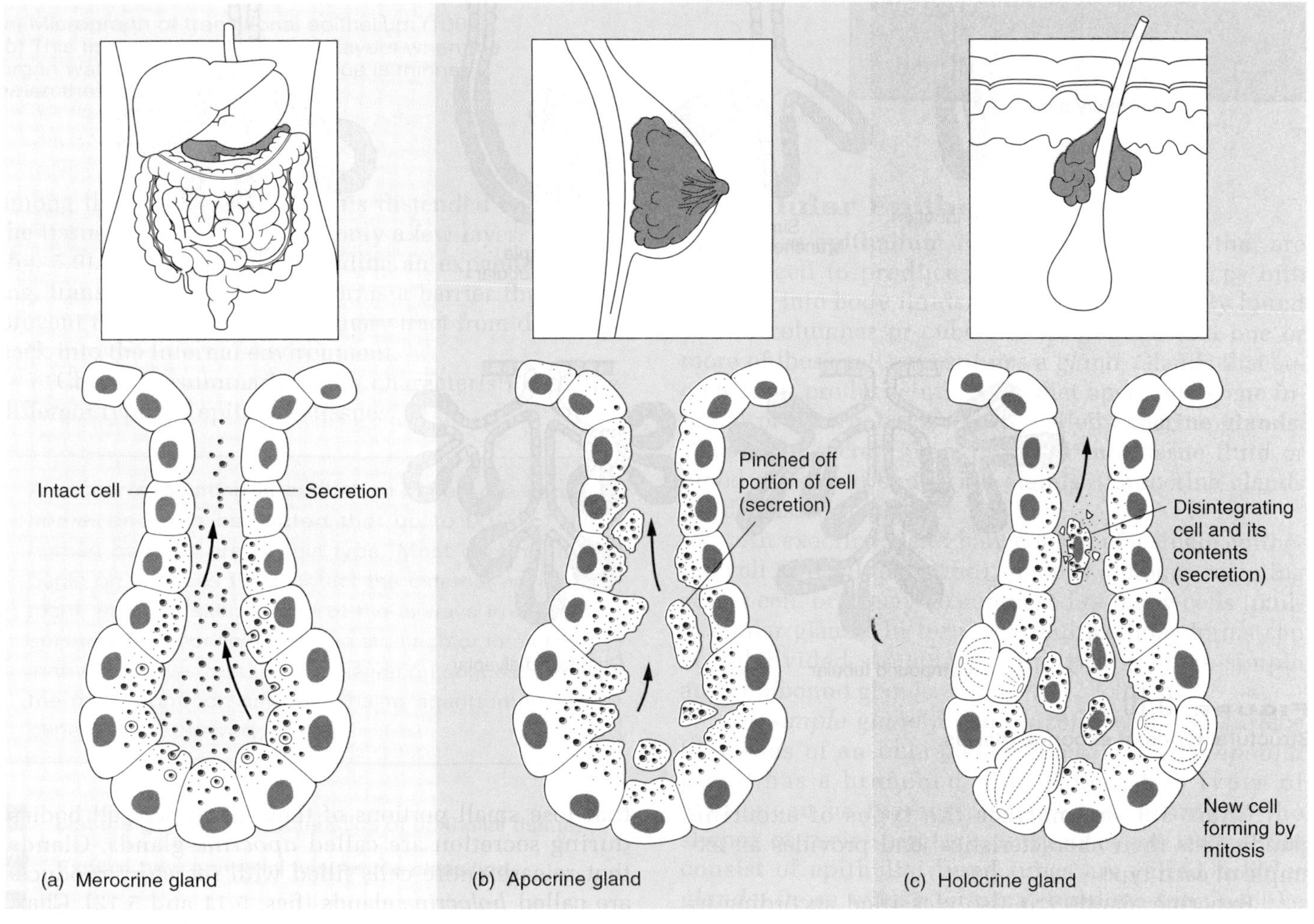


**FIGURE 5.11**
(*a*) Merocrine glands release secretions without losing cytoplasm; (*b*) apocrine glands lose small portions of their cell bodies during secretion; and (*c*) holocrine glands release entire cells filled with secretory products.

*mucous cells.* The secretion of serous cells is typically watery, has a high concentration of enzymes, and is called *serous fluid.* Such cells are common in the linings of the body cavities. Mucous cells secrete thicker *mucus.* This substance is rich in the glycoprotein *mucin* and is secreted abundantly from the inner linings of the digestive and respiratory systems.

1. Distinguish between exocrine and endocrine glands.
2. Explain how exocrine glands are classified.
3. Distinguish between a serous cell and a mucous cell.

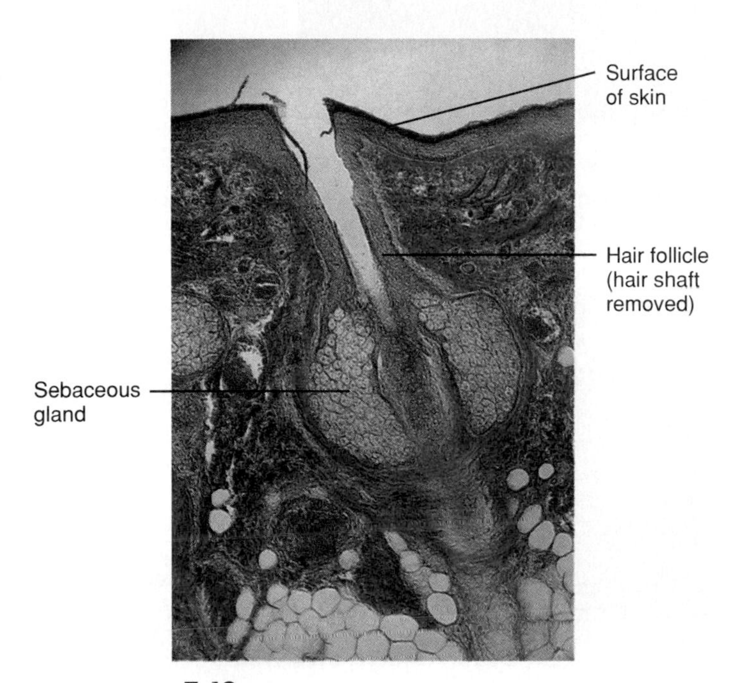


**FIGURE 5.12**
The sebaceous gland associated with a hair follicle is a simple branched alveolar gland that secretes entire cells (25×).

## CONNECTIVE TISSUES

### General Characteristics

**Connective tissues** occur throughout the body and are the most abundant type of tissue by weight. They bind structures together, provide support and protection, serve as frameworks, fill spaces, store fat, produce blood cells, protect against infections, and help repair tissue damage.

Connective tissue cells are usually farther apart than epithelial cells, and they have an abundance of intercellular material, or **matrix,** between them. This matrix consists of fibers and a ground substance whose consistency varies from fluid to semisolid to solid.

Connective tissue cells are usually able to reproduce. These tissues have varying degrees of vascularity, but in most cases, they have good blood supplies and are well nourished. Some connective tissues, such as bone and cartilage, are quite rigid. Loose connective tissue, adipose connective tissue, and fibrous connective tissue are more flexible.

### Major Cell Types

Connective tissues contain a variety of cell types. Some of them are called *resident cells* because they are usually present in relatively stable numbers. These include fibroblasts and mast cells. Other cells, such as the macrophages, are known as *wandering cells.* They appear temporarily in the tissues, usually in response to an injury or infection. The wandering cells also include several types of white blood cells.

The **fibroblast** is the most common kind of resident cell in connective tissues. It is a relatively large cell and usually is star-shaped. Fibroblasts produce fibers by secreting protein into the matrix of connective tissues (fig. 5.13).

**CHART 5.4 TYPES OF GLANDULAR SECRETIONS**

| Type | Description of Secretion | Example |
|---|---|---|
| Merocrine glands | A fluid product that is released through the cell membrane | Salivary glands, pancreatic glands, certain sweat glands of the skin |
| Apocrine glands | Cellular product and portions of the free ends of glandular cells that are pinched off during secretion | Mammary glands, certain sweat glands of the skin, ceruminous glands lining the external ear canal |
| Holocrine glands | Entire cells that are laden with secretory products | Sebaceous glands of the skin |

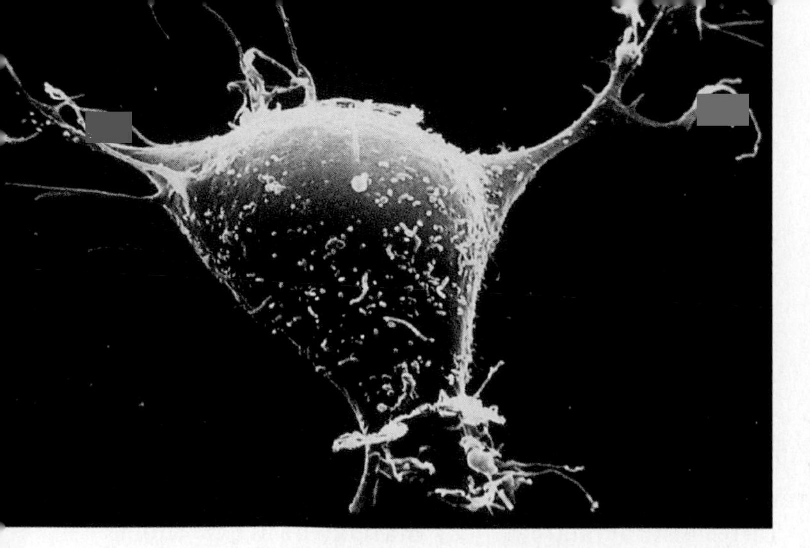

**FIGURE 5.13**
A scanning electron micrograph of a fibroblast (6,000×).

**Macrophages** (histiocytes) are almost as numerous as fibroblasts in some connective tissues. They are usually attached to fibers, but can detach and actively move about. Macrophages are specialized to carry on phagocytosis. Because they function as scavenger cells that can clear foreign particles from tissues, macrophages are an important defense against infection (fig. 5.14). They also play a role in immunity, which is discussed in chapter 16.

**Mast cells** are relatively large cells that are widely distributed in connective tissues and are usually located near blood vessels (fig. 5.15). They release *heparin,* a compound that prevents blood clotting. They also release *histamine,* a substance that promotes some of the reactions associated with inflammation and allergies, such as asthma and hay fever (see chapter 16).

## Connective Tissue Fibers

Fibroblasts produce three types of connective tissue fibers: collagenous fibers, elastic fibers, and reticular fibers.

**Collagenous fibers** are thick threads of the protein collagen, which is the major structural protein of the body. Collagen fibers are grouped in long, parallel bundles, and they are flexible but only slightly elastic (fig. 5.16). More importantly, they have great tensile strength—that is, they can resist considerable pulling force. Thus, collagenous fibers are important components of body parts that hold structures together, such as bones, ligaments (which connect bones to bones), and tendons (which connect muscles to bone).

Tissue containing abundant collagenous fibers is called *dense connective tissue.* Such tissue appears white, and for this reason collagenous fibers are sometimes called *white fibers. Loose connective tissue,* on the other hand, has sparse collagenous fibers.

Clinical Application 5.1, figure 5.17, and chart 5.5 concern disorders that result from abnormal collagen.

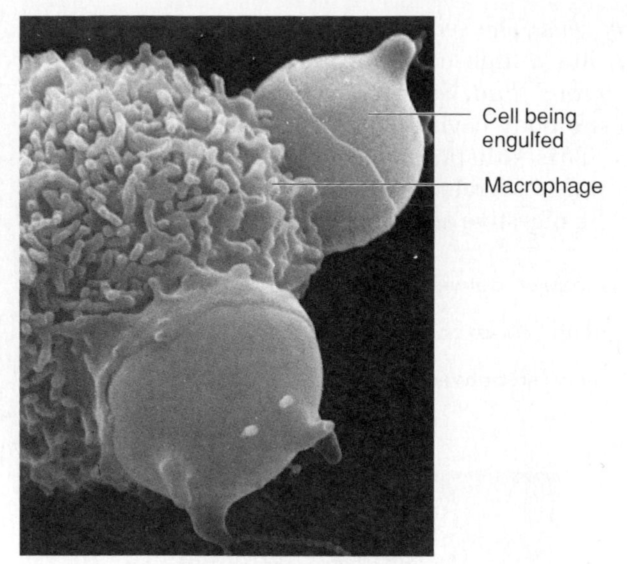


**FIGURE 5.14**
Macrophages are scavenger cells common in connective tissues. In this scanning electron micrograph, a macrophage engulfs two cells (about 5,600×).

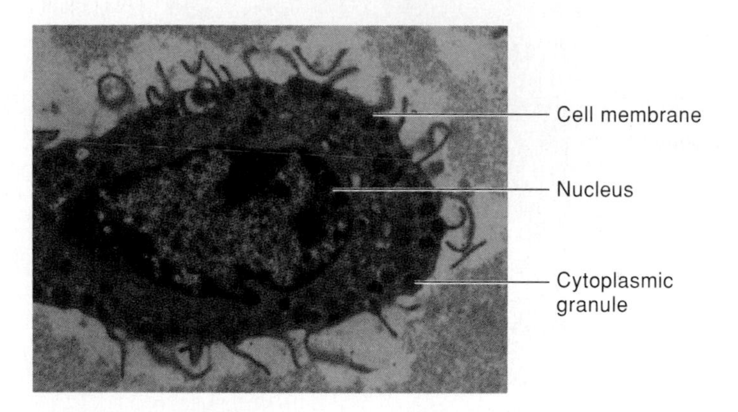


**FIGURE 5.15**
A transmission electron micrograph of a mast cell (4,000×).

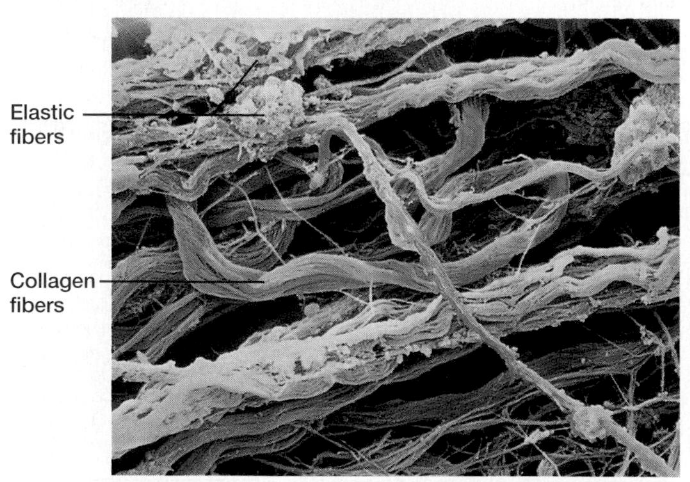


**FIGURE 5.16**
Falsely colored scanning electron micrograph of collagen fibers (here yellow) and elastic fibers (here blue) (4,100×).

CLINICAL APPLICATION

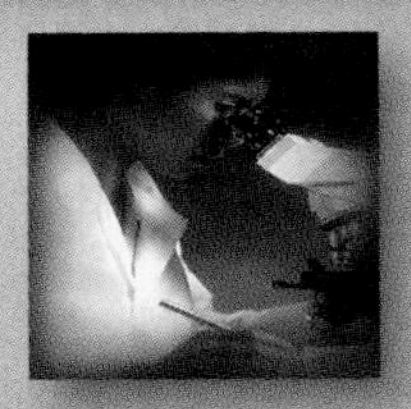

## Disorders of Orderly Collagen

Much of the human body is built of the protein collagen. It accounts for more than 60% of the protein in bone and cartilage, and provides 50–90% of the dry weight of skin, ligaments, tendons, and the dentine of teeth. Collagen is in the eyes, blood vessel linings, basement membranes, and connective tissue. It is not surprising, then, that defects in collagen lead to a variety of medical problems.

### STRUCTURE OF COLLAGEN

Collagen abnormalities are devastating, because this protein has an extremely precise conformation that is easily disrupted, even by slight alterations that might exert little noticeable effect in other proteins. Collagen is sculpted from a precursor molecule called procollagen. Three procollagen chains coil and entwine to form a very regular triple helix.

Triple helices form as the procollagen is synthesized, but once secreted from the cell, the helices are trimmed. The collagen fibrils continue to associate outside the cell, building the fibrils and networks that hold the body together. Collagen is rapidly synthesized and assembled into its rigid architecture. Many types of mutations (genetic changes) can disrupt the process, including missing procollagen chains, kinks in the triple helix, failure to cut mature collagen, and defects in aggregation outside the cell.

Knowing which specific mutations cause disorders offers a way to diagnose the condition before symptoms arise. This can be helpful if early diagnosis leads to early treatment. A woman who is told that she will develop hereditary osteoporosis, for example, might take calcium supplements or begin estrogen replacement (hormone) therapy before symptoms appear.

Aortic aneurysm is a more serious connective tissue disorder that can be diagnosed presymptomatically if the underlying mutation is detected. In aortic aneurysm, a weakened aorta (the largest blood vessel in the body, which emerges from the heart) bursts. Knowing that the mutant gene has not been inherited can ease worries—and knowing that it has been inherited can warn affected individuals to have frequent ultrasound exams so that aortic weakening can be detected early enough to surgically correct.

---

"Clinically proven, anti-aging collagen cream retexturizes skin, making it more resilient, and looking younger!" proclaims the advertisement that seems too good to be true. It is. Although any moisture on dry skin may improve its appearance temporarily, collagen molecules are far too large to actually penetrate the skin. The only thing a collagen cream is sure to affect is your wallet.

---

**Elastic fibers** are composed of bundles of microfibrils embedded in a protein called *elastin*. These fibers branch, forming complex networks in various tissues. They are weaker than collagenous fibers but very elastic. That is, they are easily stretched or deformed and will resume their original lengths and shapes when the force acting upon them is removed. Elastin fibers are common in body parts that are normally subjected to stretching, such as the vocal cords and air passages of the respiratory system. Elastic fibers are sometimes called yellow fibers, because tissues amply supplied with them appear yellowish (see fig. 5.16).

---

Surgeons use elastin in foam, powder, or sheet form to prevent scar tissue adhesions from forming at the sites of tissue removal. Elastin is produced in bacteria that are genetically engineered to contain human genes instructing them to manufacture the human protein. This is cheaper than synthesizing elastin chemically, and safer than obtaining it from cadavers.

---

## CHART 5.5 COLLAGEN DISORDERS

| Disorder | Molecular Defect | Symptoms |
|---|---|---|
| Chondrodysplasia | Collagen chains that are too wide and asymmetric due to substituted amino acids | Stunted growth; deformed joints |
| Dystrophic epidermolysis bullosa | Breakdown of collagen fibrils that attach skin layers to each other | Stretchy, easily scarred skin; lax joints |
| Hereditary osteoarthritis | Substituted amino acid in collagen chain has altered shape | Painful joints |
| Osteogenesis imperfecta type I | Too few collagen triple helices | Easily broken bones; deafness; blue sclera (whites of the eyes) |
| Stickler syndrome | Short collagen chains | Joint pain; degeneration of retina and fluid around it |

**Reticular fibers** are very thin fibers composed of collagen. They are highly branched and form delicate supporting networks in a variety of tissues.

Chart 5.6 summarizes the components of connective tissue.

1. What are the general characteristics of connective tissue?
2. What are the major types of resident cells in connective tissue?
3. What is the primary function of fibroblasts?
4. What are the characteristics of collagen and elastin?

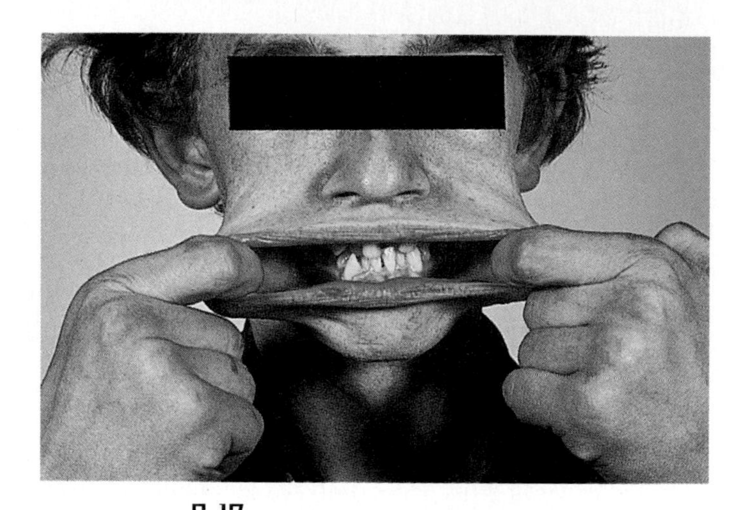

FIGURE 5.17
Abnormal collagen causes the stretchy skin of Ehlers-Danlos syndrome type I.

### Loose Fibrous Connective Tissue

**Loose fibrous connective tissue** forms delicate, thin membranes throughout the body. The cells of this tissue, mainly fibroblasts, are located some distance apart and are separated by a gel-like ground substance that contains many collagenous and elastic fibers (fig. 5.18). The fibroblasts secrete collagen and elastin.

Loose fibrous connective tissue binds the skin to the underlying organs and fills spaces between muscles. It lies beneath most layers of epithelium, where its many blood vessels nourish nearby epithelial cells.

### Adipose Tissue

**Adipose tissue,** or fat, is a specialized form of connective tissue. Certain cells within connective tissue (adipocytes) store fat in droplets within their cytoplasm. At first these cells resemble fibroblasts, but as they accumulate fat, they enlarge, and their nuclei are pushed to one side (fig. 5.19). When they become so abundant that other cell types are crowded out, they form adipose tissue. Adipose tissue is found beneath the skin, in spaces between muscles, around the kidneys, behind the eyeballs, in certain abdominal membranes, on the surface of the heart, and around certain joints.

Adipose tissue cushions joints and some organs, such as the kidneys. It also insulates beneath the skin, and it stores energy in fat molecules.

A person is born with a certain number of fat cells. Because excess food calories are likely to be converted to fat and stored, the amount of adipose tissue present in the body is usually related to a person's diet. During a period of fasting, adipose cells may lose their fat droplets, shrink in size, and become more like fibroblasts again.

## CHART 5.6 Components of Connective Tissue

| Component | Characteristic | Function |
|---|---|---|
| Fibroblasts | Widely distributed, large, star-shaped cells | Secrete proteins that become fibers |
| Macrophages | Motile cells that are sometimes attached to fibers | Clear foreign particles from tissues by phagocytosis |
| Mast cells | Large cells, usually located near blood vessels | Release substances that may help prevent blood clotting and promote inflammation |
| Collagenous fibers (white fibers) | Thick, threadlike fibers of collagen with great tensile strength | Hold structures together |
| Elastic fibers (yellow fibers) | Bundles of microfibrils composed of elastin that is very elastic | Provide elastic quality to parts that stretch |
| Reticular fibers | Thin fibers of collagen | Form supportive networks within tissues |

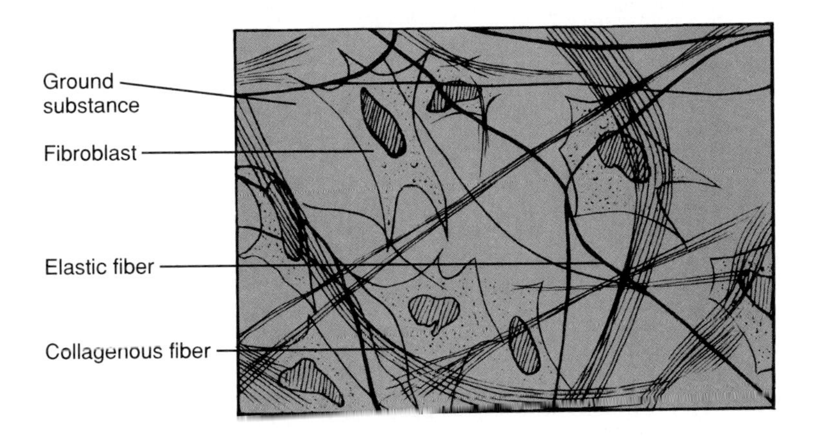


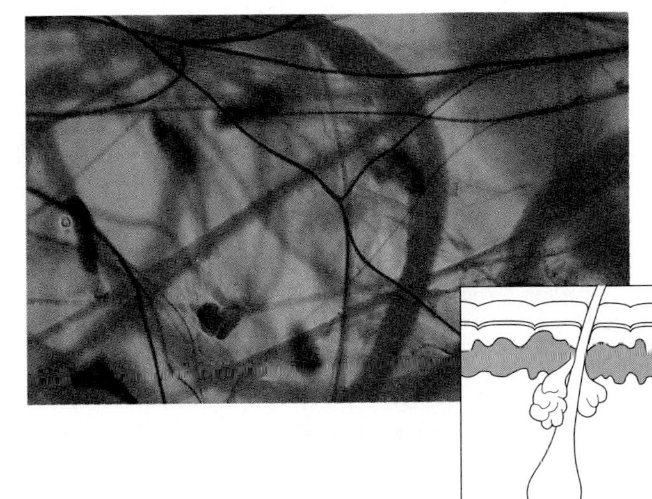

**FIGURE 5.18**
Loose fibrous connective tissue contains numerous fibroblasts that produce collagenous and elastic fibers (250×).

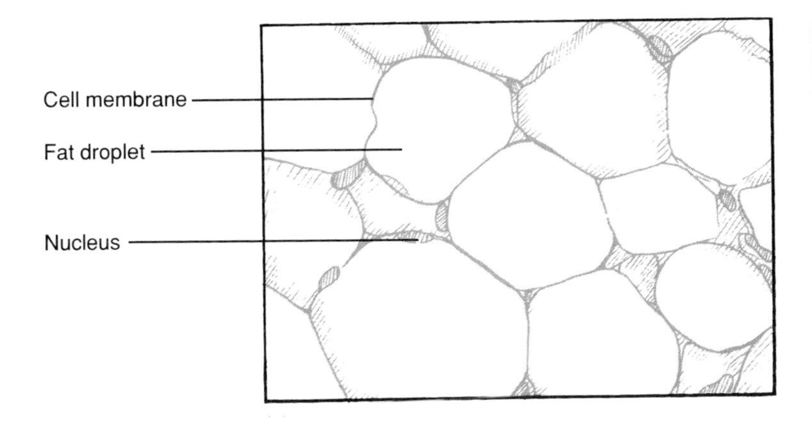


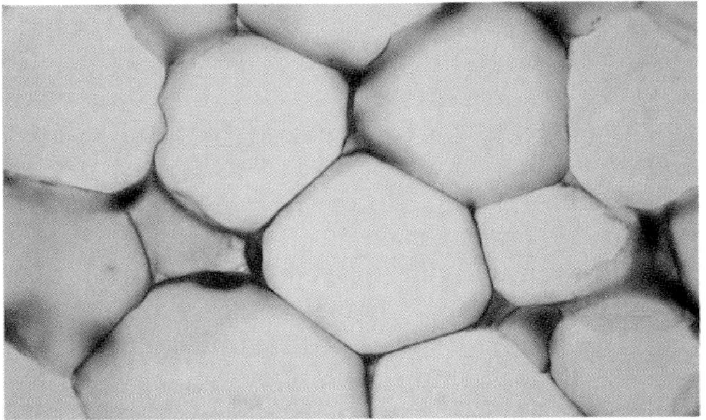

**FIGURE 5.19**
Adipose tissue cells contain large fat droplets that push the nuclei close to the cell membranes (250×).

**FIGURE 5.20**
Regular dense fibrous connective tissue consists largely of tightly packed collagenous fibers (100×).

> Infants and young children have a continuous layer of adipose tissue just beneath the skin, which gives their bodies a rounded appearance. In adults, this subcutaneous fat thins in some regions and remains thick in others. For example, in males it usually thickens in the upper back, arms, lower back, and buttocks; in females it is more likely to develop in the breasts, buttocks, and thighs.

## Dense Fibrous Connective Tissue

**Dense fibrous connective tissue** consists of many closely packed, thick, collagen fibers, a fine network of elastic fibers, and a few cells, most of which are fibroblasts. Subclasses of this tissue are regular or irregular, according to how organized the fiber patterns are.

Collagen fibers of regular tissue are very strong, enabling the tissue to withstand pulling forces (fig. 5.20). It often binds body parts together, as parts of tendons (which connect muscles to bones) and ligaments (which connect bones to bones at joints). The blood supply to regular dense fibrous connective tissue is relatively poor, so tissue repair is slow. This is why a sprain, which damages tissues surrounding a joint, may take considerable time to heal.

Fibers of irregular tissue are thicker, interwoven, and more haphazardly organized. This allows the tissue to sustain tension exerted from many different directions. Irregular dense fibrous connective tissue is found in the dermis, the inner skin layer.

## Elastic Connective Tissue

**Elastic connective tissue** consists mainly of yellow, elastic fibers in parallel strands or in branching networks. Between the fibers are collagen fibers and fibroblasts.

This tissue is found in the attachments between adjacent vertebrae of the backbone (ligamenta flava). It also occurs in the layers within the walls of certain hollow internal organs, including the larger arteries, some portions of the heart, and the larger airways, where it imparts an elastic quality (fig. 5.21).

## Reticular Connective Tissue

**Reticular connective tissue** is composed of thin, collagenous fibers arranged in a three-dimensional network. It supports the walls of certain internal organs, such as the liver, spleen, and lymphatic organs (fig. 5.22).

1. Differentiate between loose fibrous connective tissue and dense fibrous connective tissue.
2. What are the functions of adipose tissue?
3. Distinguish between fibrous, elastic, and reticular connective tissues.

## Cartilage

**Cartilage** is a rigid connective tissue. It provides support, frameworks, and attachments, protects underlying tissues, and forms structural models for many developing bones.

Cartilage matrix is abundant and is composed largely of collagenous fibers embedded in a gel-like ground substance. This ground substance is rich in a protein-polysaccharide complex (chondromucoprotein) and contains a large amount of water. Cartilage cells, or **chondrocytes,** occupy small chambers called *lacunae,* and thus are completely surrounded by matrix.

A cartilaginous structure is enclosed in a covering of fibrous connective tissue called the *perichondrium.* Although cartilage tissue lacks a direct blood supply, blood vessels are in the surrounding perichondrium.

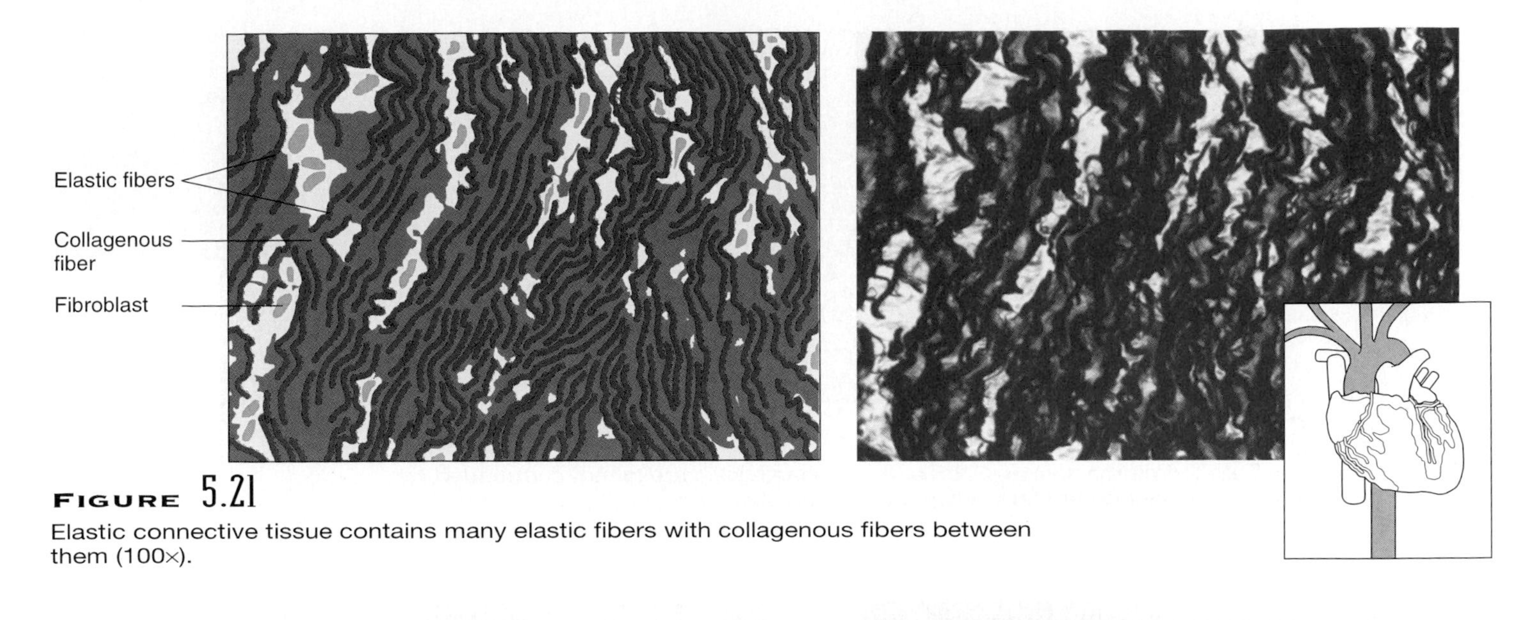


**FIGURE 5.21**
Elastic connective tissue contains many elastic fibers with collagenous fibers between them (100×).

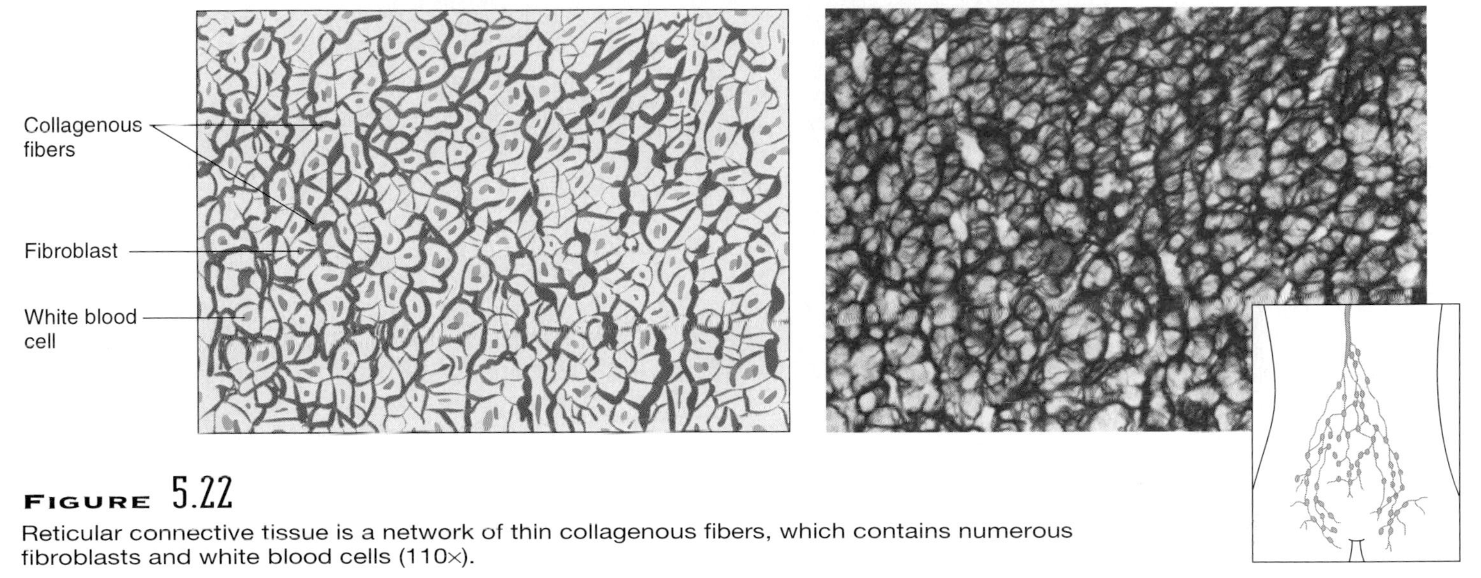


**FIGURE 5.22**
Reticular connective tissue is a network of thin collagenous fibers, which contains numerous fibroblasts and white blood cells (110×).

Cartilage cells near the perichondrium obtain nutrients from these vessels by diffusion, which is aided by the water in the matrix. This lack of a direct blood supply is why torn cartilage heals slowly, and why chondrocytes do not divide frequently.

The three types of cartilage are distinguished by their different types of intercellular material. Hyaline cartilage has very fine collagenous fibers in its matrix, elastic cartilage contains a dense network of elastic fibers, and fibrocartilage has many large collagenous fibers.

**Hyaline cartilage** (fig. 5.23), the most common type, looks somewhat like white plastic. It is found on the ends of bones in many joints, in the soft part of the nose, and in the supporting rings of the respiratory passages. Parts of an embryo's skeleton begin as hyaline cartilage "models" that bone gradually replaces. Hyaline cartilage is also important in the growth of most bones and in repair of bone fractures (see chapter 7).

**Elastic cartilage** (fig. 5.24), which is more flexible than hyaline cartilage because its matrix contains many elastic fibers, provides the framework for the external ears and parts of the larynx.

**Fibrocartilage** (fig. 5.25), a very tough tissue, contains many collagenous fibers. It is often a shock absorber for structures that are subjected to pressure. For example, fibrocartilage forms pads (intervertebral disks) between the individual parts of the backbone. It also cushions bones in the knees and in the pelvic girdle.

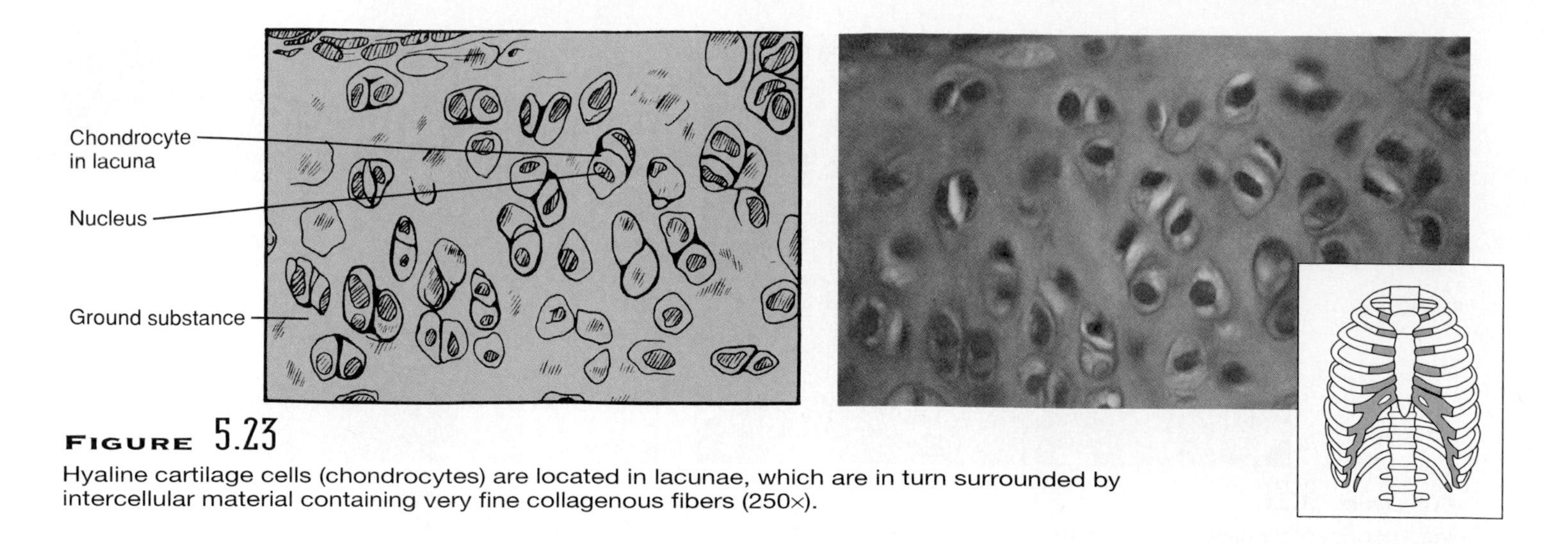


**FIGURE 5.23**

Hyaline cartilage cells (chondrocytes) are located in lacunae, which are in turn surrounded by intercellular material containing very fine collagenous fibers (250×).

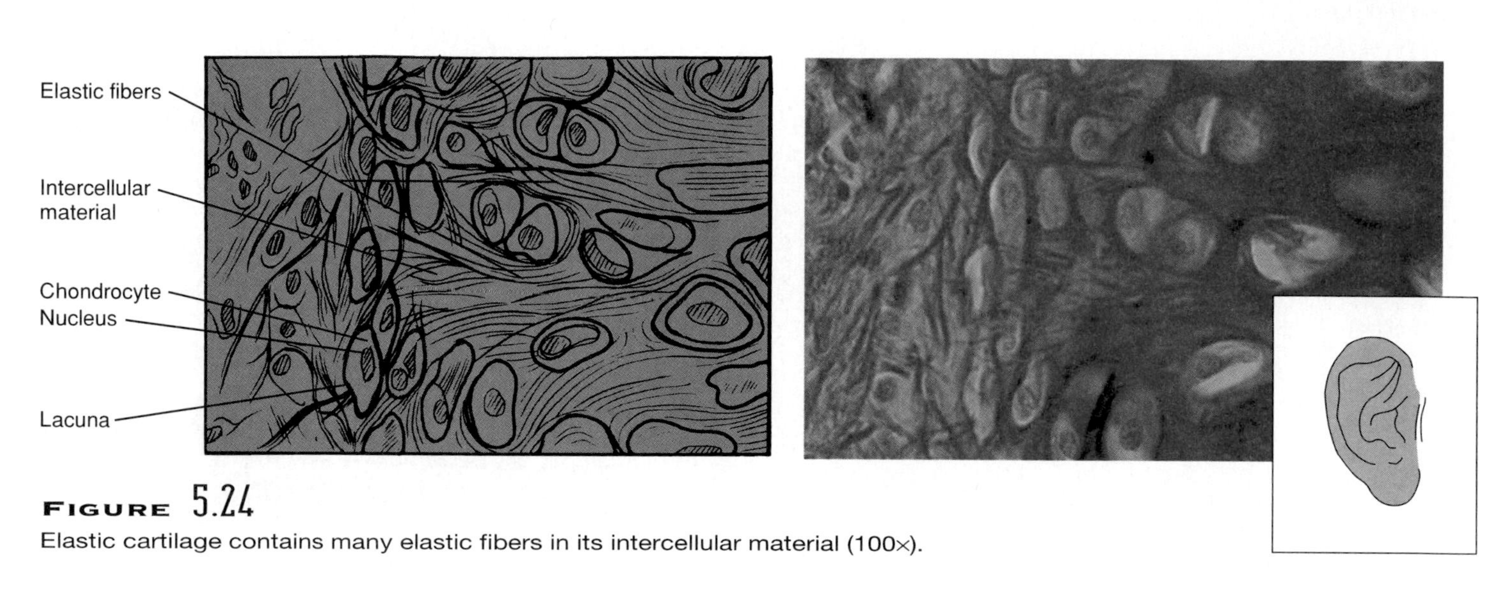


**FIGURE 5.24**

Elastic cartilage contains many elastic fibers in its intercellular material (100×).

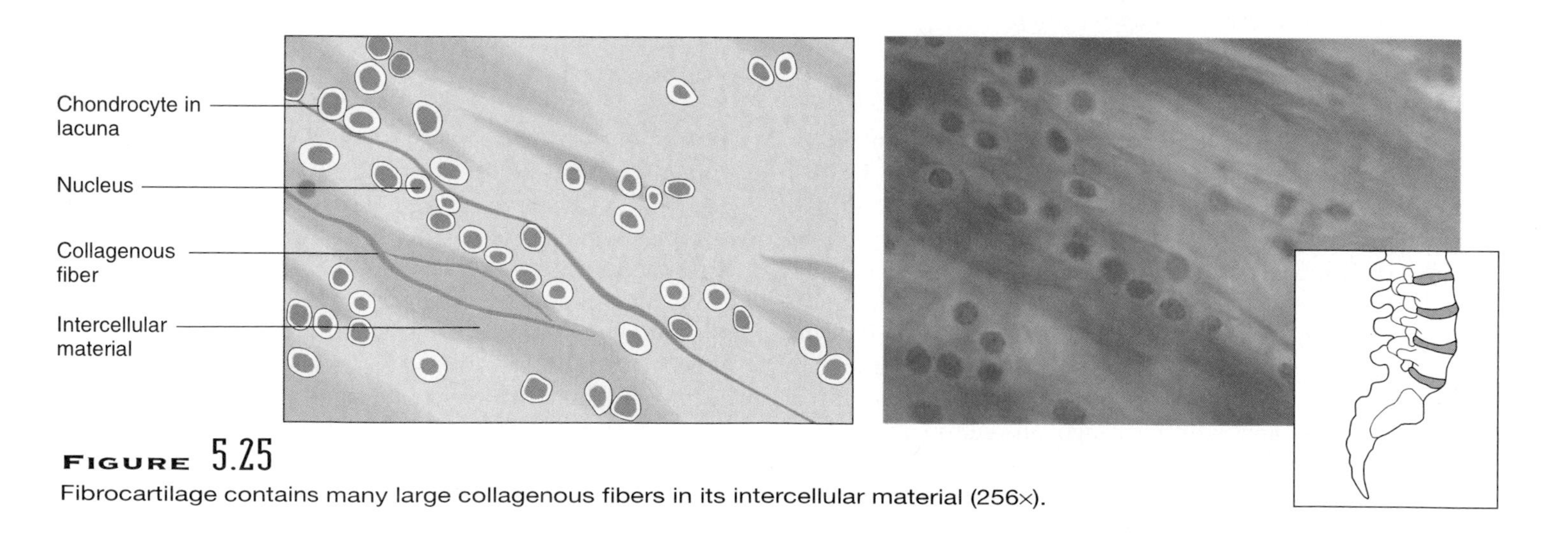


**FIGURE 5.25**

Fibrocartilage contains many large collagenous fibers in its intercellular material (256×).

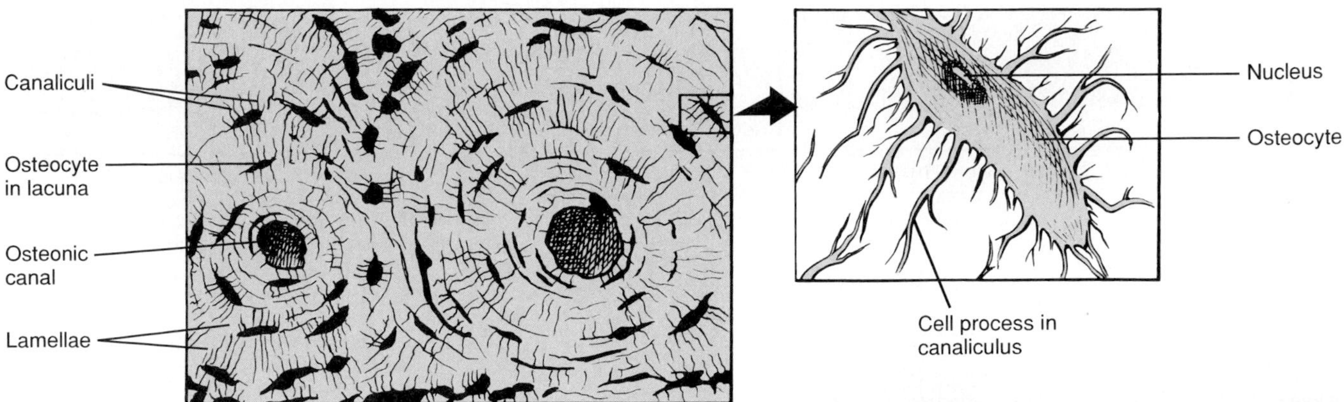


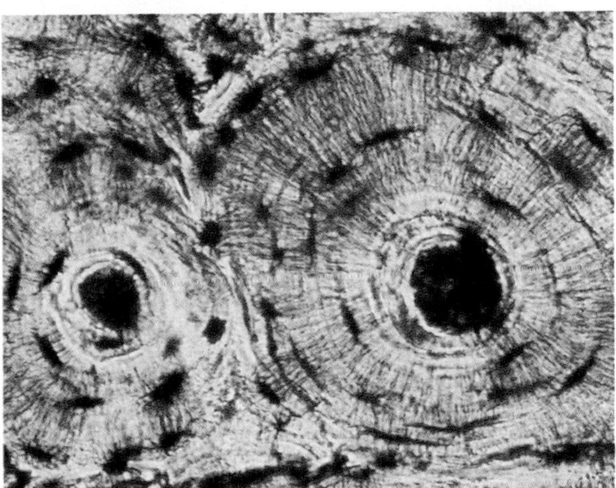

**FIGURE 5.26**
Bone matrix is deposited in concentric layers (lamellae) around osteonic canals (160×).

## Bone

**Bone** (osseous tissue) is the most rigid connective tissue. Its hardness is due largely to mineral salts, such as calcium phosphate and calcium carbonate, in its matrix. This intercellular material also contains a great amount of collagen, whose fibers flexibly reinforce the mineral components of bone.

Bone internally supports body structures. It protects vital structures in the cranial and thoracic cavities, and is an attachment for muscles. Bone also contains red marrow, which forms blood cells, and it stores various inorganic salts.

Bone matrix is deposited in thin layers called *lamellae,* which are arranged in concentric patterns around tiny longitudinal tubes called *osteonic* (*Haversian*) *canals.* Bone cells, or *osteocytes,* are located in lacunae, rather evenly spaced between the lamellae. Consequently, they too occur in patterns of concentric circles (fig. 5.26).

In a bone, the osteocytes and layers of intercellular material, which are clustered concentrically around an osteonic canal, form a cylinder-shaped unit called an **osteon** (Haversian system). Many of these units cemented together form the substance of bone (see chapter 7).

Each osteonic canal contains a blood vessel, so that every bone cell is fairly close to a nutrient supply. In addition, the bone cells have numerous cytoplasmic processes that extend outward and pass through minute tubes in the matrix called *canaliculi.* These cellular processes attach to the membranes of nearby cells by gap junctions (see chapter 3). As a result, materials can move rapidly between blood vessels and bone cells. Thus, in spite of its inert appearance, bone is a very active tissue. Injured bone heals much more rapidly than injured cartilage.

Chart 5.7 lists the characteristics of the major types of connective tissue.

## Other Connective Tissues

Other connective tissues include blood (vascular tissue) and reticuloendothelial tissue.

**Blood** is composed of cells that are suspended in a fluid intercellular matrix called *blood plasma.* These cells include red blood cells, white blood cells, and cellular fragments called *platelets.* Most blood cells form in special tissues (hematopoietic tissues) in red marrow within the hollow parts of certain bones (fig. 5.27). Blood is described in chapter 14.

Of the blood cells, only the red cells function entirely within the blood vessels. White blood cells typically migrate from the blood vessels through capillary walls. They enter connective tissues where they carry on their major activities, and they usually reside there until they die.

## CHART 5.7 CONNECTIVE TISSUES

| Type | Description | Function | Location |
|---|---|---|---|
| Loose fibrous connective tissue | Cells in fluid-gel matrix | Binds organs together, holds tissue fluids | Beneath the skin, between muscles, beneath epithelial tissues |
| Adipose tissue | Cells in fluid-gel matrix | Protects, insulates, and stores fat | Beneath the skin, around the kidneys, behind the eyeballs, on the surface of the heart |
| Dense fibrous connective tissue | Cells in fluid-gel matrix | Binds organs together | Tendons, ligaments, dermis |
| Elastic connective tissue | Cells in fluid-gel matrix | Provides elastic quality | Connecting parts of the backbone, in walls of arteries and airways |
| Reticular connective tissue | Cells in fluid-gel matrix | Supports | Walls of liver, spleen, and lymphatic organs |
| Hyaline cartilage | Cells in solid-gel matrix | Supports, protects, provides framework | Ends of bones, nose, and rings in walls of respiratory passages |
| Elastic cartilage | Cells in solid-gel matrix | Supports, protects, provides flexible framework | Framework of external ear and part of larynx |
| Fibrocartilage | Cells in solid-gel matrix | Supports, protects, absorbs shock | Between bony parts of backbone, parts of pelvic girdle, and knee |
| Bone | Cells in solid matrix | Supports, protects, provides framework | Bones of skeleton, middle ear |

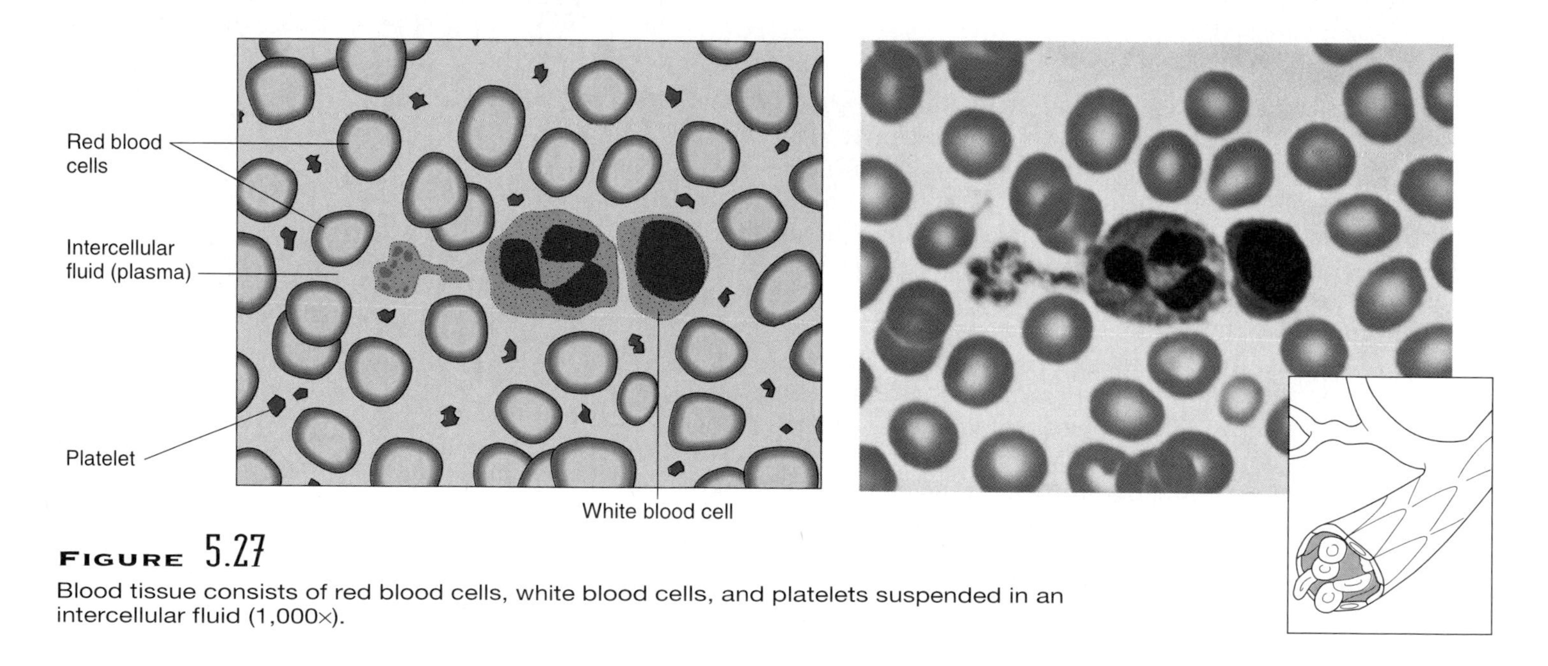


**FIGURE 5.27**
Blood tissue consists of red blood cells, white blood cells, and platelets suspended in an intercellular fluid (1,000×).

**Reticuloendothelial tissue** is composed of a variety of specialized cells that are widely scattered throughout the body. As a group, these cells are phagocytic; that is, they ingest and destroy foreign particles such as microorganisms, defending the body against infection.

Reticuloendothelial cells are found in the blood, lungs, brain, bone marrow, spleen, liver, and lymph glands. The most common ones are *macrophages.* Typically, a macrophage remains in a fixed position until it "senses" a foreign particle. Then, it moves toward the invader, and may engulf and destroy it. Once the invader has been destroyed, the macrophage becomes fixed again (see fig. 5.14). Reticuloendothelial tissue is discussed in chapter 16.

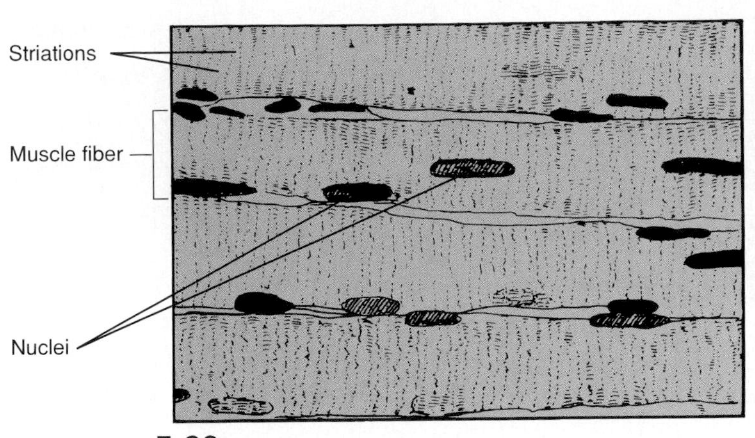


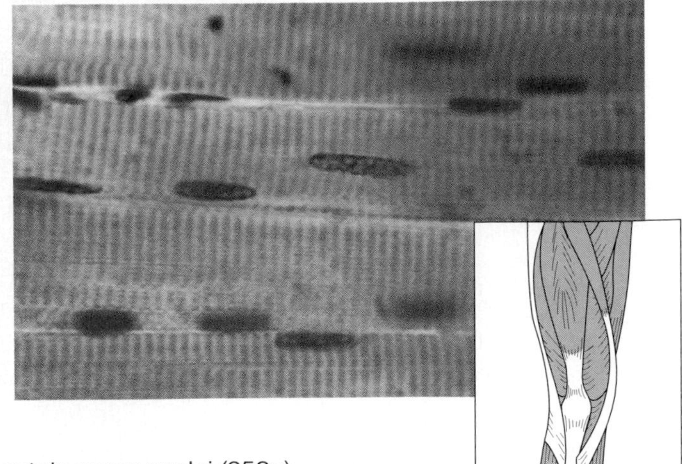

**FIGURE 5.28**
Skeletal muscle tissue is composed of striated muscle fibers that contain many nuclei (250×).

1. Describe the general characteristics of cartilage.
2. Explain why injured bone heals more rapidly than injured cartilage.
3. What are the major components of blood?
4. What do the cells of reticuloendothelial tissue have in common?

---

## MUSCLE TISSUES

### General Characteristics

**Muscle tissues** are contractile; that is, their elongated cells, or *muscle fibers,* can shorten and thicken. As they contract, muscle fibers pull at their attached ends, which moves body parts. The three types of muscle tissue, skeletal muscle, smooth muscle, and cardiac muscle, are discussed in chapter 9.

### Skeletal Muscle Tissue

**Skeletal muscle tissue** (fig. 5.28) is found in muscles that usually are attached to bones and that we control by conscious effort. For this reason, it is often called *voluntary* muscle tissue. These long and threadlike cells have alternating light and dark cross-markings called *striations.* Each cell has many nuclei. When a message from a nerve cell stimulates a muscle fiber, it contracts by protein filaments that slide past one another. Then, it relaxes. Skeletal muscles move the head, trunk, and limbs and enable us to make facial expressions, write, talk, and sing, as well as chew, swallow, and breathe.

### Smooth Muscle Tissue

**Smooth muscle tissue** (fig. 5.29) is called smooth because its cells lack striations. This tissue is found in the walls of hollow internal organs, such as the stomach, intestines, urinary bladder, uterus, and blood vessels. Unlike skeletal muscle, smooth muscle usually cannot be stimulated to contract by conscious efforts. Thus, it is a type of *involuntary* muscle tissue.

Smooth muscle cells are shorter than those of skeletal muscle, and each has a single, centrally located nucleus. The cells are spindle-shaped. Smooth muscle tissue moves food through the digestive tract, constricts blood vessels, and empties the urinary bladder.

### Cardiac Muscle Tissue

**Cardiac muscle tissue** (fig. 5.30) occurs only in the heart. Its cells, which are striated, are joined end to end. The resulting muscle fibers are branched and interconnected in complex networks. Each cell within a cardiac muscle fiber has a single nucleus. At its end, where it touches another cell, is a specialized intercellular junction called an *intercalated disk,* which is seen only in cardiac tissue.

Cardiac muscle, like smooth muscle, is controlled involuntarily and, in fact, can continue to function without being stimulated by nerve impulses. This tissue makes up the bulk of the heart and pumps blood through the heart chambers and into blood vessels.

1. List the general characteristics of muscle tissue.
2. Distinguish between skeletal, smooth, and cardiac muscle tissue.

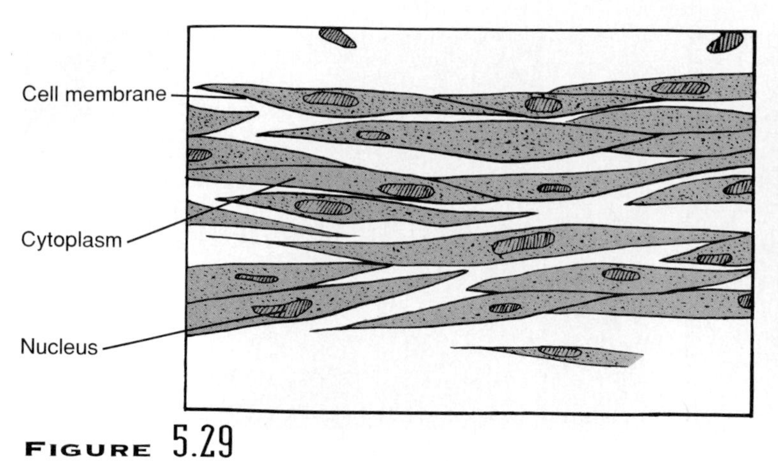

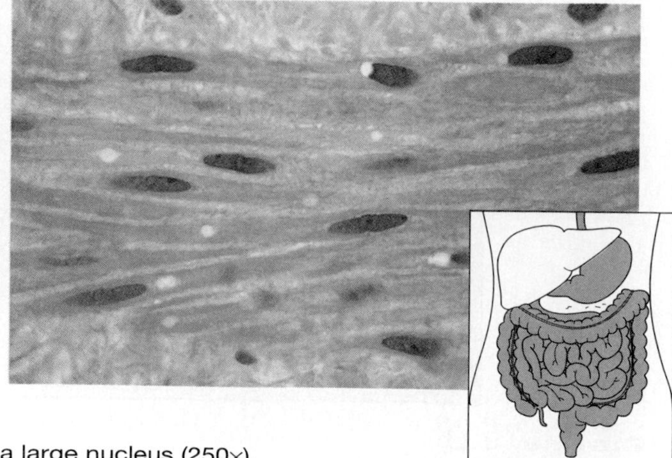

**FIGURE 5.29**
Smooth muscle tissue consists of spindle-shaped cells, each with a large nucleus (250×).

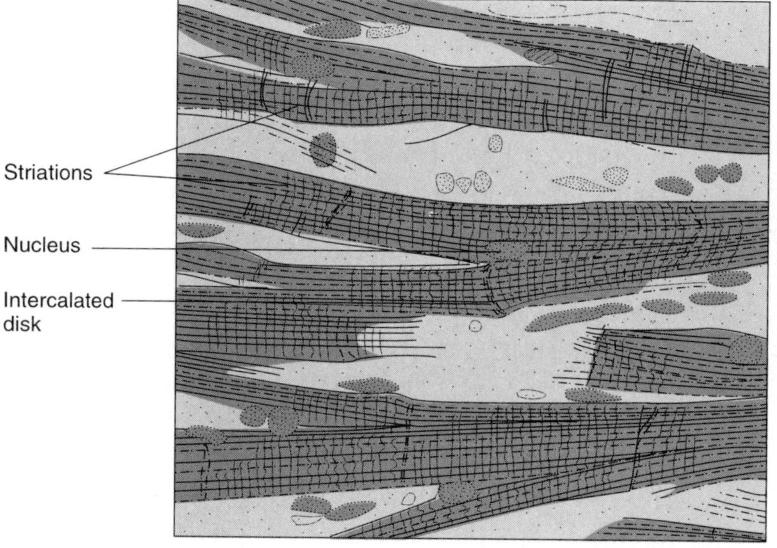

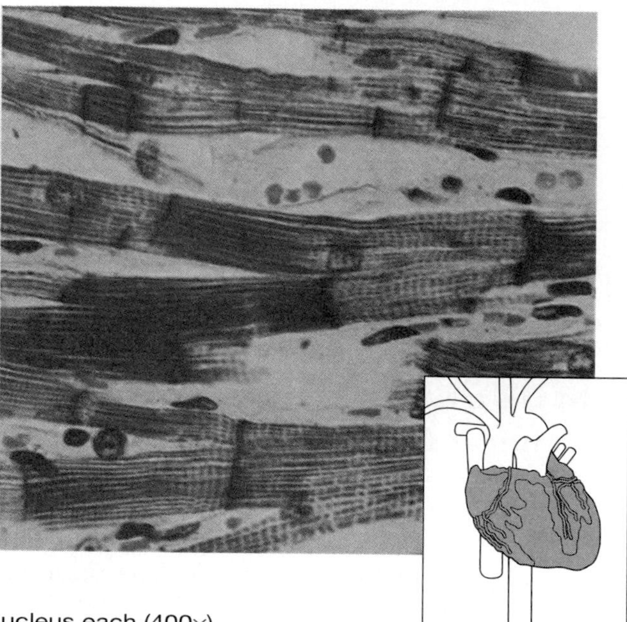

**FIGURE 5.30**
Cardiac muscle fibers are branched and interconnected, with a single nucleus each (400×).

The cells of different tissues vary greatly in their abilities to reproduce. Epithelial cells of the skin and the inner lining of the digestive tract, and the connective tissue cells that form blood cells in red bone marrow, reproduce continuously. However, striated and cardiac muscle cells and nerve cells do not usually reproduce at all after differentiating.

Fibroblasts respond rapidly to injuries by increasing in numbers and increasing fiber production. They are often the principal agents of repair in tissues that have limited abilities to regenerate themselves. For instance, cardiac muscle tissue typically degenerates in the regions damaged by a heart attack. Such tissue may be replaced by connective tissue formed by fibroblasts, which later appears as a scar.

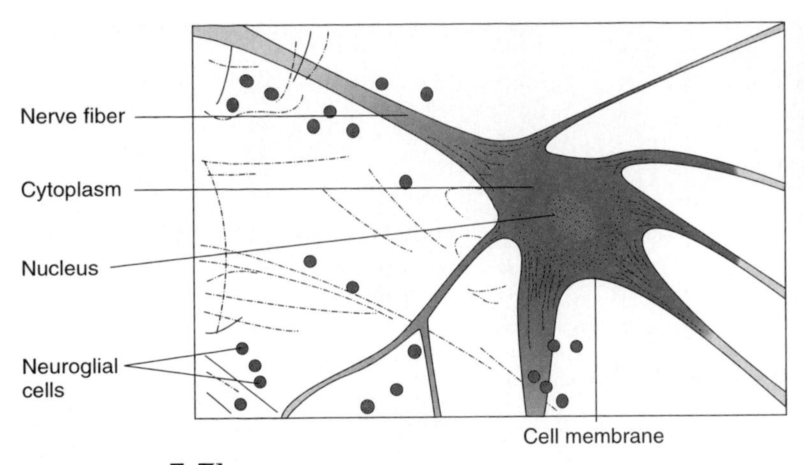

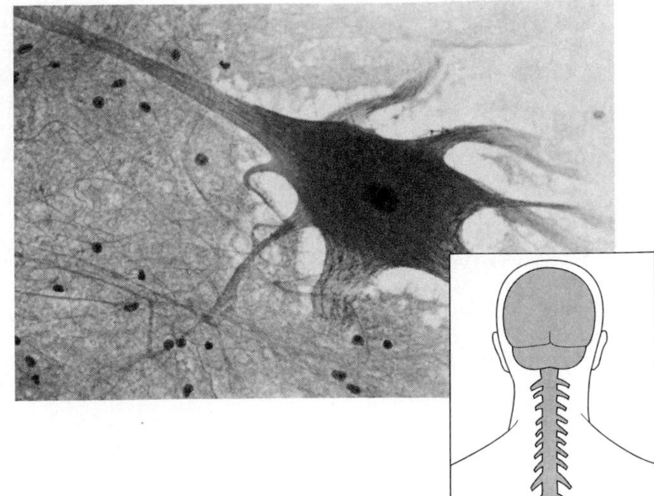

**FIGURE 5.31**
A nerve cell with nerve fibers extending into its surroundings (50×).

**CHART 5.8 MUSCLE AND NERVOUS TISSUES**

| Type | Description | Function | Location |
|---|---|---|---|
| Skeletal muscle tissue | Long, threadlike cells, striated, many nuclei | Voluntary movements of skeletal parts | Muscles usually attached to bones |
| Smooth muscle tissue | Shorter cells, single, central nucleus | Involuntary movements of internal organs | Walls of hollow internal organs |
| Cardiac muscle tissue | Branched cells, striated, single nucleus | Heart movements | Heart muscle |
| Nervous tissue | Cell with cytoplasmic extensions | Sensitivity and conduction of nerve impulses | Brain, spinal cord, and peripheral nerves |

## NERVOUS TISSUES

**Nervous tissues** are found in the brain, spinal cord, and peripheral nerves. The basic cells are called *nerve cells,* or *neurons,* and they are among the more highly specialized body cells. Neurons are sensitive to certain types of changes in their surroundings. They respond by transmitting nerve impulses along cytoplasmic extensions (nerve fibers) to other neurons or to muscles or glands (fig. 5.31). As a result of the extremely complex patterns by which neurons are connected with each other and with muscle and gland cells, they are able to coordinate, regulate, and integrate many body functions.

In addition to neurons, nervous tissue contains **neuroglial cells.** These cells support and bind the components of nervous tissue together, carry on phagocytosis, and help supply nutrients to neurons by connecting them to blood vessels. They may also play a role in cell-to-cell communications. Nervous tissue is discussed in chapter 10.

Chart 5.8 summarizes the general characteristics of muscle and nervous tissues.

Clinical Application 5.2 discusses bioengineered tissues.

1. Describe the general characteristics of nervous tissue.
2. Distinguish between neurons and neuroglial cells.

## Clinical Application 5.2

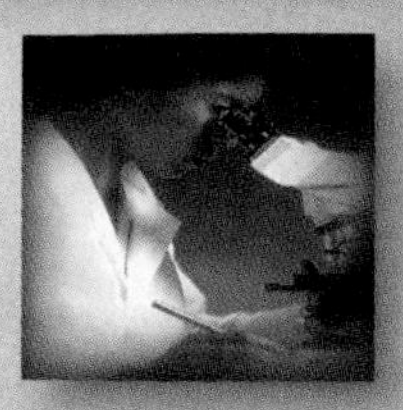

# Tissue Engineering

If an automobile or appliance part is damaged or malfunctions, replacing it is fairly simple. Not so for the human body, because the immune system presents a formidable barrier to introducing foreign tissue. Immunosuppressant drugs help, but they raise the risk of infection and cancer by dampening aspects of the immune response other than transplant rejection.

A solution to the challenge of replacing body parts is tissue engineering, which combines synthetic materials with cells. The basic recipe for a bioengineered tissue is to place cells in or on a scaffolding sculpted from a synthetic material that is accepted in the body. The cells secrete substances as they normally would, or they may be genetically altered to overproduce their natural secreted products or supply entirely different ones with therapeutic benefit, such as growth factors that might make the implant more acceptable to the body. Bioengineered tissues are just beginning to be tested in humans.

### Replacement Blood Vessels

A "GraftArtery," as one company calls its product, is a flexible yet strong tube built of collagen, a natural connective tissue (fig. 5A). An external layer of porcine collagen provides strength and handling characteristics, while an inner layer of bovine collagen provides a smooth flow surface. This biosynthetic construct does not evoke an immune response and serves as a scaffold for remodeling by the body's own cells. These bioengineered blood vessels can be manufactured to any length or width, and they withstand pressures comparable to that of the blood racing through the circulatory system. The vessels can be stitched next to natural blood vessels so smoothly that blood clots do not form nearly as easily as they do in completely synthetic grafts. If testing goes well, the vessels will be used in cardiac bypass surgery, to replace damaged arteries in the legs, and to replace brain arteries damaged by a stroke.

**Figure 5A**
This replacement blood vessel is biosynthetic. It is fabricated from collagen, a natural biomaterial. It does not evoke an immune response when used to replace a damaged blood vessel. It is remodeled by the body's own cells and functions much more like a natural blood vessel than a synthetic graft.

## New Skin and More

Burn patients may soon be helped by a bioengineered skin consisting of the patient's epidermal cells placed in sheets over dermal cells grown in culture, both layers supported by a nylon mesh framework. This semisynthetic skin may also be useful for patients who have lost a great deal of skin in surgery to remove tattoos, cancers, and moles. Bioengineered skin is already available for in vitro toxicity testing, where in many laboratories it has replaced live animals in testing cosmetic ingredients.

Other tissues on the bioengineering drawing board include liver, connective tissue, and bone marrow. A replacement cartilage similar to the skin recipe, consisting of chondrocytes in collagen, may be helpful in replacing joints destroyed by arthritis.

A scaled-down version of an engineered tissue, called a cell implant, offers a new route to drug delivery, placing cells that naturally manufacture needed substances precisely where a patient needs them. The cells are "immunoisolated" because their packaging enables them to secrete without being detected by the immune system. This is done by surrounding cells with a polymer membrane with holes small enough to allow nutrients in and the therapeutic biochemicals out, while excluding the larger molecules responsible for immune rejection.

Prime candidates for cell implants are pancreatic beta cells, which would secrete insulin to aid people with diabetes, and brain implants that would secrete dopamine, providing the biochemical that is missing in people with Parkinson's disease (fig. 5B).

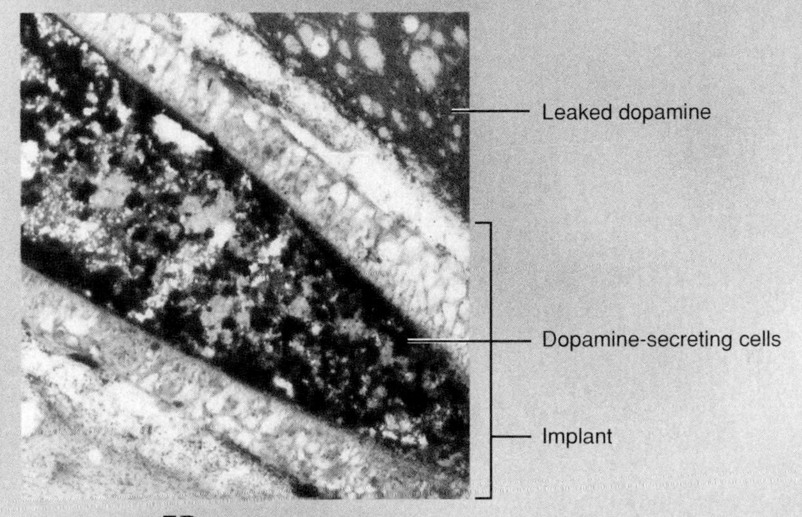


**Figure 5B**

People who have Parkinson's disease lack the neurotransmitter dopamine in certain parts of their brains, but cannot receive dopamine directly because it does not cross the blood-brain barrier. A cell implant, currently being tested in monkeys, may be able to help them. The dark, central portion of this micrograph shows a cross section of such an implant. Containing thousands of dopamine-producing cells, the implant functions like a tiny factory that continuously delivers therapeutic amounts of dopamine directly into the brain. The cells in the implant, as well as surrounding brain tissue, stain brown, indicating the presence of an enzyme required for dopamine production. The implant shown here measures approximately 1 cm long and is about the diameter of a thin pencil lead.

## Introduction (page 135)

Cells are arranged in layers or groups called tissues. Cells are separated by intercellular materials whose composition varies from solid to liquid. The four major types of human tissue are epithelial tissues, connective tissues, muscle tissues, and nervous tissues.

## Epithelial Tissues (page 135)

1. General characteristics
   a. Epithelial tissue covers all free body surfaces and is the major tissue of glands.
   b. A basement membrane anchors epithelium to connective tissue. Epithelial tissue lacks blood vessels, contains little intercellular material, and is replaced continuously.
   c. It functions in protection, secretion, absorption, and excretion.
2. Simple squamous epithelium
   a. This tissue consists of a single layer of thin, flattened cells through which substances pass relatively easily.
   b. It functions in the exchange of gases in the lungs and lines blood vessels, lymph vessels, and various membranes within the thorax and abdomen.
3. Simple cuboidal epithelium
   a. This tissue consists of a single layer of cube-shaped cells.
   b. It carries on secretion and absorption in the kidneys and various glands.
4. Simple columnar epithelium
   a. This tissue is composed of elongated cells whose nuclei are located near the basement membrane.
   b. It lines the uterus and digestive tract, where it functions in protection, secretion, and absorption.
   c. Absorbing cells often possess microvilli.
   d. This tissue usually contains goblet cells that secrete mucus.
5. Pseudostratified columnar epithelium
   a. This tissue appears stratified because the nuclei are located at two or more levels.
   b. Its cells may have cilia that move mucus or cells over the surface of the tissue.
   c. It lines tubes of the respiratory and reproductive systems.
6. Stratified squamous epithelium
   a. This tissue is composed of many layers of cells, the topmost of which are flattened.
   b. It protects underlying cells from harmful environmental effects.
   c. It covers the skin and lines the mouth, throat, vagina, and anal canal.
7. Stratified cuboidal epithelium
   a. This tissue is composed of two or three layers of cube-shaped cells.
   b. It lines the larger ducts of the sweat glands, salivary glands, and pancreas.
   c. It functions in protection.
8. Stratified columnar epithelium
   a. The top layer of cells in this tissue contains elongated columns. Cube-shaped cells make up the bottom layers.
   b. It is found in the male urethra and parts of the pharnyx.
   c. This tissue functions in protection and secretion.
9. Transitional epithelium
   a. This tissue is specialized to undergo distension.
   b. It occurs in the walls of various organs of the urinary tract.
   c. It helps prevent the contents of the urinary passageways from diffusing outward.
10. Glandular epithelium
    a. Glandular epithelium is composed of cells that are specialized to secrete substances.
    b. A gland consists of one or more cells.
       (1) Exocrine glands secrete into ducts.
       (2) Endocrine glands secrete into tissue fluid or blood.
    c. Glands are classified according to the arrangement of their cells.
       (1) Simple glands have unbranched ducts.
       (2) Compound glands have branched ducts.
       (3) Tubular glands consist of simple epithelial-lined tubes.
       (4) Alveolar glands consist of saclike dilations connected to the surface by narrowed ducts.
    d. Exocrine glands are classified according to composition of their secretions.
       (1) Merocrine glands secrete watery fluids without loss of cytoplasm. Most secretory cells are merocrine.
          (a) Serous cells secrete watery fluid with a high enzyme content.
          (b) Mucous cells secrete mucus.
       (2) Apocrine glands lose portions of their cells during secretion.
       (3) Holocrine glands release cells filled with secretory products.

## Connective Tissues (page 143)

1. General characteristics
   a. Connective tissue connects, supports, protects, provides frameworks, fills spaces, stores fat, produces blood cells, provides protection against infection, and helps repair damaged tissues.
   b. Connective tissue cells are usually some distance apart with considerable intercellular material between them.
   c. This intercellular matrix consists of fibers and a ground substance.
2. Major cell types
   a. Fibroblasts produce collagenous and elastic fibers.
   b. Macrophages are phagocytes.
   c. Mast cells may release heparin and histamine, and usually are located near blood vessels.

3. Connective tissue fibers
    a. Collagenous fibers are composed of collagen and have great tensile strength.
    b. Elastic fibers are composed of microfibrils embedded in elastin and are very elastic.
    c. Reticular fibers are very fine collagenous fibers.
4. Loose fibrous connective tissue
    a. This tissue forms thin membranes between organs and binds them together.
    b. It is found beneath the skin and between muscles.
    c. Its intercellular spaces contain tissue fluid.
5. Adipose tissue
    a. Adipose tissue is a specialized form of loose connective tissue that stores fat, provides a protective cushion, and insulates.
    b. It is found beneath the skin, in certain abdominal membranes, and around the kidneys, heart, and various joints.
6. Dense fibrous connective tissue
    a. This tissue is composed largely of strong, collagenous fibers that bind structures.
    b. Regular dense fibrous connective tissue is found in tendons and ligaments, while irregular tissue is found in the dermis.
7. Elastic connective tissue
    a. This tissue is composed mainly of elastic fibers.
    b. It imparts an elastic quality to the walls of certain hollow internal organs.
8. Reticular connective tissue
    a. This tissue consists largely of thin, branched collagenous fibers.
    b. It supports the walls of the liver, spleen, and lymphatic organs.
9. Cartilage
    a. Cartilage provides a supportive framework for various structures.
    b. Its intercellular material is composed of fibers and a gel-like ground substance.
    c. It lacks a direct blood supply and is slow to heal following an injury.
    d. Cartilaginous structures are enclosed in a perichondrium, which contains blood vessels.
    e. Major types are hyaline cartilage, elastic cartilage, and fibrocartilage.
    f. Cartilage occurs at the ends of various bones, in the ear, in the larynx, and in pads between bones of the backbone, pelvic girdle, and knees.
10. Bone
    a. The intercellular matrix of bone contains mineral salts and collagen.
    b. Its cells are usually arranged in concentric circles around osteonic canals and are interconnected by canaliculi.
    c. It is an active tissue that heals rapidly following an injury.
11. Other connective tissues
    a. Blood
        (1) Blood is composed of cells suspended in fluid.
        (2) Blood cells are formed by special tissue in the hollow parts of certain bones.
        (3) Blood transports white blood cells to connective tissues.
    b. Reticuloendothelial tissue
        (1) This tissue is composed of a variety of phagocytic cells that are widely distributed throughout the body.
        (2) It defends the body against invasion by microorganisms.

## Muscle Tissues (page 153)

1. General characteristics
    a. Muscle tissue contracts, moving structures that are attached to it.
    b. Three types are skeletal, smooth, and cardiac muscle tissues.
2. Skeletal muscle tissue
    a. Muscles containing this tissue are usually attached to bones and are controlled by conscious effort.
    b. Cells or muscle fibers are long and threadlike with alternating light and dark cross-markings.
    c. Muscle fibers contract when stimulated by nerve action, then relax immediately.
3. Smooth muscle tissue
    a. This tissue is found in walls of hollow internal organs.
    b. Usually it is controlled by involuntary activity.
4. Cardiac muscle tissue
    a. This tissue is found only in the heart.
    b. Cells are joined by intercalated disks and arranged in branched networks.
    c. Cardiac muscle tissue is controlled by involuntary activity.

## Nervous Tissues (page 155)

1. Nervous tissue is found in the brain, spinal cord, and peripheral nerves.
2. Neurons
    a. Neurons are sensitive to changes and respond by transmitting nerve impulses to other neurons or to muscles or glands.
    b. They coordinate, regulate, and integrate body activities.
3. Neuroglial cells
    a. Some forms of these cells bind and support nervous tissue.
    b. Others carry on phagocytosis.
    c. Still others connect neurons to blood vessels.
    d. Some are involved in cell-to-cell communication.

## Critical Thinking Questions

1. Joints such as the elbow, shoulder, and knee contain considerable amounts of cartilage and fibrous connective tissue. How does this relate to the fact that joint injuries are often very slow to heal?
2. Disorders of collagen are characterized by deterioration of connective tissues. Why would you expect such diseases to produce widely varying symptoms?
3. Sometimes, in response to irritants, mucous cells secrete excess mucus. What symptoms might this produce if it occurred in (a) the respiratory passageways; (b) the digestive tract?
4. As a result of continued and prolonged irritation by tobacco smoke, stratified squamous epithelium may replace the pseudostratified columnar epithelium that forms the inner lining of the upper respiratory tract. What might be the consequences of this tissue change?

## Review Exercises

1. Define *tissue.*
2. Name the four major types of tissue found in the human body.
3. Describe the general characteristics of epithelial tissues.
4. Distinguish between simple epithelium and stratified epithelium.
5. Explain how the structure of simple squamous epithelium is related to its function.
6. Name an organ in which each of the following tissues is found, and give the function of the tissue in each case:
   a. Simple squamous epithelium
   b. Simple cuboidal epithelium
   c. Simple columnar epithelium
   d. Pseudostratified columnar epithelium
   e. Stratified squamous epithelium
   f. Stratified cuboidal epithelium
   g. Stratified columnar epithelium
   h. Transitional epithelium
7. Define *gland.*
8. Distinguish between an exocrine gland and an endocrine gland.
9. Explain how glands are classified according to the structure of their ducts and the arrangement of their cells.
10. Explain how glands are classified according to the function and the nature of their cellular secretions.
11. Distinguish between a serous cell and a mucous cell.
12. Describe the general characteristics of connective tissue.
13. Describe three major types of connective tissue cells.
14. Distinguish between collagen and elastin.
15. Explain the difference between loose fibrous connective tissue and dense fibrous connective tissue.
16. Explain how the quantity of adipose tissue in the body is related to diet.
17. Distinguish between regular and irregular dense fibrous connective tissues.
18. Distinguish between elastic and reticular connective tissues.
19. Explain why injured fibrous connective tissue and cartilage are usually slow to heal.
20. Name the major types of cartilage, and describe their differences and similarities.
21. Describe how bone cells are arranged in bone tissue.
22. Explain how nutrients are supplied to bone cells.
23. Describe the solid components of blood.
24. Define *reticuloendothelial tissue.*
25. Describe the general characteristics of muscle tissues.
26. Distinguish between skeletal, smooth, and cardiac muscle tissues.
27. Describe the general characteristics of nervous tissue.
28. Distinguish between neurons and neuroglial cells.

# SKIN AND THE INTEGUMENTARY SYSTEM

## CHAPTER OBJECTIVES

AFTER YOU HAVE STUDIED THIS CHAPTER, YOU SHOULD BE ABLE TO:

1. Describe the four major types of membranes.
2. Describe the structure of the various layers of the skin.
3. List the general functions of each layer of the skin.
4. Describe the accessory organs associated with the skin.
5. Explain the functions of each accessory organ.
6. Explain how the skin regulates body temperature.
7. Summarize the factors that determine skin color.

A GLIMPSE OF FACES ON A NEW YORK CITY STREET REVEALS SOME OF THE MANY HUES OF HUMANITY.

## KEY TERMS

**conduction** (kon-duk′shun)
**convection** (kon-vek′shun)
**cutaneous membrane** (ku-ta′ne-us mem′brān)
**dermis** (der′mis)
**epidermis** (ep″i-der′mis)
**evaporation** (e-vap″o-ra′shun)
**hair follicle** (hār fol′i-kl)
**integumentary** (in-teg-u-men′tar-e)
**keratinization** (ker″ah-tin″ĭ-za′shun)
**melanin** (mel′ah-nin)
**mucous membrane** (mu′kus mem′brān)
**sebaceous gland** (se-ba′shus gland)
**serous membrane** (se′rus mem′brān)
**subcutaneous** (sub″ku-ta′ne-us)
**sweat gland** (swet gland)

On a muggy July afternoon in 1973, 26-year-old Donald C. and his father were appraising land for their real estate business. They parked near a propane gas transmission line, unaware that the line had a leak. When they finished the job, they returned to their car, and Donald's father started the ignition. The gas line blew, turning the car instantly into a huge fireball. Donald's father was fortunate, in a sense, that he died quickly. For Donald, who sustained serious burns over 68% of his body, the remainder of his life would be a hellish nightmare of agony and utter frustration. His case was written up in bioethics textbooks because he continually demanded to be allowed to die, so great was his pain.

In the weeks following the accident, Donald had many skin grafts, the remainder of his right eye was removed and the left one sewn shut to prevent infection, and his charred fingertips were amputated. But the worst part of the treatment was the daily dunkings in a tank, also to avoid infection. These seemingly drastic measures to keep his body free from infection present a painful example of the role of the integumentary system—the skin—in protecting the body.

We will return to burns at the end of the chapter. As for Donald, after many months of treatment and counseling, he was persuaded to undergo yet more surgery to restore some function to his hands and eye. He was eventually able to live at home with his mother, although his life never even approached normalcy.

---

Two or more kinds of tissues grouped together and performing specialized functions constitute an **organ.** Thus membranes, which are thin, sheetlike structures that are usually composed of epithelium and connective tissue and covering body surfaces and lining body cavities, are organs (fig. 6.1). The cutaneous membrane, together with various accessory organs, makes up the **integumentary organ system.**

## Types of Membranes

The four major types of membranes are *serous, mucous, synovial,* and *cutaneous.* Usually these structures are relatively thin. Serous, mucous, and cutaneous membranes are composed of epithelial tissue and some underlying connective tissue; synovial membranes are composed entirely of connective tissues.

**Serous membranes** line the body cavities that lack openings to the outside. They form the inner linings of the thorax and abdomen, and they cover the organs within these cavities (see fig. 1.9). A serous membrane consists of a layer of simple squamous epithelium (mesothelium) and a thin layer of loose connective tissue. Cells of a serous membrane secrete watery *serous fluid,* which helps lubricate the surfaces of the membrane.

**Mucous membranes** line the cavities and tubes that open to the outside of the body. These include the oral and nasal cavities, and the tubes of the digestive, respiratory, urinary, and reproductive systems. A mucous membrane consists of epithelium overlying a layer of loose connective tissue; however, the type of epithelium varies with the location of the membrane. For example, stratified squamous epithelium lines the oral cavity, pseudostratified columnar epithelium lines part of the nasal cavity, and simple columnar epithelium lines the small intestine. Specialized cells within a mucous membrane secrete *mucus.*

**Synovial membranes** form the inner linings of joint cavities between the ends of bones at freely movable joints (synovial joints). These membranes usually include fibrous connective tissue overlying loose connective tissue and adipose tissue. Cells of a synovial membrane secrete a thick, colorless *synovial fluid* into the joint cavity, which lubricates the ends of the bones within the joint. (Synovial joints are described in chapter 8.)

The **cutaneous membrane** is an organ of the integumentary organ system and is more commonly called *skin.* It is described in detail in this chapter.

## Skin and Its Tissues

The skin is one of the larger and more versatile organs of the body, and it is vital in maintaining homeostasis. The case of Donald C. vividly illustrates the role of the skin as a protective covering that prevents many harmful substances, including microorganisms, from entering the body. It also retards water loss by diffusion from deeper tissues. Skin helps regulate body temperature, houses sensory receptors, contains immune system cells, synthesizes various chemicals, and excretes small quantities of waste substances.

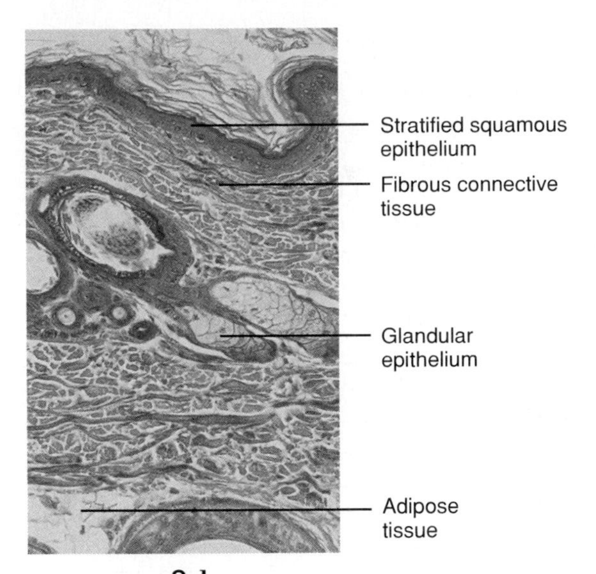


**FIGURE 6.1**
An organ, such as the skin, is composed of several kinds of tissues (30×).

The skin includes two distinct tissue layers. The outer layer, called the **epidermis,** is composed of stratified squamous epithelium. The inner layer, or **dermis,** is thicker than the epidermis, and it contains fibrous connective tissue, epithelial tissue, smooth muscle tissue, nervous tissue, and blood. A basement membrane that is anchored to the dermis by short fibrils separates the two skin layers.

An inherited condition called *epidermolysis bullosa* (EB) destroys the vital integrity of the skin's layered organization. Symptoms include very easy blistering and scarring. Different types of EB reflect the specific proteins affected. In the severe dystrophic form, the collagen fibers that anchor the dermis to the epidermis are abnormal, causing the layers to separate, forming many blisters. In EB simplex, blisters form only on the hands and feet, and usually only during warm weather. EB simplex is an abnormality in the protein keratin, which is found in epidermal cells. The basement membrane form of EB is so severe that it causes death in infancy. It is a defect in epiligrin, a protein that anchors the epidermis to the basement membrane.

Beneath the dermis, masses of loose connective and adipose tissues bind the skin to underlying organs. These tissues form the **subcutaneous layer** or **hypodermis** (fig. 6.2).

1. Name the four types of membranes, and explain how they differ.
2. List the general functions of the skin.
3. Name the tissue in the outer layer of the skin.
4. Name the tissues in the inner layer of the skin.

## Epidermis

Since the epidermis is composed of stratified squamous epithelium, it lacks blood vessels. However, the deepest layer of epidermal cells, called the stratum basale, is close to the dermis, and is nourished by dermal blood vessels. These cells can divide and grow. As the newly formed cells enlarge, they push the older epidermal cells away from the dermis toward the surface of the skin. The farther the cells travel, the poorer their nutrient supply becomes, and, in time, they die.

In psoriasis, a chronic skin disease, cells in the epidermis divide seven times more frequently than normal. Excess cells accumulate, forming bright red patches covered with silvery scales, which are keratinized cells. The anti-cancer drug methotrexate is sometimes used to treat severe cases. Five million people in the United States and 2% of all people worldwide have psoriasis.

The cell membranes of the older cells (keratinocytes) thicken and develop many desmosomes that fasten them to adjacent cells (see chapter 3). At the same time, the cells begin to harden (keratinize), with strands of tough, fibrous, waterproof keratin protein being synthesized and stored within the cell. As a result, many layers of tough, tightly packed dead cells accumulate in the outer epidermis. This outermost layer is called the *stratum corneum.* The dead cells that compose it are often rubbed away, for example, when the skin is rubbed briskly with a towel.

The structural organization of the epidermis varies from region to region. It is thickest on the palms of the hands and the soles of the feet, where it may be 0.8–1.4 mm thick. In most areas, only four layers can be distinguished. They are the *stratum basale* (stratum germinativum, or basal cell layer), which is the deepest layer; the *stratum spinosum,* a relatively thick layer; the *stratum granulosum,* a granular layer; and the *stratum corneum,* a fully keratinized layer (horny layer). An additional layer, the *stratum lucidum* (between the stratum granulosum and the stratum corneum) is in the thickened skin of the palms and soles. The cells of these layers change in shape as they are pushed toward the surface (fig. 6.3).

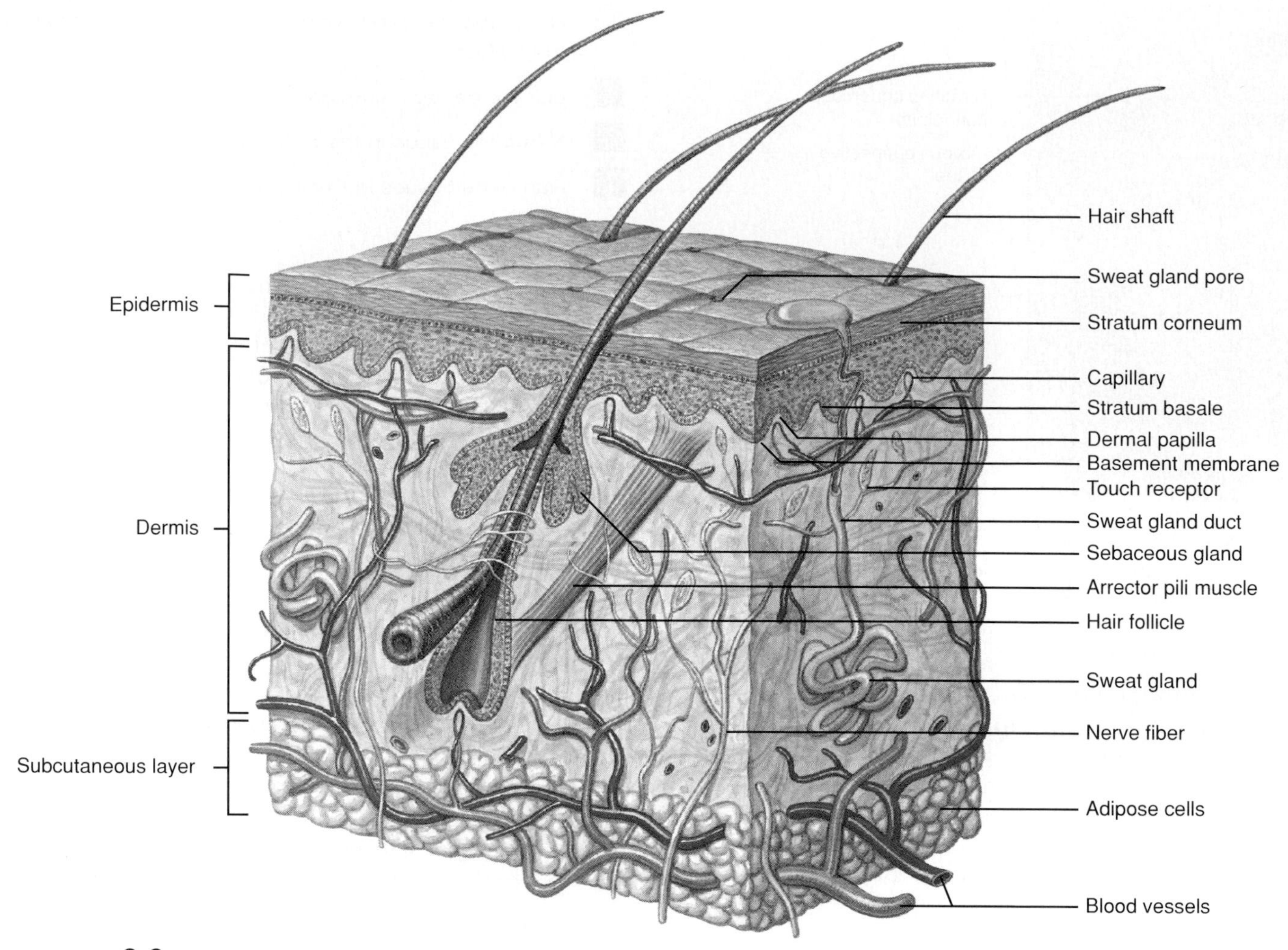


**FIGURE 6.2**
A section of skin.

Contact dermatitis is superficial inflammation (redness and swelling) or irritation of the skin. In allergic contact dermatitis, the immune system reacts to an allergen (an innocuous substance recognized as foreign), causing a red scaliness. The rash resulting from exposure to oils in poison ivy is an example of allergic contact dermatitis; 50 to 70% of people with this allergy also react to poison oak and sumac, mango peel, gingko fruit, and an oil in cashew shells. Metals in jewelry, acids in fruits, and materials in shoes also trigger allergic contact dermatitis. It is also seen among workers in certain fields, such as hairdressers, butchers, furniture makers, shrimp peelers, and bakers.

Irritant contact dermatitis is damage caused by an irritating substance, not an immune system reaction (allergy). The skin becomes red and itchy, with small, oozing blisters. Babies are famous for skin irritations—caused by everything from the perpetual drool on their faces to their wet diapers. "Dishpan hands" and reactions to cosmetics are common examples of irritant contact dermatitis. Men with outbreaks on the left sides of their necks may use an aftershave lotion that reacts with sunlight when they drive.

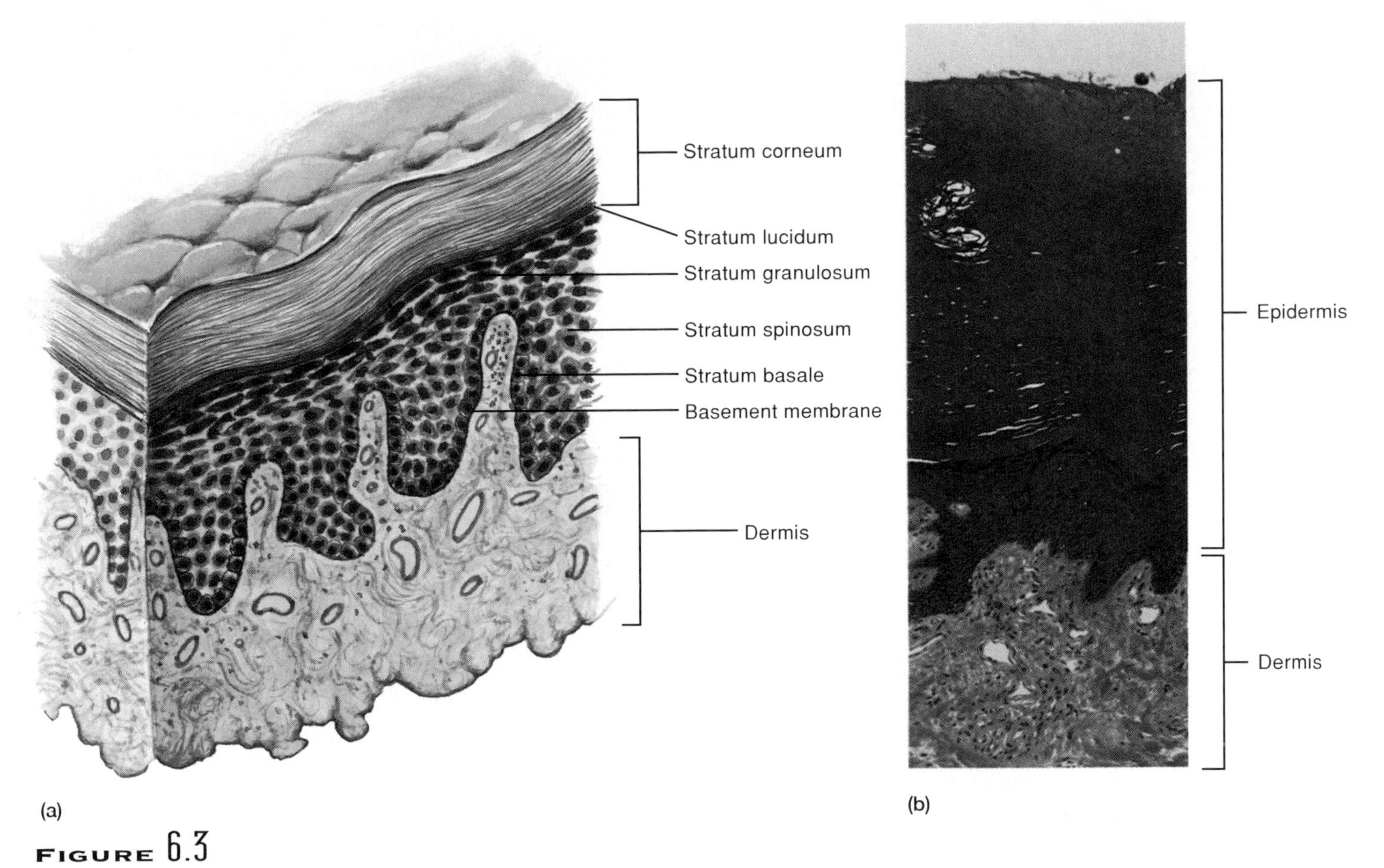


**FIGURE 6.3**
(*a*) The various layers of the epidermis are characterized by changes that occur in cells as they are pushed toward the surface of the skin. (*b*) Micrograph from the palm of the hand (50×).

In other body regions, the epidermis is usually much thinner, averaging 0.07–0.12 mm. The stratum granulosum may be missing where the epidermis is thin. Chart 6.1 describes the characteristics of each layer of the epidermis.

In healthy skin, production of epidermal cells is closely balanced with loss of dead cells from the stratum corneum, so that skin seldom wears away completely. In fact, the rate of cellular reproduction increases where the skin is rubbed or pressed regularly, causing growth of thickened areas called *calluses* on the palms and soles, and development of horny, conical masses called *corns* on the toes when poorly fitting shoes rub the skin. Other changes in the skin include the common rashes described in chart 6.2.

Specialized cells in the epidermis called **melanocytes** produce the dark pigment melanin (fig. 6.4). Melanin absorbs light energy and in this way helps protect still deeper cells from the damaging effects of ultraviolet rays of sunlight.

Because blood vessels in the dermis supply nutrients to the epidermis, any interference with blood flow may kill epidermal cells. For example, when a person lies in one position for a prolonged period, the weight of the body pressing against the bed interferes with the skin's blood supply. If cells die, the tissues begin to break down (necrosis), and a pressure ulcer (also called a decubitus ulcer or bedsore) may appear.

Pressure ulcers usually occur in the skin overlying bony projections, such as on the hip, heel, elbow, or shoulder. Changing the body position frequently or massaging the skin to stimulate blood flow in regions associated with bony prominences can prevent ulcers. In the case of a paralyzed person who cannot feel pressure or respond to it by shifting position, caretakers must turn the body often to prevent pressure ulcers.

## CHART 6.1 Layers of the Epidermis

| Layer | Location | Characteristics |
|---|---|---|
| Stratum corneum (horny layer) | Outermost layer | Many layers of keratinized, dead epithelial cells that are flattened and nonnucleated |
| Stratum lucidum | Between corneum and granulosum on soles and palms | Layer appears clear; nuclei, organelles, and cell membranes are no longer visible |
| Stratum granulosum | Beneath the stratum corneum | Three to five layers of flattened granular cells that contain shrunken fibers of keratin and shriveled nuclei |
| Stratum spinosum | Beneath the stratum granulosum | Many layers of cells with centrally located, large, oval nuclei and developing fibers of keratin; cells becoming flattened |
| Stratum basale (basal cell layer) | Deepest layer | A single row of cuboidal or columnar cells that undergo mitosis and grow; this layer also includes pigment-producing melanocytes |

Melanocytes lie in the deepest portion of the epidermis and in the underlying connective tissue of the dermis. Although they are the only cells that can produce melanin, the pigment also may be present in nearby epidermal cells. This happens because melanocytes have long, pigment-containing cellular extensions that pass upward between neighboring epidermal cells, and the extensions can transfer granules of melanin into these other cells by a process called *cytocrine secretion.* Nearby epidermal cells may contain more melanin than the melanocytes (fig. 6.5).

Clinical Application 6.1 discusses one consequence of excessive sun exposure—skin cancer.

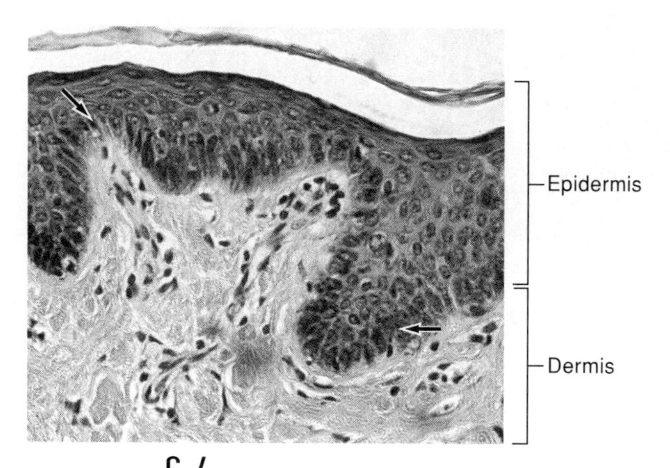


**FIGURE 6.4**
Melanocytes (arrows) that occur mainly in the stratum basale at the deepest layer of the epidermis produce the pigment melanin (160×).

1. Explain how the epidermis is formed.
2. What factors help prevent loss of body fluids through the skin?
3. What is the function of melanin?

### Dermis

The surface between the epidermis and dermis is usually uneven, because the epidermis has ridges projecting inward and the dermis has fingerlike *papillae* passing into the spaces between the ridges (see fig. 6.2).

The dermis binds the epidermis to the underlying tissues. It is composed largely of irregular dense fibrous connective tissue that includes tough collagenous fibers and elastic fibers surrounded by a gel-like ground substance. Networks of these fibers give the skin toughness and elasticity. On the average, the dermis is 1.0–2.0 mm thick; however, it may be as thin as 0.5 mm or less on the eyelids, or as thick as 3.0 mm on the soles of the feet.

Fingerprints are due to the cutaneous ridges of the dermal portion of the skin seen at the distal end of the palmar surface of a finger. Fingerprints are used for purposes of identification, because they are individually unique. Even identical twins do not have identical fingerprints. Fingerprint pattern is genetically determined, and the prints form during fetal existence. However, during a certain time early in development, when the fetus moves, the print pattern can change. Because no two fetuses move exactly alike, fingerprints of identical twins are different.

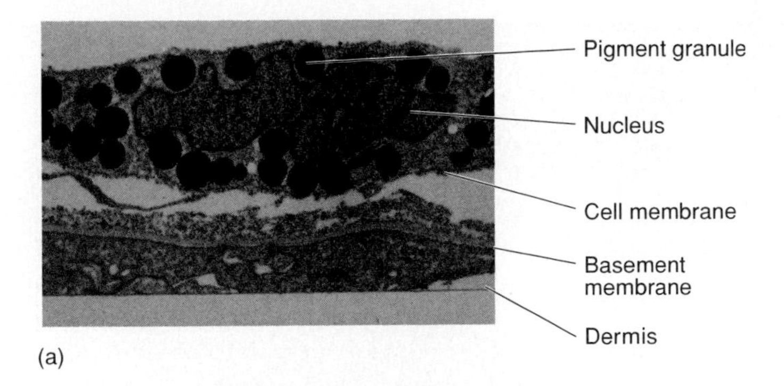


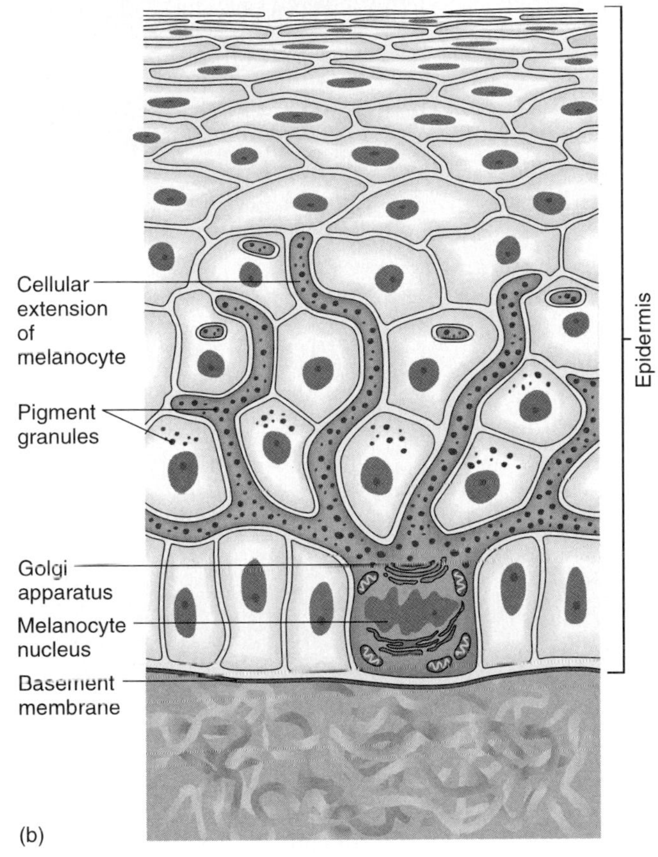


**FIGURE 6.5**

(*a*) Transmission electron micrograph of a melanocyte with pigment-containing granules (10,000×). (*b*) A melanocyte may have pigment-containing extensions that pass between neighboring epidermal cells.

> Aging skin changes, particularly in regions exposed to sunlight. The rate and degree of change vary from person to person. Typically, the skin becomes drier, scaly, thinner, and more translucent. Connective tissue fibers degenerate, leading to loss of elasticity and tensile strength, followed by wrinkling.

The dermis also contains muscle fibers. Some regions, such as the skin that encloses the testes (scrotum), contain numerous smooth muscle cells that can wrinkle the skin when they contract. Other smooth muscles in the dermis are associated with accessory organs such as hair follicles and glands. Many striated muscle fibers are anchored to the dermis in the skin of the face. They help produce the voluntary movements associated with facial expressions.

Nerve fibers are scattered throughout the dermis. Motor fibers carry impulses to dermal muscles and glands, and sensory fibers carry impulses away from specialized sensory receptors, such as touch receptors (see fig. 6.2).

One type of dermal sensory receptor (Pacinian corpuscles) is stimulated by heavy pressure, while another type (Meissner's corpuscles) is sensitive to light touch. Still other receptors are stimulated by temperature changes or by factors that can damage tissues. Sensory receptors are discussed in chapter 12.

The dermis also contains blood vessels, hair follicles, sebaceous glands, and sweat glands, which will be discussed later in the chapter.

## Subcutaneous Layer

The subcutaneous layer (hypodermis) beneath the dermis consists of loose fibrous connective and adipose tissues (see fig. 6.2). The collagenous and elastic fibers of this layer are continuous with those of the dermis. Most of these fibers run parallel to the surface of the skin, traveling in all directions. As a result, no sharp boundary separates the dermis and the subcutaneous layer.

The adipose tissue of the subcutaneous layer insulates, helping to conserve body heat and impeding the entrance of heat from the outside. The amount of adipose tissue varies greatly with each individual's nutritional condition. It also varies in thickness from one region to another. For example, adipose tissue is usually thick over the abdomen, but absent altogether in the eyelids.

The subcutaneous layer also contains the major blood vessels that supply the skin. Branches of these vessels form a network (rete cutaneum) between the dermis and the subcutaneous layer. They, in turn, give off smaller vessels that supply the dermis above and the underlying adipose tissue.

Clinical Application 6.1

## Skin Cancer

Like cigarette smoking, a deep, dark tan was once very desirable. A generation ago, a teenager might have spent hours on a beach, skin glistening with oil, maybe even using a reflecting device to concentrate sun exposure on the face. Today, as they lather on sunblock, many of these people realize that the tans of yesterday may cause cancer tomorrow.

Just four hours of unprotected sunbathing has immediate and lasting effects on the skin. The ultraviolet radiation alters collagen and elastin and dilates blood vessels in the dermis. A few days later, the outer skin layer may blister and peel. Cells that peel off have undergone programmed cell death, a protective mechanism to rid the body of cells possibly turned cancerous from environmental exposure. People with the inherited disorder xeroderma pigmentosum lack DNA repair enzymes, and can never venture outside without being completely covered by clothing or sunblock. Even the briefest exposure can provoke a sea of freckles to bloom (fig. 6A), some of which may become cancerous.

Skin cancer usually arises from nonpigmented epithelial cells within the deep layer of the epidermis or from pigmented melanocytes. Skin cancers originating from epithelial cells are called *cutaneous carcinomas* (basal cell carcinoma or squamous cell carcinoma); those arising from melanocytes are *cutaneous melanomas* (melanocarcinomas or malignant melanomas) (fig. 6B).

Cutaneous carcinomas are the most common type of skin cancer, occurring most frequently in light-skinned people over forty years of age. These cancers usually appear in persons who are exposed to sunlight regularly, such as farmers, sailors, athletes, and sunbathers.

Cutaneous carcinoma often develops from hard, dry, scaly growths (lesions) that have reddish bases. Such lesions may be either flat or raised, and they adhere firmly to the skin, appearing most often on the neck, face, and scalp. Fortunately, cutaneous carcinomas are typically slow growing and can usually be cured completely by surgical removal or radiation treatment.

Cutaneous melanomas are pigmented with melanin, often with a variety of colored areas—variegated brown, black, gray, or blue—arranged haphazardly. They usually have irregular rather than smooth outlines (fig. 6B).

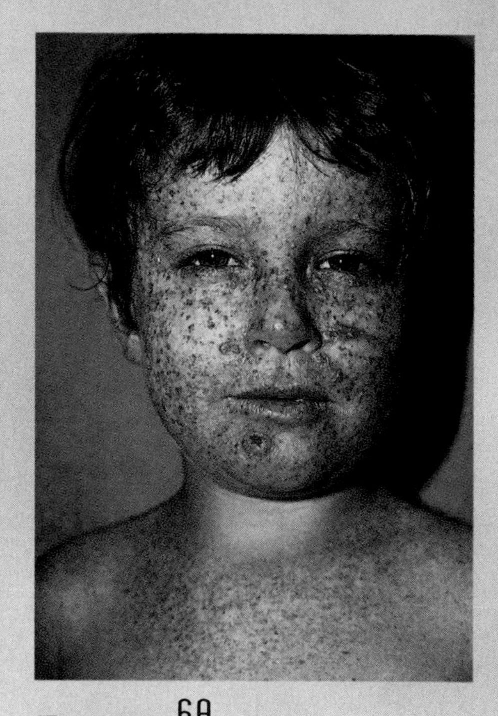

**Figure 6A**
This child has xeroderma pigmentosum. Sun exposure causes extreme freckling, and skin cancer is likely to develop because he lacks DNA repair enzymes. The large lesion on his chin is a skin cancer.

*Subcutaneous injections* are administered through a hollow needle into the subcutaneous layer beneath the skin. *Intradermal injections* are injected into tissues within the skin. Subcutaneous injections and *intramuscular injections* (administered into muscles) are sometimes called hypodermic injections.

Some substances are administered through the skin by means of an adhesive transdermal patch that includes a small reservoir containing a certain quantity of a drug. The drug passes from the reservoir through a permeable membrane at a known rate. It then diffuses into the epidermis and enters the blood vessels of the dermis. Transdermal patches are commonly used to protect against motion sickness, chest pain associated with heart disease, and elevated blood pressure. A transdermal patch that delivers nicotine is used to help people stop smoking.

Cutaneous melanomas may appear in young adults as well as in older ones, and seem to be caused by short, intermittent exposure to high-intensity sunlight. Thus, risk of melanoma increases in persons who stay indoors but occasionally sustain blistering sunburns.

Cutaneous melanomas occur most often in light-skinned people who burn rather than tan. The cancer usually appears in the skin of the trunk, especially the back, or the limbs, arising from normal-appearing skin or from a mole (nevus).

The lesion spreads through the skin horizontally, but eventually may thicken and grow downward into the skin, invading deeper tissues. If the melanoma is removed surgically while it is in its horizontal growth phase, it may be arrested. Once it thickens and spreads into deeper tissues, unfortunately, it becomes difficult to treat. The survival rate with conventional treatment is very low. However, genetic researchers are excited about a form of gene therapy in which genes injected directly into melanoma cells direct the cells to produce surface proteins that attract cancer-fighting biochemicals produced by cells of the immune system.

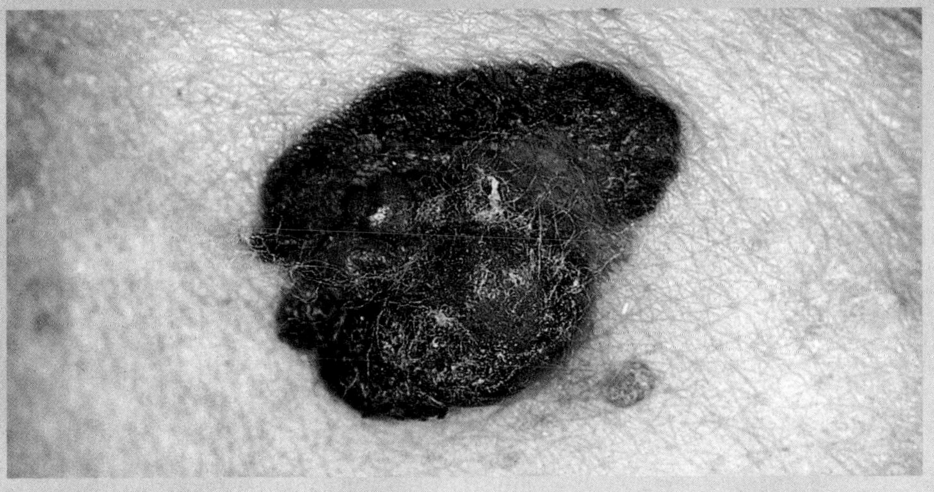

**FIGURE 6B**
A cutaneous malignant melanoma.

For reasons that are not well understood, the incidence of melanoma within the U.S. population has been increasing rapidly for the past twenty years. To reduce the chances of occurrence, avoid exposing the skin to high-intensity sunlight, use sunscreens and sunblocks, and examine the skin regularly. Report any unusual lesions—particularly those that change in color, shape, or surface texture—to a physician.

Replacements for natural suntanning may be ineffective or dangerous. "Sunless tanning agents" do not cause the skin to tan at all, but merely dye it temporarily. "Tan accelerators," according to manufacturers, are nutrients that supposedly increase melanin synthesis in the sunlight, but there is no scientific proof that they work. Tanning booths may be dangerous. Even those claiming to be safe because they emit only partial ultraviolet radiation may cause skin cancer.

Enjoy the sun—but protect yourself!

1. What kinds of tissues make up the dermis?
2. What are the functions of these tissues?
3. What are the functions of the subcutaneous layer?

## ACCESSORY ORGANS OF THE SKIN

Accessory organs of the skin extend downward from the epidermis and include hair follicles, nails, and skin glands. As long as accessory organs remain intact, severely burned or injured dermis can regenerate.

## CHART 6.2 RASHES

| Illness | Description of Rash | Cause |
|---|---|---|
| Chicken pox | Tiny pustules start on back, chest, or scalp and spread for 3 to 4 days. Pustules form blisters, then crusts, then fall away. | *Herpes varicella* |
| Fifth disease | Begins with "slapped cheek" appearance, then red spots suddenly cover entire body, lasting up to 2 days. | *Human parvovirus B19* |
| Impetigo | Thin-walled blisters and thick, crusted lesions. | *Staphylococcus aureus* *Streptococcus pyogenes* |
| Lyme disease | Large rash resembling a bull's-eye usually on thighs or trunk. | *Borrelia burgdorferi* |
| Rosacea | Flushing leads to sunburned appearance in center of face. Red pimples and then wavy red lines develop. | Unknown, but may be a microscopic mite living in hair follicles |
| Roseola infantum | Following high fever, red spots suddenly cover entire body, lasting up to 2 days. | *Herpesvirus 6* |
| Scarlet fever | Rash resembling sunburn with goose bumps begins below ears, on chest and underarms, and spreads to abdomen, limbs, and face. Skin may peel. | *Group A streptococcus* |
| Shingles | Small, clear blisters appear on inflamed skin. Blisters enlarge, become cloudy, crust, then fall off. | The virus that causes chicken pox hides in peripheral nerves, affecting the skin in a path following the nerves. |

## Hair Follicles

Hair is present on all skin surfaces except the palms, soles, lips, nipples, and parts of the external reproductive organs; however, it is not always well developed. For example, hair on the forehead is usually very fine.

Each hair develops from a group of epidermal cells at the base of a tubelike depression called a **hair follicle.** This follicle extends from the surface into the dermis and contains the hair *root.* The epidermal cells at its base are nourished from dermal blood vessels in a projection of connective tissue (hair papilla) at the deep end of the follicle. As these epidermal cells divide and grow, older cells are pushed toward the surface. The cells that move upward and away from the nutrient supply become keratinized and die. Their remains constitute the structure of a developing hair, whose *shaft* extends away from the skin surface. In other words, a hair is composed of dead epidermal cells (figs. 6.6 and 6.7).

> *Folliculitis* is an inflammation of the hair follicles in response to bacterial infection. The condition can be picked up in dirty swimming pools or hot tubs. One woman got a severe case by repeatedly using a loofah sponge containing bacteria.

Usually a hair grows for a time and then rests while it remains anchored in its follicle. Later a new hair begins to grow from the base of the follicle, and the old hair is pushed outward and drops off. Sometimes, however, the hairs are not replaced. When this occurs in the scalp, the result is baldness, described in Clinical Application 6.2.

Genes directing the type and amount of pigment produced by epidermal melanocytes determine hair color. For example, dark hair has much more melanin than blond hair. The white hair of a person with albinism lacks melanin altogether. Bright red hair contains an iron pigment (trichosiderin) that does not occur in hair of any other color. A mixture of pigmented and unpigmented hair usually appears gray.

A bundle of smooth muscle cells, forming the *arrector pili muscle* (see fig. 6.2), is attached to each hair follicle. This muscle is positioned so that a short hair within the follicle stands on end when the muscle contracts. If a person is emotionally upset or very cold, nerve impulses may stimulate the arrector pili muscles to contract, causing gooseflesh or goose bumps. Each hair follicle also has associated with it one or more sebaceous (oil-producing) glands, discussed later in the chapter.

> Some interesting hair characteristics are inherited. The direction of a cowlick is inherited, with a clockwise whorl being more common than a counterclockwise whorl. A white forelock, and hairy ears, elbows, nose tip, or palms are also inherited.

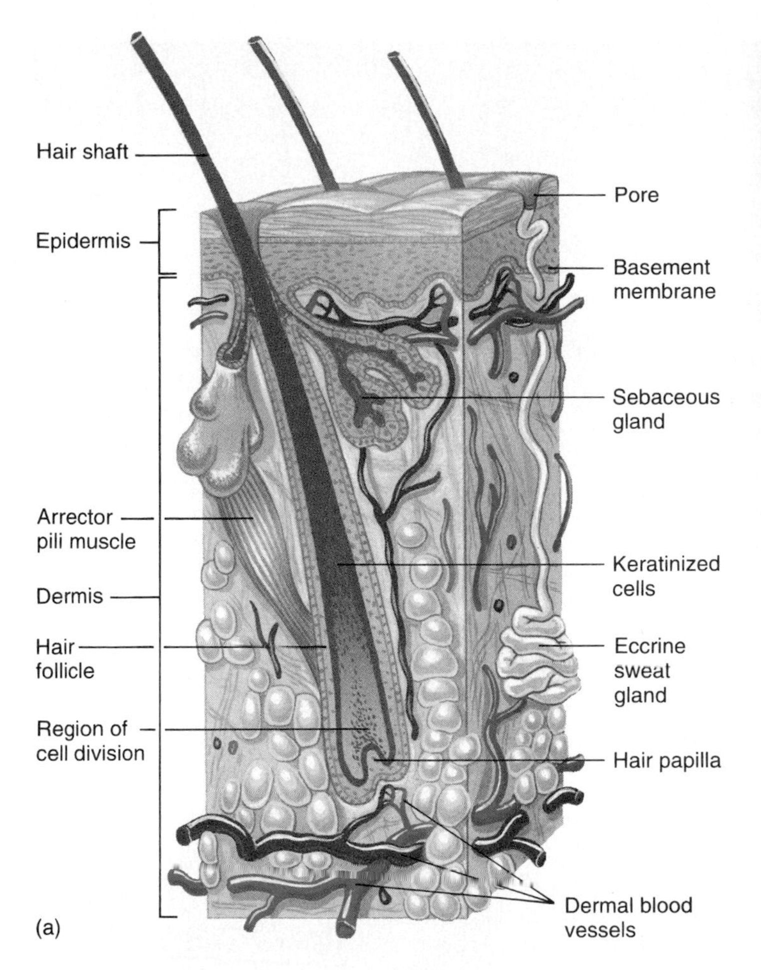


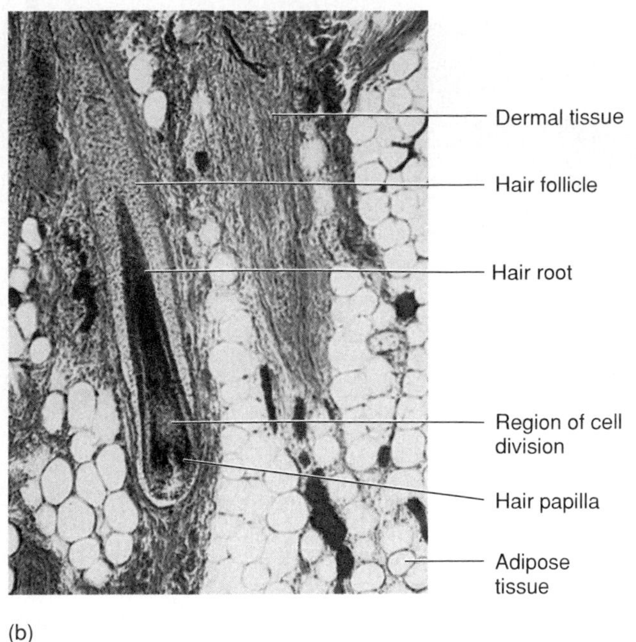


**Figure 6.6**
(*a*) A hair grows from the base of a hair follicle when epidermal cells divide and older cells move outward and become keratinized. (*b*) A light micrograph of a hair follicle (160×).

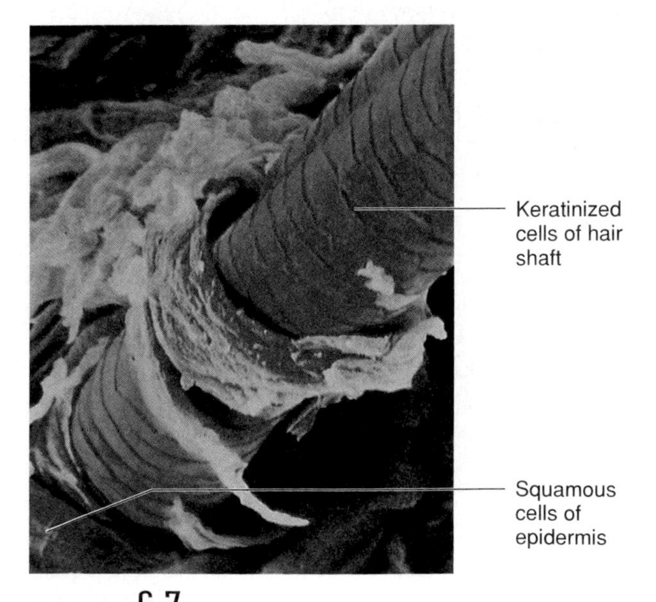


**Figure 6.7**
A scanning electron micrograph of a hair emerging from the epidermis (340×).

## Nails

**Nails** are protective coverings on the ends of the fingers and toes. Each nail consists of a *nail plate* that overlies a surface of skin called the *nail bed.* The nail plate is produced by specialized epithelial cells that are continuous with the epithelium of the skin. The whitish, thickened, half-moon-shaped region (lunula)

> Nail appearance mirrors health. Bluish nail beds may reflect a circulatory problem. A white nail bed or oval depressions in a nail can indicate anemia. A pigmented spot under a nail that isn't caused by an injury may be a melanoma. Horizontal furrows may result from a period of serious illness, or indicate malnutrition. Certain disorders of the lungs, heart, or liver may cause extreme curvature of the nails. Red streaks in noninjured nails may be traced to rheumatoid arthritis, ulcers, or hypertension.

CLINICAL APPLICATION 6.2

## Hair Loss

A healthy person loses from twenty to 100 hairs a day. This is part of the normal growth cycle of hair. A hair typically grows for two to six years, then enters a resting stage for two to three months, after which it falls out. A new hair grows in its place. At any time, 90% of the hairs are in the growth phase.

In the United States, about 57.5 million people have some degree of baldness. Pattern baldness, in which the top of the head loses hair, affects 35 million men and 20 million women. The women tend to be past menopause, when lowered amounts of the hormone estrogen contribute to hair loss, which is more even on the scalp than in men. Pattern baldness is called *androgenic alopecia,* because it is associated with testosterone, an androgenic (male) hormone. About 2.5 million people have an inherited condition called *alopecia areata,* in which the body manufactures antibodies that attack the hair follicles, resulting in oval bald spots in mild cases, to complete loss of scalp and body hair in severe cases.

Various conditions can cause temporary hair loss. Lowered estrogen levels shortly before and after giving birth may cause a woman's hair to come out in clumps. Taking birth control pills, cough medications, certain antibiotics, vitamin A derivatives, antidepressants, and many other medications can also cause temporary hair loss. A sustained high fever may prompt hair loss six weeks to three months later.

Many people losing their hair seek treatment (fig. 6C). One successful treatment is minoxidil (Rogaine), a drug originally used to lower high blood pressure. It was approved as a treatment for baldness in men in 1988 and in women in 1991. Rogaine causes new hair to grow in 10 to 14% of cases, but in 90% of people it slows hair loss. However, when a person stops taking it, any new hair falls out.

Hair transplants move hair follicles from a hairy part of a person's body to a bald part, and they are successful. Several other approaches, however, are potentially damaging—to the wallet as well as the scalp. Suturing on hair pieces often leads to scarring and infection. In 1984, the Food and Drug Administration banned hair implants of high density artificial fibers, because they too become infected easily. Products called "thinning hair supplements" are conditioners, often found in ordinary shampoo, that merely make hair feel thicker. They are generally concoctions of herbs and the carbohydrate polysorbate. Labels claim the product "releases hairs trapped in the scalp."

**FIGURE 6C**
Being bald can be beautiful, but many people with hair loss seek ways to grow hair.

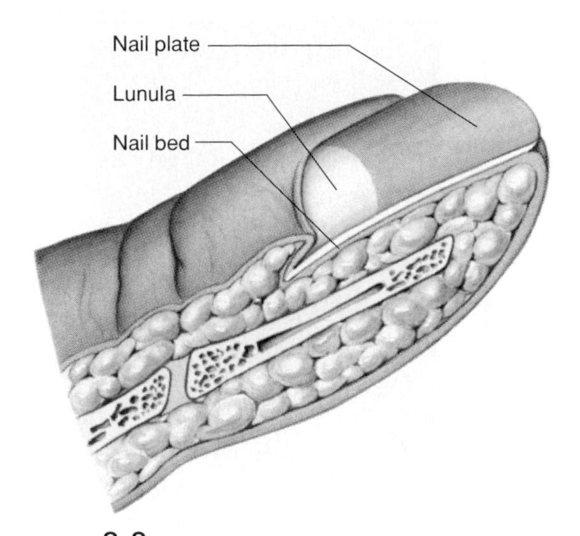


**FIGURE 6.8**
Nails grow from epithelial cells that divide and become keratinized in the lunula.

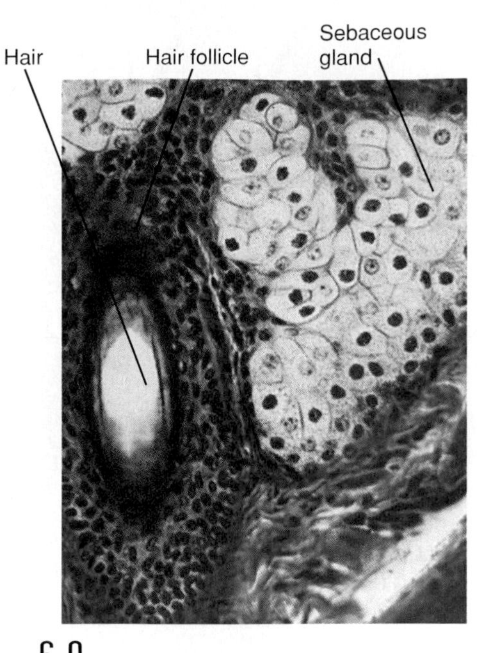


**FIGURE 6.9**
A sebaceous gland secretes sebum into a hair follicle (shown here in cross section: 175×).

at the base of a nail plate is its most active growing region. The epithelial cells in this region reproduce, and the newly formed cells are keratinized. This gives rise to tiny, horny scales that become part of the nail plate, pushing it forward over the nail bed. In time, the plate extends beyond the end of the nail bed and with normal use gradually wears away (fig. 6.8).

## Skin Glands

**Sebaceous glands** (see fig. 6.2) contain groups of specialized epithelial cells and are usually associated with hair follicles. They are holocrine glands (see chapter 5), and their cells produce globules of a fatty material that accumulate, swelling and bursting the cells. The resulting mixture of fatty material and cellular debris is called *sebum.*

Sebum is secreted into hair follicles through short ducts and helps keep the hairs and the skin soft, pliable, and relatively waterproof (fig. 6.9).

Sebaceous glands are scattered throughout the skin, but are not on the palms and soles. In some regions, such as the lips, the corners of the mouth, and parts of the external reproductive organs, sebaceous glands open directly to the surface of the skin rather than being connected to hair follicles.

> Many teens are all too familiar with a disorder of the sebaceous glands called *acne* (acne vulgaris). Overactive and inflamed glands in some body regions become plugged and surrounded by small red elevations containing blackheads (comedones) or pimples (pustules).

**Sweat glands** (sudoriferous glands) are widespread in the skin. Each gland consists of a tiny tube that originates as a ball-shaped coil in the deeper dermis or superficial subcutaneous layer. The coiled portion of the gland is closed at its deep end and is lined with sweat-secreting epithelial cells (see fig. 6.2). The most numerous sweat glands, called **eccrine glands,** respond throughout life to body temperature elevated by environmental heat or physical exercise. These glands are common on the forehead, neck, and back, where they produce profuse sweat on hot days or during intense physical activity. They also cause the moisture that appears on the palms and soles when a person is emotionally stressed.

With advancing age, sweat gland activity falls, and in the very old, fibrous tissues may replace sweat glands. As a result, the person becomes less able to control body temperature. Sebaceous gland activity also lessens with age, so that the skin of elderly people tends to be dry.

The fluid the eccrine (sweat) glands secrete is carried by a tube that opens at the surface as a *pore.* Sweat is mostly water, but it also contains small quantities of salts and wastes, such as urea and uric acid. Thus, sweating is also an excretory function.

The secretions of certain sweat glands, called **apocrine glands,** have a scent. (Although they are currently called apocrine, these glands secrete by the same mechanism as eccrine glands—see chapter 3.) Apocrine sweat glands become active at puberty, and can wet certain areas of the skin when a person is emotionally

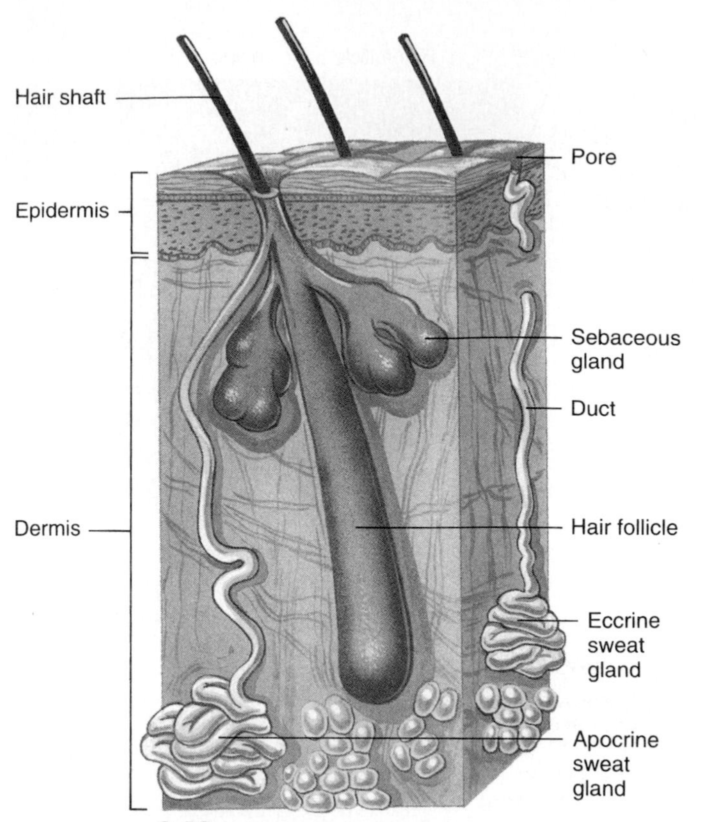


**FIGURE 6.10**
Note the difference in location of the ducts of the eccrine and apocrine sweat glands.

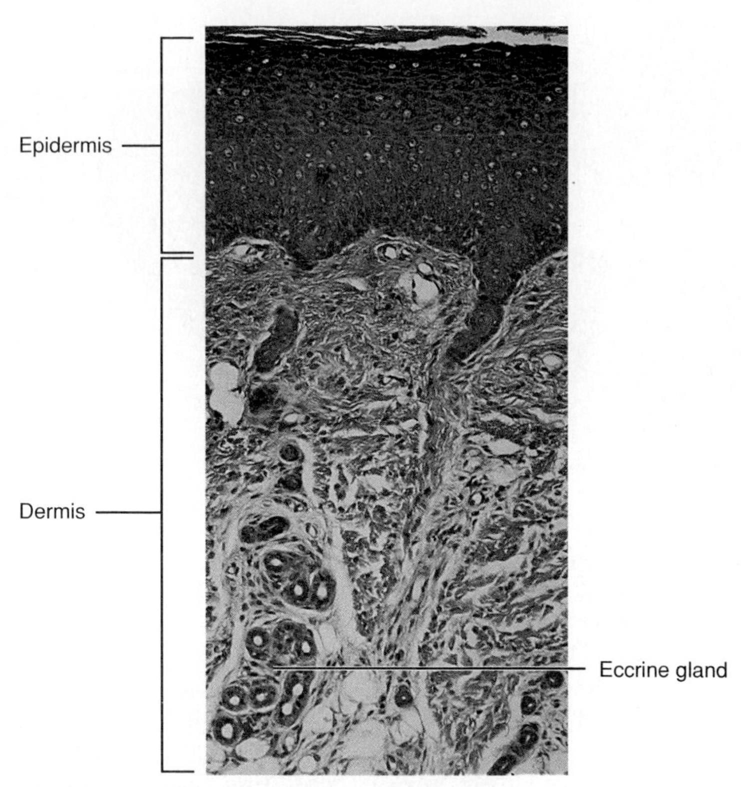


**FIGURE 6.11**
Light micrograph of an eccrine gland (30×).

upset, frightened, or in pain. Apocrine sweat glands are also active during sexual arousal. In adults, the apocrine glands are most numerous in axillary regions, the groin, and the area around the nipples. These glands are usually associated with hair follicles. Figures 6.10 and 6.11 illustrate sweat glands. Other apocrine glands are structurally and functionally modified to secrete specific fluids, such as the ceruminous glands of the external ear canal that secrete ear wax and the female mammary glands that secrete milk (see chapter 22). Chart 6.3 summarizes skin glands.

1. Explain how a hair forms.
2. What causes gooseflesh?
3. What is the function of the sebaceous glands?
4. How does the composition of a fingernail differ from that of a hair?
5. Describe the locations of the sweat glands.
6. How do the functions of eccrine sweat glands and apocrine sweat glands differ?

## Regulation of Body Temperature

The regulation of body temperature is vitally important because even slight shifts in body temperature can disrupt the rates of metabolic reactions.

Normally, the temperature of deeper body parts remains close to 37°C (98.6° F). The maintenance of a stable temperature requires that the amount of heat the body loses be balanced by the amount it produces. The skin plays a key role in the homeostatic mechanism that regulates body temperature.

### Heat Production and Loss

Because cellular metabolism releases heat, the most active cells are the major heat producers. These include skeletal and heart muscle cells, and the cells of certain glands, such as the liver.

In intense heat, nerve impulses stimulate structures in the skin and other organs to release heat. For example, during physical exercise, active muscles release heat, which the blood carries away. The warmed blood

## CHART 6.3 Skin Glands

| Type | Description | Function | Location |
|---|---|---|---|
| Sebaceous glands | Groups of specialized epithelial cells | Keep hair soft, pliable, waterproof | Near or connected to hair follicles, everywhere but on palms and soles |
| Eccrine sweat glands | Abundant sweat glands, with odorless secretion | Lower body temperature | Originate in deep dermis or subcutaneous layer and open to surface on forehead, neck, and back |
| Apocrine sweat glands | Rarer sweat glands with scented secretion | Wet skin during pain, fear, emotional upset, and sexual arousal | Near hair follicles in armpit, groin, near nipples |
| Ceruminous glands | Modified apocrine glands | Secrete earwax | External ear canal |
| Mammary glands | Modified apocrine glands | Secrete milk | Breasts |

reaches the part of the brain that contains the body's temperature set point (the hypothalamus), which signals muscles in the walls of specialized dermal blood vessels to relax. As the vessels dilate (vasodilation), more blood enters them and some of the heat the blood carries escapes to the outside. At the same time, deeper blood vessels contract (vasoconstriction), diverting blood to the surface, and the skin reddens. The heart is stimulated to beat faster, moving more blood out of the deeper regions.

The primary means of body heat loss is **radiation,** by which *infrared heat rays* escape from warmer surfaces to cooler surroundings. These rays radiate in all directions, much like those from the bulb of a heat lamp.

Conduction and convection release lesser amounts of heat. In **conduction,** heat moves from the body directly into the molecules of cooler objects in contact with its surface. For example, heat is lost by conduction into the seat of a chair when a person sits down. The heat loss continues as long as the chair is cooler than the body surface touching it.

Heat is also lost by conduction to the air molecules that contact the body. As air becomes heated, it moves away from the body, carrying heat with it, and is replaced by cooler air moving toward the body. This type of continuous circulation of air over a warm surface is **convection.**

Still another means of body heat loss is **evaporation.** When the body temperature rises above normal, the nervous system stimulates eccrine sweat glands to release sweat onto the surface of the skin. As this fluid evaporates, it carries heat away from the surface, cooling the skin.

With excessive heat loss, the brain triggers a different set of responses in skin structures. Muscles in the walls of dermal blood vessels are stimulated to contract; this decreases the flow of heat-carrying blood through the skin, which tends to lose color, and helps reduce heat loss by radiation, conduction, and convection. At the same time, sweat glands remain inactive, decreasing heat loss by evaporation. If the body temperature continues to drop, the nervous system may stimulate muscle fibers in the skeletal muscles throughout the body to contract slightly. This action requires an increase in the rate of cellular respiration and produces heat as a by-product. If this response does not raise the body temperature to normal, small groups of muscles may contract rhythmically with still greater force, causing the person to shiver, and thus generating more heat.

Figure 6.12 summarizes the body's temperature-regulating mechanism, and Clinical Application 6.3 examines two causes of elevated body temperature.

### Problems in Temperature Regulation

Unfortunately, the body's temperature-regulating mechanism does not always operate satisfactorily, and the consequences may be unpleasant and dangerous. For example, air can hold only a limited amount of water vapor, so that on a hot, humid day the air may become nearly saturated with water. At such times, the sweat glands may be activated, but the sweat cannot quickly evaporate. The skin becomes wet, but the person remains hot and uncomfortable. In addition, if the air temperature is high, heat is lost less effectively by radiation. In fact, if the air temperature is higher than the body temperature, the person may gain heat from the surroundings, and the body temperature may rise even higher.

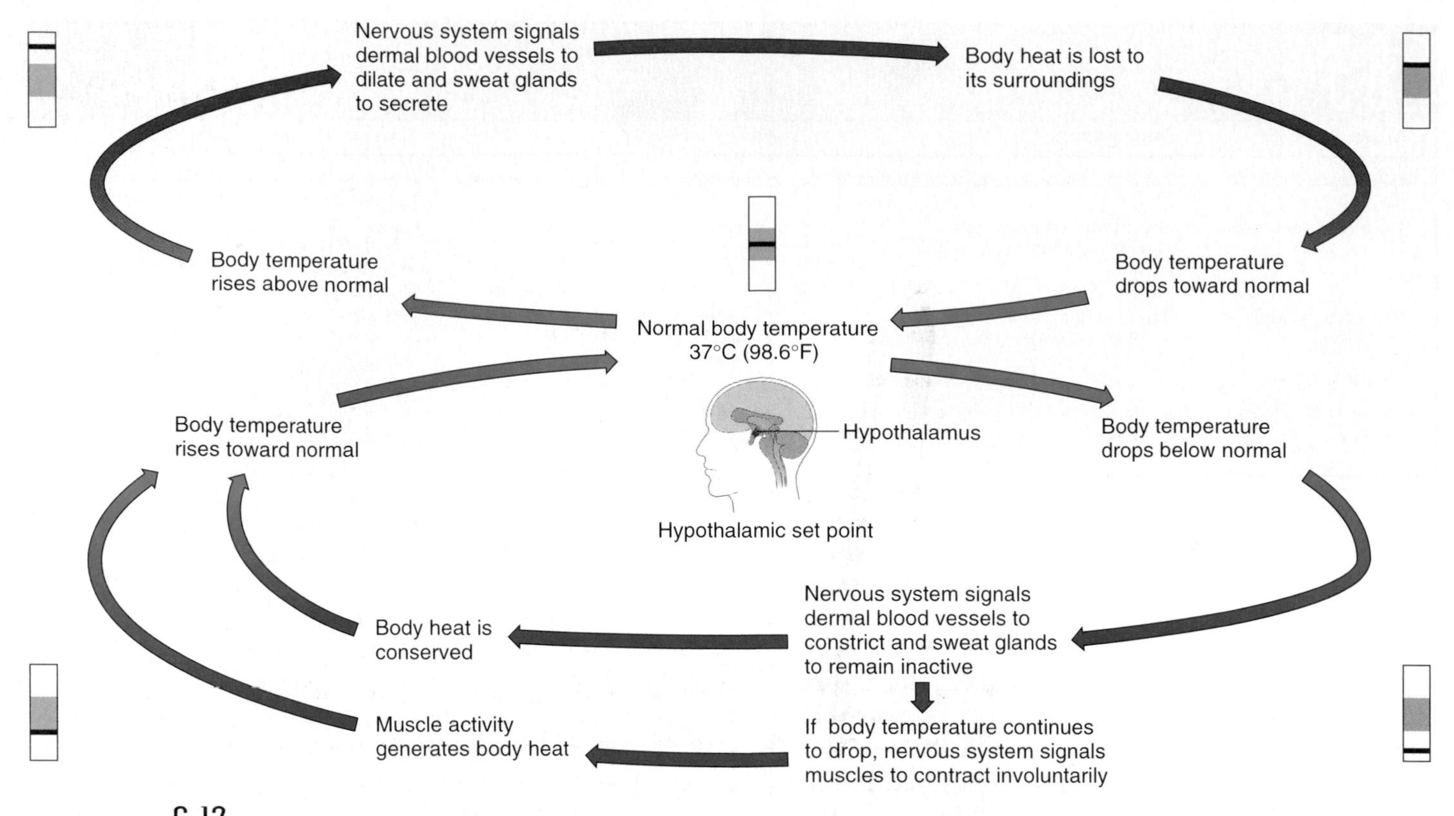


**FIGURE 6.12**
Body temperature regulation, an example of homeostasis.

> Because body temperature regulation depends, in large part, on evaporation of sweat from the skin's surface, and because high humidity hinders evaporation, athletes are advised to slow down their activities on hot, humid days. On these days, they should also stay out of the sunlight whenever possible and drink enough fluids to avoid dehydration. Such precautions can prevent the symptoms of *heat exhaustion,* which include fatigue, dizziness, headache, muscle cramps, and nausea.

Hypothermia, or lowered body temperature, can result from prolonged exposure to cold, or as part of an illness. It can be extremely dangerous. Hypothermia begins with shivering and a feeling of coldness, but if not treated progresses to mental confusion, lethargy, loss of reflexes and consciousness, and, eventually, a shutting down of major organs. If the temperature in the body's core drops just a few degrees, fatal respiratory failure or heart arrhythmia may result. However, the extremities can withstand drops of 20 to 30 degrees Fahrenheit below normal.

Certain people are at higher risk for developing hypothermia. These include the very young and the very old, very thin individuals, and the homeless. Hypothermia can be prevented by dressing appropriately and staying active in the cold. A person suffering from hypothermia must be warmed gradually so that respiratory and cardiovascular functioning remain stable.

> Hypothermia is induced intentionally during certain surgical procedures involving the heart or central nervous system (brain or spinal cord). In heart surgery, for example, body temperature may be lowered to between 78° F (26° C) and 89° F (32° C). The cooling lowers the body's metabolic rate so that less oxygen is needed. Hypothermia for surgery is accomplished by packing the patient in ice, or by removing blood, cooling it, and returning it to the body.

## Clinical Application 6.3

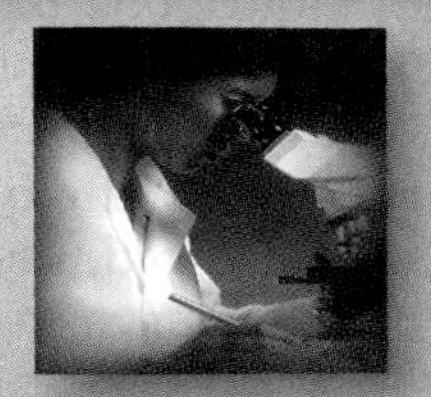

# Elevated Body Temperature

It was a warm June morning when the harried and hurried father strapped his 5-month-old son Bryan into the backseat of his car and headed for work. Tragically, the father forgot to drop his son off at the babysitter's. When his wife called him at work late that afternoon to inquire why the child was not at the sitter's, the shocked father realized his mistake and hurried down to his parked car. But it was too late—Bryan had died. Left for 10 hours in the car in the sun, all windows shut, the baby's temperature had quickly soared. Two hours after he was discovered, the child's temperature still exceeded 41° C (106° F).

Sarah L.'s case of elevated body temperature was more typical. She awoke with a fever of 40° C (104° F) and a terribly painful sore throat. Peering down the five-year-old's throat with a flashlight, her mother spotted the telltale whitish lesions that usually indicate a *Streptococcus* infection. At the doctor's office, a test revealed that Sarah did indeed have a strep throat. The fever was her body's attempt to fight the infection.

The true cases of young Bryan and Sarah illustrate two reasons why body temperature may rise—inability of the temperature homeostatic mechanism to handle an extreme environment, and an immune system response to infection.

In the case of Bryan, sustained exposure to very high heat overwhelmed the temperature-regulating mechanism resulting in hyperthermia. Body heat built up faster than it could dissipate, and body temperature rose, even though the set point of the thermostat was normal. His blood vessels dilated so greatly in an attempt to dissipate the excess heat that after a few hours his circulatory system collapsed.

In a fever, molecules on the surfaces of the infectious agents (usually bacteria or viruses) stimulate phagocytes to release a substance called interleukin-1. It is also called endogenous pyrogen, meaning "fire maker from within." The bloodstream carries IL-1 to the hypothalamus, where it raises the set point controlling temperature. In response, the brain signals skeletal muscles to increase heat production, blood flow to the skin to decrease, and sweat glands to decrease secretion. As a result, body temperature rises to the new set point, and the person has a fever. The increased body temperature helps the immune system kill the pathogens—and to make the patient quite uncomfortable for a while.

What should one do when body temperature rises? Hyperthermia in response to exposure to intense, sustained heat should be rapidly treated by administering liquids to replace lost body fluids and electrolytes, sponging the skin with water to increase cooling by evaporation, and covering the person with a refrigerated blanket. Some health professionals believe that a slightly elevated temperature should not be reduced (with medication or cold baths), because it may be part of a normal immune response.

1. Why is the regulation of body temperature so important?
2. How is body heat produced?
3. By what means is heat lost from the body when heat production is excessive?
4. How does the skin help regulate body temperature?
5. What are the dangers of hypothermia?

## Skin Color

Heredity and the environment determine skin color.

### Genetic Factors

Regardless of racial origin, all people have about the same number of melanocytes in their skin. Differences in skin color result from differences in the amount of melanin these cells produce, which is genetically controlled. The more melanin, the darker the skin. The distribution and the size of pigment granules within melanocytes also influences skin color. The granules of very dark skin are single and large; those in lighter skin occur in clusters of two to four granules and are smaller.

7

CHAPTER

# Skeletal System

The graceful movements of dancers are possible because muscles pull against bones that function as a lever system.

## Chapter Objectives

AFTER YOU HAVE STUDIED THIS CHAPTER, YOU SHOULD BE ABLE TO:

1. Classify bones according to their shapes and name an example from each group.
2. Describe the general structure of a bone and list the functions of its parts.
3. Distinguish between intramembranous and endochondral bones, and explain how such bones grow and develop.
4. Describe the effects of sunlight, nutrition, hormonal secretions, and exercise on bone development.
5. Discuss the major functions of bones.
6. Distinguish between the axial and appendicular skeletons, and name the major parts of each.
7. Locate and identify the bones and the major features of the bones that comprise the skull, vertebral column, thoracic cage, pectoral girdle, upper limb, pelvic girdle, and lower limb.

## Key Terms

**articular cartilage** (ar-tik′u-lar kar′tĭ-lij)
**compact bone** (kom′pakt bōn)
**diaphysis** (di-af′ĭ-sis)
**endochondral bone** (en′do-kon′dral bōn)
**epiphyseal disk** (ep″ĭ-fiz′e-al disk)
**epiphysis** (e-pif′ĭ-sis)
**fontanel** (fon′tah-nel)
**hematopoiesis** (hem″ah-to-poi-e′sis)
**intramembranous bone** (in″trah-mem′brah-nus bōn)
**lever** (lev′er)
**marrow** (mar′o)
**medullary cavity** (med′u-lār″e kav′ĭ-te)
**osteoblast** (os′te-o-blast)
**osteoclast** (os′te-o-klast)
**osteocyte** (os′te-o-sīt)
**osteon** (os′te-on)
**periosteum** (per″e-os′te-um)
**spongy bone** (spun′je bōn)
**trabeculae** (trah-bek′u-le)

In 1925, Australian anatomist Raymond Dart discovered the remarkably well-preserved skull of an ancient child in an outcropping of limestone in the South African village of Taung. The size and condition of the bones offered many tantalizing clues to the type of animal that had left them.

The skull of the Taung child was about 2 million years old, its owner about 6 years old when she died. The skull had several humanlike characteristics: no brow ridges, an upright vertebral column suggesting an erect stance, and teeth shaped like our own. However, a CT scan of the skull revealed the order in which the teeth had erupted from the jaw, which was more like that of apes than humans. The slower the teeth appear, the longer the period of infancy. Humans have much longer infancies than apes, and many anthropologists attribute our intelligence to this longer nurturing period. The Taung child, as her bones revealed, appeared to be a "missing link" of sorts between apes and humans.

Fossilized skeletal remains of humans and prehumans have told us much about our beginnings, enabling us to literally piece together the murky period when primates began to walk erect and venture from the forests. Today our skeletal systems remain vital to our ability to move. Experiencing a few weeks with an immobilized broken limb, or the increasing limitations of a progressive bone illness such as osteoporosis, underscores how important this system is to our enjoyment of life.

---

Nonliving material in the matrix of bone tissue makes the whole organ appear to be inert. A bone also contains very active, living tissues. An individual bone is composed of a variety of tissues: bone tissue, cartilage, fibrous connective tissue, blood, and nervous tissue.

## Bone Structure

Although the bones of the skeletal system differ greatly in size and shape, they are similar in their structure, development, and functions.

### Bone Classification

Bones are classified according to their shapes—long, short, flat, or irregular (fig. 7.1).

**Long bones** have long longitudinal axes and expanded ends. Examples are the forearm and thigh bones.

**Short bones** are somewhat cubelike, with their lengths and widths roughly equal. The bones of the wrists and ankles are examples of this type.

**Flat bones** are platelike structures with broad surfaces, such as the ribs, scapulae, and some bones of the skull.

**Irregular bones** have a variety of shapes and are usually connected to several other bones. Irregular bones include the vertebrae that comprise the backbone and many facial bones.

In addition to these four groups of bones, some authorities recognize a fifth group called *sesamoid* or round bones. These bones are usually small and nodular, and are embedded within tendons adjacent to joints, where the tendons are compressed. The kneecap (patella) is an example of a very large sesamoid bone.

### Parts of a Long Bone

The femur, a long bone in the thigh, illustrates the structure of bone (fig. 7.2). At each end of such a bone is an expanded portion called an **epiphysis** (pl. *epiphyses*), which articulates (or forms a joint) with another bone. On its outer surface, the articulating portion of the epiphysis is coated with a layer of hyaline cartilage called **articular cartilage.** The shaft of the bone, which is located between the epiphyses, is called the **diaphysis.**

Except for the articular cartilage on its ends, the bone is completely enclosed by a tough, vascular covering of fibrous tissue called the **periosteum.** This membrane is firmly attached to the bone, and the periosteal fibers are continuous with ligaments and tendons that are connected to the membrane. The periosteum also functions in the formation and repair of bone tissue.

A bone's shape is closely related to its functions. Bony projections called *processes,* for example, provide sites for attachment of ligaments and tendons; grooves and openings are passageways for blood vessels and nerves; a depression of one bone might articulate with a process of another.

The wall of the diaphysis is composed mainly of tightly packed tissue called **compact bone** (cortical bone). This type of bone is solid, strong, and resists bending (fig. 7.3*a*).

The epiphyses, on the other hand, are composed largely of **spongy bone** (cancellous bone) with thin layers of compact bone on their surfaces (fig. 7.3*b*). Spongy bone consists of numerous branching bony plates called **trabeculae.** Irregular interconnecting spaces between these plates help reduce the weight of

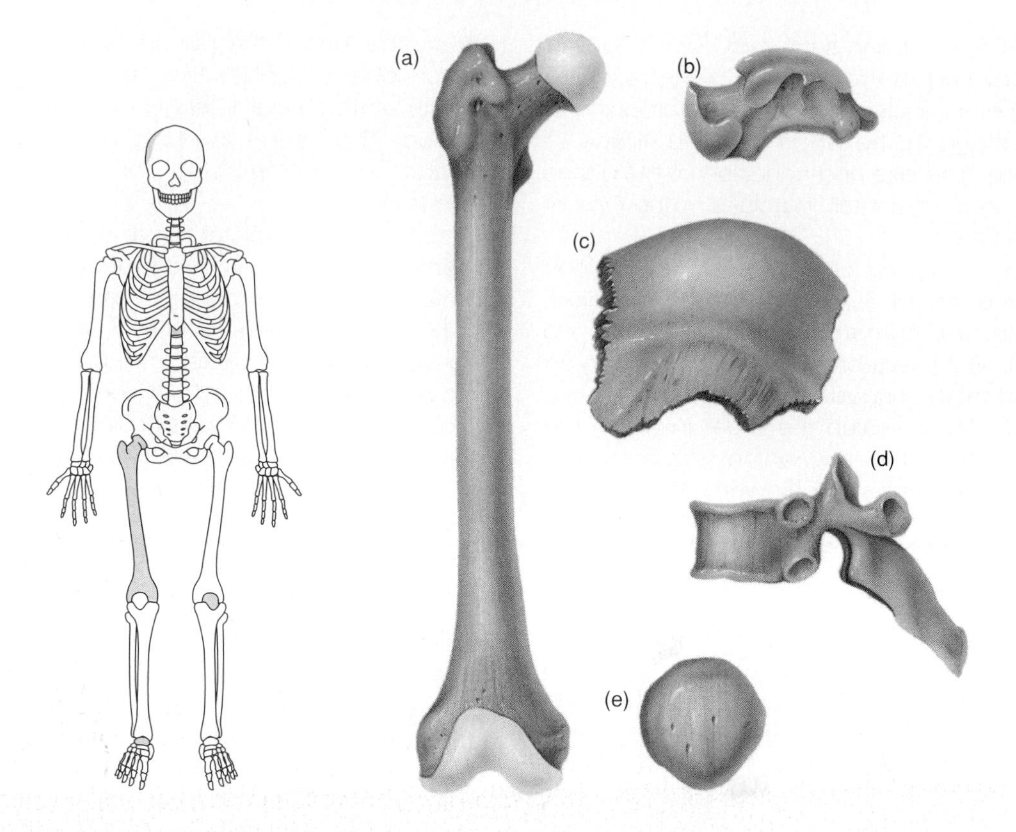


**FIGURE 7.1**
(*a*) The femur of the thigh is a long bone; (*b*) a tarsal bone of the ankle is a short bone; (*c*) a parietal bone of the skull is a flat bone; (*d*) a vertebra of the backbone is an irregular bone; and (*e*) the patella of the knee is a round bone.

the bone. Spongy bone provides strength, and its bony plates are most highly developed in the regions of the epiphyses that are subjected to compressive forces.

A bone usually has both compact and spongy tissues. Short, flat, and irregular bones typically consist of a mass of spongy bone that is either covered by a layer of compact bone or sandwiched between plates of compact bone (fig. 7.3*c*).

Compact bone in the diaphysis of a long bone forms a semirigid tube with a hollow chamber called the **medullary cavity.** This cavity is continuous with the spaces of the spongy bone. All of these areas are lined with a thin layer of squamous epithelial cells called **endosteum** and are filled with a specialized type of soft connective tissue called **marrow.** Marrow exists in two forms, red marrow and yellow marrow, described later in the chapter (see also fig. 7.2).

## Microscopic Structure

Recall from chapter 5 that bone cells (osteocytes) are located in minute, bony chambers called lacunae, which are arranged in concentric circles around osteonic canals (Haversian canals). These cells communicate with nearby cells by means of cellular processes passing through canaliculi. The intercellular material of bone tissue is largely collagen and inorganic salts. The collagen gives bone its strength and resilience, and the inorganic salts make it hard and resistant to crushing.

### Compact Bone

In compact bone, the osteocytes and layers of intercellular material clustered concentrically around an osteonic canal form a cylinder-shaped unit called an **osteon** (Haversian system). Many of these units cemented together form the substance of compact bone (fig. 7.4). The orientation of the osteons resists compressive forces.

Each osteonic canal contains one or two small blood vessels (including capillaries) and nerve fibers surrounded by some loose connective tissue. Blood in these vessels nourishes bone cells associated with the osteonic canal via gap junctions between osteocytes.

Osteonic canals travel longitudinally through bone tissue. They are interconnected by transverse *perforating canals* (Volkmann's canals). These canals contain larger blood vessels and nerves by which the vessels and nerve fibers in the osteonic canals communicate with the surface of the bone and the medullary cavity (fig. 7.5).

### Spongy Bone

Spongy bone is also composed of osteocytes and intercellular material, but the bone cells are not arranged around osteonic canals. Instead, the cells lie within the trabeculae, and they are nourished by substances diffusing into the canaliculi that lead to the surface of these thin, bony plates.

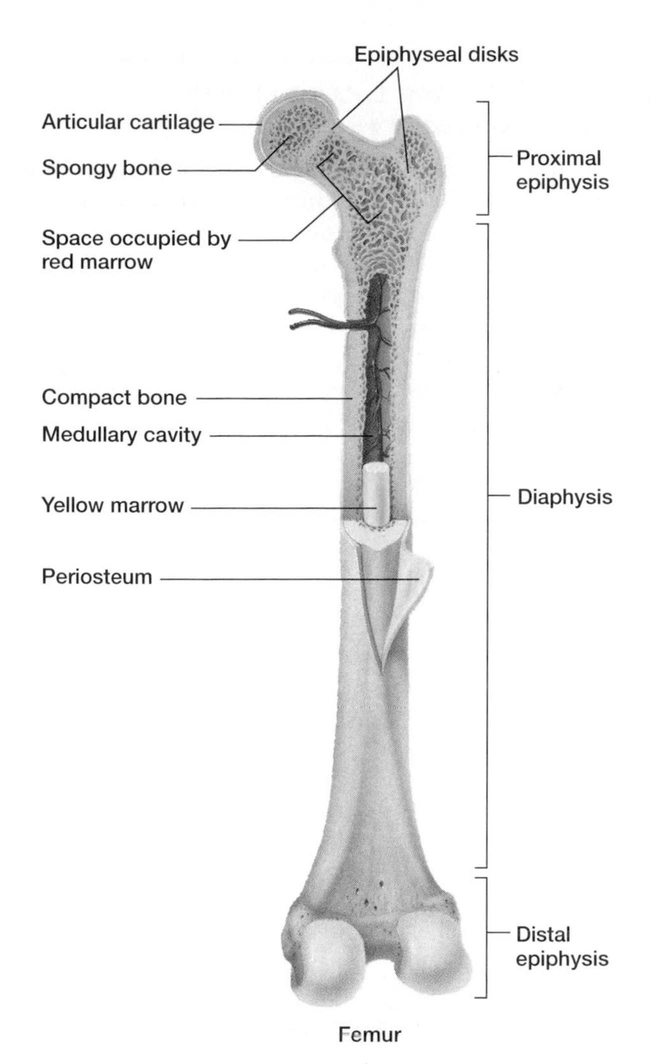


**FIGURE 7.2**
Major parts of a long bone.

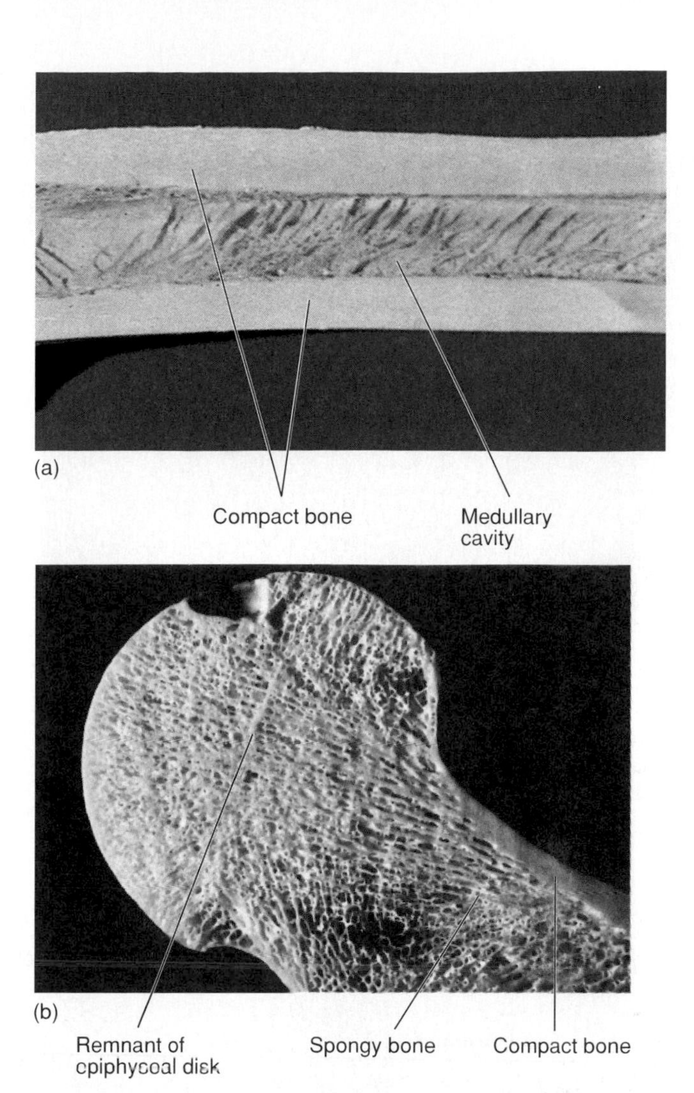


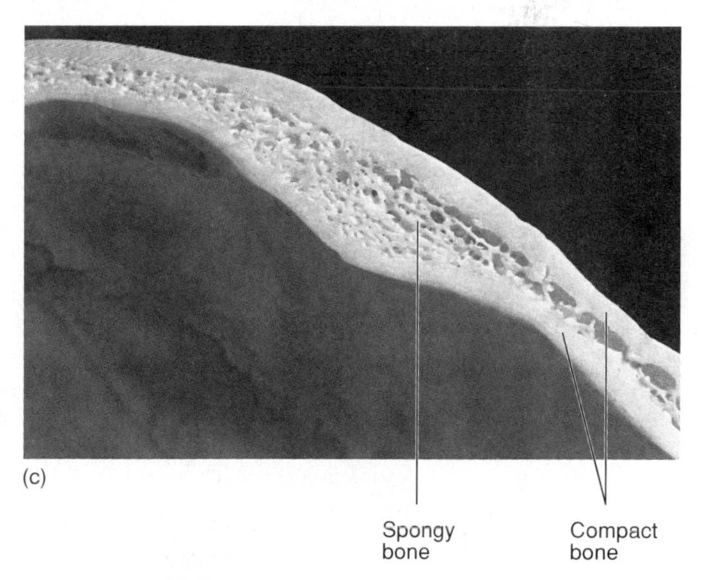


**FIGURE 7.3**
(*a*) In a femur, the wall of the diaphysis consists of compact bone. (*b*) The epiphyses of the femur contain spongy bone enclosed by a thin layer of compact bone. (*c*) This skull bone contains a layer of spongy bone sandwiched between plates of compact bone.

> Severe bone pain is a symptom of inherited sickle cell disease. Abnormal hemoglobin (an oxygen-carrying protein) bends the red blood cells that contain it into a sickle shape, which obstructs circulation. X rays can reveal blocked arterial blood flow in bones of sickle cell disease patients.

1. Explain how bones are classified.
2. List five major parts of a long bone.
3. How do compact and spongy bone differ in structure?
4. Describe the microscopic structure of compact bone.

**Figure 7.4**

Scanning electron micrograph of a single osteon in compact bone (1,300×).

*Tissues and Organs: A Text-Atlas of Scanning Electron Microscopy,* by R. G. Kessel and R. H. Kardon. © 1979 W. H. Freeman and Company.

## Bone Development and Growth

Parts of the skeletal system begin to form during the first few weeks of prenatal development, and bony structures continue to grow and develop into adulthood. Bones form by replacing existing connective tissue in one of two ways. Some bones originate between sheetlike layers of connective tissues; they are called intramembranous bones. Others begin as masses of cartilage that are later replaced by bone tissue; they are endochondral bones (fig. 7.6).

### Intramembranous Bones

The broad, flat bones of the skull are intramembranous bones. During their development (osteogenesis), membranelike layers of unspecialized, or primitive, connective tissues appear at the sites of the future bones. These layers are supplied with dense networks of blood vessels, and some of the connective tissue cells arrange around these vessels. These primitive

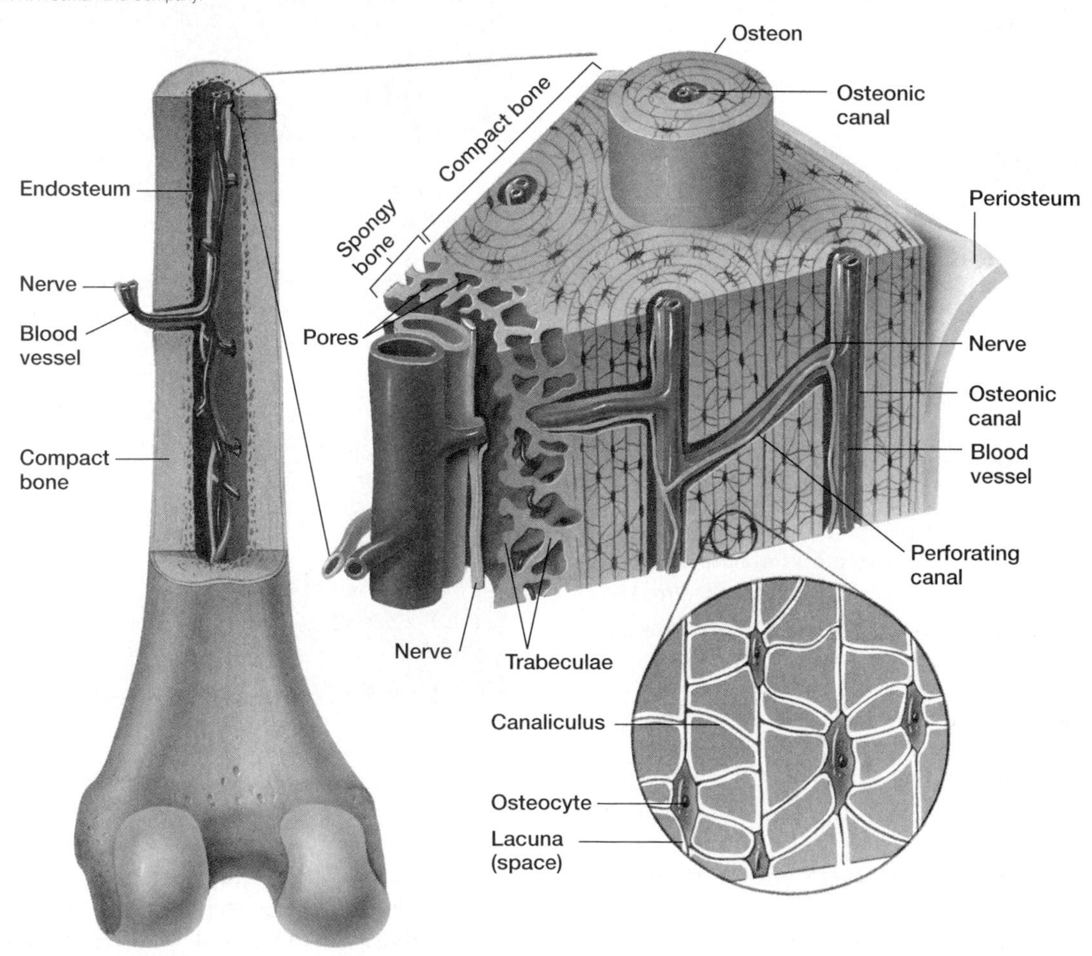


**Figure 7.5**

Compact bone is composed of osteons cemented together by bone matrix.

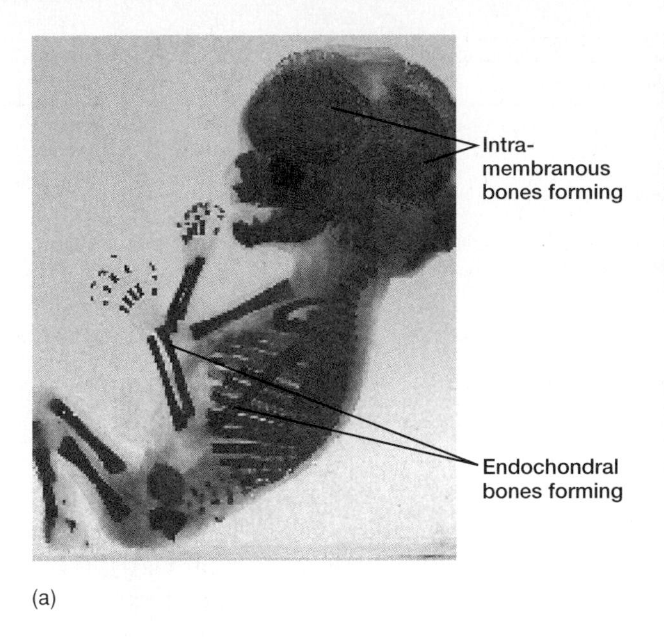


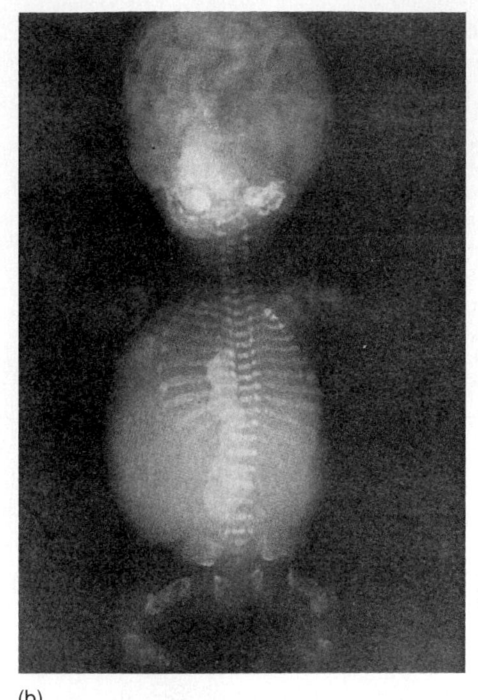


**FIGURE 7.6**

(*a*) Note the stained bones of this fourteen-week fetus. (*b*) Bones can fracture even before birth. This fetus has numerous broken bones because of an inherited defect in collagen called osteogenesis imperfecta. Often parents of such children are unfairly accused of child abuse because their children frequently break bones.

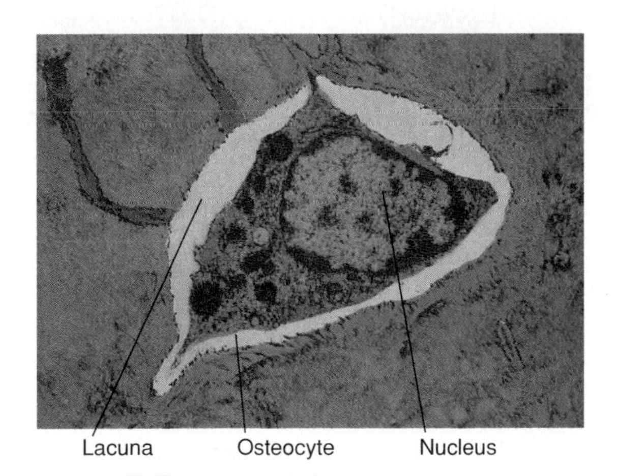


**FIGURE 7.7**

Transmission electron micrograph of an osteocyte isolated within a lacuna.

cells enlarge and differentiate into bone-forming cells called **osteoblasts,** which, in turn, deposit bony matrix around themselves. As a result, spongy bone forms in all directions along blood vessels within the layers of primitive connective tissues. Later, some spongy bone may be converted to compact bone, as spaces fill with bone matrix.

As development continues, the osteoblasts may become completely surrounded by matrix, and in this manner they become secluded within lacunae. At the same time, matrix enclosing the cellular processes of the osteoblasts gives rise to canaliculi. Once isolated in lacunae, these cells are called **osteocytes** (fig. 7.7).

Cells of the primitive connective tissue that persist outside the developing bone give rise to the periosteum. Osteoblasts on the inside of the periosteum form a layer of compact bone over the surface of the newly formed spongy bone.

This process of replacing connective tissue to form an intramembranous bone is called *intramembranous ossification.* Chart 7.1 lists the major steps of the process.

## Endochondral Bones

Most of the bones of the skeleton are **endochondral bones.** Their development proceeds from masses of hyaline cartilage with shapes similar to future bony structures. These cartilaginous models grow rapidly for a time, and then begin to extensively change. For example, cartilage cells enlarge and their lacunae grow. At the same time, the surrounding matrix breaks down, and soon the cartilage cells die and degenerate.

About the same time, a periosteum forms from connective tissue that encircles the developing structure. As the cartilage decomposes, blood vessels and undifferentiated connective tissue cells invade the disintegrating tissue. Some of the invading cells differentiate into

## CHART 7.1 MAJOR STEPS IN BONE DEVELOPMENT

| Intramembranous Ossification | Endochondral Ossification |
|---|---|
| 1. Membranelike layers of primitive connective tissue appear at sites of future bones. | 1. Masses of hyaline cartilage form models of future bones. |
| 2. Primitive connective tissue cells arrange around blood vessels in these layers. | 2. Cartilage tissue breaks down and disappears. Periosteum develops. |
| 3. Connective tissue cells differentiate into osteoblasts, which form spongy bone. | 3. Blood vessels and differentiating osteoblasts from the periosteum invade the disintegrating tissue. |
| 4. Osteoblasts become osteocytes when bony matrix completely surrounds them. | 4. Osteoblasts form spongy bone in the space occupied by cartilage. |
| 5. Connective tissue on the surface of each developing structure forms a periosteum. | 5. Osteoblasts become osteocytes when bony matrix completely surrounds them. |
| 6. Osteoblasts on the inside of the periosteum form compact bone over the spongy bone. | 6. Osteoblasts beneath the periosteum deposit compact bone around spongy bone. |

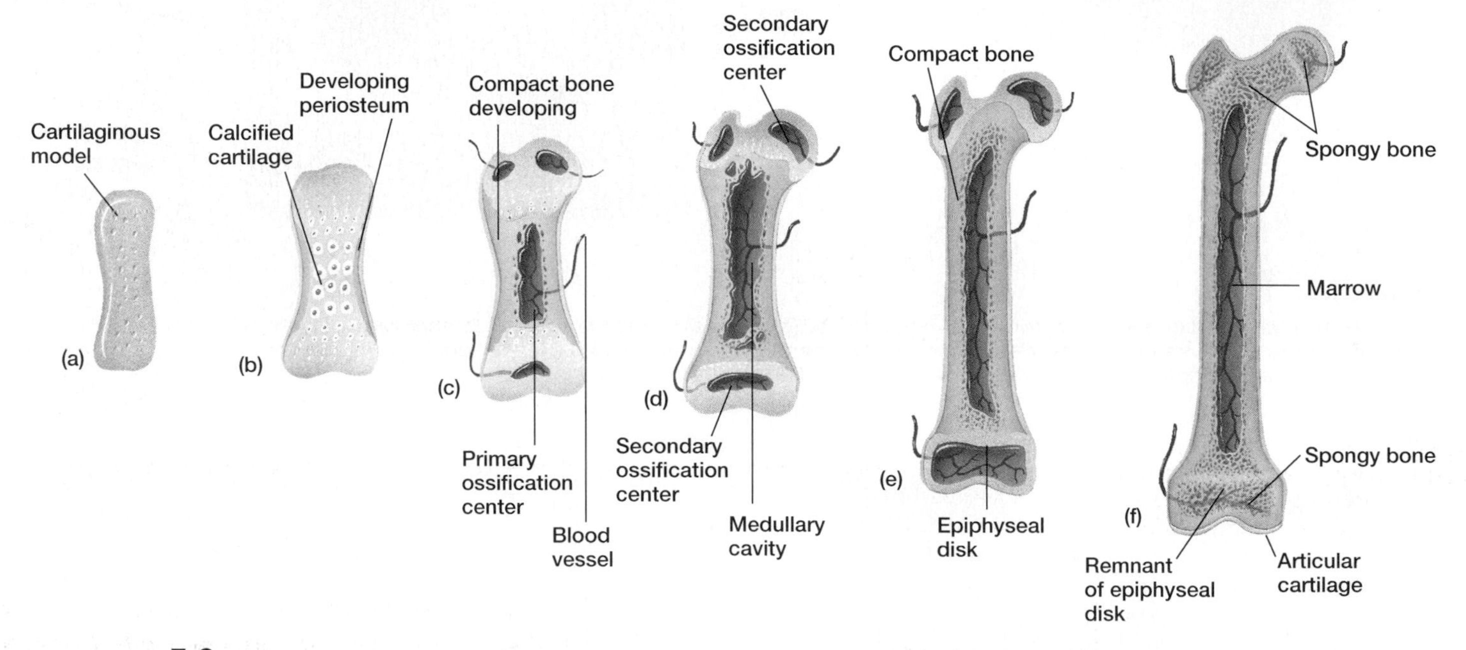


**FIGURE 7.8**
Major stages (*a–f*) in the development of an endochondral bone. (Relative bone sizes not to scale.)

osteoblasts and begin to form spongy bone in the spaces previously housing the cartilage. Once completely surrounded by the bony matrix, osteoblasts are called osteocytes. As intramembranous ossification continues, osteoblasts beneath the periosteum deposit compact bone around the spongy bone.

This process of forming an endochondral bone by the replacement of hyaline cartilage is called *endochondral ossification.* Its major steps are listed in chart 7.1 and illustrated in figure 7.8.

## Growth of an Endochondral Bone

In a long bone, replacement of hyaline cartilage by bony tissue begins in the center of the diaphysis. This region is called the *primary ossification center,* and bone develops from it toward the ends of the cartilaginous structure. Meanwhile, osteoblasts from the periosteum deposit a thin layer of compact bone around the primary ossification center by intramembranous ossification. The epiphyses of the developing bone remain cartilaginous and continue to grow. Later, *secondary ossification centers* appear in the epiphyses, and spongy bone forms in all directions from them. As spongy bone is deposited in the diaphysis and in the epiphysis, a band of cartilage, called the **epiphyseal disk** (metaphysis), remains between the two ossification centers (see figs. 7.2, 7.3*b,* and 7.8).

The cartilaginous cells of the epiphyseal disk are arranged in four layers, each of which may be several cells thick, as shown in figure 7.9. The first layer, closest to the end of the epiphysis, is composed of resting cells that do not actively participate in growth. This layer anchors the epiphyseal disk to the bony tissue of the epiphysis.

The second layer contains rows of numerous young cells undergoing mitosis. As new cells appear

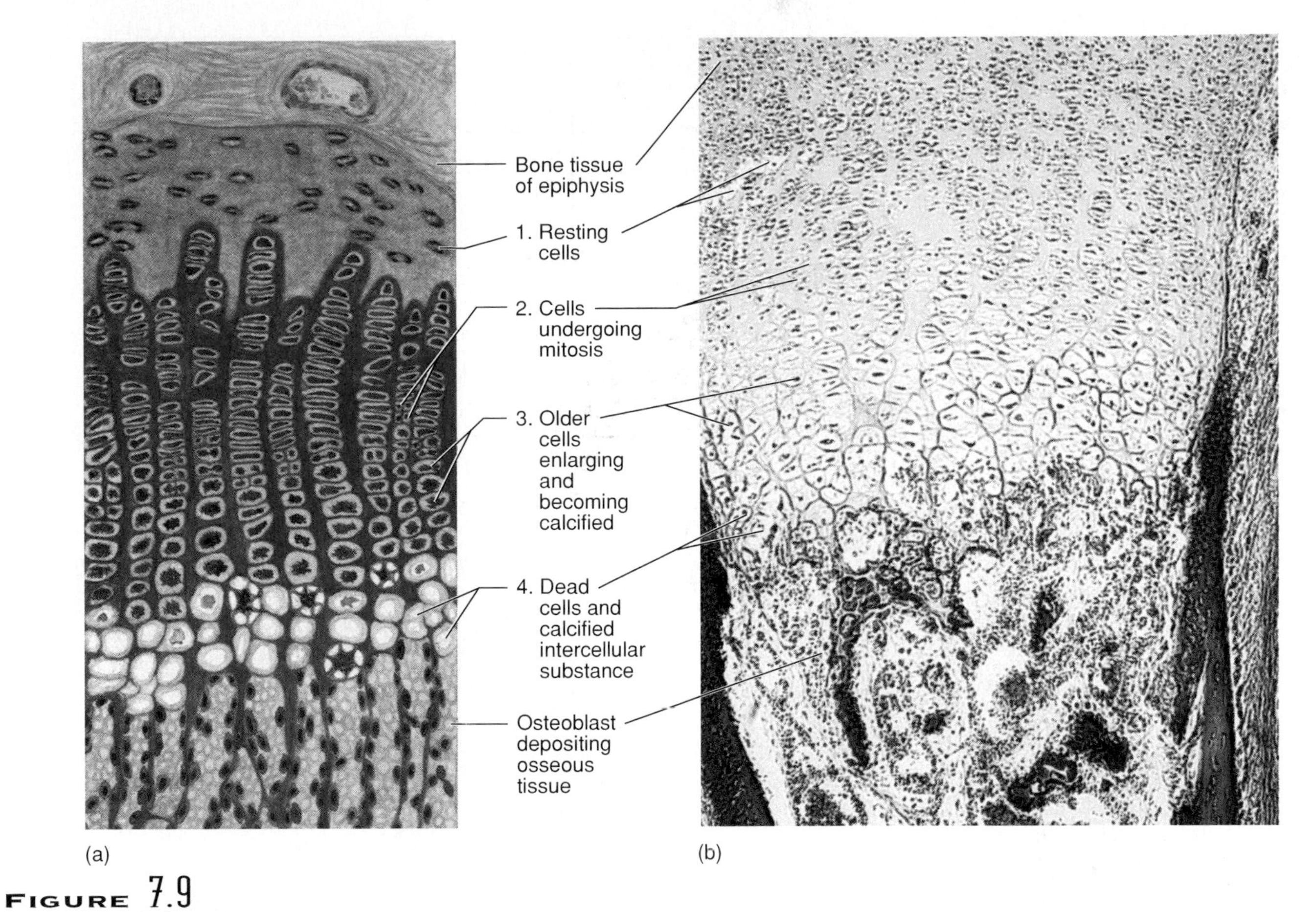


**FIGURE 7.9**

(*a*) The cartilaginous cells of an epiphyseal disk are arranged in four layers, each of which may be several cells thick. (*b*) A micrograph of an epiphyseal disk (100×).

and as intercellular material forms around them, the cartilaginous disk thickens.

The rows of older cells, which are left behind when new cells appear, form the third layer. These cells enlarge and thicken the epiphyseal disk still more. Consequently, the entire bone lengthens. At the same time, calcium salts accumulate in the intercellular matrix adjacent to the oldest cartilaginous cells, and as the matrix calcifies, the cells begin to die.

The fourth layer of the epiphyseal disk is quite thin and is composed of dead cells and calcified intercellular substance.

In time, large, multinucleated cells called **osteoclasts** break down the calcified matrix. These large cells originate by the fusion of certain single-nucleated white blood cells (monocytes) (see chapter 14). Osteoclasts secrete an acid that dissolves the inorganic component of the calcified matrix, and their lysosomal enzymes digest the organic components. After osteoclasts remove the matrix, bone-building osteoblasts invade the region and deposit bone tissue in place of the calcified cartilage.

A long bone continues to lengthen while the cartilaginous cells of the epiphyseal disks are active. However, once the ossification centers of the diaphysis and epiphyses meet and the epiphyseal disks ossify, lengthening is no longer possible in that end of the bone.

A developing bone thickens as compact bone is deposited on the outside, just beneath the periosteum. As this compact bone forms on the surface, osteoclasts erode other bone tissue on the inside (fig. 7.10). The resulting space becomes the medullary cavity of the diaphysis, which later fills with marrow.

The bone in the central regions of the epiphyses and diaphysis remains spongy, and hyaline cartilage on the ends of the epiphyses persists throughout life as articular cartilage. Chart 7.2 lists the ages at which various bones ossify.

## Homeostasis of Bone Tissue

After the intramembranous and endochondral bones form, the actions of osteoclasts and osteoblasts continually remodel them. Thus, throughout life, osteoclasts resorb bone tissue and osteoblasts replace the bone.

## CHART 7.2 Ossification Timetable

| Age | Occurrence | Age | Occurrence |
|---|---|---|---|
| Third month of prenatal development | Ossification in long bones begins. | 17 to 20 years | Bones of the upper limbs and scapulae completely ossify. |
| Fourth month of prenatal development | Most primary ossification centers have appeared in the diaphyses of bones. | 18 to 23 years | Bones of the lower limbs and coxal bones completely ossify. |
| Birth to 5 years | Secondary ossification centers appear in the epiphyses. | 23 to 25 years | Bones of the sternum, clavicles, and vertebrae completely ossify. |
| 5 to 12 years in females, or 5 to 14 years in males | Ossification spreads rapidly from the ossification centers, and certain bones are ossifying. | By 25 years | Nearly all bones completely ossify. |

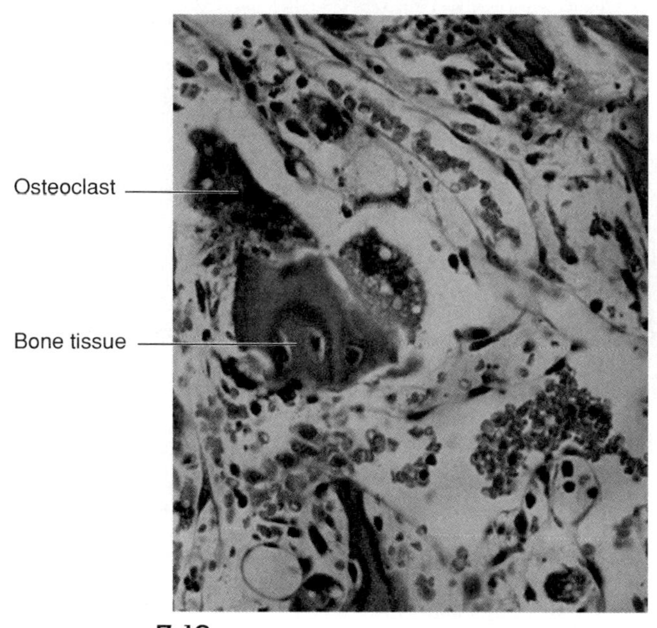


**FIGURE 7.10**
Micrograph of a bone-resorbing osteoclast.

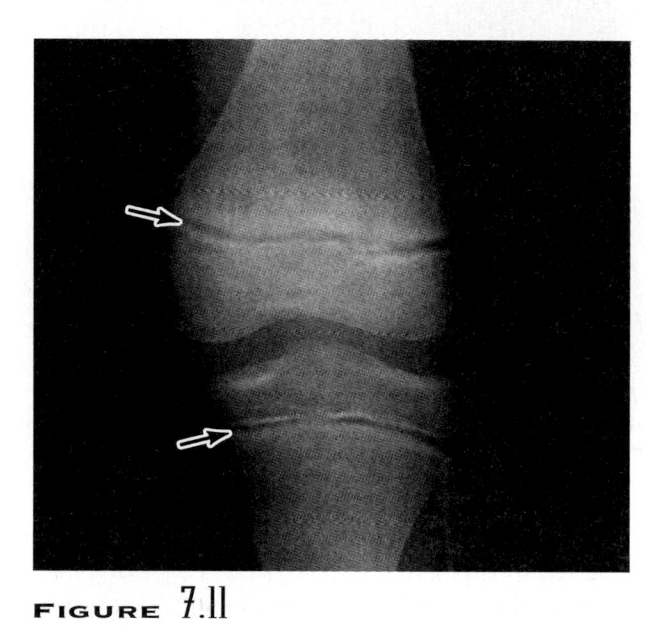

**FIGURE 7.11**
The presence of epiphyseal disks (arrows) in a child's bone indicates that the bone is still lengthening.

These opposing processes of **resorption** and **deposition** are well regulated so that the total mass of bone tissue within an adult skeleton normally remains nearly constant, even though 3% to 5% of bone calcium is exchanged each year.

> A child's long bones are still growing if an X ray shows epiphyseal disks (fig. 7.11). If a disk is damaged as a result of a fracture before it ossifies, elongation of the long bone may cease prematurely, or if growth continues, it may be uneven. For this reason, injuries to the epiphyses of a young person's bones are of special concern. On the other hand, an epiphysis is sometimes altered surgically in order to equalize growth of bones that are developing at very different rates.

> In bone cancers, abnormally active osteoclasts destroy bone tissue. Interestingly, cancer of the prostate gland can have the opposite effect. If such cancer cells reach the bone marrow, as they do in most cases of advanced prostatic cancer, they stimulate osteoblast activity, which promotes formation of new bone on the surfaces of the bony trabeculae.

## Factors Affecting Bone Development, Growth, and Repair

A number of factors influence bone development, growth, and repair, including nutrition, exposure to sunlight, hormonal secretions, and physical exercise. For example, vitamin D is necessary for proper absorption

## CLINICAL APPLICATION 7.1

### An Unusual Form of Rickets

Rickets is usually caused by vitamin D deficiency. This is why two sisters at first puzzled their doctors. At ages 7 and 3, the girls clearly suffered from the same disorder. Their too-soft leg bones bowed outwards and the ends were bearing so much strain that their knees protruded, resembling knobs. The girls' rib cages caved in. These were the unmistakable symptoms of rickets, but the children followed a healthy diet.

Even before the skeletal deformities became apparent, all was not right. As infants, the girls' growth lagged. The soft spot on the tops of their skulls did not close and harden by 18 months, as it normally does. This was the first sign of the underlying lack of vitamin D. The sisters' teeth erupted late and were soon lost to decay.

The fact that the girls' parents were blood relatives clued physicians to the possibility of a genetic explanation behind their bone disorder. A second clue came from analyzing the relative amounts of intermediate biochemicals resulting from the girls' cells using vitamin D. The diagnosis: a very rare, inherited form of vitamin D refractory rickets. Even though the girls ate foods rich in vitamin D, received vitamin supplements, and got plenty of sunshine (needed to activate vitamin D precursors in the skin), they were still sick because their bodies could not use the abundant vitamin. Vitamin D entered their cells, but it could not bind to the genetic material, a vital step in controlling production of bone proteins. ■

of calcium in the small intestine. In the absence of this vitamin, calcium is poorly absorbed, the inorganic salt portion of bone matrix lacks calcium, and bones are deformed. In children, this condition is called *rickets*, and in adults it is called *osteomalacia*. Clinical Application 7.1 describes an unusual type of rickets.

Vitamin D is relatively uncommon in natural foods, except for eggs. But it is readily available in milk and other dairy products fortified with vitamin D. Vitamin D also forms from a substance (dehydrocholesterol) produced by cells in the digestive tract or obtained in the diet. Dehydrocholesterol is carried by the blood to the skin, and when exposed to ultraviolet light from the sun, it is converted to a compound that becomes vitamin D.

Vitamins A and C are also needed for normal bone development and growth. Vitamin A is necessary for bone resorption during normal development. Thus, deficiency of vitamin A may retard bone development. Vitamin C is needed for collagen synthesis, so its lack also may inhibit bone development. In this case, osteoblasts produce less collagen in the intercellular material of the bone tissue, and the resulting bones are abnormally slender and fragile.

Hormones secreted by the pituitary gland, thyroid gland, parathyroid glands, and ovaries or testes affect bone growth and development. The pituitary gland, for instance, secretes **growth hormone,** which stimulates reproduction of cartilage cells in the epiphyseal disks. In the absence of this hormone, the long bones of the limbs fail to develop normally, and the child has pituitary dwarfism. Such a person is very short, but has normal body proportions. If excessive growth hormone is released before the epiphyseal disks are ossified, height may exceed 8 feet—a condition called *pituitary giantism.*

Pituitary dwarfism is treated with human growth hormone (HGH). Today, HGH is plentiful and pure, thanks to recombinant DNA technology. Bacteria given the human gene for HGH secrete the hormone. Previously, HGH was pooled from donors or cadavers. This introduced the risk of transmitting infection.

A controversial use of HGH is to give it to children who are of short stature, but not abnormally so, or to use it to enhance height with the goal of improving athletic ability.

In an adult, excessive secretion of growth hormone causes a condition called *acromegaly,* in which the hands, feet, and jaw enlarge (see chapter 13).

## Clinical Application 7.2

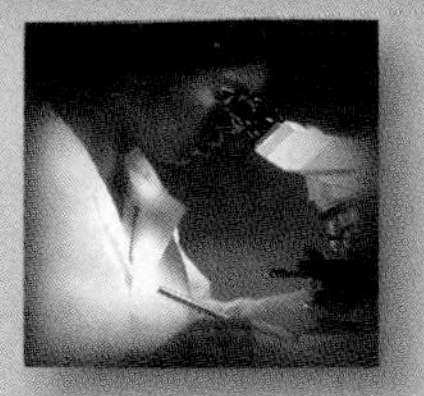

### Forcing Bones to Grow

Reza Garakani was born with achondroplasia, a hereditary form of dwarfism, and by age 14 was only 40 inches (101.6 cm) tall. He could not reach doorknobs, faucets, and elevator buttons, and he yearned to grow so that he could lead a normal life. His goal—reaching 4 feet 4 inches (132.1 cm)—would enable him to drive a car.

In April 1988, Reza began a brutally painful treatment that would literally stretch his frame—bone lengthening. He had heard about 12-year-old Anthony Terravachio, who had begun treatment in August 1985. In his legs the tibias had been stretched 7 inches (17.8 cm) and the femurs 5 inches (12.7 cm), and work was ongoing on his arms, which had already been stretched 6 inches (15.2 cm). "I can fit my hands into my pockets now!" Anthony had said excitedly on a television program. So Reza bravely underwent the procedure. The outer shells of compact bone in his legs were fractured, leaving the blood vessels and marrow undisturbed. Then pins were inserted so that they projected from the broken bone ends, like the spokes of a bicycle wheel. That was the easy part. After the surgery, Reza's mother would turn screws attached to the pins four times a day, moving the pieces of bone apart 0.039 inches (1/4 mm) each time. The separated area between the pieces would fill in with bone tissue and calcify.

It hurt terribly, but it was worth it, Reza says. For by September—just 6 months after the surgery—his thigh bones had grown 3 1/2 inches (8.9 cm) and his leg bones 7 inches (17.8 cm).

Thyroid hormone replaces cartilage in the epiphyseal disks of long bones with bone tissue. Thus, thyroid hormone can halt bone growth by causing premature ossification of the disks. Deficiency of thyroid hormone also may stunt growth, because without the normal effect of the thyroid hormone, the pituitary gland fails to secrete enough growth hormone (see chapter 13).

Both male and female sex hormones (called androgens and estrogens, respectively) from the testes, ovaries, and adrenal glands promote formation of bone tissue. Beginning at puberty, these hormones are abundant, causing the long bones to grow considerably. However, sex hormones also stimulate ossification of the epiphyseal disks, and consequently they stop bone lengthening at a relatively early age. The effect of estrogens on the disks is somewhat stronger than that of androgens. For this reason, females typically reach their maximum heights earlier than males.

Physical stress also stimulates bone growth. For example, when skeletal muscles contract, they pull at their attachments on bones, and the resulting stress stimulates the bone tissue to thicken and strengthen (hypertrophy). Conversely, with lack of exercise, the same bone tissue wastes, becoming thinner and weaker (atrophy). This is why the bones of athletes are usually stronger and heavier than those of nonathletes (fig. 7.12). It is also why fractured bones immobilized in casts may shorten. Clinical Application 7.2 discusses intentional stress on bones for a purpose—causing growth. Clinical Application 7.3 describes what happens when a bone breaks.

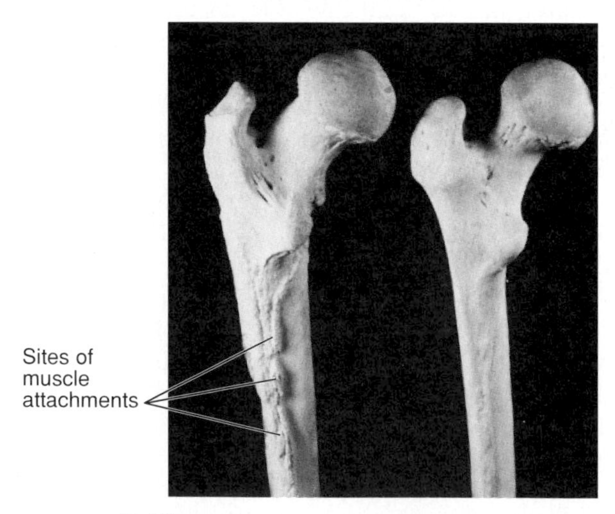


**Figure 7.12**
Note the increased amount of bone at the sites of muscle attachments in the femur on the left. The thickened bone is better able to withstand the force resulting from muscle contraction.

1. Describe the development of an intramembranous bone.
2. Explain how an endochondral bone develops.
3. List the steps in the growth of a long bone.
4. Explain how nutritional factors affect bone development.
5. What effects do hormones have on bone growth?
6. How does physical exercise affect bone structure?

## Bone Function

Bones shape, support, and protect body structures. They also act as levers that aid body movements, house tissues that produce blood cells, and store various inorganic salts.

### Support and Protection

Bones give shape to structures such as the head, face, thorax, and limbs. They also provide support and protection. For example, the bones of the lower limbs, pelvis, and backbone support the body's weight. The bones of the skull protect the eyes, ears, and brain. Those of the rib cage and shoulder girdle protect the heart and lungs, while bones of the pelvic girdle protect the lower abdominal and internal reproductive organs.

### Body Movement

Whenever limbs or other body parts move, bones and muscles interact as simple mechanical devices called *levers.* A lever has four basic components: (a) a rigid bar or rod, (b) a pivot or fulcrum on which the bar turns, (c) an object or weight (resistance) that is moved, and (d) a force that supplies energy for the movement of the bar.

A pair of scissors is a lever. The handle and blade form a rigid bar that rocks on a pivot near the center (the screw). The material to be cut by the blades represents the weight, while the person on the handle end supplies the force needed for cutting the material.

Figure 7.13 shows the three types of levers, which differ in their arrangements. A first-class lever's parts are arranged like those of a pair of scissors. Its pivot is located between the weight and the force, making the sequence of components weight–pivot–force. Other examples of first-class levers are seesaws and hemostats (devices used to clamp blood vessels).

The parts of a second-class lever are arranged in the sequence pivot–weight–force, as in a wheelbarrow. The parts of a third-class lever are arranged in the sequence weight–force–pivot. Eyebrow tweezers or forceps used to grasp an object illustrate this type of lever.

The actions of bending and straightening the upper limb at the elbow illustrate bones and muscles functioning as levers (fig. 7.14). When the upper limb bends, the forearm bones represent the rigid bar; the elbow joint is the pivot; the hand is the weight that is moved; and the force is supplied by muscles on the anterior side of the arm. One of these muscles, the *biceps brachii,* is attached by a tendon to a projection (radial tuberosity) on the *radius* bone in the forearm, a short distance below the elbow. Since the parts of this lever are arranged in the sequence weight–force–pivot, it is a third-class lever.

When the upper limb straightens at the elbow, the forearm bones again serve as the rigid bar, the hand as the weight, and the elbow joint as the pivot. However, this time the *triceps brachii,* a muscle located on the posterior side of the arm, supplies the force. A tendon of this muscle is attached to a projection (olecranon process) of the ulna bone at the point of the elbow. Since the parts of the lever are arranged weight–pivot–force, it is a first-class lever.

Levers provide a range of movements. Levers that move limbs, for example, are arranged in ways that produce rapid motions, while others, such as those that move the head, help maintain posture with minimal effort.

### Blood Cell Formation

Very early in life the process of blood cell formation, called **hematopoiesis,** occurs in the yolk sac, which lies outside the human embryo (see chapter 23). Later in development, blood cells are manufactured in the liver and spleen, and still later they form in bone marrow.

**Marrow** is a soft, netlike mass of connective tissue within the medullary cavities of long bones, in the irregular spaces of spongy bone, and in the larger osteonic canals of compact bone tissue.

There are two kinds of marrow—red marrow and yellow marrow. *Red marrow* functions in the formation of red blood cells (erythrocytes), white blood cells (leukocytes), and blood platelets (thrombocytes). It is red because of the red, oxygen-carrying pigment **hemoglobin** contained within the red blood cells.

Red marrow occupies the cavities of most bones in an infant. With increasing age, however, yellow marrow replaces much of it. *Yellow marrow* stores fat and is inactive in blood cell production.

In an adult, red marrow is found primarily in the spongy bone of the skull, ribs, sternum, clavicles, vertebrae, and pelvis. If the blood cell supply is deficient, some yellow marrow may change back into red marrow and produce blood cells. Chapter 14 discusses blood cell formation.

# CLINICAL APPLICATION 7.3

## Fractures

When 7-year-old Jacob fell from the tree limb he had been hanging from and held out his arm at an odd angle, it was obvious that he had broken a bone. An X ray at the hospital emergency room confirmed this, and Jacob spent the next six weeks with his broken arm immobilized in a cast. ■

Many of us have experienced fractured, or broken, bones. A fracture is classified by its cause and the nature of the break. For example, a break due to injury is a *traumatic* fracture, whereas one resulting from disease is a *spontaneous,* or *pathologic,* fracture.

A broken bone exposed to the outside by an opening in the skin is termed a *compound* (*open*) *fracture.* It has the added danger of infection, because microorganisms enter through the broken skin. On the other hand, a break protected by uninjured skin is a *closed fracture.* Figure 7A shows several types of traumatic fractures.

### REPAIR OF A FRACTURE

Whenever a bone breaks, blood vessels within it and its periosteum rupture, and the periosteum is likely to tear. Blood escaping from the broken vessels spreads through the damaged area and soon forms a blood clot, or *hematoma.* Vessels in surrounding tissues dilate, swelling and inflaming tissues.

Within days or weeks, developing blood vessels and large numbers of osteoblasts originating from the periosteum invade the hematoma. The osteoblasts multiply rapidly in the regions close to the new blood vessels, building spongy bone nearby. Granulation tissue develops, and in regions farther from a blood supply, fibroblasts produce masses of fibrocartilage.

Meanwhile, phagocytic cells begin to remove the blood clot as well as any dead or damaged cells in the affected area. Osteoclasts also appear and resorb bone fragments, aiding in "cleaning up" debris.

In time, fibrocartilage fills the gap between the ends of the broken bone. This mass, termed a cartilaginous *callus,* is later replaced by

A *greenstick* fracture is incomplete, and the break occurs on the convex surface of the bend in the bone.

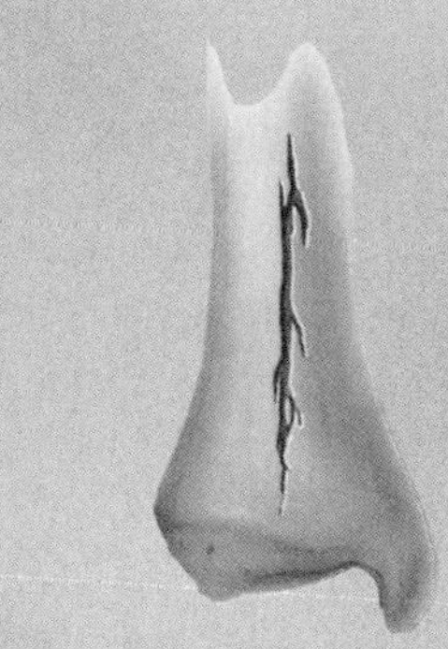

A *fissured* fracture involves an incomplete longitudinal break.

A *comminuted* fracture is complete and fragments the bone.

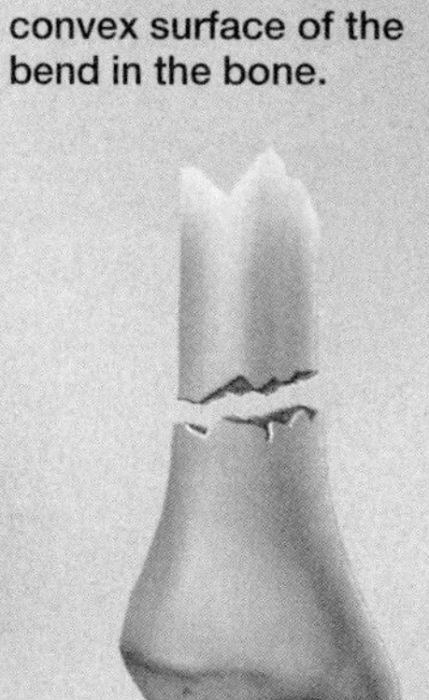

A *transverse* fracture is complete, and the break occurs at a right angle to the axis of the bone.

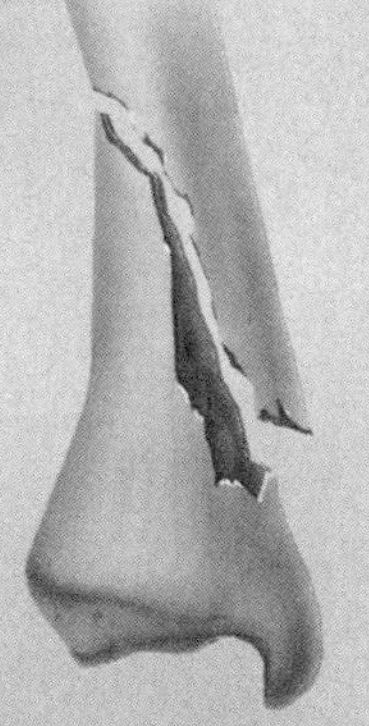

An *oblique* fracture occurs at an angle other than a right angle to the axis of the bone.

A *spiral* fracture is caused by twisting a bone excessively.

**FIGURE 7A**
Various types of fractures.

bone tissue in much the same way that the hyaline cartilage of a developing endochondral bone is replaced. That is, the cartilaginous callus breaks down, blood vessels and osteoblasts invade the area, and a bony callus fills the space.

Typically, more bone is produced at the site of a healing fracture than is needed to replace the damaged tissues. Osteoclasts remove the excess, and the final result is a bone shaped very much like the original. Figure 7B shows the steps in the healing of a fracture.

The rate of fracture repair depends upon several factors. For instance, if the ends of the broken bone are close together, healing is more rapid than if they are far apart. Setting fractured bones and using casts or metal pins to keep the broken ends together help speed healing, as well as aligning the fractured parts. Also, some bones naturally heal more rapidly than others. The long bones of the upper limbs, for example, may heal in half the time required by the long bones of the lower limbs, as Jacob was happy to discover. He also healed quickly because of his young age.

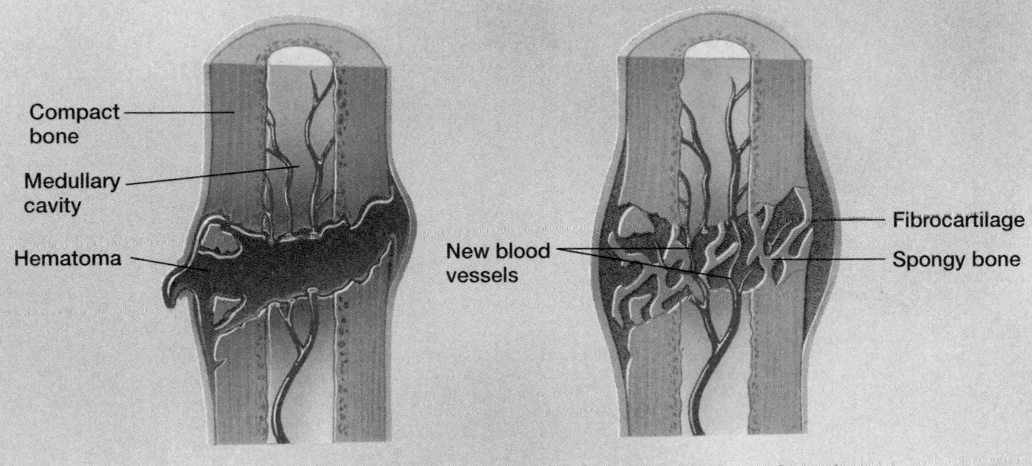


(a) Blood escapes from ruptured blood vessels and forms a hematoma.

(b) Spongy bone forms in regions close to developing blood vessels, and fibrocartilage forms in more distant regions.

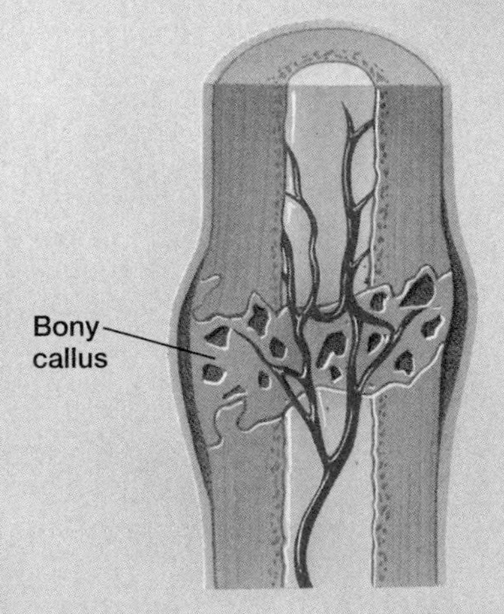


(c) A bony callus replaces fibrocartilage.

(d) Osteoclasts remove excess bony tissue, restoring new bone structure much like the original.

**FIGURE 7B**
Major steps in repair of a fracture.

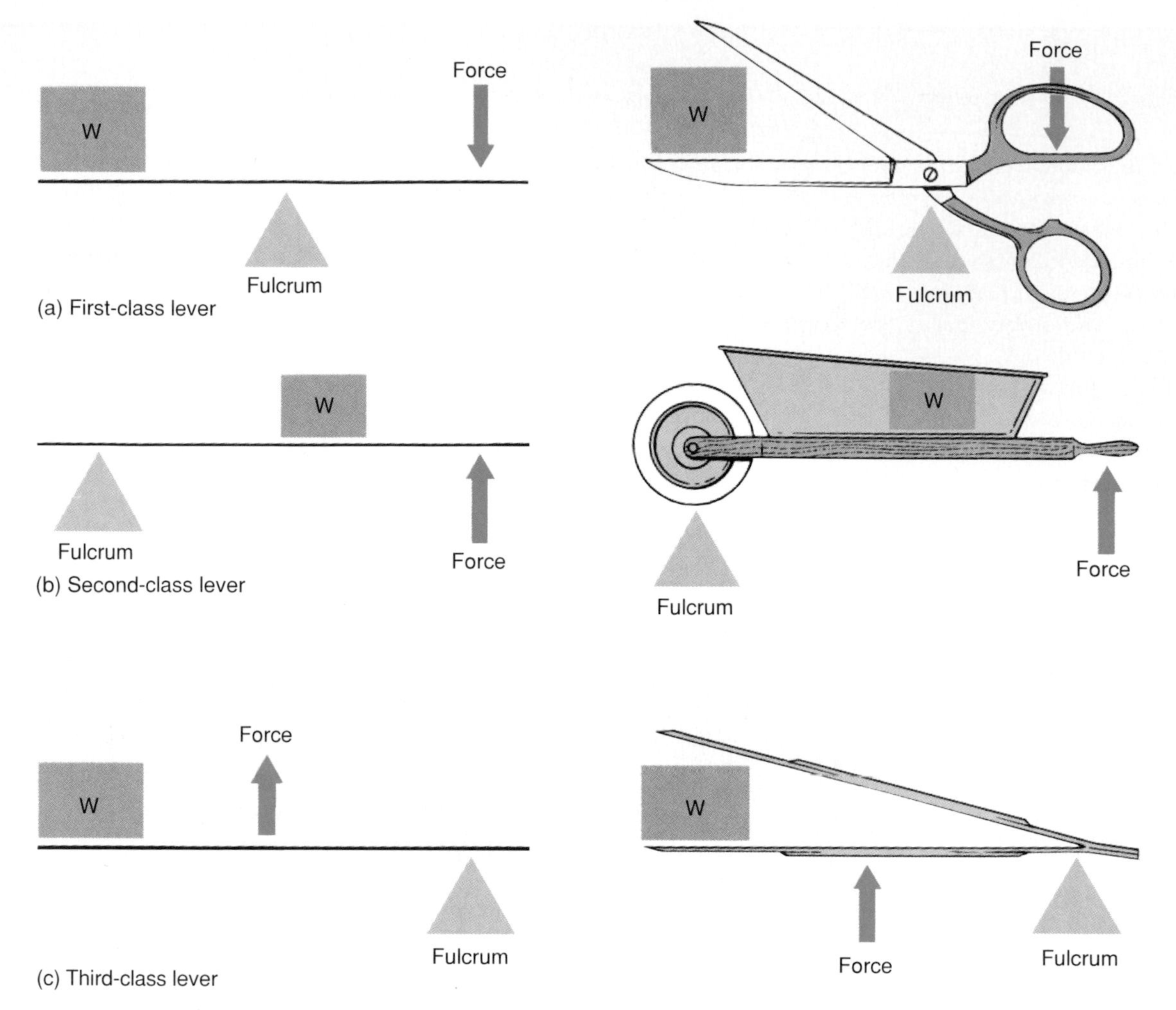


**FIGURE 7.13**
Three types of levers: (*a*) A first-class lever is used in a pair of scissors, (*b*) a second-class lever is used in a wheelbarrow, and (*c*) a third-class lever is used in a pair of forceps.

## Inorganic Salt Storage

Recall that the intercellular matrix of bone tissue contains collagen and inorganic mineral salts. The salts account for about 70% of the matrix by weight, and are mostly tiny crystals of a type of calcium phosphate called *hydroxyapatite.* Clinical Application 7.4 discusses osteoporosis, a condition that results from loss of bone mineral.

The human body requires calcium for a number of vital metabolic processes, including blood clot formation, nerve impulse conduction, and muscle cell contraction. When the blood is low in calcium, parathyroid hormone stimulates osteoclasts to break down bone tissue, releasing calcium salts from the intercellular matrix into the blood. On the other hand, very high blood calcium inhibits osteoclast activity, and calcitonin from the thyroid gland stimulates osteoblasts to form bone tissue, storing excess calcium in the matrix (fig. 7.15). The details of this homeostatic mechanism are presented in chapter 13.

In addition to storing calcium and phosphorus (as calcium phosphate), bone tissue contains lesser amounts of magnesium, sodium, potassium, and carbonate ions. Bones also accumulate certain harmful metallic elements such as lead, radium, and strontium, which are not normally present in the body but are sometimes ingested accidentally.

1. Name three major functions of bones.
2. Explain how parts of the upper limb form a first-class lever and a third-class lever.

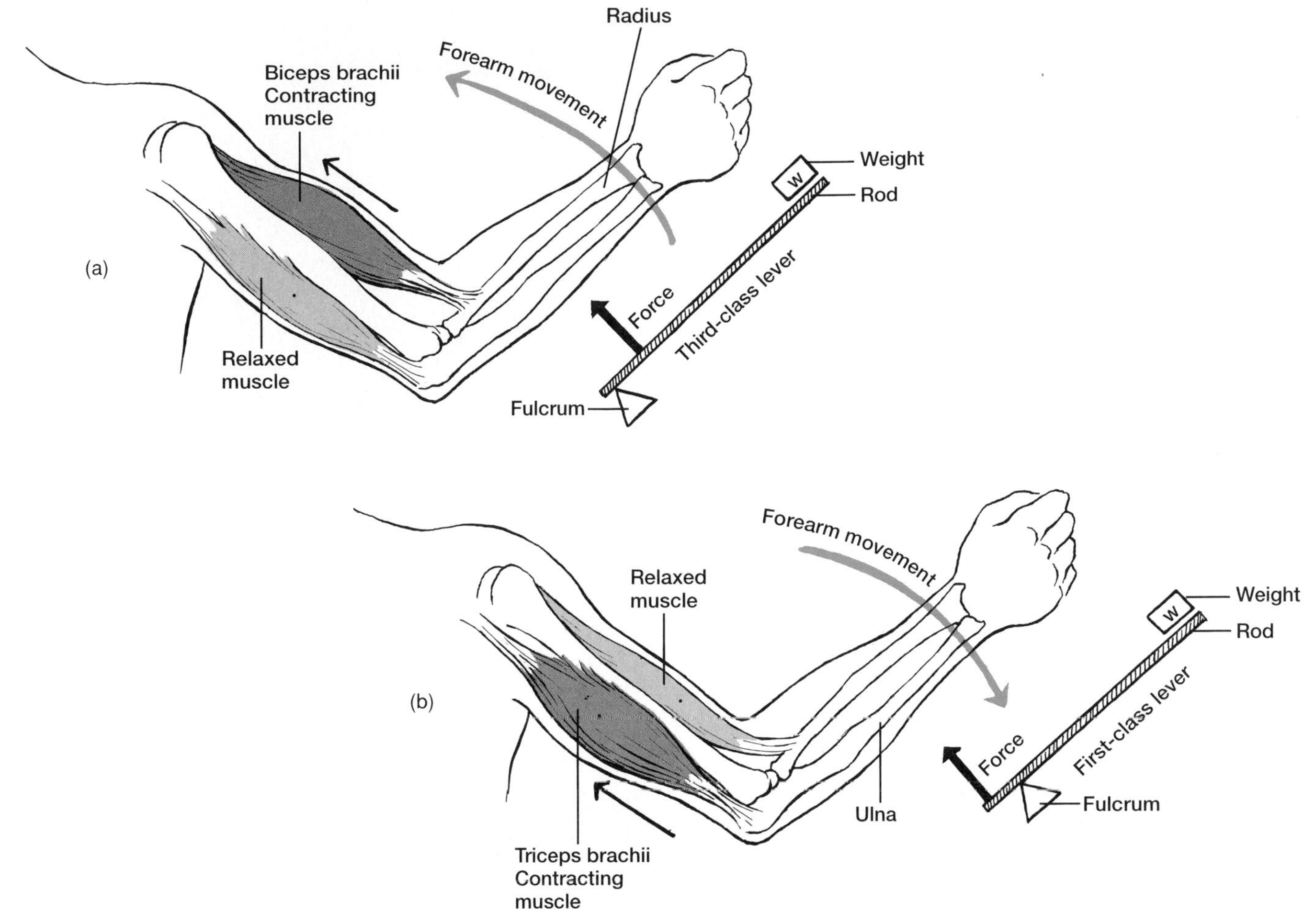


**FIGURE 7.14**

(*a*) When the upper limb bends at the elbow, bones and muscles work together as a third-class lever. (*b*) When the upper limb straightens at the elbow, bones and muscles act as a first-class lever.

**3** Explain regulation of the concentration of blood calcium.

**4** List the substances normally stored in bone tissue.

## SKELETAL ORGANIZATION

### Number of Bones

Although the number of bones in a human skeleton is often reported to be 206, the actual number varies from person to person. Some people lack certain bones, while others have extra ones. For example, the flat bones of the skull usually grow together and tightly join along irregular lines called sutures. Occasionally, extra bones called sutural bones (wormian bones) develop in these sutures (fig. 7.16). Also, extra small, round sesamoid bones may develop in tendons, where they reduce friction in places where tendons pass over bony prominences (see chart 7.3).

### Divisions of the Skeleton

For purposes of study, it is convenient to divide the skeleton into two major portions—an axial skeleton and an appendicular skeleton (fig. 7.17).

The **axial skeleton** consists of the bony and cartilaginous parts that support and protect the organs of the head, neck, and truck. These parts include the following:

1. **Skull.** The skull is composed of the *cranium* (brain case) and the *facial bones.*
2. **Hyoid bone.** The hyoid (hi′oid) bone is located in the neck between the lower jaw and the larynx (fig. 7.18). It does not articulate with any other bones but is fixed in position by muscles and ligaments. The hyoid bone supports the tongue and is an attachment for certain muscles that help move the tongue during swallowing. It can be felt approximately a finger's width above the anterior prominence of the larynx.

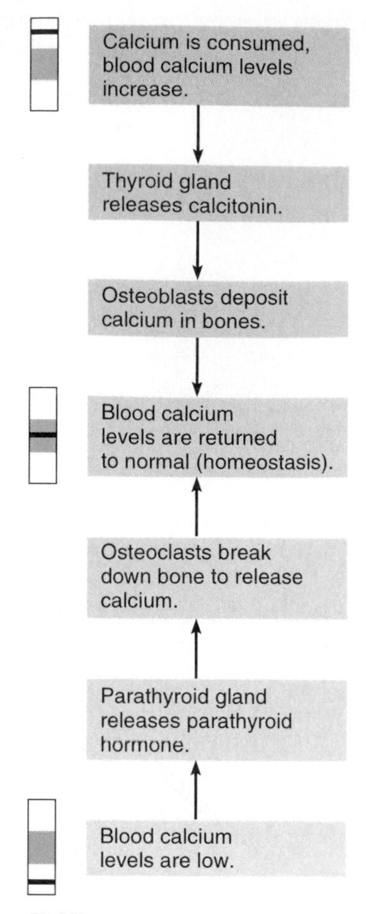


**FIGURE 7.15**
Hormonal regulation of bone calcium resorption and deposition.

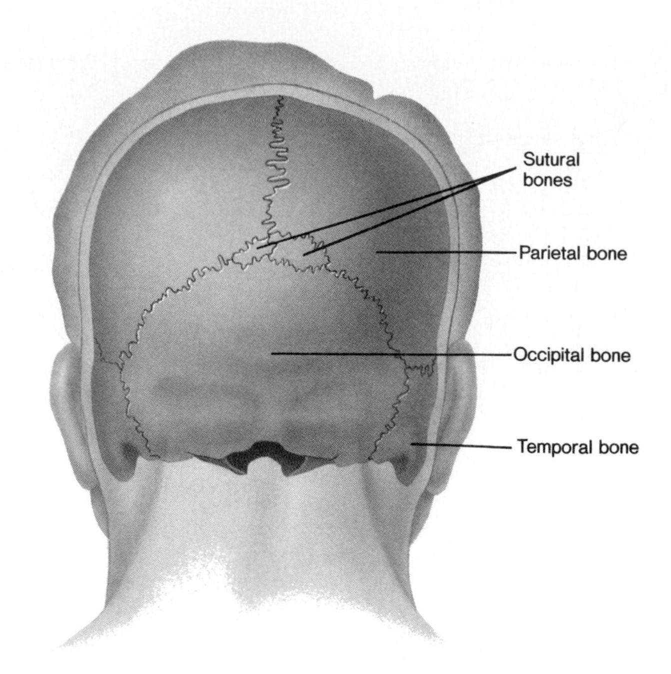


**FIGURE 7.16**
Sutural bones are extra bones that sometimes develop in sutures between the flat bones of the skull.

3. **Vertebral column.** The vertebral column, or backbone, consists of many vertebrae separated by cartilaginous *intervertebral disks.* This column forms the central axis of the skeleton. Near its distal end, several vertebrae fuse to form the **sacrum,** which is part of the pelvis. A small, rudimentary tailbone called the **coccyx** is attached to the end of the sacrum.
4. **Thorax.** The thorax protects the organs of the thoracic cavity and the upper abdominal cavity. It is composed of twelve pairs of **ribs,** which articulate posteriorly with thoracic vertebrae. It also includes the **sternum** (ster′num), or breastbone, to which most of the ribs are attached anteriorly.

The **appendicular skeleton** consists of the bones of the upper and lower limbs and the bones that anchor the limbs to the axial skeleton. It includes the following:

1. **Pectoral girdle.** The pectoral girdle is formed by a **scapula** (scap′u-lah), or shoulder blade, and a **clavicle** (klav′ĭ-k'l), or collarbone, on both sides of the body. The pectoral girdle connects the bones of the upper limbs to the axial skeleton and aids in upper limb movements.
2. **Upper limbs.** Each upper limb consists of a **humerus** (hu′mer-us), or arm bone, and two forearm bones—a **radius** (ra′de-us) and an **ulna** (ul′nah) and a hand. These three bones articulate with each other at the elbow joint. At the distal end of the radius and ulna is the hand. The hand includes the wrist, palm, and fingers. There are eight **carpals** (kar′pals) or wrist bones. The five bones of the palm are called **metacarpals,** and the fourteen finger bones are called **phalanges** (fah-lan′jēz).
3. **Pelvic girdle.** The pelvic girdle is formed by two **coxal** (kok′sal) bones or hipbones, which are attached to each other anteriorly and to the

CLINICAL APPLICATION 7.4

## Osteoporosis

It is an all-too-familiar scenario. The elderly woman pulls herself out of bed, reaches for the night table for support, and misses. She falls, landing on her hip. A younger woman would pull herself up, and maybe ache for a few minutes and develop a black and blue mark by the next day. But the 80-year-old, with weakened, brittle bones, suffers a broken hip. Each year in the United States, 200,000 senior citizens break their hips, more than 90% of the time as the result of an accident.

In **osteoporosis,** the skeletal system loses bone volume and mineral content. This disorder is associated with aging. Within affected bones, trabeculae are lost, and the bones develop spaces and canals. These enlarge and fill with fibrous and fatty tissues. Such bones easily fracture and may break spontaneously because they are no longer able to support body weight. For example, a person with osteoporosis may suffer a spontaneous fracture of the thigh bone (femur) at the hip or the collapse of sections of the backbone (vertebrae). Similarly, the distal portion of a forearm bone (radius) near the wrist may fracture as a result of a minor stress.

Osteoporosis causes many fractures in persons over forty-five years of age. Although it may affect either gender, it is most common in thin, light-complexioned females after menopause (see chapter 22).

Factors that increase the risk of osteoporosis include low intake of dietary calcium and lack of physical exercise (particularly during the early growing years). However, excessively strenuous exercise in adolescence can delay puberty, which raises the risk of developing osteoporosis later in life for both sexes.

In females, declining levels of the hormone estrogen contribute to development of osteoporosis. The ovaries produce estrogen until menopause. Evidence of the estrogen–osteoporosis link comes from studies on women who have declining estrogen levels and increased risk of osteoporosis. These include young women who have had their ovaries removed; women who have anorexia nervosa (self-starvation) that stopped their menstrual cycles; and women past menopause. Drinking alcohol, smoking cigarettes, and inheriting certain genes may also increase a person's risk of developing osteoporosis.

Fortunately, osteoporosis may be prevented if steps are taken early enough. Bone mass usually peaks at about age thirty-five. Thereafter, bone loss may exceed bone formation in both males and females. To reduce such loss, people in their mid-twenties and older should take in 1,000–1,500 milligrams of calcium daily. An 8-ounce glass of nonfat milk, for example, contains about 275 milligrams of calcium. It is also recommended that people engage in exercise regularly, especially walking or jogging, in which the bones support body weight. Additionally, postmenopausal women may require estrogen replacement therapy, which should be carried out under the supervision of a physician. As a rule, women have about 30% less bone than men; after menopause, women typically lose bone mass twice as fast as men do.

Confirming osteoporosis is sometimes difficult. An X-ray film may not reveal a decrease in bone density until 20% to 30% of the bone tissue is lost. Noninvasive diagnostic techniques, however, can detect rapid changes in bone mass. These include a *densitometer scanner* that measures the density of wrist bones, and *quantitative computed tomography,* which can visualize the density of other bones (see chapter 1).

In other cases, a physician may take a bone sample, usually from a hipbone, in order to directly assess the condition of the tissue. Such a biopsy may also be used to judge the effectiveness of treatment for bone disease.

**CHART 7.3 BONES OF THE ADULT SKELETON**

| | |
|---|---|
| 1. Axial Skeleton | |
| a. Skull | 22 bones |
| 8 cranial bones | |
| frontal 1 | |
| parietal 2 | |
| occipital 1 | |
| temporal 2 | |
| sphenoid 1 | |
| ethmoid 1 | |
| 13 facial bones | |
| maxilla 2 | |
| palatine 2 | |
| zygomatic 2 | |
| lacrimal 2 | |
| nasal 2 | |
| vomer 1 | |
| inferior nasal concha 2 | |
| 1 mandible | |
| b. Middle ear bones | 6 bones |
| malleus 2 | |
| incus 2 | |
| stapes 2 | |
| c. Hyoid | 1 bone |
| hyoid bone 1 | |
| d. Vertebral column | 26 bones |
| cervical vertebra 7 | |
| thoracic vertebra 12 | |
| lumbar vertebra 5 | |
| sacrum 1 | |
| coccyx 1 | |
| e. Thoracic cage | 25 bones |
| rib 24 | |
| sternum 1 | |
| 2. Appendicular Skeleton | |
| a. Pectoral girdle | 4 bones |
| scapula 2 | |
| clavicle 2 | |
| b. Upper limbs | 60 bones |
| humerus 2 | |
| radius 2 | |
| ulna 2 | |
| carpal 16 | |
| metacarpal 10 | |
| phalanx 28 | |
| c. Pelvic girdle | 2 bones |
| coxal bone 2 | |
| d. Lower limbs | 60 bones |
| femur 2 | |
| tibia 2 | |
| fibula 2 | |
| patella 2 | |
| tarsal 14 | |
| metatarsal 10 | |
| phalanx 28 | |
| Total | 206 bones |

sacrum posteriorly. They connect the bones of the lower limbs to the axial skeleton and, with the sacrum and coccyx, form the **pelvis,** which protects the lower abdominal and internal reproductive organs.

4. **Lower limbs.** Each lower limb consists of a **femur** (fe′mur), or thighbone, and two leg bones—a large **tibia** (tib′e-ah), or shinbone, and a slender **fibula** (fib′u-lah), or calf bone and a foot. These three bones articulate with each other at the knee joint, where the **patella** (pah-tel′ah), or kneecap, covers the anterior surface. At the distal ends of the tibia and fibula is the foot. The foot includes the ankle, instep, and toes. There are seven **tarsals** (tahr′sals) or anklebones. The five bones of the instep are called **metatarsals,** and the fourteen bones of the toes (like the fingers) are called **phalanges.** Chart 7.4 defines some terms used to describe skeletal structures.

**1** Distinguish between the axial and appendicular skeletons.

**2** List the bones of the axial skeleton; of the appendicular skeleton.

## SKULL

A human skull usually consists of twenty-two bones that, except for the lower jaw, are firmly interlocked along lines called **sutures.** Eight of these interlocked bones make up the cranium, and thirteen form the facial skeleton. The **mandible** (man′dĭ-b′l), or lower jawbone, is a movable bone held to the cranium by ligaments (figs. 7.19, 7.21, and 7.22). Some facial and cranial bones together form the orbit of the eye (fig. 7.20).

Plates 8–36 on pages 246–260 show a set of photographs of the human skull and its parts.

### Cranium

The **cranium** (kra′ne-um) encloses and protects the brain, and its surface provides attachments for muscles that make chewing and head movements possible. Some of the cranial bones contain air-filled cavities called *sinuses,* which are lined with mucous membranes and connect by passageways to the nasal cavity. Sinuses reduce the weight of the skull and increase the intensity of the voice by serving as resonant sound chambers.

The eight bones of the cranium (chart 7.5) are as follows:

1. **Frontal bone.** The frontal (frun′tal) bone forms the anterior portion of the skull above the eyes, including the forehead, the roof of the nasal cavity, and the roofs of the orbits (bony sockets) of the eyes. On the upper margin of each orbit, the frontal bone is marked by a *supraorbital foramen* through which blood vessels and nerves pass to the tissues of the forehead. Within the frontal bone are two

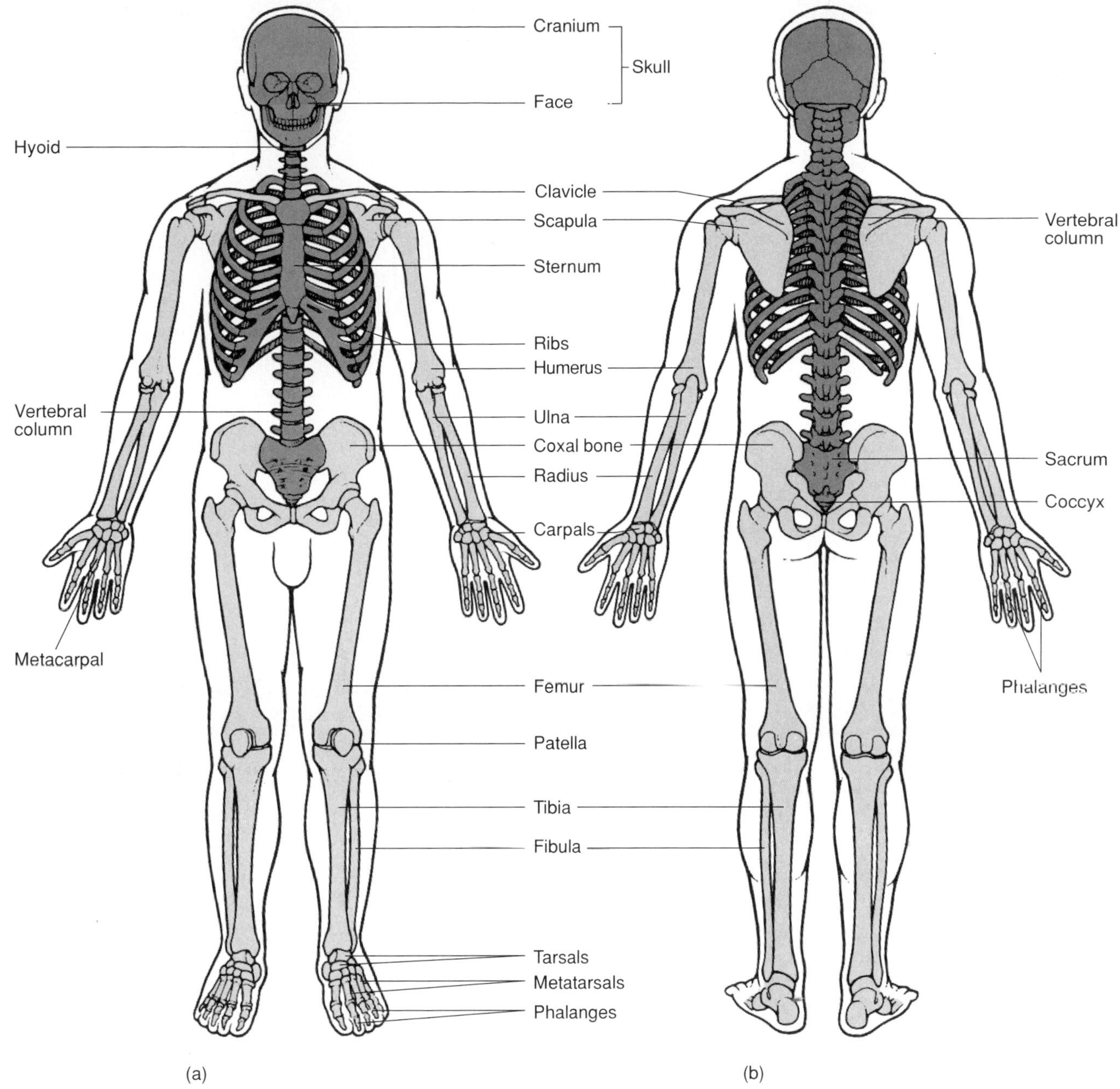


**FIGURE 7.17**

Major bones of the skeleton. (*a*) Anterior view; (*b*) posterior view. The axial portions are shown in orange and the appendicular portions are shown in yellow.

*frontal sinuses,* one above each eye near the midline. Although the frontal bone is a single bone in adults, it develops in two parts (see fig. 7.33). These halves grow together and usually completely fuse by the fifth or sixth year of life.

2. **Parietal bones.** One parietal (pah-ri′ĕ-tal) bone is located on each side of the skull just behind the frontal bone. Each is shaped like a curved plate and has four borders. Together, the parietal bones form the bulging sides and roof of the cranium. They are fused at the midline along the *sagittal suture,* and they meet the frontal bone along the *coronal suture.*

3. **Occipital bone.** The occipital (ok-sip′ĭ-tal) bone joins the parietal bones along the *lambdoidal* (lam′doid-al) *suture.* It forms the back of the skull and the base of the cranium. A large opening on its lower surface, the *foramen magnum,* houses nerve fibers from the brain that pass through and enter the vertebral canal to become part of the

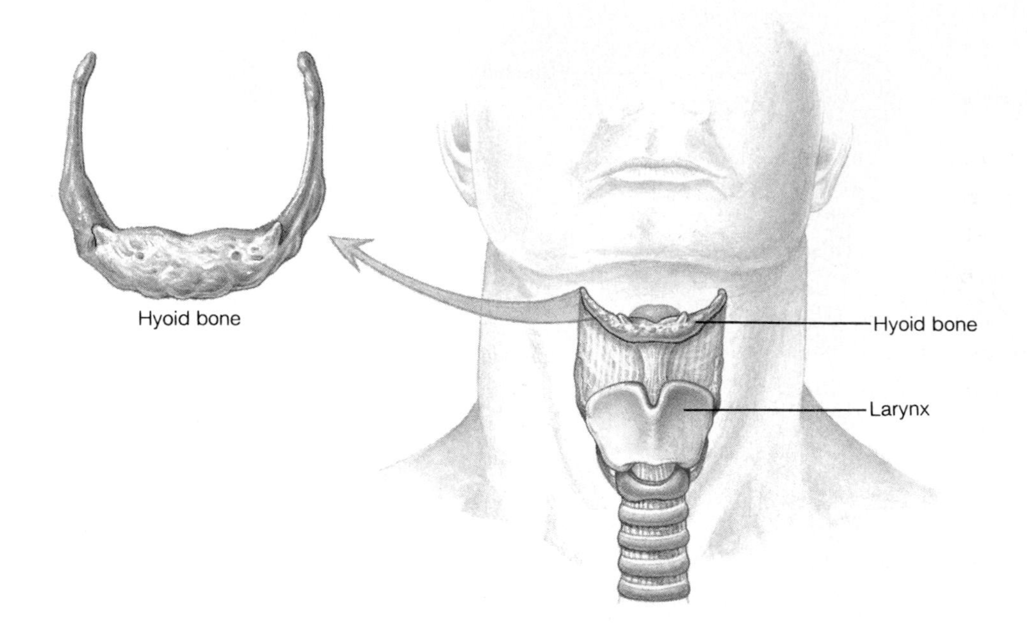


**FIGURE 7.18**

The hyoid bone supports the tongue and serves as an attachment for muscles that move the tongue and function in swallowing.

## CHART 7.4 Terms Used to Describe Skeletal Structures

| Term | Definition | Example |
|---|---|---|
| condyle (kon′dil) | A rounded process that usually articulates with another bone | Occipital condyle of the occipital bone (fig. 7.22) |
| crest (krest) | A narrow, ridgelike projection | Iliac crest of the ilium (fig. 7.50) |
| epicondyle (ep″ĭ-kon′dīl) | A projection situated above a condyle | Medial epicondyle of the humerus (fig. 7.45) |
| facet (fas′et) | A small, nearly flat surface | Facet of a thoracic vertebra (fig. 7.38) |
| fissure (fish′ūr) | A cleft or groove | Inferior orbital fissure in the orbit of the eye (fig. 7.20) |
| fontanel (fon″tah-nel′) | A soft spot in the skull where membranes cover the space between bones | Anterior fontanel between the frontal and parietal bones (fig. 7.33) |
| foramen (fo-ra′men) | An opening through a bone that usually serves as a passageway for blood vessels, nerves, or ligaments | Foramen magnum of the occipital bone (fig. 7.22) |
| fossa (fos′ah) | A relatively deep pit or depression | Olecranon fossa of the humerus (fig. 7.45) |
| fovea (fo′ve-ah) | A tiny pit or depression | Fovea capitis of the femur (fig. 7.53) |
| head (hed) | An enlargement on the end of a bone | Head of the humerus (fig. 7.45) |
| linea (lin′e-ah) | A narrow ridge | Linea aspera of the femur (fig. 7.53) |
| meatus (me-a′tus) | A tubelike passageway within a bone | External auditory meatus of the ear (fig. 7.21) |
| process (pros′es) | A prominent projection on a bone | Mastoid process of the temporal bone (fig. 7.21) |
| ramus (ra′mus) | A structure given off from another larger one | Ramus of the mandible (fig. 7.31) |
| sinus (si′nus) | A cavity within a bone | Frontal sinus of the frontal bone (fig. 7.27) |
| spine (spīn) | A thornlike projection | Spine of the scapula (fig. 7.43) |
| suture (soo′cher) | An interlocking line of union between bones | Lambdoidal suture between the occipital and parietal bones (fig. 7.21) |
| trochanter (tro-kan′ter) | A relatively large process | Greater trochanter of the femur (fig. 7.53) |
| tubercle (tu′ber-kl) | A small, knoblike process | Tubercle of a rib (fig. 7.41) |
| tuberosity (tu″bĕ-ros′ĭ-te) | A knoblike process usually larger than a tubercle | Radial tuberosity of the radius (fig. 7.46) |

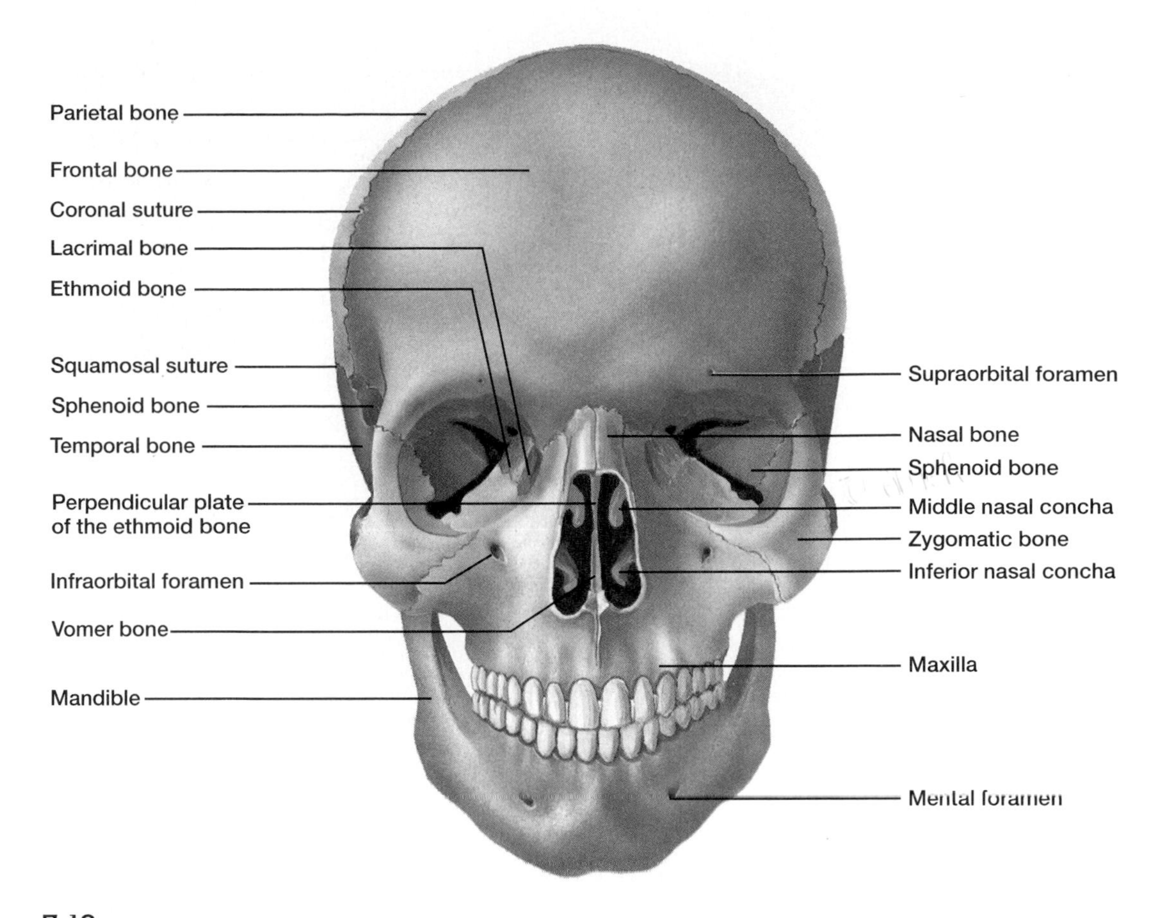


**FIGURE 7.19**
Anterior view of the skull.

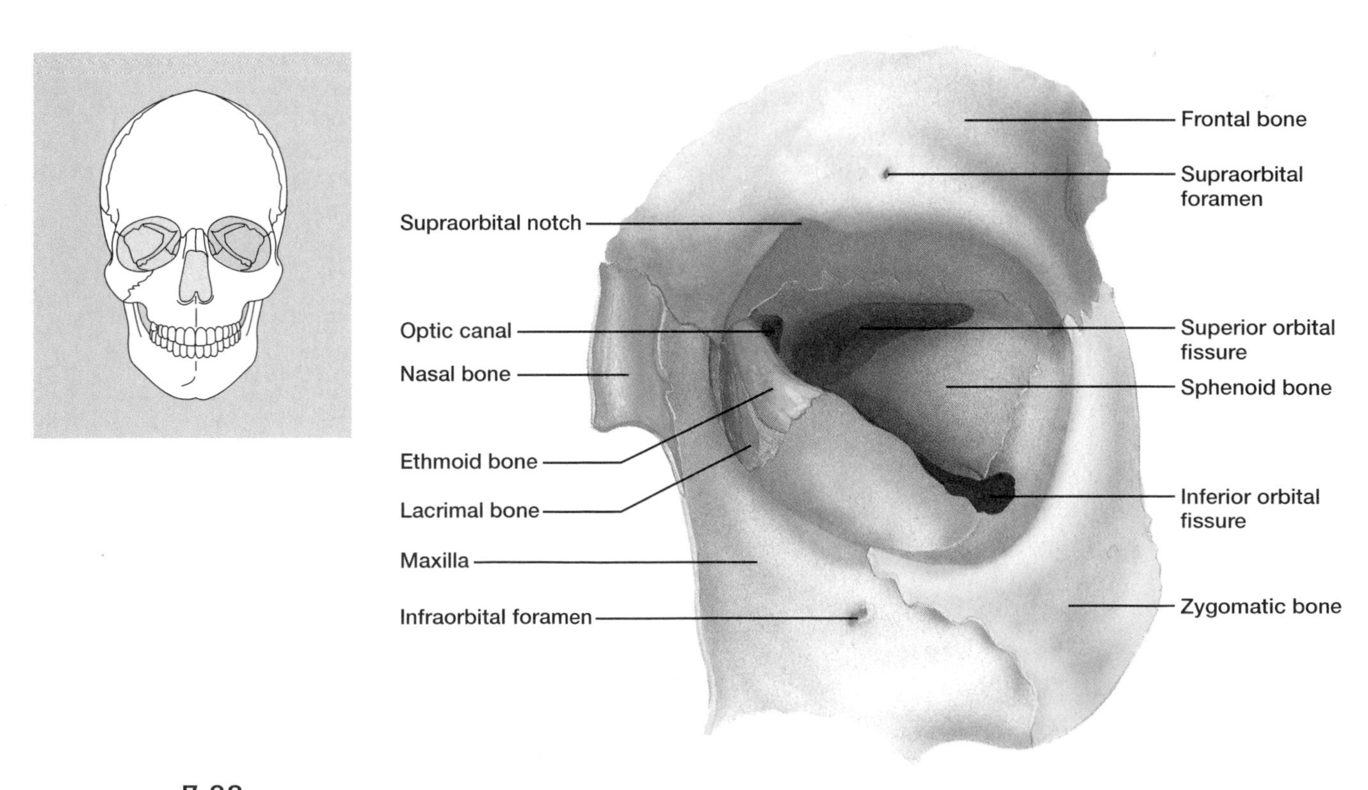


**FIGURE 7.20**
The orbit of the eye includes both cranial and facial bones.

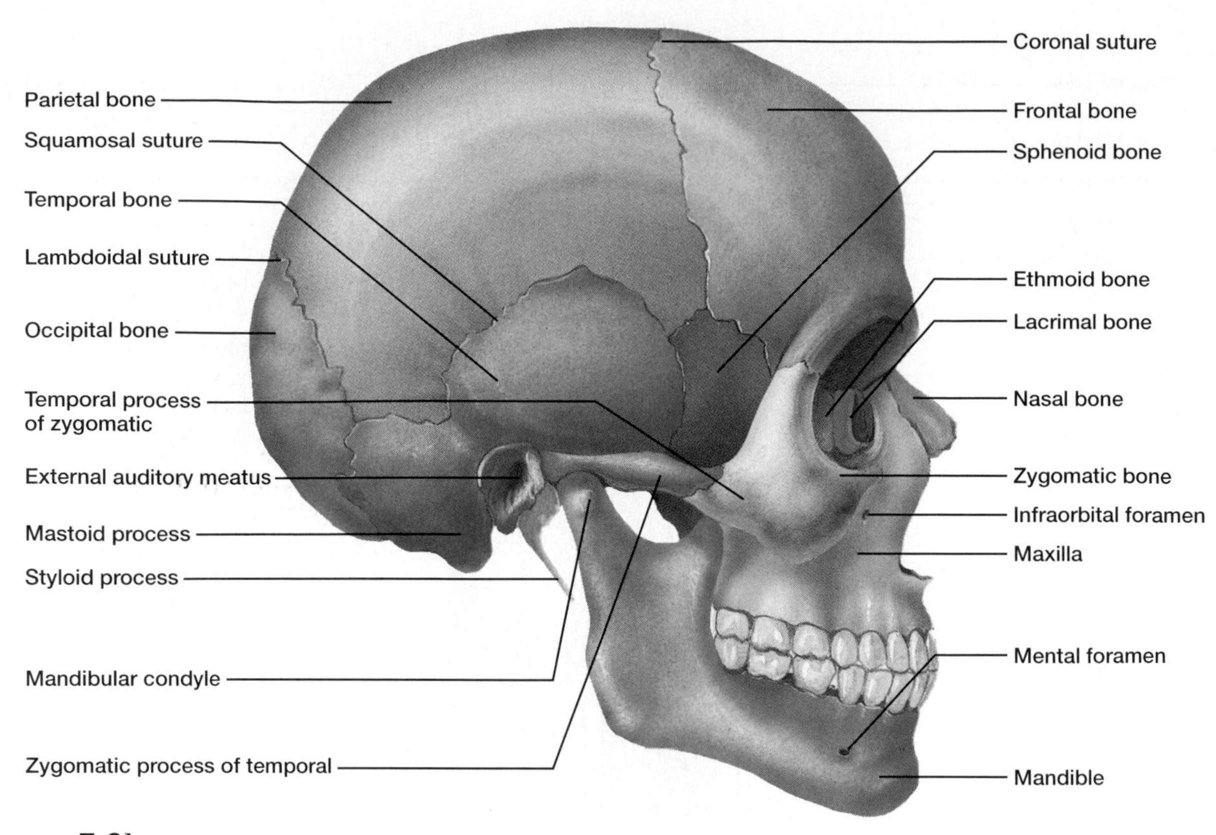


**FIGURE 7.21**
Lateral view of the skull.

**CHART 7.5 CRANIAL BONES**

| Name and Number | Description | Special Features |
|---|---|---|
| Frontal (1) | Forms forehead, roof of nasal cavity, and roofs of orbits | Supraorbital foramen, frontal sinuses |
| Parietal (2) | Form side walls and roof of cranium | Fused at midline along sagittal suture |
| Occipital (1) | Forms back of skull and base of cranium | Foramen magnum, occipital condyles |
| Temporal (2) | Form side walls and floor of cranium | External auditory meatus, mandibular fossa, mastoid process, styloid process, zygomatic process |
| Sphenoid (1) | Forms parts of base of cranium, sides of skull, and floors and sides of orbits | Sella turcica, sphenoidal sinuses |
| Ethmoid (1) | Forms parts of roof and walls of nasal cavity, floor of cranium, and walls of orbits | Cribriform plates, perpendicular plate, superior and middle nasal conchae, ethmoidal sinuses, crista galli |

spinal cord. Rounded processes called *occipital condyles,* located on each side of the foramen magnum, articulate with the first vertebra (atlas) of the vertebral column.

4. **Temporal bones.** A temporal (tem′por-al) bone on each side of the skull joins the parietal bone along a *squamosal* (skwa-mo′sal) *suture.* The temporal bones form parts of the sides and the base of the cranium. Located near the inferior margin is an opening, the *external auditory* (acoustic) *meatus,* which leads inward to parts of the ear. The temporal bones also house the internal ear structures and have depressions called the *mandibular fossae* (glenoid fossae) that articulate with condyles of the mandible. Below each external auditory meatus are two projections—a rounded *mastoid process* and a long, pointed *styloid process.* The mastoid process provides an attachment for certain muscles of the neck, while the styloid process

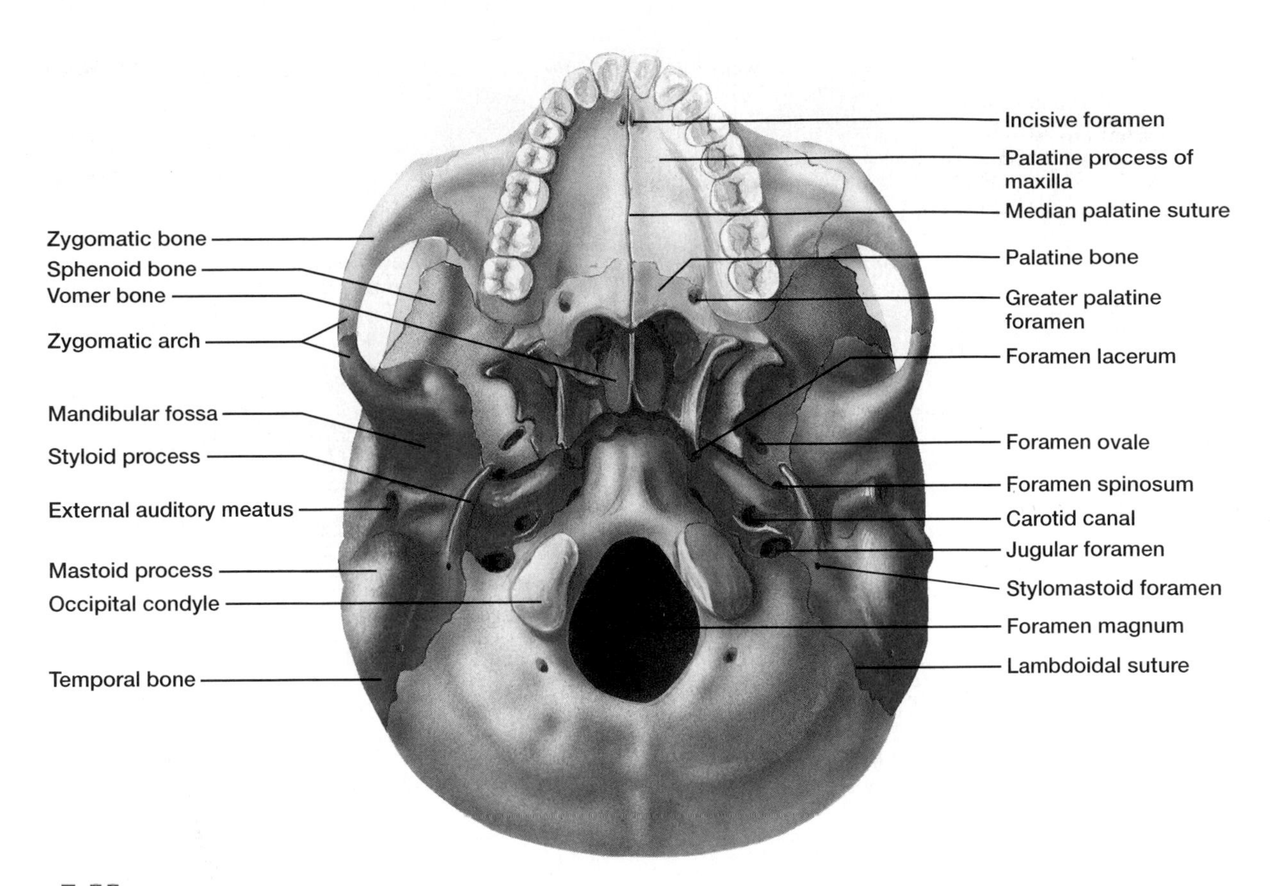


**FIGURE 7.22**
Inferior view of the skull.

anchors muscles associated with the tongue and pharynx. An opening near the mastoid process, the *carotid canal,* transmits the internal carotid artery. An opening between the temporal and occipital bones, the *jugular foramen,* accommodates the internal jugular vein (see fig. 7.22).

The mastoid process may become infected. The tissues in this region of the temporal bone contain a number of interconnected air cells lined with mucous membranes that communicate with the middle ear. These spaces sometimes become inflamed when microorganisms spread into them from an infected middle ear (*otitis media*). The resulting mastoid infection, called *mastoiditis,* is of particular concern because nearby membranes that surround the brain may become infected.

A *zygomatic process* projects anteriorly from the temporal bone in the region of the external auditory meatus. It joins the *zygomatic bone* and helps form the prominence of the cheek.

5. **Sphenoid bone.** The sphenoid (sfe′noid) bone (fig. 7.23) is wedged between several other bones in the anterior portion of the cranium. It consists of a central part and two winglike structures that extend laterally toward each side of the skull. This bone helps form the base of the cranium, the sides of the skull, and the floors and sides of the orbits. Along the midline within the cranial cavity, a portion of the sphenoid bone indents to form the saddle-shaped *sella turcica* (sel′ah tur′si-ka; Turk's saddle). In this depression lies the pituitary gland, which hangs from the base of the brain by a stalk.

   The sphenoid bone also contains two *sphenoidal sinuses,* which lie side by side and are separated by a bony septum that projects downward into the nasal cavity.

6. **Ethmoid bone.** The ethmoid (eth′moid) bone (fig. 7.24) is located in front of the sphenoid bone. It consists of two masses, one on each side of the nasal cavity, which are joined horizontally by thin *cribriform* (krib′rĭ-form) *plates.* These plates form part of the roof of the nasal cavity, and nerves associated with the sense of smell pass through tiny openings (*olfactory foramina*)

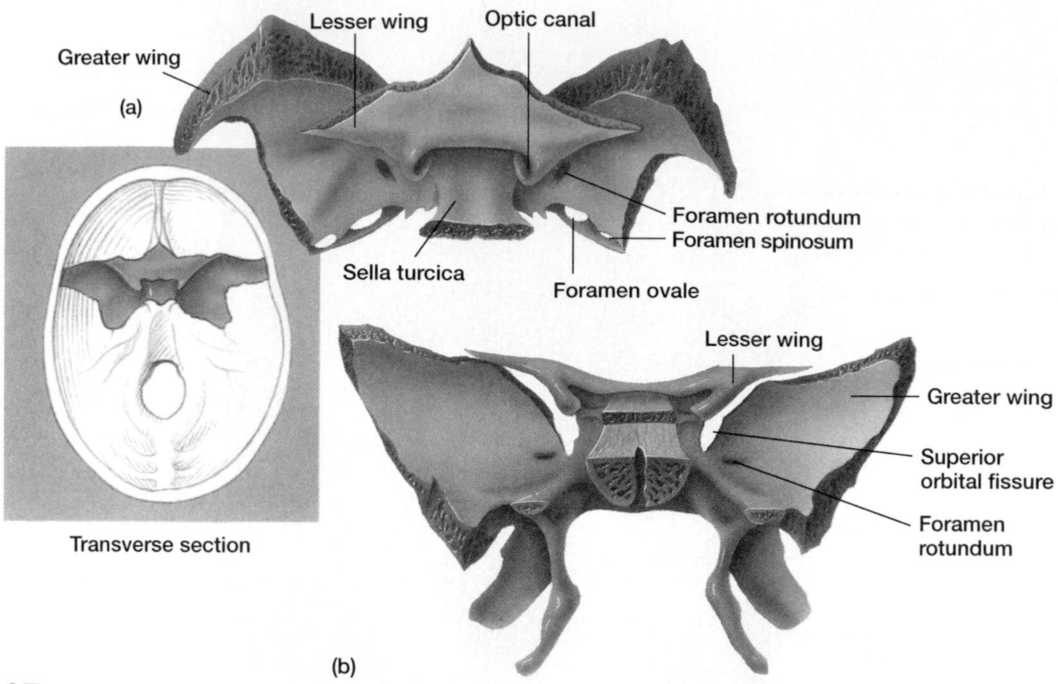


**FIGURE 7.23**

(*a*) The sphenoid bone viewed from above; (*b*) posterior view. (The sphenoidal sinuses are within the bone and are not visible in this representation.)

in them. Portions of the ethmoid bone also form sections of the cranial floor, orbital walls, and nasal cavity walls. A *perpendicular plate* projects downward in the midline from the cribriform plates to form most of the nasal septum.

Delicate, scroll-shaped plates called the *superior nasal concha* (kong′kah) and the *middle nasal concha* project inward from the lateral portions of the ethmoid bone toward the perpendicular plate. These bony plates support mucous membranes that line the nasal cavity. The mucous membranes, in turn, begin moistening, warming, and filtering air as it enters the respiratory tract. The lateral portions of the ethmoid bone contain many small air spaces, the *ethmoidal sinuses.* Figure 7.25 shows various structures in the nasal cavity.

Projecting upward into the cranial cavity between the cribriform plates is a triangular process of the ethmoid bone called the *crista galli* (kris′tă gal′li; cock's comb). Membranes that enclose the brain attach to this process. Figure 7.26 shows a view of the cranial cavity.

## Facial Skeleton

The **facial skeleton** consists of thirteen immovable bones and a movable lower jawbone. In addition to forming the basic shape of the face, these bones provide attachments for muscles that move the jaw and control facial expressions.

The bones of the facial skeleton are as follows:

1. **Maxillary bones.** The maxillary (mak′sĭ-ler″e) bones (pl. *maxillae,* mak-sĭl′e) form the upper jaw; together they form the keystone of the face, since all the other immovable facial bones articulate with them.

   Portions of these bones comprise the anterior roof of the mouth (*hard palate*), the floors of the orbits, and the sides and floor of the nasal cavity. They also contain the sockets of the upper teeth. Inside the maxillae, lateral to the nasal cavity, are *maxillary sinuses.* These spaces are the largest of the sinuses, and they extend from the floor of the orbits to the roots of the upper teeth. Figure 7.27 shows the locations of the maxillary and other sinuses. See chart 7.6 for a summary of the sinuses.

   During development, portions of the maxillary bones called *palatine processes* grow together and fuse along the midline, or median palatine suture, to form the anterior section of the hard palate (see fig. 7.22).

   The inferior border of each maxillary bone projects downward, forming an *alveolar* (al-ve′o-lar) *process.* Together these processes form a horseshoe-shaped *alveolar arch* (dental arch). Teeth occupy cavities in this arch (dental alveoli). Fibrous connective tissue binds teeth to the bony sockets (see chapter 17).

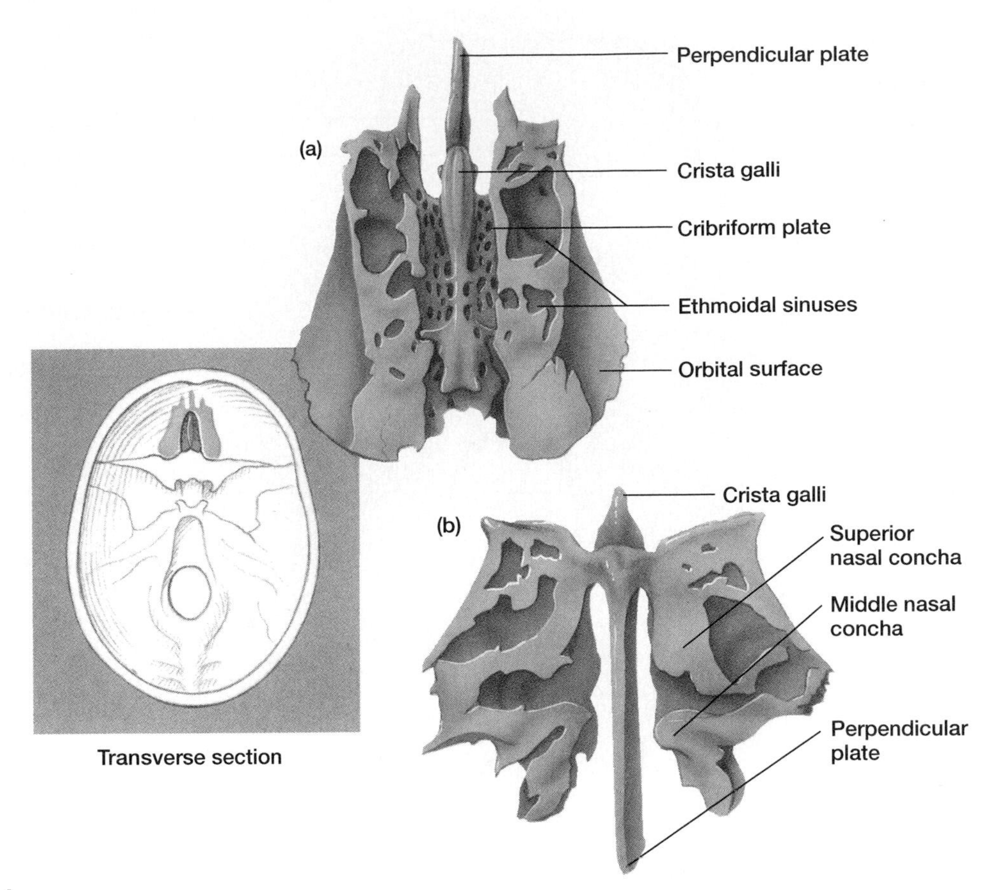


**FIGURE 7.24**
The ethmoid bone viewed (*a*) from above and (*b*) from behind.

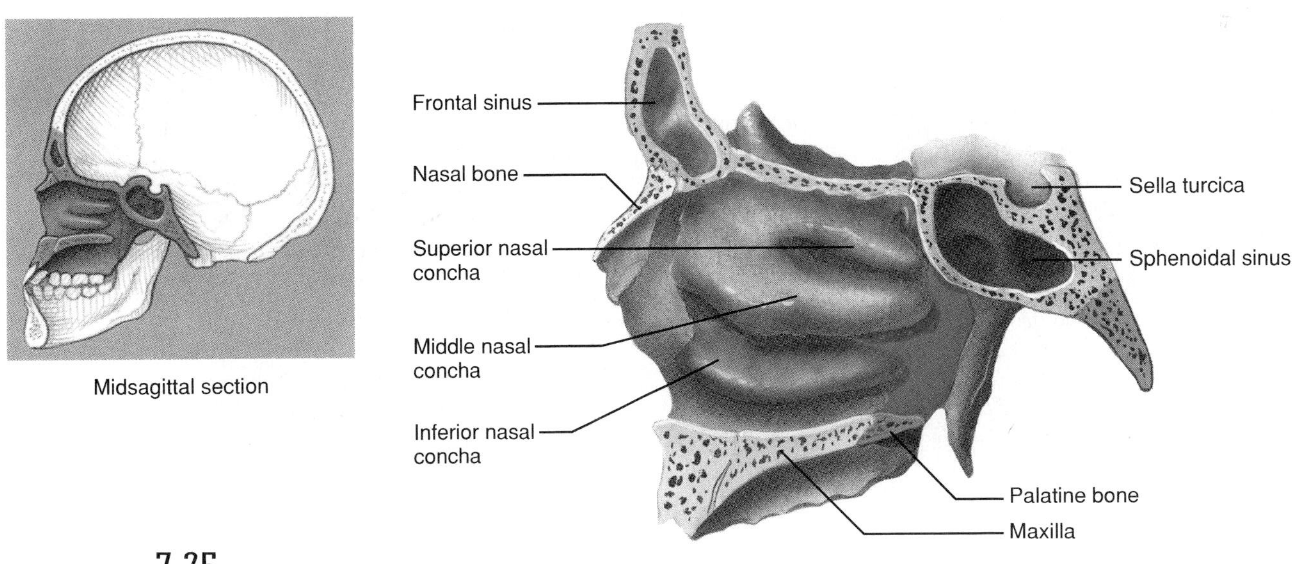


**FIGURE 7.25**
Lateral wall of the nasal cavity.

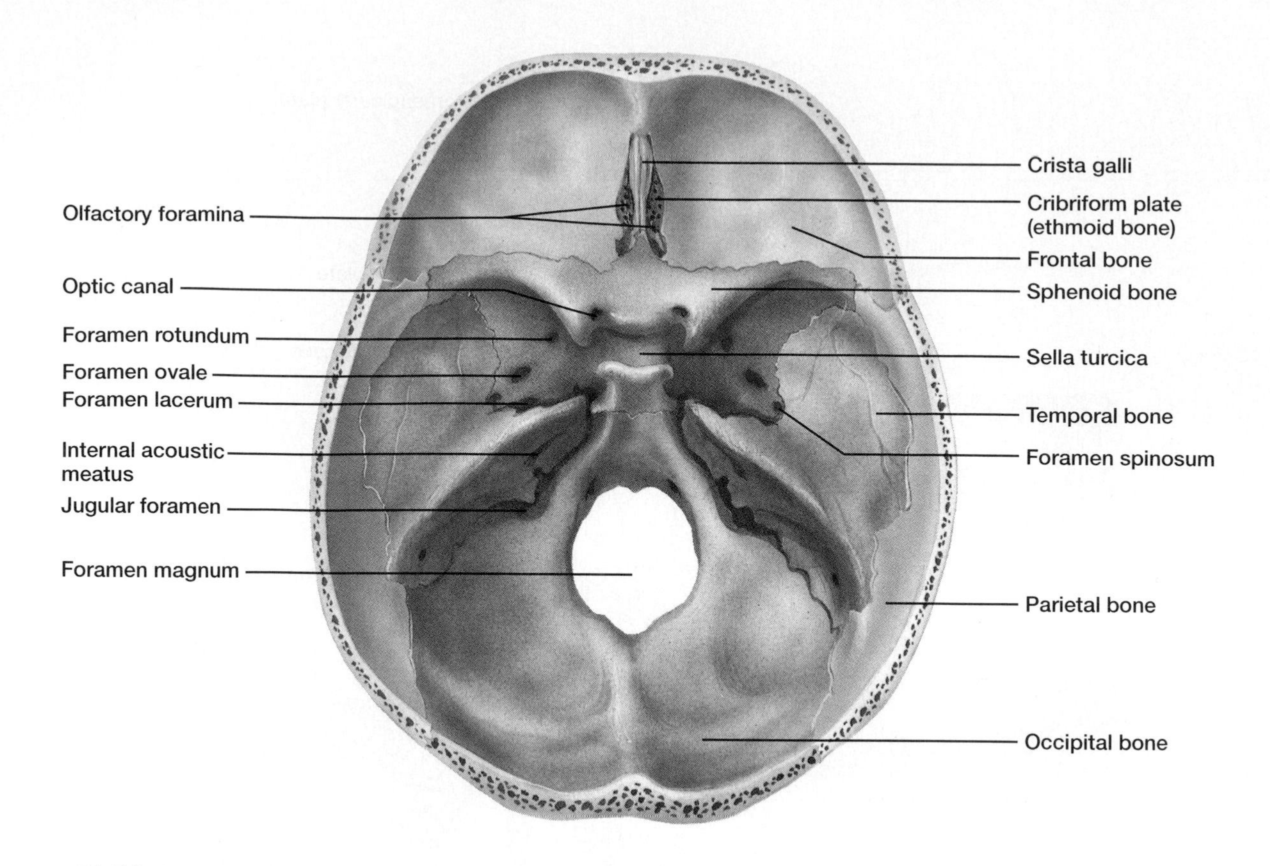


**FIGURE 7.26**
Floor of the cranial cavity viewed from above.

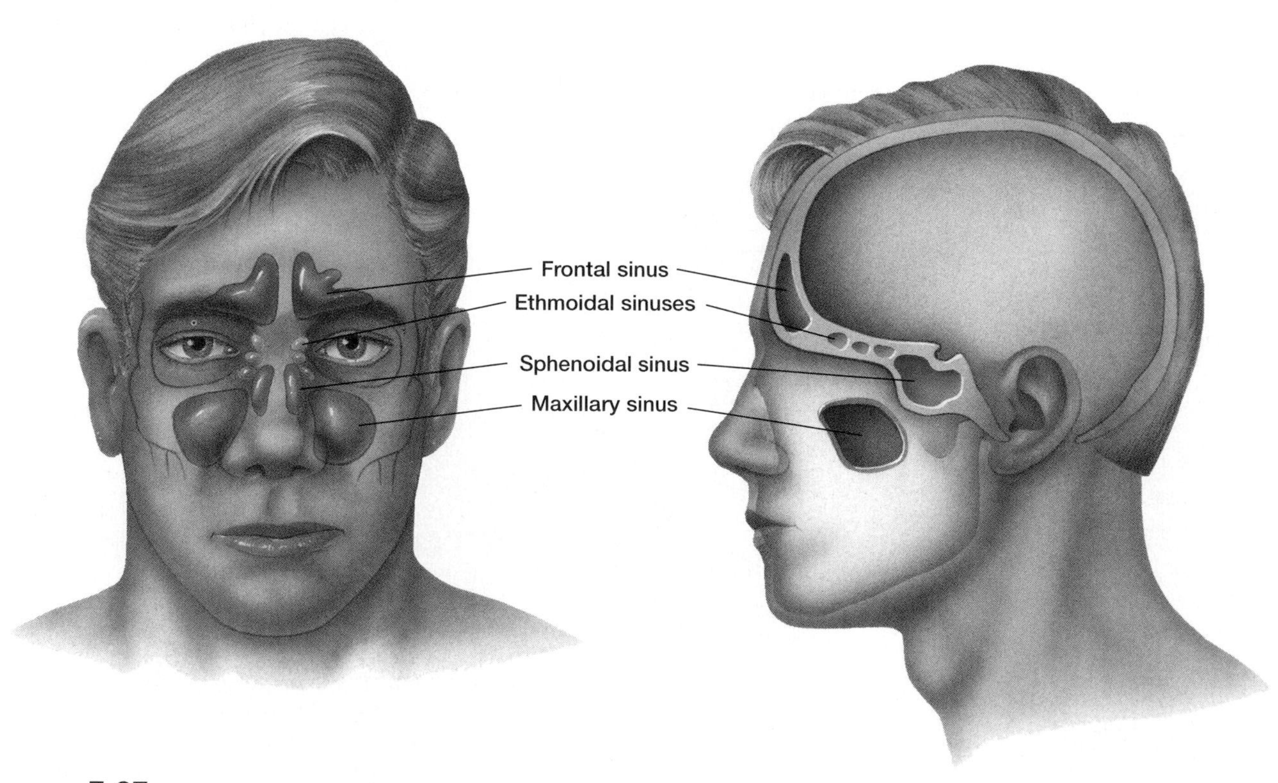


**FIGURE 7.27**
Locations of the sinuses.

## CHART 7.6 Sinuses of the Cranial and Facial Bones

| Sinuses | Number | Location |
|---|---|---|
| Frontal sinuses | 2 | Frontal bone above each eye and near the midline |
| Sphenoidal sinuses | 2 | Sphenoid bone above the posterior portion of the nasal cavity |
| Ethmoidal sinuses | 2 groups of small air cells | Ethmoid bone on either side of the upper portion of the nasal cavity |
| Maxillary sinuses | 2 | Maxillary bones lateral to the nasal cavity and extending from the floor of the orbits to the roots of the upper teeth |

A cleft palate occurs when the fusion of the palatine processes of the maxillae is incomplete at the time of birth. However, the defect in facial bone development actually occurs early in development. Infants with cleft palate may have trouble suckling because of the opening between the oral and nasal cavities. A temporary prosthetic device (artificial palate) may be inserted within the mouth, or a special type of rubber nipple can be used on bottles, until surgery can be performed.

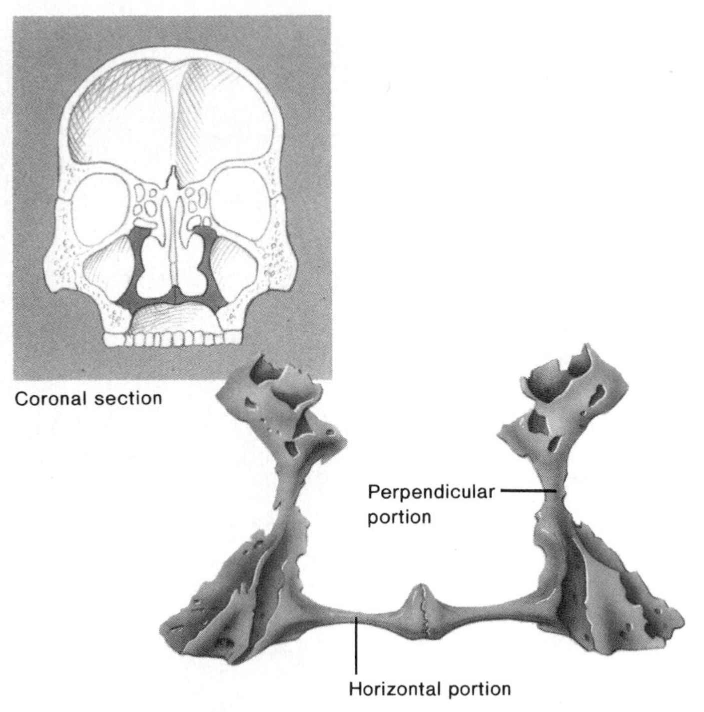


**FIGURE 7.28**
The horizontal portions of the palatine bones form the posterior section of the hard palate, and the perpendicular portions help form the lateral walls of the nasal cavity.

2. **Palatine bones.** The L-shaped palatine (pal′ah-tīn) bones (fig. 7.28) are located behind the maxillae. The horizontal portions form the posterior section of the hard palate and the floor of the nasal cavity. The perpendicular portions help form the lateral walls of the nasal cavity.
3. **Zygomatic bones.** The zygomatic (zi″go-mat′ik) bones are responsible for the prominences of the cheeks below and to the sides of the eyes. These bones also help form the lateral walls and the floors of the orbits. Each bone has a *temporal process,* which extends posteriorly to join the zygomatic process of a temporal bone. Together these processes form a *zygomatic arch* (see figs. 7.21 and 7.22).
4. **Lacrimal bones.** A lacrimal (lak′rĭ-mal) bone is a thin, scalelike structure located in the medial wall of each orbit between the ethmoid bone and the maxilla (see fig. 7.21). A groove in its anterior portion leads from the orbit to the nasal cavity, providing a pathway for a channel that carries tears from the eye to the nasal cavity.
5. **Nasal bones.** The nasal (na′zal) bones are long, thin, and nearly rectangular (see fig. 7.19). They lie side by side and are fused at the midline, where they form the bridge of the nose. These bones are attachments for the cartilaginous tissues that form the shape of the nose.
6. **Vomer bone.** The thin, flat vomer (vo′mer) bone is located along the midline within the nasal cavity. Posteriorly, it joins the perpendicular plate of the ethmoid bone, and together they form the nasal septum (figs. 7.29 and 7.30).
7. **Inferior nasal conchae.** The inferior nasal conchae (kong′ke) are fragile, scroll-shaped bones attached to the lateral walls of the nasal cavity. They are the largest of the conchae and are positioned below the superior and middle nasal conchae of the ethmoid bone (see figs. 7.19 and 7.25). Like the ethmoidal conchae, the inferior conchae support mucous membranes within the nasal cavity.
8. **Mandible.** The mandible (man′dĭ-b′l), or lower jawbone, is a horizontal, horseshoe-shaped body with a flat *ramus* projecting upward at each end. The rami are divided into a posterior *mandibular condyle* and an anterior *coronoid process* (fig. 7.31). The mandibular condyles articulate with the mandibular fossae of the temporal bones, while the coronoid processes provide

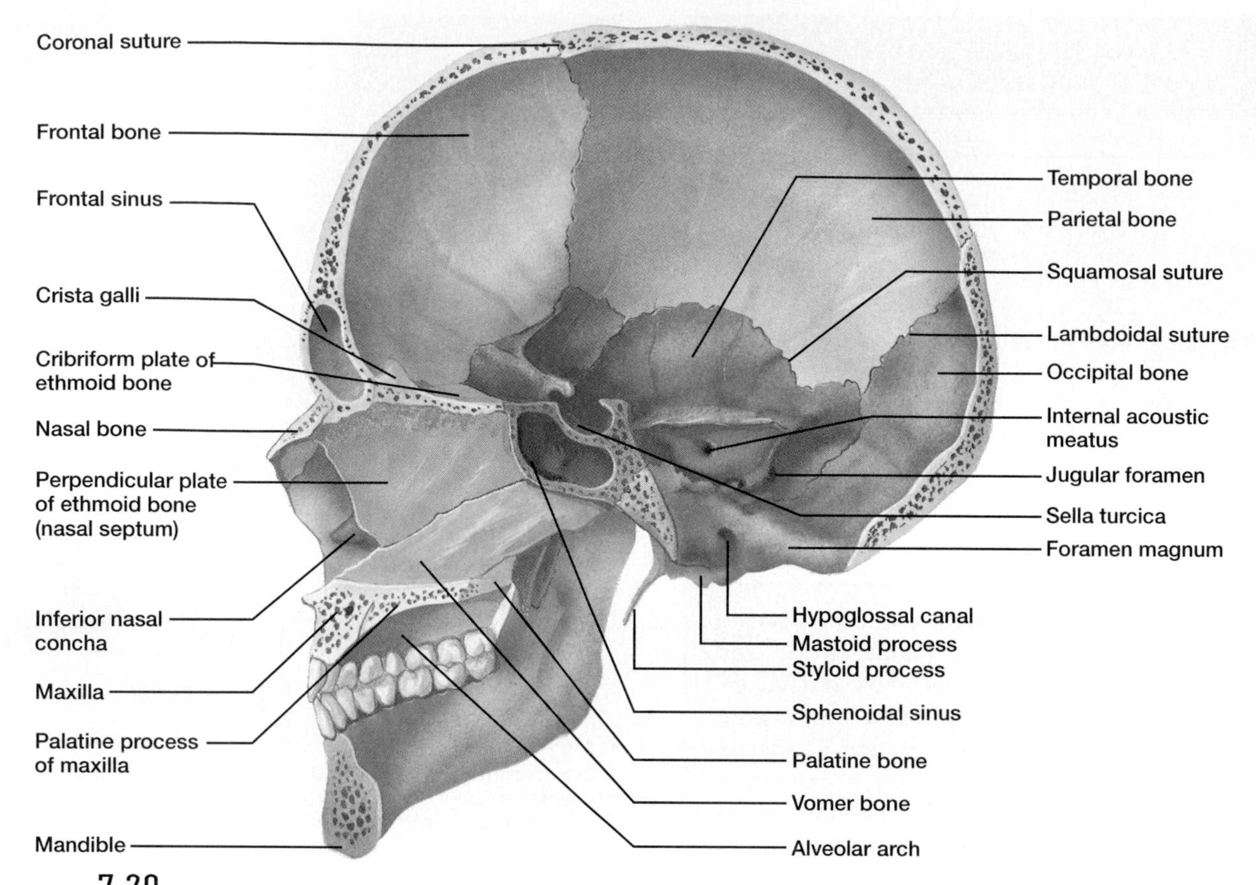


**FIGURE 7.29**
Sagittal section of the skull.

attachments for muscles used in chewing. Other large chewing muscles are inserted on the lateral surfaces of the rami.

A curved bar of bone on the superior border of the mandible, the *alveolar border,* contains the hollow sockets (dental alveoli) that bear the lower teeth.

On the medial side of the mandible, near the center of each ramus, is a *mandibular foramen.* This opening admits blood vessels and a nerve, which supply the roots of the lower teeth. Dentists inject anesthetic into the tissues near this foramen to temporarily block nerve impulse conduction and desensitize teeth on that side of the jaw. Branches of the blood vessels and the nerve emerge from the mandible through the *mental foramen,* which opens on the outside near the point of the jaw. They supply the tissues of the chin and lower lip.

Chart 7.7 describes the fourteen facial bones. Figure 7.32 shows features of these bones on X-ray films.

Chart 7.8 lists the major openings (*foramina*) and passageways through bones of the skull, as well as their general locations and the structures they transmit.

## Infantile Skull

At birth, the skull is incompletely developed, with fibrous membranes separating the cranial bones. These membranous areas are called **fontanels** (fon″tah-nel′z) or, more commonly, soft spots. They permit some movement between the bones, so that the developing skull is partially compressible and can change shape slightly. This action, called *molding,* enables an infant's skull to pass more easily through the birth canal. Eventually, the fontanels close as the cranial bones grow together. The posterior fontanel usually closes about two months after birth; the sphenoid fontanel closes at about three months; the mastoid fontanel closes near the end of the first year; and the anterior fontanel may not close until the middle or end of the second year.

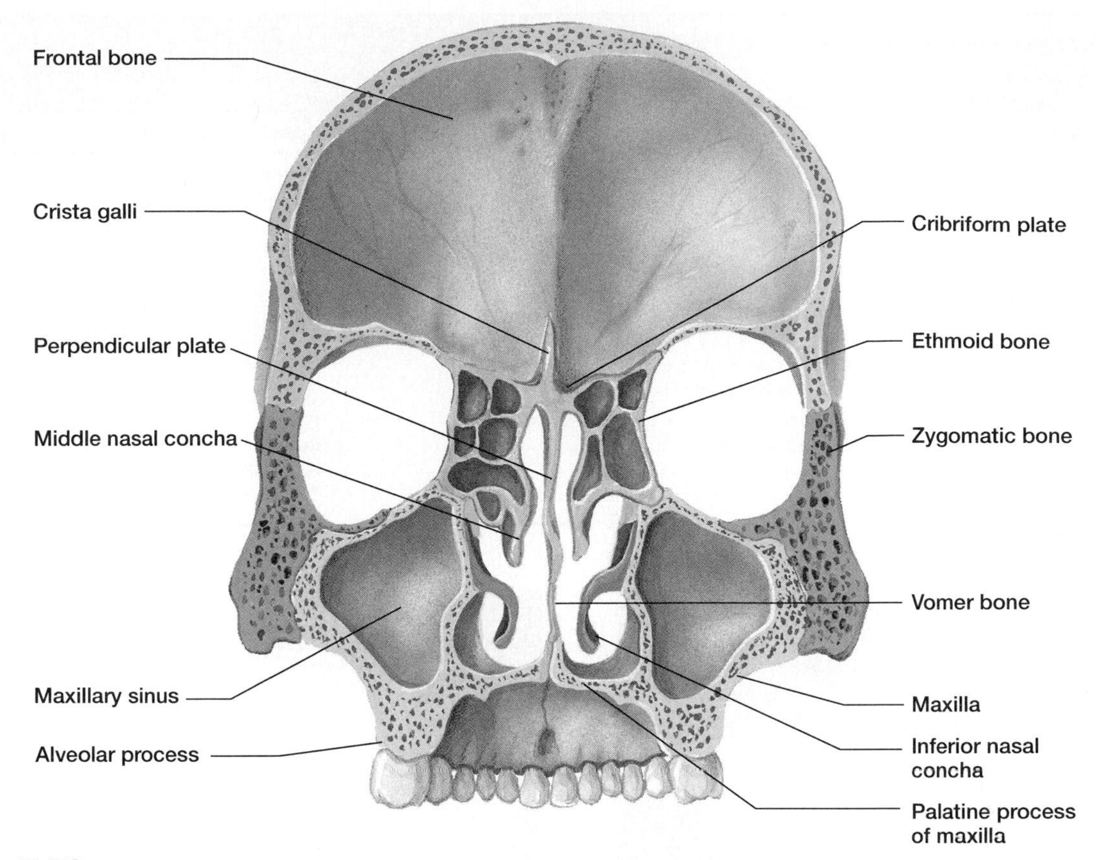


**FIGURE 7.30**
Coronal section of the skull (posterior view).

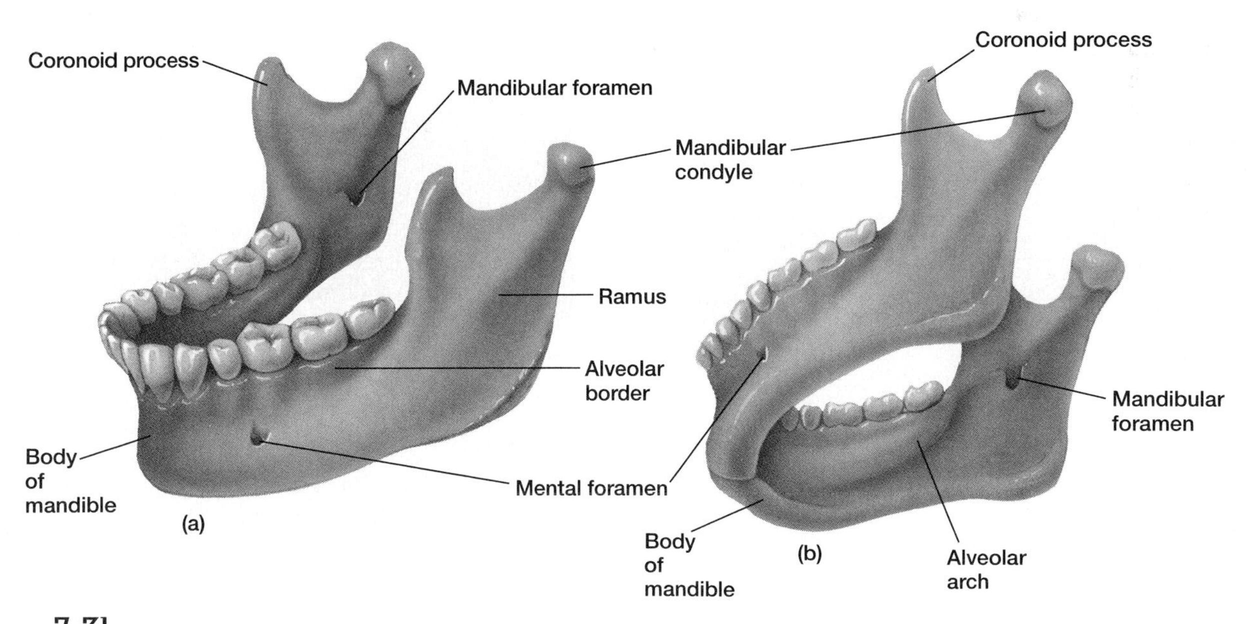


**FIGURE 7.31**
(*a*) Lateral view of the mandible; (*b*) inferior view.

## Chart 7.7 Bones of the Facial Skeleton

| Name and Number | Description | Special Features |
|---|---|---|
| Maxillary (2) | Form upper jaw, anterior roof of mouth, floors of orbits, and sides and floor of nasal cavity | Alveolar processes, maxillary sinuses, palatine process |
| Palatine (2) | Form posterior roof of mouth, and floor and lateral walls of nasal cavity | |
| Zygomatic (2) | Form prominences of cheeks, and lateral walls and floors of orbits | Temporal process |
| Lacrimal (2) | Form part of medial walls of orbits | Groove that leads from orbit to nasal cavity |
| Nasal (2) | Form bridge of nose | |
| Vomer (1) | Forms inferior portion of nasal septum | |
| Inferior nasal conchae (2) | Extend into nasal cavity from its lateral walls | |
| Mandible (1) | Forms lower jaw | Body, ramus, mandibular condyle, coronoid process, alveolar process, mandibular foramen, mental foramen |

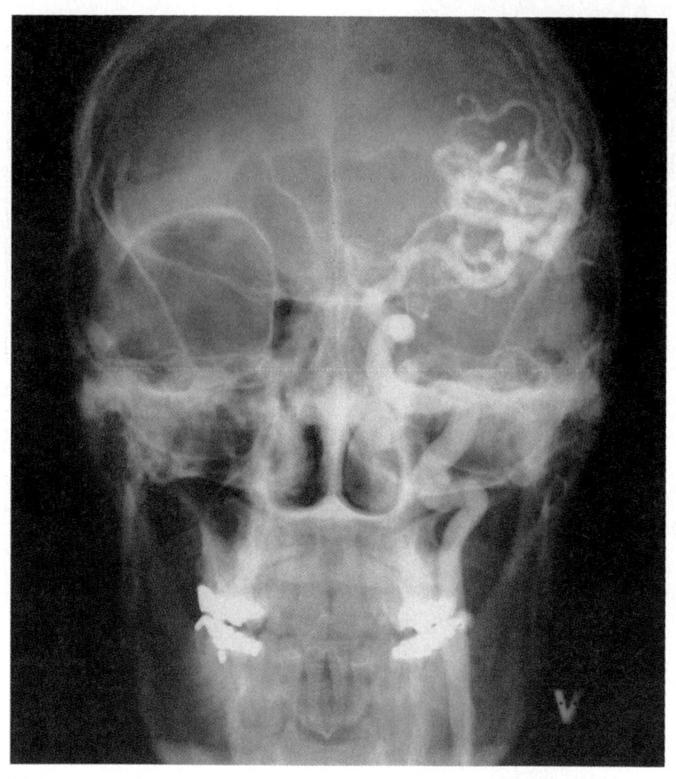

(a)

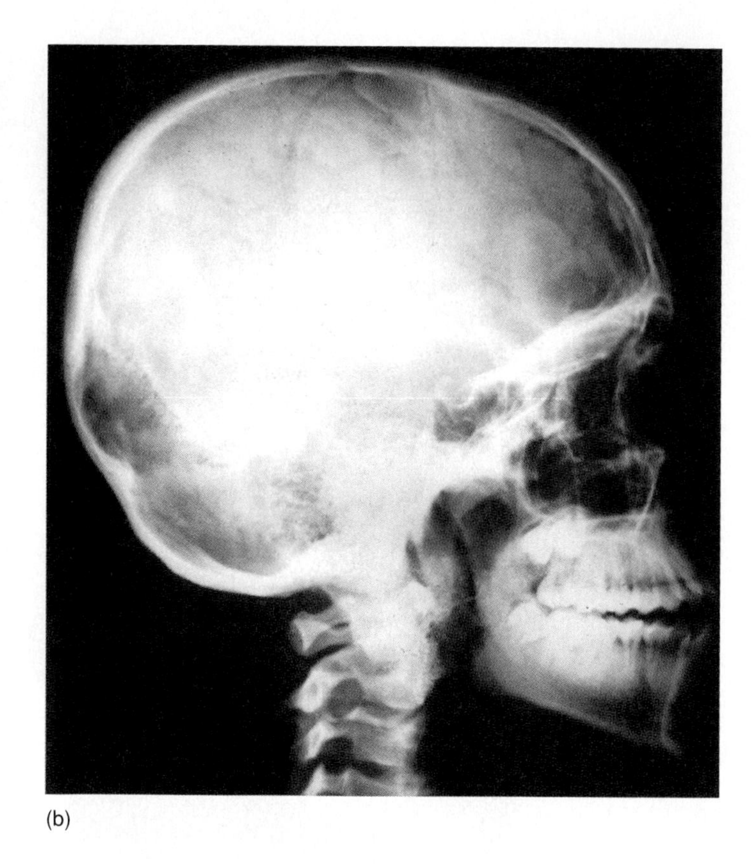

(b)

**Figure 7.32**

X-ray films of the skull (*a*) from the front and (*b*) from the side.

Other characteristics of an infantile skull (fig. 7.33) include a relatively small face with a prominent forehead and large orbits. The jaw and nasal cavity are small, the sinuses are incompletely formed, and the frontal bone is in two parts. The skull bones are thin, but they are also somewhat flexible and thus are less easily fractured than adult bones.

In the infantile skull, a frontal suture (metopic suture) separates the two parts of the developing frontal bone in the midline. This suture usually closes before the sixth year; however, in a few adults, the frontal suture remains open.

## CHART 7.8 Passageways Through Bones of the Skull

| Passageway | Location | Major Structures Transmitted |
|---|---|---|
| Carotid canal (fig. 7.22) | Inferior surface of the temporal bone | Internal carotid artery, veins, and nerves |
| Foramen lacerum (fig. 7.22) | Floor of cranial cavity between temporal and sphenoid bones | Branch of pharyngeal artery (in life opening is largely covered by fibrocartilage) |
| Foramen magnum (fig. 7.26) | Base of skull in occipital bone | Nerve fibers passing between the brain and spinal cord, and certain arteries |
| Foramen ovale (fig. 7.22) | Floor of cranial cavity in sphenoid bone | Mandibular division of trigeminal nerve and veins |
| Foramen rotundum (fig. 7.26) | Floor of cranial cavity in sphenoid bone | Maxillary division of trigeminal nerve |
| Foramen spinosum (fig. 7.26) | Floor of cranial cavity in sphenoid bone | Middle meningeal blood vessels and branch of mandibular nerve |
| Greater palatine foramen (fig. 7.22) | Posterior portion of hard palate in palatine bone | Palatine blood vessels and nerves |
| Hypoglossal canal (fig. 7.29) | Near margin of foramen magnum in occipital bone | Hypoglossal nerve |
| Incisive foramen (fig. 7.22) | Anterior portion of hard palate | Nasopalatine nerves |
| Inferior orbital fissure (fig. 7.20) | Floor of the orbit | Maxillary nerve and blood vessels |
| Infraorbital foramen (fig. 7.20) | Below the orbit in maxillary bone | Infraorbital blood vessels and nerves |
| Internal acoustic meatus (fig. 7.26) | Floor of cranial cavity in temporal bone | Branches of facial and vestibulocochlear nerves, and blood vessels |
| Jugular foramen (fig. 7.26) | Base of the skull between temporal and occipital bones | Glossopharyngeal, vagus and accessory nerves, and blood vessels |
| Mandibular foramen (fig. 7.31) | Inner surface of ramus of mandible | Inferior alveolar blood vessels and nerves |
| Mental foramen (fig. 7.31) | Near point of jaw in mandible | Mental nerve and blood vessels |
| Optic canal (fig. 7.20) | Posterior portion of orbit in sphenoid bone | Optic nerve and ophthalmic artery |
| Stylomastoid foramen (fig. 7.22) | Between styloid and mastoid processes | Facial nerve and blood vessels |
| Superior orbital fissure (fig. 7.20) | Lateral wall of orbit | Oculomotor, trochlear, and abducens nerves, and ophthalmic division of trigeminal nerve |
| Supraorbital foramen (fig. 7.19) | Upper margin or orbit in frontal bone | Supraorbital blood vessels and nerves |

1. Locate and name each of the bones of the cranium.
2. Locate and name each of the facial bones.
3. Explain how an adult skull differs from that of an infant.

## Vertebral Column

The **vertebral column** extends from the skull to the pelvis and forms the vertical axis of the skeleton (fig. 7.34). It is composed of many bony parts called **vertebrae** (ver′tĕ-bre) that are separated by masses of fibrocartilage called *intervertebral disks* and are connected to one another by ligaments. The vertebral column supports the head and the trunk of the body, yet is flexible enough to permit movements, such as bending forward, backward, or to the side, and turning or rotating on the central axis. It also protects the spinal cord, which passes through a *vertebral canal* formed by openings in the vertebrae.

An infant has thirty-three separate bones in the vertebral column. Five of these bones eventually fuse to form the sacrum, and four others join to become the coccyx. As a result, an adult vertebral column has twenty-six bones.

Normally, the vertebral column has four curvatures, which give it a degree of resiliency. The names of the curves correspond to the regions in which they occur, as shown in figure 7.34. The *thoracic* and *pelvic curvatures* are concave anteriorly and are called primary curves. The *cervical curvature* in the neck and the *lumbar curvature* in the lower back are convex anteriorly and are called secondary curves. The cervical curvature develops when a baby begins to hold up its head, and the lumbar curvature develops when the child begins to stand.

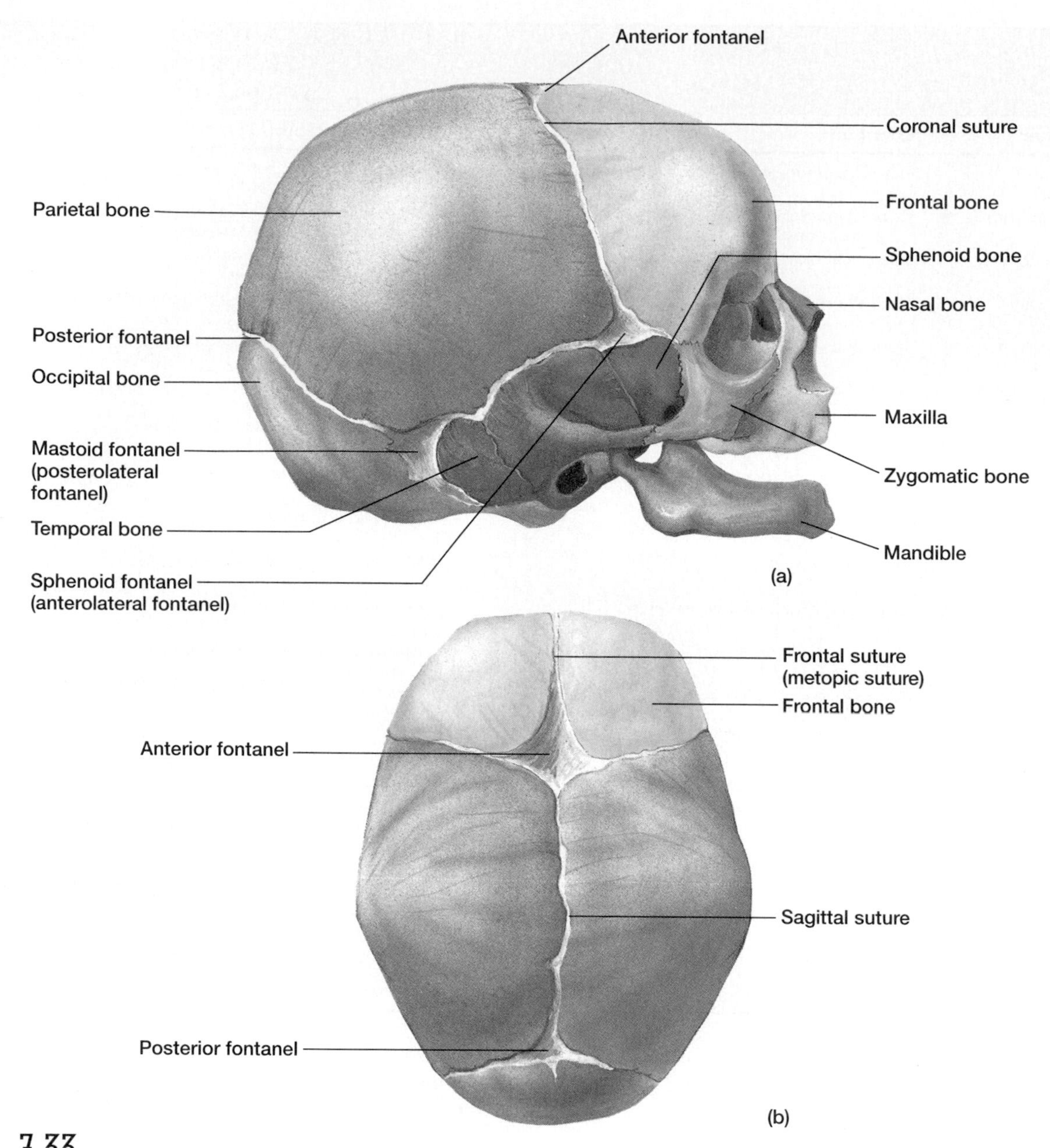


**FIGURE 7.33**
(*a*) Lateral view and (*b*) superior view of the infantile skull.

## A Typical Vertebra

Although the vertebrae in different regions of the vertebral column have special characteristics, they also have features in common. A typical vertebra (fig. 7.35) has a drum-shaped *body* (centrum), which forms the thick, anterior portion of the bone. A longitudinal row of these vertebral bodies supports the weight of the head and trunk. The intervertebral disks, which separate adjacent vertebrae, are fastened to the roughened upper and lower surfaces of the vertebral bodies. These disks cushion and soften the forces caused by such movements as walking and jumping, which might otherwise fracture vertebrae or jar the brain.

The bodies of adjacent vertebrae are joined on their anterior surfaces by *anterior ligaments* and on their posterior surfaces by *posterior ligaments.*

Projecting posteriorly from each vertebral body are two short stalks called *pedicles* (ped′ĭ-k′lz). They form the sides of the *vertebral foramen.* Two plates called *laminae* (lam′ĭ-ne) arise from the pedicles and fuse in the back to become a *spinous process.* The pedicles, laminae, and spinous process together

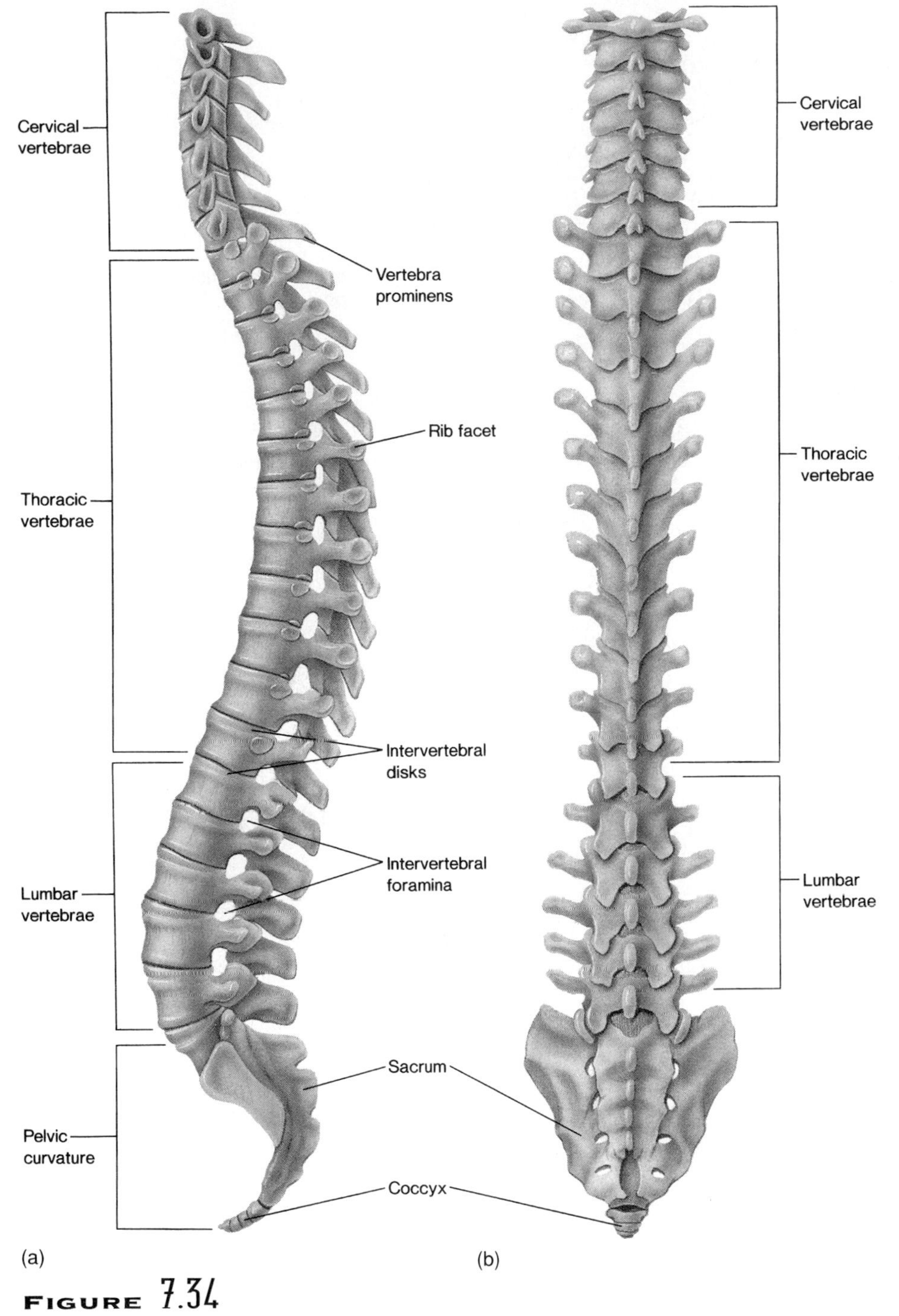


**FIGURE 7.34**

The curved vertebral column consists of many vertebrae separated by intervertebral disks. (*a*) Left lateral view; (*b*) posterior view.

complete a bony *vertebral arch* around the vertebral foramen, through which the spinal cord passes.

Between the pedicles and laminae of a typical vertebra is a *transverse process,* which projects laterally and posteriorly. Various ligaments and muscles are attached to the dorsal spinous process and the transverse processes. Projecting upward and downward from each vertebral arch are *superior* and *inferior articulating processes.* These processes bear cartilage-covered facets by which each vertebra is joined to the one above and the one below it.

On the lower surfaces of the vertebral pedicles are notches that align to help form openings called *intervertebral foramina* (in″ter-ver′tĕ-bral fo-ram′ĭ-nah). These openings provide passageways for spinal nerves that proceed between adjacent vertebrae and connect to the spinal cord.

Athletes who repeatedly land on hard surfaces, such as gymnasts, jumpers, and pole vaulters, sometimes break or crack their vertebrae, particularly the articulating processes. Such a vertebral fracture is called *spondylolysis.* It may be prevented by limiting the number of hard landings and by padding the landing areas with floor mats.

## Cervical Vertebrae

Seven **cervical vertebrae** comprise the bony axis of the neck. Although these are the smallest of the vertebrae, their bone tissues are denser than those in any other region of the vertebral column.

The transverse processes of the cervical vertebrae are distinctive because they have *transverse foramina,* which are passageways for arteries leading to the brain. Also, the spinous processes of the second through the fifth cervical vertebrae are uniquely forked (bifid). These processes provide attachments for various muscles.

The spinous process of the seventh vertebra is longer and protrudes beyond the other cervical spines. It is called the *vertebra prominens,* and because it can be felt through the skin, it is a useful landmark for locating other vertebral parts (see fig. 7.34).

Two of the cervical vertebrae, shown in figure 7.36, are of special interest. The first vertebra, or **atlas** (at′las), supports and balances the head. It has practically no body or spine and appears as a bony ring with

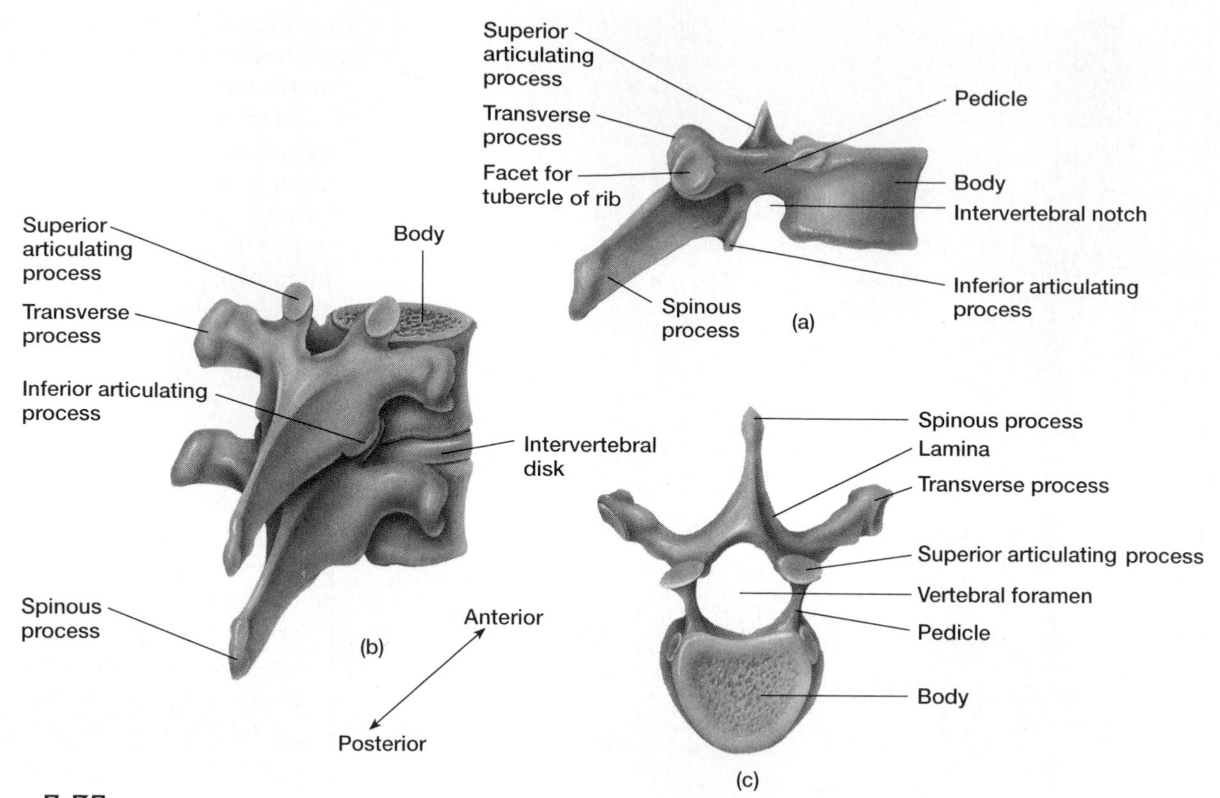


**FIGURE 7.35**

(*a*) Lateral view of a typical thoracic vertebra. (*b*) Adjacent vertebrae join at their articulating processes. (*c*) Superior view of a typical thoracic vertebra.

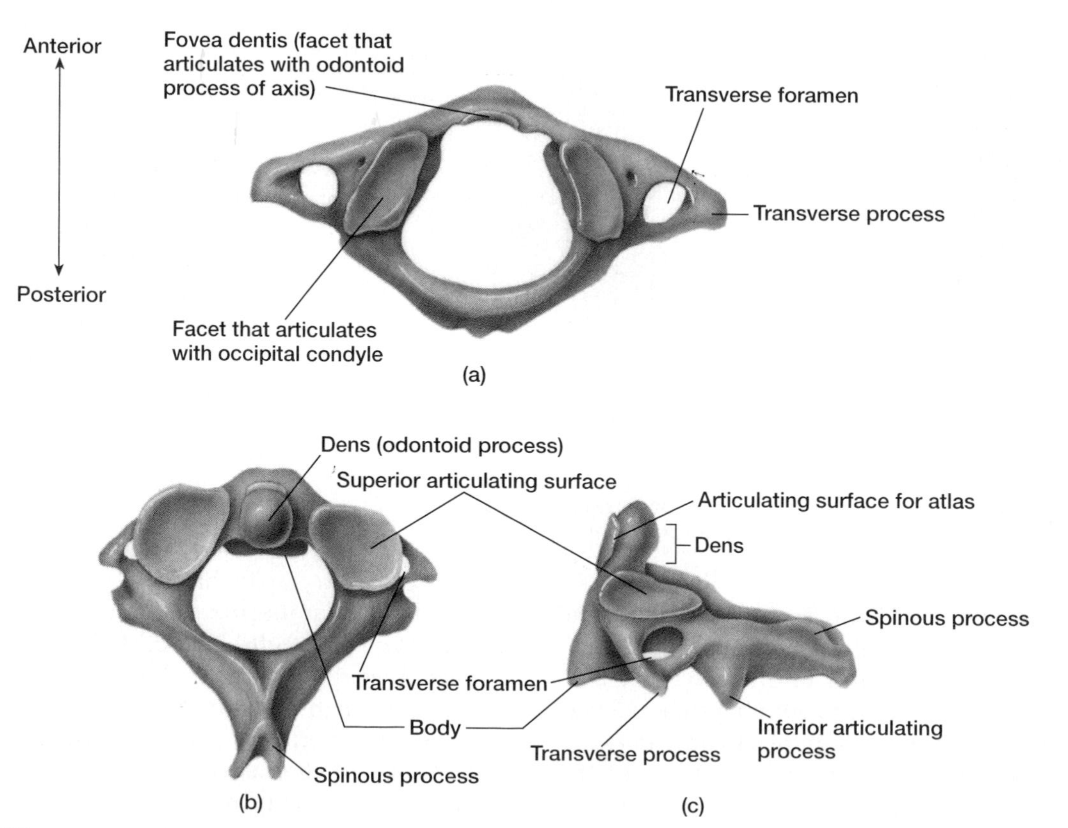


**FIGURE 7.36**

Superior view of the (*a*) atlas and (*b*) axis. (*c*) Lateral view of the axis.

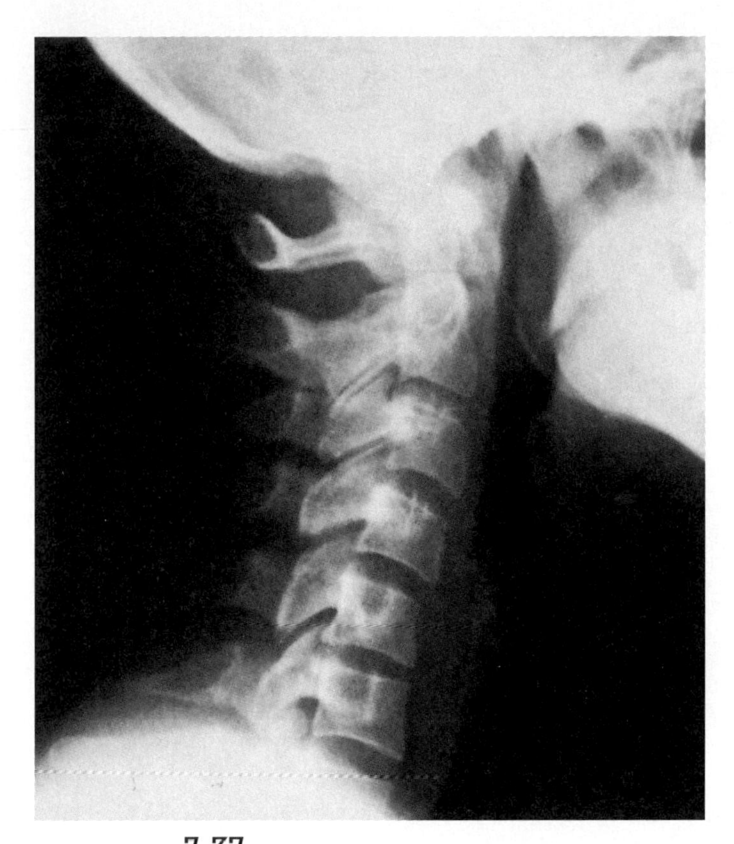

**FIGURE 7.37**
X ray of the cervical vertebrae.

two transverse processes. On its superior surface, the atlas has two kidney-shaped *facets,* which articulate with the occipital condyles of the skull.

The second cervical vertebra, or **axis** (ak′sis), bears a toothlike *dens* (odontoid process) on its body. This process projects upward and lies in the ring of the atlas. As the head is turned from side to side, the atlas pivots around the dens (see figs. 7.36 and 7.37).

### Thoracic Vertebrae

The twelve **thoracic vertebrae** are larger than those in the cervical region. Each vertebra has a long, pointed spinous process, which slopes downward, and facets on the sides of its body, which articulate with a rib.

Beginning with the third thoracic vertebra and moving inferiorly, the bodies of these bones increase in size. Thus, they are adapted to the stress placed on them by the increasing amounts of body weight they bear.

### Lumbar Vertebrae

The five **lumbar vertebrae** in the small of the back (loins) support more weight than the superior vertebrae, and have larger and stronger bodies. In comparison to the other types of vertebrae, the transverse processes of these vertebrae project posteriorly at relatively sharp angles, while their short, thick spinous processes are nearly horizontal.

Gymnasts, football players, and others who bend their vertebral columns excessively and forcefully may experience slipping of a vertebra over the one below it. This painful condition is called *spondylolisthesis*. Usually, the fifth lumbar vertebra slides forward over the body of the sacrum. Doing exercises designed to strengthen the back muscles associated with the vertebral column may prevent this condition.

Figure 7.38 compares the structures of the cervical, thoracic, and lumbar vertebrae.

### Sacrum

The **sacrum** (sa′krum) is a triangular structure at the base of the vertebral column. It is composed of five vertebrae that develop separately but gradually fuse between ages 18 and 30. The spinous processes of these fused bones form a ridge of tubercles, the *median sacral crest.* Nerves and blood vessels pass through rows of openings, called the *dorsal sacral foramina,* located to the sides of the tubercles (fig. 7.39).

The sacrum is wedged between the coxal bones of the pelvis and is united to them at its *auricular surfaces* by fibrocartilage of the *sacroiliac joints.* The pelvic girdle, at these joints, transmits the body's weight to the legs (see fig. 7.17).

The sacrum forms the posterior wall of the pelvic cavity. The upper anterior margin of the sacrum, which represents the body of the first sacral vertebra, is called the *sacral promontory* (sa′kral prom′on-to″re). During a vaginal examination, a physician can feel this projection and use it as a guide in determining the size of the pelvis. This measurement is helpful in estimating how easily an infant may be able to pass through a woman's pelvic cavity during childbirth.

The vertebral foramina of the sacral vertebrae form the *sacral canal,* which continues through the sacrum to an opening of variable size at the tip, called the *sacral hiatus* (hi-a′tus). This foramen exists because the laminae of the last sacral vertebra are not fused. On the ventral surface of the sacrum, four pairs of *pelvic sacral foramina* provide passageways for nerves and blood vessels.

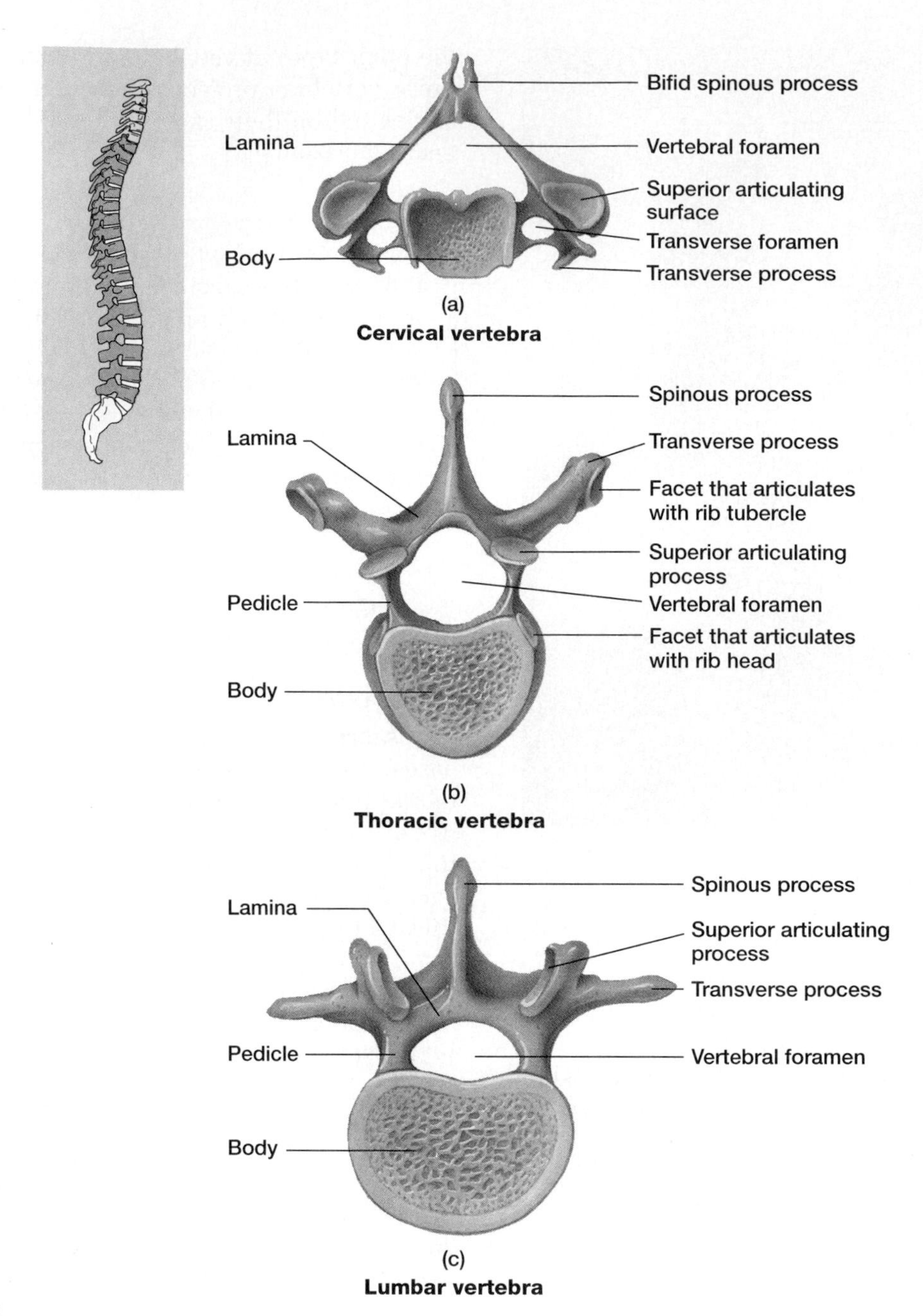


**FIGURE 7.38**

Superior view of (*a*) a cervical vertebra, (*b*) a thoracic vertebra, and (*c*) a lumbar vertebra.

## Coccyx

The **coccyx** (kok′siks), or tailbone, is the lowest part of the vertebral column and is usually composed of four vertebrae that fuse by the twenty-fifth year. Ligaments attach it to the margins of the sacral hiatus (see fig. 7.39). Sitting presses on the coccyx, and it moves forward, acting somewhat like a shock absorber. Sitting down with great force can fracture or dislocate the coccyx.

Chart 7.9 summarizes the bones of the vertebral column, and Clinical Application 7.5 discusses disorders of the vertebral column.

1. Describe the structure of the vertebral column.
2. Explain the difference between the vertebral column of an adult and that of an infant.
3. Describe a typical vertebra.
4. How do the structures of cervical, thoracic, and lumbar vertebrae differ?

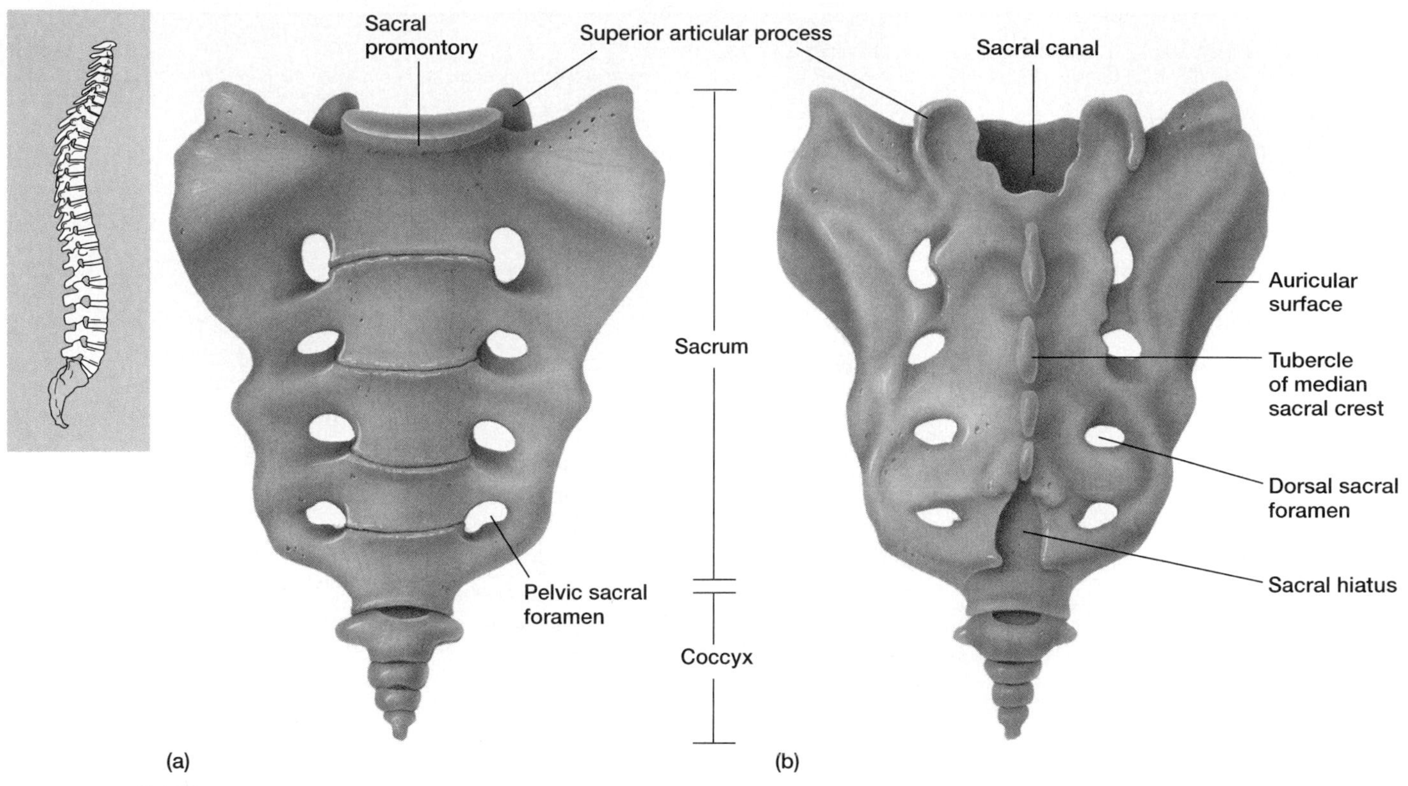


FIGURE 7.39
(*a*) Anterior view of the sacrum and coccyx; (*b*) posterior view.

CHART 7.9 BONES OF THE VERTEBRAL COLUMN

| Bones | Number | Special Features |
|---|---|---|
| Cervical vertebrae | 7 | Transverse foramina; facets of atlas articulate with occipital condyles of skull; dens of axis articulates with atlas; spinous processes of second through fifth vertebrae are bifid |
| Thoracic vertebrae | 12 | Pointed spinous processes that slope downward; facets that articulate with ribs |
| Lumbar vertebrae | 5 | Large bodies; transverse processes that project posteriorly at sharp angles; short, thick spinous processes directed nearly horizontally |
| Sacrum | 5 vertebrae fused into 1 bone | Dorsal sacral foramina, auricular surfaces, sacral promontory, sacral canal, sacral hiatus, pelvic sacral foramina |
| Coccyx | 4 vertebrae fused into 1 bone | Attached by ligaments to the margins of the sacral hiatus |

## THORACIC CAGE

The **thoracic cage** includes the ribs, the thoracic vertebrae, the sternum, and the costal cartilages that attach the ribs to the sternum. These bones support the shoulder girdle and upper limbs, protect the viscera in the thoracic and upper abdominal cavities, and play a role in breathing (fig. 7.40).

### Ribs

The usual number of rib pairs is twelve—one pair attached to each of the twelve thoracic vertebrae. Some individuals develop extra ribs associated with their cervical or lumbar vertebrae.

The first seven rib pairs, which are called the *true ribs* (vertebrosternal ribs), join the sternum directly by their costal cartilages. The remaining five

## CLINICAL APPLICATION 7.5

### Disorders of the Vertebral Column

A common vertebral problem is changes in the intervertebral disks. Each disk is composed of a tough, outer layer of fibrocartilage (annulus fibrosus) and an elastic central mass (nucleus pulposus). As a person ages, these disks degenerate, the central masses losing firmness and the outer layers thinning and weakening, and developing cracks. Extra pressure, as when a person falls or lifts a heavy object, can break the outer layers of the disks, squeezing out the central masses. Such a rupture may press on the spinal cord or on spinal nerves that branch from it. This condition, called a *ruptured,* or *herniated, disk,* may cause back pain and numbness or loss of muscular function in the parts innervated by affected spinal nerves.

A surgical procedure called a *laminectomy* may relieve the pain of a herniated disk. In this procedure, a portion of the posterior arch of a vertebra is removed to reduce the pressure on the affected nerve tissues. In other instances, a protein-digesting enzyme (chymopapain) may be injected into the injured disk to shrink it.

Sometimes problems develop in the curvatures of the vertebral column because of poor posture, injury, or disease. For example, if an exaggerated thoracic curvature appears, the person develops rounded shoulders and a hunchback. This condition, called *kyphosis,* occasionally develops in adolescents who undertake strenuous athletic activities. Unless the problem is corrected before bone growth completes, the vertebral column may be permanently deformed.

Sometimes the vertebral column develops an abnormal lateral curvature, so that one hip or shoulder is lower than the other. At the same time, the thoracic and abdominal organs may be displaced or compressed. This condition is called *scoliosis.* Although it most commonly develops without known cause in adolescent females, it also may accompany such diseases as poliomyelitis, rickets, or tuberculosis. An accentuated lumbar curvature is called *lordosis,* or swayback.

As a person ages, the intervertebral disks tend to become smaller and more rigid, and compression is more likely to fracture the vertebral bodies. Consequently, height may decrease, and the thoracic curvature of the vertebral column may be accentuated, bowing the back.

pairs are called *false ribs,* because their cartilages do not reach the sternum directly. Instead, the cartilages of the upper three false ribs (vertebrochondral ribs) join the cartilages of the seventh rib, while the last two rib pairs have no attachments to the sternum. These last two pairs (or sometimes the last three pairs) are called *floating ribs* (vertebral ribs).

A typical rib (fig. 7.41) has a long, slender shaft, which curves around the chest and slopes downward. On the posterior end is an enlarged *head* by which the rib articulates with a facet on the body of its own vertebra and with the body of the next higher vertebra. The neck of the rib is flattened, lateral to the head, where ligaments attach. Near the neck is a *tubercle,* which articulates with the transverse process of the vertebra.

The costal cartilages are composed of hyaline cartilage. They are attached to the anterior ends of the ribs and continue in line with them toward the sternum.

### Sternum

The **sternum** (ster′num), or breastbone, is located along the midline in the anterior portion of the thoracic cage. It is a flat, elongated bone that develops in three parts—an upper *manubrium* (mah-nu′bre-um), a middle *body,* and a lower *xiphoid* (zif′oid) *process* that projects downward (see fig. 7.40).

The sides of the manubrium and the body are notched where they articulate with costal cartilages. The manubrium also articulates with the clavicles by

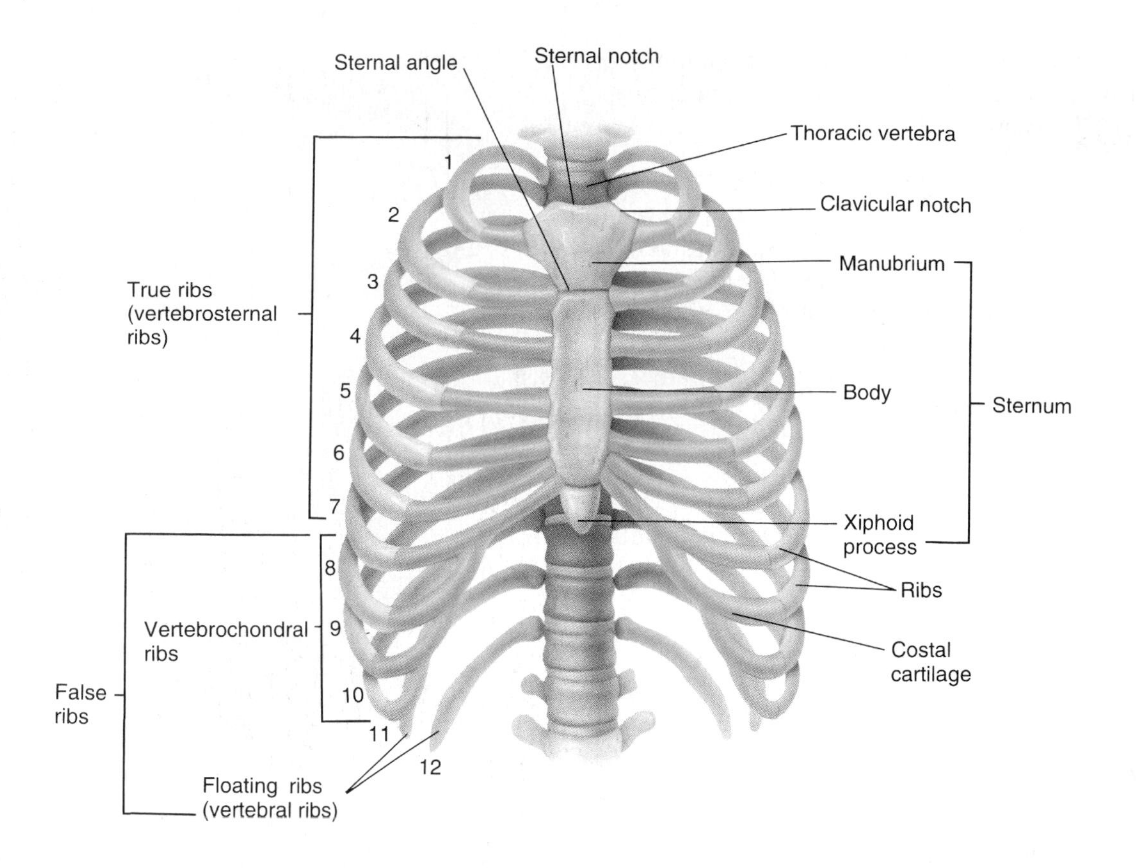


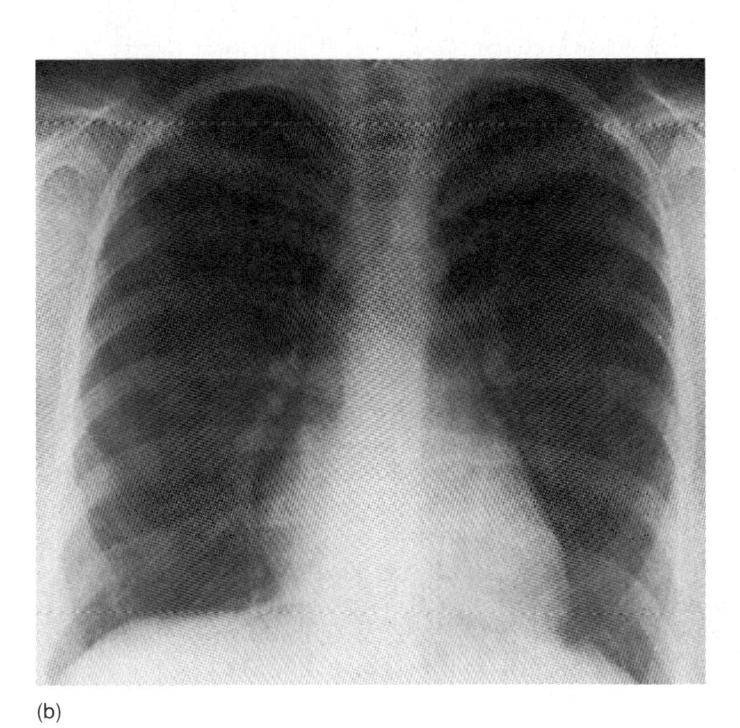


**FIGURE 7.40**

(*a*) The thoracic cage includes the thoracic vertebrae, the sternum, the ribs, and the costal cartilages that attach the ribs to the sternum. (*b*) X-ray film of the thoracic cage viewed from the front. The light region behind the sternum and above the diaphragm is the heart.

facets on its superior border. It usually remains as a separate bone until middle age or later, when it fuses to the body of the sternum.

The manubrium and body of the sternum lie in different planes, so that the line of union between them projects slightly forward. This projection, which occurs at the level of the second costal cartilage, is called the *sternal angle* (angle of Louis). It is commonly used as a clinical landmark to locate a particular rib accurately (see fig. 7.40).

The xiphoid process begins as a piece of cartilage. It slowly ossifies, and by middle age it usually fuses to the body of the sternum also.

The red marrow within the spongy bone of the sternum functions in blood cell formation into adulthood. Since the sternum has a thin covering of compact bone and is easy to reach, samples of its blood-cell-forming tissue may be removed for use in diagnosing diseases. This procedure, a *sternal puncture,* involves suctioning (aspirating) some marrow through a hollow needle. (Marrow may also be removed from the iliac crest of a coxal bone.)

1. What bones make up the thoracic cage?
2. Describe a typical rib.
3. What are the differences between true, false, and floating ribs?

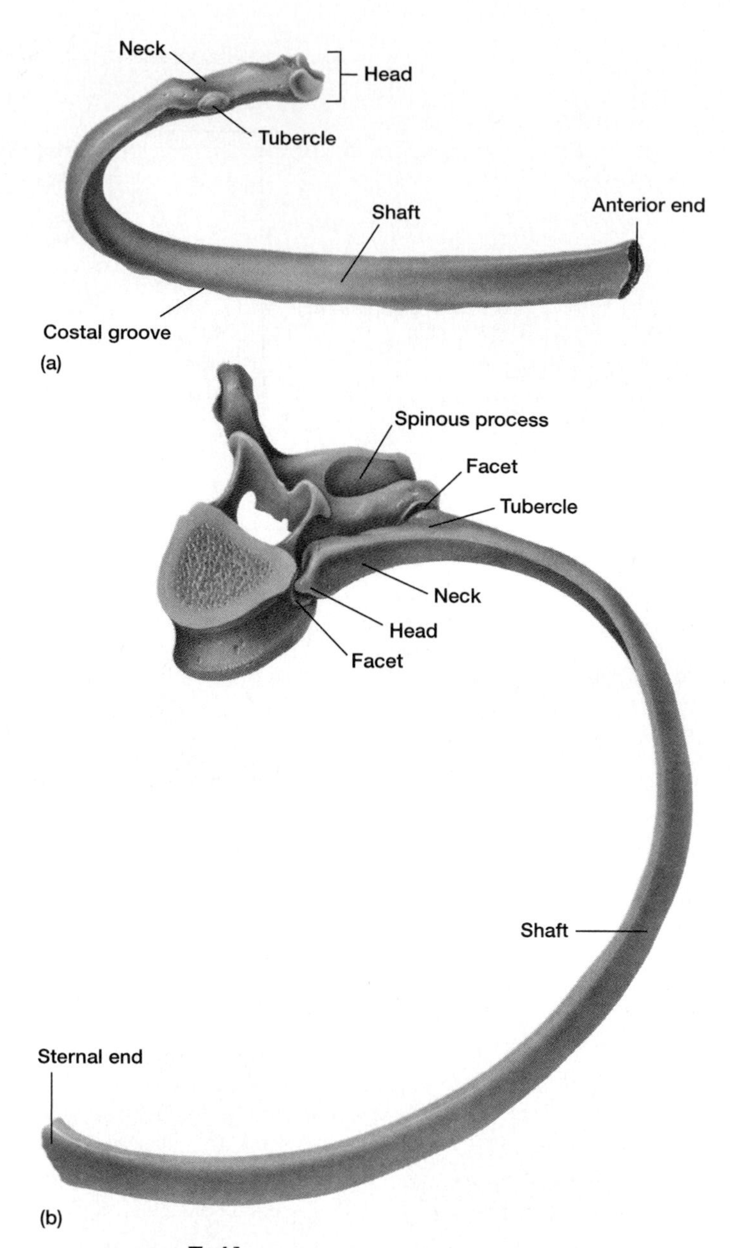


**FIGURE 7.41**
(*a*) A typical rib (posterior view); (*b*) articulations of a rib with a thoracic vertebra (superior view).

## PECTORAL GIRDLE

The **pectoral** (pek'to-ral) **girdle** (shoulder girdle) is composed of four parts—two clavicles (collarbones) and two scapulae (shoulder blades). Although the word *girdle* suggests a ring-shaped structure, the pectoral girdle is an incomplete ring. It is open in the back between the scapulae, and the sternum separates its bones in front. The pectoral girdle supports the upper limbs and is an attachment for several muscles that move them (fig. 7.42).

### Clavicles

The **clavicles** are slender, rodlike bones with elongated S-shapes (see fig. 7.42). Located at the base of the neck, they run horizontally between the sternum and the shoulders. The medial (or sternal) ends of the clavicles articulate with the manubrium, while the lateral (or acromial) ends join processes of the scapulae.

The clavicles brace the freely movable scapulae, helping to hold the shoulders in place. They also provide attachments for muscles of the upper limbs, chest, and back. Because of its elongated double curve, the clavicle is structurally weak. If compressed lengthwise due to abnormal pressure on the shoulder, it is likely to fracture.

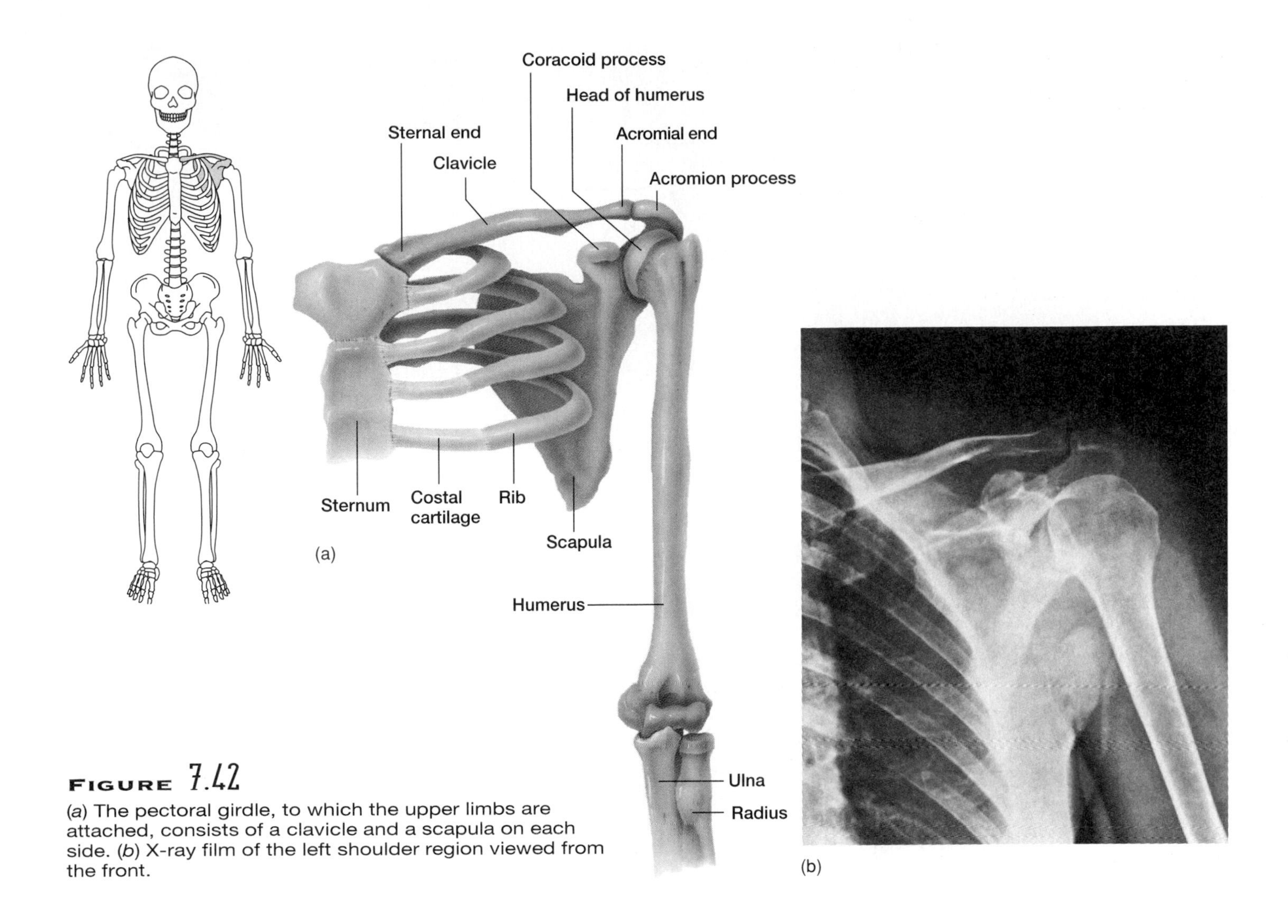


**FIGURE 7.42**

(*a*) The pectoral girdle, to which the upper limbs are attached, consists of a clavicle and a scapula on each side. (*b*) X-ray film of the left shoulder region viewed from the front.

## Scapulae

The **scapulae** are broad, somewhat triangular bones located on either side of the upper back. They have flat bodies with concave anterior surfaces. The posterior surface of each scapula is divided into unequal portions by a *spine.* Above the spine is the *supraspinous fossa,* and below the spine is the *infraspinous fossa.* This spine leads to a *head,* which bears two processes—an *acromion* (ah-kro′me-on) *process* that forms the tip of the shoulder and a *coracoid* (kor′ah-koid) *process* that curves anteriorly and inferiorly to the clavicle (fig. 7.43). The acromion process articulates with the clavicle and provides attachments for muscles of the upper limb and chest. The coracoid process also provides attachments for upper limb and chest muscles. On the head of the scapula between the processes is a depression called the *glenoid cavity.* It articulates with the head of the arm bone (humerus).

The scapula has three borders. The *superior border* is on the superior edge. The *axillary* or *lateral border* is directed toward the upper limb. The *vertebral* or *medial border* is closest to the vertebral column, about 5 cm away.

1. What bones form the pectoral girdle?
2. What is the function of the pectoral girdle?

---

# Upper Limb

The bones of the upper limb form the framework of the arm, forearm, and hand. They also provide attachments for muscles, and they function in levers that move limb parts. These bones include a humerus, a radius, an ulna, carpals, metacarpals, and phalanges (fig. 7.44).

## Humerus

The **humerus** (fig. 7.45) is a heavy bone that extends from the scapula to the elbow. At its upper end is a smooth, rounded *head* that fits into the glenoid

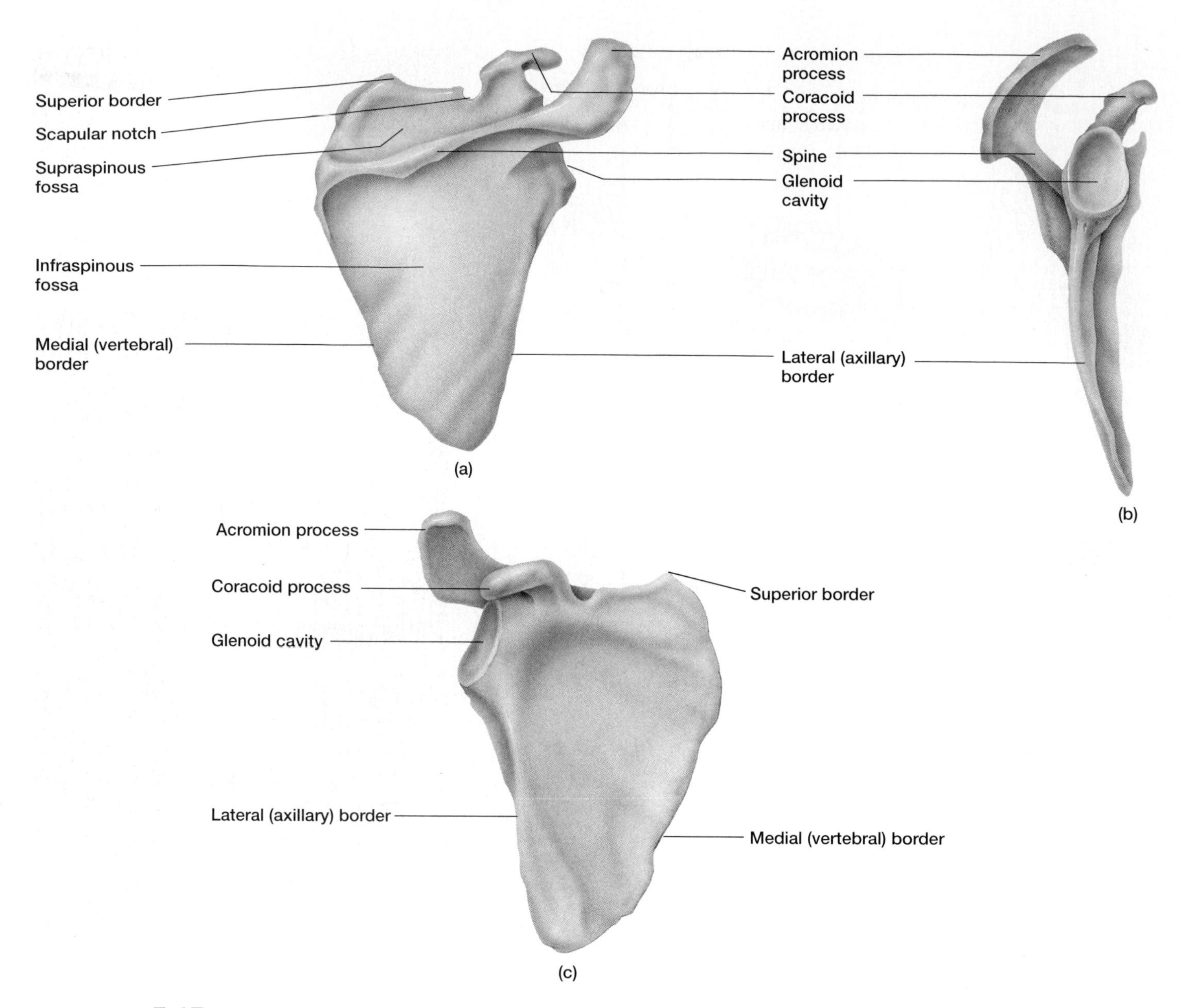


**FIGURE 7.43**

(*a*) Posterior surface of the right scapula; (*b*) lateral view showing the glenoid cavity that articulates with the head of the humerus; (*c*) anterior surface.

cavity of the scapula. Just below the head are two processes—a *greater tubercle* on the lateral side and a *lesser tubercle* on the anterior side. These tubercles provide attachments for muscles that move the upper limb at the shoulder. Between them is a narrow furrow, the *intertubercular groove,* through which a tendon passes from a muscle in the arm (biceps brachii) to the shoulder.

The narrow depression along the lower margin of the head that separates it from the tubercles is called the *anatomical neck.* Just below the head and the tubercles of the humerus is a tapering region called the *surgical neck,* so named because fractures commonly occur there. Near the middle of the bony shaft on the lateral side is a rough V-shaped area called the *deltoid tuberosity.* It provides an attachment for the muscle (deltoid) that raises the upper limb horizontally to the side.

At the lower end of the humerus are two smooth *condyles*—a knoblike *capitulum* (kah-pit′u-lum) on the lateral side and a pulley-shaped *trochlea* (trok′le-ah) on the medial side. The capitulum articulates with the radius at the elbow, while the trochlea joins the ulna.

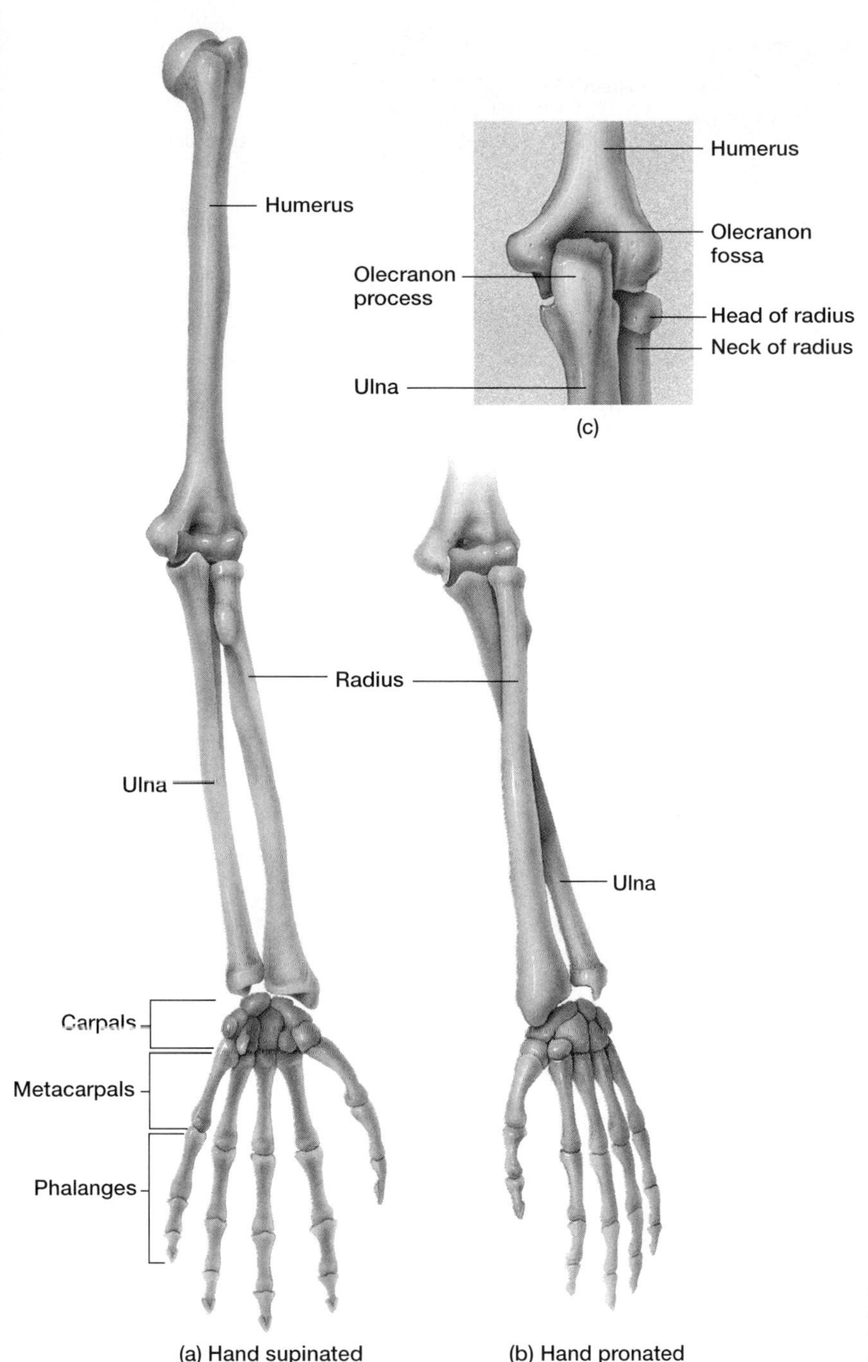


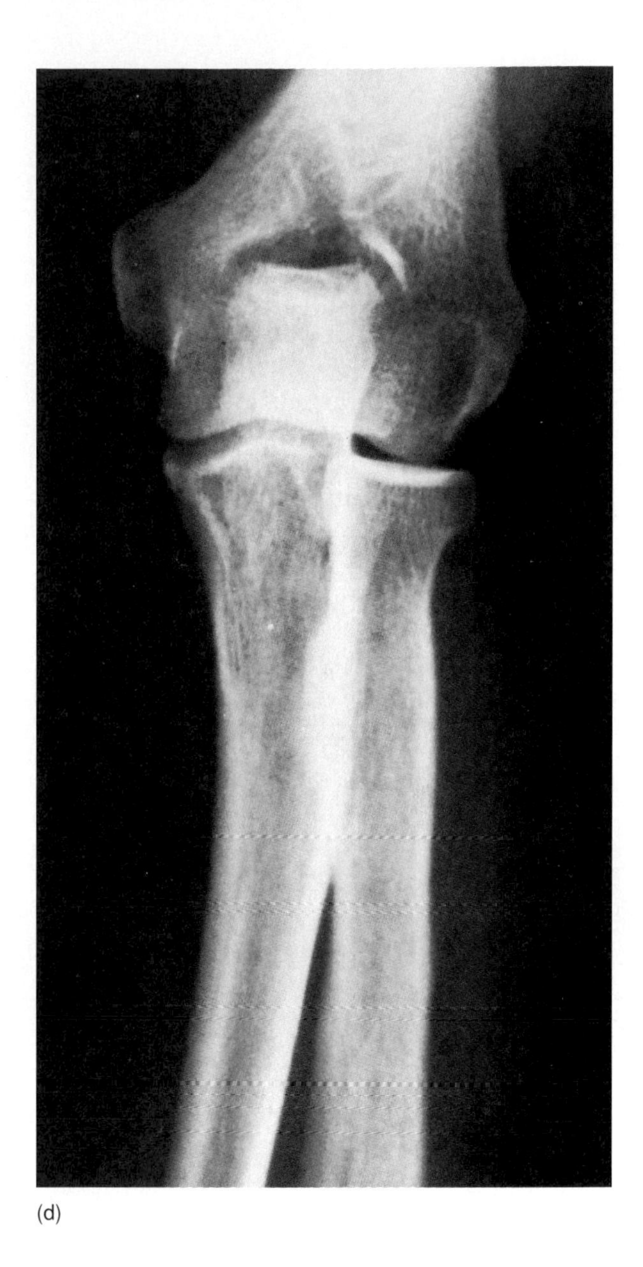


**FIGURE 7.14**

(*a*) Frontal view of the left upper limb with the hand supinated; (*b*) with the hand pronated; (*c*) posterior view of the right elbow; (*d*) X-ray film of the left elbow and forearm, viewed anteriorly.

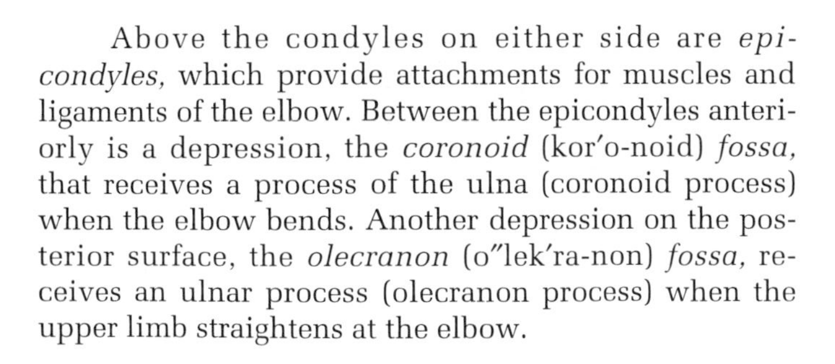

Above the condyles on either side are *epicondyles,* which provide attachments for muscles and ligaments of the elbow. Between the epicondyles anteriorly is a depression, the *coronoid* (kor′o-noid) *fossa,* that receives a process of the ulna (coronoid process) when the elbow bends. Another depression on the posterior surface, the *olecranon* (o″lek′ra-non) *fossa,* receives an ulnar process (olecranon process) when the upper limb straightens at the elbow.

Many a thirtyish parent of a young little leaguer or softball player becomes tempted to join in. But if he or she has not pitched in many years, sudden activity may break the forearm. Forearm pain while pitching is a signal that a fracture could happen. Medical specialists advise returning to the pitching mound gradually. Start with twenty pitches, five days a week, for two to three months before regular games begin. By the season's start, 120 pitches per daily practice session should be painless.

## Radius

The **radius,** located on the thumb side of the forearm, is somewhat shorter than its companion, the ulna (fig. 7.46). The radius extends from the elbow to the wrist and crosses over the ulna when the hand is turned so that the palm faces backward.

A thick, disklike *head* at the upper end of the radius articulates with the capitulum of the humerus and a notch of the ulna (radial notch). This arrangement allows the radius to rotate freely.

On the radial shaft just below the head is a process called the *radial tuberosity.* It is an attachment for a muscle (biceps brachii) that bends the upper limb at the elbow. At the distal end of the radius, a lateral *styloid* (sti'loid) *process* provides attachments for ligaments of the wrist.

## Ulna

The **ulna** is longer than the radius and overlaps the end of the humerus posteriorly. At its proximal end, the ulna has a wrenchlike opening, the *trochlear notch* (semilunar notch), that articulates with the trochlea of the humerus. A process lies on either side of this notch. The *olecranon process,* located above the trochlear notch, provides an attachment for the muscle (triceps brachii) that straightens the upper limb at the elbow. During this movement, the olecranon process of the ulna fits into the olecranon fossa of the humerus. Similarly, the *coronoid process,* just below the trochlear notch, fits into the coronoid fossa of the humerus when the elbow bends.

At the distal end of the ulna, its knoblike *head* articulates with a notch of the radius (ulnar notch) laterally and with a disk of fibrocartilage inferiorly (see fig. 7.46). This disk, in turn, joins a wrist bone (triangular). A medial *styloid process* at the distal end of the ulna provides attachments for ligaments of the wrist.

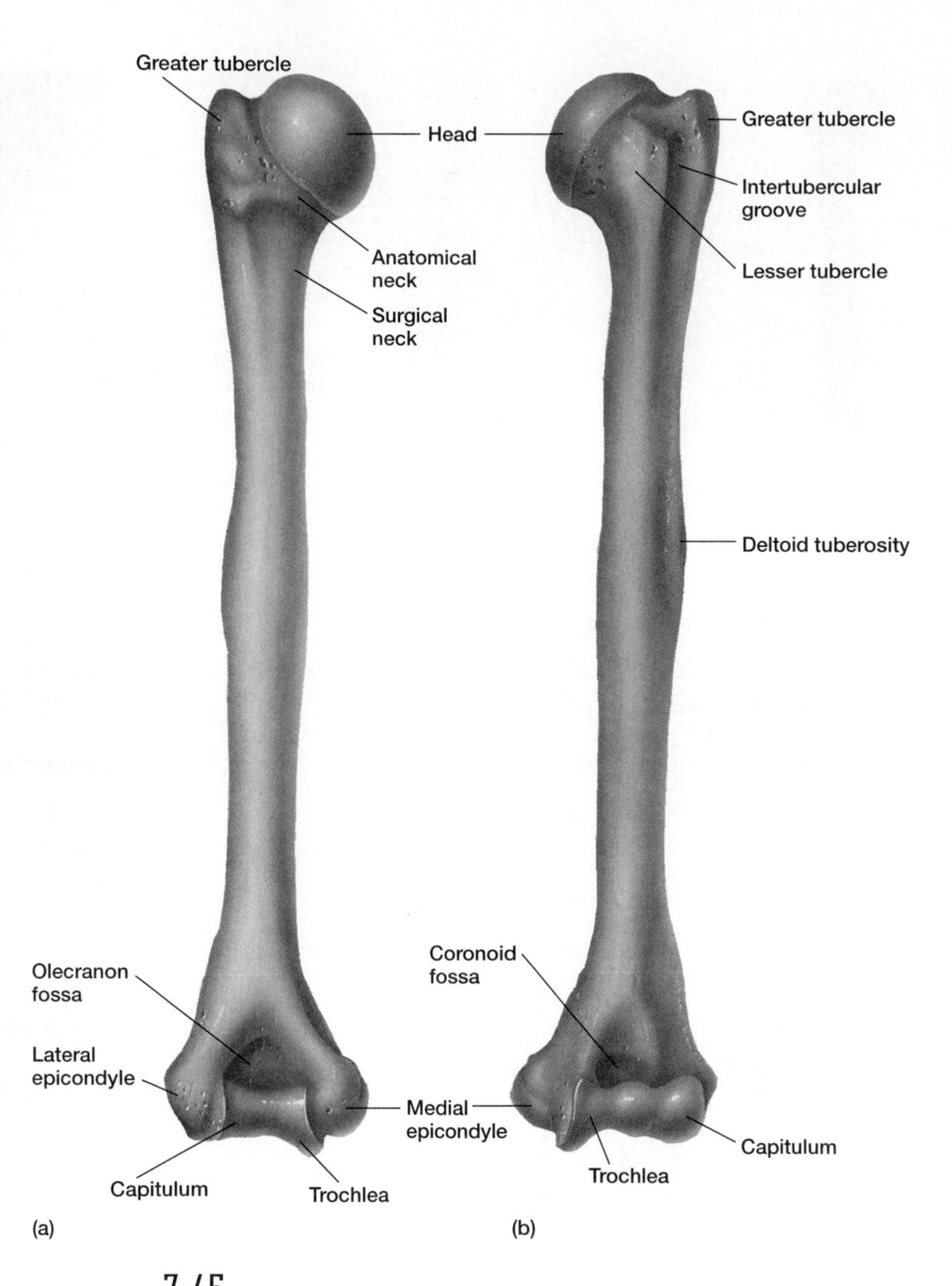


**FIGURE 7.45**

(*a*) Posterior surface and (*b*) anterior surface of the left humerus.

## Hand

The hand is composed of a wrist, a palm, and five fingers (fig. 7.47). The skeleton of the wrist consists of eight small **carpal bones** that are firmly bound in two rows of four bones each. The resulting compact mass is called a *carpus* (kar'pus).

The carpus is rounded on its proximal surface, where it articulates with the radius and with the fibrocartilaginous disk on the ulnar side. The carpus

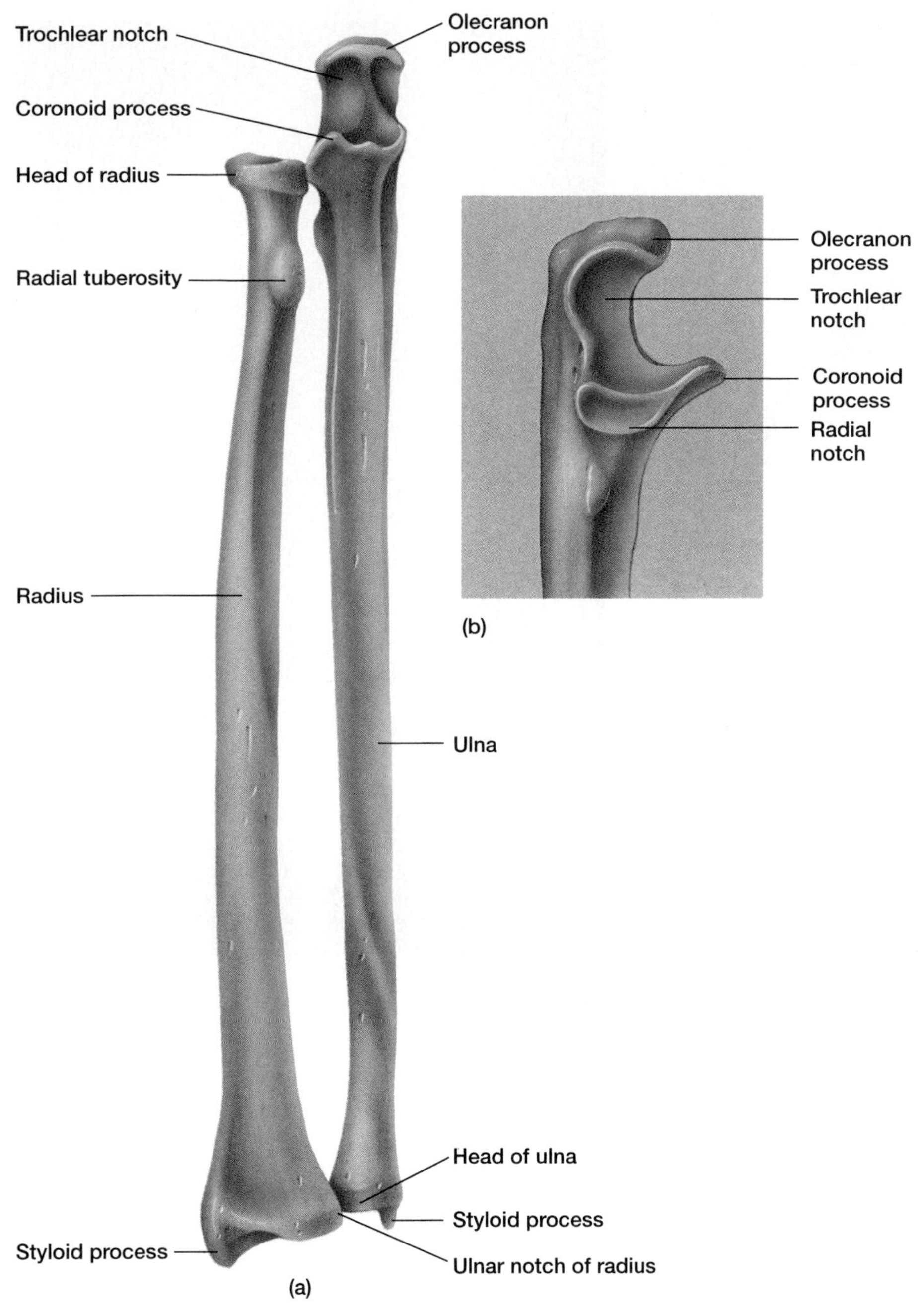


**FIGURE 7.46**

(*a*) The head of the right radius articulates with the radial notch of the ulna, and the head of the ulna articulates with the ulnar notch of the radius; (*b*) lateral view of the proximal end of the ulna.

is concave anteriorly, forming a canal through which tendons and nerves extend to the palm. Its distal surface articulates with the metacarpal bones. Figure 7.47 names the individual bones of the carpus.

Five **metacarpal** (met″ah-kar′pal) **bones,** one in line with each finger, form the framework of the palm. These bones are cylindrical, with rounded distal ends that form the knuckles of a clenched fist. The metacarpals articulate proximally with the carpals and distally with the phalanges. The metacarpal on the lateral side is the most freely movable; it permits the thumb to oppose the fingers when grasping something. These bones are numbered 1 to 5, beginning with the metacarpal of the thumb.

It is not uncommon for a baby to be born with an extra finger or toe, but since the extra digit is usually surgically removed early in life, hands like the ones in figure 7.48 are rare. Polydactyly ("many digits") is an inherited trait. It is common in cats. A lone but popular male cat brought the trait from England to colonial Boston.

The **phalanges** are the finger bones. There are three in each finger—a proximal, a middle, and a distal phalanx—and two in the thumb. (The thumb lacks a middle phalanx.) Thus, each hand has fourteen finger bones.

Chart 7.10 summarizes the bones of the pectoral girdle and upper limbs.

1. Locate and name each of the bones of the upper limb.
2. Explain how the bones of the upper limb articulate with one another.

## PELVIC GIRDLE

The **pelvic girdle** consists of the two coxal bones (hipbones), which articulate with each other anteriorly and with the sacrum posteriorly (fig. 7.49). The sacrum, coccyx, and pelvic girdle together form the ringlike *pelvis,* which supports the trunk of the body

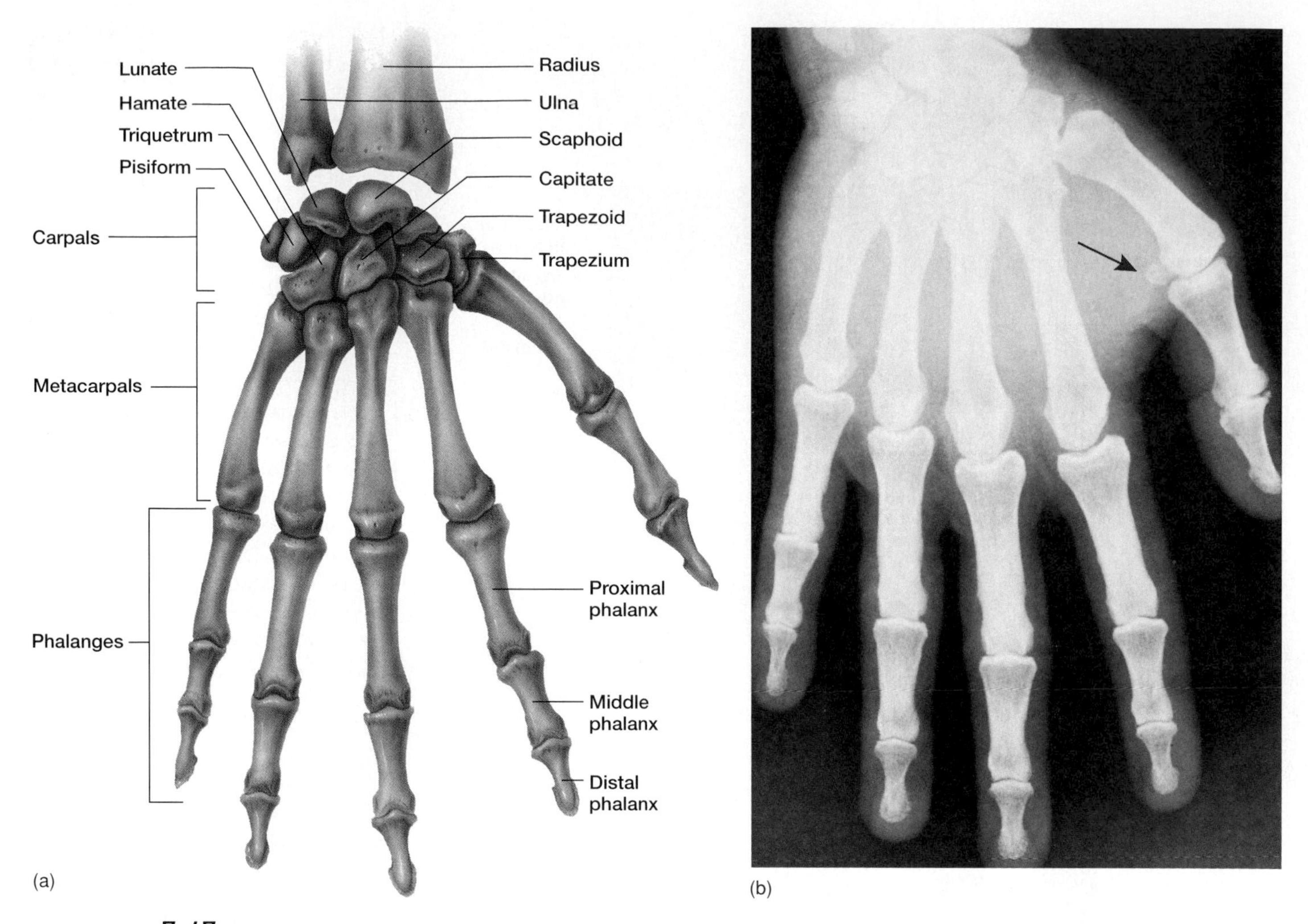


**FIGURE 7.47**
(*a*) The right hand posterior view. (*b*) X-ray film of the right hand. Note the small sesamoid bone associated with the joint at the base of the thumb (arrow).

and provides attachments for the lower limbs. The body's weight is transmitted through the pelvis to the lower limbs and then onto the ground. The pelvis also protects the urinary bladder, the distal end of the large intestine, and the internal reproductive organs.

### Coxal Bones

Each **coxal bone** develops from three parts—an ilium, an ischium, and a pubis. These parts fuse in the region of a cup-shaped cavity called the *acetabulum* (as″ĕ-tab′u-lum). This depression, on the lateral surface of the hipbone, receives the rounded head of the femur or thighbone (fig. 7.50).

The **ilium** (il′e-um), which is the largest and most superior portion of the coxal bone, flares outward, forming the prominence of the hip. The margin of this prominence is called the *iliac crest.* The smooth, concave surface on the anterior aspect of the ilium is the *iliac fossa.*

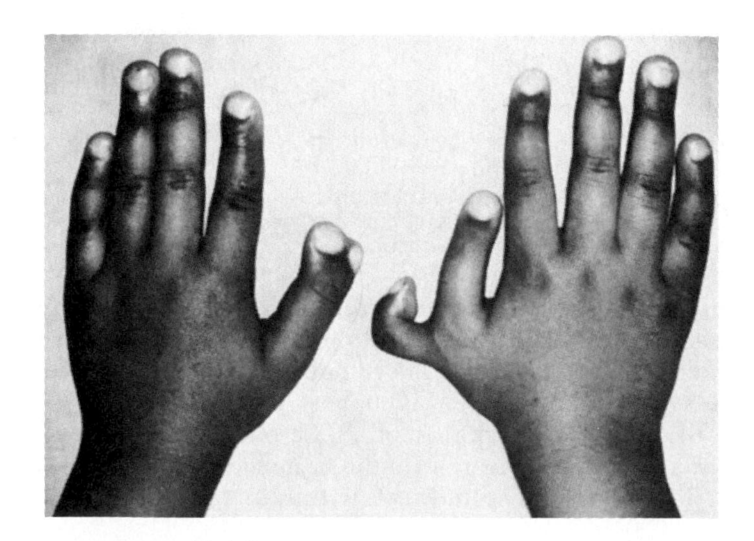

**FIGURE 7.48**
A person with polydactyly has extra digits.

## CHART 7.10 Bones of the Pectoral Girdle and Upper Limbs

| Name and Number | Location | Special Features |
|---|---|---|
| Clavicle (2) | Base of neck between sternum and scapula | Sternal end, acromial end |
| Scapula (2) | Upper back, forming part of shoulder | Body, spine, head, acromion process, coracoid process, glenoid cavity |
| Humerus (2) | Arm, between scapula and elbow | Head, greater tubercle, lesser tubercle, intertubercular groove, surgical neck, deltoid tuberosity, capitulum, trochlea, medial epicondyle, lateral epicondyle, coronoid fossa, olecranon fossa |
| Radius (2) | Lateral side of forearm, between elbow and wrist | Head, radial tuberosity, styloid process, ulnar notch |
| Ulna (2) | Medial side of forearm, between elbow and wrist | Trochlear notch, olecranon process, head, styloid process, radial notch |
| Carpal (16) | Wrist | Arranged in two rows of four bones each |
| Metacarpal (10) | Palm | One in line with each finger |
| Phalanx (28) | Finger | Three in each finger; two in each thumb |

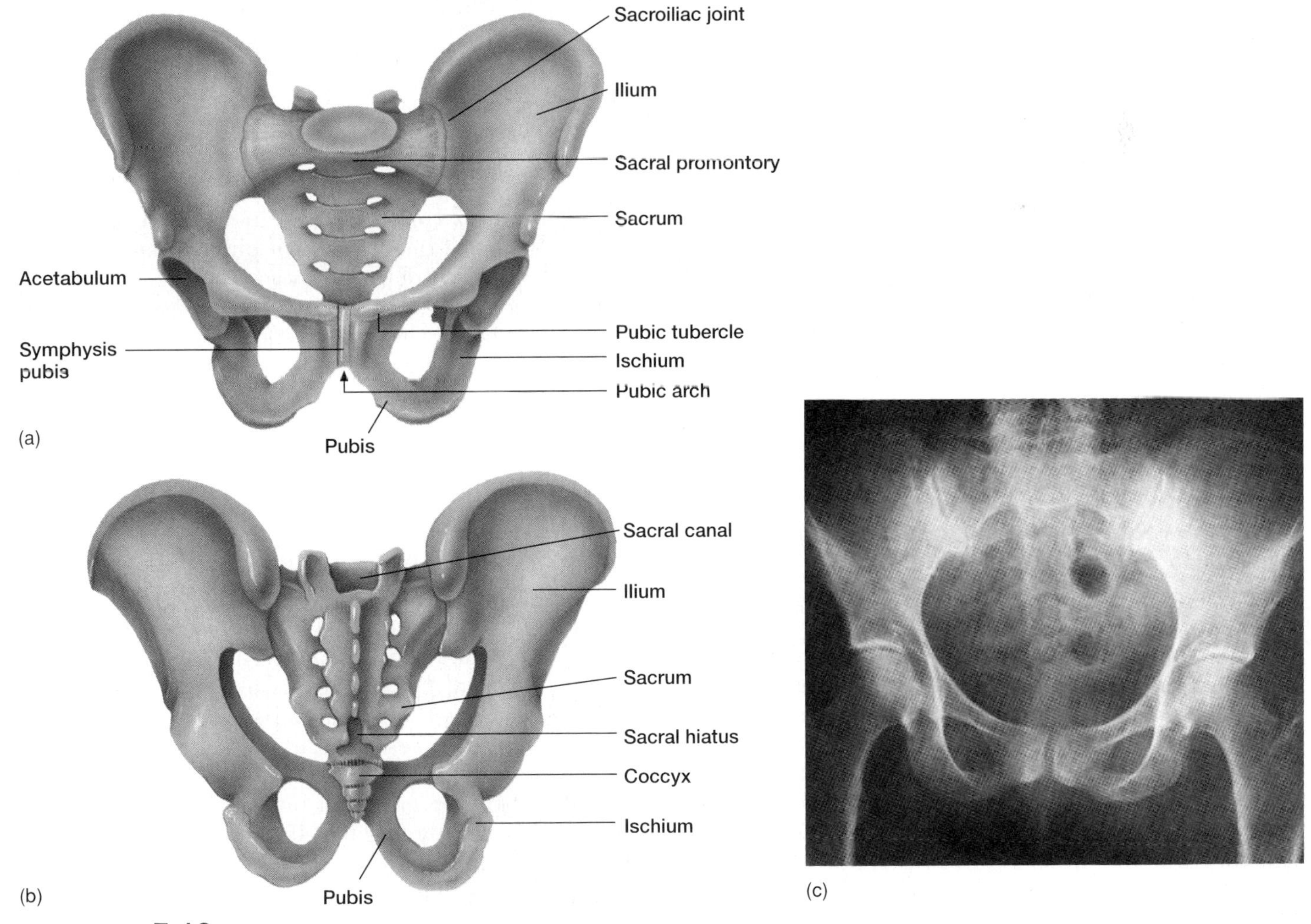


**FIGURE 7.49**

(*a*) Anterior view and (*b*) posterior view of the pelvic girdle. This girdle provides an attachment for the lower limbs, and together with the sacrum and coccyx forms the pelvis. (*c*) X-ray film of the pelvic girdle.

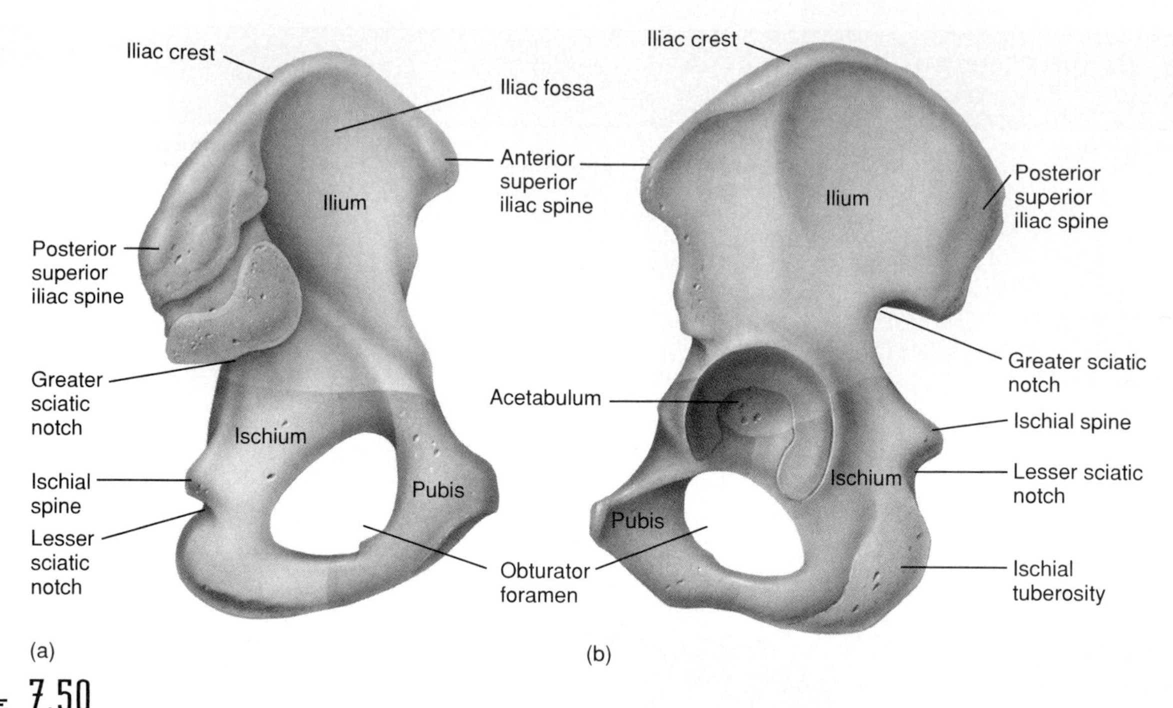


**FIGURE 7.50**
(*a*) Medial surface of the left coxal bone; (*b*) lateral view.

Posteriorly, the ilium joins the sacrum at the *sacroiliac* (sa″kro-il′e-ak) *joint.* Anteriorly, a projection of the ilium, the *anterior superior iliac spine,* can be felt lateral to the groin. This spine provides attachments for ligaments and muscles, and is an important surgical landmark.

A common injury in contact sports, such as football, is bruising the soft tissues and bone associated with the anterior superior iliac spine. Wearing protective padding can prevent this painful injury, called a *hip pointer.*

On the posterior border of the ilium is a *posterior superior iliac spine.* Below this spine is a deep indentation, the *greater sciatic notch,* through which a number of nerves and blood vessels pass.

The **ischium** (is′ke-um), which forms the lowest portion of the coxal bone, is L-shaped, with its angle, the *ischial tuberosity,* pointing posteriorly and downward. This tuberosity has a rough surface that provides attachments for ligaments and lower limb muscles. It also supports the weight of the body during sitting. Above the ischial tuberosity, near the junction of the ilium and ischium, is a sharp projection called the *ischial spine.* Like the sacral promontory this spine, which can be felt during a vaginal examination, is used as a guide for determining pelvis size. The distance between the ischial spines is the shortest diameter of the pelvic outlet.

The **pubis** (pu′bis) constitutes the anterior portion of the coxal bone. The two pubic bones come together at the midline to form a joint called the *symphysis pubis* (sim′fĭ-sis pu′bis). The angle these bones form below the symphysis is the *pubic arch.*

A portion of each pubis passes posteriorly and downward to join an ischium. Between the bodies of these bones on either side is a large opening, the *obturator foramen,* which is the largest foramen in the skeleton. An obturator membrane covers and nearly closes this foramen (see figs. 7.49 and 7.50).

## Greater and Lesser Pelves

If a line were drawn along each side of the pelvis from the sacral promontory downward and anteriorly to the upper margin of the symphysis pubis, it would mark the *pelvic brim* (linea terminalis). This margin separates the lower, or lesser (true), pelvis from the upper, or greater (false), pelvis (fig. 7.51).

The *greater pelvis* is bounded posteriorly by the lumbar vertebrae, laterally by the flared parts of the iliac bones, and anteriorly by the abdominal wall. The false pelvis helps support the abdominal organs.

The *lesser pelvis* is bounded posteriorly by the sacrum and coccyx, and laterally and anteriorly by the lower ilium, ischium, and pubis bones. This portion of the pelvis surrounds a short, canal-like cavity that has an upper inlet and a lower outlet. An infant passes through this cavity during childbirth.

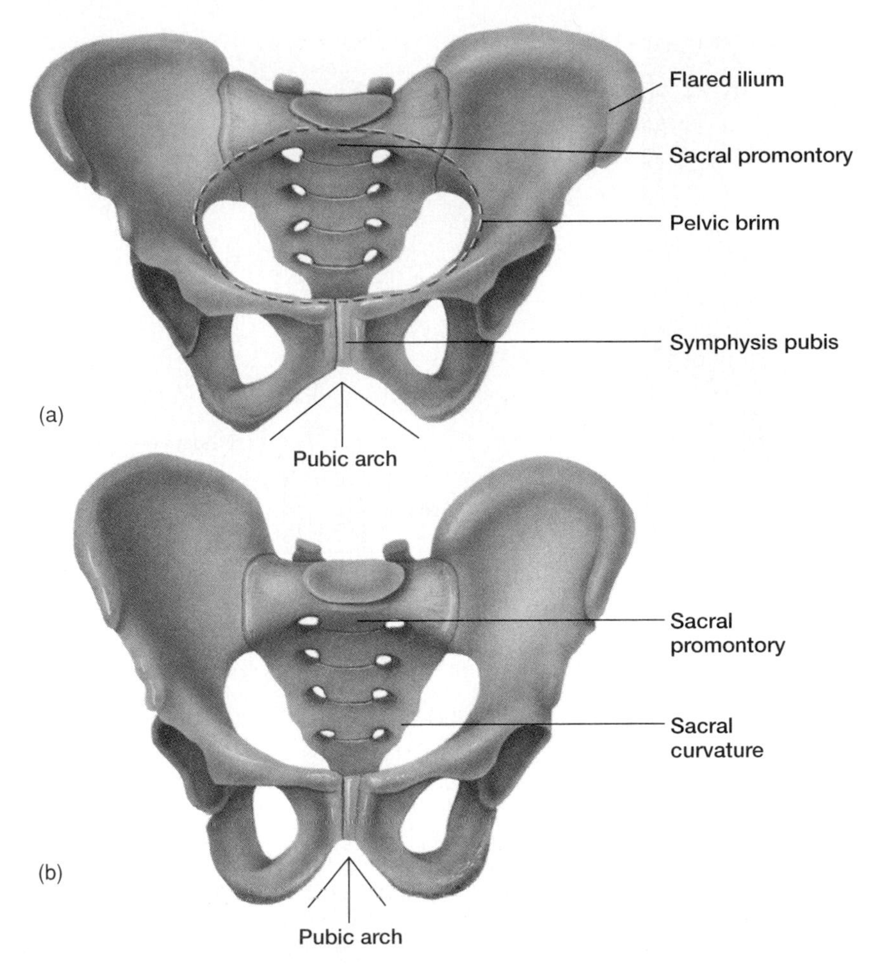


**FIGURE 7.51**
The female pelvis is usually wider in all diameters and roomier than that of the male. (*a*) Female pelvis; (*b*) male pelvis.

**CHART 7.11 DIFFERENCES BETWEEN THE MALE AND FEMALE SKELETONS**

| Part | Differences |
|---|---|
| Skull | Male skull is relatively larger and heavier, with more conspicuous muscular attachments. Male forehead is shorter, facial area is less round, jaw larger, and mastoid process more prominent than those of a female. |
| Pelvis | Male pelvic bones are heavier, thicker, and have more obvious muscular attachments. The obturator foramina and the acetabula are larger and closer together than those of a female. |
| Pelvic cavity | Male pelvic cavity is narrower in all diameters, and is longer, less roomy, and more funnel-shaped. The distances between the ischial spines and between the ischial tuberosities are lesser than in a female. |
| Sacrum | Male sacrum is relatively narrower, sacral promontory projects forward to a greater degree, and sacral curvature is bent less sharply posteriorly than in a female. |
| Coccyx | Male coccyx is less movable than that of a female. |

## Differences Between Male and Female Pelves

Some basic structural differences exist between the male and the female pelves, even though it may be difficult to find all of the "typical" characteristics in any one individual. These differences are related to the function of the female pelvis as a birth canal. Usually, the female iliac bones are more flared than those of the male, and consequently, the female hips are usually broader. The angle of the female pubic arch may be greater, there may be more distance between the ischial spines and the ischial tuberosities, and the sacral curvature may be shorter and flatter. Thus, the female pelvic cavity is usually wider in all diameters than that of the male. Also, the bones of the female pelvis are usually lighter, more delicate, and show less evidence of muscle attachments (see fig. 7.51). Chart 7.11 summarizes some of the differences between the female and male skeletons.

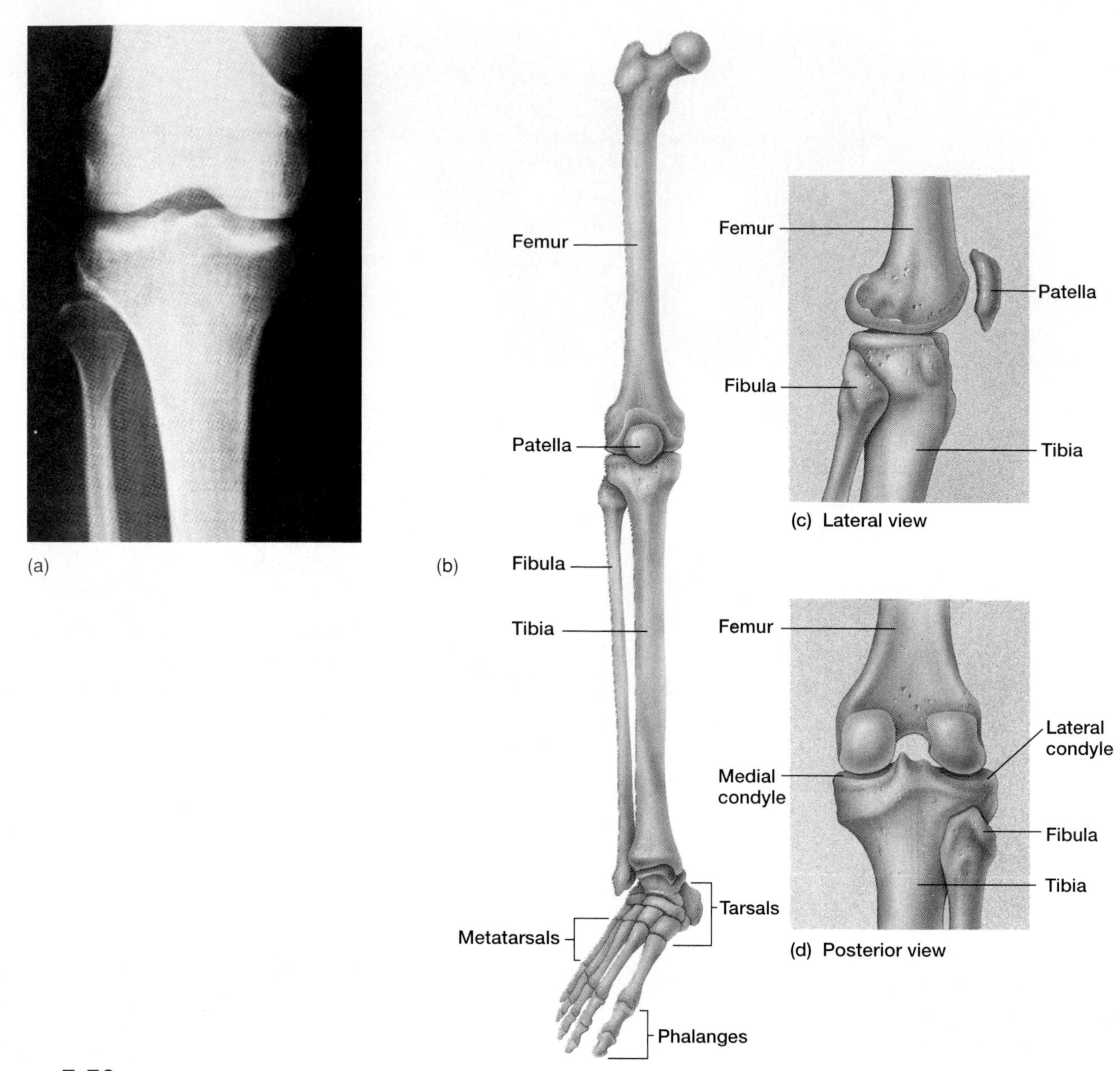


**FIGURE 7.52**

(*a*) X-ray film of the right knee, showing the ends of the femur, tibia, and fibula; (*b*) anterior view of the right lower limb; (*c*) lateral view of the right knee; (*d*) posterior view of the right knee.

1. Locate and name each pelvic bone.
2. Explain what is meant by the greater pelvis and the lesser pelvis.
3. How are male and female pelves different?

## LOWER LIMB

The bones of the lower limb form the frameworks of the thigh, leg, and foot. They include a femur, a tibia, a fibula, tarsals, metatarsals, and phalanges (fig. 7.52).

### Femur

The **femur,** or thighbone, is the longest bone in the body and extends from the hip to the knee. A large, rounded *head* at its proximal end projects medially into the acetabulum of the coxal bone. On the head, a pit called the *fovea capitis* marks the attachment of a ligament. Just below the head are a constriction, or *neck,* and two large processes—a superior, lateral *greater trochanter* and an inferior, medial *lesser trochanter.* These processes provide attachments for muscles of the legs and buttocks. On the posterior surface in the middle third of the shaft is a longitudinal crest called the *linea aspera.* This rough strip is an attachment for several muscles (fig. 7.53).

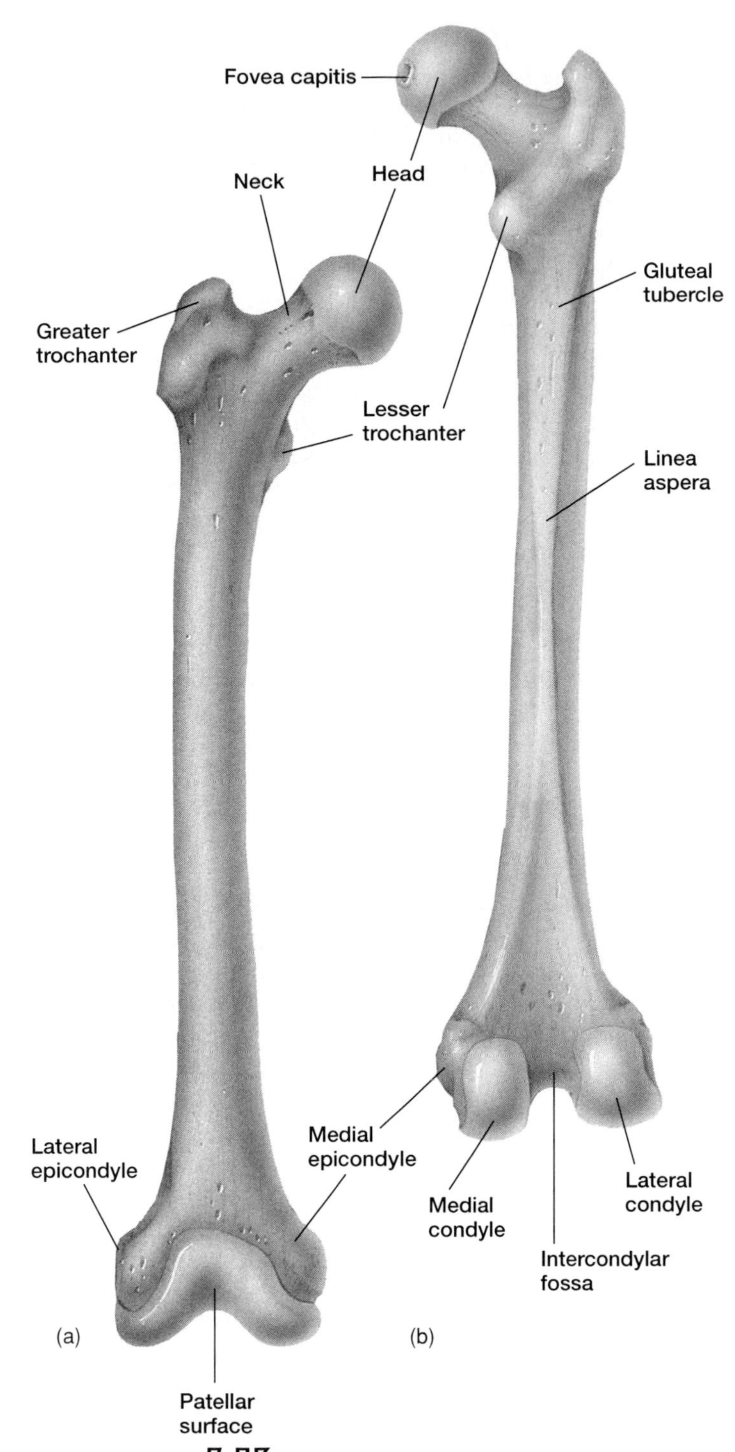


**FIGURE 7.53**

(*a*) Anterior surface and (*b*) posterior surface of the right femur.

At the distal end of the femur, two rounded processes, the *lateral* and *medial condyles,* articulate with the tibia of the leg. A patella also articulates with the femur on its distal anterior surface.

On the medial surface at its distal end is a prominent *medial epicondyle,* and on the lateral surface is a *lateral epicondyle.* These projections provide attachments for muscles and ligaments.

## Patella

The **patella,** or kneecap, is a flat sesamoid bone located in a tendon that passes anteriorly over the knee (see fig. 7.52). Because of its position, the patella controls the angle at which this tendon continues toward the tibia, and so it functions in lever actions associated with lower limb movements.

> As a result of a blow to the knee or a forceful unnatural movement of the leg, the patella sometimes slips to one side. This painful condition is called a *patellar dislocation.* Exercises that strengthen muscles associated with the knee and wearing protective padding can prevent knee displacement. Unfortunately, once the soft tissues that hold the patella in place are stretched, patellar dislocation tends to recur.

## Tibia

The **tibia,** or shinbone, is the larger of the two leg bones and is located on the medial side. Its proximal end is expanded into *medial* and *lateral condyles,* which have concave surfaces and articulate with the condyles of the femur. Below the condyles, on the anterior surface, is a process called the *tibial tuberosity,* which provides an attachment for the *patellar ligament* (a continuation of the patella-bearing tendon). A prominent *anterior crest* extends downward from the tuberosity and attaches connective tissues in the leg.

At its distal end, the tibia expands to form a prominence on the inner ankle called the *medial malleolus* (mah-le′o-lus), which is an attachment for ligaments. On its lateral side is a depression that articulates with the fibula. The inferior surface of the tibia's distal end articulates with a large bone (the talus) in the foot (fig. 7.54).

> The skeleton is particularly vulnerable to injury during the turbulent teen years, when bones grow rapidly. Athletic teens sometimes develop Osgood-Schlatter disease, painful swelling of a bony projection of the tibia below the knee. Overusing the thigh muscles to straighten the lower limb irritates the area, causing the swelling. Usually a few months of rest and no athletic activity allows the bone to heal on its own. Rarely, a cast must be used to immobilize the knee.

## Fibula

The **fibula** is a long, slender bone located on the lateral side of the tibia. Its ends are slightly enlarged into a proximal *head* and a distal *lateral malleolus.* The

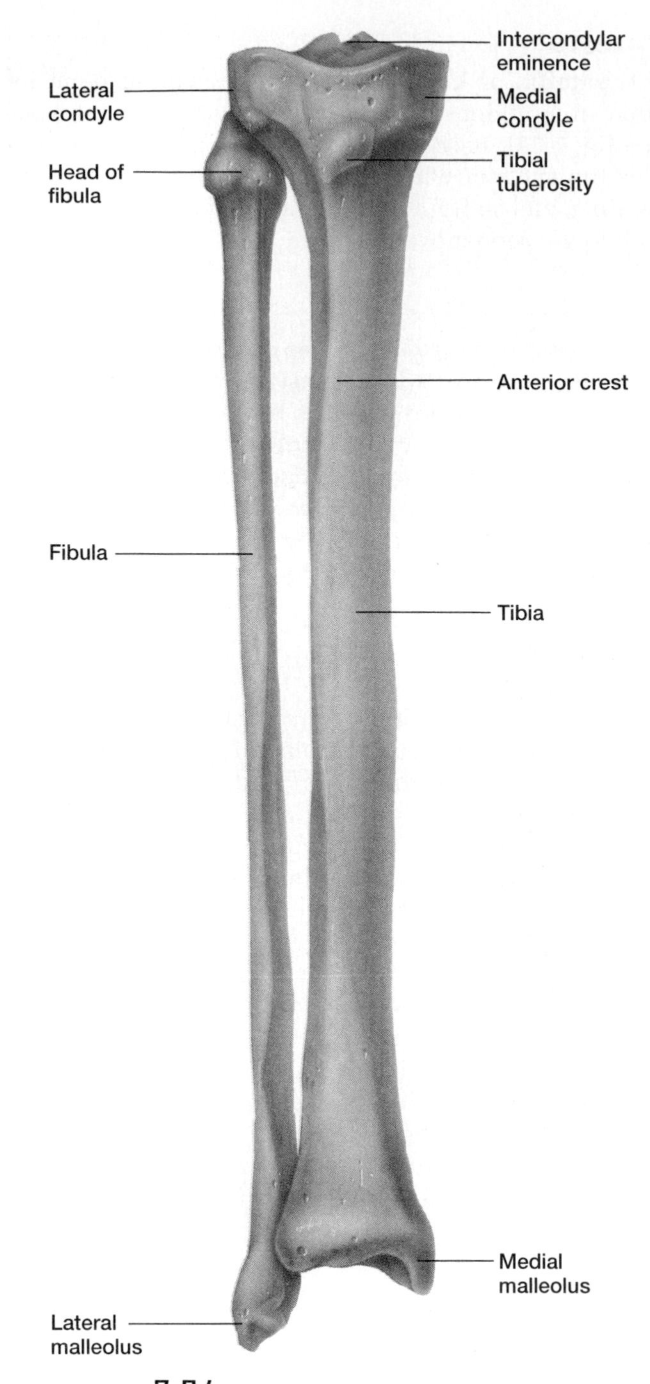


**FIGURE 7.54**
Bones of the right leg viewed from the front.

head articulates with the tibia just below the lateral condyle; however, it does not enter into the knee joint and does not bear any body weight. The lateral malleolus articulates with the ankle and forms a prominence on the lateral side (see fig. 7.54).

## Foot

The foot consists of an ankle, an instep, and five toes. The ankle is composed of seven **tarsal bones,** forming a group called the *tarsus* (tahr′sus). These bones are arranged so that one of them, the **talus** (ta′lus), can move freely where it joins the tibia and fibula. The remaining tarsal bones are bound firmly together, forming a mass supporting the talus. Figures 7.55 and 7.56 name the individual bones of the tarsus.

The largest of the anklebones, the **calcaneus** (kalka′ne-us), or heel bone, is located below the talus where it projects backward to form the base of the heel. The calcaneus helps support the weight of the body and provides an attachment for muscles that move the foot.

The instep consists of five elongated **metatarsal** (met″ah-tar′sal) **bones,** which articulate with the tarsus. They are numbered 1 to 5, beginning on the medial side (see fig. 7.56). The heads at the distal ends of these bones form the ball of the foot. The tarsals and metatarsals are arranged and bound by ligaments to form the arches of the foot. A longitudinal arch extends from the heel to the toe, and a transverse arch stretches across the foot. These arches provide a stable, springy base for the body. Sometimes, however, the tissues that bind the metatarsals weaken, producing fallen arches, or flat feet.

> An infant with two casts on her feet is probably being treated for clubfoot, a very common birth defect in which the foot twists out of its normal position, turning in, out, up, down, or some combination of these directions. Clubfoot probably results from arrested development during fetal existence, but the precise cause is not known. Clubfoot can almost always be corrected with special shoes, or surgery, followed by several months in casts to hold the feet in the correct position.

The **phalanges** of the toes are similar to those of the fingers, and align and articulate with the metatarsals. Each toe has three phalanges—a proximal, a middle, and a distal phalanx—except the great toe, which has only two because it lacks the middle phalanx (see fig. 7.56). Chart 7.12 summarizes the bones of the pelvic girdle and lower limbs.

1. Locate and name each of the bones of the lower limb.
2. Explain how the bones of the lower limb articulate with one another.
3. Describe how the foot is adapted to support the body.

## CHART 7.12 Bones of the Pelvic Girdle and Lower Limbs

| Name and Number | Location | Special Features |
|---|---|---|
| Coxal bone (2) | Hip, articulating with the other coxal bone anteriorly and with the sacrum posteriorly | Ilium, iliac crest, anterior superior iliac spine, ischium, ischial tuberosity, ischial spine, obturator foramen, acetabulum, pubis |
| Femur (2) | Thigh, between hip and knee | Head, fovea capitis, neck, greater trochanter, lesser trochanter, linea aspera, lateral condyle, medial condyle, gluteal tuberosity, intercondylar fossa |
| Patella (2) | Anterior surface of knee | A flat sesamoid bone located within a tendon |
| Tibia (2) | Medial side of leg, between knee and ankle | Medial condyle, lateral condyle, tibial tuberosity, anterior crest, medial malleolus, intercondylar eminence |
| Fibula (2) | Lateral side of leg, between knee and ankle | Head, lateral malleolus |
| Tarsal (14) | Ankle | Freely movable talus that articulates with leg bones; six other tarsal bones bound firmly together |
| Metatarsal (10) | Instep | One in line with each toe, arranged and bound by ligaments to form arches |
| Phalanx (28) | Toe | Three in each toe, two in great toe |

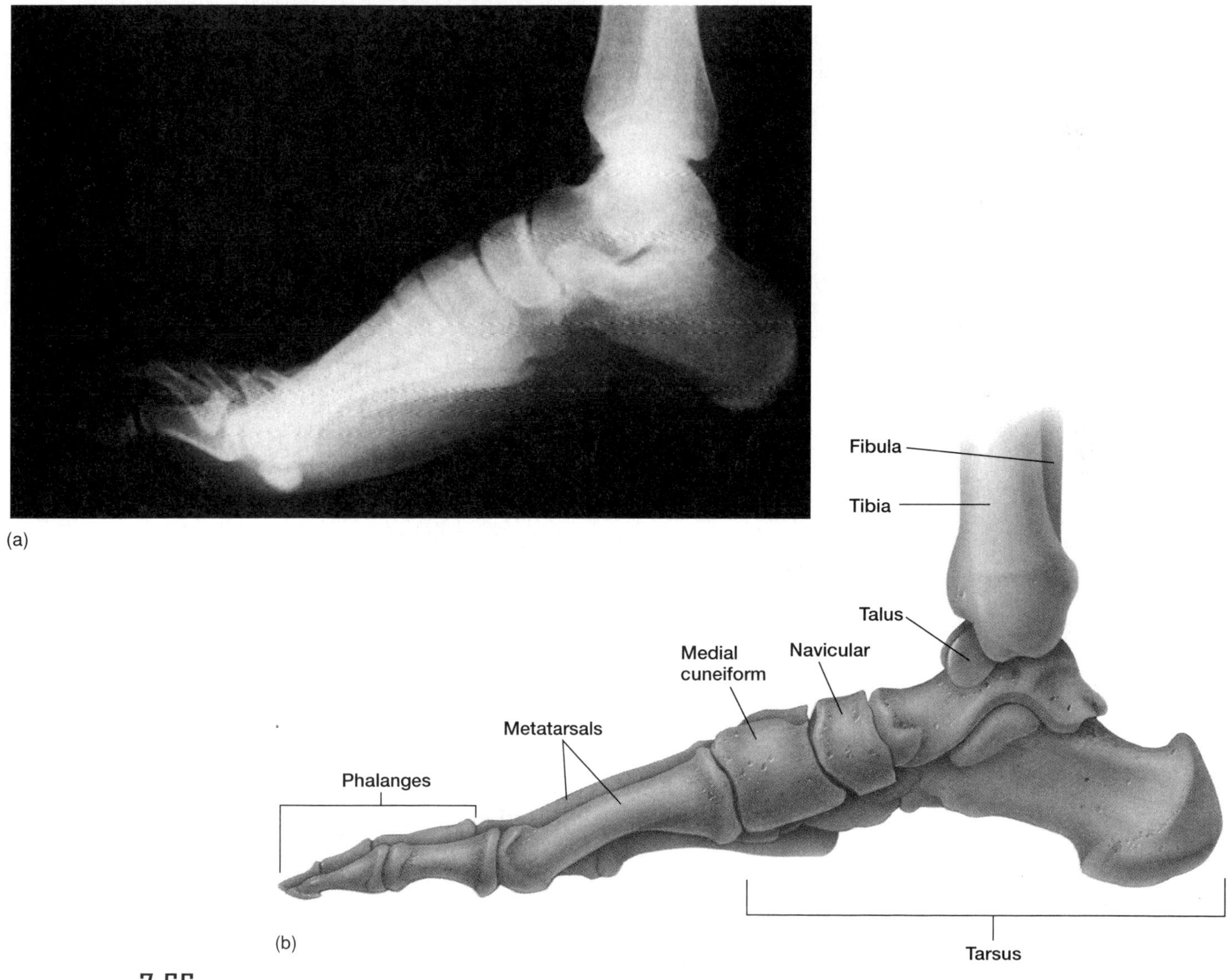


**FIGURE 7.55**
(*a*) X-ray film of the right foot viewed from the medial side; (*b*) the talus moves freely where it articulates with the tibia and fibula.

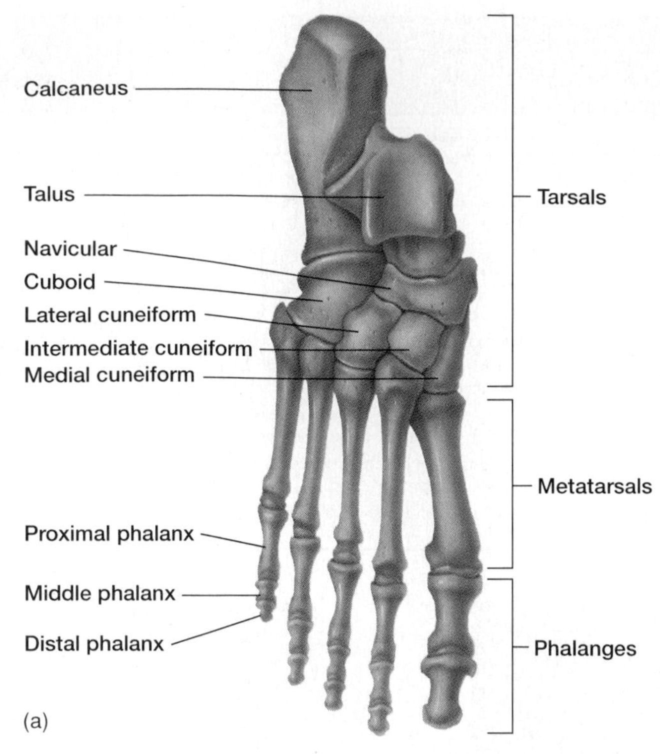


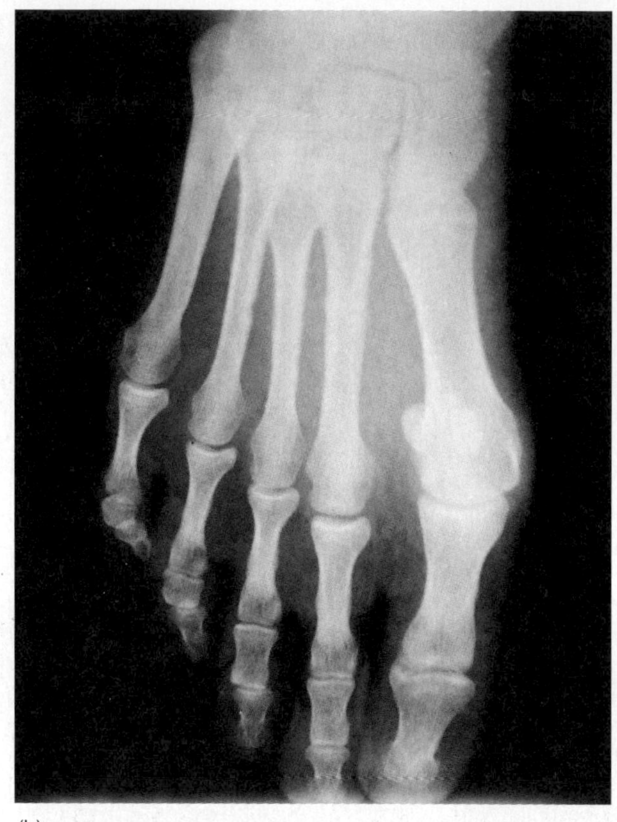


**FIGURE 7.56**
(*a*) The right foot viewed superiorly; (*b*) X-ray film of the right foot viewed superiorly.

## Clinical Terms Related to the Skeletal System

**achondroplasia** (a-kon″dro-pla′ze-ah) An inherited condition that retards formation of cartilaginous bone. The result is a type of dwarfism.

**acromegaly** (ak″ro-meg′ah-le) A condition caused by overproduction of growth hormone in adults and characterized by abnormal enlargement of facial features, hands, and feet.

**Colles fracture** (kol′ēz frak′tūre) A fracture at the distal end of the radius which posteriorly displaces the smaller fragment.

**epiphysiolysis** (ep″ĭ-fiz″e-ol′ĭ-sis) A separation or loosening of the epiphysis from the diaphysis of a bone.

**laminectomy** (lam″ĭ-nek′to-me) Surgical removal of the posterior arch of a vertebra, usually to relieve symptoms of a ruptured intervertebral disk.

**lumbago** (lum-ba′go) A dull ache in the lumbar region of the back.

**orthopedics** (or″tho-pe′diks) The science of prevention, diagnosis, and treatment of diseases and abnormalities of the skeletal and muscular systems.

**ostalgia** (os-tal′je-ah) Pain in a bone.

**ostectomy** (os-tek′to-me) Surgical removal of a bone.

**osteitis** (os″te-i′tis) Inflammation of bone tissue.

**osteochondritis** (os″te-o-kon-dri′tis) Inflammation of bone and cartilage tissues.

**osteogenesis** (os″te-o-jen′ĕ-sis) Bone development.

**osteogenesis imperfecta** (os″te-o-jen′ĕ-sis im-per-fek′ta) An inherited condition of deformed and abnormally brittle bones.

**osteoma** (os″te-o′mah) A tumor composed of bone tissue.

**osteomalacia** (os″te-o-mah-la′she-ah) A softening of adult bone due to a disorder in calcium and phosphorus metabolism, usually caused by deficiency of vitamin D.

**osteomyelitis** (os″te-o-mi″ĕ-li′tis) Bone inflammation caused by the body's reaction to bacterial or fungal infection.

**osteonecrosis** (os″te-o-ne-kro′sis) Death of bone tissue. This condition occurs most commonly in the femur head in elderly persons and may be due to obstructed arteries supplying the bone.

**osteopathology** (os″te-o-pah-thol′o-je) The study of bone diseases.

**osteopenia** (os″te-o-pe′ni-ah) Decrease in bone mass due to reduction in rate of bone tissue formation.

**osteoporosis** (os″te-o-po-ro′sis) Decreased bone mineral content.

**osteotomy** (os″te-ot′o-me) Cutting a bone.

**roentgenogram** (rent-gen′o-gram″) An image obtained on film by using X rays.

# InnerConnections
## Skeletal System

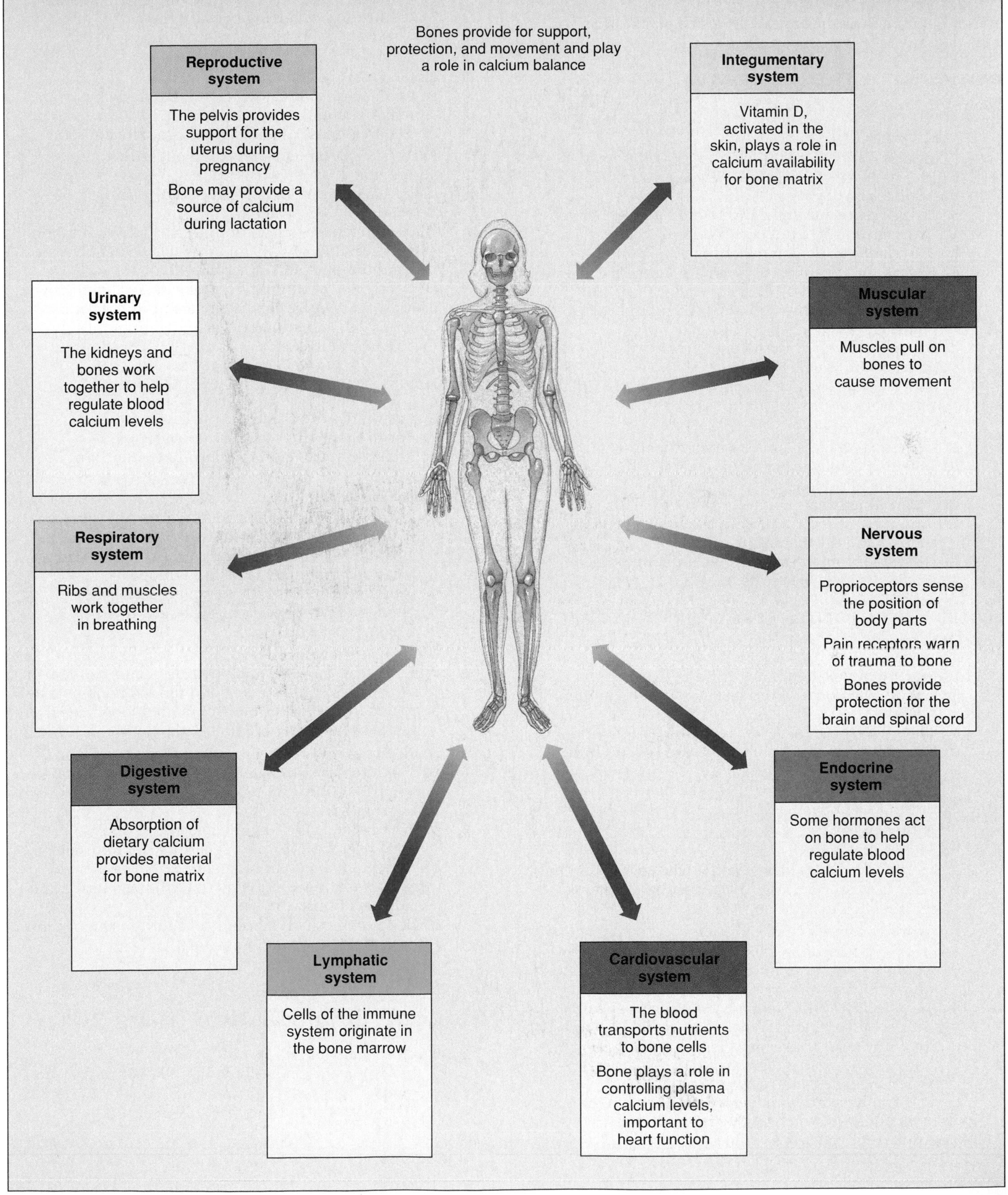

Bones provide for support, protection, and movement and play a role in calcium balance

**Reproductive system**

The pelvis provides support for the uterus during pregnancy

Bone may provide a source of calcium during lactation

**Integumentary system**

Vitamin D, activated in the skin, plays a role in calcium availability for bone matrix

**Urinary system**

The kidneys and bones work together to help regulate blood calcium levels

**Muscular system**

Muscles pull on bones to cause movement

**Respiratory system**

Ribs and muscles work together in breathing

**Nervous system**

Proprioceptors sense the position of body parts

Pain receptors warn of trauma to bone

Bones provide protection for the brain and spinal cord

**Digestive system**

Absorption of dietary calcium provides material for bone matrix

**Endocrine system**

Some hormones act on bone to help regulate blood calcium levels

**Lymphatic system**

Cells of the immune system originate in the bone marrow

**Cardiovascular system**

The blood transports nutrients to bone cells

Bone plays a role in controlling plasma calcium levels, important to heart function

# CHAPTER SUMMARY

## Introduction (page 187)

Individual bones are the organs of the skeletal system. A bone contains very active tissues.

## Bone Structure (page 187)

Bone structure reflects its function.

1. Bone classification

   Bones are grouped according to their shapes—long, short, flat, irregular, or sesamoid.
2. Parts of a long bone
   a. Epiphyses at each end are covered with articular cartilage and articulate with other bones.
   b. The shaft of a bone is called the diaphysis.
   c. Except for the articular cartilage, a bone is covered by a periosteum.
   d. Compact bone provides strength and resistance to bending.
   e. Spongy bone provides strength where needed and reduces the weight of bone.
   f. The diaphysis contains a medullary cavity filled with marrow.
3. Microscopic structure
   a. Compact bone contains osteons cemented together.
   b. Osteonic canals contain blood vessels that nourish the cells of osteons.
   c. Perforating canals connect osteonic canals transversely and communicate with the bone's surface and the medullary cavity.
   d. Diffusion from the surface of thin bony plates nourishes cells of spongy bones.

## Bone Development and Growth (page 190)

1. Intramembranous bones
   a. Certain flat bones of the skull are intramembranous bones.
   b. They develop from layers of connective tissues.
   c. Osteoblasts within the membranous layers form bone tissue.
   d. Mature bone cells are called osteocytes.
   e. Primitive connective tissue gives rise to the periosteum.
2. Endochondral bones
   a. Most of the bones of the skeleton are endochondral.
   b. They develop as hyaline cartilage that is later replaced by bone tissue.
3. Growth of an endochondral bone
   a. Primary ossification center appears in the diaphysis, while secondary ossification centers appear in the epiphyses.
   b. An epiphyseal disk remains between the primary and secondary ossification centers.
   c. An epiphyseal disk consists of layers of cells: resting cells, young reproducing cells, older enlarging cells, and dying cells.
   d. The epiphyseal disk is responsible for lengthening.
   e. Long bones continue to lengthen until the epiphyseal disks are ossified.
   f. Growth in thickness is due to intramembranous ossification occurring beneath the periosteum.
   g. The action of osteoclasts forms the medullary cavity.
4. Homeostasis of bone tissue
   a. Osteoclasts and osteoblasts continually remodel bone.
   b. The total mass of bone remains nearly constant.
5. Factors affecting bone development, growth, and repair
   a. Deficiencies of vitamin A, C, or D result in abnormal development.
   b. Insufficient secretion of pituitary growth hormone may result in dwarfism; excessive secretion may result in giantism, or acromegaly.
   c. Deficiency of thyroid hormone delays bone growth.
   d. Male and female sex hormones promote bone formation and stimulate ossification of the epiphyseal disks.

## Bone Function (page 197)

1. Support and protection
   a. Bones shape and form body structures.
   b. Bones support and protect softer, underlying tissues.
2. Body movement
   a. Bones and muscles function together as levers.
   b. A lever consists of a rod, a pivot (fulcrum), a movable weight, and a force that supplies energy.
   c. Parts of a first-class lever are arranged weight–pivot–force; of a second-class lever pivot–weight–force; of a third-class lever weight–force–pivot.
3. Blood cell formation
   a. At different ages, hematopoiesis occurs in the yolk sac, the liver, the spleen, and the red bone marrow.
   b. Red marrow houses developing red blood cells, white blood cells, and blood platelets.
4. Inorganic salt storage
   a. The intercellular material of bone tissue contains large quantities of calcium phosphate in the form of hydroxyapatite.
   b. When blood calcium ion concentration is low, osteoclasts resorb bone, releasing calcium salts.
   c. When blood calcium ion concentration is high, osteoblasts are stimulated to form bone tissue and store calcium salts.
   d. Bone stores small amounts of sodium, magnesium, potassium, and carbonate ions.
   e. Bone tissues may accumulate lead, radium, or strontium.

## Skeletal Organization (page 201)

1. Number of bones
   a. Usually a human skeleton has 206 bones, but the number may vary.
   b. Extra bones in sutures are called sutural bones.

2. Divisions of the skeleton
   a. The skeleton can be divided into axial and appendicular portions.
   b. The axial skeleton consists of the skull, hyoid bone, vertebral column, and thoracic cage.
   c. The appendicular skeleton consists of the pectoral girdle, upper limbs, pelvic girdle, and lower limbs.

## Skull (page 204)

The skull consists of twenty-two bones, which include eight cranial bones, thirteen facial bones, and one mandible.

1. Cranium
   a. The cranium encloses and protects the brain, and provides attachments for muscles.
   b. Some cranial bones contain air-filled sinuses that help reduce the weight of the skull.
   c. Cranial bones include the frontal bone, parietal bones, occipital bone, temporal bones, sphenoid bone, and ethmoid bone.
2. Facial skeleton
   a. Facial bones form the basic shape of the face and provide attachments for muscles.
   b. Facial bones include the maxillary bones, palatine bones, zygomatic bones, lacrimal bones, nasal bones, vomer bone, inferior nasal conchae, and mandible.
3. Infantile skull
   a. Incompletely developed bones, separated by fontanels, enable the infantile skull to change shape slightly during childbirth.
   b. The proportions of the infantile skull are different from those of an adult skull, and its bones are less easily fractured.

## Vertebral Column (page 217)

The vertebral column extends from the skull to the pelvis and protects the spinal cord. It is composed of vertebrae separated by intervertebral disks. An infant has thirty-three vertebral bones and an adult has twenty-six. The vertebral column has four curvatures—cervical, thoracic, lumbar, and pelvic.

1. A typical vertebra
   a. A typical vertebra consists of a body, pedicles, laminae, spinous process, transverse processes, and superior and inferior articulating processes.
   b. Notches on the upper and lower surfaces of the pedicles on adjacent vertebrae form intervertebral foramina through which spinal nerves pass.
2. Cervical vertebrae
   a. Cervical vertebrae comprise the bones of the neck.
   b. Transverse processes bear transverse foramina.
   c. The atlas (first vertebra) supports and balances the head.
   d. The dens of the axis (second vertebra) provides a pivot for the atlas when the head is turned from side to side.
3. Thoracic vertebrae
   a. Thoracic vertebrae are larger than cervical vertebrae.
   b. Their long spinous processes slope downward, and facets on the sides of bodies articulate with the ribs.
4. Lumbar vertebrae
   a. Vertebral bodies of lumbar vertebrae are large and strong.
   b. Their transverse processes project posteriorly at sharp angles, and their spinous processes are directed horizontally.
5. Sacrum
   a. The sacrum is a triangular structure that bears rows of dorsal sacral foramina.
   b. It is united with the coxal bones at the sacroiliac joints.
   c. The sacral promontory provides a guide for determining the size of the pelvis.
6. Coccyx
   a. The coccyx forms the lowest part of the vertebral column.
   b. It acts as a shock absorber when a person sits.

## Thoracic Cage (page 223)

The thoracic cage includes the ribs, thoracic vertebrae, sternum, and costal cartilages. It supports the shoulder girdle and upper limbs, protects viscera, and functions in breathing.

1. Ribs
   a. Twelve pairs of ribs are attached to the twelve thoracic vertebrae.
   b. Costal cartilages of the true ribs join the sternum directly; those of the false ribs join indirectly or not at all.
   c. A typical rib has a shaft, head, and tubercles that articulate with the vertebrae.
2. Sternum
   a. The sternum consists of a manubrium, body, and xiphoid process.
   b. It articulates with costal cartilages and clavicles.

## Pectoral Girdle (page 226)

The pectoral girdle is composed of two clavicles and two scapulae. It forms an incomplete ring that supports the upper limbs and provides attachments for muscles that move the upper limbs.

1. Clavicles
   a. Clavicles are rodlike bones that run horizontally between the sternum and shoulders.
   b. They hold the shoulders in place and provide attachments for muscles.
2. Scapulae
   a. The scapulae are broad, triangular bones with bodies, spines, heads, acromion processes, coracoid processes, glenoid cavities, supraspinous and infraspinous fossae, superior borders, axillary borders, and vertebral borders.
   b. They articulate with the humerus of each upper limb, and provide attachments for muscles of the upper limbs and chest.

## Upper Limb (page 227)

Limb bones provide the frameworks and attachments of muscles, and function in levers that move the limb and its parts.

1. Humerus
   a. The humerus extends from the scapula to the elbow.

b. It bears a head, greater tubercle, lesser tubercle, intertubercular groove, anatomical neck, surgical neck, deltoid tuberosity, capitulum, trochlea, epicondyles, coronoid fossa, and olecranon fossa.

2. Radius
   a. The radius is located on the thumb side of the forearm between the elbow and wrist.
   b. It has a head, radial tuberosity, styloid process, and ulnar notch.
3. Ulna
   a. The ulna is longer than the radius and overlaps the humerus posteriorly.
   b. It bears a trochlear notch, olecranon process, coronoid process, head, styloid process, and radial notch.
   c. It articulates with the radius laterally and with a disk of fibrocartilage inferiorly.
4. Hand
   a. The hand is composed of a wrist, palm, and five fingers.
   b. It includes eight carpals that form a carpus, five metacarpals, and fourteen phalanges.

## Pelvic Girdle (page 231)

The pelvic girdle consists of two coxal bones that articulate with each other anteriorly and with the sacrum posteriorly. The sacrum, coccyx, and pelvic girdle form the pelvis. The girdle provides support for body weight and attachments for muscles, and protects visceral organs.

1. Coxal bones

   Each coxal bone consists of an ilium, ischium, and pubis, which are fused in the region of the acetabulum.
   a. Ilium
      (1) The ilium, the largest portion of the coxal bone, joins the sacrum at the sacroiliac joint.
      (2) It bears an iliac crest with anterior and posterior superior iliac spines, and iliac fossae.
   b. Ischium
      (1) The ischium is the lowest portion of the coxal bone.
      (2) It bears an ischial tuberosity and ischial spine.
   c. Pubis
      (1) The pubis is the anterior portion of the coxal bone.
      (2) Pubis bones are fused anteriorly at the symphysis pubis.
2. Greater and lesser pelves
   a. The lesser pelvis is below the pelvic brim; the greater pelvis is above it.
   b. The lesser pelvis functions as a birth canal; the greater pelvis helps support abdominal organs.
3. Differences between male and female pelves
   a. Differences between male and female pelves are related to the function of the female pelvis as a birth canal.
   b. Usually the female pelvis is more flared; pubic arch is broader; distance between the ischial spines and the ischial tuberosities is greater; and sacral curvature is shorter.

## Lower Limb (page 236)

Bones of the lower limb provide the frameworks of the thigh, leg, and foot.

1. Femur
   a. The femur extends from the hip to the knee.
   b. It bears a head, fovea capitis, neck, greater trochanter, lesser trochanter, linea aspera, lateral condyle, and medial condyle.
2. Patella
   a. The patella is a flat sesamoid bone in the tendon that passes anteriorly over the knee.
   b. It controls the angle of this tendon and functions in lever actions associated with lower limb movements.
3. Tibia
   a. The tibia is located on the medial side of the leg.
   b. It bears medial and lateral condyles, tibial tuberosity, anterior crest, and medial malleolus.
   c. It articulates with the talus of the ankle.
4. Fibula
   a. The fibula is located on the lateral side of the tibia.
   b. It bears a head and lateral malleolus that articulates with the ankle.
5. Foot
   a. The foot consists of an ankle, an instep, and five toes.
   b. It includes seven tarsals that form the tarsus, five metatarsals, and fourteen phalanges.

# Critical Thinking Questions

1. What steps do you think should be taken to reduce the chances of a person accumulating abnormal metallic elements such as lead, radium, and strontium in bones?
2. Why do you think incomplete, longitudinal fractures of bone shafts (greenstick fractures) are more common in children than in adults?
3. When a child's bone is fractured, growth may be stimulated at the epiphyseal disk. What problems might this extra growth cause in an upper or lower limb before the growth of the other limb compensates for the difference in length?
4. Why do elderly persons often develop bowed backs and appear shorter than they were in earlier years?
5. How might the condition of an infant's fontanels be used to evaluate its development? How might the fontanels be used to estimate intracranial pressure?
6. Why are women more likely to develop osteoporosis than men? What steps can reduce the risk of developing this condition?

## Review Exercises

### Part A

1. List four groups of bones based upon their shapes, and name an example from each group.
2. Sketch a typical long bone, and label its epiphyses, diaphysis, medullary cavity, periosteum, and articular cartilages.
3. Distinguish between spongy and compact bone.
4. How are osteonic canals and perforating canals related?
5. Explain how the development of intramembranous bone differs from that of endochondral bone.
6. Distinguish between osteoblasts and osteocytes.
7. Explain the function of an epiphyseal disk.
8. Explain how a bone grows in thickness.
9. Define *osteoclast.*
10. Explain how osteoclasts and osteoblasts regulate bone mass.
11. Describe the effects of vitamin deficiencies on bone development.
12. Explain the causes of pituitary dwarfism and giantism.
13. Describe the effects of thyroid and sex hormones on bone development.
14. Explain the effects of exercise on bone structure.
15. Provide several examples to illustrate how bones support and protect body parts.
16. Describe a lever, and explain how its parts may be arranged to form first-, second-, and third-class levers.
17. Describe the functions of red and yellow bone marrow.
18. Explain the mechanism that regulates the concentration of blood calcium ions.
19. List three substances that may be stored in bone abnormally.
20. Distinguish between the axial and appendicular skeletons.
21. Name the bones of the cranium and the facial skeleton.
22. Explain the importance of fontanels.
23. Describe a typical vertebra.
24. Explain the differences between cervical, thoracic, and lumbar vertebrae.
25. Describe the locations of the sacroiliac joint, the sacral promontory, and the sacral hiatus.
26. Name the bones that comprise the thoracic cage.
27. List the bones that form the pectoral and pelvic girdles.
28. Name the bones of the upper limb.
29. Define *coxal bone.*
30. List the major differences that may occur between the male and female pelves.
31. List the bones of the lower limb.

### Part B

Match the parts listed in column I with the bones listed in column II.

| I | II |
|---|---|
| 1. Coronoid process | A. Ethmoid bone |
| 2. Cribriform plate | B. Frontal bone |
| 3. Foramen magnum | C. Mandible |
| 4. Mastoid process | D. Maxillary bone |
| 5. Palatine process | E. Occipital bone |
| 6. Sella turcica | F. Temporal bone |
| 7. Supraorbital notch | G. Sphenoid bone |
| 8. Temporal process | H. Zygomatic bone |
| 9. Acromion process | I. Femur |
| 10. Deltoid tuberosity | J. Fibula |
| 11. Greater trochanter | K. Humerus |
| 12. Lateral malleolus | L. Radius |
| 13. Medial malleolus | M. Scapula |
| 14. Olecranon process | N. Sternum |
| 15. Radial tuberosity | O. Tibia |
| 16. Xiphoid process | P. Ulna |

# HUMAN SKULL

*The following set of reference plates is presented to help you locate some of the more prominent features of the human skull. As you study these photographs, it is important to remember that individual human skulls vary in every characteristic. Also, the photographs in this set depict bones from several different skulls.*

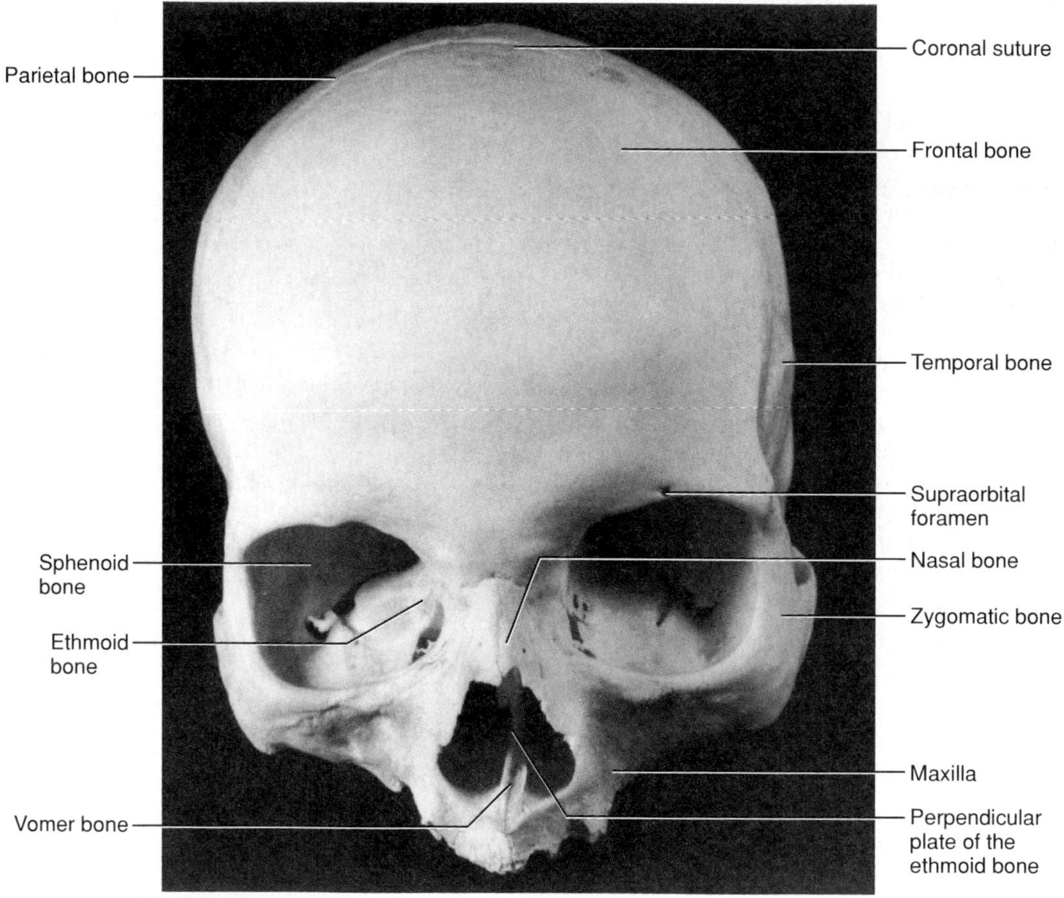


**PLATE Eight**
The skull, frontal view.

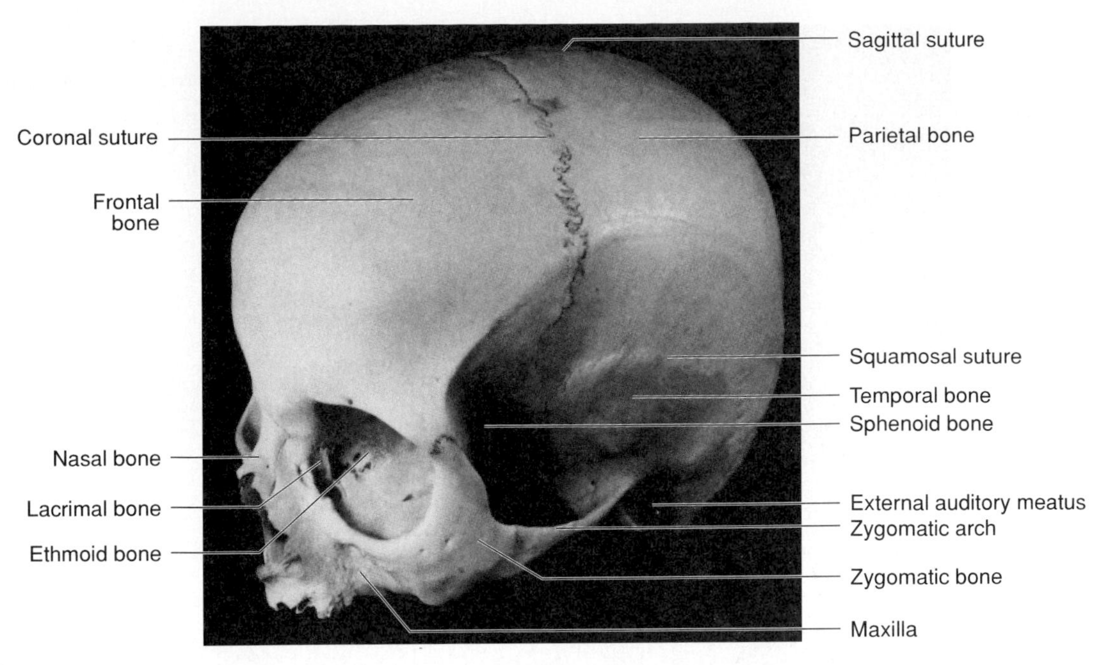


**PLATE Nine**
The skull, left lateral view.

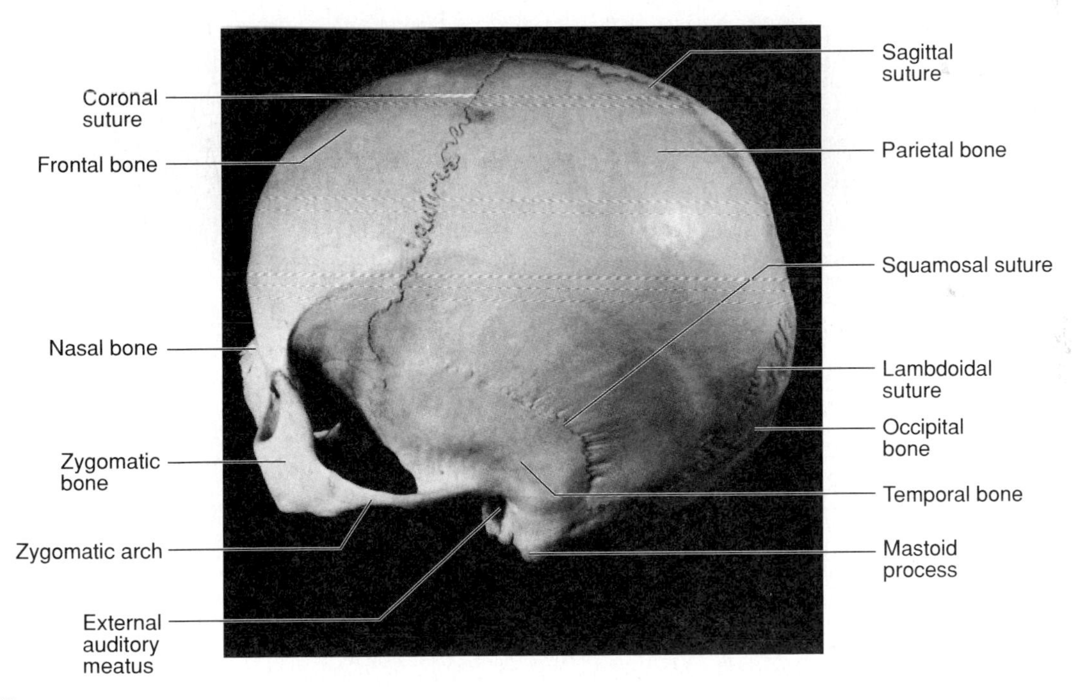


**PLATE Ten**
The skull, left posterior view.

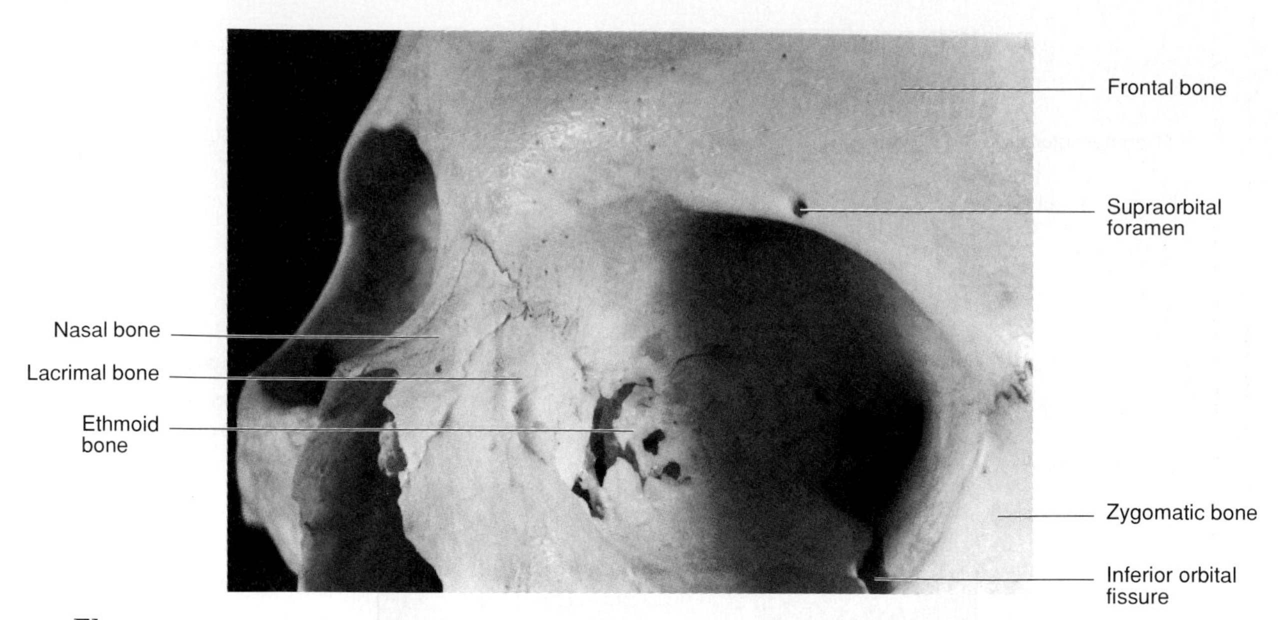


**Plate Eleven**
Bones of the left orbital region.

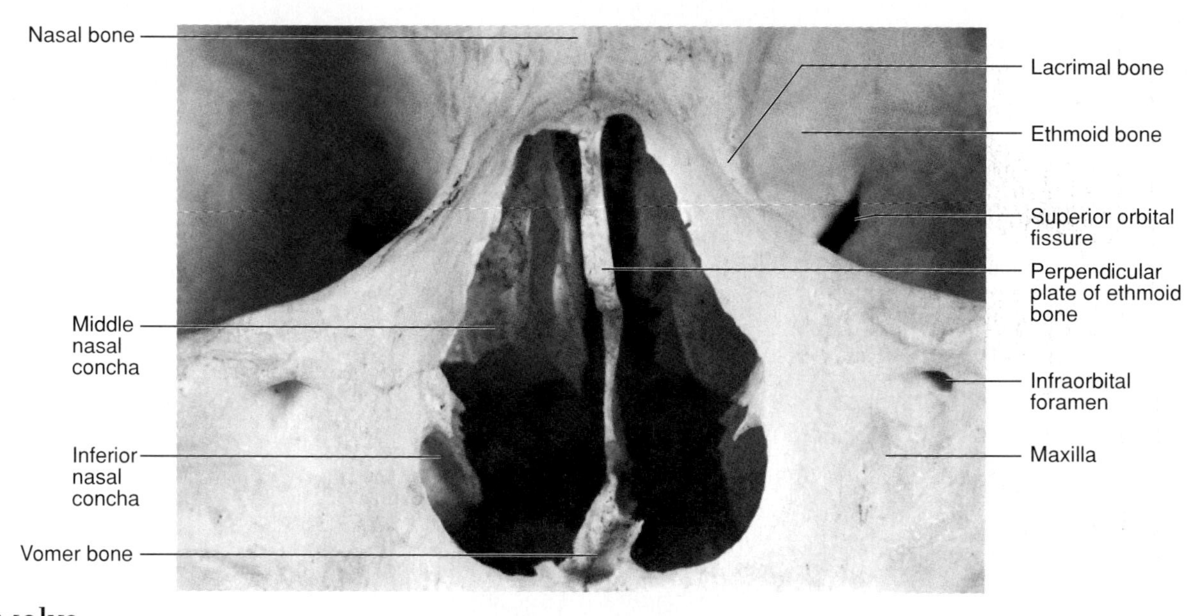


**Plate Twelve**
Bones of the anterior nasal region.

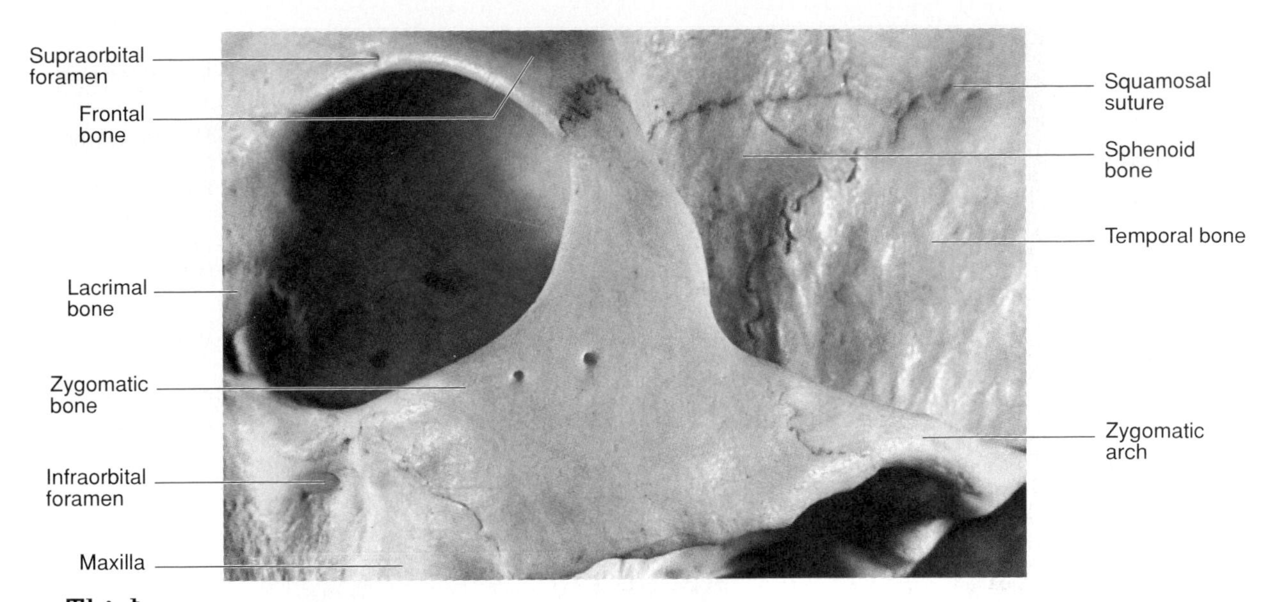


**PLATE Thirteen**

Bones of the left zygomatic region.

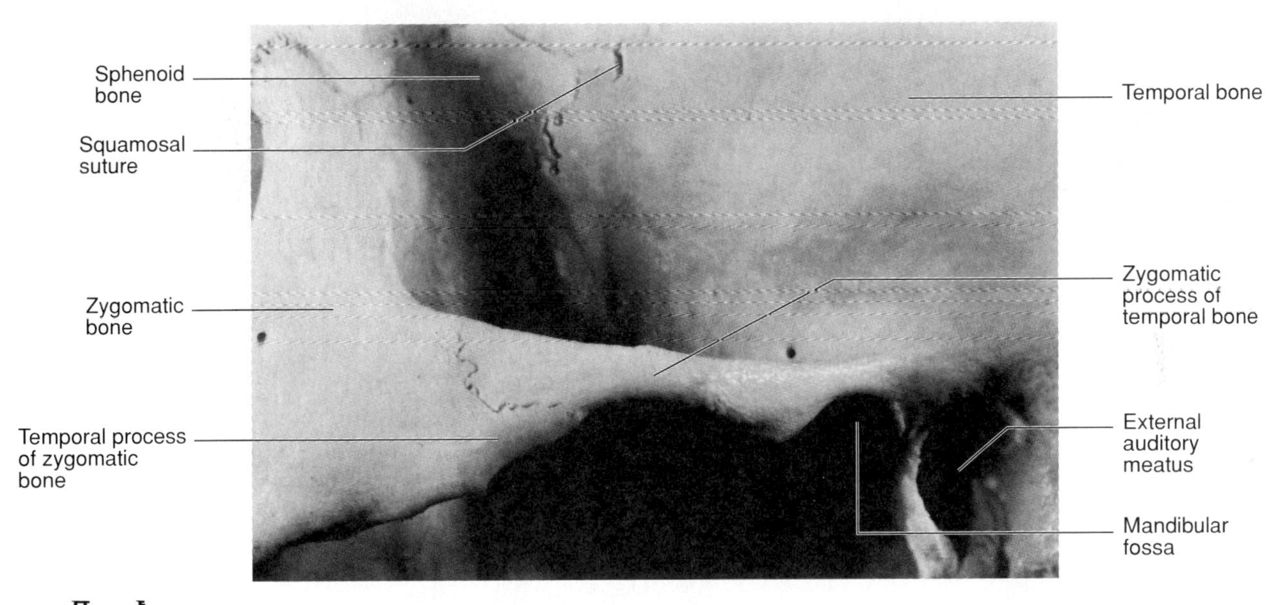


**PLATE Fourteen**

Bones of the left temporal region.

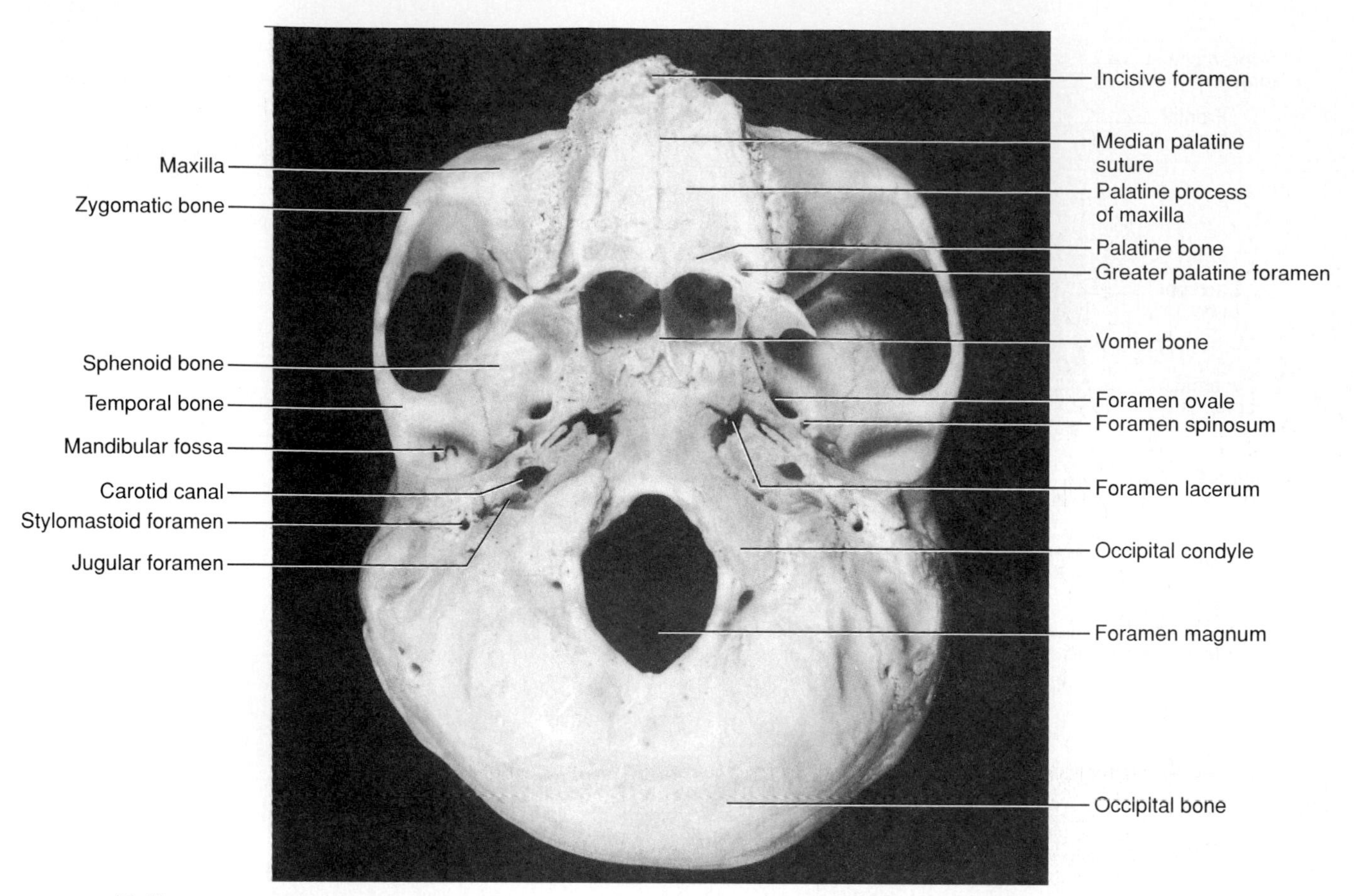


**PLATE Fifteen**
The skull, inferior view.

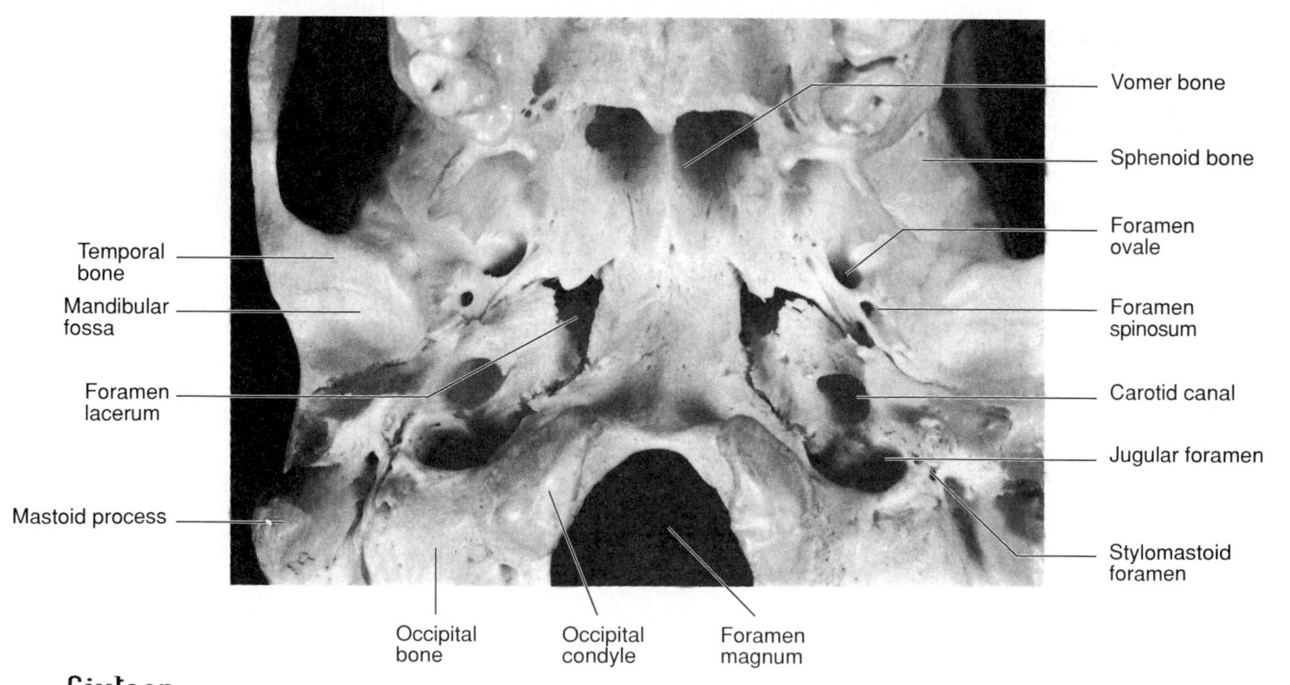


**PLATE Sixteen**
Base of the skull, sphenoidal region.

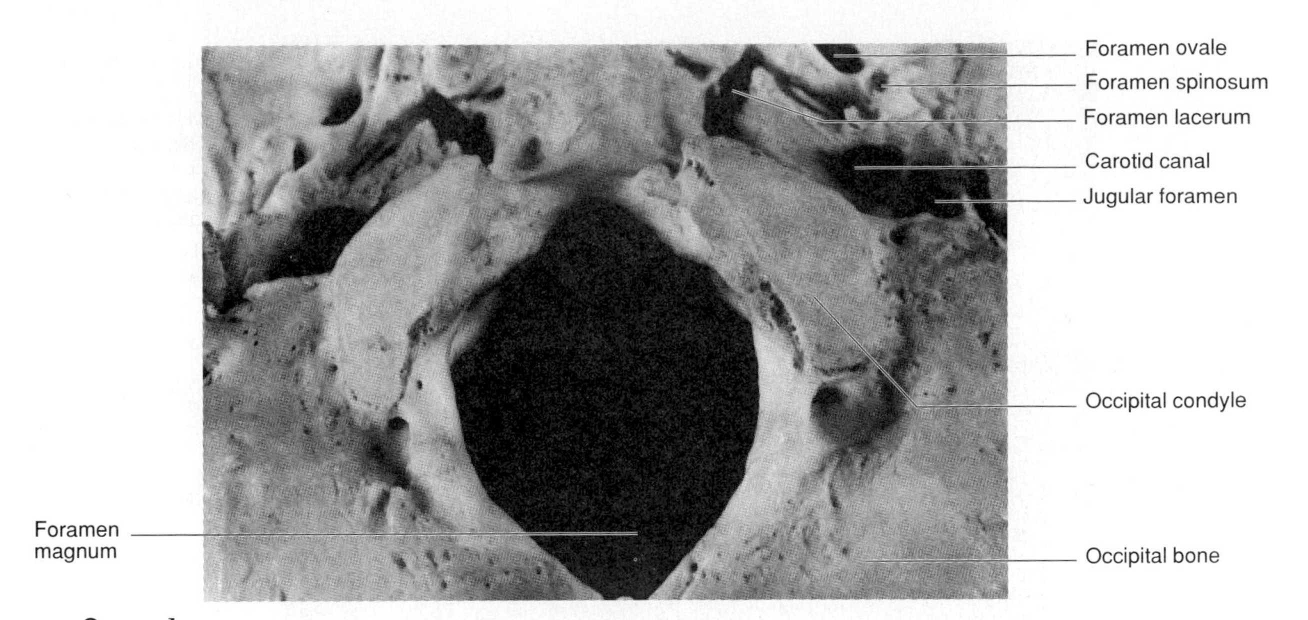


**PLATE Seventeen**

Base of the skull, occipital region.

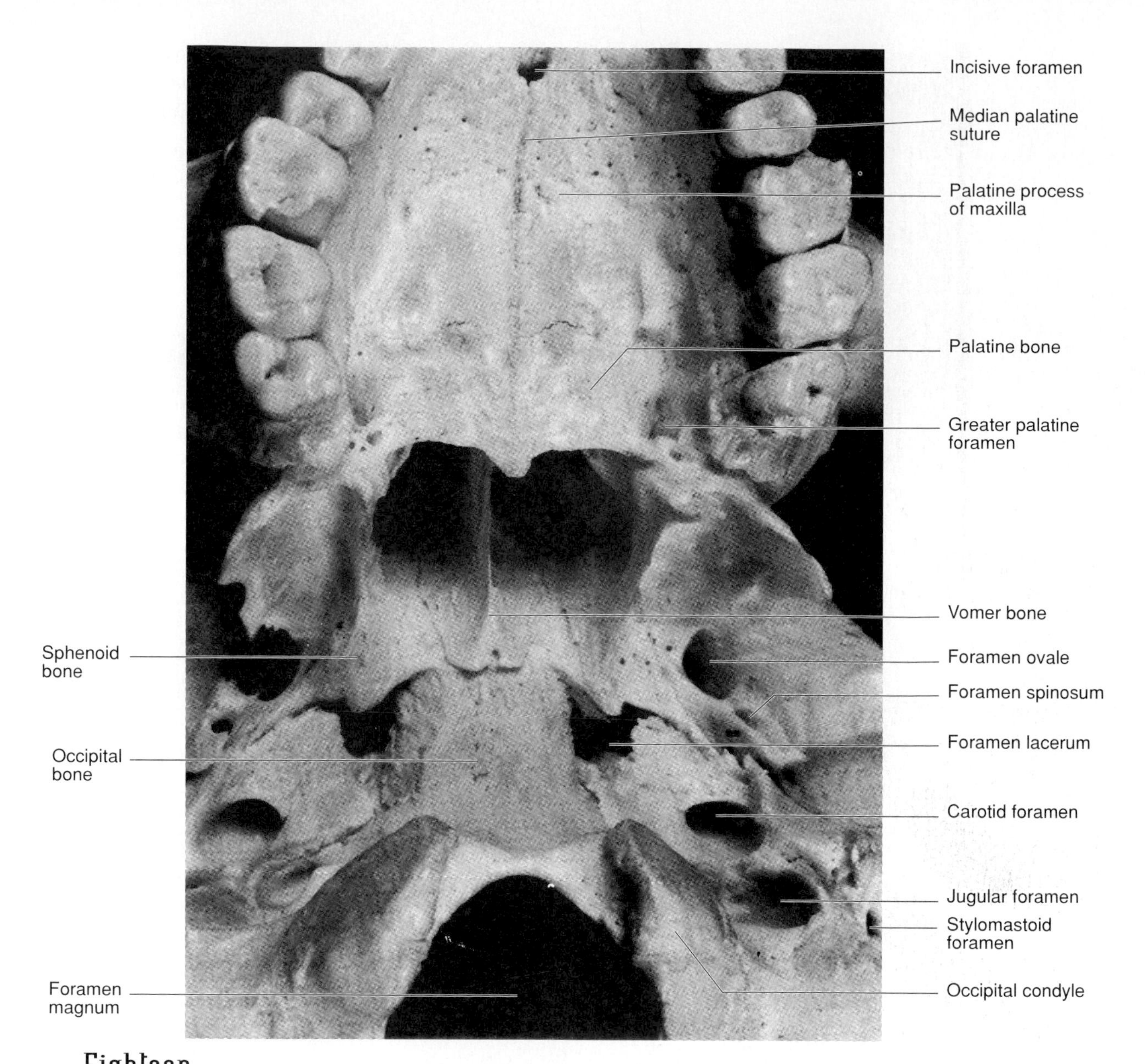


**PLATE Eighteen**
Base of the skull, maxillary region.

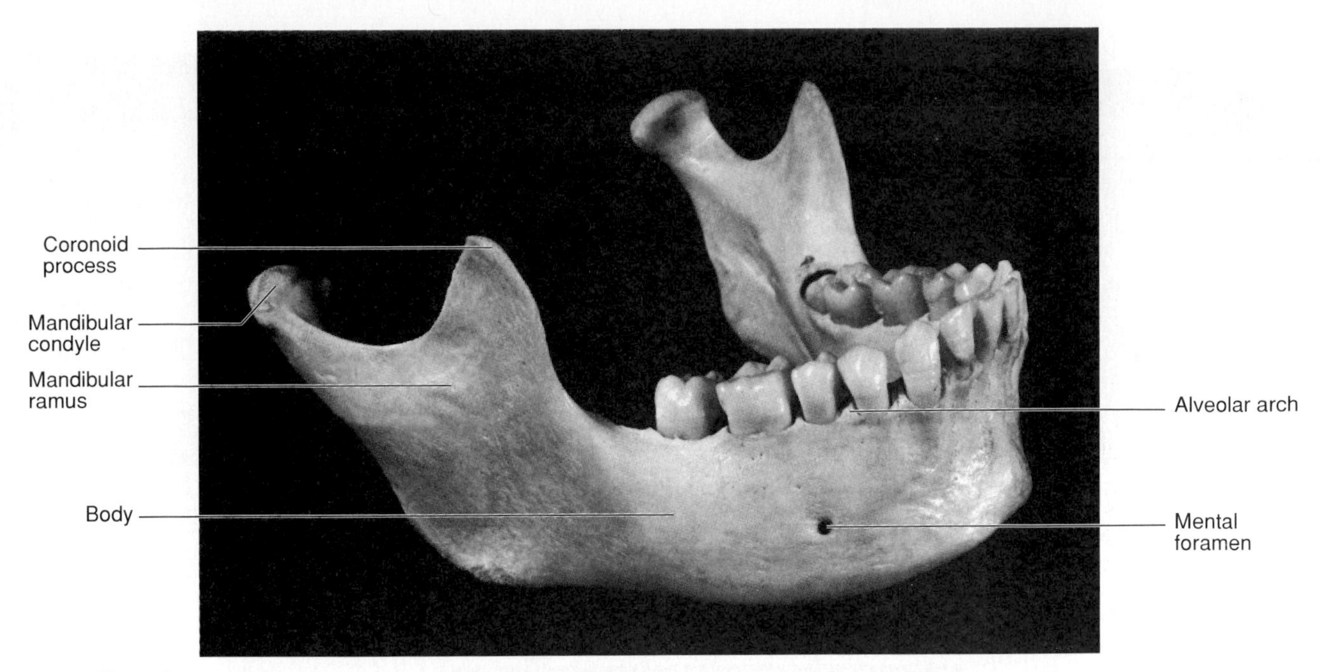


**PLATE Nineteen**
Mandible, lateral view.

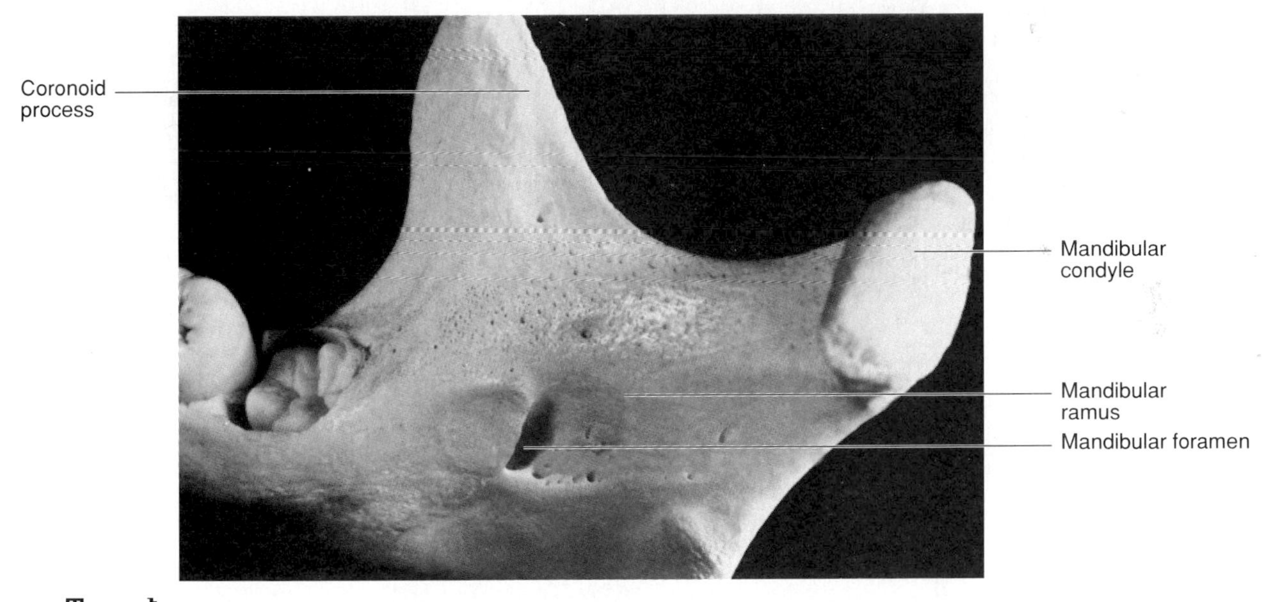


**PLATE Twenty**
Mandible, medial surface of right ramus.

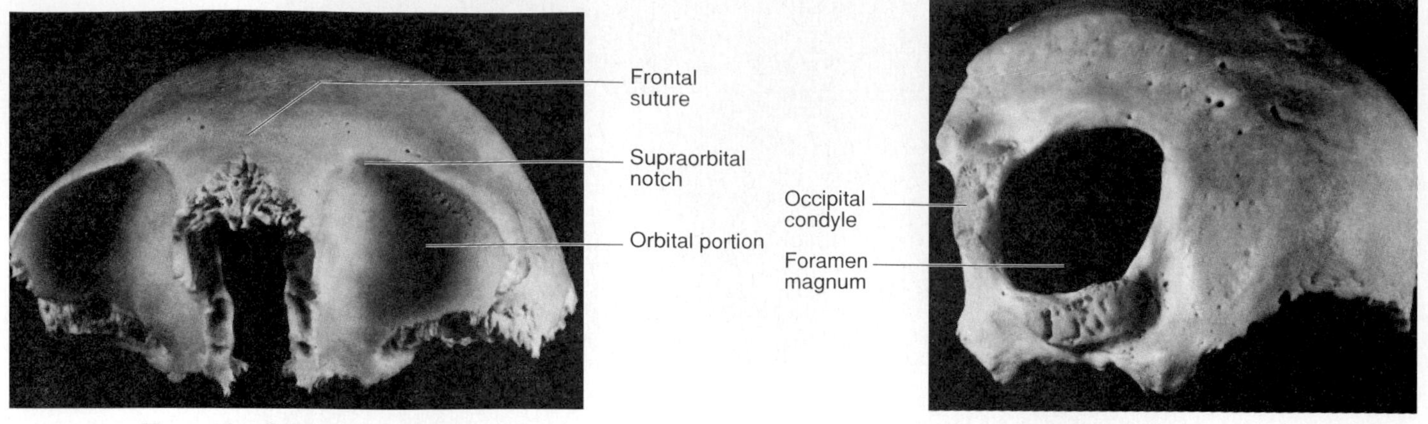


**PLATE Twenty-One**
Frontal bone, anterior view.

**PLATE Twenty-Two**
Occipital bone, inferior view.

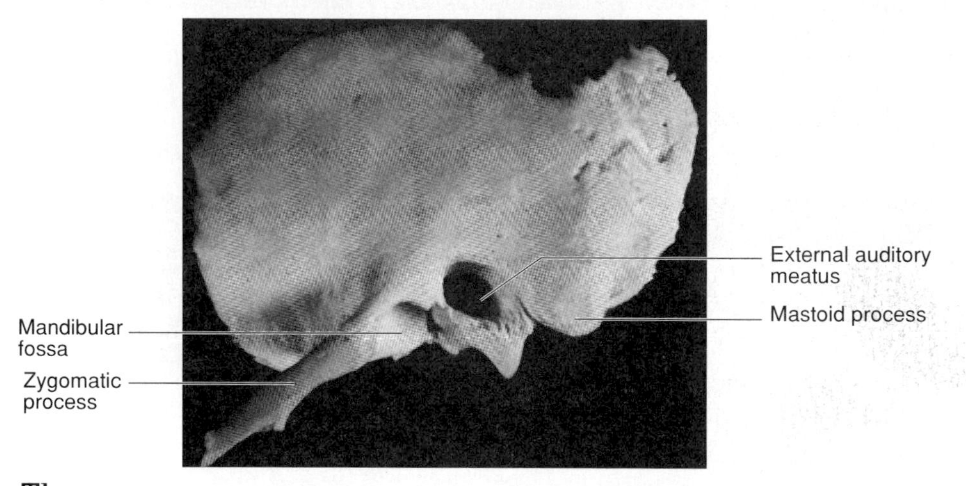


**PLATE Twenty-Three**
Temporal bone, left lateral view.

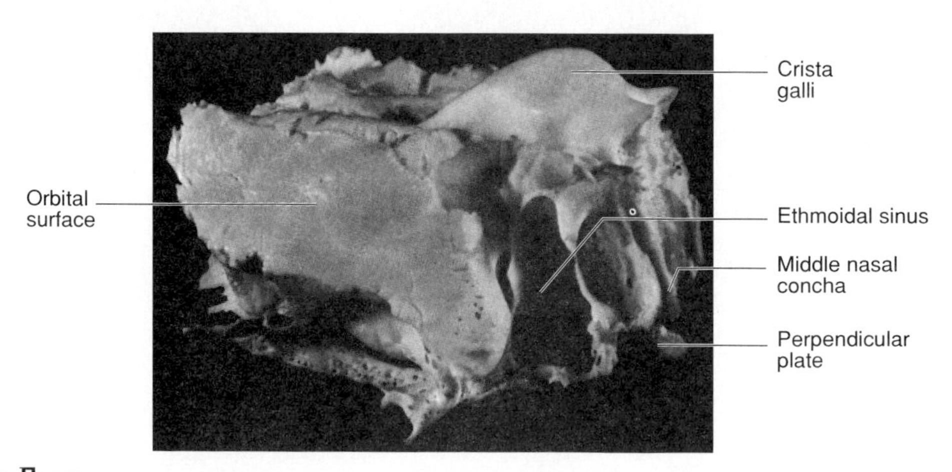


**PLATE Twenty-Four**
Ethmoid bone, right lateral view.

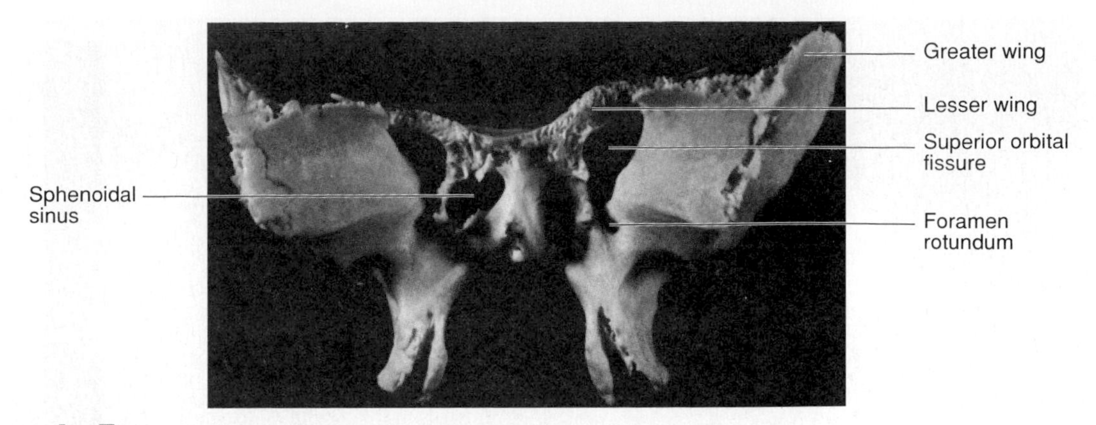


**PLATE** Twenty-Five

Sphenoid bone, anterior view.

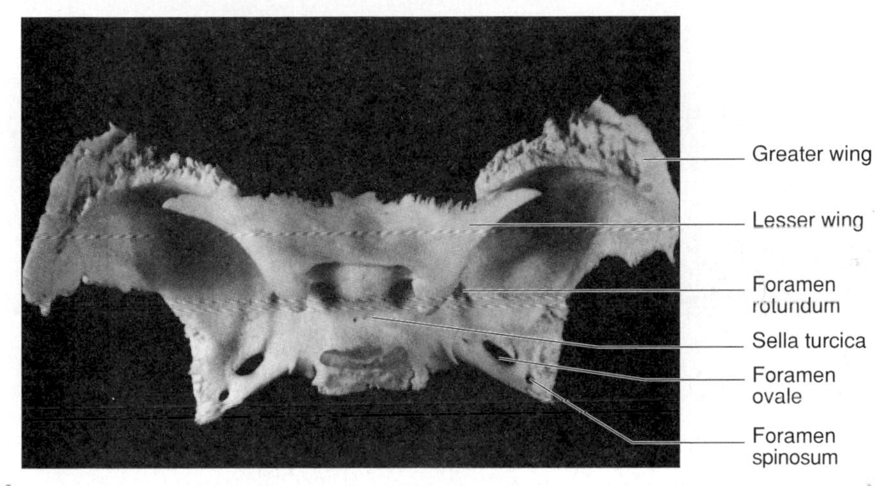


**PLATE** Twenty-Six

Sphenoid bone, posterior view.

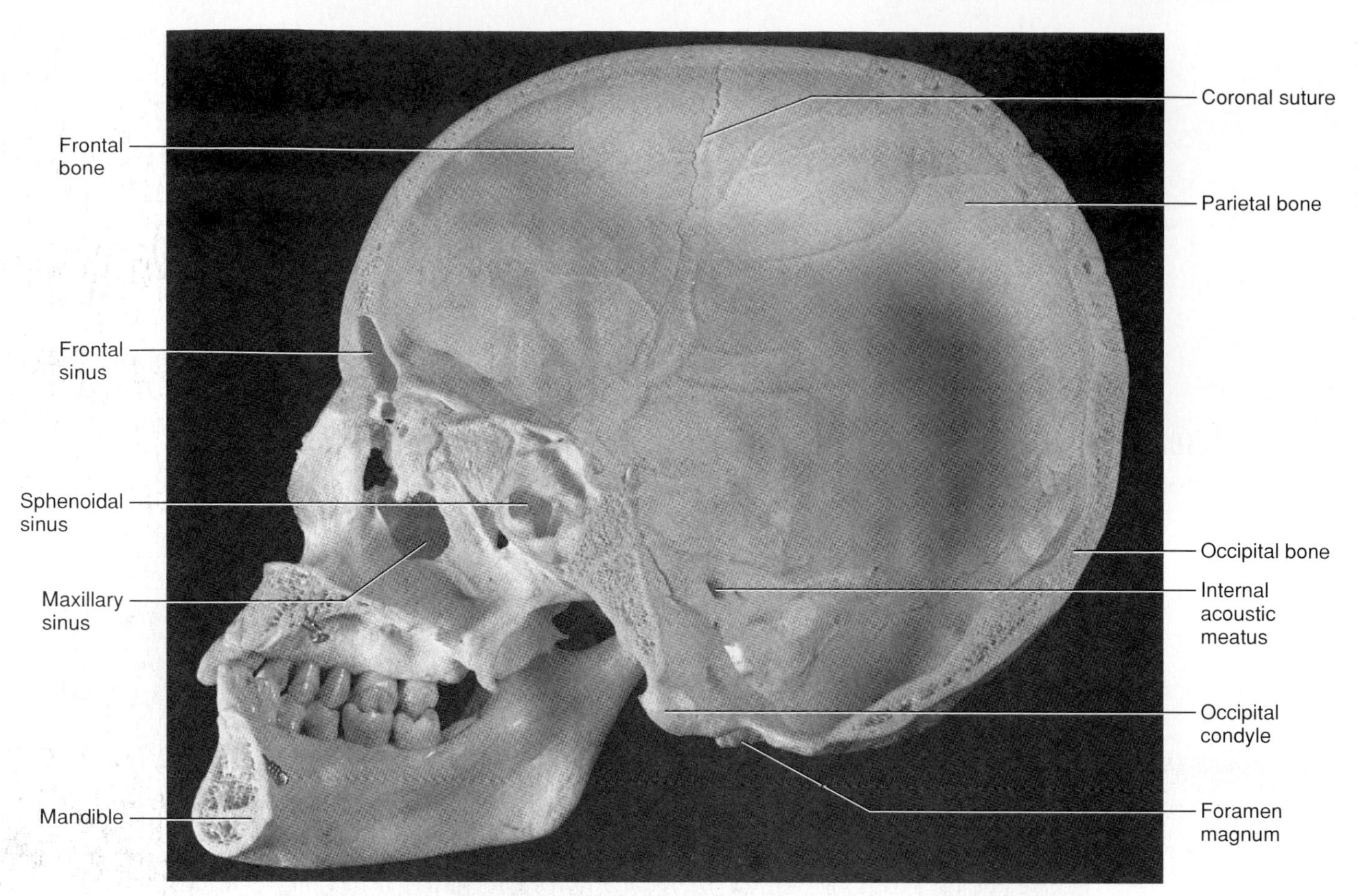


**PLATE Twenty-Seven**

The skull, sagittal section.

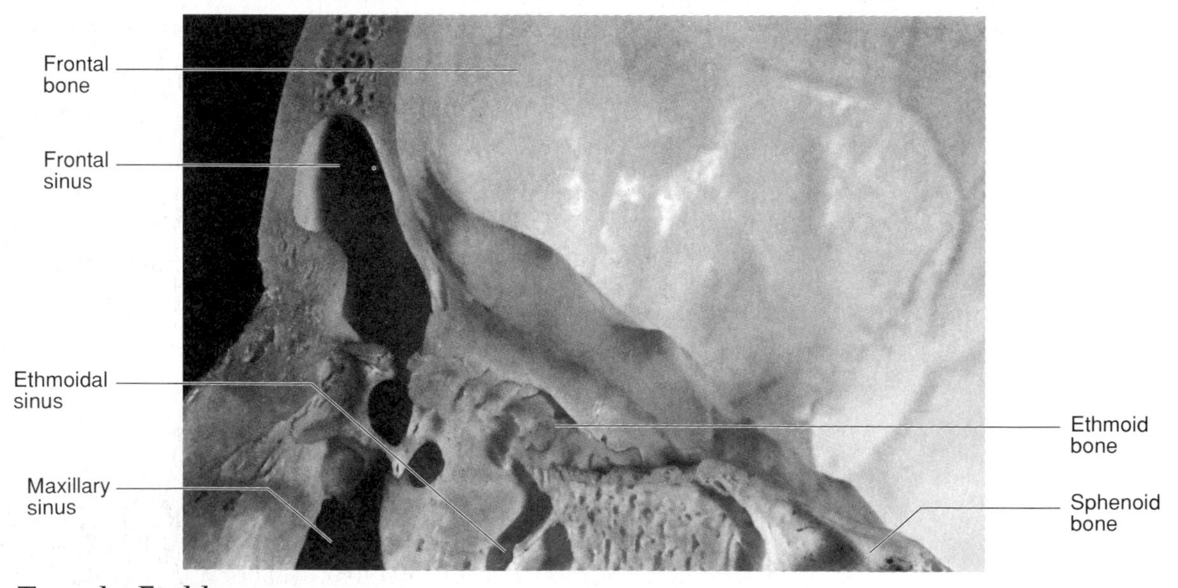


**PLATE Twenty-Eight**

Ethmoidal region, sagittal section.

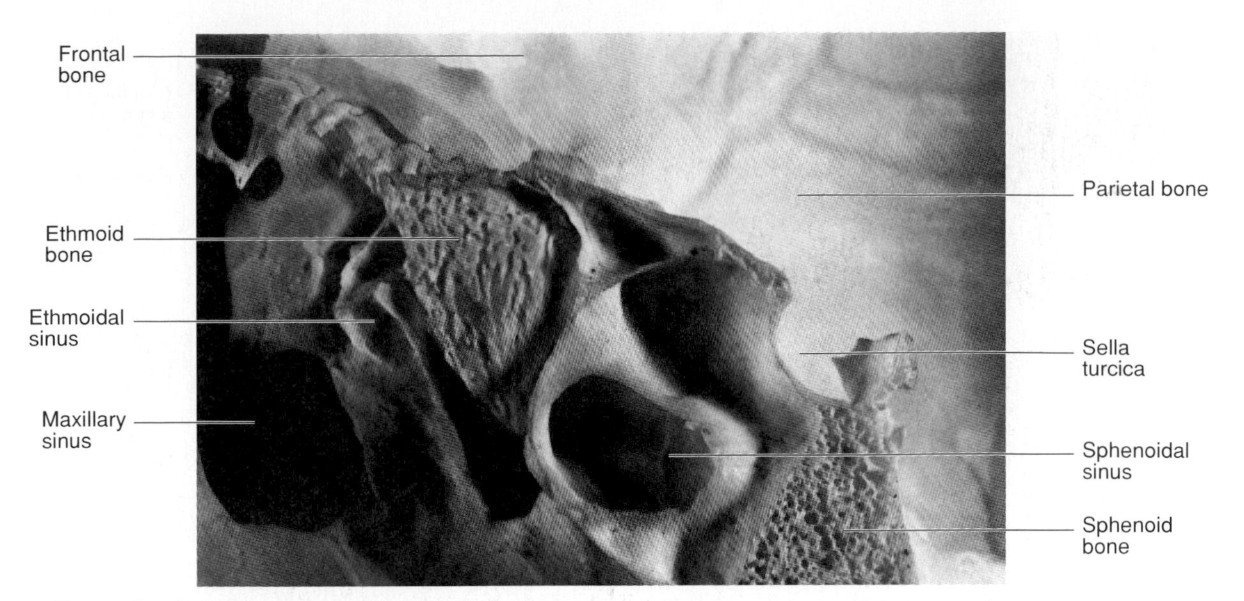


**PLATE Twenty-Nine**
Sphenoidal region, sagittal section.

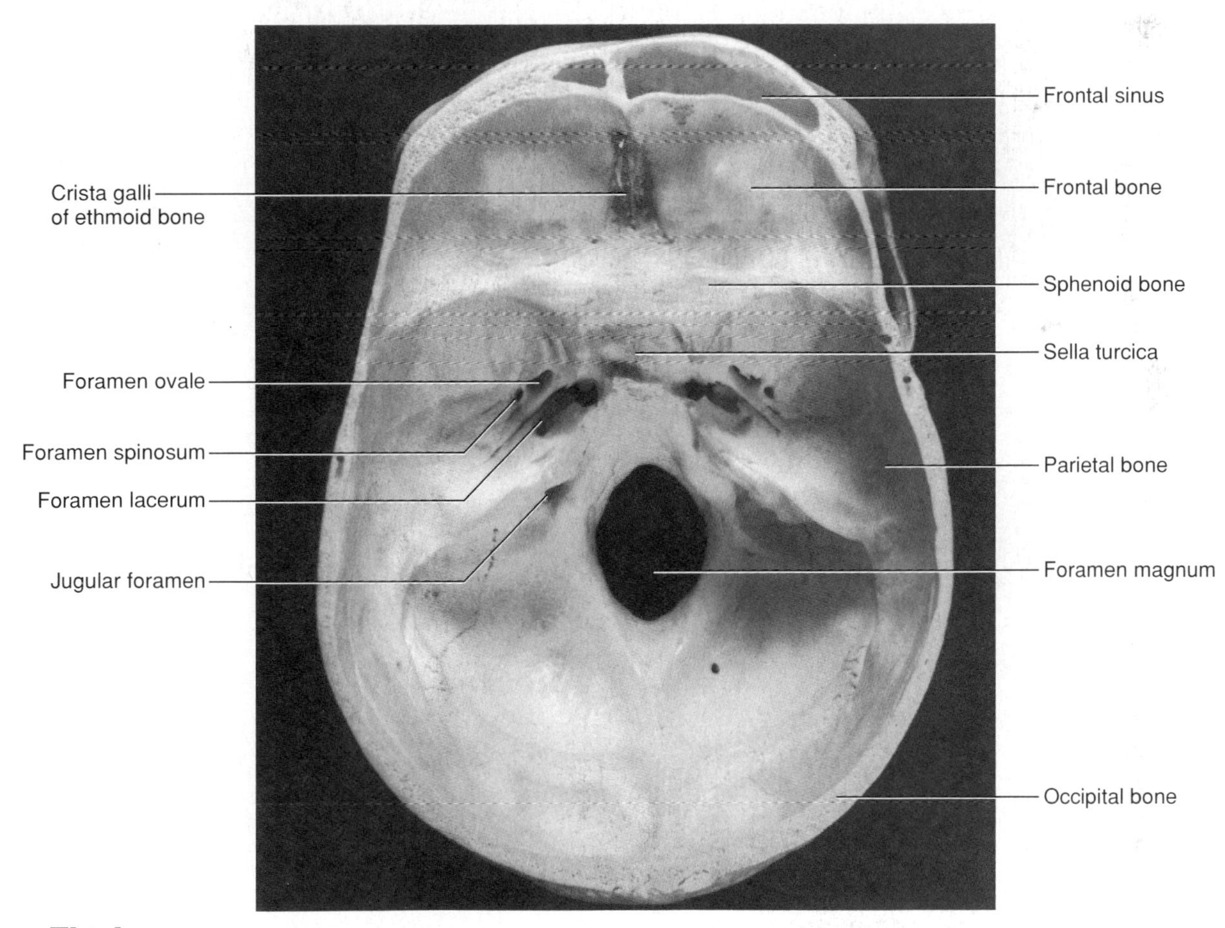


**PLATE Thirty**
The skull, floor of the cranial cavity.

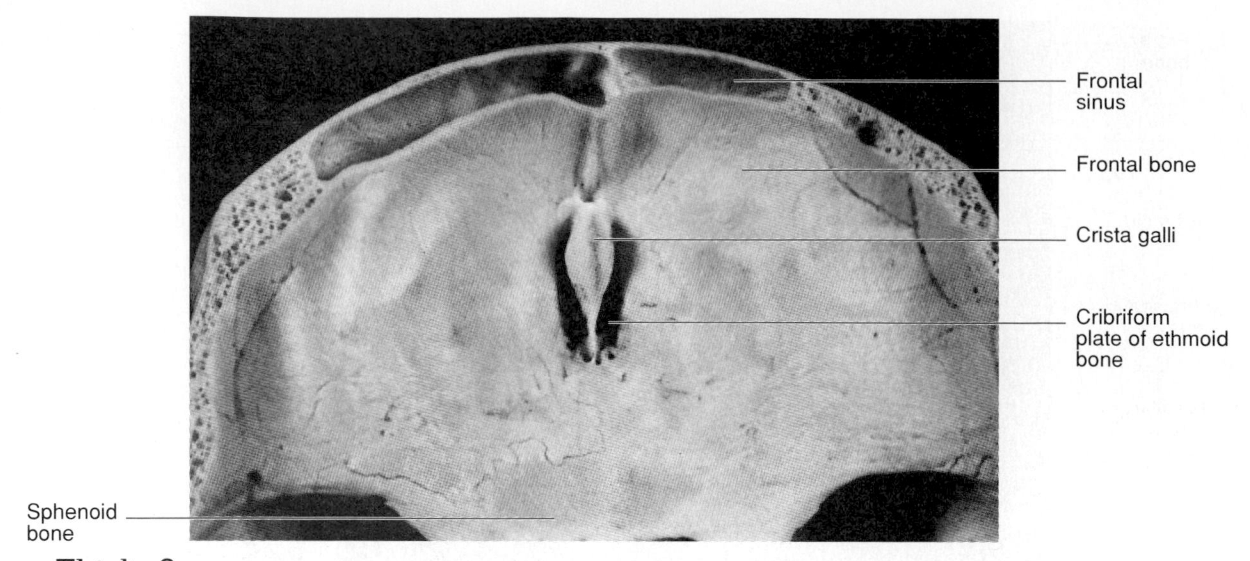


**PLATE** Thirty-One

Frontal region, transverse section.

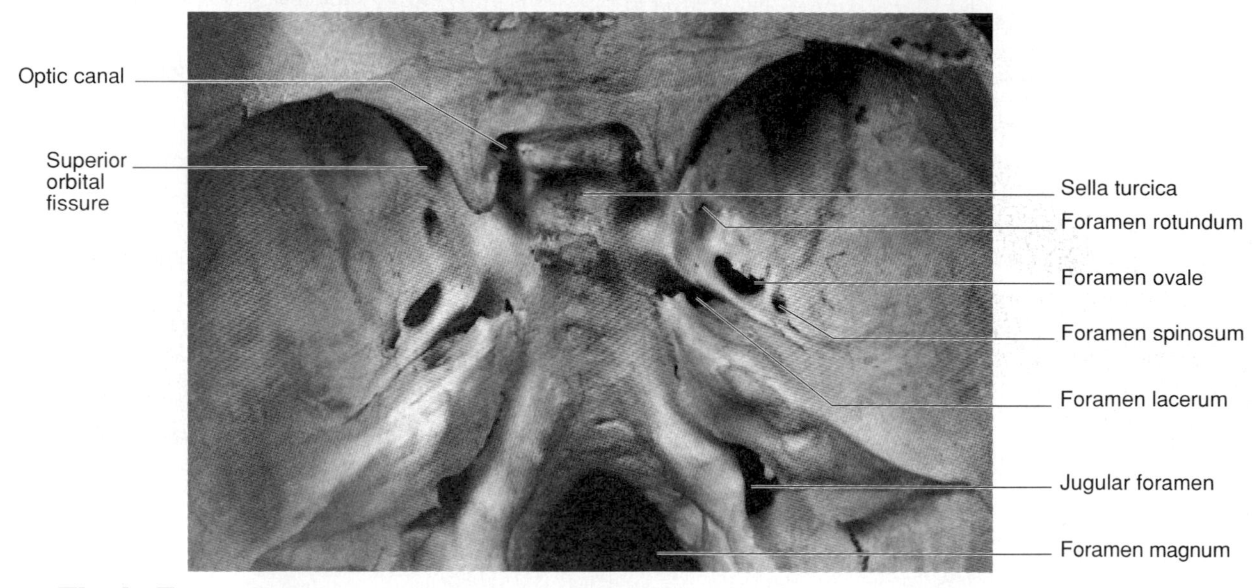


**PLATE** Thirty-Two

Sphenoidal region, floor of the cranial cavity.

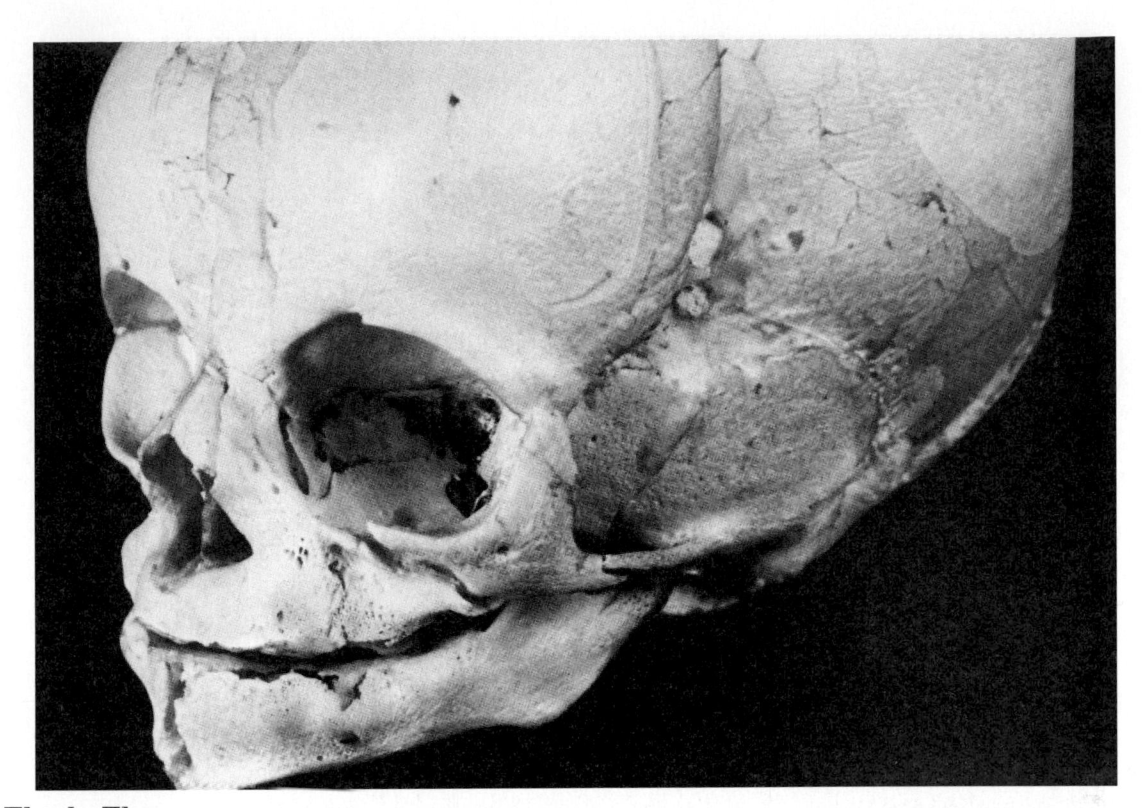

**PLATE** Thirty-Three
Skull of a fetus, left lateral view.

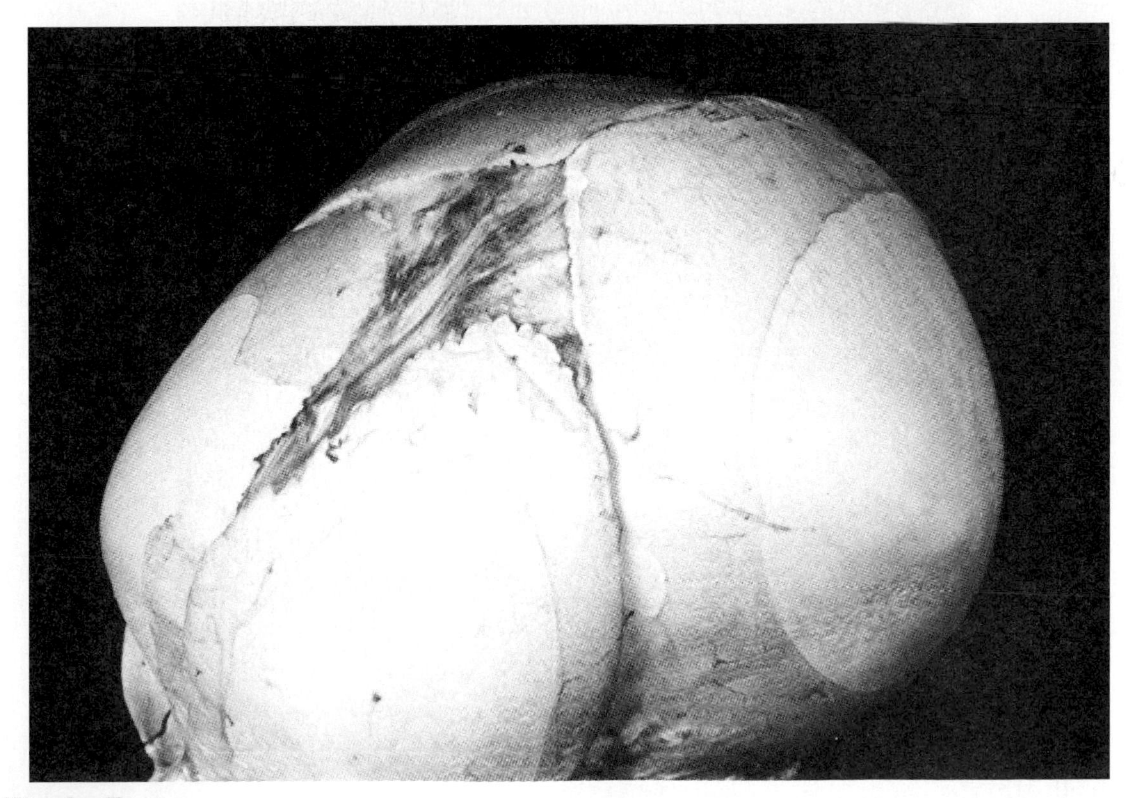

**PLATE** Thirty-Four
Skull of a fetus, left superior view.

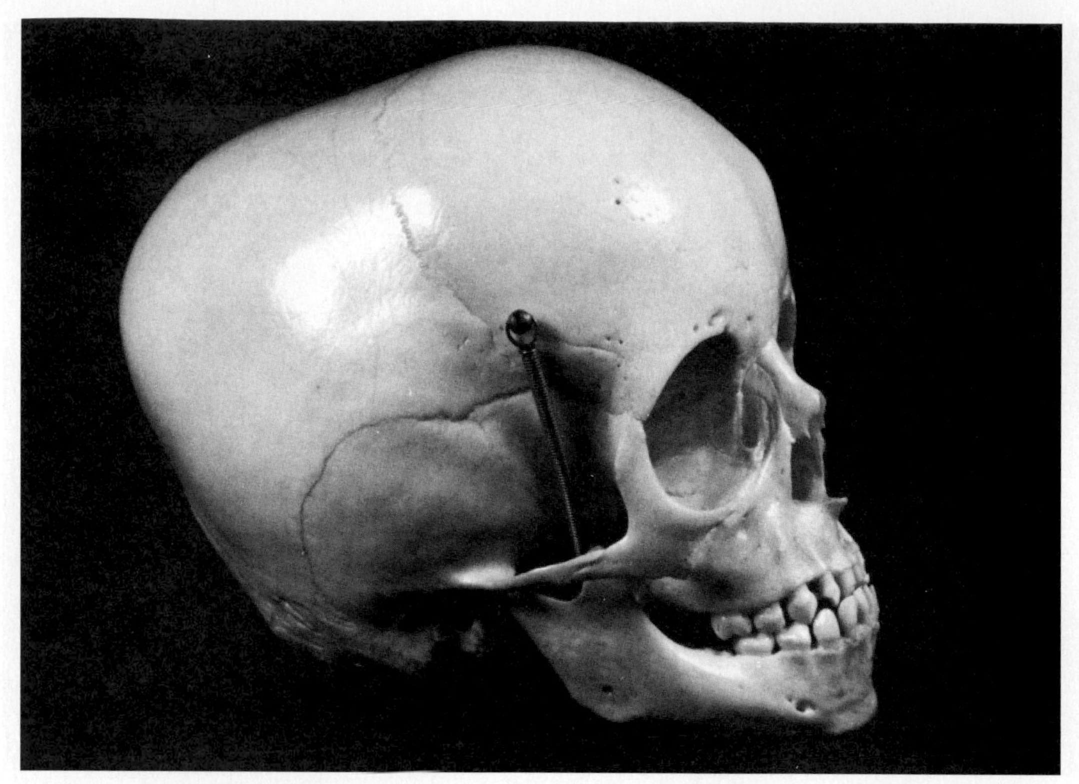

**PLATE** Thirty-Five
Skull of a child.

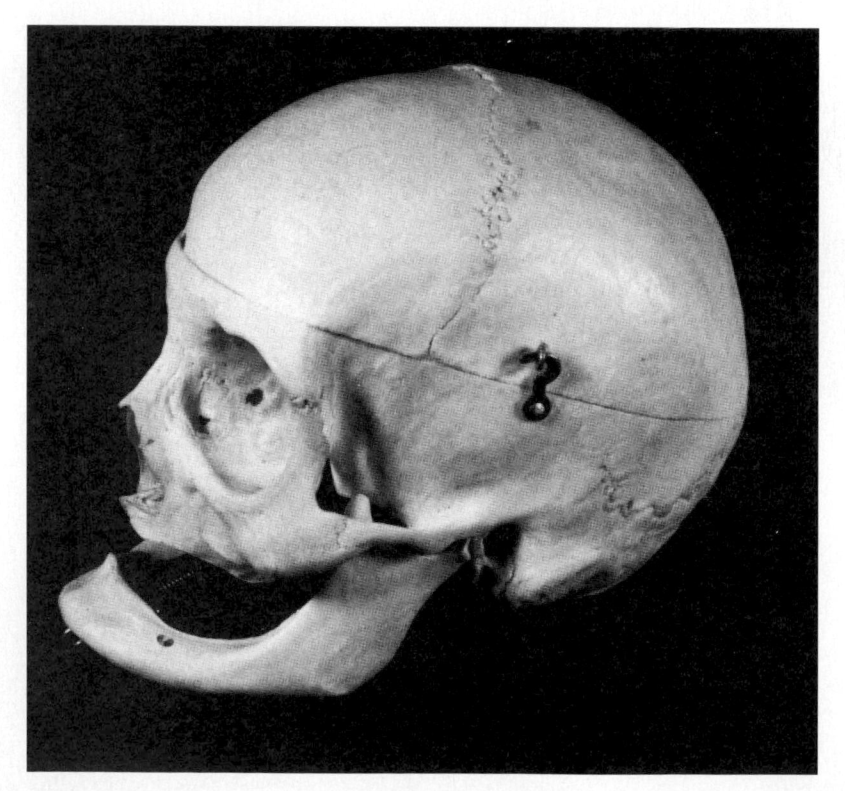

**PLATE** Thirty-Six
Skull of an aged person.

8

CHAPTER

# Joints of the Skeletal System

## Chapter Objectives

AFTER YOU HAVE STUDIED THIS CHAPTER, YOU SHOULD BE ABLE TO:

1. Explain how joints can be classified according to the type of tissue that binds the bones together.
2. Describe how bones of fibrous joints are held together.
3. Describe how bones of cartilaginous joints are held together.
4. Describe the general structure of a synovial joint.
5. List six types of synovial joints and name an example of each type.
6. Explain how skeletal muscles produce movements at joints and identify several types of joint movements.
7. Describe the shoulder joint and explain how its articulating parts are held together.
8. Describe the elbow joint and explain how its articulating parts are held together.
9. Describe the hip joint and explain how its articulating parts are held together.
10. Describe the knee joint and explain how its articulating parts are held together.

MANY MUSICIANS HAVE SUCH LAX JOINTS THAT THEY CAN STRETCH AND BEND THEIR FINGERS WITH GREAT FLEXIBILITY. PERHAPS JOINT HYPERMOBILITY CONTRIBUTES TO THEIR TALENT.

## Key Terms

**articulation** (ar-tik′u-la″shun)
**bursa** (ber′sah)
**gomphosis** (gom-fo′sis)
**ligament** (lig′ah-ment)
**meniscus** (me-nis′kus)
**suture** (su′chur)
**symphysis** (sim′fi-sis)
**synchondrosis** (sin″kon-dro′sis)
**syndesmosis** (sin″des-mo′sis)
**synovial** (si-no′ve-al)

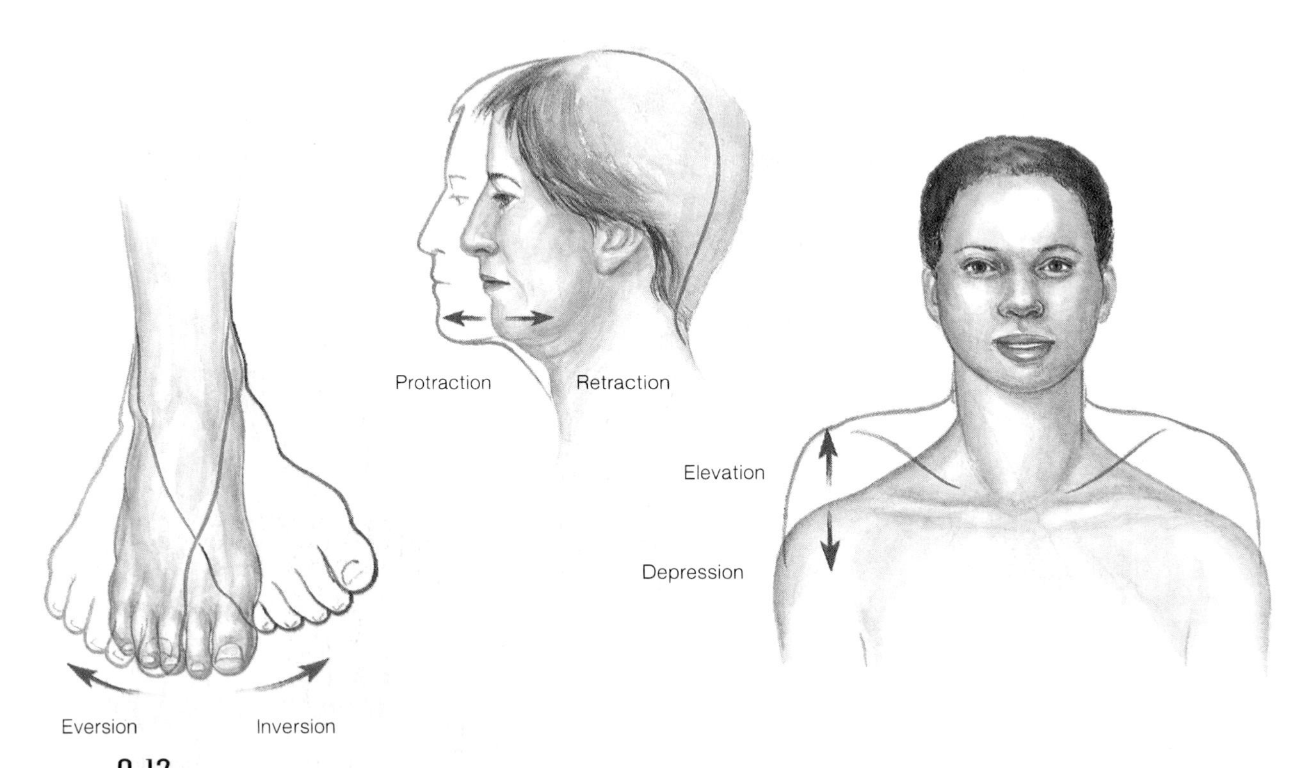


**FIGURE 8.12**
Joint movements illustrating eversion, inversion, protraction, retraction, elevation, and depression.

The ligaments that help prevent displacement of the articulating surfaces of the shoulder joint include the following (fig. 8.14):

1. **Coracohumeral** (kor″ah-ko-hu′mer-al) **ligament.** This ligament is composed of a broad band of connective tissue that connects the coracoid process of the scapula to the greater tubercle of the humerus. It strengthens the superior portion of the joint capsule.
2. **Glenohumeral** (gle″no-hu′mer-al) **ligaments.** These include three bands of fibers that appear as thickenings in the ventral wall of the joint capsule. They extend from the edge of the glenoid cavity to the lesser tubercle and the anatomical neck of the humerus.
3. **Transverse humeral ligament.** This ligament consists of a narrow sheet of connective tissue fibers that runs between the lesser and the greater tubercles of the humerus. Together with the intertubercular groove of the humerus, the ligament forms a canal (retinaculum) through which the long head of the biceps brachii muscle passes.
4. **Glenoidal labrum** (gle′noid-al la′brum). This ligament is composed of fibrocartilage. It is attached along the margin of the glenoid cavity and forms a rim with a thin, free edge that deepens the cavity.

Several bursae are associated with the shoulder joint. The major ones include the *subscapular bursa* located between the joint capsule and the tendon of the subscapularis muscle, the *subdeltoid bursa* between the joint capsule and the deep surface of the deltoid muscle, the *subacromial bursa* between the joint capsule and the undersurface of the acromion process of the scapula, and the *subcoracoid bursa* between the joint capsule and the coracoid process of the scapula. Of these, the subscapular bursa is usually continuous with the synovial cavity of the joint cavity, and although the others do not communicate with the joint cavity, they may be connected to each other (see figs. 8.13 and 8.14).

## Chart 8.2 Joints of the Body

| Joint | Location | Type of Joint | Type of Movement |
|---|---|---|---|
| Skull | Cranial and facial bones | Suture, fibrous | Immovable, synarthrotic |
| Temporomandibular | Temporal bone, mandible | Modified hinge, synovial | Elevation, depression, protraction, retraction, diarthrotic |
| Atlantooccipital | Atlas, occipital bone | Condyloid, synovial | Flexion, extension, circumduction, diarthrotic |
| Intervertebral | Between adjacent vertebral bodies | Symphysis, cartilaginous | Slight movement, amphiarthrotic |
| Intervertebral | Between articular processes | Gliding, synovial | Flexion, extension, slight rotation, diarthrotic |
| Sacroiliac | Sacrum and hipbone | Gliding, synovial | Little to no movement, diarthrotic |
| Vertebrocostal | Vertebrae and ribs | Gliding, synovial | Slight movement during breathing, diarthrotic |
| Sternoclavicular | Sternum and clavicle | Gliding, synovial | Slight movement when shrugging shoulders, diarthrotic |
| Sternocostal | Sternum and rib 1 | Synchondrosis, cartilaginous | Immovable, synarthrotic |
| Sternocostal | Sternum and ribs 2–7 | Gliding, synovial | Slight movement during breathing, diarthrotic |
| Acromioclavicular | Scapula and clavicle | Gliding, synovial | Protraction, retraction, elevation, depression, diarthrotic |
| Shoulder (glenohumeral) | Humerus and scapula | Ball-and-socket, synovial | Flexion, extension, adduction, abduction, rotation, circumduction, diarthrotic |
| Elbow | Humerus and ulna | Hinge, synovial | Flexion, extension, diarthrotic |
| Proximal radioulnar | Radius and ulna | Pivot, synovial | Rotation, diarthrotic |
| Distal radioulnar | Radius and ulna | Syndesmosis, fibrous | Slight movement, amphiarthrotic |
| Wrist (radiocarpal) | Radius and carpals | Condyloid, synovial | Flexion, extension, adduction, abduction, circumduction, diarthrotic |
| Intercarpal | Adjacent carpals | Gliding, synovial | Slight movement, diarthrotic |
| Carpometacarpal | Carpal and metacarpal 1 | Saddle, synovial | Flexion, extension, adduction, abduction, diarthrotic |
| Carpometacarpal | Carpals and metacarpals 2–5 | Condyloid, synovial | Flexion, extension, adduction, abduction, diarthrotic |
| Metacarpophalangeal | Metacarpal and proximal phalanx | Condyloid, synovial | Flexion, extension, adduction, abduction, diarthrotic |
| Interphalangeal | Adjacent phalanges | Hinge, synovial | Flexion, extension, diarthrotic |
| Symphysis pubis | Pubic bones | Symphysis, cartilaginous | Slight movement, amphiarthrotic |
| Hip | Hipbone and femur | Ball-and-socket, synovial | Flexion, extension, adduction, abduction, rotation, circumduction, diarthrotic |
| Knee (tibiofemoral) | Femur and tibia | Modified hinge, synovial | Flexion, extension, slight rotation when flexed, diarthrotic |
| Knee (femoropatellar) | Femur and patella | Gliding, synovial | Slight movement, diarthrotic |
| Proximal tibiofibular | Tibia and fibula | Gliding, synovial | Slight movement, diarthrotic |
| Distal tibiofibular | Tibia and fibula | Syndesmosis, fibrous | Slight movement, amphiarthrotic |
| Ankle (talocrural) | Talus, tibia, and fibula | Hinge, synovial | Dorsiflexion, plantar flexion, slight circumduction, diarthrotic |
| Intertarsal | Adjacent tarsals | Gliding, synovial | Inversion, eversion, diarthrotic |
| Tarsometatarsal | Tarsals and metatarsals | Gliding, synovial | Slight movement, diarthrotic |
| Metatarsophalangeal | Metatarsal and proximal phalanx | Condyloid, synovial | Flexion, extension, adduction, abduction, diarthrotic |

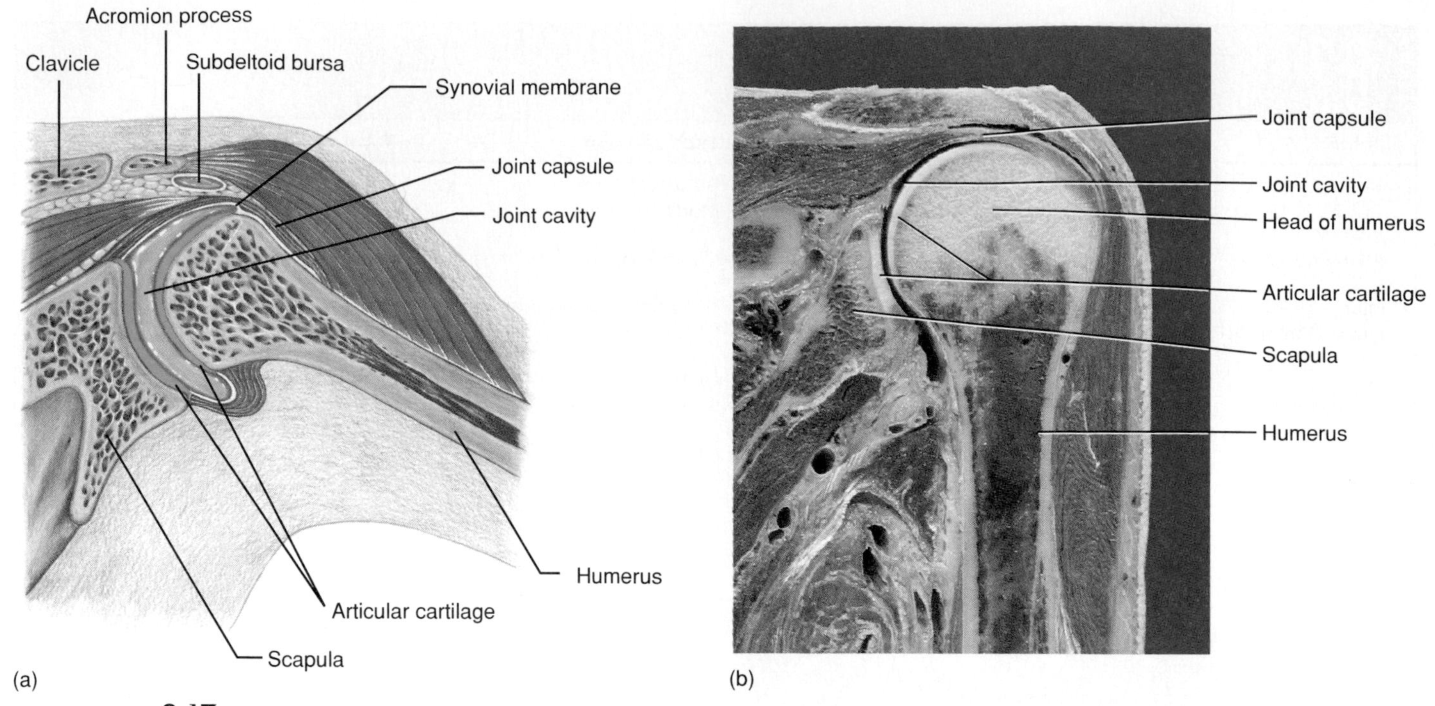


**FIGURE 8.13**

(*a*) The shoulder joint allows movements in all directions. Note that a bursa is associated with this joint. (*b*) Photograph of the shoulder joint (coronal section).

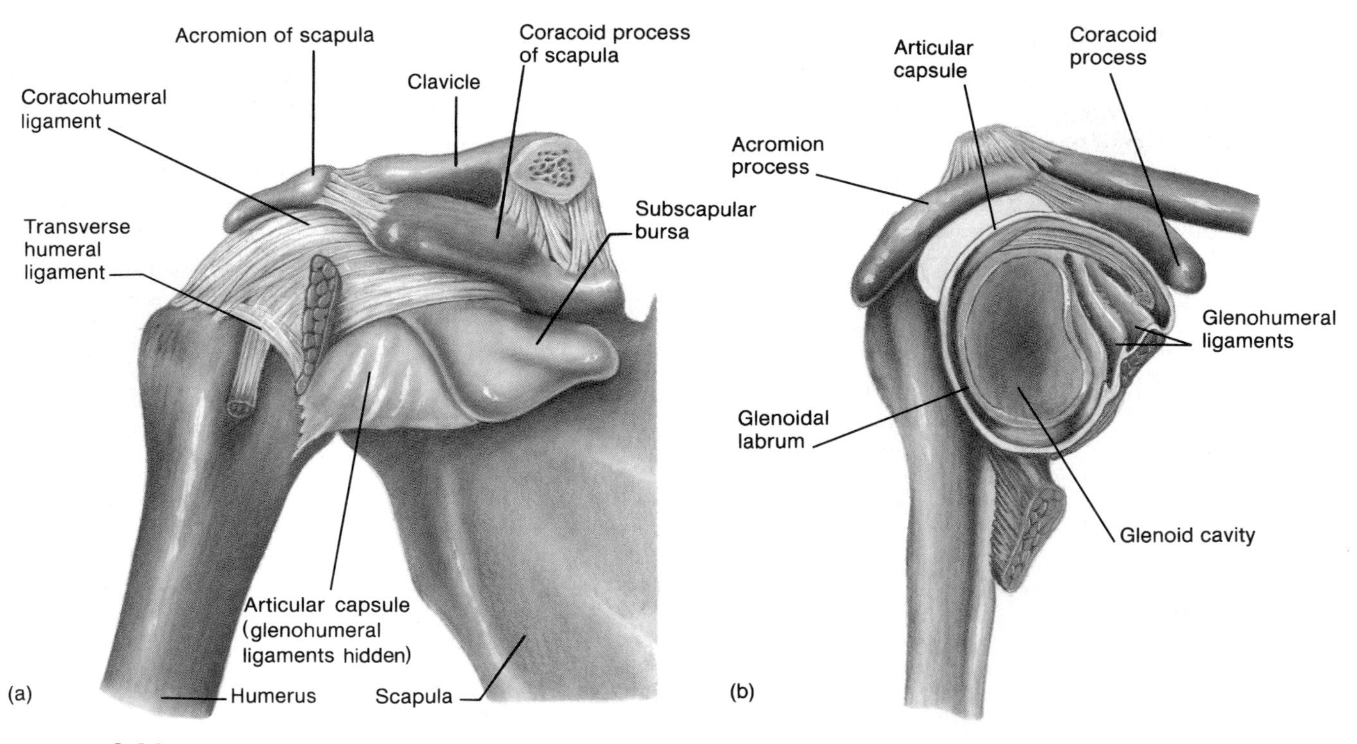


**FIGURE 8.14**

(*a*) Ligaments hold together the articulating surfaces of the shoulder. (*b*) The glenoidal labrum is a ligament composed of fibrocartilage.

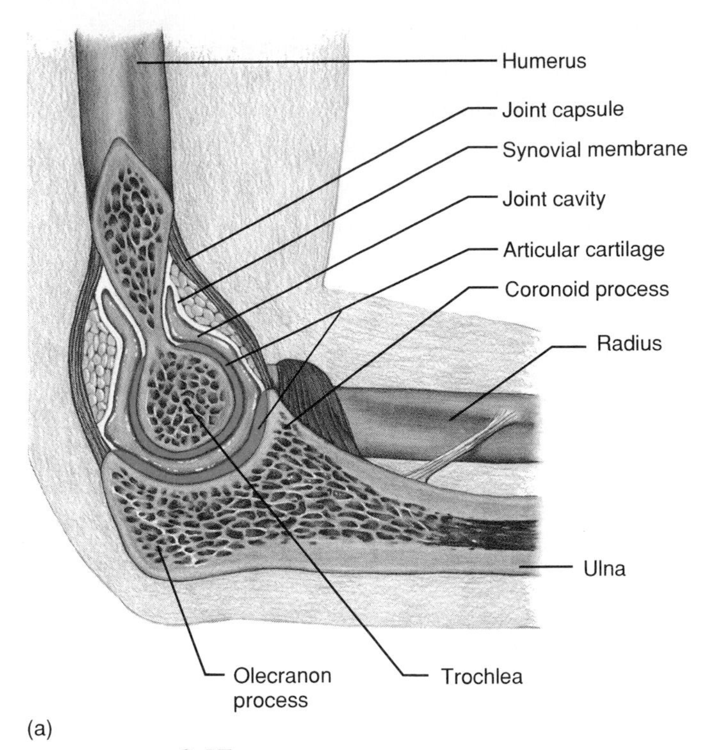


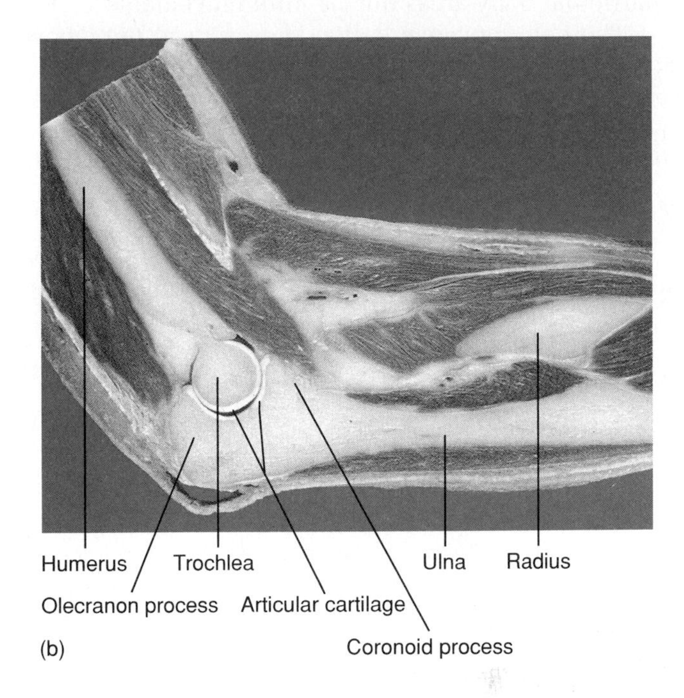


**FIGURE 8.15**

(*a*) The elbow joint allows hinge movements, as well as pronation and supination of the hand; (*b*) photograph of the elbow joint (sagittal section).

Due to the looseness of its attachments and the relatively large articular surface of the humerus compared to the shallow depth of the glenoid cavity, the shoulder joint is capable of a very wide range of movement. These movements include flexion, extension, abduction, adduction, rotation, and circumduction. Motion occurring simultaneously in the joint formed between the scapula and the clavicle may also aid such movements.

> Because the bones of the shoulder joint are held together mainly by supporting muscles rather than by bony structures and strong ligaments, the joint is somewhat weak. Consequently, the articulating surfaces may become displaced or dislocated rather easily. Such a *dislocation* most commonly occurs during forceful abduction, as when a person falls on an outstretched arm. This movement may press the head of the humerus against the lower part of the joint capsule where its wall is relatively thin and poorly supported by ligaments.

## Elbow Joint

The **elbow joint** is a complex structure that includes two articulations—a hinge joint between the trochlea of the humerus and the trochlear notch of the ulna, and a gliding joint between the capitulum of the humerus and a shallow depression (fovea) on the head of the radius. A joint capsule completely encloses and holds together these unions (fig. 8.15). Ulnar and radial collateral ligaments thicken the two joints, and fibers from a muscle (brachialis) in the arm reinforce its anterior surface (fig. 8.16).

The **ulnar collateral ligament,** which is a thick band of fibrous connective tissue, is located in the medial wall of the capsule. The anterior portion of this ligament connects the medial epicondyle of the humerus to the medial margin of the coronoid process of the ulna. Its posterior part is attached to the medial epicondyle of the humerus and to the olecranon process of the ulna.

The **radial collateral ligament,** which strengthens the lateral wall of the joint capsule, is a fibrous band extending between the lateral epicondyle of the humerus and the *annular ligament* of the radius. The annular ligament, in turn, attaches to the margin of the trochlear notch of the ulna, and it encircles the head of the radius, keeping the head in contact with the radial notch of the ulna. The elbow joint capsule encloses the resulting radioulnar joint so that its function is closely associated with the elbow (see fig. 8.16).

The *synovial membrane* that forms the inner lining of the elbow capsule projects into the joint cavity between the radius and ulna, and partially divides the joint into humerus-ulnar and humerus-radial portions. Also, varying amounts of adipose tissue form fatty pads between the synovial membrane and the fibrous layer of the joint capsule. These pads help protect nonarticular bony areas during joint movements.

The only movements that can occur at the elbow between the humerus and ulna are hinge-type movements—flexion and extension. The head of the radius, however, is free to rotate in the annular ligament. This movement allows pronation and supination of the hand.

1. What parts help keep together the articulating surfaces of the shoulder joint?
2. What factors allow an especially wide range of motion in the shoulder?
3. What structures form the hinge joint of the elbow?
4. What parts of the elbow permit pronation and supination of the hand?

Arthroscopy enables a surgeon to visualize the interior of a joint, and even perform diagnostic or therapeutic procedures, guided by the image on a video screen. An arthroscope is a thin, tubular instrument about 25 cm long containing optical fibers that transmit an image. The surgeon inserts the device through a small incision in the joint capsule. It is far less invasive than conventional surgery. Many runners have undergone arthroscopy and raced just weeks later.

Arthroscopy is combined with a genetic technique called the polymerase chain reaction (PCR) to rapidly diagnose infection. Guided by an arthroscope, the surgeon samples a small piece of the synovial membrane. PCR detects and amplifies specific DNA sequences, such as those of bacteria. For example, the technique can rapidly diagnose Lyme disease by detecting DNA from the causative bacterium *Borrelia burgdorferi.* This is valuable because a variety of bacteria can infect joints, and choosing the appropriate antibiotic, based on knowing the type of bacteria, is crucial for fast and complete recovery.

## Hip Joint

The **hip joint** is a ball-and-socket joint that consists of the head of the femur and the cup-shaped acetabulum of the coxal bone. A ligament (ligamentum capitis) attaches to a pit (fovea capitis) on the head of the femur and to connective tissue within the acetabulum. This attachment, however, seems to have little importance in holding the articulating bones together, but rather carries blood vessels to the head of the femur (fig. 8.17).

A horseshoe-shaped ring of fibrocartilage (acetabular labrum) at the rim of the acetabulum deepens the cavity of the acetabulum. It encloses the head of the femur and helps hold it securely in place. In addition, a heavy, cylindrical joint capsule that is reinforced

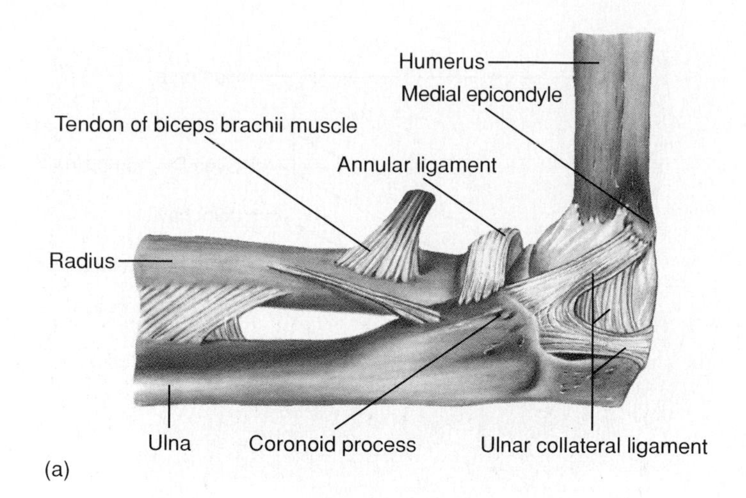


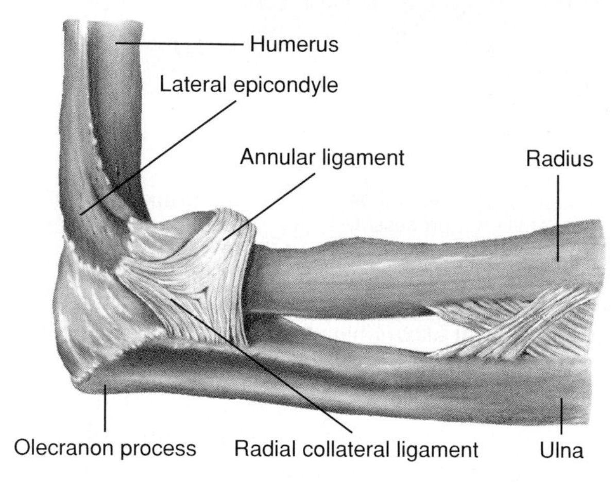


**FIGURE 8.16**
(*a*) The ulnar collateral ligament and (*b*) the radial collateral ligament strengthen the capsular wall of the elbow joint.

with still other ligaments surrounds the articulating structures and connects the neck of the femur to the margin of the acetabulum (fig. 8.18).

The major ligaments of the hip joint include the following (fig. 8.19):

1. **Iliofemoral** (il″e-o-fem′o-ral) **ligament.** This ligament consists of a Y-shaped band of very strong fibers that connects the anterior inferior iliac spine of the coxal bone to a bony line (intertrochanteric line) extending between the greater and lesser trochanters of the femur. The iliofemoral ligament is the strongest ligament in the body, and it helps prevent extension of the femur when the body is erect.
2. **Pubofemoral** (pu″bo-fem′o-ral) **ligament.** The pubofemoral ligament extends between the superior portion of the pubis and the iliofemoral ligament. Its fibers also blend with the fibers of the joint capsule.

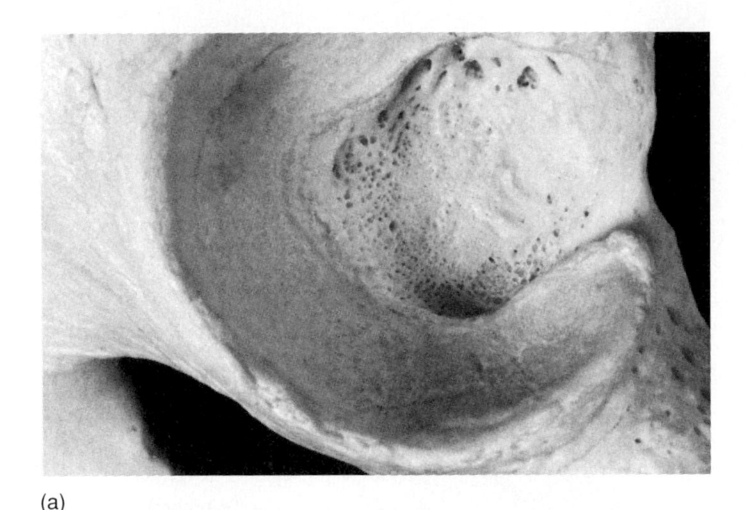
(a)

(b)

**FIGURE 8.17**
(*a*) The acetabulum provides the socket for the head of the femur in the hip joint. (*b*) The pit (fovea capitis) in the femur's head marks attachment of a ligament that carries blood vessels there.

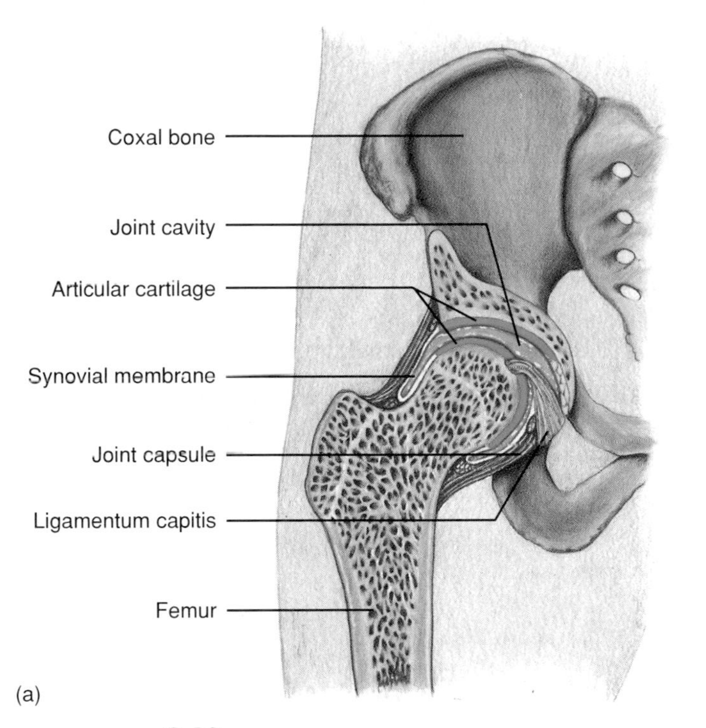

(a)

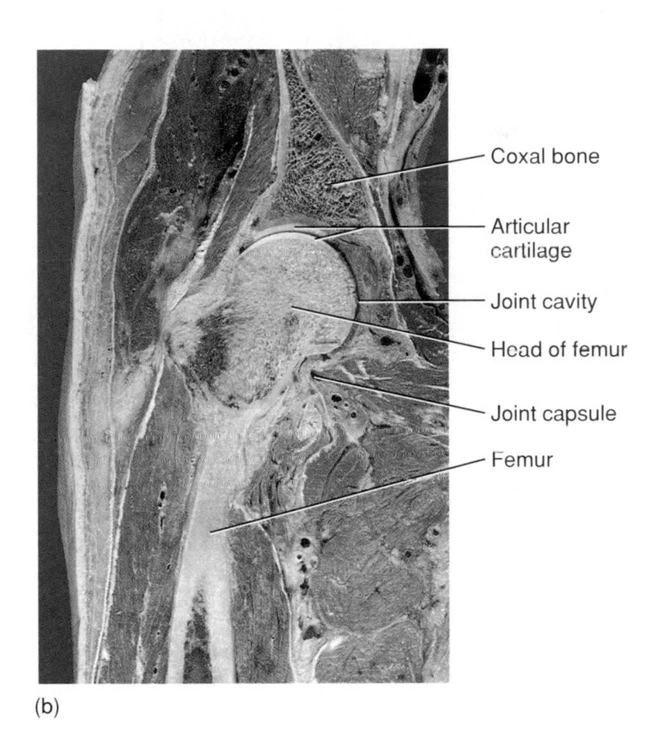

(b)

**FIGURE 8.18**
(*a*) A ring of cartilage in the acetabulum and a ligament-reinforced joint capsule hold together the hip joint; (*b*) photograph of the hip joint (coronal section).

3. **Ischiofemoral** (is″ke-o-fem′o-ral) **ligament.** This ligament consists of a band of strong fibers that originates on the ischium just posterior to the acetabulum and blends with the fibers of the joint capsule.

Muscles surround the joint capsule of the hip. The articulating parts of the hip are held more closely together than those of the shoulder, allowing considerably less freedom of movement. The structure of the hip joint, however, still permits a wide variety of movements, including extension, flexion, abduction, adduction, rotation, and circumduction. The hip is one of the joints most frequently replaced (Clinical Application 8.1).

## Knee Joint

The **knee joint** is the largest and most complex of the synovial joints. It consists of the medial and lateral condyles at the distal end of the femur, and the medial and lateral condyles at the proximal end of the tibia. In addition, the femur articulates anteriorly with the

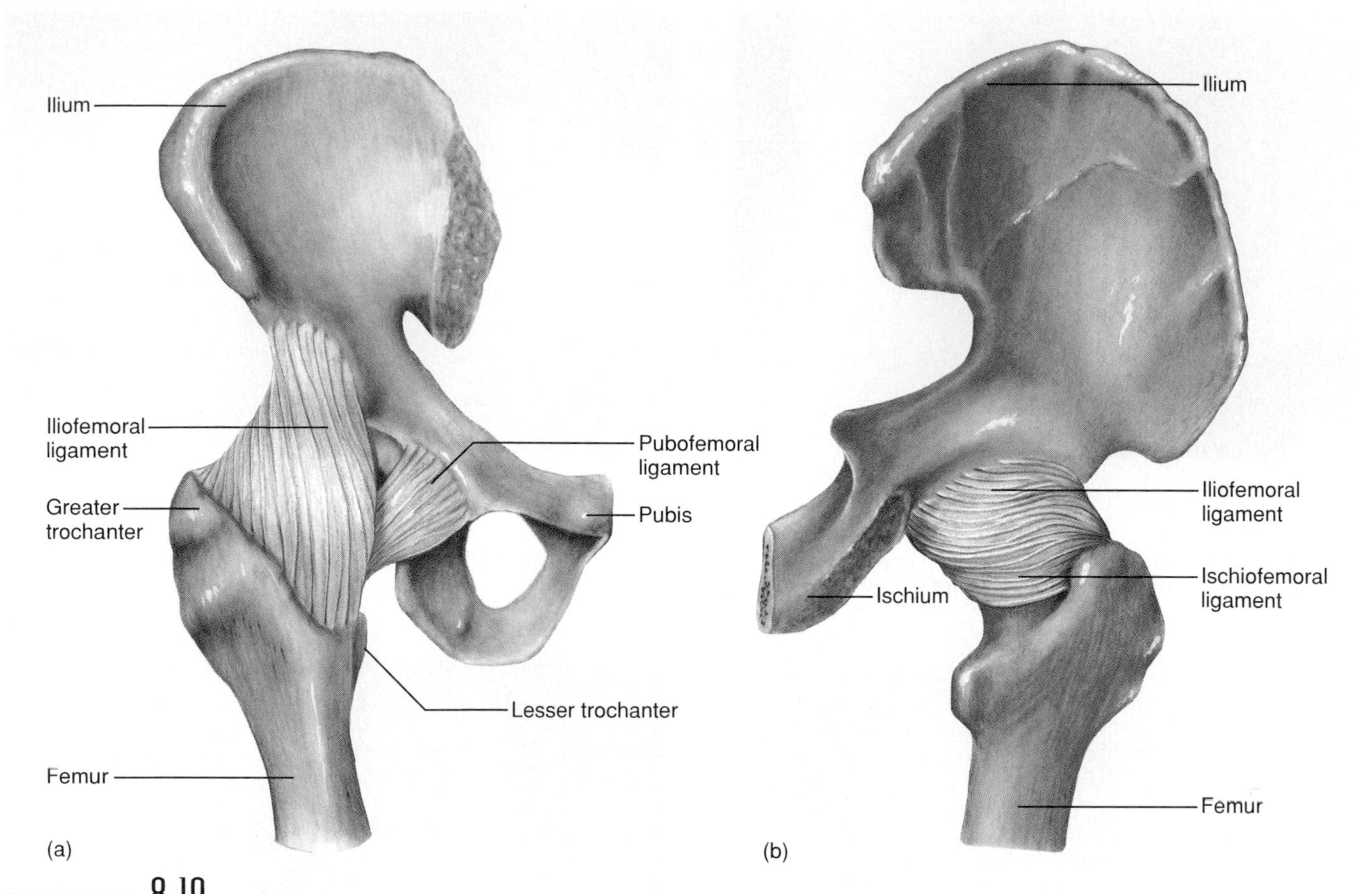


**FIGURE 8.19**
The major ligaments of the right hip joint. (*a*) Anterior view; (*b*) posterior view.

patella. Although the knee is sometimes considered a modified hinge joint, the articulations between the femur and tibia are condyloid, and the joint between the femur and patella is a gliding joint.

The *joint capsule* of the knee is relatively thin, but ligaments and the tendons of several muscles greatly strengthen it. For example, the fused tendons of several muscles in the thigh cover the capsule anteriorly. Fibers from these tendons descend to the patella, partially enclose it, and continue downward to the tibia. The capsule attaches to the margins of the femoral and tibial condyles as well as between these condyles (fig. 8.20).

The ligaments associated with the joint capsule that help keep the articulating surfaces of the knee joint in contact include the following (fig. 8.21):

1. **Patellar** (pah-tel′ar) **ligament.** This ligament is a continuation of a tendon from a large muscle group in the thigh (quadriceps femoris). It consists of a strong, flat band that extends from the margin of the patella to the tibial tuberosity.
2. **Oblique popliteal** (ŏ′blēk pop-lit′e-al) **ligament.** This ligament connects the lateral condyle of the femur to the margin of the head of the tibia.
3. **Arcuate** (ar′ku-āt) **popliteal ligament.** This ligament appears as a Y-shaped system of fibers that extends from the lateral condyle of the femur to the head of the fibula.
4. **Tibial collateral** (tib′e-al kŏ-lat′er-al) **ligament** (medial collateral ligament). This ligament is a broad, flat band of tissue that connects the medial condyle of the femur to the medial condyle of the tibia.
5. **Fibular** (fib′u-lar) **collateral ligament** (lateral collateral ligament). This ligament consists of a strong, round cord located between the lateral condyle of the femur and the head of the fibula.

In addition to the ligaments that strengthen the joint capsule, two ligaments within the joint, called **cruciate** (kroo′she-āt) **ligaments,** help prevent displacement of the articulating surfaces. These strong bands of fibrous tissue stretch upward between the tibia and the femur, crossing each other on the way. They are named according to their positions of attachment to the tibia. Thus, the *anterior cruciate ligament* originates from the anterior intercondylar area of the tibia and extends to the lateral condyle of the femur. The *posterior cruciate ligament* connects the posterior intercondylar area of the tibia to the medial condyle of the femur.

**CLINICAL APPLICATION**

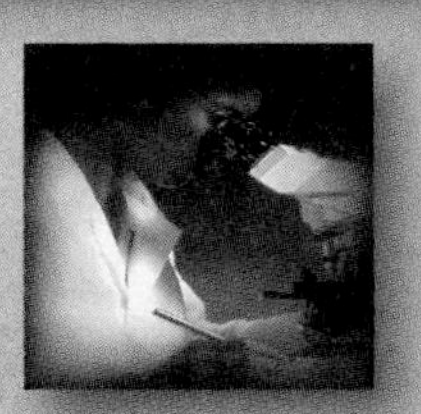

## Replacing Joints

Surgeons use a variety of synthetic materials to replace joints that are severely damaged by arthritis or injury. Metals such as cobalt-chrome and titanium alloys are used to replace larger joints, whereas silicone polymers are more commonly used to replace smaller joints. Such artificial joints must be durable yet not provoke immune system rejection. They must also allow normal healing to occur, and not move surrounding structures out of their normal positions.

Before the advent of joint replacements, surgeons removed damaged or diseased joint surfaces, hoping that scar tissue filling in the area would restore mobility. This type of surgery was rarely successful. In the 1950s, Alfred Swanson, an army surgeon in Grand Rapids, Michigan, invented the first joint implants using silicone polymers. By 1969, after much refinement, the first silicone-based joint implants hit the market. These devices provided flexible hinges for joints of the toes, fingers, and wrists. Since then, more than two dozen joint replacement models have been developed, and more than a million people have them, mostly in the hip.

A surgeon inserts a joint implant in a procedure called implant resection arthroplasty. The surgeon first removes the surface of the joint bones and excess cartilage. Next, the centers of the tips of abutting bones are hollowed out, and the stems of the implant are inserted here. The hinge part of the implant lies between the bones, aligning them yet allowing them to bend, as they would at a natural joint. Bone cement fixes the implant in place. Finally, the surgeon repairs the tendons, muscles, and ligaments. As the site of the implant heals, the patient must exercise the joint. A year of physical therapy may be necessary to fully benefit from replacement joints.

Newer joint replacements use materials that resemble natural body chemicals. Hip implants, for example, may bear a coat of hydroxylapatite, which interacts with natural bone. Instead of filling in spaces with bone cement, some investigators are testing a variety of porous coatings that allow bones to grow into the implant area.

---

The young soccer player, running at full speed, suddenly switches direction and is literally stopped in her tracks by a popping sound followed by a searing pain in her knee. Two hours after she veered toward the ball, her knee is swollen and painful, due to bleeding within the joint. She has torn the anterior cruciate ligament, a common serious knee injury.

---

Two fibrocartilaginous *menisci* separate the articulating surfaces of the femur and tibia. Each meniscus is roughly C-shaped, with a thick rim and a thinner center, and attaches to the head of the tibia. The medial and lateral menisci form depressions that fit the corresponding condyles of the femur, compensating for the differences in shapes between the surfaces of the femur and tibia (see fig. 8.20).

Several bursae are associated with the knee joint. These include a large extension of the knee joint cavity called the *suprapatellar bursa* located between the anterior surface of the distal end of the femur and the muscle group (quadriceps femoris) above it; a large *prepatellar bursa* between the patella and the skin; and a smaller *infrapatellar bursa* between the proximal end of the tibia and the patellar ligament (see fig. 8.8).

As with a hinge joint, the basic structure of the knee joint permits flexion and extension. However, when the knee is flexed, rotation is also possible.

Clinical Application 8.2 discusses some common disorders of joints.

---

Tearing or displacing a meniscus is another common knee injury, usually resulting from forcefully twisting the knee when the leg is flexed. Since the meniscus is composed of fibrocartilage, this type of injury heals very slowly. Also, a torn and displaced portion of cartilage jammed between the articulating surfaces impedes movement of the joint.

Following such a knee injury, the synovial membrane may become inflamed (acute synovitis) and secrete excess fluid, distending the joint cavity so that the knee is enlarged above and on the sides of the patella.

---

## CLINICAL APPLICATION 8.2

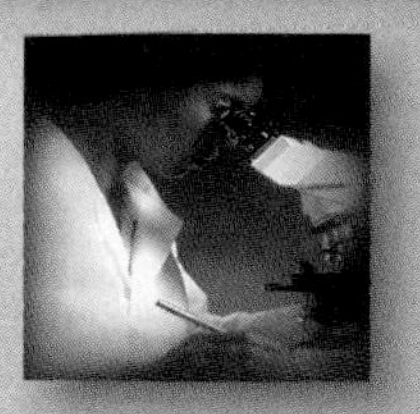

# Joint Disorders

Joints have a tough job. They must support weight, provide a great variety of body movements, and are used very frequently. In addition to this normal wear and tear, these structures are sometimes subjected to injury from overuse, infection, an immune system launching a misplaced attack, or degeneration. Here is a look at some common joint problems.

### DISLOCATION

A *dislocation* (luxation) displaces the articulating bones of a joint and usually results from a fall or other unusual body movement. The joints of the shoulders, knees, elbows, fingers, and jaw are common sites for this type of injury. A dislocation produces an obvious deformity of the joint, some loss of ability to move the articulated bones, localized pain, and swelling.

### SPRAINS

*Sprains* result from overstretching or tearing the connective tissues, ligaments, and tendons associated with a joint, but they do not dislocate the articular bones. Usually forceful wrenching or twisting sprains the wrist or ankles. For example, excessively inverting an ankle can sprain it as the ligaments on its lateral surface are stretched. Severe injuries may pull these tissues loose from their attachments.

A sprained joint is painful and swollen, restricting movement. Immediate treatment of a sprain is rest; more serious cases require medical attention. However, immobilization of a joint, even for a brief period, causes bone resorption and weakens ligaments. Consequently, exercise may help strengthen the joint.

### BURSITIS

Excessive use of a joint or stress on a bursa may cause *bursitis,* an inflammation of a bursa. The bursa between the heel bone (calcaneus) and the Achilles tendon may become inflamed as a result of a sudden increase in physical activity involving use of the feet. Similarly, a form of bursitis called tennis elbow affects the bursa between the olecranon process and the skin. Bursitis is treated with rest. Medical attention may be necessary.

### ARTHRITIS

*Arthritis* is a disease condition that causes inflamed, swollen, and painful joints. There are more than a hundred different types of arthritis, affecting 50 million people in the United States. Arthritis can also be part of other syndromes (see chart 8A). The most common causes of arthritis are discussed below.

### RHEUMATOID ARTHRITIS (RA)

This autoimmune disorder (a condition in which the immune system attacks the body's healthy tissues) is the most painful and debilitating form of arthritis. The synovial membrane of a joint becomes inflamed and thickens, forming a mass called a pannus. Then, the articular cartilage becomes damaged, and fibrous tissue infiltrates it, interfering with joint movements. In time, the joint may ossify so that the articulating bones fuse (bony ankylosis). Joints severely damaged by RA may be surgically replaced.

RA may affect many joints or only a few. It is usually a systemic illness, accompanied by fatigue, muscular atrophy, anemia, and osteoporosis, as well as changes in the skin, eyes, lungs, blood vessels, and heart. RA usually affects adults, but there is a juvenile form.

### OSTEOARTHRITIS

This degenerative disorder is the most common type of arthritis. It usually occurs with aging, but an inherited form may manifest itself as early as one's thirties. A person may first become aware of osteoarthritis when a blow to the affected joint produces pain that is much more intense than normal. Gradually, the area of the affected joint deforms (fig. 8A). For example, arthritic fingers take on a gnarled appearance, or a knee may bulge.

In osteoarthritis, articular cartilage softens and disintegrates gradually, roughening the articular surfaces. Joints become painful, with restricted movement. For example, arthritic fingers may lock into place while a person is playing the guitar or tying a shoelace. Osteoarthritis most often affects joints that are used the most over a lifetime, such as those of the fingers, hips, knees, and the lower parts of the vertebral column.

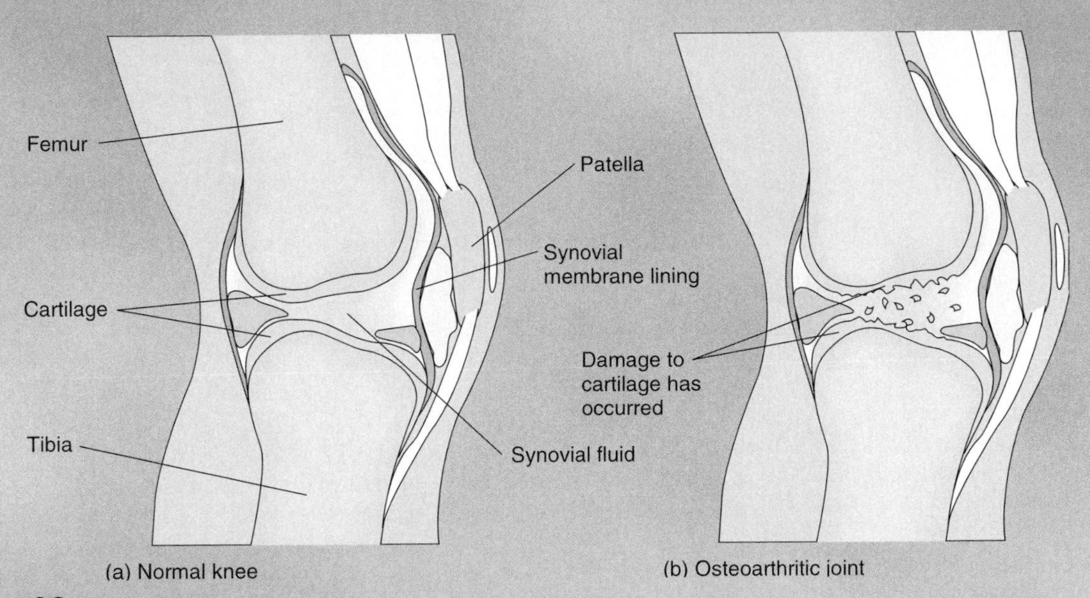


**FIGURE 8A**
An inherited defect in collagen or prolonged wear-and-tear destroys joints in osteoarthritis.

Fortunately, nonsteroidal anti-inflammatory drugs can usually control osteoarthritis symptoms. Exercise can keep stiff joints more flexible, and such simple measures as wearing gloves in the winter can alleviate symptoms.

### Gout

Gout is an "inborn error of metabolism," an inherited illness in which a defective or missing enzyme causes the chemical it normally acts on to build up. In gout, uric acid crystals accumulate in joints, most commonly the metatarsophalangeal (large toe) joint. Gout was once thought to be caused by eating too much rich food (fig. 8B). In a condition called pseudogout, a different type of crystal accumulates, usually in a knee or wrist joint.

### Lyme Arthritis

Lyme disease is a bacterial infection passed in a tick bite that causes intermittent arthritis of several joints, usually weeks after the initial symptoms of rash, fatigue, and flu-like aches and pains. Lyme arthritis was first observed in Lyme, Connecticut, where an astute woman kept a journal after noticing that many of her young neighbors had what appeared to be the very rare juvenile form of rheumatoid arthritis. Her observations led Allen Steere, a Yale University rheumatologist, to trace the illness to a tick-borne bacterial infection. Antibiotic treatment beginning as soon as the early symptoms of Lyme disease are recognized can prevent development of the associated arthritis.

Other types of bacteria can cause arthritis too. These include common *Staphylococcus* and *Streptococcus* species, *Neisseria gonorrhoeae* (which causes the sexually transmitted disease gonorrhea), and *Mycobacterium* (which causes tuberculosis). Arthritis may also be associated with AIDS, because the immunity breakdown raises the risk of infection by bacteria that can cause arthritis.

*Continued . . .*

**FIGURE 8B**

Before gout was recognized as an inborn error of metabolism, physicians attributed it to being a lazy glutton, as this cartoon illustrates. We now know that the extreme joint pain occurring in the big toe is due to accumulation of uric acid crystals.

## CHART 8A DIFFERENT TYPES OF ARTHRITIS

**Some More-Common Forms of Arthritis**

| *Type* | *Incidence in U.S.* |
|---|---|
| Osteoarthritis | 15.8 million |
| Rheumatoid arthritis | 2.9 million |
| Spondyloarthropathies | 2.5 million |

**Some Less-Common Forms of Arthritis**

| *Type* | *Incidence in U.S.* | *Age of Onset* | *Symptoms* |
|---|---|---|---|
| Gout | 1.6 million (85% male) | >40 | Sudden onset of extreme pain and swelling of a large joint |
| Juvenile rheumatoid arthritis | 250,000 | <18 | Joint stiffness, often in knee |
| Scleroderma | 300,000 | 30–50 | Skin hardens and thickens |
| Systemic lupus erythematosus | 300,000 (>90% female) | teens–50s | Fever, weakness, upper body rash, joint pain |
| Kawasaki disease | Hundreds of cases in local outbreaks | 6 months–11 years | Fever, joint pain, red rash on palms and soles, heart complications |
| Strep A infection | 100,000 | any age | Confusion, body aches, shock, low blood pressure, dizziness, arthritis, pneumonia |

Source: From "Arthritis: Modern Treatment for That Old Pain in the Joints" in *FDA Consumer*, July/August 1991.

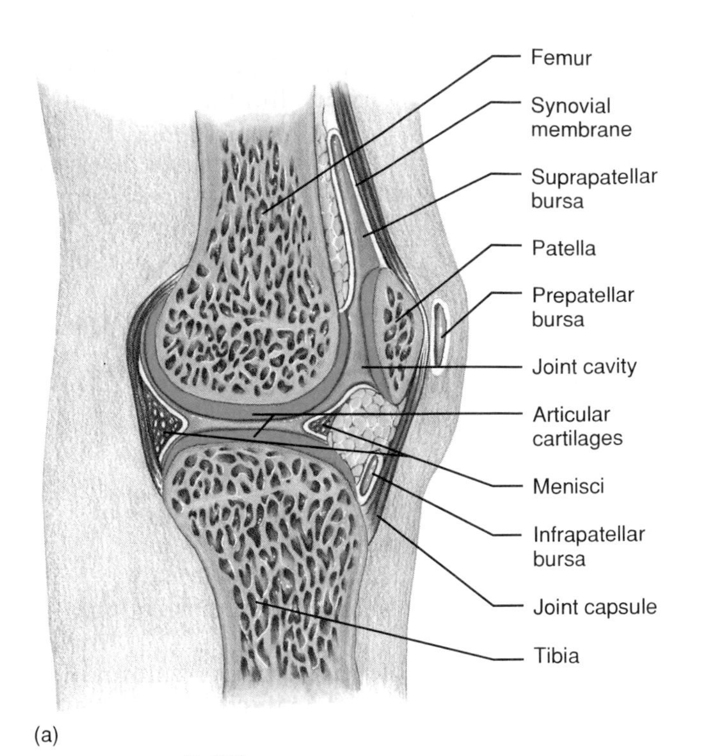


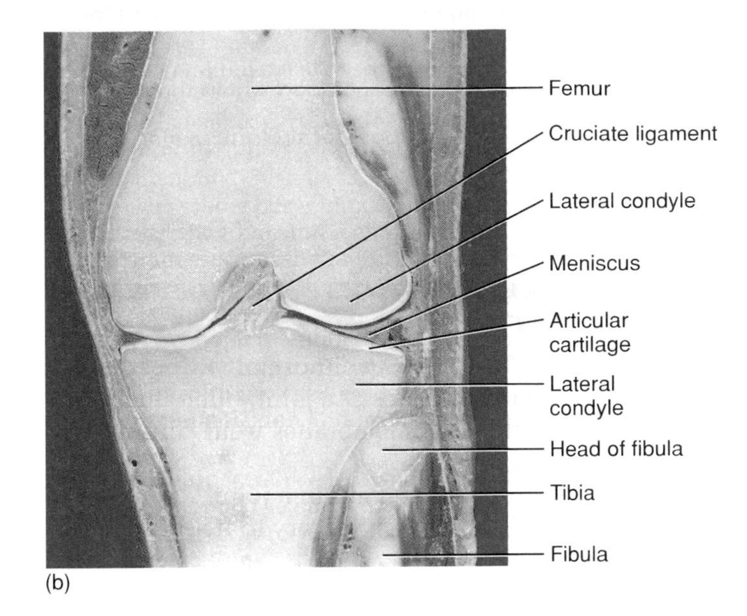


**FIGURE 8.20**

(*a*) The knee joint is the most complex of the synovial joints (sagittal section); (*b*) photograph of the knee joint (coronal section).

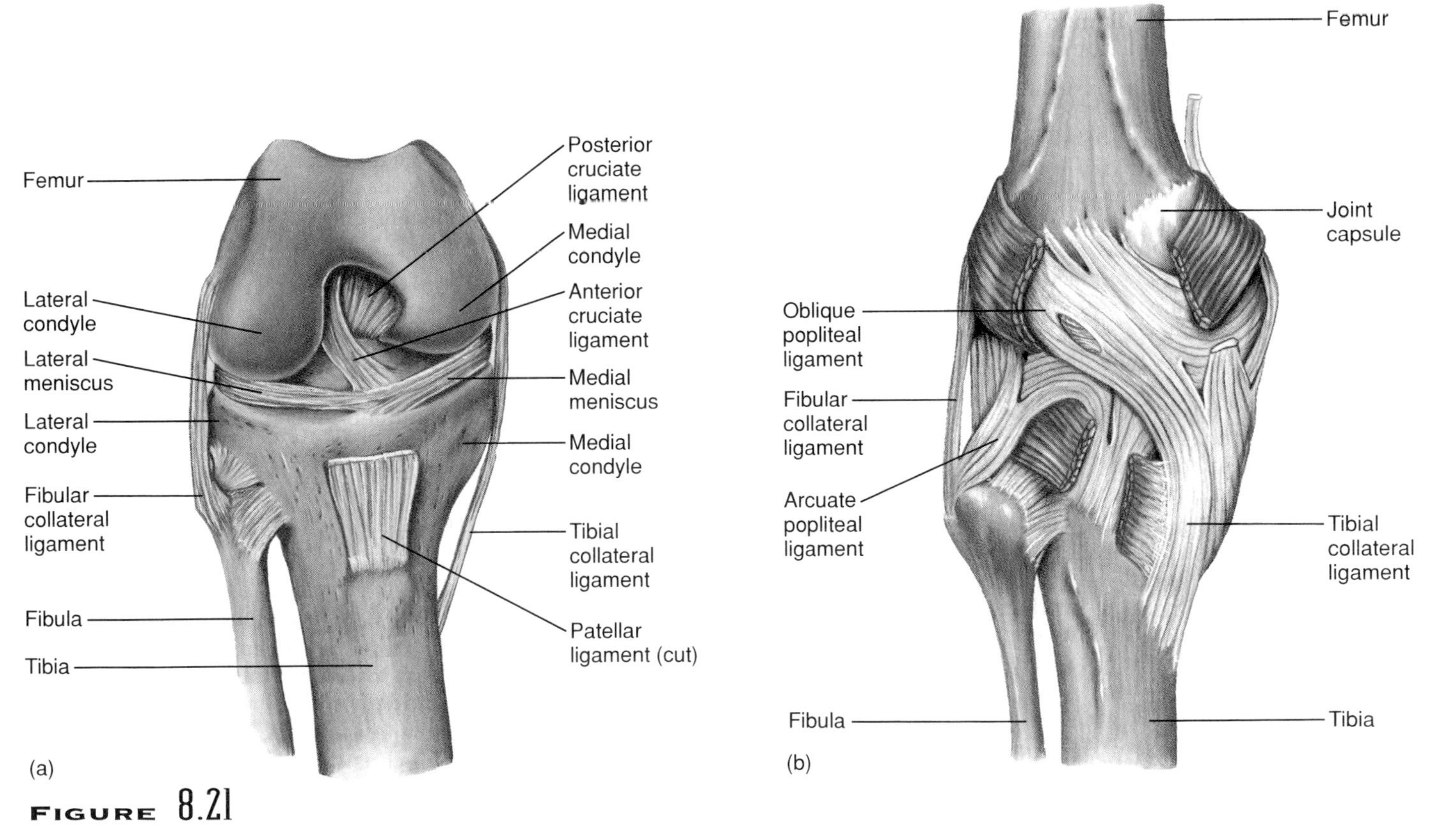


**FIGURE 8.21**

Ligaments within the knee joint help to strengthen it. (*a*) Anterior view; (*b*) posterior view.

# Review Exercises

## Part A

1. Define *joint.*
2. Explain how joints are classified.
3. Compare the structure of a fibrous joint with that of a cartilaginous joint.
4. Distinguish between a syndesmosis and a suture.
5. Describe a gomphosis, and name an example.
6. Compare the structures of a synchondrosis and a symphysis.
7. Explain how the joints between adjacent vertebrae permit movement.
8. Describe the general structure of a synovial joint.
9. Describe how a joint capsule may be reinforced.
10. Explain the function of the synovial membrane.
11. Explain the function of synovial fluid.
12. Define *meniscus.*
13. Define *bursa.*
14. List six types of synovial joints, and name an example of each type.
15. Describe the movements permitted by each type of synovial joint.
16. Name the parts that comprise the shoulder joint.
17. Name the major ligaments associated with the shoulder joint.
18. Explain why the shoulder joint permits a wide range of movements.
19. Name the parts that comprise the elbow joint.
20. Describe the major ligaments associated with the elbow joint.
21. Name the movements permitted by the elbow joint.
22. Name the parts that comprise the hip joint.
23. Describe how the articular surfaces of the hip joint are held together.
24. Explain why there is less freedom of movement in the hip joint than in the shoulder joint.
25. Name the parts that comprise the knee joint.
26. Describe the major ligaments associated with the knee joint.
27. Explain the function of the menisci of the knee.
28. Describe the locations of the bursae associated with the knee joint.

## Part B

Match the movements in column I with the descriptions in column II.

| I | II |
|---|---|
| 1. Rotation | A. Turning palm upward |
| 2. Supination | B. Decreasing angle between parts |
| 3. Extension | C. Moving part forward |
| 4. Eversion | D. Moving part around an axis |
| 5. Protraction | E. Turning sole of foot outward |
| 6. Flexion | F. Increasing angle between parts |
| 7. Pronation | G. Lowering a part |
| 8. Abduction | H. Turning palm downward |
| 9. Depression | I. Moving part away from midline |

9

CHAPTER

# Muscular System

## Chapter Objectives

AFTER YOU HAVE STUDIED THIS CHAPTER, YOU SHOULD BE ABLE TO:

1. Describe how connective tissue is included in the structure of a skeletal muscle.
2. Name the major parts of a skeletal muscle fiber and describe the function of each part.
3. Explain the major events that occur during muscle fiber contraction.
4. Explain how energy is supplied to the muscle fiber contraction mechanism, how oxygen debt develops, and how a muscle may become fatigued.
5. Distinguish between fast and slow muscles.
6. Distinguish between a twitch and a sustained contraction.
7. Describe how exercise affects skeletal muscles.
8. Explain how various types of muscular contractions produce body movements and help maintain posture.
9. Distinguish between the structures and functions of a multiunit smooth muscle and a visceral smooth muscle.
10. Compare the fiber contraction mechanisms of skeletal, smooth, and cardiac muscles.
11. Explain how the locations of skeletal muscles are related to the movements they produce and how muscles interact to produce such movements.
12. Identify and describe the locations of the major skeletal muscles of each body region and describe the action of each muscle.

MUSCLES ARE QUITE OBVIOUS IN A BODYBUILDER; BUT THEY ALSO MAKE POSSIBLE THE MORE MUNDANE MOVEMENTS OF EVERYDAY ACTIVITIES.

## Key Terms

**actin** (ak′tin)
**antagonist** (an-tag′o-nist)
**aponeurosis** (ap″o-nu-ro′sēz)
**fascia** (fash′e-ah)
**insertion** (in-ser′shun)
**motor neuron** (mo′tor nu′ron)
**motor unit** (mo′tor u′nit)
**muscle impulse** (mus′el im′puls)
**myofibril** (mi″o-fi′bril)
**myogram** (mi′o-gram)
**myosin** (mi′o-sin)
**neurotransmitter** (nu″ro-trans′mit-er)
**origin** (or′ĭ-jin)
**oxygen debt** (ok′sĭ-jen det)
**prime mover** (prim mo͞ov′er)
**recruitment** (re-kro͞ot′ment)
**sarcomere** (sar′ko-mēr)
**synergist** (sin′er-jist)
**threshold stimulus** (thresh′old stim′u-lus)

s a bodybuilder, Roger P. knows quite a bit about muscle anatomy and physiology. He knows that a used or stressed muscle increases in size and strength but that he must carefully vary weight-lifting exercises to work each muscle group. To do this, Roger takes three training approaches.

The traditional way to build muscles is called high volume training. Roger isolates body parts, then performs high-intensity repetitions of weight-lifting on one area at a time. Some evidence suggests that this approach actually stimulates muscle cells to reproduce.

In a second strategy called maximum overload, Roger lifts very heavy weights for fewer repetitions. This builds muscle size and strength by increasing the number of muscle fibers that respond to a single nerve impulse during a contraction. Finally, in a peak-power workout, Roger selects a weight that is about two-thirds of the maximum of what a particular muscle can withstand and holds it aloft for six seconds.

By mixing these approaches—a traditional workout one day on his upper body, perhaps a combination of maximum overload and peak-power on his lower body the next—Roger sculpts the overdevelopment of his muscular system.

Most of us pay less attention to our muscular systems than a bodybuilder, but we all use our muscles all the time—for enjoyment and for survival. The baseball player lunging for the ball uses dozens of muscles in a highly coordinated manner, with barely a thought, thanks to her muscular system. We use our muscles to enjoy sports, to dance, to play musical instruments, to read, and to get from one place to another.

We could not exist without the incredibly powerful muscular pump of the heart, or the smooth, contractile sheets that fold to form linings of our digestive tracts. A strong muscular contraction propels our entry into the world. Muscles provide the nuances of movement that enable us to communicate; a raised eyebrow, a shoulder shrug, and a wrinkled nose all convey unmistakable meaning. A smile that blooms instantaneously when we see something endearing or amusing requires the action of no less than fourteen different facial muscles! In short, the muscular system is important to our survival and our enjoyment of life.

---

The three types of muscle tissues are skeletal muscle, smooth muscle, and cardiac muscle, as described in chapter 5. This chapter focuses primarily on the skeletal muscles, which are attached to bones and are under conscious control.

## Structure of a Skeletal Muscle

A skeletal muscle is an organ of the muscular system and is composed of skeletal muscle tissue, nervous tissue, blood, and connective tissues.

### Connective Tissue Coverings

An individual skeletal muscle is separated from adjacent muscles and held in position by layers of fibrous connective tissue called **fascia.** This connective tissue surrounds each muscle and may project beyond the end of its muscle fibers to form a cordlike **tendon.** Fibers in a tendon intertwine with those in the periosteum of a bone, attaching the muscle to the bone. In other cases, the connective tissues associated with a muscle form broad, fibrous sheets called **aponeuroses,** which may be attached to the coverings of adjacent muscles (figs. 9.1 and 9.2).

> A tendon, or the connective tissue sheath of a tendon (tenosynovium), may become painfully inflamed and swollen following an injury or the repeated stress of athletic activity. These conditions are called *tendinitis* and *tenosynovitis,* respectively. The tendons most commonly affected are those associated with the joint capsules of the shoulder, elbow, and hip, and those involved with moving the wrist, hand, thigh, and foot.

The layer of connective tissue that closely surrounds a skeletal muscle is called the *epimysium.* Other layers of connective tissue, called the *perimysium,* extend inward from the epimysium and separate the muscle tissue into small sections. These sections contain bundles of skeletal muscle fibers called *fascicles* (fasciculi). Each muscle fiber within a fascicle (fasciculus) is surrounded by a layer of connective tissue in the form of a thin covering called *endomysium* (fig. 9.2).

All parts of a skeletal muscle, therefore, are enclosed in layers of connective tissue. This arrangement allows the parts to move somewhat independently. Also, numerous blood vessels and nerves pass through these layers (figs. 9.2 and 9.3).

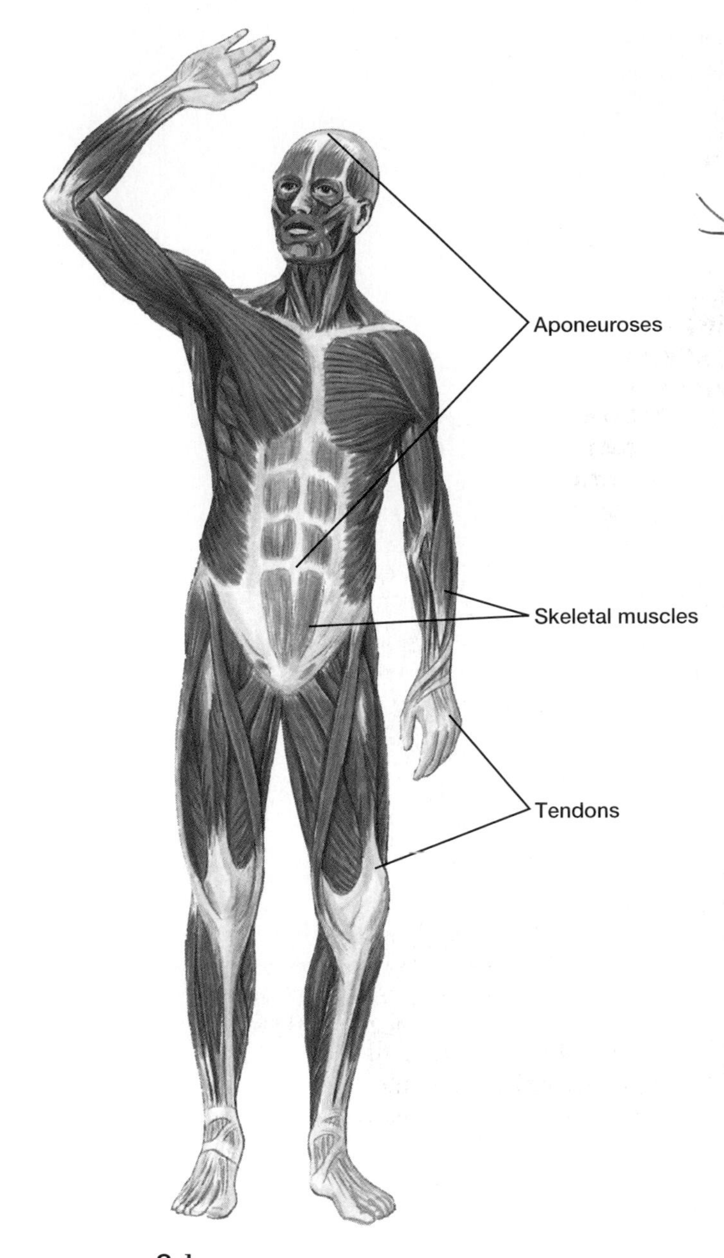


**FIGURE 9.1**
Tendons attach muscles to bones, while aponeuroses attach muscles to other muscles.

The fascia associated with the individual organs of the muscular system is part of a complex network of fasciae that extends throughout the body. The portion of the network that surrounds and penetrates the muscles is called *deep fascia.* It is continuous with the *subcutaneous fascia* that lies just beneath the skin, forming the subcutaneous layer described in chapter 6. The network is also continuous with the *subserous fascia* that forms the connective tissue layer of the serous membranes covering organs in various body cavities and lining those cavities (see chapter 6).

A *compartment* is the space occupied by a particular group of muscles, blood vessels, and nerves, all tightly enclosed by relatively inelastic fascia. There are many such compartments in the limbs. If an injury causes fluid, such as blood from an internal hemorrhage, to accumulate within a compartment, the pressure inside will rise. The increased pressure, in turn, may interfere with blood flow into the region, reducing the supply of oxygen and nutrients to the affected tissues. This condition, called *compartment syndrome,* often produces severe, unrelenting pain. Persistently elevated compartmental pressure may irreversibly damage the enclosed muscles and nerves.

Treatment for compartment syndrome may be a surgical incision through the fascia (fasciotomy) to relieve the pressure and restore circulation.

## Skeletal Muscle Fibers

Recall from chapter 5 that a skeletal muscle fiber represents a single cell of a muscle (see fig. 5.28). Each skeletal muscle fiber is a thin, elongated cylinder with rounded ends that are attached to connective tissues associated with a muscle. Just beneath its cell membrane (or *sarcolemma*), the cytoplasm (or *sarcoplasm*) of the fiber contains many small, oval nuclei and mitochondria. The sarcoplasm also contains numerous, threadlike **myofibrils** that lie parallel to one another (fig. 9.4*a*).

The myofibrils play a fundamental role in the muscle contraction mechanism. They contain two kinds of protein filaments—thick filaments composed of the protein **myosin** and thin filaments composed of the protein **actin** (fig. 9.4*c*). The arrangement of these filaments produces the characteristic alternating light and dark *striations* of skeletal muscle fiber.

The striation pattern of skeletal muscle fibers has two main parts. The first, the *I bands* (the light bands), are composed of thin actin filaments held in this arrangement by direct attachment to structures called *Z lines.*

The second part consists of the *A bands* (the dark bands), which are composed of thick myosin filaments overlapping with thin actin filaments. The myosin filaments are also held in place by the Z lines, but are

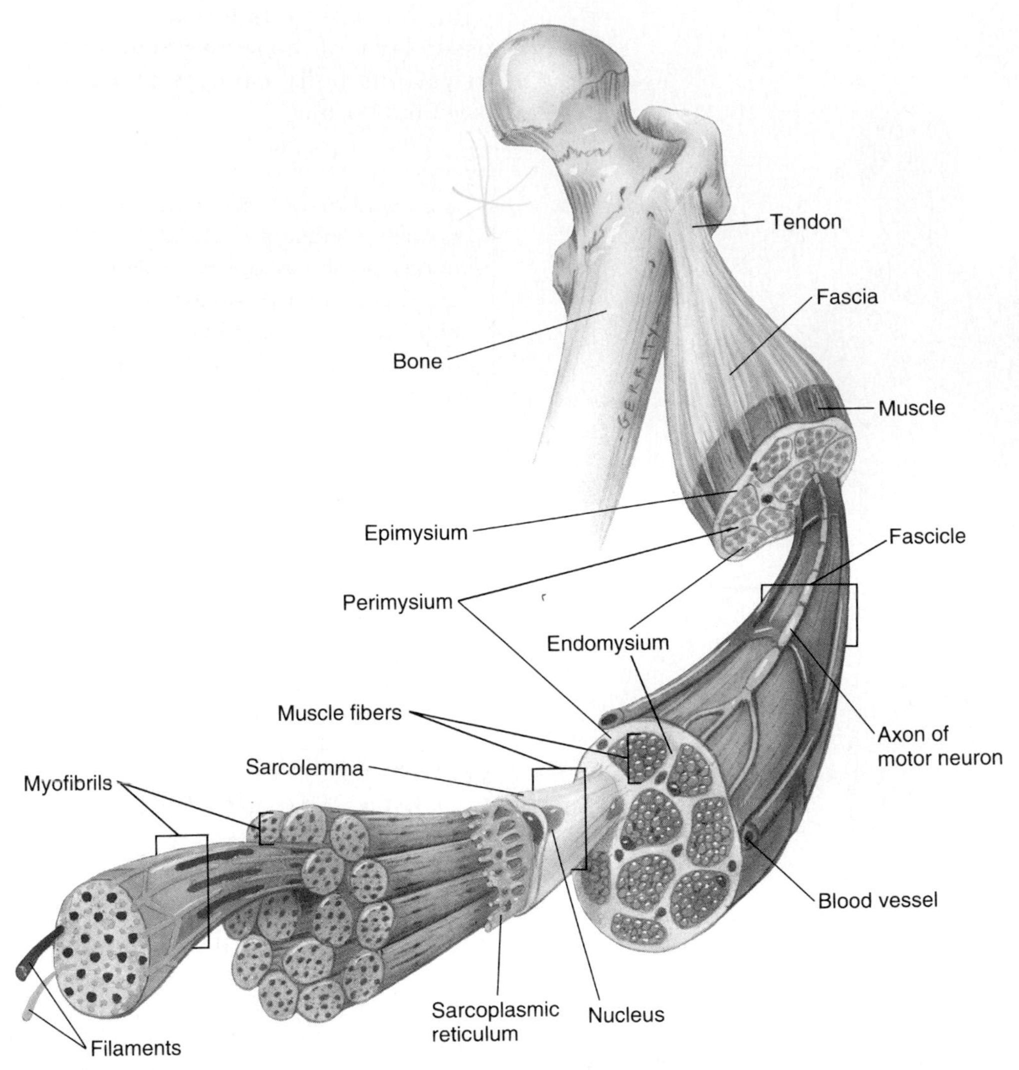


**FIGURE 9.2**

A skeletal muscle is composed of a variety of tissues, including layers of connective tissue. Fascia covers the surface of the muscle, epimysium lies beneath the fascia, and perimysium extends into the structure of the muscle where it separates muscle cells into fascicles. Endomysium separates individual muscle fibers.

attached to them by a large protein called **titin,** or connectin. Note that the A band consists not only of a region where the thick and thin filaments overlap, but also a central region (*H zone*) consisting only of thick filaments, plus a thickening known as the *M line* (fig. 9.4*b*). The segment of a myofibril that extends from one Z line to the next Z line is called a **sarcomere** (figs. 9.4 and 9.5).

Within the cytoplasm of a muscle fiber is a network of membranous channels that surrounds each myofibril and runs parallel to it. These membranes form the **sarcoplasmic reticulum,** which corresponds to the endoplasmic reticulum of other cells. Another set of membranous channels called **transverse tubules** (T-tubules) extends inward as invaginations from the fiber's membrane, and passes all the way through the fiber. Thus, each of these tubules opens to the outside of the muscle fiber and contains extracellular fluid. Furthermore, each transverse tubule lies between two enlarged portions of the sarcoplasmic reticulum called *cisternae* near the region where the actin and myosin filaments overlap. The sarcoplasmic reticulum and the transverse tubules activate muscle contraction when the fiber is stimulated (fig. 9.6).

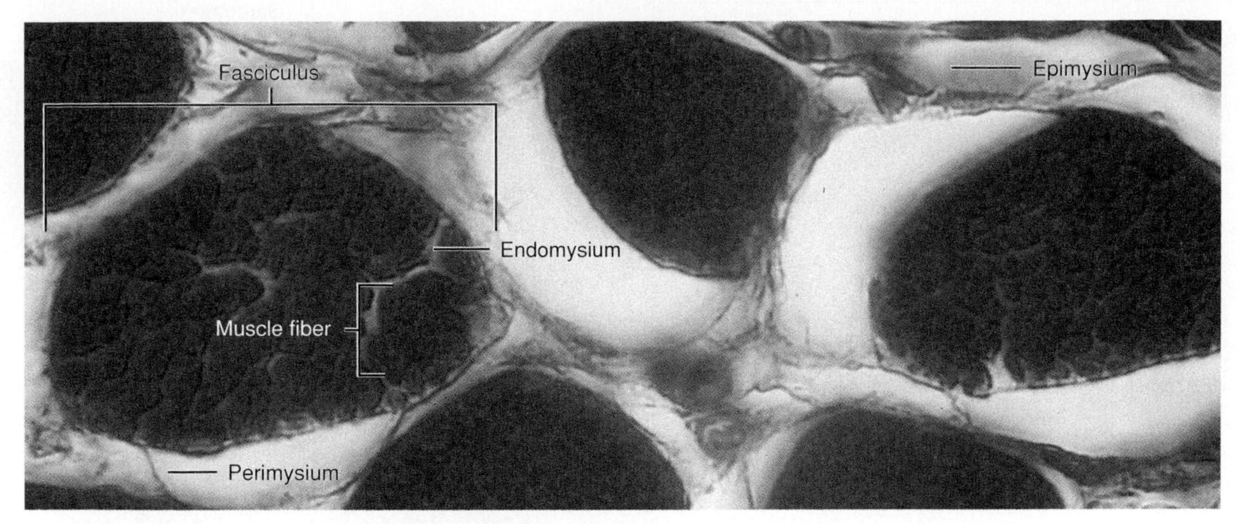


**FIGURE 9.3**

Scanning electron micrograph of a fascicle (fasciculus) surrounded by its connective tissue sheath, the perimysium. Muscle fibers within the fascicle are surrounded by endomysium (320×).

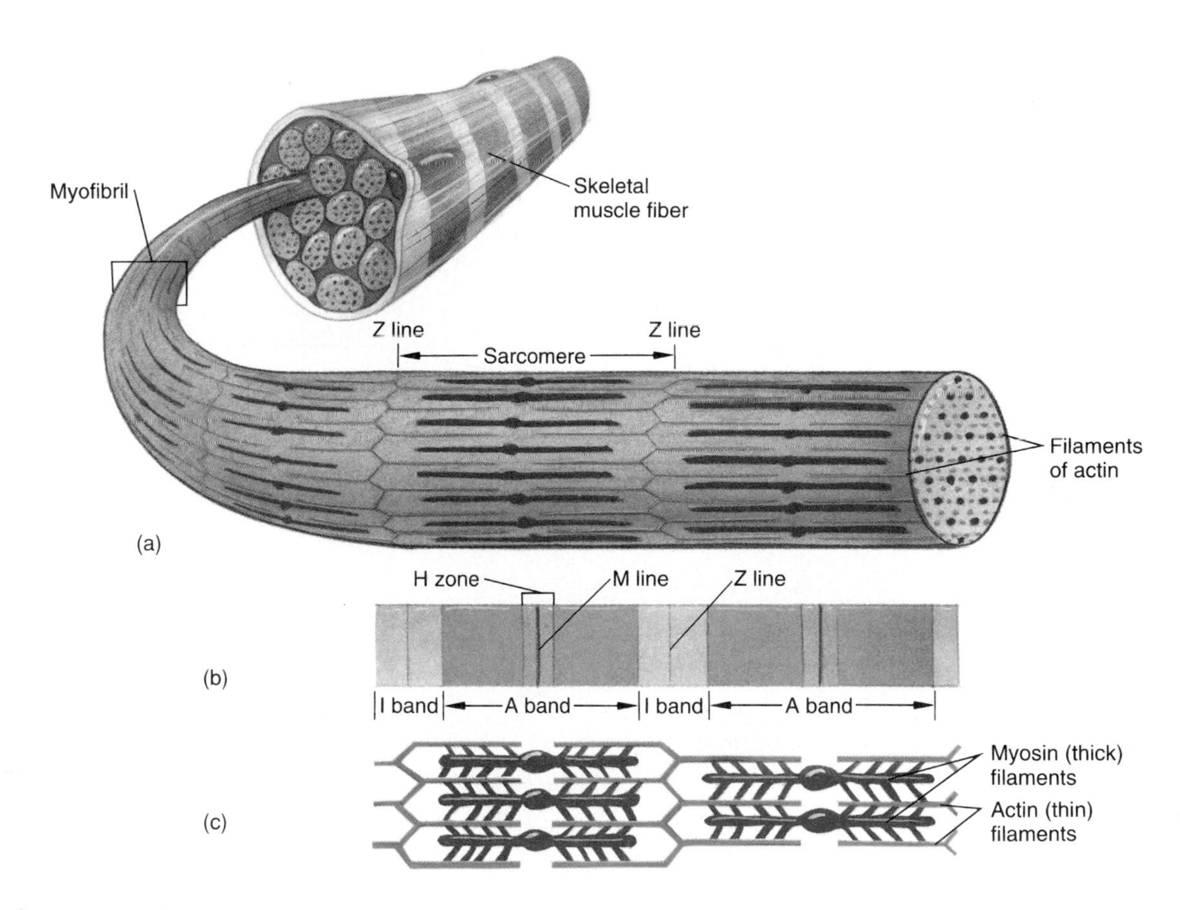


**FIGURE 9.4**

(*a*) A skeletal muscle fiber contains numerous myofibrils, each consisting of (*b*) units called sarcomeres. (*c*) The characteristic striations of a sarcomere are due to the arrangement of actin and myosin filaments.

Although muscle fibers and the connective tissues associated with them are flexible, they can tear if overstretched. This type of injury is common in athletes and is called a *muscle strain.* The seriousness of the injury depends on the degree of damage the tissues sustain. In a mild strain, only a few muscle fibers are injured, the fascia remains intact, and little function is lost. In a severe strain, many muscle fibers as well as fascia tear, and muscle function may be lost completely. A severe strain is very painful and is accompanied by discoloration and swelling of tissues due to ruptured blood vessels. Surgery may be required to reconnect the separated tissues.

1. Describe how connective tissue is associated with a skeletal muscle.
2. Describe the general structure of a skeletal muscle fiber.
3. Explain why skeletal muscle fibers appear striated.
4. Explain the relationship between the sarcoplasmic reticulum and the transverse tubules.

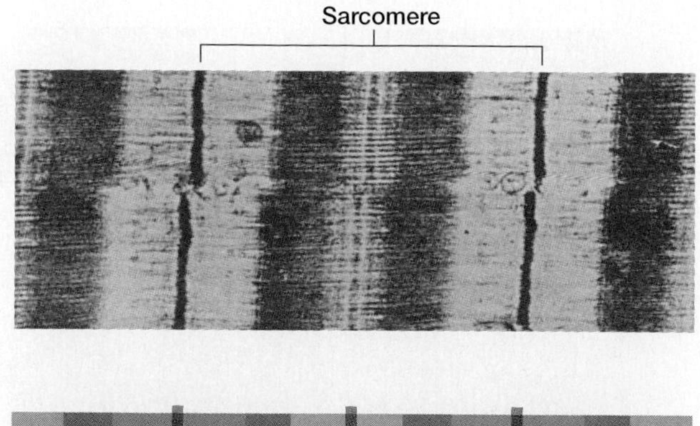


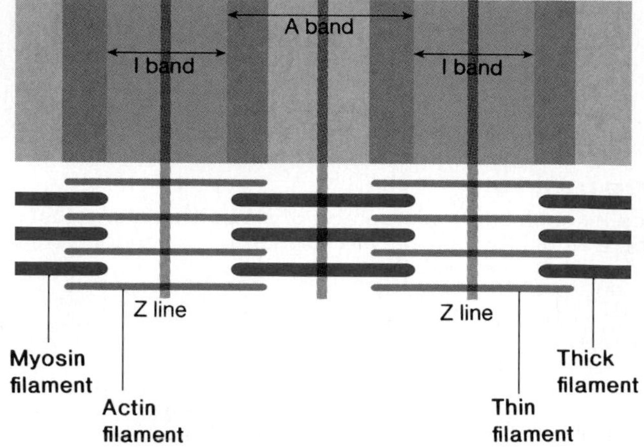


**FIGURE 9.5**
A sarcomere (80,000×).

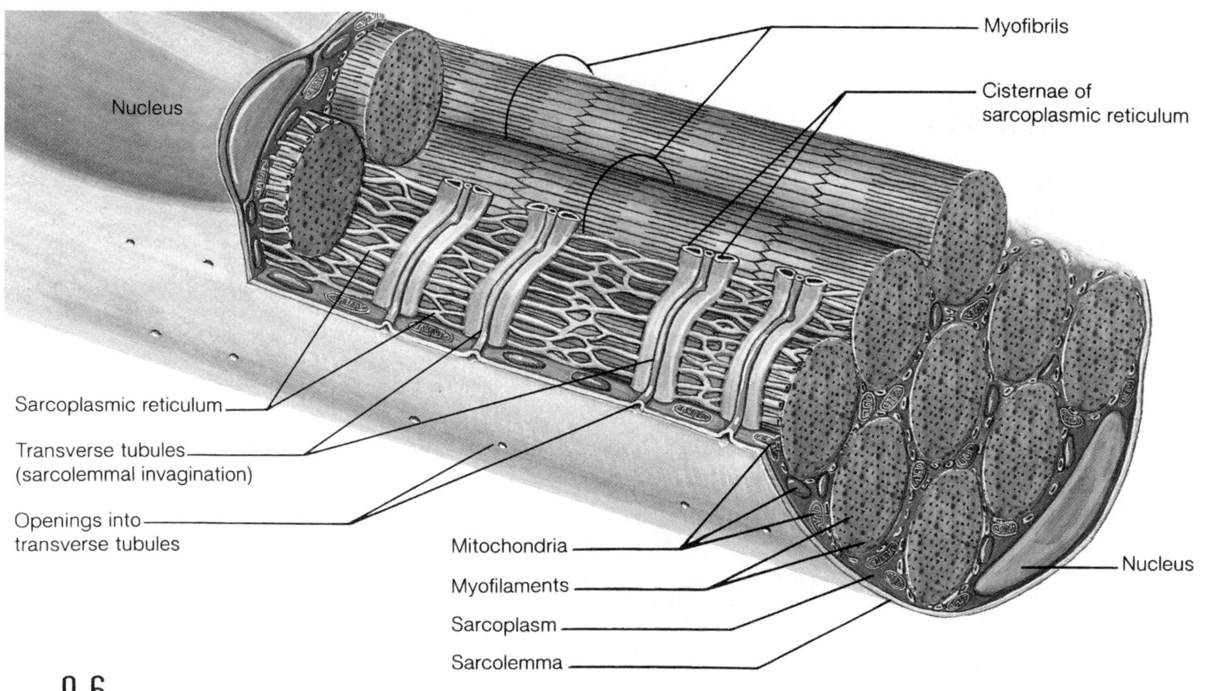


**FIGURE 9.6**
Within the sarcoplasm of a skeletal muscle fiber are a network of sarcoplasmic reticulum and a system of transverse tubules.

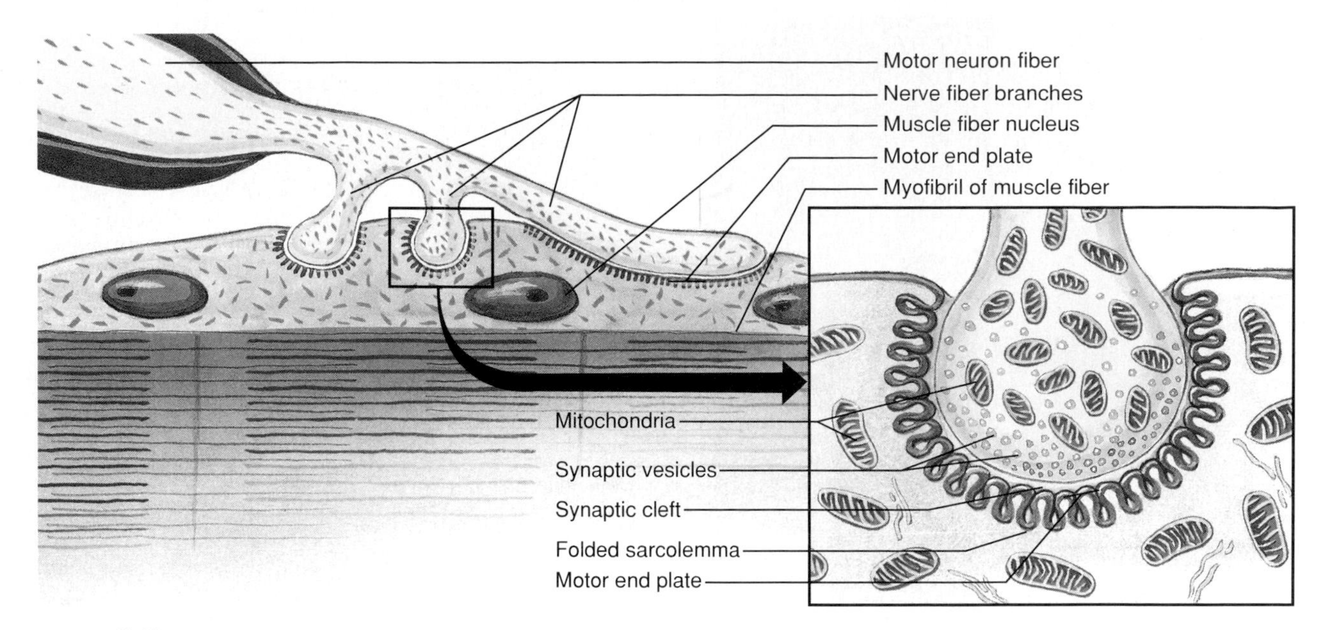


**FIGURE 9.7**

A neuromuscular junction includes the end of a motor neuron and the motor end plate of a muscle fiber.

## Neuromuscular Junction

Each skeletal muscle fiber is connected to an extension of a **motor neuron** that passes outward from the brain or spinal cord. Usually a skeletal muscle fiber contracts only upon stimulation by a motor neuron.

The site where the nerve fiber and muscle fiber meet is called a **neuromuscular junction** (myoneural junction). There, the muscle fiber membrane is specialized to form a **motor end plate** where nuclei and mitochondria are abundant, and the sarcolemma is extensively folded (fig. 9.7).

The end of the motor nerve fiber branches, projecting into recesses of the muscle fiber membrane at the motor end plate. The membrane of the nerve fiber and the membrane of the muscle fiber are separated by a small gap called the **synaptic cleft.** The cytoplasm at the distal ends of the nerve fibers is rich in mitochondria and contains many tiny vesicles (synaptic vesicles) that store chemicals called **neurotransmitters.**

When a nerve impulse traveling from the brain or spinal cord reaches the end of a motor nerve fiber, some of the vesicles release a neurotransmitter into the synaptic cleft. This action stimulates the muscle fiber to contract.

## Motor Units

A muscle fiber usually has a single motor end plate. Motor nerve fibers, however, are densely branched. By means of these fibers, one motor fiber may connect to many muscle fibers. Furthermore, when a motor nerve fiber transmits an impulse, all of the muscle fibers connected to it are stimulated to contract simultaneously. Together, a motor neuron and the muscle fibers it controls constitute a **motor unit** (fig. 9.8).

In the summer months of the early 1950s, parents in the United States lived in terror of their children contracting poliomyelitis, a viral infection that attacks nerve cells that stimulate skeletal muscles to contract. In 50% of the millions of affected children, fever, headache, and nausea rapidly progressed to a stiffened back and neck, drowsiness, and then the feared paralysis, usually of the lower limbs or muscles that control breathing or swallowing. Today, many a middle-aged person with a limp owes this slight disability to polio. Vaccines introduced in the middle 1950s ended the nightmare of polio—or so we thought.

Today, a third of the 1.6 million polio survivors in the United States are experiencing the fatigue, muscle weakness and atrophy, and difficulty breathing of post-polio syndrome. Researchers do not yet know whether this condition is a reactivation of the past infection, a new infection, or the result of aging on nerve cells damaged in childhood.

The number of muscle fibers in a motor unit varies considerably. The fewer muscle fibers in the motor units, however, the finer the movements that can

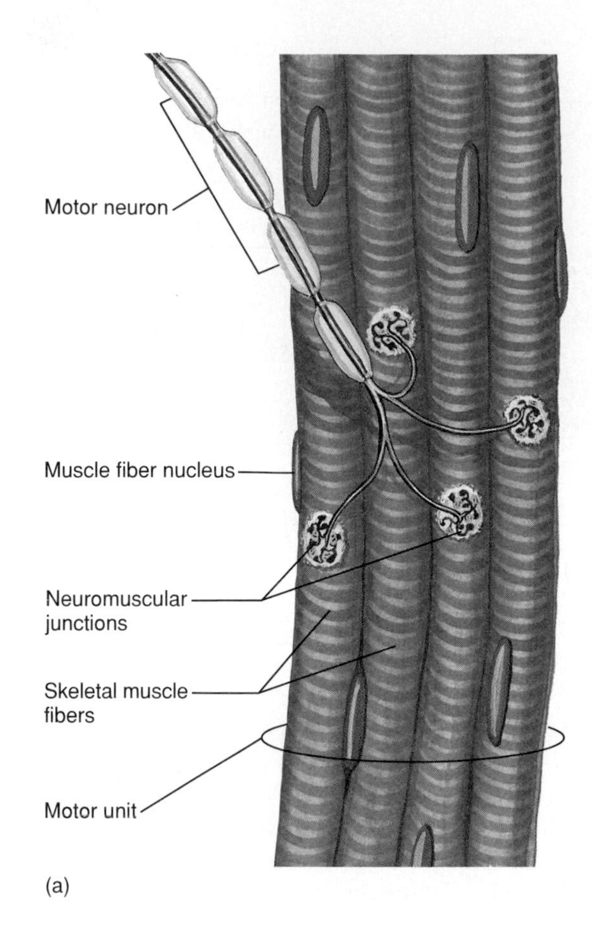


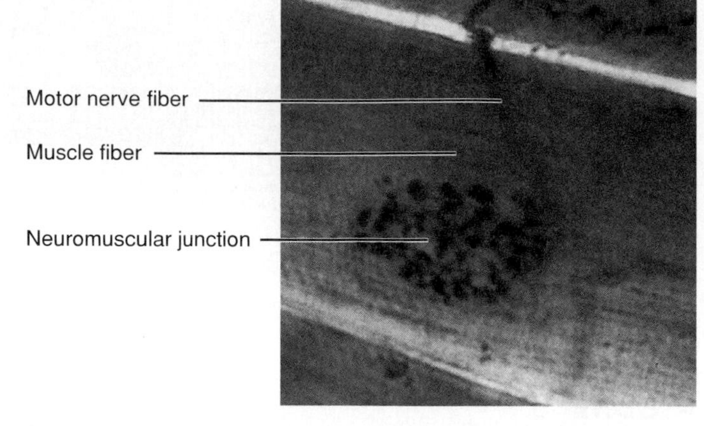


**FIGURE 9.8**
(*a*) A motor unit consists of one motor neuron and all the muscle fibers with which it communicates. (*b*) Micrograph of a neuromuscular junction (260×).

be produced in a particular muscle. For example, the motor units of the muscles that move the eyes may contain fewer than ten muscle fibers per motor unit and can produce very slight movements. Conversely, the motor units of the large muscles in the back may include a hundred or more muscle fibers. When these motor units are stimulated, the movements produced are coarse in comparison with those of the eye muscles.

## SKELETAL MUSCLE CONTRACTION

A muscle fiber contraction is a complex action involving a number of cellular and chemical constituents. The final result is a movement within the myofibrils in which the filaments of actin and myosin slide past one another, shortening sarcomeres. When this happens, the muscle fiber shortens and pulls its attachments.

### Role of Myosin and Actin

Myosin accounts for about two-thirds of the protein within skeletal muscles. A myosin molecule is composed of two twisted protein strands with globular parts called cross-bridges projecting outward along their lengths. Many of these molecules comprise a myosin filament. In the presence of calcium ions, the myosin cross-bridges react with actin filaments and form linkages with them. This reaction between the myosin and actin filaments provides the force that shortens myofibrils during muscle contraction.

When adolescents or young adults die suddenly of heart failure, the cause is often familial hypertrophic cardiomyopathy. At the molecular level, this type of inherited heart disease is caused by a defect in the myosin cross-bridge at the point where it would normally contact an actin rod. As a result, actin cannot slide past myosin quickly enough. The heart muscle overgrows (hypertrophies) to compensate for its inefficiency.

Actin accounts for about one-fourth of the total protein in skeletal muscle. An actin molecule is a globular structure with a binding site to which the cross-bridges of the myosin molecules can attach. Many of these actin molecules, arranged together in a double twisted strand (helix), form an actin filament.

Two other types of proteins, **tropomyosin** and **troponin,** associate with actin filaments. Tropomyosin molecules are rod-shaped and occupy the longitudinal

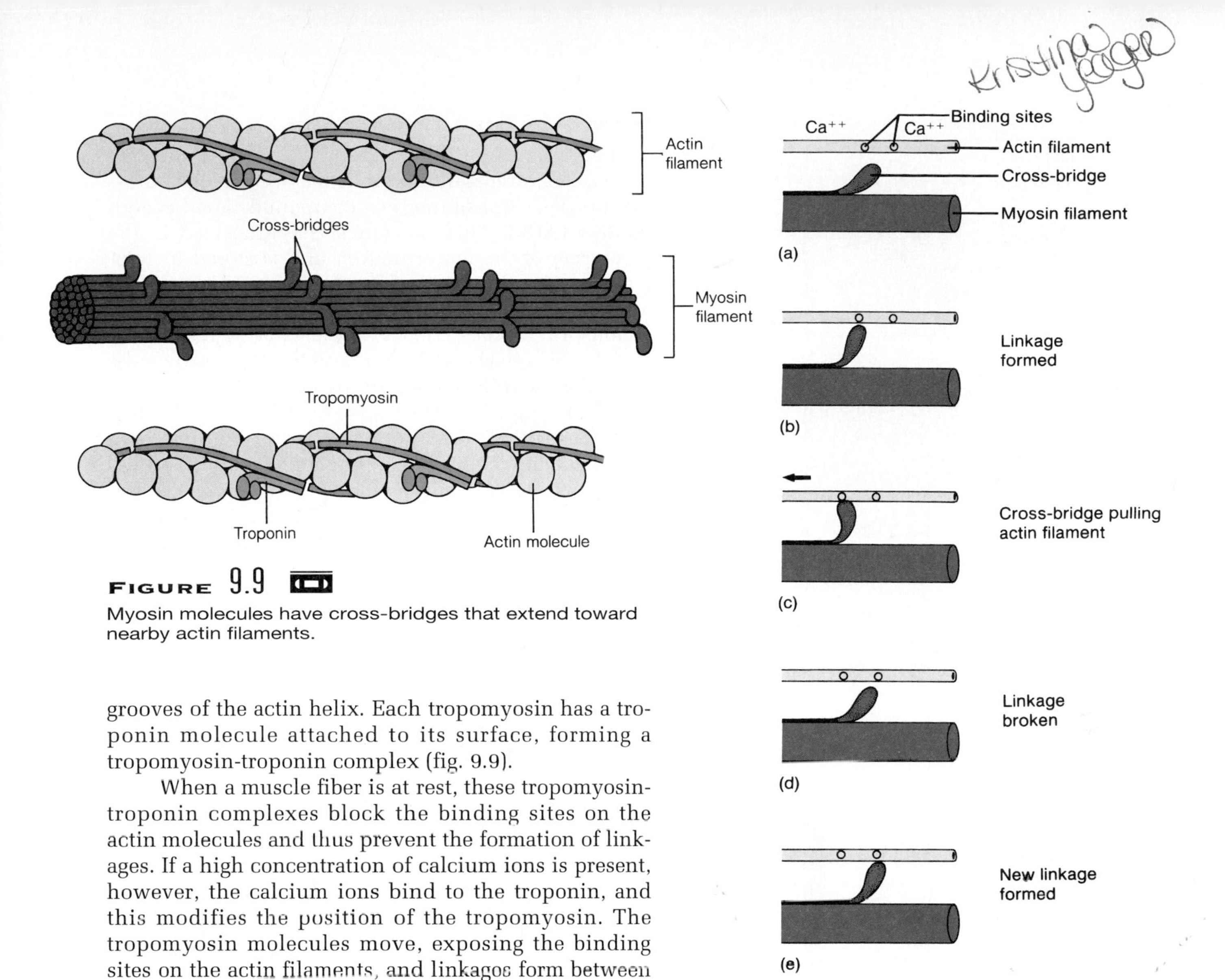


**FIGURE 9.9**

Myosin molecules have cross-bridges that extend toward nearby actin filaments.

**FIGURE 9.10**

According to the sliding filament theory, (*a*) when calcium ions are present, binding sites on an actin filament are exposed. (*b*) Cross-bridges on a myosin filament form linkages at the binding sites. (*c*) A myosin cross-bridge bends slightly, pulling an actin filament, using energy from ATP. (*d*) The linkage breaks, and (*e*) the myosin cross-bridge forms a linkage with the next binding site.

grooves of the actin helix. Each tropomyosin has a troponin molecule attached to its surface, forming a tropomyosin-troponin complex (fig. 9.9).

When a muscle fiber is at rest, these tropomyosin-troponin complexes block the binding sites on the actin molecules and thus prevent the formation of linkages. If a high concentration of calcium ions is present, however, the calcium ions bind to the troponin, and this modifies the position of the tropomyosin. The tropomyosin molecules move, exposing the binding sites on the actin filaments, and linkages form between the actin and myosin filaments.

The **sliding filament theory** of muscle contraction suggests that the head of a myosin cross-bridge can attach to an actin binding site and bend slightly, pulling the actin filament with it. Then the head can release, straighten itself, and combine with another binding site farther down the actin filament (fig. 9.10).

The cross-bridges of myosin filaments contain the enzyme **ATPase,** which catalyzes the breakdown of ATP to ADP and phosphate. This reaction releases energy (see chapter 4) that provides the force with which a cross-bridge pulls. When another ATP binds to the cross-bridge, it releases from the actin binding site before this energy-releasing reaction occurs again. Thus, according to the theory, this cycle can be repeated over and over. As the actin filaments are pulled toward the center of the sarcomere, the sarcomere shortens (fig. 9.11).

Actin, myosin, troponin, and tropomyosin are abundant in muscle cells. Scarcer proteins are also vital to muscle function. This is the case for a rod-shaped muscle protein called **dystrophin.** It accounts for only 0.002% of total muscle protein in skeletal muscle, but its absence causes the devastating inherited disorder Duchenne muscular dystrophy, a childhood disease that usually affects boys. Dystrophin binds to the inside face of muscle cell membranes, supporting them against the powerful force of contraction. Without even these minute amounts of dystrophin, muscle cells burst and die. Clinical Application 9.1 explores how genetic researchers are now attempting to restore dystrophin to boys with this illness.

## Stimulus for Contraction

A skeletal muscle fiber normally does not contract until a neurotransmitter stimulates it. In skeletal muscle, the neurotransmitter is a compound called **acetylcholine (ACh).** This substance is synthesized in the cytoplasm of the motor neuron and is stored in vesicles near the distal end of its motor nerve fiber. When a nerve impulse (or *action potential,* described in chapter 10) reaches the end of the axon, some of these vesicles release acetylcholine into the gap between the nerve fiber and the motor end plate (see fig. 9.7).

The acetylcholine diffuses rapidly across the gap, combines with certain protein molecules (receptors) in the sarcolemma, and thus stimulates the muscle fiber membrane. As a result of this stimulus, a **muscle impulse,** very much like a nerve impulse, is generated, and passes in all directions over the surface of the sarcolemma. It also travels through the transverse tubules, deep into the fiber, and reaches the sarcoplasmic reticulum.

The sarcoplasmic reticulum contains a high concentration of calcium ions compared to the sarcoplasm. This is due to active transport of calcium ions (calcium pump) in the membrane of the sarcoplasmic reticulum. In response to a muscle impulse, the membranes of the cisternae become more permeable to these ions, and the calcium ions diffuse into the sarcoplasm of the muscle fiber.

> If acetylcholine receptors at the motor end plate are decreased in number, or blocked, muscles cannot receive the signal to contract. This may occur as the result of a disease, such as myasthenia gravis, or a poison, such as nerve gas. A drug called pyridostigmine bromide is used to treat myasthenia gravis. The drug inhibits acetylcholinesterase, keeping acetylcholine around longer, stimulating muscles. It was given to veterans of the Persian Gulf War who complained of muscle aches in the months following their military service. Health officials reasoned that the drug's effect on myasthenia gravis might also help restore muscle function if the war veterans' symptoms arose from exposure to nerve gas during the military conflict. Acetylcholinesterase inhibitors are also used as insecticides. The buildup of acetylcholine causes an insect to twitch violently, then die.

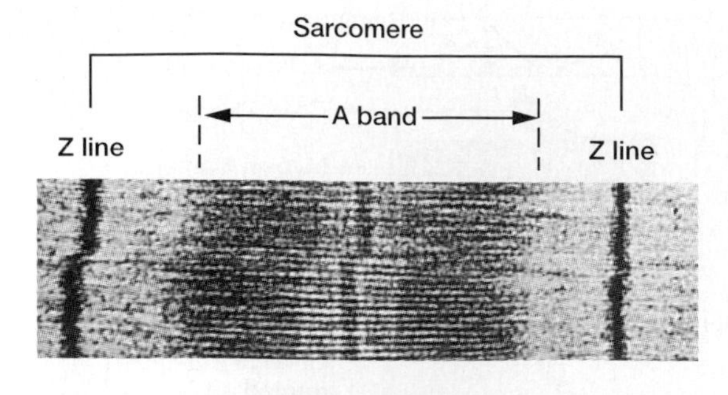


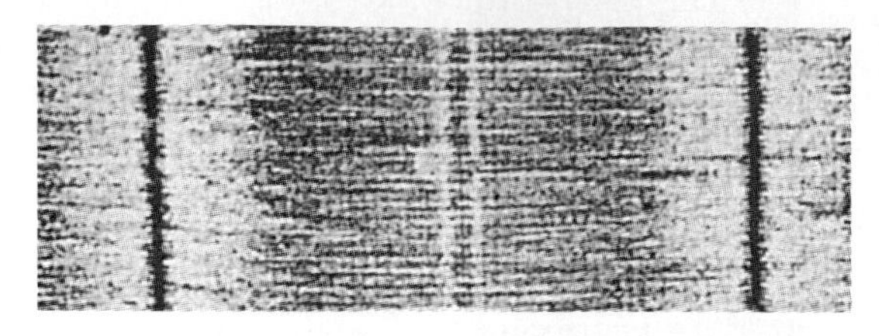

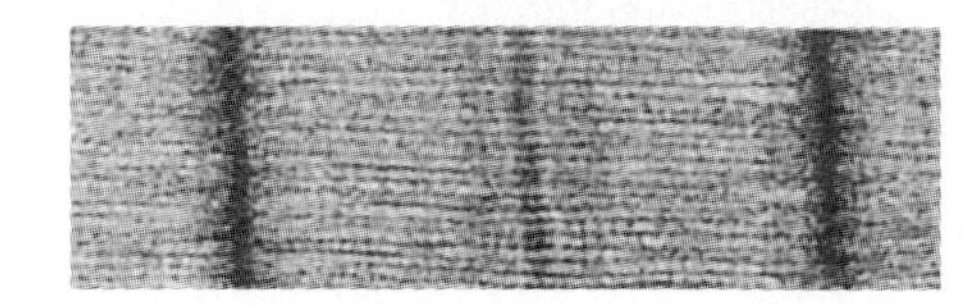

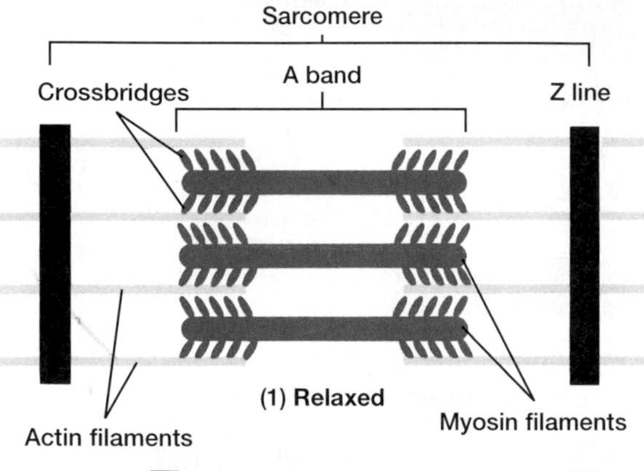


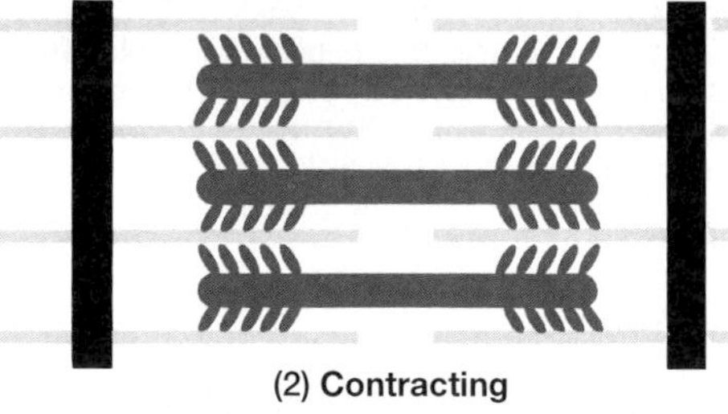


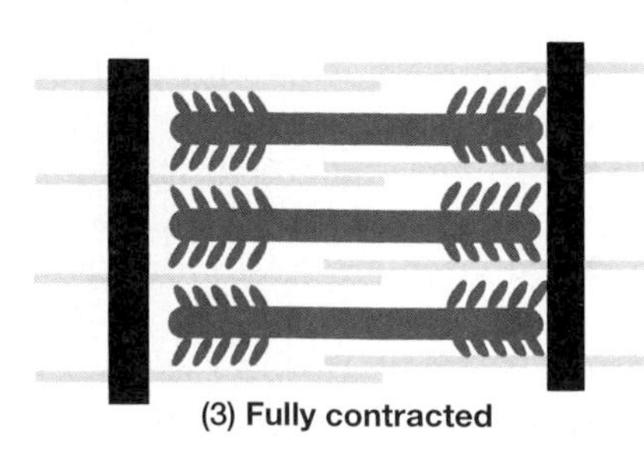


**FIGURE 9.11**

When a skeletal muscle contracts, individual sarcomeres shorten as thick and thin filaments slide past one another (about 60,000×).

CLINICAL APPLICATION 9.1

## Deficient Dystrophin Causes Duchenne Muscular Dystrophy

On a bright April morning in 1990, 9-year-old Sam Looper wiggled a toe on his left foot and made medical history. Toe wiggling is not usually cause for celebration, but in Sam's case, it was quite astounding, for he had Duchenne muscular dystrophy (DMD), and even this simple movement had been impossible for him.

It usually takes years for parents to realize that their son has DMD. The disease usually affects boys. At first the child is merely slow and clumsy, lagging behind playmates or having difficulty navigating stairs. Because his lower limb muscles appear firm and even prominent, concerned parents often do not suspect a muscular problem, yet already fat is filling in for progressively weakening muscles, making them appear deceptively robust.

A young child with DMD develops an odd but characteristic way of standing, pulling himself with his hands and gradually creeping to an erect stance. His calf muscles are too weak to power sitting and standing in the usual manner. This "Gower sign" usually alerts a physician, often during a routine physical exam, to suspect DMD. Laboratory tests confirm the diagnosis. Tests may reveal abnormal enzyme levels indicating muscle breakdown, or they may identify the abnormal gene that causes DMD. The gene is passed from the mother, who is a carrier, or arises in the boy as a spontaneous mutation (genetic change).

DMD causes progressive helplessness. The boy is typically wheelchair-bound by early adolescence and usually dies by age 20 of respiratory or heart failure, because smooth and cardiac muscle lack dystrophin too. Steroid drugs and exercise can forestall symptoms somewhat, but there is no cure.

This bleak prognosis prompted Sam Looper's family to allow him to undergo a controversial therapy. Sam received a transplant of his father's healthy, dystrophin-producing immature muscle cells (myoblasts) in four of his toes. If side effects arose, they would harm only a small part of his body. After the treatment, Sam could move his toes.

Sam's toe-wiggling ability was disappointing, however, for two reasons. First, the myoblasts do not migrate from where they are injected, so treating bodywide muscular dystrophy is difficult. Second, Sam's immune system attacked the transplanted cells, even though they came from a close relative.

To combat these problems, researchers in further trials implanted boys with myoblasts at many sites in their legs, thighs, and buttocks and gave them cyclosporine, a drug to prevent immune rejection. Scientists are not yet sure whether modest gains in strength seen in 43% of the treated muscles are due to the transplanted myoblasts or to the cyclosporine. Nor can myoblast implants in skeletal muscle treat the more life-threatening deficit in cardiac and smooth muscle. A bigger challenge is to add the needed dystrophin to muscles in and around vital organs, such as the heart, lungs, and digestive tract.

When a relatively high concentration of calcium ions is present in the sarcoplasm, linkages form between the actin and myosin filaments, and a muscle contracts. The contraction, which also requires ATP, continues as long as nerve impulses cause ACh release. When the nerve impulses cease, two events lead to muscle relaxation.

First, the acetylcholine that stimulated the muscle fiber in the first place is rapidly decomposed by action of an enzyme called **acetylcholinesterase.** This enzyme is present at the neuromuscular junction on the membranes of the motor end plate. The action of acetylcholinesterase prevents a single nerve impulse from continuously stimulating the muscle fiber.

Second, when acetylcholine breaks down, the stimulus to the sarcolemma and membranes within the muscle fiber ceases. As a result, the calcium pump quickly moves calcium ions back into the sarcoplasmic reticulum, decreasing the calcium ion concentration of the sarcoplasm. The linkages break, and, once again, the troponin and tropomyosin molecules inhibit the interaction between the filaments. Consequently, the muscle fiber relaxes.

## CHART 9.1 Major Events of Muscle Contraction and Relaxation

| Muscle Fiber Contraction | Muscle Fiber Relaxation |
|---|---|
| 1. The distal end of a motor neuron releases acetylcholine. | 1. Acetylcholinesterase decomposes acetylcholine, and the muscle fiber membrane is no longer stimulated. |
| 2. Acetylcholine diffuses across the gap at the neuromuscular junction. | 2. Calcium ions are actively transported into the sarcoplasmic reticulum. |
| 3. The sarcolemma is stimulated, and a muscle impulse travels over the surface of the muscle fiber and deep into the fiber through the transverse tubules and reaches the sarcoplasmic reticulum. | 3. Linkages between actin and myosin filaments break. |
| 4. Calcium ions diffuse from the sarcoplasmic reticulum into the sarcoplasm and bind to troponin molecules. | 4. Troponin and tropomyosin molecules inhibit the interaction between myosin and actin filaments. |
| 5. Tropomyosin molecules move and expose specific sites on actin filaments. | 5. Actin and myosin filaments slide apart. |
| 6. Actin and myosin filaments form linkages. | 6. Muscle fiber relaxes and its resting state is reestablished. |
| 7. Actin filaments are pulled inward by myosin cross-bridges. | |
| 8. Muscle fiber shortens as a contraction occurs. | |

In September 1985, two teenage tourists from Hong Kong went to the emergency room at Montreal Children's Hospital complaining of extreme nausea and weakness. Although doctors released them when they could not identify a cause of the symptoms, the girls returned that night—far sicker. Now they were becoming paralyzed and had difficulty breathing. This time, physicians recognized symptoms of botulism.

Botulism occurs when the bacterium *Clostridium botulinum* grows in an anaerobic (oxygen-poor) environment, such as in a can of food. The bacteria produce a toxin. Symptoms trace the path of the toxin: nausea, vomiting, and diarrhea (digestive tract); headache, dizziness, and blurred or double vision (nervous system); and finally, weakness, hoarseness, and difficulty swallowing and, eventually, breathing (muscular system). Fortunately, physicians can administer an antitoxin substance that binds to and inactivates botulinum toxin in the bloodstream, stemming further symptoms, although not correcting damage already done.

Prompt treatment saved the touring teens, and astute medical detective work led to a restaurant in Vancouver where the sisters and thirty-four others had eaten roast beef sandwiches. The bread had been coated with a garlic-butter spread. The garlic was bottled with soybean oil and should have been refrigerated. It was not. With bacteria that the garlic had picked up in the soil where it grew, and eight months sitting outside of the refrigerator, conditions were just right for *C. botulinum* to produce its deadly toxin.

Chart 9.1 summarizes the major events leading to muscle contraction and relaxation.

1. Describe a neuromuscular junction.
2. Define a motor unit.
3. List four proteins associated with myofibrils, and explain their relationships.
4. Explain how the filaments of a myofibril interact during muscle contraction.
5. Explain how a motor nerve impulse can trigger a muscle contraction.

## Energy Sources for Contraction

The energy used during muscle fiber contraction comes from ATP molecules. This energy makes possible the reaction between the actin and myosin filaments. However, a muscle fiber has only enough ATP to contract briefly. Therefore, when a fiber is active, ATP must be regenerated.

The initial source of energy available to regenerate ATP from ADP and phosphate is **creatine phosphate.** Like ATP, creatine phosphate contains high-energy phosphate bonds, and it is actually four to six times more abundant in muscle fibers than ATP. Creatine phosphate, however, cannot directly supply energy to a cell's energy-utilizing reactions. Instead, it stores excess energy released from mitochondria. Thus, whenever sufficient ATP is present, an enzyme in the mitochondria (creatine phosphokinase) promotes the synthesis of creatine phosphate, which stores the excess energy in its bonds (fig. 9.12).

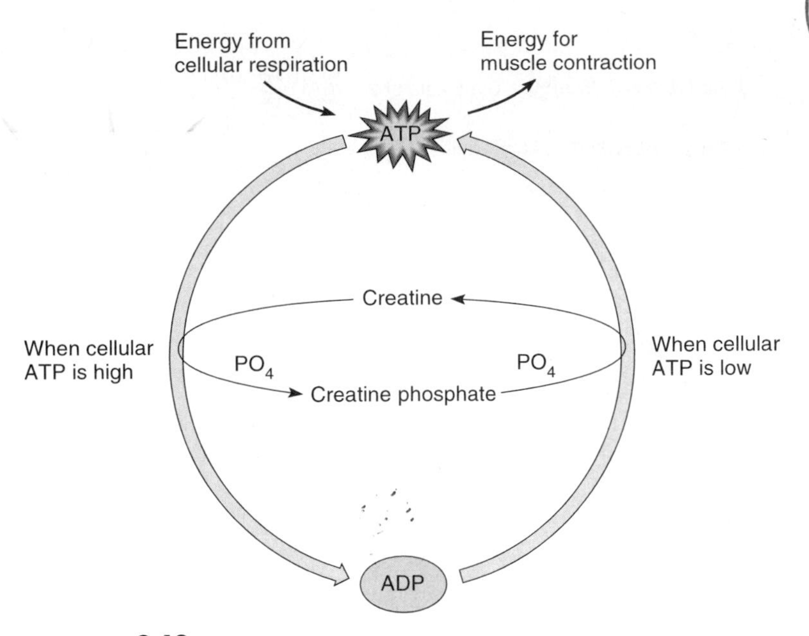


**FIGURE 9.12**
A muscle cell uses energy released in cellular respiration to synthesize ATP. ATP is then used to power muscle contraction, or to synthesize creatine phosphate. Later, creatine phosphate may be used to synthesize ATP.

As ATP is decomposed, the energy from creatine phosphate is transferred to ADP molecules, and they, in turn, are quickly converted back into ATP. The amount of ATP and creatine phosphate present in a skeletal muscle, however, is usually not sufficient to support maximal muscle activity for more than a few seconds. As a result, the muscle fibers in an active muscle soon become dependent upon cellular respiration of glucose as a source of energy for synthesizing ATP. Typically, a muscle stores glucose in the form of glycogen.

## Oxygen Supply and Cellular Respiration

Recall from chapter 4 that the early phase of cellular respiration occurs in the cytoplasm and is *anaerobic*, taking place in the absence of oxygen. This phase only partially breaks down energy-supplying molecules such as glucose, and releases only a few ATP molecules. The complete breakdown of glucose occurs in the mitochondria and is *aerobic*. This process, which includes the complex series of reactions of the *citric acid cycle*, produces many ATP molecules.

Blood carries the oxygen necessary to support aerobic respiration from the lungs to body cells. Oxygen is transported within the red blood cells, loosely bound to molecules of hemoglobin, the pigment responsible for the red color of blood. In regions of the body where the oxygen concentration is relatively low, oxygen is released from hemoglobin and becomes available for cellular respiration.

Another pigment, **myoglobin**, is synthesized in the muscle cells and is responsible for the reddish brown color of skeletal muscle tissue. Like hemoglobin, myoglobin can combine loosely with oxygen, and in fact has a greater attraction for oxygen than does hemoglobin. Myoglobin can store oxygen in muscle tissue, at least temporarily. This ability to store oxygen reduces a muscle's need for a continuous blood supply during muscular contraction. This is important because blood flow may decrease during muscular contraction, as nearby contracting muscle fibers compress blood vessels (fig. 9.13).

## Oxygen Debt

When a person is resting or moderately active, the respiratory and circulatory systems can usually supply sufficient oxygen to the skeletal muscles to support aerobic respiration. However, when skeletal muscles are used strenuously for even a minute or two, these systems usually cannot supply enough oxygen to meet the needs of aerobic respiration. The muscle fibers must depend increasingly on the anaerobic phase of respiration for energy.

Chapter 4 discussed how anaerobic respiration breaks glucose down to pyruvic acid and converts it to lactic acid, which diffuses out of the muscle fibers and is carried to the liver by the blood. The liver cells can change the lactic acid into *glucose*, but this requires energy from ATP. During strenuous exercise, available oxygen is used primarily to synthesize ATP for muscle contraction rather than to make ATP for converting lactic acid into glucose. Consequently, as lactic acid accumulates, a person develops an **oxygen debt** that must be repaid at a later time. The amount of oxygen debt equals the amount of oxygen liver cells need to convert the accumulated lactic acid into glucose, plus the amount the muscle cells need to resynthesize ATP and creatine phosphate, and return these substances to their original concentrations (fig. 9.14).

The conversion of lactic acid back into glucose is relatively slow. It may require several hours to repay an oxygen debt following strenuous exercise.

## Muscle Fatigue

A muscle exercised strenuously for a prolonged period may lose its ability to contract, a condition called *fatigue*. This condition may result from an interruption in the muscle's blood supply or, rarely, from exhaustion of

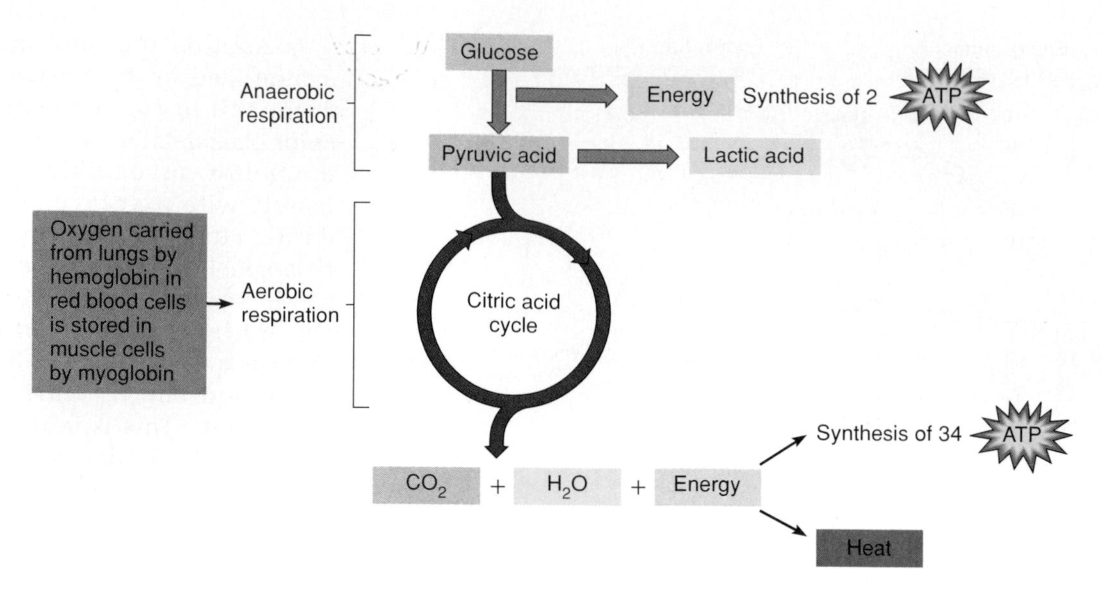


**FIGURE 9.13**

The oxygen required to support aerobic respiration is carried in the blood and stored in muscle cells in myoglobin. In the absence of sufficient oxygen, pyruvic acid is converted to lactic acid. The maximum number of ATPs generated per glucose molecule varies with cell type, and is 36 in skeletal muscle.

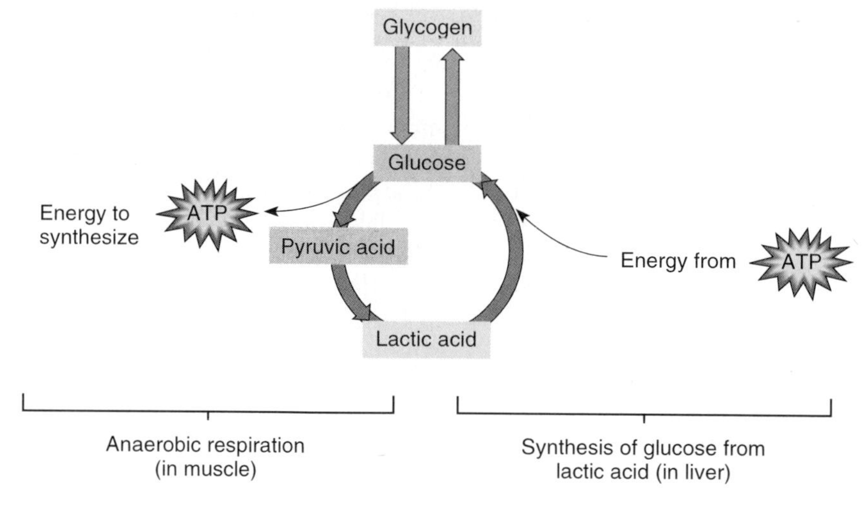


**FIGURE 9.14**

Liver cells can convert lactic acid, generated by muscles respiring anaerobically, into glucose.

the supply of acetylcholine in its motor nerve fibers. However, muscle fatigue is most likely to arise from an accumulation of lactic acid in the muscle as a result of anaerobic respiration (see chapter 4). The lowered pH from the lactic acid prevents muscle fibers from responding to stimulation.

Occasionally a muscle becomes fatigued and develops cramps at the same time. A cramp is a painful condition in which the muscle contracts spasmodically, but does not relax completely. This condition is due to a lack of ATP, which is needed to move calcium ions back into the sarcoplasmic reticulum and break the linkages between actin and myosin filaments before muscle fibers can relax.

Tolerance to lactic acid varies, so that some persons experience muscular fatigue more quickly than others. Resistance to the effects of lactic acid also varies among athletes, although in general, athletes can exercise and produce less lactic acid than nonathletes. In part, this is because the strenuous exercise of physical training stimulates new capillaries to grow within the muscles, supplying more oxygen and nutrients to the muscle fibers of athletes. Such physical training also causes muscle fibers to produce additional mitochondria, increasing their ability to carry on aerobic respiration. In addition, some muscle fibers may be more likely to accumulate lactic acid than others, as described below.

A few hours after death, the skeletal muscles partially contract, fixing the joints. This condition, called *rigor mortis,* may continue for seventy-two hours or more. It results from an increase in membrane permeability to calcium ions, which promotes cross-bridge attachment, and a decrease in availability of ATP in the muscle fibers, which prevents cross-bridge disruption. Thus, the actin and myosin filaments of the muscle fibers remain linked until the muscles begin to decompose.

## Fast and Slow Muscles

Muscles vary in contraction speed. Those that move the eyes contract about ten times faster than those responsible for maintaining posture, and the muscles that move the limbs contract at intermediate rates. Thus, the speed of contraction reflects the specialized function of the muscle.

The slow-contracting (slow-twitch) postural muscles, such as the long muscles of the back, are often called *red muscles* because most of their fibers contain the red, oxygen-storing pigment myoglobin. These fibers also are well supplied with blood, which brings oxygen. In addition, red muscle fibers contain many mitochondria, an adaptation for aerobic respiration. These fibers have a high respiratory capacity in that they can generate ATP fast enough to keep up with the ATP breakdown that occurs when they contract. For this reason, they can contract for prolonged periods without fatiguing.

Fast-contracting (fast-twitch) muscles are called *white muscles* because they contain less myoglobin and have a poorer blood supply than red muscles. They include certain hand muscles as well as those that move the eyes. Most of the fibers of these muscles have fewer mitochondria and thus have a reduced respiratory capacity. However, the white fibers have more extensive sarcoplasmic reticulum, which stores and reabsorbs calcium ions, and their ATPase has higher activity than that of red fibers. Because of these factors, white muscle fibers can contract rapidly, though they tend to fatigue as ATP and the biochemicals needed to regenerate ATP are depleted and as lactic acid accumulates.

Some muscles contain a predominate number of slow-contracting red fibers or fast-contracting white fibers, but most skeletal muscles have both types. Also, researchers have identified muscle fibers with intermediate characteristics.

Birds that migrate long distances have abundant dark, slow-twitch muscles—this is why their meat is dark. In contrast, chickens that can only flap around the barnyard have abundant fast-twitch muscles, and mostly white meat.

Alberto Salazar, who holds many first place finishes in long-distance running events, is the human equivalent of the migrating bird. His muscles contain 93% slow-twitch fibers! In some European nations, athletic coaches measure slow-twitch to fast-twitch muscle fiber ratios to predict who will excel at long-distance events and who will fare better in sprints.

## Heat Production

Heat is a by-product of cellular respiration, so all active cells generate it. Since muscle tissue represents such a large proportion of the total body mass, it is a major source of heat.

Only about 25% of the energy released in cellular respiration is available for use in metabolic processes; the rest becomes heat. Thus, active muscles release large amounts of heat. Blood transports this heat to other tissues, helping to maintain body temperature. Homeostatic mechanisms promote heat loss when heat is in excess (see chapters 1 and 6).

1. What are the sources of energy used to regenerate ATP?
2. What are the sources of the oxygen needed for aerobic respiration?
3. How are lactic acid, oxygen debt, and muscle fatigue related?
4. Distinguish between fast-contracting and slow-contracting muscles.
5. What is the relationship between cellular respiration and heat production?

# Muscular Responses

One way to observe muscle contraction is to remove a single muscle fiber from a skeletal muscle and connect it to a device that senses and records changes in the fiber's length. An electrical stimulator is usually used to promote muscle contraction.

## Threshold Stimulus

When an isolated muscle fiber is exposed to a series of stimuli of increasing strength, the fiber remains unresponsive until a certain strength of stimulation is applied. This minimal strength needed to cause a contraction is called the **threshold stimulus.** An impulse in a motor neuron normally releases enough ACh to bring the muscle fibers in its motor unit to threshold.

## All-or-None Response

A muscle fiber exposed to a stimulus of threshold strength or above responds to its fullest extent. Increasing the strength of the stimulus does not affect the degree to which the fiber contracts. In other words, a muscle fiber does not partially contract—if it contracts at all, it contracts completely, even though it may not shorten completely. This phenomenon is called the **all-or-none response.**

## Recording a Muscle Contraction

To record how a whole muscle responds to stimulation, a skeletal muscle can be removed from a frog or other small animal and mounted in a special laboratory apparatus. The muscle is then stimulated electrically, and when it contracts, it pulls on a lever. The lever's movement is recorded, and the resulting pattern is called a **myogram.**

If a muscle is exposed to a single stimulus of sufficient strength to activate some of its motor units, the muscle will contract and then relax. This action—a single contraction that lasts only a fraction of a second—is called a **twitch.** A twitch produces a myogram like that in figure 9.15. Note that there was a delay between the time the stimulus was applied and the time the muscle responded. This is the **latent period.** In a frog muscle, the latent period lasts for about 0.01 second; in a human muscle, it is even shorter. The latent period is followed by a *period of contraction* during which the muscle pulls at its attachments, and a *period of relaxation* during which it returns to its former length.

If a muscle is exposed to two stimuli (of threshold strength or above) in quick succession, it may respond with a twitch to the first stimulus but not to the second. This is because it takes an instant following a contraction for the muscle fibers to become responsive to further stimulation. Thus, for a very brief moment following stimulation, a muscle remains unresponsive. This time is called the **refractory period.**

## Staircase Effect

The force a muscle fiber exerts in a twitch may depend on whether it has recently been stimulated to contract. A muscle fiber that has been inactive can be subjected to a series of stimuli, such that it undergoes a series of twitches with complete relaxation in between (fig. 9.16*a*). However, the strength of each successive contraction increases slightly, soon reaching a maximum. This phenomenon is called the **staircase effect.**

The staircase effect seems to involve a net increase in the concentration of calcium ions available in the sarcoplasm of the muscle fibers. This increase might occur if each stimulus in the series caused the release of calcium ions and if the sarcoplasmic reticulum failed to recapture those ions immediately.

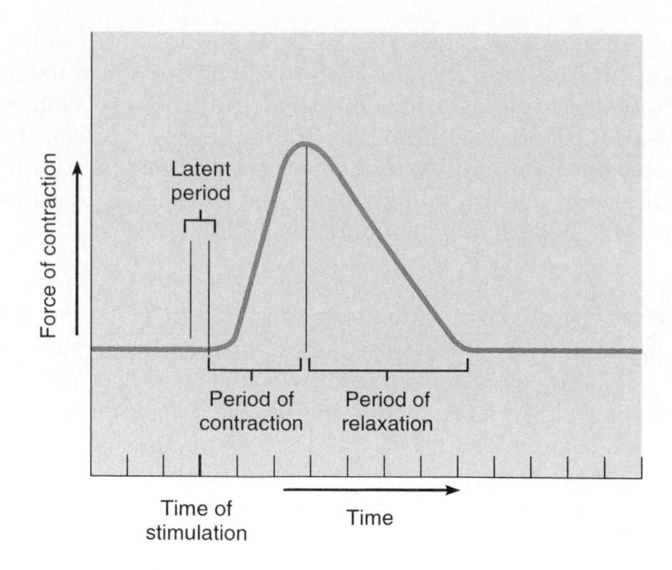


**FIGURE 9.15**
A myogram of a single muscle twitch.

## Summation

The force that a muscle fiber can generate is not limited to the maximum force of a single twitch. A muscle fiber exposed to a series of stimuli of increasing frequency reaches a point when it is unable to completely relax before the next stimulus in the series arrives. When this happens, the individual twitches begin to combine and the muscle contraction becomes sustained. In such a sustained contraction, the force of individual twitches combines by the process of **summation** (fig. 9.16*b*). When the resulting forceful, sustained contraction lacks even partial relaxation, it is called a **tetanic** (te-tan-ik) **contraction** (tetanus) (fig. 9.16*c*).

## Recruitment of Motor Units

Since the muscle fibers within a muscle are organized into motor units, and each motor unit is controlled by a single motor neuron, all the muscle fibers in a motor unit are stimulated at the same time. Therefore, a motor unit also responds in an all-or-none manner. A whole muscle, however, does not behave like this, because it is composed of many motor units controlled

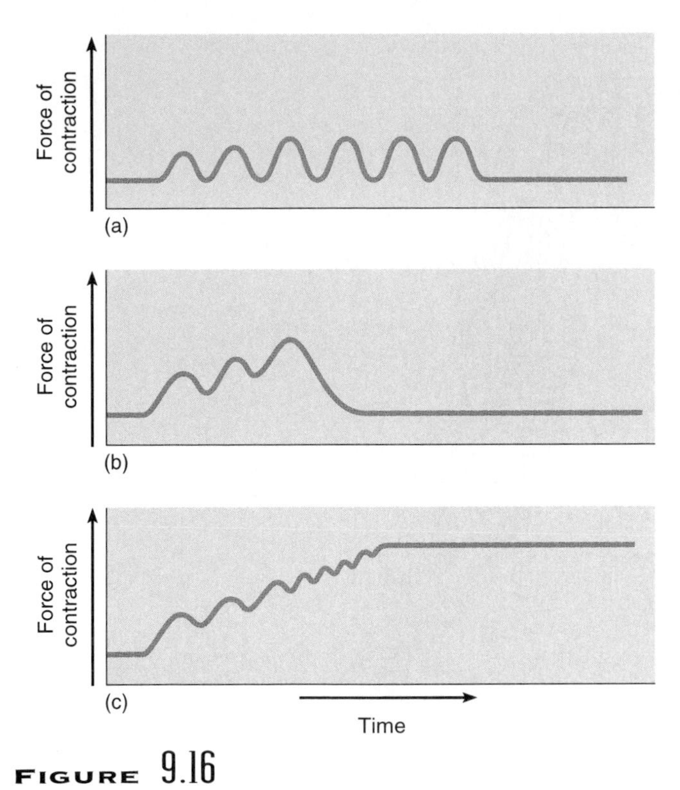


**FIGURE 9.16**

Myograms of (*a*) a series of twitches showing the staircase effect, (*b*) summation, and (*c*) a tetanic contraction. Note that stimulation frequency increases from one myogram to the next.

by different motor neurons, which respond to different thresholds of stimulation. Thus, if only the motor neurons with low thresholds are stimulated, few motor units contract. At higher intensities of stimulation, other motor neurons respond and more motor units are activated. Such an increase in the number of motor units being activated is called **recruitment.** As the intensity of stimulation increases, recruitment of motor units continues until finally all possible motor units are activated and the muscle contracts with maximal tension.

## Sustained Contractions

At the same time that twitches combine, the strength of the contractions may increase due to the recruitment of motor units. The smaller motor units, which have finer fibers, tend to respond earlier in the series of stimuli. The larger motor units, which contain thicker fibers, respond later and more forcefully. The production of a sustained contraction of increasing strength is called *multiple motor unit summation.*

Although twitches may occur occasionally in human skeletal muscles, as when an eyelid twitches, such contractions are of limited use. More commonly, muscular contractions are sustained, and they are smooth rather than irregular or jerky because a mechanism within the spinal cord stimulates contractions in different sets of motor units at different moments. Thus, while some motor units are contracting, others are relaxing.

Tetanic contractions occur frequently in skeletal muscles during everyday activities. In many cases, the condition occurs in only a portion of a muscle. For example, when a person lifts a weight or walks, sustained contractions are maintained in the upper limb or lower limb muscles for varying lengths of time. These contractions are responses to a rapid series of stimuli transmitted from the brain and spinal cord on motor neuron fibers.

Even when a muscle appears to be at rest, a certain amount of sustained contraction is occurring in its fibers. This is called **muscle tone** (tonus), and it is a response to nerve impulses originating repeatedly in the spinal cord and traveling to small numbers of muscle fibers. The result is a continuous state of partial contraction.

Muscle tone is particularly important in maintaining posture. Tautness in the muscles of the neck, trunk, and lower limbs enables a person to hold the head upright, stand, or sit. If tone is suddenly lost, such as when a person loses consciousness, the body will collapse. Although muscle tone is maintained in health, it is lost if motor nerve fibers are cut or if diseases interfere with the conduction of nerve impulses.

When skeletal muscles are contracted very forcefully, they may generate up to 50 pounds of pull for each square inch of muscle cross section. Consequently, large muscles such as those in the thigh can pull with several hundred pounds of force. Occasionally, this force is so great that the tendons of muscles are torn away from their attachments to the bones.

## Isotonic and Isometric Contractions

Sometimes muscles shorten when they contract. For example, if a person lifts an object, the tautness in the muscles remains unchanged, their attached ends pull closer together, and the object is moved. This type of contraction is termed **isotonic.**

At other times, a skeletal muscle contracts, but the parts to which it is attached do not move. This happens, for instance, when a person pushes against the wall of a building. Tension within the muscles increases, but the wall does not move, and the muscles remain the same length. Contractions of this type are called **isometric.** Isometric contractions occur continuously in postural muscles that stabilize skeletal parts and hold the body upright. Figure 9.17 illustrates these two types of contractions.

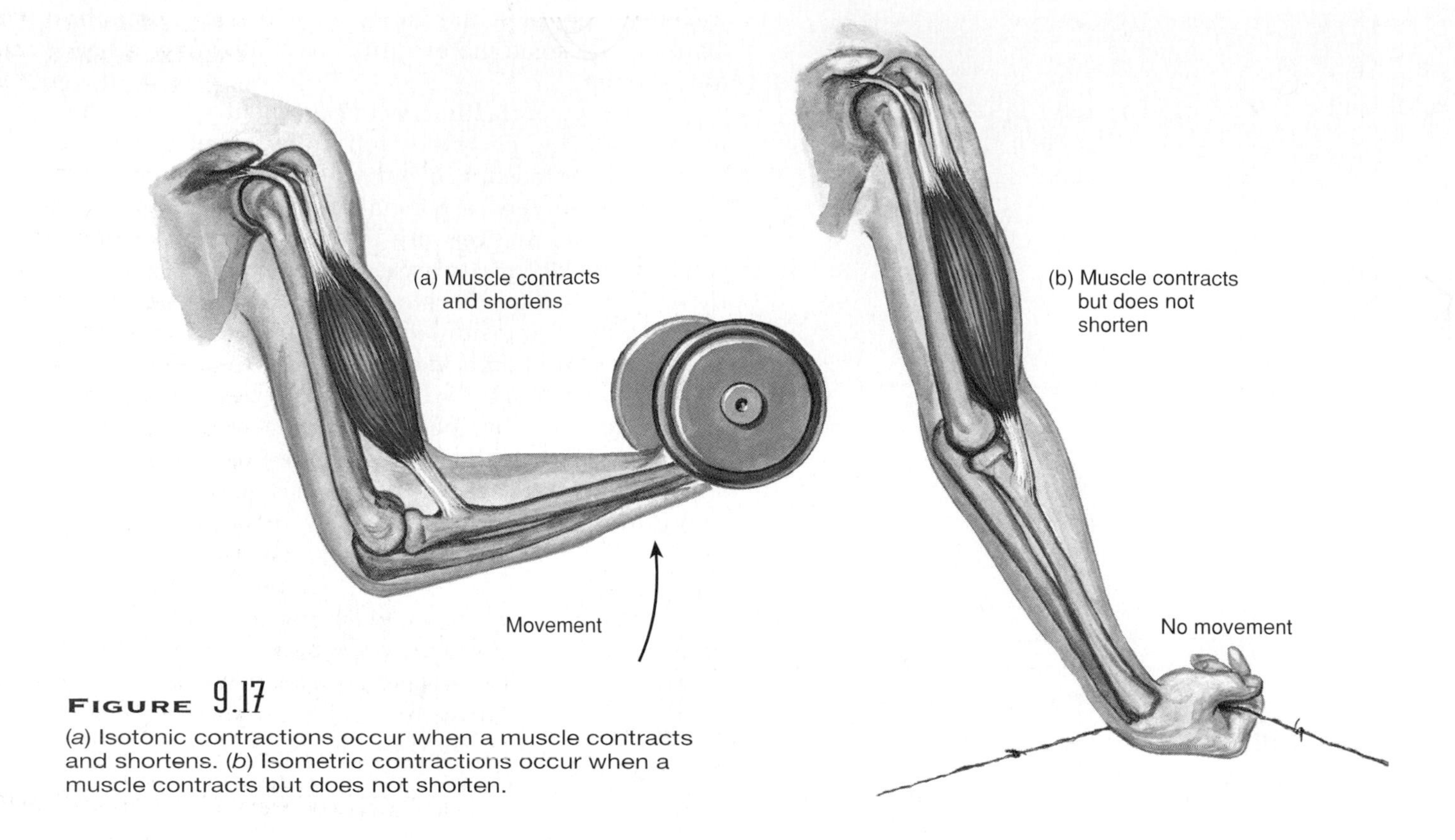


**FIGURE 9.17**
(*a*) Isotonic contractions occur when a muscle contracts and shortens. (*b*) Isometric contractions occur when a muscle contracts but does not shorten.

Skeletal muscles can contract either isotonically or isometrically, and most body actions involve both types of contraction. In walking, for instance, certain leg and thigh muscles contract isometrically and keep the limb stiff as it touches the ground, while other muscles contract isotonically, bending the limb and lifting it upward. Clinical Application 9.2 discusses very noticeable effects of muscle use and disuse.

1. Define threshold stimulus.
2. What is an all-or-none response?
3. Distinguish between a twitch and a sustained contraction.
4. Define muscle tone.
5. Explain the difference between isometric and isotonic contractions.

## SMOOTH MUSCLES

The contractile mechanisms of smooth and cardiac muscles are essentially the same as those of skeletal muscles. However, the cells of these tissues have important structural and functional differences.

### Smooth Muscle Fibers

As discussed in chapter 5, smooth muscle cells are shorter than the fibers of skeletal muscle, and they have single, centrally located nuclei. Smooth muscle cells are elongated with tapering ends and contain filaments of actin and myosin in myofibrils that extend throughout their lengths. However, the filaments are very thin and more randomly arranged than those in skeletal muscle fibers. Therefore, smooth muscle cells lack striations. They also lack transverse tubules, and their sarcoplasmic reticula are not well developed.

The two major types of smooth muscles are multiunit and visceral. In **multiunit smooth muscle,** the muscle fibers are less well organized and occur as separate fibers rather than in sheets. Smooth muscle of this type is found in the irises of the eyes and in the walls of blood vessels. Typically, multiunit smooth muscle contracts only after stimulation by motor nerve impulses or certain hormones.

**Visceral smooth muscle** is composed of sheets of spindle-shaped cells that are held in close contact by gap junctions. The thick portion of each cell is next to the thin parts of adjacent cells. Visceral smooth muscle is the more common type and is found in the walls of hollow organs, such as the stomach, intestines, urinary bladder, and uterus. Usually there are two thicknesses of smooth muscle in the walls of these organs.

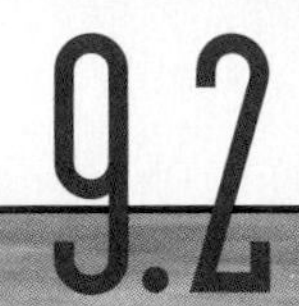

**CLINICAL APPLICATION**

## Use and Disuse of Skeletal Muscles

Skeletal muscles are very responsive to use and disuse. Those that are forcefully exercised tend to enlarge. This phenomenon is called *muscular hypertrophy*. Conversely, a muscle that is not used undergoes *atrophy*—that is, it decreases in size and strength.

The way a muscle responds to use also depends on the type of exercise. For instance, when a muscle contracts relatively weakly, as during swimming and running, its slow, fatigue-resistant red fibers are most likely to be activated. As a result, these fibers develop more mitochondria, and more extensive capillary networks. Such changes increase the fibers' abilities to resist fatigue during prolonged exercise, although their sizes and strengths may remain unchanged.

Forceful exercise, such as weightlifting, in which a muscle exerts more than 75% of its maximum tension, uses the muscle's fast, fatigable white fibers. In response, existing muscle fibers develop new filaments of actin and myosin, and as their diameters increase, the entire muscle enlarges. However, no new muscle fibers are produced during hypertrophy.

Since the strength of a contraction is directly related to the diameter of the muscle fibers, an enlarged muscle can produce stronger contractions than before. However, such a change does not increase the muscle's ability to resist fatigue during activities such as running or swimming.

If regular exercise stops, capillary networks shrink and the number of mitochondria within the muscle fibers fall. Actin and myosin filaments diminish, and the entire muscle atrophies. Injured limbs immobilized in casts or accidents or diseases that interfere with motor nerve impulses commonly cause muscle atrophy. A muscle that cannot be exercised may decrease to less than one-half its usual size within a few months.

Muscle fibers whose motor neurons are severed not only decrease in size, but also may fragment and, in time, be replaced by fat or fibrous connective tissue. However, if such a muscle is reinnervated within the first few months following an injury, function may be restored.

The fibers of the outer coats are directed longitudinally, while those of the inner coats are arranged circularly. These muscular layers are responsible for the changes in size and shape that occur in these organs as they carry on their special functions.

The fibers of visceral smooth muscles can stimulate each other. When one fiber is stimulated, the impulse moving over its surface may excite adjacent fibers that, in turn, stimulate still others. Visceral smooth muscles also display *rhythmicity*—a pattern of repeated contractions. This phenomenon is caused by self-exciting fibers that deliver spontaneous impulses periodically into surrounding muscle tissue.

These two features of visceral smooth muscle—transmission of impulses from cell to cell and rhythmicity—are largely responsible for the wavelike motion called **peristalsis** that occurs in certain tubular organs (see chapter 17). Peristalsis consists of alternate contractions and relaxations of the longitudinal and circular muscles. These movements help force the contents of a tube along its length. In the intestines, for example, peristaltic waves move masses of partially digested food and help to mix them with digestive fluids. Peristalsis in the ureters moves urine from the kidneys to the urinary bladder.

### Smooth Muscle Contraction

Smooth muscle contraction resembles skeletal muscle contraction in a number of ways. Both mechanisms reflect reactions of actin and myosin; both are triggered by membrane impulses and release of calcium ions; and both use energy from ATP molecules. There are, however, significant differences between smooth and skeletal muscle action. For example, smooth muscle fibers lack troponin, the protein that binds to calcium ions in skeletal muscle. Instead, smooth muscle uses a protein called *calmodulin,* which binds to calcium ions released when its fibers are stimulated, thus activating the actin-myosin contraction mechanism. In addition, much of the calcium necessary for smooth muscle contraction diffuses into the cell from the extracellular fluid.

Acetylcholine, the neurotransmitter in skeletal muscle, as well as *norepinephrine,* affect smooth muscle. Each of these neurotransmitters stimulates contractions in some smooth muscles and inhibits contractions in others. The discussion of the autonomic nervous system in chapter 11 describes these actions in greater detail.

Hormones affect smooth muscles by stimulating contraction in some cases and altering the amount of response to neurotransmitters in others. For example, during the later stages of childbirth, the hormone oxytocin stimulates the smooth muscles in the wall of the uterus to contract (see chapter 22).

Stretching of smooth muscle fibers can also trigger contractions. This response is particularly important to the function of visceral smooth muscle in the walls of certain hollow organs, such as the urinary bladder and the intestines. For example, when partially digested food stretches the wall of the intestine, automatic contractions move the contents away.

Smooth muscle is slower to contract and slower to relax than skeletal muscle. On the other hand, smooth muscle can maintain a forceful contraction for a longer time with the same amount of ATP. Unlike skeletal muscle, smooth muscle fibers can change length without changing tautness; because of this, smooth muscles in the stomach and intestinal walls can stretch as these organs fill, allowing the pressure inside the organs to remain constant.

1. Describe the two major types of smooth muscle.
2. What special characteristics of visceral smooth muscle make peristalsis possible?
3. How is smooth muscle contraction similar to skeletal muscle contraction?
4. How do the contraction mechanisms of smooth and skeletal muscles differ?

## Cardiac Muscle

Cardiac muscle appears only in the heart. It is composed of striated cells joined end to end, forming fibers that are interconnected in branching, three-dimensional networks. Each cell contains a single nucleus and many filaments of actin and myosin similar to those in skeletal muscle. A cardiac muscle cell also has a well-developed sarcoplasmic reticulum, a system of transverse tubules, and many mitochondria. However, the cisternae of the sarcoplasmic reticulum of a cardiac muscle fiber are less developed and store less calcium than those of a skeletal muscle fiber. On the other hand, the transverse tubules of cardiac muscle fibers are larger than those in skeletal muscle, and they release large quantities of calcium ions into the sarcoplasm in response to a single muscle impulse. This extra calcium from the transverse tubules enables cardiac muscle fibers to maintain a contraction longer than can skeletal muscle fibers.

The calcium ions in transverse tubules come from the fluid outside the muscle fiber. This extracellular calcium partially controls the strength of cardiac muscle contraction.

> Drugs called calcium channel blockers are used to stop spasms of the heart muscle. They do this by blocking ion channels that admit calcium into these cells.

The opposing ends of cardiac muscle cells are separated by cross-bands called *intercalated disks.* These bands are actually complex membrane junctions. They help hold adjacent cells together and transmit the force of contraction from cell to cell. Intercellular junctions between the fused membranes of the intercalated disks allow diffusion of ions between the cells. This makes it possible for muscle impulses to travel rapidly from cell to cell (see figs. 9.18 and 5.30).

When one portion of the cardiac muscle network is stimulated, the impulse passes to the other fibers of the network, and the whole structure contracts as a unit (a *syncytium*); that is, the network responds to stimulation in an all-or-none manner. Cardiac muscle is also self-exciting and rhythmic. Consequently, a pattern of contraction and relaxation repeats again and again, causing the rhythmic contraction of the heart. Also, unlike skeletal muscle, cardiac muscle remains refractory until a contraction ends, so that sustained or tetanic contractions do not occur in the heart muscle.

Chart 9.2 summarizes characteristics of the three types of muscles.

1. How is cardiac muscle similar to skeletal muscle?
2. How does cardiac muscle differ from skeletal muscle?
3. What is the function of intercalated disks?
4. What characteristic of cardiac muscle is responsible for the contraction of the heart as a unit?

## Skeletal Muscle Actions

Skeletal muscles are responsible for a great variety of body movements. The action of each muscle depends largely upon the kind of joint it is associated with and the way the muscle is attached on either side of that joint.

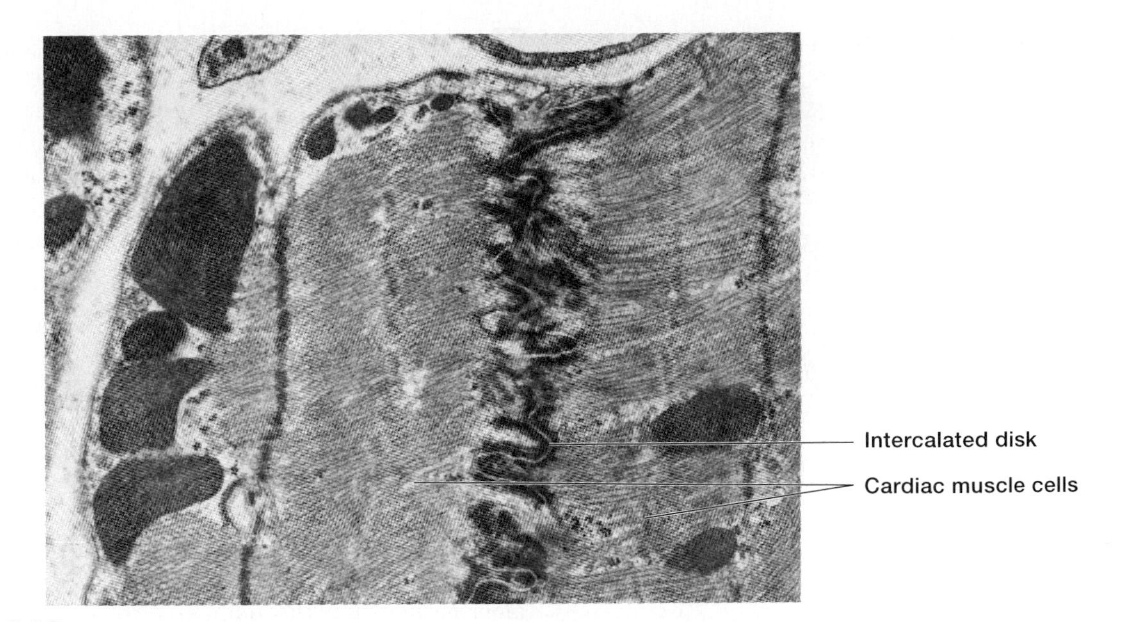


**FIGURE 9.18**
The intercalated disks of cardiac muscle, shown in this transmission electron micrograph, bind adjacent cells and allow ions to move between adjacent cells.

**CHART 9.2 CHARACTERISTICS OF MUSCLE TISSUES**

| | Skeletal | Smooth | Cardiac |
|---|---|---|---|
| *Major location* | Skeletal muscles | Walls of hollow organs | Wall of the heart |
| *Major function* | Movement of bones at joints; maintenance of posture | Movement of hollow organs; peristalsis | Pumping action of the heart |
| *Cellular characteristics* | | | |
| *Striations* | Present | Absent | Present |
| *Nucleus* | Multiple nuclei | Single nucleus | Single nucleus |
| *Special features* | Transverse tubule system is well developed | Lacks transverse tubules | Transverse tubule system is well developed; intercalated disks separate adjacent cells |
| *Mode of control* | Voluntary | Involuntary | Involuntary |
| *Contraction characteristics* | Contracts and relaxes relatively rapidly | Contracts and relaxes relatively slowly; self-exciting; rhythmic | Network of fibers contracts as a unit; self-exciting; rhythmic; remains refractory until contraction ends |

## Origin and Insertion

Recall from chapter 7 that one end of a skeletal muscle is usually fastened to a relatively immovable or fixed part, and the other end is connected to a movable part on the other side of a joint. The immovable end is called the **origin** of the muscle, and the movable end is called its **insertion.** When a muscle contracts, its insertion is pulled toward its origin (fig. 9.19).

Some muscles have more than one origin or insertion. The *biceps brachii* in the arm, for example, has two origins. This is reflected in its name *biceps,* meaning *two heads.* (Note: The head of a muscle is the part nearest its origin.) As figure 9.19 shows, one head of the muscle is attached to the coracoid process of the scapula, and the other head arises from a tubercle above the glenoid cavity of the scapula. The muscle extends along the front surface of the humerus and is inserted by a single tendon on the radial tuberosity of the radius. When the biceps brachii contracts, its insertion is pulled toward its origin, and the arm bends at the elbow.

## Interaction of Skeletal Muscles

Skeletal muscles almost always function in groups. As a result, when a particular body part moves, a person must do more than contract a single muscle; instead,

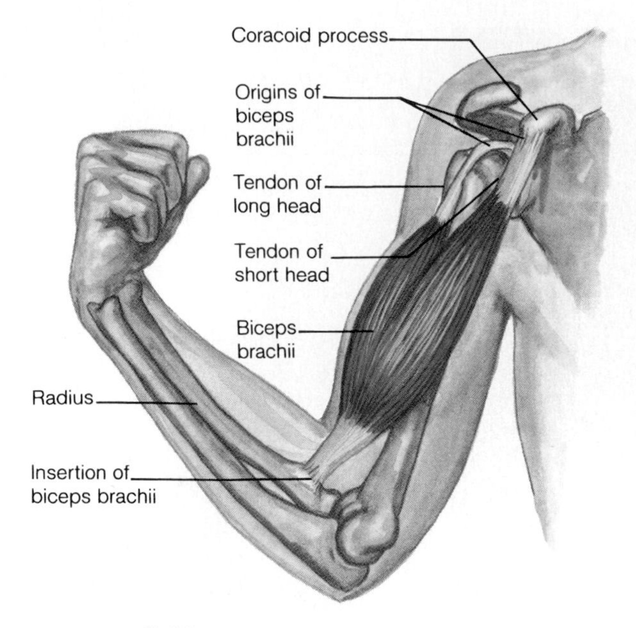


**FIGURE 9.19**
The biceps brachii has two heads that originate on the scapula. This muscle is inserted on the radius by a single tendon.

after learning to make a particular movement, the person wills the movement to occur, and the nervous system stimulates the appropriate group of muscles.

By carefully observing body movements, it is possible to determine the roles of particular muscles. For instance, lifting the upper limb horizontally away from the side requires contracting the *deltoid* muscle, which is said to be the **prime mover.** However, while a prime mover is acting, certain nearby muscles also contract. When a deltoid muscle contracts, nearby muscles help hold the shoulder steady and in this way make the action of the prime mover more effective. Muscles that contract and assist a prime mover are called **synergists.**

Still other muscles act as **antagonists** to prime movers. These muscles can resist a prime mover's action and cause movement in the opposite direction—the antagonist of the prime mover that raises the upper limb can lower the upper limb, or the antagonist of the prime mover that bends the upper limb can straighten it. If both a prime mover and its antagonist contract simultaneously, the structure they act upon remains rigid. Similarly, smooth body movements depend upon the antagonists' relaxing and giving way to the prime movers whenever the prime movers contract. Once again, the nervous system controls these complex actions, as described in chapter 11.

1. Distinguish between the origin and the insertion of a muscle.
2. Define prime mover.
3. What is the function of a synergist? An antagonist?

## MAJOR SKELETAL MUSCLES

This section concerns the locations, actions, origins, and insertions of some of the major skeletal muscles. The charts that summarize the information concerning groups of these muscles also include the names of nerves that supply the individual muscles within each group. Chapter 11 presents the origins and pathways of these nerves.

Figures 9.20 and 9.21 show the locations of superficial skeletal muscles—that is, those near the surface. Notice that the names of muscles often describe them in some way. A name may indicate a muscle's size, shape, location, action, number of attachments, or the direction of its fibers, as in the following examples:

**pectoralis major** A muscle of large size (*major*) located in the pectoral region (chest).
**deltoid** Shaped like a delta or triangle.
**extensor digitorum** Extends the digits (fingers or toes).
**biceps brachii** A muscle with two heads (*biceps*), or points of origin, located in the brachium or arm.
**sternocleidomastoid** Attached to the sternum, clavicle, and mastoid process.
**external oblique** Located near the outside, with fibers that run obliquely or in a slanting direction.

### Muscles of Facial Expression

A number of small muscles that lie beneath the skin of the face and scalp enable us to communicate feelings through facial expression. Many of these muscles are located around the eyes and mouth, and they are responsible for such expressions as surprise, sadness, anger, fear, disgust, and pain. As a group, the muscles of facial expression connect the bones of the skull to connective tissue in various regions of the overlying skin. They are shown in figure 9.22 and reference plate 61, and are listed in chart 9.3. The muscles of facial expression include the following:

| | |
|---|---|
| *Epicranius* | *Buccinator* |
| *Orbicularis oculi* | *Zygomaticus* |
| *Orbicularis oris* | *Platysma* |

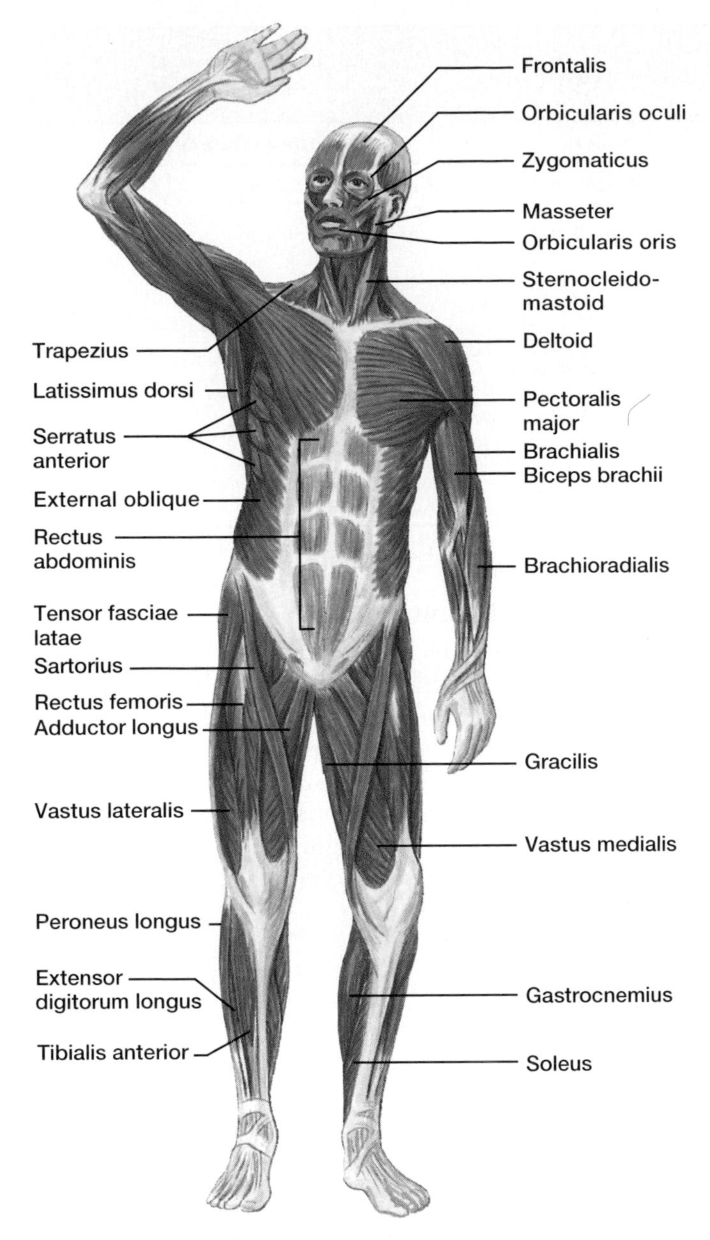


**FIGURE 9.20**
Anterior view of superficial skeletal muscles.

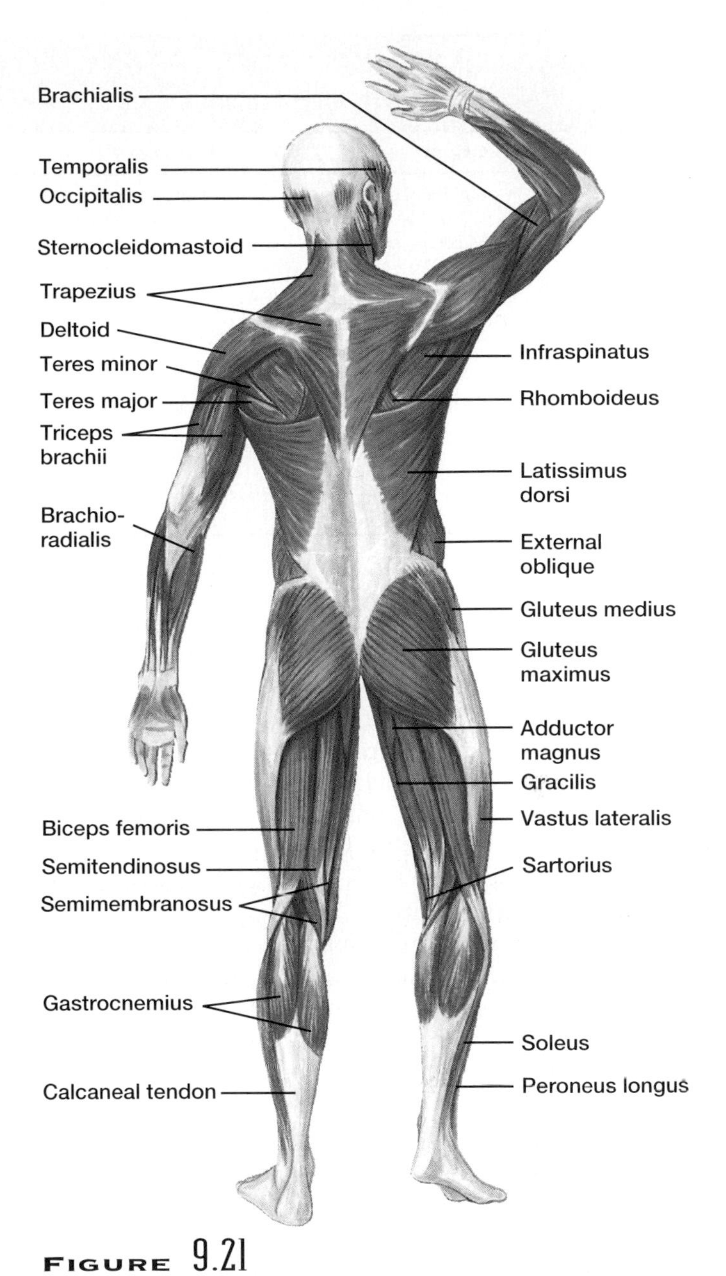


**FIGURE 9.21**
Posterior view of superficial skeletal muscles.

The **epicranius** (ep″ĭ-kra′ne-us) covers the upper part of the cranium and consists of two muscular parts—the *frontalis* (frun-ta′lis), which lies over the frontal bone, and the *occipitalis* (ok-sip″ĭ-ta′lis), which lies over the occipital bone. These muscles are united by a broad, tendinous membrane called the *epicranial aponeurosis,* which covers the cranium like a cap. Contraction of the epicranius raises the eyebrows and wrinkles the skin of the forehead horizontally, as when a person expresses surprise. Headaches often result from sustained contraction of this muscle.

The **orbicularis oculi** (or-bik′u-la-rus ok′u-li) is a ringlike band of muscle, called a *sphincter muscle,* that surrounds the eye. It lies in the subcutaneous tissue of the eyelid and closes or blinks the eye. At the same time, it compresses the nearby tear gland, or *lacrimal gland,* aiding the flow of tears over the

## CHART 9.3 MUSCLES OF FACIAL EXPRESSION

| Muscle | Origin | Insertion | Action | Nerve Supply |
|---|---|---|---|---|
| Epicranius | Occipital bone | Skin and muscles around eye | Raises eyebrow as when surprised | Facial n. |
| Orbicularis oculi | Maxillary and frontal bones | Skin around eye | Closes eye as in blinking | Facial n. |
| Orbicularis oris | Muscles near the mouth | Skin of central lip | Closes lips, protrudes lips as for kissing | Facial n. |
| Buccinator | Outer surfaces of maxilla and mandible | Orbicularis oris | Compresses cheeks inward as when blowing air | Facial n. |
| Zygomaticus | Zygomatic bone | Orbicularis oris | Raises corner of mouth as when smiling | Facial n. |
| Platysma | Fascia in upper chest | Lower border of mandible | Draws angle of mouth downward as when pouting | Facial n. |

surface of the eye. Contraction of the orbicularis oculi also causes the folds, or crow's feet, that radiate laterally from the corner of the eye.

The **orbicularis oris** (or-bik′u-la-rus o′ris) is a sphincter muscle that encircles the mouth. It lies between the skin and the mucous membranes of the lips, extending upward to the nose and downward to the region between the lower lip and chin. The orbicularis oris is sometimes called the kissing muscle because it closes and puckers the lips.

The **buccinator** (buk′sĭ-na″tor) is located in the wall of the cheek. Its fibers are directed forward from the bones of the jaws to the angle of the mouth, and when they contract, the cheek is compressed inward. This action helps hold food in contact with the teeth when a person is chewing. The buccinator also aids in blowing air out of the mouth, and for this reason, it is sometimes called the trumpeter muscle.

The **zygomaticus** (zi″go-mat′ik-us) extends from the zygomatic arch downward to the corner of the mouth. When it contracts, the corner of the mouth is drawn up, as in smiling or laughing.

The **platysma** (plah-tiz′mah) is a thin, sheetlike muscle whose fibers extend from the chest upward over the neck to the face. It pulls the angle of the mouth downward, as in pouting. The platysma also helps lower the mandible.

The muscles that move the eye are described in chapter 12.

## Muscles of Mastication

Four pairs of muscles attached to the mandible produce chewing movements. Three pairs of these muscles close the lower jaw, as in biting; the fourth pair can lower the jaw, cause side-to-side grinding motions of the mandible, and pull the mandible forward, causing it to protrude. The muscles of mastication are shown in figure 9.22 and reference plate 61, and are listed in chart 9.4. They include the following:

*Masseter*
*Temporalis*
*Medial pterygoid*
*Lateral pterygoid*

The **masseter** (mas-se′ter) is a thick, flattened muscle that can be felt just in front of the ear when the teeth are clenched. Its fibers extend downward from the zygomatic arch to the mandible. The masseter raises the jaw, but it can also control the rate at which the jaw falls open in response to gravity (fig. 9.22*a*).

The **temporalis** (tem-po-ra′lis) is a fan-shaped muscle located on the side of the skull above and in front of the ear. Its fibers, which also raise the jaw, pass downward beneath the zygomatic arch to the mandible (fig. 9.22*a* and *b*). Tensing this muscle is associated with temporomandibular joint syndrome, discussed in Clinical Application 9.3.

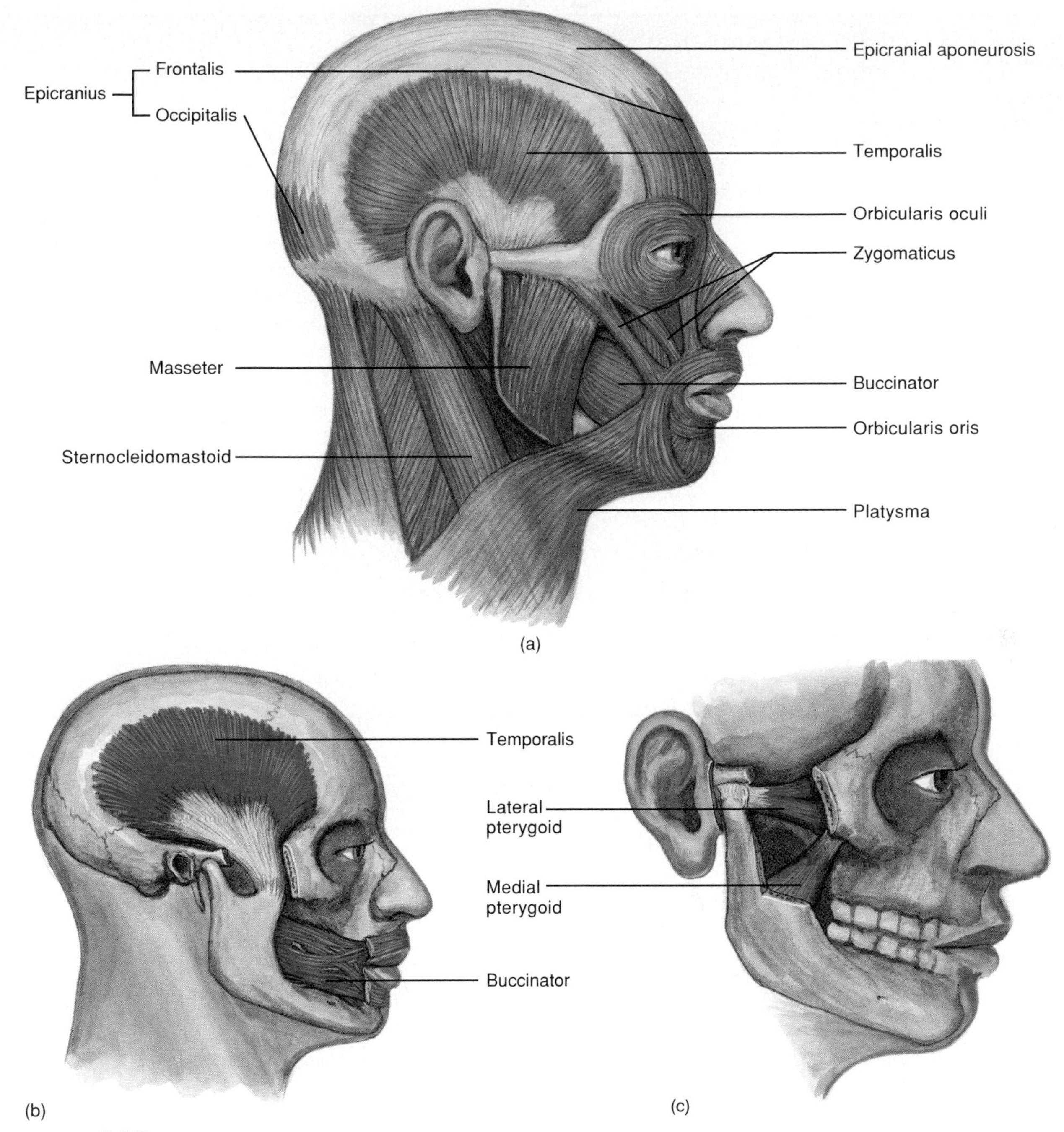


**FIGURE 9.22**

(*a*) Muscles of facial expression and mastication; isolated views of (*b*) the temporalis and buccinator muscles and (*c*) the lateral and medial pterygoid muscles.

## Clinical Application 9.3

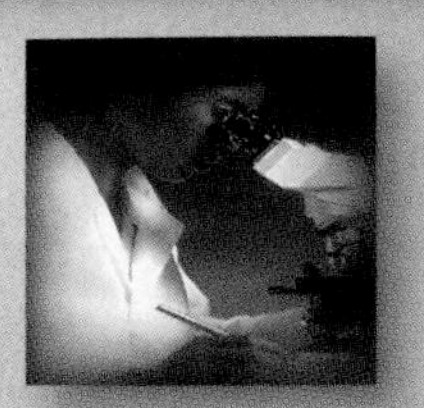

### TMJ Syndrome

Facial pain, headache, ringing in the ears, a clicking jaw, insomnia, teeth sensitive to heat or cold, backache, dizziness, and pain in front of the ears may seem to be unrelated, but these aches and pains may all result from temporomandibular joint (TMJ) syndrome. This condition is caused by a misaligned jaw or simply by a habit of grinding or clenching the teeth. This action may stress the temporomandibular joint, the articulation between the mandibular condyle of the mandible and the mandibular fossa of the temporal bone. Loss of coordination of these structures affects the nerves that pass through the neck and jaw region, causing the symptoms. When this happens, tensing a muscle in the forehead can cause a headache, or a spasm in the muscle that normally opens the auditory tubes during swallowing can produce inability to clear the ears. ■

Doctors diagnose TMJ using an electromyograph, in which electrodes record muscle activity in four pairs of head and neck muscle groups. A form of treatment is transcutaneous electrical nerve stimulation (TENS), which stimulates the facial muscles for up to an hour. Another treatment is an orthotic device fitted by a dentist. Worn for 3 to 6 months, the device fine-tunes the action of jaw muscles to form a more comfortable bite. Finally, once the correct bite is determined, a dentist can use bonding materials to alter shapes of certain teeth to offer a more permanent solution for TMJ syndrome.

## Chart 9.4 Muscles of Mastication

| Muscle | Origin | Insertion | Action | Nerve Supply |
|---|---|---|---|---|
| Masseter | Lower border of zygomatic arch | Lateral surface of mandible | Elevates mandible | Trigeminal n. |
| Temporalis | Temporal bone | Coronoid process and anterior ramus of mandible | Elevates mandible | Trigeminal n. |
| Medial pterygoid | Sphenoid, palatine, and maxillary bones | Medial surface of mandible | Elevates mandible | Trigeminal n. |
| Lateral pterygoid | Sphenoid bone | Anterior surface of mandibular condyle | Depresses and protracts mandible, and moves it from side to side | Trigeminal n. |

The **medial pterygoid** (ter′ĭ-goid) extends back and downward from the sphenoid, palatine, and maxillary bones to the ramus of the mandible. It closes the jaw (fig. 9.22*c*).

The fibers of the **lateral pterygoid** extend forward from the region just below the mandibular condyle to the sphenoid bone. This muscle can open the mouth, pull the mandible forward to make it protrude, and move the mandible from side to side (fig. 9.22*c*).

### Muscles that Move the Head and Vertebral Column

Paired muscles in the neck and back flex, extend, and rotate the head and hold the torso erect (figs. 9.23 and 9.25 and chart 9.5). They include the following:

*Sternocleidomastoid*
*Splenius capitis*
*Semispinalis capitis*
*Erector spinae*

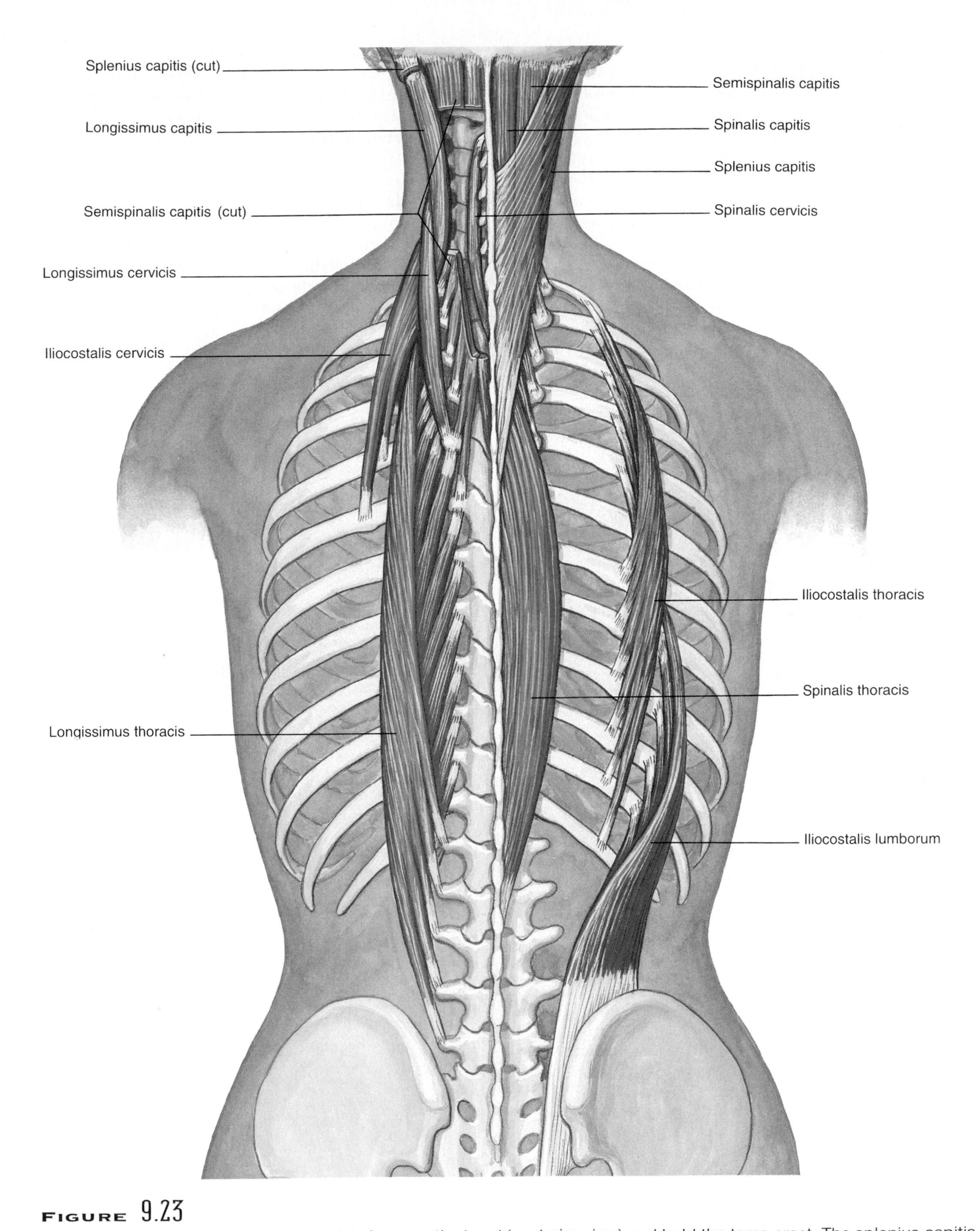


**FIGURE 9.23**
Deep muscles of the back and the neck help move the head (posterior view) and hold the torso erect. The splenius capitis and semispinalis capitis are removed on the left to show underlying muscles.

The **sternocleidomastoid** (ster″no-kli″do-mas′toid) is a long muscle in the side of the neck that extends upward from the thorax to the base of the skull behind the ear. When the sternocleido-mastoid on one side contracts, the face turns to the opposite side. When both muscles contract, the head bends toward the chest. If the head is fixed in position by other muscles, the sternocleidomastoids can raise the sternum, aiding forceful inhalation (fig. 9.25 and chart 9.5).

The **splenius capitis** (sple′ne-us kap′ĭ-tis) is a broad, straplike muscle located in the back of the neck. It connects the base of the skull to the vertebrae in the neck and upper thorax. A splenius capitis acting singly rotates the head and bends it toward one side. Acting together, these muscles bring the head into an upright position (fig. 9.23 and chart 9.5).

The **semispinalis capitis** (sem″e-spi-na′lis kap′ĭ-tis) is a broad, sheetlike muscle extending upward from the vertebrae in the neck and thorax to the occipital bone. It extends the head, bends it to one side, or rotates it (fig. 9.23 and chart 9.5).

**Erector spinae** muscles run longitudinally along the back, with origins and insertions at many places on the axial skeleton (fig. 9.23). These muscles act to extend and rotate the head, and to maintain the erect position of the spinal column. Erector spinae can be subdivided into medial, intermediate, and lateral groups (chart 9.5).

## Muscles that Move the Pectoral Girdle

The muscles that move the pectoral girdle are closely associated with those that move the arm. A number of these chest and shoulder muscles connect the scapula to nearby bones and move the scapula upward, downward, forward, and backward (figs. 9.24, 9.25, 9.26; reference plates 63, 64; chart 9.6). Muscles that move the pectoral girdle include the following:

*Trapezius*
*Rhomboideus major*
*Levator scapulae*
*Serratus anterior*
*Pectoralis minor*

The **trapezius** (trah-pe′ze-us) is a large, triangular muscle in the upper back that extends horizontally from the base of the skull and the cervical and thoracic vertebrae to the shoulder. Its fibers are arranged in three groups—upper, middle, and lower. Together these fibers rotate the scapula. The upper fibers acting alone raise the scapula and shoulder, as when the shoulders are shrugged to express a feeling of indifference. The middle fibers pull the scapula toward the vertebral column, and the lower fibers draw the scapula and shoulder downward. When other muscles fix the shoulder in position, the trapezius can pull the head backward or to one side (fig. 9.24).

The **rhomboideus** (rom-boid′-ē-us) **major** connects the upper thoracic vertebrae to the scapula. It raises the scapula and adducts it (fig. 9.24).

The **levator scapulae** (le-va′tor scap′u-lē) is a straplike muscle that runs almost vertically through the neck, connecting the cervical vertebrae to the scapula. It elevates the scapula (figs. 9.24 and 9.26).

The **serratus anterior** (ser-ra′tus an-te′re-or) is a broad, curved muscle located on the side of the chest. It arises as fleshy, narrow strips on the upper ribs and extends along the medial wall of the axilla to the ventral surface of the scapula. It pulls the scapula downward and anteriorly, and is used to thrust the shoulder forward, as when pushing something (fig. 9.25).

The **pectoralis** (pek″to-ra′lis) **minor** is a thin, flat muscle that lies beneath the larger pectoralis major. It extends laterally and upward from the ribs to the scapula and pulls the scapula forward and downward. When other muscles fix the scapula in position, the pectoralis minor can raise the ribs and thus aid forceful inhalation (fig. 9.25).

A small, triangular region, called the *triangle of auscultation,* is located in the back where the trapezius overlaps the superior border of the latissimus dorsi and the underlying rhomboideus major. This area, which is near the medial border of the scapula, enlarges when a person bends forward with the arms folded across the chest. By placing the bell of a stethoscope within the triangle of auscultation, a physician can usually hear clearly the sounds of the respiratory organs.

## Muscles that Move the Arm

The arm is one of the more freely movable parts of the body because muscles connect the humerus to regions of the pectoral girdle, ribs, and vertebral column. These muscles can be grouped according to their primary actions—flexion, extension, abduction, and rotation (figs. 9.26, 9.27, 9.28; reference plates 62, 63, 64; chart 9.7). Muscles that move the arm include the following:

**Flexors**
*Coracobrachialis*
*Pectoralis major*

**Extensors**
*Teres major*
*Latissimus dorsi*

**Abductors**
*Supraspinatus*
*Deltoid*

**Rotators**
*Subscapularis*
*Infraspinatus*
*Teres minor*

## CHART 9.5 Muscles that Move the Head and Vertebral Column

| Muscle | Origin | Insertion | Action | Nerve Supply |
|---|---|---|---|---|
| Sternocleidomastoid | Anterior surface of sternum and upper surface of clavicle | Mastoid process of temporal bone | Pulls head to one side, flexes neck or elevates sternum | Accessory, C2 and C3 cervical nerves |
| Splenius capitis | Spinous processes of lower cervical and upper thoracic vertebrae | Occipital bone | Rotates head, bends head to one side, or extends neck | Cervical nerves |
| Semispinalis capitis | Processes of lower cervical and upper thoracic vertebrae | Occipital bone | Extends head, bends head to one side, or rotates head | Cervical and thoracic spinal nerves |
| Erector spinae | | | | |
| *Iliocostalis (lateral) group* | | | | |
| Iliocostalis lumborum | Iliac crest | Lower six ribs | Extends lumbar region of vertebral column | Lumbar spinal nerves |
| Iliocostalis thoracis | Lower six ribs | Upper six ribs | Holds spine erect | Thoracic spinal nerves |
| Iliocostalis cervicis | Upper six ribs | Fourth through sixth cervical vertebrae | Extends cervical region of vertebral column | Cervical spinal nerves |
| *Longissimus (intermediate) group* | | | | |
| Longissimus thoracis | Lumbar vertebrae | Thoracic and upper lumbar vertebrae and ribs 9 and 10 | Extends thoracic region of vertebral column | Spinal nerves |
| Longissimus cervicis | Fourth and fifth thoracic vertebrae | Second through sixth cervical vertebrae | Extends cervical region of vertebral column | Spinal nerves |
| Longissimus capitis | Upper thoracic and lower cervical vertebrae | Mastoid process of temporal bone | Extends and rotates head | Cervical spinal nerves |
| *Spinalis (medial) group* | | | | |
| Spinalis thoracis | Upper lumbar and lower thoracic vertebrae | Upper thoracic vertebrae | Extends vertebral column | Spinal nerves |
| Spinalis cervicis | Ligamentum nuchae and seventh cervical vertebra | Axis | Extends vertebral column | Spinal nerves |
| Spinalis capitis | Upper thoracic and lower cervical vertebrae | Occipital bone | Extends vertebral column | Spinal nerves |

## CHART 9.6 Muscles that Move the Pectoral Girdle

| Muscle | Origin | Insertion | Action | Nerve Supply |
|---|---|---|---|---|
| Trapezius | Occipital bone and spines of cervical and thoracic vertebrae | Clavicle, spine, and acromion process of scapula | Rotates scapula; various fibers raise scapula, pull scapula medially, or pull scapula and shoulder downward | Accessory n. |
| Rhomboideus major | Spines of upper thoracic vertebrae | Medial border of scapula | Raises and adducts scapula | Dorsal scapular n. |
| Levator scapulae | Transverse processes of cervical vertebrae | Medial margin of scapula | Elevates scapula | Dorsal scapular, and cervical nerves |
| Serratus anterior | Outer surfaces of upper ribs | Ventral surface of scapula | Pulls scapula anteriorly and downward | Long thoracic n. |
| Pectoralis minor | Sternal ends of upper ribs | Coracoid process of scapula | Pulls scapula forward and downward or raises ribs | Pectoral n. |

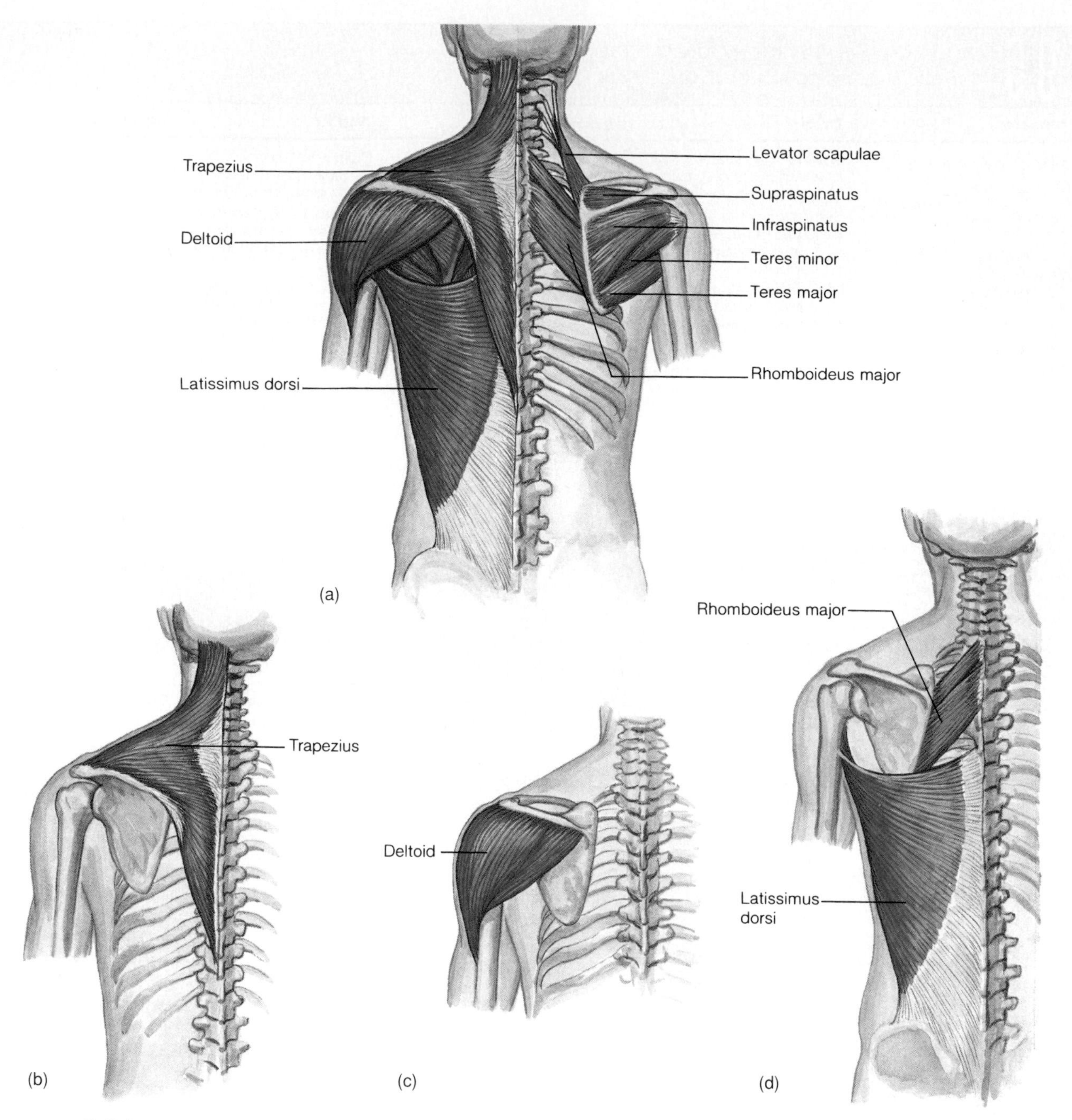


**FIGURE 9.24**

(*a*) Muscles of the posterior shoulder. The right trapezius is removed to show underlying muscles. Isolated views of (*b*) trapezius, (*c*) deltoid, and (*d*) rhomboideus and latissimus dorsi muscles.

### Flexors

The **coracobrachialis** (kor″ah-ko-bra′ke-al-is) extends from the scapula to the middle of the humerus along its medial surface. It flexes and adducts the arm (figs. 9.27 and 9.28).

The **pectoralis major** is a thick, fan-shaped muscle located in the upper chest. Its fibers extend from the center of the thorax through the armpit to the humerus. This muscle primarily pulls the arm forward and across the chest. It can also rotate the humerus medially and adduct the arm from a raised position (fig. 9.25).

### Extensors

The **teres** (te′rēz) **major** connects the scapula to the humerus. It extends the humerus and can also adduct and rotate the arm medially (figs. 9.24 and 9.26).

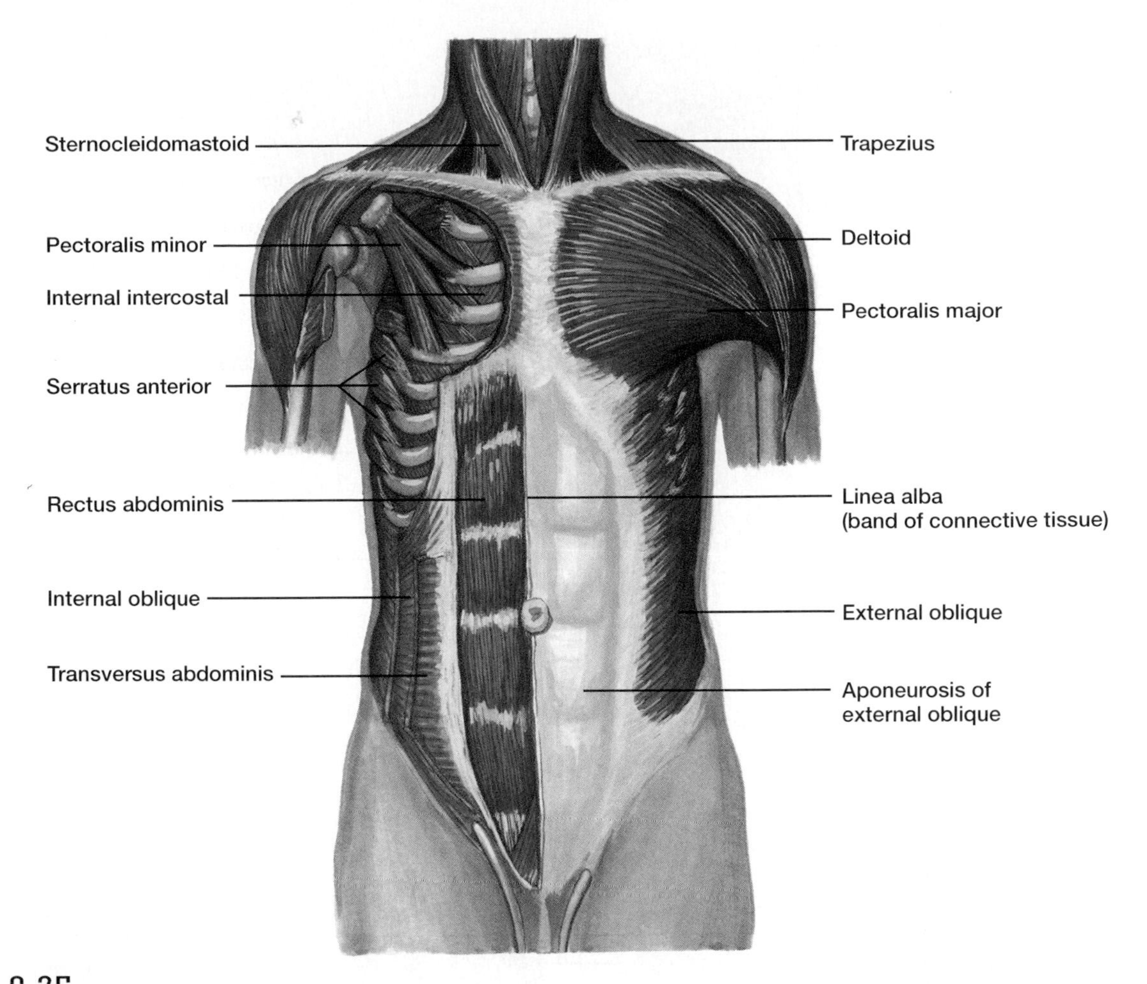


**FIGURE 9.25**
Muscles of the anterior chest and abdominal wall. The right pectoralis major is removed to show the pectoralis minor.

The **latissimus dorsi** (lah-tis′ĭ-mus dor′si) is a wide, triangular muscle that curves upward from the lower back, around the side, and to the armpit. It can extend and adduct the arm and rotate the humerus inwardly. It also pulls the shoulder downward and back. This muscle is used to pull the arm back in swimming, climbing, and rowing (figs. 9.24 and 9.27).

A humerus fractured at its surgical neck may damage the nerve that supplies the deltoid muscle (axillary nerve) (see fig. 7.45). If this occurs, the muscle is likely to shrink and weaken. In order to test the deltoid for such weakness, a physician may ask a patient to abduct the arm against some resistance and maintain that posture for a time.

### Abductors

The **supraspinatus** (su″prah-spi′na-tus) is located in the depression above the spine of the scapula on its posterior surface. It connects the scapula to the greater tubercle of the humerus and abducts the arm (figs. 9.24 and 9.26).

The **deltoid** (del′toid) is a thick, triangular muscle that covers the shoulder joint. It connects the clavicle and scapula to the lateral side of the humerus and abducts the arm. The deltoid's posterior fibers can extend the humerus, and its anterior fibers can flex the humerus (fig. 9.24).

### Rotators

The **subscapularis** (sub-scap′u-lar-is) is a large, triangular muscle that covers the anterior surface of the scapula. It connects the scapula to the humerus and rotates the arm medially (fig. 9.28).

The **infraspinatus** (in″frah-spi′na-tus) occupies the depression below the spine of the scapula on its posterior surface. The fibers of this muscle attach the scapula to the humerus and rotate the arm laterally (fig. 9.26).

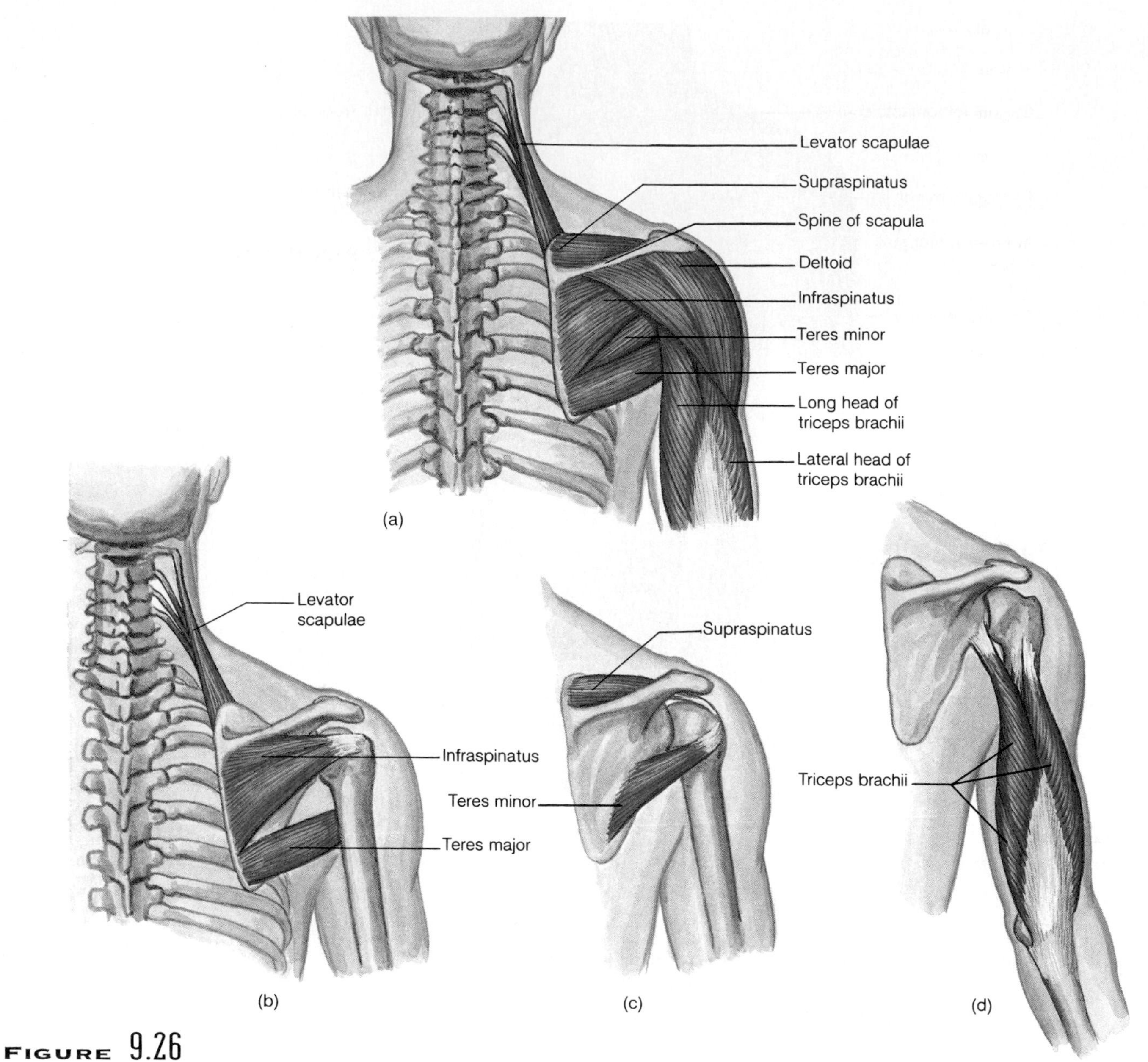


**FIGURE 9.26**

(*a*) Muscles of the posterior surface of the scapula and the arm; (*b* and *c*) muscles associated with the scapula. (*d*) Isolated view of the triceps brachii.

The **teres minor** is a small muscle connecting the scapula to the humerus. It rotates the arm laterally (figs. 9.24 and 9.26).

## Muscles that Move the Forearm

Most forearm movements are produced by muscles that connect the radius or ulna to the humerus or pectoral girdle. A group of muscles located along the anterior surface of the humerus flexes the elbow, whereas a single posterior muscle extends this joint. Other muscles cause movements at the radioulnar joint and rotate the forearm.

The muscles that move the forearm are shown in figures 9.28, 9.29, 9.30, 9.31, and in reference plates 63, 65, and are listed in chart 9.8. They include the following:

| Flexors | Extensor | Rotators |
|---|---|---|
| *Biceps brachii* | *Triceps brachii* | *Supinator* |
| *Brachialis* | | *Pronator teres* |
| *Brachioradialis* | | *Pronator quadratus* |

### Flexors

The **biceps brachii** (bi′seps bra′ke-i) is a fleshy muscle that forms a long, rounded mass on the anterior side of

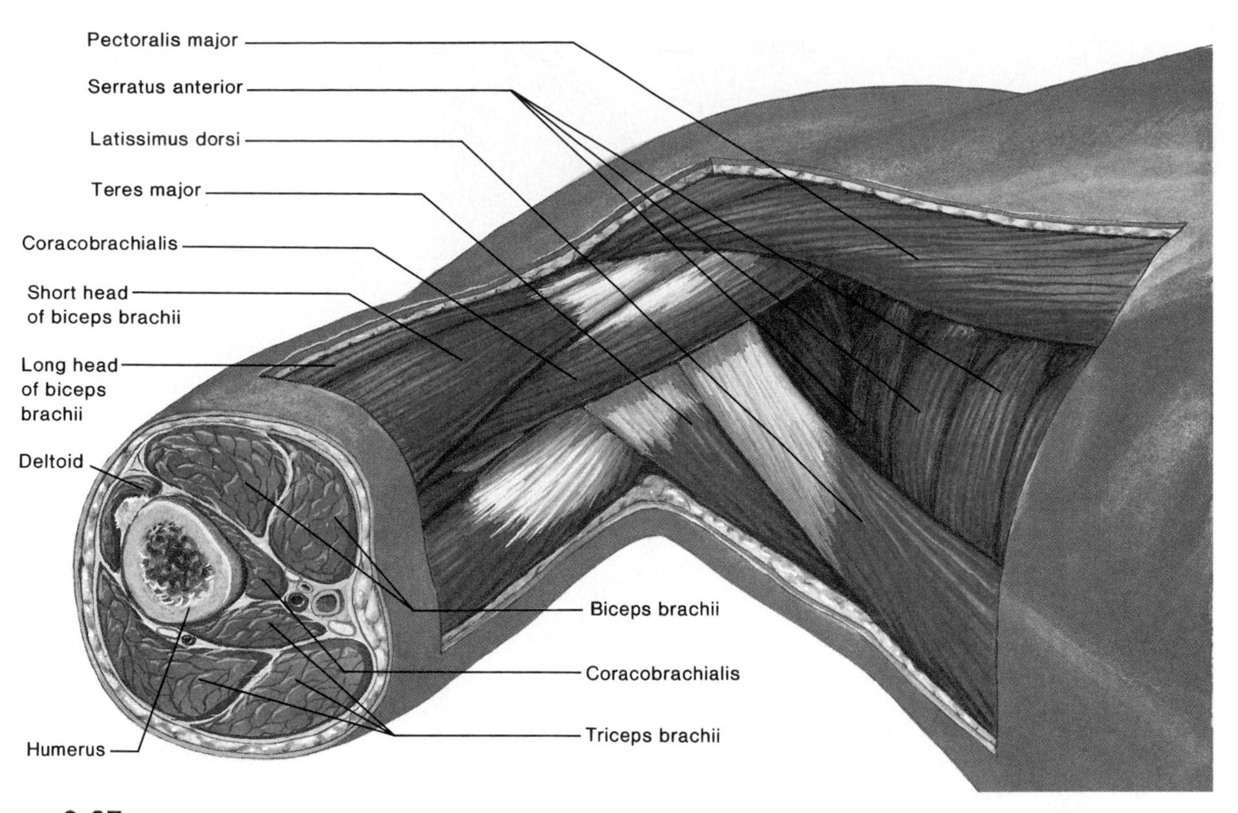


**FIGURE 9.27**
Cross section of the arm.

## CHART 9.7 Muscles that Move the Arm

| Muscle | Origin | Insertion | Action | Nerve Supply |
|---|---|---|---|---|
| Coracobrachialis | Coracoid process of scapula | Shaft of humerus | Flexes and adducts the arm | Musculocutaneus n. |
| Pectoralis major | Clavicle, sternum, and costal cartilages of upper ribs | Intertubercular groove of humerus | Flexes, adducts, and rotates arm | Pectoral n. |
| Teres major | Lateral border of scapula | Intertubercular groove of humerus | Extends, adducts, and rotates arm medially | Lower subscapular n. |
| Latissimus dorsi | Spines of sacral, lumbar, and lower thoracic vertebrae, iliac crest, and lower ribs | Intertubercular groove of humerus | Extends, adducts, and rotates the arm medially, or pulls the shoulder downward and back | Thoracodorsal n. |
| Supraspinatus | Posterior surface of scapula above spine | Greater tubercle of humerus | Abducts the arm | Suprascapular n. |
| Deltoid | Acromion process, spine of the scapula, and the clavicle | Deltoid tuberosity of humerus | Abducts, extends, and flexes arm | Axillary n. |
| Subscapularis | Anterior surface of scapula | Lesser tubercle of humerus | Rotates arm medially | Subscapular n. |
| Infraspinatus | Posterior surface of scapula below spine | Greater tubercle of humerus | Rotates arm laterally | Suprascapular n. |
| Teres minor | Lateral border of scapula | Greater tubercle of humerus | Rotates arm laterally | Axillary n. |

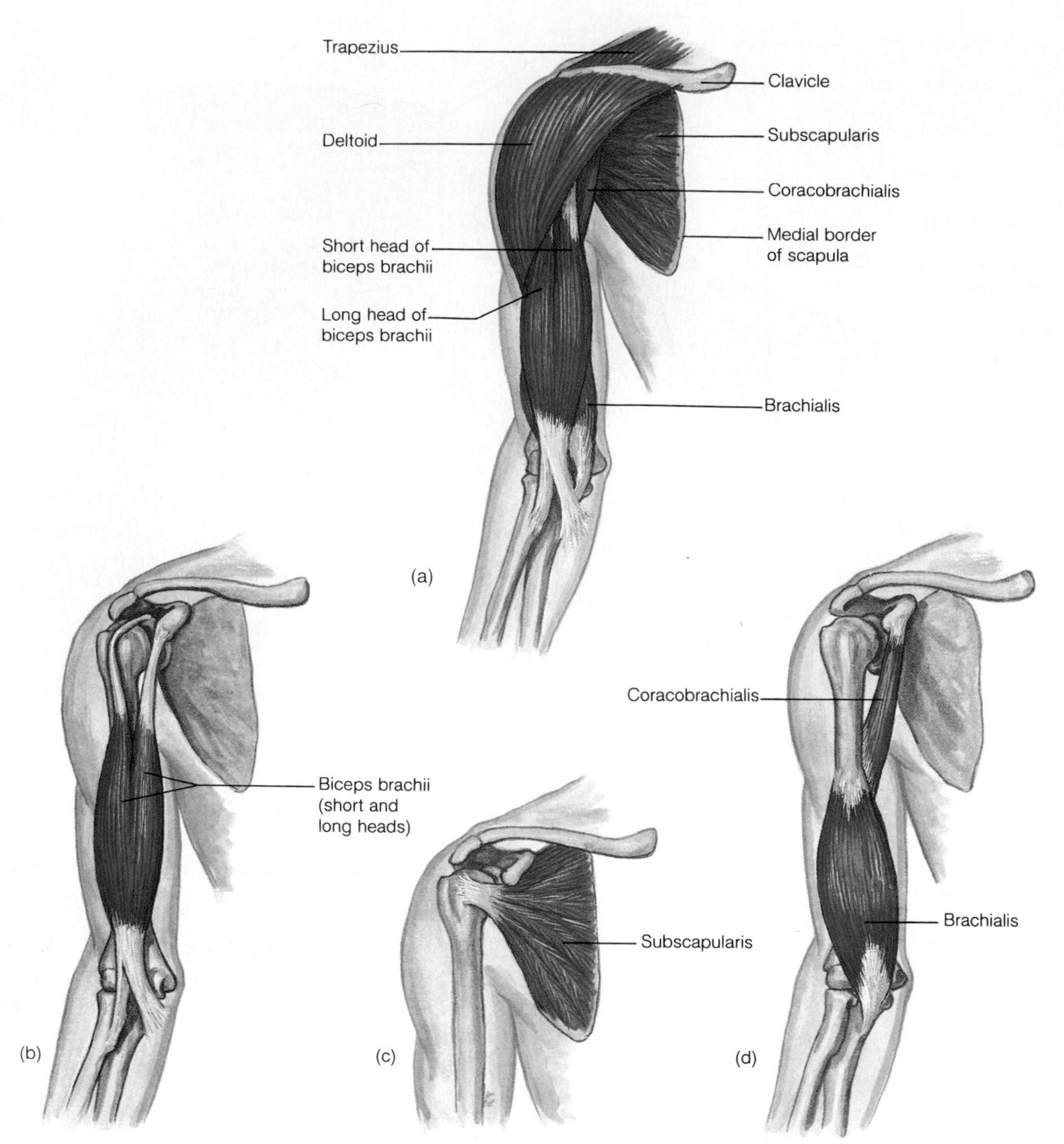


**FIGURE 9.28**

(*a*) Muscles of the anterior shoulder and the arm, with the rib cage removed. (*b, c,* and *d*) Isolated views of muscles associated with the arm.

the arm. It connects the scapula to the radius and flexes the upper limb at the elbow and rotates the hand laterally (supination), as when a person turns a doorknob or screwdriver (fig. 9.28).

The **brachialis** (bra′ke-al-is) is a large muscle beneath the biceps brachii. It connects the shaft of the humerus to the ulna and is the strongest flexor of the elbow (fig. 9.28).

The **brachioradialis** (bra″ke-o-ra″de-a′lis) connects the humerus to the radius and aids in flexing the elbow (fig. 9.29).

### Extensor

The **triceps brachii** (tri′seps bra′ke-i) has three heads and is the only muscle on the back of the arm. It connects the humerus and scapula to the ulna, and is the primary extensor of the elbow (figs. 9.26 and 9.27).

### Rotators

The **supinator** (su′pĭ-na-tor) is a short muscle whose fibers run from the ulna and the lateral end of the

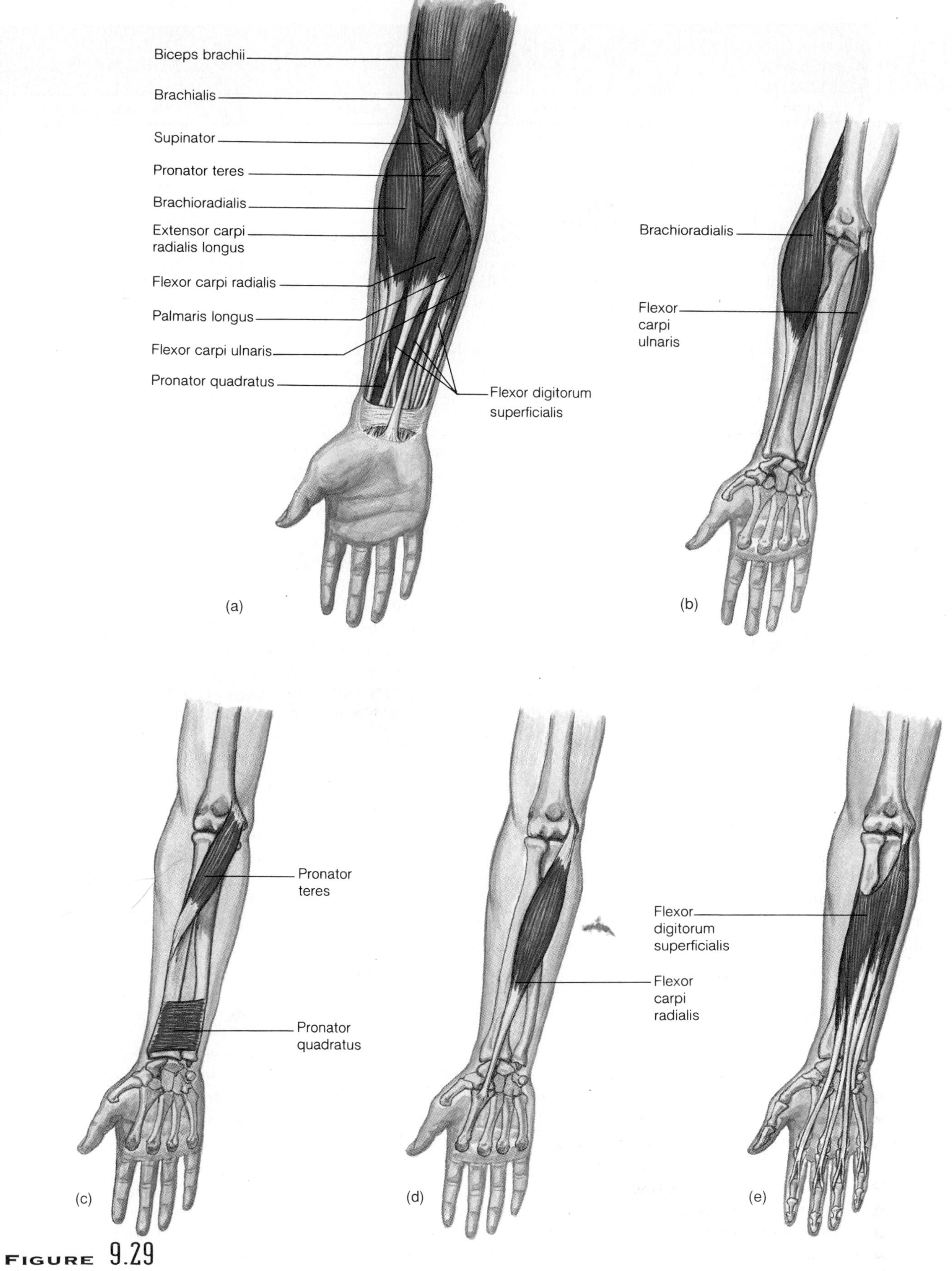


**FIGURE 9.29**

(*a*) Muscles of the anterior forearm. (*b–e*) Isolated views of muscles associated with the anterior forearm.

## CHART 9.8 Muscles that Move the Forearm

| Muscle | Origin | Insertion | Action | Nerve Supply |
|---|---|---|---|---|
| Biceps brachii | Coracoid process and tubercle above glenoid cavity of scapula | Radial tuberosity of radius | Flexes forearm at elbow and rotates hand laterally | Musculocutaneous n. |
| Brachialis | Anterior shaft of humerus | Coronoid process of ulna | Flexes forearm at elbow | Musculocutaneous, median, and radial nerves |
| Brachioradialis | Distal lateral end of humerus | Lateral surface of radius above styloid process | Flexes elbow | Radial n. |
| Triceps brachii | Tubercle below glenoid cavity and lateral and medial surfaces of humerus | Olecranon process of ulna | Extends forearm at elbow | Radial n. |
| Supinator | Lateral epicondyle of humerus and crest of ulna | Lateral surface of radius | Rotates forearm laterally | Radial n. |
| Pronator teres | Medial epicondyle of humerus and coronoid process of ulna | Lateral surface of radius | Rotates forearm medially | Median n. |
| Pronator quadratus | Anterior distal end of ulna | Anterior distal end of radius | Rotates foramen medially | Median n. |

humerus to the radius. It assists the biceps brachii in rotating the forearm laterally (supination) (fig. 9.29).

The **pronator teres** (pro-na′tor te′rēz) is a short muscle connecting the ends of the humerus and ulna to the radius. It rotates the arm medially, as when the hand is turned so the palm is facing downward (pronation) (fig. 9.29).

The **pronator quadratus** (pro-na′tor kwod-ra′tus) runs from the distal end of the ulna to the distal end of the radius. It assists the pronator teres in rotating the arm medially (fig. 9.29).

## Muscles that Move the Hand

Movements of the hand include movements of the wrist and fingers. Therefore, many muscles move the wrist, hand, and fingers. They originate from the distal end of the humerus and from the radius and ulna. The two major groups of these muscles are flexors on the anterior side of the forearm and extensors on the posterior side. Figures 9.29, 9.30, 9.31, reference plate 65, and chart 9.9 concern these muscles. The muscles that move the hand include the following:

**Flexors**

*Flexor carpi radialis*
*Flexor carpi ulnaris*
*Palmaris longus*
*Flexor digitorum profundus*
*Flexor digitorum superficialis*

**Extensors**

*Extensor carpi radialis longus*
*Extensor carpi radialis brevis*
*Extensor carpi ulnaris*
*Extensor digitorum*

### Flexors

The **flexor carpi radialis** (flek′sor kar-pi′ra″de-a′lis) is a fleshy muscle that runs medially on the anterior side of the forearm. It extends from the distal end of the humerus into the hand, where it is attached to metacarpal bones. The flexor carpi radialis flexes and abducts the hand at the wrist (fig. 9.29).

The **flexor carpi ulnaris** (flek′sor kar-pi′ ul-na′ris) is located along the medial border of the forearm. It connects the distal end of the humerus and the proximal end of the ulna to carpal and metacarpal bones. It flexes and adducts the hand at the wrist (fig. 9.29).

The **palmaris longus** (pal-ma′ris long′gus) is a slender muscle located on the medial side of the forearm between the flexor carpi radialis and the flexor carpi ulnaris. It connects the distal end of the humerus to fascia of the palm and flexes the hand at the wrist (fig. 9.29).

Some of the first signs of Parkinson's disease appear in the hands. In this disorder, certain brain cells degenerate and damage nerve cells that control muscles. Once called "shaking palsy," the disease often begins with a hand tremor that resembles the motion of rolling a marble between the thumb and forefinger. Another sign is called "cogwheel rigidity." When a doctor rotates the patient's hand in an arc, the hand resists the movement and then jerks, like the cogs in a gear.

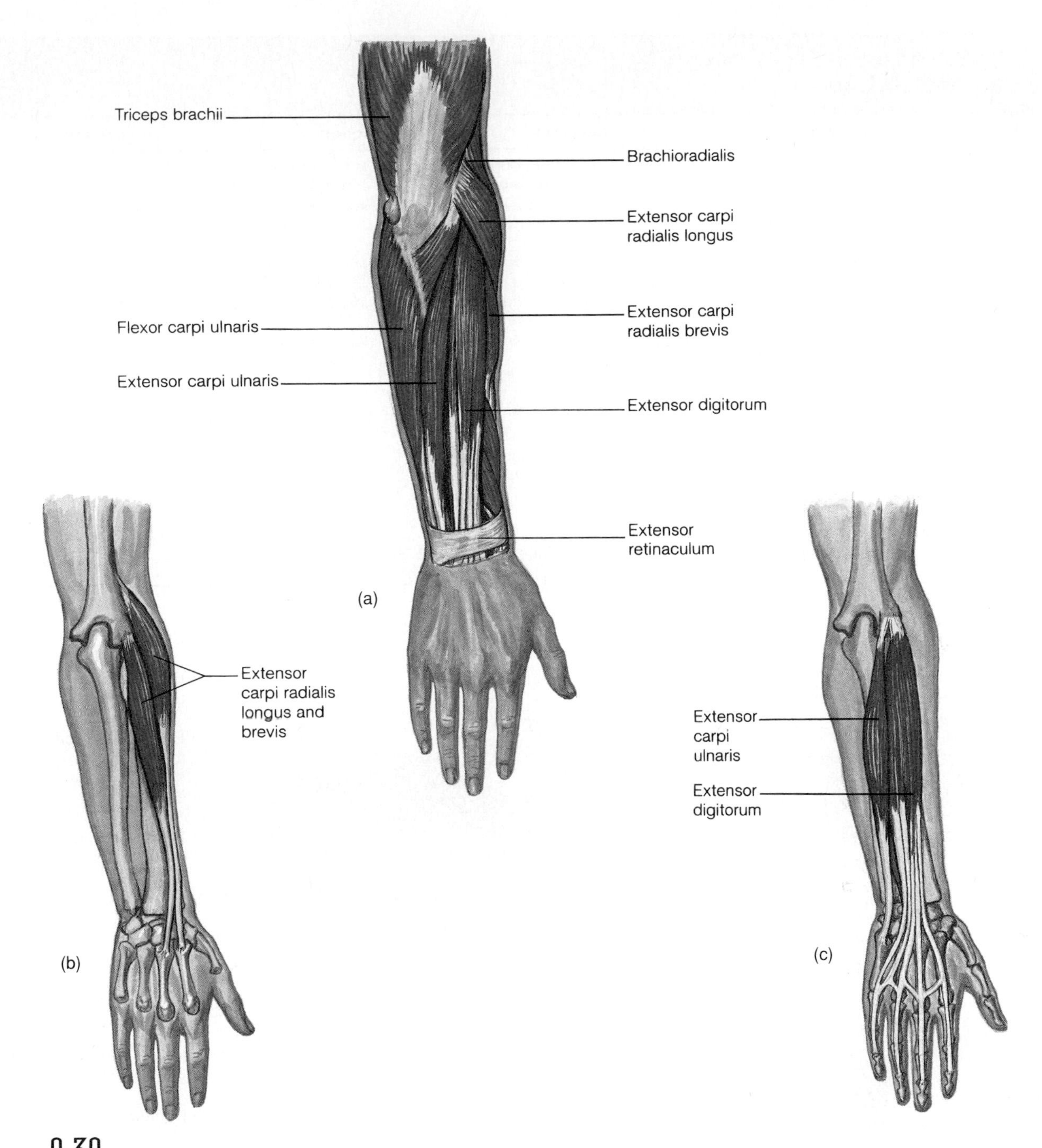


**FIGURE 9.30**
(*a*) Muscles of the posterior forearm. (*b* and *c*) Isolated views of muscles associated with the posterior forearm.

The **flexor digitorum profundus** (flek′sor dij″ĭ-to′rum pro-fun′dus) is a large muscle that connects the ulna to the distal phalanges. It flexes the distal joints of the fingers, as when a fist is made (fig. 9.31).

The **flexor digitorum superficialis** (flek′sor dij″ĭ-to′rum su″per-fish″e-a′lis) is a large muscle located beneath the flexor carpi ulnaris. It arises by three heads—one from the medial epicondyle of the humerus, one from the medial side of the ulna, and one from the radius. It is inserted in the tendons of the fingers and flexes the fingers and, by a combined action, flexes the hand at the wrist (fig. 9.29).

### Extensors

The **extensor carpi radialis longus** (eks-ten′sor kar-pi′ ra″de-a′lis long′gus) runs along the lateral side of the forearm, connecting the humerus to the hand. It extends the hand at the wrist and assists in abducting the hand (figs. 9.30 and 9.31).

## Chart 9.9 Muscles That Move the Hand

| Muscle | Origin | Insertion | Action | Nerve Supply |
|---|---|---|---|---|
| Flexor carpi radialis | Medial epicondyle of humerus | Base of second and third metacarpals | Flexes and abducts hand at the wrist | Median n. |
| Flexor carpi ulnaris | Medial epicondyle of humerus and olecranon process | Carpal and metacarpal bones | Flexes and adducts hand at the wrist | Ulnar n. |
| Palmaris longus | Medial epicondyle of humerus | Fascia of palm | Flexes hand at the wrist | Median n. |
| Flexor digitorum profundus | Anterior surface of ulna | Bases of distal phalanges in fingers 2–5 | Flexes distal joints of fingers | Median and ulnar nerves |
| Flexor digitorum superficialis | Medial epicondyle of humerus, coronoid process of ulna, and radius | Tendons of fingers | Flexes fingers and hand | Median n. |
| Extensor carpi radialis longus | Distal end of humerus | Base of second metacarpal | Extends and abducts hand at the wrist | Radial n. |
| Extensor carpi radialis brevis | Lateral epicondyle of humerus | Base of second and third metacarpals | Extends and abducts hand at the wrist | Radial n. |
| Extensor carpi ulnaris | Lateral epicondyle of humerus | Base of fifth metacarpal | Extends and adducts hand at the wrist | Radial n. |
| Extensor digitorum | Lateral epicondyle of humerus | Posterior surface of phalanges in fingers 2–5 | Extends fingers | Radial n. |

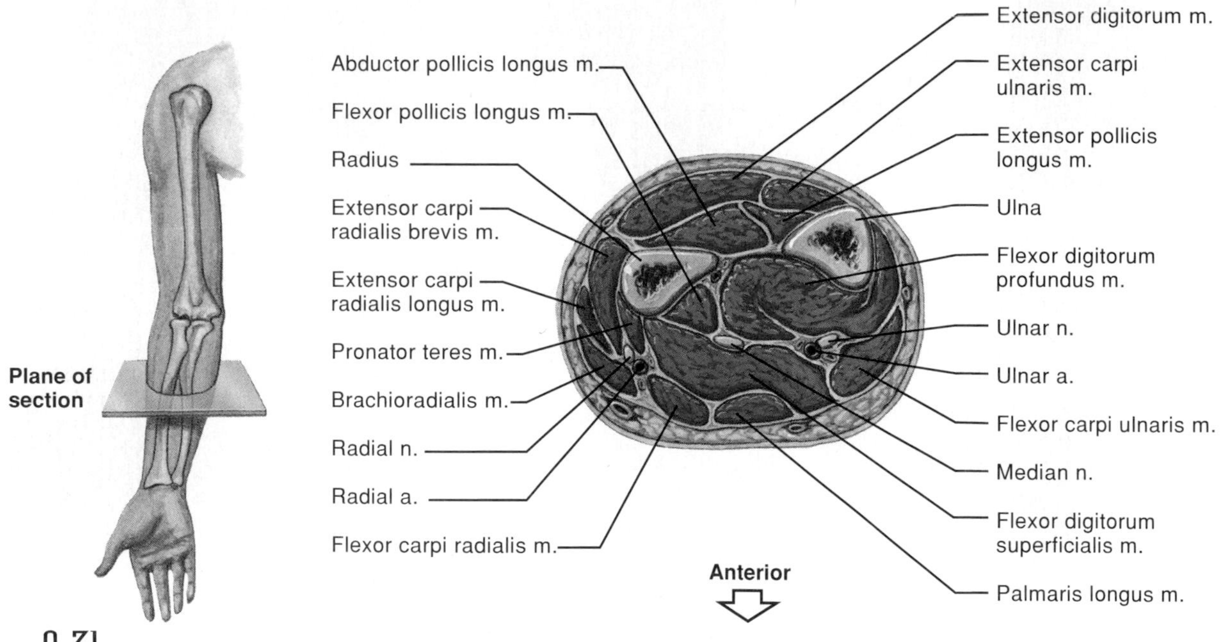


**Figure 9.31**
A cross section of the forearm.

The **extensor carpi radialis brevis** (eks-ten′sor kar-pi′ ra″de-a′lis brev′ ĭs) is a companion of the extensor carpi radialis longus and is located medially to it. This muscle runs from the humerus to metacarpal bones and extends the hand at the wrist. It also assists in abducting the hand (figs. 9.30 and 9.31).

The **extensor carpi ulnaris** (eks-ten′sor kar-pi′ ul-na′ris) is located along the posterior surface of the ulna and connects the humerus to the hand. It extends the hand at the wrist and assists in adducting it (figs. 9.30 and 9.31).

The **extensor digitorum** (eks-ten′sor dij″ĭ-to′rum) runs medially along the back of the forearm. It connects the humerus to the posterior surface of the phalanges and extends the fingers (figs. 9.30 and 9.31).

A structure called the *extensor retinaculum* consists of a group of heavy connective tissue fibers in the fascia of the wrist (fig. 9.30). It connects the lateral margin of the radius with the medial border of the styloid process of the ulna and certain bones of the wrist. The retinaculum gives off branches of connective tissue to the underlying wrist bones, creating a series of sheathlike compartments through which the tendons of the extensor muscles pass to the wrist and fingers.

## Muscles of the Abdominal Wall

The walls of the chest and pelvic regions are supported directly by bone, but those of the abdomen are not. Instead, the anterior and lateral walls of the abdomen are composed of layers of broad, flattened muscles. These muscles connect the rib cage and vertebral column to the pelvic girdle. A band of tough connective tissue, called the **linea alba,** extends from the xiphoid process of the sternum to the symphysis pubis. It is an attachment for some of the abdominal wall muscles.

Contraction of these muscles decreases the size of the abdominal cavity and increases the pressure inside. This action helps press air out of the lungs during forceful exhalation, and also aids in defecation, urination, vomiting, and childbirth.

The abdominal wall muscles are shown in figure 9.32 and reference plate 62, and are listed in chart 9.10. They include the following:

*External oblique*
*Internal oblique*
*Transversus abdominis*
*Rectus abdominis*

The **external oblique** (eks-ter′nal ŏ-blēk) is a broad, thin sheet of muscle whose fibers slant downward from the lower ribs to the pelvic girdle and the linea alba. When this muscle contracts, it tenses the abdominal wall and compresses the contents of the abdominal cavity.

Similarly, the **internal oblique** (in-ter′nal ŏ-blēk) is a broad, thin sheet of muscle located beneath the external oblique. Its fibers run up and forward from the pelvic girdle to the lower ribs. Its function is similar to that of the external oblique.

The **transversus abdominis** (trans-ver′sus ab-dom′ĭ-nis) forms a third layer of muscle beneath the external and internal obliques. Its fibers run horizontally from the lower ribs, lumbar vertebrae, and ilium to the linea alba and pubic bones. It functions in the same manner as the external and internal obliques.

The **rectus abdominis** (rek′tus ab-dom′ĭ-nis) is a long, straplike muscle that connects the pubic bones to the ribs and sternum. Three or more fibrous bands cross the muscle transversely, giving it a segmented appearance. The muscle functions with other abdominal wall muscles to compress the contents of the abdominal cavity, and it also helps to flex the vertebral column.

## Muscles of the Pelvic Outlet

Two muscular sheets span the outlet of the pelvis—a deeper **pelvic diaphragm** and a more superficial **urogenital diaphragm.** The pelvic diaphragm forms the floor of the pelvic cavity, and the urogenital diaphragm fills the space within the pubic arch. Figure 9.33 and chart 9.11 show the muscles of the male and female pelvic outlets. They include the following:

**Pelvic diaphragm**
*Levator ani*
*Coccygeus*
**Urogenital diaphragm**
*Superficial transversus perinei*
*Bulbospongiosus*
*Ischiocavernosus*
*Sphincter urethrae*

### Pelvic Diaphragm

The **levator ani** (le-va′tor ah-ni′) muscles form a thin sheet across the pelvic outlet. They are connected at the midline posteriorly by a ligament that extends from the tip of the coccyx to the anal canal. Anteriorly, they are separated in the male by the urethra and the anal canal, and in the female by the urethra, vagina, and anal canal. These muscles help support the pelvic viscera and provide sphincterlike action in the anal canal and vagina.

An *external anal sphincter* that is under voluntary control and an *internal anal sphincter* that is formed of involuntary muscle fibers of the intestine encircle the anal canal and keep it closed.

The **coccygeus** (kok-sij′e-us) is a fan-shaped muscle that extends from the ischial spine to the coccyx and sacrum. It aids the levator ani.

### Urogenital Diaphragm

The **superficial transversus perinei** (su″per-fish′al trans-ver′sus per″ĭ-ne′i) consists of a small bundle of muscle fibers that passes medially from the ischial tuberosity along the posterior border of the urogenital diaphragm. It assists other muscles in supporting the pelvic viscera.

In males, the **bulbospongiosus** (bul″bo-spon″je-o′sus) muscles are united surrounding the base of the penis. They assist in emptying the urethra. In females, these muscles are separated medially by the vagina and constrict the vaginal opening. They can

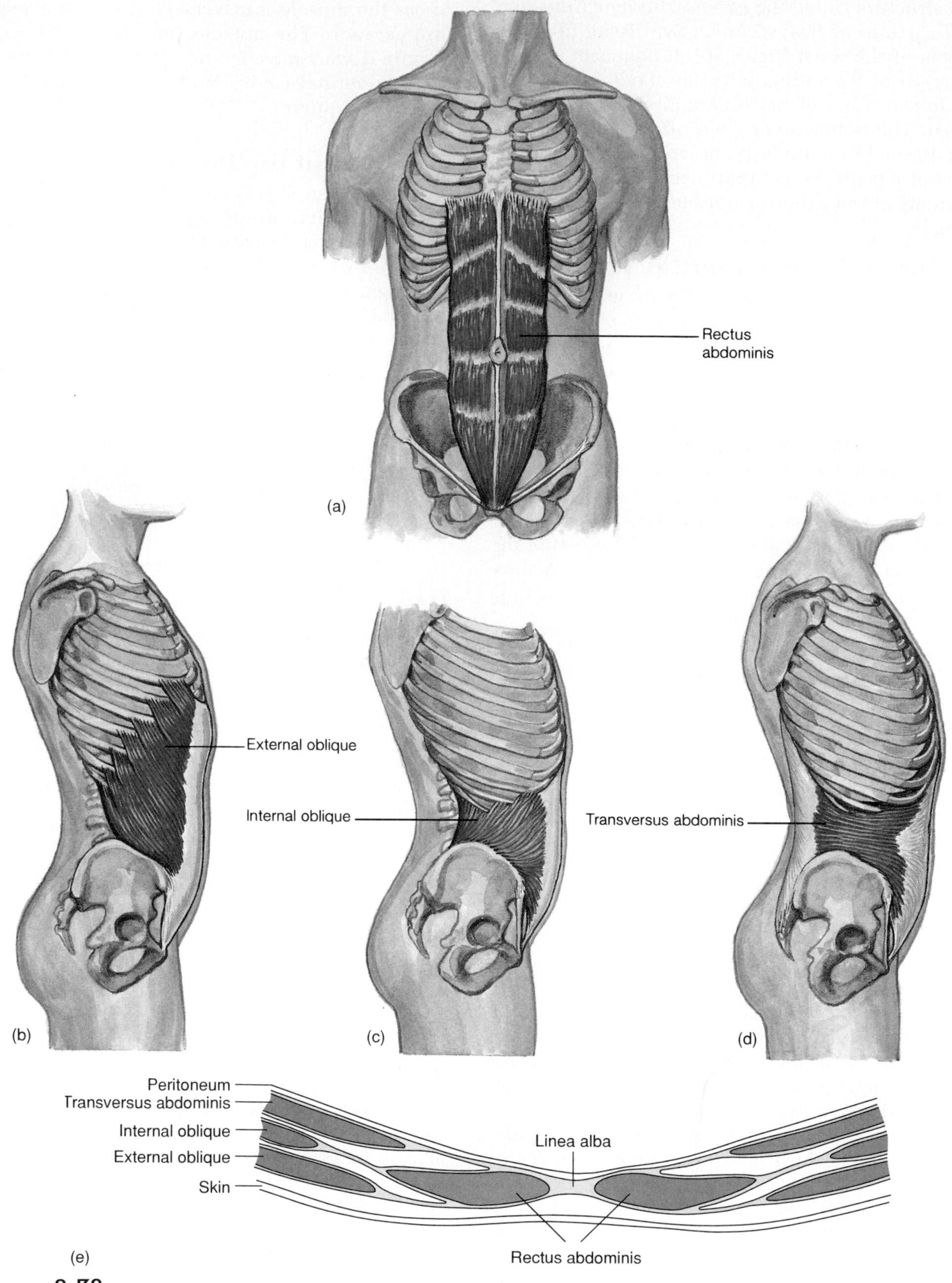


**FIGURE 9.32**

(*a–d*) Isolated muscles of the abdominal wall; (*e*) transverse section through the abdominal wall.

## CHART 9.10 Muscles of the Abdominal Wall

| Muscle | Origin | Insertion | Action | Nerve Supply |
|---|---|---|---|---|
| External oblique | Outer surfaces of lower ribs | Outer lip of iliac crest and linea alba | Tenses abdominal wall and compresses abdominal contents | Intercostal nerves 7–12 |
| Internal oblique | Crest of ilium and inguinal ligament | Cartilages of lower ribs, linea alba, and crest of pubis | Same as above | Intercostal nerves 7–12 |
| Transversus abdominis | Costal cartilages of lower ribs, processes of lumbar vertebrae, lip of iliac crest, and inguinal ligament | Linea alba and crest of pubis | Same as above | Intercostal nerves 7–12 |
| Rectus abdominis | Crest of pubis and symphysis pubis | Xiphoid process of sternum and costal cartilages | Same as above; also flexes vertebral column | Intercostal nerves 7–12 |

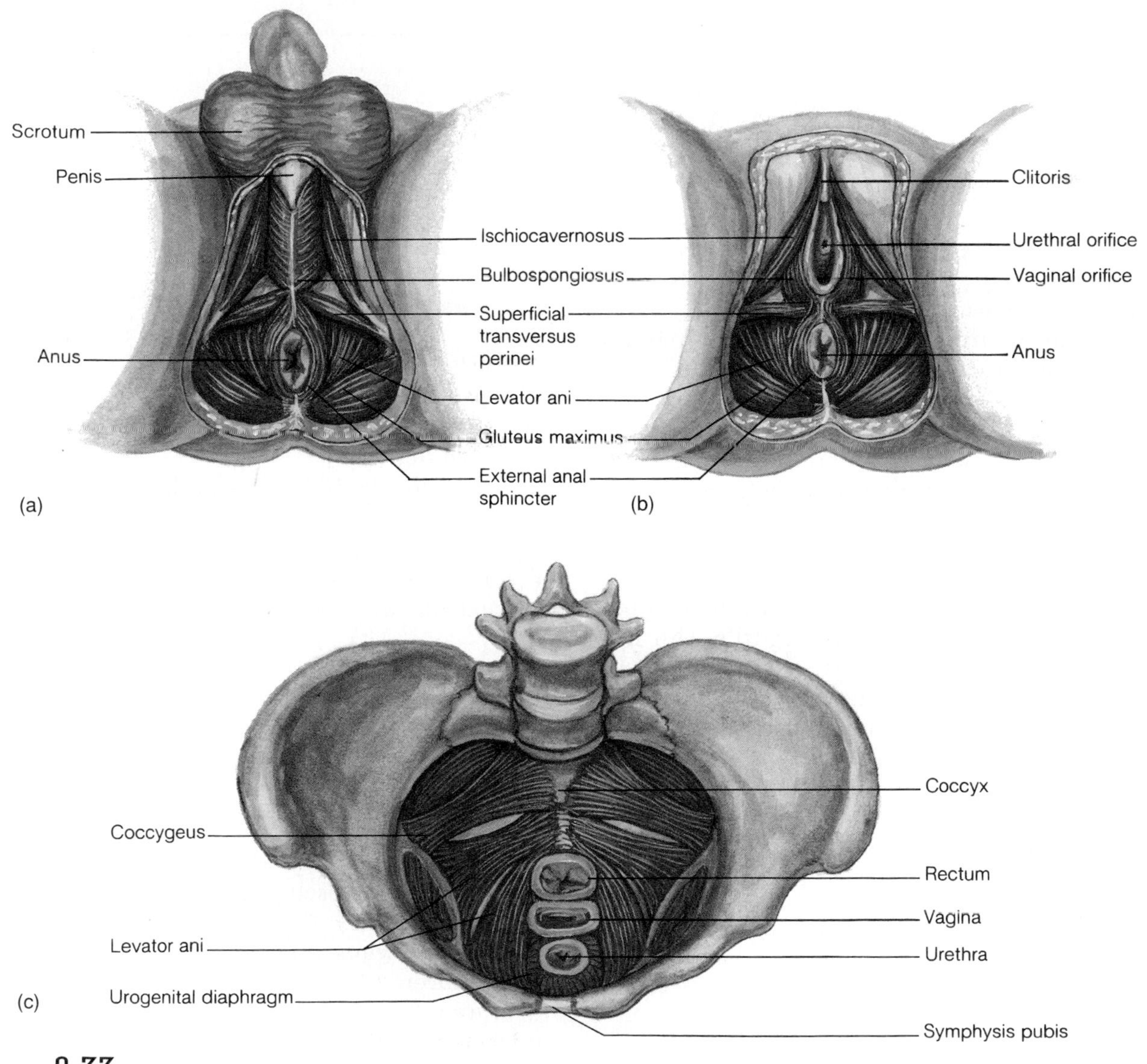


**FIGURE 9.33**
External view of muscles of (*a*) the male pelvic outlet, and (*b*) the female pelvic outlet. (*c*) Internal view of pelvic and urogenital diaphragms.

## CHART 9.11 Muscles of the Pelvic Outlet

| Muscle | Origin | Insertion | Action | Nerve Supply |
|---|---|---|---|---|
| Levator ani | Pubic bone and ischial spine | Coccyx | Supports pelvic viscera, and provides sphincterlike action in anal canal and vagina | Pudendal n. |
| Coccygeus | Ischial spine | Sacrum and coccyx | Same as above | S4 and S5 nerves |
| Superficial transversus perinei | Ischial tuberosity | Central tendon | Supports pelvic viscera | Pudendal n. |
| Bulbospongiosus | Central tendon | Males: Urogenital diaphragm and fascia of penis | Males: Assists emptying of urethra | Pudendal n. |
| | | Females: Pubic arch and root of clitoris | Females: Constricts vagina | |
| Ischiocavernosus | Ischial tuberosity | Pubic arch | Assists function of bulbospongiosus | Pudendal n. |
| Sphincter urethrae | Margins of pubis and ischium | Fibers of each unite with those from other side | Opens and closes urethra | Pudendal n. |

also retard the flow of blood in veins, which helps maintain an erection in the penis of the male and in the clitoris of the female.

The **ischiocavernosus** (is″ke-o-kav″er-no′sus) muscle is a tendinous structure that extends from the ischial tuberosity to the margin of the pubic arch. It assists the bulbospongiosus muscle.

The **sphincter urethrae** (sfingk′ter u-re′thrē) are muscles that arise from the margins of the pubic and ischial bones. Each arches around the urethra and unites with the one on the other side. Together they act as a sphincter that closes the urethra by compression and opens it by relaxation, thus helping control the flow of urine.

## Muscles that Move the Thigh

The muscles that move the thigh are attached to the femur and to some part of the pelvic girdle. (An important exception is the sartorius, described later.) They can be separated into anterior and posterior groups. The muscles of the anterior group primarily flex the thigh; those of the posterior group extend, abduct, or rotate it. The muscles in these groups are shown in figures 9.34, 9.35, 9.36, 9.37, and in reference plates 66 and 67, and are listed in chart 9.12. Muscles that move the thigh include the following:

**Anterior group**
- *Psoas major*
- *Iliacus*

**Posterior group**
- *Gluteus maximus*
- *Gluteus medius*
- *Gluteus minimus*
- *Tensor fasciae latae*

Still another group of muscles, attached to the femur and pelvic girdle, adducts the thigh. This group includes:

- *Adductor longus*
- *Adductor magnus*
- *Gracilis*

### Anterior Group

The **psoas** (so′as) **major** is a long, thick muscle that connects the lumbar vertebrae to the femur. It flexes the thigh (fig. 9.34).

The **iliacus** (il′e-ak-us), a large, fan-shaped muscle, lies along the lateral side of the psoas major. The iliacus and the psoas major are the primary flexors of the thigh, and they advance the lower limb in walking movements (fig. 9.34).

### Posterior Group

The **gluteus maximus** (gloo′te-us mak′si-mus) is the largest muscle in the body and covers a large part of each buttock. It connects the ilium, sacrum, and coccyx to the femur by fascia of the thigh and extends the thigh. The gluteus maximus helps to straighten the lower limb at the hip when a person walks, runs, or climbs. It is also used to raise the body from a sitting position (fig. 9.35).

The **gluteus medius** (gloo′te-us me′de-us) is partly covered by the gluteus maximus. Its fibers extend from the ilium to the femur, and they abduct the thigh and rotate it medially (fig. 9.35).

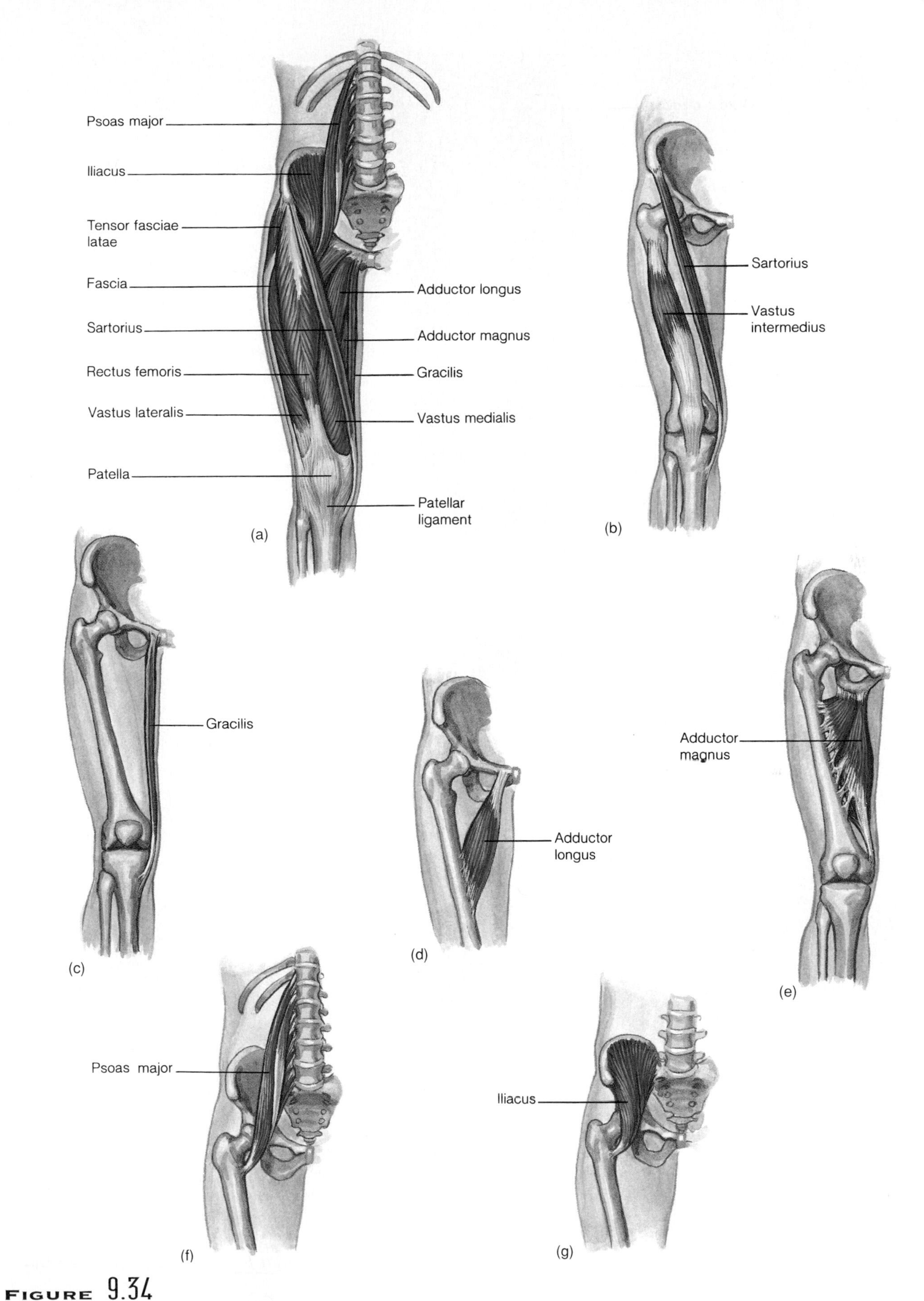


**FIGURE 9.34**
(*a*) Muscles of the anterior right thigh. Isolated views of (*b*) the vastus intermedius; (*c–e*) adductors of the thigh; (*f–g*) flexors of the thigh.

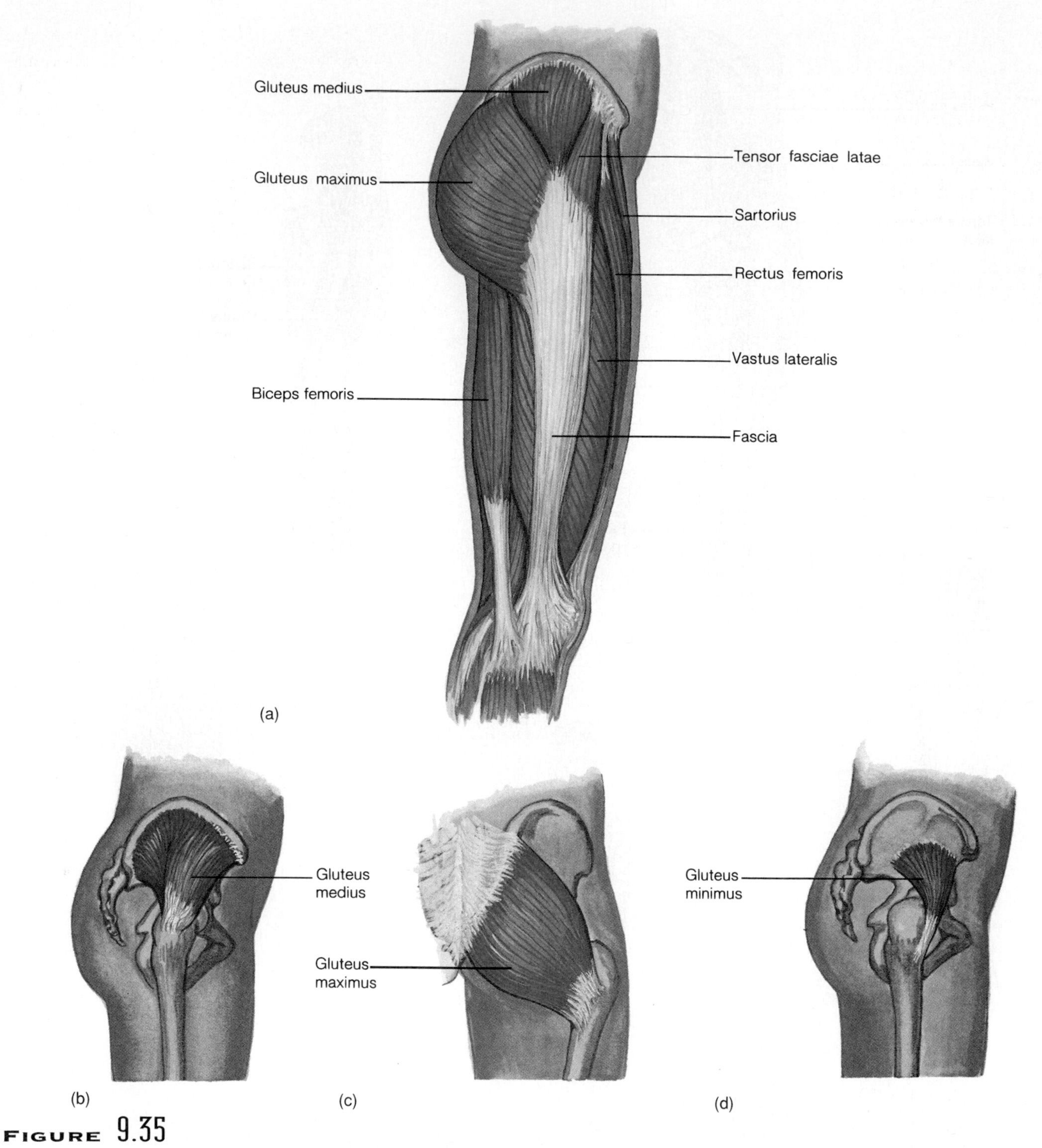


**FIGURE 9.35**

(*a*) Muscles of the lateral right thigh. (*b–d*) Isolated views of the gluteal muscles.

The **gluteus minimus** (gloo′te-us min′ĭ-mus) lies beneath the gluteus medius and is its companion in attachments and functions (fig. 9.35).

The **tensor fasciae latae** (ten′sor fash′e-e lah-tē) connects the ilium to the fascia of the thigh, which continues downward to the tibia. This muscle abducts and flexes the thigh and rotates it medially (fig. 9.35).

The gluteus medius and gluteus minimus help support and maintain the normal position of the pelvis. If these muscles are paralyzed as a result of injury or disease, the pelvis tends to drop to one side whenever the foot on that side is raised. Consequently, the person walks with a waddling limp called the *gluteal gait.*

### Thigh Adductors

The **adductor longus** (ah-duk′tor long′gus) is a long, triangular muscle that runs from the pubic bone to the femur. It adducts the thigh and assists in flexing and rotating it laterally (fig. 9.34).

The **adductor magnus** (ah-duk′tor mag′nus) is the largest adductor of the thigh. It is a triangular muscle that connects the ischium to the femur. It adducts the thigh and assists in extending and rotating it laterally (fig. 9.34).

The **gracilis** (gras′il-is) is a long, straplike muscle that passes from the pubic bone to the tibia. It adducts the thigh and flexes the leg at the knee (fig. 9.34).

## Muscles that Move the Leg

The muscles that move the leg connect the tibia or fibula to the femur or to the pelvic girdle. They fall into two major groups—those that flex the knee and those that extend it. The muscles of these groups are shown in figures 9.34, 9.35, 9.36, 9.37, in reference plates 66 and 67, and are listed in chart 9.13. Muscles that move the leg include the following:

**Flexors**

*Biceps femoris*
*Semitendinosus*
*Semimembranosus*
*Sartorius*

**Extensor**

*Quadriceps femoris group*

### Flexors

As its name implies, the **biceps femoris** (bi′seps fem′or-is) has two heads, one attached to the ischium and the other attached to the femur. This muscle passes along the back of the thigh on the lateral side and connects to the proximal ends of the fibula and tibia. The biceps femoris is one of the hamstring muscles, and its tendon (hamstring) can be felt as a lateral ridge behind the knee. This muscle flexes and rotates the leg laterally and extends the thigh (figs. 9.35 and 9.36).

The **semitendinosus** (sem″e-ten′dĭ-no-sus) is another hamstring muscle. It is a long, bandlike muscle on the back of the thigh toward the medial side, connecting the ischium to the proximal end of the tibia. The semitendinosus is so named because it becomes tendinous in the middle of the thigh, continuing to its insertion as a long, cordlike tendon. It flexes and rotates the leg medially, and extends the thigh (fig. 9.36).

The **semimembranosus** (sem″e-mem′brah-no-sus) is the third hamstring muscle and is the most medially located muscle in the back of the thigh. It connects the ischium to the tibia, and flexes and rotates the leg medially and extends the thigh (fig. 9.36).

The **sartorius** (sar-to′re-us) is an elongated, straplike muscle that passes obliquely across the front of the thigh and then descends over the medial side of the knee. It connects the ilium to the tibia, and flexes the leg and the thigh. It can also abduct the thigh and rotate it laterally (figs. 9.34 and 9.35).

> The tendinous attachments of the hamstring muscles to the ischial tuberosity are sometimes torn as a result of strenuous running or kicking motions. This painful injury is commonly called "pulled hamstrings," and is usually accompanied by internal bleeding from damaged blood vessels that supply the muscles.

### Extensor

The large, fleshy muscle group called the **quadriceps femoris** (kwod′rĭ-seps fem′or-is) occupies the front and sides of the thigh, and is the primary extensor of the knee. It is composed of four parts—*rectus femoris, vastus lateralis, vastus medialis,* and *vastus intermedius* (figs. 9.34 and 9.37). These parts connect the ilium and femur to a common *patellar tendon,* which passes over the front of the knee and attaches to the patella. This tendon then continues as the *patellar ligament* to the tibia.

> Occasionally, as a result of traumatic injury in which a muscle such as the quadriceps femoris is compressed against an underlying bone, new bone tissue may begin to develop within the damaged muscle. This condition is called *myositis ossificans.* When the bone tissue matures several months following the injury, the newly formed bone can be removed surgically.

## Muscles that Move the Foot

Movements of the foot include movements of the ankle and toes. Therefore, a number of muscles that move the foot are located in the leg. They attach the femur, tibia, and fibula to bones of the foot and are responsible for moving the foot upward (dorsiflexion) or downward (plantar flexion), and turning the sole of the foot inward (inversion) or outward (eversion). These muscles

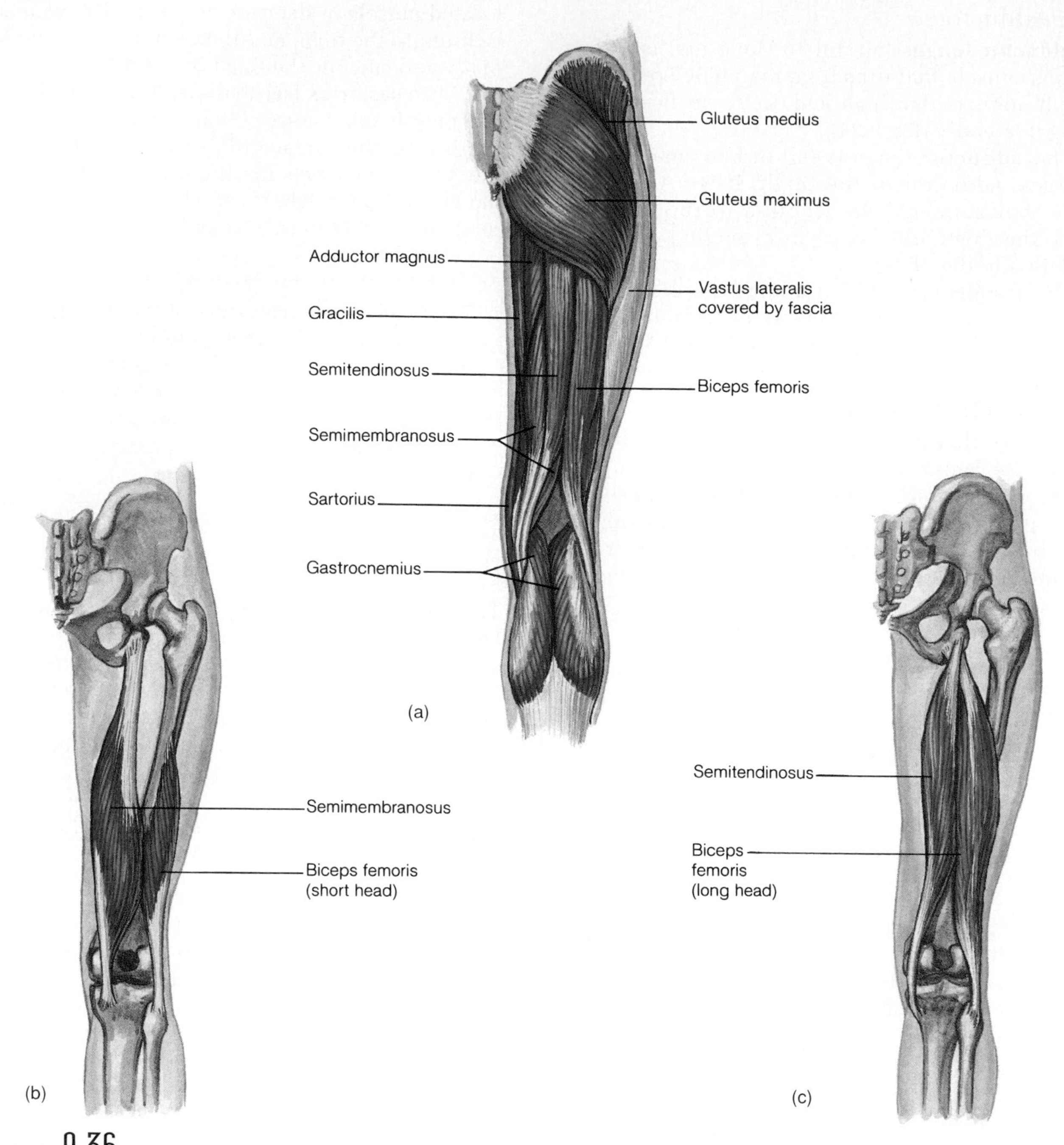


**FIGURE 9.36**
(*a*) Muscles of the posterior right thigh. (*b* and *c*) Isolated views of muscles that flex the leg at the knee.

are shown in figures 9.38, 9.39, 9.40, 9.41, in reference plates 68, 69, 70, and are listed in chart 9.14. Muscles that move the foot include the following:

**Dorsal flexors**
*Tibialis anterior*
*Peroneus tertius*
*Extensor digitorum longus*

**Plantar flexors**
*Gastrocnemius*
*Soleus*
*Flexor digitorum longus*

**Invertor**
*Tibialis posterior*

**Evertor**
*Peroneus longus*

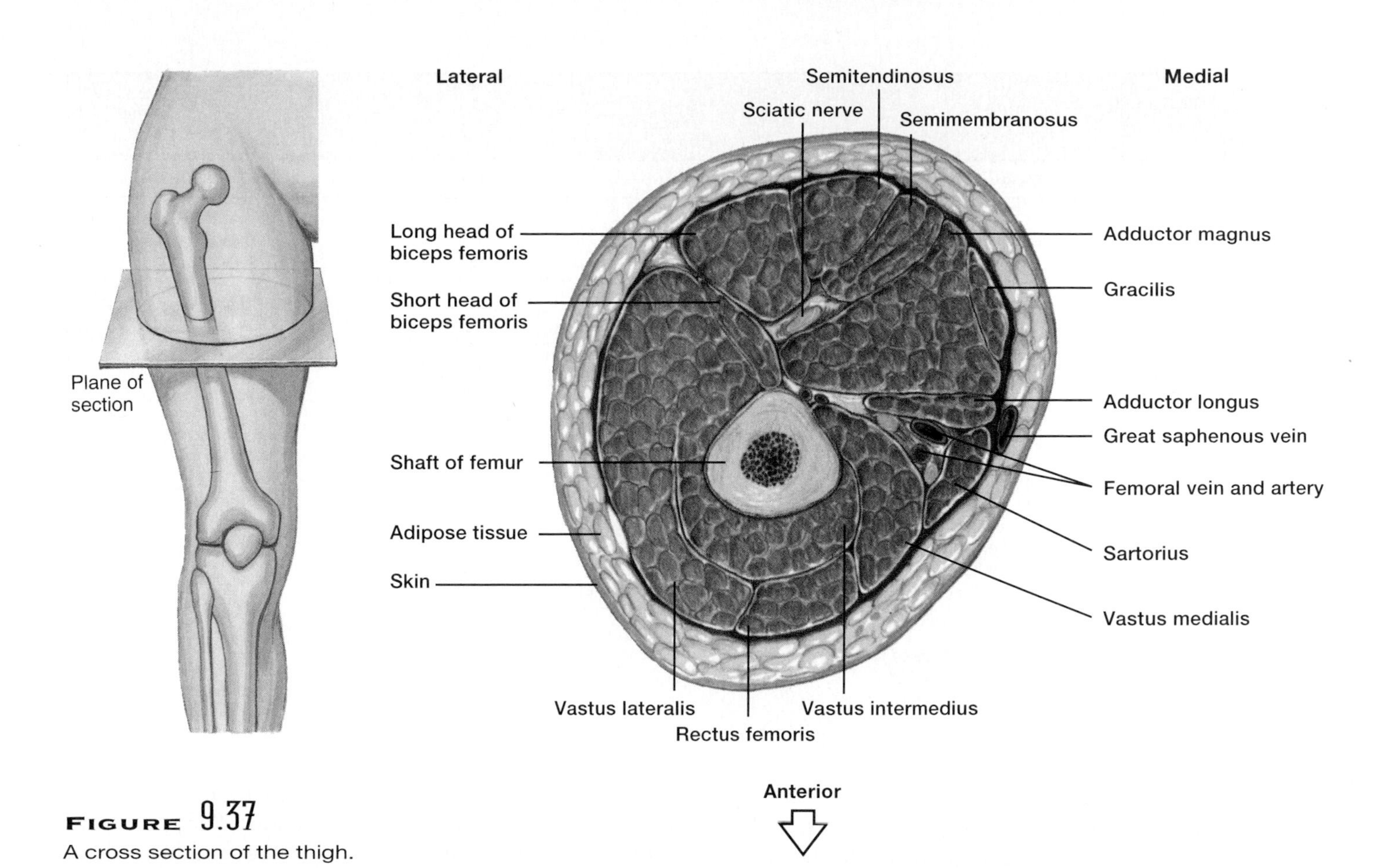


**FIGURE 9.37**
A cross section of the thigh.

**CHART 9.12 MUSCLES THAT MOVE THE THIGH**

| Muscle | Origin | Insertion | Action | Nerve Supply |
|---|---|---|---|---|
| Psoas major | Lumbar intervertebral disks; bodies and transverse processes of lumbar vertebrae | Lesser trochanter of femur | Flexes thigh | Branches of L1–3 nerves |
| Iliacus | Iliac fossa of ilium | Lesser trochanter of femur | Flexes thigh | Femoral n. |
| Gluteus maximus | Sacrum, coccyx, and posterior surface of ilium | Posterior surface of femur and fascia of thigh | Extends thigh at hip | Inferior gluteal n. |
| Gluteus medius | Lateral surface of ilium | Greater trochanter of femur | Abducts and rotates thigh medially | Superior gluteal n. |
| Gluteus minimus | Lateral surface of ilium | Greater trochanter of femur | Same as gluteus medius | Superior gluteal n. |
| Tensor fasciae latae | Anterior iliac crest | Fascia of thigh | Abducts, flexes, and rotates thigh medially | Superior gluteal n. |
| Adductor longus | Pubic bone near symphysis pubis | Posterior surface of femur | Adducts, flexes, and rotates thigh laterally | Obturator n. |
| Adductor magnus | Ischial tuberosity | Posterior surface of femur | Adducts, extends, and rotates thigh laterally | Obturator and branch of sciatic n. |
| Gracilis | Lower edge of symphysis pubis | Medial surface of tibia | Adducts thigh and flexes leg at the knee | Obturator n. |

CHART 9.13 MUSCLES THAT MOVE THE LEG

| Muscle | Origin | Insertion | Action | Nerve Supply |
|---|---|---|---|---|
| ***Hamstring group*** | | | | |
| Biceps femoris | Ischial tuberosity and linea aspera of femur | Head of fibula and lateral condyle of tibia | Flexes and rotates leg laterally and extends thigh | Tibial n. |
| Semitendinosus | Ischial tuberosity | Medial surface of tibia | Flexes and rotates leg medially and extends thigh | Tibial n. |
| Semimembranosus | Ischial tuberosity | Medial condyle of tibia | Flexes and rotates leg medially and extends thigh | Tibial n. |
| ***Sartorius*** | Anterior superior iliac spine | Medial surface of tibia | Flexes leg and thigh, abducts and rotates thigh laterally | Femoral n. |
| ***Quadriceps femoris group*** | | | | |
| Rectus femoris | Spine of ilium and margin of acetabulum | | | |
| Vastus lateralis | Greater trochanter and posterior surface of femur | Patella by common tendon, which continues as patellar ligament to tibial tuberosity | Extends leg at knee | Femoral n. |
| Vastus medialis | Medial surface of femur | | | |
| Vastus intermedius | Anterior and lateral surfaces of femur | | | |

### Dorsal Flexors

The **tibialis anterior** (tib″e-a′lis an-te′re-or) is an elongated, spindle-shaped muscle located on the front of the leg. It arises from the surface of the tibia, passes medially over the distal end of the tibia, and attaches to bones of the ankle and foot. Contraction of the tibialis anterior causes dorsiflexion and inversion of the foot (fig. 9.38).

The **peroneus tertius** (per″o-ne′us ter′shus) is a muscle of variable size that connects the fibula to the lateral side of the foot. It functions in dorsiflexion and eversion of the foot (fig. 9.38).

The **extensor digitorum longus** (eks-ten′sor dij″ĭ-to′rum long′gus) is situated along the lateral side of the leg just behind the tibialis anterior. It arises from the proximal end of the tibia and the shaft of the fibula. Its tendon divides into four parts as it passes over the front of the ankle. These parts continue over the surface of the foot and attach to the four lateral toes. The actions of the extensor digitorum longus include dorsiflexion of the foot, eversion of the foot, and extension of the toes (figs. 9.38 and 9.39).

### Plantar Flexors

The **gastrocnemius** (gas″trok-ne′me-us) on the back of the leg forms part of the calf. It arises by two heads from the femur. The distal end of this muscle joins the strong *calcaneal tendon* (Achilles tendon), which descends to the heel and attaches to the calcaneus. The gastrocnemius is a powerful plantar flexor of the foot that aids in pushing the body forward when a person walks or runs. It also flexes the leg at the knee (figs. 9.39 and 9.40).

Strenuous athletic activity may partially or completely tear the calcaneal (Achilles) tendon. This injury occurs most frequently in middle-aged athletes who run or play sports that involve quick movements and directional changes. A torn calcaneal tendon usually requires surgical treatment.

The **soleus** (so′le-us) is a thick, flat muscle located beneath the gastrocnemius, and together these two muscles form the calf of the leg. The soleus arises from the tibia and fibula, and it extends to the heel by way of the calcaneal tendon. It acts with the gastrocnemius to cause plantar flexion of the foot (figs. 9.39 and 9.40).

The **flexor digitorum longus** (flek′sor dij″ĭ-to′rum long′gus) extends from the posterior surface of the tibia to the foot. Its tendon passes along the plantar surface of the foot. There it divides into four parts that attach to the terminal bones of the four lateral toes.

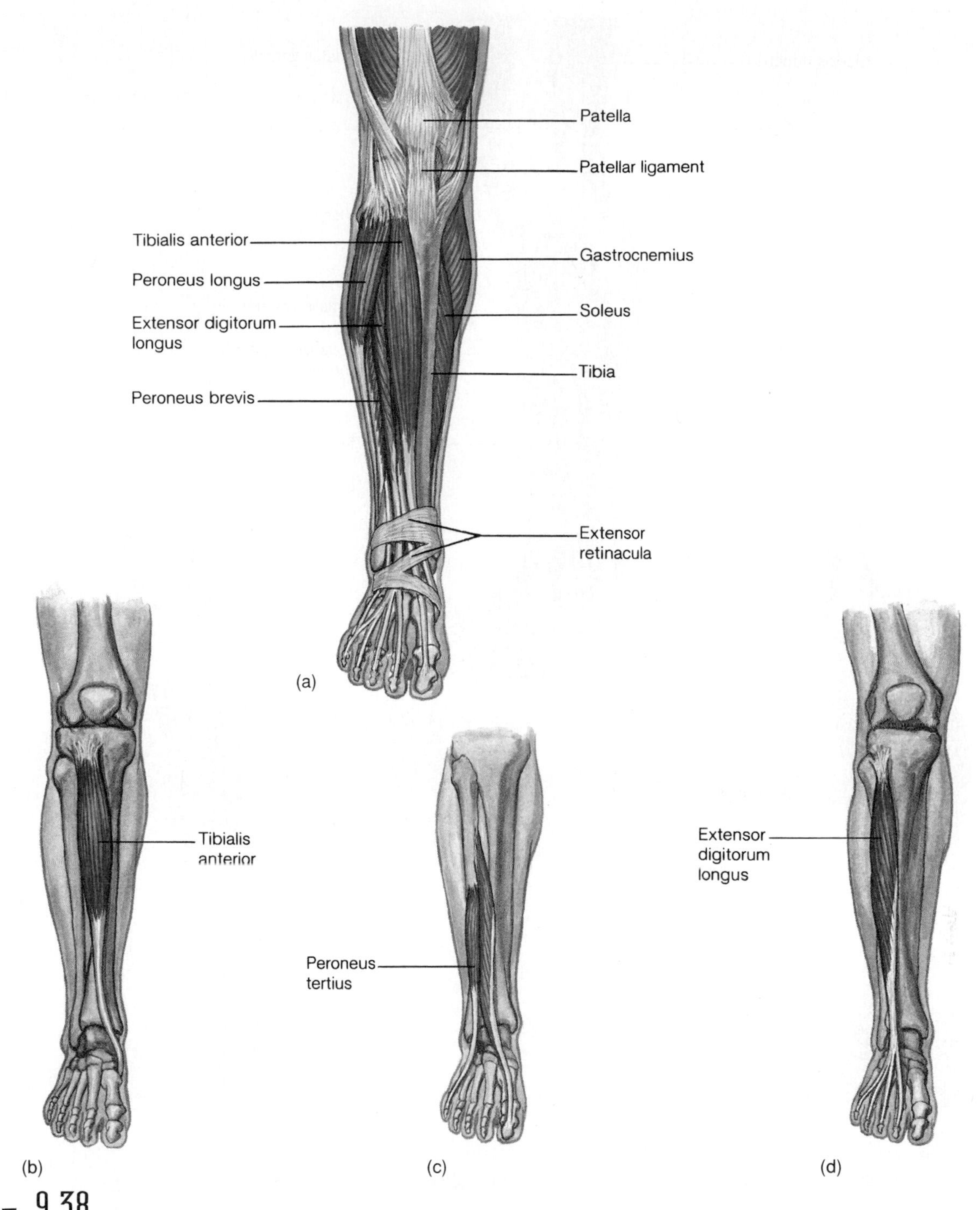


**FIGURE 9.38**
(*a*) Muscles of the anterior right leg. (*b–d*) Isolated views of muscles associated with the anterior leg.

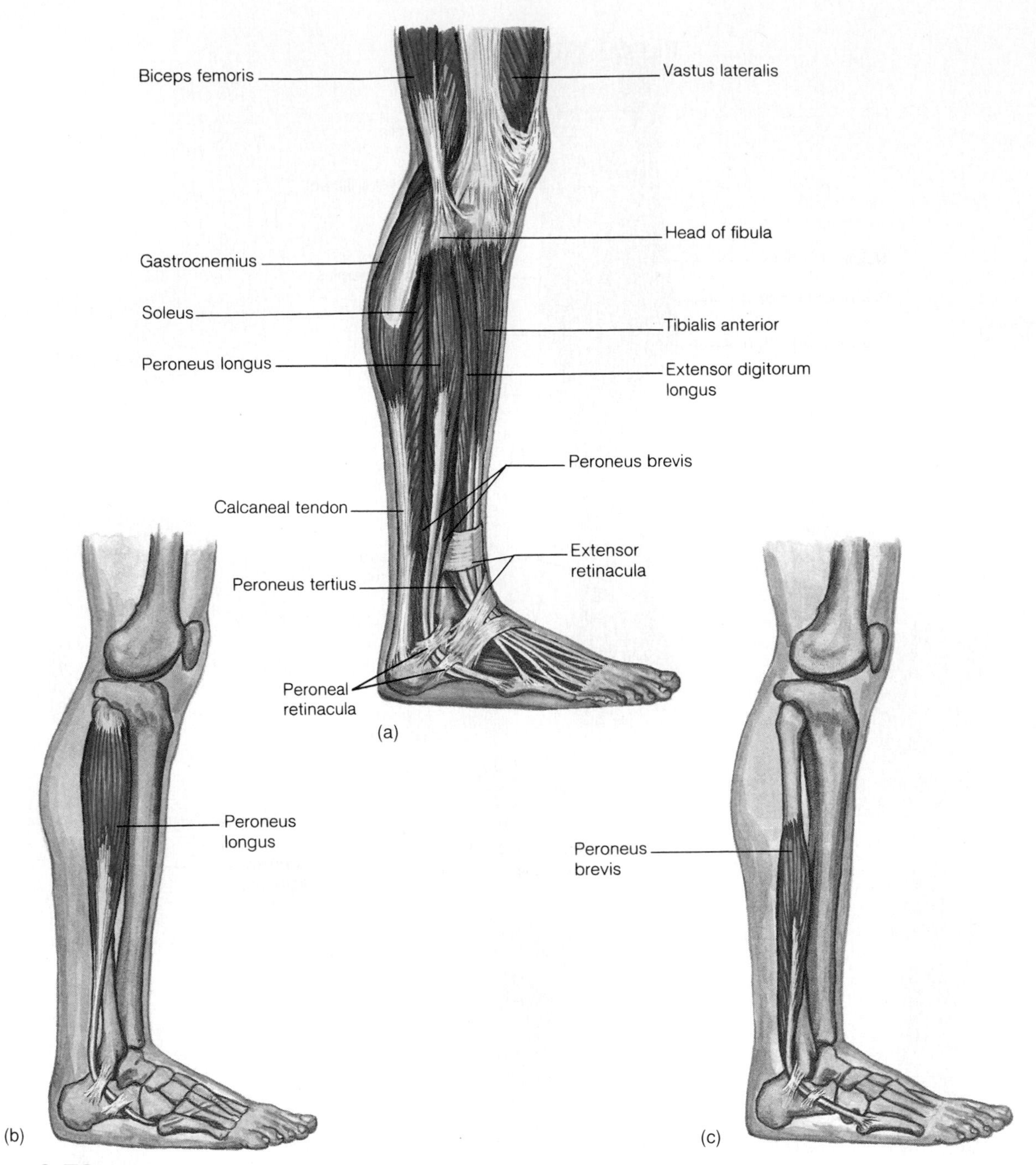


**FIGURE 9.39**
(*a*) Muscles of the lateral right leg. Isolated views of (*b*) peroneus longus and (*c*) peroneus brevis.

This muscle assists in plantar flexion of the foot, flexion of the four lateral toes, and inversion of the foot (fig. 9.40).

### Invertor

The **tibialis posterior** (tib″e-a′lis pos-tēr′e-or) is the deepest of the muscles on the back of the leg. It connects the fibula and tibia to the ankle bones by means of a tendon that curves under the medial malleolus. This muscle assists in inversion and plantar flexion of the foot (fig. 9.40).

### Evertor

The **peroneus longus** (per″o-ne′us long′gus) is a long, straplike muscle located on the lateral side of the leg. It connects the tibia and the fibula to the foot by means

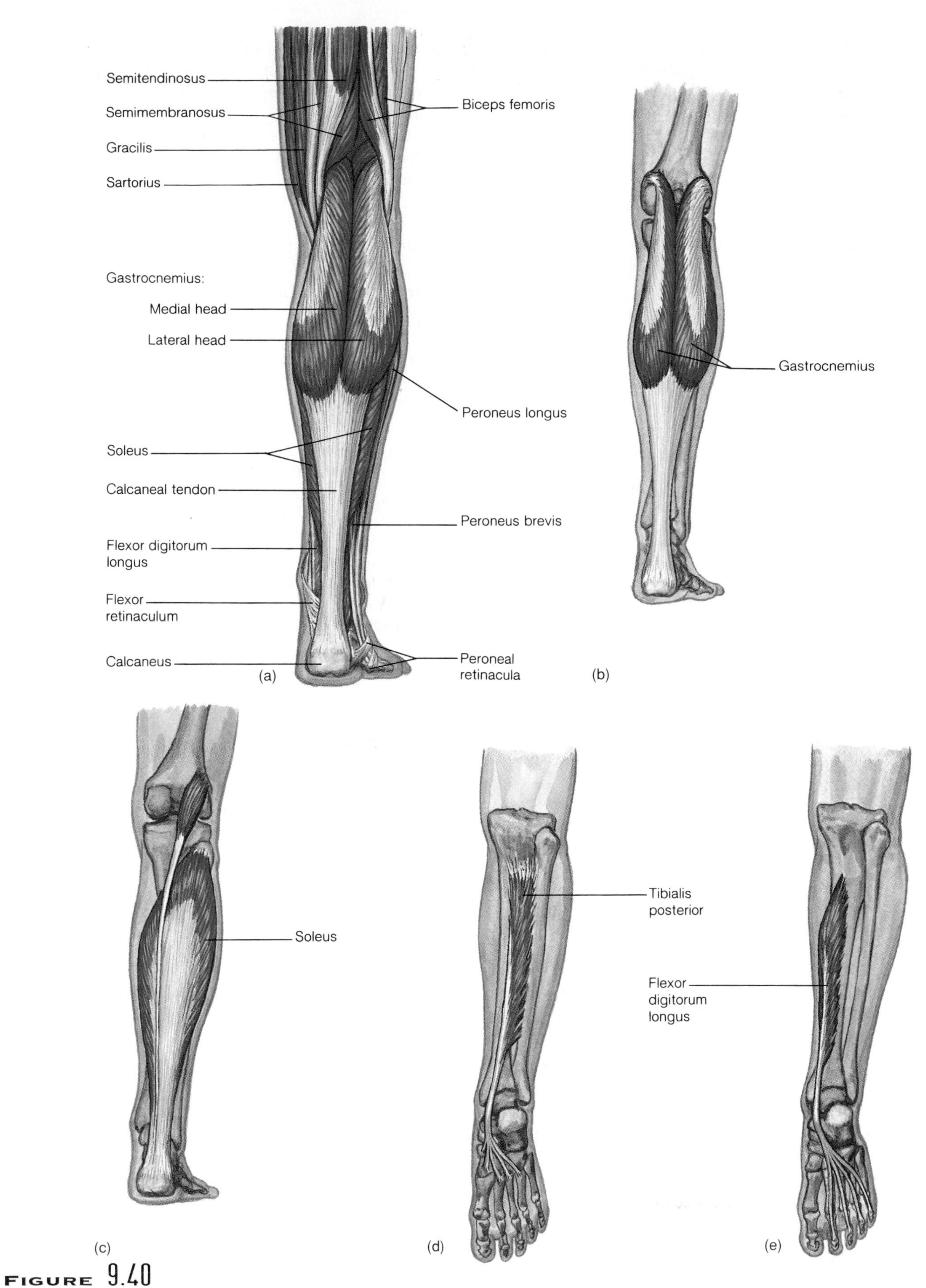


**FIGURE 9.40**
(*a*) Muscles of the posterior right leg. (*b–e*) Isolated views of muscles associated with the posterior right leg.

of a stout tendon that passes behind the lateral malleolus. It everts the foot, assists in plantar flexion, and helps support the arch of the foot (figs. 9.39 and 9.41).

As in the wrist, fascia in various regions of the ankle thicken to form retinacula. Anteriorly, for example, *extensor retinacula* connect the tibia and fibula as well as the calcaneus and fascia of the sole. These retinacula form sheaths for tendons crossing the front of the ankle (fig. 9.39).

Posteriorly, on the inside, a *flexor retinaculum* runs between the medial malleolus and the calcaneus, and forms sheaths for tendons passing beneath the foot (fig. 9.40). *Peroneal retinacula* connect the lateral malleolus and the calcaneus, providing sheaths for tendons on the lateral side of the ankle (fig. 9.39).

## Clinical Terms Related to the Muscular System

**contracture** (kon-trak′tūr) A condition in which there is great resistance to the stretching of a muscle.

**convulsion** (kun-vul′shun) A series of involuntary contractions of various voluntary muscles.

**electromyography** (e-lek″tro-mi-og′rah-fe) A technique for recording the electrical changes that occur in muscle tissues.

**fibrillation** (fi″bri-la′shun) Spontaneous contractions of individual muscle fibers, producing rapid and uncoordinated activity within a muscle.

**fibrosis** (fi-bro′sis) A degenerative disease in which fibrous connective tissue replaces skeletal muscle tissue.

**fibrositis** (fi″bro-si′tis) An inflammation of fibrous connective tissues, especially in the muscle fascia. This disease is also called muscular rheumatism.

**muscular dystrophy** (mus′ku-lar dis′tro-fe) Progressive muscle weakness and atrophy caused by deficient dystrophin protein.

**myalgia** (mi-al′je-ah) Pain resulting from any muscular disease or disorder.

**myasthenia gravis** (mi″as-the′ne-ah grav′is) A chronic disease characterized by muscles that are weak and easily fatigued. It results from the immune system's attack on neuromuscular junctions so that stimuli are not transmitted from motor neurons to muscle fibers.

**myokymia** (mi″o-ki′me-ah) Persistent quivering of a muscle.

**myology** (mi-ol′o-je) The study of muscles.

**myoma** (mi-o′mah) A tumor composed of muscle tissue.

**myopathy** (mi-op′ah-the) Any muscular disease.

**myositis** (mi″o-si′tis) An inflammation of skeletal muscle tissue.

**myotomy** (mi-ot′o-me) The cutting of muscle tissue.

**myotonia** (mi″o-to′ne-ah) A prolonged muscular spasm.

**paralysis** (pah-ral′ĭ-sis) The loss of ability to move a body part.

**paresis** (pah-re′sis) A partial or slight paralysis of the muscles.

**shin splints** (shin′ splints) A soreness on the front of the leg due to straining the flexor digitorum longus, often as a result of walking up and down hills.

**torticollis** (tor″tĭ-kol′is) A condition in which the neck muscles, such as the sternocleidomastoids, contract involuntarily. It is more commonly called wryneck.

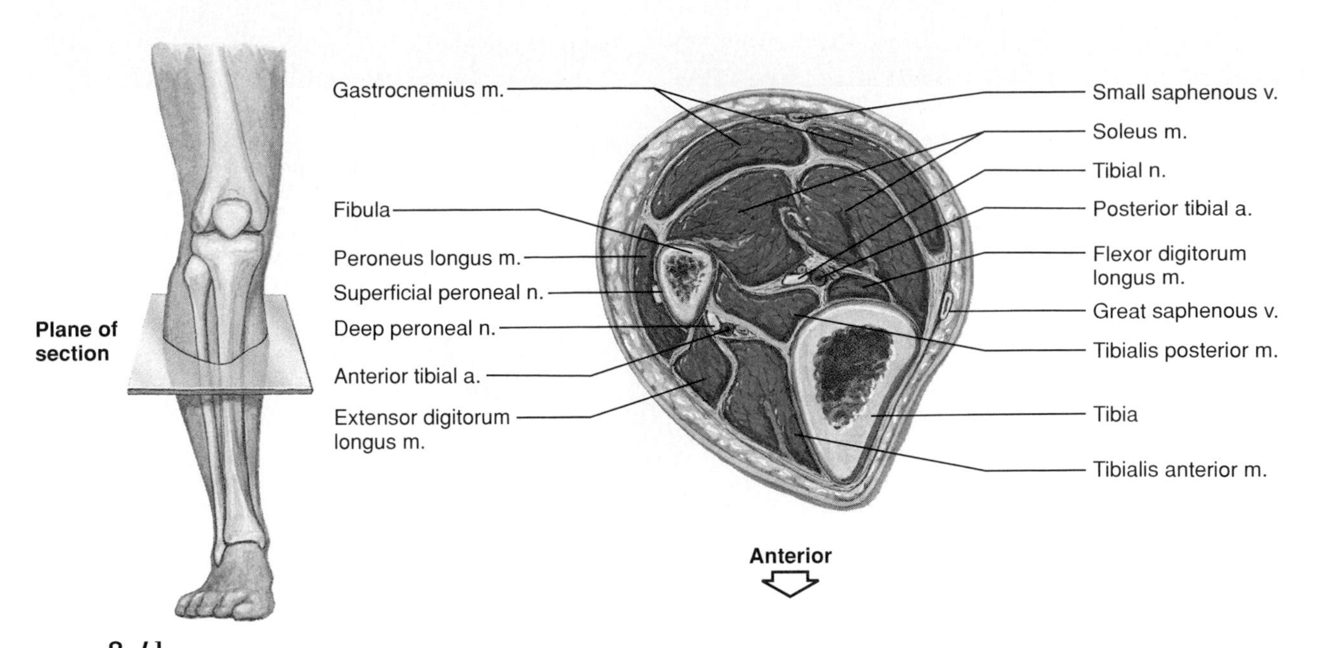


**FIGURE 9.41**
A cross section of the leg (superior view).

## CHART 9.14 Muscles that Move the Foot

| Muscle | Origin | Insertion | Action | Nerve Supply |
|---|---|---|---|---|
| Tibialis anterior | Lateral condyle and lateral surface of tibia | Tarsal bone (cuneiform) and first metatarsal | Dorsiflexion and inversion of foot | Deep peroneal n. |
| Peroneus tertius | Anterior surface of fibula | Dorsal surface of fifth metatarsal | Dorsiflexion and eversion of foot | Deep peroneal n. |
| Extensor digitorum longus | Lateral condyle of tibia and anterior surface of fibula | Dorsal surfaces of second and third phalanges of four lateral toes | Dorsiflexion and eversion of foot and extension of toes | Deep peroneal n. |
| Gastrocnemius | Lateral and medial condyles of femur | Posterior surface of calcaneus | Plantar flexion of foot and flexion of leg at knee | Tibial n. |
| Soleus | Head and shaft of fibula and posterior surface of tibia | Posterior surface of calcaneus | Plantar flexion of foot | Tibial n. |
| Flexor digitorum longus | Posterior surface of tibia | Distal phalanges of four lateral toes | Plantar flexion and inversion of foot, and flexion of four lateral toes | Tibial n. |
| Tibialis posterior | Lateral condyle and posterior surface of tibia, and posterior surface of fibula | Tarsal and metatarsal bones | Plantar flexion and inversion of foot | Tibial n. |
| Peroneus longus | Lateral condyle of tibia, and head and shaft of fibula | Tarsal and metatarsal bones | Plantar flexion and eversion of foot; also supports arch | Superficial peroneal n. |

# InnerConnections
## Muscular System

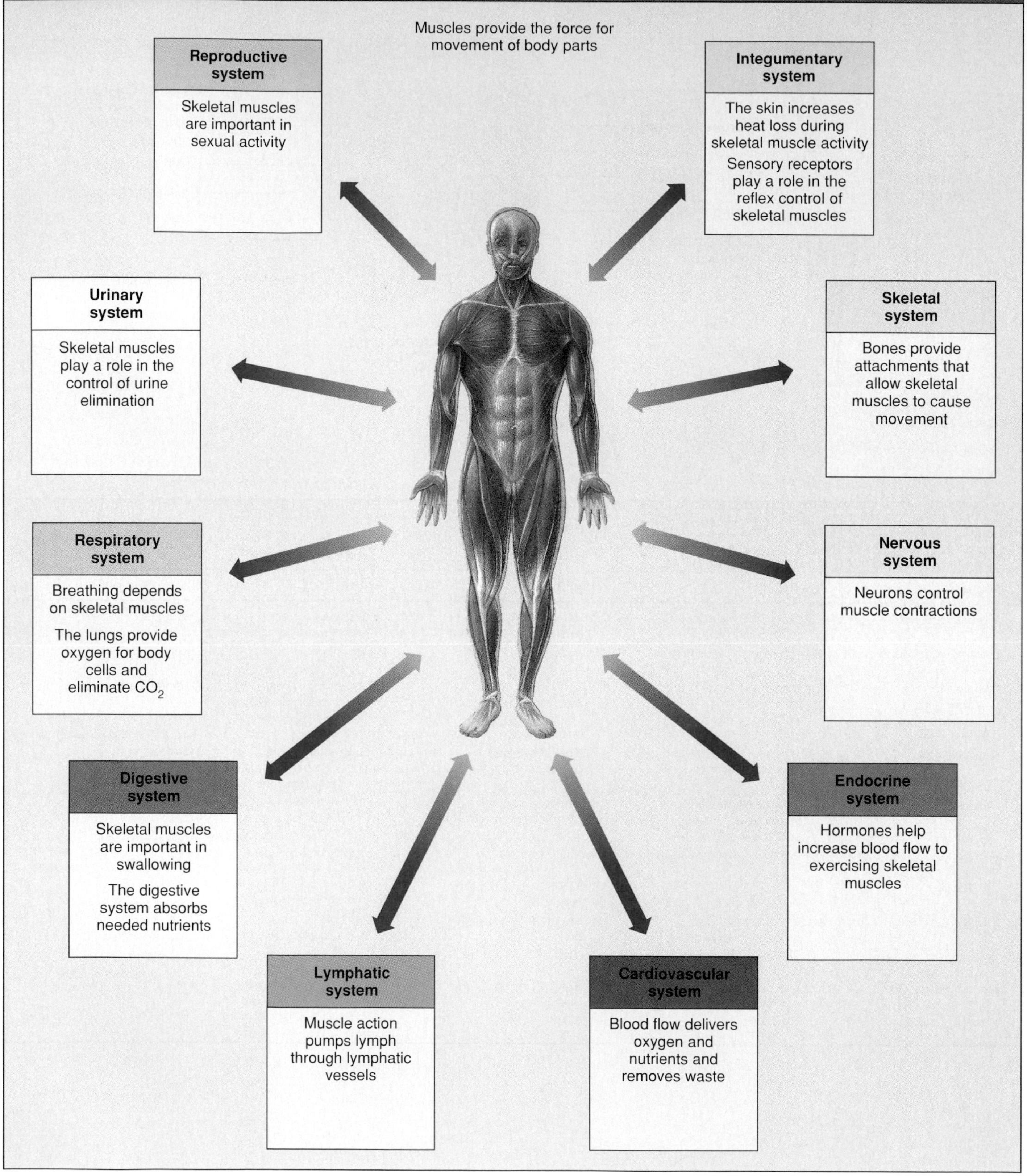

Muscles provide the force for movement of body parts

**Reproductive system**
Skeletal muscles are important in sexual activity

**Integumentary system**
The skin increases heat loss during skeletal muscle activity
Sensory receptors play a role in the reflex control of skeletal muscles

**Urinary system**
Skeletal muscles play a role in the control of urine elimination

**Skeletal system**
Bones provide attachments that allow skeletal muscles to cause movement

**Respiratory system**
Breathing depends on skeletal muscles
The lungs provide oxygen for body cells and eliminate $CO_2$

**Nervous system**
Neurons control muscle contractions

**Digestive system**
Skeletal muscles are important in swallowing
The digestive system absorbs needed nutrients

**Endocrine system**
Hormones help increase blood flow to exercising skeletal muscles

**Lymphatic system**
Muscle action pumps lymph through lymphatic vessels

**Cardiovascular system**
Blood flow delivers oxygen and nutrients and removes waste

## CHAPTER SUMMARY

### Introduction (page 288)

The three types of muscle tissue are skeletal, smooth, and cardiac.

### Structure of a Skeletal Muscle (page 288)

Skeletal muscles are composed of nervous, vascular, and various connective tissues, as well as skeletal muscle tissue.

1. Connective tissue coverings
   a. Fascia cover skeletal muscles.
   b. Other connective tissues surround cells and groups of cells within the muscle's structure.
   c. Fascia is part of a complex network of connective tissue that extends throughout the body.
2. Skeletal muscle fibers
   a. Each skeletal muscle fiber is a single muscle cell, which is the unit of contraction.
   b. Muscle fibers are cylindrical cells with many nuclei.
   c. The cytoplasm contains mitochondria, sarcoplasmic reticulum, and myofibrils of actin and myosin.
   d. The arrangement of the actin and myosin filaments causes striations.
   e. Transverse tubules extend from the cell membrane into the cytoplasm and are associated with the cisternae of the sarcoplasmic reticulum.
3. Neuromuscular junction
   a. Motor neurons stimulate muscle fibers to contract.
   b. The motor end plate of a muscle fiber lies on one side of a neuromuscular junction.
   c. In response to a nerve impulse, the end of a motor nerve fiber secretes a neurotransmitter, which diffuses across the junction and stimulates the muscle fiber.
4. Motor units
   a. One motor neuron and the muscle fibers associated with it constitute a motor unit.
   b. Muscles whose motor units contain small numbers of muscle fibers produce finer movements.

### Skeletal Muscle Contraction (page 294)

Muscle fiber contraction results from a sliding movement of actin and myosin filaments.

1. Role of myosin and actin
   a. Cross-bridges of myosin filaments form linkages with actin filaments.
   b. The reaction between actin and myosin filaments provides the force of contraction.
   c. When a fiber is at rest, troponin and tropomyosin molecules interfere with linkage formation.
   d. Calcium ions remove the inhibition.
2. Stimulus for contraction
   a. Muscle fiber is usually stimulated by acetylcholine released from the end of a motor nerve fiber.
   b. Acetylcholinesterase decomposes acetylcholine.
   c. Stimulation causes muscle fiber to conduct an impulse that travels over the surface of the sarcolemma and reaches the deep parts of the fiber by means of the transverse tubules.
   d. A muscle impulse signals the sarcoplasmic reticulum to release calcium ions.
   e. Linkages form between myosin and actin, and the actin filaments move inward, shortening the sarcomere.
   f. The muscle fiber relaxes when calcium ions are transported back into the sarcoplasmic reticulum.
3. Energy sources for contraction
   a. ATP supplies the energy for muscle fiber contraction.
   b. Creatine phosphate stores energy that can be used to synthesize ATP as it is decomposed.
   c. Active muscles depend upon cellular respiration for energy.
4. Oxygen supply and cellular respiration
   a. Anaerobic respiration yields few ATP molecules, while aerobic respiration provides many ATP molecules.
   b. Hemoglobin in red blood cells carries oxygen from the lungs to body cells.
   c. Myoglobin in muscle cells stores some oxygen temporarily.
5. Oxygen debt
   a. During rest or moderate exercise, oxygen is sufficient to support aerobic respiration.
   b. During strenuous exercise, oxygen deficiency may develop, and lactic acid may accumulate as a result of anaerobic respiration.
   c. The amount of oxygen needed to convert accumulated lactic acid to glucose and to restore supplies of ATP and creatine phosphate is called oxygen debt.
6. Muscle fatigue
   a. A fatigued muscle loses its ability to contract.
   b. Muscle fatigue is usually due to the effects of accumulation of lactic acid
   c. Athletes usually produce less lactic acid than nonathletes because of their increased ability to supply oxygen and nutrients to muscles.
7. Fast and slow muscles
   a. The speed of contraction is related to a muscle's specific function.
   b. Slow-contracting, or red, muscles can generate ATP fast enough to keep up with ATP breakdown and can contract for long periods.
   c. Fast-contracting, or white, muscles have reduced ability to carry on aerobic respiration and tend to fatigue relatively rapidly.
8. Heat production
   a. Muscles represent an important source of body heat.
   b. Most of the energy released by cellular respiration is lost as heat.

### Muscular Responses (page 301)

1. Threshold stimulus is the minimal stimulus needed to elicit a muscular contraction.
2. All-or-none response
   a. If a muscle fiber contracts at all, it will contract completely.
   b. Motor units respond in an all-or-none manner.
3. Recording a muscle contraction
   a. A myogram is a recording of an electrically stimulated isolated muscle pulling a lever.

b. A twitch is a single, short contraction reflecting stimulation of some motor units in a muscle.
c. The latent period is the time between stimulus and responding muscle contraction.
d. During the refractory period immediately following contraction, a muscle cannot respond.
4. Staircase effect
a. An inactive muscle undergoes a series of contractions of increasing strength when subjected to a series of stimuli.
b. This staircase effect seems to be due to failure to remove calcium ions from the sarcoplasm rapidly enough.
5. Summation
a. A rapid series of stimuli may produce summation of twitches and sustained contraction.
b. Forceful, sustained contraction without relaxation is a tetanic contraction.
6. Recruitment of motor units
a. At low intensity of stimulation, relatively small numbers of motor units contract.
b. At increasing intensities of stimulation, other motor units are recruited until the muscle contracts with maximal tension.
7. Sustained contractions
a. When contractions fuse, the strength of contraction may increase due to recruitment of fibers.
b. Even when a muscle is at rest, its fibers usually maintain tone—that is, remain partially contracted.
8. Isotonic and isometric contractions
a. When a muscle contracts and its ends are pulled closer together, the contraction is called isotonic.
b. When a muscle contracts but its attachments do not move, the contraction is called isometric.
c. Most body movements involve both isometric and isotonic contractions.

## Smooth Muscles (page 304)

The contractile mechanisms of smooth and cardiac muscles are similar to those of skeletal muscle.

1. Smooth muscle fibers
a. Smooth muscle cells contain filaments of myosin and actin.
b. They lack transverse tubules, and the sarcoplasmic reticula are not well developed.
c. Types include multiunit smooth muscle and visceral smooth muscle.
d. Visceral smooth muscle displays rhythmicity.
e. Peristalsis aids movement of material through hollow organs.
2. Smooth muscle contraction
a. In smooth muscles, calmodulin binds to calcium ions and activates the contraction mechanism.
b. Both acetylcholine and norepinephrine are neurotransmitters for smooth muscles.
c. Hormones and stretching affect smooth muscle contractions.
d. With a given amount of energy, smooth muscle can maintain a contraction for a longer time than can skeletal muscle.
e. Smooth muscles can change lengths without changing tautness.

## Cardiac Muscle (page 306)

1. Cardiac muscle contracts for a longer time than skeletal muscle because transverse tubules supply extra calcium ions.
2. Intercalated disks separate the ends of adjacent cardiac muscle cells and hold the cells together.
3. A network of fibers contracts as a unit and responds to stimulation in an all-or-none manner.
4. Cardiac muscle is self-exciting, rhythmic, and remains refractory until a contraction is completed.

## Skeletal Muscle Actions (page 306)

1. Origin and insertion
a. The movable end of a skeletal muscle is its insertion, and the immovable end is its origin.
b. Some muscles have more than one origin or insertion.
2. Interaction of skeletal muscles
a. Skeletal muscles function in groups.
b. A prime mover is responsible for most of a movement; synergists aid prime movers; antagonists can resist the movement of a prime mover.
c. Smooth movements depend upon antagonists giving way to the actions of prime movers.

## Major Skeletal Muscles (page 308)

Muscle names often describe sizes, shapes, locations, actions, number of attachments, or direction of fibers.

1. Muscles of facial expression
a. These muscles lie beneath the skin of the face and scalp and are used to communicate feelings through facial expression.
b. They include the epicranius, orbicularis oculi, orbicularis oris, buccinator, zygomaticus, and platysma.
2. Muscles of mastication
a. These muscles are attached to the mandible and are used in chewing.
b. They include the masseter, temporalis, medial pterygoid, and lateral pterygoid.
3. Muscles that move the head and vertebral column
a. Muscles in the neck and back move the head.
b. They include the sternocleidomastoid, splenius capitis, semispinalis capitis, and erector spinae.
4. Muscles that move the pectoral girdle
a. Most of these muscles connect the scapula to nearby bones and are closely associated with muscles that move the arm.
b. They include the trapezius, rhomboideus major, levator scapulae, serratus anterior, and pectoralis minor.

5. Muscles that move the arm
   a. These muscles connect the humerus to various regions of the pectoral girdle, ribs, and vertebral column.
   b. They include the coracobrachialis, pectoralis major, teres major, latissimus dorsi, supraspinatus, deltoid, subscapularis, infraspinatus, and teres minor.
6. Muscles that move the forearm
   a. These muscles connect the radius and ulna to the humerus and pectoral girdle.
   b. They include the biceps brachii, brachialis, brachioradialis, triceps brachii, supinator, pronator teres, and pronator quadratus.
7. Muscles that move the hand
   a. These muscles arise from the distal end of the humerus and from the radius and ulna.
   b. They include the flexor carpi radialis, flexor carpi ulnaris, palmaris longus, flexor digitorum profundus, flexor digitorum superficialis, extensor carpi radialis longus, extensor carpi radialis brevis, extensor carpi ulnaris, and extensor digitorum.
   c. An extensor retinaculum forms sheaths for tendons of the extensor muscles.
8. Muscles of the abdominal wall
   a. These muscles connect the rib cage and vertebral column to the pelvic girdle.
   b. They include the external oblique, internal oblique, transversus abdominis, and rectus abdominis.
9. Muscles of the pelvic outlet
   a. These muscles form the floor of the pelvic cavity and fill the space of the pubic arch.
   b. They include the levator ani, coccygeus, superficial transversus perinei, bulbospongiosus, ischiocavernosus, and sphincter urethrae.
10. Muscles that move the thigh
    a. These muscles are attached to the femur and to some part of the pelvic girdle.
    b. They include the psoas major, iliacus, gluteus maximus, gluteus medius, gluteus minimus, tensor fasciae latae, adductor longus, adductor magnus, and gracilis.
11. Muscles that move the leg
    a. These muscles connect the tibia or fibula to the femur or pelvic girdle.
    b. They include the biceps femoris, semitendinosus, semimembranosus, sartorius, and the quadriceps femoris group.
12. Muscles that move the foot
    a. These muscles attach the femur, tibia, and fibula to various bones of the foot.
    b. They include the tibialis anterior, peroneus tertius, extensor digitorum longus, gastrocnemius, soleus, flexor digitorum longus, tibialis posterior, and peroneus longus.
    c. Retinacula form sheaths for tendons passing to the foot.

**Explorations in Human Anatomy and Physiology CD-ROM**
The module accompanying Chapter Nine is #4 Muscle Contraction.

## Critical Thinking Questions

1. Why do you think athletes generally perform better if they warm up by exercising lightly before a competitive event?
2. Following childbirth, a woman may lose urinary control (incontinence) when sneezing or coughing. What muscles of the pelvic floor should be strengthened by exercise to help control this problem?
3. What steps might be taken to minimize atrophy of skeletal muscles in patients who are confined to bed for prolonged times?
4. As lactic acid and other substances accumulate in an active muscle, they stimulate pain receptors, and the muscle may feel sore. How might the application of heat or substances that dilate blood vessels help relieve such soreness?
5. Several important nerves and blood vessels course through the muscles of the gluteal region. In order to avoid the possibility of damaging such parts, intramuscular injections into this region are usually made into the lateral, superior portion of the gluteus medius. What landmarks would help you locate this muscle in a patient?
6. Following an injury to a nerve, the muscles it supplies with motor nerve fibers may become paralyzed. How would you explain to a patient the importance of moving the disabled muscles passively or contracting them with electrical stimulation?

# Review Exercises

## Part A

1. List the three types of muscle tissue.
2. Distinguish between a tendon and an aponeurosis.
3. Describe the connective tissue coverings of a skeletal muscle.
4. Distinguish between deep fascia, subcutaneous fascia, and subserous fascia.
5. List the major parts of a skeletal muscle fiber, and describe the function of each part.
6. Describe a neuromuscular junction.
7. Explain the function of a neurotransmitter substance.
8. Define *motor unit*, and explain how the number of fibers within a unit affects muscular contractions.
9. Describe the major events that occur when a muscle fiber contracts.
10. Explain how ATP and creatine phosphate are related and how these substances function in the muscle fiber contraction mechanism.
11. Describe how oxygen is supplied to skeletal muscles.
12. Describe how an oxygen debt may develop.
13. Explain how muscles may become fatigued and how a person's physical condition may affect tolerance to fatigue.
14. Distinguish between fast-contracting and slow-contracting muscles.
15. Explain how the maintenance of body temperature is related to the actions of skeletal muscles.
16. Define *threshold stimulus*.
17. Explain what is meant by an *all-or-none response*.
18. Explain what is meant by *motor unit recruitment*.
19. Describe the staircase effect.
20. Explain how a skeletal muscle can be stimulated to produce a sustained contraction.
21. Distinguish between a tetanic contraction and muscle tone.
22. Distinguish between isometric and isotonic contractions, and explain how each is used in body movements.
23. Compare the structures of smooth and skeletal muscle fibers.
24. Distinguish between multiunit and visceral smooth muscles.
25. Define *peristalsis* and explain its function.
26. Compare the characteristics of smooth and skeletal muscle contractions.
27. Compare the structures of cardiac and skeletal muscle fibers.
28. Compare the characteristics of cardiac and skeletal muscle contractions.
29. Distinguish between a muscle's origin and its insertion.
30. Define *prime mover, synergist,* and *antagonist*.

## Part B

Match the muscles in column I with the descriptions and functions in column II.

| I | II |
|---|---|
| 1. Buccinator | A. Inserted on the coronoid process of the mandible. |
| 2. Epicranius | B. Draws the corner of the mouth upward. |
| 3. Lateral pterygoid | C. Can raise and adduct the scapula. |
| 4. Platysma | D. Can pull the head into an upright position. |
| 5. Rhomboideus major | E. Consists of two parts—the frontalis and the occipitalis. |
| 6. Splenius capitis | F. Compresses the cheeks. |
| 7. Temporalis | G. Extends over the neck from the chest to the face. |
| 8. Zygomaticus | H. Pulls the jaw from side to side. |

| I | II |
|---|---|
| 9. Biceps brachii | I. Primary extensor of the elbow. |
| 10. Brachialis | J. Pulls the shoulder back and downward. |
| 11. Deltoid | K. Abducts the arm. |
| 12. Latissimus dorsi | L. Rotates the arm laterally. |
| 13. Pectoralis major | M. Pulls the arm forward and across the chest. |
| 14. Pronator teres | N. Rotates the arm medially. |
| 15. Teres minor | O. Strongest flexor of the elbow. |
| 16. Triceps brachii | P. Strongest supinator of the forearm. |

| I | II |
|---|---|
| 17. Biceps femoris | Q. Inverts the foot. |
| 18. External oblique | R. A member of the quadriceps femoris group. |
| 19. Gastrocnemius | S. A plantar flexor of the foot. |
| 20. Gluteus maximus | T. Compresses the contents of the abdominal cavity. |
| 21. Gluteus medius | U. Largest muscle in the body. |
| 22. Gracilis | V. A hamstring muscle. |
| 23. Rectus femoris | W. Adducts the thigh. |
| 24. Tibialis anterior | X. Abducts the thigh. |

## Part C

What muscles can you identify in the bodies of these models whose muscles are enlarged by exercise?

# Surface Anatomy

■ *The following set of reference plates is presented to help you locate some of the more prominent surface features in various regions of the body. For the most part, the labeled structures are easily seen or palpated through the skin. As a review, you may want to locate as many of these features as possible on your own body.*

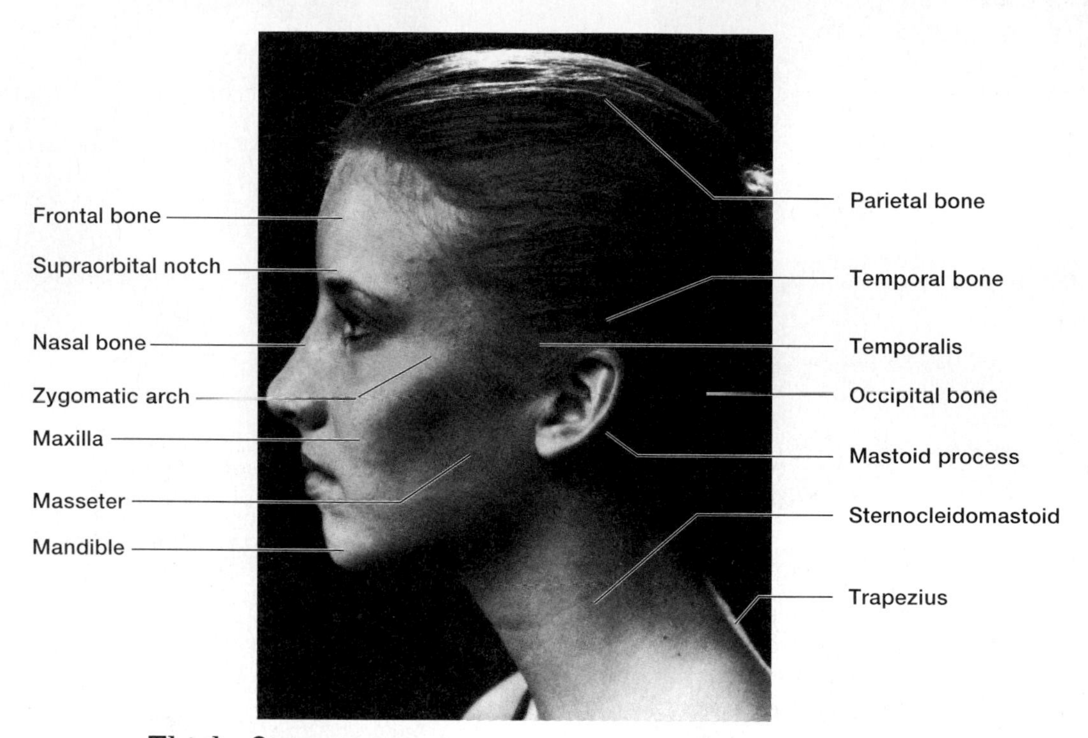


**PLATE Thirty-Seven**
Surface anatomy of head and neck, lateral view.

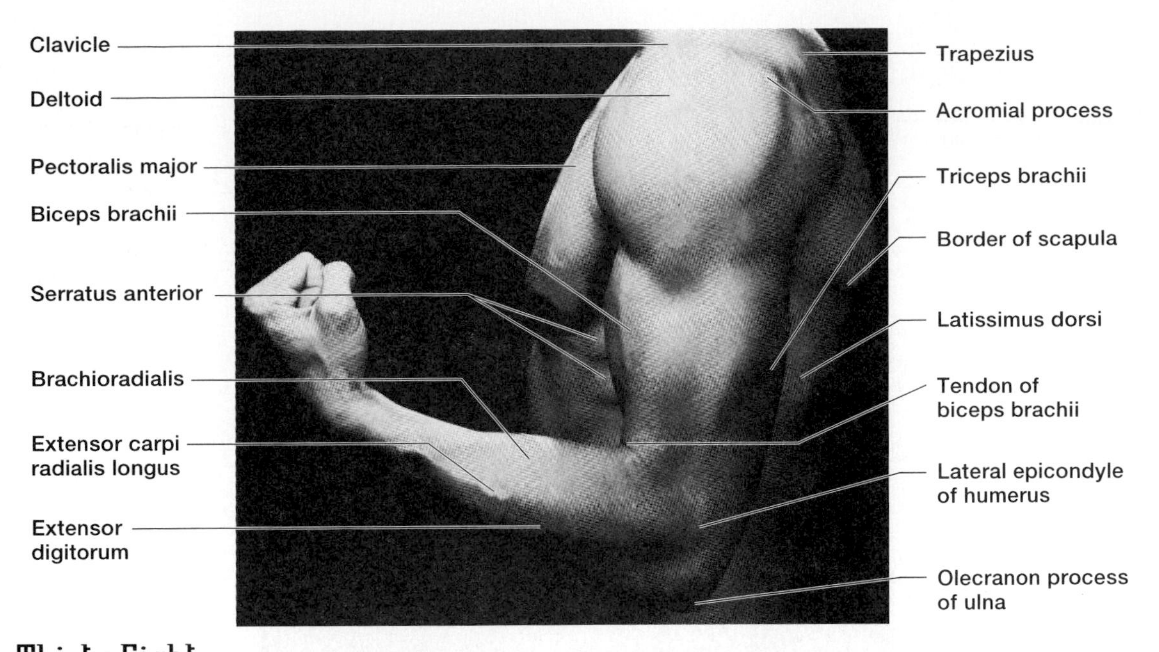


**PLATE Thirty-Eight**
Surface anatomy of upper limb and thorax, lateral view.

**PLATE Thirty-Nine**
Surface anatomy of back and upper limbs, posterior view.

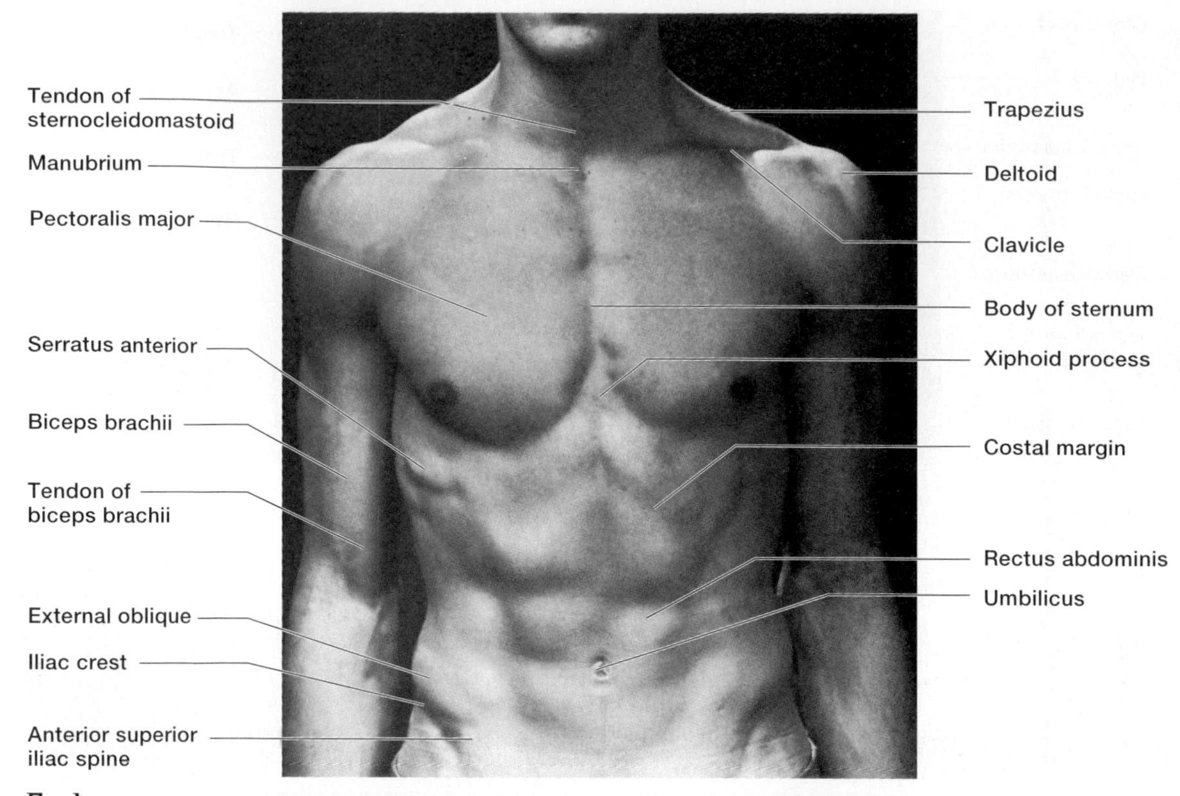


**PLATE Forty**

Surface anatomy of torso and arms, anterior view.

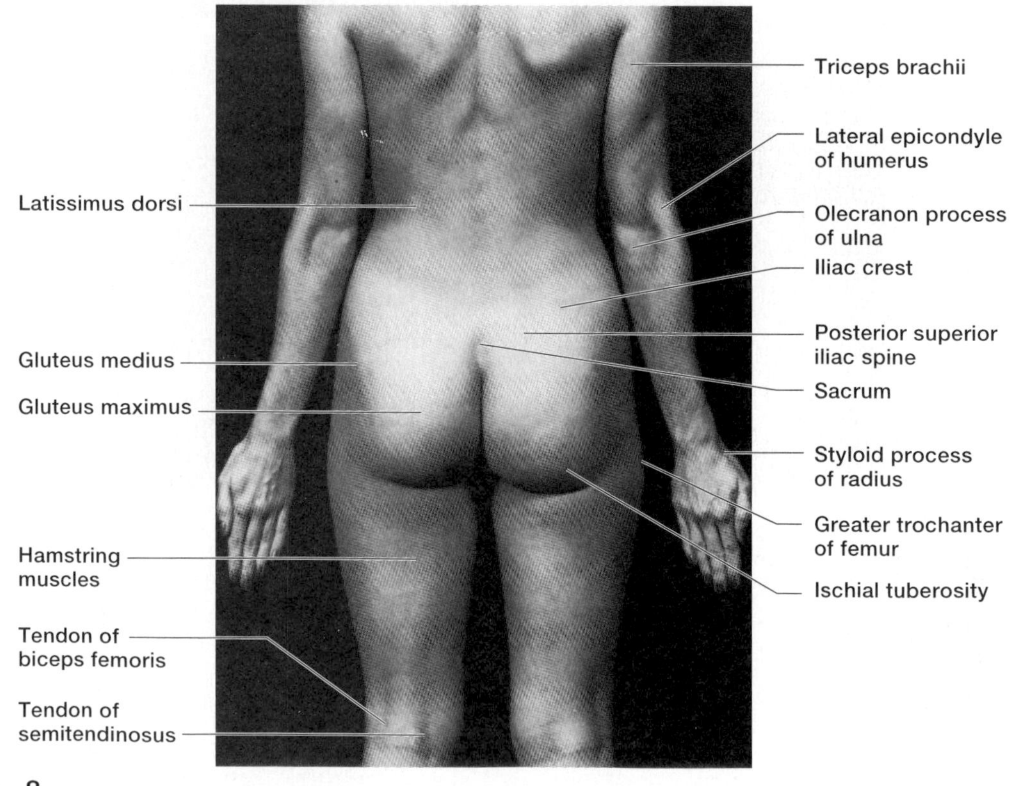


**PLATE Forty-One**

Surface anatomy of torso and thighs, posterior view.

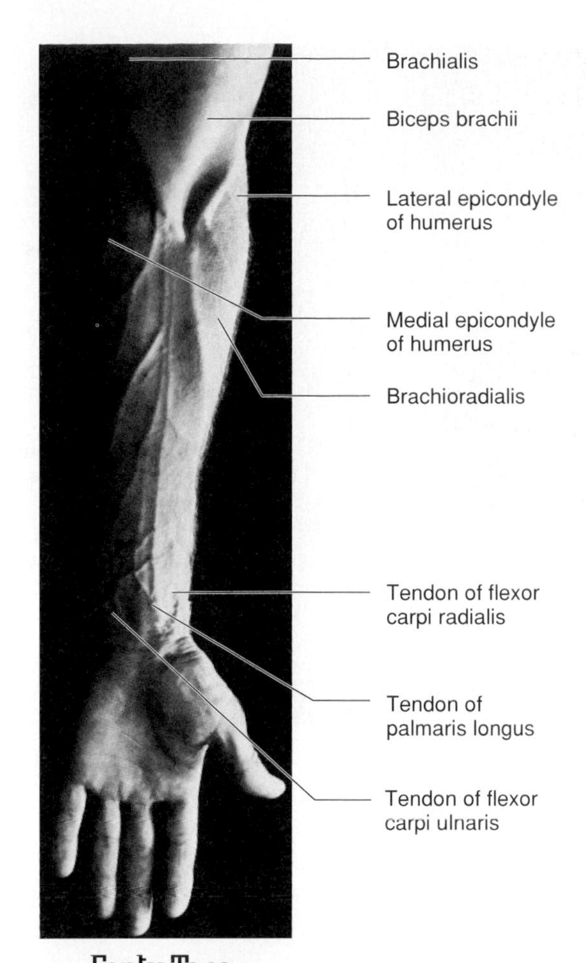


**PLATE Forty-Two**
Surface anatomy of forearm, anterior view.

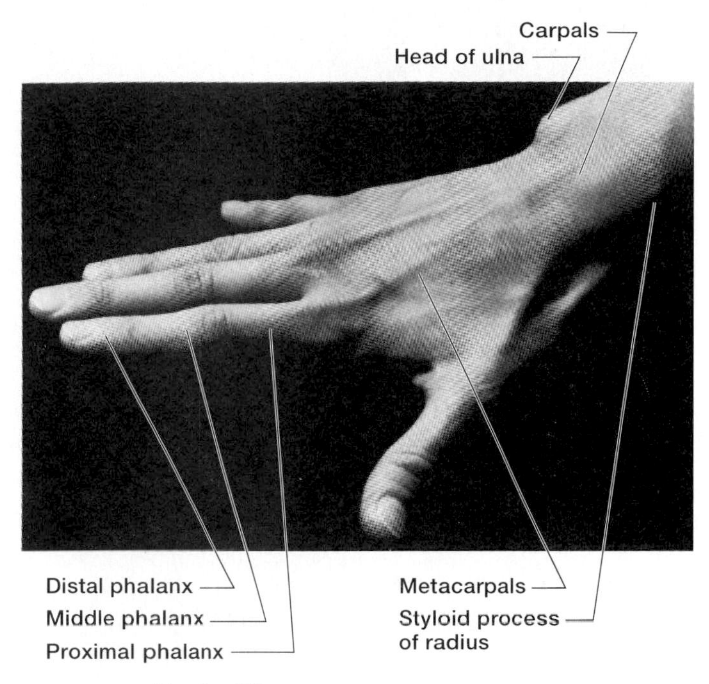


**PLATE Forty-Three**
Surface anatomy of the hand.

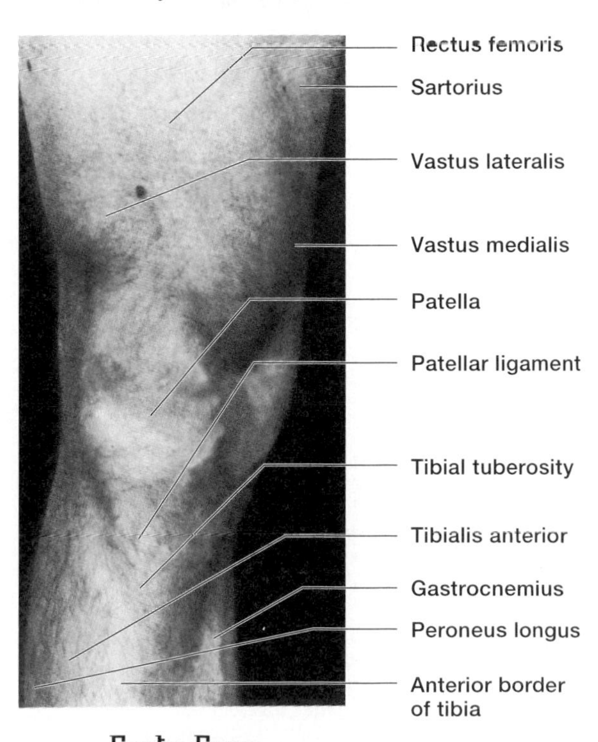


**PLATE Forty-Four**
Surface anatomy of knee and surrounding area, anterior view.

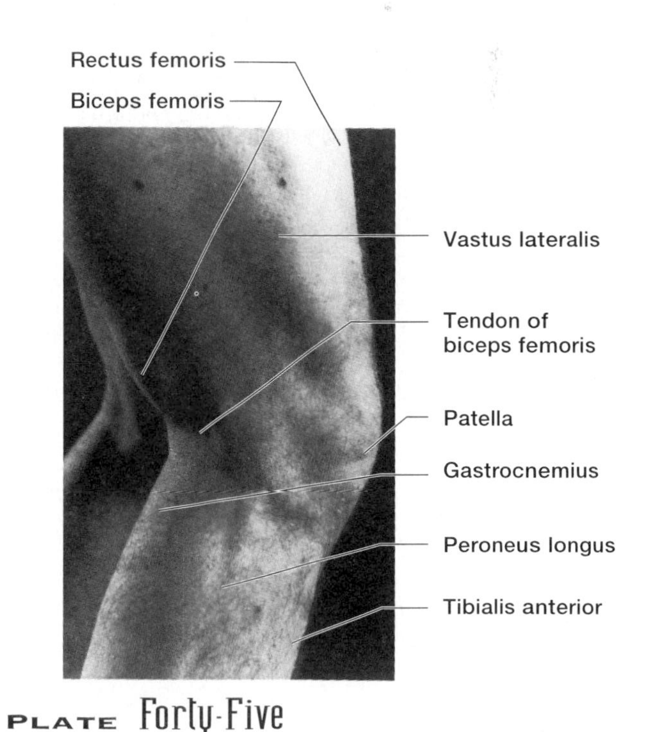


**PLATE Forty-Five**
Surface anatomy of knee and surrounding area, lateral view.

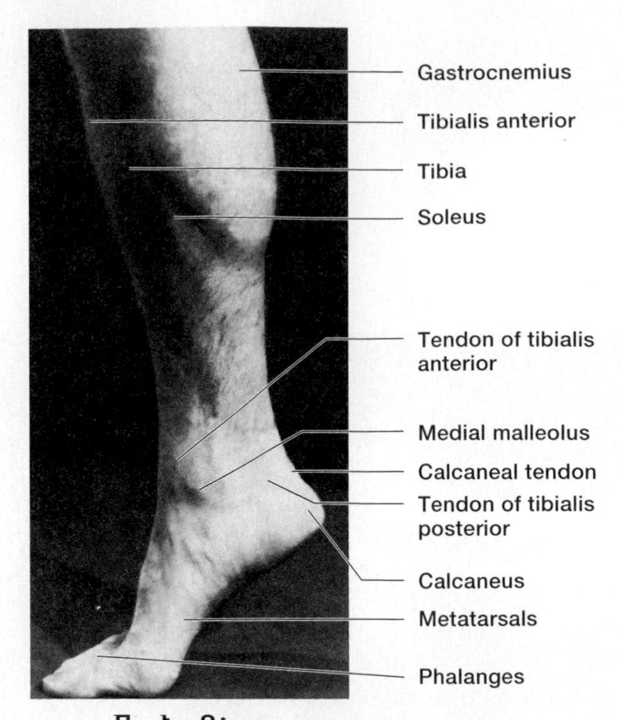


**PLATE Forty-Six**
Surface anatomy of ankle and leg, medial view.

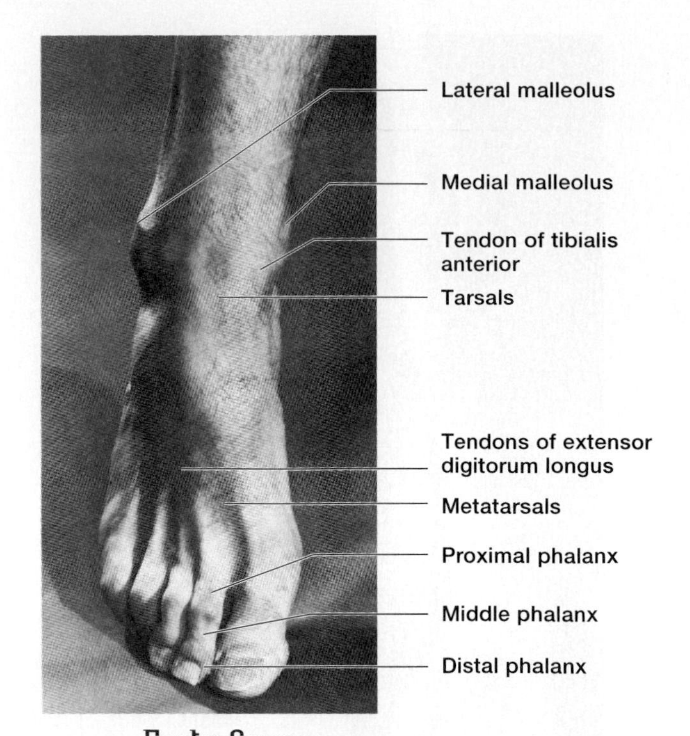


**PLATE Forty-Seven**
Surface anatomy of ankle and foot.

10

CHAPTER

# NERVOUS SYSTEM I

## Basic Structure and Function

### CHAPTER OBJECTIVES

AFTER YOU HAVE STUDIED THIS CHAPTER, YOU SHOULD BE ABLE TO:

1. Explain the general functions of the nervous system.
2. Describe the general structure of a neuron.
3. Name four types of neuroglial cells and describe the functions of each.
4. Explain how an injured nerve fiber may regenerate.
5. Explain how a membrane becomes polarized.
6. Describe the events that lead to the conduction of a nerve impulse.
7. Explain how a nerve impulse is transmitted from one neuron to another.
8. Distinguish between excitatory and inhibitory postsynaptic potentials.
9. Explain two ways impulses are processed in neuronal pools.
10. Explain how neurons are classified.
11. Describe how nerve fibers in peripheral nerves are classified.
12. Describe a reflex arc.
13. Explain what is meant by reflex behavior.

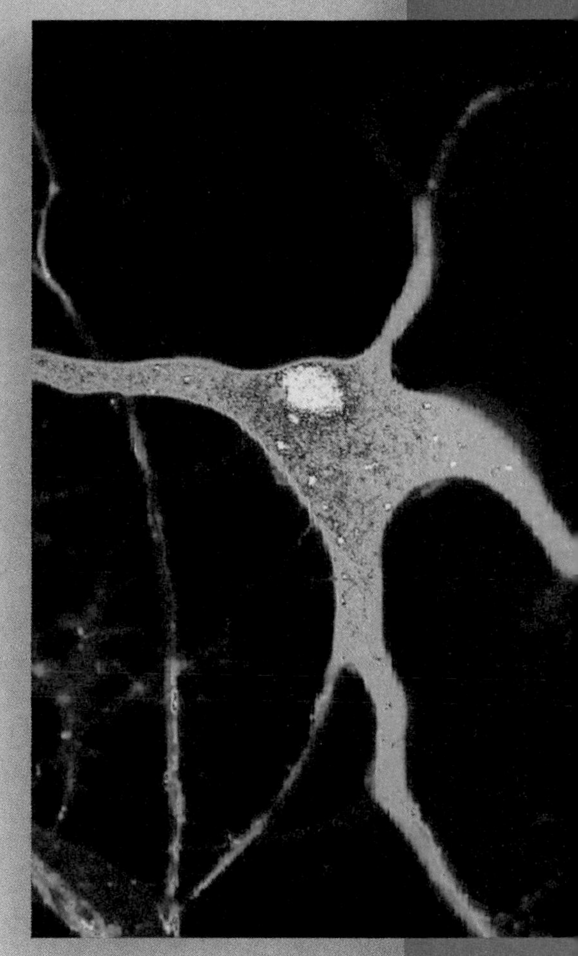

THIS NEURON IS FILLING WITH CALCIUM IONS (YELLOW) AFTER EXPOSURE IN LABORATORY CULTURE TO A PROTEIN FRAGMENT THOUGHT TO CAUSE ALZHEIMER'S DISEASE. IDENTIFYING THE CAUSE OF THIS DEVASTATING DISORDER AT THE MOLECULAR AND CELLULAR LEVELS WILL AID IN DEVELOPING TREATMENTS (500×).

### KEY TERMS

**action potential** (ak′shun po-ten′shal)
**axon** (ak′son)
**central nervous system** (sen′tral ner′vus sis′tem)
**convergence** (kon-ver′jens)
**dendrite** (den′drit)
**divergence** (di-ver′jens)
**effector** (e-fek′tor)
**facilitation** (fah-sil″ĭ-ta′shun)
**myelin** (mi′ĕ-lin)
**neurilemma** (nu″ri-lem′mah)
**neuroglia** (nu-rog′le-ah)
**neuron** (nu′ron)
**neurotransmitter** (nu″ro-trans-mit′er)
**Nissl body** (nis′l bod′e)
**peripheral nervous system** (pĕ-rif′er-al ner′vus sis′tem)
**receptor** (re-sep′tor)
**reflex** (re′fleks)
**summation** (sum-ma′shun)
**synapse** (sin′aps)
**threshold** (thresh′old)

UNIT THREE

When the five living American presidents gathered at the funeral of former president Richard M. Nixon in April 1994, Gerald Ford, Jimmy Carter, George Bush, and Bill Clinton knew that all was not right with their compatriot Ronald Reagan. The former president was forgetful, responded inappropriately to questions, and, in the words of Gerald Ford, seemed "hollowed out." Reagan's memory continued to fade in and out, and 6 months later he penned a moving letter to the public confirming that he has Alzheimer's disease. The spells of forgetfulness and cloudy reasoning would come and go over the next several years, eventually increasing in frequency and severity to the point where he would not even recognize his wife Nancy.

Because Alzheimer's disease affects 4 million Americans and their families, finding a treatment or cure is a primary health care objective. Doing so requires pinpointing just what goes awry in the nervous system to trigger the cascade of destruction that ultimately strangles the brain in gummy protein plaques and tangles. Research is focusing on replenishing the messenger molecule that is deficient in the brain of a person with Alzheimer's disease, and supplying other biochemicals that stimulate nervous tissue to grow. But as is often the case in medical science, new discoveries often come from an astute researcher who makes an intriguing mental connection. This may be the case for a test to diagnose Alzheimer's disease.

Diagnosis of Alzheimer's disease is typically based on performance on tests of mental function, and exclusion of other possible causes of forgetfulness and confusion, such as thyroid disorders or depression. Definitive diagnosis is possible only at autopsy, which reveals the characteristic plaques and tangles in the brain—not a very useful approach. However, a researcher in Boston may have discovered a simple way to diagnose and possibly predict Alzheimer's disease.

The researcher worked with people who have Down syndrome, a condition caused by extra genetic material. These people invariably develop Alzheimer's disease, if they live long enough. Noting that the eyes of people with Down syndrome respond much more strongly to a drug that causes the pupils to dilate than do those of healthy people, the researcher tried the eye drug on people with Alzheimer's disease. Their pupils too widened almost completely. Two individuals who were healthy at the time and whose pupils dilated greatly developed symptoms of Alzheimer's disease months later. The drug, called tropicamide, blocks the effects of a biochemical (acetylcholine) normally produced by the brain cells that degenerate in Alzheimer's disease. The test is currently being tested on many more people, to see if the association holds up.

## General Functions of the Nervous System

The nervous system is composed of some blood vessels and connective tissue but predominantly neural tissue, which consists of two cell types: nerve cells or **neurons,** and **neuroglia,** sometimes called simply *glia.* Neurons, the structural and functional units of the nervous system, are specialized to react to physical and chemical changes in their surroundings. They transmit information in the form of electrochemical changes, called **nerve impulses,** along **nerve fibers** to other neurons and to cells outside the nervous system (fig. 10.1). **Nerves** are bundles of nerve fibers. Neuroglia were once thought only to fill spaces and surround or support neurons. Today we know that they have many other functions, nourishing neurons and perhaps even sending and receiving messages.

Perhaps the most important part of the nervous system at the cellular level is not a cell at all but the small spaces between neurons, called **synapses.** Much of the effort of the nervous system centers on sending and receiving electrochemical messages from neuron to neuron at synapses. The actual carriers of this information are biological messenger molecules called **neurotransmitters.**

The organs of the nervous system can be divided into two groups. One group, consisting of the brain and spinal cord, forms the **central nervous system** (CNS), and the other, composed of the nerves (peripheral nerves) that connect the central nervous system to other body parts, is called the **peripheral nervous system** (PNS) (fig. 10.2). Together these systems provide three general functions—sensory, integrative, and motor.

Structures called **sensory receptors** at the ends of peripheral neurons provide the sensory function of the nervous system (see chapter 11). These receptors gather information by detecting changes inside and outside the body. They monitor external environmental factors such as light and sound intensities as well as the temperature, oxygen concentration, and other conditions of the body's internal environment.

Sensory receptors convert their information into nerve impulses, which are then transmitted over peripheral nerves to the central nervous system. There the signals are integrated—that is, they are brought together, creating sensations (perceptions), adding to memory, or helping produce thoughts. Following integration, conscious or subconscious decisions are made and then acted upon by means of motor functions.

The motor functions of the nervous system employ peripheral neurons that carry impulses from the

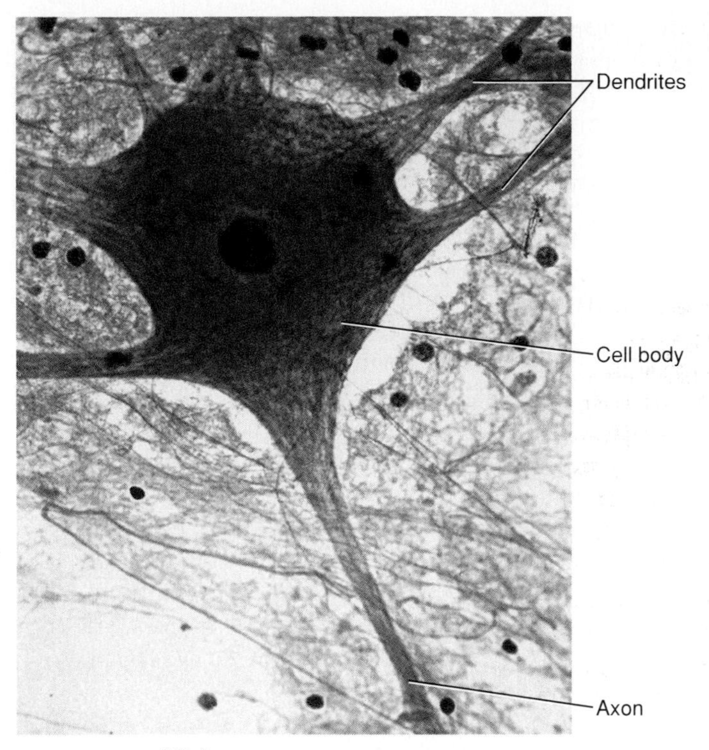


**FIGURE 10.1**
Neurons are the structural and functional units of the nervous system (50×). Nuclei of neuroglial cells surround the neuron, appearing as dark dots. Note the location of nerve fibers (dendrites and a single axon).

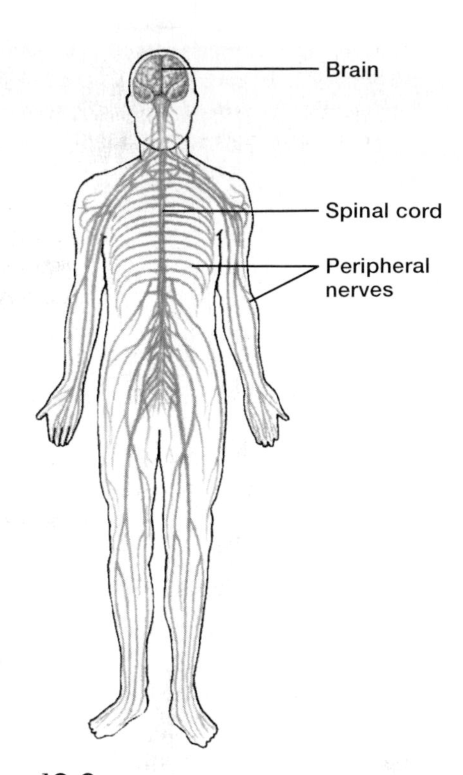


**FIGURE 10.2**
The nervous system consists of the brain and spinal cord (central nervous system) and peripheral nerves (peripheral nervous system).

central nervous system to responsive structures called *effectors*. These effectors are outside the nervous system and include muscles that contract in response to nerve impulse stimulation, and glands that secrete when they are stimulated.

Thus, the nervous system can detect changes in the body, make decisions on the basis of the information received, and stimulate muscles or glands to respond. Typically, these responses counteract the effects of the changes, and in this way, the nervous system helps maintain homeostasis.

Clinical Application 10.1 discusses a common medical problem attributed to the nervous system that actually involves its blood supply.

## Neuron Structure

Although neurons vary considerably in size and shape, they have certain features in common. For example, every neuron has a **cell body** and nerve fibers (tubular membrane-bound processes containing cytoplasm) that conduct nerve impulses to or from the cell body. Figure 10.3 shows some of the other structures common to neurons.

A neuron's cell body (soma or perikaryon) contains granular cytoplasm, mitochondria, lysosomes, a Golgi apparatus, and many microtubules. A network of fine threads called **neurofibrils** extends into the fibers and supports them. Scattered throughout the cytoplasm are many membranous sacs called **Nissl bodies,** which consist of rough endoplasmic reticulum. Cytoplasmic inclusions common in neurons contain glycogen, lipids, or pigments such as melanin.

> Neurons are notoriously difficult to culture in the laboratory, but cell biologists can do so by coaxing other cell types to express the genes that cause them to specialize as neurons. Researchers at one biotechnology company begin by culturing human olfactory epithelium, cells that provide our sense of smell. Under certain culture conditions, these cells gradually become immature neurons. By carefully manipulating the chemicals in the culture medium, researchers stimulate the cells to specialize into neurons characteristic of those in the central nervous system. This technique may one day provide replacement cell implants for treating such brain disorders as Parkinson's disease and Alzheimer's disease. But first, the cultured neurons have to show that they can release the same neurotransmitters and form the same connections to other neurons as do the brain neurons that they would replace.

## Clinical Application 10.1

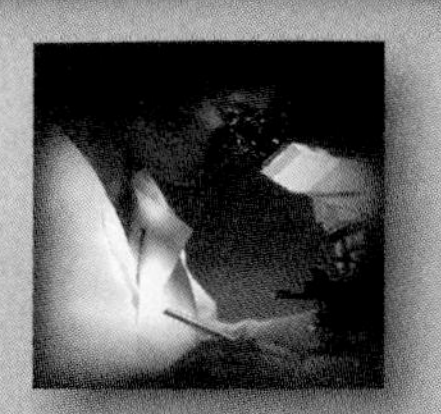

### Migraine

Heather L. knew a migraine was coming on within minutes. Her head began to pound on one side and waves of nausea washed over her. She ran into a windowless bathroom, swallowed two ibuprofen tablets, shut off the light, and huddled in a corner, eyes tightly closed. Thankfully, this one was short, just a few hours. By now Heather had learned precisely what to do. All it took was a strong shaft of sunlight into her room as she slept to trigger an attack.

Although migraine is considered a type of headache, it is actually a response to changes in the diameters of blood vessels in the face, head, and neck. Constriction followed by dilation of these vessels causes head pain (usually on one side), nausea and perhaps vomiting, and sensitivity to light. There are two variants—10 to 15% of sufferers experience classic migraine, which lasts 4 to 6 hours and is preceded by an "aura" of light in the peripheral vision. Common migraine usually lacks an aura but may last for 3 to 4 days.

On a cellular level, abnormal levels of the neurotransmitter serotonin in the brain are thought to constrict the blood vessels that cause migraines. On a practical level, many sufferers link migraine attacks to eating a particular food (chocolate, red wine, nuts, and processed meats top the list), lack of sleep, excess stress, glaring lights, high altitude, and stormy weather.

Nearly all migraine sufferers can be helped. Over-the-counter pain relievers help some people, as do the prescription beta blockers and a drug called Imitrex. All of these constrict blood vessels in the head. Tricyclic antidepressant drugs also help by keeping serotonin available longer, thereby offsetting a deficit that may cause this painful disorder.

Near the center of the neuron cell body is a large, spherical nucleus with a conspicuous nucleolus. Mature neurons do not divide.

Two kinds of nerve fibers, **dendrites** and **axons,** extend from the cell bodies of most neurons. A neuron may have many dendrites but only one axon.

Dendrites are usually highly branched, providing receptive surfaces to which processes from other neurons communicate. (In some kinds of neurons, the cell body itself provides such a receptive surface.) Often the dendrites have tiny, thornlike spines (dendritic spines) on their surfaces, which are contact points for other neurons.

The axon, which often arises from a slight elevation of the cell body (axonal hillock), is a slender, cylindrical process with a nearly smooth surface and uniform diameter. It is specialized to conduct nerve impulses away from the cell body. Its cytoplasm includes many mitochondria, microtubules, and neurofibrils. An axon begins as a single fiber but may give off branches, called *collaterals.* Near its end, it may have many fine extensions, each with a specialized ending called a *presynaptic terminal,* which terminates very close to the receptive surface of another cell, separated only by the synaptic cleft.

In addition to conducting nerve impulses, an axon conveys biochemicals that are produced in the neuron cell body, which can be quite a task in these very long cells. This process, called *axonal transport,* involves vesicles, mitochondria, ions, nutrients, and neurotransmitters that travel from the cell body to the ends of the axon.

Sheaths composed of many neuroglial cells called **Schwann cells** commonly enclose larger axons of peripheral neurons. These cells wind tightly around these fibers, somewhat like a bandage wrapped around a finger. As a result, such fibers are coated with many layers of cell membranes that have little or no cytoplasm

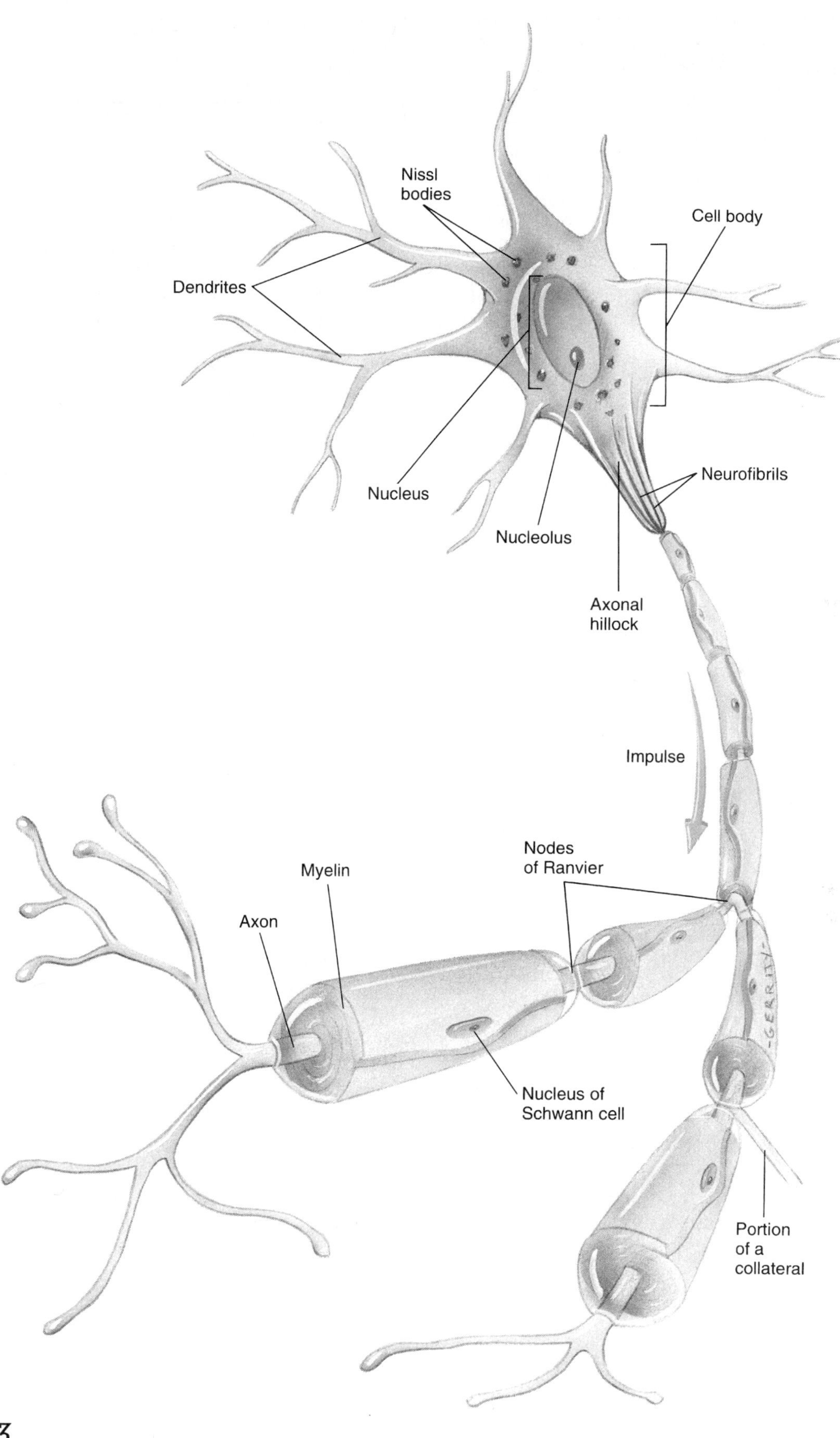


**FIGURE 10.3**
A common neuron.

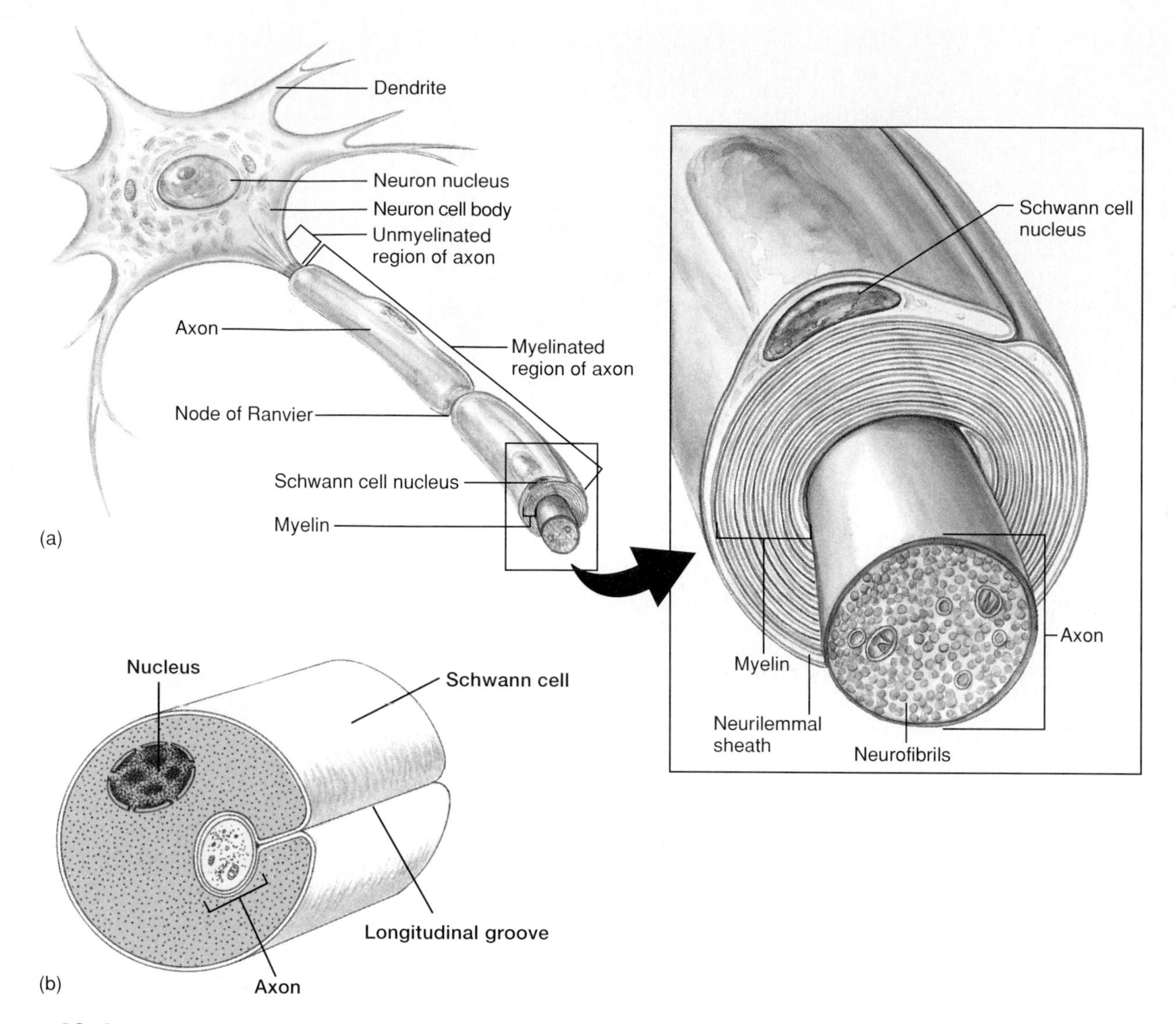


**FIGURE 10.4**

(*a*) The portion of a Schwann cell that winds tightly around an axon forms the myelin sheath. The cytoplasm and nucleus of the Schwann cell, remaining on the outside, form the neurilemmal sheath. (*b*) An axon lying in a longitudinal groove of a Schwann cell lacks a myelin sheath.

between them. These membranes, composed largely of a lipoprotein called **myelin,** have a higher proportion of lipid than other surface membranes. Myelin forms a *myelin sheath* on the outside of an axon. In addition, the portions of the Schwann cells that contain most of the cytoplasm and the nuclei remain outside the myelin sheath and comprise a **neurilemma** or *neurilemmal sheath,* which surrounds the myelin sheath (fig. 10.4). Narrow gaps in the myelin sheath between adjacent Schwann cells are called **nodes of Ranvier** (fig. 10.4).

Schwann cells also enclose the smallest axons of peripheral neurons, but the Schwann cells are not wound around these fibers. Consequently, such fibers lack myelin sheaths. Instead, the fiber or a group of fibers may lie in a longitudinal groove of Schwann cells.

Fibers that possess myelin sheaths are called *myelinated* (medullated) nerve fibers, and those that lack these sheaths are *unmyelinated* nerve fibers (fig. 10.5). Groups of myelinated fibers appear white. Masses of such fibers impart color to the *white matter* in the brain and spinal cord, but here myelin is produced by another kind of neuroglial cell (oligodendrocyte) rather than by Schwann cells. Myelinated nerve fibers in the brain and spinal cord lack neurilemmal sheaths.

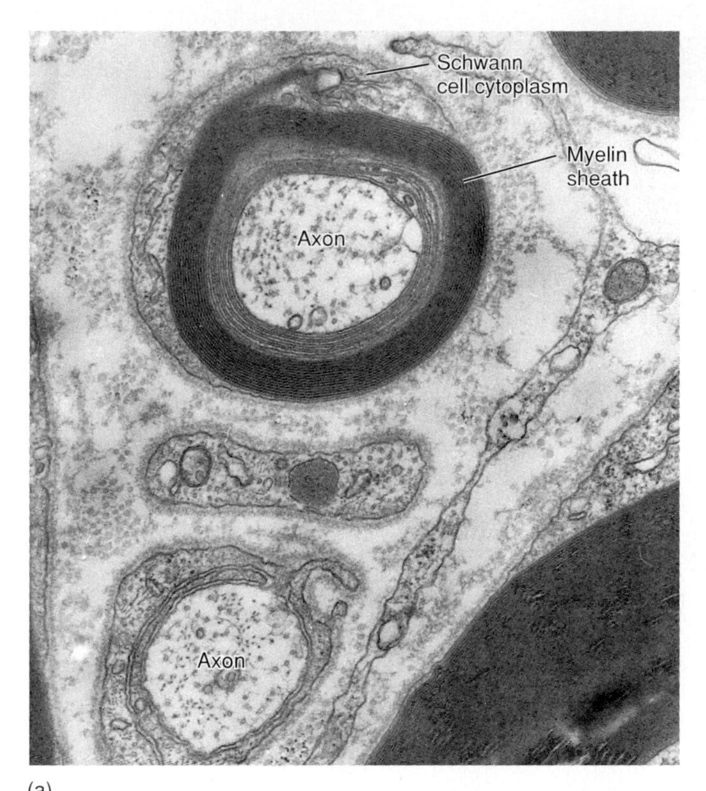


(a)

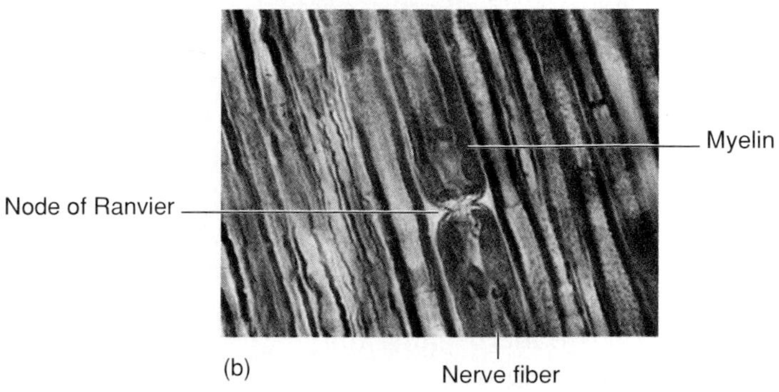


(b)

**FIGURE 10.5**

(*a*) A transmission electron micrograph of myelinated and unmyelinated axons in cross section (51,200×); (*b*) light micrograph of a myelinated nerve fiber (longitudinal section) (300×).

Unmyelinated nerve tissue appears gray. Thus, the *gray matter* within the brain and spinal cord contains an abundance of unmyelinated nerve fibers and neuron cell bodies. Clinical Application 10.2 discusses a disorder in which peripheral neurons lose their myelin.

1. List the general functions of the nervous system.
2. Describe a neuron.
3. Explain how a fiber in the peripheral nervous system becomes myelinated.

## Classification of Neurons and Neuroglia

Neurons and their processes vary in size and shape. They may differ in the length and size of their axons and dendrites, and in the number of synaptic knobs by which they communicate with other neurons.

Neurons also vary in function. Some carry impulses into the brain or spinal cord; others carry impulses out from the brain or spinal cord; and still others conduct impulses from neuron to neuron within the brain or spinal cord.

### Classification of Neurons

On the basis of *structural differences,* neurons can be classified into three major groups as figure 10.6 shows. Each type of neuron is specialized to send a nerve impulse in one direction, originating at a sensitive region called the **trigger zone.**

1. **Bipolar neurons.** The cell body of a bipolar neuron has only two nerve fibers, one arising from either end. Although these fibers are similar in structure, one is an axon and the other is a dendrite. Such neurons are found within specialized parts of the eyes, nose, and ears.
2. **Unipolar neurons.** Each unipolar neuron has a single nerve fiber extending from its cell body. A short distance from the cell body, this fiber divides into two branches: One branch (peripheral process) approaches a peripheral body part and serves as a dendrite, and the other (central process) enters the brain or spinal cord and serves as an axon. The cell bodies of some unipolar neurons aggregate in specialized masses of nerve tissue called *ganglia,* which are located outside the brain and spinal cord.

CLINICAL APPLICATION

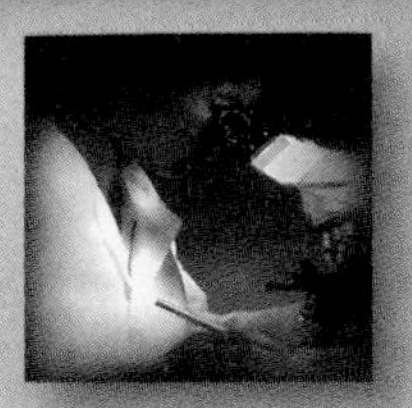

## Multiple Sclerosis

In 1964, at age 20, skier Jimmie Huega won the Olympic bronze medal in the slalom. In 1967, his vision blurred and then his legs became slightly numb. Jimmie ignored these intermittent symptoms, and after a while they disappeared.

Three years later, the symptoms returned, and this time Jimmie sought medical help. On the basis of his symptoms, which affected more than one body part and occurred sporadically, physicians diagnosed multiple sclerosis (MS). Today, diagnosis also includes a magnetic resonance imaging scan, which can detect brain lesions.

Jimmie can still ski and cycle and do situps and pushups, but he never knows when the on-again off-again disorder will strike. For many of the 300,000 people in the United States with MS, the progressive deterioration causes permanent paralysis. Fortunately, taking the immune system biochemical beta interferon as a drug can prevent flare-ups and mitigate symptoms.

In MS, the myelin coating in various sites through the brain and spinal cord forms hard scars called scleroses. The neurons that they surround can no longer transmit messages. Muscles that no longer receive input from motor neurons stop contracting. Without stimulation, muscles atrophy. Symptoms depend upon which neurons are affected. Short-circuiting in one part of the brain may affect fine coordination in one hand; if another brain part is affected, vision may be altered.

What might be responsible for the destruction of myelin in MS? A virus may cause the body's immune system to attack the cells producing myelin. This would happen if viruses lay latent in nerve cells, then emerged years later bearing proteins also found on nerve cells. The immune system, interpreting the proteins as foreign, would attack the viruses as well as the neurons.

A virus is suspected for a few reasons: viral infections are known to strip neurons of their myelin sheaths; viral infections can cause repeated bouts of symptoms; and most compelling, MS is far more common in some geographical regions (the temperate zones of Europe, South America, and North America) than others, suggesting a pattern of infection. A possible culprit is the virus that causes measles.

3. **Multipolar neurons.** Multipolar neurons have many nerve fibers arising from their cell bodies. Only one fiber of each neuron is an axon; the rest are dendrites. Most neurons whose cell bodies lie within the brain or spinal cord are of this type. The neuron illustrated in figure 10.3 is multipolar.

Neurons can also be classified by *functional differences* into the following groups:

1. **Sensory neurons** (afferent neurons) carry nerve impulses from peripheral body parts into the brain or spinal cord. These neurons have specialized *receptor ends* at the tips of their dendrites, or they have dendrites that are closely associated with *receptor cells* located in the skin or in certain sensory organs.

   Changes that occur inside or outside the body are likely to stimulate receptor ends or receptor cells, triggering sensory nerve impulses. The impulses travel along the sensory neuron fibers that lead to the brain or spinal cord, where they are processed by other neurons. Most sensory neurons are unipolar, although some are bipolar.

2. **Interneurons** (also called association or internuncial neurons) lie within the brain or spinal cord. They are multipolar and form links between other neurons. Interneurons transmit impulses from one part of the brain or spinal cord to another. That is, they may direct incoming sensory impulses to appropriate regions for processing and interpreting. Other incoming impulses are transferred to motor neurons.

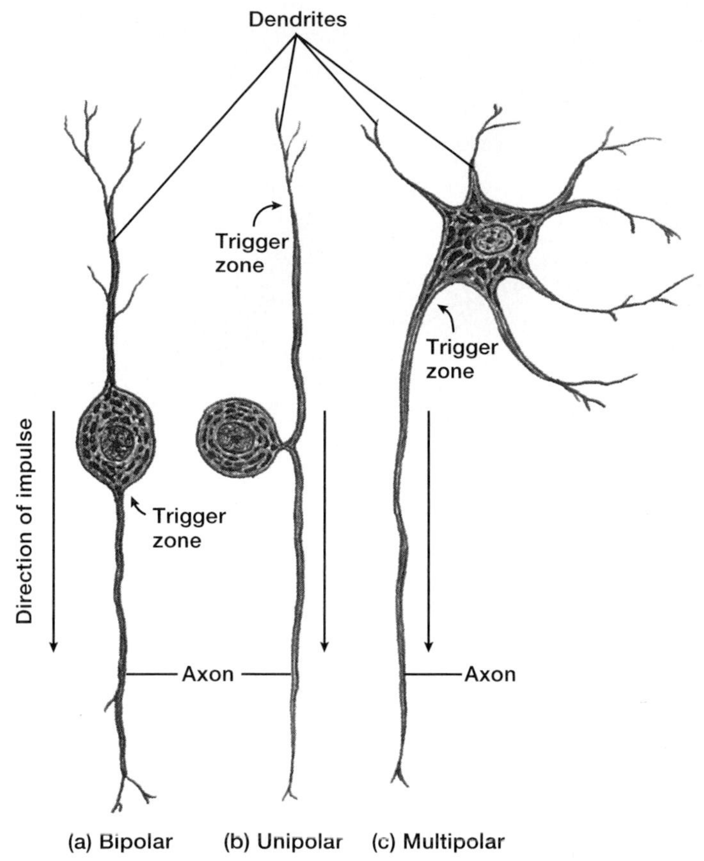


**FIGURE 10.6**

Structural types of neurons include (*a*) the bipolar neuron, (*b*) the unipolar neuron, and (*c*) the multipolar neuron.

3. **Motor neurons** (efferent neurons) are multipolar and carry nerve impulses out of the brain or spinal cord to effectors—structures that respond, such as muscles or glands. For example, when motor impulses reach muscles, they contract; when motor impulses reach glands, they release secretions.

Two specialized groups of motor neurons, accelerator and inhibitory neurons, innervate smooth and cardiac muscles. *Accelerator neurons* increase muscular activities, while *inhibitory neurons* decrease such actions.

Chart 10.1 summarizes the classification of neurons.

## Classification of Neuroglia

Neuroglia (glial cells) were once thought to be mere bystanders to neural function, filling spaces, providing scaffolding, and controlling the sites at which neurons contact one another (figs. 10.7 and 10.8). These important cells have additional functions. In the embryo, neuroglia guide neurons to their positions and may stimulate them to specialize. Neuroglia also produce the growth factors that nourish neurons, and remove ions and neurotransmitters that accumulate between neurons, enabling them to continue transmitting information. Experiments in which neuroglia grow in culture suggest that some types of these cells may play a direct role in communicating with neurons.

**CHART 10.1 TYPES OF NEURONS**

**A. Classified by Structure**

| *Type* | *Structural characteristics* | *Location* |
|---|---|---|
| **1.** Bipolar neuron | Cell body with a nerve fiber arising from each end | In specialized parts of the eyes, nose, and ears |
| **2.** Unipolar neuron | Cell body with a single nerve fiber that divides into two branches | In ganglia outside the brain or spinal cord |
| **3.** Multipolar neuron | Cell body with many nerve fibers, one of which is an axon | Most common type of neuron in the brain and spinal cord |

**B. Classified by Function**

| *Type* | *Functional characteristics* | *Structural characteristics* |
|---|---|---|
| **1.** Sensory neuron | Conducts nerve impulses from receptors in peripheral body parts into the brain or spinal cord | Most unipolar; some bipolar |
| **2.** Interneuron | Transmits nerve impulses between neurons within the brain and spinal cord | Multipolar |
| **3.** Motor neuron | Conducts nerve impulses from the brain or spinal cord out to effectors—muscles or glands | Multipolar |

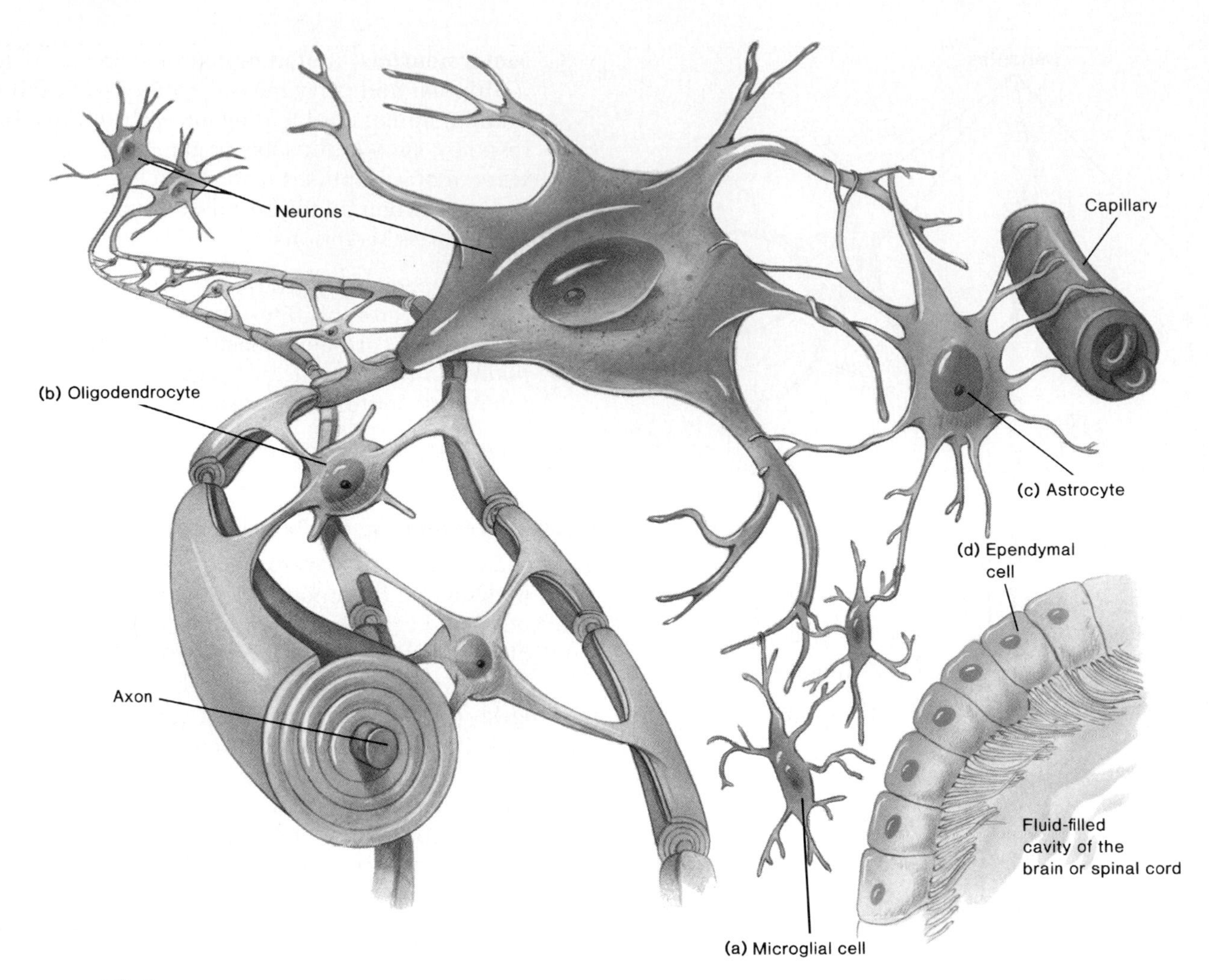


**FIGURE 10.7**

Types of neuroglial cells in the central nervous system include (*a*) microglial cell, (*b*) oligodendrocyte, (*c*) astrocyte, and (*d*) ependymal cell.

Schwann cells are the neuroglia of the peripheral nervous system. The central nervous system contains the following types of neuroglia:

1. **Astrocytes.** As their name implies, astrocytes are star-shaped cells. They are commonly found between neurons and blood vessels, where they provide support and hold structures together by means of numerous cellular processes. Astrocytes aid metabolism of certain substances, such as glucose, and they may help regulate the concentrations of important ions, such as potassium ions, within the interstitial space of nervous tissue. Astrocytes also respond to injury of brain tissue and form scar tissue, which fills spaces and closes gaps. These multifunctional cells may also have a nutritive function, transporting substances from blood vessels to neurons and bathing nearby neurons in growth factors. Gap junctions link astrocytes

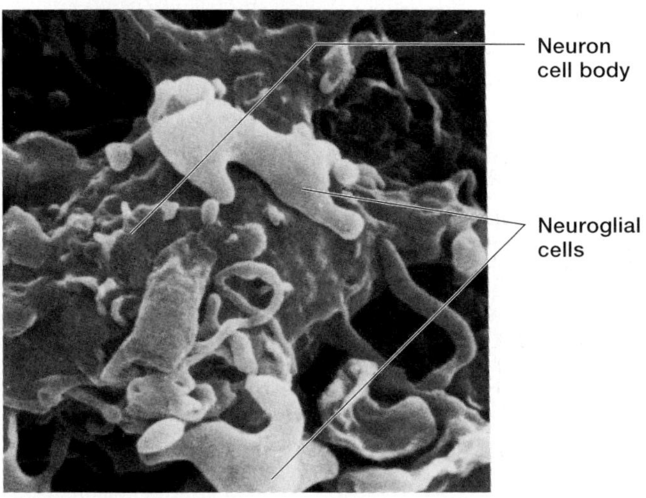


**FIGURE 10.8**

A scanning electron micrograph of a neuron cell body and some of the neuroglial cells associated with it (10,000×).

## CHART 10.2 Types of Neuroglial Cells of the Central Nervous System

| Type | Characteristics | Functions |
|---|---|---|
| Astrocytes | Star-shaped cells found between neurons and blood vessels | Structural support, formation of scar tissue, transport of substances between blood vessels and neurons, communicate with one another and with neurons, mop up excess ions and transmitters |
| Oligodendrocytes | Shaped like astrocytes, but with fewer cellular processes, and arranged in rows along nerve fibers | Form myelin sheaths within the brain and spinal cord; produce nerve growth factors |
| Microglia | Small cells with few cellular processes and found throughout the CNS | Structural support and phagocytosis (immune protection) |
| Ependyma | Cuboidal and columnar cells in the inner lining of the ventricles of the brain and the central canal of the spinal cord | Form a porous layer through which substances diffuse between the interstitial fluid of the brain and spinal cord, and the cerebrospinal fluid |

to one another, forming protein-lined channels through which calcium ions travel, possibly stimulating neurons.

2. **Oligodendrocytes.** Oligodendrocytes resemble astrocytes but are smaller and have fewer processes. They are commonly arranged in rows along myelinated nerve fibers, and they function in forming myelin within the brain and spinal cord.

   Unlike the Schwann cells of the peripheral nervous system, oligodendrocytes can send out a number of cellular processes, each of which forms a myelin sheath around a nearby axon. In this way, a single oligodendrocyte may provide myelin for many axons. However, these cells do not form neurilemmal sheaths.

3. **Microglia.** Microglial cells are relatively small and have fewer processes than other types of neuroglial cells. These cells are scattered throughout the central nervous system, where they help support neurons and phagocytize bacterial cells and cellular debris. They usually increase in number whenever the brain or spinal cord is inflamed because of injury or disease.

4. **Ependyma.** Ependymal cells are cuboidal or columnar in shape and may have cilia. They form the inner lining of the *central canal* that extends downward through the spinal cord. They also form an epithelial-like membrane that is one cell thick and covers the inside of spaces within the brain called *ventricles* (see chapter 11), and they cover the specialized capillaries called *choroid plexuses* that are associated with the ventricles of the brain.

   Gap junctions join ependymal cells to one another. They form a porous layer through which substances diffuse freely between the interstitial fluid of the brain tissues and the fluid (cerebrospinal fluid) within the ventricles.

Neuroglia form more than half of the volume of the brain. Chart 10.2 summarizes characteristics of neuroglial cells.

Abnormal neuroglia are associated with certain disorders. Most brain tumors, for example, consist of neuroglia dividing too often. Gene therapy is being tested to treat such tumors. Researchers add genes that instruct the cancerous glia to bear cell surface proteins that attract the immune system or render them more sensitive to cancer-fighting drugs.

Another experimental medical approach utilizing neuroglia is to construct implants consisting of certain types of neuroglia. Once implanted, these cells would secrete substances that may

- replace neurochemicals whose absence causes degenerative diseases of the nervous system, such as Parkinson's disease, Alzheimer's disease, multiple sclerosis, and amyotrophic lateral sclerosis,
- repair damaged spinal cords,
- halt damage to delicate nervous tissues caused by AIDS or cancer chemotherapy.

## Regeneration of Nerve Fibers

Injury to the cell body usually kills the neuron; however, a damaged axon may regenerate. For example, if injury or disease separates an axon in a peripheral nerve from its cell body, the distal portion of the axon and its myelin sheath deteriorate within a few weeks. Macrophages remove the fragments of myelin and

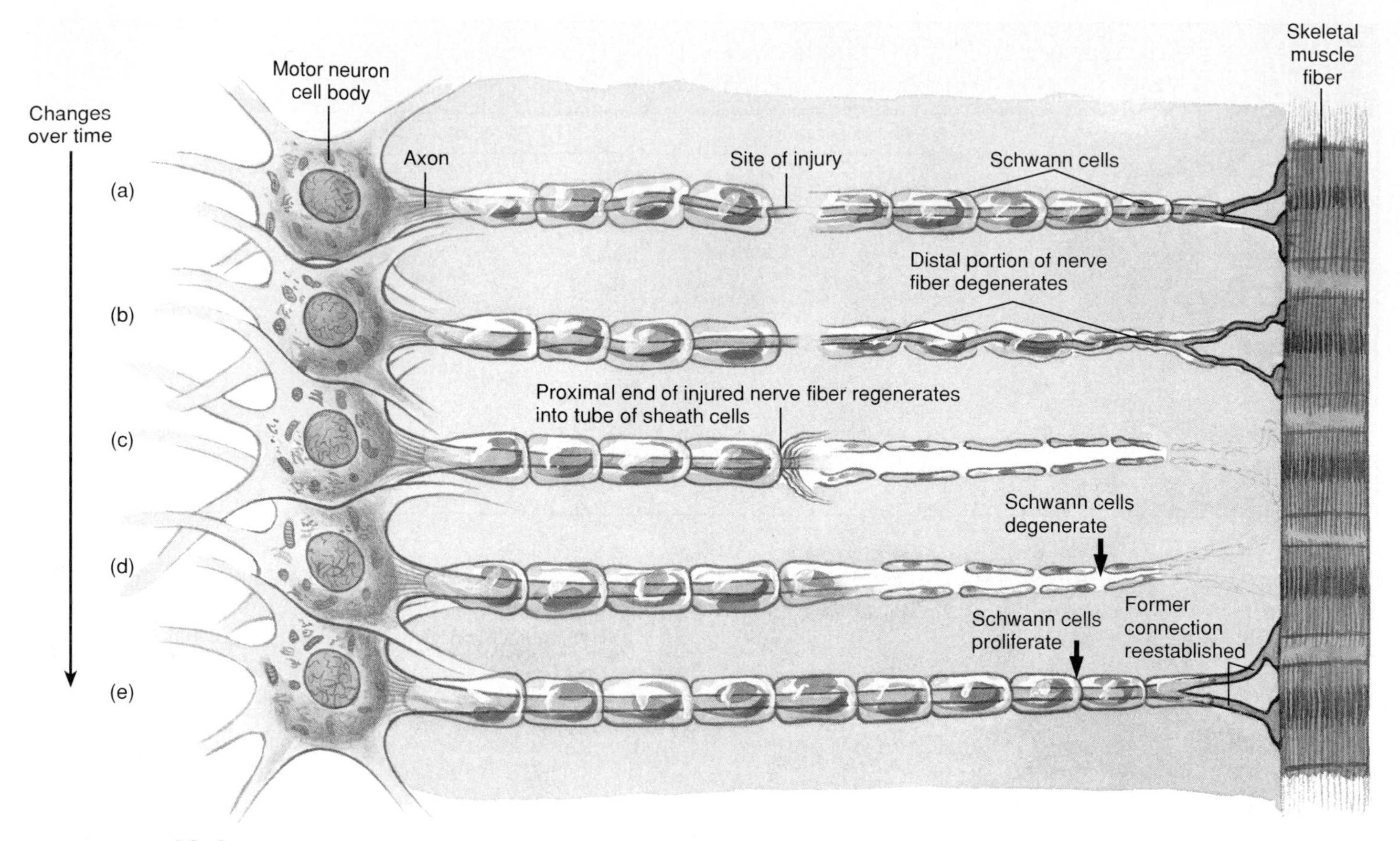


**FIGURE 10.9**

If a myelinated axon is injured, the following events may occur over several weeks to months: (*a*) The proximal portion of the fiber may survive, but (*b*) the portion distal to the injury degenerates. (*c* and *d*) In time, the proximal portion may develop extensions that grow into the tube of basement membrane and connective tissue cells that the fiber previously occupied, and (*e*) possibly reestablish the former connection. Nerve growth factors that neuroglial cells secrete assist in the regeneration process.

> Myelin begins to form on nerve fibers during the fourteenth week of prenatal development. By the time of birth, many nerve fibers are not completely myelinated. All myelinated fibers have begun to develop sheaths by the time a child starts to walk, and myelination continues into adolescence.
>
> Excess myelin seriously impairs nervous system functioning. In Tay-Sachs disease, an inherited defect in a lysosomal enzyme causes myelin to accumulate, burying neurons in fat. The affected child begins to show symptoms by 6 months of age, gradually losing sight, hearing, and muscle function until death occurs by age 4. Thanks to genetic screening among people of eastern European descent who are most likely to carry this gene, fewer than ten children are born in the United States with this condition each year.

other cellular debris. The proximal end of the injured axon develops sprouts shortly after the injury. Influenced by nerve growth factors nearby glia secrete, one of these sprouts may grow into a tube formed by remaining basement membrane and connective tissue. At the same time, any remaining Schwann cells proliferate along the length of the degenerating fiber, and form new myelin around the growing axon.

Growth of a regenerating fiber is slow (3 to 4 millimeters per day), but eventually the new fiber may reestablish the former connection (fig. 10.9). Nerve growth factors, secreted by glial cells, may help direct the growing fiber to the proper place.

If an axon of a nerve fiber within the central nervous system is separated from its cell body, the distal portion of the axon will degenerate, but more slowly than a separated axon in the peripheral nervous system. However, axons within the central nervous system lack a neurilemma, and the myelin-producing oligodendrocytes fail to proliferate following an

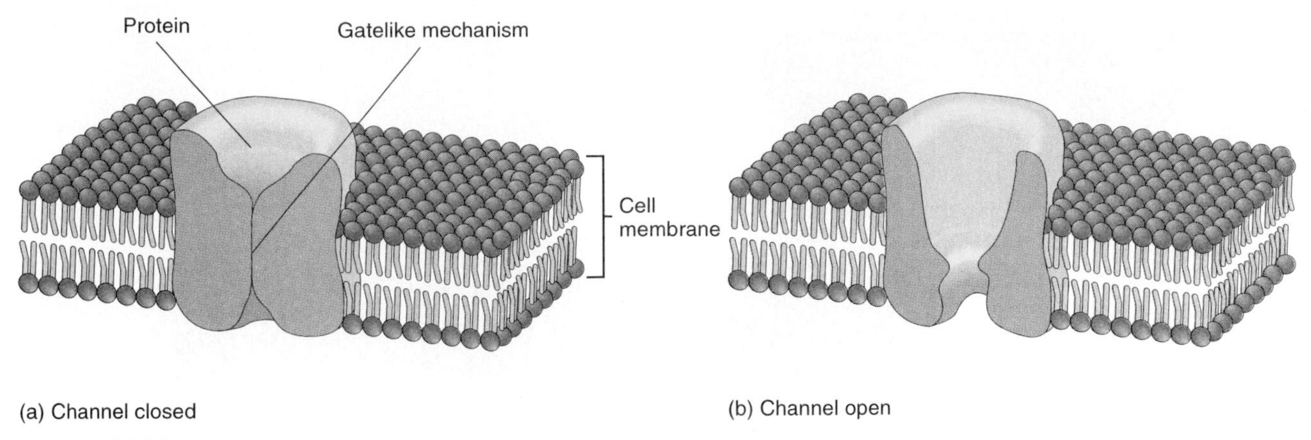


**FIGURE 10.10**
A gatelike mechanism can close (*a*) or open (*b*) some of the channels in cell membranes through which ions pass.

injury. Consequently, if the proximal end of a damaged axon begins to grow, there is no tube of sheath cells to guide it. Therefore regeneration is unlikely.

> If a peripheral nerve is severed, it is very important that the two cut ends be closely connected as soon as possible so that the regenerating sprouts of the nerve fibers can more easily reach the tubes formed by the basement membranes and connective tissues on the distal side of the gap.
>
> When the gap exceeds 3 millimeters, the regenerating fibers may form a tangled mass called a *neuroma*. It is composed of sensory nerve fibers, and is painfully sensitive to pressure. Neuromas sometimes complicate a patient's recovery following limb amputation.

1. What is a neuroglial cell?
2. Name and describe four types of neuroglial cells.
3. What are some functions of neuroglia?
4. Explain how an injured peripheral nerve fiber might regenerate.
5. Explain why functionally significant regeneration is unlikely in the central nervous system.

## Cell Membrane Potential

A cell membrane is usually electrically charged, or *polarized,* so that the inside is negatively charged with respect to the outside. This polarization is due to an unequal distribution of ions on either side of the membrane, and it is particularly important in the conduction of muscle and nerve impulses.

### Distribution of Ions

The distribution of the ions inside and outside of cell membranes is determined in part by the presence of pores or channels in those membranes, as discussed in chapter 3. These channels, formed by membrane proteins, can be quite selective; that is, a particular channel may allow one kind of ion to pass through and exclude all other ions of different size and charge. Furthermore, some channels are always open, whereas others may be either open or closed, somewhat like a gate. As we shall see, both chemical and electrical factors can affect the opening and closing of particular channels (fig. 10.10).

As a consequence of the selectivity of specific ion channels, potassium ions (the major intracellular positive ion, or **cation**) tend to pass through resting cell membranes much more easily than sodium ions (the major extracellular cation). The relative ease with which potassium ions diffuse through membranes makes them a major contributor to membrane polarization. Calcium ions are less able to cross the resting cell membrane than either sodium ions or potassium ions, and have a special role in nerve function, described later.

### Resting Potential

Because of the active transport of sodium and potassium ions, cells throughout the body have a relatively greater concentration of sodium ions ($Na^+$) outside their membranes and a relatively greater concentration of potassium ions ($K^+$) inside their membranes. In the cytoplasm of these cells are many negatively charged ions called **anions,** including those of phosphate ($PO_4^=$), sulfate ($SO_4^=$), and protein, that cannot diffuse through the cell membranes (fig. 10.11*a*).

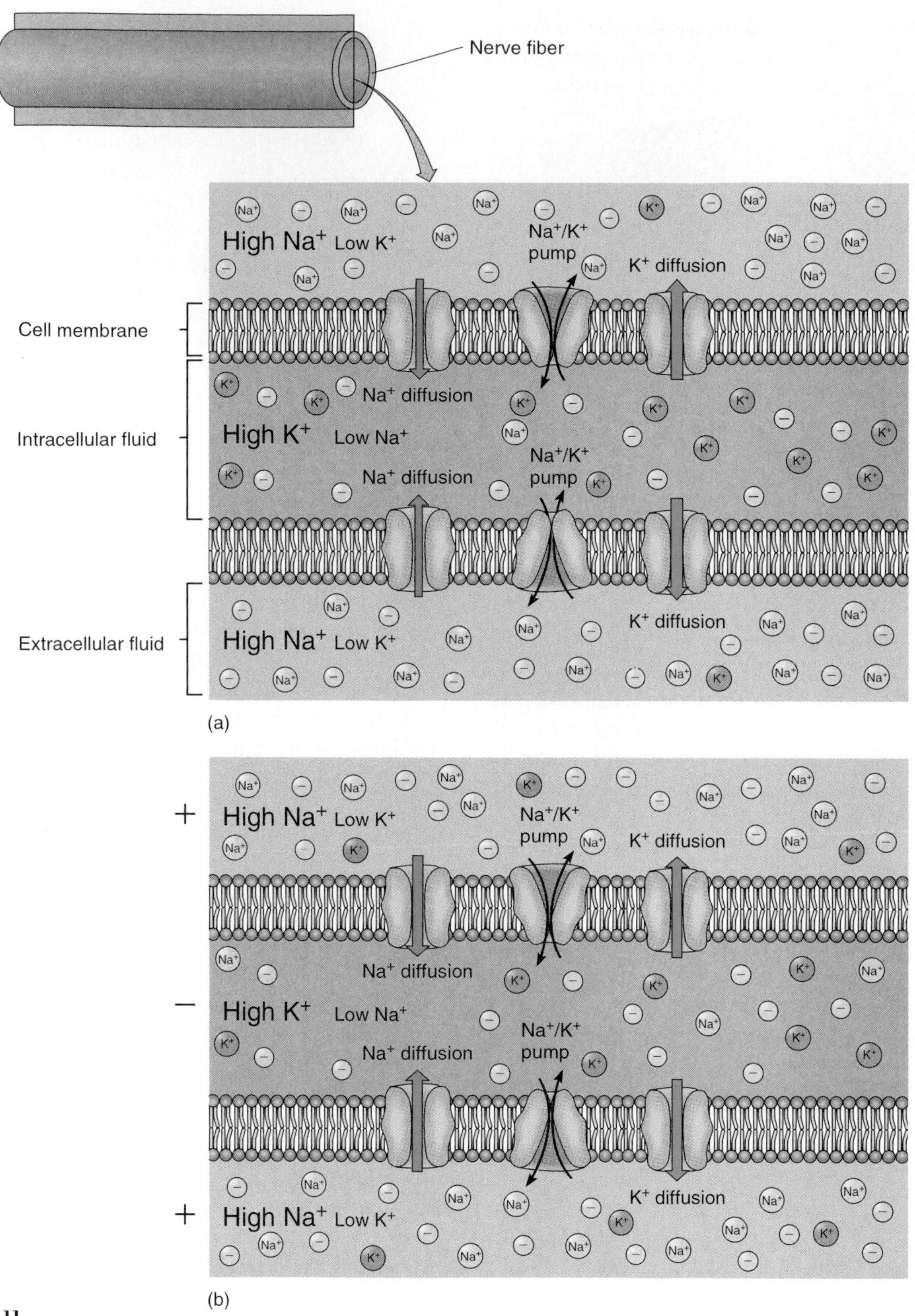


**FIGURE 10.11**

Development of the resting membrane potential. (*a*) Active transport creates a concentration gradient across the cell membrane for sodium ions ($Na^+$) and potassium ions ($K^+$); $K^+$ diffuses out of the cell rather slowly but nonetheless faster than $Na^+$ can diffuse in; (*b*) this unequal diffusion results in a net loss of positive charge and a resultant excess of negative charge inside the membrane.

Sodium and potassium ions follow the laws of diffusion stated earlier and show a net movement from high concentration to low concentration as permeabilities permit. Because a resting cell membrane is more permeable to potassium ions than to sodium ions, potassium ions tend to diffuse out of the cell more rapidly than sodium ions can diffuse in (fig. 10.11*a*). This means that, every millisecond, more positive charges leave the cell by diffusion than enter it. As a result, the outside of the cell membrane gains a slight surplus of positive charges and the inside is left with a slight surplus of impermeable negative charges. At the same time, the cell continues to expend energy to actively transport sodium and potassium ions in the opposite directions, thus maintaining the concentration gradients for these ions responsible for their diffusion in the first place (fig. 10.11*b*).

The difference in electrical charge between two points is called the *potential difference* because it represents stored electrical energy that can be used to do work. It is measured in units called *volts.* In the case of a resting nerve cell, the potential difference between the inside and the outside of the cell membrane is about –70 millivolts (negative because of the excess negative charges on the inside of the cell membrane). This separation of charge is called the **resting potential,** and the work it may do is to send the nerve impulse or **action potential.** To understand how neurons initiate action potentials, we must first understand how nerve cells respond to signals called **stimuli.**

### Local Potential Changes

Nerve cells are excitable; that is, they can respond to changes in their surroundings. Some nerve cells, for example, detect changes in temperature, light, or pressure occurring outside the body, while others respond to signals coming from inside the body, often from other neurons. In either case, such changes or stimuli usually affect the resting potential in a particular local region of the cell membrane. If, in response to a stimulus, the membrane's resting potential becomes more negative (moving away from zero), the membrane is *hyperpolarizing;* if the resting potential becomes less negative (moving toward zero), the membrane is *depolarizing.*

Changes in the local potential of a membrane are *graded.* This means the amount of change in potential is directly related to the intensity of the stimulation. If a sufficiently strong depolarization occurs, a level called the *threshold potential* will be reached. In many cases, a single stimulus is not sufficient to reach threshold potential.

If another stimulus of the same type is received before the effect of the first one subsides, the change in local membrane potential is greater. This additive phenomenon is called *summation,* and as a result of summated potentials, the threshold potential may be reached. Thus, many subthreshold potential changes may combine to reach threshold. At threshold, an action potential is produced in a nerve fiber.

### Action Potentials

Once the threshold potential is reached, the portion of the cell membrane being stimulated undergoes a sudden change in permeability. Channels, in the trigger zone that are highly selective for sodium ions, open, allowing the sodium ions in. This movement is aided by the fact that the sodium ions are attracted by the negative electrical condition on the inside of the membrane.

As the sodium ions rush inward, the membrane potential changes from its resting value (fig. 10.12*a*) and momentarily becomes positive on the inside (this is still considered depolarization). At the peak of the action potential, membrane potential may reach +30mV (fig. 10.12*b*). At almost the same time, channels open in the membrane that allow some of the potassium ions to pass through, and as they diffuse outward, the inside of the membrane becomes negatively charged once more. The membrane then is said to become **repolarized,** and it remains in this state until it is stimulated again (fig. 10.12*c*).

This rapid sequence of changes, involving depolarization and repolarization, takes about 1/1,000th second or less—this is the action potential (fig. 10.13). Actually, only a small proportion of the sodium and potassium ions present move through the membrane during an action potential, so that many action potentials could occur before the concentrations of sodium ions and potassium ions on either side of the membrane would change significantly. The active transport mechanism in the membrane works to maintain the original concentrations of sodium and potassium. Thus, the resting potential is quickly reestablished.

When an action potential occurs at the trigger zone, it causes an electric current to flow a short distance down the fiber. This local current stimulates the adjacent membrane to its threshold level, triggering another action potential. This in turn causes another electric current farther down the fiber. This sequence of events results in a wave of action potentials moving down the fiber, traveling to the end of the fiber without decreasing in amplitude, even if branches occur. This propagation of action potentials along a fiber constitutes a **nerve impulse** (fig. 10.14).

A nerve impulse is similar to the muscle impulse mentioned in chapter 9. In the muscle fiber, stimulation at the motor end plate triggers an impulse to travel over the surface of the fiber and down into its transverse tubules. See chart 10.3 for a summary of the events leading to the conduction of a nerve impulse.

### Refractory Period

For a very short time following the passage of a nerve impulse, a threshold stimulus will not trigger another impulse on a nerve fiber. This brief period, called the **refractory period,** has two parts. During the *absolute refractory period,* which lasts about 1/2,500 of a second, the fiber's membrane is changing in sodium permeability and cannot be stimulated. This is followed by a *relative refractory period,* during which the membrane is reestablishing its resting potential. While the membrane is in the relative refractory period, even though polarization is incomplete, a stimulus of high intensity may trigger an impulse.

As time passes, the intensity of stimulation needed to trigger an impulse decreases until the fiber's original excitability is restored. This return to the resting state usually takes from 10 to 30 milliseconds.

Because of the refractory period, a nerve fiber cannot be stimulated continuously. Thus, the refractory period limits the rate at which nerve impulses can be conducted. The time between impulses cannot be less than the absolute refractory period, and the maximum rate of nerve impulses is therefore about one impulse per millisecond.

## All-or-None Response

Like muscle fiber contraction, nerve impulse conduction is an all-or-none response. In other words, if a nerve fiber responds at all, it responds completely. Thus, a nerve impulse is conducted whenever a stimulus of threshold intensity or above is applied to a nerve fiber, and all impulses carried on that fiber will be of the same strength. A greater intensity of stimulation produces not a stronger impulse but rather more impulses per second.

## Impulse Conduction

An unmyelinated nerve fiber conducts an impulse over its entire surface. A myelinated fiber functions differently. Myelin contains a high proportion of

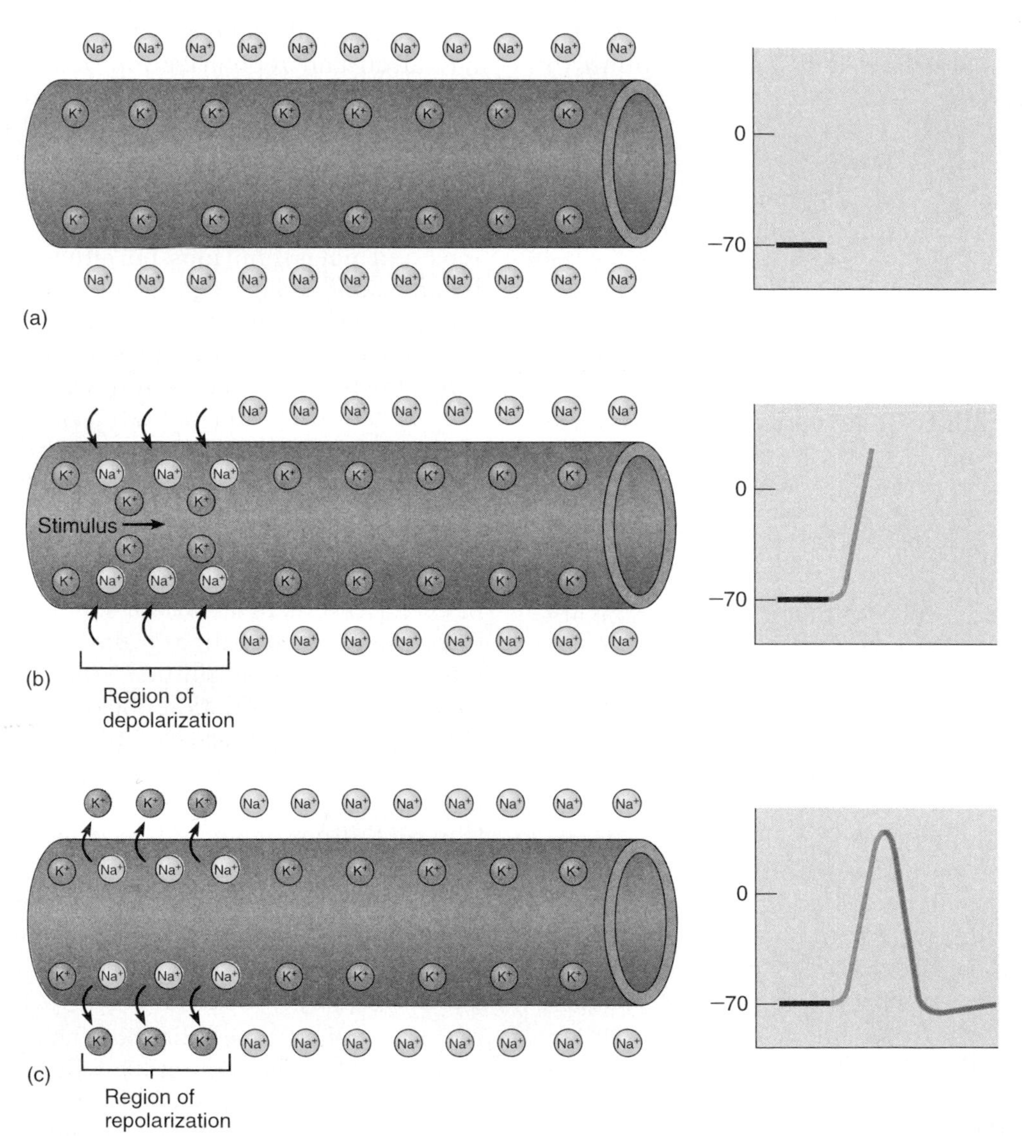


**FIGURE 10.12**

At rest (*a*), the membrane potential is about −70 millivolts. When the membrane reaches threshold (*b*), sodium channels open, some $Na^+$ diffuses inward, and the membrane is depolarized. Soon afterward (*c*), potassium channels open, $K^+$ diffuses out, and the membrane is repolarized.

lipid that excludes water and water-soluble substances. Thus, it serves as an insulator and prevents almost all flow of ions through the membrane that is enclosed in myelin.

Considering this, it might seem that the myelin sheath would prevent the conduction of a nerve impulse altogether, and this would be true if the sheath were continuous. It is, however, interrupted by nodes of Ranvier between adjacent Schwann cells (see fig. 10.3). At these nodes, the nerve fiber membrane contains channels for sodium and potassium ions that open during a threshold depolarization.

When such a myelinated nerve fiber is stimulated to threshold, an action potential occurs at the trigger zone. This causes an electric current to flow

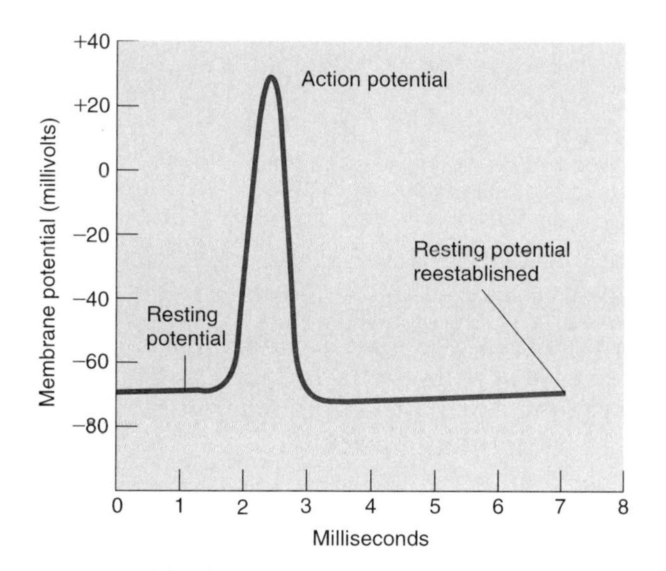


**FIGURE 10.13**
An oscilloscope records an action potential.

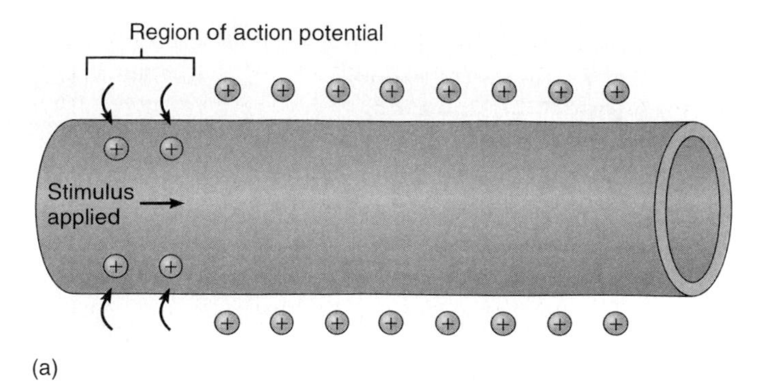


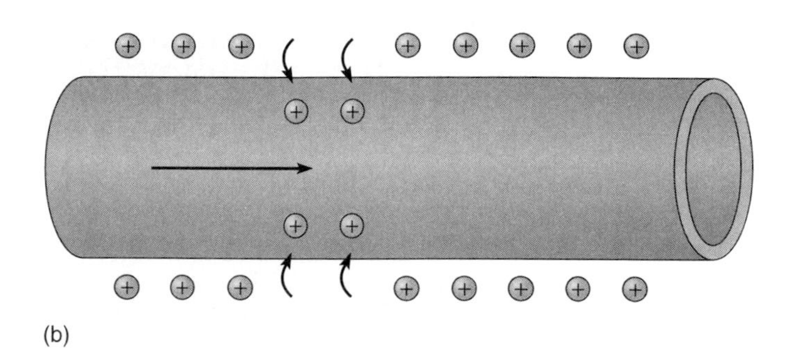


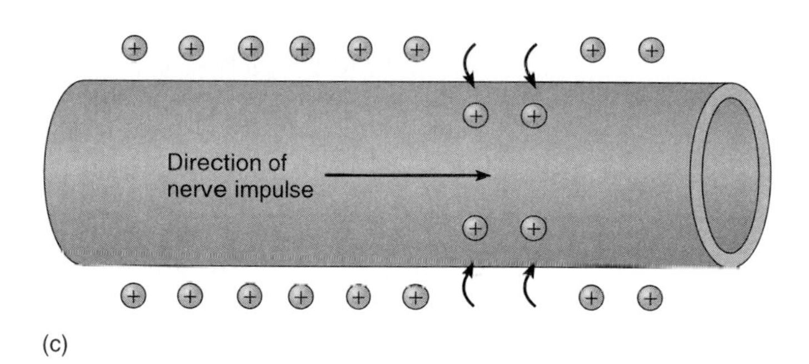


**FIGURE 10.14**
(*a*) An action potential in one region stimulates the adjacent region, and (*b* and *c*) a wave of action potentials (a nerve impulse) moves along the fiber.

## CHART 10.3 EVENTS LEADING TO NERVE IMPULSE CONDUCTION

1. Nerve fiber membrane maintains resting potential by diffusion of $Na^+$ and $K^+$ down their concentration gradients as the cell pumps them up the gradients.
2. Neurons receive stimulation, causing local potentials, which may sum to reach threshold.
3. Sodium channels in a local region of the membrane open.
4. Sodium ions diffuse inward, depolarizing the membrane.
5. Potassium channels in the membrane open.
6. Potassium ions diffuse outward, repolarizing the membrane.
7. The resulting action potential causes an electric current that stimulates adjacent portions of the membrane.
8. Wave of action potentials travels the length of the nerve fiber as a nerve impulse.

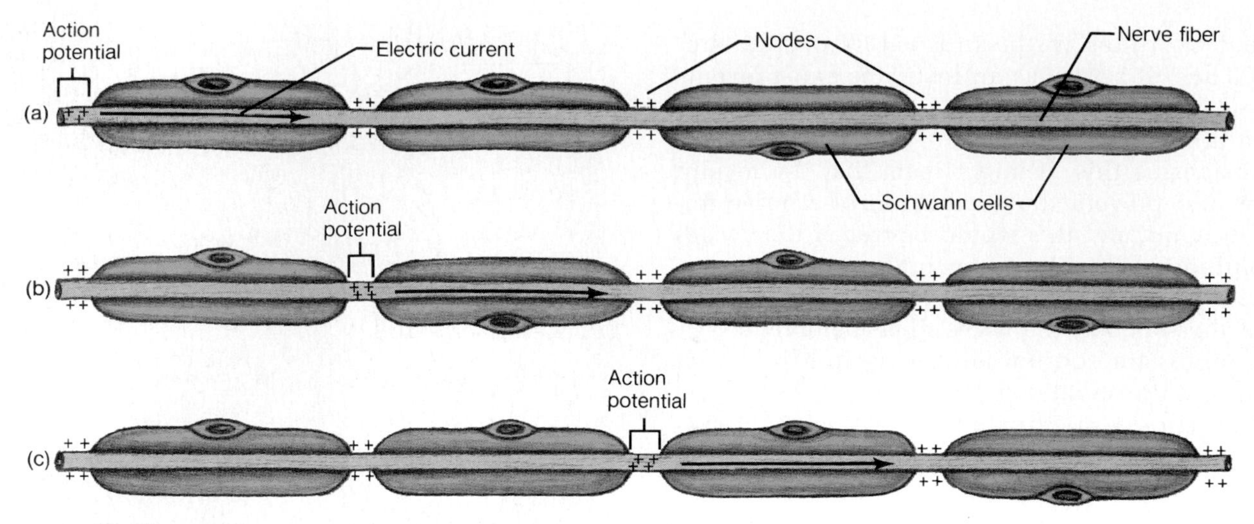


**FIGURE 10.15**

On a myelinated fiber, a nerve impulse appears to jump from node to node.

away from the trigger zone through the cytoplasm of the axon. As this local current reaches the first node, it stimulates the membrane to its threshold level. An action potential occurs there sending an electric current to the next node. Consequently, a nerve impulse traveling along a myelinated nerve fiber appears to jump from node to node. This type of impulse conduction, called **saltatory conduction,** is many times faster than conduction on an unmyelinated nerve fiber (fig. 10.15).

The speed of nerve impulse conduction is also related to the diameter of the fiber—the greater the diameter, the faster the impulse. For example, an impulse on a relatively thick, myelinated nerve fiber, such as a motor fiber associated with a skeletal muscle, might travel 120 meters per second, while an impulse on a thin, unmyelinated nerve fiber, such as a sensory fiber associated with the skin, might move only 0.5 meter per second. Clinical Application 10.3 discusses some factors that influence nerve impulse conduction.

1. Summarize how a resting potential is achieved.
2. Explain how a polarized nerve fiber responds to stimulation.
3. List the major events that occur during an action potential.
4. Define refractory period.
5. Explain how impulse conduction differs in myelinated and unmyelinated nerve fibers.

## THE SYNAPSE

Nerve impulses may pass from neuron to neuron at synapses along complex nerve pathways (fig. 10.16). A **presynaptic** and **postsynaptic neuron** surround the gap, or **synaptic cleft** (fig. 10.17).

### Synaptic Transmission

A nerve impulse travels from a dendrite or cell body along the axon to the presynaptic terminal at its end. The process by which the impulse in the presynaptic neuron signals the postsynaptic neuron is called **synaptic transmission.**

Axons usually have several rounded *synaptic knobs* at their presynaptic terminals, which dendrites lack. These knobs contain numerous membranous sacs, called *synaptic vesicles.* When a nerve impulse reaches a synaptic knob, some of the synaptic vesicles respond by releasing a neurotransmitter.

## Clinical Application 10.3

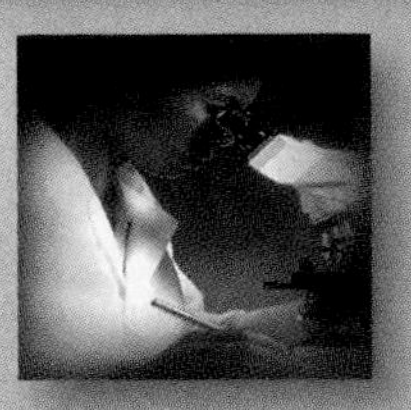

### Factors Affecting Impulse Conduction

A number of substances alter nerve fiber membrane permeability to ions. For example, calcium ions are needed to close sodium channels in nerve fiber membranes during an action potential. Consequently, if calcium is deficient, sodium channels remain open, and sodium ions diffuse through the membrane again and again so that impulses are transmitted repeatedly. If these spontaneous impulses travel along nerve fibers to skeletal muscle fibers, the muscles continuously spasm (tetanus or tetany). This can occur in women during pregnancy as the developing fetus uses maternal calcium. Tetanic contraction may also occur when the diet lacks calcium or vitamin D, or when prolonged diarrhea depletes the body of calcium.

A small increase in the concentration of extracellular potassium ions causes the resting potential of nerve fibers to be less negative (partially depolarized). As a result, the threshold potential is reached with a less intense stimulus than usual. The affected fibers are very excitable, and the person may experience convulsions.

If the extracellular potassium ion concentration is greatly decreased, the resting potentials of the nerve fibers may remain so negative that action potentials cannot occur. In this case, impulses are not triggered, and muscles become paralyzed.

Certain anesthetic drugs, such as procaine, produce special effects by decreasing membrane permeability to sodium ions. These drugs present in the tissue fluids surrounding a nerve fiber prevent impulses from passing through the affected region. Consequently, these drugs keep impulses from reaching the brain, preventing the perception of touch and pain.

At least twenty types of proteins jut from the presynaptic and postsynaptic membranes, participating in as yet unknown ways in synaptic transmission.

### Synaptic Potentials

Released neurotransmitter diffuses across the synaptic cleft and reacts with specific receptors in or on the postsynaptic neuron membrane. Effects of neurotransmitters may vary. Some open ion channels, and others close them. Neurotransmitter molecules binding to postsynaptic receptors create local potentials, called **synaptic potentials,** which enable one neuron to influence another.

Synaptic potentials are graded and can depolarize or hyperpolarize the receiving cell membrane. For example, if a neurotransmitter binds to a postsynaptic receptor and opens sodium ion channels, the ions diffuse inward, depolarizing the membrane, possibly triggering an action potential. This type of membrane change is called an **excitatory postsynaptic potential** (EPSP), and it lasts for about 15 milliseconds.

If a different neurotransmitter binds other receptors and increases membrane permeability to potassium ions, these ions diffuse outward, hyperpolarizing the membrane. Since an action potential is now less likely to occur, this change is called an **inhibitory postsynaptic potential** (IPSP).

Within the brain and spinal cord, each neuron may receive the synaptic knobs of a thousand or more nerve fibers on its dendrites and cell body. Furthermore, at any moment, some of the postsynaptic potentials are excitatory on a particular neuron, while others are inhibitory (fig. 10.18).

The integrated sum of the EPSPs and IPSPs determines whether an action potential results. That is, if the net effect is more excitatory than inhibitory, threshold may be reached, and an action potential triggered. Conversely, if the net effect is inhibitory, no impulse will be transmitted.

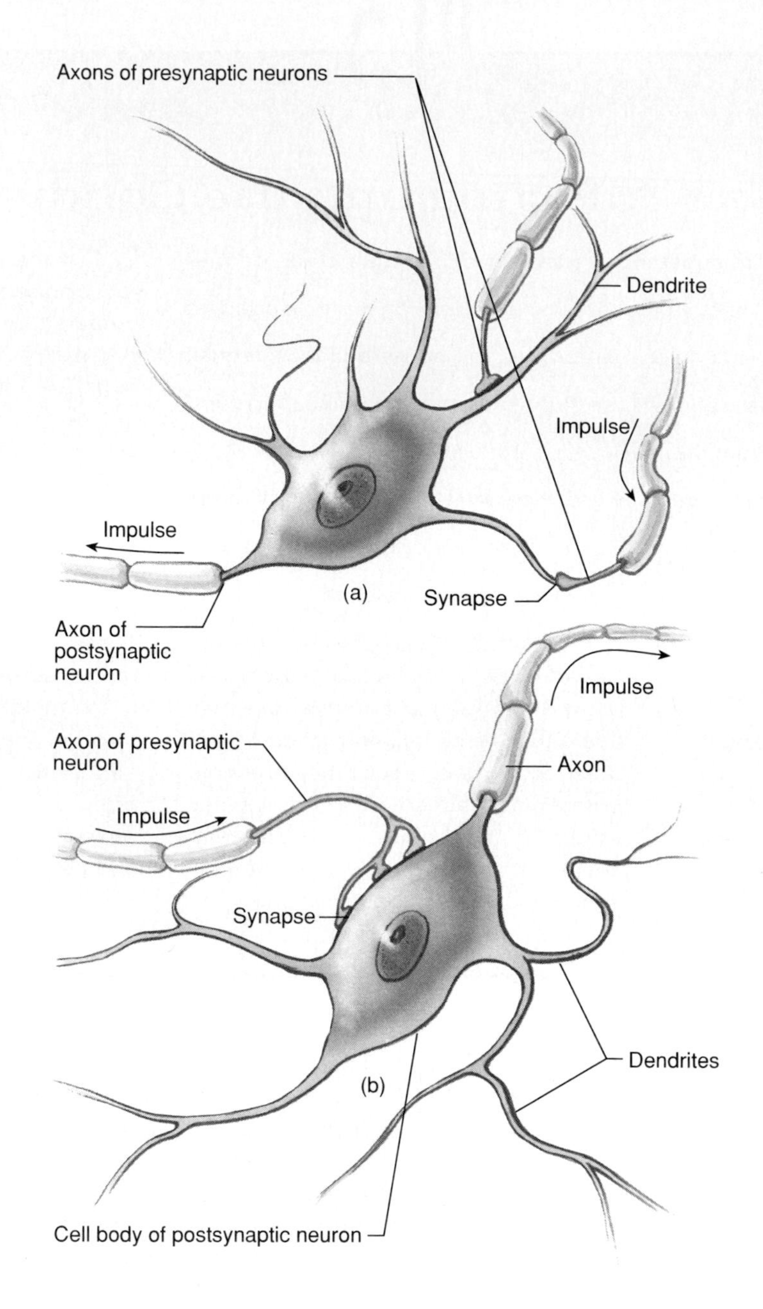


**FIGURE 10.16**

For an impulse to continue from one neuron to another, it must cross the synaptic cleft at a synapse. A synapse usually occurs (*a*) between an axon and a dendrite, or (*b*) between an axon and a cell body.

This summation of the excitatory and inhibitory effects of the postsynaptic potentials commonly takes place at the trigger zone, usually in a proximal region of the axon, but found in some dendrites as well. This region has an especially low threshold for triggering an action potential; thus, it serves as a decision-making part of the neuron.

1. Describe a synapse.
2. Explain the function of a neurotransmitter.
3. Distinguish between EPSP and IPSP.
4. Describe the net effects of EPSPs and IPSPs.

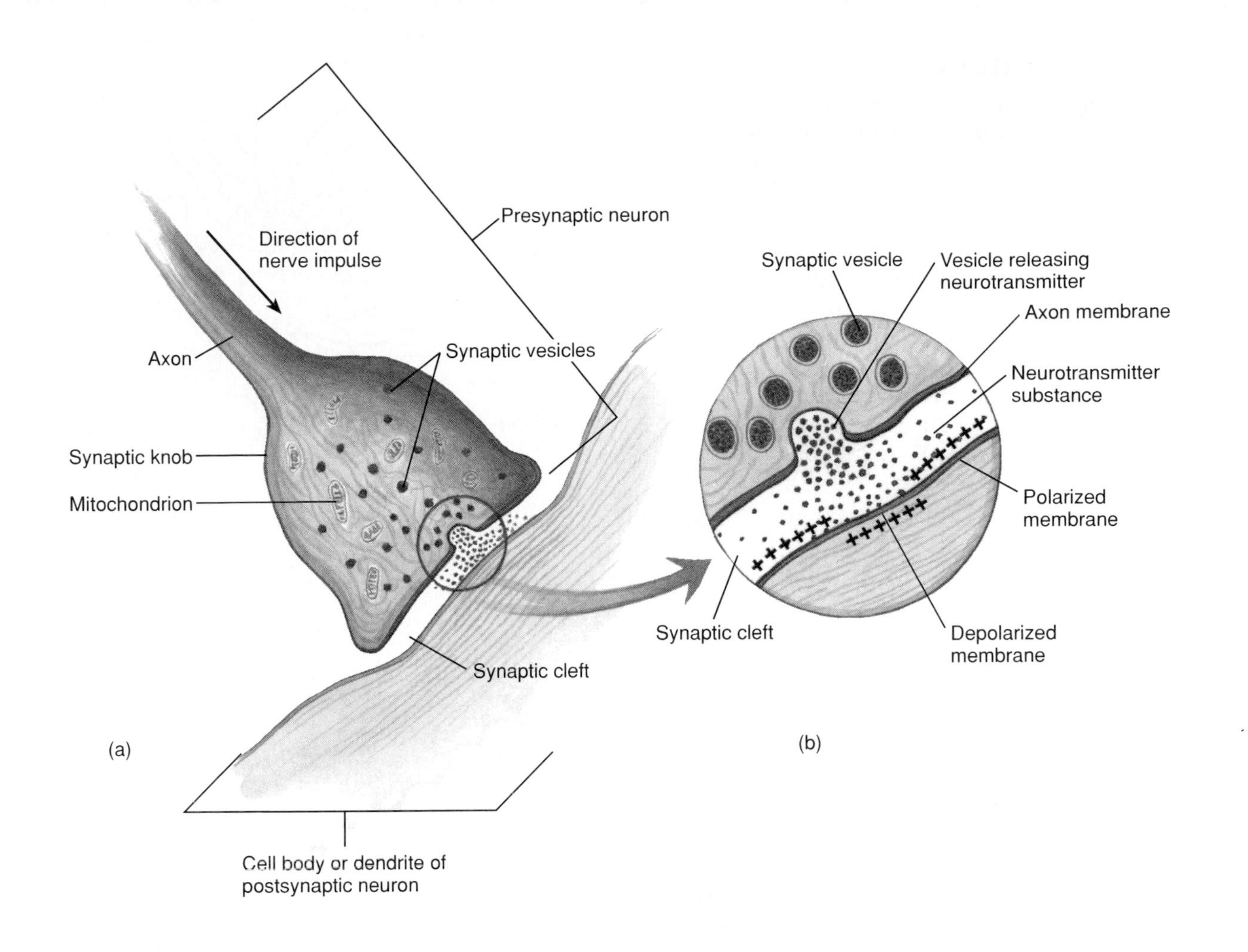


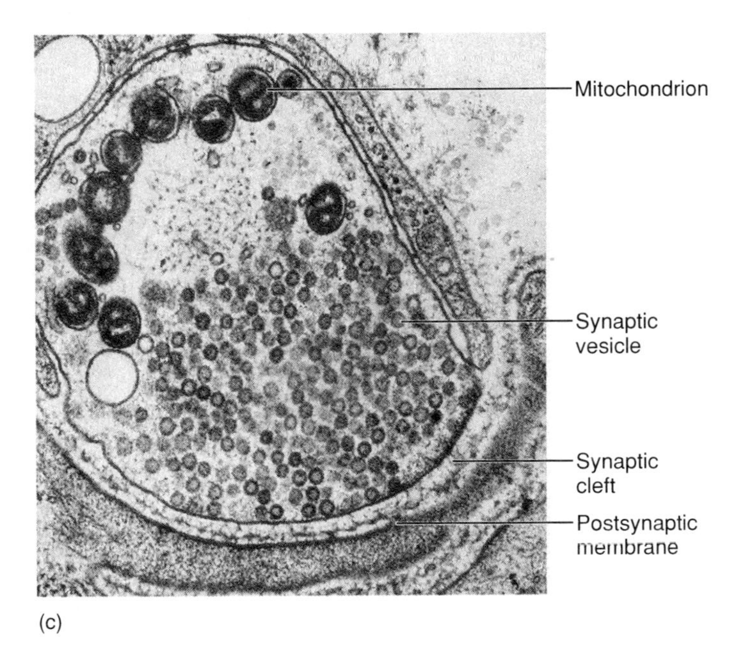


**FIGURE 10.17**

(*a*) When a nerve impulse reaches the synaptic knob at the end of an axon, (*b*) synaptic vesicles release a neurotransmitter substance that diffuses across the synaptic cleft. (*c*) A transmission electron micrograph of a synaptic knob filled with synaptic vesicles.

## Neurotransmitters

The nervous system produces at least thirty different neurotransmitters. Some neurons release only one kind, while other neurons produce two or three kinds. Neurotransmitters include *acetylcholine,* which stimulates skeletal muscle contractions (see chapter 9 and Clinical Application 10.4); a group of compounds called *monoamines* (such as epinephrine, norepinephrine, dopamine, and serotonin), which are formed by modifying amino acid molecules; a group of unmodified *amino acids* (such as glycine, glutamic acid, aspartic acid, and gamma-aminobutyric acid—GABA); and a large group of *peptides* (such as enkephalins and substance P), which are short chains of amino acids.

Most types of neurotransmitters are synthesized in the cytoplasm of the synaptic knobs and stored in the synaptic vesicles. When an action potential passes along the membrane of a synaptic knob, it increases the membrane's permeability to calcium ions by opening its calcium ion channels. Calcium ions diffuse inward, and in response, some of the synaptic vesicles fuse with the presynaptic membrane and release their contents by exocytosis into the synaptic cleft. The quantity of neurotransmitter that vesicles release is related directly to the quantity of calcium that enters the synaptic knob. Clinical Application 10.4 discusses myasthenia gravis, an autoimmune disorder that affects neuromuscular junctions. Charts 10.4 and 10.5 list disorders and drugs that alter neurotransmitter levels.

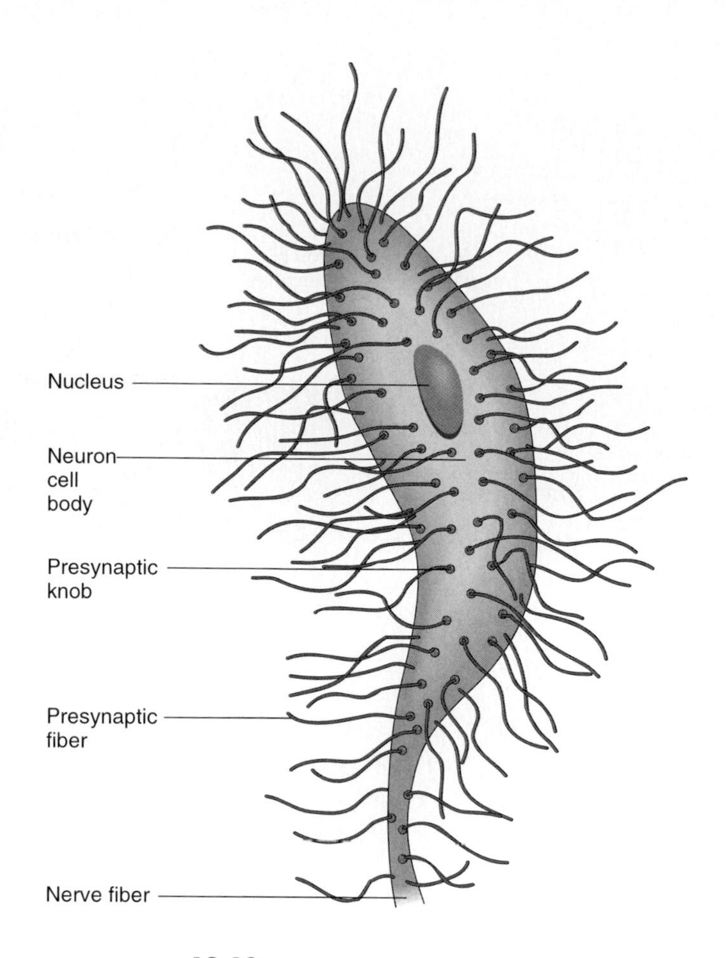


**Figure 10.18**
The synaptic knobs of many axons may communicate with the cell body of a neuron.

**Chart 10.4 Disorders Associated with Neurotransmitter Imbalances**

| Condition | Symptoms | Imbalance of Neurotransmitter in Brain |
|---|---|---|
| Alzheimer's disease | Memory loss, depression, disorientation, dementia, hallucinations, death | Deficient acetylcholine |
| Clinical depression | Debilitating, inexplicable sadness | Deficient norepinephrine and/or serotonin |
| Epilepsy | Seizures, loss of consciousness | Excess GABA leads to excess norepinephrine and dopamine |
| Huntington disease | Personality changes, loss of coordination, uncontrollable dancelike movements, death | Deficient GABA |
| Hypersomnia | Excessive sleeping | Excess serotonin |
| Insomnia | Inability to sleep | Deficient serotonin |
| Mania | Elation, irritability, overtalkativeness, increased movements | Excess norepinephrine |
| Myasthenia gravis | Progressive muscular weakness | Deficient acetylcholine receptors at neuromuscular junctions |
| Parkinson's disease | Tremors of hands, slowed movements, muscle rigidity | Deficient dopamine |
| Schizophrenia | Inappropriate emotional responses, hallucinations | Deficient GABA leads to excess dopamine |
| Sudden infant death syndrome ("crib death") | Baby stops breathing, dies if unassisted | Excess dopamine |
| Tardive dyskinesia | Uncontrollable movements of facial muscles | Deficient dopamine |

## Clinical Application 10.4

# Myasthenia Gravis

An autoimmune disorder in which the immune system attacks part of the nervous system is myasthenia gravis (MG). The target in this disorder is receptors for acetylcholine on muscle cells at neuromuscular junctions. (Recall that a neuromuscular junction is the place where a neuron meets a muscle cell.) People with MG have one-third the normal number of acetylcholine receptors at these junctions. On a whole-body level, this translates into muscle weakness and easily fatigued muscles.

MG affects 25,000 people in the United States, usually women beginning in their twenties or thirties and men in their sixties and seventies. The specific symptoms depend upon the site of attack. For 85% of patients, the disease causes generalized muscle weakness. Many people develop a characteristic flat smile and nasal voice and have difficulty chewing and swallowing due to affected facial and neck muscles. Many have limb weakness. About 15% of patients experience the illness only in the muscles surrounding their eyes. The disease reaches crisis level when respiratory muscles are affected, requiring a ventilator to breathe. MG does not affect sensation or reflexes.

Until 1958, MG was a serious threat to health, with a third of patients dying, a third worsening, and a third maintaining or improving. Today, most people with MG can live near-normal lives, thanks to a combination of the following treatments:

- Drugs that disable the enzyme that normally breaks down acetylcholine, increasing availability of the neurotransmitter.
- Removing the thymus gland, which oversees much of the immune response.
- Immunosuppressant drugs such as cyclosporine, prednisone, and azathioprine.
- Intravenous antibodies to bind the ones causing the damage.
- Plasma exchange, which rapidly removes the damaging antibodies from the circulation. This helps people in crisis.

## Chart 10.5 Drugs That Alter Neurotransmitter Levels

| Drug | Neurotransmitter Affected | Mechanism of Action | Effect |
|---|---|---|---|
| Tryptophan | Serotonin | Stimulates neurotransmitter synthesis | Sleepiness |
| Reserpine | Norepinephrine | Packaging neurotransmitter into vesicles | Limb tremors |
| Curare | Acetylcholine | Decreases neurotransmitter in synaptic cleft | Muscle paralysis |
| Valium | GABA | Enhances receptor binding | Decreases anxiety |
| Nicotine | Acetylcholine | Stimulates synthesis of enzyme that degrades neurotransmitter | Increases alertness |
| Cocaine | Norepinephrine | Blocks reuptake | Euphoria |
| Tricyclic antidepressants | Norepinephrine | Blocks reuptake | Mood elevation |
| Monoamine oxidase inhibitors | Norepinephrine | Blocks enzymatic degradation of transmitter in presynaptic cell | Mood elevation |
| Prozac and related drugs | Serotonin | Blocks reuptake | Mood elevation |

## CHART 10.6 Events Leading to Neurotransmitter Release

1. Action potential passes along a nerve fiber and over the surface of its synaptic knob.
2. Synaptic knob membrane becomes more permeable to calcium ions, and they diffuse inward.
3. In the presence of calcium ions, synaptic vesicles fuse to synaptic knob membrane.
4. Synaptic vesicles release their neurotransmitter by exocytosis into the synaptic cleft.
5. Synaptic vesicles reenter the cytoplasm of the nerve fiber and pick up more neurotransmitter.

After a vesicle releases its neurotransmitter, it eventually returns to the cytoplasm, where it picks up more neurotransmitter. Damaged or lost vesicles are replaced by new ones formed in the neuron cell body and transported through the axon to the synaptic knob. Chart 10.6 summarizes this activity.

In order to keep signal duration short, enzymes in synaptic clefts rapidly decompose some neurotransmitters. Other neurotransmitters are transported back into the synaptic knob of the presynaptic neuron or into nearby neurons or neuroglial cells, a process called *reuptake.* The enzyme acetylcholinesterase, for example, decomposes acetylcholine. It is present on postsynaptic membranes. Similarly, the enzyme monoamine oxidase inactivates the monoamine neurotransmitters epinephrine and norepinephrine after reuptake. This enzyme is found in mitochondria in the synaptic knob. Destruction or removal of the neurotransmitter prevents a continuous stimulation of the postsynaptic neuron.

### Neuropeptides

Neurons in the brain or spinal cord synthesize **neuropeptides.** These peptides act as neurotransmitters or as *neuromodulators*—substances that alter a neuron's response to a neurotransmitter or block the release of a neurotransmitter.

Among the neuropeptides are the *enkephalins* that occur throughout the brain and spinal cord. Each enkephalin molecule consists of five amino acids in a chain. Synthesis of enkephalins seems to increase during periods of painful stress, and they bind to the same receptors in the brain (opiate receptors) as the narcotic morphine. Enkephalins relieve pain sensations, and probably have other functions as well (Clinical Application 10.5).

Another morphinelike peptide, called *beta endorphin,* is found in the brain and spinal cord. It remains active longer than enkephalins and is a much more potent pain reliever.

Still another neuropeptide, which consists of eleven amino acids and is widely distributed throughout the nervous system, is *substance P.* It functions as a neurotransmitter (or perhaps as a neuromodulator) in the neurons that transmit pain impulses into the spinal cord and on to the brain. Enkephalins and endorphins may relieve pain by inhibiting the release of substance P from pain-transmitting neurons.

## Impulse Processing

The way the nervous system processes nerve impulses and acts upon them reflects, in part, the organization of neurons and their nerve fibers within the brain and spinal cord.

### Neuronal Pools

The neurons within the central nervous system are organized into groups called **neuronal pools** that have varying numbers of cells. Each neuronal pool receives impulses from input (afferent) nerve fibers. These impulses are processed according to the special characteristics of the pool, and any resulting impulses are conducted away on output (efferent) fibers.

Each input fiber divides many times as it enters, and its branches spread over a certain region of the neuronal pool. The branches give off smaller branches, and their terminals form hundreds of synapses with the dendrites and cell bodies of the neurons in the pool.

### Facilitation

As a result of incoming impulses and neurotransmitter release, a particular neuron of the neuronal pool may receive excitatory and inhibitory stimulation. If the net effect of the stimulation is excitatory, threshold may be reached, and an outgoing impulse is triggered. If the net effect is excitatory but subthreshold, an impulse is not triggered, but the neuron is more excitable to incoming stimulation, a state called **facilitation.**

### Convergence

Any single neuron in a neuronal pool may receive impulses from two or more incoming fibers. Fibers originating from different parts of the nervous system leading to the same neuron exhibit **convergence.**

CLINICAL APPLICATION

## Opiates in the Human Body

Opiate drugs, such as morphine, heroin, codeine, and opium, are potent painkillers derived from the poppy plant. These drugs alter pain perception, making it easier to tolerate, and elevate mood.

The human body produces its own opiates, called endorphins (for "endogenous morphine"), that are peptides. Like the poppy-derived opiates that they structurally resemble, endorphins influence mood and perception of pain.

The discovery of endorphins began in 1971 in research laboratories at Stanford University and the Johns Hopkins School of Medicine, where researchers exposed pieces of brain tissue from experimental mammals to morphine. The morphine was radioactively labeled (some of the atoms were radioactive isotopes) so researchers could follow its destination in the brain.

The morphine indeed bound to receptors on the membranes of certain nerve cells, particularly in the neurons that transmit pain. Why, the investigators wondered, would an animal's brain contain receptors for a chemical made by a poppy? Could a mammal's body manufacture its own opiates? The opiate receptor, then, would normally bind the body's own opiates (the endorphins) but would also be able to bind the chemically similar compounds made by the poppy. Over the next few years, researchers identified several types of endorphins in the human brain and associated their release with situations involving pain relief, such as acupuncture and analgesia to mother and child during childbirth.

The existence of endorphins explains why some people who are addicted to opiate drugs such as heroin experience withdrawal pain when they stop taking the drug. Initially, the body interprets the frequent binding of heroin to its endorphin receptors as an excess of endorphins. To bring the level down, the body slows its own production of endorphins. Then, when the addict stops taking the heroin, the body is caught short of opiates (heroin and endorphins). The result is pain.

Convergence makes it possible for a neuron to summate impulses from different sources. For example, if a neuron receives subthreshold stimulation from one input fiber, it may reach threshold if it receives additional stimulation from a second input fiber. Thus, an output impulse triggered from this neuron reflects summation of impulses from two different sources. Such an output impulse may travel to a particular effector and evoke a response (fig. 10.19*a*).

Incoming impulses often represent information from various sensory receptors that detect changes. Convergence allows the nervous system to collect, process, and respond to information.

### Divergence

Impulses leaving a neuron of a neuronal pool often exhibit **divergence** by passing into several other output fibers. For example, an impulse from one neuron may stimulate two others; each of these, in turn, may stimulate several others, and so forth. Such an arrangement of diverging nerve fibers can amplify an impulse—that is, spread it to increasing numbers of neurons within the pool (fig. 10.19*b*).

As a result of divergence, an impulse originating from a single neuron in the central nervous system may be amplified so that enough impulses reach the motor units within a skeletal muscle to cause forceful contraction.

Similarly, an impulse originating from a sensory receptor may diverge and reach several different regions of the central nervous system, where the resulting impulses can be processed and acted upon.

1. Define neuropeptide.
2. What is a neuronal pool?
3. Define facilitation.
4. What is convergence?
5. What is the relationship between divergence and amplification?

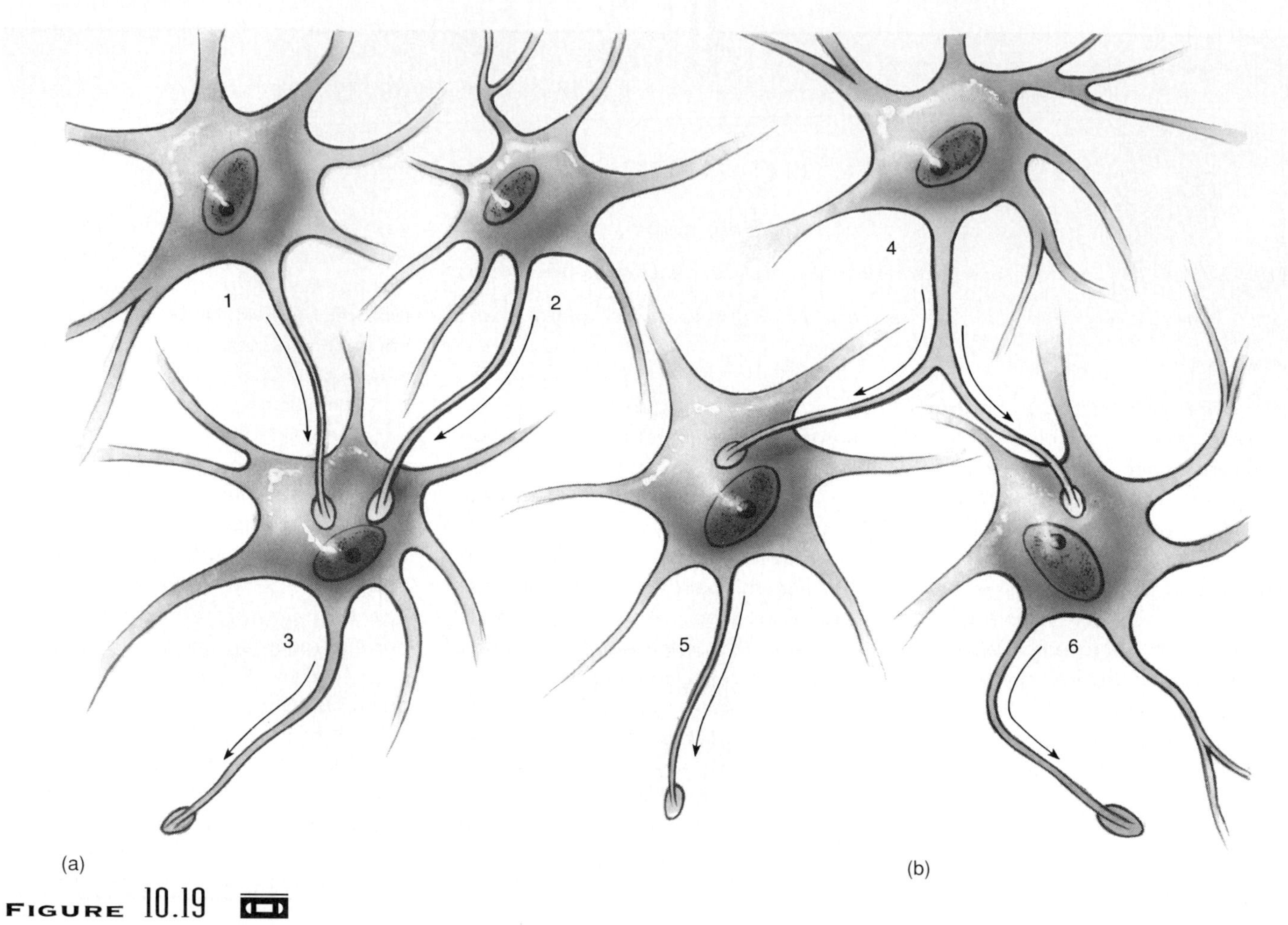


**FIGURE 10.19**

(*a*) Nerve fibers of neurons 1 and 2 converge to the cell body of neuron 3. (*b*) The nerve fiber of neuron 4 diverges to the cell bodies of neurons 5 and 6.

## Nerves and Nerve Fiber Classification

Although a nerve fiber is an extension of a neuron, a nerve is a cordlike bundle (or group of bundles) of nerve fibers held together by layers of connective tissue (fig. 10.20). Chapter 11 describes nerve structure.

Like nerve fibers, nerves that conduct impulses into the brain or spinal cord are called **sensory nerves,** and those that carry impulses to muscles or glands are termed **motor nerves.** Most nerves, however, include both sensory and motor fibers, and they are called **mixed nerves.**

Nerves originating from the spinal cord that communicate with other body parts are called **spinal nerves,** while those originating from the brain that communicate with other body parts are called **cranial nerves.** The nerve fibers within these structures can be subdivided further into four groups as follows:

1. **General somatic efferent fibers** carry motor impulses outward from the brain or spinal cord to skeletal muscles and stimulate them to contract.
2. **General visceral efferent fibers** carry motor impulses outward to various smooth muscles and glands associated with internal organs, causing certain muscles to contract or glands to secrete.
3. **General somatic afferent fibers** carry sensory impulses inward to the brain or spinal cord from receptors in the skin and skeletal muscles.

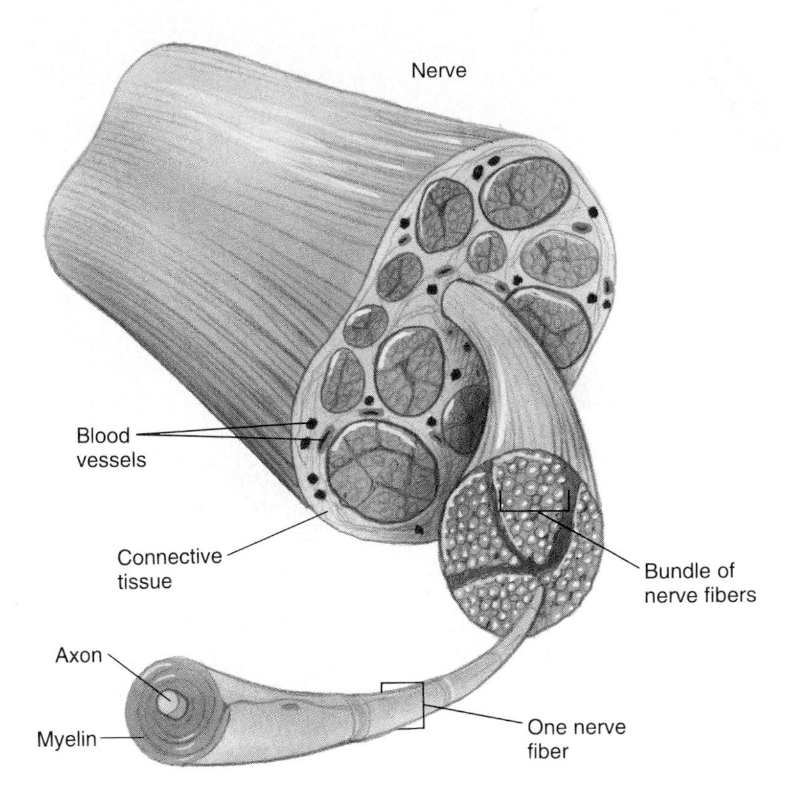


**FIGURE 10.20**
Connective tissue holds together a bundle of nerve fibers, forming a nerve.

4. **General visceral afferent fibers** carry sensory impulses to the central nervous system from blood vessels and internal organs.

The term *general* in each of these categories indicates that the fibers are associated with general structures such as the skin, skeletal muscles, glands, and viscera. Three other groups of fibers, found only in cranial nerves, are associated with more specialized, or *special,* structures:

1. **Special visceral efferent fibers** carry motor impulses outward from the brain to the muscles used in chewing, swallowing, speaking, and forming facial expressions.
2. **Special visceral afferent fibers** carry sensory impulses inward to the brain from the olfactory and taste receptors.
3. **Special somatic afferent fibers** carry sensory impulses inward from the receptors of sight, hearing, and equilibrium.

1. Explain how neurons are classified according to structure and according to function.
2. How is a neuron related to a nerve?
3. What is a mixed nerve?
4. Explain how cranial and spinal nerve fibers are grouped.

## NERVE PATHWAYS

Nerve impulses follow nerve pathways as they travel through the nervous system. The simplest of these pathways, including only a few neurons, constitutes a **reflex arc.** Reflex arcs carry out the simplest responses—reflexes.

### Reflex Arcs

A reflex arc begins with a receptor at the end of a sensory nerve fiber. This fiber usually leads to several interneurons within the central nervous system, which

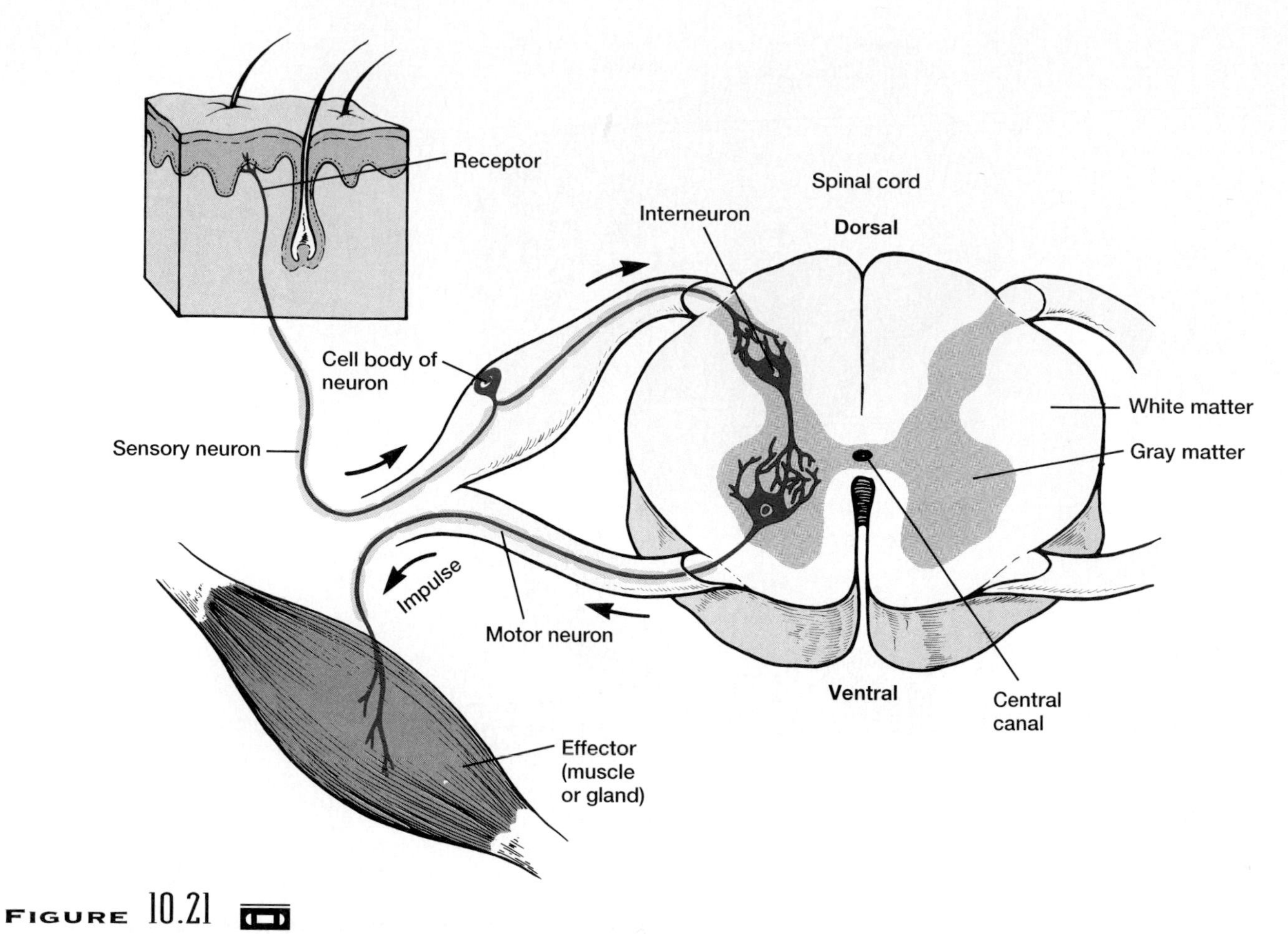


**FIGURE 10.21**

A reflex arc usually includes a receptor, a sensory neuron, interneurons, a motor neuron, and an effector.

serve as a processing center, or *reflex center.* Fibers from these interneurons may connect with interneurons in other parts of the nervous system. They also communicate with motor neurons, whose fibers pass outward from the central nervous system to effectors (fig. 10.21).

## Reflex Behavior

**Reflexes** are automatic, subconscious responses to changes (stimuli) within or outside the body. They help maintain homeostasis by controlling many involuntary processes such as heart rate, breathing rate, blood pressure, and digestion. Reflexes also carry out the automatic actions of swallowing, sneezing, coughing, and vomiting.

The *knee-jerk reflex* (patellar tendon reflex) is an example of a simple reflex using only two neurons—a sensory neuron communicating directly to a motor neuron. Striking the patellar ligament just below the patella initiates this reflex. The quadriceps femoris muscle group, which is attached to the patella by a tendon, is pulled slightly, stimulating stretch receptors within the muscle group. These receptors, in turn, trigger impulses that pass along the fiber of a sensory neuron into the lumbar region of the spinal cord. Within the spinal cord, the sensory axon synapses with a motor neuron. The impulse then continues along the axon of the motor neuron and travels back to the quadriceps femoris. The muscles respond by contracting, and the reflex is completed as the leg extends (fig. 10.22).

This reflex helps maintain an upright posture. For example, if a person is standing still and the knee begins to bend as a result of gravity, the quadriceps femoris is stretched, the reflex is triggered, and the leg straightens again.

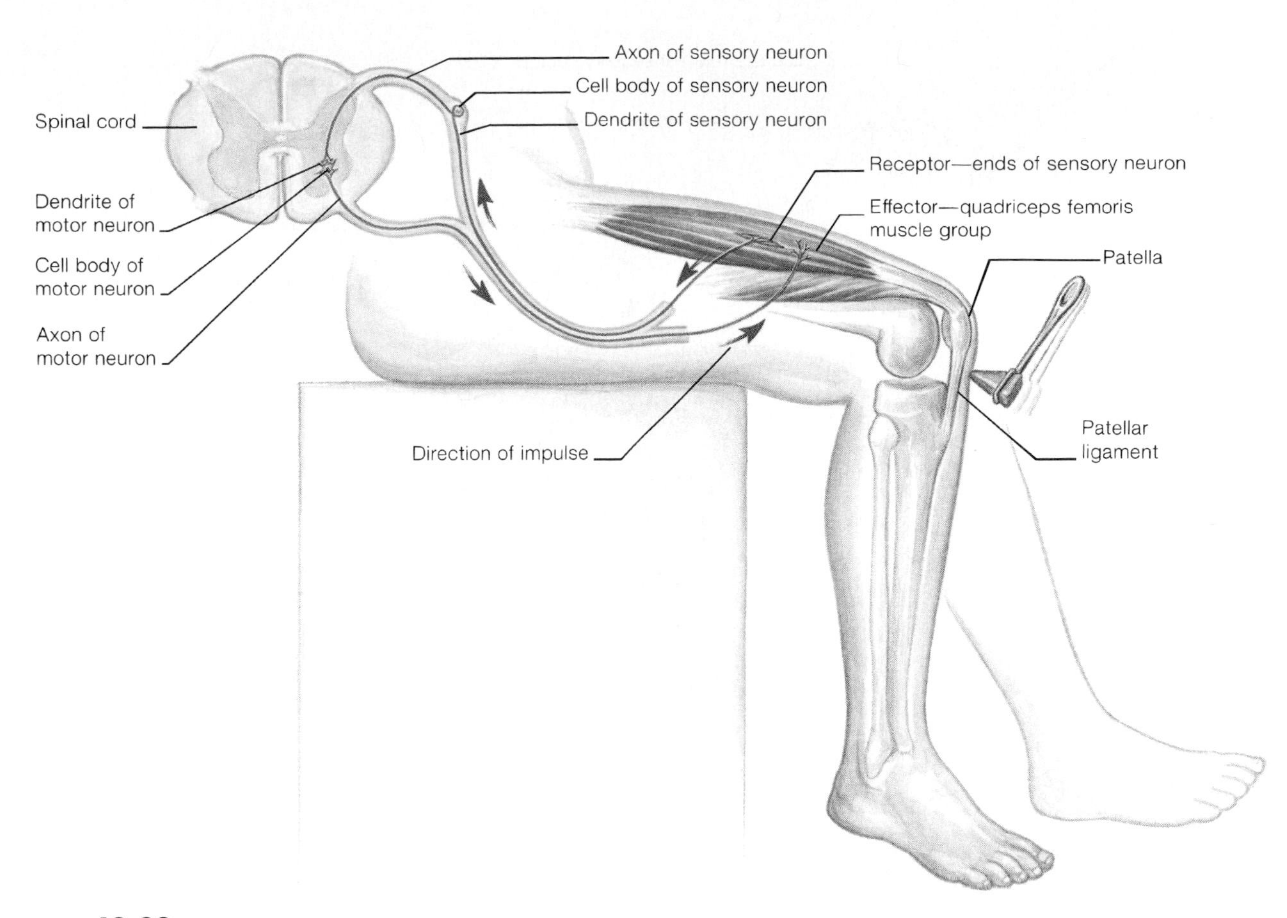


**FIGURE 10.22**
The knee-jerk reflex involves two neurons—a sensory neuron and a motor neuron. It is an example of a monosynaptic reflex.

Another type of reflex, called a *withdrawal reflex* (fig. 10.23), occurs when a person unexpectedly touches a finger to something painful, activating skin receptors and sending sensory impulses to the spinal cord. There the impulses pass on to interneurons of a reflex center and are directed to motor neurons. The motor neurons transmit signals to the flexor muscles in the arm, which contract in response, pulling the part away from the painful stimulus.

At the same time, some of the incoming impulses stimulate interneurons that inhibit the action of the antagonistic extensor muscles (reciprocal innervation). This inhibition allows flexor muscles to effectively withdraw the hand.

While flexor muscles of the stimulated side (ipsilateral side) contract, the flexor muscles of the other arm (contralateral side) are inhibited. Furthermore, the extensor muscles on this other side contract. This phenomenon, called a *crossed extensor reflex,* is due to interneuron pathways within the reflex center of the spinal cord that allow sensory impulses arriving on one side of the cord to pass across to the other side and produce an opposite effect (fig. 10.24).

Concurrent with the withdrawal reflex, other interneurons in the spinal cord carry sensory impulses upward to the brain. The person becomes aware of the experience and may feel pain.

A withdrawal reflex is protective because it prevents or limits tissue damage when a body part touches something potentially harmful. Chart 10.7 summarizes the components of a reflex arc. Clinical Application 10.6 discusses some familiar reflexes.

1. What is a nerve pathway?
2. Describe a reflex arc.
3. Define a reflex.
4. Describe the actions that occur during a withdrawal reflex.

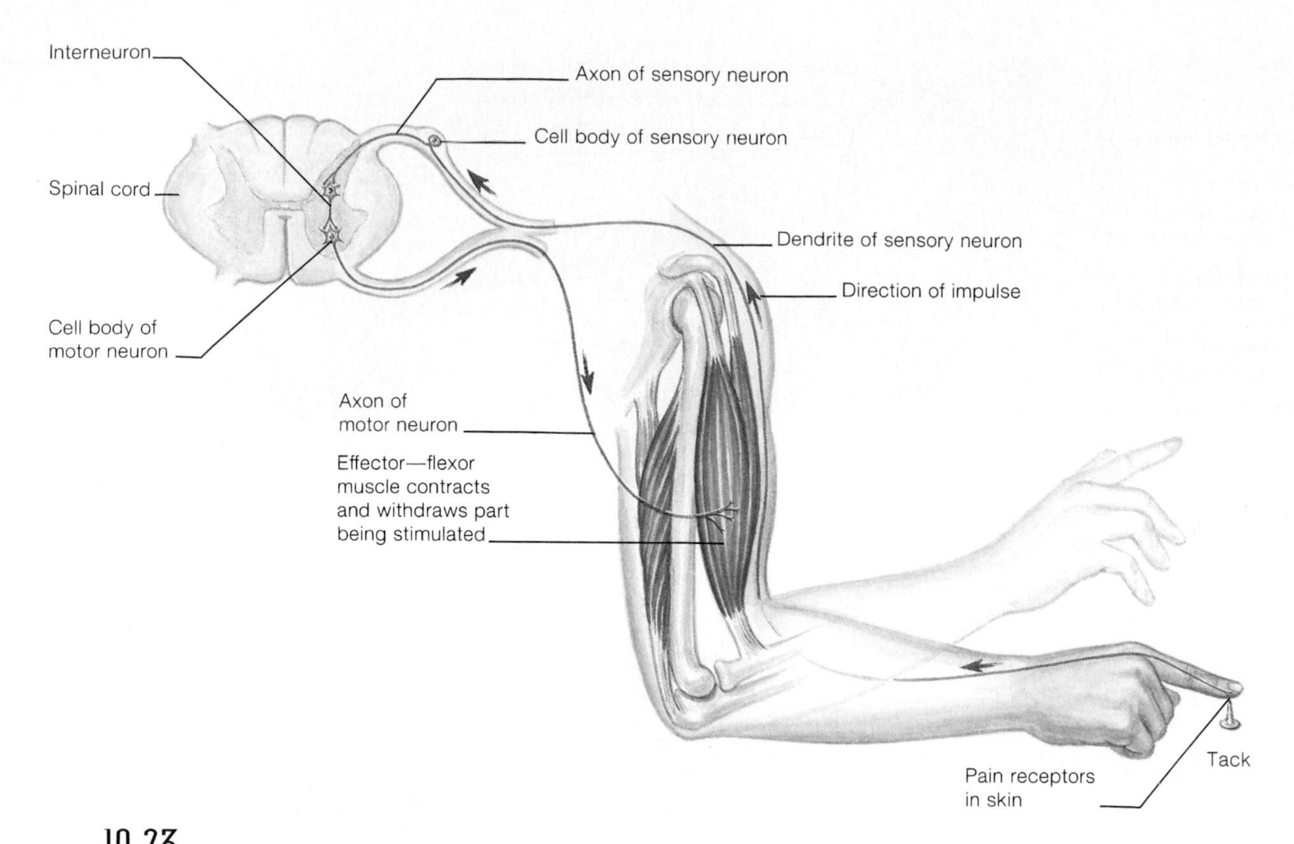


**FIGURE 10.23**
A withdrawal reflex involves a sensory neuron, an interneuron, and a motor neuron.

## CHART 10.7 Parts of a Reflex Arc

| Part | Description | Function |
|---|---|---|
| Receptor | The receptor end of a dendrite or a specialized receptor cell in a sensory organ | Sensitive to a specific type of internal or external change |
| Sensory neuron | Dendrite, cell body, and axon of a sensory neuron | Transmits nerve impulse from the receptor into the brain or spinal cord |
| Interneuron | Dendrite, cell body, and axon of a neuron within the brain or spinal cord | Serves as processing center; conducts nerve impulse from the sensory neuron to a motor neuron |
| Motor neuron | Dendrite, cell body, and axon of a motor neuron | Transmits nerve impulse from the brain or spinal cord out to an effector |
| Effector | A muscle or gland | Responds to stimulation by the motor neuron and produces the reflex or behavioral action |

CLINICAL APPLICATION

## Uses of Reflexes

Since normal reflexes depend on normal neuron functions, reflexes are commonly used to obtain information concerning the condition of the nervous system. An anesthesiologist, for instance, may try to initiate a reflex in a patient who is being anesthetized in order to determine how the anesthetic drug is affecting nerve functions. Also, in the case of injury to some part of the nervous system, observing reflexes may reveal the location and extent of damage.

Injury to any component of a reflex arc alters its function. For example, a *plantar reflex* is normally initiated by stroking the sole of the foot, and the usual response is flexion of the foot and toes. However, damage to certain nerve pathways (corticospinal tract) may trigger an abnormal response called the *Babinski reflex,* which is a dorsiflexion, extending the great toe

upward and fanning apart the smaller toes. If the injury is minor, the response may consist of plantar flexion with failure of the great toe to flex, or plantar flexion followed by dorsiflexion. The Babinski reflex is present normally in infants up to the age of twelve months and is thought to reflect immaturity in their corticospinal tracts.

Other reflexes that may be tested during a neurological examination include the following:

1. *Biceps-jerk reflex.* Bending a person's arm at the elbow elicits this reflex. The examiner's finger is placed on the inside of the bent elbow over the tendon of the biceps muscle, and the finger is tapped. The biceps contracts in response, and the forearm flexes at the elbow.
2. *Triceps-jerk reflex.* Flexing a person's arm at the elbow and tapping the short tendon of the triceps muscle close to its insertion near the tip of the elbow elicits this reflex. The muscle contracts in response, and the forearm extends slightly.
3. *Abdominal reflexes.* These reflexes occur when the examiner strokes the skin of the abdomen. For example, a dull pin drawn from the sides of the abdomen upward toward the midline and above the umbilicus causes the abdominal muscles underlying the skin to contract, and the umbilicus moves toward the stimulated region.
4. *Ankle-jerk reflex.* Tapping the calcaneal tendon just above its insertion on the calcaneus elicits this reflex. The response is plantar flexion, produced by contraction of the gastrocnemius and soleus muscles.
5. *Cremasteric reflex.* This reflex is elicited in males by stroking the upper inside of the thigh. In response, the testis on the same side is elevated by contracting muscles.

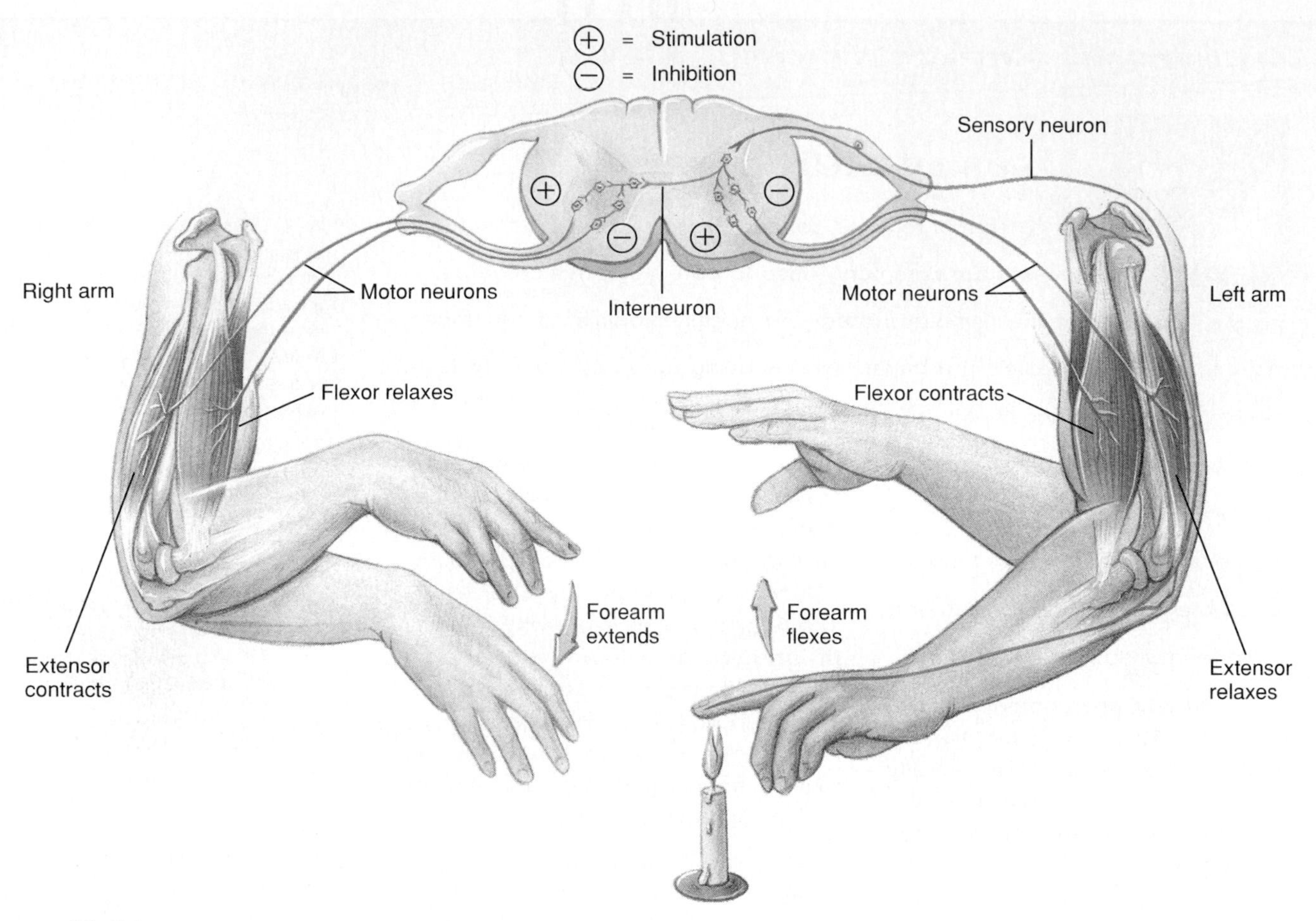


**Figure 10.24**

When the flexor muscle on one side is stimulated to contract in a withdrawal reflex, the extensor muscle on the opposite side also contracts. This helps to maintain balance.

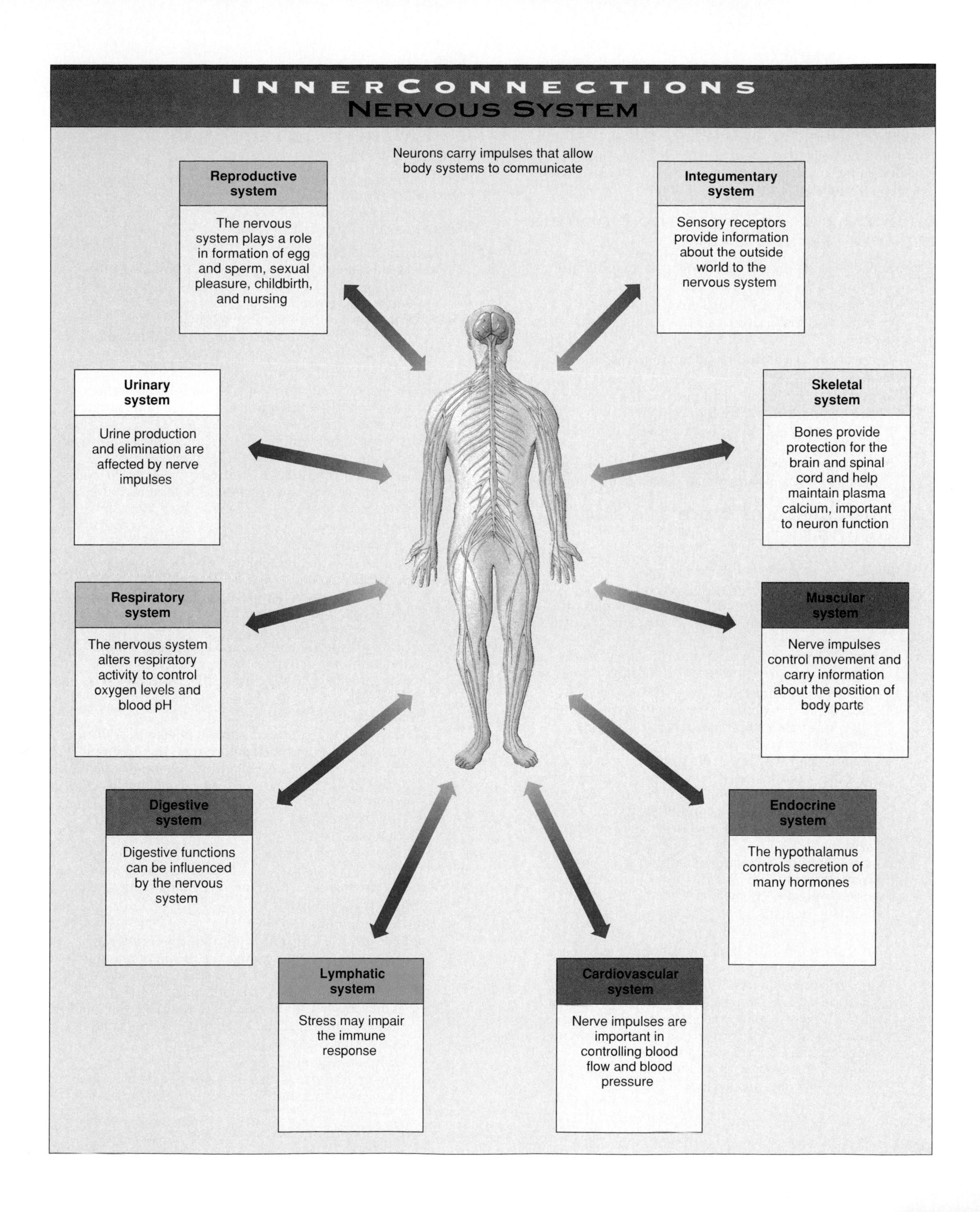
InnerConnections
Nervous System
Neurons carry impulses that allow body systems to communicate
Reproductive system
The nervous system plays a role in formation of egg and sperm, sexual pleasure, childbirth, and nursing
Integumentary system
Sensory receptors provide information about the outside world to the nervous system
Urinary system
Urine production and elimination are affected by nerve impulses
Skeletal system
Bones provide protection for the brain and spinal cord and help maintain plasma calcium, important to neuron function
Respiratory system
The nervous system alters respiratory activity to control oxygen levels and blood pH
Muscular system
Nerve impulses control movement and carry information about the position of body parts
Digestive system
Digestive functions can be influenced by the nervous system
Endocrine system
The hypothalamus controls secretion of many hormones
Lymphatic system
Stress may impair the immune response
Cardiovascular system
Nerve impulses are important in controlling blood flow and blood pressure

## Introduction (page 352)

Organs of the nervous system are divided into the central and peripheral nervous systems. These divisions provide sensory, integrative, and motor functions.

## General Functions of the Nervous System (page 352)

1. Sensory receptors detect changes in internal and external body conditions.
2. Integrative functions bring sensory information together and make decisions that motor functions act upon.
3. Motor impulses stimulate effectors to respond.
4. Neuron structure
   a. A neuron includes a cell body, cell processes, and the organelles usually found in cells.
   b. Dendrites and the cell body provide receptive surfaces.
   c. A single axon arises from the cell body and may be enclosed in a myelin sheath and a neurilemma.

## Classification of Neurons and Neuroglia (page 357)

Neurons differ in structure and function.

1. Classification of neurons
   a. On the basis of structure, neurons are classified as bipolar neurons, unipolar neurons, and multipolar neurons.
   b. On the basis of function, neurons are classified as sensory neurons, interneurons, or motor neurons.
2. Classification of neuroglia
   a. Neuroglial cells make up a large portion of the nervous system and have several functions.
   b. They fill spaces, support neurons, hold nervous tissue together, play a role in the metabolism of glucose, help regulate potassium ion concentration, produce myelin, carry on phagocytosis, rid synapses of excess ions and neurotransmitters, nourish neurons, and may directly communicate with neurons.
   c. They include Schwann cells in the peripheral nervous system and astrocytes, oligodendrocytes, microglia, and ependymal cells in the central nervous system.
3. Regeneration of nerve fibers
   a. If a neuron cell body is injured, the neuron is likely to die.
   b. If a peripheral nerve fiber is severed, its distal portion will die, but under the influence of nerve growth factors, the proximal portion may regenerate and reestablish its former connections, provided a tube of connective tissue guides it.
   c. Significant regeneration is unlikely to take place in the central nervous system.

## Cell Membrane Potential (page 363)

A cell membrane is usually polarized as a result of an unequal distribution of ions on either side. Pores and channels in membranes that allow passage of some ions but not others control ion distribution.

1. Distribution of ions
   a. Membrane pores and channels, formed by proteins, may be always open or sometimes open and sometimes closed.
   b. Potassium ions pass more readily through resting neuron cell membranes than do sodium and calcium ions.
2. Resting potential
   a. A high concentration of sodium ions is on the outside of the membrane and a high concentration of potassium ions is on the inside.
   b. Large numbers of negatively charged ions are inside the cell.
   c. In a resting cell, more positive ions leave the cell than enter it, so the outside of the cell membrane develops a positive charge with respect to the inside.
3. Local potential changes
   a. Stimulation of a membrane affects its resting potential in a local region.
   b. The membrane is depolarized if it becomes less negative; it is hyperpolarized if it becomes more negative.
   c. Local potential changes are graded and subject to summation.
   d. Reaching threshold potential triggers an action potential.
4. Action potentials
   a. At threshold, sodium channels open and sodium ions diffuse inward, depolarizing the membrane.
   b. About the same time, potassium channels open and potassium ions diffuse outward, repolarizing the membrane.
   c. This rapid change in potential is an action potential.
   d. Many action potentials can occur before active transport reestablishes the original resting potential.
   e. The propagation of action potentials along a nerve fiber is an impulse.
5. Refractory period
   a. The refractory period is a brief time following passage of a nerve impulse when the membrane is unresponsive to an ordinary stimulus.
   b. During the absolute refractory period, the membrane cannot be stimulated; during the relative refractory period, the membrane can be stimulated with a high intensity stimulus.
6. All-or-none response
   a. A nerve impulse is an all-or-none response to a stimulus of threshold intensity applied to a fiber.

b. All the impulses conducted on a fiber are the same strength.
7. Impulse conduction
   a. Unmyelinated fibers conduct impulses that travel over their entire surfaces.
   b. Myelinated fibers conduct impulses that travel from node to node.
   c. Impulse conduction is more rapid on myelinated fibers with large diameters.

## The Synapse (page 368)

A synapse is a junction between two neurons. A synaptic cleft is the gap between parts of two neurons at a synapse.

1. Synaptic transmission
   a. Impulses usually travel from dendrite or cell body, then along the axon to a synapse.
   b. Axons have synaptic knobs at their distal ends that secrete neurotransmitters.
   c. The neurotransmitter is released when a nerve impulse reaches the end of an axon, and the neurotransmitter diffuses across the synaptic cleft.
   d. A neurotransmitter reaching the nerve fiber or cell body on the distal side of the cleft triggers a nerve impulse.
2. Synaptic potentials
   a. Some neurotransmitters can depolarize postsynaptic membranes, triggering an action potential. This is an excitatory postsynaptic potential (EPSP).
   b. Others hyperpolarize the membranes, inhibiting action potentials. This is an inhibitory postsynaptic potential (IPSP).
   c. EPSPs and IPSPs are summed in a trigger zone of the neuron.
3. Neurotransmitters
   a. The nervous system produces at least thirty types of neurotransmitters.
   b. Calcium ions diffuse into synaptic knobs in response to action potentials, releasing neurotransmitters.
   c. Neurotransmitters are quickly decomposed or removed from synaptic clefts.
4. Neuropeptides
   a. Neuropeptides are chains of amino acids.
   b. Some neuropeptides are neurotransmitters or neuromodulators.
   c. They include enkephalins, endorphins, and substance P.

## Impulse Processing (page 374)

The way impulses are processed reflects the organization of neurons in the brain and spinal cord.

1. Neuronal pools
   a. Neurons are organized into pools within the central nervous system.
   b. Each pool receives, processes, and conducts away impulses.
2. Facilitation
   a. Each neuron in a pool may receive excitatory and inhibitory stimuli.
   b. A neuron is facilitated when it receives subthreshold stimuli and becomes more excitable.
3. Convergence
   a. Impulses from two or more incoming fibers may converge on a single neuron.
   b. Convergence enables a neuron to summate impulses from different sources.
4. Divergence
   a. Impulses leaving a pool may diverge by passing onto several output fibers.
   b. Divergence amplifies impulses.

## Nerves and Nerve Fiber Classification (page 376)

Nerves are cordlike bundles of nerve fibers.

1. Nerves can be classified as sensory nerves, motor nerves, or mixed nerves, depending on which type of fibers they contain.
2. Nerve fibers within the central nervous system can be subdivided into groups with general and special functions.

## Nerve Pathways (page 377)

A nerve pathway is the route an impulse follows through the nervous system.

1. Reflex arcs
   a. A reflex arc usually includes a sensory neuron, a reflex center composed of interneurons, and a motor neuron.
   b. The reflex arc is the behavioral unit of the nervous system.
2. Reflex behavior
   a. Reflexes are automatic, subconscious responses to changes.
   b. They help maintain homeostasis.
   c. The knee-jerk reflex employs only two neurons.
   d. Withdrawal reflexes are protective actions.

**Explorations in Human Anatomy and Physiology CD-ROM**
The modules accompanying Chapter Ten are #8 Nerve Conduction and #9 Synaptic Transmission.

## Critical Thinking Questions

1. A drug called tacrine slows breakdown of acetylcholine in synaptic clefts. What illness discussed in the chapter might tacrine theoretically treat?
2. How would you explain the following observations?
   a. When motor nerve fibers in the leg are severed, the muscles they innervate become paralyzed; however, in time, control over the muscles often returns.
   b. When motor nerve fibers in the spinal cord are severed, the muscles they control become paralyzed permanently.
3. People who inherit familial periodic paralysis often develop very low blood potassium concentrations. How would you explain the fact that the paralysis may disappear quickly when potassium ions are administered intravenously?
4. What might be deficient in the diet of a pregnant woman who is complaining of leg muscle cramping? How would you explain this to her?
5. Why are rapidly growing cancers that originate in nervous tissue more likely to be composed of neuroglial cells than of neurons?
6. The biceps-jerk reflex employs motor neurons that exit from the spinal cord in the 5th spinal nerve (C5), that is, fifth from the top of the cord. The triceps-jerk reflex involves motor neurons in the 7th spinal nerve (C7). How might these reflexes be used to help locate the site of damage in a patient with a neck injury?
7. How are multiple sclerosis and Tay-Sachs disease opposite one another?

## Review Exercises

1. Explain the relationship between the central nervous system and the peripheral nervous system.
2. List three general functions of the nervous system.
3. Distinguish between neurons and neuroglial cells.
4. Describe the generalized structure of a neuron.
5. Define *myelin.*
6. Distinguish between myelinated and unmyelinated nerve fibers.
7. Describe how an injured nerve fiber may regenerate.
8. Discuss the functions of each type of neuroglial cell.
9. Explain how a membrane may become polarized.
10. Define *resting potential.*
11. Distinguish between depolarizing and hyperpolarizing.
12. List the changes that occur during an action potential.
13. Explain how action potentials are related to nerve impulses.
14. Define *refractory period.*
15. Define *saltatory conduction.*
16. Define *synapse.*
17. Explain how a nerve impulse is transmitted from one neuron to another.
18. Explain the role of calcium in the release of neurotransmitters.
19. Define *neuropeptide.*
20. Distinguish between excitatory and inhibitory postsynaptic potentials.
21. Explain what is meant by the "trigger zone" of a neuron.
22. Describe the relationship between an input nerve fiber and its neuronal pool.
23. Define *facilitation.*
24. Distinguish between convergence and divergence.
25. Explain how nerve impulses are amplified.
26. Explain how neurons are classified on the basis of their structure.
27. Explain how neurons are classified on the basis of their function.
28. Distinguish between sensory, motor, and mixed nerves.
29. List four general types of nerve fibers.
30. Describe a reflex arc.
31. Define *reflex.*
32. Describe a withdrawal reflex.

CHAPTER 11

# NERVOUS SYSTEM II

## Divisions of the Nervous System

### CHAPTER OBJECTIVES

AFTER YOU HAVE STUDIED THIS CHAPTER, YOU SHOULD BE ABLE TO:

1. Describe the coverings of the brain and spinal cord.
2. Describe the structure of the spinal cord and its major functions.
3. Name the major parts of the brain and describe the functions of each.
4. Distinguish among motor, sensory, and association areas of the cerebral cortex.
5. Explain what is meant by hemisphere dominance.
6. Explain the stages in memory storage.
7. Describe the formation and function of cerebrospinal fluid.
8. Explain the functions of the limbic system and the reticular formation.
9. List the major parts of the peripheral nervous system.
10. Describe the structure of a peripheral nerve.
11. Name the cranial nerves and list their major functions.
12. Explain how spinal nerves are named.
13. Explain the function of a spinal nerve.
14. Describe the general characteristics of the autonomic nervous system.
15. Distinguish between the sympathetic and the parasympathetic divisions of the autonomic nervous system.
16. Describe a sympathetic and a parasympathetic nerve pathway.
17. Explain how the autonomic neurotransmitters differently affect visceral effectors.

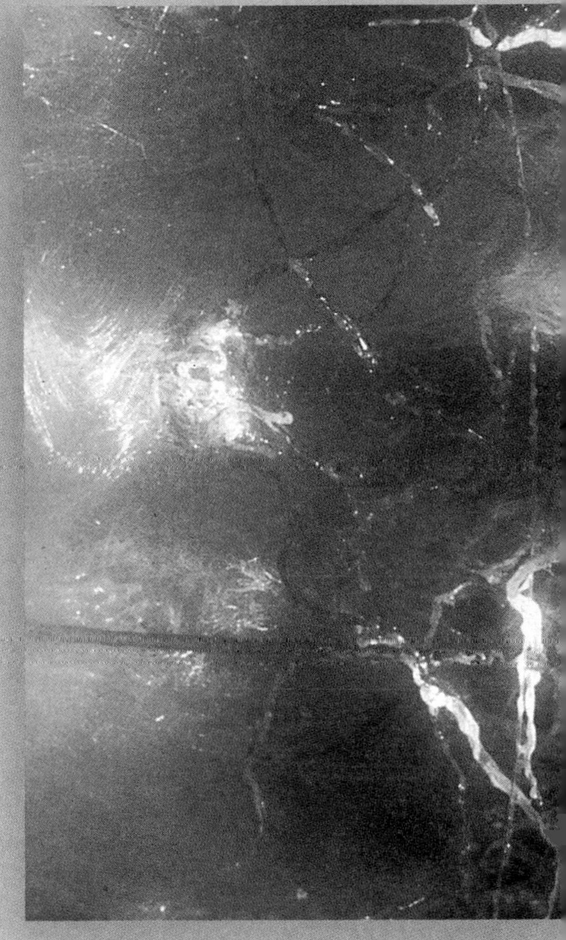

THE HUMAN NERVOUS SYSTEM IS A COMPLEX NETWORK OF NERVE CELLS AND THEIR ASSOCIATED GLIA. (ARTIST'S RENDERING OF NERVE CELLS FIRING)

### KEY TERMS

**adrenergic** (ad″ren-er′jik)

**autonomic nervous system** (aw″to-nom′ik ner′vus sis′tem)

**brain stem** (brān stem)

**cerebellum**(ser″ĕ-bel′um)

**cerebral cortex** (ser′ĕ-bral kor′teks)

**cerebral hemisphere** (ser′ĕ-bral hem′i-sfĕr)

**cerebrospinal fluid** (ser″ĕ-bro-spi′nal floo′id)

**cerebrum** (ser′ē-brum)

**cholinergic** (ko″lin-er′jik)

**choroid plexus** (ko′roid plek′sus)

**diencephalon** (di″en-sef′ah-lon)

**hypothalamus** (hi″po-thal′ah-mus)

**medulla oblongata** (mĕ-dul′ah ob″long-ga′tah)

**meninges** (mĕ-nin′jēz)

**midbrain** (mid′brān)

**parasympathetic** (par″ah-sim″pah-thet′ik)

**postganglionic** (pōst″gang-gle-on′ik)

**preganglionic** (pre″gang-gle-on′ik)

**reticular formation** (rĕ-tik′u-lar fōr-ma′shun)

**sympathetic** (sim″pah-thet′ik)

**thalamus** (thal′ah-mus)

**ventricle** (ven′trĭ-kl)

September 13, 1848, was a momentous day for Phineas Gage, a young man who worked in Vermont evening out terrain for railroad tracks. To blast away rock, he would drill a hole, fill it with gunpowder, cover that with sand, insert a fuse, and then press down with an iron rod called a tamping iron. The explosion would go down into the rock.

But on that fateful September day, Gage began pounding on the tamping iron before his coworker had put down the sand. The gunpowder exploded outward, slamming the inch-thick, 40-inch-long iron rod straight through Gage's skull. It pierced his brain like an arrow propelled through a soft melon, shooting out the other side of his head. Curiously, Gage stood up just a few moments later, fully conscious and apparently unharmed by the hole just blasted through his head.

As it turned out, Gage was harmed by the freak accident, but in ways so subtle that they were not at first evident. His friends reported that "Gage was no longer Gage." Although retaining his intellect and abilities to move, speak, learn, and remember, Gage underwent a profound personality transformation. Once a trusted, honest, and dedicated worker, the 25-year-old became irresponsible, shirking work, cursing, and pursuing what his doctor termed "animal propensities."

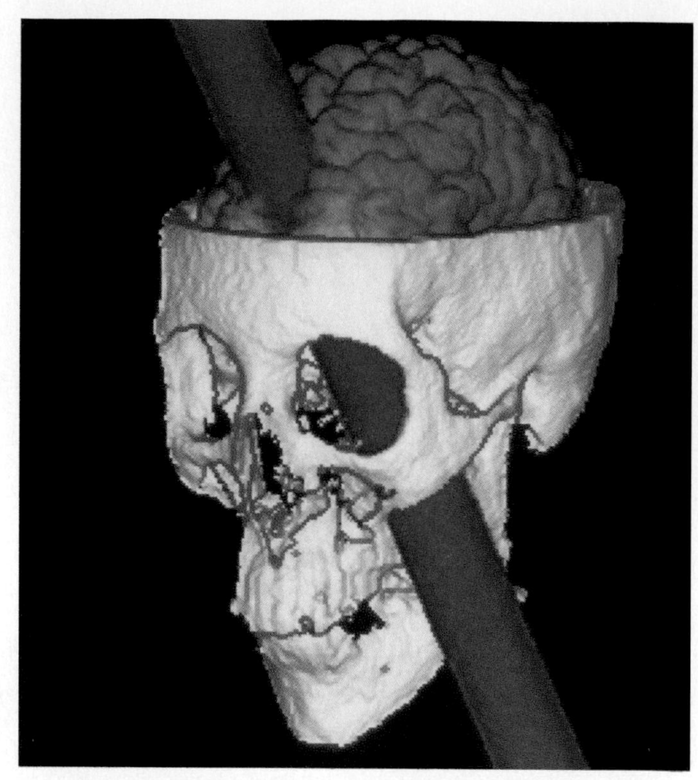

A rod blasted through the head of a young railway worker has taught us much about the biology of personality.

Researchers as long ago as 1868 hypothesized that the tamping iron had ripped out a part of Gage's brain controlling personality. In 1994, computer analysis more precisely pinpointed the damage to the famous brain of Phineas Gage, which, along with the tamping iron, wound up in a museum at Harvard University. Researchers reconstructed the trajectory of the tamping iron, localizing two small areas in the front of the brain that control rational decision making and processing of emotion.

More than a hundred years after Gage's accident, in 1975, 21-year-old Karen Ann Quinlan drank alcohol after taking a prescription sedative, and her heart and lungs stopped working. When she was found, Karen had no pulse, was not breathing, had dilated pupils, and was unresponsive. Cardiopulmonary resuscitation restored her pulse, but once at the hospital she was placed on a ventilator. Within 12 hours, some functions returned—her pupils constricted, she moved, gagged, grimaced, and even opened her eyes. Within a few months, she could even breathe unaided for short periods.

Because Karen's responses were random and not purposeful, and she was apparently unaware of herself and her environment, she was said to be in a *persistent vegetative state.* Her basic life functions were intact, but she had to be fed and given water intravenously. Fourteen months after Karen took the fateful pills and alcohol, her parents made a request that was to launch the right-to-die movement. They asked that Karen be taken off of life support. Doctors removed Karen's ventilator, and a nursing home accepted her, where she lived for 9 more years before dying of infection, never regaining awareness.

Throughout Karen and her family's ordeal, researchers tried to fathom what had happened to her. A CAT scan performed 5 years after the accident showed atrophy in two major brain regions, the cerebrum and the cerebellum. But when researchers analyzed Karen Ann Quinlan's brain in 1993, they were surprised. The most severely damaged part of her brain was the thalamus, an area thought to function merely as a relay station to higher brain structures. Karen's tragic case revealed that the thalamus is also important in processing thoughts, in providing the awareness and responsiveness that makes a person a conscious being.

The cases of Phineas Gage and Karen Ann Quinlan dramatically illustrate the function of the human brain by revealing what can happen when it is damaged. Nearly every aspect of our existence depends upon the brain and other parts of the nervous system, from thinking and feeling; to sensing, perceiving, and responding to the environment; to carrying out vital functions such as breathing and heartbeat. This chapter describes how the billions of neurons comprising the nervous system interact to enable us to survive and to enjoy the world around us.

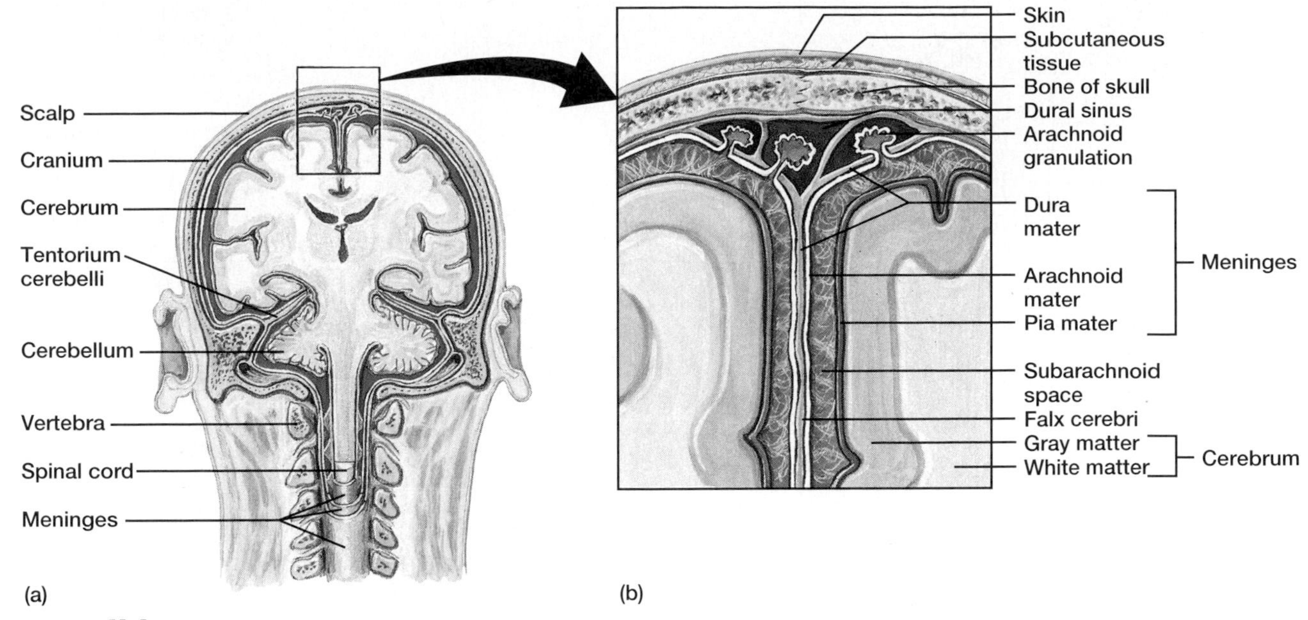


**FIGURE 11.1**

(*a*) Membranes called meninges enclose the brain and spinal cord. (*b*) The meninges include three layers: dura mater, arachnoid mater, and pia mater.

(*Tissues and Organs: A Text-Atlas of Scanning Electron Microscopy,* by R. G. Kessel and R. H. Kardon. © 1979 W. H. Freeman and Company.)

Bones, membranes, and fluid surround the organs of the central nervous system (CNS). More specifically, the brain lies within the cranial cavity of the skull, while the spinal cord, which is continuous with the brain, occupies the vertebral canal within the vertebral column. Beneath these bony coverings, membranes called meninges, located between the bone and the soft tissues of the nervous system, protect the brain and spinal cord (fig. 11.1*a*).

## Meninges

The **meninges** (singular, *meninx*) have three layers—dura mater, arachnoid mater, and pia mater (fig. 11.1*b*).

The **dura mater** is the outermost layer. It is composed primarily of tough, white fibrous connective tissue and contains many blood vessels and nerves. It is attached to the inside of the cranial cavity and forms the internal periosteum of the surrounding skull bones (see reference plate 53).

In some regions, the dura mater extends inward between lobes of the brain and forms supportive and protective partitions (chart 11.1). In other areas, the dura mater splits into two layers, forming channels called *dural sinuses,* shown in figure 11.1*b.* Venous blood flows through these channels as it returns from the brain to vessels leading to the heart.

**CHART 11.1 Partitions of the Dura Mater**

| Partition | Location |
|---|---|
| Falx cerebelli | Separates the right and left cerebellar hemispheres |
| Falx cerebri | Extends downward into the longitudinal fissure, and separates the right and left cerebral hemispheres (fig. 11.1*b*) |
| Tentorium cerebelli | Separates the occipital lobes of the cerebrum from the cerebellum (fig. 11.1*a*) |

The dura mater continues into the vertebral canal as a strong, tubular sheath that surrounds the spinal cord. It is attached to the cord at regular intervals by a band of pia mater (denticulate ligaments) that extends the length of the spinal cord on either side. The dural sheath terminates as a blind sac at the level of the second sacral vertebra, below the end of the spinal cord. The sheath around the spinal cord is not attached directly to the vertebrae but is separated by an *epidural space,* which lies between the dural sheath and the bony walls (fig. 11.2). This space contains blood vessels, loose connective tissue, and adipose tissue that provide a protective pad around the spinal cord.

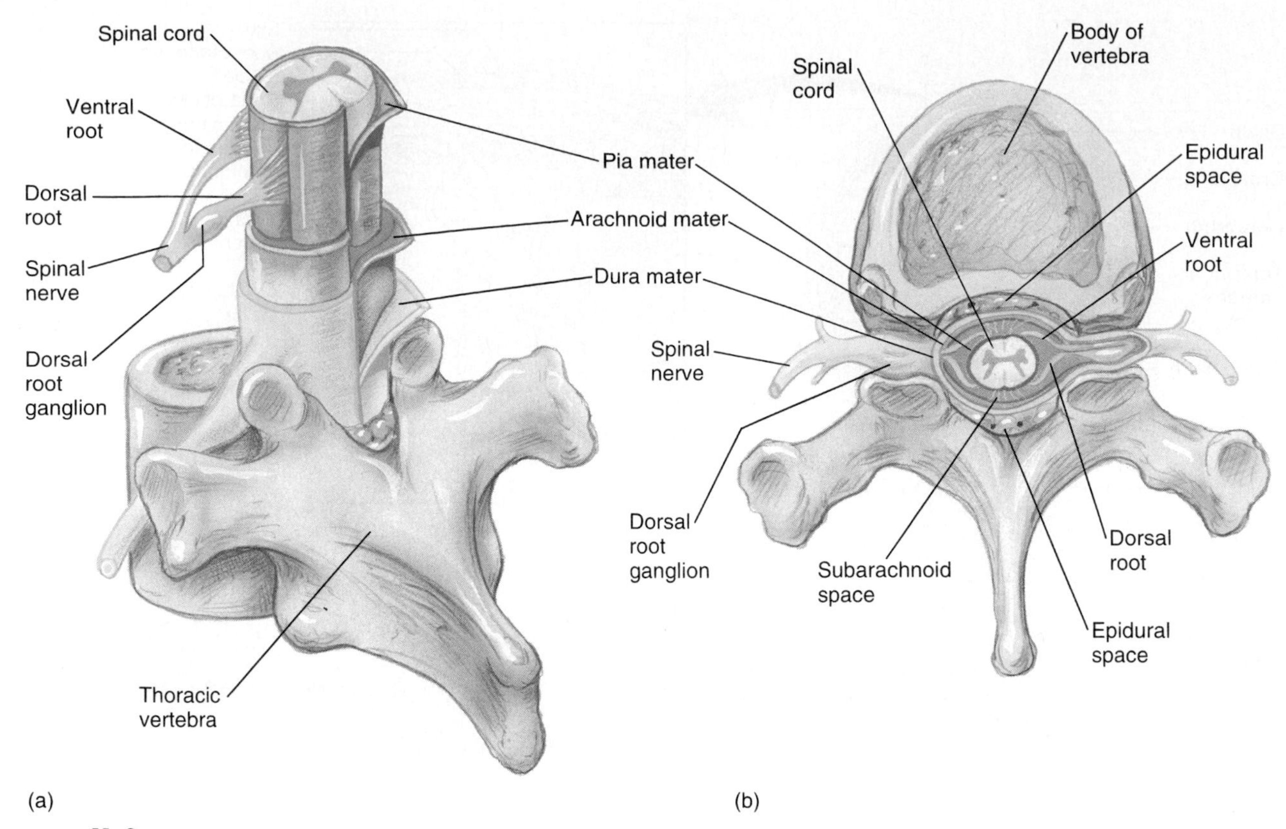


**FIGURE 11.2**

(*a*) The dura mater ensheaths the spinal cord. (*b*) Tissues forming a protective pad around the cord fill the epidural space between the dural sheath and the bone of the vertebra.

A blow to the head may rupture some blood vessels associated with the brain, and the escaping blood may collect in the space beneath the dura mater. This condition, called *subdural hematoma,* can create increasing pressure between the rigid bones of the skull and the soft tissues of the brain. Unless the accumulating blood is evacuated promptly, the resulting compression of the brain may lead to functional losses or even death.

The **arachnoid mater** is a thin, weblike membrane that lacks blood vessels and is located between the dura and pia maters. It spreads over the brain and spinal cord but generally does not dip into the grooves and depressions on their surfaces. Many thin strands extend from its undersurface and are attached to the pia mater.

Between the arachnoid and pia maters is a *subarachnoid space,* which contains the clear, watery **cerebrospinal fluid.**

The **pia mater** is very thin and contains many nerves, as well as blood vessels that nourish the underlying cells of the brain and spinal cord. The pia mater is attached to the surfaces of these organs and follows their irregular contours, passing over the high areas and dipping into the depressions.

An inflammation of the meninges is called *meningitis.* Bacteria or viruses that invade the cerebrospinal fluid are the usual causes of this condition. Although meningitis may affect the dura mater, it is more commonly limited to the arachnoid and pia maters.

Meningitis occurs most often in infants and children and is considered a serious childhood infection. Possible complications of this disease include loss of vision or hearing, paralysis, mental retardation, or death.

1. Describe the meninges.
2. Name the layers of the meninges.
3. Explain where cerebrospinal fluid is located.

## Spinal Cord

The **spinal cord** is a slender nerve column that passes downward from the brain into the vertebral canal. Although it is continuous with the brain, the spinal cord is said to begin where nervous tissue leaves the cranial cavity at the level of the foramen magnum (see reference plate 55). The cord tapers to a point and terminates near the intervertebral disk that separates the first and second lumbar vertebrae (fig. 11.3*a*).

### Structure of the Spinal Cord

The spinal cord consists of thirty-one segments, each of which gives rise to a pair of **spinal nerves.** These nerves branch to various body parts and connect them with the central nervous system.

In the neck region, a thickening in the spinal cord, called the *cervical enlargement,* supplies nerves to the upper limbs. A similar thickening in the lower back, the *lumbar enlargement,* gives off nerves to the lower limbs. Just inferior to the lumbar enlargement, the spinal cord narrows to a sharp tip called the *conus medullaris.* From this tip, a thin cord of connective tissue descends to the upper surface of the coccyx. This cord is called the *filum terminale* (fig. 11.3*b*).

Two grooves, a deep *anterior median fissure* and a shallow *posterior median sulcus,* extend the length of the spinal cord, dividing it into right and left halves. A cross section of the cord (fig. 11.4) reveals that it consists of white matter surrounding a core of gray matter. The pattern the gray matter produces roughly resembles a butterfly with its wings outspread. The upper and lower wings of gray matter are called the *posterior horns* and the *anterior horns,* respectively. Between them on either side is a protrusion of gray matter called the *lateral horn.* Neurons with relatively large cell bodies in the anterior horns (anterior horn cells) give rise to motor fibers that pass out through spinal nerves to various skeletal muscles. However, the majority of neurons in the gray matter are interneurons (see chapter 10).

A horizontal bar of gray matter in the middle of the spinal cord, the *gray commissure,* connects the wings of the gray matter on the right and left sides. This bar surrounds the **central canal,** which is continuous with the ventricles of the brain and contains cerebrospinal fluid. The central canal is prominent during embryonic development, but it becomes almost microscopic in an adult.

The gray matter divides the white matter of the spinal cord into three regions on each side—the *anterior, lateral,* and *posterior funiculi.* Each funiculus consists of longitudinal bundles of myelinated nerve fibers that comprise major nerve pathways called **nerve tracts** (fig. 11.4).

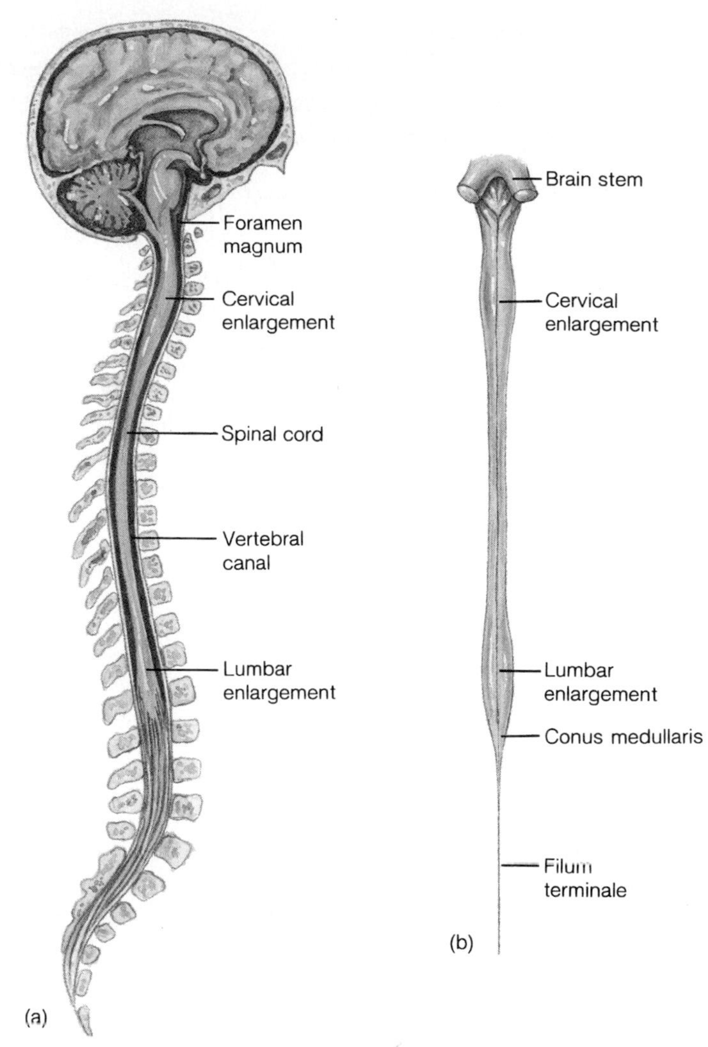


**Figure 11.3**

(*a*) The spinal cord begins at the level of the foramen magnum. (*b*) Posterior view of the spinal cord with the spinal nerves removed.

### Functions of the Spinal Cord

The spinal cord has two major functions: it conducts nerve impulses and it serves as a center for spinal reflexes.

The nerve tracts of the spinal cord, together with the spinal nerves, provide a two-way communication system between the brain and body parts outside the nervous system. The tracts that conduct sensory impulses to the brain are called **ascending tracts;** those that conduct motor impulses from the brain to motor neurons reaching muscles and glands are called **descending tracts.**

The nerve fibers within these tracts are axons. Typically, all the axons within a given tract originate from neuron cell bodies located in the same part of the nervous system and end together in some other part. The names that identify nerve tracts often reflect these

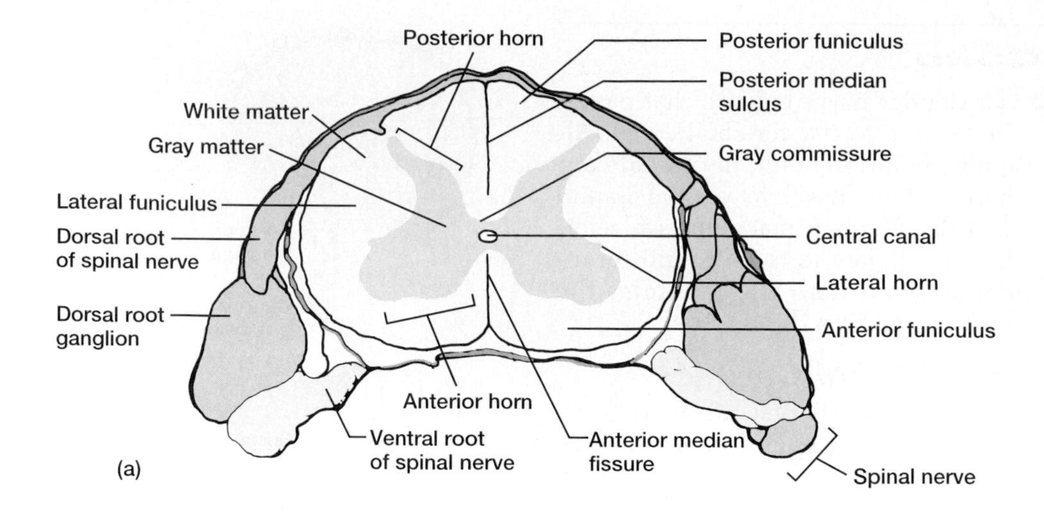


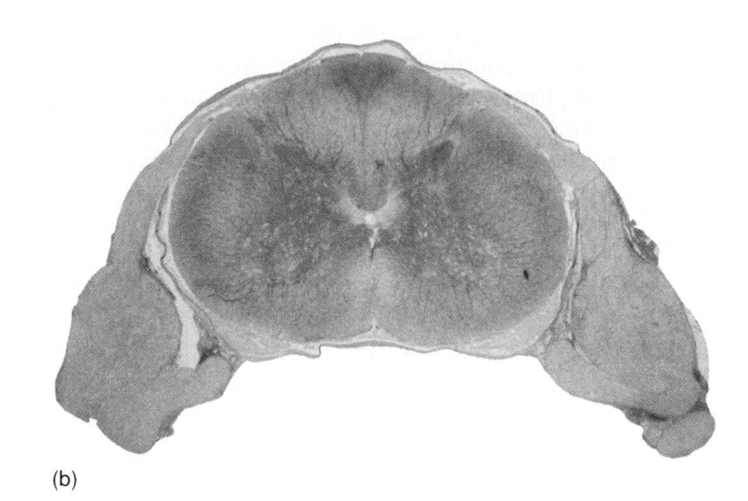


**FIGURE 11.4**

(*a*) A cross section of the spinal cord. (*b*) Identify the parts of the spinal cord in this micrograph (7.5×).

common origins and terminations. For example, a *spinothalamic tract* begins in the spinal cord and carries sensory impulses associated with the sensations of pain and touch to the thalamus of the brain. A *corticospinal tract* originates in the cortex of the brain and carries motor impulses downward through the spinal cord and spinal nerves. These impulses control skeletal muscle movements.

## Ascending Tracts

Among the major ascending tracts of the spinal cord are the following (fig. 11.5):

1. **Fasciculus gracilis** (fah-sik′u-lus gras′il-is) and **fasciculus cuneatus** (ku′ne-at-us). These tracts are located in the posterior funiculi of the spinal cord. Their fibers conduct sensory impulses from the skin, muscles, tendons, and joints to the brain, where they are interpreted as sensations of touch, pressure, and body movement.

   At the base of the brain in an area called the *medulla oblongata,* most of these fibers cross over (decussate) from one side to the other—that is, those ascending on the left side of the spinal cord pass across to the right side and vice versa. As a result, the impulses originating from sensory receptors on the left side of the body reach the right side of the brain, and those originating on the right side of the body reach the left side of the brain (fig. 11.6).

2. **Spinothalamic** (spi″no-thah-lam′ik) **tracts.** The *lateral* and *anterior spinothalamic tracts* are located in the lateral and anterior funiculi, respectively. Impulses in these tracts also cross over in the medulla. The lateral tracts conduct impulses from various body regions to the brain and give rise to sensations of pain and temperature. Impulses carried on fibers of the anterior tracts are interpreted as touch and pressure.

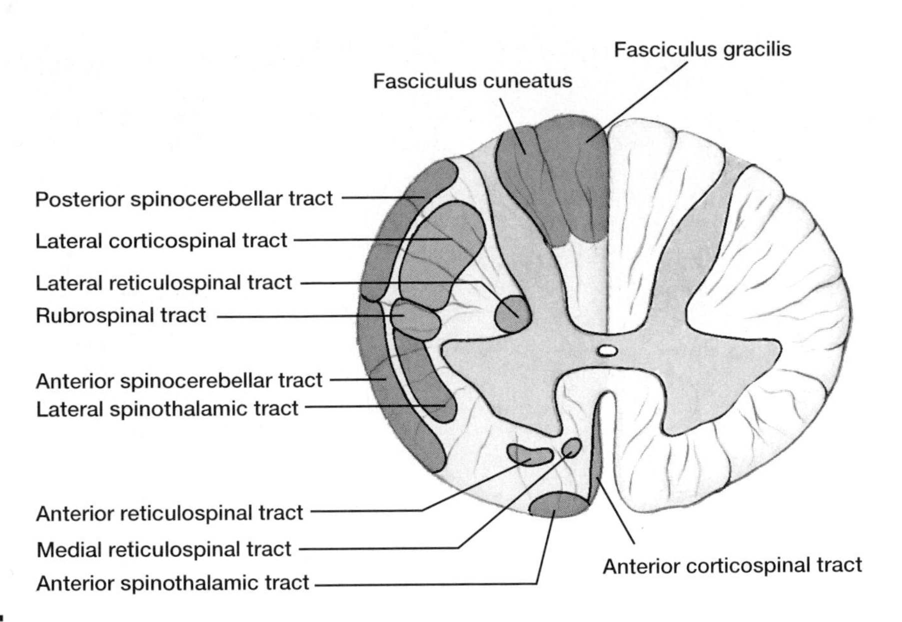


**FIGURE 11.5**
Major ascending and descending tracts within a cross section of the spinal cord. Ascending tracts are in orange, descending tracts in tan. (Tracts are shown only on one side.)

> As a last resort to relieve severe, unremitting pain, a person may undergo a surgical procedure called a *cordotomy*. Fibers in the lateral spinothalamic tracts on the side opposite the affected area are severed above the level of the stimulated pain receptors.

3. **Spinocerebellar** (spi″no-ser″ĕ-bel′ar) **tracts.** The *posterior* and *anterior spinocerebellar tracts* lie near the surface in the lateral funiculi of the spinal cord. Fibers in the posterior tracts remain uncrossed, while those in the anterior tracts cross over in the medulla. Impulses conducted on their fibers originate in the muscles of the lower limbs and trunk and then travel to the cerebellum of the brain. These impulses coordinate muscular movements.

## Descending Tracts

The major descending tracts of the spinal cord are shown in figure 11.5. They include the following:

1. **Corticospinal** (kor″tĭ-ko-spi′nal) **tracts.** The *lateral* and *anterior corticospinal tracts* occupy the lateral and anterior funiculi, respectively. Most of the fibers of the lateral tracts cross over in the lower portion of the medulla oblongata. Those of the anterior tracts descend uncrossed. The corticospinal tracts conduct motor impulses from the brain to spinal nerves and outward to various skeletal muscles. Thus, they help control voluntary movements (fig. 11.7).

   The corticospinal tracts are also called *pyramidal tracts* after the pyramid-shaped regions in the medulla oblongata through which they pass. The other descending tracts are called *extrapyramidal tracts,* and they include the reticulospinal and rubrospinal tracts.

2. **Reticulospinal** (rĕ-tik″u-lo-spi′nal) **tracts.** The *lateral reticulospinal tracts* are located in the lateral funiculi, while the *anterior* and *medial reticulospinal tracts* are in the anterior funiculi. Some fibers in the lateral tracts cross over, while others remain uncrossed. Those of the anterior and medial tracts remain uncrossed. Motor impulses transmitted on the reticulospinal tracts originate in the brain and control muscular tone and activity of sweat glands.

3. **Rubrospinal** (roo″bro-spi′nal) **tracts.** The fibers of the rubrospinal tracts cross over in the brain and pass through the lateral funiculi. They carry motor impulses from the brain to skeletal muscles, and coordinate muscles and control posture.

Chart 11.2 summarizes the nerve tracts of the spinal cord.

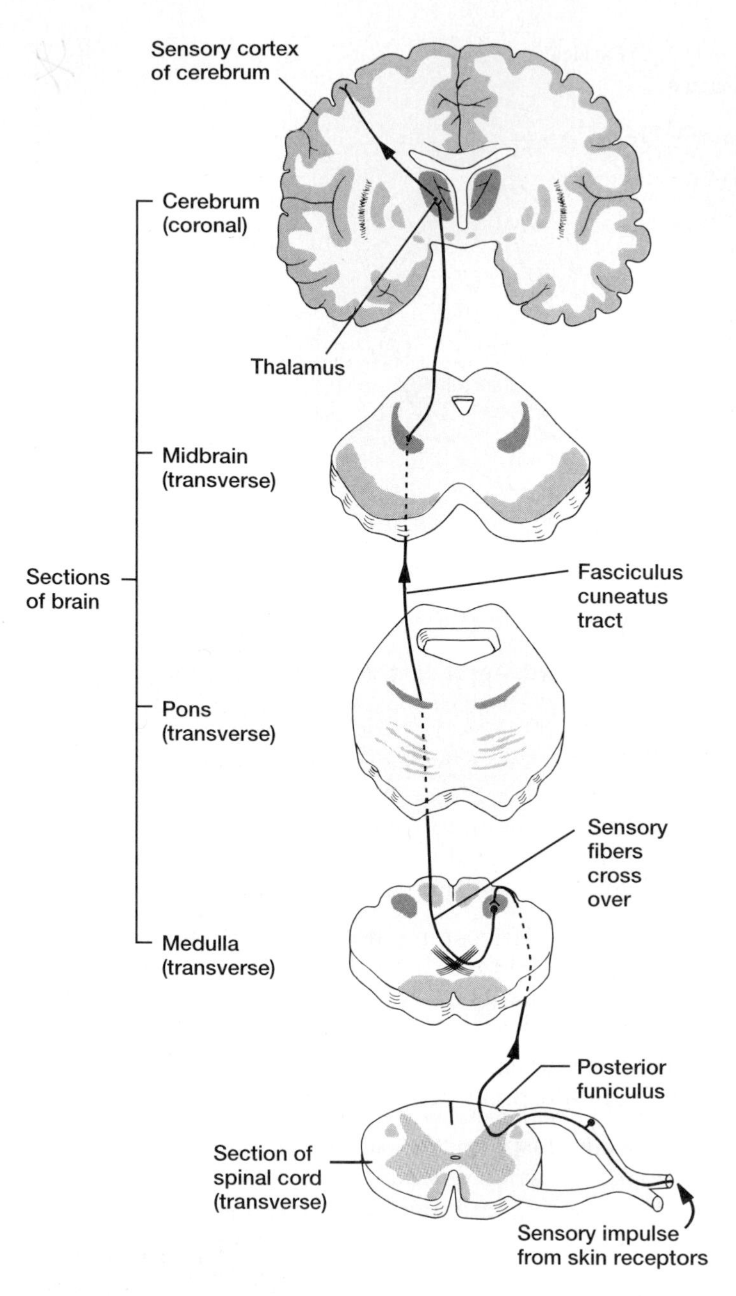


**FIGURE 11.6**

Sensory impulses originating in skin receptors ascend in the fasciculus cuneatus tract and cross over in the medulla of the brain.

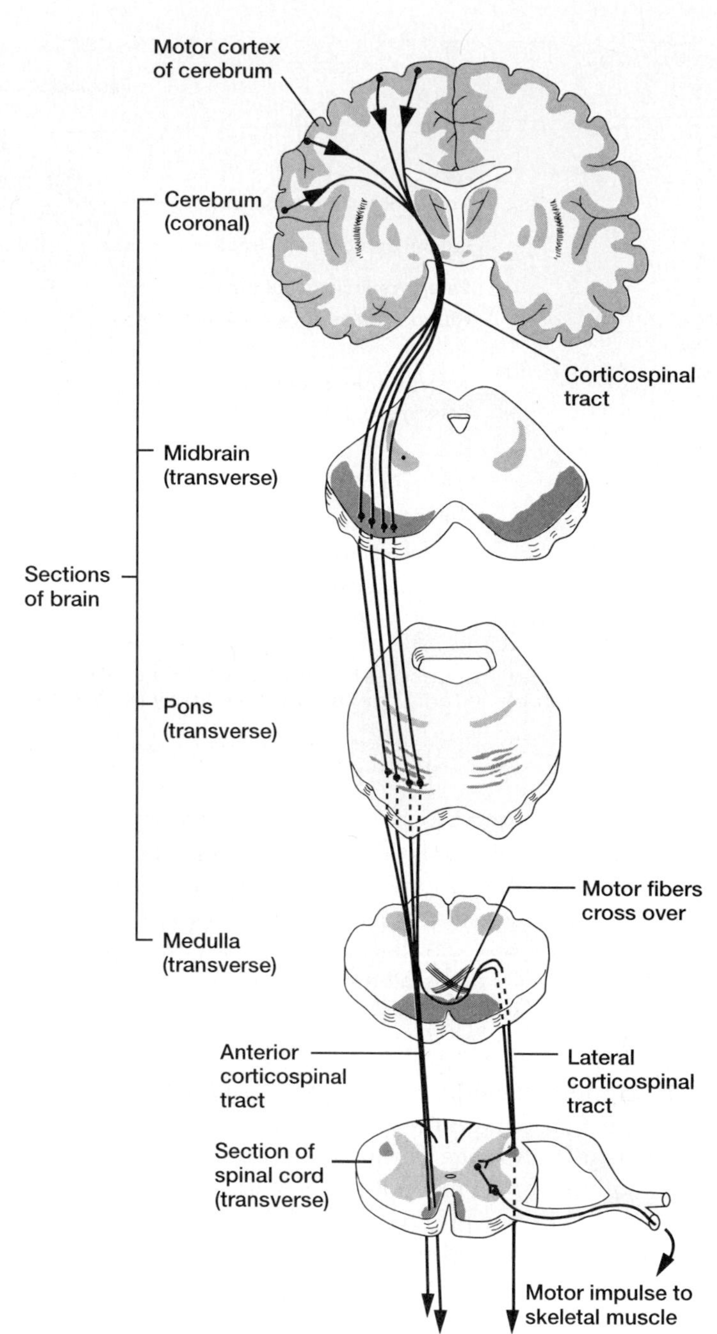


**FIGURE 11.7**

Motor fibers of the corticospinal tract begin in the cerebral cortex, cross over in the medulla, and descend in the spinal cord, where they synapse with neurons whose fibers lead to spinal nerves supplying skeletal muscles.

In addition to providing a pathway for various nerve tracts, the spinal cord functions in many reflexes, like the knee-jerk and withdrawal reflexes described in chapter 10. Such reflexes are called **spinal reflexes,** because their reflex arcs pass through the spinal cord.

Clinical Application 11.1 describes injuries to the spinal cord.

1. Describe the structure of the spinal cord.
2. What is meant by ascending and descending tracts?
3. What is the consequence of fibers crossing over?
4. Name the major tracts of the spinal cord, and list the kinds of impulses each conducts.

## CHART 11.2 NERVE TRACTS OF THE SPINAL CORD

| Tract | Location | Function |
|---|---|---|
| ***Ascending Tracts*** | | |
| **1.** Fasciculus gracilis and fasciculus cuneatus | Posterior funiculi | Conduct sensory impulses associated with the senses of touch, pressure, and body movement from skin, muscles, tendons, and joints to the brain |
| **2.** Spinothalamic tracts (anterior and posterior) | Lateral and anterior funiculi | Conduct sensory impulses associated with the senses of pain, temperature, touch, and pressure from various body regions to the brain |
| **3.** Spinocerebellar tracts (posterior and anterior) | Lateral funiculi | Conduct sensory impulses needed for the coordination of muscle movements from muscles of the lower limbs and trunk to the cerebellum |
| ***Descending Tracts*** | | |
| **1.** Corticospinal tracts (lateral and anterior) | Lateral and anterior funiculi | Conduct motor impulses associated with voluntary movements from the brain to various skeletal muscles |
| **2.** Reticulospinal tracts (lateral, anterior, and medial) | Lateral and anterior funiculi | Conduct motor impulses associated with the maintenance of muscle tone and the activity of sweat glands from the brain |
| **3.** Rubrospinal tracts | Lateral funiculi | Conduct motor impulses associated with muscular coordination and the maintenance of posture from the brain |

*Amyotrophic lateral sclerosis* (ALS), also known as Lou Gehrig's disease, begins with slight stiffening and weakening of the upper and lower limbs, loss of finger dexterity, wasting hand muscles, severe muscle cramps, and difficulty swallowing. Muscle function declines throughout the body, and usually the person dies within 5 years from respiratory muscle paralysis. Some people, however, live many years with ALS, such as noted astronomer and author Stephen Hawking.

In ALS, motor neurons degenerate within the spinal cord, brain stem, and cerebral cortex. Fibrous tissue replaces them. By studying some of the 10% of ALS patients who inherit the disorder, researchers traced the cause to an abnormal form of an enzyme, superoxide dismutase, which normally dismantles oxygen-free radicals, which are toxic byproducts of metabolism. Although the disorder remains devastating, investigators now know where to look to conquer it.

## BRAIN

The **brain** is the largest and most complex part of the nervous system. It occupies the cranial cavity and is composed of about one hundred billion ($10^{11}$) multipolar neurons and innumerable nerve fibers, by which these neurons communicate with one another and with neurons in other parts of the system.

The brain contains nerve centers associated with sensory functions and is responsible for sensations and perceptions. It issues motor commands to skeletal muscles and carries on higher mental functions, such as memory and reasoning. It also contains centers that coordinate muscular movements, as well as centers and nerve pathways that regulate visceral activities. In addition to overseeing the function of the entire body, the brain also provides characteristics like personality, as evidenced by the strange case of Phineas Gage discussed at the chapter's start.

### Brain Development

The basic structure of the brain reflects the way it forms during early (embryonic) development. It begins as the neural tube that gives rise to the central nervous system. The portion that becomes the brain has three major cavities, or vesicles, at one end—the *forebrain* (prosencephalon), *midbrain* (mesencephalon), and *hindbrain* (rhombencephalon) (fig. 11.8). Later, the forebrain divides into anterior and posterior portions (telencephalon and diencephalon, respectively), and the hindbrain partially divides into two parts (metencephalon and myelencephalon). The resulting five cavities persist in the mature brain as fluid-filled spaces called **ventricles** and the tubes that connect them. The tissue surrounding the spaces differentiates into various structural and functional regions of the brain.

The wall of the anterior portion of the forebrain gives rise to the *cerebrum* and *basal ganglia,* while the posterior portion forms a section of the brain called the

CLINICAL APPLICATION 11.1

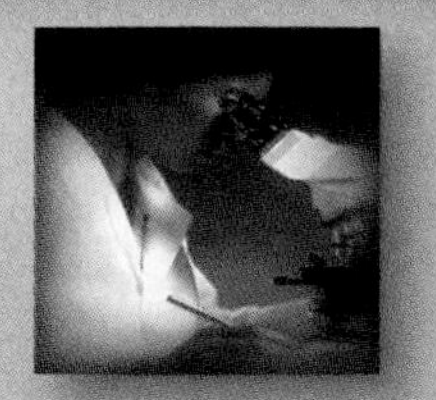

## Spinal Cord Injuries

Injuries to the spinal cord may be caused indirectly, as by a blow to the head or a fall, or they may be caused by forces applied directly to the cord. The consequences depend on the amount of damage the cord sustains and where damage occurs.

Normal spinal reflexes depend on two-way communication between the spinal cord and the brain. Injuring nerve pathways depresses the cord's reflex activities in sites below the injury. At the same time, sensations and muscular tone in the parts the affected fibers innervate lessen. This condition, *spinal shock,* may last for days or weeks, although normal reflex activity may return eventually. However, if nerve fibers are severed, some of the cord's functions are likely to be permanently lost.

Less severe injuries to the spinal cord, as from a blow to the head, whiplash in an automobile accident, or rupture of an intervertebral disk, compress or distort the cord. Pain, weakness, and muscular atrophy in the regions the damaged nerve fibers supply often accompany such injuries.

Among the more common causes of severe direct injury to the spinal cord are gunshot wounds, stabbings, and fractures or dislocations of vertebrae during vehicular accidents (fig. 11A). Regardless of the cause, if nerve fibers in ascending tracts are cut, sensations arising from receptors below the level of the injury are lost. If descending tracts are damaged, the result is a loss of motor functions.

For example, if the right lateral corticospinal tract is severed in the neck near the first cervical vertebra, control of the voluntary muscles in the right upper and lower limbs is lost, paralyzing them (hemiplegia). Problems of this type in fibers of the descending tracts produce *upper motor neuron syndrome,* characterized by *spastic paralysis* in which muscle tone increases, with very little atrophy of the muscles. However, uncoordinated reflex activity (hyperreflexia) usually occurs, during which the flexor and extensor muscles of affected limbs alternately spasm.

Injury to motor neurons or their fibers in the horns of the spinal cord results in *lower motor neuron syndrome.* It is characterized by *flaccid paralysis,* a total loss of muscle tone and reflex activity, and the muscles atrophy.

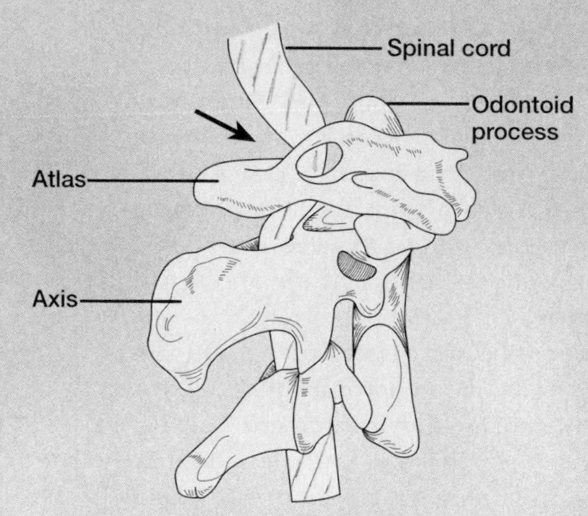


FIGURE 11A
A dislocation of the atlas may cause a compression injury to the spinal cord.

*diencephalon.* The region the midbrain produces continues to be called the *midbrain* in the adult structure, and the hindbrain gives rise to the *cerebellum, pons,* and *medulla oblongata* (fig. 11.9 and chart 11.3). Together, the midbrain, pons, and medulla oblongata comprise the **brain stem,** which attaches the brain to the spinal cord.

On a cellular level, the brain develops as specific neurons attract others by secreting growth hormones. Excess neural connections are destroyed by apoptosis, a form of programmed cell death.

### Structure of the Cerebrum

The cerebrum, which develops from the anterior portion of the forebrain, is the largest part of the mature brain. It consists of two large masses of **cerebral hemispheres,** which are essentially mirror images of each other (fig. 11.10 and reference plate 49). A deep bridge of nerve fibers called the **corpus callosum** connects the cerebral hemispheres. A layer of dura mater called the *falx cerebri* separates them.

The expectant couple was staggered when they saw the ultrasound image of their fetus. The defect was so great that it was obvious even to their untrained eyes: there was no head. The fetus had *anencephaly;* it had a face and lower brain structures but lacked most higher brain structures. The woman miscarried a few days later. Rarely, an infant with anencephaly survives to be born but then almost always dies within 24 hours. Sometimes the parents donate its organs.

Anencephaly is a type of birth defect called a neural tube defect (NTD). It occurs on the twenty-eighth day of prenatal development, when a sheet of tissue that normally folds to form a neural tube, which develops into the central nervous system, remains open at the top. A less serious NTD is *spina bifida,* in which an opening occurs farther down the neural tube, causing a lesion in the spine and paralysis from that point downward. Sometimes spina bifida can be improved or even corrected with surgery.

The precise cause of neural tube defects is not known, but it somehow involves folic acid; taking supplements of this vitamin cuts the recurrence risk sharply among women who have had an affected child. Most pregnant women take a blood test at the fifteenth week of pregnancy to detect fluid leaking from an NTD.

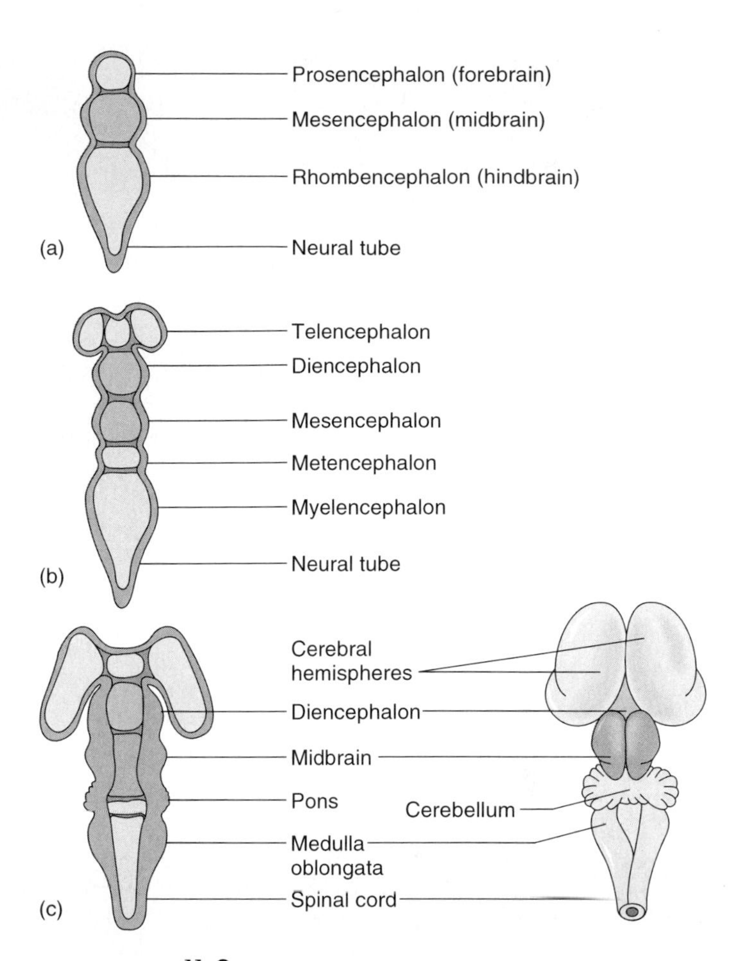


**FIGURE 11.8**

(*a*) The brain develops from a tubular structure with three cavities. (*b*) The cavities persist as the ventricles and their interconnections. (*c*) The wall of the tube gives rise to various regions of the brain, brain stem, and spinal cord.

Numerous ridges or **convolutions** (sing. *gyrus;* pl. *gyri*), separated by grooves, mark the cerebrum's surface. A shallow groove is called a **sulcus,** and a very deep groove is called a **fissure.** Although the arrangement of these elevations and depressions is complex, they form distinct patterns in all normal brains. For example, a *longitudinal fissure* separates the right and left cerebral hemispheres; a *transverse fissure* separates the cerebrum from the cerebellum; and various sulci divide each hemisphere into lobes (figs. 11.9 and 11.10).

In a disorder called *lissencephaly* ("smooth brain"), a newborn has a smooth cerebral cortex, completely lacking the characteristic convolutions. Absence of a protein early in prenatal development prevents certain neurons from migrating within the brain, which somehow blocks formation of convolutions. The child is profoundly mentally retarded, with frequent seizures and other neurological problems.

The lobes of the cerebral hemispheres (fig. 11.10) are named after the skull bones that they underlie. They include the following:

1. **Frontal lobe.** The frontal lobe forms the anterior portion of each cerebral hemisphere. It is bordered posteriorly by a *central sulcus* (fissure of Rolando), which passes out from the longitudinal fissure at a right angle, and inferiorly by a *lateral sulcus* (fissure of Sylvius), which exits the undersurface of the brain along its sides.
2. **Parietal lobe.** The parietal lobe is posterior to the frontal lobe and is separated from it by the central sulcus.
3. **Temporal lobe.** The temporal lobe lies below the frontal and parietal lobes and is separated from them by the lateral sulcus.
4. **Occipital lobe.** The occipital lobe forms the posterior portion of each cerebral hemisphere and is separated from the cerebellum by a shelflike

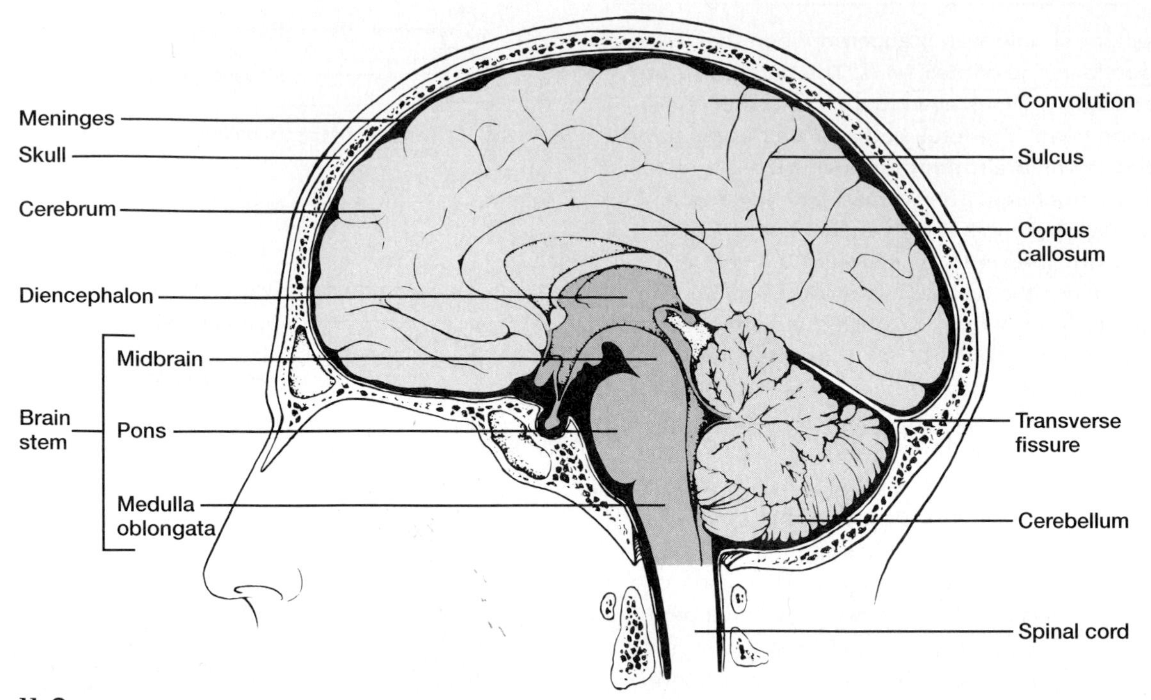


**FIGURE 11.9**
The major portions of the brain include the cerebrum, the diencephalon, the cerebellum, and the brain stem.

### CHART 11.3 STRUCTURAL DEVELOPMENT OF THE BRAIN

| Embryonic Vesicle | Spaces Produced | Regions of the Brain Produced |
|---|---|---|
| Forebrain (prosencephalon) | | |
| Anterior portion (telencephalon) | Lateral ventricles | Cerebrum |
| | | Basal ganglia |
| Posterior portion (diencephalon) | Third ventricle | Thalamus |
| | | Hypothalamus |
| | | Posterior pituitary gland |
| | | Pineal gland |
| Midbrain (mesencephalon) | Cerebral aqueduct | Midbrain |
| Hindbrain (rhombencephalon) | | |
| Anterior portion (metencephalon) | Fourth ventricle | Cerebellum |
| | | Pons |
| Posterior portion (myelencephalon) | Fourth ventricle | Medulla oblongata |

extension of dura mater called the *tentorium cerebelli.* The occipital lobe and the parietal and temporal lobes have no distinct boundary.

5. **Insula.** The insula (island of Reil) is located deep within the lateral sulcus and is covered by parts of the frontal, parietal, and temporal lobes. A *circular sulcus* separates it from them.

A thin layer of gray matter (2 to 5 millimeters thick) called the **cerebral cortex** constitutes the outermost portion of the cerebrum. It covers the convolutions, dipping into the sulci and fissures. The cerebral cortex contains nearly 75% of all the neuron cell bodies in the nervous system.

Just beneath the cerebral cortex is a mass of white matter that makes up the bulk of the cerebrum. This mass contains bundles of myelinated nerve fibers that connect neuron cell bodies of the cortex with other parts of the nervous system. Some of these fibers pass from one cerebral hemisphere to the other by way of the corpus callosum, and others

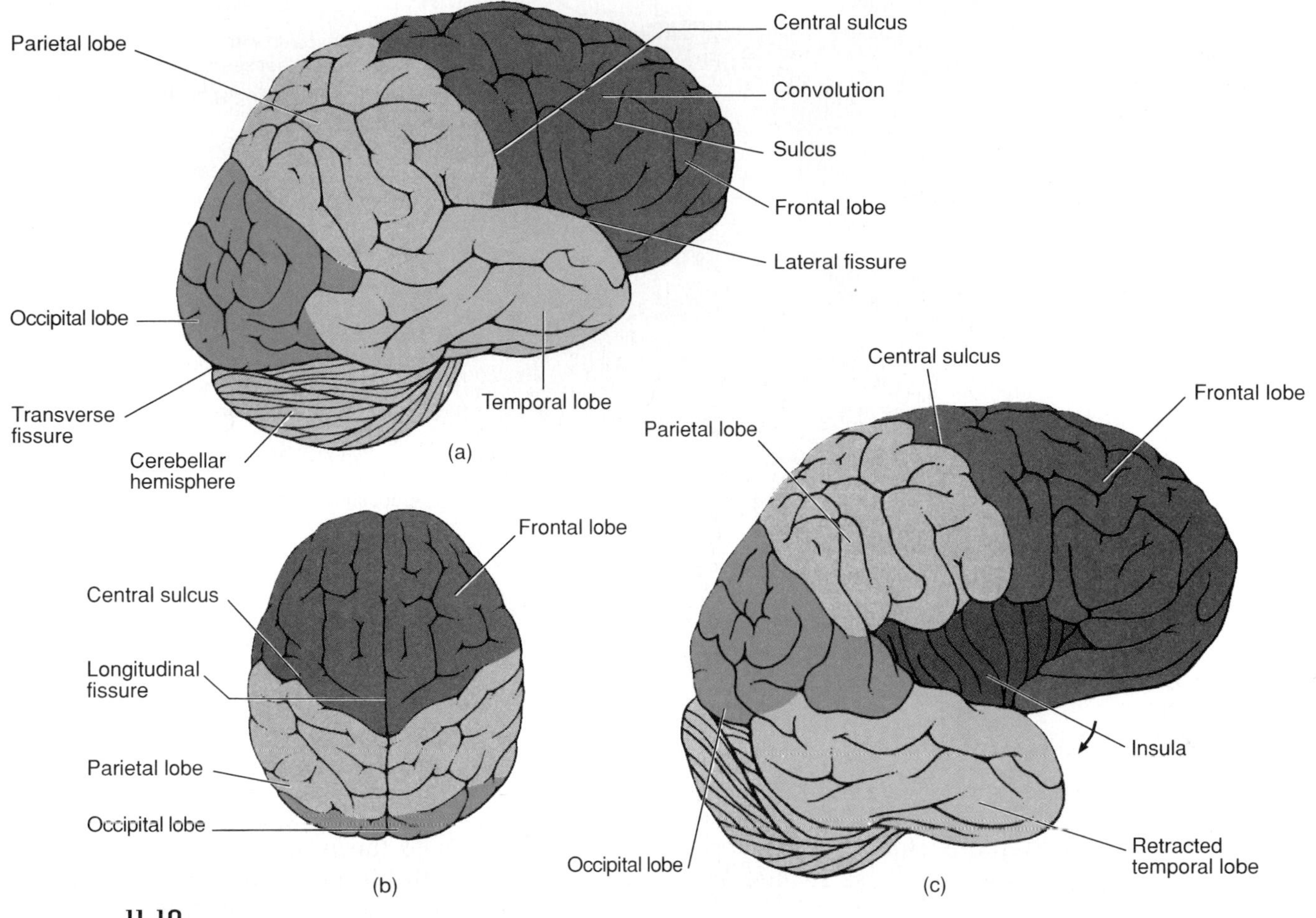


**FIGURE 11.10**

Color distinguishes the lobes of the right cerebral hemisphere. (*a*) Lateral view of the right hemisphere; (*b*) hemispheres viewed from above; (*c*) lateral view of the right hemisphere with the insula exposed.

carry sensory or motor impulses from the cortex to nerve centers in the brain or spinal cord.

> Chess players helped researchers identify a part of the cerebral cortex in the frontal lobes used for planning an action. PET scans, which highlight brain activity, were taken as the players answered questions about a chessboard displayed on a computer screen. When the hardest question was posed (Does one move remain until checkmate?), only then did a specific part of the cortex light up on the scan, relating structure to function.

## Functions of the Cerebrum

The cerebrum provides higher brain functions: interpreting impulses from sense organs, initiating voluntary muscular movements, storing information as memory, and retrieving this information in reasoning. The cerebrum is also the seat of intelligence and personality.

### Functional Regions of the Cortex

The regions of the cerebral cortex that perform specific functions have been located using a variety of techniques. Persons who have suffered brain disease or injury, such as Karen Ann Quinlan and Phineas Gage, or have had portions of their brains removed surgically, have provided clues as to the functions of the impaired brain regions.

In other studies, areas of cortices have been exposed surgically and stimulated mechanically or electrically, with researchers observing the responses in certain muscles or the specific sensations that result. Such investigations have enabled investigators to map the functional regions of the cerebral cortex. The cortex is divided into sections known as motor, sensory, and association areas. They overlap somewhat.

### Motor Areas

The *primary motor areas* of the cerebral cortex lie in the frontal lobes just in front of the central sulcus (precentral gyrus), and in the anterior wall of this sulcus

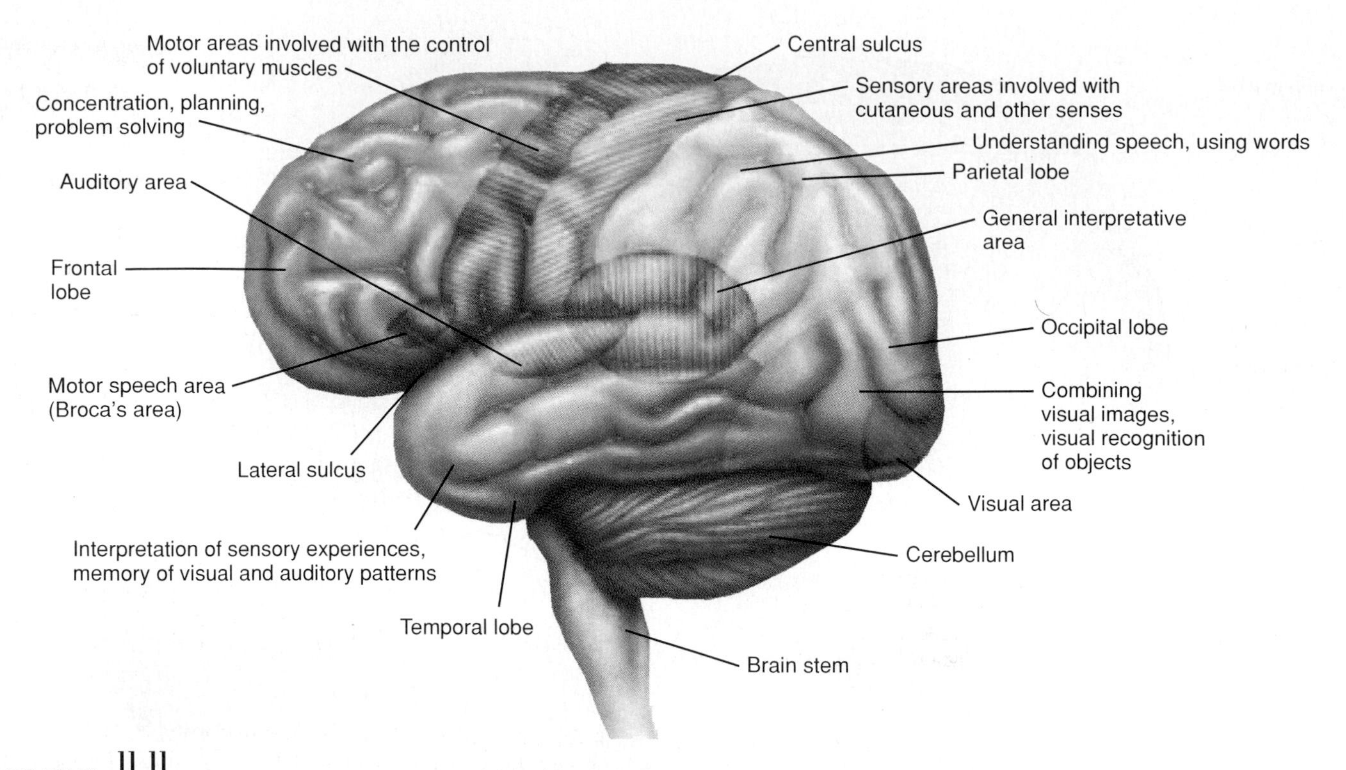


**FIGURE 11.11**
Some motor, sensory, and association areas of the left cerebral cortex.

(fig. 11.11). The nervous tissue in these regions contains numerous, large *pyramidal cells,* so named because of their pyramid-shaped cell bodies.

Impulses from these pyramidal cells travel downward through the brain stem and into the spinal cord on the *corticospinal tracts.* Most of the nerve fibers in these tracts cross over from one side of the brain to the other within the brain stem and descend as lateral corticospinal tracts (see fig. 11.7).

Within the spinal cord, the corticospinal fibers synapse with motor neurons in the gray matter of the anterior horns. Axons of the motor neurons lead outward through peripheral nerves to voluntary muscles. Impulses transmitted on these pathways in special patterns and frequencies are responsible for fine movements in skeletal muscles. More specifically, as figure 11.12 shows, cells in the upper portions of the motor areas send impulses to muscles in the thighs and legs; those in the middle portions control muscles in the arms and forearms; and those in lower portions activate muscles of the head, face, and tongue.

The *reticulospinal* and *rubrospinal tracts* coordinate and control motor functions that maintain balance and posture. Many of these fibers pass into the basal ganglia on the way to the spinal cord. Some of the impulses conducted on these pathways normally inhibit muscular actions.

In addition to the primary motor areas, certain other regions of the frontal lobe control motor functions. For example, a region called *Broca's area* is just anterior to the primary motor cortex and superior to the lateral sulcus (see fig. 11.11), usually in the left cerebral hemisphere. It coordinates the complex muscular actions of the mouth, tongue, and larynx, which make speech possible. A person with an injury to this area may be able to understand spoken words, but may be unable to speak.

Above Broca's area is a region called the *frontal eye field.* The motor cortex in this area controls voluntary movements of the eyes and eyelids. Nearby is the cortex responsible for movements of the head that direct the eyes. Another region just in front of the primary motor area controls the muscular movements of the hands and fingers that make such skills as writing possible (see fig. 11.11).

An injury to the motor system may result in a loss of the ability to produce purposeful muscular movements. Such a condition that impairs use of the upper and lower limbs, head, or eyes is called *apraxia.* Apraxia affecting the speech muscles, impairing speaking ability, is called *aphasia.*

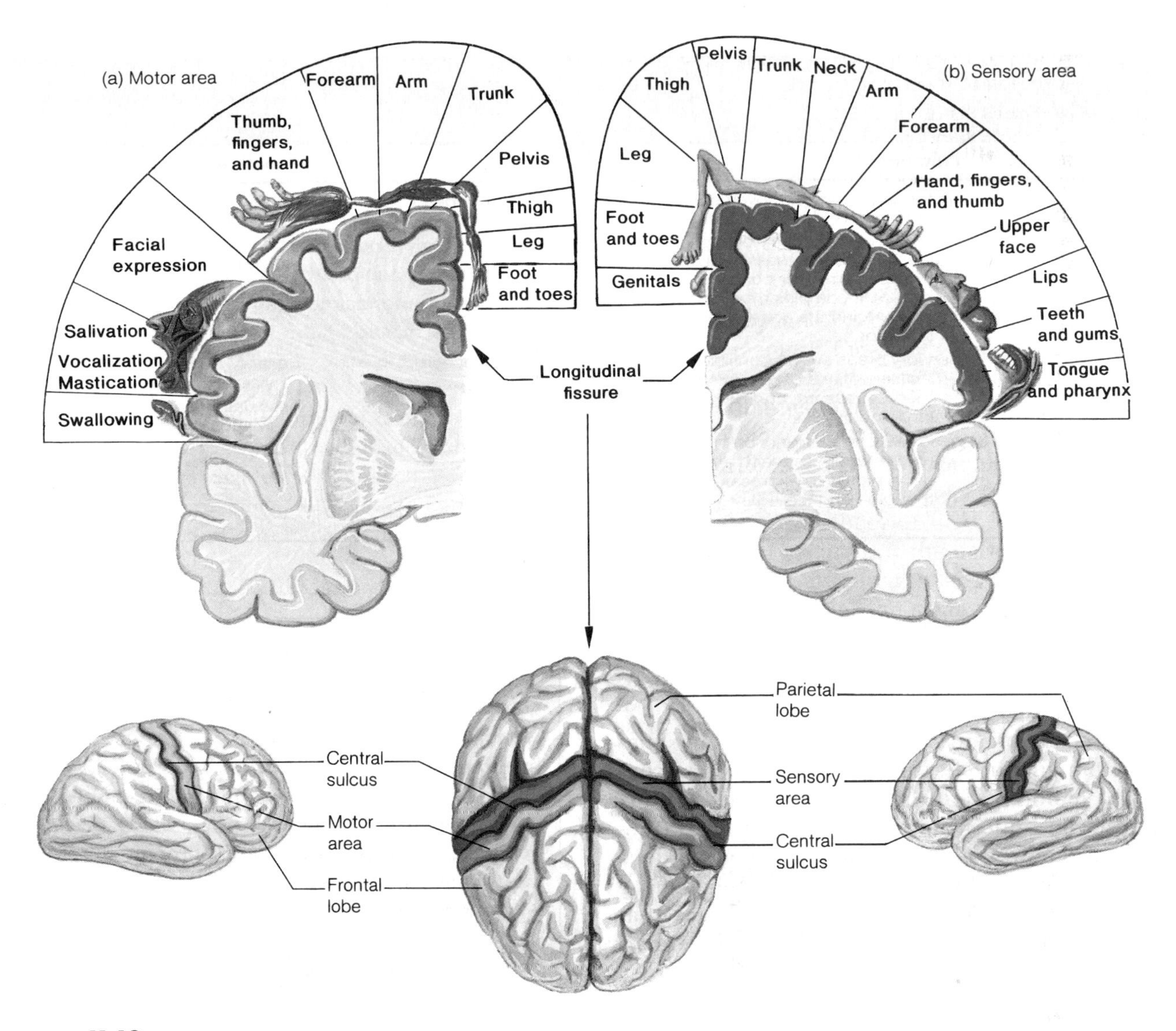


**FIGURE 11.12**
(*a*) Motor areas that control voluntary muscles; (*b*) sensory areas involved with cutaneous and other senses.

## Sensory Areas

Sensory areas, which occur in several lobes of the cerebrum, interpret impulses arriving from sensory receptors, producing feelings or sensations. For example, the sensations of temperature, touch, pressure, and pain in the skin arise in the anterior portions of the parietal lobes along the central sulcus (postcentral gyrus) and in the posterior wall of this sulcus (see fig. 11.11). The posterior parts of the occipital lobes are concerned with vision, while the posterior, dorsal portions of the temporal lobes contain the centers for hearing. The sensory areas for taste are located near the bases of the central sulci along the lateral sulci, and the sense of smell arises from centers deep within the cerebrum.

Like motor fibers, sensory fibers, such as those in the *fasciculus cuneatus tract*, cross over (see fig. 11.6). Thus, the centers in the right central hemisphere interpret impulses originating from the left side of the body, and vice versa. However, the sensory areas concerned with vision receive impulses from both eyes, and those concerned with hearing receive impulses from both ears.

## Association Areas

Association areas are regions of the cerebral cortex that are not primarily sensory or motor in function, and interconnect with each other and with other brain structures. These areas occupy the anterior portions of the frontal lobes and are widespread in the lateral

CLINICAL APPLICATION 11.2

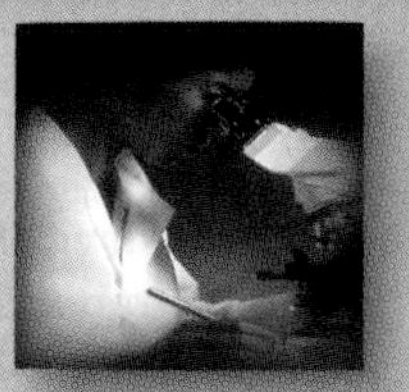

## Cerebral Injuries and Abnormalities

The specific symptoms associated with a cerebral injury or abnormality depend upon the areas and extent of damage. A person with damage to the association areas of the frontal lobes may have difficulty concentrating on complex mental tasks, appearing disorganized and easily distracted.

If the general interpretative area of the dominant hemisphere is injured, the person may be unable to interpret sounds as words or to understand ideas presented in writing. However, the dominance of one hemisphere usually does not become established until after five or six years of age. Consequently, if the general interpretative area is destroyed in a child, the corresponding region of the other side of the brain may be able to take over the functions, and the child's language abilities may develop normally. If such an injury occurs in an adult, the nondominant hemisphere may develop only limited interpretative functions, producing a severe intellectual disability. Following are three common cerebral abnormalities.

- The last thing the young man recalled before waking up in a hospital room was riding his bike around a sharp curve. Witnesses said that he smashed into a parked car. When he regained consciousness, he found that a gash on his face had been sewn up and his shoulder was dislocated. But the primary reason for keeping him in the hospital was a *concussion.* He had lost consciousness when slamming into the car jarred his brain. He recovered consciousness just a few minutes after the accident, witnesses reported, but he did not recall anything prior to waking up in the hospital. Gradually, he became more aware but felt mentally fuzzy. He tried to watch television but would forget a statement as soon as it was spoken. For several days he had difficulty concentrating and remembering, and he had a fierce headache. Soon, however, he completely recovered.
- *Cerebral palsy* (CP) is motor impairment at birth, stemming from a brain anomaly occurring early in development. Until recently, most cases of CP were blamed on "birth trauma," but recently researchers determined that the most common cause is a blocked cerebral blood vessel, which leads to atrophy of the brain region deprived of its blood supply. Birth trauma and brain infection cause some cases.

  CP affects about 2 in every 2,000 births and is especially prevalent among premature babies. One-half to two-thirds of affected babies improve and can even outgrow the condition by age 7. Sometimes seizures or learning disabilities are also present. Clinicians classify CP by the number of limbs and the types of neurons affected.
- In a "stroke," or *cerebrovascular accident* (CVA), a sudden interruption in blood flow in a vessel supplying brain tissues damages the cerebrum. The affected blood vessel may rupture, bleeding into the brain, or be blocked by a clot. In either case, brain tissues downstream from the vascular accident die or permanently lose function. Temporary interruption in cerebral blood flow, perhaps by a clot that quickly breaks apart, produces a much less serious *transient ischemic attack* (TIA).

the forebrain (fig. 11.13). The neuron cell bodies that the basal ganglia contain relay motor impulses originating in the cerebral cortex and passing into the brain stem and spinal cord. The basal ganglia produce most of the inhibitory neurotransmitter *dopamine.* Impulses from the basal ganglia normally inhibit motor functions, thus controlling certain muscular activities.

1. What does hemisphere dominance mean?
2. What are the functions of the nondominant hemisphere?
3. Distinguish between short-term and long-term memory.
4. What is the function of the basal ganglia?

### Ventricles and Cerebrospinal Fluid

A series of interconnected cavities called *ventricles* are located within the cerebral hemispheres and brain stem (fig. 11.14 and reference plates 53 and 54). These spaces are continuous with the central canal of the spinal cord and also are filled with cerebrospinal fluid.

The largest ventricles are the *lateral ventricles,* which are the first and second ventricles (the first ventricle in the left cerebral hemisphere and the second ventricle in the right cerebral hemisphere). They extend into the cerebral hemispheres and occupy portions of the frontal, temporal, and occipital lobes.

A narrow space that constitutes the *third ventricle* is located in the midline of the brain, beneath the

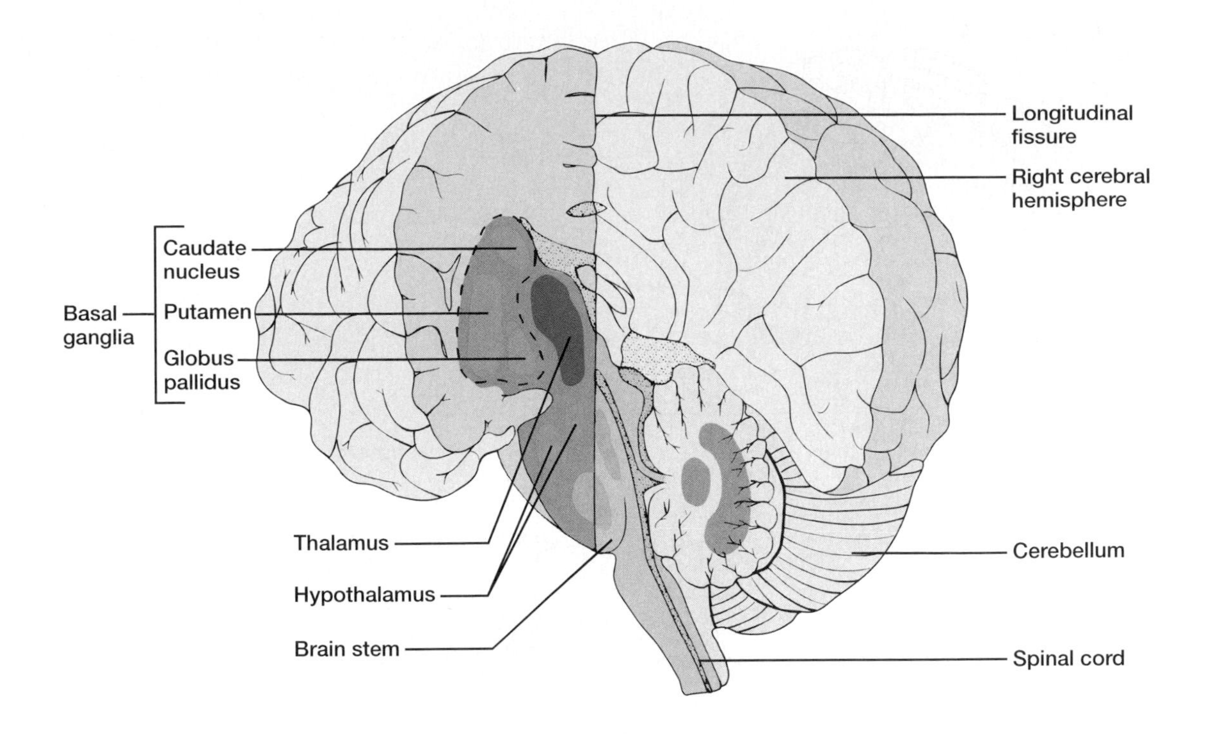


**FIGURE 11.13**
A coronal section of the left cerebral hemisphere reveals some of the basal ganglia.

Movement disorders often reflect an abnormality in the basal ganglia. In *Parkinson's disease,* neurons that synthesize dopamine in the basal ganglia degenerate, producing slow movements, difficulty in initiating voluntary muscular actions, and tremors. Giving a patient L-dopa, a dopamine precursor that can cross the blood-brain barrier, stems symptoms for a short time because it is converted to dopamine inside the brain. However, L-dopa eventually produces a condition called tardive dyskinesia, in which the patient has uncontrollable facial tics. A newer treatment for Parkinson's disease is to transplant a patient's adrenal tissue, or fetal tissue, to the basal ganglia. The transplanted cells secrete dopamine.

The movements of *Huntington disease* are quite distinct from those of Parkinson's disease. This inherited disorder was once called Huntington's chorea, referring to the dancelike, seemingly choreographed movements. It begins usually in middle age with small uncontrollable motions, such as turning the hands. Gait is affected early on, and sometimes people are thought to be intoxicated. Gradually, the movements become more pronounced. Personality changes, particularly great anger, are also part of Huntington disease. Some people may become demented, but many, such as folksinger Woody Guthrie, maintain their intellects as they slowly lose control of their bodies. Just before he died, Guthrie could communicate only by blinking.

Researchers discovered the gene causing Huntington disease in 1993. The mutant gene actually expands with each generation. Persons with the disorder have a larger-than-normal gene. The gene encodes a protein called huntingtin, but we still do not know how it causes the disorder.

corpus callosum. This ventricle communicates with the lateral ventricles through openings (*interventricular foramina*) in its anterior end.

*The fourth ventricle* is located in the brain stem just in front of the cerebellum. A narrow canal, the *cerebral aqueduct* (aqueduct of Sylvius), connects it to the third ventricle and passes lengthwise through the brain stem. This ventricle is continuous with the central canal of the spinal cord and has openings in its roof that lead into the subarachnoid space of the meninges.

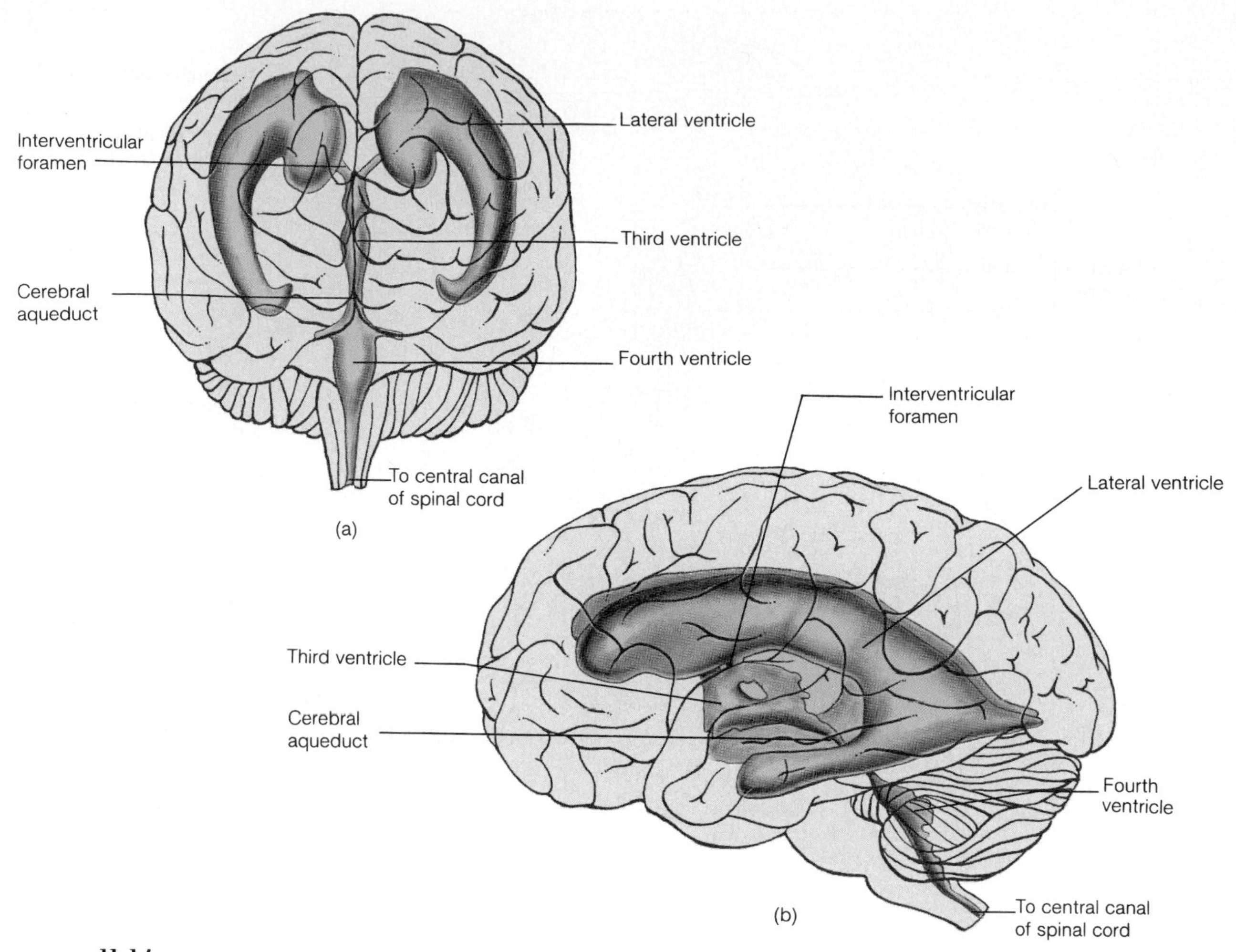


**FIGURE 11.14**

(*a*) Anterior view of the ventricles within the cerebral hemispheres and brain stem; (*b*) lateral view.

Tiny, reddish cauliflowerlike masses of specialized capillaries from the pia mater, called **choroid plexuses,** secrete *cerebrospinal fluid* (CSF). These structures project into the cavities of the ventricles (fig. 11.15). A single layer of specialized ependymal cells (see chapter 10) joined closely by tight junctions covers the choroid plexuses. These cells block passage of water-soluble substances between the blood and the cerebrospinal fluid. At the same time, the cells selectively transfer certain substances from the blood into the cerebrospinal fluid by facilitated diffusion and transfer other substances by active transport (see chapter 3), thus regulating the composition of the cerebrospinal fluid.

Most of the cerebrospinal fluid arises in the lateral ventricles, from where it circulates slowly into the third and fourth ventricles and into the central canal of the spinal cord. It also enters the subarachnoid space of the meninges by passing through the wall of the fourth ventricle near the cerebellum.

Humans secrete nearly 500 milliliters of cerebrospinal fluid daily. However, only about 140 milliliters are in the nervous system at any time, because cerebrospinal fluid is continuously reabsorbed into the blood. The CSF is reabsorbed through tiny, fingerlike structures called *arachnoid granulations* that project from the subarachnoid space into the blood-filled dural sinuses (see fig. 11.15).

Cerebrospinal fluid is a clear, somewhat viscid liquid that differs in composition from the fluid that leaves the capillaries in other parts of the body. Specifically, it contains a greater concentration of sodium and lesser concentrations of glucose and potassium than do other extracellular fluids. Its function is nutritive as well as protective. Cerebrospinal fluid helps maintain a stable ionic concentration in the central nervous system and provides a pathway to the blood for waste. The cerebrospinal fluid may also supply information about the internal environment to autonomic centers in the hypothalamus and brain stem, because the fluid forms from blood plasma and therefore its composition reflects changes in body fluids. Clinical Application 11.3 discusses the pressure that cerebrospinal fluid generates.

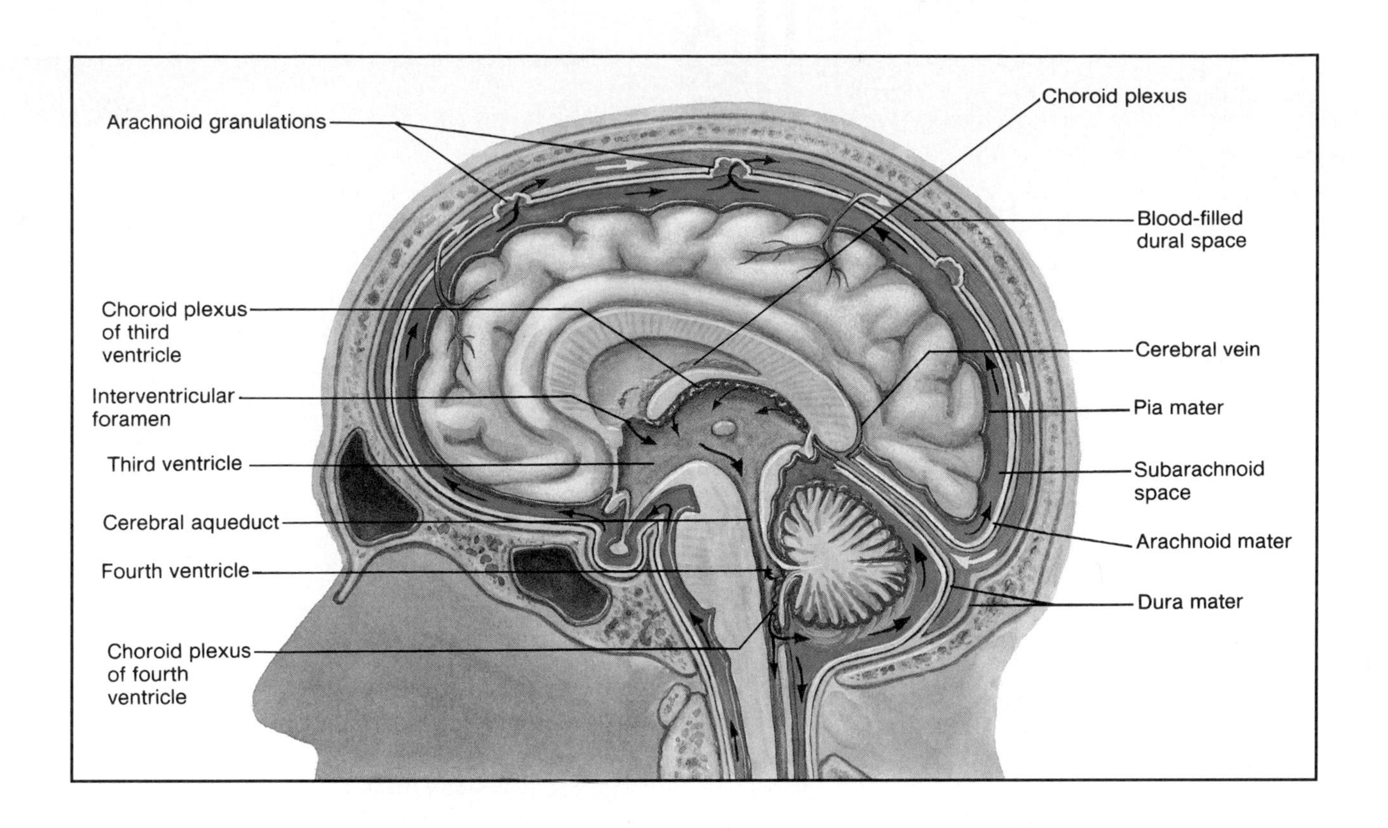


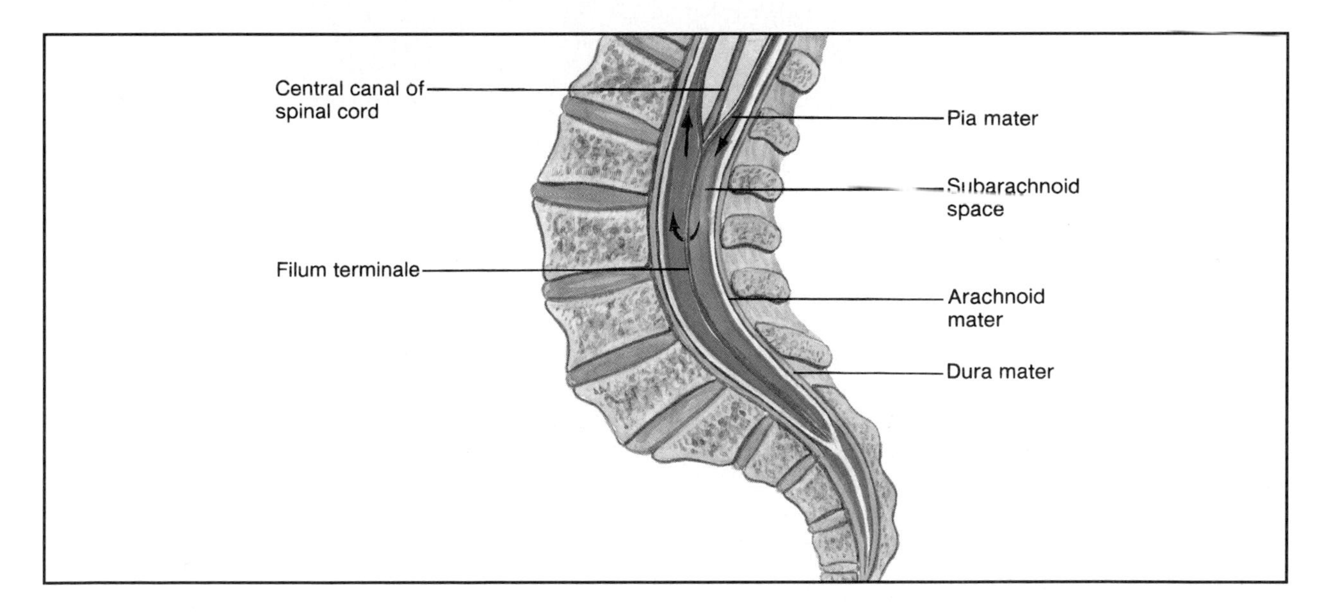


**FIGURE 11.15**
Choroid plexuses in ventricle walls secrete cerebrospinal fluid. The fluid circulates through the ventricles and central canal, enters the subarachnoid space, and is reabsorbed into the blood of the dural sinuses through arachnoid granulations.

Because it occupies the subarachnoid space of the meninges, cerebrospinal fluid completely surrounds the brain and spinal cord. In effect, these organs float in the fluid. The CSF protects them by absorbing forces that might otherwise jar and damage their delicate tissues.

1. Where are the ventricles of the brain located?
2. How does cerebrospinal fluid form?
3. Describe the pattern of cerebrospinal fluid circulation.

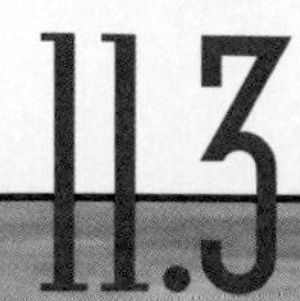

Clinical Application 11.3

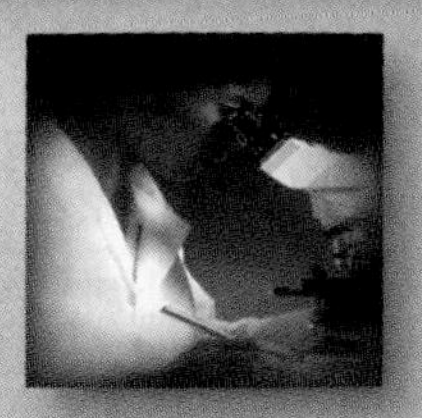

## Cerebrospinal Fluid Pressure

Because cerebrospinal fluid is secreted and reabsorbed continuously, the fluid pressure in the ventricles remains relatively constant. However, infection, a tumor, or a blood clot can interfere with the fluid's circulation, increasing pressure within the ventricles (intracranial pressure). This can collapse cerebral blood vessels, retarding blood flow. Brain tissues may be injured by being forced against the skull.

A *lumbar puncture* (spinal tap) can measure cerebrospinal fluid pressure. A physician inserts a fine, hollow needle into the subarachnoid space between the third and fourth or between the fourth and fifth lumbar vertebrae—below the end of the spinal cord (fig. 11B). An instrument called a

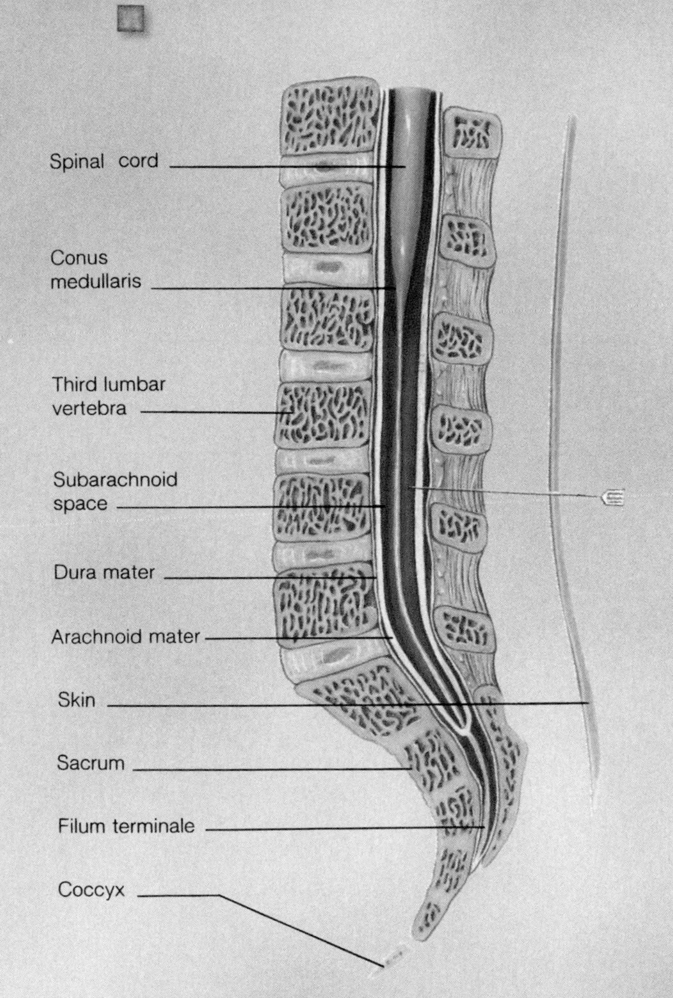


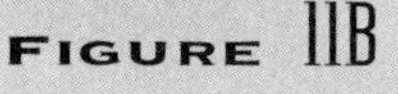

**Figure 11B**
A lumbar puncture is performed by inserting a fine needle between the third and fourth lumbar vertebrae and withdrawing a sample of cerebrospinal fluid from the subarachnoid space.

### Diencephalon

The **diencephalon** develops from the posterior forebrain and is located between the cerebral hemispheres and above the brain stem (figs. 11.9 and 11.16). It surrounds the third ventricle and is composed largely of gray matter. Within the diencephalon, a dense mass, called the **thalamus,** bulges into the third ventricle from each side. Another region of the diencephalon that includes many nuclei is the **hypothalamus.** It lies below the thalamic nuclei and forms the lower walls and floor of the third ventricle (see reference plates 49 and 53).

Other parts of the diencephalon include (a) the **optic tracts** and the **optic chiasma** that is formed by the optic nerve fibers crossing over; (b) the **infundibulum,** a conical process behind the optic chiasma to which the pituitary gland is attached; (c) the **posterior**

*manometer* measures the pressure of the fluid, which is usually about 130 millimeters of water (10 millimeters of mercury). At the same time, samples of cerebrospinal fluid may be withdrawn and tested for the presence of abnormal constituents. Red blood cells, for example, may indicate a hemorrhage in the central nervous system.

A temporary drain inserted into the subarachnoid space between the fourth and fifth lumbar vertebrae can relieve excess pressure. In a fetus or infant whose cranial sutures have not yet united, increasing intracranial pressure (ICP) may cause an enlargement of the cranium called *hydrocephalus,* or "water on the brain." A shunt to relieve hydrocephalus drains fluid away from the cranial cavity and into the digestive tract, where it is either reabsorbed into the blood or excreted (fig. 11C).

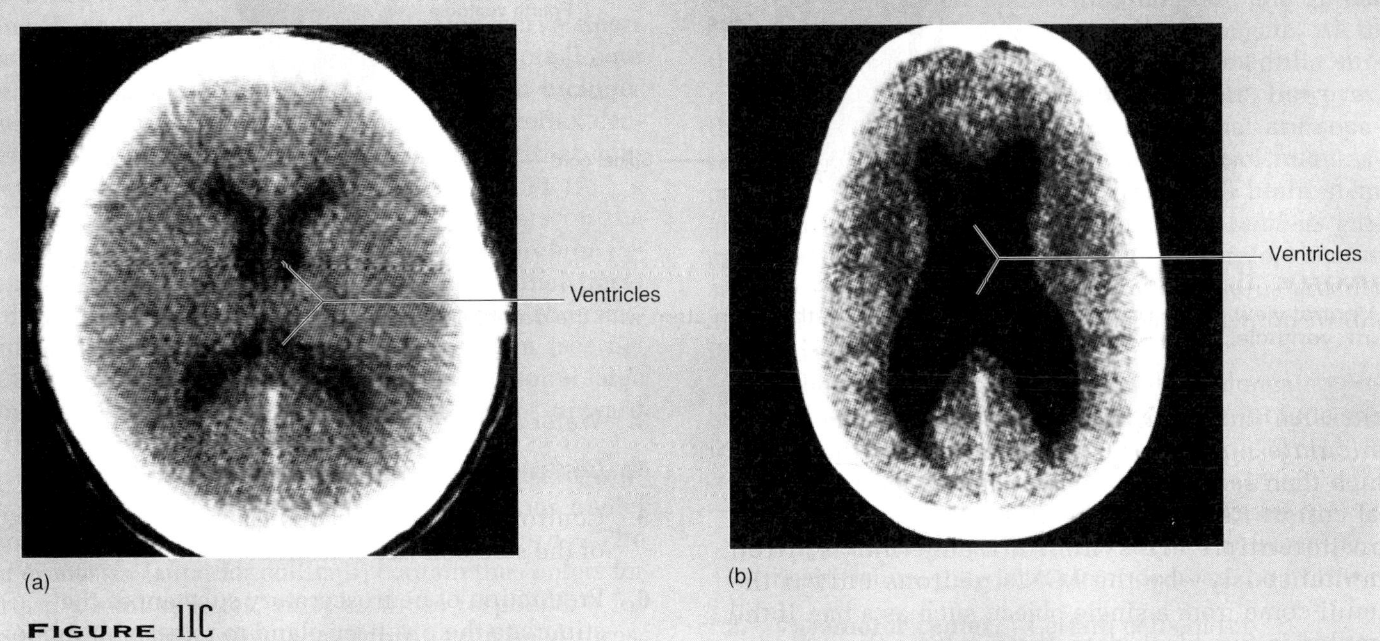


**FIGURE 11C**
CT scans of the human brain. (*a*) Normal ventricles; (*b*) ventricles enlarged by accumulated fluid.

**pituitary gland,** which hangs from the floor of the hypothalamus; (d) the **mammillary** (mam′ĭ-ler″e) **bodies,** which are two rounded structures behind the infundibulum; and (e) the **pineal gland,** which forms as a cone-shaped evagination from the roof of the diencephalon (see chapter 13).

The thalamus is a selective gateway for sensory impulses ascending from other parts of the nervous system to the cerebral cortex. It receives all sensory impulses (except those associated with the sense of smell) and channels them to appropriate regions of the cortex for interpretation. In addition, all regions of the cerebral cortex can communicate with the thalamus by means of descending fibers.

The thalamus seems to transmit sensory information by synchronizing action potentials. Consider

Thus, the cerebellum integrates sensory information concerning the position of body parts and coordinates skeletal muscle activity and maintains posture. It receives sensory impulses from receptors in muscles, tendons, and joints (proprioceptors) and from special sense organs, such as the eyes and ears (see chapter 12). For example, the cerebellum uses sensory information from the semicircular canals of the inner ears concerning the motion and position of the head to help maintain equilibrium. Damage to the cerebellum is likely to result in tremors, inaccurate movements of voluntary muscles, loss of muscle tone, a reeling walk, and loss of equilibrium (fig. 11.19).

Chart 11.5 summarizes the characteristics and functions of the major parts of the brain. Clinical Application 11.4 discusses how brain waves reflect brain activity.

1. Where is the cerebellum located?
2. What are the major functions of the cerebellum?
3. What kinds of receptors provide information to the cerebellum?

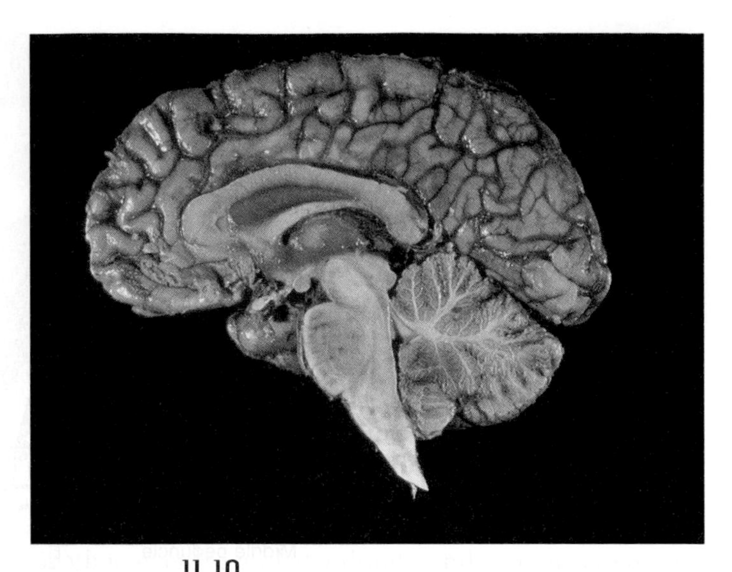

**FIGURE 11.19**
A sagittal section of the brain showing the cerebrum, diencephalon, brain stem, and cerebellum.

### CHART 11.5 MAJOR PARTS OF THE BRAIN

| Part | Characteristics | Functions |
|---|---|---|
| **1.** Cerebrum | Largest part of the brain; consists of two hemispheres connected by the corpus callosum | Controls higher brain functions, including interpreting sensory impulses, initiating muscular movements, storing memory, reasoning, and determining intelligence |
| **2.** Basal ganglia | Masses of gray matter located deep within the cerebral hemispheres | Act as relay stations for motor impulses originating in the cerebral cortex and passing into the brain stem and spinal cord |
| **3.** Diencephalon | Includes masses of gray matter called the thalamus and hypothalamus | The thalamus serves as a relay station for sensory impulses ascending from other parts of the nervous system to the cerebral cortex; the hypothalamus helps maintain homeostasis by regulating various visceral activities and by linking the nervous and endocrine systems |
| **4.** Brain stem | Region that connects the cerebrum to the spinal cord | |
| **a.** Midbrain | Contains masses of gray matter and bundles of nerve fibers that join the spinal cord to higher regions of the brain | Contains reflex centers that move the eyes and head, and maintains posture |
| **b.** Pons | A bulge on the underside of the brain stem that contains masses of gray matter and nerve fibers | Relays nerve impulses to and from the medulla oblongata and cerebrum; helps regulate rate and depth of breathing |
| **c.** Medulla oblongata | An enlarged continuation of the spinal cord that extends from the foramen magnum to the pons, and contains masses of gray matter and nerve fibers | Conducts ascending and descending impulses between the brain and spinal cord; contains cardiac, vasomotor, and respiratory control centers and various nonvital reflex control centers |
| **5.** Cerebellum | A large mass of tissue located below the cerebrum and posterior to the brain stem; includes two lateral hemispheres connected by the vermis | Communicates with other parts of the CNS by nerve tracts; integrates sensory information concerning the position of body parts; and coordinates muscle activities and maintains posture |

CLINICAL APPLICATION 11.4

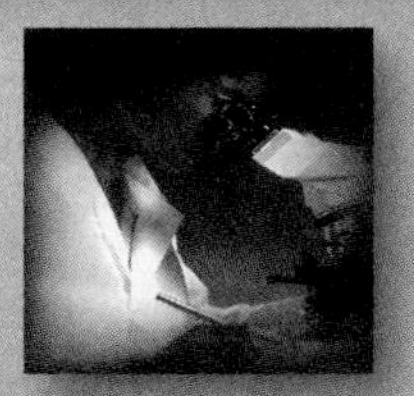

## Brain Waves

*Brain waves* are recordings of fluctuating electrical changes in the brain. To obtain such a recording, electrodes are positioned on the surface of a surgically exposed brain (an electrocorticogram, ECoG) or on the outer surface of the head (an electroencephalogram, EEG). These electrodes detect electrical changes in the extracellular fluid of the brain in response to changes in potential among large groups of neurons. The resulting signals from the electrodes are amplified and recorded. Brain waves originate from the cerebral cortex, but also reflect activities in other parts of the brain that influence the cortex, such as the reticular formation. Because the intensity of electrical changes is directly related to the degree of neuronal activity, brain waves vary markedly in amplitude and frequency between sleep and wakefulness.

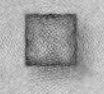

Brain waves are classified as alpha, beta, theta, and delta waves. *Alpha waves* are recorded most easily from the posterior regions of the head and have a frequency of 8–13 cycles per second. They occur when a person is awake but resting, with the eyes closed. These waves disappear during sleep, and if the wakeful person's eyes open, higher frequency beta waves replace the alpha waves.

*Beta waves* have a frequency of more than 13 cycles per second and are usually recorded in the anterior region of the head. They occur when a person is actively engaged in mental activity or is under tension.

*Theta waves* have a frequency of 4–7 cycles per second and occur mainly in the parietal and temporal regions of the cerebrum. They are normal in children, but do not usually occur in adults. However, some adults produce theta waves in early stages of sleep or at times of emotional stress.

*Delta waves* have a frequency below 4 cycles per second and occur during sleep. They originate from the cerebral cortex when it is not being activated by the reticular formation (fig. 11D).

Brain wave patterns can be useful for diagnosing disease conditions, such as distinguishing types of seizure disorders (epilepsy) and to locate brain tumors. Brain waves are also used to determine when *brain death* has occurred. Brain death, characterized by the cessation of neuronal activity, may be verified by an EEG that lacks waves (isoelectric EEG). However, drugs that greatly depress brain functions must be excluded as the cause of the flat EEG pattern before brain death can be confirmed.

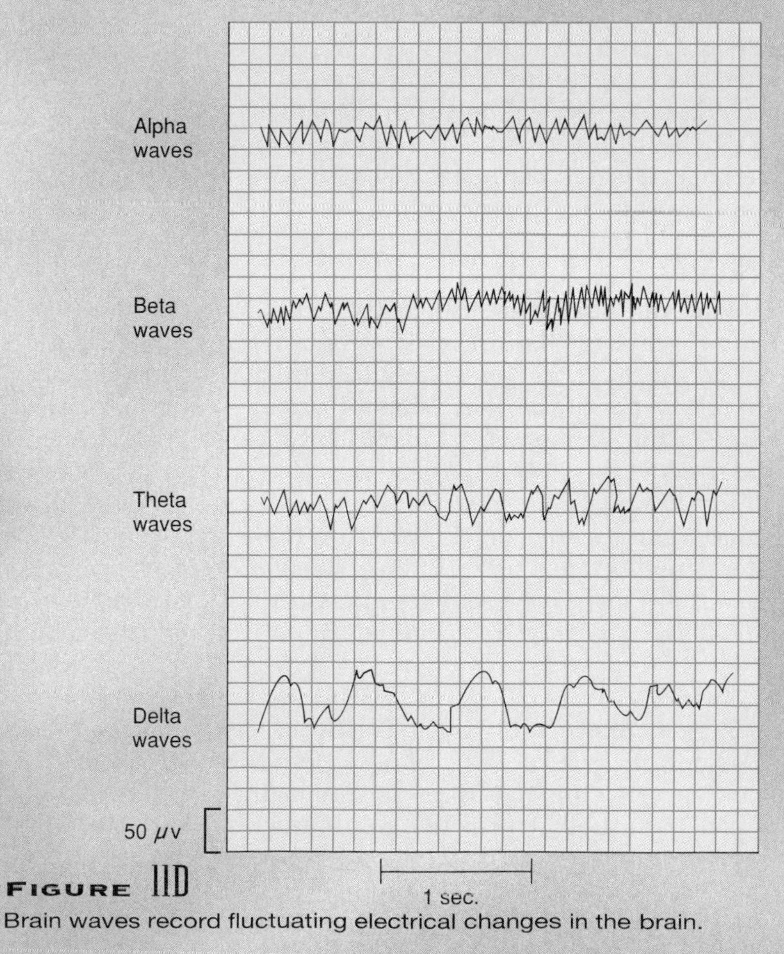


**FIGURE 11D**
Brain waves record fluctuating electrical changes in the brain.

## Peripheral Nervous System

The **peripheral nervous system** (PNS) consists of the nerves that branch from the central nervous system (CNS), connecting it to other body parts. The PNS includes the *cranial nerves* that arise from the brain and the *spinal nerves* that arise from the spinal cord.

The peripheral nervous system can also be subdivided into somatic and autonomic nervous systems. Generally, the **somatic nervous system** consists of the cranial and spinal nerve fibers that connect the CNS to the skin and skeletal muscles, so it oversees conscious activities. The **autonomic nervous system** includes fibers that connect the CNS to viscera such as the heart, stomach, intestines, and various glands. Thus, the autonomic nervous system controls unconscious actions. Chart 11.6 outlines the subdivisions of the nervous system.

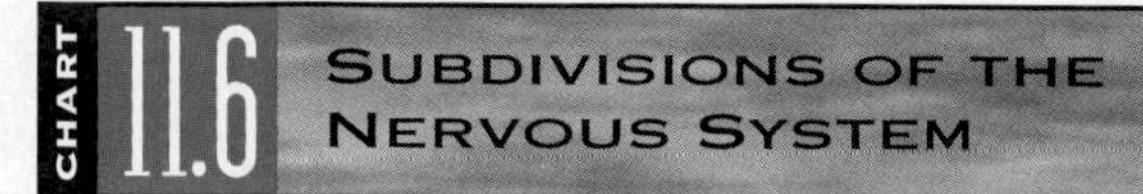


1. Central nervous system (CNS)
   a. Brain
   b. Spinal cord
2. Peripheral nervous system (PNS)
   a. Cranial nerves arising from the brain
      (1) Somatic fibers connecting to the skin and skeletal muscles
      (2) Autonomic fibers connecting to viscera
   b. Spinal nerves arising from the spinal cord
      (1) Somatic fibers connecting to the skin and skeletal muscles
      (2) Autonomic fibers connecting to viscera

### Structure of Peripheral Nerves

As described in chapter 10, a peripheral nerve consists of connective tissue surrounding bundles of nerve fibers. The outermost layer of the connective tissue, called the *epineurium,* is dense and includes many collagenous fibers. Each bundle of nerve fibers (fascicle) is, in turn, enclosed in a sleeve of looser connective tissue called the *perineurium.* A small amount of loose connective tissue called *endoneurium* surrounds individual nerve fibers (figs. 11.20 and 11.21). Blood vessels in the epineurium and perineurium give rise to a network of capillaries in the endoneurium.

### Cranial Nerves

Twelve pairs of **cranial nerves** arise from the underside of the brain. Except for the first pair, which begins within the cerebrum, these nerves originate from the brain stem. They pass from their sites of origin through foramina of the skull and lead to areas of the head, neck, and trunk.

Although most cranial nerves are mixed nerves, some of those associated with special senses, such as smell and vision, contain only sensory fibers. Others that innervate muscles and glands are composed primarily of motor fibers and have only limited sensory functions.

Neuron cell bodies to which the sensory fibers in the cranial nerves attach are located outside the brain and are usually in groups called *ganglia* (sing. *ganglion*). On the other hand, motor neuron cell bodies are typically located within the gray matter of the brain.

Cranial nerves are designated by numbers or name. The numbers indicate the order in which the nerves arise from the brain, from anterior to posterior. The names describe primary functions or the general distribution of their fibers (fig. 11.22).

The first pair of cranial nerves, the **olfactory nerves (I),** are associated with the sense of smell and contain only sensory neurons. These neurons synapse with bipolar neurons, located in the lining of the upper nasal cavity, that serve as *olfactory receptor cells.* Axons from these receptors pass upward through the cribriform plates of the ethmoid bone. The synapses occur in the *olfactory bulbs,* which are extensions of the cerebral cortex, located just beneath the frontal lobes. Sensory impulses travel from the olfactory bulbs along *olfactory tracts* to cerebral centers where they are interpreted. The result of this interpretation is the sensation of smell.

The second pair, the **optic nerves (II),** lead from the eyes to the brain and are associated with vision. The sensory cell bodies of these nerve fibers form ganglion cell layers within the eyes, and their axons pass through the *optic foramina* of the orbits and continue into the visual nerve pathways of the brain (chapter 12).

The third pair, the **oculomotor** (ok″u-lo-mo′tor) **nerves (III),** arise from the midbrain and pass into the orbits of the eyes. One component of each nerve connects to a number of voluntary muscles, including those that raise the eyelids and most of the muscles that move the eyes.

A second portion of each oculomotor nerve is part of the autonomic nervous system, supplying involuntary muscles inside the eyes. These muscles help adjust the amount of light that enters the eyes and help focus the lenses of the eyes.

The fibers of the oculomotor nerves are primarily motor, with a few sensory fibers. These transmit sensory information to the brain concerning the condition of various muscles.

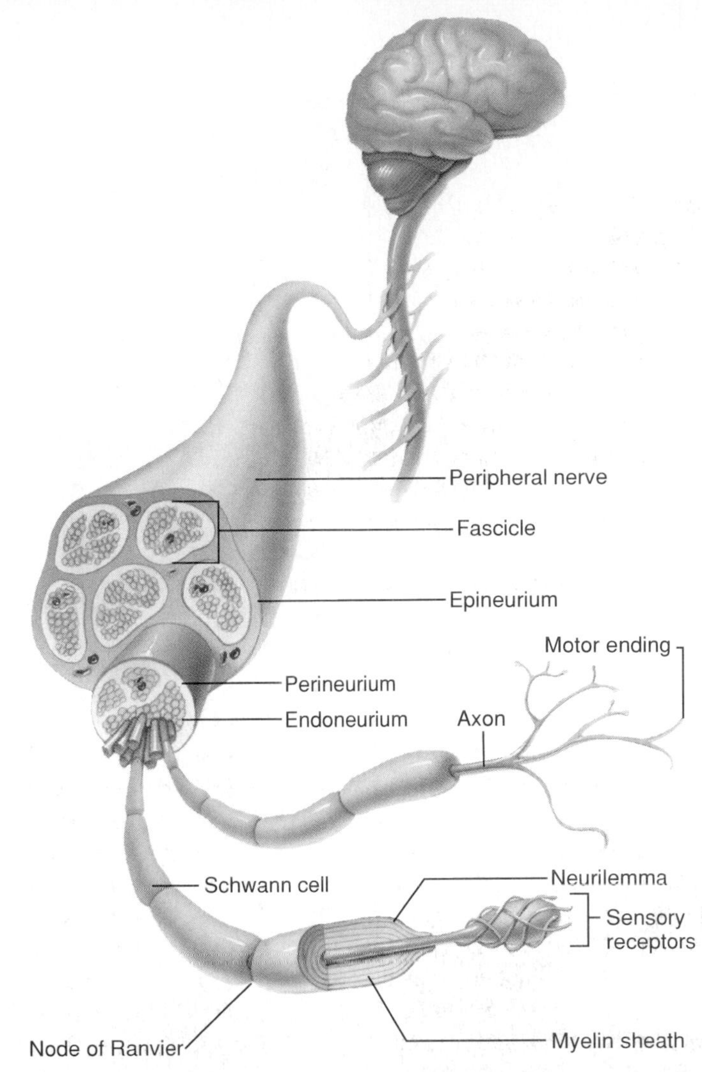


**FIGURE 11.20**
The structure of a peripheral mixed nerve.

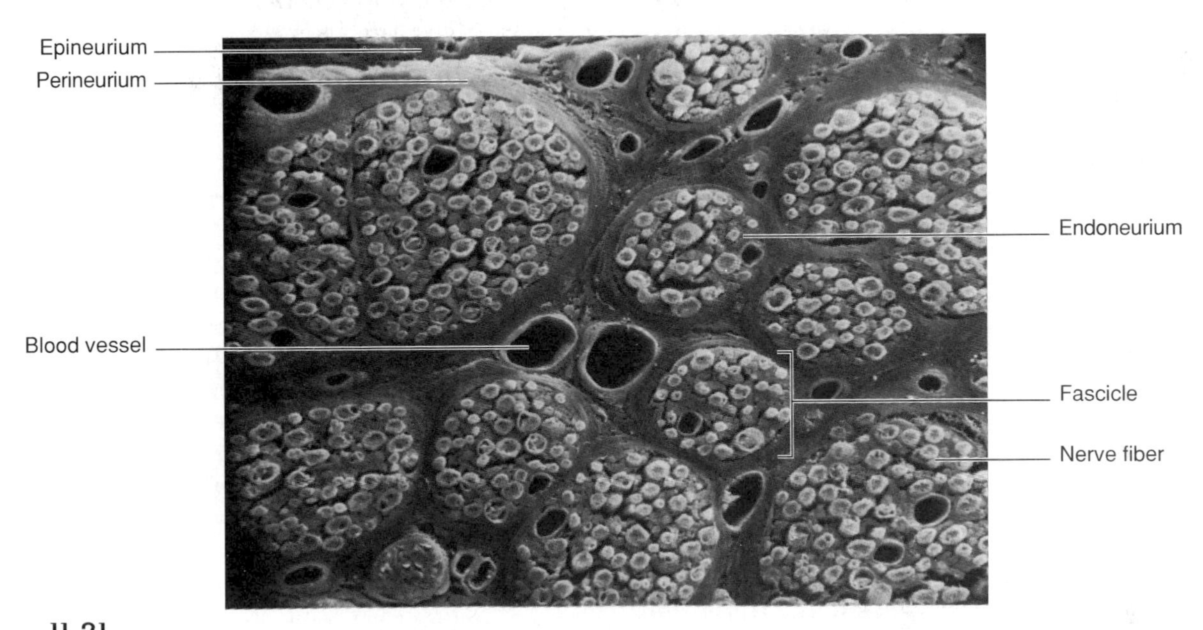


**FIGURE 11.21**
Scanning electron micrograph of a peripheral nerve in cross section (1,000×). Note the bundles or fascicles of nerve fibers.

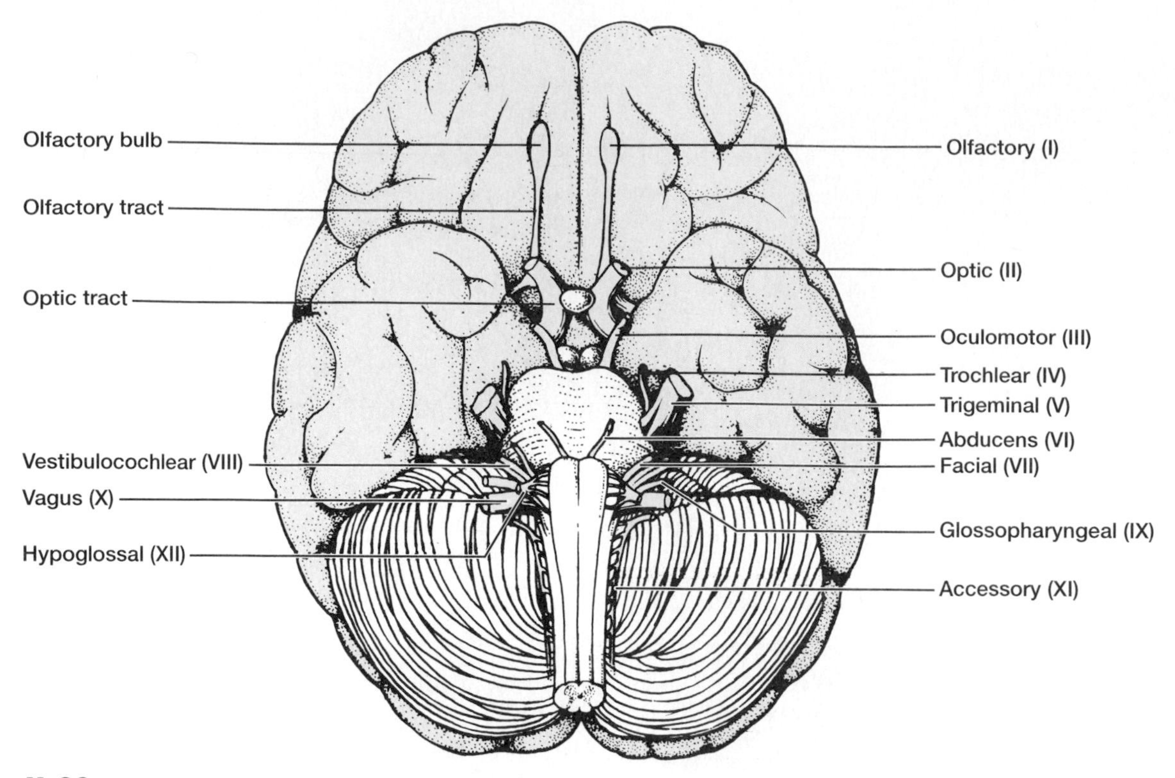


**FIGURE 11.22**
The cranial nerves, except for the first pair, arise from the brain stem. They are identified either by numbers indicating their order, or by their function or the distribution of their fibers.

The fourth pair, the **trochlear** (trok′le-ar) **nerves (IV),** are the smallest cranial nerves. They arise from the midbrain and carry motor impulses to a pair of external eye muscles, the *superior oblique muscles,* which are not supplied by the oculomotor nerves. The trochlear nerves also contain sensory fibers that transmit information about the condition of certain muscles.

The fifth pair, the **trigeminal** (tri-jem′i-nal) **nerves (V),** are the largest of the cranial nerves and arise from the pons. They are mixed nerves, with the sensory portions more extensive than the motor portions. Each sensory component includes three large branches, called the ophthalmic, maxillary, and mandibular divisions (fig. 11.23).

The *ophthalmic division* consists of sensory fibers that bring impulses to the brain from the surface of the eye, the tear gland, and the skin of the anterior scalp, forehead, and upper eyelid. The fibers of the *maxillary division* carry sensory impulses from the upper teeth, upper gum, and upper lip, as well as from the mucous lining of the palate and facial skin. The *mandibular division* includes both motor and sensory fibers. The sensory branches transmit impulses from the scalp behind the ears, the skin of the jaw, the lower teeth, the lower gum, and the lower lip. The motor branches supply the muscles of mastication and certain muscles in the floor of the mouth.

The sixth pair, the **abducens** (ab-du′senz) **nerves (VI),** are quite small, and originate from the pons near the medulla oblongata. They enter the orbits of the eyes and supply motor impulses to a pair of external eye muscles, the *lateral rectus muscles.* The sensory fibers of these nerves provide the brain with information concerning the condition of certain muscles.

A disorder of the trigeminal nerve called *trigeminal neuralgia* (tic douloureux) causes severe recurring pain in the face and forehead on the affected side. If drugs cannot control the pain, surgery may be used to sever the sensory portion of the nerve. Although this procedure may relieve the pain, the patient loses sensations in other body regions that the sensory branch supplies. Consequently, after such surgery, care must be taken when eating or drinking hot foods or liquids, and the mouth inspected daily for the presence of food particles or damage to the cheeks from biting.

The seventh pair, the **facial** (fa′shal) **nerves (VII),** arise from the lower part of the pons and emerge on the sides of the face. Their sensory branches are associated with taste receptors on the anterior two-thirds

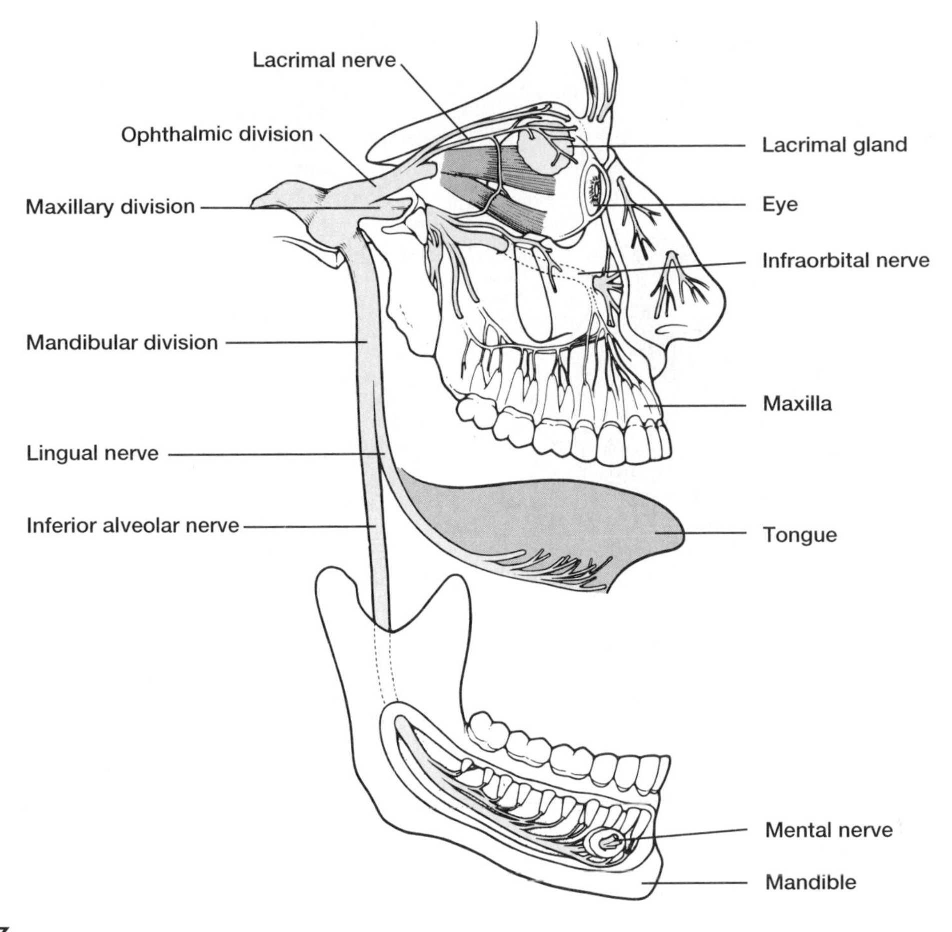


**FIGURE 11.23**

Each trigeminal nerve has three large branches that supply various regions of the head and face: the ophthalmic division, the maxillary division, and the mandibular division.

of the tongue, and some of their motor fibers transmit impulses to muscles of facial expression. Still other motor fibers of these nerves function in the autonomic nervous system by stimulating secretions from tear glands and certain salivary glands (submandibular and sublingual glands) (fig. 11.24).

The eighth pair, the **vestibulocochlear** (ves-tib″u-lo-kok′le-ar) **nerves (VIII,** acoustic, or auditory, nerves), are sensory nerves that arise from the medulla oblongata. Each of these nerves has two distinct parts—a vestibular branch and a cochlear branch.

The neuron cell bodies of the *vestibular branch* fibers are located in ganglia near the vestibule and semicircular canals of the inner ear. These structures contain receptors that sense changes in the position of the head, and in response initiate and send impulses to the cerebellum, where they are used in reflexes that maintain equilibrium.

The neuron cell bodies of the *cochlear branch* fibers are located in a ganglion of the cochlea, a part of the inner ear that houses the hearing receptors. Impulses from this branch pass through the medulla oblongata and midbrain on their way to the temporal lobe, where they are interpreted.

The ninth pair, the **glossopharyngeal** (glos″o-fah-rin′je-al) **nerves (IX),** are associated with the tongue and pharynx. These nerves arise from the medulla oblongata, and, although they are mixed nerves, their predominant fibers are sensory. These fibers carry impulses from the lining of the pharynx, tonsils, and posterior third of the tongue to the brain. Fibers in the motor component of the glossopharyngeal nerves innervate constrictor muscles in the wall of the pharynx that function in swallowing.

The tenth pair, the **vagus** (va′gus) **nerves (X),** originate in the medulla oblongata and extend downward through the neck into the chest and abdomen. These nerves are mixed, containing both somatic and autonomic branches, with the autonomic fibers predominant.

Among the somatic components of the vagus nerves are motor fibers that carry impulses to muscles of the larynx. These fibers are associated with speech and swallowing reflexes that employ muscles in the soft palate and pharynx. Vagal sensory fibers carry impulses from the linings of the pharynx, larynx, and esophagus, and from the viscera of the thorax and abdomen to the brain.

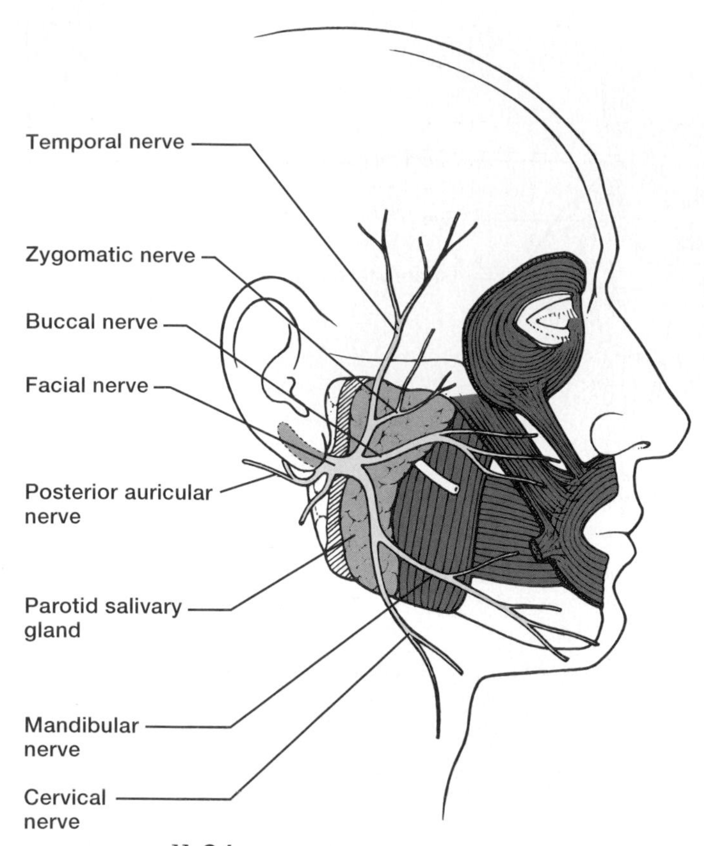


**FIGURE 11.24**
The facial nerves are associated with taste receptors on the tongue and with muscles of facial expression.

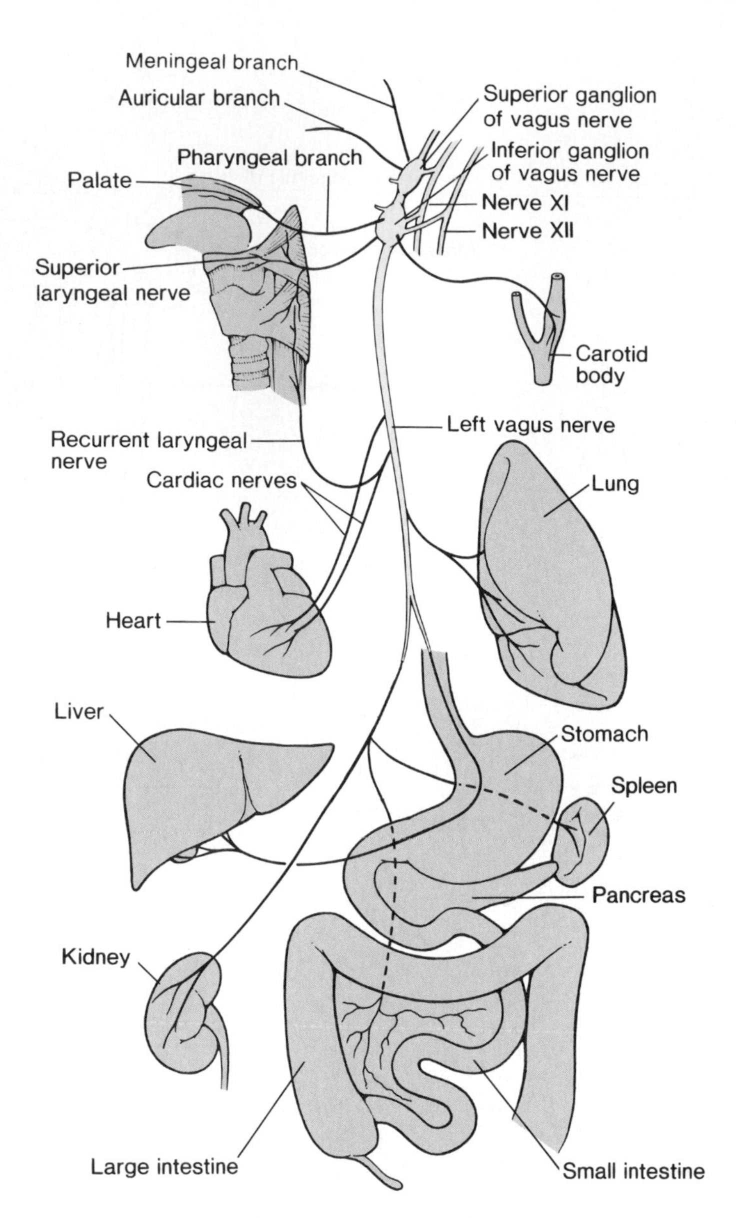


**FIGURE 11.25**
The vagus nerves extend from the medulla oblongata downward into the chest and abdomen to supply many organs.

Autonomic motor fibers of the vagus nerves supply the heart and many smooth muscles and glands in the viscera of the thorax and abdomen (fig. 11.25).

The eleventh pair, the **accessory** (ak-ses′o-re) **nerves (XI,** spinal accessory), originate in the medulla oblongata and the spinal cord; thus, they have both cranial and spinal branches.

Each *cranial branch* of an accessory nerve joins a vagus nerve and carries impulses to muscles of the soft palate, pharynx, and larynx. The *spinal branch* descends into the neck and supplies motor fibers to the trapezius and sternocleidomastoid muscles.

The twelfth pair, the **hypoglossal** (hi″po-glos′al) **nerves (XII),** arise from the medulla oblongata and pass into the tongue. They consist primarily of motor fibers that carry impulses to muscles that move the tongue in speaking, chewing, and swallowing.

Chart 11.7 summarizes the functions of the cranial nerves.

1 Define *peripheral nervous system.*

2 Distinguish between somatic and autonomic nerve fibers.

3 Describe the structure of a peripheral nerve.

4 Name the cranial nerves, and list the major functions of each.

## Spinal Nerves

Thirty-one pairs of spinal nerves originate from the spinal cord. They are mixed nerves, and they provide two-way communication between the spinal cord and parts of the upper and lower limbs, neck, and trunk.

## CHART 11.7 Functions of Cranial Nerves

| Nerve | | Type | Function |
|---|---|---|---|
| I | Olfactory | Sensory | Sensory fibers transmit impulses associated with the sense of smell. |
| II | Optic | Sensory | Sensory fibers transmit impulses associated with the sense of vision. |
| III | Oculomotor | Primarily motor | Motor fibers transmit impulses to muscles that raise the eyelids, move the eyes, adjust the amount of light entering the eyes, and focus the lenses. |
| | | | Some sensory fibers transmit impulses associated with the condition of muscles. |
| IV | Trochlear | Primarily motor | Motor fibers transmit impulses to muscles that move the eyes. |
| | | | Some sensory fibers transmit impulses associated with the condition of muscles. |
| V | Trigeminal | Mixed | |
| | Ophthalmic division | | Sensory fibers transmit impulses from the surface of the eyes, tear glands, scalp, forehead, and upper eyelids. |
| | Maxillary division | | Sensory fibers transmit impulses from the upper teeth, upper gum, upper lip, lining of the palate, and skin of the face. |
| | Mandibular division | | Sensory fibers transmit impulses from the scalp, skin of the jaw, lower teeth, lower gum, and lower lip. |
| | | | Motor fibers transmit impulses to muscles of mastication and to muscles in the floor of the mouth. |
| VI | Abducens | Primarily motor | Motor fibers transmit impulses to muscles that move the eyes. |
| | | | Some sensory fibers transmit impulses associated with the condition of muscles. |
| VII | Facial | Mixed | Sensory fibers transmit impulses associated with taste receptors of the anterior tongue. |
| | | | Motor fibers transmit impulses to muscles of facial expression, tear glands, and salivary glands. |
| VIII | Vestibulocochlear | Sensory | |
| | Vestibular branch | | Sensory fibers transmit impulses associated with the sense of equilibrium. |
| | Cochlear branch | | Sensory fibers transmit impulses associated with the sense of hearing. |
| IX | Glossopharyngeal | Mixed | Sensory fibers transmit impulses from the pharynx, tonsils, posterior tongue, and carotid arteries. |
| | | | Motor fibers transmit impulses to salivary glands and to muscles of the pharynx used in swallowing. |
| X | Vagus | Mixed | Somatic motor fibers transmit impulses to muscles associated with speech and swallowing; autonomic motor fibers transmit impulses to the viscera of the thorax and abdomen. |
| | | | Sensory fibers transmit impulses from the pharynx, larynx, esophagus, and viscera of the thorax and abdomen. |
| XI | Accessory | Primarily motor | |
| | Cranial branch | | Motor fibers transmit impulses to muscles of the soft palate, pharynx, and larynx. |
| | Spinal branch | | Motor fibers transmit impulses to muscles of the neck and back. |
| XII | Hypoglossal | Primarily motor | Motor fibers transmit impulses to muscles that move the tongue. |

Spinal nerves are not named individually, but are grouped by the level from which they arise, with each nerve numbered in sequence (fig. 11.26). Thus, there are eight pairs of *cervical nerves* (numbered C1 to C8), twelve pairs of *thoracic nerves* (numbered T1 to T12), five pairs of *lumbar nerves* (numbered L1 to L5), five pairs of *sacral nerves* (numbered S1 to S5), and one pair of *coccygeal nerves* (Co).

The nerves arising from the superior part of the spinal cord pass outward almost horizontally, while those from the inferior portions of the spinal cord descend at sharp angles. This arrangement is a consequence of growth. In early life, the spinal cord extends the entire length of the vertebral column, but with age, the column grows more rapidly than the cord. Thus, the adult spinal cord ends at the level between the first and second lumbar vertebrae, so the lumbar, sacral, and coccygeal nerves descend to their exits beyond the end of the cord. These descending nerves form a structure called the *cauda equina* (horse's tail) (see fig. 11.26).

Each spinal nerve emerges from the cord by two short branches, or *roots,* which lie within the vertebral column. The **dorsal root** (posterior, or sensory, root) can be identified by an enlargement called the *dorsal root ganglion.* This ganglion contains the cell bodies of the sensory neurons whose dendrites conduct impulses inward from the peripheral body parts. The axons of these neurons extend through the dorsal root and into the spinal cord, where they form synapses with dendrites of other neurons.

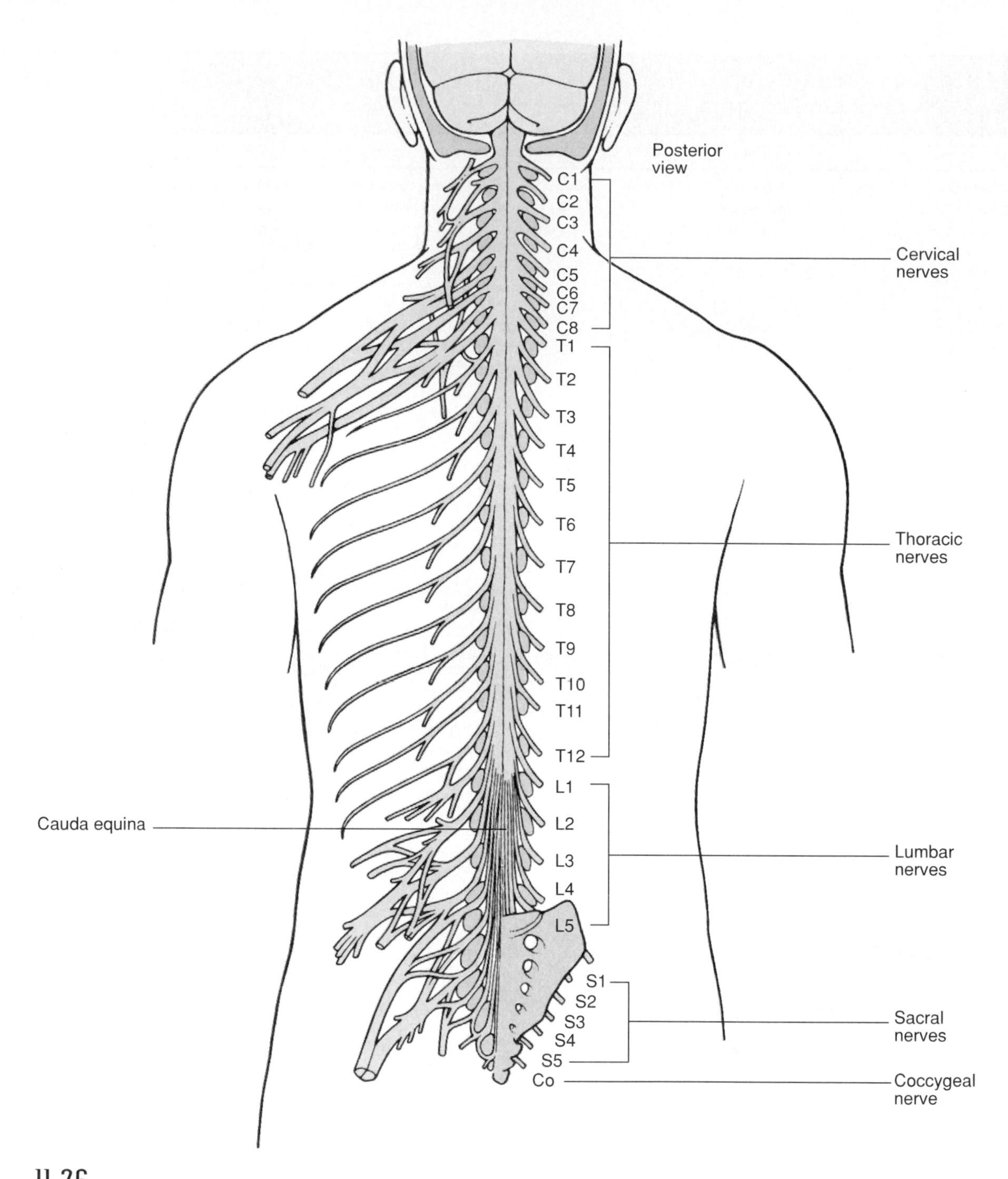


**FIGURE 11.26**

The 31 pairs of spinal nerves are grouped according to the level from which they arise, and are numbered in sequence.

> An area of skin that the sensory nerve fibers of a particular spinal nerve innervate is called a *dermatome*. Dermatomes are highly organized, but they vary considerably in size and shape, as figure 11.27 indicates. A map of the dermatomes is often useful in localizing the sites of injuries to dorsal roots or the spinal cord.

The **ventral root** (anterior, or motor, root) of each spinal nerve consists of axons from the motor neurons whose cell bodies are located within the gray matter of the cord.

A ventral root and a dorsal root unite to form a spinal nerve, which extends outward from the vertebral canal through an *intervertebral foramen*. Just beyond its foramen, each spinal nerve branches.

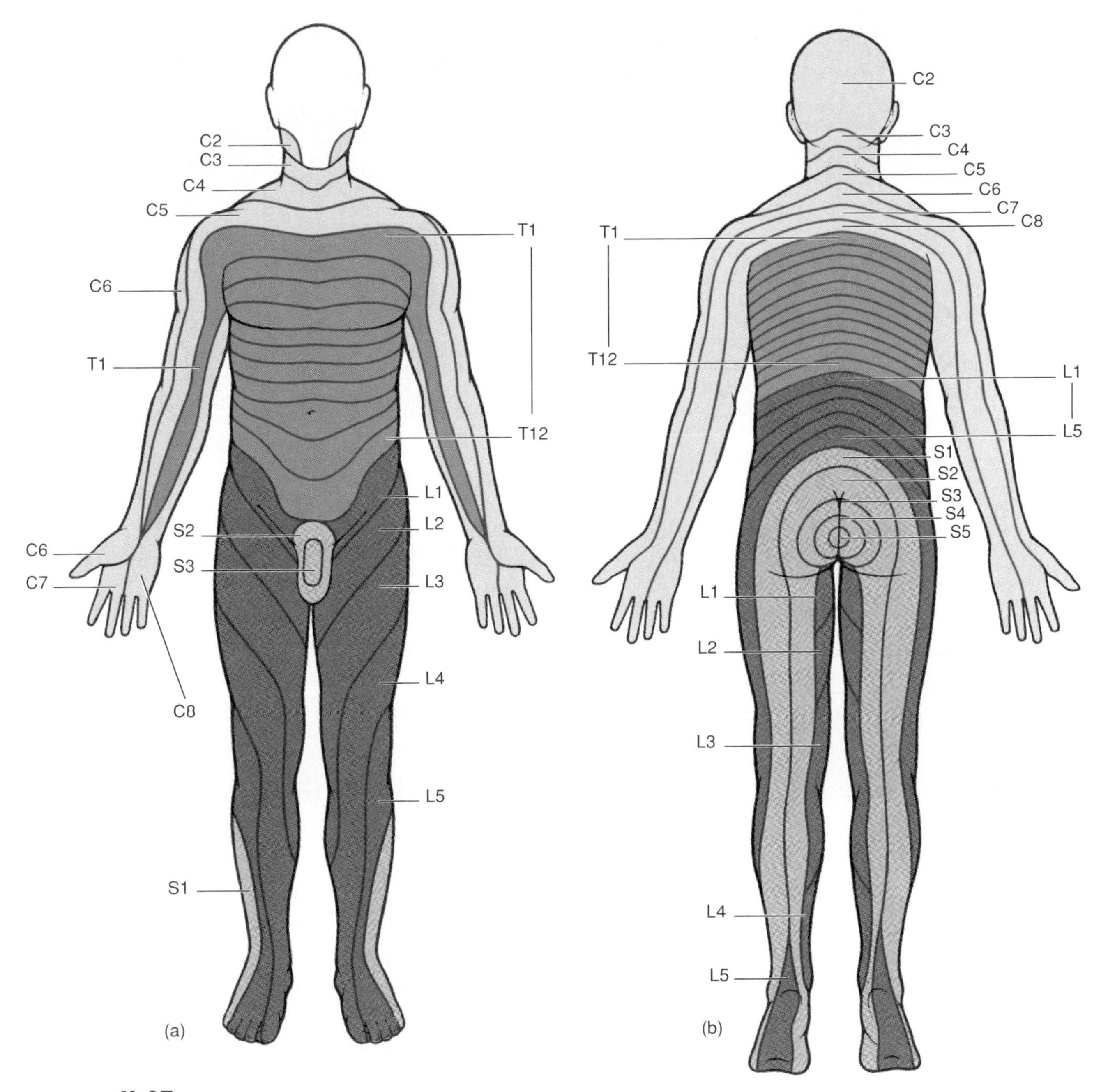


**FIGURE 11.27**

(*a*) Dermatomes on the anterior body surface and (*b*) on the posterior surface. Note that spinal nerve C1 does not supply any skin area.

One of these parts, the small *meningeal branch,* reenters the vertebral canal through the intervertebral foramen and supplies the meninges and blood vessels of the cord, as well as the intervertebral ligaments and the vertebrae.

As figure 11.28 shows, a *posterior branch* (posterior ramus) of each spinal nerve turns posteriorly and innervates the muscles and skin of the back. The main portion of the nerve, the *anterior branch* (anterior ramus), continues forward to supply muscles and skin on the front and sides of the trunk and limbs.

The spinal nerves in the thoracic and lumbar regions have a fourth, or *visceral branch,* which is part of the autonomic nervous system.

Except in the thoracic region, anterior branches of the spinal nerves combine to form complex networks called **plexuses** instead of continuing directly to the peripheral body parts. In a plexus, the fibers of various spinal nerves are sorted and recombined, so that fibers associated with a particular peripheral body part reach it in the same nerve, even though the fibers originate from different spinal nerves (fig. 11.29).

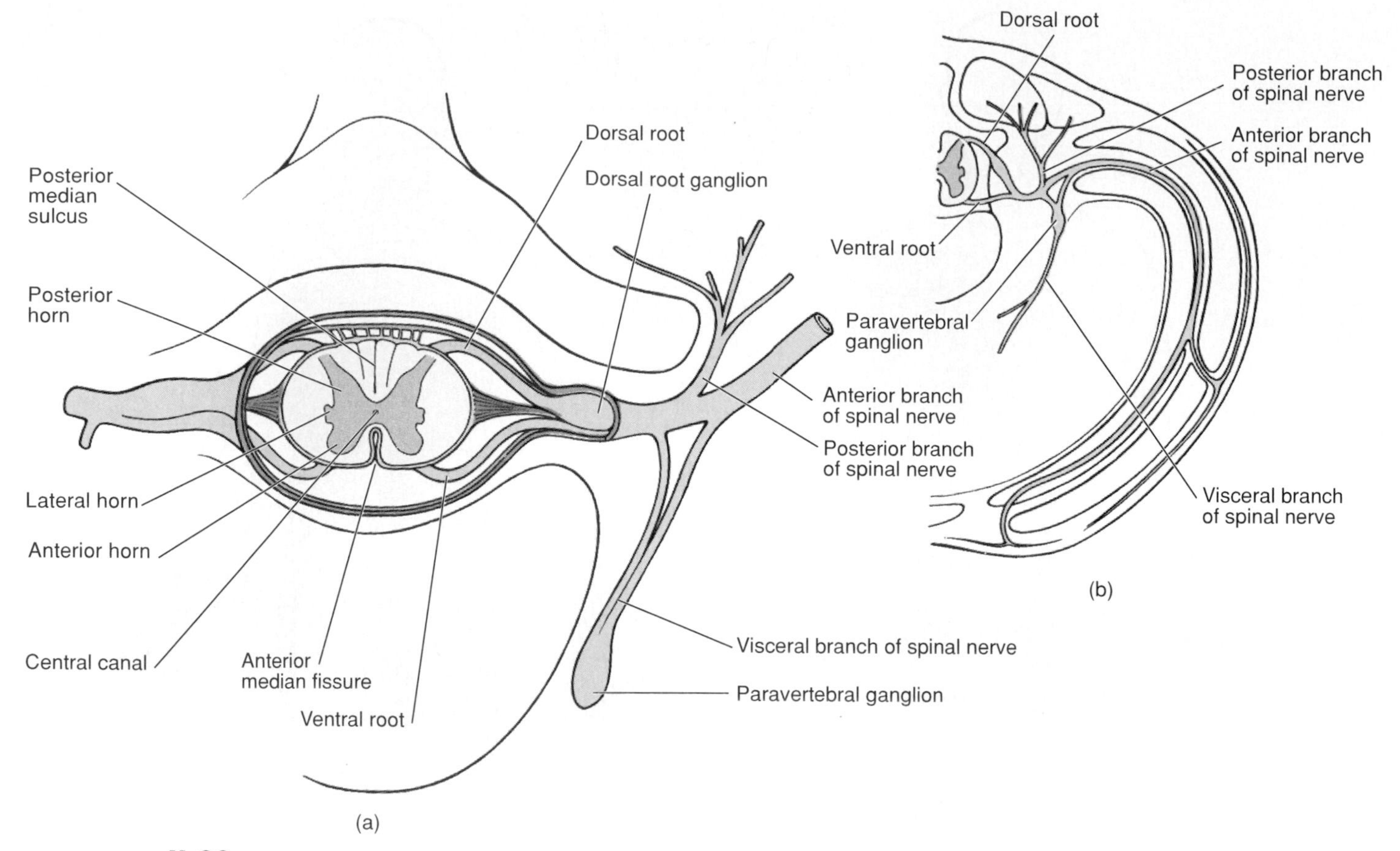


**FIGURE 11.28**

(*a*) Each spinal nerve has a posterior and an anterior branch. (*b*) The thoracic and lumbar spinal nerves also have a visceral branch.

## Cervical Plexuses

The **cervical plexuses** lie deep in the neck on either side. They are formed by the anterior branches of the first four cervical nerves. Fibers from these plexuses supply the muscles and skin of the neck. In addition, fibers from the third, fourth, and fifth cervical nerves pass into the right and left **phrenic** (fren′ik) **nerves,** which conduct motor impulses to the muscle fibers of the diaphragm.

## Brachial Plexuses

The anterior branches of the lower four cervical nerves and the first thoracic nerve give rise to **brachial plexuses.** These networks of nerve fibers are located deep within the shoulders between the neck and the axillae (armpits). The major branches emerging from the brachial plexuses include the following (fig. 11.30):

1. *Musculocutaneous nerves* supply muscles of the arms on the anterior sides, and the skin of the forearms.
2. *Ulnar nerves* supply muscles of the forearms and hands, and the skin of the hands.
3. *Median nerves* supply muscles of the forearms, and muscles and skin of the hands.
4. *Radial nerves* supply muscles of the arms on the posterior sides, and the skin of the forearms and hands.
5. *Axillary nerves* supply muscles and skin of the superior, lateral, and posterior regions of the arm.

Other nerves associated with the brachial plexus that innervate various skeletal muscles include:

1. The *lateral* and *medial pectoral nerves* supply the pectoralis major and pectoralis minor muscles.
2. The *dorsal scapular nerve* supplies the rhomboideus major and levator scapulae muscles.

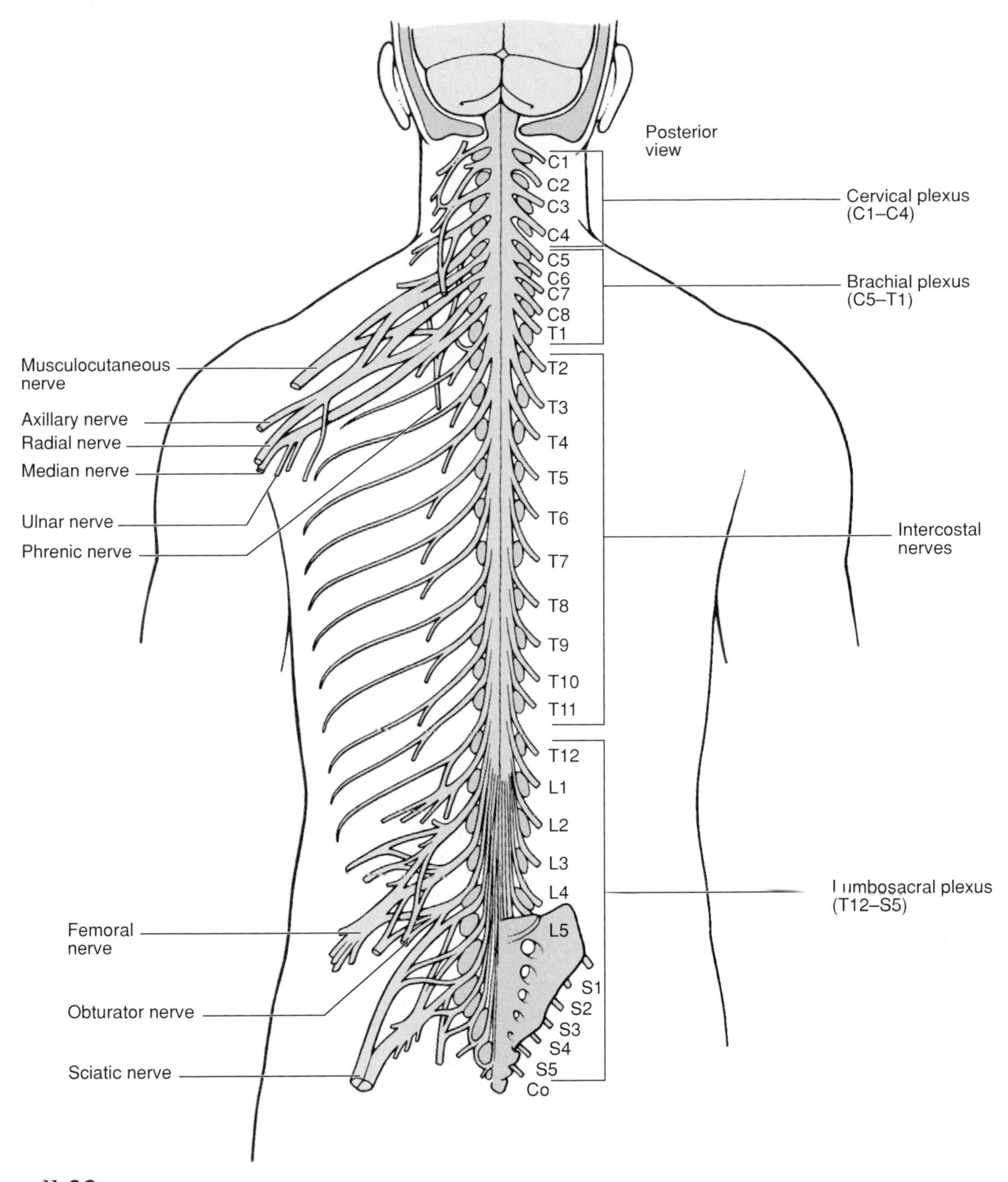


**FIGURE 11.29**
The anterior branches of the spinal nerves in the thoracic region give rise to intercostal nerves. Those in other regions combine to form complex networks called plexuses.

3. The *lower subscapular nerve* supplies the subscapularis and teres major muscles.
4. The *thoracodorsal nerve* supplies the latissimus dorsi muscle.
5. The *suprascapular nerve* supplies the supraspinatus and infraspinatus muscles.

## Lumbosacral Plexuses

The **lumbosacral** (lum″bo-sa′kral) **plexuses** are formed by the last thoracic nerve and the lumbar, sacral, and coccygeal nerves. These networks of nerve fibers extend from the lumbar region of the back into the pelvic cavity, giving rise to a number of motor and sensory fibers associated with the lower abdominal

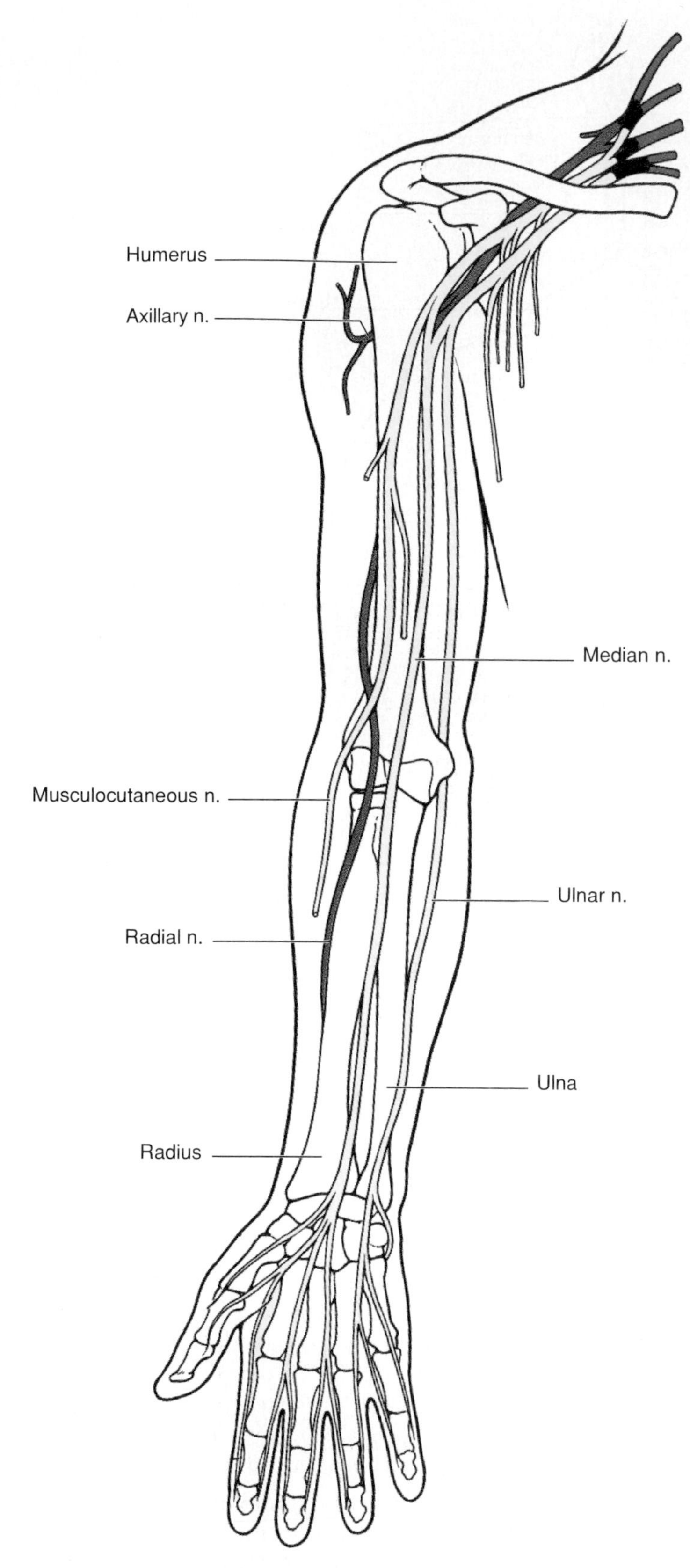


**FIGURE 11.30**
Nerves of the brachial plexus.

wall, external genitalia, buttocks, thighs, legs, and feet. The major branches of these plexuses include the following (fig. 11.31):

1. The *obturator nerves* supply the adductor muscles of the thighs.
2. The *femoral nerves* divide into many branches, supplying motor impulses to muscles of the thighs and legs, and receiving sensory impulses from the skin of the thighs and legs.
3. The *sciatic nerves* are the largest and longest nerves in the body. They pass downward into the buttocks and descend into the thighs, where they divide into *tibial* and *common peroneal nerves.* The many branches of these nerves supply muscles and skin in the thighs, legs, and feet.

Other nerves associated with the lumbosacral plexus that innervate various skeletal muscles include:

1. The *pudendal nerve* supplies the muscles of the perineum.
2. The *inferior* and *superior gluteal nerves* supply the gluteal muscles and the tensor fasciae latae muscle.

The anterior branches of the thoracic spinal nerves do not enter a plexus. Instead, they travel into spaces between the ribs and become **intercostal** (in″ter-kos′tal) **nerves.** These nerves supply motor impulses to the intercostal muscles and the upper abdominal wall muscles. They also receive sensory impulses from the skin of the thorax and abdomen.

Clinical Application 11.5 discusses injuries to the spinal nerves.

1. How are spinal nerves grouped?
2. Describe how a spinal nerve joins the spinal cord.
3. Name and locate the major nerve plexuses.

---

## AUTONOMIC NERVOUS SYSTEM

The autonomic nervous system is the part of the peripheral nervous system that functions independently (autonomously) and continuously, without conscious effort. This system controls visceral activities by regulating the actions of smooth muscles, cardiac muscles, and various glands. It oversees heart rate, blood pressure, breathing rate, body temperature, and other visceral activities that aid in maintaining homeostasis.

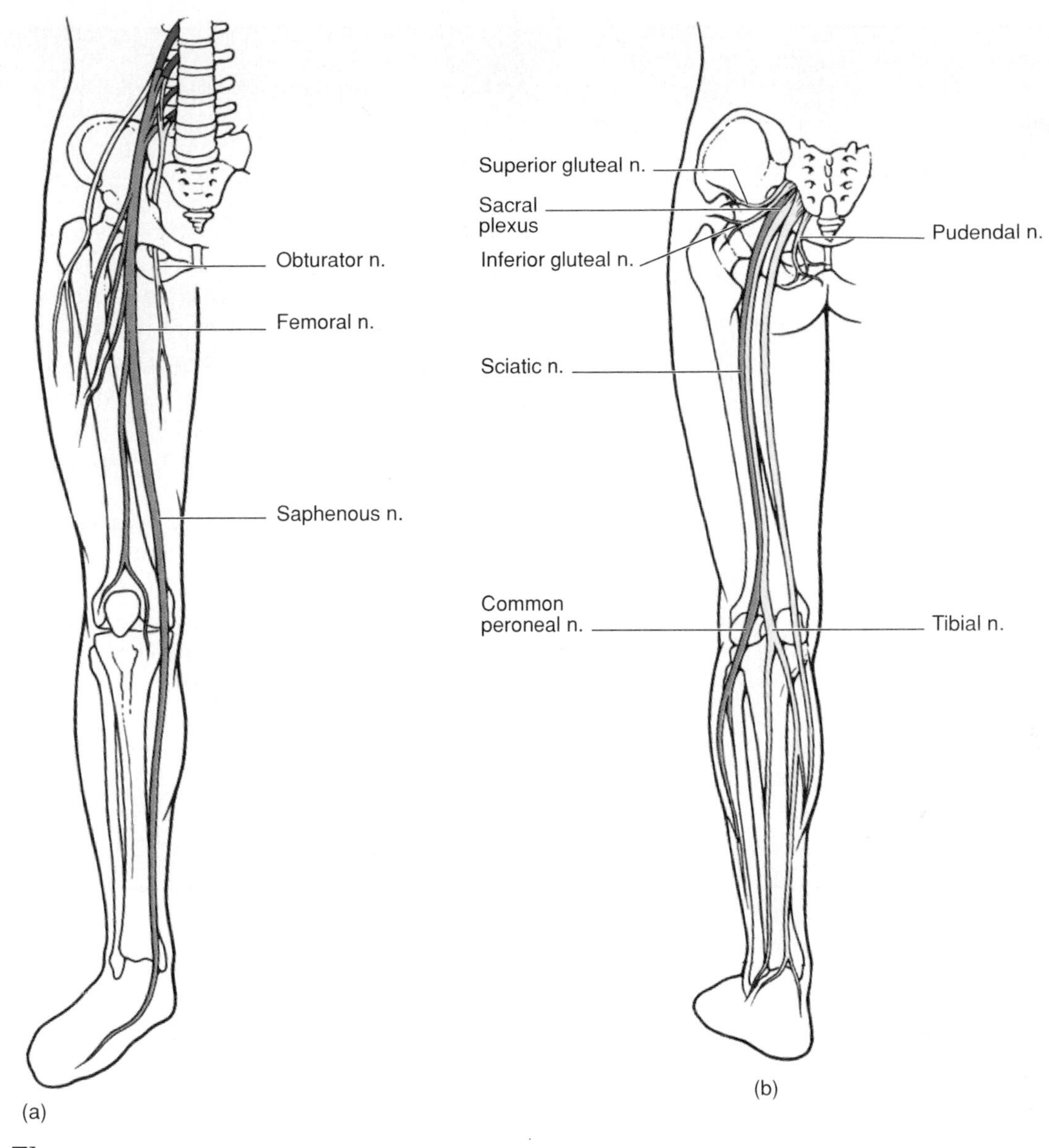


**FIGURE 11.31**
Nerves of the lumbosacral plexus. (*a*) Anterior view; (*b*) posterior view.

Portions of the autonomic nervous system also respond during times of emotional stress and prepare the body to meet the demands of strenuous physical activity.

## General Characteristics

Reflexes in which sensory signals originate from receptors within the viscera and the skin regulate autonomic activities. Afferent nerve fibers transmit these signals to nerve centers within the brain or spinal cord. In response, motor impulses travel out from these centers on efferent nerve fibers within cranial and spinal nerves.

Typically, these efferent fibers lead to ganglia outside the central nervous system. The impulses they carry are integrated within the ganglia and are relayed to various organs (muscles or glands) that respond by contracting, secreting, or being inhibited. The integrative function of the ganglia provides the autonomic nervous system with some degree of independence from the brain and spinal cord, and the visceral efferent nerve fibers associated with these ganglia comprise the autonomic nervous system.

The autonomic nervous system includes two divisions, called the **sympathetic** and **parasympathetic**

## Clinical Application 11.5

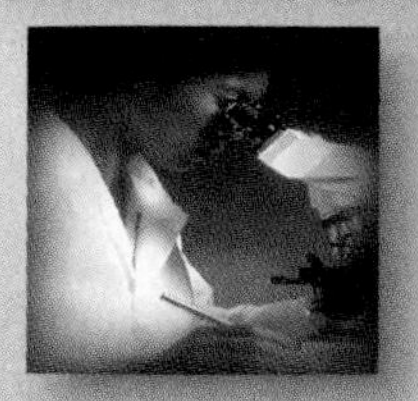

### Spinal Nerve Injuries

Birth injuries, dislocations, vertebral fractures, stabs, gunshot wounds, and pressure from tumors can all injure spinal nerves. Suddenly bending the neck, called *whiplash*, can compress the nerves of the cervical plexuses, causing persistent headache and pain in the neck and skin, which the cervical nerves supply. If a broken or dislocated vertebra severs or damages the phrenic nerves associated with the cervical plexuses, partial or complete paralysis of the diaphragm may result.

Intermittent or constant pain in the neck, shoulder, or upper limb may result from prolonged abduction of the upper limb, as in painting or typing. This is due to excessive pressure on the brachial plexus. This condition, called *thoracic outlet syndrome*, may also result from a congenital skeletal malformation that compresses the plexus during upper limb and shoulder movements.

Degenerative changes may compress an intervertebral disk in the lumbar region, producing a condition called *sciatica*, causing pain in the lower back and gluteal region that can radiate to the thigh, calf, ankle, and foot. Sciatica is most common in middle-aged people, particularly distance runners. It usually compresses spinal nerve roots between L2 and S1, some of which contain fibers of the sciatic nerve. Rest, drugs, or surgery are used to treat sciatica.

In *carpal tunnel syndrome*, repeated hand movements, such as typing, inflame the tendons that pass through the carpal tunnel, which is a space between bones in the wrist. The swollen tendons compress the median nerve in the wrist, causing pain to shoot up the upper limb. Surgery or avoiding repetitive hand movements can relieve symptoms.

**divisions,** that act in conjunction with one another in various situations. For example, some organs have nerve fibers from each of the divisions. Impulses on one set of fibers may activate an organ, while impulses on the other set inhibit it. Thus, the divisions may act antagonistically, regulating the actions of some organs by alternately activating or inhibiting them.

The functions of the autonomic divisions are varied; that is, each activates some organs and inhibits others. However, the divisions have important functional differences. The sympathetic division is concerned primarily with preparing the body for energy-expending, stressful, or emergency situations. Conversely, the parasympathetic division is most active under ordinary, restful conditions. It also counterbalances the effects of the sympathetic division, and restores the body to a resting state following a stressful experience. For example, during an emergency, the sympathetic division increases heart and breathing rates; following the emergency, the parasympathetic division decreases these activities.

## Autonomic Nerve Fibers

All of the nerve fibers of the autonomic nervous system are efferent, or motor, fibers. In the motor pathways of the somatic nervous system, a single neuron typically links the brain or spinal cord and a skeletal muscle. In the autonomic system, motor pathways include two neurons, as figure 11.32 shows. The cell body of one neuron is located in the brain or spinal cord. Its axon, the **preganglionic fiber,** leaves the CNS and synapses with one or more nerve fibers whose cell bodies are housed within an autonomic ganglion. The axon of such a second neuron is called a **postganglionic fiber,** and it extends to a visceral effector.

### Sympathetic Division

Within the sympathetic division (thoracolumbar division), the preganglionic fibers originate from neurons within the lateral horn of the spinal cord. These neurons are found in all of the thoracic segments and in the upper two or three lumbar segments of the cord. Their axons exit through the ventral roots of spinal nerves along with various somatic motor fibers.

After traveling a short distance, preganglionic fibers leave the spinal nerves through branches called *white rami* (sing. *ramus*) and enter sympathetic ganglia. Two groups of such ganglia, called **paravertebral ganglia,** are located in chains along the sides of the vertebral column. These ganglia, together with the fibers that connect them, comprise the **sympathetic trunks** (fig. 11.33).

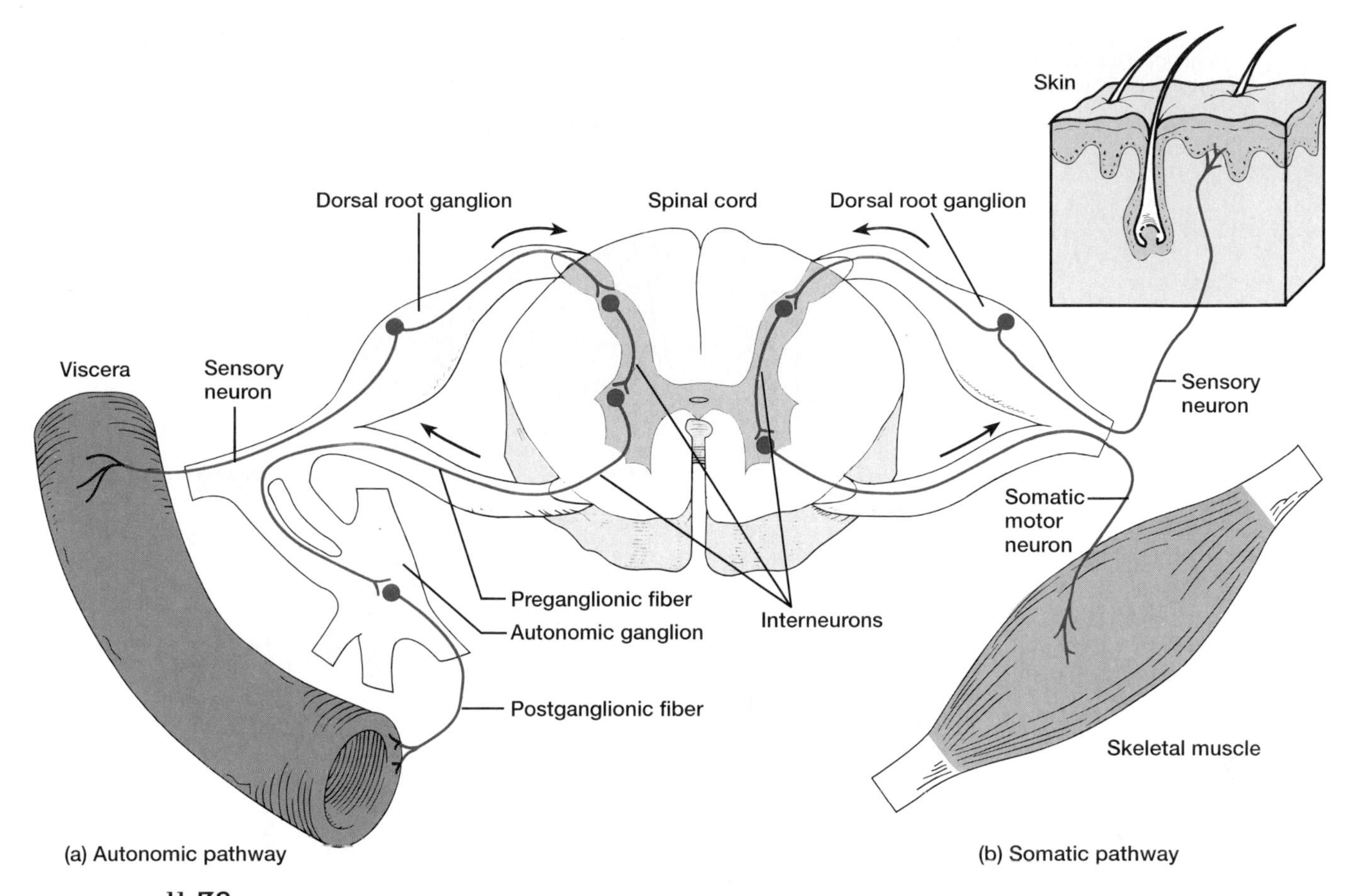


**FIGURE 11.32**

(*a*) Autonomic pathways include two neurons between the central nervous system and an effector. (*b*) Somatic pathways usually have a single neuron between the central nervous system and an effector.

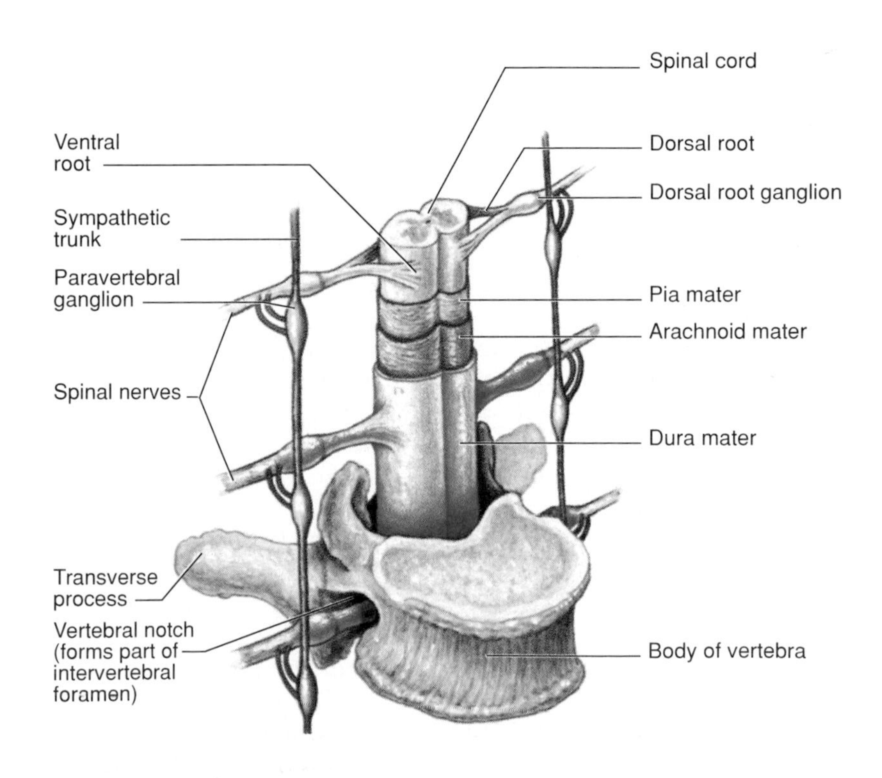


**FIGURE 11.33**

A chain of paravertebral ganglia extends along each side of the vertebral column.

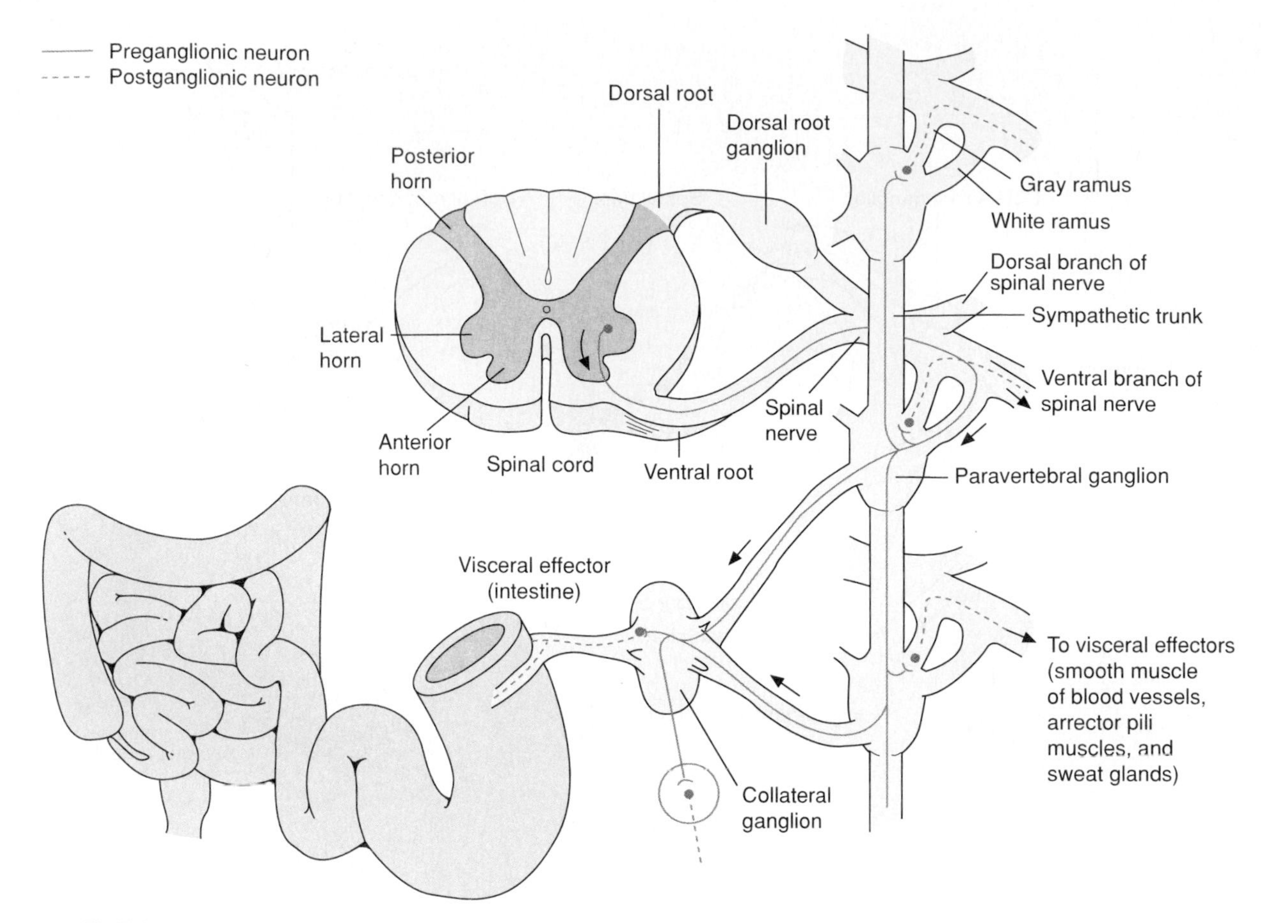


**FIGURE 11.34**
Sympathetic fibers leave the spinal cord in the ventral roots of spinal nerves, enter sympathetic ganglia, and synapse with other neurons that extend to visceral effectors.

The paravertebral ganglia lie just beneath the parietal pleura in the thorax and beneath the parietal peritoneum in the abdomen (see chapter 1). Although these ganglia are located some distance from the viscera they help control, other sympathetic ganglia are nearer to the viscera. The *collateral ganglia,* for example, are found within the abdomen, closely associated with certain large blood vessels (fig. 11.34).

Some of the preganglionic fibers that enter paravertebral ganglia synapse with neurons within these ganglia. Other fibers extend through the ganglia and pass up or down the sympathetic trunk and synapse with neurons in ganglia at higher or lower levels within the chain. Still other fibers pass through to collateral ganglia before they synapse. Typically, a preganglionic axon will synapse with several other neurons within a sympathetic ganglion.

The axons of the second neurons in sympathetic pathways, the postganglionic fibers, extend out from the sympathetic ganglia to visceral effectors. Those leaving paravertebral ganglia usually pass through branches called **gray rami** and return to a spinal nerve before proceeding to an effector (see fig. 11.34). These branches appear gray because the postganglionic axons generally are unmyelinated, whereas the preganglionic axons in the white rami are nearly all myelinated. An important exception to the usual arrangement of sympathetic fibers occurs in a set of preganglionic fibers that pass through the sympathetic ganglia and extend out to the medulla of each adrenal gland. These fibers terminate within the glands on special hormone-secreting cells that release **norepinephrine** (20%) and **epinephrine** (80%) when they are stimulated. The functions of the adrenal medulla and its hormones are discussed in chapter 13. The sympathetic division is shown in figure 11.35.

## Parasympathetic Division

The preganglionic fibers of the parasympathetic division (craniosacral division) arise from neurons in the midbrain, pons, and medulla oblongata of the brain stem and from the sacral region of the spinal cord (fig. 11.36). From there, they lead outward on cranial or sacral nerves to ganglia located near or within various organs (terminal ganglia). The relatively short postganglionic fibers continue from the ganglia to specific

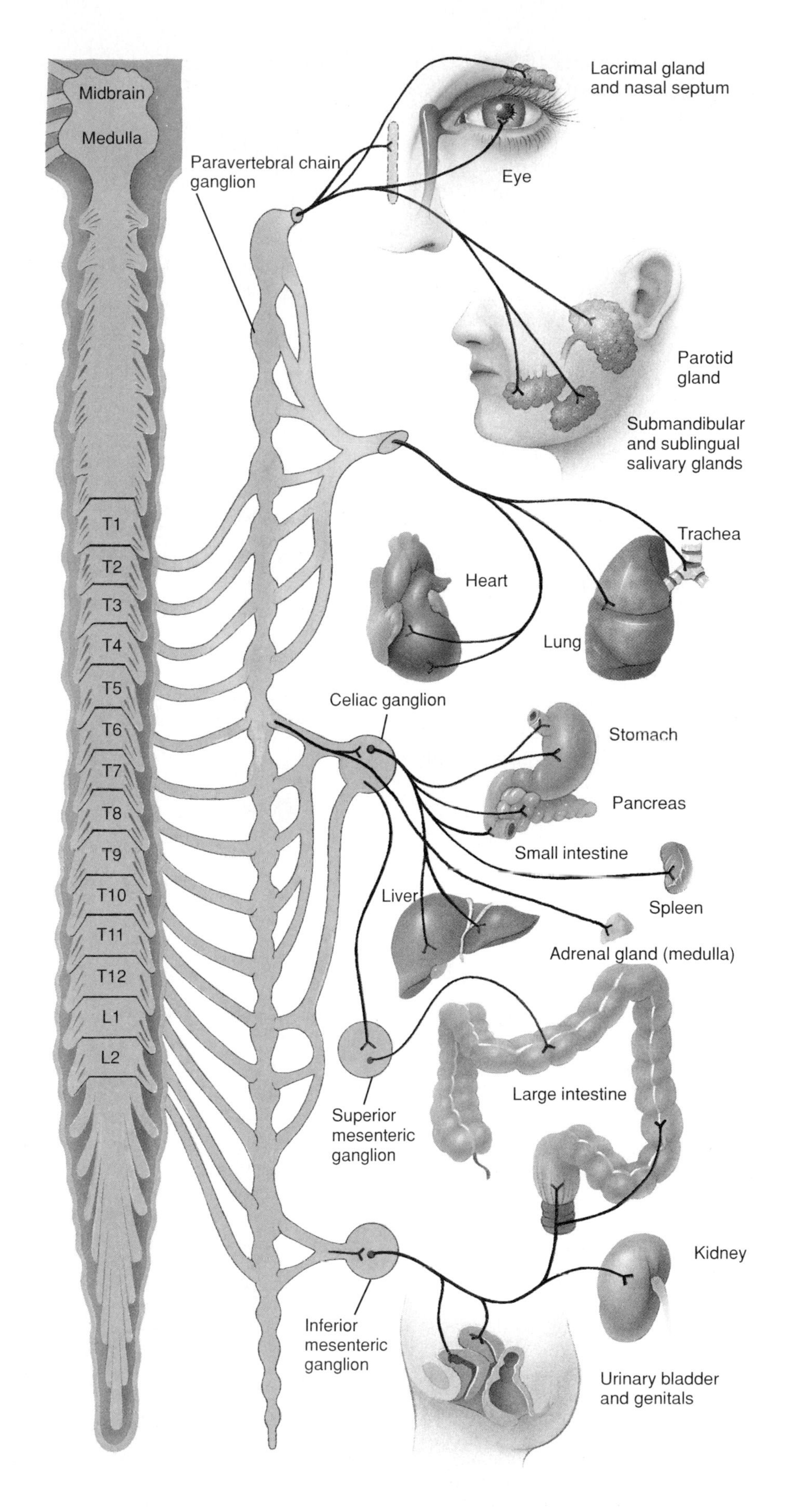


**FIGURE 11.35**

The preganglionic fibers of the sympathetic division of the autonomic nervous system arise from the thoracic and lumbar regions of the spinal cord. Note that the adrenal medulla is innervated directly by a preganglionic fiber, as described in the text.

muscles or glands within these organs (fig. 11.37). Parasympathetic preganglionic axons are usually myelinated, and the parasympathetic postganglionic fibers are unmyelinated.

The parasympathetic preganglionic fibers associated with parts of the head are included in the oculomotor, facial, and glossopharyngeal nerves. Those that innervate organs of the thorax and upper abdomen are parts of the vagus nerves. (The vagus nerves carry about 75% of all parasympathetic fibers.) Preganglionic fibers arising from the sacral region of the spinal cord lie within the branches of the second through the fourth sacral spinal nerves, and they carry impulses to viscera within the pelvic cavity (see fig. 11.36).

1. What is the general function of the autonomic nervous system?
2. How are the divisions of the autonomic system distinguished?
3. Describe a sympathetic nerve pathway and a parasympathetic nerve pathway.

## Autonomic Neurotransmitters

The preganglionic fibers of the sympathetic and parasympathetic divisions all secrete acetylcholine, and for this reason they are called **cholinergic fibers.** The parasympathetic postganglionic fibers are also cholinergic fibers. Most sympathetic postganglionic fibers, however, secrete norepinephrine (noradrenalin) and are called **adrenergic fibers** (see fig. 11.37). Exceptions to this include the sympathetic postganglionic fibers which stimulate sweat glands and a few sympathetic fibers to blood vessels in skeletal muscles and skin (which cause vasodilation); these fibers secrete acetylcholine and therefore are cholinergic (adrenergic sympathetic fibers to blood vessels cause vasoconstriction).

The different postganglionic neurotransmitters (mediators) are responsible for the different effects that the sympathetic and parasympathetic divisions have on organs.

Although each division can activate some effectors and inhibit others, one division primarily controls most organs. In other words, the divisions usually are

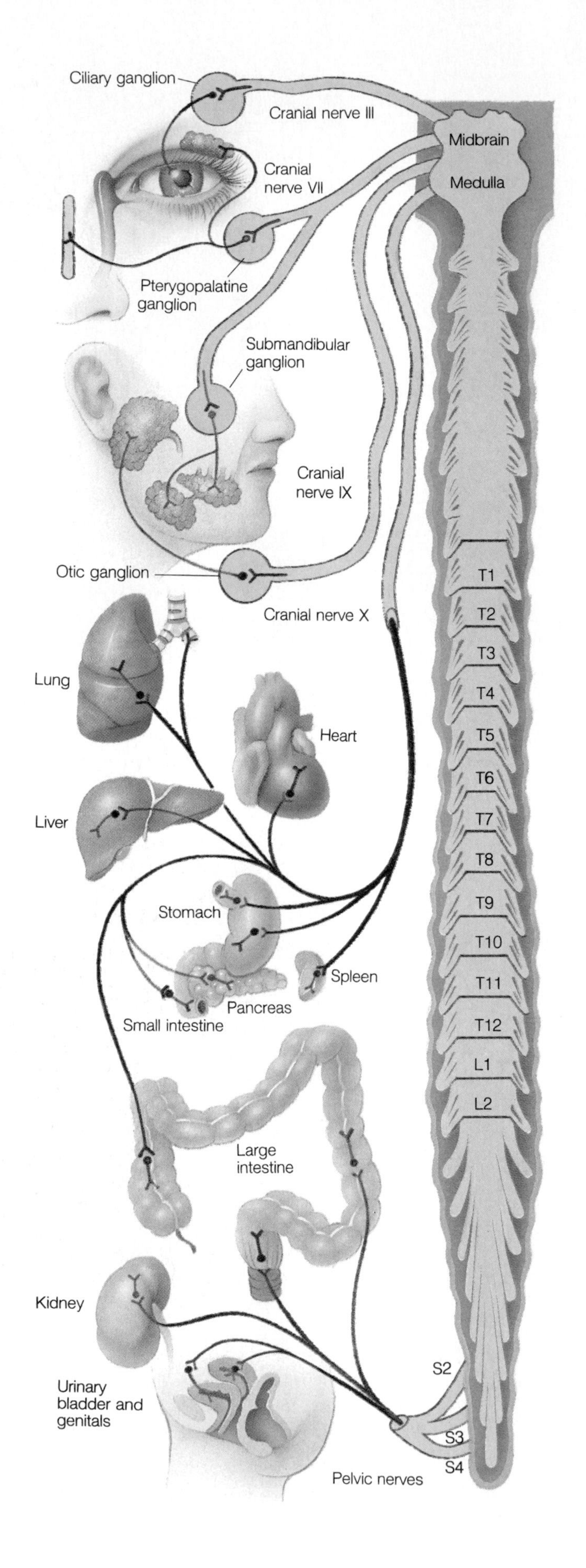


**FIGURE 11.36**
The preganglionic fibers of the parasympathetic division of the autonomic nervous system arise from the brain and sacral region of the spinal cord.

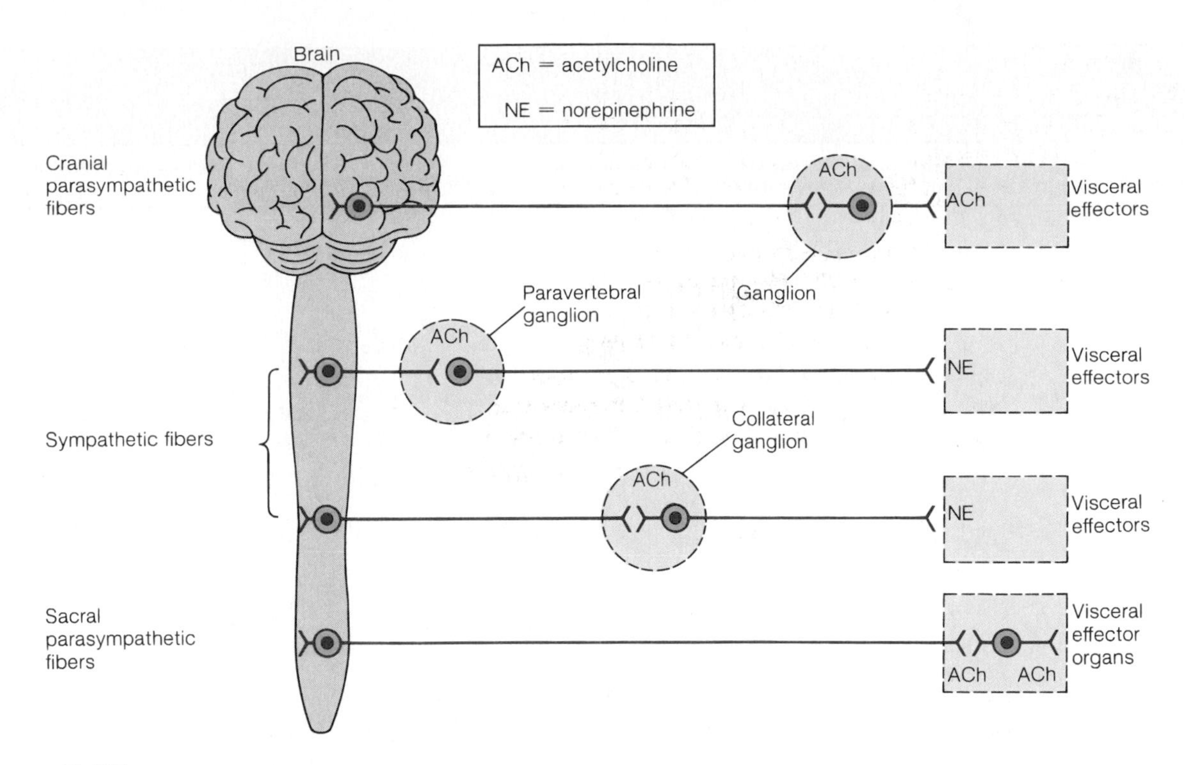


**FIGURE 11.37**
Most sympathetic fibers are adrenergic and secrete norepinephrine at the ends of the postganglionic fibers; parasympathetic fibers are cholinergic and secrete acetylcholine at the ends of the postganglionic fibers. The two arrangements of parasympathetic postganglionic fibers are seen in both cranial and sacral portions. Similarly, paravertebral and collateral ganglia are seen in both the thoracic and lumbar portions of the nervous system.

not actively antagonistic. For example, the diameters of most blood vessels, which lack parasympathetic innervation, are regulated by the sympathetic division. Smooth muscles in the walls of these vessels are continuously stimulated by sympathetic impulses; thus, they are maintained in a state of partial contraction called *sympathetic tone.* Decreasing sympathetic stimulation allows the muscular walls of such blood vessels to relax, increasing their diameters (vasodilation). Conversely, increasing sympathetic stimulation vasoconstricts vessels.

Similarly, the parasympathetic division is dominant in controlling movement in the digestive system. Parasympathetic impulses stimulate stomach and intestinal motility, and when these impulses decrease, movement is reduced.

Chart 11.8 summarizes the effects of adrenergic and cholinergic fibers on various visceral effectors.

## Actions of Autonomic Neurotransmitters

As in the case of stimulation at neuromuscular junctions (see chapter 9) and synapses (see chapter 10), the actions of autonomic neurotransmitters result from their binding to protein receptors in the membranes of effector cells. Receptor binding alters the membrane. For example, the membrane's permeability to certain ions may increase, and in smooth muscle cells, an action potential followed by muscular contraction may result. Similarly, a gland cell may respond to a change in its membrane by secreting a product.

Acetylcholine can combine with two types of cholinergic receptors, called *muscarinic receptors* and *nicotinic receptors.* These receptor names come from *muscarine,* a toxin from a fungus that can activate muscarinic receptors, and *nicotine,* the toxin of tobacco that can activate nicotinic receptors. The muscarinic receptors are located in the membranes of effector cells at the ends of all postganglionic parasympathetic nerve fibers and at the ends of the cholinergic sympathetic fibers. Responses from these receptors are excitatory and occur relatively slowly. The nicotinic receptors occur in the synapses between the preganglionic and postganglionic neurons of the parasympathetic and sympathetic pathways. They produce rapid, excitatory responses (fig. 11.38). (Receptors at neuromuscular junctions of skeletal muscles are nicotinic.)

Epinephrine and norepinephrine are the two chemical mediators of the sympathetic nervous system. The adrenal gland releases both as hormones, but only norepinephrine is released as a neurotransmitter

## CHART 11.8 Effects of Autonomic Stimulation on Various Effectors

| Effector Location | Response to Sympathetic Stimulation | Response to Parasympathetic Stimulation |
|---|---|---|
| Integumentary system | | |
| Apocrine glands | Increased secretion | No action |
| Eccrine glands | Increased secretion (cholinergic effect) | No action |
| Special senses | | |
| Iris of eye | Dilation | Constriction |
| Tear gland | Slightly increased secretion | Greatly increased secretion |
| Endocrine system | | |
| Adrenal cortex | Increased secretion | No action |
| Adrenal medulla | Increased secretion | No action |
| Digestive system | | |
| Muscle of gallbladder wall | Relaxation | Contraction |
| Muscle of intestinal wall | Decreased peristaltic action | Increased peristaltic action |
| Muscle of internal anal sphincter | Contraction | Relaxation |
| Pancreatic glands | Reduced secretion | Greatly increased secretion |
| Salivary glands | Reduced secretion | Greatly increased secretion |
| Respiratory system | | |
| Muscles in walls of bronchioles | Dilation | Constriction |
| Cardiovascular system | | |
| Blood vessels supplying muscles | Constriction (alpha adrenergic)<br>Dilation (beta adrenergic)<br>Dilation (cholinergic) | No action |
| Blood vessels supplying skin | Constriction | No action |
| Blood vessels supplying heart (coronary arteries) | Dilation (beta adrenergic)<br>Constriction (alpha adrenergic) | Dilation |
| Muscles in wall of heart | Increased contraction rate | Decreased contraction rate |
| Urinary system | | |
| Muscle of bladder wall | Relaxation | Contraction |
| Muscle of internal urethral sphincter | Contraction | Relaxation |
| Reproductive systems | | |
| Blood vessels to clitoris and penis | No action | Dilation leading to erection of clitoris and penis |
| Muscles associated with internal reproductive organs | Male ejaculation, female orgasm | |

by the sympathetic nervous system. These biochemicals can then combine with adrenergic receptors of effector cells.

The two major types of adrenergic receptors are termed *alpha* and *beta.* Exciting them elicits different responses in the effector organs. For example, stimulation of the alpha receptors in vascular smooth muscle causes vasoconstriction, whereas stimulation of the beta receptors in bronchial smooth muscle causes relaxation leading to bronchodilation. Furthermore, although norepinephrine has a somewhat stronger effect on alpha receptors, both of these mediators can stimulate both kinds of receptors. Consequently, the way each of these adrenergic substances influences effector cells depends on the relative numbers of alpha and beta receptors in the cell membranes.

The enzyme acetylcholinesterase rapidly decomposes the acetylcholine that cholinergic fibers release. Recall that this decomposition also occurs in the neuromuscular junctions of skeletal muscle. Thus, acetylcholine usually produces an effect on the postsynaptic membrane for only a fraction of a second.

Much of the norepinephrine released from adrenergic fibers is removed by active transport back into the nerve endings. The enzyme monoamine oxidase found in mitochondria then inactivates norepinephrine. This may take a few seconds, during which some molecules may diffuse into nearby tissues, where other enzymes decompose them. On the other hand, some norepinephrine molecules may enter the blood and remain active until they diffuse into tissues containing the inactivating enzymes. For these reasons, norepinephrine is likely to produce a more prolonged effect than acetylcholine. In fact, when the adrenal medulla releases norepinephrine and epinephrine into the blood in response to sympathetic

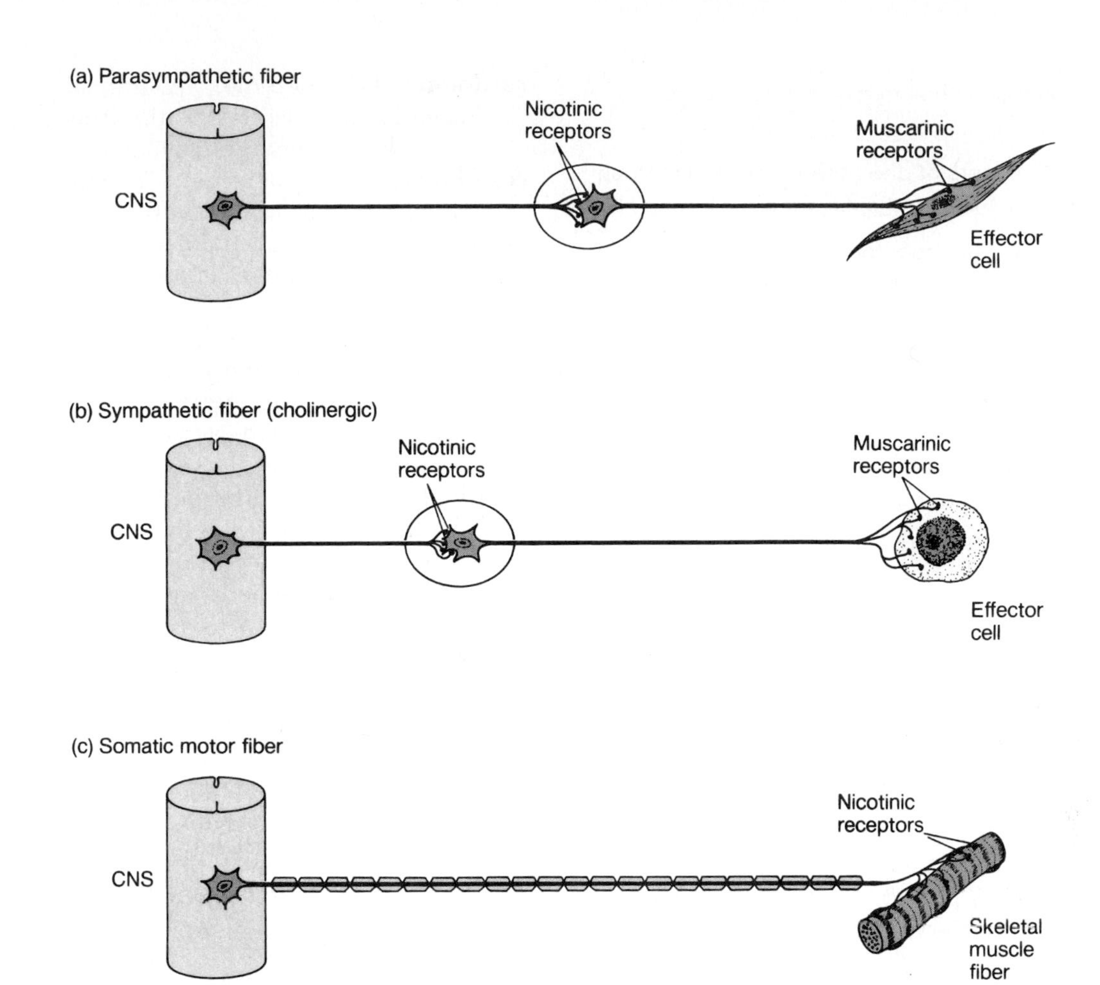


**FIGURE 11.38**

(*a* and *b*) Muscarinic receptors occur in the membranes of effector cells at the ends of autonomic cholinergic fibers. (*c*) Nicotinic receptors are found in the membranes of skeletal muscle fibers.

stimulation, these substances may trigger sympathetic responses in organs throughout the body that last up to thirty seconds.

> Many drugs influence autonomic functions. Some, like ephedrine, enhance sympathetic effects by stimulating release of norepinephrine from postganglionic sympathetic nerve endings. Others, like reserpine, inhibit sympathetic activity by preventing norepinephrine synthesis. Another group of drugs, which includes pilocarpine, produce parasympathetic effects, and some, like atropine, block the action of acetylcholine in visceral effectors.

## Control of Autonomic Activity

Although the autonomic nervous system has some independence resulting from impulse integration within its ganglia, it is controlled largely by the brain and spinal cord. For example, recall the control centers in the medulla oblongata for cardiac, vasomotor, and respiratory activities. These reflex centers receive sensory impulses from viscera by means of vagus nerve fibers, and use autonomic nerve pathways to stimulate motor responses in various muscles and glands. Thus, they control the autonomic nervous system. Similarly, the hypothalamus helps regulate body temperature, hunger, thirst, and water and electrolyte balance by influencing autonomic pathways.

Still higher levels within the brain, including the limbic system and the cerebral cortex, control the autonomic nervous system during emotional stress. In this way, the autonomic pathways regulate emotional expression and behavior.

Subsequent chapters that deal with individual organs and organ systems discuss regulation of particular organs.

1. Distinguish between cholinergic and adrenergic fibers.
2. Explain how the fibers of one autonomic division can control the actions of a particular organ.
3. What neurotransmitter substances are used in the autonomic nervous system?
4. Describe two types of cholinergic receptors and two types of adrenergic receptors.

## Clinical Terms Related to the Nervous System

**analgesia** (an″al-je′ze-ah) Loss or reduction in the ability to sense pain, but without loss of consciousness.
**analgesic** (an″al-je′sik) A pain-relieving drug.
**anesthesia** (an″es-the′ze-ah) A loss of feeling.
**aphasia** (ah-fa′ze-ah) A disturbance or loss in the ability to use words or to understand them, usually due to damage to cerebral association areas.
**apraxia** (ah-prak′se-ah) An impairment in a person's ability to make correct use of objects.
**ataxia** (ah-tak′se-ah) A partial or complete inability to coordinate voluntary movements.
**cordotomy** (kor-dot′o-me) A surgical procedure severing a nerve tract within the spinal cord, usually to relieve intractable pain.
**craniotomy** (kra″ne-ot′o-me) A surgical procedure opening a part of the skull.
**encephalitis** (en″sef-ah-li′tis) Inflammation of the brain and meninges characterized by drowsiness and apathy.
**epilepsy** (ep′ĭ-lep″se) A disorder of the central nervous system that is characterized by temporary disturbances in normal brain impulses; it may be accompanied by convulsive seizures and loss of consciousness.
**Huntington disease** (hunt′ing-tun di-zez′) A hereditary disorder of the brain producing progressively worsening, uncontrollable dancelike movements and personality changes.
**laminectomy** (lam″i-nek′to-me) Surgical removal of the posterior arch of a vertebra, usually to relieve symptoms of a ruptured intervertebral disk that is pressing on a spinal nerve.
**neuralgia** (nu-ral′je-ah) A sharp, recurring pain associated with a nerve, usually caused by inflammation or injury.
**neuritis** (nu-ri′tis) Inflammation of a nerve.
**vagotomy** (va-got′o-me) Surgical severing of a vagus nerve.

## Chapter Summary

### Introduction (page 389)

Bone and protective membranes called meninges surround the brain and spinal cord.

### Meninges (page 389)

1. The meninges consist of a dura mater, arachnoid mater, and pia mater.
2. Cerebrospinal fluid occupies the space between the arachnoid and pia maters.

### Spinal Cord (page 391)

The spinal cord is a nerve column that extends from the brain into the vertebral canal. It terminates at the level between the first and second lumbar vertebrae.

1. Structure of the spinal cord
   a. The spinal cord is composed of thirty-one segments, each of which gives rise to a pair of spinal nerves.
   b. It is characterized by a cervical enlargement, a lumbar enlargement, and two deep longitudinal grooves that divide it into right and left halves.
   c. White matter surrounds a central core of gray matter.
   d. The white matter is composed of bundles of myelinated nerve fibers.
2. Functions of the spinal cord
   a. The cord provides a two-way communication system between the brain and structures outside the nervous system.
   b. Ascending tracts carry sensory impulses to the brain; descending tracts carry motor impulses to muscles and glands.
   c. Many of the fibers in the ascending and descending tracts cross over in the spinal cord or brain.

### Brain (page 395)

The brain is the largest and most complex part of the nervous system. It contains nerve centers that are associated with sensations. The brain issues motor commands and carries on higher mental functions.

1. Brain development
   a. Brain structure reflects the way it forms.
   b. The brain develops from a neural tube with three cavities—the forebrain, midbrain, and hindbrain.
   c. The cavities persist as ventricles, and the walls give rise to structural and functional regions.
2. Structure of the cerebrum
   a. The cerebrum consists of two cerebral hemispheres connected by the corpus callosum.

b. Its surface is marked by ridges and grooves; sulci divide each hemisphere into lobes.
c. The cerebral cortex is a thin layer of gray matter near the surface.
d. White matter consists of myelinated nerve fibers that interconnect neurons within the nervous system and communicate with other body parts.

3. Functions of the cerebrum
   a. The cerebrum is concerned with higher brain functions, such as thought, reasoning, interpretation of sensory impulses, control of voluntary muscles, and memory storage.
   b. The cerebral cortex has sensory, motor, and association areas.
   c. The primary motor regions lie in the frontal lobes near the central sulcus. Other areas of the frontal lobes control special motor functions.
   d. Areas that interpret sensory impulses from the skin are located in the parietal lobes near the central sulcus; other specialized sensory areas are in the temporal and occipital lobes.
   e. Association areas analyze and interpret sensory impulses and provide memory, reasoning, verbalizing, judgment, and emotions.
   f. One cerebral hemisphere usually dominates for certain intellectual functions.
   g. Short-term memory is probably electrical. Long-term memory is thought to be encoded in patterns of synaptic connections.
4. Basal ganglia
   a. Basal ganglia are masses of gray matter located deep within the cerebral hemispheres.
   b. They relay motor impulses originating in the cerebral cortex, and aid in controlling motor activities.
5. Ventricles and cerebrospinal fluid
   a. Ventricles are interconnected cavities within the cerebral hemispheres and brain stem.
   b. Cerebrospinal fluid fills the ventricles.
   c. Choroid plexuses in the walls of the ventricles secrete cerebrospinal fluid.
   d. Ependymal cells of the choroid plexus regulate the composition of cerebrospinal fluid.
   e. Cerebrospinal fluid circulates through the ventricles and is reabsorbed into the blood of the dural sinuses.
6. Diencephalon
   a. The diencephalon contains the thalamus and hypothalamus.
   b. The thalamus selects incoming sensory impulses and relays them to the cerebral cortex
   c. The hypothalamus is important in maintaining homeostasis.
   d. The limbic system produces emotional feelings and modifies behavior.
7. Brain stem
   a. The brain stem extends from the base of the brain to the spinal cord.
   b. The brain stem consists of the midbrain, pons, and medulla oblongata.
   c. The midbrain contains reflex centers associated with eye and head movements.
   d. The pons transmits impulses between the cerebrum and other parts of the nervous system and contains centers that help regulate rate and depth of breathing.
   e. The medulla oblongata transmits all ascending and descending impulses and contains several vital and nonvital reflex centers.
   f. The reticular formation filters incoming sensory impulses, arousing the cerebral cortex into wakefulness in response to meaningful impulses.
   g. Normal sleep results from decreasing activity of the reticular formation, and paradoxical sleep occurs when activating impulses are received by some parts of the brain, but not by others.
8. Cerebellum
   a. The cerebellum consists of two hemispheres connected by the vermis.
   b. A thin cortex of gray matter surrounds the white matter of the cerebellum.
   c. The cerebellum functions primarily as a reflex center, coordinating skeletal muscle movements and maintaining equilibrium.

## Peripheral Nervous System (page 416)

The peripheral nervous system consists of cranial and spinal nerves that branch out from the brain and spinal cord to all body parts. It can be subdivided into somatic and autonomic portions.

1. Structure of peripheral nerves
   a. A nerve consists of a bundle of nerve fibers surrounded by connective tissues.
   b. The connective tissues form an outer epineurium, a perineurium enclosing bundles of nerve fibers, and an endoneurium surrounding each fiber.
2. Cranial nerves
   a. Twelve pairs of cranial nerves connect the brain to parts in the head, neck, and trunk.
   b. Although most cranial nerves are mixed, some are pure sensory, and others are primarily motor.
   c. The names of cranial nerves indicate their primary functions or the general distributions of their fibers.
   d. Some cranial nerve fibers are somatic and others are autonomic.
3. Spinal nerves
   a. Thirty-one pairs of spinal nerves originate from the spinal cord.
   b. These mixed nerves provide a two-way communication system between the spinal cord and the upper limbs, lower limbs, neck, and trunk.
   c. Spinal nerves are grouped according to the levels from which they arise, and they are numbered sequentially.
   d. Each nerve emerges by a dorsal and a ventral root.
      (1) A dorsal root contains sensory fibers and has a dorsal root ganglion.
      (2) A ventral root contains motor fibers.
   e. Just beyond its foramen, each spinal nerve divides into several branches.
   f. Most spinal nerves combine to form plexuses that direct nerve fibers to a particular body part.

## Autonomic Nervous System (page 426)

The autonomic nervous system functions without conscious effort. It is concerned primarily with regulating visceral activities that maintain homeostasis.

1. General characteristics
   a. Autonomic functions are reflexes controlled from centers in the hypothalamus, brain stem, and spinal cord.

b. Autonomic nerve fibers are associated with ganglia where impulses are integrated before distribution to effectors.
c. The integrative function of the ganglia provides a degree of independence from the central nervous system.
d. The autonomic nervous system consists of the visceral efferent fibers associated with these ganglia.
e. The autonomic nervous system is subdivided into two divisions—sympathetic and parasympathetic.
f. The sympathetic division prepares the body for stressful and emergency conditions.
g. The parasympathetic division is most active under ordinary conditions.

2. Autonomic nerve fibers
   a. The autonomic fibers are efferent, or motor.
   b. Sympathetic fibers leave the spinal cord and synapse in ganglia.
      (1) Preganglionic fibers pass through white rami to reach paravertebral ganglia.
      (2) Paravertebral ganglia and interconnecting fibers comprise the sympathetic trunks.
      (3) Preganglionic fibers synapse within paravertebral or collateral ganglia.
      (4) Postganglionic fibers usually pass through gray rami to reach spinal nerves before passing to effectors.
      (5) A special set of sympathetic preganglionic fibers passes through ganglia and extends to the adrenal medulla.
   c. Parasympathetic fibers begin in the brain stem and sacral region of the spinal cord, and synapse in ganglia near various organs or in the organs themselves.
3. Autonomic neurotransmitters
   a. Sympathetic and parasympathetic preganglionic fibers secrete acetylcholine.
   b. Most sympathetic postganglionic fibers secrete norepinephrine and are adrenergic; postganglionic parasympathetic fibers secrete acetylcholine and are cholinergic.
   c. The different effects of the autonomic divisions are due to different transmitter substances the postganglionic fibers release.
4. Actions of autonomic neurotransmitters
   a. Neurotransmitters combine with receptors and alter cell membranes.
   b. There are two types of cholinergic receptors and two types of adrenergic receptors.
   c. How cells respond to neurotransmitters depends upon the number and type of receptors in their membranes.
   d. Acetylcholine acts very briefly; norepinephrine and epinephrine may have more prolonged effects.
5. Control of autonomic activity
   a. The central nervous system largely controls the autonomic nervous system.
   b. The medulla oblongata uses autonomic fibers to regulate cardiac, vasomotor, and respiratory activities.
   c. The hypothalamus uses autonomic fibers in regulating visceral functions.
   d. The limbic system and cerebral cortex control emotional responses through the autonomic nervous system.

## Critical Thinking Questions

1. If a physician plans to obtain a sample of spinal fluid from a patient, what anatomical site can be safely used, and how should the patient be positioned to facilitate this procedure?
2. What functional losses would you expect to observe in a patient who has suffered injury to the right occipital lobe of the cerebral cortex? To the right temporal lobe?
3. The Brown-Sequard syndrome is due to an injury on one side of the spinal cord. It is characterized by paralysis below the injury and on the same side as the injury, and by loss of sensations of temperature and pain on the opposite side. How would you explain these symptoms?
4. Substances used by intravenous drug abusers are sometimes obtained in tablet form and are crushed and dissolved before they are injected. Such tablets may contain fillers, such as talc or cornstarch, that may obstruct tiny blood vessels in the cerebrum. What problems might these obstructions create?
5. In planning treatment for a patient who has had a cerebrovascular accident (CVA), why would it be important to know whether the CVA was caused by a ruptured or obstructed blood vessel?
6. What symptoms might the sympathetic division of the autonomic nervous system produce in a patient who is experiencing stress?
7. How would you distinguish between a patient in a coma and one in a persistent vegetative state?

## Review Exercises

1. Name the layers of the meninges, and explain their functions.
2. Describe the location of cerebrospinal fluid within the meninges.
3. Describe the structure of the spinal cord.
4. Name the major ascending and descending tracts of the spinal cord, and list the functions of each.
5. Explain the consequences of nerve fibers crossing over.
6. Describe how the brain develops.
7. Describe the structure of the cerebrum.
8. Define *cerebral cortex.*
9. Describe the location and function of the primary motor areas of the cortex.
10. Describe the location and function of Broca's area.
11. Describe the location and function of the sensory areas of the cortex.
12. Explain the function of the association areas of the lobes of the cerebrum.
13. Define *hemisphere dominance.*
14. Explain the function of the corpus callosum.
15. Distinguish between short-term and long-term memory.
16. Describe the location and function of the basal ganglia.
17. Describe the location of the ventricles of the brain.
18. Explain how cerebrospinal fluid is produced and how it functions.
19. Name the parts of the diencephalon, and describe the general functions of each.
20. Define the limbic system, and explain its functions.
21. Name the parts of the midbrain, and describe the general functions of each.
22. Describe the pons and its functions.
23. Describe the medulla oblongata and its functions.
24. Describe the location and function of the reticular formation.
25. Distinguish between normal and paradoxical sleep.
26. Describe the functions of the cerebellum.
27. Distinguish between the somatic and autonomic nervous systems.
28. Describe the structure of a peripheral nerve.
29. Name, locate, and describe the major functions of each pair of cranial nerves.
30. Explain how the spinal nerves are grouped and numbered.
31. Define *cauda equina.*
32. Describe the structure of a spinal nerve.
33. Define *plexus,* and locate the major plexuses of the spinal nerves.
34. Distinguish between the sympathetic and parasympathetic divisions of the autonomic nervous system.
35. Explain how autonomic ganglia provide a degree of independence from the central nervous system.
36. Distinguish between a preganglionic fiber and a postganglionic fiber.
37. Define *paravertebral ganglion.*
38. Trace a sympathetic nerve pathway through a ganglion to an effector.
39. Explain why the effects of the sympathetic and parasympathetic autonomic divisions differ.
40. Distinguish between cholinergic and adrenergic nerve fibers.
41. Define *sympathetic tone.*
42. Explain how autonomic transmitters influence the actions of effector cells.
43. Distinguish between alpha adrenergic and beta adrenergic receptors.
44. Describe three examples in which the central nervous system employs autonomic nerve pathways.

12

CHAPTER

# Somatic and Special Senses

We use many of our senses to "experience" an apple.

## Chapter Objectives

After you have studied this chapter, you should be able to:

1. Name five kinds of receptors and explain the function of each.
2. Explain how receptors stimulate sensory impulses.
3. Explain how a sensation is produced.
4. Distinguish between somatic and special senses.
5. Describe the receptors associated with the senses of touch and pressure, temperature, and pain.
6. Describe how the sense of pain is produced.
7. Explain the importance of stretch receptors in muscles and tendons.
8. Explain the relationship between the senses of smell and taste.
9. Name the parts of the ear and explain the function of each part.
10. Distinguish between static and dynamic equilibrium.
11. Name the parts of the eye and explain the function of each part.
12. Explain how light is refracted by the eye.
13. Explain how depth and distance are perceived.
14. Describe the visual nerve pathway.

## Key Terms

**accommodation** (ah-kom″o-da′shun)
**ampulla** (am-pul′lah)
**auditory** (aw′di-to″re)
**chemoreceptor** (ke″mo-re-sep′tor)
**cochlea** (kok′le-ah)
**cornea** (kor′ne-ah)
**dynamic equilibrium** (di-nam′ik e″kwĭ-lib′re-um)
**labyrinth** (lab′i-rinth)
**macula** (mak′u-lah)
**mechanoreceptor** (mek″ah-no-re-sep′tor)
**olfactory** (ol-fak′to-re)
**optic** (op′tik)
**photoreceptor** (fo″to-re-sep′tor)
**projection** (pro-jek′shun)
**proprioceptor** (pro″pre-o-sep′tor)
**referred pain** (re-furd′ pān)
**refraction** (re-frak′shun)
**retina** (ret′i-nah)
**rhodopsin** (ro-dop′sin)
**sclera** (skle′rah)
**sensory adaptation** (sen′so-re ad″ap-ta′shun)
**static equilibrium** (stat′ik e″kwĭ-lib′re-um)
**thermoreceptor** (ther″mo-re-sep′tor)

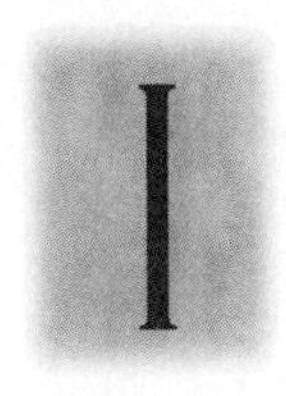

In Herman Wouk's classic novel *The Winds of War,* a young woman, Natalie, is in a railroad car packed with starving people. She is en route to a concentration camp in Europe during the Second World War. Natalie hasn't eaten anything in many days. Suddenly, a door opens, and a kind-hearted farmer tosses apples into the car. Natalie is thrown to the floor in the ensuing scramble, and miraculously, lands right on top of an apple! In the second before she consumes it, her starving senses are afire. She cherishes the round hardness hidden in her hands, revels in the shiny red skin and the tempting fresh odor. When she bites into the fruit, the taste washes over her, so wonderful that she nearly forgets her terror. Wouk's descriptions of the morsel, and of Natalie's plight, are so vivid that the reader can almost taste the apple. Our special senses enable us to experience an apple—and everything else in the outside world.

What is an apple? It is a fresh, pleasant odor, a sweet taste, a crunchy sound, and a shiny, smooth orb of yellow, red, or green. Biting into a crisp apple may also be enjoyable because it evokes memories—perhaps of a childhood picnic or just images of early fall—that we may not even be aware of. Our senses enable us to experience the environment from multiple viewpoints, melding them into a meaningful perception.

CHART 12.1 INFORMATION FLOW FROM THE ENVIRONMENT THROUGH THE NERVOUS SYSTEM

| Information Flow | Smell | Taste | Sight | Hearing |
|---|---|---|---|---|
| Sense receptors | Olfactory cells in nose | Taste bud receptor cells | Rods and cones in retina | Hair cells in cochlea |
| ↓ | ↓ | ↓ | ↓ | ↓ |
| Stimulation of nerve fibers | Olfactory nerve fibers | Sensory fibers in various cranial nerves | Optic nerve fibers | Auditory nerve fibers |
| ↓ | ↓ | ↓ | ↓ | ↓ |
| Impulse transmission to CNS | Cortex | Cortex | Midbrain and visual cortex | Midbrain and cortex |
| ↓ | ↓ | ↓ | ↓ | ↓ |
| Sensation (memory, experience) | A pleasant smell | A sweet taste | A small, round, red object | A crunching sound |
| ↓ | ↓ | ↓ | ↓ | ↓ |
| Perception | The smell of an apple | The taste of an apple | The sight of an apple | Biting into an apple |

All senses work in basically the same way. Special sensory receptors collect information from the environment and stimulate neurons to send a message to the brain. There the cerebral cortex forms a perception, a person's particular view of the stimulus. Chart 12.1 outlines the pathways from sensation to perception that describe an apple.

## Receptors and Sensations

Sensory receptors are diverse, but share certain features. For example, each type of receptor is particularly sensitive to a distinct kind of environmental change and is much less sensitive to other forms of stimulation.

### Receptor Types

Five general groups of sensory receptors are known, based on their sensitivities to changes in one of the following factors:

1. *Chemical concentration.* Receptors that are stimulated by changes in the concentration of chemical substances are called **chemoreceptors.** Receptors associated with the senses of smell and taste are of this type. Chemoreceptors in internal organs, for example, can detect changes in the blood concentrations of substances including oxygen, hydrogen ions, and glucose.
2. *Tissue damage.* Tissue damage stimulates **pain receptors** (nociceptors). Triggering factors include exposure to excessive mechanical, electrical, thermal, or chemical energy.
3. *Temperature change.* Receptors sensitive to temperature change are called **thermoreceptors.**
4. *Mechanical forces.* A number of sensory receptors sense mechanical forces, such as changes in the pressure or movement of fluids. As a rule, these **mechanoreceptors** detect changes that deform the receptors. For example, **proprioceptors** sense changes in the tensions of muscles and tendons, whereas **baroreceptors** (pressoreceptors) in certain blood vessels can detect changes in blood pressure. Similarly, **stretch receptors** in the lungs sense degree of inflation.

5. *Light intensity.* Light receptors, or **photoreceptors,** occur only in the eyes, and they respond whenever they are exposed to light energy of sufficient intensity.

## Sensory Impulses

Sensory receptors can be ends of nerve fibers or other kinds of cells located close to them. In either case, stimulation causes local changes in their membrane potentials (receptor potentials), generating a graded electric current that reflects the intensity of stimulation (see chapter 10).

If a receptor is a nerve fiber and the change in membrane potential reaches threshold, an action potential is generated, and a sensory impulse travels on the nerve fiber. However, if the receptor is another type of cell, its receptor potential must be transferred to a nerve fiber to trigger an action potential. Peripheral nerves transmit such sensory impulses to the central nervous system, where they are analyzed and interpreted within the brain (see chapter 11).

## Sensations

A **sensation** (perception) is a feeling that occurs when the brain interprets sensory impulses. Because all the nerve impulses that travel away from sensory receptors into the central nervous system are alike, the resulting sensation depends on what region of the cerebral cortex receives the impulse. For example, impulses reaching one region are always interpreted as sounds, and those reaching another portion are always sensed as touch.

It makes little difference how receptors are stimulated. Pain receptors, for example, can be stimulated by heat, cold, or pressure, but the sensation is always the same because in each case the same part of the brain interprets the resulting nerve impulses. Similarly, factors other than light, such as a sharp blow to the head, may trigger nerve impulses in visual receptors. When this happens, the person may "see" lights, even though no light is entering the eye, since any impulses reaching the visual cortex are interpreted as light. Normally receptors only respond to specific stimuli, so the brain creates the correct sensation.

At the same time that a sensation forms, the cerebral cortex interprets it to seem to come from the receptors being stimulated. This process is called **projection,** because the brain projects the sensation back to its apparent source. Projection allows a person to pinpoint the region of stimulation. Thus, we are aware that the eyes see an apple, the nose smells it, and the ears hear the teeth crunch into it.

## Sensory Adaptation

When sensory receptors are continuously stimulated, many of them undergo an adjustment called **sensory adaptation.** As the receptors adapt, impulses leave them at decreasing rates, until finally these receptors may completely fail to send signals. Once receptors adapt, impulses are triggered only if the strength of the stimulus changes.

Imagine a person walking into a fish market. At first the fishy odor seems intense, but it becomes less and less noticeable as the smell (olfactory) receptors adapt. Soon, the once offensive odor is barely noticeable, but if the person leaves and reenters, the stench returns full force.

1 List the five general types of sensory receptors.

2 What do all types of receptors have in common?

3 Explain how a sensation occurs.

4 What is sensory adaptation?

# Somatic Senses

Receptors associated with the skin, muscles, joints, and viscera provide somatic senses. These senses can be divided into three groups:

1. Senses associated with changes at the body surface (exteroceptive senses), which include the senses of touch, pressure, and temperature.
2. Senses associated with changes in muscles and tendons and in body position (proprioceptive senses).
3. Senses associated with changes in viscera (visceroceptive senses). Except for visceral pain, these senses are discussed in subsequent chapters.

## Touch and Pressure Senses

The senses of touch and pressure employ three kinds of receptors (fig. 12.1). As a group, these receptors sense mechanical forces that deform or displace tissues. The touch and pressure receptors include:

1. **Sensory nerve fibers.** These receptors are common in epithelial tissues where their free ends lie between epithelial cells. They are associated with sensations of touch and pressure.
2. **Meissner's corpuscles.** These are small, oval masses of flattened connective tissue cells in

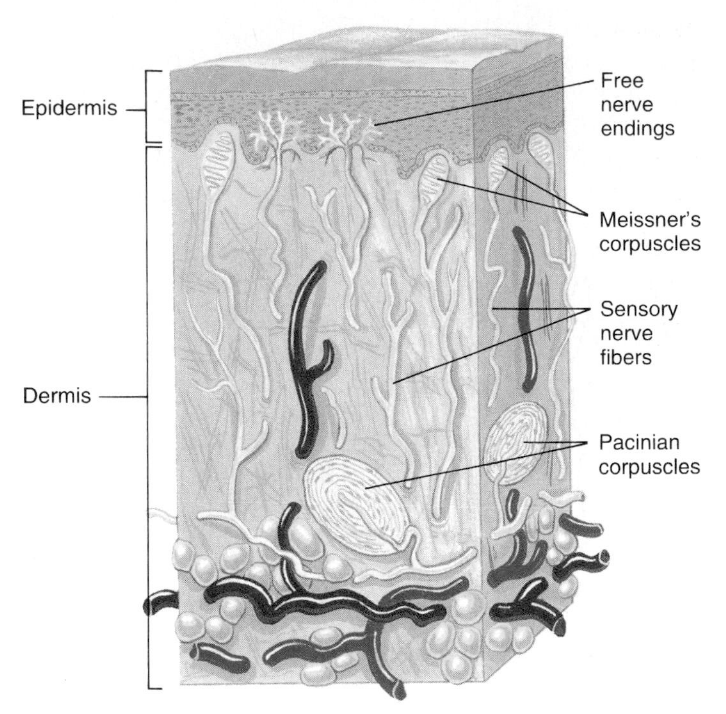


**FIGURE 12.1**
Touch and pressure receptors include free ends of sensory nerve fibers, Meissner's corpuscles, and Pacinian corpuscles.

connective tissue sheaths. Two or more sensory nerve fibers branch into each corpuscle and end within it as tiny knobs.

Meissner's corpuscles are abundant in the hairless portions of the skin, such as the lips, fingertips, palms, soles, nipples, and external genital organs. They sense motion of objects that barely contact the skin, interpreting impulses from them as the sensation of light touch. They are also used when a person touches something to judge its texture (fig. 12.2*a*).

3. **Pacinian corpuscles.** These sensory bodies are relatively large, ellipsoidal structures composed of connective tissue fibers and cells. They are common in the deeper subcutaneous tissues of the hands, feet, penis, clitoris, urethra, and breasts, and also in tendons of muscles and ligaments of joints (fig. 12.2*b*).

   Heavy pressure stimulates Pacinian corpuscles, which sense deep pressure. They may also detect vibrations in tissues.

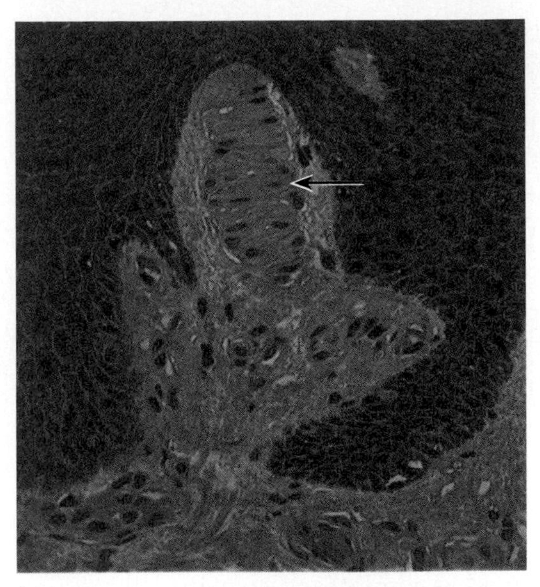

(a)

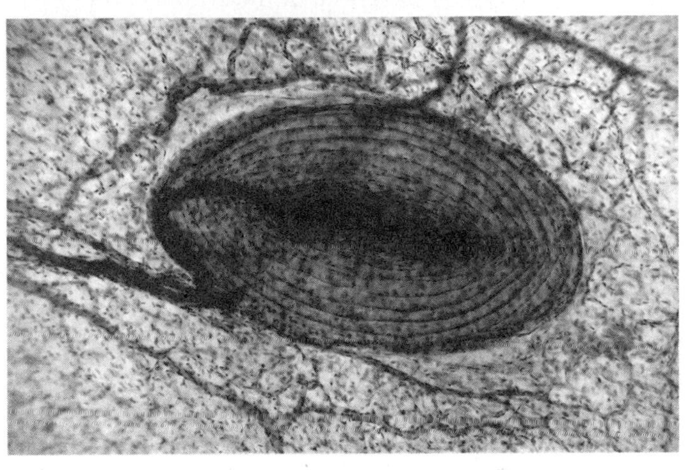

(b)

**FIGURE 12.2**
(*a*) Light micrograph of a Meissner's corpuscle from the skin of the palm (400×); (*b*) light micrograph of a Pacinian corpuscle (25×).

## Temperature Senses

Temperature receptors (thermoreceptors) include two types of *free nerve endings* located in the skin. Those which respond to warmer temperatures are called *heat receptors* and those which respond to colder temperatures are called *cold receptors*. The heat receptors are most sensitive to temperatures above 25°C (77°F) and become unresponsive at temperatures above 45°C (113°F). As 45°C is approached, pain receptors are also triggered, producing a *burning sensation*.

Cold receptors are most sensitive to temperatures between 10°C (50°F) and 20°C (68°F). If the temperature drops below 10°C, pain receptors are stimulated, and the person feels a *freezing sensation*.

At intermediate temperatures, the brain interprets sensory input from different combinations of these receptors as a particular temperature sensation.

Both heat and cold receptors rapidly adapt, so that within about a minute of continuous stimulation, the sensation of heat or cold begins to fade. Imagine stepping into a bathtub containing "hot" water that soon feels pleasantly warm, or diving into a freezing pool of water that soon becomes refreshingly cool. These are familiar temperature adaptations.

## Sense of Pain

Other receptors that consist of *free nerve endings* sense pain. These receptors are widely distributed throughout the skin and internal tissues, except in the nervous tissue of the brain, which lacks pain receptors.

### Pain Receptors

The pain receptors are protective in that they are stimulated when tissues are damaged. Pain sensation is usually perceived as unpleasant, signaling that action be taken to remove the source of the stimulation.

Although most pain receptors can be stimulated by more than one type of change, some are most sensitive to mechanical damage while others are particularly sensitive to extremes in temperature. Still other pain receptors are most responsive to chemicals, such as hydrogen ions, potassium ions, or specific breakdown products of proteins, histamine, and acetylcholine. A deficiency of blood (ischemia) and thus a deficiency of oxygen (hypoxia) in a tissue, or stimulation of mechanical-sensitive receptors, also triggers the pain sensation. The pain elicited during a muscle cramp, for example, is related to interruption of blood flow that occurs as the sustained contraction squeezes capillaries, as well as to stimulation of mechanical-sensitive pain receptors. Also, when blood flow is interrupted, pain-stimulating chemicals accumulate. Increasing blood flow through the sore tissue may relieve the resulting pain, and this is why heat is sometimes applied to reduce muscle soreness. The heat dilates blood vessels and thus promotes blood flow, which helps reduce the concentration of the pain-stimulating substances. In some conditions, accumulating chemicals lower the thresholds of pain receptors, making inflamed tissues more sensitive to heat or pressure than before.

Pain receptors adapt very little, if at all. Once such a receptor is activated, even by a single stimulus, it may continue to send impulses into the central nervous system for some time.

### Visceral Pain

As a rule, pain receptors are the only receptors in viscera whose stimulation produces sensations. Pain receptors in these organs respond differently to stimulation than those associated with surface tissues. For example, localized damage to intestinal tissue during surgical procedures may not elicit any pain sensations, even in a conscious person. However, when visceral tissues are subjected to more widespread stimulation, as when intestinal tissues are stretched or when the smooth muscles in the intestinal walls undergo spasms, a strong pain sensation may follow. Once again, the resulting pain is related to stimulation of mechanical-sensitive receptors and to decreased blood flow accompanied by a lower tissue oxygen concentration and accumulation of pain-stimulating chemicals.

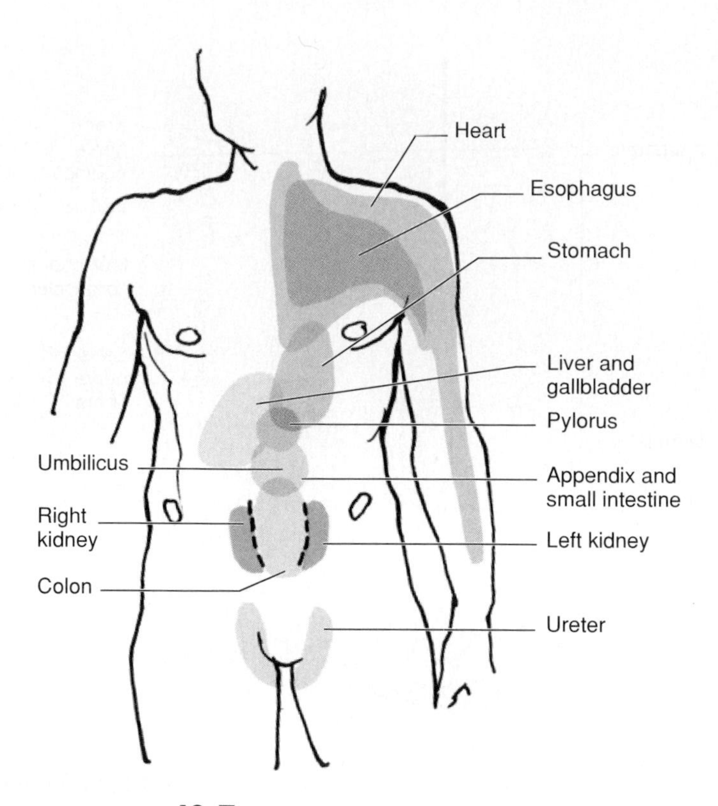


**FIGURE 12.3**
Surface regions to which visceral pain may be referred.

Visceral pain may feel as if it is coming from some part of the body other than the part being stimulated—a phenomenon called **referred pain.** For example, pain originating in the heart may be referred to the left shoulder or the inside of the left arm. Pain from the lower esophagus, stomach, or small intestine may seem to be coming from the upper central (epigastric) region of the abdomen. Pain from the urogenital tract may be referred to the lower central (hypogastric) region of the abdomen or to the sides between the ribs and the hip (fig. 12.3).

Referred pain may derive from *common nerve pathways* that sensory impulses coming both from skin areas and from internal organs use (see chapter 11).

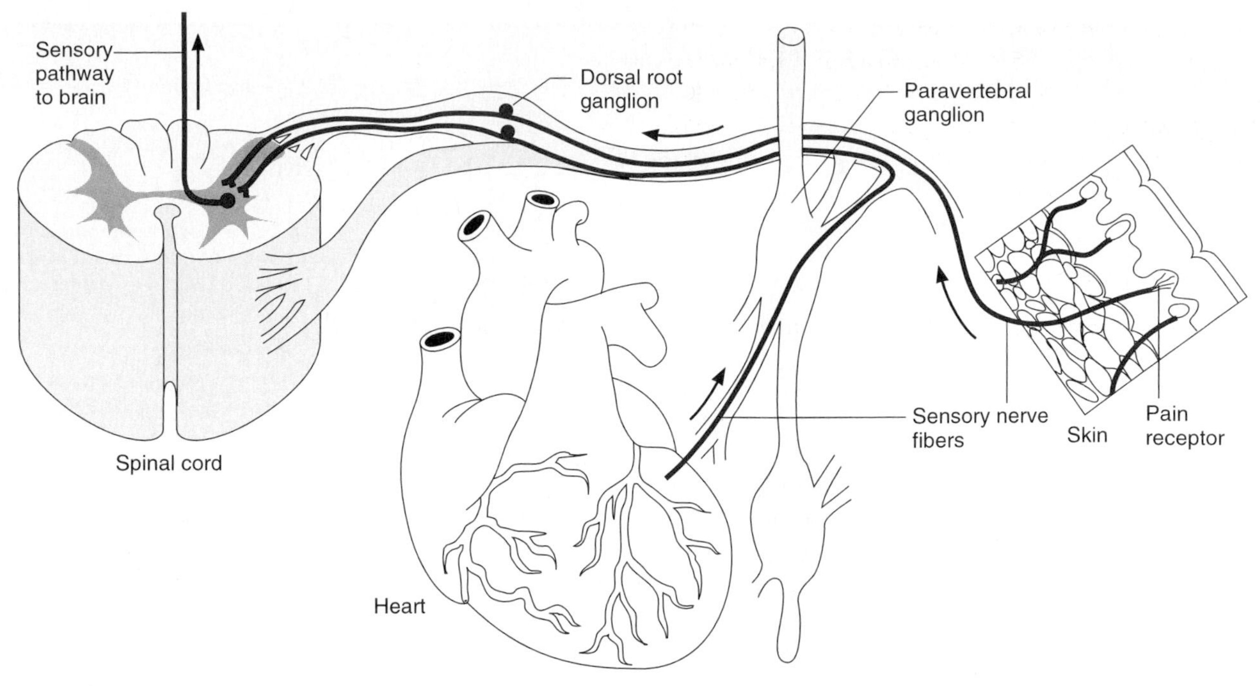


**FIGURE 12.4**
Pain originating in the heart may feel as if it is coming from the skin because sensory impulses from those two regions follow common nerve pathways to the brain.

Pain impulses from the heart seem to be conducted over the same nerve pathways as those from the skin of the left shoulder and the inside of the left arm, as shown in figure 12.4. Consequently, the cerebral cortex may incorrectly interpret the source of the impulses as the shoulder and the inside of the left arm, rather than the heart.

Pain originating in the parietal layers of thoracic and abdominal membranes—parietal pleura, parietal pericardium, or parietal peritoneum—is usually not referred; instead, such pain is felt directly over the area being stimulated.

## Pain Nerve Pathways

The nerve fibers that conduct impulses away from pain receptors are of two main types: acute pain fibers and chronic pain fibers.

The *acute pain fibers* (also known as A-delta fibers) are relatively thin, myelinated nerve fibers. They conduct nerve impulses rapidly, at velocities up to 30 meters per second. These impulses are associated with the sensation of sharp pain, which typically seems to originate in the skin and is restricted to a local area. This type of pain seldom continues after the pain-producing stimulus is discontinued.

The *chronic pain fibers* (C fibers) are thin, unmyelinated nerve fibers. They conduct impulses more slowly than acute pain fibers, at velocities up to 2 meters per second. These impulses cause the dull, aching pain sensation that may be widespread and difficult to pinpoint. Such pain may continue for some time after the original stimulus is eliminated. Although acute pain is usually sensed as coming from the surface, chronic pain is likely to be felt in deeper tissues as well as in the skin. Visceral pain impulses are usually carried on C fibers. Clinical Application 12.1 discusses treatments for severe pain.

Commonly, an event that stimulates pain receptors will trigger impulses on both types of pain fibers. This causes a dual sensation—a sharp, pricking pain, then a dull, aching one. The aching pain is usually more intense and may worsen over time. Chronic pain that is resistant to relief and control can be debilitating.

Pain impulses that originate from tissues of the head reach the brain on sensory fibers of the fifth, seventh, ninth, and tenth cranial nerves. All other pain impulses travel on sensory fibers of spinal nerves, and they pass into the spinal cord by way of the dorsal roots of these spinal nerves.

Upon reaching the spinal cord, pain impulses enter the gray matter of the dorsal horn, where they are processed (see chapters 10 and 11). The fast-conducting fibers synapse with long nerve fibers that cross over to the opposite side of the spinal cord and ascend in the lateral spinothalamic tracts.

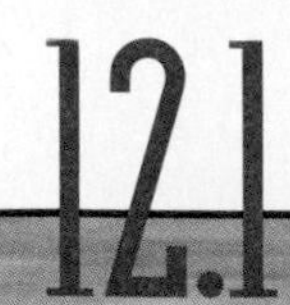

**Clinical Application 12.1**

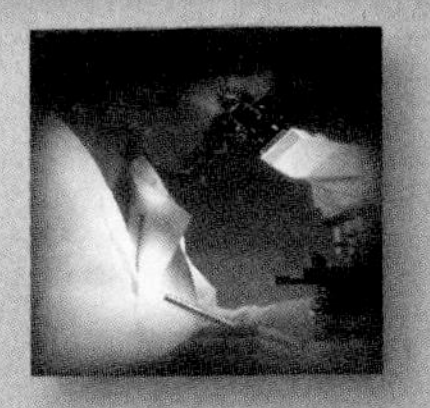

## Cancer Pain and Chronic Pain

One of the shortcomings of modern medicine is that many patients suffering from the pain of advanced cancer do not receive adequate medication to ease their final days. Studies show that 50% of people with cancer are in pain at the time of their diagnosis, and 90% of those in advanced stages are in pain.

Doctors are beginning to advocate pain relief for all people with cancer who require it, not only for humanitarian reasons but for clinical ones. Studies on rats with lung cancer show that tumors spread more rapidly in animals not given pain medication. Plus, the widespread belief that giving narcotics would addict cancer patients has been shown to be unfounded. Narcotics are much more likely to be addicting when they are taken to provide euphoria than when they are taken to relieve severe pain.

Drugs used to help cancer patients include:

- Anti-inflammatory agents such as aspirin and ibuprofen.
- Weak narcotics such as codeine.
- Strong narcotics such as morphine.
- Opiates delivered directly to the spine via an implanted reservoir.

Many patients are offered the option of patient-controlled analgesia, in which they determine their own dosage schedule.

Chronic pain is of three types: lower back pain, migraine, and myofascial syndrome (inflammation of muscles and their fascia). A variety of approaches is used to treat chronic pain:

- Biofeedback. A mechanical device detects and amplifies an autonomic body function, such as blood pressure or heart rate. A person made aware of the measurement can concentrate to attempt to alter it to within a normal range.
- Anti-inflammatory drugs.
- Stretching exercises.
- Trigger point injections of local anesthetic drugs into cramping muscles.
- The antidepressant amitriptyline raises serotonin levels in the central nervous system, relieving some chronic pain.
- Transcutaneous electrical nerve stimulation (TENS) places electrodes on nerves causing pain. The patient feels a tingling sensation, then pain relief.
- A nerve block is invasive and interrupts a pain signal by freezing (cryotherapy) or introducing an anesthetic drug (neurolysis). Side effects include numbness or paralysis.
- A dorsal column stimulator consists of electrodes implanted near the spinal cord.

The impulses carried on the slow-conducting fibers pass through one or more interneurons before reaching the long fibers that cross over and ascend to the brain.

Within the brain, most of the pain fibers terminate in the reticular formation (see chapter 11), and from there are conducted on fibers of still other neurons to the thalamus, hypothalamus, and cerebral cortex.

### Regulation of Pain Impulses

Awareness of pain seems to occur when pain impulses reach the level of the thalamus—that is, even before they reach the cerebral cortex. However, the cerebral cortex must judge the intensity of pain and locate its source. The cerebral cortex is also responsible for emotional and motor responses to pain.

Still other parts of the brain regulate the flow of pain impulses from the spinal cord, including areas of gray matter in the midbrain, pons, and medulla oblongata. Impulses from special neurons in these areas descend in the lateral funiculus (see chapter 11) to various levels within the spinal cord. The impulses stimulate the ends of certain nerve fibers to release biochemicals that can block pain signals by inhibiting presynaptic nerve fibers in the posterior horn of the spinal cord.

Among the inhibiting substances released in the posterior horn are neuropeptides called *enkephalins* and the monoamine *serotonin* (see chapter 10). Enkephalins can suppress both acute and chronic pain impulses; thus they can relieve relatively strong pain sensations, much as morphine and other opiate drugs do. In fact, enkephalins were discovered because they bind to the same receptor sites on neuron membranes as does morphine. Serotonin stimulates other neurons to release enkephalins.

Another group of neuropeptides with pain-suppressing, morphinelike actions are the *endorphins.* They occur in the pituitary gland and in regions of the nervous system, such as the hypothalamus, that transmit pain impulses. Enkephalins and endorphins are released in response to extreme pain impulses, providing natural pain control.

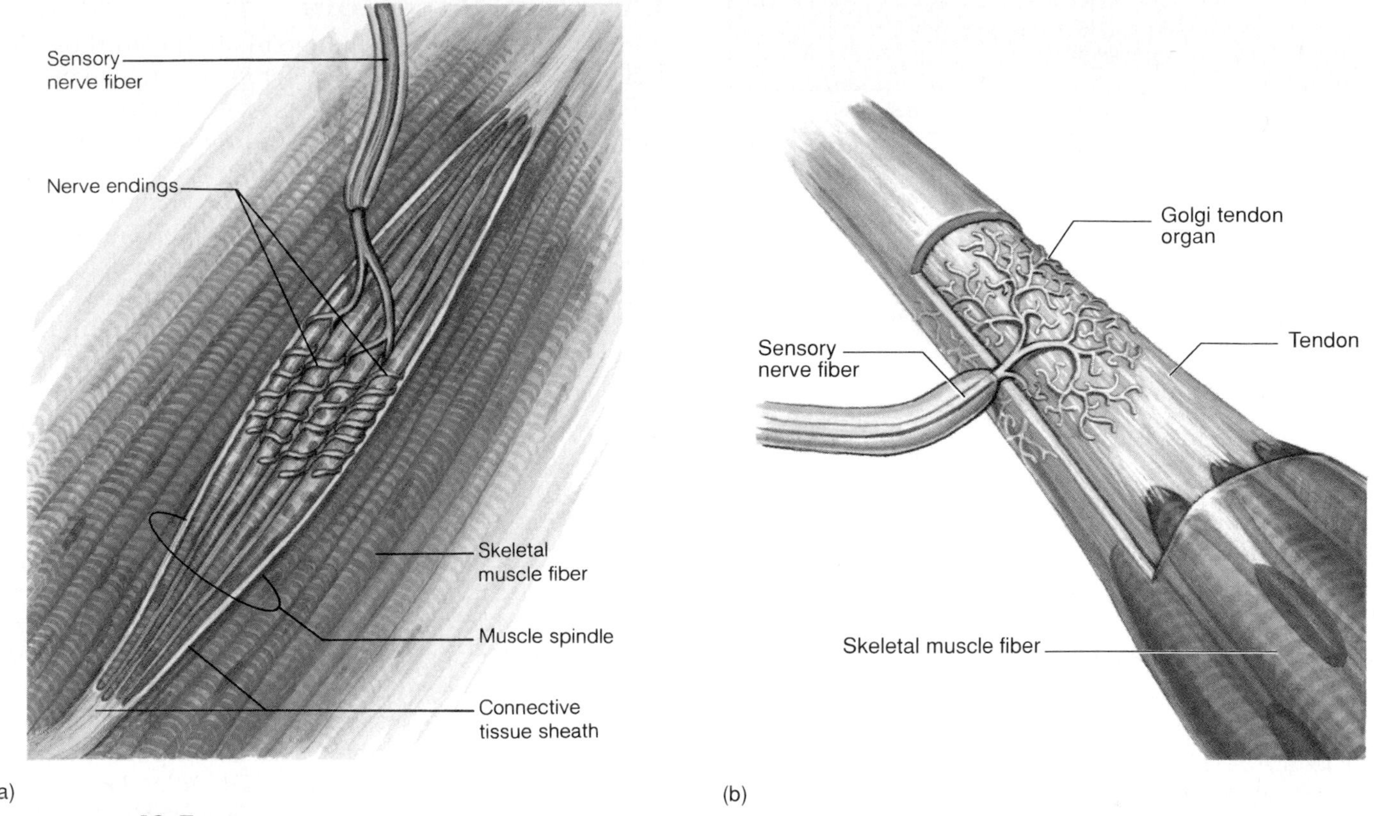


**FIGURE 12.5**

(*a*) Changes in muscle length stimulate muscle spindles, which are modified muscle fibers; (*b*) Golgi tendon organs occupy tendons, where they inhibit muscle contraction.

1. Describe three types of touch and pressure receptors.
2. Describe thermoreceptors.
3. What types of stimuli excite pain receptors?
4. What is referred pain?
5. Explain how neuropeptides control pain.

## Stretch Receptors

*Stretch receptors* are proprioceptors that send information to the spinal cord and brain concerning the lengths and tensions of muscles. The two main kinds of stretch receptors are muscle spindles and Golgi tendon organs; however, no sensation results when they are stimulated.

**Muscle spindles** are located in skeletal muscles near their junctions with tendons. Each spindle consists of one or more small, modified skeletal muscle fibers (intrafusal fibers) enclosed in a connective tissue sheath. Near its center, each fiber has a specialized nonstriated region with the end of a sensory nerve fiber wrapped around it (fig. 12.5*a*).

If striated portions of the muscle fiber contract, they relax the spindle. However, if the whole muscle relaxes and is stretched longer than usual, the muscle spindle is also stretched, triggering sensory nerve impulses on its nerve fiber. Such sensory impulses travel into the spinal cord and onto motor fibers leading back to the same muscle, contracting it. This action, called a **stretch reflex,** opposes the lengthening of the muscle and helps maintain the desired position of a limb in spite of gravitational or other forces tending to move it (see chapter 10).

**Golgi tendon organs** are found in tendons close to their attachments to muscles. Each is connected to a set of skeletal muscle fibers and is innervated by a sensory neuron (fig. 12.5*b*).

These receptors have relatively high thresholds and are stimulated by increased tension. Sensory impulses from them produce a reflex that inhibits contraction of the muscle whose tendon they occupy. Thus, the Golgi tendon organs stimulate a reflex with an effect opposite that of a stretch reflex. This reflex also helps maintain posture, and it protects muscle attachments from being pulled

## CHART 12.2 Somatic Receptors

| Type | Function | Sensation |
|---|---|---|
| Free nerve endings (mechanoreceptors) | Detect changes in pressure | Touch, pressure |
| Meissner's corpuscles (mechanoreceptors) | Detect objects moving over the skin | Touch, texture |
| Pacinian corpuscles (mechanoreceptors) | Detect changes in pressure | Deep pressure, vibrations |
| Free nerve endings (thermoreceptors) | Detect changes in temperature | Heat, cold |
| Free nerve endings (pain receptors) | Detect tissue damage | Pain |
| Muscle spindle (mechanoreceptors) | Detect changes in muscle length | None |
| Golgi tendon organ (mechanoreceptors) | Detect changes in muscle tension | None |

away from their insertions by excessive tension. Chart 12.2 summarizes the somatic receptors and their functions.

1. Describe a muscle spindle.
2. Explain how muscle spindles help maintain posture.
3. Where are Golgi tendon organs located?
4. What is the function of Golgi tendon organs?

## Special Senses

Special senses are those whose sensory receptors are within relatively large, complex sensory organs in the head. These organs and their respective senses include the following:

- olfactory organs → smell
- taste buds → taste
- ears → hearing
- organs of equilibrium → static equilibrium, dynamic equilibrium
- eyes → sight

## Sense of Smell

The ability to detect the strong scent of a fish market, the antiseptic odor of a hospital, the aroma of a ripe melon—and thousands of other smells—is possible thanks to a yellowish patch of tissue the size of a quarter high up in the nose. This fabric of sensation is actually a layer of 12 million specialized cells.

### Olfactory Receptors

Smell, or olfactory, receptors are similar to those for taste in that they are chemoreceptors sensitive to chemicals dissolved in liquids. These two chemical senses function closely together and aid in food selection, because we smell food at the same time we taste it. In fact, it is often difficult to tell what part of a food sensation is due to smell and what part is due to taste. For this reason, an onion tastes quite different when sampled with the nostrils closed, because much of the usual onion sensation is due to odor. Similarly, if copious mucous secretions from an upper respiratory infection (such as a cold) cover the olfactory receptors, food may seem to lose its taste. Clinical Application 12.2 discusses an unusual and disturbing type of sensory abnormality.

### Olfactory Organs

The **olfactory organs,** which contain the olfactory receptors, also include epithelial supporting cells. These organs appear as yellowish brown masses surrounded by pinkish mucous membrane. They cover the upper parts of the nasal cavity, the superior nasal conchae, and a portion of the nasal septum (fig. 12.6).

The **olfactory receptor cells** are bipolar neurons surrounded by columnar epithelial cells. These neurons have knobs at the distal ends of their dendrites covered with hairlike cilia. The cilia project into the nasal cavity and are thought to be the sensitive portions of the receptors (fig. 12.7). A person's 12 million olfactory receptor cells each have ten to twenty cilia.

Chemicals that stimulate olfactory receptors enter the nasal cavity as gases, but they must dissolve at least partially in the watery fluids that surround the cilia before they can be detected. In some cases, these chemicals bind to specific plasma membrane receptors on the olfactory receptor cells, depolarizing them and thereby generating a nerve impulse.

The expert nose of the bloodhound is due to its 4 billion olfactory cells. A bloodhound's olfactory epithelium spread out covers 59 square inches, compared to 1 1/2 square inches in a human. Still, the human sense of smell is nothing to sneeze at—people can detect one molecule of green pepper smell in a gaseous sea of 3 trillion other molecules. Our 12 million smell cells and their many million more receptors allow us to discern some 10,000 scents. But, without air, we cannot smell, as early astronauts could attest. In the vacuum of space, odorant molecules could not reach the astronauts' senses, and eating in space was a rather tasteless—and some would say joyless—experience. Currently, astronauts eat in pressurized cabins and are not affected by the vacuum in which they travel.

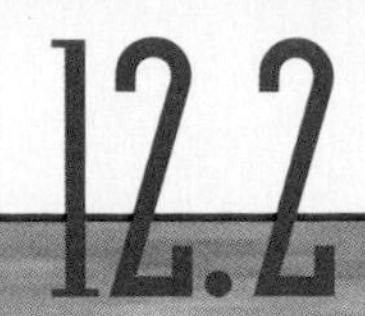

CLINICAL APPLICATION 12.2

## Mixed-up Senses—Synesthesia

"His name was purple."
"The song was full of shimmering green triangles."
"The paint smelled red."
"The sunset was salty."

To 1 in 500,000 of us, senses become unusually associated in a little-understood condition called synesthesia. A sensation involving one sense is perceived in terms of another, so that, most commonly, visions take on characteristic smells or sounds are associated with particular colors. Although synesthetic people are unlikely to share the same associations, they do share some aspects of their talents—their particular quirks are involuntary, remain constant over their lifetimes (a certain name or song is always a certain color), and they are absolutely convinced that what they perceive is reality, not something in their "mind's eye."

We do not know what causes synesthesia, although it does seem to be inherited. The condition has been sheepishly reported to psychologists and physicians for at least 200 years. Various theories (all unproven) have attributed the condition to an immature nervous system that cannot sort out sensory stimuli; altered brain circuitry; and simply an exaggerated use of metaphor—taking such descriptions as a sharp flavor, warm color, and sweet person too literally.

In 1980, measurements of cerebral blood flow linked synesthesia to a brain abnormality. A person who tasted in geometric shapes inhaled a harmless radioactive chemical, xenon-133, which her tissues absorbed. The rate at which the chemical left certain parts of her brain indicated how metabolically active that part was. When brain activity was so assessed during a synesthetic experience, it was found that blood flow in her left hemisphere, particularly in the temporal lobe, plummeted 18%—a decrease seen only when tissue dies as a result of a stroke. But the woman was perfectly healthy. The left hemisphere is the site of the language center. Another clue is that people with temporal lobe epilepsy are often synesthetic. Researchers believe that synesthesia reflects a breakdown in the translation of a perception into language—but we have much to learn about this fascinating mixing of the senses.

Sensory receptors are not the same as membrane receptors. Sensory receptors may be as small as individual cells or as large as complex organs such as the eye or ear. They respond to sensory stimuli. Membrane receptors are molecules such as proteins and glycoproteins located on the plasma membranes of cells. They allow cells, such as neurons and olfactory receptor cells, to respond to specific molecules reaching their membranes. Thus, the sensory receptors known as taste buds respond to chemical stimuli, but they depend on plasma membrane receptors to do so.

### Olfactory Nerve Pathways

Once olfactory receptors are stimulated, nerve impulses travel along the axons of the receptor cells through tiny openings in the cribriform plates of the ethmoid bone and lead to neurons located in the enlargements of the **olfactory bulbs.** These structures lie on either side of the crista galli of the ethmoid bone (see figs. 7.26 and 12.6). Within the olfactory bulbs, the sensory impulses are analyzed, and as a result, additional impulses travel along the **olfactory tracts** to portions of the limbic system (see chapter 11). The main interpreting areas for the olfactory impulses (olfactory cortex) are located deep within the temporal lobes and at the bases of the frontal lobes, anterior to the hypothalamus.

### Olfactory Stimulation

Biologists are not certain how stimulated receptors encode specific smells, but one hypothesis is that each odor stimulates a distinct subset of receptor subtypes. The brain then recognizes the combination of receptors as an *olfactory code.* For example, perhaps there are ten types of odor receptors. Banana might stimulate receptors 2, 4, and 7; garlic receptors 1, 5, and 9.

Because the olfactory organs are high in the nasal cavity above the usual pathway of inhaled air, sniffing and forcing air over the receptor areas may be necessary to smell a faint odor. Olfactory receptors undergo sensory adaptation rather rapidly, so the intensity of a smell drops about 50% within a second following the stimulation. Within a minute, the receptors may

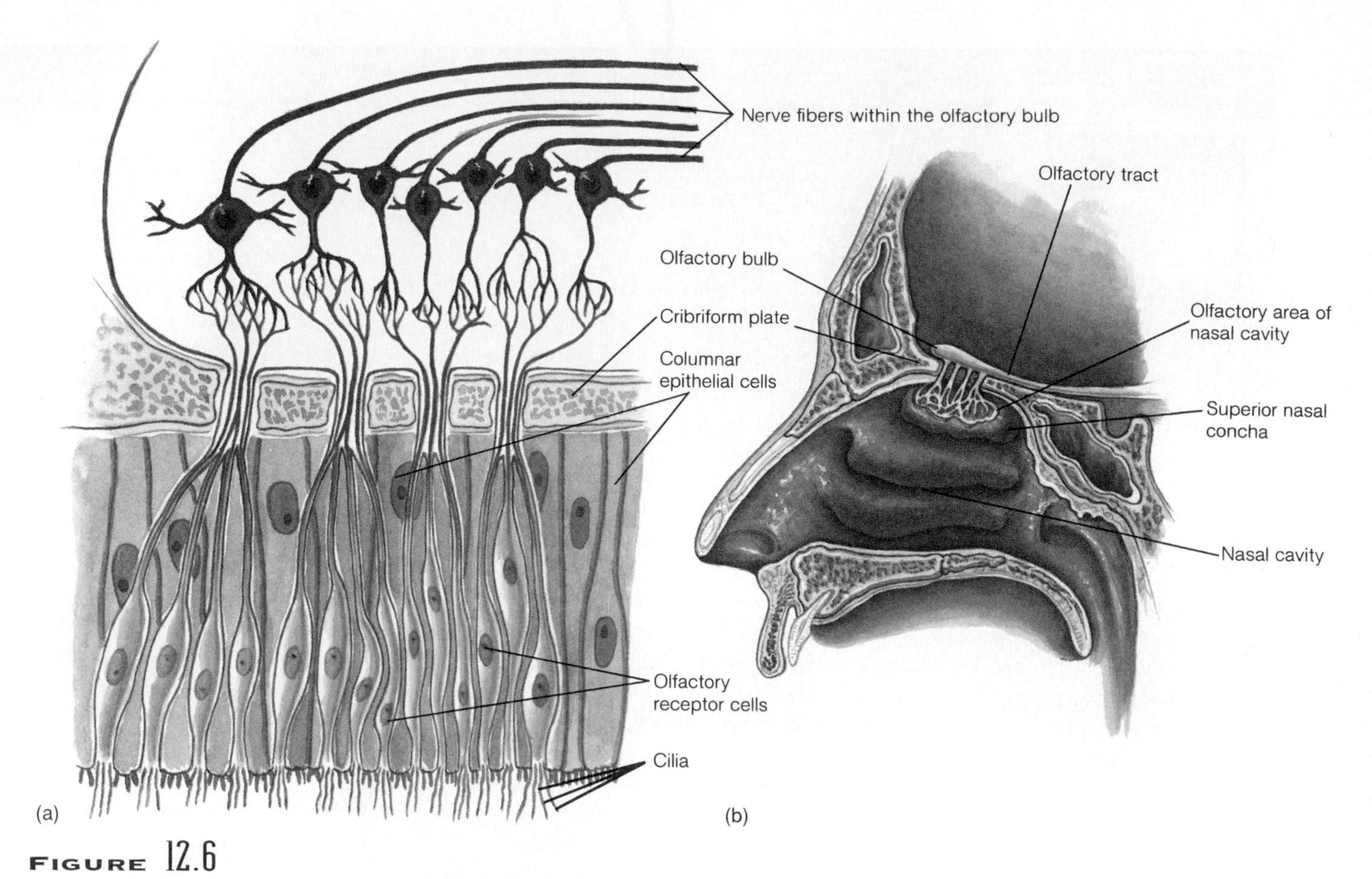


**FIGURE 12.6**

(*a*) Columnar epithelial cells support olfactory receptor cells, which have cilia at their distal ends. (*b*) The olfactory area is associated with the superior nasal concha.

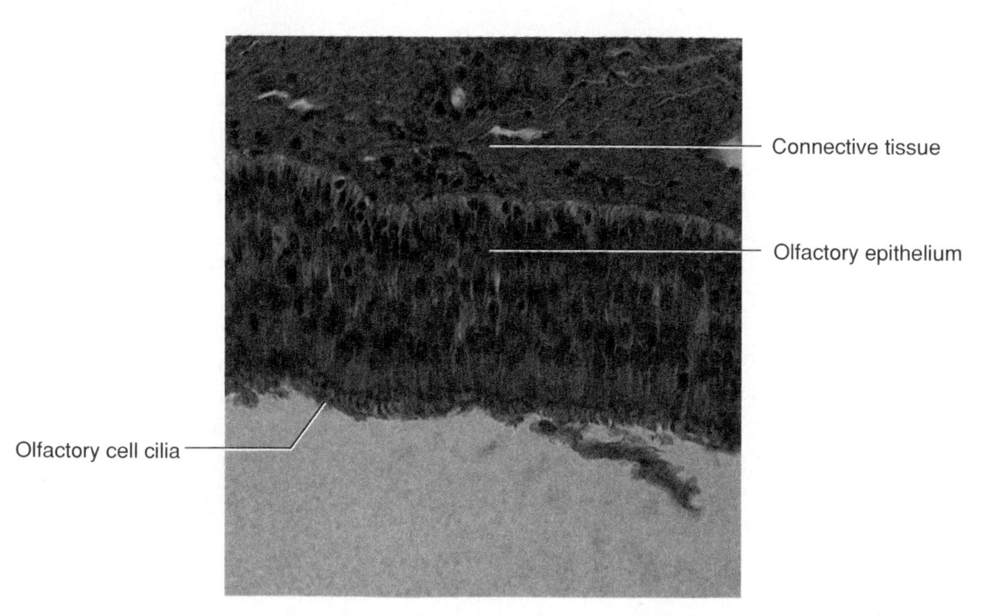


**FIGURE 12.7**

Light micrograph of the olfactory epithelium (250×).

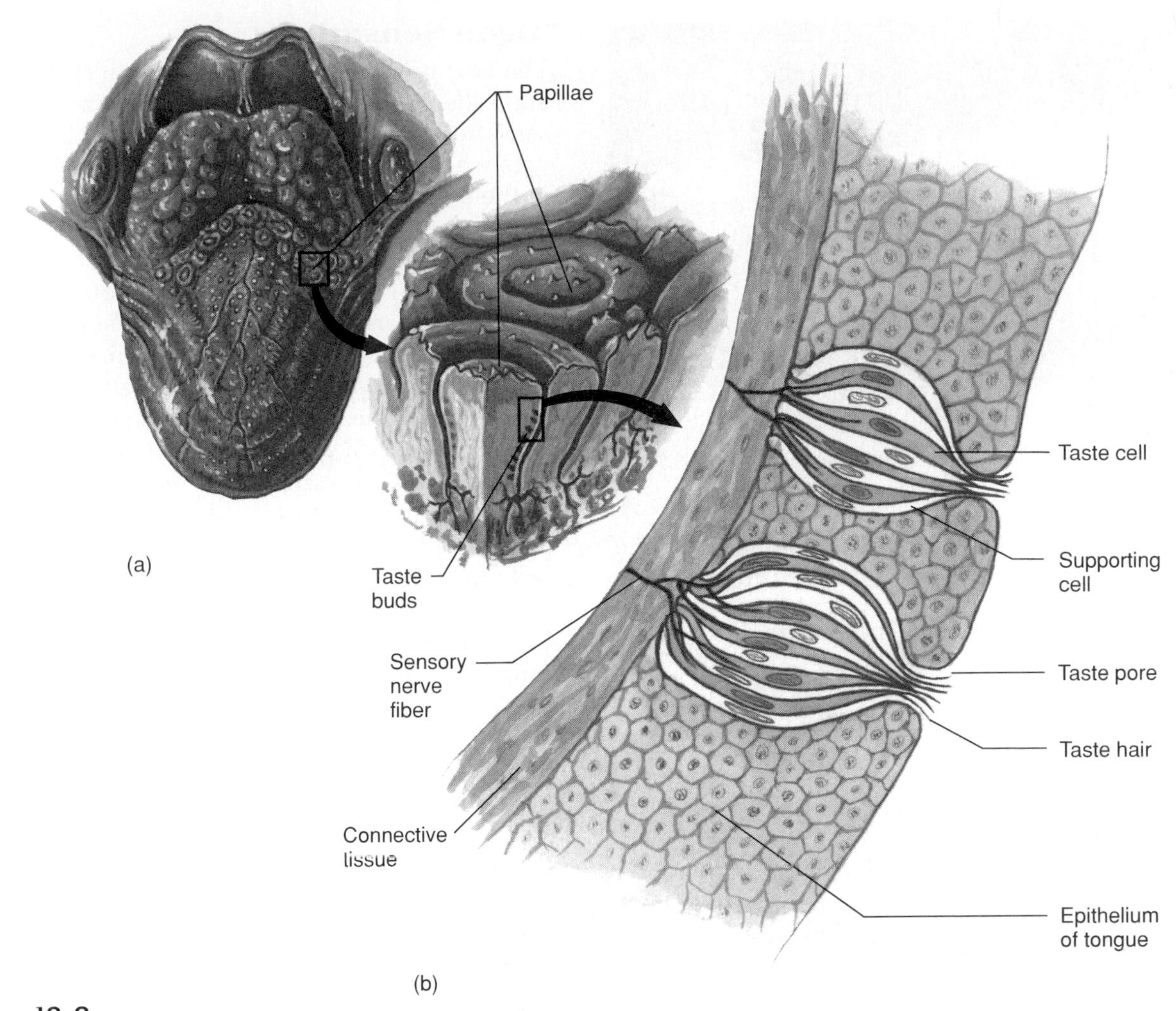


**FIGURE 12.8**

(*a*) Taste buds on the surface of the tongue are associated with nipplelike elevations called papillae. (*b*) A taste bud contains taste cells and has an opening, the taste pore, at its free surface.

become almost insensitive to a given odor, but even though they have adapted to one scent, their sensitivity to other odors remains unchanged.

The olfactory receptor neurons are the only parts of the nervous system that are in direct contact with the outside environment. Because of their exposed positions, these neurons are subject to damage; and because damaged neurons usually are not replaced, progressive diminishing of olfactory sense usually occurs with age. In fact, some investigators estimate that about 1% of the olfactory receptors are lost every year.

1. Where are the olfactory receptors located?
2. Trace the pathway of an olfactory impulse from a receptor to the cerebrum.
3. Why is the sense of smell likely to decrease with age?

One person enjoys liver and onions; another finds them distasteful. Of the 68% of women who can detect armpit odor (a chemical called androsterone), 72% report disliking it; of the 57% of men who can smell androsterone, only 50% dislike it. What accounts for these individual palates and noses? To some extent, what you taste or smell is in your genes. For example, the ability to smell a squashed skunk or freesia flowers is inherited.

## SENSE OF TASTE

**Taste buds** are the special organs of taste. They resemble orange sections and are located primarily on the surface of the tongue where they are associated with tiny elevations called **papillae** (figs. 12.8 and

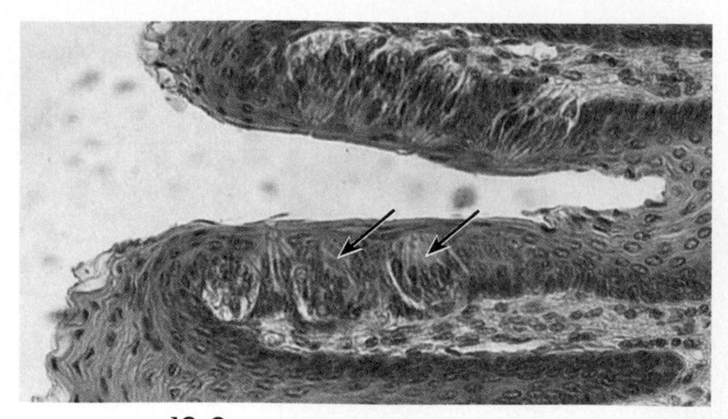

**FIGURE 12.9**
A light micrograph of some taste buds (arrows) (400×).

12.9). They also are found in smaller numbers in the roof of the mouth, lining the cheeks, and in the walls of the pharynx.

## Taste Receptors

Each taste bud includes a group of modified epithelial cells, which are the **taste cells** (gustatory cells) that function as receptors. Each of our 10,000 taste buds houses 60 to 100 taste cells. The taste bud also incudes a number of epithelial supporting cells. The entire structure is somewhat spherical, with an opening, the **taste pore,** on its free surface. Tiny projections (microvilli), called **taste hairs,** protrude from the outer ends of the taste cells and jut out through the taste pore. These taste hairs are the sensitive parts of the receptor cells.

Interwoven among and wrapped around the taste cells is a network of nerve fibers. The ends of these fibers closely contact the receptor cell membranes. A stimulated receptor cell triggers an impulse on a nearby nerve fiber, which travels into the brain.

A chemical to be tasted must dissolve in the watery fluid surrounding the taste buds. The salivary glands supply this fluid. To demonstrate its importance, blot your tongue and try to taste some dry food; then repeat the test after moistening your tongue with saliva.

As is the case for smell, the mechanism of tasting is not well understood, but may involve combinations of chemicals binding specific receptor sites on taste hair surfaces, altering membrane polarization, and thereby generating sensory impulses on nearby nerve fibers. The amount of change is directly proportional to the concentration of the stimulating substance.

Although the taste cells in all taste buds appear very much alike microscopically, there are at least four types, corresponding to a particular kind of chemical stimulus. This produces at least four primary taste (gustatory) sensations.

## Taste Sensations

The four *primary taste sensations* are *sweet, sour, salty,* and *bitter.* Some investigators recognize two other taste sensations, *alkaline* and *metallic.* Each of the many flavors we experience daily is believed to result from one of the primary sensations or from some combination of two or more of them. The way we experience flavors may also reflect the concentration of chemicals as well as the sensations of smell, texture (touch), and temperature. Furthermore, chemicals in some foods—chili peppers and ginger, for instance—may stimulate pain receptors that cause a burning sensation.

The tongue is often depicted as having regions in which taste buds specialize in a particular sensation—the tip tastes sweetness, the front saltiness, the sides sour, and the back bitter. Although taste buds with all specificities are scattered everywhere, and a single taste bud can have receptors for all four general types of tastes, each of the four major types of taste receptors is somewhat concentrated in certain regions of the tongue's surface (fig. 12.10). For example, *sweet receptors* are most plentiful near the tip of the tongue. This is why a child may prefer to lick a lollipop rather than chew it. The chemicals that stimulate these receptors are usually carbohydrates, but a few inorganic substances, including some salts of lead and beryllium, also elicit sweet sensations.

*Sour receptors* occur primarily along the margins of the tongue, and acids stimulate them. The intensity of a sour sensation is roughly proportional to the concentration of the hydrogen ions in the substance being tasted.

*Salt receptors* are most common in the tip and the upper front portion of the tongue. They are stimulated mainly by ionized inorganic salts. The quality of the sensation that each produces depends upon the kind of positively charged ion, such as $Na^+$ from table salt, that it releases into solution.

For many years biology teachers have given students strips of paper impregnated with a chemical called PTC to study a "bitterness tasting" gene. Typically, seven in ten students make a horrible face upon tasting the paper, because they taste PTC as extremely bitter. The other students look around in amazement, for they cannot taste a thing! PTC-tasters also find artificial sweeteners offensive and may taste cheese as bitter rather than pleasantly tart. Researchers hypothesize that some infants who fail to gain weight may simply not like the taste of milk.

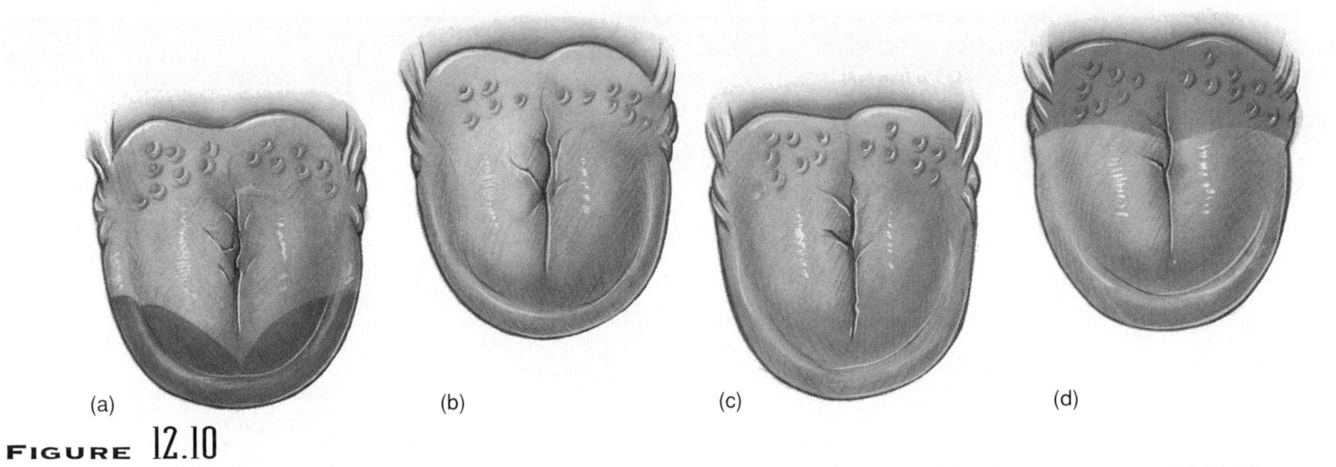


**FIGURE 12.10**
Color indicates patterns of taste receptors: (*a*) sweet receptors; (*b*) sour receptors; (*c*) salt receptors; and (*d*) bitter receptors.

*Bitter receptors* are located toward the back of the tongue. A variety of chemical substances stimulates them, including many organic compounds. Inorganic salts of magnesium and calcium produce bitter sensations, too.

One group of bitter compounds of particular interest are the *alkaloids,* which include a number of poisons such as strychnine, nicotine, and morphine. The fact that we spit out bitter substances may be a protective mechanism to avoid poisonous alkaloids in foods.

Taste receptors, like olfactory receptors, undergo sensory adaptation relatively rapidly. The resulting loss of taste can be avoided by moving bits of food over the surface of the tongue to stimulate different receptors at different moments.

Although taste cells are located very close to the surface of the tongue and are therefore exposed to environmental wear and tear, the sense of taste is not as likely to diminish with age as is the sense of smell. This is because taste cells reproduce continually, so that any one of these cells functions for only about three days before it is replaced.

### Taste Nerve Pathways

Sensory impulses from taste receptors located in the anterior two-thirds of the tongue travel on fibers of the facial nerve (VII); impulses from receptors in the posterior one-third of the tongue and the back of the mouth pass along the glossopharyngeal nerve (IX); and impulses from receptors at the base of the tongue and the pharynx travel on the vagus nerve (X) (see chapter 11).

These cranial nerves conduct the impulses into the medulla oblongata. From there, the impulses ascend to the thalamus and are directed to the gustatory cortex of the cerebrum, located in the parietal lobe along a deep portion of the lateral sulcus.

Sensory information from olfactory receptors is also conducted to the limbic system, a brain center for memory and emotions. This is why we may become nostalgic over a scent from the past. A whiff of the perfume that grandma used to wear may bring back a flood of memories. The input to the limbic system also explains why odors can alter mood so easily. For example, the scent of new-mown hay or rain on a summer's morning generally makes us feel good.

Clinical Application 12.3 and chart 12.3 discuss disorders of smell and taste.

1. Why is saliva necessary to taste?
2. Name the four primary taste sensations.
3. What characteristic of taste receptors helps maintain a sense of taste with age?
4. Trace a sensory impulse from a taste receptor to the cerebral cortex.

## SENSE OF HEARING

The organ of hearing, the **ear,** has external, middle, and inner sections. In addition to making hearing possible, the ear provides the sense of equilibrium.

### External Ear

The external ear consists of two parts: an outer, funnel-like structure called the **auricle** (pinna), and an S-shaped tube, the **external auditory meatus** (external auditory canal) that leads inward for about 2.5 centimeters (fig. 12.11).

The external auditory meatus passes into the temporal bone. Near its opening, hairs guard the tube. It is lined with skin that contains many modified sweat glands called *ceruminous glands,* which secrete wax (cerumen). The hairs and wax help keep relatively large foreign objects, such as insects, out of the ear.

Clinical Application 12.3

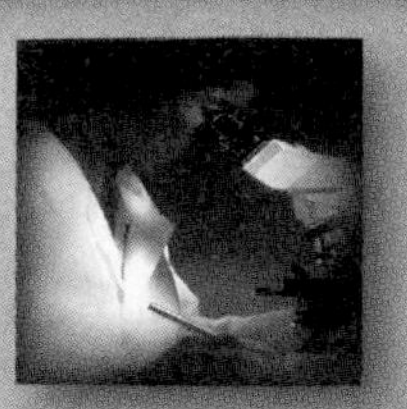

## Smell and Taste Disorders

Imagine a spicy slice of pizza, or freshly brewed coffee, and your mouth waters in anticipation. But for 2 million people in the United States, the senses of smell and taste are dulled, distorted, or gone altogether. Many more of us get some idea of their plight when a cold temporarily stifles these senses.

Compared to the loss of hearing or sight, being unable to taste or smell normally may seem more an oddity than an illness. But those with such ailments would probably disagree. In some situations, a poor or lacking sense of smell can even be dangerous. One patient died in a house fire because he did not smell the smoke in time to escape.

The direct connection between the outside environment and the brain makes the sense of smell very vulnerable to damage. Smell and taste disorders can be triggered by colds and flu, allergies, nasal polyps (swollen mucous membranes inside the nose), a head injury, chemical exposure, a nutritional or metabolic problem, or a disease. In many cases, a cause cannot be identified.

Drugs can alter taste and smell in many ways, affecting cell turnover, the neural conduction system, the status of receptors, and changes in nutritional status. Consider what happened to twelve hikers touring Peru and Bolivia. A day before a long hike, three of them had begun taking acetazolamide (Diamox), a drug that prevents acute mountain sickness. The night after the climb, the group went out for beer. To three of the people, the brew tasted unbearably bitter, and a drink of cola to wash away the taste was equally offensive. At fault: acetazolamide.

Drugs containing sulfur atoms are notorious for squelching taste. They include the anti-inflammatory drug penicillamine, the anti-hypertensive drug captopril (Capoten), and transdermal (patch) nitroglycerin to treat chest pain. The antibiotic tetracycline and the antiprotozoan metronidazole (Flagyl) cause a metallic taste. Cancer chemotherapy and radiation treatment often alter taste and smell.

Exposure to toxic chemicals can affect taste and smell, too. A 45-year-old woman from Altoona, Pennsylvania, suddenly found that once-pleasant smells had become offensive. Her doctor traced her problem to inhaling a paint stripper. Hydrocarbon solvents in the product—toluene, methanol, and methylene chloride—were the culprits responsible for her *cacosmia,* the association of an odor of decay with normally inoffensive stimuli.

Source: From "When Smell and Taste Go Awry" in *FDA Consumer,* November 1991, pp. 29–33.

### Chart 12.3 Types of Smell and Taste Disorders

Doctors use these terms when discussing taste and smell disorders:

| | Smell | Taste |
|---|---|---|
| Loss of sensation | Anosmia | Ageusia |
| Diminished sensation | Hyposmia | Hypogeusia |
| Heightened sensation | Hyperosmia | Hypergeusia |
| Distorted sensation | Dysosmia | Dysgeusia |

Source: From "When Smell and Taste Go Awry" in *FDA Consumer,* November 1991, page 33.

Vibrations are transmitted through matter as sound waves. Just as the sounds of some musical instruments are produced by vibrating strings or reeds, the sounds of the human voice are caused by vibrating vocal folds in the larynx. The auricle of the ear helps collect sound waves traveling through air and directs them into the external auditory meatus.

After entering the meatus, the sound waves pass to the end of the tube and alter the pressure on the **tympanic membrane** (eardrum). The tympanic membrane moves back and forth in response, reproducing the vibrations of the sound wave source.

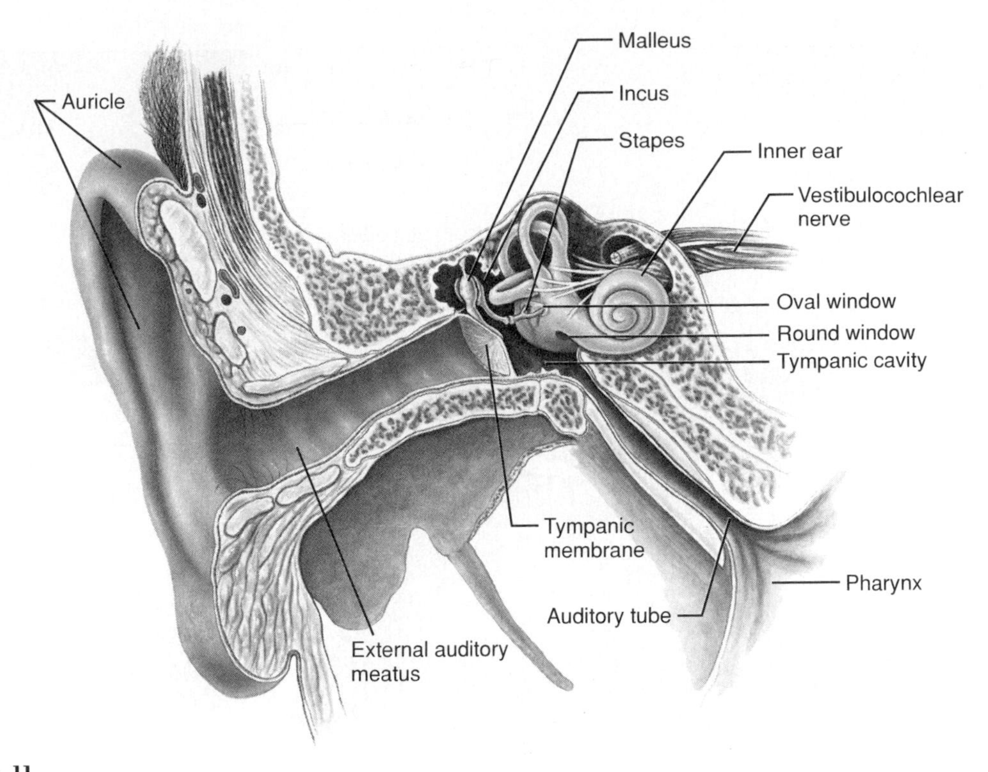


**FIGURE 12.11**
Major parts of the ear.

## Middle Ear

The tympanic membane is thought of as the boundary between the outer ear and the middle ear, but it is technically part of the middle ear. The middle ear includes the tympanic membrane, the tympanic cavity, and three small bones called auditory ossicles.

The **tympanic cavity** is an air-filled space in the temporal bone that separates the external and internal ears.

The tympanic membrane is a semitransparent membrane covered by a thin layer of skin on its outer surface and by mucous membrane on the inside. It has an oval margin and is cone-shaped, with the apex of the cone directed inward. One of the auditory ossicles (malleus) maintains its cone shape.

The three **auditory ossicles,** called the *malleus,* the *incus,* and the *stapes,* are attached to the wall of the tympanic cavity by tiny ligaments and are covered by mucous membrane. These bones bridge the eardrum and the inner ear, transmitting vibrations between these parts. Specifically, the malleus is attached to the eardrum, and when the eardrum vibrates, the malleus vibrates in unison with it. The malleus vibrates the incus, and the incus passes the movement onto the stapes. Ligaments hold the stapes to an opening in the wall of the tympanic cavity called the **oval window.** Vibration of the stapes, which acts like a piston at the oval window, moves a fluid within the inner ear. These vibrations of the fluid stimulate the hearing receptors (fig. 12.11).

In addition to transmitting vibrations, the auditory ossicles form a lever system that helps increase (amplify) the force of the vibrations as they pass from the eardrum to the oval window. Also, because the ossicles transmit vibrations from the relatively large surface of the eardrum to a much smaller area at the oval window, the vibrational force concentrates as it travels from the external to the inner ear. As a result of these two factors, the pressure (per square millimeter) that the stapes applies at the oval window is about twenty-two times greater than that which sound waves exert on the eardrum.

The middle ear also contains two small skeletal muscles that are attached to the auditory ossicles and

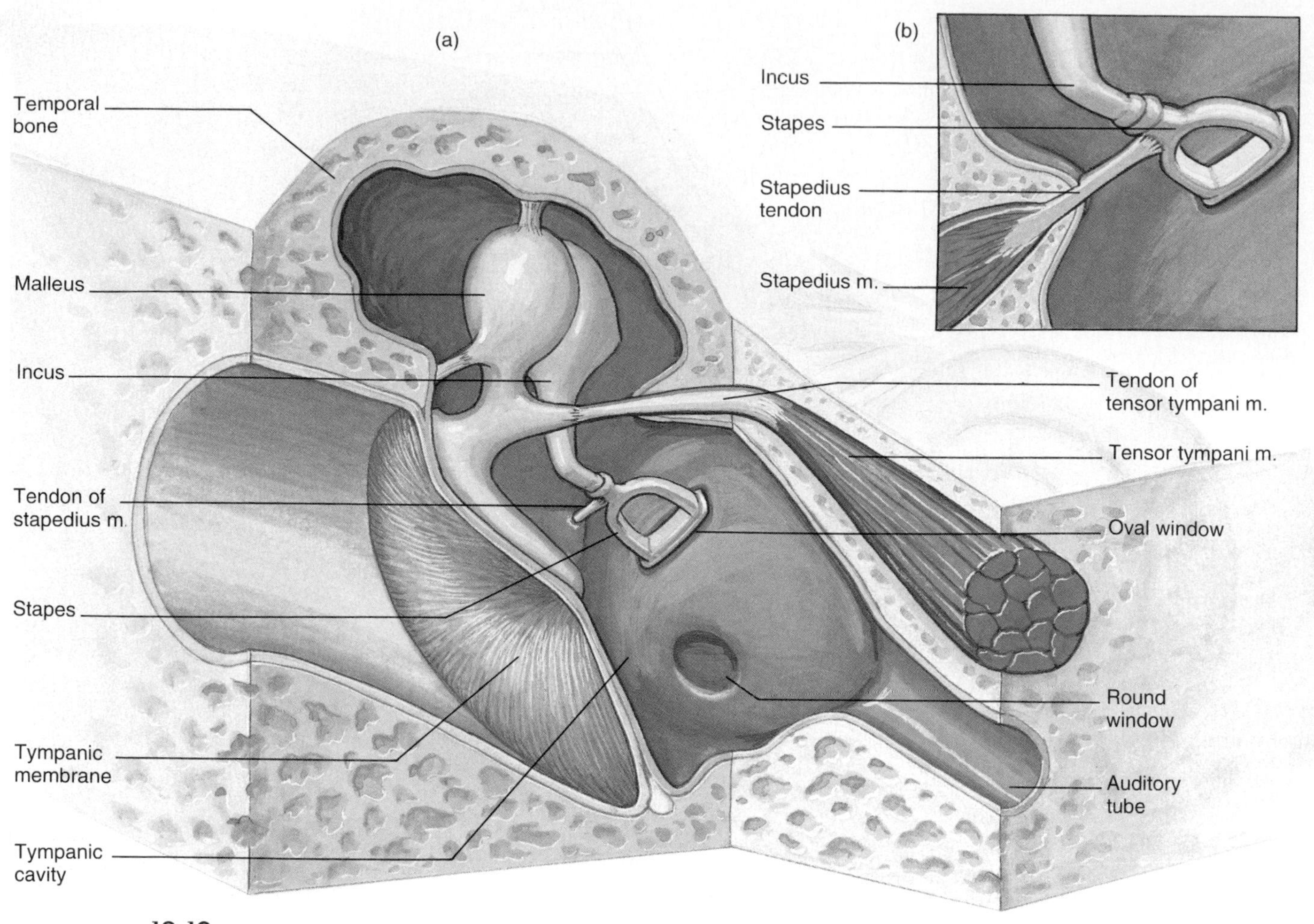


**FIGURE 12.12**

Two small muscles attached to the (*a*) malleus and (*b*) stapes, the tensor tympani and the stapedius, are effectors in the tympanic reflex. Figure 12.11 does not show these muscles.

are controlled involuntarily. One of them, the *tensor tympani,* is inserted on the medial surface of the malleus and is anchored to the cartilaginous wall of the auditory tube. When it contracts, it pulls the malleus inward. The other muscle, the *stapedius,* is attached to the posterior side of the stapes and the inner wall of the tympanic cavity. It pulls the stapes outward (fig. 12.12). These muscles are the effectors in the **tympanic reflex,** which is elicited in about one-tenth second following a long, external sound. When the reflex occurs, the muscles contract, and the malleus and stapes move. As a result, the bridge of ossicles in the middle ear becomes more rigid, reducing its effectiveness in transmitting vibrations to the inner ear.

The tympanic reflex reduces pressure from loud sounds that might otherwise damage the hearing receptors. The tympanic reflex is also elicited by ordinary vocal sounds, as when a person speaks or sings, and this action muffles the lower frequencies of such sounds, improving the hearing of higher frequencies, which are common in human vocal sounds. In addition, the tensor tympani muscle also steadily pulls on the eardrum. This is important because a loose tympanic membrane would not be able to transmit vibrations effectively to the auditory ossicles.

The middle ear muscles take 100 to 200 milliseconds to contract. For this reason, the tympanic reflex cannot protect the hearing receptors from the effects of loud sounds that occur very rapidly, such as those from an explosion or a gunshot. On the other hand, this protective mechanism can reduce the effects of intense sounds that arise relatively slowly, such as the roar of thunder.

## Auditory Tube

An **auditory tube** (eustachian tube) connects each middle ear to the throat. This tube allows air to pass between the tympanic cavity and the outside of the body by way of the throat (nasopharynx) and mouth. It helps maintain equal air pressure on both sides of the eardrum, which is necessary for normal hearing (see fig. 12.11).

The function of the auditory tube becomes noticeable during rapid change in altitude. As a person moves from a high altitude to a lower one, the air pressure on the outside of the eardrum steadily increases. As a result, the eardrum may be pushed inward, out of its normal position, impairing hearing.

When the air pressure difference is great enough, some air may force its way up through the auditory tube into the middle ear. This equalizes the pressure on both sides of the eardrum, which moves back into its regular position, causing a popping sound as normal hearing returns. A reverse movement of air ordinarily occurs when a person moves from a low altitude to a higher one.

The auditory tube is usually closed by valvelike flaps in the throat, which may inhibit air movements into the middle ear. Swallowing, yawning, or chewing aid in opening the valves, and can hasten equalization of air pressure.

> Most parents sooner or later learn to recognize the signs of an ear infection in their infant or toddler—irritability, screaming incessantly for no apparent reason, perhaps fever or tugging on the affected ear. A doctor viewing the painful ear with an instrument called an otoscope sees a red and bulging eardrum. The diagnosis: *otitis media,* or a middle ear infection.
>
> Ear infections occur because the mucous membranes that line the auditory tubes are continuous with the linings of the middle ears, creating a conduit for bacteria infecting the throat or nasal passages to travel to the ear. This route to infection is greater in young children because their auditory tubes are shorter than they are in adults. Half of all children in the United States have an ear infection by the first birthday, and 90% have one by age six.
>
> Physicians treat acute otitis media with antibiotics. Because recurrent infections may cause hearing loss and interfere with learning, children with recurrent otitis media are often fitted with tympanostomy tubes, which are inserted into affected ears during a brief surgical procedure. The tubes form a small tunnel through the eardrum so the ears can drain. By the time the tubes fall out, the child has usually outgrown the susceptibility to ear infections.

## Inner Ear

The inner ear is a complex system of intercommunicating chambers and tubes called a **labyrinth.** Each ear has two such regions—the osseous labyrinth and the membranous labyrinth.

The *osseous labyrinth* is a bony canal in the temporal bone; the *membranous labyrinth* is a tube that lies within the osseous labyrinth and has a similar shape (fig. 12.13). Between the osseous and membranous labyrinths is a fluid called *perilymph,* which cells in the wall of the bony canal secrete. The membranous labyrinth contains a slightly different fluid called *endolymph.*

The parts of the labyrinths include a **cochlea** that functions in hearing, and three **semicircular canals** that provide a sense of equilibrium. A bony chamber called the **vestibule,** which is located between the cochlea and the semicircular canals, houses membranous structures that serve both hearing and equilibrium.

The cochlea is shaped like a coiled snail shell. Inside, it contains a bony core (modiolus) and a thin, bony shelf (spiral lamina) that entwines the core like the threads of a screw. The shelf divides the bony labyrinth of the cochlea into upper and lower compartments. The upper compartment, called the *scala vestibuli,* leads from the oval window to the apex of the spiral. The lower compartment, the *scala tympani,* extends from the apex of the cochlea to a membrane-covered opening in the wall of the inner ear called the **round window.** These compartments constitute the bony labyrinth of the cochlea, and they are filled with perilymph. At the apex of the cochlea, the fluids in the chambers can flow together through a small opening (helicotrema).

A portion of the membranous labyrinth within the cochlea, called the *cochlear duct* (scala media), is filled with endolymph and lies between the two bony compartments. The cochlear duct ends as a closed sac at the apex of the cochlea. The duct is separated from the scala vestibuli by a *vestibular membrane* (Reissner's membrane) and from the scala tympani by a *basilar membrane* (fig. 12.14).

The basilar membrane extends from the bony shelf of the cochlea and forms the floor of the cochlear duct. It contains many thousands of stiff, elastic fibers whose lengths vary, becoming progressively longer from the base of the cochlea to its apex. Vibrations entering the perilymph at the oval window travel along the scala vestibuli and pass through the vestibular membrane to enter the endolymph of the cochlear duct, where they move the basilar membrane.

After passing through the basilar membrane, the vibrations enter the perilymph of the scala tympani,

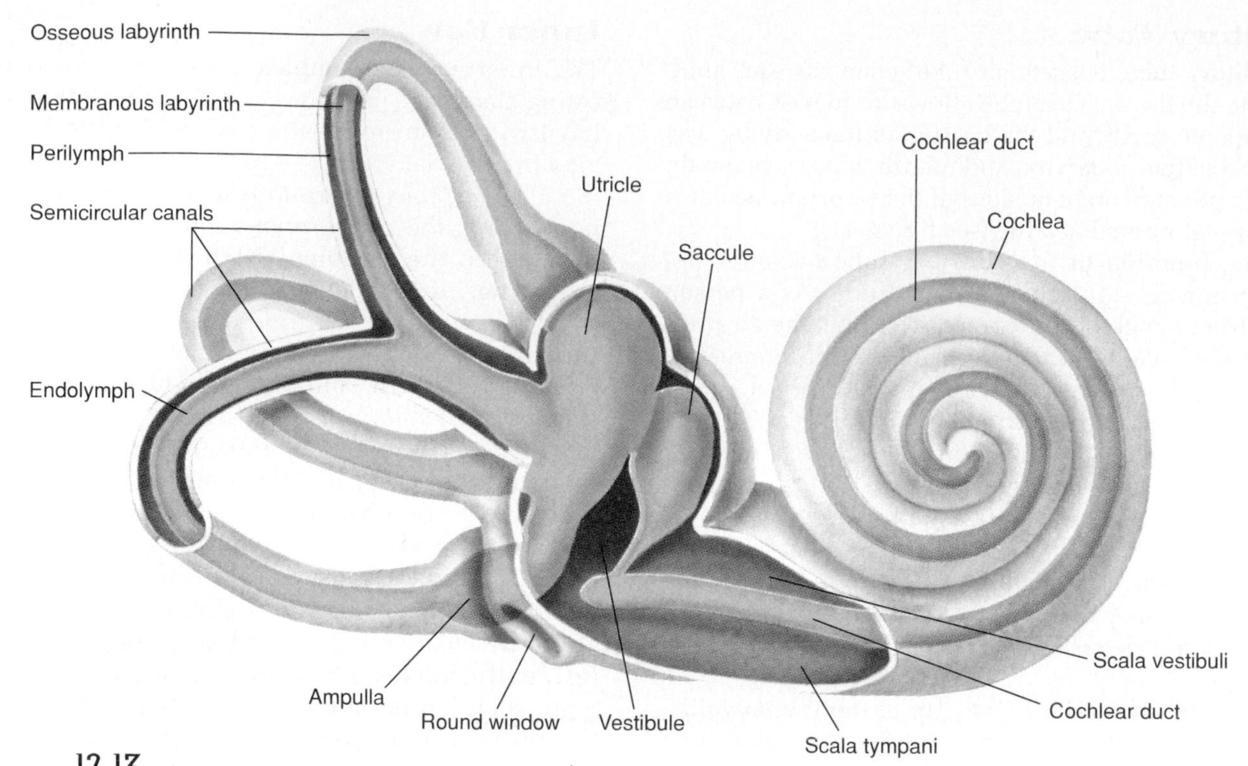


**FIGURE 12.13**

Perilymph separates the osseous labyrinth of the inner ear from the membranous labyrinth, which contains endolymph.

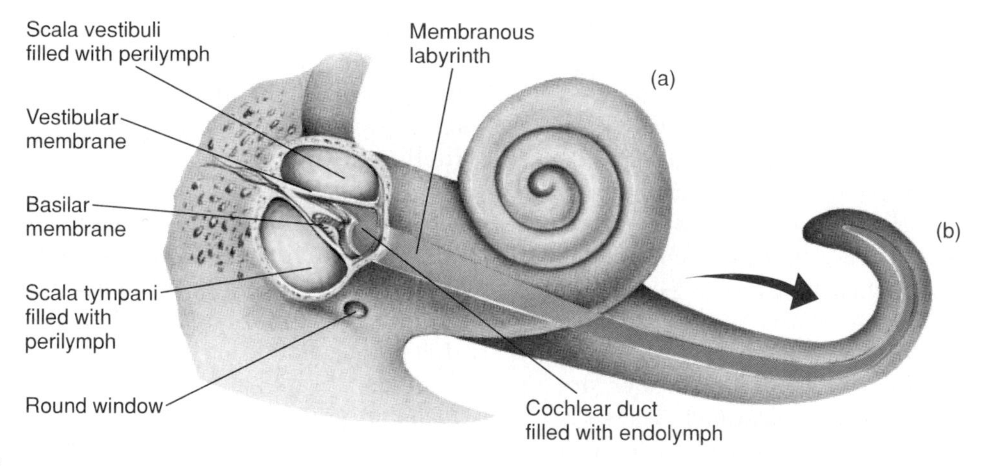


**FIGURE 12.14**

(*a*) The cochlea is a coiled, bony canal with a membranous tube inside. (*b*) If the cochlea could be unwound, the membranous tube would be seen ending as a closed sac at the apex of the bony canal.

and their forces are dissipated into the air in the tympanic cavity by movement of the membrane covering the round window.

The **organ of Corti,** which contains about 16,000 hearing receptor cells, is located on the upper surface of the basilar membrane and stretches from the apex to the base of the cochlea. The receptor cells, called **hair cells,** are arranged in four parallel rows, and they possess numerous hairlike processes (stereocilia) that extend into the endolymph of the cochlear duct. Above these hair cells is a *tectorial membrane,* which is attached to the bony shelf of the cochlea and passes like a roof over the receptor cells, contacting the tips of their hairs (figs. 12.15 and 12.16).

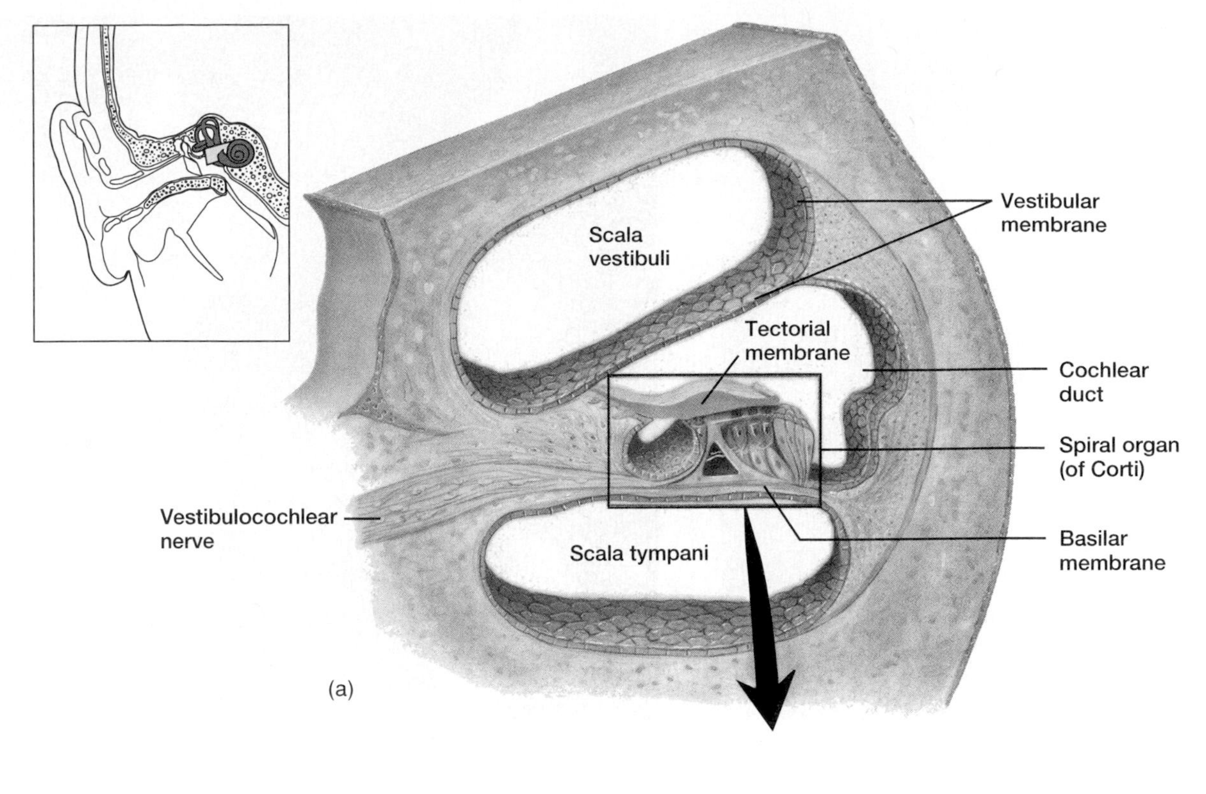


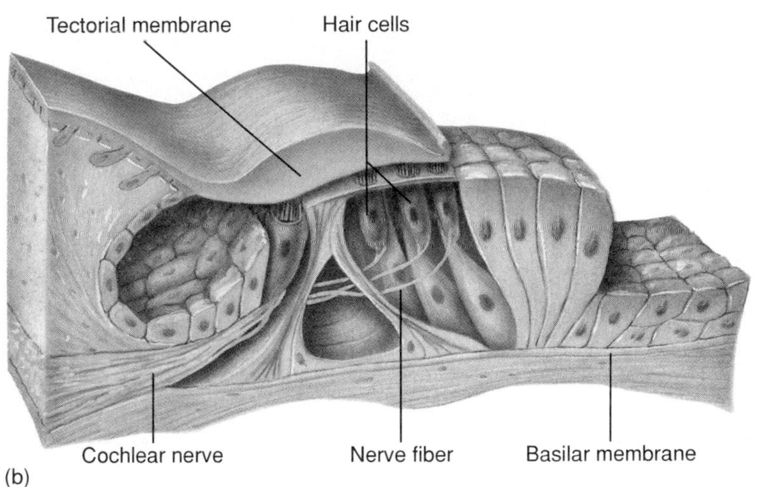


**FIGURE 12.15**

(*a*) Cross section of the cochlea; (*b*) organ of Corti and the tectorial membrane.

Because of its structure, different portions of the basilar membrane are moved by different frequencies of vibration. Thus, a particular sound frequency causes the hairs of a specific group of receptor cells to shear back and forth against the tectorial membrane more intensely, while each of the other frequencies deflects another set of receptor cells. There is some evidence that movements originating in some hair cells may contribute to this process.

These receptor cells are epithelial cells, but they respond to stimuli somewhat like neurons (see chapter 10). For example, when a receptor cell is at rest, its membrane is polarized. When its hairs move in a certain direction, selective ion channels open, and its cell membrane depolarizes. The membrane then becomes more permeable specifically to calcium ions. The receptor cell has no axon or dendrites, but it does have neurotransmitter-containing

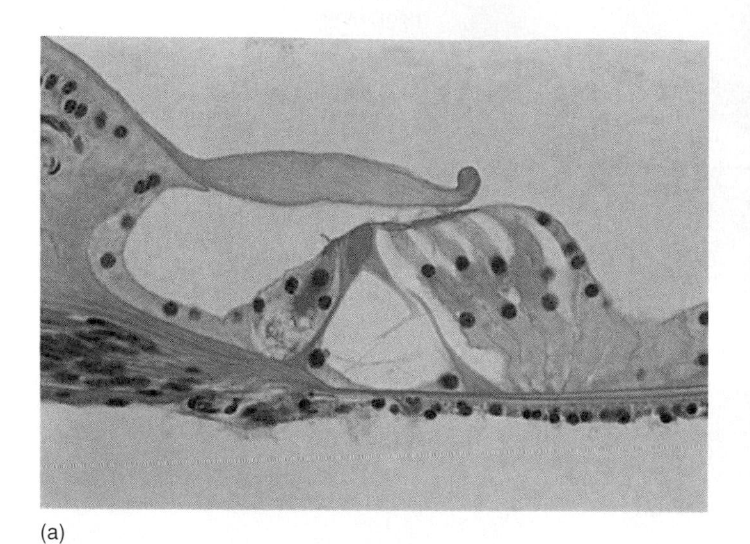

(a)

(b)

**FIGURE 12.16**

(*a*) A micrograph of the organ of Corti and the tectorial membrane (75×). (*b*) A scanning electron micrograph of hair cells in the organ of Corti (6,700×).

vesicles in the cytoplasm near its base. In the presence of calcium ions, some of these vesicles fuse with the cell membrane and release neurotransmitter to the outside. The neurotransmitter stimulates the ends of nearby sensory nerve fibers, and in response they transmit nerve impulses along the cochlear branch of the vestibulocochlear nerve to the brain.

Although the ear of a young person with normal hearing is able to detect sound waves with frequencies varying from about 20 to 20,000 or more vibrations per second, the range of greatest sensitivity is between 2,000 and 3,000 vibrations per second (fig. 12.17).

## Auditory Nerve Pathways

The cochlear branches of the vestibulocochlear nerves enter the auditory nerve pathways that extend into the medulla oblongata and proceed through the midbrain to the thalamus. From there they pass into the auditory cortices of the temporal lobes of the cerebrum, where they are interpreted. On the way, some of these fibers cross over, so that impulses arising from each ear are interpreted on both sides of the brain. Consequently, damage to a temporal lobe on one side of the brain is not necessarily accompanied by complete hearing loss in the ear on that side (fig. 12.18).

Chart 12.4 summarizes the pathway of vibrations through the parts of the middle and inner ears. Clinical Application 12.4 examines types of hearing loss.

> Units called *decibels* (dB) measure sound intensity. The decibel scale begins at 0 dB, which is the intensity of the sound that is least perceptible by a normal human ear. The decibel scale is logarithmic, so that a sound of 10 dB is 10 times as intense as the least perceptible sound; a sound of 20 dB is 100 times as intense; and a sound of 30 dB is 1,000 times as intense.
>
> On this scale, a whisper has an intensity of about 40 dB, normal conversation measures 60–70 dB, and heavy traffic produces about 80 dB. A sound of 120 dB, such as that commonly produced by the amplified sound at a rock concert, produces discomfort, and a sound of 140 dB, such as that emitted by a jet plane at takeoff, causes pain. Frequent or prolonged exposure to sounds with intensities above 90 dB is likely to cause permanent hearing loss. Many rock stars of the 1960s are now hearing impaired, due to their persistent exposure to very loud sounds.

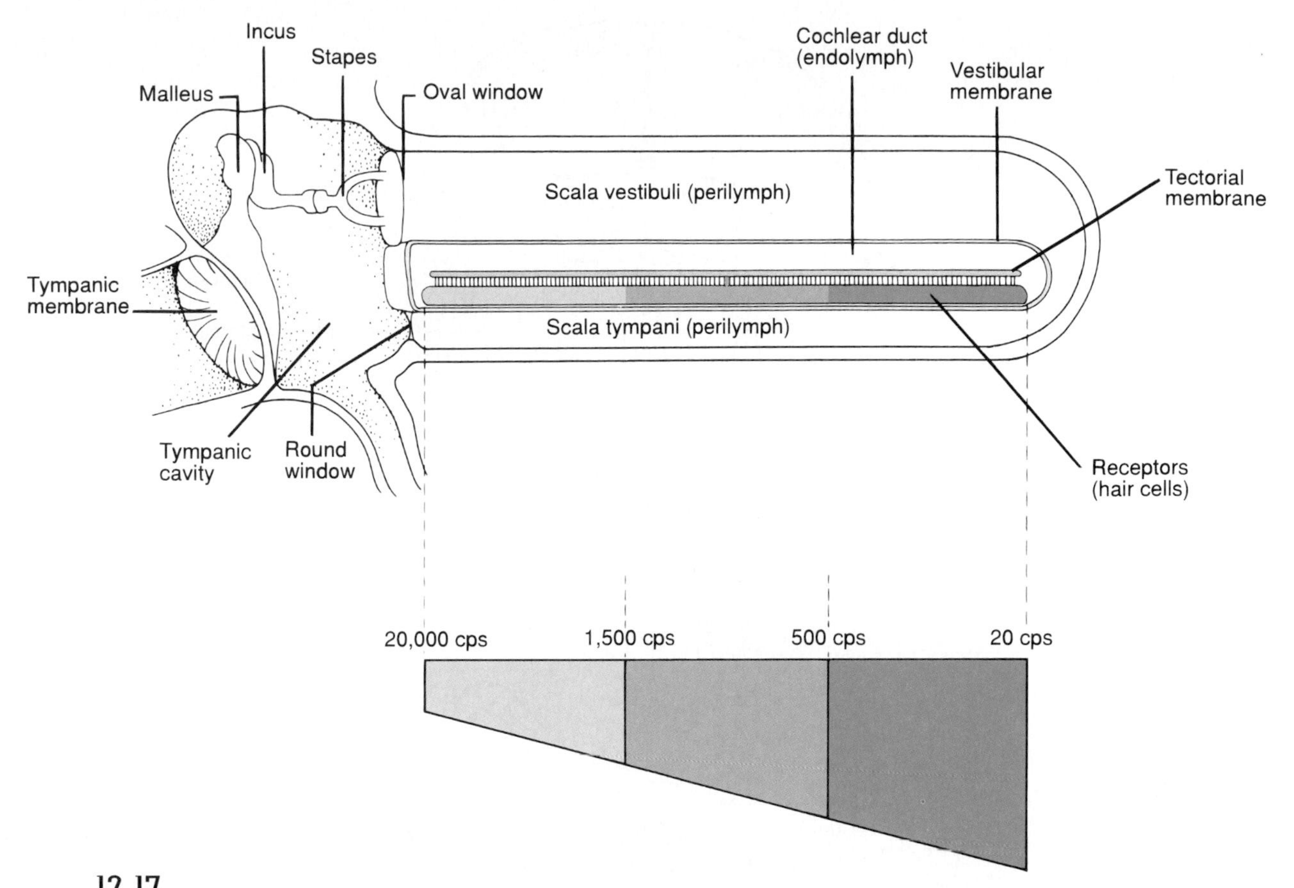


**FIGURE 12.17**
Receptors in regions of the cochlear duct sense different frequencies of vibration, expressed in cycles per second (cps).

## CHART 12.4 STEPS IN THE GENERATION OF SENSORY IMPULSES FROM THE EAR

1. Sound waves enter the external auditory meatus.
2. Waves of changing pressures cause the eardrum to reproduce the vibrations coming from the sound wave source.
3. Auditory ossicles amplify and transmit vibrations to the end of the stapes.
4. Movement of the stapes at the oval window transmits vibrations to the perilymph in the scala vestibuli.
5. Vibrations pass through the vestibular membrane and enter the endolymph of the cochlear duct.
6. Different frequencies of vibration in endolymph stimulate different sets of receptor cells.
7. A receptor cell becomes depolarized; its membrane becomes more permeable to calcium ions.
8. In the presence of calcium ions, vesicles at the base of the receptor cell release neurotransmitter.
9. Neurotransmitter stimulates the ends of nearby sensory neurons.
10. Sensory impulses are triggered on fibers of the cochlear branch of the vestibulocochlear nerve.
11. The auditory cortex of the temporal lobe interprets the sensory impulses.

1. Describe the external, middle, and inner ears.
2. Explain how sound waves are transmitted through the parts of the ear.
3. Describe the tympanic reflex.
4. Distinguish between the osseous and membranous labyrinths.
5. Explain the function of the organ of Corti.

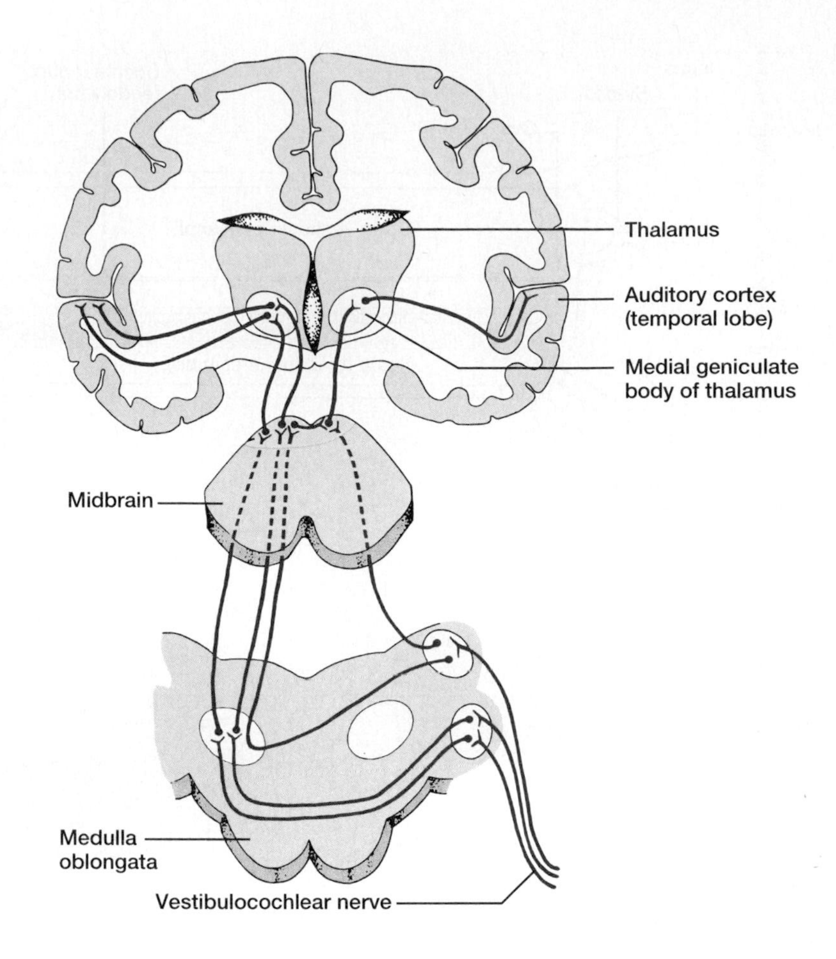


**FIGURE 12.18**
The auditory nerve pathway extends into the medulla oblongata, proceeds through the midbrain to the thalamus, and passes into the auditory cortex of the cerebrum.

## Sense of Equilibrium

The feeling of equilibrium derives from two senses—*static equilibrium* and *dynamic equilibrium.* Different sensory organs provide these two components of equilibrium. The organs associated with static equilibrium sense the position of the head, maintaining stability and posture when the head and body are still. When the head and body suddenly move or rotate, the organs of dynamic equilibrium detect such motion and aid in maintaining balance.

### Static Equilibrium

The organs of **static equilibrium** are located within the vestibule, a bony chamber between the semicircular canals and the cochlea. More specifically, the membranous labyrinth inside the vestibule consists of two expanded chambers—a **utricle** and a **saccule.** The larger utricle communicates with the saccule and the membranous portions of the semicircular canals; the saccule, in turn, communicates with the cochlear duct (fig. 12.19).

Each of these chambers has a small patch of hair cells and supporting cells called a **macula** on its wall. When the head is upright, the hairs of the macula in the utricle project vertically, while those in the saccule project horizontally. In each case, the hairs contact a sheet of gelatinous material (otolithic membrane) that has crystals of calcium carbonate (otoliths) embedded on its surface. These particles add weight to the gelatinous sheet, making it more responsive to changes in position. The hair cells, which serve as sensory receptors, have nerve fibers wrapped around their bases. These fibers are associated with the vestibular portion of the vestibulocochlear nerve.

The usual stimulus to the hair cells occurs when the head bends forward, backward, or to one side. Such movements tilt the gelatinous mass of one or more maculae, and as the gelatinous material sags in response to gravity, the hairs projecting into it bend. This action stimulates the hair cells, and they signal the nerve fibers associated with them in a manner similar to that

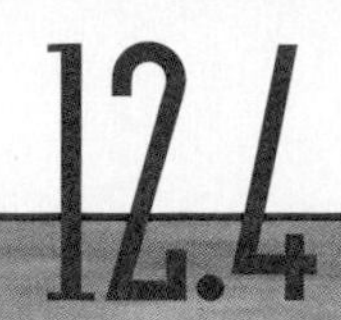

**CLINICAL APPLICATION 12.4**

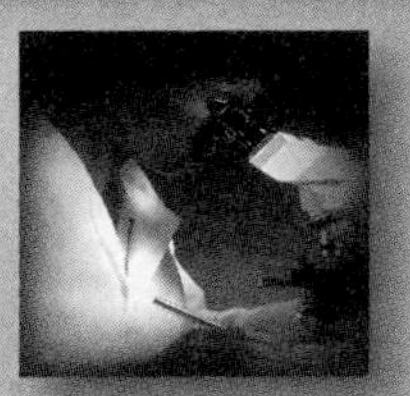

## Hearing Loss

Several factors can impair hearing, including interference with transmission of vibrations to the inner ear (*conductive deafness*) or damage to the cochlea or the auditory nerve and its pathways (*sensorineural deafness*). Disease, injury, and heredity all can impair hearing. There are more than 100 forms of inherited deafness, many of which are part of other syndromes. In the U.S., 8% of the population, or 21 million people, have hearing loss.

About 95% of cases of hearing loss are conductive. One cause is accumulated dry wax or a foreign object in the ear, which plugs the auditory meatus. Changes in the eardrum or auditory ossicles can also block hearing. The eardrum may harden as a result of disease, becoming less responsive to sound waves, or an injury may tear or perforate it.

A common disorder of the auditory ossicles is *otosclerosis,* in which new bone is deposited abnormally around the base of the stapes. This interferes with the ossicles' movement, which is necessary to transmit vibrations to the inner ear. Surgery often can restore some hearing to a person with otosclerosis by chipping away the bone that holds the stapes fixed in position, or replacing the stapes with a wire or plastic substitute.

Two tests used to diagnose conductive deafness are the Weber test and the Rinne test. In the Weber test, the handle of a vibrating tuning fork is pressed against the forehead. A person with normal hearing perceives the sound coming from directly in front, while a person with sound conduction blockage in one middle ear hears the sound coming from the impaired side.

In the Rinne test, a vibrating tuning fork is held against the bone behind the ear. After the sound is no longer heard by conduction through the bones of the skull, the fork is moved to just in front of the external auditory meatus. In middle ear conductive deafness, the vibrating fork can no longer be heard, but a normal ear will continue to hear its tone.

Very loud sounds can cause sensorineural deafness. If exposure is brief, hearing loss may be temporary, but when exposure is repeated and prolonged, such as occurs in foundries, near jackhammers, or on a firing range, impairment may be permanent. Pete Townshend, of the rock group The Who, suffers from hearing loss from many years of performing loud concerts.

Other causes of sensorineural deafness include tumors in the central nervous system, brain damage as a result of vascular accidents, and the use of certain drugs.

Because hearing loss and other ear problems can begin gradually, it is important to be aware of their signs, which may include:

- difficulty hearing people talking softly
- inability to understand speech when there is background noise
- ringing in the ears
- dizziness
- loss of balance

New parents should notice whether their infant responds to sounds in a way that indicates normal hearing. Before 1993, 50% of hearing impaired infants were not diagnosed until age 2. Since then, the federal government has advised hearing exams as part of a well-baby visit to a doctor. If the baby's responses indicate a possible problem, the next step is to see an audiologist, who identifies and measures hearing loss.

Often a hearing aid can help people with conductive hearing loss. A hearing aid has a tiny microphone that picks up sound waves and converts them to electrical signals, which are then amplified so that the person can hear them. An ear mold holds the device in place, either behind the outer ear, in the outer ear, or in the ear canal.

A cochlear implant enables people with sensorineural hearing loss to detect some sounds, although it usually remains difficult to discern distinct words. The device converts sound waves to electrical signals, which stimulate neurons in the cochlea.

Many persons over age 65 lose the ability to hear high-frequency sounds and can no longer discriminate speech sounds as well as they once did—a condition called *presbycusis.* Thus, it is important to use lower voice tones in speaking to an older person. Also, the speaker should face the listener so that lip-reading is possible.

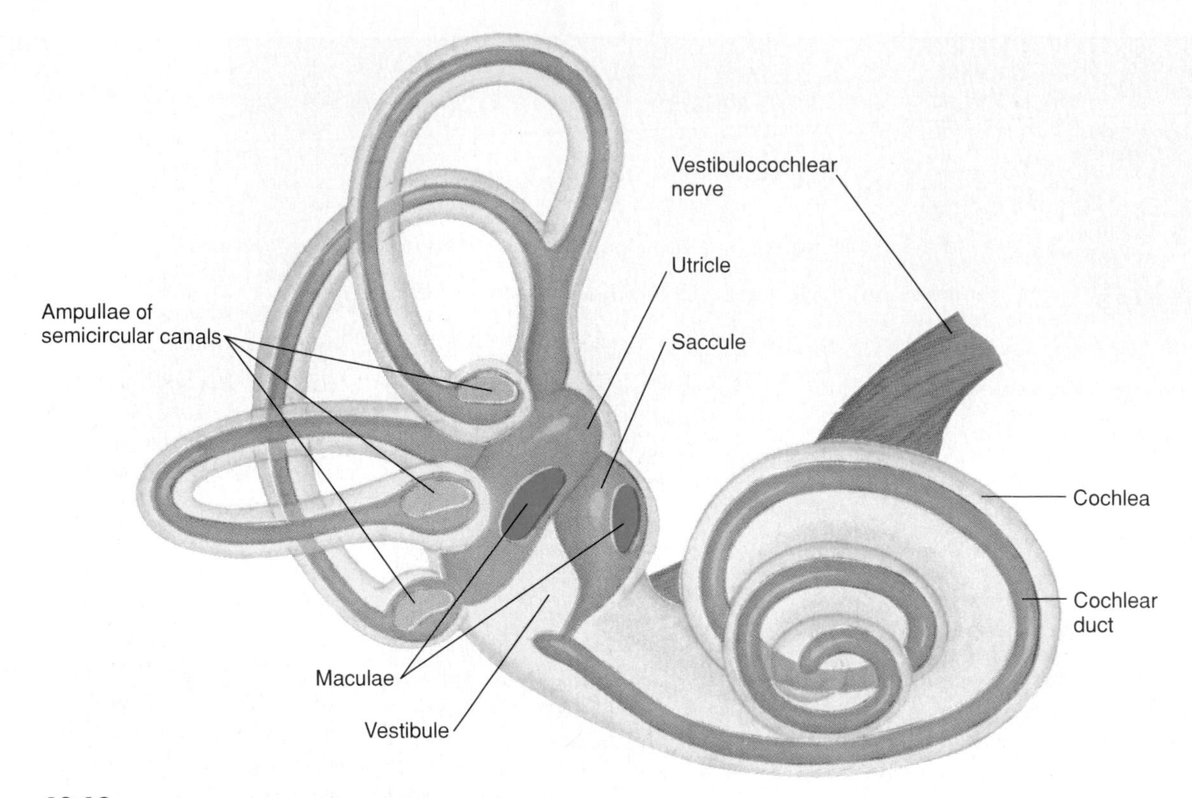


**FIGURE 12.19**
The saccule and utricle, which are expanded portions of the membranous labyrinth, are located within the bony chamber of the vestibule.

of the hearing receptors. The resulting nerve impulses travel into the central nervous system by means of the vestibular branch of the vestibulocochlear nerve, informing the brain of the head's position. The brain responds to this information by sending motor impulses to skeletal muscles, and they may contract or relax appropriately to maintain balance (figs. 12.20 and 12.21).

The maculae also participate in the sense of dynamic equilibrium. For example, if the head or body is thrust forward or backward abruptly, the gelatinous mass of the maculae lags slightly behind, and the hair cells are stimulated. In this way, the maculae aid the brain in detecting movements such as falling, and in maintaining posture while walking.

## Dynamic Equilibrium

Each semicircular canal follows a circular path about 6 millimeters in diameter. The three bony semicircular canals lie at right angles to each other and occupy three different planes in space. Two of them, the *superior canal* and the *posterior canal,* stand vertically, while the third, the *lateral canal,* is horizontal. Their orientations closely approximate the three body planes (see chapter 1).

Suspended in the perilymph of each bony canal is a membranous semicircular canal that ends in a swelling called an **ampulla.** The ampullae communicate with the utricle of the vestibule.

An ampulla contains a septum that crosses the tube and houses a sensory organ. Each of these organs, called a **crista ampullaris,** contains a number of sensory hair cells and supporting cells. As in the maculae, the hairs of the hair cells extend upward into a dome-shaped gelatinous mass called the *cupula.* Also, the hair cells are connected at their bases to nerve fibers that make up part of the vestibular branch of the vestibulocochlear nerve (fig. 12.22).

Rapid turns of the head or body stimulate the hair cells of the crista. At such times, the semicircular canals move with the head or body, but the fluid inside the membranous canals tends to remain stationary because of inertia. This bends the cupula in one or more of the canals in a direction opposite that of the head or body movement, and the hairs embedded in it also bend. This bending of the hairs stimulates the hair cells to signal their associated nerve fibers, and, as a result, impulses travel to the brain (fig. 12.23).

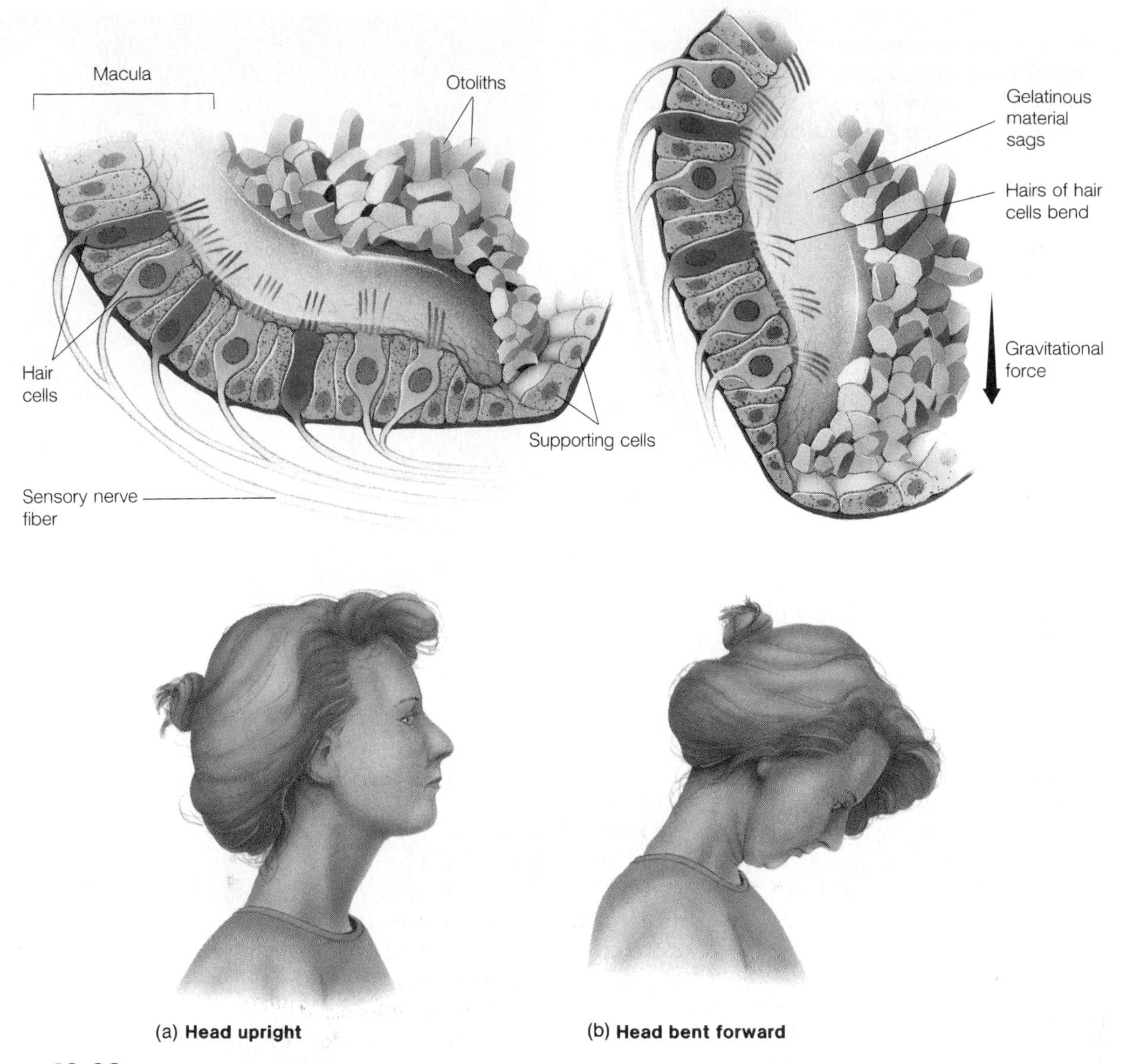


**FIGURE 12.20**

The macula responds to changes in position of the head. (*a*) Macula with the head in an upright position; (*b*) macula with the head bent forward.

Parts of the cerebellum are particularly important in interpreting impulses from the semicircular canals. Analysis of such information allows the brain to predict the consequences of rapid body movements, and by modifying signals to appropriate skeletal muscles, the cerebellum can maintain balance.

Other sensory structures aid in maintaining equilibrium. For example, various proprioceptors, particularly those associated with the joints of the neck, inform the brain about the position of body parts. In addition, the eyes detect changes in posture that result from body movements. Such visual information is so important that even if the organs of equilibrium are damaged, keeping the eyes open and moving slowly is sufficient to maintain normal balance.

1. Distinguish between the senses of static and dynamic equilibrium.
2. What structures provide the sense of static equilibrium? Of dynamic equilibrium?
3. How does sensory information from other receptors help maintain equilibrium?

Motion sickness is a disturbance of the inner ear's sensation of balance. Nine out of 10 people have experienced this nausea and vomiting, usually when riding in a car or on a boat. Astronauts began reporting a form of motion sickness called space adaptation syndrome in 1968, when spacecraft began to be made roomy enough for astronauts to move about while in flight.

Although the cause of motion sickness is not known, one theory is that it results when visual information contradicts the inner ear's sensation that one is motionless. Consider a woman riding in a car. Her inner ears tell her that she is not moving, but the passing scenery tells her eyes that she is moving. The problem is compounded if she tries to read. The brain reacts to these seemingly contradictory sensations by signaling a "vomiting center" in the medulla oblongata.

**FIGURE 12.21**
Scanning electron micrograph of hairs of hair cells, such as those found in the utricle and saccule (about 7,000×).

## SENSE OF SIGHT

A number of *accessory organs* assist the visual receptors, which are in the eyes. These include the eyelids and lacrimal apparatus that help protect the eyes, and a set of extrinsic muscles that move them.

### Visual Accessory Organs

Each eye, lacrimal gland, and associated extrinsic muscles are housed within the pear-shaped orbital cavity of the skull. The orbit, which is lined with the periosteums of various bones, also contains fat, blood vessels, nerves, and connective tissues.

Each **eyelid** (palpebra) is composed of four layers—skin, muscle, connective tissue, and conjunctiva. The skin of the eyelid, which is the thinnest skin of the body, covers the lid's outer surface and fuses with its inner lining near the margin of the lid (fig. 12.24).

The muscles that move the eyelids include the *orbicularis oculi* and the *levator palpebrae superioris*. Fibers of the orbicularis oculi encircle the opening between the lids and spread out onto the cheek and forehead. This muscle acts as a sphincter that closes the lids when it contracts.

Fibers of the levator palpebrae superioris muscle arise from the roof of the orbit and are inserted in the connective tissue of the upper lid. When these fibers contract, the upper lids are raised and the eye opens.

The connective tissue layer of the eyelid, which helps give it form, contains many modified sebaceous glands (tarsal glands). Ducts carry the oily secretions of

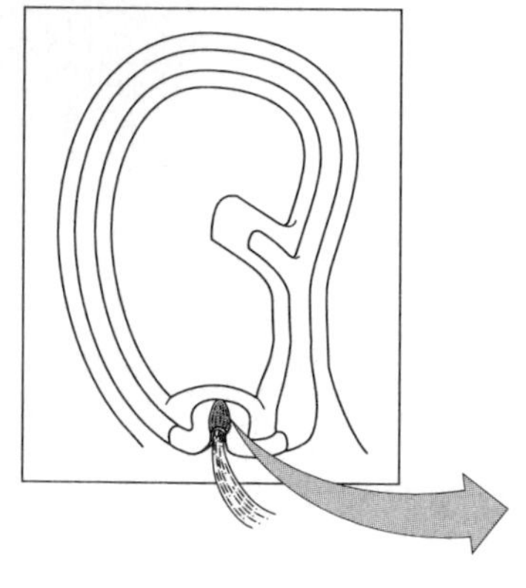


**FIGURE 12.22**
A crista ampullaris is located within the ampulla of each semicircular canal.

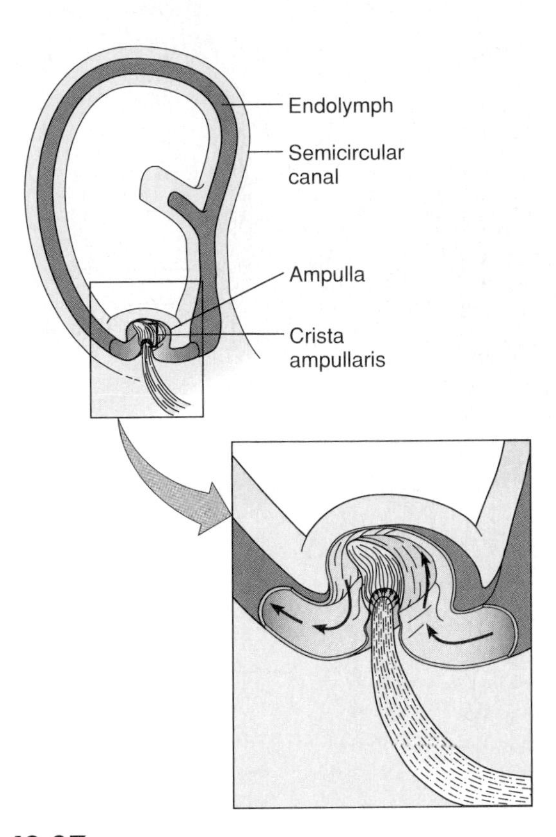


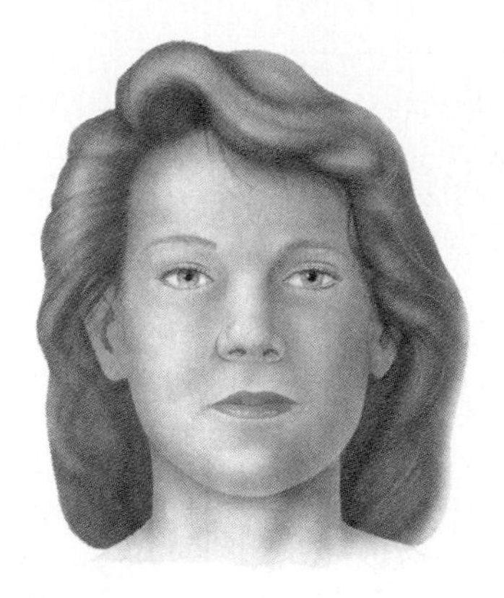

(a) Head in still position

(b) Head rotating

**FIGURE 12.23**

(*a*) When the head is stationary, the cupula of the crista ampullaris remains upright. (*b*) When the head is moving rapidly, the cupula is bent opposite the motion of the head, stimulating sensory receptors.

these glands to openings along the borders of the lids. This secretion helps keep the lids from sticking together.

The **conjunctiva** is a mucous membrane that lines the inner surfaces of the eyelids and folds back to cover the anterior surface of the eyeball, except for its central portion (cornea). Although the tissue that lines the eyelids is relatively thick, the conjunctiva that covers the eyeball is very thin. It is also freely movable and quite transparent, so that blood vessels are clearly visible beneath it.

> A child in school with "pinkeye" is usually sent straight home. Bacteria cause this highly contagious form of inflammation of the conjunctiva, or *conjunctivitis.* Viral conjunctivitis is not usually contagious. Allergy or exposure to an irritating chemical may also cause conjunctivitis.

The *lacrimal apparatus* consists of the **lacrimal gland,** which secretes tears, and a series of *ducts,* which carry the tears into the nasal cavity (fig. 12.25). The gland is located in the orbit, above and to the lateral side of the eye. It secretes tears continuously, and they pass out through tiny tubules and flow downward and medially across the eye.

Two small ducts (superior and inferior canaliculi) collect tears, and their openings (puncta) can be seen on the medial borders of the eyelids. From these ducts, the fluid moves into the *lacrimal sac,* which lies in a deep groove of the lacrimal bone, and then into the *nasolacrimal duct,* which empties into the nasal cavity.

Glandular cells of the conjunctiva also secrete a tearlike liquid that, together with the secretion of the lacrimal gland, moistens and lubricates the surface of the eye and the lining of the lids. Tears contain an enzyme, called *lysozyme,* that has antibacterial properties, reducing the chance of eye infections.

Tear glands secrete excessively when a person is upset, or the conjunctiva is irritated. Tears spill over the edges of the eyelids, and the nose fills with fluid. When a person cries, parasympathetic nerve fibers carry motor impulses to the lacrimal glands.

The **extrinsic muscles** of the eye arise from the bones of the orbit and are inserted by broad tendons on

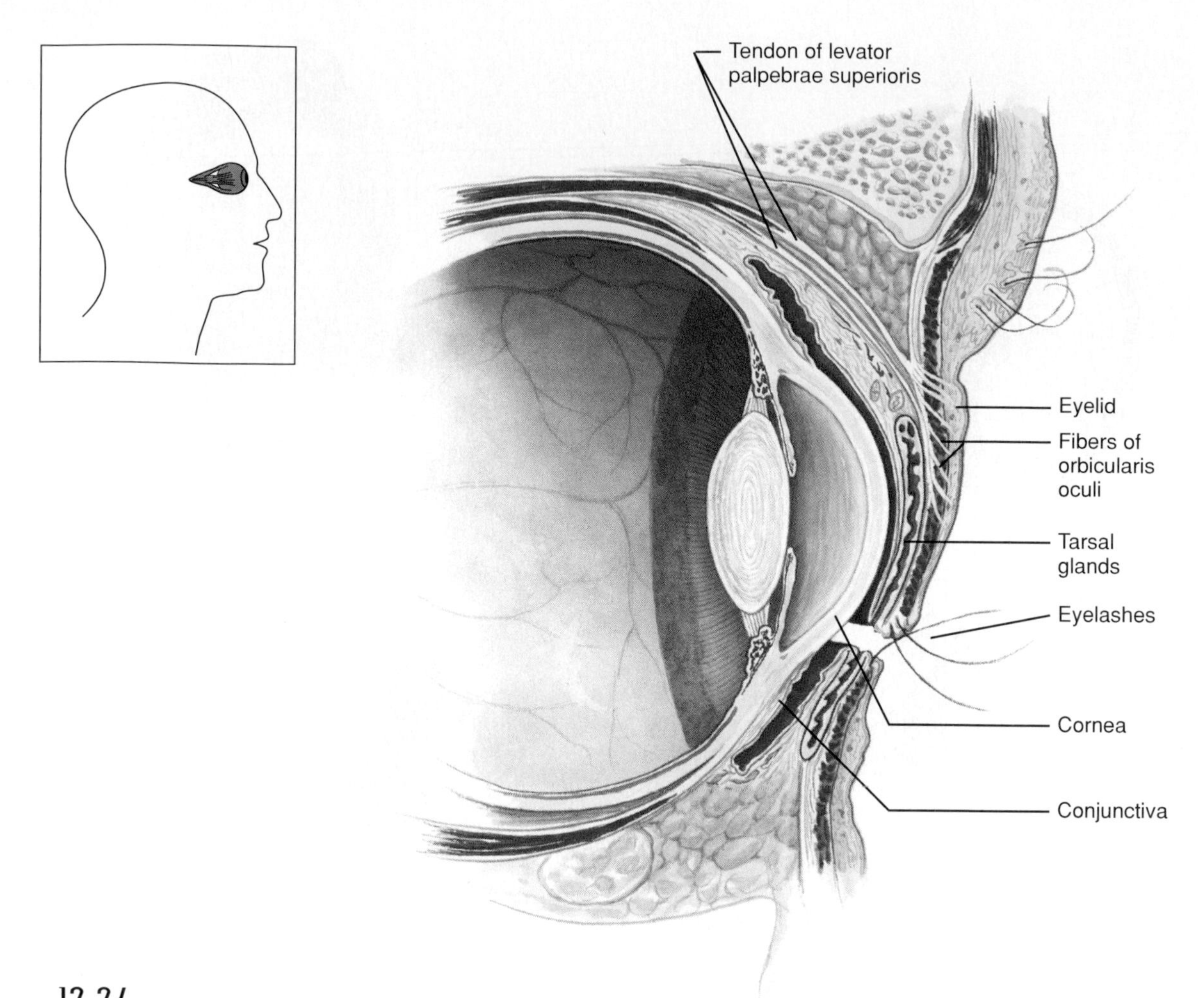


**Figure 12.24**
Sagittal section of the closed eyelids and the anterior portion of the eye.

the eye's tough outer surface. Six such muscles move the eye in various directions (fig. 12.26). Although any given eye movement may use more than one of them, each muscle is associated with one primary action, as follows:

1. **Superior rectus**—rotates the eye upward and toward the midline.
2. **Inferior rectus**—rotates the eye downward and toward the midline.
3. **Medial rectus**—rotates the eye toward the midline.
4. **Lateral rectus**—rotates the eye away from the midline.
5. **Superior oblique**—rotates the eye downward and away from the midline.
6. **Inferior oblique**—rotates the eye upward and away from the midline.

The motor units of the extrinsic eye muscles contain the fewest muscle fibers (five to ten) of any mus-

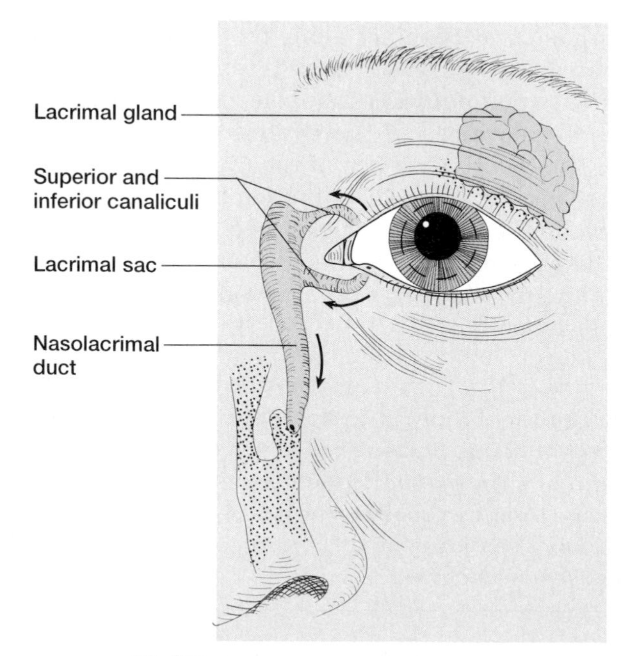


**Figure 12.25**
The lacrimal apparatus consists of a tear-secreting gland and a series of ducts.

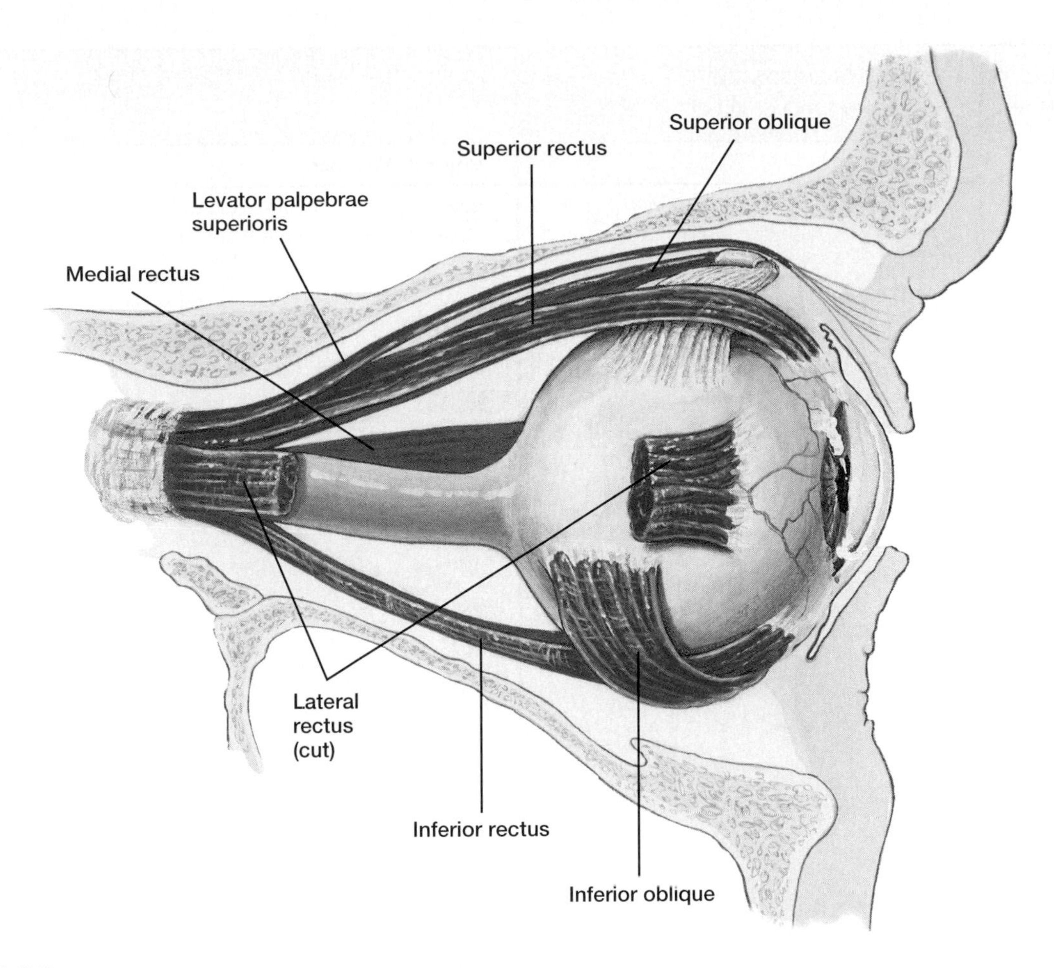


**FIGURE 12.26**
The extrinsic muscles of the right eye (lateral view).

cles in the body, enabling the eyes to move with great precision. Also, the eyes move together so that they align when looking at something. Such alignment is the result of complex motor adjustments that contract certain eye muscles while relaxing their antagonists. For example, when the eyes move to the right, the lateral rectus of the right eye and the medial rectus of the left eye must contract. At the same time, the medial rectus of the right eye and the lateral rectus of the left eye must relax. A person whose eyes are not coordinated well enough to align has *strabismus.*

Chart 12.5 summarizes the muscles associated with the eyelids and eye.

1. Explain how the eyelid is moved.
2. Describe the conjunctiva.
3. What is the function of the lacrimal apparatus?
4. Describe the function of each extrinsic eye muscle.

When one eye deviates from the line of vision, the person has double vision (diplopia). If this condition persists, the brain may eventually suppress the image from the deviated eye. As a result, the turning eye may become blind (suppression amblyopia). Such monocular blindness can often be prevented if the eye deviation is treated early in life with exercises, eyeglasses, and surgery. For this reason, vision screening programs for preschool children are very important.

## Structure of the Eye

The eye is a hollow, spherical structure about 2.5 centimeters in diameter. Its wall has three distinct layers—an outer *fibrous tunic,* a middle *vascular tunic,* and an inner *nervous tunic.* The spaces within the eye are filled with fluids that support its wall and internal structures, and help maintain its shape. Figure 12.27 shows the major parts of the eye.

## CHART 12.5 Muscles Associated with the Eyelids and Eyes

**Skeletal Muscles**

| Name | Innervation | Function |
|---|---|---|
| Muscles of the eyelids | | |
| Orbicularis oculi | Facial nerve (VII) | Closes eye |
| Levator palpebrae superioris | Oculomotor nerve (III) | Opens eye |
| Extrinsic muscles of the eyes | | |
| Superior rectus | Oculomotor nerve (III) | Rotates eye upward and toward midline |
| Inferior rectus | Oculomotor nerve (III) | Rotates eye downward and toward midline |
| Medial rectus | Oculomotor nerve (III) | Rotates eye toward midline |
| Lateral rectus | Abducens nerve (VI) | Rotates eye away from midline |
| Superior oblique | Trochlear nerve (IV) | Rotates eye downward and away from midline |
| Inferior oblique | Oculomotor nerve (III) | Rotates eye upward and away from midline |

**Smooth Muscles**

| Name | Innervation | Function |
|---|---|---|
| Ciliary muscles | Oculomotor nerve (III) parasympathetic fibers | Relax suspensory ligaments |
| Iris, circular muscles | Oculomotor nerve (III) parasympathetic fibers | Constrict pupil |
| Iris, radial muscles | Sympathetic fibers | Dilate pupil |

### The Outer Tunic

The anterior sixth of the outer tunic bulges forward as the transparent **cornea,** which is the window of the eye and helps focus entering light rays. It is composed largely of connective tissue with a thin layer of epithelium on its surface. The cornea is transparent because it contains relatively few cells and no blood vessels. Also, the cells and collagenous fibers are arranged in unusually regular patterns.

On the other hand, the cornea is well supplied with nerve fibers that enter its margin and radiate toward its center. These fibers are associated with many pain receptors that have very low thresholds. Cold receptors are also abundant in the cornea, but heat and touch receptors are not.

Along its circumference, the cornea is continuous with the **sclera,** the white portion of the eye. The sclera makes up the posterior five-sixths of the outer tunic and is opaque due to the presence of many large, haphazardly arranged collagenous and elastic fibers. The sclera protects the eye and is an attachment for the extrinsic muscles.

In the back of the eye, the **optic nerve** and blood vessels pierce the sclera. The dura mater that encloses these structures attaches to the sclera.

### The Middle Tunic

The middle, or vascular, tunic of the eyeball (uveal layer) includes the choroid coat, the ciliary body, and the iris.

The **choroid coat,** in the posterior five-sixths of the globe of the eye, is loosely joined to the sclera. Blood vessels pervade the choroid coat and nourish surrounding tissues. The choroid coat also contains numerous pigment-producing melanocytes that give it a brownish black appearance. The melanin of these cells absorbs excess light and helps keep the inside of the eye dark.

In 1905, doctors transplanted the cornea of an 11-year-old boy who lost his eye in an accident into a man whose cornea had been destroyed by a splash of a caustic chemical, marking one of the first successful human organ transplants. Today, corneal transplants are commonly used to treat corneal disease, the most common cause of blindness worldwide. In this procedure, called a *penetrating keratoplasty,* a piece of donor cornea replaces the central two-thirds of the defective cornea. These transplants are highly successful because the cornea lacks blood vessels, and therefore the immune system does not have direct access to the new, "foreign" tissue. Unfortunately, as is the case for many transplantable body parts, donor tissue is in short supply.

The **ciliary body,** which is the thickest part of the middle tunic, extends forward from the choroid coat and forms an internal ring around the front of the eye.

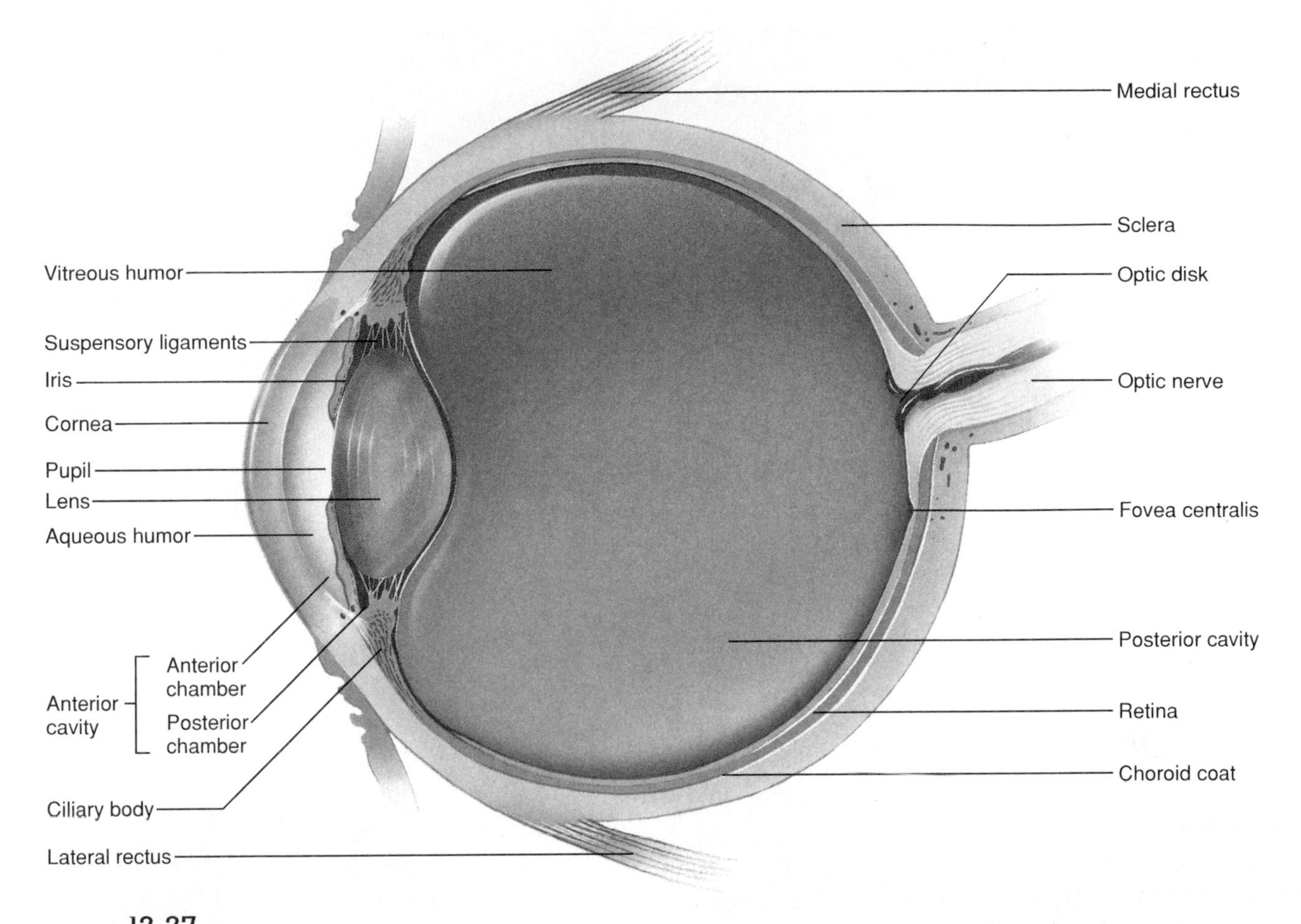


**FIGURE 12.27**
Transverse section of the left eye (superior view).

Within the ciliary body are many radiating folds called *ciliary processes* and two distinct groups of muscle fibers that constitute the *ciliary muscles.* Figure 12.28 shows these structures.

Many strong but delicate fibers, called *suspensory ligaments* (zonular fibers), extend inward from the ciliary processes and hold the transparent lens in position. The distal ends of these fibers are attached along the margin of a thin capsule that surrounds the lens. The body of the lens, which lacks blood vessels, lies directly behind the iris and pupil and is composed of specialized epithelial cells.

These epithelial cells originate from a single layer of epithelium located beneath the anterior portion of the lens capsule. The cells divide, and the newly formed cells on the surface of the lens capsule differentiate into columnar cells called *lens fibers.* These lens fibers constitute the substance of the lens. Lens fiber production continues slowly throughout life, thickening the lens from front to back. Simultaneously, the deeper lens fibers are compressed toward the center of the structure (fig. 12.29).

The lens capsule is a clear, membranelike structure composed largely of intercellular material. It is quite elastic, a quality that keeps it under constant tension. As a result, the lens can assume a globular shape. However, the suspensory ligaments attached to the margin of the capsule are also under tension, and they pull outward, flattening the capsule and the lens (fig. 12.30).

If the tension on the suspensory ligaments relaxes, the elastic capsule rebounds, and the lens surface becomes more convex. This change occurs in the lens when the eye focuses to view a close object and is called **accommodation.**

The ciliary muscles relax the suspensory ligaments during accommodation. One set of these muscle fibers forms a circular sphincterlike structure around the ciliary processes. The fibers of the other set extend back from fixed points in the sclera to the choroid coat. When the circular muscle fibers contract, the diameter of the ring formed by the ciliary processes decreases; when the other fibers contract, the choroid coat is pulled forward and the ciliary body shortens. Both of

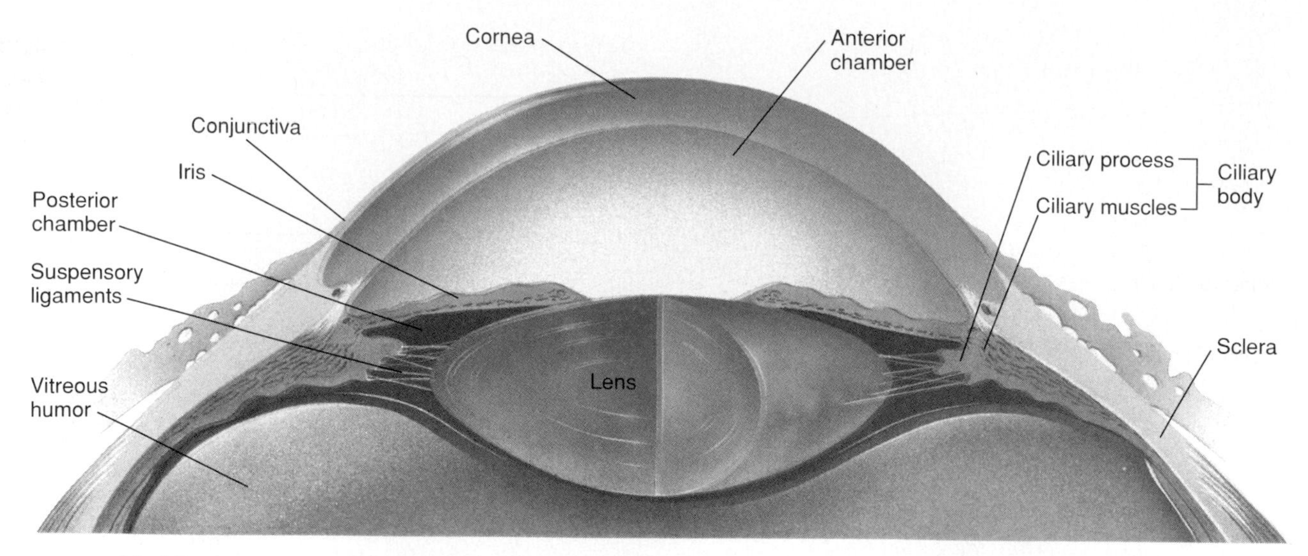


**FIGURE 12.28**
Anterior portion of the eye.

**FIGURE 12.29**
A scanning electron micrograph of the long, flattened lens fibers (4,800×). Note the fingerlike junctions where one fiber joins another.

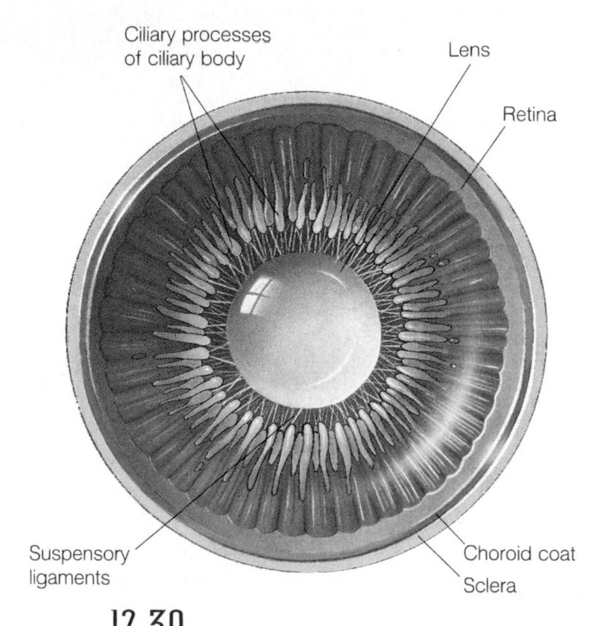


**FIGURE 12.30**
The lens and ciliary body viewed from behind.

these actions relax the suspensory ligaments, thickening the lens. In this thickened state, the lens is focused for viewing closer objects than before (fig. 12.31).

To focus on a distant object, the ciliary muscles relax, increasing tension on the suspensory ligaments. The lens thins again.

A relatively common eye disorder, particularly in older people, is *cataract.* The lens or its capsule slowly becomes cloudy, opaque, and discolored. As a result, clear images cannot focus on the retina. In time, the person may become blind.

Cataract is often treated by surgically removing the lens and replacing it with an artificial one (intraocular lens implant). Sometimes eyeglasses or contact lenses can correct the loss of refractive power following removal of the lens.

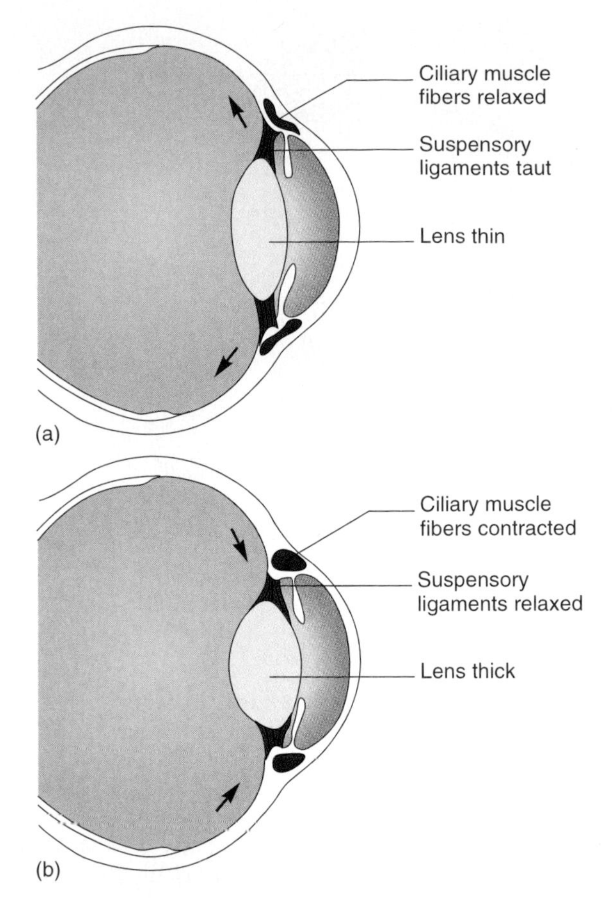


**FIGURE 12.31**

(*a*) The lens thins as the ciliary muscle fibers relax. (*b*) The lens thickens as the ciliary muscle fibers contract.

1. Describe the outer and middle tunics of the eye.
2. What factors contribute to the transparency of the cornea?
3. How does the shape of the lens change during accommodation?

The **iris** is a thin diaphragm composed mostly of connective tissue and smooth muscle fibers. It is seen from the outside as the colored portion of the eye. The iris extends forward from the periphery of the ciliary body and lies between the cornea and the lens. It divides the space separating these parts, which is called the *anterior cavity,* into an *anterior chamber* (between the cornea and the iris) and a *posterior chamber* (between the iris and the vitreous humor, and occupied by the lens).

The epithelium on the inner surface of the ciliary body continuously secretes a watery fluid called **aqueous humor** into the posterior chamber. The fluid circulates from this chamber through the **pupil,** a circular opening in the center of the iris, and into the anterior chamber (fig. 12.32). Aqueous humor fills the space between the cornea and the lens, providing nutrients and maintaining the shape of the front of the eye. It subsequently leaves the anterior chamber through veins and a special drainage canal, the scleral venous sinus (canal of Schlemm), located in its wall.

> A disorder called *glaucoma* sometimes develops in the eyes as a person ages when the rate of aqueous humor formation exceeds the rate of its removal. Fluid accumulates in the anterior chamber of the eye, and the fluid pressure rises.
>
> Because liquids cannot be compressed, the increasing pressure from the anterior chamber is transmitted to all parts of the eye, and, in time, the blood vessels that supply the receptor cells of the retina may squeeze shut. If this happens, cells that fail to receive needed nutrients and oxygen may die, and permanent blindness may result.
>
> Drugs, laser therapy, or surgery to promote the outflow of aqueous humor can treat glaucoma if it is diagnosed early. However, since glaucoma in its early stages typically produces no symptoms, discovery of the condition usually depends on measuring the intraocular pressure using an instrument called a *tonometer.*

The smooth muscle fibers of the iris form two groups, a *circular set* and a *radial set.* These muscles control the size of the pupil, which is the opening that light passes through as it enters the eye. The circular set of muscle fibers acts as a sphincter, and when it contracts, the pupil gets smaller (constricts) and the intensity of the light entering decreases. When the radial muscle fibers contract, the diameter of the pupil increases (dilates) and the intensity of the light entering increases.

The sizes of the pupils change constantly in response to pupillary reflexes that are triggered by such factors as light intensity, gaze, accommodation, and variations in emotional state. For example, bright light elicits a reflex, and impulses travel along parasympathetic nerve fibers to the *circular muscles* of the irises. The pupils constrict in response. Conversely, in dim light, impulses travel on sympathetic nerve fibers to the *radial muscles* of the irises, and the pupils dilate (fig. 12.33).

The amount and distribution of melanin in the irises and the density of the tissue within the body of

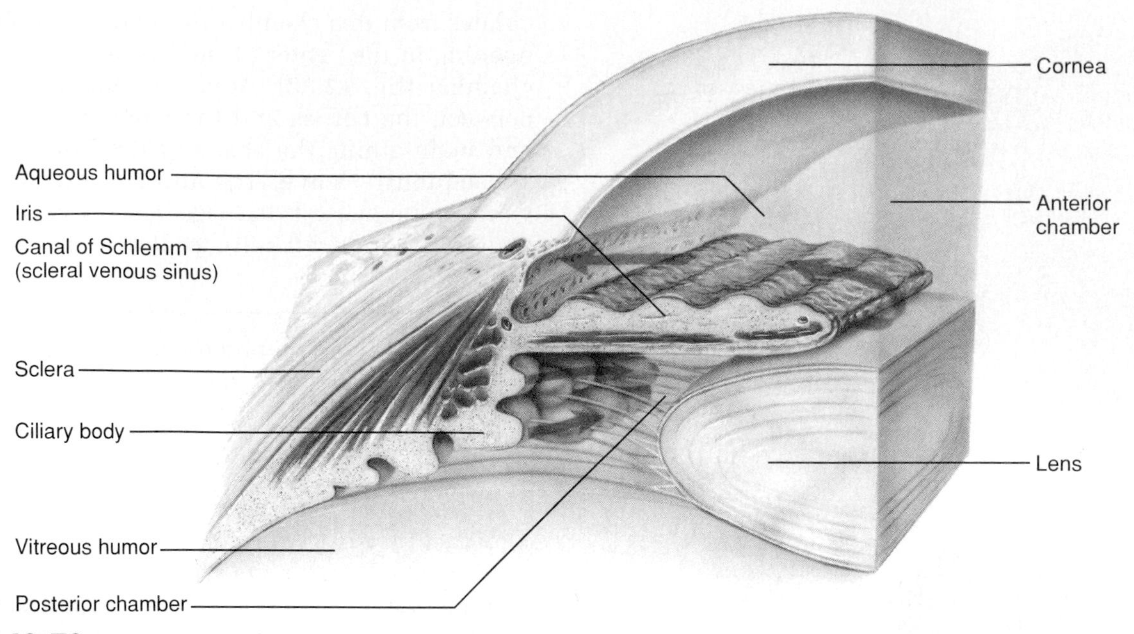


**FIGURE 12.32**
Aqueous humor (blue arrows), which is secreted into the posterior chamber, circulates into the anterior chamber and leaves it through the canal of Schlemm (scleral venous sinus).

the iris determine eye color. If melanin is present only in the epithelial cells that cover an iris's posterior surface, the iris appears blue. When this condition exists together with denser than usual tissue within the body of the iris, it looks gray. When melanin is present within the body of the iris as well as in the epithelial covering, the iris appears brown.

## The Inner Tunic

The inner tunic of the eye consists of the **retina,** which contains the visual receptor cells (photoreceptors). This nearly transparent sheet of tissue is continuous with the optic nerve in the back of the eye and extends forward as the inner lining of the eyeball. It ends just behind the margin of the ciliary body.

The retina is thin and delicate, but its structure is quite complex. It has a number of distinct layers, including pigmented epithelium, neurons, nerve fibers, and limiting membranes (figs. 12.34 and 12.35).

There are five major groups of retinal neurons. The nerve fibers of three of these groups—the *receptor cells, bipolar neurons,* and *ganglion cells*—provide a direct pathway for impulses triggered in the receptors to the optic nerve and brain. The nerve fibers of the other two groups of retinal cells, called *horizontal cells* and *amacrine cells,* pass laterally between retinal cells (see fig. 12.34). The horizontal and amacrine cells modify the impulses transmitted on the fibers of the direct pathway.

In the central region of the retina is a yellowish spot called the **macula lutea** that occupies about 1 square millimeter. A depression in its center, called the **fovea centralis,** is in the region of the retina that produces the sharpest vision.

Just medial to the fovea centralis is an area called the **optic disk** (fig. 12.36). Here the nerve fibers from the retina leave the eye and become parts of the optic nerve. A central artery and vein also pass through at the optic disk. These vessels are continuous with capillary networks of the retina, and together with vessels in the underlying choroid coat, they supply blood to the cells of the inner tunic. Because the optic disk lacks receptor cells, it is commonly referred to as the *blind spot* of the eye.

The space enclosed by the lens, ciliary body, and retina is the largest compartment of the eye and is called the *posterior cavity.* It is filled with a transparent, jellylike fluid called **vitreous humor,** which together with some collagenous fibers comprise the **vitreous body.** The vitreous body supports the internal structures of the eye and helps maintain its shape.

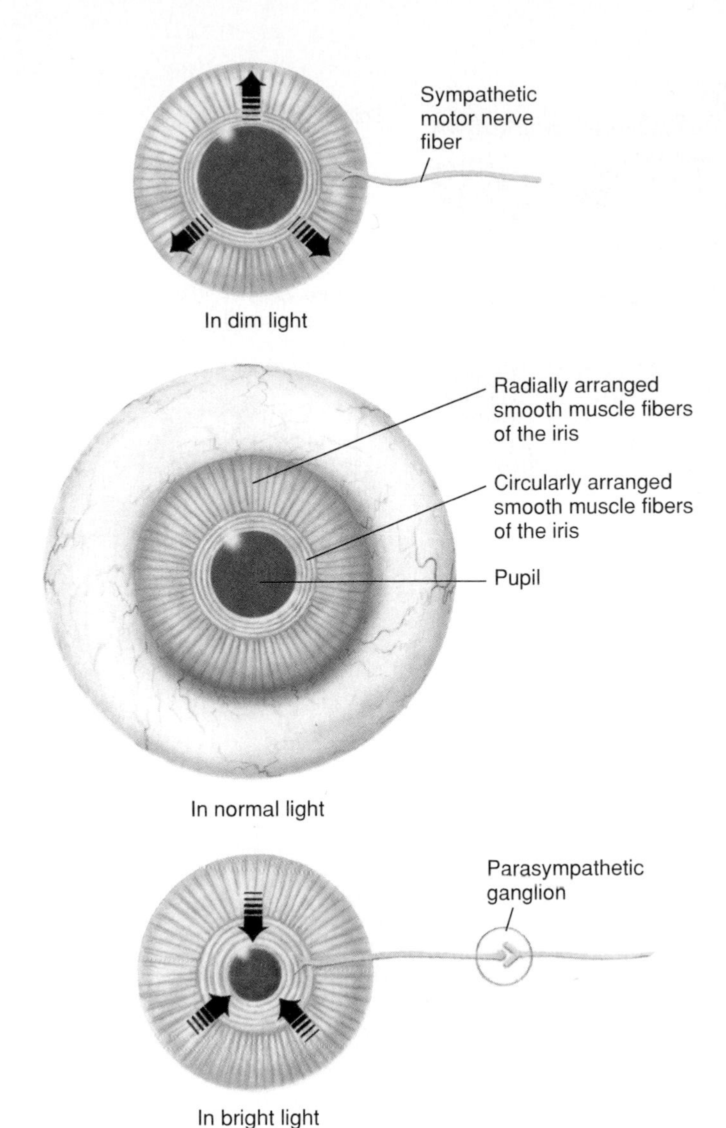


**FIGURE 12.33**
In dim light, the radial muscles of the iris are stimulated to contract, and the pupil dilates. In bright light, the circular muscles of the iris are stimulated to contract, and the pupil constricts.

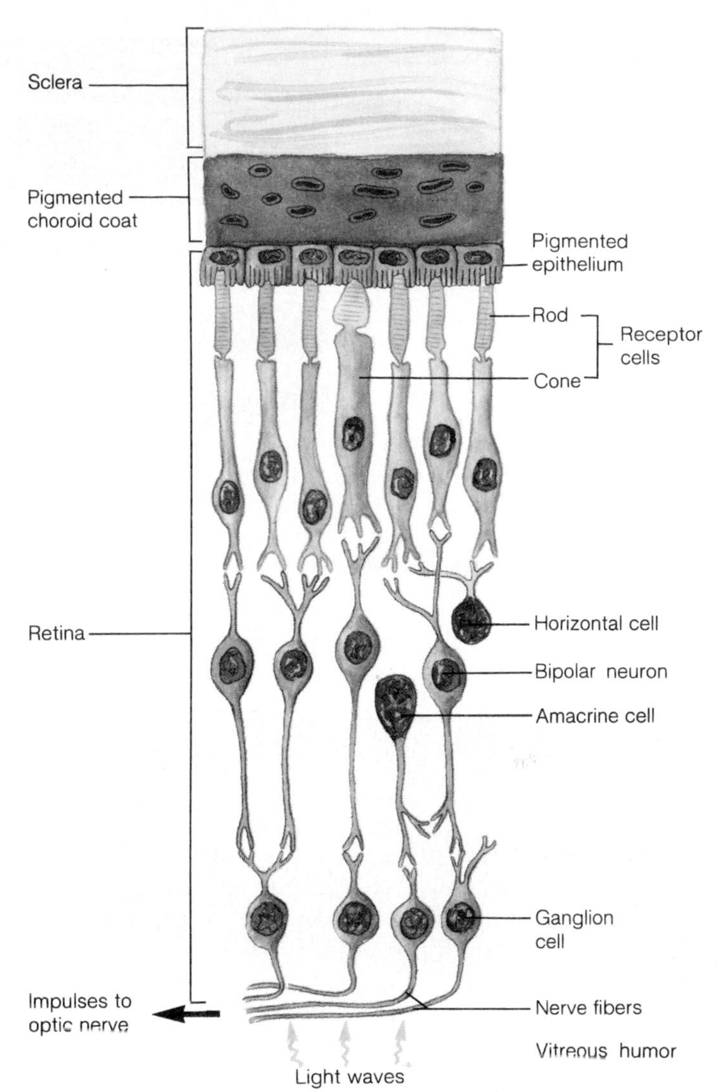


**FIGURE 12.34**
The retina consists of several cell layers.

In summary, light waves entering the eye must pass through the cornea, aqueous humor, lens, vitreous humor, and several layers of the retina before they reach the photoreceptors (see fig. 12.34). Chart 12.6 summarizes the layers of the eye.

1. Explain the origin of aqueous humor and trace its path through the eye.
2. How is the size of the pupil regulated?
3. Describe the structure of the retina.

With age, tiny dense clumps of gel or deposits of crystal-like substances form in the vitreous humor. These clumps may cast shadows on the retina, and as a result, the person sees small moving specks in the field of vision. These specks, or *floaters,* are most apparent when looking at a plain background, such as the sky or a blank wall.

Also with age, the vitreous humor may shrink and pull away from the retina. This may mechanically stimulate receptor cells of the retina, and the person may see flashes of light. The presence of floaters or light flashes usually does not indicate a serious eye problem; however, if they develop suddenly or seem to be increasing in number or frequency, an ophthalmologist should examine the eye.

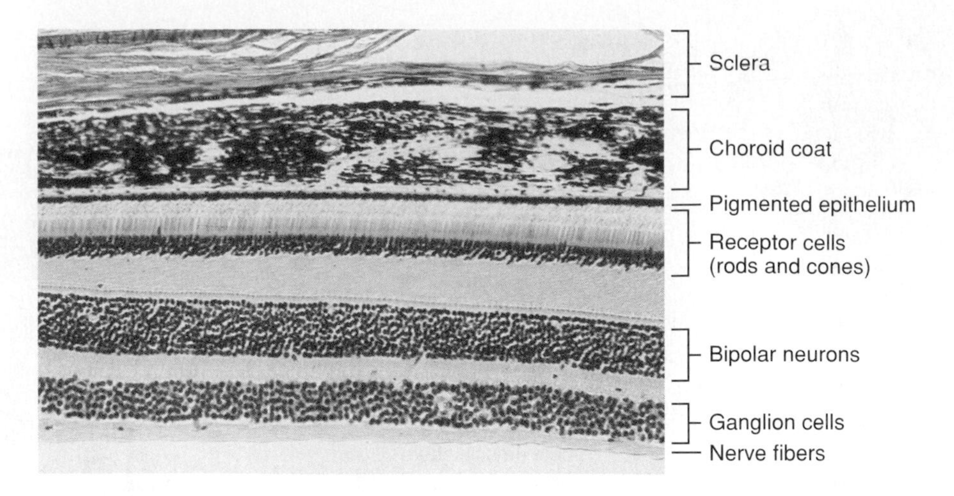


**FIGURE 12.35**

Note the layers of cells and nerve fibers in this light micrograph of the retina (200×).

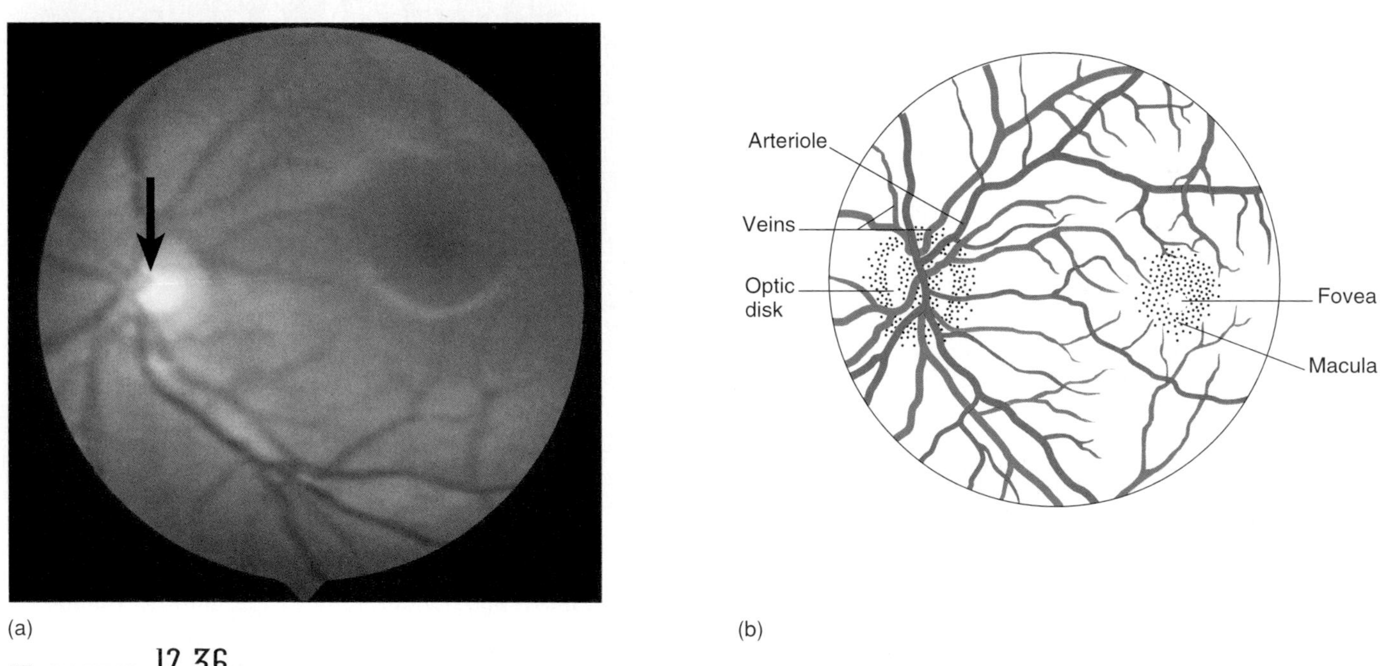


**FIGURE 12.36**

(*a*) Nerve fibers leave the eye in the area of the optic disk (arrow) to form the optic nerve (53×); (*b*) major features of the retina.

## Light Refraction

When a person sees something, either the object is giving off light, or light waves are reflected from it. These light waves enter the eye, and an image of what is seen focuses upon the retina. The light rays must bend to be focused, a phenomenon called **refraction.**

Refraction occurs when light waves pass at an oblique angle from a medium of one optical density into a medium of a different optical density. For example, as figure 12.37 shows, when light passes obliquely from a less dense medium such as air into a denser medium such as glass, or from air into the cornea of the eye, the light is bent toward a line perpendicular to the surface between these substances. When the surface between such refracting media is curved, a lens is formed. A lens with a *convex* surface causes light waves to converge, and a lens with a *concave* surface causes light waves to diverge (fig. 12.38). Clinical Application 12.5 discusses some familiar problems with refraction.

The convex surface of the cornea refracts light waves from objects outside the eye, providing about 75% of the total refractive power of the eye. The light is refracted again by the convex surface of the lens and to a lesser extent by the surfaces of the fluids within the eye chambers.

If the shape of the eye is normal, light waves are focused sharply upon the retina, much as a motion picture image is focused on a screen for viewing. Unlike the motion picture image, however, the one formed on the retina is upside down and reversed from left to

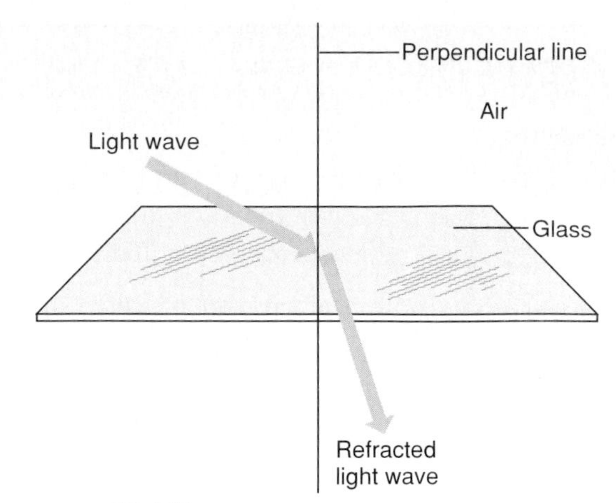


**FIGURE 12.37**
When light passes at an oblique angle from air into glass, the light waves bend toward a line perpendicular to the surface of the glass.

**CHART 12.6 LAYERS OF THE EYE**

| Layer/Tunic | Posterior Portion | Function | Anterior Portion | Function |
|---|---|---|---|---|
| Outer layer | Sclera | Protection | Cornea | Light transmission and refraction |
| Middle layer | Choroid coat | Blood supply, pigment prevents reflection | Ciliary body, iris | Accommodation; controls light intensity |
| Inner layer | Retina | Photoreception, impulse transmission | None | |

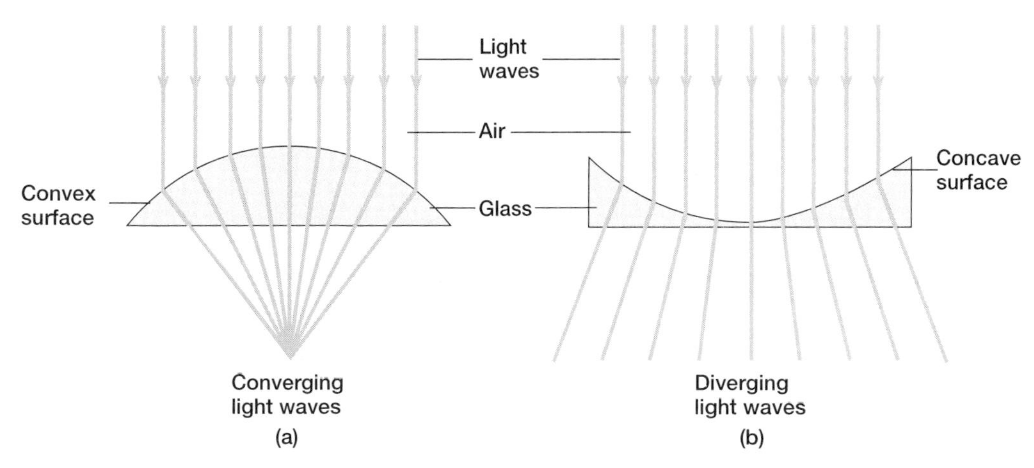


**FIGURE 12.38**
(*a*) A lens with a convex surface causes light waves to converge; (*b*) a lens with a concave surface causes them to diverge.

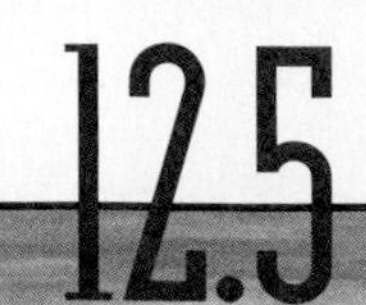

CLINICAL APPLICATION 12.5

## Refraction Disorders

The elastic quality of the lens capsule tends to lessen with time. People over 45 years of age are often unable to accommodate sufficiently to read the fine print in books and newspapers or on medicine bottles. Their eyes remain focused for distant vision. This condition is termed *presbyopia,* or farsightedness of age. Eyeglasses or contact lenses can usually make up for the eye's loss of refracting power.

Other visual problems result from eyeballs that are too short or too long for sharp focusing. If an eye is too short, light waves are not focused sharply on the retina because their point of focus lies behind it. A person with this condition may be able to bring the image of distant objects into focus by accommodation, but this requires contraction of the ciliary muscles at times when these muscles are at rest in a normal eye. Still more accommodation is needed to view closer objects, and the person may suffer from ciliary muscle fatigue, pain, and headache when doing close work.

Since people with short eyeballs are usually unable to accommodate enough to focus on the very close objects, they are said to be *farsighted.* Eyeglasses or contact lenses with *convex* surfaces can remedy this condition (hyperopia) by focusing images closer to the front of the eye.

If an eyeball is too long, light waves tend to be focused in front of the retina, and the image produced on it is blurred. In other words, the refracting power of the eye, even when the lens is flattened, is too great. Although a person with this problem may be able to focus on close objects by accommodation, distance vision is invariably poor. For this reason, the person is said to be *nearsighted.* Eyeglasses or contact lenses with *concave* surfaces that focus images farther from the front of the eye treat nearsightedness (myopia) (figs. 12A and 12B).

Still another refraction problem, *astigmatism,* reflects a defect in the curvature of the cornea or the lens. The normal cornea has a spherical curvature, like the inside of a ball; an astigmatic cornea usually has an elliptical curvature, like the bowl of a spoon. As a result, some portions of an image are in focus on the retina, but other portions are blurred, and vision distorts.

Without corrective lenses, astigmatic eyes tend to accommodate back and forth reflexly in an attempt to sharpen focus. The consequence of this continual action is often ciliary muscle fatigue and headache.

right (fig. 12.39). When the visual cortex of the cerebrum interprets such an image, it somehow corrects this, and things are seen in their proper positions.

Light waves coming from objects more than 20 feet away are traveling in nearly parallel lines, and they are focused on the retina by the cornea and by the lens in its more flattened or "at rest" condition. Light waves arriving from objects less than 20 feet away, however, reach the eye along more divergent lines—in fact, the closer the object, the more divergent the lines.

Divergent light waves tend to focus behind the retina unless something increases the refracting power of the eye. Accommodation accomplishes this increase,

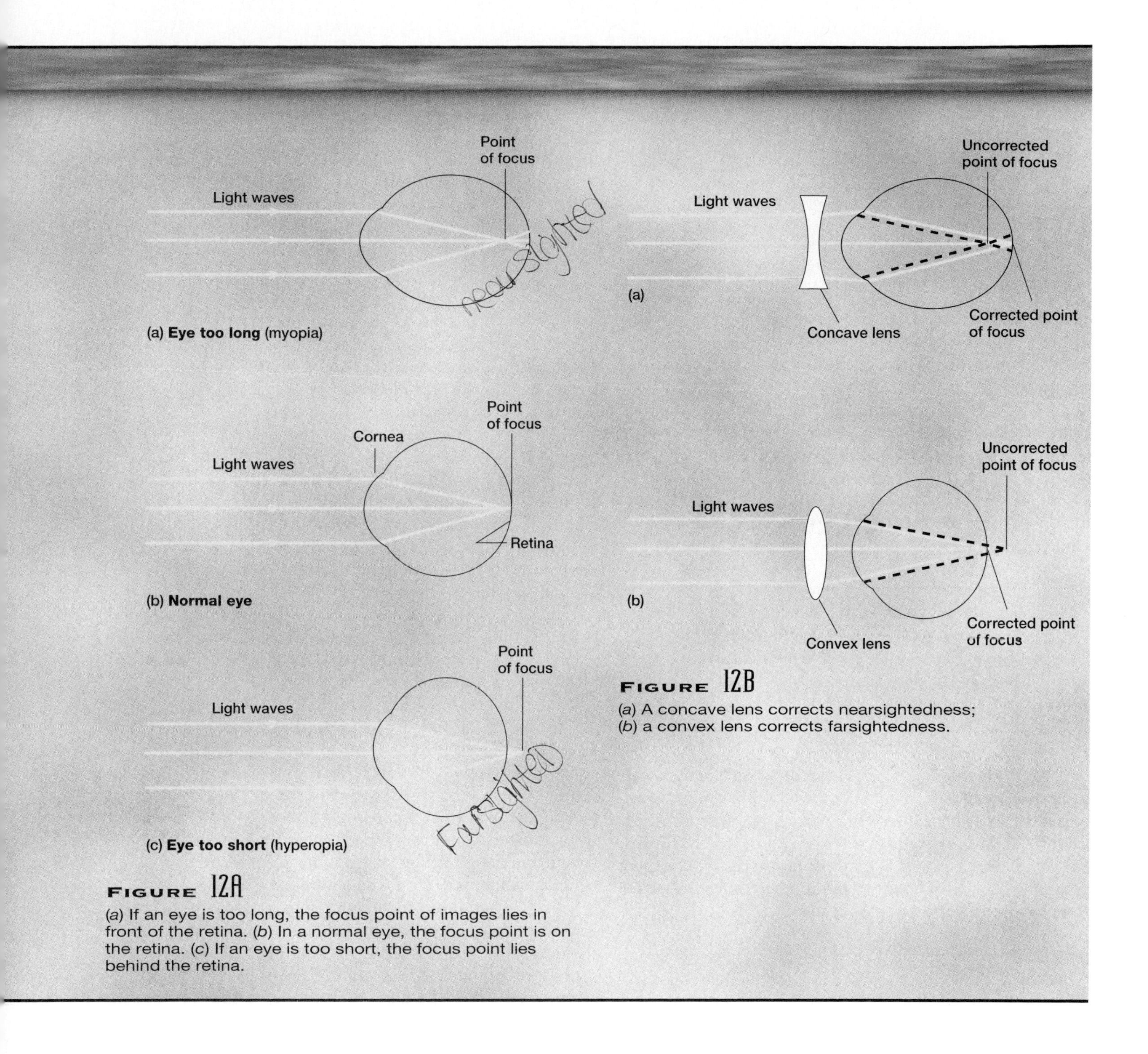


**FIGURE 12A**

(*a*) If an eye is too long, the focus point of images lies in front of the retina. (*b*) In a normal eye, the focus point is on the retina. (*c*) If an eye is too short, the focus point lies behind the retina.

**FIGURE 12B**

(*a*) A concave lens corrects nearsightedness; (*b*) a convex lens corrects farsightedness.

thickening the lens. As the lens thickens, light waves converge more strongly, so that diverging light waves coming from close objects focus on the retina.

1. What is refraction?
2. What parts of the eye provide refracting surfaces?
3. Why is it necessary to accommodate for viewing close objects?

## Visual Receptors

The photoreceptors of the eye are modified neurons of two distinct kinds. One group of receptor cells, called **rods,** have long, thin projections at their terminal ends. The cells of the other group, called **cones,** have short, blunt projections. The retina contains about 100 million rods and three million cones.

Instead of being located in the surface layer of the retina, rods and cones are found in a deep portion,

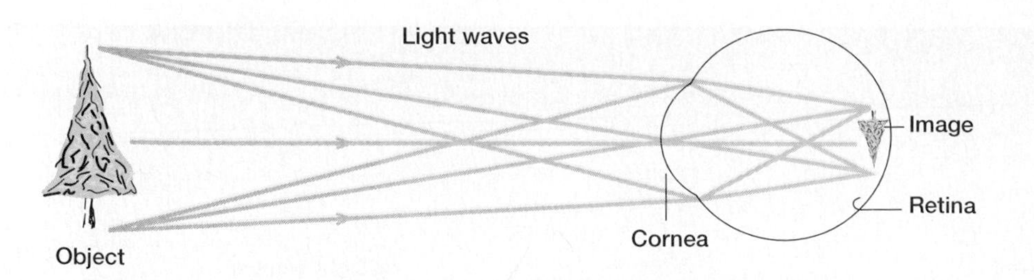


**FIGURE 12.39**
The image of an object forms on the retina upside down.

closely associated with a layer of pigmented epithelium. The projections from the receptors extend into the pigmented layer and contain light-sensitive visual pigments.

The epithelial pigment of the retina absorbs light waves that are not absorbed by the receptor cells, and together with the pigment of the choroid coat, it keeps light from reflecting off the surfaces inside the eye. The pigment layer also stores vitamin A, which the receptor cells use to synthesize visual pigments.

> Researchers can grow retinal epithelial cells in laboratory cultures, and the cells retain their pigment. This means that someday scientists may be able to grow tissue that can be implanted into a person's eye to treat some forms of blindness.

The visual receptors are only stimulated when light reaches them. Thus, when a light image is focused on an area of the retina, some receptors are stimulated, and impulses travel from them to the brain. However, the impulse leaving each activated receptor provides only a fragment of the information needed for the brain to interpret a total scene.

> *Albinism* is an inherited condition in which a missing enzyme blocks pigment synthesis, causing very pale, highly sun-sensitive skin. More severe forms of albinism also affect the eyes, making vision blurry and intolerant to light. A person may squint even in very faint light. The extra sensitivity is due to the fact that light reflects inside the lenses, excessively stimulating visual receptors. The eyes of many people with albinism also dart about uncontrollably, a condition called *nystagmus.*

Rods and cones function differently. Rods are hundreds of times more sensitive to light than are cones, and as a result, rods provide vision in dim light. In addition, rods produce colorless vision, while cones can detect colors.

Still another difference is that cones provide sharp images, while rods produce more general outlines of objects. This difference is due to the fact that nerve fibers from many rods may converge, and their impulses may be transmitted to the brain on the same nerve fiber (see chapter 10). Thus, if a point of light stimulates a rod, the brain cannot tell which one of many receptors has actually been stimulated. Such a convergence of impulses occurs to a much lesser degree among cones, so when a cone is stimulated, the brain is able to pinpoint the stimulation more accurately (fig. 12.40).

The area of sharpest vision, the fovea centralis in the macula lutea, lacks rods, but contains densely packed cones with few or no converging fibers. Also, the overlying layers of the retina, as well as the retinal blood vessels, are displaced to the sides in the fovea, which more fully exposes the receptors to incoming light. Consequently, to view something in detail, a person moves the eyes so that the important part of an image falls upon the fovea centralis.

The concentration of cones decreases in areas farther away from the macula lutea, whereas the concentration of rods increases in these areas. Also, the degree of convergence among the rods and cones increases toward the periphery of the retina. As a result, the visual sensations from images focused on the sides of the retina tend to be blurred compared with those focused on the central portion of the retina.

### Visual Pigments

Both rods and cones contain light-sensitive pigments that decompose when they absorb light energy. The light-sensitive pigment in rods is called **rhodopsin,** or visual purple, and it is embedded in membranous

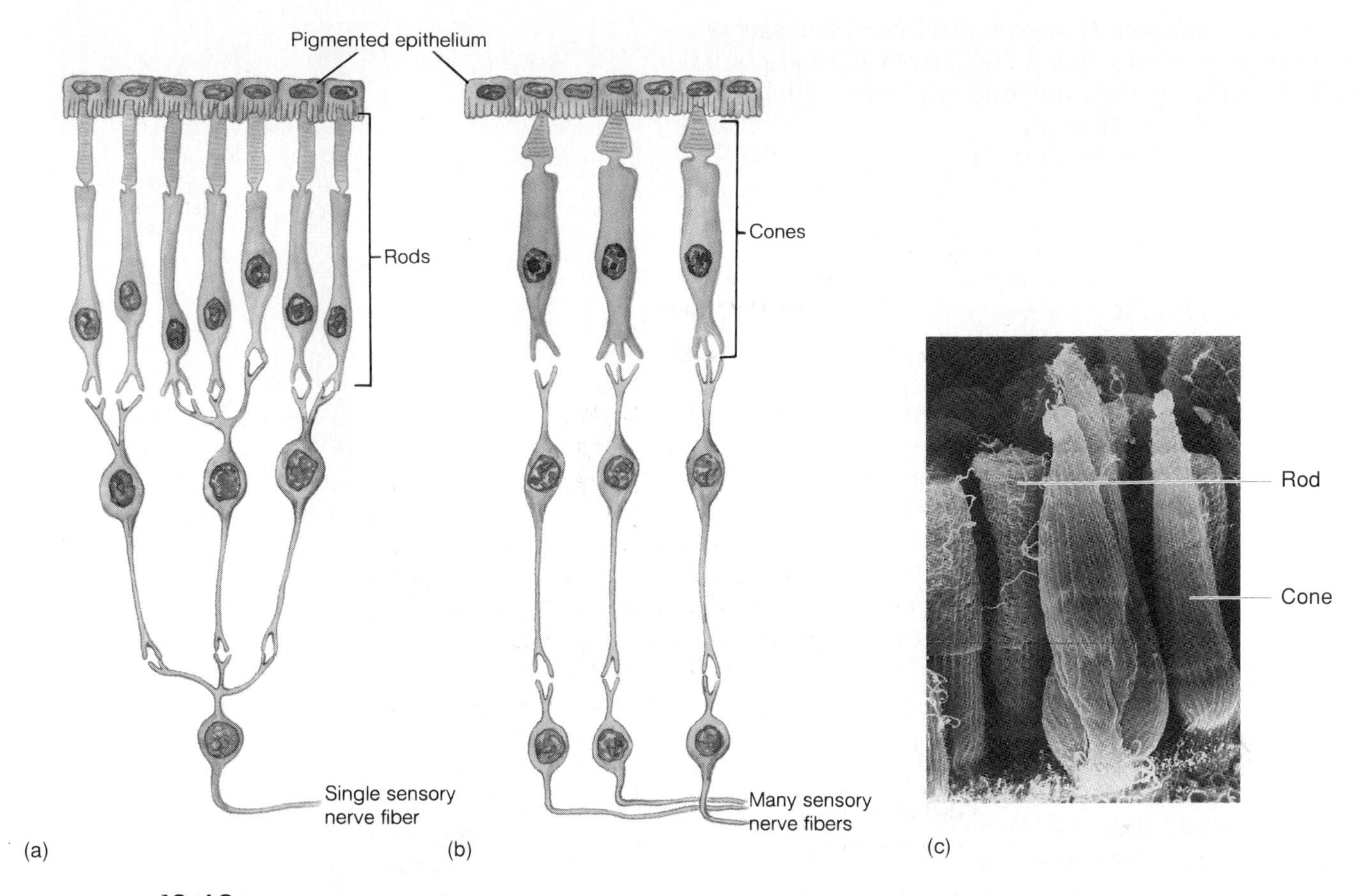


**FIGURE 12.40**
(*a*) A single sensory nerve fiber transmits impulses from several rods to the brain. (*b*) Separate sensory nerve fibers transmit impulses from cones to the brain. (*c*) A scanning electron micrograph of rods and cones.

disks that are stacked within these receptor cells (fig. 12.41). In the presence of light, rhodopsin molecules break down into molecules of a colorless protein called *opsin* and a yellowish organic molecule called *retinal* (retinene) that is synthesized from vitamin A.

Although the bony orbit usually protects the eye, sometimes a forceful blow causes damage. If the eye is jarred sufficiently, some of its contents may be displaced. The suspensory ligaments may be torn and the lens dislocated into the posterior cavity, or the retina may pull away from the underlying vascular choroid coat. Once the retina detaches, photoreceptor cells may die because of lack of oxygen and nutrients. Unless a *detached retina* is repaired surgically, this injury may cause visual loss or blindness.

In darkness, sodium channels in portions of the receptor cell membranes are kept open by a nucleotide called *cyclic guanosine monophosphate* (cGMP). When rhodopsin molecules absorb light, they release opsin, which becomes an active enzyme. This enzyme activates another enzyme (transducin), which in turn activates still another one (phosphodiesterase). The third enzyme of this series breaks down cGMP, and as the concentration of cGMP decreases, sodium channels close, and the receptor cell membrane hyperpolarizes (see chapter 10). The degree of hyperpolarization is directly proportional to the intensity of the light stimulating the receptor cells.

The hyperpolarization reaches the synaptic end of the cell, inhibiting release of neurotransmitter. Through a complex mechanism, decreased release of neurotransmitter by photoreceptor cells either stimulates or inhibits nerve impulses (action potentials) in nearby retinal neurons. Consequently, complex patterns of nerve impulses travel away from the retina, through the optic nerve, and into the brain, where they are interpreted as vision.

In bright light, nearly all of the rhodopsin in the rods decomposes, sharply reducing the sensitivity of these receptors. The cones continue to function, however, and in bright light we therefore see in color. In dim light, rhodopsin can be regenerated from opsin and retinal faster than it is broken down. This regeneration requires cellular energy, which ATP provides (see chapter 4). Under these conditions, the rods continue to function and the cones remain unstimulated. Hence, we see only shades of gray in dim light.

The light sensitivity of an eye whose rods have converted the available opsin and retinal to rhodopsin increases about 100,000 times, and the eye is said to be *dark adapted.* A person needs a dark-adapted eye to see in dim light. For example, when going from daylight into a darkened theater, it may be difficult to see well enough to locate a seat, but soon the eyes adapt to the dim light and vision improves. Later, leaving the theater and entering the sunlight may cause discomfort or even pain. This occurs at the moment that most of the rhodopsin decomposes in response to the bright light. At the same time, the light sensitivity of the eyes decreases greatly, and they become *light adapted.*

The light-sensitive pigments of cones, called *iodopsins,* are similar to rhodopsin in that they are composed of retinal combined with a protein; the protein, however, differs from the protein in the rods. In fact, there are three sets of cones within the retina, each containing an abundance of one of three different visual pigments.

> Too little vitamin A in the diet reduces the quantity of retinal, impairing rhodopsin production and sensitivity of the rods. The result is night blindness, or poor vision in dim light.

The wavelength of a particular kind of light determines the color perceived from it. For example, the shortest wavelengths of visible light are perceived as violet, while the longest wavelengths of visible light are seen as red. One type of cone pigment (erythrolabe) is most sensitive to red light waves, another (chlorolabe) to green light waves, and a third (cyanolabe) to blue light waves. The sensitivities of these pigments do overlap somewhat. For example, both red and green light pigments are sensitive to orange light waves. On the other hand, red pigment absorbs orange light waves more effectively.

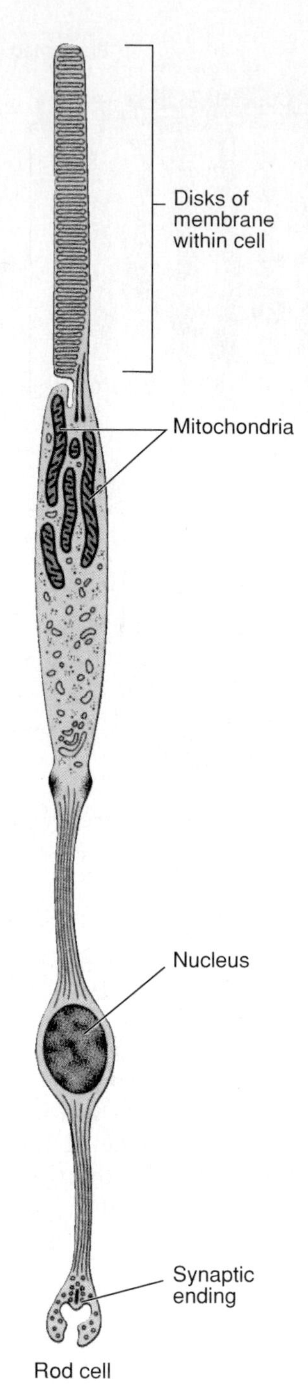


**FIGURE 12.41**
Rhodopsin is embedded in disks of membrane that are stacked within the rod cells.

The color perceived depends upon which sets of cones the light in a given image stimulates. If all three types of sets of cones are stimulated, the light is perceived as white, and if none are stimulated, it is seen as black.

Color blindness is an inherited trait reflecting absence of cone pigments. In "green weakness" color blindness, the most common type, red appears brown and green appears tan. Less common is "red weakness," in which colors appear washed out. Red and green weakness are carried on the X chromosome, which means they are usually passed from a mother who is a carrier to a son, who is color-blind. Thus, these forms of color blindness are more common in males. A rare type of color blindness, affecting blue pigment, is not carried on the X chromosome and so affects males and females with equal frequency.

## Stereoscopic Vision

**Stereoscopic vision** (stereopsis) simultaneously perceives distance, depth, height, and width of objects. Such vision is possible because the pupils are 6–7 centimeters apart. Consequently, objects that are relatively close (less than 20 feet away) produce slightly different retinal images. That is, the right eye sees a little more of one side of an object, while the left eye sees a little more of the other side. These two images are somehow superimposed and interpreted by the visual cortex of the brain. The result is the perception of a single object in three dimensions (fig. 12.42).

Because this type of depth perception depends on vision with two eyes (binocular vision), it follows that a one-eyed person is less able to judge distance and depth accurately. To compensate, a person with one eye can use the relative sizes and positions of familiar objects as visual clues.

## Visual Nerve Pathways

As mentioned in chapter 11, the axons of the ganglion cells in the retina leave the eyes to form the *optic nerves.* Just anterior to the pituitary gland, these nerves give rise to the X-shaped *optic chiasma,* and within the chiasma some of the fibers cross over. More specifically, the fibers from the nasal (medial) half of each retina cross over, while those from the temporal (lateral) sides do not. Thus, fibers from the nasal half of the left eye and the temporal half of the right eye form the right *optic tract;* and fibers from the nasal half of the right eye and the temporal half of the left eye form the left optic tract.

The nerve fibers continue in the optic tracts, and just before they reach the thalamus, a few of them leave to enter nuclei that function in various visual reflexes. Most of the fibers, however, enter the thalamus and synapse in its posterior portion (lateral geniculate body). From this region the visual impulses enter nerve pathways called *optic radiations,* and the pathways lead to the visual cortex of the occipital lobes (fig. 12.43).

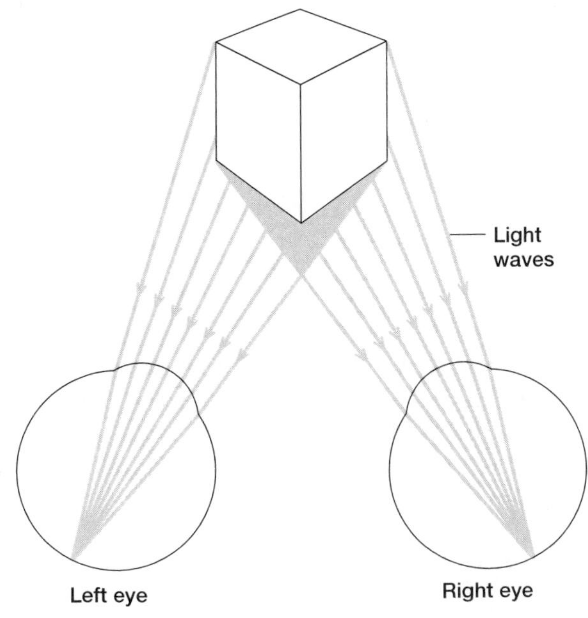


**FIGURE 12.42**
Stereoscopic vision results from formation of two slightly different retinal images.

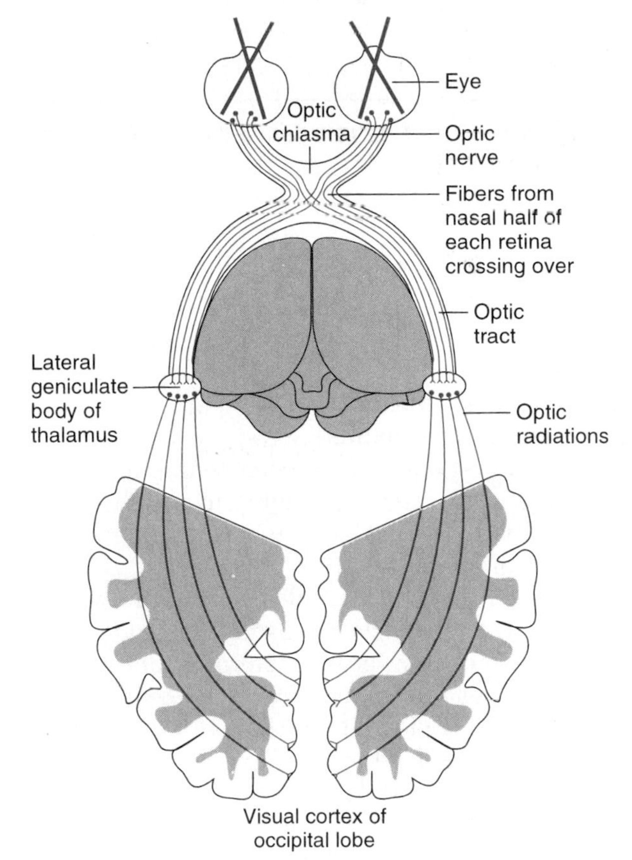


**FIGURE 12.43**
The visual pathway includes the optic nerve, optic chiasma, optic tract, and optic radiations.

Since each visual cortex receives impulses from each eye, a person may develop partial blindness in both eyes if either visual cortex is injured. For example, if the right visual cortex (or the right optic tract) is injured, sight may be lost in the temporal side of the right eye and the nasal side of the left eye. Similarly, damage to the central portion of the optic chiasma, where fibers from the nasal sides of the eyes cross over, blinds the nasal sides of both eyes.

Other fibers conducting visual impulses pass downward into various regions of the brain stem. These impulses are important for controlling head and eye movements associated with tracking an object visually, for controlling the simultaneous movements of both eyes, and for controlling certain visual reflexes, such as those that move the iris muscles.

1. Distinguish between the rods and the cones of the retina.
2. Explain the roles of visual pigments.
3. What factors make stereoscopic vision possible?
4. Trace the pathway of visual impulses from the retina to the occipital cortex.

## Clinical Terms Related to the Senses

**amblyopia** (am″ble-o′pe-ah) Dim vision due to a cause other than a refractive disorder or lesion.
**anopia** (an-o′pe-ah) Absence of an eye.
**audiometry** (aw″de-om′ĕ-tre) The measurement of auditory acuity for various frequencies of sound waves.
**blepharitis** (blef″ah-ri′tis) An inflammation of the margins of the eyelids.
**causalgia** (kaw-zal′je-ah) A persistent, burning pain usually associated with injury to a limb.
**conjunctivitis** (kon-junk″tĭ-vi′tis) An inflammation of the conjunctiva.
**diplopia** (di-plo′pe-ah) Double vision, or the sensation of seeing two objects when only one is viewed.
**emmetropia** (em″ĕ-tro′pe-ah) Normal condition of the eyes; eyes with no refractive defects.
**enucleation** (e-nu″kle-a′shun) Removal of the eyeball.
**exophthalmos** (ek″sof-thal′mos) Condition in which the eyes protrude abnormally.
**hemianopsia** (hem″e-an-op′se-ah) Defective vision affecting half of the visual field.
**hyperalgesia** (hi″per-al-je′ze-ah) An abnormally increased sensitivity to pain.
**iridectomy** (ir″ĭ-dek′to-me) The surgical removal of part of the iris.
**iritis** (i-ri′tis) An inflammation of the iris.
**keratitis** (ker″ah-ti′tis) An inflammation of the cornea.
**labyrinthectomy** (lab″i-rin-thek′to-me) The surgical removal of the labyrinth.
**labyrinthitis** (lab″i-rin-thi′tis) An inflammation of the labyrinth.
**Ménière's disease** (men″e-ārz′ dĭ-zez) An inner ear disorder characterized by ringing in the ears, increased sensitivity to sounds, dizziness, and loss of hearing.
**neuralgia** (nu-ral′je-ah) Pain resulting from inflammation of a nerve or a group of nerves.
**neuritis** (nu-ri′tis) An inflammation of a nerve.
**nystagmus** (nis-tag′mus) An involuntary oscillation of the eyes.
**otitis media** (o-ti′tis me′de-ah) An inflammation of the middle ear.
**otosclerosis** (o″to-skle-ro′sis) A formation of spongy bone in the inner ear, which often causes deafness by fixing the stapes to the oval window.
**pterygium** (tĕ-rij′e-um) An abnormally thickened patch of conjunctiva that extends over part of the cornea.
**retinitis pigmentosa** (ret″ĭ-ni′tis pig″men-to′sa) An inherited progressive retinal sclerosis characterized by deposits of pigment in the retina and by retinal atrophy.
**retinoblastoma** (ret″i-no-blas-to′mah) An inherited, highly malignant tumor arising from immature retinal cells.
**tinnitus** (tĭ-ni′tus) A ringing or buzzing noise in the ears.
**tonometry** (to-nom′e-tre) The measurement of fluid pressure within the eyeball.
**trachoma** (trah-ko′mah) A bacterial disease of the eye, characterized by conjunctivitis, which may lead to blindness.
**tympanoplasty** (tim″pah-no-plas′te) The surgical reconstruction of the middle ear bones and establishment of continuity from the tympanic membrane to the oval window.
**uveitis** (u″ve-i′tis) An inflammation of the uvea, the region of the eye that includes the iris, the ciliary body, and the choroid coat.
**vertigo** (ver′tĭ-go) A sensation of dizziness.

# Chapter Summary

## Introduction (page 441)

Sensory receptors are sensitive to environmental changes and initiate impulses to the brain and spinal cord.

## Receptors and Sensations (page 441)

1. Receptor types
   a. Each type of receptor is sensitive to a distinct type of stimulus.
   b. The major types of receptors include the following:
      (1) Chemoreceptors, sensitive to changes in chemical concentration.
      (2) Pain receptors, sensitive to tissue damage.
      (3) Thermoreceptors, sensitive to temperature changes.
      (4) Mechanoreceptors, sensitive to mechanical forces.
      (5) Photoreceptors, sensitive to light.
2. Sensory impulses
   a. When receptors are stimulated, changes occur in their membrane potentials.
   b. Receptor potentials are transferred to nerve fibers, triggering action potentials.
3. Sensations
   a. Sensations are feelings resulting from sensory stimulation.
   b. A particular part of the sensory cortex interprets every impulse that reaches it in the same way.
   c. The cerebral cortex projects a sensation back to the region of stimulation.
4. Sensory adaptations are adjustments of sensory receptors to continuous stimulation. Impulses are triggered at slower and slower rates.

## Somatic Senses (page 442)

Somatic senses receive information from receptors in skin, muscles, joints, and viscera. They can be grouped as exteroceptive, proprioceptive, and visceroceptive senses.

1. Touch and pressure senses
   a. Free ends of sensory nerve fibers are the receptors for the sensations of touch and pressure.
   b. Meissner's corpuscles are the receptors for the sensations of light touch.
   c. Pacinian corpuscles are the receptors for the sensations of heavy pressure and vibrations.
2. Thermoreceptors include two sets of free nerve endings that are heat and cold receptors.
3. Sense of pain
   a. Pain receptors
      (1) Pain receptors are free nerve endings that tissue damage stimulates.
      (2) Pain receptors provide protection, do not adapt rapidly, and can be stimulated by changes in temperature, mechanical force, and chemical concentration.
   b. The only receptors in viscera that provide sensations are pain receptors. These receptors are most sensitive to certain chemicals and lack of blood flow. The sensations they produce feel as if they come from some other part of the body (referred pain).
   c. Pain nerve pathways
      (1) The two main types of pain fibers are acute pain fibers and chronic pain fibers.
      (2) Acute pain fibers are fast conducting; chronic pain fibers are slower conducting.
      (3) Pain impulses are processed in the dorsal horn of the spinal cord, and they ascend in the spinothalamic tracts.
      (4) Within the brain, pain impulses pass through the reticular formation before being conducted to the cerebral cortex.
   d. Regulation of pain impulses
      (1) Awareness of pain occurs when impulses reach the thalamus.
      (2) The cerebral cortex judges the intensity of pain and locates its source.
      (3) Impulses descending from the brain cause neurons to release pain-relieving substances, such as enkephalins and serotonin.
      (4) Endorphin is a pain-relieving biochemical produced in the brain.
   e. Certain neuropeptides synthesized in the brain and spinal cord inhibit pain impulses.
4. Stretch receptors
   a. Stretch receptors provide information about the condition of muscles and tendons.
   b. Muscle spindles are stimulated when a muscle is relaxed, and they initiate a reflex that contracts the muscle.
   c. Golgi tendon organs are stimulated when muscle tension increases, and they initiate a reflex that relaxes the muscle.

## Special Senses (page 448)

Special senses are those whose receptors occur in relatively large, complex sensory organs of the head.

## Sense of Smell (page 448)

1. Olfactory receptors
   a. Olfactory receptors are chemoreceptors that chemicals dissolved in liquid stimulate.
   b. Olfactory receptors function together with taste receptors and aid in food selection.
2. Olfactory organs
   a. The olfactory organs consist of receptors and supporting cells in the nasal cavity.
   b. Olfactory receptors are neurons with cilia that sense lipid-soluble chemicals.
3. Olfactory nerve pathways.
   a. Nerve impulses travel from the olfactory receptors through the olfactory nerves, olfactory bulbs, and olfactory tracts.
   b. They go to interpreting centers in the limbic system of the brain.
4. Olfactory stimulation
   a. Olfactory impulses may result when various gaseous molecules combine with specific sites on the cilia of the receptor cells.
   b. Olfactory receptors adapt rapidly.
   c. Olfactory receptors are often damaged by environmental factors, but are not replaced.

## Sense of Taste (page 451)

1. Taste receptors
   a. Taste buds consist of receptor cells and supporting cells.

b. Taste cells have taste hairs that are sensitive to particular chemicals dissolved in water.
c. Taste hair surfaces have receptor sites to which chemicals combine and trigger impulses to the brain.
d. Each of the four primary kinds of taste cells is particularly sensitive to a certain group of chemicals.

2. Taste sensations
   a. The four primary taste sensations are sweet, sour, salty, and bitter.
   b. Various taste sensations result from the stimulation of one or more sets of taste receptors.
   c. Sweet receptors are most plentiful near the tip of the tongue, sour receptors along the margins, salt receptors in the tip and upper front, and bitter receptors toward the back.
3. Taste nerve pathways
   a. Sensory impulses from taste receptors travel on fibers of the facial, glossopharyngeal, and vagus nerves.
   b. These impulses are carried to the medulla and ascend to the thalamus, and then to the gustatory cortex in the parietal lobes.

## Sense of Hearing (page 453)

1. The external ear collects sound waves created by vibrating objects.
2. Middle ear
   a. Auditory ossicles of the middle ear conduct sound waves from the tympanic membrane to the oval window of the inner ear. They also increase the force of these waves.
   b. Skeletal muscles attached to the auditory ossicles provide the tympanic reflex, which protects the inner ear from the effects of loud sounds.
3. Auditory tubes connect the middle ears to the throat and help maintain equal air pressure on both sides of the eardrums.
4. Inner ear
   a. The inner ear consists of a complex system of interconnected tubes and chambers—the osseous and membranous labyrinths. It includes the cochlea, which in turn houses the organ of Corti.
   b. The organ of Corti contains the hearing receptors that vibrations in the fluids of the inner ear stimulate.
   c. Different frequencies of vibrations stimulate different sets of receptor cells; the human ear can detect sound frequencies from about 20 to 20,000 vibrations per second.
5. Auditory nerve pathways
   a. The nerve fibers from hearing receptors travel in the cochlear branch of the vestibulocochlear nerves.
   b. Auditory impulses travel into the medulla oblongata, midbrain, and thalamus, and are interpreted in the temporal lobes of the cerebrum.

## Sense of Equilibrium (page 462)

1. Static equilibrium maintains the stability of the head and body when they are motionless. The organs of static equilibrium are located in the vestibule.
2. Dynamic equilibrium balances the head and body when they are moved or rotated suddenly. The organs of this sense are located in the ampullae of the semicircular canals.
3. Other structures that help maintain equilibrium include the eyes and the proprioceptors associated with certain joints.

## Sense of Sight (page 466)

1. Visual accessory organs include the eyelids and lacrimal apparatus that protect the eye, and the extrinsic muscles that move the eye.
2. Structure of the eye
   a. The wall of the eye has an outer, a middle, and an inner tunic that function as follows:
      (1) The outer layer (sclera) is protective, and its transparent anterior portion (cornea) refracts light entering the eye.
      (2) The middle layer (choroid coat) is vascular and contains pigments that help keep the inside of the eye dark.
      (3) The inner layer (retina) contains the visual receptor cells.
   b. The lens is a transparent, elastic structure. The ciliary muscles control its shape.
   c. The iris is a muscular diaphragm that controls the amount of light entering the eye; the pupil is an opening in the iris.
   d. Spaces within the eye are filled with fluids (aqueous and vitreous humors) that help maintain its shape.
3. Light refraction
   a. Light waves are refracted primarily by the cornea and lens to focus an image on the retina.
   b. The lens must thicken to focus on close objects.
4. Visual receptors
   a. The visual receptors are rods and cones.
   b. Rods are responsible for colorless vision in relatively dim light, and cones provide color vision.
5. Visual pigments
   a. A light-sensitive pigment in rods (rhodopsin) decomposes in the presence of light and triggers a complex series of reactions that initiate nerve impulses on the optic nerve.
   b. Three sets of cones provide color vision. Each set contains a different light-sensitive pigment, and each set is sensitive to a different wavelength of light; the color perceived depends on which set or sets of cones are stimulated.
6. Stereoscopic vision
   a. Stereoscopic vision provides perception of distance and depth.
   b. Stereoscopic vision occurs because of the formation of two slightly different retinal images that the brain superimposes and interprets as one image in three dimensions.
   c. A one-eyed person uses relative sizes and positions of familiar objects to judge distance and depth.
7. Visual nerve pathways
   a. Nerve fibers from the retina form the optic nerves.
   b. Some fibers cross over in the optic chiasma.
   c. Most of the fibers enter the thalamus and synapse with others that continue to the visual cortex of the occipital lobes.
   d. Other impulses pass into the brain stem and function in various visual reflexes.

## Critical Thinking Questions

1. How would you interpret the following observation? A person enters a tub of water and reports that it is uncomfortably warm, yet a few moments later says the water feels comfortable, even though the water temperature is unchanged.
2. Why are some serious injuries, such as those produced by a bullet entering the abdomen, relatively painless, while others, such as those that crush the skin, are quite painful?
3. Labyrinthitis is an inflammation of the tissues of the inner ear. What symptoms would you expect to observe in a patient with this disorder?
4. Sometimes, as a result of an injury to the eye, the retina detaches from its pigmented epithelium. Assuming that the retinal tissues remain functional, what is likely to happen to the person's vision if the retina moves unevenly toward the interior of the eye?
5. The auditory tubes of a child are shorter and directed more horizontally than those of an adult. How might this explain the greater prevalence of middle ear infections in children compared to adults?
6. A patient with heart disease experiences pain at the base of the neck and in the left shoulder and arm after exercise. How would you explain to the patient the origin of this pain?

## Review Exercises

1. List five groups of sensory receptors, and name the kind of change to which each is sensitive.
2. Explain how sensory receptors stimulate sensory impulses.
3. Define *sensation.*
4. Explain what is meant by the projection of a sensation.
5. Define *sensory adaptation.*
6. Explain how somatic senses can be grouped.
7. Describe the functions of free nerve endings, Meissner's corpuscles, and Pacinian corpuscles.
8. Explain how thermoreceptors function.
9. Compare pain receptors with other types of somatic receptors.
10. List the factors that are likely to stimulate visceral pain receptors.
11. Define *referred pain.*
12. Explain how neuropeptides relieve pain.
13. Distinguish between muscle spindles and Golgi tendon organs.
14. Explain how the senses of smell and taste function together to create the flavors of foods.
15. Describe the olfactory organ and its function.
16. Trace a nerve impulse from the olfactory receptor to the interpreting centers of the brain.
17. Explain how an olfactory code distinguishes odor stimuli.
18. Explain how the salivary glands aid the taste receptors.
19. Name the four primary taste sensations, and describe the patterns in which the taste receptors are roughly distributed on the tongue.
20. Explain why taste sensation is less likely to diminish with age than olfactory sensation.
21. Trace the pathway of a taste impulse from the receptor to the cerebral cortex.
22. Distinguish between the external, middle, and inner ears.
23. Trace the path of a sound vibration from the tympanic membrane to the hearing receptors.
24. Describe the functions of the auditory ossicles.
25. Describe the tympanic reflex, and explain its importance.
26. Explain the function of the auditory tube.
27. Distinguish between the osseous and the membranous labyrinths.
28. Describe the cochlea and its function.
29. Describe a hearing receptor.
30. Explain how a hearing receptor stimulates a sensory neuron.
31. Trace a nerve impulse from the organ of Corti to the interpreting centers of the cerebrum.
32. Describe the organs of static and dynamic equilibrium and their functions.
33. Explain how the sense of vision helps maintain equilibrium.
34. List the visual accessory organs, and describe the functions of each.
35. Name the three layers of the eye wall, and describe the functions of each.
36. Describe how accommodation is accomplished.
37. Explain how the iris functions.
38. Distinguish between aqueous humor and vitreous humor.
39. Distinguish between the macula lutea and the optic disk.
40. Explain how light waves focus on the retina.
41. Distinguish between rods and cones.
42. Explain why cone vision is generally more acute than rod vision.
43. Describe the function of rhodopsin.
44. Explain how the eye adapts to light and dark.
45. Describe the relationship between light wavelengths and color vision.
46. Define *stereoscopic vision.*
47. Explain why a person with binocular vision is able to judge distance and depth of close objects more accurately than a one-eyed person.
48. Trace a nerve impulse from the retina to the visual cortex.

13

CHAPTER

# Endocrine System

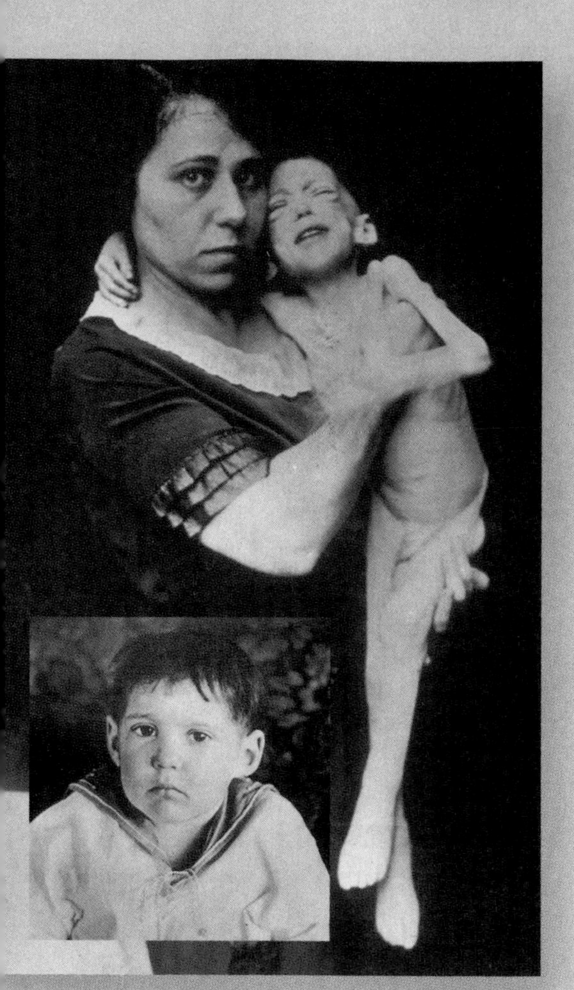

Before and after insulin treatment. The boy in his mother's arms is 3 years old but weighs only 15 pounds because of diabetes mellitus. The inset shows the same child after just two months of receiving insulin.

## Chapter Objectives

After you have studied this chapter, you should be able to:

1. Distinguish between endocrine and exocrine glands.
2. Describe how hormones can be classified according to their chemical composition.
3. Explain how steroid and nonsteroid hormones produce effects on target cells.
4. Discuss how negative feedback mechanisms regulate hormonal secretions.
5. Explain how the nervous system controls hormonal secretions.
6. Name and describe the locations of the major endocrine glands of the body, and list the hormones they secrete.
7. Describe the general functions of the hormones secreted by the endocrine glands.
8. Explain how the secretion of each hormone is regulated.
9. Distinguish between physical and psychological stress.
10. Describe the general stress response.

## Key Terms

**adenylate cyclase** (ah-den′ĭ-lat si′klās)

**adrenal cortex** (ah-dre′nal kor′teks)

**adrenal medulla** (ah-dre′nal me-dul′ah)

**anterior pituitary** (an-ter′e-or pĭ-tu′ĭ-tar″e)

**cyclic AMP** (sik′lik AMP)

**hormone** (hor′mōn)

**kinase** (ki′nās)

**metabolic rate** (met″ah-bol′ik rāt)

**negative feedback** (neg′ah-tiv fēd′bak)

**pancreas** (pan′kre-as)

**parathyroid gland** (par″ah-thi′roid gland)

**pineal gland** (pin′e-al gland)

**posterior pituitary** (pos-ter′e-or pĭ-tu′ĭ-tar″e)

**prostaglandin** (pros″tah-glan′din)

**steroid** (ste′roid)

**target cell** (tar′get sel)

**thymus gland** (thi′mus gland)

**thyroid gland** (thi′roid gland)

The sweet-smelling urine that is the hallmark of diabetes mellitus was noted as far back as an Egyptian papyrus from 1500 B.C. In A.D. 96 in Greece, Aretaeus of Cappadocia described the condition as a "melting down of limbs and flesh into urine." A look at the 3-year-old boy on the previous page illustrates how apt a description that was. In December 1922, the boy weighed only 15 pounds. But he was far luckier than people who had suffered with diabetes before him. He was among the first to receive insulin, a hormone that his body could not produce. As the inset shows, the boy rapidly improved, doubling his weight in just two months of insulin treatment.

Insulin and the gland that produces it—the pancreas—are familiar components of the endocrine system. Understanding diabetes provides a fascinating glimpse into the evolution of medical technology that continues today.

In 1921, Canadian physiologists Sir Frederick Grant Banting and Charles Herbert Best discovered the link between lack of insulin and diabetes. They induced diabetes symptoms in a dog by removing its pancreas, then cured it by administering insulin from another dog's healthy pancreas. Just a year later, people with diabetes—such as the starving 3-year-old—began to receive insulin extracted from pigs or cows.

And so it went until 1982, when pure human insulin became available by genetically engineering bacteria to produce the human protein, helping those who were allergic to the animal product. Today, people with diabetes receive insulin in one or two injections a day, or with an infusion pump inserted beneath the skin. A pump more closely approximates the pancreas's natural release of insulin, but it is not perfect. Alternatives include a pancreas transplant or an implant of insulin-producing cell clusters.

Although a person with diabetes today is certainly far better off than the boy on the brink of the insulin discovery, we still cannot exactly duplicate the function of the pancreas in secreting its hormone product, insulin. Better understanding of the endocrine system will lead to better treatment of diabetes and other hormonal disorders.

## General Characteristics of the Endocrine System

The **endocrine system** includes the cells, tissues, and organs that secrete hormones into body fluids (the internal environment). In contrast, **exocrine** structures secrete into tubes or ducts, which lead to internal or external body surfaces. The thyroid and parathyroid glands, for example, secrete hormones into the blood and are therefore endocrine (ductless) glands, but sweat glands and sebaceous (oil) glands in the skin are exocrine (fig. 13.1). Two other patterns of secretion are **paracrine,** which affects neighboring cells, and **autocrine,** which affects the secreting cell itself.

The interrelationships of the glands comprising the endocrine system, although not well understood, are vividly obvious in families that have an inherited cancer syndrome called *multiple endocrine neoplasia* (MEN). One family member might have a tumor of the adrenal glands called pheochromocytoma; another might have thyroid cancer; yet a third relative might have parathyroid hyperplasia, a precancerous condition.

Endocrine glands and their hormones help regulate metabolic processes. They control the rates of certain chemical reactions, aid in transporting substances through membranes, and help regulate water balance, electrolyte balance, and blood pressure. They also play vital roles in reproduction, development, and growth.

Specialized small groups of cells produce some hormones. However, the larger endocrine glands—the pituitary gland, thyroid gland, parathyroid glands, adrenal glands, and pancreas—are the subject of this chapter (fig. 13.2). Subsequent chapters discuss several other hormone-secreting glands and tissues.

## Hormone Action

A **hormone** is a biochemical secreted by a cell that affects the functions of another cell. Hormones are released into the extracellular spaces surrounding secretory cells. Paracrine hormones travel only short distances and affect nearby cells. Other hormones are carried by the blood to all parts of the body and may produce general effects. In either case, the physiological action of a particular hormone is restricted to its *target cells*—those cells that possess specific receptors for the hormone molecules. In other words, a hormone's target cells have receptors that other cells lack.

### Chemistry of Hormones

Chemically, most hormones are either steroids (or steroidlike substances) that are synthesized from cholesterol (see chapter 2), or they are amines, peptides, proteins, or glycoproteins that are synthesized from amino acids. Hormones are organic compounds. They can stimulate changes in target cells even if present in extremely low concentrations.

*Steroids* are lipids that include complex rings of carbon and hydrogen atoms. Steroids differ by the types and numbers of atoms attached to these rings

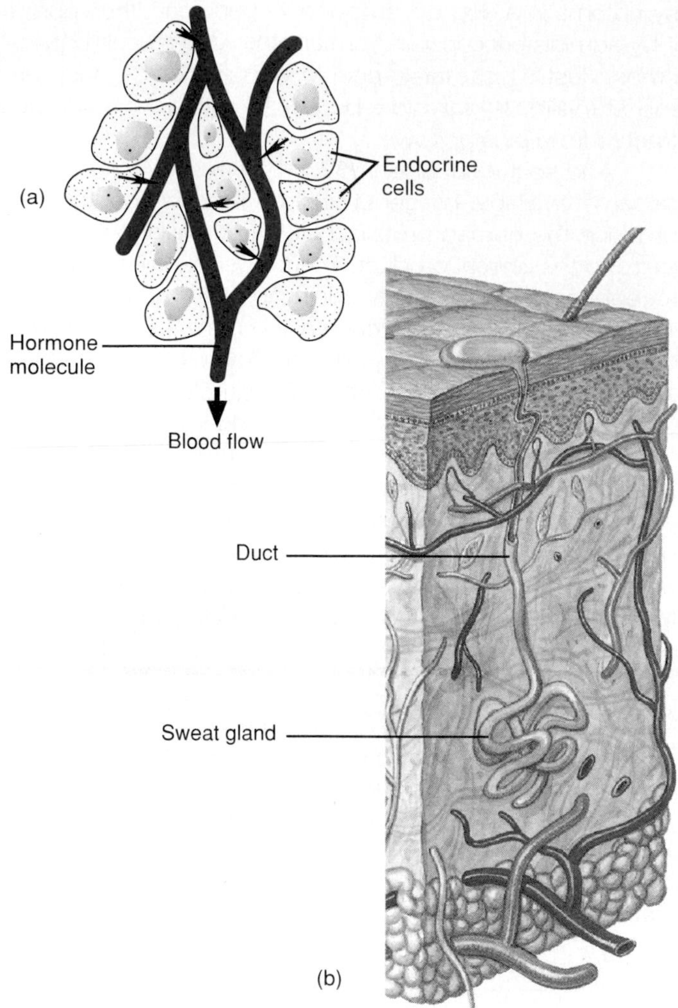


**FIGURE 13.1**
(*a*) Endocrine glands, such as the thyroid gland, release hormones into the internal environment (body fluids).
(*b*) Exocrine glands, such as sweat glands, secrete to the outside environment, through ducts that lead to body surfaces.

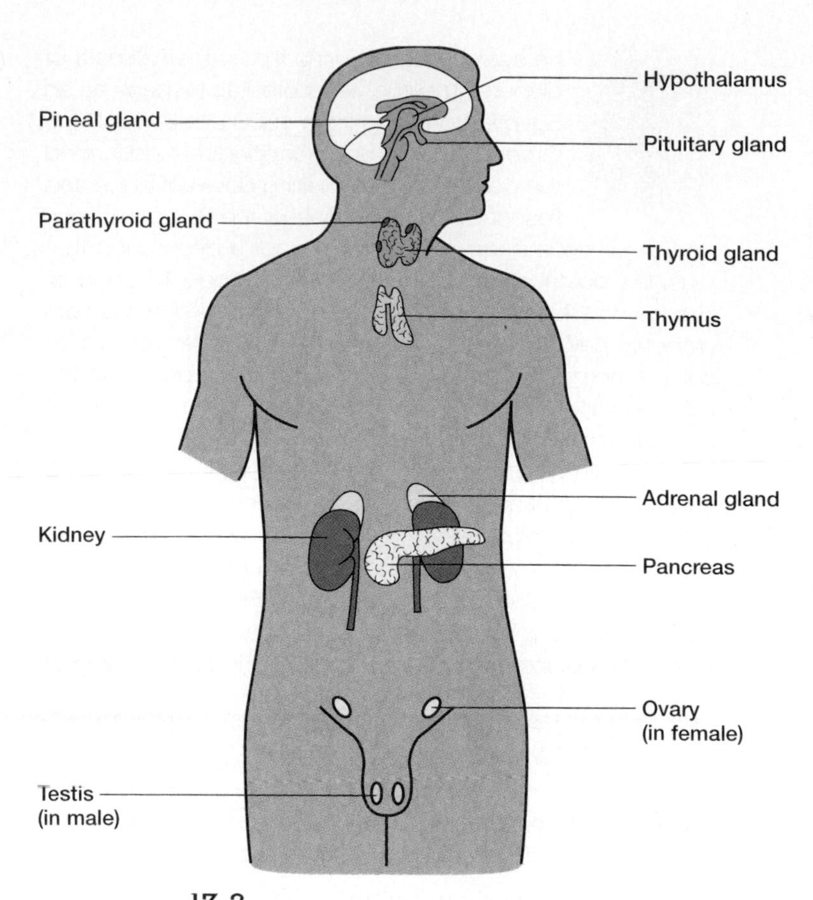


**FIGURE 13.2**
Locations of major endocrine glands.

and the ways they are joined (see chapter 2). The steroid hormones include sex hormones (such as estrogens and testosterone) and those the adrenal cortex secretes (including aldosterone and cortisol). Vitamin D is a modified steroid and can be converted into a hormone, as is discussed later in this chapter (see also chapter 18).

Some neurons produce hormones that are *amines,* including norepinephrine and epinephrine. These hormones are also synthesized in the adrenal medulla (the inner portion of the adrenal gland) from the amino acid tyrosine (see chapter 11).

The *peptide* hormones are short chains of amino acids. This group includes hormones associated with the posterior pituitary gland and some produced in the hypothalamus.

*Protein* hormones are composed of long chains of amino acids, linked to form intricate molecular structures (see chapter 2). They include the hormone secreted by the parathyroid gland and some of those secreted by the anterior pituitary gland (fig. 13.3). Certain other hormones from the anterior pituitary gland are *glycoproteins,* which consist of a protein joined to a carbohydrate.

Another group of compounds, called prostaglandins, are usually considered paracrine substances rather than true hormones. They have regulating effects on neighboring cells. Prostaglandins are lipids (20-carbon fatty acids that include 5-carbon rings) and are synthesized from a fatty acid (arachidonic acid) in cell membranes. Prostaglandins occur in a wide variety of cells, including those of the liver, kidneys, heart, lungs, thymus gland, pancreas, brain, and various reproductive organs. (Their cousins, the leukotrienes, affect white blood cells and are discussed in chapter 16.)

Chart 13.1 lists the names and abbreviations of some of the hormones discussed in this chapter. Chart 13.2 and fig. 13.3 summarize the chemical composition of various hormones.

(a) Cortisol

(b) Norepinephrine

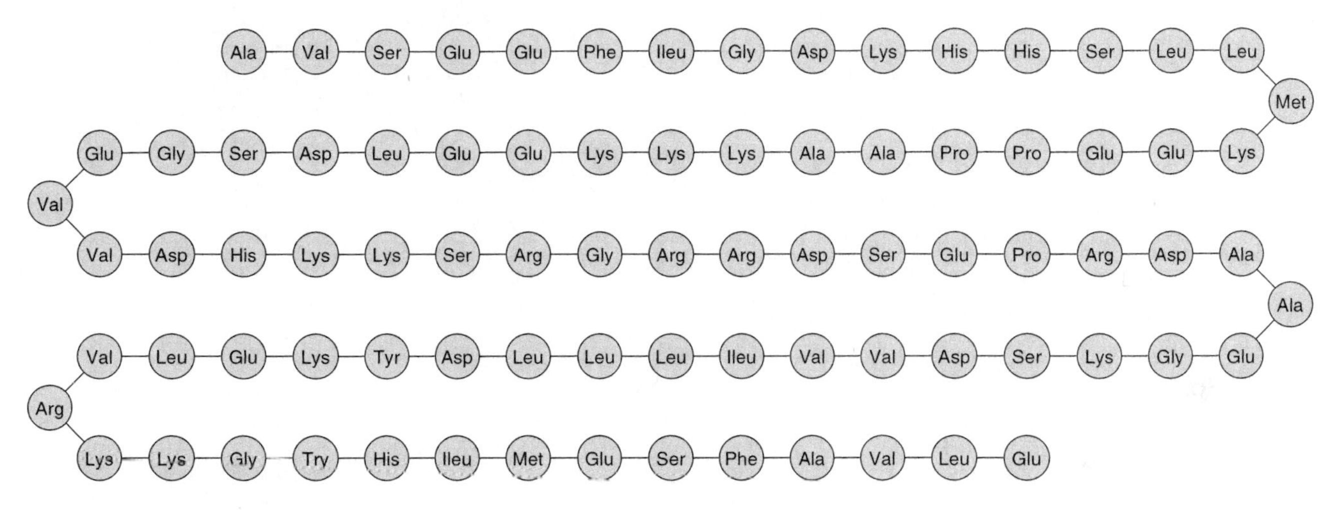


(c) Parathyroid hormone (PTH)

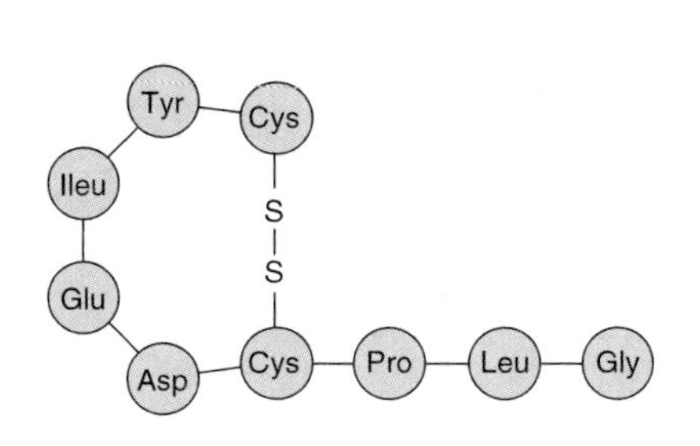


(d) Oxytocin

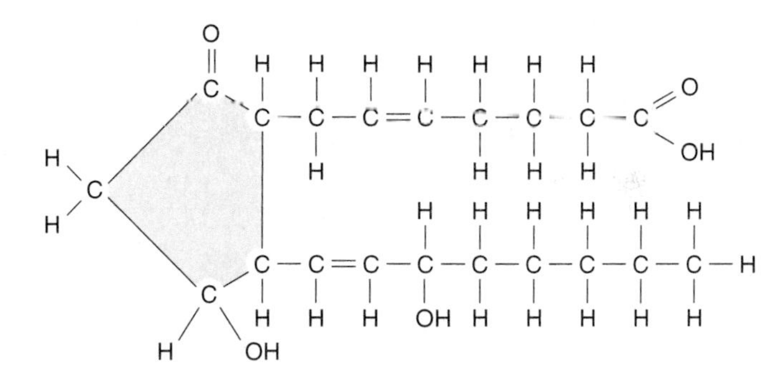


(e) Prostaglandin $PGE_2$

**FIGURE 13.3**

(*a* and *b*) Structural formulas of a steroid hormone (cortisol) and an amine hormone (norepinephrine). (*c* and *d*) Amino acid sequence of a protein hormone (PTH) and a peptide hormone (oxytocin). (*e*) A prostaglandin ($PGE_2$).

**1** What is the difference between an endocrine gland and an exocrine gland?

**2** What is a hormone?

**3** How are hormones chemically classified?

## Actions of Hormones

Hormones exert their effects by altering metabolic processes. For example, a hormone might alter activity of an enzyme necessary for synthesizing a particular substance, or alter the rate at which particular chemicals are transported through cell membranes.

Hormones bind to specific receptors in order to exert effects on their target cells. Each hormone receptor is a protein or glycoprotein molecule that has a *binding site* for a specific hormone. A hormone delivers

## CHART 13.1 Hormone Names and Abbreviations

| Source | Name | Abbreviation | Synonym |
|---|---|---|---|
| Hypothalamus | Corticotropin-releasing hormone | CRH | |
| | Gonadotropin-releasing hormone | GnRH | Luteinizing hormone-releasing hormone (LHRH) |
| | Somatostatin | SS | Growth hormone release-inhibiting hormone (GIH) |
| | Growth hormone-releasing hormone | GHRH | |
| | Prolactin release-inhibiting hormone | PIH | |
| | Prolactin-releasing hormone | PRH | |
| | Thyrotropin-releasing hormone | TRH | |
| Anterior pituitary gland | Adrenocorticotropic hormone | ACTH | Corticotropin |
| | Follicle-stimulating hormone | FSH | Follitropin |
| | Growth hormone | GH | Somatotropin (STH) |
| | Luteinizing hormone | LH | Lutropin, interstitial cell-stimulating hormone (ICSH) |
| | Prolactin | PRL | |
| | Thyroid-stimulating hormone | TSH | Thyrotropin |
| Posterior pituitary gland | Antidiuretic hormone | ADH | Vasopressin |
| | Oxytocin | OT | |
| Thyroid gland | Calcitonin | | |
| | Thyroxine | $T_4$ | |
| | Triiodothyronine | $T_3$ | |
| Parathyroid gland | Parathyroid hormone | PTH | Parathormone |
| Adrenal medulla | Epinephrine | EPI | Adrenalin |
| | Norepinephrine | NE | Noradrenalin |
| Adrenal cortex | Aldosterone | | |
| | Cortisol | | Hydrocortisone |
| Pancreas | Glucagon | | |
| | Insulin | | |
| | Somatostatin | SS | |

## CHART 13.2 Types of Hormones

| Type of Compound | Formed from | Examples |
|---|---|---|
| Amines | Amino acids | Norepinephrine, epinephrine |
| Peptides | Amino acids | ADH, OT, TRH, SS, GnRH |
| Proteins | Amino acids | PTH, GH, PRL |
| Glycoproteins | Protein and carbohydrate | FSH, LH, TSH |
| Steroids | Cholesterol | Estrogen, testosterone, aldosterone, cortisol |

its message to a cell by uniting with the binding site of its receptor. The more receptors the hormone binds, the greater the cellular response.

### Steroid Hormones

Steroid hormones are soluble in the lipids that make up the bulk of cell membranes. These hormones can enter target cells relatively easily by diffusion. Once inside a target cell, they combine with specific protein molecules—the receptors. The *hormone-receptor complex* thus formed binds within the nucleus to a particular region of the target cell's DNA and activates specific genes. The activated genes, in turn, are transcribed into messenger RNA (mRNA), which enters the cytoplasm, where it directs manufacture of specific proteins. These proteins bring about the cellular changes associated with the particular hormone (fig. 13.4, chart 13.3, and Clinical Application 13.1).

## CHART 13.3 Sequence of Steroid Hormone Action

1. Endocrine gland secretes steroid hormone.
2. Steroid hormone enters target cell by diffusing through the cell membrane and then enters the nucleus.
3. Hormone combines with a receptor molecule.
4. Steroid hormone-receptor complex binds to DNA and promotes the synthesis of messenger RNA.
5. Messenger RNA enters the cytoplasm and directs protein synthesis.
6. Newly synthesized proteins produce hormone's specific effects.

CLINICAL APPLICATION 13.1

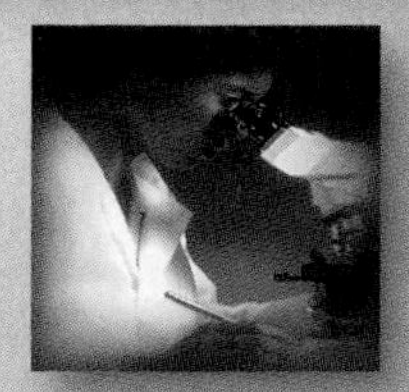

## Steroids and Athletes—An Unhealthy Combination

Canadian track star Ben Johnson flew past his competitors in the 100-meter run at the 1988 Summer Olympics in Seoul (fig. 13A). But 72 hours later, the gold medal awarded for his record-breaking time of 9.79 seconds was rescinded after traces of the drug stanozolol were detected in his urine.

Stanozolol is one of several synthetic versions of the steroid hormone testosterone. Like testosterone, these drugs promote signs of masculinity (their androgenic effect) and increased synthesis of muscle proteins (their anabolic effect). Used in the past to treat a handful of medical conditions—anemia and breast cancer among them—steroids are used by professional and amateur athletes alike, all looking to build muscle tissue easily.

Steroid users may improve their performances and physiques in the short term, but in the long run they may suffer for it. Steroids hasten adulthood, stunting height and causing early baldness, and in males lead to breast development and in females to a deepened voice, hairiness, and a male physique. The kidney, liver, and heart may be damaged, and atherosclerosis may develop because steroids raise LDL and lower HDL—the opposite of a healthy cholesterol profile. In males, the body mistakes the synthetic steroids for the natural hormone and lowers its own production of testosterone. The price of athletic prowess today may be infertility later.

In 1976 the International Olympic Committee banned the use of steroids by athletes and required urine tests. The Anabolic Steroids Act of 1990 made possession with intent to distribute the drugs a federal offense in the United States.

Ben Johnson was caught in his tracks by a urine test able to detect part-per-billion traces of synthetic steroids even weeks after they are taken. Johnson at first claimed the stanozolol in his urine was the result of a drink spiked with an approved anti-inflammatory drug used on his ankle, but a test of his natural testosterone showed it to be only 15% of normal—a sure sign that this athlete had been taking steroids for a long time.

Unfortunately, athletes' use of drugs to pump up their muscles continues. At the 1992 Summer Olympics in Barcelona, athletes from Germany, the United States, China, Britain, and the former Soviet Union were turned away because of evidence they had used nonsteroid drugs that they thought would have steroidlike effects.

**FIGURE 13A**
Canadian track star Ben Johnson ran away with the gold medal in the 100-meter race at the 1988 Summer Olympics—but then had to return the award when traces of a steroid drug showed up in his urine.

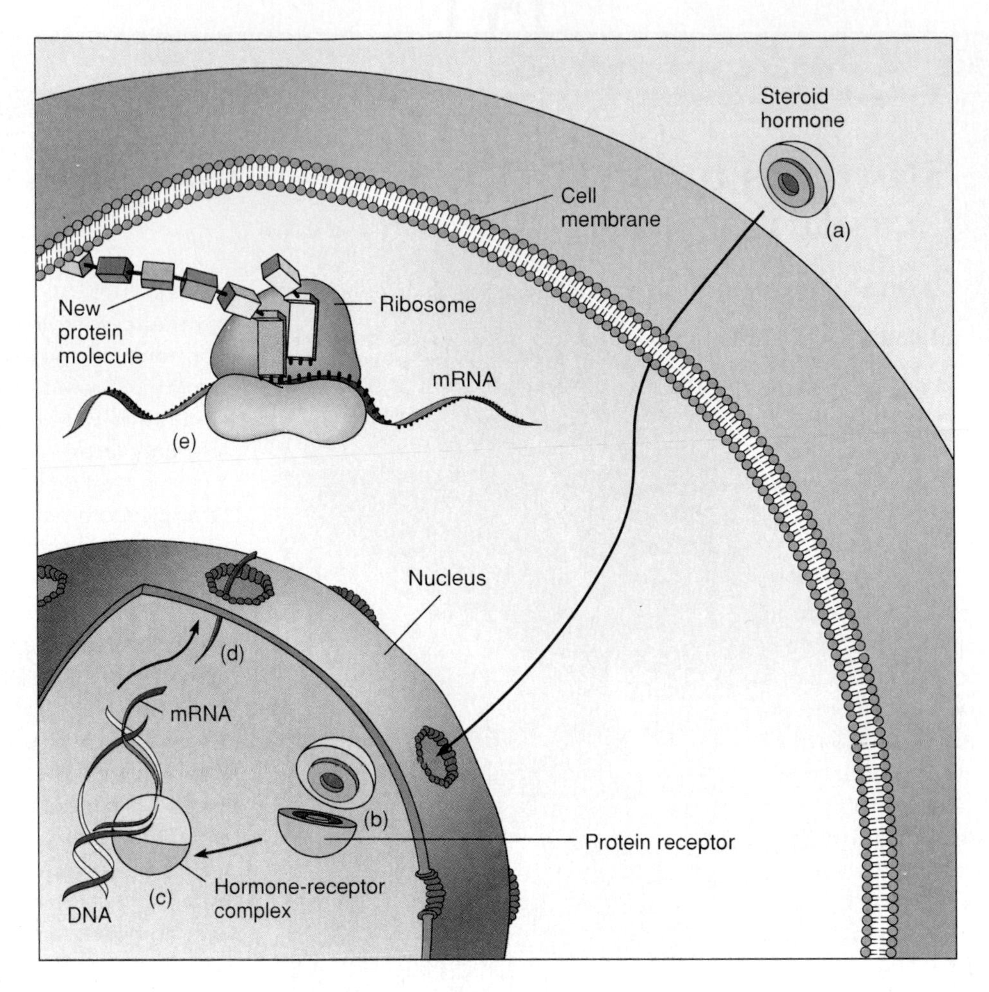


**FIGURE 13.4**

(*a*) A steroid hormone passes through a cell membrane and (*b*) combines with a protein receptor, usually in the nucleus. (*c*) The hormone-receptor complex activates messenger RNA synthesis. (*d*) The mRNA leaves the nucleus and (*e*) guides protein synthesis.

## Nonsteroid Hormones

A nonsteroid hormone, such as an amine, peptide, or protein, contacts its receptor at a binding site on the cell membrane. This stimulates the receptor's activity site to interact with other membrane proteins. This receptor binding may alter the function of enzymes or membrane transport mechanisms, changing the concentrations of still other cellular components. The hormone that triggers this cascade of biochemical activity may be considered a *first messenger.* The biochemicals in the cell that induce the changes that are recognized as responses to the hormone's binding are thus called *second messengers.*

Many hormones use **cyclic adenosine monophosphate** (cyclic AMP, or cAMP) as a second messenger. In this mechanism, a hormone binds to its receptor, and the resulting hormone-receptor complex activates a protein called a G protein, which then activates an enzyme called **adenylate cyclase** that is bound to the inside of the cell membrane. The activated enzyme removes two phosphates from ATP and circularizes it, forming cyclic AMP (fig. 13.5). Cyclic AMP, in turn, activates another set of enzymes called **protein kinases.** Protein kinases transfer phosphate groups from ATP molecules to protein substrate molecules. This phosphorylation alters the shapes of the substrate molecules and converts some of them from inactive forms into active ones. The activated protein molecules then induce changes in various cellular processes (fig. 13.6). Thus, the response of any particular cell to such a hormone is determined not only by the type of membrane receptors present, but also by the kinds of protein substrate molecules the cell contains. Chart 13.4 summarizes these actions.

Cellular responses to second messenger activation include altering membrane permeabilities, activating enzymes, promoting synthesis of certain proteins, stimulating or inhibiting specific metabolic pathways, promoting cellular movements, and initiating secretion of hormones and other substances.

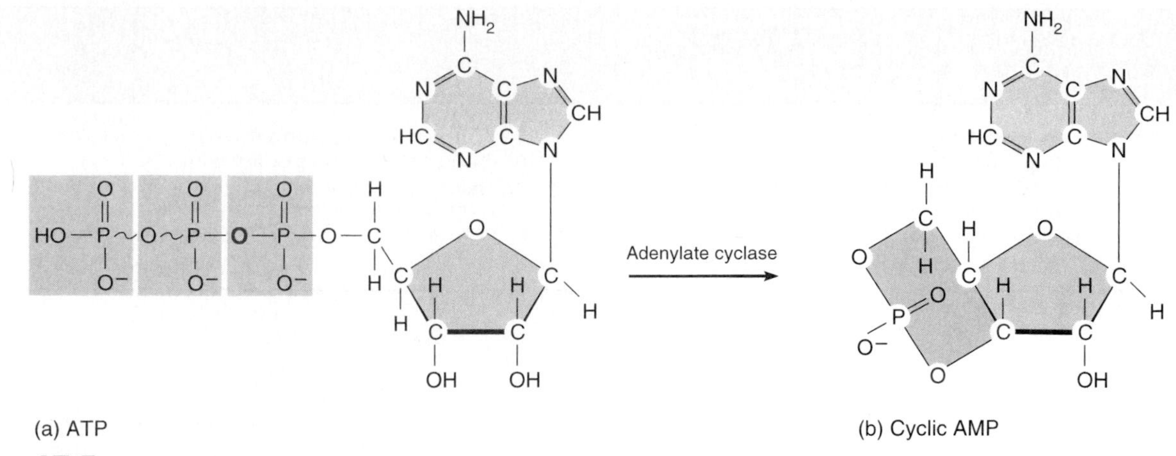


**FIGURE 13.5**

Adenylate cyclase catalyzes conversion of ATP molecules (*a*) into cyclic AMP (*b*). The atoms forming the new bond are shown in red.

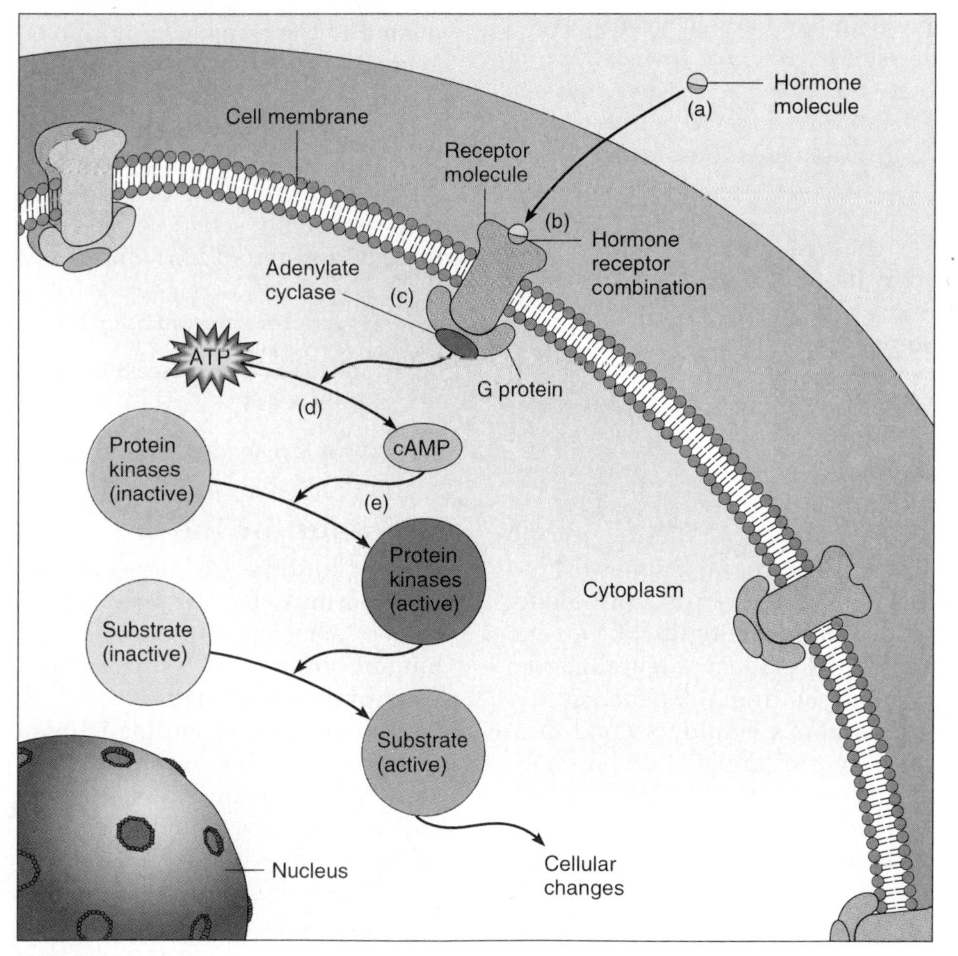


**FIGURE 13.6**

(*a*) Body fluids carry nonsteroid hormone molecules to the target cell, where they combine (*b*) with receptor sites on the cell membrane. (*c*) This may activate molecules of adenylate cyclase, which (*d*) catalyze conversion of ATP into cyclic AMP. (*e*) Cyclic AMP promotes a series of reactions leading to the cellular changes associated with the hormone's action.

## CHART 13.4 SEQUENCE OF NONSTEROID HORMONE ACTIONS

1. Endocrine gland secretes hormone.
2. Body fluid carries hormone to its target cell.
3. Hormone combines with receptor site on membrane of its target cell, thereby activating G protein.
4. Adenylate cyclase molecules are activated within target cell's membrane.
5. Adenylate cyclase converts ATP into cyclic AMP.
6. Cyclic AMP activates protein kinases.
7. These enzymes activate protein substrates in the cell that change metabolic processes.
8. Cellular changes are recognized as the hormone's action.

Because many hormones utilize the cAMP-mediated second messenger system to exert their effects, an abnormality in part of the second messenger signaling system can lead to symptoms stemming from many endocrine glands. In McCune-Albright syndrome, for example, a defect in the G protein that activates adenylate cyclase results in conversion of ATP to cAMP even without hormonal stimulation. As a result, cells in the pituitary, thyroid, gonads, and adrenal glands secrete hormones in excess. These conditions are inherited because the components of the second messenger system are proteins.

Another enzyme, phosphodiesterase, quickly inactivates cAMP, so that its action is short-lived. For this reason, a continuing response in a target cell depends upon a continuing signal produced by hormone molecules combining with receptors in the target-cell membrane. Since hormones themselves are continuously degraded by enzymes, a sustained response depends on continued hormone secretion.

Hormones whose actions depend upon cyclic AMP include releasing hormones from the hypothalamus; TSH, ACTH, FSH, and LH from the anterior pituitary gland; ADH from the posterior pituitary gland; PTH from the parathyroid glands; norepinephrine and epinephrine from the adrenal glands; calcitonin from the thyroid gland; and glucagon from the pancreas.

Other nonsteroid hormones employ second messengers other than cAMP. For example, a second messenger called diacylglycerol (DAG), like cAMP, activates a protein kinase leading to a cellular response.

In another mechanism, a hormone binding to its receptor increases calcium ion concentration within the target cell. Such a hormone may stimulate transport of calcium ions inward through the cell membrane or induce release of calcium ions from cellular storage sites via a second messenger called inositol triphosphate (IP3). The calcium ions combine with the protein *calmodulin* (see chapter 9), altering its molecular structure in a way that activates the molecule. Activated calmodulin can then interact with enzymes, altering their activities and thus producing a variety of responses.

Still another mechanism uses *cyclic guanosine monophosphate* (cyclic GMP, or cGMP). Like cAMP, cGMP is a nucleotide derivative, and functions in much the same manner as a second messenger.

Cellular response to a steroid hormone is closely related to the number of hormone-receptor complexes formed. In contrast, response to a hormone operating through a second messenger is greatly amplified. This is possible because many second messenger molecules can be activated in response to just a few hormone-receptor complexes. Because of such amplification, cells are highly sensitive to changes in the concentrations of nonsteroid hormones.

1. How does a steroid hormone act on its target cells?
2. How does a nonsteroid hormone act on its target cells?
3. What is a second messenger?

## Prostaglandins

Prostaglandins are paracrine substances, affecting neighboring cells, that are very potent and are present in very small quantities. They are not stored in cells, but are synthesized just before they are released. They are rapidly inactivated.

Some prostaglandins regulate cellular responses to hormones. For example, different prostaglandins can either activate or inactivate adenylate cyclase in cell membranes, thereby controlling production of cAMP and altering the cell's response to a hormone.

Prostaglandins produce a variety of effects. For example, some can relax smooth muscle in the airways of lungs and in the blood vessels dilating these passageways. Other prostaglandins can contract smooth muscle in the walls of the uterus, causing menstrual cramps and labor contractions. They stimulate secretion of hormones from the adrenal cortex and inhibit secretion of hydrochloric acid from the wall of the stomach. Prostaglandins also influence the movements of sodium ions and water in the kidneys, help regulate blood pressure, and have powerful effects on both male and female reproductive physiology. When tissues are injured, prostaglandins promote inflammation (see chapter 5).

Understanding prostaglandin function has medical applications. Drugs such as aspirin and certain steroids that relieve the joint pain of rheumatoid arthritis inhibit production of prostaglandins in the synovial fluid of affected joints. Daily doses of aspirin may reduce the risk of heart attack by altering prostaglandin activity. Prostaglandins may be used as drugs to dilate constricted blood vessels to relieve hypertension.

1. What are prostaglandins?
2. Describe one possible function of prostaglandins.
3. What kinds of effects do prostaglandins produce?

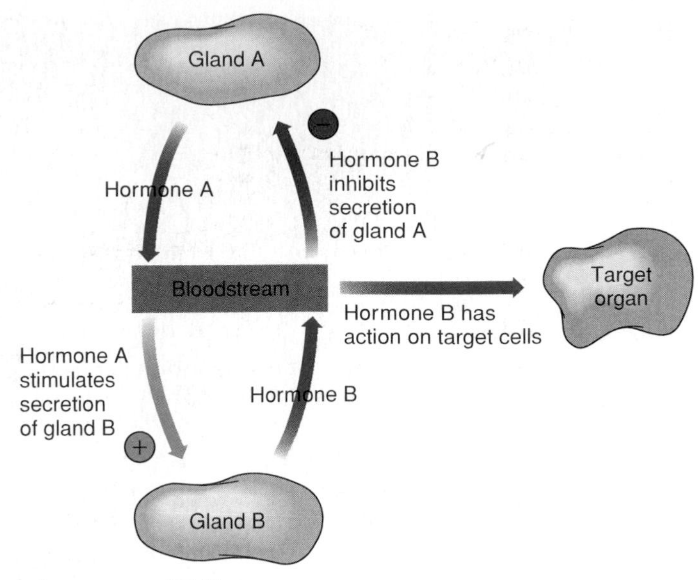


**FIGURE 13.7**
An example of a negative feedback system: (1) Gland A secretes a hormone that stimulates gland B to increase secretion of another hormone. (2) The hormone from gland B alters its target cells and inhibits activity of gland A.

## Control of Hormonal Secretions

Hormone secretion is under precise regulation.

### Negative Feedback Systems

Commonly, **negative feedback systems** control hormonal secretions (see chapter 1). In such a system, an endocrine gland or the system controlling it is sensitive to the concentration of a substance the gland regulates or to the concentration of a product from a process it controls. Whenever this concentration reaches a certain level, the endocrine gland is inhibited (a negative effect), and its secretory activity decreases (fig. 13.7). Then, as the concentration of the gland's hormone decreases, the concentration of the regulated substance decreases too, and the inhibition of the gland ceases. When the gland is no longer inhibited, it begins to secrete its hormone again. Such negative feedback systems keep the concentrations of some hormones relatively stable (fig. 13.8).

### Control Sources

Hormone secretion must be precisely controlled so that these biochemicals can effectively maintain the internal environment. Hormone control occurs in three ways:

- The hypothalamus controls the anterior pituitary gland's release of **tropic hormones,** which then stimulate other endocrine glands to release hormones. The hypothalamus constantly receives information about the internal environment from neural connections and cerebrospinal fluid, made possible by its location near the thalamus and the third ventricle (fig. 13.9).
- Another group of glands responds directly to changes in the composition of the internal environment. For example, when the blood glucose level rises, the pancreas secretes insulin, and when the blood glucose level falls, it secretes glucagon, as we shall see later in the chapter.
- The nervous system stimulates some glands directly. The adrenal medulla, for example, secretes its hormones (epinephrine and norepinephrine) in response to preganglionic sympathetic nerve impulses. The secretory cells replace the postganglionic sympathetic neurons, which would normally secrete norepinephrine alone as a neurotransmitter.

1. What is a feedback system?
2. How does a negative feedback system control hormonal secretions?
3. How does the nervous system help regulate hormonal secretions?

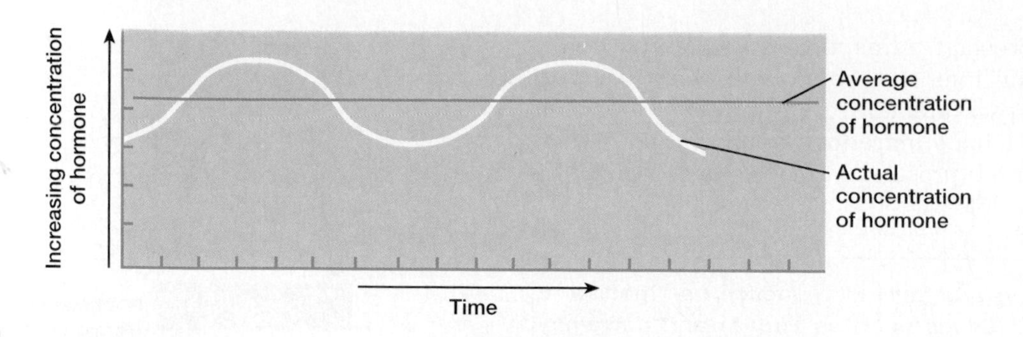


**FIGURE 13.8**
As a result of negative feedback, some hormone concentrations remain relatively stable, although they may fluctuate slightly above and below average concentrations.

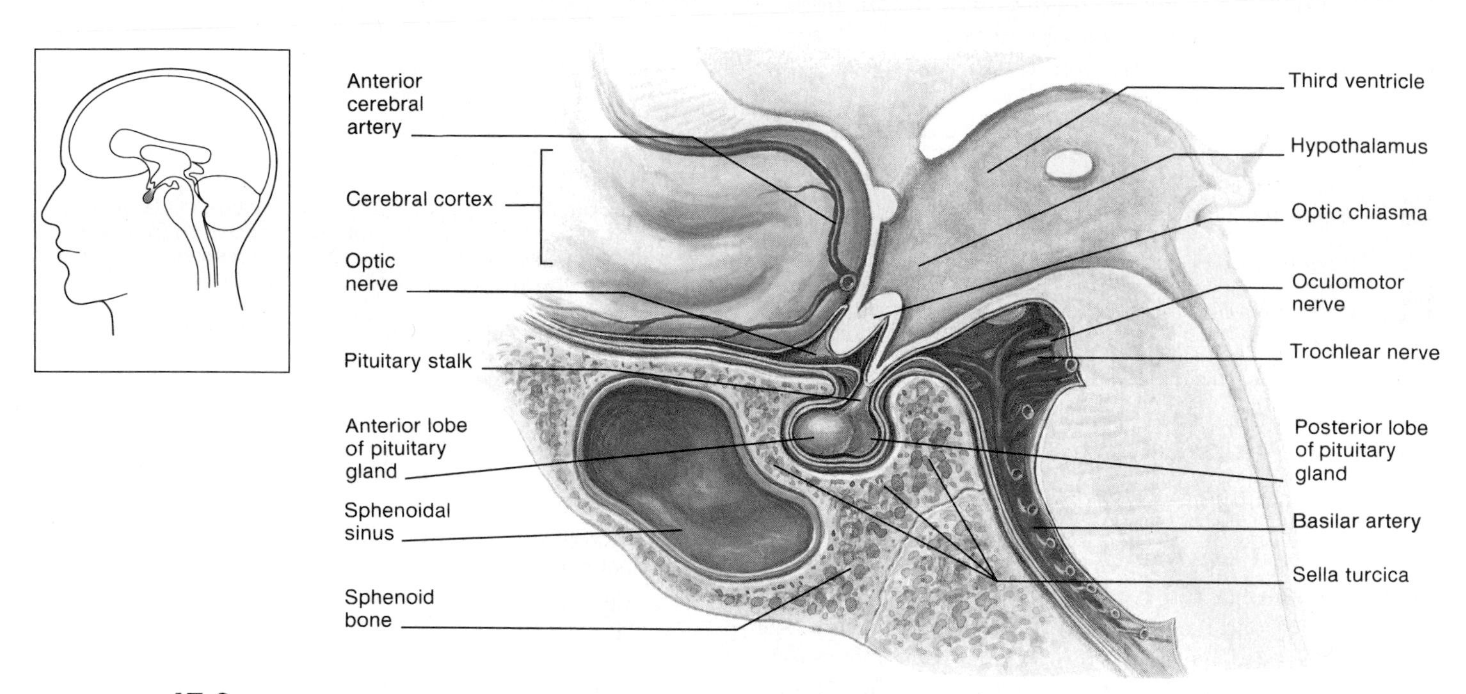


**FIGURE 13.9**
The pituitary gland is attached to the hypothalamus and lies in the sella turcica of the sphenoid bone.

## PITUITARY GLAND

The **pituitary gland** (hypophysis) is about 1 centimeter in diameter and is located at the base of the brain. It is attached to the hypothalamus by the pituitary stalk, or *infundibulum,* and lies in the sella turcica of the sphenoid bone, as shown in figure 13.9.

The pituitary gland consists of two distinct portions: an *anterior lobe* (adenohypophysis) and a *posterior lobe* (neurohypophysis). The anterior lobe secretes a number of hormones, including **growth hormone** (GH), **thyroid-stimulating hormone** (TSH), **adrenocorticotropic hormone** (ACTH), **follicle-stimulating hormone** (FSH), **luteinizing hormone** (LH), and **prolactin** (PRL). Although the cells of the posterior lobe (pituicytes) do not synthesize any hormones, two important hormones, **antidiuretic hormone** (ADH) and **oxytocin** (OT), are secreted by special neurons called **neurosecretory cells** whose nerve endings secrete into the bloodstream within the posterior lobe. The cell bodies of these neurosecretory cells are located in the hypothalamus.

During fetal development, a narrow region appears between the anterior and posterior lobes of the pituitary gland. Called the *intermediate lobe* (pars intermedia), this region produces melanocyte-stimulating hormone (MSH), which regulates the formation of melanin—the pigment found in the skin, and in portions of the eyes and brain. The region atrophies during prenatal development, and appears only as a vestige in adults.

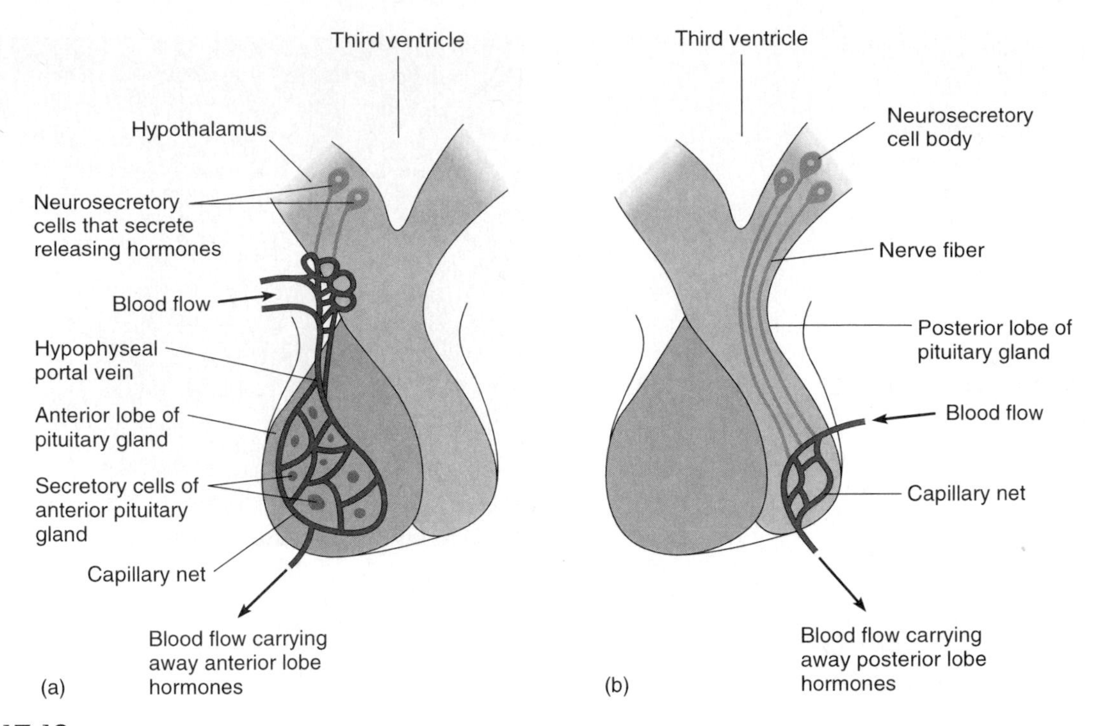


**FIGURE 13.10**

(*a*) Hypothalamic releasing hormones stimulate cells of the anterior lobe to secrete hormones. (*b*) Nerve impulses originating in the hypothalamus stimulate nerve endings in the posterior lobe of the pituitary gland to release hormones.

The brain controls most of the pituitary gland's activities (fig. 13.10). For example, the pituitary gland's posterior lobe releases hormones into the bloodstream in response to nerve impulses from the hypothalamus. However, releasing hormones from the hypothalamus primarily control secretions from the anterior lobe. These releasing hormones are carried in the blood via a capillary net associated with the hypothalamus. The vessels merge to form the **hypophyseal portal veins** that pass downward along the pituitary stalk and give rise to a capillary net in the anterior lobe. Thus, substances released into the blood from the hypothalamus are carried directly to the anterior lobe. The hypothalamus, therefore, is an endocrine gland itself, yet it also controls other endocrine glands. This is also true of the anterior pituitary.

Upon reaching the anterior lobe of the pituitary, each of the hypothalamic releasing hormones acts on a specific population of cells in the anterior pituitary. Some of the resulting actions are inhibitory (prolactin inhibitory hormone and somatostatin), but most stimulate the anterior pituitary to release hormones that stimulate the secretions of peripheral endocrine glands. In many of these cases, important negative feedback relationships regulate hormone levels in the bloodstream. Figure 13.11 shows this general relationship.

The following sections discuss the hormones of the anterior pituitary as a group. The hormones of the peripheral endocrine glands will then be discussed individually.

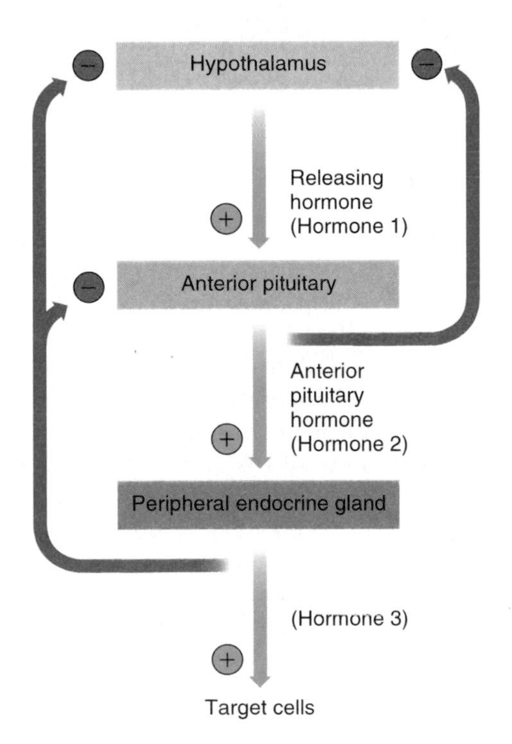


**FIGURE 13.11**

Hypothalamic control of the peripheral endocrine glands may utilize as many as three types of hormones, with multiple negative feedback controls, indicated by ⊖.

1. Where is the pituitary gland located?
2. List the hormones that the anterior and posterior lobes of the pituitary gland secrete.
3. Explain how the hypothalamus controls the actions of the pituitary gland.

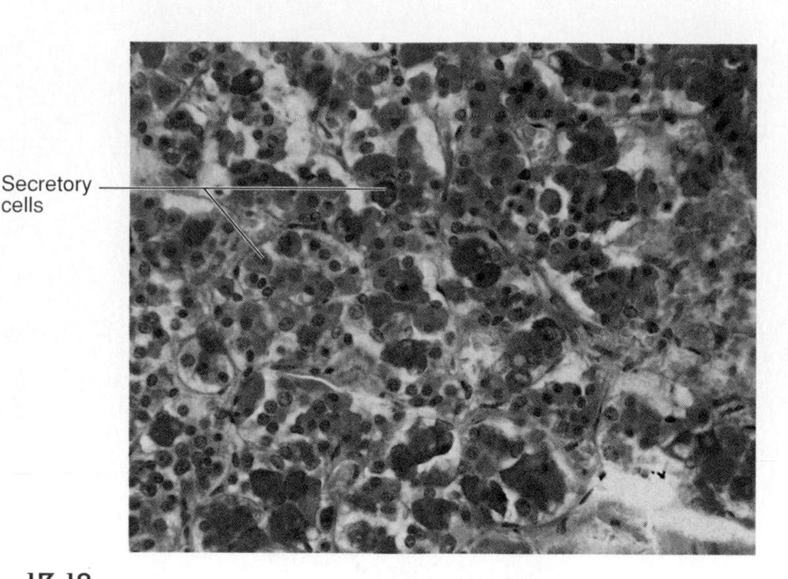


**FIGURE 13.12**
Light micrograph of the anterior pituitary gland (100×).

## Anterior Pituitary Hormones

The anterior lobe of the pituitary gland is enclosed in a dense capsule of collagenous connective tissue and consists largely of epithelial tissue arranged in blocks around many thin-walled blood vessels. Within the epithelial tissue are five types of secretory cells. They are *somatotropes* that secrete GH, *mammatropes* that secrete PRL, *thyrotropes* that secrete TSH, *corticotropes* that secrete ACTH, and *gonadotropes* that secrete FSH and LH (figs. 13.12 and 13.13). In males, LH (luteinizing hormone) is known as ICSH (interstitial cell-stimulating hormone) because it affects the interstitial cells of the testes (see chapter 22).

**Growth hormone**, which is also called *somatotropin* (STH), is a protein that stimulates cells to increase in size and more rapidly divide. It enhances the movement of amino acids through the cell membranes and increases the rate of protein synthesis. GH also decreases the rate at which cells utilize carbohydrates and increases the rate at which they use fats.

Growth hormone is secreted in pulses, especially during sleep. Two biochemicals from the hypothalamus control its secretion. They are released alternately, exerting opposite effects. *Growth hormone-releasing hormone* (GHRH) stimulates secretion of GH, and *somatostatin* (SS) inhibits secretion.

Nutritional state seems to play a role in control of GH. For example, more GH is released during periods of protein deficiency and abnormally low blood glucose concentration. Conversely, when blood protein and glucose concentrations increase, growth hormone secretion decreases. Apparently, the hypothalamus is able to sense changes in the concentrations of certain blood nutrients, and it releases (GHRH) in response to some of them.

Growth hormone can stimulate elongation of bone tissue directly, but its effect on cartilage requires a mediator substance. The liver releases this biochemical, called *somatomedin* (insulin-like growth factor), in response to GH. Somatomedin promotes growth of cartilage.

Clinical Application 13.2 discusses some clinical uses of growth hormone.

**Prolactin** is a protein, and as its name suggests, it promotes milk production. In males, PRL decreases secretion of luteinizing hormone (LH). Because LH is necessary for production of male sex hormones (androgens), excess prolactin secretion may cause a male to produce too few sex hormones and become infertile (see chapter 22).

Two biochemicals from the hypothalamus regulate prolactin secretion. One of these, *prolactin release-inhibiting hormone* (PIH), restrains secretion of prolactin, while the other, *prolactin-releasing hormone* (PRH), stimulates its secretion.

**Thyroid-stimulating hormone,** also called *thyrotropin,* is a glycoprotein. It controls secretion of certain hormones from the thyroid gland. TSH can also stimulate growth of the gland, and abnormally high TSH levels may lead to an enlarged thyroid gland, or *goiter.*

The hypothalamus partially regulates TSH secretion by producing thyrotropin-releasing hormone (TRH). Circulating thyroid hormones help regulate TSH secretion by inhibiting release of TRH and TSH; therefore, as the blood concentration of thyroid hormones increases, secretions of TRH and TSH decline (fig. 13.14).

External factors that influence release of TRH and TSH include exposure to extreme cold, which is accompanied by increased hormonal secretion, and emotional stress, which sometimes increases hormonal secretion and other times decreases secretion.

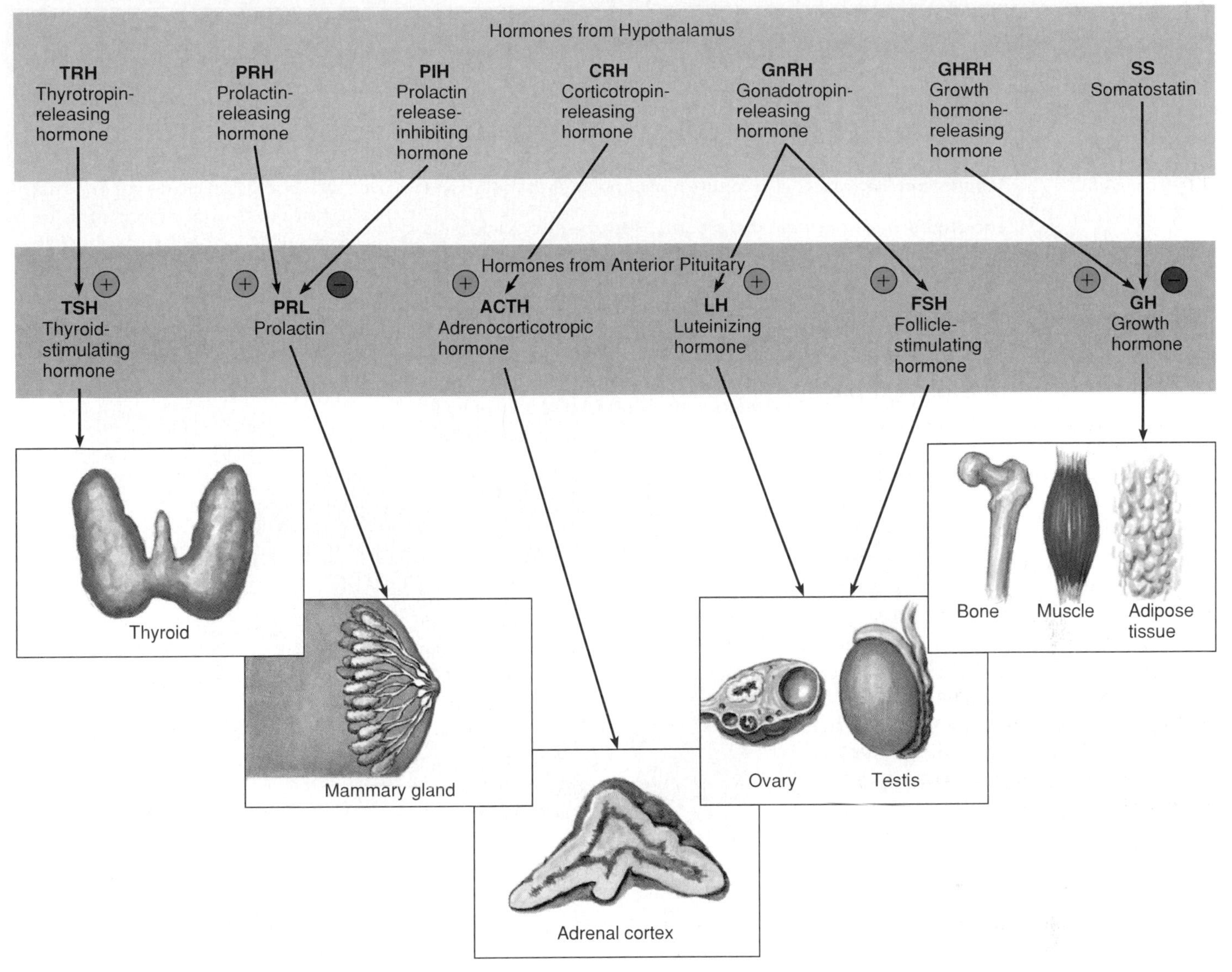


**FIGURE 13.13**

Hormones released from the hypothalamus, the corresponding hormones released from the anterior lobe of the pituitary gland, and their target organs.

1. How does growth hormone affect the cellular metabolism of carbohydrates, fats, and proteins?
2. What are the functions of prolactin in females? In males?
3. How is TSH secretion regulated?

**Adrenocorticotropic hormone** is a peptide that controls the manufacture and secretion of certain hormones from the outer layer (cortex) of the adrenal gland. The secretion of ACTH is regulated in part by *corticotropin-releasing hormone* (CRH), which is released from the hypothalamus in response to decreased concentrations of adrenal cortical hormones. Stress can increase secretion of ACTH by stimulating release of CRH.

Both **follicle-stimulating hormone** and **luteinizing hormone** are glycoproteins and are called *gonadotropins,* which means they exert their actions on the gonads or reproductive organs. FSH, for example, is responsible for growth and development of egg-cell-containing follicles in the ovaries. It also stimulates the follicular cells to secrete a group of female sex hormones, collectively called *estrogen.*

In males, FSH stimulates the initial production of sperm cells in the testes at puberty. LH promotes secretion of sex hormones in both males and females and is essential for release of egg cells from the ovaries. Other functions of the gonadotropins and their interactions are discussed in chapter 22.

The mechanism that regulates secretion of gonadotropins is not well understood. However, it

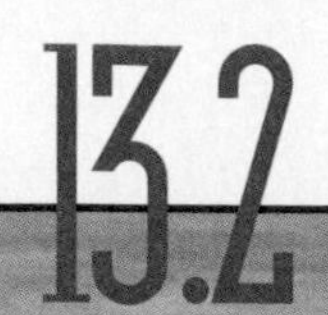

CLINICAL APPLICATION 13.2

## Growth Hormone Ups and Downs

Insufficient secretion of growth hormone during childhood produces *hypopituitary dwarfism*. Body proportions and mental development are normal, but because secretions of other anterior pituitary hormones are also below normal amounts, additional hormone deficiency symptoms may appear. For example, a child with this condition often fails to develop adult sexual features unless he or she receives hormone therapy.

Human growth hormone, manufactured using recombinant DNA technology, is a valuable drug in treating pituitary dwarfism, although treatment must begin before the bones completely ossify. The hormone also has some controversial uses. Some people want to use it to increase height in children who are short, but not abnormally so. A few years ago, growth hormone was given experimentally to older individuals to see if it would slow aging. While muscle tone improved and the participants reported feeling well, side effects arose, and the value of such treatment was not confirmed. Bovine growth hormone is given to dairy cows to increase their milk production.

Oversecretion of growth hormone in childhood may result in *gigantism,* in which height may eventually exceed 8 feet. Gigantism is usually accompanied by a tumor of the pituitary gland, which causes pituitary hormones and GH to be secreted excessively, so that a giant often suffers from other metabolic disturbances.

If growth hormone is oversecreted in an adult after the epiphyses of the long bones have ossified, the person does not grow taller. The soft tissues, however, continue to enlarge and the bones thicken, producing a large tongue, nose, hands and feet, and a protruding jaw. This condition, *acromegaly,* is also often associated with a pituitary tumor (fig. 13B).

(a) (b)

(c)

(d)

**FIGURE 13B**
Oversecretion of growth hormone in adulthood causes acromegaly. Note the changes in this woman's facial features at ages (*a*) 9, (*b*) 16, (*c*) 33, and (*d*) 52.

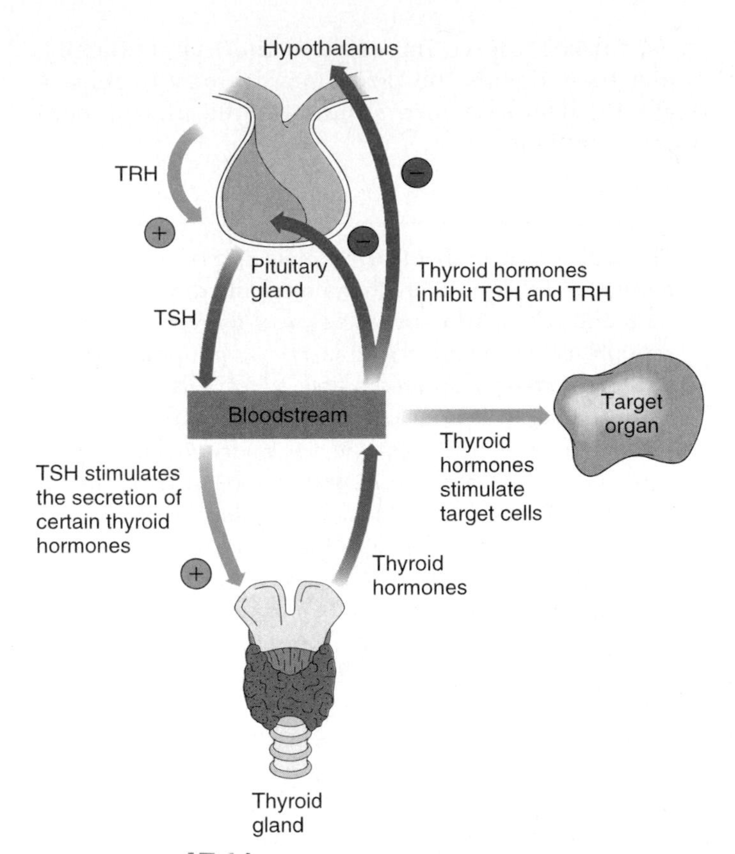


**FIGURE 13.14**

TRH from the hypothalamus stimulates the anterior pituitary gland to release TSH, which stimulates the thyroid gland to release hormones. These thyroid hormones reduce the secretion of TSH and TRH.

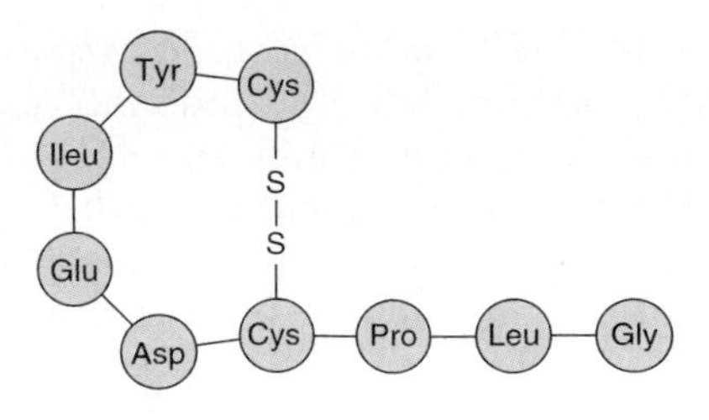


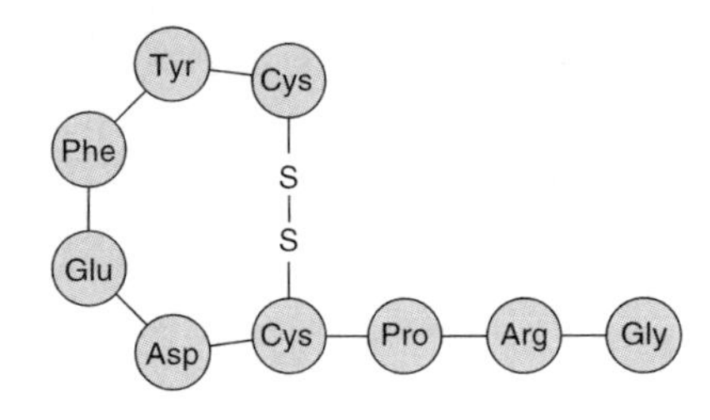


**FIGURE 13.15**

The structure of oxytocin differs from that of ADH by only two amino acids, yet they function differently.

is known that the hypothalamus secretes a *gonadotropin-releasing hormone* (GnRH). The hypothalamus does not secrete this hormone in significant amounts until puberty, because gonadotropins are virtually absent in the body fluids of infants and children.

1. What is the function of ACTH?
2. Describe the functions of FSH and LH in a male and in a female.
3. What is a gonadotropin?

## Posterior Pituitary Hormones

Unlike the anterior lobe of the pituitary gland, which is composed primarily of glandular epithelial cells, the posterior lobe consists largely of nerve fibers and neuroglial cells (*pituicytes*). The neuroglial cells support the nerve fibers that originate in the hypothalamus. The hypothalamic cells that give rise to these fibers are called neurosecretory cells because their secretions function not as neurotransmitters but as hormones.

Specialized neurons in the hypothalamus produce the two hormones associated with the posterior pituitary—**antidiuretic hormone** and **oxytocin** (see fig. 13.10*b*). These hormones travel down axons through the pituitary stalk to the posterior pituitary and are stored in vesicles (secretory granules) near the ends of the axons. The hormones are released into the blood in response to nerve impulses coming from the hypothalamus.

Antidiuretic hormone and oxytocin are short polypeptides with similar sequences (fig. 13.15). A *diuretic* is a substance that increases urine production. An *antidiuretic,* then, is a chemical that decreases urine formation. ADH produces its antidiuretic effect by causing the kidneys to reduce the amount of water they excrete. In this way, ADH plays an important role in regulating the concentration of body fluids (see chapter 20).

> Drinking alcohol is often followed by frequent and copious urination. This is because alcohol (ethyl alcohol) inhibits ADH secretion. A person must replace the lost body fluid to maintain normal water balance. Although it seems counterintuitive, excessive beer drinking can lead to dehydration because the body loses more fluid than it replaces.

ADH present in sufficient concentrations contracts certain smooth muscles, including those in the walls of blood vessels. As a result, vascular resistance may increase and blood pressure may increase as well.

For this reason, ADH is also called *vasopressin.* Although ADH is seldom present in quantities sufficient to elevate blood pressure, it may be released following severe blood loss. Blood pressure may drop as a consequence of excessive bleeding, and in this situation ADH's vasopressor effect may help return blood pressure toward normal.

ADH's two effects—vasoconstriction and water retention—are possible because the hormone binds two different receptors on target cells. V1 receptors, when ADH binds, increase the concentration of a compound called inositol triphosphate, which increases calcium ion concentration, leading to vasoconstriction. The second receptor, V2, is found on parts of the kidneys' microscopic tubules called collecting ducts. There ADH binding activates the cAMP second messenger system, which ultimately causes collecting duct cells to reabsorb water that would otherwise be excreted as urine.

The hypothalamus regulates secretion of ADH. Certain neurons in this part of the brain, called *osmoreceptors,* sense changes in the concentration of body fluids. For example, if a person is dehydrating due to a lack of water intake, the solutes in blood become more and more concentrated. The osmoreceptors, sensing the resulting increase in osmotic pressure, signal the posterior pituitary to release ADH, which causes the kidneys to retain water.

On the other hand, if a person drinks a large amount of water, body fluids become more dilute, inhibiting the release of ADH. Consequently, the kidneys excrete more dilute urine until the concentration of body fluids returns to normal.

Blood volume also affects ADH secretion. Increased blood volume stretches the walls of certain blood vessels, stimulating volume receptors that signal the hypothalamus to inhibit release of ADH. However, if hemorrhage decreases blood volume, these receptors are stretched less, and therefore send fewer inhibiting impulses. As a result, ADH secretion increases, and as before, ADH causes the kidneys to conserve water. This helps prevent further volume loss.

Oxytocin also has an antidiuretic action, but less so than ADH. In addition, oxytocin can contract smooth muscles in the uterine wall, playing a role in the later stages of childbirth. The uterus does become more and more sensitive to oxytocin's effects during pregnancy. Stretching of uterine and vaginal tissues late in pregnancy, caused by the growing fetus, initiates nerve impulses to the hypothalamus, which then signals the posterior pituitary to release oxytocin, which in turn stimulates the uterine contractions of labor.

The baby first started to display symptoms at 5 months of age—he drank huge amounts of water. By 13 months, he had become severely dehydrated, although he continued to drink nearly continuously, with a seemingly unquenchable thirst. His parents were constantly changing his wet diapers. Doctors finally diagnosed a form of the condition *diabetes insipidus,* which impairs ADH regulation of water balance. The boy was drinking sufficient fluids, but his kidneys could not absorb them. The specific defect was defective ADH V2 receptors on the kidney tubules' collecting ducts. The hormone binds, but the receptor fails to trigger cAMP formation. The boy's ADH was still able to constrict blood vessels because the V1 receptors were unaffected. A high-calorie diet and providing lots of water preserved the boy's mental abilities, but he remained small for his age. Tumors and injury affecting the hypothalamus and posterior pituitary can also cause diabetes insipidus.

In the breasts, oxytocin contracts certain cells associated with the milk-producing glands and their ducts. In lactating breasts, this action forces liquid from the milk glands into the milk ducts and ejects the milk from the breasts.

The mechanical stimulation provided by suckling the nipple initiates nerve impulses that travel to the mother's hypothalamus. The hypothalamus responds by signaling the posterior pituitary to release oxytocin, which in turn stimulates milk release. Thus, milk is normally not ejected from the milk glands and ducts until it is needed. Oxytocin has no known function in males, although it is present in the male posterior pituitary.

Chart 13.5 reviews the hormones of the pituitary gland.

If the uterus is not contracting sufficiently to expel a fully developed fetus, oxytocin is sometimes given intravenously to stimulate uterine contractions, thus inducing labor. Oxytocin is also administered to the mother following childbirth to ensure that the uterine muscles contract enough to squeeze broken blood vessels closed, minimizing the danger of hemorrhage.

## CHART 13.5 HORMONES OF THE PITUITARY GLAND

| Anterior Lobe | | |
|---|---|---|
| *Hormone* | *Action* | *Source of Control* |
| Growth hormone (GH) | Stimulates increase in size and rate of reproduction of body cells; enhances movement of amino acids through membranes; promotes growth of long bones | Growth hormone-releasing hormone (GHRH) and somatostatin (SS) from the hypothalamus |
| Prolactin (PRL) | Sustains milk production after birth; amplifies effect of LH in males | Secretion restrained by prolactin release-inhibiting hormone (PIH) and stimulated by prolactin-releasing hormone (PRH) from the hypothalamus |
| Thyroid-stimulating hormone (TSH) | Controls secretion of hormones from the thyroid gland | Thyrotropin-releasing hormone (TRH) from the hypothalamus |
| Adrenocorticotropic hormone (ACTH) | Controls secretion of certain hormones from the adrenal cortex | Corticotropin-releasing hormone (CRH) from the hypothalamus |
| Follicle-stimulating hormone (FSH) | Responsible for development of egg-containing follicles in ovaries; stimulates follicular cells to secrete estrogen; in males, stimulates production of sperm cells | Gonadotropin-releasing hormone (GnRH) from the hypothalamus |
| Luteinizing hormone (LH or ICSH in males) | Promotes secretion of sex hormones; plays role in release of egg cell in females | Gonadotropin-releasing hormone (GnRH) from the hypothalamus |

| Posterior Lobe | | |
|---|---|---|
| *Hormone* | *Action* | *Source of Control* |
| Antidiuretic hormone (ADH) | Causes kidneys to reduce water excretion; in high concentration, raises blood pressure | Hypothalamus in response to changes in blood water concentration and blood volume |
| Oxytocin (OT) | Contracts muscles in uterine wall and those associated with milk-secreting glands | Hypothalamus in response to stretch in uterine and vaginal walls, and stimulation of breasts |

1. What is the function of ADH?
2. How is the secretion of ADH controlled?
3. What effects does oxytocin produce in females?

## THYROID GLAND

The **thyroid gland,** shown in figure 13.16, is a very vascular structure that consists of two large lateral lobes connected by a broad isthmus. It is located just below the larynx on either side and anterior to the trachea. It has a special ability to remove iodine from the blood.

### Structure of the Gland

A capsule of connective tissue covers the thyroid gland, which is made up of many secretory parts called *follicles.* The cavities within these follicles are lined with a single layer of cuboidal epithelial cells, and are filled with a clear, viscous glycoprotein called *colloid.* The follicular cells produce and secrete hormones that may either be stored in the colloid or released into the blood of nearby capillaries (fig. 13.17). Other hormone-secreting cells, called *extrafollicular cells* (C cells), lie outside the follicles.

### Thyroid Hormones

The thyroid gland produces three important hormones. The follicular cells synthesize two of these, which have marked effects on the metabolic rates of body cells. The extrafollicular cells produce the third type of hormone, which influences blood concentrations of calcium and phosphate ions.

The two important thyroid hormones that affect cellular metabolic rates are **thyroxine,** or tetraiodothyronine (also called $T_4$, because it includes four atoms of iodine), and **triiodothyronine** (also called $T_3$, because it includes three atoms of iodine). These hormones help regulate the metabolism of carbohydrates, lipids, and proteins. For example, they increase the rate at which cells release energy from carbohydrates, enhance the rate of protein synthesis, and stimulate breakdown and mobilization of lipids. These hormones are essential for normal growth and development and for maturation of the nervous system (fig. 13.18).

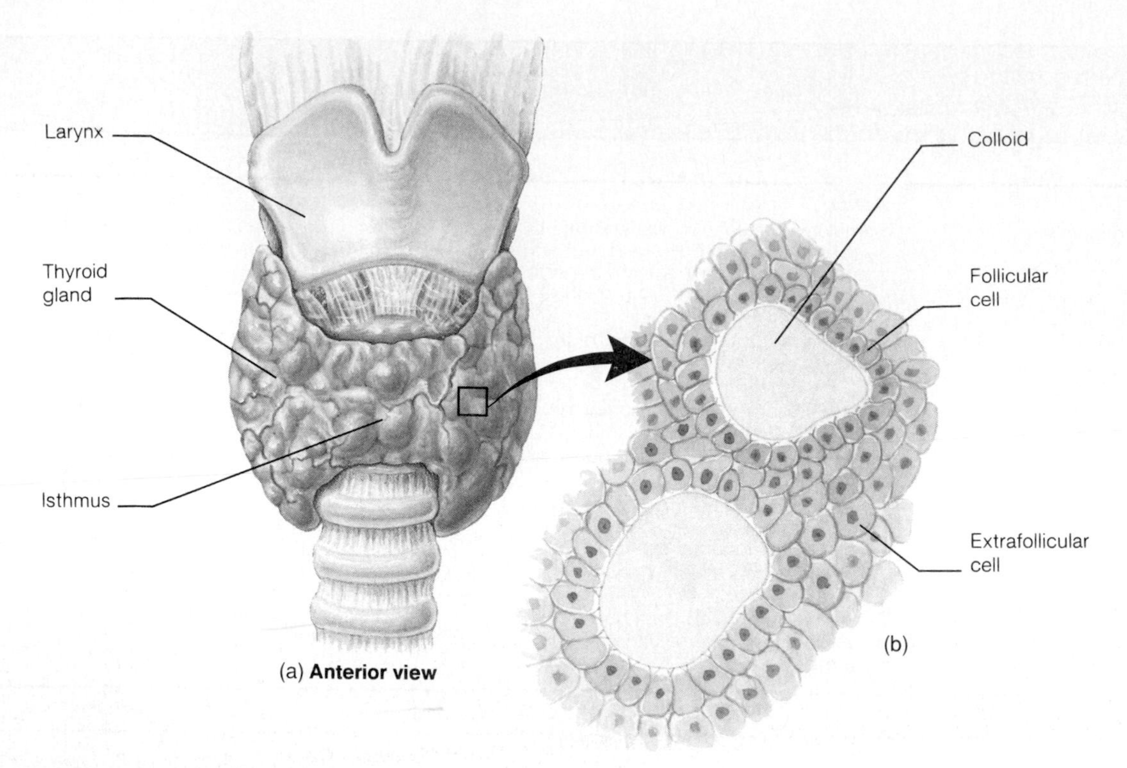


**FIGURE 13.16**
(*a*) The thyroid gland consists of two lobes connected anteriorly by an isthmus. (*b*) Follicular cells secrete thyroid hormones.

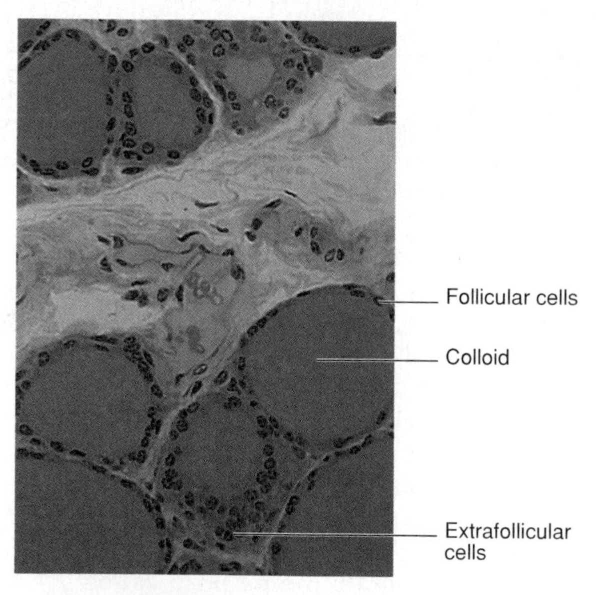


**FIGURE 13.17**
A light micrograph of thyroid gland tissue (40×). The open spaces that follicular cells surround are filled with colloid.

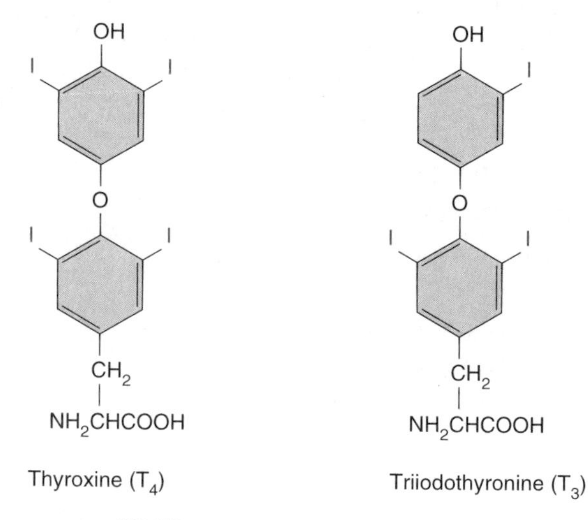


**FIGURE 13.18**
The hormones thyroxine and triiodothyronine have very similar molecular structures.

Follicular cells require iodine salts (iodides) to produce thyroxine and triiodothyronine. Such salts are normally obtained from foods, and after they have been absorbed from the intestine, the blood carries some of them to the thyroid gland. An efficient active transport mechanism called the *iodine pump* moves the iodides into the follicular cells, where they are concentrated. The iodides, together with an amino acid (tyrosine), are used to synthesize these thyroid hormones.

Follicular cells also secrete a protein called *thyroglobulin,* which is the main ingredient of thyroid colloid. Thyroglobulin stores thyroid hormones whenever they are produced in excess. The stored hormones are bound to the thyroglobulin until the

## CHART 13.6 Hormones of the Thyroid Gland

| Hormone | Action | Source of Control |
|---|---|---|
| Thyroxine ($T_4$) | Increases rate of energy release from carbohydrates; increases rate of protein synthesis; accelerates growth; stimulates activity in the nervous system | TSH from the anterior pituitary gland |
| Triiodothyronine ($T_3$) | Same as above, but five times more potent than thyroxine | Same as above |
| Calcitonin | Lowers blood calcium and phosphate ion concentrations by inhibiting release of calcium and phosphate ions from bones, and by increasing the rate at which calcium and phosphate ions are deposited in bones | Elevated blood calcium ion concentration, digestive hormones |

hormone concentration of the body fluids drops below a certain level, then enzymes release the hormones from the colloid, and they diffuse into the blood. Once in the blood, thyroid hormones combine with blood proteins (alpha globulins) and are transported to body cells. Triiodothyronine is nearly five times more potent, but thyroxine accounts for at least 95% of circulating thyroid hormones.

Although produced by cells in the thyroid gland, **calcitonin** is sometimes not considered to be a thyroid hormone because it is synthesized by the C cells, which are distinct from the gland's follicles. Calcitonin influences blood calcium and phosphate ion concentrations. It helps lower concentrations of calcium and phosphate ions by inhibiting the rate at which they leave the bones and enter extracellular fluids by action of osteoclasts (see chapter 7). At the same time, calcitonin increases the rate at which calcium and phosphate ions are deposited in bone matrix by stimulating activity of osteoblasts. It also increases the excretion of calcium ions and phosphate ions by the kidneys.

Calcitonin secretion is stimulated by a high blood calcium ion concentration, as may occur following absorption of calcium ions from a recent meal. Certain hormones also prompt its secretion, such as gastrin, which is released from active digestive organs. Calcitonin helps prevent prolonged elevation of blood calcium ion concentration after eating.

Chart 13.6 summarizes the actions and sources of control of the thyroid hormones. Clinical Application 2.1, chart 13.7, and figures 13.19, 13.20, and 13.21 discuss disorders of the thyroid gland.

1. Where is the thyroid gland located?
2. What hormones of the thyroid gland affect carbohydrate metabolism and protein synthesis?
3. What substance is essential for the production of thyroxine and triiodothyronine?
4. How does calcitonin influence the concentrations of blood calcium and phosphate ions?

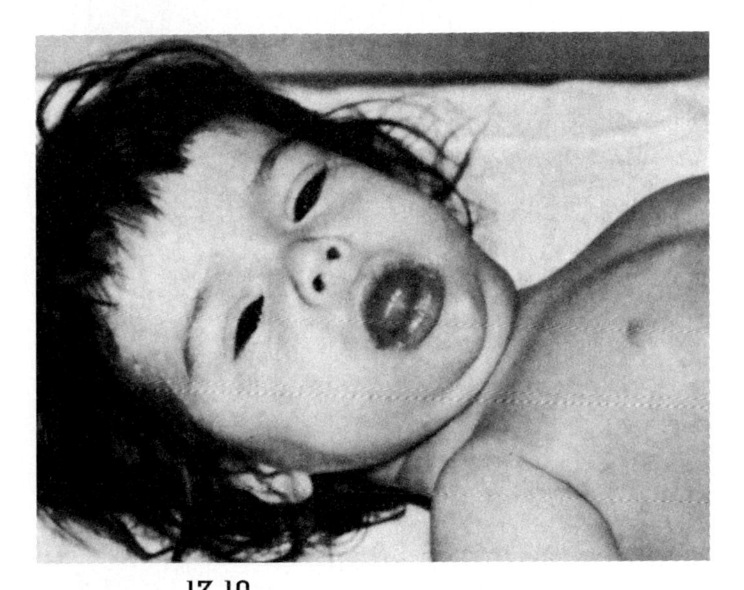

FIGURE 13.19
Cretinism is due to an underactive thyroid gland during infancy and childhood.

On a spring morning in 1991, U.S. President George Bush set out on his daily run. After only a few hundred yards, he stopped, winded and experiencing an irregular heartbeat. Over the preceding two months, his wife, Barbara, had inexplicably lost 18 pounds, and her eyesight had become blurry and her eyes dry. Both Bushes were feeling unusually fidgety. Blood tests found that they had hyperthyroidism—and so did their dog, Millie. Because of the three cases, the White House environment was carefully studied, but no cause was ever identified.

## PARATHYROID GLANDS

The **parathyroid glands** are located on the posterior surface of the thyroid gland, as shown in figure 13.22. Usually there are four of them—a superior and an inferior gland associated with each of the thyroid's

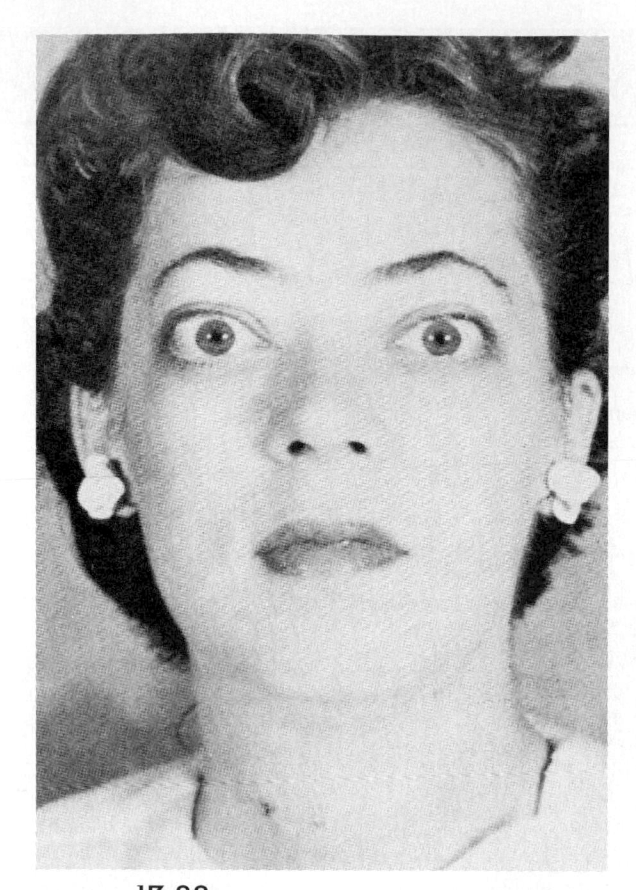

**Figure 13.20**
Hyperthyroidism may cause the eyes to protrude.

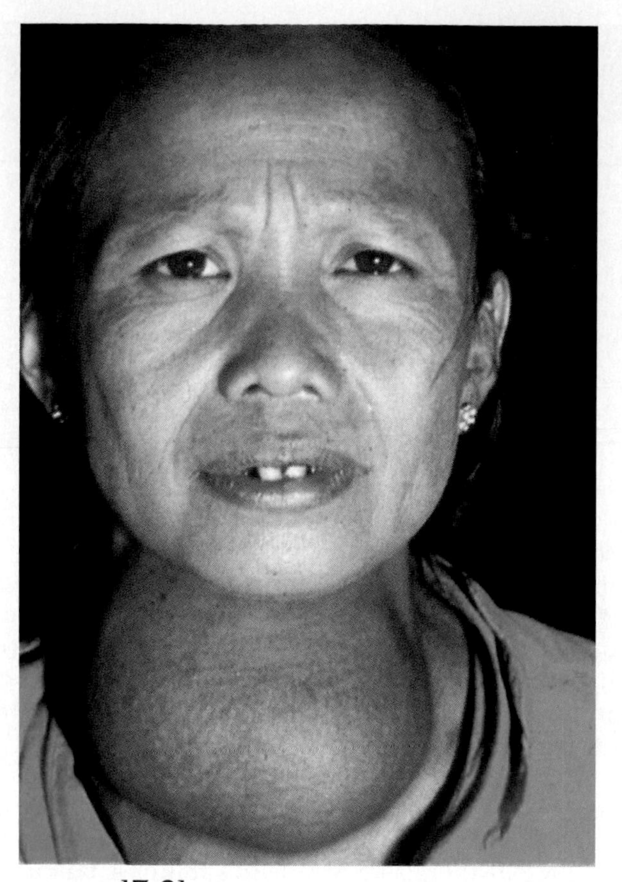

**Figure 13.21**
An iodine deficiency causes simple (endemic) goiter, and results in high levels of TSH.

**Chart 13.7 Disorders of the Thyroid Gland**

| Condition | Mechanism/Symptoms |
|---|---|
| Grave's disease | Autoantibodies bind TSH receptors on thyroid cell membranes, mimicking action of TSH, overstimulating gland (hyperthyroidism); this is an exothalmic goiter |
| Hashimoto's disease | Autoantibodies (against self) attack thyroid cells, producing hypothyroidism |
| Hyperthyroidism | High metabolic rate, sensitivity to heat, restlessness, hyperactivity, weight loss, protruding eyes, goiter |
| Hypothyroidism (infantile) | Cretinism—stunted growth, abnormal bone formation, mental retardation, low body temperature, sluggishness |
| Hypothyroidism (adult) | Myxedema—low metabolic rate, sensitivity to cold, sluggishness, poor appetite, swollen tissues, mental dullness |
| Simple goiter | Deficiency of thyroid hormones due to iodine deficiency; because no thyroid hormones inhibit pituitary release of TSH, thyroid is overstimulated and enlarges, but functions below normal (hypothyroidism) |

lateral lobes. The parathyroid glands secrete a hormone that regulates the concentrations of calcium and phosphate ions in the blood.

### Structure of the Glands

Each parathyroid gland is a small, yellowish brown structure covered by a thin capsule of connective tissue. The body of the gland consists of numerous tightly packed secretory cells that are closely associated with capillary networks (fig. 13.23).

### Parathyroid Hormone

The parathyroid glands secrete a protein, **parathyroid hormone** (PTH) or *parathormone* (see fig. 13.3), which increases blood calcium ion concentration and decreases blood phosphate ion concentration through actions in the bones, kidneys, and intestines.

The intercellular matrix of bone tissue contains a considerable amount of calcium phosphate and calcium carbonate. PTH stimulates bone resorption by osteocytes and osteoclasts, and inhibits the activity of osteoblasts

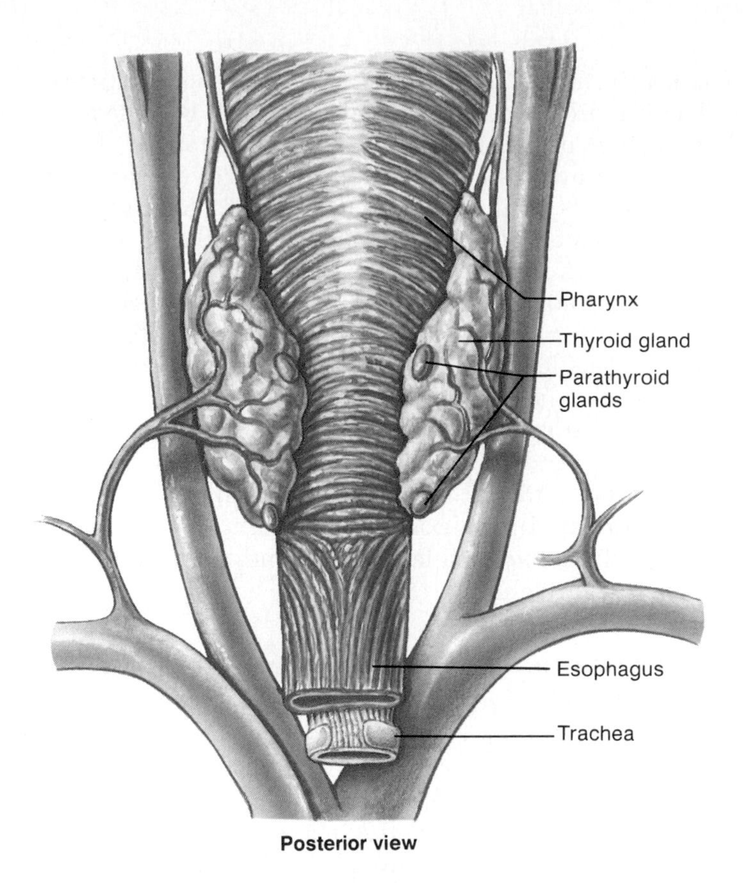


**FIGURE 13.22**
The parathyroid glands are embedded in the posterior surface of the thyroid gland.

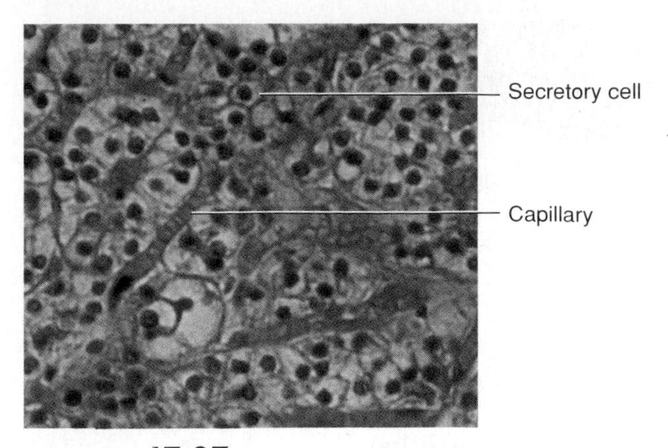


**FIGURE 13.23**
Light micrograph of the parathyroid gland.

(see chapter 7). As bone resorption increases, calcium and phosphate ions are released into the blood. At the same time, PTH causes the kidneys to conserve blood calcium ions and to excrete more phosphate ions in the urine. It also indirectly stimulates absorption of calcium ions from food in the intestine by influencing metabolism of vitamin D.

Vitamin D (cholecalciferol) is synthesized from dietary cholesterol, which intestinal enzymes convert into provitamin D (7-dehydrocholesterol). This provitamin is stored largely in the skin, and exposure to the ultraviolet wavelengths of sunlight changes it to vitamin D. Vitamin D is also obtained from some foods.

Vitamin D is changed by the liver to hydroxycholecalciferol, which is carried in the bloodstream or is stored in tissues. When PTH is present, hydroxycholecalciferol can be changed in the kidneys into an active form of vitamin D (dihydroxycholecalciferol). This active form controls absorption of calcium ions from the intestine (fig. 13.24).

A negative feedback mechanism operating between the parathyroid glands and the blood calcium ion concentration regulates secretion of PTH (fig. 13.25). As the concentration of blood calcium ions rises, less PTH is secreted; as the concentration of blood calcium ions drops, more PTH is released.

> Cancer cells in nonendocrine tissues sometimes secrete hormones, producing symptoms of an *endocrine paraneoplastic syndrome.* A patient with an unusual such syndrome in 1994 led to discovery of a new hormone. In a middle-aged woman with oncogenic osteomalacia, a small tumor in the soft tissue of her thigh disrupted phosphate homeostasis, causing her bones to demineralize, leading to her symptoms of bone aching. To the surprise of medical researchers, the tumor was not producing parathyroid hormone but a new hormone, tentatively named phosphaturic factor, which increases excretion of phosphate in urine, weakening the bones. The woman recovered when her tumor was removed, but it grew back and symptoms returned.

The homeostasis of calcium ions, maintained by the opposite effects of calcitonin and PTH, is important in a number of physiological processes. For example, as the blood calcium ion concentration drops (hypocalcemia), the nervous system becomes abnormally excitable, and impulses may be triggered spontaneously. As a result, muscles, including the respiratory muscles, may undergo tetanic contractions, and the person may die due to a failure to breathe. An abnormally high concentration of blood calcium ions (hypercalcemia) depresses the nervous system. Consequently, muscle contractions are weak and reflexes are sluggish. Chart 13.8 lists parathyroid disorders.

1. Where are the parathyroid glands located?
2. How does parathyroid hormone help regulate the concentrations of blood calcium and phosphate ions?
3. How does the negative feedback system of the parathyroid glands differ from that of the thyroid gland?

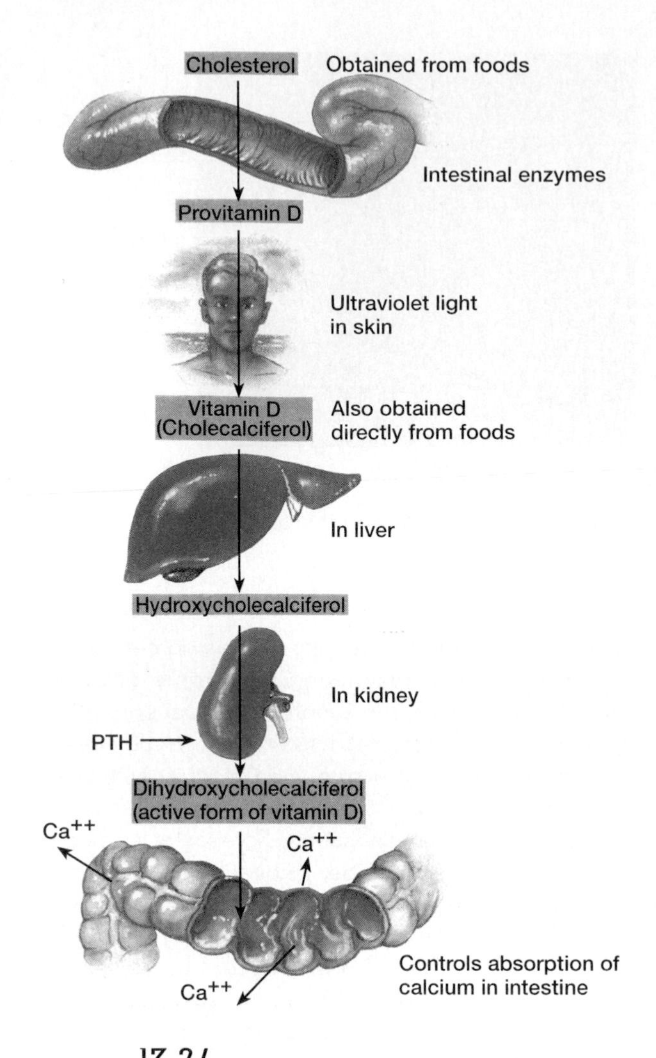


**Figure 13.24**
Mechanism by which PTH promotes calcium absorption in the intestine.

## Adrenal Glands

The **adrenal glands** (suprarenal glands) are closely associated with the kidneys. A gland sits atop each kidney like a cap and is embedded in the mass of adipose tissue that encloses the kidney.

### Structure of the Glands

The adrenal glands are shaped like pyramids. Each adrenal gland is very vascular and consists of two parts (fig. 13.26). The central portion is the adrenal medulla, and the outer part is the adrenal cortex. These regions are not sharply divided, but they are distinct glands that secrete different hormones.

The **adrenal medulla** consists of irregularly shaped cells arranged in groups around blood vessels. These cells are intimately connected with the sympathetic division of the autonomic nervous system. In fact, these adrenal medullary cells are modified postganglionic neurons, and preganglionic autonomic nerve fibers lead to them directly from the central nervous system (see chapter 11).

The **adrenal cortex** makes up the bulk of the adrenal gland. It is composed of closely packed masses of epithelial layers that form an outer, a middle, and an inner zone of the cortex—the zona glomerulosa, the zona fasciculata, and the zona reticularis, respectively (fig. 13.27).

### Hormones of the Adrenal Medulla

The cells of the adrenal medulla (chromaffin cells) produce, store, and secrete two closely related hormones, **epinephrine** (adrenalin) and **norepinephrine** (noradrenalin). Both of these substances are a type of amine called a *catecholamine,* and they have similar molecular structures and physiological functions (fig. 13.28). In fact, epinephrine is synthesized from norepinephrine.

The synthesis of these hormones begins with the amino acid *tyrosine.* In the first step of the process, an enzyme (tyrosine hydroxylase) in the secretory cells converts tyrosine into a substance called *dopa.* A second enzyme (dopa decarboxylase) converts dopa to dopamine, and a third enzyme (dopamine betahydroxylase) converts dopamine to norepinephrine. Still another enzyme (phenylethanolamine N-methyltransferase) converts the norepinephrine to epinephrine. About 15% of the norepinephrine is not converted but is stored unchanged. The hormones are stored in tiny vesicles (chromaffin granules), much like neurotransmitters are stored in neurons.

The effects of the adrenal medullary hormones generally resemble those that result when sympathetic nerve fibers stimulate their effectors. These effects include increased heart rate and increased force of cardiac muscle contraction, elevated blood pressure, increased breathing rate, and decreased activity in the digestive system (see chart 11.8). The hormonal effects last up to ten times longer than the neurotransmitter effects because the hormones are removed from the tissues relatively slowly.

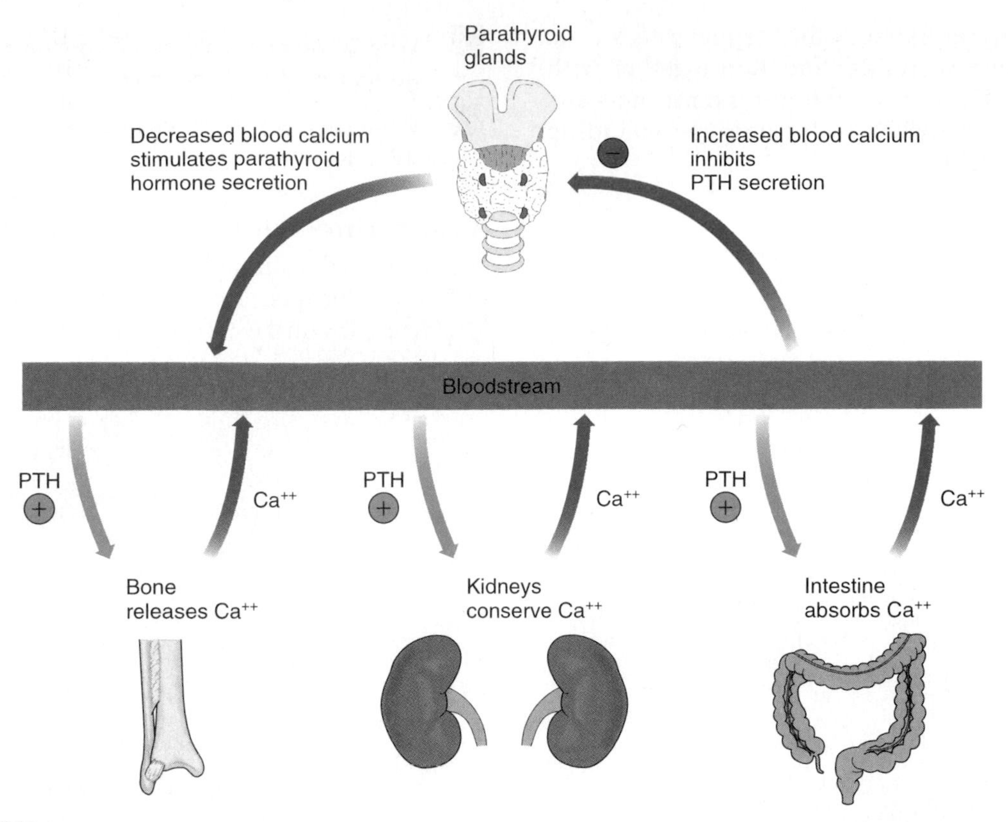


**FIGURE 13.25**
Parathyroid hormone stimulates bone to release calcium and the kidneys to conserve calcium. It indirectly stimulates the intestine to absorb calcium. The resulting increase in blood calcium concentration inhibits secretion of parathyroid hormone.

### CHART 13.8 DISORDERS OF THE PARATHYROID GLANDS

| Condition | Symptoms/Mechanism | Cause | Treatment |
|---|---|---|---|
| Hyperparathyroidism | Fatigue, muscular weakness, painful joints, altered mental functions, depression, weight loss, bone weakening. Increased PTH secretion overstimulates osteoclasts. | Tumor | Remove tumor, correct bone deformities |
| Hypoparathyroidism | Muscle cramps and seizures. Decreased PTH secretion reduces osteoclast activity, diminishing blood calcium ion concentration. | Inadvertent surgical removal; injury | Calcium salt injections, massive doses of vitamin D |

The ratio of the two hormones in the adrenal medullary secretion varies with different physiological conditions, but usually the secretion is about 80% epinephrine and 20% norepinephrine. Although these hormones' effects are generally similar, certain effectors respond differently to them. These differences are due to the relative numbers of alpha and beta receptors in the membranes of the effector cells. As described in chapter 11, both hormones can stimulate both classes of receptors, although norepinephrine has a greater effect on alpha receptors. Chart 13.9 compares some of the differences in the effects of these hormones.

Impulses arriving by way of sympathetic nerve fibers stimulate the adrenal medulla to release its hormones at the same time other effectors are stimulated by sympathetic impulses. As a rule, these impulses originate in the hypothalamus in response to stress. Thus, the medullary secretions function together with the sympathetic division of the autonomic nervous system in preparing the body for energy-expending action—"fight or flight."

1. Describe the location and structure of the adrenal glands.
2. Name the hormones the adrenal medulla secretes.
3. What general effects do hormones secreted by the adrenal medulla produce?
4. What usually stimulates release of hormones from the adrenal medulla?

## Hormones of the Adrenal Cortex

The cells of the adrenal cortex produce more than thirty different steroids, including several hormones (corticosteroids). Unlike the adrenal medullary hormones,

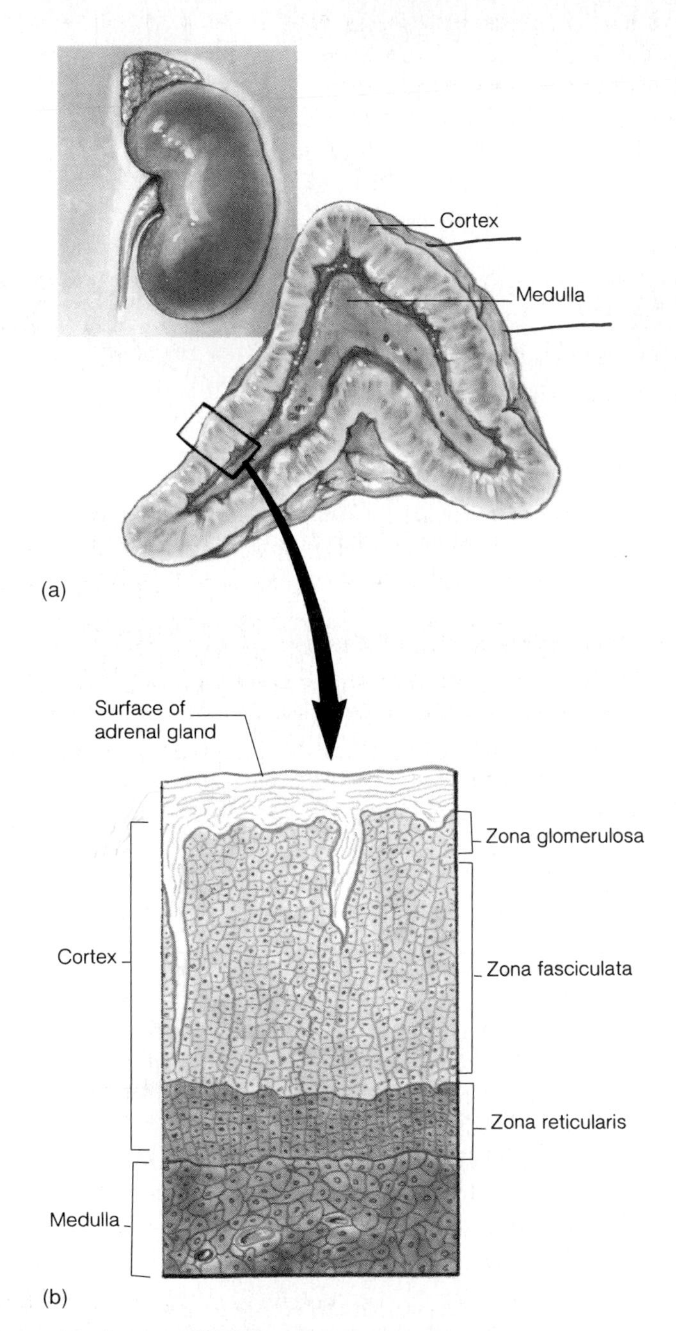


**FIGURE 13.26**
(*a*) An adrenal gland consists of an outer cortex and an inner medulla. (*b*) The cortex has three layers, or zones, of cells.

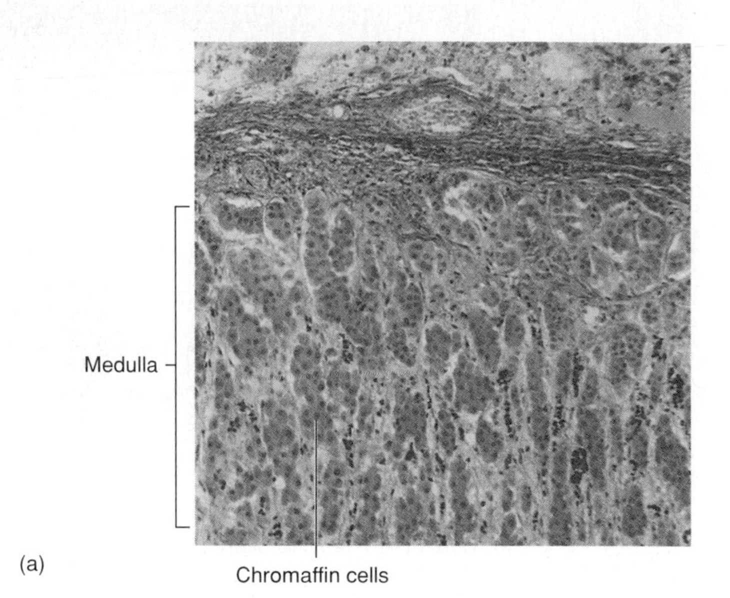


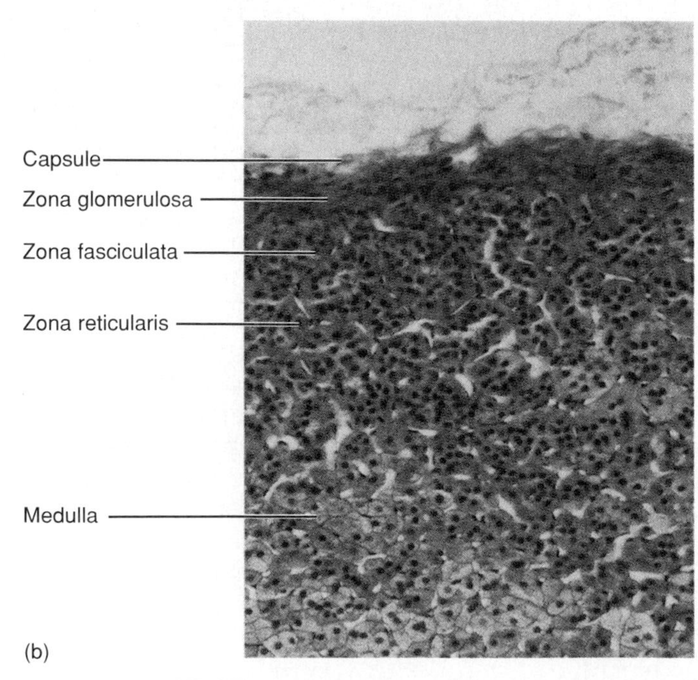


**FIGURE 13.27**
Light micrograph of (*a*) the adrenal medulla (45×) and (*b*) the adrenal cortex (30×).

without which a person can survive, some of those released by the cortex are vital. In fact, in the absence of adrenal cortical secretions, a person usually dies within a week without extensive electrolyte therapy. The most important adrenal cortical hormones are aldosterone, cortisol, and certain sex hormones.

### Aldosterone

Cells in the outer zone (zona glomerulosa) of the adrenal cortex synthesize **aldosterone.** This hormone is called a *mineralocorticoid* because it helps regulate the concentration of mineral electrolytes, such as sodium and potassium ions. More specifically, aldosterone causes the kidney to conserve sodium ions and to excrete potassium ions. The cells that secrete aldosterone respond directly to changes in the composition of blood plasma. However, while an increase in plasma potassium strongly stimulates these cells, a decrease in plasma sodium only slightly stimulates them. Control of aldosterone secretion is linked indirectly to plasma sodium level by action of groups of specialized kidney cells (juxtaglomerular cells) that respond to changes in blood pressure and plasma sodium ion concentration. If either of these factors decreases, the cells release an enzyme called **renin,** which decomposes a blood protein called **angiotensinogen,** releasing a peptide called **angiotensin I.** Another enzyme (angiotensin-converting enzyme, or ACE) in the lungs converts angiotensin I into another form, **angiotensin II,** which is carried in the bloodstream. When angiotensin II reaches the adrenal cortex, it stimulates release of aldosterone. ACTH also stimulates secretion of aldosterone, and is necessary for aldosterone secretion to respond to other stimuli.

Aldosterone, in conserving sodium ions, retains water indirectly by osmosis. This helps maintain blood sodium ion concentration and blood volume. Angiotensin II helps maintain systemic blood pressure by constricting blood vessels (fig. 13.29).

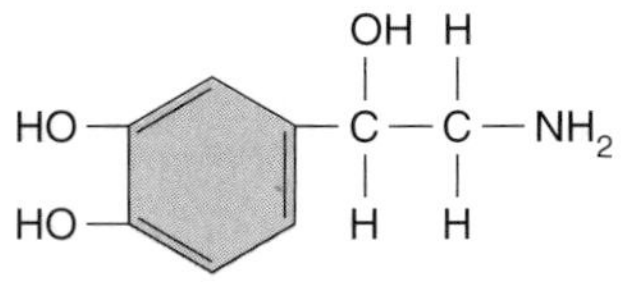

> ACE inhibitors are a class of drugs used to treat some forms of high blood pressure (hypertension). They work by competing with angiotensin-converting enzyme, blocking formation of angiotensin II and preventing inactivation of a substance called bradykinin, a vasodilator. Both effects lead to dilation of blood vessels, lowering blood pressure.

### Cortisol

**Cortisol** (hydrocortisone) is a *glucocorticoid,* which means it affects glucose metabolism. It is produced in the middle zone (zona fasciculata) of the adrenal cortex and has a molecular structure similar to aldosterone

Norepinephrine

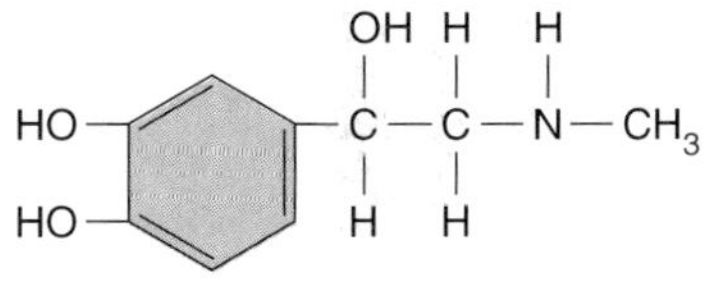

Epinephrine

**FIGURE 13.28**

Epinephrine and norepinephrine have similar molecular structures and similar functions.

**CHART 13.9 Comparative Effects of Epinephrine and Norepinephrine**

| Structure or Function Affected | Epinephrine | Norepinephrine |
|---|---|---|
| Heart | Rate increases<br>Force of contraction increases | Rate increases<br>Force of contraction increases |
| Blood vessels | Vessels in skeletal muscle vasodilate, decreasing resistance to blood flow | Blood flow to skeletal muscles increases, resulting from constriction of blood vessels in skin and viscera |
| Systemic blood pressure | Some increase due to increased cardiac output | Great increase due to vasoconstriction |
| Airways | Dilated | Some dilation |
| Reticular formation of brain | Activated | Little effect |
| Liver | Promotes breakdown of glycogen to glucose, increasing blood sugar level | Little effect on blood sugar |
| Metabolic rate | Increases | Increases |

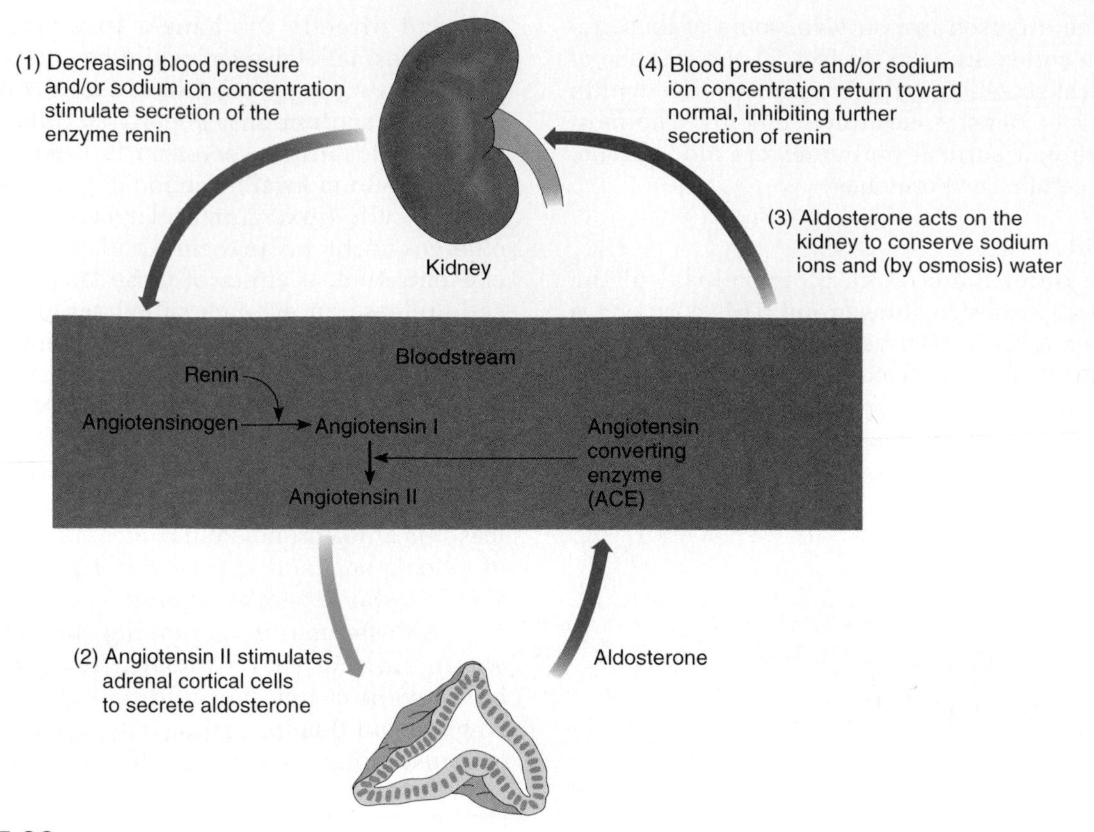


**FIGURE 13.29**

Aldosterone increases blood volume and pressure by promoting conservation of sodium ions and water (steps 1–4).

(fig. 13.30). In addition to affecting glucose, cortisol influences protein and fat metabolism. Among the more important actions of cortisol are the following:

1. It inhibits the synthesis of protein in various tissues, increasing blood concentration of amino acids.
2. It promotes the release of fatty acids from adipose tissue, increasing the use of fatty acids as an energy source and decreasing the use of glucose as an energy source.
3. It stimulates liver cells to synthesize glucose from noncarbohydrates (gluconeogenesis), such as circulating amino acids and glycerol, thus increasing blood glucose concentration.

Cortisol's actions help keep the blood glucose concentration within the normal range between meals. These actions are important because just a few hours without food can exhaust liver glycogen, another major source of glucose.

A negative feedback mechanism much like that controlling the thyroid hormones $T_3$ and $T_4$ regulates cortisol release. It involves the hypothalamus, anterior pituitary gland, and adrenal cortex. The hypothalamus secretes CRH (corticotropin-releasing hormone) into

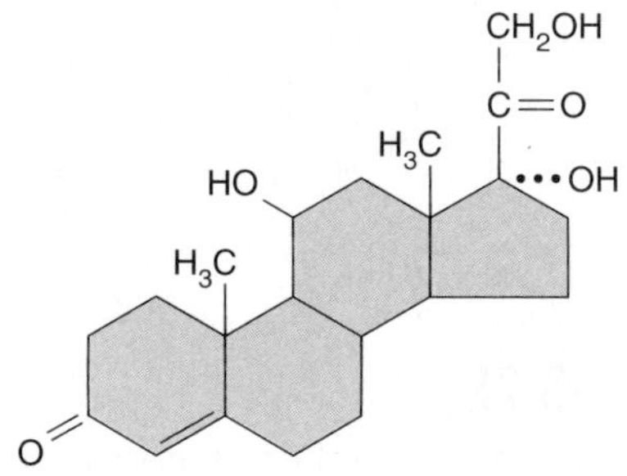

$CH_2OH$, C=O, ···OH, $H_3C$, HO, $H_3C$, O

Cortisol

$CH_2OH$, H, C=O, O=C, HO, $H_3C$, O

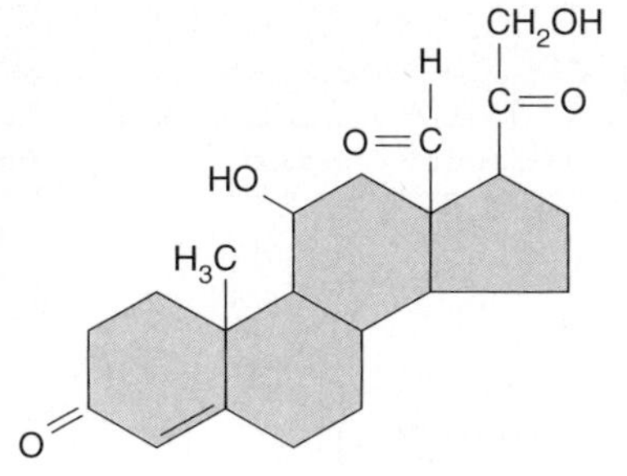

Aldosterone

**FIGURE 13.30**

Cortisol and aldosterone are steroids with similar molecular structures.

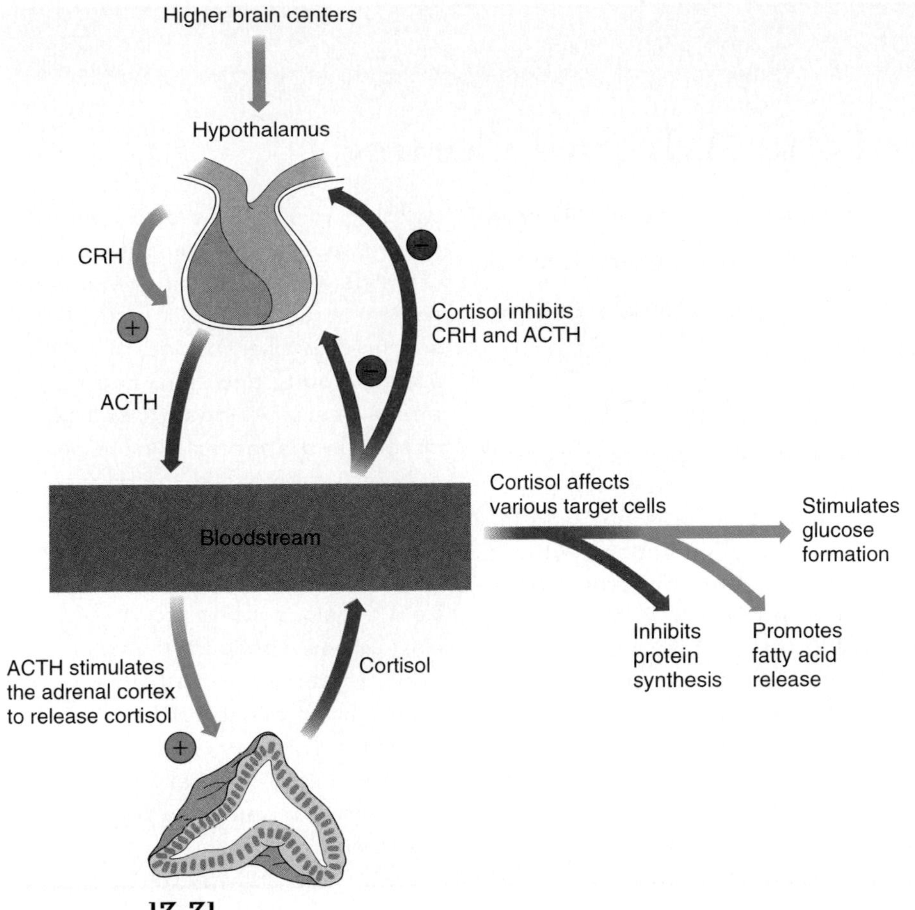


**FIGURE 13.31**
Negative feedback regulates cortisol secretion, similar to the regulation of thyroid hormone secretion (see fig. 13.14).

the hypophyseal portal veins, which carry the CRH to the anterior pituitary gland, stimulating it to secrete ACTH. In turn, ACTH stimulates the adrenal cortex to release cortisol. Cortisol inhibits release of both CRH and ACTH. As concentration of these substances falls, cortisol production drops.

Cortisol and related compounds are often used as drugs to reduce painful inflammation. They provide relief by:

- decreasing permeability of capillaries, preventing leakage of fluids that swell surrounding tissues
- stabilizing lysosomal membranes, preventing release of their enzymes, which destroy tissue
- inhibiting prostaglandin synthesis

Unfortunately, the concentration of cortisol compounds necessary to stifle inflammation is toxic, so these drugs can only be used for a short time. They are used to treat autoimmune disorders, allergies, asthma, and patients who have received organ transplants or tissue grafts.

The set point of the feedback loop controlling cortisol secretion changes from time to time, enabling hormone output to alter to meet the demands of changing conditions. For example, under stress—injury, disease, extreme temperature, or emotional upset—nerve impulses send the brain information concerning the stressful condition. In response, brain centers signal the hypothalamus to release more CRH, leading to a higher concentration of cortisol until the stress subsides (fig. 13.31).

### Sex Hormones

Cells in the inner zone (zona reticularis) of the adrenal cortex produce sex hormones. These hormones are male (adrenal androgens), but some of them are converted into female hormones (estrogens) by the skin, liver, and adipose tissues. These hormones may supplement the supply of sex hormones from the gonads and stimulate early development of the reproductive organs. Also, adrenal androgens may play a role in controlling the female sex drive. Chart 13.10 summarizes the actions of the cortical hormones.

Clinical Application 13.3 discusses some of the effects of a malfunctioning adrenal gland on health.

1. Name the most important hormones of the adrenal cortex.
2. What is the function of aldosterone?
3. What does cortisol do?
4. How are blood concentrations of aldosterone and cortisol regulated?

## PANCREAS

The **pancreas** contains two major types of secretory tissues, reflecting its dual function as an exocrine gland that secretes digestive juice through a duct, and an endocrine gland that releases hormones into body fluids.

### Structure of the Gland

The pancreas is an elongated, somewhat flattened organ that is posterior to the stomach and behind the parietal

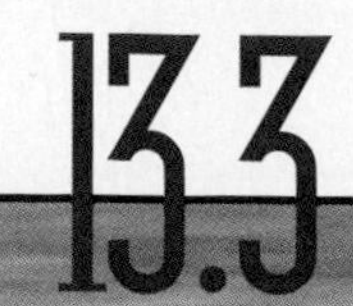

CLINICAL APPLICATION 13.3

## Disorders of the Adrenal Cortex

John F. Kennedy's beautiful bronze complexion may have resulted not from sunbathing, but from a disorder of the adrenal glands. When he ran for the presidency in 1960, Kennedy knew he had *Addison's disease,* but his staff kept his secret, for fear it would affect his career. Kennedy had almost no adrenal tissue but was able to function by receiving mineralocorticoids and glucocorticoids, the standard treatment.

In Addison's disease, the adrenal cortex does not secrete hormones sufficiently, due to immune system attack (autoimmunity) or an infection such as tuberculosis. Signs and symptoms include decreased blood sodium, increased blood potassium, low blood glucose level (hypoglycemia), dehydration, low blood pressure, and increased skin pigmentation. Without treatment, death comes within days from severe disturbances in electrolyte balance.

An adrenal tumor or oversecretion of ACTH by the anterior pituitary causes hypersecretion of glucocorticoids (primarily cortisol), resulting in *Cushing's syndrome.* Tissue protein level plummets, accompanied by wasting of muscles and loss of bone tissue. Blood glucose level remains elevated, and excess sodium is retained. As a result, tissue fluid increases, and the skin appears puffy and thin, and may bruise easily. Adipose tissue deposited in the face and back produce a characteristic "moon face" and "buffalo hump." At the same time, an increase in adrenal sex hormone secretion may masculinize a female, causing growth of facial hair and a deepening voice.

Treatment of Cushing's syndrome attempts to reduce ACTH secretion. This may entail removing a tumor in the pituitary gland or partially or completely removing the adrenal glands.

CHART 13.10 HORMONES OF THE ADRENAL CORTEX

| Hormone | Action | Factors Regulating Secretion |
|---|---|---|
| Aldosterone | Helps regulate the concentration of extracellular electrolytes by conserving sodium ions and excreting potassium ions | Electrolyte concentrations in body fluids, and renin-angiotensin mechanism |
| Cortisol | Decreases protein synthesis, increases fatty acid release, and stimulates glucose synthesis from noncarbohydrates | CRH from the hypothalamus, and ACTH from the anterior pituitary gland |
| Adrenal androgens | Supplement sex hormones from the gonads, may be converted into estrogens | |

peritoneum (fig. 13.32). It is attached to the first section of the small intestine (duodenum) by a duct, which transports its digestive juice into the intestine. The digestive functions of the pancreas are discussed in chapter 17.

The endocrine portion of the pancreas consists of cells arranged in groups that are closely associated with blood vessels. These groups, called *islets of Langerhans,* include three distinct types of hormone-secreting cells—*alpha cells,* which secrete glucagon; *beta cells,* which secrete insulin; and *delta cells,* which secrete somatostatin (fig. 13.33).

### Hormones of the Islets of Langerhans

**Glucagon** is a protein that stimulates the liver to break down glycogen into glucose (glycogenolysis) and to convert noncarbohydrates, such as amino acids, into glucose (gluconeogenesis). Glucagon also stimulates breakdown of fats into fatty acids and glycerol.

In a negative feedback system, a low concentration of blood sugar stimulates release of glucagon from the alpha cells. When blood sugar concentration returns toward normal, glucagon secretion decreases. This mechanism prevents hypoglycemia from occurring at times

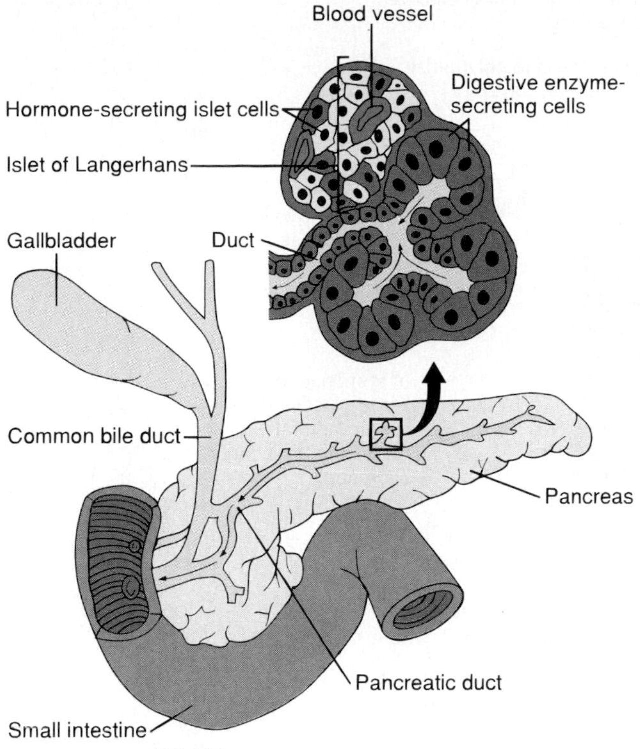


**FIGURE 13.32**

The hormone-secreting cells of the pancreas are grouped in clusters, or islets, that are closely associated with blood vessels. Other pancreatic cells secrete digestive enzymes into ducts.

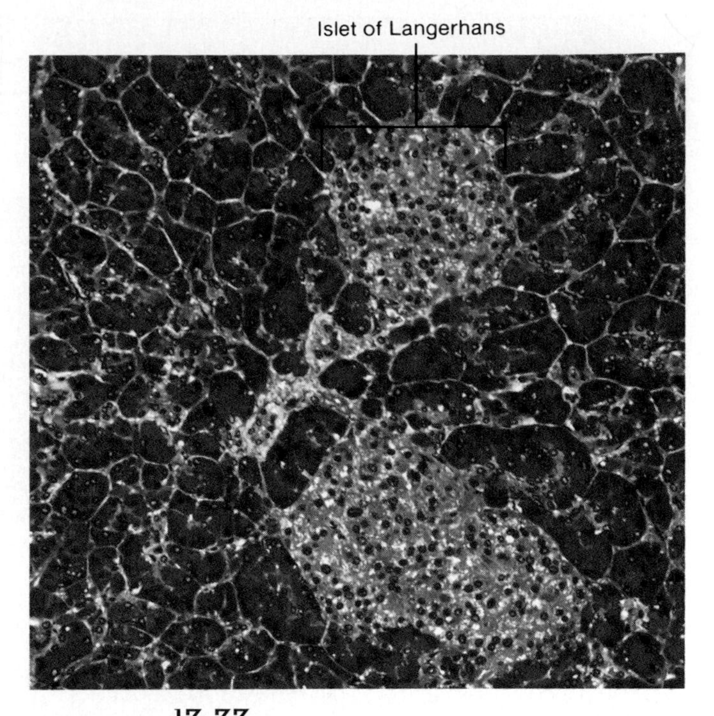


**FIGURE 13.33**

Light micrograph of an islet of Langerhans within the pancreas (50×).

when glucose concentration is relatively low, such as between meals, or when glucose is being used rapidly—during periods of exercise, for example.

> An enzyme called glucokinase enables beta cells to "sense" glucose level, important information in determining rates of synthesis of glucagon and insulin. A rare type of diabetes mellitus, maturity-onset diabetes of the young (MODY), is caused by a mutation in a gene encoding glucokinase—the beta cells cannot accurately assess when they need to produce insulin. MODY is treated with drugs or dietary modification.

The hormone **insulin** is also a protein, and its main effect is exactly opposite that of glucagon. Insulin prompts the liver to stimulate formation of glycogen from glucose and to inhibit conversion of noncarbohydrates into glucose. Insulin also has the special effect of promoting the facilitated diffusion (see chapter 3) of glucose through the membranes of cells bearing insulin receptors. These cells include those of the cardiac muscles, adipose tissues, and resting skeletal muscles (glucose uptake by exercising skeletal muscles is not dependent on insulin). Insulin action decreases the concentration of blood glucose, promotes transport of amino acids into cells, and increases protein synthesis. It also stimulates adipose cells to synthesize and store fat.

A negative feedback system sensitive to the concentration of blood glucose regulates insulin secretion. When glucose concentration is relatively high, as may occur following a meal, the beta cells release insulin. By promoting formation of glycogen in the liver and entrance of glucose into adipose and muscle cells, insulin helps prevent excessive rise in the blood glucose concentration (hyperglycemia). Then, when the glucose concentration falls, between meals or during the night, insulin secretion decreases (fig. 13.34).

As insulin concentration decreases, less and less glucose enters the adipose and muscle cells, and the glucose remaining in the blood is available for use by cells that lack insulin receptors, such as nerve cells. Neurons readily use a continuous supply of glucose for ATP production.

At the same time that insulin concentration is decreasing, glucagon secretion is increasing. Therefore, these hormones function together to maintain a relatively constant blood glucose concentration, despite great variations in the amounts of ingested carbohydrates.

Nerve cells, including those of the brain, obtain glucose by a facilitated-diffusion mechanism that is

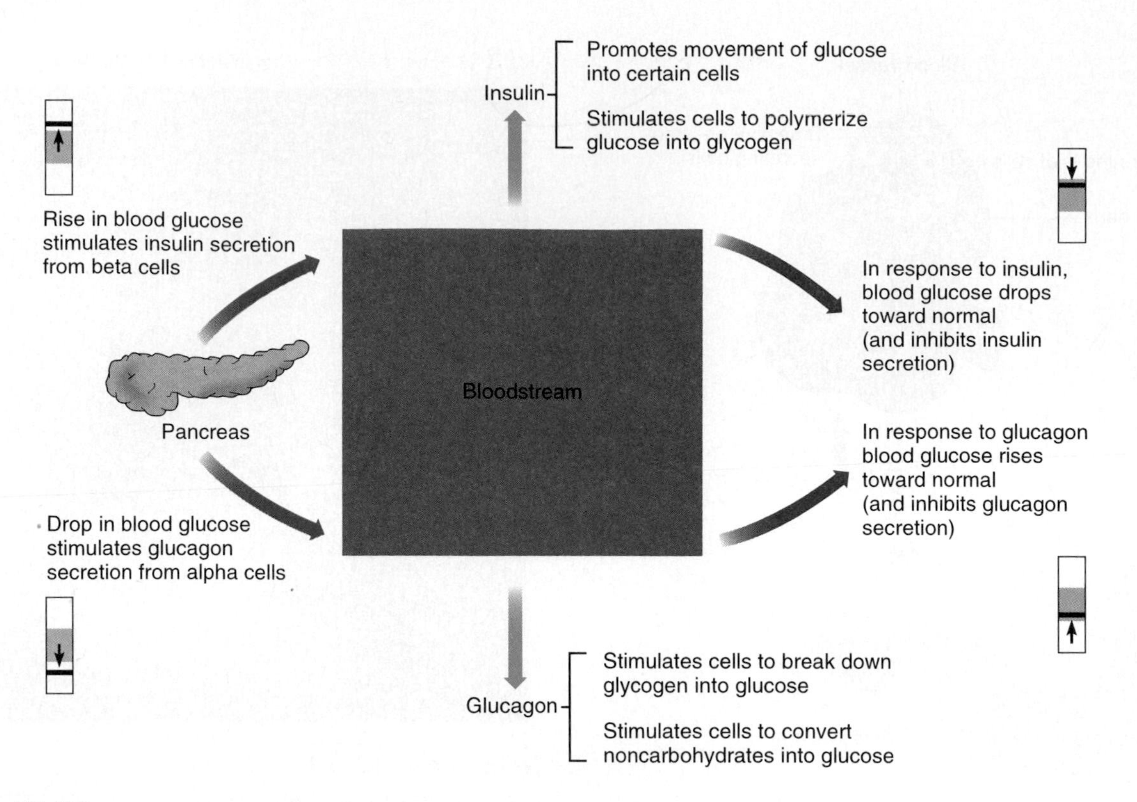


**FIGURE 13.34**
Insulin and glucagon function together to help maintain a relatively stable blood glucose concentration. Negative feedback responding to blood glucose concentration controls the levels of both hormones.

### CHART 13.11 HORMONES OF THE ISLETS OF LANGERHANS

| Hormone | Action | Source of Control |
|---|---|---|
| Glucagon | Stimulates the liver to break down glycogen and convert noncarbohydrates into glucose; stimulates breakdown of fats | Blood glucose concentration |
| Insulin | Promotes formation of glycogen from glucose, inhibits conversion of noncarbohydrates into glucose, and enhances movement of glucose through adipose and muscle cell membranes, decreasing blood glucose concentration; promotes transport of amino acids into cells; enhances synthesis of proteins and fats | Blood glucose concentration |
| Somatostatin | Helps regulate carbohydrates | |

not dependent on insulin. For this reason, nerve cells are particularly sensitive to changes in the blood glucose concentration, and conditions that cause such changes—excessive insulin secretion, for example—are likely to affect brain functions.

**Somatostatin** (similar to the hypothalamic hormone), which the delta cells release, helps regulate glucose metabolism by inhibiting secretion of glucagon and insulin.

Chart 13.11 summarizes the hormones of the islets of Langerhans, and Clinical Application 13.4 discusses diabetes mellitus, a derangement of the control of glucose metabolism.

1. What is the name of the endocrine portion of the pancreas?
2. What is the function of glucagon?

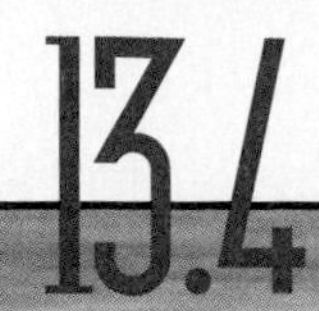

**Clinical Application 13.4**

## Diabetes Mellitus

Life for a person with *diabetes mellitus* means constant awareness of the illness—one or two insulin shots a day, frequent finger punctures to monitor blood glucose level, a restrictive diet, and concern over complications, which include loss of vision, leg ulcers, and kidney damage. The many symptoms of diabetes reflect disturbances in carbohydrate, protein, and fat metabolism.

*Diabetes* in Latin means increased urine output, and *mellitus* means honey, referring to urine's sugar content. Lack of insulin decreases movement of glucose into skeletal muscle and adipose cells, inhibiting glycogen formation. As a result, blood sugar concentration rises (hyperglycemia). When blood sugar reaches a certain level, the kidneys begin to excrete the excess, and glucose appears in the urine (glycosuria). The osmotic pressure of water follows the glucose by osmosis, causing excess urinary water loss. As a result, the person becomes dehydrated and very thirsty.

Diabetes mellitus decreases protein synthesis, causing tissues to waste away as glucose-starved cells use protein as an energy source. Weight falls and wounds cannot heal. Fatty acids accumulate in the blood as a result of decreased fat synthesis and storage. Ketone bodies, a by-product of fat metabolism, also increase in the blood. They are excreted in the urine as sodium salts, and large quantities of water follow by osmosis, intensifying dehydration and lowering sodium ion concentration in the blood. Accumulation of ketones and loss of sodium ions lead to metablic acidosis, a condition that lowers the pH of body fluids. Acidosis and dehydration adversely affect brain neurons. Without treatment (insulin replacement), the person becomes disoriented and may enter a diabetic coma and die.

There are two common forms of diabetes mellitus. Insulin-dependent diabetes mellitus (IDDM, also called type I, or juvenile onset diabetes) usually appears before age 20. About 15% of the 11 million Americans with diabetes mellitus have this form. It is an autoimmune disorder attacking pancreatic beta cells, ultimately destroying them so that insulin secretion halts. Treatment is administering enough insulin to control carbohydrate metabolism. Insulin cannot be taken in oral form because it is a peptide and digestive enzymes break it down.

The milder, other common form of diabetes, non-insulin-dependent diabetes mellitus (NIDDM), begins gradually, in people over 40. Usually in NIDDM cells lose insulin receptors, and therefore cannot respond to insulin, even if it is abundant. Heredity and a lifestyle of overeating and underexercising are risk factors for developing NIDDM. People who develop it are often overweight. Treatment for NIDDM includes careful control of diet to avoid foods that stimulate insulin production, exercising, and maintaining desirable body weight.

The glucose-tolerance test is used to diagnose diabetes mellitus. The patient ingests a known amount of glucose, and blood glucose concentration is measured at intervals to determine glucose utilization. If the person has diabetes, blood glucose concentration rises excessively and remains elevated for several hours. In a healthy person, glucose rise is less dramatic and the level returns to normal in about one and a half hours.

3. What is the function of insulin?
4. How are the secretions of glucagon and insulin controlled?
5. Why are nerve cells particularly sensitive to changes in blood glucose concentration?

## Other Endocrine Glands

Other organs that produce hormones and, therefore, are parts of the endocrine system include the pineal gland, thymus gland, reproductive glands, and certain glands of the digestive tract, heart, and kidneys.

Daniel R. was terrified by his symptoms. Several times a day, he would become incredibly weak and shaky and would break out into a drenching sweat. The attacks left him extremely anxious. A doctor diagnosed hypoglycemia, which, despite its notoriety some years ago, is very rare. Daniel's blood glucose level was low, indicating excess insulin in his bloodstream. Now that he knows the problem he can control his symptoms by following a diet of frequent, small meals that are low in carbohydrates and high in protein, which prevents surges of insulin that lower blood sugar level. Hypoglycemia is most often seen when a person with diabetes injects too much insulin, but it can also reflect a tumor of the insulin-producing cells of the pancreas, or may occur transiently following very strenuous exercise.

## Pineal Gland

The **pineal gland** is a small, oval structure located deep between the cerebral hemispheres, where it attaches to the upper portion of the thalamus near the roof of the third ventricle. It consists largely of specialized *pineal cells* and supportive *neuroglial cells* (see fig. 11.16*b*).

The pineal gland secretes a hormone, **melatonin,** which is synthesized from serotonin. Varying patterns of light and dark outside the body control the gland, with environmental information reaching the gland by means of nerve impulses. In the presence of light, nerve impulses originating in the retinas of the eyes travel to the hypothalamus. From the hypothalamus, they enter the reticular formation and then pass downward into the spinal cord. In the spinal cord, the impulses travel along sympathetic nerve fibers back into the brain, and finally they reach the pineal gland. In response to impulses that light triggers, melatonin secretion from the pineal gland decreases.

The role of the pineal gland in humans is not well understood, possibly because we alter natural light-dark cycles with artificial lighting. Some researchers hypothesize that mood swings are linked to abnormal melatonin secretion patterns, particularly a form of depression called seasonal affective disorder (SAD). Exposing such individuals to additional hours of daylight elevates their mood.

In the absence of light, nerve impulses from the eyes are decreased, and secretion of melatonin increases. Melatonin secretion is believed to be involved in the regulation of **circadian rhythms,** which are patterns of repeated activity associated with the environmental cycles of day and night. Circadian rhythms responding to light and dark include sleep/wake rhythms and seasonal cycles of fertility seen in many mammals.

Although the precise mechanism of melatonin secretion is poorly understood, the hormone seems to inhibit secretion of gonadotropins from the anterior pituitary gland and helps regulate the female reproductive cycle (menstrual cycle). It may also control onset of puberty (see chapter 22).

In 1992, physicians in Spain studied a 21-year-old man who had not yet matured sexually. He had low levels of male sex hormones but also had very high levels of melatonin. Within the next two years, his melatonin levels spontaneously declined to normal and he matured, even fathering a child. This report, although anecdotal, links melatonin production to control of puberty onset.

## Thymus Gland

The **thymus gland,** which lies in the mediastinum posterior to the sternum and between the lungs, is large in young children but diminishes in size with age. This gland secretes a group of hormones, called **thymosins,** that affect production of certain white blood cells (T lymphocytes). This gland plays an important role in immunity and is discussed in chapter 16.

## Reproductive Glands

The reproductive organs that secrete important hormones include the **ovaries,** which produce estrogens and progesterone; the **placenta,** which produces estrogens, progesterone, and a gonadotropin, and the **testes,** which produce testosterone. These glands and their secretions are discussed in chapter 22.

## Digestive Glands

The digestive glands that secrete hormones are generally associated with the linings of the stomach and small intestine. These structures and their secretions are described in chapter 17.

## Other Hormone-producing Organs

Other organs that produce hormones include the heart, which secretes *atrial natriuretic peptide* (chapter 15), and the kidneys, which secrete a red blood cell growth hormone called *erythropoietin* (chapter 14).

1. Where is the pineal gland located?
2. What seems to be the function of the pineal gland?
3. Where is the thymus gland located?

## Stress and Its Effects

Because survival depends upon maintaining homeostasis, factors that change the body's internal environment are potentially life threatening. When such dangers are sensed, nerve impulses are directed to the hypothalamus, triggering physiological responses that resist a loss of homeostasis. These responses often include increased activity in the sympathetic division of the autonomic nervous system and increased secretion of adrenal hormones. A factor capable of stimulating such a response is called a **stressor,** and the condition it produces in the body is called **stress.**

### Types of Stress

Stressors may be physical, psychological, or a combination of both.

*Physical stress* threatens tissues. This includes extreme heat or cold, decreased oxygen concentration, infections, injuries, prolonged heavy exercise, and loud sounds. Often physical stress is accompanied by unpleasant or painful sensations.

*Psychological stress* results from thoughts about real or imagined dangers, personal losses, unpleasant social interactions (or lack of social interactions), or any factors that threaten a person. Feelings of anger, fear, grief, anxiety, depression, and guilt cause psychological stress. In other instances, psychological stress may stem from pleasant stimuli, such as friendly social contact, feelings of joy or happiness, or sexual arousal. The factors that produce psychological stress vary greatly from person to person. A situation that is stressful to one person may not affect another, and what is stressful at one time may not be at another time.

### Responses to Stress

The hypothalamus controls response to stress, termed the *general stress* (or *general adaptation*) *syndrome.* This response, evoked to stress of any kind, maintains homeostasis.

> Inflammation is the immune system's generalized response to limit the effects of injury or infection. However, inflammation is painful and possibly destructive. The endocrine system keeps the immune system in check by increasing secretion by the pituitary and adrenal glands to temper inflammation. This is an example of how homeostasis operates between organ systems as well as within them.

Recall that the hypothalamus receives information from nearly all body parts, including visceral receptors, the cerebral cortex, the reticular formation, and limbic system. At times of stress, the hypothalamus responds to incoming impulses by activating the "fight or flight" response. More specifically, sympathetic impulses from the hypothalamus raise blood glucose concentration, the level of blood glycerol and fatty acids, heart rate, blood pressure, increase breathing rate, and dilate the air passages. The response also shunts blood from the skin and digestive organs into the skeletal muscles, and increases secretion of epinephrine from the adrenal medulla. The epinephrine, in turn, intensifies these sympathetic responses and prolongs their effects (fig. 13.35).

At the same time, the hypothalamus's release of corticotropin-releasing hormone (CRH) stimulates the anterior pituitary gland to secrete ACTH, which increases the adrenal cortex's secretion of cortisol. Cortisol supplies cells with amino acids and extra energy sources, and diverts glucose from skeletal muscles to brain tissue (fig. 13.36). Stress can also trigger release of glucagon from the pancreas, growth hormone (GH) from the anterior pituitary, and antidiuretic hormone (ADH) from the posterior pituitary gland. The secretion of renin from the kidney may also be stimulated.

Glucagon and growth hormone help mobilize energy sources, such as glucose, glycerol, and fatty acids, and stimulate cells to take up amino acids, facilitating repair of injured tissues. ADH stimulates the kidneys to retain water. This action decreases urine output and helps to maintain blood volume—particularly important if a person is bleeding excessively or sweating heavily. Renin, by increasing angiotensin II levels, helps stimulate the kidneys to retain sodium (through aldosterone), and through the vasoconstrictor action of angiotension II contributes to maintaining blood pressure. Chart 13.12 summarizes the body's reactions to stress.

> The 17-year-old was visiting her family physician for the third time. She had recurrent stomach pains and seemed to have a constant respiratory infection. Unlike her previous visits, this time the woman seemed noticeably upset. Suspecting that her physical symptoms stemmed from a struggle to deal with a stressful situation, the doctor looked for signs—increased heart and respiratory rate, elevated blood pressure, and excessive sweating. He took a blood sample to measure levels of epinephrine and cortisol, while prodding her to talk about whatever was bothering her.
>
> The young woman's cortisol was indeed elevated, which could account for her gastrointestinal pain and high blood pressure, as well as her impaired immunity. On the doctor's advice, she began seeing a psychologist. Her symptoms began to abate when she was able to discuss the source of her stress—a family member's illness and anxiety about beginning college.

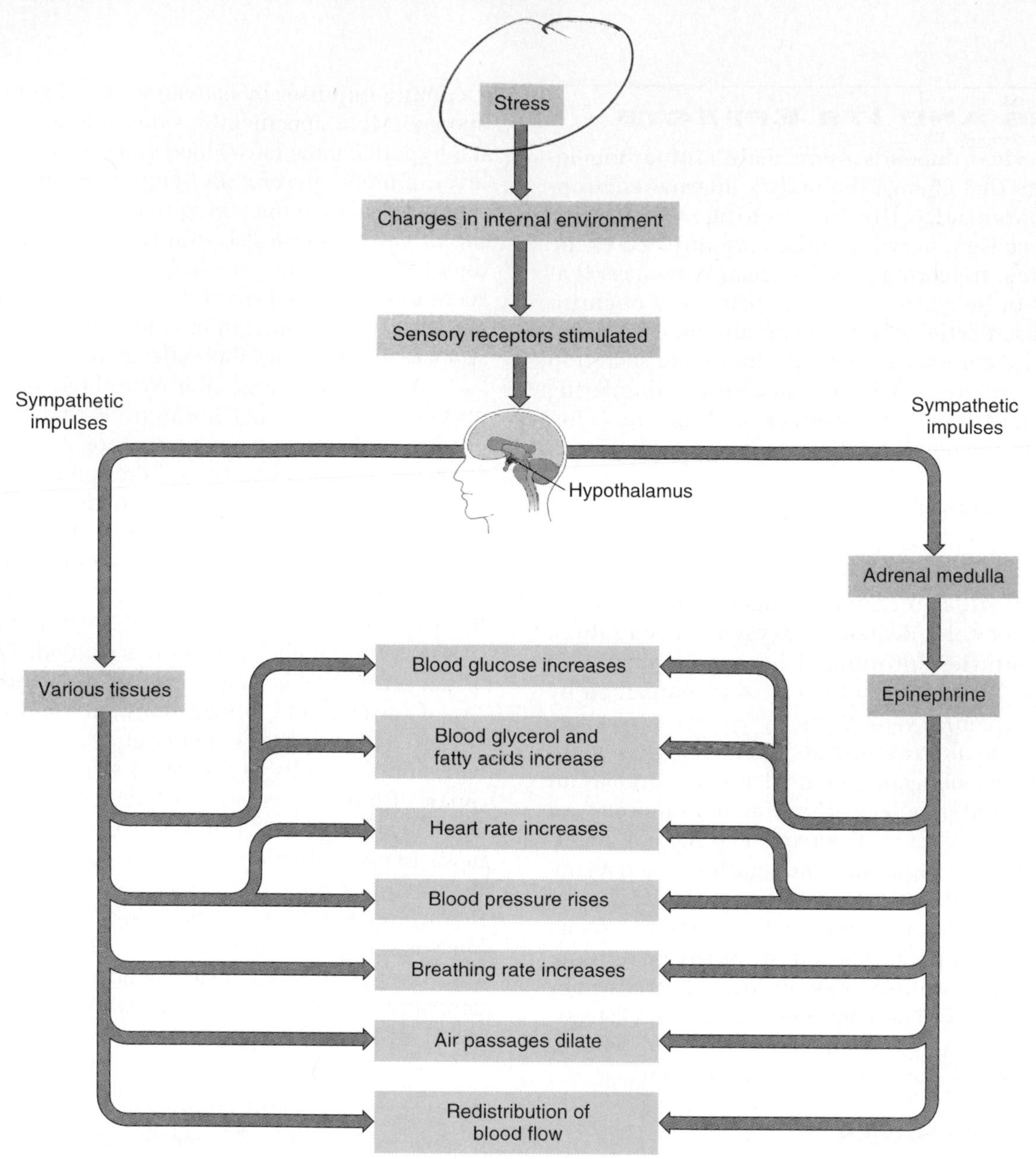


**FIGURE 13.35**

During stress, the hypothalamus helps prepare the body for fight or flight by triggering sympathetic impulses to various organs. It also stimulates release of epinephrine, intensifying the sympathetic responses.

1. What is stress?
2. Distinguish between physical stress and psychological stress.
3. Describe the general stress syndrome.

## C[illegible]INICAL TERMS RELATED [illegible]OCRINE SYSTEM

[illegible]″nah-lek′to-me) Surgical [illegible]renal glands.

**adrenogenital syndrome** (ah-dre″no-jen′ĭ-tal sin′drōm) A group of symptoms associated with changes in sexual characteristics as a result of increased secretion of adrenal androgens.

**exophthalmos** (ek″sof-thal′mos) An abnormal protrusion of the eyes.

**hirsutism** (her′sut-izm) Excessive growth of hair, especially in women.

**hypercalcemia** (hi″per-kal-se′me-ah) An excess of blood calcium.

**hyperglycemia** (hi″per-gli-se′me-ah) An excess of blood glucose.

**hypocalcemia** (hi″po-kal-se′me-ah) A deficiency of blood calcium.

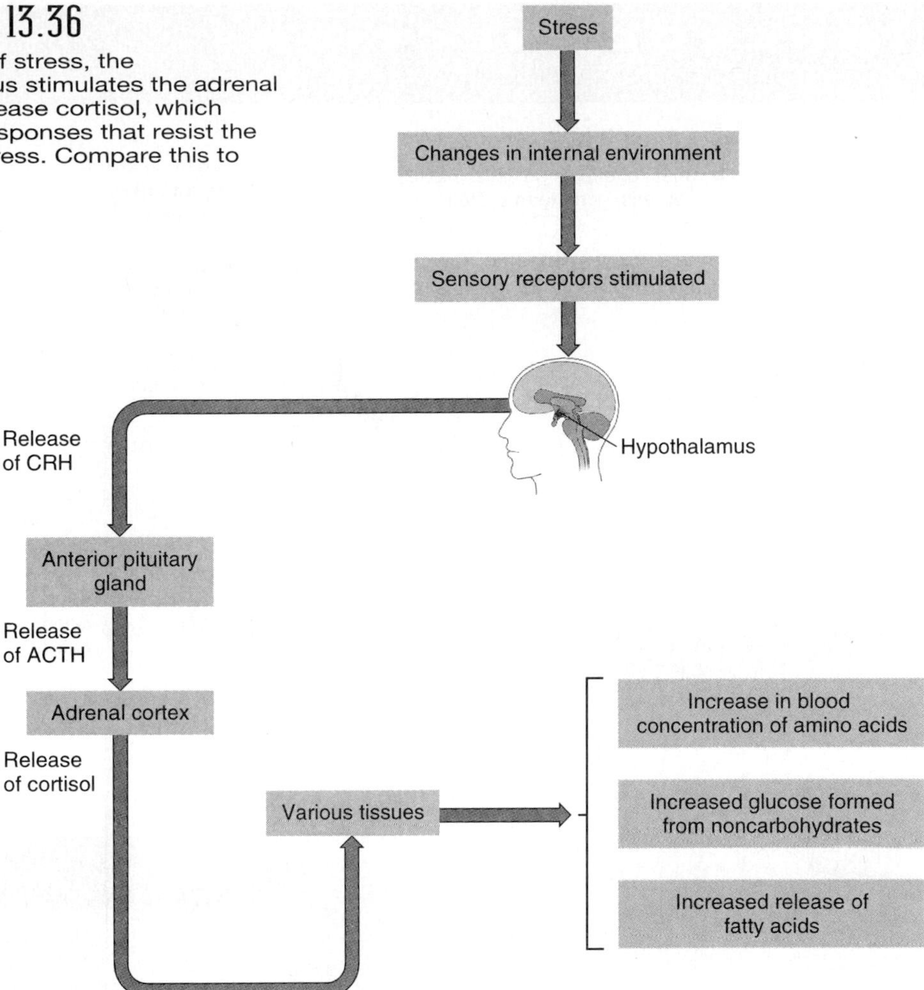


**FIGURE 13.36**
As a result of stress, the hypothalamus stimulates the adrenal cortex to release cortisol, which promotes responses that resist the effects of stress. Compare this to figure 13.31.

## CHART 13.12 Major Events in the General Stress Syndrome

1. As a result of stress, nerve impulses are transmitted to the hypothalamus.
2. Sympathetic impulses arising from the hypothalamus increase blood glucose concentration, blood glycerol concentration, blood fatty acid concentration, heart rate, blood pressure, and breathing rate. They dilate air passages, shunt blood into skeletal muscles, and increase secretion of epinephrine from the adrenal medulla.
3. Epinephrine intensifies and prolongs sympathetic actions.
4. The hypothalamus secretes CRH, which stimulates secretion of ACTH by the anterior pituitary gland.
5. ACTH stimulates release of cortisol by the adrenal cortex.
6. Cortisol increases the concentration of blood amino acids, releases fatty acids, and forms glucose from noncarbohydrate sources.
7. Secretion of glucagon from the pancreas and growth hormone from the anterior pituitary increase.
8. Glucagon and growth hormone aid mobilization of energy sources and stimulate uptake of amino acids by cells.
9. Secretion of ADH from the posterior pituitary increases.
10. ADH promotes the retention of water by the kidneys, which increases blood volume.
11. Renin increases blood levels of angiotensin II, which acts as a vasoconstrictor and also stimulates aldosterone secretion by the adrenal cortex.
12. Aldosterone stimulates sodium retention by the kidneys.

**hypophysectomy** (hi-pof″ĭ-sek′to-me) Surgical removal of the pituitary gland.
**parathyroidectomy** (par″ah-thi″roi-dek′to-me) Surgical removal of the parathyroid glands.
**pheochromocytoma** (fe-o-kro″mo-si-to′mah) A type of tumor found in the adrenal medulla and usually accompanied by high blood pressure.
**polyphagia** (pol″e-fa′je-ah) Excessive eating.
**thymectomy** (thi-mek′to-me) Surgical removal of the thymus gland.
**thyroidectomy** (thi″roi-dek′to-me) Surgical removal of the thyroid gland.
**thyroiditis** (thi″roi-di′tis) Inflammation of the thyroid gland.

# InnerConnections
## Endocrine System

Glands secrete hormones that have a variety of effects on cells, tissues, organs, and organ systems

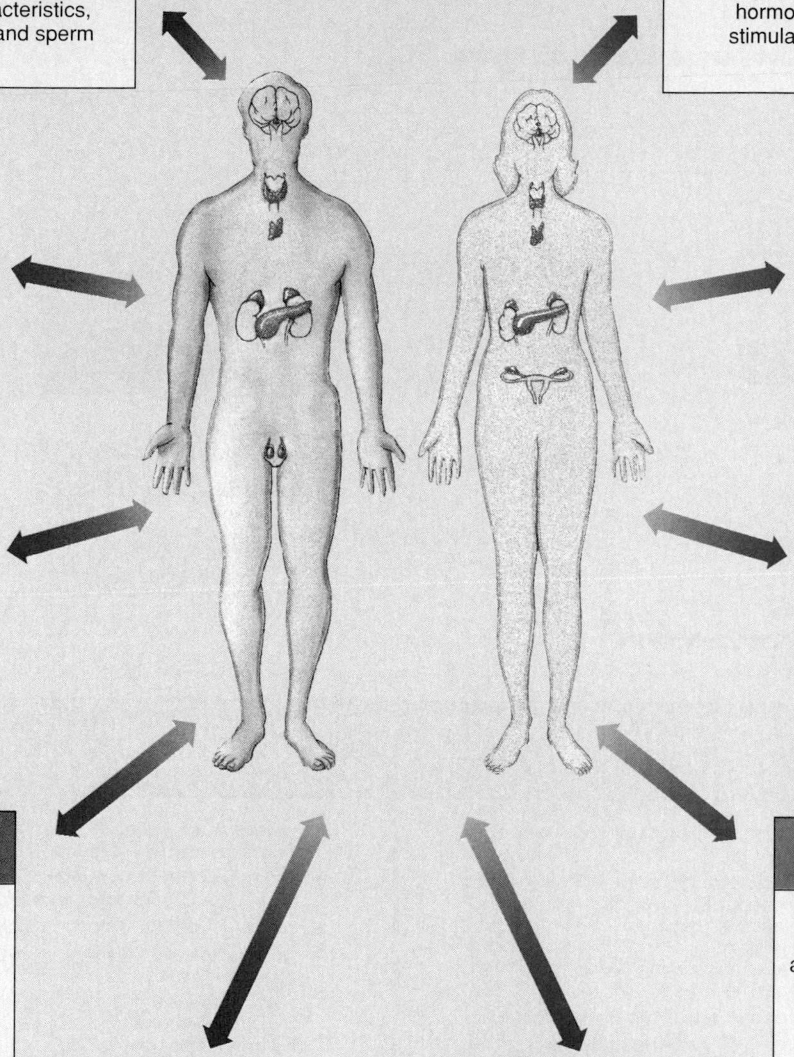

**Reproductive system**
Sex hormones play a major role in development of secondary sex characteristics, egg, and sperm

**Integumentary system**
Melanocytes produce skin pigment in response to hormonal stimulation

**Urinary system**
Hormones act on the kidneys to help control water and electrolyte balance

**Skeletal system**
Hormones act on bones to control calcium balance

**Respiratory system**
Decreased oxygen causes hormonal stimulation of red blood cell production; red blood cells transport oxygen and carbon dioxide

**Muscular system**
Hormones help increase blood flow to exercising muscles

**Digestive system**
Hormones play a role in controlling digestive secretions and motility

**Nervous system**
Neurons control the secretions of the anterior and posterior pituitary glands and the adrenal medulla

**Lymphatic system**
Hormones stimulate lymphocyte production

**Cardiovascular system**
Hormones are carried in the bloodstream; some have direct actions on the heart and blood vessels

## Chapter Summary

### Introduction (page 489)

Endocrine glands secrete their products into body fluids (the internal environment); exocrine glands secrete their products into ducts that lead to the outside of the body.

### General Characteristics of the Endocrine System (page 489)

As a group, endocrine glands regulate metabolic processes.

### Hormone Action (page 489)

Endocrine glands secrete hormones that affect target cells possessing specific receptors.

1. Chemistry of hormones
   a. Each kind of hormone has a special molecular structure and is very potent.
   b. Chemically, hormones are steroids, amines, peptides, proteins, or glycoproteins.
2. Actions of Hormones
   a. Steroid Hormones
      (1) Steroid hormones enter target cells and combine with receptors to form complexes.
      (2) These complexes activate specific genes in the nucleus, which direct synthesis of specific proteins.
      (3) The degree of cellular response is proportional to the number of hormone-receptor complexes formed.
   b. Nonsteroid hormones
      (1) Nonsteroid hormones combine with receptors in the target-cell membrane.
      (2) A hormone-receptor complex stimulates membrane proteins, such as adenylate cyclase, to induce the formation of second messenger molecules.
      (3) A second messenger, such as cAMP, activates protein kinases.
      (4) Protein kinases activate certain protein substrate molecules, which, in turn, change cellular processes.
      (5) The cellular response to a nonsteroid hormone is amplified because the enzymes induced by a small number of hormone-receptor complexes can catalyze formation of a large number of second messenger molecules.
3. Prostaglandins
   a. Prostaglandins are paracrine substances present in small quantities that have powerful hormonelike effects.
   b. Prostaglandins modulate hormones that regulate formation of cyclic AMP.

### Control of Hormonal Secretions (page 497)

The concentration of each hormone in the body fluids is regulated precisely.

1. Negative feedback systems
   a. In a negative feedback system, a gland is sensitive to the concentration of a substance it regulates.
   b. When the concentration of the regulated substance reaches a certain concentration, it inhibits the gland.
   c. As the gland secretes less hormone, the controlled substance also decreases.
2. Control sources
   a. Some endocrine glands secrete in response to nerve impulses.
   b. Other glands secrete hormones in response to releasing hormones the hypothalamus secretes.

### Pituitary Gland (page 498)

The pituitary gland, which is attached to the base of the brain, has an anterior lobe and a posterior lobe. Releasing hormones from the hypothalamus control most pituitary secretions.

1. Anterior pituitary hormones
   a. The anterior pituitary consists largely of epithelial cells, and it secretes GH, PRL, TSH, ACTH, FSH, and LH.
   b. Growth hormone (GH)
      (1) Growth hormone stimulates body cells to grow and reproduce.
      (2) Growth hormone-releasing hormone and somatostatin from the hypothalamus control GH secretion.
   c. Prolactin (PRL)
      (1) PRL promotes breast development and stimulates milk production.
      (2) In males, prolactin decreases secretion of LH (ICSH).
      (3) Prolactin release-inhibiting hormone from the hypothalamus restrains secretion of prolactin, while prolactin-releasing hormone promotes its secretion.
   d. Thyroid-stimulating hormone (TSH)
      (1) TSH controls secretion of hormones from the thyroid gland.
      (2) The hypothalamus, by secreting thyrotropin-releasing hormone, regulates TSH secretion.
   e. Adrenocorticotropic hormone (ACTH)
      (1) ACTH controls the secretion of certain hormones from the adrenal cortex.
      (2) The hypothalamus, by secreting corticotropin-releasing hormone, regulates ACTH secretion.
   f. Follicle-stimulating hormone (FSH) and luteinizing hormone (LH) are gonadotropins that influence the reproductive organs.
2. Posterior pituitary hormones
   a. The posterior lobe of the pituitary gland consists largely of neuroglial cells and nerve fibers that originate in the hypothalamus.
   b. The two hormones of the posterior pituitary are produced in the hypothalamus.
   c. Antidiuretic hormone (ADH)
      (1) ADH causes the kidneys to reduce the amount of water they excrete.
      (2) In high concentration, ADH constricts blood vessel walls, raising blood pressure.
      (3) The hypothalamus regulates ADH secretion.

d. Oxytocin (OT)
   (1) Oxytocin has an antidiuretic effect and can contract muscles in the uterine wall.
   (2) OT also contracts certain cells associated with production and ejection of milk from the milk glands of the breasts.

## Thyroid Gland (page 505)

The thyroid gland is located in the neck and consists of two lateral lobes.

1. Structure of the gland
   a. The thyroid gland consists of many hollow secretory parts called follicles.
   b. The follicles are fluid filled and store the hormones the follicle cells secrete.
   c. Extrafollicular cells secrete calcitonin.
2. Thyroid hormones
   a. Thyroxine and triiodothyronine
      (1) These hormones increase the rate of metabolism, enhance protein synthesis, and stimulate lipid breakdown.
      (2) These hormones are needed for normal growth and development, and for maturation of the nervous system.
   b. Calcitonin
      (1) Calcitonin lowers blood calcium and phosphate ion concentrations.
      (2) This hormone prevents prolonged elevation of calcium after a meal.

## Parathyroid Glands (page 507)

The parathyroid glands are located on the posterior surface of the thyroid.

1. Structure of the glands
   a. Each gland is small and yellow-brown, within a thin connective tissue capsule.
   b. Each gland consists of secretory cells that are well supplied with capillaries.
2. Parathyroid hormone (PTH)
   a. PTH increases blood calcium ion concentration and decreases blood phosphate ion concentration.
   b. PTH stimulates resorption of bone tissue, causes the kidneys to conserve calcium ions and excrete phosphate ions, and indirectly stimulates absorption of calcium ions from the intestine.
   c. A negative feedback mechanism operating between the parathyroid glands and the blood regulates these glands.

## Adrenal Glands (page 510)

The adrenal glands are located atop the kidneys.

1. Structure of the glands
   a. Each adrenal gland consists of a medulla and a cortex.
   b. The adrenal medulla and adrenal cortex are distinct glands that secrete different hormones.
2. Hormones of the adrenal medulla
   a. The adrenal medulla secretes epinephrine and norepinephrine.
   b. These hormones are synthesized from tyrosine and are closely related to each other chemically.
   c. These hormones produce effects similar to those of the sympathetic nervous system.
   d. Sympathetic impulses originating from the hypothalamus stimulate secretion of these hormones.
3. Hormones of the adrenal cortex
   a. The cortex produces several types of steroids that include hormones.
   b. Aldosterone
      (1) It causes the kidneys to conserve sodium ions and water, and to excrete potassium ions.
      (2) It is secreted in response to increased potassium ion concentration, or presence of angiotensin II.
      (3) By conserving sodium ions and water, it helps maintain blood volume and pressure.
   c. Cortisol
      (1) It inhibits protein synthesis, releases fatty acids, and stimulates glucose formation from noncarbohydrates.
      (2) A negative feedback mechanism involving secretion of CRH from the hypothalamus and ACTH from the anterior pituitary gland controls its level.
   d. Adrenal sex hormones
      (1) These hormones are of the male type although some can be converted into female hormones.
      (2) They supplement the sex hormones produced by the gonads.

## Pancreas (page 515)

The pancreas secretes digestive juices as well as hormones.

1. Structure of the gland
   a. The pancreas is posterior to the stomach and is attached to the small intestine.
   b. The endocrine portion, which is called the islets of Langerhans, secretes glucagon, insulin, and somatostatin.
2. Hormones of the islets of Langerhans
   a. Glucagon stimulates the liver to produce glucose, increasing concentration of blood glucose. It also breaks down fat.
   b. Insulin moves glucose through cell membranes, stimulates its storage, promotes protein synthesis, and stimulates fat storage.
   c. Nerve cells lack insulin receptors and depend upon diffusion for a glucose supply.
   d. Somatostatin inhibits insulin and glucagon release.

## Other Endocrine Glands (page 519)

1. Pineal gland
   a. The pineal gland is attached to the thalamus near the roof of the third ventricle.
   b. Postganglionic sympathetic nerve fibers innervate it.
   c. It secretes melatonin, which inhibits secretion of gonadotropins from the anterior pituitary gland.
   d. It may help regulate the female reproductive cycle.
2. Thymus gland
   a. The thymus gland lies posterior to the sternum and between the lungs.
   b. Its size diminishes with age.
   c. It secretes thymosin, which affects the production of certain lymphocytes that, in turn, provide immunity.

3. Reproductive glands
   a. The ovaries secrete estrogens and progesterone.
   b. The placenta secretes estrogens, progesterone, and a gonadotropin.
   c. The testes secrete testosterone.
4. The digestive glands include certain glands of the stomach and small intestine that secrete hormones.
5. Other hormone-producing organs include the heart and kidneys.

## Stress and Its Effects (page 521)

Stress occurs when the body responds to stressors that threaten the maintenance of homeostasis. Stress responses include increased activity of the sympathetic nervous system and increased secretion of adrenal hormones.

1. Types of stress
   a. Physical stress results from environmental factors that are harmful or potentially harmful to tissues.
   b. Psychological stress results from thoughts about real or imagined dangers.
   c. Factors that produce psychological stress vary with the individual and the situation.
2. Responses to stress
   a. Responses to stress maintain homeostasis.
   b. The hypothalamus controls a general stress syndrome.

**Explorations in Human Anatomy and Physiology CD-ROM**
The module accompanying Chapter Thirteen is #11 Hormone Action.

## Critical Thinking Questions

1. Based on your understanding of the actions of glucagon and insulin, would a person with diabetes mellitus be likely to require more insulin or more sugar following strenuous exercise? Why?
2. What problems might result from the prolonged administration of cortisol to a person with a severe inflammatory disease?
3. How might the environment of a patient with hyperthyroidism be modified to minimize the drain on body energy resources?
4. What hormones should be administered to an adult whose anterior pituitary gland has been removed? Why?
5. A patient who has lost a relatively large volume of blood will excessively secrete aldosterone from the adrenal cortex. What effect will this increased secretion have on the patient's blood concentrations of sodium and potassium ions?
6. Both growth hormone and growth hormone-releasing hormone have been successfully used to promote growth in children with short statures. How would you explain the difference in the ways these hormones produce their effects?

## Review Exercises

1. What is an *endocrine gland?*
2. Define *hormone* and *target cell.*
3. Explain how hormones can be grouped on the basis of their chemical composition.
4. Explain how steroid hormones influence cells.
5. Distinguish between the binding site and the activity site of a receptor molecule.
6. Explain how nonsteroid hormones may function through the formation of cAMP.
7. Explain how nonsteroid hormones may function through an increase in intracellular calcium ion concentration.
8. Explain how the cellular response to a hormone operating through a second messenger is amplified.
9. Define *prostaglandins,* and explain their general function.
10. Describe a negative feedback system.
11. Define *releasing hormone,* and provide an example of one.
12. Describe the location and structure of the pituitary gland.
13. List the hormones the anterior pituitary gland secretes.
14. Explain how the brain controls pituitary gland activity.
15. Explain how growth hormone produces its effects.
16. List the major factors that affect growth hormone secretion.
17. Summarize the functions of prolactin.
18. Describe regulation of concentrations of circulating thyroid hormones.
19. Explain the control of secretion of ACTH.
20. List the major gonadotropins, and explain the general functions of each.
21. Compare the cellular structures of the anterior and posterior lobes of the pituitary gland.
22. Name the hormones associated with the posterior pituitary, and explain their functions.
23. Explain how the release of ADH is regulated.
24. Describe the location and structure of the thyroid gland.

25. Name the hormones the thyroid gland secretes, and list the general functions of each.
26. Define *iodine pump*.
27. Describe the location and structure of the parathyroid glands.
28. Explain the general functions of parathyroid hormone.
29. Describe mechanisms that regulate the secretion of parathyroid hormone.
30. Distinguish between the adrenal medulla and the adrenal cortex.
31. List the hormones produced by the adrenal medulla, and describe their general functions.
32. List the steps in the synthesis of adrenal medullary hormones.
33. Name the most important hormones of the adrenal cortex, and describe the general functions of each.
34. Describe the regulation of the secretion of aldosterone.
35. Describe how the secretion of cortisol is regulated.
36. Describe the location and structure of the pancreas.
37. List the hormones the islets of Langerhans secrete, and describe the general functions of each.
38. Summarize how the secretion of hormones from the pancreas is regulated.
39. Describe the location and general function of the pineal gland.
40. Describe the location and general function of the thymus gland.
41. Distinguish between a stressor and stress.
42. List several factors that cause physical and psychological stress.
43. Describe the general stress syndrome.

CHAPTER 14

# BLOOD

## CHAPTER OBJECTIVES

AFTER YOU HAVE STUDIED THIS CHAPTER, YOU SHOULD BE ABLE TO:

1. Describe the general characteristics of the blood and discuss its major functions.
2. Distinguish between the various types of blood cells.
3. Explain how blood cell counts are made and how they are used.
4. Discuss the life cycle of a red blood cell.
5. Explain how red blood cell production is controlled.
6. List the major components of blood plasma and describe the functions of each.
7. Define *hemostasis,* and explain the mechanisms that help to achieve it.
8. Review the major steps in blood coagulation.
9. Explain how coagulation can be prevented.
10. Explain the basis for blood typing.
11. Describe how blood reactions may occur between the fetal and maternal tissues.

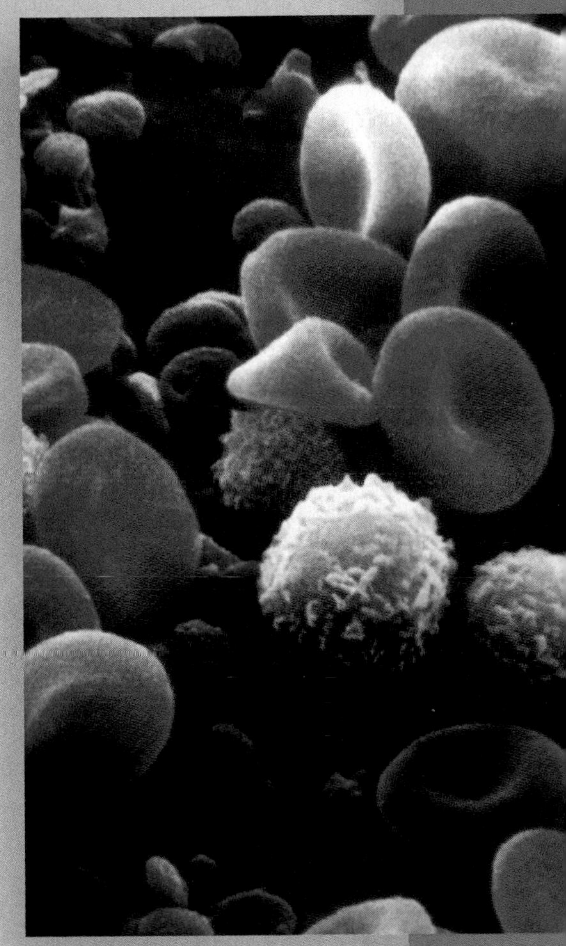

BLOOD IS A COMPLEX MIXTURE OF FORMED ELEMENTS (RED BLOOD CELLS, WHITE BLOOD CELLS, AND PLATELETS) WITHIN A WATERY MATRIX CALLED PLASMA (1,000×).

## KEY TERMS

**albumin** (al-bu′min)
**antibody** (an′ti-bod″e)
**antigen** (an′-ti-jen)
**basophil** (ba′-so-fil)
**coagulation** (ko-ag″u-la′-shun)
**embolus** (em′-bo-lus)
**eosinophil** (e″o-sin′o-fil)
**erythrocyte** (ĕ-rith′ro-sīt)
**erythropoietin** (e-rith″ro-poi′ĕ-tin)
**fibrinogen** (fi-brin′o-jen)
**globulin** (glob′u-lin)
**hemostasis** (he″mo-sta′sis)
**leukocyte** (lu′ko-sīt)
**lymphocyte** (lim′fo-sīt)
**macrophage** (mak′ro-fāj)
**monocyte** (mon′o-sīt)
**neutrophil** (nu′tro-fil)
**plasma** (plaz′mah)
**platelet** (plāt′let)
**thrombus** (throm′bus)

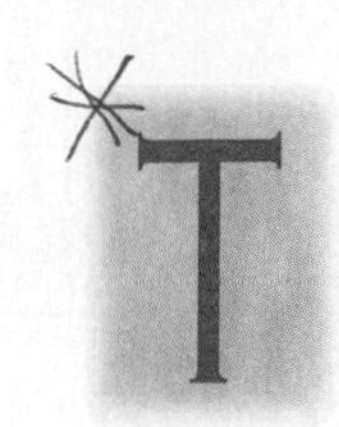

They are known in medical journals simply as "the erythrocytosis family," some 102 individuals in Denmark descended from a couple born in the mid-1850s, one of whom had a most unusual gene. Over the years, thirty family members have inherited the gene. But unlike most inherited conditions, this one confers superb good health. For in erythrocytosis, precursor red blood cells are extra sensitive to a hormone, *erythropoietin* (EPO), that stimulates their maturation. As a result, these individuals produce about 25% more red blood cells than normal. Their tissues receive more oxygen, powering muscles and providing great physical endurance.

Members of the erythrocytosis family have no symptoms whatsoever. But researchers are wondering if this inborn ability to withstand vigorous exercise explains the successful athletic career of one particular relative—a man, now in his mid-fifties, who has won three Olympic gold medals and two world championships in cross-country skiing. Sports medicine physicians now fear that EPO may become a drug of abuse among some athletes, with a normal response to EPO, seeking ways to build endurance.

Having too little EPO is a far more serious medical matter. Very small premature babies often have severe anemia (too few red blood cells) because they have too little EPO. Fortunately, human EPO manufactured in bacteria helps these infants, as it does older people with anemia from kidney failure.

The different effects of excess or deficient EPO illustrate the importance of balance among the components of that most complex mixture, blood.

Blood is often considered a type of connective tissue whose cells are suspended in a liquid intercellular material. It is vital in transporting substances between the body cells and the external environment, thereby aiding in maintaining a stable cellular environment.

## Blood and Blood Cells

Whole blood is slightly heavier and three to four times more viscous than water. Its cells, which are formed mostly in red bone marrow, include red blood cells and white blood cells. The blood also contains cellular fragments called blood platelets (fig. 14.1).

### Blood Volume and Composition

Blood volume varies with body size, changes in fluid and electrolyte concentrations, and the amount of adipose tissue. However, an average-sized adult has a blood volume of about 5 liters.

If a blood sample is allowed to stand in a tube for a while and prevented from clotting, the cells separate from the liquid portion of the blood and settle to the bottom. Centrifuging the sample packs the cells quickly into the lower part of the centrifuge tube, as figure 14.2 shows. The percentage of cells and liquid in the blood sample can then be calculated.

A blood sample is usually about 45% cells. This percentage is called the **hematocrit** (HCT), or **packed cell volume** (PCV). The remaining 55% of a blood sample is clear, straw-colored **plasma.** Appendix C lists values for the hematocrit and other common blood tests in healthy individuals.

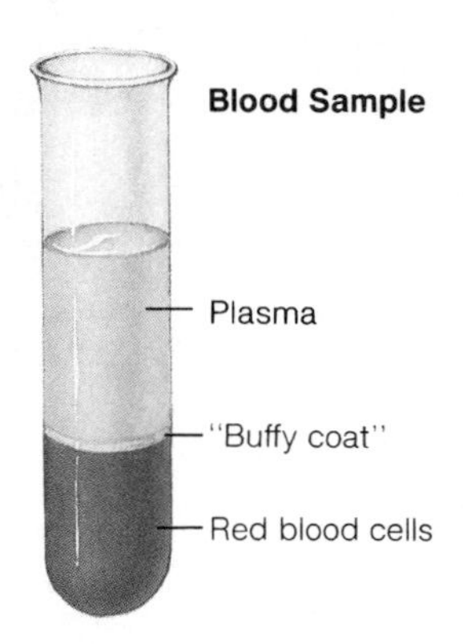


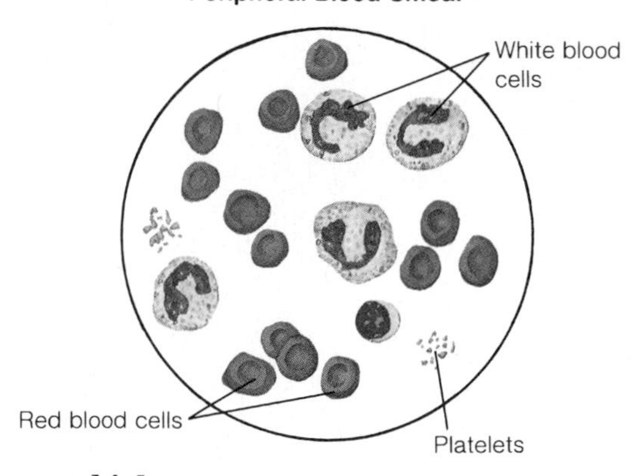


**Figure 14.1**
Blood consists of a liquid portion called plasma and a solid portion that includes red blood cells, white blood cells, and platelets. When blood components are separated, the white blood cells and platelets form a thin layer, called the "buffy coat," between the plasma and the red blood cells.

In addition to red blood cells, which comprise over 99% of the blood cells, the formed elements of the blood include white blood cells and blood platelets. The plasma is composed of a complex mixture that includes water, amino acids, proteins, carbohydrates, lipids, vitamins, hormones, electrolytes, and cellular wastes.

## The Origin of Blood Cells

Blood cells originate in bone marrow from **hemocytoblasts,** a type of stem cell (fig. 14.3). A stem cell can differentiate into any of a number of specialized cell types. As hemocytoblasts divide, the new cells respond to different secreted growth factors, called **colony stimulating factors,** that turn on some genes and turn off others, ultimately sculpting the distinctive formed elements of blood, including the cellular components of the immune system. A protein called *thrombopoietin* (TPO) stimulates large cells called megakaryocytes to proliferate, much in the same way that erythropoietin (EPO) stimulates red blood cell proliferation.

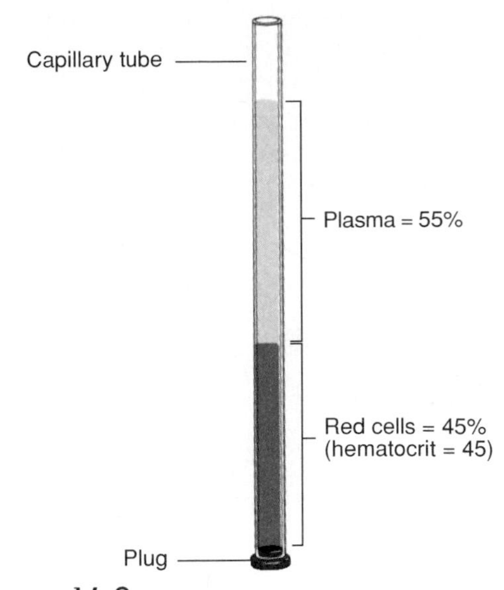


**FIGURE 14.2**

If a blood-filled capillary tube is centrifuged, the red cells pack in the lower portion, and the percentage of red cells (hematocrit) can be determined.

> Because stem cells can produce blood cells, they are a very valuable medical resource. The challenge is in obtaining them, because they comprise less than a tenth of a percent of the billions of cells in the bone marrow. The rarity of stem cells is why a bone marrow transplant delivers some 40 billion cells, in the hope of giving a patient with a life-threatening blood disorder stem cells to rebuild his or her supplies.
>
> Because bone marrow transplants are very dangerous, researchers are developing new ways to harness stem cells. A colony stimulating factor can be given as a drug. It sends stem cells into a person's bloodstream, where they can be extracted using a device called a cell sorter. In the future, bone marrow transplants will not be necessary, because we will be able to store umbilical cord blood, which is an especially rich source of these valuable cells.

1. What are the major components of blood?
2. What factors affect blood volume?
3. How is hematocrit determined?
4. How do blood cells form?

## Characteristics of Red Blood Cells

**Red blood cells,** or **erythrocytes,** are tiny, biconcave disks that are thin near their centers and thicker around their rims (fig. 14.4). This special shape is an adaptation for the red cell's function of transporting gases that provides an increased surface area through which gases can diffuse. The shape also places the cell membrane closer to oxygen-carrying *hemoglobin* within the cell. Because of its shape, a red blood cell can also deform readily as it squeezes through the narrow passages of capillaries.

Each red blood cell is about one-third hemoglobin by volume. This protein is responsible for the color of the blood. The rest of the cell consists mainly of membrane, water, electrolytes, and enzymes. When the hemoglobin combines with oxygen, the resulting *oxyhemoglobin* is bright red, and when the oxygen is released, the resulting *deoxyhemoglobin* is darker. Blood rich in deoxyhemoglobin may appear bluish when it is viewed through blood vessel walls.

> A person experiencing prolonged oxygen deficiency (hypoxia) may become *cyanotic.* The skin and mucous membranes appear bluish due to an abnormally high concentration of deoxyhemoglobin in the blood. Cyanosis may also occur as a result of exposure to low temperature. In this case, the superficial blood vessels constrict, blood flow slows, and more oxygen than usual is removed from the blood flowing through the vessels.

Red blood cells have nuclei during their early stages of development, but extrude them as the cells mature, providing more space for hemoglobin. Since they lack nuclei, red blood cells cannot synthesize protein or reproduce. Because they also lack mitochondria, red blood cells produce ATP through glycolysis only, and use none of the oxygen they carry.

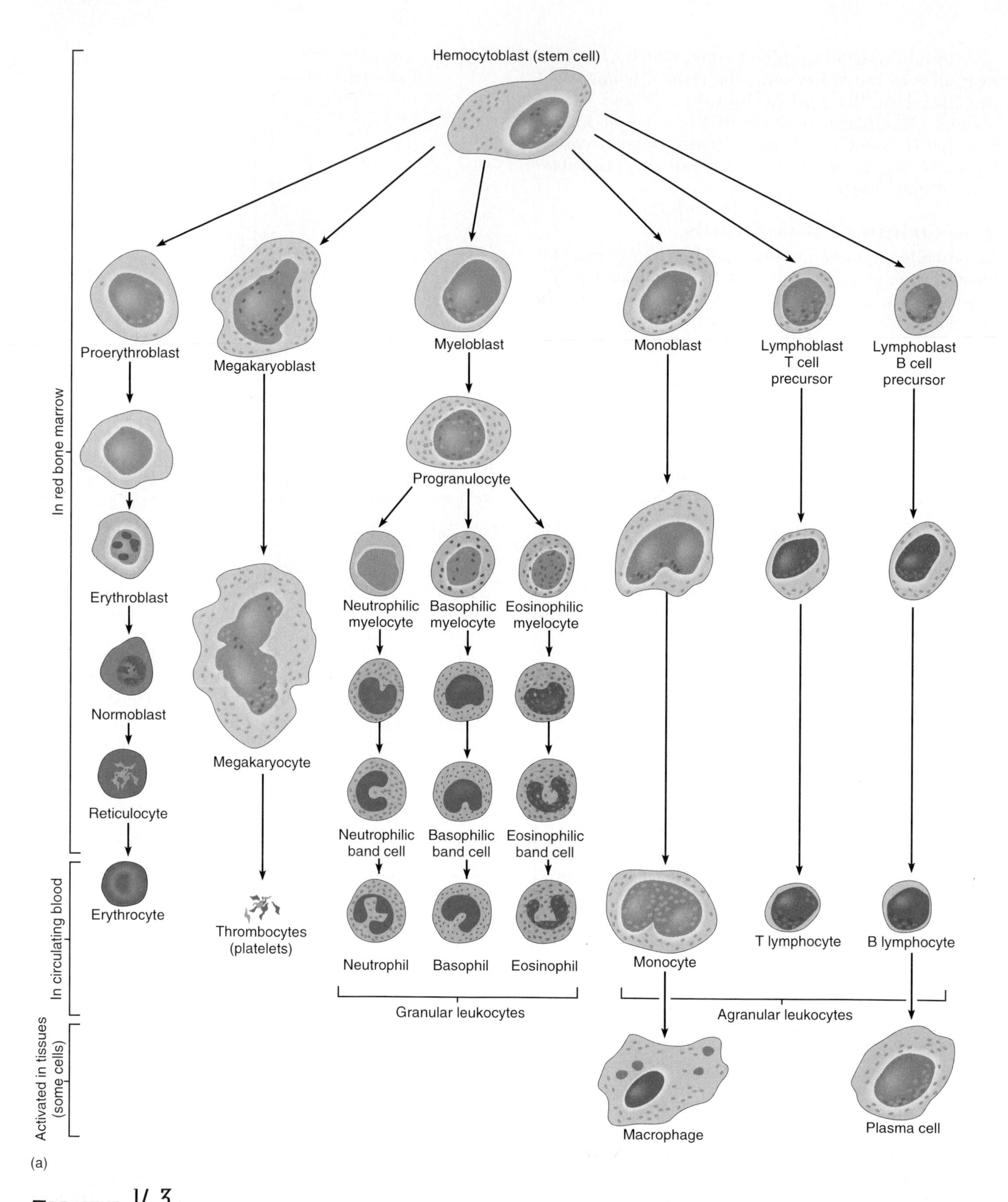


**FIGURE 14.3**

(*a*) Origin and development of blood cells from a hemocytoblast (stem cell) in bone marrow. (*b*) (next page) Light micrograph of a hemocytoblast in red bone marrow (500×).

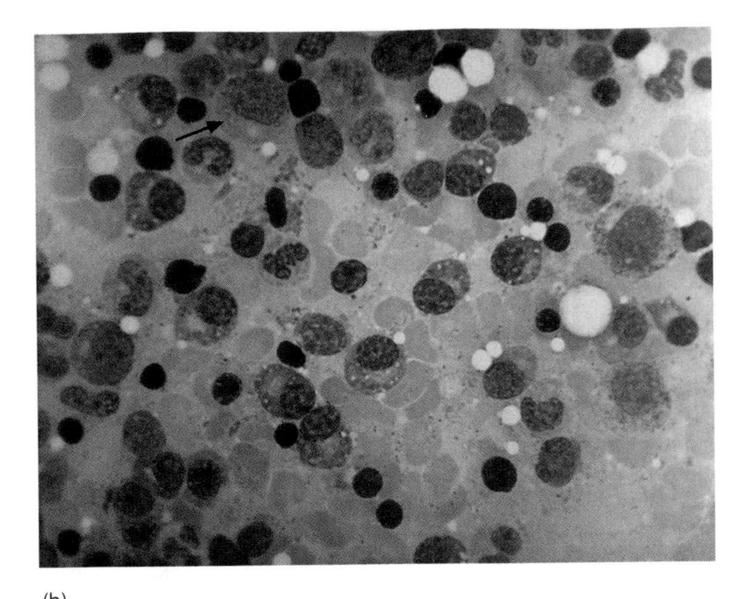

(b)

**FIGURE 14.3** *Continued*

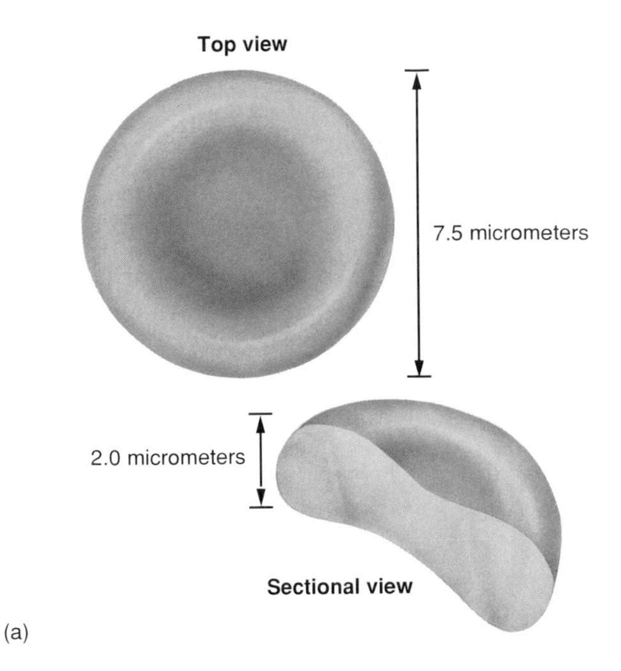


(a)

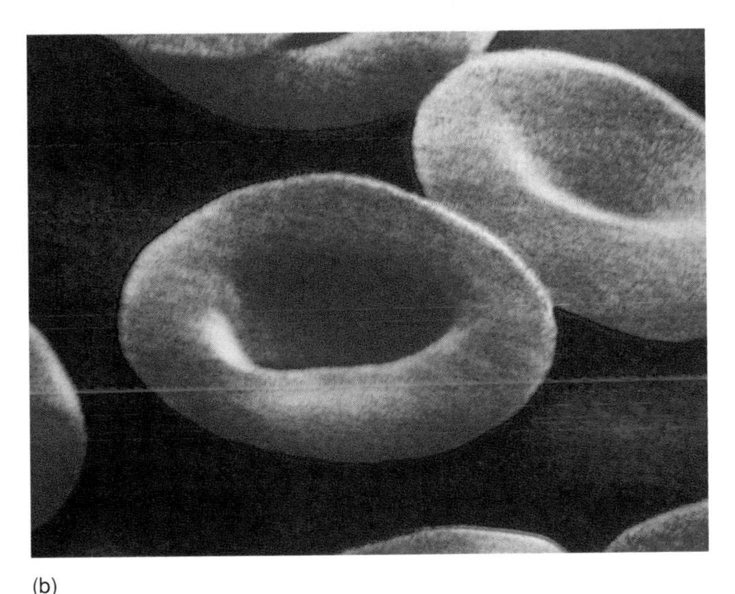

(b)

**FIGURE 14.4**

(*a*) Note the biconcave shape of a red blood cell.
(*b*) Scanning electron micrograph of human red blood cells (falsely colored) (5,600×).

As long as cytoplasmic enzymes remain functional, these can carry on vital energy-releasing processes. With time, however, red blood cells become less and less active.

> In *sickle cell disease,* a single DNA base change causes hemoglobin to crystallize in a low-oxygen environment. This bends the red blood cells containing the hemoglobin into a sickle shape, which blocks circulation in small vessels, causing excruciating joint pain and damaging many organs. As the spleen works overtime to recycle the short-lived red blood cells, infection becomes likely.
>
> Most children with sickle cell disease are diagnosed at birth and are given antibiotics to prevent infection. Hospitalization for blood transfusions may be necessary if the person experiences painful sickling "crises" of blocked circulation. An experimental approach to treating the disorder uses drugs to activate genes that normally produce a slightly different form of hemoglobin in the fetus. Unlike the mature, sickle hemoglobin, the fetal hemoglobin is functional. A bone marrow transplant can completely cure sickle cell disease, but it has a mortality rate of 15%.

1. Describe a red blood cell.
2. How is the biconcave shape of a red blood cell related to its function?
3. What is the function of hemoglobin?
4. What changes occur in a red blood cell as it matures?

## Red Blood Cell Counts

The number of red blood cells in a cubic millimeter ($mm^3$) of blood is called the *red blood cell count* (RBCC or RCC). Although this number varies from time to time even in healthy individuals, the typical range for adult males is 4,600,000–6,200,000 cells per $mm^3$, and that for adult females is 4,200,000–5,400,000 cells per $mm^3$. For children, the average range is 4,500,000–5,100,000 cells per $mm^3$. These values may vary slightly with the hospital, physician, and type of equipment used to make blood cell counts. The

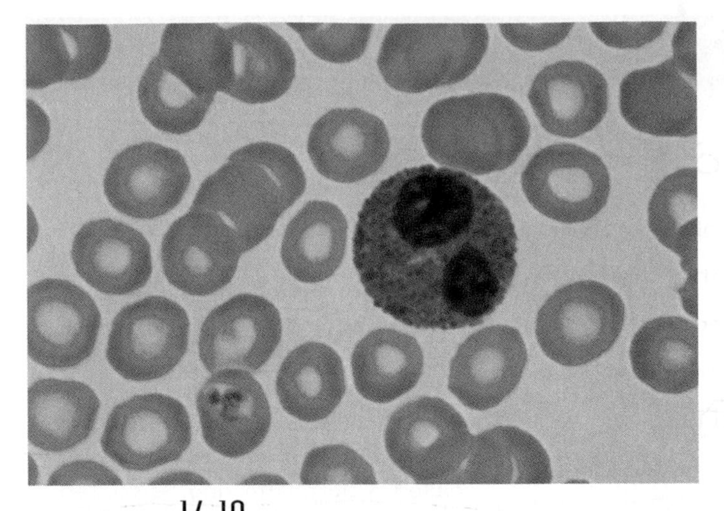

**FIGURE 14.10**
An eosinophil has red-staining cytoplasmic granules (500×).

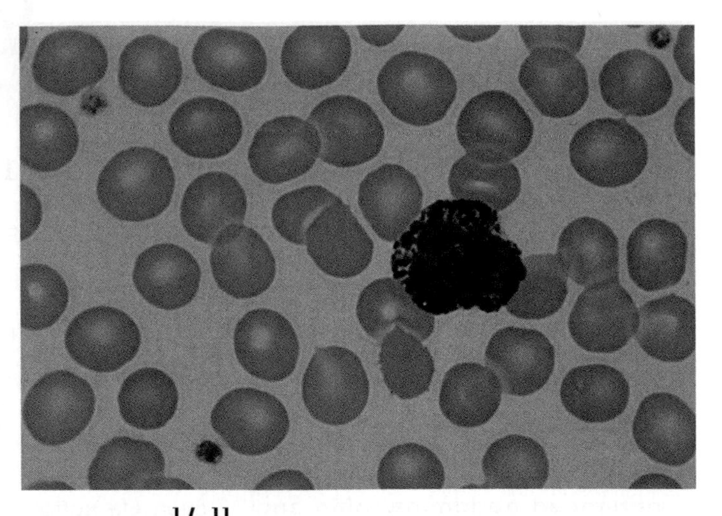

**FIGURE 14.11**
A basophil has cytoplasmic granules that stain deep blue (400×).

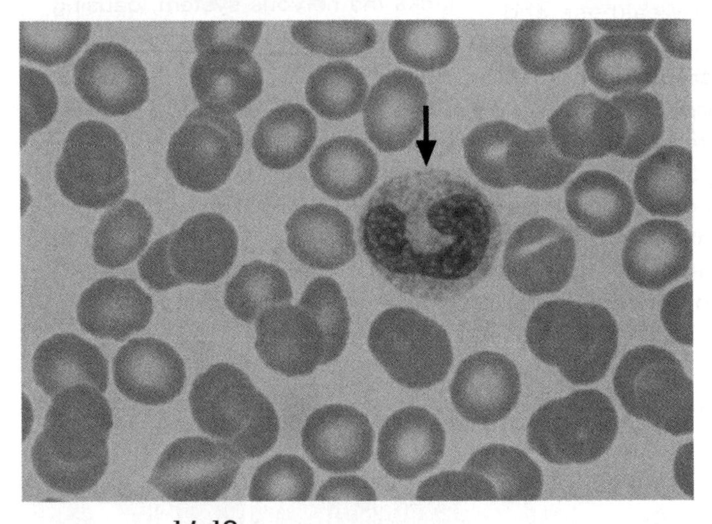

**FIGURE 14.12**
A monocyte is the largest type of blood cell (400×).

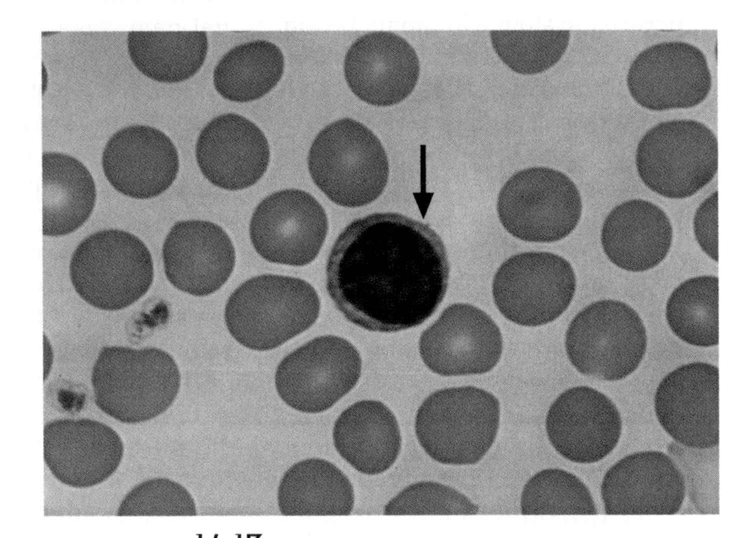

**FIGURE 14.13**
The lymphocyte (arrow) contains a large, round nucleus (500×).

The leukocytes of the agranulocyte group include **monocytes** and **lymphocytes.** Monocytes generally arise from red bone marrow. Lymphocytes are formed in the organs of the lymphatic system as well as in the red bone marrow.

**Monocytes** are the largest cells found in the blood, two to three times greater in diameter than red blood cells. Their nuclei vary in shape and are round, kidney-shaped, oval, or lobed. They usually make up 3% to 9% of the leukocytes in a blood sample and live for several weeks or even months (fig. 14.12).

**Lymphocytes** are usually only slightly larger than erythrocytes. A typical lymphocyte contains a relatively large, round nucleus surrounded by a thin rim of cytoplasm. These cells account for 25% to 33% of the circulating leukocytes. They have relatively long life spans that may extend for years (fig. 14.13).

1. Distinguish between granulocytes and agranulocytes.
2. List five types of white blood cells, and explain how they differ from one another.

## White Blood Cell Counts

The procedure used to count white blood cells is similar to that used for counting red cells. However, before a *white cell count* is made the red cells in the blood sample are destroyed so they will not be mistaken for white cells. Normally, a cubic millimeter of blood includes 5,000 to 10,000 white cells.

The total number and percentages of different white blood cell types are of clinical interest. A rise in the number of circulating white cells may indicate infection. A total number of white cells exceeding

10,000 per $mm^3$ of blood constitutes **leukocytosis,** indicating acute infection, such as appendicitis. It may also appear following vigorous exercise, emotional disturbances, or great loss of body fluids.

A total white cell count below 5,000 per $mm^3$ of blood is called **leukopenia.** Such a deficiency may accompany typhoid fever, influenza, measles, mumps, chicken pox, AIDS, or poliomyelitis. It may also result from anemia or from lead, arsenic, or mercury poisoning.

A *differential white blood cell count* (DIFF) lists percentages of the types of leukocytes in a blood sample. This test is useful because the relative proportions of white cells may change in particular diseases. Neutrophils, for instance, usually decrease during bacterial infections, and eosinophils may increase during certain parasitic infections and allergic reactions. In HIV infection and AIDS, the numbers of a type of lymphocyte called helper T cells plummet.

Chart 14.5 lists some disorders that alter the numbers of particular types of white blood cells.

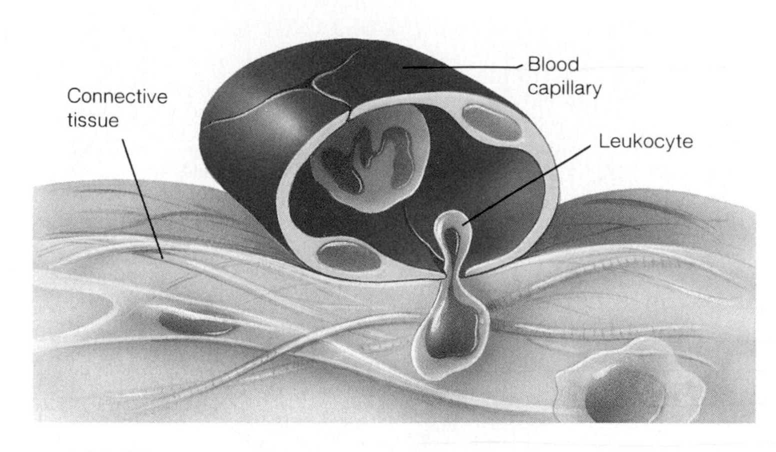


**FIGURE 14.14**
Leukocytes can squeeze between the cells of a capillary wall and enter the tissue space outside the blood system.

**CHART 14.5 WHITE BLOOD CELL ALTERATIONS**

| White Blood Cell Population Change | Illness |
|---|---|
| Elevated lymphocytes | Hairy cell leukemia, tuberculosis, whooping cough |
| Elevated eosinophils | Tapeworm infection, hookworm infection |
| Elevated monocytes | Typhoid fever, malaria, mononucleosis |
| Too few helper T cells | AIDS |

1. What is the normal human white blood cell count?
2. Distinguish between leukocytosis and leukopenia.
3. What is a differential white blood cell count?

## Functions of White Blood Cells

Leukocytes protect against infection in various ways. Some leukocytes phagocytize bacterial cells in the body, and others produce antibody proteins that destroy or disable foreign particles.

Leukocytes can squeeze between the cells that form blood vessel walls. This movement, called **diapedesis,** allows the white cells to leave the circulation (fig. 14.14). Once outside the blood, they move through interstitial spaces using a form of self-propulsion called *ameboid motion.*

The most mobile and active phagocytic leukocytes are neutrophils and monocytes. Although neutrophils are unable to ingest particles much larger than bacterial cells, monocytes can engulf relatively large objects. Monocytes contain numerous lysosomes, which are filled with digestive enzymes that break down organic molecules in captured bacteria. Neutrophils and monocytes often become so engorged with digestive products and bacterial toxins that they also die.

When microorganisms invade human tissues, certain body cells respond by releasing biochemicals that dilate local blood vessels. One such substance, **histamine,** dilates arterioles, allowing more blood to flood the capillaries. The tissues redden and copious fluids leak into the interstitial spaces. The swelling this *inflammatory reaction* produces delays the spread of invading microorganisms into other regions (see chapter 5). At the same time, damaged cells release chemicals that attract leukocytes. This phenomenon is called **positive chemotaxis,** and when combined with diapedesis brings many white blood cells into inflamed areas quickly (fig. 14.15).

> As bacteria, leukocytes, and damaged cells accumulate in an inflamed area, a thick fluid called *pus* often forms, and remains while the invading microorganisms are active. If the pus cannot escape to the outside of the body or into a body cavity, it may remain walled in the tissues for some time. Eventually surrounding cells absorb it.

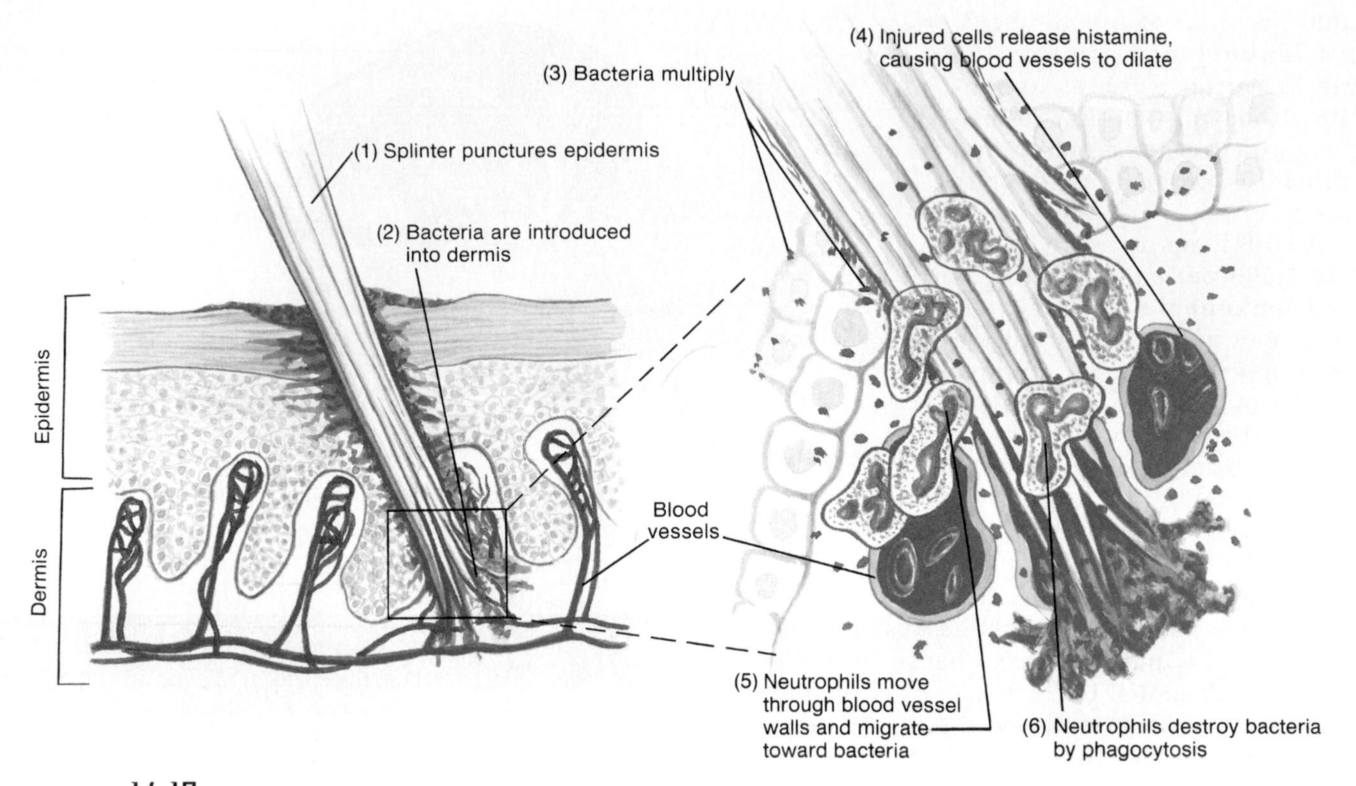


**FIGURE 14.15**
When bacteria invade the tissues, leukocytes migrate into the region and destroy the microbes by phagocytosis.

Eosinophils are only weakly phagocytic, but they are attracted to and can kill certain parasites. Eosinophils also help control inflammation and allergic reactions by removing biochemicals associated with these reactions.

Some of the cytoplasmic granules of basophils contain a blood clot inhibiting substance called *heparin,* while other granules contain histamine. Basophils may help prevent intravascular blood clot formation by releasing heparin, and may increase blood flow to injured tissues by releasing histamine. Basophils also play major roles in certain allergic reactions (see chapter 16).

Lymphocytes are important in *immunity.* They include B lymphocytes (B cells) that produce antibodies, and T lymphocytes (T cells) that produce biochemicals called cytokines. Chapter 16 discusses lymphocytes in greater detail.

Clinical Application 14.2 examines leukemia, cancer of white blood cells.

1. What are the primary functions of white blood cells?
2. How do white blood cells reach microorganisms that are outside blood vessels?
3. Which white blood cells are the most active phagocytes?
4. What are the functions of eosinophils and basophils?

## Blood Platelets

**Platelets,** or **thrombocytes,** are not complete cells. They arise from very large cells in the red bone marrow, called **megakaryocytes,** that fragment a little like a shattered plate, releasing small sections of cytoplasm—platelets—into the circulation. The larger fragments of the megakaryocytes shrink and become platelets as they pass through the blood vessels of the lungs.

Each platelet is a round disk that lacks a nucleus and is less than half the size of a red blood cell. It is capable of ameboid movement and may live for about ten days. In normal blood, the *platelet count* varies from 130,000 to 360,000 platelets per $mm^3$.

Platelets help repair damaged blood vessels by sticking to broken surfaces. They release **serotonin,** which contracts smooth muscles in the vessel walls, reducing blood flow. In addition, platelets initiate the formation of blood clots, explained later in this chapter.

Chart 14.6 summarizes the characteristics of blood cells and platelets.

## CHART 14.6 CELLULAR COMPONENTS OF BLOOD

| Component | Description | Number Present | Function |
|---|---|---|---|
| Red blood cell (erythrocyte) | Biconcave disk without a nucleus, about one-third hemoglobin | 4,200,000 to 6,200,000 per $mm^3$ | Transports oxygen and carbon dioxide |
| White blood cell (leukocyte) | | 5,000 to 10,000 per $mm^3$ | Destroys pathogenic microorganisms and parasites and removes worn cells |
| *Granulocytes* | About twice the size of red cells; cytoplasmic granules are present | | |
| **1.** Neutrophil | Nucleus with two to five lobes; cytoplasmic granules stain pink in neutral stain | 54%–62% of white cells present | Phagocytizes small particles |
| **2.** Eosinophil | Nucleus bilobed; cytoplasmic granules stain red in acid stain | 1%–3% of white cells present | Kills parasites and helps control inflammation and allergic reaction |
| **3.** Basophil | Nucleus lobed; cytoplasmic granules stain blue in basic stain | Less than 1% of white cells present | Releases anticoagulant, heparin, and histamine |
| *Agranulocytes* | Cytoplasmic granules are absent | | |
| **1.** Monocyte | Two to three times larger than a red cell; nuclear shape varies from round to lobed | 3%–9% of white cells present | Phagocytizes large particles |
| **2.** Lymphocyte | Only slightly larger than a red cell; its nucleus nearly fills cell | 25%–33% of white cells present | Provides immunity |
| Platelet (thrombocyte) | Cytoplasmic fragment | 130,000 to 360,000 per $mm^3$ | Helps control blood loss from broken vessels |

1. What is the function of blood platelets?
2. What is the normal human blood platelet count?

## BLOOD PLASMA

**Plasma** is the clear, straw-colored, liquid portion of the blood in which the cells and platelets are suspended. It is approximately 92% water and contains a complex mixture of organic and inorganic biochemicals. Functions of plasma constituents include transporting nutrients, gases, and vitamins; helping to regulate fluid and electrolyte balance; and maintaining a favorable pH. Figure 14.16 shows the chemical makeup of plasma.

### Plasma Proteins

**Plasma proteins** are the most abundant dissolved substances (solutes) in plasma. These proteins remain in the blood and interstitial fluids, and ordinarily are not used as energy sources. The three main plasma protein groups are **albumins, globulins,** and **fibrinogen.** The groups differ in chemical composition and physiological function.

**Albumins** account for about 60% of the plasma proteins, but are the smallest. They are synthesized in the liver, and because they are so plentiful, albumins are an important determinant of the *osmotic pressure* of the blood.

As explained in chapter 3, whenever the concentration of solutes changes on either side of a cell membrane, water moves through the membrane toward the region where the solutes are in higher concentration. For this reason, it is important that the concentration of solutes in the plasma remains relatively stable. Otherwise, water tends to leave the blood and enter the tissues, or leave the tissues and enter the blood, by osmosis. Because the presence of albumins (and other plasma proteins) adds to the osmotic pressure of the plasma, albumins aid in regulating the water balance between the blood and the tissues. In doing so, they help control blood volume, which, in turn, is directly related to blood pressure. Albumins also bind and transport certain molecules, such as bilirubin and free fatty acids.

If concentration of plasma proteins falls, tissues swell, a condition called *edema.* This may result from starvation or a protein-deficient diet, in which the body must use protein to obtain energy, or it may result if an impaired liver cannot synthesize plasma proteins. As blood concentration of plasma proteins drops, so does the osmotic pressure of blood, sending fluids into the intercellular spaces.

The **globulins,** which make up about 36% of the plasma proteins, can be further subdivided into *alpha, beta,* and *gamma globulins.* The liver synthesizes

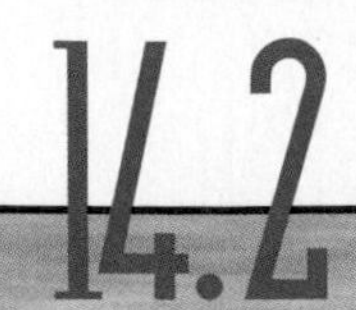

CLINICAL APPLICATION 14.2

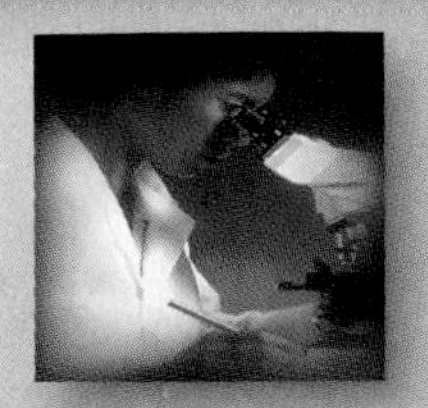

## Leukemia

The young woman had noticed symptoms for several months before she finally went to the doctor. At first it was just fatigue and headaches, which she attributed to studying for final exams. She had frequent colds and bouts of fever, chills, and sweats that she thought were just minor infections. When she developed several bruises and bone pain and noticed that her blood did not clot very quickly after minor cuts and scrapes, she consulted her physician, who examined her and took a blood sample. One glance at a blood smear under a microscope alarmed the doctor—there were far too few red blood cells and platelets, and too many white blood cells. She sent the sample to a laboratory to diagnose the type of *leukemia,* or cancer of the white blood cells, that was causing her patient's symptoms.

The young woman had *myeloid leukemia.* Her red bone marrow was producing too many granulocytes, but they were immature cells, unable to fight infection as they normally do. This explained the frequent illnesses. The leukemic cells were crowding out red blood cells and their precursors in the red marrow, causing her anemia and resulting fatigue. Platelet deficiency (thrombocytopenia) led to increased tendency to bleed. Finally, spread of the cancer cells outside the marrow painfully weakened the surrounding bone. Eventually, if she wasn't treated, the cancer cells would spread outside the circulatory system, causing other tissues that would normally not produce white blood cells to do so.

A second type of leukemia, distinguished by the source of the cancer cells, is *lymphoid leukemia.* These cancer cells are lymphocytes, produced in lymph nodes. Many of the symptoms are similar to those of myeloid leukemia. Sometimes a person has no leukemia symptoms at all, and a routine blood test detects the condition (fig. 14A).

Leukemia is also classified as acute or chronic. An acute condition appears suddenly, symptoms progress rapidly, and death occurs in a few months without treatment. Chronic forms begin more slowly and may remain undetected for months or even years or, in rare cases, decades. Without treatment, life expectancy is about 3 years. With treatment, 50% to

alpha and beta globulins. They have a variety of functions, including transport of lipids and fat-soluble vitamins. Lymphatic tissues produce the gamma globulins, which are a type of antibody (see chapter 16).

**Fibrinogen,** which constitutes about 4% of the plasma protein, plays a primary role in blood coagulation. Synthesized in the liver, it is the largest of the plasma proteins. The function of fibrinogen is discussed later in this chapter.

Chart 14.7 summarizes the characteristics of the plasma proteins.

1. List three types of plasma proteins.
2. How does albumin help maintain water balance between the blood and the tissues?
3. Which of the globulins functions in immunity?
4. What is the role of fibrinogen?

### Nutrients and Gases

The *plasma nutrients* include amino acids, simple sugars, nucleotides, and lipids absorbed from the digestive tract. For example, plasma transports glucose from the

80% of patients enter remission, a period of stability that may become a cure. Chemotherapy may be necessary for a year or two to increase the chances of long remission.

Leukemia treatment includes correcting symptoms by giving blood transfusions and treating infections, and using drugs that kill cancer cells. Several drugs in use for many years have led to spectacular increases in cure rates, particularly for acute lymphoid leukemia in children. Some newer treatments offer hope for other types of leukemia too. For example, alpha interferon treats a chronic form called hairy cell leukemia that mostly affects adult males, and a type of retinoic acid treats acute promyelocytic leukemia. A bone marrow transplant can cure leukemia, but it is a very risky procedure.

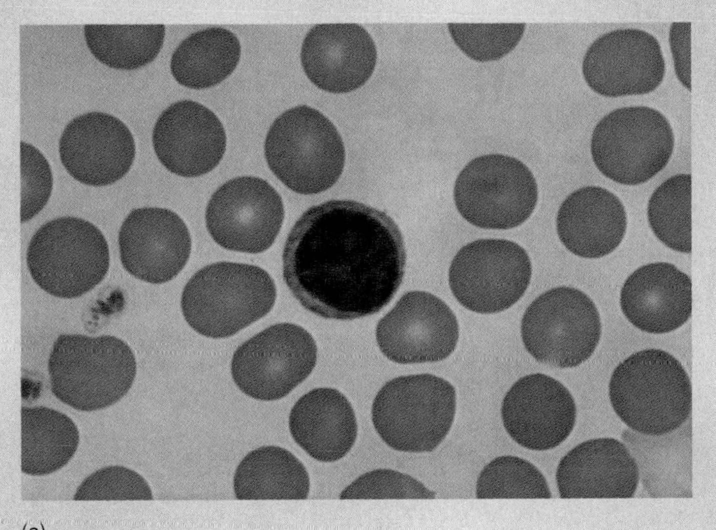

(a)

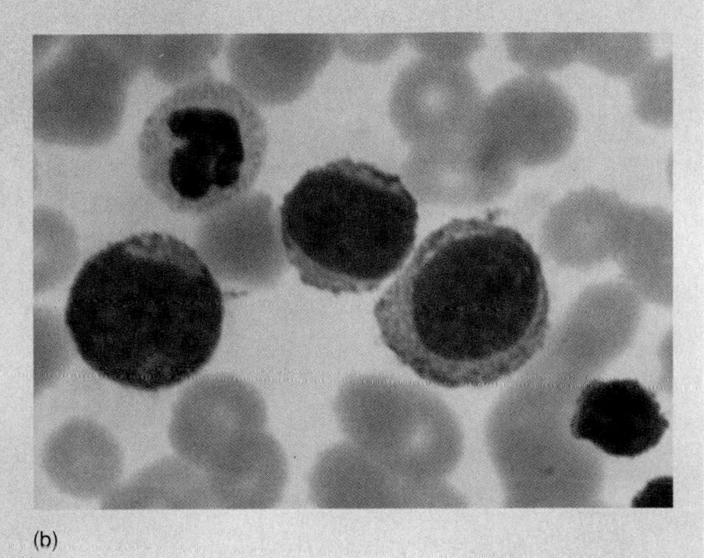

(b)

**FIGURE 14A**

(*a*) Normal blood cells (500×); (*b*) blood cells from a person with lymphoid leukemia (500×). Note the increased number of leukocytes.

small intestine to the liver, where it may be stored as glycogen or altered to form fat. If blood glucose concentration drops below the normal range, glycogen may be broken down into glucose, as described in chapter 13.

Recently absorbed amino acids are also carried to the liver, where they may be used to manufacture proteins and used as an energy source (see chapter 4).

Plasma lipids include fats (triglycerides), phospholipids, and cholesterol. Because lipids are not water soluble and plasma is almost 92% water, these lipids combine with proteins in **lipoprotein** complexes. Lipoprotein molecules are relatively large and consist of a surface layer of phospholipid, cholesterol, and protein surrounding a triglyceride core. *Apoproteins* in the outer layer can combine with receptors on the membranes of specific target cells. Lipoprotein molecules vary in the proportions of the lipids they contain.

Because fats are less dense than proteins, as the proportion of triglycerides in a lipoprotein increases, the density of the particle decreases. Conversely, as the proportion of triglycerides decreases, the density increases. Lipoproteins are classified on the basis of their densities, which reflect their composition.

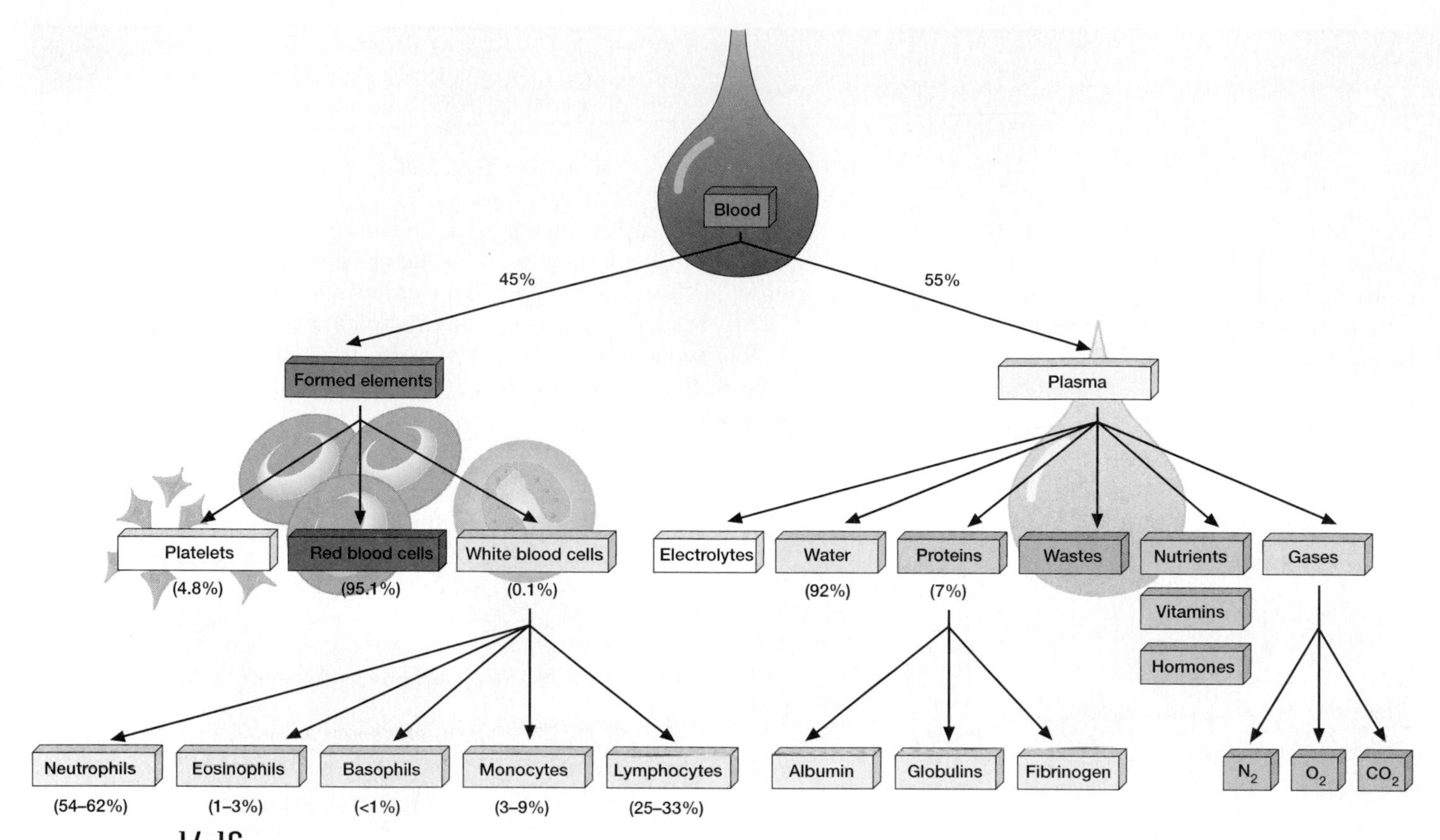


**FIGURE 14.16**
Blood composition.

**CHART 14.7 PLASMA PROTEINS**

| Protein | Percentage of Total | Origin | Function |
|---|---|---|---|
| Albumin | 60% | Liver | Helps maintain blood osmotic pressure |
| Globulin | 36% | | |
| Alpha globulins | | Liver | Transports lipids and fat-soluble vitamins |
| Beta globulins | | Liver | Transports lipids and fat-soluble vitamins |
| Gamma globulins | | Lymphatic tissues | Constitute the antibodies of immunity |
| Fibrinogen | 4% | Liver | Plays a key role in blood coagulation |

*Chylomicrons* consist mainly of triglycerides absorbed from the small intestine (see chapter 17). *Very low-density lipoproteins* (VLDL) have a relatively high concentration of triglycerides. *Low-density lipoproteins* (LDL) have a relatively high concentration of cholesterol and are the major cholesterol-carrying lipoproteins. *High-density lipoproteins* (HDL) have a relatively high concentration of protein and a lower concentration of lipids.

Many residents of Tangier Island in the Chesapeake Bay have inherited a deficiency in a particular blood lipoprotein from the original settlers, who arrived in 1686. Low blood levels of alpha lipoprotein, which normally transports cholesterol in the blood, leads to low blood cholesterol. However, excess cholesterol accumulates elsewhere, such as in the thymus gland, and in scavenging macrophages.

Chylomicrons, discussed further in chapter 17, transport dietary fats to muscle and adipose cells. Similarly, very low-density lipoproteins, produced in

## CHART 14.8 PLASMA LIPOPROTEINS

| Lipoprotein | Characteristics | Functions |
| --- | --- | --- |
| Chylomicron | High concentration of triglycerides | Transports dietary fats to muscle and adipose cells |
| Very low-density lipoprotein (VLDL) | Relatively high concentration of triglycerides; produced in the liver | Transports triglycerides synthesized in the liver from carbohydrates to adipose cells |
| Low-density lipoprotein (LDL) | Relatively high concentration of cholesterol; formed from remnants of VLDL molecules that have given up their triglycerides | Delivers cholesterol to various cells, including liver cells |
| High-density lipoprotein (HDL) | Relatively high concentration of protein and low concentration of lipid | Transports to the liver remnants of chylomicrons that have given up their triglycerides |

the liver, transport triglycerides synthesized from excess dietary carbohydrates. After VLDL molecules deliver their loads of triglycerides to adipose cells, an enzyme, lipoprotein lipase, converts their remnants to low-density lipoproteins. Because most of the triglycerides have been removed, LDL molecules have a relatively higher cholesterol content than do the original VLDL molecules. Various cells, including liver cells, have surface receptors that combine with apoproteins associated with LDL molecules. These cells slowly remove LDL from plasma by receptor-mediated endocytosis, supplying cells with cholesterol (see chapter 3).

After chylomicrons deliver their triglycerides to cells, their remnants are transferred to high-density lipoproteins. These HDL molecules, which form in the liver and small intestine, transport chylomicron remnants to the liver, where they enter cells rapidly by receptor-mediated endocytosis. The liver disposes of the cholesterol it obtains in this manner by secreting it into bile or by using it to synthesize bile salts.

Chart 14.8 summarizes the characteristics and functions of these lipoproteins.

Much of the cholesterol and bile salts in bile are later reabsorbed by the small intestine and transported back to the liver, and the secretion-reabsorption cycle repeats. During each cycle, some of the cholesterol and bile salts escape reabsorption, reach the large intestine, and are eliminated with the feces.

The most important *blood gases* are oxygen and carbon dioxide. Plasma also contains a considerable amount of dissolved nitrogen, which ordinarily has no physiological function. The blood gases and their transport will be discussed in chapter 21.

As a rule, blood gases are evaluated using a fresh sample of whole blood obtained from an artery. This blood is cooled to decrease the rate of metabolic reactions, and clotting is prevented by adding an anticoagulant. In the laboratory, the partial pressures of oxygen and carbon dioxide of the blood are determined, the blood pH is measured, and the plasma bicarbonate concentration is calculated. Such information is commonly used to diagnose and treat disorders of circulation, respiration, and electrolyte balance. Appendix C lists average values for these laboratory tests.

1. What nutrients are in blood plasma?
2. How are triglycerides transported in plasma?
3. How is cholesterol eliminated from the liver?
4. What gases are in plasma?

### Nonprotein Nitrogenous Substances

Molecules that contain nitrogen atoms but are not proteins comprise a group called **nonprotein nitrogenous substances.** In plasma, this group includes amino acids, urea, uric acid, creatine, and creatinine. Amino acids come from protein digestion and amino acid absorption. Urea and uric acid are products of protein and nucleic acid catabolism, respectively, and creatinine results from the metabolism of creatine. As discussed in chapter 9, creatine occurs as **creatine phosphate** in muscle and brain tissues as well as in the blood, where it stores high-energy phosphate bonds, much like those of ATP molecules.

Normally, the concentration of nonprotein nitrogenous (NPN) substances remains relatively stable because protein intake and utilization are balanced with excretion of nitrogenous wastes. Because about half of the NPN is urea, which the kidneys ordinarily excrete, a rise in the blood urea nitrogen (BUN) may suggest a kidney disorder. It may also result from excessive protein catabolism or infection.

### Plasma Electrolytes

Blood plasma contains a variety of *electrolytes.* Recall that electrolytes release ions when dissolved in water. Plasma electrolytes are absorbed from the intestine or released as by-products of cellular metabolism. They include sodium, potassium, calcium, magnesium,

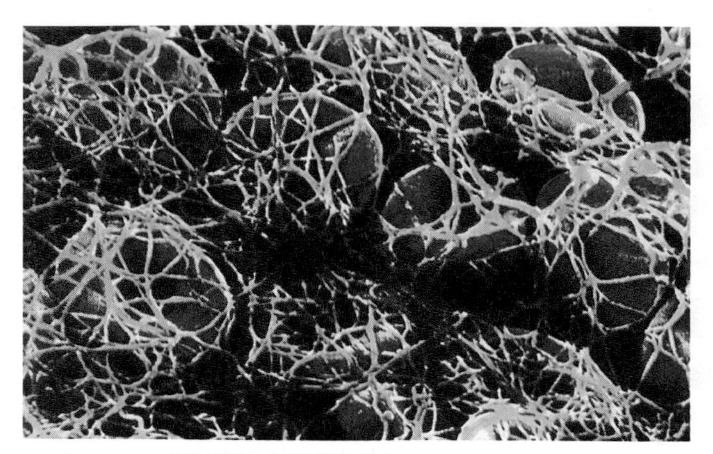

**FIGURE 14.19**
A scanning electron micrograph of fibrin threads (2,000×).

form, it promotes still more clotting, because thrombin also acts directly on blood clotting factors other than fibrinogen, causing prothrombin to form still more thrombin. This type of self-initiating action is an example of a **positive feedback system,** in which the original action stimulates more of the same type of action. Such a mechanism produces unstable conditions and can operate for only a short time in a living system, because life depends on the maintenance of a stable internal environment (see chapter 1).

Normally, blood flow throughout the body prevents formation of a massive clot within the cardiovascular system by rapidly carrying excess thrombin away and keeping its concentration too low to enhance further clotting. Also, a naturally occurring substance called *antithrombin,* present in the blood and on the surfaces of endothelial cells that line blood vessels, limits thrombin formation. Consequently, blood coagulation is usually limited to blood that is standing still (or moving relatively slowly), and clotting ceases where a clot contacts circulating blood.

> In *disseminated intravascular clotting,* coagulation is abnormally activated in widespread regions of the circulatory system. This condition is usually associated with bacteria or bacterial toxins in the blood or with a disorder causing widespread tissue damage. As a result, many small clots may appear and obstruct blood flow into various tissues and organs, particularly the kidneys. As plasma clotting factors and platelets are depleted, the patient may develop serious bleeding.

## Intrinsic Clotting Mechanism

Unlike extrinsic clotting, all of the components necessary for intrinsic clotting are in the blood. Activation of a substance called the *Hageman factor* (factor XII) initiates intrinsic clotting. This happens when blood is exposed to a foreign surface such as collagen or when blood is stored in a glass container. In the presence of calcium ions, the activated factor triggers a complex series of changes utilizing several clotting factors and leads to formation of prothrombin activator. The subsequent steps of blood clot formation are the same as those described for the extrinsic mechanism (fig. 14.20). Chart 14.10 compares extrinsic and intrinsic clotting mechanisms.

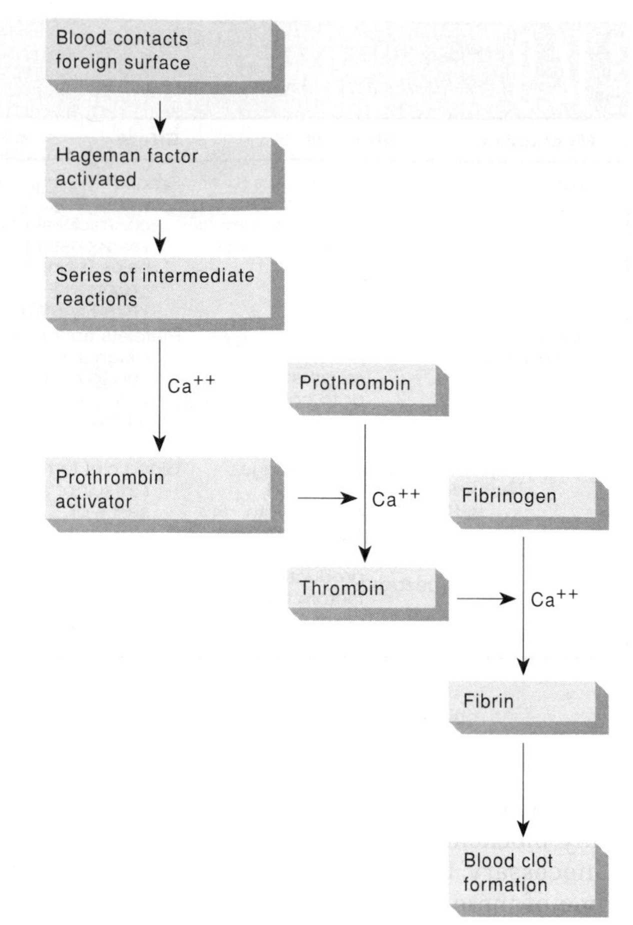


**FIGURE 14.20**
The intrinsic blood clotting mechanism may begin when blood contacts a foreign surface.

> Laboratory tests commonly used to evaluate the blood coagulation mechanisms include *prothrombin time* (PT) and *partial thromboplastin time* (PTT). Both of these tests measure the time it takes for fibrin threads to form in a sample of blood plasma. However, the prothrombin time test checks the extrinsic clotting mechanism, whereas the partial thromboplastin test evaluates intrinsic clotting.

## CHART 14.10 Blood Coagulation

| Steps | Extrinsic Clotting Mechanism | Intrinsic Clotting Mechanism |
|---|---|---|
| Trigger | Damage to vessel or tissue | Blood contacts foreign surface |
| Initiation | Tissue thromboplastin | Hageman factor |
| Series of reactions involving several clotting factors and calcium ions ($Ca^{++}$) lead to the production of | Prothrombin activator | Prothrombin activator |
| Prothrombin activator and calcium ions cause the conversion of | Prothrombin to thrombin | Prothrombin to thrombin |
| Thrombin causes fragmentation, then joining of | Fibrinogen to fibrin | Fibrinogen to fibrin |

### Fate of Blood Clots

After a blood clot forms, it soon begins to retract as the tiny processes extending from the platelet membranes adhere to strands of fibrin within the clot, and contract. The blood clot shrinks, pulling the edges of the broken vessel closer together and squeezing a fluid called **serum** from the clot. Serum is essentially plasma minus all of its fibrinogen and most of the other clotting factors. Platelets associated with a blood clot also release *platelet-derived growth factor* (PDGF), which stimulates smooth muscle cells and fibroblasts to repair damaged blood vessel walls.

Fibroblasts soon invade blood clots that form in ruptured vessels, producing fibrous connective tissue throughout the clots, which helps strengthen and seal vascular breaks. Many clots, including those that form in tissues as a result of blood leakage (hematomas), disappear in time. In clot dissolution, fibrin threads absorb a plasma protein called *plasminogen* (profibrinolysin). Then a substance called plasminogen activator released from the lysosomes of damaged tissue cells converts plasminogen to *plasmin.* Plasmin is a protein-splitting enzyme that can digest fibrin threads and other proteins associated with blood clots. Plasmin formation may dissolve a whole clot; however, clots that fill large blood vessels are seldom removed naturally.

A blood clot forming in a vessel abnormally is a **thrombus.** If the clot dislodges or if a fragment of it breaks loose and is carried away by the blood flow, it is called an **embolus.** Generally, emboli continue to move until they reach narrow places in vessels where they lodge and may interfere with blood flow.

Such abnormal clot formations are often associated with conditions that change the endothelial linings of vessels. For example, in *atherosclerosis,* accumulations of fatty deposits change arterial linings, sometimes initiating inappropriate clotting. This is the most common cause of thrombosis in medium-sized arteries (fig. 14.21).

Coagulation may also occur in blood that is flowing too slowly. The concentration of clot-promoting substances may increase to a critical level instead of being carried away by more rapidly moving blood, and a clot may form. This mechanism is the usual cause of thrombosis in veins.

A blood clot forming in a vessel that supplies a vital organ, such as the heart (coronary thrombosis) or the brain (cerebral thrombosis), kills tissues the vessel serves (*infarction*), and may be fatal. A blood clot that travels and then blocks a vessel that supplies a vital organ, such as the lungs (pulmonary embolism), affects the portion of the organ supplied by the blocked blood vessel. Plasminogen activators are biochemicals that dissolve blood clots. They can be very useful in breaking up abnormal blood clots.

Drugs based on "clot-busting" biochemicals can be lifesavers. *Tissue plasminogen activator* (tPA) has had remarkable success in restoring blocked coronary circulation if given within 4 hours of a heart attack. It was heralded as a wonder drug in the mid-1980s, but many clinical trials have given conflicting results—in some studies, a drug derived from bacteria called *streptokinase* has been just as successful, for a fraction of the cost. Another plasminogen activator used as a drug is *urokinase,* an enzyme produced in certain human kidney cells.

1. Distinguish between extrinsic and intrinsic clotting mechanisms.
2. What is the basic event in blood clot formation?
3. What factors initiate the formation of fibrin?
4. What prevents the formation of massive clots throughout the cardiovascular system?
5. Distinguish between a thrombus and an embolus.
6. How might atherosclerosis promote the formation of blood clots?

### Prevention of Coagula

In a healthy vascular system, sp formation is prevented in part b the blood vessels. This smooth li accumulation of platelets and dothelial cells also produce a

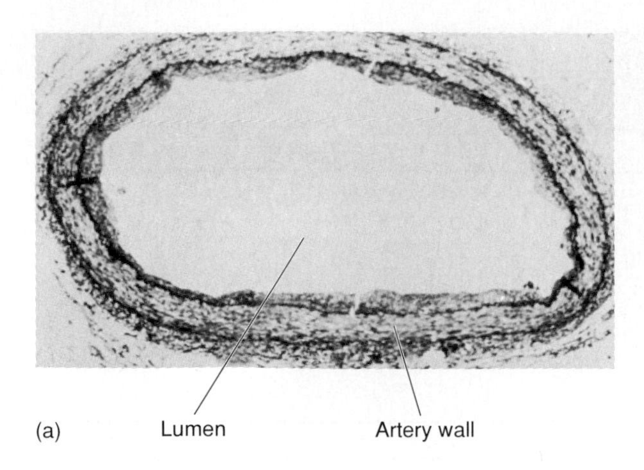


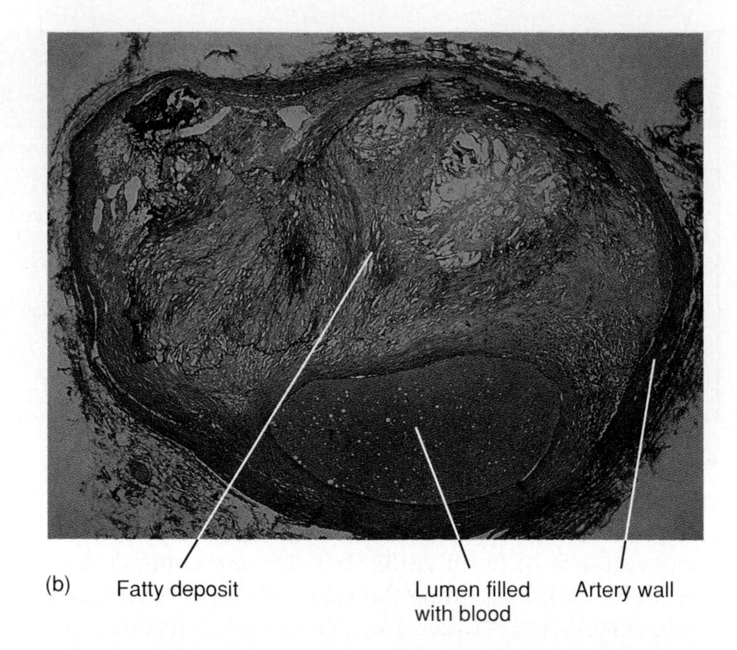


**FIGURE 14.21**
(*a*) Light micrograph of a normal artery (60×); (*b*) the inner wall of an artery changed as a result of atherosclerosis (60×).

**CHART 14.11 FACTORS THAT INHIBIT BLOOD CLOT FORMATION**

| Factor | Action |
| --- | --- |
| Smooth lining of blood vessel | Prevents activation of intrinsic blood clotting mechanism |
| Prostacyclin | Inhibits adherence of platelets to blood vessel wall |
| Fibrin threads | Adsorbs thrombin |
| Antithrombin in plasma | Interferes with the action of thrombin |
| Heparin from mast cells and basophils | Interferes with the formation of prothrombin activator |

chapter 13) called *prostacyclin* ($PGI_2$), which inhibits the adherence of platelets to the inner surface of healthy blood vessel walls.

When a clot is forming, fibrin threads adsorb thrombin, thus helping prevent the spread of the clotting reaction. A plasma alpha globulin, *antithrombin,* inactivates additional thrombin by binding to it and blocking its action on fibrinogen.

In addition, basophils and mast cells in the connective tissue surrounding capillaries secrete the anticoagulant *heparin.* This substance interferes with formation of prothrombin activator, prevents the action of thrombin on fibrinogen, and promotes removal of thrombin by antithrombin and fibrin adsorption.

Heparin-secreting cells are particularly abundant in the liver and lungs, where capillaries trap small blood clots that commonly form in the slow-moving blood of veins. These cells secrete heparin continually, preventing additional clotting in the circulatory system. Chart 14.11 summarizes clot-inhibiting factors. Clinical Application 14.3 discusses an ancient anticlotting treatment becoming popular again—medicinal leeches. Clinical Application 14.4 discusses clotting disorders.

*Thrombocytopenia* is a tendency to bleed because of an abnormally low platelet count, dropping below 100,000 platelets per cubic millimeter of blood. Symptoms include bleeding easily, capillary hemorrhages throughout the body, and small, bruiselike spots on the skin. Thrombocytopenia is a common side effect of cancer chemotherapy and radiation treatments, and can also develop as a complication of pregnancy, leukemia, bone marrow transplantation, infectious disease, cardiac surgery, or anemia.

Transfusion of platelets is the conventional treatment for thrombocytopenia, but recently researchers isolated a substance called thrombopoietin (TPO), which stimulates formation and maturation of megakaryocytes. It may be useful as a drug to boost platelet levels.

1. How does the lining of a blood vessel help prevent blood clot formation?
2. What is the function of antithrombin?
3. How does heparin help prevent blood clot formation?

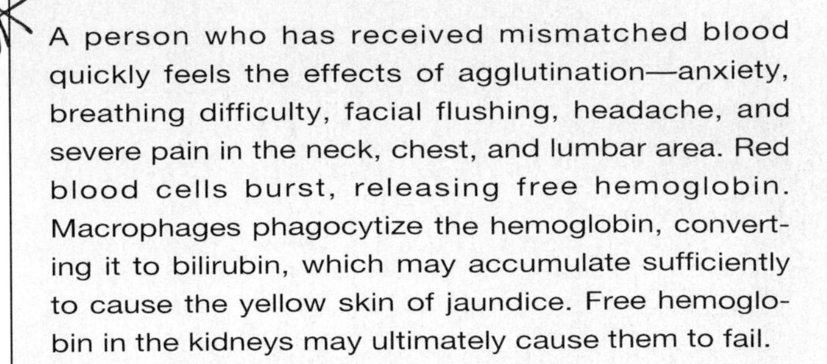
A person who has received mismatched blood quickly feels the effects of agglutination—anxiety, breathing difficulty, facial flushing, headache, and severe pain in the neck, chest, and lumbar area. Red blood cells burst, releasing free hemoglobin. Macrophages phagocytize the hemoglobin, converting it to bilirubin, which may accumulate sufficiently to cause the yellow skin of jaundice. Free hemoglobin in the kidneys may ultimately cause them to fail.

## Blood Groups and Transfusions

Experiments transfusing blood, which date from the late 1600s, first used lamb blood. By the 1800s, human blood was being used. Results were unpredictable—some recipients were cured, but some were killed when their kidneys failed under the strain of handling red blood cells that clumped when blood types were incompatible. So poor was the success rate that, by the late 1800s, many nations banned transfusions.

Around this time, Austrian physician Karl Landsteiner began investigating why transfusions sometimes worked and sometimes did not. In 1900, he determined that blood was of differing types, and that only certain combinations of them were compatible. In 1910, identification of the ABO blood antigen gene explained the observed blood type incompatibilities. Today, twenty different genes are known to contribute to the surface features of red blood cells, which determine compatibility between blood types.

### Antigens and Antibodies

The clumping of red blood cells following a transfusion reaction is called **agglutination.** This phenomenon is due to a reaction between red blood cell surface molecules **antigens** (formerly called *agglutinogens*), and protein **antibodies** (formerly called *agglutinins*) carried in the plasma. Although many different antigens are associated with human erythrocytes, only a few of them are likely to produce serious transfusion reactions. These include the antigens of the ABO group and those of the Rh group.

Avoiding the mixture of certain kinds of antigens and antibodies prevents adverse transfusion reactions.

### ABO Blood Group

The *ABO blood group* is based on the presence (or absence) of two major antigens in red cell membranes—antigen *A* and antigen *B*. A person's erythrocytes contain one of four antigen combinations: only A, only B, both A and B, or neither A nor B.

A person with only antigen A has *type A blood;* a person with only antigen B has *type B blood;* one with both antigen A and B has *type AB blood;* and one with neither antigen A nor B has *type O blood.*

Certain antibodies are synthesized in the plasma about two to eight months following birth. Specifically, whenever antigen A is absent in the red blood cells, an antibody called *anti-A* develops, and whenever antigen B is absent, an antibody called *anti-B* develops. Therefore, persons with type A blood also have antibody anti-B in their plasma; those with type B blood have antibody anti-A; those with type AB blood have neither antibody; and those with type O blood have both anti-A and anti-B (fig. 14.22 and chart 14.12). The antibodies anti-A and anti-B are large and do not cross the placenta. Thus, the pregnant woman and the fetus may be of different ABO blood types and agglutination within the fetus will not occur.

The percentage of blood types in human populations reflects history and migration patterns. Type B blood is found in 5% to 10% of the English and Irish, but gradually increases eastward, reaching 25% to 30% in the former Soviet Union. Some Native American populations in the United States today do not have type B or AB blood, yet the Asian populations from whom they descend do. This suggests that the original group of settlers lacked individuals with type B or AB blood.

Because an antibody of one kind will react with an antigen of the same kind and clump red blood cells, such combinations must be avoided. The major concern in blood transfusion procedures is that the cells in the donated blood not clump due to antibodies present in the recipient's plasma. For this reason, a person with type A (anti-B) blood must not receive blood of type B or AB, either of which would clump in the presence of anti-B in the recipient's type A blood. Likewise, a person with type B (anti-A) blood must not be given type [illegible] and a person with type O (anti-A an[illegible] must not be given type A, B, or AB bl[illegible]

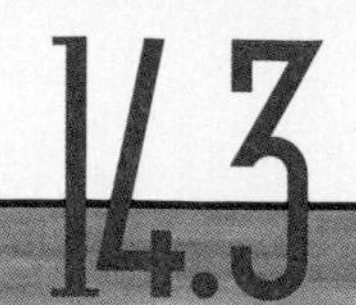

Clinical Application 14.3

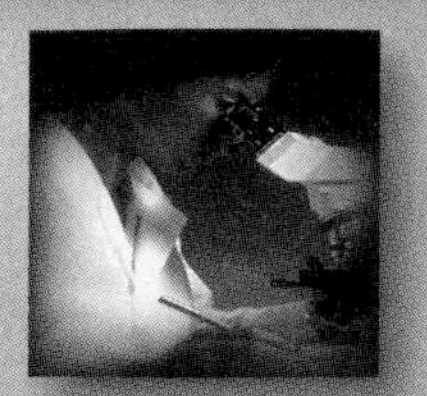

## The Return of the Medicinal Leech

It had taken surgeon Joseph Upton ten hours to sew the 5-year-old's ear back on, after a dog had bitten it off. At first the operation appeared to be a success, but after four days, trouble began. Blood flow in the ear was blocked. Close examination showed that the arteries that the surgeon had repaired were fine, but the smaller veins were becoming congested. So Dr. Upton tried an experimental technique—he applied twenty-four leeches to the wound area.

The leeches latched on for up to an hour each, drinking the boy's blood. Leech saliva contains several biochemicals, one of which is a potent anticoagulant called hirudin in honor of its source, the medicinal leech *Hirudo medicinalis.* Unlike conventional anticlotting agents such as heparin, which are short-acting, hirudin works for up to 24 hours after the leech has drunk its fill and dropped off. Hirudin blocks thrombin specifically in veins. The long-acting leech biochemical gave the boy's ear time to heal.

Leeches have long been part of medical practice, with references hailing back to the ancient Egyptians

**Figure 14B**
For centuries, bloodletting with leeches was believed to cure many ills. This woman in seventeenth-century Belgium applies a medicinal leech to her arm.

Because type AB blood lacks both anti-A and anti-B antibodies, an AB person can receive a transfusion of blood of any other type. For this reason, type AB persons are sometimes called *universal recipients.* However, type A (anti-B) blood, type B (anti-A) blood, and type O (anti-A and anti-B) blood still contain antibodies (either anti-A and/or anti-B) that could agglutinate type AB cells. Consequently, even for AB individuals, it is always best to use donor blood of the same type as the recipient blood. If the matching type is not available and type A, B, or O is used, it should be transfused slowly so that the donor blood is well diluted by the recipient's larger blood volume. This precaution usually avoids serious reactions between the donor's antibodies and the recipient's antigens.

Similarly, because type O blood lacks antigens A and B, this type could theoretically be transfused into persons with blood of any other type. Therefore, persons with type O blood are sometimes called *universal donors.* Type O blood, however, does contain both anti-A and anti-B antibodies and if it is given to a person with blood type A, B, or AB, it too should be transfused slowly to minimize the chance of an adverse reaction.

2,500 years ago (fig. 14B). The leech's popularity peaked in Europe in the nineteenth century, when French physicians alone used more than a billion a year, to drain "bad humours" from the body to cure nearly every ill. Use of leeches fell in the latter half of the nineteenth century. They were rediscovered by Yugoslav plastic surgeons in 1960 and by French microsurgeons in the early 1980s. In 1985 Dr. Upton made headlines and brought leeches into the limelight by saving the boy's ear at Children's Hospital in Boston.

A leech's bite does not hurt, patients say. But for those unwilling to have one or more 3-inch long, slimy green-gray invertebrates picnicking on a wound, hirudin is being developed as a drug produced by recombinant DNA technology. It will be called hirulog (fig. 14C).

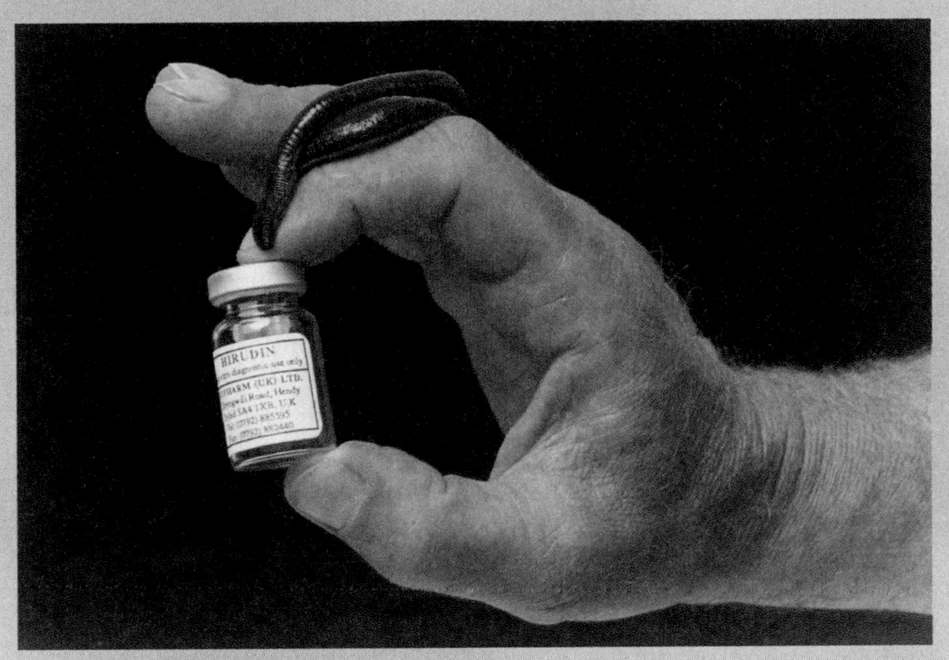


**FIGURE 14C**

Microsurgeons sometimes use leeches to help maintain blood flow through veins in patients after reattaching severed ears or digits. An anticoagulant in the leech's saliva keeps the blood thin enough to flow.



Chart 14.13 summarizes preferred blood types for normal transfusions and permissible blood types for emergency transfusions.

1. Distinguish between antigens and antibodies.
2. What is a blood type?
3. What is the main concern when blood is transfused from one individual to another?
4. Why is a type AB person called a universal recipient?

When is type O blood not really type O blood? When a person has two copies of a rare gene, called the "h" gene. Such a person lacks an enzyme that inserts a particular sugar onto red blood cell surfaces. Without that sugar, the A and B antigens cannot bind. The result—blood that tests as O (because it lacks A and B antigens) but can genetically be of any ABO type—A, B, AB, or O. Although this does not affect health, it can lead to questions when a child's ABO type cannot be derived from those of the parents.

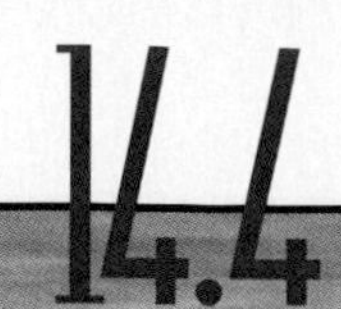

CLINICAL APPLICATION 14.4

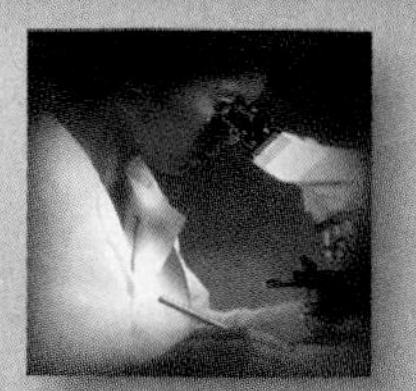

## Coagulation Disorders

### HEMOPHILIA

In 1962, 5-year-old Bob Massie developed uncontrollable bleeding in his left knee, a symptom of his *hemophilia A,* an inherited clotting disorder. It took thirty transfusions of plasma over the next three months to stop the bleeding. Because the knee joint had swelled and locked into place during that time, Bob was unable to walk for the next seven years. Today, Bob still suffers from painful joint bleeds, but he injects himself with factor VIII, the coagulation protein that his body cannot make. The factor VIII soon controls the bleed.

Hemophilia has left its mark on history. One of the earliest descriptions is in the Talmud, a second century B.C. Jewish document, which reads "If she circumcised her first child and he died, and a second one also died, she must not circumcise her third child." Queen Victoria (1819–1901) passed the hemophilia gene to several of her children, eventually spreading the condition to the royal families of England, Russia, Germany, and Spain. Hemophilia achieved notoriety when factor VIII pooled from blood donations was discovered to transmit HIV in 1985. Ninety percent of people with severe hemophilia who used such pooled factor VIII prior to then have contracted AIDS.

Different forms of hemophilia are caused by abnormalities of different clotting factors, but hemophilia A is by far the most common. Symptoms of the hemophilias include tendency to hemorrhage severely following minor injuries, frequent nosebleeds, large intramuscular hematomas, and blood in the urine.

### VON WILLEBRAND DISEASE

Easy bruising and a tendency to bleed easily are signs of *von Willebrand disease,* another inherited clotting disorder that is usually far less severe than hemophilia. Affected persons lack a plasma protein, von Willebrand factor, that is secreted by endothelial cells lining blood vessels and enables platelets to adhere to damaged blood vessel walls, a key step preceding actual clotting. Sometimes the condition can cause spontaneous bleeding from the mucous membranes of the gastrointestinal and urinary tracts.

## Rh Blood Group

The *Rh blood group* was named after the rhesus monkey in which it was first studied. In humans, this group includes several Rh antigens (factors). The most important of these is *antigen D;* however, if any of the antigen D and other Rh antigens are present in the red cell membranes, the blood is said to be *Rh positive.* Conversely, if the red cells lack the Rh antigens, the blood is called *Rh negative.*

As in the case of antigens A and B, the presence (or absence) of Rh antigens is an inherited trait. Unlike anti-A and anti-B, antibodies for Rh (*anti-Rh*) do not appear spontaneously. Instead, they form only in Rh-negative persons in response to special stimulation.

If an Rh-negative person receives a transfusion of Rh-positive blood, the recipient's antibody-producing cells are stimulated by the presence of the Rh antigens and will begin producing *anti-Rh antibodies.* Generally, no serious consequences result from this initial transfusion, but if the Rh-negative person—who is now sensitized to Rh-positive blood—receives another transfusion of Rh-positive blood some months later, the donor's red cells are likely to agglutinate.

A related condition may occur when an Rh-negative woman is pregnant with an Rh-positive fetus for the first time. Such a pregnancy may be uneventful; however, at the time of this infant's birth (or if a miscarriage occurs), the placental membranes that separated the maternal blood from the fetal blood during the pregnancy tear, and some of the infant's Rh-positive blood cells may enter the maternal circulation. These Rh-positive cells may then stimulate the maternal tissues to begin producing anti-Rh antibodies (fig. 14.24).

If a woman who has already developed anti-Rh antibodies becomes pregnant with a second Rh-positive fetus, these anti-Rh antibodies, called hemolysins, pass through the placental membrane and destroy

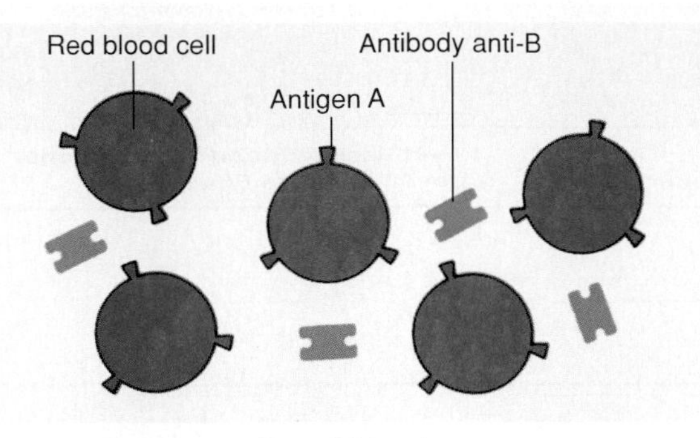


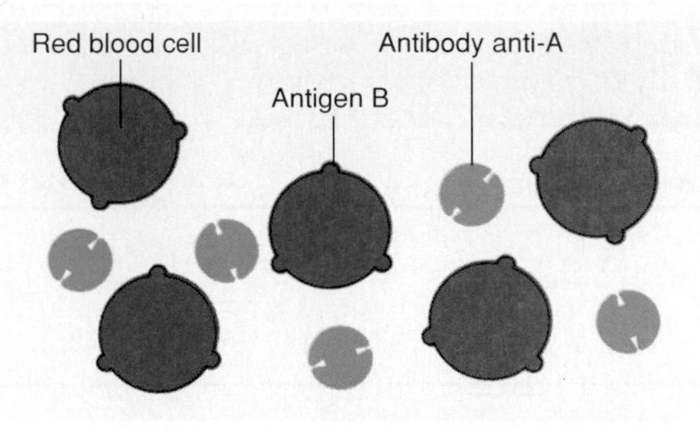


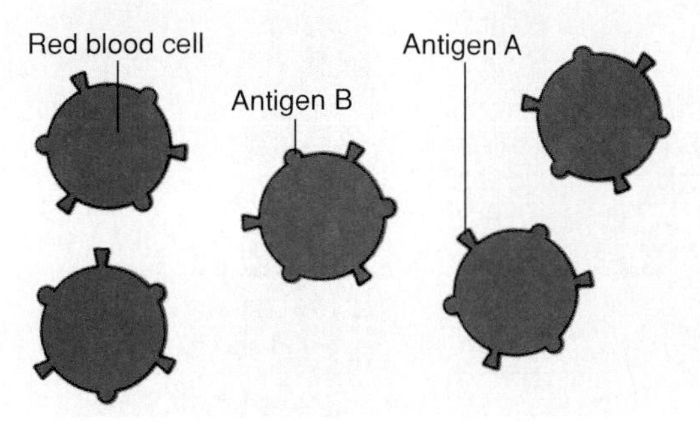


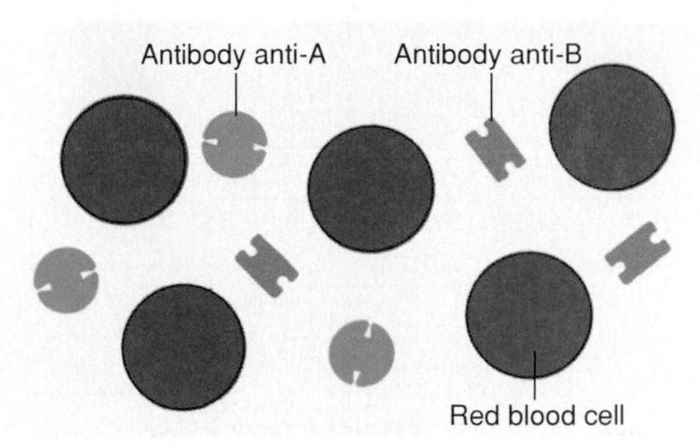


**FIGURE 14.22**
Each ABO blood type is distinguished by a different set of antigens and antibodies.

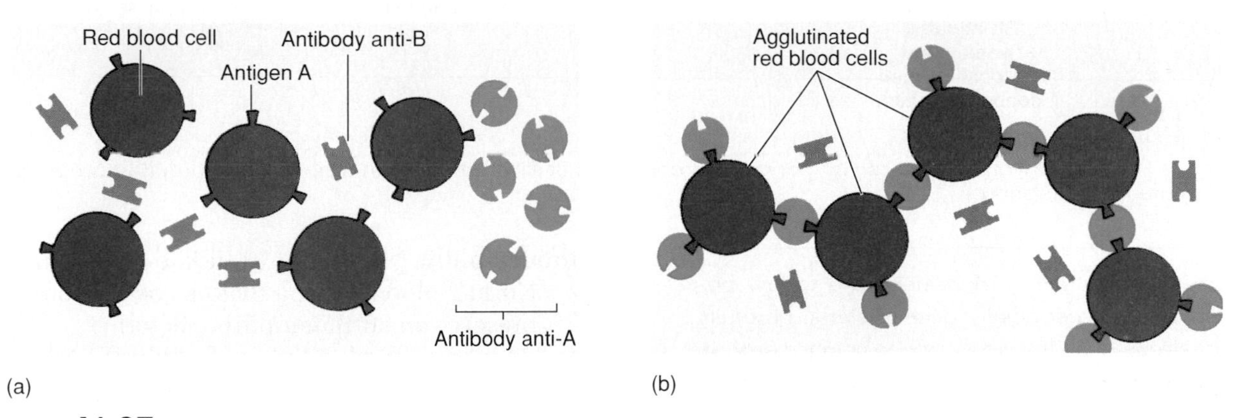


**FIGURE 14.23**
(*a*) If red blood cells with antigen A are added to blood containing antibody anti-A, (*b*) the antibodies react with the antigens, causing clumping (agglutination).

the fetal red cells. The fetus then develops a condition called **erythroblastosis fetalis** (hemolytic disease of the newborn).

Clinical Application 14.5 discusses attempts to simulate human blood.

1. What is the Rh blood group?
2. What are two ways that Rh incompatibility can arise?

**CHART 14.12 Antigens and Antibodies of the ABO Blood Group**

| Blood Type | Antigen | Antibody |
|---|---|---|
| A | A | anti-B |
| B | B | anti-A |
| AB | A and B | Neither anti-A nor anti-B |
| O | Neither A nor B | Both anti-A and anti-B |

## CHART 14.13 Preferred and Permissible Blood Types for Transfusions

| Blood Type of Recipient | Preferred Blood Type of Donor | Permissible Blood Type of Donor (in an Extreme Emergency) |
|---|---|---|
| A | A | A,O |
| B | B | B,O |
| AB | AB | AB, A, B, O |
| O | O | O |

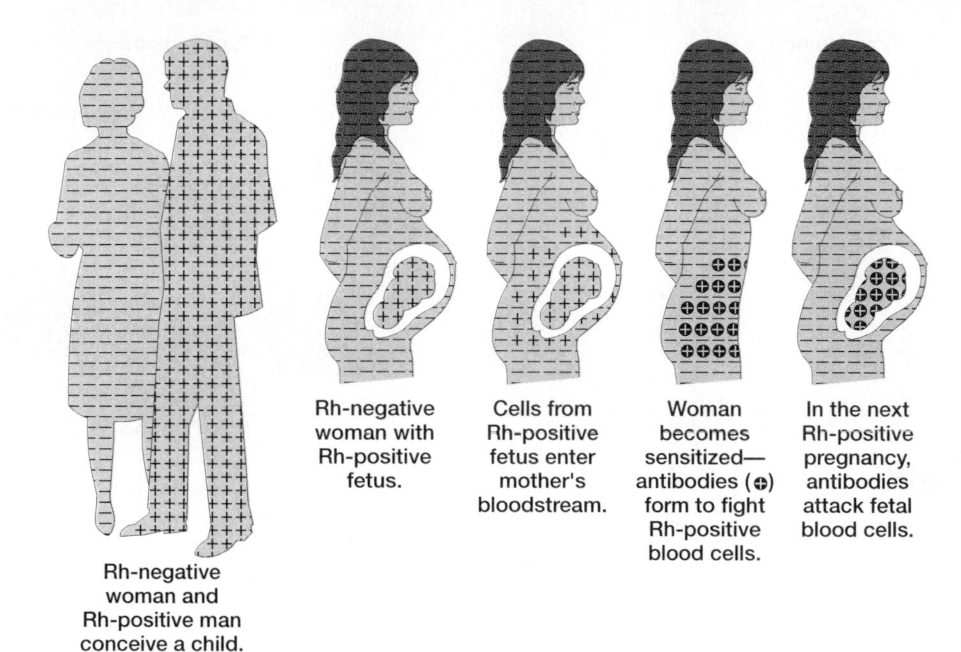


**FIGURE 14.24**

If a man who is $Rh^+$ and a woman who is $Rh^-$ conceive a child who is $Rh^+$, the woman's body may manufacture antibodies that attack future $Rh^+$ offspring.

Erythroblastosis fetalis is extremely rare today, because physicians, especially obstetricians, carefully track Rh status. An $Rh^-$ woman who might carry an $Rh^+$ fetus is given an injection of a drug called Rhogam. This is actually anti-Rh antibodies, which bind to and shield any $Rh^+$ fetal cells that might contact the woman's cells, sensitizing her immune system. Rhogam must be given within 72 hours of possible contact with $Rh^+$ cells—such situations include giving birth, terminating a pregnancy, miscarrying, or undergoing amniocentesis (a prenatal test in which a needle is inserted into the uterus).

## Clinical Terms Related to the Blood

**anisocytosis** (an-i″so-si-to′sis) A condition characterized by an abnormal variation in the size of erythrocytes.

**antihemophilic plasma** (an″ti-he″mo-fil′ik plaz′mah) Normal blood plasma that has been processed to preserve an antihemophilic factor.

**citrated whole blood** (sit′rāt-ed hōl blud) Normal blood to which a solution of acid citrate has been added to prevent coagulation.

**dried plasma** (drīd plaz′mah) Normal blood plasma that has been vacuum dried to prevent the growth of microorganisms.

**hemorrhagic telangiectasia** (hem″o-raj′ik tel-an″je-ek-ta′ze-ah) A hereditary disorder characterized by a tendency to bleed from localized lesions of the capillaries.

**heparinized whole blood** (hep′er-i-nīzed″ hōl blud) Normal blood to which a solution of heparin has been added to prevent coagulation.

**macrocytosis** (mak″ro-si-to′sis) A condition characterized by the presence of abnormally large erythrocytes.

**microcytosis** (mi″kro-si-to′sis) A condition characterized by the presence of abnormally small erythrocytes.

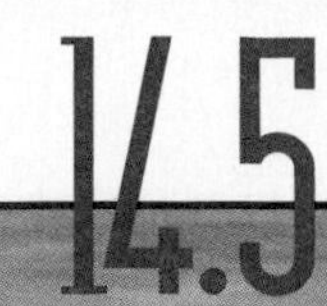

CLINICAL APPLICATION 14.5

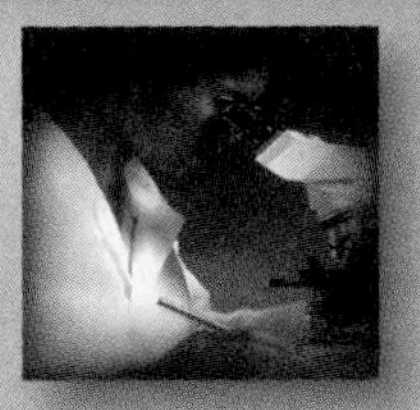

## Replacing Blood

In the past, a blood transfusion used whole blood. Today, whole blood is often separated into its component parts, and only those needed are used.

A patient with anemia or an acute blood loss might receive concentrated (packed) red blood cells; a person with too few platelets might be given platelets; a person with cancer whose treatment has depleted the white blood cell count might be given a white blood cell preparation. Similarly, blood plasma might be used to replace lost blood volume, or to provide clotting factors.

Before donor blood is used to obtain blood components, it is tested for particular viruses, including those that cause hepatitis B and AIDS (see chapter 16). Then the blood components must be stored properly. For example, packed red blood cells can be stored for several years if they are frozen, but can be used for only about a month if they are not frozen. After thawing, however, such cells must be used within about a day. Platelet preparations must be used within five days, and white blood cell concentrates must be used immediately.

Synthetic bloods—all still experimental—try to duplicate blood's oxygen-carrying ability. They are variations on chemicals that bind oxygen, mimicking the hemoglobin carried in red blood cells.

The first such synthetic blood was silicone oil. A mouse immersed into a beaker of silicone oil, for example, can "breathe" the oxygen in the oil (fig. 14D). Researchers have tried related chemicals containing fluorine and carbon, which bind oxygen similarly to silicone oil. In 1990, the first red blood cell substitute was approved for use to maintain localized blood flow during certain surgical procedures. The product consists of two fluorine compounds, a mild detergent, and lipid from egg yolk.

In another approach, hemoglobin molecules are linked together and administered alone, rather than in cells as occurs in the circulatory system. Yet another type of artificial blood consists of hemoglobin packaged into synthetic microcapsules, which are tiny fat bubbles currently used in carbonless carbon paper and paper that releases an odor when scratched. However, these "neohemocytes" so far provoke an immune attack from the animals in which they have been tested. Artificial bloods based on hemoglobin do not offer the freedom from contamination possible with a completely synthetic product.

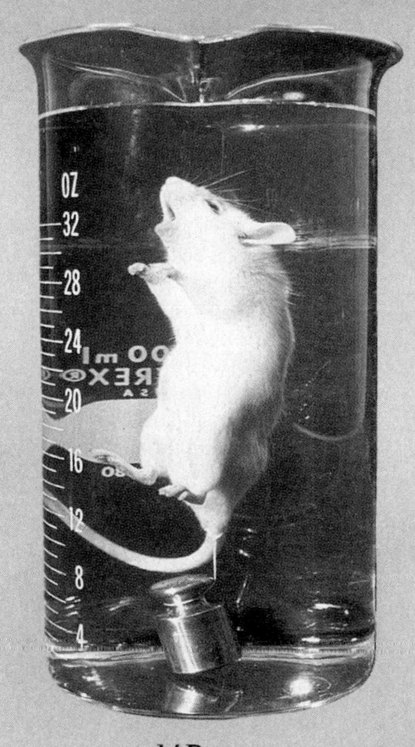

FIGURE 14D
This mouse is "breathing" a fluorocarbon liquid that is saturated with oxygen. Such a compound may one day be useful in humans as a blood substitute.

**neutrophilia** (nu″tro-fil′e-ah) A condition in which the number of circulating neutrophils has increased.

**packed red cells** A concentrated suspension of red blood cells from which the plasma has been removed.

**pancytopenia** (pan″si-to-pe′ne-ah) A condition characterized by an abnormal depression of all the cellular components of blood.

**poikilocytosis** (poi″ki-lo-si-to′sis) A condition in which the erythrocytes are irregularly shaped.

**purpura** (per′pu-rah) A disease characterized by spontaneous bleeding into the tissues and through the mucous membrane.

**septicemia** (sep″ti-se′me-ah) A condition in which disease-causing microorganisms or their toxins are present in the blood.

**spherocytosis** (sfēr″o-si-to′sis) Hemolytic anemia caused by defective proteins supporting the cell membranes of red blood cells. The cells are abnormally spherical.

**thalassemia** (thal″ah-se′me-ah) A group of hereditary hemolytic anemias characterized by the presence of very thin, fragile erythrocytes.

## CHAPTER SUMMARY

### Introduction (page 530)

Blood is often considered a type of connective tissue whose cells are suspended in a liquid intercellular material. It transports substances between the body cells and the external environment, and helps maintain a stable cellular environment.

### Blood and Blood Cells (page 530)

Blood contains red blood cells, white blood cells, and platelets.

1. Blood volume and composition
   a. Blood volume varies with body size, fluid and electrolyte balance, and adipose tissue content.
   b. Blood can be separated into formed elements and liquid portions.
      (1) The formed elements portion is mostly red blood cells.
      (2) The liquid plasma includes water, nutrients, hormones, electrolytes, and cellular wastes.
2. The origin of blood cells
   a. Blood cells develop from hemocytoblasts in bone marrow.
   b. Cells descended from hemocytoblasts respond to colony stimulating factors to specialize.
   c. Thrombopoietin stimulates megakaryocytes to give rise to platelets, much as erythropoietin stimulates formation of red blood cells.
3. Characteristics of red blood cells
   a. Red blood cells are biconcave disks with shapes that provide increased surface area and place their cell membranes close to internal structures.
   b. They contain hemoglobin, which combines loosely with oxygen.
   c. The mature form lacks nuclei, but contain enzymes needed for energy-releasing processes.
4. Red blood cell counts
   a. The red blood cell count equals the number of cells per $mm^3$ of blood.
   b. The average count may range from approximately 4,000,000 to 6,000,000 cells per $mm^3$.
   c. Red cell count is related to the oxygen-carrying capacity of the blood and is used in diagnosing and evaluating the courses of diseases.
5. Destruction of red blood cells
   a. Red cells are fragile and are damaged while moving through capillaries.
   b. Macrophages in the liver and spleen phagocytize damaged red blood cells.
   c. Hemoglobin molecules are decomposed, and the iron they contain is recycled.
   d. Biliverdin and bilirubin are pigments released from hemoglobin.
6. Red blood cell production and its control
   a. During fetal development, red cells form in the yolk sac, liver, and spleen; later, red cells are produced by the red bone marrow.
   b. The number of red blood cells remains relatively stable.
   c. A negative feedback mechanism involving erythropoietin from the kidneys and liver controls rate of red cell production.
      (1) Erythropoietin is released in response to low oxygen levels.
      (2) High altitude, loss of blood, or chronic lung disease can lower oxygen concentration.
7. Dietary factors affecting red blood cell production
   a. The availability of vitamin $B_{12}$, iron, and folic acid affect red blood cell production.
   b. The rate of iron absorption varies with the amount of iron in the body.
8. Types of white blood cells
   a. White blood cells control infection.
   b. Granulocytes include neutrophils, eosinophils, and basophils.
   c. Agranulocytes include monocytes and lymphocytes.
9. White blood cell counts
   a. Normal total white blood cell counts vary from 5,000 to 10,000 cells per $mm^3$ of blood.
   b. The number of white cells may change in abnormal conditions such as infections, emotional disturbances, or excessive loss of body fluids.
   c. A differential white cell count indicates the percentages of various types of leukocytes present.
10. Functions of white blood cells
    a. Neutrophils and monocytes phagocytize foreign particles.
    b. Chemicals released by damaged cells attract and stimulate leukocytes.
    c. Eosinophils kill parasites and help control inflammation and allergic reactions.
    d. Basophils release heparin, which inhibits blood clotting.
    e. Lymphocytes produce antibodies that attack specific foreign antigens.
11. Blood platelets
    a. Blood platelets are fragments of giant cells that detach and enter the circulation.
    b. The normal count varies from 130,000 to 360,000 platelets per $mm^3$.
    c. They help close breaks in blood vessels.

### Blood Plasma (page 543)

Plasma is the liquid part of the blood that is composed of water and a mixture of organic and inorganic substances. It transports nutrients and gases, helps regulate fluid and electrolyte balance, and helps maintain stable pH.

1. Plasma proteins
   a. Plasma proteins remain in blood and interstitial fluids, and are not normally used as energy sources.
   b. Three major groups exist.
      (1) Albumins help maintain the osmotic pressure of blood.
      (2) Globulins include antibodies. They provide immunity and transport lipids and fat-soluble vitamins.
      (3) Fibrinogen functions in blood clotting.
2. Nutrients and gases
   a. Plasma nutrients include amino acids, simple sugars, and lipids.
      (1) Glucose is stored in the liver as glycogen and is released whenever the blood glucose concentration falls.

(2) Amino acids are used to synthesize proteins and are deaminated for use as energy sources.
(3) Lipoproteins function in the transport of lipids.
b. Gases in plasma include oxygen, carbon dioxide, and nitrogen.
3. Nonprotein nitrogenous substances
a. Nonprotein nitrogenous substances are composed of molecules that contain nitrogen atoms but are not proteins.
b. They include amino acids, urea, uric acid, creatine, and creatinine.
(1) Urea and uric acid are products of catabolic metabolism.
(2) Creatinine results from the metabolism of creatine.
c. These substances usually remain stable; an increase may indicate a kidney disorder.
4. Plasma electrolytes
a. Plasma electrolytes are obtained by absorption from the intestines and are released as by-products of cellular metabolism.
b. They include ions of sodium, potassium, calcium, magnesium, chloride, bicarbonate, phosphate, and sulfate.
c. They are important in the maintenance of osmotic pressure and pH.

## Hemostasis (page 548)

Hemostasis refers to the stoppage of bleeding. Hemostatic mechanisms are most effective in controlling blood loss from small vessels.

1. Blood vessel spasm
a. Smooth muscles in walls of arterioles and arteries contract reflexly following injury.
b. Platelets release serotonin that stimulates vasoconstriction and helps maintain vessel spasm.
2. Platelet plug formation
a. Platelets adhere to rough surfaces and exposed collagen.
b. Platelets stick together at the sites of injuries and form platelet plugs in broken vessels.
3. Blood coagulation
a. Blood clotting is the most effective means of hemostasis and may be initiated by extrinsic or intrinsic mechanisms.
b. Clot formation depends on the balance between clotting factors that promote clotting and those that inhibit clotting.
c. The basic event of coagulation is the conversion of soluble fibrinogen into insoluble fibrin.
d. Biochemicals that promote clotting include prothrombin activator, prothrombin, and calcium ions.
e. After forming, the clot retracts and pulls the edges of a broken vessel closer together.
f. A thrombus is a blood clot in a vessel; an embolus is a clot or fragment of a clot that has moved in a vessel.
g. Fibroblasts invade a clot, forming connective tissue throughout.
h. Protein-splitting enzymes may eventually destroy a clot.
4. Prevention of coagulation
a. The smooth lining of blood vessels discourages the accumulation of platelets.
b. As a clot forms, fibrin adsorbs thrombin and prevents the reaction from spreading.
c. Antithrombin interferes with the action of excessive thrombin.
d. Some cells secrete heparin, an anticoagulant.

## Blood Groups and Transfusions (page 553)

Blood can be typed on the basis of the surface structures of its cells.

1. Antigens and antibodies
a. Red blood cell membranes may contain antigens and blood plasma may contain antibodies.
b. Blood typing involves identifying the antigens present in the red cell membranes.
2. ABO blood group
a. Blood can be grouped according to the presence or absence of antigens A and B.
b. Whenever antigen A is absent, antibody anti-A is present; whenever antigen B is absent, antibody anti-B is present.
c. Adverse transfusion reactions are avoided by preventing the mixing of red cells that contain an antigen with plasma that contains the corresponding antibody.
d. Adverse reactions involve agglutination (clumping) of the red blood cells.
3. Rh blood group
a. Rh antigens are present in the red cell membranes of Rh-positive blood; they are absent in Rh-negative blood.
b. If an Rh-negative person is exposed to Rh-positive blood, anti-Rh antibodies are produced in response.
c. Mixing Rh-positive red cells with plasma that contains anti-Rh antibodies agglutinates the positive cells.
d. If an Rh-negative female is pregnant with an Rh-positive fetus, some of the positive cells may enter the maternal blood at the time of birth and stimulate the maternal tissues to produce anti-Rh antibodies.
e. Anti-Rh antibodies in maternal blood may pass through the placental tissues and react with the red cells of an Rh-positive fetus.

## Critical Thinking Questions

1. What changes would you expect to occur in the hematocrit of a person who is dehydrated? Why?
2. If a patient with an inoperable cancer is treated using a drug that reduces the rate of cell division, what changes might occur in the patient's white blood cell count? How might the patient's environment be modified to compensate for the effects of these changes?
3. Hypochromic (iron-deficiency) anemia is relatively common among aging persons who are admitted to hospitals for other conditions. What environmental and sociological factors might promote this form of anemia?
4. How would you explain to a patient with leukemia, who has a greatly elevated white blood cell count, the importance of avoiding bacterial infections?
5. If a woman whose blood is Rh negative and contains anti-Rh antibodies is carrying a fetus with Rh negative blood, will the fetus be in danger of developing erythroblastosis fetalis? Why?
6. In the United States, between 1977 and 1985, more than 10,000 men with hemophilia contracted the human immunodeficiency virus (HIV) from contaminated factor VIII that they received to treat the hemophilia. What are two abnormalities in the blood of these men?

## Review Exercises

1. List the major components of blood.
2. Define *hematocrit,* and explain how it is determined.
3. Describe a red blood cell.
4. Distinguish between oxyhemoglobin and deoxyhemoglobin.
5. Explain what is meant by a *red blood cell count.*
6. Describe the life cycle of a red blood cell.
7. Distinguish between biliverdin and bilirubin.
8. Define *erythropoietin,* and explain its function.
9. Explain how vitamin $B_{12}$ and folic acid deficiencies affect red blood cell production.
10. List two sources of iron that can be used for the synthesis of hemoglobin.
11. Distinguish between granulocytes and agranulocytes.
12. Name five types of leukocytes, and list the major functions of each type.
13. Explain the significance of white blood cell counts as aids to diagnosing diseases.
14. Describe a blood platelet, and explain its functions.
15. Name three types of plasma proteins, and list the major functions of each type.
16. Define *lipoprotein.*
17. Define *apoprotein.*
18. Distinguish between low-density lipoprotein and high-density lipoprotein.
19. Name the sources of VLDL, LDL, HDL, and chylomicrons.
20. Describe how lipoproteins are removed from plasma.
21. Explain how cholesterol is eliminated from plasma and from the body.
22. Define *nonprotein nitrogenous substances,* and name those commonly present in plasma.
23. Name several plasma electrolytes.
24. Define *hemostasis.*
25. Explain how blood vessel spasms are stimulated following an injury.
26. Explain how a platelet plug forms.
27. List the major steps leading to the formation of a blood clot.
28. Distinguish between fibrinogen and fibrin.
29. Provide an example of a positive feedback system that operates during blood clotting.
30. Define *serum.*
31. Distinguish between a thrombus and an embolus.
32. Explain how a blood clot may be removed naturally from a blood vessel.
33. Describe how blood coagulation may be prevented.
34. State a vitamin required for blood clotting.
35. Distinguish between antigen and antibody.
36. Explain the basis of ABO blood types.
37. Explain why a person with blood type AB is sometimes called a universal recipient.
38. Explain why a person with blood type O is sometimes called a universal donor.
39. Distinguish between Rh-positive and Rh-negative blood.
40. Describe how a person may become sensitized to Rh-positive blood.
41. Define *erythroblastosis fetalis,* and explain how this condition may develop.

15

CHAPTER

# Cardiovascular System

Blood vessels branch, bringing sustenance to all parts of the body.

## Chapter Objectives

After you have studied this chapter, you should be able to:

1. Name the organs of the cardiovascular system and discuss their functions.
2. Name and describe the locations of the major parts of the heart and discuss the function of each part.
3. Trace the pathway of the blood through the heart and the vessels of the coronary circulation.
4. Discuss the cardiac cycle and explain how it is controlled.
5. Compare the structures and functions of the major types of blood vessels.
6. Describe the mechanisms that aid in returning venous blood to the heart.
7. Explain how blood pressure is produced and controlled.
8. Compare the pulmonary and systemic circuits of the cardiovascular system.
9. Identify and locate the major arteries and veins of the pulmonary and systemic circuits.

## Key Terms

**arteriole** (ar-te′re-ōl)
**atrium** (a′tre-um)
**cardiac conduction system** (kar′de-ak kon-duk′shun sis′tem)
**cardiac cycle** (kar′de-ak si′kl)
**cardiac output** (kar′de-ak owt′poot)
**diastolic pressure** (di″ah-stol′ik presh′ur)
**electrocardiogram** (e-lek″tro-kar′de-o-gram″)
**endocardium** (en″do-kar′de-um)
**epicardium** (ep″i-kar′de-um)
**functional syncytium** (funk′shun-al sin-sish′e-um)
**myocardium** (mi″o-kar′de-um)
**pacemaker** (pās′māk-er)
**pericardium** (per″i-kar′de-um)
**peripheral resistance** (pe-rif′er-al re-zis′tans)
**pulmonary circuit** (pul′mo-ner″e sur′kit)
**sphygmomanometer** (sfig″mo-mah-nom′ĕ-ter)
**systemic circuit** (sis-tem′ik sur′kit)
**systolic pressure** (sis-tol′ik presh′ur)
**vasoconstriction** (vas″o-kon-strik′shun)
**vasodilation** (vas″o-di-la′shun)
**ventricle** (ven′tri-kl)
**venule** (ven′ūl)
**viscosity** (vis-kos′i-te)

Massachusetts physician John T. Harrington's history of heart trouble dates back to when he was 8 years old and suffered a bout of rheumatic fever. He was ill for several weeks but recovered completely—or so he thought for many years.

Rheumatic heart disease is an inflammation that follows an infection by *Streptococcus* bacteria. Because the cell surfaces of these bacteria bear striking resemblances to those of heart valve cells, the alerted immune system sometimes attacks the valve cells, mistaking them for bacteria. (The valves are tissue flaps that route blood through the heart.) The result—valve damage that can take many years to produce symptoms.

Harrington's aortic valve—the one leading from his heart—was indeed damaged. It kept him out of sports in high school, then out of Vietnam, and became an object of much curiosity and amusement to his medical school buddies. Although the valve damage produced an interesting sound and abnormalities in various diagnostic tests, he felt no ill effects until 1983.

At that time, Harrington began jogging, just a little, a half mile down a hill from his house, and the half mile back. But late that year, he started feeling a pressure in his chest upon completing the workout, which would vanish when he rested. Finally, in early 1984, at age 47, he saw a cardiologist, who said he needed a "valve job."

Harrington's first replacement valve was bioprosthetic, meaning that it came from an animal—specifically, the outer heart muscle layer of a cow. It wouldn't last more than 10 years, but the human body is less likely to react to a bioprosthetic valve by forming dangerous blood clots than it is to a valve made of synthetic materials.

The cow valve lasted 8 years. Just after Christmas in 1992, while sitting at the breakfast table with his young grandsons, Harrington experienced crushing chest pain. The valve had torn. Waking up in the hospital some time later, Harrington learned that he now had a synthetic valve. This one should last about 30 years.

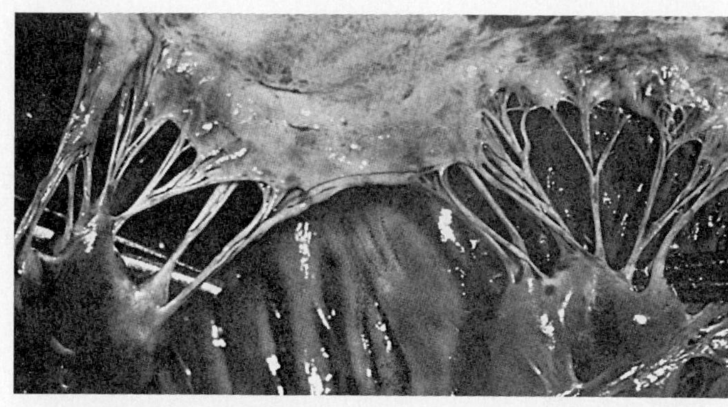

(a)

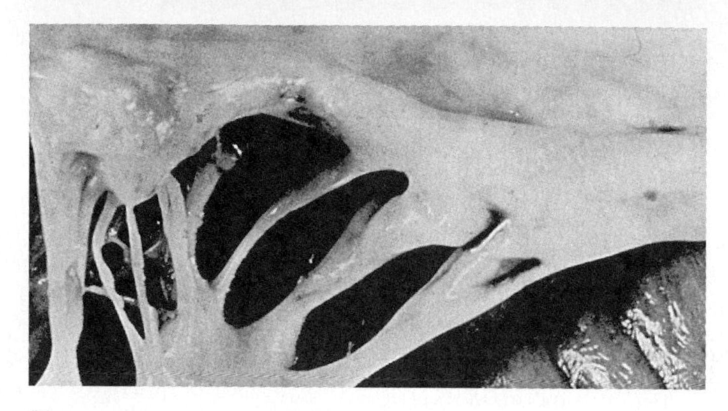

(b)

Rheumatic fever progresses to rheumatic heart disease when heart valves become scarred. The valve on the top is healthy (*a*); the thickened and scarred valve leaflets on the bottom are the result of rheumatic heart disease, a misplaced immune attack triggered by a bacterial infection (*b*).

A heart valve is only a tiny component of the vast cardiovascular system, but size has nothing to do with importance, as Harrington's reliance on a replacement heart valve indicates—an experience he shares with 75,000 more people each year.

A functional cardiovascular system is vital for survival, because without circulation the tissues lack a supply of oxygen and nutrients, and wastes accumulate. Under such conditions the cells soon begin irreversible change, which quickly leads to death. Figure 15.1 shows the general pattern of the cardiovascular system.

## Structure of the Heart

The heart is a hollow, cone-shaped, muscular pump located within the mediastinum of the thorax and resting upon the diaphragm.

### Size and Location of the Heart

Although heart size varies with body size, an average adult's heart is generally about 14 centimeters long and 9 centimeters wide (fig. 15.2).

The heart is bordered laterally by the lungs, posteriorly by the backbone, and anteriorly by the sternum (fig. 15.3 and reference plates 50, 56, 71, and 72). Its *base*, which is attached to several large blood vessels, lies beneath the second rib. Its distal end extends downward and to the left, terminating as a bluntly pointed *apex* at the level of the fifth intercostal space. For this reason, it is possible to sense the *apical heartbeat* by

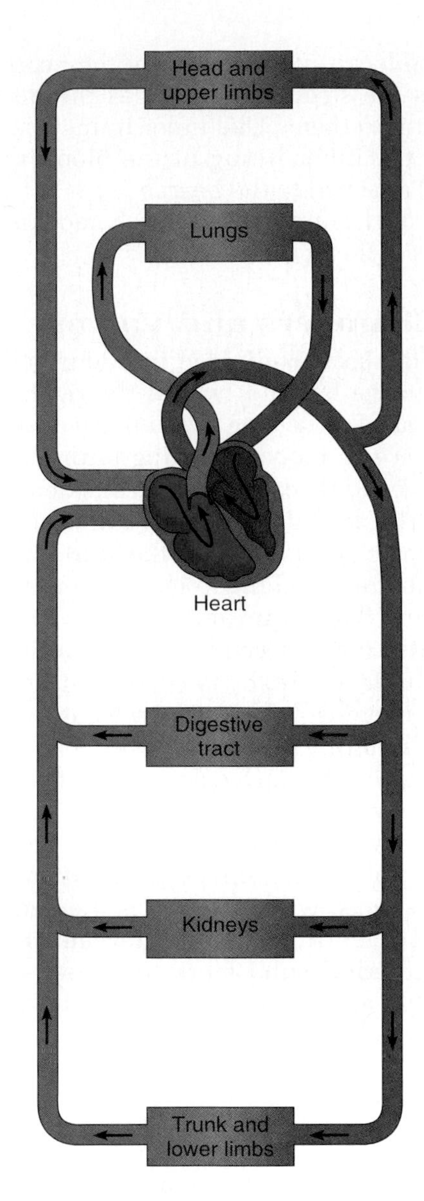


**FIGURE 15.1**

The cardiovascular system transports blood between the body cells and organs such as the lungs, intestines, and kidneys that communicate with the external environment.

feeling or listening to the chest wall between the fifth and sixth ribs, about 7.5 centimeters to the left of the midline.

## Coverings of the Heart

The **pericardium** encloses the heart and the proximal ends of the large blood vessels to which it attaches. The pericardium consists of an outer fibrous bag, the *fibrous pericardium,* that surrounds a more delicate, double-layered sac. The inner layer of this sac, the *visceral pericardium* (epicardium), covers the heart. At the base of the heart, the visceral pericardium turns back upon itself to become the *parietal pericardium.* The parietal pericardium, in turn, forms the inner lining of the fibrous pericardium.

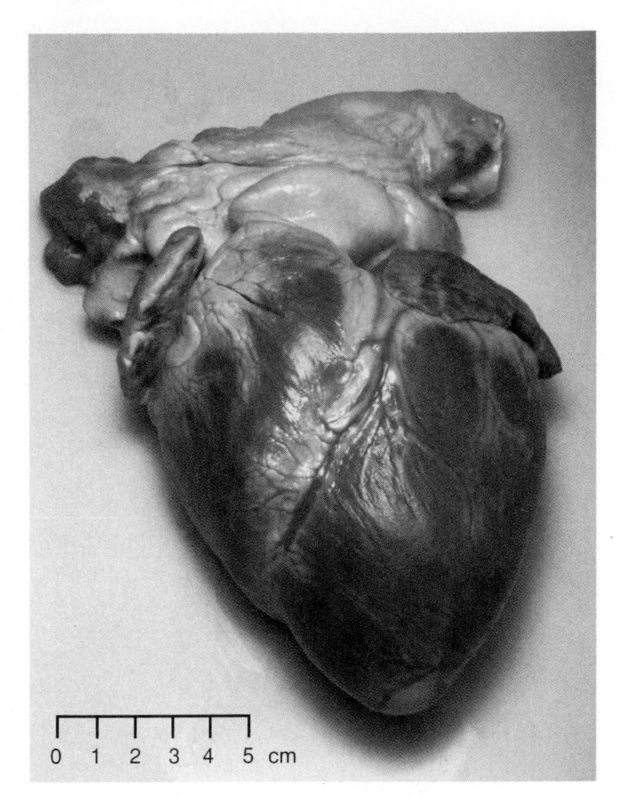


**FIGURE 15.2**

Anterior view of a human heart.

The fibrous pericardium is a tough, protective sac composed largely of white fibrous connective tissue. It is attached to the central portion of the diaphragm, the posterior of the sternum, the vertebral column, and the large blood vessels emerging from the heart (see figs. 1.9, 15.4 and reference plates 56 and 57). Between the parietal and visceral layers of the pericardium is a space, the *pericardial cavity,* that contains a small amount of serous fluid that the pericardial membranes secrete. This fluid reduces friction between the pericardial membranes as the heart moves within them.

In *pericarditis,* inflammation of the pericardium due to viral or bacterial infection produces adhesions that stick together the layers of the pericardium. This is very painful and interferes with heart movements.

1. Where is the heart located?
2. Where would you listen to hear the apical heartbeat?
3. Distinguish between the visceral pericardium and the parietal pericardium.
4. What is the function of the fluid in the pericardial cavity?

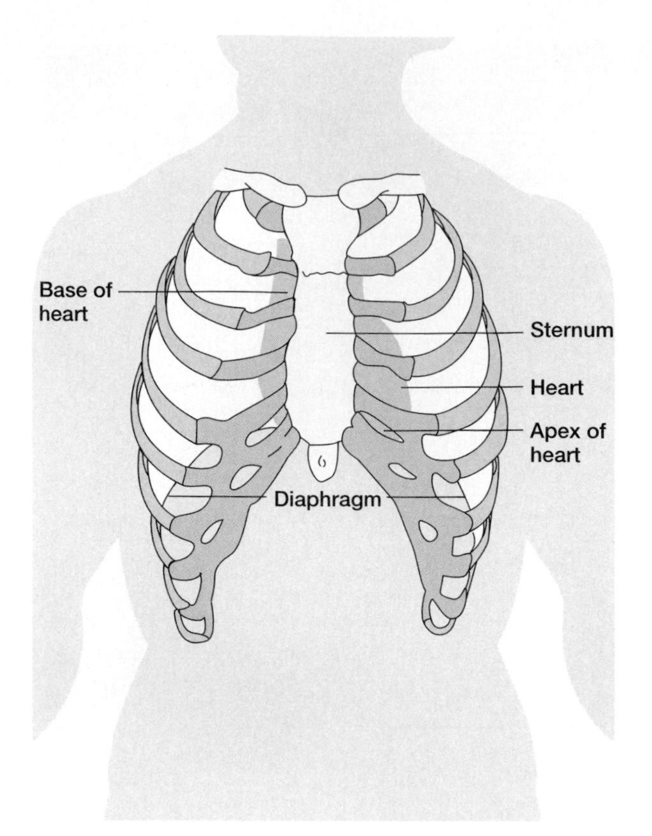


**FIGURE 15.3**
The heart is located posterior to the sternum, where it lies upon the diaphragm.

## Wall of the Heart

The wall of the heart is composed of three distinct layers: an outer **epicardium,** a middle **myocardium,** and an inner **endocardium** (fig. 15.5).

The epicardium, which corresponds to the visceral pericardium, is protective. It is a serous membrane that consists of connective tissue covered by epithelium, and it includes blood capillaries, lymph capillaries, and nerve fibers. The deeper portion of the epicardium often contains fat, particularly along the paths of coronary arteries and cardiac veins that provide blood flow through the myocardium.

The middle layer, or myocardium, is relatively thick and consists largely of the cardiac muscle tissue that forces blood out of the heart chambers. The muscle fibers are arranged in planes, separated by connective tissues that are richly supplied with blood capillaries, lymph capillaries, and nerve fibers.

The inner layer, or endocardium, consists of epithelium and connective tissue that contains many elastic and collagenous fibers. The endocardium also contains blood vessels and some specialized cardiac muscle fibers called *Purkinje fibers,* described later in this chapter.

The endocardium lines all of the heart chambers and covers the structures, such as the heart valves, that project into them. This inner lining is also continuous with the inner lining of the blood vessels (endothelium) attached to the heart.

Chart 15.1 summarizes the characteristics of the three layers of the heart wall.

## Heart Chambers and Valves

Internally, the heart is divided into four hollow chambers, two on the left and two on the right. The upper chambers, called **atria** (sing. *atrium*), have relatively thin walls and receive blood returning to the heart. Small, ear-like projections, called **auricles,** extend anteriorly from the atria (see fig. 15.4). The lower chambers, the **ventricles,** force the blood out of the heart into arteries.

A structure called the *interatrial septum* separates the right from the left atrium. An *interventricular septum* separates the two ventricles. The atrium on each side communicates with its corresponding ventricle through an opening called the **atrioventricular orifice,** guarded by an *atrioventricular valve* (*A-V valve*).

Grooves on the surface of the heart mark the divisions between its chambers, and they also contain major blood vessels that supply the heart tissues. The deepest of these grooves is the **atrioventricular** (coronary) **sulcus,** which encircles the heart between the atria and ventricles. Two **interventricular** (anterior and posterior) **sulci** mark the septum that separates the right and left ventricles (see fig. 15.4).

When increasing blood volume stretches muscle cells associated with the atria, the cells secrete a peptide hormone called *atrial natriuretic peptide* (ANP). ANP inhibits release of renin from the kidneys and of aldosterone from the adrenal cortex. The overall result is increased excretion of sodium ions and water from the kidneys and lowered blood volume and pressure. Researchers are investigating use of ANP to treat high blood pressure.

1. Describe the layers of the heart wall.
2. Name and locate the four chambers of the heart.
3. Name the orifices between the upper and the lower chambers of the heart.
4. Name the structure that separates the right and left sides of the heart.

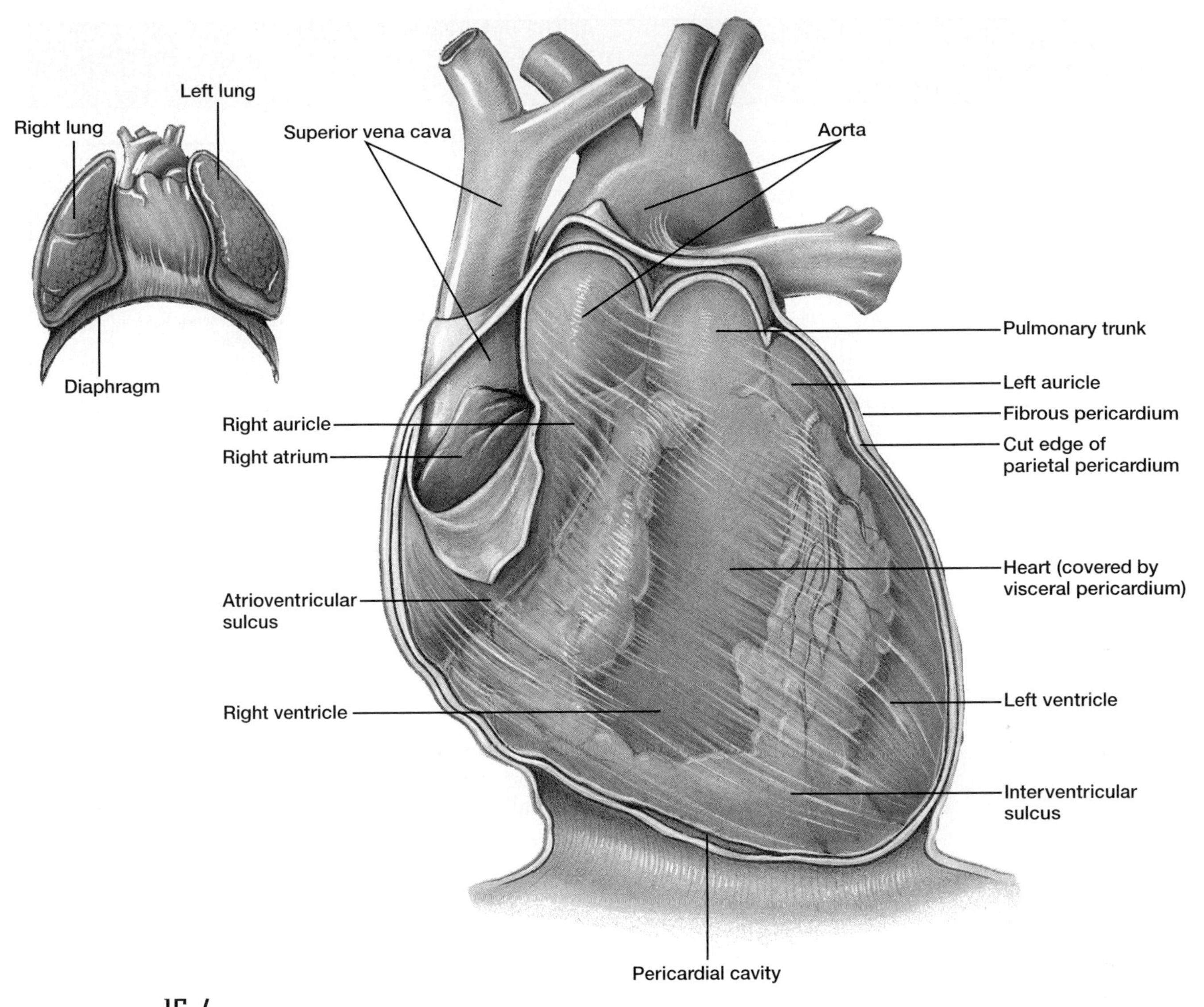


**FIGURE 15.4**
The heart is within the mediastinum and is enclosed by a layered pericardium.

The right atrium receives blood from two large veins: the *superior vena cava* and the *inferior vena cava.* These veins return blood that is low in oxygen from tissues. A smaller vein, the *coronary sinus,* also drains blood into the right atrium from the wall of the heart.

A large **tricuspid valve** guards the atrioventricular orifice between the right atrium and the right ventricle. It is composed of three leaflets, or cusps, as its name implies. This valve permits the blood to move from the right atrium into the right ventricle and prevents it from moving in the opposite direction. The cusps fold passively out of the way against the ventricular wall when the blood pressure is greater on the atrial side, and they close passively when the pressure is greater on the ventricular side (figs. 15.6, 15.7, 15.8, and 15.9).

Strong, fibrous strings, called *chordae tendineae,* are attached to the cusps on the ventricular side. These strings originate from small mounds of cardiac muscle tissue, the **papillary muscles,** that project inward from the walls of the ventricle. When the tricuspid valve closes, the chordae tendineae and papillary muscles prevent the cusps from swinging back into the atrium because they contract when the ventricle contracts.

The right ventricle has a thinner muscular wall than the left ventricle. This right chamber pumps the blood a fairly short distance to the lungs against a relatively low resistance to blood flow. The left ventricle, on the other hand, must force the blood to all the other parts of the body against a much greater resistance to flow.

## Chart 15.1 Wall of the Heart

| Layer | Composition | Function |
|---|---|---|
| Epicardium (visceral pericardium) | Serous membrane of connective tissue covered with epithelium and including blood capillaries, lymph capillaries, and nerve fibers | Forms a protective outer covering; secretes serous fluid |
| Myocardium | Cardiac muscle tissue separated by connective tissues and including blood capillaries, lymph capillaries, and nerve fibers | Contracts to force blood from the heart chambers |
| Endocardium | Membrane of epithelium and connective tissues, including elastic and collagenous fibers, blood vessels, and specialized muscle fibers | Forms a protective inner lining of the chambers and valves |

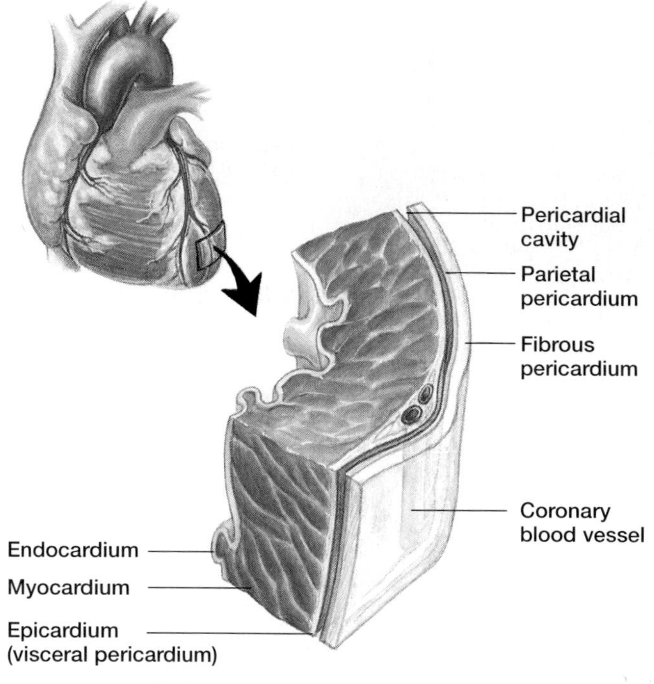


**Figure 15.5**
The wall of the heart consists of three layers: an endocardium, a myocardium, and an epicardium.

When the muscular wall of the right ventricle contracts, the blood inside its chamber is put under increasing pressure, and the tricuspid valve closes passively. As a result, the only exit is through the *pulmonary trunk,* which divides to form the left and right *pulmonary arteries.* At the base of this trunk is a **pulmonary valve** (pulmonary semilunar valve), which consists of three cusps (see fig. 15.8). This valve opens as the right ventricle contracts. However, when the ventricular muscles relax, the blood begins to back up in the pulmonary trunk. This closes the pulmonary valve, preventing a return flow into the ventricle.

The left atrium receives the blood from the lungs through four *pulmonary veins*—two from the right lung and two from the left lung. The blood passes from the left atrium into the left ventricle through the atrioventricular orifice, which a valve guards. This valve consists of two leaflets, and it is appropriately named the **bicuspid,** or **mitral** (shaped like a miter, a turbanlike headdress), **valve.** It prevents the blood from flowing back into the left atrium from the ventricle. As with the tricuspid valve, the papillary muscles and the chordae tendineae prevent the cusps of the bicuspid valve from swinging back.

When the left ventricle contracts, the bicuspid valve closes passively, and the only exit is through a large artery called the *aorta.* Its branches distribute blood to all parts of the body.

At the base of the aorta is an **aortic valve** (aortic semilunar valve) that consists of three cusps (see fig. 15.8). It opens and allows blood to leave the left ventricle as it contracts. When the ventricular muscles relax, this valve closes and prevents blood from backing up into the ventricle. Chart 15.2 summarizes the heart valves.

*Mitral valve prolapse* (MVP) affects up to 6% of the U.S. population. In this condition, one (or both) of the cusps of the mitral valve stretches, so that it bulges into the left atrium during ventricular contraction. Although the valve usually continues to function adequately, sometimes blood regurgitates into the left atrium. In a stethoscope, a regurgitating mitral valve prolapse sounds like a click at the end of ventricular contraction, then a murmur as blood goes back through the valve into the left atrium.

Symptoms of MVP include chest pain, palpitations, fatigue, and anxiety. These people are particularly susceptible to *endocarditis,* an inflammation of the endocardium due to bacterial infection. It appears as a plantlike growth on the valve. Individuals with MVP must take antibiotics before undergoing dental work, to prevent infection by *Streptococcus* bacteria in the mouth.

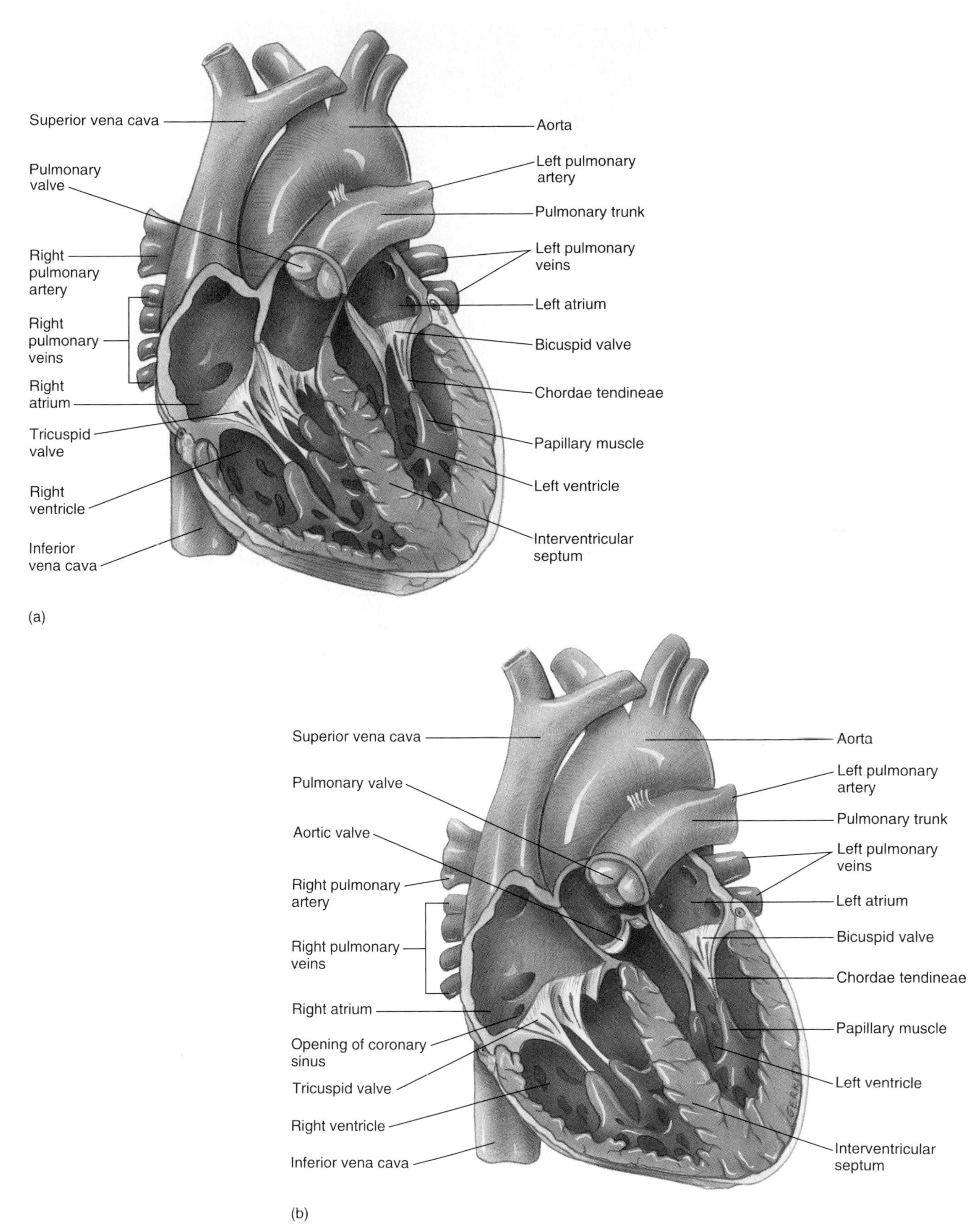


**FIGURE 15.6**

Coronal sections of the heart (*a*) showing the connection between the right ventricle and the pulmonary trunk and (*b*) showing the connection between the left ventricle and the aorta.

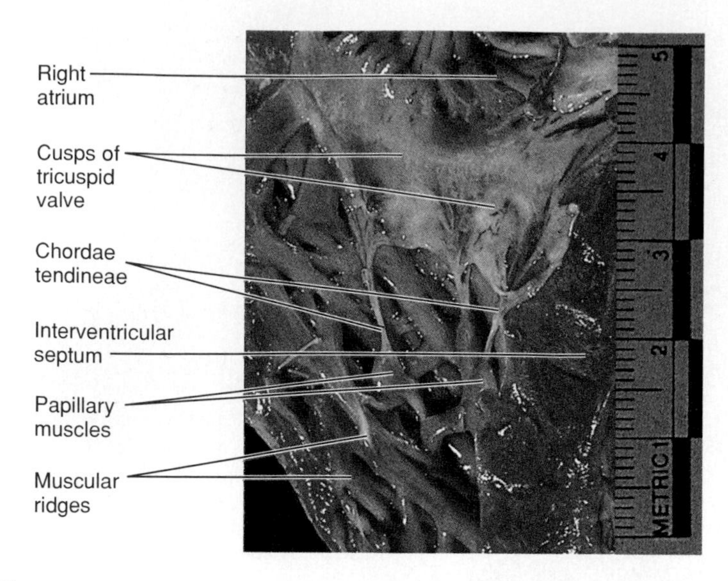

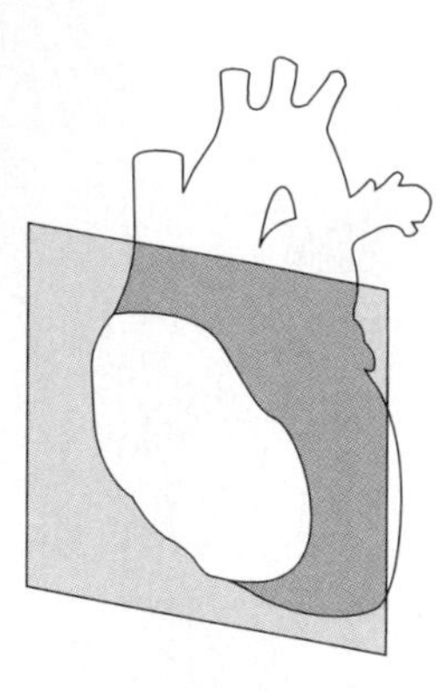

**FIGURE 15.7**
Photograph of a human tricuspid valve.

**CHART 15.2 Valves of the Heart**

| Valve | Location | Function | Valve | Location | Function |
|---|---|---|---|---|---|
| Tricuspid valve | Right atrioventricular orifice | Prevents blood from moving from right ventricle into right atrium during ventricular contraction | Bicuspid (mitral) valve | Left atrioventricular orifice | Prevents blood from moving from left ventricle into left atrium during ventricular contraction |
| Pulmonary valve | Entrance to pulmonary trunk | Prevents blood from moving from pulmonary trunk into right ventricle during ventricular relaxation | Aortic valve | Entrance to aorta | Prevents blood from moving from aorta into left ventricle during ventricular relaxation |

1. Which blood vessels carry blood into the right atrium?
2. Where does the blood go after it leaves the right ventricle?
3. Which blood vessels carry blood into the left atrium?
4. What prevents blood from flowing back into the ventricles when they are relaxed?

## Skeleton of the Heart

Rings of dense fibrous connective tissue surround the pulmonary trunk and aorta at their proximal ends. These rings are continuous with others that encircle the atrioventricular orifices. They provide firm attachments for the heart valves and for muscle fibers, and prevent the outlets of the atria and ventricles from dilating during contraction. The fibrous rings, together with other masses of dense fibrous tissue in the upper portion of the septum between the ventricles (interventricular septum), constitute the *skeleton of the heart* (see fig. 15.9).

## Path of Blood Through the Heart

Blood that is relatively low in oxygen concentration and relatively high in carbon dioxide concentration enters the right atrium through the venae cavae and the coronary sinus. As the right atrial wall contracts, the blood passes through the right atrioventricular orifice and enters the chamber of the right ventricle (fig. 15.10).

When the right ventricular wall contracts, the tricuspid valve closes the right atrioventricular orifice, and the blood moves into the pulmonary trunk and its branches (pulmonary arteries). From these vessels, blood enters the capillaries associated with the alveoli of the lungs. Gas exchanges occur between

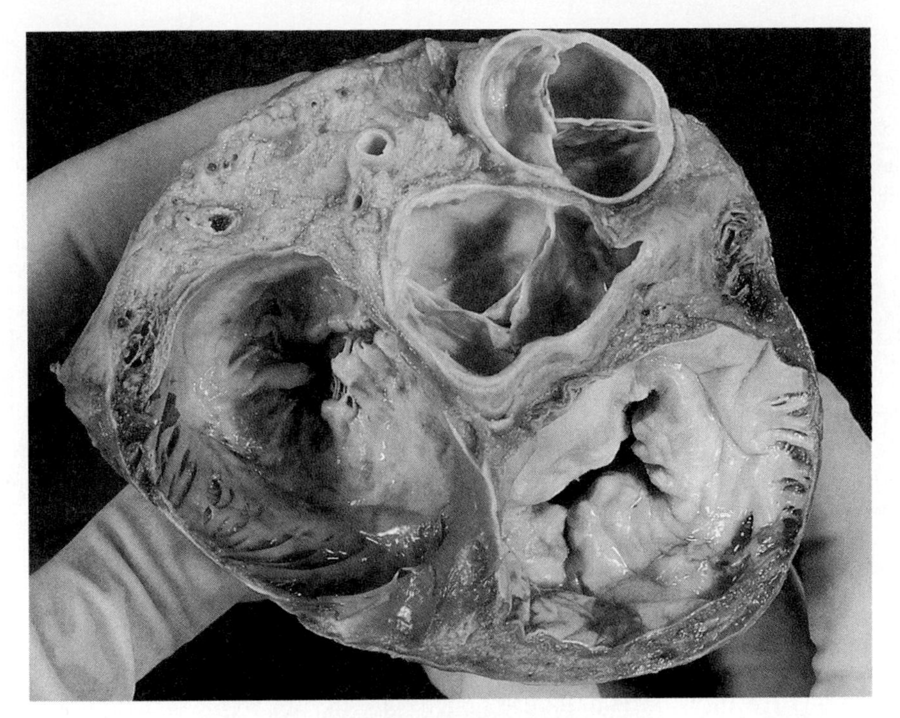

**FIGURE 15.8**
Photograph of the pulmonary and aortic valves of the heart (superior view).

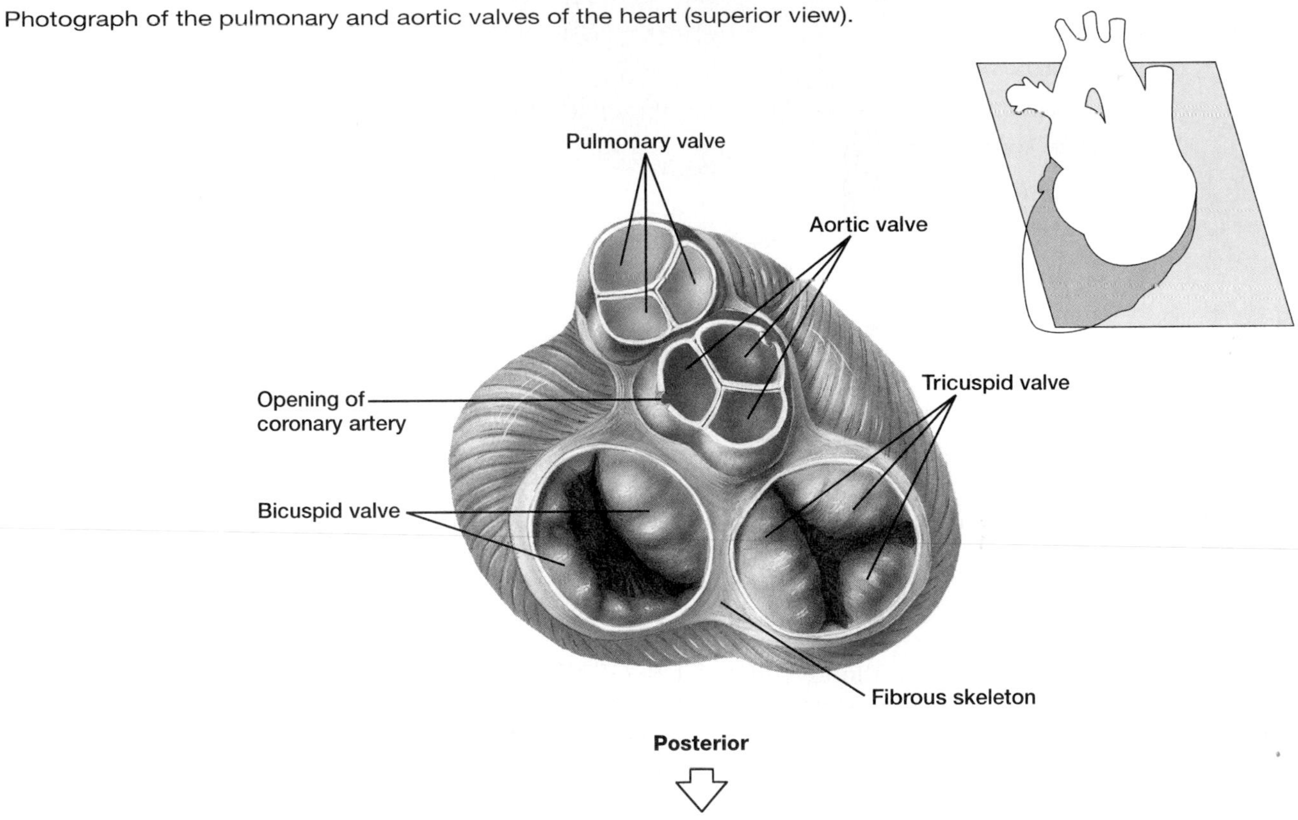


**FIGURE 15.9**
The skeleton of the heart consists of fibrous rings to which the heart valves are attached (superior view).

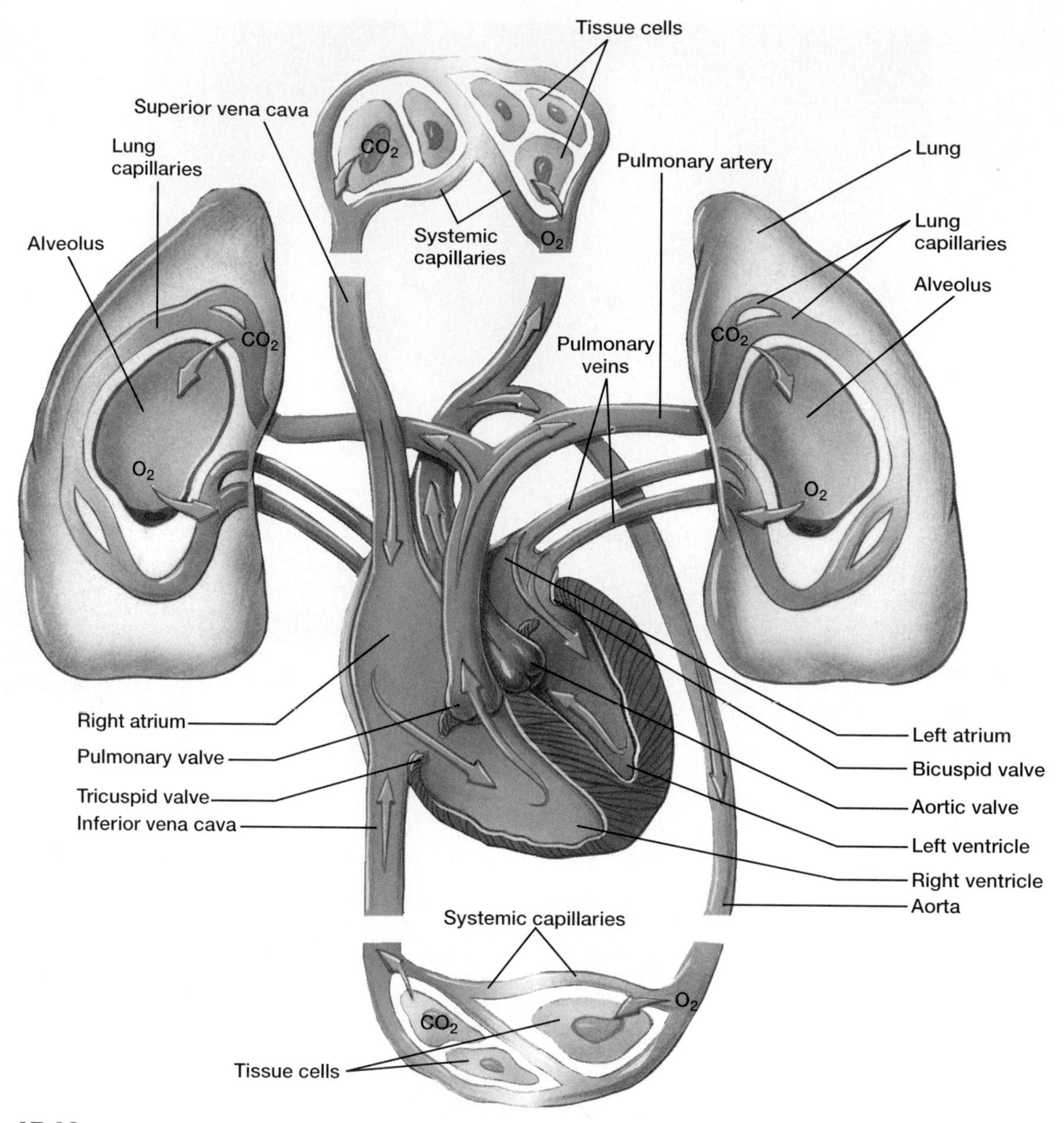


**FIGURE 15.10**

The right ventricle forces blood to the lungs, while the left ventricle forces blood to all other body parts (structures are not drawn to scale).

the blood in the capillaries and the air in the alveoli. The freshly oxygenated blood, which is now relatively low in carbon dioxide, returns to the heart through the pulmonary veins that lead to the left atrium.

The left atrial wall contracts, and the blood moves through the left atrioventricular orifice and into the ch... of the left ventricle. When the left ven-tr... ontracts, the bicuspid valve closes the ...ular orifice, and the blood passes into ... branches. Figure 15.11 summarizes ... takes as it moves through the heart ... circuit.

*Magnetic resonance imaging* (MRI) can noninvasively image coronary arteries. Blood flow appears as a bright signal, and areas of diminished or absent blood flow, or blood turbulence, appear as blank areas. This approach, still experimental, is less invasive than the standard procedure of *coronary angiography,* in which a catheter is snaked through a blood vessel into the heart and a contrast image is used to show heart structure.

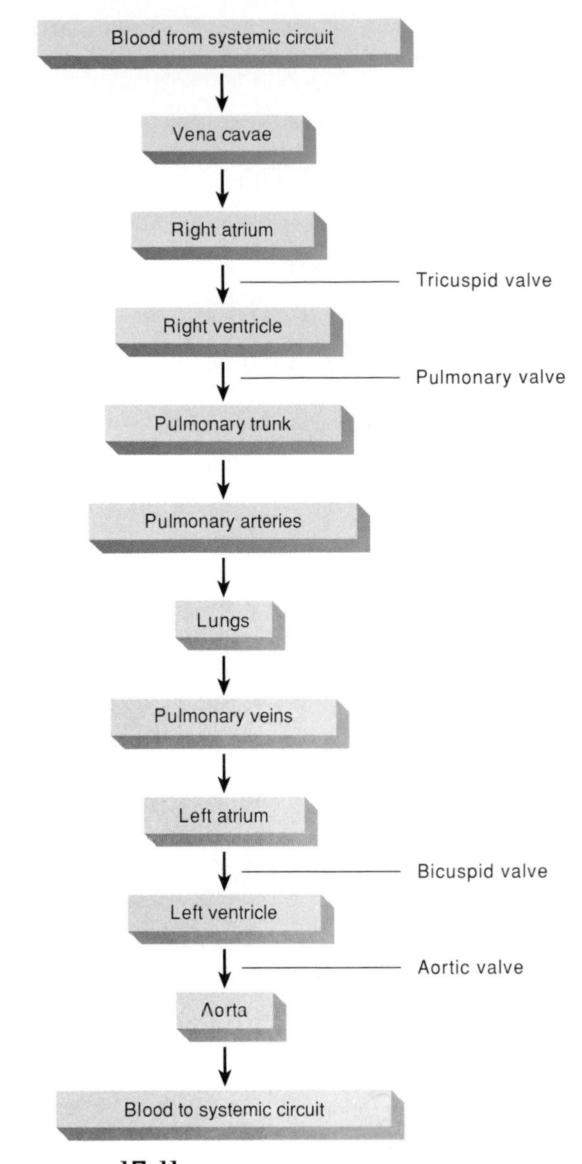


**FIGURE 15.11**
Path of blood through the heart and pulmonary circuit.

## Blood Supply to the Heart

Blood is supplied to the tissues of the heart by the first two branches of the aorta, called the right and left **coronary arteries.** Their openings lie just beyond the aortic valve (figs. 15.12 and 15.13).

One branch of the left coronary artery, the *circumflex artery,* follows the atrioventricular sulcus between the left atrium and the left ventricle. Its branches supply blood to the walls of the left atrium and the left ventricle. Another branch of the left coronary artery, the *anterior interventricular artery* (or *left anterior descending artery*), travels in the anterior interventricular sulcus, and its branches supply the walls of both ventricles.

The right coronary artery passes along the atrioventricular sulcus between the right atrium and the right ventricle. It gives off two major branches—a *posterior interventricular artery,* which travels along the posterior interventricular sulcus and supplies the walls of both ventricles, and a *marginal artery,* which passes along the lower border of the heart. Branches of the marginal artery supply the walls of the right atrium and the right ventricle (figs. 15.12, 15.13, and 15.14).

Because the heart must beat continually to supply blood to the tissues, myocardial cells require a constant supply of freshly oxygenated blood. The myocardium contains many capillaries fed by branches of the coronary arteries. The smaller branches of these arteries usually have connections (anastomoses) between vessels that provide alternate pathways for blood, called collateral circulation.

In most body parts, blood flow in arteries peaks during ventricular contraction. However, blood flow in the vessels of the myocardium is poorest during ventricular contraction. This is because the muscle fibers of the myocardium compress nearby vessels as they contract, interfering with blood flow. Also, the openings into the coronary arteries are partially blocked as the flaps of the aortic valve open. Conversely, during ventricular relaxation, the myocardial vessels are no longer compressed and the orifices of the coronary arteries are not blocked by the aortic valve. This increases blood flow into the myocardium.

The blood that has passed through the capillaries of the myocardium is drained by branches of the **cardiac veins,** whose paths roughly parallel those of the coronary arteries. As figure 15.13*b* shows, these veins join the **coronary sinus,** an enlarged vein on the posterior surface of the heart in the atrioventricular sulcus. The coronary sinus empties into the right atrium. Figure 15.15 summarizes the path of the blood that supplies the tissues of the heart.

Blockage or narrowing of a branch of a coronary artery by a thrombus or embolus deprives the myocardial cells it supplies of oxygen, producing ischemia and a painful condition called *angina pectoris.* The pain usually occurs during physical activity, when the myocardial cells' oxygen requirements exceed the blood oxygen supply. Pain lessens with rest. Emotional disturbance may also trigger angina pectoris.

Angina pectoris may cause a sensation of heavy pressure, tightening, or squeezing in the chest. The pain is usually felt in the region behind the sternum or in the anterior portion of the upper thorax, but may radiate to the neck, jaw, throat, arm, shoulder, elbow, back, or upper abdomen. Other symptoms include profuse perspiration (diaphoresis), difficulty breathing (dyspnea), nausea, or vomiting.

A blood clot completely obstructing a coronary artery or one of its branches (coronary thrombosis) kills part of the heart. This is a *myocardial infarction* (MI), more commonly known as a heart attack.

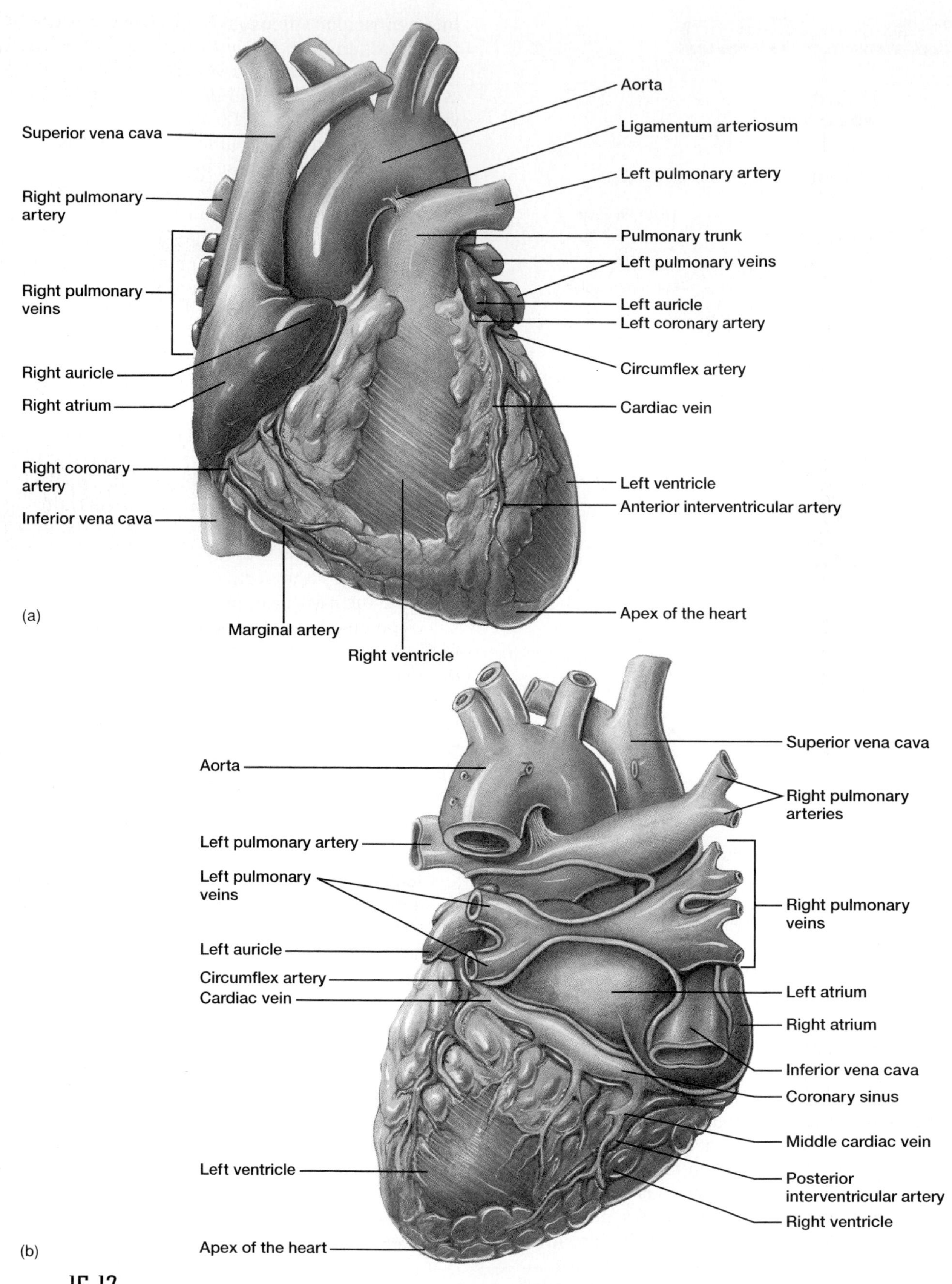


**FIGURE 15.12**
Blood vessels associated with the surface of the heart. (*a*) Anterior view; (*b*) posterior view.

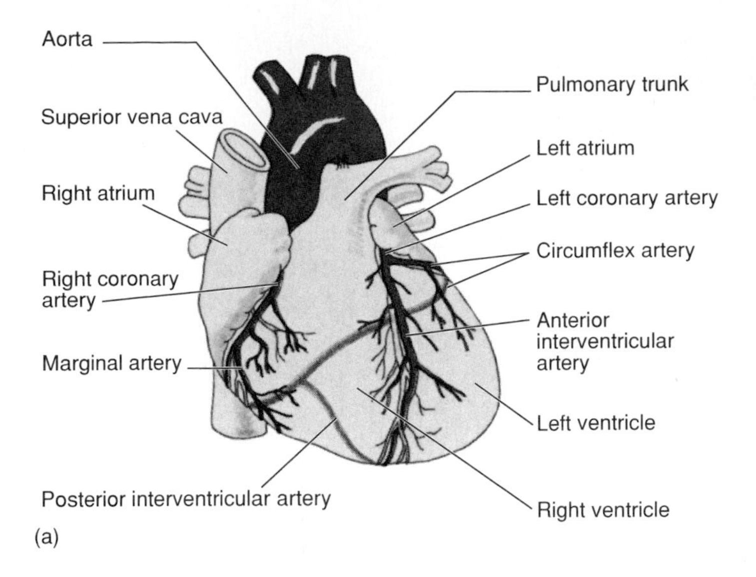


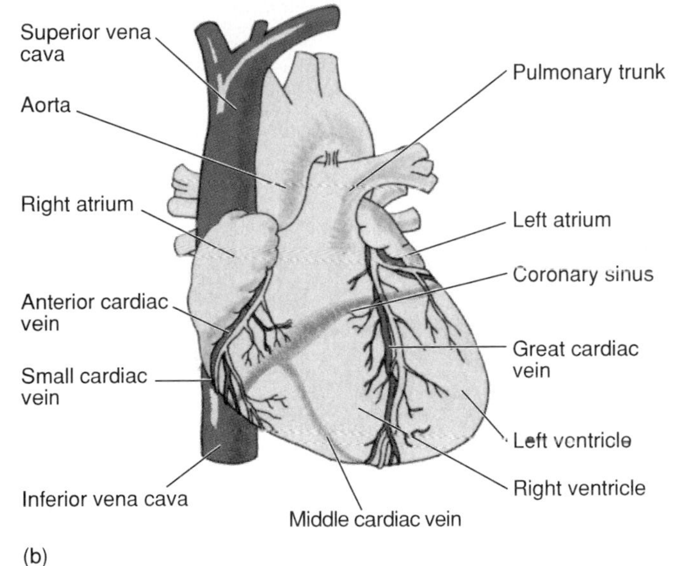


**FIGURE 15.13**

(*a*) Branches of the coronary arteries supply blood to heart tissues. (*b*) Branches of the cardiac veins drain blood from heart tissues. (Faded color indicates vessels on posterior surface.)

1. What structures make up the skeleton of the heart?
2. Review the path of blood through the heart.
3. How does blood composition differ in the right and left ventricle?
4. What vessels supply blood to the myocardium?
5. How does blood return from the cardiac tissues to the right atrium?

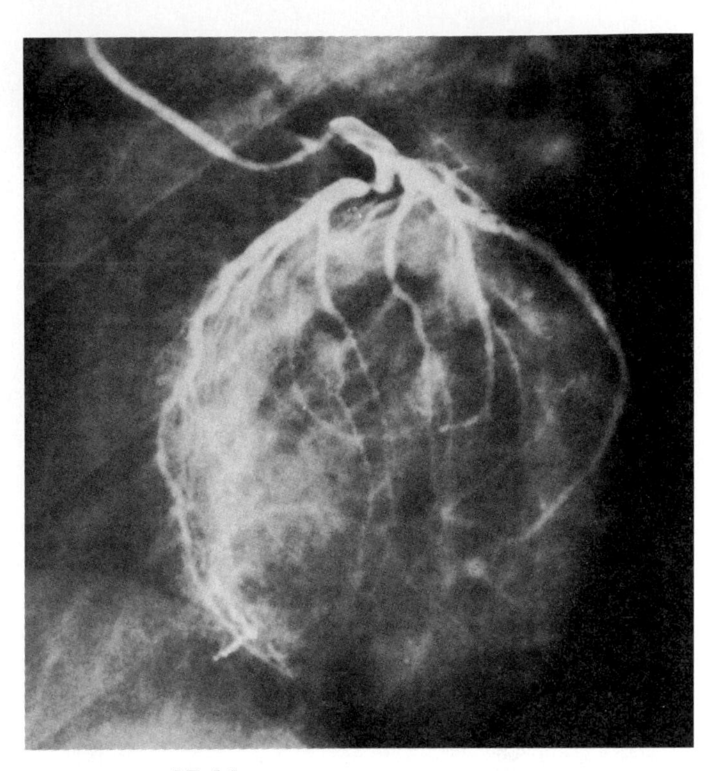

**FIGURE 15.14**

An angiogram (X-ray film) of the coronary arteries is a diagnostic procedure used to examine specific blood vessels.

In *heart transplantation*, the recipient's failing heart is removed, except for the posterior walls of the right and left atria and their connections to the venae cavae and pulmonary veins. The donor heart is prepared similarly and is attached to the atrial cuffs remaining in the recipient's thorax. Finally, the recipient's aorta and pulmonary arteries are connected to those of the donor heart.

A study of more than 8,000 recipients of heart transplants since 1985 shows that likelihood of survival is greatly increased if the pattern of a person's cell surfaces (HLA antigens) closely matches that of the cells of the donor heart. If three to six HLA genes do not match, the chance of success is 71%; if one or none do not match, the chance of success is 83%. But HLA typing takes time. Since no more than 4 hours can elapse between availability of a heart and the transplant, HLA typing, given current technology, is not usually possible.

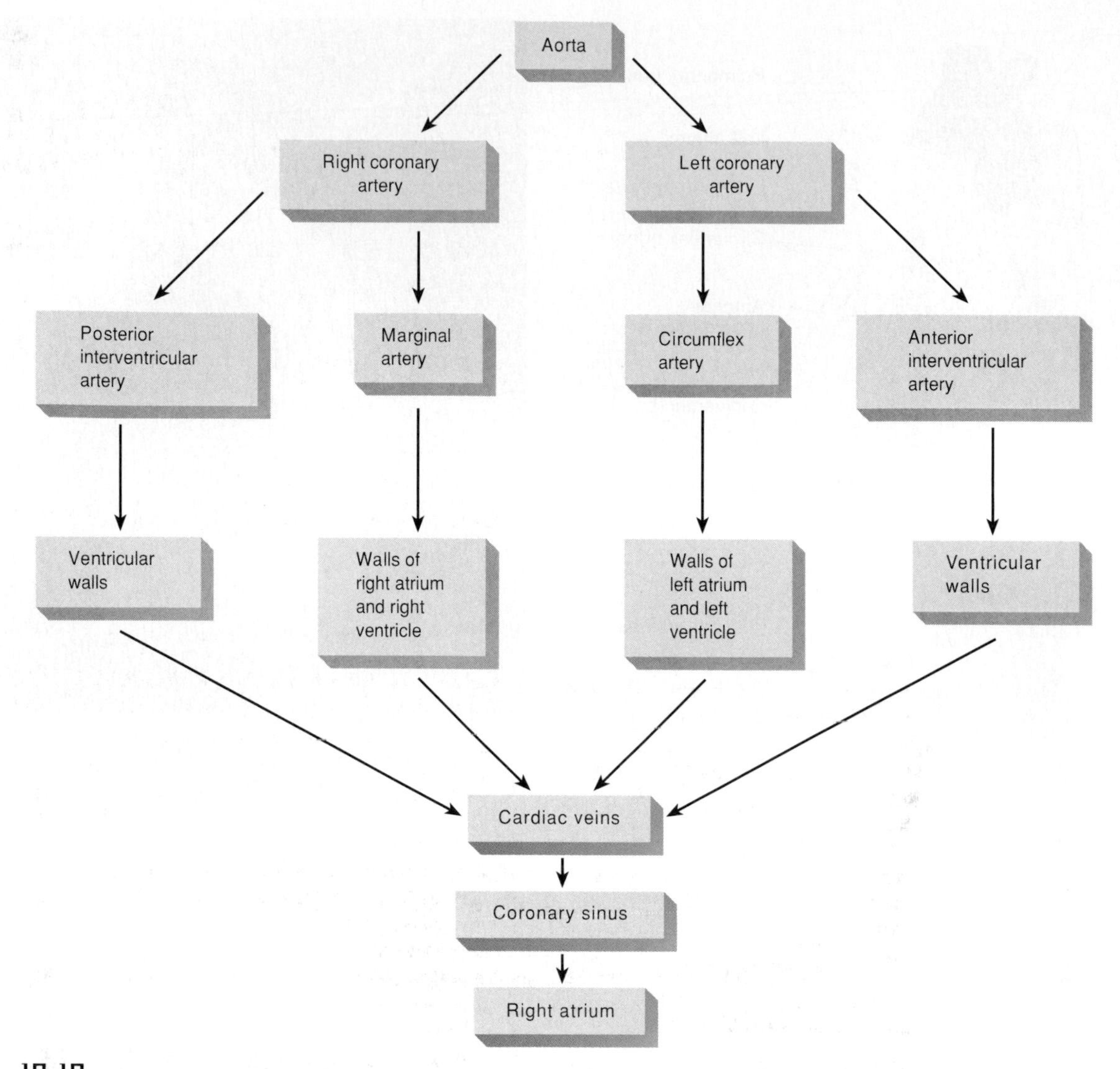


**FIGURE 15.15**

Path of blood through the coronary circulation.

## HEART ACTIONS

Heart chambers function in a coordinated, interacting manner, so that the atrial walls contract while the ventricular walls relax, and ventricular walls contract while the atrial walls relax. Then, the atria and the ventricles relax for a moment. Such a series of events constitutes a complete heartbeat, or **cardiac cycle.**

### Cardiac Cycle

During a cardiac cycle, the pressure within the chambers of the heart rises and falls. For example, when the atria are relaxed, the blood flows into them from the large, attached veins. As these chambers fill, the pressure inside gradually increases, forcing the A-V valves open. About 70% of the entering blood flows directly into the ventricles through the atrioventricular orifices before the atrial walls contract. Then, during atrial contraction (atrial systole), the atrial pressure rises suddenly, forcing the remaining 30% of the atrial contents into the ventricles. This is followed by atrial relaxation (atrial diastole) (fig. 15.16).

As the ventricles contract (ventricular systole), the A-V valves guarding the atrioventricular orifices close passively and begin to bulge back into the atria, increasing atrial pressure sharply. At the same time, the papillary muscles contract, and by pulling on the chordae tendineae, they prevent the cusps of the A-V valves from bulging too far into the atria. The atrial pressure soon falls, however, as blood flows out of the ventricles into the arteries. During ventricular contraction, the A-V valves remain closed, and the atrial pressure gradually increases as the atria fill with blood. When the ventricles relax (ventricular diastole), the A-V valves open passively, the blood flows through them into the ventricles, and the atrial pressure drops to a low point.

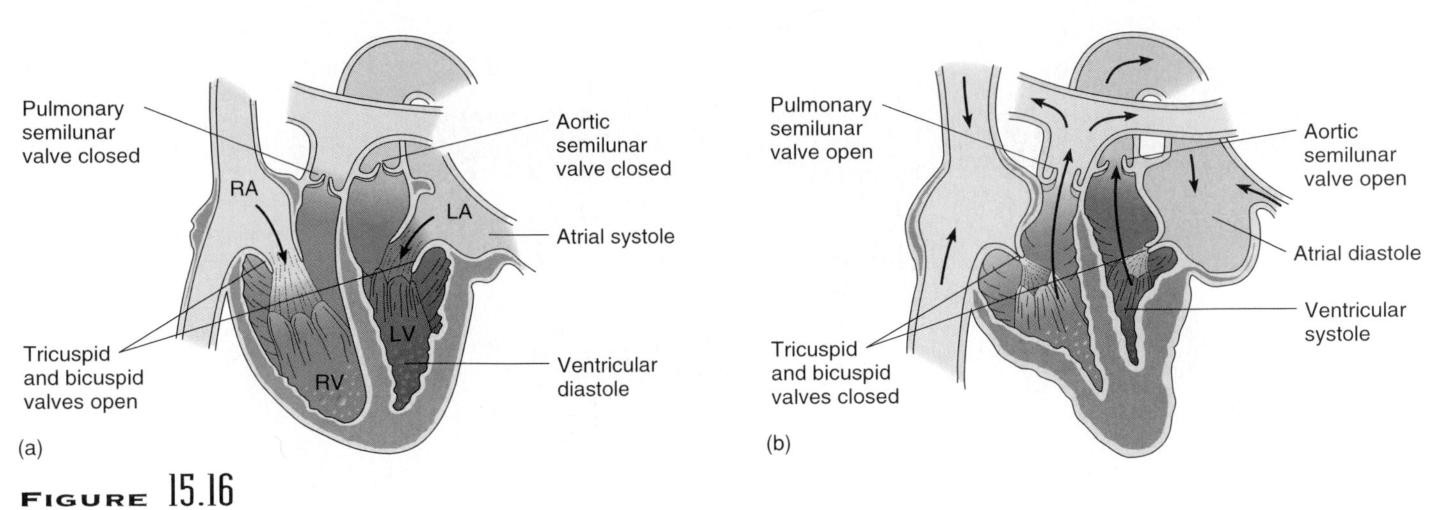


**FIGURE 15.16**
(*a*) The atria empty during atrial systole and (*b*) fill with blood during atrial diastole.

Pressure in the ventricles is low while they are filling, but when the atria contract, the ventricular pressure increases slightly. Then, as the ventricles contract, the ventricular pressure rises sharply, and as soon as the pressure exceeds that in the atria, the A-V valves close. The ventricular pressure continues to increase until it exceeds the pressure in the pulmonary trunk and aorta. Then, the pulmonary and aortic valves open, and blood is ejected from the ventricles into these arteries. When the ventricles are nearly empty, the ventricular pressure begins to drop, and it continues to drop as the ventricles relax. When the ventricular pressure is less than that in the arteries, the pulmonary and aortic valves are closed by arterial blood flowing back toward the ventricles. As soon as the ventricular pressure falls below that of the atria, the A-V valves open, and the ventricles begin to fill once more. The graph in figure 15.17 summarizes some of the changes that occur in the left ventricle during a cardiac cycle.

## Heart Sounds

A heartbeat heard with a stethoscope sounds like "lub–dup." These sounds are due to vibrations in the heart tissues produced as the blood flow is suddenly speeded or slowed with the contraction and relaxation of the heart chambers, and with the opening and closing of the valves.

The first part of a heart sound (*lub*) occurs during the ventricular contraction, when the A-V valves are closing. The second part (*dup*) occurs during ventricular relaxation, when the pulmonary and aortic valves are closing (see fig. 15.17).

Sometimes during inspiration, the interval between the closure of the pulmonary and the aortic valves is long enough that a sound related to each of these events can be heard. In this case, the second heart sound is said to be *split.*

Heart sounds are of particular interest because they provide information concerning the condition of the heart valves. For example, inflammation of the endocardium (endocarditis) may change the shapes of the valvular cusps (valvular stenosis). Then, the cusps may close incompletely, and some blood may leak back through the valve. This produces an abnormal sound called a *murmur.* The seriousness of a murmur depends on the extent of valvular damage. Many heart murmurs are harmless. Fortunately for those who have serious problems, it is often possible to repair the damaged valves or to replace them, as happened to the doctor described at the chapter's beginning. Clinical Application 15.1 describes more severe heart problems.

Using a stethoscope, it is possible to hear sounds associated with the aortic and pulmonary valves by listening from the second intercostal space on either side of the sternum. The *aortic sound* is heard on the right, and the *pulmonic sound* is heard on the left. The sound associated with the bicuspid (mitral) valve can be heard from the fifth intercostal space at the nipple line on the left. The sound of the tricuspid valve can be heard at the tip of the sternum (fig. 15.18).

## Cardiac Muscle Fibers

Recall that cardiac muscle fibers function like those of skeletal muscles, but the fibers connect in branching networks. Stimulation to any part of the network sends impulses throughout the heart, which contracts as a unit.

A mass of merging cells that act as a unit is called a **funtional syncytium.** Two such structures are

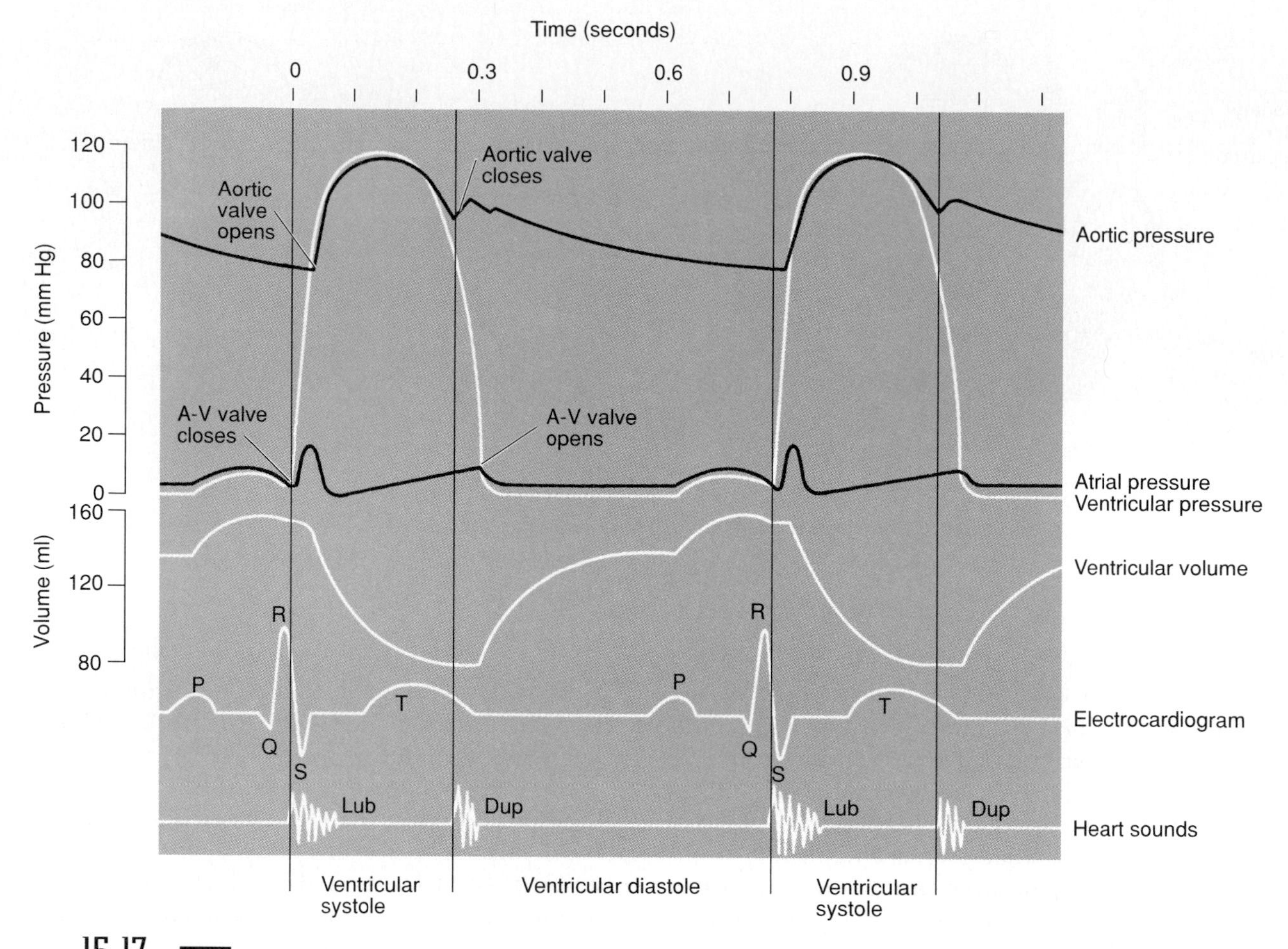


**FIGURE 15.17**

A graph of some of the changes that occur in the left ventricle during a cardiac cycle.

in the heart—in the atrial walls and in the ventricular walls. These masses of cardiac muscle fibers are separated from each other by portions of the heart's fibrous skeleton, except for a small area in the right atrial floor. In this region, the *atrial syncytium* and the *ventricular syncytium* are connected by fibers of the **cardiac conduction system.**

1. Describe the pressure changes that occur in the atria and ventricles during a cardiac cycle.
2. What causes heart sounds?
3. What is a functional syncytium?
4. Where are the functional syncytia of the heart located?

## Cardiac Conduction System

Throughout the heart are clumps and strands of specialized cardiac muscle tissue whose fibers contain only a few myofibrils. Instead of contracting, these areas initiate and distribute impulses (cardiac impulses) throughout the myocardium. They comprise the cardiac conduction system, which coordinates the events of the cardiac cycle.

A key portion of this conduction system is the **sinoatrial node** (S-A node), a small, elongated mass of specialized cardiac muscle tissue just beneath the epicardium. It is located in the right atrium near the opening of the superior vena cava, and its fibers are continuous with those of the atrial syncytium.

The membranes of the nodal cells contact one another, and these cells can excite themselves. That is, without being stimulated by nerve fibers or any other outside agents, the nodal cells initiate impulses that spread into the surrounding myocardium and stimulate the cardiac muscle fibers to contract.

S-A node activity is rhythmic. The S-A node initiates one impulse after another, seventy to eighty times a minute in an adult. Thus, it is responsible for the rhythmic contractions of the heart and is often called the **pacemaker.** From the SA node, bundles of atrial muscle, called *internodal atrial muscle,* preferentially conduct impulses along tracts to specific regions of the heart.

As a cardiac impulse travels from the S-A node into the atrial syncytium, it travels from cell to cell via gap junctions. The right and left atria contract almost simultaneously. Instead of passing directly into the ventricular syncytium, which is separated from the

CLINICAL APPLICATION 15.1

## Heart Transplants

When Tina Orbacz was pregnant with her second child in 1990, she attributed her increasing fatigue to her pregnant state. But a month after her son's birth, she was even more exhausted. Plus, she had lost weight during her pregnancy, not gained it. Finally, cardiologists discovered that she was suffering from heart failure due to a birth defect, called an *atrial septal defect,* that weakened the tissue between the atria. Because her heart was failing, and a complication of this disorder is lung failure, Tina Orbacz, at the age of 24 years, found herself waiting for a heart-lung transplant.

Tina was lucky. Her condition stabilized, and she was able to survive the three-year wait until a 16-year-old who died in an accident donated a heart and lungs. With youth and a good HLA match on her side, Tina has done extraordinarily well. Today, she exercises regularly and leads a normal life.

Because of the shortage of donor hearts, and the severity of the illnesses that lead to heart failure, many potential recipients die waiting for a transplant. A mechanical half-heart, called a *left ventricular assist device* (LVAD), can often maintain cardiac function long enough for a donor heart to become available.

This happened to Mike Dorsey, a 41-year-old father of six. Although he had to stay in the hospital to wear his LVAD, the device worked so well that he was able to exercise and help with office work while awaiting his transplant. The LVAD enabled him to increase his physical fitness, which contributed to the success of his eventual heart transplant six months later.

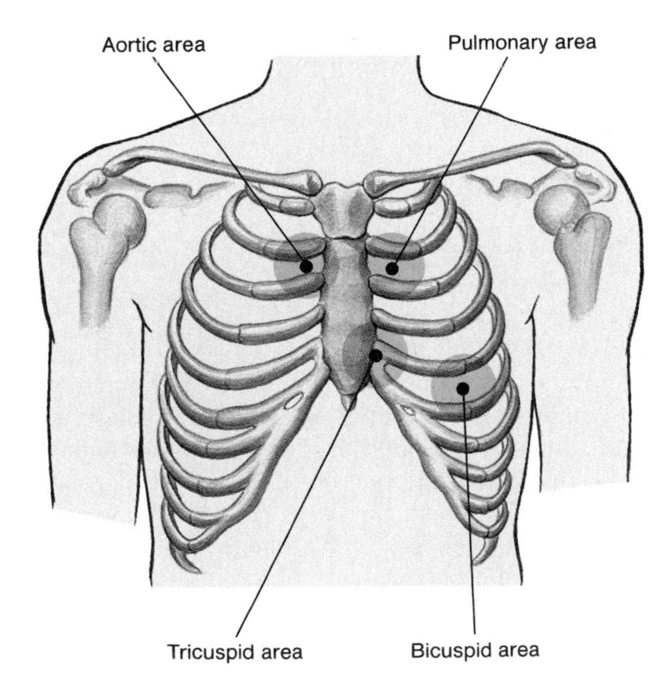


FIGURE 15.18
Thoracic regions where the sounds of each heart valve are heard most easily.

atrial syncytium by the fibrous skeleton of the heart, the cardiac impulse passes along fibers of the conduction system that are continuous with atrial muscle fibers. These conducting fibers lead to a mass of specialized cardiac muscle tissue called the **atrioventricular node** (A-V node). This node is located in the inferior portion of the septum that separates the atria (interatrial septum) and just beneath the endocardium. It provides the only normal conduction pathway between the atrial and ventricular syncytia, because the fibrous skeleton does not conduct the impulse.

The fibers that conduct the cardiac impulse into the A-V node (junctional fibers) have very small diameters, and because small fibers conduct impulses slowly, they delay transmission of the impulse. The impulse is delayed still more as it travels through the A-V node, allowing time for the atria to completely empty and the ventricles to fill with blood prior to ventricular contraction.

Once the cardiac impulse reaches the distal side of the A-V node, it passes into a group of large fibers that make up the **A-V bundle** (atrioventricular bundle or bundle of His), and the impulse moves rapidly through them. The A-V bundle enters the upper part of the interventricular septum, and divides into right and left branches that lie just beneath the endocardium. About halfway down the septum, the branches give rise to enlarged **Purkinje fibers.** These larger fibers carry the impulse to distant regions of the ventricular myocardium much faster than cell-to-cell conduction could. Thus, the massive ventricular myocardium will also contract as a functioning unit.

The base of the aorta, which contains the aortic valves, is enlarged and protrudes somewhat into the interatrial septum close to the A-V bundle. Consequently, inflammatory conditions, such as bacterial endocarditis affecting the aortic valves (aortic valvulitis), may also affect the A-V bundle.

If a portion of the bundle is damaged, it may no longer conduct impulses normally. As a result, cardiac impulses may reach the two ventricles at different times so that they fail to contract together. This condition is called a *bundle branch block.*

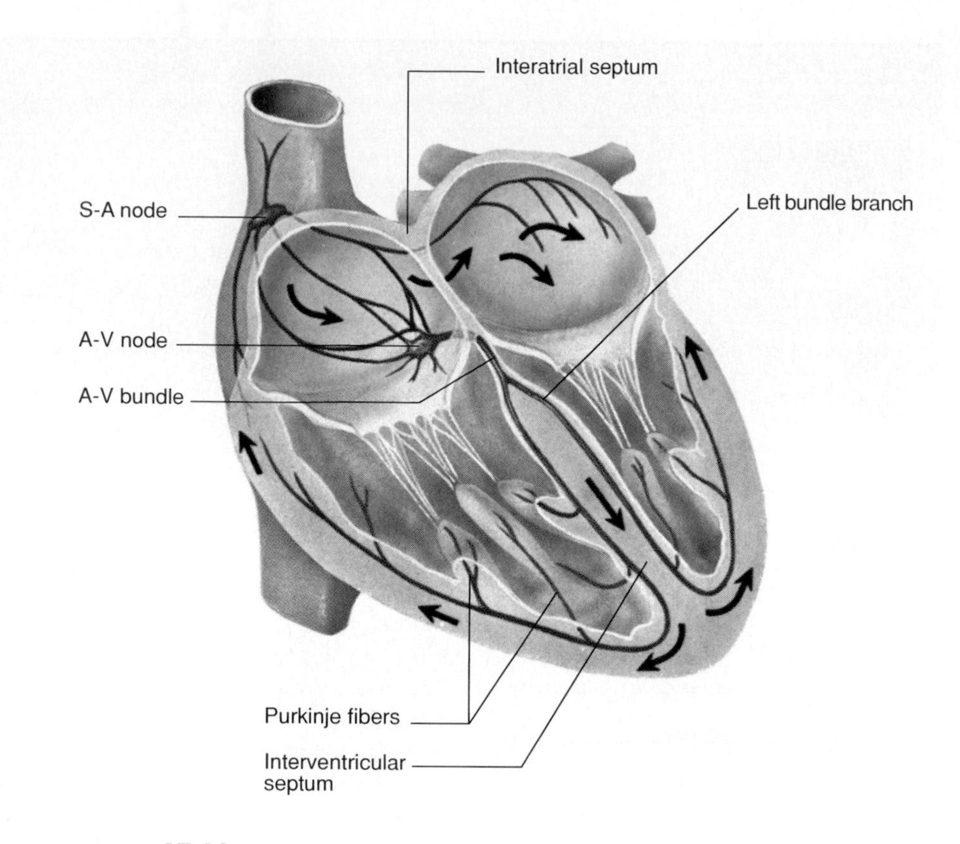


**FIGURE 15.19**
The cardiac conduction system.

The Purkinje fibers spread from the interventricular septum into the papillary muscles, which project inward from the ventricular walls, and then continue downward to the apex of the heart. There they curve around the tips of the ventricles and pass upward over the lateral walls of these chambers. Along the way, the Purkinje fibers give off many small branches, which become continuous with cardiac muscle fibers. These parts of the conduction system are shown in figure 15.19 and are summarized in figure 15.20.

The muscle fibers in the ventricular walls occur in irregular whorls, so that when impulses on the Purkinje fibers stimulate them, the ventricular walls contract with a twisting motion (fig. 15.21). This action squeezes blood out of the ventricular chambers and forces it into the arteries.

Another property of the conduction system is that the Purkinje fibers transmit the impulse to the apex of the heart first. As a result, contraction begins at the apex and pushes the blood superiorly toward the aortic and pulmonary semilunar valves, rather than having the impulse begin superiorly and push blood toward the apex, as it would if the impulse traveled from cell to cell.

1. What is the function of the cardiac conduction system?
2. What kinds of tissues make up the cardiac conduction system?
3. How is a cardiac impulse initiated?
4. How is this impulse transmitted from the right atrium to the other heart chambers?

## Regulation of the Cardiac Cycle

The quantity of blood pumped changes to accommodate cellular requirements. For example, during strenuous exercise, the amount of blood the skeletal muscles require increases greatly, and the rate of the heartbeat increases in response to this need. Since the S-A node normally controls heart rate, changes in this rate often involve factors that affect the pacemaker, such as the motor impulses carried on the parasympathetic and sympathetic nerve fibers (see figs. 11.35, 11.36, 15.22, and 15.37).

The parasympathetic fibers that innervate the heart arise from neurons in the medulla oblongata and make up parts of the *vagus nerves.* Most of these fibers branch to the S-A and A-V nodes. When the nerve impulses

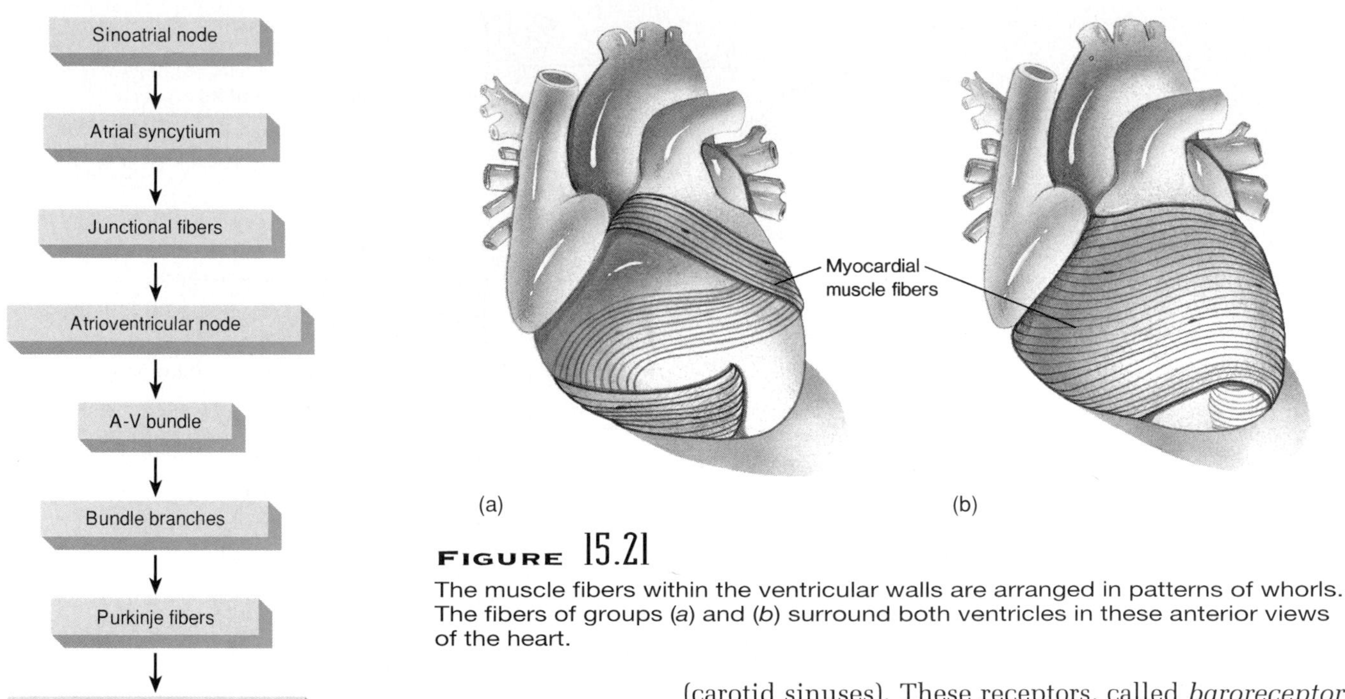


**FIGURE 15.21**

The muscle fibers within the ventricular walls are arranged in patterns of whorls. The fibers of groups (*a*) and (*b*) surround both ventricles in these anterior views of the heart.

**FIGURE 15.20**

Components of the cardiac conduction system.

reach their endings, these fibers secrete acetylcholine, which decreases S-A and A-V nodal activity. As a result, heart rate decreases.

The vagus nerves carry impulses continually to the S-A and A-V nodes, imposing a braking action on the heart. Consequently, parasympathetic activity can change heart rate in either direction. An increase in the impulses slows the heart, and a decrease in the impulses releases the parasympathetic brake and increases heart rate.

Sympathetic fibers reach the heart by means of the *accelerator nerves,* whose branches join the S-A and A-V nodes as well as other areas of the atrial and ventricular myocardium. The endings of these fibers secrete norepinephrine in response to nerve impulses, which increases the rate and force of myocardial contractions.

The *cardiac control center* of the medulla oblongata maintains balance between the inhibitory effects of the parasympathetic fibers and the excitatory effects of the sympathetic fibers. In this region of the brain, masses of neurons function as *cardioinhibitor* and *cardioaccelerator reflex centers.* These centers receive sensory impulses from throughout the circulatory system and relay motor impulses to the heart in response.

For example, receptors that are sensitive to stretch are located in certain regions of the aorta (aortic sinus and aortic arch) and in the carotid arteries (carotid sinuses). These receptors, called *baroreceptors* (pressoreceptors), can detect changes in blood pressure. Rising pressure stretches the receptors, and they signal the cardioinhibitor center in the medulla. In response, the medulla sends parasympathetic motor impulses to the heart, decreasing the heart rate and force of contraction. This action also helps to decrease blood pressure toward normal (see fig. 15.22).

Another regulatory reflex involves stretch receptors in the venae cavae near the entrances to the right atrium. If venous blood pressure increases abnormally in these vessels, the receptors signal the cardioaccelerator center, and sympathetic impulses flow to the heart. As a result, heart rate and force of contraction increase, and the venous pressure is reduced.

Impulses from the cerebrum or hypothalamus also influence the cardiac control center. Such impulses may decrease heart rate, as occurs when a person faints following an emotional upset, or they may increase heart rate during a period of anxiety.

Two other factors that influence heart rate are temperature change and certain ions. Rising body temperature increases heart action, which is why heart rate usually increases during fever. On the other hand, abnormally low body temperature decreases heart action. As a result, a patient's body temperature is sometimes deliberately lowered (hypothermia) to slow the heart during surgery.

Of the ions that influence heart action, the most important are potassium ($K^+$) and calcium ($Ca^{++}$). Potassium affects the electrical potential of the cell membrane, altering its ability to reach the threshold for conducting an impulse (see chapter 10). Calcium affects the cardiac

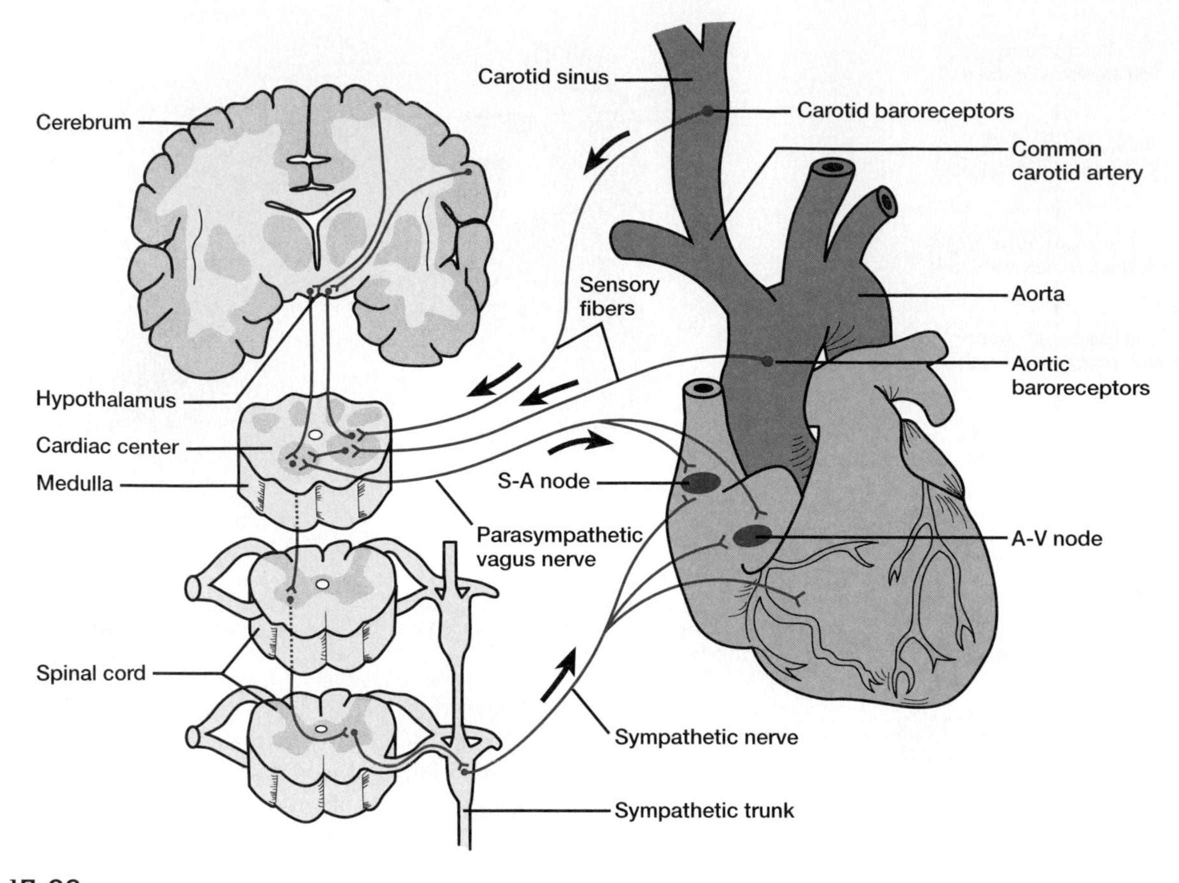


**FIGURE 15.22**
Autonomic nerve impulses alter the activities of the S-A and A-V nodes.

muscle fiber because its sarcoplasmic reticulum contains less calcium reserve than skeletal muscle, and therefore is more dependent on extracellular (blood-borne) calcium (see chapter 9). Although homeostatic mechanisms normally maintain the concentrations of these ions within narrow ranges, these mechanisms sometimes fail, and the consequences can be serious or even fatal.

An excess of potassium ions (hyperkalemia) alters the usual polarized state of the cardiac muscle fibers, and the result is a decrease in the rate and force of contractions. In fact, very high potassium ion concentration may block conduction of cardiac impulses, and heart action may suddenly stop (cardiac arrest). Conversely, if the potassium concentration drops below normal (hypokalemia), the heart may develop a serious abnormal rhythm (arrhythmia), which can also be life threatening.

Excess calcium ions (hypercalcemia) increase heart action, introducing danger that the heart will undergo a prolonged contraction. Conversely, low calcium ion concentration (hypocalcemia) depresses heart action because these ions help initiate muscle contraction.

Clinical Application 15.2 discusses the electrical activity of the heart, and Clinical Application 15.3 examines abnormal heart rhythms.

1. Which nerves supply parasympathetic fibers to the heart? Which nerves supply sympathetic fibers?
2. How do parasympathetic and sympathetic impulses help control heart rate?
3. How do changes in body temperature affect heart rate?

## BLOOD VESSELS

The systemic blood vessels are organs of the cardiovascular system, and they form a closed circuit of tubes that carry blood from the heart to the body cells and back again. These systemic vessels include **arteries, arterioles, capillaries, venules,** and **veins.** The arteries and arterioles conduct blood away from the ventricles of the heart and lead to the capillaries. The capillaries are sites of exchange of substances between the blood and the body cells, and the venules and veins return blood from the capillaries to the atria. (Figure 5A shows a semisynthetic blood vessel substitute.)

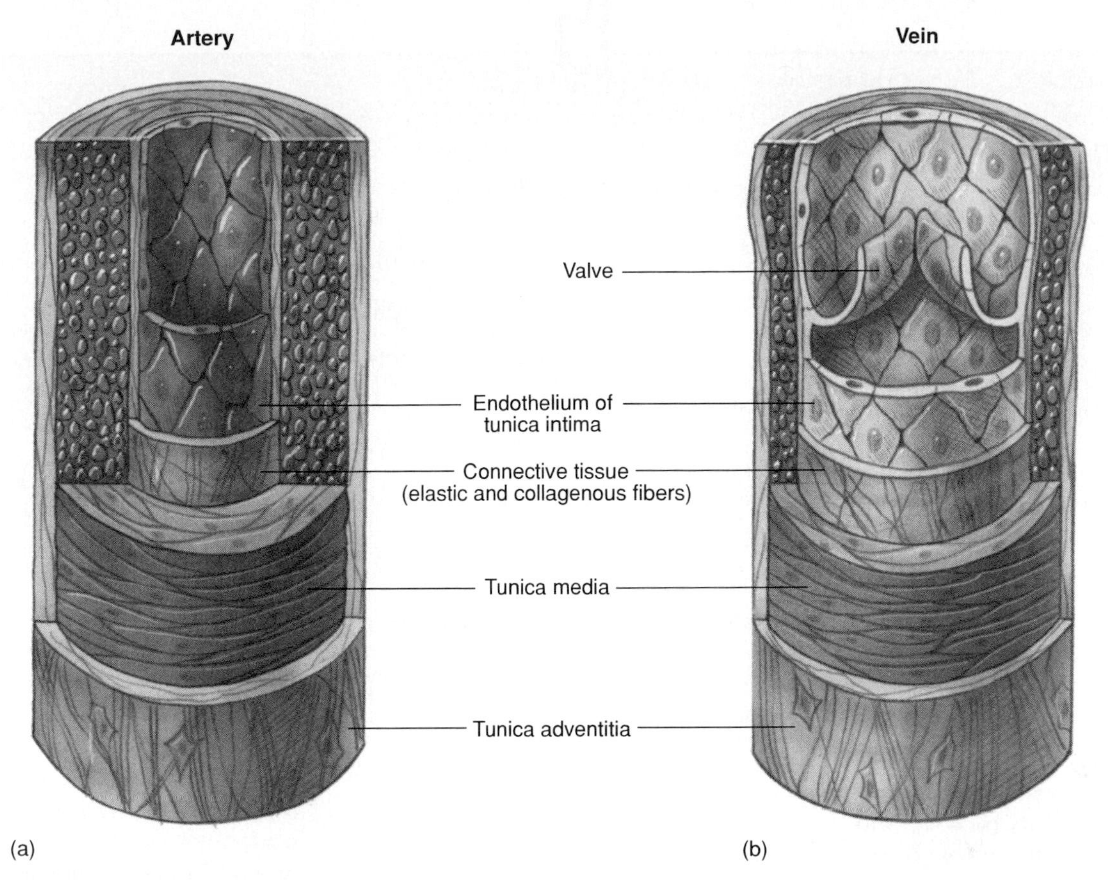


**FIGURE 15.23**
(*a*) The wall of an artery and (*b*) the wall of a vein.

## Arteries and Arterioles

**Arteries** are strong, elastic vessels that are adapted for carrying the blood away from the heart under relatively high pressure. These vessels subdivide into progressively thinner tubes and eventually give rise to fine branches called **arterioles.**

The wall of an artery consists of three distinct layers, or *tunics,* shown in figure 15.23. The innermost layer (tunica intima) is composed of a layer of simple squamous epithelium, called *endothelium,* resting on a connective tissue membrane that is rich in elastic and collagenous fibers.

In addition to separating the flowing blood from the blood vessel wall, endothelium helps prevent blood clotting by secreting biochemicals that inhibit platelet aggregation (see chapter 14). Endothelium also may help regulate local blood flow by secreting substances that dilate or constrict blood vessels.

The middle layer (tunica media) makes up the bulk of the arterial wall. It includes smooth muscle fibers, which encircle the tube, and a thick layer of elastic connective tissue.

The outer layer (tunica adventitia) is relatively thin and consists chiefly of connective tissue with irregularly arranged elastic and collagenous fibers. This layer attaches the artery to the surrounding tissues. It also contains minute vessels (vasa vasorum) that give rise to capillaries and provide blood to the more external cells of the artery wall.

The endothelial lining of an artery provides a smooth surface that allows blood cells and platelets to flow through without being damaged. The connective tissues give the vessel a tough elasticity that enables it to withstand the force of blood pressure and, at the same time, to stretch and accommodate the sudden increase in blood volume that accompanies each ventricular contraction.

The sympathetic branches of the autonomic nervous system innervate smooth muscle in artery and arteriole walls. Impulses on these *vasomotor fibers* contract the smooth muscles, reducing the diameter of the vessel. This is called **vasoconstriction.** If vasomotor impulses are inhibited, the muscle fibers relax and the diameter of the vessel increases. This is called **vasodilation.** Changes in the diameters of arteries and arterioles greatly influence blood flow and pressure.

Although the walls of the larger arterioles have three layers similar to those of arteries, these walls thin as the arterioles approach the capillaries. The wall of a very small arteriole consists only of an endothelial lining

Clinical Application 15.2

## The Electrocardiogram

An *electrocardiogram* (ECG) is a recording of the electrical changes that occur in the myocardium during a cardiac cycle. These changes result from depolarization and repolarization as action potentials occur in contracting cardiac muscle fibers. Because body fluids can conduct electrical currents, such changes can be detected on the surface of the body.

To record an ECG, electrodes are placed on the skin and connected by wires to an instrument that responds to very weak electrical changes by moving a pen or stylus on a moving strip of paper (fig. 15A). Up-and-down movements of the pen correspond to electrical changes within the body that reflect myocardial activity. Because the paper moves past the pen at a known rate, the distance between pen deflections indicates time elapsing between phases of the cardiac cycle.

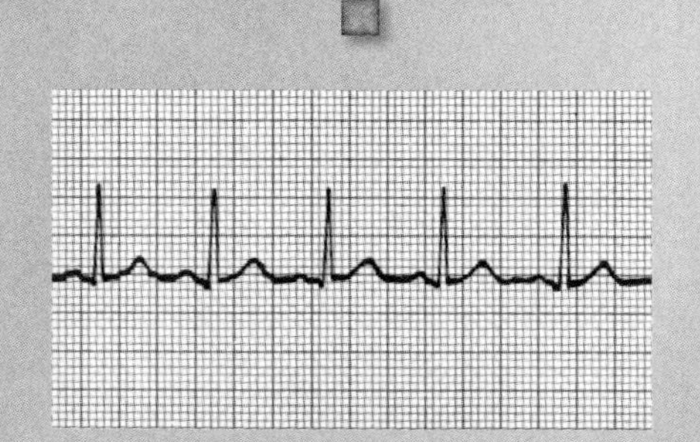

**Figure 15B**
A normal ECG.

As figure 15B illustrates, a normal ECG pattern includes several deflections, or *waves,* during each cardiac cycle. Between cycles, the muscle fibers remain polarized, with no detectable electrical changes. Consequently, the pen does not move and simply marks along the baseline as the paper passes through the instrument. However, when the S-A node triggers a cardiac impulse, the atrial fibers depolarize, producing an electrical change. The pen moves, and when the electrical change completes, the pen returns to the base position. This first pen movement produces a *P wave,* corresponding to depolarization of the atrial fibers just before they contract.

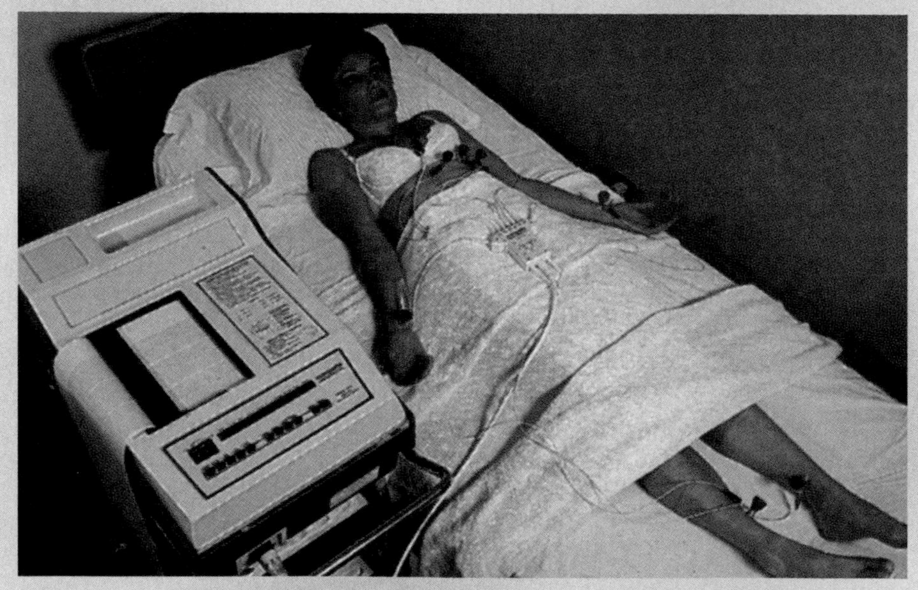

**Figure 15A**
This instrument is used to record an ECG.

When the cardiac impulse reaches the ventricular fibers, they rapidly depolarize. Because the ventricular walls are of much greater mass than those of the atria, the amount of electrical change is greater, and the pen is deflected to a greater degree than before. When the electrical change finishes, the pen returns to the baseline, leaving a mark called the *QRS complex,* which usually consists of a *Q wave,* an *R wave,* and an *S wave.* This complex appears due to depolarization of the ventricular fibers just prior to the contraction of the ventricular walls.

The electrical changes occurring as the ventricular muscle fibers repolarize slowly produces a *T wave* as the pen deflects again, ending the ECG pattern. The record of the atrial repolarization is missing from the pattern because the atrial fibers repolarize at the same time that the ventricular fibers depolarize. Thus, the QRS complex obscures the recording of the atrial repolarization (figs. 15C and 15D).

ECG patterns allow a physician to assess the heart's ability to conduct impulses. For example, the time period between the beginning of a P wave and the beginning of a QRS complex is the *P-Q interval* (*P-R interval* if the initial portion of the QRS wave is upright). This indicates how long it takes for the cardiac impulse to travel from the S-A node through the

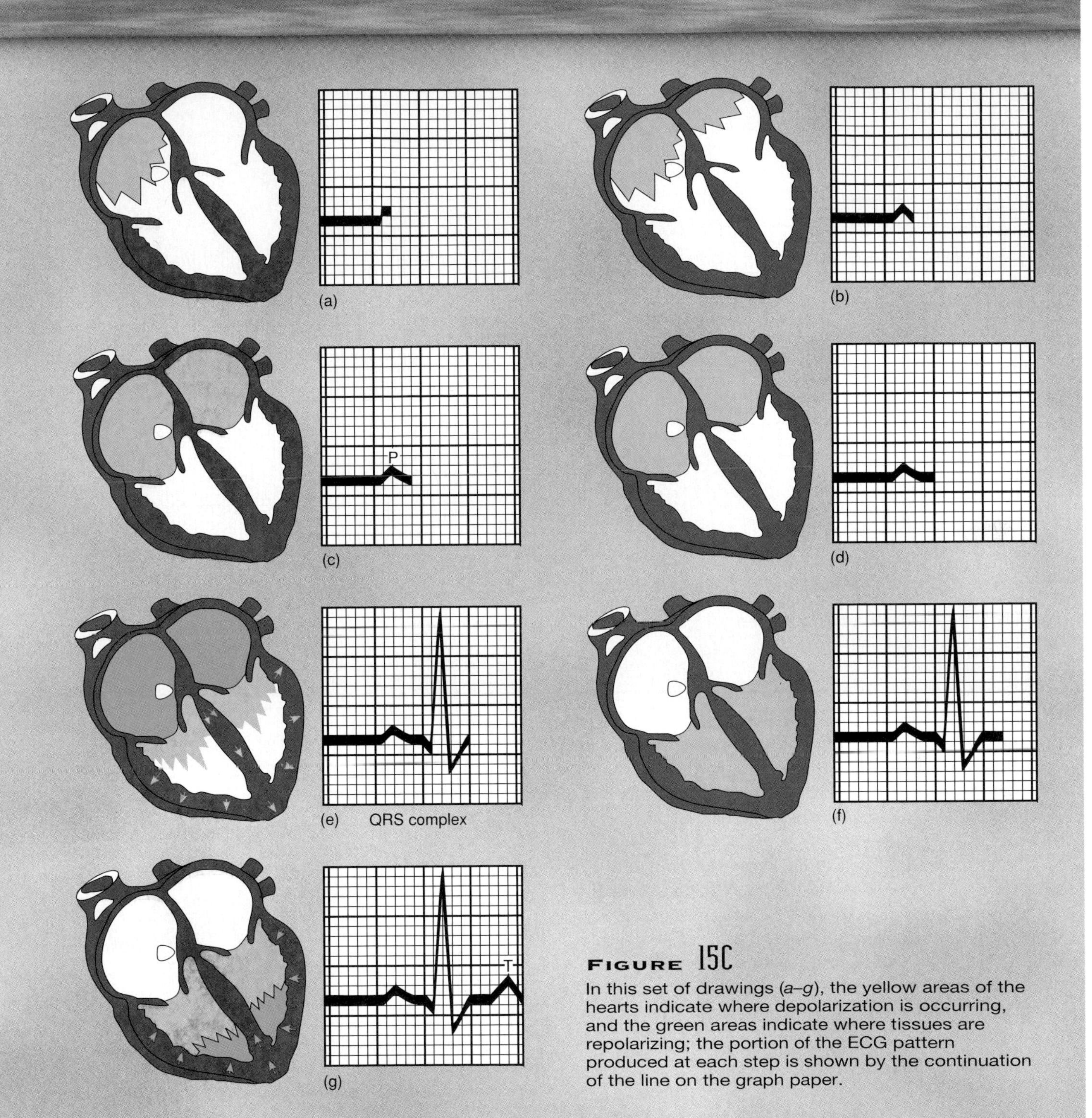


**FIGURE 15C**

In this set of drawings (*a–g*), the yellow areas of the hearts indicate where depolarization is occurring, and the green areas indicate where tissues are repolarizing; the portion of the ECG pattern produced at each step is shown by the continuation of the line on the graph paper.

A-V node. Ischemia or other problems affecting the fibers of the A-V conduction pathways can increase this P-Q interval. Similarly, injury to the A-V bundle can increase duration of the QRS complex, because it may take longer for an impulse to spread throughout the ventricular walls (fig. 15E).

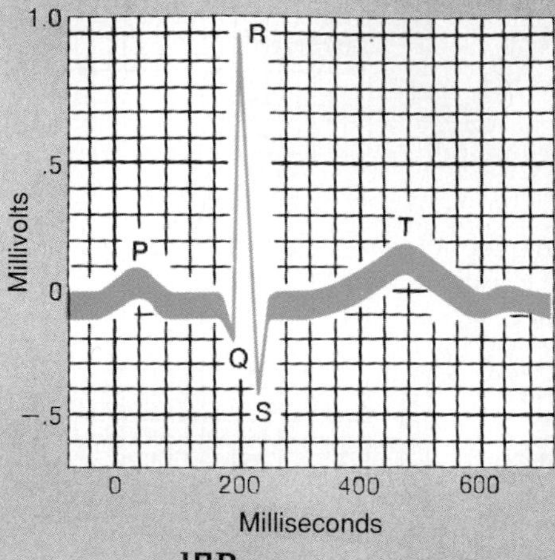


**FIGURE 15D**
In an ECG pattern, the P wave results from depolarization of the atria; the QRS complex results from depolarization of the ventricles (and repolarization of the atria); and the T wave results from repolarization of the ventricles.

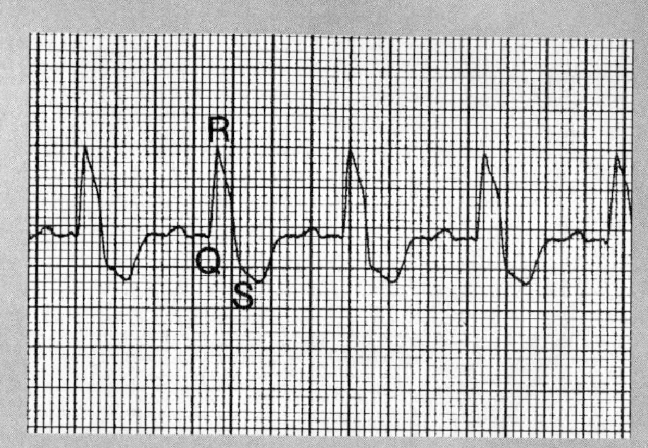


**FIGURE 15E**
A prolonged QRS complex may result from damage to the A-V bundle fibers.

and some smooth muscle fibers, surrounded by a small amount of connective tissue (figs. 15.24 and 15.25). Arterioles, which are microscopic continuations of arteries, give off branches called *metarterioles* that, in turn, join capillaries.

The arteriole and metarteriole walls are adapted for vasoconstriction and vasodilation in that their muscle fibers respond to impulses from the autonomic nervous system by contracting or relaxing. Thus, these vessels help control the flow of blood into the capillaries.

Sometimes metarterioles connect directly to venules, and the blood entering them can bypass the capillaries. These connections between arteriole and venous pathways, shown in figure 15.26, are called *arteriovenous shunts.*

1. Describe the wall of an artery.
2. What is the function of the smooth muscle in the arterial wall?
3. How is the structure of an arteriole different from that of an artery?

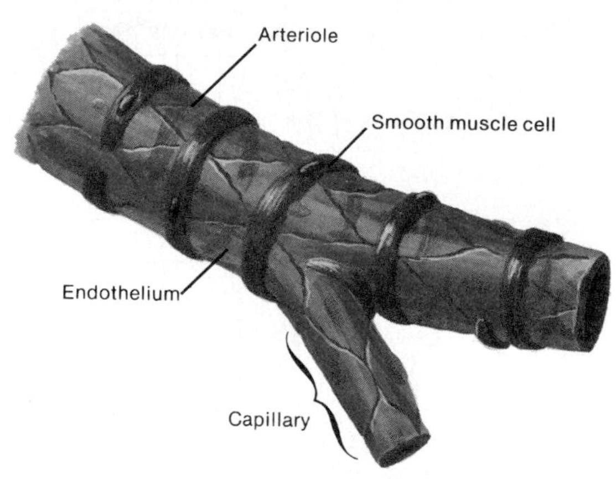


**FIGURE 15.24**
Small arterioles have smooth muscle fibers in their walls; capillaries lack these fibers.

## Capillaries

**Capillaries** are the smallest blood vessels. They connect the smallest arterioles and the smallest venules. Capillaries are extensions of the inner linings of these larger

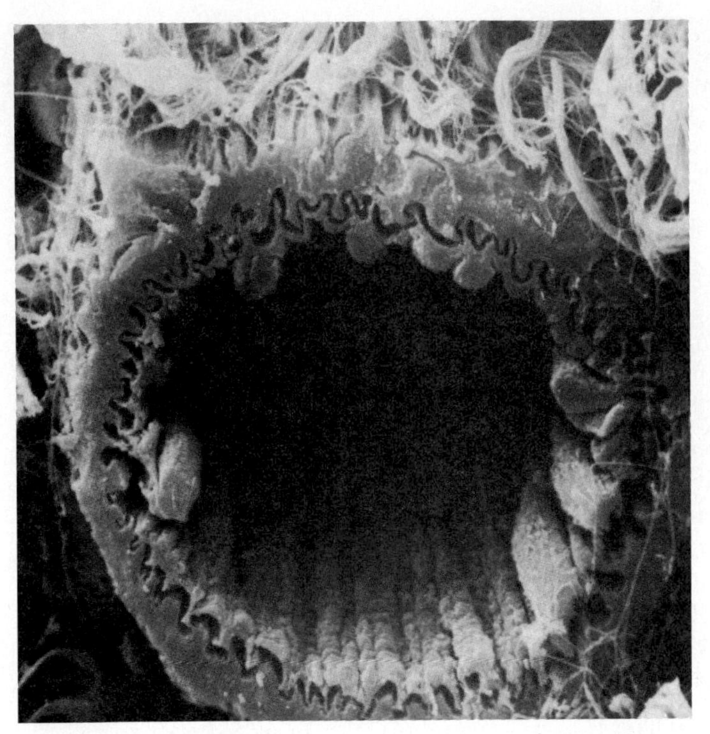

**FIGURE 15.25**
Scanning electron micrograph of an arteriole cross section (3,900×).

(*Tissues and Organs: A Text-Atlas of Scanning Electron Microscopy*, by R. G. Kessel and R. H. Kardon. © 1979 W. H. Freeman and Company.)

vessels in that their walls are endothelium—a single layer of squamous epithelial cells (fig. 15.27*a*). These thin walls form the semipermeable membranes through which substances in the blood are exchanged for substances in the tissue fluid surrounding body cells.

### Capillary Permeability

The openings or intercellular channels in the capillary walls are thin slits occurring where two adjacent endothelial cells overlap. The sizes of these openings, and consequently the permeability of the capillary wall, vary from tissue to tissue. For example, the openings are relatively small in the capillaries of smooth, skeletal, and cardiac muscle, whereas those in capillaries associated with endocrine glands, kidneys, and the lining of the small intestine are larger.

Capillaries with the largest openings include those of the liver, spleen, and red bone marrow. These capillaries are discontinuous and the distance between their cells appears as little cavities (sinusoids) in the organ. Discontinuous capillaries allow large proteins and even intact cells to pass through as they enter or leave the circulation (fig. 15.27*b* and *c*). Clinical Application 3.1 discusses the blood-brain barrier, the protective tight arrangement of capillaries in the brain. The barrier is not present in the pituitary and pineal glands and parts of the hypothalamus.

### Capillary Arrangement

The density of capillaries within tissues varies directly with the tissues' rates of metabolism. Muscle and nerve tissues, which use large quantities of oxygen and nutrients, are richly supplied with capillaries; cartilaginous tissues, the epidermis, and the cornea, where metabolism is slow, lack capillaries.

> If the capillaries of an adult were unwound and spread end to end, they would cover from 25,000 to 60,000 miles.

The patterns of capillary arrangement also differ in various body parts. For example, some capillaries pass directly from arterioles to venules, but others lead to highly branched networks (fig. 15.28). Such arrangements make it possible for the blood to follow different pathways through a tissue and meet the varying demands of its cells. During exercise, for example, blood can be directed into the capillary networks of the skeletal muscles, where the cells have increased need for oxygen and nutrients. At the same time, the blood can bypass some of the capillary nets in the tissues of the digestive tract, where demand for blood is less critical. Conversely, when a person is relaxing after a meal, blood can be shunted from the inactive skeletal muscles into the capillary networks of the digestive organs.

### Regulation of Capillary Blood Flow

The distribution of blood in the various capillary pathways is regulated mainly by the smooth muscles that encircle the capillary entrances. As figure 15.26 shows, these muscles form *precapillary sphincters*, which may close a capillary by contracting, or open it by relaxing. The precapillary sphincters seem to respond to the demands of the cells. When the cells have low concentrations of oxygen and nutrients, the sphincter relaxes; and when cellular requirements have been met, the sphincter may contract again.

1. Describe a capillary wall.
2. What is the function of a capillary?
3. What controls blood flow into capillaries?

**CLINICAL APPLICATION 15.3**

## Arrhythmias

Each year, 400,000 people in the United States die from a fast or irregular heartbeat. These are types of altered heart rhythm called *arrhythmia.* John Thomas was almost one of them—several times.

In a crowded Boston train in December 1990, 36-year-old Thomas lost consciousness for about 8 seconds, during which his heart *fibrillated.* In this state, small areas of the myocardium contract in an uncoordinated fashion (fig. 15F). As a result, the myocardium fails to contract as a whole, and blood is no longer pumped. Atrial fibrillation is not life threatening because the ventricles still pump blood, but ventricular fibrillation, which struck Thomas, is often deadly. Ventricular fibrillation can be caused by an obstructed coronary artery, toxic drug exposure, electric shock, or traumatic injury to the heart or chest wall.

Fortunately, Thomas had had a device called a defibrillator surgically implanted a year earlier, after he suffered a cardiac arrest. The device sends out an electrical jolt during fibrillation, which interrupts the abnormal heart rhythm, allowing the heart to resume a normal beat. Thomas' defibrillator saves his life several times a year.

Fibrillation is cardiac chaos. An abnormally fast heartbeat, usually more than one hundred beats per minute, is called *tachycardia*. Increase in body temperature, nodal stimulation by sympathetic fibers, certain drugs or hormones, heart disease, excitement, exercise, anemia, or shock can all cause tachycardia. Figure 15G shows the ECG of a tachycardic heart.

Bradycardia means a slow heart rate, usually fewer than sixty beats per minute. Decreased body temperature, nodal stimulation by parasympathetic impulses, or certain drugs may cause bradycardia. It also may occur during sleep. Figure 15H shows the ECG of a bradycardic heart. Athletes sometimes have unusually slow heartbeats because their hearts have developed the ability to pump a greater than normal volume of blood with each beat. The slowest heartbeat recorded in a healthy athlete was 25 beats per minute!

A *premature beat* occurs before it is expected in a normal series of cardiac cycles. Cardiac impulses originating from unusual (ectopic) regions of the heart probably cause a premature beat. That is, the impulse originates from a site other than the S-A node. Cardiac impulses may arise from ischemic tissues or from muscle fibers that disease or drugs irritate.

A heart chamber *flutters* when it contracts regularly, but very rapidly, such as 250–350 times per minute. Although normal hearts may flutter occasionally, this condition is more likely to be due to damage to the myocardium (fig. 15I).

Any interference or block in cardiac impulse conduction may cause arrhythmia, the type varying with the location and extent of the block. Such arrhythmias are related to the fact that certain cardiac tissues other than the S-A node can function as pacemakers.

The S-A node usually initiates seventy to eighty heartbeats per minute, called a sinus rhythm. If the S-A node is damaged, impulses originating in the A-V node may travel upward into the atrial myocardium and downward into the ventricular walls, stimulating them to contract. Under the influence of the A-V node acting as a *secondary pacemaker,* the heart may continue to pump blood, but at a rate of forty to sixty beats per minute, called a nodal rhythm. Similarly, the

### Exchanges in the Capillaries

The vital function of exchanging gases, nutrients, and metabolic by-products between the blood and the tissue fluid surrounding body cells occurs in the capillaries. The biochemicals exchanged move through the capillary walls by diffusion, filtration, and osmosis (see chapter 3). Diffusion is the most important means of transfer.

Because blood entering certain capillaries carries relatively high concentrations of oxygen and nutrients, these substances diffuse through the capillary walls and enter the tissue fluid. Conversely, the concentrations of carbon dioxide and other wastes are generally greater in such tissues, and the wastes tend to diffuse into the capillary blood.

Purkinje fibers can initiate cardiac impulses, contracting the heart fifteen to forty times per minute.

An *artificial pacemaker* can treat a disorder of the cardiac conduction system. This device includes an electrical pulse generator and a lead wire that communicates with a portion of the myocardium. The pulse generator contains a permanent battery that provides energy and a microprocessor that can sense the cardiac rhythm and signal the heart to alter its contraction rate.

An artificial pacemaker is surgically implanted beneath the patient's skin in the shoulder. An external programmer adjusts its functions from the outside. The first pacemakers, made in 1958, were crude. Today, thanks to telecommunications advances, a physician can check a patient's pacemaker over the phone! In 1993, a device called a pacemaker-cardioverter-defibrillator became available, which can correct both abnormal heart rhythm and cardiac arrest.

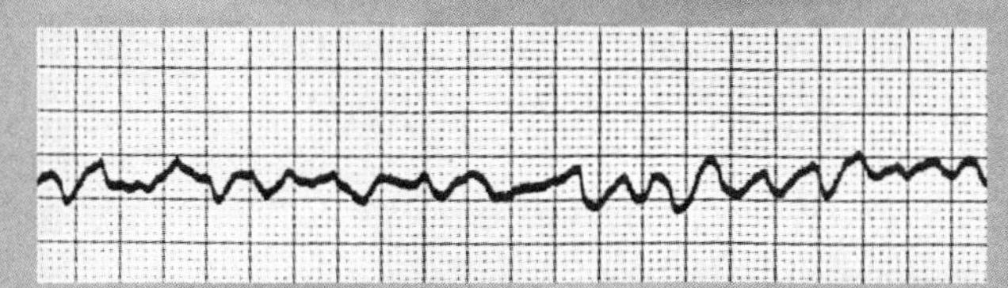

**FIGURE 15F**
Ventricular fibrillation is rapid, uncoordinated depolarizations of the ventricles.

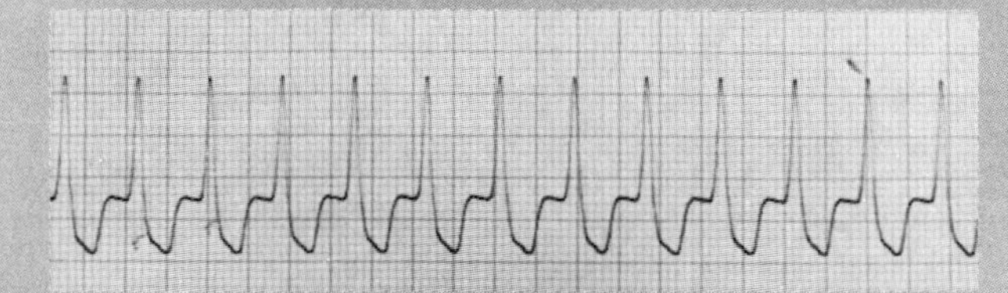

**FIGURE 15G**
Tachycardia is a rapid heartbeat.

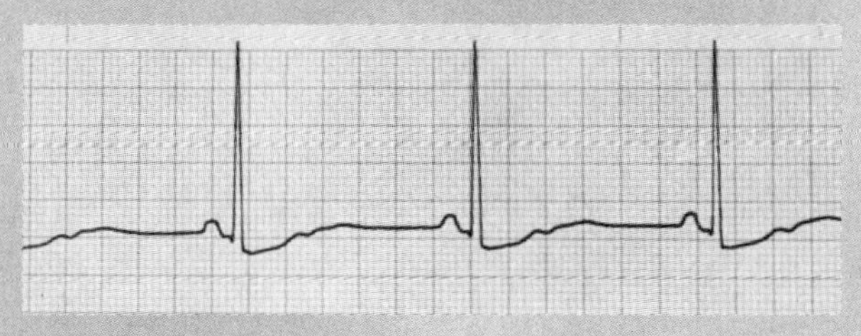

**FIGURE 15H**
Bradycardia is a slow heartbeat.

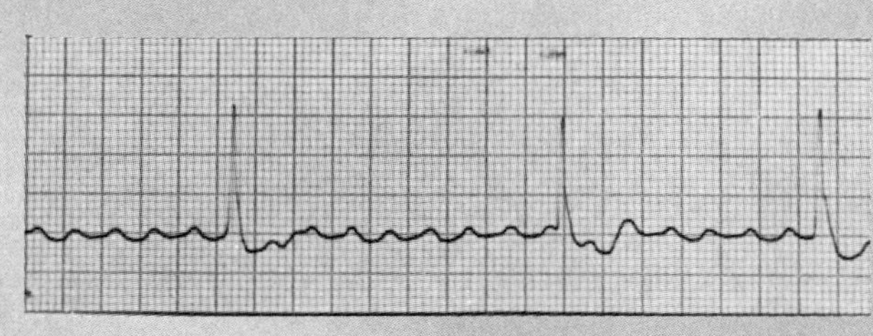

**FIGURE 15I**
Atrial flutter is an abnormally rapid rate of atrial depolarization.

The paths these substances follow depend primarily on their solubilities in lipids. Those that are soluble in lipid, such as oxygen, carbon dioxide, and fatty acids, can diffuse through most areas of the cell membranes that make up the capillary wall because the membranes are largely lipid. Lipid-insoluble substances, such as water, sodium ions, and chloride ions, diffuse through pores in the cell membranes and through the slitlike openings between the endothelial cells that form the capillary wall (see fig. 15.27).

Plasma proteins generally remain in the blood because they are not soluble in the lipid portions of the capillary membranes, and they are too large to diffuse through the membrane pores or slitlike openings between the endothelial cells of most capillaries.

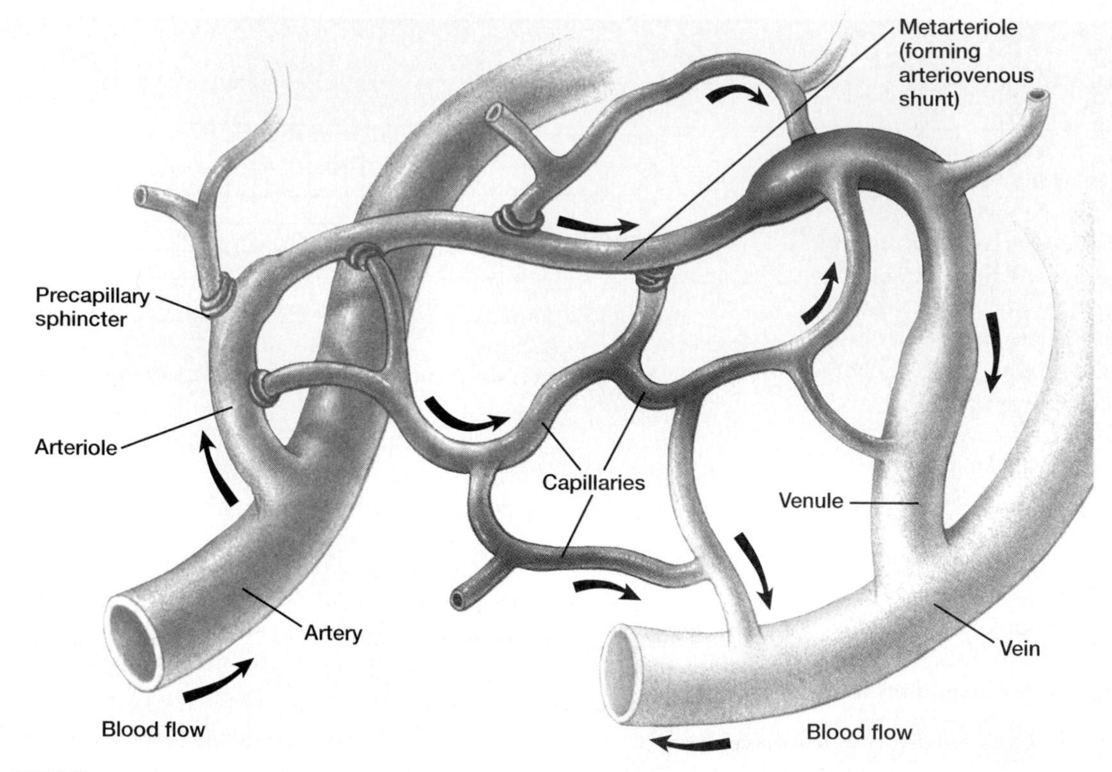


**FIGURE 15.26**
Some metarterioles provide arteriovenous shunts by connecting arterioles directly to venules.

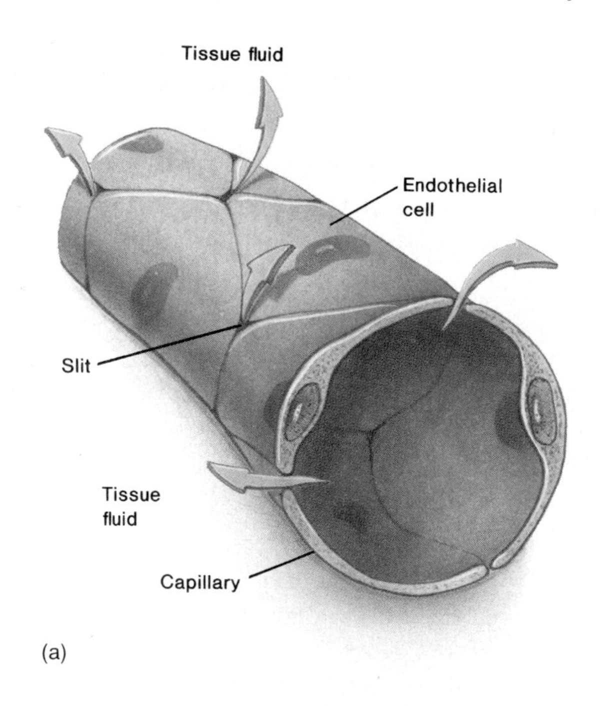


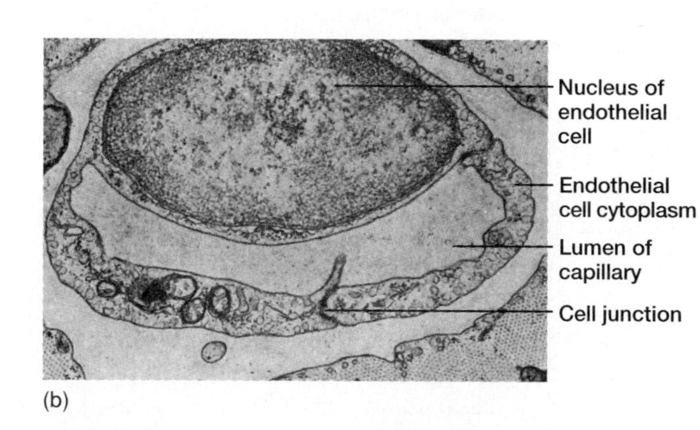


(c)

**FIGURE 15.27**
(*a*) Substances are exchanged between the blood and tissue fluid through openings separating adjacent endothelial cells. (*b*) Transmission electron micrograph of a capillary cross section. (*c*) Note the narrow slit-like openings at the cell junctions (arrow).

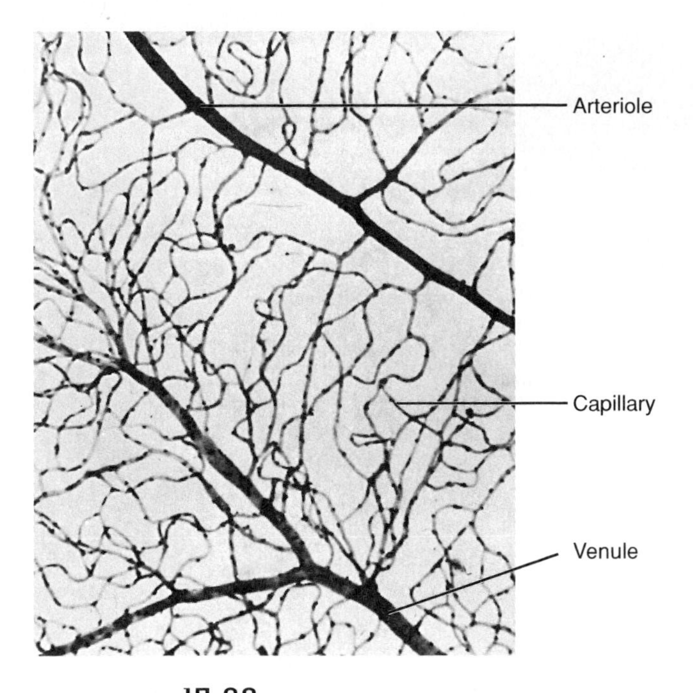


**FIGURE 15.28**
Light micrograph of a capillary network (100×).

Recall from chapter 3 that in *filtration, hydrostatic pressure* forces molecules through a membrane. In the capillaries, the blood pressure generated when ventricle walls contract provides the force for filtration.

Blood pressure also moves blood through the arteries and arterioles. This pressure decreases as distance from the heart increases, because friction (peripheral resistance) between the blood and the vessel walls slows the flow. For this reason, blood pressure is greater in the arteries than in the arterioles, and greater in the arterioles than in the capillaries. It is similarly greater at the arteriole end of a capillary than at the venule end.

The walls of arteries and arterioles are too thick to allow blood components to pass through. However, the hydrostatic pressure of the blood pushes small molecules through capillary walls by filtration. This effect occurs primarily at the arteriole ends of capillaries whereas diffusion takes place along their entire lengths.

Plasma proteins, which remain in the capillaries, help make the *osmotic pressure* of the blood greater (hypertonic) than that of the tissue fluid. Although capillary blood has a greater osmotic attraction for water than does tissue fluid, the greater force of the blood pressure overcomes this attraction at the arteriolar end of the capillary. As a result, the net movement of water and dissolved substances is outward at the arteriole end of the capillary by filtration.

More specifically, the force, including hydrostatic pressure, that moves fluid outward at the arteriole end of a capillary is 41.3 millimeters of mercury (mm Hg), while the net osmotic force, which tends to move fluid inward, is only 28 mm Hg. Consequently, the net movement of water and dissolved substances is outward. However, blood pressure decreases as the blood moves through the capillary, and, at the venule end, the outward force equals 21.3 mm Hg, whereas the osmotic pressure of the blood remains unchanged at 28 mm Hg. Thus, there is a net movement of water and dissolved materials into the venule end of the capillary by diffusion. Figure 15.29 shows this process.

Normally, more fluid leaves the capillaries than returns to them. *Lymphatic vessels* collect the excess fluid and return it to the venous circulation. This mechanism is discussed in chapter 16.

Sometimes unusual events increase blood flow to capillaries, and excess fluid enters spaces between tissue cells (interstitial spaces). This may occur, for instance, in response to certain chemicals such as *histamine* that vasodilate the metarteriole and increase capillary permeability. Enough fluid may leak out of the capillaries to overwhelm lymphatic drainage, and affected tissues become swollen (edematous) and painful.

1. What forces are responsible for the exchange of substances between the blood and the tissue fluid?
2. Why is the fluid movement out of a capillary greater at its arteriole end than at its venule end?
3. Since more fluid leaves the capillary than returns to it, how is the remainder returned to the vascular system?

If the right ventricle of the heart is unable to pump blood out as rapidly as it enters, other parts of the body may develop edema, because the blood backs up into the veins, venules, and capillaries, increasing blood pressure in these vessels. As a result of this increased *back pressure,* osmotic pressure of the blood in the venule ends of the capillaries is less effective in attracting water from tissue fluid, and the tissues swell. This is true particularly in the lower extremities if the person is upright, or in the back if the person is supine. In the terminal stages of heart failure, edema is widespread, and fluid accumulates in the peritoneal cavity of the abdomen. This condition is called *ascites.*

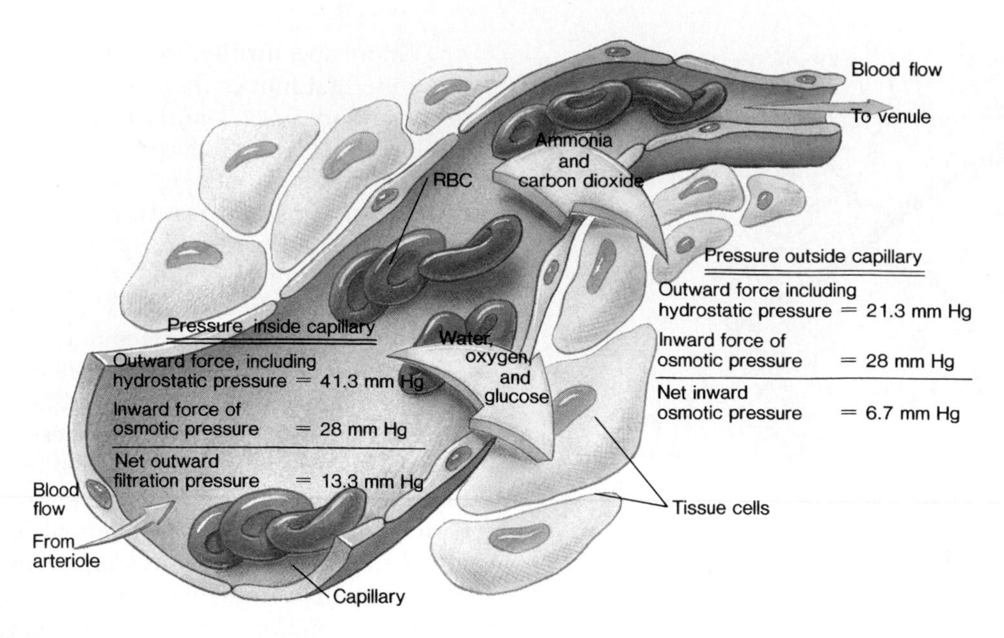


**FIGURE 15.29**

Water and other substances leave the capillaries because of a net outward filtration pressure at the arteriole end of the capillaries. Water enters at the venule end of the capillaries because of a net inward force of osmotic pressure. Substances move in and out along the length of the capillaries according to their respective diffusion gradients.

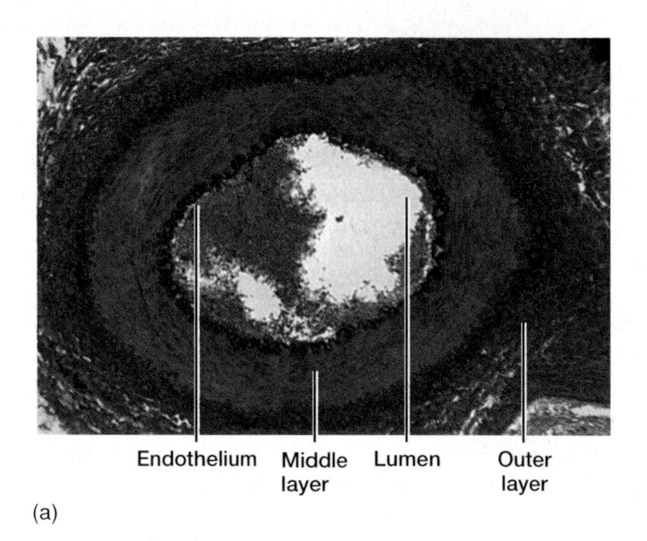


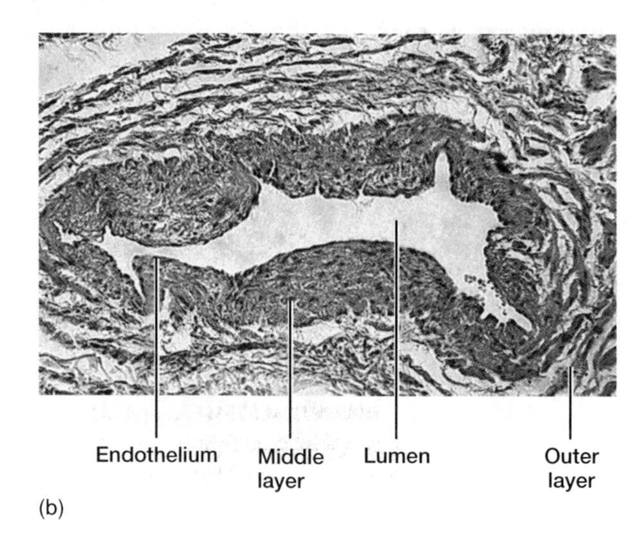


**FIGURE 15.30**

Note the structural differences in these cross sections of (*a*) an artery (100×) and (*b*) a vein (160×).

## Venules and Veins

**Venules** are the microscopic vessels that continue from the capillaries and merge to form **veins.** The veins, which carry blood back to the atria, follow pathways that roughly parallel those of the arteries.

The walls of veins are similar to those of arteries in that they are composed of three distinct layers. However, the middle layer of the venous wall is poorly developed. Consequently, veins have thinner walls that contain less smooth muscle and less elastic tissue than those of comparable arteries (figs. 15.23 and 15.30).

Many veins, particularly those in the upper and lower limbs, contain flaplike *valves,* which project inward from their linings. Valves, shown in figure 15.31, are usually composed of two leaflets that close if the blood begins to back up in a vein. These valves aid in returning blood to the heart because they open as long as the flow is toward the heart, but close if it is in the opposite direction.

Veins also function as *blood reservoirs,* useful in times of need. For example, if a hemorrhage accompanied by a drop in arterial blood pressure occurs, the

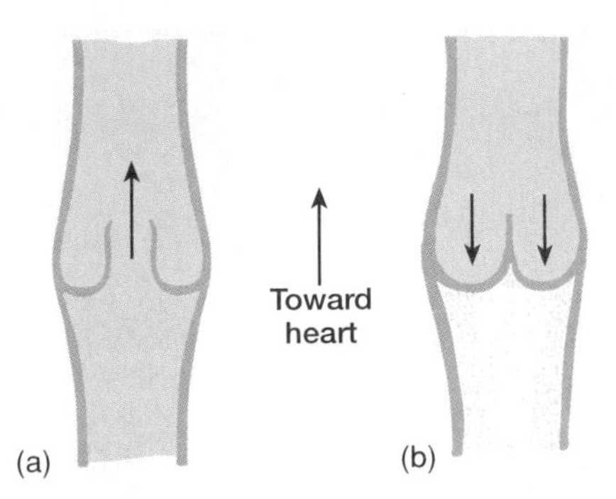


**FIGURE 15.31**

(*a*) Venous valves allow blood to move toward the heart, but (*b*) prevent blood from moving backward away from the heart.

muscular walls of the veins are stimulated reflexly by sympathetic nerve impulses. The resulting venous constrictions help maintain blood pressure by returning more blood to the heart. This mechanism ensures a nearly normal blood flow even when as much as 25% of the blood volume is lost. Figure 15.32 illustrates the relative volumes of blood in the veins and other blood vessels.

Chart 15.3 summarizes the characteristics of blood vessels. Clinical Application 15.4 examines disorders of blood vessels.

1. How does the structure of a vein differ from that of an artery?
2. What are the functions of veins and venules?

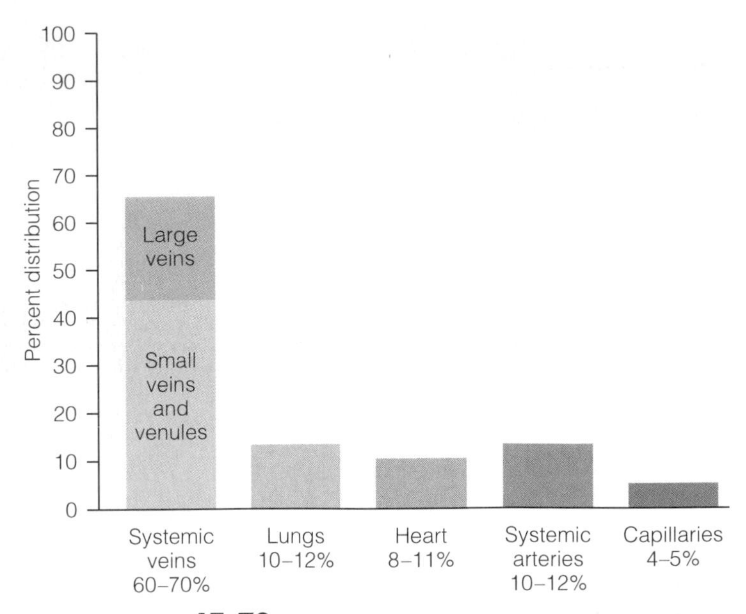


**FIGURE 15.32**

Most of the blood volume is contained within the veins and venules.

## Blood Pressure

Blood pressure is the force the blood exerts against the inner walls of the blood vessels. Although this force occurs throughout the vascular system, the term *blood pressure* most commonly refers to pressure in arteries supplied by branches of the aorta (systemic arteries).

### Arterial Blood Pressure

The arterial blood pressure rises and falls in a pattern corresponding to the phases of the cardiac cycle. That is, when the ventricles contract (ventricular systole), their walls squeeze the blood inside their chambers and force it into the pulmonary trunk and aorta. As a result, the pressures in these arteries increase sharply. The maximum pressure achieved during ventricular contraction is called the **systolic pressure.** When the ventricles relax (ventricular diastole), the arterial pressure drops, and the lowest pressure that remains in the arteries before the next ventricular contraction is termed the **diastolic pressure.**

The surge of blood entering the arterial system during a ventricular contraction distends the elastic walls of the arteries, but the pressure drops almost immediately as the contraction ends, and the arterial walls recoil. This alternate expanding and recoiling of the arterial wall can be felt as a *pulse* in an artery that runs close to the surface. Figure 15.33 shows several sites where a pulse can be detected. The radial artery, for example, courses near the surface at the wrist and is commonly used to sense a person's radial pulse.

The radial pulse rate is equal to the rate at which the left ventricle contracts, and for this reason, it can be used to determine heart rate. A pulse can also reveal something about blood pressure, because an elevated pressure produces a pulse that feels full, while a low pressure produces a pulse that is easily compressed.

Clinical Application 15.5 describes how to measure arterial blood pressure.

1. Distinguish between systolic and diastolic blood pressure.
2. What cardiac event is responsible for the systolic pressure? For the diastolic pressure?
3. What causes a pulse in an artery?

### Factors that Influence Arterial Blood Pressure

Arterial pressure depends on a variety of factors, including heart action, blood volume, resistance to flow, and blood viscosity (fig. 15.34).

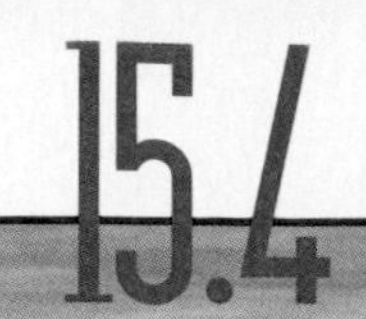

CLINICAL APPLICATION 15.4

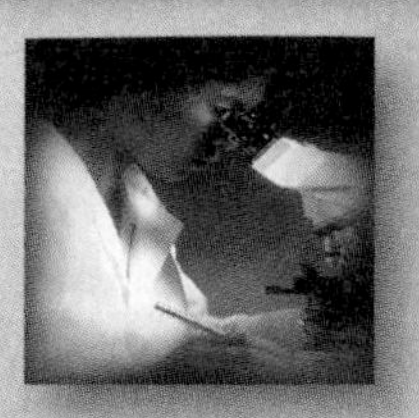

## Blood Vessel Disorders

Nearly half of all deaths in the United States are due to the arterial disease *atherosclerosis,* in which soft masses of fatty materials, particularly cholesterol, accumulate on the inside of the arterial walls. Such deposits, called *plaque,* protrude into the lumens of the vessels and interfere with blood flow (fig. 15J). Furthermore, plaque often forms a surface that can initiate formation of a blood clot, increasing the risk of developing thrombi or emboli that cause blood deficiency (*ischemia*) or tissue death (*necrosis*) downstream from the obstruction.

The walls of affected arteries also tend to degenerate, losing their elasticity and becoming hardened or *sclerotic.* This stage of the disease, called *arteriosclerosis,* introduces the danger that a sclerotic vessel will rupture under the force of blood pressure.

Atherosclerosis is often associated with a fatty diet, elevated blood pressure, cigarette smoking, obesity, and lack of physical exercise. Emotional and genetic factors may also increase susceptibility to atherosclerosis (see chapter 18).

If atherosclerosis so weakens the wall of an artery that blood pressure dilates a region of it, a pulsating sac called an *aneurysm* may form. Aneurysms tend to grow. If the resulting sac develops by a longitudinal splitting of the middle layer of the arterial wall, it is called a *dissecting aneurysm.* An aneurysm may cause symptoms by pressing on nearby organs, or it may rupture and produce a great loss of blood.

Aneurysms may also result from trauma, high blood pressure, infections, inherited disorders such as Marfan syndrome, or congenital defects in blood vessels. Common sites of aneurysms include the thoracic and abdominal aorta, and an arterial circle at the base of the brain (circle of Willis).

*Phlebitis,* or inflammation of a vein, is relatively common. It may occur in association with an injury or infection or after surgery, or it may develop for no apparent reason.

If inflammation is restricted to a superficial vein, such as the greater or lesser saphenous veins, blood flow may be rechanneled through other vessels. But if it occurs in a deep vein, such as the tibial, peroneal, popliteal, or femoral veins, the consequences can be quite serious, particularly if the blood within the affected vessel clots and blocks normal circulation. This condition is called *thrombophlebitis.* There is now a risk that a blood clot within a vein will detach, move with the venous blood, pass through the heart, and lodge in the pulmonary arterial system within a lung. Such an obstruction is called a *pulmonary embolism.*

*Varicose veins* are abnormal and irregular dilations in superficial veins, particularly in the legs. This condition is usually associated with prolonged, increased back pressure within the affected vessels due to gravity, as occurs when a person stands. Crossing the legs or sitting in a chair so that its edge presses against the area behind the knee can obstruct venous blood flow and aggravate varicose veins.

Increased venous back pressure stretches the veins and increases their diameters. Because the valves within these vessels do not change size, they soon lose their abilities to block the backward flow of blood, and blood tends to accumulate in the enlarged regions.

### Heart Action

In addition to producing blood pressure by forcing blood into the arteries, heart action determines how much blood enters the arterial system with each ventricular contraction, as well as the rate of this fluid output.

The volume of blood discharged from the ventricle with each contraction is called the **stroke volume** and equals about 70 milliliters at rest. The volume discharged from the ventricle per minute is called the **cardiac output.** It is calculated by multiplying the stroke volume by the heart rate in beats per minute. (Cardiac output = stroke volume × heart rate.) Thus, if the stroke volume is 70 milliliters and the heart rate is 72 beats per minute, the cardiac output is 5,040 milliliters per minute.

Blood pressure varies with the cardiac output. If either the stroke volume or the heart rate increases, so

Increased venous pressure is also accompanied by rising pressure within the venules and capillaries that supply the veins. Consequently, tissues in affected regions typically become edematous and painful.

Heredity, pregnancy, obesity, and standing for long periods raise the risk of developing varicose veins. Elevating the legs above the level of the heart or putting on support hosiery before arising in the morning can relieve discomfort. Intravenous injection of a substance that destroys veins (a sclerosing agent) or surgical removal of the affected veins may be necessary.

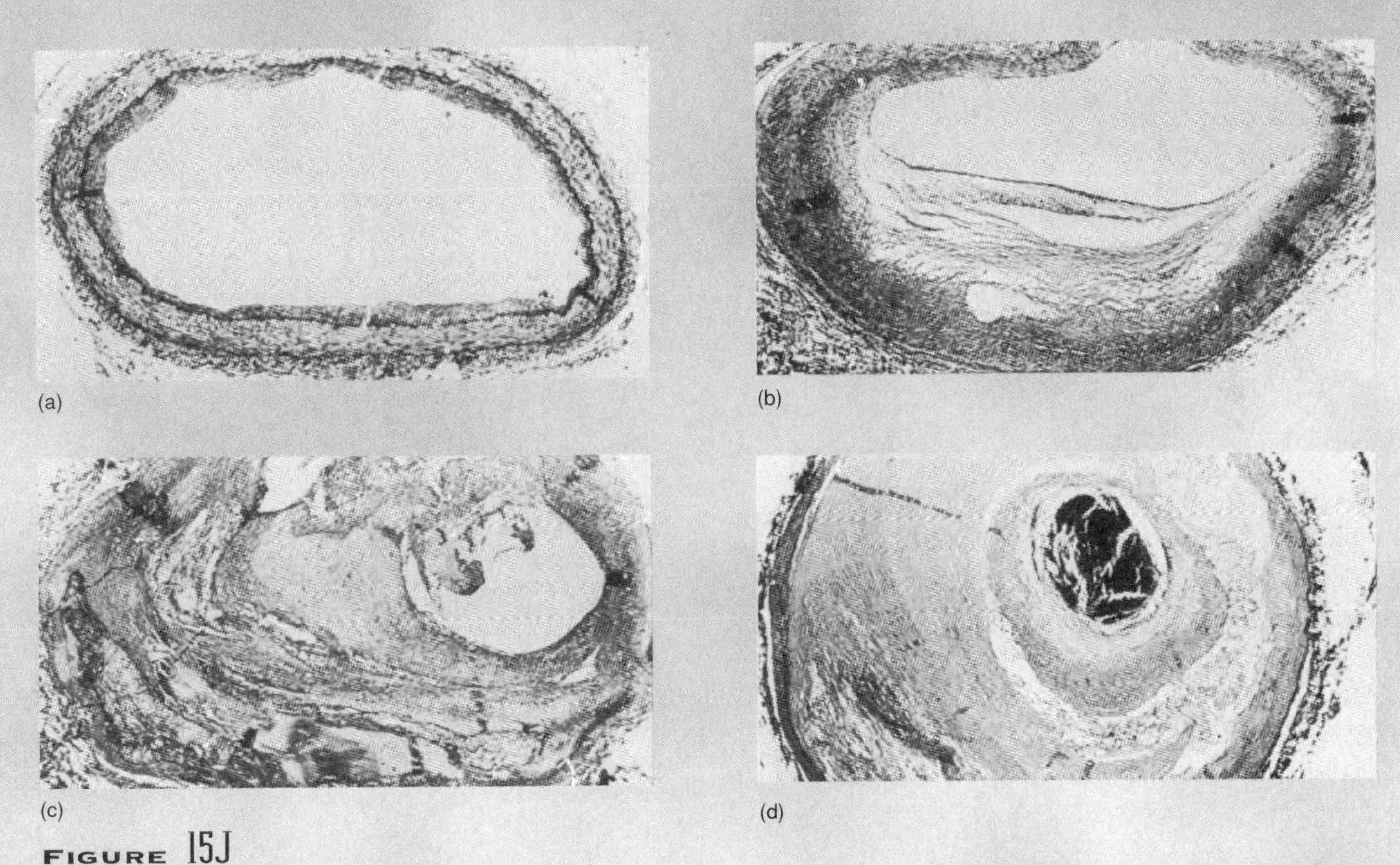


**FIGURE 15J**

As atherosclerosis develops, masses of fatty materials accumulate beneath the inner linings of certain arteries and arterioles and protrude into their lumens. (*a*) Normal arteriole; (*b,c,* and *d*) accumulation of plaque on the inner wall of the arteriole.

does the cardiac output, and, as a result, blood pressure rises. Conversely, if the stroke volume or the heart rate decreases, the cardiac output and blood pressure decrease also.

## Blood Volume

**Blood volume** equals the sum of the formed elements and plasma volumes in the vascular system. Although the blood volume varies somewhat with age, body size, and sex, it is usually about 5 liters for adults.

> Blood volume can be determined by injecting a known volume of an indicator, such as radioactive iodine, into the blood. After a time that allows for thorough mixing, a blood sample is withdrawn, and the concentration of the indicator measured. The total blood volume is calculated using the formula: blood volume = amount of indicator injected/concentration of indicator in blood sample.

## CHART 15.3 Characteristics of Blood Vessels

| Vessel | Type of Wall | Function |
|---|---|---|
| Artery | Thick, strong wall with three layers—an endothelial lining, a middle layer of smooth muscle and elastic tissue, and an outer layer of connective tissue | Carries relatively high pressure blood from the heart to arterioles |
| Arteriole | Thinner wall than an artery, but with three layers; smaller arterioles have an endothelial lining, some smooth muscle tissue, and a small amount of connective tissue | Connects an artery to a capillary, helps control the blood flow into a capillary by vasoconstricting or vasodilating |
| Capillary | Single layer of squamous epithelium | Provides a membrane through which nutrients, gases, and wastes are exchanged between the blood and tissue cells; connects an arteriole to a venule |
| Venule | Thinner wall, less smooth muscle and elastic tissue than in an arteriole | Connects a capillary to a vein |
| Vein | Thinner wall than an artery but with similar layers; the middle layer is more poorly developed; some with flaplike valves | Carries relatively low pressure blood from a venule to the heart; valves prevent a backflow of blood; serves as blood reservoir |

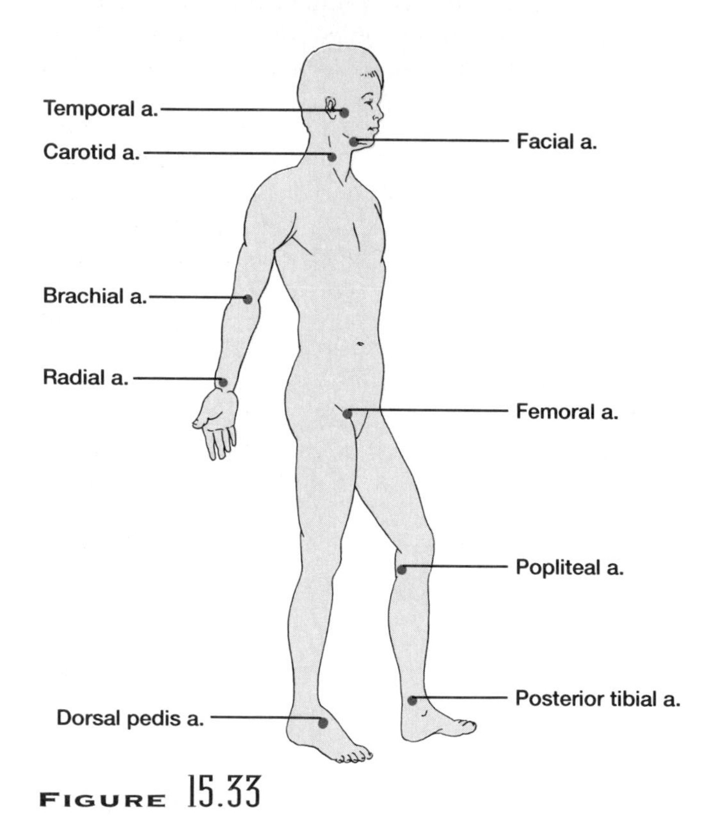


**FIGURE 15.33**
Sites where an arterial pulse is most easily detected (a. stands for *artery*).

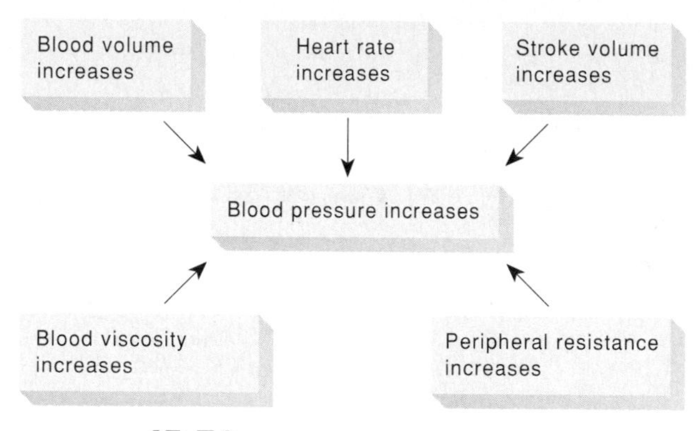


**FIGURE 15.34**
Some of the factors that influence arterial blood pressure.

Blood pressure is normally directly proportional to the volume of the blood within the cardiovascular system. Thus, any changes in the blood volume alter the blood pressure. For example, if a hemorrhage reduces blood volume, blood pressure drops. If a transfusion restores normal blood volume, normal pressure may be reestablished. Blood volume can also fall if the fluid balance is upset, as happens in dehydration. Fluid replacement can reestablish normal blood volume and pressure.

### Peripheral Resistance

Friction between the blood and the walls of the blood vessels produces a force called **peripheral resistance,** which hinders blood flow. Blood pressure must overcome this force if the blood is to continue flowing. Therefore, factors that alter the peripheral resistance change blood pressure.

For example, smooth muscles in the walls of arterioles contracting increases the peripheral resistance of these constricted vessels. Blood tends to back up into the arteries supplying the arterioles, and the arterial pressure rises. Dilation of the arterioles has the opposite effect—peripheral resistance lessens, and the arterial blood pressure drops in response (fig. 15.35).

Because arterial walls are quite elastic, when the ventricles discharge a surge of blood, the arteries swell. Almost immediately, the elastic tissues recoil, and the vessel walls press against the blood inside. This action helps force the blood onward against the peripheral resistance in arterioles and capillaries. It also tends to convert the intermittent flow of blood, which is characteristic of the arterial system, into a more continuous movement through the capillaries.

### Viscosity

The viscosity of a fluid is a physical property related to the ease with which its molecules flow past one another. The greater the viscosity, the greater the resistance to flowing.

Blood cells and plasma proteins increase blood viscosity. Since the greater the blood's resistance to flowing, the greater the force needed to move it through the vascular system, it is not surprising that blood pressure rises as blood viscosity increases, and drops as viscosity decreases.

Although the viscosity of blood normally remains relatively stable, any condition that alters the concentrations of blood cells or plasma proteins may alter viscosity. For example, anemia and hemorrhage may decrease viscosity and consequently lower blood pressure. Excess red blood cells increase viscosity and blood pressure.

1. How is cardiac output calculated?
2. What is the relationship between cardiac output and blood pressure?
3. How does blood volume affect blood pressure?
4. What is the relationship between peripheral resistance and blood pressure? Between viscosity and blood pressure?

## Control of Blood Pressure

Blood pressure (BP) is determined by cardiac output (CO) and peripheral resistance (PR) according to this relationship: $BP = CO \times PR$. Maintenance of normal blood pressure therefore involves the regulation of these two factors.

*Cardiac output* depends on the volume of blood discharged from the left ventricle with each contraction (stroke volume) and heart rate. Mechanical, neural, and chemical factors affect these actions.

For example, the volume of blood entering the ventricle affects the stroke volume. As the blood enters, myocardial fibers in the ventricular wall are mechanically stretched. Within limits, the greater the length of these fibers, the greater the force with which they contract. This relationship between fiber length (due to stretching of the cardiac muscle cell just before contraction) and force of contraction is called *Starling's law of the heart.* Because of it, the heart can respond to the immediate demands placed on it by the varying quantities of blood that return from the venous system. In other words, the more blood that enters the heart from the veins, the greater the ventricular distension, the stronger the ventricular contraction, the greater the stroke volume, and the greater the cardiac output (fig. 15.36). The less blood that returns from the veins, the lesser the ventricular distension, the weaker the ventricular contraction, the lesser the stroke volume, and the lesser the cardiac output.

This mechanism ensures that the volume of blood discharged from the heart is equal to the volume entering its chambers. Consequently, the volume of blood that enters the right atrium from the venae cavae is normally equal to the volume that leaves the left ventricle and enters the aorta.

Recall that baroreceptors in the walls of the aortic arch and carotid sinuses sense changes in blood pressure. If arterial pressure increases, nerve impulses travel from the receptors to the *cardiac center* of the medulla oblongata. This center relays parasympathetic impulses to the S-A node in the heart, and heart rate decreases in response. As a result of this *cardioinhibitor reflex,* cardiac output falls, and blood pressure decreases toward the normal level. Figure 15.37 summarizes this mechanism.

Conversely, decreasing arterial blood pressure initiates the *cardioaccelerator reflex,* which involves sympathetic impulses to the S-A node. As a result, the heart beats faster. This response increases cardiac output, increasing arterial pressure.

Recall that epinephrine increases heart rate, and consequently alters cardiac output and blood pressure. Other factors that increase heart rate and blood pressure include emotional responses, such as fear and anger; physical exercise; and a rise in body temperature.

Changes in arteriole diameters regulate peripheral resistance. Because blood vessels with smaller diameters offer a greater resistance to blood flow, factors that cause arteriole vasoconstriction increase peripheral resistance, and factors causing vasodilation decrease resistance.

The *vasomotor center* of the medulla oblongata continually sends sympathetic impulses to the smooth muscles in the arteriole walls, keeping them in a state

**CLINICAL APPLICATION**

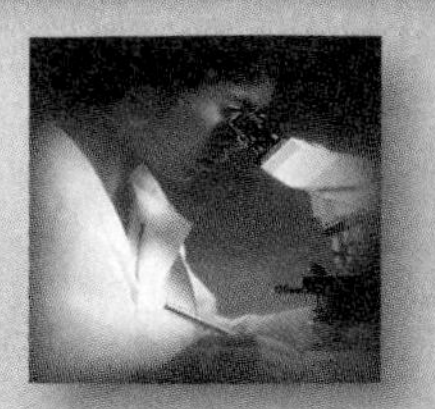

## Measurement of Arterial Blood Pressure

Systemic arterial blood pressure usually is measured using an instrument called a *sphygmomanometer* (fig. 15K). This device consists of an inflatable rubber cuff connected by tubing to a compressible bulb and a glass tube containing a column of mercury. The bulb is used to pump air into the cuff, and a rise in the mercury column indicates the pressure produced. Thus, the pressure in the cuff can be expressed in millimeters of mercury (mm Hg). A pressure of 100 mm Hg, for example, would be enough to force the mercury column upward for a distance of 100 mm.

To measure arterial blood pressure, the cuff of the sphygmomanometer is usually wrapped around the arm so that it surrounds the brachial artery. Air is pumped into the cuff until the cuff pressure exceeds the pressure in that artery. As a result, the vessel is squeezed closed, and its blood flow stopped. At this moment, if the diaphragm of a stethoscope is placed over the brachial artery at the distal border of the cuff, no sounds can be heard from the vessel because the blood flow is interrupted. As air is slowly released from the cuff, the air pressure inside it decreases. When the cuff pressure is approximately equal to the systolic blood pressure within the brachial artery, the artery opens enough for a small amount of blood to spurt through. This movement produces a sharp sound (Korotkoff's sound) that can be heard through the stethoscope. The height of the mercury column when this first tapping sound is heard represents the *arterial systolic pressure* (SP).

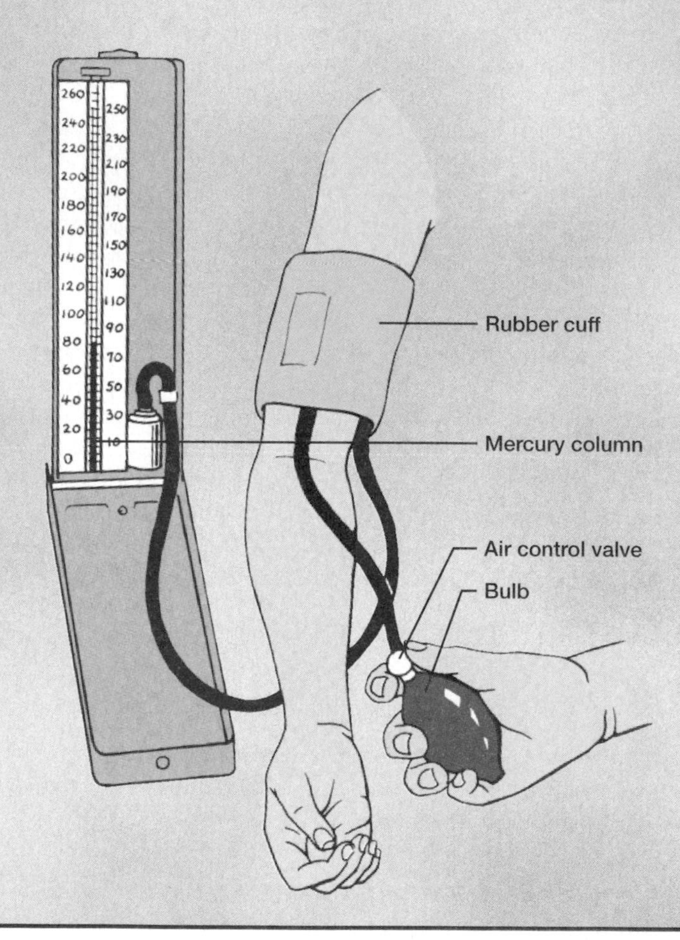


**FIGURE 15K**
A sphygmomanometer is used to measure arterial blood pressure.

of tonic contraction, which helps maintain the peripheral resistance associated with normal blood pressure. Because the vasomotor center responds to changes in blood pressure, it can increase peripheral resistance by increasing its outflow of sympathetic impulses, or it can decrease such resistance by decreasing its sympathetic outflow. In the latter case, the vessels undergo vasodilation as sympathetic stimulation falls.

For instance, as figure 15.38 illustrates, whenever arterial blood pressure suddenly increases, baroreceptors in the aortic arch and carotid sinuses signal the vasomotor center, and the sympathetic outflow to the arteriole walls falls. The resulting vasodilation decreases peripheral resistance, and blood pressure decreases toward the normal level.

As the cuff pressure continues to drop, a series of increasingly louder sounds can be heard. Then, when the cuff pressure is approximately equal to that within the fully opened artery, the sounds become abruptly muffled and disappear. The height of the mercury column when this happens represents the *arterial diastolic pressure* (DP).

The results of a blood pressure measurement are reported as a fraction, such as 120/80. In this notation, the upper number indicates the systolic pressure in mm Hg (SP), and the lower number indicates the diastolic pressure in mm Hg (DP). Figure 15L shows how these pressures decrease as distance from the left ventricle increases.

The difference between the systolic and diastolic pressures (SP–DP), which is called the *pulse pressure* (PP), is generally about 40 mm Hg.

The average pressure in the arterial system is also of interest because it represents the force that is effective throughout the cardiac cycle for driving blood to the tissues. This force, called the *mean arterial pressure,* is approximated by adding the diastolic pressure and one-third of the pulse pressure (DP + 1/3PP).

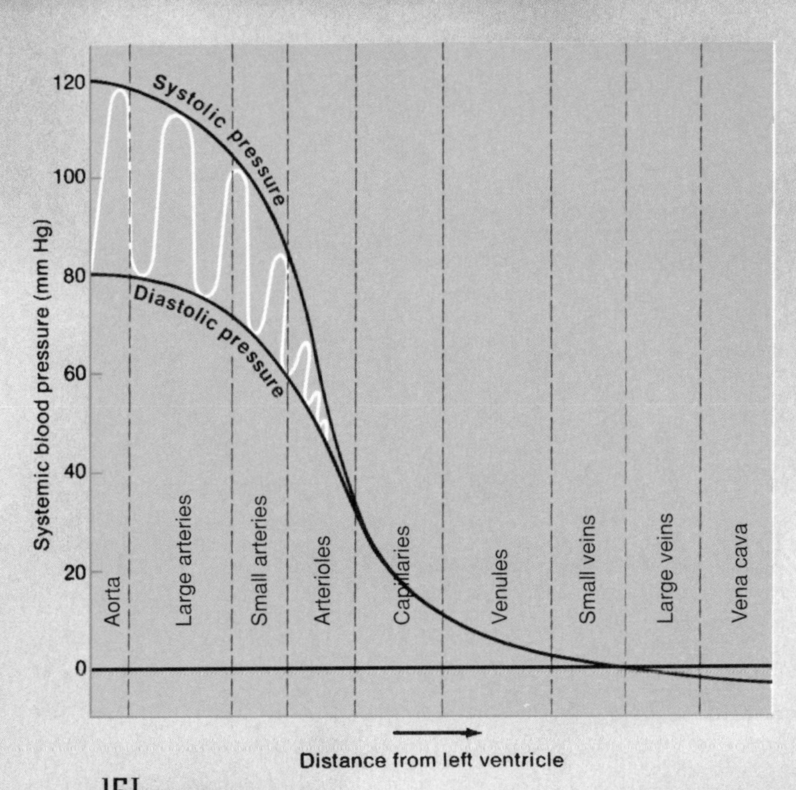


**FIGURE 15L**
Blood pressure decreases as the distance from the left ventricle increases.

The vasomotor center's control of vasoconstriction and vasodilation is especially important in the arterioles of the *abdominal viscera* (splanchnic region). These vessels, if fully dilated, could accept nearly all the blood of the body and cause the arterial pressure to approach zero. Thus, control of their diameters is essential in regulating normal peripheral resistance.

Certain chemicals, including carbon dioxide, oxygen, and hydrogen ions, also influence peripheral resistance by affecting precapillary sphincters and smooth muscles in arteriole and metarteriole walls. For example, increasing carbon dioxide, decreasing oxygen, and lowering pH relaxes these muscles in the systemic circulation. This increases local blood flow

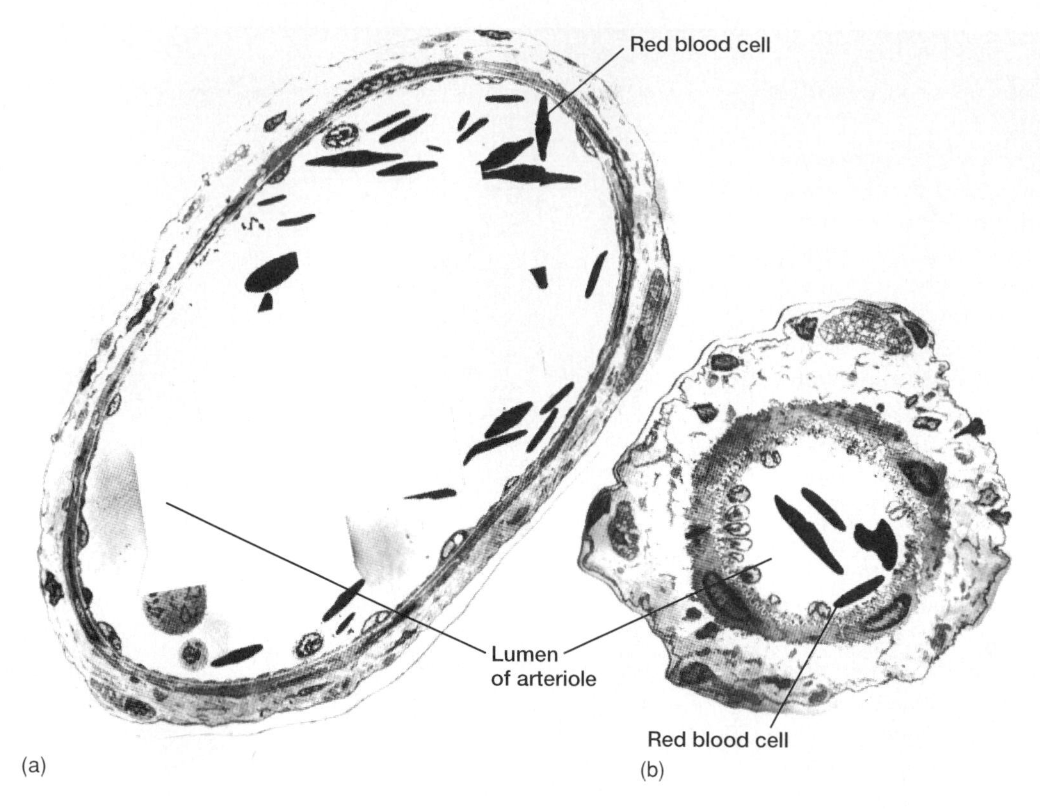


**FIGURE 15.35**

(*a*) Relaxation of smooth muscle in the arteriole wall produces dilation, while (*b*) contraction of the smooth muscle causes constriction (*a* and *b* 1,100×).

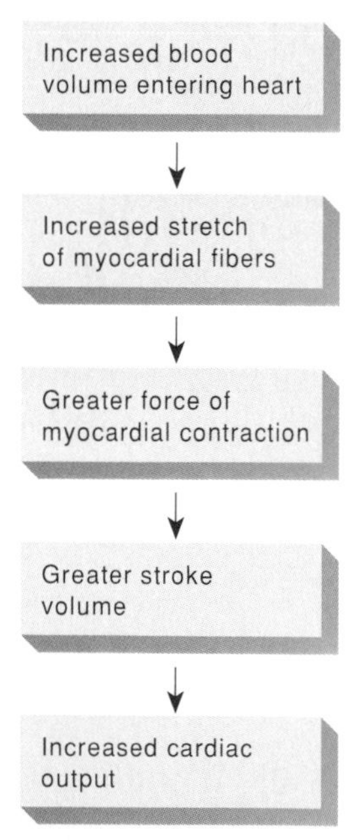


**FIGURE 15.36**

Cardiac output is related to the volume of blood entering the heart.

to tissues with high metabolic rates, such as exercising skeletal muscles. In addition, epinephrine and norepinephrine vasoconstrict many systemic vessels, increasing peripheral resistance, even though epinephrine vasodilates vessels within the skeletal muscles.

Clinical Application 15.6 discusses high blood pressure.

1. What factors affect cardiac output?
2. Define Starling's law of the heart.
3. What is the function of the baroreceptors in the walls of the aortic arch and carotid sinuses?
4. How does the vasomotor center control the diameter of arterioles?

## Venous Blood Flow

Blood pressure decreases as the blood moves through the arterial system and into the capillary networks, so that little pressure remains at the venule ends of capillaries (see fig. 15L). Instead,

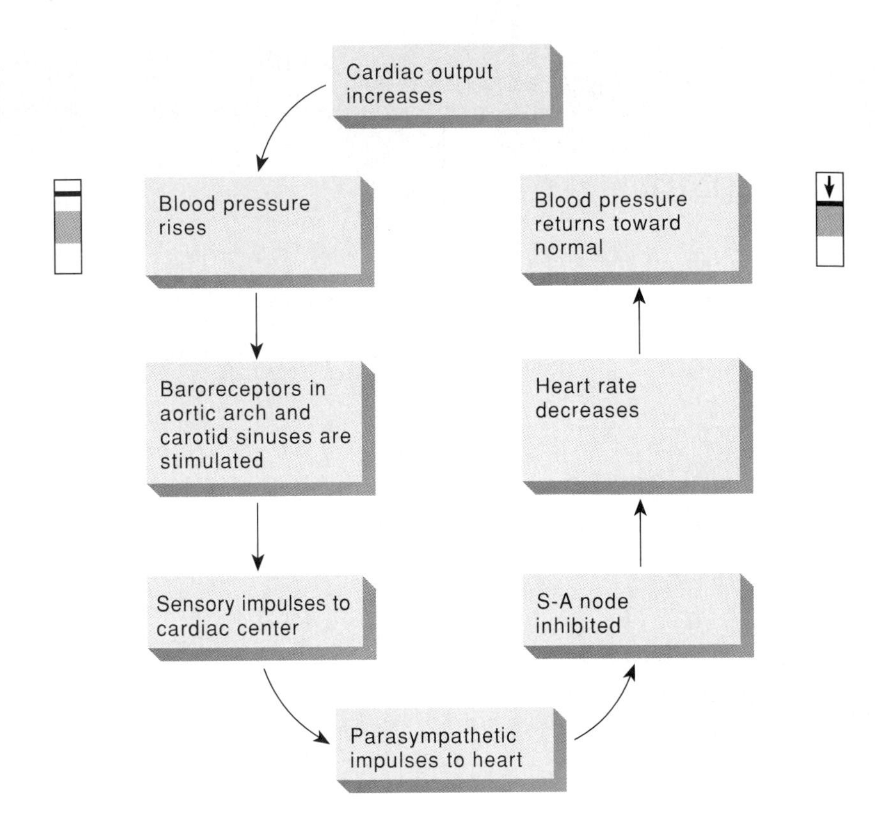


**FIGURE 15.37**
This mechanism helps regulate blood pressure by inhibiting the S-A node.

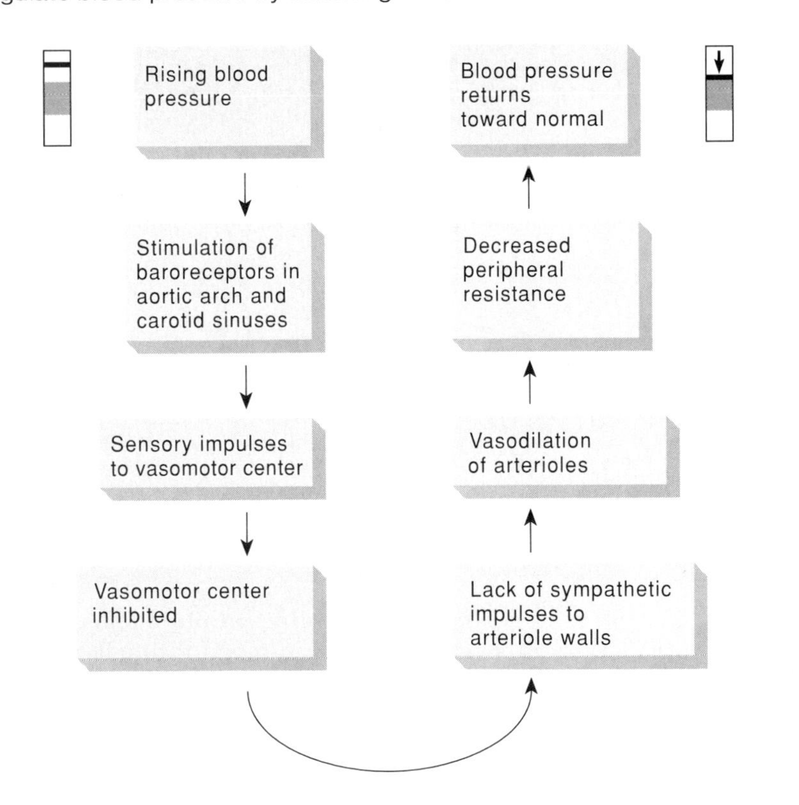


**FIGURE 15.38**
This mechanism helps regulate blood pressure by dilating arterioles.

CLINICAL APPLICATION 15.6

## Hypertension

Hypertension, or high blood pressure, is persistently elevated arterial pressure. It is one of the more common diseases of the cardiovascular system.

High blood pressure with unknown cause is called *essential* (also primary or idiopathic) *hypertension.* Elevated pressure related to another problem, such as arteriosclerosis or kidney disease, is *secondary hypertension.*

Arteriosclerosis is accompanied by decreased elasticity of the arterial walls and narrowed vessel lumens, which contribute to increased blood pressure. Kidney diseases often produce changes that interfere with blood flow to kidney cells. In response, the affected tissues may release an enzyme called *renin* that leads to the production of *angiotensin II,* a powerful vasoconstrictor that increases peripheral resistance in the arterial system, raising arterial pressure. Angiotensin II also stimulates the adrenal cortex to release *aldosterone,* which stimulates the kidneys to retain sodium ions and water. The resulting increase in blood volume contributes to increased blood pressure (fig. 15M).

Normally, this mechanism ensures that a decrease in blood flow to the kidneys is followed by an increase in arterial pressure, which, in turn, restores blood flow to the kidneys. If the decreased blood flow is the result of disease, such as atherosclerosis, the mechanism may cause high blood pressure and promote further deterioration of the arterial system.

In some individuals, high sodium intake leads to vasoconstriction, raising blood pressure. Obesity also is a risk factor for hypertension because it tends to increase peripheral resistance. Psychological stress, which activates sympathetic nerve impulses that cause generalized vasoconstriction, may also lead to hypertension. Yet another cause of hypertension may be an inability of endothelium to respond to a relaxing factor, leading to vasoconstriction.

Hypertension is called a "silent killer" because it may not have direct symptoms yet can set the stage for serious cardiovascular complications.

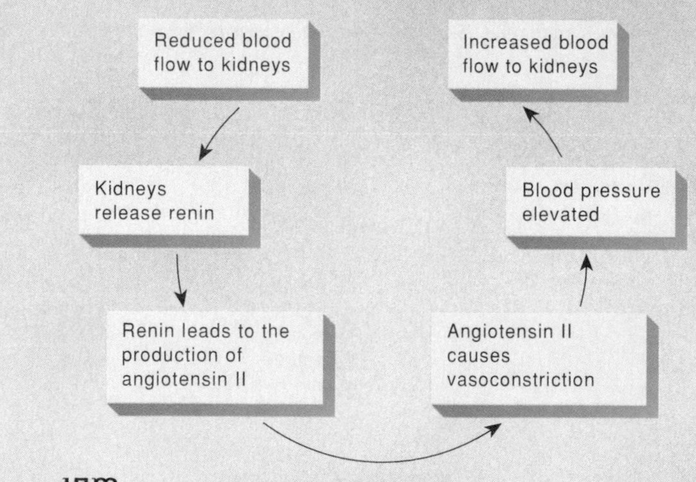


**FIGURE 15M**
This mechanism elevates blood pressure.

blood flow through the venous system is only partly the direct result of heart action and depends on other factors, such as skeletal muscle contraction, breathing movements, and vasoconstriction of veins.

For example, contracting skeletal muscles thicken and press on nearby vessels, squeezing the blood inside. As skeletal muscles exert pressure on veins with valves (called semilunar valves), some blood moves from one valve section to another. This massaging action of contracting skeletal muscles helps push the blood through the venous system toward the heart (fig. 15.39).

Respiratory movements also move venous blood. During inspiration, the pressure within the thoracic cavity is reduced as the diaphragm contracts and the rib cage moves upward and outward. At the same time, the pressure within the abdominal cavity is increased as the diaphragm presses downward on the

For example, as the left ventricle works overtime to pump enough blood, the myocardium thickens, enlarging the heart. If the coronary blood vessels cannot support this overgrowth, parts of the heart muscle die and become replaced with fibrous tissue. Eventually, the enlarged and weakened heart dies.

Hypertension also contributes to the development of atherosclerosis. As arteries accumulate plaque, *a coronary thrombosis* or a *coronary embolism* may occur. Similar changes in the arteries of the brain increase the chances of a *cerebral vascular accident* (CVA), or stroke, due to a cerebral thrombosis, embolism, or hemorrhage.

When an embolus or hemorrhage causes a stroke, paralysis and other functional losses appear suddenly. A thrombus-caused stroke is slower. It may begin with clumsiness, progress to partial visual loss, then affect speech. One arm becomes paralyzed, then a day later, perhaps an entire side of the body is affected. Chart 15A lists risk factors for a stroke.

A *transient ischemic attack* (TIA, or "ministroke") is a temporary block in a small artery. Symptoms include difficulty in speaking or understanding speech; numbness or weakness in the face, upper limb, lower limb, or one side; dizziness; falling; an unsteady gait; blurred vision; or blindness. These symptoms typically resolve within 24 hours with no lasting effects but may be a warning of an impending, more serious stroke.

Treatment of hypertension varies among patients and may include exercising regularly, controlling body weight, reducing stress, and limiting the diet to foods that are low in sodium. Drugs, such as diuretics and/or inhibitors of sympathetic nerve activity, may help control blood pressure. Diuretics increase urinary excretion of sodium and water, reducing the volume of body fluids. Sympathetic inhibitors block the synthesis of neurotransmitters, such as norepinephrine, or block receptor sites of effector cells. Chart 15B summarizes drugs that treat hypertension.

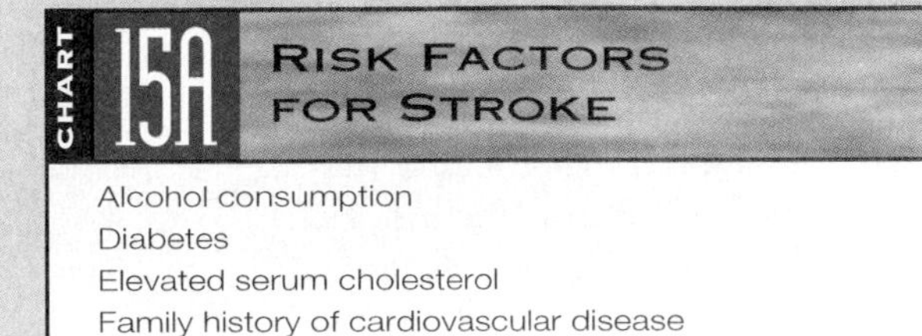

CHART 15A RISK FACTORS FOR STROKE

Alcohol consumption
Diabetes
Elevated serum cholesterol
Family history of cardiovascular disease
Hypertension
Smoking
Transient ischemic attacks

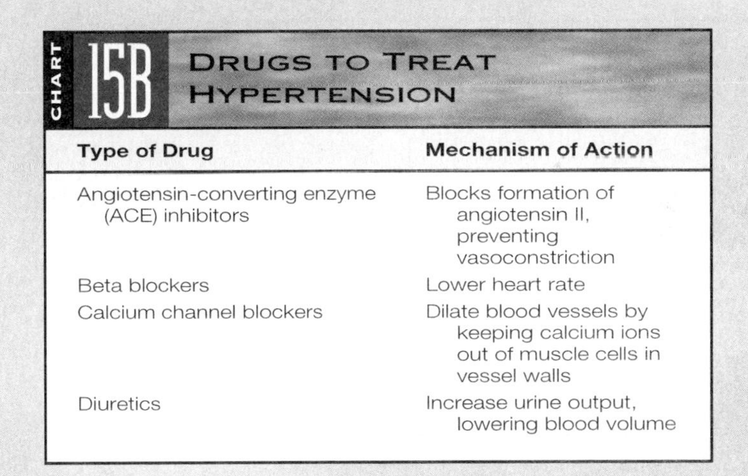

CHART 15B DRUGS TO TREAT HYPERTENSION

| Type of Drug | Mechanism of Action |
| --- | --- |
| Angiotensin-converting enzyme (ACE) inhibitors | Blocks formation of angiotensin II, preventing vasoconstriction |
| Beta blockers | Lower heart rate |
| Calcium channel blockers | Dilate blood vessels by keeping calcium ions out of muscle cells in vessel walls |
| Diuretics | Increase urine output, lowering blood volume |

abdominal viscera. Consequently, the blood tends to be squeezed out of the abdominal veins and forced into thoracic veins. During exercise, these respiratory movements act with skeletal muscle contractions to increase return of venous blood to the heart.

Vein constriction also returns venous blood to the heart. When venous pressure is low, sympathetic reflexes stimulate smooth muscles in the walls of veins to contract. The veins also provide a blood reservoir that can adapt its capacity to changes in blood volume (see fig. 15.32). If some blood is lost and blood pressure falls, venoconstriction can force blood out of this reservoir. In both of these examples, *venoconstriction* helps to maintain blood pressure by forcing more blood toward the heart.

## Central Venous Pressure

Because all the veins, except those of the pulmonary circuit, drain into the right atrium, the pressure within this heart chamber is called *central venous pressure*. This pressure is of special interest because it affects the pressure within the peripheral veins. For example, if the heart is beating weakly, the central venous pressure increases, and blood backs up in the venous network, raising its pressure too. However, if the heart is beating forcefully, the central venous pressure and the pressure within the venous network decrease.

Other factors that increase the flow of blood into the right atrium, and thus elevate the central venous pressure, include increase in blood volume or widespread venoconstriction. An increase in central venous pressure can lead to peripheral edema because the resulting higher capillary hydrostatic pressure favors movement of fluid into the tissues (see chapter 21).

As a result of disease or injury, blood or tissue fluid may accumulate in the pericardial cavity where it increases pressure. This condition, called *acute cardiac tamponade,* can be life threatening. As the pressure around the heart increases, it may compress the heart, interfere with the flow of blood into its chambers, and prevent pumping action. An early symptom of acute cardiac tamponade may be increased central venous pressure, with visible engorgement of the veins in the neck.

Clinical Application 15.7 discusses the effects of exercise on the heart and blood vessels.

1. What is the function of the venous valves?
2. How do skeletal muscles affect venous blood flow?
3. How do respiratory movements affect venous blood flow?
4. What factors stimulate venoconstriction?

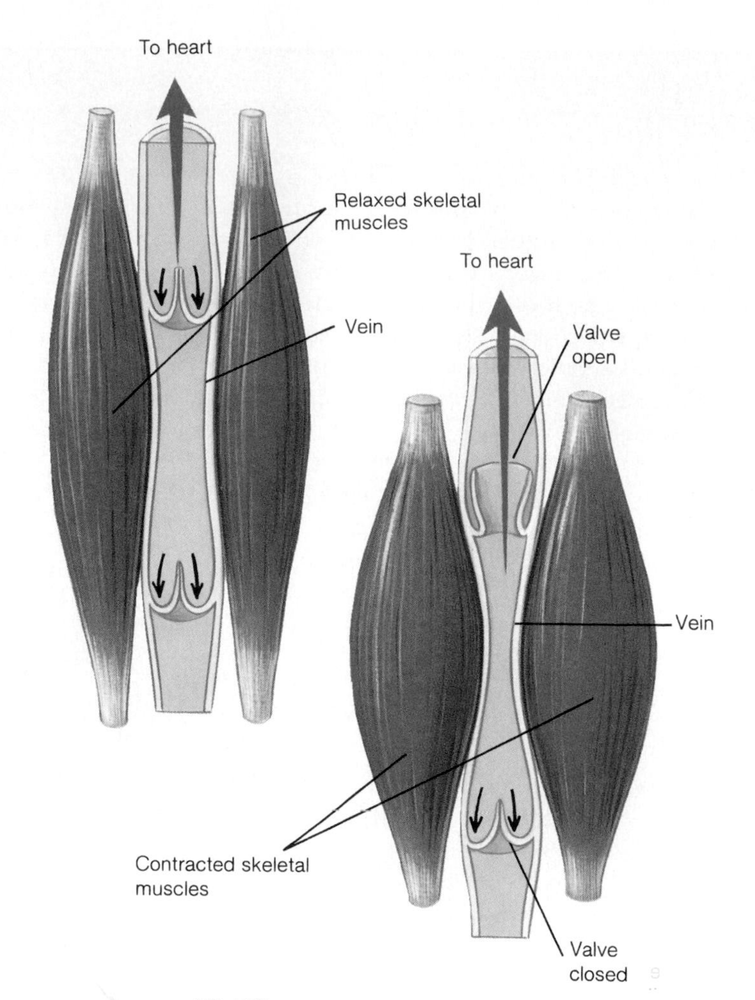


**FIGURE 15.39**
The massaging action of skeletal muscles helps move blood through the venous system toward the heart.

## PATHS OF CIRCULATION

The blood vessels can be divided into two major pathways. The **pulmonary circuit** consists of vessels that carry blood from the heart to the lungs and back to the heart. The **systemic circuit** carries blood from the heart to all other parts of the body and back again (fig. 15.40).

The circulatory pathways described in the following sections are those of an adult. The fetal pathways, which are somewhat different, are described in chapter 23.

### Pulmonary Circuit

The blood enters the pulmonary circuit as it leaves the right ventricle through the pulmonary trunk. The pulmonary trunk extends upward and posteriorly from the heart, and about 5 centimeters above its origin, it divides into the right and left pulmonary arteries. These branches penetrate the right and left lungs, respectively. Within the lungs, they divide into *lobar branches* (three on the right side and two on the left) that accompany the main divisions of the bronchi into the lobes of the lungs. After repeated divisions, the lobar branches give rise to arterioles that continue into the capillary networks associated with the walls of the alveoli (fig. 15.41).

The blood in the arteries and arterioles of the pulmonary circuit has a relatively low concentration of oxygen and a relatively high concentration of carbon dioxide. Gases are exchanged between the blood and the air as the blood moves through the *pulmonary capillaries,* discussed in chapter 19.

Because the right ventricle contracts with less force than the left ventricle, the arterial pressure in the pulmonary circuit is less than that in the systemic circuit. Therefore, the pulmonary capillary pressure is relatively low.

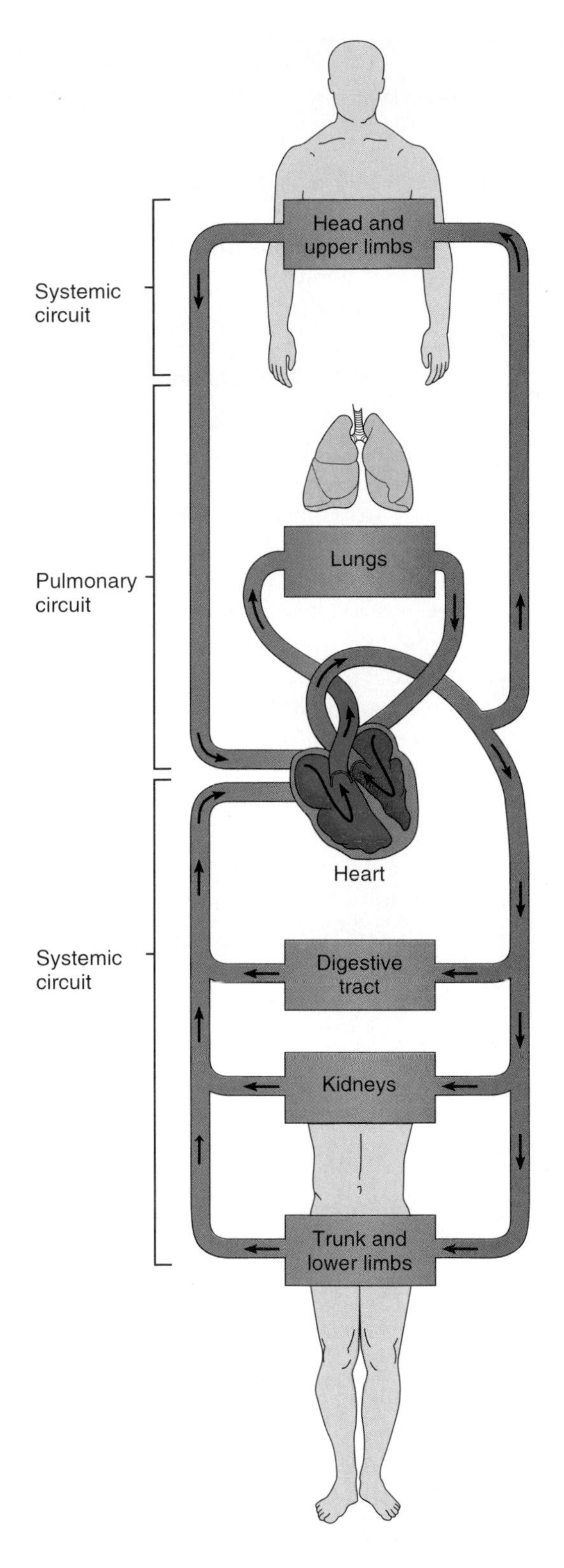


**FIGURE 15.40**

The pulmonary circuit consists of the vessels that carry blood between the heart and the lungs; all other vessels are included in the systemic circuit.

The force that moves fluid out of a pulmonary capillary is 23 mm Hg; the force pulling fluid into it is 22 mm Hg. Thus, such a capillary has a net filtration pressure of 1 mm Hg. This pressure causes a slight, continuous flow of fluid into the narrow interstitial space between the pulmonary capillary and the alveolus.

The epithelial cells of the alveolar membranes are so tightly joined that sodium, chloride, and potassium ions, as well as glucose and urea, enter the interstitial space but usually fail to enter the alveoli. This helps maintain a relatively high osmotic pressure in the interstitial fluid. Consequently, osmosis rapidly moves any water that gets into the alveoli back into the interstitial space. Although the alveolar surface must be moist to allow diffusion of oxygen and carbon dioxide, this mechanism prevents excess water from entering the alveoli and helps keep the alveoli from filling with fluid (fig. 15.42).

Fluid in the interstitial space may be drawn back into the pulmonary capillaries by osmotic pressure of the blood, or it may be returned to the circulation by means of lymphatic vessels (see chapter 16).

As a result of the gas exchanges between the blood and the alveolar air, blood entering the venules of the pulmonary circuit is rich in oxygen and low in carbon dioxide. These venules merge to form small veins, and they, in turn, converge to form still larger ones. Four *pulmonary veins,* two from each lung, return blood to the left atrium, and this completes the vascular loop of the pulmonary circuit.

The ductus arteriosus is a fetal blood vessel that joins the aorta and the pulmonary artery. It normally closes shortly after birth, forcing blood to circulate through the lungs, which it did not do during fetal existence. Failure of this vessel to close, a condition called *patent ductus arteriosus,* can result in heart failure.

In a condition called *hypoplastic left heart syndrome,* physicians keep the ductus arteriosus open. This severe underdevelopment of the left side of the heart is responsible for 8% of fatal heart problems before the age of a year. Sometimes surgery can correct the condition, but in some cases a heart transplant is the only option. In order to transplant a heart, surgeons must keep the ductus arteriosus open. They do this by giving the small patient a prostaglandin drug, or by inserting a mold, called a stent, that keeps the vessel open. Several years ago, researchers attempted to treat hypoplastic left heart syndrome by transplanting a chimpanzee heart into a newborn named Baby Fae, to keep her alive until a human heart could be found. The experiment did not work.

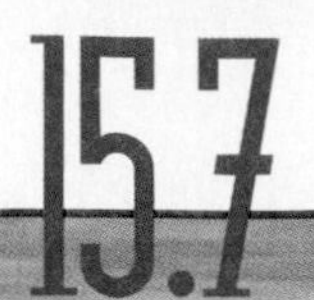

CLINICAL APPLICATION 15.7

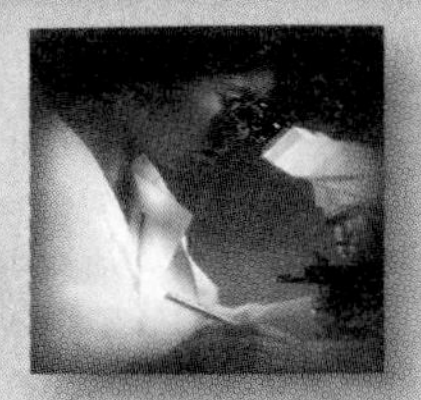

## Exercise and the Cardiovascular System

We all know that exercise is good for the heart. Yet each year, a few individuals die of sudden cardiac arrest while shoveling snow, running, or engaging in some other strenuous activity. The explanation for this apparent paradox is that exercise *is* good for the heart—but only if it is a regular part of life.

### EFFECTS DURING A REGULAR WORKOUT

Physiological responses to intense aerobic exercise generally increase blood flow, and therefore oxygen delivery, to active muscles. In muscles, sympathetic nerve impulses stimulate vasodilation, opening more capillaries. At the same time, vasoconstriction diminishes blood flow where it is not immediately needed, such as to the digestive tract. Blood flow, however, is maintained in the brain and kidneys, which need a steady stream of oxygen and nutrients to function. Respiratory rate rises, increasing the rate of venous return to the heart. Heart rate increases in response to decreased parasympathetic stimulation of the S-A and A-V nodes and to sympathetic reflexes triggered by stimulation of proprioceptors in the skeletal muscles and of stretch receptors in the lungs. As the volume of blood returning to the heart increases, ventricular walls stretch, stimulating them to contract with greater force.

### LONG-TERM CHANGES

The cardiovascular system adapts to exercise as a way of life. The conditioned athlete experiences increases in heart pumping efficiency, blood volume, blood hemoglobin concentration, and the number of mitochondria in muscle fibers. All of these adaptations improve oxygen delivery to and utilization by muscle tissue.

An athlete's heart typically changes in response to these increased demands, enlarging as much as 40% or more. Myocardial mass increases, the ventricular cavities expand, and the ventricle walls thicken. Stroke volume increases

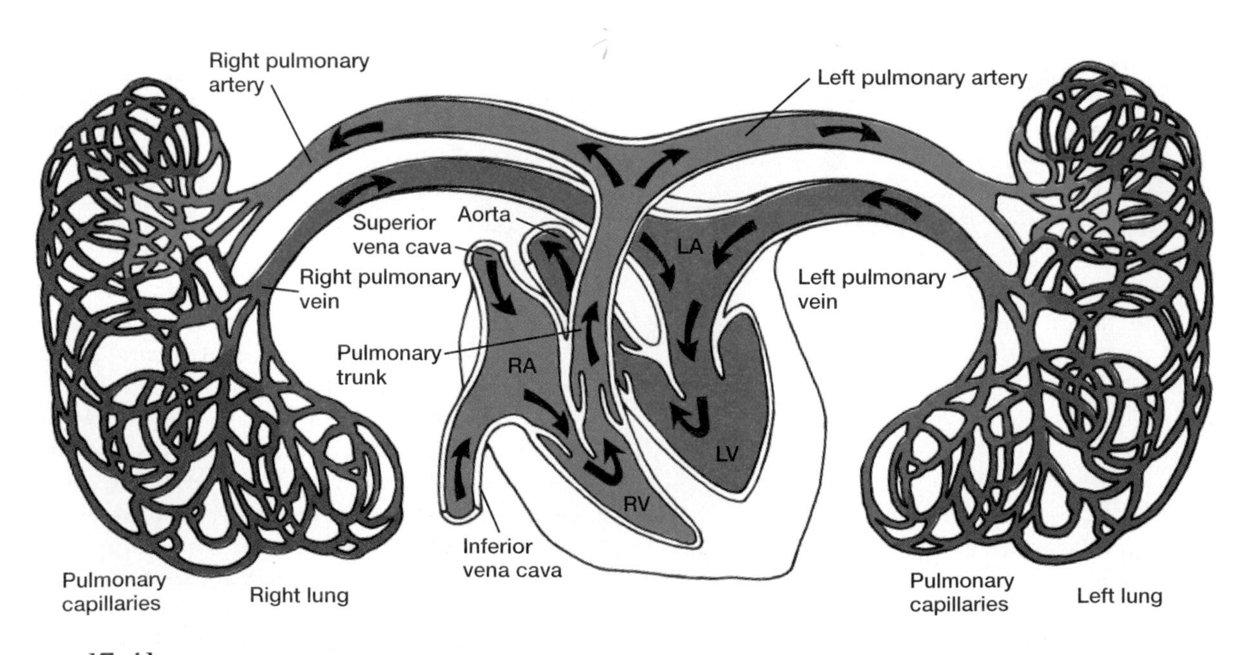


FIGURE 15.41
Blood is carried to the lungs through branches of the pulmonary arteries, and it returns to the heart through pulmonary veins.

and heart rate decreases, as does blood pressure. To a physician unfamiliar with a conditioned cardiovascular system, a trained athlete may appear to be abnormal!

### Overstressing the System

The cardiovascular system responds beautifully to a slow, steady buildup in exercise frequency and intensity. It does not react well to sudden demands—such as a person who never exercises suddenly shoveling snow or running 3 miles. A recent study confirmed age-old anecdotal reports of unaccustomed exercise causing heart failure.

The study was simple, but telling. Researchers in the United States and Germany each interviewed about 1,000 patients hospitalized for heart attacks, asking them what they were doing in the hour before the attack and what their exercise habits were before the attack. They also questioned the same number of people who had not had heart attacks about their activities during the same hours as the ill people. The people with heart attacks were much more likely to have been engaging in unaccustomed strenuous exercise. But the study also turned up good news for those who exercise regularly: Although sedentary people have a two- to sixfold increased risk of cardiac arrest while exercising than when not, people in shape have little or no excess risk while exercising.

How much exercise is enough to benefit the circulatory system? To achieve the benefits of exercise, the heart rate must be elevated to 70% to 85% of its "theoretical maximum" for at least half an hour three times a week. You can calculate your theoretical maximum by subtracting your age from 220. If you are 18 years old, your theoretical maximum is 202 beats per minute. Seventy to 85% of this value is 141 to 172 beats per minute. Some good activities for raising the heart rate are tennis, skating, skiing, handball, vigorous dancing, hockey, basketball, biking, and fast walking.

It is wise to consult a physician before starting an exercise program. People over the age of 30 are advised to have a stress test, which is an electrocardiogram taken while the subject is exercising. (The standard electrocardiogram is taken at rest.) An arrhythmia that appears only during exercise may indicate heart disease that has not yet produced symptoms.

*Pulmonary edema,* in which lungs fill with fluid, can accompany a failing left ventricle or a damaged bicuspid valve. A weak left ventricle may be unable to move the normal volume of blood into the systemic circuit. Blood backing up into the pulmonary circuit increases pressure in the pulmonary capillaries, flooding the interstitial spaces with fluid. Increasing pressure in the interstitial fluid may rupture the alveolar membranes, and fluid may enter the alveoli more rapidly than it can be removed. This reduces the alveolar surface available for gas exchange, and the person may suffocate.

### Systemic Circuit

Contraction of the left ventricle forces the freshly oxygenated blood entering the left atrium into the systemic circuit. This circuit includes the aorta and its branches that lead to all of the body tissues, as well as the companion system of veins that returns blood to the right atrium.

1. Distinguish between the pulmonary and systemic circuits of the cardiovascular system.
2. Trace a drop of blood through the pulmonary circuit from the right ventricle.
3. Explain why the alveoli normally do not fill with fluid.

## Arterial System

The **aorta** is the largest artery in the body. It extends upward from the left ventricle, arches over the heart to the left, and descends just anterior and to the left of the vertebral column.

### Principal Branches of the Aorta

The first portion of the aorta is called the *ascending aorta.* Located at its base are the three cusps of the aortic valve, and opposite each cusp is a swelling in the aortic wall called an **aortic sinus.** The right and left *coronary arteries* spring from two of these sinuses.

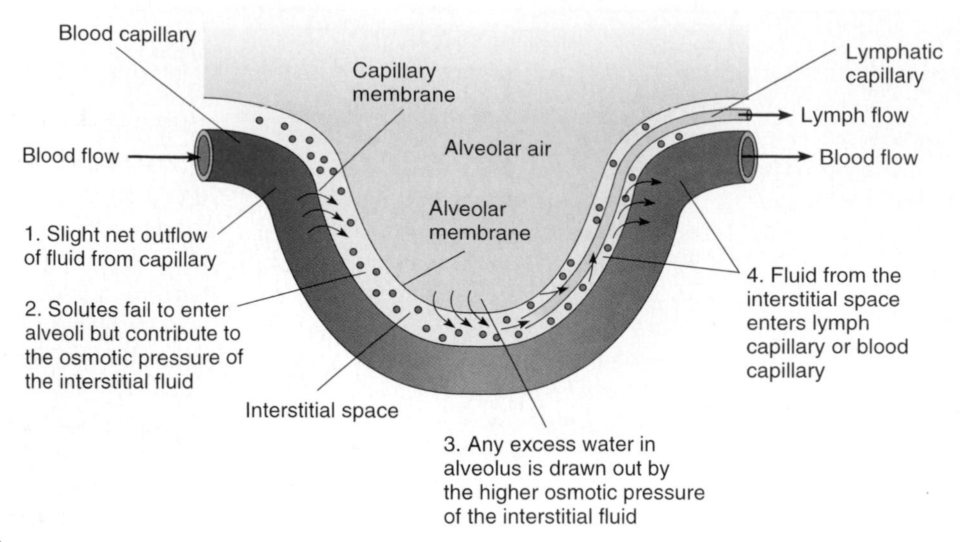


**FIGURE 15.42**

Cells of the alveolar wall are tightly joined. The relatively high osmotic pressure of the interstitial fluid draws water out of them.

Blood flow into these arteries is intermittent and is driven by the elastic recoil of the aortic wall following contraction of the left ventricle.

Recall that several small structures called **aortic bodies** are located within the epithelial lining of the aortic sinuses. These bodies contain baroreceptors that detect changes in blood pressure, and chemoreceptors that sense blood concentrations of oxygen and carbon dioxide.

Three major arteries originate from the *arch of the aorta* (aortic arch). They are the brachiocephalic (innominate) artery, the left common carotid artery, and the left subclavian artery.

The **brachiocephalic** (brak″e-o-sĕ-fal′ik) **artery** supplies blood to the tissues of the upper limb and head, as its name suggests. It is the first branch from the aortic arch, and rises upward through the mediastinum to a point near the junction of the sternum and the right clavicle. There it divides, giving rise to the right **common carotid** (kah-rot′id) **artery,** which carries blood to the right side of the neck and head, and the right **subclavian** (sub-kla′ve-an) **artery,** which leads into the right arm. Branches of the subclavian artery also supply blood to parts of the shoulder, neck, and head.

The left *common carotid artery* and the left *subclavian artery* are respectively the second and third branches of the aortic arch. They supply blood to regions on the left side of the body corresponding to those supplied by their counterparts on the right (see fig. 15.43 and reference plates 71, 72, and 73).

Although the upper part of the *descending aorta* is positioned to the left of the midline, it gradually moves medially and finally lies directly in front of the vertebral column at the level of the twelfth thoracic vertebra. The portion of the descending aorta above the diaphragm is the **thoracic** (tho-ras′ik) **aorta,** and it gives off numerous small branches to the thoracic wall and the thoracic visceral organs. These branches, the *bronchial, pericardial,* and *esophageal arteries,* supply blood to the structures for which they were named. Other branches become *mediastinal arteries,* supplying various tissues within the mediastinum, and *posterior intercostal arteries* that pass into the thoracic wall.

Below the diaphragm, the descending aorta becomes the **abdominal aorta,** and it gives off branches to the abdominal wall and various abdominal organs. These branches include the following:

1. **Celiac** (se′le-ak) **artery.** This single vessel gives rise to the left *gastric, splenic,* and *hepatic arteries,* which supply upper portions of the digestive tract, the spleen, and the liver, respectively. (Note: The hepatic artery supplies the liver with about one-third of its blood flow, and this blood is oxygen-rich. The remaining two-thirds of the liver's blood flow arrives by means of the portal vein and is oxygen-poor.)
2. **Phrenic** (fren′ik) **arteries.** These paired arteries supply blood to the diaphragm.
3. **Superior mesenteric** (mes″en-ter′ik) **artery.** The superior mesenteric is a large, unpaired artery that branches to many parts of the intestinal tract, including the jejunum, ileum, cecum, ascending colon, and transverse colon.
4. **Suprarenal arteries.** This pair of vessels supplies blood to the adrenal glands.

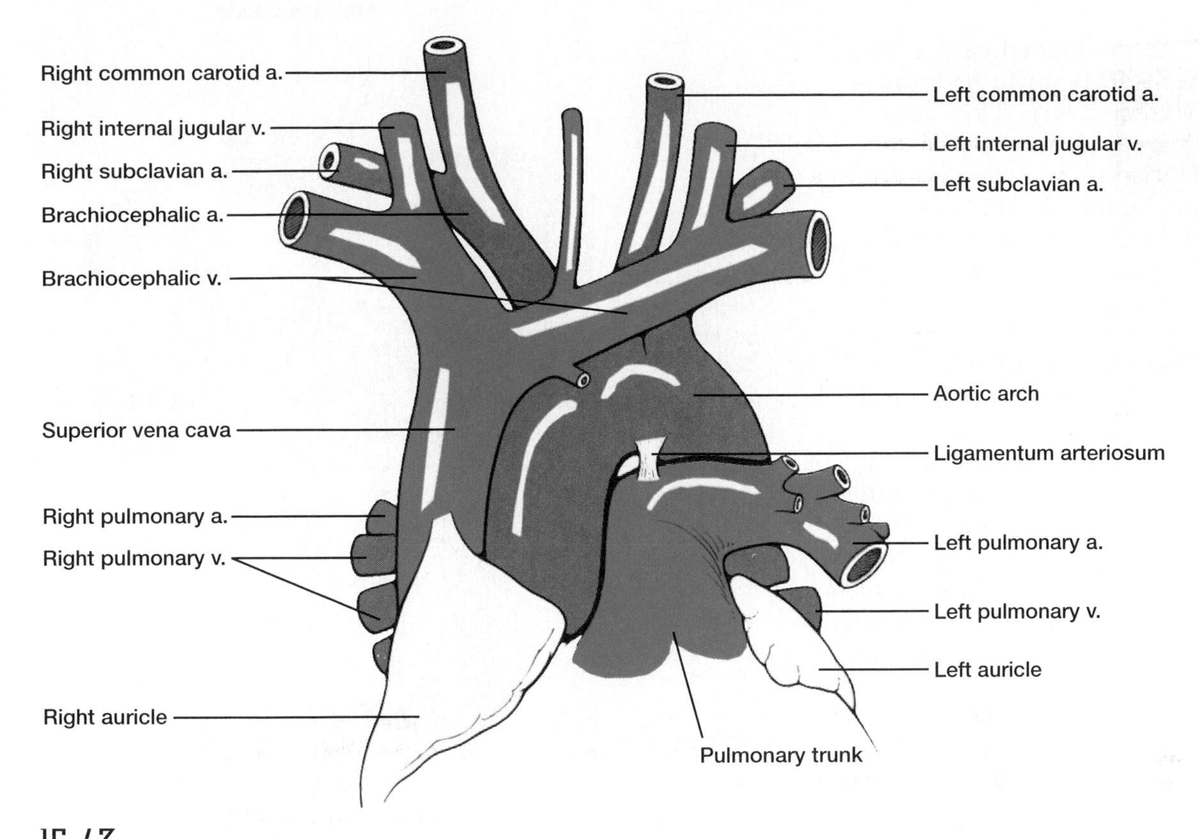


**FIGURE 15.43**
The major blood vessels associated with the heart.

5. **Renal** (re′nal) **arteries.** The renal arteries pass laterally from the aorta into the kidneys. Each artery then divides into several lobar branches within the kidney tissues.
6. **Gonadal** (go′nad-al) **arteries.** In a female, paired *ovarian arteries* arise from the aorta and pass into the pelvis to supply the ovaries. In a male, *spermatic arteries* originate in similar locations. They course downward and pass through the body wall by way of the *inguinal canal* to supply the testes.
7. **Inferior mesenteric artery.** Branches of this single artery lead to the descending colon, the sigmoid colon, and the rectum.
8. **Lumbar arteries.** Three or four pairs of lumbar arteries arise from the posterior surface of the aorta in the region of the lumbar vertebrae. These arteries supply muscles of the skin and the posterior abdominal wall.
9. **Middle sacral** (sa′kral) **artery.** This small, single vessel descends medially from the aorta along the anterior surfaces of the lower lumbar vertebrae. It carries blood to the sacrum and coccyx.

The abdominal aorta terminates near the brim of the pelvis, where it divides into right and left *common iliac arteries.* These vessels supply blood to lower regions of the abdominal wall, the pelvic organs, and the lower extremities (see fig. 15.44).

Chart 15.4 summarizes the main branches of the aorta.

### Arteries to the Neck, Head, and Brain

Branches of the subclavian and common carotid arteries supply blood to structures within the neck, head, and brain (figs. 15.45 and 15.46). The main divisions of the subclavian artery to these regions are the vertebral, thyrocervical, and costocervical arteries. The common carotid artery communicates with these regions by means of the internal and external carotid arteries.

The **vertebral** (ver′te-bral) **arteries** arise from the subclavian arteries in the base of the neck near the tips of the lungs. They pass upward through the foramina of the transverse processes of the cervical vertebrae and enter the skull by way of the foramen magnum. Along their paths, these vessels supply blood to vertebrae and to their associated ligaments and muscles.

Within the cranial cavity, the vertebral arteries unite to form a single *basilar artery.* This vessel passes along the ventral brain stem and gives rise to branches leading to the pons, midbrain, and cerebellum. The basilar artery terminates by dividing into two *posterior cerebral arteries* that supply portions of the occipital and temporal lobes of the cerebrum. The posterior cerebral arteries also help form an arterial circle at the base of the brain, the **circle of Willis,** which connects the vertebral artery and internal carotid artery systems (fig. 15.47). The union of these systems provides alternate pathways through which blood can reach brain tissues in the event of an arterial occlusion.

The **thyrocervical** (thi″ro-ser′vĭ-kal) **arteries** are short vessels that give off branches to the thyroid gland, parathyroid glands, larynx, trachea, esophagus, and pharynx, as well as to various muscles in the neck, shoulder, and back.

The **costocervical** (kos″to-ser′vĭ-kal) **arteries,** which are the third vessels to branch from the subclavians, carry blood to muscles in the neck, back, and thoracic wall.

The left and right *common carotid arteries* ascend deeply within the neck on either side. At the level of the upper laryngeal border, they divide to form the internal and external carotid arteries.

The **external carotid artery** courses upward on the side of the head, giving off branches to structures in the neck, face, jaw, scalp, and base of the skull. The main vessels that originate from this artery include the following:

1. *Superior thyroid artery* to the hyoid bone, larynx, and thyroid gland.
2. *Lingual artery* to the tongue, muscles of the tongue, and salivary glands beneath the tongue.
3. *Facial artery* to the pharynx, palate, chin, lips, and nose.
4. *Occipital artery* to the scalp on the back of the skull, the meninges, the mastoid process, and various muscles in the neck.
5. *Posterior auricular artery* to the ear and the scalp over the ear.

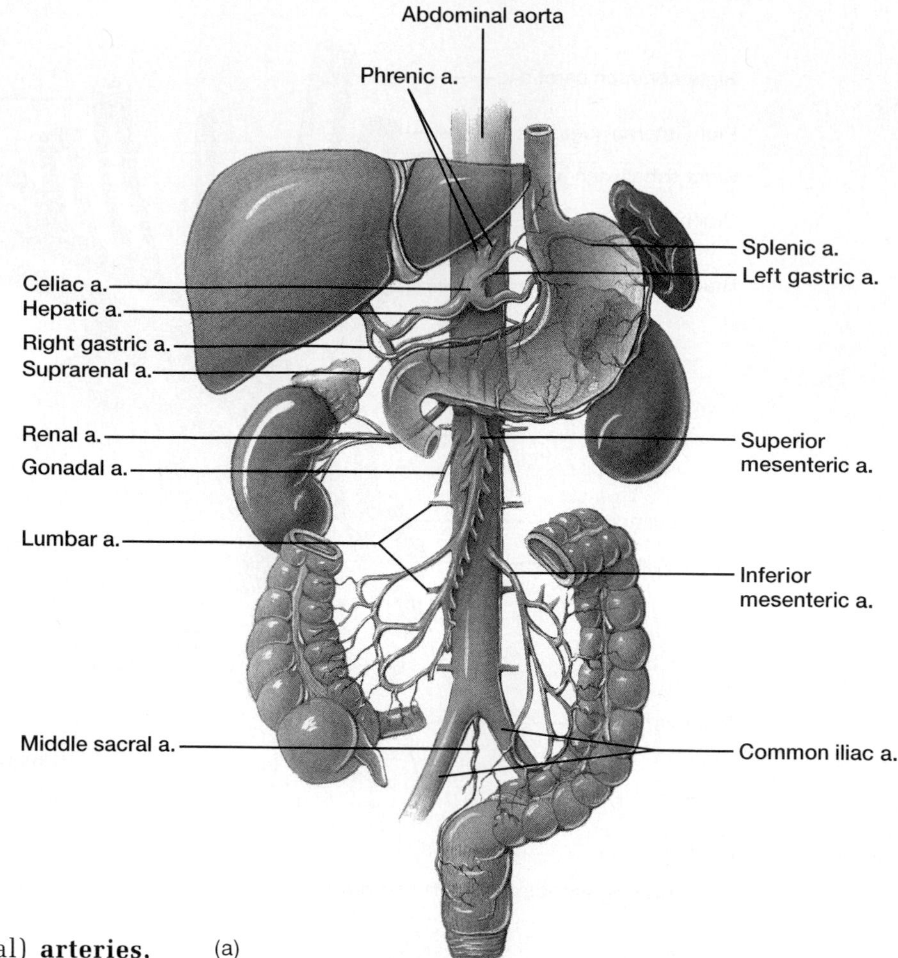


(a)

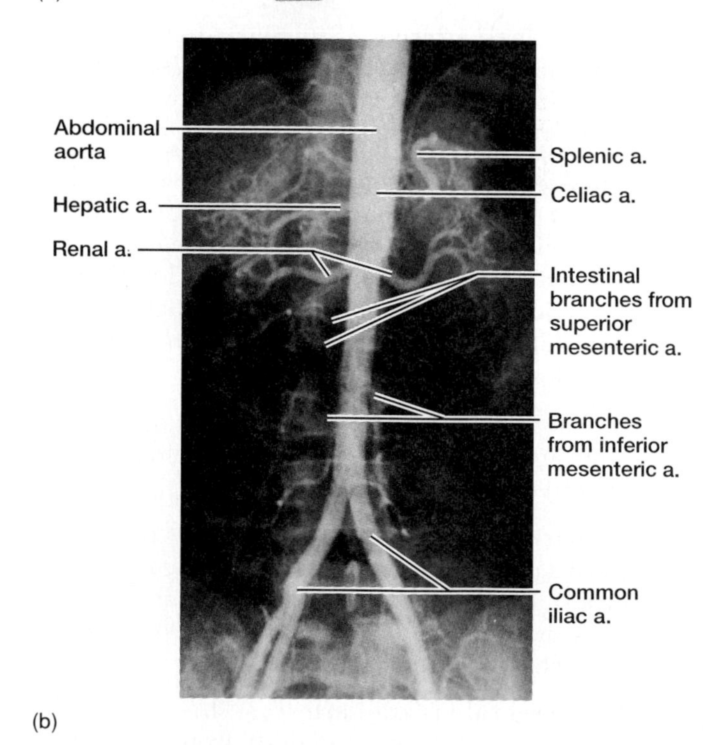


(b)

**Figure 15.44**

(*a*) Abdominal aorta and its major branches; (*b*) angiogram (X-ray film) of the abdominal aorta.

## CHART 15.4 The Aorta and Its Principal Branches

| Portion of Aorta | Major Branch | General Regions or Organs Supplied |
|---|---|---|
| Ascending aorta | Right and left coronary arteries | Heart |
| Arch of aorta | Brachiocephalic artery | Right upper limb, right side of head |
| | Left common carotid artery | Left side of head |
| | Left subclavian artery | Left upper limb |
| Descending aorta | | |
| Thoracic aorta | Bronchial artery | Bronchi |
| | Pericardial artery | Pericardium |
| | Esophageal artery | Esophagus |
| | Mediastinal artery | Mediastinum |
| | Posterior intercostal artery | Thoracic wall |
| Abdominal aorta | Celiac artery | Organs of upper digestive tract |
| | Phrenic artery | Diaphragm |
| | Superior mesenteric artery | Portions of small and large intestines |
| | Suprarenal artery | Adrenal gland |
| | Renal artery | Kidney |
| | Gonadal artery | Ovary or testis |
| | Inferior mesenteric artery | Lower portions of large intestine |
| | Lumbar artery | Posterior abdominal wall |
| | Middle sacral artery | Sacrum and coccyx |
| | Common iliac artery | Lower abdominal wall, pelvic organs, and lower limb |

**FIGURE 15.45**
The main arteries of the head and neck. (Note that the clavicle has been removed.)

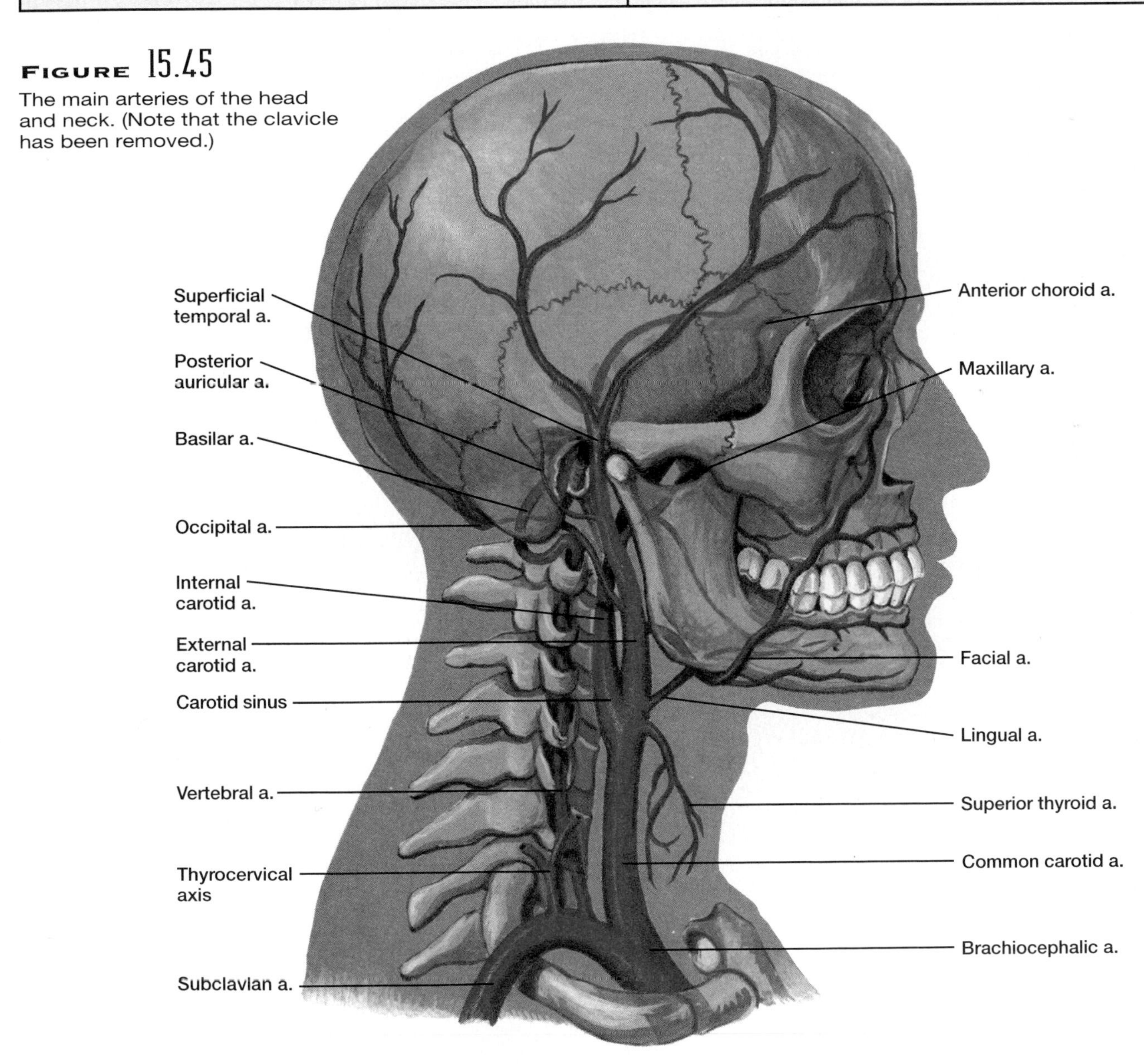

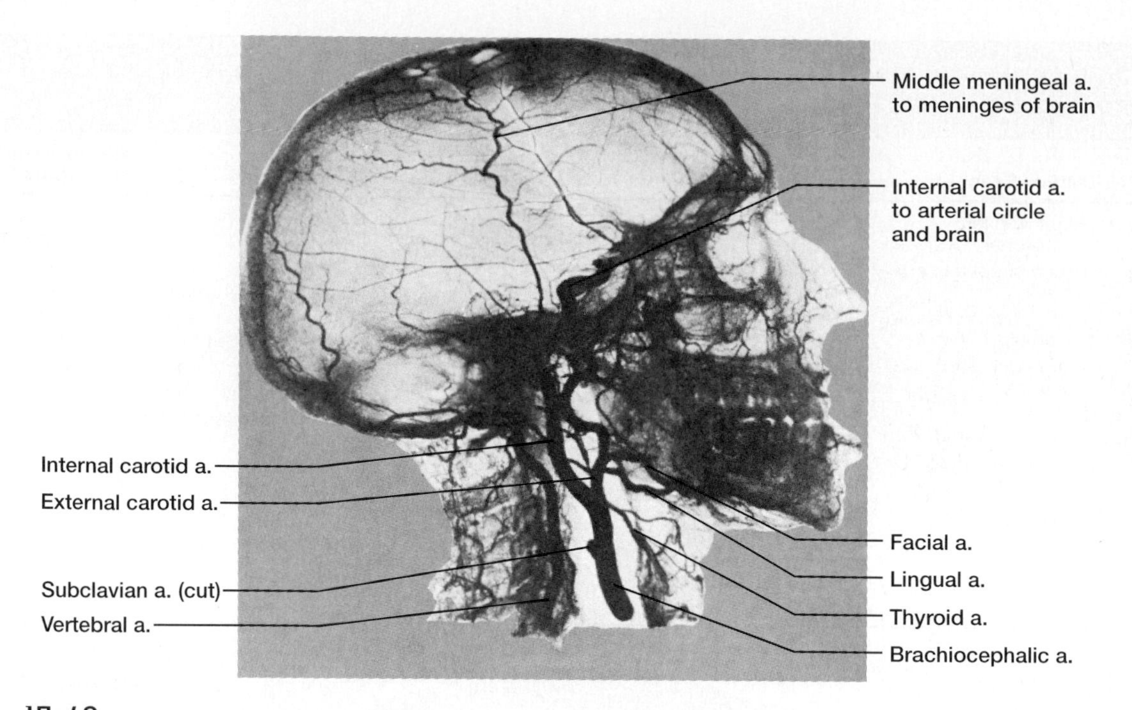


**FIGURE 15.46**
An angiogram of the arteries associated with the head.

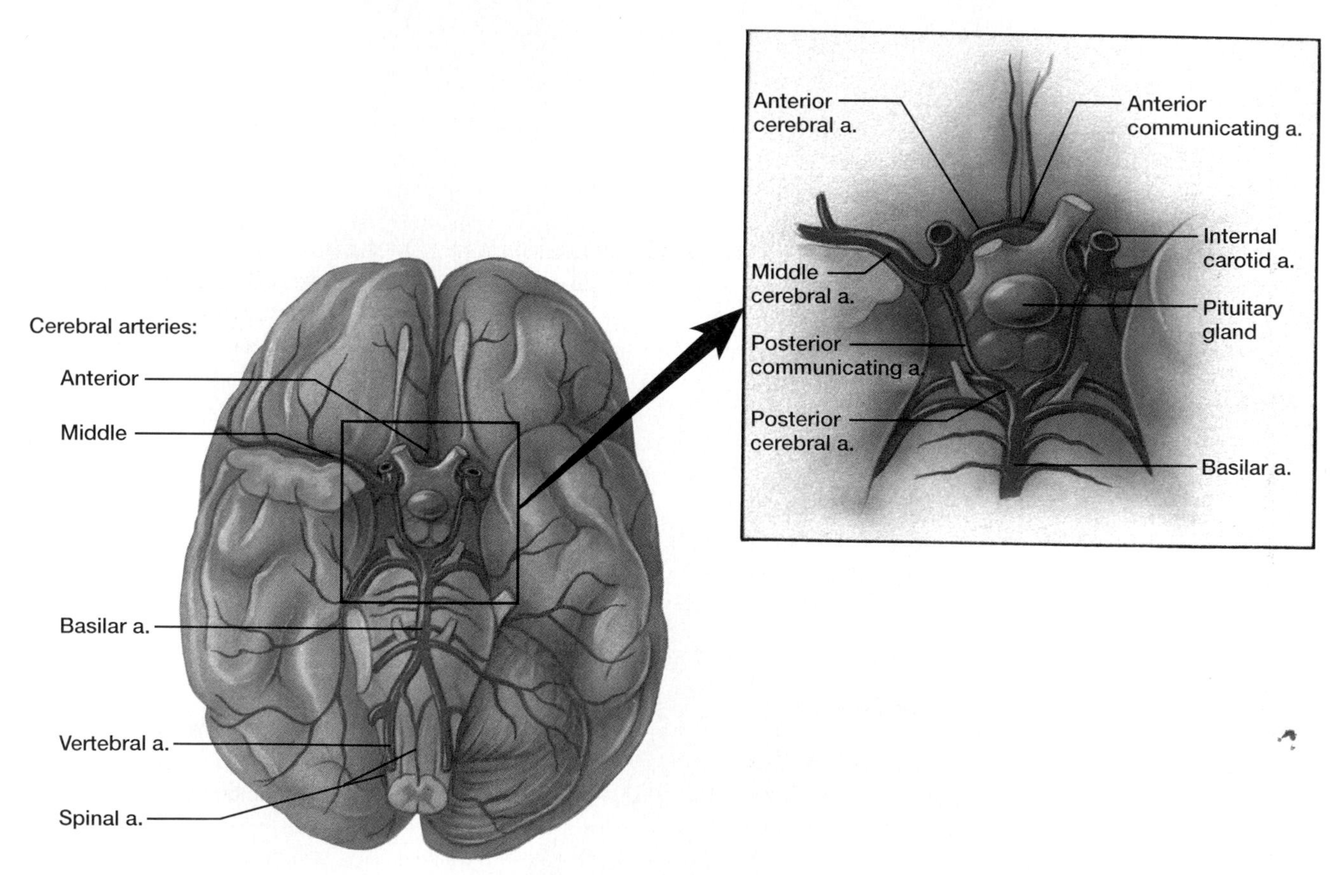


**FIGURE 15.47**
View of inferior surface of the brain. The circle of Willis is formed by the anterior cerebral arteries, which are connected by the anterior communicating artery, and by the posterior vertebral arteries, which are connected to the internal carotid arteries by the posterior communicating arteries.

The external carotid artery terminates by dividing into *maxillary* and *superficial temporal arteries.* The maxillary artery supplies blood to the teeth, gums, jaws, cheek, nasal cavity, eyelids, and meninges. The temporal artery extends to the parotid salivary gland and to various surface regions of the face and scalp.

The **internal carotid artery** follows a deep course upward along the pharynx to the base of the skull. Entering the cranial cavity, it provides the major blood supply to the brain. Its major branches include the following:

1. *Ophthalmic artery* to the eyeball and to various muscles and accessory organs within the orbit.
2. *Posterior communicating artery* that forms part of the circle of Willis.
3. *Anterior choroid artery* to the choroid plexus within the lateral ventricle of the brain and to nerve structures within the brain.

The internal carotid artery terminates by dividing into *anterior* and *middle cerebral arteries.* The middle cerebral artery passes through the lateral sulcus and supplies the lateral surface of the cerebrum, including the primary motor and sensory areas of the face and upper limbs, the optic radiations, and the speech area (see chapter 11). The anterior cerebral artery extends anteriorly between the cerebral hemispheres and supplies the medial surface of the brain.

Near the base of each internal carotid artery is an enlargement called a **carotid sinus.** Like the aortic sinuses, these structures contain baroreceptors that control blood pressure. A number of small epithelial masses, called **carotid bodies,** also occur in the wall of the carotid sinus. These bodies are very vascular and contain chemoreceptors that act with those of the aortic bodies to regulate circulation and respiration.

## Arteries to the Shoulder and Upper Limb

The subclavian artery, after giving off branches to the neck, continues into the arm (fig. 15.48). It passes between the clavicle and the first rib, and becomes the axillary artery.

The **axillary artery** supplies branches to structures in the axilla and the chest wall, including the skin of the shoulder, part of the mammary gland, the upper end of the humerus, the shoulder joint, and muscles in the back, shoulder, and chest. As this vessel leaves the axilla, it becomes the brachial artery.

The **brachial artery** courses along the humerus to the elbow. It gives rise to a *deep brachial artery* that curves posteriorly around the humerus and supplies the triceps muscle. Shorter branches pass into the muscles on the anterior side of the arm, while others descend on each side to the elbow and connect with arteries in the forearm. The resulting arterial network allows blood to reach the forearm even if a portion of the distal brachial artery becomes obstructed.

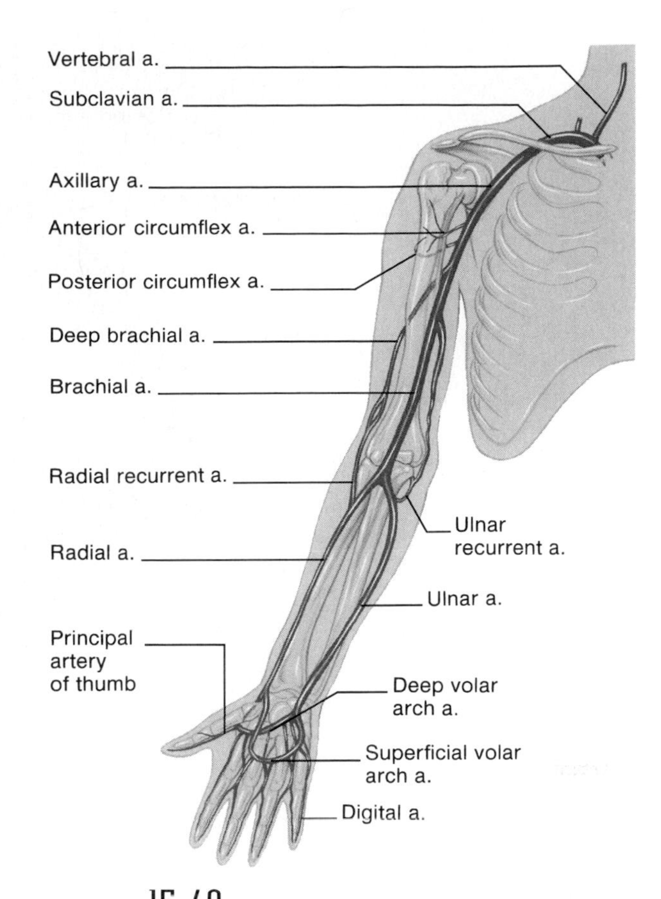


**FIGURE 15.48**
The main arteries to the shoulder and upper limb.

Within the elbow, the brachial artery divides into an ulnar artery and a radial artery. The **ulnar** (ul′nar) **artery** leads downward on the ulnar side of the forearm to the wrist. Some of its branches join the anastomosis around the elbow joint, while others supply blood to flexor and extensor muscles in the forearm.

The **radial** (ra′de-al) **artery,** a continuation of the brachial artery, travels along the radial side of the forearm to the wrist. As it nears the wrist, it comes close to the surface and provides a convenient vessel for taking the pulse (radial pulse). Branches of the radial artery join the anastomosis of the elbow and supply the lateral muscles of the forearm.

At the wrist, the branches of the ulnar and radial arteries join to form a network of vessels. Arteries arising from this network supply blood to structures in the wrist, hand, and fingers.

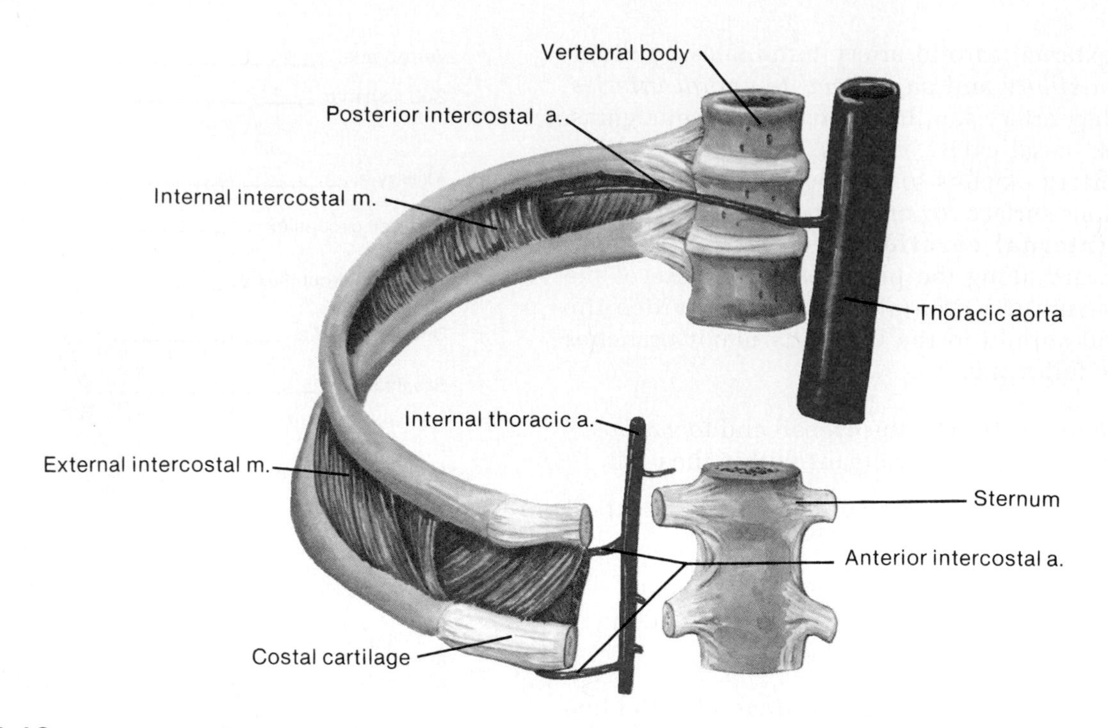


**FIGURE 15.49**
Arteries that supply the thoracic wall (a. stands for *artery;* m. for *muscle*).

## Arteries to the Thoracic and Abdominal Walls

Blood reaches the thoracic wall through several vessels, including branches from the subclavian artery and the thoracic aorta (fig. 15.49).

The subclavian artery contributes to this supply through a branch called the **internal thoracic artery.** This vessel originates in the base of the neck and passes downward on the pleura and behind the cartilages of the upper six ribs. It gives off two *anterior intercostal arteries* to each of the upper six intercostal spaces; these two arteries supply the intercostal muscles, other intercostal tissues, and the mammary glands.

The *posterior intercostal arteries* arise from the thoracic aorta and enter the intercostal spaces between the third through the eleventh ribs. These arteries give off branches that supply the intercostal muscles, the vertebrae, the spinal cord, and deep muscles of the back.

Branches of the *internal thoracic* and *external iliac arteries* provide blood to the anterior abdominal wall. Paired vessels originating from the abdominal aorta, including the *phrenic* and *lumbar arteries,* supply blood to structures in the posterior and lateral abdominal wall.

## Arteries to the Pelvis and Lower Limb

The abdominal aorta divides to form the **common iliac arteries** at the level of the pelvic brim, and these vessels provide blood to the pelvic organs, gluteal region, and lower limbs.

Each common iliac artery descends a short distance and divides into an internal (hypogastric) branch and an external branch. The **internal iliac artery** gives off numerous branches to various pelvic muscles and visceral structures, as well as to the gluteal muscles and the external genitalia. Figure 15.50 shows important branches of this vessel, including the following:

1. *Iliolumbar artery* to the ilium and muscles of the back.
2. *Superior* and *inferior gluteal arteries* to the gluteal muscles, pelvic muscles, and skin of the buttocks.
3. *Internal pudendal artery* to muscles in the distal portion of the alimentary canal, the external genitalia, and the hip joint.
4. *Superior* and *inferior vesical arteries* to the urinary bladder. In males, these vessels also supply the seminal vesicles and the prostate gland.

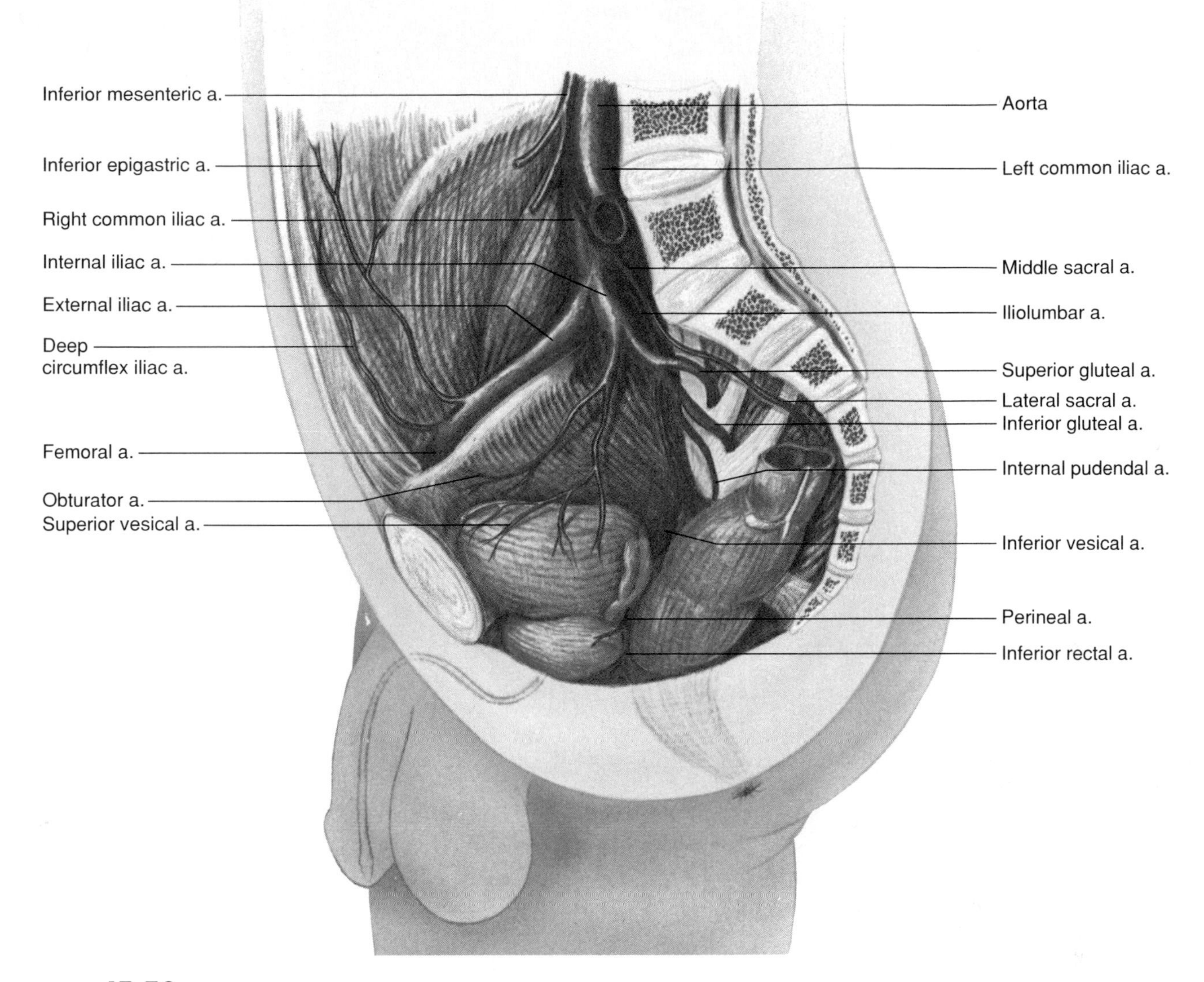


**FIGURE 15.50**
Arteries that supply the pelvic region.

5. *Middle rectal artery* to the rectum.
6. *Uterine artery* to the uterus and vagina.

The **external iliac artery** provides the main blood supply to the lower limbs (fig. 15.51). It passes downward along the brim of the pelvis and gives off two large branches—an *inferior epigastric artery* and a *deep circumflex iliac artery.* These vessels supply the muscles and skin in the lower abdominal wall.

Midway between the symphysis pubis and the anterior superior iliac spine of the ilium, the external iliac artery becomes the femoral artery.

The **femoral** (fem′or-al) **artery,** which passes fairly close to the anterior surface of the upper thigh, gives off many branches to muscles and superficial tissues of the thigh. These branches also supply the skin of the groin and the lower abdominal wall. Important subdivisions of the femoral artery include the following:

1. *Superficial circumflex iliac artery* to the lymph nodes and skin of the groin.
2. *Superficial epigastric artery* to the skin of the lower abdominal wall.

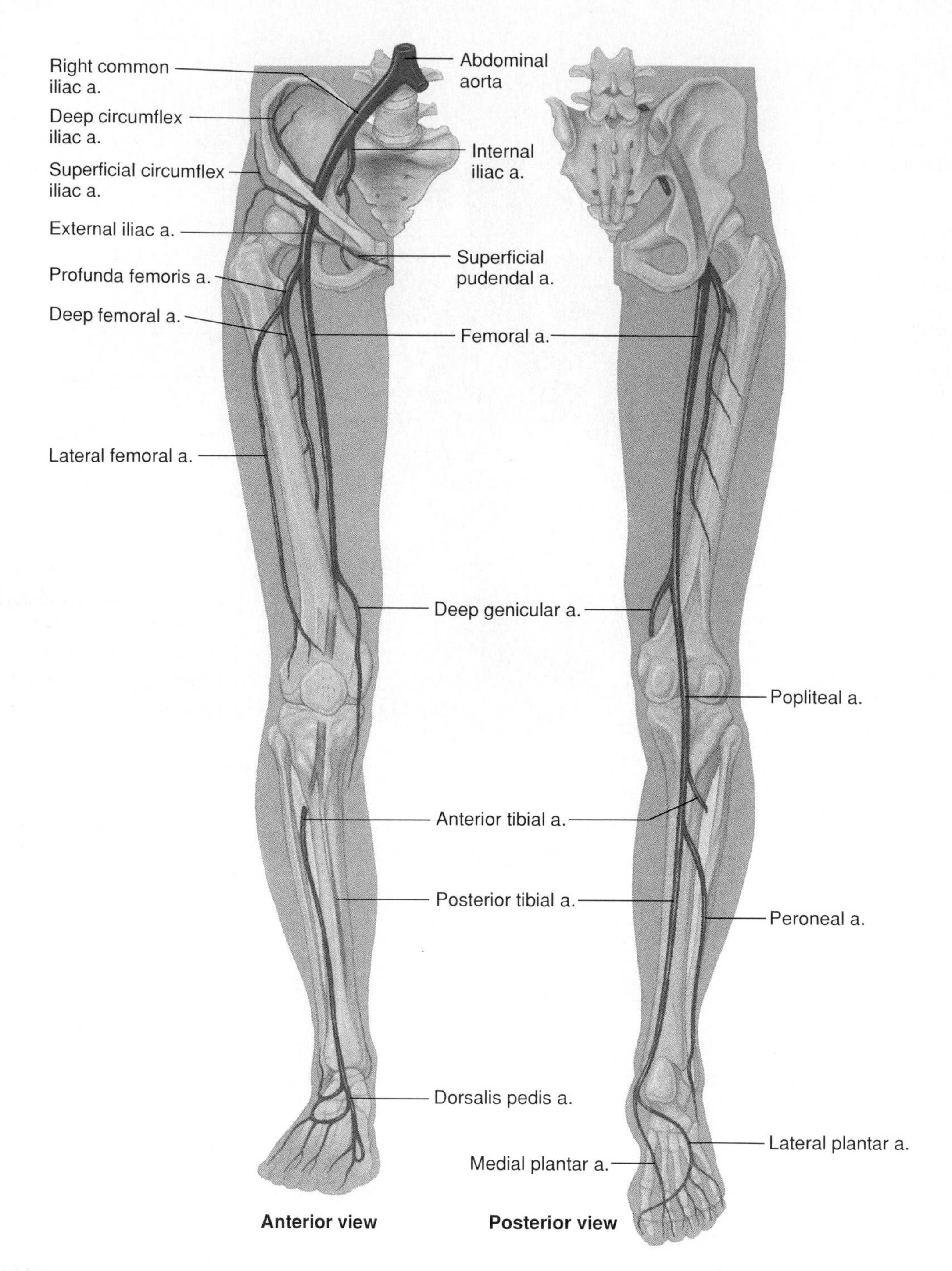


**FIGURE 15.51**
Main branches of the external iliac artery.

3. *Superficial* and *deep external pudendal arteries* to the skin of the lower abdomen and external genitalia.
4. *Profunda femoris artery* (the largest branch of the femoral artery) to the hip joint and muscles of the thigh.
5. *Deep genicular artery* to distal ends of thigh muscles and to an anastomosis around the knee joint.

As the femoral artery reaches the proximal border of the space behind the knee (popliteal fossa), it becomes the **popliteal** (pop-lit′e-al) **artery.** Branches of this artery supply blood to the knee joint and to certain muscles in the thigh and calf. Also, many of its branches join the anastomosis of the knee and help provide alternate pathways for blood in the case of arterial obstructions. At the lower border of the popliteal fossa, the popliteal artery divides into the anterior and posterior tibial arteries.

The **anterior tibial** (tib′e-al) **artery** passes downward between the tibia and the fibula, giving off branches to the skin and muscles in anterior and lateral regions of the leg. It also communicates with the anastomosis of the knee and with a network of arteries around the ankle. This vessel continues into the foot as the *dorsalis pedis artery,* which supplies blood to the foot and toes.

The **posterior tibial artery,** the larger of the two popliteal branches, descends beneath the calf muscles, giving off branches to the skin, muscles, and other tissues of the leg along the way. Some of these vessels join the anastomoses of the knee and ankle. As it passes between the medial malleolus and the heel, the posterior tibial artery divides into the *medial* and *lateral plantar arteries.* Branches from these arteries supply blood to tissues of the heel, foot, and toes.

The largest branch of the posterior tibial artery is the *peroneal artery,* which travels downward along the fibula and contributes to the anastomosis of the ankle.

The major vessels of the arterial system are shown in figure 15.52.

1. Describe the structure of the aorta.
2. Name the vessels that arise from the aortic arch.
3. Name the branches of the thoracic and abdominal aorta.
4. Which vessels supply blood to the head? To the upper limb? To the abdominal wall? To the lower limb?

## Venous System

Venous circulation returns blood to the heart after gases, nutrients, and wastes are exchanged between the blood and body cells.

### Characteristics of Venous Pathways

The vessels of the venous system begin with the merging of capillaries into venules, venules into small veins, and small veins into larger ones. Unlike the arterial pathways, however, those of the venous system are difficult to follow. This is because the vessels commonly connect in irregular networks, so that many unnamed tributaries may join to form a relatively large vein.

On the other hand, the larger veins typically parallel the courses of named arteries, and these veins often have the same names as their companions in the arterial system. For example, the renal vein parallels the renal artery, and the common iliac vein accompanies the common iliac artery.

The veins that carry the blood from the lungs and myocardium back to the heart have already been described. The veins from all the other parts of the body converge into two major pathways, the *superior* and *inferior venae cavae,* which lead to the right atrium.

### Veins from the Head, Neck, and Brain

The **external jugular** (jug′u-lar) **veins** drain blood from the face, scalp, and superficial regions of the neck. These vessels descend on either side of the neck, passing over the sternocleidomastoid muscles and beneath the platysma. They empty into the *right* and *left subclavian veins* in the base of the neck (fig. 15.53).

The **internal jugular veins,** which are somewhat larger than the external jugular veins, arise from numerous veins and venous sinuses of the brain and from deep veins in various parts of the face and neck. They pass downward through the neck beside the common carotid arteries and also join the subclavian veins. These unions of the internal jugular and subclavian veins form large **brachiocephalic** (innominate) **veins** on each side. These vessels then merge in the mediastinum and give rise to the **superior vena cava,** which enters the right atrium.

A lung cancer, enlarged lymph node, or an aortic aneurysm can compress the superior vena cava, interfering with return of blood from the upper body to the heart. This produces pain, shortness of breath, distension of veins draining into the superior vena cava, and swelling of tissues in the face, head, and lower limbs. Restriction of blood flow to the brain may threaten life.

### Veins from the Upper Limb and Shoulder

A set of deep veins and a set of superficial ones drain the upper limb. The deep veins generally parallel the arteries in each region and are given similar names, such as the *radial vein, ulnar vein, brachial vein,* and *axillary vein.* The superficial veins connect in complex networks just beneath the skin. They also communicate with the deep vessels of the upper limb, providing many alternate pathways through which the blood can leave the tissues (fig. 15.54).

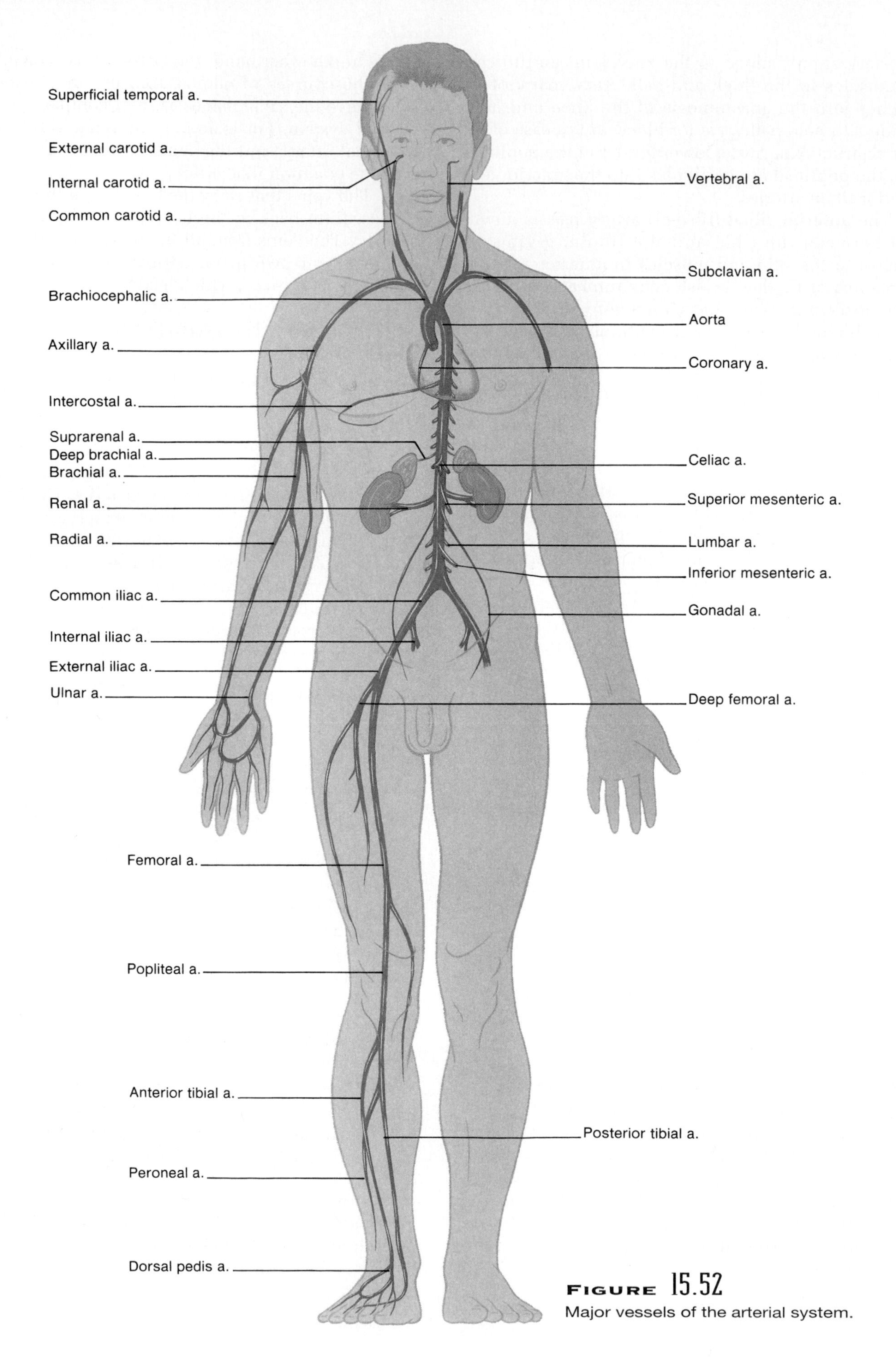


**FIGURE 15.52**
Major vessels of the arterial system.

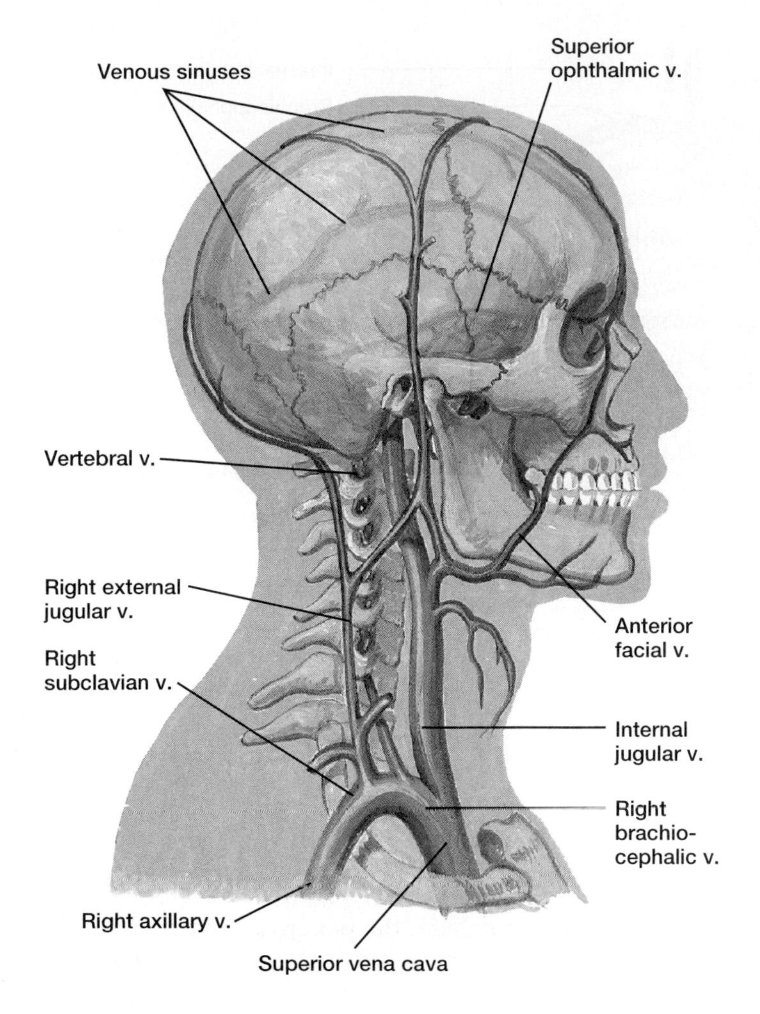


**FIGURE 15.53**
The main veins of the head and neck (v. stands for *vein*). (Note that the clavicle has been removed.)

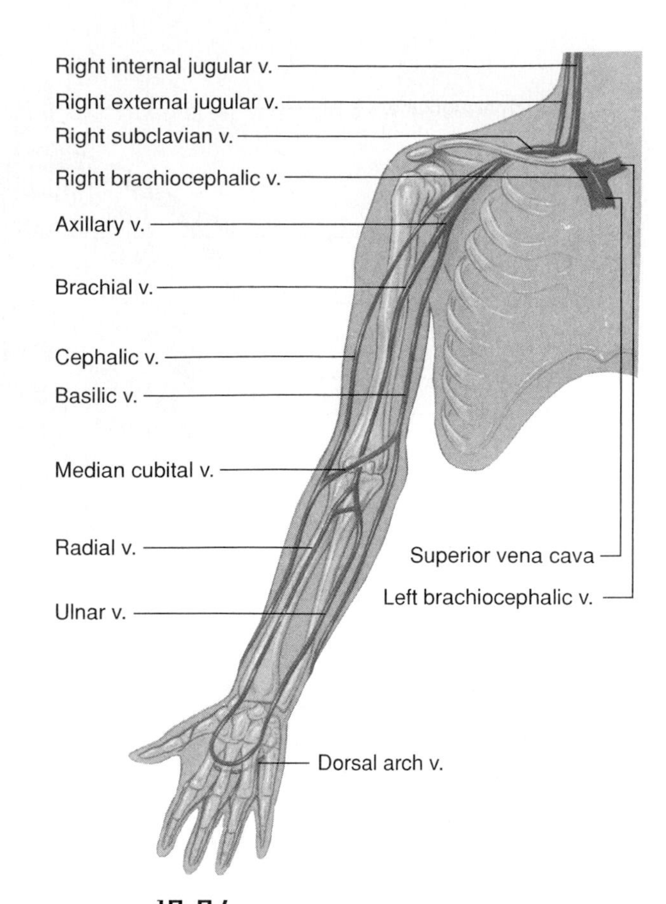


**FIGURE 15.54**
The main veins of the upper limb and shoulder.

The main vessels of the superficial network are the basilic and cephalic veins. They arise from anastomoses in the hand and wrist on the ulnar and radial sides, respectively.

The **basilic vein** passes along the back of the forearm on the ulnar side for a distance and then curves forward to the anterior surface below the elbow. It continues ascending on the medial side until it reaches the middle of the arm. There it penetrates the tissues deeply and joins the *brachial vein.* As the basilic and brachial veins merge, they form the *axillary vein.*

The **cephalic vein** courses upward on the lateral side of the upper limb from the hand to the shoulder. In the shoulder it pierces the tissues and empties into the axillary vein. Beyond the axilla, the axillary vein becomes the *subclavian vein.*

In the bend of the elbow, a *median cubital vein* ascends from the cephalic vein on the lateral side of the forearm to the basilic vein on the medial side. This relatively large vein is usually visible. It is often used as a site for *venipuncture,* when it is necessary to remove a sample of blood for examination or to add fluids to the blood.

## Veins from the Abdominal and Thoracic Walls

Tributaries of the brachiocephalic and azygos veins drain the abdominal and thoracic walls. For example, the *brachiocephalic vein* receives blood from the *internal thoracic vein,* which generally drains the tissues the internal thoracic artery supplies. Some *intercostal veins* also empty into the brachiocephalic vein (fig. 15.55).

The **azygos** (az′ĭ-gos) **vein** originates in the dorsal abdominal wall and ascends through the mediastinum on the right side of the vertebral column to join the superior vena cava. It drains most of the muscular tissue in the abdominal and thoracic walls.

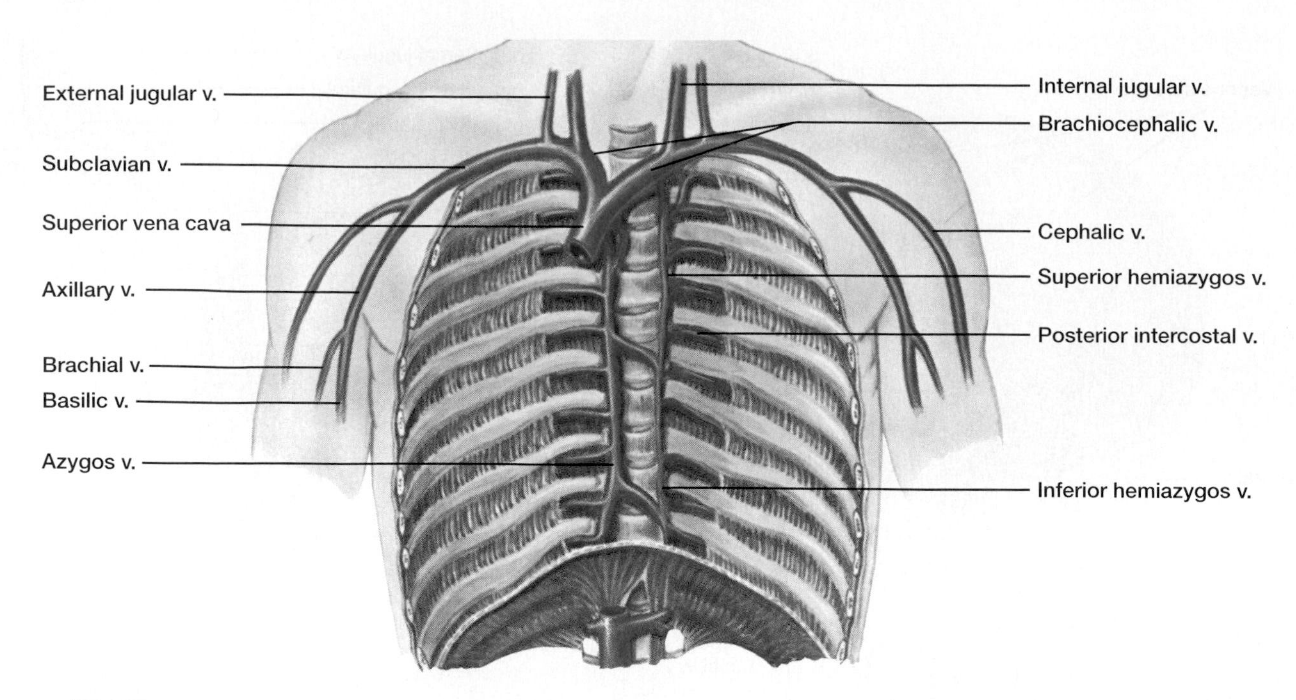


**FIGURE 15.55**
Veins that drain the thoracic wall.

Tributaries of the azygos vein include the *posterior intercostal veins* on the right side, which drain the intercostal spaces, and the *superior* and *inferior hemiazygos veins,* which receive blood from the posterior intercostal veins on the left. The right and left *ascending lumbar veins,* with tributaries that include vessels from the lumbar and sacral regions, also connect to the azygos system.

## Veins from the Abdominal Viscera

Although veins usually carry the blood directly to the atria of the heart, those that drain the abdominal viscera are exceptions (fig. 15.56). They originate in the capillary networks of the stomach, intestines, pancreas, and spleen, and carry blood from these organs through a **portal** (por′tal) **vein** to the liver. There the blood enters capillarylike **hepatic sinusoids** (hĕ-pat′ik si′nŭ-soids). This unique venous pathway is called the **hepatic portal system** (fig. 15.57).

The tributaries of the portal vein include the following vessels:

1. Right and left *gastric veins* from the stomach.
2. *Superior mesenteric vein* from the small intestine, ascending colon, and transverse colon.
3. *Splenic vein* from a convergence of several veins draining the spleen, the pancreas, and a portion of the stomach. Its largest tributary, the *inferior mesenteric vein,* brings blood upward from the descending colon, sigmoid colon, and rectum.

About 80% of the blood flowing to the liver in the hepatic portal system comes from the capillaries in the stomach and intestines and is oxygen-poor but rich in nutrients. As discussed in chapter 17, the liver handles these nutrients in a variety of ways. It regulates blood glucose concentration by converting excess glucose into glycogen for storage, or by converting glycogen into glucose when blood glucose concentration drops below normal.

Similarly, the liver helps regulate blood concentrations of recently absorbed amino acids and lipids by modifying their molecules into forms cells can use, by oxidizing them, or by changing them into storage forms. The liver also stores certain vitamins and detoxifies harmful substances.

Blood in the portal vein nearly always contains bacteria that have entered through intestinal capillaries. Large *Kupffer cells* lining the hepatic sinusoids phagocytize these microorganisms, removing them from the portal blood before it leaves the liver.

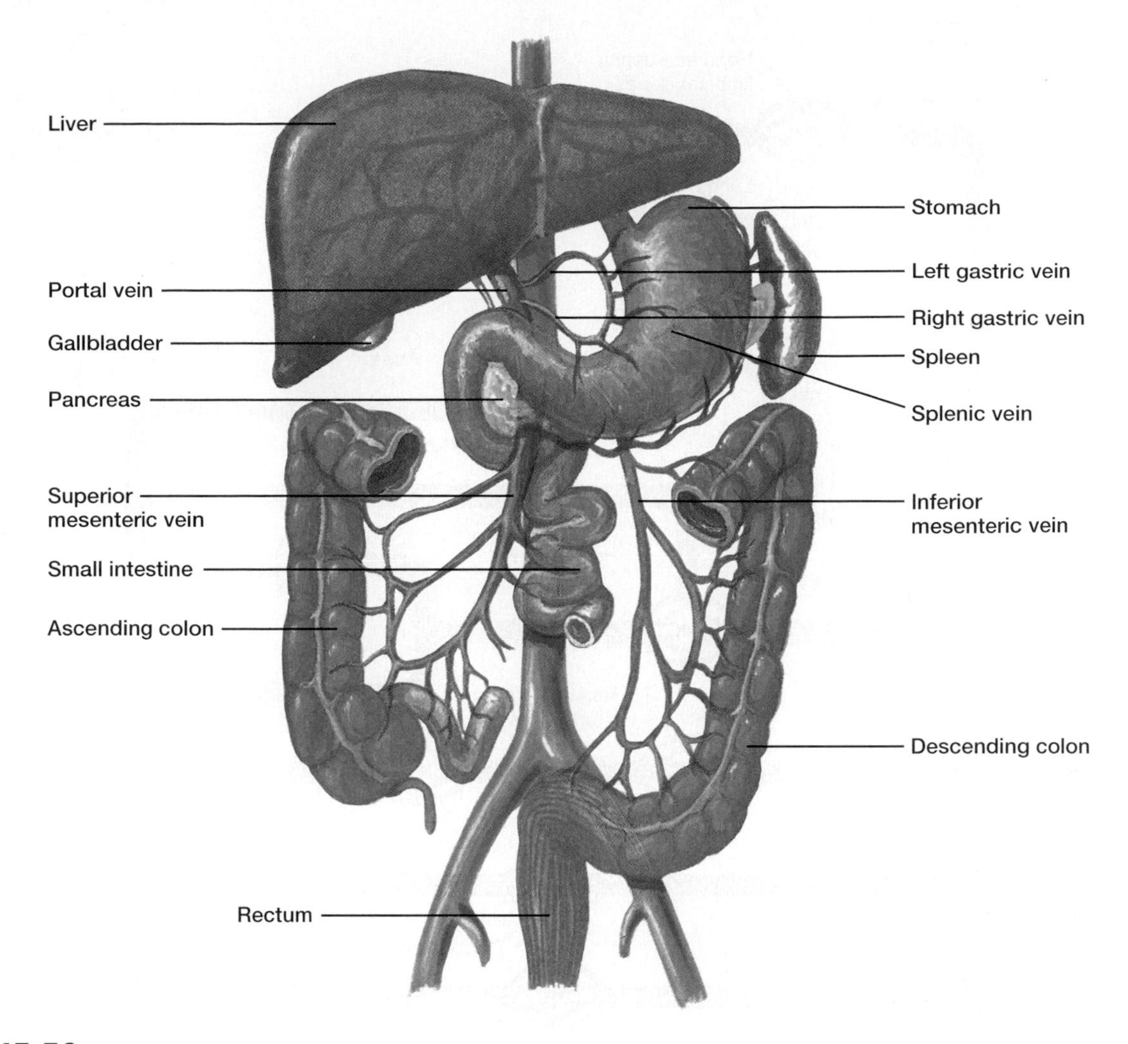


**FIGURE 15.56**
Veins that drain the abdominal viscera.

After passing through the hepatic sinusoids of the liver, the blood in the hepatic portal system travels through a series of merging vessels into **hepatic veins.** These veins empty into the *inferior vena cava,* returning the blood to the general circulation.

Other veins empty into the inferior vena cava as it ascends through the abdomen. They include the *lumbar, gonadal, renal, suprarenal,* and *phrenic veins.* These vessels drain regions that arteries with corresponding names supply.

### Veins from the Lower Limb and Pelvis

As in the upper limb, veins that drain the blood from the lower limb can be divided into deep and superficial groups (fig. 15.58).

The deep veins of the leg, such as the *anterior* and *posterior tibial veins,* have names that correspond to the arteries they accompany. At the level of the knee, these vessels form a single trunk, the **popliteal vein.** This vein continues upward through the thigh as the **femoral vein,** which, in turn, becomes the **external iliac vein.**

The superficial veins of the foot, leg, and thigh connect to form a complex network beneath the skin. These vessels drain into two major trunks: the small and great saphenous veins.

The **small saphenous** (sah-fe′nus) **vein** begins in the lateral portion of the foot and passes upward behind the lateral malleolus. It ascends along the back of the calf, enters the popliteal fossa, and joins the popliteal vein.

The **great saphenous vein,** which is the longest vein in the body, originates on the medial side of the

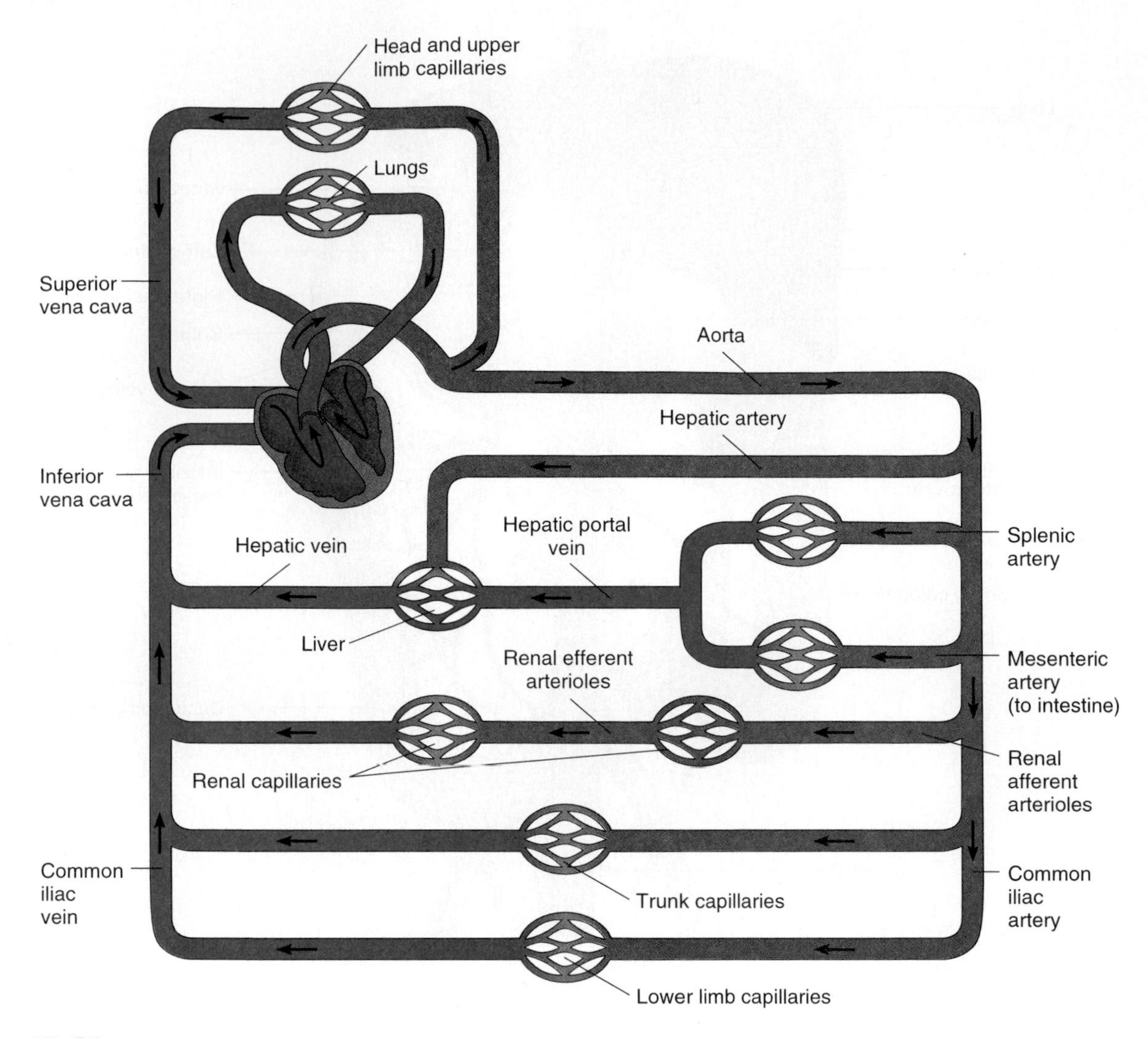


**FIGURE 15.57**
In this schematic drawing of the circulatory system, note how the hepatic portal vein drains one set of capillaries and leads to another set. A similar relationship exists in the kidneys.

foot. It ascends in front of the medial malleolus and extends upward along the medial side of the leg and thigh. In the thigh just below the inguinal ligament, it penetrates deeply and joins the femoral vein. Near its termination, the great saphenous vein receives tributaries from a number of vessels that drain the upper thigh, groin, and lower abdominal wall.

In addition to communicating freely with each other, the saphenous veins communicate extensively with the deep veins of the leg and thigh. Blood can thus return to the heart from the lower extremities by several routes.

In the pelvic region, vessels leading to the **internal iliac vein** carry blood away from organs of the reproductive, urinary, and digestive systems. This vein is formed by tributaries corresponding to the branches of the internal iliac artery, such as the *gluteal, pudendal, vesical, rectal, uterine,* and *vaginal veins.* Typically, these veins have many connections and form complex networks (plexuses) in the regions of the rectum, urinary bladder, and prostate gland (in the male) or uterus and vagina (in the female).

The internal iliac veins originate deep within the pelvis and ascend to the pelvic brim. There they unite with the right and left external iliac veins to form the **common iliac veins.** These vessels, in turn, merge to produce the *inferior vena cava* at the level of the fifth lumbar vertebra. Figure 15.59 shows the major vessels of the venous system.

The chapter concludes with Clinical Application 15.8 which looks at molecular explanations of certain cardiovascular disorders, and Clinical Application 15.9, which discusses coronary artery disease.

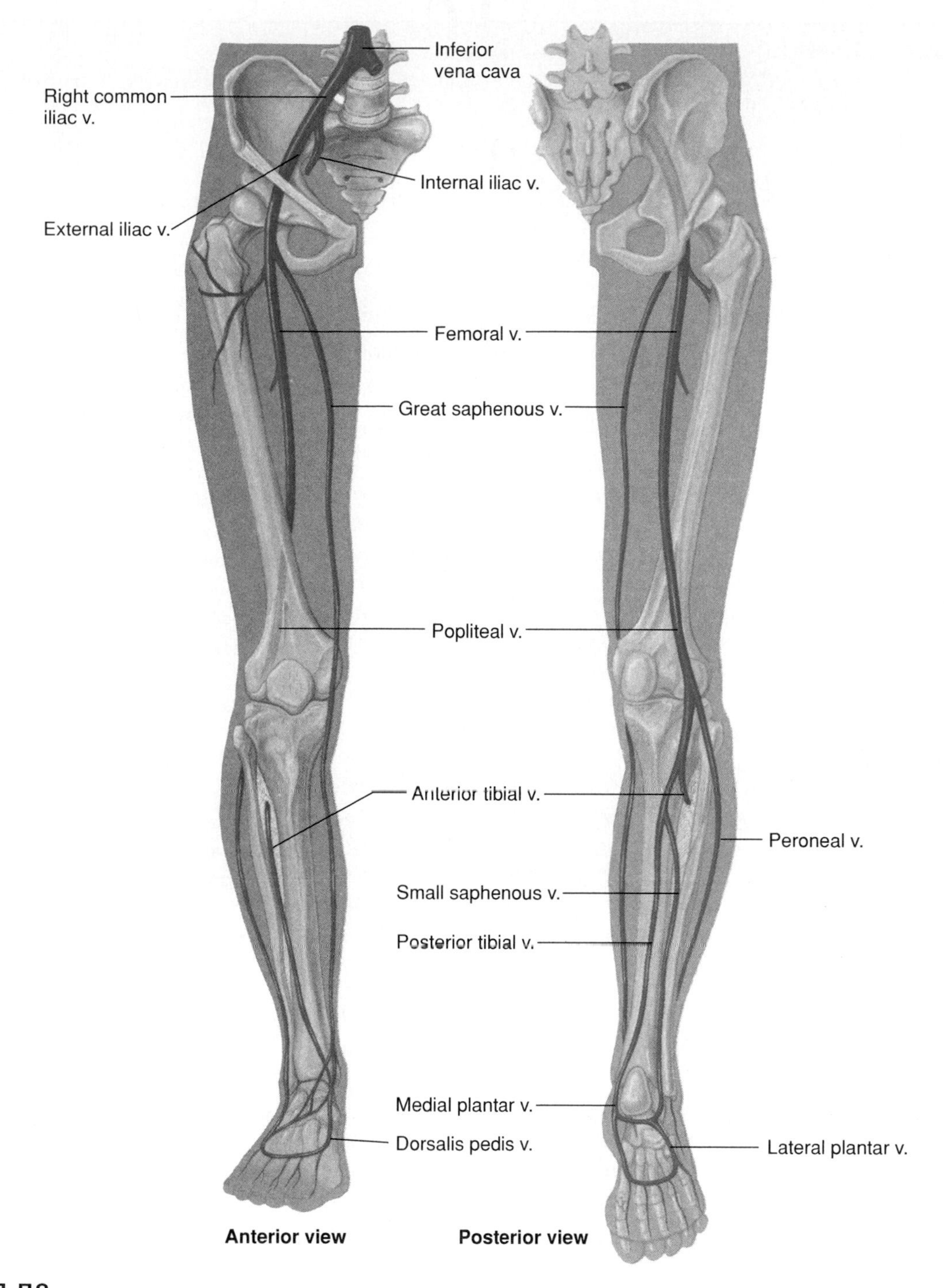


**FIGURE 15.58**
The main veins of the lower limb and pelvis.

**1** Name the veins that return the blood to the right atrium.

**2** Which major veins drain the blood from the head? From the upper limb? From the abdominal viscera? From the lower limb?

CLINICAL APPLICATION

# Molecular Underpinnings of Cardiovascular Disease

## A Connective Tissue Defect

In January 1986, volleyball champion Flo Hyman left the court during a game in Japan, collapsed, and very suddenly died. Her aorta, the largest blood vessel of the body, which leads out of the heart, had burst. Death was instant. Hyman had *Marfan syndrome,* an inherited condition that also caused the characteristics that led her to excel in her sport—her great height and long fingers (fig. 15N).

In Marfan syndrome, an abnormal form of a connective tissue protein called fibrillin weakens the aorta wall. After Flo died, her siblings were examined, and her brother Michael was found to have a weakened aorta. By surgically repairing his aorta and placing him on drugs to control his blood pressure and heart rate, physicians enabled him to avoid the sudden death that claimed his famous sister. Testing for the causative gene can alert physicians to affected individuals even before the dangerous swelling in the aorta begins.

## A Myosin Defect

Each year, one or two seemingly healthy young people die suddenly during a sports event, usually basketball. The culprit is *familial hypertrophic cardiomyopathy,* an inherited overgrowth of the heart muscle. The defect in this disorder is different than that behind Marfan syndrome. It is an abnormality in one of the myosin chains that comprise cardiac muscle. Again, detecting the responsible gene can alert affected individuals to their increased risk of sudden death. They can adjust the type of exercise they do to avoid stressing the cardiovascular system.

## A Metabolic Glitch

Sometimes inherited heart disease strikes very early in life. Jim D. died at 4 days of age, two days after suffering cardiac arrest. Two years later, his parents had another son. Like Jim, Kerry seemed normal at birth, but at 36 hours of age his heart rate plummeted, he had a seizure, and he stopped breathing. He was resuscitated. A blood test revealed excess long-chain fatty acids, indicating a metabolic disorder, an inability to utilize fatty acids. Lack of food triggered the symptoms because the boys could not use fatty acids for energy, as healthy people do. Kerry was able to survive for three years by following a diet low in fatty acids and eating frequently. Once he became comatose because he missed a meal. Eventually, he died of respiratory failure.

Kerry and Jim had inherited a deficiency of a mitochondrial enzyme that processes long-chain fatty acids. Because this is a primary energy source for cardiac muscle, their tiny hearts ultimately failed. There are several other types of inborn errors of cardiac energy metabolism.

The human genome project—an ongoing worldwide effort to describe all human genes—is revealing the genetic underpinnings of many illnesses that affect the cardiovascular system. Medical geneticists begin by gathering members of families with a specific illness, then identifying a particular DNA sequence that ill relatives have in common but healthy individuals lack. Finding this gene then leads to finding the protein it encodes. The protein abnormality or absence may explain the illness—at the molecular, cellular, tissue, organ, and organismal levels.

(a)

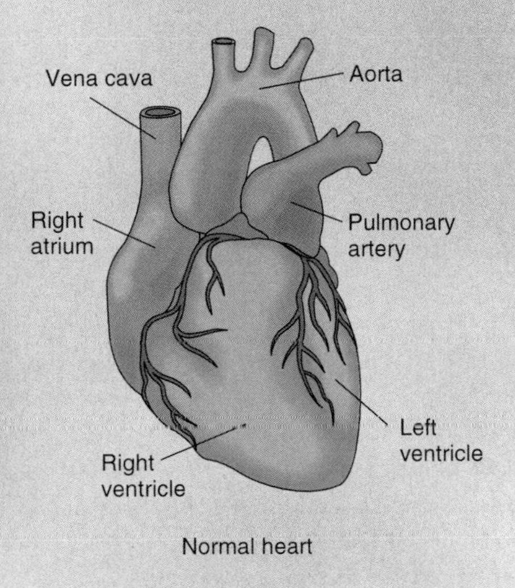


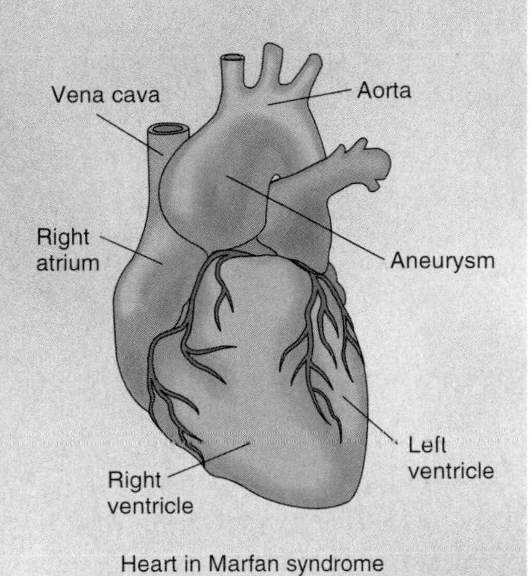

(b)

**FIGURE 15N**

(*a*) Two years after she led the 1984 U.S. women's volleyball team to a silver medal in the Olympics, Flo Hyman died suddenly when her aorta burst, a symptom of Marfan syndrome. (*b*) Note the swelling (aneurysm) of the aorta in the heart on the right. A burst aneurysm is fatal.

CLINICAL APPLICATION 15.9

## Coronary Artery Disease

Dave R., a 52-year-old overweight accountant, had been having chest pains occasionally for several months. The mild pain occurred during his usual weekend tennis match, and he attributed it to overeating or indigestion. The discomfort almost always diminished after the game. Recently, however, the pain seemed more severe and was lasting longer. Dave asked his physician about the problem.

The physician explained that Dave was probably experiencing *angina pectoris*, a symptom of *coronary artery disease* (CAD), and suggested that he undergo an *exercise stress test.* Dave walked on a treadmill, whose speed and incline were increased while he exercised. During the test, an ECG recorded continuously, and Dave's blood pressure was monitored. Near the end of the test, when Dave's heart had reached the desired rate, a small quantity of radioactive thallium-201 was injected into a vein. A *scintillation counter* scanned Dave's heart to determine if the blood carrying the thallium was uniformly distributed to the myocardium by branches of his coronary arteries (see fig. 15.13*a*).

The test revealed that Dave was developing CAD. In addition, he had hypertension and high blood cholesterol.

Dave was advised to stop smoking, to reduce his intake of foods high in saturated fats, cholesterol, and sodium, and to exercise regularly, rather than on weekends only. He was given medications to reduce his blood pressure and to relieve the pain of angina. Dave was also cautioned to avoid stressful situations and to lose weight.

Six months later, in spite of faithful compliance with medical advice, Dave suffered a heart attack—a sign that blood flow to part of his myocardium had been obstructed, producing oxygen deficiency (ischemia). He was at home at the time of the attack, which began as severe, crushing chest pain accompanied by shortness of breath and sweating. Paramedics stabilized Dave's condition and transported him to a hospital.

At the hospital, a cardiologist concluded from an ECG that Dave's heart attack (acute myocardial infarction) was caused by a blood clot obstructing a coronary artery (occlusive coronary thrombosis). The cardiologist intravenously administered a "clot-busting" (thrombolytic) drug, tissue plasminogen activator (t-PA).

After some time, the ECG showed that the blood vessel remained partially obstructed, so the cardiologist ordered a *coronary angiogram.* In this X-ray procedure, which was conducted in the cardiac

## CLINICAL TERMS RELATED TO THE CARDIOVASCULAR SYSTEM

**anastomosis** (ah-nas″to-mo′sis) An interconnection between two blood vessels, sometimes produced surgically.

**angiospasm** (an′je-o-spazm″) A muscular spasm in the wall of a blood vessel.

**arteriography** (ar″te-re-og′rah-fe) The injection of radiopaque solution into the vascular system for an X-ray examination of arteries.

**asystole** (a-sis′to-le) A condition in which the myocardium fails to contract.

**cardiac tamponade** (kar′de-ak tam″po-nād′) Compression of the heart by an accumulation of fluid within the pericardial cavity.

**congestive heart failure** (kon-jes′tiv hart fāl′yer) Inability of the heart to pump an adequate amount of blood to the body cells.

**cor pulmonale** (kor pul-mo-na′le) A heart-lung disorder of pulmonary hypertension and hypertrophy of the right ventricle.

**embolectomy** (em″bo-lek′to-me) Removal of an embolus through an incision in a blood vessel.

catheterization laboratory, a thin plastic catheter was passed through a guiding sheath inserted into the femoral artery of Dave's right inguinal area. From there, the catheter was pushed into the aorta until it reached the region of the opening to the left coronary artery, and then near the opening to the right coronary artery.

The progress of the catheter was monitored with *X-ray fluoroscopy*. Each time the catheter was in proper position, a radiopaque dye (contrast medium) was released from its distal end into the blood. X-ray images that revealed the path of the dye as it entered a coronary artery and its branches were recorded on videotape and on motion picture film, which were later analyzed "frame by frame." A single severe narrowing was discovered near the origin of Dave's left anterior descending artery. The cardiologist decided to perform *percutaneous transluminal coronary angioplasty* (PTCA) in order to enlarge the opening (lumen) of that vessel.

The PTCA was performed by passing another plastic catheter through the guiding sheath used for the angiogram. This second tube had a tiny deflated balloon at its tip, and when the balloon was located in the region of the arterial narrowing, it was inflated for a short time with relatively high pressure. The inflating balloon compressed the atherosclerotic plaque (atheroma), which was responsible for the obstruction, against the arterial wall. It also stretched the blood vessel wall, thus increasing the diameter of its lumen (recanalization). The blood flow to the myocardial tissue downstream from the obstruction improved immediately.

About 50% of the time, a vessel opened with PTCA becomes occluded again, because the underlying disease is still present. To prevent this *restenosis,* the doctor inserted a *coronary stent,* which is an expandable tube or coil that literally holds the vessel wall open. The cardiologist had two other options that have a slightly higher risk of causing damage. He might have vaporized the plaque obstructing the vessel with an excimer laser pulse delivered along optical fibers threaded through the catheter. Or, he could have performed atherectomy, in which a cutting device attached to the balloon inserted into the catheter spins, removing plaque that is withdrawn on the catheter tip.

Should the coronary stent fail, or an obstruction block another heart vessel, Dave might benefit from *coronary bypass surgery.* A portion of his internal mammary artery inside his chest wall would be removed and stitched between the aorta and the blocked coronary artery at a point beyond the obstruction.

Once again, Dave was advised to avoid saturated fats, cholesterol, and sodium in his diet, to exercise regularly, to give up smoking, to take medication to control hypertension, to reduce stress when possible, and to maintain a desirable body weight, all in the hope of retarding the progress of his coronary artery disease. Dave was also advised to undergo exercise stress tests periodically to monitor the progress of his disease.

**endarterectomy** (en″dar-ter-ek′to-me) Removal of the inner wall of an artery to reduce an arterial occlusion.

**palpitation** (pal″pi-ta′shun) Awareness of a heartbeat that is unusually rapid, strong, or irregular.

**pericardiectomy** (per″i-kar″de-ek′to-me) An excision of the pericardium.

**phlebitis** (flĕ-bi′tis) An inflammation of a vein, usually in the lower limbs.

**phlebotomy** (flĕ-bot′o-me) An incision of a vein for the purpose of withdrawing blood.

**sinus rhythm** (si′nus rithm) The normal cardiac rhythm regulated by the S-A node.

**thrombophlebitis** (throm″bo-fle-bi′tis) The formation of a blood clot in a vein in response to inflammation of the venous wall.

**valvotomy** (val-vot′o-me) An incision of a valve.

**venography** (ve-nog′rah-fe) An injection of radiopaque solution into the vascular system for X-ray examination of veins.

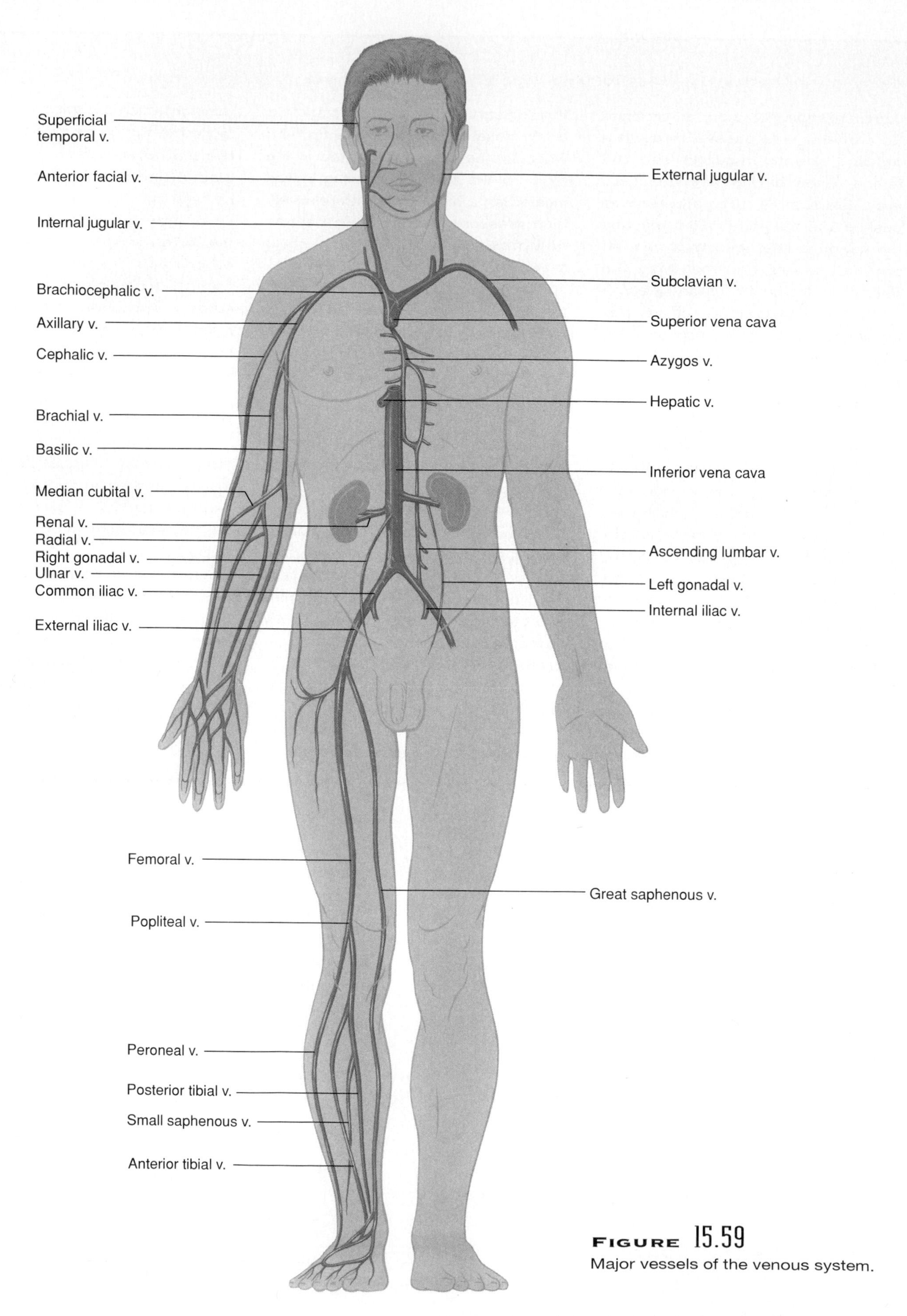


**FIGURE 15.59**
Major vessels of the venous system.

# InnerConnections
## Cardiovascular System

The heart pumps blood through approximately 25,000–60,000 miles of blood vessels, reaching all body cells to deliver nutrients and remove wastes

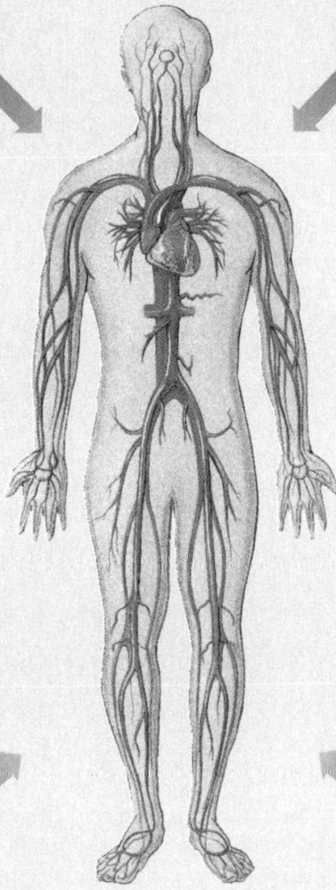

**Reproductive system**
Blood pressure is important in normal function of the sex organs

**Integumentary system**
Changes in skin blood flow are important in temperature control

**Urinary system**
The kidneys clear the blood of wastes and excess substances
The kidneys play an important role in controlling blood pressure and blood volume

**Skeletal system**
Bones play a role in controlling plasma calcium levels

**Respiratory system**
The respiratory system oxygenates the blood and removes carbon dioxide
Respiratory movements help the blood circulate

**Muscular system**
Blood flow increases to exercising skeletal muscle, delivering oxygen and nutrients, and removing wastes
Muscle actions help the blood circulate

**Digestive system**
The digestive system breaks down nutrients into forms readily absorbed by the bloodstream

**Nervous system**
The brain is especially dependent on blood flow for survival
The nervous system is important in controlling blood flow and blood pressure

**Lymphatic system**
Tissue fluids are returned to the bloodstream by the lymphatic system

**Endocrine system**
Hormones are carried in the bloodstream
Many hormones have direct effects on the heart and blood vessels

## CHAPTER SUMMARY

### Introduction (page 564)

The cardiovascular system provides oxygen and nutrients to tissues, and removes wastes.

### Structure of the Heart (page 564)

1. Size and location of the heart
   a. The heart is about 14 centimeters long and 9 centimeters wide.
   b. It is located within the mediastinum and rests on the diaphragm.
2. Coverings of the heart
   a. A layered pericardium encloses the heart.
   b. The pericardial cavity is a space between the visceral and parietal layers of the pericardium.
3. Wall of the heart
   a. The wall of the heart has three layers.
   b. These layers include an epicardium, a myocardium, and an endocardium.
4. Heart chambers and valves
   a. The heart is divided into four chambers—two atria and two ventricles—that communicate through atrioventricular orifices on each side.
   b. Right chambers and valves
      (1) The right atrium receives blood from the venae cavae and coronary sinus.
      (2) The tricuspid valve guards the right atrioventricular orifice.
      (3) The right ventricle pumps blood into the pulmonary trunk.
      (4) A pulmonary valve guards the base of the pulmonary trunk.
   c. Left chambers and valves
      (1) The left atrium receives blood from the pulmonary veins.
      (2) The bicuspid valve guards the left atrioventricular orifice.
      (3) The left ventricle pumps blood into the aorta.
      (4) An aortic valve guards the base of the aorta.
5. Skeleton of the heart
   a. The skeleton of the heart consists of fibrous rings that enclose the bases of the pulmonary artery, aorta, and atrioventricular orifices.
   b. The fibrous rings provide attachments for valves and muscle fibers, and prevent the orifices from dilating excessively during ventricular contractions.
6. Path of blood through the heart
   a. Blood that is relatively low in oxygen concentration and high in carbon dioxide concentration enters the right side of the heart from the vena cavae and is pumped into the pulmonary circulation.
   b. After the blood is oxygenated in the lungs and some of its carbon dioxide is removed, it returns to the left side of the heart through the pulmonary veins.
   c. From the left ventricle, it moves into the aorta.
7. Blood supply to the heart
   a. The coronary arteries supply blood to the myocardium.
   b. It is returned to the right atrium through the cardiac veins and coronary sinus.

### Heart Actions (page 576)

1. Cardiac cycle
   a. The atria contract while the ventricles relax; the ventricles contract while the atria relax.
   b. Pressure within the chambers rises and falls in repeated cycles.
2. Heart sounds
   a. Heart sounds can be described as *lub-dup*.
   b. Heart sounds are due to the vibrations the blood and valve movements produce.
   c. The first part of the sound occurs as A-V valves close, and the second part is associated with the closing of pulmonary and aortic valves.
3. Cardiac muscle fibers
   a. Cardiac muscle fibers connect to form a functional syncytium.
   b. If any part of the syncytium is stimulated, the whole structure contracts as a unit.
   c. Except for a small region in the floor of the right atrium, the fibrous skeleton separates the atrial syncytium from the ventricular syncytium.
4. Cardiac conduction system
   a. This system, composed of specialized cardiac muscle tissue, initiates and conducts depolarization waves through the myocardium.
   b. Impulses from the S-A node pass slowly to the A-V node; impulses travel rapidly along the A-V bundle and Purkinje fibers.
   c. Muscle fibers in the ventricular walls are arranged in whorls that squeeze blood out of the contracting ventricles.
5. Regulation of the cardiac cycle
   a. Physical exercise, body temperature, and concentration of various ions affect heartbeat.
   b. Branches of sympathetic and parasympathetic nerve fibers innervate the S-A and A-V nodes.
      (1) Parasympathetic impulses decrease heart action; sympathetic impulses increase heart action.
      (2) The cardiac center in the medulla oblongata regulates autonomic impulses.

### Blood Vessels (page 582)

The blood vessels form a closed circuit of tubes that transport blood between the heart and body cells. The tubes include arteries, arterioles, capillaries, venules, and veins.

1. Arteries and arterioles
   a. The arteries are adapted to carry relatively high pressure blood away from the heart.
   b. The arterioles are branches of arteries.
   c. The walls of arteries and arterioles consist of layers of endothelium, smooth muscle, and connective tissue.
   d. Autonomic fibers that can stimulate vasoconstriction or vasodilation innervate smooth muscles in vessel walls.
2. Capillaries
   Capillaries connect arterioles and venules. The capillary wall is a single layer of cells that forms a semipermeable membrane.

a. Capillary permeability
   (1) Openings in the capillary walls are thin slits between adjacent endothelial cells.
   (2) The sizes of the openings vary from tissue to tissue.
   (3) Endothelial cells of brain capillaries are tightly fused, forming a blood-brain barrier through which substances move by facilitated diffusion.
b. Capillary arrangement
   Capillary density varies directly with tissue metabolic rates.
c. Regulation of capillary blood flow
   (1) Precapillary sphincters regulate capillary blood flow.
   (2) Precapillary sphincters open when cells are low in oxygen and nutrients, and close when cellular needs are met.

3. Exchanges in the capillaries
   a. Gases, nutrients, and metabolic by-products are exchanged between the capillary blood and the tissue fluid.
   b. Diffusion provides the most important means of transport.
   c. Diffusion pathways depend on lipid solubilities.
   d. Plasma proteins generally remain in the blood.
   e. Filtration, which is due to the hydrostatic pressure of blood, causes a net outward movement of fluid at the arterial end of a capillary.
   f. Osmosis causes a net inward movement of fluid at the venule end of a capillary.
   g. Some factors cause fluids to accumulate in the tissues.
4. Venules and veins
   a. Venules continue from capillaries and merge to form veins.
   b. Veins carry blood to the heart.
   c. Venous walls are similar to arterial walls, but are thinner and contain less muscle and elastic tissue.

## Blood Pressure (page 593)

Blood pressure is the force blood exerts against the insides of blood vessels.

1. Arterial blood pressure
   a. The arterial blood pressure is produced primarily by heart action; it rises and falls with phases of the cardiac cycle.
   b. Systolic pressure occurs when the ventricle contracts; diastolic pressure occurs when the ventricle relaxes.
2. Factors that influence arterial blood pressure
   a. Heart action, blood volume, resistance to flow, and blood viscosity influence arterial blood pressure.
   b. Arterial pressure increases as cardiac output, blood volume, peripheral resistance, or blood viscosity increases.
3. Control of blood pressure
   a. Blood pressure is controlled in part by the mechanisms that regulate cardiac output and peripheral resistance.
   b. Cardiac output depends on the volume of blood discharged from the ventricle with each beat and on the rate of heartbeat.
      (1) The more blood that enters the heart, the stronger the ventricular contraction, the greater the stroke volume, and the greater the cardiac output.
      (2) The cardiac center of the medulla oblongata regulates heart rate.
   c. Changes in the diameter of arterioles, controlled by the vasomotor center of the medulla oblongata, regulates peripheral resistance.
4. Venous blood flow
   a. Venous blood flow is not a direct result of heart action; it depends on skeletal muscle contraction, breathing movements, and venoconstriction.
   b. Many veins contain flaplike valves that prevent blood from backing up.
   c. Venous constriction can increase venous pressure and blood flow.
5. Central venous pressure
   a. Central venous pressure is the pressure in the right atrium.
   b. Factors that influence it alter the flow of blood into the right atrium.
   c. It affects pressure within the peripheral veins.

## Paths of Circulation (page 604)

1. Pulmonary circuit
   a. The pulmonary circuit consists of vessels that carry blood from the right ventricle to the lungs, pulmonary capillaries, and vessels that lead back to the left atrium.
   b. Pulmonary capillaries exert less pressure than those of the systemic circuit.
   c. Tightly joined epithelial cells of alveoli walls prevent most substances from entering the alveoli.
   d. Osmotic pressure rapidly draws water out of alveoli into the interstitial fluid, so alveoli do not fill with fluid.
2. Systemic circuit
   a. The systemic circuit is composed of vessels that lead from the heart to the body cells and back to the heart.
   b. It includes the aorta and its branches as well as the system of veins that return blood to the right atrium.

## Arterial System (page 607)

1. Principal branches of the aorta
   a. The branches of the ascending aorta include the right and left coronary arteries.
   b. The branches of the aortic arch include the brachiocephalic, left common carotid, and left subclavian arteries.
   c. The branches of the descending aorta include the thoracic and abdominal groups.
   d. The abdominal aorta terminates by dividing into right and left common iliac arteries.
2. Arteries to the neck, head, and brain include branches of the subclavian and common carotid arteries.
3. Arteries to the shoulder and upper limb
   a. The subclavian artery passes into the arm, and in various regions is called the axillary and brachial artery.
   b. Branches of the brachial artery include the ulnar and radial arteries.

4. Arteries to the thoracic and abdominal walls
   a. Branches of the subclavian artery and thoracic aorta supply the thoracic wall.
   b. Branches of the abdominal aorta and other arteries supply the abdominal wall.
5. Arteries to the pelvis and lower limb
   The common iliac artery supplies the pelvic organs, gluteal region, and lower limb.

## Venous System (page 617)

1. Characteristics of venous pathways
   a. The veins return blood to the heart.
   b. Larger veins usually parallel the paths of major arteries.
2. Veins from the head, neck, and brain
   a. The jugular veins drain these regions.
   b. Jugular veins unite with subclavian veins to form the brachiocephalic veins.
3. Veins from the upper limb and shoulder
   a. Sets of superficial and deep veins drain the upper limb.
   b. The major superficial veins are the basilic and cephalic veins.
   c. The median cubital vein in the bend of the elbow is often used as a site for venipuncture.
4. Veins from the abdominal and thoracic walls include tributaries of the brachiocephalic and azygos veins drain these walls.
5. Veins from the abdominal viscera
   a. The blood from the abdominal viscera generally enters the hepatic portal system and is carried to the liver.
   b. The blood in the portal system is rich in nutrients.
   c. The liver helps regulate the blood concentrations of glucose, amino acids, and lipids.
   d. Phagocytic cells in the liver remove bacteria from the portal blood.
   e. From the liver, hepatic veins carry blood to the inferior vena cava.
6. Veins from the lower limb and pelvis
   a. Sets of deep and superficial veins drain these regions.
   b. The deep veins include the tibial veins, and the superficial veins include the saphenous veins.

**Explorations in Human Anatomy and Physiology CD-ROM**
The module accompanying Chapter Fifteen is #5 Evolution of the Human Heart.

## Critical Thinking Questions

1. Based on your understanding of the way capillary blood flow is regulated, do you think it is wiser to rest or to exercise following a heavy meal? Give a reason for your answer.
2. If a patient develops a blood clot in the femoral vein of the left lower limb, and a portion of the clot breaks loose, where is the blood flow likely to carry the embolus? What symptoms is this condition likely to produce?
3. When a person strains to lift a heavy object, intrathoracic pressure increases. What do you think will happen to the rate of venous blood returning to the heart during such lifting? Why?
4. Why is a ventricular fibrillation more likely to be life threatening than an atrial fibrillation?
5. Cirrhosis of the liver, a disease commonly associated with alcoholism, obstructs blood flow through the hepatic blood vessels. As a result, the blood backs up, and the capillary pressure greatly increases in the organs drained by the hepatic portal system. What effects might this increasing capillary pressure produce, and which organs would it affect?
6. If a cardiologist inserted a catheter into a patient's right femoral artery, which arteries would the tube have to pass through in order to reach the entrance of the left coronary artery?

## Review Exercises

1. Describe the general structure, function, and location of the heart.
2. Describe the pericardium.
3. Compare the layers of the cardiac wall.
4. Identify and describe the locations of the chambers and the valves of the heart.
5. Describe the skeleton of the heart, and explain its function.
6. Trace the path of the blood through the heart.
7. Trace the path of the blood through the coronary circulation.
8. Describe a cardiac cycle.
9. Describe the pressure changes that occur in the atria and ventricles during a cardiac cycle.
10. Explain the origin of heart sounds.
11. Describe the arrangement of the cardiac muscle fibers.
12. Distinguish between the S-A node and the A-V node.
13. Explain how the cardiac conduction system functions in controlling the cardiac cycle.
14. Discuss how the nervous system functions in the regulation of the cardiac cycle.
15. Describe two factors other than the nervous system that affect the cardiac cycle.
16. Describe and explain the normal ECG pattern.

17. Distinguish between an artery and an arteriole.
18. Explain how vasoconstriction and vasodilation are controlled.
19. Describe the structure and function of a capillary.
20. Describe the function of the blood-brain barrier.
21. Explain how the blood flow through a capillary is controlled.
22. Explain how diffusion functions in the exchange of substances between blood plasma and tissue fluid.
23. Explain why water and dissolved substances leave the arteriole end of a capillary and enter the venule end.
24. Describe the effect of histamine on a capillary.
25. Distinguish between a venule and a vein.
26. Explain how veins function as blood reservoirs.
27. Distinguish between systolic and diastolic blood pressures.
28. Name several factors that influence the blood pressure, and explain how each produces its effect.
29. Describe how blood pressure is controlled.
30. List the major factors that promote the flow of venous blood.
31. Define *central venous pressure.*
32. Distinguish between the pulmonary and systemic circuits of the cardiovascular system.
33. Trace the path of blood through the pulmonary circuit.
34. Explain why the alveoli normally do not fill with fluid.
35. Describe the aorta, and name its principal branches.
36. Describe the relationship between the major venous pathways and the major arterial pathways.

16

CHAPTER

# Lymphatic System and Immunity

ON THE AIDS QUILT, EACH SQUARE REPRESENTS A LIFE LOST TO THE DISEASE.

## Chapter Objectives

AFTER YOU HAVE STUDIED THIS CHAPTER, YOU SHOULD BE ABLE TO:

1. Describe the general functions of the lymphatic system.
2. Describe the location of the major lymphatic pathways.
3. Describe how tissue fluid and lymph form and explain the function of lymph.
4. Explain how lymphatic circulation is maintained and describe the consequence of lymphatic obstruction.
5. Describe a lymph node and its major functions.
6. Describe the location of the major chains of lymph nodes.
7. Discuss the functions of the thymus and spleen.
8. Distinguish between specific and nonspecific immunity, and provide examples of each.
9. Explain how two major types of lymphocytes are formed and how they function in immune mechanisms.
10. Name the major types of immunoglobulins and discuss their origins and actions.
11. Distinguish between primary and secondary immune responses.
12. Distinguish between active and passive immunity.
13. Explain how allergic reactions, tissue rejection reactions, and autoimmunity are related to immune mechanisms.

## Key Terms

**allergen** (al′er-jen)
**antibody** (an′ti-bod″e)
**antigen** (an′ti-jen)
**clone** (klōn)
**complement** (kom′plĕ-ment)
**hapten** (hap′ten)
**immunity** (i-mu′ni-te)
**immunoglobulin** (im″u-no-glob′u-lin)
**interferon** (in″ter-fēr′on)
**lymph** (limf)
**lymphatic pathway** (lim-fat′ik path′wa)
**lymph node** (limf nōd)
**lymphocyte** (lim′fo-sīt)
**macrophage** (mak′ro-fāj)
**pathogen** (path′o-jen)
**reticuloendothelial tissue** (rĕ-tik″u-lo-en″do-the′le-al tish′u)
**spleen** (splēn)
**thymus** (thi′mus)
**vaccine** (vak′sēn)

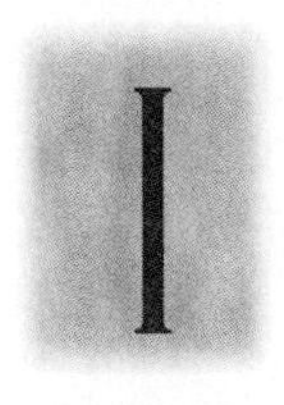

In 1974, a Danish physician working in a poor village in Zaire, Africa, came down with the first inklings of a mystifying disease. For the first 2 years, the illness manifested itself mostly as inexplicable, profound fatigue, weight loss, and persistent diarrhea. But by the time the woman returned to Denmark in 1977, her body was wracked by infections that ravaged her mouth, bloodstream, and lungs. After dozens of medical tests proved fruitless, just before Christmas, Margrethe Rask became one of the first to die of what we know today as acquired immunodeficiency syndrome (AIDS). She and thousands of others since then have experienced the gradual breakdown of a most vital body system—the lymphatic system, which protects the body against infectious disease and cancer.

The lymphatic system is closely associated with the cardiovascular system because it includes a network of vessels that assist in circulating body fluids. Lymphatic vessels transport excess fluid away from interstitial spaces between cells in most tissues and return it to the bloodstream (fig. 16.1). The organs of the lymphatic system also help defend the body against invasion by disease-causing agents.

## Lymphatic Pathways

The lymphatic pathways begin as lymphatic capillaries that merge to form lymphatic vessels, which, in turn, lead to larger vessels that unite with the veins in the thorax.

### Lymphatic Capillaries

**Lymphatic capillaries** are microscopic, closed-ended tubes. They extend into the interstitial spaces, forming complex networks that parallel the networks of the blood capillaries (fig. 16.2). The walls of the lymphatic capillaries, like those of the blood capillaries, consist of a single layer of squamous epithelial cells. This thin wall makes it possible for tissue fluid (interstitial fluid) from the interstitial space to enter the lymphatic capillary. Fluid inside a lymphatic capillary is called **lymph.**

The villi of the small intestine contain specialized lymphatic capillaries called *lacteals,* described in chapter 17. These vessels transport recently absorbed fats away from the digestive tract.

### Lymphatic Vessels

The walls of **lymphatic vessels** are similar to those of veins, but thinner. That is, their walls are composed of three layers: an endothelial lining, a middle layer of smooth muscle and elastic fibers, and an outer layer of connective tissue. Also like veins, the lymphatic vessels have semilunar valves, which help prevent the backflow of lymph. Figure 16.3 shows one of these valves.

The larger lymphatic vessels lead to specialized organs called **lymph nodes.** After leaving the nodes, the vessels merge to form still larger lymphatic trunks.

### Lymphatic Trunks and Collecting Ducts

The **lymphatic trunks,** which drain lymph, are named for the regions they serve. For example, the *lumbar trunk* drains lymph from the lower limbs, lower abdominal wall, and pelvic organs; the *intestinal trunk* drains the abdominal viscera; the *intercostal* and *bronchomediastinal trunks* drain lymph from portions of the thorax; the *subclavian trunk* drains the upper limb; and the *jugular trunk* drains portions of the neck and head. These lymphatic trunks then join one of two **collecting ducts**—the thoracic duct or the right lymphatic duct. Figure 16.4 shows the location of the major lymphatic trunks and collecting ducts, and figure 16.5 shows a lymphangiogram, or X-ray film, of some lymphatic vessels and lymph nodes.

The **thoracic duct** is the larger and longer of the two collecting ducts. It begins in the abdomen, passes upward through the diaphragm beside the aorta, ascends in front of the vertebral column through the mediastinum, and empties into the left subclavian vein near the junction of the left jugular vein. This duct drains lymph from the intestinal, lumbar, and intercostal trunks, as well as from the left subclavian, left jugular, and left bronchomediastinal trunks.

The **right lymphatic duct** originates in the right thorax at the union of the right jugular, right subclavian, and right bronchomediastinal trunks. It empties into the right subclavian vein near the junction of the right jugular vein.

After leaving the two collecting ducts, lymph enters the venous system and becomes part of the plasma just before the blood returns to the right atrium.

Thus, lymph from the lower body regions, the left upper limb, and the left side of the head and neck

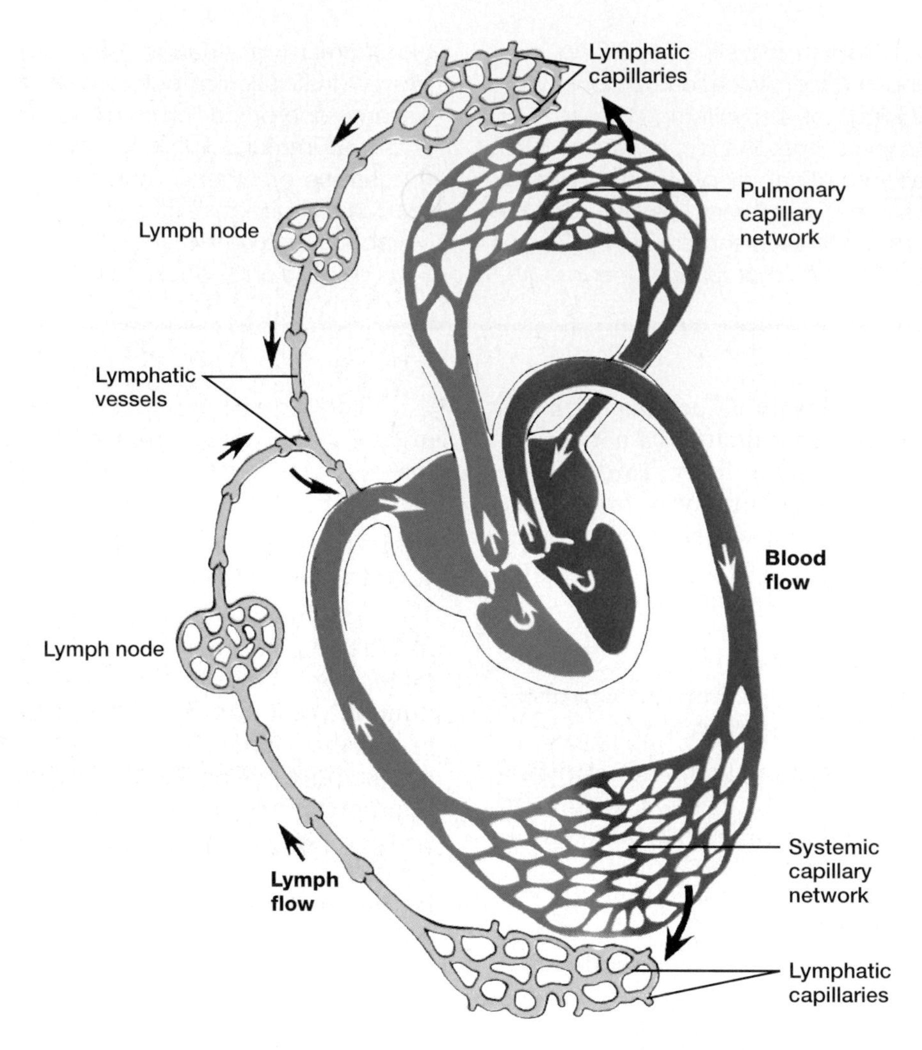


**FIGURE 16.1**
Lymphatic vessels transport fluid from interstitial spaces to the bloodstream.

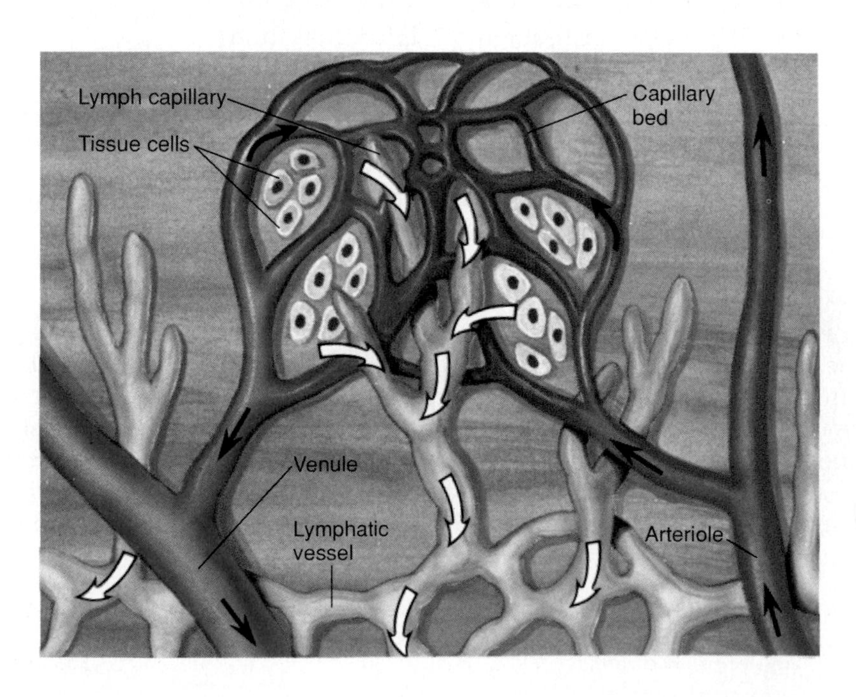


**FIGURE 16.2**
Lymph capillaries are microscopic, closed-ended tubes that begin in the interstitial spaces of most tissues.

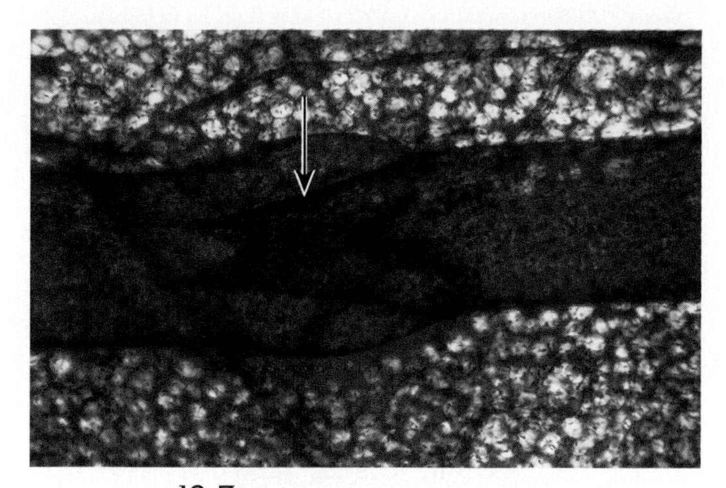

**FIGURE 16.3**
A light micrograph of the flaplike valve (arrow) within a lymphatic vessel (25×).

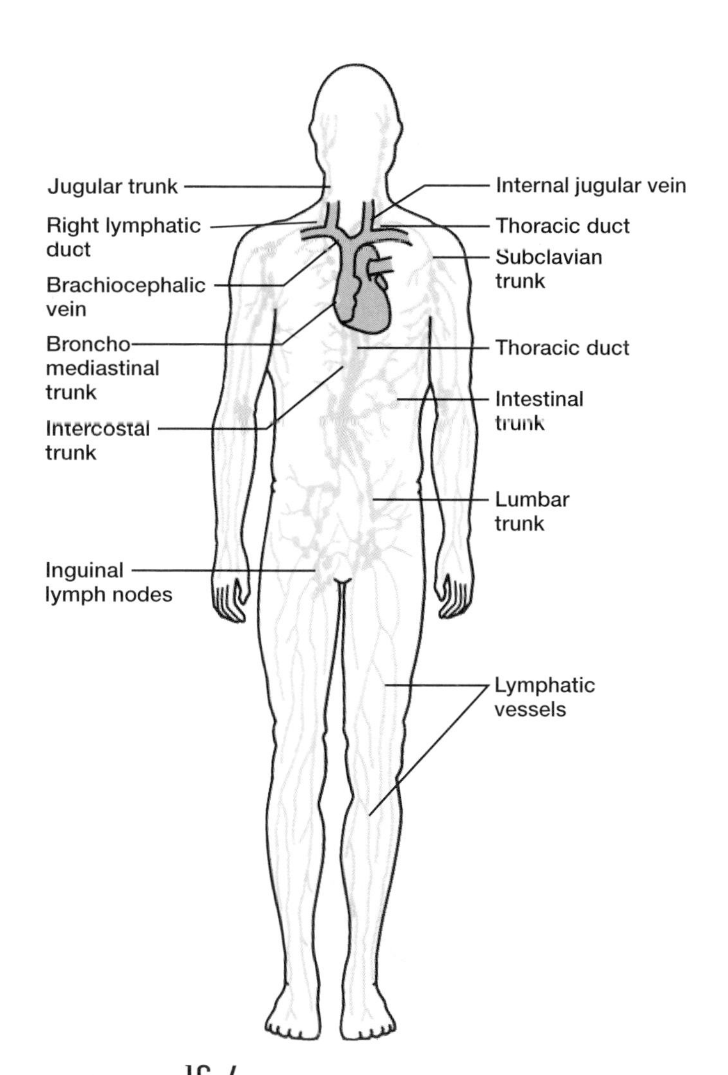


**FIGURE 16.4**
Lymphatic vessels merge into larger lymphatic trunks, which in turn drain into collecting ducts.

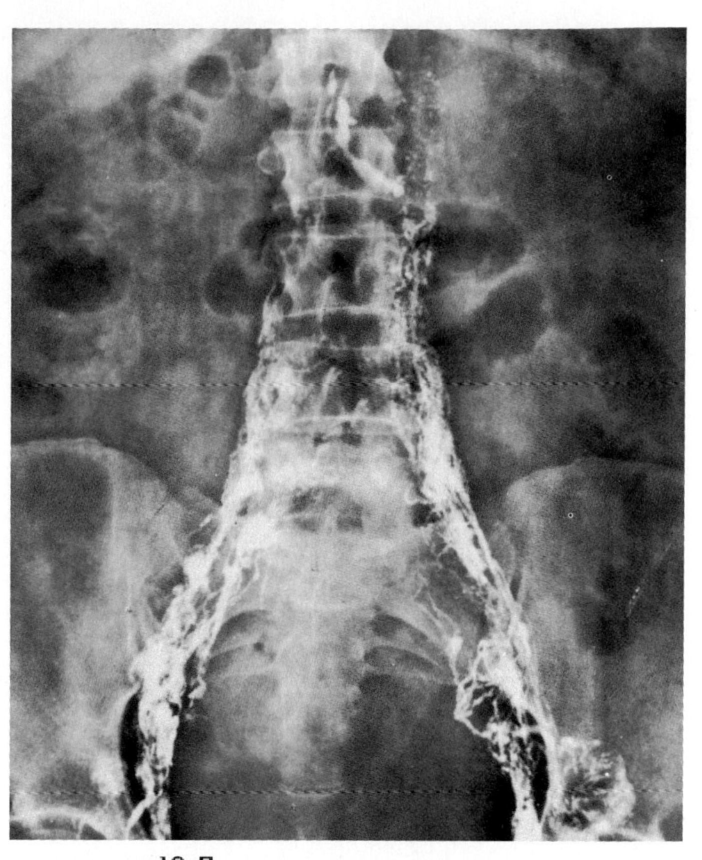

**FIGURE 16.5**
A lymphangiogram (X-ray film) of the lymphatic vessels and lymph nodes of the pelvic region.

enters the thoracic duct; lymph from the right side of the head and neck, the right upper limb, and the right thorax enters the right lymphatic duct (fig. 16.6). Figure 16.7 summarizes the lymphatic pathway.

> The skin is richly supplied with lymphatic capillaries. Consequently, if the skin is broken, or if something is injected into it (such as venom from a stinging insect), foreign substances are likely to enter the lymphatic system relatively rapidly.

1. What is the general function of the lymphatic system?
2. Through which lymphatic vessels would lymph pass in traveling from a lower limb to the bloodstream?

## TISSUE FLUID AND LYMPH

Lymph is essentially tissue fluid that has entered a lymphatic capillary. Thus, lymph formation is closely associated with tissue fluid formation.

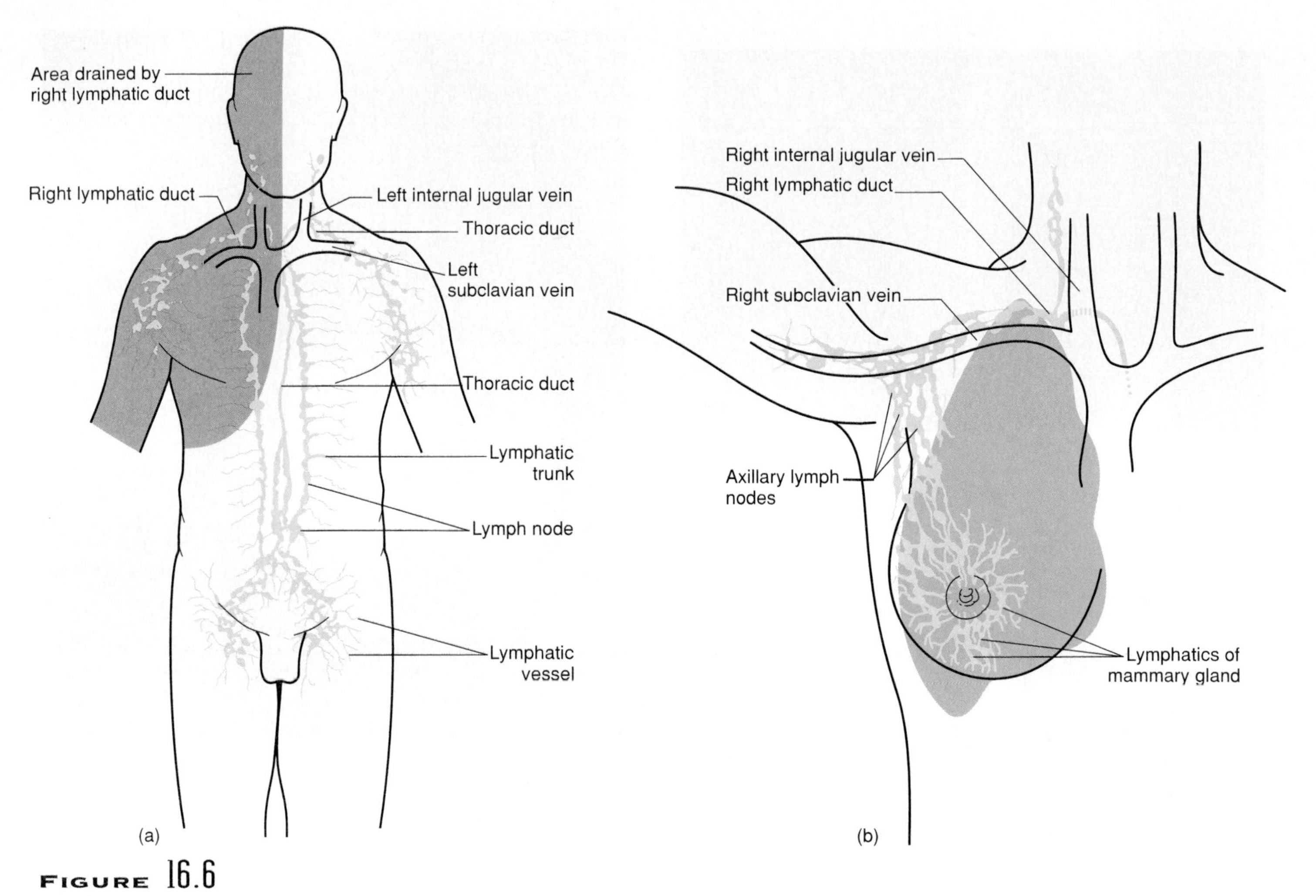


**FIGURE 16.6**

(*a*) The right lymphatic duct drains lymph from the upper right side of the body, whereas the thoracic duct drains lymph from the rest of the body. (*b*) Lymph drainage of the right breast.

## Tissue Fluid Formation

Recall from chapter 15 that tissue fluid originates from blood plasma. Tissue fluid is composed of water and dissolved substances that leave the blood capillaries by diffusion and filtration.

Tissue fluid contains nutrients and gases found in plasma and lacks large proteins. Some smaller proteins do leak out of the blood capillaries and enter the interstitial spaces. Usually, these smaller proteins are not reabsorbed when water and dissolved substances move back into the venule ends of these capillaries by diffusion and osmosis. As a result, the protein concentration of the tissue fluid tends to rise, raising the *osmotic pressure* of the fluid.

## Lymph Formation

Rising osmotic pressure of the tissue fluid interferes with the blood capillaries' osmotic reabsorption of water. The volume and pressure of fluid in the interstitial spaces then tends to increase. This increasing interstitial pressure forces some of the tissue fluid into the lymphatic capillaries, where it becomes lymph (see fig. 16.2).

## Lymph Function

Lymph returns to the bloodstream most of the proteins that leak out of the blood capillaries. At the same time, lymph transports foreign particles, such as bacteria or viruses, to lymph nodes.

Although these proteins and foreign particles cannot easily enter blood capillaries, the lymphatic capillaries are adapted to receive them. Specifically, the epithelial cells that form the walls of lymphatic vessels are arranged so that the edge of one cell overlaps the edge of an adjacent cell, but is not attached to it. This configuration, shown in figure 16.8, creates flaplike valves in the lymphatic capillary wall, which are pushed inward when the pressure is greater on the outside of the capillary, but close when the pressure is greater on the inside.

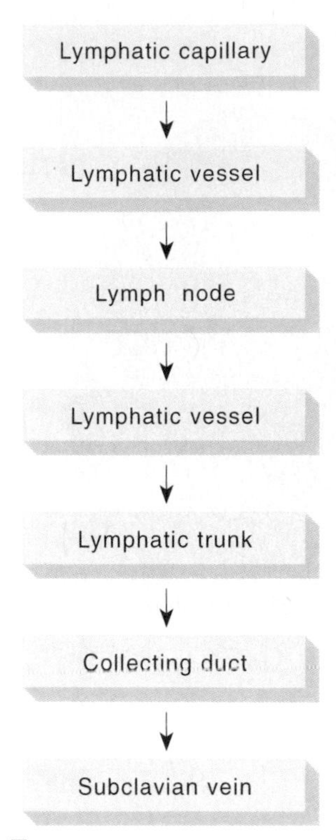


**FIGURE 16.7**
The lymphatic pathway.

The epithelial cells of the lymphatic capillary wall are also attached to surrounding connective tissue cells by thin filaments. As a result, the lumen of a lymphatic capillary remains open even when the outside pressure increases.

1. What is the relationship between tissue fluid and lymph?
2. How does the presence of protein in tissue fluid affect lymph formation?
3. What are the major functions of lymph?

## Lymph Movement

The osmotic pressure of tissue fluid influences the entry of lymph into lymphatic capillaries. However, muscular activity largely influences the movement of lymph through the lymphatic vessels.

### Lymph Flow

Lymph, like venous blood, is under relatively low hydrostatic pressure and may not flow readily through the lymphatic vessels without outside help. These forces include contraction of the skeletal muscles,

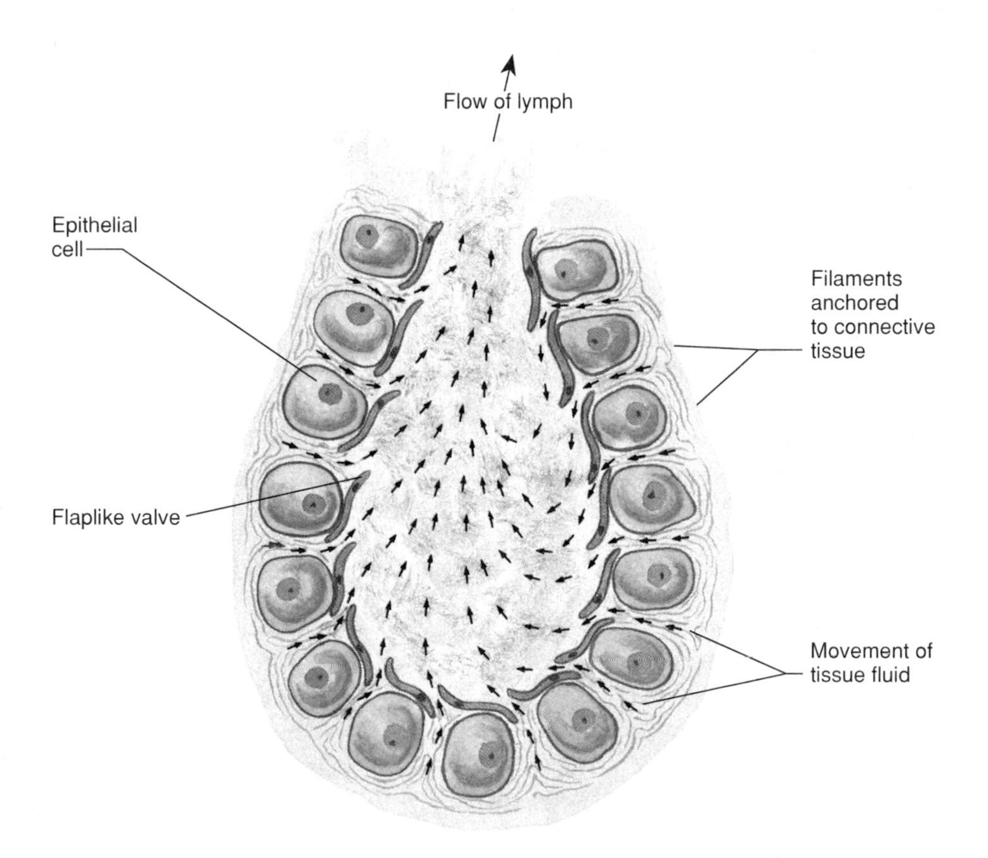


**FIGURE 16.8**
Tissue fluid enters lymphatic capillaries through flaplike valves between adjacent epithelial cells.

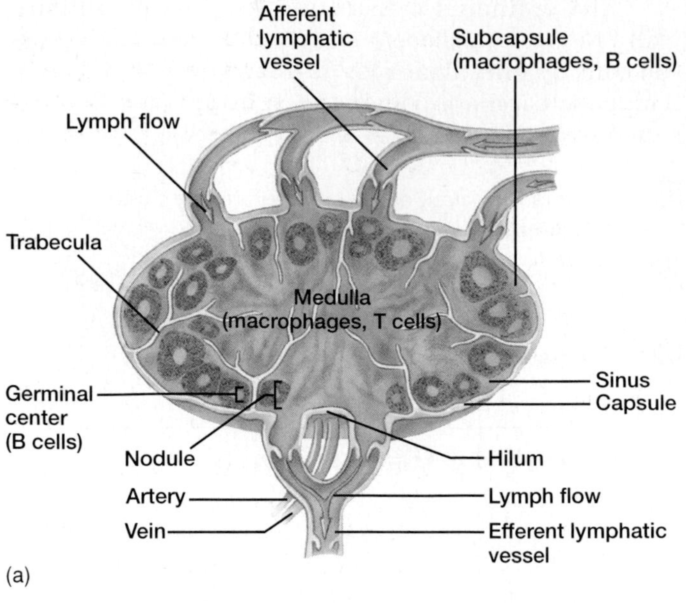


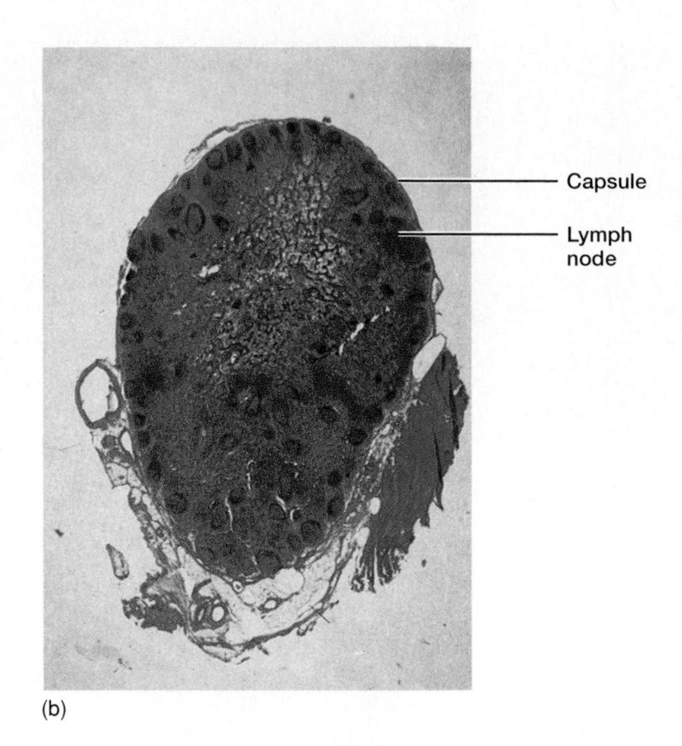


**FIGURE 16.9**
(*a*) A section of a lymph node. (*b*) Light micrograph of a lymph node (2.5×).

pressure changes due to the action of the breathing muscles, and contraction of smooth muscles in the walls of the larger lymphatic trunks.

Contracting skeletal muscles compress lymphatic vessels. This squeezing action moves the lymph inside a vessel, but because the lymphatic vessels contain valves that prevent backflow, the lymph can only move toward a collecting duct. Similarly, the smooth muscles in the walls of the larger lymphatic trunks may contract and compress the lymph inside, forcing the fluid onward.

The breathing muscles aid lymph circulation by creating a relatively low pressure in the thorax during inhalation. At the same time, the contracting diaphragm increases the pressure in the abdominal cavity. Consequently, lymph is squeezed out of the abdominal vessels and forced into the thoracic vessels. Once again, valves within the lymphatic vessels prevent lymph backflow.

Lymph flow peaks during physical exercise, due to the actions of skeletal and breathing muscles.

### Obstruction of Lymph Movement

The continuous movement of fluid from interstitial spaces into blood capillaries and lymphatic capillaries stabilizes the volume of fluid in these spaces. Conditions that interfere with lymph movement cause tissue fluids to accumulate in the interstitial spaces, producing *edema*. This may happen when surgery removes lymphatic tissue, obstructing certain lymphatic vessels. For example, surgery to remove a cancerous breast also usually removes nearby axillary lymph nodes, because associated lymphatic vessels can transport cancer cells to other sites (metastasis). Removing the lymphatic tissue can obstruct drainage from the upper limb, causing edema.

1. What factors promote lymph flow?
2. What is the consequence of lymphatic obstruction?

## Lymph Nodes

Lymph nodes (lymph glands) are located along the lymphatic pathways. They contain large numbers of *lymphocytes* and *macrophages,* which fight invading microorganisms.

### Structure of a Lymph Node

Lymph nodes vary in size and shape, but are usually less than 2.5 centimeters long and are somewhat bean-shaped. Figure 16.9 illustrates a section of a typical lymph node.

The indented region of a bean-shaped node is called the **hilum,** and it is the portion through which the blood vessels and nerves connect with the structure. The lymphatic vessels leading to a node (afferent vessels) enter separately at various points on its convex surface, but the lymphatic vessels leaving the node (efferent vessels) exit from the hilum.

A capsule of white fibrous connective tissue encloses each lymph node. The capsule extends into the

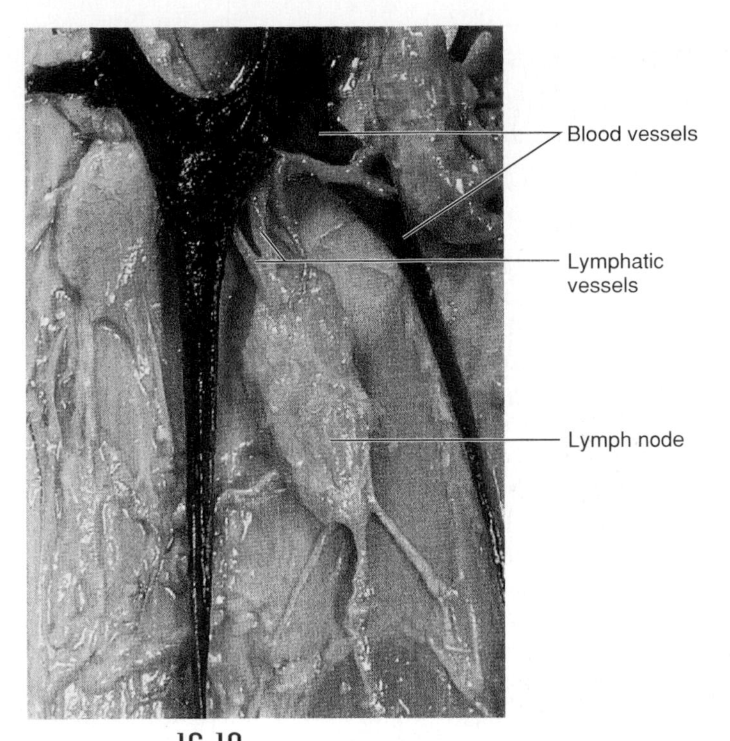


**FIGURE 16.10**
Lymph enters and leaves a lymph node through lymphatic vessels.

node and partially subdivides it into compartments that contain dense masses of actively dividing lymphocytes and macrophages, called germinal centers. These masses, or **nodules,** are the structural units of the lymph node.

The spaces within the node, called **lymph sinuses,** provide a complex network of chambers and channels through which lymph circulates as it passes through. Lymph enters a lymph node through an *afferent lymphatic vessel,* moves slowly through the lymph sinuses, and leaves through an *efferent lymphatic vessel* (fig. 16.10).

> Superficial lymphatic vessels inflamed by bacterial infection appear as red streaks beneath the skin, a condition called *lymphangitis.* Inflammation of the lymph nodes, called *lymphadenitis,* often follows. Affected nodes enlarge and may be quite painful.

Nodules also occur singly or in groups associated with the mucous membranes of the respiratory and digestive tracts. The *tonsils,* described in chapter 17, are composed of partially encapsulated lymph nodules. Also, aggregations of nodules called *Peyer's patches* are scattered throughout the mucosal lining of the ileum of the small intestine.

## Locations of Lymph Nodes

The lymph nodes generally occur in groups or chains along the paths of the larger lymphatic vessels throughout the body, but are absent in the central nervous system. The major locations of the lymph nodes, shown in figure 16.11, are as follows:

1. **Cervical region.** Nodes in the cervical region occur along the lower border of the mandible, in front of and behind the ears, and deep within the neck along the paths of the larger blood vessels. These nodes are associated with the lymphatic vessels that drain the skin of the scalp and face, as well as the tissues of the nasal cavity and pharynx.
2. **Axillary region.** Nodes in the underarm region receive lymph from vessels that drain the upper limbs, the wall of the thorax, the mammary glands (breasts), and the upper wall of the abdomen.
3. **Inguinal region.** Nodes in the inguinal region receive lymph from the lower limbs, the external genitalia, and the lower abdominal wall.
4. **Pelvic cavity.** Within the pelvic cavity, nodes occur primarily along the paths of the iliac blood vessels. They receive lymph from the lymphatic vessels of the pelvic viscera.
5. **Abdominal cavity.** Nodes within the abdominal cavity occur in chains along the main branches of the mesenteric arteries and the abdominal aorta. These nodes receive lymph from the abdominal viscera.
6. **Thoracic cavity.** Nodes of the thoracic cavity occur within the mediastinum and along the trachea and bronchi. They receive lymph from the thoracic viscera and from the internal wall of the thorax.
7. **Supratrochlear region.** These nodes are located superficially on the medial side of the elbow. They often enlarge in children as a result of infections from cuts and scrapes on the hands.

> The illness described as "swollen glands" actually refers to enlarged cervical lymph nodes associated with throat or respiratory infection.

## Functions of Lymph Nodes

Lymph nodes have two primary functions: filtering potentially harmful particles from lymph before returning it to the bloodstream, and immune surveillance,

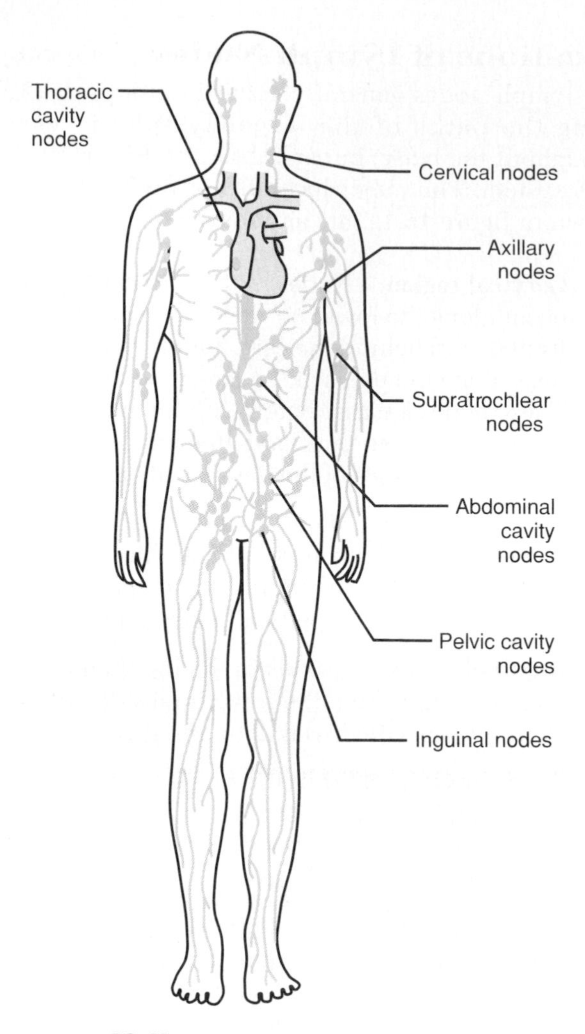


**FIGURE 16.11**
Major locations of lymph nodes.

provided by lymphocytes and macrophages. Along with the red bone marrow, the lymph nodes are centers for lymphocyte production. These cells attack infecting viruses, bacteria, and other parasitic cells brought to the nodes in lymphatic vessels. Macrophages in the nodes engulf and destroy foreign substances, damaged cells, and cellular debris.

1. Distinguish between a lymph node and a lymph nodule.
2. What factors promote the flow of lymph through a node?
3. In what body regions are lymph nodes most abundant?
4. What are the major functions of lymph nodes?

## THYMUS AND SPLEEN

Two other lymphatic organs, whose functions are closely related to those of the lymph nodes, are the **thymus** and the **spleen.**

### Thymus

The thymus gland is a soft, bilobed structure enclosed in a connective tissue capsule (fig. 16.12). It is located within the mediastinum, in front of the aortic arch and behind the upper part of the sternum, extending from the root of the neck to the pericardium. The thymus varies in size from person to person, and it is usually relatively large during infancy and early childhood. After puberty, it shrinks, and in an adult, it may be quite small. In elderly persons, adipose and connective tissues replace the normal lymphatic tissue.

Connective tissues extend inward from the thymus's surface, subdividing it into lobules (fig. 16.13). The lobules contain large numbers of lymphocytes that developed from precursor cells originating in the bone marrow. The majority of these cells (thymocytes) remain inactive; however, some of them mature into **T lymphocytes,** which leave the thymus and provide immunity.

Epithelial cells within the thymus secrete a protein hormone called *thymosin,* which stimulates maturation of T lymphocytes after they leave the thymus and migrate to other lymphatic tissues.

### Spleen

The spleen is the largest of the lymphatic organs. It is located in the upper left portion of the abdominal cavity, just beneath the diaphragm and behind the stomach (see fig. 16.12 and reference plates 4, 5, and 6).

The spleen resembles a large lymph node in that it is enclosed in connective tissue extending inward from the surface and partially subdividing the organ into chambers, or lobules. It also has a hilum on one surface through which blood vessels and nerves enter. However, unlike the sinuses of a lymph node, the spaces (venous sinuses) within the chambers of the spleen are filled with blood instead of lymph.

The tissues within the lobules of the spleen are of two types. The *white pulp* is distributed throughout the spleen in tiny islands. This tissue is composed of nodules (splenic nodules), which are similar to those found in lymph nodes and contain many lymphocytes. The *red pulp,* which fills the remaining spaces of the lobules, surrounds the venous sinuses. This pulp contains numerous red blood cells, which impart its color, plus many lymphocytes and macrophages (figs. 16.14 and 16.15).

Blood capillaries within the red pulp are quite permeable. Red blood cells can squeeze through the pores in these capillary walls and enter the venous sinuses. The older, more fragile red blood cells may rupture as they make this passage, and the resulting cellular debris is removed by phagocytic macrophages located within the splenic sinuses.

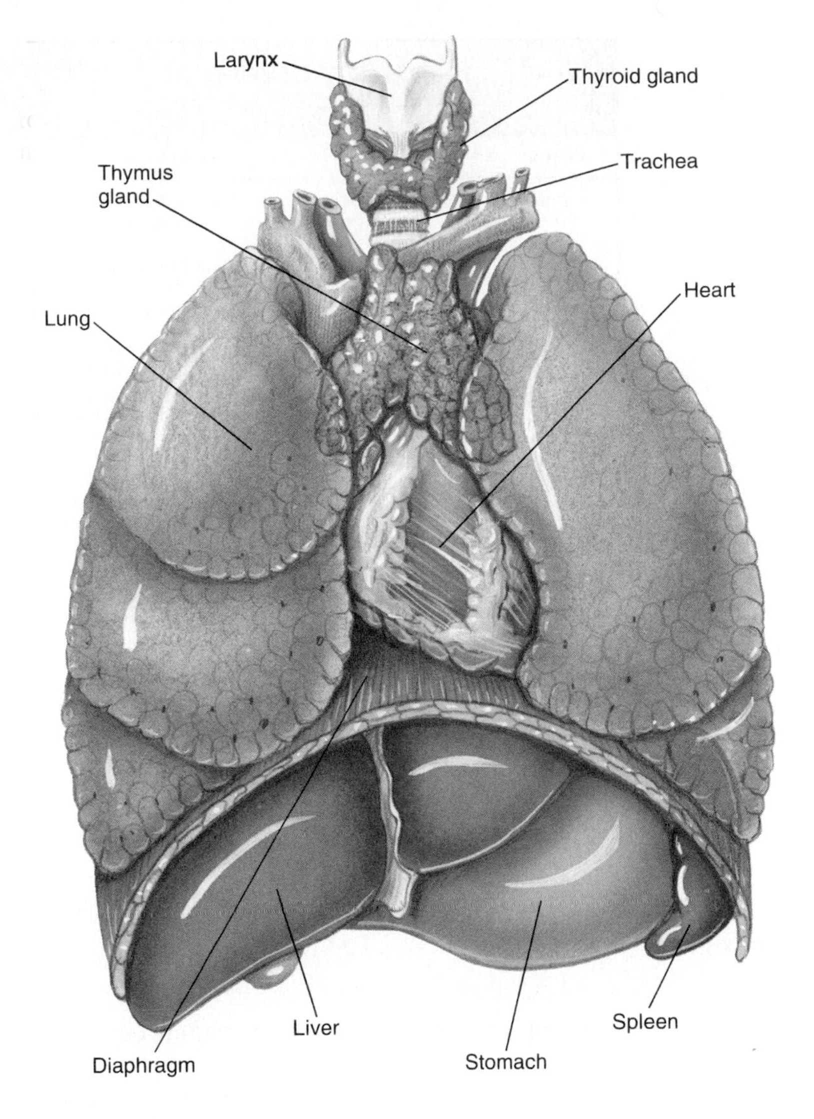


**FIGURE 16.12**
The thymus gland is bilobed and located between the lungs and above the heart. The spleen is located beneath the diaphragm and behind the stomach.

> During fetal development, pulp cells of the spleen produce blood cells, much as red bone marrow cells do in later life. As the time of birth approaches, this splenic function ceases. However, in certain diseases, such as *erythroblastosis fetalis,* in which large numbers of red blood cells are destroyed, the splenic pulp cells may resume their hematopoietic activity.

The macrophages also engulf and destroy foreign particles, such as bacteria, that may be carried in the blood as it flows through the sinuses.

The lymphocytes of the spleen, like those of the thymus, lymph nodes, and nodules, also help defend the body against infections. Thus, the spleen filters blood much as the lymph nodes filter lymph.

Chart 16.1 summarizes the characteristics of the major organs of the lymphatic system.

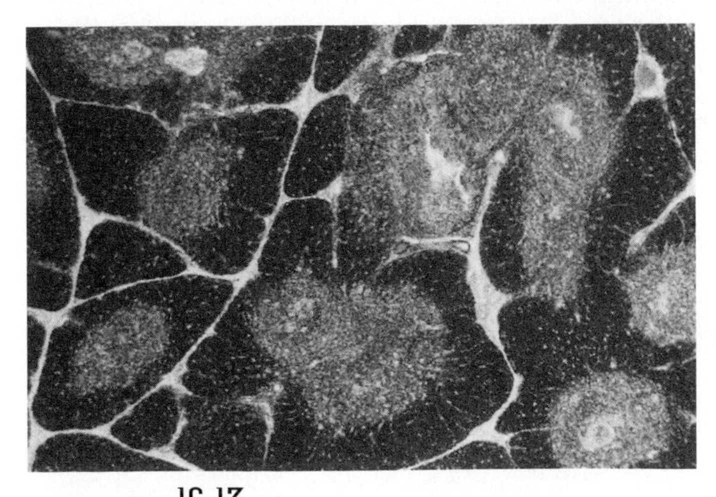

**FIGURE 16.13**
A cross section of the thymus gland (10×). Note how the gland is subdivided into lobules.

1. Why are the thymus and spleen considered organs of the lymphatic system?
2. What are the major functions of the thymus and the spleen?

## Body Defenses Against Infection

The presence and multiplication of a disease-causing agent, or **pathogen,** causes an **infection.** Pathogens include simple microorganisms (bacteria), complex

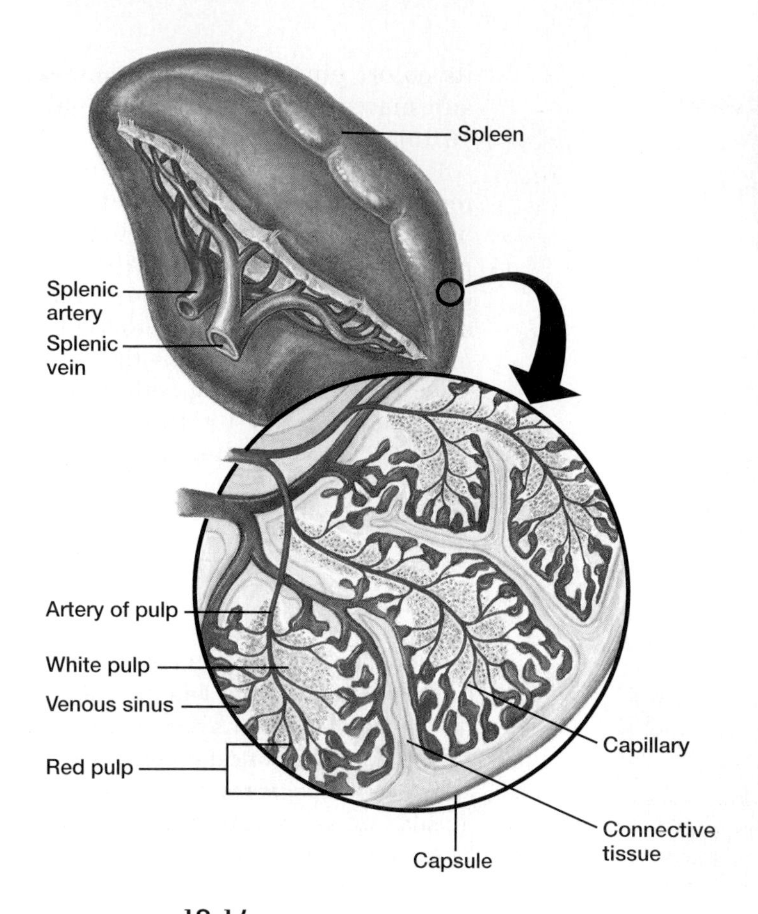


**Figure 16.14**
The spleen resembles a large lymph node.

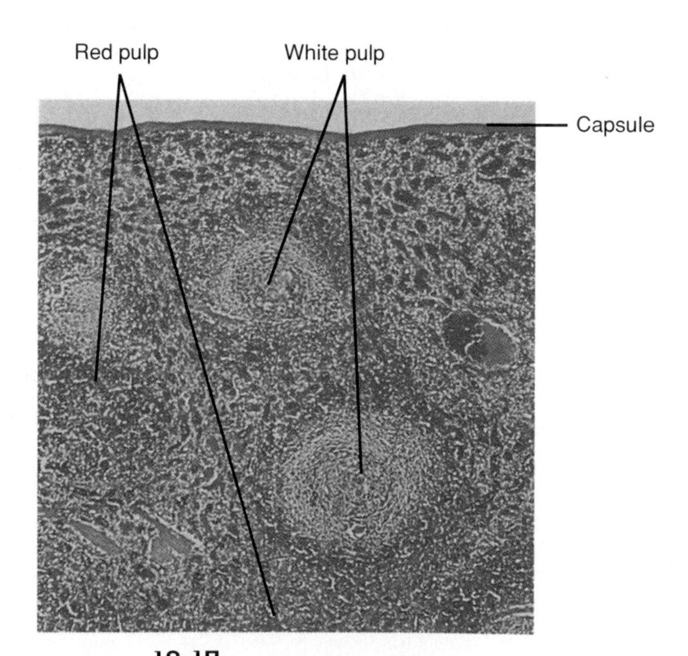


**Figure 16.15**
Light micrograph of the spleen (40×).

### Chart 16.1 Major Organs of the Lymphatic System

| Organ | Location | Function |
|---|---|---|
| Lymph nodes | In groups or chains along the paths of larger lymphatic vessels | Filter foreign particles and debris from lymph; produce and house lymphocytes that destroy foreign particles in lymph; house macrophages that engulf and destroy foreign particles and cellular debris carried in lymph |
| Thymus | Within the mediastinum behind the upper portion of the sternum | Houses lymphocytes; differentiates thymocytes into T lymphocytes |
| Spleen | In upper left portion of abdominal cavity beneath the diaphragm and behind the stomach | Houses macrophages that remove foreign particles, damaged red blood cells, and cellular debris from the blood; contains lymphocytes |

microorganisms such as protozoa, and even spores of multicellular organisms, such as fungi. Viruses are pathogens, but they are considered to be infectious agents and not organisms, because their structure is far simpler than that of a living cell. An infection may be present even though an individual feels well. People who are infected with the human immunodeficiency virus (HIV), which causes AIDS, often live for many years in good health before becoming ill.

The human body can prevent entry of pathogens, or destroy them if they gain entrance. Some mechanisms are quite general, protecting against many types of pathogens, providing **nonspecific defense.** These mechanisms include species resistance, mechanical barriers, enzyme action, interferon, fever, inflammation, and phagocytosis. Other defense mechanisms are very precise, targeting specific pathogens. These mechanisms provide **specific defense,** or immunity. They are carried out by specialized lymphocytes that recognize foreign molecules (nonself antigens) in the body and respond to them.

## Nonspecific Defenses

### Species Resistance

**Species resistance** refers to the fact that a given kind of organism, or species (such as the human species, *Homo sapiens*), develops diseases that are unique to it. At the same time, a species may be resistant to diseases that affect other species, because its tissues somehow fail to provide the temperature or chemical environment that a particular pathogen needs. For example, humans are subject to infections by the microorganisms that cause measles, mumps, gonorrhea, and syphilis, but other animal species are not. Similarly, humans are resistant to certain forms of malaria and tuberculosis that affect birds.

> In San Francisco in 1982, Simon Guzman became one of the very first recorded individuals to succumb to AIDS. He was *the* first whose death was attributed to a parasitic infection previously seen only in sheep—*cryptosporidiosis.* This infection, which causes relentless diarrhea, illustrates one of the most bizarre effects of AIDS: alteration of species resistance.

### Mechanical Barriers

The skin and mucous membranes lining the passageways of the respiratory, digestive, urinary, and reproductive systems create **mechanical barriers** that prevent the entrance of some infectious agents. As long as these barriers remain intact, many pathogens are unable to penetrate them. The epidermis also sloughs off, removing superficial bacteria with it. In addition, the mucus-coated ciliated epithelium, described in chapter 19, that lines the respiratory passages entraps particles and sweeps them out of the airways and into the pharynx, where they are swallowed.

### Chemical Barriers

Enzymes in body fluids provide a **chemical barrier** to pathogens. Gastric juice, for example, contains the protein-splitting enzyme pepsin, and has a low pH due to the presence of hydrochloric acid. The combined effect of these substances is lethal to many pathogens that enter the stomach. Similarly, tears contain the enzyme lysozyme, which has an antibacterial action against certain pathogens that may get onto the surfaces of the eyes. The accumulation of salt from perspiration also kills certain bacteria on the skin.

**Interferons** are hormonelike peptides produced by certain cells, including lymphocytes and fibroblasts, in response to viruses or certain tumor cells. The effect of interferon is nonspecific. It blocks viral replication, stimulates phagocytosis, and enhances the activity of certain other cells that help to resist infections and the growth of tumors.

### Fever

A **fever** offers powerful protection. A viral or bacterial infection stimulates lymphocytes to proliferate, producing cells that secrete a substance called *endogenous pyrogen* ("fire maker from within"). Endogenous pyrogen raises the thermoregulatory set point in the brain's hypothalamus to maintain a higher body temperature. Fever counters microbial growth indirectly because higher body temperature reduces the level of iron in the blood. Since bacteria and fungi need more iron as the temperature rises, their growth is squelched in a fever-ridden body. Plus, phagocytic cells attack more vigorously when the temperature rises.

### Inflammation

**Inflammation** is a tissue response to pathogen invasion (see chapter 5), producing localized redness, swelling, heat, and pain. The redness is a result of blood vessel dilation and the consequent increase of blood volume within the affected tissues (hyperemia). This effect, coupled with an increase in the permeability of the nearby capillaries, causes tissue swelling (edema). The heat comes from blood from deeper body parts, which is generally warmer than that near the surface, and the pain results from stimulation of nearby pain receptors.

White blood cells tend to accumulate at the sites of inflammation, and some of these cells help control pathogens by phagocytosis. In bacterial infections, the resulting mass of white blood cells, bacterial cells, and damaged tissue may form a thick fluid called *pus.*

Body fluids (exudate) also collect in inflamed tissues. These fluids contain fibrinogen and other clotting factors. As a result of clotting, a network of fibrin threads may develop within the affected region. Later, fibroblasts may appear and form fibers around the area until it is enclosed in a sac of fibrous connective tissue. This action inhibits the spread of pathogens and toxic substances to adjacent tissues.

## CHART 16.2 Major Actions That May Occur During an Inflammation Response

***Blood vessels dilate.***

***Capillary permeability increases.***
Tissues become red, swollen, warm, and painful.

***White blood cells invade the region.***
Pus may form as white blood cells, bacterial cells, and cellular debris accumulate.

***Body fluids seep into the area.***
A clot containing threads of fibrin may form.

***Fibroblasts appear.***
A connective tissue sac may form around the injured tissues.

***Phagocytes are active.***
Bacteria, dead cells, and other debris are removed.

***Cells reproduce.***
Newly formed cells replace injured ones.

Once an infection is controlled, phagocytic cells remove dead cells and other debris from the site of inflammation. At the same time mitosis replaces lost cells. Chart 16.2 summarizes the process of inflammation.

### Phagocytosis

Recall from chapter 14 that the most active phagocytic cells of the blood are *neutrophils* and *monocytes.* These wandering cells can leave the bloodstream by squeezing between the cells of the blood vessel walls (diapedesis). Chemicals released from injured tissues attract these cells (chemotaxis). Neutrophils engulf and digest smaller particles, while monocytes phagocytize somewhat larger ones.

Monocytes give rise to *macrophages* (histiocytes), which become fixed in various tissues and attached to the inner walls of blood and lymphatic vessels. These relatively nonmotile, phagocytic cells, which can divide and produce new macrophages, are found in lymph nodes, the spleen, liver, and lungs. A macrophage can engulf up to one hundred bacteria, compared to the twenty or so bacteria that a neutrophil can engulf. This diffuse group of phagocytic cells constitutes the **reticuloendothelial tissue** (reticuloendothelial system, or RES, also called the mononuclear phagocyte system).

As a result of reticuloendothelial activities, foreign particles are removed from the lymph as it moves from the interstitial spaces to the bloodstream. Particles reaching the blood are usually removed by phagocytes in the vessels and tissues of the spleen, liver, or bone marrow.

Chart 16.3 summarizes the types of nonspecific defenses.

1 What is an infection?

2 Explain six nonspecific defense mechanisms.

3 Define reticuloendothelial tissue.

## Specific Defenses (Immunity)

**Immunity** is resistance to particular pathogens or to their toxins or metabolic by-products. Lymphocytes and macrophages that recognize specific foreign molecules (nonself antigens) carry out several types of immune responses.

### Lymphocyte Origins

During fetal development, red bone marrow releases undifferentiated lymphocytes into the circulation. About half of these cells reach the thymus, where they remain for a time. Here, these thymocytes specialize into T lymphocytes, or **T cells.** (T refers to *thymus-derived* lymphocytes.) Later, the blood transports T cells, where they comprise 70% to 80% of the circulating lymphocytes. T cells reside in lymphatic organs and are particularly abundant in the lymph nodes, the thoracic duct, and the white pulp of the spleen.

Other lymphocytes are thought to remain in the red bone marrow until they differentiate into B lymphocytes or **B cells.** Historically, the B stands for *bursa of Fabricius,* an organ in the chicken where these cells were first identified. Ironically, the B can now be thought of as referring to bone marrow.

The blood distributes B cells, which constitute 20% to 30% of circulating lymphocytes. B cells settle in lymphatic organs along with T cells and are abundant in lymph nodes, the spleen, bone marrow, intestinal lining, and reticuloendothelial tissues (figs. 16.16 and 16.17).

### Antigens

Before birth, cells inventory the proteins and other large molecules in the body, learning to distinguish these as "self." The lymphatic system responds to

## CHART 16.3 Types of Nonspecific Defenses

| Type | Description |
|---|---|
| Species resistance | A species or organism is resistant to certain diseases to which other species are susceptible. |
| Mechanical barriers | Unbroken skin and mucous membranes prevent the entrance of some infectious agents. |
| Chemical barriers | Enzymes in various body fluids kill pathogens. pH and high salt concentration also harm pathogens. Interferon blocks reproduction of viruses, stimulates phagocytosis, and enhances the activity of cells that resist infections and the growth of tumors. |
| Fever | Elevated body temperature aids recovery from bacterial infections. |
| Inflammation | A tissue response to injury that helps prevent the spread of infectious agents into nearby tissues. |
| Phagocytosis | Neutrophils, monocytes, and macrophages engulf and destroy foreign particles and cells. |

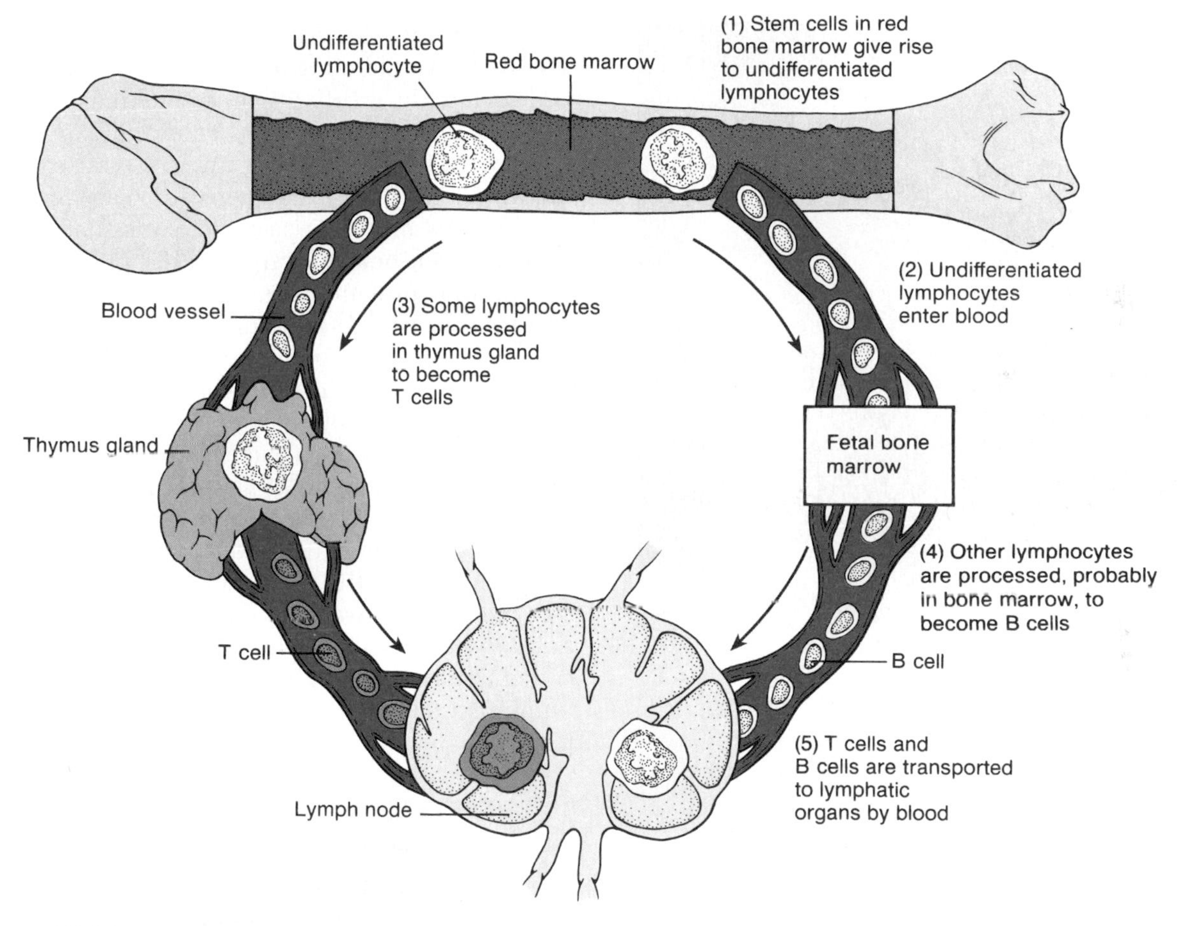


**FIGURE 16.16**
Bone marrow releases undifferentiated lymphocytes, which after processing become T lymphocytes (T cells) or B lymphocytes (B cells). Note that the diaphysis contains red marrow in the fetus.

nonself, or foreign antigens, but not normally to self antigens. Receptors on T and B cell surfaces enable the cells to recognize foreign antigens.

**Antigens** may be proteins, polysaccharides, or glycolipids. The antigens that are most effective in eliciting an immune response are large and complex, with few repeating parts. Sometimes, a smaller molecule that cannot by itself stimulate an immune response combines with a larger one, which makes it able to do so (antigenic). Such a small molecule is called a **hapten.** Stimulated lymphocytes react either to the hapten or to the larger molecule of the combination. Haptens are found in certain drugs, such as penicillin; in household and industrial chemicals; in dust particles; and in products of animal skins (dander).

CHART 16.4 A Comparison of T Cells and B Cells

| Characteristic | T Cells | B Cells |
|---|---|---|
| Origin of undifferentiated cell | Red bone marrow | Red bone marrow |
| Site of differentiation | Thymus | Probably the red bone marrow |
| Primary locations | Lymphatic tissues, 70%–80% of the circulating lymphocytes | Lymphatic tissues, 20%–30% of the circulating lymphocytes |
| Primary functions | Responsible for cell-mediated immunity in which T cells interact directly with antigen-bearing agents | Responsible for antibody-mediated immunity in which B cells interact indirectly with antigen-bearing agents by producing antibodies. |

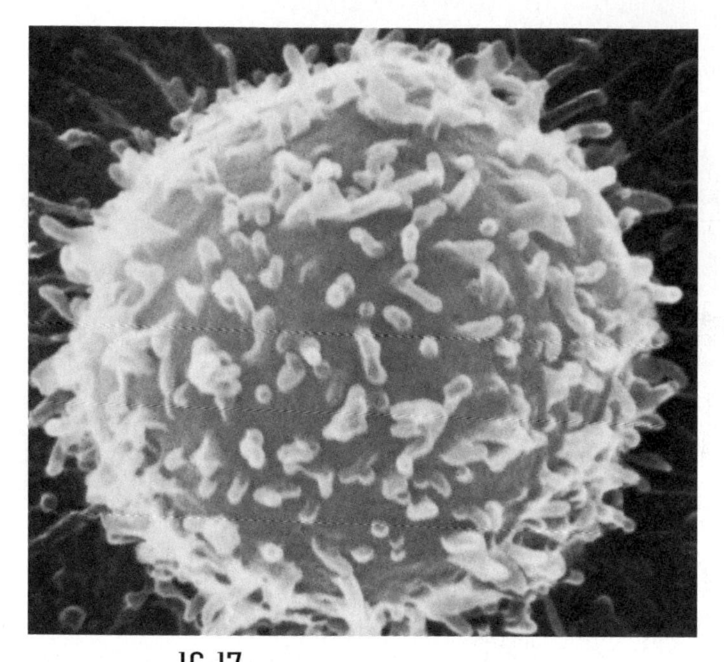

FIGURE 16.17
Scanning electron micrograph of a human circulating lymphocyte (36,000×).

1 What is immunity?

2 How do T cells and B cells originate?

3 What is the difference between an antigen and a hapten?

## Lymphocyte Functions

T cells and B cells respond to the antigens they recognize in different ways. T cells attach to foreign, antigen-bearing cells, such as bacterial cells, and interact directly—that is, by cell-to-cell contact. This type of response is called **cell-mediated immunity** (CMI).

T cells (and some macrophages) also synthesize and secrete polypeptides called **cytokines** (or more specifically, lymphokines) that enhance certain cellular responses to antigens. For example, *interleukin-1* and *interleukin-2* stimulate synthesis of several cytokines from other T cells. In addition, interleukin-1 helps activate T cells, while interleukin-2 causes T cells to proliferate and activates a certain type of T cell (cytotoxic T cells). Other cytokines called *colony stimulating factors* (CSFs) stimulate production of leukocytes in the red bone marrow, cause growth and maturation of B cells, and activate macrophages.

T cells may also secrete toxins that are lethal to their antigen-bearing target cells: growth-inhibiting factors that prevent target-cell growth, or interferon that inhibits proliferation of viruses and tumor cells.

B cells attack foreign antigens in a different way. They differentiate into *plasma* cells, which produce and secrete large globular proteins called **antibodies,** or immunoglobulins. A plasma cell is a virtual antibody factory, as evidenced by its characteristically huge Golgi apparatus. At the peak of an infection, a plasma cell may produce and secrete 2,000 antibody molecules a second! Body fluids carry antibodies, which then react in various ways to destroy specific antigens or antigen-bearing particles. This type of response is called **antibody-mediated immunity** (AMI), or humoral immunity. ("Humoral" refers to fluid.) Chart 16.4 compares the characteristics of T cells and B cells.

Each year, a few children are born defenseless against infection due to *severe combined immune deficiency* (*SCID*), in which neither T nor B cells function. David Vetter was one such youngster. Born in Texas in 1971, he had no thymus gland and spent the 12 years of his life in a vinyl bubble, awaiting a treatment that never came. As David reached adolescence, he wanted to leave his bubble. An experimental bone marrow transplant was unsuccessful—soon afterwards David began vomiting and developed diarrhea, signs of infection. David left the bubble but died within days of a massive infection. Today some children with SCID caused by an inherited enzyme deficiency have more pleasant treatment options, including enzyme replacement therapy and gene therapy.

1. Define immunity.
2. Explain how T cells and B cells originate.
3. Explain the difference between an antigen and a hapten.
4. What are the functions of a T cell? Of a B cell?

## Lymphocyte Clones

A person has millions of varieties of T and B cells. The members of each variety originate from a single early cell, so that the members are all alike, forming a **clone** of cells. The members of each variety have a particular type of antigen receptor on their cell membranes that is capable of responding only to a specific antigen.

> The human body can manufacture an apparently limitless number of different antibodies, but we have a limited number of antibody genes. How can this be? The great diversity of antibody types is possible because different antibody protein subunits combine. During the early development of B cells, sections of their antibody genes randomly move to other chromosomal locations, creating new genetic instructions for antibodies. Antibody diversity is like using the limited number of words in a language to compose an infinite variety of stories.

## B Cell Activation

A lymphocyte must be activated before it can respond to an antigen. A B cell, for example, may become activated when it encounters an antigen whose molecular shape fits the shape of the B cell's antigen receptors. In response to the receptor-antigen combination, the B cell divides repeatedly, expanding its clone. However, most antigens require T cell "help" to activate B cells. Some members of the activated B cell's clone differentiate further into plasma cells, which secrete antibodies. These antibodies are similar in structure to the antigen-receptor molecules present on the original B cell's surface. Thus, antibodies can combine with the antigen-bearing agent that has invaded the body, and react against it.

A single type of B cell carries information to produce a single type of antibody. However, different B cells respond to different antigens on a pathogen's surface. Therefore, an immune response may include several types of antibodies manufactured against a single microbe or virus. This is called a **polyclonal response.** Clinical Application 16.1 discusses how researchers use clones of single B cells to produce single, or monoclonal, antibodies.

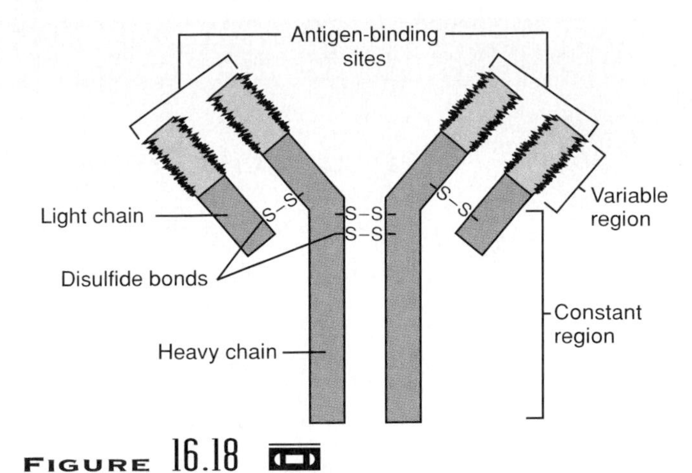


**FIGURE 16.18** An immunoglobulin molecule consists basically of two identical light chains of amino acids and two identical heavy chains of amino acids.

## Antibody Molecules

Antibodies are soluble, globular proteins that constitute the *gamma globulin* fraction of plasma proteins (see chapter 14). They are also called **immunoglobulins.**

Each immunoglobulin molecule is composed of four chains of amino acids that are linked together by pairs of sulfur atoms (disulfide bonds). Two of these chains are identical **light chains** (L-chains), and two are identical **heavy chains** (H-chains). The heavy chains contain about twice as many amino acids as the light chains. There are five major types of immunoglobulin molecules, each containing a particular kind of heavy chain.

As with other proteins, the sequence of amino acids within the peptide chains confers the unique, three-dimensional structure of each immunoglobulin molecule. This special conformation, in turn, imparts the physiological properties of the molecule. For example, one end of each of the heavy and light chains consists of variable sequences of amino acids (variable regions). These regions are specialized to react with the shape of a specific antigen molecule.

Antibodies can bind to certain antigens because of the three-dimensional shape of the variable regions. The antibody contorts to form a pocket around the antigen. These specialized ends of the antibody molecule are called **antigen-binding sites,** and the particular parts that actually bind the antigen are called **idiotypes.**

The remaining portions of the chains are termed constant regions because their amino acid sequences are relatively unchanged from molecule to molecule. Constant regions are responsible for other properties of the immunoglobulin molecule, such as its ability to bond to cellular structures or to combine with certain chemicals (fig. 16.18).

## Types of Immunoglobulins

Of the five major types of immunoglobulins, three constitute the bulk of the circulating antibodies. They are called immunoglobulin G, which accounts for about

CLINICAL APPLICATION 16.1

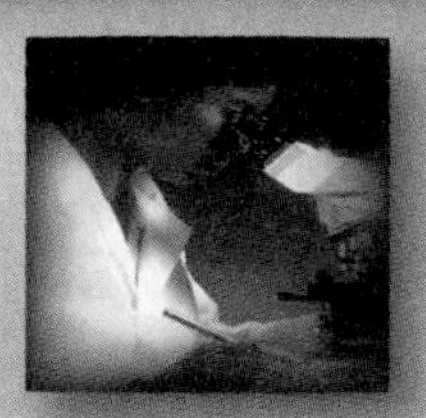

## Immunotherapy

At the turn of the last century, German bacteriologist Paul Ehrlich developed the concept of the "magic bullet"—a substance that could enter the body and destroy diseased cells, yet spare the healthy ones. The biochemicals and cells of the immune system, with their great specificity for attacking nonself tissue, would be ideal magic bullets. *Immunotherapy* uses immune system components to fight disease—both humoral immunity (antibodies) and cell-mediated immunity (cytokines).

### Monoclonal Antibodies—Targeting Immunity

Tapping the specificity of a single B cell, and using its single type, or *monoclonal,* antibody to target a specific antigen (such as on a cancer or bacterial cell) awaited finding a way to entice the normally short-lived mature B cells into persisting in culture. In 1975, British researchers Cesar Milstein and Georges Kohler devised *monoclonal antibody* (MAb) technology, an ingenious way to capture the antibody-making capacity of a single B cell.

Milstein and Kohler injected a mouse with red blood cells from a sheep. They then isolated a single B cell from the mouse's spleen, and fused it with a cancerous white blood cell from a mouse. The result was a fused cell, or *hybridoma,* with a valuable pair of talents: like the B cell, it produces large amounts of a single antibody type; like the cancer cell, it divides continuously (fig. 16A).

MAbs are used in basic research, veterinary and human health care, and agriculture. Cell biologists use pure antibodies to localize and isolate proteins. Diagnostic MAb "kits" detect tiny amounts of a single molecule. Most kits consist of a paper strip impregnated with a MAb, to which the user adds a body fluid. For example, a woman who suspects she is pregnant places drops of her urine onto the paper. A color change ensues if the MAb binds to human chorionic gonadotropin, indicating pregnancy.

MAbs can highlight cancer before it can be detected by other means. The MAb is attached to a radioactive chemical, which is then detected when the MAb binds an antigen unique to the cancer cell surfaces. Detecting a cancer's recurrence with a MAb requires only an injection followed by a painless imaging procedure—compared to surgery.

In cancer treatment, MAbs hold the promise of being "magic bullets" that ferry conventional cancer treatments to where they are needed and limit their toxicity. Drugs or radioactive chemicals are attached to MAbs that are attracted to antigens on cancer cells. When injected into a patient, the MAb and its cargo are engulfed by the cancer cells, which are destroyed. Because many people have allergic reactions to MAbs derived from mice, geneticists have engineered mice and bacteria to secrete human antibodies.

### Cytokines

Immunotherapy experiments were difficult to do in the late 1960s because the needed biochemicals could only be obtained in small amounts from cadavers. In the 1970s, however, the advent of recombinant DNA and monoclonal antibody technologies provided the ability to make unlimited amounts of pure proteins—just as the

80% of the antibodies; immunoglobulin A, which makes up about 13%; and immunoglobulin M, which is responsible for about 6%. Immunoglobulin D and immunoglobulin E account for the remainder of the antibodies.

*Immunoglobulin G* (IgG) is in plasma and tissue fluids, and is particularly effective against bacteria, viruses, and toxins. IgG also activates a group of enzymes called **complement,** which is described in the following section.

*Immunoglobulin A* (IgA) is commonly found in exocrine gland secretions. It is in breast milk, tears, nasal fluid, gastric juice, intestinal juice, bile, and urine.

A newborn does not yet have its own antibodies, but does retain IgG that passed through the placenta from the mother. These maternal antibodies protect the infant against some illnesses to which the mother is immune. Just as the maternal antibody supply falls, the infant begins to manufacture its own. The newborn receives IgA from colostrum, a substance secreted from the mother's breasts for the first few days after birth. Antibodies in colostrum protect against certain digestive and respiratory infections.

first inklings of the AIDS epidemic were making it essential to find a purer source than cadavers!

Interferon was the first cytokine to be tested on a large scale. Although interferon did not live up to early expectations of being a wonder drug, it is effective against a dozen or so conditions, including a type of leukemia, multiple sclerosis, hepatitis, and genital warts.

In another cytokine-based cancer treatment, certain T cells are removed from tumor samples, and incubated with interleukin-2, which the patient also receives intravenously. The removed T cells, activated by the cytokine, are injected into the patient. Interleukin-2 is also used to treat kidney cancer. In another approach, colony stimulating factors can boost the white blood cell supply in people whose immune systems are temporarily suppressed, such as those receiving drugs to treat cancer or AIDS.

Cancer treatment may soon consist of combinations of immune system cells and biochemicals, plus standard therapies. Immunotherapy can enable a patient to withstand higher doses of a conventional drug, or destroy cancer cells remaining after standard treatment.

The idea of using the immune system to treat cancer is not new. In 1893, a New York surgeon, William Coley, intentionally infected cancer patients with killed *Streptococcus* bacteria, after noticing that some patients with bacterial infections spontaneously recovered from the cancer. Sometimes "Coley's toxins" worked, although he did not understand precisely how. Today we know that an immune system biochemical called *tumor necrosis factor* is often responsible for the body's success in overcoming cancer. However, researchers do not yet understand all of this intriguing molecule's effects sufficiently to use it as an anticancer drug safely.

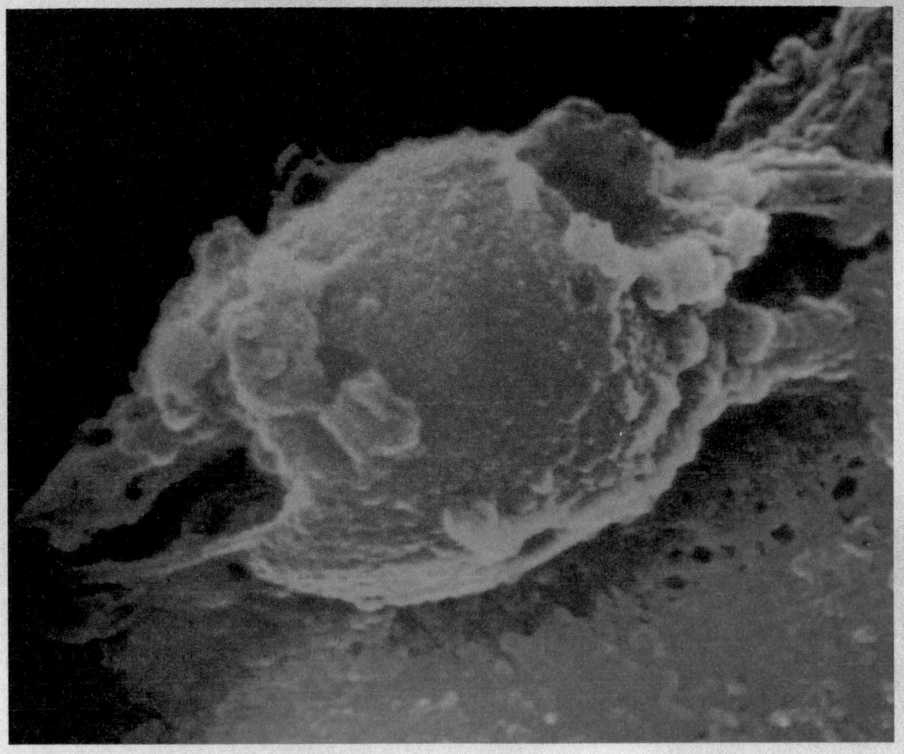

**FIGURE 16A**
Monoclonal antibodies are produced by a type of artificial cell combination called a hybridoma. It consists of a cancer cell (the flat blue cell) fused with a B cell (the round green cell). The cancer cell contributes rapid and continuous division; the B cell secretes a single antibody type.

*Immunoglobulin M* (IgM) is a type of antibody that develops in the blood plasma, in response to contact with certain antigens in foods or bacteria. The antibodies anti-A and anti-B, described in chapter 14, are examples of IgM. IgM also activates complement.

*Immunoglobulin D* (IgD) is found on the surfaces of most B cells, especially those of infants. IgD is important in activating B cells (see figure 16.20).

*Immunoglobulin E* (IgE) appears in exocrine secretions along with IgA. It is associated with allergic reactions that are described in a subsequent section of this chapter.

Chart 16.5 summarizes the major immunoglobulins and their functions.

1. What is an immunoglobulin?
2. Describe the structure of an immunoglobulin molecule.
3. Name the five major types of immunoglobulins.

## CHART 16.5 Characteristics of Major Immunoglobulins

| Type | Occurrence | Major Function |
|---|---|---|
| IgG | Tissue fluid and plasma | Acts against bacterial cells, viruses, and toxins; activates complement |
| IgA | Exocrine gland secretions | Acts against bacterial cells and viruses |
| IgM | Plasma | Reacts with antigens occurring naturally on some red blood cell membranes following certain blood transfusions; activates complement |
| IgD | Surface of most B lymphocytes | B cell activation |
| IgE | Exocrine gland secretions | Promotes allergic reactions |

### Antibody Actions

In general, antibodies react to antigens by attacking them directly, by activating a set of enzymes (complement) that attack the antigens, or by stimulating changes in local areas that help prevent spread of the antigens.

In a direct attack, antibodies combine with antigens and cause them to clump together (agglutinate) or to form insoluble substances (precipitation). Such actions make it easier for phagocytic cells to engulf the antigen-bearing agents and eliminate them. In other instances, antibodies cover the toxic portions of antigen molecules and neutralize their effects (neutralization). However, under normal conditions, complement activation is more important in protecting against infection than direct antibody attack.

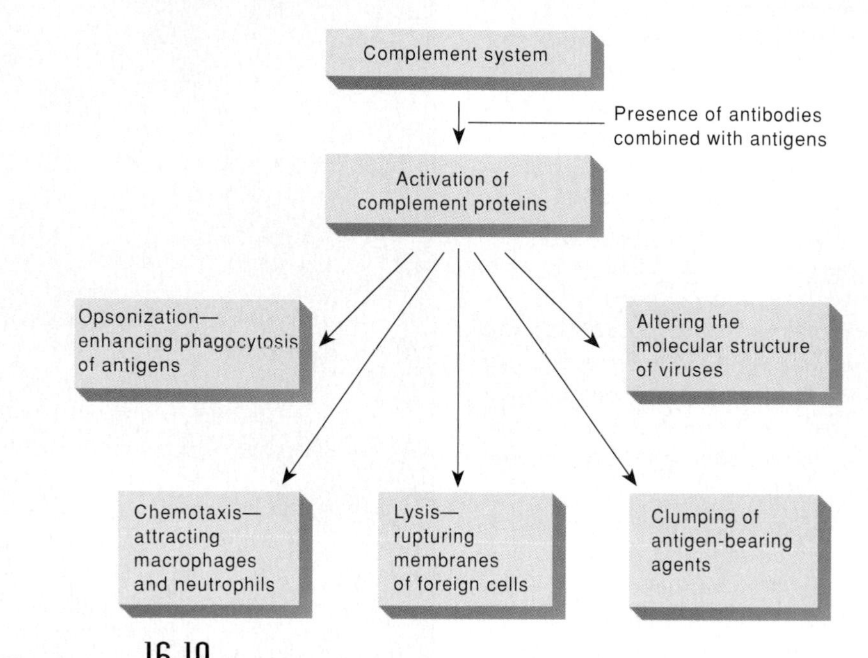


FIGURE 16.19
Actions of the complement system.

**Complement** is a group of proteins (complement system) in plasma and other body fluids. Certain IgG or IgM antibodies combining with antigens expose reactive sites on the constant regions of antibody molecules. This triggers a series of reactions leading to activation of the complement proteins, which in turn produce a variety of effects, including: coating the antigen-antibody complexes (opsonization), making them more susceptible to phagocytosis; attracting macrophages and neutrophils into the region (chemotaxis); clumping antigen-bearing agents together; rupturing membranes of foreign cells; and altering the molecular structure of viruses, making them harmless (fig. 16.19). Other proteins promote inflammation, which helps prevent the spread of infectious agents.

Immunoglobulin E promotes inflammation that may be so intense it damages tissues. This antibody is usually attached to the membranes of widely distributed mast cells (see chapter 5). When antigens combine with the antibodies, the resulting antigen-antibody complexes stimulate mast cells to release biochemicals, such as histamine, that cause the changes associated with inflammation, such as vasodilation and edema.

Chart 16.6 summarizes the actions of antibodies.

1. In what general ways do antibodies act?
2. What is the function of complement?
3. How is complement activated?

**CHART 16.6 ACTIONS OF ANTIBODIES**

| General Action | Type of Effect | Description |
|---|---|---|
| *Direct attack* | | |
| | Agglutination | Antigens clump |
| | Precipitation | Antigens become insoluble |
| | Neutralization | Antigens lose toxic properties |
| *Activation of complement* | | |
| | Opsonization | Alters cell membranes so cells are more susceptible to phagocytosis |
| | Chemotaxis | Attracts macrophages and neutrophils into region |
| | Inflammation | Promotes local tissue changes that help prevent spread of antigens |
| | Lysis | Cell membranes rupture |

## T Cell Types

T cell activation requires the presence of processed fragments of antigen attached to the surface of another kind of cell, called an **antigen-presenting cell** (or accessory cell). Macrophages, B cells, and several other cell types can serve as antigen-presenting cells.

There are several types of T cells. The three major types are helper cells, cytotoxic cells, and suppressor cells.

*Helper T cells* mobilize the immune system to stop a bacterial infection through a complex series of steps. First, a macrophage phagocytizes a bacterium, digesting it in its lysosomes. Some bacterial antigens exit the lysosomes and move to the macrophage's surface. Here, they are displayed on the outside membrane near certain protein molecules that are part of the "self" topography of the cell's surface. These self proteins belong to a group called the *major histocompatibility complex* (MHC). Then, a helper T cell contacts a displayed foreign antigen. If the displayed antigen fits and combines with the helper T cell's antigen receptors, the helper cell becomes activated.

When an activated helper T cell encounters a B cell that has already combined with an identical foreign antigen, the helper cell releases certain cytokines. These cytokines stimulate the B cell to proliferate, thus enlarging its clone of antibody-producing cells (figs. 16.20 and 16.21). The cytokines also attract macrophages and leukocytes into inflamed tissues and help retain them there. Chart 16.7 summarizes the steps leading to antibody production as a result of B and T cell activities.

A type of helper T cell called a CD4 cell is the prime target of HIV, the virus that causes AIDS. (CD4 stands for the "cluster-of-differentiation" antigen it bears that enables it to recognize a macrophage displaying a foreign antigen.) Considering the role of CD4 helper T cells as key players in establishing immunity—they stimulate B cells and secrete cytokines—it is no wonder that harming them destroys immunity.

Learning how different subtypes of T cells act is helping researchers learn how HIV cripples the immune system, causing AIDS. Differences in T cell subpopulations among individuals may explain why some people infected with the virus develop AIDS quickly and die, while others remain healthy for many years, perhaps never becoming ill. One such clue may be the predominance of one type of CD4 helper T cell over another.

Specifically, cells called Th1 seem to protect against AIDS, whereas predominance of another type, called Th2, increases susceptibility to AIDS. The two types of CD4 helper T cell secrete different combinations of cytokines. Th1 T cells secrete interleukin-2 and gamma interferon, which are products of cell-mediated immunity. In contrast, Th2 T cells secrete interleukin-4 and interleukin-5, which stimulate antibody production and therefore control humoral immunity.

Perhaps HIV infection somehow shifts the balance of T cell subtypes from Th1 predominating to Th2, setting the stage for AIDS. Although this theory is far from proven, it at least suggests new therapeutic routes to explore in treating AIDS: giving a person the cytokines that Th1 cells normally secrete, or suppressing Th2 cell activity.

A second major type of T cell is a *cytotoxic T cell,* which recognizes nonself antigens that cancerous cells or virally infected cells display on their surfaces near certain MHC proteins. A cytotoxic T cell becomes activated when it fits and combines with an antigen that fits its receptors. Then, the T cell proliferates, enlarging its clone. Cytotoxic T cells then bind to the surfaces of antigen-bearing cells, where they release a protein called **perforin** that cuts porelike openings, destroying these cells. In this way, cytotoxic T cells continually monitor the body's cells, recognizing and eliminating tumor cells and cells infected with viruses. Another type of cell, called a **natural killer**

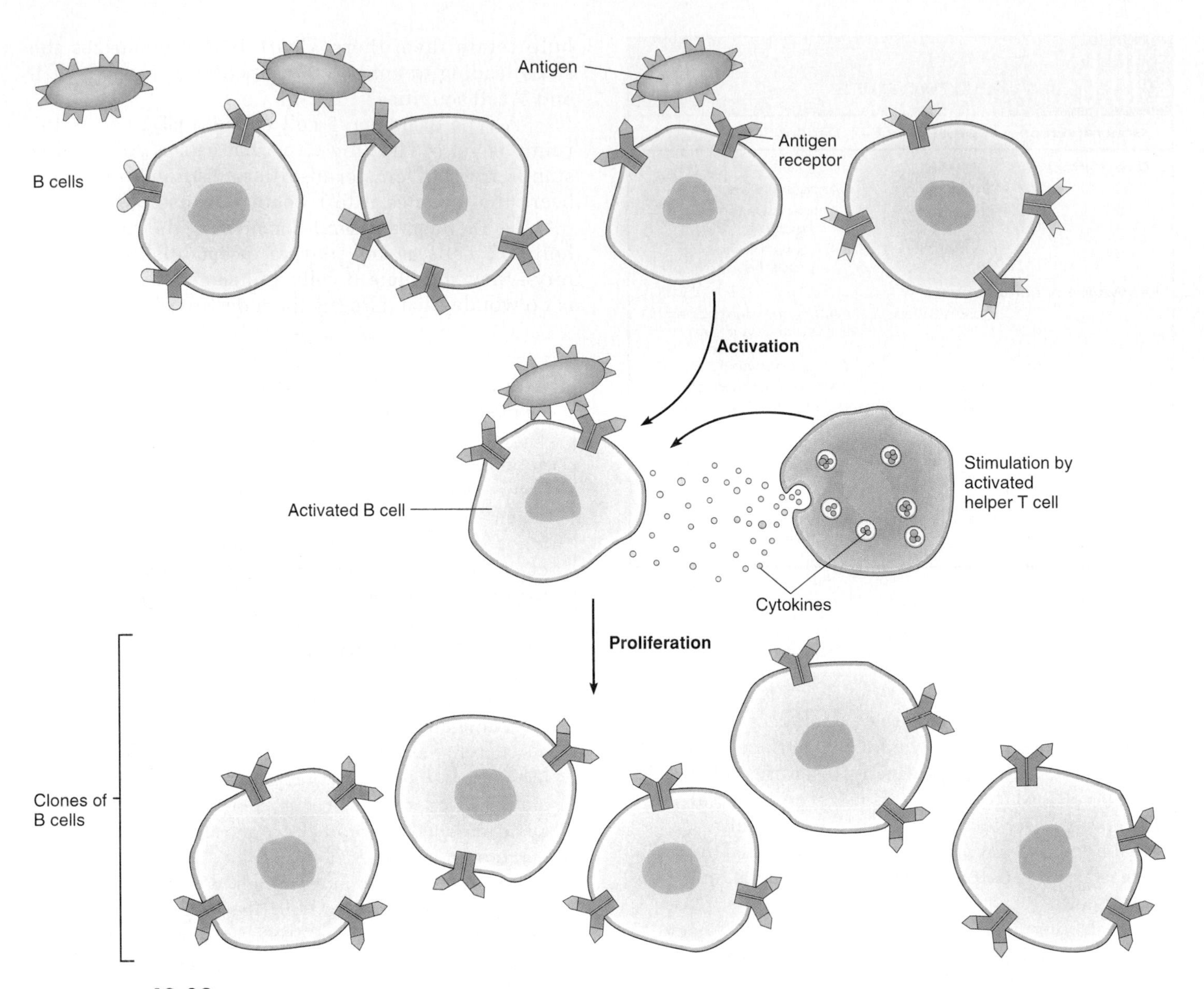


**FIGURE 16.20**

When a B cell encounters an antigen that fits its antigen receptor, it becomes activated and proliferates, thus enlarging its clone. Note that all cells in the clone have the correct antigen receptor (an IgD molecule).

**cell,** also uses perforin to destroy tumor cells. Both cytotoxic T cells and natural killer cells have other ways to lyse cells too.

Immunologists have hypothesized the existence of a third type of T cell, a *suppressor T cell* that inhibits the activities of both other T cells and B cells once an infection is controlled. However, such cells have not been isolated. Instead, researchers now believe that certain cytokines shut off the immune response.

## Immune Responses

When B cells or T cells become activated after first encountering the antigens for which they are specialized to react, their actions constitute a **primary immune response.** During such a response, plasma cells release antibodies (IgM, followed by IgG) into the lymph (fig. 16.22). The antibodies are transported to the blood and then throughout the body, where they help destroy antigen-bearing agents. Production and release of antibodies continues for several weeks.

Following a primary immune response, some of the B cells produced during proliferation of the clone remain dormant and serve as *memory cells* (fig. 16.22). If the identical antigen is encountered in the future, the clones of these memory cells enlarge, and they can respond rapidly with IgG to the antigen to which they were previously sensitized. Such a subsequent reaction is called a **secondary immune response.** Cells in

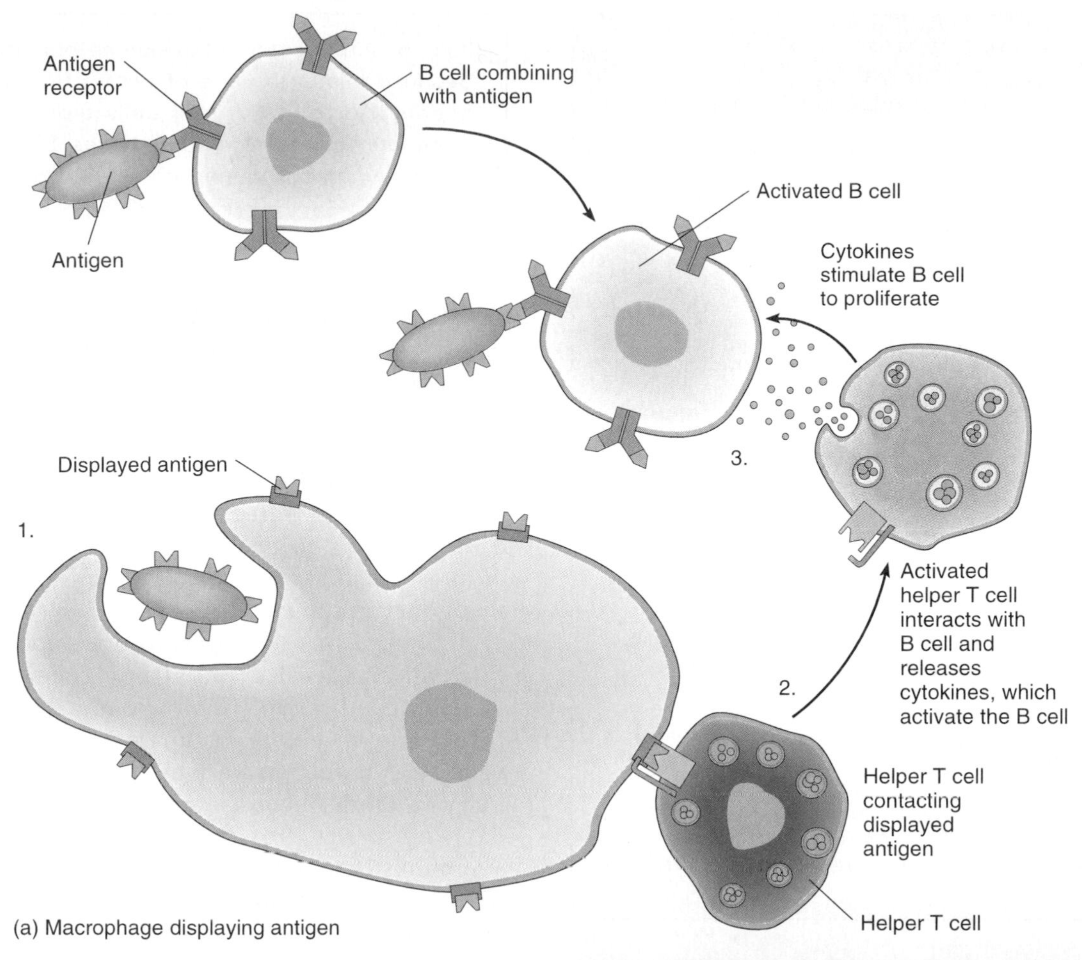


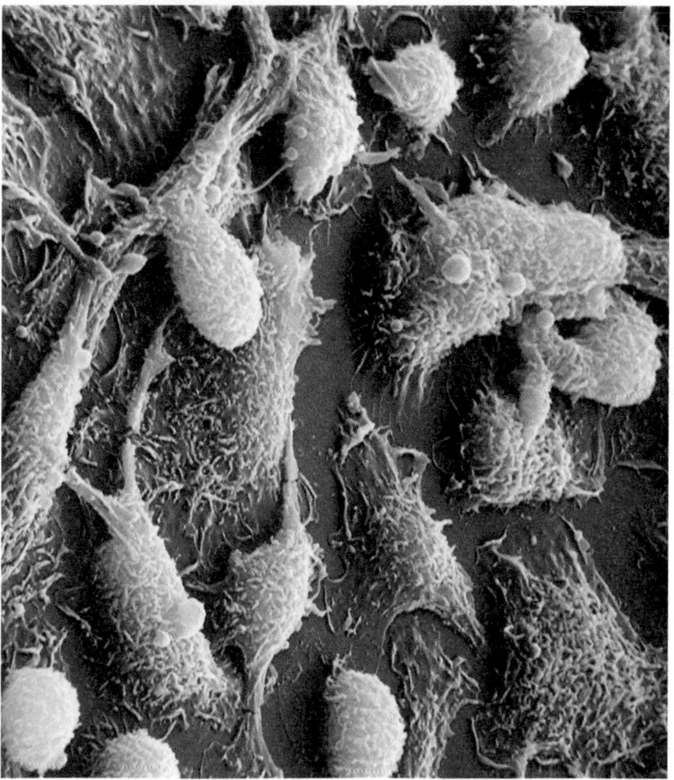

(b)

**FIGURE 16.21**

(*a*) 1. After digesting antigen-bearing agents, a macrophage displays antigens on its surface. 2. Helper T cells become activated when they contact displayed antigens that fit their antigen receptors. 3. An activated helper T cell interacts with a B cell that has combined with an identical antigen and causes the B cell to proliferate. (*b*) Macrophages bind to lymphocytes. During an infection, macrophages bind to helper T cells, activating them to trigger other immune defenses. Here, the round cells are helper T cells and the cells bearing projections are macrophages (1,040×).

## CHART 16.7 Steps in Antibody Production

**B Cell Activity**

1. Antigen-bearing agents enter tissues.
2. B cell becomes activated when it encounters an antigen that fits its antigen receptors, either alone or in conjunction with helper T cells.
3. Activated B cell proliferates, enlarging its clone.
4. Some of the newly formed B cells differentiate further to become plasma cells.
5. Plasma cells synthesize and secrete antibodies whose molecular structure is similar to the activated B cell's antigen receptors.
6. Antibodies combine with antigen-bearing agents, helping to destroy them.

**T Cell Activity**

1. Antigen-bearing agents enter tissues.
2. Accessory cell, such as a macrophage, phagocytizes antigen-bearing agent, and the macrophage's lysosomes digest the agent.
3. Antigens from the digested antigen-bearing agents are displayed on the surface membrane of the accessory cell.
4. Helper T cell becomes activated when it encounters a displayed antigen that fits its antigen receptors.
5. Activated helper T cell releases cytokines when it encounters a B cell that has previously combined with an identical antigen-bearing agent.
6. Cytokines stimulate the B cell to proliferate.
7. Some of the newly formed B cells differentiate into antibody-secreting plasma cells.
8. Antibodies combine with antigen-bearing agents, helping to destroy them.

lymph nodes called **follicular dendritic cells** may help memory by harboring and slowly releasing viral antigens after an initial infection. This constantly stimulates memory B cells, which present the antigens to T cells, maintaining immunity.

As a result of a primary immune response, detectable concentrations of antibodies usually appear in the body fluids within five to ten days following an exposure to antigens. If the identical antigen is encountered some time later, a secondary immune response may produce additional antibodies within a day or two (fig. 16.23). Although newly formed antibodies may persist in the body for only a few months, or perhaps a few years, memory cells live much longer. Consequently, the ability to produce a secondary immune response may be long-lasting.

1. How do T cells become activated?
2. What is the function of cytokines?
3. How do cytotoxic T cells destroy antigen-bearing cells?
4. What is the difference between a primary and a secondary immune response?

It is not surprising that the skin harbors many T cells, because this is the site of entry of many infectious agents. Additionally, skin cells (keratinocytes) secrete interleukin-1, which activates T cells. The skin may also contain antigen-presenting cells.

## Types of Immunity

A generation ago, it was common at certain times of the year for grade school classrooms to be nearly empty, due to several infectious "childhood diseases," including measles, mumps, rubella, and chicken pox. However, each child usually suffered each illness only once, thanks to *naturally acquired active immunity.* This follows a primary immune response and occurs as a result of exposure to a live pathogen and development of symptoms.

Today, most children in developed countries do not contract measles, mumps, or rubella because they develop another type of active immunity, produced in response to receiving a preparation called a **vaccine.** A vaccine contains an antigen that can stimulate a primary immune response against a particular disease-causing agent, but does not produce the severe symptoms of that disease.

A vaccine might contain bacteria or viruses that have been killed or attenuated (weakened) so that they cannot cause a serious infection, or it may contain a toxoid (toxin of an infectious organism that has been chemically altered to destroy its toxic effects). A vaccine may even consist of a single glycoprotein or similar large molecule from the pathogen's surface. This may be sufficient evidence of a foreign antigen to alert the immune system. A vaccine causes a person to develop *artificially acquired active immunity.*

Vaccines are available to stimulate active immunity against a variety of diseases, including typhoid fever, cholera, whooping cough, diphtheria, tetanus, polio, chicken pox, measles (rubeola), German measles (rubella), mumps, influenza, hepatitis B, and bacterial pneumonia. Vaccines have virtually eliminated smallpox from the world, and greatly diminished the number of cases of poliomyelitis and infection by *Haemophilus influenzae* type B, which causes meningitis in children. Unfortunately, distribution of vaccines is not equitable worldwide. Many thousands of people living in underdeveloped countries die of infectious diseases for which vaccines are widely available in other nations.

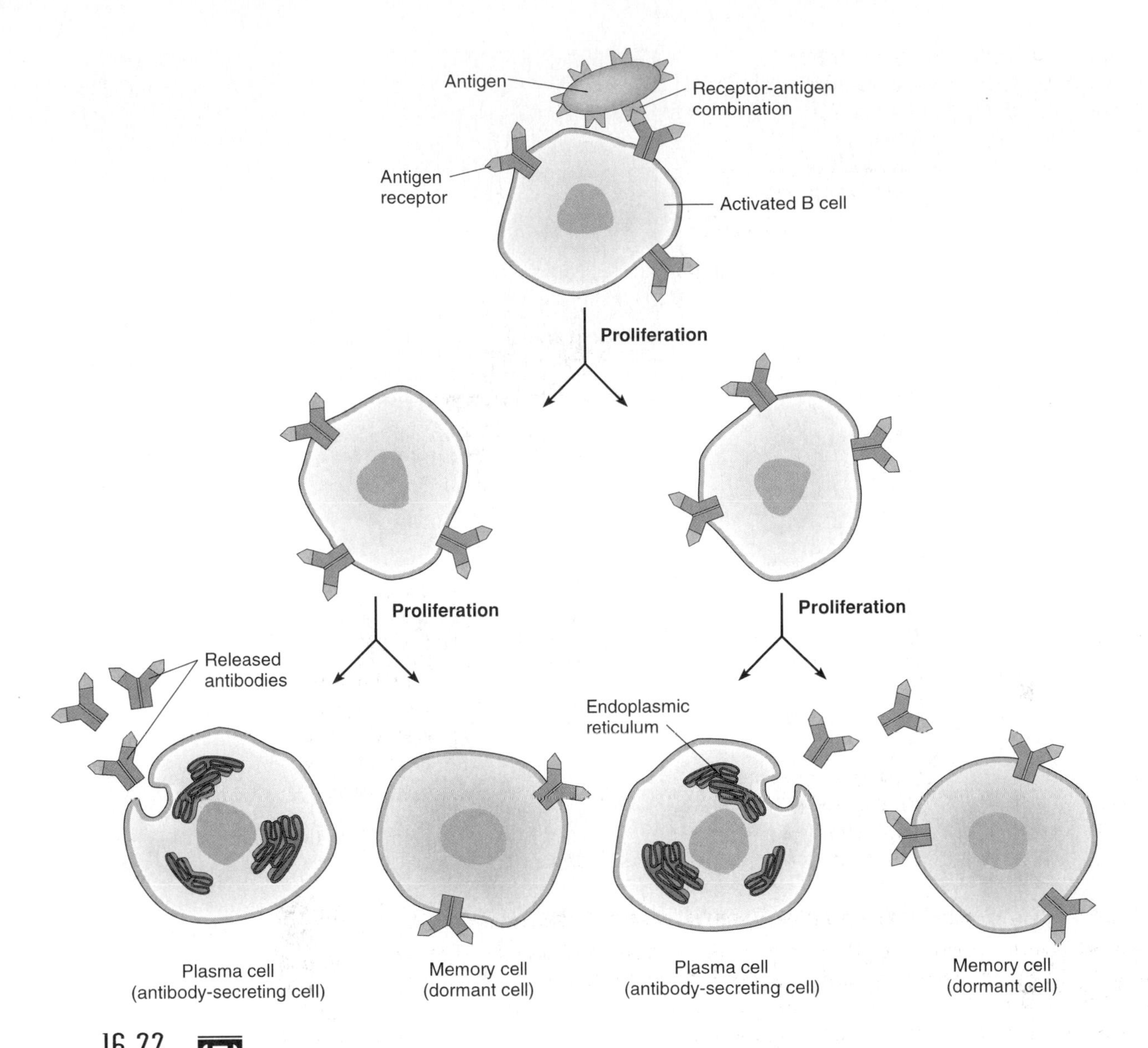


**FIGURE 16.22**
Activated B cells proliferate, giving rise to antibody-secreting plasma cells and dormant memory cells.

Viruses that rapidly mutate (change) present a great challenge to vaccine development. It is a little like fighting an enemy who is constantly changing disguises. For this reason, a new vaccine against influenza must be developed each year. HIV, the virus that causes AIDS, is particularly changeable, which has severely hampered efforts to produce a vaccine.

Sometimes a person who has been exposed to infection needs protection against a disease-causing microorganism, but lacks the time needed to develop active immunity. This happens with hepatitis, a viral infection of the liver. In such a case, it may be possible to inject the person with antiserum (ready-made antibodies). These antibodies may be obtained from gamma globulin separated from the blood plasma of persons who have already developed immunity against the particular disease.

An injection of gamma globulin provides *artificially acquired passive immunity.* This type of immunity is called passive because the recipient's cells do not produce the antibodies. Such immunity is relatively short-term, seldom lasting more than a few weeks. Furthermore, because the recipient's lymphocytes were not involved in fighting against the pathogens for which protection was needed, susceptibility to those pathogens persists in the future.

During pregnancy, certain antibodies (IgG) pass from the maternal blood into the fetal bloodstream.

Receptor-mediated endocytosis (see chapter 3) utilizing receptor sites on cells of the fetal yolk sac accomplishes the transfer (see chapter 23). These receptor sites bind to a region common to the structure of IgG molecules. After entering the fetal cells, the antibodies are secreted into the fetal blood. As a result, the fetus acquires a limited amount of immunity against the pathogens for which the pregnant woman has developed active immunities. The fetus thus has *naturally acquired passive immunity,* which may remain effective for six months to a year after birth.

Chart 16.8 summarizes the types of immunity.

Competence of the immune system tends to decline with advancing age. Consequently, elderly persons have a higher incidence of tumors and an increased susceptibility to infections, such as influenza, pneumonia, and tuberculosis. This decline in effectiveness is primarily due to a loss of T cells. B cell activity, on the other hand, changes little with age.

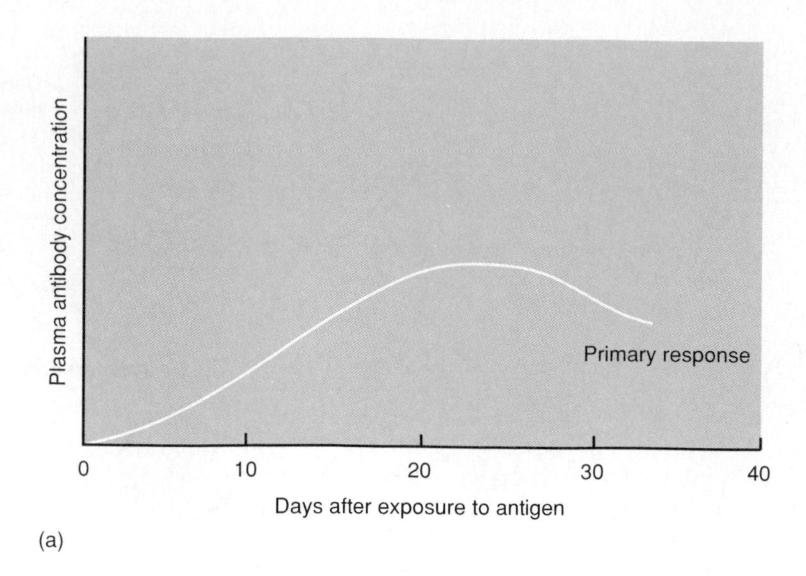


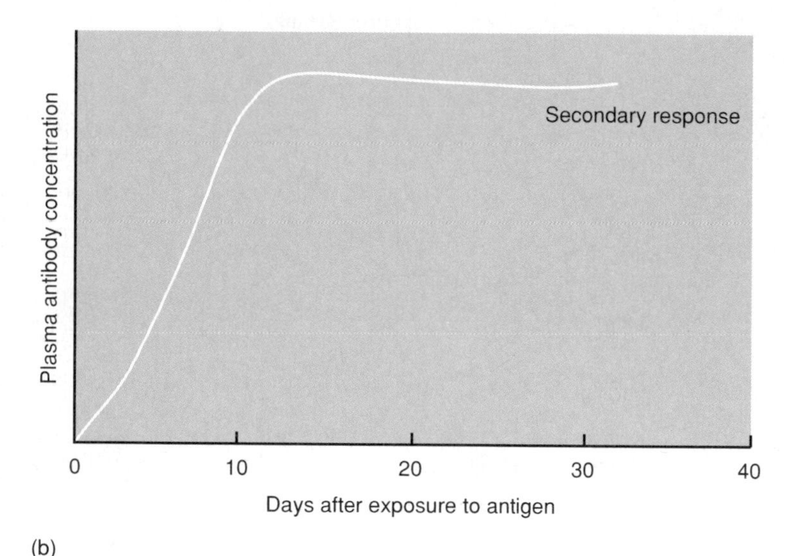


**FIGURE 16.23**

(*a*) A primary immune response produces a lesser concentration of antibodies than does (*b*) a secondary immune response.

## Allergic Reactions

Allergic reactions are closely related to immune responses in that both may involve the sensitizing of lymphocytes or the combining of antigens with antibodies. Allergic reactions, however, can damage tissues. An allergy is also called a hypersensitivity reaction. One form of allergic reaction can occur in almost anyone, but another form affects only those people who have inherited an ability to produce exaggerated immune responses. The antigens that trigger allergic responses are called **allergens.**

A *delayed-reaction allergy* (type IV) may occur in anyone. It results from repeated exposure of the skin to certain chemicals—commonly, household or industrial chemicals or some cosmetics. After repeated contacts, the presence of the foreign substance activates T cells, a large number of which collect in the skin. The T cells and the macrophages they attract release chemical factors, which, in turn, cause eruptions and inflammation of the skin (dermatitis). This reaction is called *delayed* because it usually takes about 48 hours to occur.

Hypersensitivities that take 1 to 3 hours to develop include *antibody-dependent cytotoxic reactions* (type II) and *immune complex reactions* (type III). Antibody-dependent cytotoxic reactions are those in which an antigen binds to a specific body cell, stimulating phagocytosis and complement-mediated lysis of

## CHART 16.8 TYPES OF IMMUNITY

| Type | Stimulus | Result |
|---|---|---|
| Naturally acquired active immunity | Exposure to live pathogens | Symptoms of a disease and stimulation of an immune response |
| Artificially acquired active immunity | Exposure to a vaccine containing weakened or dead pathogens or their components | Stimulation of an immune response without the severe symptoms of a disease |
| Artificially acquired passive immunity | Injection of gamma globulin containing antibodies | Immunity for a short time without stimulating an immune response |
| Naturally acquired passive immunity | Antibodies passed to fetus from pregnant woman with active immunity | Short-term immunity for infant, without stimulating an immune response |

the antigen. A transfusion reaction to mismatched blood is a type II hypersensitivity reaction. In an immune complex reaction (type III), widespread antigen-antibody complexes cannot be cleared. **Autoimmunity,** the loss of the ability to tolerate self antigens, illustrates this type of hypersensitivity reaction.

The *tuberculin skin test* is used to detect individuals who have tuberculosis (TB) or who have had it (or a closely related infection) or been exposed to it in the past. The test uses a tuberculin preparation called *purified protein derivative* (PPD), which is introduced into the superficial layers of the test subject's skin (Mantoux test). If the person's T cells have been sensitized to the antigens of the mycobacteria that cause tuberculosis, an allergic reaction (positive test result) occurs within 48 to 72 hours. In a positive reaction, a localized region of the skin and subcutaneous tissue hardens (indurates). The absence of this reaction (negative result) signifies that the person's T cells have not previously been exposed to the mycobacterial antigens.

In other cases, an allergic reaction may occur within minutes after contact with a nonself substance. Persons with this *immediate-reaction allergy* (type I or anaphylactic) have an inherited ability to overproduce IgE antibodies in response to certain antigens. (IgE normally comprises a minute fraction of plasma proteins.) In this instance, the allergic reaction activates B cells.

In an immediate-reaction allergy, the B cells become sensitized when the allergen is first encountered, and subsequent exposures trigger allergic reactions. In the initial exposure, IgE attaches to the membranes of widely distributed mast cells and basophils. When a subsequent allergen-antibody reaction occurs, these cells release allergy mediators such as *histamine, prostaglandin $D_2$,* and *leukotrienes* (fig. 16.24). These substances cause a variety of physiological effects, including dilation of blood vessels, increased vascular permeability that swells tissues, contraction of bronchial and intestinal smooth muscles, and increased mucus production. The result is a severe inflammation reaction that is responsible for the symptoms of the allergy, such as hives, hay fever, asthma, eczema, or gastric disturbances.

Anaphylactic shock is a severe form of immediate-reaction allergy, in which mast cells release allergy mediators throughout the body. The person may at first feel an inexplicable apprehension, and then suddenly the entire body itches and breaks out in red hives. Vomiting and diarrhea may follow. The face, tongue, and larynx begin to swell, and breathing becomes difficult. Unless the person receives an injection of epinephrine (adrenaline) and sometimes a tracheotomy (an incision into the windpipe so that breathing is restored), he or she will lose consciousness and may die within 5 minutes. Anaphylactic shock most often results from an allergy to penicillin or insect stings. Fortunately, thanks to prompt medical attention and people who know they have allergies avoiding the allergens, fewer than 100 people a year actually die of anaphylactic shock.

Anaphylactic shock seems to be a severe response to something as innocuous as an insect bite or penicillin. One theory of the origin of allergies, particularly anaphylactic shock, is that they evolved at a time when insect bites and the natural substances from which antibiotics such as penicillin are made threatened human survival. The observation that IgE protects against roundworm and flatworm infections, in addition to taking part in allergic reactions, supports the idea that this antibody class is a holdover from times past, when immunity requirements might have been different from what they are today.

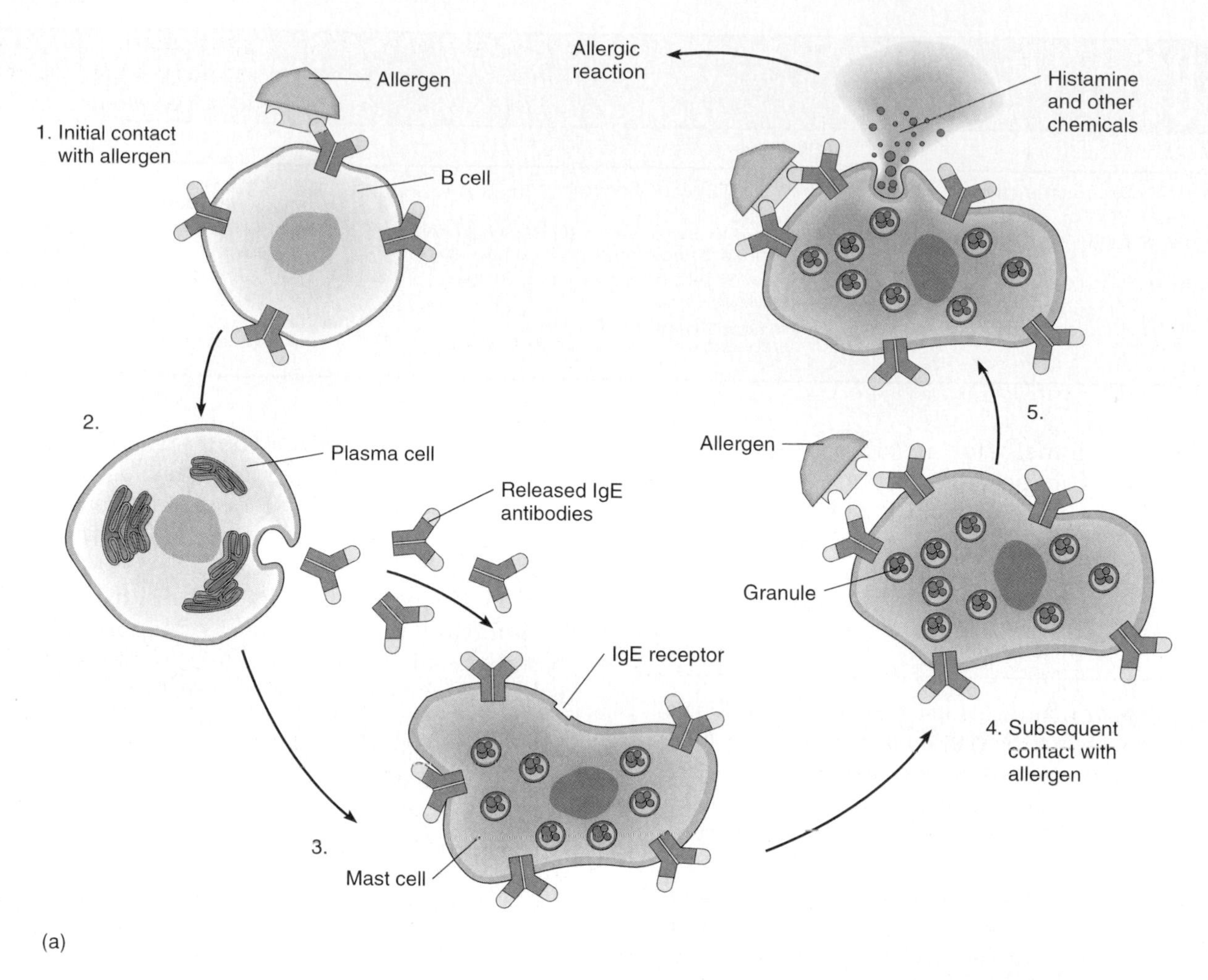


(a)

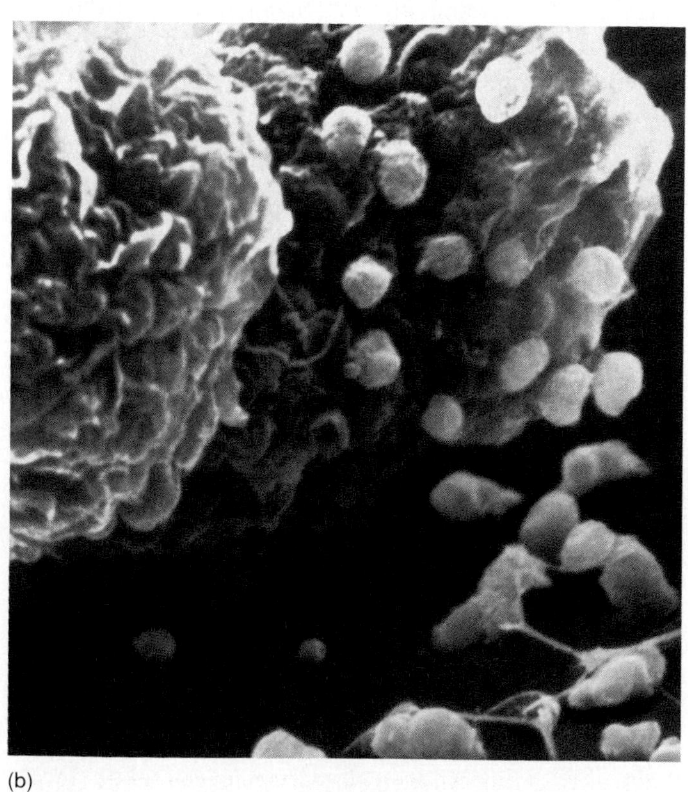

(b)

**FIGURE 16.24**

(*a*) Immediate-reaction allergy. 1. B cells are activated when they contact an allergen. 2. An activated B cell differentiates into an antibody-secreting plasma cell. 3. Antibodies attach to mast cells. 4. When allergens are encountered, they combine with the antibodies on the mast cells. 5. The mast cells release allergy mediators, which cause the symptoms of the allergy attack. (*b*) A mast cell releases histamine granules (3,000×).

## Transplantation and Tissue Rejection

When a car breaks down, replacing the damaged or malfunctioning part often fixes the trouble. The same is sometimes true for the human body. Transplanted tissues and organs include corneas, kidneys, pieces of skin, livers, and even hearts. The danger the immune system poses to transplanted tissue is that the recipient's cells may recognize the donor's tissues as foreign and attempt to destroy the transplanted tissue. Such a response is called a **tissue rejection reaction.**

Tissue rejection resembles the cell-mediated response against a nonself antigen. The greater the antigenic difference between the cell surface molecules of the recipient tissues and the donor tissues, the more rapid and severe the rejection reaction. Matching donor and recipient tissues can minimize the rejection reaction. This means locating a donor whose tissues are antigenically similar to those of the person needing a transplant—a procedure much like matching the blood of a donor with that of a recipient before giving a blood transfusion.

The four major varieties of grafts (transplant tissue) include:

- *Isograft.* Tissue is taken from a genetically identical twin.
- *Autograft.* Tissue is taken from elsewhere in a person's body, as in the case of skin grafts.
- *Allograft.* Tissue comes from an individual who is not genetically identical to the recipient, but of the same species.
- *Xenograft.* Tissue comes from a different species.

*Major histocompatibility complex* is a general term for cell surface proteins in many species. The human version of the MHC produces human leukocyte antigens, or HLA types, which are patterns of proteins on white blood cells. HLA antigens are "typed" to determine the chances of acceptance of transplanted tissue. HLA typing is also used in forensics to match the pattern of cell surface proteins between a suspect and tissue left at a crime scene, and to establish biological relationships, such as between a child and several men who may be the father.

*Immunosuppressive drugs* are used to reduce rejection of transplanted tissues. These drugs interfere with the recipient's immune response by suppressing formation of antibodies or production of T cells, thereby reducing the humoral and cellular responses. Unfortunately, the use of immunosuppressive drugs leaves the recipient relatively unprotected against infections. It is not uncommon for a patient to survive a transplant but die of infection because of a weakened immune system.

Less drastic than an organ transplant is a cell implant, consisting of small pieces of tissue. Implants of liver cells may treat cirrhosis; pancreas cells may treat diabetes; skeletal muscle cells may replace heart muscle damaged in a heart attack or treat muscular dystrophy; adrenal gland cells may treat Parkinson's disease; and brain cells may treat Alzheimer's and Huntington diseases.

Several biotechnology companies produce skin, blood vessel, and other tissue replacements that consist of biocompatible polymers seeded with cells that secrete tissue-specific biochemicals. Antigens are removed from the cells so that they do not stimulate the immune system. In another approach, a person can donate part of his or her liver to a relative in need of this vital organ. Many people already store their own blood for future use. Stem cells taken from a newborn's umbilical cord can be stored and then cultured later to provide healthy bone marrow stem cells, either to the donor or a compatible recipient.

Fetal tissue, which is less likely to evoke immune rejection than adult tissue, is a possible source of transplant material. A source of fetal tissues is from aborted healthy fetuses, but there is concern that use of fetal tissue in medicine will encourage abortion. Fetuses lost in miscarriage are very often chromosomally abnormal, and therefore not suitable as donor tissue.

An alternative approach is to use cells from the yolk sac surrounding an embryo. These cells can be coaxed to develop as any of several cell types and are not rejected.

## Autoimmunity

Sometimes the immune system backfires, manufacturing **autoantibodies** that attack the body's own cells, causing autoimmune disorders. The specific nature of an autoimmune disorder depends upon the cell types that are the target of the immune attack. Chart 16.9 lists some autoimmune disorders.

Why might the immune system attack body tissues? Perhaps a virus, while replicating within a human cell, "borrows" proteins from the host cell's surface and incorporates them onto its own surface. When the immune system "learns" the surface of the virus to destroy it, it also learns to attack the human cells that normally bear the particular protein. Another explanation of autoimmunity is that somehow T cells escape their "education" in the thymus, never learning to distinguish self from nonself.

## CHART 16.9 AUTOIMMUNE DISORDERS

| Disorder | Symptoms | Antibodies against |
|---|---|---|
| Glomerulonephritis | Lower back pain | Kidney cell antigens that resemble streptococcal bacteria antigens |
| Grave's disease | Restlessness, weight loss, irritability, increased heart rate and blood pressure | Thyroid gland antigens near thyroid-stimulating hormone receptor, causing overactivity |
| Juvenile diabetes | Thirst, hunger, weakness, emaciation | Pancreatic beta cells |
| Hemolytic anemia | Fatigue and weakness | Red blood cells |
| Myasthenia gravis | Muscle weakness | Receptors for nerve messages on skeletal muscle |
| Pernicious anemia | Fatigue and weakness | Binding site for vitamin B on cells lining stomach |
| Rheumatic fever | Weakness, shortness of breath | Heart cell antigens that resemble streptococcal bacteria antigens |
| Rheumatoid arthritis | Joint pain and deformity | Cells lining joints |
| Scleroderma | Thick, hard, pigmented skin patches | Connective tissue cells |
| Systemic lupus erythematosus | Red rash on face, prolonged fever, weakness, kidney damage | DNA, neurons, blood cells |
| Ulcerative colitis | Lower abdominal pain | Colon cells |

A third possible route of autoimmunity is when a nonself antigen coincidentally resembles a self antigen. This may explain juvenile diabetes, which is a deficiency of insulin, which is needed to transport glucose in the blood into cells that use it. Juvenile diabetes strikes 1 in 500 people under the age of 20—they must inject insulin several times a day.

Part of a protein on insulin-producing cells matches part of bovine serum albumin (BSA), which is a protein in cow's milk. Might children with an allergy to cow's milk develop antibodies against BSA, which later on attack the similar-appearing pancreas cells, causing juvenile diabetes? An ongoing study is tracking the health of children with family histories of juvenile diabetes who have avoided drinking cow's milk, compared to similarly high-risk children who have drunk cow's milk.

Some 2 million to 5 million people in the United States suffer from a poorly understood immune system imbalance called chronic fatigue syndrome. The condition begins suddenly, producing fatigue so great that getting out of bed is an effort. Chills, fever, sore throat, swollen glands, muscle and joint pain, and headaches are also symptoms.

The various disabling aches and pains reflect an overactive immune system. These people have up to forty times the normal amount of interleukin-2, and too many cytotoxic T cells, yet too little interferon. It is as if the immune system mounts a defense, and then doesn't know when to shut it off. The culprit behind chronic fatigue syndrome seems to be a virus, but none has yet been identified that is common to all sufferers.

Clinical Application 16.2 discusses the immunity breakdown known as AIDS.

1. Explain the difference between active and passive immunities.
2. In what ways is an allergic reaction related to an immune reaction?
3. In what ways is a tissue rejection reaction related to an immune response?
4. How is autoimmunity an abnormal functioning of the immune response?

## CLINICAL TERMS RELATED TO THE LYMPHATIC SYSTEM AND IMMUNITY

**asplenia** (ah-sple′ne-ah) The absence of a spleen.

**immunocompetence** (im″u-no-kom′pe-tens) The ability to produce an immune response to the presence of antigens.

**immunodeficiency** (im″u-no-de-fish′en-se) A lack of the ability to produce an immune response.

**lymphadenectomy** (lim-fad″ĕ-nek′to-me) The surgical removal of lymph nodes.

**lymphadenopathy** (lim-fad″ĕ-nop′ah-the) Enlargement of the lymph nodes.

**lymphadenotomy** (lim-fad″ĕ-not′o-me) An incision of a lymph node.

**lymphocytopenia** (lim″fo-si″to-pe′ne-ah) An abnormally low concentration of lymphocytes in the blood.

**lymphocytosis** (lim″fo-si″to′sis) An abnormally high concentration of lymphocytes in the blood.

## InnerConnections
## Lymphatic System

The lymphatic system is an important link between the interstitial fluid and the plasma; it also plays a major role in the response to infection

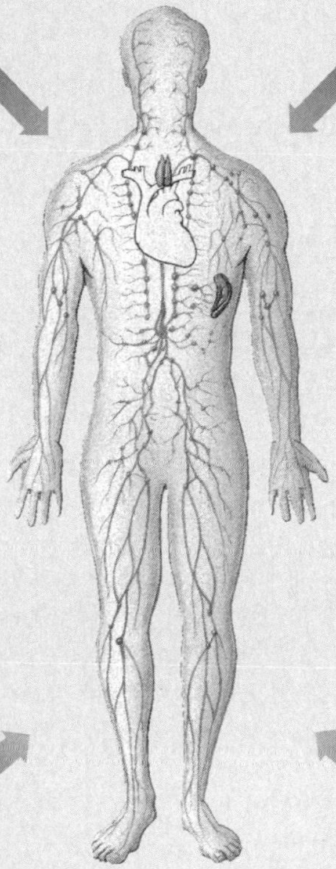

**Reproductive system**

Special mechanisms inhibit the female immune system in its attack of sperm as foreign invaders

**Integumentary system**

The skin provides a first line of defense against infection

**Urinary system**

The kidneys control the volume of extracellular fluid, including lymph

**Skeletal system**

Cells of the immune system originate in the bone marrow

**Respiratory system**

Cells of the immune system patrol the respiratory system to defend against infection

**Muscular system**

Muscle action helps pump lymph through the lymphatic vessels

**Digestive system**

Lymph plays a major role in the absorption of fats

**Nervous system**

Stress may impair the immune response

**Cardiovascular system**

Tissue fluids are returned to the bloodstream by the lymphatic system

Lymph originates as interstitial fluid, formed by the action of blood pressure

**Endocrine system**

Hormones stimulate lymphocyte production

**CLINICAL APPLICATION** 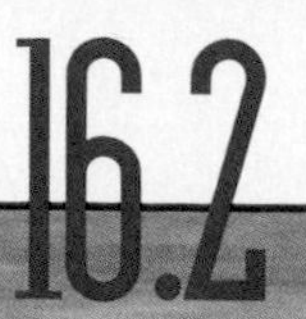

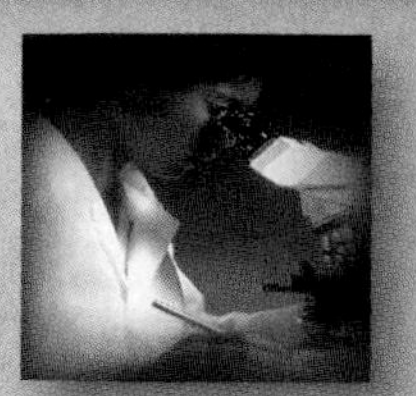

# Immunity Breakdown: AIDS

## NATURAL HISTORY OF A MODERN PLAGUE

In late 1981 and early 1982, physicians from large cities in the United States began reporting to the Centers for Disease Control and Prevention cases of formerly rare infections in otherwise healthy young men. Some of the infections were prevalent in the general population, such as herpes simplex and cytomegalovirus, but in these young men were unusually severe. Oddly, some of the infections were caused by organisms known to infect only nonhuman animals. Other infections, particularly pneumonia caused by the microorganism *Pneumocystis carinii* and a cancer, Kaposi's sarcoma, were known only in individuals whose immune systems were suppressed (fig. 16B).

The bodies of the sick young men had become nesting places for all types of infectious agents, including viruses, bacteria, protozoans, and fungi. The infections were *opportunistic,* which means that they take advantage of a weakened immune system.

As the infections spread, a portrait of a lethal disease emerged. *Acquired immunodeficiency syndrome,* or AIDS, starts with recurrent fever, weakness, and weight loss. Then usually after a relatively healthy period, infections begin. Many patients die within 2 or 3 years of symptom onset, but the virus that causes AIDS can be present for a decade or longer efore a person feels ill. Five percent infected people have remained for more than 15 years .

ientists and physicians entify the source of the outbreak of infections. Attention turned first to the sufferers. What did they have in common? All had had blood contact with another AIDS patient. Many in the United States and Europe were homosexual men who transferred the infections through fragile rectal tissues during anal intercourse. Intravenous drug users spread AIDS by sharing needles. A growing AIDS population includes children who acquired the disease before or during birth from infected mothers. Today, worldwide, the most common mode of transmission is heterosexual intercourse.

Some people acquired AIDS from HIV-tainted blood transfused before 1985, when stricter blood banking precautions began. Many people with hemophilia were infected when they received blood products pooled from many donors. The toll on people with hemophilia has been staggering—90% of those who used clotting factors several times a month have AIDS, as do 70% of those who have ever received clotting factors. Today, half of all people with hemophilia are infected with HIV. Blood samples meticulously stored from people with hemophilia since the mid-1970s have greatly aided researchers in tracing the birth of the epidemic.

## HIV

By 1983, researchers linked the human immunodeficiency virus, HIV, to AIDS. It targets lymphocytes in lymph nodes, and CD4 helper T cells. In most cases, HIV is transmitted through blood or semen.

HIV is a *retrovirus,* which means that its genetic material is RNA. The virus attaches to CD4 receptors on helper T cells, and sends in its RNA. A viral enzyme, *reverse transcriptase,* catalyzes construction of a DNA strand complementary to the viral RNA. The initial viral DNA strand replicates to form a DNA double helix, which enters the T cell's nucleus.

Using the invaded cell's protein synthetic machinery, the virus replicates, filling the cell with HIV RNA and proteins. Not only can the dying T cell no longer release cytokines or stimulate B cells to manufacture antibodies, but it bursts, unleashing many new HIV particles. HIV replicates at an astounding rate, producing 3 billion new particles a day, from the start of infection. Immune system cells can remove only 100 million to a billion per day, so soon the system becomes overwhelmed. Opportunistic infections set in. A danger sign in HIV infected people is a falling CD4 T cell count, signaling collapse of the coordination of the immune response (chart 16A).

## TREATMENT

AIDS treatment includes treating individual opportunistic infections and cancers as they arise, and using palliative measures to make life more comfortable. A drug called azidothymidine (AZT) can delay symptom onset, but once symptoms begin they tend to be more severe. AZT does not extend survival, and its side effects may be intolerable. However, it does sharply decrease rate of transmission from a pregnant woman to a fetus. AZT is normally given to people who are infected with HIV and whose CD4 helper T cell counts dip below 500/mm$^3$. It inhibits the ability of HIV to replicate. Many other AIDS drugs are in development and testing stages, including existing drugs such as acyclovir, interferon, thalidomide, and amphotericin B, as well as many new compounds. Because HIV mutates into drug-resistant forms rapidly, from the very beginning of infection, a combination of drugs may be the best approach to treatment.

The key to conquering AIDS may lie in the long-term survivors and in a few children whose HIV infections have cleared. These children had HIV genetic material present during infancy, but do not have it several years later, and are healthy. By learning how their immune systems function, and how the virus is impaired, we may finally be able to defeat this foe.

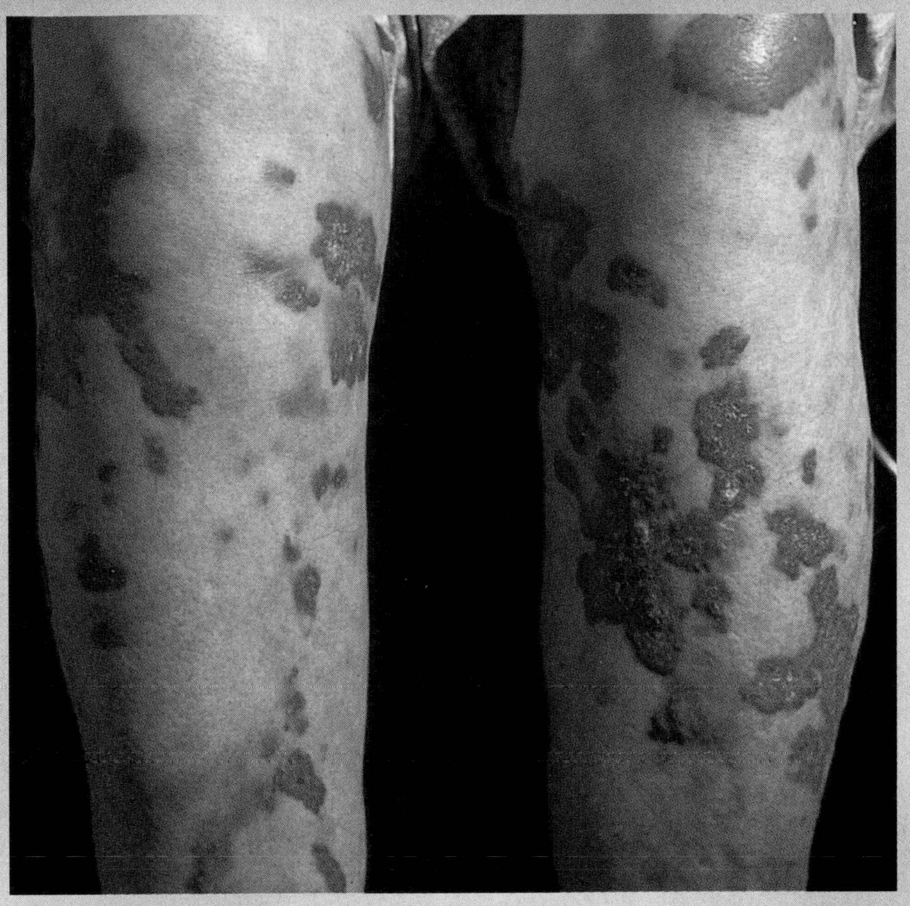

**FIGURE 16B**
Prior to the appearance of AIDS, Kaposi's sarcoma was a rare cancer seen only in elderly Jewish and Italian men and in people with suppressed immune systems. In these groups it produces purplish patches on the lower limbs, but in AIDS patients, Kaposi's sarcoma patches appear all over the body and sometimes internally too. These lower limbs display characteristic lesions.

**CHART 16A WALTER REED CLASSIFICATION* OF HIV INFECTION**

| Stage | Characteristics | Approximate Time For Progression to Next Stage |
|---|---|---|
| 0 | Exposure to HIV | Months to years |
| 1 | Documented presence of HIV | 6 weeks–1 year |
| 2 | Chronically swollen lymph nodes (lymphadenopathy) | 3–5 years |
| 3 | Lymphadenopathy may disappear; helper T cell count falls below 400/$mm^3$ | 18 months |
| 4 | Impaired cell-mediated immunity, as indicated by failure to respond to three of four skin tests measuring delayed hypersensitivity | 1 year |
| 5 | Total lack of delayed hypersensitivity; thrush and other viral or fungal infections of skin and mucous membranes | 7 months |
| 6 | Systemic immune deficiency, as indicated by rampant opportunistic infections; neurological impairment | 1 year |

*Since the Walter Reed classification was developed in 1986, several cases have arisen of individuals who remain in the early stages for extended periods, perhaps even permanently. Most individuals, however, follow the stages outlined here.

**lymphoma** (lim-fo′mah) A tumor composed of lymphatic tissue.
**lymphosarcoma** (lim″fo-sar-ko′mah) A cancer within the lymphatic tissue.
**splenectomy** (sple-nek′to-me) The surgical removal of the spleen.
**splenitis** (sple-ni′tis) An inflammation of the spleen.
**splenomegaly** (sple″no-meg′ah-le) An abnormal enlargement of the spleen.
**splenotomy** (sple-not′o-me) An incision of the spleen.
**thymectomy** (thi-mek′to-me) The surgical removal of the thymus.
**thymitis** (thi-mi′tis) An inflammation of the thymus.

## Chapter Summary

### Introduction (page 635)

The lymphatic system is closely associated with the cardiovascular system. It transports excess fluid to the bloodstream and helps defend the body against disease-causing agents.

### Lymphatic Pathways (page 635)

1. Lymphatic capillaries
   a. Lymphatic capillaries are microscopic, closed-ended tubes that extend into interstitial spaces.
   b. They receive lymph through their thin walls.
   c. Lacteals are lymphatic capillaries in the villi of the small intestine.
2. Lymphatic vessels
   a. Lymphatic vessels are formed by the merging of lymphatic capillaries.
   b. They have walls similar to veins, only thinner, and possess valves that prevent backflow of lymph.
   c. Larger lymphatic vessels lead to lymph nodes and then merge into lymphatic trunks.
3. Lymphatic trunks and collecting ducts
   a. Lymphatic trunks drain lymph from relatively large body regions.
   b. Trunks lead to two collecting ducts within the thorax.
   c. Collecting ducts join the subclavian veins.

### Tissue Fluid and Lymph (page 637)

1. Tissue fluid formation
   a. Tissue fluid originates from blood plasma and includes water and dissolved substances that have passed through the capillary wall.
   b. It generally lacks large proteins, but some smaller proteins leak into interstitial spaces.
   c. As the protein concentration of tissue fluid increases, osmotic pressure increases.
2. Lymph formation
   a. Rising osmotic pressure in tissue fluid interferes with return of water to the blood capillaries.
   b. Increasing pressure within interstitial spaces forces some tissue fluid into lymphatic capillaries, and this fluid becomes lymph.
3. Lymph function
   a. Lymph returns protein molecules to the bloodstream.
   b. It transports foreign particles to the lymph nodes.

### Lymph Movement (page 639)

1. Lymph flow
   a. Lymph is under low pressure and may not flow readily without external aid.
   b. Lymph is moved by the squeezing of skeletal muscles and low pressure in the thorax created by breathing movements.
2. Obstruction of lymph movement
   a. Any condition that interferes with the flow of lymph results in edema.
   b. Obstruction of lymphatic vessels due to surgery causes edema.

### Lymph Nodes (page 640)

1. Structure of a lymph node
   a. Lymph nodes are usually bean-shaped, with blood vessels, nerves, and efferent lymphatic vessels attached to the indented region: afferent lymphatic vessels enter at points on the convex surface.
   b. Lymph nodes are enclosed in connective tissue that extends into the nodes and subdivides them into nodules.
   c. Nodules contain masses of lymphocytes and macrophages, as well as spaces through which lymph flows.
2. Locations of lymph nodes
   a. Lymph nodes aggregate in groups or chains along the paths of larger lymphatic vessels.
   b. They occur primarily in cervical, axillary, and inguinal regions, and within the pelvic, abdominal, and thoracic cavities.
3. Functions of lymph nodes
   a. Lymph nodes filter potentially harmful foreign particles from the lymph before it is returned to the bloodstream.
   b. Lymph nodes are centers for production of lymphocytes that act against foreign particles.
   c. They contain macrophages that remove foreign particles from lymph.

### Thymus and Spleen (page 642)

1. Thymus
   a. The thymus is a soft, bilobed organ located within the mediastinum.
   b. It slowly shrinks after puberty.
   c. It is composed of lymphatic tissue subdivided into lobules.

d. Lobules contain lymphocytes, most of which are inactive, that develop from precursor cells in bone marrow.
e. Some lymphocytes leave the thymus and provide immunity.
f. The thymus secretes thymosin, which stimulates lymphocytes that have migrated to other lymphatic tissues.

2. Spleen
   a. The spleen is located in the upper left portion of the abdominal cavity.
   b. It resembles a large lymph node that is encapsulated and subdivided into lobules by connective tissue.
   c. Spaces within lobules are filled with blood.
   d. The spleen contains many macrophages and lymphocytes, which filter foreign particles and damaged red blood cells from the blood.

## Body Defenses Against Infection (page 643)

The presence and multiplication of pathogens causes infection. Pathogens include bacteria, complex single-celled organisms, fungi, and viruses. An infection may be present without immediately causing symptoms. The body has nonspecific and specific defenses against infection.

## Nonspecific Defenses (page 645)

1. Species resistance
   Each species is resistant to certain diseases that may affect other species but is susceptible to diseases other species may resist.
2. Mechanical barriers
   a. Mechanical barriers include skin and mucous membranes.
   b. Intact mechanical barriers prevent entrance of some pathogens.
3. Chemical barriers
   a. Enzymes in gastric juice and tears kill some pathogens.
   b. Low pH in the stomach prevents growth of some bacteria.
   c. High salt concentration in perspiration kills some bacteria.
   d. Interferons block proliferation of viruses, stimulate phagocytosis, and enhance activity of cells that help resist infections and stifle tumor growth.
4. Fever
   a. Viral or bacterial infection stimulates certain lymphocytes to secrete endogenous pyrogen, which temporarily raises body temperature.
   b. Higher body temperature and the resulting decrease in blood iron level and increased phagocytic activity hamper infection.
5. Inflammation
   a. Inflammation is a tissue response to damage, injury, or infection.
   b. The response includes localized redness, swelling, heat, and pain.
   c. Chemicals released by damaged tissues attract white blood cells to the site.
   d. Clotting may occur in body fluids that accumulate in affected tissues.
   e. Fibrous connective tissue may form a sac around the injured tissue and thus prevent the spread of pathogens.
6. Phagocytosis
   a. The most active phagocytes in blood are neutrophils and monocytes; monocytes give rise to macrophages, which remain fixed in tissues.
   b. Phagocytic cells associated with the linings of blood vessels in the bone marrow, liver, spleen, and lymph nodes constitute the reticuloendothelial tissue.
   c. Phagocytes remove foreign particles from tissues and body fluids.

## Specific Defenses (Immunity) (page 646)

1. Lymphocyte origins
   a. Lymphocytes originate in red bone marrow and are released into the blood before they differentiate.
   b. Some reach the thymus where they mature into T cells.
   c. Others, the B cells, mature in the bone marrow.
   d. Both T cells and B cells reside in lymphatic tissues and organs.
2. Antigens
   a. Before birth, body cells inventory "self" proteins and other large molecules.
   b. After inventory, lymphocytes develop receptors that allow them to differentiate between foreign and self antigens.
   c. Nonself (foreign) antigens combine with T cell and B cell surface receptors and stimulate these cells to cause an immune reaction.
   d. Haptens are small molecules that can combine with larger ones, becoming antigenic.
3. Lymphocyte functions
   a. Some T cells interact with antigen-bearing agents directly, providing cell-mediated immunity.
   b. T cells secrete cytokines, such as interleukins, that enhance cellular responses to antigens.
   c. T cells may also secrete substances that are toxic to their target cells.
   d. B cells interact with antigen-bearing agents indirectly, providing antibody-mediated immunity (or humoral immunity).
4. Lymphocyte clones
   a. There are millions of varieties of T and B cells.
   b. The members of each variety respond only to a specific antigen.
   c. As a group, the members of each variety form a clone.
5. B cell activation
   a. A B cell is activated when it encounters an antigen that fits its antigen receptors.
   b. An activated B cell proliferates, enlarging its clone.
   c. Some activated B cells specialize into antibody-producing plasma cells.
   d. Antibodies react against the antigen-bearing agent that stimulated their production.
   e. An individual's diverse B cells defend against a very large number of pathogens.

6. Antibody molecules
    a. Antibodies are proteins called immunoglobulins.
    b. They constitute the gamma globulin fraction of plasma.
    c. Each immunoglobulin molecule consists of four chains of amino acids linked together.
    d. Variable regions at the ends of these chains are specialized into antigen binding sites to react with antigens.
7. Types of immunoglobulins
    a. The five major types of immunoglobulins are IgG, IgA, IgM, IgD, and IgE.
    b. IgG, IgA, and IgM make up most of the circulating antibodies.
8. Antibody actions
    a. Antibodies attack antigens directly, activate complement, or stimulate local tissue changes that are unfavorable to antigen-bearing agents.
    b. Direct attacks occur by means of agglutination, precipitation, or neutralization.
    c. Activated proteins of complement attract phagocytes, alter cells so they become more susceptible to phagocytosis, and cause lysis by rupturing foreign cell membranes.
9. T cell types
    a. T cells are activated when an antigen-presenting cell displays a foreign antigen.
    b. When a macrophage acts as an accessory cell, it phagocytizes an antigen-bearing agent, digests the agent, and displays the antigens on its surface membrane in association with certain MHC proteins.
    c. A helper T cell becomes activated when it encounters displayed antigens for which it is specialized to react.
    d. CD4 helper T cells stimulate antibody-mediated and cell-mediated immunity. HIV cripples these cells.
    e. An activated helper T cell contacts a B cell that carries the foreign antigen which the T cell previously encountered on an antigen-presenting cell. In response, the T cell secretes cytokines, stimulates B cell proliferation, and attracts macrophages.
    f. Cytotoxic T cells recognize foreign antigens on tumor cells and cells whose surfaces indicate that they are infected by viruses. Stimulated cytotoxic T cells secrete perforin to destroy its target cells.
10. Immune responses
    a. When B cells or T cells first encounter an antigen for which they are specialized to react, the reaction is called a primary immune response.
        (1) During this response, antibodies are produced for several weeks.
        (2) Some B cells remain dormant as memory cells, aided by follicular dendritic cells.
    b. A secondary immune response occurs rapidly if the same antigen is encountered later.
11. Types of immunity
    a. A person who encounters a pathogen and has a primary immune response develops naturally acquired active immunity.
    b. A person who receives a vaccine containing a dead or weakened pathogen, or part of it, develops artificially acquired active immunity.
    c. A person who receives an injection of antibodies has artificially acquired passive immunity.
    d. When antibodies pass through a placental membrane from a pregnant woman to her fetus, the fetus develops naturally acquired passive immunity.
    e. Active immunity lasts much longer than passive immunity.
12. Allergic reactions
    a. Allergic or hypersensitivity reactions are excessive immune responses and may damage tissue.
    b. Delayed-reaction allergy, which can occur in anyone and inflame the skin, results from repeated exposure to antigens.
    c. Antibody-dependent cytotoxic allergic reactions occur when blood transfusions are mismatched.
    d. Immune complex allergic reactions involve autoimmunity, which is an immune reaction against self antigens.
    e. Immediate-reaction allergy is an inborn ability to overproduce IgE.
    f. Allergic reactions result from mast cells bursting and releasing allergy mediators such as histamine and serotonin.
    g. In anaphylactic shock, allergy mediators flood the body, causing severe symptoms.
13. Transplantation and tissue rejection
    a. A transplant recipient's immune system may react against the donated tissue in a tissue rejection reaction.
    b. Matching donor and recipient tissues and using immunosuppressive drugs can minimize tissue rejection.
    c. Immunosuppressive drugs may increase susceptibility to infection.
    d. Transplants may take place between genetically identical twins, from one body part to another, between unrelated individuals of the same species, or between individuals of different species.
14. Autoimmunity
    a. In autoimmune disorders, the immune system manufactures autoantibodies that attack body tissues.
    b. Autoimmune disorders may result from a previous viral infection, faulty T cell development, or reaction to a nonself antigen that resembles a self antigen.

**Explorations in Human Anatomy and Physiology CD-ROM**
The modules accompanying Chapter Sixteen are #12 Immune Response and #13 AIDS.

## Critical Thinking Questions

1. Based on your understanding of the functions of lymph nodes, how would you explain the fact that enlarged nodes are often removed for microscopic examination as an aid to diagnosing certain diseases?
2. Why is it true that an injection into the skin is, to a large extent, an injection into the lymphatic system?
3. Explain why vaccination provides long-lasting protection against a disease, while gamma globulin provides only short-term protection.
4. When a breast is removed surgically for the treatment of breast cancer, the lymph nodes in the nearby axillary region are sometimes excised also. Why is this procedure likely to be followed by swelling of the upper limb on the treated side?
5. If an infant was found to be lacking a thymus because of a developmental disorder, what could be predicted about the infant's susceptibility to infections? Why?
6. A patient sees a doctor because of fatigue, night sweats (awaking drenched in perspiration), and weight loss.
   a. What questions should the physician ask this patient?
   b. What other symptoms might the physician look for or ask about?
   c. When the physician orders a blood test, what, specifically, should be examined?

## Review Exercises

1. Explain how the lymphatic system is related to the cardiovascular system.
2. Trace the general pathway of lymph from the interstitial spaces to the bloodstream.
3. Identify and describe the locations of the major lymphatic trunks and collecting ducts.
4. Distinguish between tissue fluid and lymph.
5. Describe the primary functions of lymph.
6. Explain why physical exercise promotes lymphatic circulation.
7. Explain how a lymphatic obstruction leads to edema.
8. Describe the structure of a lymph node, and list its major functions.
9. Locate the major body regions occupied by lymph nodes.
10. Describe the structure and functions of the thymus.
11. Describe the structure and functions of the spleen.
12. Distinguish between specific and nonspecific body defenses against infection.
13. Explain what is meant by *species resistance.*
14. Name two mechanical barriers to infection.
15. Describe how enzymatic actions function as defense mechanisms.
16. Define *interferon,* and explain its action.
17. List the major symptoms of inflammation, and explain why each occurs.
18. Identify the major phagocytic cells in the blood and other tissues.
19. Explain where the reticuloendothelial tissue is located.
20. Review the origin of T cells and B cells.
21. Distinguish between an antigen and an antibody.
22. Define *hapten.*
23. Explain what is meant by *cell-mediated immunity.*
24. Define *cytokine.*
25. Explain what is meant by *antibody-mediated immunity.*
26. Define *clone of lymphocytes.*
27. Explain how a B cell is activated.
28. Explain the function of plasma cells.
29. Describe an immunoglobulin molecule.
30. Distinguish between the variable region and the constant region of an immunoglobulin molecule.
31. List the major types of immunoglobulins, and describe their main functions.
32. Explain four mechanisms by which antibodies may attack antigens directly.
33. Explain the function of complement.
34. Describe how T cells become activated.
35. Distinguish between a primary and a secondary immune response.
36. Distinguish between active and passive immunity.
37. Define *vaccine.*
38. Explain how a vaccine produces its effect.
39. Describe how a fetus may obtain antibodies from the maternal blood.
40. Explain the relationship between an allergic reaction and an immune response.
41. Distinguish between an antigen and an allergen.
42. List the major events leading to a delayed-reaction allergic response.
43. Describe how an immediate-reaction allergic response may occur.
44. Explain the relationship between a tissue rejection and an immune response.
45. Describe two methods used to reduce the severity of a tissue rejection reaction.
46. Explain the relationship between autoimmunity and an immune response.
47. How do vaccines augment immunity, and monoclonal antibodies target immunity?
48. How do immunosuppressant drugs increase the likelihood of success of a transplant, yet place a patient at a higher risk of developing infections?

17

CHAPTER

# Digestive System

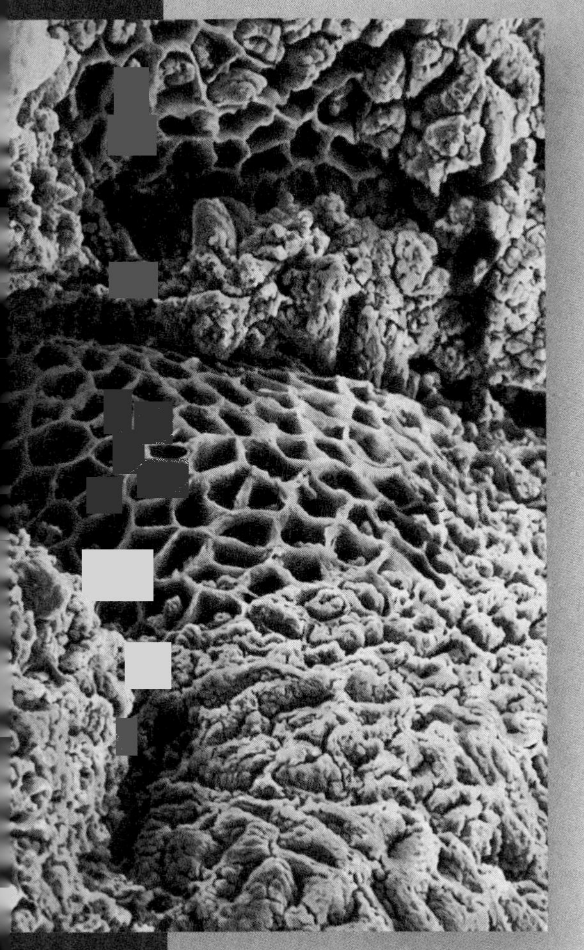

False-colored scanning electron micrograph of stomach lining (greenish) surrounding eroded ulcerated regions (pink) (200×).

## Chapter Objectives

After you have studied this chapter, you should be able to:

1. Name and describe the locations of the organs of the digestive system and their major parts.
2. Describe the general functions of each digestive organ.
3. Describe the structure of the wall of the alimentary canal.
4. Explain how the contents of the alimentary canal are mixed and moved.
5. List the enzymes the various digestive organs and glands secrete and describe the function of each.
6. Describe how digestive secretions are regulated.
7. Explain how digestive reflexes control movement of material through the alimentary canal.
8. Describe the mechanisms of swallowing, vomiting, and defecating.
9. Explain how the products of digestion are absorbed.

## Key Terms

**absorption** (ab-sorp′shun)
**accessory organ** (ak-ses′o-re or′gan)
**alimentary canal** (al″i-men′tar-e kah-nal′)
**bile** (bīl)
**chyme** (kīm)
**circular muscle** (ser′ku-lar mus′el)
**deciduous** (de-sid′u-us)
**feces** (fe′sēz)
**gastric juice** (gas′trik jo͞os)
**intestinal juice** (in-tes′ti-nal jo͞os)
**intrinsic** (in-trin′sik)
**longitudinal muscle** (lon″ji-tu′di-nal mus′el)
**mesentery** (mes′en-ter″e)
**mucous membrane** (mu′kus mem′brān)
**pancreatic juice** (pan″kre-at′ik jo͞os)
**peristalsis** (per″i-stal′sis)
**serous layer** (se′rus la′er)
**sphincter muscle** (sfingk′ter mus′el)
**villi; sing., villus** (vil′i, vil′us)

Unit Five

In late 1982, two young Australian physicians, Barry J. Marshall and J. Robin Warren, performed a daring experiment. They had been studying gastritis (stomach inflammation) and peptic ulcers (sores in the lining of the stomach or small intestine) and had proposed that a bacterium, *Helicobacter pylori,* causes these problems. But because the medical community had for many decades been convinced that ulcers were the result of excess stomach acid due to stress, Marshall and Warren's idea was not very popular.

So they drank a culture of the suspect bacteria—and developed gastritis.

In the decade following Marshall and Warren's publication of the bacteria–gastritis link in early 1983, abundant evidence accumulated that the brash young doctors were right about gastritis. The medical community soon realized that this odd microbe that thrives in the highly acidic environs of the human stomach is also responsible for a majority of peptic ulcers.

By 1994, treatment for gastritis and ulcers finally began to change. Instead of taking drugs for many years to lower stomach acid, sufferers can now often take a combination of already available antibiotic drugs for just two weeks. A once debilitating digestive problem may now, for many people, be a short-lived pain and inconvenience.

**Digestion** is the mechanical and chemical breakdown of foods into forms that cell membranes can absorb. Mechanical digestion breaks large pieces into smaller ones without altering their chemical composition. Chemical digestion breaks food into simpler chemicals. The organs of the **digestive system** carry out these processes. The digestive system consists of the **alimentary canal** extending from the mouth to the anus, and several accessory organs, which release secretions into the canal. The alimentary canal includes the mouth, pharynx, esophagus, stomach, small intestine, large intestine, and anal canal. The accessory organs include the salivary glands, liver, gallbladder, and pancreas. Figure 17.1 shows the major organs of the digestive system.

The digestive system originates from the middle layer (endoderm) of the embryo, which folds to form the tube of the alimentary canal. The accessory organs appear as buds from the tube.

## General Characteristics of the Alimentary Canal

The alimentary canal is a muscular tube about 9 meters long that passes through the body's ventral cavity. The structure of its wall, the method by which it moves food, and its type of innervation are similar throughout its length (fig. 17.2).

### Structure of the Wall

The wall of the alimentary canal consists of four distinct layers that are developed to different degrees from region to region. Although the four-layered structure persists throughout the alimentary canal, certain

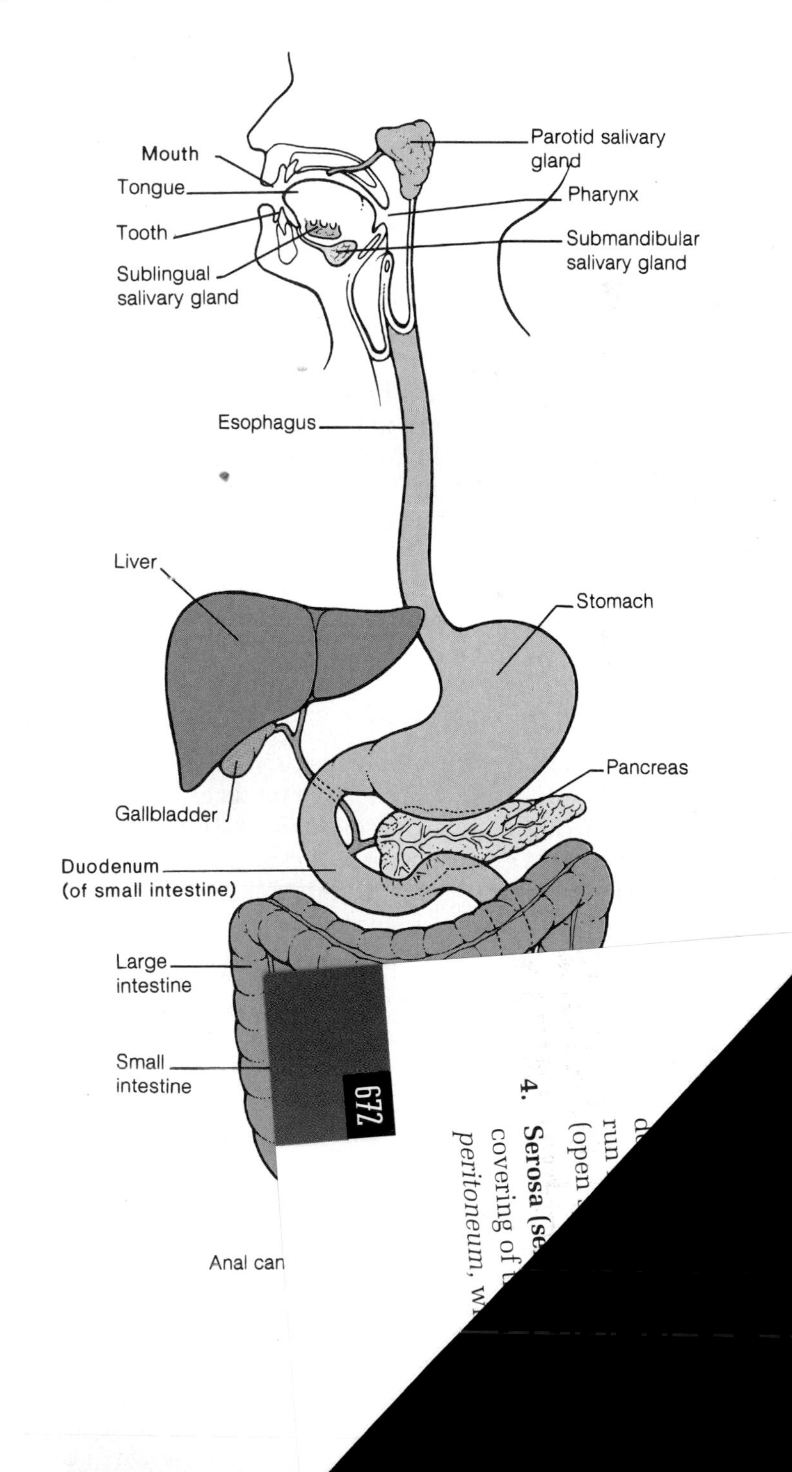


**Figure 17.1**
Major organs of the digestive system.

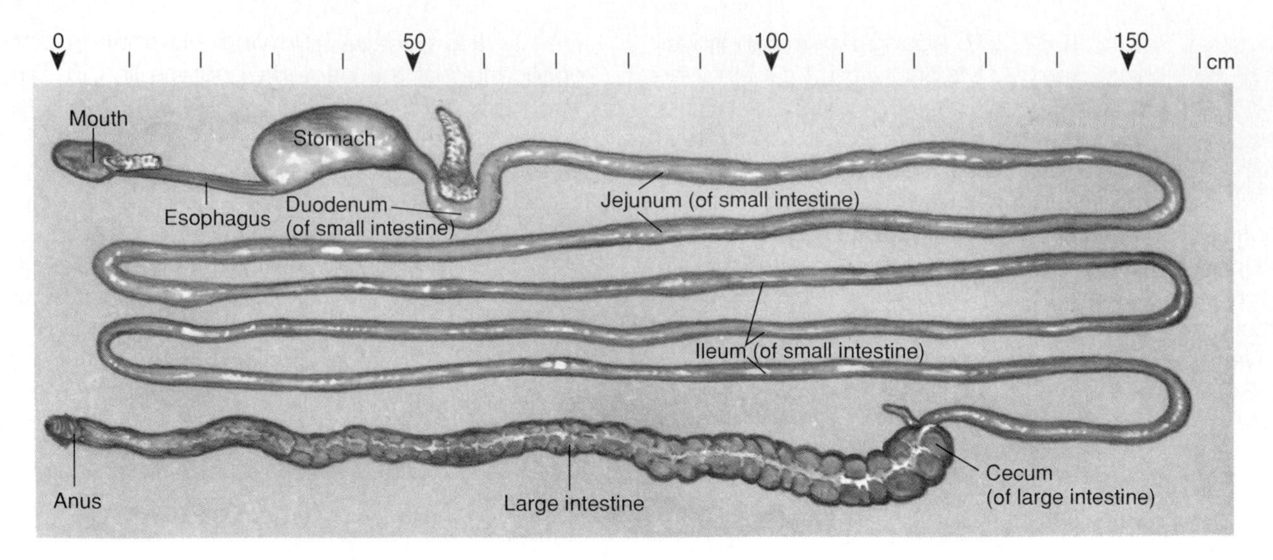


**FIGURE 17.2**
The alimentary canal is a muscular tube about 9 meters long.

regions are specialized for particular functions. Beginning with the innermost tissues, these layers, shown in figure 17.3, include the following:

1. **Mucosa (mucous membrane).** This layer is formed of surface epithelium, underlying connective tissue (lamina propria), and a small amount of smooth muscle. In some regions, it develops folds and tiny projections, which extend into the lumen of the digestive tube and increase its absorptive surface area. It may also contain glands that are tubular invaginations into which the lining cells secrete mucus and digestive enzymes. The mucosa protects the tissues beneath it and carries on absorption and secretion.
2. **Submucosa.** The submucosa contains considerable loose connective tissue as well as glands, blood vessels, lymphatic vessels, and nerves. Its vessels nourish the surrounding tissues and carry away absorbed materials.
3. **Muscular layer.** This layer, which is responsible for the movements of the tube, consists of two coats of smooth muscle tissue. The fibers of the inner coat encircle the tube. When these *circular fibers* (they are actually closed spirals) contract, the diameter of the tube ecreases. The fibers of the outer muscular coat lengthwise. When these *longitudinal fibers* spirals) contract, the tube shortens.
4. rous layer). The serous layer, or outer he tube, is composed of the *visceral* hich is formed of epithelium on the outside and connective tissue beneath. The cells of the serosa protect underlying tissues and secrete serous fluid, which moistens and lubricates the tube's outer surface so that the organs within the abdominal cavity slide freely against one another.

Chart 17.1 summarizes the characteristics of these layers.

## Movements of the Tube

The motor functions of the alimentary canal are of two basic types—*mixing movements* and *propelling movements* (fig. 17.4). Mixing occurs when smooth muscles in relatively small segments of the tube contract rhythmically. For example, when the stomach is full, waves of muscular contractions move along its wall from one end to the other. These waves occur every twenty seconds or so, and their action mixes foods with the digestive juices the mucosa secretes.

Propelling movements include a wavelike motion called **peristalsis.** When peristalsis occurs, a ring of contraction appears in the wall of the tube. At the same time, the muscular wall just ahead of the ring relaxes—a phenomenon called *receptive relaxation.* As the wave moves along, it pushes the tubular contents ahead of it. Peristalsis begins when food expands the tube. It causes the sounds that can be heard through a stethoscope applied to the abdominal wall.

## Innervation of the Tube

Branches of the sympathetic and parasympathetic divisions of the autonomic nervous system extensively innervate the alimentary canal. These nerve fibers,

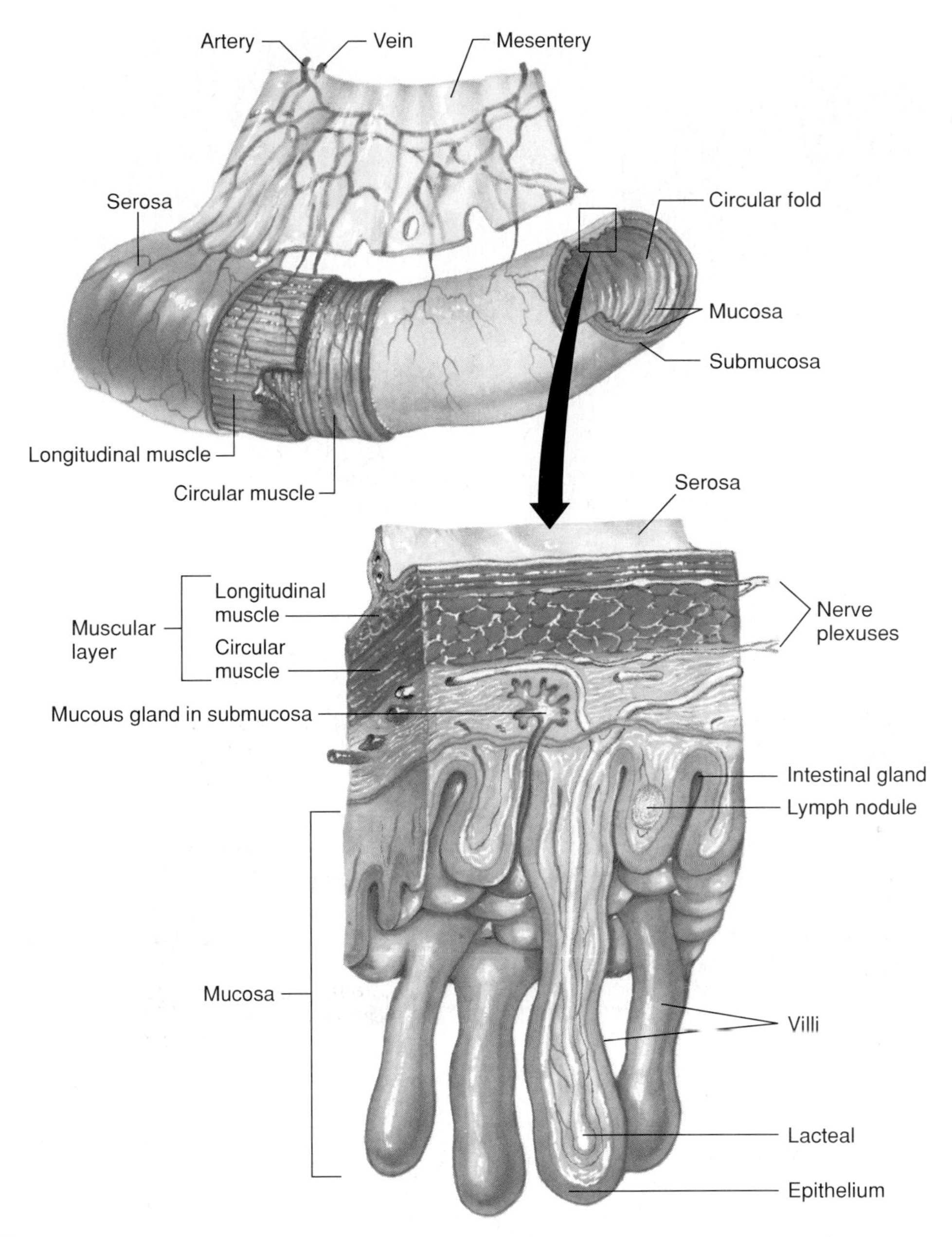


**FIGURE 17.3**

The wall of the small intestine, as in other portions of the alimentary canal, includes four layers: an inner mucosa, a submucosa, a muscular layer, and an outer serosa.

**CHART 17.1 LAYERS OF THE WALL OF THE ALIMENTARY CANAL**

| Layer | Composition | Function |
|---|---|---|
| Mucosa | Epithelium, connective tissue, smooth muscle | Protection, absorption, secretion |
| Submucosa | Loose connective tissue, blood vessels, lymphatic vessels, nerves | Nourishes surrounding tissues, transports absorbed materials |
| Muscular layer | Smooth muscle fibers arranged in circular and longitudinal groups | Movements of the tube and its contents |
| Serosa | Epithelium, connective tissue | Protection, lubrication |

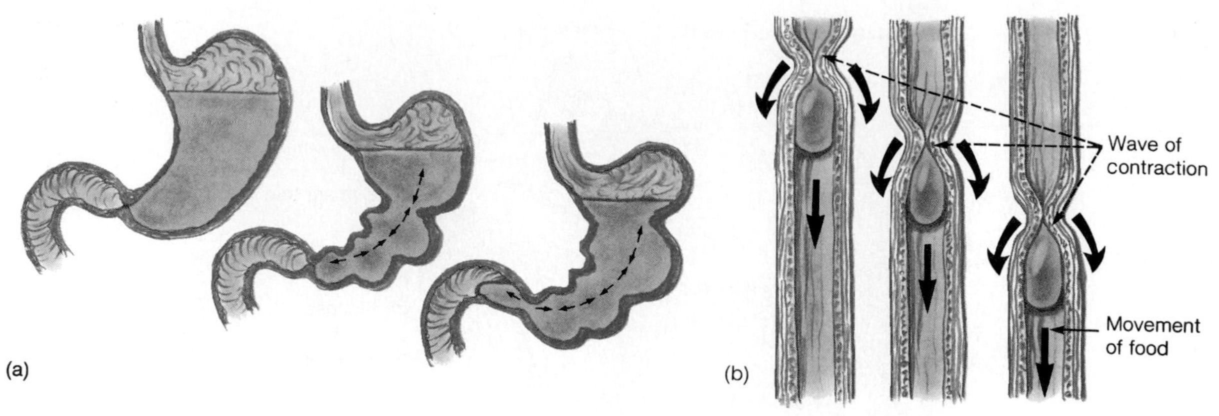


**FIGURE 17.4**

(*a*) Mixing movements occur when small segments of the muscular wall of the alimentary canal rhythmically contract. (*b*) Peristaltic waves move the contents along the canal.

associated mainly with the tube's muscular layer, maintain muscle tone and regulate the strength, rate, and velocity of muscular contractions. Many of the postganglionic fibers synapse with a network or plexus of neurons within the wall of the canal (see figure 17.3).

Parasympathetic impulses generally increase the activities of the digestive system. Some of these impulses originate in the brain and are conducted through branches of the vagus nerves to the esophagus, stomach, pancreas, gallbladder, small intestine, and proximal half of the large intestine. Other parasympathetic impulses arise in the sacral region of the spinal cord and supply the distal half of the large intestine.

Sympathetic nerve impulses' effects on digestive actions usually are opposite those of the parasympathetic division. That is, sympathetic impulses inhibit various digestive actions. For example, such impulses contract sphincter muscles in the wall of the alimentary canal, blocking movement of materials through the tube.

1. What are the general functions of the digestive system?
2. Which organs constitute the digestive system?
3. Describe the wall of the alimentary canal.
4. Name the two types of movements that occur in the alimentary canal.
5. How do parasympathetic nerve impulses affect digestive actions? What effect do sympathetic nerve impulses have?

## MOUTH

The **mouth,** which is the first portion of the alimentary canal, receives food and begins digestion by mechanically reducing the size of solid particles and mixing them with saliva (mastication). The mouth also functions as an organ of speech and sensory reception. It is surrounded by the lips, cheeks, tongue, and palate, and includes a chamber between the palate and tongue called the *oral cavity,* as well as a narrow space between the teeth, cheeks, and lips called the *vestibule* (fig. 17.5 and reference plates 49 and 55).

### Cheeks and Lips

The **cheeks** form the lateral walls of the mouth. They consist of outer layers of skin, pads of subcutaneous fat, muscles associated with expression and chewing, and inner linings of moist, stratified squamous epithelium.

> Because cheek cells are easily removed, many biology laboratory classes have students collect them and use them to study cell structure. A similar "cheekbrush test" is used to identify people who are carriers of the inherited disorder cystic fibrosis. The patient simply swishes a brush on the inside of the cheek, then the doctor sends the brush to a laboratory. Here, cheek cells are removed from the brush, and the DNA is extracted and analyzed for the presence of the gene variant that causes cystic fibrosis.

The **lips** are highly mobile structures that surround the mouth opening. They contain skeletal muscles and sensory receptors useful in judging the temperature and texture of foods. Their normal reddish color is due to the many blood vessels near their surfaces. The external borders of the lips mark the boundaries between the skin of the face and the mucous membrane that lines the alimentary canal.

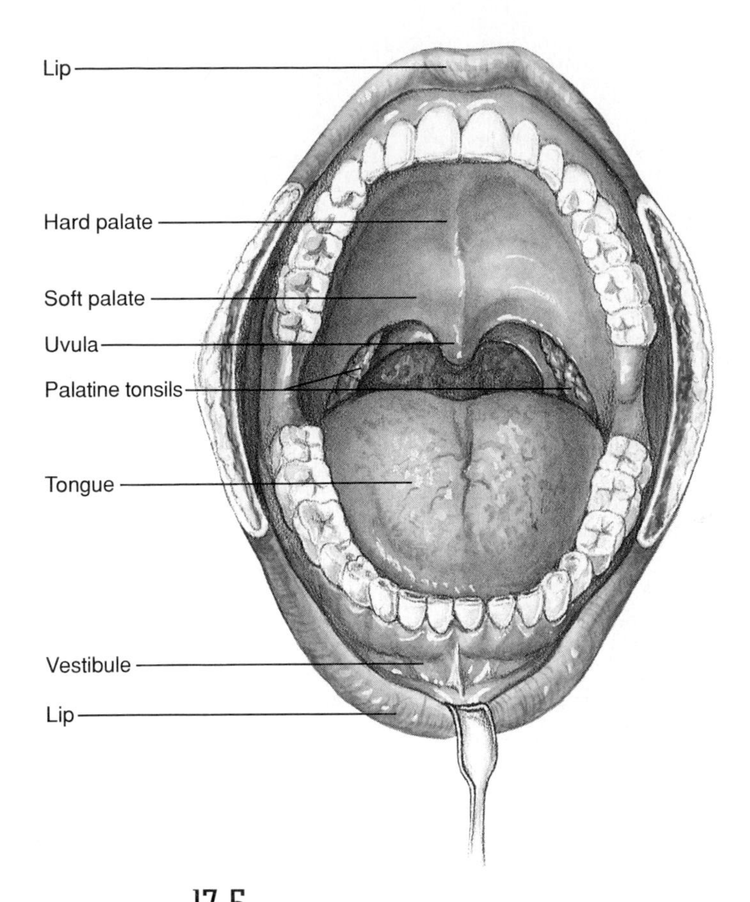


**FIGURE 17.5**
The mouth is adapted for ingesting food and preparing it for digestion.

## Tongue

The **tongue** is a thick, muscular organ that occupies the floor of the mouth and nearly fills the oral cavity when the mouth is closed. It is covered by mucous membrane and is connected in the midline to the floor of the mouth by a membranous fold called the **frenulum.**

The *body* of the tongue is composed largely of skeletal muscle fibers running in several directions. These muscles mix food particles with saliva during chewing and move food toward the pharynx during swallowing. Rough projections, called **papillae,** on the surface of the tongue provide friction, which is useful in handling food. Between these papillae are most of the taste buds. Some taste buds are in the papillae (fig. 17.6).

The posterior region, or *root,* of the tongue is anchored to the hyoid bone and is covered with rounded masses of lymphatic tissue called **lingual tonsils.**

## Palate

The **palate** forms the roof of the oral cavity and consists of a hard anterior part and a soft posterior part. The *hard palate* is formed by the palatine processes of the

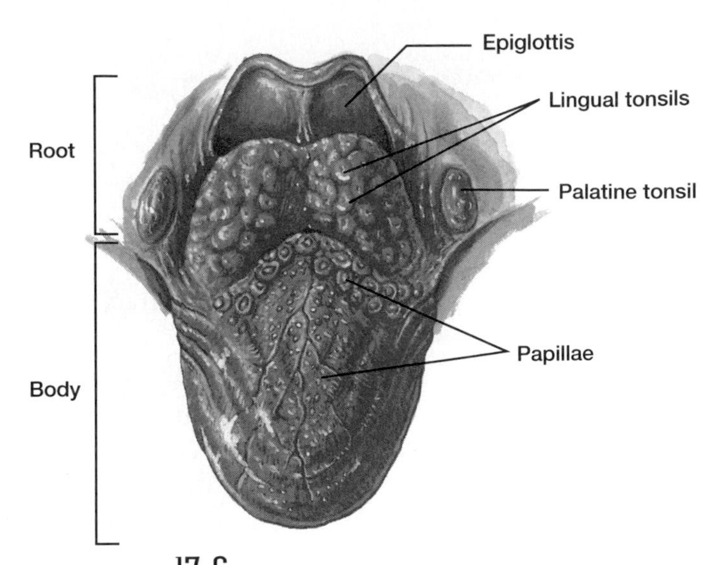


**FIGURE 17.6**
The surface of the tongue, viewed from above.

maxillary bones in front and the horizontal portions of the palatine bones in back. The *soft palate* forms a muscular arch, which extends posteriorly and downward as a cone-shaped projection called the **uvula.**

During swallowing, muscles draw the soft palate and the uvula upward. This action closes the opening between the nasal cavity and the pharynx, preventing food from entering the nasal cavity.

In the back of the mouth, on either side of the tongue and closely associated with the palate, are masses of lymphatic tissue called **palatine tonsils.** These structures lie beneath the epithelial lining of the mouth and, like other lymphatic tissues, help protect the body against infections (see chapter 16).

> The palatine tonsils themselves are common sites of infections, and if they become inflamed, the condition is termed *tonsillitis.* Infected tonsils may become so swollen that they block the passageways of the pharynx and interfere with breathing and swallowing. Because the mucous membranes of the pharynx, auditory tubes, and middle ears are continuous, such an infection may travel from the throat into the middle ears (otitis media).
>
> When tonsillitis occurs repeatedly and does not respond to antibiotic treatment, the tonsils are sometimes removed. This surgical procedure is called *tonsillectomy.* Removing the tonsils is far less common today than it was a generation ago, because we are aware now of their role in immunity.

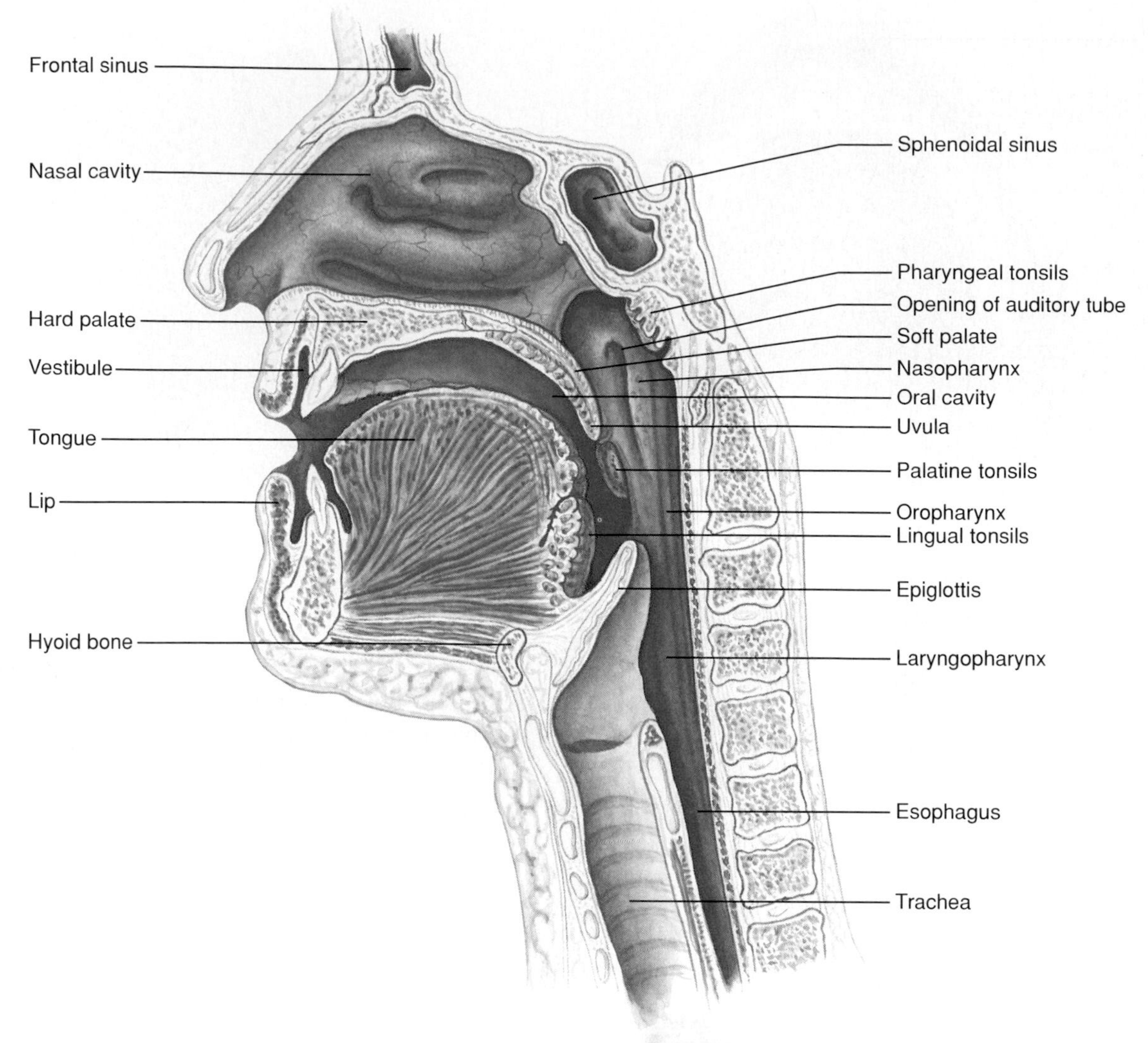


**FIGURE 17.7**
A sagittal section of the mouth, nasal cavity, and pharynx.

Still other masses of lymphatic tissue, called **pharyngeal tonsils,** or *adenoids,* are on the posterior wall of the pharynx, above the border of the soft palate. If the adenoids enlarge and block the passage between the pharynx and the nasal cavity, they also may be removed surgically (fig. 17.7).

1. What is the function of the mouth?
2. How does the tongue function as part of the digestive system?
3. What is the role of the soft palate in swallowing?
4. Where are the tonsils located?

## Teeth

The **teeth** are the hardest structures in the body, but they are not part of the skeletal system. They develop in sockets within the alveolar processes of the mandibular and maxillary bones. Teeth are unique structures in that two sets form during development (fig. 17.8). The members of the first set, the *primary teeth* (deciduous teeth), usually erupt through the gums (gingiva) at regular intervals between the ages of 6 months and 2 to 4 years. The ten primary teeth in each jaw are located from the midline toward the sides in the following sequence: central incisor, lateral incisor, cuspid (canine), first molar, and second molar.

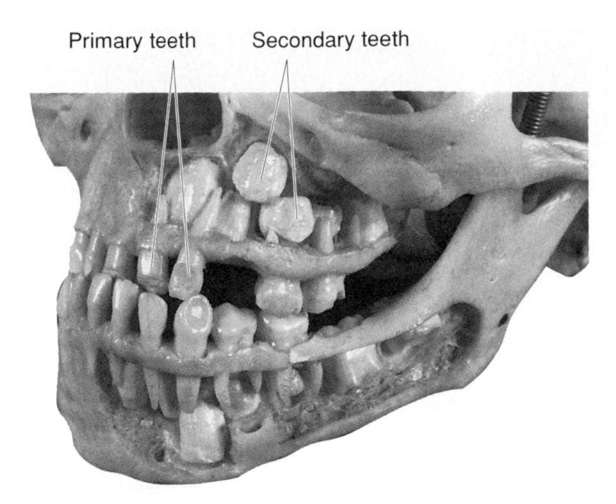


**FIGURE 17.8**
This child's skull reveals the primary and secondary teeth developing in the maxilla and mandible.

**CHART 17.2 PRIMARY AND SECONDARY TEETH**

| Primary Teeth (Deciduous) | | Secondary Teeth (Permanent) | |
|---|---|---|---|
| *Type* | *Number* | *Type* | *Number* |
| Incisor | | Incisor | |
| Central | 4 | Central | 4 |
| Lateral | 4 | Lateral | 4 |
| Cuspid | 4 | Cuspid | 4 |
| | | Bicuspid | |
| | | First | 4 |
| | | Second | 4 |
| Molar | | Molar | |
| First | 4 | First | 4 |
| Second | 4 | Second | 4 |
| | | Third | 4 |
| Total | 20 | Total | 32 |

The primary teeth are usually shed in the same order they appeared. Before this happens, though, their roots are resorbed. The teeth are then pushed out of their sockets by pressure from the developing *secondary teeth* (permanent teeth). This secondary set consists of thirty-two teeth—sixteen in each jaw—and they are arranged from the midline as follows: central incisor, lateral incisor, cuspid (canine), first bicuspid (premolar), second bicuspid (premolar), first molar, second molar, and third molar (fig. 17.9).

Chart 17.2 summarizes the types and numbers of primary and secondary teeth.

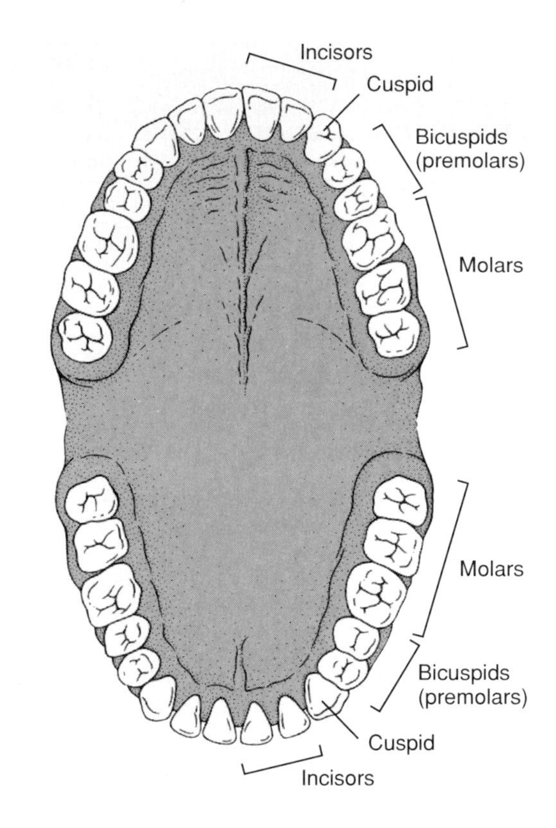


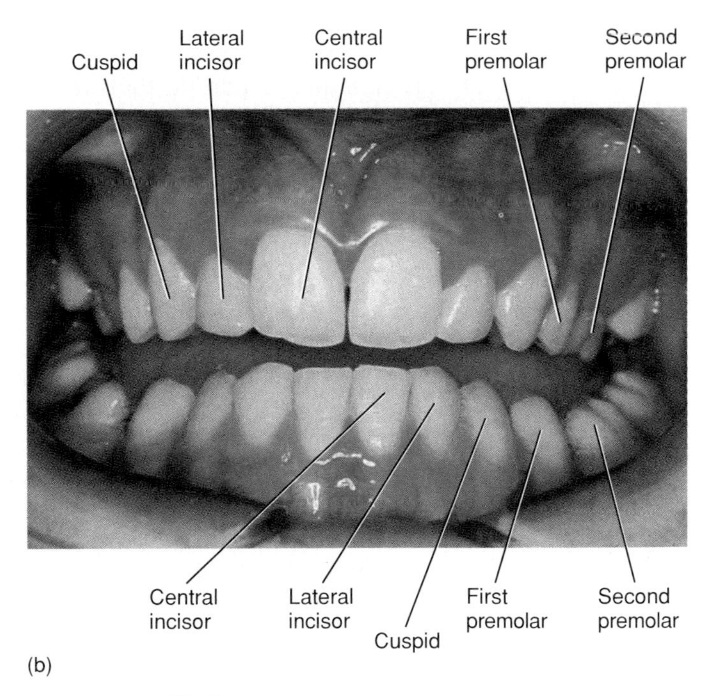


**FIGURE 17.9**
(*a*) The secondary teeth of the upper and lower jaws; (*b*) anterior view of the secondary teeth.

## CHART 17.3 Mouth Parts and Their Functions

| Part | Location | Function | Part | Location | Function |
|---|---|---|---|---|---|
| Cheeks | Form lateral walls of mouth | Hold food in mouth; muscles function in chewing | Tongue | Occupies floor of mouth | Aids in mixing food with saliva; moves food toward pharynx; contains taste receptors |
| Lips | Surround mouth opening | Contain sensory receptors used to judge characteristics of foods | Palate | Forms roof of mouth | Holds food in mouth; directs food to pharynx |
| | | | Teeth | In sockets of mandibular and maxillary bones | Break food particles into smaller pieces; help mix food with saliva during chewing |

The permanent teeth usually begin to appear at 6 years of age, but the set may not be completed until the third molars appear between 17 and 25 years of age. Sometimes these third molars, which are also called wisdom teeth, become wedged in abnormal positions within the jaws and fail to erupt. Such teeth are said to be *impacted.*

The teeth break pieces of food into smaller pieces, the beginning of mechanical digestion. This action increases the surface area of the food particles and thus makes it possible for digestive enzymes to react more effectively with the food molecules.

Different teeth are adapted to handle food in different ways. The *incisors* are chisel-shaped, and their sharp edges bite off relatively large pieces of food. The *cuspids* are cone-shaped, and they grasp and tear food. The *bicuspids* and *molars* have somewhat flattened surfaces and are specialized for grinding food particles.

Each tooth consists of two main portions—the *crown,* which projects beyond the gum, and the *root,* which is anchored to the alveolar process of the jaw. The region where these portions meet is called the *neck* of the tooth. The crown is covered by glossy, white *enamel.* Enamel consists mainly of calcium salts and is the hardest substance in the body. Unfortunately, if enamel is damaged by abrasive action or injury, it is not replaced. It also tends to wear away with age.

The bulk of a tooth beneath the enamel is composed of *dentin,* a substance much like bone, but somewhat harder. The dentin, in turn, surrounds the tooth's central cavity (pulp cavity), which contains blood vessels, nerves, and connective tissue (pulp). The blood vessels and nerves reach this cavity through tubular *root canals,* which extend upward into the root. Loss of teeth is most often associated with diseases of the gums (gingivitis) and the dental pulp (endodontitis).

The root is enclosed by a thin layer of bonelike material called *cementum,* which is surrounded by a *periodontal ligament* (periodontal membrane). This ligament contains bundles of thick collagenous fibers, which pass between the cementum and the bone of the alveolar process, firmly attaching the tooth to the jaw. It also contains blood vessels and nerves near the surface of the cementum-covered root (fig. 17.10).

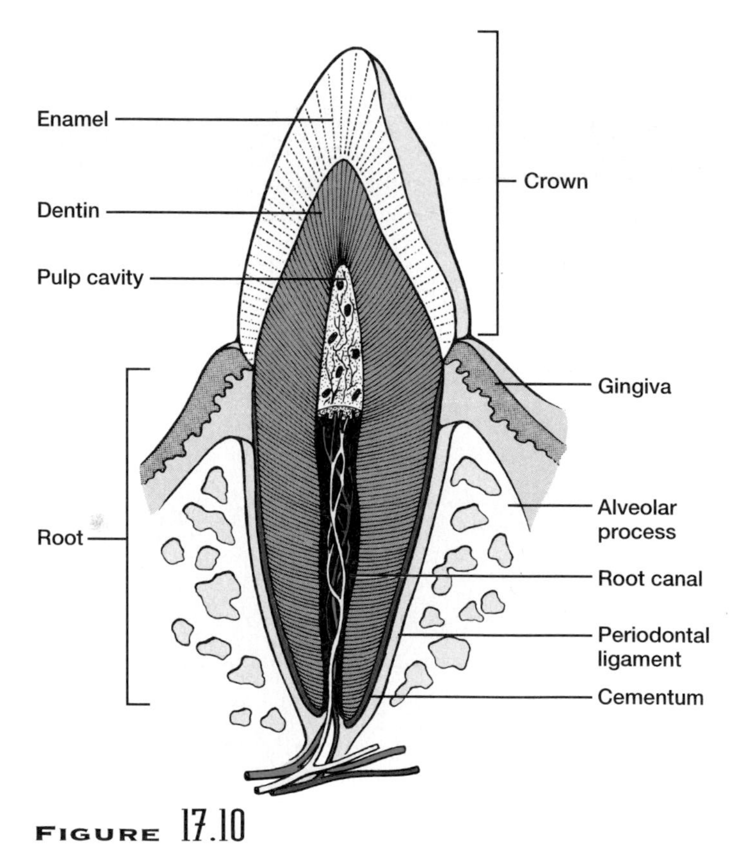


**FIGURE 17.10**
A section of a cuspid tooth.

The mouth parts and their functions are summarized in chart 17.3. Clinical Application 17.1 describes the effect of bacteria on teeth.

**CLINICAL APPLICATION** 

## Dental Caries

You are at the movies and eat a sticky candy bar. Inevitably, some of the caramel lodges between your teeth, or in the crevices of your molars. The sticky snack feeds not only you, but bacteria in your mouth. Species include *Actinomyces, Streptococcus mutans,* and *Lactobacillus.* These microbes metabolize carbohydrates in the food, producing acid by-products that destroy tooth enamel and dentin (fig. 17A). The bacteria also produce sticky substances that hold them in place.

If you eat candy bars but do not brush your teeth soon afterward, the actions of the acid-forming bacteria will produce *dental caries.* Unless a dentist cleans and fills the resulting cavity that forms where enamel is destroyed, the damage will spread to the underlying dentin. The tooth becomes very sensitive.

You can prevent dental caries in several ways:

1. Brush and floss teeth regularly.
2. See the dentist regularly.
3. Drink fluoridated water or receive a fluoride treatment. Fluoride is actually incorporated into the enamel's chemical structure, strengthening it.
4. Apply a sealant to children and adolescents' teeth that have crevices that might hold onto decay-causing bacteria. The sealant is a coating that keeps acids from eating away at tooth enamel.

One dental researcher took an unconventional approach to preventing dental caries that, understandably, was never commercialized. He invented a mouthwash consisting of mutant bacteria that would replace *Streptococcus mutans* but would not decay enamel. Consumer acceptance of a mutant bacterial brew was an obstacle!

**FIGURE 17A**
*Actinomyces* bacteria clinging to teeth release acids that decay tooth enamel (1,250×).

1. How do primary teeth differ from secondary teeth?
2. How are types of teeth adapted to provide specialized functions?
3. Describe the structure of a to
4. Explain how a tooth is attache of the jaw.

## Salivary Glands

The **salivary glands** secrete saliva. This fluid moistens food particles, helps bind them together, and begins the chemical digestion of carbohydrates. Saliva is also a solvent, dissolving foods so that they can be tasted, and it helps cleanse the mouth and teeth. Bicarbonate ions ($HCO_3^-$) in saliva help buffer its acid concentration so that the pH of saliva usually remains near neutral, between 6.5 and 7.5. This is a favorable range for the action of the salivary enzyme and protects the teeth from dissolving in a highly acidic environment.

Many minor salivary glands are scattered throughout the mucosa of the tongue, palate, and cheeks. They secrete fluid continuously so that the lining of the mouth remains moist. There are also three pairs of major salivary glands: the parotid glands, the submandibular glands, and the sublingual glands.

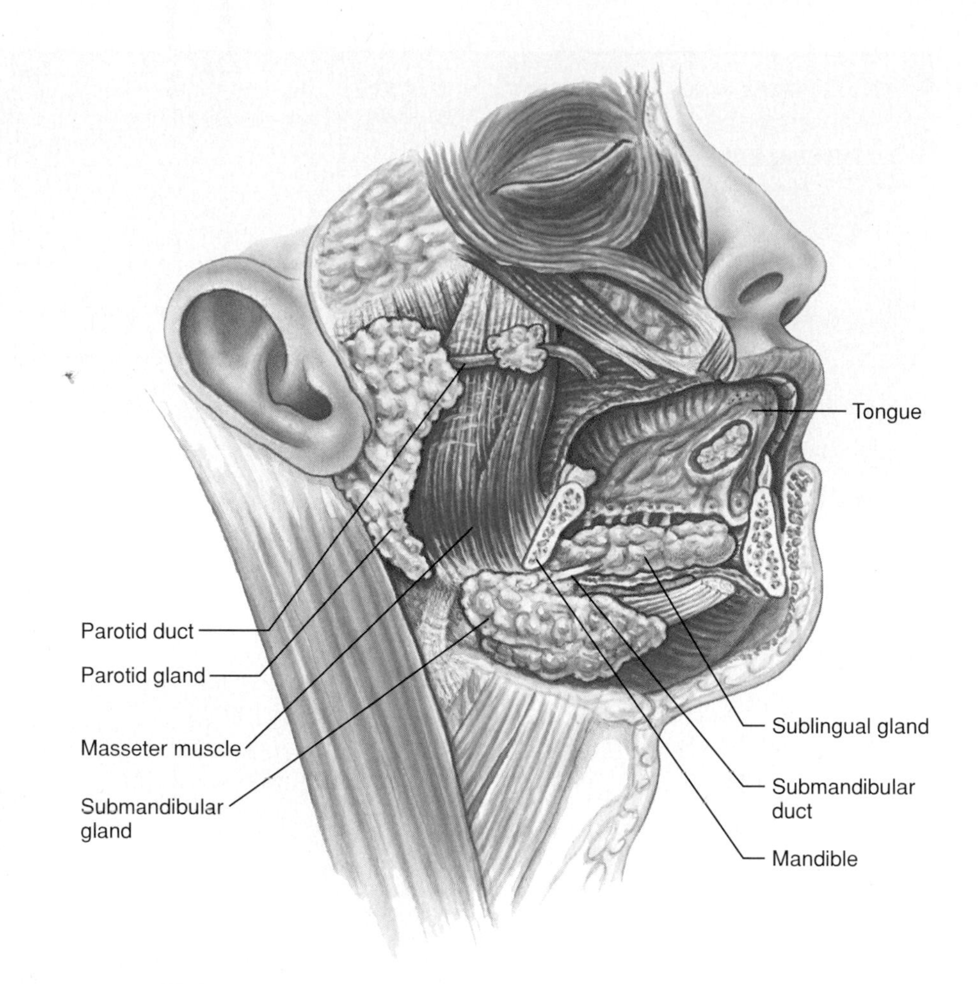


**Figure 17.11**
Locations of the major salivary glands.

### Salivary Secretions

Within a salivary gland are two types of secretory cells, *serous cells* and *mucous cells*. These cells are present in varying proportions within different glands. Serous cells produce a watery fluid that contains a digestive enzyme called **amylase.** This enzyme splits starch and glycogen molecules into disaccharides—the first step in the chemical digestion of carbohydrates. Mucous cells secrete a thick liquid called **mucus,** which binds food particles together and acts as a lubricant during swallowing.

Like other digestive structures, the salivary glands are innervated by branches of both sympathetic and parasympathetic nerves. Impulses arriving on sympathetic fibers stimulate the gland cells to secrete a small quantity of viscous saliva. Parasympathetic impulses, on the other hand, elicit the secretion of a large volume of watery saliva. Such parasympathetic impulses are activated reflexly when a person sees, smells, tastes, or even thinks about pleasant foods. Conversely, if food looks, smells, or tastes unpleasant, parasympathetic activity is inhibited, so that less saliva is produced and swallowing may become difficult.

### Major Salivary Glands

The **parotid glands** are the largest of the major salivary glands. Each gland lies in front of and somewhat below each ear, between the skin of the cheek and the masseter muscle. A *parotid duct* (Stensen's duct) passes from the gland inward through the buccinator muscle, entering the mouth just opposite the upper second molar on either side of the jaw. The parotid glands secrete a clear, watery fluid that is rich in amylase (fig. 17.11).

The **submandibular glands** are located in the floor of the mouth on the inside surface of the lower jaw. The secretory cells of these glands are predominantly serous, with a few mucous cells. The submandibular glands secrete a more viscous fluid than the parotid glands. The ducts of the submandibular glands (Wharton's ducts) open under the tongue, near the frenulum (see fig. 17.11).

The **sublingual glands** are the smallest of the major salivary glands. They are found on the floor of the mouth under the tongue. Because their cells are

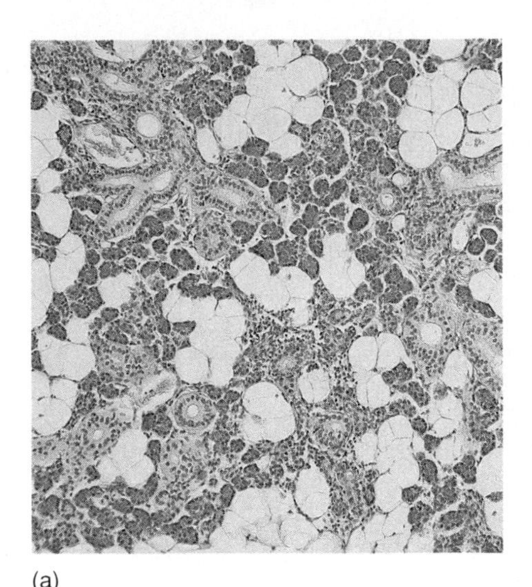

(a)

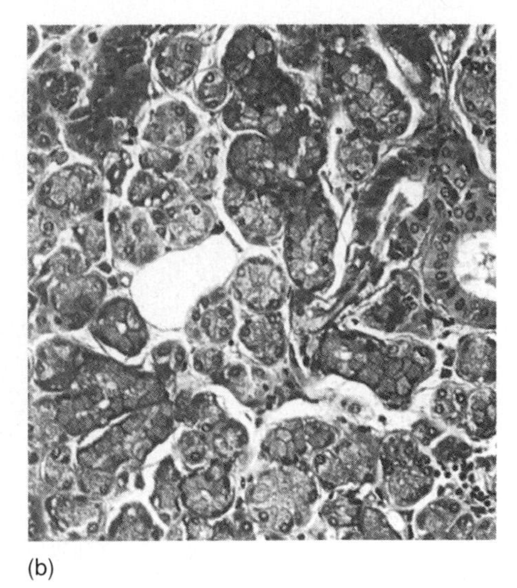

(b)

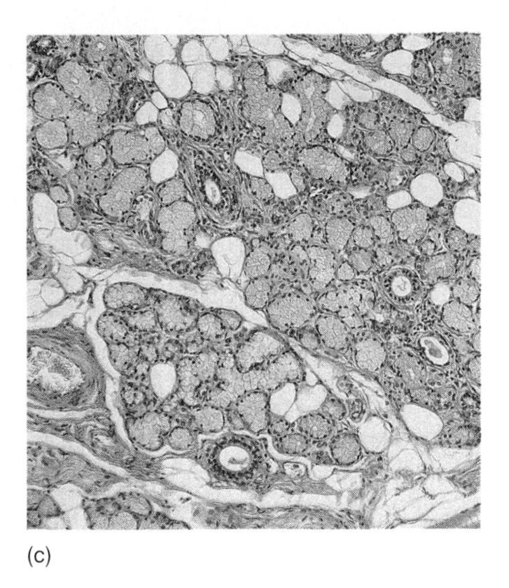

(c)

**FIGURE 17.12**
Light micrographs of (*a*) the parotid salivary gland (200×), (*b*) the submandibular salivary gland (150×), and (*c*) the sublingual salivary gland (200×).

**CHART 17.4 THE MAJOR SALIVARY GLANDS**

| Gland | Location | Duct | Type of Secretion |
|---|---|---|---|
| Parotid glands | In front of and somewhat below the ears between the skin of the cheeks and the masseter muscles | Parotid ducts pass through the buccinator muscles and enter the mouth opposite the upper second molars | Clear, watery serous fluid, rich in amylase |
| Submandibular glands | In the floor of the mouth on the inside surface of the mandible | Ducts open beneath the tongue near the frenulum | Primarily serous fluid, but with some mucus; more viscous than parotid secretion |
| Sublingual glands | In the floor of the mouth beneath the tongue | Many separate ducts | Primarily thick, stringy mucus |

primarily the mucous type, their secretions, which enter the mouth through many separate ducts (Rivinus's ducts) tend to be thick and stringy (see fig. 17.12).

Chart 17.4 summarizes the characteristics of the major salivary glands.

**1** What is the function of saliva?

**2** What stimulates the salivary glands to secrete saliva?

**3** Where are the major salivary glands located?

## PHARYNX AND ESOPHAGUS

The pharynx is a cavity behind the mouth from which the tubular esophagus leads to the stomach. The pharynx and the esophagus do not digest food, but both are important passageways, and their muscular walls function in swallowing.

### Structure of the Pharynx

The **pharynx** connects the nasal and oral cavities with the larynx and esophagus (see fig. 17.7). It can be divided into the following parts:

1. The **nasopharynx** is located above the soft palate. It communicates with the nasal cavity and provides a passageway for air during breathing. The auditory tubes, which connect the pharynx with the middle ears, open through the walls of the nasopharynx (see chapter 12).
2. The **oropharynx** is behind the mouth. It opens behind the soft palate into the nasopharynx and projects downward to the upper border of the epiglottis. This portion is a passageway for food moving downward from the mouth, and for air moving to and from the nasal cavity.
3. The **laryngopharynx** is located just below the oropharynx. It extends from the upper border of

the epiglottis downward to the lower border of the cricoid cartilage of the larynx, where it is continuous with the esophagus. This portion is a passageway for food only; it is not a respiratory tube.

The muscles in the walls of the pharynx are arranged in inner circular and outer longitudinal groups (fig. 17.13). The circular muscles, called *constrictor muscles,* pull the walls inward during swallowing. The *superior constrictor muscles,* which are attached to bony processes of the skull and mandible, curve around the upper part of the pharynx. The *middle constrictor muscles* arise from projections on the hyoid bone and fan around the middle of the pharynx. The *inferior constrictor muscles* originate from cartilage of the larynx and pass around the lower portion of the pharyngeal cavity. Some of the lower inferior constrictor muscle fibers contract most of the time, which prevents air from entering the esophagus during breathing.

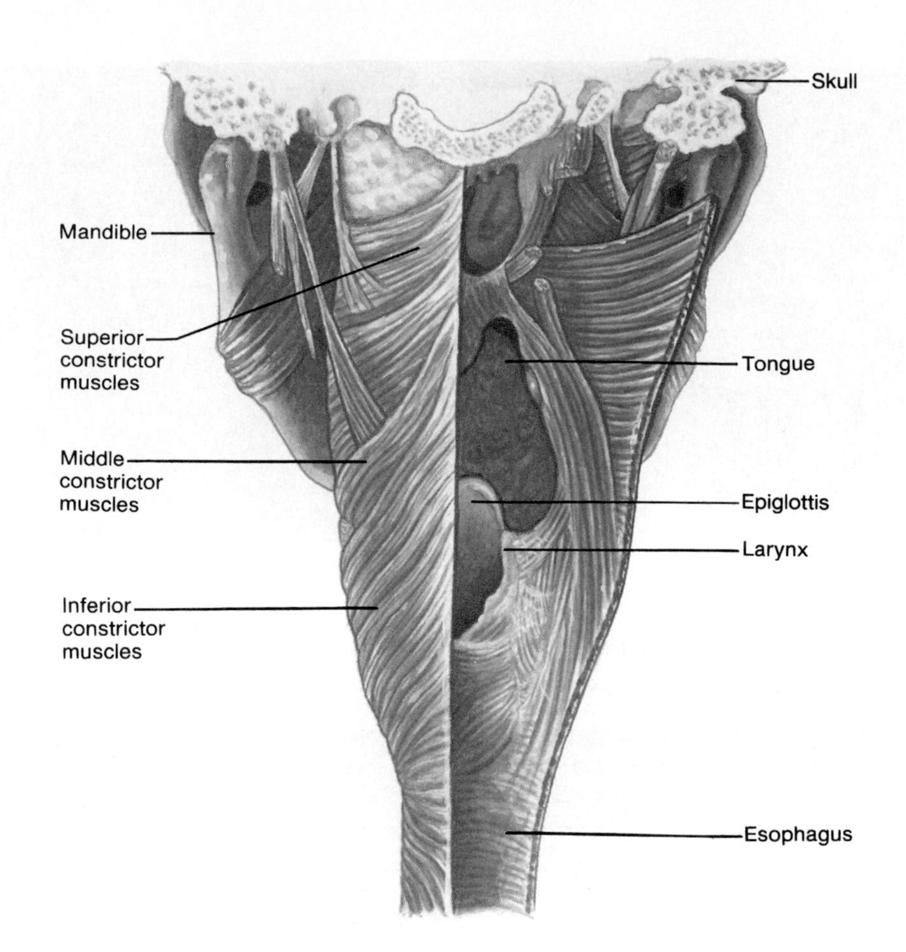


FIGURE 17.13
Muscles of the pharyngeal wall, posterior view.

Although the pharyngeal muscles are skeletal muscles, they are under voluntary control, only in the sense that swallowing (deglutition) can be voluntarily initiated. The precise actions of these muscles during swallowing is controlled by complex reflexes.

## Swallowing Mechanism

Swallowing reflexes can be divided into three stages. In the first stage, which is initiated voluntarily, food is chewed and mixed with saliva. Then, the tongue rolls this mixture into a mass (bolus) and forces it into the pharynx. The second stage begins as food reaches the pharynx and stimulates sensory receptors around the pharyngeal opening. This triggers the swallowing reflexes, illustrated in figure 17.14, which includes the following actions:

1. The soft palate raises, preventing food from entering the nasal cavity.
2. The hyoid bone and the larynx are elevated; the epiglottis closes off the top of the trachea so that food is less likely to enter.
3. The tongue is pressed against the soft palate, sealing off the oral cavity from the pharynx.
4. The longitudinal muscles in the pharyngeal wall contract, pulling the pharynx upward toward the food.
5. The lower portion of the inferior constrictor muscles relaxes, opening the esophagus.
6. The superior constrictor muscles contract, stimulating a peristaltic wave to begin in other pharyngeal muscles. This wave forces the food into the esophagus.

As the swallowing reflexes occur, breathing is momentarily inhibited. Then, during the third stage of swallowing, peristalsis transports the food in the esophagus to the stomach.

## Esophagus

The **esophagus** is a straight, collapsible tube about 25 centimeters long. It provides a passageway for food from the pharynx to the stomach. The esophagus descends through the thorax behind the trachea, passing through the mediastinum. It penetrates the diaphragm through an opening, the *esophageal hiatus,* and is

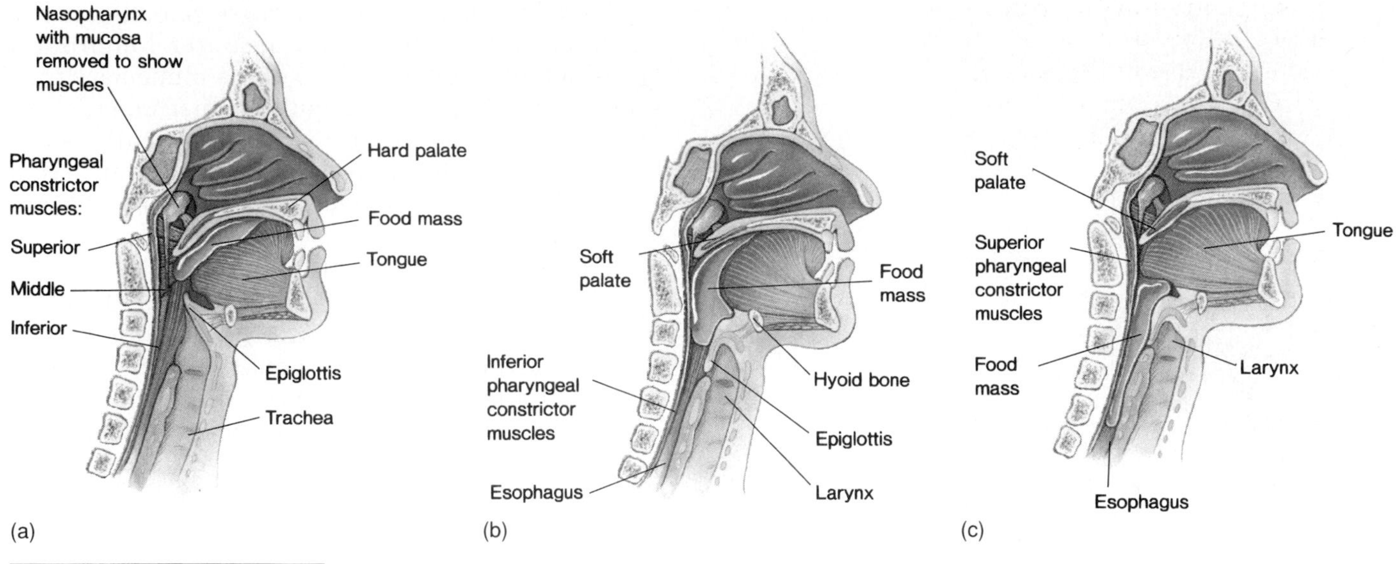


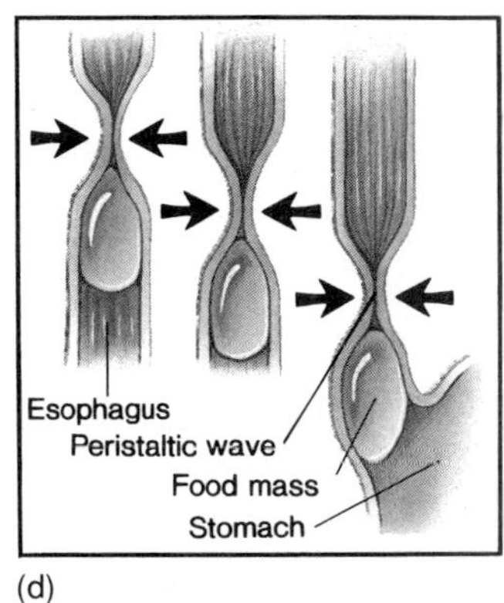


**FIGURE 17.14**

Steps in the swallowing reflex. (*a*) The tongue forces food into the pharynx. (*b*) The soft palate, hyoid bone, and larynx are raised, the tongue is pressed against the palate, the epiglottis closes, and the inferior constrictor muscles relax so that the esophagus opens. (*c*) Superior constrictor muscles contract and force food into the esophagus. (*d*) Peristaltic waves move food through the esophagus to the stomach.

continuous with the stomach on the abdominal side of the diaphragm (figs. 17.15, 17.16, and reference plates 57, 73).

Mucous glands are scattered throughout the submucosa of the esophagus. Their secretions moisten and lubricate the inner lining of the tube.

> In a *hiatal hernia,* a portion of the stomach or large intestine protrudes through a weakened area of the diaphragm, through the esophageal hiatus and into the thorax. Gastric juice from the stomach entering the esophagus as a result of a hiatal hernia or from regurgitation (reflux) may inflame the esophageal mucosa, causing heartburn, difficulty in swallowing, or ulceration and blood loss. In response to the destructive action of gastric juice, columnar epithelium may replace the squamous epithelium that normally lines the esophagus (see chapter 5). This condition, called *Barrett's esophagus,* increases the risk of developing esophageal cancer.

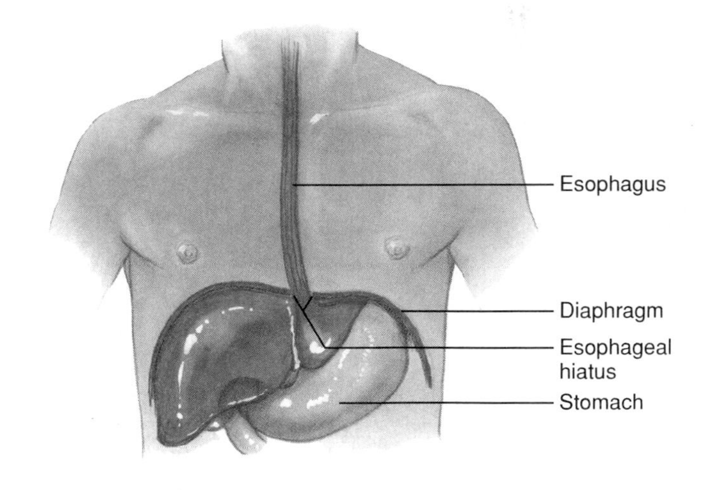


**FIGURE 17.15**

The esophagus functions as a passageway between the pharynx and the stomach.

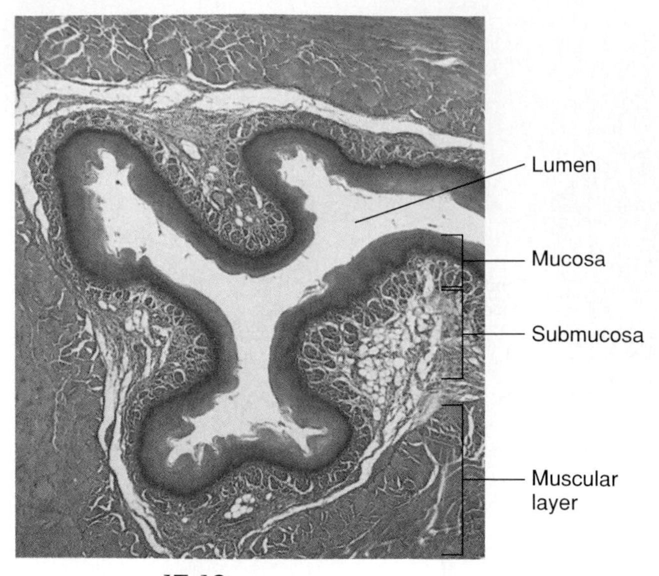


**FIGURE 17.16**
This cross section of the esophagus shows its muscular wall (10×).

Just above the point where the esophagus joins the stomach, some of the circular muscle fibers have increased sympathetic muscle tone, forming the **lower esophageal sphincter** (cardiac sphincter) (fig. 17.17). These fibers usually remain contracted and they close the entrance to the stomach. In this way, they help prevent regurgitation of the stomach contents into the esophagus.

When peristaltic waves reach the stomach, the muscle fibers that guard its entrance relax and allow the swallowed food to enter.

1. Describe the regions of the pharynx.
2. List the major events that occur during swallowing.
3. What is the function of the esophagus?

## STOMACH

The **stomach** is a J-shaped, pouchlike organ, about 25–30 centimeters long, which hangs under the diaphragm in the upper left portion of the abdominal cavity. It has a capacity of about one liter or more, and its inner lining is marked by thick folds (rugae) of the mucosal and submucosal layers that tend to disappear when its wall is distended. The stomach receives food from the esophagus, mixes it with gastric juice, initiates the digestion of proteins, carries on a limited amount of absorption, and moves food into the small intestine.

In addition to the two layers of smooth muscle—an inner circular layer and an outer longitudinal layer—found in other regions of the alimentary canal, some parts of the stomach have another inner layer of oblique fibers. This third innermost muscular layer is most highly developed near the opening of the esophagus and in the body of the stomach (see fig. 17.17).

### Parts of the Stomach

The stomach, shown in figures 17.17 and 17.18 and reference plate 51, can be divided into the cardiac, fundic, body, and pyloric regions. The *cardiac region* is a small area near the esophageal opening (cardia). The *fundic region,* which balloons above the cardiac portion, is a temporary storage area and sometimes fills with swallowed air. This produces a gastric air bubble, which may be used as a landmark on an X-ray film of the abdomen. The dilated *body region,* which is the main part of the stomach, is located between the fundic and pyloric portions. The *pyloric region* (antrum) is a funnel-shaped portion which narrows and becomes the *pyloric canal* as it approaches the small intestine.

At the end of the pyloric canal, the circular layer of fibers in its muscular wall thickens, forming a powerful muscle, the **pyloric sphincter** (pylorus). This muscle is a valve that prevents regurgitation of food from the intestine back into the stomach, and also prevents excessive release of gastric acids into the duodenum.

*Hypertrophic pyloric stenosis* is a birth defect in which muscle overgrowth blocks the pyloric canal. The newborn vomits, with increasing force. To diagnose the condition, an X ray is taken of the area after the infant drinks formula containing a radiopaque barium compound. Surgical splitting of the muscle blocking the passageway from stomach to small intestine is necessary to enable the infant to eat normally. Pyloric stenosis can occur later in life as a result of ulcers or cancer.

### Gastric Secretions

The mucous membrane that forms the inner lining of the stomach is relatively thick, and its surface is studded with many small openings. These openings, called *gastric pits,* are located at the ends of tubular **gastric glands** (oxyntic glands) (fig. 17.19). Although their structure and the composition of their secretions vary in different parts of the stomach, gastric glands

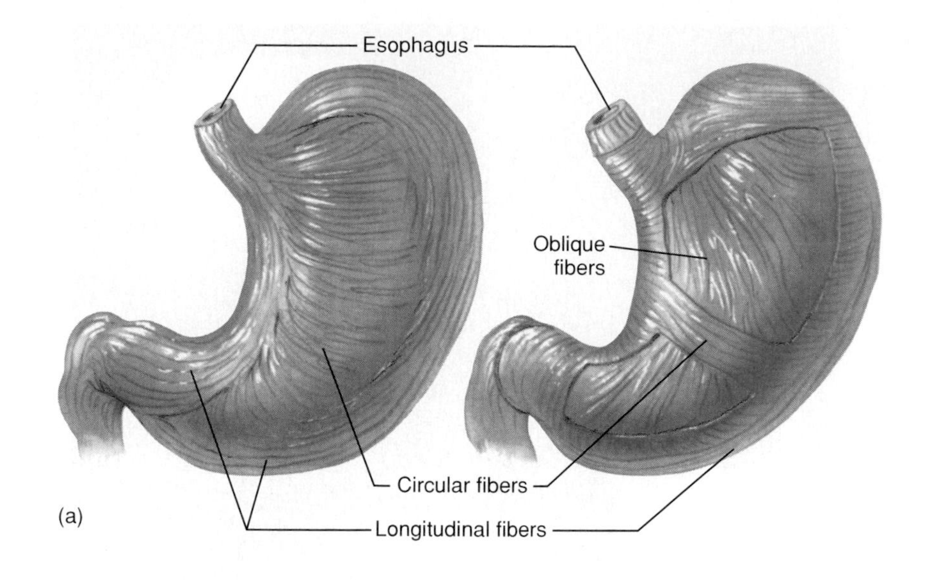


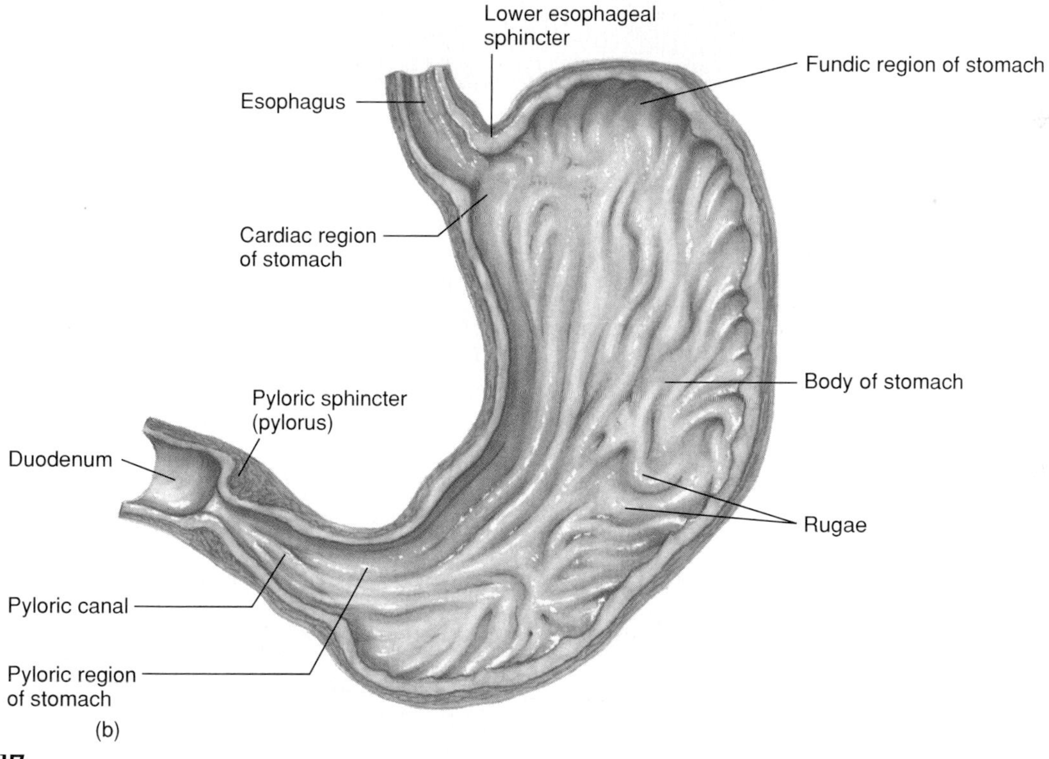


**FIGURE 17.17**

(*a*) Some parts of the stomach have three layers of muscle fibers. (*b*) Major regions of the stomach.

generally contain three types of secretory cells. One type, the *mucous cell* (goblet cell), occurs in the necks of the glands near the openings of the gastric pits. The other types, *chief cells* (peptic cells) and *parietal cells* (oxyntic cells), are found in the deeper parts of the glands (figs. 17.19 and 17.20). The chief cells secrete digestive enzymes, and the parietal cells release a solution containing hydrochloric acid. The products of the mucous cells, chief cells, and parietal cells together form **gastric juice.**

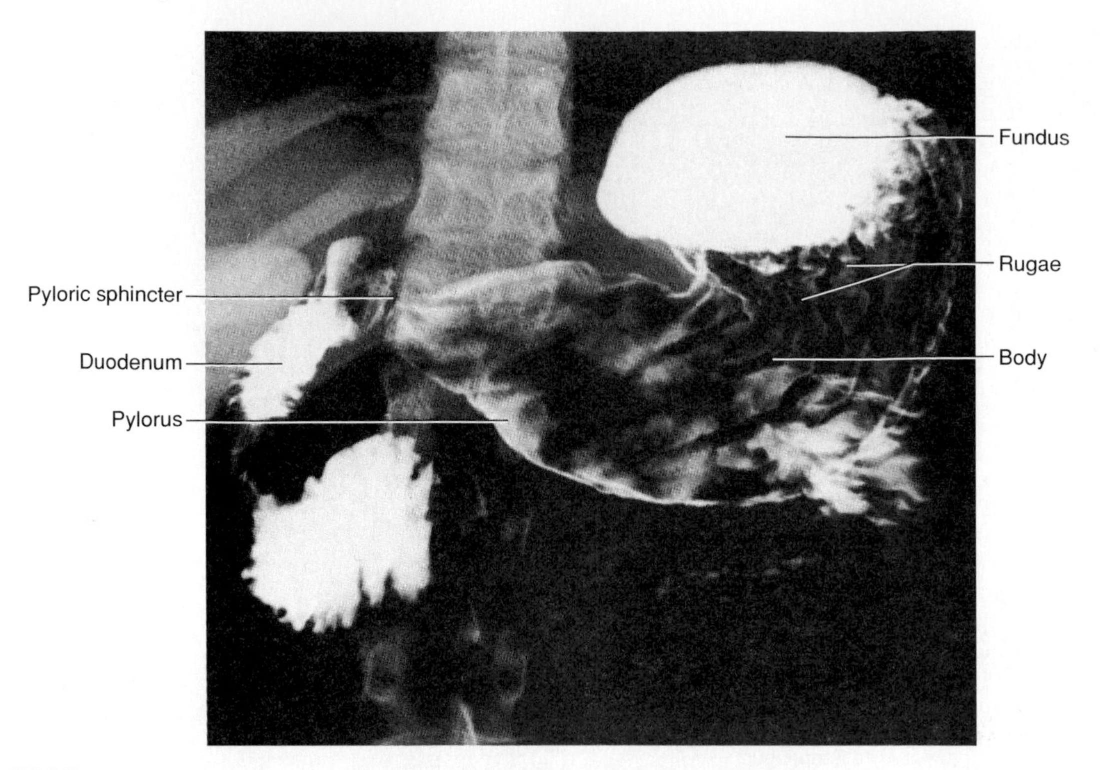


**FIGURE 17.18**
X-ray film of a stomach. (Note: A radiopaque compound the patient swallowed appears white in the X-ray film.)

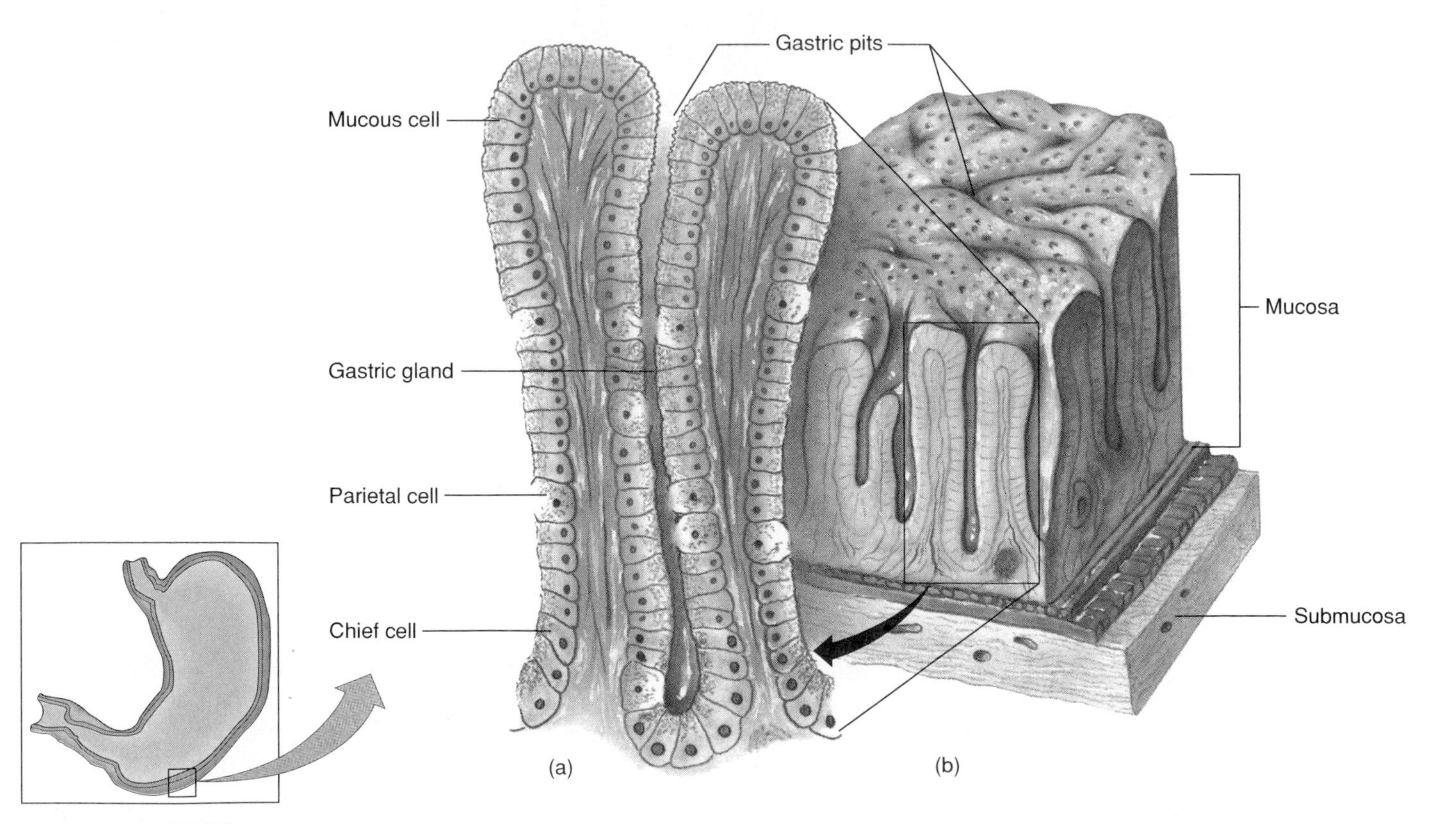


**FIGURE 17.19**
(*a*) Gastric glands include mucous cells, parietal cells, and chief cells. (*b*) The mucosa of the stomach is studded with gastric pits that are the openings of the gastric glands.

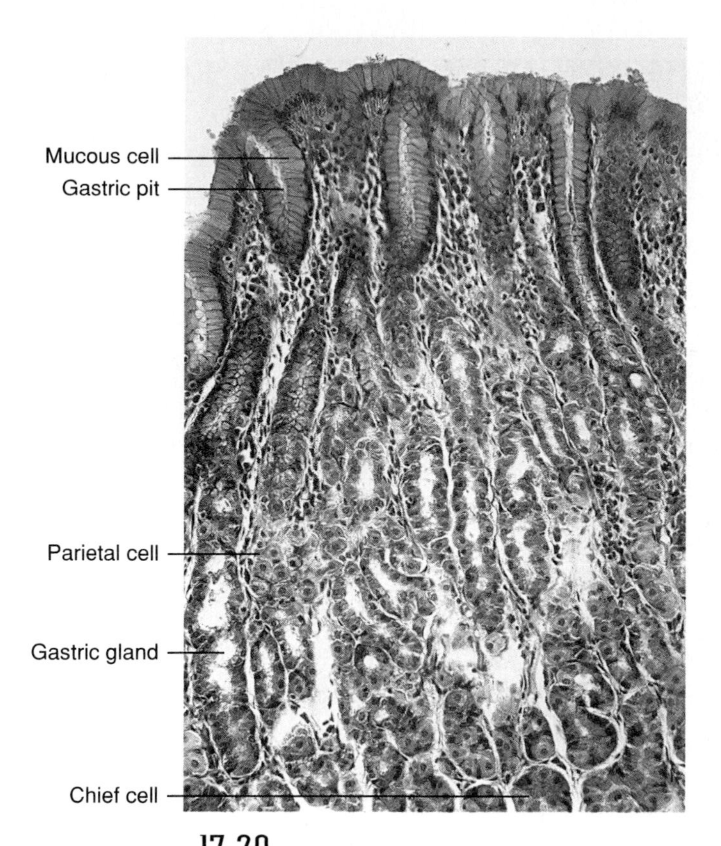


**FIGURE 17.20**
A light micrograph of cells associated with the gastric glands (50×).

Much of what we know about the stomach's functioning comes from a French-Canadian explorer, Alexis St. Martin, who in 1822 accidentally shot himself in the abdomen. His extensive injuries eventually healed, but a hole, called a fistula, was left, allowing observers to look at his stomach in action. A U.S. Army surgeon, William Beaumont, spent eight years watching food digesting in the stomach, noting how the stomach lining changed in response to stress.

Of the several digestive enzymes in gastric juice, **pepsin** is by far the most important. The chief cells secrete it as an inactive, nonerosive enzyme precursor called **pepsinogen.** When pepsinogen contacts the hydrochloric acid from the parietal cells, however, it changes rapidly into pepsin. Pepsin, in turn, can also convert pepsinogen into more pepsin.

Pepsin begins the digestion of nearly all types of dietary protein. This enzyme is most active in an acid environment, which the hydrochloric acid in gastric juice provides.

Gastric juice also contains small quantities of a fat-splitting enzyme, *gastric lipase.* However, its action is relatively weak due in part to the low pH of gastric juice. Gastric lipase acts mainly on butterfat.

The mucous cells of the gastric glands secrete large quantities of thin mucus. In addition, the cells of the mucous membrane, associated with the inner lining of the stomach and between the gastric glands, release a more viscous and alkaline secretion, which coats the inside of the stomach wall. This coating is especially important because pepsin can digest the proteins of stomach tissues, as well as those in foods. Thus, the coating normally prevents the stomach from digesting itself.

Still another component of gastric juice is **intrinsic factor.** The parietal cells of the gastric glands secrete intrinsic factor, which aids in the absorption of vitamin $B_{12}$ from the small intestine.

Chart 17.5 summarizes the components of gastric juice.

1. Where is the stomach located?
2. What are the secretions of the chief cells and parietal cells?
3. Which is the most important digestive enzyme in gastric juice?
4. Why doesn't the stomach digest itself?

## Regulation of Gastric Secretions

Gastric juice is produced continuously, but the rate varies considerably from time to time and is controlled neurally and hormonally. More specifically, within the gastric glands specialized cells closely associated with the parietal cells secrete the hormone *somatostatin,* which inhibits acid secretion. However, acetylcholine released from nerve fibers in response to parasympathetic impulses arriving on the vagus nerves suppresses the secretion of somatostatin and stimulates the gastric glands to secrete large amounts of gastric juice, which is rich in hydrochloric acid and pepsinogen. These parasympathetic impulses also stimulate certain stomach cells, mainly in the pyloric region, to release a peptide hormone called **gastrin** that stimulates the gastric glands to secrete (fig. 17.21). Furthermore, parasympathetic impulses and gastrin promote release of *histamine* from gastric mucosal cells, which, in turn, stimulates additional gastric secretion.

Histamine is very effective in promoting secretion of gastric acid. Consequently, drugs that block the histamine receptors of gastric mucosal cells ($H_2$-blockers) are often used to inhibit gastric secretions in patients with excess gastric acid.

Three stages of gastric secretion are recognized—the cephalic, gastric, and intestinal phases.

The *cephalic phase* of gastric secretion begins before any food reaches the stomach and possibly even

## CHART 17.5 MAJOR COMPONENTS OF GASTRIC JUICE

| Component | Source | Function |
|---|---|---|
| Pepsinogen | Chief cells of the gastric glands | Inactive form of pepsin |
| Pepsin | Formed from pepsinogen in the presence of hydrochloric acid | A protein-splitting enzyme that digests nearly all types of dietary protein |
| Hydrochloric acid | Parietal cells of the gastric glands | Provides the acid environment needed for the conversion of pepsinogen into pepsin and for the action of pepsin |
| Mucus | Goblet cells and mucous glands | Provides a viscous, alkaline protective layer on the stomach wall |
| Intrinsic factor | Parietal cells of the gastric glands | Aids in absorption of vitamin $B_{12}$ |

before eating. In this stage, parasympathetic reflexes operating through the vagus nerves stimulate gastric secretion whenever a person tastes, smells, sees, or even thinks about food. Furthermore, the hungrier the person is, the greater the amount of gastric secretion. The cephalic phase of secretion is responsible for 30%–50% of the secretory response to a meal.

The *gastric phase* of gastric secretion, which accounts for 40%–50% of the secretory activity, starts when food enters the stomach. The presence of food and the distension of the stomach wall trigger the stomach to release gastrin, which stimulates production of still more gastric juice.

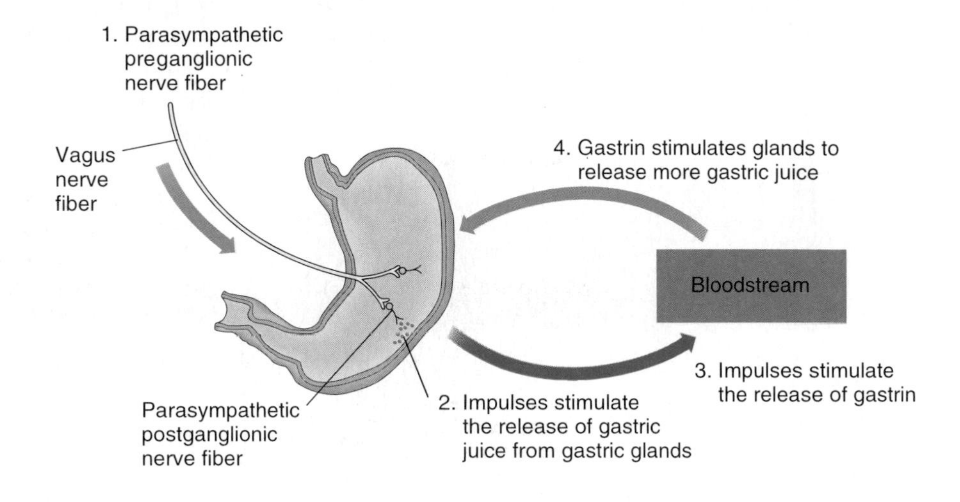


**FIGURE 17.21**
The secretion of gastric juice is regulated in part by parasympathetic nerve impulses that stimulate the release of gastric juice and gastrin.

As food enters the stomach and mixes with gastric juice, the pH of the contents rises, which enhances gastrin secretion. Consequently, the pH of the stomach contents drops. As the pH approaches 3.0, secretion of gastrin is inhibited. When the pH reaches 1.5, gastrin secretion ceases.

Gastrin also stimulates cell growth in the mucosa of the stomach and intestines, except where gastrin is produced. This effect helps repair mucosal cells damaged as a result of disease or medical treatments.

For the stomach to secrete hydrochloric acid, hydrogen ions are removed from the blood, and an equivalent amount of alkaline bicarbonate ions is released into the blood. Consequently, following a meal, the blood concentration of bicarbonate ions increases, and the urine excretes excess bicarbonate ions. This phenomenon is called the *alkaline tide.*

The *intestinal phase* of gastric secretion, which accounts for about 5% of the total secretory response to a meal, begins when food leaves the stomach and enters the small intestine. When food first contacts the intestinal wall, it stimulates intestinal cells to release a hormone that again enhances gastric gland secretion. Many investigators believe this hormone is identical to gastrin, and they call it *intestinal gastrin.*

As more food moves into the small intestine, secretion of gastric juice from the stomach wall is inhibited, due to a sympathetic reflex triggered by acid in the upper part of the small intestine. Also, the presence of proteins and fats in this region of the intestine causes the release of the peptide hormone *cholecystokinin* from the intestinal wall, which decreases gastric motility. Similarly, fats in the small intestine stimulate intestinal cells to release *intestinal somatostatin,* which inhibits release of gastric juice. Overall, these actions decrease gastric secretion and motility, as the small intestine fills with food.

## CLINICAL APPLICATION 17.2

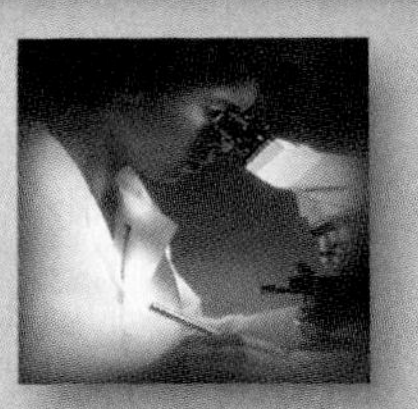

### Oh, My Aching Stomach!

The food at the barbecue was so terrific that Perry W. didn't notice how much he was packing away—two burgers, three hot dogs, beans in a spicy sauce, loads of chips, several beers, topped off with ice cream for dessert. Shortly after going home, tired from all that eating, he lay down. And then the pain started.

At first Perry just felt nicely full, but the fullness seemed to grow. Then he felt pain in his abdomen, and later in the evening, heartburn, as some of his stomach contents backed up into his esophagus.

Fortunately, Perry found relief with an over-the-counter antacid product. Antacids raise the pH of the stomach within minutes. They usually include a compound containing either sodium, calcium, magnesium, or aluminum. Another ingredient in some products is simethicone, which breaks up gas bubbles in the digestive tract. If antacids do not bring some relief within a few minutes, or they are used for longer than two weeks, a doctor should be consulted. The problem may be more serious than overeating.

Avoiding acid indigestion and heartburn is a more healthful approach than gorging and then reaching for the antacid bottle. Some tips:

- Avoid large meals. The more food, the more stomach acid is produced.
- Eat slowly, so that stomach acid secretion is more gradual.
- Do not lie down immediately after eating. In an upright stance, gravity helps food along the alimentary canal.
- If you are prone to indigestion or heartburn, avoid caffeine, which increases stomach acid secretion.
- Cigarettes and alcohol irritate the stomach lining and lower the pressure at the junction between the stomach and the esophagus. This makes it easier for food to return to the esophagus, causing heartburn.
- Do not eat acidic foods, such as citrus fruits and tomatoes, unless it is at least three hours before bedtime.
- Use a pillow that elevates the head six to eight inches above the stomach.

**CHART 17.6 PHASES OF GASTRIC SECRETION**

| Phase | Action |
|---|---|
| Cephalic phase | The sight, taste, smell, or thought of food triggers parasympathetic reflexes. Gastric juice is secreted in response. |
| Gastric phase | Food in stomach chemically and mechanically stimulates release of gastrin, which, in turn, stimulates secretion of gastric juice; reflex responses also stimulate gastric juice secretion. |
| Intestinal phase | As food enters the small intestine, it stimulates intestinal cells to release intestinal gastrin, which, in turn, promotes the secretion of gastric juice from the stomach wall. |

Chart 17.6 summarizes the phases of gastric secretion, and Clinical Application 17.2 discusses a common problem—indigestion.

1. What controls gastric juice secretion?
2. Distinguish between the cephalic, gastric, and intestinal phases of gastric secretion.
3. What is the function of cholecystokinin?

### Gastric Absorption

Although gastric enzymes begin breaking down proteins, the stomach wall is not well adapted to absorb digestive products. However, the stomach absorbs small quantities of water, glucose, certain salts, alcohol, and some lipid-soluble drugs.

> Most nutrients are absorbed in the small intestine. Alcohol, which is not a nutrient, is absorbed in the stomach. This is why the intoxicating effects of alcohol are felt soon after consuming alcoholic beverages.

### Mixing and Emptying Actions

Food entering the stomach stretches the smooth muscles in its wall. The stomach may enlarge, but its muscles maintain their tone, and internal pressure of the stomach normally is unchanged. A person may eat more than the stomach can comfortably hold, and when this happens, the internal pressure may rise enough to stimulate pain receptors. The result is a stomachache.

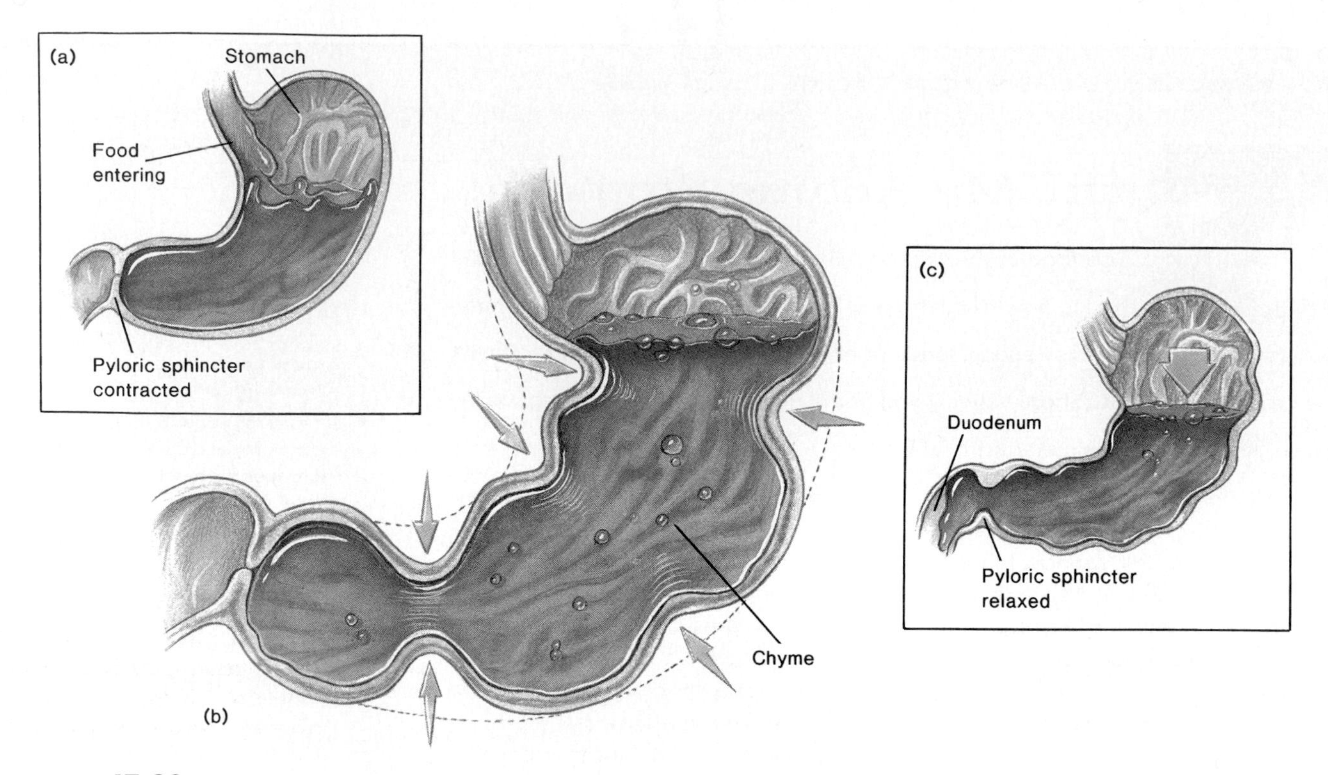


**FIGURE 17.22**

Stomach movements. (*a*) As the stomach fills, its muscular wall stretches, but the pyloric sphincter remains closed. (*b*) Mixing movements combine food and gastric juice, creating chyme. (*c*) Peristaltic waves move the chyme toward the pyloric sphincter, which relaxes and admits some chyme into the duodenum.

Following a meal, the mixing movements of the stomach wall aid in producing a semifluid paste of food particles and gastric juice called **chyme.** Peristaltic waves push the chyme toward the pyloric region of the stomach, and as chyme accumulates near the pyloric sphincter, this muscle begins to relax. The muscular pyloric region then pumps the chyme a little (5–15 milliliters) at a time into the small intestine. Most of the chyme is forced backward into the stomach by these waves of contraction, where it is mixed some more. Figure 17.22 illustrates this process.

The rate at which the stomach empties depends on the fluidity of the chyme and the type of food present. For example, liquids usually pass through the stomach quite rapidly, but solids remain until they are well mixed with gastric juice. Fatty foods may remain in the stomach three to six hours; foods high in proteins tend to move through more quickly; and carbohydrates usually pass through more rapidly than either fats or proteins.

When the duodenum fills with chyme, its internal pressure increases, and the intestinal wall is stretched. This action stimulates sensory receptors in the wall, triggering an **enterogastric reflex.** The name of this reflex, like those of other digestive reflexes, describes the origin and termination of reflex impulses. Thus, the enterogastric reflex begins in the small intestine (*entero*) and ends in the stomach (*gastro*).

As a result of the enterogastric reflex, fewer parasympathetic impulses arrive at the stomach, inhibiting peristalsis (fig. 17.23). Consequently, the intestine fills less rapidly. Also, if the chyme entering the intestine is fatty, the intestinal wall releases the hormone **cholecystokinin,** which inhibits peristalsis even more.

As chyme enters the duodenum (the first portion of the small intestine), accessory organs add their secretions. These organs include the pancreas, liver, and gallbladder.

*Vomiting* results from a complex reflex that empties the stomach another way. Irritation or distension in the stomach or intestines can trigger vomiting. Sensory impulses travel from the site of stimulation to the *vomiting center* in the medulla oblongata, and a number of motor responses follow. These include taking a deep breath, raising the soft palate and thus closing the nasal cavity, closing the opening to the trachea (glottis), relaxing the circular muscle fibers at the base of the esophagus, contracting the diaphragm so it moves downward over the stomach, and contracting the abdominal wall muscles so that pressure inside the abdominal cavity increases. As a result, the stomach is squeezed from all sides, forcing its contents upward and out through the esophagus, pharynx, and mouth.

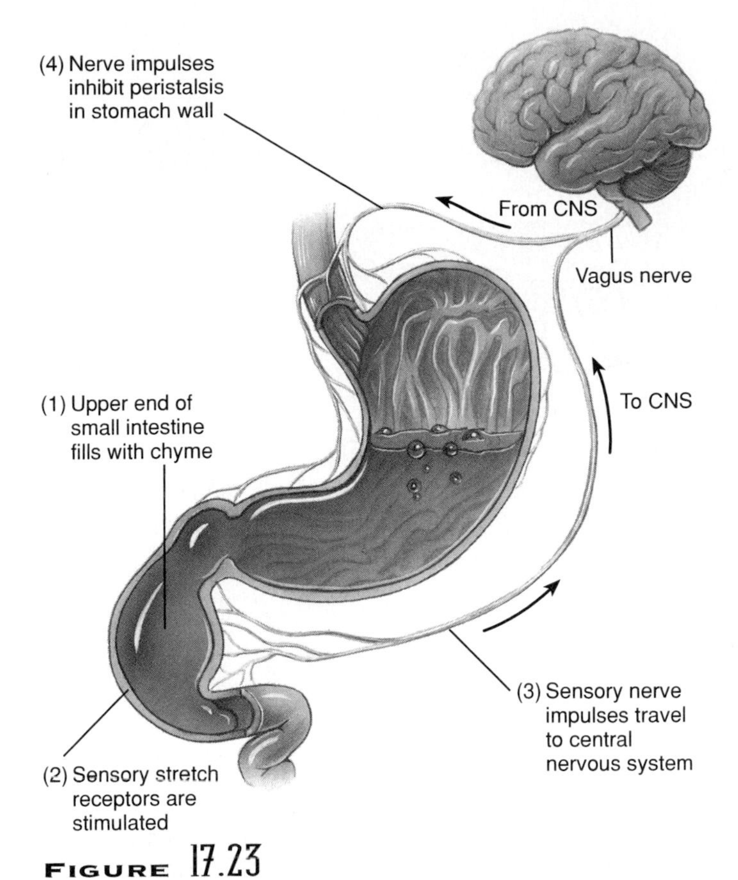


**FIGURE 17.23**
The enterogastric reflex partially regulates the rate at which chyme leaves the stomach.

Activity in the vomiting center can be stimulated by drugs (emetics), by toxins in contaminated foods, and sometimes by rapid changes in body motion. In this last situation, sensory impulses from the labyrinths of the inner ears reach the vomiting center, and can produce motion sickness. The vomiting center can also be activated by stimulation of higher brain centers through sights, sounds, odors, tastes, emotions, or mechanical stimulation of the back of the pharynx.

*Nausea* emanates from activity in the vomiting center or in nerve centers closely associated with it. During nausea, stomach movements usually are diminished or absent, and duodenal contents may move back into the stomach.

1. How is chyme produced?
2. What factors influence how quickly chyme leaves the stomach?
3. Describe the enterogastric reflex.
4. Describe the vomiting reflex.
5. What factors may stimulate the vomiting reflex?

## PANCREAS

The pancreas, discussed as an endocrine gland in chapter 13, also has an exocrine function—secretion of a digestive juice called **pancreatic juice.**

### Structure of the Pancreas

The pancreas is closely associated with the small intestine and is located behind the parietal peritoneum. It extends horizontally across the posterior abdominal wall, with its head in the C-shaped curve of the duodenum and its tail against the spleen (fig. 17.24 and reference plate 59).

The cells that produce pancreatic juice, called *pancreatic acinar cells,* make up the bulk of the pancreas. These cells cluster around tiny tubes, into which they release their secretions. The smaller tubes unite to form larger ones, which, in turn, give rise to a *pancreatic duct* extending the length of the pancreas. The pancreatic duct usually connects with the duodenum at the same place where the bile duct from the liver and gallbladder joins the duodenum (see figs. 13.32 and 17.24).

The pancreatic and bile ducts join at a short, dilated tube called the *hepatopancreatic ampulla* (ampulla of Vater). A band of smooth muscle, called the *hepatopancreatic sphincter* (sphincter of Oddi), surrounds this ampulla.

### Pancreatic Juice

Pancreatic juice contains enzymes that digest carbohydrates, fats, proteins, and nucleic acids.

The carbohydrate-digesting enzyme, **pancreatic amylase,** splits molecules of starch or glycogen into double sugars (disaccharides). The fat-digesting enzyme, **pancreatic lipase,** breaks triglyceride molecules into fatty acids and monoglycerides. (A monoglyceride molecule consists of one fatty acid bound to glycerol.)

The protein-splitting (proteolytic) enzymes are **trypsin, chymotrypsin,** and **carboxypeptidase.** Each of these enzymes splits the bonds between particular combinations of amino acids in proteins. Because no single enzyme can split all of the possible combinations of amino acids, several enzymes are necessary for the complete digestion of protein molecules.

The protein-splitting enzymes are stored in inactive forms within tiny cellular structures called *zymogen granules.* These enzymes, like gastric pepsin, are secreted in inactive forms and must be activated by other enzymes after they reach the small intestine. For example, the pancreatic cells release inactive **trypsinogen,** which is activated to trypsin when it contacts an enzyme called *enterokinase,* which the mucosa of the small intestine secretes. Chymotrypsin and carboxypeptidase are

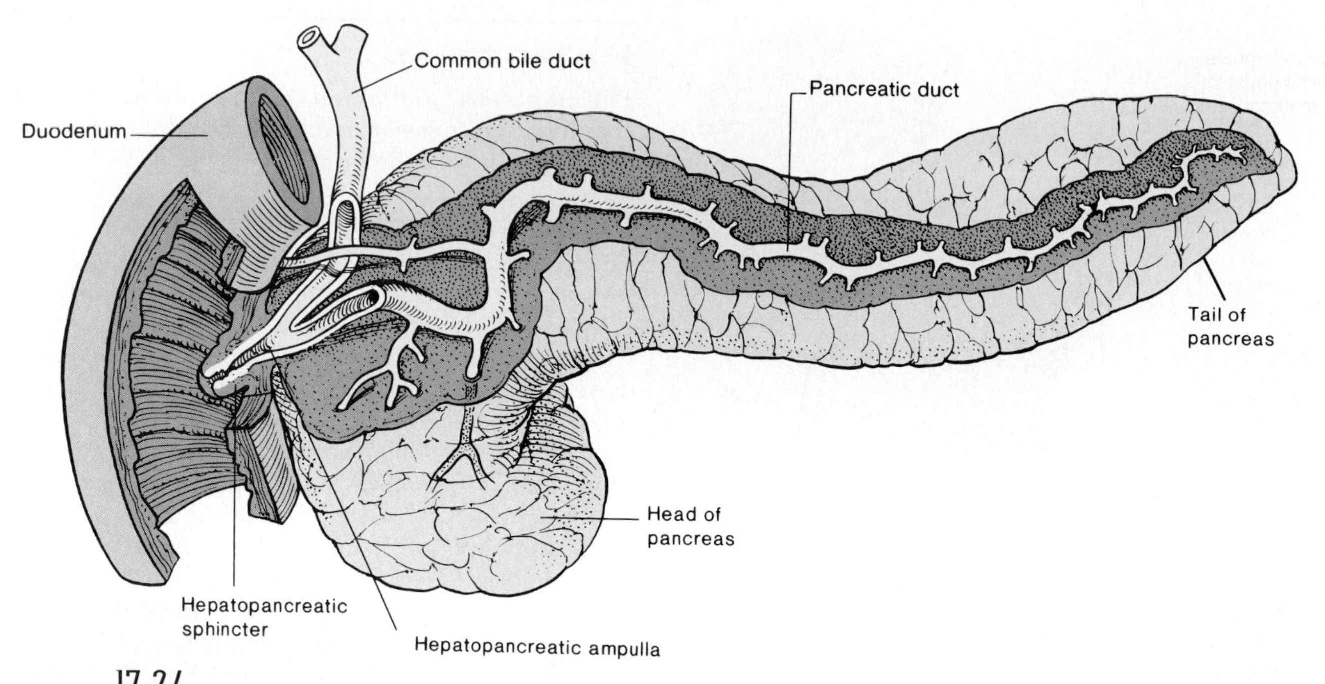


**FIGURE 17.24**
The pancreas is closely associated with the duodenum.

activated, in turn, by trypsin. This mechanism prevents enzymatic digestion of proteins within the secreting cells and the pancreatic ducts.

A painful condition called *acute pancreatitis* results from a blockage in the release of pancreatic juice. Trypsinogen, activated as pancreatic juice builds up, digests parts of the pancreas. Pancreatitis may be caused by alcoholism, gallstones, certain infections, a traumatic injury, or as a side effect of some drugs.

Pancreatic juice also contains two **nucleases,** which are enzymes that break nucleic acid molecules into nucleotides, as well as a high concentration of bicarbonate ions that makes it alkaline. The alkaline content of pancreatic juice provides a favorable environment for the actions of the digestive enzymes and helps neutralize the acidic chyme as it arrives from the stomach. At the same time, the alkaline condition in the small intestine blocks the action of pepsin, which might otherwise damage the duodenal wall.

## Regulation of Pancreatic Secretion

As with gastric and small intestinal secretions, the nervous and endocrine systems regulate release of pancreatic juice. For example, during the cephalic and gastric phases of gastric secretion, parasympathetic impulses reach the pancreas and stimulate it to release digestive enzymes. A peptide hormone, **secretin,** stimulates the pancreas to secrete a large quantity of fluid when acidic chyme enters the duodenum. Secretin is released into the blood from the duodenal mucous membrane in response to the acid in chyme. The pancreatic juice secreted at this time contains few, if any, digestive enzymes, but has a high concentration of bicarbonate ions that neutralizes the acid in chyme (fig. 17.25).

In cystic fibrosis, abnormal chloride channels in cells in various tissues cause water to be drawn into the cells from interstitial spaces. This dries out secretions in the lungs and pancreas, leaving a very sticky mucus that impairs the functioning of these organs. When the pancreas is plugged with this mucus, its secretions, containing digestive enzymes, cannot reach the duodenum. The person must take digestive enzyme supplements—usually as a powder mixed with a soft food such as applesauce—to prevent malnutrition.

The presence of proteins and fats in chyme within the duodenum also stimulates the release of cholecystokinin from the intestinal wall. As in the case of secretin, cholecystokinin reaches the pancreas

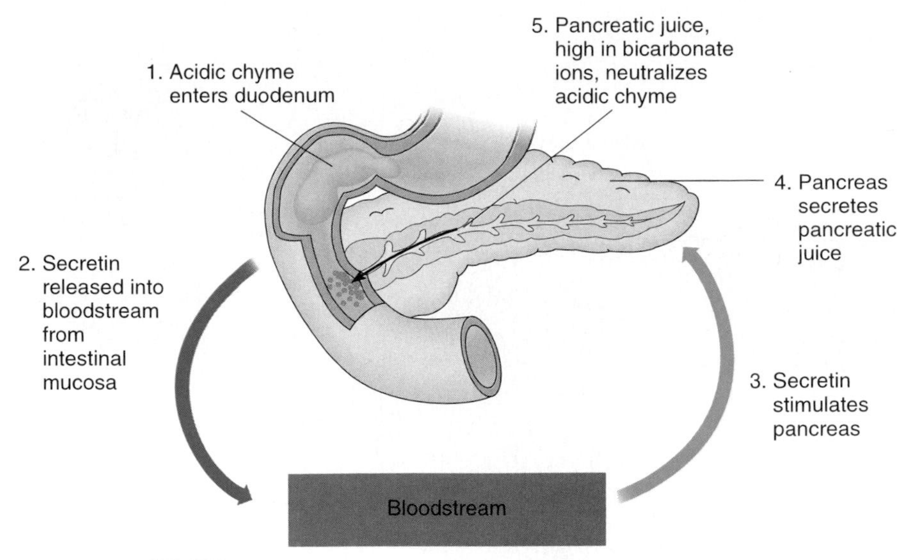


**FIGURE 17.25**
Acidic chyme entering the duodenum from the stomach stimulates release of secretin, which in turn stimulates release of pancreatic juice.

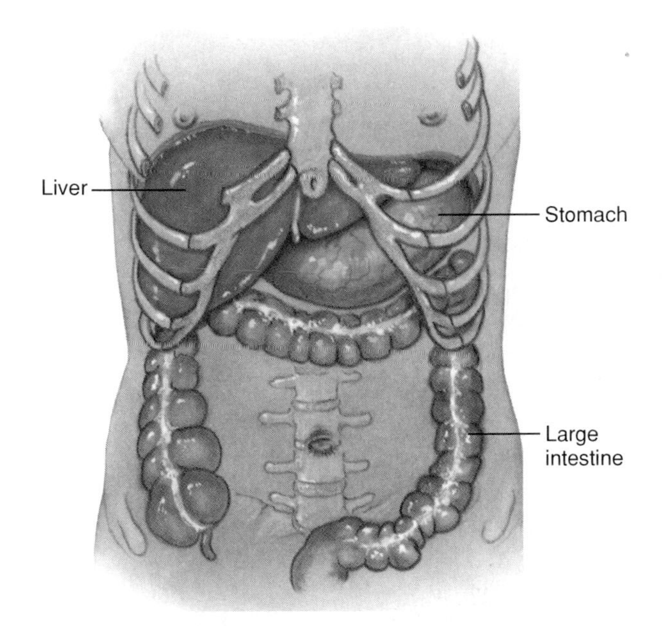


**FIGURE 17.26**
The ribs partially surround the liver.

by way of the bloodstream. Pancreatic juice secreted in response to cholecystokinin has a high concentration of digestive enzymes.

1. Where is the pancreas located?
2. List the enzymes found in pancreatic juice.
3. What are the functions of the enzymes in pancreatic juice?
4. What regulates secretion of pancreatic juice?

## LIVER

The **liver,** the largest internal organ, is located in the upper right and central portions of the abdominal cavity, just below the diaphragm. It is partially surrounded by the ribs, and extends from the level of the fifth intercostal space to the lower margin of the ribs. It is reddish brown in color and well supplied with blood vessels (figs. 17.26, 17.27, 17.28 and reference plates 51, 57, 74).

### Liver Functions

The liver carries on many important metabolic activities. Recall from chapter 13 that the liver plays a key role in carbohydrate metabolism by helping maintain the normal concentration of blood glucose. Liver cells responding to hormones such as insulin and glucagon decrease blood glucose by polymerizing glucose to glycogen and increase blood glucose by breaking down glycogen to glucose or by converting noncarbohydrates into glucose.

The liver's effects on lipid metabolism include oxidizing fatty acids at an especially high rate (see chapter 4); synthesizing lipoproteins, phospholipids, and cholesterol; and converting portions of carbohydrate and protein molecules into fat molecules. The blood transports fats synthesized in the liver to adipose tissue for storage.

The most vital liver functions are probably those related to protein metabolism. They include deaminating amino acids; forming urea (see chapter 20); synthesizing blood proteins, such as clotting factors (see chapter 14); and converting certain amino acids to other amino acids.

> Bacteria in the intestine produce ammonia, which is carried in the blood to the liver, where it is converted to urea. When this liver function fails, concentration of blood ammonia sharply rises, causing *hepatic coma,* a condition that can lead to death.

The liver also stores many substances, including glycogen, iron, and vitamins A, D, and $B_{12}$. Extra iron from the blood combines with a protein (apoferritin) in liver cells, forming *ferritin.* The iron is stored in this form until blood iron concentration lowers, when some of the iron is released. Thus, the liver is important in iron homeostasis.

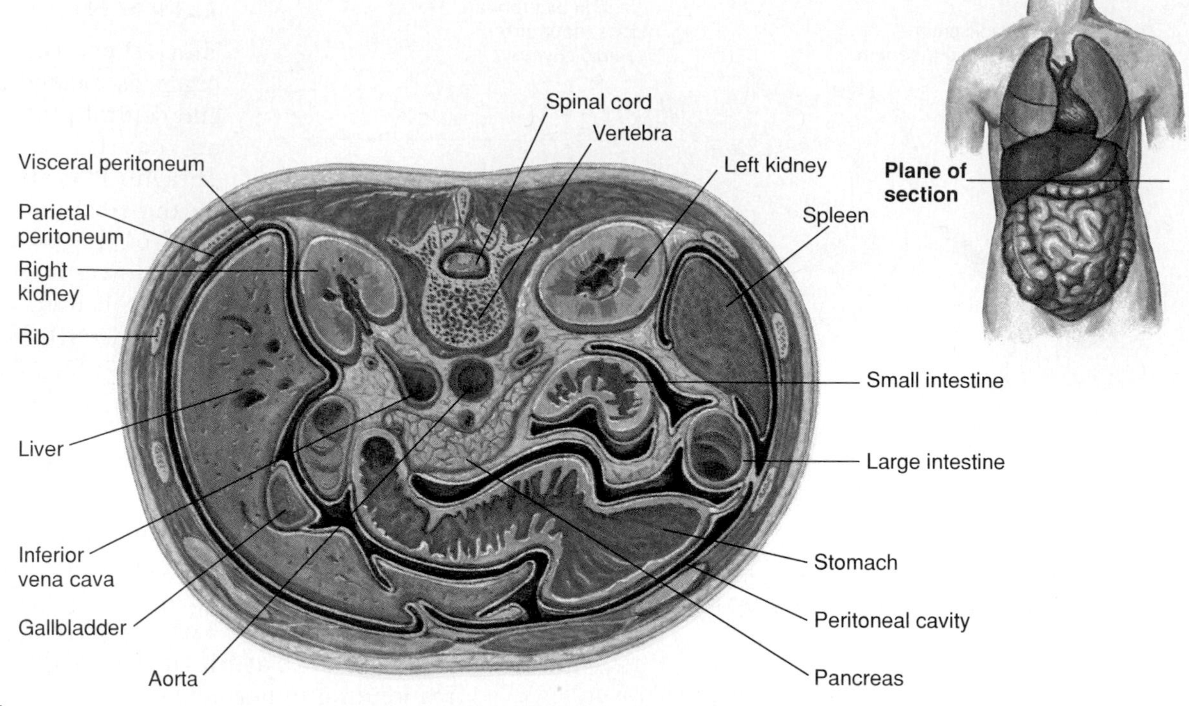


**FIGURE 17.27**
This transverse section of the abdomen reveals the liver and other organs within the upper portion of the abdominal cavity.

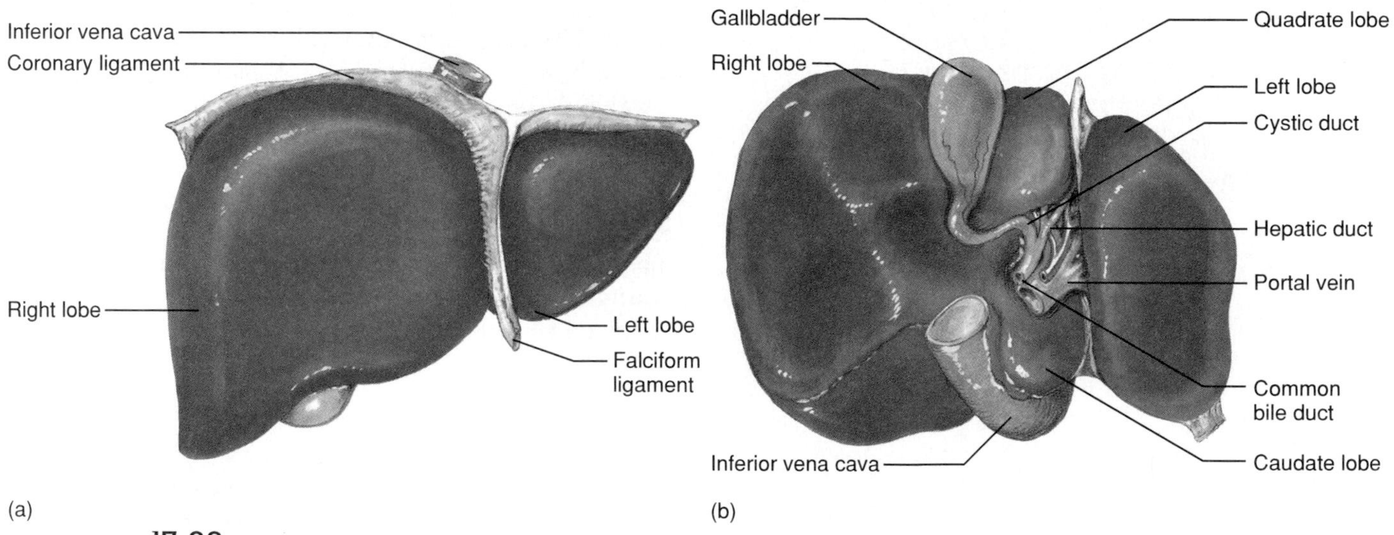


**FIGURE 17.28**
Lobes of the liver, viewed (*a*) from the front and (*b*) from below.

Liver cells help destroy damaged red blood cells and phagocytize foreign antigens. The liver removes toxic substances such as alcohol from the blood (detoxification). The liver can also serve as a blood reservoir, storing 200 to 400 milliliters of blood. The liver's role in digestion is to secrete bile, discussed later in the chapter. Chart 17.7 summarizes the major functions of the liver.

## Liver Structure

A fibrous capsule encloses the liver, and connective tissue divides the organ into a large *right lobe* and a smaller *left lobe*. The *falciform ligament* is a fold of visceral peritoneum that separates the lobes and fastens the liver to the abdominal wall anteriorly. The liver also has two minor lobes, the *quadrate lobe* near the gallbladder, and the *caudate lobe* close to the vena cava (see fig. 17.28).

## Chart 17.7 Major Functions of the Liver

| General Function | Specific Function |
|---|---|
| Carbohydrate metabolism | Polymerizes glucose to glycogen, breaks down glycogen to glucose, and converts noncarbohydrates to glucose |
| Lipid metabolism | Oxidizes fatty acids; synthesizes lipoproteins, phospholipids, and cholesterol; converts portions of carbohydrate and protein molecules into fats |
| Protein metabolism | Deaminates amino acids; forms urea; synthesizes blood proteins; interconverts amino acids |
| Storage | Stores glycogen, vitamins A, D, and $B_{12}$, iron, and blood |
| Blood filtering | Removes damaged red blood cells and foreign substances by phagocytosis |
| Detoxification | Removes toxins from the blood |
| Secretion | Secretes bile |

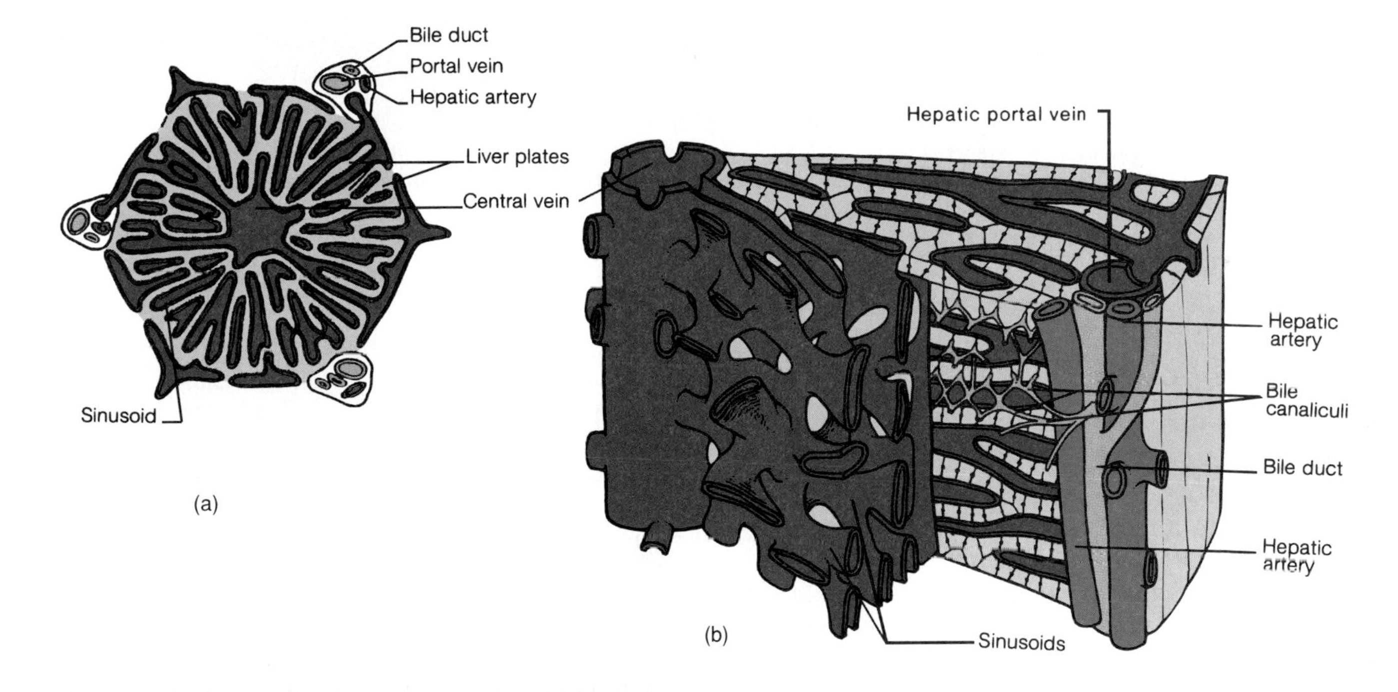


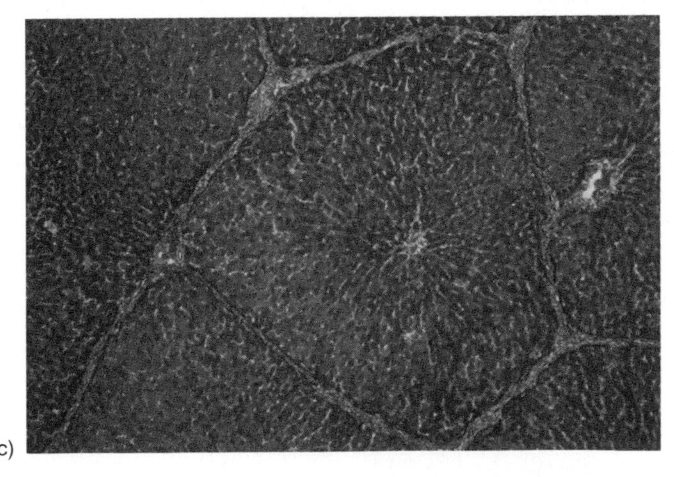


**Figure 17.29**
(*a*) Cross section of a hepatic lobule; (*b*) enlarged longitudinal section of a hepatic lobule; (*c*) light micrograph of hepatic lobules (cross section) (160×).

A fold of visceral peritoneum called the *coronary ligament* attaches the liver to the diaphragm on its superior surface. Each lobe is separated into many tiny **hepatic lobules,** which are the functional units of the gland (fig. 17.29). A lobule consists of many *hepatic cells* that radiate outward from a *central vein.* Vascular channels called **hepatic sinusoids** separate platelike groups of these cells from each other. Blood from the

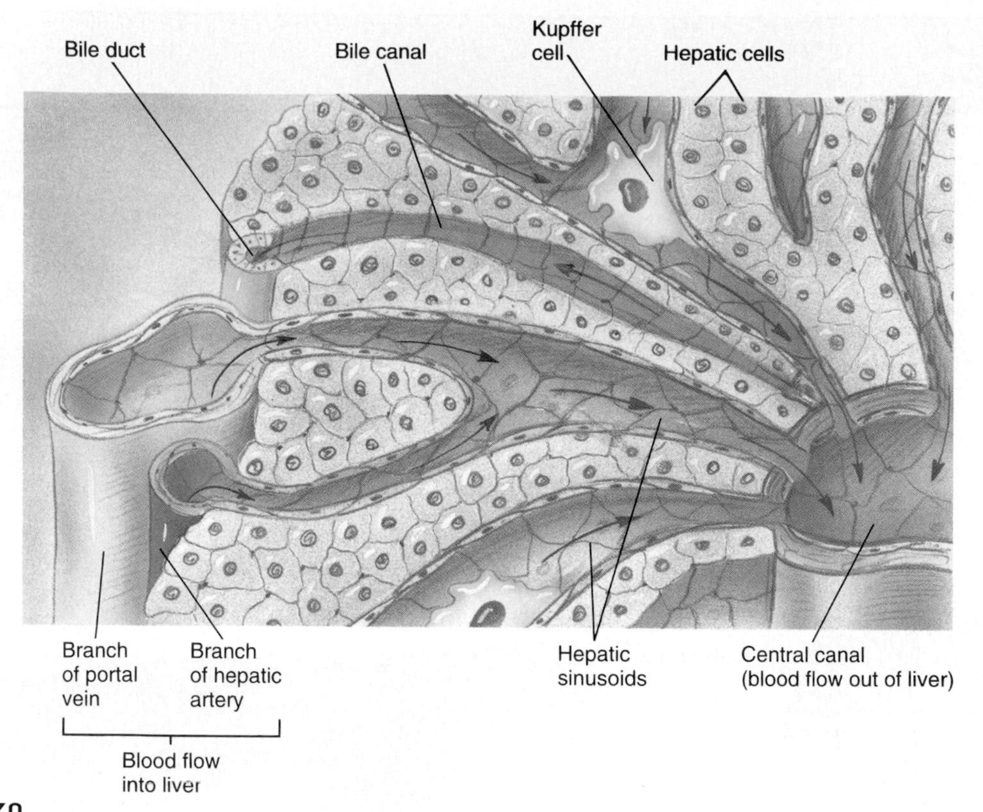


**FIGURE 17.30**
The paths of blood and bile within a hepatic lobule.

digestive tract, which is carried in the hepatic *portal vein* (see chapter 15), brings newly absorbed nutrients into the sinusoids (fig. 17.30). At the same time, oxygenated blood from the hepatic artery mixes freely with the blood containing nutrients, then flows through the liver sinusoids and nourishes the hepatic cells.

Often blood in the portal veins contains some bacteria that have entered through the intestinal wall. However, large **Kupffer cells,** which are fixed to the inner lining (endothelium) of the hepatic sinusoids, remove most of the bacteria from the blood by phagocytosis. Then the blood passes into the *central veins* of the hepatic lobules and moves out of the liver.

Within the liver lobules are many fine *bile canals* (or canaliculi), which receive secretions from the hepatic cells. The canals of neighboring lobules unite to form larger ducts, and then converge to become the **hepatic ducts.** These ducts merge, in turn, to form the *common hepatic duct.*

Clinical Application 17.3 discusses hepatitis, an inflammation of the liver.

1 Describe the location of the liver.

2 Review the functions of the liver.

3 Which liver function is directly related to digestion?

4 Describe a hepatic lobule.

## Composition of Bile

**Bile** is a yellowish green liquid that hepatic cells secrete continuously. In addition to water, it contains *bile salts, bile pigments, cholesterol,* and *electrolytes.* Of these, bile salts are the most abundant. They are the only bile substances that have a digestive function.

Hepatic cells use cholesterol to produce bile salts, and in secreting these salts, they release some cholesterol into the bile. Cholesterol has no special function in bile or in the alimentary canal.

Bile pigments (bilirubin and biliverdin) are breakdown products of hemoglobin from red blood cells. These pigments are normally excreted in the bile (see chapter 14). The yellowish skin, sclerae, and mucous membranes of jaundice result from excess deposition of bile pigments.

> Jaundice can have several causes. In *obstructive jaundice,* bile ducts are blocked. In *hepatocellular jaundice,* the liver is diseased. In *hemolytic jaundice,* red blood cells are destroyed too rapidly.

CLINICAL APPLICATION 17.3

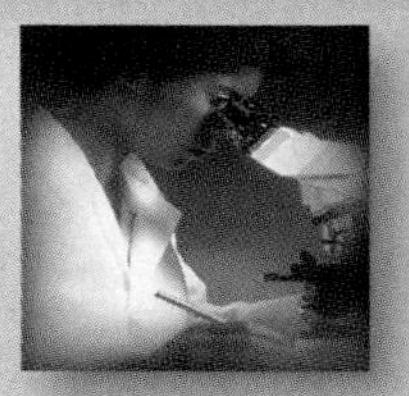

## Hepatitis

Hepatitis is an inflammation of the liver. It has several causes, but the various types have similar symptoms.

For the first few days, hepatitis may resemble the flu, producing mild headache, low fever, fatigue, lack of appetite, nausea and vomiting, and perhaps stiff joints. Between days 3 and 7, more distinctive symptoms appear: a rash, pain in the upper right quadrant of the chest, dark and foamy urine, and pale feces. About this time the skin and sclera of the eyes begin to turn yellow from accumulating bile pigments. (This is a form of jaundice.) Great fatigue may continue for two or three weeks, and then gradually the person begins to feel better.

This is hepatitis in its most common, least dangerous acute guise. It is not always so. About half a million people develop hepatitis in the United States each year, and only 6,000 die. In a rare form, called *fulminant hepatitis,* symptoms occur suddenly and severely, along with altered behavior and personality. Medical attention is necessary to prevent kidney or liver failure, or coma. Hepatitis that persists for more than 6 months is termed chronic. Perhaps as many as 300 million people worldwide are carriers of hepatitis. They do not have symptoms, but can infect others. Five percent of carriers develop liver cancer.

Rarely, hepatitis results from alcoholism, autoimmunity, or use of certain drugs. It is usually caused by one of five types of viruses, designated A through E in order of their discovery. The type of virus causing a particular case of hepatitis is often suspected by the route of infection, and a diagnosis is confirmed by detecting a surface molecule specific to a particular virus type. Distinctions between the viral types are discussed below:

*Hepatitis A* This virus is spread by contact with food or objects that have been contaminated with virus-containing feces. In day-care centers, for example, the infection is frequently spread through diaper changing. Infected food handlers can spread hepatitis A to many people. In one Missouri restaurant, 110 sick customers ate lettuce on the particular day that a worker who did not yet realize that she had hepatitis A shredded the lettuce. The course of hepatitis A is short and mild.

*Hepatitis B* This form of the illness is spread by contact with virus-containing body fluids, such as blood, saliva, or semen. It may be transmitted by blood transfusions, hypodermic needles, or sexual activity. Recently, twenty patients who had spent time on the same ward in a California hospital came down with hepatitis B. All of the patients had diabetes, and the hepatitis infection was traced to a spring-loaded fingerstick device used to sample blood from their fingers. Although nurses routinely discarded the lancet portion of the device between patients, they did not replace the platform on which the lancet rests. This was the source of the hepatitis B virus, which can live for up to a week outside the body.

*Hepatitis C* About half of all cases of hepatitis are now believed to be caused by the hepatitis C virus. It is primarily transmitted in blood—from sharing razors or needles, from pregnant woman to fetus, or in blood transfusions or use of blood products. For example, 112 cases of hepatitis C recently reported to U.S. health officials were traced to intravenous gamma globulin each person had received. The tainted treatment came from blood donated to the American Red Cross. As many as 60% of individuals infected with the hepatitis C virus suffer chronic symptoms.

*Hepatitis D* This form of hepatitis occurs in people already infected with the hepatitis B virus. It is blood borne, and associated with blood transfusions and intravenous use of drugs. About 20% of individuals infected with this virus die.

*Hepatitis E* The hepatitis E virus is usually transmitted in water contaminated with feces in developing nations—not to residents, who are immune, but most often to visitors. One woman contracted hepatitis E from crushed ice in her margaritas on a trip to Mexico. Several college students developed hepatitis E after swimming in the Ganges river while visiting India.

Because hepatitis is most often caused by a virus, antibiotic drugs, which are effective against bacteria, are not helpful. Usually the person must just wait out the symptoms. Hepatitis C, however, sometimes responds to a form of interferon, an immune system biochemical.

### Gallbladder

The **gallbladder** is a pear-shaped sac located in a depression on the inferior surface of the liver. It is connected to the **cystic duct,** which, in turn, joins the hepatic duct (see fig. 17.28 and The gallbladder has a capacity lined with columnar epithelial muscular layer in its wall. I



tion

meals, concentrates bile by reabsorbing water, and releases bile into the duodenum when stimulated by *cholecystokinin* from the small intestine.

The **common bile duct** is formed by the union of the common hepatic and cystic ducts. It leads to the duodenum, where the hepatopancreatic sphincter muscle guards its exit (see fig. 17.24). This sphincter normally remains contracted, so that bile collects in the common bile duct and backs up into the cystic duct. When this happens, the bile flows into the gallbladder, where it is stored.

While the bile is in the gallbladder, bile salts, bile pigments, and cholesterol become increasingly concentrated as the gallbladder lining reabsorbs some of the water and electrolytes. Although the cholesterol normally remains in solution, under certain conditions it may precipitate and form solid crystals. If cholesterol continues to come out of solution, these crystals become larger and larger, forming *gallstones* (fig. 17.31).

Gallstones may form if the bile is too concentrated, if the hepatic cells secrete too much cholesterol, or if the gallbladder is inflamed (cholecystitis). If such stones get into the bile duct, they may block the flow of bile, causing obstructive jaundice and considerable pain.

Clinical Application 17.4 discusses disorders of the gallbladder.

> A famous photograph of former U.S. president Lyndon Johnson depicts him displaying the 8-inch scar on his abdomen from where his gallbladder was removed to treat gallstones. Today, many such procedures—called *cholecystectomies*—are performed using a laser, which leaves four tiny cuts, each a quarter to a half inch long. Recovery from the laser procedure is much faster than from traditional surgery.

## Regulation of Bile Release

Normally, bile does not enter the duodenum until cholecystokinin stimulates the gallbladder to contract. The intestinal mucosa releases this hormone in response to the presence of proteins and fats in the small intestine. The hepatopancreatic sphincter usually remains contracted until a peristaltic wave in the duodenal wall approaches it. Then, just before the wave reaches it, the sphincter relaxes and a squirt of bile enters the duodenum (fig. 17.32).

The hormones that help control digestive functions are summarized in chart 17.8.

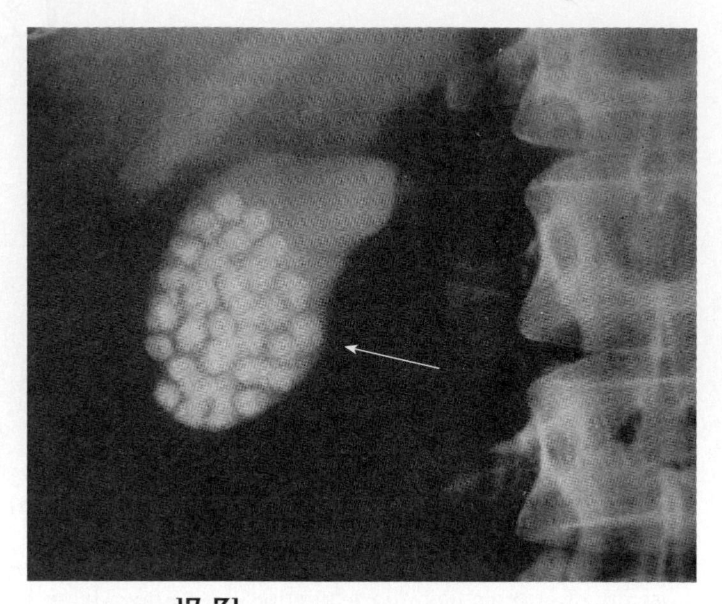

**FIGURE 17.31**
X-ray film of a gallbladder that contains gallstones (arrow).

## Functions of Bile Salts

Bile salts aid digestive enzymes and enhance absorption of fatty acids and certain fat-soluble vitamins. Molecules of fats tend to clump into *fat globules.* Bile salts affect fat globules much like a soap or detergent would affect them. That is, bile salts reduce surface tension and break fat globules into smaller droplets, an action called **emulsification.** Monoglycerides that form from the action of pancreatic lipase on triglyceride molecules aid emulsification. This emulsification greatly increases the total surface area of the fatty substance, and the tiny droplets mix with water. Lipases can then digest the fat molecules more effectively.

Bile salts aid in absorption of fatty acids and cholesterol by forming complexes (micelles) that are very soluble in chyme and that epithelial cells can more easily absorb. The fat-soluble vitamins A, D, E, and K are also absorbed. Lack of bile salts results in poor lipid absorption and vitamin deficiencies.

The mucous membrane of the small intestine reabsorbs nearly all of the bile salts, along with fatty acids. The bile salts are carried in the blood to the liver, where hepatic cells resecrete them into the bile ducts. The small quantities that are lost in the feces are replaced by synthesis of bile salts in liver cells.

1. Explain how bile originates.
2. Describe the function of the gallbladder.
3. How is secretion of bile regulated?
4. How does bile function in digestion?

## Clinical Application 17.4

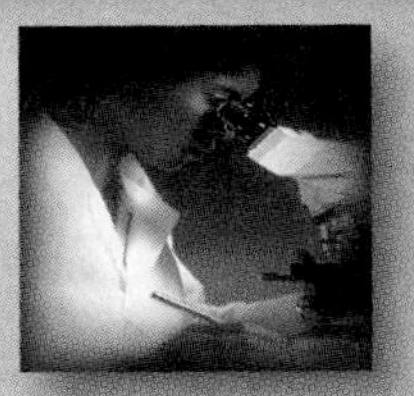

### Gallbladder Disease

Molly G., an overweight, 47-year-old college administrator and mother of four, had been feeling healthy until recently. Then she regularly began to feel pain in the upper right quadrant of her abdomen (see fig. 1.21). Sometimes the discomfort seemed to radiate around to her back and move upward into her right shoulder. Most commonly, she felt this pain after her evening meal; occasionally it also occurred during the night, awakening her. After an episode of particularly severe pain accompanied by sweating (diaphoresis) and nausea, Molly approached her physician.

During an examination of Molly's abdomen, the physician discovered tenderness in the epigastric region (see fig. 1.20). She decided that Molly might be experiencing the symptoms of *acute cholecystitis*—an inflammation of the gallbladder. The physician recommended that Molly have a *cholecystogram*—an X ray of the gallbladder.

Molly took tablets containing a contrast medium the night before the X-ray procedure. This schedule allowed time for the small intestine to absorb the substance, which was carried to the liver and excreted into the bile. Later, the bile and contrast medium would be stored and concentrated in the gallbladder and would make the contents of the gallbladder opaque to X rays.

Molly's cholecystogram (see fig. 17.31) revealed several stones (calculi) in her gallbladder, a condition called *cholelithiasis.* Because Molly's symptoms of gallbladder disease were worsening, her physician recommended that she consult with an abdominal surgeon about undergoing a *cholecystectomy*—surgical removal of the gallbladder.

During the surgical procedure, an incision was made in Molly's right subcostal region. Her gallbladder was excised from the liver. Then the cystic duct (see fig. 17.28) and hepatic ducts were explored for the presence of stones, but none were found.

Unfortunately, following her recovery from surgery, Molly's symptoms persisted. Her surgeon ordered a *cholangiogram*—an X-ray series of the bile ducts. This study showed a residual stone at the distal end of Molly's common bile duct (see fig. 17.24).

The surgeon extracted the residual stone using a *fiberoptic endoscope,* a long, flexible tube that can be passed through the patient's esophagus and stomach and into the duodenum. This instrument enables a surgeon to observe features of the gastrointestinal tract by viewing them directly through the eyepiece of the endoscope or by watching an image on the screen of a television monitor. A surgeon can also perform manipulations using specialized tools that are passed through the endoscope to its distal end.

In Molly's case, the surgeon performed an *endoscopic papillotomy*—an incision of the hepatopancreatic sphincter (see fig. 17.24). This was done by applying an electric current to a wire extending from the end of the endoscope. He then removed the exposed stone by manipulating a tiny basket at the tip of the endoscope.

## Small Intestine

The **small intestine** is a tubular organ that extends from the pyloric sphincter to the beginning of the large intestine. With its many loops and coils, it fills much of the abdominal cavity. Although it is 5.5–6.0 meters (18–20 feet) long in a cadaver when the muscular wall lacks tone, the small intestine may be only half this long in a living person.

The small intestine receives secretions from the pancreas and liver. It also completes digestion of the nutrients in chyme, absorbs the products of digestion, and transports the remaining residues to the large intestine.

### Parts of the Small Intestine

The small intestine, shown in figures 17.33 and 17.34, and in reference plates 51, 58, 74, and 75, consists of three portions: the duodenum, the jejunum, and the ileum.

The **duodenum,** which is about 25 centimeters long and 5 centimeters in diameter, lies behind the parietal peritoneum (retroperitoneal). It is the shortest and most fixed portion of the small intestine. The duodenum follows a C-shaped path as it passes in front of the right kidney and the upper three lumbar vertebrae.

The remainder of the small intestine is mobile and lies free in the peritoneal cavity. The proximal

## CHART 17.8 Hormones of the Digestive Tract

| Hormone | Source | Function |
|---|---|---|
| Gastrin | Gastric cells, in response to the presence of food | Causes gastric glands to increase their secretory activity |
| Intestinal gastrin | Cells of small intestine, in response to the presence of chyme | Causes gastric glands to increase their secretory activity |
| Somatostatin | Gastric cells | Inhibits secretion of acid by parietal cells |
| Intestinal somatostatin | Intestinal wall cells in response to the presence of fats | Inhibits secretion of acid by parietal cells |
| Cholecystokinin | Intestinal wall cells, in response to the presence of proteins and fats in the small intestine | Causes gastric glands to decrease their secretory activity and inhibits gastric motility; stimulates pancreas to secrete fluid with a high digestive enzyme concentration; stimulates gallbladder to contract and release bile |
| Secretin | Cells in the duodenal wall, in response to acidic chyme entering the small intestine | Stimulates pancreas to secrete fluid with a high bicarbonate ion concentration |

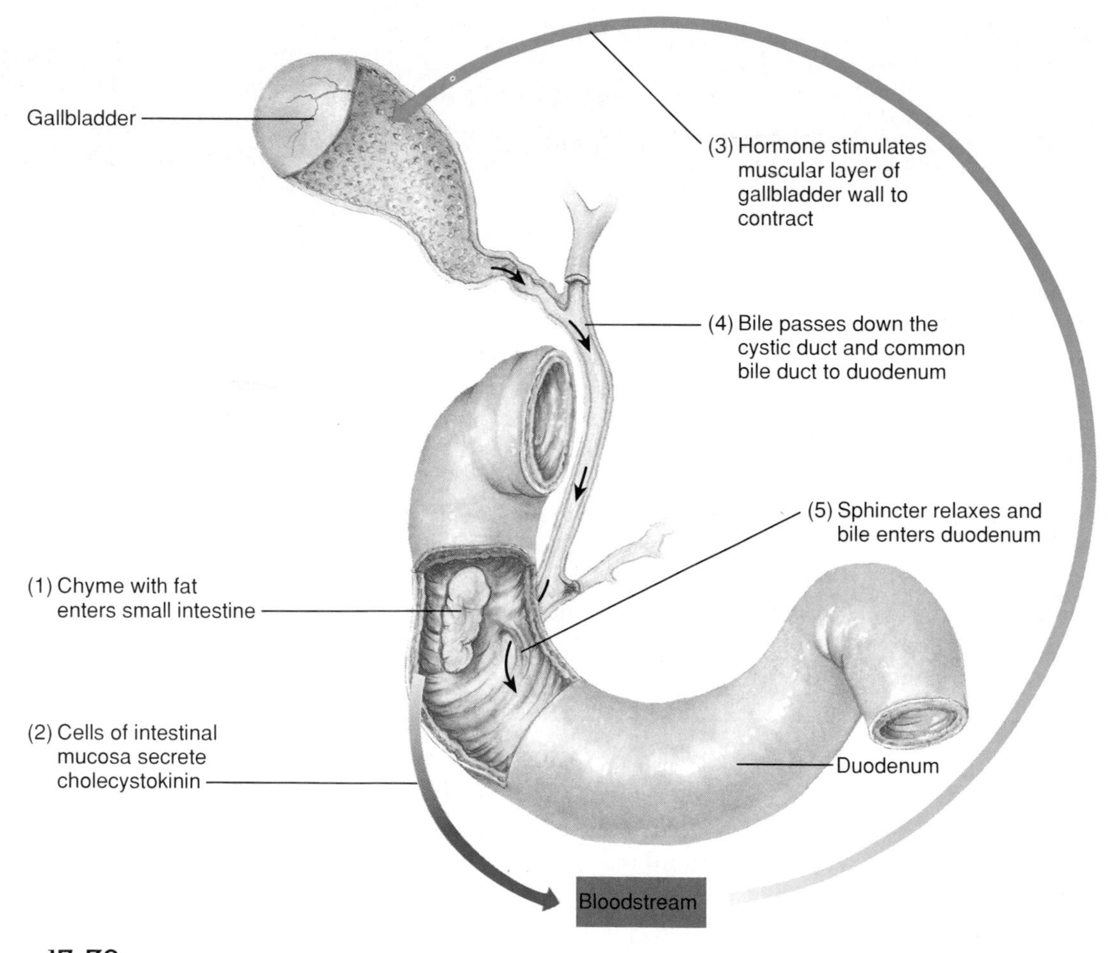


**FIGURE 17.32**
Fatty chyme entering the duodenum stimulates the gallbladder to release bile.

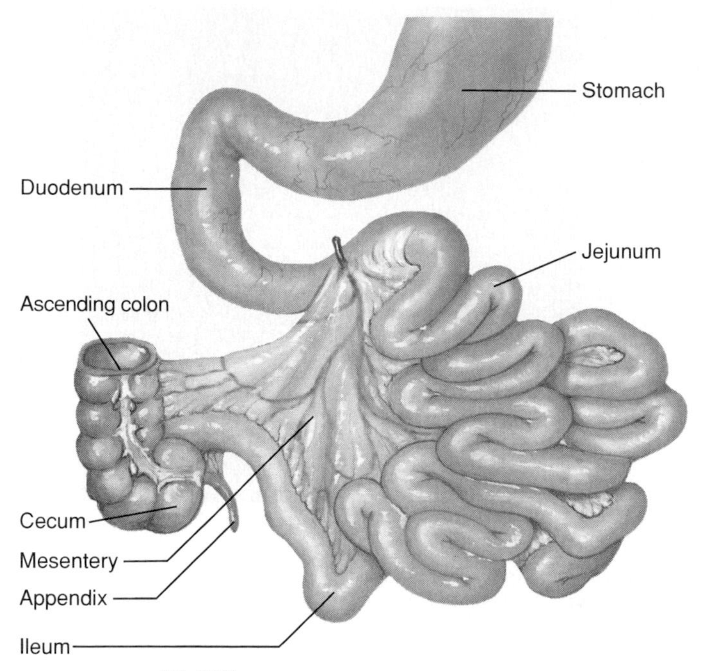


**FIGURE 17.33**
The three parts of the small intestine are the duodenum, the jejunum, and the ileum.

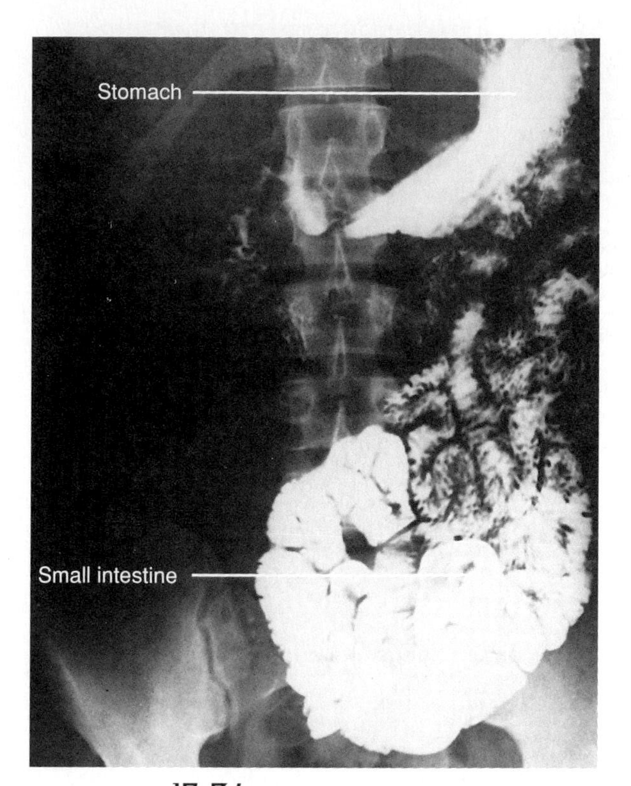


**FIGURE 17.34**
X-ray film showing a normal small intestine.

two-fifths of this portion is the **jejunum,** and the remainder is the **ileum.** There is no distinct separation between the jejunum and ileum, but the diameter of the jejunum is usually greater, and its wall is thicker, more vascular, and more active than that of the ileum.

The jejunum and ileum are suspended from the posterior abdominal wall by a double-layered fold of peritoneum called **mesentery** (fig. 17.35). This supporting tissue contains the blood vessels, nerves, and lymphatic vessels that supply the intestinal wall.

A filmy, double fold of peritoneal membrane called the *greater omentum* drapes like an apron from the stomach over the transverse colon and the folds of the small intestine. If infections occur in the wall of the alimentary canal, cells from the omentum may adhere to the inflamed region and help wall it off so that the infection is less likely to enter the peritoneal cavity (fig. 17.36).

## Structure of the Small Intestinal Wall

Throughout its length, the inner wall of the small intestine has a velvety appearance. This is due to innumerable tiny projections of mucous membrane called **intestinal villi** (figs. 17.37 and 17.38). These structures

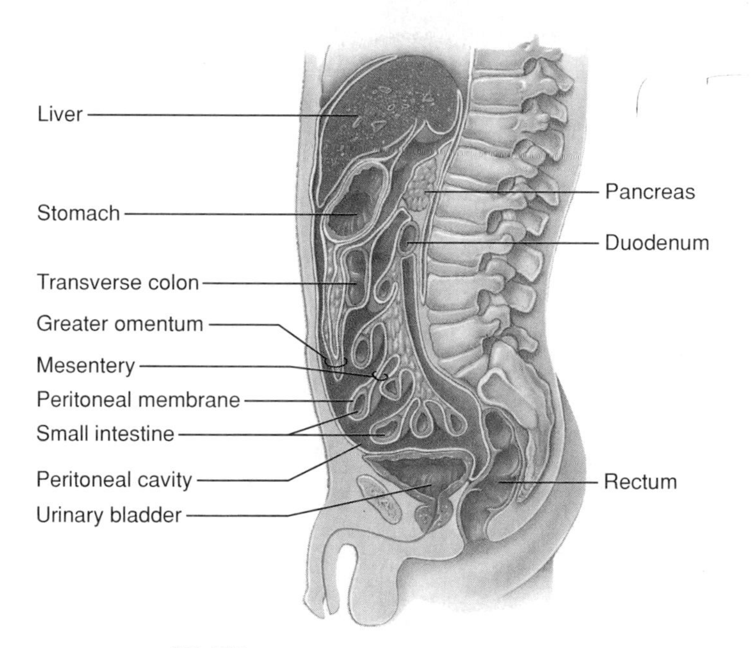


**FIGURE 17.35**
Portions of the small intestine are suspended from the posterior abdominal wall by mesentery formed by folds of the peritoneal membrane.

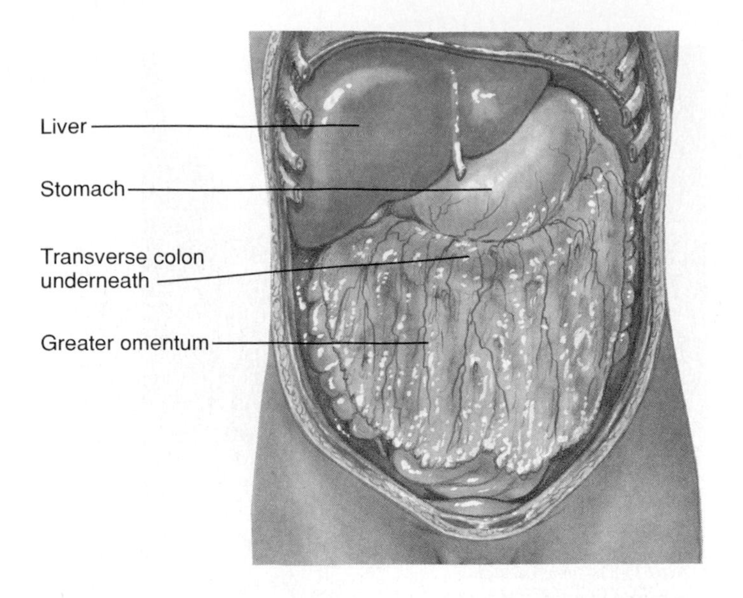


**FIGURE 17.36**
The greater omentum hangs like an apron over the abdominal organs.

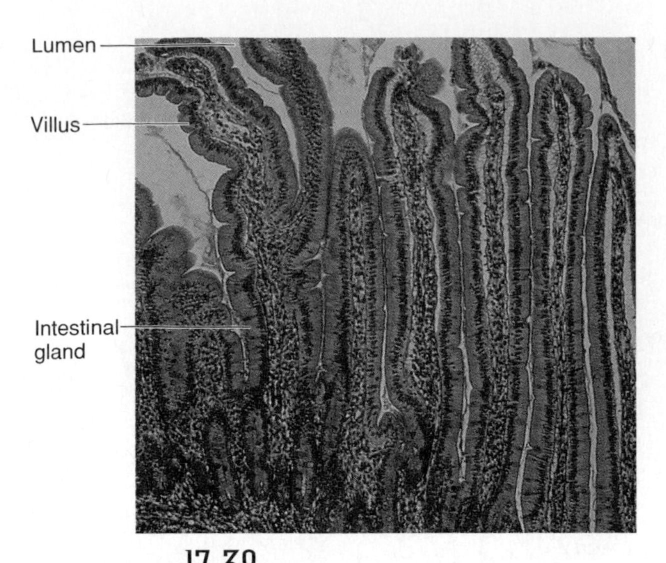


**FIGURE 17.38**
Light micrograph of intestinal villi from the wall of the duodenum (100×).

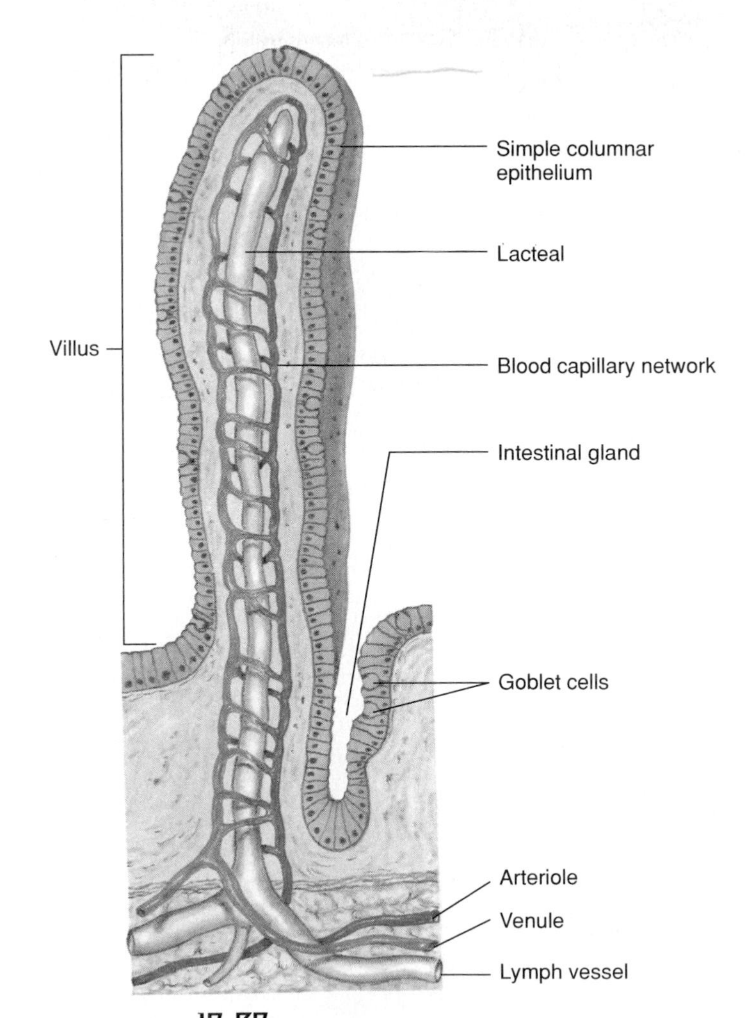


**FIGURE 17.37**
Structure of a single intestinal villus.

are most numerous in the duodenum and the proximal portion of the jejunum. They project into the passageway, or **lumen,** of the alimentary canal, contacting the intestinal contents. Villi increase the surface area of the intestinal lining, aiding absorption of digestive products.

Each villus consists of a layer of simple columnar epithelium and a core of connective tissue containing blood capillaries, a lymphatic capillary called a **lacteal,** and nerve fibers. At their free surfaces, the epithelial cells possess many fine extensions called *microvilli* that form a brushlike border and greatly increase the surface area of the intestinal cells, enhancing absorption (fig. 17.39). The blood and lymph capillaries carry away absorbed nutrients, and impulses transmitted by the nerve fibers can stimulate or inhibit activities of the villus.

Between the bases of adjacent villi are tubular **intestinal glands** (crypts of Lieberkühn), which extend downward into the mucous membrane. The deeper layers of the small intestinal wall are much like those of other parts of the alimentary canal in that they include a submucosa, a muscular layer, and a serosa.

The lining of the small intestine also has many circular folds of mucosa, called *plicae circulares,* that are especially well developed in the lower duodenum and upper jejunum. Together with the villi and microvilli, these folds help increase the surface area of the intestinal lining (fig. 17.40).

## Secretions of the Small Intestine

In addition to the mucus-secreting goblet cells, which are abundant throughout the mucosa of the small intestine, many specialized *mucus-secreting glands* (Brunner's glands) are in the submucosa within the

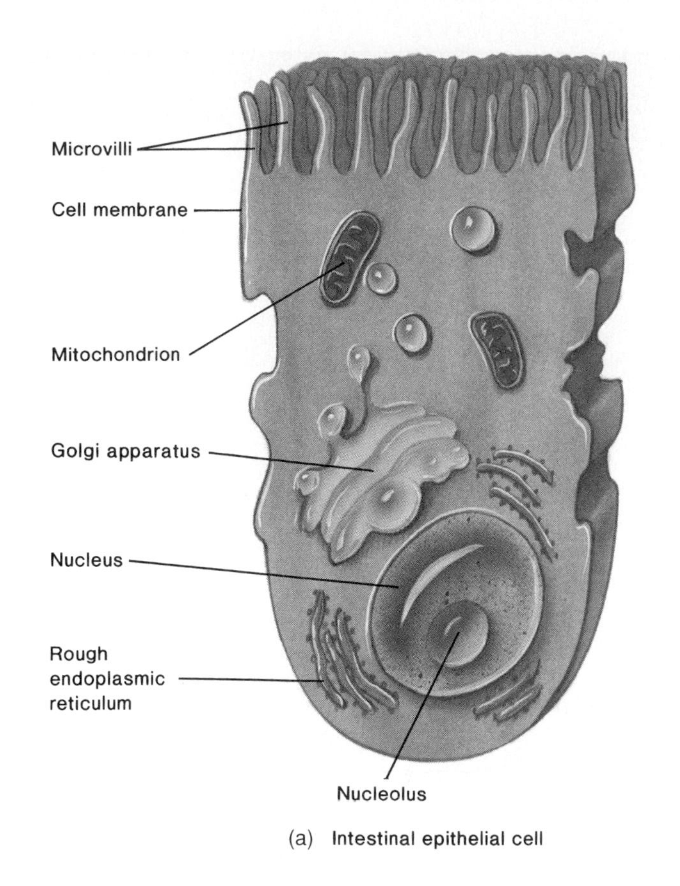


(a) Intestinal epithelial cell

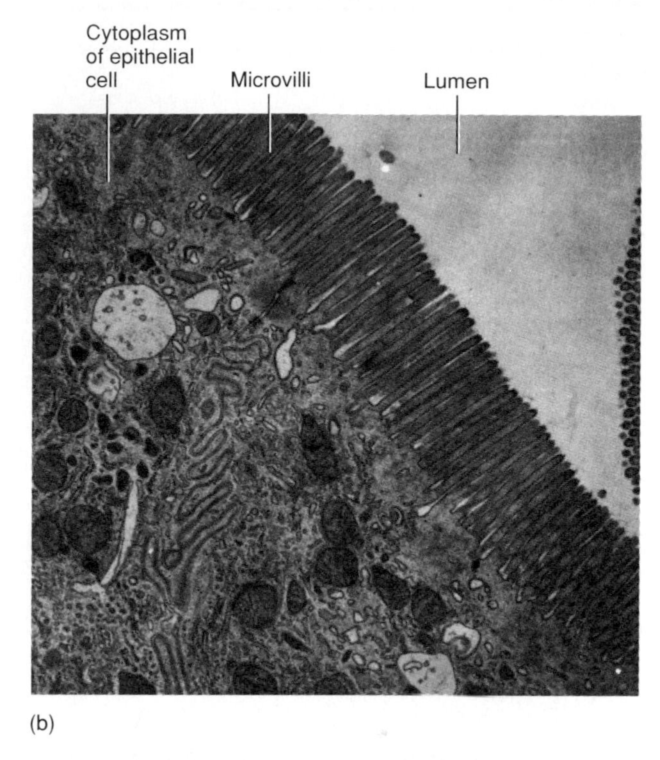


(b)

**FIGURE 17.39**
(*a*) Microvilli increase the surface area of intestinal epithelial cells; (*b*) transmission electron micrograph of microvilli.

proximal portion of the duodenum. These glands secrete large quantities of viscid, alkaline mucus in response to certain stimuli.

> The epithelial cells that form the lining of the small intestine are continually being replaced. The new cells form within the intestinal glands by mitosis, and they migrate outward onto the surface of a villus. When the migrating cells reach the tip of the villus, they are shed. As a result of this process, called *cellular turnover,* the epithelial lining of the small intestine is renewed every three to six days.

The intestinal glands at the bases of the villi secrete relatively large amounts of a watery fluid (see fig. 17.37). The villi rapidly reabsorb this fluid, and it provides a vehicle for moving digestive products into the villi. The fluid secreted by the intestinal glands has a pH that is nearly neutral (6.5–7.5), and it seems to lack digestive enzymes. However, the epithelial cells of the intestinal mucosa have digestive enzymes embedded in the surfaces of their microvilli. These enzymes break down food molecules just before absorption takes place. They include **peptidases,** which split peptides into their constituent amino acids; **sucrase, maltase,** and **lactase,** which split the double sugars (disaccharides) sucrose, maltose, and lactose into the simple sugars (monosaccharides) glucose, fructose, and galactose; and **intestinal lipase,** which splits fats into fatty acids and glycerol.

Chart 17.9 summarizes the sources and actions of the major digestive enzymes.

> Many adults do not produce sufficient lactase to adequately digest lactose, or milk sugar. In this *lactose intolerance,* lactose remains undigested, increasing osmotic pressure of the intestinal contents and drawing water into the intestines. At the same time, intestinal bacteria metabolize undigested sugar, producing organic acids and gases. The overall result of lactose intolerance is bloating, intestinal cramps, and diarrhea. To avoid these unpleasant symptoms, people with lactose intolerance can take lactase in pill form before eating dairy products. Infants with lactose intolerance can drink formula based on soybeans rather than milk.

## CHART 17.9 SUMMARY OF THE MAJOR DIGESTIVE ENZYMES

| Enzyme | Source | Digestive Action |
|---|---|---|
| ***Salivary enzyme*** | | |
| Amylase | Salivary glands | Begins carbohydrate digestion by breaking down starch and glycogen to disaccharides |
| ***Gastric enzymes*** | | |
| Pepsin | Gastric glands | Begins protein digestion |
| Lipase | Gastric glands | Begins butterfat digestion |
| ***Pancreatic enzymes*** | | |
| Amylase | Pancreas | Breaks down starch and glycogen into disaccharides |
| Lipase | Pancreas | Breaks down fats into fatty acids and glycerol |
| Proteinases<br>a. Trypsin<br>b. Chymotrypsin | Pancreas | Breaks down proteins or partially digested proteins into peptides |
| c. Carboxypeptidase | | Breaks down peptides into amino acids |
| Nucleases | Pancreas | Breaks down nucleic acids into nucleotides |
| ***Intestinal enzymes*** | | |
| Peptidase | Mucosal cells | Breaks down peptides into amino acids |
| Sucrase,<br>maltase,<br>lactase | Mucosal cells | Breaks down disaccharides into monosaccharides |
| Lipase | Mucosal cells | Breaks down fats into fatty acids and glycerol |
| Enterokinase | Mucosal cells | Breaks down trypsinogen into trypsin |

## Regulation of Small Intestinal Secretions

Since mucus protects the intestinal wall in the same way it protects the stomach lining, it is not surprising that mucus secretion increases in response to mechanical stimulation and the presence of irritants, such as gastric juice. Consequently, stomach contents entering the small intestine stimulate the duodenal mucous glands to release large quantities of mucus.

Goblet cells and intestinal glands secrete when they are stimulated by direct contact with chyme, which provides both chemical and mechanical stimuli, and by parasympathetic reflexes triggered as the intestinal wall distends.

1. Describe the parts of the small intestine.
2. What is the function of an intestinal villus?
3. Distinguish between intestinal villi and microvilli.
4. How is surface area maximized in the small intestine?
5. What is the function of the intestinal glands?
6.  l digestive enzymes.

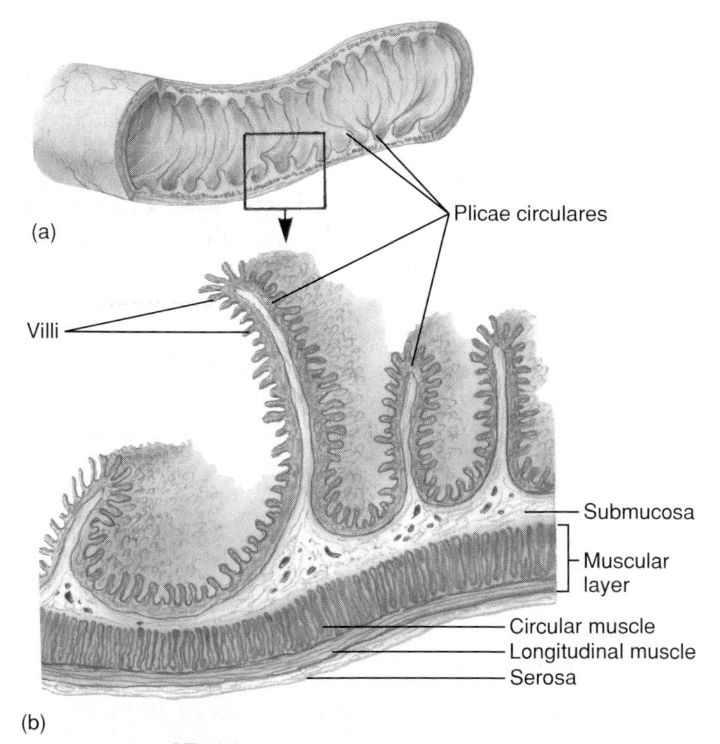


**FIGURE 17.40**

(*a*) The inner lining of the small intestine contains many circular folds, the plicae circulares; (*b*) a longitudinal section through some of these folds.

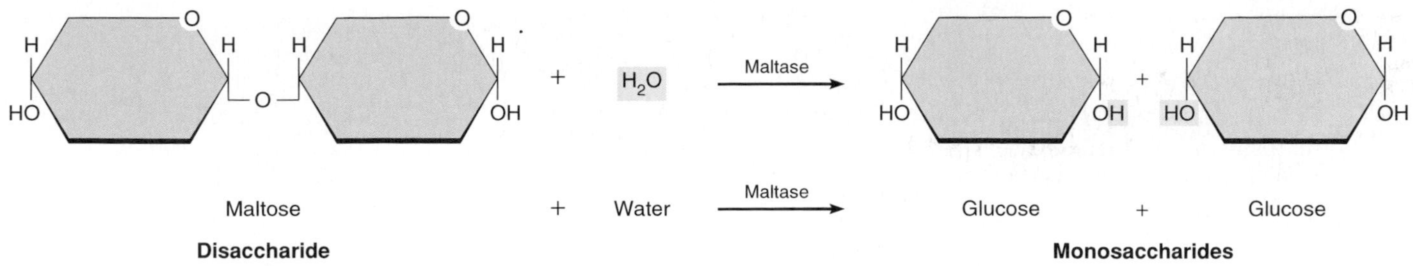

Maltose + Water $\xrightarrow{\text{Maltase}}$ Glucose + Glucose

Disaccharide | Monosaccharides

**FIGURE 17.41**
Digestion breaks down complex carbohydrates into disaccharides, which are then broken down into monosaccharides, which are small enough for intestinal villi to absorb. The monosaccharides then enter the bloodstream.

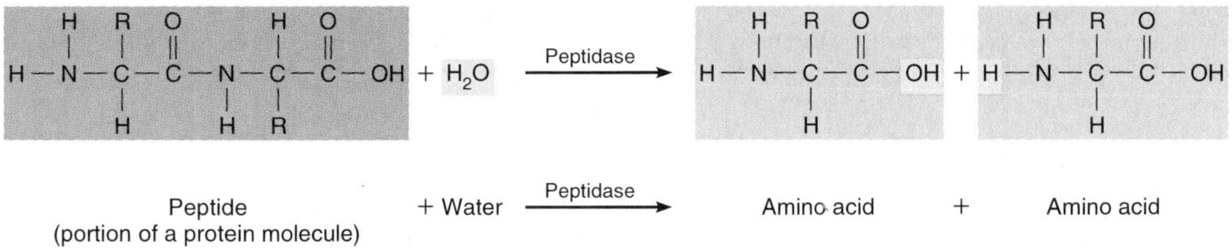

Peptide (portion of a protein molecule) + Water $\xrightarrow{\text{Peptidase}}$ Amino acid + Amino acid

**FIGURE 17.42**
The amino acids that result from dipeptide are absorbed by intestinal villi and enter the blood.

## Absorption in the Small Intestine

Because villi greatly increase the surface area of the intestinal mucosa, the small intestine is the most important absorbing organ of the alimentary canal. In fact, the small intestine is so effective in absorbing digestive products, water, and electrolytes, that very little absorbable material reaches its distal end.

Carbohydrate digestion begins in the mouth with the activity of salivary amylase, and is completed in the small intestine by enzymes from the intestinal mucosa and pancreas. The resulting monosaccharides are absorbed by the villi and enter blood capillaries (fig. 17.41). Simple sugars are absorbed by active transport or facilitated diffusion (see chapter 3).

Protein digestion begins in the stomach as a result of pepsin activity, and is completed in the small intestine by enzymes from the intestinal mucosa and the pancreas. During this process, large protein molecules are broken down into amino acids, which are then absorbed into the villi by active transport and are carried away by the blood (fig. 17.42).

Fat molecules are digested almost entirely by enzymes from the intestinal mucosa and pancreas (fig. 17.43). The resulting fatty acid molecules are absorbed in the following steps: (1) The fatty acid molecules dissolve in the epithelial cell membranes of the villi and diffuse through them. (2) The endoplasmic reticula of the cells use the fatty acids to resynthesize fat molecules similar to those previously digested. (3) These fats collect in clusters that become encased in protein. (4) The resulting large molecules of lipoprotein are called *chylomicrons,* and they make their way to the lacteal of the villus. (5) Periodic contractions of smooth muscles in the villus help empty the lacteal into the cysterna chyli, which empties into the thoracic duct. The lymph carries the chylomicrons to the bloodstream (fig. 17.44).

Chylomicrons are transported by the blood to capillaries of muscle and adipose tissues. A specific type of protein (apoprotein) associated with the surface of chylomicrons activates an enzyme (lipoprotein lipase) that is attached to the inner lining of such capillaries. As a result of the enzyme's action, fatty acids and monoglycerides are released from the chylomicrons, and they enter muscle or adipose cells for use as energy sources or to be stored. The remnants of the chylomicrons travel in the blood to the liver, where they bind to receptors on the surface of liver cells. The remnants quickly enter liver cells by receptor-mediated endocytosis (see chapter 3), and lysosomes degrade them.

Some fatty acids with relatively short carbon chains may be absorbed directly into the blood capillary of the villus without being converted back into fat.

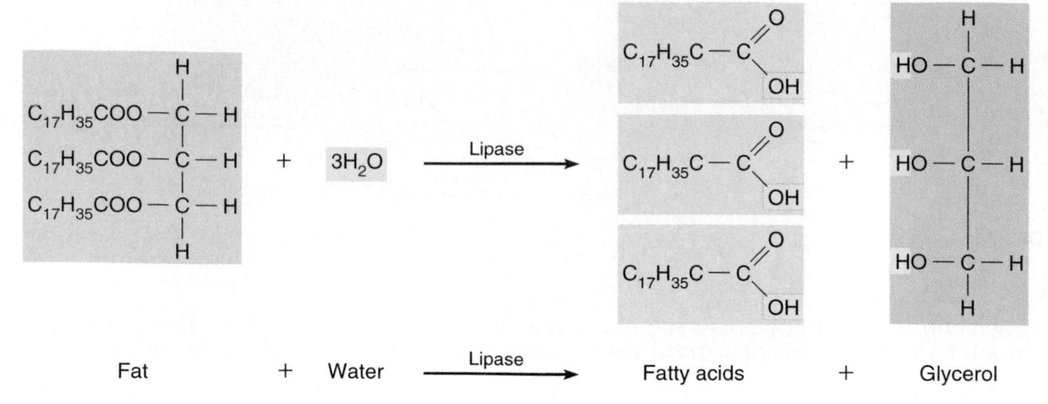


**FIGURE 17.43**

Fatty acids and glycerol result from fat digestion. Intestinal villi absorb them, and most are resynthesized into fat molecules before they enter the blood or lymph.

Be sure to read the labels on cakes claiming to be fat free. Often these products contain diglycerides, rather than the triglycerides found in dietary fats. Once diglycerides are broken down to their constituent fatty acids during digestion and enter the small intestine epithelium, they are re-formed into fat. So yes, these cakes are technically no-fat, but the result after digestion may be the same as eating the real thing.

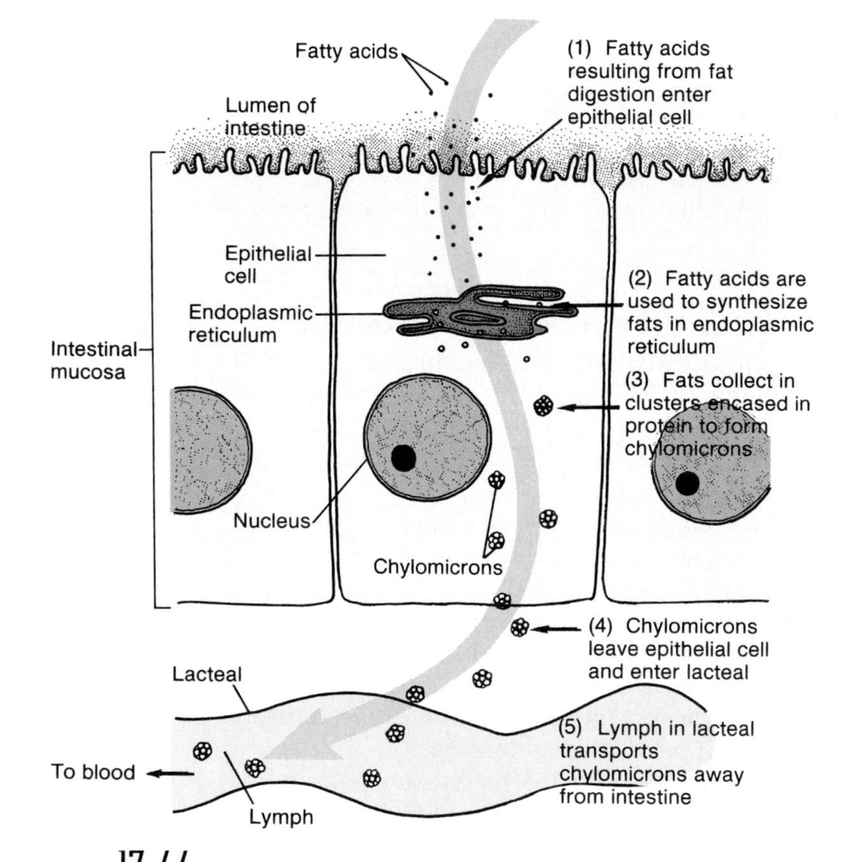


**FIGURE 17.44**

Fatty acid absorption has several steps.

In addition to absorbing the products of carbohydrate, protein, and fat digestion, the intestinal villi absorb electrolytes and water. Certain ions, such as those of sodium, potassium, chloride, nitrate, and bicarbonate, are readily absorbed; but others, including ions of calcium, magnesium, and sulfate, are poorly absorbed.

Electrolytes are usually absorbed by active transport, and water by osmosis. Thus, even though the intestinal contents may be hypertonic to the epithelial cells at first, as nutrients and electrolytes are absorbed, they become hypotonic to the cells. Then, water follows the nutrients and electrolytes into the villi by osmosis.

The absorption process is summarized in chart 17.10.

1. Which substances resulting from the digestion of carbohydrate, protein, and fat molecules does the small intestine absorb?
2. Which ions does the small intestine absorb?
3. What transport mechanisms do intestinal villi use?
4. Describe how fatty acids are absorbed and transported.

## CHART 17.10 INTESTINAL ABSORPTION OF NUTRIENTS

| Nutrient | Absorption Mechanism | Means of Transport |
|---|---|---|
| Monosaccharides | Facilitated diffusion and active transport | Blood in capillaries |
| Amino acids | Active transport | Blood in capillaries |
| Fatty acids and glycerol | Diffusion into cells | |
| | (a) Most fatty acids are converted back into fats and incorporated in chylomicrons for transport. | Lymph in lacteals |
| | (b) Some fatty acids with relatively short carbon chains are transported without being converted back into fats. | Blood in capillaries |
| Electrolytes | Diffusion and active transport | Blood in capillaries |
| Water | Osmosis | Blood in capillaries |

In *malabsorption,* the small intestine digests, but does not absorb, some nutrients. Causes of malabsorption include surgical removal of a portion of the small intestine, obstruction of lymphatic vessels due to a tumor, or interference with the production and release of bile as a result of liver disease.

Another cause of malabsorption is a reaction to *gluten,* found in certain grains, especially wheat and rye. This condition is called *celiac disease.* Microvilli are damaged and, in severe cases, villi may be destroyed. Both of these effects reduce the absorptive surface of the small intestine, preventing absorption of some nutrients. Symptoms of malabsorption include diarrhea, weight loss, weakness, vitamin deficiencies, anemia, and demineralization of the bones.

### Movements of the Small Intestine

Like the stomach, the small intestine carries on mixing movements and peristalsis. The major mixing movement is called *segmentation,* in which small, ringlike contractions occur periodically, cutting the chyme into segments and moving it back and forth. Segmentation also slows the movement of chyme through the small intestine.

Peristaltic waves propel chyme through the small intestine. These waves are usually weak, and they stop after pushing the chyme a short distance. Consequently, chyme moves relatively slowly through the small intestine, taking from three to ten hours to travel its length.

As might be expected, parasympathetic impulses enhance both mixing and peristaltic movements, and sympathetic impulses inhibit them. Reflexes involving parasympathetic impulses to the small intestine sometimes originate in the stomach. For example, as the stomach fills with food and its wall becomes distended, a reflex (gastroenteric reflex) is triggered, and peristaltic activity in the small intestine is greatly increased. Another reflex is initiated when the duodenum is filled with chyme and its wall is stretched. This reflex speeds movement through the small intestine.

If the small intestine wall becomes overdistended or irritated, a strong *peristaltic rush* may pass along the entire length of the organ, sweeping chyme into the large intestine so quickly that water, nutrients, and electrolytes that would normally be absorbed are not. The result is *diarrhea,* a condition in which defecation becomes more frequent and the stools become watery. Prolonged diarrhea causes imbalances in water and electrolyte concentrations.

A sphincter muscle called the **ileocecal valve** joins the small intestine's ileum to the large intestine's cecum. Normally, this sphincter remains constricted, preventing the contents of the small intestine from entering the large intestine, and at the same time preventing the contents of the large intestine from backing up into the ileum. However, after a meal, a gastroileal reflex is elicited that increases peristalsis in the ileum, forcing some of the contents of the small intestine into the cecum.

1. Describe the movements of the small intestine.
2. How are the movements of the small intestine initiated?
3. What is a peristaltic rush?
4. What stimulus relaxes the ileocecal valve?

## LARGE INTESTINE

The **large intestine** is so named because its diameter is greater than that of the small intestine. This portion of the alimentary canal is about 1.5 meters long, and it begins in the lower right side of the abdominal cavity where the ileum joins the cecum. From there, the large intestine travels upward on the right side, crosses obliquely to the left, and descends into the pelvis. At its distal end, it opens to the outside of the body as the anus.

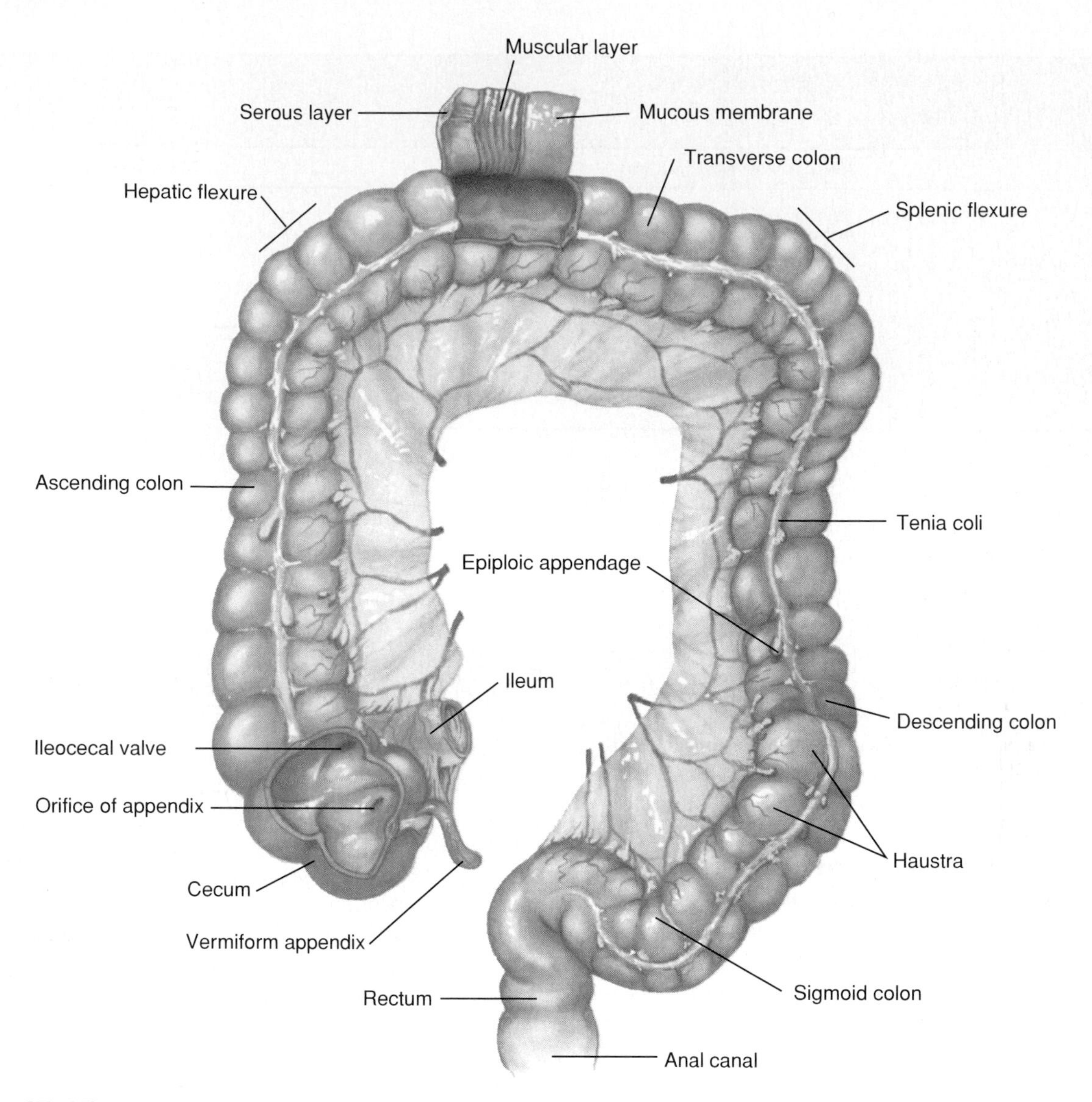


**FIGURE 17.45**
Parts of the large intestine (anterior view).

The large intestine reabsorbs water and electrolytes from the chyme remaining in the alimentary canal. It also reabsorbs and recycles water and remnants of digestive secretions. The large intestine also forms and stores feces.

## Parts of the Large Intestine

The large intestine, shown in figures 17.45 and 17.46 and in reference plates 51, 52, 58, and 75, consists of the cecum, the colon, the rectum, and the anal canal.

The **cecum,** which represents the beginning of the large intestine, is a dilated, pouchlike structure that hangs slightly below the ileocecal opening. Projecting downward from it is a narrow tube with a closed end called the **vermiform** (wormlike) **appendix.** The human appendix has no known digestive function. It contains lymphatic tissue.

In *appendicitis,* the appendix becomes inflamed and infected. Surgery is necessary to prevent it from rupturing. If it does break open, the contents of the large intestine may enter the abdominal cavity and cause a serious infection of the peritoneum called *peritonitis.*

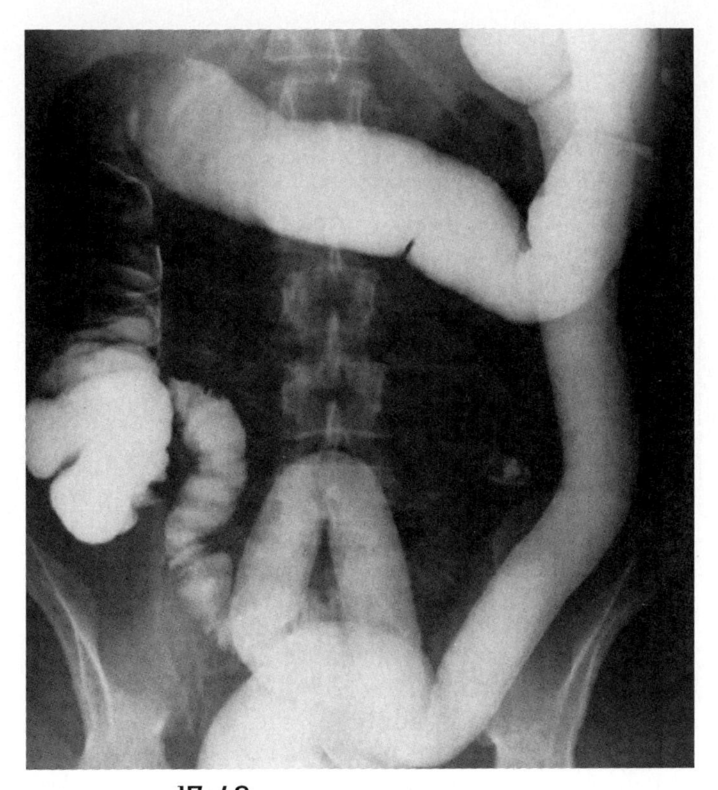

**FIGURE 17.46**
X-ray film of the large intestine.

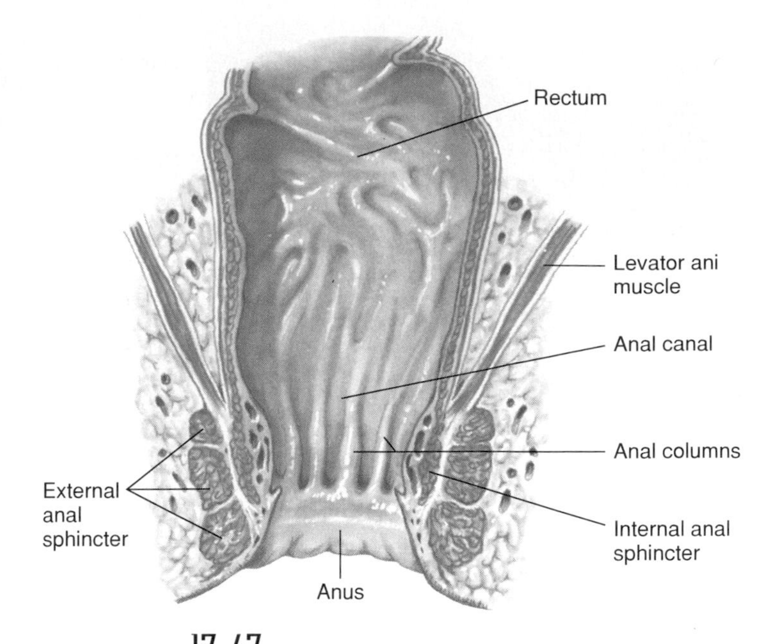


**FIGURE 17.47**
The rectum and the anal canal are located at the distal end of the alimentary canal.

The **colon** is divided into four portions—the ascending, transverse, descending, and sigmoid colons. The **ascending colon** begins at the cecum and travels upward against the posterior abdominal wall to a point just below the liver. There it turns sharply to the left (as the right colic, or hepatic, flexure) and becomes the **transverse colon.** The transverse colon is the longest and most movable part of the large intestine. It is suspended by a fold of peritoneum and sags in the middle below the stomach. As the transverse colon approaches the spleen, it turns abruptly downward (as the left colic, or splenic, flexure) and becomes the **descending colon.** At the brim of the pelvis, the descending colon makes an S-shaped curve, called the **sigmoid colon,** and then becomes the rectum.

The **rectum** lies next to the sacrum and generally follows its curvature. It is firmly attached to the sacrum by the peritoneum, and it ends about 5 centimeters below the tip of the coccyx, where it becomes the anal canal (fig. 17.47).

The **anal canal** is formed by the last 2.5 to 4.0 centimeters of the large intestine. The mucous membrane in the canal is folded into a series of six to eight longitudinal *anal columns.* At its distal end, the canal opens to the outside as the **anus.** Two sphincter muscles guard the anus—an *internal anal sphincter muscle* composed of smooth muscle under involuntary control, and an *external anal sphincter muscle* composed of skeletal muscle under voluntary control.

> *Hemorrhoids* are, literally, a pain in the rear. Enlarged and inflamed branches of the rectal vein in the anal columns causes intense itching, sharp pain, and sometimes bright red bleeding. The hemorrhoids may be internal (which do not produce symptoms) or bulge out of the anus. Causes of hemorrhoids include anything that puts prolonged pressure on the delicate rectal tissue, including obesity, pregnancy, constipation, diarrhea, and liver disease.
>
> Preventing or curing hemorrhoids is usually possible by eating more fiber-rich foods and drinking lots of water. Warm soaks in the tub, cold packs, and careful wiping of painful areas also helps, as do external creams and ointments. Surgery—with a scalpel or a laser—can remove severe hemorrhoids.

1. What is the general function of the large intestine?
2. Describe the parts of the large intestine.
3. Distinguish between the internal sphincter muscle and the external sphincter muscle of the anus.

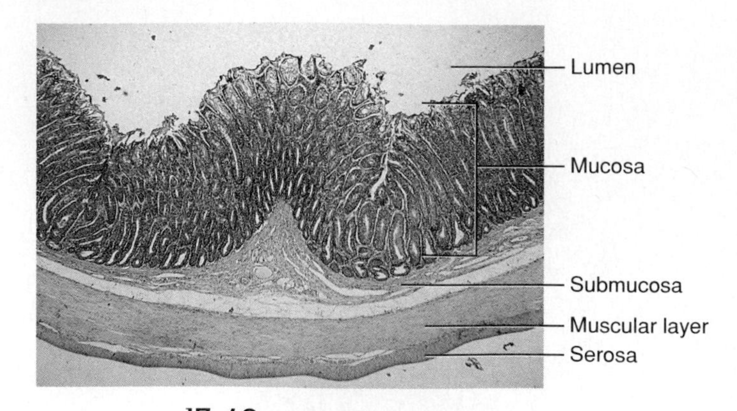


**FIGURE 17.48**
Light micrograph of the large intestinal wall (10×).

## Structure of the Large Intestinal Wall

Although the wall of the large intestine includes the same types of tissues found in other parts of the alimentary canal, it has some unique features. For example, it lacks the villi and plicae circularis that are characteristic of the small intestine. Also, the layer of longitudinal muscle fibers does not cover its wall uniformly; the fibers are arranged in three distinct bands (teniae coli) that extend the entire length of the colon. These bands exert tension on the wall, creating a series of pouches (haustra). The large intestinal wall also has small collections of fat (epiploic appendages) in the serosa on its outer surface (fig. 17.48).

## Functions of the Large Intestine

Unlike the small intestine, which secretes digestive enzymes and absorbs the products of digestion, the large intestine has little or no digestive function. However, the mucous membrane that forms the inner lining of the large intestine contains many tubular glands. Structurally, these glands are similar to those of the small intestine, but they are composed almost entirely of goblet cells. Consequently, mucus is the only significant secretion of this portion of the alimentary canal (fig. 17.49).

The rate of mucus secretion is controlled by mechanical stimulation from chyme and by parasympathetic impulses. In both cases, the goblet cells respond by increasing mucus production, which, in turn, protects the intestinal wall against the abrasive action of materials passing through it. Mucus also holds particles of fecal matter together, and, because it is alkaline, mucus helps control the pH of the large intestinal contents. This is important because acids are sometimes released from the feces as a result of bacterial activity.

> Aspirin may reduce the death rate from cancers of the esophagus, stomach, colon, and rectum by up to 40%. These digestive cancers kill 81,000 people a year. Researchers do not know why aspirin seems to protect against these cancers. One suggestion is that it causes bleeding of the alimentary canal, as evidenced by rectal bleeding, which may send people to the doctor earlier, leading to earlier diagnosis, when treatment is more likely to be successful. Alternatively, aspirin may protect against these cancers by interfering with prostaglandin activity.

The chyme entering the large intestine usually has few nutrients remaining in it, consisting mostly of materials not digested or absorbed in the small intestine. It also contains water, electrolytes, mucus, and bacteria.

Absorption in the large intestine is normally limited to water and electrolytes, and this usually takes place in the proximal half of the tube. Electrolytes, such as sodium ions, can be absorbed by active transport, while the water follows, passively entering the mucosa by osmosis. As a result, about 90% of the water that enters the large intestine is absorbed, and little sodium or water is lost in the feces.

The many bacteria that normally inhabit the large intestine break down some of the molecules that escape the actions of human digestive enzymes. These bacteria are colorfully called *intestinal flora.* For instance, cellulose, a complex carbohydrate found in food of plant origin, passes through the alimentary canal almost unchanged, but colon bacteria can break down cellulose and use it as an energy source. This is why we can utilize dietary fiber. At the same time, these bacteria synthesize certain vitamins, such as K, $B_{12}$, thiamine, and riboflavin. The intestinal mucosa absorbs these vitamins. Bacterial actions in the large intestine also may give rise to intestinal gas (flatus).

1. How does the structure of the large intestine differ from that of the small intestine?
2. What substances does the large intestine absorb?
3. What useful substances do bacteria inhabiting the large intestine produce?

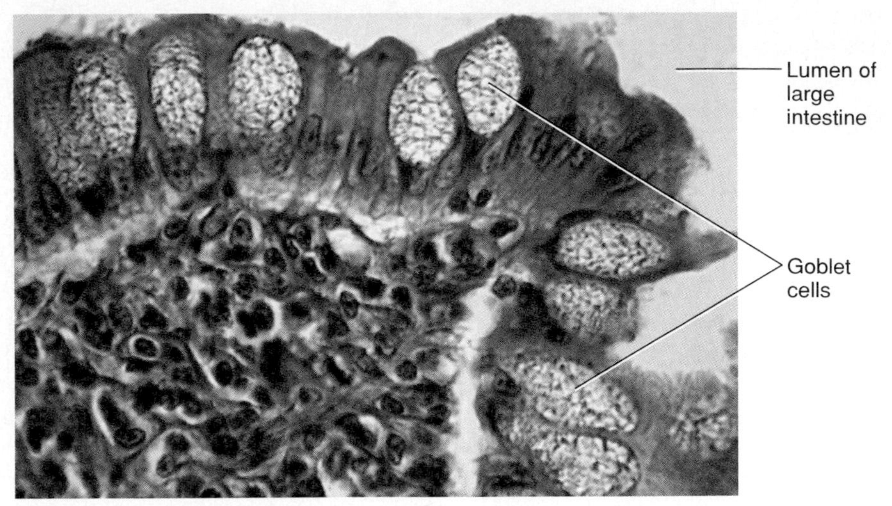


**FIGURE 17.49**
Light micrograph of the large intestinal mucosa (250×).

## Movements of the Large Intestine

The movements of the large intestine—mixing and peristalsis—are similar to those of the small intestine, although they are usually more sluggish. The mixing movements break the fecal matter into segments and turn it so that all portions are exposed to the intestinal mucosa. This helps in water and electrolyte absorption.

The peristaltic waves of the large intestine are different from those of the small intestine. Instead of occurring frequently, they happen only two or three times each day. These waves produce *mass movements* in which a relatively large section of the intestinal wall constricts vigorously, forcing the intestinal contents to move toward the rectum. Typically, mass movements occur following a meal, as a result of the gastrocolic reflex. Abnormal irritations of the intestinal mucosa can also trigger such movements. For instance, a person suffering from an inflamed colon (colitis) may experience frequent mass movements.

When it is appropriate to defecate, a person usually can initiate a *defecation reflex* by holding a deep breath and contracting the abdominal wall muscles. This action increases the internal abdominal pressure and forces feces into the rectum. As the rectum fills, its wall is distended and the defecation reflex is triggered, stimulating peristaltic waves in the descending colon, and the internal anal sphincter relaxes. At the same time, other reflexes involving the sacral region of the spinal cord strengthen the peristaltic waves, lower the diaphragm, close the glottis, and contract the abdominal wall muscles. These actions additionally increase the internal abdominal pressure and assist in squeezing the rectum. The external anal sphincter is signaled to relax, and the feces are forced to the outside. A person can inhibit defecation voluntarily by contracting the external anal sphincter.

## Feces

**Feces** are composed of materials that were not digested or absorbed, together with water, electrolytes, mucus, and bacteria. Usually the feces are about 75% water, and their color derives from bile pigments that have been altered somewhat by bacterial action.

The pungent odor of the feces results from a variety of compounds produced by bacteria. These compounds include phenol, hydrogen sulfide, indole, skatole, and ammonia.

Clinical Application 17.5 examines conditions affecting the large intestine.

1. How does peristalsis in the large intestine differ from peristalsis in the small intestine?
2. List the major events that occur during defecation.
3. Describe the composition of feces.

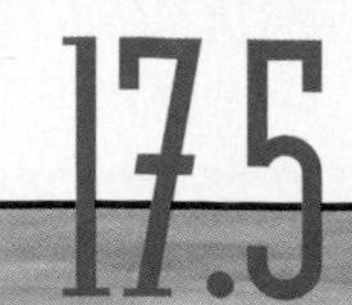

CLINICAL APPLICATION 17.5

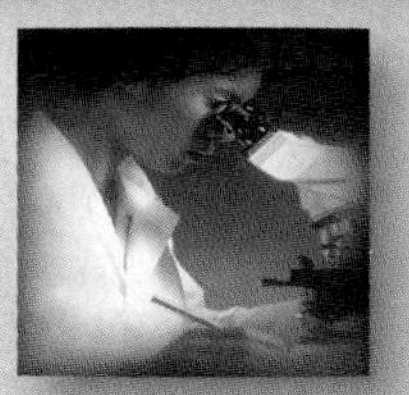

# Disorders of the Large Intestine

The large intestine is the source of many medical problems, from such familiar digestive discomforts as gas to more serious disorders.

## INTESTINAL GAS

People do not often talk about intestinal gas, but this common evidence of digestion is a source of pain and sometimes embarrassment to many. What exactly is in intestinal gas? Most of it is nitrogen and oxygen gulped in while breathing and eating. Undigested food fermented by bacteria contributes methane ($CH_4$), carbon dioxide ($CO_2$), and hydrogen. These five gases account for 99% of intestinal gas. The other 1% comes from compounds also produced by intestinal bacteria, and these impart foul odors. Intestinal gas can be minimized by eating slowly, avoiding milk if you are lactose intolerant, and not eating gas-inducing foods—beans, bagels, bran, broccoli, brussels sprouts, cabbage, cauliflower, and onions.

## DIARRHEA

Because the large intestine absorbs water from material within it, the rate of movement through it determines the consistency of feces. *Diarrhea,* the frequent and too-rapid passage of loose feces, results when material moves along so quickly that too little water is absorbed. The condition may reflect poisoning, infection, a diet too high in fiber, or nervousness. When the large intestine is the site of an infection or chemical irritation, diarrhea protects by flushing toxins out of the body.

## CONSTIPATION

*Constipation,* the infrequent passage of hard feces, is caused by abnormally slow movement of fecal matter through the large intestine. Because the feces remain in the large intestine longer than usual, excess water is absorbed. Constipation can be caused by a failure of the sensory cells in the rectum to signal the spinal cord to defecate or by the conscious suppression of defecation, both of which can be a result of emotional stress. A diet low in fiber can also cause constipation by slowing fecal movement through the large intestine. Eating foods high in fiber, drinking at least eight 8-ounce glasses of water a day, and regular exercise can prevent constipation.

## MORE SERIOUS DISORDERS

In *diverticulosis,* parts of the intestinal wall weaken, and the inner mucous membrane protrudes through (fig. 17B). Many times there are no symptoms, but if the outpouching becomes blocked with chyme and then infected (a condition called divertic-

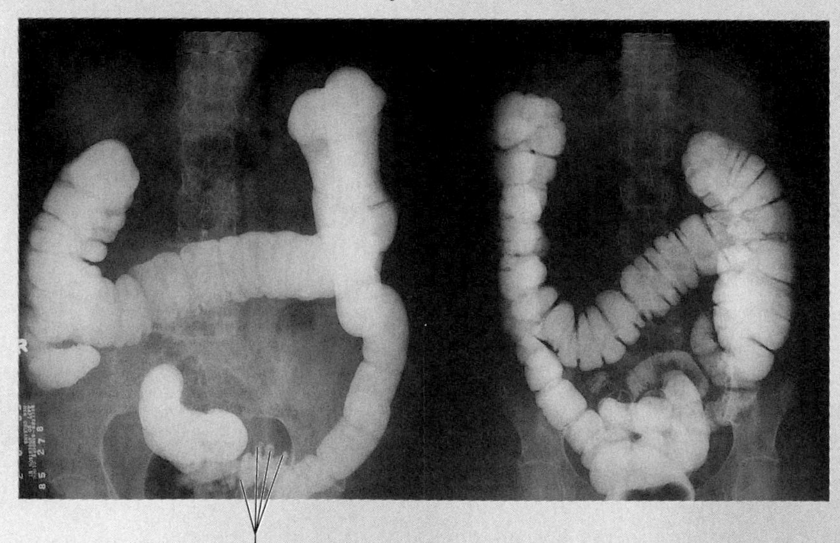


FIGURE 17B
Diverticulosis of the large intestine. The large intestine on the right is healthy. The organ on the left shows diverticula on the left side.

ulitis), antibiotics or surgery may be required. More than half of all Americans develop diverticulosis by age 60. It is not known exactly what causes diverticulosis, but the fact that it is nonexistent in populations whose diets are high in fiber and that it was not known in the United States before refined foods were introduced earlier this century suggests that lack of dietary fiber may be a cause. Fiber speeds the movement of material through the digestive system, hastening exit of toxic substances.

In *ulcerative colitis,* inflammation of the inner lining of the colon and rectum produces abdominal and rectal pain, bloody diarrhea, and weight loss. Drugs can often treat the symptoms, but sometimes removal of the colon is necessary. Again, the cause of this disorder is not known.

Cancer of the large intestine and rectum, known as *colorectal cancer,* is the second most prevalent cancer in the United States, with 152,000 new cases and nearly 60,000 deaths yearly. Symptoms include a change in the frequency or consistency of bowel movements, bloody feces, and abdominal pain. A home test kit called a *hemoccult* (hidden blood) *test* can detect intestinal bleeding that may signal the presence of cancer. Blood in feces is often black and not visible. Follow-up at a doctor's office entails use of a fiber-optic colonoscope that searches and samples colorectal tissue for cancer. Soon, physicians will be able to diagnose colon cancer by performing genetic tests on cells in the feces.

Cancer or other severe maladies may require that the large intestine be removed. A new opening for feces to exit the body is then needed. The free end of the intestine can be surgically attached to an opening created through the skin of the abdomen, and a bag attached to the opening to collect the fecal matter. This procedure is called a *colostomy.*

Some colon cancers may take many years to develop. A sequence of genetic changes cause the cells lining the large intestine first to divide more frequently than normal, then to enter a precancerous stage. Next, the growths form benign polyps, then, perhaps years later, cancer. Still other genetic changes control the cancer's spread. The entire process may be triggered by a susceptibility gene. By understanding the steps that lead to inherited types of colon cancer, researchers may learn how other colon malignancies form and progress as well (fig. 17C).

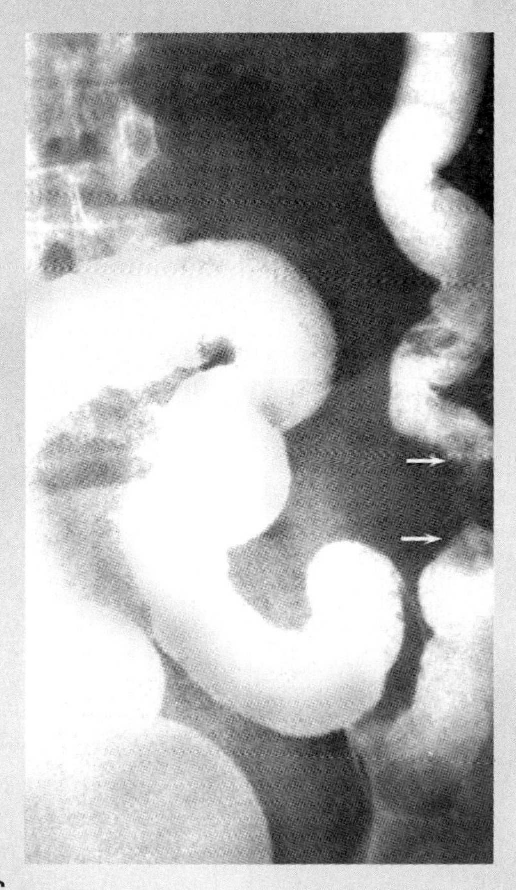

**FIGURE 17C**
To detect colon cancer, a patient receives an enema containing barium. The barium highlights the lower digestive tract, revealing an obstruction caused by a tumor.

## Clinical Terms Related to the Digestive System

**achalasia** (ak″ah-la′ze-ah) Failure of the smooth muscle to relax at some junction in the digestive tube, such as that between the esophagus and stomach.

**achlorhydria** (ah″klōr-hi′dre-ah) A lack of hydrochloric acid in gastric secretions.

**aphagia** (ah-fa′je-ah) Inability to swallow.

**cholecystitis** (ko″le-sis-ti′tis) Inflammation of the gallbladder.

**cholelithiasis** (ko″le-li-thi′ah-sis) The presence of stones in the gallbladder.

**cholestasis** (ko″le-sta′sis) Condition in which the flow of bile from the gallbladder is prevented.

**cirrhosis** (si-ro′sis) A liver condition in which the hepatic cells degenerate and the surrounding connective tissues thicken.

**diverticulitis** (di″ver-tik″u-li′tis) Inflammation of small pouches (diverticula) that sometimes form in the lining and wall of the colon.

**dumping syndrome** (dum′ping sin′drōm) A set of symptoms, including diarrhea, that often occur following a gastrectomy.

**dysentery** (dis′en-ter″e) An intestinal infection, caused by viruses, bacteria, or protozoans, that is accompanied by diarrhea and cramps.

**dyspepsia** (dis-pep′se-ah) Indigestion; difficulty in digesting a meal.

**dysphagia** (dis-fa′je-ah) Difficulty in swallowing.

**enteritis** (en″tĕ-ri′tis) Inflammation of the intestine.

**esophagitis** (e-sof″ah-ji′tis) Inflammation of the esophagus.

**gastrectomy** (gas-trek′to-me) Partial or complete removal of the stomach.

**gastrostomy** (gas-tros′to-me) The creation of an opening in the stomach wall through which food and liquids may be administered when swallowing is not possible.

**glossitis** (glŏ-si′tis) Inflammation of the tongue.

**ileitis** (il″e-i′tis) Inflammation of the ileum.

**ileus** (il′e-us) An obstruction of the intestine due to an inhibition of motility or a mechanical cause.

**pharyngitis** (far″in-ji′tis) Inflammation of the pharynx.

**pylorospasm** (pi-lor′o-spazm) A spasm of the pyloric portion of the stomach or of the pyloric sphincter.

**pyorrhea** (pi″o-re′ah) Inflammation of the dental periosteum, accompanied by the formation of pus.

**stomatitis** (sto″mah-ti′tis) Inflammation of the lining of the mouth.

# InnerConnections
## Digestive System

The digestive system is responsible for the ingestion, digestion, and absorption of nutrients for use by all body cells

**Reproductive system**

Adequate availability of nutrients, including fat, is essential for conception and normal development

**Integumentary system**

Vitamin D activated in the skin is important in absorption of calcium from the digestive tract

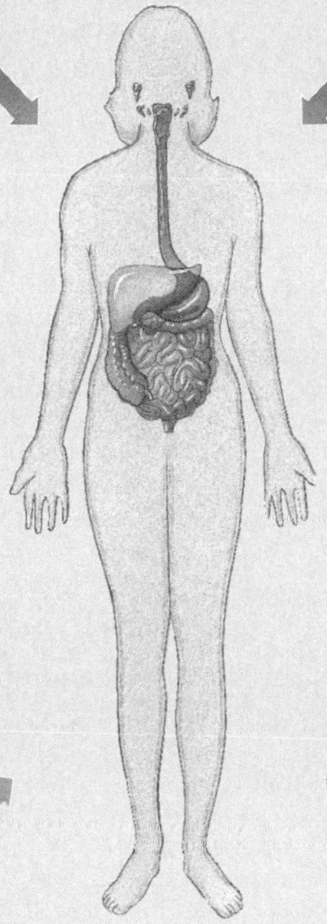

**Urinary system**

The kidney and liver work together to activate vitamin D

**Skeletal system**

Bones play an important role in mastication

**Respiratory system**

The digestive system and the respiratory system share common anatomical structures

**Muscular system**

Muscles are important in mastication, swallowing, and the mixing and moving of digestion products through the gastrointestinal tract

**Lymphatic system**

The lymphatic system plays a major role in the absorption of fats

**Nervous system**

The nervous system can influence the activity of the digestive system

**Cardiovascular system**

Absorbed nutrients are carried to all body cells by the bloodstream

**Endocrine system**

Hormones can influence the activity of the digestive system

# Chapter Summary

## Introduction (page 671)

Digestion is the process of mechanically and chemically breaking down foods so that they can be absorbed. The digestive system consists of an alimentary canal and several accessory organs.

## General Characteristics of the Alimentary Canal (page 671)

Regions of the alimentary canal perform specific functions.

1. Structure of the wall
   a. The wall consists of four layers.
   b. These layers include the mucosa, submucosa, muscular layer, and serosa.
2. Movements of the tube
   a. Motor functions include mixing and propelling movements.
   b. Peristalsis is responsible for propelling movements.
   c. The wall of the tube undergoes receptive relaxation just ahead of a peristaltic wave.
3. Innervation of the tube
   a. The tube is innervated by branches of the sympathetic and parasympathetic divisions of the autonomic nervous system.
   b. Parasympathetic impulses generally increase digestive activities; sympathetic impulses generally inhibit digestive activities.
   c. Sympathetic impulses contract certain sphincter muscles, controlling movement through the alimentary canal.

## Mouth (page 674)

The mouth is adapted to receive food and begin preparing it for digestion. It also serves as an organ of speech and sensory perception.

1. Cheeks and lips
   a. Cheeks form the lateral walls of the mouth.
   b. Lips are highly mobile and possess a variety of sensory receptors useful in judging the characteristics of food.
2. Tongue
   a. The tongue is a thick, muscular organ that mixes food with saliva and moves it toward the pharynx.
   b. The rough surface of the tongue handles food and contains taste buds.
   c. Lingual tonsils are located on the root of the tongue.
3. Palate
   a. The palate comprises the roof of the mouth and includes hard and soft portions.
   b. The soft palate closes the opening to the nasal cavity during swallowing.
   c. Palatine tonsils are located on either side of the tongue in the back of the mouth.
   d. Tonsils consist of lymphatic tissues.
4. Teeth
   a. Two sets of teeth develop in sockets of the mandibular and maxillary bones.
   b. There are twenty primary and thirty-two secondary teeth.
   c. Teeth mechanically break food into smaller pieces, increasing the surface area exposed to digestive actions.
   d. Different kinds of teeth are adapted to handle foods in different ways, such as biting, grasping, or grinding.
   e. Each tooth consists of a crown and root, and is composed of enamel, dentin, pulp, nerves, and blood vessels.
   f. A tooth is attached to the alveolar process by collagenous fibers of the periodontal ligament.

## Salivary Glands (page 680)

Salivary glands secrete saliva, which moistens food, helps bind food particles together, begins chemical digestion of carbohydrates, makes taste possible, helps cleanse the mouth, and regulates pH in the mouth.

1. Salivary secretions
   a. Salivary glands include serous cells that secrete digestive enzymes and mucous cells that secrete mucus.
   b. Parasympathetic impulses stimulate the secretion of serous fluid.
2. Major salivary glands
   a. The parotid glands are the largest, and they secrete saliva that is rich in amylase.
   b. The submandibular glands in the floor of the mouth produce viscid saliva.
   c. The sublingual glands in the floor of the mouth primarily secrete mucus.

## Pharynx and Esophagus (page 681)

The pharynx and esophagus serve as passageways.

1. Structure of the pharynx
   a. The pharynx is divided into a nasopharynx, oropharynx, and laryngopharynx.
   b. The muscular walls of the pharynx contain fibers arranged in circular and longitudinal groups.
2. Swallowing mechanism
   a. Swallowing occurs in three stages.
      (1) Food is mixed with saliva and forced into the pharynx.
      (2) Involuntary reflex actions move the food into the esophagus.
      (3) Food is transported to the stomach.
3. Esophagus
   a. The esophagus passes through the mediastinum and penetrates the diaphragm.
   b. Circular muscle fibers at the distal end of the esophagus help prevent regurgitation of food from the stomach.

## Stomach (page 684)

The stomach receives food, mixes it with gastric juice, carries on a limited amount of absorption, and moves food into the small intestine.

1. Parts of the stomach
   a. The stomach is divided into cardiac, fundic, body, and pyloric regions.
   b. The lower esophageal sphincter serves as a valve between the esophagus and the stomach.
   c. The pyloric sphincter serves as a valve between the stomach and the small intestine.

2. Gastric secretions
   a. Gastric glands secrete gastric juice.
   b. Gastric juice contains pepsin, hydrochloric acid, lipase, and intrinsic factor.
3. Regulation of gastric secretions
   a. Parasympathetic impulses and the hormone gastrin enhance gastric secretion.
   b. The three stages of gastric secretion are the cephalic, gastric, and intestinal phases.
   c. The presence of food in the small intestine reflexly inhibits gastric secretions.
4. Gastric absorption
   a. The stomach is not well adapted for absorption.
   b. A few substances such as water and other small molecules may be absorbed through the stomach wall.
5. Mixing and emptying actions
   a. As the stomach fills, its wall stretches, but its internal pressure remains unchanged.
   b. Mixing movements aid in producing chyme; peristaltic waves move the chyme into the pyloric region.
   c. The muscular wall of the pyloric region pumps chyme into the small intestine.
   d. The rate of emptying depends on the fluidity of the chyme and the type of food present.
   e. The upper part of the small intestine fills, and an enterogastric reflex inhibits peristalsis in the stomach.
   f. Vomiting results from a complex reflex that has many stimuli.

## Pancreas (page 691)

1. Structure of the pancreas
   a. The pancreas is closely associated with the duodenum.
   b. It produces pancreatic juice that is secreted into a pancreatic duct.
   c. The pancreatic duct leads to the duodenum.
2. Pancreatic juice
   a. Pancreatic juice contains enzymes that can split carbohydrates, proteins, fats, and nucleic acids.
   b. Pancreatic juice has a high bicarbonate ion concentration that helps neutralize chyme and causes the intestinal contents to be alkaline.
3. Regulation of pancreatic secretion
   a. Secretin from the duodenum stimulates the release of pancreatic juice that contains few digestive enzymes but has a high bicarbonate ion concentration.
   b. Cholecystokinin from the intestinal wall stimulates the release of pancreatic juice that has a high concentration of digestive enzymes.

## Liver (page 693)

1. Liver functions
   a. The liver is located in the upper right and central portion of the abdominal cavity.
   b. The liver has many functions. It metabolizes carbohydrates, lipids, and proteins; stores some substances; filters blood; destroys toxic chemicals; and secretes bile.
   c. Bile is the only liver secretion that directly affects digestion.
2. Liver structure
   a. The liver is a highly vascular organ, enclosed in a fibrous capsule, and divided into lobes.
   b. Each lobe consists of hepatic lobules, the functional units of the liver.
   c. Bile from the lobules is carried by bile canals to hepatic ducts that unite to form the common bile duct.
3. Composition of bile
   a. Bile contains bile salts, bile pigments, cholesterol, and electrolytes.
   b. Only the bile salts have digestive functions.
   c. Bile pigments are products of red blood cell breakdown.
4. Gallbladder
   a. The gallbladder stores bile between meals.
   b. A sphincter muscle controls release of bile from the common bile duct.
   c. Gallstones may sometimes form within the gallbladder.
5. Regulation of bile release
   a. Cholecystokinin from the small intestine stimulates bile release.
   b. The sphincter muscle at the base of the common bile duct relaxes as a peristaltic wave in the duodenal wall approaches.
6. Functions of bile salts
   a. Bile salts emulsify fats and aid in the absorption of fatty acids, cholesterol, and certain vitamins.
   b. Bile salts are reabsorbed in the small intestine.

## Small Intestine (page 699)

The small intestine extends from the pyloric sphincter to the large intestine. It receives secretions from the pancreas and liver, completes digestion of nutrients, absorbs the products of digestion, and transports the residues to the large intestine.

1. Parts of the small intestine
   a. The small intestine consists of the duodenum, jejunum, and ileum.
   b. The small intestine is suspended from the posterior abdominal wall by mesentery.
2. Structure of the small intestinal wall
   a. The wall is lined with villi that increase the surface area and aid in mixing and absorption.
   b. Microvilli on the free ends of epithelial cells greatly increase the surface area.
   c. Intestinal glands are located between the villi.
   d. Circular folds in the lining of the intestinal wall also increase its surface area.
3. Secretions of the small intestine
   a. Intestinal glands secrete a watery fluid that lacks digestive enzymes but provides a vehicle for moving chyme to the villi.
   b. Digestive enzymes embedded in the surfaces of microvilli split molecules of sugars, proteins, and fats.
4. Regulation of small intestinal secretions
   Secretion is stimulated by gastric juice, chyme, and reflexes stimulated by distension of the small intestinal wall.
5. Absorption in the small intestine
   a. Villi absorb monosaccharides, amino acids, fatty acids, and glycerol.
   b. Villi also absorb water and electrolytes.

c. Fat molecules with longer chains of carbon atoms enter the lacteals of the villi; fatty acids with relatively short carbon chains enter the blood capillaries of the villi.

6. Movements of the small intestine
   a. Movements include mixing by segmentation and peristalsis.
   b. Overdistension or irritation may stimulate a peristaltic rush and result in diarrhea.
   c. The ileocecal valve controls movement of the intestinal contents from the small intestine into the large intestine.

## Large Intestine (page 707)

The large intestine reabsorbs water and electrolytes, and forms and stores feces.

1. Parts of the large intestine
   a. The large intestine consists of the cecum, colon, rectum, and anal canal.
   b. The colon is divided into ascending, transverse, descending, and sigmoid portions.
2. Structure of the large intestinal wall
   a. The large intestine wall resembles the wall in other parts of the alimentary canal.
   b. The large intestine wall has a unique layer of longitudinal muscle fibers arranged in distinct bands.
3. Functions of the large intestine
   a. The large intestine has little or no digestive function, although it secretes mucus.
   b. Mechanical stimulation and parasympathetic impulses control the rate of mucus secretion.
   c. The large intestine absorbs water and electrolytes.
   d. Many bacteria inhabit the large intestine, where they break down certain undigestible substances and synthesize certain vitamins.
4. Movements of the large intestine
   a. Movements are similar to those in the small intestine.
   b. Mass movements occur two to three times each day.
   c. Defecation is stimulated by a reflex.
5. Feces
   a. Feces are formed and stored in the large intestine.
   b. Feces consist largely of water, undigested material, mucus, and bacteria.
   c. The color of feces is due to bile salts that have been altered by bacterial action.

## Critical Thinking Questions

1. If a patient has 95% of the stomach removed (subtotal gastrectomy) as treatment for severe ulcers or cancer, how would the digestion and absorption of foods be affected? How would the patient's eating habits have to be altered? Why?
2. Why may a person with inflammation of the gallbladder (cholecystitis) also develop an inflammation of the pancreas (pancreatitis)?
3. What effect is a before-dinner cocktail likely to have on digestion? Why are such beverages inadvisable for persons with ulcers?
4. What type of acid-alkaline disorder is likely to develop if the stomach contents are lost repeatedly by vomiting over a prolonged period? What acid-alkaline disorder is likely to develop as a result of prolonged diarrhea?
5. Several years ago, an extract from kidney beans was sold in health food stores as a "starch blocker." Advertisements claimed that one could eat a plate of spaghetti, yet absorb none of it, because starch-digesting enzyme function would be blocked. The kidney bean product indeed kept salivary amylase from functioning. However, people who took the starch blocker developed abdominal pain, bloating, and gas. Suggest a reason for the ill effects of the supposed starch blocker.

## Review Exercises

1. List and describe the locations of the major parts of the alimentary canal.
2. List and describe the locations of the accessory organs of the digestive system.
3. Name the four layers of the wall of the alimentary canal.
4. Distinguish between mixing movements and propelling movements.
5. Define *peristalsis.*
6. Explain the relationship between peristalsis and receptive relaxation.
7. Describe the general effects of parasympathetic and sympathetic impulses on the alimentary canal.
8. Discuss the functions of the mouth and its parts.
9. Distinguish between lingual, palatine, and pharyngeal tonsils.
10. Compare the primary and secondary teeth.
11. Explain how the various types of teeth are adapted to perform specialized functions.
12. Describe the structure of a tooth.
13. Explain how a tooth is anchored in its socket.
14. List and describe the locations of the major salivary glands.
15. Explain how the secretions of the salivary glands differ.
16. Discuss the digestive functions of saliva.
17. Name and locate the three major regions of the pharynx.
18. Describe the mechanism of swallowing.
19. Explain the function of the esophagus.
20. Describe the structure of the stomach.
21. List the enzymes in gastric juice, and explain the function of each enzyme.
22. Explain how gastric secretions are regulated.
23. Describe the mechanism that controls the emptying of the stomach.
24. Describe the enterogastric reflex.
25. Explain the mechanism of vomiting.
26. Describe the locations of the pancreas and the pancreatic duct.
27. List the enzymes found in pancreatic juice, and explain the function of each enzyme.
28. Explain how pancreatic secretions are regulated.
29. List the major functions of the liver.
30. Describe the structure of the liver.
31. Describe the composition of bile.
32. Trace the path of bile from a bile canal to the small intestine.
33. Explain how gallstones form.
34. Define *cholecystokinin.*
35. Explain the functions of bile salts.
36. List and describe the locations of the parts of the small intestine.
37. Name the enzymes of the intestinal mucosa, and explain the function of each enzyme.
38. Explain how the secretions of the small intestine are regulated.
39. Describe the functions of the intestinal villi.
40. Summarize how each major type of digestive product is absorbed.
41. Explain how the movement of the intestinal contents is controlled.
42. List and describe the locations of the parts of the large intestine.
43. Explain the general functions of the large intestine.
44. Describe the defecation reflex.

18

CHAPTER

# Nutrition and Metabolism

## Chapter Objectives

AFTER YOU HAVE STUDIED THIS CHAPTER, YOU SHOULD BE ABLE TO:

1. Define *nutrition, nutrients,* and *essential nutrients.*
2. List the major sources of carbohydrates, lipids, and proteins.
3. Describe how carbohydrates are utilized by cells.
4. Describe how lipids are utilized by cells.
5. Describe how amino acids are utilized by cells.
6. Define *nitrogen balance.*
7. Explain how the energy values of foods are determined.
8. Explain how various factors affect an individual's energy needs.
9. Define *energy balance.*
10. Explain what is meant by desirable weight.
11. List the fat-soluble and water-soluble vitamins.
12. Describe the general functions of each vitamin.
13. Distinguish between a vitamin and a mineral.
14. List the major minerals and trace elements.
15. Describe the general functions of each mineral and trace element.
16. Describe an adequate diet.
17. Distinguish between primary and secondary malnutrition.

A VARIED DIET HELPS MAINTAIN HEALTH.

## Key Terms

**acetyl coenzyme A** (as′ĕ-til ko-en′zīm)

**antioxidant** (an″te-ok′sĭ-dant)

**basal metabolic rate** (ba′sal met″ah-bol′ik rāt)

**calorie** (kal′o-re)

**calorimeter** (kal″o-rim′ĕ-ter)

**dynamic equilibrium** (di-nam′ik e″kwĭ-lib′re-um)

**energy balance** (en′er-je bal′ans)

**malnutrition** (mal″nu-trish′un)

**mineral** (min′er-al)

**nitrogen balance** (ni′tro-jen bal′ans)

**nutrient** (nu′tre-ent)

**triglyceride** (tri-glis′er-īd)

**vitamin** (vi′tah-min)

When gymnast Christy Henrich was buried on a Friday morning in July 1994, she weighed 61 pounds. Three weeks earlier, she had weighed an unbelievable 47 pounds. Between those two dates, she celebrated her twenty-second birthday.

Christy suffered from *anorexia nervosa,* a psychological disorder common among professional athletes in which the patient, usually a girl, perceives herself as fat—even though she is obviously not. A year before her death, Christy told a newspaper reporter, "My life is a horrifying nightmare. It feels like there's a beast inside me, like a monster."

Christy's decline into self-starvation began in 1988, when a judge told her she'd have to lose weight if she wanted to make the U.S. Olympic team. Christy weighed 90 pounds. From then on, with only a few respites, she ate nearly nothing, exercised many hours each day, and used laxatives to drop her weight lower and lower. Finally, it dropped so low that her vital organs could not function.

World-class gymnast Christy Henrich died of complications of the self-starvation eating disorder anorexia nervosa in July 1994. In this photo, taken 11 months before her death, she weighed under 60 pounds. Concern over weight gain propelled her down the path of this deadly nutritional illness.

Anorexia nervosa is a killer. We will return to it, other eating disorders, and starvation at the chapter's end. First, we consider the human body's nutritional requirements.

## Why We Eat

The human body requires a daily input of calories, distributed among food substances needed in large amounts as well as some needed in small amounts. The **macronutrients,** needed in bulk, include carbohydrates, proteins, and fats. **Micronutrients,** needed in small daily doses, include vitamins and minerals. The body also requires water.

In countries with adequate food supplies, most healthy individuals can obtain nourishment by eating a variety of foods and limiting fat intake. People who do not eat meat products can also receive adequate nutrition but must pay more attention to food choices to avoid developing nutrient deficiencies. For example, eliminating red meat also means eliminating an excellent source of iron, copper, zinc, and vitamin $B_{12}$. The fiber that often makes up much of a vegetarian's diet, although very healthful in many ways, also decreases absorption of iron. Therefore, a vegetarian must be careful to obtain sufficient iron from non-meat sources. This is easily done. Fortified foods, green leafy vegetables, and especially whole grains provide many of the nutrients also present in meat. Chart 18.1 lists the different types of vegetarian diets.

Nutrients are obtained from foods, and digestion breaks them down to sizes that can be absorbed and transported in the bloodstream. **Metabolism** refers to the ways that nutrients are altered chemically and used in anabolism (building up or synthesis) a
tabolism (breaking down) to suppor
life. Chapter 4 discusses the meta
carbohydrates, lipids, and proteins i

Nutrients that human cells cannot synthesize, such as certain amino acids, are particularly important and are therefore called **essential nutrients.**

## Carbohydrates

**Carbohydrates** are organic compounds, and include the sugars and starches. The energy held in their chemical bonds is used to power cellular processes.

### Carbohydrate Sources

Carbohydrates are ingested in a variety of forms. Complex carbohydrates include starch from grains and vegetables, and glycogen from meats. Foods containing starch and glycogen usually have many other nutrients, including valuable vitamins and minerals. The simple carbohydrates include *disaccharides* from cane sugar, beet sugar, and molasses, and *monosaccharides* from honey and fruits. During digestion, complex carbohydrates are broken down to monosaccharides, which are small enough to be absorbed.

Cellulose is a complex carbohydrate that is abundant in our food—it provides the crunch to celery and the crispness to lettuce. However, we cannot digest cellulose, and most of it passes through the alimentary canal largely unchanged. However, cellulose provides bulk (also called fiber or roughage) against which the muscular wall of the digestive system can push, facilitating the movement of food. *Hemicellulose, pectin,* and *lignin* are other plant carbohydrates that provide fiber.

### Carbohydrate Utilization

The monosaccharides that are absorbed from the digestive tract include *fructose, galactose,* and *glucose.* Liver enzymes convert fructose and galactose into glucose (fig. 18.1). Recall that glucose is the form of carbohydrate that is most commonly oxidized in glycolysis for cellular fuel.

Sugar substitutes provide concentrated sweetness, so that a food can be sweetened with fewer calories. Aspartame, which is a dipeptide (two joined amino acids), is two hundred times as sweet as table sugar (sucrose). This means that just a pinch of it equals the sweetness of a teaspoonful or more of sugar. Saccharin, another sugar substitute, is three hundred times as sweet as sugar.

Some excess glucose is polymerized to *glycogen* (glycogenesis) and stored as a glucose reserve in the liver and muscles. Glucose can be mobilized rapidly om glycogen (glycogenolysis) when it is needed to ply energy. However, only a certain amount of glycogen can be stored. Excess glucose beyond what can be stored as glycogen is usually converted into fat and stored in adipose tissue (fig. 18.2).

Cells also use carbohydrates as starting materials for such vital biochemicals as the 5-carbon sugars *ribose* and *deoxyribose,* which are needed for the production of the nucleic acids RNA and DNA, and the

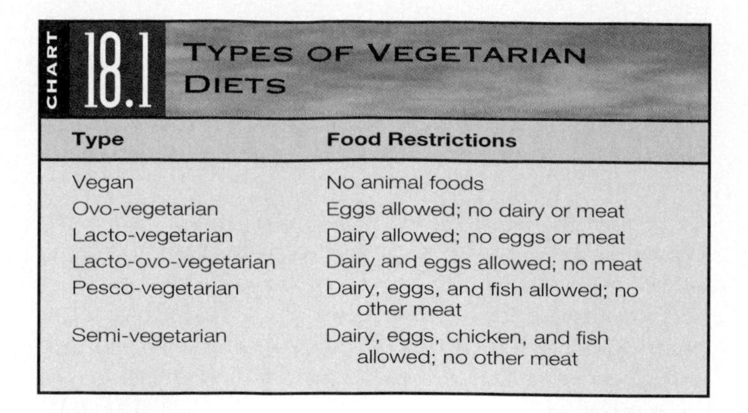

**Chart 18.1 Types of Vegetarian Diets**

| Type | Food Restrictions |
|---|---|
| Vegan | No animal foods |
| Ovo-vegetarian | Eggs allowed; no dairy or meat |
| Lacto-vegetarian | Dairy allowed; no eggs or meat |
| Lacto-ovo-vegetarian | Dairy and eggs allowed; no meat |
| Pesco-vegetarian | Dairy, eggs, and fish allowed; no other meat |
| Semi-vegetarian | Dairy, eggs, chicken, and fish allowed; no other meat |

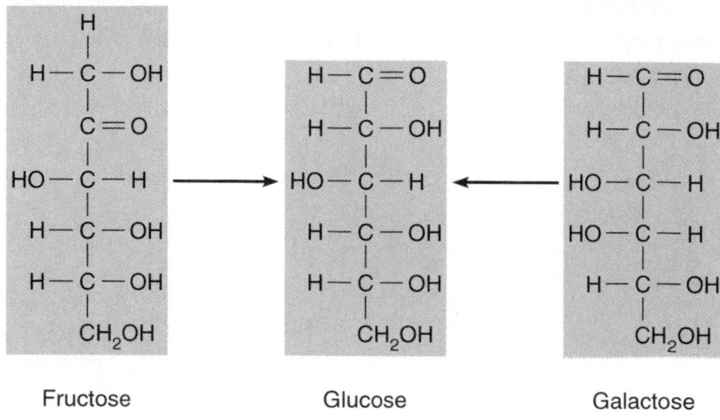


**Figure 18.1**
The liver converts the monosaccharides fructose and galactose into glucose.

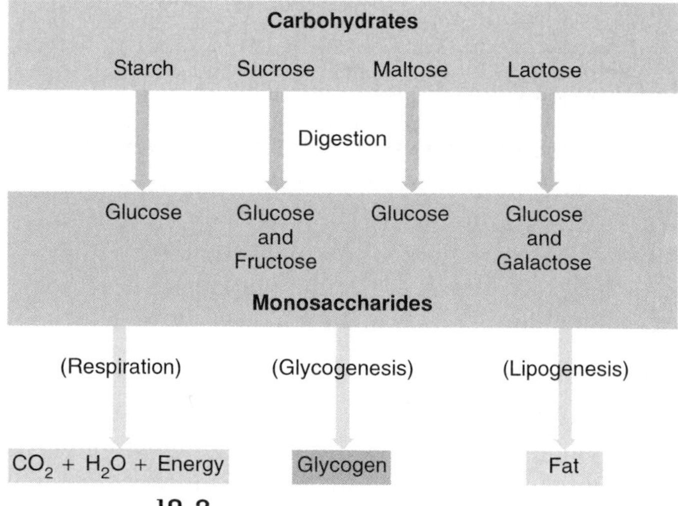


**Figure 18.2**
Monosaccharides from foods are used for energy, stored as glycogen, or converted to fat.

disaccharide *lactose* (milk sugar), which is synthesized when the breasts are actively secreting milk.

Many cells can also obtain energy by oxidizing fatty acids. However, some cells, such as neurons, normally depend on a continuous supply of glucose for survival. (Under some conditions, such as prolonged starvation, neurons may use other fuel sources.) Even a temporary decrease in the glucose supply may seriously impair nervous system function. Consequently, the body requires a minimum amount of carbohydrates. If an adequate supply of carbohydrates is not received in foods, the liver may convert some noncarbohydrates, such as amino acids from proteins or glycerol from fats, into glucose—a process called *gluconeogenesis*. Thus, the need for glucose has physiological priority over the need to synthesize certain other substances, such as proteins, from the available amino acids.

1. List several common sources of carbohydrates.
2. In what form are carbohydrates utilized as a cellular fuel?
3. Explain what happens to excess glucose in the body.
4. Name two uses of carbohydrates other than supplying energy.
5. How does the body obtain glucose when its food supply of carbohydrates is insufficient?

An adult liver stores about 100 grams of glycogen, and muscle tissue stores another 200 grams, providing enough reserve to meet energy demands for about 12 hours when the person is resting. Whether these stores are filled depends, of course, on diet. People live on widely varying amounts of carbohydrates, often reflecting economic conditions. In the United States, a typical adult's diet supplies about 50% of total body energy from carbohydrates. Because carbohydrate foods are generally inexpensive, people in lower economic groups tend to exceed this percentage of carbohydrates in the diet. This is not dangerous, as long as they also obtain essential amino acids, vitamins, and minerals.

### Carbohydrate Requirements

Because carbohydrates provide the primary source of fuel for cellular processes, the need for carbohydrates varies with individual energy requirements. Therefore, physically active individuals require more fuel than those who are sedentary. The minimal requirement for carbohydrates in the human diet is unknown. It is estimated, however, that an intake of at least 125–175 grams daily is necessary to spare protein (that is, to avoid protein breakdown) and to avoid metabolic disorders resulting from excess fat utilization. Most people in the United States include 200–300 grams of carbohydrates in their daily diets.

1. Why do the daily requirements for carbohydrates vary from person to person?
2. What is the daily minimum requirement of carbohydrates?

## LIPIDS

**Lipids** are organic compounds that include fats, oils, and fatlike substances (see chapter 2). They supply energy for cellular processes and to build structures, such as cell membranes. Although lipids include fats, phospholipids, and cholesterol, the most common dietary lipids are the fats called *triglycerides* (see fig. 2.11).

### Lipid Sources

Triglycerides are found in plant- and animal-based foods. Saturated fats (which should comprise no more than 10% of the diet) are found mainly in foods of animal origin, such as meat, eggs, milk, and lard, as well as in palm and coconut oil. Unsaturated fats are contained in seeds, nuts, and plant oils.

Cholesterol is abundant in liver and egg yolk, and to a lesser extent in whole milk, butter, cheese, and meats. It is generally not present in foods of plant origin. This is why a label on a plant-based food stating that it is "cholesterol-free" may be meaningless.

### Lipid Utilization

Digestion breaks triglycerides down into fatty acids and glycerol. After being absorbed, these products are transported in the lymph and blood to tissues. In the cells, the process of beta oxidation breaks fatty acids down into two-carbon units, which are further catabolized in the citric acid cycle to release the energy held in their bonds. Glycerol becomes an intermediate within the glycolytic pathway for further ATP production. The liver and adipose tissues control metabolism of lipids (fig. 18.3).

The liver can convert fatty acids from one form to another. However, it cannot synthesize certain of the fatty acids, such as *linoleic acid*. This **essential fatty acid** is needed for the synthesis of phospholipids, which, in turn, are necessary for the formation of cell membranes and the transport of circulating lipids. Good sources of linoleic acid include corn oil, cottonseed oil, and soy oil. Other essential fatty acids are *linolenic acid* and *arachidonic acid*.

The liver uses free fatty acids to synthesize triglycerides, phospholipids, and lipoproteins that may then be released into the blood. Thus, the liver regulates circulating lipids (see fig. 18.3). It also controls the total amount of cholesterol in the body by synthesizing cholesterol and releasing it into the blood, or by removing cholesterol from the blood and excreting it into the bile. The liver uses cholesterol to produce bile salts. Cholesterol is not used as an energy source, but it does provide structural material for cell and organelle membranes, and it furnishes starting materials for the synthesis of certain sex hormones and hormones produced by the adrenal cortex.

Excess triglycerides are stored in adipose tissue. If the blood lipid concentration drops (in response to fasting, for example), some of these triglycerides are hydrolyzed into free fatty acids and glycerol, and then released into the blood.

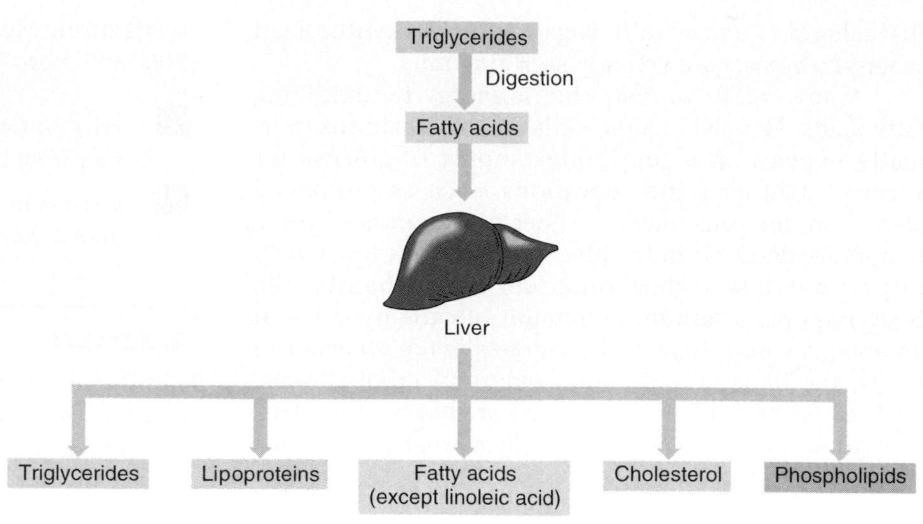


**FIGURE 18.3**

The liver uses fatty acids to synthesize a variety of lipids.

### Lipid Requirements

As with carbohydrates, the lipid content of human diets varies widely. A person who lives on burgers, fries, and shakes may consume 50% or more of total daily calories from fat. For a vegetarian, the percentage may be far lower. The American Heart Association advises that the diet not exceed 30% of total daily calories from fat.

The amounts and types of fats needed for health are unknown. However, linoleic acid is an essential fatty acid, and to prevent deficiency conditions from developing, nutritionists recommend that infants fed formula receive 3% of the energy intake in the form of linoleic acid. A typical adult diet consisting of a variety of foods usually provides an adequate supply of fats.

Fat intake must also supply the required amounts of fat-soluble vitamins.

> What tastes like fat, feels like fat, but isn't fat? Fake fat, or more formally, a fat substitute. Because it is really a carbohydrate or protein, a fat substitute offers half the calories of fat.
>
> The first fake fat, introduced in 1990, was made from egg protein (albumin) and milk protein (casein), mixed and broken down into tiny particles that impart the texture of fat. A variation on this theme that can withstand heating and is therefore good for baked products is made from whey protein. Other fake fats introduced since 1990 are made from carbohydrates.

> Be wary of "lite" and similar markings on prepared food packages. Before 1993, "lite" could, and did, mean almost anything. A bottle of vegetable oil labeled "lite" referred to the product's color; a "lite" cheesecake referred to its texture! Now, "lite," or any similar spelling, has a distinct meaning: the product must have a third fewer calories than the "real thing," or half the fat calories.
>
> Also beware of claims that a food product is "99% fat-free." This usually refers to percentage by weight—not calories, which is what counts. A creamy concoction that is 99% fat-free may be largely air and water, and therefore in that form, fat comprises very little of it. But when the air is compressed and the water absorbed, as happens in the stomach, the fat percentage may skyrocket.

1. What foods commonly supply lipids?
2. Which fatty acid is an essential nutrient?
3. What is the role of the liver in the utilization of lipids?
4. What is the function of cholesterol?

## PROTEINS

**Proteins** are polymers of amino acids with a wide variety of functions. Proteins include enzymes that control the rates of metabolic reactions; clotting factors; the keratin of skin and hair; elastin and collagen of connective tissue; plasma proteins that regulate water balance; the muscle components actin and myosin; certain hormones; and the antibodies that protect against infection (fig. 18.4).

Amino acids are also potential sources of energy. They are oxidized to release energy when they are present in excess or when supplies of carbohydrates and fats are insufficient. (Fig. 4.16 describes how the three macronutrients enter the energy pathways.) When

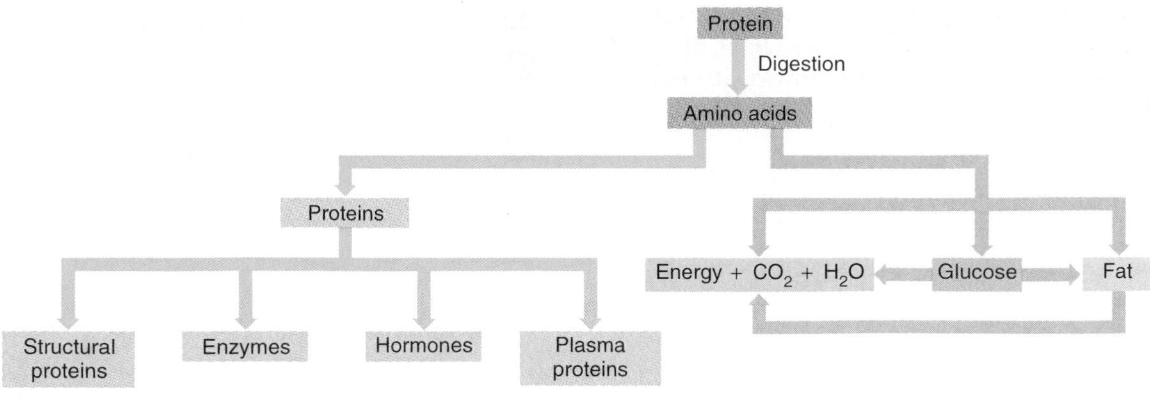


**FIGURE 18.4**
Proteins are digested to their constituent amino acids. These amino acids are then linked together, following genetic instructions, to build new proteins. Free amino acids are also used to supply energy under certain conditions.

carbohydrate intake is sufficient, proteins remain available for tissue building and repair, rather than being converted into carbohydrates for use as an energy source. Using structural proteins to generate energy causes the tissue wasting characteristic of starvation.

## Protein Sources

Foods rich in proteins include meats, fish, poultry, cheese, nuts, milk, eggs, and cereals. Legumes, including beans and peas, contain lesser amounts of protein. Digestion breaks proteins down into their component amino acids, and these smaller molecules are absorbed by intestinal cells and transported in the blood.

The human body can synthesize many amino acids (nonessential amino acids). However, eight amino acids the adult body needs (ten needed by growing children) cannot be synthesized sufficiently or at all. They are called **essential amino acids** because they are essential in the diet.

All of the amino acids must be present in the body at the same time for growth and tissue repair to occur. In other words, if one essential amino acid is missing from the diet, protein synthesis cannot take place. Since essential amino acids are not stored, those present but not used in protein synthesis are oxidized as energy sources or are converted into carbohydrates or fats.

> Plant proteins typically contain too little of one or more essential amino acids to provide adequate nutrition for a person. Combining appropriate plant foods can provide a diversity of dietary amino acids. For example, beans are low in methionine, but have enough lysine. Rice lacks lysine, but has enough methionine. A meal of beans and rice provides enough of both types of amino acids.

**CHART 18.2 AMINO ACIDS FOUND IN FOODS**

| | |
|---|---|
| Alanine | Leucine (e) |
| Arginine (ch) | Lysine (e) |
| Asparagine | Methionine (e) |
| Aspartic acid | Phenylalanine (e) |
| Cysteine | Proline |
| Glutamic acid | Serine |
| Glutamine | Threonine (e) |
| Glycine | Tryptophan (e) |
| Histidine (ch) | Tyrosine |
| Hydroxyproline | Valine (e) |
| Isoleucine (e) | |

Eight essential amino acids (e) cannot be synthesized by human cells and must be provided in the diet. Two additional amino acids (ch) are essential in growing children.

Chart 18.2 lists the amino acids found in foods and indicates those that are essential.

Proteins are classified as complete or incomplete on the basis of the amino acid types they provide. The **complete proteins,** which include those available in milk, meats, and eggs, contain adequate amounts of the essential amino acids to maintain human body tissues and promote normal growth and development. **Incomplete proteins,** such as *zein* in corn, which has too little of the essential amino acids tryptophan and lysine, are unable by themselves to maintain human tissues or to support normal growth and development.

> Genetic engineers can alter certain plants so that their protein is more "complete." For example, genetic instructions for producing the amino acid tryptophan are inserted into corn cells, compensating for the low levels of this nutrient normally found in corn.

A protein in wheat called *gliadin* is an example of a *partially complete protein.* Although it does not contain enough lysine to promote growth, it contains enough to maintain life.

1. Which foods provide rich sources of protein?
2. Why are some amino acids called essential?
3. Distinguish between a complete protein and an incomplete protein.

## Nitrogen Balance

In a healthy adult, proteins are continuously built up and broken down. This occurs at different rates in different tissues, but the overall gain of body proteins equals the loss, producing a state of *dynamic equilibrium.* Since proteins contain a relatively high percentage of nitrogen, dynamic equilibrium also brings **nitrogen balance**—a condition in which the amount of nitrogen taken in is equal to the amount excreted.

A person who is starving, however, has a *negative nitrogen balance* because the amount of nitrogen excreted as a result of amino acid oxidation exceeds the amount the diet replaces. A growing child, a pregnant woman, or an athlete in training is likely to have a *positive nitrogen balance,* since the amount of protein being built into new tissue probably exceeds the amount being used for energy.

## Protein Requirements

In addition to supplying essential amino acids, proteins provide nitrogen and other elements for the synthesis of nonessential amino acids and certain nonprotein nitrogenous substances. Consequently, the amount of protein individuals require varies according to body size, metabolic rate, and nitrogen balance condition.

For an average adult, nutritionists recommend a daily protein intake of about 0.8 grams per kilogram (0.4 grams per pound) of body weight. For a pregnant woman, who needs to maintain a positive nitrogen balance, the recommendation is increased by an additional 30 grams of protein per day. Similarly, a nursing mother requires an additional 20 grams of protein per day to maintain a high level of milk production.

In addition to tissue wasting, protein deficiency may also decrease the level of plasma proteins, which decreases the osmotic pressure of the blood. As a result, fluids collect in the tissues, producing a condition called *nutritional edema.* The swollen bellies of children suffering from protein starvation illustrate nutritional edema.

Chart 18.3 summarizes the sources, uses, and requirements for protein, lipid, and carbohydrate nutrients.

1. What are the physiological functions of proteins?
2. What is a negative nitrogen balance? A positive nitrogen balance?
3. How can inadequate nutrition cause edema?

# Energy Expenditures

Carbohydrates, fats, and proteins supply energy. Since energy is required for all metabolic processes, its supply is of prime importance to cell survival. If the diet is deficient in energy-supplying nutrients, structural molecules may gradually be consumed, leading to death.

## Energy Values of Foods

The amount of potential energy contained in a food can be expressed as **calories,** which are units of heat.

Although a calorie is defined as the amount of heat needed to raise the temperature of a gram of water by 1 degree Celsius (° C), the calorie used to measure food energy is 1,000 times greater. This *large calorie* (Cal.) is equal to the amount of heat needed to raise the temperature of a kilogram (1,000 grams) of water by 1° C (actually from 14.5° C to 15.5° C). This unit is also called a *kilocalorie,* but it is customary in nutritional studies to refer to it simply as a calorie.

Figure 18.5 shows a bomb calorimeter, which is used to measure the caloric contents of foods. It consists of a metal chamber submerged in a known volume of water. A food sample is dried, weighed, and placed in a nonreactive dish inside the chamber. The chamber is filled with oxygen gas and submerged in the water. Then, the food is ignited and allowed to oxidize completely. Heat released from the food raises the temperature of the surrounding water, and the change in temperature is measured. Since the volume of the water is known, the amount of heat released from the food can be calculated in calories.

Caloric values determined this way are somewhat higher than the amount of energy actually released by metabolic oxidation, because nutrients generally are not completely absorbed from the digestive tract. Also, the body does not completely oxidize amino acids but excretes parts of them in urea or transforms them into other nitrogenous substances. When such losses are taken into account, cellular oxidation yields on the average about 4.1 calories from 1 gram of carbohydrate, 4.1 calories from 1 gram of protein, and 9.5 calories from 1 gram of fat.

The fact that more than twice as much energy is derived from equal amounts by weight of fats as from either proteins or carbohydrates is why avoiding fatty foods helps lead to weight loss.

## CHART 18.3 PROTEIN, LIPID, AND CARBOHYDRATE NUTRIENTS

| Nutrient | Sources and RDA* for Adults | Calories per Gram | Utilization | Conditions Associated with: Excesses | Conditions Associated with: Deficiencies |
|---|---|---|---|---|---|
| *Protein* | Meats, cheese, nuts, milk, eggs, cereals, legumes 125–175 g | 4.1 | Production of protein molecules used to build cell structure and to function as enzymes or hormones; used in the transport of oxygen, regulation of water balance, control of pH, formation of antibodies; amino acids may be broken down and oxidized for energy or converted to carbohydrates or fats for storage | Obesity | Extreme weight loss, wasting, anemia, growth retardation |
| *Lipid* | Meats, eggs, milk, lard, plant oils 80–100 g | 9.5 | Oxidized for energy; production of triglycerides, phospholipids, lipoproteins, and cholesterol, stored in adipose tissue; glycerol portions of fat molecules may be used to synthesize glucose | Obesity, increased serum cholesterol, increased risk of heart disease | Weight loss, skin lesions |
| *Carbohydrate* | Primarily from starch and sugars in foods of plant origin, and from glycogen in meats .8g/kg body weight | 4.1 | Oxidized for energy; used in production of ribose, deoxyribose, and lactose; stored in liver and muscles as glycogen; converted to fats and stored in adipose tissue | Obesity, dental caries, nutritional deficits | Metabolic acidosis |

*RDA = recommended dietary allowance.

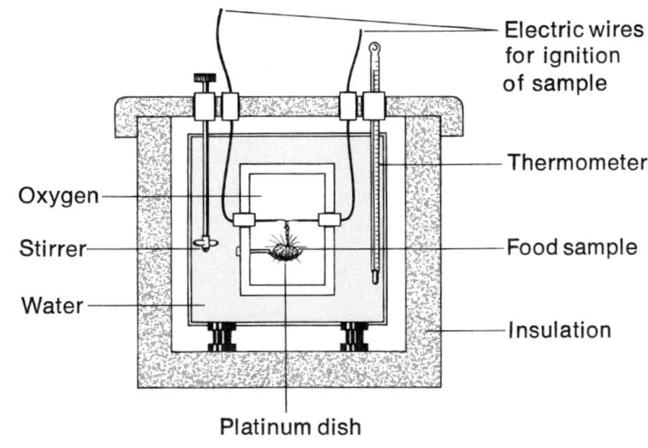


**FIGURE 18.5**

A bomb calorimeter measures the caloric content of a food sample.

1. What term designates the potential energy in a food substance?
2. How can the energy value of a food be determined?
3. What is the energy value of a gram of carbohydrate? A gram of protein? A gram of fat?

## Energy Requirements

The amount of energy required to support metabolic activities for 24 hours varies from person to person. The factors that influence energy needs include the individual's basal metabolic rate, degree of muscular activity, body temperature, and rate of growth.

The **basal metabolic rate** (BMR) is a measurement of the rate at which the body expends energy under *basal conditions*—when a person is awake and at rest, after an overnight fast, and in a comfortable, controlled environment. Tests of thyroid function can be used to estimate a person's BMR.

The amount of oxygen the body consumes is directly proportional to the amount of energy cellular respiration releases. The BMR, therefore, reveals the total amount of energy expended in a given time period to support the activities of such organs as the brain, heart, lungs, liver, and kidneys.

Although the average adult basal metabolic rate indicates a need for approximately one calorie of energy per hour for each kilogram of body weight, this need varies with such factors as sex, body size, body temperature, and level of endocrine gland activity. For

CHART 18.4 CALORIES USED DURING VARIOUS ACTIVITIES

| Activity | Calories (per Hour) |
|---|---|
| Walking up stairs | 1,100 |
| Running (jogging) | 570 |
| Swimming | 500 |
| Vigorous exercise | 450 |
| Slow walking | 200 |
| Dressing and undressing | 118 |
| Sitting at rest | 100 |

example, since heat loss is directly proportional to the body surface area, and a smaller person has a relatively larger surface area, such a person will have a higher BMR. Males tend to have higher metabolic rates than females. As body temperature increases, the BMR increases, and as the blood level of thyroxine or epinephrine increases, so does the BMR.

Maintaining the basal metabolic rate usually requires the body's greatest expenditure of energy. The energy required to support voluntary muscular activity comes next, though this amount varies greatly with the type of activity (chart 18.4). For example, energy needed to maintain posture while sitting at a desk might amount to 100 Cal. per hour above the basal need, whereas running or swimming might require 500–600 Cal. per hour.

Maintenance of body temperature may require additional energy expenditure, particularly in cold weather. In this case, extra energy may be released by involuntary muscular contractions, such as shivering, or through voluntary muscular actions, such as walking.

Growing children and pregnant women, because their bodies are actively producing new tissues, also require increased caloric intake, because tissue building uses energy.

## Energy Balance

A state of **energy balance** exists when caloric intake in the form of foods equals caloric output resulting from the basal metabolic rate and muscular activities. Under these conditions, body weight remains constant, except perhaps for slight variations due to changes in water content.

If, on the other hand, the caloric intake exceeds the output, a *positive energy balance* occurs, and the excess nutrients are stored in the tissues. At the same time, body weight increases, since an excess of 3,500 Cal. can be stored as a pound of fat. Conversely, if caloric output exceeds input, the energy balance is negative, and stored materials are mobilized from the tissues for oxidation, causing body weight loss.

1. What is basal metabolic rate?
2. What factors influence the BMR?
3. What is *energy balance?*

## Desirable Weight

The most obvious and common nutritional disorders reflect calorie imbalances, which may result from societal and geographic factors. Obesity is a common problem in the United States because of plentiful and diverse food supplies. In many African nations, however, natural famines combined with political unrest cause mass starvation. Obesity and starvation are considered later in the chapter.

It is difficult to determine a desirable body weight. In the past, weight standards were based on average weights and heights within a certain population, and the degrees of underweight and overweight were expressed as percentage deviations from these averages. These standards reflected the gradual gain in weight that usually occurs in members of the United States population as they grow older. Later, medical researchers recognized that such an increase in weight after the age of 25 to 30 years is not necessary and may not be conducive to health. Consequently, standards of *desirable weights* were prepared. More recent height-weight guidelines are based upon the characteristics of people who live the longest. These weights are somewhat more lenient than those in the desirable weight charts.

*Overweight* can be defined as exceeding desirable weight by 10% to 20%. A person who exceeds this standard by more than 20% is considered *obese,* though **obesity** is more correctly defined as excess adipose tissue. In other words, being overweight and being obese are not the same. For example, as figure 18.6 shows, an athlete or a person whose work involves heavy muscular activity may be overweight, but not obese. Clinical Application 18.1 discusses obesity.

When a person needs to gain weight, the diet can be altered to include more calories and to emphasize particular macronutrients. For example, a person recovering from a debilitating illness might stress carbohydrates, whereas a bodybuilder might consume excess protein to spur muscle development. An infant also needs to gain weight rapidly, which is best accomplished by consumption of human milk.

1. What is meant by desirable weight?
2. Distinguish between overweight and obesity.
3. Under what conditions is weight gain desirable?

(a)

(b)

**FIGURE 18.6**

(*a*) An obese person is overweight and has excess adipose tissue. (*b*) An athlete may be overweight due to muscle overgrowth but is not considered obese. In fact, many athletes have very low percentages of body fat.

## VITAMINS

**Vitamins** are organic compounds (other than carbohydrates, lipids, and proteins) that must be present in small amounts for normal metabolic processes, but cannot be synthesized in adequate amounts by body cells. Thus, they are essential nutrients that must be supplied in foods.

Vitamins can be classified on the basis of solubility, since some are soluble in fats (or fat solvents) and others are soluble in water. Those that are *fat-soluble* include vitamins A, D, E, and K; the *water-soluble* group includes the B vitamins and vitamin C.

> Different species have different vitamin requirements. For example, ascorbic acid is a vitamin (C) in humans, guinea pigs, and Indian fruit bats, but not in other animals, which can manufacture their own ascorbic acid.

Chart 18.5 lists, and corrects, some common misconceptions about vitamins.

**CHART 18.5 VITAMIN FALLACIES AND FACTS**

| Fallacy | Fact |
|---|---|
| Vitamins from foods are superior to those in tablet form | A vitamin molecule is chemically the same, whether it comes from a bean or a bottle |
| The more vitamins, the better | Too much of a water-soluble vitamin results in excretion of the vitamin through urination; too much of a fat-soluble vitamin can harm one's health |
| A varied diet provides all needed vitamins | Many people do need vitamin supplements, particularly pregnant and breast-feeding women |
| Vitamins provide energy | Vitamins do not directly supply energy; they aid in the release of energy from carbohydrates, fats, and proteins |

### Fat-Soluble Vitamins

Since the fat-soluble vitamins dissolve i[n] ... asso-ciate with lipids and are influenced by ... that affect lipid absorption. For example ... bile salts in the intestine promotes the ab...

CLINICAL APPLICATION 18.1

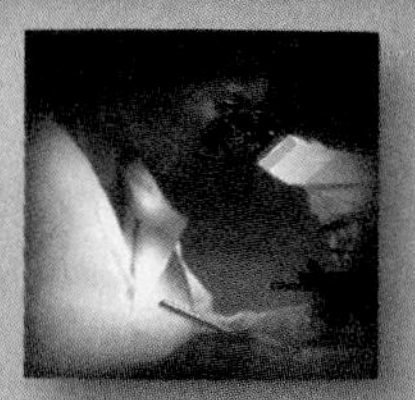

## Obesity

The National Institutes of Health considers obesity a "killer disease," and for good reason. A person who is *obese*—defined as 20% above "ideal" weight based on population statistics considering age, sex, and build—is at higher risk for diabetes, digestive disorders, heart disease, kidney failure, hypertension, stroke, and cancers of the female reproductive organs and the gallbladder. The body is enormously strained to support the extra weight—miles of blood vessels are needed to nourish the additional pounds.

Obesity refers specifically to extra pounds of fat. The proportion of fat in a human body ranges from 5% to more than 50%, with "normal" for males falling between 12% and 23% and for females between 16% and 28%. An elite athlete may have a body fat level as low as 4%. Fat distribution also affects health. Excess poundage above the waist is linked to increased risk of heart disease, diabetes, hypertension, and lipid disorders. Figure 18A shows a do-it-yourself way to estimate obesity using a measurement called body mass index.

Both heredity and the environment contribute to obesity. We inherit genes that control metabolism, but the fact that identical twins reared in different households can grow into adults of vastly different weights indicates that environment influences weight too. Studies comparing body mass index between adult identical twins reared apart indicate that weight is about 70% influenced by genes and 30% by the environment.

A safe goal for weight loss is 1 pound of fat per week. A pound of fat contains 3,500 calories of energy, so that pound can be shed by an appropriate combination of food restriction and exercise. This might mean eating 500 calories less per day or exercising off 500 calories each day. Actually more than a pound of weight will drop because water is lost as well as fat.

Calorie cutting should apply to the energy-providing nutrients (carbohydrates, proteins, and fats) but never to the vitamins and minerals. A rule of thumb is to leave the proportion of protein calories about the same or slightly increased, cut fat calories in half, and cut carbohydrates by a third. Choose foods that you like, and distribute them into three or four balanced meals of 250 to 500 calories each.

Diet plans abound. Many are simply variations on the preceding suggestions, with a gimmick added, such as a cup of chili or an ice cream cone each day. Other diets emphasize a particular food, such as bananas, pineapple, or rice. Avoid diets that are based on erroneous but impressive-sounding scientific principles, such as the "immune power" diet that predicts that the effects of certain food extracts on isolated

vitamins. As a group, the fat-soluble vitamins are stored in moderate quantities within various tissues, and because they are fairly resistant to the effects of heat, they are usually not destroyed by cooking or food processing.

1. What are vitamins?
2. Distinguish between fat-soluble vitamins and water-soluble vitamins.
3. How do bile salts affect the absorption of vitamins?

**Vitamin A** occurs in several forms, including retinol and retinal (retinene). Body cells synthesize this vitamin from a group of yellowish plant pigments called *carotenes*. Excess vitamin A or its precursors are stored mainly in the liver, which regulates their concentration in the body (fig. 18.7). An adult's liver stores enough vitamin A to supply body needs for a year. Infants and children usually lack such reserves, and are therefore more likely to develop vitamin A deficiencies if their diets are inadequate.

Though relatively stable to the effects of heat, acids, and alkalis, vitamin A is readily destroyed by oxidation and is unstable in light.

Vitamin A is important in vision. Retinal is used in the synthesis of *rhodopsin* (visual purple) in the rods of the retina. Vitamin A is thought to be required for the production of light-sensitive pigments in the cones as well. The vitamin also functions in the synthesis of mucoproteins and mucopolysaccharides of white blood cells will be echoed in the body, or diets that suggest that the order in which foods are eaten influences how many calories are absorbed. Also avoid diets very low in carbohydrates (these deplete energy), diets high in protein (these strain the kidneys), and diets with less than 1,200 calories per day, which is starvation level.

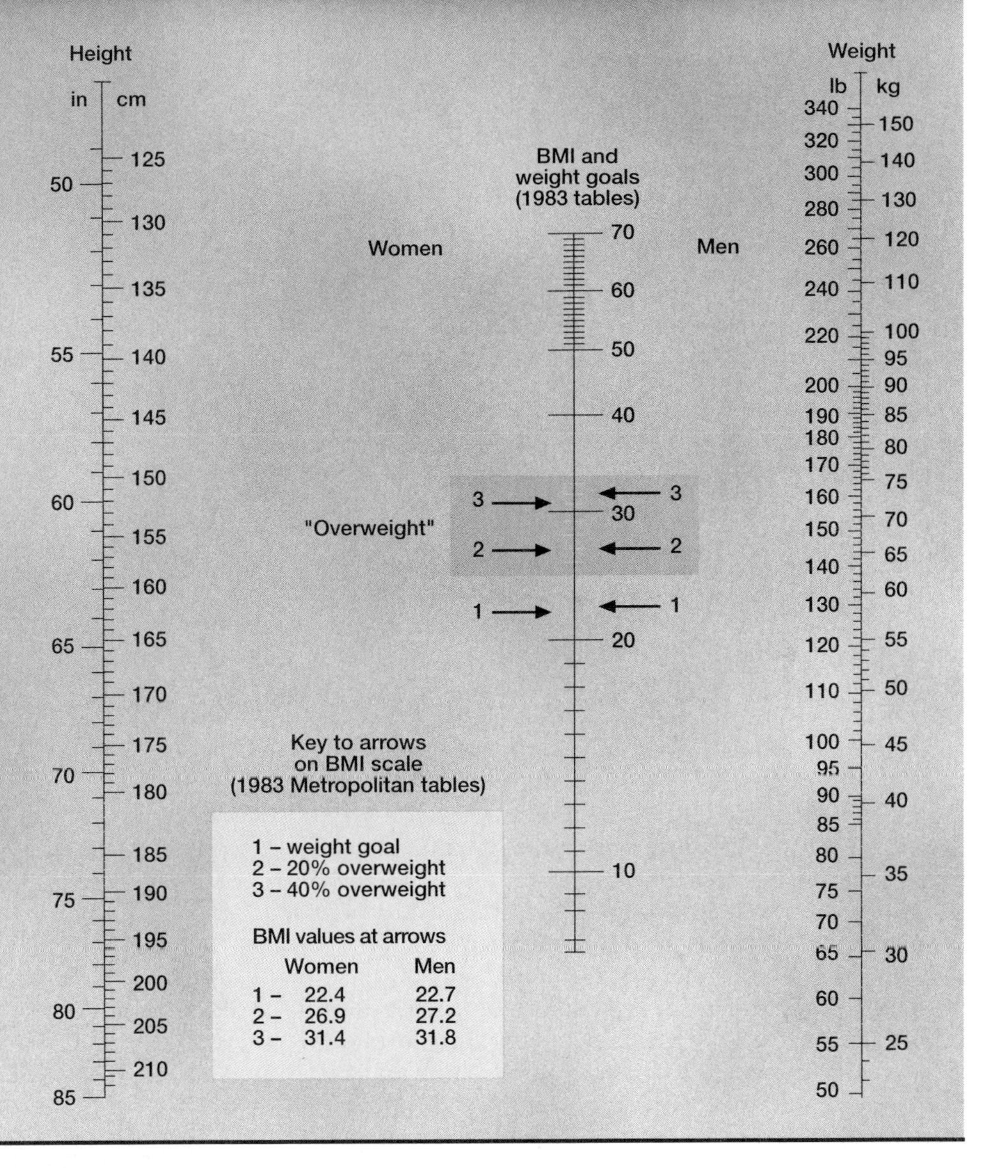


**FIGURE 18A**

Body mass index (BMI) is a measurement that estimates obesity. BMI equals weight in kilograms divided by height in meters, squared (BMI = weight [kg]/height [m]$^2$). To calculate your BMI, draw a line from your height on the left side of the figure to your weight on the right side. The point where the line crosses the scale in the middle is your BMI. A value above 27.2 for men and 26.9 for women indicates obesity. A value above 30 is cause for concern; a value above 35 may be life threatening.

mucus, in the development of normal bones and teeth, and in the maintenance of epithelial cells in skin and mucous membranes.

Only foods of animal origin such as liver, fish, whole milk, butter, and eggs are sources of vitamin A. However, its precursor, carotene, is widespread in leafy green vegetables and in yellow or orange vegetables and fruits.

Clinical Application 18.2 examines the ongoing controversy concerning the value of vitamins A and E in preventing illness.

Excess vitamin A produces peeling skin, hair loss, nausea, headache, and dizziness, a condition called *hypervitaminosis A*. Chronic overdoses of the vitamin may inhibit growth and cause the bones and joints to degenerate. "Megadosing" on fat-soluble vitamins is particularly dangerous during pregnancy. Excess vitamin A, for example, can cause birth defects.

A deficiency of vitamin A causes *night blindness*, in which a person cannot see normally in dim light. Vitamin A deficiency also causes degenerative changes in certain epithelial tissues, and the body becomes more susceptible to infection.

1. What biochemical do body cells use to synthesize vitamin A?
2. What conditions destroy vitamin A?
3. Which foods are good sources of vitamin A?

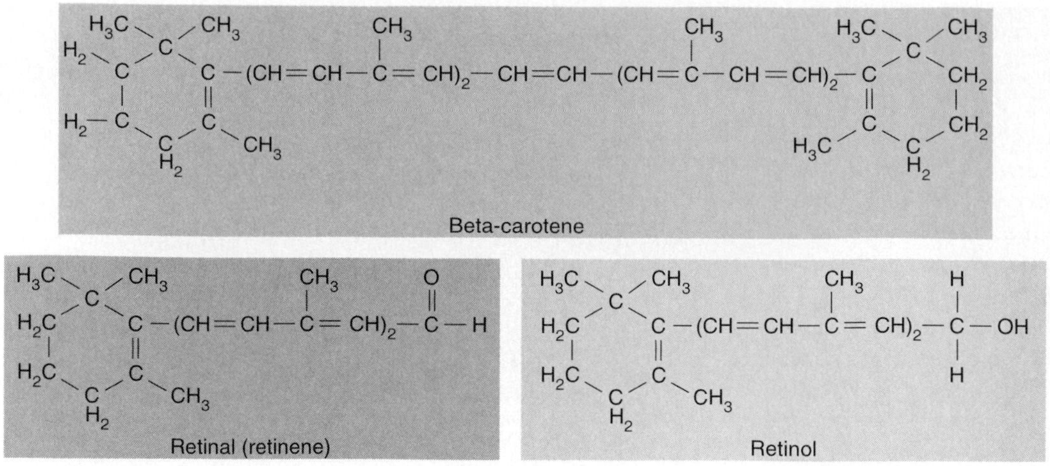

Beta-carotene

Retinal (retinene)

Retinol

**FIGURE 18.7**

A molecule of beta-carotene can be converted into two molecules of retinal, which in turn can be changed into retinol.

**Vitamin D** is a group of steroids that have similar properties. One of these substances, vitamin $D_3$ (cholecalciferol), is found in foods such as milk, egg yolk, and fish liver oils. Vitamin $D_2$ (ergocalciferol) is produced commercially by exposing a steroid obtained from yeasts (ergosterol) to ultraviolet light.

Vitamin D can also be synthesized from dietary cholesterol that has been converted to provitamin D by intestinal enzymes, then stored in the skin and exposed to ultraviolet light (see chapter 13).

Like other fat-soluble vitamins, vitamin D resists the effects of heat, oxidation, acids, and alkalis. It is stored primarily in the liver and is found to a lesser extent in the skin, brain, spleen, and bones.

As it is needed, vitamin D stored in the form of hydroxycholecalciferol is released into the blood. When parathyroid hormone is present, this form of vitamin D is converted in the kidneys into an active form of the vitamin (dihydroxycholecalciferol). This substance, in turn, is carried as a hormone in the blood to the intestine. Here, it promotes absorption of calcium and phosphorus, ensuring that adequate amounts of these minerals are available in the blood for tooth and bone formation and metabolic processes.

Because natural foods are often poor sources of vitamin D, it is often added to food during processing. For example, homogenized, nonfat, and evaporated milk are typically fortified with vitamin D. *Fortified* means essential nutrients have been added to a food where they originally were absent or scarce. *Enriched* means essential nutrients have been partially replaced in a food that has lost nutrients as a result of processing.

Excess vitamin D, or *hypervitaminosis D,* produces diarrhea, nausea, and weight loss and, over time, calcification of certain soft tissues and irreversible kidney damage.

**FIGURE 18.8**

Vitamin D deficiency causes rickets, in which the bones and teeth do not develop normally.

In children, vitamin D deficiency results in *rickets,* in which the bones and teeth fail to develop normally (fig. 18.8). In adults or in the elderly who have little exposure to sunlight, such a deficiency may lead to *osteomalacia,* in which the bones decalcify and weaken due to disturbances in calcium and phosphorus metabolism.

## Clinical Application 18.2

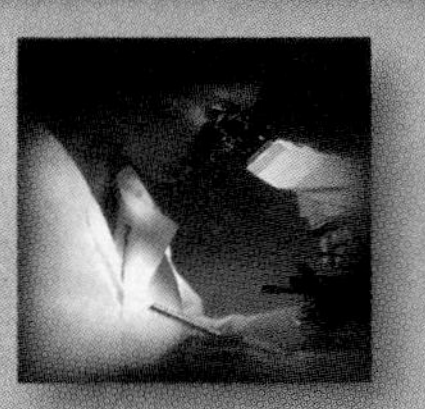

### Do Vitamins Protect Against Heart Disease and Cancer?

Vitamins A and E are popularly believed to protect against cardiovascular disease and cancer. While studies have shown that people who eat diets rich in fruits and vegetables containing these vitamins have slightly lower incidences of these disorders, these investigations do not prove that the vitamins, and not something else in these foods, are responsible for any protective effect. A more meaningful study would carefully follow the result of taking a specific vitamin supplement. The Finnish Alpha-Tocopherol, Beta-Carotene Cancer Prevention Study did just that—with surprising results.

Researchers monitored the health of 29,000 adult male smokers over a 6-year period. Each man received vitamin A or E, both vitamins, or neither. The "endpoint" of the study was to determine which men developed lung cancer. Not only did these vitamins fail to protect against this specific cancer, but the men who received beta-carotene actually had an increased incidence of lung cancer! In addition, neither vitamin lowered incidences of death from stroke or coronary heart disease.

The Finnish study has several limitations. Some critics claim that it was not conducted over a long enough time period. Also, the study considers only one type of person who engages in a dangerous activity—smoking—and may not account for all possible contributing factors to disease development. Unraveling the connections between specific nutrients and health is extremely complex, because of the many variables. Overall, a diet rich in fruits and vegetables is healthful. But the Finnish study shows that claims that certain vitamins protect against cancer or heart disease may be premature.

1. Where is vitamin D stored?
2. What are the functions of vitamin D?
3. Which foods are good sources of vitamin D?
4. What are symptoms of vitamin D excess and deficiency?

**Vitamin E** includes a group of compounds, the most active of which is *alphatocopherol.* This vitamin is resistant to the effects of heat, acids, and visible light, but is unstable in alkalis and in the presence of ultraviolet light or oxygen. In fact, it combines so readily with oxygen that it may prevent oxidation of other compounds. For this reason, it is termed an **antioxidant.**

Vitamin E is found in all tissues, but is stored primarily in the muscles and adipose tissue. It is also highly concentrated in the pituitary and adrenal glands.

The precise functions of vitamin E are unknown, but it apparently acts as an antioxidant by inhibiting the oxidation of vitamin A in the digestive tract and of polyunsaturated fatty acids in the tissues. It may also help maintain the stability of cell membranes. Antioxidants are thought to prevent the formation of free radicals that may play a role in disease and aging.

While vitamin E is widely distributed among foods, its richest sources are oils from cereal seeds such as wheat germ oil. Other good sources are salad oils, margarine, shortenings, fruits, nuts, and vegetables. Since this vitamin is so easily obtained, deficiency conditions are rare.

1. Where is vitamin E stored?
2. What are the functions of vitamin E?
3. Which foods are good sources of vitamin E?

**Vitamin K,** like the other fat-soluble vitamins, occurs in several chemical forms. One of these, vitamin $K_1$ (phylloquinone), is found in foods, while another, vitamin $K_2$, is produced by bacteria (*Escherichia coli*) that normally inhabit the human intestinal tract.

These vitamins resist the effects of heat, but are destroyed by oxidation or by exposure to acids, alkalis, or light. The liver stores them to a limited degree.

CHART 18.6 FAT-SOLUBLE VITAMINS

| Vitamin | Characteristics | Functions | Sources and RDA* for Adults | Conditions Associated with | |
|---|---|---|---|---|---|
| | | | | *Excesses* | *Deficiencies* |
| *Vitamin A* | Occurs in several forms; synthesized from carotenes; stored in liver; stable in heat, acids, and alkalis; unstable in light | Necessary for synthesis of visual pigments, mucoproteins, and mucopolysaccharides; for normal development of bones and teeth; and for maintenance of epithelial cells | Liver, fish, whole milk, butter, eggs, leafy green vegetables, yellow and orange vegetables and fruits 4,000–5,000 IU** | Nausea, headache, dizziness, hair loss | Night blindness, degeneration of epithelial tissues |
| *Vitamin D* | A group of steroids; resistant to heat, oxidation, acids, and alkalis; stored in liver, skin, brain, spleen, and bones | Promotes absorption of calcium and phosphorus; promotes development of teeth and bones | Produced in skin exposed to ultraviolet light; in milk, egg yolk, fish liver oils, fortified foods 400 IU | Diarrhea, calcification of soft tissues, renal damage | Rickets, bone decalcification and weakening |
| *Vitamin E* | A group of compounds; resistant to heat and visible light; unstable in presence of oxygen and ultraviolet light; stored in muscles and adipose tissue | An antioxidant; prevents oxidation of vitamin A and polyunsaturated fatty acids; may help maintain stability of cell membranes | Oils from cereal seeds, salad oils, margarine, shortenings, fruits, nuts, and vegetables 30 IU | Hypertension | Rare, uncertain effects |
| *Vitamin K* | Occurs in several forms; resistant to heat, but destroyed by acids, alkalis, and light; stored in liver | Needed for synthesis of prothrombin which functions in blood clotting | Leafy green vegetables, egg yolk, pork liver, soy oil, tomatoes, cauliflower 55–70 μg | None known | Easy bruising and bleeding |

*RDA = recommended dietary allowance.
**IU = international unit.

Vitamin K functions primarily in the liver, where it is necessary for the formation of several proteins needed for blood clotting, including *prothrombin* (see chapter 14). Consequently, deficiency of vitamin K causes prolonged blood clotting time and a tendency to hemorrhage.

The richest sources of vitamin K are leafy green vegetables. Other good sources are egg yolk, pork liver, soy oil, tomatoes, and cauliflower.

Chart 18.6 summarizes the fat-soluble vitamins and their properties.

About 1 in every 200 to 400 newborns develops vitamin K deficiency because of an immature liver, poor transfer of vitamin K through the placenta, or lack of intestinal bacteria that can synthesize this vitamin. This "hemorrhagic disease of the newborn" typically appears during the second to fifth day of life, producing abnormal bleeding. To prevent it, infants in the United States routinely receive injections of vitamin K shortly after birth. Adults may develop vitamin K deficiency if they are treated with antibiotic drugs that kill the intestinal bacteria that manufacture the vitamin. People with cystic fibrosis may develop the deficiency because they cannot digest fats well.

1. Where in the body is vitamin K synthesized?
2. What is the function of vitamin K?
3. Which foods are good sources of vitamin K?

## Water-Soluble Vitamins

The water-soluble vitamins include the B vitamins and vitamin C. The **B vitamins** are several compounds that are essential for normal cellular metabolism. They help oxidize carbohydrates, lipids, and proteins. Since the B vitamins often occur together in foods, they are usually referred to as the *vitamin B complex.* Members of this group differ chemically and functionally.

The B-complex vitamins include the following:

1. **Thiamine,** or **vitamin $B_1$.** In its pure form, thiamine is a crystalline compound called thiamine hydrochloride. It is destroyed by exposure to heat and oxygen, especially in alkaline environments. Figure 18.14 shows its molecular structure.

   Thiamine is part of a coenzyme called *cocarboxylase,* which oxidizes carbohydrates. More specifically, thiamine is required for pyruvic acid to enter the citric acid cycle (see chapter 4); in the absence of this vitamin, pyruvic acid accumulates in the blood. Thiamine also functions as a coenzyme in the synthesis of the sugar ribose, which is part of the nucleic acid RNA.

Thiamine is absorbed primarily through the wall of the duodenum and is transported by the blood to body cells. Only small amounts are stored in the tissues, and excesses are excreted in the urine.

Since this vitamin functions in the oxidation of carbohydrates, the quantity cells require varies with caloric intake. It is recommended that an adult diet contain 0.5 milligram (mg) of thiamine for every 1,000 calories ingested daily.

Good sources of thiamine are lean pork, other lean meats, liver, eggs, whole-grain cereals, leafy green vegetables, and legumes.

> A mild deficiency of thiamine produces loss of appetite, fatigue, and nausea. Prolonged deficiency leads to a disease called *beriberi,* which causes gastrointestinal disturbances, mental confusion, muscular weakness and paralysis, and enlargement of the heart. In severe cases, the heart may fail.
>
> In the United States, beriberi occurs mainly in chronic alcoholics who have substituted alcohol for needed foods. Moreover, since thiamine is required for the metabolic oxidation of alcohol, alcoholics are particularly likely to develop a thiamine deficiency.

2. **Riboflavin,** or **vitamin $B_2$.** Riboflavin is a yellowish-brown crystalline substance that is relatively stable to the effects of heat, acids, and oxidation, but is destroyed by exposure to bases and ultraviolet light. This vitamin is part of several enzymes and coenzymes that are known as *flavoproteins.* One such coenzyme, FAD, is an electron carrier in the citric acid cycle and electron transport chain of aerobic respiration. Flavoproteins are essential for the oxidation of glucose and fatty acids, and for cellular growth. The absorption of riboflavin seems to be regulated by an active transport system that controls the amount taken in by the intestinal mucosa. It is carried in the blood combined with blood proteins called *albumins.* Excess riboflavin in the blood is excreted in the urine, and any that remains unabsorbed in the intestine is lost in the feces.

   The amount of riboflavin the body requires varies with caloric intake. About 0.6 mg of riboflavin per 1,000 calories is sufficient to meet daily cellular needs.

   Riboflavin is widely distributed in foods, and rich sources include meats and dairy products. Leafy green vegetables, whole-grain cereals, and enriched cereals provide lesser amounts.

   Vitamin $B_2$ deficiency produces dermatitis and blurred vision.

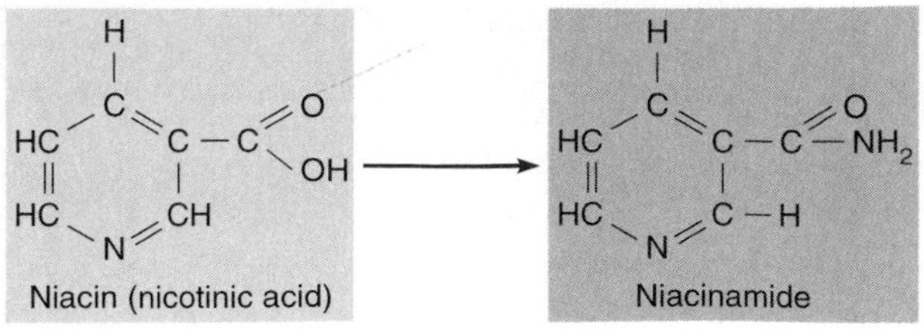

Niacin (nicotinic acid) → Niacinamide

**FIGURE 18.9**

Body cells convert niacin from foods into physiologically active niacinamide.

3. **Niacin.** Niacin, which is also known as *nicotinic acid,* occurs in plant tissues, and is stable in the presence of heat, acids, and bases. After ingestion, it is converted to a physiologically active form called *niacinamide* (fig. 18.9). Niacinamide is the form of niacin that is present in foods of animal origin.

   Niacin functions as part of two coenzymes (coenzyme I, also called NAD [fig. 18.10], and coenzyme II, called NADP) that play essential roles in the oxidation of glucose, acting as electron carriers in glycolysis, the citric acid cycle, and the electron transport chain, as well as in the synthesis of proteins and fats. These coenzymes are also needed for the synthesis of the sugars used in the production of nucleic acids.

> Niacin in large doses is prescribed as a drug to lower serum cholesterol.

Niacin is readily absorbed from foods, and human cells synthesize it from the essential amino acid *tryptophan.* Consequently, the daily requirement for niacin varies with tryptophan intake. Nutritionists recommend a daily niacin (or niacin equivalent) intake of 6.6 mg per 1,000 calories.

Rich sources of niacin (and tryptophan) include liver, lean meats, peanut butter, and legumes. Milk is a poor source of niacin, but a good source of tryptophan.

Historically, niacin deficiencies have been associated with diets consisting largely of corn and corn products, which are very low in niacin and lack adequate tryptophan. Such a deficiency causes a disease called *pellagra* that produces dermatitis, inflammation of the digestive tract, diarrhea, and mental disorders.

Although pellagra is relatively rare in the United States today, it was a serious problem in the rural South in the early 1900s. It sometimes occurs in chronic alcoholics who have substituted alcohol for needed foods.

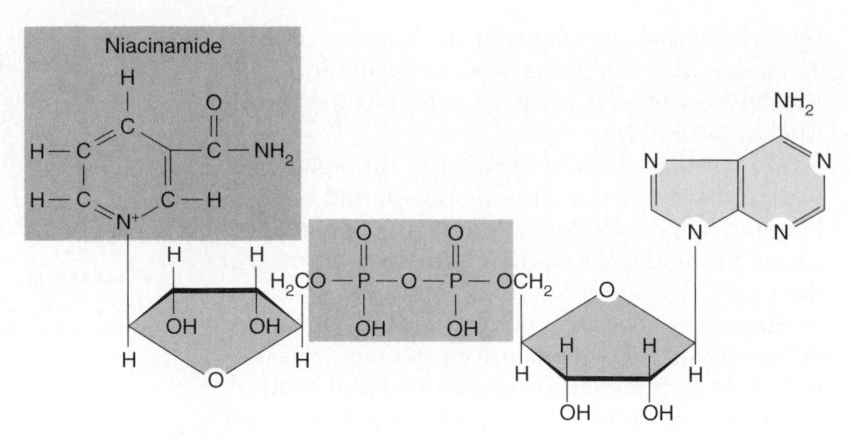


**FIGURE 18.10**
Niacinamide is incorporated into molecules of coenzyme I.

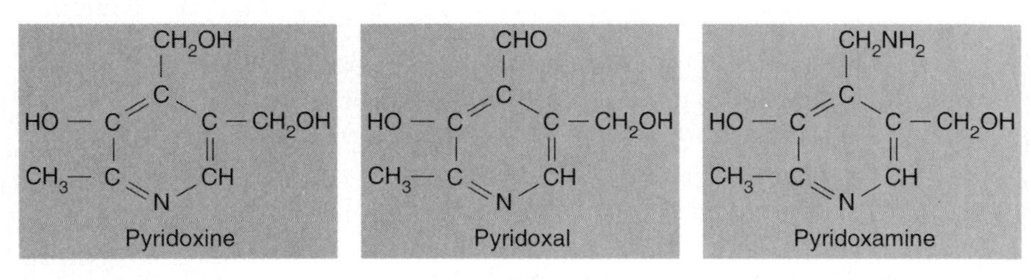


**FIGURE 18.11**
Vitamin $B_6$ includes three closely related chemical compounds.

4. **Vitamin $B_6$.** Vitamin $B_6$ is a group of three compounds that are closely related chemically, as figure 18.11 shows. They are called *pyridoxine, pyridoxal,* and *pyridoxamine.* These biochemicals have similar actions and are fairly stable in the presence of heat and acids; but they are destroyed by oxidation, or by exposure to bases or ultraviolet light.

   The vitamin $B_6$ compounds function as coenzymes that are essential in a wide variety of metabolic pathways, including those that synthesize proteins, amino acids, antibodies, and nucleic acids, as well as the conversion of tryptophan to niacin.

   Since vitamin $B_6$ functions in the metabolism of nitrogen-containing substances, the requirement for this vitamin varies with the protein content of the diet rather than with caloric intake.

   The recommended daily allowance of vitamin $B_6$ is 2.0 mg, but because it is so widespread in foods, deficiency conditions are quite rare.

   Good sources of vitamin $B_6$ include liver, meats, bananas, avocados, beans, peanuts, whole-grain cereals, and egg yolk.

5. **Pantothenic acid.** Pantothenic acid is a yellowish oil that is destroyed by heat, acids, and alkalis. It functions as part of a complex molecule called *coenzyme A,* which, in turn, reacts with intermediate products of carbohydrate and fat metabolism to become *acetyl coenzyme A,* which enters the citric acid cycle. Pantothenic acid is therefore essential to cellular energy release.

   A daily adult intake of 4–7 mg seems to be adequate. Most diets apparently provide sufficient amounts, since deficiencies are rare, and no clearly defined set of deficiency symptoms is known.

   Good sources of pantothenic acid include meats, whole-grain cereals, legumes, milk, fruits, and vegetables.

6. **Cyanocobalamin,** or **vitamin $B_{12}$.** Cyanocobalamin has a complex molecular structure that contains a single atom of the element *cobalt* (fig. 18.12). In its pure form, this vitamin is red. It is stable to the effects of heat, but is inactivated by exposure to light or strong acids or bases.

   Secretion of *intrinsic factor* from the parietal cells of the gastric glands regulates cyanocobalamin absorption. Intrinsic factor is thought to combine with cyanocobalamin and to facilitate its transfer through the epithelial lining of the small intestine and into the blood. Calcium ions must be present for the process to take place.

   Cyanocobalamin is stored in various tissues, particularly those of the liver. An average adult has a reserve sufficient to supply cellular needs for three to five years.

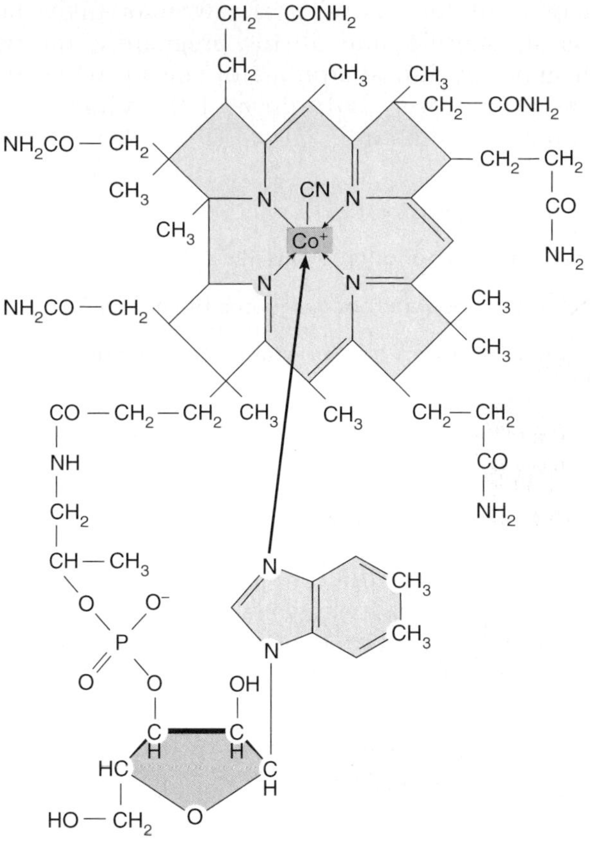

Vitamin $B_{12}$ (cyanocobalamin)

**FIGURE 18.12**
Vitamin $B_{12}$, which has the most complex molecular structure of the vitamins, contains cobalt (Co).

This vitamin is essential for the functions of all cells. It is part of coenzymes needed for the synthesis of nucleic acids and the metabolism of carbohydrates and fats. It also seems to help form myelin in the central nervous system.

Cyanocobalamin is found only in foods of animal origin; good sources include liver, meats, milk, cheese, and eggs. Because persons in the United States consume relatively large amounts of such foods, a dietary lack of this vitamin seldom occurs, although strict vegetarians (vegans) may develop a deficiency.

> When the gastric glands of some individuals fail to secrete adequate amounts of intrinsic factor, cyanocobalamin is poorly absorbed. This leads to *pernicious anemia,* in which abnormally large red blood cells called macrocytes are produced when bone marrow cells do not divide properly because of defective DNA synthesis.

7. **Folacin,** or **folic acid.** Folacin is a yellow crystalline compound that exists in several forms. It is easily oxidized in an acid environment and is destroyed by heat in alkaline solutions; consequently, this vitamin may be lost in foods that are stored or cooked.

   Folacin is readily absorbed from the digestive tract and is stored in the liver, where it is converted to a physiologically active substance called *folinic acid.*

   Folinic acid functions as a coenzyme that is necessary for the metabolism of certain amino acids and for the synthesis of DNA. It also acts together with cyanocobalamin in promoting the production of normal red blood cells.

   Good sources of folacin include liver, leafy green vegetables, whole-grain cereals, and legumes.

   A deficiency of folacin leads to *megaloblastic anemia,* which is characterized by a reduced number of normal red blood cells and the presence of large, nucleated red cells. Folacin deficiency has been linked to neural tube defects, in which the tube that becomes the central nervous system in a fetus fails to close entirely. Taking folacin supplements just before and during pregnancy can greatly reduce the risk of a neural tube defect.

8. **Biotin.** Biotin is a relatively simple compound that is stable to the effects of heat, acids, and light, but may be destroyed by oxidation or bases. Figure 18.14 shows the molecular structure of biotin.

   Biotin is a coenzyme in metabolic pathways for amino acids and fatty acids. It also plays a role in the synthesis of certain organic bases (purines) which are essential in the synthesis of nucleic acids.

   Small quantities of biotin are stored in the tissues of metabolically active organs such as the brain, liver, and kidneys. Bacteria that inhabit the intestinal tract synthesize biotin.

   Biotin is widely distributed in foods, and dietary deficiencies are relatively rare. Good sources include liver, egg yolk, nuts, legumes, and mushrooms.

1. Which biochemicals comprise the vitamin B complex?
2. Which foods are good sources of vitamin B complex?
3. Which of the B-complex vitamins can be synthesized from tryptophan?
4. What is the general function of each member of the B complex?

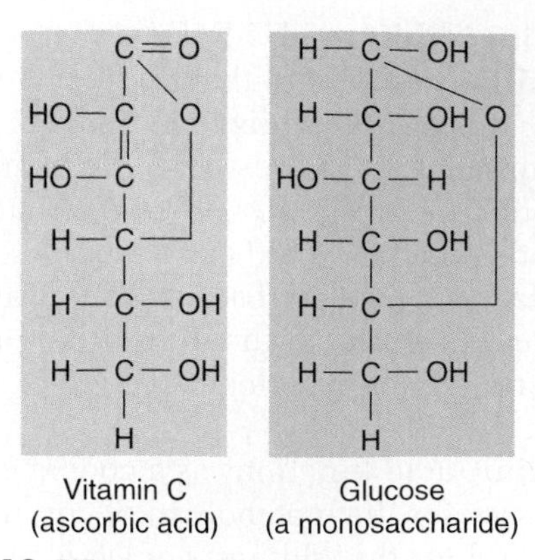


**FIGURE 18.13**

Vitamin C is closely related chemically to some six-carbon monosaccharides.

**Vitamin C,** or **ascorbic acid.** Ascorbic acid is a crystalline compound that contains six carbon atoms. Chemically, it is closely related to the monosaccharides (fig. 18.13). It is one of the least stable of the vitamins, in that it can be destroyed by oxidation, heat, light, or bases. It is fairly stable in acids.

Ascorbic acid is necessary for the production of the connective tissue protein *collagen,* for conversion of folacin to folinic acid, and in the metabolism of certain amino acids. It also promotes iron absorption and synthesis of certain hormones from cholesterol.

Although vitamin C is not stored in any great amount, tissues of the adrenal cortex, pituitary gland, and intestinal glands contain relatively high concentrations. Excess quantities are either excreted in the urine or oxidized.

Individual needs for ascorbic acid may vary. Ten mg per day is sufficient to prevent deficiency symptoms, and 80 mg per day will saturate the tissues within a few weeks. Many nutritionists recommend a daily adult intake of 60 mg, which is enough to replenish normal losses and to provide a satisfactory level for cellular needs.

Ascorbic acid is fairly widespread in plant foods; particularly high concentrations are found in citrus fruits and tomatoes. Leafy green vegetables are also good sources.

Chart 18.7 summarizes the water-soluble vitamins and their characteristics.

A prolonged deficiency of ascorbic acid leads to *scurvy,* which occurs more frequently in infants and children. Scurvy produces abnormal bone development and swollen, painful joints. Because of a tendency for cells to pull apart, the gums may swell and bleed easily, resistance to infection is lowered, and wounds heal slowly. If a woman takes large doses of ascorbic acid during pregnancy, the newborn may develop symptoms of scurvy when it receives a far lower daily dose of the vitamin after birth than it did before.

1. What factors destroy vitamin C?
2. What is the function of vitamin C?
3. Which foods are good sources of vitamin C?
4. What are symptoms of vitamin C deficiency?

## MINERALS

Carbohydrates, lipids, proteins, and vitamins are all organic compounds. Dietary **minerals** are inorganic elements that are essential in human metabolism. These elements are usually extracted from the soil by plants. Humans, in turn, obtain them from plant foods or from animals that have eaten plants.

### Characteristics of Minerals

Minerals are responsible for about 4% of the body weight and are most concentrated in the bones and teeth. The minerals *calcium* and *phosphorus,* which are very abundant in these tissues, account for nearly 75% of the body's minerals.

Minerals are usually incorporated into organic molecules. For example, phosphorus is found in phospholipids, iron in hemoglobin, and iodine in thyroxine. However, some minerals are part of inorganic compounds, such as the calcium phosphate of bone. Other minerals are free ions, such as the sodium, chloride, and calcium ions in the blood.

Minerals are present in all body cells, where they comprise parts of the structural materials. They are also portions of enzyme molecules, contribute to the osmotic pressure of body fluids, and play vital roles in the conduction of nerve impulses, the contraction of muscle fibers, the coagulation of blood, and the maintenance of pH. The physiologically active form of minerals is the ionized form, such as $Ca^{++}$.

Homeostatic mechanisms regulate the concentrations of minerals in body fluids, ensuring that excretion matches intake.

1. How do minerals differ from other nutrients?
2. What are the major functions of minerals?
3. Which are the most abundant minerals in the body?

## CHART 18.7 WATER-SOLUBLE VITAMINS

| Vitamin | Characteristics | Functions | Sources and RDA* for Adults | Conditions Associated with Excesses | Conditions Associated with Deficiencies |
|---|---|---|---|---|---|
| *Thiamine* (Vitamin $B_1$) | Destroyed by heat and oxygen, especially in alkaline environment | Part of coenzyme needed for oxidation of carbohydrates; coenzyme needed in synthesis of ribose | Lean meats, liver, eggs, whole-grain cereals, leafy green vegetables, legumes 1.5 mg | None known | Beriberi, muscular weakness, heart enlarges |
| *Riboflavin* (Vitamin $B_2$) | Stable to heat, acids, and oxidation; destroyed by bases and ultraviolet light | Part of enzymes and coenzymes, such as FAD, needed for oxidation of glucose and fatty acids and for cellular growth | Meats, dairy products, leafy green vegetables, whole-grain cereals 1.7 mg | None known | Dermatitis, blurred vision |
| *Niacin* (Nicotinic Acid) | Stable to heat, acids, and alkalis; converted to niacinamide by cells; synthesized from tryptophan | Part of coenzymes NAD and NADP needed for oxidation of glucose and synthesis of proteins, fats, and nucleic acids | Liver, lean meats, peanuts, legumes 20 mg | Hyperglycemia, vasodilation, gout | Pellagra, photosensitive dermatitis, diarrhea, mental disorders |
| *Vitamin $B_6$* | Group of three compounds; stable to heat and acids; destroyed by oxidation, bases, and ultraviolet light | Coenzyme needed for synthesis of proteins and various amino acids, for conversion of tryptophan to niacin, for production of antibodies, and for synthesis of nucleic acids | Liver, meats, bananas, avocados, beans, peanuts, whole-grain cereals, egg yolk 2 mg | Numbness | Rare, convulsions, vomiting, seborrhea lesions |
| *Pantothenic Acid* | Destroyed by heat, acids, and bases | Part of coenzyme A needed for oxidation of carbohydrates and fats | Meats, whole-grain cereals, legumes, milk, fruits, vegetables 10 mg | None known | Rare, loss of appetite, mental depression, muscle spasms |
| *Cyanocobalamin* (Vitamin $B_{12}$) | Complex, cobalt-containing compound; stable to heat; inactivated by light, strong acids, and strong bases; absorption regulated by intrinsic factor from gastric glands; stored in liver | Part of coenzyme needed for synthesis of nucleic acids and for metabolism of carbohydrates; plays role in synthesis of myelin | Liver, meats, milk, cheese, eggs 3–6 μg | None known | Pernicious anemia |
| *Folacin* (Folic Acid) | Occurs in several forms; destroyed by oxidation in acid environment or by heat in alkaline environment; stored in liver where it is converted into folinic acid | Coenzyme needed for metabolism of certain amino acids and for synthesis of DNA; promotes production of normal red blood cells | Liver, leafy green vegetables, whole-grain cereals, legumes 0.4 mg | None known | Megaloblastic anemia |
| *Biotin* | Stable to heat, acids, and light; destroyed by oxidation and bases | Coenzyme needed for metabolism of amino acids and fatty acids, and for synthesis of nucleic acids | Liver, egg yolk, nuts, legumes, mushrooms 0.3 mg | None known | Rare, elevated blood cholesterol, nausea, fatigue, anorexia |
| *Ascorbic Acid* (Vitamin C) | Closely related to monosaccharides; stable in acids but destroyed by oxidation, heat, light, and bases | Needed for production of collagen, conversion of folacin to folinic acid, and metabolism of certain amino acids; promotes absorption of iron and synthesis of hormones from cholesterol | Citrus fruits, tomatoes, potatoes, leafy green vegetables, 60 mg | Exacerbates gout and kidney stone formation | Scurvy, lowered resistance to infection, wounds heal slowly |

*RDA = recommended dietary allowance.

## Major Minerals

Calcium and phosphorus account for nearly 75% of the mineral elements in the body; thus, they are **major minerals** (macrominerals). Other major minerals, each of which accounts for 0.05% or more of the body weight, include potassium, sulfur, sodium, chlorine, and magnesium. Descriptions of the major minerals follow:

1. **Calcium.** Calcium (Ca) is widely distributed in cells and body fluids, even though 99% of the body's supply is in the inorganic salts of the bones and teeth. It is essential for nerve impulse conduction, muscle fiber contraction, and blood coagulation. It also decreases the permeability of cell membranes and activates certain enzymes.

   The amount of calcium absorbed varies with a number of factors. For example, the proportion of calcium absorbed increases as the body's need for calcium increases. Vitamin D and high protein intake promote calcium absorption; increased motility of the digestive tract or an excess intake of fats decreases absorption. Consequently, the amount of calcium needed to supply cells may vary. However, a daily intake of 800 mg is sufficient to cover adult needs in spite of variations in absorption.

   Only a few foods contain significant amounts of calcium. Milk and milk products and fish with bones, such as salmon or sardines, are the richest sources. Leafy green vegetables, such as mustard greens, turnip greens, and kale, are good sources. Since these vegetables are not very popular foods in the United States, it is difficult for most people to maintain an adequate intake of calcium unless they regularly consume milk or milk products, such as cheese, cottage cheese, and ice cream.

   Calcium deficiency in children causes stunted growth, misshapen bones, and enlarged wrists and ankles. In adults, such a deficiency may remove calcium from the bones, thinning them and raising risk of fracture. Because calcium is required to close the sodium channels in nerve cell membranes, too little calcium (hypocalcemia) can cause tetany. Extra calcium demands in pregnancy can cause cramps.

2. **Phosphorus.** Phosphorus (P) is responsible for about 1% of total body weight, and most of it is incorporated in the calcium phosphate of bones and teeth. The remainder serves as structural components and plays important roles in nearly all metabolic reactions. Phosphorus is a constituent of nucleic acids, many proteins, some enzymes, and some vitamins. It is also found in the phospholipids of cell membranes, in the energy-carrying molecule ATP, and in the phosphates of body fluids that regulate pH. (Review the molecular structure of ATP in fig. 4.7.)

   The recommended daily adult intake of phosphorus is 800 mg, and since this mineral is abundant in protein foods, diets adequate in proteins are also adequate in phosphorus. Phosphorus-rich foods include meats, cheese, nuts, whole-grain cereals, milk, and legumes.

1 What are the functions of calcium?

2 What are the functions of phosphorus?

3 Which foods are good sources of calcium and phosphorus?

3. **Potassium.** Potassium (K) is widely distributed throughout the body and tends to be concentrated inside cells rather than in extracellular fluids. On the other hand, sodium, which has similar chemical properties, tends to be concentrated outside the cells. The ratio of potassium to sodium within a cell is 10:1, while the ratio outside the cell is 1:28.

   Potassium helps maintain the intracellular osmotic pressure and pH. It promotes reactions of carbohydrate and protein metabolism, and plays a vital role in the membrane potential that occurs in nerve impulse conduction and muscle fiber contraction.

   Nutritionists recommend a daily adult intake of 2.5 grams (2,500 mg) of potassium. Because this mineral is widely distributed in foods, a typical adult diet provides between 2 and 6 grams each day. Dietary potassium deficiency is rare, but it may occur for other reasons. For example, when a person has diarrhea, the intestinal contents may pass through the digestive tract so rapidly that potassium absorption is greatly reduced. Vomiting or using diuretic drugs may also deplete potassium. The consequences of such losses may include muscular weakness, cardiac abnormalities, and edema.

   Foods rich in potassium are avocados, dried apricots, meats, milk, peanut butter, potatoes, and bananas. Citrus fruits, apples, carrots, and tomatoes provide lesser amounts.

1 How is potassium distributed in the body?

2 What is the function of potassium?

3 Which foods are good sources of potassium?

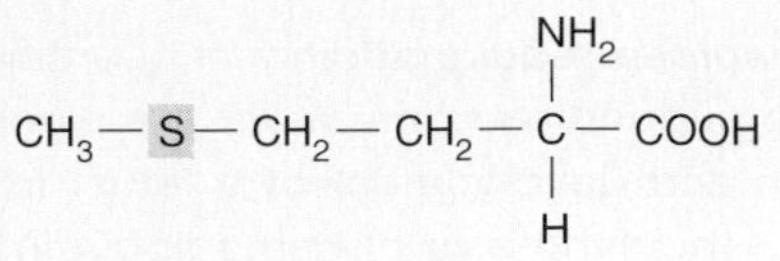

Methionine

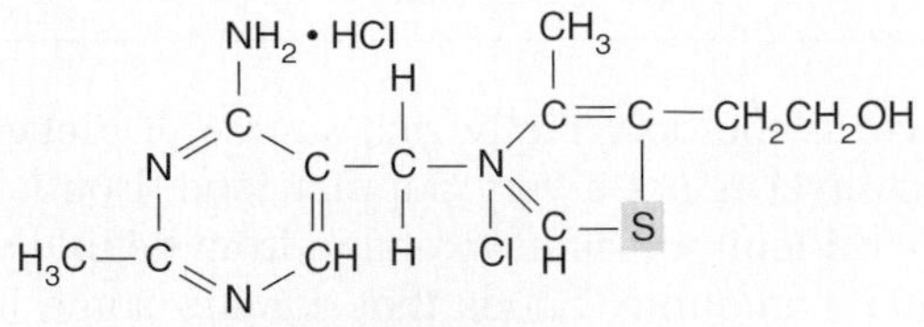

Thiamine hydrochloride (vitamin $B_1$)

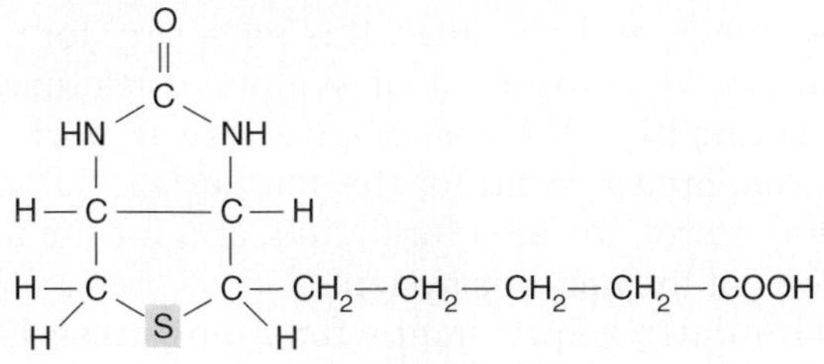

Biotin

**FIGURE 18.14**

Three examples of essential sulfur-containing nutrients.

4. **Sulfur.** Sulfur (S) is responsible for about 0.25% of body weight and is widely distributed through tissues. It is particularly abundant in skin, hair, and nails. Most sulfur is part of the amino acids *methionine* and *cysteine.* Other sulfur-containing compounds include thiamine, insulin, and biotin (see fig. 18.14). In addition, sulfur is a constituent of mucopolysaccharides found in cartilage, tendons, and bones, and of sulfolipids that are in the liver, kidneys, salivary glands, and brain.

   No daily requirement for sulfur has been established. It is thought, however, that a diet providing adequate amounts of protein will also meet the body's need for sulfur. Good food sources of this mineral include meats, milk, eggs, and legumes.

5. **Sodium.** About 0.15% of adult body weight is due to sodium (Na), which is widely distributed throughout the body. Only about 10% of this mineral is inside the cells, and about 40% is within the extracellular fluids. The remainder is bound to the inorganic salts of bones.

   Sodium is readily absorbed from foods by active transport, and the blood concentration of this element is regulated by the kidneys under the influence of the adrenal cortical hormone *aldosterone,* which causes the kidneys to reabsorb sodium while expelling potassium.

   Sodium makes a major contribution to the solute concentration of extracellular fluids and thus regulates water movement between cells and their surroundings. It is necessary for nerve impulse conduction and muscle fiber contraction, and plays a role in moving substances, such as chloride ions, through cell membranes (see chapter 21).

   The usual human diet probably provides more than enough sodium to meet the body's needs. Sodium may be lost as a result of diarrhea, vomiting, kidney disorders, sweating, or using diuretics. Such losses may cause a variety of symptoms including nausea, muscular cramps, and convulsions.

   The amount of sodium naturally present in foods varies greatly, and it is commonly added to foods in the form of table salt (sodium chloride). In some geographic regions, drinking water contains significant concentrations of sodium.

   Foods that have relatively high sodium contents include cured ham, sauerkraut, cheese, and graham crackers.

**1** In what compounds and tissues of the body is sulfur found?

**2** Which hormone regulates the blood concentration of sodium?

**3** What are the functions of sodium?

6. **Chlorine.** Chlorine (Cl) in the form of chloride ions is closely associated with sodium. Like sodium, it is widely distributed throughout the body, although it is most highly concentrated in cerebrospinal fluid and in gastric juice.

   Together with sodium, chlorine helps to maintain the solute concentration of extracellular fluids, regulate pH, and maintain electrolyte balance. It is also essential for the formation of hydrochloric acid in gastric juice, and it functions in the transport of carbon dioxide by red blood cells.

   Chlorine and sodium are usually obtained together in the form of table salt (sodium chloride), and as in the case of sodium, an ordinary diet usually provides considerably more chlorine than the body needs. Vomiting, diarrhea, kidney disorders, sweating, or using diuretics can deplete chlorine in the body.

7. **Magnesium.** Magnesium (Mg) is responsible for about 0.05% of body weight and is found in all cells. It is particularly abundant in bones in the form of phosphates and carbonates.

Magnesium is important in ATP-forming reactions in mitochondria, as well as in breaking down ATP to ADP. Therefore, it is important in providing energy for cellular processes.

Magnesium absorption in the intestinal tract adapts to dietary intake of the mineral. When the intake of magnesium is high, a smaller percentage is absorbed from the intestinal tract, and when the intake is low, a larger percentage is absorbed. Absorption increases as protein intake increases, and decreases as calcium and vitamin D intake increase. Bone tissue stores a reserve supply of magnesium, and excess amounts are excreted in the urine.

The recommended daily allowance of magnesium is 300 mg for females and 350 mg for males. A typical diet usually provides only about 120 mg of magnesium for every 1,000 calories, barely meeting the body's needs.

Good sources of magnesium include milk and dairy products (except butter), legumes, nuts, and leafy green vegetables.

Chart 18.8 summarizes the major minerals.

1. Where are chloride ions most highly concentrated in the body?
2. Where is magnesium stored?
3. What factors influence the absorption of magnesium from the intestinal tract?

## Trace Elements

**Trace elements** (microminerals) are essential minerals found in minute amounts, each making up less than 0.005% of adult body weight. They include iron, manganese, copper, iodine, cobalt, zinc, fluorine, selenium, and chromium.

**Iron** (Fe) is most abundant in the blood, is stored in the liver, spleen, and bone marrow, and is found to some extent in all cells. Iron enables *hemoglobin* molecules in red blood cells to carry oxygen (fig. 18.15). Iron is also part of *myoglobin,* which stores oxygen in muscle cells. In addition, iron assists in vitamin A synthesis, is incorporated into a number of enzymes, and is included in the cytochrome molecules that participate in ATP-generating reactions.

An adult male requires from 0.7 to 1 mg of iron daily, and a female needs 1.2 to 2 mg. A typical diet supplies about 10 to 18 mg of iron each day, but only 2% to 10% of the iron is absorbed. For some people, this may not be enough iron. Eating foods rich in vitamin C along with iron-containing foods can increase absorption of this important mineral.

Pregnant women require additional quantities of iron in order to support the formation of a placenta, and the growth and development of a fetus. Iron is also needed for the synthesis of hemoglobin in the fetus as well as in the pregnant woman, whose blood volume increases by a third during pregnancy.

Liver is the only really rich source of dietary iron, and since liver is not a very popular food, iron is one of the more difficult nutrients to obtain from natural sources in adequate amounts. Foods that contain some iron include lean meats; dried apricots, raisins, and prunes; enriched whole-grain cereals; legumes; and molasses.

**Manganese** (Mn) is most concentrated in the liver, kidneys, and pancreas. It is necessary for normal growth and development of skeletal structures and other connective tissues. Manganese is part of enzymes that are essential for the synthesis of fatty acids and cholesterol, for urea formation, and for the normal functions of the nervous system.

The daily requirement for manganese is 2.5–5 mg. The richest sources include nuts, legumes, and whole-grain cereals; leafy green vegetables and fruits are good sources.

A compulsive disorder that may result from mineral deficiency is *pica,* in which people consume huge amounts of nondietary substances such as ice chips, dirt, sand, laundry starch, clay and plaster, and even such strange things as hair, toilet paper, matchheads, inner tubes, mothballs, and charcoal. The condition is named for the magpie bird, *Pica pica,* which eats a range of odd things. Pica affects people of all cultures and was noted as early as 40 B.C. The connection to dietary deficiency stems from the observation that slaves suffering from pica in colonial America recovered when their diets improved, particularly when they were given iron supplements. But pica is largely a medical mystery—trace mineral deficiencies apparently both cause it and result from it.

1. What is the primary function of iron?
2. Why does the usual diet provide only a narrow margin of safety in supplying iron?
3. How is manganese utilized?
4. Which foods are good sources of manganese?

**Copper** (Cu) is found in all body tissues, but is most highly concentrated in the liver, heart, and brain. It is essential for hemoglobin synthesis, bone development, melanin production, and formation of myelin within the nervous system.

## CHART 18.8 Major Minerals

| Mineral | Distribution | Functions | Sources and RDA* for Adults | Conditions Associated with Excesses | Conditions Associated with Deficiencies |
|---|---|---|---|---|---|
| *Calcium* (Ca) | Mostly in the inorganic salts of bones and teeth | Structure of bones and teeth; essential for nerve impulse conduction, muscle fiber contraction, and blood coagulation; increases permeability of cell membranes; activates certain enzymes | Milk, milk products, leafy green vegetables 800 mg | Kidney stones | Stunted growth, misshapen bones, fragile bones |
| *Phosphorus* (P) | Mostly in the inorganic salts of bones and teeth | Structure of bones and teeth; component in nearly all metabolic reactions; constituent of nucleic acids, many proteins, some enzymes, and some vitamins; occurs in cell membrane, ATP, and phosphates of body fluids | Meats, cheese, nuts, whole-grain cereals, milk, legumes 800 mg | None known | Stunted growth |
| *Potassium* (K) | Widely distributed; tends to be concentrated inside cells | Helps maintain intracellular osmotic pressure and regulate pH; promotes metabolism; needed for nerve impulse conduction and muscle fiber contraction | Avocados, dried apricots, meats, nuts, potatoes, bananas 2,500 mg | None known | Muscular weakness, cardiac abnormalities, edema |
| *Sulfur* (S) | Widely distributed; abundant in skin, hair, and nails | Essential part of various amino acids, thiamine, insulin, biotin, and mucopolysaccharides | Meats, milk, eggs, legumes None established | None known | None known |
| *Sodium* (Na) | Widely distributed; large proportion occurs in extracellular fluids and bonded to inorganic salts of bone | Helps maintain osmotic pressure of extracellular fluids and regulate water movement; needed for conduction of nerve impulses and contraction of muscle fibers; aids in regulation of pH and in transport of substances across cell membranes | Table salt, cured ham, sauerkraut, cheese, graham crackers 2,500 mg | Hypertension, edema | Nausea, muscle cramps, convulsions |
| *Chlorine* (Cl) | Closely associated with sodium; most highly concentrated in cerebrospinal fluid and gastric juice | Helps maintain osmotic pressure of extracellular fluids, regulate pH, and maintain electrolyte balance; essential in formation of hydrochloric acid; aids transport of carbon dioxide by red blood cells | Same as for sodium None established | Vomiting | Muscle cramps |
| *Magnesium* (Mg) | Abundant in bones | Needed in metabolic reactions that occur in mitochondria and are associated with the production of ATP; plays role in the breakdown of ATP to ADP | Milk, dairy products, legumes, nuts, leafy green vegetables 300–350 mg | Diarrhea | Neuromuscular disturbances |

*RDA = recommended dietary allowance.

A daily intake of 2 mg of copper is sufficient to provide the needs of body cells. Because a typical adult diet supplies about 2–5 mg of this mineral, adults seldom develop copper deficiencies.

Foods rich in copper include liver, oysters, crabmeat, nuts, whole-grain cereals, and legumes.

**Iodine** (I) is found in minute quantities in all tissues, but is highly concentrated within the thyroid gland. Its only known function is as an essential component of thyroid hormones. (The molecular structures of two of these hormones, thyroxine and triiodothyronine, are shown in figure 13.18.)

A daily intake of 1 microgram (.001 mg) of iodine per kilogram of body weight is adequate for most adults. Since the iodine content of foods varies with the iodine content of soils in different geographic regions, many people use *iodized* table salt to season foods to prevent deficiencies.

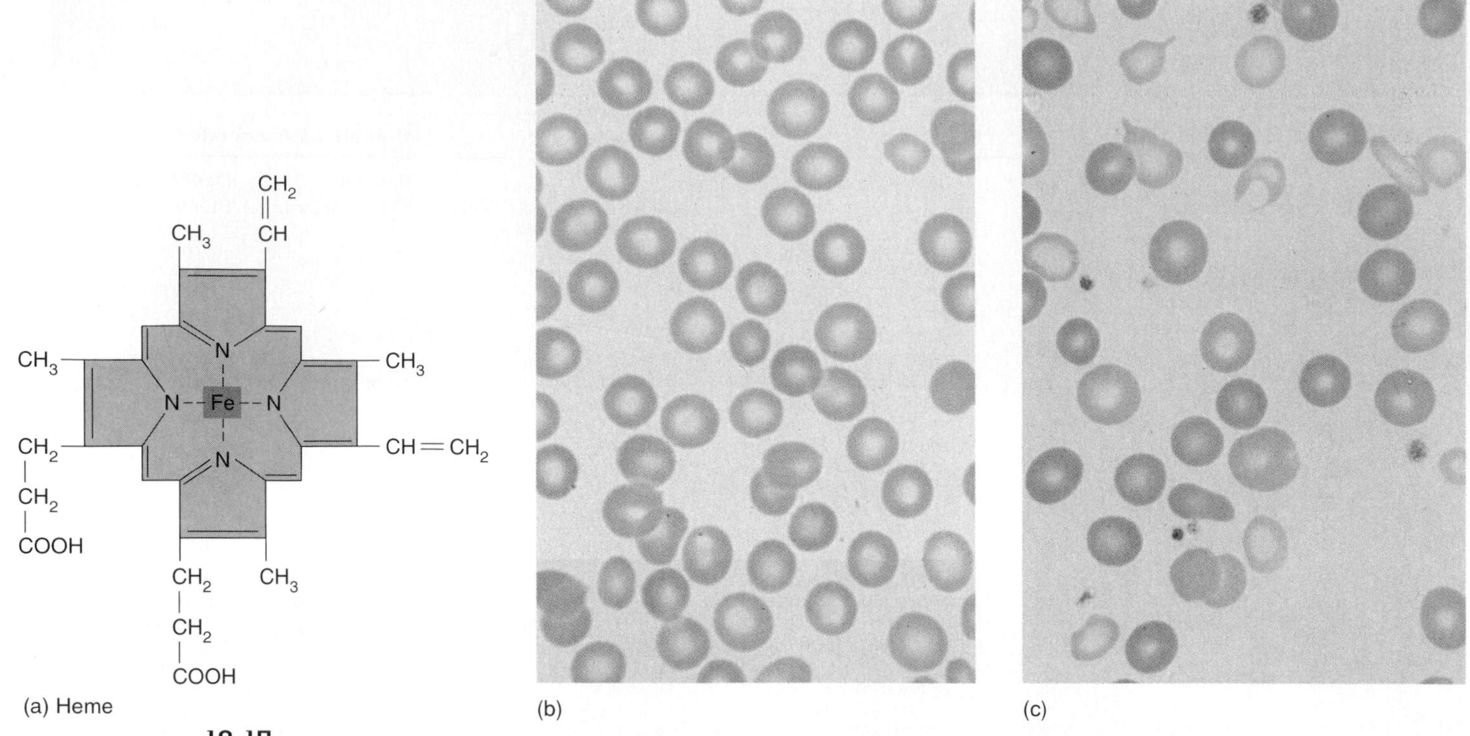


**FIGURE 18.15**

(*a*) A hemoglobin molecule contains four heme groups, each of which houses a single iron atom (Fe) that can combine with oxygen. Iron deficiency anemia can result from a diet poor in iron-containing foods. The blood cells in (*b*) are normal (250×), but many of those in (*c*) are small and pale (250×). They contain too little hemoglobin, because iron is lacking in the diet. Vegetarians must be especially careful to consume sufficient iron.

**Cobalt** (Co) is widely distributed in the body, because it is an essential part of cyanocobalamin molecules (vitamin $B_{12}$). It is also necessary for the synthesis of several important enzymes.

The amount of cobalt required in the daily diet is unknown. This mineral is found in a great variety of foods, and the quantity in the average diet is apparently sufficient to meet the body's need. Good sources of cobalt include liver, lean meats, and milk.

**Zinc** (Zn) is most concentrated in the liver, kidneys, and brain. It is part of many enzymes involved in digestion, respiration, and bone and liver metabolism. It is also necessary for normal wound healing and for maintaining the integrity of the skin.

The daily requirement for zinc is about 15 mg, and most diets provide 10–15 mg. Since only a portion of this amount may be absorbed, zinc deficiencies may be relatively common.

The richest sources of zinc are meats; cereals, legumes, nuts, and vegetables provide lesser amounts.

**Fluorine** (F), as part of the compound fluoroapatite, replaces hydroxyapatite in teeth, strengthening the enamel and preventing dental caries. **Selenium** (Se) is stored in the liver and kidneys. It is a constituent of certain enzymes and participates in heart function. This mineral is found in lean meats, whole-grain cereals, and onions. **Chromium** (Cr) is widely distributed throughout the body and regulates glucose utilization. It is found in liver, lean meats, yeast, and pork kidneys.

Chart 18.9 summarizes the characteristics of trace elements.

1. How is copper used?
2. What is the function of iodine?
3. Why are zinc deficiencies thought to be relatively common?

## HEALTHY EATING—THE FOOD PYRAMID AND READING LABELS

An adequate diet provides sufficient energy (calories), essential fatty acids, essential amino acids, vitamins, and minerals to support optimal growth and to maintain and repair body tissues. However, because individual needs for nutrients vary greatly with age, sex, growth rate, amount of physical activity, and

## CHART 18.9 Trace Elements

| Trace Element | Distribution | Functions | Sources and RDA* for Adults | Conditions Associated with | |
|---|---|---|---|---|---|
| | | | | *Excesses* | *Deficiencies* |
| *Iron* (Fe) | Primarily in blood; stored in liver, spleen, and bone marrow | Part of hemoglobin molecule; catalyzes formation of vitamin A; incorporated into a number of enzymes | Liver, lean meats, dried apricots, raisins, enriched whole-grain cereals, legumes, molasses<br>10–18 mg | Liver damage | Anemia |
| *Manganese* (Mn) | Most concentrated in liver, kidneys, and pancreas | Occurs in enzymes needed for synthesis of fatty acids and cholesterol, formation of urea, and normal functioning of the nervous system | Nuts, legumes, whole-grain cereals, leafy green vegetables, fruits<br>2.5–5 mg | None known | None known |
| *Copper* (Cu) | Most highly concentrated in liver, heart, and brain | Essential for synthesis of hemoglobin, development of bone, production of melanin, and formation of myelin | Liver, oysters, crabmeat, nuts, whole-grain cereals, legumes<br>2–3 mg | Rare | Rare |
| *Iodine* (I) | Concentrated in thyroid gland | Essential component for synthesis of thyroid hormones | Food content varies with soil content in different geographic regions; iodized table salt<br>.15 mg | Decreased synthesis of thyroid hormones | Goiter |
| *Cobalt* (Co) | Widely distributed | Component of cyanocobalamin; needed for synthesis of several enzymes | Liver, lean meats, milk<br>None established | Heart disease | Pernicious anemia |
| *Zinc* (Zn) | Most concentrated in liver, kidneys, and brain | Constituent of several enzymes involved in digestion, respiration, bone metabolism, liver metabolism; necessary for normal wound healing and maintaining integrity of the skin | Meats, cereals, legumes, nuts, vegetables<br>15 mg | Slurred speech, problems walking | Depressed immunity, loss of taste and smell, learning difficulties |
| *Fluorine* (F) | Primarily in bones and teeth | Component of tooth structure | Fluoridated water<br>1.5–4 mg | Mottled teeth | None known |
| *Selenium* (Se) | Concentrated in liver and kidneys | Occurs in enzymes | Lean meats, fish, cereals<br>0.05–2 mg | Vomiting, fatigue | None known |
| *Chromium* (Cr) | Widely distributed | Essential for use of carbohydrates | Liver, lean meats, wine<br>0.05–2 mg | None known | None known |

*RDA = recommended dietary allowance.

level of stress, as well as with genetic and environmental factors, it is not possible to design a diet that is adequate for everyone. However, nutrients are so widely distributed in foods that satisfactory amounts and combinations can usually be obtained in spite of individual food preferences.

As the chapter so far shows, it is very difficult to keep track of the different nutrients in a diet and be certain that an adequate amount of each is consumed daily. Nutritionists have devised several ways to help consumers make healthy food choices.

Most familiar is the RDA guideline that has appeared on several charts in the chapter. *RDA* stands for United States Recommended Daily Allowance. An RDA is actually the upper limit of another measurement, called the Recommended Dietary Allowance, which lists optimal calorie intake for each sex at various ages, and the amounts of vitamins and minerals needed to avoid deficiency or excess conditions. The RDA values on food packages are set high, ensuring that most people who follow them receive sufficient amounts of each nutrient. Government panels meet every 5 years to evaluate the RDAs in light of new data.

Placing foods into groups is an easier way to follow a healthy diet. Figure 18.16 depicts the **food pyramid** system, which shows at a glance that complex carbohydrates should comprise the bulk of the diet. The four food group plan that the pyramid replaces, devised

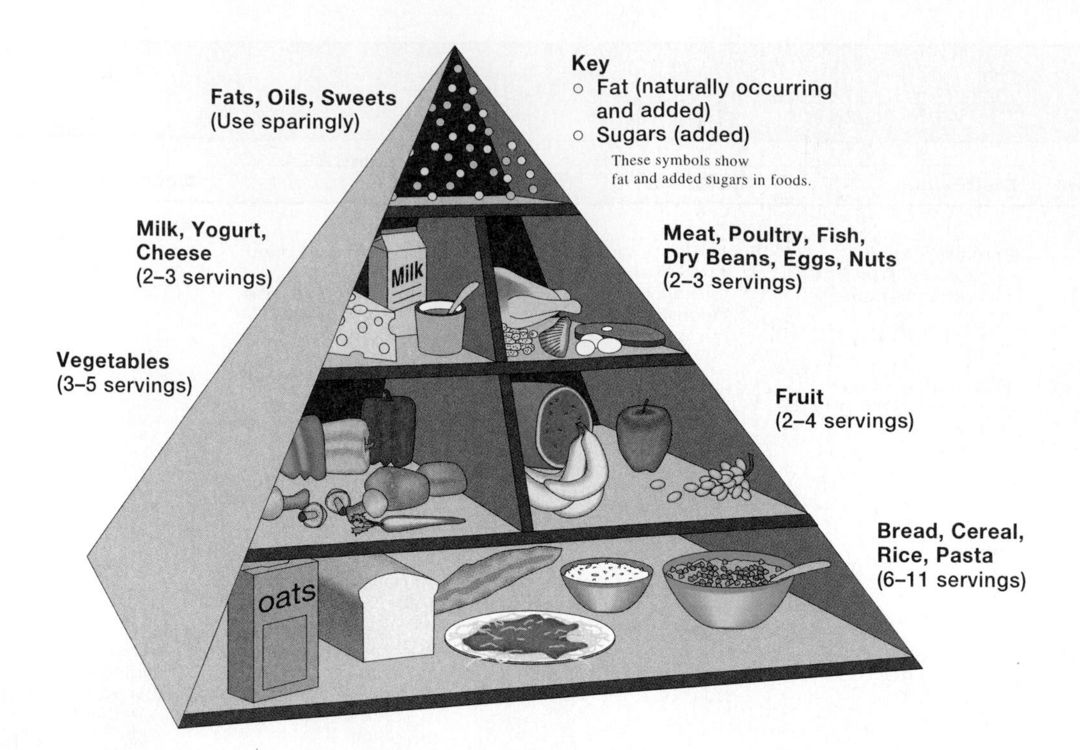


**FIGURE 18.16**

The U.S. Department of Agriculture introduced the food pyramid in 1992. The chart is a guideline to healthy eating. In contrast to former food group plans, the pyramid gives an instant idea of which foods should make up the bulk of the diet—whole grains, fruits, and vegetables.

in the 1950s, put meat on equal footing with grains, dairy products, and fruits and vegetables in terms of number of servings. A food group plan from the 1940s had eight food groups, including separate groups for butter and margarine, and for eggs—foods now associated with the development of heart disease. In the 1920s, an entire food group was devoted to sweets!

When making individual food choices, it helps to read and understand food labels, which became much more informative in 1994. Disregard claims such as "light" and "low fat" and skip right to the calories of different ingredients; ideally, fat percentage should be below 30, and that of saturated fats below 10. Ingredients are listed in descending order by weight. Figure 18.17 shows the information on a food label.

## Malnutrition

**Malnutrition** is poor nutrition that results from a lack of essential nutrients or a failure to utilize them. It may result from *undernutrition* and produce the symptoms of deficiency diseases, or it may be due to *overnutrition* arising from excess nutrient intake.

The factors leading to malnutrition vary. For example, a deficiency condition may stem from lack of availability or poor quality of food. On the other hand, malnutrition may result from overeating or taking too many vitamin supplements. Malnutrition from diet alone is called *primary malnutrition.*

The Food and Drug Administration allows only the following specific food and health claims:

- Dietary calcium decreases risk of osteoporosis (a bone-thinning condition).
- A low-fat diet lowers risk of some cancers.
- A diet low in saturated fat and cholesterol lowers risk of coronary heart disease.
- Fiber, fruits and vegetables, and whole grains reduce the risk of some cancers and coronary heart disease.
- Lowering sodium intake lowers blood pressure.
- Folic acid lowers the risk of neural tube defects.

Take any other health claims with a grain of salt!

*Secondary malnutrition* occurs when an individual's characteristics make a normally adequate diet insufficient. For example, a person who secretes low

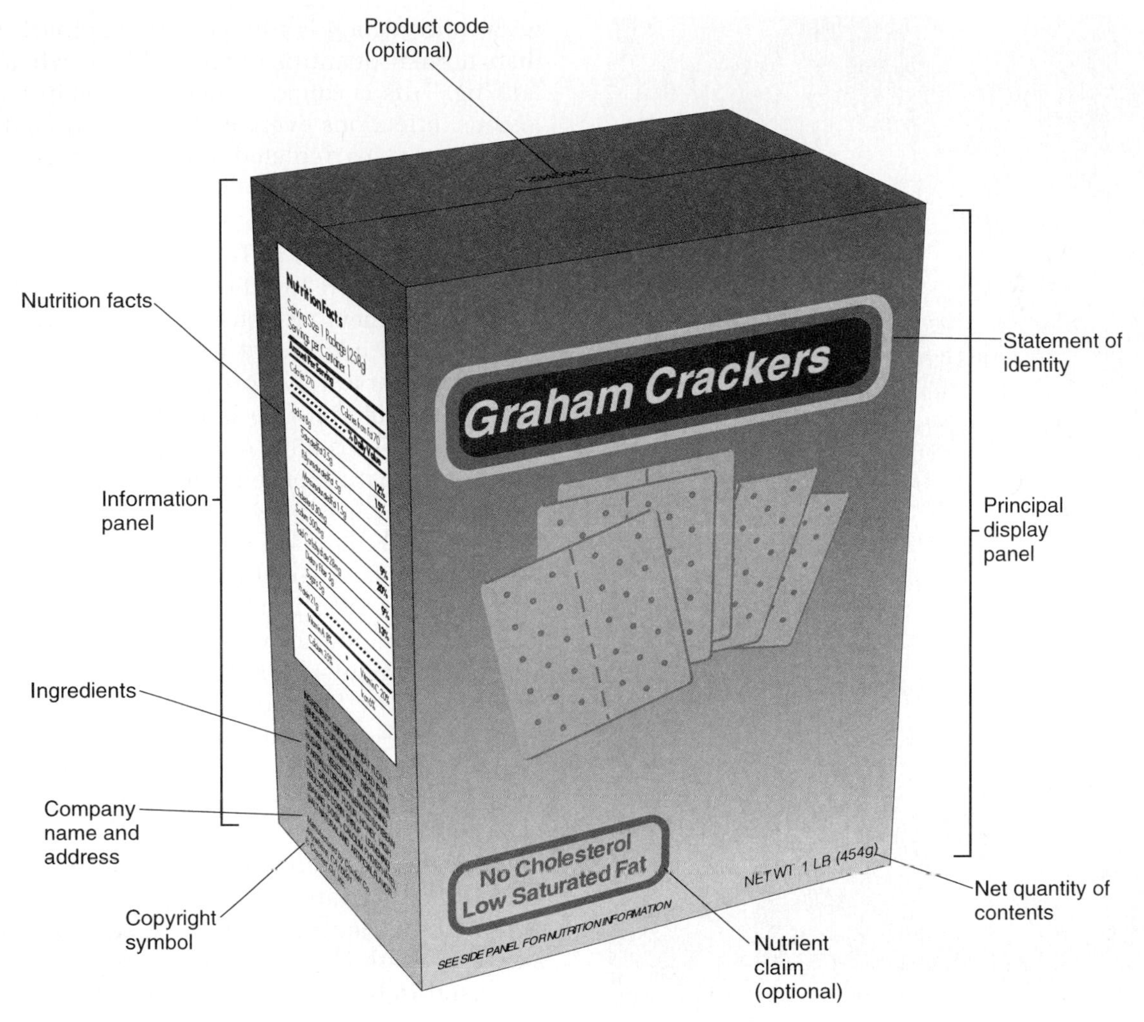


**FIGURE 18.17**

A key to healthy eating is to become familiar with the "Nutrition Facts" panel on food packages. Look for products with fewer than 30% of calories from fat.

quantities of bile salts may develop a deficiency of fat-soluble vitamins because bile salts promote absorption of fats. Likewise, severe and prolonged emotional stress may lead to secondary malnutrition, because stress can change hormonal concentrations, and such changes may result in amino acid breakdown or excretion of nutrients.

## Starvation

A healthy human can stay alive for 50 to 70 days without food. In prehistoric times, this margin allowed survival during seasonal famines. In some areas of Africa today, famine is not a seasonal event but a constant condition, and many millions have starved to death. Starvation is also seen in hunger strikers, in inmates of concentration camps, and in sufferers of psychological eating disorders such as **anorexia nervosa** and **bulimia.**

Whatever the cause, the starving human body begins to digest itself. After only 1 day without eating, the body's reserves of sugar and starch are gone. Next, the body extracts energy from fat and then from muscle protein. By the third day, hunger ceases as the body uses energy from fat reserves. Gradually, metabolism slows to conserve energy, blood pressure drops, the pulse slows, and chills set in. Skin becomes dry and hair falls out as the proteins in these structures are broken down to release amino acids that are used for the more vital functioning of the brain, heart, and lungs. When the immune system's antibody proteins are dismantled for their amino acids, protection against infection declines. Mouth sores and anemia develop, the heart beats irregularly, and bone begins to degenerate. After several weeks without food, coordination is gradually lost. Near the end, the starving human is blind, deaf, and emaciated (fig. 18.18).

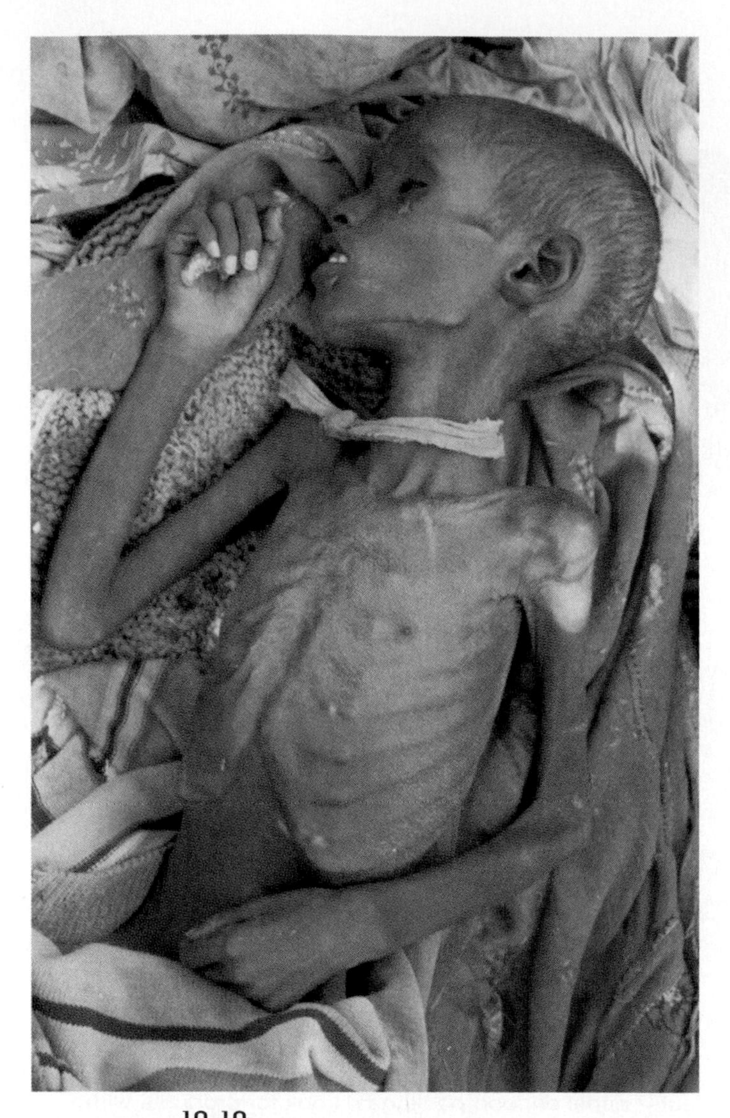

FIGURE 18.18
A child dying of starvation.

## Marasmus and Kwashiorkor

Lack of nutrients is called **marasmus,** and it causes people to resemble living skeletons (fig. 18.19). Children under the age of two with marasmus often die of measles or other infections, their immune systems too weakened to fight off normally mild viral illnesses.

Some starving children do not look skeletal but have protruding bellies. These youngsters suffer from a form of protein starvation called **kwashiorkor,** which in the language of Ghana means "the evil spirit which infects the first child when the second child is born." Kwashiorkor typically appears in a child who has recently been weaned from the breast, usually because of the birth of a sibling. The switch from protein-rich breast milk to the protein-poor gruel that is the staple of many developing nations is the source of this protein deficiency. The children's bellies swell with fluid, which is filtered from capillaries in greater than normal quantities that build up when protein is lacking. This is called **ascites.** Their skin may develop lesions. Infections overwhelm the body as the immune system becomes depleted of its protective antibodies.

## Anorexia Nervosa

**Anorexia nervosa** is self-imposed starvation. The condition is reported to affect 1 out of 250 adolescents, and 95% of them are female. The sufferer, typically a well-behaved adolescent girl from an affluent family, perceives herself to be overweight and eats barely enough to survive. She is terrified of gaining weight and usually loses 25% of her original body weight. In addition to eating only small amounts of low-calorie foods, she further loses weight by vomiting, by taking laxatives and diuretics, or by intense exercise. Her eating behavior is often ritualized. She may meticulously arrange her meager meal on her plate or consume only a few foods. She develops low blood pressure, a slowed or irregular heartbeat, constipation, and constant chilliness. She stops menstruating as her body fat level plunges. Like any starving person, the hair becomes brittle and the skin dries out. She may develop soft, pale, fine body hair called lanugo, normally seen only on a developing fetus, to preserve body heat.

When the anorectic reaches an obviously emaciated state, her parents usually have her hospitalized, where she is fed intravenously so that she does not starve to death or die suddenly of heart failure due to a mineral imbalance. She also receives psychotherapy and nutritional counseling. Despite these efforts, 15% to 21% of people with anorexia die.

Anorexia nervosa has no known physical cause. One hypothesis is that the person is rebelling against approaching womanhood. Indeed, her body is astonishingly childlike, with a flat chest and slim hips, and she has often ceased to menstruate. She typically has low self-esteem and believes that others, particularly her parents, are controlling her life. Her weight is something that she can control. Anorexia can be a one-time, short-term experience—little more than a weight-loss diet gone out of control—or a lifelong obsession.

## Bulimia

A person suffering from **bulimia** is often of normal weight. She eats whatever she wants, often in huge amounts, but she then rids her body of the thousands of extra calories by vomiting, taking laxatives, or exercising frantically. For an estimated one in five college students, the great majority of them female, "binging and purging" appears to be a way of coping with stress.

Sometimes a bulimic's dentist is the first to spot her problem by observing teeth decayed from frequent vomiting. The backs of her hands may bear telltale scratches from efforts to induce vomiting. Her throat is

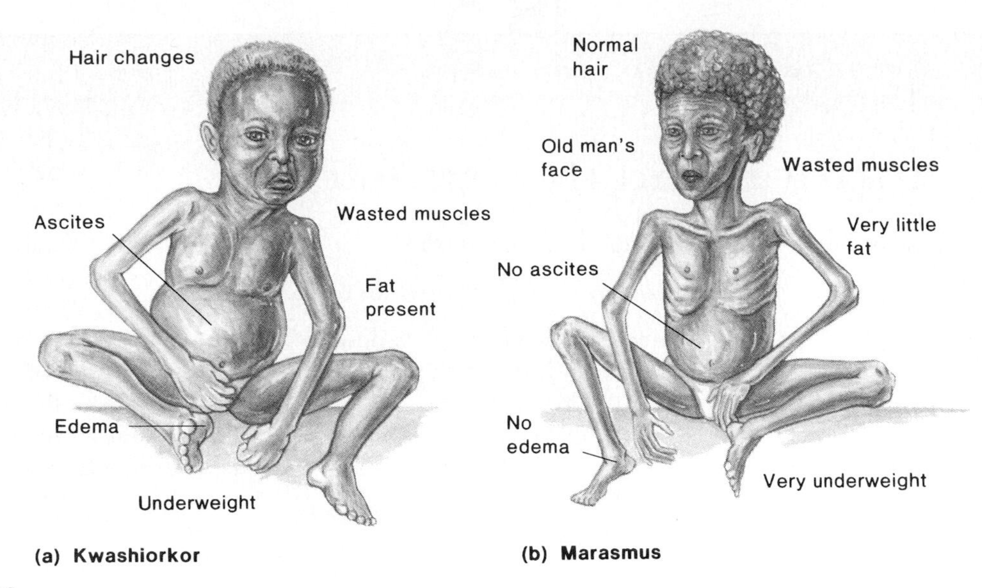


**FIGURE 18.19**

Two types of starvation in the young. (*a*) The child on the left suffers from kwashiorkor. Although he may have received adequate nourishment from breast milk early in life, he became malnourished when his diet switched to a watery, white extract from cassava that looks like milk but has very little protein. The lack of protein in the diet causes edema and the ascites that swells his belly. (*b*) The child on the right, suffering from marasmus, did not even have adequate nutrition as an infant.

raw and her stomach lining ulcerated from the stomach acid forced forward by vomiting. The binge and purge cycle is very hard to break, even with psychotherapy and nutritional counseling.

> A person with bulimia tends to eat soft foods that can be consumed in large amounts quickly with minimal chewing. Reveals one athletic young woman, "For me a binge consists of a pound of cottage cheese, a head of lettuce, a steak, a loaf of Italian bread, a 10-ounce serving of broccoli, spinach or a head of cabbage, a cake, an 18-ounce pie, with a quart or half gallon of ice cream. When my disease is at its worst, I eat raw oatmeal with butter, laden with mounds of sugar, or a loaf of white bread with butter and syrup poured over it." She follows her "typical" 20,000-calorie binge with hours of bicycle riding, running, and swimming.

Clinical Application 18.3 discusses some ways that understanding nutrition can help athletic performance.

1. What is an adequate diet?
2. What factors influence individual needs for nutrients?
3. How can the food pyramid plan help consumers make wise food choices?
4. What is primary malnutrition? Secondary malnutrition?
5. What happens to the body during starvation?
6. How do marasmus and kwashiorkor differ?
7. How do anorexia nervosa and bulimia differ?

## Clinical Terms Related to Nutrients and Nutrition

**cachexia** (kah-kek′se-ah) A state of chronic malnutrition and physical wasting.

**casein** (ka′se-in) The primary protein found in milk.

**celiac disease** (se′le-ak dĭ-zēz′) A digestive disorder characterized by the inability to digest or use fats and carbohydrates.

**emaciation** (e-ma″se-a′shun) Extreme leanness due to tissue wasting.

**hyperalimentation** (hi″per-al″ĭ men-ta′shun) Prolonged intravenous nutrition provided for patients with severe digestive disorders.

**hypercalcemia** (hi″-per-kal-se′me-ah) Excess calcium in the blood.

**hypercalciuria** (hi″per-kal-se-u′re-ah) Excess excretion of calcium in the urine.

**hyperglycemia** (hi″per-gli-se′me-ah) Excess glucose in the blood.

**hyperkalemia** (hi″per-kah-le′me-ah) Excess potassium in the blood.

**hypernatremia** (hi″per-nah-tre′me-ah) Excess sodium in the blood.

**hypoalbuminemia** (hi″po-al-bu″mĭ-ne′me-ah) A low level of albumin in the blood.

**hypoglycemia** (hi″po-gli-se′me-ah) A low level of glucose in the blood.

CLINICAL APPLICATION 18.3

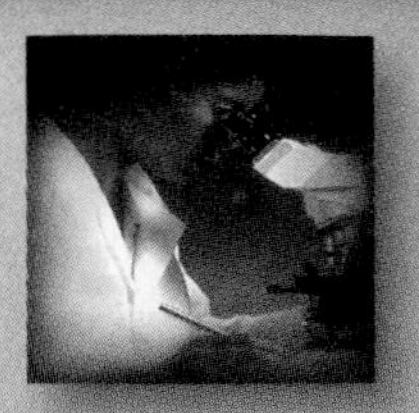

## Nutrition and the Athlete

Can a marathoner, cross-country skier, weight lifter, or competitive swimmer eat to win? A diet of 60% or more carbohydrate, 18% protein, and 22% fat should be adequate to support frequent, strenuous activity.

### Macronutrients

As the source of immediate energy, carbohydrates are the athlete's best friend. Athletes should get the bulk of their carbohydrates from vegetables and grains in frequent meals, because the muscles can store only 1,800 calories worth of glycogen.

Many people erroneously believe that an athlete needs protein supplements. Excess dietary protein, however, can strain the kidneys in ridding the body of the excess nitrogen, dehydrating the athlete as more water is used in urine. The American Dietetic Association suggests that athletes eat 1 gram of protein per kilogram of weight per day, compared to 0.8 grams for nonathletes. Athletes should not rely solely on meat for protein, because these foods can be high in fat. Supplements are only necessary for young athletes at the start of training, under a doctor's supervision. Too little protein in an athlete is linked to "sports anemia," in which hemoglobin levels decline and blood may appear in the urine.

The body stores 140,000 calories of fat, so it is clear why no one needs to replenish that constantly with fatty foods. Athletes should use low-fat milk and meats.

### Water

A sedentary person loses a quart of water a day as sweat; an athlete may lose 2 to 4 quarts of water an hour! To stay hydrated, athletes should drink 3 cups of cold water 2 hours before an event, then 2 more cups 15 minutes before the event, and small amounts every 15 minutes during the event. They should drink afterward too. Another way to determine water needs is to weigh in before and after training. For each pound lost, athletes should drink a pint of water. They should also avoid sugary fluids, which slow water's trip through the digestive system, and alcohol, which increases fluid loss.

### Vitamins and Minerals

If an athlete eats an adequate, balanced diet, vitamin supplements are not needed. Supplements of sodium and potassium are usually not needed either, because the active body naturally conserves these nutrients. To be certain of enough sodium, athletes may want to salt their food; to get enough potassium, they can eat bananas, dates, apricots, oranges, or raisins.

A healthy pregame meal should be eaten 2 to 5 hours before the game, provide 500 to 1,500 kilocalories, include 4 or 5 cups of fluid, and be high in carbohydrates, which taste good, provide energy, and are easy to digest.

**hypokalemia** (hi″po-kah-le′me-ah) A low level of potassium in the blood.
**hyponatremia** (hi″po-nah-tre′me-ah) A low level of sodium in the blood.
**isocaloric** (i″so-kah-lo′rik) Containing equal amounts of heat energy.
**lipogenesis** (lip″o-jen′e-sis) The formation of fat.
**nyctalopia** (nik″tah-lo′pe-ah) Night blindness.
**polyphagia** (pol″e-fa′je-ah) Overeating.
**proteinuria** (pro″te-i-nu′re-ah) Protein in the urine.
**provitamin** (pro-vi′tah-min) A biochemical precursor of a vitamin.

## Chapter Summary

### Why We Eat (page 721)

Nutrients include carbohydrates, lipids, proteins, vitamins, and minerals. The ways nutrients are used to support life processes constitute metabolism. Essential nutrients are needed for health and cannot be synthesized by body cells. Macronutrients include carbohydrates, lipids, and proteins. Micronutrients are vitamins and minerals. Water is also essential.

### Carbohydrates (page 722)

Carbohydrates are organic compounds that are used primarily to supply cellular energy.

1. Carbohydrate sources
   a. Carbohydrates are ingested in a variety of forms.
   b. Starch, glycogen, disaccharides, and monosaccharides are carbohydrates.
   c. Cellulose is a polysaccharide that human enzymes cannot digest, but it is important in providing bulk that facilitates movement of intestinal contents.
2. Carbohydrate utilization
   a. Carbohydrates are absorbed as monosaccharides.
   b. The liver converts fructose and galactose to glucose.
   c. Oxidation releases energy from glucose.
   d. Excess glucose is stored as glycogen or converted to fat.
3. Carbohydrate requirements
   a. Most carbohydrates are used to supply energy, although some are used to produce sugars.
   b. Some cells depend on a continuous supply of glucose to survive.
   c. If glucose is scarce, amino acids may be converted to glucose.
   d. Humans survive with a wide range of carbohydrate intakes.
   e. Poor nutritional status is usually related to low intake of nutrients other than carbohydrates.

### Lipids (page 723)

Lipids are organic compounds that supply energy and are used to build cell structures. They include fats, phospholipids, and cholesterol.

1. Lipid sources
   a. Triglycerides are obtained from foods of plant and animal origins.
   b. Cholesterol is mostly obtained in foods of animal origin.
2. Lipid utilization
   a. The liver and adipose tissue control triglyceride metabolism.
   b. The liver can alter the molecular structures of fatty acids.
   c. Linoleic acid is an essential fatty acid.
   d. The liver regulates the amount of cholesterol by synthesizing or excreting it.
3. Lipid requirements
   a. Humans survive with a wide range of lipid intakes.
   b. The amounts and types of lipids needed for health are unknown.
   c. Fat intake must be sufficient to carry fat-soluble vitamins.

### Proteins (page 724)

Proteins are organic compounds that serve as structural materials, act as enzymes, and provide energy. Amino acids are incorporated into various structural and functional proteins, including enzymes. During starvation, tissue proteins may be used as energy sources; thus, the tissues waste away.

1. Protein sources
   a. Proteins are obtained mainly from meats, dairy products, cereals, and legumes.
   b. During digestion, proteins are broken down into amino acids.
   c. Eight amino acids are essential for adults, while ten are essential for growing children.
   d. All essential amino acids must be present at the same time in order for growth and repair of tissues to take place.
   e. Complete proteins contain adequate amounts of all the essential amino acids needed to maintain the tissues and promote growth.
   f. Incomplete proteins lack adequate amounts of one or more essential amino acids.
2. Nitrogen balance
   a. In healthy adults, the gain of protein equals the loss of protein, and a nitrogen balance exists.
   b. A starving person has a negative nitrogen balance; a growing child, a pregnant woman, or an athlete in training usually has a positive nitrogen balance.
3. Protein requirements
   a. Proteins and amino acids are needed to supply essential amino acids and nitrogen for the synthesis of nitrogen-containing molecules.
   b. The consequences of protein deficiencies are particularly severe among growing children.

### Energy Expenditures (page 726)

Energy is of prime importance to survival and may be obtained from carbohydrates, fats, or proteins.

1. Energy values of foods
   a. The potential energy values of foods are expressed in calories.
   b. When energy losses due to incomplete absorption and incomplete oxidation are taken into account, 1 gram of carbohydrate or 1 gram of protein yields about 4 calories, while 1 gram of fat yields about 9 calories.
2. Energy requirements
   a. The amount of energy required varies from person to person.
   b. Factors that influence energy requirements include basal metabolic rate, muscular activity, body temperature, and nitrogen balance.
3. Energy balance
   a. Energy balance exists when caloric intake equals caloric output.
   b. If energy balance is positive, body weight increases; if energy balance is negative, body weight decreases.

4. Desirable weight
   a. The most common nutritional disorders involve caloric imbalances.
   b. Average weights of persons 25–30 years of age are desirable for older persons as well.
   c. Height-weight guidelines are based on longevity.
   d. A person who exceeds the desirable weight by 10%–20% is called overweight.
   e. A person whose body contains an excess of fatty tissue is obese.

## Vitamins (page 729)

Vitamins are organic compounds (other than carbohydrates, lipids, and proteins) that are essential for normal metabolic processes and cannot be synthesized by body cells in adequate amounts.

1. Fat-soluble vitamins
   a. General characteristics
      (1) Fat-soluble vitamins are carried in lipids and are influenced by the same factors that affect lipid absorption.
      (2) They are fairly resistant to the effects of heat; thus, they are not destroyed by cooking or food processing.
   b. Vitamin A
      (1) Vitamin A occurs in several forms, is synthesized from carotenes, and is stored in the liver.
      (2) It functions in the production of pigments necessary for vision.
   c. Vitamin D
      (1) Vitamin D is a group of related steroids.
      (2) It is found in certain foods and is produced commercially; it can also be synthesized in the skin.
      (3) When needed, vitamin D is converted by the kidneys to an active form that functions as a hormone and promotes the intestine's absorption of calcium and phosphorus.
   d. Vitamin E
      (1) Vitamin E is a group of compounds that are antioxidants.
      (2) It is stored in muscles and adipose tissue.
      (3) Its precise functions are unknown, but it seems to prevent oxidation of vitamin A and polyunsaturated fatty acids, and to stabilize cell membranes.
   e. Vitamin K
      (1) Vitamin $K_1$ occurs in foods; vitamin $K_2$ is produced by certain bacteria that normally inhabit the intestinal tract.
      (2) It is stored to a limited degree in the liver.
      (3) It is used in the production of prothrombin that is needed for normal blood clotting.
2. Water-soluble vitamins
   a. General characteristics
      (1) Water-soluble vitamins include the B vitamins and vitamin C.
      (2) B vitamins make up a group called the vitamin B complex and oxidize carbohydrates, lipids, and proteins.
   b. Vitamin B complex
      (1) Thiamine
         (a) Thiamine functions as part of coenzymes that oxidize carbohydrates and synthesize essential sugars.
         (b) Small amounts are stored in the tissues; excess is excreted in the urine.
         (c) Quantities needed vary with caloric intake.
      (2) Riboflavin
         (a) Riboflavin functions as part of several enzymes and coenzymes that are essential to the oxidation of glucose and fatty acids.
         (b) Its absorption is regulated by an active transport system; excess is excreted in the urine.
         (c) Quantities needed vary with caloric intake.
      (3) Niacin
         (a) Niacin functions as part of coenzymes needed for the oxidation of glucose and for the synthesis of proteins and fats.
         (b) It can be synthesized from tryptophan; daily requirement varies with the tryptophan intake.
      (4) Vitamin $B_6$
         (a) Vitamin $B_6$ is a group of compounds that function as coenzymes in metabolic pathways that synthesize proteins, certain amino acids, antibodies, and nucleic acids.
         (b) Its requirement varies with protein intake.
      (5) Pantothenic acid
         (a) Pantothenic acid functions as part of coenzyme A; thus, it is essential for energy-releasing mechanisms.
         (b) Most diets provide sufficient amounts; deficiencies are rare.
      (6) Cyanocobalamin
         (a) The cyanocobalamin molecule contains cobalt.
         (b) Its absorption is regulated by the secretion of intrinsic factor from the gastric glands.
         (c) It functions as part of coenzymes needed for the synthesis of nucleic acids and for the metabolism of carbohydrates and fats.
      (7) Folacin
         (a) Folacin is converted by the liver to physiologically active folinic acid.
         (b) It is a coenzyme needed for the metabolism of certain amino acids, the synthesis of DNA, and the normal production of red blood cells.
      (8) Biotin
         (a) Biotin is a coenzyme needed for the metabolism of amino acids and fatty acids, and for the synthesis of nucleic acids.
         (b) It is stored in metabolically active organs.
   c. Ascorbic acid (vitamin C)
      (1) Vitamin C is closely related chemically to monosaccharides.
      (2) It is needed for production of collagen, the metabolism of certain amino acids, and iron absorption.
      (3) It is not stored in large amounts; excess is excreted in the urine.

## Minerals (page 738)

1. Characteristics of minerals
   a. Minerals are responsible for about 4% of body weight.
   b. About 75% of the minerals are found in bones and teeth as calcium and phosphorus.

c. Minerals are usually incorporated into organic molecules, although some occur in inorganic compounds or as free ions.
d. They comprise structural materials, function in enzymes, and play vital roles in various metabolic processes.
e. Mineral concentrations are generally regulated by homeostatic mechanisms.
f. The physiologically active form of minerals is the ionized form.

2. Major minerals
   a. Calcium
      (1) Calcium is essential for the formation of bones and teeth, the conduction of nerve impulses, the contraction of muscle fibers, the coagulation of blood, and the activation of various enzymes.
      (2) Its absorption is affected by existing calcium concentration, vitamin D, protein intake, and motility of the digestive tract.
   b. Phosphorus
      (1) Phosphorus is incorporated for the most part into the salts of bones and teeth.
      (2) It plays roles in nearly all metabolic reactions as a constituent of nucleic acids, proteins, enzymes, and some vitamins.
      (3) It also occurs in the phospholipids of cell membranes, in ATP, and in phosphates of body fluids.
   c. Potassium
      (1) Potassium tends to be concentrated inside cells.
      (2) It functions in maintenance of osmotic pressure, regulation of pH, metabolism of carbohydrates and proteins, conduction of nerve impulses, and contraction of muscle fibers.
   d. Sulfur
      (1) Sulfur is incorporated for the most part into the molecular structures of certain amino acids.
      (2) It is also included in thiamine, insulin, biotin, and mucopolysaccharides.
   e. Sodium
      (1) Most sodium occurs in extracellular fluids or is bound to the inorganic salts of bone.
      (2) The blood concentration of sodium is regulated by the kidneys under the influence of aldosterone.
      (3) Sodium helps maintain solute concentration and regulates water balance.
      (4) It is essential for the conduction of nerve impulses, the contraction of muscle fibers, and the movement of substances through cell membranes.
   f. Chlorine
      (1) Chlorine is closely associated with sodium in the form of chloride ions.
      (2) It acts with sodium to help maintain osmotic pressure, regulate pH, and maintain electrolyte balance.
      (3) It is essential for the formation of hydrochloric acid and for the transport of carbon dioxide by red blood cells.
   g. Magnesium
      (1) Magnesium is particularly abundant in the bones as phosphates and carbonates.
      (2) It functions in the production of ATP and in the breakdown of ATP to ADP.
      (3) A reserve supply of magnesium is stored in the bones; excesses are excreted in the urine.

3. Trace elements
   a. Iron
      (1) Iron occurs primarily in hemoglobin of red blood cells and in myoglobin of muscles.
      (2) A reserve supply of iron is stored in the liver, spleen, and bone marrow.
      (3) It is needed to catalyze the formation of vitamin A; it is also incorporated into various enzymes and the cytochrome molecules.
   b. Manganese
      (1) Most manganese is concentrated in the liver, kidneys, and pancreas.
      (2) It is necessary for normal growth and development of skeletal structures and other connective tissues; it is essential for the synthesis of fatty acids, cholesterol, and urea.
   c. Copper
      (1) Most copper is concentrated in the liver, heart, and brain.
      (2) It is needed for synthesis of hemoglobin, development of bones, production of melanin, and formation of myelin.
   d. Iodine
      (1) Iodine is most highly concentrated in the thyroid gland.
      (2) It provides an essential component for the synthesis of thyroid hormones.
      (3) It is often added to foods in the form of iodized table salt.
   e. Cobalt
      (1) Cobalt is widely distributed throughout the body.
      (2) It is an essential part of cyanocobalamin and is probably needed for the synthesis of several important enzymes.
   f. Zinc
      (1) Zinc is most concentrated in the liver, kidneys, and brain.
      (2) It is a constituent of several enzymes involved with digestion, respiration, bone metabolism, and liver metabolism.
   g. Fluorine
      (1) The teeth concentrate fluorine.
      (2) It is incorporated into enamel and prevents dental caries.
   h. Selenium
      (1) The liver and kidneys store selenium.
      (2) It is a constituent of certain enzymes.
   i. Chromium
      (1) Chromium is widely distributed throughout the body.
      (2) It regulates glucose utilization.

## Healthy Eating—The Food Pyramid and Reading Labels (page 744)

1. An adequate diet provides sufficient energy and essential nutrients to support optimal growth, as well as maintenance and repair, of tissues.
2. Individual needs vary so greatly that it is not possible to design a diet that is adequate for everyone.

3. Devices to help consumers make healthy food choices include Recommended Daily Allowances, Recommended Dietary Allowances, food group plans, the food pyramid, and food labels.
4. Malnutrition
   a. Poor nutrition is due to lack of foods or failure to make the best use of available foods.
   b. Primary malnutrition is due to poor diet.
   c. Secondary malnutrition is due to an individual characteristic that makes a normal diet inadequate.
5. Starvation
   a. A person can survive 50 to 70 days without food.
   b. A starving body digests itself, starting with carbohydrates, then proteins, then fats.
   c. Symptoms include low blood pressure, slow pulse, chills, dry skin, hair loss, and poor immunity. Finally, vital organs cease to function.
   d. Marasmus is lack of all nutrients.
   e. Kwashiorkor is protein starvation.
   f. Anorexia nervosa is a self-starvation eating disorder.
   g. Bulimia is an eating disorder characterized by binging and purging.

**Explorations in Human Anatomy and Physiology CD-ROM**
The module accompanying Chapter Eighteen is #7 Diet and Weight Loss.

## Critical Thinking Questions

1. For each of the following diets, indicate how the diet is nutritionally unsound (if it is ) and why it would be easy or difficult to follow.
   a. The bikini diet consists of 500 calories per day, including 45% to 55% protein, 40% to 50% carbohydrate, and 4% to 6% fat.
   b. The Cambridge diet is a powder mixed with water and drunk three times a day. One day's intake equals 33 grams of protein, 40 grams of carbohydrate, and 3 grams of fat.
   c. For the first 10 days of the Beverly Hills diet, nothing but fruit is eaten. On day 10 you can eat a bagel and butter, and then it is back to fruit only until day 19, when you can eat steak or lobster. The cycle repeats, adding more meat. This diet is based on "conscious combining"—the idea that eating certain combinations of foods leads to weight loss.
   d. The Weight Loss Clinic diet consists of 800 calories per day, with 46.1% protein, 35.2% carbohydrate, and 18.7% fat.
   e. The macrobiotic diet includes 10% to 20% protein, 70% carbohydrate, and 10% fat, with a half hour of walking each day. Most familiar foods are forbidden, but you can eat many unusual foods—such as rice cakes, seaweed, barley stew, pumpkin soup, rice gruel, kasha and onions, millet balls, wheat berries, and parsnip chips.
   f. The No Aging diet maintains that eating foods rich in nucleic acids (RNA and DNA) can prolong life, since these are the genetic materials. Recommended foods include sardines, salmon, calves liver, lentils, and beets.
2. Why does the blood sugar concentration of a person whose diet is relatively low in carbohydrates remain stable?
3. A young man takes several vitamin supplements each day, claiming that they give him energy. Is he correct? Why or why not?
4. A soccer coach advises his players to eat a hamburger and french fried potatoes about 2 hours before a game. Suggest a more sensible pregame meal.
5. Anorexia nervosa is a form of starvation. If it is a nutritional problem, then why should treatment include psychotherapy?
6. Why do starving children often die of infections that are usually mild in well-nourished children?
7. With the aid of nutrient tables available in reference books, calculate the carbohydrate, lipid, and protein content of your diet in grams for a 24-hour period. Also calculate the total calories represented by these foods. Assuming that this 24-hour sample is representative of your normal eating habits, what improvements could be made in its composition?
8. Examine the label information on the packages of a variety of dry breakfast cereals. Which types of cereals provide the best sources of vitamins and minerals? Which major nutrients are lacking in these cereals?
9. If a person decided to avoid eating meat and other animal products, such as milk, cheese, and eggs, what foods might be included in the diet to provide essential amino acids?
10. How might a diet be modified in order to limit the intake of cholesterol?
11. How do you think the nutritional requirements of a healthy 12-year-old boy, a 24-year-old pregnant woman, and a healthy 60-year-old man would differ?

## Review Exercises

1. Define *essential nutrient.*
2. List some common sources of carbohydrates.
3. Explain the importance of cellulose in the diet.
4. Explain what happens to excess glucose in the body.
5. Explain why a temporary drop in the blood glucose concentration may impair nervous system functioning.
6. List some of the factors that affect an individual's need for carbohydrates.
7. Define *triglyceride.*
8. List some common sources of lipids.
9. Describe the role of the liver in fat metabolism.
10. Discuss the functions of cholesterol.
11. List some common sources of protein.
12. Distinguish between essential and nonessential amino acids.
13. Explain why all of the essential amino acids must be present before growth can occur.
14. Distinguish between complete and incomplete proteins.
15. Review the major functions of amino acids.
16. Define *nitrogen balance.*
17. Explain why a protein deficiency may be accompanied by edema.
18. Define *calorie.*
19. Explain how the caloric values of foods are determined.
20. Define *basal metabolic rate.*
21. List some of the factors that affect the BMR.
22. Define *energy balance.*
23. Explain what is meant by *desirable weight.*
24. Distinguish between overweight and obesity.
25. Discuss the general characteristics of fat-soluble vitamins.
26. List the fat-soluble vitamins, and describe the major functions of each vitamin.
27. List some good sources for each of the fat-soluble vitamins.
28. Explain what is meant by the *vitamin B complex.*
29. List the water-soluble vitamins, and describe the major functions of each vitamin.
30. List some good sources for each of the water-soluble vitamins.
31. Discuss the general characteristics of the mineral nutrients.
32. List the major minerals, and describe the major functions of each mineral.
33. List some good sources for each of the major minerals.
34. Distinguish between a major mineral and a trace element.
35. List the trace elements, and describe the major functions of each trace element.
36. List some good sources of each of the trace elements.
37. Define *adequate diet.*
38. Define *malnutrition.*
39. Distinguish between primary and secondary malnutrition.
40. Discuss bodily changes during starvation.
41. Distinguish between marasmus, kwashiorkor, anorexia nervosa, and bulimia.

19

CHAPTER

# RESPIRATORY SYSTEM

## CHAPTER OBJECTIVES

AFTER YOU HAVE STUDIED THIS CHAPTER, YOU SHOULD BE ABLE TO:

1. List the general functions of the respiratory system.
2. Name and describe the locations of the organs of the respiratory system.
3. Describe the functions of each organ of the respiratory system.
4. Explain how inspiration and expiration are accomplished.
5. Name and define each of the respiratory air volumes and capacities.
6. Explain how the alveolar ventilation rate is calculated.
7. List several nonrespiratory air movements and explain how each occurs.
8. Locate the respiratory center and explain how it controls normal breathing.
9. Discuss how various factors affect the respiratory center.
10. Describe the structure and function of the respiratory membrane.
11. Explain how oxygen and carbon dioxide are transported in the blood.
12. Review the major events that occur during cellular respiration.
13. Explain how cells use oxygen.

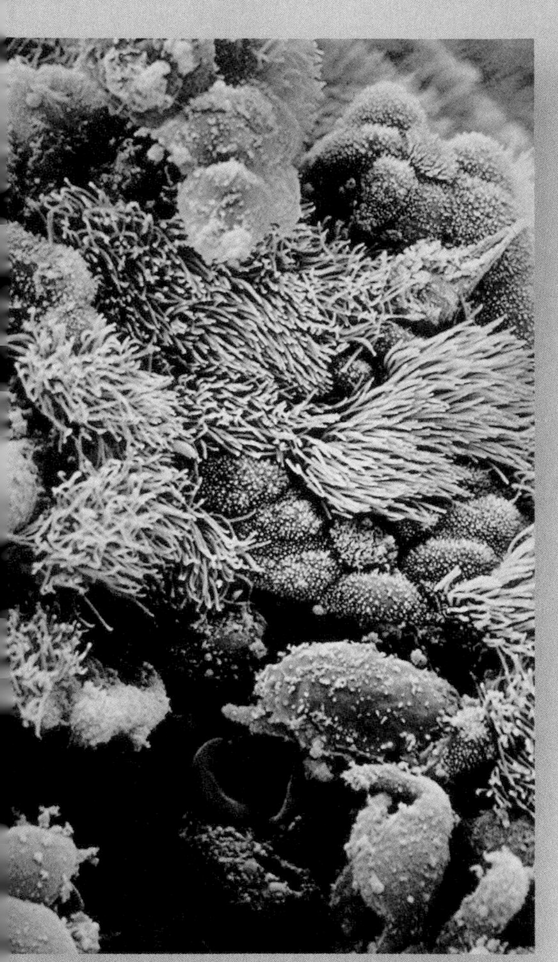

THE INNER LININGS OF THE BRONCHIAL TUBES ARE FRINGED WITH CILIA, WHICH MOVE IRRITANTS OUT OF THE RESPIRATORY SYSTEM (2,000X).

## KEY TERMS

**alveolus** (al-ve′o-lus)

**bronchial tree** (brong′ke-al trē)

**carbaminohemoglobin** (kar-bam″ ĭ-no-he″mo-glo′bin)

**carbonic anhydrase** (kar-bon′ik an-hi′drās)

**cellular respiration** (sel′u-lar res″pĭ-ra′ shun)

**citric acid cycle** (sit′rik as′id si′kl)

**expiration** (ek″spi-ra′shun)

**glottis** (glot′is)

**hemoglobin** (he″mo-glo′bin)

**hyperventilation** (hi″per-ven″tĭ-la′shun)

**inspiration** (in″spi-ra′shun)

**oxyhemoglobin** (ok″se-he″mo-glo′bin)

**partial pressure** (par′shil presh′ur)

**pleural cavity** (ploo′ral kav′ĭ-te)

**respiratory center** (re-spi′rah-to″re sen′ter)

**respiratory membrane** (re-spi′rah-to″re mem′brān)

**respiratory volume** (re-spi′rah-to″re vol′ūm)

**surface tension** (ser′fas ten′shun)

**surfactant** (ser-fak′tant)

The young man was talking excitedly to his lunch partner while gobbling a burger. Suddenly, he began to choke and gestured wildly at his throat. Fortunately, his friend knew just what to do.

First, she asked him if he could speak, and when he frantically motioned that he couldn't, she ran behind him, enclosed her right fist in her left hand, and pushed inward and upward under his rib cage from the center of his abdomen. The force of her thrust created a sudden burst of pressure that pushed air out of his lungs, dislodging the blockage in the air passage. Out popped the piece of hamburger that was blocking the man's airway. The Heimlich maneuver saved his life.

Each year in the United States, 8,000 to 10,000 people are not as lucky as the young man lunching with a friend who knew the Heimlich maneuver—they choke to death. A total block of the respiratory system is quickly deadly.

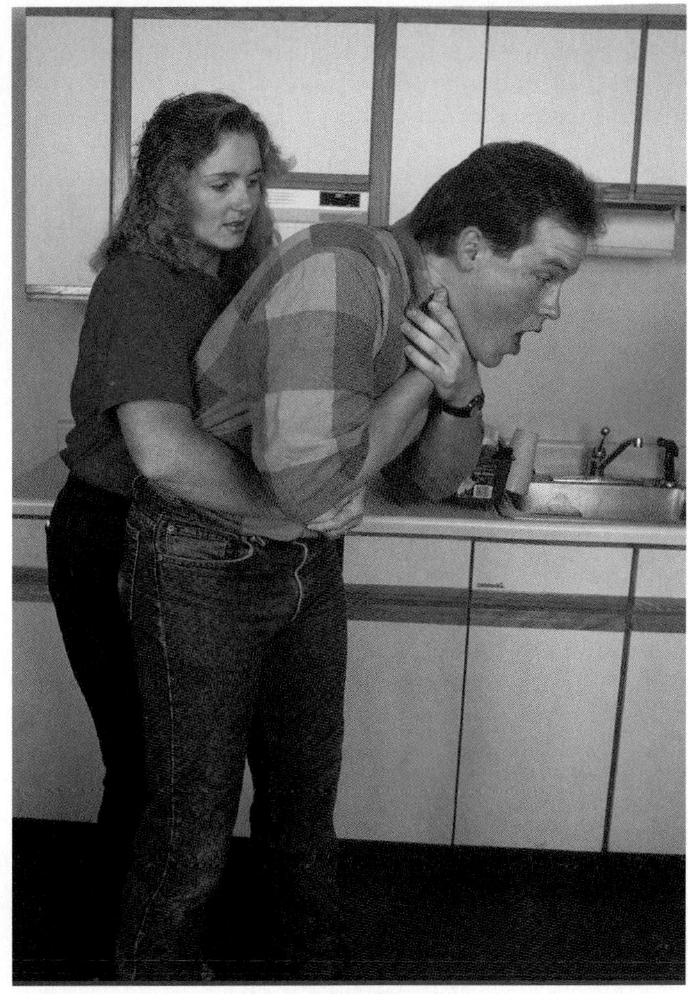

The Heimlich maneuver is used to expel food stuck in the trachea.

The respiratory system consists of a group of passages that filter incoming air and transport it into the body, into the lungs, and to the many microscopic air sacs where gases are exchanged. The entire process of exchanging gases between the atmosphere and body cells is called **respiration.** It involves several events:

- Movement of air in and out of the lungs, commonly called breathing or *ventilation.* This part of the process is sometimes called *external respiration.*
- Exchange of gases between the air in the lungs and the blood, sometimes called *internal respiration.*
- Transport of gases by the blood between the lungs and body cells.
- Exchange of gases between the blood and the body cells.

## Why We Breathe

Respiration occurs on a macroscopic level—a function provided by an organ system. However, the reason that body cells need to exchange gases—that is, take up oxygen and rid themselves of carbon dioxide—is apparent at the cellular and molecular levels. Oxygen utilization and production of carbon dioxide by body cells is part of the process of **cellular respiration.**

The gas exchange that respiration makes possible enables cells to harness the energy held in the chemical bonds of nutrient molecules. Recall from chapter 4 that energy is slowly liberated from food molecules by stripping off electrons and channeling them through a series of electron carriers. Without oxygen as a final electron acceptor, much energy remains locked in nutrient molecules. Recall also that in the reactions that remove electrons, carbons are cleaved, combined with oxygen, and released as carbon dioxide, a metabolic waste. Cellular respiration, then, explains why we need to obtain oxygen and get rid of carbon dioxide.

## Organs of the Respiratory System

The organs of the respiratory system include the nose, nasal cavity, sinuses, pharynx, larynx, trachea, bronchial tree, and lungs.

The parts of the respiratory system, shown in figure 19.1, can be divided into two groups, or tracts. Those organs outside the thorax constitute the *upper respiratory tract,* and those within the thorax comprise the *lower respiratory tract.*

### Nose

The nose is covered with skin and is supported internally by bone and cartilage. Its two *nostrils* (external nares) provide openings through which air can enter and leave the nasal cavity. Many internal hairs guard these openings, preventing entry of large particles carried in the air.

### Nasal Cavity

The **nasal cavity,** a hollow space behind the nose, is divided medially into right and left portions by the **nasal septum.** This cavity is separated from the cranial cavity by the cribriform plate of the ethmoid bone and from the mouth by the hard palate.

> The nasal septum is usually straight at birth, but may be bent during birth. It remains straight throughout early childhood, but then may bend toward one side or the other. Such a *deviated septum* may obstruct the nasal cavity, making breathing difficult.

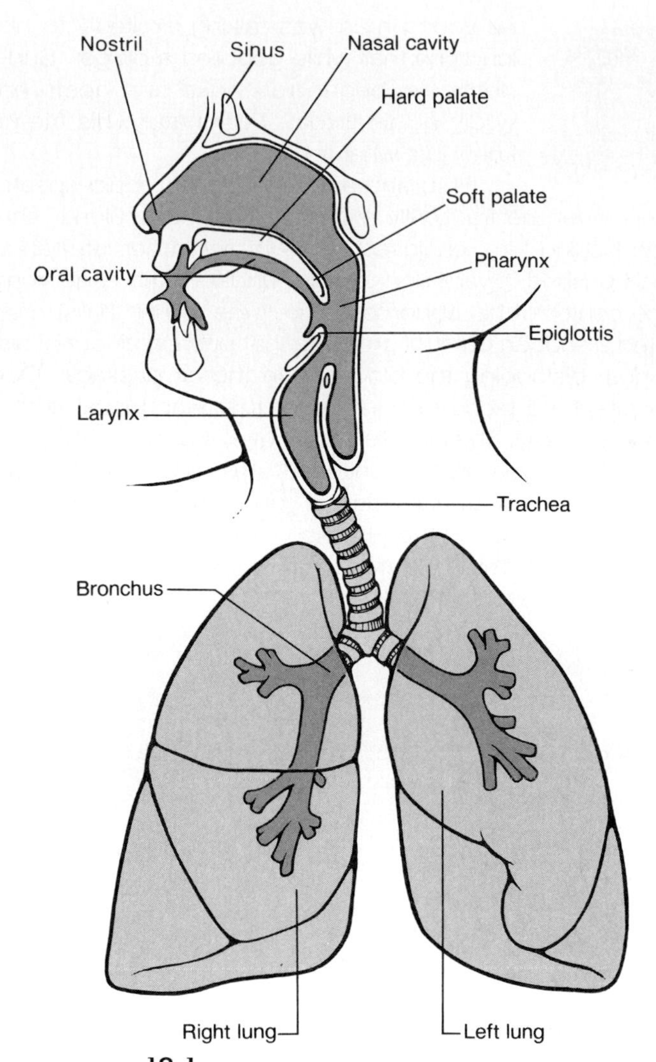


**Figure 19.1**
Organs of the respiratory system.

As figure 19.2 shows, **nasal conchae** (turbinate bones) curl out from the lateral walls of the nasal cavity on each side, dividing the cavity into passageways called the *superior, middle,* and *inferior meatuses* (see chapter 7). They also support the mucous membrane that lines the nasal cavity and help increase its surface area.

The upper posterior portion of the nasal cavity, below the cribriform plate, is slitlike, and its lining contains the olfactory receptors that provide the sense of smell. The remainder of the cavity conducts air to and from the nasopharynx.

The mucous membrane lining the nasal cavity contains pseudostratified ciliated epithelium that is rich in mucus-secreting goblet cells (see chapter 5). It also includes an extensive network of blood vessels, and normally appears pinkish. As air passes over the membrane, heat radiates from the blood and warms the air, adjusting its temperature to that of the body. In addition, evaporation of water from the mucous lining moistens the air. The sticky mucus the mucous membrane secretes entraps dust and other small particles entering with the air.

As the cilia of the epithelial cells move, a thin layer of mucus and any entrapped particles are pushed toward the pharynx (fig. 19.3). When the mucus reaches the pharynx, it is swallowed. In the stomach, any microorganisms in the mucus, including disease-causing forms, are likely to be destroyed by the action of gastric juice. Thus, the filtering mechanism provided by the mucous membrane not only prevents particles from reaching the lower air passages, but also helps prevent respiratory infections. Clinical Application 19.1 discusses how cigarette smoking impairs the respiratory system, beginning with the cleansing mucus and cilia.

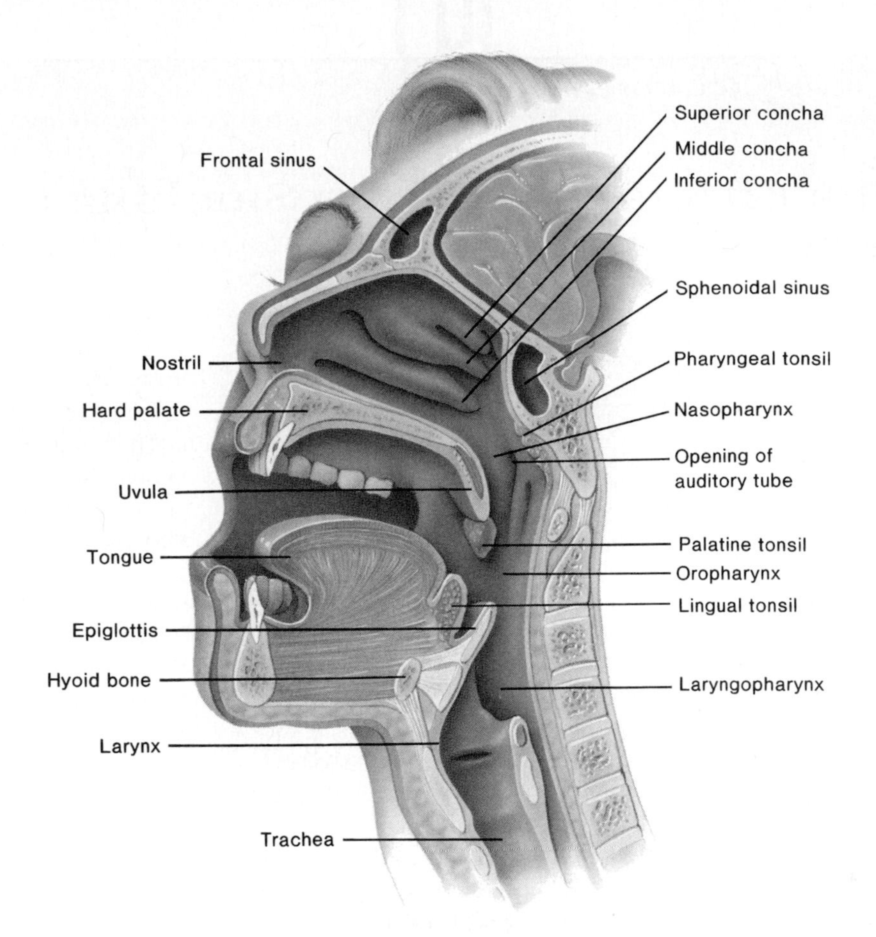


**FIGURE 19.2**
Major features of the upper respiratory tract.

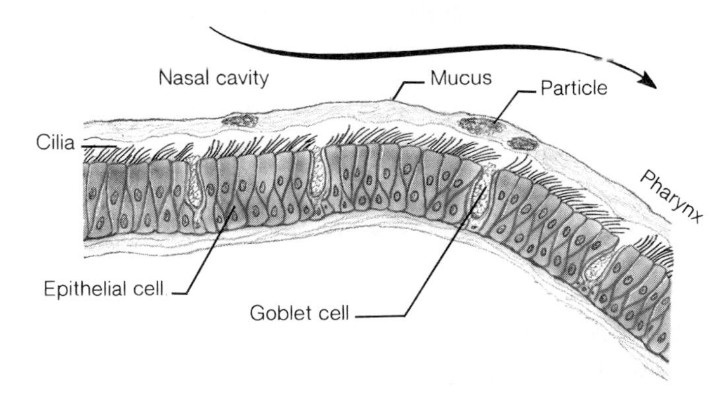


**FIGURE 19.3**
Cilia move mucus and trapped particles from the nasal cavity to the pharynx.

1. What is respiration?
2. What organs constitute the respiratory system?
3. What is the function of the mucous membrane that lines the nasal cavity?
4. What is the function of the cilia in the cells that line the nasal cavity?

## Sinuses

Recall from chapter 7 that the **sinuses** (paranasal sinuses) are air-filled spaces located within the *maxillary, frontal, ethmoid,* and *sphenoid bones* of the skull (fig. 19.4). These spaces open into the nasal cavity and are lined with mucous membranes that are continuous with the lining of the nasal cavity. Consequently, mucus secretions drain from the sinuses into the nasal cavity. Membranes that are inflamed and swollen because of nasal infections or allergic reactions (sinusitis) may block this drainage, causing increasing pressure within a sinus and a painful headache.

The sinuses reduce the weight of the skull, and serve as a resonant chamber that affects the quality of the voice.

> It is possible to illuminate a person's frontal sinus in a darkened room by holding a small flashlight just beneath the eyebrow. Similarly, the maxillary sinuses can be illuminated by holding the flashlight in the mouth.

CLINICAL APPLICATION 19.1

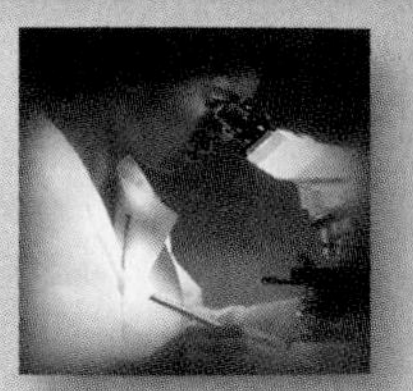

## The Effects of Cigarette Smoking on the Respiratory System

Damage to the respiratory system from cigarette smoking is slow, progressive, and deadly. A healthy respiratory system is continuously cleansed. The mucus produced by the respiratory tubules traps dirt and disease-causing organisms, which cilia sweep toward the mouth, where it can be eliminated. Smoking greatly impairs this housekeeping. With the very first inhalation of smoke, the beating of the cilia slows. With time, the cilia become paralyzed and, eventually, disappear altogether. The loss of cilia leads to the development of smoker's cough. The cilia no longer effectively remove mucus, so the individual must cough it up. Coughing is usually worse in the morning because mucus has accumulated during sleep.

To make matters worse, excess mucus is produced and accumulates, clogging the air passageways. Pathogenic organisms that are normally removed now have easier access to the respiratory surfaces and the resulting lung congestion favors their growth. This is why smokers are sick more often than nonsmokers. In addition, a lethal chain reaction begins. Smoker's cough leads to chronic bronchitis, caused by destroyed respiratory cilia. Mucus production increases and the lining of the bronchioles thickens, making breathing difficult. The bronchioles lose elasticity and are no longer able to absorb the pressure changes accompanying coughing. As a result, a cough can increase the air pressure within the alveoli (microscopic air sacs) enough to rupture the delicate alveolar walls; this condition is the hallmark of smoking-induced *emphysema.* The burst alveoli cause worsening of the cough, fatigue, wheezing, and impaired breathing. Emphysema is fifteen times more common among individuals who smoke a pack of cigarettes a day than among nonsmokers.

Simultaneous with the structural changes progressing to emphysema may be cellular changes leading to lung cancer. First, cells in the outer border of the bronchial lining begin to divide more rapidly than usual. Eventually, these displace the ciliated cells. Their nuclei begin to resemble those of cancerous cells—large and distorted with abnormal numbers of chromosomes. Up to this point, the damage can be repaired if smoking ceases. However, if smoking continues, these cells may eventually break through the basement membrane and begin dividing within the lung tissue, forming a tumor with the potential of spreading throughout lung tissue (figs. 19A and 19B). Eighty percent of lung cancer cases are due to cigarette smoking. Only 13% of lung cancer patients live as long as 5 years after the initial diagnosis.

1. Where are the sinuses located?
2. What are the functions of the sinuses?

### Pharynx

The **pharynx** (throat) is located behind the oral cavity and between the nasal cavity and the larynx. It is a passageway for food traveling from the oral cavity to the esophagus and for air passing between the nasal cavity and the larynx (see fig. 19.2). It also aids in producing the sounds of speech. The subdivisions of the pharynx—the nasopharynx, oropharynx, and laryngopharynx—are described in chapter 17.

### Larynx

The **larynx** is an enlargement in the airway at the top of the trachea and below the pharynx. It is a passageway for air moving in and out of the trachea and prevents foreign objects from entering the trachea. It also houses the *vocal cords* (see reference plates 49 and 71).

The larynx is composed of a framework of muscles and cartilages bound together by elastic tissue. The largest of the cartilages (shown in fig. 19.5) are the thyroid, cricoid, and epiglottic cartilages. These structures are single. The other laryngeal cartilages—the arytenoid, corniculate, and cuneiform cartilages—are paired.

It pays to quit. Much of the damage to the respiratory system can be repaired. Cilia are restored and the thickening of alveolar walls due to emphysema can be reversed. But ruptured alveoli are gone forever. The nicotine in tobacco smoke causes a powerful dependency by binding to certain receptors on brain cells.

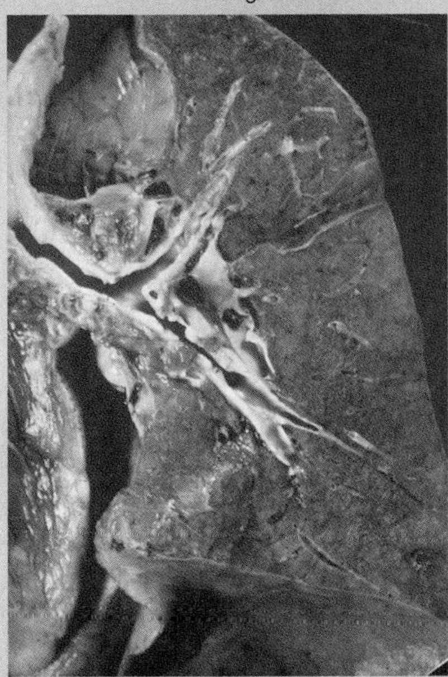


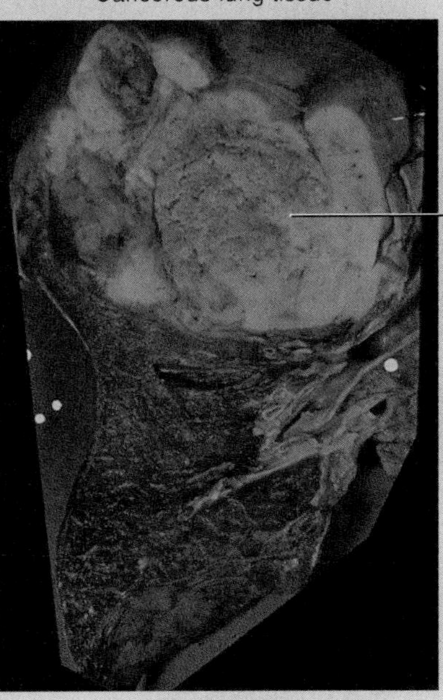


**FIGURE 19A**
The lung on the left is healthy. A cancerous tumor has invaded the lung on the right, taking up nearly half of the lung space.

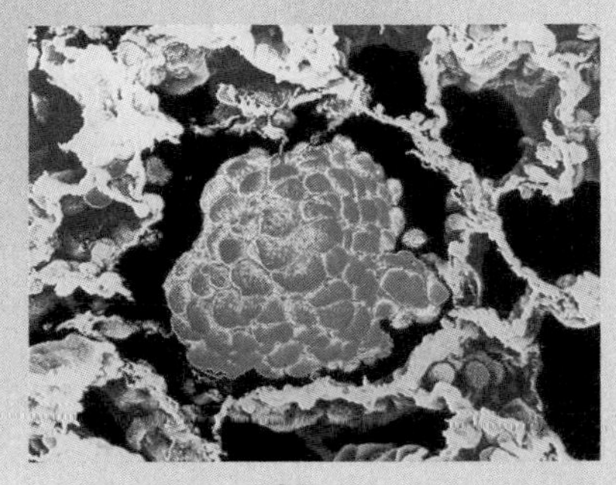

**FIGURE 19B**
Lung cancer may begin as a tiny tumor growing in an alveolus, a microscopic air sac (260×).

The **thyroid cartilage** was named for the thyroid gland that covers its lower area. This cartilage is the shieldlike structure that protrudes in the front of the neck and is sometimes called the Adam's apple. The protrusion typically is more prominent in males than in females because of an effect of male sex hormones on the development of the larynx.

The **cricoid cartilage** lies below the thyroid cartilage and marks the lowermost portion of the larynx.

The **epiglottic cartilage,** the only one of the laryngeal cartilages that is elastic, not hyaline, cartilage, is attached to the upper border of the thyroid cartilage and supports a flaplike structure called the **epiglottis.** The epiglottis usually stands upright and allows air to enter the larynx. During swallowing, however, muscular contractions raise the larynx, and the base of the tongue presses the epiglottis downward. As a result, the epiglottis partially covers the opening into the larynx, helping prevent foods and liquids from entering the air passages.

The pyramid-shaped **arytenoid cartilages** are located above and on either side of the cricoid cartilage. Attached to the tips of the arytenoid cartilages are the tiny, conelike **corniculate cartilages.** These cartilages are attachments for muscles that help regulate tension on the vocal cords during speech and aid in closing the larynx during swallowing.

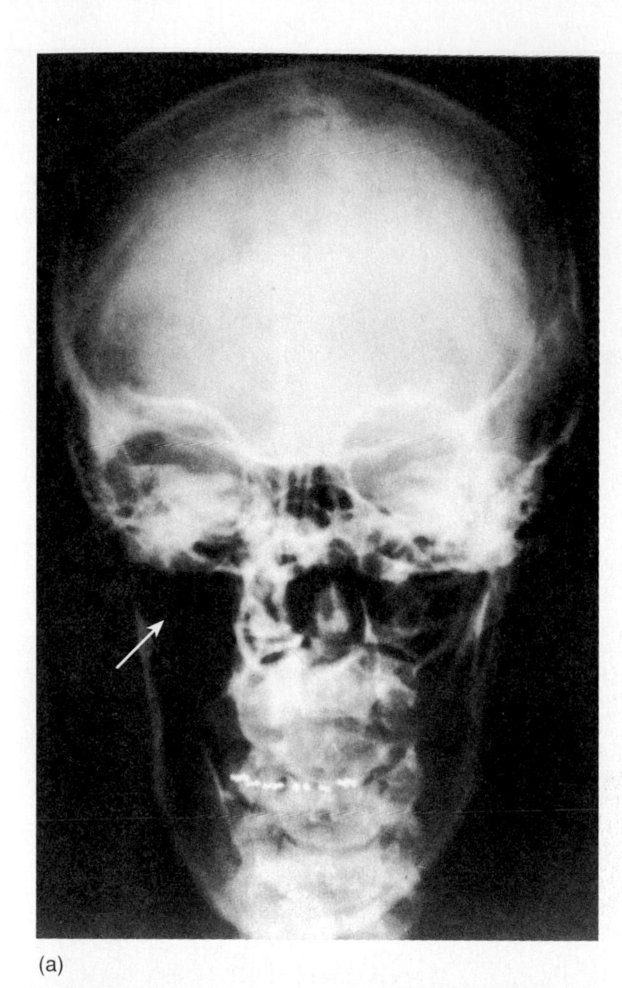

(a)

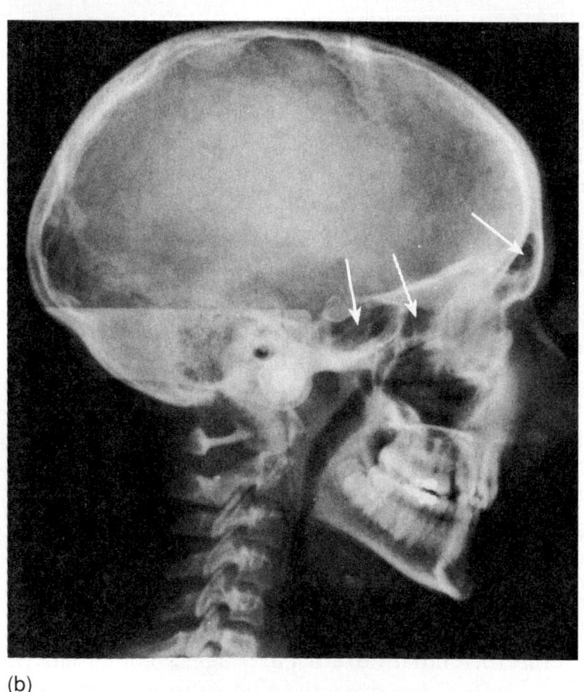

(b)

**Figure 19.4**

(*a*) X-ray film of a skull from the anterior and (*b*) from the lateral view, showing air-filled sinuses within the bones (arrows).

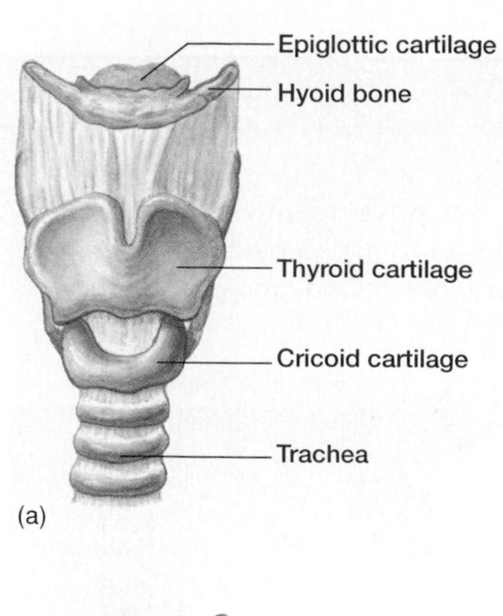


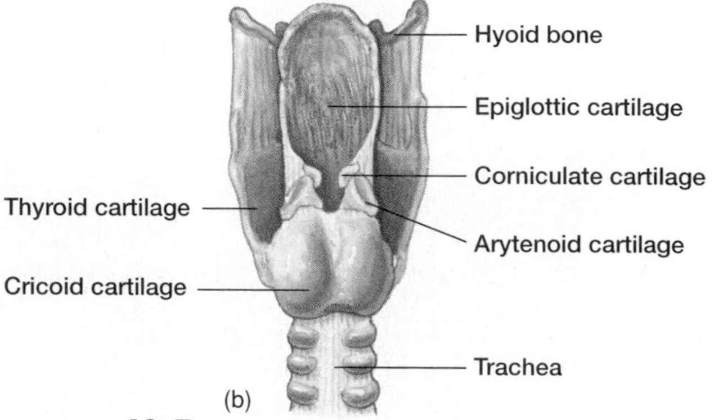


**Figure 19.5**

(*a*) Anterior view and (*b*) posterior view of the larynx.

On December 13, 1799, George Washington spent the day walking on his estate in a freezing rain. The next day, he had trouble breathing and swallowing. Several doctors were called in. One suggested a tracheostomy, in which a hole would be cut in the throat so that the president could breathe. He was voted down. The other physicians suggested bleeding the patient, plastering his throat with bran and honey, and placing blister beetles on his legs to produce blisters. Within a few hours, Washington's voice became muffled, breathing was more labored, and he was restless. For a short time he seemed euphoric, and then he died.

Today, Washington's problem is recognized as *epiglottitis,* in which the epiglottis swells to ten times its normal size. A tracheostomy might have saved his life.

The **cuneiform cartilages** are small, cylindrical structures in the mucous membrane between the epiglottic and the arytenoid cartilages. They stiffen the soft tissues in this region.

Inside the larynx, two pairs of horizontal folds in the mucous membrane extend inward from the lateral

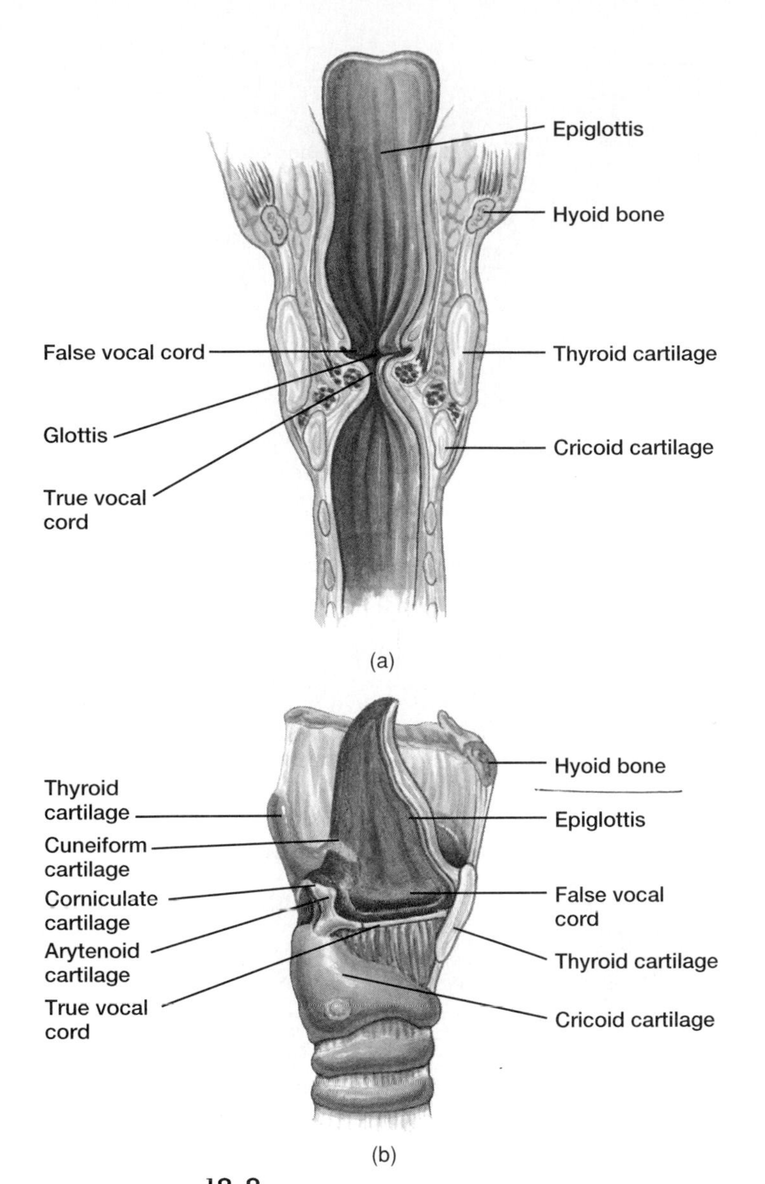


**FIGURE 19.6**

(*a*) Coronal section and (*b*) sagittal section of the larynx.

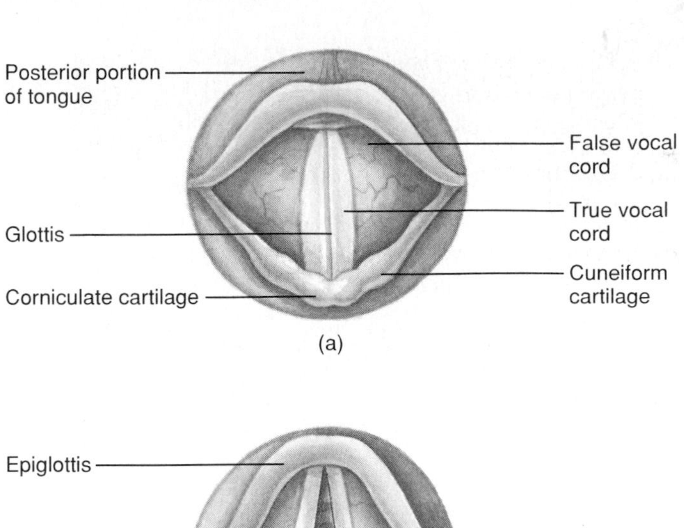


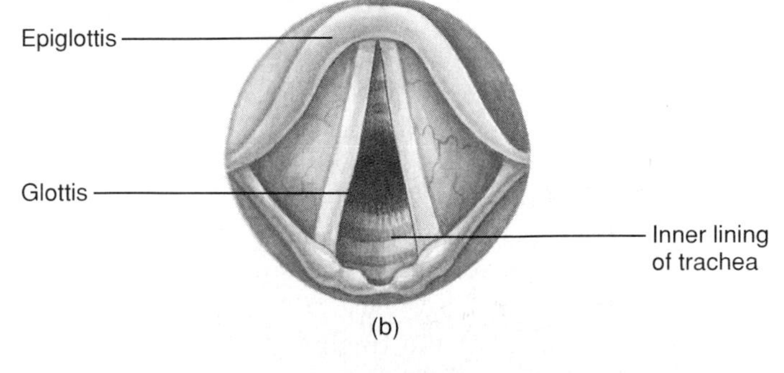


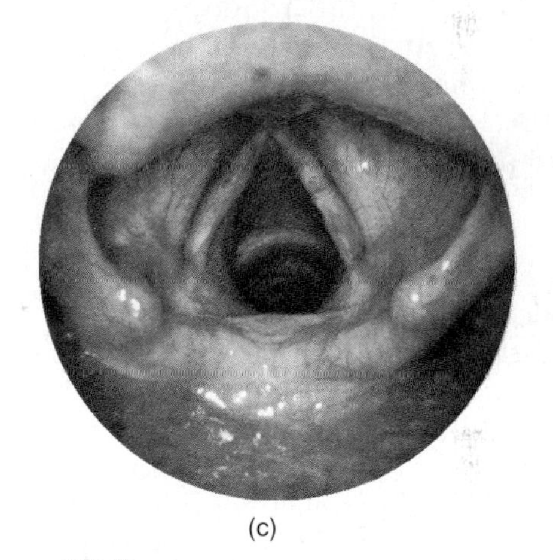


**FIGURE 19.7**

Vocal cords as viewed from above (*a*) with the glottis closed and (*b*) with the glottis open. (*c*) A photograph of the glottis and vocal folds.

walls. The upper folds (vestibular folds) are called *false vocal cords* because they do not produce sounds. Muscle fibers within these folds help close the larynx during swallowing.

The lower folds are the *true vocal cords.* They contain elastic fibers and are responsible for vocal sounds, which are created when air is forced between these folds, causing them to vibrate from side to side. This action generates sound waves, which can be formed into words by changing the shapes of the pharynx and oral cavity and by using the tongue and lips. Figure 19.6 shows both pairs of folds.

Changing the tension on the vocal cords controls the *pitch* (musical tone) of the voice. Contracting or relaxing laryngeal muscles controls pitch. Increasing the tension produces a higher pitch, and decreasing the tension creates a lower pitch.

The *intensity* (loudness) of a vocal sound is related to the force of the air passing over the vocal cords. Stronger blasts of air result in greater vibration of the vocal cords and louder sound.

During normal breathing, the vocal cords remain relaxed, and the opening between them, called the **glottis,** is a triangular slit. However, when food or liquid is swallowed, muscles close the glottis within the false vocal folds. This action assists closing of the epiglottis, which prevents food or liquid from entering the trachea (fig. 19.7).

The mucous membrane that lines the larynx continues to filter incoming air by entrapping particles and moving them toward the pharynx by ciliary action.

1. What part of the respiratory tract is shared with the alimentary canal?
2. Describe the structure of the larynx.
3. How do the vocal cords produce sounds?
4. What is the function of the glottis? Of the epiglottis?

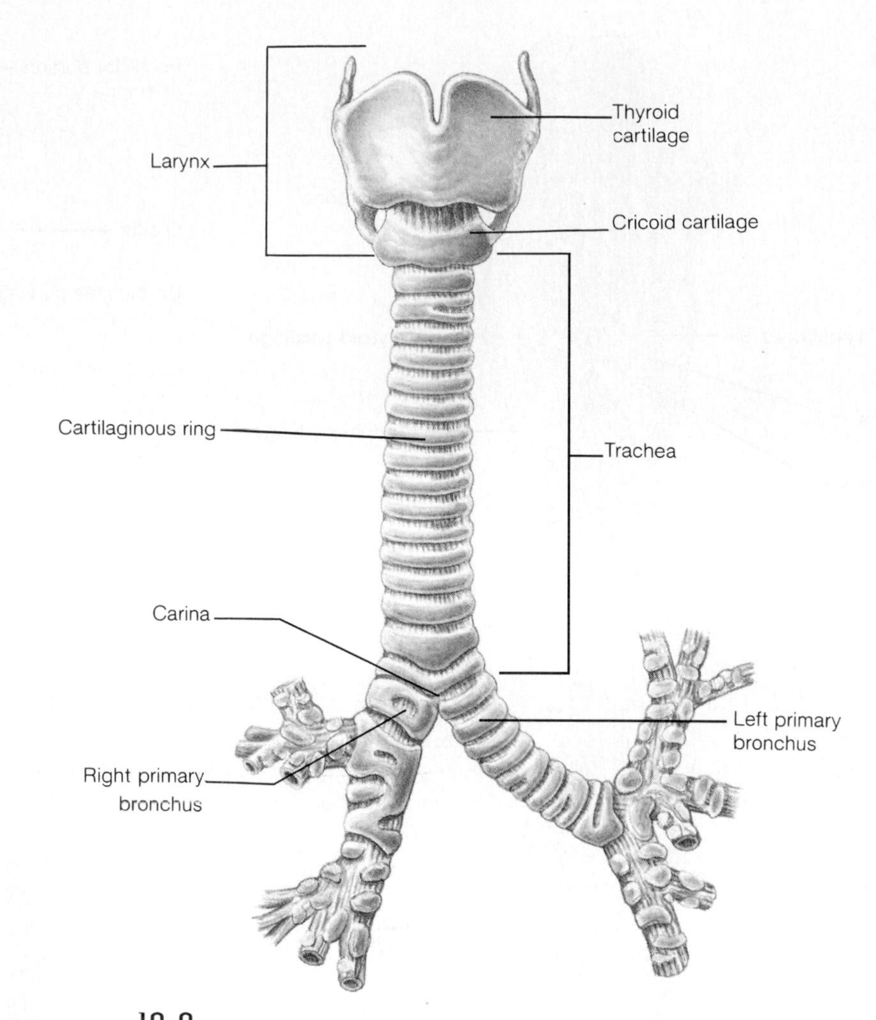


**FIGURE 19.8**
The trachea transports air between the larynx and the bronchi.

## Trachea

The **trachea** (windpipe) is a flexible cylindrical tube about 2.5 centimeters in diameter and 12.5 centimeters in length. It extends downward in front of the esophagus and into the thoracic cavity, where it splits into right and left bronchi (fig. 19.8 and reference plate 50).

The inner wall of the trachea is lined with a ciliated mucous membrane that contains many goblet cells. This membrane continues to filter the incoming air and to move entrapped particles upward into the pharynx where the mucus can be swallowed.

Within the tracheal wall are about twenty C-shaped pieces of hyaline cartilage, one above the other. The open ends of these incomplete rings are directed posteriorly, and the gaps between their ends are filled with smooth muscle and connective tissues (figs. 19.9 and 19.10). These cartilaginous rings prevent the trachea from collapsing and blocking the airway. At the same time, the soft tissues that complete the rings in the back allow the nearby esophagus to expand as food moves through it on the way to the stomach.

A blocked trachea can cause asphyxiation in minutes. If swollen tissues, excess secretions, or a foreign object obstruct the trachea, making a temporary, external opening in the tube so that air can bypass the obstruction is lifesaving. This procedure, shown in figure 19.11, is called a *tracheostomy*.

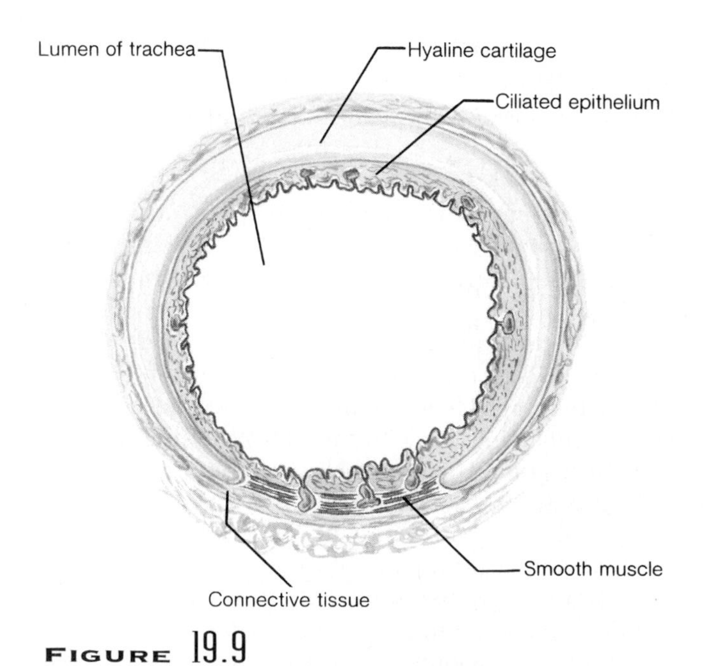


**FIGURE 19.9**
Cross section of the trachea. Note the C-shaped rings of hyaline cartilage in the wall.

## Bronchial Tree

The **bronchial tree** includes the trachea and the branched airways leading from it to the microscopic air sacs in the lungs. Its branches begin with the right and left **primary bronchi,** which arise from the trachea at the level of the fifth thoracic vertebrae. The openings of the primary bronchi are separated by a ridge of

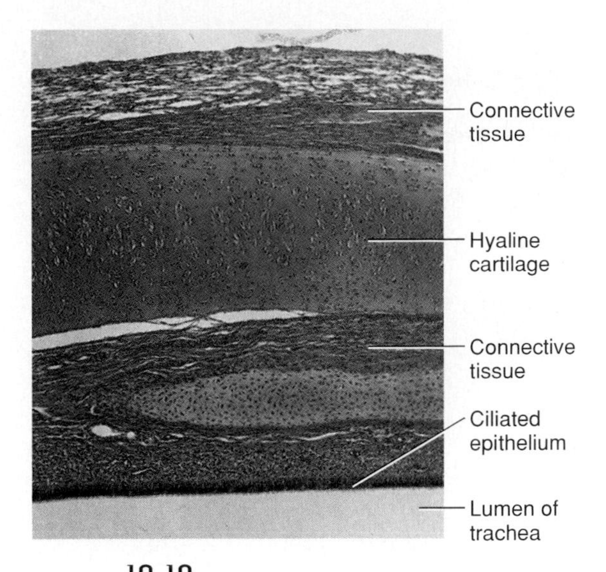


**FIGURE 19.10**
Light micrograph of a section of the tracheal wall (63x).

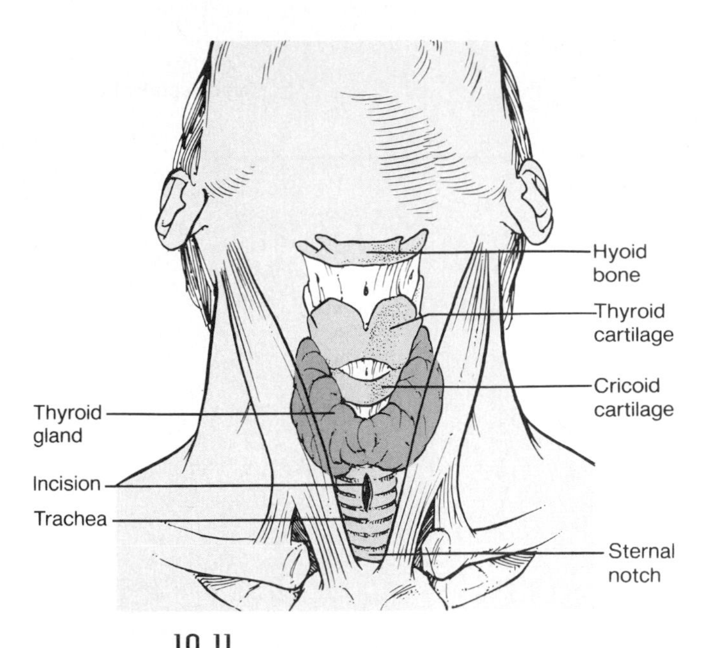


**FIGURE 19.11**
A tracheostomy may be performed to allow air to bypass an obstruction within the larynx.

cartilage called the *carina* (see fig. 19.8). Each bronchus, accompanied by large blood vessels, enters its respective lung.

## Branches of the Bronchial Tree

A short distance from its origin, each primary bronchus divides into **secondary,** or **lobar, bronchi** (two on the left and three on the right) that, in turn, branch again and again into finer and finer tubes (figs. 19.12 and 19.13). When stripped of their associated blood vessels and tissues, the airways appear as an upside down tree. The successive divisions of these branches from the lobar bronchus to the microscopic air sacs follow:

1. **Tertiary,** or **segmental, bronchi.** Each of these branches supplies a portion of the lung called a *bronchopulmonary segment.* Usually there are ten such segments in the right lung and eight in the left lung.
2. **Intralobular bronchioles.** These small branches of the segmental bronchi enter the basic units of the lung—the *lobules.*
3. **Terminal bronchioles.** These tubes branch from a bronchiole. There are fifty to eighty terminal bronchioles within a lobule of the lung.
4. **Respiratory bronchioles.** Two or more respiratory bronchioles branch from each terminal bronchiole. They are relatively short and about 0.5 millimeter in diameter. They are called "respiratory" because a few air sacs bud from their sides, making them the first structures in the sequence that can take part in gas exchange.
5. **Alveolar ducts.** Two to ten long, branching alveolar ducts extend from each respiratory bronchiole (fig. 19.14).
6. **Alveolar sacs.** Alveolar sacs are thin-walled, closely packed outpouchings of the alveolar ducts.
7. **Alveoli.** Alveoli are thin-walled, microscopic air sacs that open only on the side communicating with an alveolar sac. Thus, air can diffuse freely from the alveolar ducts, through the alveolar sacs, and into the alveoli (fig. 19.15).

Dust particles, asbestos fibers, and other pollutants travel at speeds of 200 centimeters per second in the trachea, but slow to 1 centimeter per second when deep in the lungs. Such particles are deposited by gravity, particularly at branchpoints in the respiratory tree, like traffic backing up at an exit to a highway.

## Structure of the Respiratory Tubes

The structure of a bronchus is similar to that of the trachea, but the C-shaped cartilaginous rings are replaced with cartilaginous plates where the bronchus enters the lung. These plates are irregularly shaped, and completely surround the tube. However, as finer and finer branch tubes appear, the amount of cartilage decreases, and it finally disappears in the bronchioles, which have diameters of about 1 millimeter.

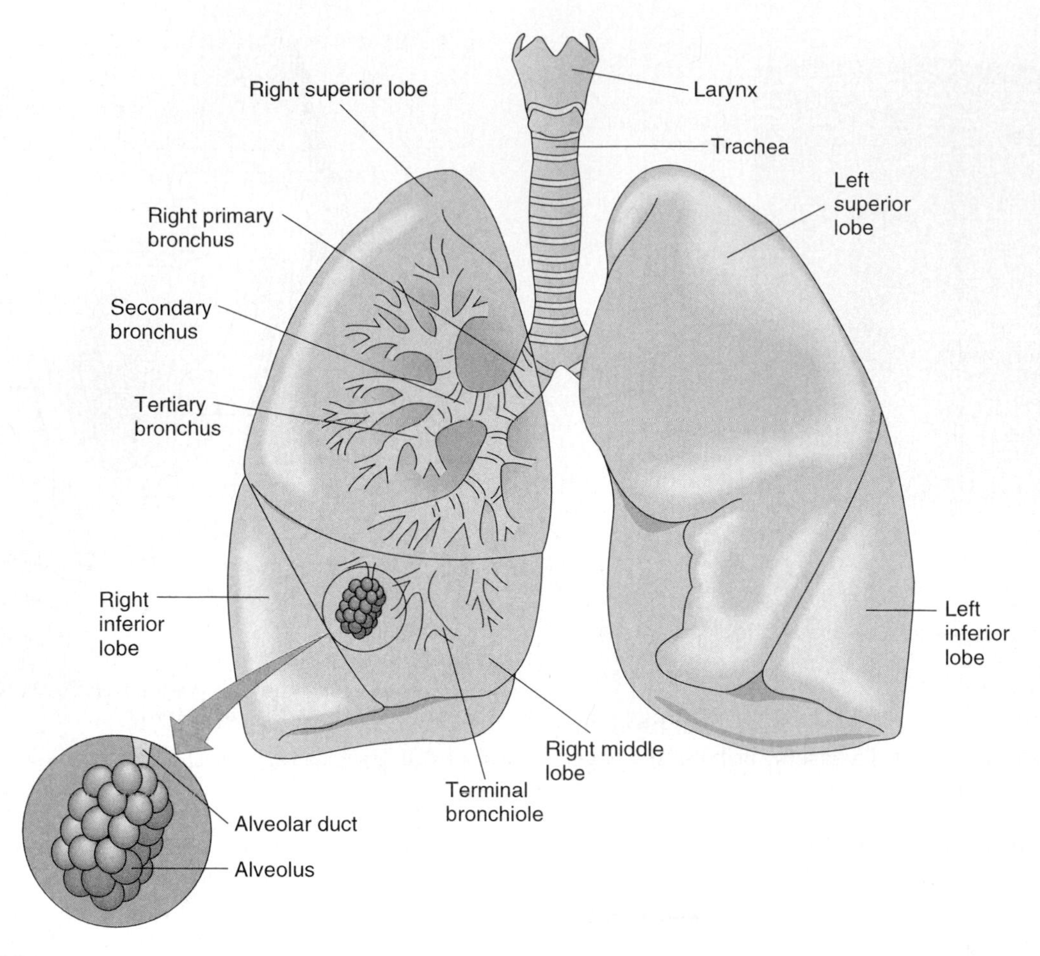


**FIGURE 19.12**

The bronchial tree consists of the passageways that connect the trachea and the alveoli. The alveolar duct and alveoli are enlarged to show their location.

As the amount of cartilage decreases, a layer of smooth muscle that surrounds the tube just beneath the mucosa becomes more prominent. This muscular layer remains in the wall to the ends of the respiratory bronchioles, and only a few muscle fibers are in the walls of the alveolar ducts.

Elastic fibers are scattered among the smooth muscle cells and are abundant in the connective tissue that surrounds the respiratory tubes. These fibers play an important role in breathing, as is explained later in this chapter.

As the tubes become smaller in diameter, a change occurs in the type of cells that line them. The lining of the larger tubes consists of pseudostratified, ciliated columnar epithelium and mucus-secreting goblet cells. However, along the way the number of goblet cells and the height of the other epithelial cells decline, and cilia become scarcer. In the finer tubes, beginning with the respiratory bronchioles, the lining is cuboidal epithelium; and in the alveoli, it is simple squamous epithelium closely associated with a dense network of capillaries. The mucous lining gradually thins, until none appears in the alveoli.

> The trachea and bronchial tree can be examined directly using a flexible optical instrument called a *fiberoptic bronchoscope.* This procedure (bronchoscopy) is sometimes used in diagnosing tumors or other pulmonary diseases. It may also be used to locate and remove aspirated foreign bodies in the air passages.

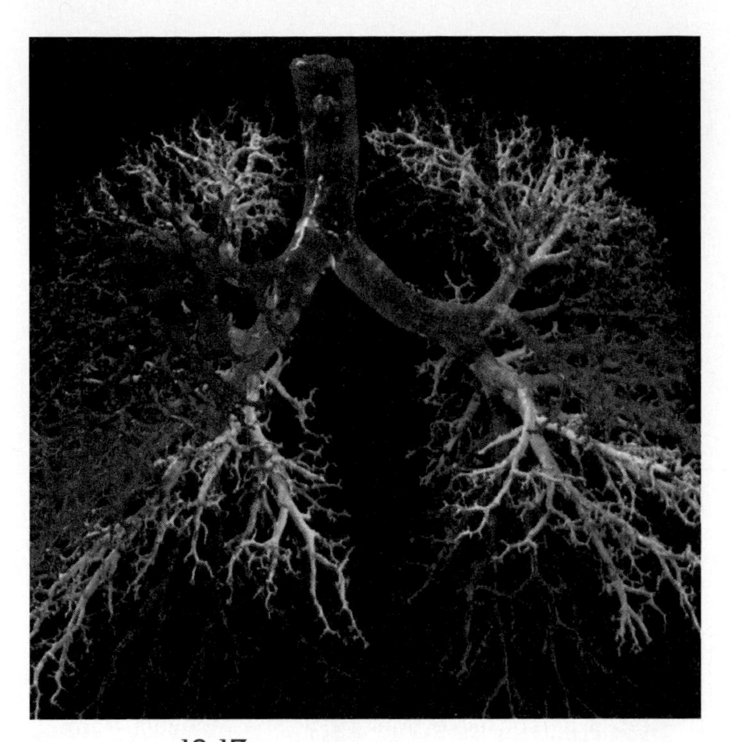

**Figure** 19.13
A plastic cast of the bronchial tree.

## Functions of the Respiratory Tubes and Alveoli

The branches of the bronchial tree are air passages, which continue to filter the incoming air and distribute it to the alveoli in all parts of the lungs. The alveoli, in turn, provide a large surface area of thin epithelial cells through which gas exchanges can occur (fig. 19.16). If the 300 million alveoli in the human lung were spread out, they would cover an area the size of half a tennis court—between 70 and 80 square meters.

During these exchanges, oxygen diffuses through the alveolar walls and enters the blood in nearby capillaries, and carbon dioxide diffuses from the blood through these walls and enters the alveoli (figs. 19.17 and 19.18).

> In the inherited illness *cystic fibrosis,* airways become clogged with thick, sticky mucus, which attracts bacteria. As damaged white blood cells accumulate at the infection site, their DNA may leak out and further clog the area. A new treatment that moderately eases breathing is deoxyribonuclease (DNase), an enzyme normally found in the body that degrades the accumulating extracellular DNA.

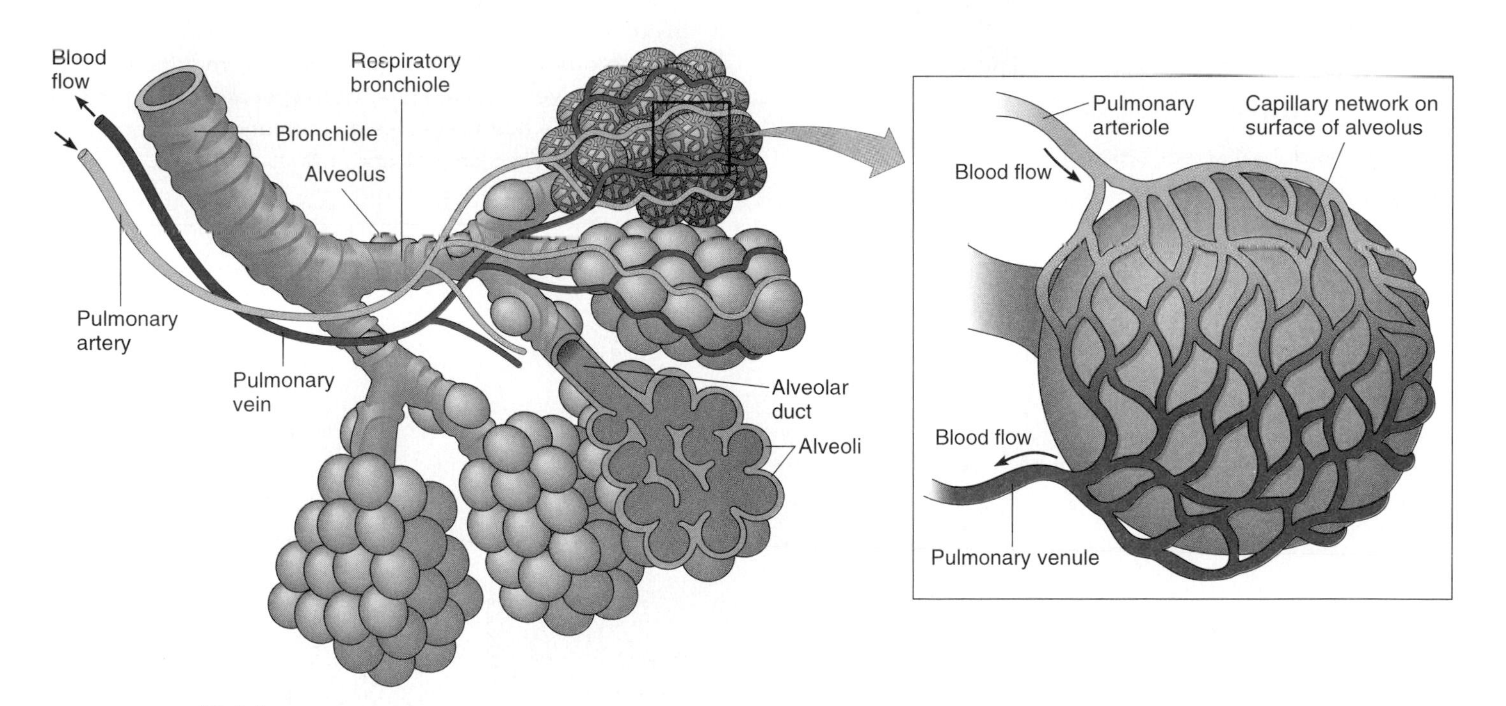


**Figure** 19.14
The respiratory tubes end in tiny alveoli, each of which is surrounded by a capillary network.

1. What is the function of the cartilaginous rings in the tracheal wall?
2. How do the right and left bronchi differ in structure?
3. List the branches of the bronchial tree.
4. Describe the changes in structure that occur in the respiratory tubes as they become smaller and smaller.
5. How are gases exchanged in the alveoli?

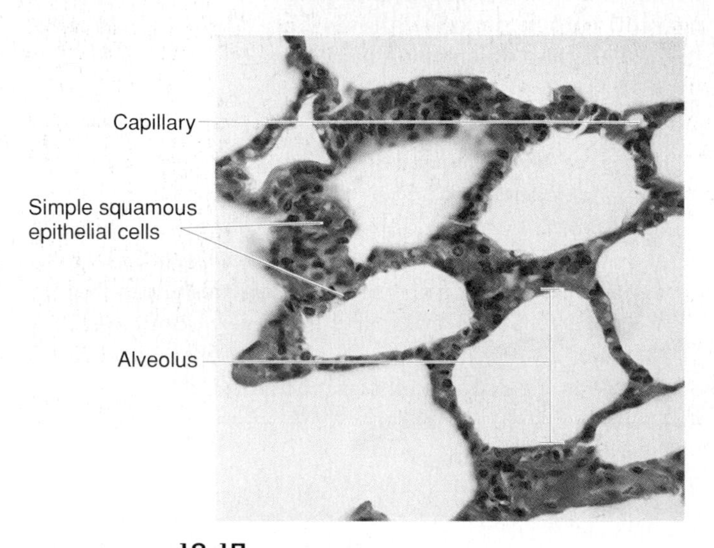


**FIGURE 19.15**
Light micrograph of alveoli (250x).

## Lungs

The lungs are soft, spongy, cone-shaped organs located in the thoracic cavity. The right and left lungs are separated medially by the heart and the mediastinum, and they are enclosed by the diaphragm and the thoracic cage (see figs. 1.7, 19.19 and reference plates 56, 57, and 71).

Because the respiratory organs are vital, malfunction can be deadly. Temporarily maintaining respiratory function can be lifesaving, and can be accomplished in several ways.

In *artificial respiration,* a person blows into the mouth of a person who has stopped breathing. The oxygen in the rescuer's exhaled breath can keep the victim alive.

In *extracorporeal membrane oxygenation*, blood is pumped out of the body and across a gas-permeable membrane that adds oxygen and removes carbon dioxide, simulating lung function. Such a devise can keep a person alive until he or she recovers from other problems but is too costly and cumbersome to maintain life indefinitely.

A lung assist device, called an *intravascular oxygenator,* is in clinical trials. It consists of hundreds of tiny porous hair-thin fibers surgically implanted in the inferior vena cava. Here, deoxygenated blood returning to the heart receives oxygen and is rid of carbon dioxide—but only at about 30% the capacity of a healthy respiratory system.

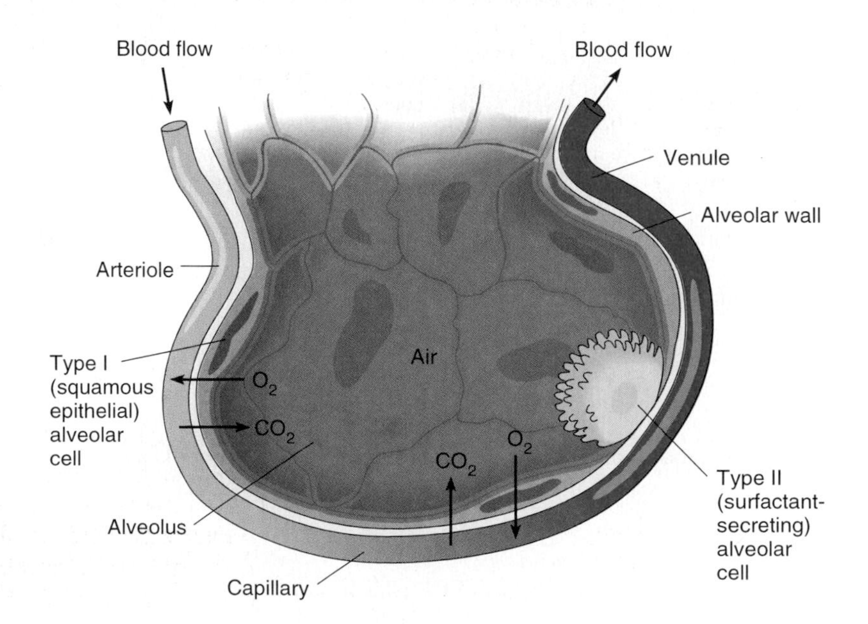


**FIGURE 19.16**
Oxygen diffuses from the air within the alveolus into the capillary, while carbon dioxide diffuses from the blood within the capillary into the alveolus.

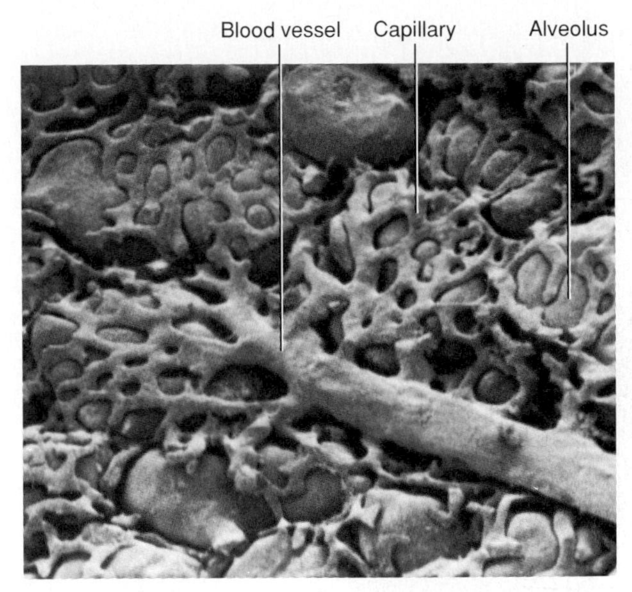


**FIGURE 19.17**
Scanning electron micrograph of casts of alveoli and associated capillary networks. These casts were prepared by filling the alveoli and blood vessels with resin and later removing the soft tissues by digestion, leaving only the resin casts (420x).

(*Tissues and Organs: A Text-Atlas of Scanning Electron Microscopy*, by Richard D. Kessel and Randy Kardon. © 1979 W. H. Freeman and Company.)

**FIGURE 19.18**
Scanning electron micrograph of lung alveoli and a bronchiole (70×).

Each lung occupies most of the thoracic space on its side and is suspended in the cavity by a bronchus and some large blood vessels. These tubular structures enter the lung on its medial surface through a region called the **hilus.** A layer of serous membrane, the **visceral pleura,** is firmly attached to the surface of each lung, and this membrane folds back at the hilus to become the **parietal pleura.** The parietal pleura, in turn, forms part of the mediastinum and lines the inner wall of the thoracic cavity.

There is no significant space between the visceral and parietal pleurae, but the potential space between them is called the **pleural cavity.** It contains a thin film of serous fluid that lubricates the adjacent pleural surfaces, reducing friction as they move against one another during breathing. This fluid also helps hold the pleural membranes together.

The right lung is larger than the left lung, and it is divided into three parts, called the superior, middle, and inferior lobes. The left lung consists of two parts, a superior and an inferior lobe.

A lobar bronchus of the bronchial tree supplies each lobe. A lobe also has connections to blood and lymphatic vessels, and is enclosed by connective tissues. Connective tissue further subdivides a lobe into **lobules,** each of which contains terminal bronchioles together with their alveolar ducts, alveolar sacs, alveoli, nerves, and associated blood and lymphatic vessels. Chart 19.1 summarizes the characteristics of the major parts of the respiratory system.

1. Where are the lungs located?
2. What is the function of the serous fluid within the pleural cavity?
3. How does the structure of the right lung differ from that of the left lung?
4. What kinds of structures make up a lung?

## Breathing Mechanism

Breathing, which is also called *ventilation,* is the movement of air from outside the body into the bronchial tree and alveoli, followed by a reversal of this air movement. The actions responsible for these air movements are termed **inspiration** (inhalation) and **expiration** (exhalation).

### Inspiration

Atmospheric pressure due to the weight of the air is the force that moves air into the lungs. At sea level, this pressure is sufficient to support a column of mercury about 760 millimeters high in a tube (fig. 19.20). Thus, normal air pressure equals 760 millimeters (mm) of mercury (Hg).

Air pressure is exerted on all surfaces in contact with the air, and because people breathe air, the inside

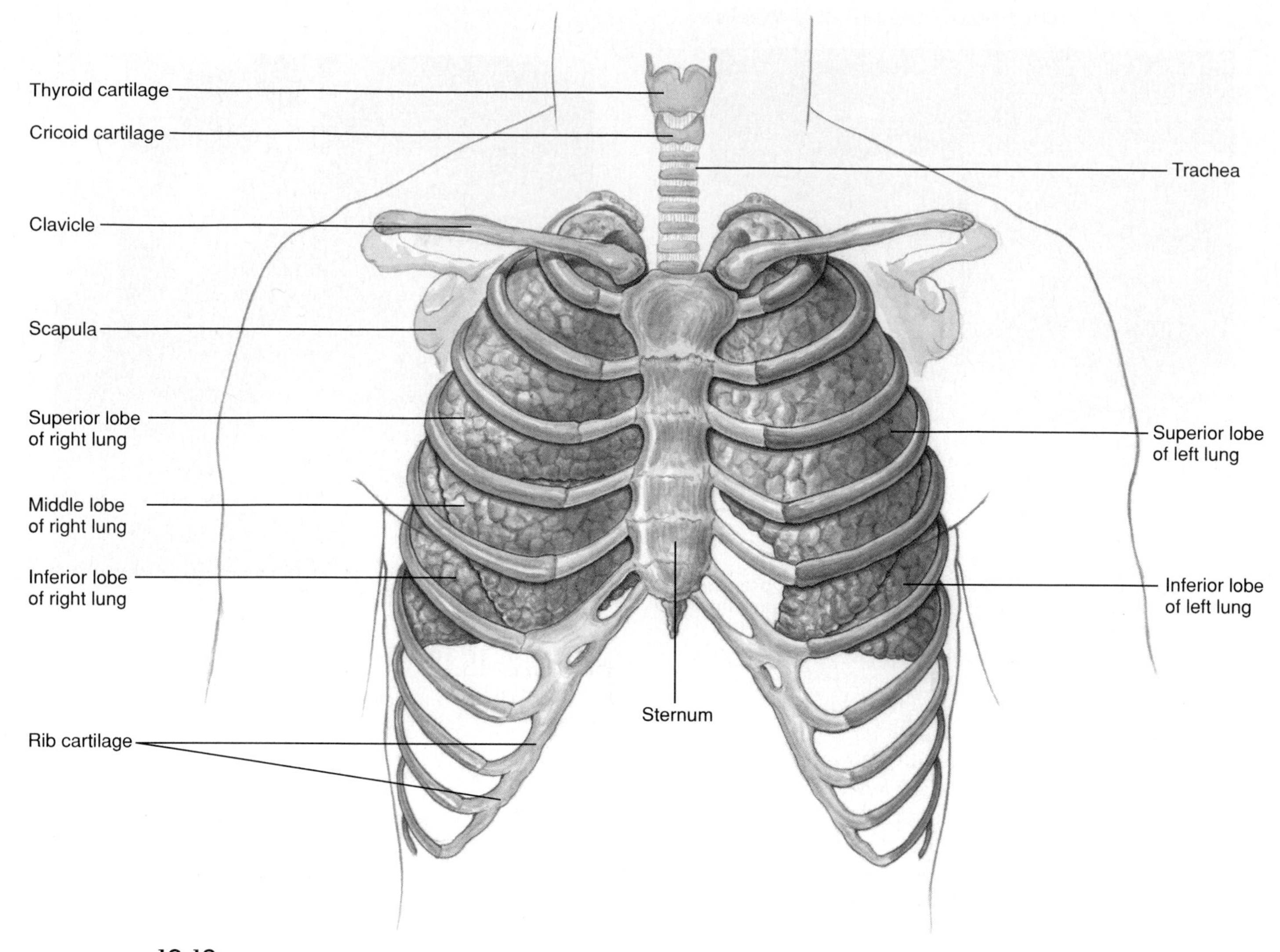


**FIGURE 19.19**
Locations of the lungs within the thoracic cavity.

surfaces of their lungs are also subjected to pressure. In other words, the pressures on the inside of the lungs and alveoli and on the outside of the thoracic wall are about the same (fig. 19.21).

If the pressure inside the lungs and alveoli (intra-alveolar pressure) decreases, outside air will then be pushed into the airways by atmospheric pressure. This is what happens during normal inspiration, and it involves the action of muscle fibers within the dome-shaped *diaphragm.*

The diaphragm is located just below the lungs. It consists of an anterior group of skeletal muscle fibers (costal fibers), which originate from the ribs and sternum, and a posterior group (crural fibers), which originate from the vertebrae. Both groups of muscle fibers are inserted on a tendinous central portion of the diaphragm (reference plate 71).

The muscle fibers of the diaphragm are stimulated to contract by impulses carried by the *phrenic nerves,* which are associated with the cervical plexuses. When this occurs, the diaphragm moves downward, the thoracic cavity enlarges, and the intra-alveolar pressure falls about 2 mm Hg below that of atmospheric pressure. In response to this decreased pressure, air is forced into the airways by atmospheric pressure (fig. 19.22).

While the diaphragm is contracting and moving downward, the *external* (*inspiratory*) *intercostal muscles* and certain thoracic muscles may be stimulated to contract. This action raises the ribs and elevates the sternum, increasing the size of the thoracic cavity even more. As a result, the intra-alveolar pressure falls further, and the relatively greater atmospheric pressure forces more air into the airways.

## CHART 19.1 Parts of the Respiratory System

| Part | Description | Function |
|---|---|---|
| Nose | Part of face centered above the mouth and below the space between the eyes | Nostrils provide entrance to nasal cavity; internal hairs begin to filter incoming air |
| Nasal cavity | Hollow space behind nose | Conducts air to pharynx; mucous lining filters, warms, and moistens incoming air |
| Sinuses | Hollow spaces in various bones of the skull | Reduce weight of the skull; serve as resonant chambers |
| Pharynx | Chamber behind mouth cavity and between nasal cavity and larynx | Passageway for air moving from nasal cavity to larynx and for food moving from mouth cavity to esophagus |
| Larynx | Enlargement at the top of the trachea | Passageway for air; prevents foreign objects from entering trachea; houses vocal cords |
| Trachea | Flexible tube that connects larynx with bronchial tree | Passageway for air; mucous lining continues to filter air |
| Bronchial tree | Branched tubes that lead from the trachea to the alveoli | Conducts air to the alveoli; mucous lining continues to filter incoming air |
| Lungs | Soft, cone-shaped organs that occupy a large portion of the thoracic cavity | Contain the air passages, alveoli, blood vessels, connective tissues, lymphatic vessels, and nerves of the lower respiratory tract |

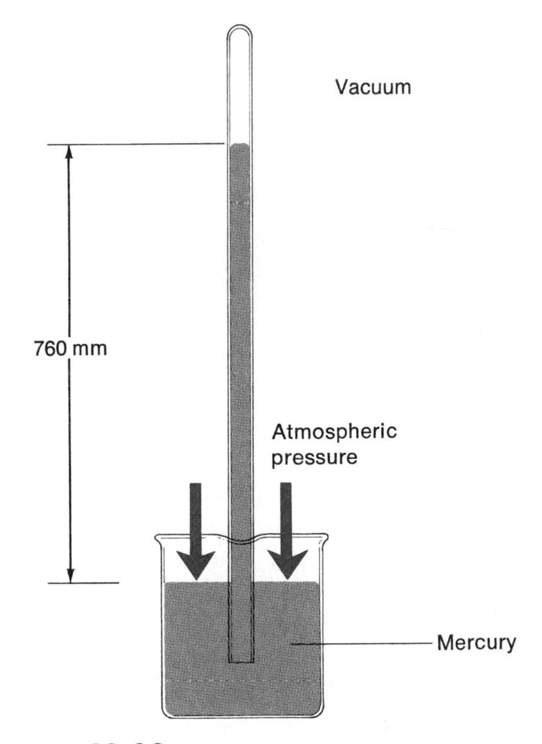


**FIGURE 19.20**
Atmospheric pressure at sea level is sufficient to support a column of mercury 760 millimeters high at sea level.

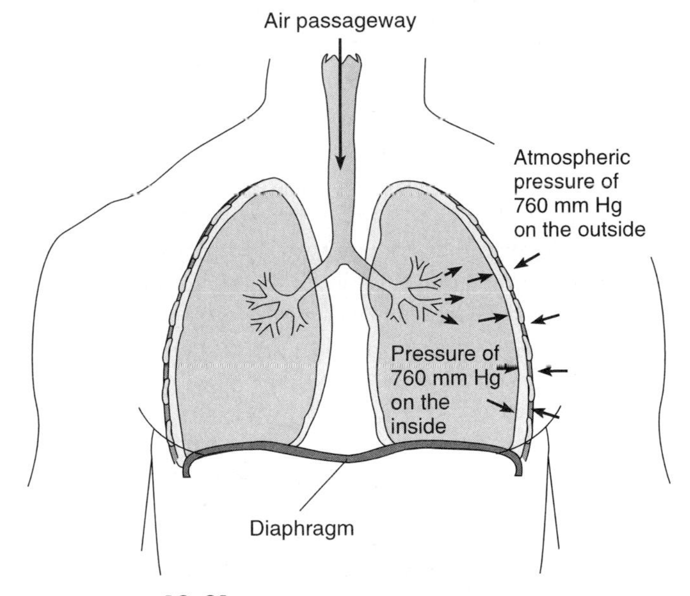


**FIGURE 19.21**
When the lungs are at rest, the pressure on the inside of the lungs is equal to the pressure on the outside of the thorax.

Lung expansion in response to movements of the diaphragm and chest wall depends on movements of the pleural membranes as follows. The parietal pleura, on the inner wall of the thoracic cavity, and the visceral pleura, attached to the surface of the lungs, are separated only by a thin film of serous fluid. The water molecules in this fluid greatly attract one another, creating a force called **surface tension** that holds the moist surfaces of the pleural membranes tightly together. In addition, any separation of the pleural membranes will decrease pressure in the intrapleural space, also tending to hold these membranes together. Consequently, when the intercostal muscles move the thoracic wall upward and outward, the parietal pleura moves too, and the visceral pleura follows it. This helps expand the lungs in all directions. Chart 19.2 summarizes the steps in inspiration.

The surface tension between the adjacent moist membranes would be sufficient to collapse the alveoli,

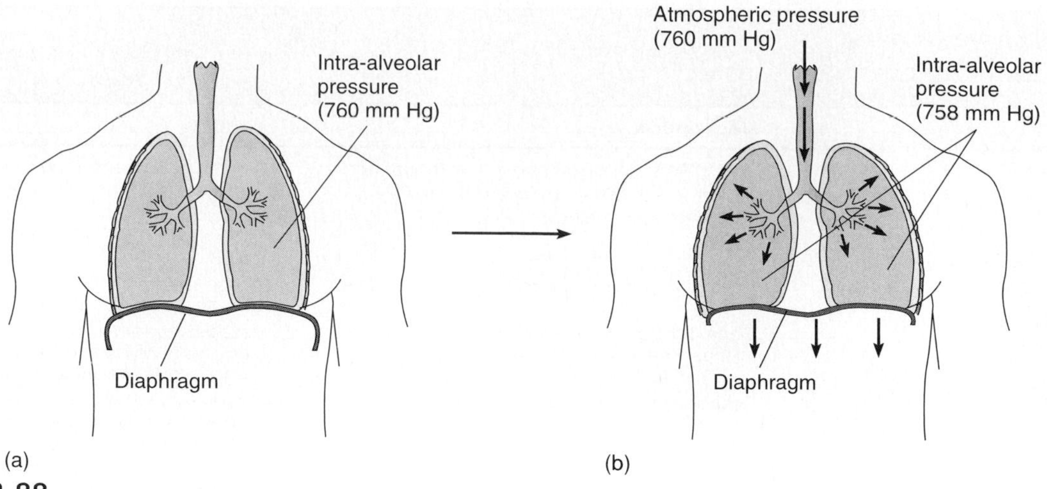


**FIGURE 19.22**

(*a*) Prior to inspiration, the intra-alveolar pressure is 760 mm Hg. (*b*) The intra-alveolar pressure decreases to about 758 mm Hg as the thoracic cavity enlarges, and atmospheric pressure forces air into the airways.

which have moist inner surfaces. However, certain alveolar cells (alveolar type II cells) synthesize a mixture of lipoproteins, called **surfactant** (see figure 19.16). Surfactant, which is secreted continuously into the alveolar air spaces, reduces the surface tension and thus decreases the tendency of the alveoli to collapse when the lung volume is low.

When Benjamin McClatchey was born prematurely in 1990, he weighed only 2 pounds, 13 ounces. Like many of the 380,000 preemies born each year in the United States, Benjamin had *respiratory distress syndrome* (RDS). His lungs were too immature to produce sufficient surfactant, and as a result, they could not overcome the force of surface tension enough to inflate.

Just a decade ago, Benjamin might not have survived RDS. But with the help of a synthetic surfactant dripped into his lungs through an endotracheal tube, and a ventilator machine designed to assist breathing in premature infants, he survived. Unlike conventional ventilators, which force air into the lungs at pressures that could damage delicate newborn lungs, the high-frequency ventilator used on preemies delivers the lifesaving oxygen in tiny, gentle puffs.

If a person needs to take a deeper than normal breath, the diaphragm and external intercostal muscles may contract to an even greater extent. Additional muscles, such as the pectoralis minors and sternocleidomastoids, can also be used to pull the thoracic cage further upward and outward, enlarging the thoracic cavity, decreasing alveolar pressure (fig. 19.23).

The ease with which the lungs can be expanded as a result of pressure changes occurring during breathing is called *compliance* (distensibility). In a normal lung, compliance decreases as lung volume increases, because an inflated lung is more difficult to expand than a lung at rest. Conditions that obstruct air passages, destroy lung tissue, or impede lung expansion in other ways also decrease compliance.

### CHART 19.2 Major Events in Inspiration

1. Nerve impulses travel on phrenic nerves to muscle fibers in the diaphragm, contracting them.
2. As the dome-shaped diaphragm moves downward, the thoracic cavity expands.
3. At the same time, the external intercostal muscles may contract, raising the ribs and increasing the size of the thoracic cavity still more.
4. The intra-alveolar pressure decreases.
5. Atmospheric pressure, which is relatively greater on the outside, forces air into the respiratory tract through the air passages.
6. The lungs fill with air.

## Expiration

The forces responsible for normal resting expiration come from *elastic recoil* of tissues and from surface tension. The lungs and thoracic wall contain a considerable amount of elastic tissue, and lung expansion during inspiration stretches these tissues. When the diaphragm lowers, the abdominal organs beneath it are compressed. As the diaphragm and the external intercostal muscles relax following inspiration, the elastic tissues cause the lungs and thoracic cage to recoil, and they return to their original shapes. Similarly, elastic tissues within the abdominal organs cause them to spring back into their previous shapes, pushing the diaphragm upward. At the same time, surface tension

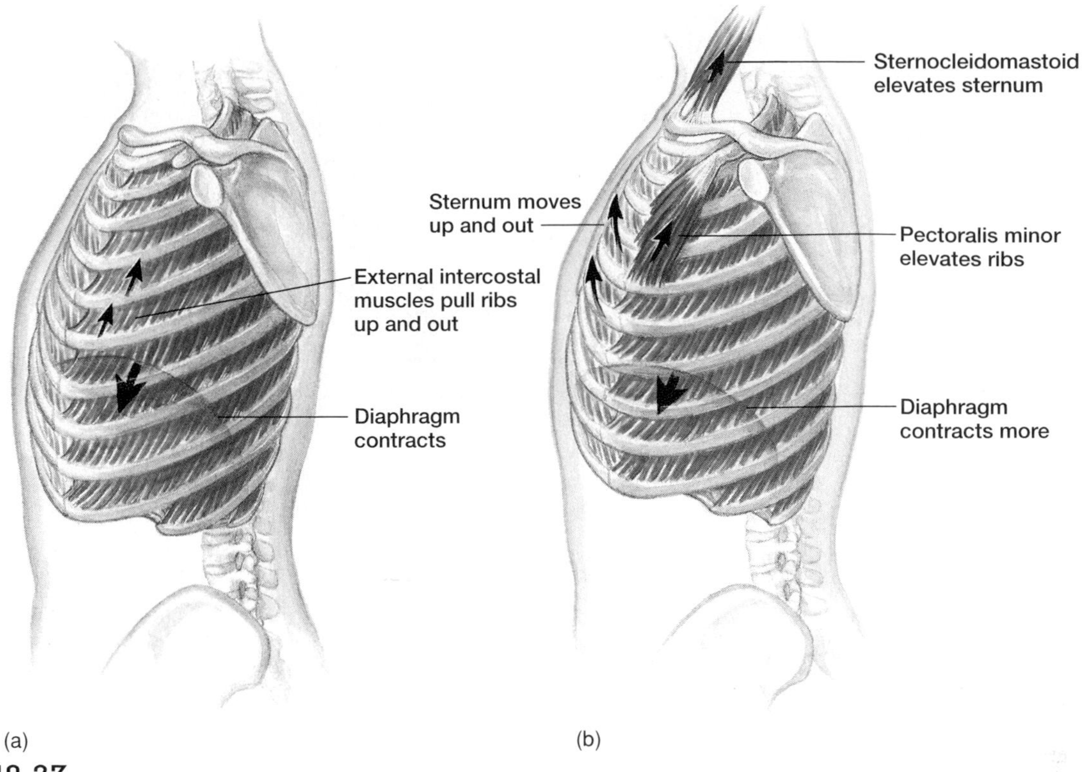


**FIGURE 19.23**

(*a*) Shape of the thorax at the end of normal inspiration. (*b*) Shape of the thorax at the end of maximal inspiration, aided by contraction of the sternocleidomastoid and pectoralis minor muscles.

that develops between the moist surfaces of the alveolar linings tends to shrink alveoli. Each of these factors increases the intra-alveolar pressure about 1 mm Hg above atmospheric pressure, so that the air inside the lungs is forced out through the respiratory passages. Thus, normal resting expiration is a passive process.

The recoil of the elastic fibers within the lung tissues reduces pressure in the pleural cavity. Consequently, the pressure between the pleural membranes (intrapleural pressure) is usually about 4 mm Hg less than atmospheric pressure.

> Because surface tension adheres the visceral and parietal pleural membranes, no significant space normally exists in the pleural cavity between them. However, if the thoracic wall is punctured, atmospheric air may enter the pleural cavity and create a substantial space between the membranes. This condition is called *pneumothorax,* and when it occurs, the lung on the affected side may collapse because of its elasticity.
>
> Pneumothorax may be treated by covering the chest wound with an impermeable bandage, passing a tube (chest tube) through the thoracic wall into the pleural cavity, and applying suction to the tube. In response to the suction, negative pressure is reestablished within the cavity, and the collapsed lung expands.

If a person needs to exhale more air than normal, the posterior *internal* (*expiratory*) *intercostal muscles* can be contracted. These muscles pull the ribs and sternum downward and inward, increasing the pressure in the lungs. Also, the *abdominal wall muscles,* including the external and internal obliques, the transversus abdominis, and the rectus abdominis, can be used to squeeze the abdominal organs inward. Thus, the abdominal wall muscles can increase pressure in the abdominal cavity and force the diaphragm still higher against the lungs, squeezing additional air out of the lungs (fig. 19.24).

Chart 19.3 summarizes the steps in expiration.

1. Describe the events in inspiration.
2. How does surface tension aid in expanding the lungs during inspiration?
3. What forces are responsible for normal expiration?

## Respiratory Volumes and Capacities

Different degrees of effort in breathing move different volumes of air in or out of the lungs. The measurement of such air volumes is called *spirometry,* and it describes four distinct *respiratory volumes.* The amount

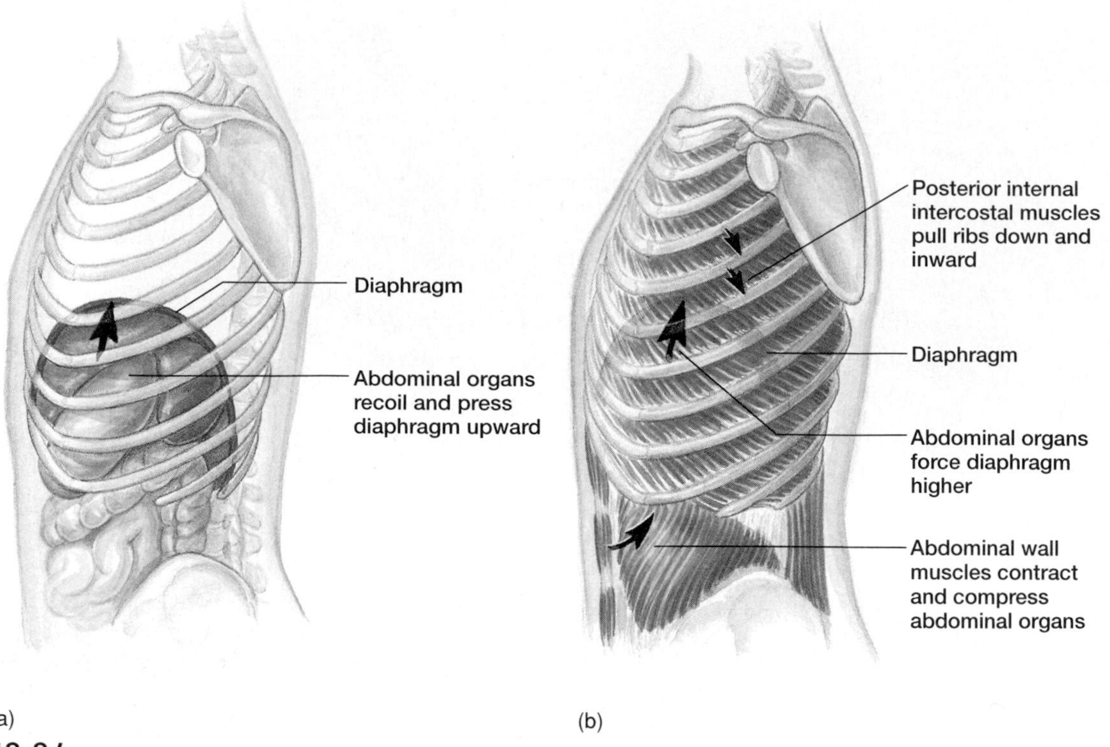


**FIGURE 19.24**

(*a*) Normal resting expiration is due to elastic recoil of the lung tissues, the thoracic wall, and the abdominal organs.
(*b*) Contraction of the abdominal wall muscles and posterior internal intercostal muscles aids maximal expiration.

### CHART 19.3 MAJOR EVENTS IN EXPIRATION

1. The diaphragm and external respiratory muscles relax.
2. Elastic tissues of the lungs and thoracic cage, which were stretched during inspiration, suddenly recoil, and surface tension collapses alveolar walls.
3. Tissues recoiling around the lungs increase the intra-alveolar pressure.
4. Air is squeezed out of the lungs.

of air that enters the lungs during a normal, resting inspiration is about 500 milliliters (ml). On the average, the same amount leaves during a normal resting expiration. One inspiration plus the following expiration is called a **respiratory cycle.** This amount of air that enters or leaves during a respiratory cycle is termed the **tidal volume** (fig. 19.25).

During forced maximal inspiration, a quantity of air in addition to the resting tidal volume enters the lungs. This additional volume is called the **inspiratory reserve volume** (complemental air) and it equals about 3,000 ml.

During a maximal forced expiration, about 1,100 ml of air in addition to the resting tidal volume can be expelled from the lungs. This quantity is called the **expiratory reserve volume** (supplemental air). However, even after the most forceful expiration, about 1,200 ml of air remains in the lungs. This is called the **residual volume.**

Residual air remains in the lungs at all times, and consequently, newly inhaled air is always mixed with air that is already in the lungs. This prevents the oxygen and carbon dioxide concentrations in the lungs from fluctuating greatly with each breath.

Once the respiratory volumes are known, four *respiratory capacities* can be calculated by combining two or more of the volumes. Thus, if the inspiratory reserve volume (3,000 ml) is combined with the tidal volume (500 ml) and the expiratory reserve volume (1,100 ml) the total is termed the **vital capacity** (4,600 ml). This capacity is the maximum amount of air a person can exhale after taking the deepest breath possible.

The tidal volume (500 ml) plus the inspiratory reserve volume (3,000 ml) gives the **inspiratory capacity** (3,500 ml), which is the maximum volume of air a person can inhale following a resting expiration. Similarly, the expiratory reserve volume (1,100 ml) plus the residual volume (1,200 ml) equals the **functional residual capacity** (2,300 ml), which is the volume of air that remains in the lungs following a resting expiration.

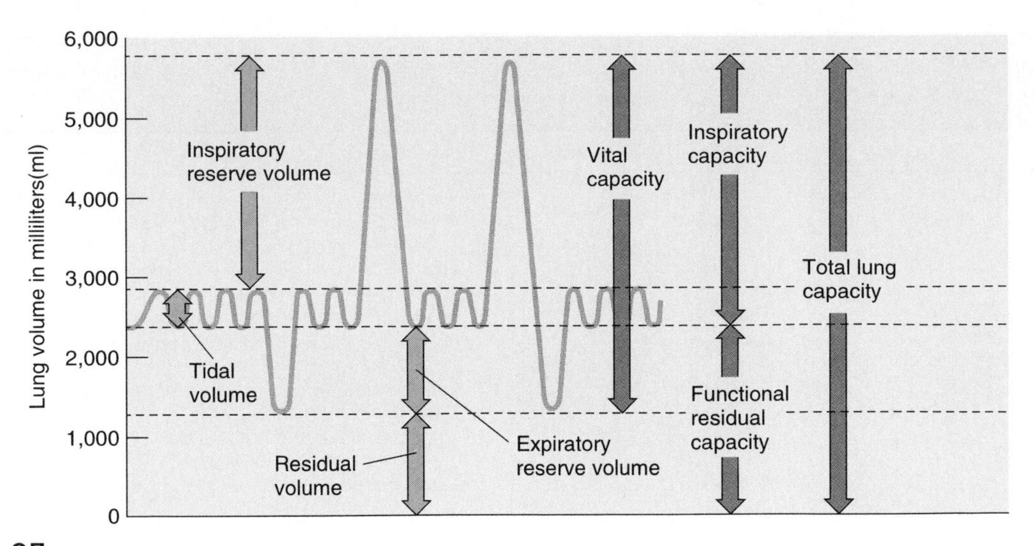


**FIGURE 19.25**
Respiratory volumes and capacities.

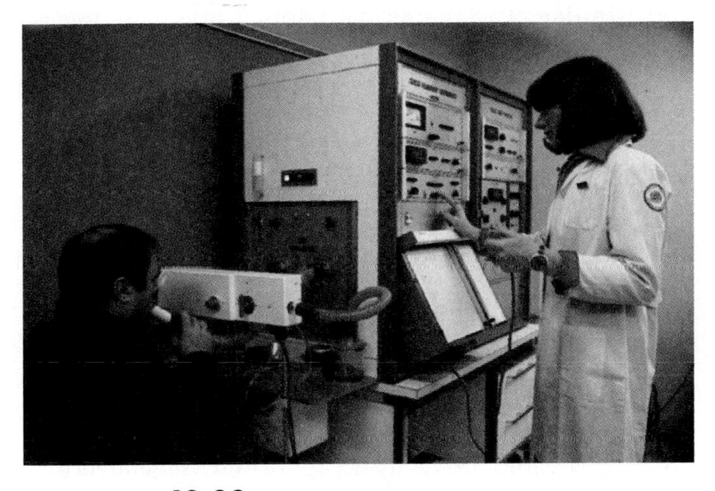

**FIGURE 19.26**
A spirometer can be used to measure respiratory air volumes.

The vital capacity plus the residual volume equals the **total lung capacity** (about 5,800 ml). This total varies with age, sex, and body size.

Some of the air that enters the respiratory tract during breathing fails to reach the alveoli. This volume (about 150 ml) remains in the passageways of the trachea, bronchi, and bronchioles. Since gas exchanges do not occur through the walls of these passages, this air is said to occupy *anatomic dead space.*

Occasionally, air sacs in some regions of the lungs are nonfunctional due to poor blood flow in the adjacent capillaries. This creates *alveolar dead space.* The anatomic and alveolar dead space volumes combined equal *physiologic dead space.* In a normal lung the anatomic and physiologic dead spaces are essentially the same (about 150 ml).

A spirometer (fig. 19.26) is used to measure respiratory air volumes (except the residual volume). These measurements are used to evaluate the course of respiratory illnesses, such as emphysema, pneumonia, lung cancer, and bronchial asthma.

Chart 19.4 summarizes the respiratory air volumes and capacities.

1. What is tidal volume?
2. Distinguish between inspiratory and expiratory reserve volumes.
3. How is vital capacity measured?
4. How is the total lung capacity calculated?

## Alveolar Ventilation

The amount of new atmospheric air that is moved into the respiratory passages each minute is called the *minute ventilation.* It equals the tidal volume multiplied by the breathing rate. Thus, if the tidal volume is 500 ml and the breathing rate is 12 breaths per minute, the minute respiratory volume is 500 ml × 12, or 6,000 ml per minute. However, much of the new air remains in the physiologic dead space.

The volume of new air that does reach the alveoli and is available for gas exchange is calculated by subtracting the physiologic dead space (150 ml) from the tidal volume (500 ml). The resulting volume (350 ml) multiplied by the breathing rate (12 breaths per minute), is the *alveolar ventilation rate* (4,200 ml per minute). The alveolar ventilation rate is a major factor affecting the concentrations of oxygen and carbon dioxide in the alveoli.

## CHART 19.4 Respiratory Air Volumes and Capacities

| Name | Volume (Average) | Description |
|---|---|---|
| Tidal volume (TV) | 500 ml | Volume moved in or out of the lungs during a respiratory cycle |
| Inspiratory reserve volume (IRV) | 3,000 ml | Volume that can be inhaled during forced breathing in addition to tidal volume |
| Expiratory reserve volume (ERV) | 1,100 ml | Volume that can be exhaled during forced breathing in addition to tidal volume |
| Residual volume (RV) | 1,200 ml | Volume that remains in the lungs at all times |
| Inspiratory capacity (IC) | 3,500 ml | Maximum volume of air that can be inhaled following exhalation of tidal volume: IC = TV + IRV |
| Functional residual capacity (FRC) | 2,300 ml | Volume of air that remains in the lungs following exhalation of tidal volume: FRC = ERV + RV |
| Vital capacity (VC) | 4,600 ml | Maximum volume of air that can be exhaled after taking the deepest breath possible: VC = TV + IRV + ERV |
| Total lung capacity (TLC) | 5,800 ml | Total volume of air that the lungs can hold: TLC = VC + RV |

### Nonrespiratory Air Movements

Air movements that occur in addition to breathing are called *nonrespiratory movements.* They are used to clear air passages, as in coughing and sneezing, or to express emotional feelings, as in laughing and crying.

Nonrespiratory movements usually result from *reflexes,* although sometimes they are initiated voluntarily. A cough, for example, can be produced through conscious effort or may be triggered by the presence of a foreign object in an air passage.

The act of *coughing* involves taking a deep breath, closing the glottis, and forcing air upward from the lungs against the closure. Then the glottis is suddenly opened, and a blast of air is forced upward from the lower respiratory tract. Usually this rapid rush of air will remove the substance that triggered the reflex.

> The most sensitive areas of the air passages are in the larynx and in regions near the branches of the major bronchi. The distal portions of the bronchioles (respiratory bronchioles), alveolar ducts, and alveoli lack a nerve supply. Consequently, before any material in these parts can trigger a cough reflex, it must be moved into the larger passages of the respiratory tract.

A *sneeze* is much like a cough, but it clears the upper respiratory passages rather than the lower ones. This reflex is usually initiated by a mild irritation in the lining of the nasal cavity, and, in response, a blast of air is forced up through the glottis. This time, the air is directed into the nasal passages by depressing the uvula, thus closing the opening between the pharynx and the oral cavity.

*Laughing* involves taking a breath and releasing it in a series of short expirations. *Crying* consists of very similar movements, and sometimes it is necessary to note a person's facial expression in order to distinguish laughing from crying.

A *hiccup* is caused by sudden inspiration due to a spasmodic contraction of the diaphragm while the glottis is closed. Air striking the vocal folds causes the sound of the hiccup. We do not know the function, if any, of hiccups.

*Yawning* may aid respiration by providing an occasional deep breath. During normal, quiet breathing, not all of the alveoli are ventilated and some blood may pass through the lungs without becoming well oxygenated. This low blood oxygen concentration somehow triggers the yawn reflex, prompting a very deep breath that ventilates a larger proportion of the alveoli.

Chart 19.5 summarizes the characteristics of the nonrespiratory air movements. Clinical Application 19.2 discusses respiratory problems that affect ventilation.

1. How is the minute ventilation calculated? The alveolar ventilation rate?
2. What nonrespiratory air movements help clear the air passages?
3. What nonrespiratory air movements are used to express emotions?
4. What seems to be the function of a yawn?

## Control of Breathing

Normal breathing is a rhythmic, involuntary act that continues when a person is unconscious.

## CHART 19.5 NONRESPIRATORY AIR MOVEMENTS

| Air Movement | Mechanism | Function |
|---|---|---|
| Coughing | Deep breath is taken, glottis is closed, and air is forced against the closure; suddenly the glottis is opened and a blast of air passes upward | Clears lower respiratory passages |
| Sneezing | Same as coughing, except air moving upward is directed into the nasal cavity by depressing the uvula | Clears upper respiratory passages |
| Laughing | Deep breath is released in a series of short expirations | Expresses happiness |
| Crying | Same as laughing | Expresses sadness |
| Hiccuping | Diaphragm contracts spasmodically while glottis is closed | No useful function known |
| Yawning | Deep breath is taken | Ventilates a larger proportion of the alveoli and aids oxygenation of the blood |
| Speech | Air is forced through the larynx, causing vocal cords to vibrate; words are formed by lips, tongue, and soft palate | Vocal communication |

### Respiratory Center

Groups of neurons in the brain stem comprise the **respiratory center,** which controls breathing. This center periodically initiates impulses that travel on cranial and spinal nerves to breathing muscles, causing inspiration and expiration. The respiratory center also adjusts the rate and depth of breathing to meet cellular needs for a supply of oxygen and removal of carbon dioxide, even during strenuous physical exercise.

The components of the respiratory center are widely scattered throughout the pons and medulla oblongata. However, two areas of the respiratory center are of special interest. They are the rhythmicity area of the medulla and the pneumotaxic area of the pons (fig. 19.27).

The **medullary rhythmicity area** includes two groups of neurons that extend throughout the length of the medulla oblongata. They are called the dorsal respiratory group and the ventral respiratory group.

The *dorsal respiratory group* is responsible for the basic rhythm of breathing. The neurons of this group emit bursts of impulses that signal the diaphragm and other inspiratory muscles to contract. The impulses of each burst begin weakly, increase in strength for about two seconds, and cease abruptly. The breathing muscles that contract in response to the impulses cause the volume of air entering the lungs to increase steadily. The neurons remain inactive while expiration occurs passively, and then they emit another burst of inspiratory impulses so that the inspiration–expiration cycle repeats.

The *ventral respiratory group* is quiescent during normal breathing. However, when more forceful breathing is necessary, the neurons in this group generate impulses that increase inspiratory movement. Other neurons in the group activate the muscles associated with forceful expiration as well.

Each year, 10,000 babies in the United States die because they stop breathing in their sleep, a condition called *sleep apnea.* Babies who have difficulty breathing just after birth are often sent home with monitors, which sound an alarm when the child stops breathing, alerting parents to resuscitate the infant. The position in which the baby sleeps seems to affect the risk of sleep apnea—sleeping on the back or side is safest.

Six million adults in the United States may have sleep apnea, although many of them don't know it. Each night, they cease breathing for 10 to 20 seconds, hundreds of times. People who sleep with them may be aware of the problem because the frequent cessation in breathing causes snoring. The greatest danger of adult sleep apnea is the fatigue, headache, depression, and drowsiness that the person experiences during waking hours.

Sleep apnea is diagnosed in a sleep lab, where breathing during slumber is monitored. There is a treatment for sleep apnea, but many people find it too uncomfortable to use consistently. Called *nasal continuous positive airway pressure,* the treatment is a device strapped onto the nose at night that regulates the amount of air sent into and out of the respiratory system.

The neurons in the **pneumotaxic area** transmit impulses to the dorsal respiratory group continuously and regulate the duration of inspiratory bursts

CLINICAL APPLICATION 

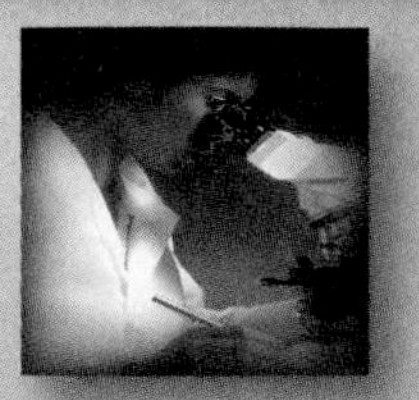

## Respiratory Disorders that Decrease Ventilation

Injuries to the respiratory center or to spinal nerve tracts that transmit motor impulses may paralyze breathing muscles. Paralysis may also be due to a disease, such as *poliomyelitis,* that affects parts of the central nervous system and injures motor neurons. The consequences of such paralysis depend on which muscles are affected. Sometimes, by increasing their responses, other muscles are able to compensate for functional losses of a paralyzed muscle. Otherwise, mechanical ventilation is necessary.

*Bronchial asthma* is usually an allergic reaction to foreign antigens in the respiratory tract, such as plant pollen that enters with inhaled air. The walls of the small bronchioles become edematous, the cells lining the respiratory tubes secrete abnormally large amounts of thick mucus, and the smooth muscles in these tubes contract, constricting the bronchioles and reducing the diameters of the air passages. Breathing becomes increasingly difficult, and it produces a characteristic wheezing sound as air moves through narrowed passages.

A person with asthma usually finds it harder to force air out of the lungs than to bring it in. This is because inspiration utilizes powerful breathing muscles, and, as they contract, the lungs expand, opening the air passages. Expiration, on the other hand, is a passive process due to elastic recoil of stretched tissues. Also, it compresses the tissues and constricts the bronchioles, further impairing air movement through the narrowed air passages.

*Emphysema* is a progressive, degenerative disease that destroys many alveolar walls. As a result, clusters of small air sacs merge to form larger chambers, decreasing the total surface area of the alveolar walls. At the same time, the alveolar walls lose their elasticity, and the capillary networks associated with the alveoli become less abundant (fig. 19C).

Because of the loss of tissue elasticity, a person with emphysema finds it increasingly difficult to force air out of the lungs. Abnormal muscular efforts are required to produce the elastic recoil of inflated tissues needed for expiration. Exposure to respiratory irritants, such as those in tobacco smoke and polluted air, is believed to cause emphysema.

(a)

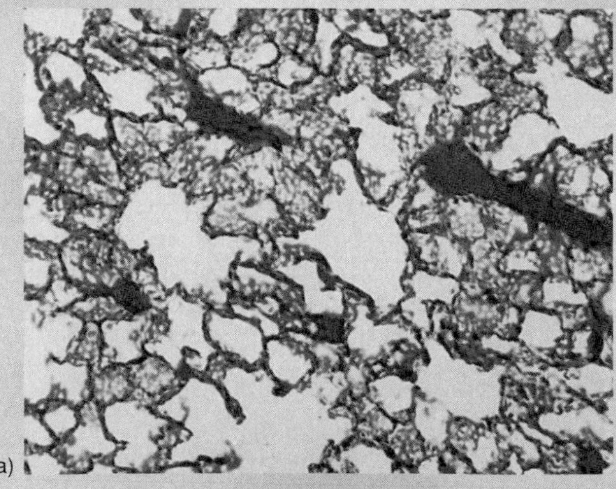

(b)

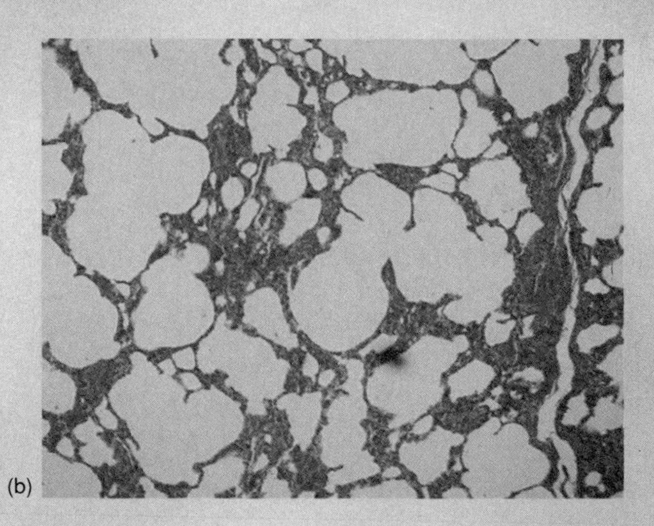

**FIGURE 19C**

(*a*) Normal lung tissue (200×). (*b*) As emphysema develops, the alveoli tend to merge, forming larger chambers (200×).

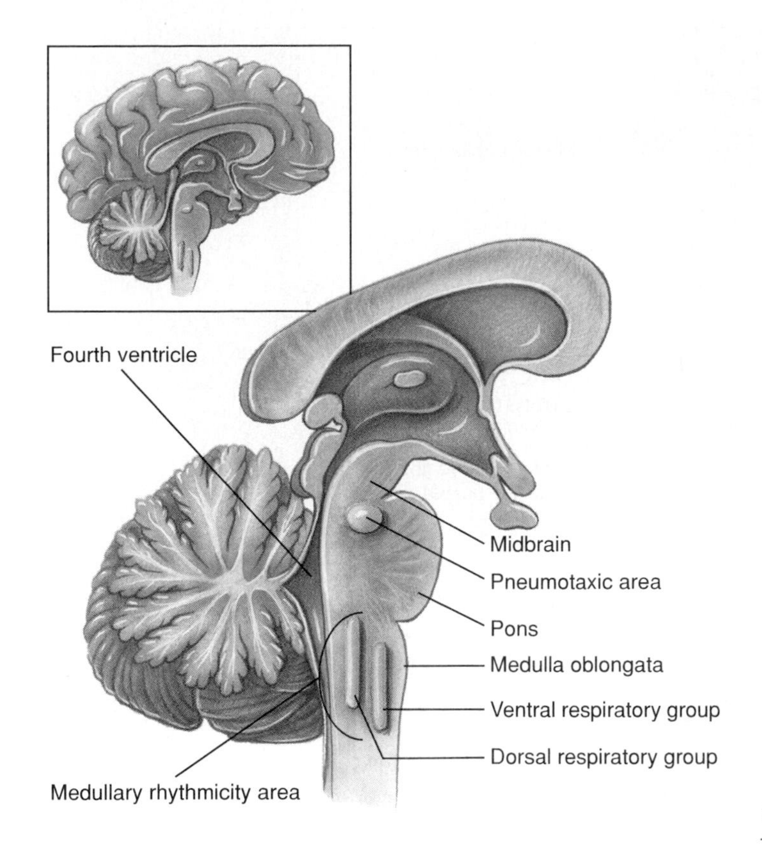


**FIGURE 19.27**
The respiratory center is located in the pons and the medulla oblongata.

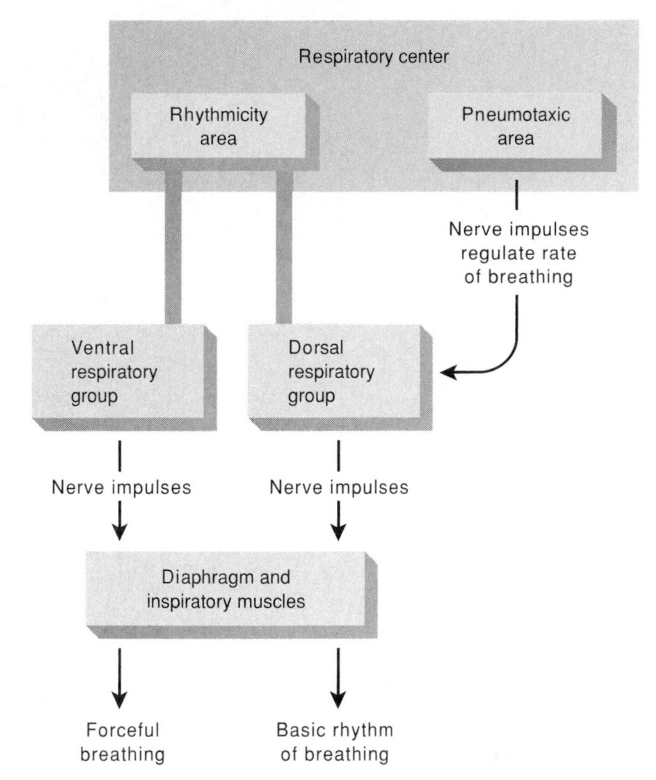


**FIGURE 19.28**
The medullary rhythmicity and pneumotaxic areas of the respiratory center control breathing.

originating from the dorsal group. In this way the pneumotaxic neurons control the rate of breathing. More specifically, when the pneumotaxic signals are strong, the inspiratory bursts are shorter, and the rate of breathing increases; when the pneumotaxic signals are weak, the inspiratory bursts are longer, and the rate of breathing decreases (fig. 19.28).

1. Where is the respiratory center located?
2. Describe how the respiratory center maintains a normal breathing pattern.
3. Explain how the breathing pattern may be changed.

## Factors Affecting Breathing

Factors in addition to the respiratory center influence breathing rate and depth. These include certain chemicals in body fluids, the degree to which lung tissues are stretched, and emotional state. For example, *chemosensitive areas* within the respiratory center, located in the ventral portion of the medulla oblongata near the origins of the vagus nerves, are very sensitive to changes in the blood concentration of carbon dioxide and hydrogen ions. If carbon dioxide or hydrogen ion concentration rises, the chemosensitive areas signal the respiratory center, and breathing rate increases.

The similarity of the effects of carbon dioxide and hydrogen ions is related to the fact that carbon dioxide combines with water in the cerebrospinal fluid to form carbonic acid ($H_2CO_3$):

$$CO_2 + H_2O \rightarrow H_2CO_3$$

The carbonic acid thus formed soon becomes ionized, releasing hydrogen ions ($H^+$) and bicarbonate ions ($HCO_3^-$):

$$H_2CO_3 \rightarrow H^+ + HCO_3^-$$

It is the presence of hydrogen ions rather than the carbon dioxide that influences the chemosensitive areas. In any event, breathing rate and tidal volume increase when a person inhales air rich in carbon dioxide or when body cells produce excess carbon dioxide or hydrogen ions. These changes increase alveolar ventilation. As a result, more carbon dioxide is exhaled, and the blood concentrations of carbon dioxide and hydrogen ions return toward normal.

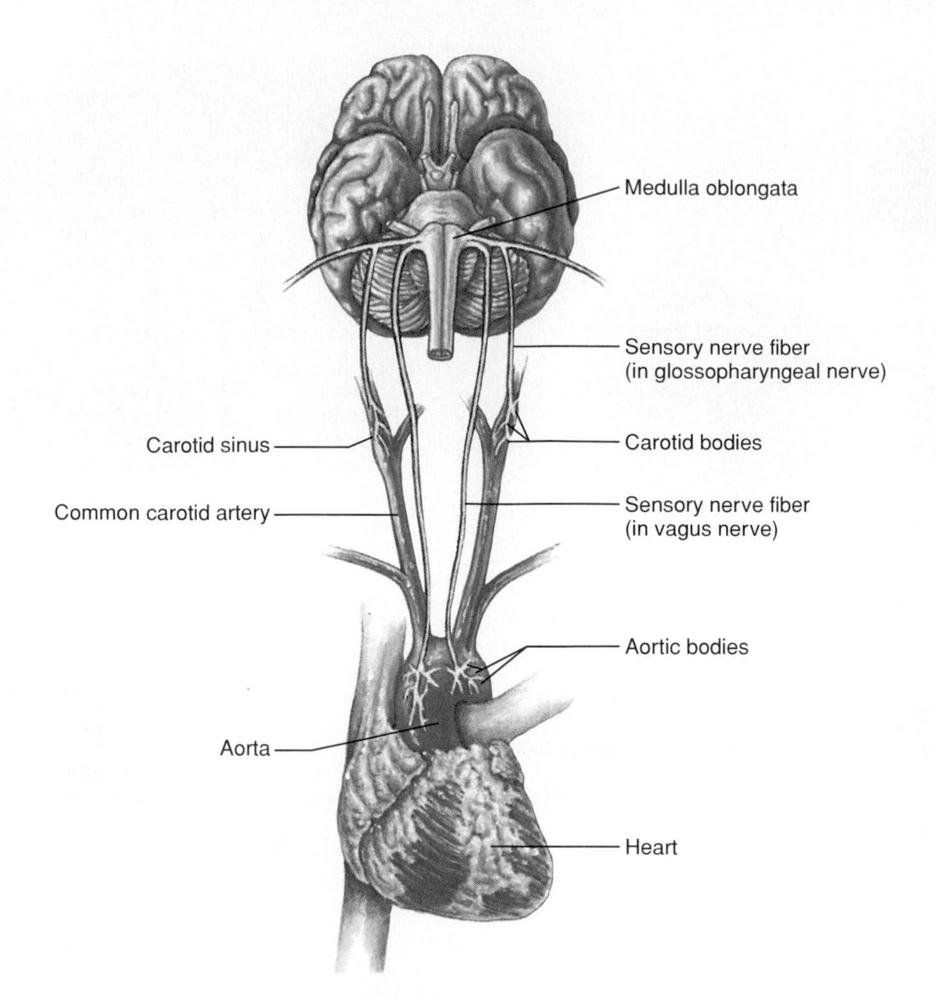


**FIGURE 19.29**
Decreased blood oxygen concentration stimulates chemoreceptors in the carotid and aortic bodies.

> Adding carbon dioxide to air can stimulate the rate and depth of breathing. Ordinary air is about 0.04% carbon dioxide. If a patient inhales air containing 4% carbon dioxide, breathing rate usually doubles.

Low blood oxygen concentration seems to have little direct effect on the chemosensitive areas associated with the respiratory center. Instead, changes in the blood oxygen concentration are sensed by *chemoreceptors* in specialized structures called the *carotid bodies* and *aortic bodies,* which are located in the walls of the carotid arteries and aorta (fig. 19.29). When decreased oxygen concentration stimulates these receptors, impulses are transmitted to the respiratory center, and the breathing rate and tidal volume increase, thus increasing alveolar ventilation. This mechanism is usually not triggered until the blood oxygen concentration reaches a very low level; thus, oxygen seems to play only a minor role in the control of normal respiration.

The chemoreceptors of the carotid and aortic bodies are also stimulated by changes in the blood concentrations of carbon dioxide and hydrogen ions. However, these substances have a much more powerful effect when they act on the chemosensitive areas of the respiratory center. The effects of carbon dioxide and hydrogen ions on the carotid and aortic bodies are relatively unimportant.

> An exception to the normal pattern of chemical control may occur in patients who have chronic obstructive pulmonary diseases (COPD), such as asthma, bronchitis, and emphysema. Over a period of time, these patients seem to adapt to high concentrations of carbon dioxide, and for them, low oxygen concentrations may serve as a necessary respiratory stimulus. Paradoxically, when such a patient is placed on 100% oxygen, breathing may stop.

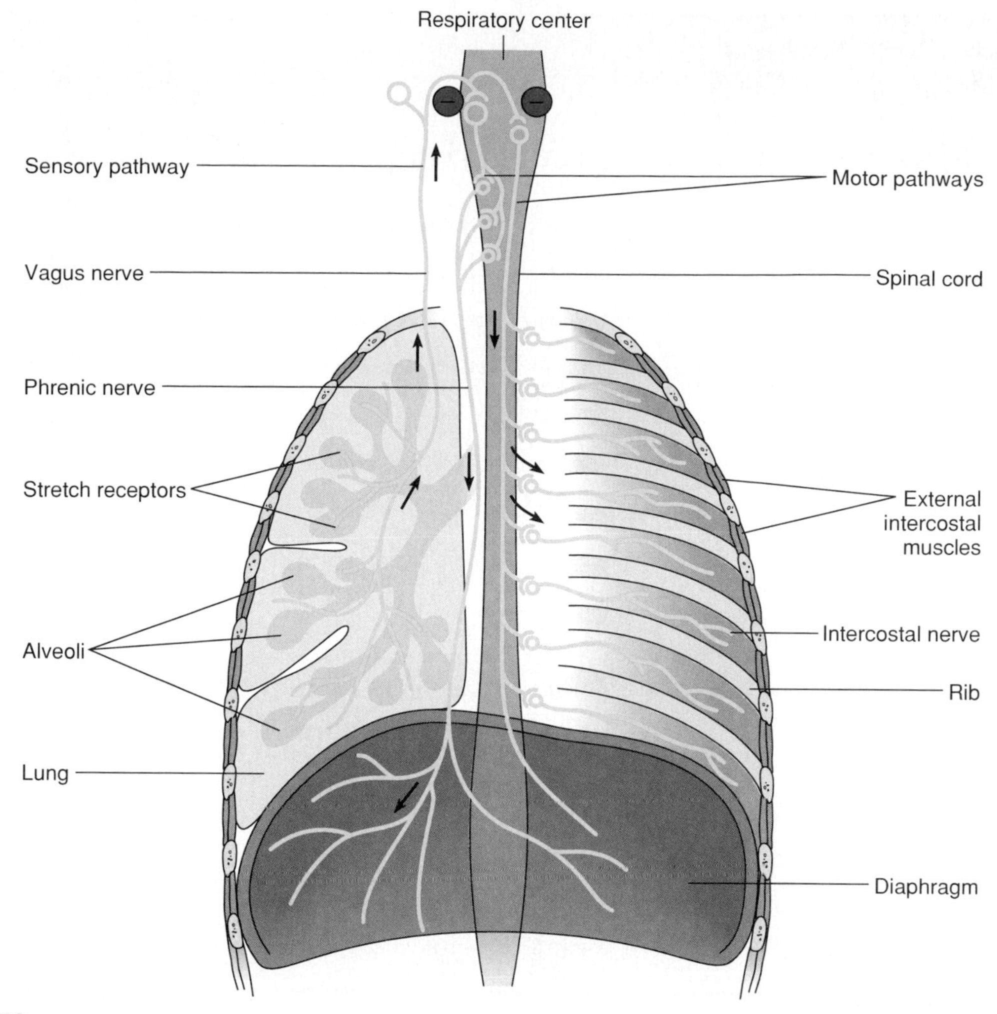


**FIGURE 19.30**

In the process of inspiration, motor impulses travel from the respiratory center to the diaphragm and external intercostal muscles, which contract and cause the lungs to expand. This expansion stimulates stretch receptors in the lungs to send inhibiting impulses to the respiratory center, thus preventing overinflation.

An *inflation reflex* (Hering-Breuer reflex) helps regulate the depth of breathing. This reflex occurs when stretch receptors in the visceral pleura, bronchioles, and alveoli are stimulated as lung tissues are stretched. The sensory impulses of the reflex travel via the vagus nerves to the pneumotaxic area of the respiratory center, and shorten the duration of inspiratory movements. This action prevents overinflation of the lungs during forceful breathing (fig. 19.30).

Emotional upset or strong sensory stimulation may alter the normal breathing pattern. Gasping and rapid breathing are familiar responses to fear, anger, shock, excitement, horror, surprise, sexual stimulation, or even the chill of stepping into a cold shower. Because control of the respiratory muscles is voluntary, we can alter breathing pattern consciously, or even stop it altogether for a short time. During childbirth, for example, women often concentrate on controlling their breathing, which distracts them from the pain.

If a person decides to stop breathing, the blood concentrations of carbon dioxide and hydrogen ions begin to rise, and the concentration of oxygen falls. These changes (primarily the increased $CO_2$) stimulate the respiratory center, and soon the need to inhale overpowers the desire to hold the breath—much to the relief of parents when young children threaten to hold their breaths until they turn blue! On the other hand, a person can increase the breath-holding time by breathing rapidly and deeply in advance. This action, termed **hyperventilation,** lowers the blood carbon dioxide concentration. Following hyperventilation, it takes longer than usual for the carbon dioxide concentration to reach the level needed to override the conscious effort of breath holding.

Chart 19.6 discusses factors affecting breathing. Clinical Application 19.3 focuses on one influence on breathing—exercise.

CLINICAL APPLICATION

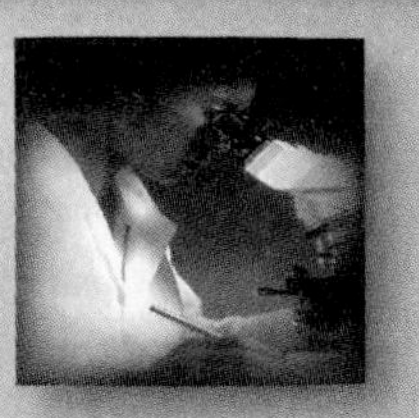

## Exercise and Breathing

Moderate to heavy physical exercise greatly increases the amount of oxygen skeletal muscles use. For example, a young man at rest will utilize about 250 milliliters of oxygen per minute, but he may require 3,600 milliliters per minute during maximal exercise. While oxygen utilization is increasing, the volume of carbon dioxide produced increases also. Since decreased blood oxygen and increased blood carbon dioxide concentration stimulate the respiratory center, it is not surprising that exercise is accompanied by an increased breathing rate. However, studies have revealed that blood oxygen and carbon dioxide concentrations usually remain nearly unchanged during exercise—a reflection of the respiratory system's effectiveness in obtaining oxygen and releasing carbon dioxide to the outside.

What other factors might increase breathing rate during exercise? The cerebral cortex and the proprioceptors associated with muscles and joints seem to be involved (see chapter 12). Specifically, the cortex transmits stimulating impulses to the respiratory center whenever it signals the skeletal muscles to contract. At the same time, muscular movements stimulate the proprioceptors, triggering a *joint reflex*. In this reflex, sensory impulses are transmitted from the proprioceptors to the respiratory center, and breathing rate increases.

The increase in breathing rate during exercise requires an increase in blood flow to meet the needs of skeletal muscles. Thus, physical exercise increases demand on both the respiratory and the circulatory systems. If either of these systems fails to keep up with cellular demands, the person will begin to feel out of breath. This sensation however, is usually due to the inability of the heart and circulatory system to move enough blood between the lungs and the body cells, rather than to the inability of the respiratory system to provide enough air.

CHART 19.6 FACTORS AFFECTING BREATHING

| Factor | Receptors Stimulated | Response | Effect |
|---|---|---|---|
| Stretch of tissues | Stretch receptors in visceral pleura, bronchioles, and alveoli | Inhibits inspiration | Prevents overinflation of lungs during forceful breathing |
| Low blood oxygen | Chemoreceptors in carotid and aortic bodies | Increases alveolar ventilation | Increases blood oxygen concentration |
| High blood carbon dioxide | Chemosensitive areas of the respiratory center | Increases alveolar ventilation | Decreases blood carbon dioxide concentration and CSF hydrogen ion concentration |
| High CSF hydrogen ion | Chemosensitive areas of the respiratory center | Increases breathing rate and alveolar ventilation | Decreases blood carbon dioxide concentration and CSF hydrogen ion concentration |

Sometimes a person who is emotionally upset may hyperventilate, become dizzy, and lose consciousness. This is due to a lowered carbon dioxide concentration followed by a rise in pH (respiratory alkalosis) and a localized vasoconstriction of cerebral arterioles, decreasing blood flow to nearby brain cells. Hampered oxygen supply to the brain causes fainting.

Hyperventilation should never be used to help hold the breath while swimming, because the person may lose consciousness under water and drown.

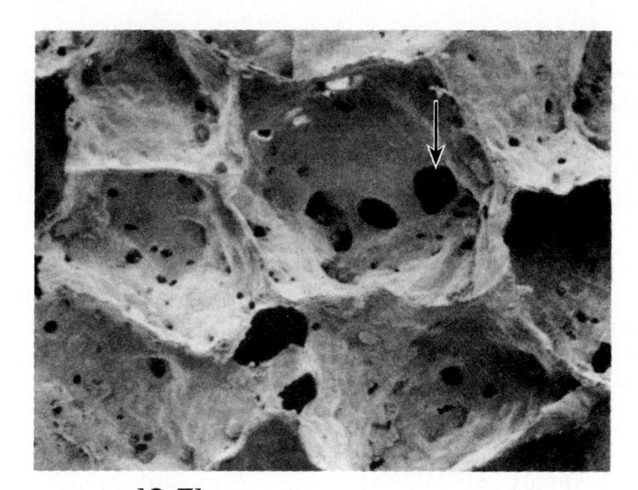

**FIGURE 19.31**
Alveolar pores (arrow) allow air to pass from one alveolus to another (300×).

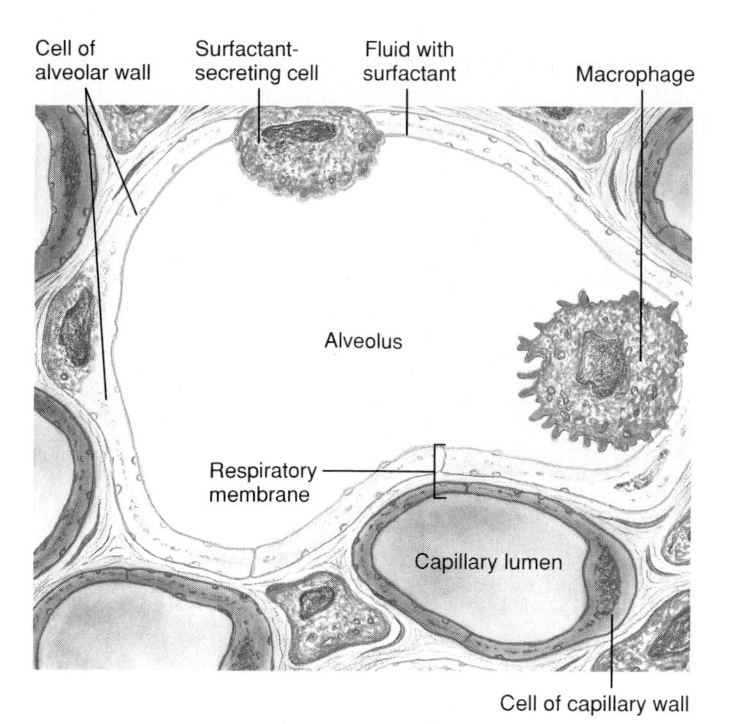


**FIGURE 19.32**
The respiratory membrane consists of the walls of the alveolus and the capillary.

1. Describe the inflation reflex.
2. What chemical factors affect breathing?
3. How does hyperventilation decrease respiratory rate?

## ALVEOLAR GAS EXCHANGES

While other parts of the respiratory system conduct air and move it in and out of the air passages, the alveoli carry on the vital process of exchanging gases between the air and the blood.

### Alveoli

The **alveoli** are microscopic air sacs clustered at the distal ends of the finest respiratory tubes—the alveolar ducts. Each alveolus consists of a tiny space surrounded by a thin wall that separates it from adjacent alveoli. Minute openings, called **alveolar pores,** in the walls of some alveoli, may permit air to pass from one alveolus to another (fig. 19.31). This arrangement provides alternate air pathways if the passages in some portions of the lung become obstructed.

Phagocytic cells called *alveolar macrophages* are found in alveoli and in the pores connecting the air sacs. These macrophages phagocytize airborne agents, including bacteria, thereby keeping the alveoli clean.

### Respiratory Membrane

The wall of an alveolus consists of an inner lining of simple squamous epithelium (type I cells, see fig. 19.16) and a dense network of capillaries, which are also lined with simple squamous epithelial cells (fig. 19.32). Thin basement membranes separate the layers of these flattened cells, and in the spaces between them are elastic and collagenous fibers that help support the alveolar wall. At least two thicknesses of epithelial cells and basement membranes separate the air in an alveolus and the blood in a capillary. These layers make up the **respiratory membrane** (alveolar-capillary membrane), through which gas exchange occurs between the alveolar air and the blood (figs. 19.33 and 19.34).

### Diffusion Through the Respiratory Membrane

Recall from chapter 3 that molecules diffuse from regions where they are in higher concentration toward regions where they are in lower concentration. In the case of gases, it is more useful to think of diffusion occurring from regions of higher pressure toward regions of lower pressure. The pressure of a gas determines the rate at which it will diffuse from one region to another.

Measured by volume, ordinary air is about 78% nitrogen, 21% oxygen, and 0.04% carbon dioxide. Air also contains small amounts of other gases that have little or no physiological importance.

In a mixture of gases, such as air, each gas is responsible for a portion of the total pressure the mixture produces. The amount of pressure each gas contributes is called the **partial pressure,** and this pressure is directly related to the concentration of the gas in the mixture. For example, because air is 21% oxygen, this gas is responsible for 21% of the atmospheric pressure. Because 21% of 760 mm Hg is equal to 160 mm Hg, the partial pressure of oxygen, symbolized $PO_2$, in atmospheric air is said to be 160 mm Hg. Similarly, the partial pressure of carbon dioxide ($PCO_2$) in air is 0.3 mm Hg.

When a mixture of gases dissolves in the blood, each gas exerts its own partial pressure in proportion to its dissolved concentration. Furthermore, each gas will diffuse between the blood and its surroundings from areas of higher partial pressure to areas of lower partial pressure. As figure 19.34 shows, this movement will tend to equalize partial pressures in the two regions.

For example, the $P_{CO_2}$ in capillary blood is 45 mm Hg, but the $P_{CO_2}$ in alveolar air is 40 mm Hg. As a consequence of the difference between these partial pressures, carbon dioxide diffuses from the blood, where its partial pressure is higher, through the respiratory membrane, and into the alveolar air. When the blood leaves the lungs, its $P_{CO_2}$ has equilibrated at 40 mm Hg, which is the same as the $P_{CO_2}$ of the alveolar air.

Similarly, the $P_{O_2}$ of capillary blood is 40 mm Hg, but that of alveolar air is 104 mm Hg. Thus, oxygen diffuses from the alveolar air into the blood, and the blood leaves the lungs with a $P_{O_2}$ of 104 mm Hg.

Clinical Application 19.4 examines illnesses that result from impaired gas exchange.

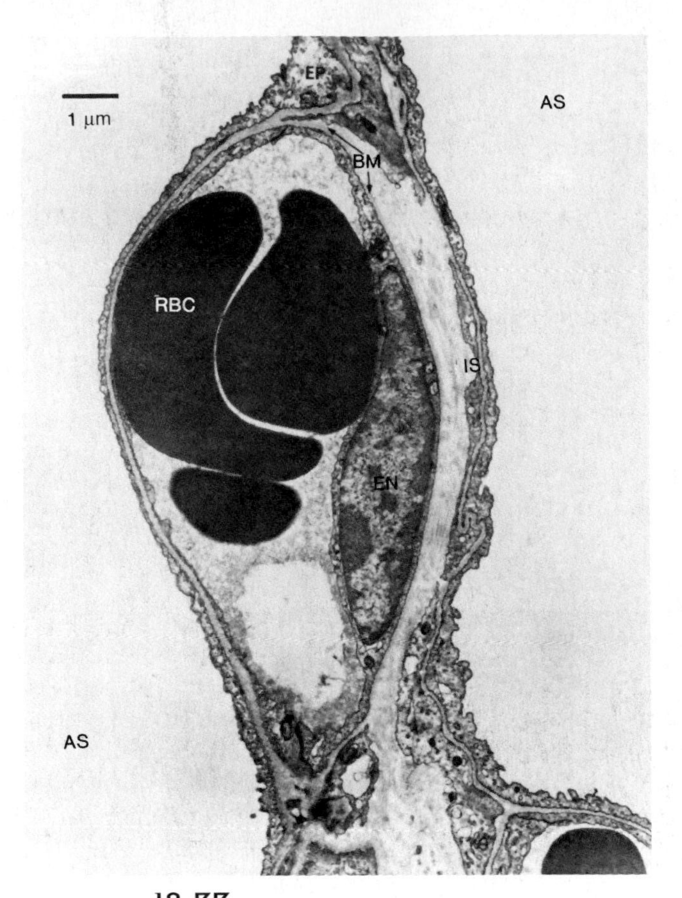


**FIGURE 19.33**
Electron micrograph of a capillary located between two adjacent alveoli. (AS, alveolar space; RBC, red blood cell; BM, basement membrane; IS, interstitial connective tissue; EN, epithelial nucleus.)

1. Describe the structure of the respiratory membrane.
2. What is partial pressure?
3. What causes oxygen and carbon dioxide to move across the respiratory membrane?

> Exposure to high oxygen concentration (hyperoxia) for a prolonged time may damage lung tissues, particularly those that form the capillary walls. As a result of such damage, excess fluid may escape and flood the alveolar air spaces, interfering with gas exchange. This may lead to death. Similarly, hyperoxia can damage the retinal capillaries of premature infants causing *retrolental fibroplasia* (RLF), a condition that may lead to blindness.

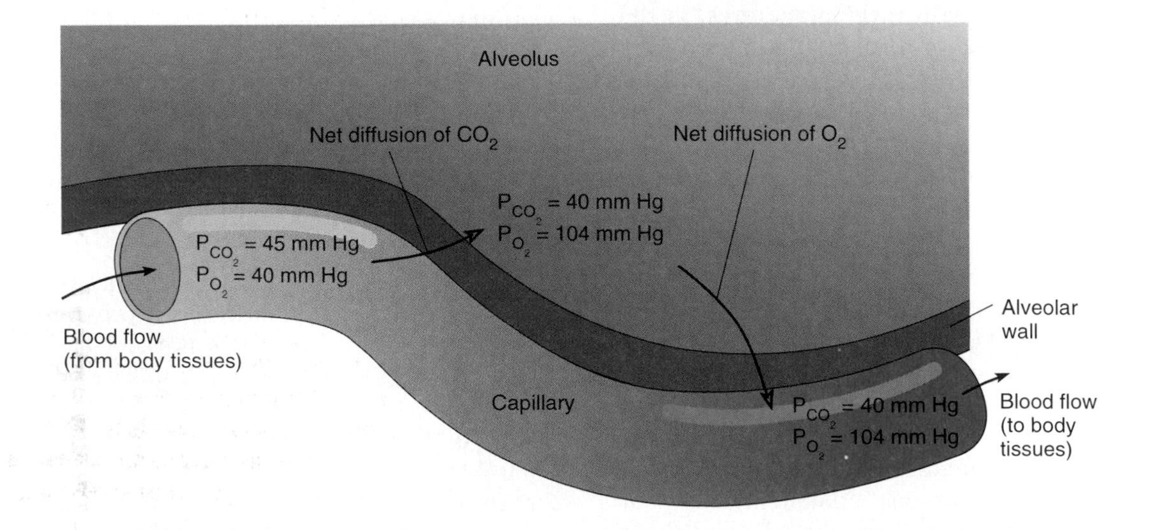


**FIGURE 19.34**
Gas exchanges occur between the air of the alveolus and the blood of the capillary as a result of differences in partial pressures.

## GAS TRANSPORT

The blood transports oxygen and carbon dioxide between the lungs and the body cells. As these gases enter the blood, they dissolve in the liquid portion, the plasma, or combine chemically with other atoms or molecules.

### Oxygen Transport

Almost all the oxygen (over 98%) is carried in the blood bound to the protein **hemoglobin** in red blood cells. The iron in hemoglobin provides the color of these blood cells. The remainder of the oxygen is dissolved in the blood plasma.

Hemoglobin consists of two types of components called *heme* and *globin* (see chapter 14). Globin is a protein that contains 574 amino acids arranged in four polypeptide chains. Each chain is associated with a heme group, and each heme group contains an atom of iron. Each iron atom can combine loosely with an oxygen molecule. As oxygen dissolves in blood, it combines rapidly with hemoglobin, forming a new compound called **oxyhemoglobin.** Each hemoglobin molecule can combine with a maximum of four oxygen atoms.

The $Po_2$ determines the amount of oxygen that combines with hemoglobin. The greater the $Po_2$, the more oxygen will combine with hemoglobin, until the hemoglobin molecules are saturated with oxygen (fig. 19.35).

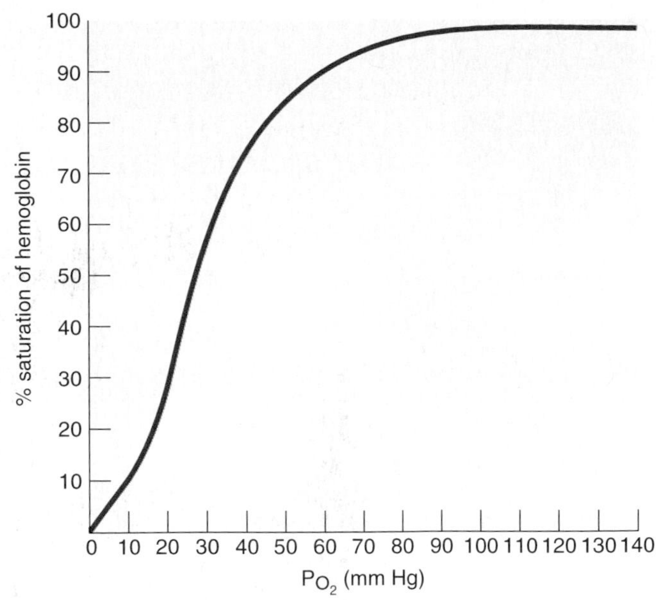


**FIGURE 19.35**

Hemoglobin is completely saturated at normal systemic arterial $Po_2$, but releases oxygen readily at the $Po_2$ of the body tissues.

> Each year, about 100,000 mountain climbers and high-altitude exercisers experience varying degrees of altitude sickness. This is because at high altitudes, the proportion of oxygen in air remains the same (about 21%), but the $Po_2$ decreases. Consequently, if a person breathes high altitude air, oxygen diffuses less rapidly from the alveoli into the blood, and the hemoglobin is less likely to become saturated with oxygen. The body's efforts to get more oxygen—increased breathing and heart rate and enhanced red blood cell and hemoglobin production—cannot keep pace with the plummeting amount of oxygen available.
>
> Such a person may develop the symptoms of oxygen deficiency (hypoxia). For example, at 8,000 feet a person may experience anxiety, restlessness, an increased breathing rate, and a rapid pulse. At 12,500 feet, where the $Po_2$ is only 100 mm Hg rather than 160 mm Hg at sea level, the symptoms may include drowsiness, mental fatigue, headache, and nausea. Above 13,000 feet, the brain swells (cerebral edema), producing severe headache, vomiting, loss of coordination, and hallucinations. Finally, the person loses consciousness and may die at about 23,000 feet.

The chemical bonds that form between oxygen and hemoglobin molecules are relatively unstable, and as the $Po_2$ decreases, oxygen is released from oxyhemoglobin molecules (fig. 19.35). This happens in tissues, where cells have used oxygen in their respiratory processes and the free oxygen diffuses from the blood into nearby cells, as figure 19.36 shows.

Increasing blood concentration of carbon dioxide ($Pco_2$), acidity, and temperature all increase the amount of oxygen that oxyhemoglobin releases (figs. 19.37, 19.38, and 19.39). These influences explain why more oxygen is released from the blood to the skeletal muscles during periods of exercise. The increased muscular activity accompanied by increased oxygen use increases the $Pco_2$, decreases the pH, and raises the local temperature. At the same time, less active cells receive less oxygen.

### Carbon Monoxide

**Carbon monoxide** (CO) is a toxic gas produced in gasoline engines and some stoves as a result of incomplete combustion of fuels. It is also a component of tobacco smoke. The toxic effect of carbon monoxide occurs because it combines with hemoglobin more effectively than does oxygen. Furthermore, carbon monoxide does not dissociate readily from hemoglobin. Thus, when a person breathes carbon monoxide, less hemoglobin is available for oxygen transport, and the body cells soon begin to suffer from oxygen deficiency. The effects of carbon monoxide on hemoglobin may be responsible for the lower average birth weights of infants born to women who smoked while pregnant.

Clinical Application 19.4

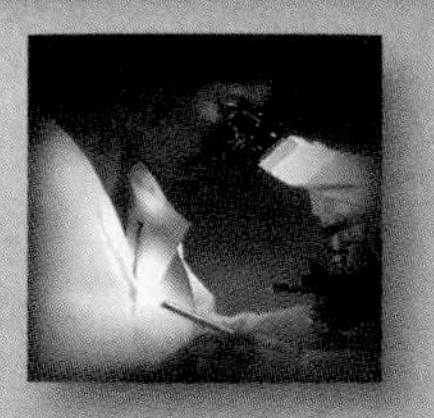

## Disorders Impairing Gas Exchange

Five-year-old Carly had what her parents at first thought was just a "bug" that was passing through the family. But after 12 hours of flu-like symptoms, Carly's temperature shot up to 105°F, her chest began to hurt, and her breathing became rapid and shallow. Later that day, a chest X ray confirmed what the doctor suspected—Carly had *pneumonia.* Apparently, the bacteria that had caused a mild upper respiratory infection in her parents and sisters had taken a detour in her body, infecting her lower respiratory structures instead.

Carly's bacterial pneumonia was successfully treated with antibiotics. Pneumonia can also be caused by a viral infection, or, as is often the case in people with AIDS, by *Pneumocystis carinii* infection. Whatever the cause, the events within the infected lung are similar: alveolar linings swell with edema and become abnormally permeable, allowing fluids and white blood cells to accumulate in the air sacs. As the alveoli fill, the surface area available for gas exchange diminishes. Breathing becomes difficult. Untreated, pneumonia can kill.

*Tuberculosis* is a different type of lung infection, caused by the bacterium *Mycobacterium tuberculosis* (fig. 19D). In this condition, fibrous connective tissue develops around the sites of infection, forming structures called *tubercles.* By walling off the bacteria, the tubercles help inhibit their spread. Sometimes this protective mechanism fails, and the bacteria flourish throughout the lungs, and may even spread to other organs. In the later stages of infection, other types of bacteria may cause secondary infections. As lung tissue is destroyed, the surface area for gas exchange decreases. In addition, the widespread fibrous tissue increases the thickness of the respiratory membrane, further restricting gas exchange. A variety of drugs can treat tuberculosis, but in recent years, strains resistant to drugs have arisen, and these can be swiftly deadly.

Another type of condition that impairs gas exchange is *atelectasis.* This is the collapse of a lung, or some part of it, together with the collapse of the blood vessels that supply the affected region. Obstruction of a respiratory tube, such as by an inhaled foreign object or excess mucus secretion, may cause atelectasis. The air in the alveoli beyond the obstruction is absorbed, and as the air pressure in the alveoli decreases, their elastic walls collapse, and they can no longer function. Fortunately, after a portion of a lung collapses, the functional re-

Treatment for carbon monoxide poisoning is to administer oxygen in high concentration to replace some of the carbon monoxide bound to hemoglobin molecules. Carbon dioxide is usually administered simultaneously to stimulate the respiratory center which, in turn, increases breathing rate. Rapid breathing helps reduce the concentration of carbon monoxide in the alveoli.

1. How is oxygen transported from the lungs to the body cells?
2. What factors affect the release of oxygen from oxyhemoglobin?
3. Why is carbon monoxide toxic?

### Carbon Dioxide Transport

Blood flowing through the capillaries of the body tissues gains carbon dioxide because the tissues have a relatively high $PCO_2$. This carbon dioxide is transported to the lungs in one of three forms: as carbon dioxide dissolved in plasma, as part of a compound formed by bonding to hemoglobin, or as part of a bicarbonate ion (fig. 19.40).

The amount of carbon dioxide that dissolves in plasma is determined by its partial pressure. The higher the $PCO_2$ of the tissues, the more carbon dioxide will go into solution. However, only about 7% of the carbon dioxide is transported in this form.

Unlike oxygen, which combines with the iron atoms of hemoglobin molecules, carbon dioxide bonds with the amino groups ($—NH_2$) of these molecules. Consequently, oxygen and carbon dioxide do not compete for binding sites—a hemoglobin molecule can transport both gases at the same time.

gions that remain are often able to carry on enough gas exchange to sustain the body cells.

*Adult respiratory distress syndrome* (ARDS) is a special form of atelectasis in which alveoli collapse. It has a variety of causes, all of which damage lung tissues. These include pneumonia and other infections, near drowning, aspiration of stomach acid into the respiratory system, or physical trauma to the lungs from an injury or surgical procedure. This damage disrupts the respiratory membrane that separates the air in the alveoli from the blood in the pulmonary capillaries, allowing protein-rich fluid to escape from the capillaries and flood the alveoli. They collapse in response, and surfactant is nonfunctional. Blood vessels and airways narrow, greatly elevating blood pressure in the lungs. Delivery of oxygen to tissues is seriously impaired. ARDS is fatal about 60% of the time.

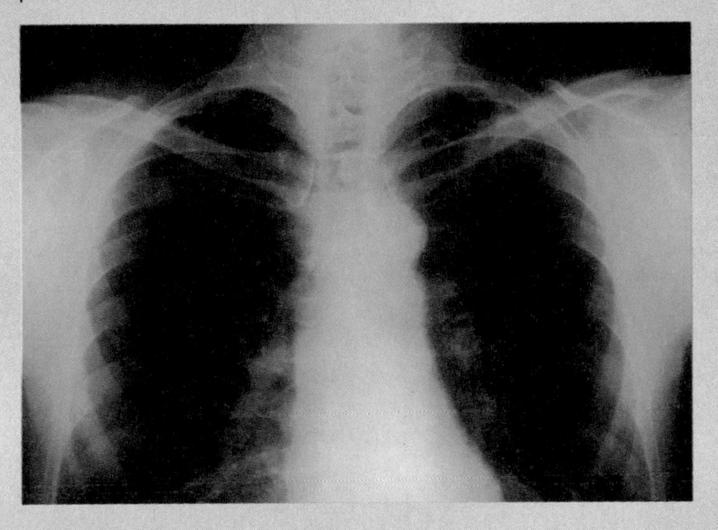

Healthy lungs

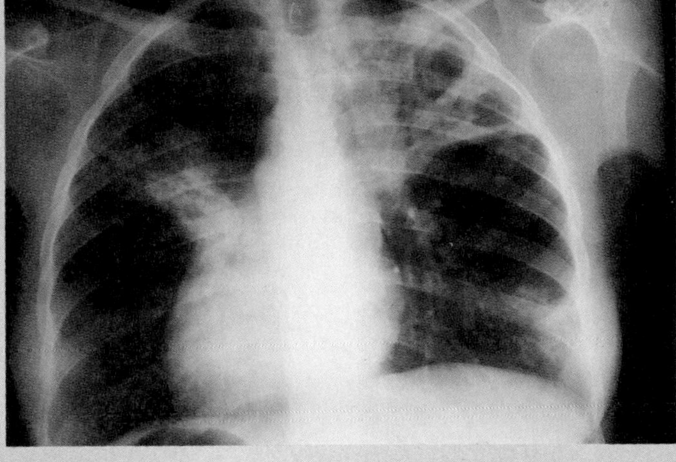

Tuberculosis

**FIGURE 19D**
Healthy lungs appear dark and clear on X-ray film. Lungs with tuberculosis have cloudy areas where fibrous tissue grows to wall off infected areas.

Carbon dioxide combining with hemoglobin forms a loosely bound compound called **carbaminohemoglobin.** This molecule decomposes readily in regions where the $P_{CO_2}$ is low, and thus releases its carbon dioxide. Although this method of transporting carbon dioxide is theoretically quite effective, carbaminohemoglobin forms relatively slowly. Only about 15%–25% of the total carbon dioxide is carried this way.

The most important carbon dioxide transport mechanism involves the formation of **bicarbonate ions** ($HCO_3^-$). Recall that carbon dioxide reacts with water to form carbonic acid ($H_2CO_3$). This reaction occurs slowly in the blood plasma, but much of the carbon dioxide diffuses into the red blood cells. These cells contain an enzyme, **carbonic anhydrase,** which speeds the reaction between carbon dioxide and water.

The resulting carbonic acid dissociates almost immediately, releasing hydrogen ions ($H^+$) and bicarbonate ions ($HCO_3^-$):

$$H_2CO_3 \rightarrow H^+ + HCO_3^-$$

These new hydrogen ions might be expected to lower blood pH, but this reaction occurs in the systemic capillaries, where deoxyhemoglobin is generated. Deoxyhemoglobin is an excellent buffer because hydrogen ions combine readily with it. The bicarbonate ions diffuse out of the red blood cells and enter the blood plasma. At least 70% of the carbon dioxide transported in the blood is carried in this form.

As the bicarbonate ions leave the red blood cells and enter the plasma, *chloride ions,* which also have negative charges, are repelled electrically, and they move

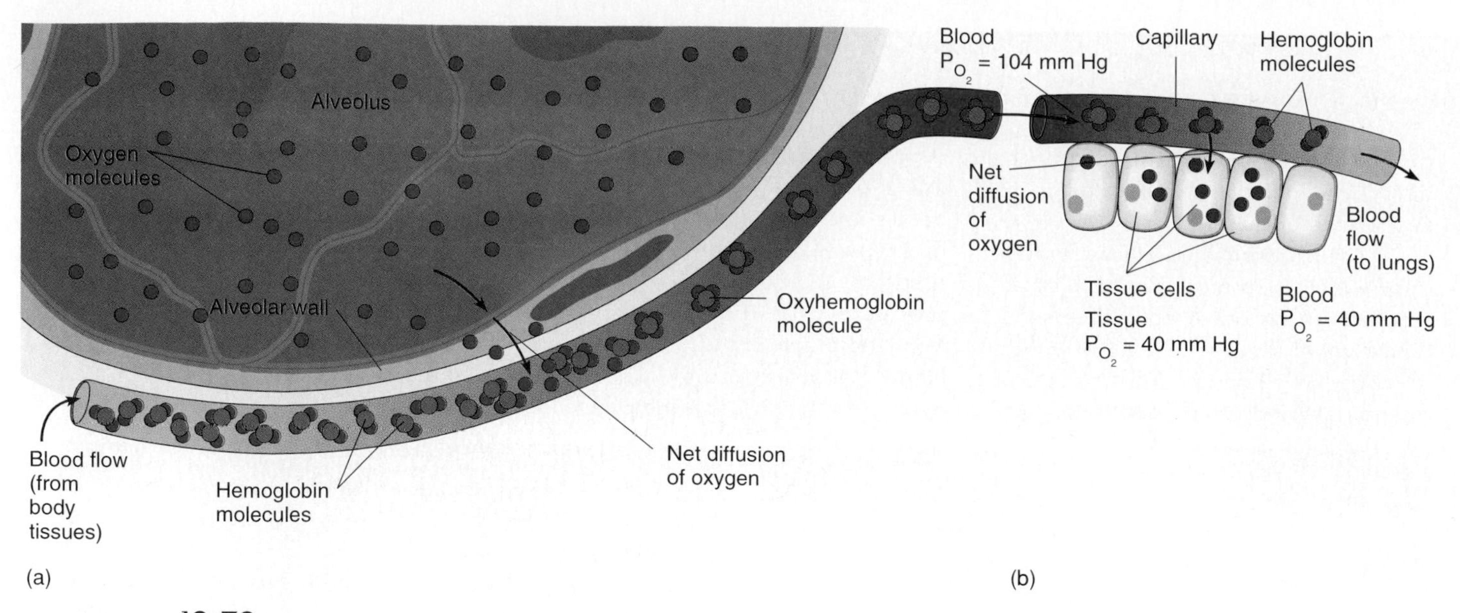


**FIGURE 19.36**

(*a*) Oxygen molecules, entering the blood from the alveolus, bond to hemoglobin, forming oxyhemoglobin. (*b*) In the regions of the body cells, oxyhemoglobin releases oxygen. Note that some oxygen is still bound to hemoglobin at the $P_{O_2}$ of systemic venous blood (see fig. 19.35).

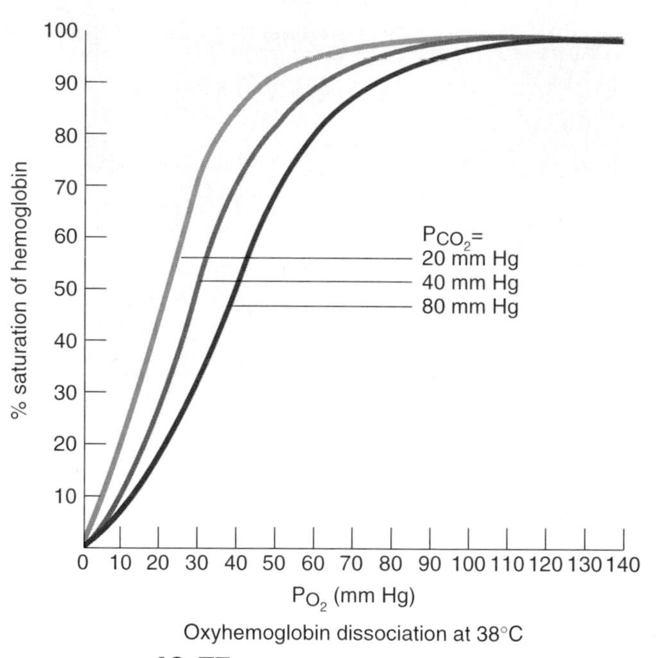


**FIGURE 19.37**

The amount of oxygen released from oxyhemoglobin increases as the $P_{CO_2}$ increases.

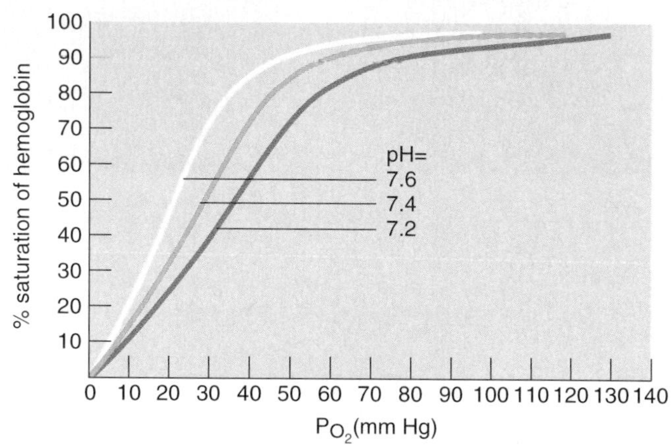


**FIGURE 19.38**

The amount of oxygen released from oxyhemoglobin increases as the blood pH decreases.

from the plasma into the red blood cells. This exchange in position of the two negatively charged ions, shown in figure 19.41, maintains the ionic balance between the red blood cells and the plasma. It is termed the **chloride shift.**

When blood passes through the capillaries of the lungs, its dissolved carbon dioxide diffuses into the alveoli, in response to the relatively low $P_{CO_2}$ of the alveolar air. As the plasma $P_{CO_2}$ drops, hydrogen ions and bicarbonate ions in the red blood cells recombine to form carbonic acid, and under the influence of carbonic anhydrase, the carbonic acid quickly yields new carbon dioxide and water:

$$H^+ + HCO_3^- \rightarrow H_2CO_3 \rightarrow CO_2 + H_2O$$

Carbaminohemoglobin also releases its carbon dioxide, and both of these events contribute to the $P_{CO_2}$ of the alveolar capillary blood. Carbon dioxide diffuses out of the blood until an equilibrium is established between the $P_{CO_2}$ of the blood and the $P_{CO_2}$ of the alveolar air. Figure 19.42 summarizes this process, and chart 19.7 summarizes the transport of blood gases.

1. Describe three ways carbon dioxide can be transported from body cells to the lungs.
2. How can hemoglobin carry oxygen and carbon dioxide at the same time?
3. How do bicarbonate ions help buffer the blood (maintain its pH)?
4. What is meant by the chloride shift?
5. How is carbon dioxide released from the blood into the lungs?

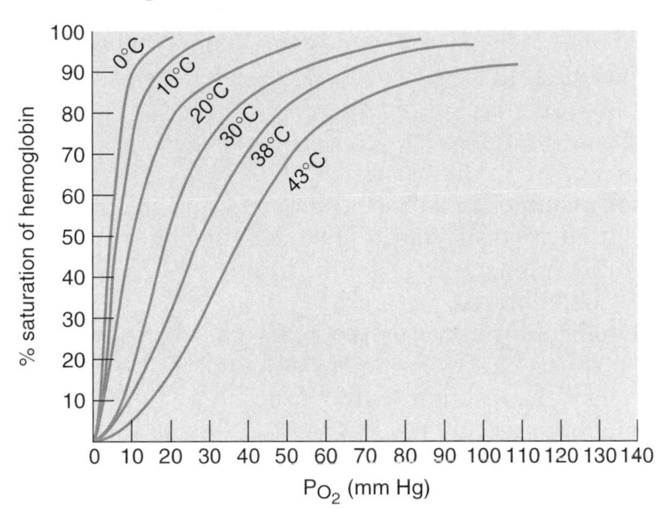


**FIGURE 19.39**
The amount of oxygen released from oxyhemoglobin increases as the blood temperature increases.

## Clinical Terms Related to the Respiratory System

**anoxia** (ah-nok′se-ah) An absence or a deficiency of oxygen within tissues.
**asphyxia** (as-fik′se-ah) Deficiency of oxygen and excess carbon dioxide in the blood and tissues.
**atelectasis** (at″e-lek′tah-sis) The collapse of a lung or some portion of it.
**bradypnea** (brad″e-ne′ah) Abnormally slow breathing.
**bronchitis** (brong-ki′tis) An inflammation of the bronchial lining.
**Cheyne-Stokes respiration** (chān stōks res″pi-ra′shun) Irregular breathing pattern of a series of shallow breaths that increase in depth and rate, followed by breaths that decrease in depth and rate.
**dyspnea** (disp′ne-ah) Difficulty in breathing.
**eupnea** (up-ne′ah) Normal breathing.
**hemothorax** (he″mo-tho′raks) Blood in the pleural cavity.
**hypercapnia** (hi″per-kap′ne-ah) Excess carbon dioxide in the blood.
**hyperoxia** (hi″per-ok′se-ah) Excess oxygenation of the blood.
**hyperpnea** (hi″perp-ne′ah) Increase in the depth and rate of breathing.
**hyperventilation** (hi″per-ven″ti-la′shun) Prolonged, rapid, and deep breathing.

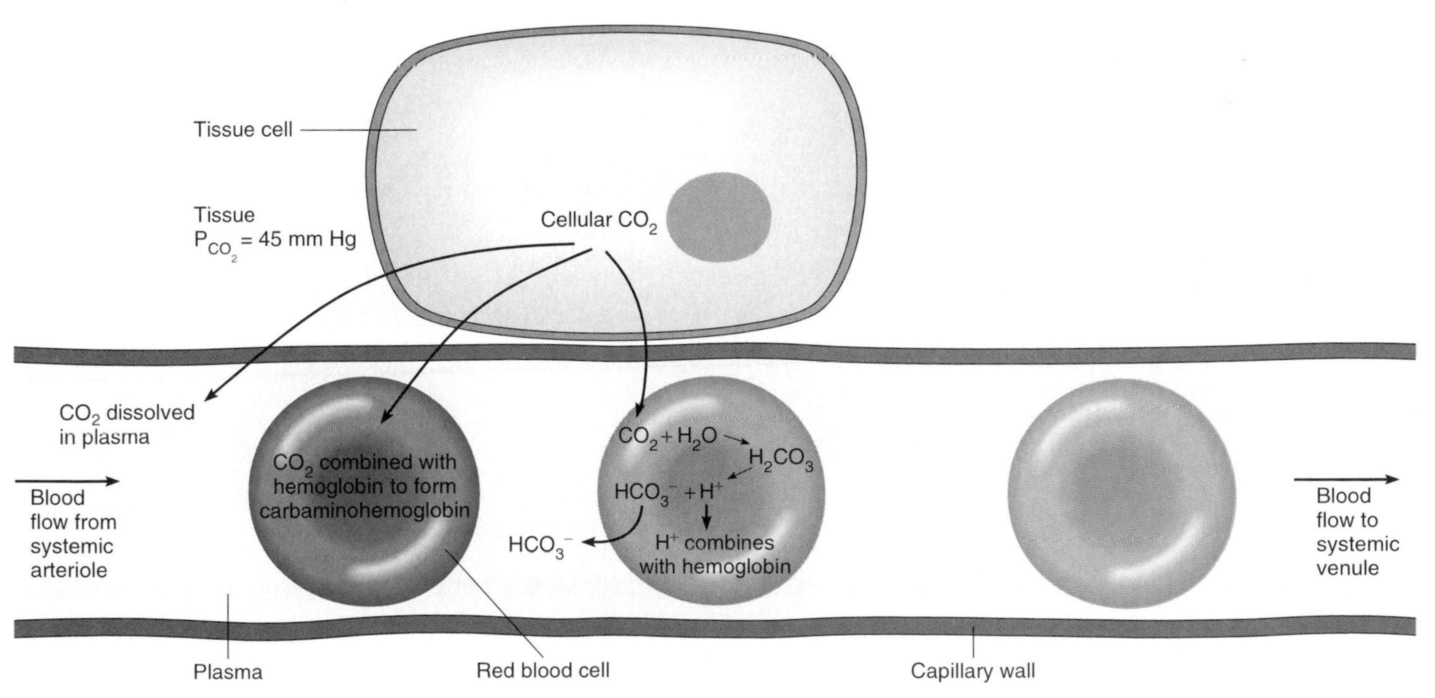


**FIGURE 19.40**
Carbon dioxide produced by tissue cells is transported in the blood in a dissolved state, combined with hemoglobin or in the form of bicarbonate ions ($HCO_3^-$).

## CHART 19.7 GASES TRANSPORTED IN BLOOD

| Gas | Reaction Involved | Substance Transported |
|---|---|---|
| Oxygen | Combines with iron atoms of hemoglobin molecules | Oxyhemoglobin |
| Carbon dioxide | About 7% dissolves in plasma | Carbon dioxide |
| | About 23% combines with the amino groups of hemoglobin molecules | Carbaminohemoglobin |
| | About 70% reacts with water to form carbonic acid; the carbonic acid then dissociates to release hydrogen ions and bicarbonate ions | Bicarbonate ions |

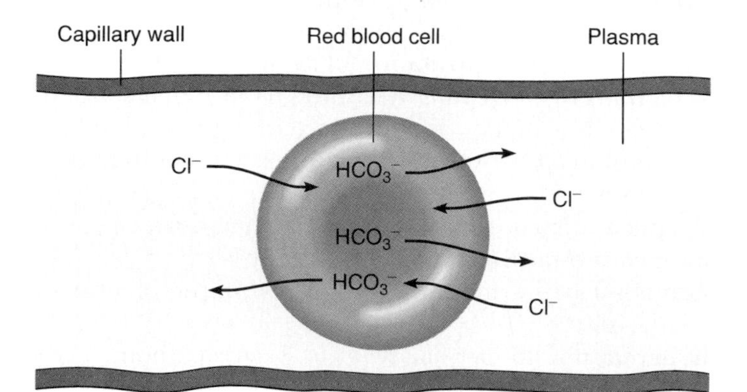


**FIGURE 19.41**

As bicarbonate ions ($HCO_3^-$) diffuse out of the red blood cell, chloride ions ($Cl^-$) from the plasma diffuse into the cell, thus maintaining the electrical balance between ions. This exchange of ions is called the chloride shift.

**hypoxemia** (hi″pok-se′me-ah) A deficiency in the oxygenation of the blood.

**hypoxia** (hi-pok′se-ah) A diminished availability of oxygen in the tissues.

**lobar pneumonia** (lo′ber nu-mo′ne-ah) Pneumonia that affects an entire lobe of a lung.

**pleurisy** (ploo′rĭ-se) An inflammation of the pleural membranes.

**pneumoconiosis** (nu″mo-ko″ne-o′sis) Accumulation of particles from the environment in the lungs and the reaction of the tissues.

**pneumothorax** (nu″mo-tho′raks) The entrance of air into the space between the pleural membranes, followed by collapse of the lung.

**rhinitis** (ri-ni′tis) An inflammation of the nasal cavity lining.

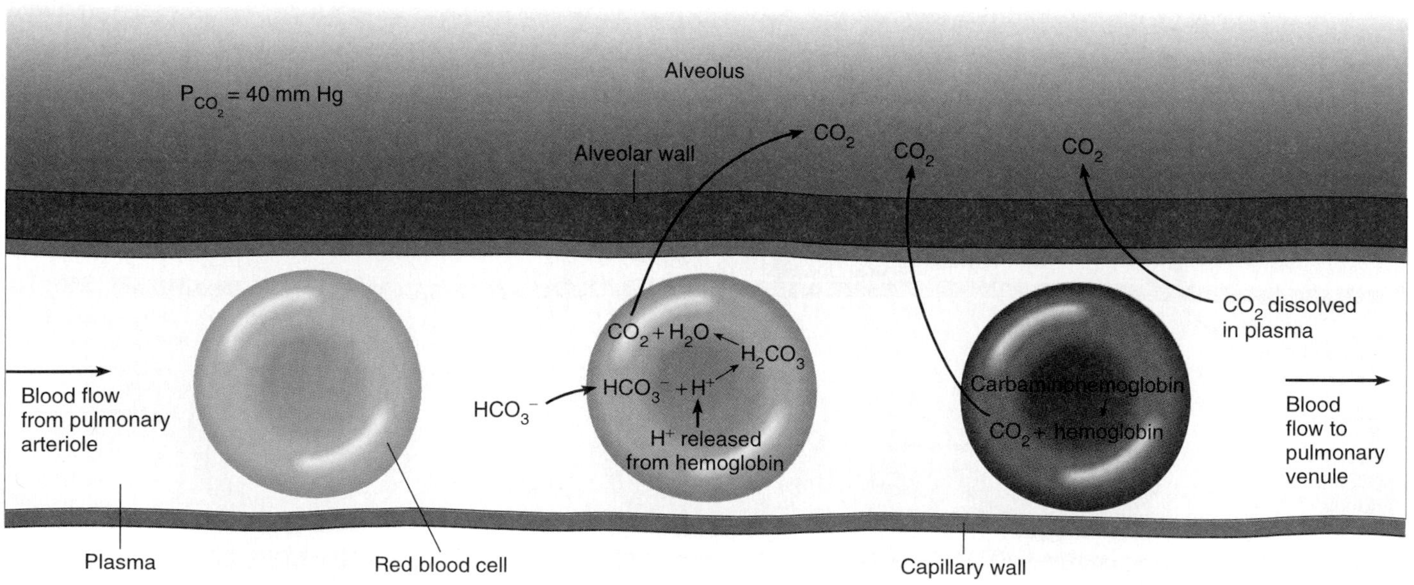


**FIGURE 19.42**

In the lungs, carbon dioxide diffuses from the blood into the alveoli.

# InnerConnections
## Respiratory System

The respiratory system provides oxygen for the internal environment and excretes carbon dioxide

**Reproductive system**

Respiration increases during sexual activity

Fetal "respiration" begins before birth

**Integumentary system**

Stimulation of skin receptors may alter respiratory rate

**Urinary system**

The kidneys and the respiratory system work together to maintain blood pH

The kidneys compensate for water lost through breathing

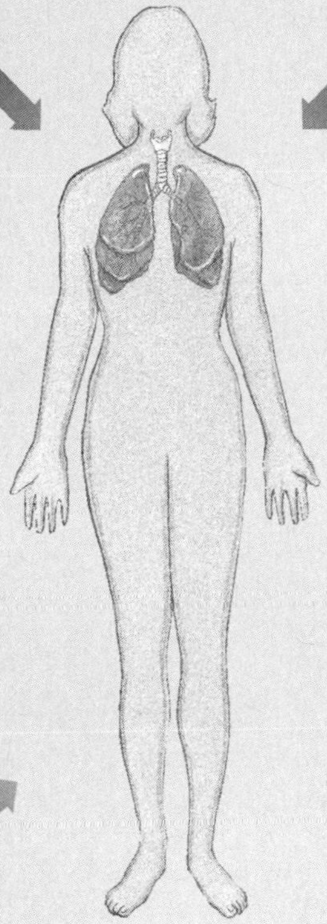

**Skeletal system**

Bones provide attachments for muscles involved in breathing

**Digestive system**

The digestive system and respiratory system share openings to the outside

**Muscular system**

The respiratory system eliminates carbon dioxide produced by exercising muscles

**Lymphatic system**

Cells of the immune system patrol the lungs and defend against infection

**Nervous system**

The brain controls the respiratory system

The respiratory system helps control pH of the internal environment

**Cardiovascular system**

As blood is pumped through the lungs by the heart, the lungs oxygenate the blood and excrete carbon dioxide

**Endocrine system**

Hormone-like substances control the production of red blood cells that transport oxygen and carbon dioxide

**sinusitis** (si″nu-si′tis) An inflammation of the sinus cavity lining.

**tachypnea** (tak″ip-ne′ah) Rapid, shallow breathing.

**tracheotomy** (tra″ke-ot′o-me) An incision in the trachea for exploration or the removal of a foreign object.

## Chapter Summary

### Introduction (page 757)

The respiratory system includes the passages that transport air to and from the lungs, and the air sacs in which gas exchanges occur. Respiration is the entire process by which gases are exchanged between the atmosphere and the body cells. Respiration is necessary because of cellular respiration. Cells require oxygen to extract maximal energy from nutrient molecules, and to rid themselves of carbon dioxide, a metabolic waste.

### Organs of the Respiratory System (page 758)

The respiratory system includes the nose, nasal cavity, sinuses, pharynx, larynx, trachea, bronchial tree, and lungs. The upper respiratory tract includes the respiratory organs outside the thorax; the lower respiratory tract includes those respiratory organs within the thorax.

1. Nose
   a. Bone and cartilage support the nose.
   b. Nostrils provide entrances for air.
2. Nasal cavity
   a. The nasal cavity is a space behind the nose.
   b. The nasal septum divides it medially.
   c. Nasal conchae divide the cavity into passageways and help increase the surface area of the mucous membrane.
   d. Mucous membrane filters, warms, and moistens incoming air.
   e. Particles trapped in the mucus are carried to the pharynx by ciliary action and are swallowed.
3. Sinuses
   a. Sinuses are spaces in the bones of the skull that open into the nasal cavity.
   b. They are lined with mucous membrane that is continuous with the lining of the nasal cavity.
4. Pharynx
   a. The pharynx is located behind the mouth and between the nasal cavity and the larynx.
   b. It provides a common passage for air and food.
   c. It aids in creating vocal sounds.
5. Larynx
   a. The larynx is an enlargement at the top of the trachea.
   b. It is a passageway for air and helps prevent foreign objects from entering the trachea.
   c. It is composed of muscles and cartilages; some of these cartilages are single while others are paired.
   d. It contains the vocal cords, which produce sounds by vibrating as air passes over them.
      (1) The pitch of a sound is related to the tension on the cords.
      (2) The intensity of a sound is related to the force of the air passing over the cords.
   e. The glottis and epiglottis help prevent food and liquid from entering the trachea.
6. Trachea
   a. The trachea extends into the thoracic cavity in front of the esophagus.
   b. It divides into the right and left bronchi.
   c. The mucous lining continues to filter incoming air.
   d. Incomplete cartilaginous rings support the wall.
7. Bronchial tree
   a. The bronchial tree consists of branched air passages that lead through the trachea to the air sacs.
   b. The branches of the bronchial tree include primary bronchi, lobar bronchi, segmental bronchi, intralobular bronchioles, terminal bronchioles, respiratory bronchioles, alveolar ducts, alveolar sacs, and alveoli.
   c. Structure of the respiratory tubes
      (1) As tubes branch, the amount of cartilage in the walls decreases and the muscular layer becomes more prominent.
      (2) Elastic fibers in the walls aid breathing.
      (3) The epithelial lining changes from pseudostratified and ciliated to cuboidal and simple squamous as the tubes become progressively smaller.
   d. Functions of the respiratory tubes include distribution of air and exchange of gases between the alveolar air and the blood.
8. Lungs
   a. The left and right lungs are separated by the mediastinum and are enclosed by the diaphragm and the thoracic cage.
   b. The visceral pleura is attached to the surface of the lungs; parietal pleura lines the thoracic cavity.
   c. The right lung has three lobes and the left lung has two.
   d. Each lobe is composed of lobules that contain alveolar ducts, alveolar sacs, alveoli, nerves, blood vessels, lymphatic vessels, and connective tissues.

### Breathing Mechanism (page 769)

Inspiration and expiration movements are accompanied by changes in the size of the thoracic cavity.

1. Inspiration
   a. Atmospheric pressure forces air into the lungs.
   b. Inspiration occurs when the intra-alveolar pressure is reduced.
   c. The intra-alveolar pressure is reduced when the diaphragm moves downward, and the thoracic cage moves upward and outward.
   d. Surface tension holding the pleural membranes together aids lung expansion.
   e. Surfactant reduces surface tension within the alveoli.
2. Expiration
   a. The forces of expiration come from the elastic recoil of tissues and from surface tension within the alveoli.
   b. Expiration can be aided by thoracic and abdominal wall muscles that pull the thoracic cage downward and inward, and compress the abdominal organs inward and upward.

3. Respiratory volumes and capacities
   a. One inspiration followed by one expiration is called a respiratory cycle.
   b. The amount of air that normally moves in and out during a respiratory cycle is the tidal volume.
   c. Additional air that can be inhaled is the inspiratory reserve volume; additional air that can be exhaled is the expiratory reserve volume.
   d. Residual air remains in the lungs and is mixed with newly inhaled air.
   e. The inspiratory capacity is the maximum volume of air a person can inhale following exhalation of the tidal volume.
   f. The functional residual capacity is the volume of air that remains in the lungs following the exhalation of the tidal volume.
   g. The vital capacity is the maximum amount of air a person can exhale after taking the deepest breath possible.
   h. The total lung capacity is equal to the vital capacity plus the residual air volume.
   i. Air in the anatomic and alveolar dead spaces is not available for gas exchange.
4. Alveolar ventilation
   a. Minute ventilation is calculated by multiplying the tidal volume by the breathing rate.
   b. Alveolar ventilation rate is calculated by subtracting the physiologic dead space from the tidal volume and multiplying the result by the breathing rate.
   c. The alveolar ventilation rate is a major factor affecting the exchange of gases between the alveolar air and the blood.
5. Nonrespiratory air movements
   a. Nonrespiratory air movements are air movements other than breathing.
   b. They include coughing, sneezing, laughing, crying, hiccuping, and yawning.

## Control of Breathing (page 776)

Normal breathing is rhythmic and involuntary, although the respiratory muscles can be controlled voluntarily.

1. Respiratory center
   a. The respiratory center is located in the brain stem, and includes parts of the medulla oblongata and pons.
   b. The medullary rhythmicity area includes two groups of neurons.
      (1) The dorsal respiratory group is responsible for the basic rhythm of breathing.
      (2) The ventral respiratory group increases inspiratory and expiratory movements during forceful breathing.
   c. The pneumotaxic area regulates the rate of breathing.
2. Factors affecting breathing
   a. Chemicals, lung tissue stretching, and emotional state affect breathing.
   b. Chemosensitive areas are associated with the respiratory center.
      (1) Carbon dioxide combines with water to form carbonic acid, which, in turn, releases hydrogen ions in the CSF.
      (2) Stimulation of these areas increases alveolar ventilation.
   c. Chemoreceptors are in the carotid bodies and aortic bodies of certain arteries.
      (1) These chemoreceptors sense low oxygen concentration.
      (2) When oxygen concentration is low, alveolar ventilation increases.
   d. Stretching the lung tissues triggers an inflation reflex.
      (1) This reflex reduces the duration of inspiratory movements.
      (2) This prevents overinflation of the lungs during forceful breathing.
   e. Hyperventilation decreases carbon dioxide concentration, but *this is very dangerous when associated with breath holding during underwater swimming.*

## Alveolar Gas Exchanges (page 783)

Gas exchanges between the air and the blood occur within the alveoli.

1. Alveoli
   a. The alveoli are tiny sacs clustered at the distal ends of the alveolar ducts.
   b. Some alveoli open into adjacent air sacs that provide alternate pathways for air when passages are obstructed.
2. Respiratory membrane
   a. The respiratory membrane consists of the alveolar and capillary walls.
   b. Gas exchanges take place through these walls.
3. Diffusion through the respiratory membrane
   a. The partial pressure of a gas is determined by the concentration of that gas in a mixture of gases or the concentration of gas dissolved in a liquid.
   b. Gases diffuse from regions of higher partial pressure toward regions of lower partial pressure.
   c. Oxygen diffuses from the alveolar air into the blood; carbon dioxide diffuses from the blood into the alveolar air.

## Gas Transport (page 785)

Blood transports gases between the lungs and the body cells.

1. Oxygen transport
   a. Oxygen is mainly transported in combination with hemoglobin molecules.
   b. The resulting oxyhemoglobin is relatively unstable and releases its oxygen in regions where the $Po_2$ is low.
   c. More oxygen is released as the blood concentration of carbon dioxide increases, as the blood becomes more acidic, and as the blood temperature increases.
2. Carbon monoxide
   a. Carbon monoxide forms as a result of incomplete combustion of fuels.
   b. It combines with hemoglobin more readily than oxygen and forms a stable compound.
   c. Carbon monoxide is toxic because the hemoglobin with which it combines is no longer available for oxygen transport.

3. Carbon dioxide transport
   a. Carbon dioxide may be carried in solution, either as dissolved $CO_2$, $CO_2$ bound to hemoglobin, or as a bicarbonate ion.
   b. Most carbon dioxide is transported in the form of bicarbonate ions.
   c. Carbonic anhydrase, an enzyme, speeds the reaction between carbon dioxide and water to form carbonic acid.
   d. Carbonic acid dissociates to release hydrogen ions and bicarbonate ions.

## Critical Thinking Questions

1. If the upper respiratory passages are bypassed with a tracheostomy, how might the air entering the trachea be different from air normally passing through this canal? What problems might this cause for the patient?
2. Certain respiratory disorders, such as emphysema, reduce the capacity of the lungs to recoil elastically. Which respiratory air volumes will this condition affect? Why?
3. What changes would you expect to occur in the relative concentrations of blood oxygen and carbon dioxide in a patient who breathes rapidly and deeply for a prolonged time? Why?
4. If a person has stopped breathing and is receiving pulmonary resuscitation, would it be better to administer pure oxygen or a mixture of oxygen and carbon dioxide? Why?
5. The air pressure within the passenger compartment of a commercial aircraft may be equivalent to an altitude of 8,000 feet. What problem might this create for a person with a serious respiratory disorder?
6. Patients experiencing asthma attacks are often advised to breathe through pursed (puckered) lips. How might this help reduce the symptoms of the asthma?

## Review Exercises

1. Describe the general functions of the respiratory system.
2. Distinguish between the upper and lower respiratory tracts.
3. Explain how the nose and nasal cavity filter incoming air.
4. Name and describe the locations of the major sinuses, and explain how a sinus headache may occur.
5. Distinguish between the pharynx and the larynx.
6. Name and describe the locations and functions of the cartilages of the larynx.
7. Distinguish between the false vocal cords and the true vocal cords.
8. Compare the structure of the trachea with the structure of the branches of the bronchial tree.
9. List the successive branches of the bronchial tree, from the primary bronchi to the alveoli.
10. Describe how the structure of the respiratory tube changes as the branches become finer.
11. Explain the functions of the respiratory tubes.
12. Distinguish between visceral pleura and parietal pleura.
13. Name and describe the locations of the lobes of the lungs.
14. Explain how normal inspiration and forced inspiration are accomplished.
15. Define *surface tension,* and explain how it aids the breathing mechanism.
16. Define *surfactant,* and explain its function.
17. Define *compliance.*
18. Explain how normal expiration and forced expiration are accomplished.
19. Distinguish between vital capacity and total lung capacity.
20. Distinguish between anatomic, alveolar, and physiologic dead spaces.
21. Distinguish between minute respiratory volume and alveolar ventilation rate.
22. Compare the mechanisms of coughing and sneezing, and explain the function of each.
23. Explain the function of yawning.
24. Describe the location of the respiratory center, and name its major components.
25. Describe how the basic rhythm of breathing is controlled.
26. Explain the function of the pneumotaxic area of the respiratory center.
27. Explain why increasing blood concentrations of carbon dioxide and hydrogen ions have similar effects on the respiratory center.
28. Describe the function of the chemoreceptors in the carotid and aortic bodies of certain arteries.
29. Describe the inflation reflex.
30. Discuss the effects of emotions on breathing.
31. Define *hyperventilation,* and explain how it affects the respiratory center.
32. Define *respiratory membrane,* and explain its function.
33. Explain the relationship between the partial pressure of a gas and its rate of diffusion.
34. Summarize the gas exchanges that occur through the respiratory membrane.
35. Describe how oxygen is transported in blood.
36. List three factors that increase release of oxygen from the blood.
37. Explain why carbon monoxide is toxic.
38. List three ways that carbon dioxide is transported in blood.
39. Explain the function of carbonic anhydrase.
40. Define *chloride shift.*

20

CHAPTER

# URINARY SYSTEM

## CHAPTER OBJECTIVES

AFTER YOU HAVE STUDIED THIS CHAPTER, YOU SHOULD BE ABLE TO:

1. Name the organs of the urinary system and list their general functions.
2. Describe the locations of the kidneys and the structure of a kidney.
3. List the functions of the kidneys.
4. Trace the pathway of blood through the major vessels within a kidney.
5. Describe a nephron and explain the functions of its major parts.
6. Explain how glomerular filtrate is produced and describe its composition.
7. Explain how various factors affect the rate of glomerular filtration and how this rate is regulated.
8. Discuss the role of tubular reabsorption in urine formation.
9. Explain why the osmotic concentration of the glomerular filtrate changes as it passes through a renal tubule.
10. Describe a countercurrent mechanism and explain how it helps concentrate urine.
11. Define tubular secretion and explain its role in urine formation.
12. Describe the structure of the ureters, urinary bladder, and urethra.
13. Discuss the process of micturition and explain how it is controlled.

CRYSTALS OF URIC ACID (40×).

## KEY TERMS

**afferent arteriole** (af′er-ent ar-te′re-ōl)

**autoregulation** (aw″to-reg″u-la′shun)

**countercurrent mechanism** (kown′ter-kur′ent mek′ah-nizm)

**destrusor muscle** (de-truz′or mus′l)

**efferent arteriole** (ef′er-ent ar-te′re-ōl)

**glomerulus** (glo-mer′u-lus)

**juxtaglomerular apparatus** (juks″tah-glo-mer′u-lar ap″ah-ra′tus)

**micturition** (mik″tu-rish′un)

**nephron** (nef′ron)

**peritubular capillary** (per″i-tū′bu-lar kap′ i-ler″e)

**renal corpuscle** (re′nal kor′pusl)

**renal cortex** (re′nal kor′teks)

**renal medulla** (re′nal mĕ-dul′ah)

**renal plasma threshold** (re′nal plaz′mah thresh′old)

**renal tubule** (re′nal tu′būl)

**retroperitoneal** (re″tro-per″ĭ-to-ne′al)

The first time Miguel woke up to extreme pain in his right great toe, he attributed it to the long hike he'd taken the day before. After he hobbled about for two days, the discomfort ebbed away. Miguel forgot about it. Then, four months later, he again awoke with agonizing pain in the same toe. Because this time the pain lasted longer and was accompanied by headache and fever, Miguel consulted a physician.

Miguel was astonished to learn that his painful toe was actually caused by a failure of his kidneys to excrete uric acid, a by-product of nucleic acid catabolism. Instead of exiting the body in the urine, the uric acid was deposited as crystals in the joint of his large toe. The ultimate source of Miguel's condition, called *gout,* was his genes—he found that several male relatives also had mysterious joint pains.

To confirm her suspicion of gout, based on the painful toe, the physician aspirated fluid from the affected toe and found the telltale needle-shaped sodium urate crystals in the synovial fluid. The next step was to treat Miguel. He found relief with drugs that increase the kidneys' excretion of uric acid and block an enzyme in the biosynthetic pathway for uric acid. He also limited his intake of foods that are sources of uric acid, including organ meats, anchovies, and sardines. He was encouraged to lose some weight and to drink more fluids to dilute his urine, which would enhance uric acid excretion. The physician also warned Miguel that he was at high risk of uric acid crystals depositing in the urinary system as painful stones.

Gout is an illness with a long history in medicine. Hippocrates mentioned it, and in 1793, English physician Alfred Baring Garrod isolated and implicated uric acid from the blood of a patient with gout, and noted that people with this condition often had relatives suffering from it too. At that time, gout was thought to be the result of being a lazy glutton! Today we know that it is an inborn error of metabolism affecting the urinary system, but producing symptoms in the joints.

---

The urinary system consists of a pair of glandular *kidneys,* which remove substances from the blood, form urine, and help regulate certain metabolic processes; a pair of tubular *ureters,* which transport urine from the kidneys; a saclike *urinary bladder,* which is a urine reservoir; and a tubular *urethra,* which conveys urine to the outside of the body. Figures 20.1 and 20.2 show these organs.

## KIDNEYS

A **kidney** is a reddish brown, bean-shaped organ with a smooth surface. It is about 12 centimeters long, 6 centimeters wide, and 3 centimeters thick in an adult, and it is enclosed in a tough, fibrous capsule (tunic fibrosa).

### Location of the Kidneys

The kidneys lie on either side of the vertebral column in a depression high on the posterior wall of the abdominal cavity. Although the positions of the kidneys may vary slightly with changes in posture and with breathing movements, their upper and lower borders are generally at the levels of the twelfth thoracic and third lumbar vertebrae, respectively. The left kidney is usually about 1.5 to 2 centimeters higher than the right one.

The kidneys are positioned *retroperitoneally,* which means they are behind the parietal peritoneum and against the deep muscles of the back. Connective tissue (renal fascia) and masses of adipose tissue (renal fat) surrounding the kidneys hold them in place (see fig. 20.3 and reference plates 58, 59).

### Kidney Structure

The lateral surface of each kidney is convex, but its medial side is deeply concave. The resulting medial depression leads into a hollow chamber called the **renal sinus.** Through the entrance to this sinus, termed the *hilum,* pass blood vessels, nerves, lymphatic vessels, and the ureter (see fig. 20.1).

The superior end of the ureter expands to form a funnel-shaped sac called the **renal pelvis,** which is located inside the renal sinus. The pelvis is subdivided into two or three tubes, called *major calyces* (sing. *calyx*), and they, in turn, are subdivided into eight to fourteen *minor calyces* (fig. 20.4).

A series of small projections called *renal papillae* project into the renal sinus from its wall. Tiny openings that lead into a minor calyx pierce each projection.

Two common inherited kidney abnormalities are *polycystic kidney disease,* which affects adults, and *Wilms tumor,* which affects young children. In polycystic kidney disease, cysts present in the kidneys since childhood or adolescence begin to produce symptoms in the thirties, including abdominal pain, bloody urine, and elevated blood pressure. Eventually the condition can lead to renal failure. In Wilms tumor, pockets of cells within a child's kidney remain as they were in the embryo—they are unspecialized and divide rapidly, forming a cancerous tumor. Loss of a tumor suppressor gene causes Wilms tumor.

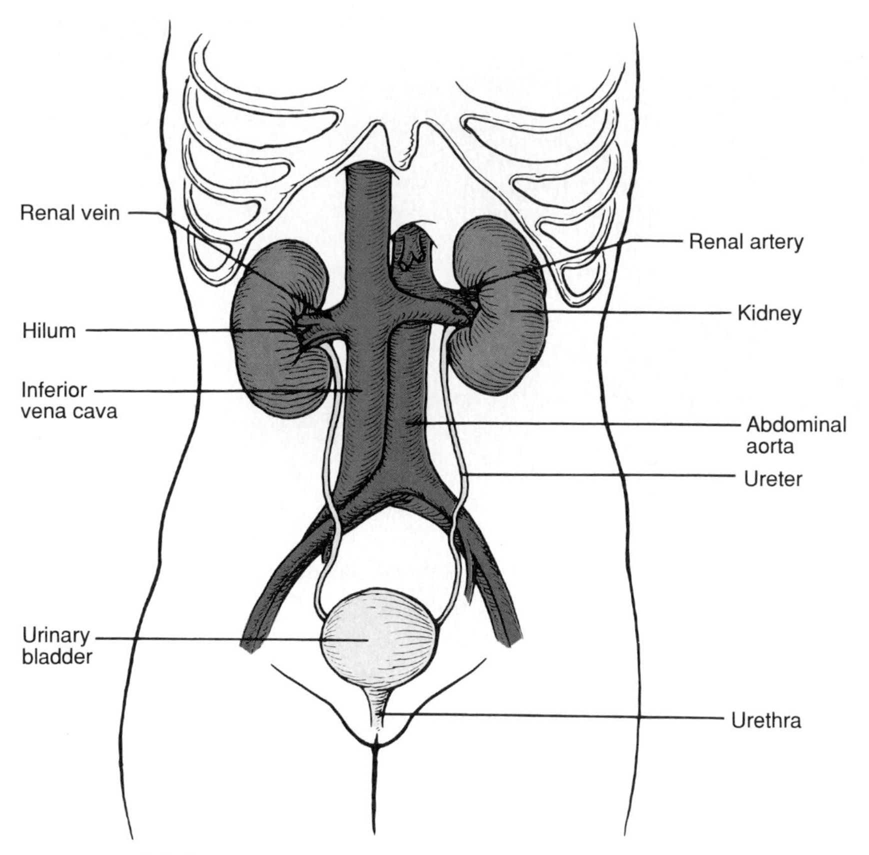


**FIGURE 20.1**

The urinary system includes the kidneys, ureters, urinary bladder, and urethra. Notice the relationship of these structures to the major blood vessels.

The kidney includes two distinct regions: an inner medulla and an outer cortex. The **renal medulla** is composed of conical masses of tissue called *renal pyramids,* the bases of which are directed toward the convex surface of the kidney, and the apexes of which form the renal papillae. The tissue of the medulla appears striated because it consists of microscopic tubules leading from the cortex to the renal papillae.

The **renal cortex,** which appears somewhat granular, forms a shell around the medulla. Its tissue dips into the medulla between adjacent renal pyramids, forming *renal columns.* The granular appearance of the cortex is due to the random arrangement of tiny tubules associated with the **nephrons,** the functional units of the kidney.

## Functions of the Kidneys

The main function of the kidneys is to regulate the volume, composition, and pH of body fluids. In the process, the kidneys remove metabolic wastes from the blood and excrete them to the outside. They also help control the rate of red blood cell formation by secreting the hormone *erythropoietin* (see chapter 14), regulate the blood pressure by secreting the enzyme *renin* (see chapter 15), and regulate the absorption of calcium ions by activating *vitamin D* (see chapter 13).

Medical technology can take over the role of a kidney. In *hemodialysis,* a person's blood is rerouted across an artificial membrane that "cleanses" it, removing substances that would normally be excreted in the urine. A patient usually must use this artificial kidney three times a week, for several hours each time. Hemodialysis is discussed further in Clinical Application 20.1.

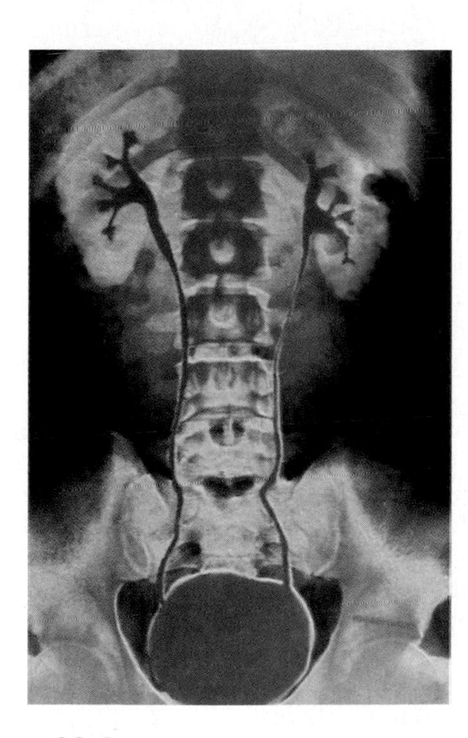

**FIGURE 20.2**

Structures of the urinary system are visible in this false-color X-ray film.

About two in every ten patients with renal failure can use a procedure that can be done at home called *continuous ambulatory peritoneal dialysis* instead of hemodialysis. The patient infuses a solution into the abdominal cavity through a permanently implanted tube. The solution stays in for 4 to 8 hours, while it takes up substances that would normally be excreted into urine. Then the patient drains the waste-laden solution out of the tube, replacing it with clean fluid.

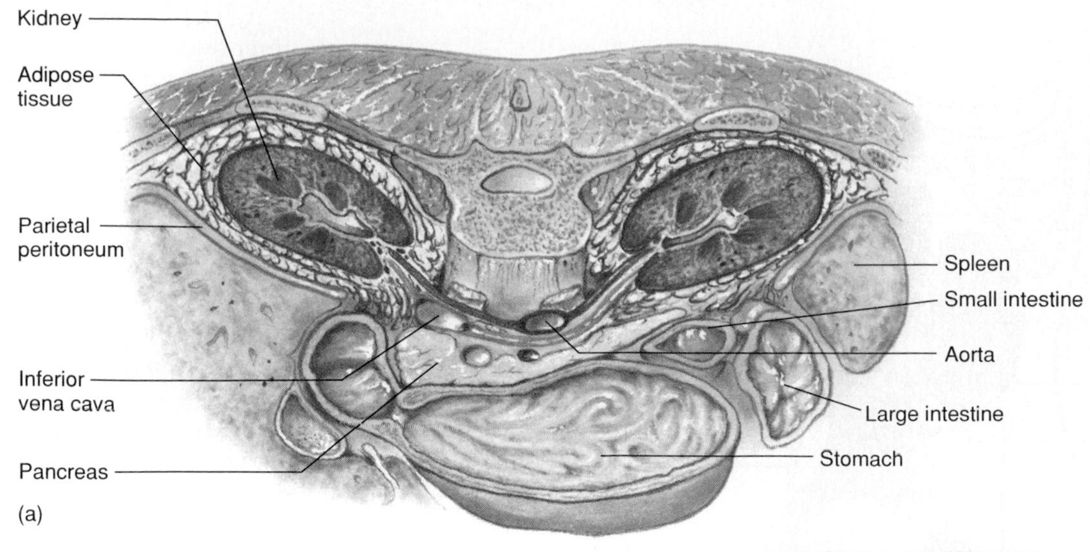


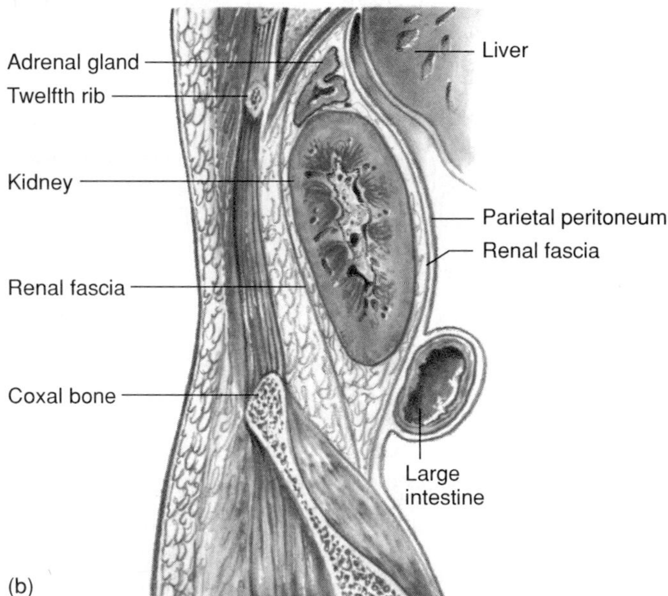


**FIGURE 20.3**
(*a*) Transverse sections of the kidneys, which are located behind the parietal peritoneum. Adipose and other connective tissue surround and support the kidneys. (*b*) Sagittal section through the posterior abdominal cavity showing the kidney.

1. Where are the kidneys located?
2. Describe the structure of a kidney.
3. Name the functional unit of the kidney.
4. What are the general functions of the kidneys?

## Renal Blood Vessels

The **renal arteries,** which arise from the abdominal aorta, supply blood to the kidneys (fig. 20.5). These arteries transport a large volume of blood. When a person is at rest, the renal arteries usually carry from 15% to 30% of the total cardiac output into the kidneys although the kidneys account for only 1% of body weight.

A renal artery enters a kidney through the hilum and gives off several branches, called the *interlobar arteries,* which pass between the renal pyramids. At the junction between the medulla and the cortex, the interlobar arteries branch to form a series of incomplete arches, the *arciform arteries* (arcuate arteries), which, in turn, give rise to *interlobular arteries.* The final branches of the interlobular arteries, called **afferent arterioles,** lead to the nephrons, the functional units of the kidneys.

Venous blood is returned through a series of vessels that correspond generally to the arterial pathways. For example, the venous blood passes through interlobular, arciform, interlobar, and renal veins. The **renal vein** then joins the inferior vena cava as it courses through the abdominal cavity. (Figs. 20.6 and 20.7 show branches of the renal arteries and veins.)

## Nephrons

### Structure of a Nephron

A kidney contains about one million nephrons, each consisting of a **renal corpuscle** and a **renal tubule** (see fig. 20.4).

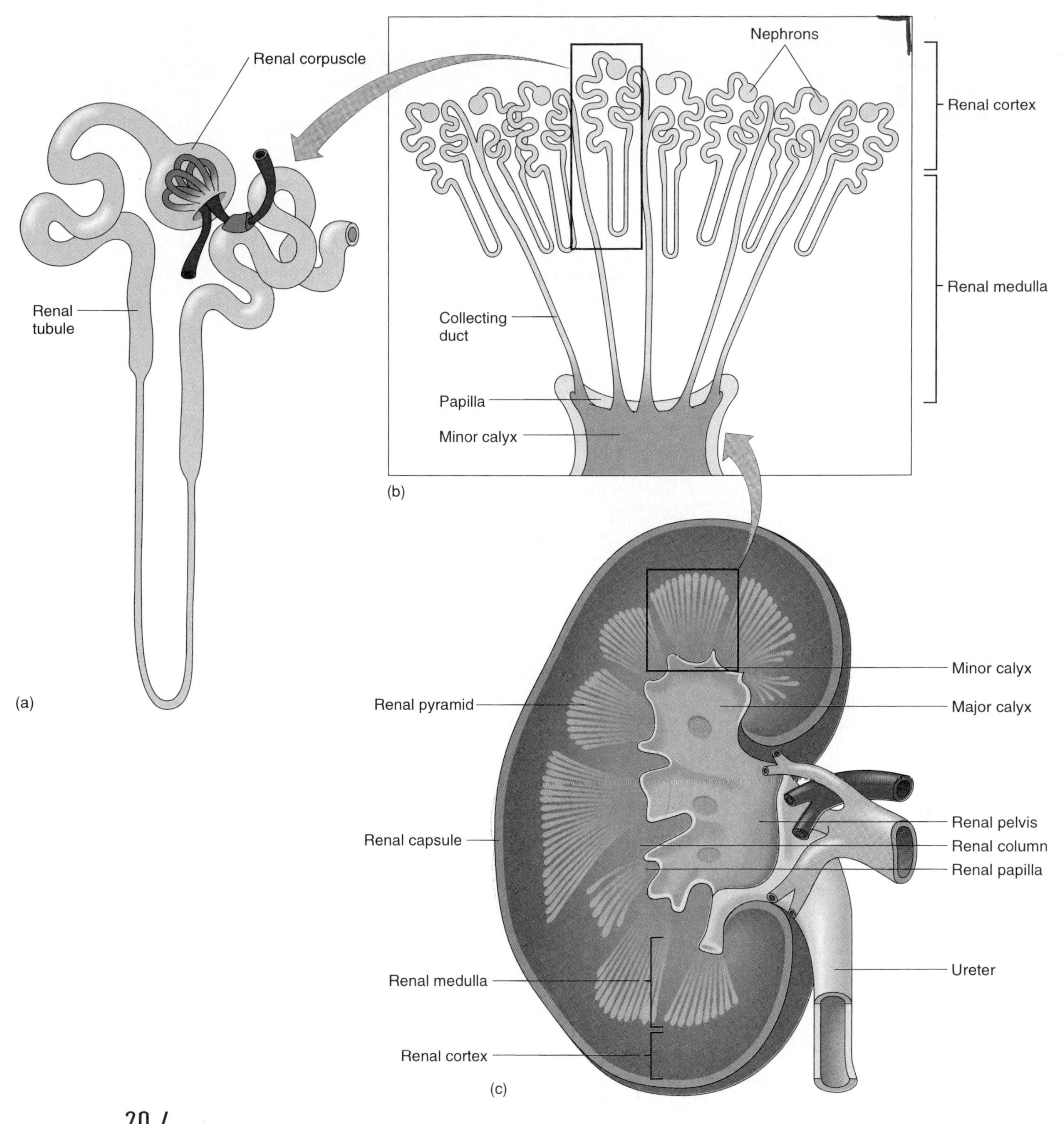


**FIGURE 20.4**
(*a*) A single nephron; (*b*) a renal pyramid containing nephrons; (*c*) longitudinal section of a kidney.

A renal corpuscle is composed of a filtering unit consisting of a tangled cluster of blood capillaries called a **glomerulus** and a surrounding thin-walled, saclike structure called a **glomerular capsule.** Afferent arterioles give rise to these capillaries, which lead to **efferent arterioles** (fig. 20.7).

The glomerular capsule is an expansion at the end of a renal tubule that receives the fluid filtered through the glomerulus. It is composed of two layers of squamous epithelial cells: a visceral layer that closely covers the glomerulus, and an outer parietal layer that is continuous with the

**CLINICAL APPLICATION** 

## Chronic Kidney Failure

Charles B., a 43-year-old muscular construction worker, had been feeling unusually tired for several weeks. Occasionally he had felt dizzy and found it increasingly difficult to sleep. More recently he had noticed a burning pain in his lower back, just below his rib cage, and his urine had darkened.

In addition, his feet, ankles, and face seemed swollen. His wife suggested that he consult their family physician about these symptoms.

The physician found that Charles had elevated blood pressure (hypertension) and that the regions of his kidneys were sensitive to pressure. A urinalysis revealed excess protein (proteinuria) and blood (hematuria). Blood tests indicated elevated blood urea nitrogen (BUN), elevated serum creatinine, and decreased serum protein (hypoproteinemia) concentrations.

The physician told Charles that he probably had *chronic glomerulonephritis,* an inflammation of the capillaries within the glomeruli of the renal nephrons (see fig. 20.9), and that this was a progressive degenerative disease lacking a cure. Examination of a tiny sample of kidney tissue (biopsy) examined microscopically later confirmed the diagnosis (see fig. 20.10).

In spite of medical treatment and careful attention to his diet, Charles's condition deteriorated rapidly. When it appeared that most of his kidney function had been lost (end-stage renal disease, or ESRD), he was offered artificial kidney treatments (hemodialysis).

To prepare Charles for hemodialysis, a vascular surgeon created a fistula in his left forearm by surgically connecting an artery to a vein. The greater pressure of the blood in the artery that now flowed directly into the vein swelled the vein, making it more accessible.

During hemodialysis treatment, a hollow needle was inserted into the vein of the fistula near its arterial connection. This allowed the blood to flow, with the aid of a blood pump, through a tube leading to the blood compartment of a dialysis machine. Within this compartment, the blood passed over a semipermeable membrane. On the opposite side of the membrane was a dialysate solution with a controlled composition. Negative pressure on the dialysate side of the membrane, created by a vacuum pump, increased the movement of fluid through the membrane. At the same time, waste and excess electrolytes diffused from the blood through the membrane and entered the dialysate solution. The blood was then returned through a tube to the vein of the fistula.

In order to maintain favorable blood concentrations of waste, electrolytes, and water, Charles had to undergo hemodialysis three times per week, with each treatment lasting three to four hours. During the treatments, he was given an anticoagulant to prevent blood clotting, an antibiotic drug to control infections, and an antihypertensive drug to reduce his blood pressure.

Charles was advised to carefully control his intake of water, sodium, potassium, proteins, and total calories between treatments. He was also asked to consider another option for the treatment of ESRD—a kidney transplant—which could free him from the time-consuming dependence on hemodialysis.

In this surgical procedure, a kidney from a living donor or a cadaver, whose tissues are antigenically similar (histocompatible) to those of the recipient, is placed in the depression on the medial surface of the right or left ilium (iliac fossa). The renal artery and vein of the donor kidney are connected to the recipient's iliac artery and vein, respectively, and the kidney's ureter is attached to the dome of the recipient's urinary bladder. The patient must then remain on immunosuppressant drugs to prevent rejection of the transplant.

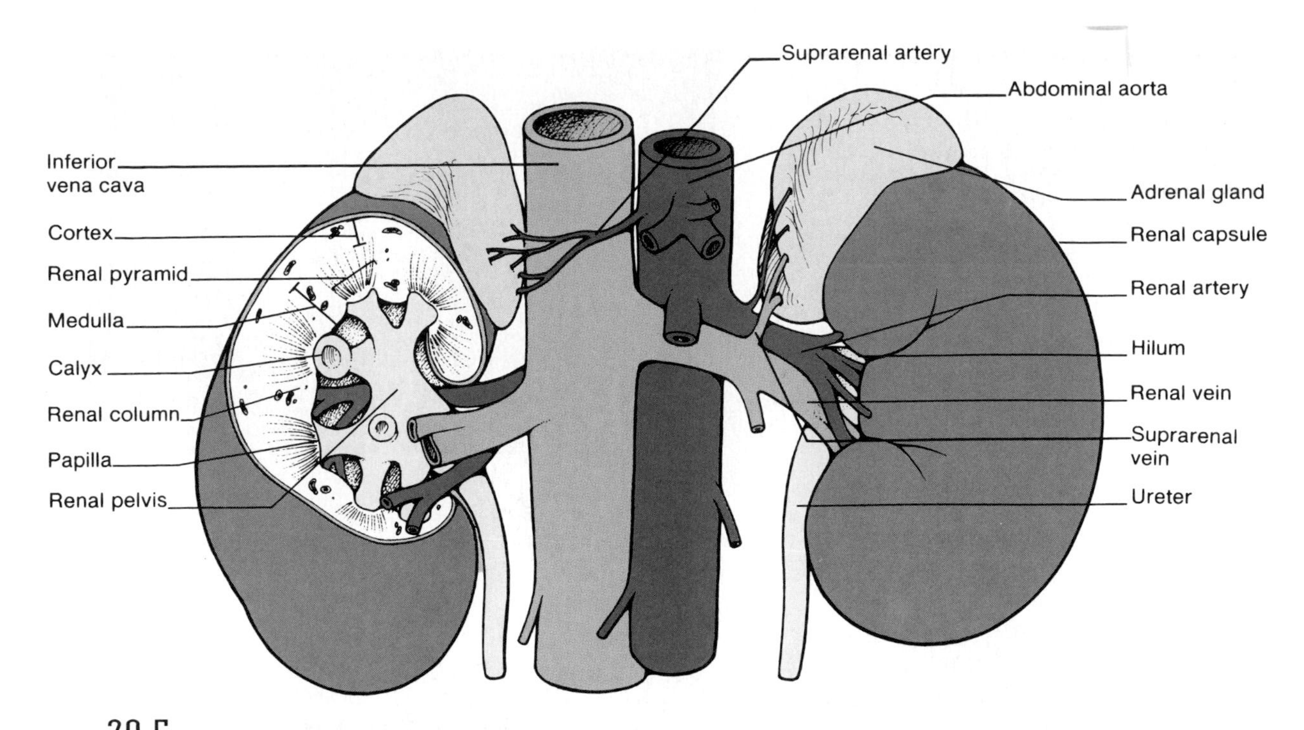


**FIGURE 20.5**

Blood vessels associated with the kidneys and adrenal glands. Note their relationship with the renal pelvis and ureters.

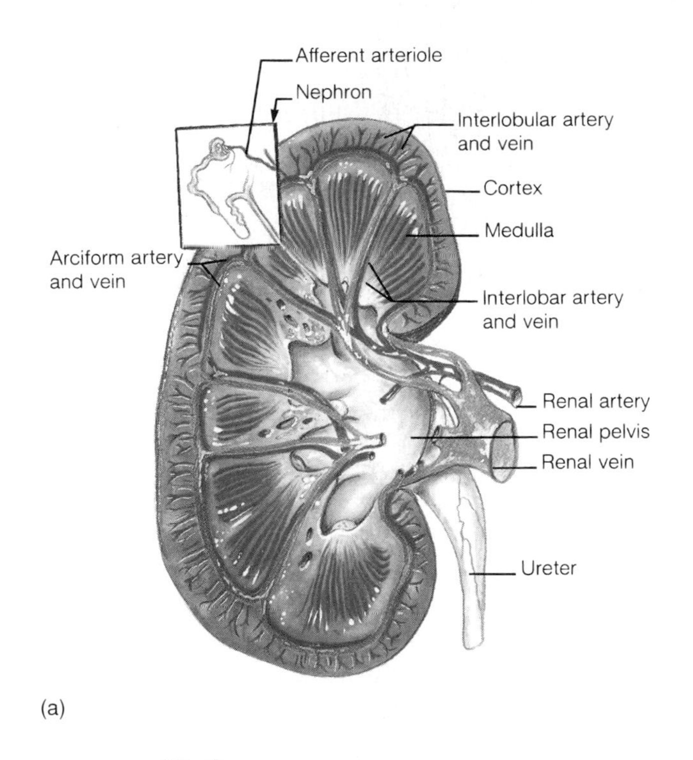


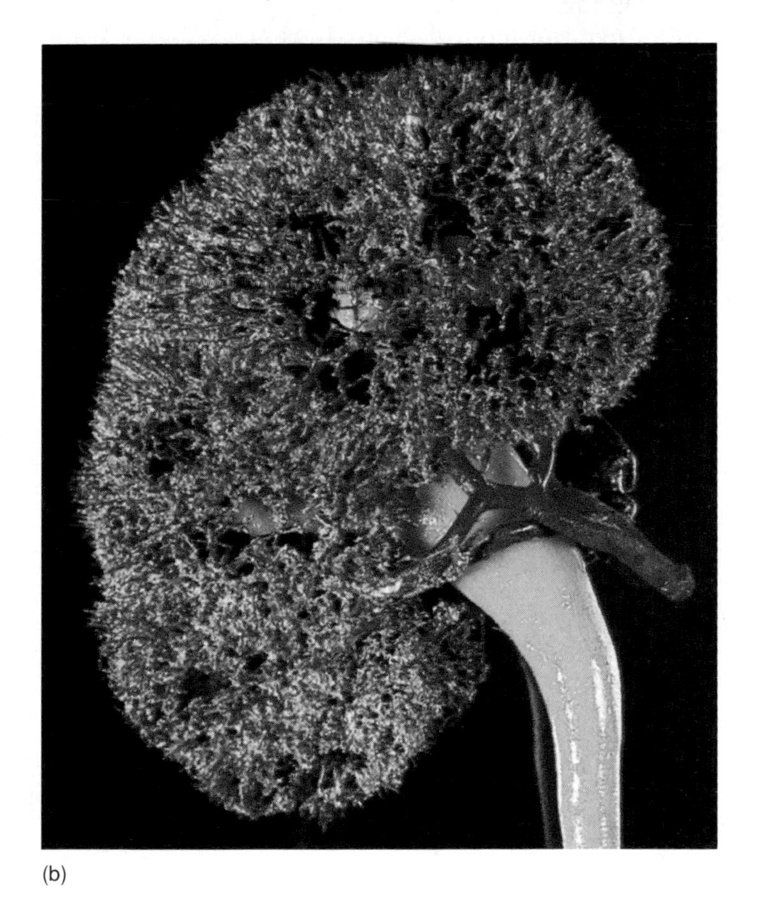


**FIGURE 20.6**

(*a*) Main branches of the renal artery and vein; (*b*) corrosion cast of the renal arterial system. Not all blood vessels associated with the nephron are shown.

visceral layer and with the wall of the renal tubule.

The cells of the parietal layer are typical squamous epithelial cells; however, those of the visceral layer are highly modified epithelial cells called *podocytes.* Each podocyte has several primary processes extending from its cell body, and these processes, in turn, bear numerous secondary processes, or *pedicels.* The pedicels of each cell interdigitate with those of adjacent podocytes, and the clefts between them form a complicated system of *slit pores* (fig. 20.8).

The renal tubule leads away from the glomerular capsule and becomes highly coiled. This coiled portion of the tubule is named the *proximal convoluted tubule.*

Following the proximal convoluted tubule is a structure known as the **nephron loop.** The proximal convoluted tubule dips toward the renal pelvis to become the *descending limb of the nephron loop.* The tubule then curves back toward its renal corpuscle and forms the *ascending limb of the nephron loop.*

The ascending limb returns to the region of the renal corpuscle, where it becomes highly coiled again and is called the *distal convoluted tubule.* This distal portion tends to be shorter and less convoluted than the proximal tubule.

Several distal convoluted tubules merge in the renal cortex to form a *collecting duct* (collecting tubule) which, in turn, passes into the renal medulla, becoming larger and larger in diameter as it joins other collecting ducts. The resulting tube empties into a minor calyx through an opening in a renal papilla. Figures 20.9 and 20.10 show the parts of a nephron. Clinical Application 20.2 examines glomerulonephritis, an inflammation of the glomeruli.

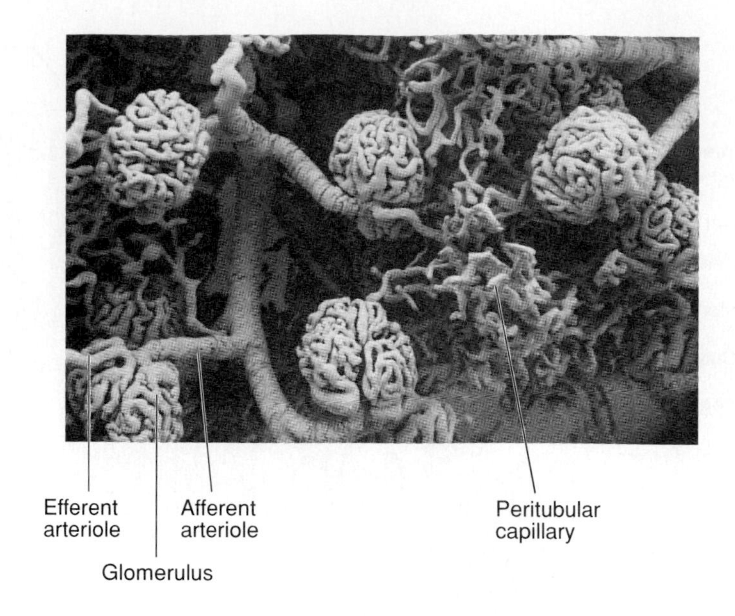


(a)

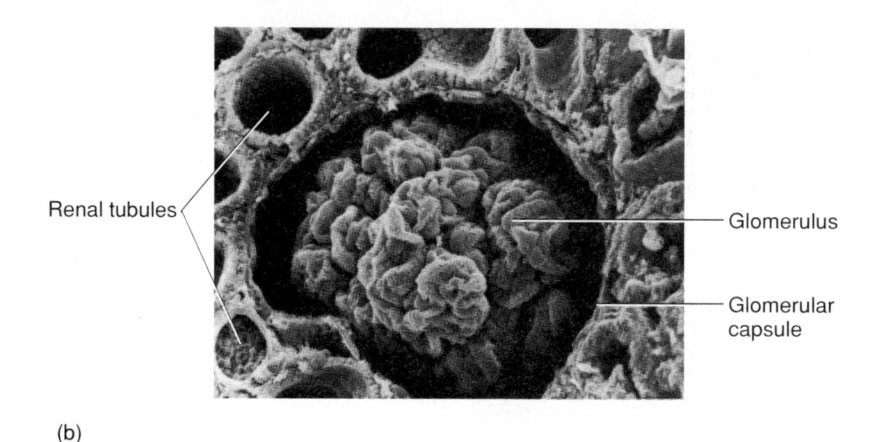


(b)

**FIGURE 20.7**

(*a*) A scanning electron micrograph of a cast of the renal blood vessels associated with the glomeruli (260×). (*b*) Scanning electron micrograph of a glomerular capsule surrounding a glomerulus.

(*Tissues and Organs: A Text-Atlas of Scanning Electron Microscopy,* by R. G. Kessel and R. H. Kardon. © 1979 W. H. Freeman and Company.)

## Juxtaglomerular Apparatus

Near its origin, the distal convoluted tubule passes between the afferent and efferent arterioles and contacts them. At the point of contact, the epithelial cells of the distal tubule are quite narrow and densely packed. These cells comprise a structure called the *macula densa.*

Close by, in the wall of the afferent arteriole near its attachment to the glomerulus, are some large, vascular smooth muscle cells. They are called *juxtaglomerular cells,* and together with the cells of the macula densa, they constitute the **juxtaglomerular apparatus** (complex). This structure plays an important role in regulating the secretion of renin (see chapter 13) (fig. 20.11).

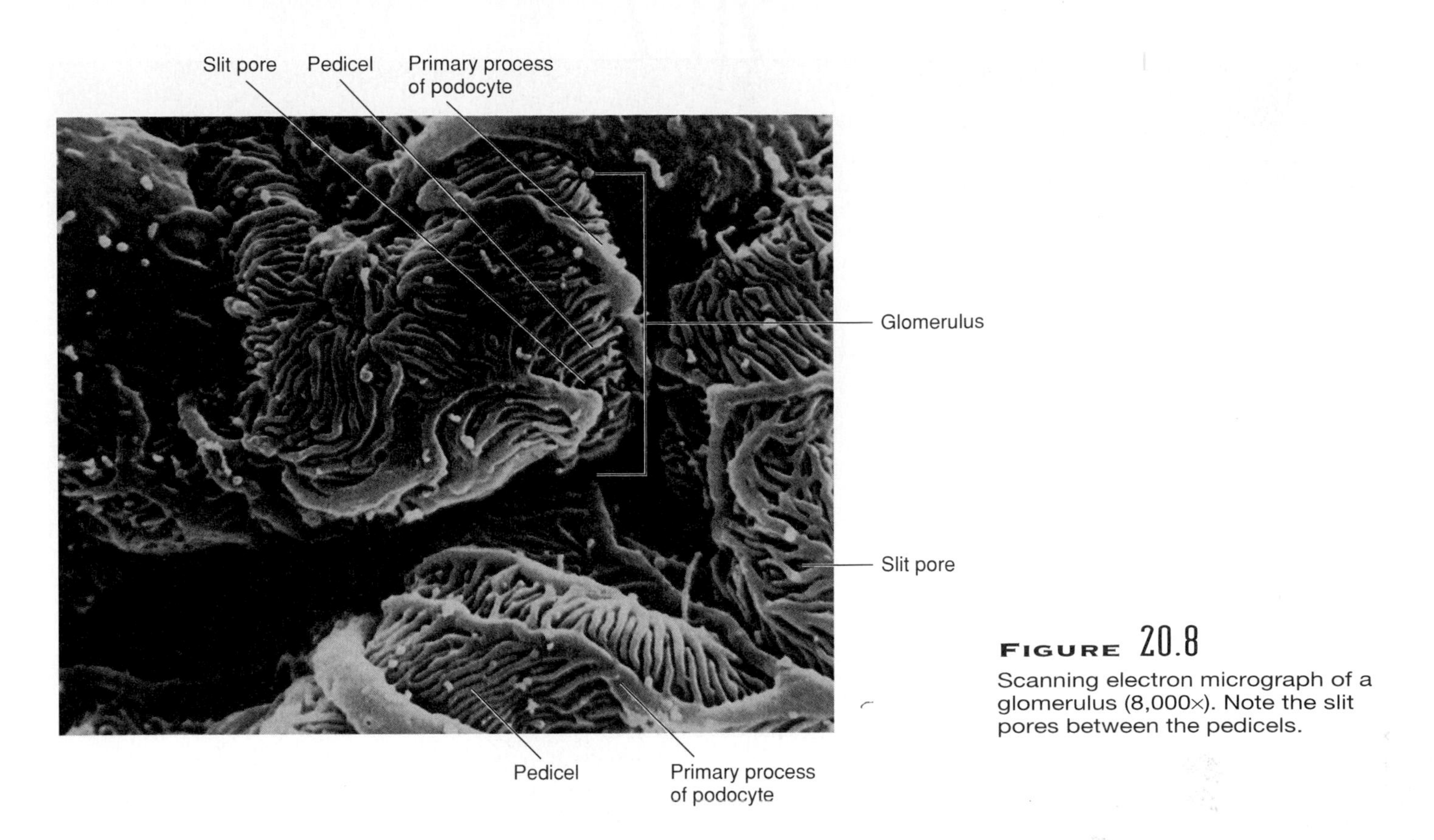


**FIGURE 20.8**
Scanning electron micrograph of a glomerulus (8,000×). Note the slit pores between the pedicels.

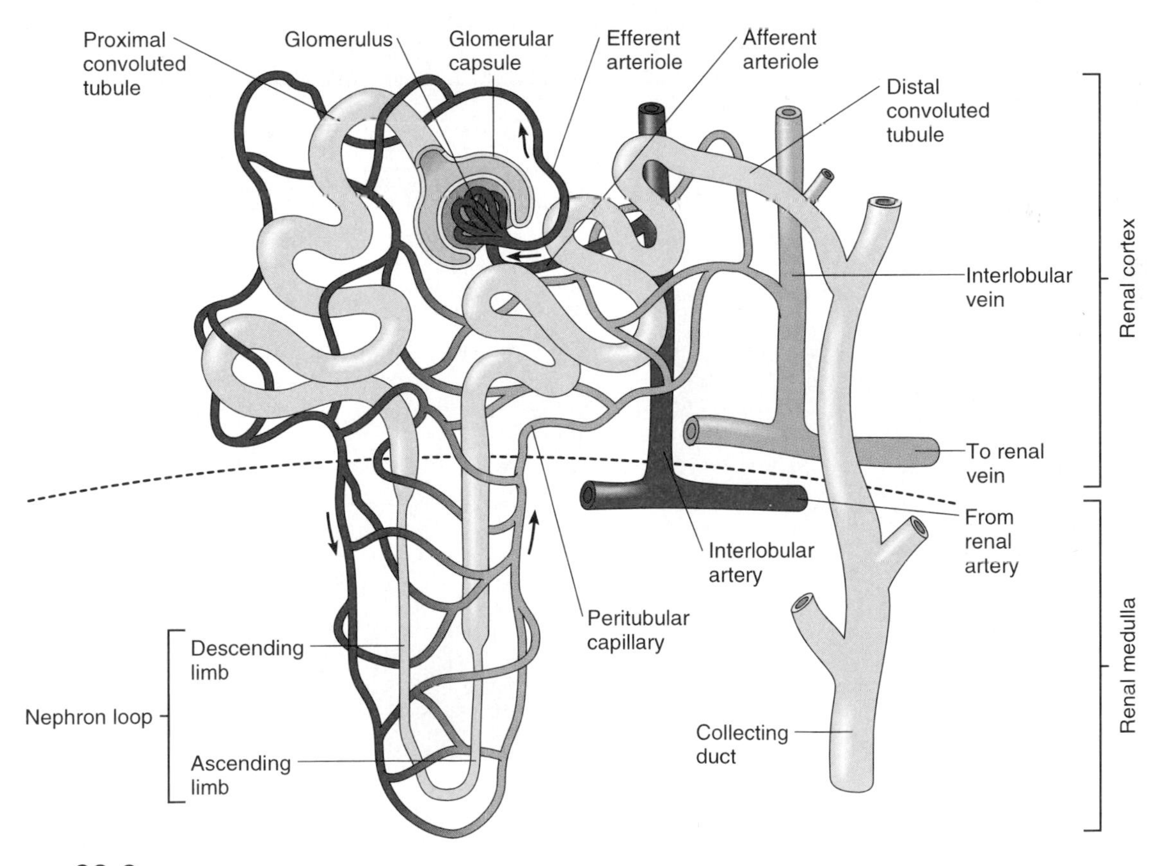


**FIGURE 20.9**
Structure of a nephron and the blood vessels associated with it.

20.2

CLINICAL APPLICATION

## Glomerulonephritis

*Nephritis* refers to an inflammation of the kidney; *glomerulonephritis* refers to an inflammation affecting the glomeruli. This latter condition may be acute or chronic, and can lead to renal failure.

*Acute glomerulonephritis* (AGN) usually results from an abnormal immune reaction that develops one to three weeks following bacterial infection by beta-hemolytic streptococci. As a rule, the infection occurs in some other part of the body and does not affect the kidneys directly. Instead, bacterial antigens trigger an immune reaction. Antibodies are produced against these antigens, forming insoluble immune complexes (see chapter 16) that are carried by the blood to the kidneys. The antigen-antibody complexes are deposited in and block the glomerular capillaries.

At the same time, inflammation sends large numbers of white blood cells to the region, blocking the capillaries still more. Those capillaries remaining open may become abnormally permeable, sending plasma proteins and red blood cells into the urine.

Most glomerulonephritis patients eventually regain normal kidney function; however, in severe cases, renal functions may fail completely, and without treatment, the person is likely to die within a week or so.

*Chronic glomerulonephritis* is a progressive disease in which increasing numbers of nephrons are slowly damaged until finally the kidneys are unable to function. This condition is usually associated with certain diseases other than streptococcal infections, and it also involves formation of antigen-antibody complexes that precipitate and accumulate in the glomeruli. The resulting inflammation is prolonged, and it is accompanied by fibrous tissue replacing glomerular membranes. As this happens, the functions of the nephrons are permanently lost, and eventually the kidneys fail.

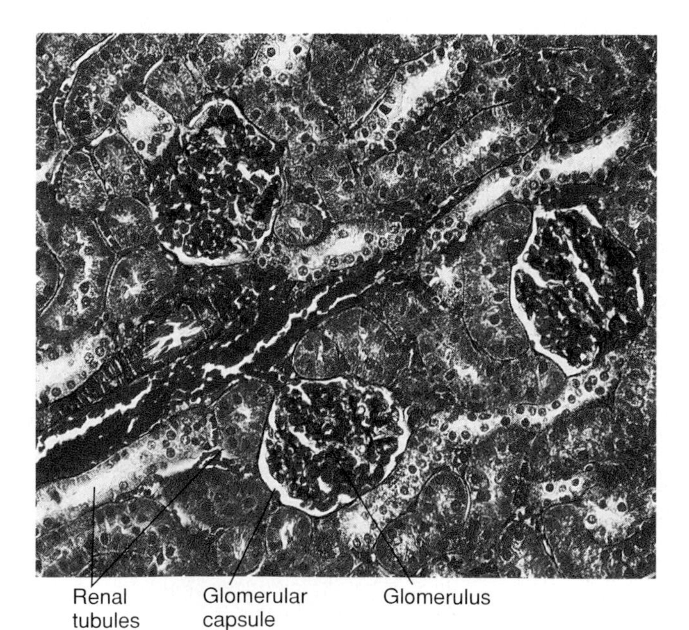


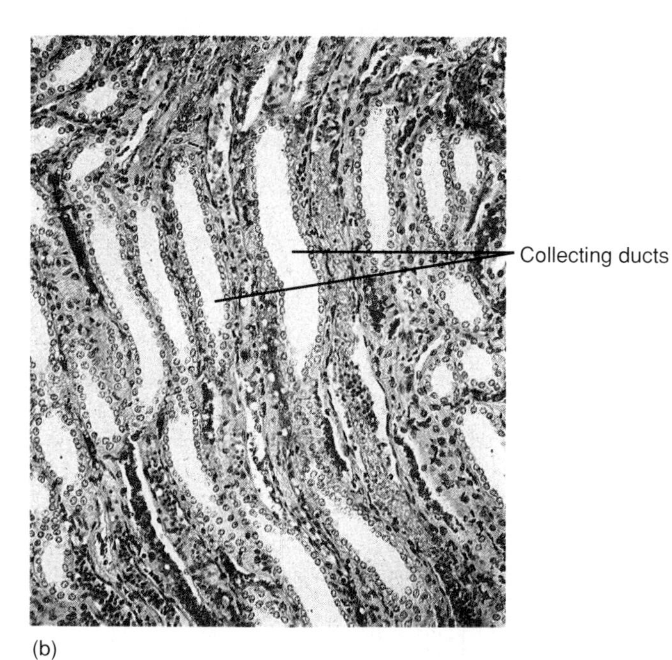


**FIGURE 20.10**
(*a*) Light micrograph of a section of the human renal cortex (240×). (*b*) Light micrograph of the renal medulla (80×).

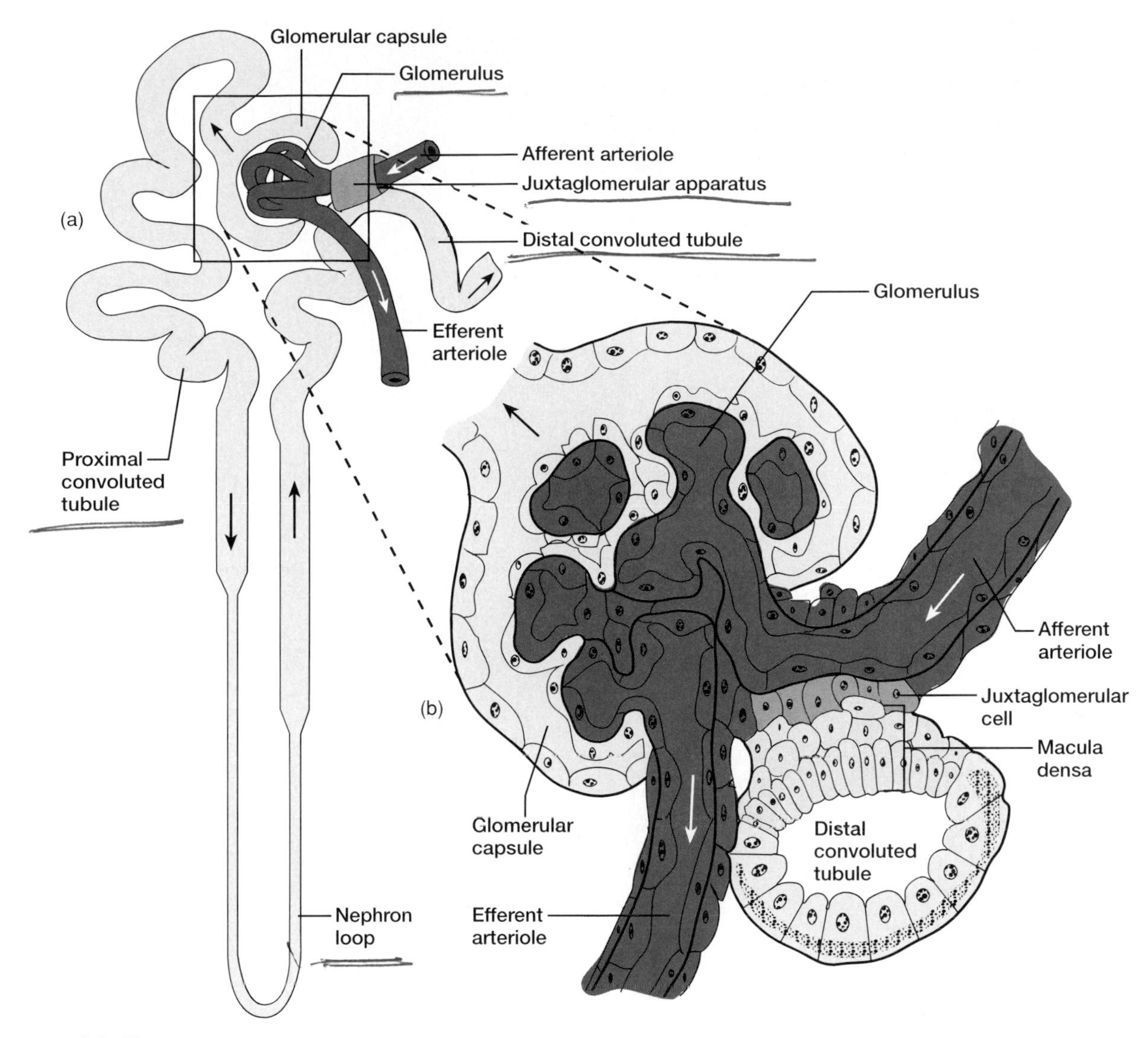


**FIGURE 20.11**

(*a*) Location of the juxtaglomerular apparatus. (*b*) Enlargement of a section of the juxtaglomerular apparatus, which consists of the macula densa and the juxtaglomerular cells. White arrows indicate direction of blood flow. Black arrows indicate flow of glomerular filtrate and tubular fluid.

## Cortical and Juxtamedullary Nephrons

Most nephrons have corpuscles located in the renal cortex near the surface of the kidney. These are called *cortical nephrons,* and they have relatively short nephron loops that usually do not reach the renal medulla.

Another group, called *juxtamedullary nephrons,* have corpuscles close to the renal medulla, and their nephron loops extend deep into the medulla. Although they represent only about 20% of the total, these nephrons play an important role in regulating water balance (fig. 20.12).

## Blood Supply of a Nephron

The cluster of capillaries forming a glomerulus arises from an afferent arteriole whose diameter is greater than that of other arterioles. Blood passes through the capillaries of the glomerulus, then enters an efferent arteriole (rather than a venule), whose diameter is smaller than that of the afferent arteriole. The greater resistance to blood flow of the efferent arteriole causes blood to back up into the glomerulus. This results in a relatively high pressure in the glomerular capillaries.

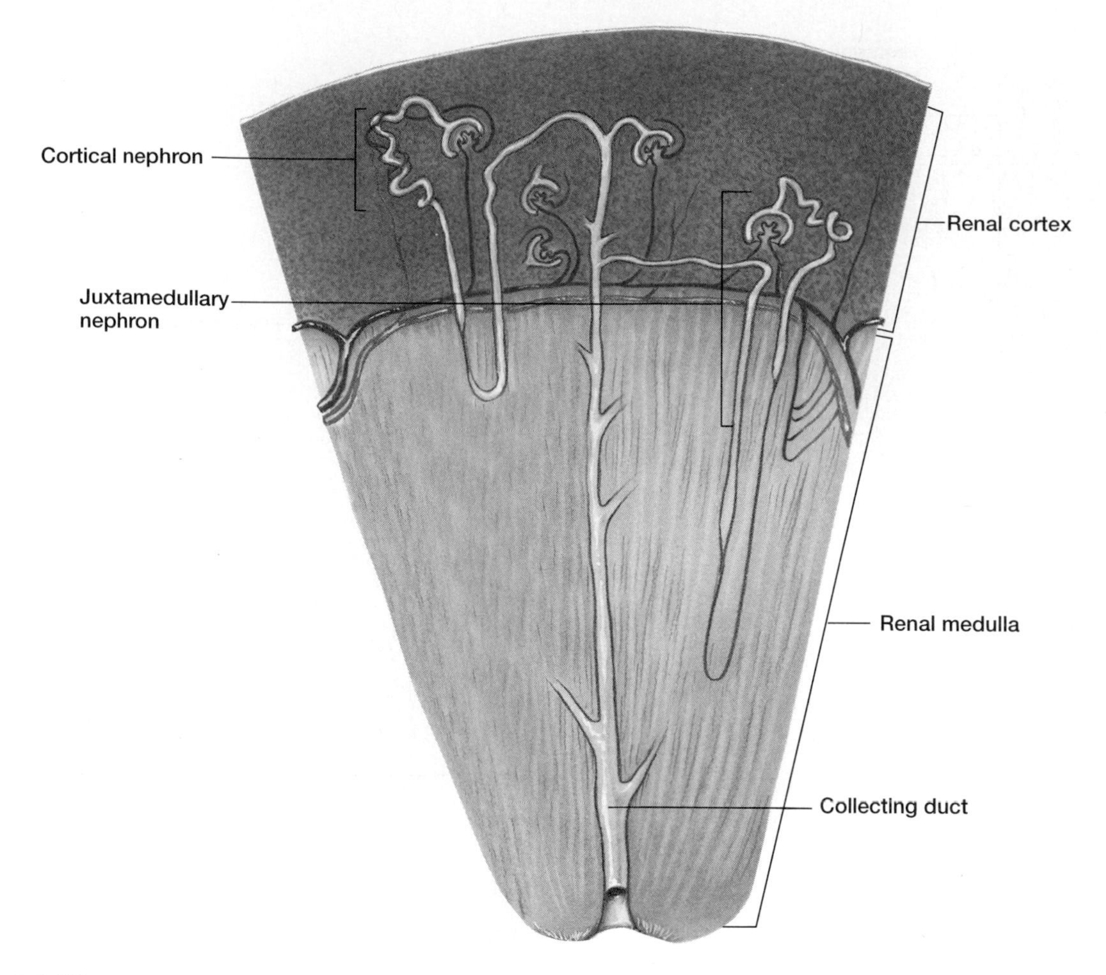


**FIGURE 20.12**
Cortical nephrons are close to the surface of a kidney; juxtamedullary nephrons are near the renal medulla.

The efferent arteriole branches into a complex network of capillaries surrounding the renal tubule called the **peritubular capillary system.** Blood in it is under relatively low pressure (see fig. 20.9). Branches of this system that receive blood primarily from the efferent arterioles of the juxtamedullary nephrons form capillary loops called *vasa recta.* These loops dip into the renal medulla and are closely associated with the loops of the juxtamedullary nephrons (fig. 20.13).

After flowing through the vasa recta, blood returns to the renal cortex, where it joins blood from other branches of the peritubular capillary system and enters the venous system of the kidney.

Figure 20.14 summarizes the pathway the blood follows as it passes through the blood vessels of the kidney and nephron.

1. Describe the system of vessels that supplies blood to the kidney.
2. Name the parts of a nephron.
3. What structures comprise the juxtaglomerular apparatus?
4. Distinguish between a cortical nephron and a juxtamedullary nephron.
5. Describe the blood supply of a nephron.

## URINE FORMATION

The main function of the nephrons is to control the composition of body fluids and remove wastes from the blood. The end product is **urine,** which is excreted from the body. It contains wastes and excess water and electrolytes.

Urine formation involves three processes. In *glomerular filtration,* certain chemicals move from the glomerular capillaries into the renal tubules. In *tubular reabsorption,* some of these chemicals move back into the blood plasma. In *tubular secretion,* certain

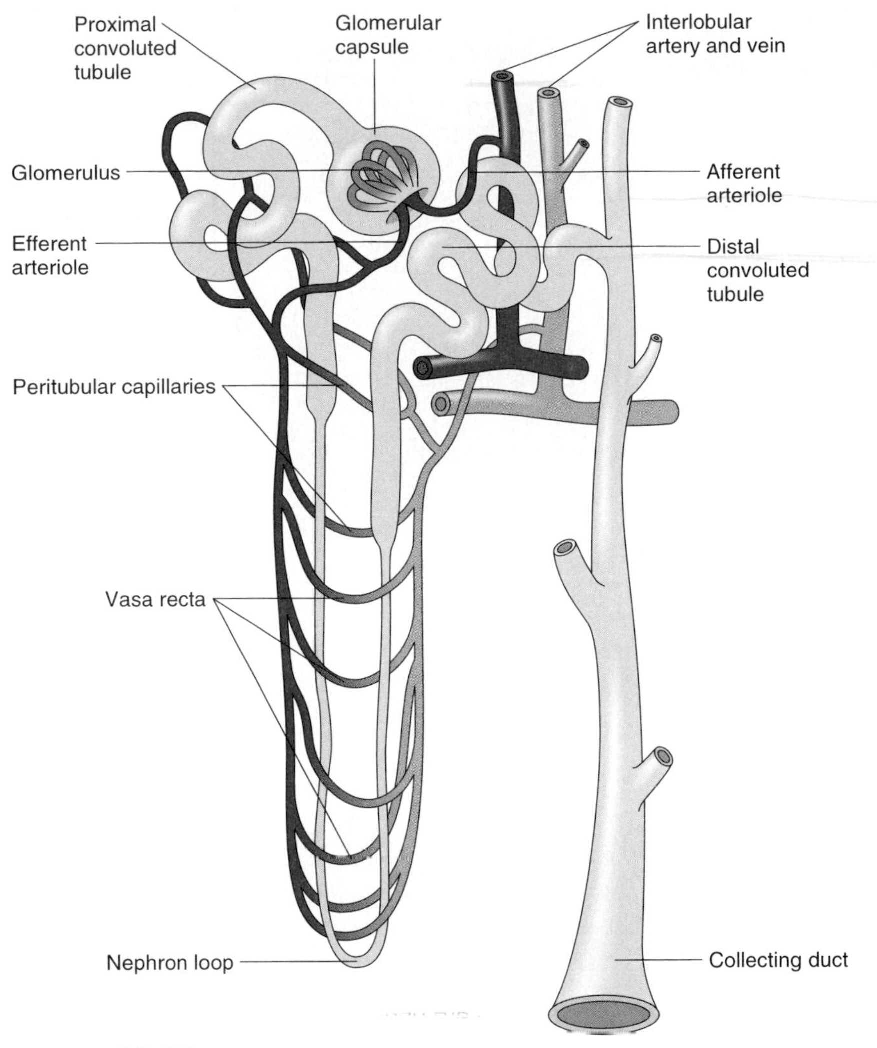


**FIGURE 20.13**
The capillary loop of the vasa recta is closely associated with the nephron loop of a juxtamedullary nephron.

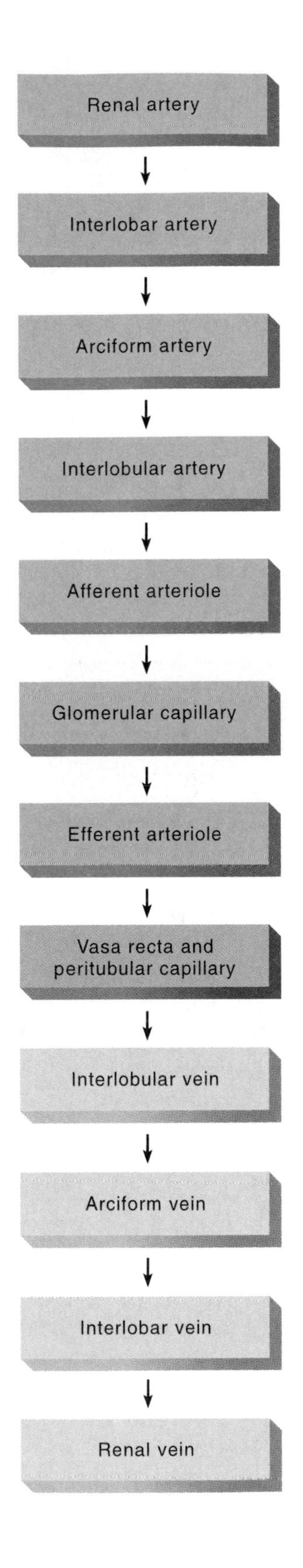


**FIGURE 20.14**
Pathway of blood through the blood vessels of the kidney and nephron.

other chemicals move from the peritubular capillaries into the renal tubules. In other words, the following relationship determines the amount of substance excreted in the urine:

$$\text{urinary excretion} = \text{glomerular filtration} + \text{tubular secretion} - \text{tubular reabsorption}$$

## Glomerular Filtration

Urine formation begins when water and other small dissolved molecules and ions are filtered out of the glomerular capillary plasma and into the glomerular capsules. The filtration of these materials through the capillary walls is much like the filtration that occurs at the arteriole ends of other capillaries throughout the body. The glomerular capillaries, however, are

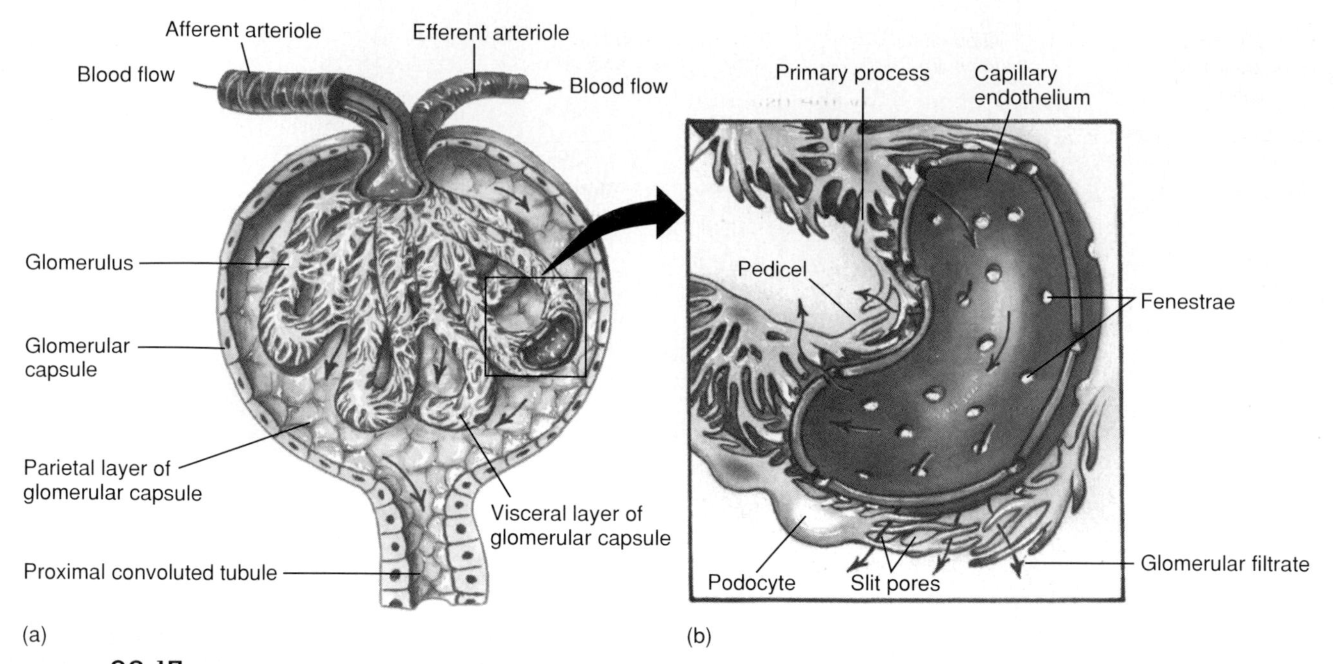


**FIGURE 20.15**

(*a*) The first step in urine formation is filtration of substances through the glomerular membrane into the glomerular capsule. (*b*) The glomerular filtrate passes through fenestrae of the capillary endothelium.

**CHART 20.1 RELATIVE CONCENTRATIONS OF PLASMA, GLOMERULAR FILTRATE, AND URINE COMPONENTS**

| | Concentrations (mEq/l) | | |
|---|---|---|---|
| *Substance* | *Plasma* | *Glomerular Filtrate* | *Urine* |
| Sodium ($Na^+$) | 142 | 142 | 128 |
| Potassium ($K^+$) | 5 | 5 | 60 |
| Calcium ($Ca^{+2}$) | 4 | 4 | 5 |
| Magnesium ($Mg^{+2}$) | 3 | 3 | 15 |
| Chloride ($Cl^-$) | 103 | 103 | 134 |
| Bicarbonate ($HCO_3^-$) | 27 | 27 | 14 |
| Sulfate ($SO_4^{-2}$) | 1 | 1 | 33 |
| Phosphate ($PO_4^{-3}$) | 2 | 2 | 40 |

| | Concentrations (mg/100 ml) | | |
|---|---|---|---|
| *Substance* | *Plasma* | *Glomerular Filtrate* | *Urine* |
| Glucose | 100 | 100 | 0 |
| Urea | 26 | 26 | 1,820 |
| Uric acid | 4 | 4 | 53 |
| Creatinine | 1 | 1 | 196 |

many times more permeable to small molecules than the capillaries in other tissues, due to the many tiny openings (fenestrae) in their walls (fig. 20.15).

The glomerular capsule receives the resulting **glomerular filtrate,** which has about the same composition as the filtrate that becomes tissue fluid elsewhere in the body. That is, glomerular filtrate is mostly water and the same solutes as in blood plasma, except for the larger protein molecules. More specifically, glomerular filtrate contains water, glucose, amino acids, urea, uric acid, creatine, creatinine, and sodium, chloride, potassium, calcium, bicarbonate, phosphate, and sulfate ions. Chart 20.1 shows the relative concentrations of some of the substances in the blood plasma, glomerular filtrate, and urine.

## Filtration Pressure

As in the case of other capillaries, the main force responsible for moving substances through the glomerular capillary wall is the hydrostatic pressure of the blood inside. (Recall that glomerular capillary pressure is relatively high compared to other capillaries.) Glomerular filtration is also influenced by the osmotic pressure of the blood plasma in the glomerulus and by the hydrostatic pressure inside the glomerular capsule.

Due to the plasma proteins, the osmotic pressure of the plasma is always higher than that of the glomerular filtrate (except in some kinds of kidney disease). This tends to draw water back into the glomerular capillaries, thus opposing filtration. Any increase in glomerular capsule hydrostatic pressure would also oppose filtration.

The net pressure that forces substances out of the glomerulus is called **filtration pressure,** determined by the above factors. It can be calculated as follows:

Filtration pressure =

force favoring filtration – forces opposing filtration

| (glomerular capillary hydrostatic pressure) | (capsular hydrostatic pressure and glomerular capillary osmotic pressure) |
|---|---|

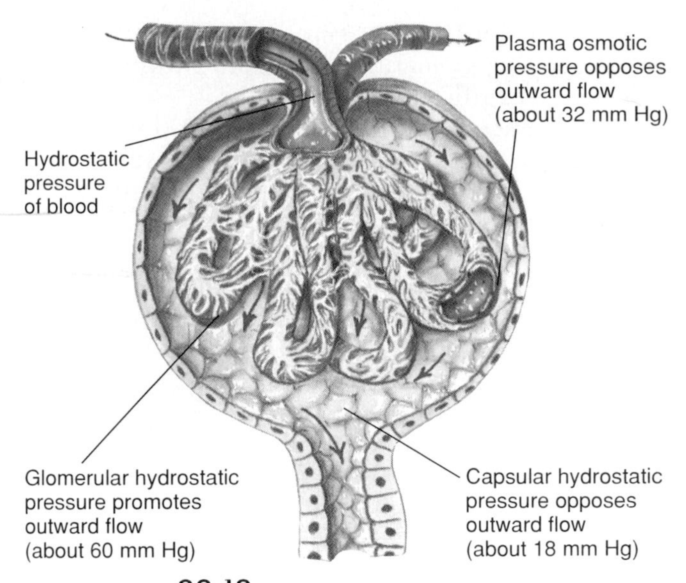


**FIGURE 20.16**

The rate of glomerular filtration is affected by the hydrostatic and osmotic pressure of the plasma, and the hydrostatic pressure of the fluid in the glomerular capsule.

> The concentrations of certain components of the blood plasma can be used to evaluate kidney functions. For example, if the kidneys are functioning inadequately, the plasma concentrations of urea (as indicated by a blood urea nitrogen test) and of creatinine may increase as much as tenfold above normal.

> If arterial blood pressure drops drastically, as may occur during *shock,* the glomerular hydrostatic pressure may fall below the level required for filtration, leading to acute renal failure. At the same time, the epithelial cells of the renal tubules may fail to receive sufficient nutrients to maintain their high rates of metabolism. As a result, cells may die (tubular necrosis), and renal functions may be lost permanently, resulting in chronic renal failure.

## Filtration Rate

The glomerular filtration rate (GFR) is directly proportional to the filtration pressure described in figure 20.16. Consequently, the factors that affect the glomerular hydrostatic pressure, glomerular plasma osmotic pressure, or hydrostatic pressure in the glomerular capsule will also affect the rate of filtration.

Normally, glomerular hydrostatic pressure is the most important factor determining GFR. Since the glomerular capillary is located between two arterioles—the afferent and efferent arterioles—any change in the diameters of these vessels is likely to change glomerular hydrostatic pressure, causing a change in glomerular filtration rate. The afferent arteriole, through which the blood enters the glomerulus, may constrict in response to mild stimulation by sympathetic nerve impulses. If this occurs, blood flow diminishes, glomerular hydrostatic pressure decreases, and filtration rate drops. If, on the other hand, the efferent arteriole (through which the blood leaves the glomerulus) constricts, blood backs up into the glomerulus, glomerular hydrostatic pressure increases, and filtration rate rises. Vasodilation of these vessels produces opposite effects.

Recall that the osmotic pressure of the glomerular plasma also influences filtration pressure and the rate of filtration. In capillaries, the blood hydrostatic pressure, forcing water and dissolved substances outward, is opposed by the plasma osmotic pressure that attracts water inward. Actually, as filtration occurs through the capillary wall, proteins remaining in the plasma raise the osmotic pressure within the glomerular capillary. When this pressure reaches a certain high level, filtration ceases. Conversely, conditions that lower plasma osmotic pressure, such as a decrease in plasma protein concentration, increase filtration rate.

The hydrostatic pressure in the glomerular capsule is another factor that may affect filtration pressure and rate. This capsular pressure sometimes changes as a result of an obstruction, such as a stone in a ureter or an enlarged prostate gland pressing on the urethra. If this occurs, fluids back up into the renal tubules and raise the hydrostatic pressure in the glomerular capsules. Because any increase in capsular pressure opposes glomerular filtration, filtration rate may decrease significantly.

In an average adult, the glomerular filtration rate for the nephrons of both kidneys is about 125 milliliters per minute, or 180,000 milliliters (180 liters) in 24 hours. Assuming that the blood plasma volume is about 3 liters, the production of 180 liters of filtrate in 24 hours means that all of the plasma must be filtered through the glomeruli about 60 times each day (fig. 20.17). Since this 24-hour volume is nearly 45 gallons, it is obvious that not all of it is excreted as urine. Instead, most of the fluid that passes through the renal tubules is reabsorbed and reenters the plasma.

The volume of plasma the kidneys filter is also related to the amount of *surface area* within the glomerular capillaries. This surface area is estimated to be about 2 square meters—approximately equal to the surface area of an adult's skin.

An injury to a kidney can be more dangerous than an injury to another organ. An injured kidney produces a protein called *transforming growth factor beta*. This protein causes scars to form. The scars further damage the kidney, impairing its function.

1. What processes occur in urine formation?
2. How is filtration pressure calculated?
3. What factors influence the rate of glomerular filtration?

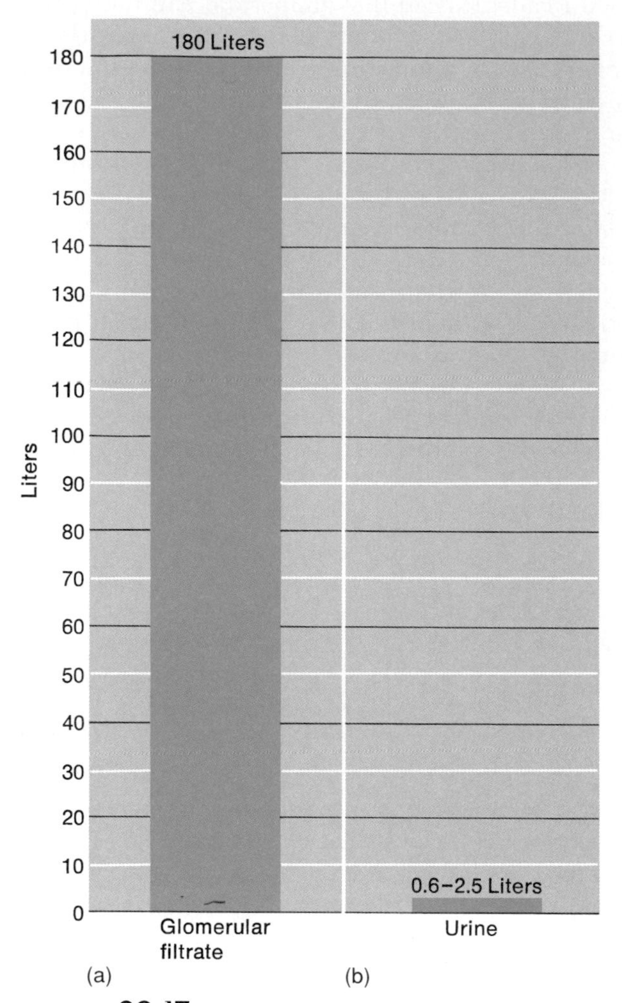


**FIGURE 20.17**
Relative amounts of (*a*) glomerular filtrate and (*b*) urine formed in 24 hours.

## Regulation of Filtration Rate

In general, glomerular filtration rate remains relatively constant through a process called **autoregulation.** However, certain conditions override autoregulation. GFR may increase, for example, when body fluids are in excess, and decrease when the body needs to conserve fluid.

Recall from chapter 10 that sympathetic nervous system fibers synapse with the vascular smooth muscle of arterioles. Reflexes responding to changes in blood pressure and volume control the activity of these sympathetic fibers. If blood pressure and volume drop, vasoconstriction of the afferent arterioles results, decreasing GFR. The result is an appropriate decrease in the rate of urine formation when the body must conserve water. If receptors detect excess body fluids, vasodilation of the afferent arteriole results, increasing GFR. A second control of GFR is a hormone-like system called the **renin-angiotensin system,** whose primary function seems to be regulation of renal sodium excretion.

Renin is an enzyme the juxtaglomerular cells of the afferent arterioles secrete in response to stimulation from sympathetic nerves and pressure sensitive cells called **renal baroreceptors** in the afferent arteriole. These factors stimulate renin secretion if blood pressure drops. The macula densa also controls renin secretion. Somehow the cells of the macula sense the concentration of sodium, potassium, and chloride ions in the distal renal tubule. Decreasing levels of these ions stimulates renin secretion.

Once in the bloodstream, renin reacts with the plasma protein *angiotensinogen* to form *angiotensin I.* An enzyme, *angiotensin converting enzyme* (ACE), present on capillary endothelial cells (particularly in the lungs), rapidly converts angiotensin I to *angiotensin II.*

Angiotensin II has a number of renal effects which help maintain sodium balance, water balance, and blood pressure (fig. 20.18). As a vasoconstrictor, it affects both the afferent and efferent arterioles. Although

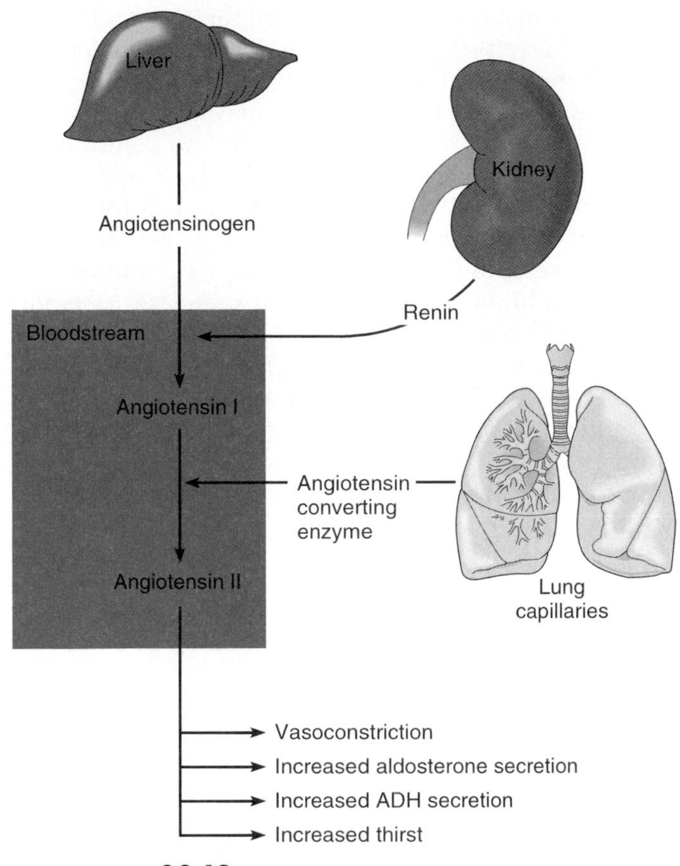


**FIGURE 20.18**
The formation of angiotensin II in the bloodstream involves several organs, and has multiple actions that conserve sodium and water.

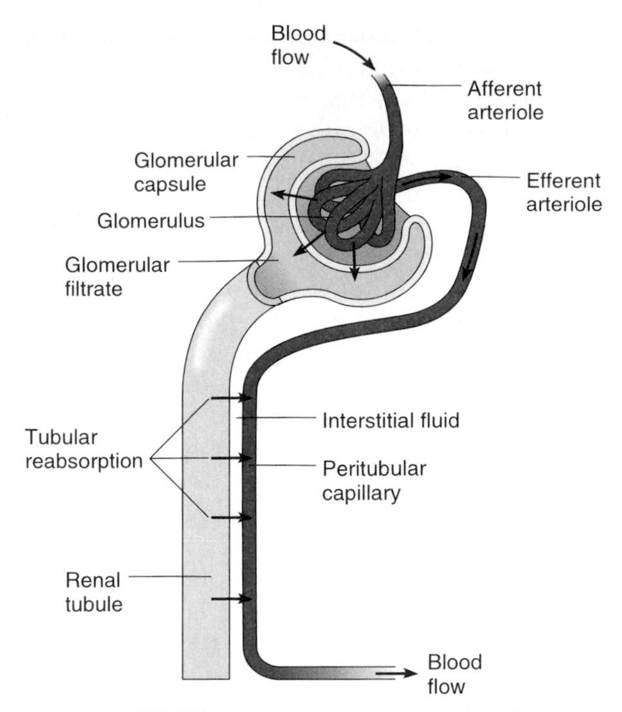


**FIGURE 20.19**
Reabsorption is the process by which substances are transported from the glomerular filtrate into the blood of the peritubular capillary.

afferent arteriolar constriction decreases GFR, efferent arteriolar constriction tends to minimize the decrease, thus contributing to autoregulation. Angiotensin II has a major affect on the kidneys through the adrenal cortical hormone aldosterone, which stimulates sodium reabsorption in the distal tubule. By stimulating aldosterone secretion, angiotensin II helps to reduce the amount of sodium excreted in the urine.

The hormone **atrial natriuretic peptide** (ANP) also controls GFR. ANP secretion increases when the atria of the heart stretch due to increased blood volume. ANP stimulates sodium excretion by a number of mechanisms, including increasing GFR.

## Tubular Reabsorption

If the composition of the glomerular filtrate entering the renal tubule is compared with that of the urine leaving the tubule, it is clear that changes occur as the fluid passes through the tubule (see chart 20.1). For example, glucose is present in the filtrate, but absent in the urine. Also, urea and uric acid are considerably more concentrated in urine than they are in the glomerular filtrate. Such changes in fluid composition are largely the result of **tubular reabsorption,** the process by which substances are transported out of the tubular fluid, through the epithelium of the renal tubule, and into the interstitial fluid. These substances then diffuse into the peritubular capillaries (fig. 20.19).

Peritubular capillary blood is under relatively low pressure because it has already passed through two arterioles. Also, the wall of the peritubular capillary is more permeable than that of other capillaries. Finally, the relatively high rate of glomerular filtration has increased the protein concentration of the peritubular capillary plasma. All of these factors enhance the rate of fluid reabsorption from the renal tubule.

Although tubular reabsorption occurs throughout the renal tubule, most of it occurs in the proximal convoluted portion. The epithelial cells in this portion have many *microvilli* that form a "brush border" on their free surfaces facing the tubular lumen (see chapter 17). These tiny extensions greatly increase the surface area exposed to the glomerular filtrate and enhance reabsorption (see fig. 17.39).

Segments of the renal tubule are adapted to reabsorb specific substances, using particular modes of transport. Glucose reabsorption, for example, occurs through the walls of the proximal tubule by active transport. Water also is reabsorbed rapidly through the epithelium of the proximal tubule by osmosis; however, portions of the distal tubule and collecting duct are almost impermeable to water. This characteristic of the distal tubule is important in the regulation of urine concentration and volume, as we will soon see.

As described in chapter 3, active transport depends on the presence of carrier proteins in a cell membrane. Exactly how these carrier molecules work is not known, but it appears to involve a binding site on the carrier and the ability of proteins to alter their shapes. The molecule to be transported binds to the carrier; the carrier alters its shape, releases the transported molecule on the other side of the cell membrane, and then returns to its original position to repeat the process. Such a mechanism has a *limited transport capacity;* that is, it can only transport a certain number of molecules in a given amount of time because the number of carriers is limited.

Usually all of the glucose in the glomerular filtrate is reabsorbed, because there are enough carrier molecules to transport it. When plasma glucose concentration increases to a critical level, called the *renal plasma threshold,* more glucose molecules are in the filtrate than the active transport mechanism can handle. As a result, some glucose remains in the filtrate and is excreted in the urine.

Any increase in urine volume is called *diuresis.* Glucose in the filtrate will draw water into the renal tubule by osmosis, thus increasing urine volume. Such an increase is called an *osmotic diuresis.*

Glucose in the urine is called *glucosuria.* It may follow intravenous administration of glucose, epinephrine, or eating a candy bar, or it may occur in a person with insulin-dependent (type I) diabetes mellitus. In the case of diabetes, blood glucose concentration rises because of insufficient insulin from the pancreas.

One in three people who have diabetes mellitus sustains kidney damage as part of their condition, as evidenced by protein in their urine. Following a low-protein diet can slow the loss of kidney function.

*Amino acids* also enter the glomerular filtrate and are reabsorbed in the proximal convoluted tubule. Three different active transport mechanisms are thought to reabsorb different groups of amino acids, whose members have similar structures. As a result, normally only a trace of amino acids remains in the urine.

Although the glomerular filtrate is nearly free of protein, a number of smaller protein molecules, such as albumin, somehow squeeze through the glomerular capillaries. These proteins are reabsorbed by *pinocytosis* through the brush border of epithelial cells lining the proximal convoluted tubule. Once they are inside an epithelial cell, the proteins are degraded to amino acids and moved into the blood of the peritubular capillary.

The epithelium of the proximal convoluted tubule also reabsorbs creatine; lactic, citric, uric, and ascorbic (vitamin C) acids; and phosphate, sulfate, calcium, potassium, and sodium ions. Active transport mechanisms with limited transport capacities reabsorb all of these chemicals. Such a substance will appear in the urine in greater amounts when its concentration in the glomerular filtrate exceeds its particular renal plasma threshold. Clinical Application 20.3 discusses how plasma proteins in the urine result from the nephrotic syndrome.

## Sodium and Water Reabsorption

Water reabsorption occurs passively by osmosis, primarily in the proximal convoluted tubule, and is closely associated with the active reabsorption of sodium ions. In the proximal tubule, if sodium reabsorption increases, water reabsorption increases; if sodium reabsorption decreases, water reabsorption decreases also.

Much of the sodium ion reabsorption occurs in the proximal segment of the renal tubule by active transport (sodium pump mechanism). When the positively charged sodium ions ($Na^+$) are moved through the tubular wall, negatively charged ions, including chloride ions ($Cl^-$), phosphate ions ($PO_4^{-3}$), and bicarbonate ions ($HCO_3^-$), accompany them. This movement of negatively charged ions is due to the electrochemical attraction between particles of opposite electrical charge. Although it depends on active transport of sodium, this movement of negatively charged ions is considered a passive process because it does not require a direct expenditure of cellular energy. Some negative ions (such as $HCO_3^-$) are also reabsorbed by active transport.

As more and more sodium ions are reabsorbed into the peritubular capillary along with negatively charged ions, the concentration of solutes within the peritubular blood might be expected to increase. However, since water diffuses through cell membranes from regions of lesser solute concentration (hypotonic) toward regions of greater solute concentration (hypertonic), water moves by osmosis from the renal tubule into the peritubular capillary.

The proximal tubule reabsorbs about 70% of the filtered sodium, other ions, and water. By the end of the proximal tubule, osmotic equilibrium is reached, and the remaining tubular fluid is isotonic (fig. 20.20).

Active transport continues to reabsorb sodium ions as the tubular fluid moves through the nephron loop, the distal convoluted tubule, and the collecting

CLINICAL APPLICATION 20.3

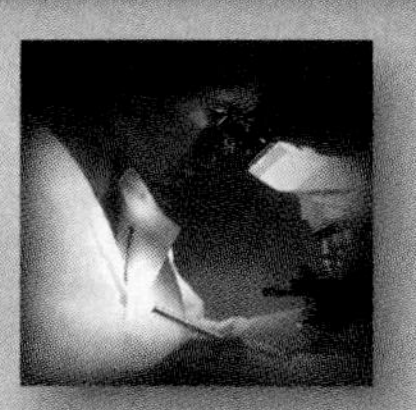

## The Nephrotic Syndrome

The *nephrotic syndrome* is a set of symptoms that often appears in patients with renal diseases. It causes considerable loss of plasma proteins into the urine (proteinuria), widespread edema, and increased susceptibility to infections.

Plasma proteins are lost into the urine because of increased permeability of the glomerular membranes, which accompanies renal disorders such as glomerulonephritis. As a consequence of a decreasing plasma protein concentration (hypoproteinemia), the plasma osmotic pressure falls, increasing filtration pressure in capillaries throughout the body. This may lead to widespread, severe edema as a large volume of fluid accumulates in the interstitial spaces within the tissues and in body spaces such as the abdominal cavity, pleural cavity, pericardial cavity, and joint cavities.

Also, as edema develops, blood volume decreases and blood pressure drops. These changes may activate the renin-angiotensin system, leading to the release of aldosterone from the adrenal cortex (see chapter 13), which, in turn, stimulates the kidneys to conserve sodium ions and water. This action reduces the urine output and may aggravate the edema.

The nephrotic syndrome sometimes appears in young children who have *lipoid nephrosis*. The cause of this condition is unknown, but it alters the epithelial cells of the glomeruli so that the glomerular membranes enlarge and distort, allowing proteins to leak through.

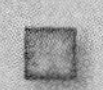

duct. Consequently, almost all of the sodium and water (97%–99%) that enters the renal tubules as part of the glomerular filtrate may be reabsorbed before the urine is excreted. However, aldosterone controls sodium reabsorption and antidiuretic hormone controls water reabsorption, and reabsorption of these substances can change with conditions in the body fluids. The next chapter discusses this further.

Recall that the kidneys filter an extremely large volume of fluid (180 liters) each day. Thus, if 99% of the glomerular filtrate is reabsorbed, the remaining 1% excreted includes a relatively large amount of sodium and water. Furthermore, if sodium and water reabsorption decrease to 97% of the amount filtered, the amount excreted triples! Therefore, small changes in the tubular reabsorption of water and sodium result in large changes in urinary excretion.

1. How is the peritubular capillary adapted for reabsorption?
2. What substances present in the glomerular filtrate are not normally present in the urine?
3. What mechanisms reabsorb solutes from the glomerular filtrate?
4. Define the renal plasma threshold.
5. Describe the role of passive transport in urine formation.

### Regulation of Urine Concentration and Volume

In contrast to conditions in the proximal tubule, the tubular fluid reaching the distal convoluted tubule is hypotonic to its surroundings due to changes that occur through the loop segment of the nephron. The cells lining the distal tubule and the collecting duct that follows continue to reabsorb sodium and chloride ions under the influence of aldosterone, which the adrenal cortex secretes (see chapter 13). In addition, the interstitial fluid surrounding the collecting ducts is hypertonic, particularly in the medulla. These might seem to be ideal conditions for water reabsorption as well. However, the cells lining the later portion of the distal convoluted tubule and the collecting duct are impermeable to water. Thus, water tends to accumulate inside the tubule and may be excreted as dilute urine.

As discussed in chapter 13, neurosecretory cells in the hypothalamus produce antidiuretic hormone (ADH). The posterior lobe of the pituitary gland releases ADH in response to decreasing concentration of water in the body fluids or to decreasing blood volume and blood pressure. When ADH reaches the kidney, it increases permeability of the epithelial cell linings of the distal convoluted tubule and the collecting duct; consequently, water moves rapidly out of these segments by osmosis, especially where the collecting duct passes through the extremely hypertonic

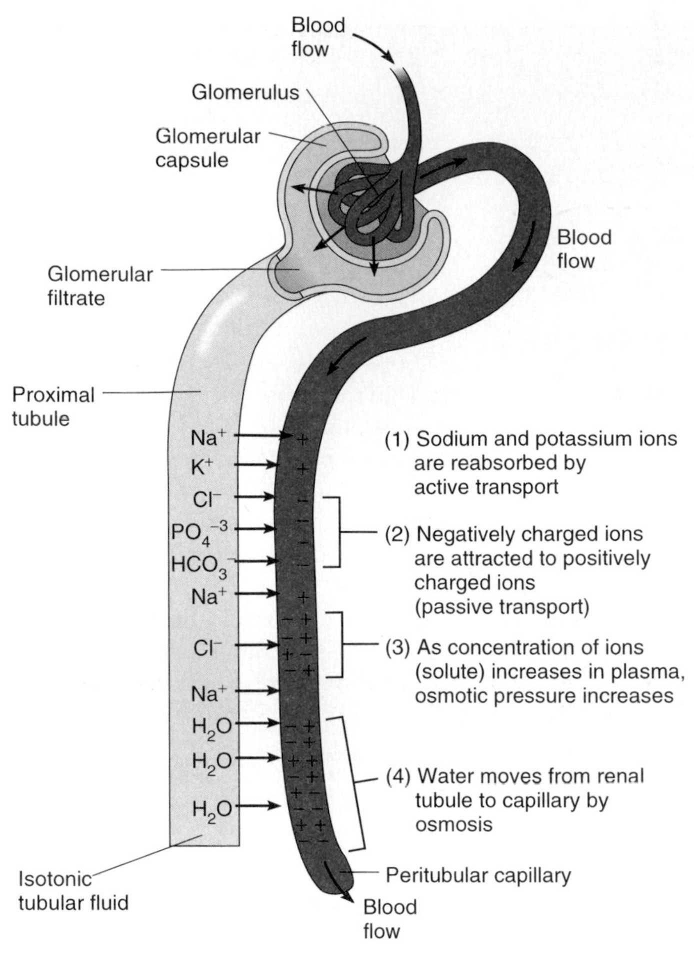


**FIGURE 20.20**

Water reabsorption by osmosis occurs in response to the reabsorption of sodium and other solutes by active transport in the proximal tubule.

medulla. Thus, the urine becomes more concentrated (hypertonic), and water is retained in the internal environment (fig. 20.21).

Examining the **countercurrent mechanism,** which involves the nephron loops, particularly of the juxtamedullary nephrons, reveals how the medullary interstitial fluid becomes so hypertonic. The descending and ascending limbs of these U-shaped structures lie parallel and are very close to one another. This mechanism is named partly for the fact that fluid moving down the descending limb creates a current that is counter to that of the fluid moving up in the ascending limb.

The different parts of the nephron loop have important functional differences. For example, the epithelial lining in the thick upper portion of the ascending limb (thick segment) is relatively impermeable to water. However, the epithelium does actively reabsorb sodium and chloride ions (some potassium is actively reabsorbed as well). As these solutes accumulate in the interstitial fluid outside the ascending limb, it becomes hypertonic, while the tubular fluid inside becomes hypotonic because it is losing its solute.

In contrast, the epithelium of the descending limb (thin segment) is quite permeable to water, but relatively impermeable to solutes. Because this segment is surrounded by hypertonic fluid created by the ascending limb, water tends to leave the descending limb by osmosis. Thus, the contents of the descending limb become more and more concentrated, or hypertonic (fig. 20.22).

This very concentrated tubular fluid now moves into the ascending limb, and sodium chloride (NaCl) is again actively reabsorbed into the medullary interstitial fluid, raising the interstitial NaCl concentration even more. With the increased interstitial fluid solute concentration, even more water diffuses out of the descending limb, further increasing the salt concentration of the tubular fluid. Each time this circuit is completed, the concentration of NaCl increases, or multiplies. For this reason, the mechanism is called a *countercurrent multiplier.* In humans, this mechanism creates a tubular fluid solute concentration near the tip of the loop that is more than four times the solute concentration of plasma (fig. 20.23*a*).

The solute concentration of the tubular fluid decreases progressively toward the renal cortex. Since the descending limb of the loop is permeable to water, the interstitial fluid at any level of the loop is essentially in equilibrium with the fluid in the tubule. Thus, the concentration gradient in the loop is also found in the interstitial fluid (fig. 20.23*b*).

The vasa recta is another countercurrent mechanism that maintains the NaCl concentration gradient in the renal medulla. Blood flows relatively slowly down the descending portion of the vasa recta, and NaCl enters it by diffusion. Then, as the blood moves back up toward the renal cortex, most of the NaCl diffuses from the blood and reenters the medullary interstitial fluid. Consequently, the bloodstream carries little NaCl away from the renal medulla, preserving the gradient (fig. 20.24).

To summarize, the countercurrent multiplier creates a large concentration gradient for water reabsorption in the interstitial fluid surrounding the distal convoluted tubules and the collecting ducts of the nephron. Although the epithelial lining of these structures is impermeable to water, in the presence of ADH it becomes permeable. The higher the blood levels of ADH, the more permeable it becomes. In this way, ADH stimulates the production of concentrated urine,

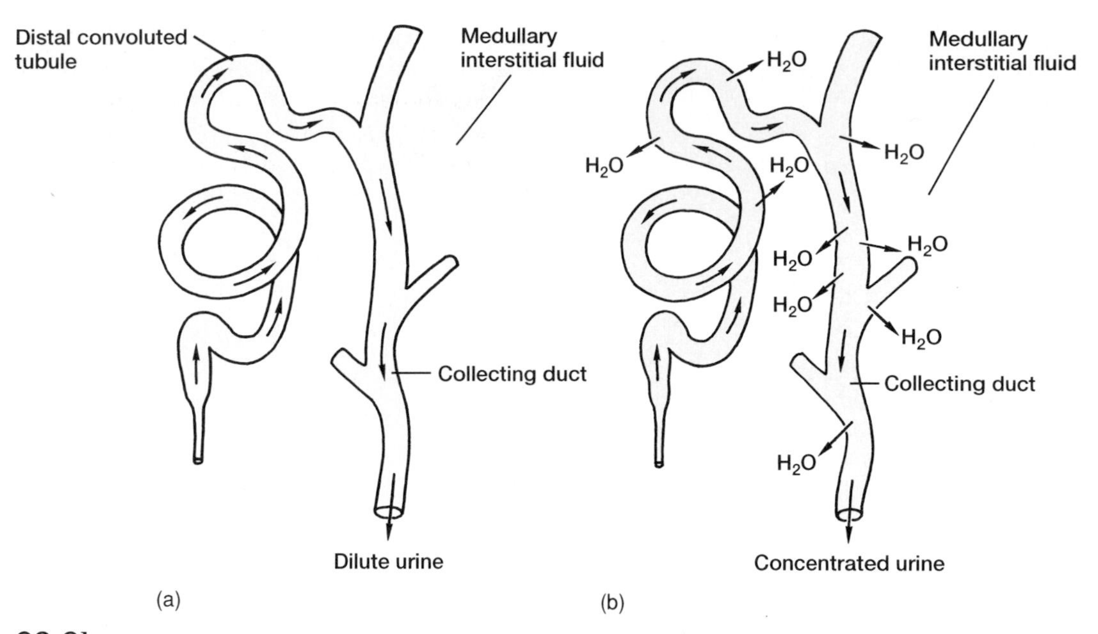


FIGURE 20.21

(*a*) The distal convoluted tubule and collecting duct are impermeable to water, so water may be excreted as dilute urine. (*b*) If ADH is present, however, these segments become permeable, and water is reabsorbed by osmosis into the hypertonic medullary interstitial fluid.

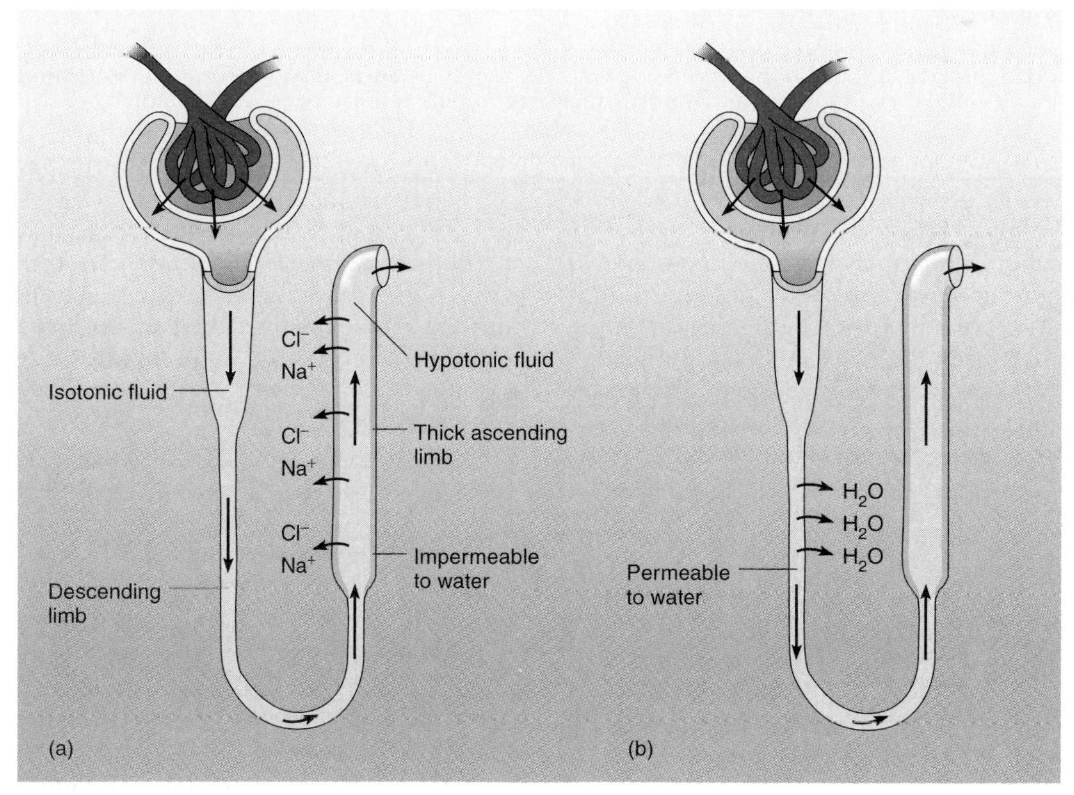


FIGURE 20.22

(*a*) Fluid in the ascending limb of the nephron loop becomes hypotonic as solute is reabsorbed. (*b*) Fluid in the descending limb becomes hypertonic as it loses water by osmosis.

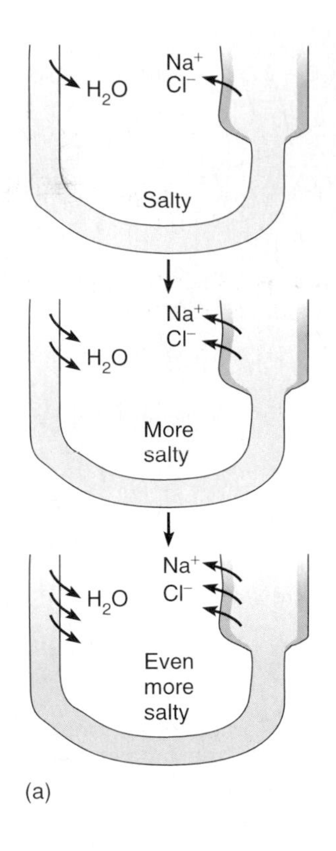


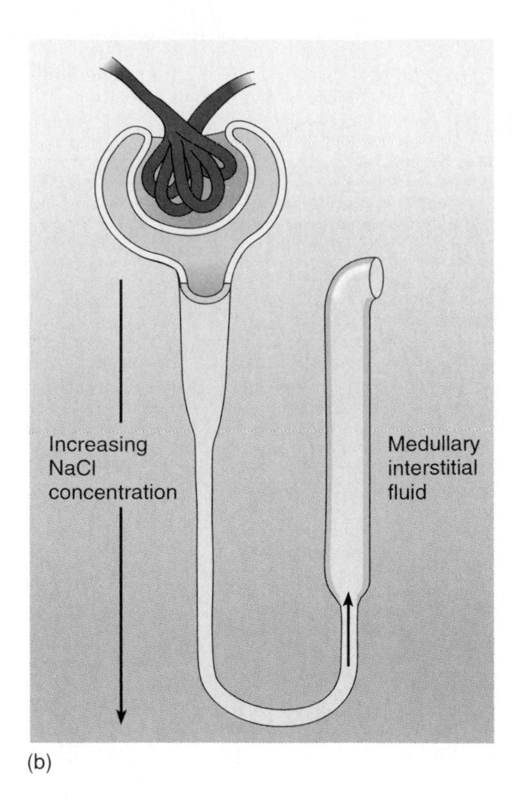


**FIGURE 20.23**

(*a*) As NaCl completes the countercurrent circuit again and again, more water exits the descending limb, leaving more concentrated solute in the tubular fluid in the ascending limb. This, in turn, transports even more solute into the interstitial fluid. (*b*) As a result, an NaCl concentration gradient is established in the medullary interstitial fluid.

which contains soluble wastes and other substances in a minimum of water, thus minimizing the loss of body water when dehydration is a threat. If the body fluids become too dilute, ADH secretion is decreased, and the epithelial linings of the distal segment and the collecting duct become less permeable to water. Thus less water is reabsorbed, and the urine will be more dilute.

Chart 20.2 summarizes the role of ADH in urine production.

## Urea and Uric Acid Excretion

Because **urea** is a by-product of amino acid catabolism made in the liver, its plasma concentration is directly related to the amount of protein in the diet. Urea enters the renal tubule by filtration. About 50% of it is reabsorbed (passively) by diffusion, but the remainder is excreted in the urine.

Urea moves within the kidney in other ways, including the countercurrent multiplier mechanism that concentrates the medullary interstitial fluid. As a result, urea contributes to the reabsorption of water from the collecting duct.

**Uric acid,** which results from the metabolism of certain organic bases (purines) in nucleic acids, is reabsorbed by active transport. About 10% of the amount filtered is excreted in the urine. It is apparently secreted into the renal tubule.

1. Describe a countercurrent mechanism.
2. What role does the hypothalamus play in regulating urine concentration and volume?
3. Explain how urea and uric acid are excreted.

## Tubular Secretion

In **tubular secretion,** certain substances move from the plasma of the peritubular capillary into the fluid of the renal tubule. As a result, the amount of a particular chemical excreted in the urine may be greater than the amount filtered from the plasma in the glomerulus (fig. 20.25).

Active transport mechanisms similar to those that function in reabsorption secrete some substances. However, the *secretory mechanisms* transport substances in the opposite direction. For example, the epithelium of

## CHART 20.2 Role of ADH in Regulating Urine Concentration and Volume

1. Concentration of water in the blood decreases.
2. Increase in the osmotic pressure of body fluids stimulates osmoreceptors in the hypothalamus.
3. Hypothalamus signals the posterior pituitary gland to release ADH.
4. Blood carries ADH to the kidneys.
5. ADH causes the distal convoluted tubules and collecting ducts to increase water reabsorption by osmosis.
6. Urine becomes more concentrated and the urine volume decreases.

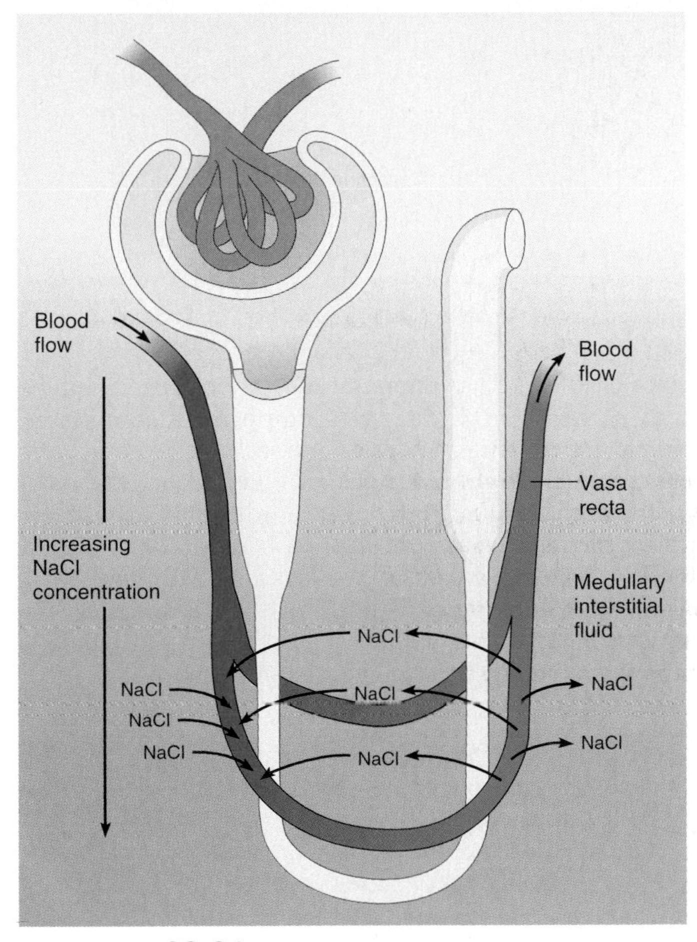


**FIGURE 20.24**
A countercurrent mechanism in the vasa recta helps maintain the NaCl concentration gradient in the medullary interstitial fluid (see fig. 20.13).

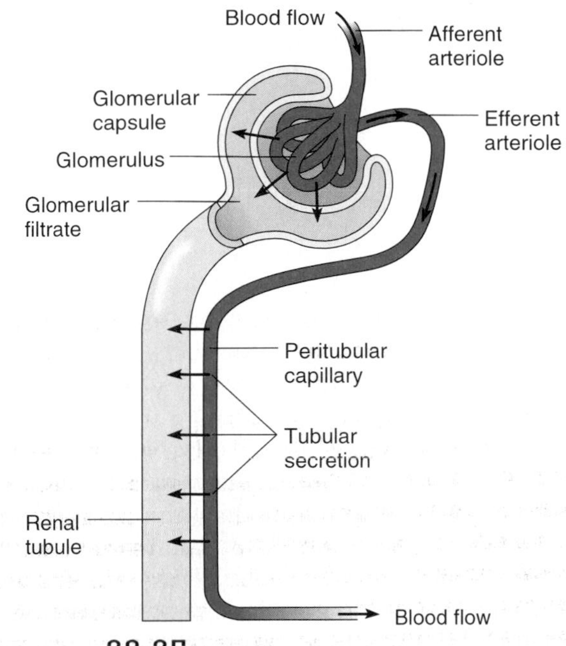


**FIGURE 20.25**
Secretory mechanisms move substances from the plasma of the peritubular capillary into the fluid of the renal tubule.

the proximal convoluted segment actively secretes certain organic compounds, including penicillin and histamine, into the tubular fluid.

Hydrogen ions are also actively secreted throughout the entire renal tubule. As a result, the urine is usually acidic by the time it is excreted, although the urinary pH can vary considerably. The secretion of hydrogen ions plays an important role in regulating the pH of body fluids, as is explained in chapter 21.

> For the 20,000 people in the United States diagnosed with kidney cancer each year, surgery to remove the affected kidney is the first treatment. However, in half the cases, cancer returns, usually in the lungs. A promising new treatment for this metastatic kidney cancer is the immune system protein interleukin-2. It is administered intravenously in cycles in a hospital setting because of sometimes severe side effects. Interleukin-2 presumably stimulates the immune system to attack tumor cells. In about 15% of patients, this drug shrinks lung tumors.

Although most of the potassium ions in the glomerular filtrate are actively reabsorbed in the proximal convoluted tubule, some may be secreted in the distal segment and collecting duct. During this process, the active reabsorption of sodium ions out of

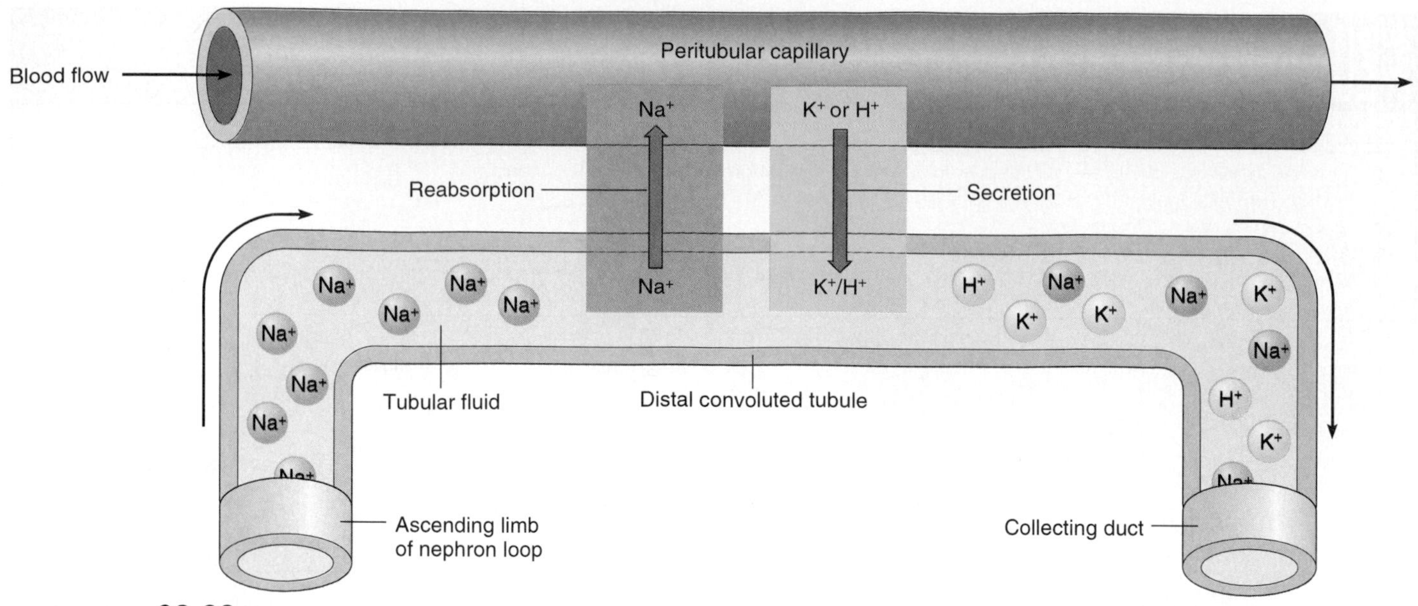


**FIGURE 20.26**

Passive secretion of potassium ions (or hydrogen ions) may occur in response to the active reabsorption of sodium ions.

the tubular fluid under the influence of aldosterone produces a negative electrical charge within the tube. Because positively charged potassium ions ($K^+$) are attracted to regions that are negatively charged, these ions move passively through the tubular epithelium and enter the tubular fluid. Potassium ions are also secreted by active processes (fig. 20.26).

To summarize, urine forms as a result of the following:

- Glomerular filtration of materials from blood plasma.
- Reabsorption of substances including glucose; water; urea; proteins; creatine; amino, lactic, citric, and uric acids; and phosphate, sulfate, calcium, potassium, and sodium ions.
- Secretion of substances including penicillin, histamine, phenobarbital, hydrogen ions, ammonia, and potassium ions (chart 20.3).

1 Define *tubular secretion.*

2 What substances are actively secreted? Passively secreted?

3 How does the reabsorption of sodium affect the secretion of potassium?

## Urine Composition

Urine composition reflects the amounts of water and solutes that the kidneys must eliminate from the body or retain in the internal environment to maintain homeostasis. It varies considerably from time to time because of variations in dietary intake and physical activity. In addition to containing about 95% water, urine usually contains urea from the catabolic metabolism of amino acids, uric acid from the catabolic metabolism of nucleic acids, and creatinine from the metabolism of creatine. It may also contain a trace of amino acids, as well as electrolytes whose concentrations vary directly with the amounts included in the diet (see chart 20.1). Appendix C lists the normal concentrations of urine components.

> Abnormal constituents of urine may not indicate illness. Glucose in the urine may reflect a sugary meal or may occur toward the end of pregnancy; protein may appear in the urine following vigorous physical exercise; ketones appear in the urine during a prolonged fast or during a very low calorie diet.

The volume of urine produced usually varies between 0.6 and 2.5 liters per day. The exact volume is influenced by such factors as fluid intake, environmental temperature, relative humidity of the surrounding air, and a person's emotional condition, respiratory rate, and body temperature. An output of 50–60 milliliters of urine per hour is considered normal, and an output of less than 30 milliliters per hour may indicate kidney failure.

Clinical Application 20.4 discusses renal clearance, which is a measure of kidney efficiency.

CLINICAL APPLICATION 20.4

## Renal Clearance

The rate at which a particular chemical is removed from the plasma indicates kidney efficiency. This rate of removal is called *renal clearance.*

Tests of renal clearance can detect glomerular damage or judge the progress of renal disease. One such test, the *inulin clearance test,* uses *inulin* (not to be confused with insulin), a complex polysaccharide found in certain plant roots. In the test, a known amount of inulin is infused into the blood at a constant rate. The inulin passes freely through the glomerular membranes, so that its concentration in the glomerular filtrate equals that of the plasma. In the renal tubule, inulin is not reabsorbed to any significant degree, nor is it secreted. Consequently, the rate at which it appears in the urine can be used to calculate the rate of glomerular filtration.

Similarly, the kidneys remove creatinine, which is produced at a constant rate as a result of muscle metabolism, from the blood. Like inulin, creatinine is filtered, but neither reabsorbed nor secreted by the kidneys. Thus, the *creatinine clearance test,* which compares a patient's blood and urine creatinine concentrations, can also be used to calculate the GFR. A significant advantage is that the bloodstream normally has a constant level of creatinine. Therefore, a single measurement of plasma creatinine levels provides a rough index of kidney function. For example, significantly elevated plasma creatinine levels suggest that GFR is greatly reduced. Because nearly all of the creatinine the kidneys filter normally appears in the urine, a change in the rate of creatinine excretion may reflect a renal disorder.

Another plasma clearance test uses *para-aminohippuric acid* (PAH), a substance that filters freely through the glomerular membranes. However, unlike inulin, any PAH remaining in the peritubular capillary plasma after filtration is secreted into the proximal convoluted tubules. Therefore, essentially all PAH passing through the kidneys appears in the urine. For this reason, the rate of PAH clearance can be used to calculate the rate of plasma flow through the kidneys. Then, if the hematocrit is known (see chapter 14), the rate of total blood flow through the kidneys can be calculated.

CHART 20.3 FUNCTIONS OF NEPHRON COMPONENTS

| Part | Function |
|---|---|
| ***Renal Corpuscle*** | |
| Glomerulus | Filtration of water and dissolved substances from the plasma |
| Glomerular capsule | Receives the glomerular filtrate |
| ***Renal Tubule*** | |
| Proximal convoluted tubule | Reabsorption of glucose; amino acids; creatine; lactic, citric, uric, and ascorbic acids; phosphate, sulfate, calcium, potassium, and sodium ions by active transport |
| | Reabsorption of proteins by pinocytosis |
| | Reabsorption of water by osmosis |
| | Reabsorption of chloride ions and other negatively charged ions by electrochemical attraction |
| | Active secretion of substances such as penicillin, histamine, creatinine, and hydrogen ions |
| Descending limb of nephron loop | Reabsorption of water by osmosis |
| Ascending limb of nephron loop | Reabsorption of sodium, potassium, and chloride ions by active transport |
| Distal convoluted tubule | Reabsorption of sodium ions by active transport |
| | Reabsorption of water by osmosis |
| | Active secretion of hydrogen ions |
| | Secretion of potassium ions both actively and by electrochemical attraction |
| Collecting duct | Reabsorption of water by osmosis |

Parents of infants may be startled when a physician hospitalizes their child for an illness that in an adult might be considered mild—a day or two of vomiting and diarrhea. Because the kidneys of infants and young children are unable to concentrate urine and conserve water as effectively as those of adults, they can lose water rapidly, which may lead to dehydration. A 20-pound infant can lose a pound in just a day of an acute viral illness, and this is a sufficiently significant proportion of body weight to warrant hospitalization, where intravenous fluids are given to restore water and electrolyte balance (see chapter 21).

1. List the normal constituents of urine.
2. What is the normal hourly output of urine? The minimal hourly output?

## Elimination of Urine

After forming along the nephrons, urine passes from the collecting ducts through openings in the renal papillae and enters the major and minor calyces of the kidney. From there it passes through the renal pelvis, into a ureter, and into the urinary bladder. The urethra delivers urine to the outside.

### Ureters

Each **ureter** is a tubular organ about 25 centimeters long, which begins as the funnel-shaped renal pelvis. It extends downward behind the parietal peritoneum and parallel to the vertebral column. Within the pelvic cavity, it courses forward and medially to join the urinary bladder from underneath.

The wall of a ureter is composed of three layers. The inner layer, or *mucous coat,* includes several thicknesses of transitional epithelial cells and is continuous with the linings of the renal tubules and the urinary bladder. The middle layer, or *muscular coat,* consists largely of smooth muscle fibers arranged in circular and longitudinal bundles. The outer layer, or *fibrous coat,* is composed of connective tissue (fig. 20.27).

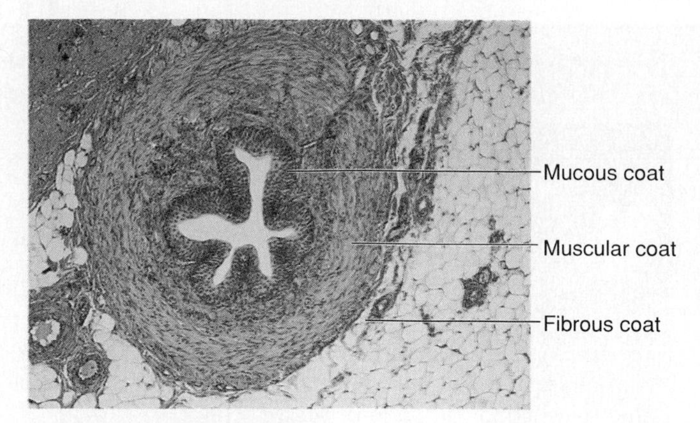


**Figure 20.27**
Cross section of a ureter (160×).

Because the linings of the ureters and the urinary bladder are continuous, bacteria may ascend from the bladder into the ureters, causing infection. An inflammation of the bladder, called *cystitis,* is more common in women than in men because the female urethral pathway is shorter. Inflammation of the ureter is called *ureteritis.*

Although the ureter is simply a tube leading from the kidney to the urinary bladder, its muscular wall helps move the urine. Muscular peristaltic waves, originating in the renal pelvis, force the urine along the length of the ureter. The presence of urine in the renal pelvis initiates these waves, whose frequency is related to the rate of urine formation. If this rate is high, a peristaltic wave may occur every few seconds; if the rate is low, a wave may occur every few minutes.

When such a peristaltic wave reaches the urinary bladder, it causes a jet of urine to spurt into the bladder. The opening through which the urine enters is covered by a flaplike fold of mucous membrane. This fold acts as a valve, allowing urine to enter the bladder from the ureter but preventing it from backing up from the bladder into the ureter.

If a ureter becomes obstructed, as when a small kidney stone (renal calculus) is present in its lumen, strong peristaltic waves are initiated in the proximal portion of the tube, which may help move the stone into the bladder. At the same time, the presence of a stone usually stimulates a sympathetic reflex (ureterorenal reflex) that constricts the renal arterioles and reduces urine production in the affected kidney.

1. Describe the structure of a ureter.
2. How is urine moved from the renal pelvis to the urinary bladder?
3. What prevents urine from backing up from the urinary bladder into the ureters?
4. How does an obstruction in a ureter affect urine production?

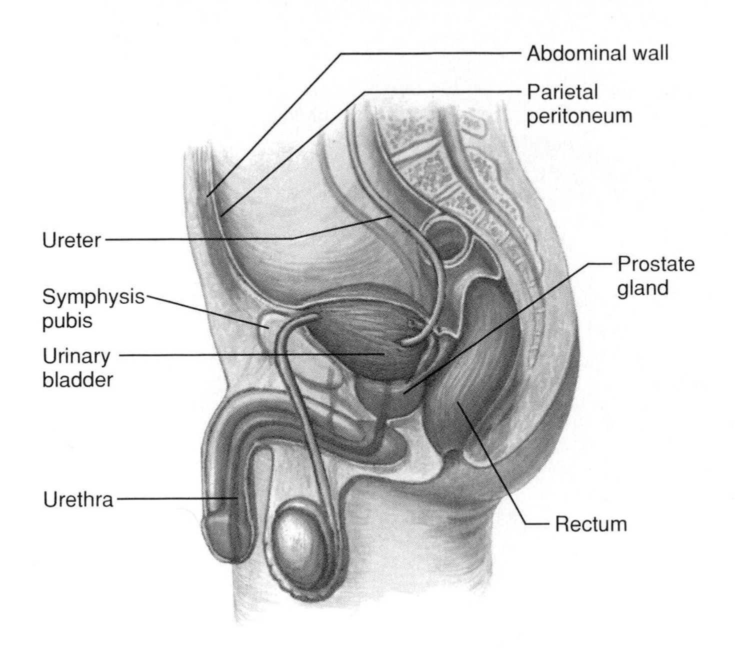


**FIGURE 20.28**
The urinary bladder is located within the pelvic cavity and behind the symphysis pubis. In a male, it lies against the rectum.

Kidney stones, which are usually composed of calcium oxalate, calcium phosphate, uric acid, or magnesium phosphate, sometimes form in the renal pelvis. If such a stone passes into a ureter, it may produce severe pain, beginning in the region of the kidney and radiating into the abdomen, pelvis, and lower limbs. Nausea and vomiting may also occur.

About 60% of kidney stone patients pass their stones spontaneously. The others must have the stones removed. In the past, such removal required surgery or tubular instruments that could be passed through the tubes of the urinary tract and used to capture or crush the stones. Today, shock waves applied from outside the body are used to fragment kidney stones. This procedure, called *extracorporeal shock-wave lithotripsy* (ESWL), focuses high energy shock waves through water (either in a tub or in a water-filled sack placed against the patient). The shock waves break the stones into fragments small enough to be eliminated with the urine.

## Urinary Bladder

The **urinary bladder** is a hollow, distensible, muscular organ. It is located within the pelvic cavity, behind the symphysis pubis, and beneath the parietal peritoneum (fig. 20.28 and reference plate 52). In a male, the bladder lies against the rectum posteriorly, and in a female, it contacts the anterior walls of the uterus and vagina.

The pressure of surrounding organs alters the spherical shape of the bladder. When the bladder is empty, its inner wall forms many folds, but as it fills with urine, the wall becomes smoother. At the same time, the superior surface of the bladder expands upward into a dome.

When it is greatly distended, the bladder pushes above the pubic crest and into the region between the abdominal wall and the parietal peritoneum. The dome can reach the level of the umbilicus and press against the coils of the small intestine.

The internal floor of the bladder includes a triangular area called the *trigone,* which has an opening at each of its three angles (fig. 20.29). Posteriorly, at the base of the trigone, the openings are those of the ureters. Anteriorly, at the apex of the trigone, is a short, funnel-shaped extension called the *neck* of the bladder, which contains the opening into the urethra. The trigone generally remains in a fixed position, even though the rest of the bladder distends and contracts.

The wall of the urinary bladder consists of four layers. The inner layer, or *mucous coat,* includes several thicknesses of transitional epithelial cells, similar to those lining the ureters and the upper portion of the urethra (see chapter 5). The thickness of this tissue changes as the bladder expands and contracts. Thus, during distension, the tissue appears to be only two or three cells thick, but during contraction, it appears to be five or six cells thick (see fig. 5.9).

Former vice president Hubert Humphrey died in 1978 of bladder cancer. He had been diagnosed in 1976, and an exam in 1973 had found a "borderline malignancy." In 1967, he had experienced his first symptom—blood in the urine. In 1994, researchers used the polymerase chain reaction to mass-produce a known cancer-causing gene (called p53) from preserved samples of the vice president's bladder tissue. The test revealed a tiny genetic glitch affecting only one DNA base, which was responsible for the cancer.

Can detecting a genetic change in cancer cells be clinically useful? Yes. It can permit early diagnosis, perhaps even before symptoms appear, when treatment is more likely to be successful. Absence of the causative gene in cells bordering an excised tumor indicates that sufficient tissue has been removed to stop the cancer. Finally, a genetic test can detect a recurrence of cancer before symptoms arise.

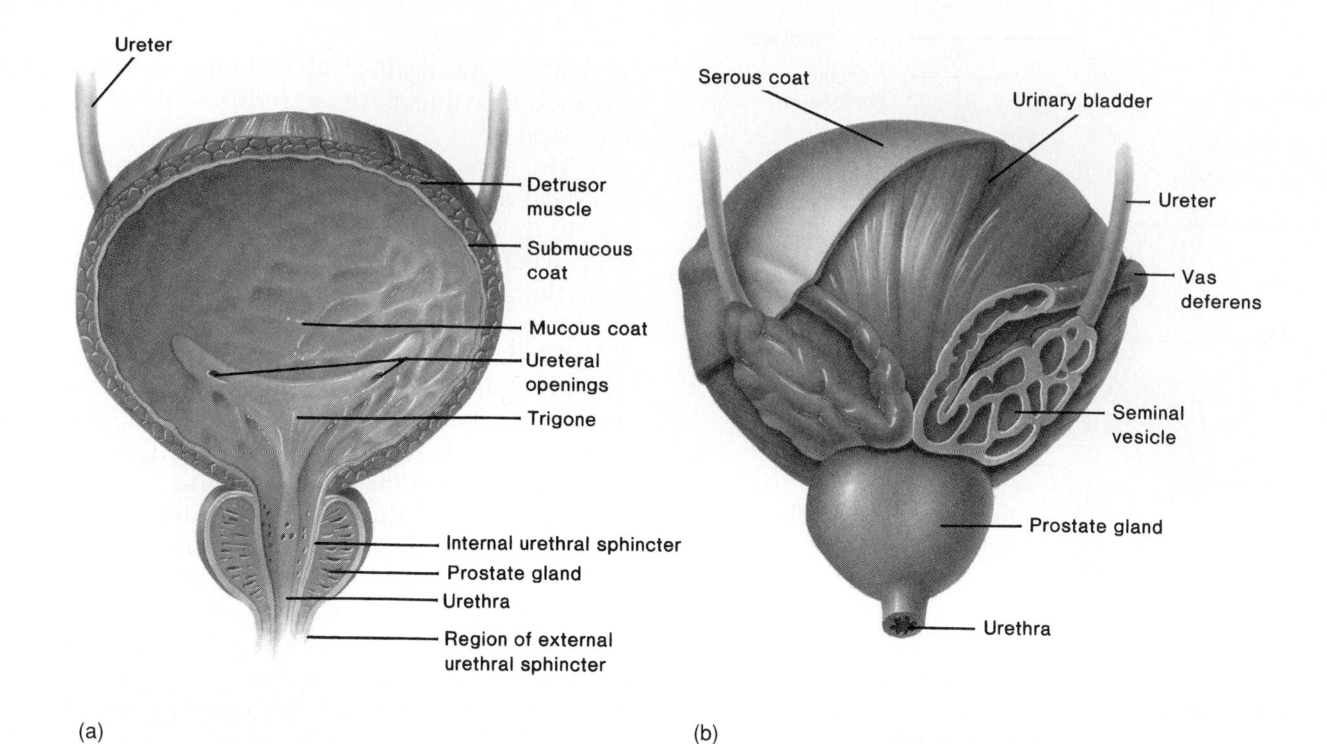


**FIGURE 20.29**
A male urinary bladder. (*a*) Coronal section; (*b*) posterior view.

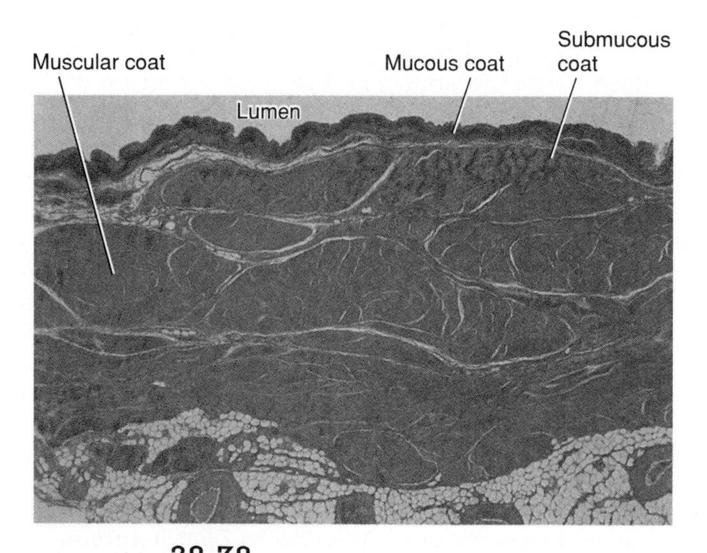


**FIGURE 20.30**
Light micrograph of the human urinary bladder wall (6×).

The second layer of the wall is the *submucous coat.* It consists of connective tissue and contains many elastic fibers.

The third layer of the bladder wall, the *muscular coat,* is composed primarily of coarse bundles of smooth muscle fibers. These bundles are interlaced in all directions and at all depths, and together they comprise the **detrusor muscle.** The portion of the detrusor muscle that surrounds the neck of the bladder forms an *internal urethral sphincter.* Sustained contraction of this sphincter muscle prevents the bladder from emptying until the pressure within it increases to a certain level. The detrusor muscle has parasympathetic nerve fibers that function in the micturition reflex that passes urine.

The outer layer of the wall, the *serous coat,* consists of the parietal peritoneum. It is found only on the upper surface of the bladder. Elsewhere, the outer coat is composed of fibrous connective tissue (fig. 20.30).

1. Describe the trigone of the urinary bladder.
2. Describe the structure of the bladder wall.
3. What kind of nerve fibers supply the detrusor muscle?

## Micturition

Urine leaves the urinary bladder by the **micturition** (urination) reflex. The detrusor muscle contracts, and contractions of muscles in the abdominal wall and pelvic floor may help, as well as fixation of the thoracic wall and diaphragm. Micturition also involves relaxation of the *external urethral sphincter.* This muscle, which is part of the urogenital diaphragm (see chapter 9), surrounds the urethra about 3 centimeters from the bladder and is composed of voluntary skeletal muscle tissue.

Distension of the bladder wall as it fills with urine stimulates the need to urinate. The wall expands, stimulating stretch receptors, which triggers the micturition reflex.

Clinical Application

## Urinalysis: Clues to Health

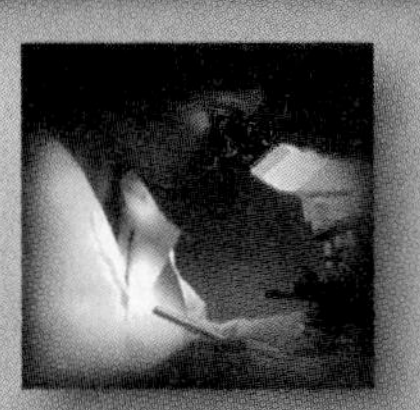

Urine has long fascinated medical minds. As a folk remedy, urine has been used as a mouthwash, toothache treatment, and a cure for sore eyes. Hippocrates (460–377 B.C.) was the first to observe that the condition of the urine can reflect health, noting that frothy urine denoted kidney disease. During the Middle Ages, health practitioners consulted charts in which certain urine colors were matched to certain diseases. In the seventeenth century, British physicians diagnosed diabetes by having their medical students taste sugar in patients' urine. Today, urine composition is still used as a window on health, and also to check for illicit drug use.

Certain inherited disorders can alter urine quite noticeably. The name *maple syrup urine disease* vividly describes what this inborn error of metabolism does to the urine. This condition, which causes mental retardation, results from a block in the breakdown pathways for certain amino acids. In *alkaptonuria,* one of the first inborn errors to be described, urine turns black on standing. This condition also produces painful arthritis and blackened ear tips. People with *Wilson disease* have an inherited inability to excrete copper. If they are properly diagnosed and given the drug penicillamine, they excrete some of the built-up copper, and their urine takes on a coppery appearance.

Other genetic conditions alter urine without causing health problems. People with *beeturia* excrete dark pink urine after they eat beets. The problem for people with *urinary excretion of odoriferous component of asparagus* is obvious. Parents of newborns who have inherited *blue diaper syndrome* are in for a shock when they change their child's first diaper. Due to a defect in transport of the amino acid tryptophan in the small intestine, the infant's urine turns blue when it hits the air. Bacteria degrade the partially digested tryptophan, producing a compound that turns blue on contact with oxygen.

The *micturition reflex center* is located in the sacral portion of the spinal cord. When sensory impulses from the stretch receptors signal the reflex center, parasympathetic motor impulses travel out to the detrusor muscle, which contracts rhythmically in response. A sensation of urgency accompanies this action.

The urinary bladder may hold as much as 600 milliliters of urine. The desire to urinate usually appears when it contains about 150 milliliters. Then, as urine volume increases to 300 milliliters or more, the sensation of fullness becomes increasingly uncomfortable.

As the bladder fills with urine, and its internal pressure increases, contractions of its wall become more and more powerful. When these contractions become strong enough to force the internal urethral sphincter to open, another reflex signals the external urethral sphincter to relax, and the bladder may empty.

However, because the external urethral sphincter is composed of skeletal muscle, it is under conscious control. Thus, the sphincter muscle ordinarily remains contracted until a decision is made to urinate. Nerve centers in the brain stem and cerebral cortex that inhibit the micturition reflex aid this control. When a person decides to urinate, the external urethral sphincter relaxes, and the micturition reflex is no longer inhibited. Nerve centers within the pons and the hypothalamus heighten the micturition reflex. Consequently, the detrusor muscle contracts, and urine is excreted through the urethra. Within a few moments, the neurons of the micturition reflex tire, the detrusor muscle relaxes, and the bladder begins to fill with urine again.

Chart 20.4 outlines the micturition process, and Clinical Application 20.5 discusses inherited conditions that affect urine composition.

> Damage to the spinal cord above the sacral region may abolish voluntary control of urination. However, if the micturition reflex center and its sensory and motor fibers are uninjured, micturition may continue to occur reflexively. In this case, the bladder collects urine until its walls stretch enough to trigger a micturition reflex, and the detrusor muscle contracts in response. This condition is called an *automatic bladder.*

CHART 20.4 MAJOR EVENTS OF MICTURITION

1. Urinary bladder distends as it fills with urine.
2. Stretch receptors in the bladder wall are stimulated and they signal the micturition center in the sacral spinal cord.
3. Parasympathetic nerve impulses travel to the detrusor muscle, which responds by contracting rhythmically.
4. The need to urinate is sensed as urgent.
5. Voluntary contraction of the external urethral sphincter and inhibition of the micturition reflex by impulses from the brain stem and the cerebral cortex prevent urination.
6. Following the decision to urinate, the external urethral sphincter is relaxed, and impulses from the pons and the hypothalamus facilitate the micturition reflex.
7. The detrusor muscle contracts, and urine is expelled through the urethra.
8. Neurons of the micturition reflex center fatigue, the detrusor muscle relaxes, and the bladder begins to fill with urine again.

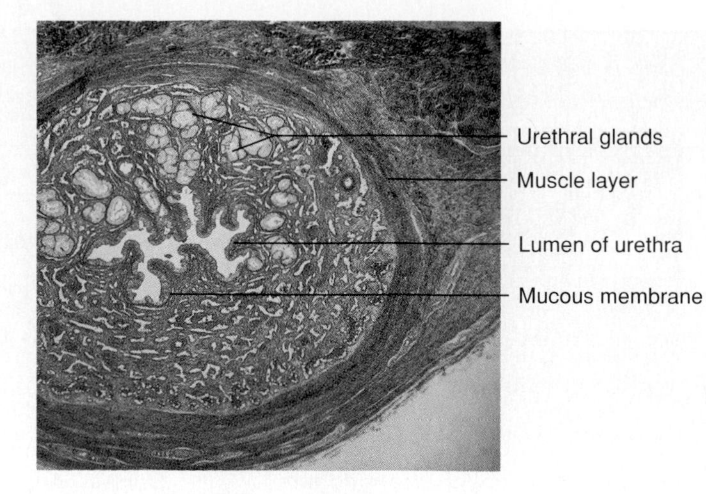


FIGURE 20.31
Cross section of the urethra (10×).

## Urethra

The **urethra** is a tube that conveys urine from the urinary bladder to the outside of the body. Its wall is lined with mucous membrane and contains a relatively thick layer of longitudinal smooth muscle fibers. It also contains numerous mucous glands, called *urethral glands,* which secrete mucus into the urethral canal (fig. 20.31).

In a female, the urethra is about 4 centimeters long. It passes forward from the bladder, courses below the symphysis pubis, and empties between the labia minora. Its opening, the *external urethral orifice* (urinary meatus), is located anterior to the vaginal opening and about 2.5 centimeters posterior to the clitoris (fig. 20.32).

In a male, the urethra, which functions both as a urinary canal and a passageway for cells and secretions from the reproductive organs, can be divided into three sections: the prostatic urethra, the membranous urethra, and the penile urethra (see fig. 20.32 and reference plate 60).

The **prostatic urethra** is about 2.5 centimeters long and passes from the urinary bladder through the *prostate gland,* which is located just below the bladder. Ducts from reproductive structures join the urethra in this region.

The **membranous urethra** is about 2 centimeters long. It begins just distal to the prostate gland, passes through the urogenital diaphragm, and is surrounded by the fibers of the external urethral sphincter muscle.

The **penile urethra** is about 15 centimeters long and passes through the corpus spongiosum of the penis, where erectile tissue surrounds it. This portion of the urethra terminates with the *external urethral orifice* at the tip of the penis.

1. Describe micturition.
2. How is it possible to consciously inhibit the micturition reflex?
3. Describe the structure of the urethra.
4. How does the urethra of a male differ from that of a female?

## CLINICAL TERMS RELATED TO THE URINARY SYSTEM

**anuria** (ah-nu′re-ah) Absence of urine due to failure of kidney function or to an obstruction in a urinary pathway.

**bacteriuria** (bak-te″re-u′re-ah) Bacteria in the urine.

**cystectomy** (sis-tek′to-me) The surgical removal of the urinary bladder.

**cystitis** (sis-ti′tis) Inflammation of the urinary bladder.

**cystoscope** (sis′to-skop) An instrument to visually examine the interior of the urinary bladder.

**cystotomy** (sis-tot′o-me) An incision of the wall of the urinary bladder.

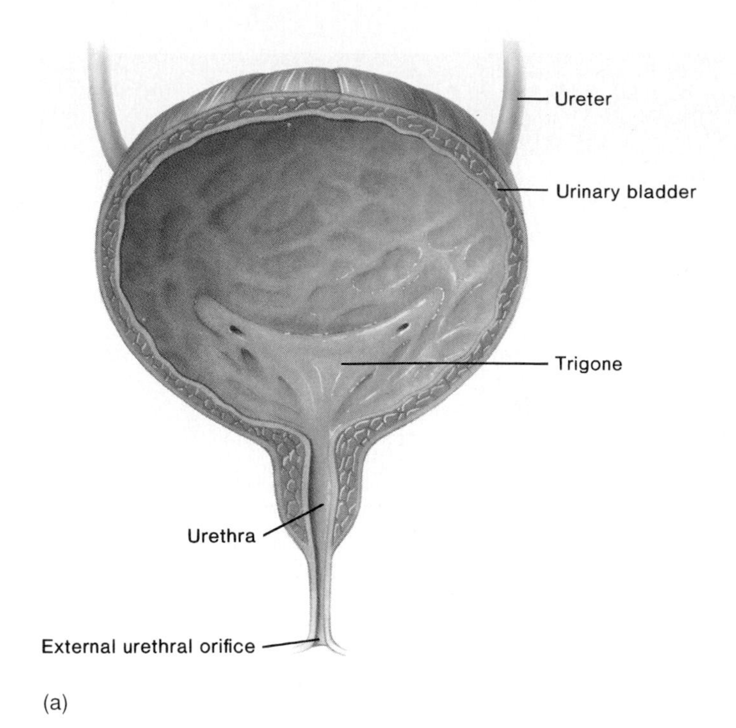


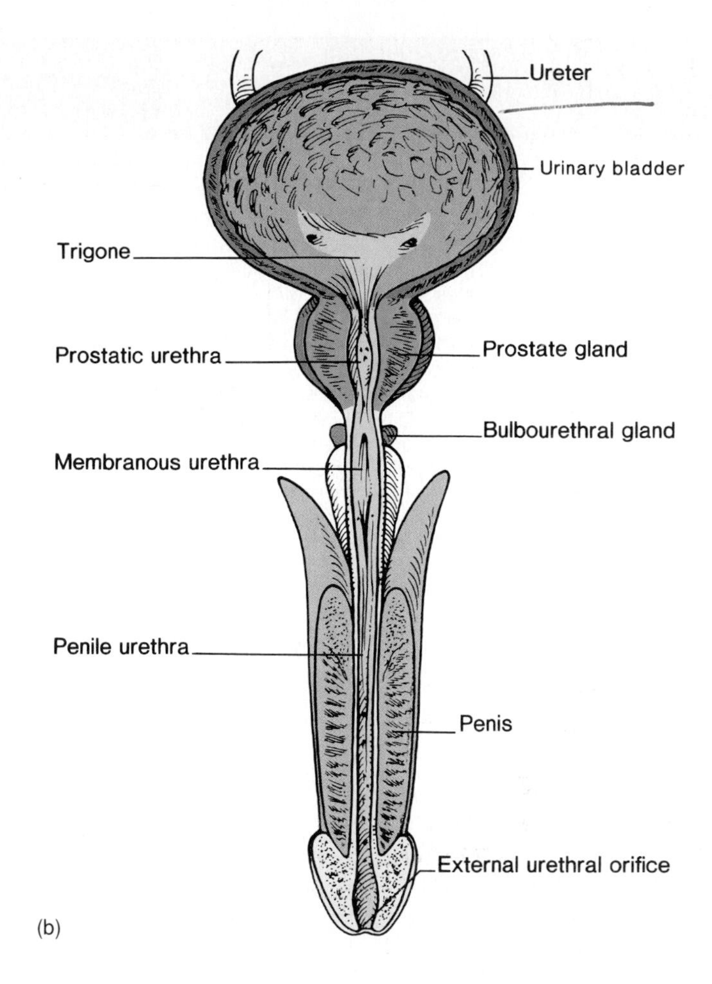


**FIGURE 20.32**
(*a*) Longitudinal section of the female urinary bladder and urethra; (*b*) longitudinal section of the male urinary bladder and urethra.

**diuresis** (di″u-re′sis) Increased production of urine.
**diuretic** (di″u-ret′ik) A substance that increases urine production.
**dysuria** (dis-u′re-ah) Painful or difficult urination.
**enuresis** (en″u-re′sis) Uncontrolled urination.
**hematuria** (hem″ah-tu′re-ah) Blood in the urine.
**incontinence** (in-kon′ti-nens) An inability to control urination and/or defecation reflexes.
**nephrectomy** (ne-frek′to-me) Surgical removal of a kidney.
**nephrolithiasis** (nef″ro-li-thi′ah-sis) Kidney stones.
**nephroptosis** (nef″rop-to′sis) A movable or displaced kidney.
**oliguria** (ol″i-gu′re-ah) A scanty output of urine.
**polyuria** (pol″e-u′re-ah) Excess urine.
**pyelolithotomy** (pi″ĕ-lo-li-thot′o-me) The removal of a stone from the renal pelvis.
**pyelonephritis** (pi″ĕ-lo-ne-fri′tis) Inflammation of the renal pelvis.
**pyelotomy** (pi″ĕ-lot′o-me) An incision into the renal pelvis.
**pyuria** (pi-u′re-ah) Pus (excess white blood cells) in the urine.
**uremia** (u-re′me-ah) Condition in which substances ordinarily excreted in the urine accumulate in the blood.
**ureteritis** (u-re″ter-i′tis) Inflammation of the ureter.
**urethritis** (u″re-thri′tis) Inflammation of the urethra.

# InnerConnections
## Urinary System

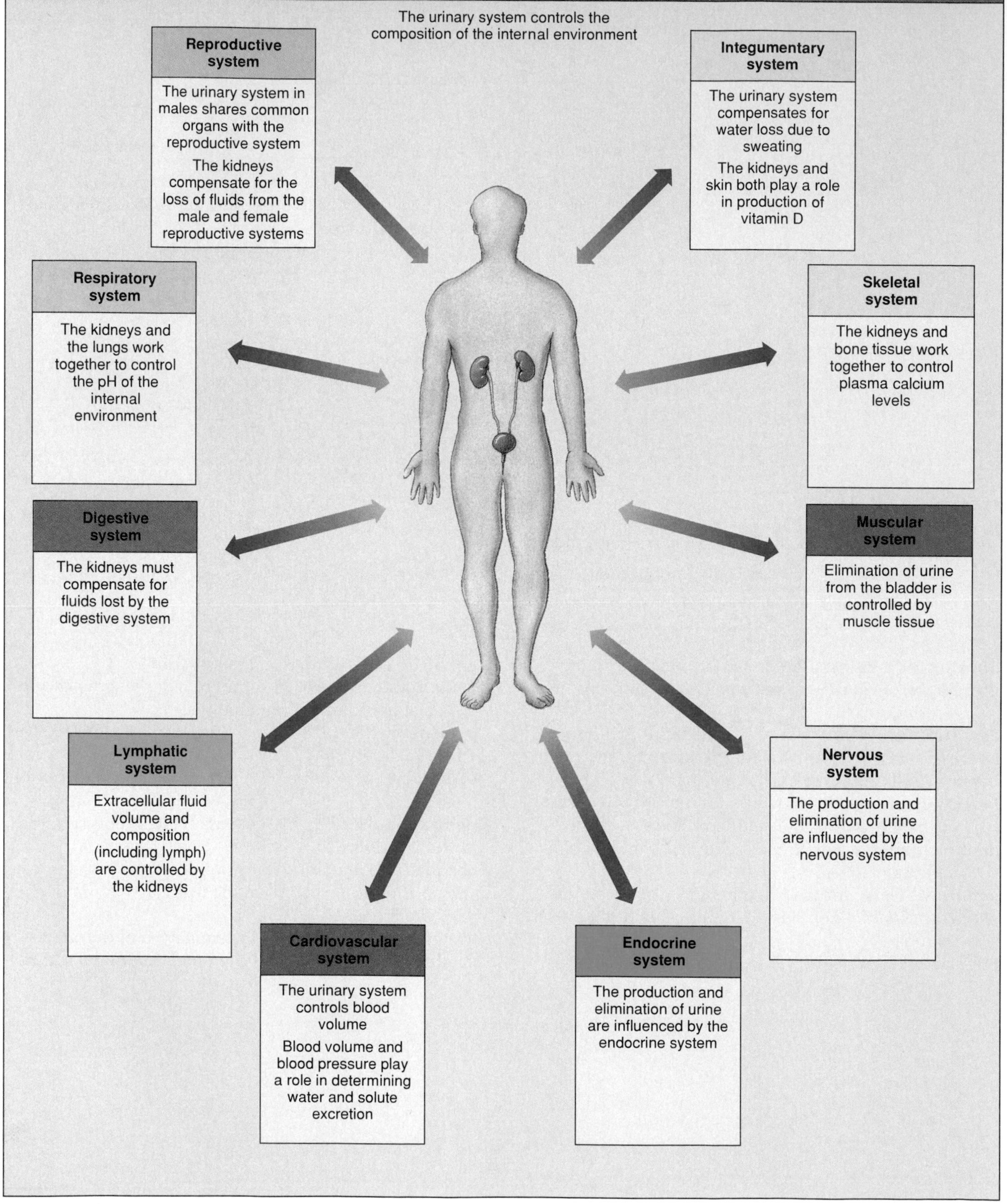

The urinary system controls the composition of the internal environment

**Reproductive system**

The urinary system in males shares common organs with the reproductive system

The kidneys compensate for the loss of fluids from the male and female reproductive systems

**Integumentary system**

The urinary system compensates for water loss due to sweating

The kidneys and skin both play a role in production of vitamin D

**Respiratory system**

The kidneys and the lungs work together to control the pH of the internal environment

**Skeletal system**

The kidneys and bone tissue work together to control plasma calcium levels

**Digestive system**

The kidneys must compensate for fluids lost by the digestive system

**Muscular system**

Elimination of urine from the bladder is controlled by muscle tissue

**Lymphatic system**

Extracellular fluid volume and composition (including lymph) are controlled by the kidneys

**Nervous system**

The production and elimination of urine are influenced by the nervous system

**Cardiovascular system**

The urinary system controls blood volume

Blood volume and blood pressure play a role in determining water and solute excretion

**Endocrine system**

The production and elimination of urine are influenced by the endocrine system

## Chapter Summary

### Introduction (page 796)

The urinary system consists of the kidneys, ureters, urinary bladder, and urethra.

### Kidneys (page 796)

1. Location of the kidneys
   a. The kidneys are on either side of the vertebral column, high on the posterior wall of the abdominal cavity.
   b. They are positioned behind the parietal peritoneum, and held in place by adipose and connective tissue.
2. Kidney structure
   a. A kidney contains a hollow renal sinus.
   b. The ureter expands into the renal pelvis, which, in turn, is divided into major and minor calyces.
   c. Renal papillae project into the renal sinus.
   d. Kidney tissue is divided into a medulla and a cortex.
3. Functions of the kidneys
   a. The kidneys remove metabolic wastes from the blood and excrete them to the outside.
   b. They also help regulate red blood cell production, blood pressure, calcium ion absorption, and the volume, composition, and pH of the blood.
4. Renal blood vessels
   a. Arterial blood flows through the renal artery, interlobar arteries, arciform arteries, interlobular arteries, afferent arterioles, glomerular capillaries, efferent arterioles, and peritubular capillaries.
   b. Venous blood returns through a series of vessels that correspond to those of the arterial pathways.
5. Nephrons
   a. Structure of a nephron
      (1) A nephron is the functional unit of the kidney.
      (2) It consists of a renal corpuscle and a renal tubule.
         (a) The corpuscle consists of a glomerulus and a glomerular capsule.
         (b) Portions of the renal tubule include the proximal convoluted tubule, the nephron loop (ascending and descending limbs), the distal convoluted tubule, and the collecting duct.
      (3) The collecting duct empties into the minor calyx of the renal pelvis.
   b. Juxtaglomerular apparatus
      (1) The juxtaglomerular apparatus is located at the point of contact between the distal convoluted tubule and the afferent and efferent arterioles.
      (2) It consists of the macula densa and the juxtaglomerular cells.
   c. Cortical and juxtamedullary nephrons
      (1) Cortical nephrons are the most numerous and have corpuscles near the surface of the kidney.
      (2) Juxtamedullary nephrons have corpuscles near the medulla.
   d. Blood supply of a nephron
      (1) The glomerular capillary receives blood from the afferent arteriole and passes it to the efferent arteriole.
      (2) The efferent arteriole gives rise to the peritubular capillary system, which surrounds the renal tubule.
      (3) Capillary loops, called vasa recta, dip down into the medulla.

### Urine Formation (page 806)

Nephrons remove wastes from the blood and regulate water and electrolyte concentrations. Urine is the end product of these functions. Kidney function includes filtration, reabsorption, and secretion of substances from renal tubules.

1. Glomerular filtration
   a. Urine formation begins when water and dissolved materials are filtered out of the glomerular capillary.
   b. The glomerular capillaries are much more permeable than the capillaries in other tissues.
2. Filtration pressure
   a. Filtration is due mainly to hydrostatic pressure inside the glomerular capillaries.
   b. The osmotic pressure of the blood plasma and hydrostatic pressure in the glomerular capsule also affect filtration.
   c. Filtration pressure is the net force acting to move material out of the glomerulus and into the glomerular capsule.
   d. The composition of the filtrate is similar to that of tissue fluid.
3. Filtration rate
   a. The rate of filtration varies with the filtration pressure.
   b. Filtration pressure changes with the diameters of the afferent and efferent arterioles.
   c. As the osmotic pressure in the glomerulus increases, filtration decreases.
   d. Filtration rate varies with the rate of blood flow through the glomerulus.
   e. As the hydrostatic pressure in a glomerular capsule increases, the filtration rate decreases.
   f. The kidneys produce about 125 milliliters of glomerular fluid per minute, most of which is reabsorbed.
   g. The volume of filtrate varies with the surface area of the glomerular capillary.
4. Regulation of filtration rate
   a. Glomerular filtration rate (GFR) remains relatively constant, but may be increased or decreased when the need arises. Increased sympathetic nerve activity can decrease GFR.
   b. When tubular fluid NaCl concentration decreases, the macula densa causes the juxtaglomerular cells to release renin. This triggers a series of changes leading to vasoconstriction, which may affect GFR, and secretion of aldosterone, which stimulates tubular sodium reabsorption.
   c. Autoregulation is the ability of an organ or tissue to maintain a constant blood flow under certain conditions when the arterial blood pressure is changing.

5. Tubular reabsorption
   a. Substances are selectively reabsorbed from the glomerular filtrate.
   b. The peritubular capillary is adapted for reabsorption.
      (1) It carries low pressure blood.
      (2) It is very permeable.
   c. Most reabsorption occurs in the proximal tubule, where the epithelial cells possess microvilli.
   d. Different modes of transport reabsorb various substances in particular segments of the renal tubule.
      (1) Glucose and amino acids are reabsorbed by active transport.
      (2) Water is reabsorbed by osmosis.
      (3) Proteins are reabsorbed by pinocytosis.
   e. Active transport mechanisms have limited transport capacities.
   f. If the concentration of a substance in the filtrate exceeds its renal plasma threshold, the excess is excreted in the urine.
   g. Substances that remain in the filtrate are concentrated as water is reabsorbed.
   h. Sodium ions are reabsorbed by active transport.
      (1) As positively charged sodium ions are transported out of the filtrate, negatively charged ions accompany them.
      (2) Water is passively reabsorbed by osmosis as sodium ions are actively reabsorbed.
6. Regulation of urine concentration and volume
   a. Most of the sodium ions are reabsorbed before the urine is excreted.
   b. Sodium ions are concentrated in the renal medulla by the countercurrent mechanism.
      (1) Chloride ions are actively reabsorbed in the ascending limb, and sodium ions follow them passively.
      (2) Tubular fluid in the ascending limb becomes hypotonic as it loses solutes.
      (3) Water leaves the descending limb by osmosis, and NaCl enters this limb by diffusion.
      (4) Tubular fluid in the descending limb becomes hypertonic as it loses water and gains NaCl.
      (5) As NaCl repeats this circuit, its concentration in the medulla increases.
   c. The vasa recta countercurrent mechanism helps maintain the NaCl concentration in the medulla.
   d. The distal tubule and collecting duct are impermeable to water, which therefore is excreted in urine.
   e. ADH from the posterior pituitary gland increases the permeability of the distal tubule and collecting duct, promoting water reabsorption.
7. Urea and uric acid excretion
   a. Urea is a by-product of amino acid metabolism.
      (1) It is reabsorbed passively by diffusion.
      (2) About 50% of the urea is excreted in urine.
      (3) A countercurrent mechanism helps in the excretion of urea.
   b. Uric acid results from the metabolism of nucleic acids.
      (1) Most is reabsorbed by active transport.
      (2) Some is secreted into the renal tubule.
8. Tubular secretion
   a. Tubular secretion transports certain substances from the plasma to the tubular fluid.
   b. Some substances are secreted actively.
      (1) These include various organic compounds and hydrogen ions.
      (2) The proximal and distal segments of the renal tubule secrete hydrogen ions.
   c. Potassium ions are secreted both actively and passively in the distal tubule and collecting duct.
9. Urine composition
   a. Urine is about 95% water, and it usually contains urea, uric acid, and creatinine.
   b. It may contain a trace of amino acids and varying amounts of electrolytes, depending upon dietary intake.
   c. The volume of urine varies with the fluid intake and with certain environmental factors.

## Elimination of Urine (page 820)

1. Ureters
   a. The ureter is a tubular organ that extends from each kidney to the urinary bladder.
   b. Its wall has mucous, muscular, and fibrous layers.
   c. Peristaltic waves in the ureter force urine to the bladder.
   d. Obstruction in the ureter stimulates strong peristaltic waves and a reflex that decreases urine production.
2. Urinary bladder
   a. The urinary bladder is a distensible organ that stores urine and forces it into the urethra.
   b. The openings for the ureters and urethra are located at the three angles of the trigone in the floor of the urinary bladder.
   c. Muscle fibers in the wall form the detrusor muscle.
   d. A portion of the detrusor muscle forms an internal urethral sphincter.
3. Micturition
   a. Micturition is the process of expelling urine.
   b. It involves contraction of the detrusor muscle and relaxation of the external urethral sphincter.
   c. Micturition reflex
      (1) Distension stimulates stretch receptors in the bladder wall.
      (2) The micturition reflex center in the sacral portion of the spinal cord sends parasympathetic motor impulses to the detrusor muscle.
      (3) As the bladder fills, its internal pressure increases, forcing the internal urethral sphincter open.
      (4) A second reflex relaxes the external urethral sphincter, unless its contraction is voluntarily controlled.
      (5) Nerve centers in the brain stem and cerebral cortex aid control of urination.
4. Urethra
   a. The urethra conveys urine from the bladder to the outside.
   b. In females, it empties between the labia minora.
   c. In males, it conveys products of reproductive organs as well as urine.
      (1) Three portions of the male urethra are prostatic, membranous, and penile.
      (2) The urethra empties at the tip of the penis.

## Critical Thinking Questions

1. If an infant is born with narrowed renal arteries, what effect would this condition have on the volume of urine produced? Explain your answer.
2. Why are people with nephrotic syndrome, in which plasma proteins are lost into the urine, more susceptible to infections?
3. If a patient who has had major abdominal surgery receives intravenous fluids equal to the volume of blood lost during surgery, would you expect the volume of urine produced to be greater or less than normal? Why?
4. If a physician prescribed oral penicillin therapy for a patient with an infection of the urinary bladder, how would you describe for the patient the route by which the drug would reach the bladder?
5. If the blood pressure of a patient who is in shock as a result of a severe injury decreases greatly, how would you expect the volume of urine to change? Why?
6. Inflammation of the urinary bladder is more common in women than in men. How might this observation be related to the anatomy of the female and male urethras?

## Review Exercises

1. Name the organs of the urinary system, and list their general functions.
2. Describe the external and internal structure of a kidney.
3. List the functions of the kidneys.
4. Name the vessels the blood passes through as it travels from the renal artery to the renal vein.
5. Distinguish between a renal corpuscle and a renal tubule.
6. Name the structures fluid passes through as it travels from the glomerulus to the collecting duct.
7. Describe the location and structure of the juxtaglomerular apparatus.
8. Distinguish between cortical and juxtamedullary nephrons.
9. Distinguish between filtration, reabsorption, and secretion as they relate to urine formation.
10. Define *filtration pressure.*
11. Compare the composition of the glomerular filtrate with that of the blood plasma.
12. Explain how the diameters of the afferent and efferent arterioles affect the rate of glomerular filtration.
13. Explain how changes in the osmotic pressure of the blood plasma may affect the rate of glomerular filtration.
14. Explain how the hydrostatic pressure of a glomerular capsule affects the rate of glomerular filtration.
15. Describe two mechanisms by which the body regulates the filtration rate.
16. Define *autoregulation.*
17. Discuss how tubular reabsorption is a selective process.
18. Explain how the peritubular capillary is adapted for reabsorption.
19. Explain how the epithelial cells of the proximal convoluted tubule are adapted for reabsorption.
20. Explain why active transport mechanisms have limited transport capacities.
21. Define *renal plasma threshold,* and explain its significance in tubular reabsorption.
22. Describe the renal defects associated with polycystic kidney disease, Wilms tumor, and alkaptonuria.
23. Explain how amino acids and proteins are reabsorbed.
24. Describe the effect of sodium reabsorption on the reabsorption of negatively charged ions.
25. Explain how sodium ion reabsorption affects water reabsorption.
26. Explain how hypotonic tubular fluid is produced in the ascending limb of the nephron loop.
27. Explain why fluid in the descending limb of the nephron loop is hypertonic.
28. Describe the function of ADH.
29. Explain how urine may become concentrated as it moves through the collecting duct.
30. Compare the processes by which urea and uric acid are reabsorbed.
31. Explain how the renal tubule is adapted to secrete hydrogen ions.
32. Explain how potassium ions may be secreted passively.
33. List the more common substances found in urine and their sources.
34. List some of the factors that affect the volume of urine produced each day.
35. Describe the structure and function of a ureter.
36. Explain how the muscular wall of the ureter aids in moving urine.
37. Discuss what happens if a ureter becomes obstructed.
38. Describe the structure and location of the urinary bladder.
39. Define *detrusor muscle.*
40. Distinguish between the internal and external urethral sphincters.
41. Describe the micturition reflex.
42. Explain how the micturition reflex can be voluntarily controlled.
43. Compare the urethra of a female with that of a male.

21

CHAPTER

# Water, Electrolyte, and Acid-Base Balance

ATHLETES ENGAGING IN VIGOROUS, SUSTAINED EXERCISE, ESPECIALLY IN HOT WEATHER, MUST DRINK BEYOND THE POINT OF SATISFYING THIRST SO THAT THEIR TISSUES REMAIN SUFFICIENTLY HYDRATED.

## Chapter Objectives

AFTER YOU HAVE STUDIED THIS CHAPTER, YOU SHOULD BE ABLE TO:

1. Explain what is meant by water and electrolyte balance, and discuss the importance of this balance.
2. Describe how the body fluids are distributed within compartments, how fluid composition differs between compartments, and how fluids move from one compartment to another.
3. List the routes by which water enters and leaves the body, and explain how water input and output are regulated.
4. Explain how electrolytes enter and leave the body, and how the input and output of electrolytes are regulated.
5. Explain what is meant by acid-base balance.
6. Describe how hydrogen ion concentrations are expressed.
7. List the major sources of hydrogen ions in the body.
8. Distinguish between strong and weak acids and bases.
9. Explain how changing pH values of the body fluids are minimized by chemical buffer systems, the respiratory center, and the kidneys.

## Key Terms

**acid** (as′id)
**acidosis** (as″i-do′sis)
**alkalosis** (al″kah-lo′sis)
**base** (bās)
**buffer system** (buf′er sis′tem)
**electrolyte balance** (e-lek′tro-līt bal′ans)
**extracellular** (ek″strah-sel′u-lar)
**intracellular** (in″trah sel′u-lar)
**osmoreceptor** (oz″mo-re-sep′tor)
**transcellular** (trans-sel′u-lar)
**water balance** (wot′er bal′ans)

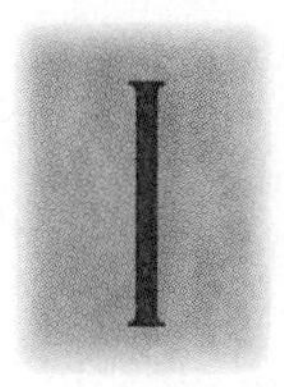

t was Rachel's first road race. She had been running daily since early April, and by this sunny July morning, felt she was ready for the 10 kilometer (6.2 mile) run. She was accustomed to running 4 or 5 miles just after sunrise. Today's race was at noon.

Rachel drank two glasses of water an hour before the race, which she thought would be sufficient to get her through the 45-minute run. On the race course, she wondered why cups of water had been placed on tables at every mile marker. Not wanting to waste precious seconds by grabbing a drink, as the other runners did, Rachel plowed on.

By sheer strength of will, she finished, but at a much slower pace than she had expected. More surprising was that she practically stumbled for the last half mile, then almost collapsed from weakness just after passing the finish line. Although she had sweated profusely during the event, she didn't seem to be sweating at all now that the race was over. And instead of feeling exhilarated at meeting her goal, Rachel was nauseous, dizzy, and felt the beginnings of a terrible headache.

Rachel would learn later, and never forget, that her mistake had not been lack of training or an impending viral infection, as she initially suspected, but simply failing to drink enough water. Losing a few seconds every few miles to replenish water lost in sweat would have gone a long way toward making her first race a much more pleasant, and healthy, experience.

Runners, bikers, and other outdoor athletes can lose massive amounts of water through sweat. In the 1984 Olympic marathon, Alberto Salazar lost 12 pounds of body weight in 5 quarts of sweat! Although sweating is a very effective cooling mechanism—without it, hyperthermia could begin after only 20 minutes of vigorous exercise in the heat—water loss exceeding 2% of body weight impairs function by raising body temperature and decreasing blood flow from the heart to the body's tissues. Rachel learned this firsthand.

Researchers in sports medicine laboratories have duplicated Rachel's experience by comparing people running on a treadmill under controlled conditions. Runners who drink water every 20 minutes have greater endurance and keep heart rate and body temperature lower than runners who do not drink. Sports physiologists suggest that an athlete drink 200 to 400 milliliters (½ to 1 cup) every 20 minutes during exercise. The athlete feels the effects of the water quickly—within 15 minutes, about two-thirds of it is absorbed in the small intestine and is well on the way to hydrating tissues. The athlete must drink about a third more volume of water than usual—beyond the point of satisfying thirst.

What is the best drink for an athlete in action? Plain old water is best. Electrolyte solutions aren't necessary, because relatively few electrolytes are lost in sweat, compared to the amounts in other body fluids. If a vigorous workout slightly depletes sodium, calcium, potassium, magnesium, zinc, or calcium ion supplies, the next meal usually replaces them. Similarly, salt tablets are unnecessary because a trained athlete's body excretes less salt in sweat and urine. After just a week or so of working out daily, a person's sweat contains a third less salt.

Rachel ran another 10 kilometer race in mid-August. This time, she took the paper cups offered her. Not only was her finish place much improved, but she felt terrific. Rachel had learned the importance of water, electrolyte, and acid-base balance in her body.

The term *balance* suggests a state of equilibrium, and in the case of water and electrolytes, it means that the quantities entering the body equal the quantities leaving it. Maintaining such a balance requires mechanisms to ensure that lost water and electrolytes will be replaced and that any excesses will be excreted. As a result, the levels of water and electrolytes in the body remain relatively stable at all times.

It is important to remember that water balance and electrolyte balance are interdependent because electrolytes are dissolved in the water of body fluids. Consequently, anything that alters the concentrations of the electrolytes will necessarily alter the concentration of the water by adding solutes to it or by removing solutes from it. Likewise, anything that changes the concentration of the water will change the concentrations of the electrolytes by making them either more concentrated or more dilute.

## Distribution of Body Fluids

Body fluids are not uniformly distributed throughout the tissues. Instead, they occur in regions, or *compartments,* of different volumes that contain fluids of varying compositions. The movement of water and electrolytes between these compartments is regulated so that their distribution and the composition of body fluids remain relatively stable.

### Fluid Compartments

The body of an average adult female is about 52% water by weight, and that of an average male is about 63% water. The differences between the sexes is due to the fact that females generally have more adipose tissue, which has little water, than do males. Water in the body (about 40 liters), together with its dissolved electrolytes, is distributed into two major compartments: an intracellular fluid compartment and an extracellular fluid compartment (fig. 21.1).

The **intracellular fluid compartment** includes all the water and electrolytes that cell membranes enclose. In other words, intracellular fluid is the fluid within the cells, and, in an adult, it represents about 63% by volume of the total body water.

The **extracellular fluid compartment** includes all the fluid outside the cells—within the tissue spaces (interstitial fluid), the blood vessels (plasma), and the lymphatic vessels (lymph). Epithelial layers separate a specialized fraction of the extracellular fluid from other extracellular fluids. This **transcellular fluid** includes *cerebrospinal fluid* of the central nervous system, *aqueous* and *vitreous humors* of the eyes, *synovial fluid* of the joints, *serous fluid* within the various body cavities, and fluid *secretions* of the exocrine glands. The fluids of the extracellular compartment constitute about 37% by volume of the total body water (fig. 21.2).

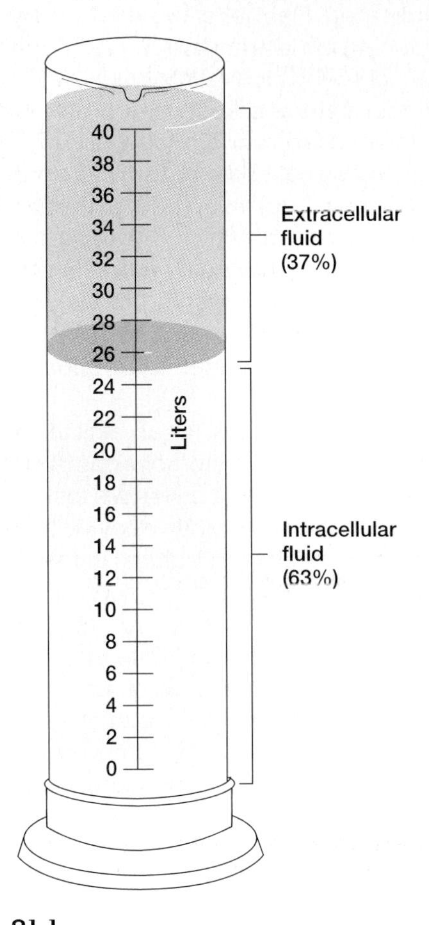


**FIGURE 21.1**
Of the 40 liters of water in the body of an average adult male, about 63% is intracellular and 37% is extracellular.

## Body Fluid Composition

*Extracellular fluids* generally have similar compositions, including relatively high concentrations of sodium, chloride, and bicarbonate ions, and lesser concentrations of potassium, calcium, magnesium, phosphate, and sulfate ions. The blood plasma fraction of extracellular fluid contains considerably more protein than do either interstitial fluid or lymph.

*Intracellular fluid* contains relatively high concentrations of potassium, phosphate, and magnesium ions. It includes a greater concentration of sulfate ions, and lesser concentrations of sodium, chloride, and bicarbonate ions than does extracellular fluid. Intracellular fluid also has a greater concentration of protein than plasma. Figure 21.3 shows these relative concentrations.

1. How are fluid balance and electrolyte balance interdependent?
2. Describe the normal distribution of water within the body.

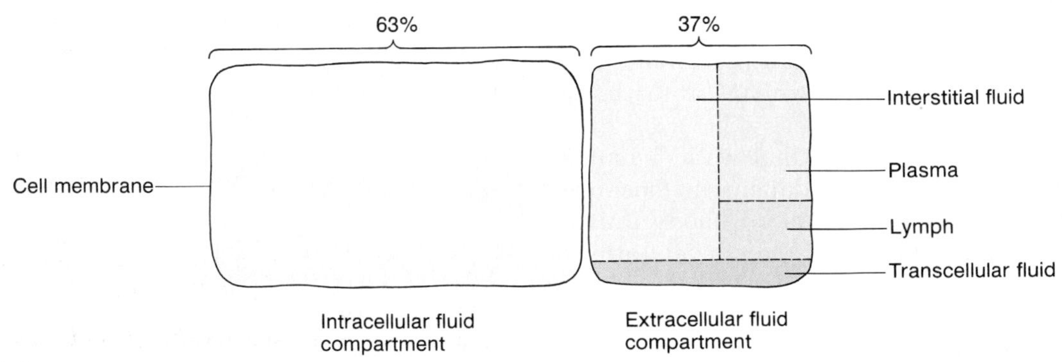


**FIGURE 21.2**
Cell membranes separate fluid in the intracellular compartment from fluid in the extracellular compartment.

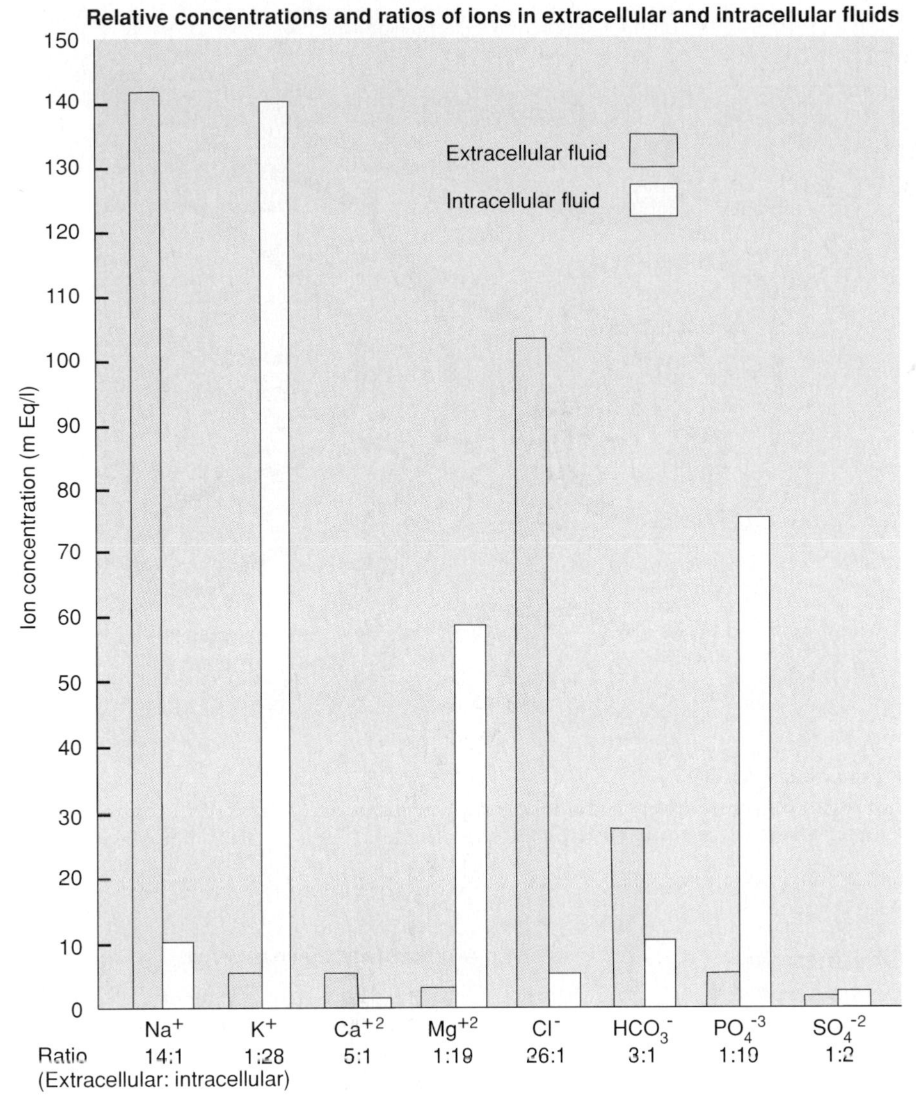


**FIGURE 21.3**

Extracellular fluid has relatively high concentrations of sodium, chloride, and bicarbonate ions. Intracellular fluid has relatively high concentrations of potassium, magnesium, and phosphate ions.

3 Which electrolytes are in higher concentrations in extracellular fluids? In intracellular fluid?

4 How does the concentration of protein vary in the various body fluids?

## Movement of Fluid Between Compartments

Two major factors regulate the movement of water and electrolytes from one fluid compartment to another: hydrostatic pressure and osmotic pressure. For example, as explained in chapter 15, fluid leaves the plasma at the arteriole ends of capillaries and enters the interstitial spaces because of the net outward force of *hydrostatic pressure* (blood pressure). Fluid returns to the plasma from the interstitial spaces at the venule ends of capillaries because of the net inward force of *osmotic pressure*. Likewise, as mentioned in chapter 16, fluid leaves the interstitial spaces and enters the lymph capillaries due to the hydrostatic pressure that develops within these spaces. As a result of the circulation of lymph, interstitial fluid returns to the plasma.

Such pressures similarly control the movement of fluid between the intracellular and extracellular compartments. However, because hydrostatic pressure within the cells and surrounding interstitial fluid is ordinarily equal and remains stable, any net fluid movement that occurs is likely to be the result of changes in osmotic pressure (fig. 21.4). Thus, although the composition of body fluids varies from one compartment to another, the total solute concentrations and the water concentrations are normally equal. A net gain or loss of water will therefore cause shifts affecting both the intracellular and extracellular fluids.

For example, because the sodium ion concentration in the extracellular fluids is especially high, a decrease in this concentration will cause a net movement of water from the extracellular compartment into the intracellular compartment by osmosis. As a consequence, the cells will swell. Conversely, if the concentration of sodium ions in the interstitial fluid increases, the net movement of water will be outward from the intracellular compartment, and the cells will shrink as they lose water.

1 What factors control the movement of water and electrolytes from one fluid compartment to another?

2 How does the sodium ion concentration within body fluids affect the net movement of water between the compartments?

# Water Balance

**Water balance** exists when the total intake of water equals the total loss of water.

## Water Intake

The volume of water gained each day varies from individual to individual. An average adult living in a moderate environment takes in about 2,500 milliliters. Of this amount, probably 60% is obtained from drinking water or beverages, while another 30% comes from moist foods. The remaining 10% is a by-product of the oxidative metabolism of nutrients, which is called **water of metabolism** (fig. 21.5).

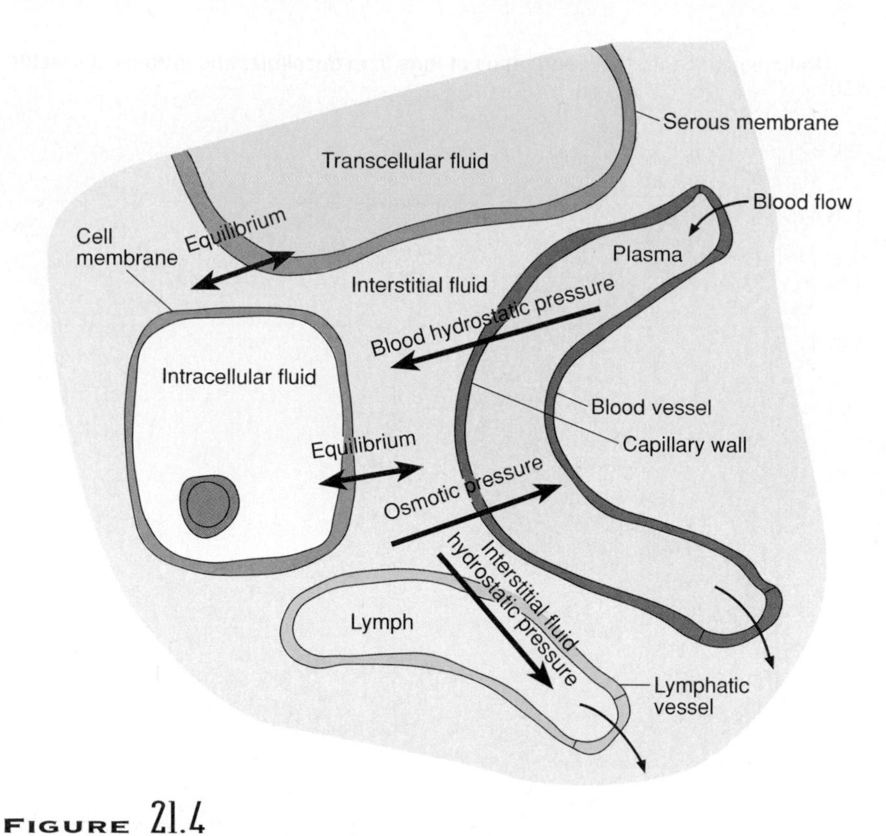


**FIGURE 21.4**
Net movements of fluids between compartments result from differences in hydrostatic and osmotic pressures.

## Regulation of Water Intake

The primary regulator of water intake is thirst. The intense feeling of thirst derives from the osmotic pressure of extracellular fluids and a *thirst center* in the hypothalamus of the brain.

As the body loses water, the osmotic pressure of the extracellular fluids increases. Such a change is thought to stimulate *osmoreceptors* in the thirst center, and as a result, the hypothalamus causes the person to feel thirsty and to seek water. A thirsty person usually has a dry mouth, which is related to the loss of extracellular water and a consequent decreased flow of saliva.

The thirst mechanism is normally triggered whenever the total body water is decreased by as little as 1%. As a thirsty person drinks water, the act of drinking and the resulting distension of the stomach wall seem to trigger nerve impulses that inhibit the thirst mechanism. Thus, drinking stops long before the swallowed water is absorbed. This inhibition helps prevent the person from drinking more than is required to replace the quantity lost, avoiding development of an imbalance of the opposite type. Chart 21.1 summarizes the steps in this mechanism.

1. What is water balance?
2. Where is the thirst center located?
3. What stimulates fluid intake? What inhibits it?

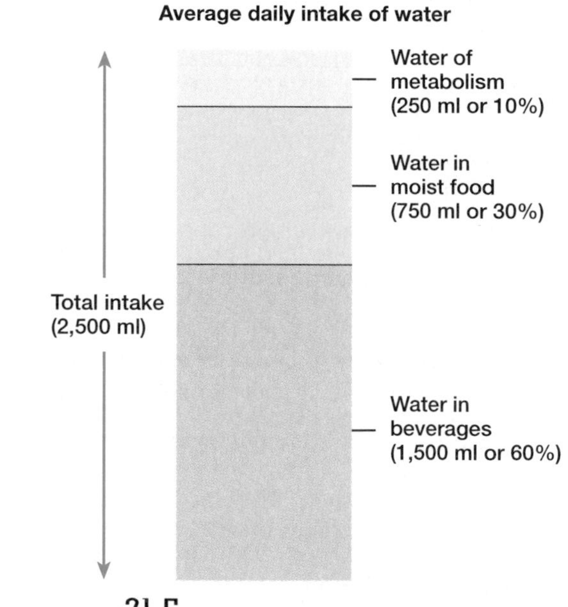


**FIGURE 21.5**
Major sources of body water.

## CHART 21.1 Regulation of Water Intake

1. The body loses as little as 1% of its water.
2. An increase in the osmotic pressure of extracellular fluid due to water loss stimulates osmoreceptors in the thirst center.
3. Activity in the hypothalamus causes the person to feel thirsty and to seek water.
4. Drinking and the resulting distension of the stomach by water stimulate nerve impulses that inhibit the thirst center.
5. Water is absorbed through the walls of the stomach and small intestine.
6. The osmotic pressure of extracellular fluid decreases.

## Water Output

Water normally enters the body only through the mouth, but it can be lost by a variety of routes. These include obvious losses in urine, feces, and sweat (sensible perspiration), as well as less obvious losses, which occur by evaporation of water from the skin (insensible perspiration) and from the lungs during breathing.

If an average adult takes in 2,500 milliliters of water each day, then 2,500 milliliters must be eliminated to maintain water balance. Of this volume, perhaps 60% will be lost in urine, 6% in feces, and 6% in sweat. About 28% will be lost by evaporation from the skin and lungs (fig. 21.6). These percentages vary with such environmental factors as temperature and relative humidity, and with physical exercise.

Water lost by sweating is a necessary part of the body's temperature control mechanism; water lost in feces accompanies the elimination of undigested food materials; and water lost by evaporation is largely unavoidable. Therefore, the primary means of regulating water output is urine production.

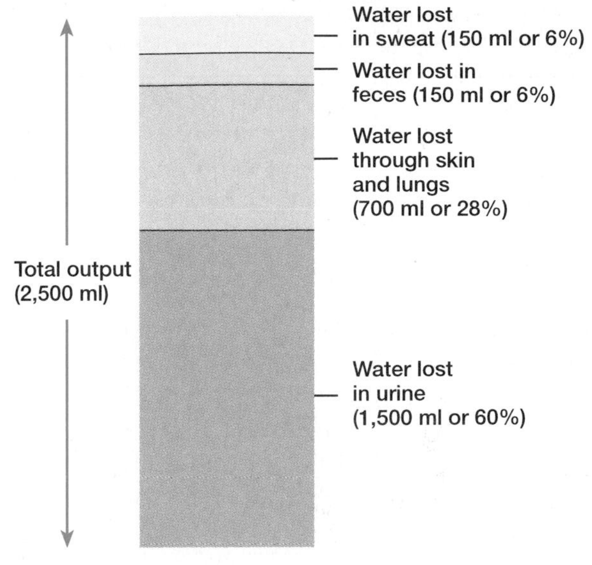


**FIGURE 21.6**

Routes by which the body loses water. Regulation of urinary water loss is most important in the control of water balance.

Proteins called *aquaporins* form water-selective membrane channels that enable body cells, including red blood cells and cells in the proximal renal tubules and descending limbs of the nephron tubules, to admit water. A mutation in one aquaporin gene (which instructs cells to manufacture a type of aquaporin protein) causes a form of *diabetes insipidus,* in which the renal tubules fail to reabsorb water. Interestingly, rare individuals have been identified who lack certain other aquaporin genes, and they apparently have no symptoms. This suggests that cells have more than one way to admit water.

## Regulation of Water Balance

The distal convoluted tubules and collecting ducts of the nephrons regulate the volume of water excreted in the urine. The epithelial linings of these segments of the renal tubule remain relatively impermeable to water unless antidiuretic hormone (ADH) is present.

Recall from chapter 13 that osmoreceptors in the hypothalamus help control release of ADH. If the blood plasma becomes more concentrated because of excessive water loss, these osmoreceptors dehydrate and shrink. This change triggers impulses that signal the posterior pituitary gland to release ADH. The ADH released into the bloodstream reaches the kidneys, where it increases the permeability of the distal tubule and collecting duct. Consequently, water reabsorption increases, and water is conserved. This action resists further osmotic change in the plasma. In fact, the *osmoreceptor-ADH mechanism* can reduce a normal urine production of 1,500 milliliters per day to about 500 milliliters per day when the body is dehydrating.

On the other hand, if a person drinks too much water, the plasma becomes less concentrated, and the osmoreceptors swell as they receive extra water by osmosis. In this instance, ADH release is inhibited, and the distal segment and collecting duct remain impermeable to water. Consequently, less water is reabsorbed and more urine is produced. Chart 21.2 summarizes the steps in this mechanism.

## CHART 21.2 EVENTS IN REGULATION OF WATER BALANCE

| Dehydration | Excess Water Intake |
|---|---|
| 1. Extracellular fluid becomes more concentrated.<br>2. Osmoreceptors in the hypothalamus are stimulated by a change in the osmotic pressure of body fluids.<br>3. The hypothalamus signals the posterior pituitary gland to release ADH into the blood.<br>4. Blood carries ADH to the kidneys.<br>5. ADH causes the distal convoluted tubules and collecting ducts to increase water reabsorption.<br>6. Urine output decreases, and water is conserved. | 1. Extracellular fluid becomes osmotically less concentrated.<br>2. This change stimulates osmoreceptors in the hypothalamus.<br>3. The posterior pituitary gland decreases ADH release.<br>4. Renal tubules decrease water reabsorption.<br>5. Urine output increases, and excess water is excreted. |

*Diuretics* are substances that promote urine production. A number of common substances, such as caffeine in coffee and tea, have diuretic effects, as do a variety of drugs used to reduce the volume of body fluids.

Diuretics produce their effects in different ways. Some, such as alcohol and certain narcotic drugs, promote urine formation by inhibiting ADH release. Certain other substances, such as caffeine, inhibit the reabsorption of sodium ions or other solutes in portions of the renal tubules. As a consequence, the osmotic pressure of the tubular fluid increases, reducing osmotic reabsorption of water and increasing urine volume.

Clinical Application 21.1 discusses disorders resulting from water imbalance.

1. By what routes does the body lose water?
2. What is the primary regulator of water loss?
3. What types of water loss are unavoidable?
4. What role does the hypothalamus play in the regulation of water balance?

## ELECTROLYTE BALANCE

An **electrolyte balance** exists when the quantities of electrolytes (molecules that release ions in water) the body gains equal those lost (fig. 21.7).

### Electrolyte Intake

The electrolytes of greatest importance to cellular functions release sodium, potassium, calcium, magnesium, chloride, sulfate, phosphate, bicarbonate, and hydrogen ions. These electrolytes are obtained primarily from foods, but they may also be found in drinking water and other beverages. In addition, some electrolytes are by-products of metabolic reactions.

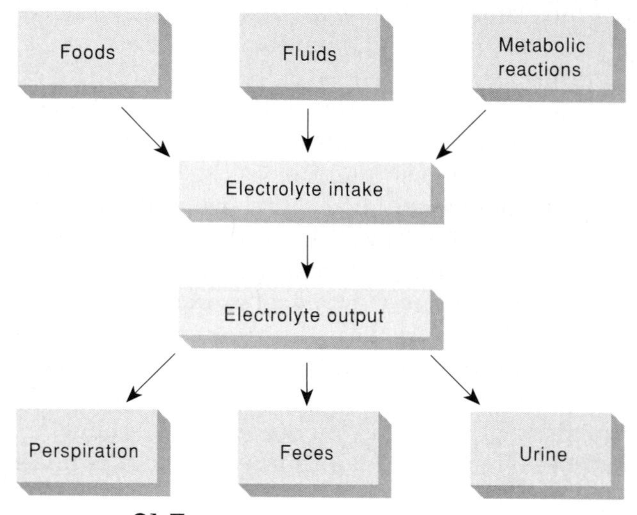


FIGURE 21.7
Electrolyte balance exists when the intake of electrolytes from all sources equals the output of electrolytes.

### Regulation of Electrolyte Intake

Ordinarily, a person obtains sufficient electrolytes by responding to hunger and thirst. However, a severe electrolyte deficiency may cause *salt craving,* which is a strong desire to eat salty foods.

### Electrolyte Output

The body loses some electrolytes by perspiring (sweat has about half the solute concentration of plasma). The quantities of electrolytes leaving vary with the amount of perspiration produced. More electrolytes are lost in sweat on warmer days and during times of strenuous exercise. Varying amounts of electrolytes are lost in the feces. The greatest electrolyte output occurs as a result of kidney function and urine production. The kidneys alter renal electrolyte losses to maintain the proper composition of body fluids.

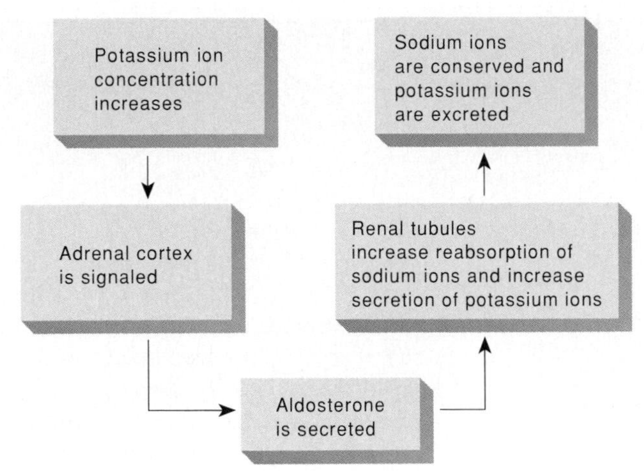


**FIGURE 21.8**
If the concentration of potassium ions increases, the kidneys conserve sodium ions and excrete potassium ions.

1. What electrolytes are most important to cellular functions?
2. What mechanisms ordinarily regulate electrolyte intake?
3. By what routes does the body lose electrolytes?

## Regulation of Electrolyte Balance

The concentrations of positively charged ions, such as sodium ($Na^+$), potassium ($K^+$), and calcium ($Ca^{+2}$), are particularly important. For example, certain concentrations of these ions are necessary for nerve impulse conduction, muscle fiber contraction, and maintenance of cell membrane permeability. Thus, regulating their concentrations is vital.

*Sodium ions* account for nearly 90% of the positively charged ions in extracellular fluids. The kidneys and the hormone aldosterone provide the primary mechanism regulating these ions. Aldosterone, which the adrenal cortex secretes, increases sodium ion reabsorption in the distal tubules and collecting ducts of the nephrons.

Aldosterone also regulates *potassium ions.* In fact, the most important stimulus for aldosterone secretion is a rising potassium ion concentration, which directly stimulates cells of the adrenal cortex. This hormone enhances the reabsorption of sodium ions and, at the same time, causes the renal tubules to excrete potassium ions (fig. 21.8).

Recall from chapter 13 that the parathyroid gland as well as calcitonin from the thyroid gland regulate the concentration of *calcium ions* in extracellular fluids. Calcium ion concentration dropping below normal stimulates the parathyroids directly to secrete parathyroid hormone.

Parathyroid hormone increases activity in bone-resorbing cells (osteocytes and osteoclasts), increasing the concentrations of both calcium and phosphate ions in the extracellular fluids. Parathyroid hormone also indirectly stimulates calcium absorption from the intestine (see chapter 13). Concurrently, this hormone causes the kidneys to conserve calcium ions (through increased tubular reabsorption) and increases the urinary excretion of phosphate ions. Thus, the net effect of the hormone is to increase the *calcium ion* concentration of the extracellular fluid, but to maintain a normal *phosphate ion* concentration (fig. 21.9).

Abnormal increases in blood calcium (hypercalcemia) sometimes result from hyperparathyroidism, in which excess secretion of PTH increases bone resorption (see chapter 13). Hypercalcemia may also be caused by cancers, particularly those originating in the bone marrow, breasts, lungs, or prostate gland. Usually the increase in calcium occurs when bone tissue being invaded and destroyed by cancerous growths releases ions. In other cases, however, the blood calcium concentration increases when cancer cells produce biochemicals that have physiological effects similar to parathyroid hormone. This most often occurs in lung cancer. Symptoms of cancer-induced hypercalcemia include weakness and fatigue, impaired mental function, headache, nausea, increased urine volume (polyuria), and increased thirst (polydipsia).

Abnormal decreases in blood calcium (hypocalcemia) may result from a reduced availability of PTH following surgical removal of the parathyroid glands, or from vitamin D deficiency, which may result from decreased absorption following gastrointestinal surgery or excess excretion due to kidney disease. Hypocalcemia may be life-threatening because it may produce muscle spasms within the airways and cardiac arrhythmias. Administering calcium salts and high doses of vitamin D to promote calcium absorption can correct this condition.

Generally, the regulatory mechanisms that control positively charged ions secondarily control the concentrations of negatively charged ions. For example, chloride ions ($Cl^-$), the most abundant negatively charged ions in the extracellular fluids, are passively reabsorbed from the renal tubules in response to the active reabsorption of sodium ions. That is, the negatively charged chloride ions are electrically attracted to the positively charged sodium ions and accompany them as they are reabsorbed.

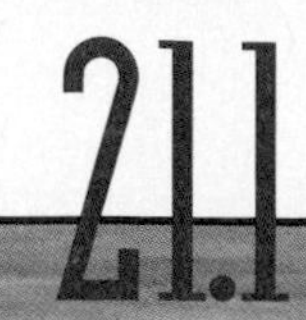

CLINICAL APPLICATION 21.1

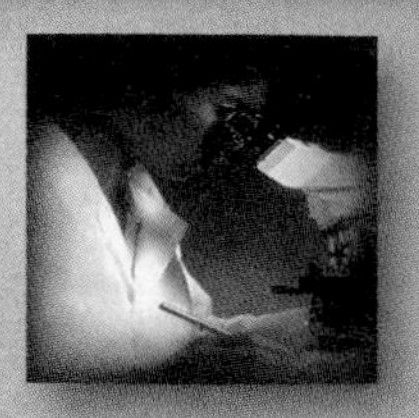

# Water Balance Disorders

Among the more common disorders involving an imbalance in the water of body fluids are dehydration, water intoxication, and edema.

### Dehydration

In 1994, the world watched in horror as thousands of starving people died in the African nation of Rwanda. It wasn't the lack of food that killed most of these people, but cholera, a bacterial infection that cripples the ability of intestinal lining cells to reabsorb water. The severe diarrhea that develops can kill in days, sometimes even hours. Dehydration is deadly.

*Dehydration* is a deficiency condition that occurs when the output of water exceeds the intake. This condition may develop following excessive sweating, or as a result of prolonged water deprivation accompanied by continued water output. In either case, as water is lost, the extracellular fluid becomes increasingly more concentrated, and water tends to leave cells by osmosis (fig. 21A). Dehydration may also accompany illnesses in which prolonged vomiting or diarrhea depletes body fluids. During dehydration, the skin and mucous membranes of the mouth feel dry, and body weight drops. Also, severe hyperthermia may develop as the body temperature-regulating mechanism becomes less effective due to a lack of water needed for sweating. In severe cases, as waste products accumulate in the extracellular fluid, symptoms of cerebral disturbances, including mental confusion, delirium, and coma, may develop.

Because the kidneys of infants are less able to conserve water than are those of adults, infants are more likely to become dehydrated. Elderly people are also especially susceptible to developing water imbalances because the sensitivity of their thirst mechanisms decreases with age, and physical disabilities may make it difficult for them to obtain adequate fluids.

The treatment for dehydration is to replace the lost water and electrolytes. But if only water is replaced, the extracellular fluid will become more dilute than normal. This may produce a condition called water intoxication.

### Water Intoxication

Babies rushed to emergency rooms because they are having seizures sometimes are suffering from drinking too much water, a rare condition called *water intoxication*. This can occur when a baby under six months of age is given several bottles of water a day, or very dilute infant formula. The hungry infant gobbles down the water, and soon its tissues swell with the excess fluid. When the serum sodium level drops, the eyes begin to flutter and a seizure occurs. As extracellular fluid becomes hypotonic, water enters the cells rapidly by osmosis (fig. 21B). Coma resulting from swelling brain tissues may follow unless water intake is restricted and hypertonic salt solutions given. Usually, recovery is complete within a few days.

Water intoxication in infancy sometimes occurs among poverty-stricken families who dilute formula to make it last longer. Another source of water intoxication in infants is bottled water products that are sold alongside infant formula. Their placement on grocery shelves leads some parents to believe that these products are adequate nutritional supplements. Used this way, they can be dangerous.

### Edema

*Edema* is an abnormal accumulation of extracellular fluid within the interstitial spaces. A variety of factors can cause

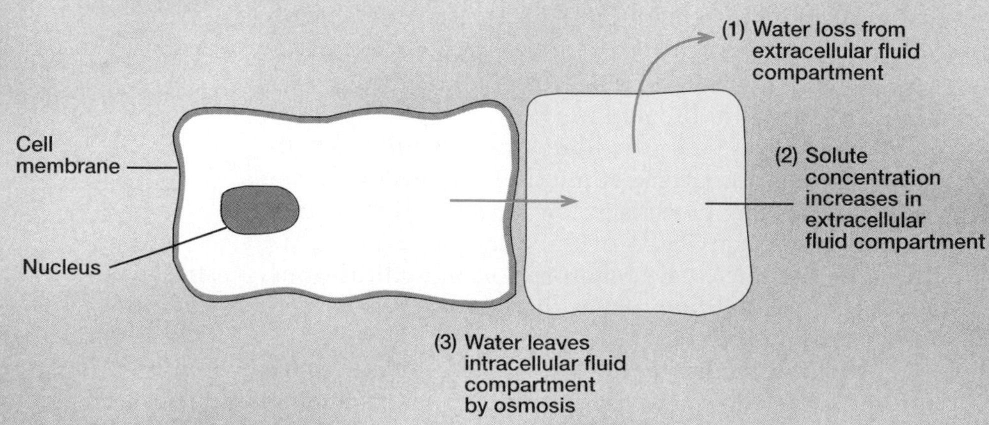


FIGURE 21A
If excess extracellular fluids are lost, cells dehydrate by osmosis.

it, including decrease in the plasma protein concentration (hypoproteinemia), obstructions in lymphatic vessels, increased venous pressure, and increased capillary permeability.

*Hypoproteinemia* may result from liver disease causing failure to synthesize plasma proteins: kidney disease (glomerulonephritis) that damages glomerular capillaries, allowing proteins to escape into the urine; or starvation, in which the intake of amino acids is insufficient to support synthesis of plasma proteins.

In each of these instances, the plasma protein concentration is decreased, which decreases plasma osmotic pressure, reducing the normal return of tissue fluid to the venule ends of capillaries. Tissue fluid consequently accumulates in the interstitial spaces.

As discussed in chapter 16, *lymphatic obstructions* may result from surgery or from parasitic infections of lymphatic vessels. Back pressure develops in the lymphatic vessels, interfering with the normal movement of tissue fluid into them. At the same time, proteins that the lymphatic circulation ordinarily removes accumulate in the interstitial spaces, raising osmotic pressure of the interstitial fluid. This effect attracts still more fluid into the interstitial spaces.

If the outflow of blood from the liver into the inferior vena cava is blocked, the venous pressure within the liver and portal blood vessels increases greatly. As a result, fluid with a high concentration of protein is exuded from the surfaces of the liver and intestine into the peritoneal cavity. This causes a rise in the osmotic pressure of the abdominal fluid, which, in turn, attracts more water into the peritoneal cavity by osmosis. This condition, called *ascites*, distends the abdomen. It is quite painful.

Edema may also result from increased capillary permeability accompanying *inflammation.* Recall that inflammation is a response to tissue damage and usually releases chemicals such as histamine from damaged cells. Histamine causes vasodilation and increased capillary permeability, so that excess fluid leaks out of the capillary and enters the interstitial spaces.

Chart 21A summarizes the factors that result in edema.

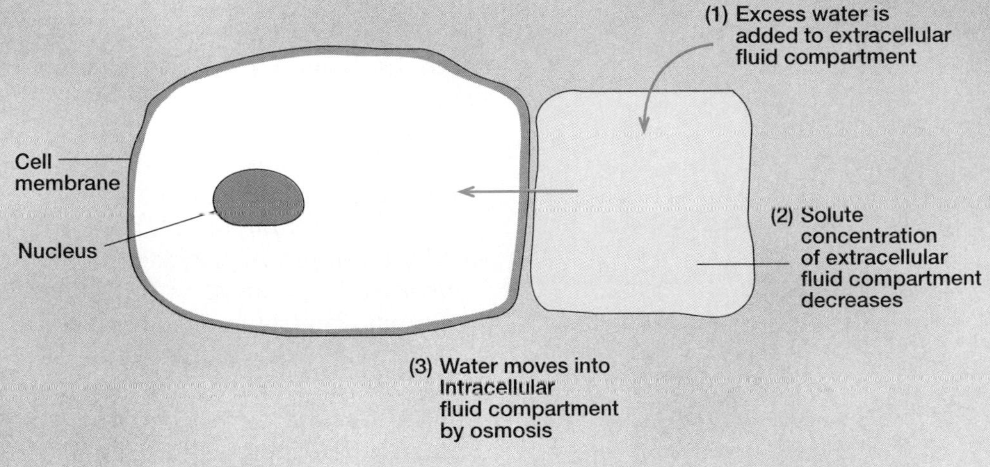


**FIGURE 21B**
If excess water is added to the extracellular fluid compartment, cells gain water by osmosis.

## CHART 21A FACTORS ASSOCIATED WITH EDEMA

| Factor | Cause | Effect |
|---|---|---|
| Low plasma protein concentration | Liver disease and failure to synthesize proteins; kidney disease and loss of proteins in urine; lack of proteins in diet due to starvation | Plasma osmotic pressure decreases, less fluid enters venule ends of capillaries by osmosis |
| Obstruction of lymph vessels | Surgical removal of portions of lymphatic pathways; certain parasitic infections | Back pressure in lymph vessels interferes with movement of fluid from interstitial spaces into lymph capillaries |
| Increased venous pressure | Venous obstructions or faulty venous valves | Back pressure in veins interferes with return of fluid from interstitial spaces into venule ends of capillaries |
| Inflammation | Tissue damage | Capillaries become abnormally permeable; fluid leaks from plasma into interstitial spaces |

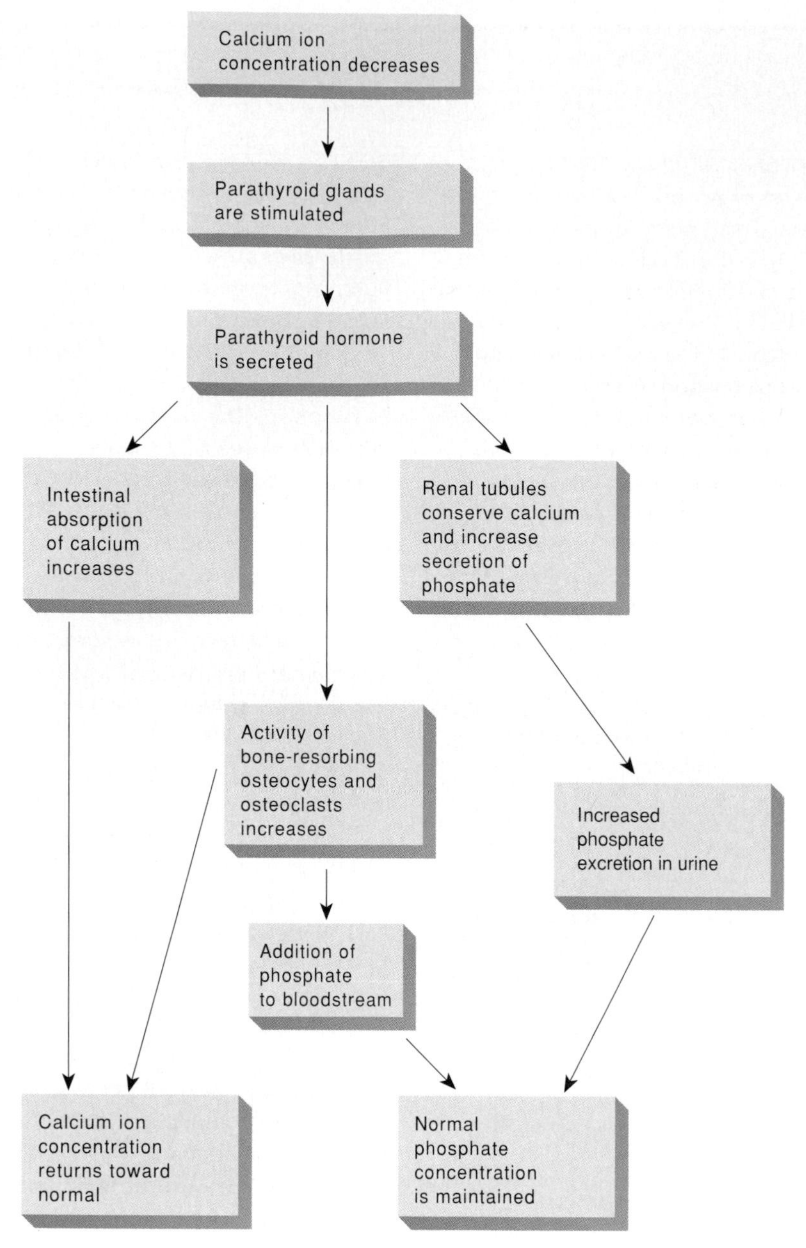


**FIGURE 21.9**

If the concentration of calcium ions decreases, parathyroid hormone increases calcium ion concentration. Increased urinary phosphate excretion offsets bone resorption (which increases blood phosphate levels) to maintain a normal concentration of phosphate ions.

Some negatively charged ions, such as phosphate ions ($PO_4^{-3}$) and sulfate ions ($SO_4^{-2}$), also are regulated partially by active transport mechanisms that have limited transport capacities. Thus, if the extracellular phosphate ion concentration is low, the phosphate ions in the renal tubules are conserved. On the other hand, if the renal plasma threshold is exceeded, the excess phosphate will be excreted in the urine.

Clinical Application 21.2 discusses symptoms associated with sodium and potassium imbalances.

1. What is the action of aldosterone?
2. How are the concentrations of sodium and potassium ions controlled?
3. How is calcium regulated?
4. What mechanism regulates the concentrations of most negatively charged ions?

## Acid-Base Balance

As discussed in chapter 2, electrolytes that ionize in water and release hydrogen ions are called *acids*. Substances that combine with hydrogen ions are called *bases*. Maintenance of homeostasis depends on control of concentrations of acids and bases within body fluids. Regulation of hydrogen ions is very important, because slight changes in hydrogen ion concentrations can alter the rates of enzyme-controlled metabolic reactions, shift the distribution of other ions, or modify hormone actions. Thus, acid-base balance is primarily concerned with the regulation of hydrogen ion concentrations, usually measured in pH units, of the body fluids.

### Sources of Hydrogen Ions

Most of the hydrogen ions in body fluids originate as by-products of metabolic processes, although the digestive tract may directly absorb small quantities.

The major metabolic sources of hydrogen ions include the following:

1. **Aerobic respiration of glucose.** This process produces carbon dioxide and water. Carbon dioxide diffuses out of the cells and reacts with water in the extracellular fluids to form *carbonic acid:*

$$CO_2 + H_2O \rightarrow H_2CO_3$$

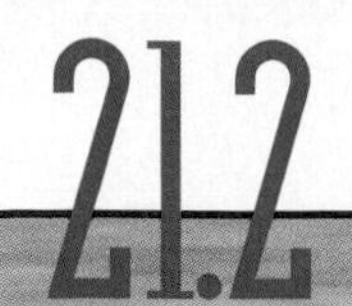

CLINICAL APPLICATION

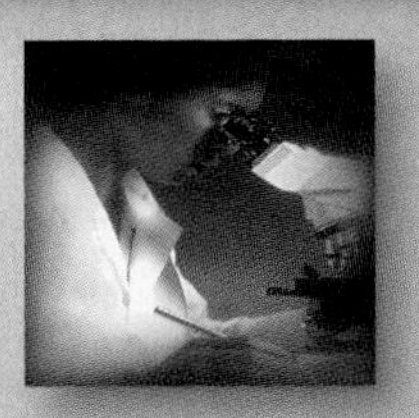

## Sodium and Potassium Imbalances

Extracellular fluids usually have high sodium ion concentrations, and intracellular fluid usually has high potassium ion concentrations. The renal regulation of sodium is closely related to that of potassium because active reabsorption of sodium (under the influence of aldosterone) is accompanied by secretion (and excretion) of potassium. Thus, it is not surprising that conditions involving sodium ion imbalance are closely related to those involving potassium ion imbalance.

Such disorders can be summarized as follows:

1. *Low sodium concentration* (hyponatremia). Possible causes of sodium deficiencies include prolonged sweating, vomiting, or diarrhea; renal disease in which sodium is reabsorbed inadequately; adrenal cortex disorders in which aldosterone secretion is insufficient to promote the reabsorption of sodium (Addison's disease); and drinking too much water.

   Possible effects of hyponatremia include the development of extracellular fluid that is hypotonic and promotes the movement of water into the cells by osmosis. This is accompanied by the symptoms of water intoxication previously described.
2. *High sodium concentration* (hypernatremia). Possible causes of elevated sodium concentration include excessive water loss by evaporation and diffusion, as may occur during high fever, or increased water loss accompanying diabetes insipidus, in one form of which ADH secretion is insufficient to maintain water conservation by the renal tubules.

   Possible effects of hypernatremia include disturbances of the central nervous system, such as confusion, stupor, and coma.
3. *Low potassium concentration* (hypokalemia). Possible causes of potassium deficiency include excessive release of aldosterone by the adrenal cortex (Cushing's syndrome), which increases renal excretion of potassium; use of diuretic drugs that promote potassium excretion; kidney disease; and prolonged vomiting or diarrhea.

   Possible effects of hypokalemia include muscular weakness or paralysis, respiratory difficulty, and severe cardiac disturbances, such as atrial or ventricular arrhythmias.
4. *High potassium concentration* (hyperkalemia). Possible causes of elevated potassium concentration include renal disease, which decreases potassium excretion; use of drugs that promote renal conservation of potassium; insufficient secretion of aldosterone by the adrenal cortex (Addison's disease); or a shift of potassium from the intracellular fluid to the extracellular fluid, a change that accompanies an increase in hydrogen ion concentration (acidosis).

   Possible effects of hyperkalemia include paralysis of the skeletal muscles and severe cardiac disturbances, such as cardiac arrest.

---

The resulting carbonic acid then ionizes to release hydrogen ions and bicarbonate ions:

$$H_2CO_3 \rightarrow H^+ + HCO_3^-$$

2. **Anaerobic respiration of glucose.** Glucose metabolized anaerobically produces *lactic acid,* which adds hydrogen ions to body fluids.
3. **Incomplete oxidation of fatty acids.** The incomplete oxidation of fatty acids produces *acidic ketone bodies,* which increase hydrogen ion concentration.
4. **Oxidation of amino acids containing sulfur.** The oxidation of sulfur-containing amino acids yields *sulfuric acid* ($H_2SO_4$), which ionizes to release hydrogen ions.
5. **Breakdown (hydrolysis) of phosphoproteins and nucleic acids.** Phosphoproteins and nucleic acids contain phosphorus. Their oxidation produces *phosphoric acid* ($H_3PO_4$), which ionizes to release hydrogen ions.

The acids resulting from metabolism vary in strength. Thus, their effects on the hydrogen ion concentration of body fluids vary (fig. 21.10).

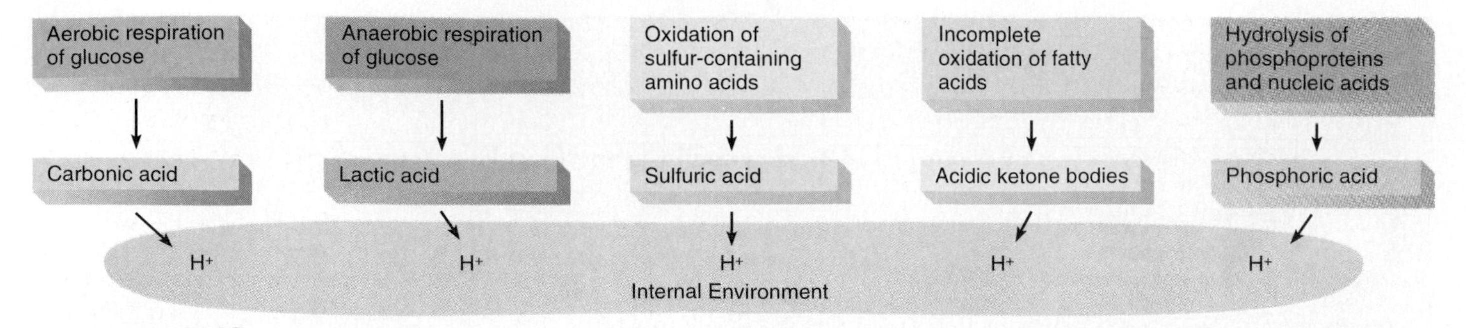


**FIGURE 21.10**
Some of the metabolic processes that provide hydrogen ions.

**1** Explain why the regulation of hydrogen ion concentration is so important.

**2** What are the major sources of hydrogen ions in the body?

## Strengths of Acids and Bases

Acids that ionize more completely are *strong acids,* and those that ionize less completely are *weak acids.* For example, the hydrochloric acid (HCl) of gastric juice is a strong acid, but the carbonic acid ($H_2CO_3$) produced when carbon dioxide reacts with water is weak.

Bases are substances that, like hydroxyl ions ($OH^-$), combine with hydrogen ions. *Chloride ions* ($Cl^-$) and *bicarbonate ions* ($HCO_3^-$) are also bases. Furthermore, since chloride ions combine less readily with hydrogen ions, they are *weak* bases, but the bicarbonate ions, which combine more readily with hydrogen ions, are *strong bases.*

## Regulation of Hydrogen Ion Concentration

The concentration of hydrogen ions in the body fluid, measured by pH, is regulated primarily by acid-base buffer systems, the respiratory center in the brain stem, and the nephrons in the kidneys. Normal metabolic reactions generally produce more acid than base. Consequently, the maintenance of that acid-base balance usually entails eliminating acid. Strong acids and bases pose a greater threat to homeostasis than weak acids do.

### Acid-Base Buffer Systems

**Acid-base buffer systems** occur in all the body fluids and are usually composed of sets of two or more chemicals that combine with acids or bases when in excess. More specifically, the chemical components of a buffer system can convert strong acids, which tend to release large quantities of hydrogen ions, into weak acids, which release fewer hydrogen ions. Likewise, these buffers can combine with strong bases and change them into weak bases. Such activity helps minimize the pH changes in body fluids.

The three most important acid-base buffer systems in body fluids are these:

1. **Bicarbonate buffer system.** The bicarbonate buffer system, which is present in intracellular and extracellular body fluids, consists of carbonic acid ($H_2CO_3$) and sodium bicarbonate ($NaHCO_3$). A strong acid, like hydrochloric acid, reacts with the sodium bicarbonate. The products of the reaction are carbonic acid, which is a weaker acid, and sodium chloride. This minimizes an increase in the hydrogen ion concentration in the body fluid:

$$\underset{\text{(strong acid)}}{HCl} + NaHCO_3 \rightarrow \underset{\text{(weak acid)}}{H_2CO_3} + NaCl$$

On the other hand, a strong base like sodium hydroxide (NaOH) reacts with the carbonic acid. The products are sodium bicarbonate ($NaHCO_3$), which is a weaker base, and water. This minimizes a shift toward a more basic (alkaline) state:

$$\underset{\text{(strong base)}}{NaOH} + H_2CO_3 \rightarrow \underset{\text{(weak base)}}{NaHCO_3} + H_2O$$

Recall that water is neutral:

$$H_2O \rightarrow H^+ + OH^-$$

2. **Phosphate buffer system.** The phosphate acid-base buffer system is also present in intracellular and extracellular body fluids. It is particularly important as a regulator of the

hydrogen ion concentrations in the tubular fluid of the nephrons and in urine. This buffer system consists of two phosphate compounds—sodium monohydrogen phosphate ($Na_2HPO_4$) and sodium dihydrogen phosphate ($NaH_2PO_4$).

If a strong acid is present, it reacts with the sodium monohydrogen phosphate to produce a weaker acid (sodium dihydrogen phosphate) and sodium chloride:

$$\underset{\text{(strong acid)}}{HCl} + Na_2HPO_4 \rightarrow \underset{\text{(weak acid)}}{NaH_2PO_4} + NaCl$$

If a strong base is present, it reacts with the sodium dihydrogen phosphate, and again these products are a weak base (sodium monohydrogen phosphate) and water:

$$\underset{\text{(strong base)}}{NaOH} + NaH_2PO_4 \rightarrow \underset{\text{(weak base)}}{Na_2HPO_4} + H_2O$$

3. **Protein buffer system.** The protein acid-base buffer system consists of the plasma proteins, such as albumins, and certain proteins within the cells, including the hemoglobin of red blood cells.

As described in chapter 2, proteins are chains of amino acids. Some of these amino acids have freely exposed groups of atoms, called *carboxyl groups*. Under certain conditions, a carboxyl group (–COOH) can become ionized, releasing a hydrogen ion:

$$-COOH \rightarrow -COO^- + H^+$$

Some of the amino acids within a protein molecule also contain freely exposed *amino groups* ($-NH_2$). Under certain conditions, these amino groups can accept hydrogen ions:

$$-NH_2 + H^+ \rightarrow -NH_3^+$$

Thus, protein molecules can function as acids by releasing hydrogen ions from their carboxyl groups, or as bases by accepting hydrogen ions into their amino groups. This special property allows protein molecules to operate as an acid-base buffer system. In the presence of excess hydrogen ions, the $-COO^-$ portions of the protein molecules accept hydrogen ions and become –COOH groups again. This action decreases the number of free hydrogen ions in body fluid and minimizes the pH change.

In the presence of excess hydroxyl ions ($OH^-$), the $-NH_3^+$ groups within protein molecules give up hydrogen ions and become $-NH_2$ groups again. These hydrogen ions then combine with the hydroxyl ions to form water molecules. Once again, the pH change is reduced to a minimum.

*Hemoglobin* is an example of a protein that buffers hydrogen ions. As explained in chapter 19, carbon dioxide, produced by cellular oxidation of glucose, diffuses through the capillary wall and enters the plasma and red blood cells. The red cells contain an enzyme, *carbonic anhydrase,* that speeds the reaction between carbon dioxide and water, producing carbonic acid:

$$CO_2 + H_2O \rightarrow H_2CO_3$$

The carbonic acid quickly dissociates, releasing hydrogen ions and bicarbonate ions ($HCO_3^-$):

$$H_2CO_3 \rightarrow H^+ + HCO_3^-$$

Hemoglobin molecules accept the hydrogen ions, thus minimizing the pH change that would otherwise occur.

Individual amino acids in body fluids can also function as acid-base buffers by accepting or giving up hydrogen ions because every amino acid has an amino group ($-NH_2$) and a carboxyl group (–COOH).

To summarize, acid-base buffer systems take up hydrogen ions when body fluids are becoming more acidic, and give up hydrogen ions when the fluids are becoming more basic (alkaline). Buffer systems accomplish this by converting stronger acids into weaker ones or by converting stronger bases into weaker ones, as chart 21.3 summarizes.

In addition to minimizing pH fluctuations, acid-base buffer systems in body fluids buffer each other. Consequently, whenever the hydrogen ion concentration begins to change, the chemical balances within all of the buffer systems change too, resisting the drift in pH.

Neurons are particularly sensitive to changes in the pH of body fluids. For example, if the interstitial fluid becomes more alkaline than normal (alkalosis), neurons become more excitable and seizures may result. Conversely, if conditions become more acidic (acidosis), neuron activity is depressed, and the level of consciousness may decrease.

## CHART 21.3 Chemical Acid-Base Buffer Systems

| Buffer System | Constituents | Actions |
|---|---|---|
| Bicarbonate system | Sodium bicarbonate ($NaHCO_3$) | Converts a strong acid into a weak acid |
| | Carbonic acid ($H_2CO_3$) | Converts a strong base into a weak base |
| Phosphate system | Sodium monohydrogen phosphate ($Na_2HPO_4$) | Converts a strong acid into a weak acid |
| | Sodium dihydrogen phosphate ($NaH_2PO_4$) | Converts a strong base into a weak base |
| Protein system (and amino acids) | $-COO^-$ group of an amino acid or protein | Accepts hydrogen ions in the presence of excess acid |
| | $-NH_3^+$ group of an amino acid or protein | Releases hydrogen ions in the presence of excess base |

**1** What is the difference between a strong acid or base and a weak acid or base?

**2** How does a chemical buffer system help regulate the pH?

**3** List the major buffer systems of the body.

Chemical buffer systems only temporarily solve the problem of acid-base balance. Ultimately, the body must eliminate excess acid or base. This task becomes the job of the lungs (controlled by the respiratory center) and the kidneys.

### The Respiratory Center

The **respiratory center** in the brain stem helps regulate hydrogen ion concentrations in the body fluids by controlling the rate and depth of breathing. Specifically, if the body cells increase their production of carbon dioxide, as occurs during periods of physical exercise, carbonic acid production increases. As the carbonic acid dissociates, the concentration of hydrogen ions increases, and the pH of the internal environment drops (see chapter 19). Such an increasing concentration of carbon dioxide in the central nervous system and the subsequent increase in hydrogen ion concentration in the cerebrospinal fluid stimulates chemosensitive areas within the respiratory center.

In response, the respiratory center increases the depth and rate of breathing, so that the lungs excrete more carbon dioxide. At the same time, hydrogen ion concentration in body fluids returns toward normal, because the released carbon dioxide comes from carbonic acid (fig. 21.11):

$$H_2CO_3 \rightarrow CO_2 + H_2O$$

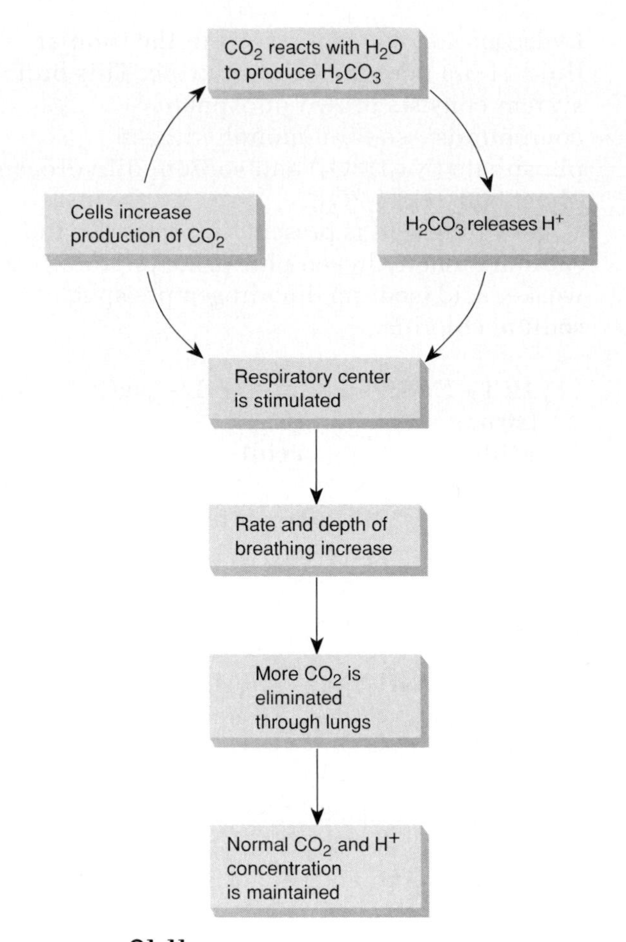


**FIGURE 21.11**
An increase in carbon dioxide elimination follows an increase in carbon dioxide production.

Conversely, if body cells are less active, concentrations of carbon dioxide and hydrogen ions in body fluids remain relatively low. As a result, breathing rate and depth fall. In time, this may cause accumulation of carbon dioxide in the body fluids, and as pH drops, the respiratory center is stimulated to increase the rate and depth of breathing once again.

Thus, the activity of the respiratory center changes in response to shifts in the pH of the body fluids, reducing these shifts to a minimum. Because most of the hydrogen ions in the body fluids originate from carbonic acid produced when carbon dioxide reacts with water, the respiratory regulation of hydrogen ion concentration is of considerable importance.

### The Kidneys

Nephrons help regulate the hydrogen ion concentration of the body fluids by excreting hydrogen ions in the urine. Recall from chapter 20 that the epithelial cells that line the proximal and distal convoluted tubules and the collecting ducts secrete these ions into the tubular fluid. The kidneys also regulate the concentration of bicarbonate ions in body fluids. These mechanisms are

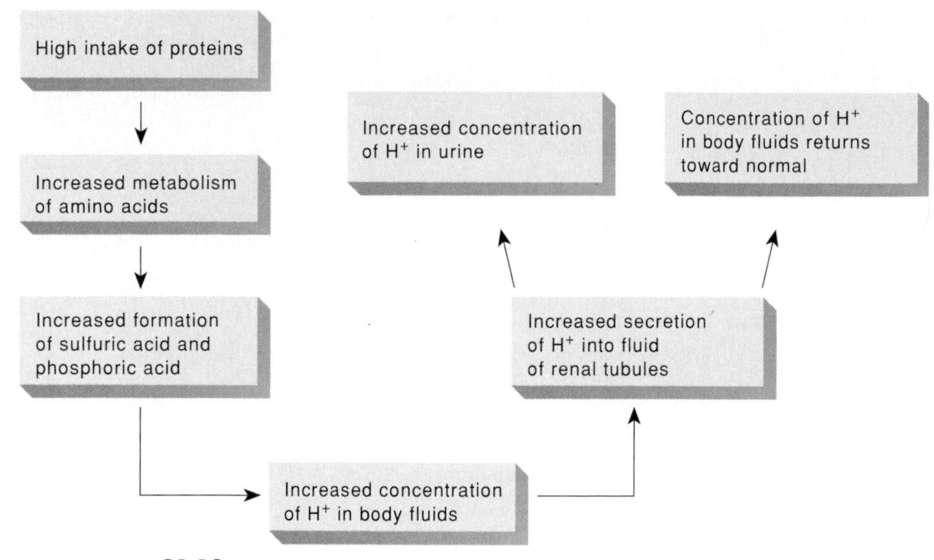


**FIGURE 21.12**

If the concentration of hydrogen ions in body fluids increases, the renal tubules increase their secretion of hydrogen ions into the urine.

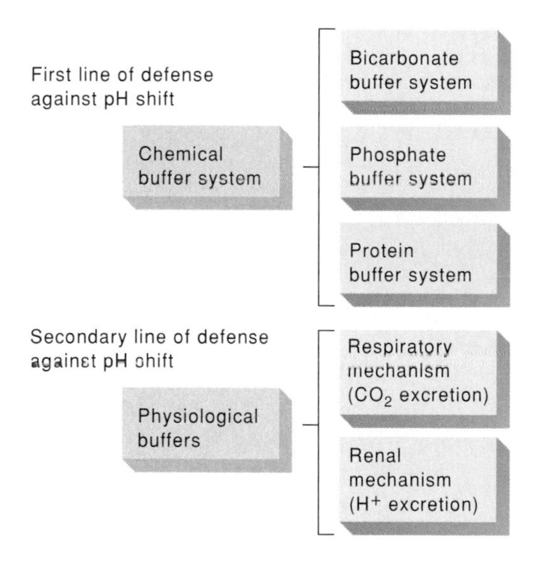


**FIGURE 21.13**

Chemical buffers act rapidly, while physiological buffers may require several minutes to several days to begin resisting a change in pH.

important in balancing the quantities of sulfuric acid, phosphoric acid, and various organic acids that appear in body fluids as by-products of metabolic processes.

The metabolism of certain amino acids, for example, produces sulfuric and phosphoric acids. Consequently, a diet high in proteins may trigger excess acid formation. The kidneys compensate for such gains in acids by altering the rate of hydrogen ion secretion, thus resisting a shift in the pH of body fluids (fig. 21.12).

Once hydrogen ions are secreted, phosphates that were filtered into the fluid of the renal tubule buffer them. Ammonia aids in this buffering action.

Through deamination of certain amino acids, the cells of the renal tubules produce ammonia ($NH_3$). It diffuses readily through the cell membranes and enters the urine. Because ammonia is a weak base, it can accept hydrogen ions and *ammonium ions* ($NH_4^+$) result:

$$H^+ + NH_3 \rightarrow NH_4^+$$

Cell membranes are quite impermeable to ammonium ions. Consequently, these ions are trapped in the urine as they form, and they are excreted from the body with the urine. When increase in the hydrogen ion concentration of body fluids is prolonged, the renal tubules increase ammonia production. This mechanism helps to transport excess hydrogen ions to the outside and helps prevent the urine from becoming too acidic.

The various regulators of hydrogen ion concentration operate at different rates. Acid-base buffers, for example, function rapidly and can convert strong acids or bases into weak acids or bases almost immediately. For this reason, these chemical buffer systems are sometimes called the body's *first line of defense* against shifts in pH.

Physiological buffer systems, such as the respiratory and renal mechanisms, function more slowly, and constitute *secondary defenses.* The respiratory mechanism may require several minutes to begin resisting a change in pH, and the renal mechanism may require one to three days to regulate a changing hydrogen ion concentration (fig. 21.13).

Clinical Application 21.3 examines the effects of acid-base imbalances.

1. How does the respiratory system help regulate acid-base balance?
2. How do the kidneys respond to excess hydrogen ions?
3. How do the rates at which chemical and physiological buffer systems act differ?

CLINICAL APPLICATION

# Acid-Base Imbalances

Ordinarily, chemical and physiological buffer systems maintain the hydrogen ion concentration of body fluids within very narrow pH ranges. Abnormal conditions may disturb the acid-base balance. For example, the pH of arterial blood is normally 7.35–7.45. A value below 7.35 produces *acidosis*. A pH above 7.45 produces *alkalosis*. Such shifts in the pH of body fluids may be life threatening. In fact, a person usually cannot survive if the pH drops to 6.8 or rises to 8.0 for more than a few hours (fig. 21C).

Acidosis results from an accumulation of acids or a loss of bases, both of which cause abnormal increases in the hydrogen ion concentrations of body fluids. Conversely, alkalosis results from a loss of acids or an accumulation of bases accompanied by a decrease in hydrogen ion concentrations (fig. 21D).

The two major types of acidosis are *respiratory acidosis* and *metabolic acidosis*. Factors that increase carbon dioxide, also increasing the concentration of carbonic acid (the respiratory acid), cause respiratory acidosis. Metabolic acidosis is due to an abnormal accumulation of any other acids in the body fluids or to a loss of bases, including bicarbonate ions.

Similarly, the two major types of alkalosis are *respiratory alkalosis* and *metabolic alkalosis*. Excessive loss of carbon dioxide and consequent loss of carbonic acid cause respiratory alkalosis. Metabolic alkalosis is due to excess loss of hydrogen ions or gain of bases.

Since in respiratory acidosis carbon dioxide accumulates, it can result from factors that hinder pulmonary ventilation (fig. 21E). These include the following:

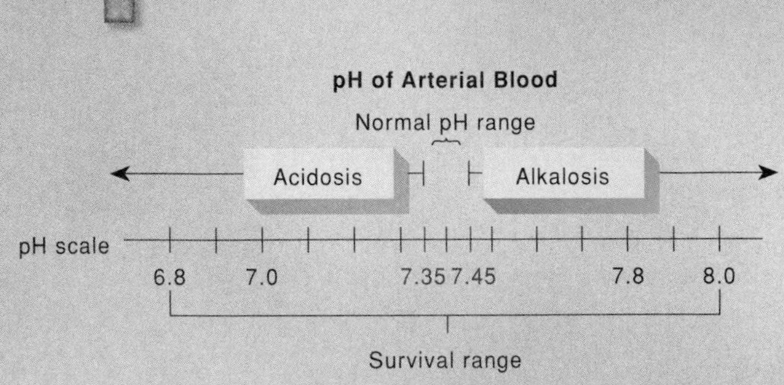


FIGURE 21C
If the pH of the arterial blood drops to 6.8 or rises to 8.0 for more than a few hours, the person usually cannot survive.

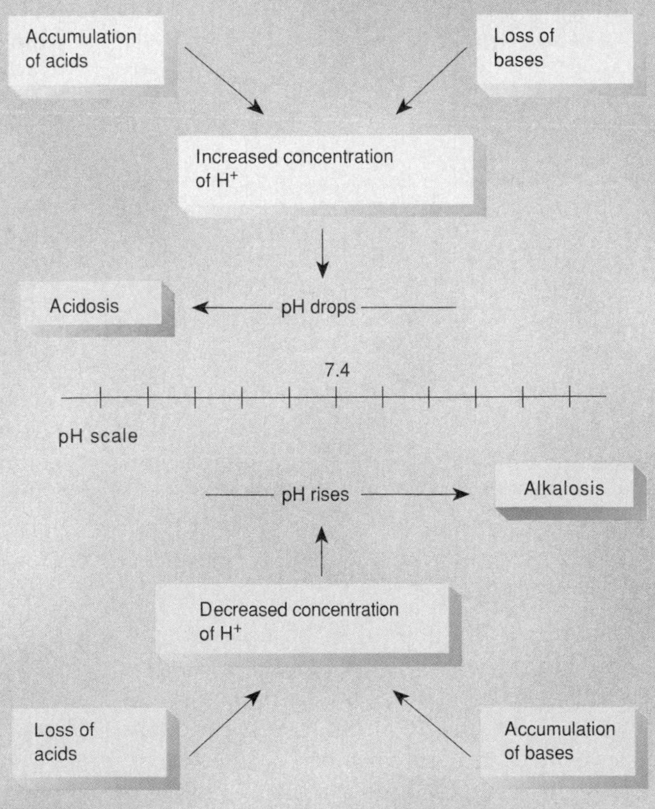


FIGURE 21D
Acidosis results from an accumulation of acids or a loss of bases. Alkalosis results from a loss of acids or an accumulation of bases.

1. Injury to the respiratory center of the brain stem, decreasing rate and depth of breathing.
2. Obstructions in air passages that interfere with air movement into the alveoli.
3. Diseases that decrease gas exchanges, such as pneumonia, or those that reduce surface area of the respiratory membrane, such as emphysema.

Any of these conditions can increase the level of carbonic acid and hydrogen ions in body fluids, lowering pH. Chemical buffers, such as hemoglobin, may resist this shift in pH. At the same time, increasing concentrations of carbon dioxide and hydrogen ions stimulate the respiratory center, increasing breathing rate and depth and thereby lowering carbon dioxide concentration. Also, the kidneys may begin to excrete more hydrogen ions.

Eventually, thanks to these chemical and physiological buffers, the pH of the body fluids may return to normal. When this happens, the acidosis is said to be *compensated.*

The symptoms of respiratory acidosis result from depression of central nervous system function, and include drowsiness, disorientation, and stupor. Evidence of respiratory insufficiency, such as labored breathing and cyanosis, is usually also evident. In *uncompensated acidosis,* the person may become comatose and die.

Metabolic acidosis is due to either accumulation of nonrespiratory acids or loss of bases (fig. 21F). Factors that may lead to this condition include:

1. Kidney disease that reduces glomerular filtration and fails to excrete the acids produced in metabolism (uremic acidosis).
2. Prolonged vomiting that loses the alkaline contents of the upper intestine and stomach contents. (Losing only the stomach contents produces metabolic alkalosis.)
3. Prolonged diarrhea in which excess alkaline intestinal secretions are lost (especially in infants).
4. Diabetes mellitus, in which some fatty acids are converted into ketone bodies. These ketone bodies include *acetoacetic acid, beta-hydroxybutyric acid,* and *acetone.* Normally these molecules are produced in relatively small quantities, and cells oxidize them as energy sources. However, if fats are being utilized at an abnormally high rate, as may occur in diabetes mellitus, ketone bodies may accumulate faster

*Continued . . .*

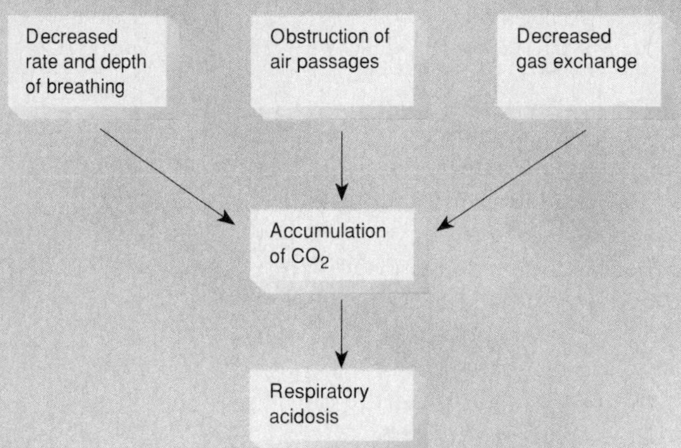


**FIGURE 21E**
Some of the factors that lead to respiratory acidosis.

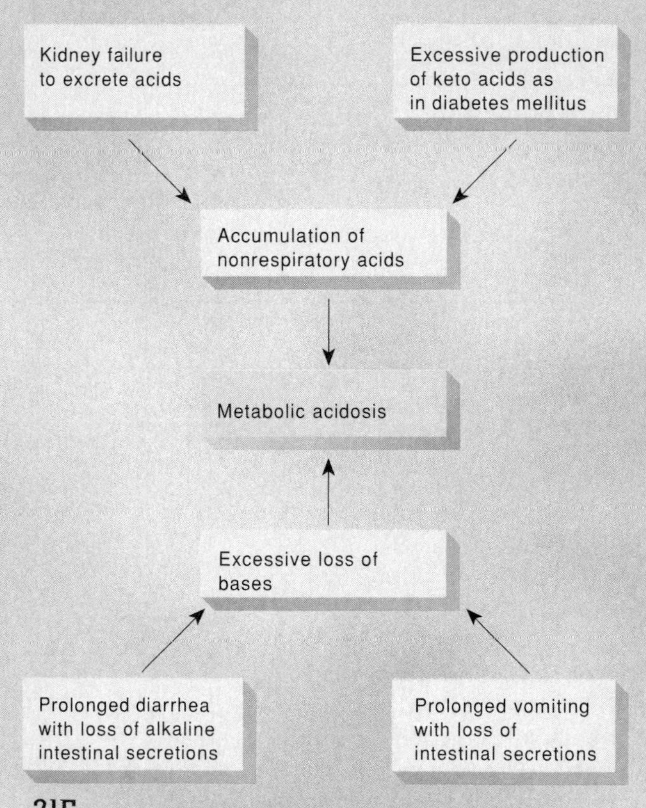


**FIGURE 21F**
Some of the factors that lead to metabolic acidosis.

than they can be oxidized. At such times, these compounds may be excreted in the urine (ketonuria); in addition, acetone, which is volatile, may be excreted by the lungs and impart a fruity odor to the breath. More seriously, the accumulation of acetoacetic acid and beta-hydroxybutyric acid may lower pH (ketonemic acidosis).

These acids may also combine with bicarbonate ions in the urine. As a result, excess bicarbonate ions are excreted, interfering with the function of the bicarbonate acid-base buffer system.

In each case, the pH tends to shift toward lower values. However, the following factors resist this shift: chemical buffer systems, which accept excess hydrogen ions; the respiratory center, which increases the breathing rate and depth; and the kidneys, which excrete more hydrogen ions.

Respiratory alkalosis develops as a result of *hyperventilation*, described in chapter 19. Hyperventilation is accompanied by too great a loss of carbon dioxide and consequent decreases in carbonic acid and hydrogen ion concentrations (fig. 21G).

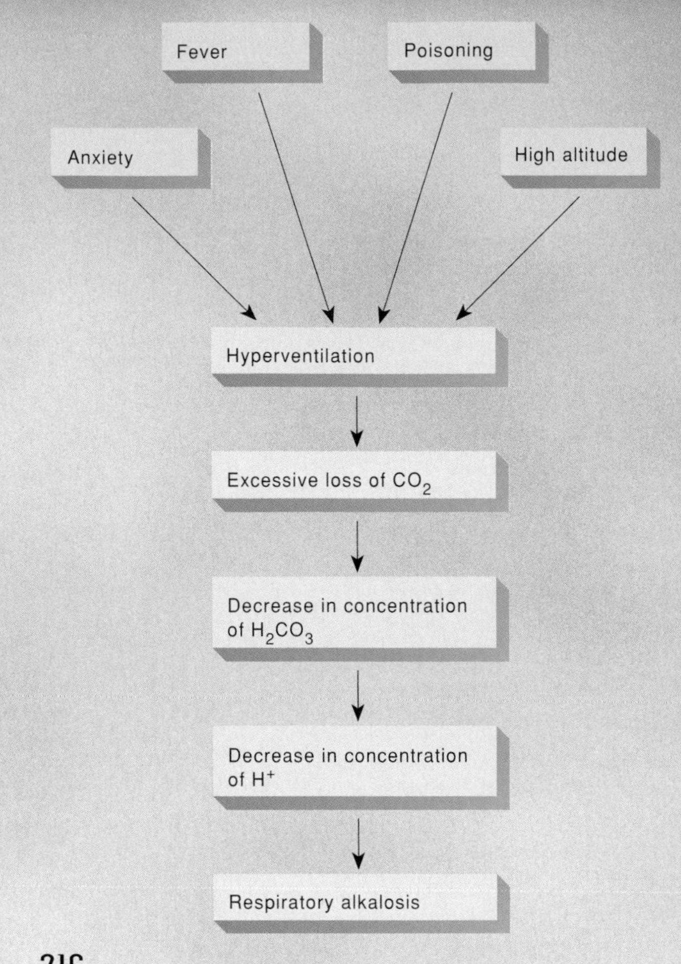


**Figure 21G**
Some of the factors that lead to respiratory alkalosis.

## Clinical Terms Related to Water and Electrolyte Balance

**acetonemia** (as″e-to-ne′me-ah) Abnormal amounts of acetone in the blood.
**acetonuria** (as″e-to-nu′re-ah) Abnormal amounts of acetone in the urine.
**albuminuria** (al-bu″mĭ-nu′re-ah) Albumin in the urine.
**anasarca** (an″ah-sar′kah) A widespread accumulation of tissue fluid.
**antacid** (ant-as′id) A substance that neutralizes an acid.
**anuria** (ah-nu′re-ah) The absence of urine excretion.
**azotemia** (az″o-te′me-ah) Accumulation of nitrogenous wastes in the blood.
**diuresis** (di″u-re′sis) Increased production of urine.
**glucosuria** (gli″ko-su′re-ah) Excess sugar in the urine.
**hyperglycemia** (hi″per-gli-se′me-ah) An abnormally high concentration of blood sugar.
**hyperkalemia** (hi″per-kah-le′me-ah) Excess potassium in the blood.
**hypernatremia** (hi″per-na-tre′me-ah) Excess sodium in the blood.
**hyperuricemia** (hi″per-u″rĭ-se′me-ah) Excess uric acid in the blood.
**hypoglycemia** (hi″po-gli-se′me-ah) An abnormally low concentration of blood sugar.
**ketonuria** (ke″to-nu′re-ah) Ketone bodies in the urine.
**ketosis** (ke″to′sis) Acidosis due to excess ketone bodies in the body fluids.
**proteinuria** (pro″te-ĭ-nu′re-ah) Protein in the urine.
**uremia** (u-re′me-ah) A toxic condition resulting from nitrogenous wastes in the blood.

Hyperventilation may occur during periods of anxiety, although it may also accompany fever or poisoning from salicylates, such as aspirin. At high altitudes, hyperventilation may be a response to low oxygen pressure. Also, musicians, such as bass tuba players, who must provide a large volume of air when playing sustained passages, sometimes hyperventilate. In each case, rapid, deep breathing depletes carbon dioxide and the pH of body fluids increases.

Chemical buffers, such as hemoglobin, that release hydrogen ions resist this pH change. Also, the lower concentrations of carbon dioxide and hydrogen ions stimulate the respiratory center to a lesser degree. This inhibits hyperventilation, thus reducing further carbon dioxide loss. At the same time, the kidneys decrease their secretion of hydrogen ions, and the urine becomes alkaline as bases are excreted.

The symptoms of respiratory alkalosis include lightheadedness, agitation, dizziness, and tingling sensations. In severe cases, impulses may be triggered spontaneously on peripheral nerves, and muscles may respond with tetanic contractions (see chapter 9).

Metabolic alkalosis results from a great loss of hydrogen ions or from a gain in bases, both of which are accompanied by a rise in the pH of the blood (alkalemia) (fig. 21.H).

This condition may occur following gastric drainage (lavage), prolonged vomiting in which only the stomach contents are lost, or the use of certain diuretic drugs. Because gastric juice is very acidic, its loss leaves the body fluids with a net increase of basic substances and a pH shift toward alkaline values. Metabolic alkalosis may also develop as a result of ingesting too much antacid, such as sodium bicarbonate, medication to relieve the symptoms of indigestion. The symptoms of metabolic alkalosis include a decrease in the breathing rate and depth, which, in turn, results in an increased concentration of carbon dioxide in the blood.

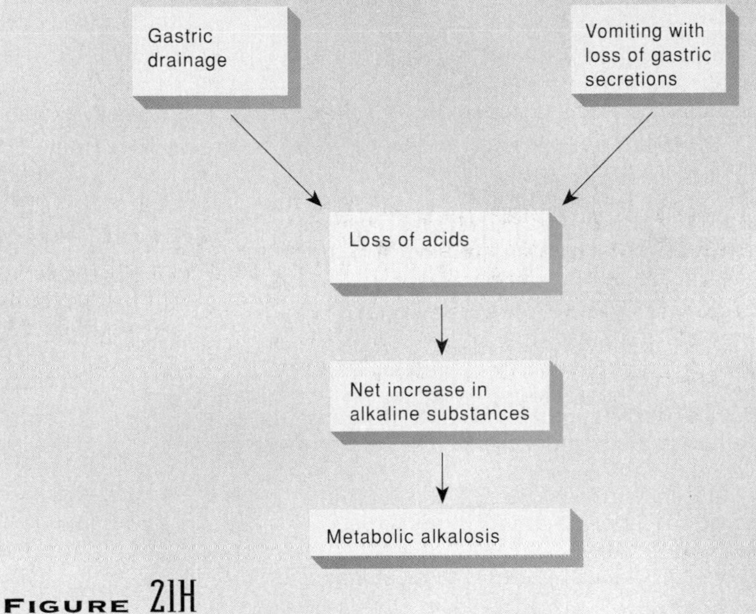


**FIGURE 21H**
Some of the factors that lead to metabolic alkalosis.

## CHAPTER SUMMARY

### Introduction (page 831)

The maintenance of water and electrolyte balance requires that the quantities of these substances entering the body equal the quantities leaving it. Altering the water balance necessarily affects the electrolyte balance.

### Distribution of Body Fluids (page 831)

1. Fluid compartments
   a. The intracellular fluid compartment includes the fluids and electrolytes cell membranes enclose.
   b. The extracellular fluid compartment includes all fluids and electrolytes outside cell membranes.
      (1) Interstitial fluid within tissue spaces
      (2) Plasma within blood
      (3) Lymph within lymphatic vessels
      (4) Transcellular fluid within body cavities
2. Body fluid composition
   a. Extracellular fluids
      (1) Extracellular fluids have high concentrations of sodium, chloride, and bicarbonate ions, with less potassium, calcium, magnesium, phosphate, and sulfate ions.
      (2) Plasma contains more protein than does either interstitial fluid or lymph.
   b. Intracellular fluid contains relatively high concentrations of potassium, phosphate, and magnesium ions; it also contains a greater concentration of sulfate ions and lesser concentrations of sodium, chloride, and bicarbonate ions than does extracellular fluid.

3. Movement of fluid between compartments
    a. Hydrostatic and osmotic pressure regulate fluid movements.
        (1) Fluid leaves plasma because of hydrostatic pressure and returns to plasma because of osmotic pressure.
        (2) Osmotic pressure drives fluid into lymph vessels.
        (3) Osmotic pressure regulates fluid movement in and out of cells.
    b. Sodium ion concentrations are especially important in fluid movement regulation.

## Water Balance (page 834)

1. Water intake
    a. The volume of water taken in varies from person to person.
    b. Most water comes from consuming liquid or moist foods.
    c. Oxidative metabolism produces some water.
2. Regulation of water intake
    a. The thirst mechanism is the primary regulator of water intake.
    b. Drinking and the resulting stomach distension inhibit the thirst mechanism.
3. Water output
    a. Water is lost in a variety of ways.
        (1) It is excreted in the urine, feces, and sweat.
        (2) Insensible loss occurs through evaporation from the skin and lungs.
    b. Urine production regulates water output.
4. Regulation of water balance
    a. The distal convoluted tubules and collecting ducts of the nephrons regulate water balance.
        (1) ADH from the hypothalamus and posterior pituitary gland stimulates water reabsorption in these segments.
        (2) The mechanism involving ADH can reduce normal output of 1,500 milliliters to 500 milliliters per day.
    b. If excess water is taken in, the ADH mechanism is inhibited.

## Electrolyte Balance (page 836)

1. Electrolyte intake
    a. The most important electrolytes in the body fluids are those that release ions of sodium, potassium, calcium, magnesium, chloride, sulfate, phosphate, and bicarbonate.
    b. These ions are obtained in foods and beverages or as by-products of metabolic processes.
2. Regulation of electrolyte intake
    a. Electrolytes are usually obtained in sufficient quantities in response to hunger and thirst mechanisms.
    b. In a severe electrolyte deficiency, a person may experience a salt craving.
3. Electrolyte output
    a. Electrolytes are lost through perspiration, feces, and urine.
    b. Quantities lost vary with temperature and physical exercise.
    c. The greatest electrolyte loss occurs as a result of kidney functions.
4. Regulation of electrolyte balance
    a. Concentrations of sodium, potassium, and calcium ions in the body fluids are particularly important.
    b. The regulation of sodium ions involves the secretion of aldosterone from the adrenal glands.
    c. The regulation of potassium ions also involves aldosterone.
    d. Calcitonin from the thyroid gland and parathyroid hormone from the parathyroid glands regulate calcium ion concentration.
    e. The mechanisms that control positively charged ions secondarily regulate negatively charged ions.
        (1) Chloride ions are passively reabsorbed in renal tubules as sodium ions are actively reabsorbed.
        (2) Some negatively charged ions, such as phosphate ions, are reabsorbed partially by limited active transport mechanisms.

## Acid-Base Balance (page 840)

Acids are electrolytes that release hydrogen ions. Bases combine with hydrogen ions.

1. Sources of hydrogen ions
    a. Aerobic respiration of glucose
        (1) Aerobic respiration of glucose produces carbon dioxide, which reacts with water to form carbonic acid.
        (2) Carbonic acid dissociates to release hydrogen and bicarbonate ions.
    b. Anaerobic respiration of glucose produces lactic acid.
    c. Incomplete oxidation of fatty acids releases acidic ketone bodies.
    d. Oxidation of sulfur-containing amino acids produces sulfuric acid.
    e. Hydrolysis of phosphoproteins and nucleic acids gives rise to phosphoric acid.
2. Strengths of acids and bases
    a. Acids vary in the extent to which they ionize.
        (1) Strong acids, such as hydrochloric acid, ionize more completely.
        (2) Weak acids, such as carbonic acid, ionize less completely.
    b. Bases vary in strength also.
        (1) Strong bases, such as hydroxyl ions, combine readily with hydrogen ions.
        (2) Weak bases, such as chloride ions, combine with hydrogen ions less readily.
3. Regulation of hydrogen ion concentration
    a. Acid-base buffer systems
        (1) Buffer systems are composed of sets of two or more chemicals.
        (2) They convert strong acids into weaker acids, or strong bases into weaker bases.
        (3) They include the bicarbonate buffer system, phosphate buffer system, and protein buffer system.
        (4) Buffer systems minimize pH changes.
    b. Respiratory center
        (1) The respiratory center is located in the brain stem.
        (2) It helps regulate pH by controlling the rate and depth of breathing.
        (3) Increasing carbon dioxide and hydrogen ion concentrations stimulate chemosensitive areas

associated with the respiratory center; breathing rate and depth increase, and carbon dioxide concentration decreases.
   (4) If the carbon dioxide and hydrogen ion concentrations are low, the respiratory center inhibits breathing.
c. Kidneys
   (1) Nephrons secrete hydrogen ions to regulate pH.
   (2) Phosphates buffer hydrogen ions in urine.
   (3) Ammonia produced by renal cells helps transport hydrogen ions to the outside of the body.
d. Chemical buffers act rapidly; physiological buffers act more slowly.

## Critical Thinking Questions

1. An elderly, semiconscious patient is tentatively diagnosed as having acidosis. What components of the arterial blood will be most valuable in determining if the acidosis is of respiratory origin?
2. Some time ago, several newborn infants died due to an error in which sodium chloride was substituted for sugar in their formula. What symptoms would this produce? Why do you think infants are more prone to the hazard of excess salt intake than adults?
3. Explain the threat to fluid and electrolyte balance in the following situation: A patient is being nutritionally maintained on concentrated solutions of hydrolyzed protein that are administered through a gastrostomy tube.
4. Describe what might happen to the plasma pH of a patient as a result of:
   a. prolonged diarrhea
   b. suction of the gastric contents
   c. hyperventilation
   d. hypoventilation
5. Radiation therapy may damage the mucosa of the stomach and intestines. What effect might this have on the patient's electrolyte balance?
6. If the right ventricle of a patient's heart is failing, increasing the venous pressure, what changes might occur in the patient's extracellular fluid compartments?

## Review Exercises

1. Explain how water balance and electrolyte balance are interdependent.
2. Name the body fluid compartments, and describe their locations.
3. Explain how the fluids within these compartments differ in composition.
4. Describe how fluid movements between the compartments are controlled.
5. Prepare a list of sources of normal water gain and loss to illustrate how the input of water equals the output of water.
6. Define *water of metabolism*.
7. Explain how water intake is regulated.
8. Explain how the nephrons function in the regulation of water output.
9. List the most important electrolytes in the body fluids.
10. Explain how electrolyte intake is regulated.
11. List the routes by which electrolytes leave the body.
12. Explain how the adrenal cortex functions in the regulation of electrolyte output.
13. Describe the role of the parathyroid glands in regulating electrolyte balance.
14. Describe the role of the renal tubule in regulating electrolyte balance.
15. Distinguish between an acid and a base.
16. List five sources of hydrogen ions in the body fluids, and name an acid that originates from each source.
17. Distinguish between a strong acid and a weak acid, and name an example of each.
18. Distinguish between a strong base and a weak base, and name an example of each.
19. Explain how an acid-base buffer system functions.
20. Describe how the bicarbonate buffer system resists changes in pH.
21. Explain why a protein has acidic as well as basic properties.
22. Describe how a protein functions as a buffer system.
23. Describe the function of hemoglobin as a buffer of carbonic acid.
24. Explain how the respiratory center functions in the regulation of the acid-base balance.
25. Explain how the kidneys function in the regulation of the acid-base balance.
26. Describe the role of ammonia in the transport of hydrogen ions to the outside of the body.
27. Distinguish between a chemical buffer system and a physiological buffer system.

22

CHAPTER

# REPRODUCTIVE SYSTEMS

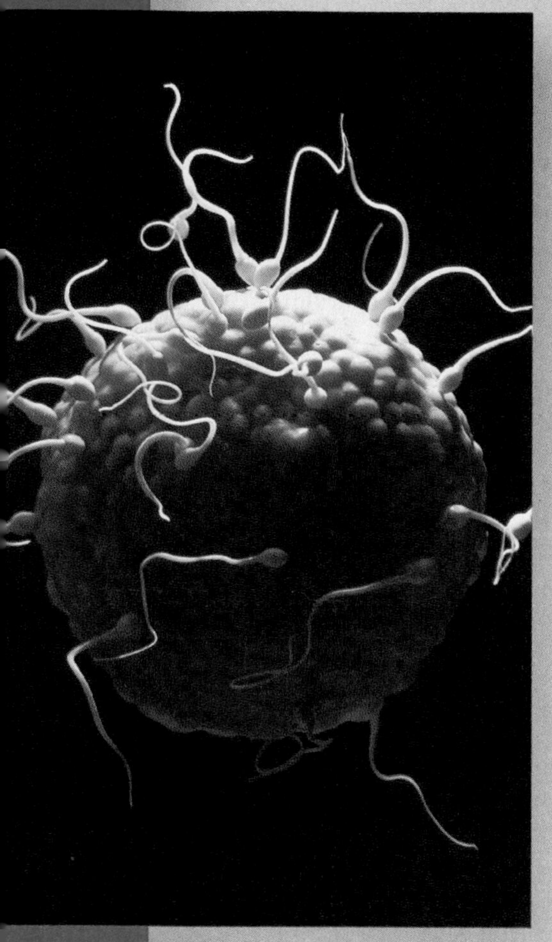

SPERM AROUND AN EGG.

## CHAPTER OBJECTIVES

AFTER YOU HAVE STUDIED THIS CHAPTER, YOU SHOULD BE ABLE TO:

1. State the general functions of the male reproductive system.
2. Name the parts of the male reproductive system and describe the general functions of each part.
3. Outline the process of spermatogenesis.
4. Trace the path followed by sperm cells from their site of formation to the outside.
5. Describe the structure of the penis and explain how its parts function to produce an erection.
6. Explain how hormones control the activities of the male reproductive organs and the development of male secondary sexual characteristics.
7. State the general functions of the female reproductive system.
8. Name the parts of the female reproductive system and describe the general functions of each part.
9. Outline the process of oogenesis.
10. Describe how hormones control the activities of the female reproductive system and the development of female secondary sexual characteristics.
11. Describe the major events that occur during a menstrual cycle.
12. Describe the hormonal changes that occur in the maternal body during pregnancy.
13. Describe the birth process and explain the role of hormones in this process.
14. List several methods of birth control and describe the relative effectiveness of each method.

## KEY TERMS

**androgen** (an′dro-jen)
**cleavage** (klēv′ij)
**clitoris** (klit′o-ris)
**coitus** (ko′ĭ-tus)
**contraception** (kon″trah-sep′shun)
**ejaculation** (e-jak″u-la′shun)
**emission** (e-mish′un)
**epididymis** (ep″ĭ-did′ĭ-mis)
**estrogen** (es′tro jen)
**fertilization** (fer″ti-li-za′shun)
**fimbriae** (fim′bre)
**follicle** (fol′ĭ-kl)
**gonadotropin** (go-nad″o-trōp′in)
**gubernaculum** (gu″ber-nak′u-lum)
**implantation** (im plan-ta′shun)
**infundibulum** (in″fun-dib′u-lum)
**inguinal** (ing′gwĭ-nal)
**meiosis** (mi-o′sis)
**menopause** (men′o-pawz)
**menstrual cycle** (men′stroo-al si′kl)
**oogenesis** (o″o-jen′ĕ-sis)
**orgasm** (or′gazm)
**ovulation** (o″vu-la′shun)
**placenta** (plah-sen′tah)
**pregnancy** (preg′nan-se)
**progesterone** (pro-jes′tĕ-rōn)
**puberty** (pu′ber-te)
**semen** (se′men)
**spermatogenesis** (sper″mah-to-jen′ĕ-sis)
**testosterone** (tes-tos′te-rōn)
**vas deferens** (vas def′er-enz)
**zygote** (zi′gōt)

UNIT SIX

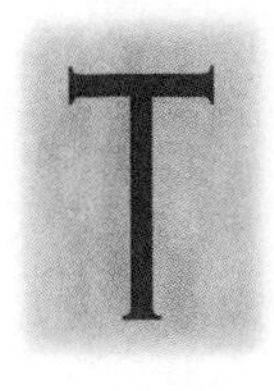

The two cells approach each other, one 90,000 times the volume of the other. Although they look about as different as two cells can, they are alike in that each carries a half set of genetic instructions from each individual parent. The two cells meet and merge, launching a new life.

Each of us harbors many of these sex cells—sperm or eggs—that have the potential to participate in forming a new life. These very special cells are generated and nurtured in our reproductive systems, which are similarly organized in the male and female. Each system has paired structures where sperm or eggs are manufactured, a network of tubes to transport these cells, and hormones and exocrine secretions that oversee the entire process.

Our reproductive systems profoundly affect our lives, even when we are not parents. Many a teen has anguished over facial blemishes, turbulent and disturbing feelings, and a body seemingly out of control—all attributed to the hormonal upheaval that reflects a maturing reproductive system. We use our reproductive systems not only to make babies but also to intimately interact with another individual. But with sexual activity comes responsibility—to avoid pregnancy when one is not ready for parenthood, and to protect ourselves and our partners from infections of the reproductive system.

## Organs of the Male Reproductive System

The organs of the male reproductive system are specialized to produce and maintain the male sex cells, or *sperm cells;* to transport these cells, together with supporting fluids, to the female reproductive tract; and to secrete male sex hormones.

The *primary sex organs* (gonads) of this system are the two testes in which the sperm cells (spermatozoa) and the male sex hormones are formed. The other structures of the male reproductive system are termed *accessory sex organs* (secondary sex organs). They include the internal reproductive organs and the external reproductive organs (fig. 22.1).

## Testes

The **testes** are ovoid structures about 5 centimeters in length and 3 centimeters in diameter. Each testis is suspended by a spermatic cord and is contained within the cavity of the saclike *scrotum* (see fig. 22.1 and reference plate 52).

### Descent of the Testes

In a male fetus, the testes originate from masses of tissue behind the parietal peritoneum, near the developing kidneys. Usually a month or two before birth, these organs descend to the lower abdominal cavity and pass through the abdominal wall into the scrotum.

The male sex hormone *testosterone,* which is secreted by the developing testes, stimulates the testes to descend. A fibromuscular cord called the **gubernaculum** aids movement of the testes. This cord is attached to each developing testis and extends into the inguinal region of the abdominal cavity. It passes through the abdominal wall and is fastened to the skin on the outside of the scrotum. The testis descends, guided by the gubernaculum, passing through the **inguinal canal** of the abdominal wall and entering the scrotum, where it remains anchored by the gubernaculum. Each testis carries a developing *vas deferens,* blood vessels, and nerves. These structures later form parts of the **spermatic cord** by which the testis is suspended in the scrotum (fig. 22.2).

If the testes fail to descend into the scrotum, they will not produce sperm cells because the temperature in the abdominal cavity is too high. If this condition, called *cryptorchidism,* is left untreated the cells that normally produce sperm cells degenerate, and the male is infertile.

During the descent of a testis, a pouch of peritoneum, called the *vaginal process,* moves through the inguinal canal and into the scrotum. In about one-quarter of males, this process remains open, providing a potential passageway through which a loop of intestine may be forced by great abdominal pressure, producing an *indirect inguinal hernia.* If the protruding intestinal loop is so tightly constricted within the inguinal canal that its blood supply stops, the condition is called a *strangulated hernia.* Without prompt treatment, the strangulated tissues may die.

1. What are the primary sex organs of the male reproductive system?
2. Describe the descent of the testes.
3. What is the function of the gubernaculum, both during and after the descent of the testes?
4. What happens if the testes fail to descend into the scrotum?

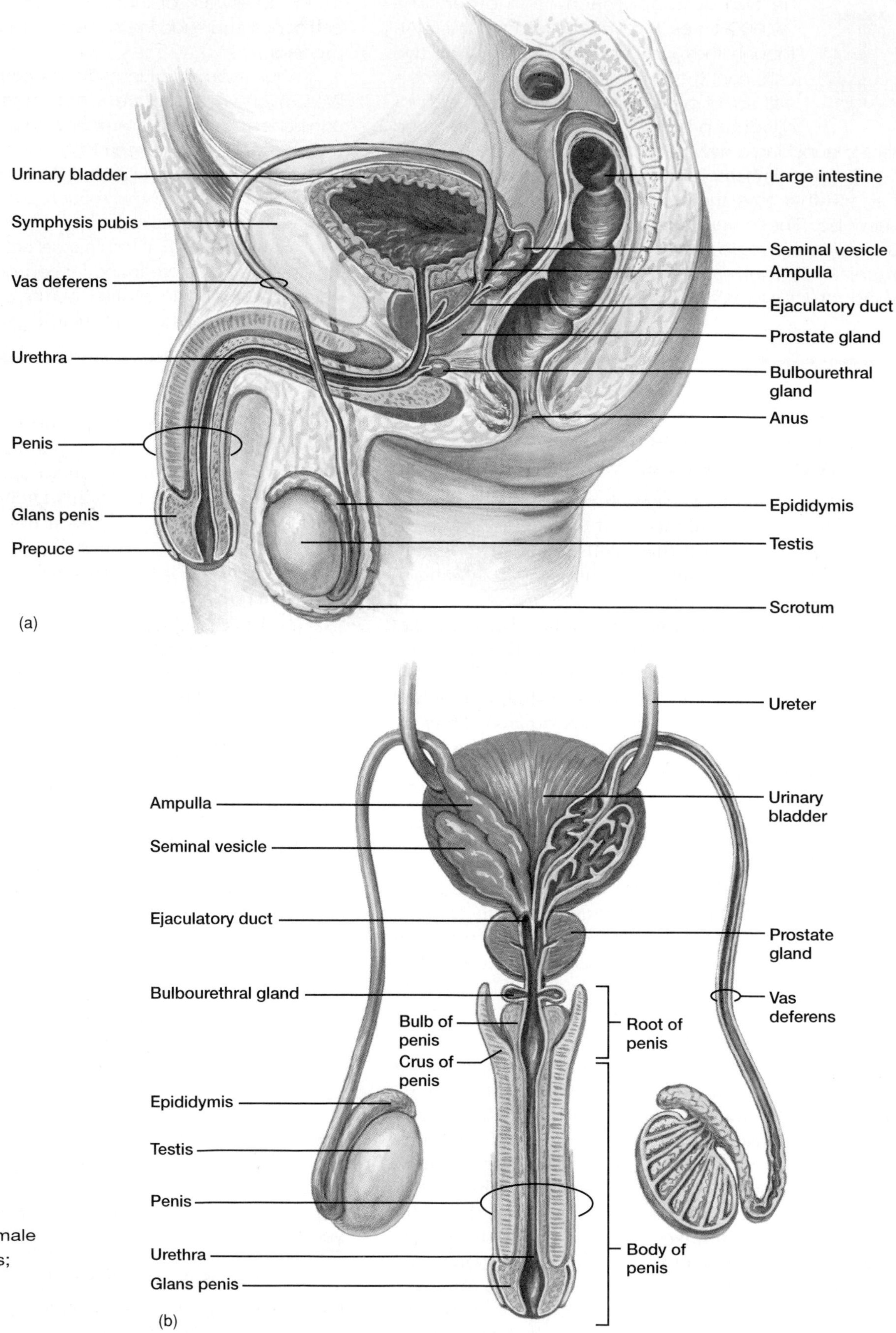


**FIGURE 22.1**

(*a*) Sagittal view of male reproductive organs; (*b*) posterior view.

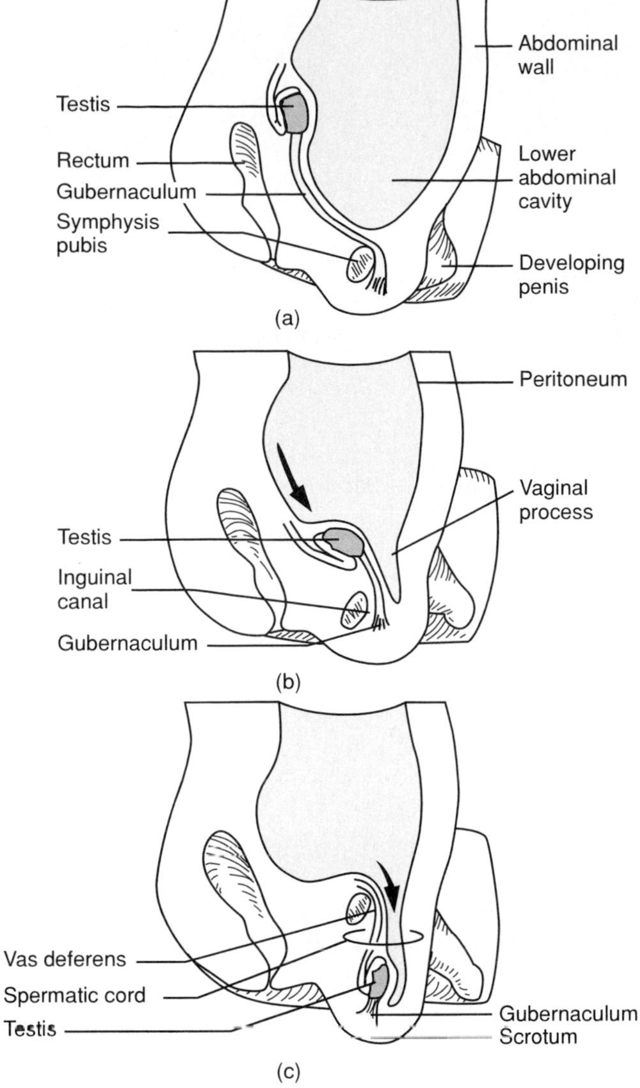


**FIGURE 22.2**

During fetal development, each testis descends through an inguinal canal and enters the scrotum (*a–c*).

## Structure of the Testes

A tough, white, fibrous capsule called the *tunica albuginea* encloses the testis. Along its posterior border, the connective tissue thickens and extends into the organ, forming a mass called the *mediastinum testis.* From this structure, thin layers of connective tissue, called *septa,* pass into the testis and subdivide it into about 250 *lobules.*

A lobule contains one to four highly coiled, convoluted **seminiferous tubules,** each of which is approximately 70 centimeters long when uncoiled. These tubules course posteriorly and unite to form a complex network of channels called the *rete testis.* The rete testis is located within the mediastinum testis and gives rise to several ducts that join a tube called the *epididymis.* The epididymis, in turn, is coiled on the outer surface of the testis.

The seminiferous tubules are lined with a specialized stratified epithelium, which includes the spermatogenic cells that give rise to the sperm cells. Other specialized cells, called **interstitial cells** (cells of Leydig), lie in the spaces between the seminiferous tubules. Interstitial cells produce and secrete male sex hormones (figs. 22.3, 22.4, and 22.5*a*).

The epithelial cells of the seminiferous tubules can give rise to *testicular cancer,* one of the more common types of cancer in young men. In most cases, the first sign of this condition is a painless enlargement of a testis or a scrotal mass that seems to be attached to a testis. When testicular cancer is suspected, a tissue sample is usually removed (biopsied) and examined microscopically. If cancer cells are present, the affected testis is surgically removed (orchiectomy). Depending upon the type of cancerous tissue present and the extent of the disease, a patient may be treated with radiation or chemotherapy. Testicular cancer is often curable.

1. Describe the structure of a testis.
2. Where are the sperm cells produced within the testes?
3. What cells produce male sex hormones?

## Formation of Sperm Cells

The epithelium of the seminiferous tubules consists of supporting cells (*sustentacular cells,* or Sertoli's cells) and spermatogenic cells. The sustentacular cells are tall, columnar cells that extend the full thickness of the epithelium from its base to the lumen of the seminiferous tubule. Many thin processes project from these cells, filling the spaces between nearby spermatogenic cells. They support, nourish, and regulate the *spermatogenic cells,* which give rise to sperm cells (spermatozoa).

In the male embryo, the spermatogenic cells are undifferentiated and are called *spermatogonia.* Each of these cells contains 46 chromosomes in its nucleus, which is the usual number for human cells (see fig. 22.5).

Beginning during embryonic development, hormones stimulate the spermatogonia to become active. Some of the cells undergo mitosis, giving rise to new spermatogonia and providing a reserve supply of these undifferentiated cells. Others enlarge and become *primary spermatocytes.* Sperm production (spermatogenesis) (see fig. 22.5) arrests at this stage. At puberty, testosterone secretion rises, and the primary spermatocytes then divide by a special type of cell division called **meiosis,** which is described in detail in chapter 24.

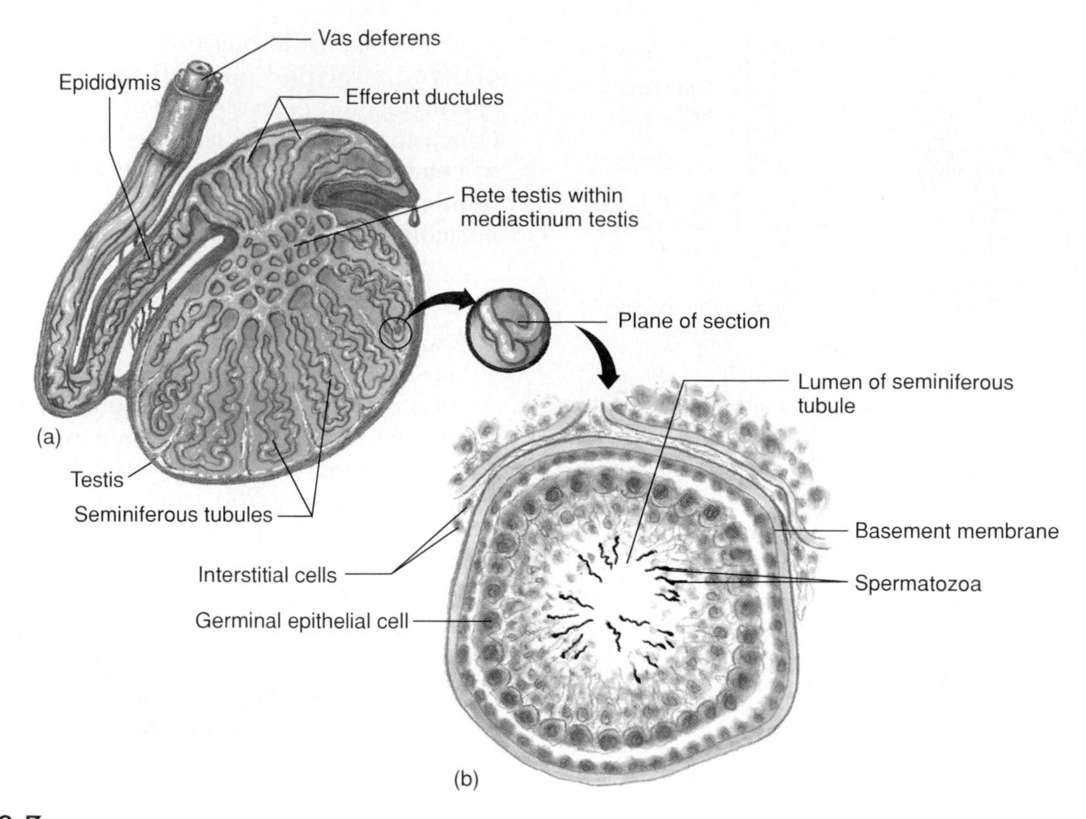


**FIGURE 22.3**
(*a*) Sagittal section of a testis; (*b*) cross section of a seminiferous tubule.

**FIGURE 22.4**
Scanning electron micrograph of cross sections of human seminiferous tubules (100×).

(*Tissues and Organs: A Text-Atlas of Scanning Electron Microscopy,* by R. G. Kessel and R. H. Kardon. © 1979 W. H. Freeman and Company.)

During meiosis, the primary spermatocytes each divide to form two *secondary spermatocytes.* Each of these cells, in turn, divides to form two *spermatids,* which mature into sperm cells. Meiosis also reduces the number of chromosomes in each cell by one-half. Consequently, for each primary spermatocyte that undergoes meiosis, four sperm cells with 23 chromosomes in each of their nuclei are formed. The chromosome number is halved so that when a sperm and egg join, the new individual has a complete set of 23 pairs of chromosomes.

The spermatogonia are located near the base of the germinal epithelium. As spermatogenesis occurs, cells in more advanced stages are pushed along the sides of sustentacular cells toward the lumen of the seminiferous tubule.

Near the base of the epithelium, membranous processes from adjacent sustentacular cells fuse by specialized junctions (occluding junctions) into complexes that divide the tissue into two layers. The spermatogonia are on one side of this barrier, and the cells in more advanced stages are on the other side. This membranous complex helps maintain a favorable environment for development of sperm cells by

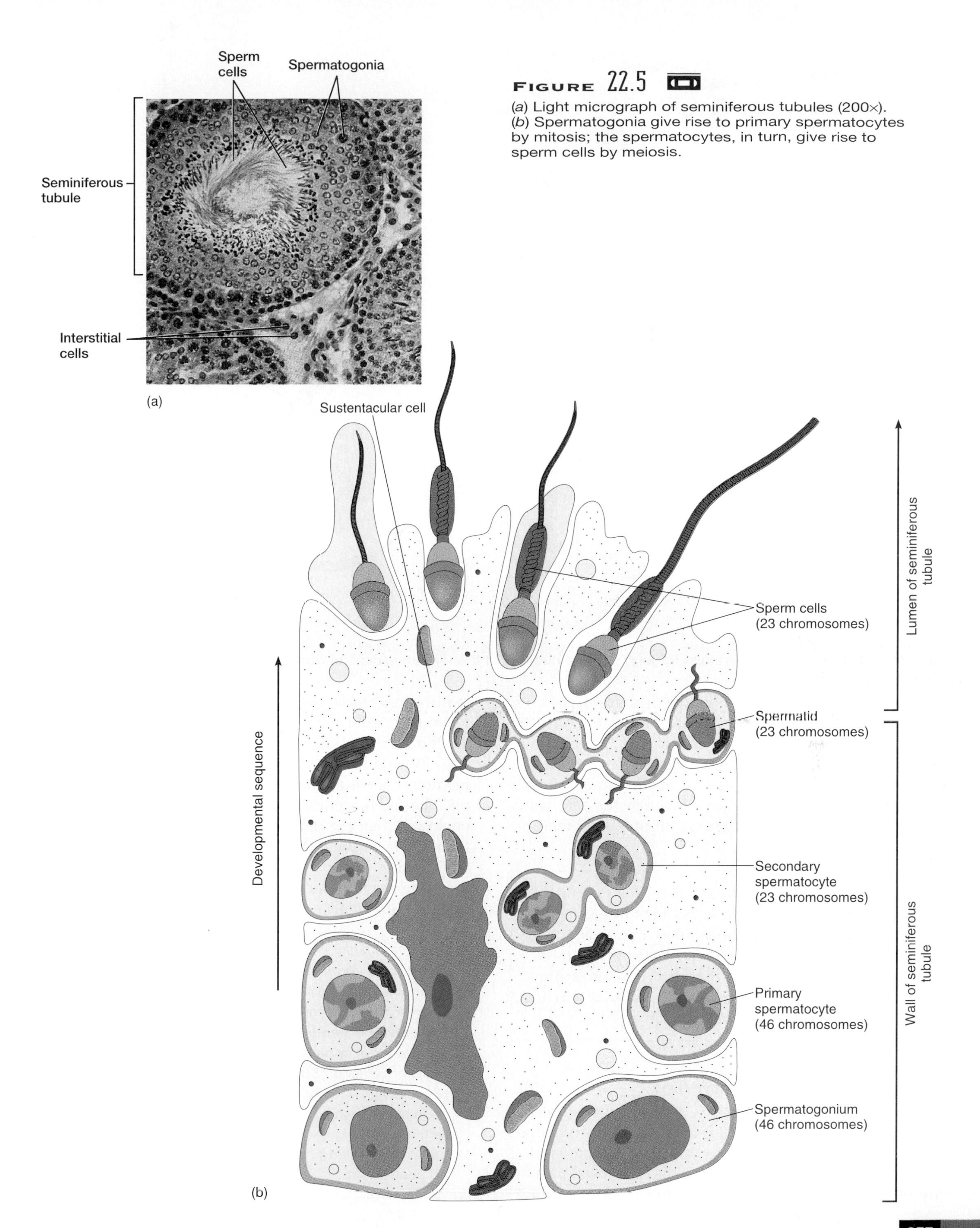


**FIGURE 22.5**

(*a*) Light micrograph of seminiferous tubules (200×). (*b*) Spermatogonia give rise to primary spermatocytes by mitosis; the spermatocytes, in turn, give rise to sperm cells by meiosis.

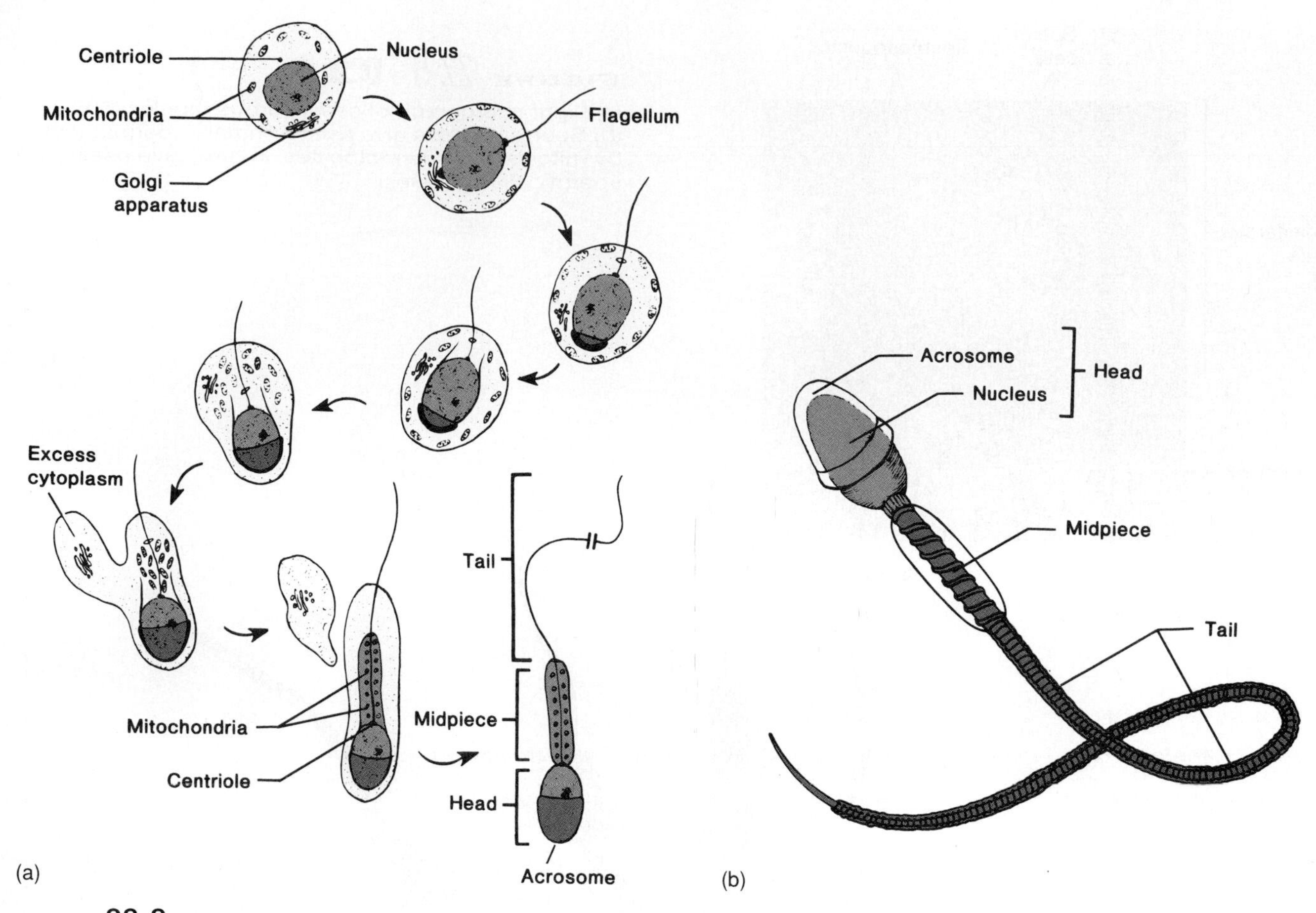


**FIGURE 22.6**
(*a*) The head of the sperm develops largely from the nucleus of the formative cell. (*b*) Parts of a mature sperm cell.

preventing certain large molecules from moving from the interstitial fluid of the basal epithelium into the region of the differentiating cells.

Spermatogenesis occurs continually in a male. The resulting sperm cells collect in the lumen of each seminiferous tubule, then pass through the rete testis to the epididymis, where they accumulate and mature.

Sperm have fascinated biologists for centuries. Anton van Leeuwenhoek was the first to view human sperm under a microscope in 1678, concluding that they were parasites in semen. By 1685, he had modified his view, writing that sperm contain a preformed human being and are seeds of sorts, requiring nurturing in a female to start a new life.

## Structure of a Sperm Cell

A mature sperm cell is a tiny, tadpole-shaped structure about 0.06 millimeters long. It consists of a flattened head, a cylindrical body, and an elongated tail.

The oval *head* of a sperm cell is composed primarily of a nucleus, and contains highly compacted chromatin of 23 chromosomes. A small protrusion at its anterior end, called the *acrosome,* contains enzymes that aid the sperm cell in penetrating an egg cell (fig. 22.6).

The midpiece of a sperm contains a central, filamentous core and many mitochondria arranged in a spiral. The *tail* (flagellum) consists of several microtubules enclosed in an extension of the cell membrane. The mitochondria provide ATP for the lashing movement of the tail that propels the sperm cell through fluid. The scanning electron micrograph in figure 22.7 shows a few mature sperm cells.

Many toxic chemicals that affect sperm hamper their ability to swim, so the cells cannot transmit the toxin to an egg. One notable exception is cocaine, which attaches to thousands of binding sites on human sperm cells, without apparently harming the cells or impeding their movements. Sperm can ferry cocaine to an egg, but scientists do not know what harm, if any, the drug causes the prenatal human that develops from it once fertilized. We do know that fetuses exposed to cocaine in the uterus may suffer a stroke, or, as infants, be unable to react normally to their surroundings.

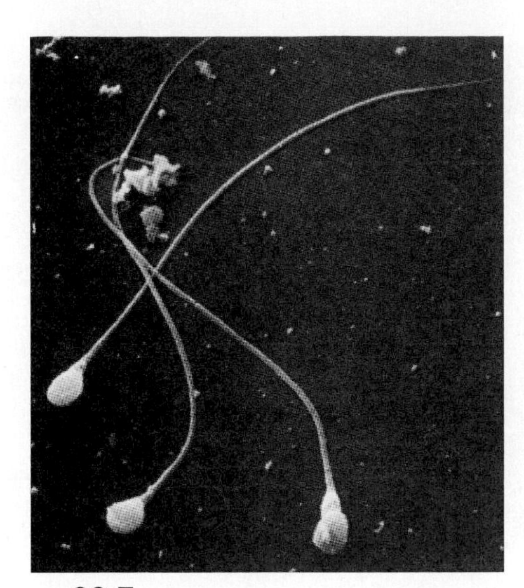

**FIGURE 22.7**
Scanning electron micrograph of human sperm cells (4,000×).

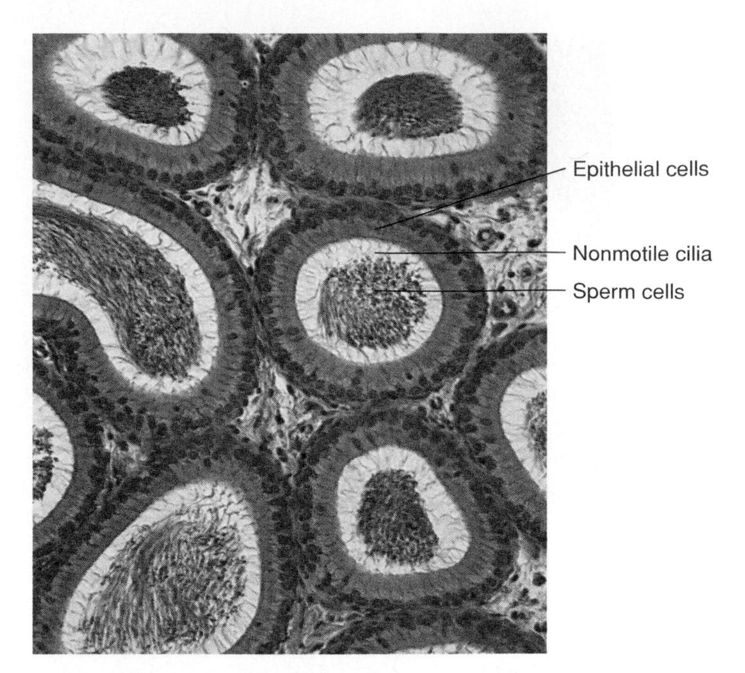


**FIGURE 22.8**
Cross section of a human epididymis (50×).

1. Explain the function of the sustentacular cells in the seminiferous tubules.
2. Describe spermatogenesis.
3. Describe the structure of a sperm cell.

## Male Internal Accessory Organs

The internal accessory organs of the male reproductive system include the epididymides, vasa deferentia, ejaculatory ducts, and urethra, as well as the seminal vesicles, prostate gland, and bulbourethral glands.

### Epididymis

Each **epididymis** (pl. *epididymides*) is a tightly coiled, threadlike tube about 6 meters long (see figs. 22.1, 22.8, and reference plate 52). The epididymis is connected to ducts within a testis. It emerges from the top of the testis, descends along its posterior surface, and then courses upward to become the vas deferens.

The inner lining of the epididymis is composed of pseudostratified columnar cells that bear nonmotile cilia. These cells secrete glycogen and other substances which support stored sperm cells and promote their maturation.

When immature sperm cells reach the epididymis, they are nonmotile. However, as they travel through the epididymis, as a result of rhythmic peristaltic contractions, they undergo maturation. Following this aging process, the sperm cells can move independently and fertilize egg cells (ova). However, they usually do not "swim" until after ejaculation.

### Vas Deferens

Each **vas deferens** (pl. *vasa deferentia*), also called ductus deferens, is a muscular tube about 45 centimeters long that is lined with pseudostratified columnar epithelium (fig. 22.9). It begins at the lower end of the epididymis and passes upward along the medial side of a testis to become part of the spermatic cord. It passes through the inguinal canal, enters the abdominal cavity outside the parietal peritoneum, and courses over the pelvic brim. From there, it extends backward and medially into the pelvic cavity, where it ends behind the urinary bladder.

Near its termination, the vas deferens dilates into a portion called the *ampulla*. Just outside the prostate gland, the tube becomes slender again and unites with the duct of a seminal vesicle. The fusion of these two ducts forms an **ejaculatory duct,** which passes through the prostate gland and empties into the urethra through a slitlike opening (see fig. 22.1).

### Seminal Vesicle

A **seminal vesicle** (see fig. 22.1) is a convoluted, saclike structure about 5 centimeters long that is attached to the vas deferens near the base of the urinary bladder.

The glandular tissue lining the inner wall of the seminal vesicle secretes a slightly alkaline fluid. This fluid helps regulate the pH of the tubular contents as sperm cells travel to the outside. The secretion of the seminal vesicle also contains *fructose,* a monosaccharide that provides energy to the sperm cells, and

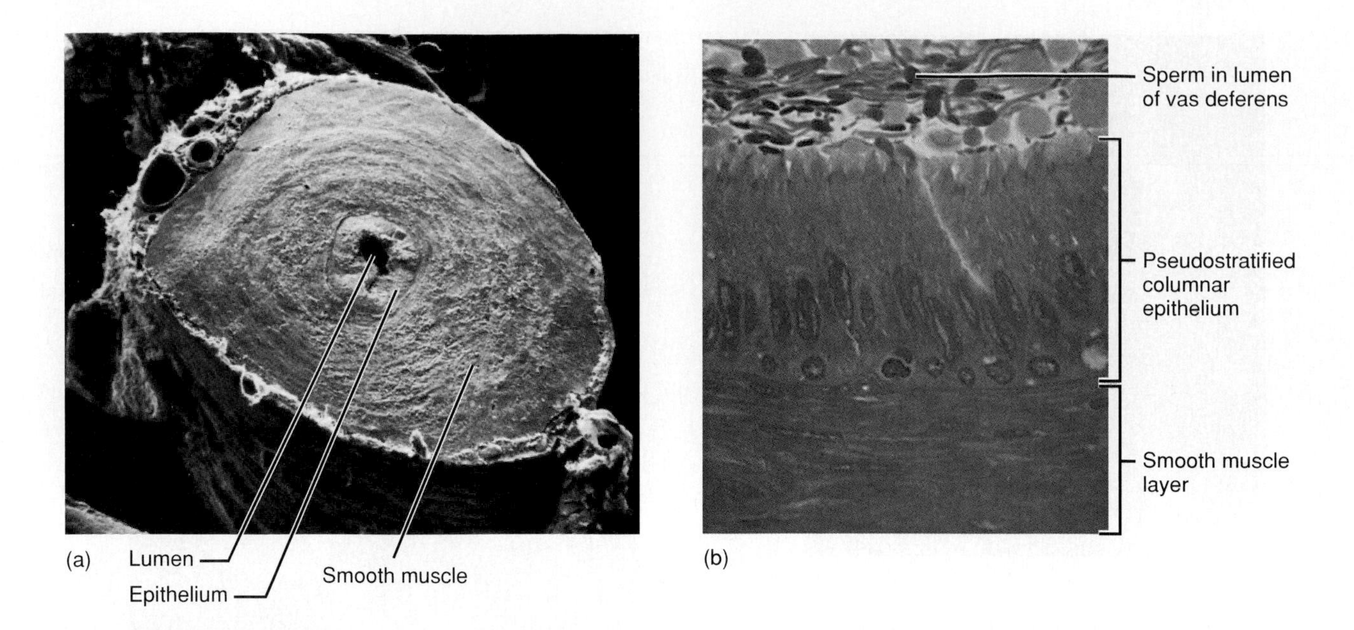


**FIGURE 22.9**

(*a*) Scanning electron micrograph of a cross section of the vas deferens (70×); (*b*) light micrograph of the wall of the vas deferens (250×).

(*Tissues and Organs: A Text-Atlas of Scanning Electron Microscopy,* by R. G. Kessel and R. H. Kardon. © 1979. W. H. Freeman and Company.)

*prostaglandins,* which stimulate muscular contractions within the female reproductive organs, aiding the movement of sperm cells toward the egg cell.

At emission, the contents of the seminal vesicles empty into the ejaculatory ducts, greatly increasing the volume of the fluid discharged from the vas deferens.

1. Describe the structure of the epididymis.
2. Trace the path of the vas deferens.
3. What is the function of a seminal vesicle?

## Prostate Gland

The **prostate gland** (see figs. 22.1 and 22.10) is a chestnut-shaped structure about 4 centimeters across and 3 centimeters thick that surrounds the beginning of the urethra, just below the urinary bladder. It is composed of many branched tubular glands enclosed in connective tissue. Septa of connective tissue and smooth muscle extend inward from the capsule, separating the tubular glands. The ducts of these glands open into the urethra.

The prostate gland secretes a thin, milky fluid. This alkaline secretion neutralizes the sperm cell-containing fluid, which is acidic from accumulation of metabolic wastes produced by the stored sperm cells. Prostate fluid also enhances the motility of sperm cells, which remain relatively nonmotile in the acidic contents of the epididymis. In addition, the prostatic fluid helps neutralize the acidic secretions of the vagina, thus helping to sustain sperm cells that enter the female reproductive tract.

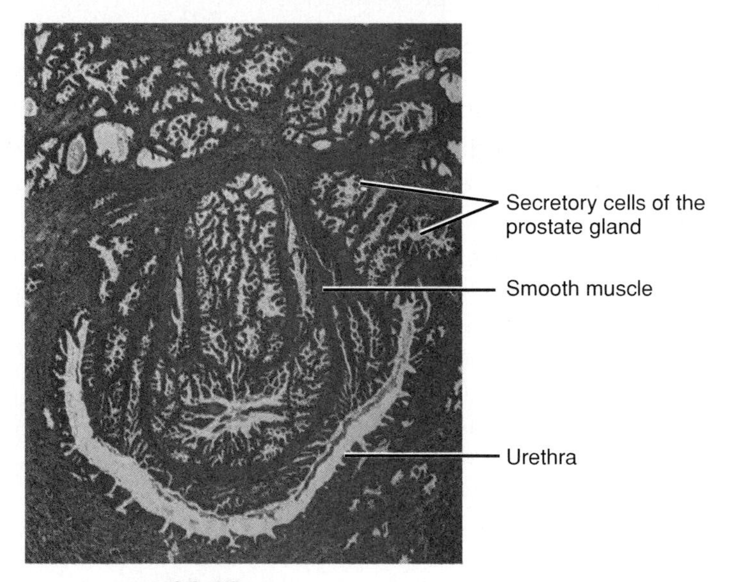


**FIGURE 22.10**

Light micrograph of the prostate gland (100×).

The prostate gland releases its secretions into the urethra as a result of smooth muscle contractions in its capsular wall. As this release occurs the contents of the vas deferens and the seminal vesicles enter the urethra, increasing the volume of the fluid.

Clinical Application 22.1 discusses the effects of prostate enlargement.

## CLINICAL APPLICATION 22.1

### Prostate Enlargement

The prostate gland is relatively small in boys, beginning to grow in early adolescence and reaching adult size a few years later. Usually the gland does not grow again until age 50, when in half of all men it enlarges enough to press on the urethra. This produces a feeling of pressure on the bladder because it cannot empty completely, and thus frequent urination. Retained urine can lead to infection and inflammation, bladder stones, or kidney disease.

Medical researchers do not know what causes prostate enlargement. Risk factors include a fatty diet, having had a vasectomy, and possibly occupational exposure to batteries or the metal cadmium. The enlargement may be benign or cancerous. Because prostate cancer is nearly 100% treatable if detected early, it is imperative that men have their prostates examined regularly.

Diagnostic tests include a rectal exam as well as a blood test to detect prostate specific antigen (PSA), a cell surface protein normally found on prostate cells. Elevated PSA levels indicate an enlarged prostate, which may be caused by a benign or cancerous growth. Ultrasound may provide further information. Chart 22A summarizes treatments for an enlarged prostate.

#### CHART 22A MEDICAL APPROACHES TO TREATING AN ENLARGED PROSTATE GLAND

Surgical removal of prostate
Radiation
Drugs to block the growth-stimulating effect of testosterone on the prostate (finasteride)
Microwave energy delivered through a probe inserted into the urethra or rectum
Balloon inserted into the urethra and inflated with liquid
Tumor frozen with liquid nitrogen delivered by a probe through the skin
Device (stent) inserted between lobes of prostate to relieve pressure on the urethra

## Bulbourethral Glands

The **bulbourethral glands** (Cowper's glands) are two small structures, each about a centimeter in diameter, that are located below the prostate gland lateral to the membranous urethra and enclosed by fibers of the external urethral sphincter muscle (see fig. 22.1).

These glands are composed of many tubes whose epithelial linings secrete a mucuslike fluid. This fluid is released in response to sexual stimulation and lubricates the end of the penis in preparation for sexual intercourse (coitus). Females secrete most of the lubricating fluid for intercourse, however.

## Semen

The fluid conveyed by the urethra to the outside during ejaculation is called **semen.** It consists of sperm cells from the testes and secretions of the seminal vesicles, prostate gland, and bulbourethral glands. Semen is slightly alkaline (pH about 7.5) and its contents include prostaglandins and nutrients.

The volume of semen released at one time varies from 2 to 6 milliliters, and the average number of sperm cells present in the fluid is about 120 million per milliliter.

Sperm cells remain nonmotile while they are in the ducts of the testis and epididymis, but become activated as they mix with the secretions of accessory glands. However, sperm cells cannot fertilize an egg cell until they enter the female reproductive tract. The development of this ability is called *capacitation,* and it involves changes that weaken the acrosomal membranes of the sperm cells. When sperm cells are placed with egg cells in a laboratory dish to achieve fertilization—a technique called in vitro fertilization, discussed in the next chapter—chemicals are added to simulate capacitation.

Although sperm cells can live for many weeks in the ducts of the male reproductive tract, they usually survive only a day or two after being expelled to the outside, even when they are maintained at body

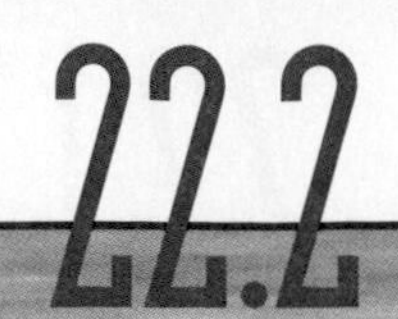

# Clinical Application 22.2

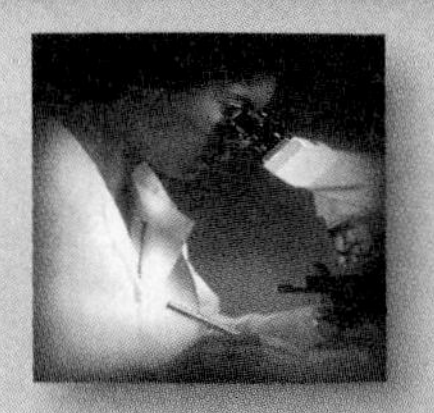

## Male Infertility

Male infertility—the inability of sperm to fertilize an egg—has several causes. If during fetal development the testes do not descend into the scrotum, the higher temperature of the abdominal cavity or inguinal canal causes the developing sperm cells in the seminiferous tubules to degenerate. Certain diseases, such as mumps, may inflame the testes (orchitis), producing infertility by destroying cells in the seminiferous tubules.

Both the quality and quantity of sperm cells are essential factors in the ability of a man to father a child. If a sperm head is misshapen, if a sperm cannot swim, or if there are simply too few sperm cells, completing the arduous journey to the well-protected female cell may be impossible. Sometimes even a sperm that enters an egg is unsuccessful because it lacks the microtubules necessary to attract and merge the nuclei of the two cells.

Until recently, sperm analysis was rather subjective, based on a person's viewing the cells under a microscope. Now, computer-aided sperm analysis (CASA) is standardizing and expanding criteria for normalcy in human male seminal fluid and the sperm cells it contains.

To analyze sperm, a man abstains from intercourse for 2 to 3 days, then provides a sperm sample, which must be examined within the hour. The man must also provide information about his reproductive history and possible exposure to toxins. The sperm sample is placed on a slide under a microscope, and then technology intervenes. A video camera sends an image to a videocassette recorder,

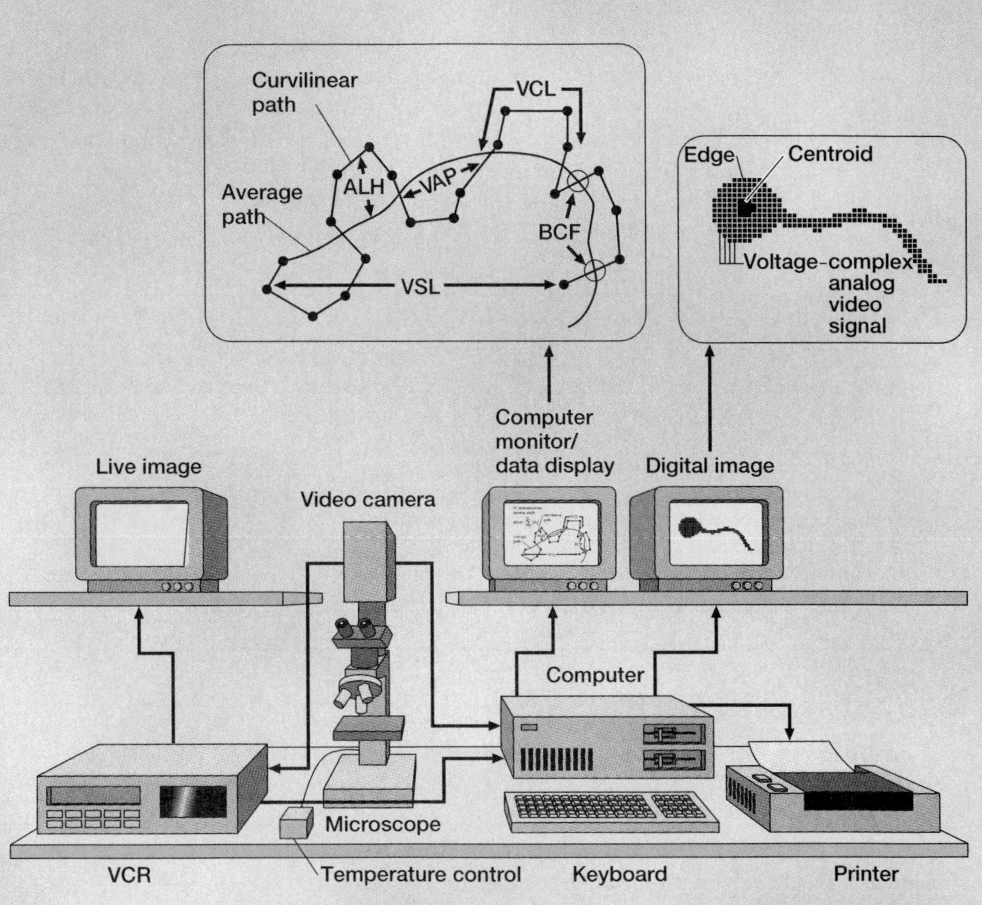


**Figure 22A**
Computer analysis improves the consistency and accuracy of describing sperm motility, morphology, and abundance.

temperature. On the other hand, sperm cells can be stored and kept viable for years if they are frozen at a temperature below –100° C.

Clinical Application 22.2 describes some causes of male infertility.

1. Where is the prostate gland located?
2. What is the function of the prostate gland's secretion?
3. What is the function of the bulbourethral glands?
4. What are the components of semen?

which projects a live or digitized image. The camera also sends the image to a computer, which traces sperm trajectories and displays them on a monitor or prints a hard copy. Figures 22A and 22B show a CASA of normal sperm, depicting how the swimming pattern alters as they travel.

Hundreds of CASA systems are now in use, mostly at fertility clinics. The devices are also helpful in using sperm as "biomarkers" of exposure to toxins. For example, in one study, the sperm of men who work in the dry cleaning industry and are exposed to the solvent perchloroethylene (believed to damage sperm) were compared with sperm from men who work in the laundry industry and are exposed to many of the same chemicals except this one. CASA showed a difference in sperm motility that was directly related to level of exposure, as measured by exhalation of the chemical. This result supported the reproductive evidence: Although the men in both groups had the same numbers of children, the dry cleaners' wives took far longer to conceive than did the launderers' wives.

Chart 22B lists the components of a sperm analysis.

**CHART 22B Semen Analysis**

| Characteristic | Normal Value |
|---|---|
| Volume | 2–6 milliliters/ejaculate |
| Sperm density | 120 million cells/milliliter |
| Percent motile | > 40% |
| Motile sperm density | > 8 million/milliliter |
| Average velocity | > 20 micrometers/second |
| Motility | > 8 micrometers/second |
| Percent abnormal morphology | > 40% |
| White blood cells | > 5 million/milliliter |

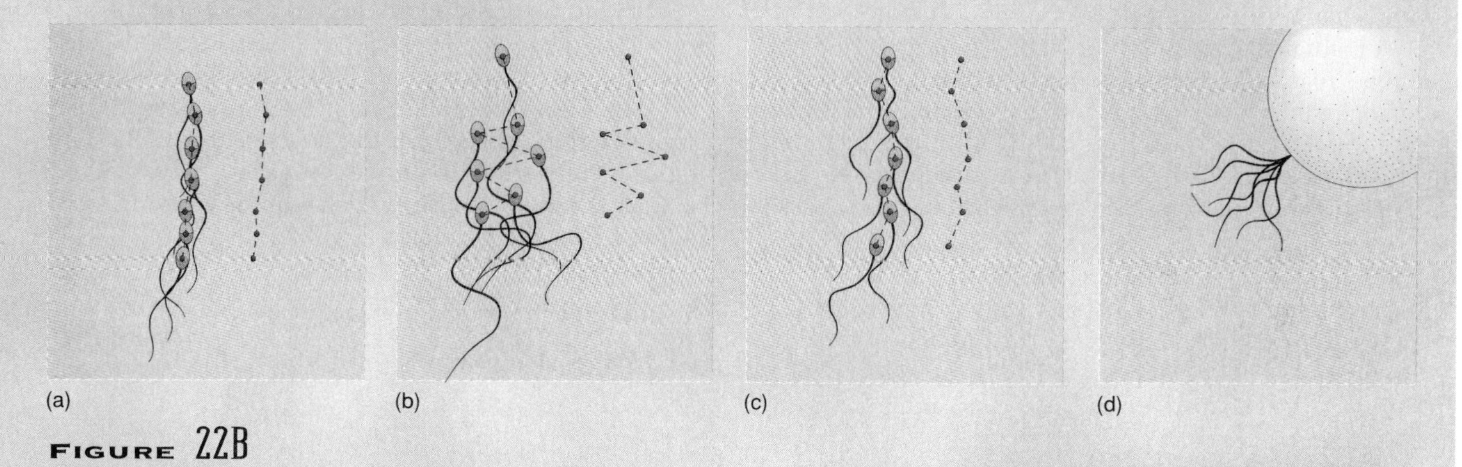


**FIGURE 22B**

A computer tracks sperm movements. In semen, sperm swim in a straight line (*a*), but as they are activated by biochemicals normally found in the woman's body, their trajectories widen (*b*). The sperm in (*c*) are in the mucus of a woman's cervix, and the sperm in (*d*) are attempting to penetrate the structures surrounding an oocyte.

## Male External Reproductive Organs

The male external reproductive organs are the scrotum, which encloses the testes, and the penis, through which the urethra passes.

### Scrotum

The **scrotum** is a pouch of skin and subcutaneous tissue that hangs from the lower abdominal region behind the penis.

The subcutaneous tissue of the scrotal wall lacks fat but contains a layer of smooth muscle fibers that

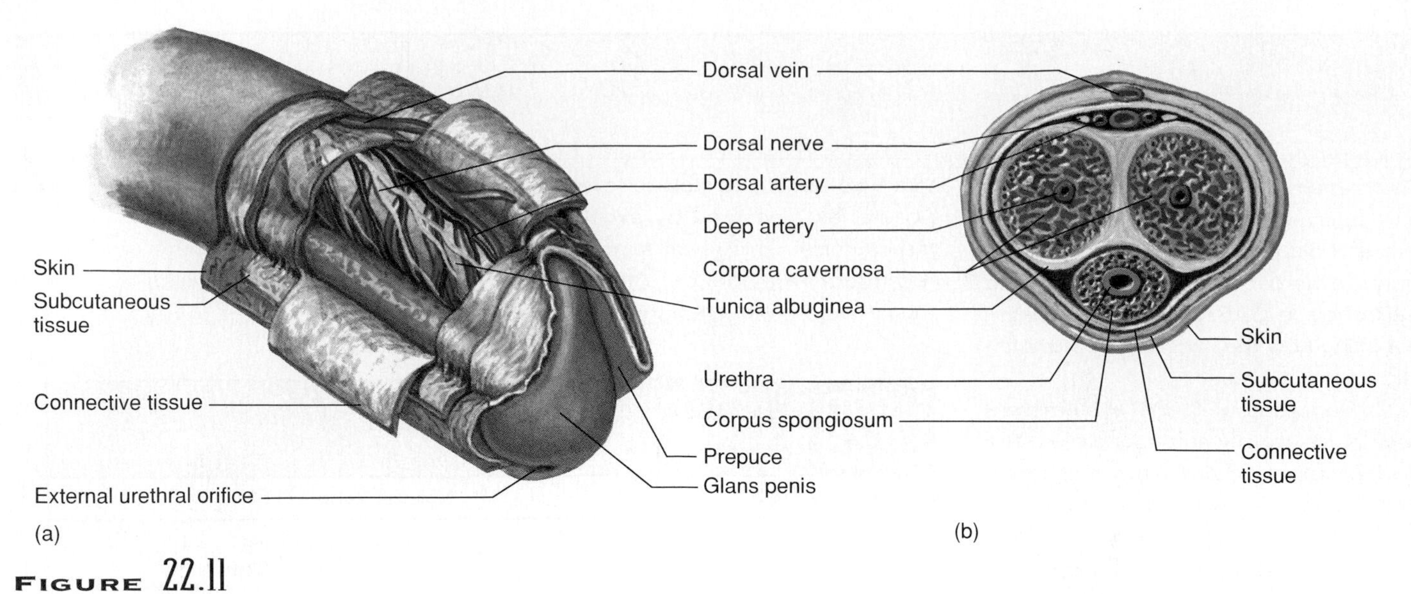


**FIGURE 22.11**

(*a*) Interior structure of the penis; (*b*) cross section of the penis.

constitute the *dartos muscle.* When these muscle fibers contract, the scrotal skin becomes wrinkled and is held close to the testes; when the fibers relax, the scrotum hangs more loosely.

A medial septum subdivides the scrotum into two chambers, each of which encloses a testis. Each chamber also contains a serous membrane, which covers the front and sides of the testis and the epididymis, helping to ensure that the testis and epididymis move smoothly within the scrotum (see fig. 22.1).

## Penis

The **penis** is a cylindrical organ that conveys urine and semen through the urethra to the outside. It is also specialized to enlarge and stiffen by a process called erection, so that it can insert into the vagina during sexual intercourse.

The *body,* or shaft, of the penis is composed of three columns of erectile tissue, which include a pair of dorsally located *corpora cavernosa* and a single *corpus spongiosum* below. The penis is enclosed by skin, a thin layer of subcutaneous tissue, and a layer of connective tissue. In addition, a tough capsule of white fibrous connective tissue called a *tunica albuginea* surrounds each column (fig. 22.11).

The corpus spongiosum, through which the urethra extends, is enlarged at its distal end to form a sensitive, cone-shaped **glans penis.** This glans covers the ends of the corpora cavernosa and bears the urethral opening—the *external urethral orifice.* The skin of the glans is very thin, hairless, and contains sensory receptors for sexual stimulation. Also, a loose fold of skin called the *prepuce* (foreskin) begins just behind the glans and extends forward to cover it as a sheath. The prepuce is sometimes removed by a surgical procedure called *circumcision.*

At the *root* of the penis, the columns of erectile tissue separate. The corpora cavernosa diverge laterally in the perineum and are firmly attached to the inferior surface of the pubic arch by connective tissue. These diverging parts form the *crura* of the penis. The single corpus spongiosum is enlarged between the crura as the *bulb* of the penis, which is attached to membranes of the perineum.

1. Describe the structure of the penis.
2. What is circumcision?
3. How is the penis attached to the perineum?

## Erection, Orgasm, and Ejaculation

During sexual stimulation, parasympathetic nerve impulses pass from the sacral portion of the spinal cord to the arteries leading into the penis, causing them to dilate. At the same time, the increasing pressure of the arterial blood entering the vascular spaces of the erectile tissue compresses the veins of the penis, reducing flow of venous blood away from the penis. Consequently, blood accumulates in the erectile tissues, and the penis swells and elongates, producing an **erection** (fig. 22.12).

The culmination of sexual stimulation is **orgasm,** a pleasurable feeling of physiological and psychological release. Orgasm in the male is accompanied by emission and ejaculation.

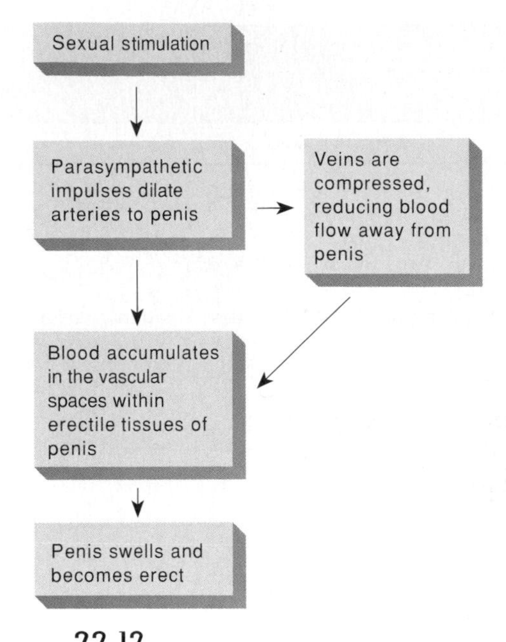


**FIGURE 22.12**
Mechanism of penile erection.

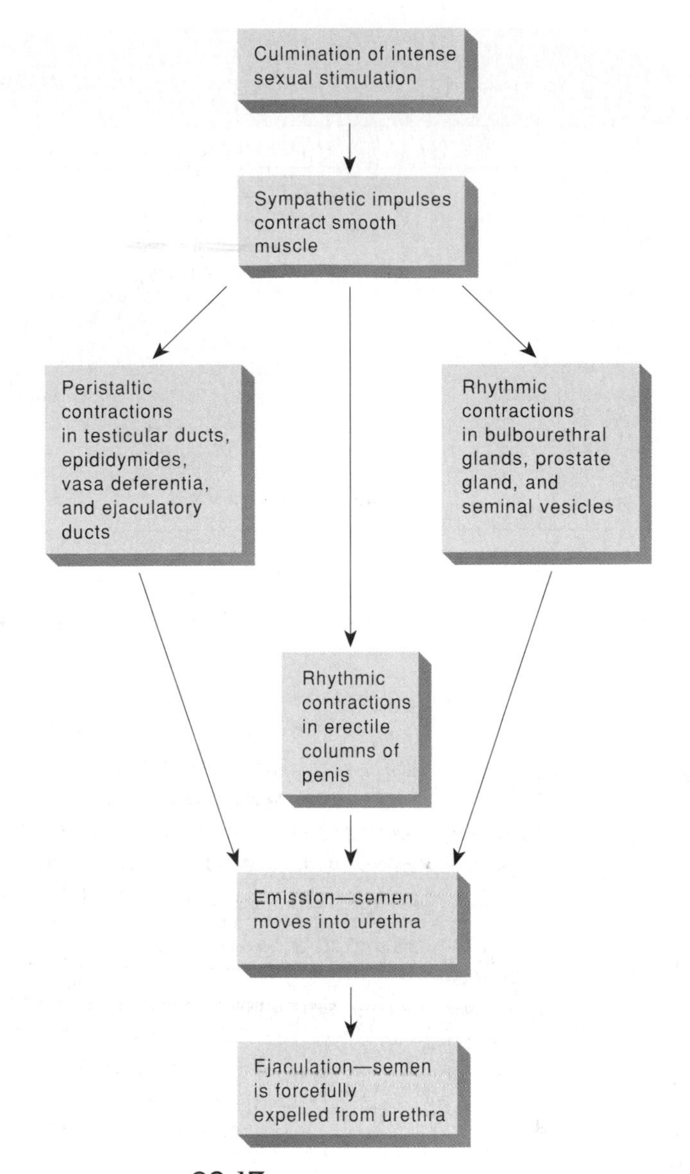


**FIGURE 22.13**
Mechanism of emission and ejaculation.

**Emission** is the movement of sperm cells from the testes and secretions from the prostate gland and seminal vesicles into the urethra, where they mix to form semen. Emission occurs in response to sympathetic nerve impulses from the spinal cord, which stimulate peristaltic contractions in smooth muscles within the walls of the testicular ducts, epididymides, vasa deferentia, and ejaculatory ducts. At the same time, other sympathetic impulses stimulate rhythmic contractions of the seminal vesicles and prostate gland.

As the urethra fills with semen, sensory impulses are stimulated and pass into the sacral portion of the spinal cord. In response, motor impulses are transmitted from the cord to certain skeletal muscles at the base of the erectile columns of the penis, causing them to contract rhythmically. This increases the pressure within the erectile tissues and aids in forcing the semen through the urethra to the outside—a process called **ejaculation.**

The sequence of events during emission and ejaculation is regulated so that the fluid from the bulbourethral glands is expelled first. This is followed by the release of fluid from the prostate gland, the passage of the sperm cells, and finally, the ejection of fluid from the seminal vesicles (fig. 22.13).

Immediately after ejaculation, sympathetic impulses constrict the arteries that supply the erectile tissue, reducing the inflow of blood. Smooth muscles within the walls of the vascular spaces partially contract again, and the veins of the penis carry the excess blood out of these spaces. The penis gradually returns to its flaccid state, and usually another erection and ejaculation cannot be triggered for a period of 10 to 30 minutes or longer.

Chart 22.1 summarizes the functions of the male reproductive organs.

Spontaneous emissions and ejaculations commonly occur in adolescent males during sleep. Changes in hormonal concentrations that accompany adolescent development and sexual maturation cause these *nocturnal emissions.*

## CHART 22.1 Functions of the Male Reproductive Organs

| Organ | Function |
|---|---|
| Testis | |
| Seminiferous tubules | Production of sperm cells |
| Interstitial cells | Production and secretion of male sex hormones |
| Epididymis | Storage and maturation of sperm cells; conveys sperm cells to vas deferens |
| Vas deferens | Conveys sperm cells to ejaculatory duct |
| Seminal vesicle | Secretes an alkaline fluid containing nutrients and prostaglandins; fluid helps neutralize acidic semen |
| Prostate gland | Secretes an alkaline fluid that helps neutralize acidic semen and enhances motility of sperm cells |
| Bulbourethral gland | Secretes fluid that lubricates end of the penis |
| Scrotum | Encloses and protects testes |
| Penis | Conveys urine and semen to outside of body; inserted into the vagina during sexual intercourse; the glans penis is richly supplied with sensory nerve endings associated with feelings of pleasure during sexual stimulation |

1. What controls blood flow into the erectile tissues of the penis?
2. Distinguish between orgasm, emission, and ejaculation.
3. Review the events associated with emission and ejaculation.

## Hormonal Control of Male Reproductive Functions

Hormones secreted by the *hypothalamus,* the *anterior pituitary gland,* and the *testes* control male reproductive functions. These hormones initiate and maintain sperm cell production and oversee the development and maintenance of male sexual characteristics.

### Hypothalamic and Pituitary Hormones

Prior to 10 years of age, a boy's body is reproductively immature. During this period, the body is childlike, and the spermatogenic cells of the testes are undifferentiated. Then a series of changes leads to development of a reproductively functional adult. The hypothalamus controls many of these changes.

Recall from chapter 13 that the hypothalamus secretes gonadotropin-releasing hormone (GnRH), which enters the blood vessels leading to the anterior pituitary gland. In response, the anterior pituitary gland secretes the **gonadotropins** called *luteinizing hormone* (LH) and *follicle-stimulating hormone* (FSH). LH, which in males is called interstitial cell-stimulating hormone (ICSH), promotes development of the interstitial cells (cells of Leydig) of the testes, and they, in turn, secrete male sex hormones. FSH stimulates the sustentacular cells of the seminiferous tubules to proliferate, grow, mature, and become responsive to the effects of the male sex hormone testosterone. Then, in the presence of FSH and testosterone, these cells stimulate the spermatogenic cells to undergo spermatogenesis, giving rise to sperm cells (fig. 22.14). The sustentacular cells also secrete a hormone called *inhibin,* which inhibits the anterior pituitary gland by negative feedback, and thus prevents oversecretion of FSH.

### Male Sex Hormones

Male sex hormones are termed **androgens.** The interstitial cells of the testes produce most of them, but small amounts are synthesized in the adrenal cortex (see chapter 13).

The hormone **testosterone** is the most abundant androgen. It is secreted and transported in the blood loosely attached to plasma proteins. Like other steroid hormones, testosterone combines with receptor molecules usually in the nuclei of its target cells (see chapter 13). However, in many target cells, such as those in the prostate gland, seminal vesicles, and male external accessory organs, testosterone is first converted to another androgen called **dihydrotestosterone,** which stimulates the cells of these organs.

Androgen molecules that do not reach receptors in target cells are usually changed by the liver into forms that can be excreted in bile or urine.

Although secretion of testosterone begins during fetal development and continues for a few weeks following birth, it nearly ceases during childhood. Between the ages of 13 and 15, a young man's androgen production usually increases rapidly. This phase in development, when an individual becomes reproductively functional, is **puberty.** After puberty, testosterone secretion continues throughout the life of a male.

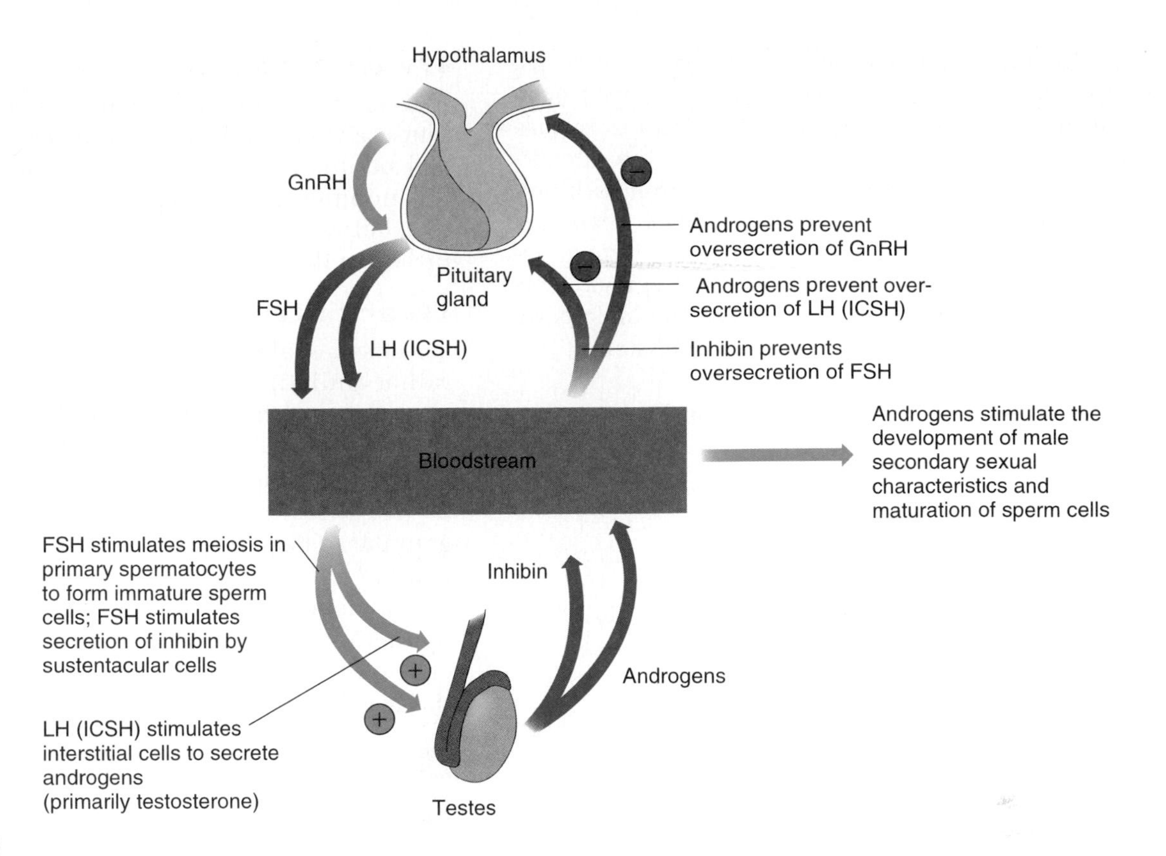


**FIGURE 22.14**
The hypothalamus controls maturation of sperm cells and development of male secondary sexual characteristics. A negative feedback mechanism operating between the hypothalamus, the anterior lobe of the pituitary gland, and the testes controls the concentration of testosterone.

## Actions of Testosterone

Testosterone is first produced by cells of the embryonic testes after about eight weeks of development. It stimulates the formation of the male reproductive organs, including the penis, scrotum, prostate gland, seminal vesicles, and ducts. Still later, it causes the testes to descend into the scrotum.

During puberty, testosterone stimulates enlargement of the testes (the primary male sexual characteristic) and accessory organs of the reproductive system, as well as development of male *secondary sexual characteristics,* which are special features associated with the adult male body. Secondary sexual characteristics in the male include:

1. Increased growth of body hair, particularly on the face, chest, axillary region, and pubic region, but sometimes accompanied by decreased growth of hair on the scalp.
2. Enlargement of the larynx and thickening of the vocal folds, accompanied by lowering of the pitch of the voice.
3. Thickening of the skin.
4. Increased muscular growth accompanied by broadening shoulders and narrowing of the waist.
5. Thickening and strengthening of the bones.

Testosterone also increases the rate of cellular metabolism and production of red blood cells (by stimulating erythropoietin, see chapter 19), so that the average number of red blood cells in a cubic millimeter of blood is usually greater in males than in females. Testosterone stimulates sexual activity by affecting certain portions of the brain.

## Regulation of Male Sex Hormones

The extent to which male secondary sexual characteristics develop is directly related to the amount of testosterone that the interstitial cells secrete. A negative feedback system involving the hypothalamus regulates the quantity of testosterone (see fig. 22.14).

As the concentration of testosterone in the blood increases, the hypothalamus becomes inhibited, and its stimulation of the anterior pituitary gland by GnRH decreases. As the pituitary's secretion of LH (ICSH) falls in response, the amount of testosterone released by the interstitial cells is reduced.

As the blood concentration of testosterone drops, the hypothalamus becomes less inhibited, and it once again stimulates the anterior pituitary gland to release LH. The increasing secretion of LH causes the interstitial cells to release more testosterone, and its blood concentration increases. Testosterone level decreases somewhat during and after the *male climacteric,* a decline in sexual function that occurs with aging. Overall, the concentration of testosterone in the male body is regulated so that it remains relatively constant.

1. What hormone initiates the changes associated with male sexual maturity?
2. Describe several of the male secondary sexual characteristics.
3. List the functions of testosterone.
4. Explain the regulation of secretion of male sex hormones.

## Organs of the Female Reproductive System

The organs of the female reproductive system are specialized to produce and maintain the female sex cells, or *egg cells;* to transport these cells to the site of fertilization; to provide a favorable environment for a developing offspring; to move the offspring to the outside; and to produce female sex hormones.

The *primary sex organs* (gonads) of this system are the ovaries, which produce the female sex cells and sex hormones. The other parts of the system comprise the internal and external *accessory organs.*

## Ovaries

The **ovaries** are solid, ovoid structures measuring about 3.5 centimeters in length, 2 centimeters in width, and 1 centimeter in thickness. An individual ovary is located in a shallow depression (ovarian fossa) on each side in the lateral wall of the pelvic cavity (fig. 22.15).

### Ovary Attachments

Several ligaments help hold each ovary in position. The largest of these, formed by a fold of peritoneum, is called the *broad ligament.* It is also attached to the uterine tubes and the uterus.

A small fold of peritoneum, called the *suspensory ligament,* holds the ovary at its upper end. This ligament also contains the ovarian blood vessels and nerves. At its lower end, the ovary is attached to the uterus by a rounded, cordlike thickening of the broad ligament called the *ovarian ligament* (fig. 22.16).

### Ovary Descent

Like the testes in a male fetus, the ovaries in a female fetus originate from masses of tissue behind the parietal peritoneum, near the developing kidneys. During development, these structures descend to locations just below the pelvic brim, where they remain attached to the lateral pelvic wall.

### Ovary Structure

The tissues of an ovary can be subdivided into two rather indistinct regions, an inner *medulla* and an outer *cortex.* The ovarian medulla is composed mostly of loose connective tissue, and contains many blood vessels, lymphatic vessels, and nerve fibers. The ovarian cortex consists of more compact tissue, and has a granular appearance due to tiny masses of cells called *ovarian follicles.*

A layer of cuboidal epithelium cells (germinal epithelium) covers the free surface of the ovary. Just beneath this epithelium is a layer of dense connective tissue called the *tunica albuginea.*

1. What are the primary sex organs of the female?
2. Describe the descent of the ovary.
3. Describe the structure of an ovary.

### Primordial Follicles

During prenatal development (before birth), small groups of cells in the outer region of the ovarian cortex form several million **primordial follicles.** Each of these structures consists of a single, large cell called a *primary oocyte,* which is closely surrounded by a layer of flattened epithelial cells called *follicular cells.*

Early in development, the primary oocytes begin to undergo meiosis, but the process soon halts and does not continue until puberty. Once the primordial follicles appear, no new ones form. Instead, the number of oocytes in the ovary steadily declines, as many of the oocytes degenerate. Of the several million oocytes formed originally, only a million or so remain at the time of birth, and perhaps 400,000 are present at puberty. Of these, probably fewer than 400 or 500 will be released from the ovary during the reproductive life of a female. Probably fewer than ten will go on to form a new individual!

> Scientists think that the reason for the increased incidence of chromosome defects in children of older mothers is that the eggs, having been present for several decades, had time to be exposed extensively to damaging agents, such as radiation, viruses, and toxins.

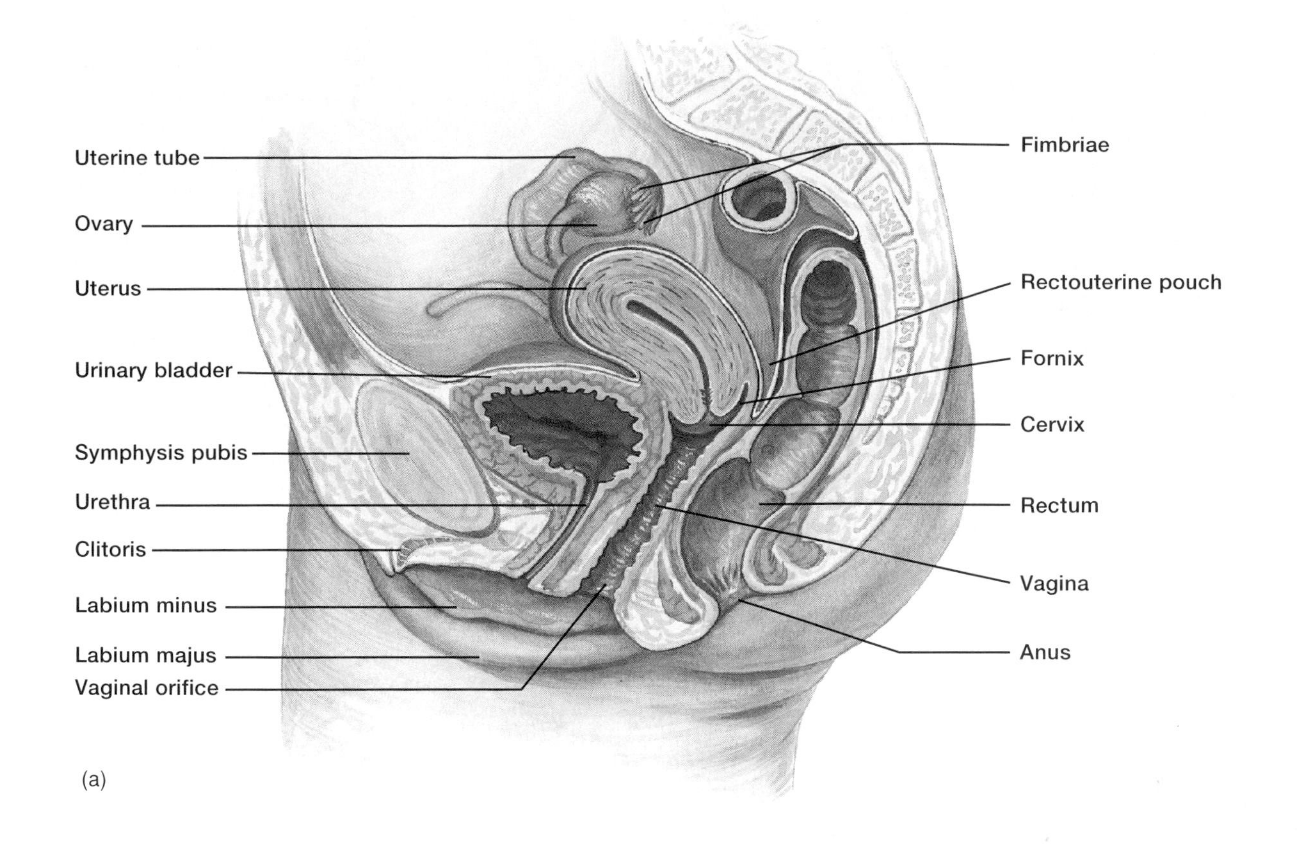


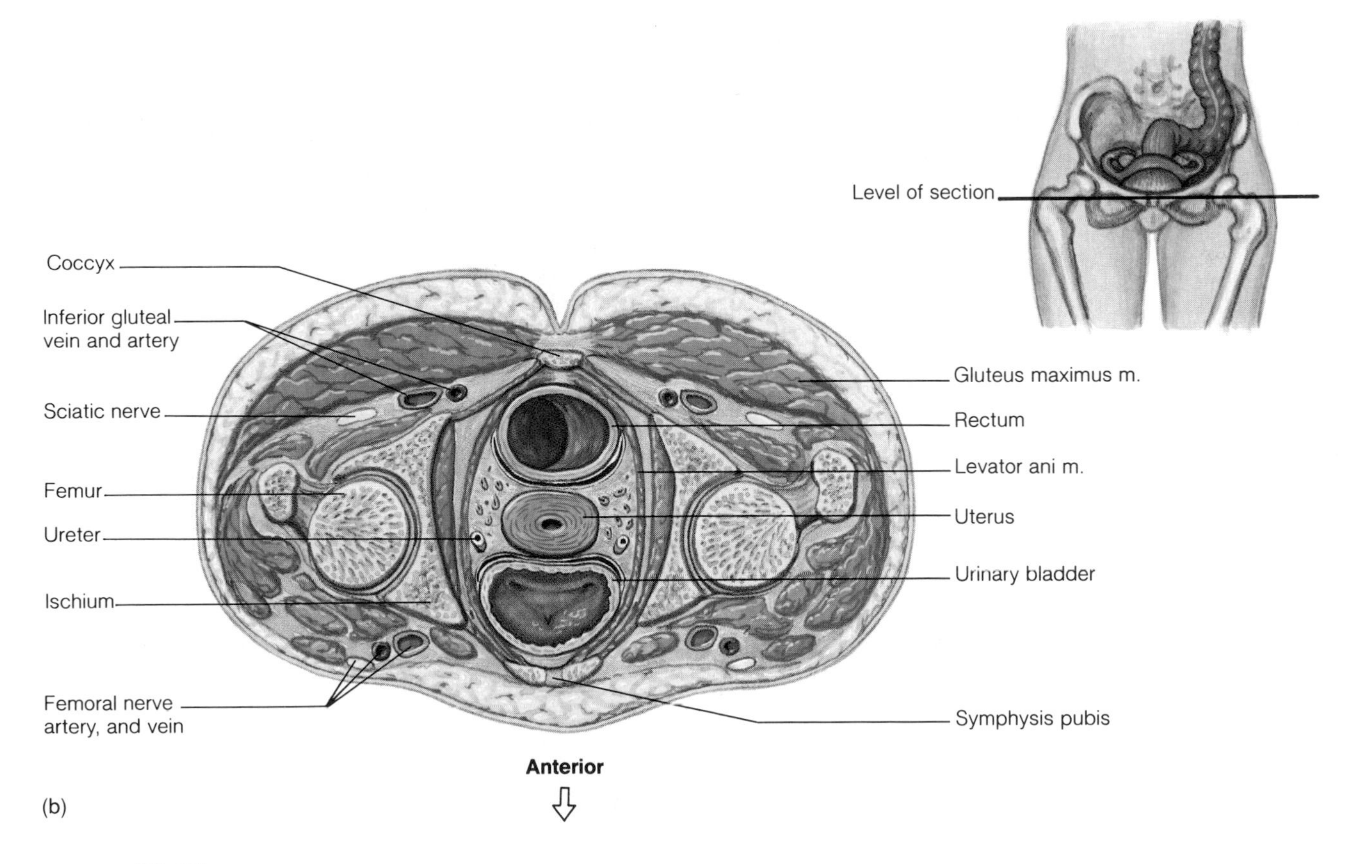


**FIGURE 22.15**
(*a*) Sagittal view of the female reproductive organs; (*b*) transverse section of the female pelvic cavity.

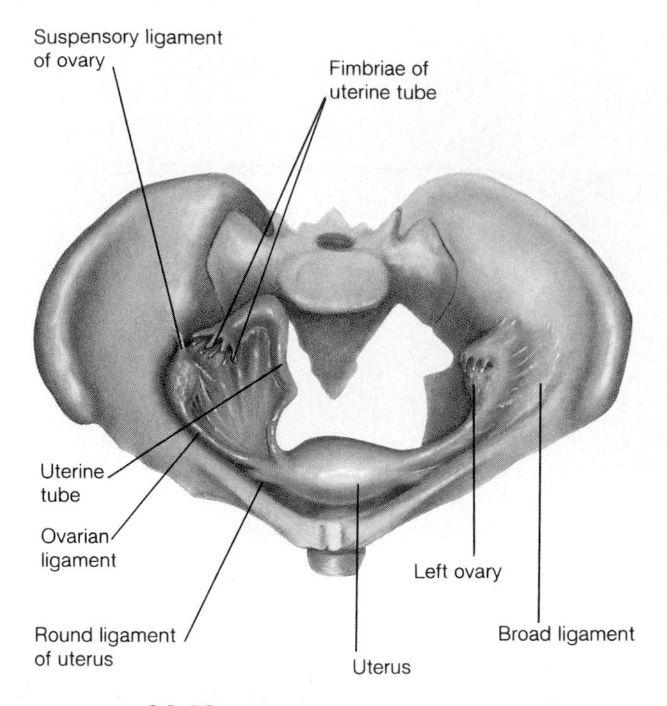


**FIGURE 22.16**
The ovaries are located on each side against the lateral walls of the pelvic cavity. The right uterine tube is retracted to reveal the ovarian ligament.

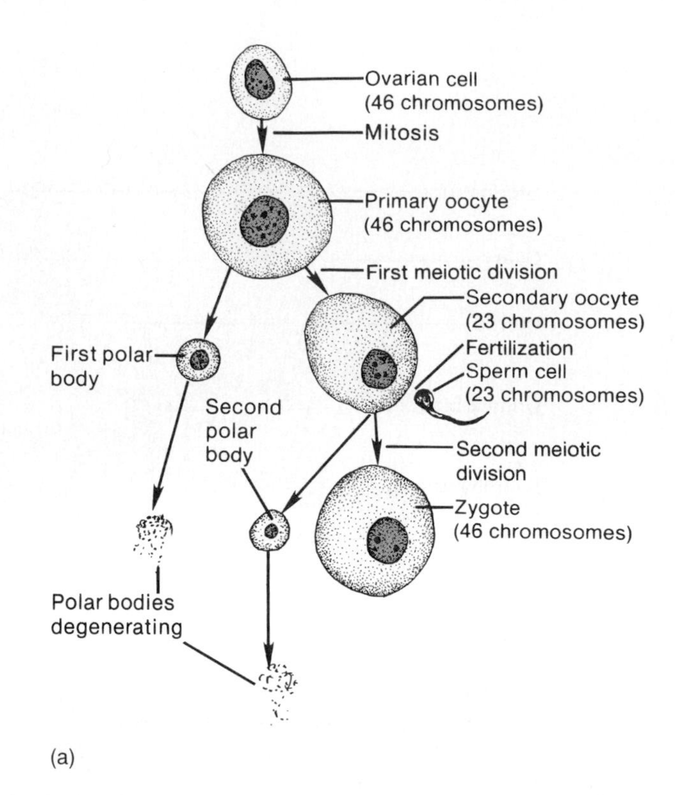


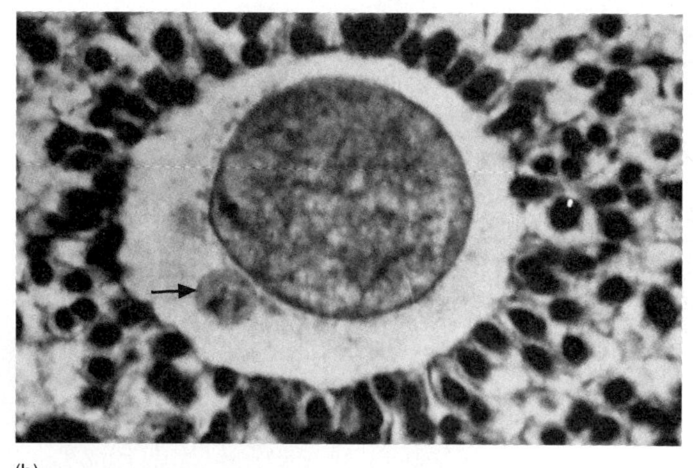


**FIGURE 22.17**
(*a*) During oogenesis, a single egg cell (secondary oocyte) results from the meiosis of a primary oocyte. If the egg cell is fertilized, it forms a second polar body and becomes a zygote; (*b*) light micrograph of a secondary oocyte and a polar body (arrow).

## Oogenesis

Beginning at puberty, some primary oocytes are stimulated to continue meiosis. As in the case of sperm cells, the resulting cells have one-half as many chromosomes (23) in their nuclei as their parent cells (see chapter 24). Egg formation is called **oogenesis.**

When a primary oocyte divides, the distribution of the cytoplasm is unequal. One of the resulting cells, called a *secondary oocyte,* is relatively large, and the other, called the *first polar body,* is very small.

The large secondary oocyte represents a future *egg cell* (ovum) that can be fertilized by uniting with a sperm cell. If this happens, the oocyte divides unequally to produce a tiny *second polar body* and a relatively large fertilized egg cell, or **zygote** that can give rise to an embryo (fig. 22.17). The polar bodies have no further function, and they soon degenerate.

Formation of polar bodies may appear wasteful, but it has an important biological function—it produces an egg cell that has the massive amounts of cytoplasm and abundant organelles needed to carry a zygote through the first few cell divisions. After that, its own genes take over the tasks of supplying and processing raw materials.

1. Describe the major events of oogenesis.
2. How is polar body formation beneficial to an egg?

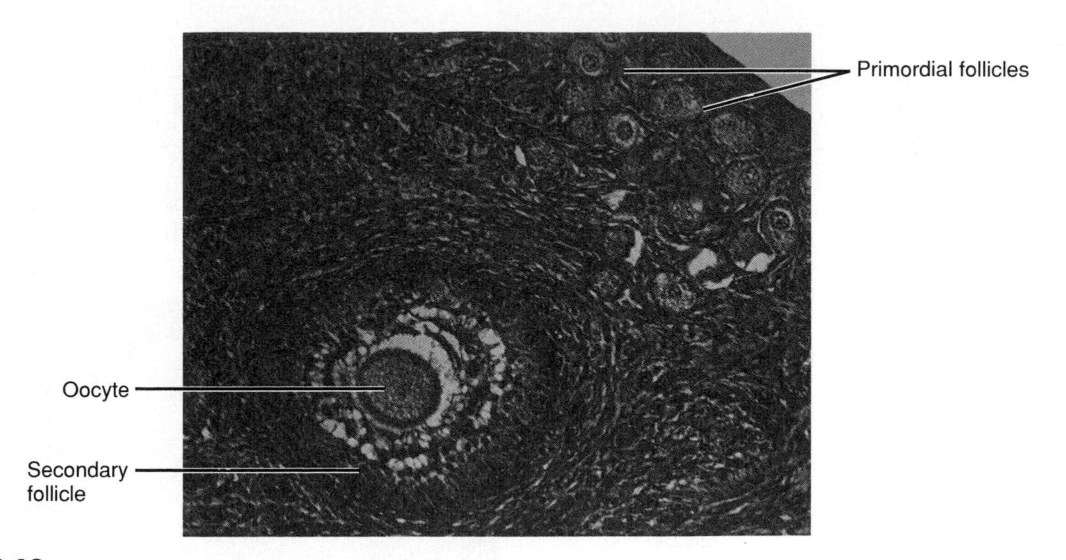


**FIGURE 22.18**
Light micrograph of the surface of a mammalian ovary (50×).

An experimental procedure called polar body biopsy allows couples to select an egg that does not carry a disease-causing gene that the woman carries. First, oocytes with attached first polar bodies are removed from the woman and cultured in a laboratory dish. Then the polar bodies are screened with a "DNA probe," which is a piece of genetic material that homes in on and signals the presence of a specific disease-causing gene, such as one that causes sickle cell disease or cystic fibrosis.

In polar body biopsy, bad news is really good news. Because of the laws of inheritance (discussed in chapter 24), if the defective gene is in a polar body, it is *not* in the egg cell that it is physically attached to. Researchers can then fertilize the egg with sperm in the laboratory and implant it in the woman who donated it, with some confidence that the disorder carried in the family will not pass to this particular future child.

## Follicle Maturation

At puberty, the anterior pituitary gland secretes increased amounts of FSH, and the ovaries enlarge in response. At the same time, some of the primordial follicles mature (fig. 22.18).

During maturation, the oocyte of a follicle enlarges, and the surrounding follicular cells proliferate by mitosis, giving rise to a stratified epithelium composed of *granulosa cells.* A layer of glycoprotein, called the *zona pellucida,* gradually separates the oocyte from the granulosa cells; at this stage, the structure is called a *primary follicle.*

Meanwhile, the ovarian cells outside the follicle organize into two cellular layers—an *inner vascular layer* (theca interna), composed largely of steroid-secreting cells, plus some loose connective tissue and blood vessels, and an *outer fibrous layer* (theca externa), composed of tightly packed connective tissue cells.

The follicular cells continue to proliferate, and when there are six to twelve layers of cells, irregular, fluid-filled spaces appear among them. These spaces soon join to form a single cavity (antrum), and the oocyte is pressed to one side of the follicle. At this stage, the follicle has a diameter of about 0.2 millimeters and is called a *secondary follicle.*

Maturation takes ten to fourteen days. The *mature follicle* (preovulatory, or Graafian, follicle) is about 10 millimeters or more across, and its fluid-filled cavity bulges outward on the surface of the ovary, like a blister. The oocyte within the mature follicle is a large, spherical cell, surrounded by a relatively thick zona pellucida, attached to a mantle of follicular cells (corona radiata). Processes from these follicular cells extend through the zona pellucida and supply nutrients to the oocyte (fig. 22.19).

Although as many as twenty primary follicles may begin maturing at any one time, one follicle (dominant follicle) usually outgrows the others. Typically, only the dominant follicle fully develops, and the other follicles degenerate (fig. 20.20).

Certain drugs used to treat female infertility, such as Clomid (clomiphene), may cause a woman to "superovulate." More than one follicle grows, more than one oocyte is released, and the result is twins—or even triplets!

## Ovulation

As a follicle matures, its primary oocyte undergoes oogenesis, giving rise to a secondary oocyte and a first polar body. A process called **ovulation** releases these cells from the follicle.

Hormones from the anterior pituitary gland stimulate ovulation, causing the mature follicle to swell rapidly and its wall to weaken. Eventually the wall ruptures, and the follicular fluid, accompanied by the oocyte, oozes outward from the surface of the ovary. Figure 22.21 shows expulsion of a mammalian oocyte.

After ovulation, the oocyte and one or two layers of follicular cells surrounding it are usually propelled

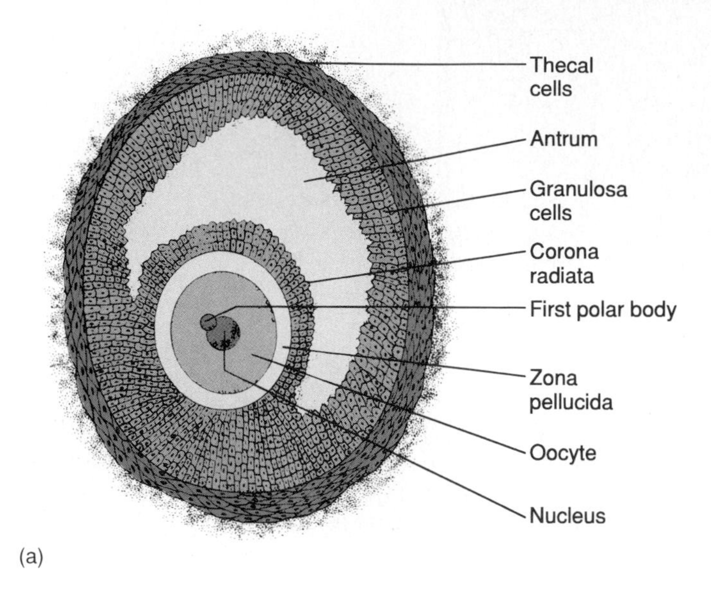


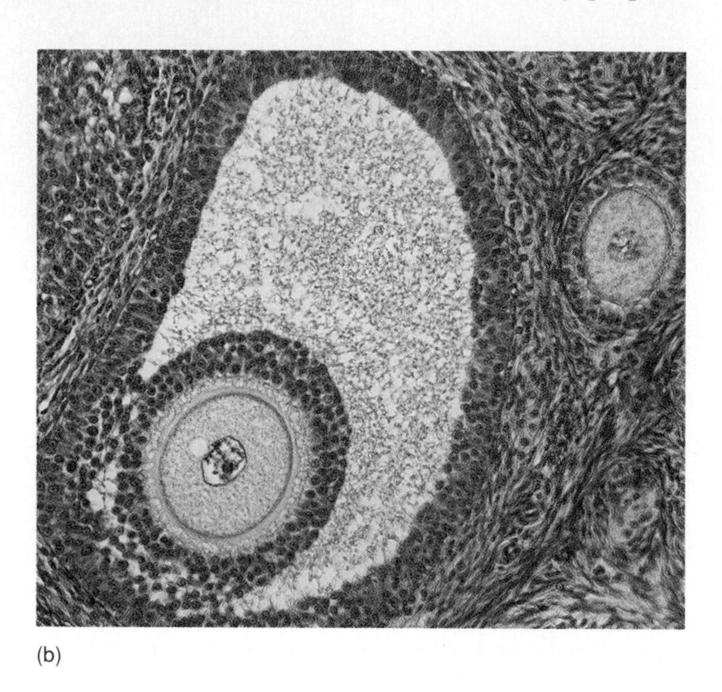


**FIGURE 22.19**

(*a*) Structure of a mature (Graafian) follicle. (*b*) Light micrograph of a Graafian follicle (250×). The first polar body has just separated and has not been expelled from the cell.

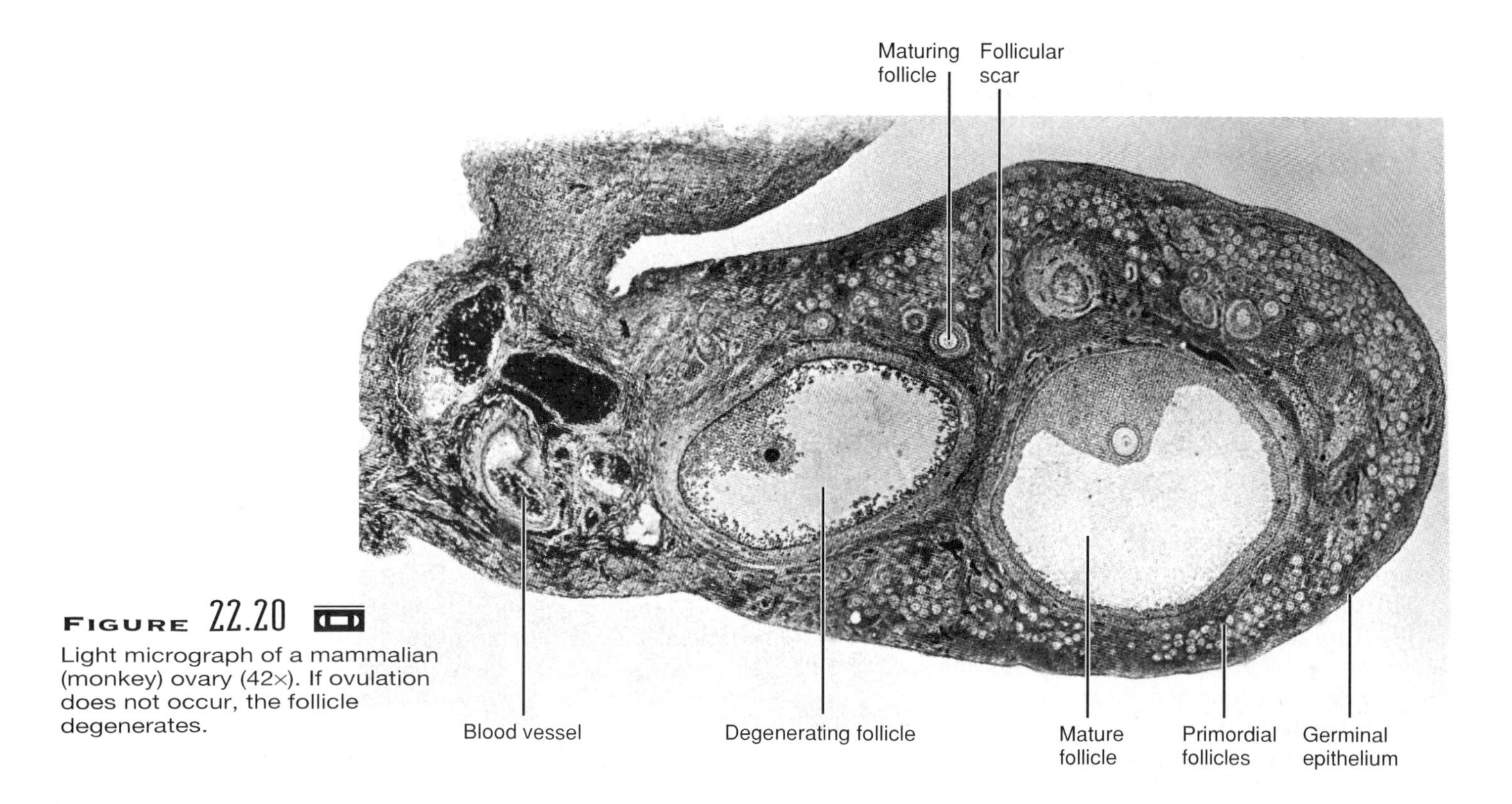


**FIGURE 22.20**

Light micrograph of a mammalian (monkey) ovary (42×). If ovulation does not occur, the follicle degenerates.

to the opening of a nearby **uterine tube** (fallopian tube, or oviduct). If the oocyte is not fertilized within a relatively short time, it degenerates. Figure 22.22 illustrates maturation of a follicle and the release of an oocyte.

1. What changes occur in a follicle and its oocyte during maturation?
2. What causes ovulation?
3. What happens to an oocyte following ovulation?

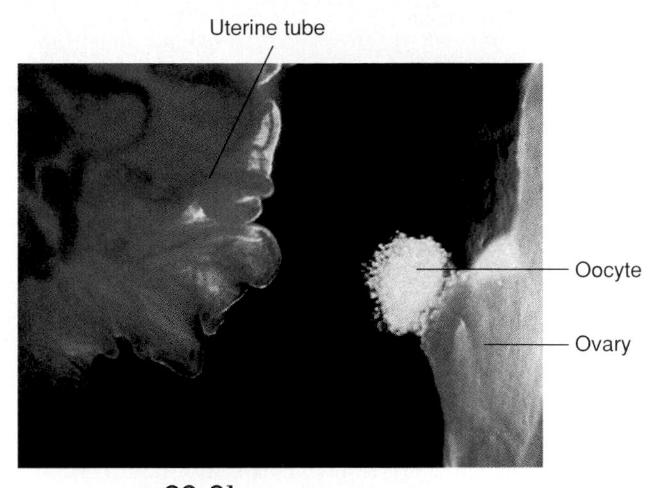


**FIGURE 22.21**
Light micrograph of a follicle during ovulation.

## FEMALE INTERNAL ACCESSORY ORGANS

The internal accessory organs of the female reproductive system include a pair of uterine tubes, a uterus, and a vagina.

### Uterine Tubes

The uterine tubes are suspended by portions of the broad ligament, and open near the ovaries. Each tube, which is about 10 centimeters long and 0.7 centimeters in diameter, passes medially to the uterus, penetrates its wall, and opens into the uterine cavity.

Near each ovary, a uterine tube expands to form a funnel-shaped *infundibulum,* which partially encircles the ovary medially. On its margin, the infundibulum bears a number of irregular, branched extensions called *fimbriae* (fig. 22.23). Although the infundibulum generally does not touch the ovary, one of the larger extensions (ovarian fimbria) connects directly to the ovary.

The wall of a uterine tube consists of an inner mucosal layer, a middle muscular layer, and an outer covering of peritoneum. The mucosal layer is drawn into many longitudinal folds and is lined with simple columnar epithelial cells, some of which are *ciliated* (fig. 22.24). The epithelium secretes mucus, and the

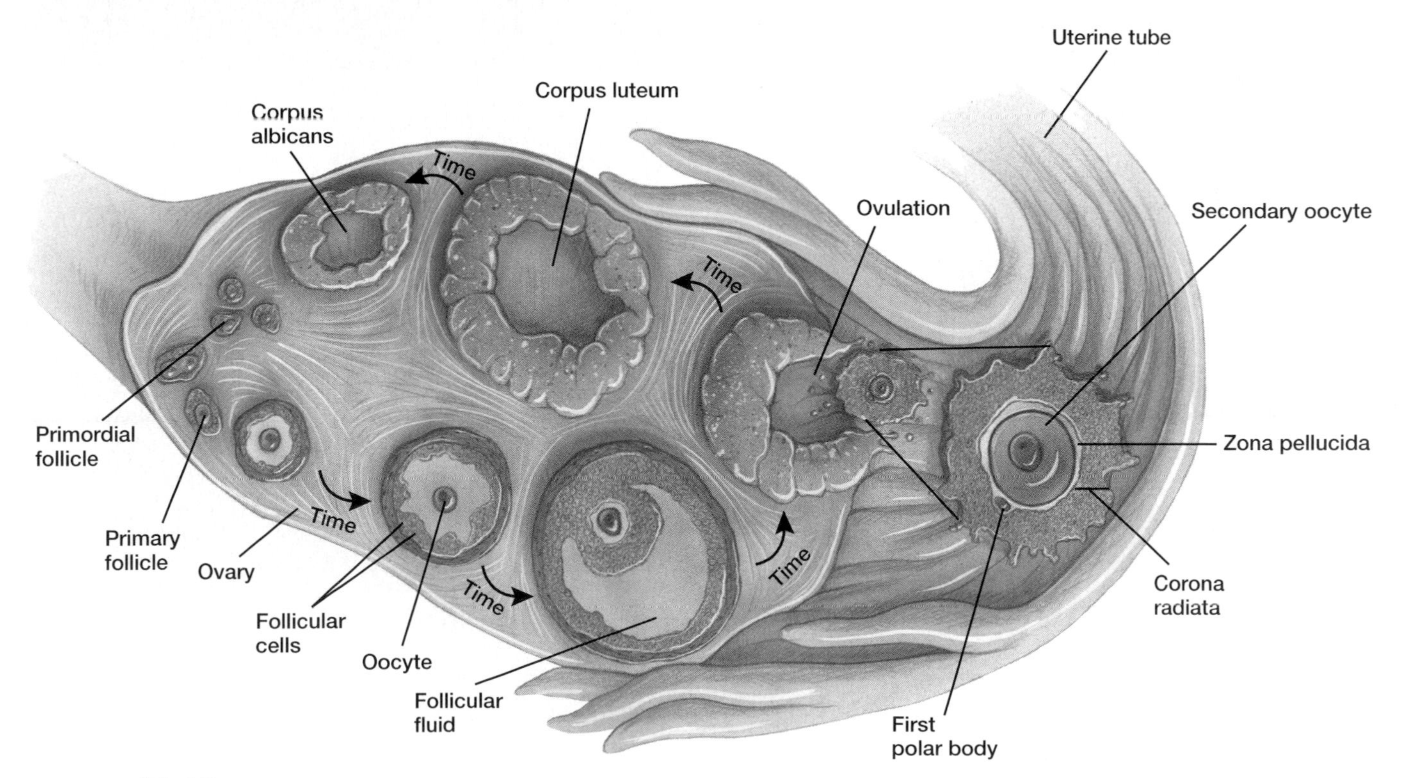


**FIGURE 22.22**
As a follicle matures, the egg cell enlarges and becomes surrounded by follicular cells and fluid. Eventually, the mature follicle ruptures, releasing the egg cell.

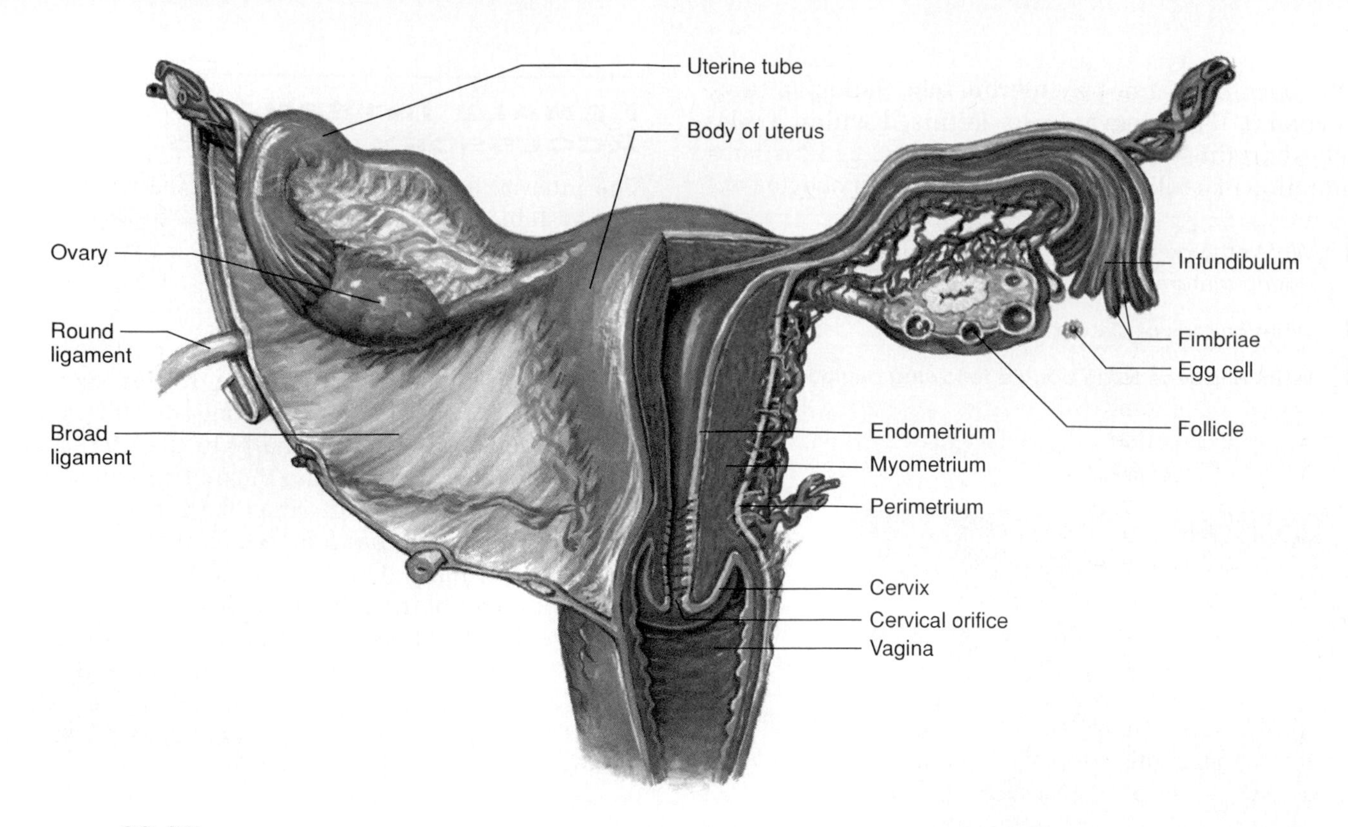


**FIGURE 22.23**
The funnel-shaped infundibulum of the uterine tube partially encircles the ovary.

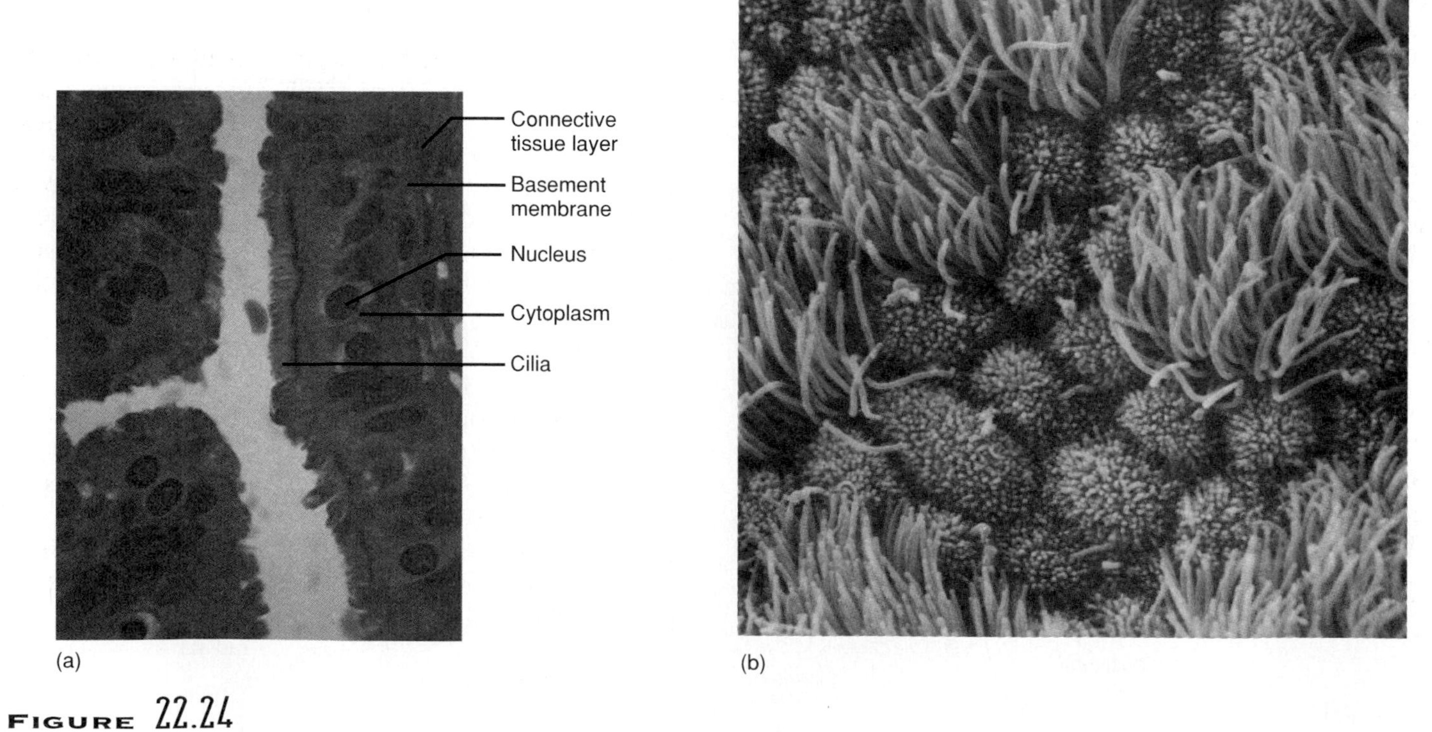


**FIGURE 22.24**
(*a*) Light micrograph of a uterine tube (250×). (*b*) Scanning electron micrograph of ciliated cells that line the uterine tube (650×).

cilia beat toward the uterus. These actions help draw the egg cell and expelled follicular fluid into the infundibulum following ovulation.

Ciliary action and peristaltic contractions of the tube's muscular layer aid transport of the egg down the uterine tube.

## Uterus

The **uterus** receives the embryo that develops from an egg cell that has been fertilized in the uterine tube, and sustains its development. It is a hollow, muscular organ, shaped somewhat like an inverted pear.

The *broad ligament,* which also attaches to the ovaries and uterine tubes, extends from the lateral walls of the uterus to the pelvic walls and floor, creating a drape across the top of the pelvic cavity (see fig. 22.23). A flattened band of tissue within the broad ligament, called the *round ligament,* connects the upper end of the uterus to the anterior pelvic wall (see figs. 22.16 and 22.23).

The size of the uterus changes greatly during pregnancy. In its nonpregnant, adult state, it is about 7 centimeters long, 5 centimeters wide (at its broadest point), and 2.5 centimeters in diameter. The uterus is located medially within the anterior portion of the pelvic cavity, above the vagina, and is usually bent forward over the urinary bladder.

The upper two-thirds, or *body,* of the uterus has a dome-shaped top, called the *fundus,* and is joined by the uterine tubes, which enter its wall at its broadest part.

The lower one-third of the uterus is called the **cervix.** This tubular part extends downward into the upper portion of the vagina. The cervix surrounds the opening called the *cervical orifice* (ostium uteri), through which the uterus connects with the vagina.

The uterine wall is thick and composed of three layers (fig. 22.25). The **endometrium** is the inner mucosal layer lining the uterine cavity. It is covered with columnar epithelium and contains abundant tubular glands. The **myometrium,** a very thick, muscular layer, consists largely of bundles of smooth muscle fibers in longitudinal, circular, and spiral patterns, and is interlaced with connective tissues. During the monthly female reproductive cycles and during pregnancy, the endometrium and myometrium extensively change. The **perimetrium** consists of an outer serosal layer, which covers the body of the uterus and part of the cervix.

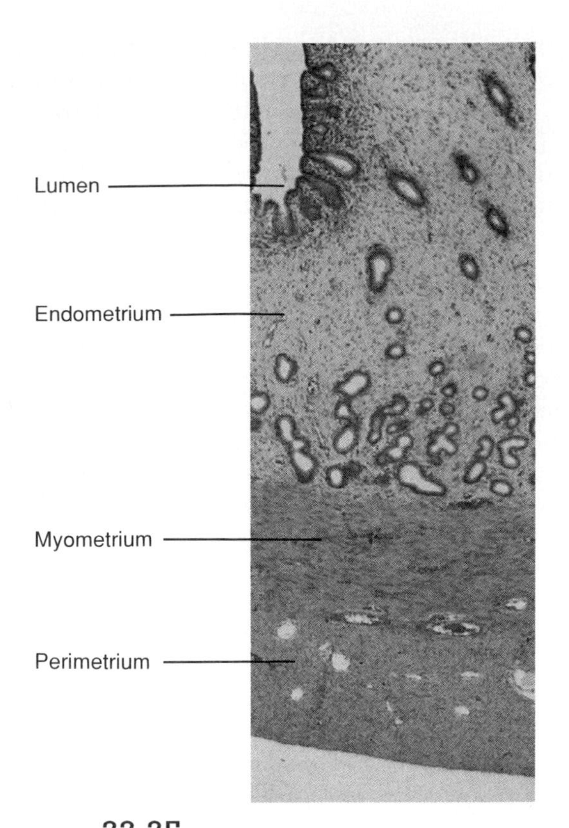


FIGURE 22.25
Light micrograph of the uterine wall (10.5×).

## Vagina

The **vagina** is a fibromuscular tube, about 9 centimeters long, extending from the uterus to the outside. It conveys uterine secretions, receives the erect penis during sexual intercourse, and provides the open channel for the offspring during birth.

The vagina extends upward and back into the pelvic cavity. It is posterior to the urinary bladder and urethra, anterior to the rectum, and attached to these structures by connective tissues. The upper one-fourth of the vagina is separated from the rectum by a pouch (rectouterine pouch). The tubular vagina also surrounds the end of the cervix, and the recesses between the vaginal wall and the cervix are termed *fornices* (sing. *fornix*).

> The fornices are clinically important because they are relatively thin-walled and allow the physician to palpate the internal abdominal organs during a physical examination. Also, the posterior fornix, which is somewhat longer than the others, provides a surgical access to the peritoneal cavity through the vagina.

The *vaginal orifice* is partially closed by a thin membrane of connective tissue and stratified squamous epithelium called the **hymen.** A central opening of varying size allows uterine and vaginal secretions to pass to the outside.

The vaginal wall consists of three layers. The inner *mucosal layer* is stratified squamous epithelium and is drawn into many longitudinal and transverse ridges (vaginal rugae). This layer lacks mucous glands; the mucus found in the lumen of the vagina comes from the glands of the cervix and the vestibular glands at the mouth of the vagina.

> Daughters of women who took the drug DES (diethylstilbestrol) while pregnant with them may develop a benign condition called *adenosis.* It arises when secretory columnar epithelium, resembling normal cells of the uterine lining, grow in the wrong place—in the vagina, up near the cervix. It is a little as if the lining of the mouth were to grow onto the face.
>
> Adenosis may produce a slight vaginal discharge. It is detected with a procedure called the *Pap* (Papanicolaou) *smear test.* A doctor or nurse scrapes off a tiny sample of cervical tissue, smears the sample on a glass slide, stains it, and sends it to a laboratory, where cytotechnologists examine it for the presence of abnormal cells.
>
> If the Pap smear is abnormal, the doctor follows up with a direct observation with a special type of microscope called a *colposcope.* The physician paints the patient's cervix with acetic acid (vinegar), which stains a purplish blue in the presence of carbohydrate in the discharge. The abnormally placed tissue can be removed painlessly with a laser.

The middle *muscular layer* consists mainly of smooth muscle fibers in longitudinal and circular patterns. At the lower end of the vagina is a thin band of striated muscle. This band helps close the vaginal opening; however, a voluntary muscle (bulbospongiosus) is primarily responsible for closing this orifice.

The outer *fibrous layer* consists of dense fibrous connective tissue interlaced with elastic fibers. It attaches the vagina to surrounding organs.

1. What factors aid the movement of an egg cell into the infundibulum following ovulation?
2. How is an egg cell moved along a uterine tube?
3. Describe the structure of the uterus.
4. What is the function of the uterus?
5. Describe the structure of the vagina.

## Female External Reproductive Organs

The *external accessory organs* of the female reproductive system include the labia majora, the labia minora, the clitoris, and the vestibular glands. These structures that surround the openings of the urethra and vagina compose the **vulva** (fig. 22.26).

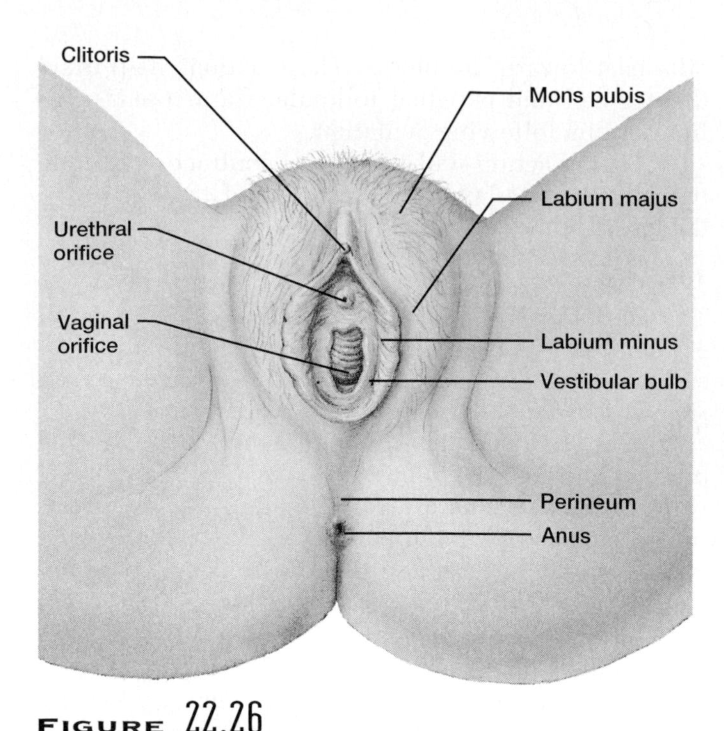


**Figure 22.26**
Female external reproductive organs and vestibular bulbs.

### Labia Majora

The **labia majora** (sing. *labium majus*) enclose and protect the other external reproductive organs. They correspond to the scrotum of the male, and are composed of rounded folds of adipose tissue and a thin layer of smooth muscle, covered by skin. On the outside, this skin includes hairs, sweat glands, and sebaceous glands, while on the inside, it is thinner and hairless.

The labia majora lie closely together and are separated longitudinally by a cleft (pudendal cleft), which includes the urethral and vaginal openings. At their anterior ends, the labia merge to form a medial, rounded elevation of adipose tissue called the *mons pubis,* which overlies the symphysis pubis. At their posterior ends, the labia are somewhat tapered, and they merge into the perineum near the anus.

### Labia Minora

The **labia minora** (sing. *labium minus*) are flattened longitudinal folds between the labia majora, extending along either side of the vestibule. They are composed of connective tissue richly supplied with blood vessels, causing a pinkish appearance. This tissue is covered with stratified squamous epithelium.

Posteriorly, the labia minora merge with the labia majora, while anteriorly, they converge to form a hoodlike covering around the clitoris.

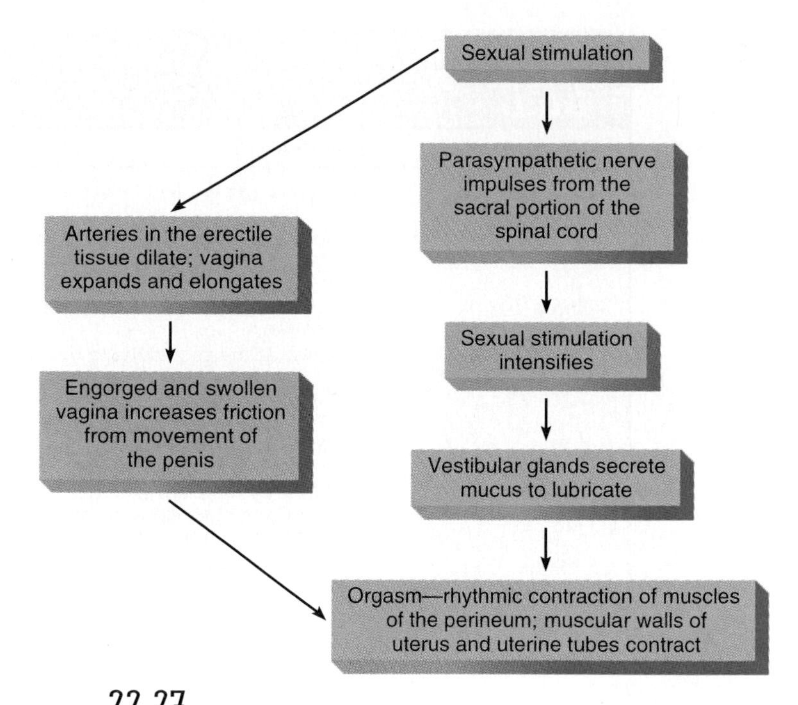


**FIGURE 22.27**
Mechanism of erection, lubrication, and orgasm in the human female.

## Clitoris

The **clitoris** is a small projection at the anterior end of the vulva between the labia minora. It is usually about 2 centimeters long and 0.5 centimeters in diameter, including a portion embedded in surrounding tissues. The clitoris corresponds to the penis and has a similar structure. It is composed of two columns of erectile tissue called *corpora cavernosa.* These columns are separated by a septum and are surrounded by a covering of dense fibrous connective tissue.

At the root of the clitoris, the corpora cavernosa diverge to form *crura,* which, in turn, attach to the sides of the pubic arch. At its anterior end, a small mass of erectile tissue forms a **glans,** which is richly supplied with sensory nerve fibers.

## Vestibule

The labia minora enclose a space called the **vestibule.** The vagina opens into the posterior portion of the vestibule, and the urethra opens in the midline, just in front of the vagina and about 2.5 centimeters behind the glans of the clitoris.

A pair of **vestibular glands** (Bartholin's glands), corresponding to the bulbourethral glands, lie on either side of the vaginal opening. Their ducts open into the vestibule near the lateral margins of the vaginal orifice.

Beneath the mucosa of the vestibule on either side is a mass of vascular erectile tissue. These structures, shown in figure 22.26, are called the *vestibular bulbs.* They are separated from each other by the vagina and the urethra, and they extend forward from the level of the vaginal opening to the clitoris.

1. What is the male counterpart of the labia majora? Of the clitoris?
2. What structures are located within the vestibule?

## Erection, Lubrication, and Orgasm

Erectile tissues located in the clitoris and around the vaginal entrance respond to sexual stimulation. Following such stimulation, parasympathetic nerve impulses pass out from the sacral portion of the spinal cord, dilating the arteries associated with the erectile tissues. As a result, inflow of blood increases, and the erectile tissues swell. At the same time, the vagina begins to expand and elongate.

If sexual stimulation is sufficiently intense, parasympathetic impulses stimulate the vestibular glands to secrete mucus into the vestibule. This secretion moistens and lubricates the tissues surrounding the vestibule and the lower end of the vagina, facilitating insertion of the penis into the vagina. Mucus secretion continuing during sexual intercourse helps prevent irritation of tissues that might occur if the vagina remained dry.

The clitoris is abundantly supplied with sensory nerve fibers, which are especially sensitive to local stimulation. The culmination of such stimulation is orgasm, the pleasurable sense of physiological and psychological release.

Just prior to orgasm, the tissues of the outer third of the vagina engorge with blood and swell. This action increases the friction on the penis during intercourse. Orgasm initiates a series of reflexes involving the sacral and lumbar portions of the spinal cord. In response to these reflexes, the muscles of the perineum and the walls of the uterus and uterine tubes contract rhythmically. These contractions help transport sperm cells through the female reproductive tract toward the upper ends of the uterine tubes (fig. 22.27).

Following orgasm, the flow of blood into the erectile tissues slackens, and the muscles of the perineum and reproductive tract relax. Consequently, the organs return to a state similar to that prior to sexual stimulation.

Chart 22.2 summarizes the functions of the female reproductive organs.

1. What events result from parasympathetic stimulation of the female reproductive organs?
2. What changes take place in the vagina just prior to and during female orgasm?
3. How do the uterus and the uterine tubes respond to orgasm?

## Hormonal Control of Female Reproductive Functions

Hormones secreted by the *hypothalamus,* the *anterior pituitary gland,* and the *ovaries* control development and maintenance of female secondary sexual characteristics, maturation of female sex cells, and changes that occur during the monthly reproductive cycle.

### Female Sex Hormones

A girl's body is reproductively immature until about 10 years of age. Then, the hypothalamus begins to secrete increasing amounts of gonadotropin-releasing hormone (GnRH), which, in turn, stimulate the anterior pituitary gland to release the gonadotropins FSH and LH. These hormones play primary roles in controlling female sex cell maturation and in producing female sex hormones.

Several tissues, including the ovaries, the adrenal cortices, and the placenta (during pregnancy), secrete female sex hormones belonging to two major groups, called **estrogen** and **progesterone.**

The primary source of estrogen (in a nonpregnant female) is the ovaries, although some estrogen is also synthesized in adipose tissue from adrenal androgens. At puberty, under the influence of the anterior pituitary gland, the ovaries secrete increasing amounts of the hormone. Estrogen stimulates enlargement of accessory organs including the vagina, uterus, uterine tubes, and ovaries, as well as the external structures, and is also responsible for the development and maintenance of female *secondary sexual characteristics.* These are listed in figure 22.28 and include the following:

1. Development of the breasts and the ductile system of the mammary glands within the breasts.
2. Increased deposition of adipose tissue in the subcutaneous layer generally and in the breasts, thighs, and buttocks particularly.
3. Increased vascularization of the skin.

**CHART 22.2 Functions of the Female Reproductive Organs**

| Organ | Function |
|---|---|
| Ovary | Produces egg cells and female sex hormones |
| Uterine tube | Conveys egg cell toward uterus; site of fertilization; conveys developing embryo to uterus |
| Uterus | Protects and sustains embryo during pregnancy |
| Vagina | Conveys uterine secretions to outside of body; receives erect penis during sexual intercourse; provides open channel for offspring during birth process |
| Labia majora | Enclose and protect other external reproductive organs |
| Labia minora | Form margins of vestibule; protect openings of vagina and urethra |
| Clitoris | Glans is richly supplied with sensory nerve endings associated with feeling of pleasure during sexual stimulation |
| Vestibule | Space between labia minora that includes vaginal and urethral openings |
| Vestibular glands | Secrete fluid that moistens and lubricates vestibule |

The ovaries are also the primary source of progesterone (in a nonpregnant female). This hormone promotes changes that occur in the uterus during the female reproductive cycle, affects the mammary glands, and helps regulate secretion of gonadotropins from the anterior pituitary gland.

Certain other changes that occur in females at puberty seem to be related to *androgen* (male sex hormone) concentrations. For example, increased growth of hair in the pubic and axillary regions is due to androgen secreted by the adrenal cortices. Conversely, development of the female skeletal configuration, which includes narrow shoulders and broad hips, seems to be related to a low concentration of androgen.

Female athletes who train for endurance events, such as the marathon, typically maintain about 6% body fat. Male endurance athletes usually have about 4% body fat. This difference of 50% in proportion of body fat is related to the actions of sex hormones. The male hormone testosterone promotes deposition of protein throughout the body and especially in skeletal muscles; the female hormone estrogen increases deposition of adipose tissue in the breasts, thighs, buttocks, and the subcutaneous layer of the skin.

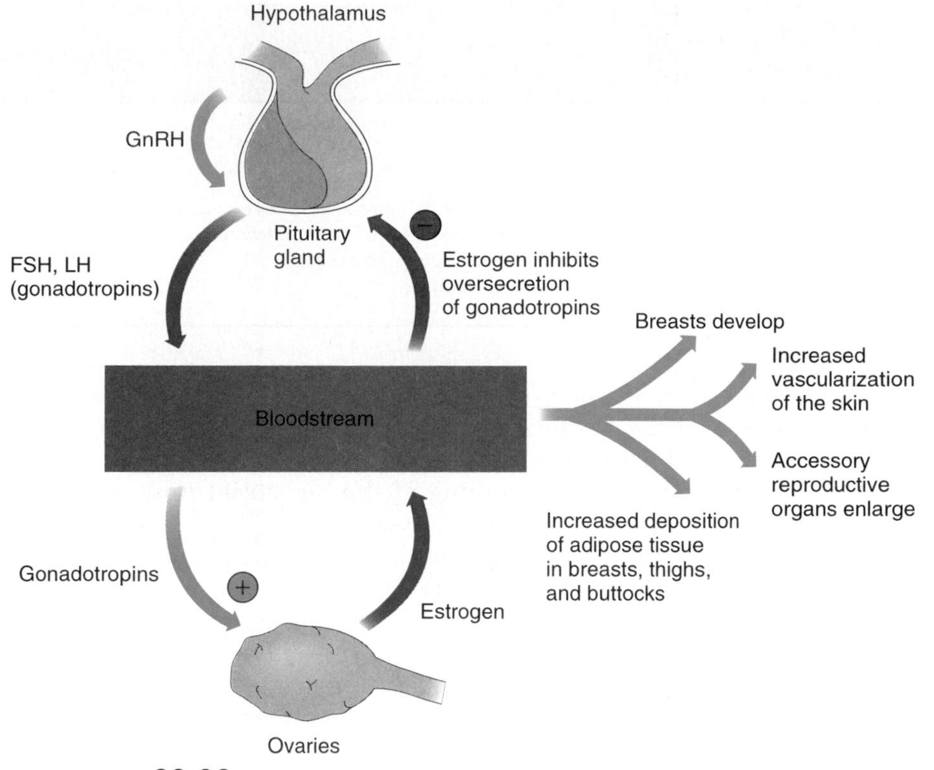


**FIGURE 22.28**

Control of female secondary sexual development. Estrogen inhibits LH and FSH during most of the menstrual cycle.

1. What factors initiate sexual maturation in a female?
2. Name the two major female sex hormones.
3. What is the function of estrogen?
4. What is the function of androgen in a female?

## Female Reproductive Cycle

The female reproductive cycle, or **menstrual cycle,** is characterized by regular, recurring changes in the uterine lining, which culminate in menstrual bleeding (menses). Such cycles usually begin near the thirteenth year of life and continue into middle age, then cease.

> Women athletes sometimes experience disturbances in their menstrual cycles, ranging from diminished menstrual flow (oligomenorrhea) to its complete stoppage (amenorrhea). The incidence of menstrual disorders generally increases with the intensity and duration of exercise periods, occurring most commonly in athletes who perform the most strenuous activities and who follow the most intense training schedules. This effect is related to a loss of adipose tissue and a consequent decline in estrogen, which is synthesized in these tissues from adrenal androgens.

A female's first menstrual cycle (menarche) occurs after the ovaries and other organs of the female reproductive control system mature and respond to certain hormones. Then, the hypothalamic secretion of gonadotropin-releasing hormone (GnRH) stimulates the anterior pituitary gland to release threshold levels of FSH (follicle-stimulating hormone) and LH (luteinizing hormone). As its name implies, FSH stimulates maturation of an ovarian *follicle.* The granulosa cells of the follicle produce increasing amounts of estrogen and some progesterone. LH also stimulates certain ovarian cells (theca interna) to secrete precursor molecules (testosterone) used to produce estrogen.

In a young female, estrogen stimulates development of various secondary sexual characteristics. Estrogen secreted during subsequent menstrual cycles continues development of these traits and maintains them. Chart 22.3 summarizes the hormonal control of female secondary sexual characteristics.

Increasing concentration of estrogen during the first week or so of a menstrual cycle changes the uterine lining, thickening the glandular endometrium (proliferative phase). Meanwhile, the developing follicle completes maturation, and by the fourteenth day of the cycle, the follicle appears on the surface of the ovary as a blisterlike bulge.

Within the follicle, the granulosa cells, which surround the oocyte and connect it to the inner wall, loosen. Follicular fluid accumulates rapidly.

While the follicle matured, estrogen that it secreted inhibited the release of LH from the anterior pituitary gland, but allowed LH to be stored in the gland. Estrogen also made the anterior pituitary cells more sensitive to the action of GnRH, which is released from the hypothalamus in rhythmic pulses about 90 minutes apart.

Near the fourteenth day of follicular development, the anterior pituitary cells finally respond to the pulses of GnRH and to the increasing quantity of progesterone released from the follicle. In response, the anterior pituitary cells release the LH that they have stored. The resulting surge in LH concentration, which lasts for about 36 hours, weakens and ruptures the bulging follicular wall. At the same time, the oocyte and follicular fluid escape from the ovary (ovulation).

## CHART 22.3 HORMONAL CONTROL OF FEMALE SECONDARY SEXUAL CHARACTERISTICS

1. The hypothalamus releases GnRH, which stimulates the anterior pituitary gland.
2. The anterior pituitary gland secretes FSH and LH.
3. FSH stimulates the maturation of a follicle.
4. Granulosa cells of the follicle produce and secrete estrogen; LH stimulates certain cells to secrete estrogen precursor molecules.
5. Estrogen is responsible for the development and maintenance of most of the female secondary sexual characteristics.
6. Concentrations of androgen affect other secondary sexual characteristics, including skeletal growth and growth of hair.
7. Progesterone, secreted by the ovaries, affects cyclical changes in the uterus and mammary glands.

Following ovulation, the remnants of the follicle and the theca interna within the ovary rapidly change. The space occupied by the follicular fluid fills with blood, which soon clots, and under the influence of LH, the follicular and thecal cells enlarge greatly to form a temporary glandular structure within the ovary, called a **corpus luteum** (see fig. 22.22).

Follicular cells secrete some progesterone during the first part of the menstrual cycle. However, corpus luteum cells secrete abundant progesterone and estrogen during the last half of the cycle. Consequently, as a corpus luteum is established, the blood concentration of progesterone increases sharply.

Progesterone causes the endometrium to become more vascular and glandular. It also stimulates the uterine glands to secrete more glycogen and lipids (secretory phase). As a result, the endometrial tissues fill with fluids containing nutrients and electrolytes, which provide a favorable environment for the development of an embryo.

Estrogen and progesterone inhibit the release of LH and FSH from the anterior pituitary gland. Consequently, no other follicles are stimulated to develop when the corpus luteum is active. However, if the oocyte released at ovulation is not fertilized by a sperm cell, the corpus luteum begins to degenerate (regress) about the twenty-fourth day of the cycle. Eventually, fibrous connective tissue replaces it. The remnant of such a corpus luteum is called a *corpus albicans* (see fig. 22.22).

When the corpus luteum ceases to function, concentrations of estrogen and progesterone decline rapidly, and in response, blood vessels in the endometrium constrict. This reduces the supply of oxygen and nutrients to the thickened endometrium, and these lining tissues (decidua) soon disintegrate and slough off. At the same time, blood escapes from damaged capillaries, creating a flow of blood and cellular debris, which passes through the vagina as the *menstrual flow* (menses). This flow usually begins about the twenty-eighth day of the cycle and continues for three to five days, while the estrogen concentration is relatively low.

The beginning of the menstrual flow marks the end of a menstrual cycle and the beginning of a new cycle. This cycle is diagrammed in figure 22.29 and summarized in chart 22.4.

Because the blood concentrations of estrogen and progesterone are low at the beginning of the menstrual cycle, the hypothalamus and anterior pituitary gland are no longer inhibited. Consequently, the concentrations of FSH and LH soon increase, and a new follicle is stimulated to mature. As this follicle secretes estrogen, the uterine lining undergoes repair, and the endometrium begins to thicken again.

A woman who becomes extremely irritable, fatigued, bloated, or easily upset at the same time each month is likely to have *premenstrual syndrome* (PMS). This condition may occur only during some months. The cause of PMS is unknown, but it is believed to involve hormonal imbalance.

## Menopause

After puberty, menstrual cycles continue at regular intervals into the late forties or early fifties, when they usually become increasingly irregular. Then, in a few months or years, the cycles cease altogether. This period in life is called **menopause** (female climacteric).

The cause of menopause is aging of the ovaries. After about 35 years of cycling, few primary follicles remain to be stimulated by pituitary gonadotropins. Consequently, the follicles no longer mature, ovulation does not occur, and the blood concentration of estrogen plummets, although many women continue to synthesize some estrogen from adrenal androgens.

As a result of reduced estrogen concentration and lack of progesterone, the female secondary sexual

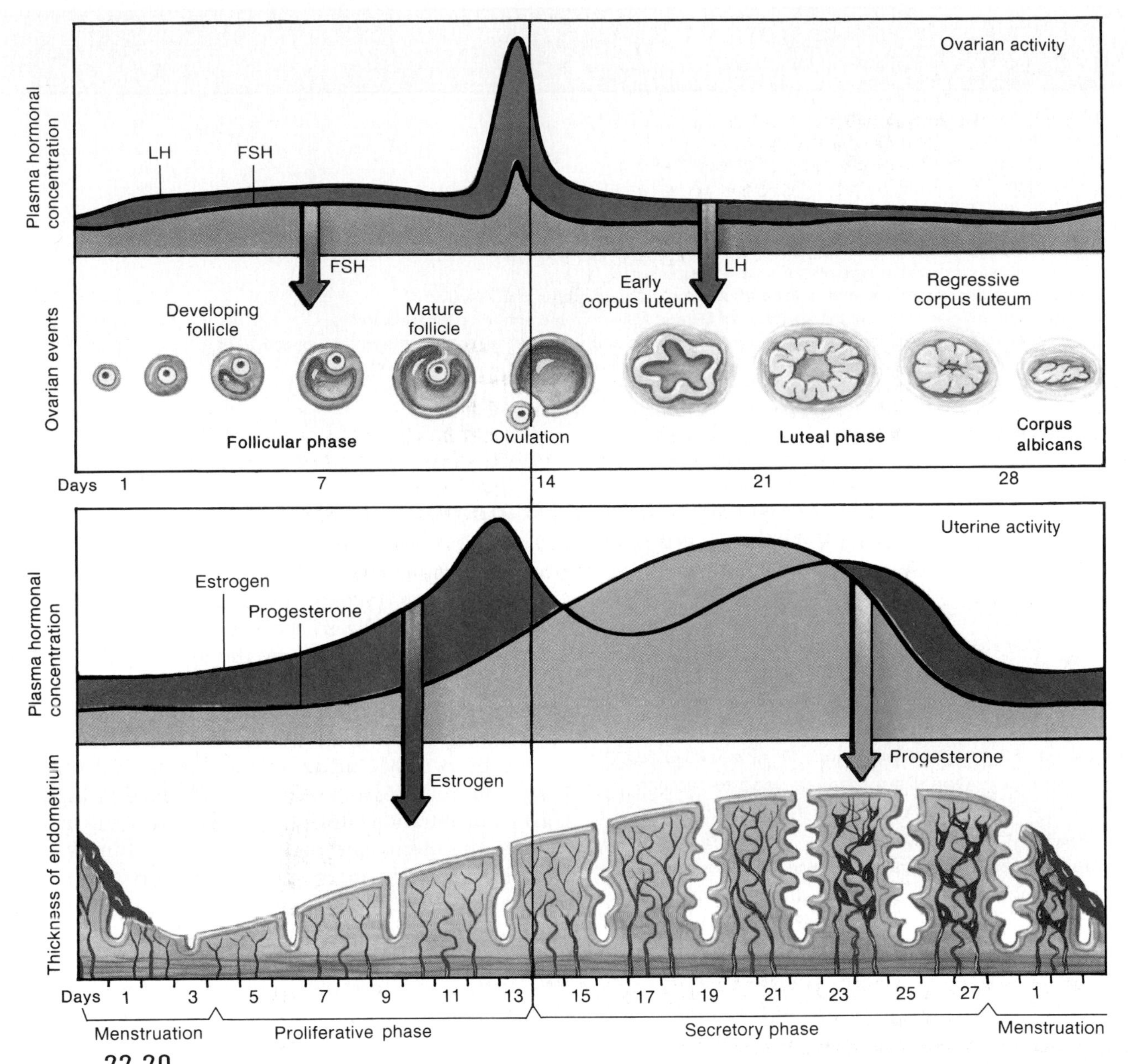


**FIGURE 22.29**
Major events in the female ovarian and menstrual cycles.

characteristics may change. The breasts, vagina, uterus, and uterine tubes may shrink, and the pubic and axillary hair may thin. The epithelial linings associated with urinary and reproductive organs may thin. There may be increased loss of bone matrix (osteoporosis) and thinning of the skin. Because the pituitary secretions of FSH and LH are no longer inhibited, these hormones may be released continuously for some time.

About 50% of women reach menopause by age 50, and 85% reach it by age 52. Of these, perhaps 20% have no unusual symptoms—they simply stop menstruating. However, about 50% of menopausal women experience unpleasant vasomotor symptoms, including sensations of heat in the face, neck, and upper body called "hot flashes." Such a sensation may last for 30 seconds to 5 minutes and may be accompanied by chills and sweating. Menopausal women may also experience headache, backache, and fatigue. These vasomotor symptoms may involve changes in the rhythmic secretion of GnRH by the hypothalamus in response to declining concentrations of

CHART 22.4 MAJOR EVENTS IN A MENSTRUAL CYCLE

1. The anterior pituitary gland secretes FSH and LH.
2. FSH stimulates maturation of a follicle.
3. Granulosa cells of the follicle produce and secrete estrogen.
   a. Estrogen maintains secondary sexual traits.
   b. Estrogen causes the uterine lining to thicken.
4. The anterior pituitary gland releases a surge of LH, which stimulates ovulation.
5. Follicular and thecal cells become corpus luteum cells, which secrete estrogen and progesterone.
   a. Estrogen continues to stimulate uterine wall development.
   b. Progesterone stimulates the uterine lining to become more glandular and vascular.
   c. Estrogen and progesterone inhibit secretion of LH and FSH from the anterior pituitary gland.
6. If the egg cell is not fertilized, the corpus luteum degenerates and no longer secretes estrogen and progesterone.
7. As the concentrations of luteal hormones decline, blood vessels in the uterine lining constrict.
8. The uterine lining disintegrates and sloughs off, producing a menstrual flow.
9. The anterior pituitary gland, which is no longer inhibited, again secretes FSH and LH.
10. The menstrual cycle repeats.

sex hormones. To reduce the unpleasant side effects of menopause, women are often treated with *estrogen replacement therapy* (ERT).

At age 53, Mary Shearing is the mother of newborn twins, Amy Leigh and Kelly Ann. In the last stages of menopause, Mary already had three grown children from a previous marriage, and two grandchildren. She and her 32-year-old husband, Don, decided to try for pregnancy. Don's sperm was used to fertilize oocytes in a laboratory dish, and the fertilized ova were implanted in Mary's uterus. The Shearings' success showed that it is not the condition of the uterine lining in an older woman that makes conceiving, carrying, and delivering a healthy baby difficult, but the age of the oocyte. Women in their sixties have given birth.

Most women who have the treatment that Shearing underwent are under 45 years of age, and have lost ovaries to disease. Often the donor is a sister.

1. Trace the events of the female menstrual cycle.
2. What effect does progesterone have on the endometrium?
3. What causes the menstrual flow?
4. What are some changes that may occur at menopause?

## PREGNANCY

**Pregnancy** is the presence of a developing offspring within the uterus. It results from the union of the genetic packages of an egg cell and a sperm cell—an event called **fertilization.**

### Transport of Sex Cells

Ordinarily, before fertilization can occur, a secondary oocyte must be ovulated and enter a uterine tube. During sexual intercourse, the male deposits semen containing sperm cells in the vagina near the cervix. To reach the secondary oocyte, the sperm cells must then move upward through the uterus and uterine tube. Lashing of sperm tails and muscular contractions within the walls of the uterus and uterine tube, stimulated by prostaglandins in the semen, aid the sperm cells' journey. Also, under the influence of high estrogen concentrations during the first part of the menstrual cycle, the uterus and cervix contain a thin, watery secretion that promotes sperm transport and survival. Conversely, during the latter part of the cycle, when the progesterone concentration is relatively high, the female reproductive tract secretes a viscous fluid that hampers sperm transport and survival (fig. 22.30).

Sperm movement is relatively inefficient. Even though as many as 300 million to 500 million sperm cells may be deposited in the vagina by a single ejaculation, only a few hundred ever reach a secondary oocyte.

Sperm cells are thought to reach the upper portions of the uterine tube within an hour following sexual intercourse. Although many sperm cells may reach a secondary oocyte, only one sperm cell actually fertilizes it (fig. 22.31).

About one in a million births produces a severely deformed child who has inherited three sets of chromosomes. This may be the result of two sperm cells entering a single egg cell.

A secondary oocyte may survive for only 12 to 24 hours following ovulation, while sperm cells may live up to 72 hours within the female reproductive tract.

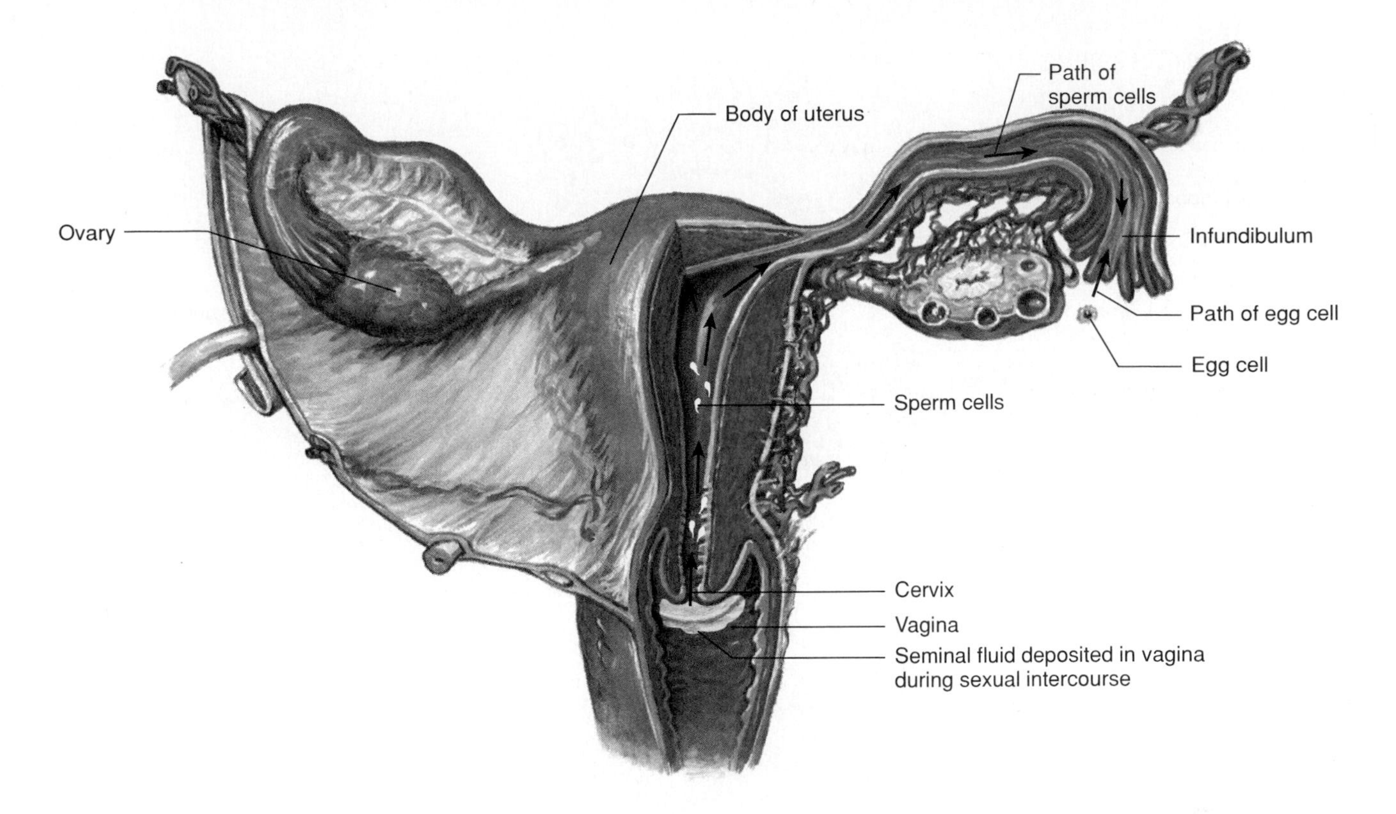


**FIGURE 22.30**
The paths of the egg and sperm cells through the female reproductive tract.

Consequently, sexual intercourse probably must occur no more than 72 hours before ovulation or 24 hours following ovulation if fertilization is to take place.

### Fertilization

When a sperm reaches a secondary oocyte, it invades the follicular cells that adhere to the oocyte's surface (corona radiata) and binds to the **zona pellucida** that surrounds the oocyte's cell membrane. The acrosome of a sperm cell attached to the zona pellucida releases an enzyme that helps the motile sperm penetrate the zona pellucida (fig. 22.32).

> In "zona blasting," an experimental procedure to aid certain infertile men, an egg cell cultured in a laboratory dish is chemically stripped of its zona pellucida. The more vulnerable egg now presents less of a barrier to a sperm.

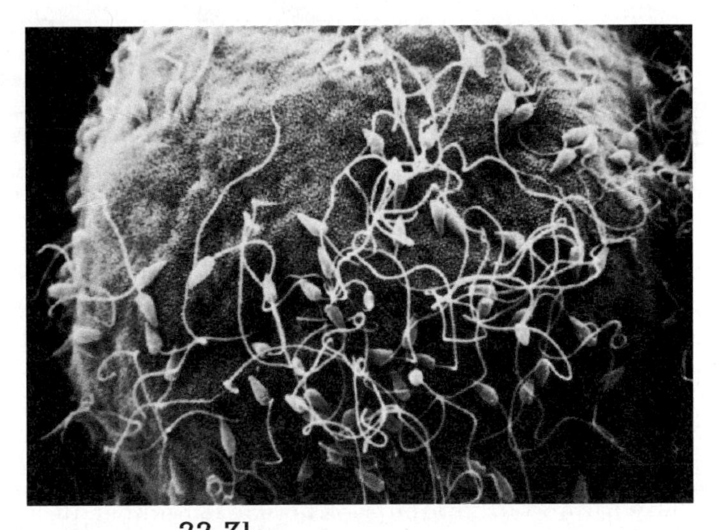

**FIGURE 22.31**
Scanning electron micrograph of sperm cells on the surface of an egg cell (1,200×).

The sperm cell's plasma membrane fuses with that of the secondary oocyte, and sperm movement ceases. The sperm's tail does not enter. At the same time, the oocyte cell membrane becomes unresponsive to other sperm cells. The union of the oocyte and sperm cell membranes also triggers some lysosome-like granules (cortical granules) just beneath the oocyte cell membrane to release enzymes that harden the zona pellucida. This reduces the chance that other sperm cells will penetrate, and it forms a protective layer around the newly formed fertilized egg cell.

Once the sperm enters the oocyte's cytoplasm, the nucleus within its head swells. The secondary oocyte then divides unequally to form a relatively large cell and a tiny second polar body, which is later expelled. Therefore, female meiosis completes only after the

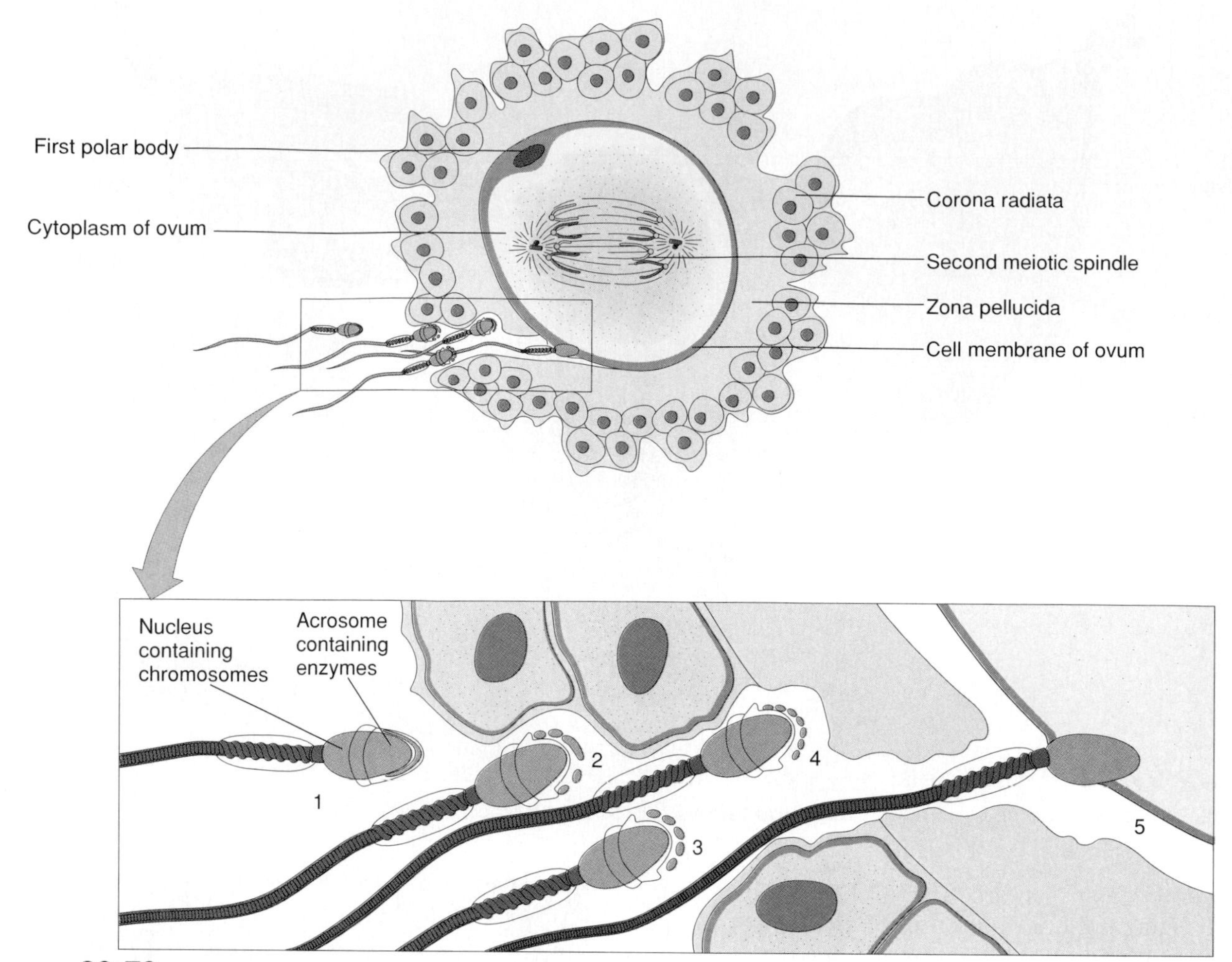


**FIGURE 22.32**

Steps in fertilization: (1) Sperm cell reaches corona radiata surrounding the egg cell. (2) Acrosome of sperm cell releases protein-digesting enzyme. (3 and 4) Sperm cell penetrates zona pellucida surrounding egg cell. (5) Sperm cell's plasma membrane fuses with egg cell membrane.

sperm enters the egg. Next, the nuclei of the male and female cells unite. Their nuclear membranes disassemble, and their chromosomes mingle, completing the process of fertilization, diagrammed in figure 22.32.

Because the sperm cell and the egg cell each provide 23 chromosomes, the product of fertilization is a cell with 46 chromosomes—the usual number in a human cell. This cell, called a **zygote,** is the first cell of the future offspring.

1. What factors aid the movements of the egg and sperm cells through the female reproductive tract?
2. Where in the female reproductive system does fertilization normally take place?
3. List the events that occur during fertilization.

## Early Embryonic Development

About 30 hours after forming, the zygote undergoes *mitosis,* giving rise to two cells (blastomeres). These cells, in turn, divide into four cells, which divide into eight cells, and so forth. With each subsequent division, the resulting cells are smaller and smaller. This phase of development is termed **cleavage.**

Meanwhile, the tiny mass of cells moves through the uterine tube to the uterine cavity, aided by the beating of cilia of the tubular epithelium and by weak peristaltic contractions of smooth muscles in the tubular wall. Secretions from the epithelial lining may provide the developing organism with nutrients.

The trip to the uterus takes about three days, and by then, the structure consists of a solid ball **(morula)** of about sixteen cells. The morula remains free within

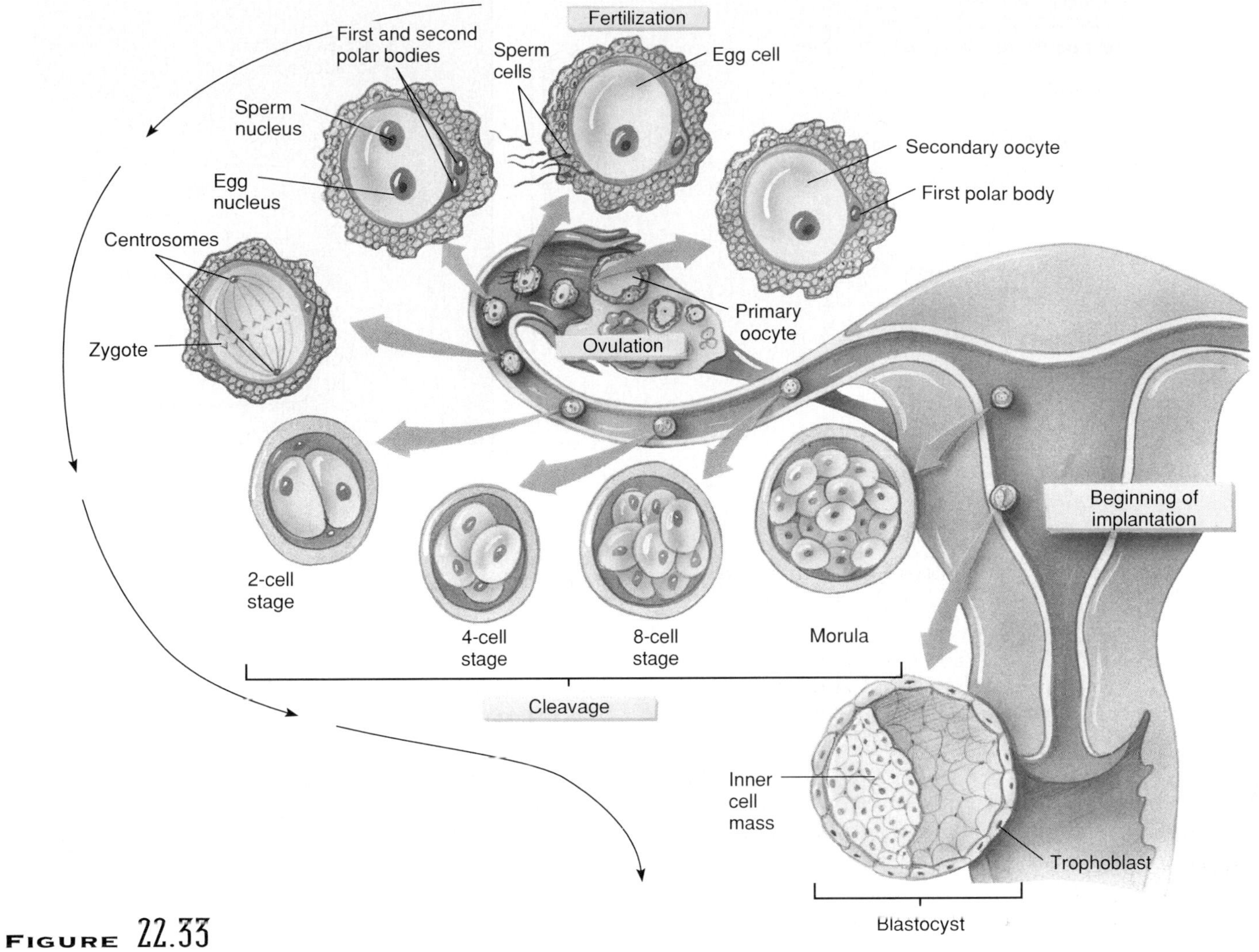


**FIGURE 22.33**
Stages in early human development.

the uterine cavity for about three days. During this stage, the zona pellucida of the original egg cell degenerates, and the structure, which now consists of a hollow ball of cells **(blastocyst),** drops into one of the tubules in the endometrium. By the end of the first week of development, it superficially *implants* in the endometrium (fig. 22.33).

About the time of implantation, certain cells within the blastocyst organize into a group, called the inner cell mass, that will give rise to the body. This marks the beginning of the *embryonic period* of development. The offspring is termed a preembryo for two weeks, then an **embryo** until the end of the eighth week, after which it is called a **fetus** up to the time of birth.

The "abortion pill" RU486 interferes with prenatal development at the implantation stage. Taken early in pregnancy, the drug prevents progesterone from binding to cells in the uterine lining, where it is necessary for a blastocyst to implant. A prostaglandin drug given two days later contracts the uterus and opens the cervix. The combined action of RU486 and the prostaglandin expels the blastocyst 97% of the time. Protocols using RU486 alone are being developed in France, where the pill was developed in the 1980s. In the United States, a combination of a prostaglandin and a cancer drug is being studied as an abortion-causing regimen.

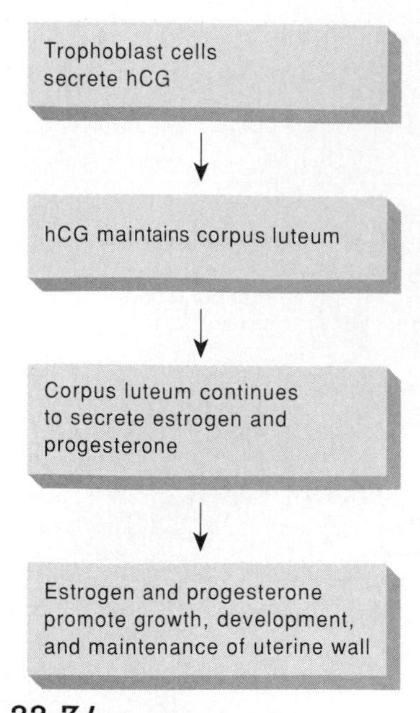


**FIGURE 22.34**
Mechanism that preserves the uterine lining during early pregnancy.

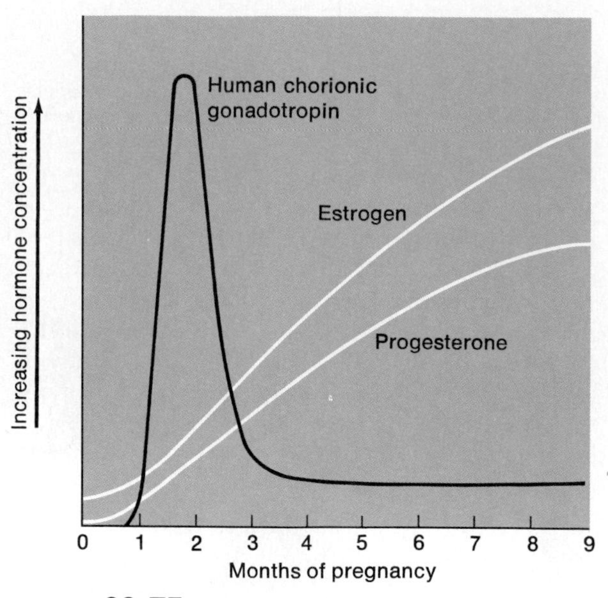


**FIGURE 22.35**
Relative concentrations of three hormones in the blood during pregnancy.

Eventually, the outer cells of the embryo, together with cells of the endometrium, form a complex vascular structure called the **placenta.** This organ attaches the embryo to the uterine wall and exchanges nutrients, gases, and wastes between the maternal blood and the embryonic blood. The placenta also secretes hormones. The placenta is described in more detail in chapter 23.

Occasionally, a blastocyst implants in tissues outside the uterus, such as those of a uterine tube, an ovary, the cervix, or an organ in the abdominal cavity. The result is called an *ectopic pregnancy.* If a fertilized egg implants within the uterine tube, it is called a *tubal pregnancy.* A tubal pregnancy is dangerous to a pregnant woman and the developing offspring because the tube usually ruptures as the embryo enlarges, and is accompanied by severe pain and heavy vaginal bleeding. Treatment is prompt surgical removal of the embryo and repair or removal of the damaged uterine tube.

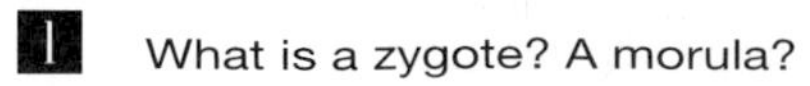

1 What is a zygote? A morula?

2 How does an embryo implant in the uterine wall?

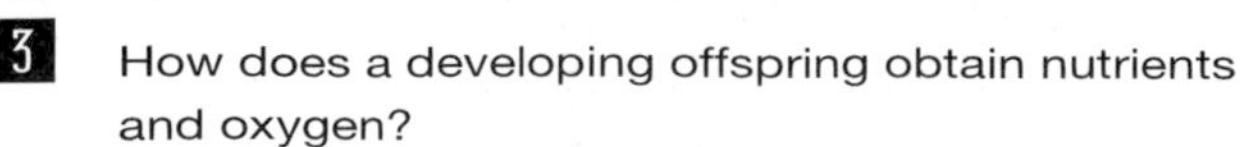

3 How does a developing offspring obtain nutrients and oxygen?

## Hormonal Changes during Pregnancy

During a typical menstrual cycle, the corpus luteum degenerates about two weeks after ovulation. Consequently, estrogen and progesterone concentrations decline rapidly, the uterine lining is no longer maintained, and the endometrium sloughs off as menstrual flow. If this occurs following implantation, the embryo is lost (spontaneously aborted).

The mechanism that normally prevents such a termination of pregnancy involves a hormone called hCG **(human chorionic gonadotropin).** A layer of embryonic cells called a **trophoblast** (see fig. 22.33) that surrounds the developing embryo and later helps form the placenta secretes hCG (see chapter 23). This hormone has properties similar to those of LH, and it maintains the corpus luteum, which continues secreting estrogen and progesterone. Thus, the uterine wall continues to grow and develop. At the same time, release of FSH and LH from the anterior pituitary gland is inhibited, so that normal menstrual cycles cease (fig. 22.34).

Secretion of hCG continues at a high level for about two months, then declines to a relatively low level by the end of four months. Although the corpus luteum persists throughout pregnancy, its function as a source of hormones becomes less important after the first three months (first trimester), when the placenta secretes sufficient estrogen and progesterone (fig. 22.35).

The excited woman takes a drop of her first morning's urine and places it in a tube that came with an early pregnancy test kit. She and her partner anxiously watch for a few minutes. They are looking for a sign of hCG, the hormone that a zygote begins secreting shortly after fertilization. Many such pregnancy tests utilize a monoclonal antibody to hCG, which is linked to a molecule that produces a color change when the antibody binds hCG. The couple learns of their impending parenthood just two days after the woman expected her menstrual period. Some tests detect hCG eight to ten days following fertilization. When the woman visits her physician to begin prenatal care, her blood is tested for hCG to confirm the home test.

The level of hCG in a pregnant woman's body fluids peaks at 50 to 60 days of gestation, then falls to a much lower level for the remainder of pregnancy. Later on, measuring hCG has other uses. If a woman miscarries but her blood still shows hCG, fetal tissue may remain in her uterus, and it must be removed. At the fifteenth week of pregnancy, most women have a blood test that measures levels of three substances produced by the fetus—alpha fetoprotein, estriol (a form of estrogen), and hCG. If alpha fetoprotein and estriol are low but hCG is high, the fetus may have an extra chromosome, with severe effects on health.

For the remainder of the pregnancy, *placental estrogen* and *placental progesterone* maintain the uterine wall. The placenta also secretes a hormone called **placental lactogen** that may stimulate breast development and prepare the mammary glands for milk secretion, with the aid of placental estrogen and progesterone. Placental progesterone and a polypeptide hormone called *relaxin* from the corpus luteum inhibit the smooth muscles in the myometrium, suppressing uterine contractions until the birth process begins.

The high concentration of placental estrogen during pregnancy enlarges the vagina and the external reproductive organs. Also, relaxin relaxes the ligaments holding the symphysis pubis and sacroiliac joints together. This action, which usually occurs during the last week of pregnancy, allows for greater movement at these joints, aiding passage of the fetus through the birth canal.

Other hormonal changes that occur during pregnancy include increased secretions of aldosterone from the adrenal cortex and of parathyroid hormone from the parathyroid glands. Aldosterone promotes renal reabsorption of sodium, leading to fluid retention. Parathyroid hormone helps to maintain a high concentration of maternal blood calcium, since fetal demand for calcium can cause hypocalcemia, which promotes cramps (see chapter 13).

CHART 22.5 HORMONAL CHANGES DURING PREGNANCY

1. Following implantation, the embryonic cells begin to secrete hCG.
2. hCG maintains the corpus luteum, which continues secreting estrogen and progesterone.
3. As the placenta develops, it secretes large quantities of estrogen and progesterone.
4. Placental estrogen and progesterone
   a. stimulate the uterine lining to continue development.
   b. maintain the uterine lining.
   c. inhibit secretion of FSH and LH from the anterior pituitary gland.
   d. stimulate development of the mammary glands.
   e. inhibit uterine contractions (progesterone).
   f. enlarge the reproductive organs (estrogen).
5. Relaxin from the corpus luteum also inhibits uterine contractions and relaxes the pelvic ligaments.
6. The placenta secretes placental lactogen that stimulates breast development.
7. Aldosterone from the adrenal cortex promotes reabsorption of sodium.
8. Parathyroid hormone from the parathyroid glands helps maintain a high concentration of maternal blood calcium.

Chart 22.5 summarizes the hormonal changes of pregnancy.

1 What mechanism maintains the uterine wall during pregnancy?

2 What is the source of hCG during the first few months of pregnancy?

3 What is the source of the hormones that sustain the uterine wall during pregnancy?

4 What other hormonal changes occur during pregnancy?

## Other Changes during Pregnancy

A number of other changes occur in a woman's body as a result of the increased demands of a growing fetus. As the fetus increases in size, the uterus enlarges greatly, and instead of being confined to its normal location in the pelvic cavity, it extends upward and may eventually reach the level of the ribs. At the same time, the abdominal organs are displaced upward and compressed against the diaphragm. The enlarging uterus also presses on the urinary bladder.

A pregnant woman is well aware of the effects of her expanding uterus. She can no longer eat large meals, develops heartburn often as stomach contents are pushed up into the esophagus, and frequently visits the bathroom as her uterus presses on her bladder.

As the placenta grows and develops, it requires more blood, and as the fetus enlarges, it needs more oxygen and produces greater amounts of wastes that must be excreted. The pregnant woman's blood volume, cardiac output, breathing rate, and urine production all increase in response to fetal demands.

The pregnant woman must eat more to meet the fetal need for more nutrients. Her intake must supply adequate vitamins, minerals, and proteins for herself and the fetus. The fetal tissues have a greater capacity to capture available nutrients than do the maternal tissues. Consequently, if the pregnant woman's diet is inadequate, her body will usually show symptoms of a deficiency condition before fetal growth is adversely affected.

## Birth

Pregnancy usually continues for forty weeks (280 days, or ten average menstrual cycles), or about nine calendar months (ten lunar months), if it is measured from the beginning of the last menstrual cycle. Pregnancy terminates with the *birth process* (parturition).

Birth is a complex, little understood process. Progesterone seems to play a major role in beginning parturition. During pregnancy, this hormone suppresses uterine contractions. As the placenta ages, the concentration of progesterone within the uterus declines, which may also stimulate synthesis of a prostaglandin that promotes uterine contractions.

Another stimulant beginning the birth process is stretching of the uterine and vaginal tissues late in pregnancy. This may initiate nerve impulses to the hypothalamus which, in turn, signals the posterior pituitary gland to release the hormone **oxytocin.**

Oxytocin stimulates powerful uterine contractions. Combined with the greater excitability of the myometrium due to the decline in progesterone secretion, oxytocin aids labor in its later stages.

During *labor,* muscular contractions force the fetus through the birth canal. Rhythmic contractions that begin at the top of the uterus and travel down its length force the contents of the uterus toward the cervix.

Since the fetus is usually positioned head downward, labor contractions force the head against the cervix. This action stretches the cervix, which elicits a reflex that stimulates still stronger labor contractions. Thus, a positive feedback system operates in which uterine contractions produce more intense uterine contractions until a maximum effort is achieved (fig. 22.36). At the same time, dilation of the cervix reflexly stimulates an increased release of oxytocin from the posterior pituitary gland.

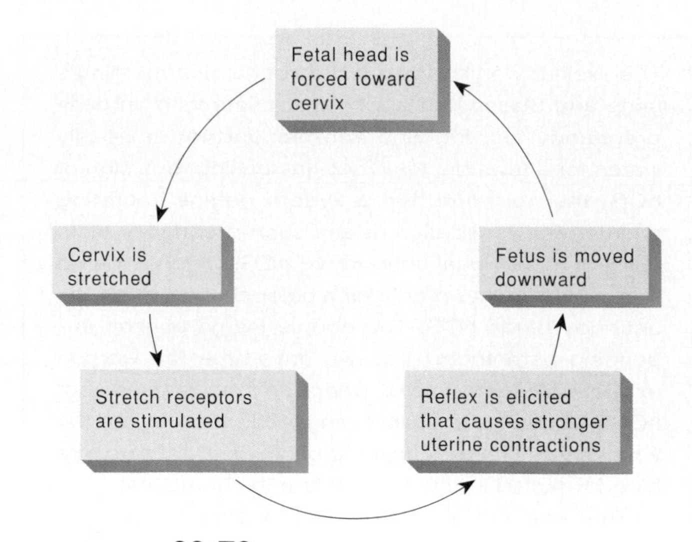


**FIGURE 22.36**
The birth process involves this positive feedback mechanism.

> During childbirth, the tissues between the vulva and anus (perineum) are sometimes torn by the stretching that occurs as the infant passes through the birth canal. For this reason, an incision may be made along the midline of the perineum from the vestibule to within 1.5 centimeters of the anus before the birth is completed. This procedure, called an *episiotomy,* ensures that the perineal tissues are cut cleanly rather than torn.

As labor continues, positive feedback stimulates abdominal wall muscles to contract. These muscles also aid in forcing the fetus through the cervix and vagina to the outside.

Chart 22.6 summarizes some of the factors promoting labor, and figure 22.37 illustrates the steps of the birth process.

Following birth of the fetus, the placenta, which remains inside the uterus, separates from the uterine wall and is expelled by uterine contractions through the birth canal. This expulsion, termed the *afterbirth,* is accompanied by bleeding, because vascular tissues are damaged in the process. However, the loss of blood is usually minimized by continued contraction of the uterus that compresses the bleeding vessels. The action of oxytocin stimulates this contraction.

For several weeks following childbirth, the uterus shrinks by a process called *involution.* Also, its

## CHART 22.6 Factors Contributing to the Labor Process

1. As the time of birth approaches, secretion of progesterone declines, and its inhibiting effect on uterine contractions may lessen.
2. Decreasing progesterone concentration may stimulate synthesis of prostaglandins, which may initiate labor.
3. Stretching uterine tissues stimulates release of oxytocin from the posterior pituitary gland.
4. Oxytocin may stimulate uterine contractions and aid labor in its later stages.
5. As the fetal head stretches the cervix, a positive feedback mechanism results in stronger and stronger uterine contractions and a greater release of oxytocin.
6. Positive feedback stimulates abdominal wall muscles to contract with greater and greater force.
7. The fetus is forced through the birth canal to the outside.

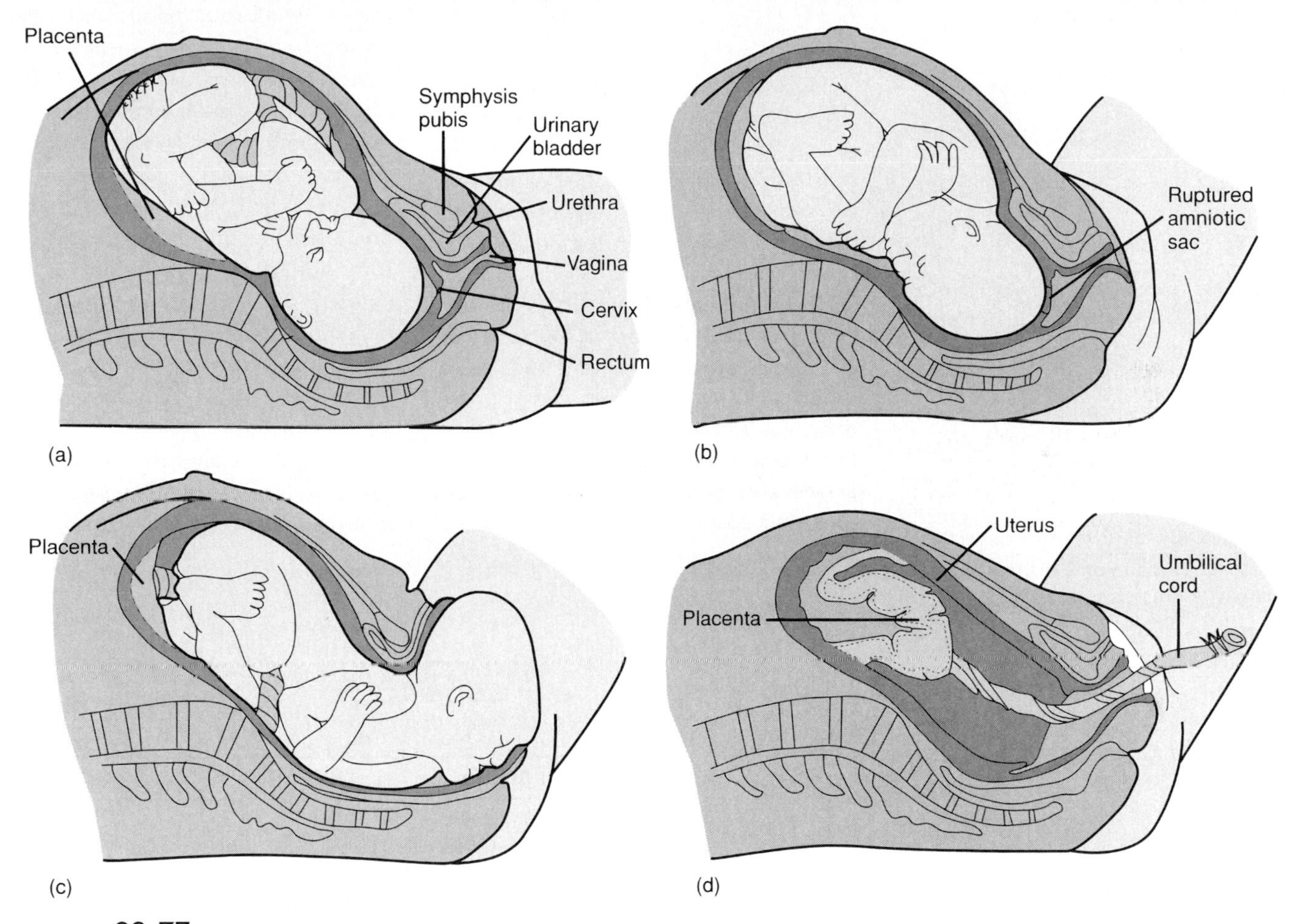


**FIGURE 22.37**
Stages in birth. (*a*) Fetal position before labor; (*b*) dilation of the cervix; (*c*) expulsion of the fetus; (*d*) expulsion of the placenta.

endometrium sloughs off and is discharged through the vagina. The new mother passes a bloody and then yellowish discharge from the vagina for a few weeks. This is followed by the return of an epithelial lining characteristic of a nonpregnant female.

Clinical Application 22.3 addresses some causes of infertility in the female.

1. List some of the physiological changes that occur in a woman's body during pregnancy.
2. Describe the role of progesterone in initiating labor.
3. Explain how dilation of the cervix affects labor.
4. How is bleeding controlled naturally after the placenta is expelled?

## Clinical Application 22.3

# Female Infertility

For one out of six couples in the United States, trying for parenthood is not a time of carefree joy, but one of increasing worry and despair, as pregnancy remains elusive. Physicians define infertility as the inability to conceive after a year of trying. A physical cause of the problem is found in 90% of cases, and 60% of the time, the abnormality lies in the female's reproductive system.

One of the more common causes of female infertility is hyposecretion of gonadotropic hormones from the anterior pituitary gland, followed by failure to ovulate (anovulation). This type of anovulatory cycle can sometimes be detected by testing the female's urine for the presence of *pregnanediol,* a product of progesterone metabolism. Since the concentration of progesterone normally rises following ovulation, no increase in pregnanediol in the urine during the latter part of the menstrual cycle suggests lack of ovulation.

Fertility specialists can treat absence of ovulation due to too little secretion of gonadotropic hormones by administering hCG (obtained from human placentas) or another ovulation-stimulating biochemical, human menopausal gonadotropin (hMG), which contains LH and FSH and is obtained from urine of women who are past menopause. However, either hCG or hMG may overstimulate the ovaries and cause many follicles to release egg cells simultaneously, resulting in multiple births later.

Another cause of female infertility is *endometriosis,* in which tissue resembling the inner lining of the uterus (endometrium) is present abnormally in the abdominal cavity. Small pieces of the endometrium may move up through the uterine tubes during menses and implant in the abdominal cavity. Here it undergoes changes similar to those that take place in the uterine lining during the menstrual cycle. However, when the tissue begins to break down at the end of the cycle, it cannot be expelled to the outside. Instead, its products remain in the abdominal cavity where they may irritate its lining (peritoneum) and cause considerable abdominal pain. These products also tend to stimulate formation of fibrous tissue (fibrosis), which, in turn, may encase the ovary, preventing ovulation or may obstruct the uterine tubes.

Some women become infertile as a result of infections, such as gonorrhea, which may inflame and obstruct the uterine tubes, or may stimulate production of viscous mucus that can plug the cervix and prevent entry of sperm.

The first step in finding the right treatment for a particular patient is to determine the cause of the infertility. Chart 22C describes diagnostic tests that a woman with a fertility problem may undergo.

### Chart 22C Tests to Assess Female Infertility

| Test | What It Checks |
|---|---|
| Hormone levels | If ovulation occurs |
| Ultrasound | Placement and appearance of reproductive organs and structures |
| Postcoital test | Cervix examined soon after unprotected intercourse to see if mucus is thin enough to allow sperm through |
| Endometrial biopsy | Small piece of uterine lining sampled and viewed under microscope to see if it can support an embryo |
| Hysterosalpingogram | Dye injected into fallopian tube and followed with scanner shows if tube is clear or blocked |
| Laparoscopy | Small, lit optical device inserted near navel to detect scar tissue blocking tubes, which could be missed in ultrasound |
| Laparotomy | Scar tissue in tubes removed through incision made for laparoscopy |

(Lewis, *Human Genetics* Table 20.1)

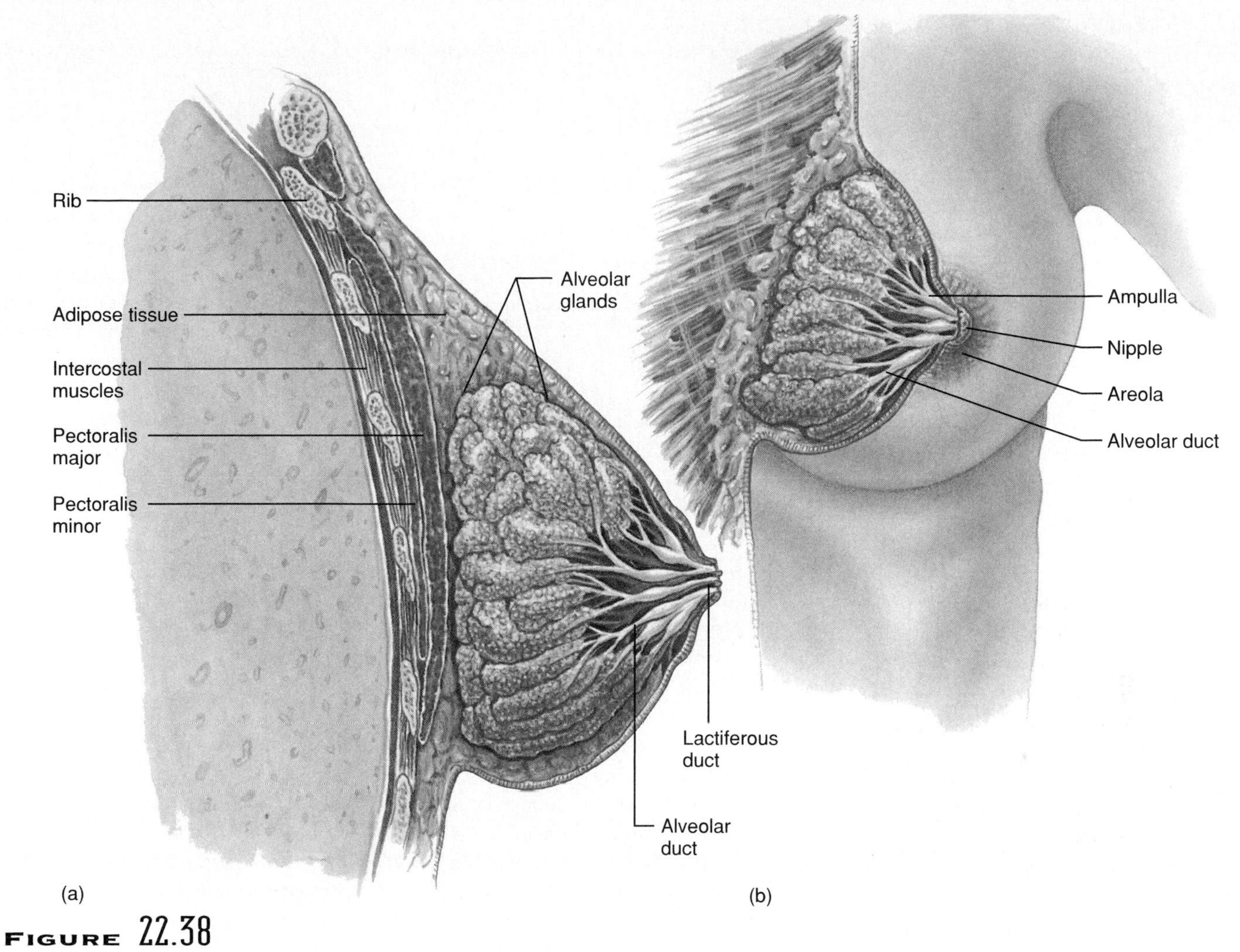


**FIGURE 22.38**
Structure of the breast. (*a*) Sagittal section; (*b*) anterior view.

## MAMMARY GLANDS

The **mammary glands** are accessory organs of the female reproductive system that are specialized to secrete milk following pregnancy.

### Location of the Glands

The mammary glands are located in the subcutaneous tissue of the anterior thorax within the hemispherical elevations called *breasts.* The breasts overlie the *pectoralis major* muscles and extend from the second to the sixth ribs and from the sternum to the axillae.

A *nipple* is located near the tip of each breast at about the level of the fourth intercostal space, and it is surrounded by a circular area of pigmented skin called the *areola* (fig. 22.38).

### Structure of the Glands

A mammary gland is composed of fifteen to twenty irregularly shaped lobes. Each lobe contains glands (alveolar glands) and a duct (lactiferous duct) that leads to the nipple and opens to the outside. The lobes are separated by dense connective and adipose tissues. These tissues also support the glands and attach them to the fascia of the underlying pectoral muscles. Other connective tissue, which forms dense strands called *suspensory ligaments,* extends inward from the dermis of the breast to the fascia, helping support the weight of the breast.

Clinical Application 22.4 discusses a pressing problem concerning breast health—cancer.

CLINICAL APPLICATION 22.4

# Breast Cancer Update

## FINDING THE LUMP

Few discoveries are as terrifying to a woman as finding a lump in her breast. The woman may first notice the irregularity as a small area of thickening, or, she may detect a dimple, a change in contour, or a nipple that is flatter than usual, points in an unusual direction, or produces a discharge.

The next step is having a thorough physical exam. The doctor palpates the breast and performs a mammogram, which is an X-ray scan that pinpoints the location and approximate extent of abnormal tissue (fig. 22C). An ultrasound scan might be done to distinguish between a cyst and a tumor. If an area might be cancerous, the next step is a biopsy, in which a very thin needle samples the affected tissue.

Eighty percent of the time, a breast lump is a sign of fibrocystic breast disease, which is benign. It may be a fluid-filled sac of glandular tissue (a cyst) or a solid, fibrous mass of connective tissue (a fibroadenoma). Treatment includes taking vitamin E or synthetic androgens under a doctor's care, or lowering caffeine intake and examining unusual lumps with mammograms or biopsies, because women with fibrocystic breasts are 1.6 times more likely to develop breast cancer than are other women.

## TREATMENT

If biopsied breast cells are cancerous, treatment is usually surgical. A *lumpectomy* removes a small tumor and some surrounding tissue; a *modified mastectomy* removes an entire breast; and a *radical mastectomy* removes the breast and surrounding lymph and muscle tissue. Follow-up treatment varies depending upon the extent of the tumor. If abnormal tissue extends to nearby lymph nodes, chemotherapy and possibly radiation therapy will follow surgery. Chemotherapy is also indicated if a bone scan or other test reveals that the cancer has already spread. This happens in 7% of newly diagnosed cases of breast cancer.

The types of estrogen and progesterone receptors in the cancer cells determine which drugs are used. Some drugs block estrogen or progesterone receptors, which blocks signals that tell the cells to divide. Tumor cells that lack hormone receptors are associated with a poor prognosis. Women with these tumors receive some type of chemotherapy even if the cancer has not spread to adjacent lymph nodes.

## CAUSES OF BREAST CANCER

A woman who has a close relative who has had breast cancer is at a higher risk of developing the condition than a woman without a family history of breast cancer. Experimental genetic tests can tell whether a woman with a strong family history will develop the inherited form of the illness. This was the case for a young woman whose mother and sister had died of breast cancer. Another sister had recently been diagnosed. Just before she was to undergo breast removal to avoid her feared fate, a genetic test showed that she had not inherited the responsible gene. The woman's cousin, who thought she could not inherit breast cancer because her father was related to the affected family, found that she had indeed inherited the gene. A mammogram revealed a tiny tumor, and surgery saved her life.

Ten percent of all breast cancers are inherited, with the two major responsible genes—called BRCA 1 and BRCA 2—discovered in 1993 and 1994. Researchers are still searching for the environmental triggers of the remaining 90% of cases. One candidate is prolonged exposure to estrogen. The estrogen link was first recognized in the 1970s, when researchers realized that young women who had had their ovaries removed very rarely developed breast cancer. Did their lack of estrogen protect against breast cancer? It appears so.

Prolonged exposure to estrogen can occur in a variety of ways:

- Early menarche (first period) and late menopause (last period).
- Pesticide residues and other pollutants. These contaminants stimulate cell division like estrogens do. Wildlife populations exposed to environmental estrogens demonstrate reproductive problems, such as infertility, undersized genitalia, and thin eggshells.
- Having no children, or having a first child after the age of 30.
- Not breast-feeding. Women who do not breast-feed have a higher breast cancer risk than women who do.

## BREAST CANCER STATISTICS

Nearly 3 million women in the United States currently have breast cancer. People frequently misunderstand the oft-quoted figure that a woman's lifetime risk of developing breast cancer is 1 in 8, believing that at any given time 1 in 8 U.S. women have breast cancer. This figure, however, refers to the lifetime risk for women who live to be 95 (chart 22D).

Statistics indicate that breast cancer is on the rise. The lifetime risk of 1 in 8 was only 1 in 16 in the 1940s. It has increased by 1% a year since then, and by 4% a year since 1987. At least some of that rise is due to more widespread and earlier diagnosis, a result of public health programs to educate women about breast self-exam and mammography. Thanks to these efforts, most women will recover from breast cancer. However, the illness is the second most common cancer in women after lung cancer and causes 46,000 deaths in the United States each year.

### Prevention

Health agencies advise mammograms once every two years for women aged 40 to 49, and yearly tests after that. A single "baseline" mammogram should be taken between age 35 and 40, as a basis of comparison. For a woman with a family history of breast cancer, this timetable is moved up. A mammogram can spot a tumor two years before a woman can feel it.

Thanks to the U.S. government's Women's Health Initiative, begun in 1993, we may soon learn enough about breast cancer to more effectively prevent and treat it. Some 70,000 women are participating in one or both of two investigations: The first study will test the ability of a diet low in fat and high in fruits and vegetables to prevent breast cancer. In the second study, healthy women with family histories of breast cancer are taking daily a drug called tamoxifen. This drug plugs up estrogen receptors in breast cells, preventing cell division.

Another candidate for breast cancer prevention is RU486, the drug used as an abortion pill because it blocks binding of progesterone to its cellular receptors. This action may also prevent breast cancer. Tamoxifen and RU486 are currently being tested as breast cancer treatments too. Tamoxifen, for example, decreases risk of cancer recurrence by 30%.

In addition, at least two dozen substances are being tested for their ability to direct the immune system to fight breast cancer cells. And the Women's Health Initiative is attempting to sort through the various purported environmental causes of breast cancer, including alcohol, pesticides, birth control pills, and electromagnetic fields.

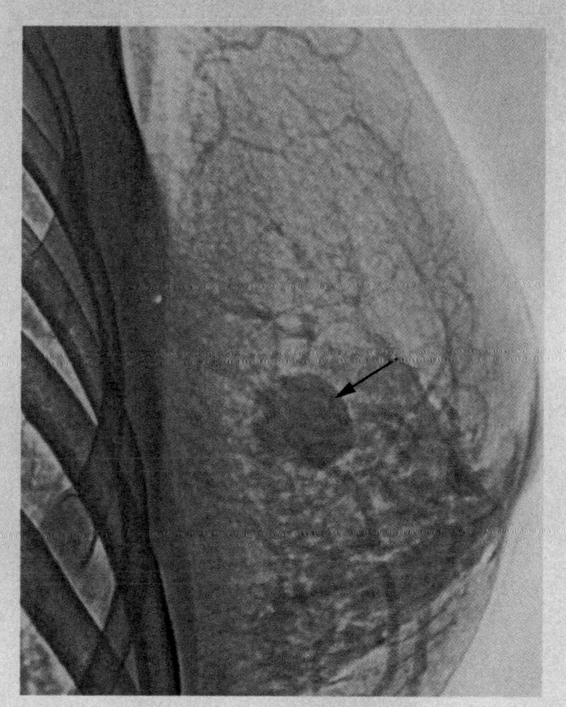

**Figure 22C**
Mammogram (X-ray film) of a breast with a tumor (arrow).

**Chart 22D Breast Cancer Risk**

| By Age | Odds | By Age | Odds |
|---|---|---|---|
| 25 | 1 in 19,608 | 60 | 1 in 24 |
| 30 | 1 in 2,525 | 65 | 1 in 17 |
| 35 | 1 in 622 | 70 | 1 in 14 |
| 40 | 1 in 217 | 75 | 1 in 11 |
| 45 | 1 in 93 | 80 | 1 in 10 |
| 50 | 1 in 50 | 85 | 1 in 9 |
| 55 | 1 in 33 | 95 or older | 1 in 8 |

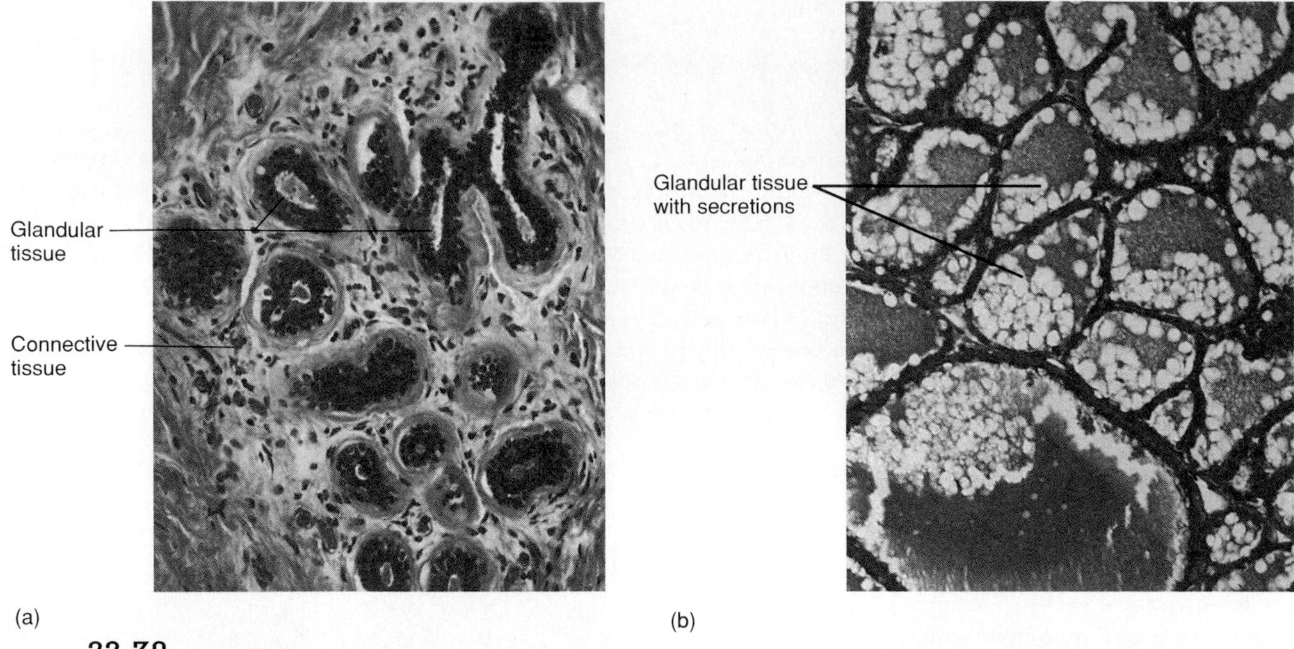


**FIGURE 22.39**
(*a*) Light micrograph of a mammary gland in a nonpregnant woman (63×); (*b*) light micrograph of an active (lactating) mammary gland (63×).

## Development of the Breasts

The mammary glands of boys and girls are similar. As children reach *puberty,* the male glands fail to develop, while ovarian hormones stimulate the female glands to develop. As a result, the alveolar glands and ducts enlarge, and fat is deposited, so that the breasts become surrounded by adipose tissue, except for the region of the areola.

During pregnancy, placental estrogen and progesterone stimulate further development of the mammary glands. Estrogen causes the ductile systems to grow and branch, and deposits large quantities of fat around them. Progesterone stimulates the development of the alveolar glands at the ends of the ducts. Placental lactogen also promotes these changes.

Because of hormonal activity, the breasts may double in size during pregnancy. At the same time, glandular tissue replaces the adipose tissue of the breasts. Beginning about the fifth week of pregnancy, the anterior pituitary gland releases increasing amounts of *prolactin.* Prolactin is synthesized from early pregnancy throughout gestation, peaking at the time of birth. However, milk secretion does not begin until after birth. This is because during pregnancy, placental progesterone inhibits milk production, and placental lactogen blocks the action of prolactin. Consequently, even though the mammary glands can secrete milk, none is produced. The micrographs in figure 22.39 compare the mammary gland tissues of a nonpregnant woman with those of a lactating woman.

## Milk Production and Secretion

Following childbirth and the expulsion of the placenta, the maternal blood concentrations of placental hormones decline rapidly, and the action of prolactin is no longer inhibited.

Prolactin stimulates the mammary glands to secrete large quantities of milk. This hormonal effect does not occur until two or three days following birth, and in the meantime, the glands secrete a thin, watery fluid called *colostrum.* Although colostrum is relatively rich in proteins, particularly protective antibodies, its concentrations of carbohydrates and fats are lower than those of milk.

Milk does not flow readily through the ductile system of the mammary gland, but must be actively ejected by contraction of specialized *myoepithelial cells* surrounding the alveolar glands. A reflex action controls this process and is elicited when the breast is suckled or the nipple or areola is otherwise mechanically stimulated (fig. 22.40). Then, impulses from sensory receptors within the breasts travel to the hypothalamus, which signals the posterior pituitary gland to release oxytocin. The oxytocin reaches the breasts by means of the blood and stimulates the myoepithelial cells to contract (in both breasts). As a result, milk is ejected into a suckling infant's mouth in about thirty seconds (fig. 22.41).

Sensory impulses triggered by mechanical stimulation of the nipples also signal the hypothalamus to continue secreting prolactin. Thus, prolactin is released as long as milk is removed from the breasts.

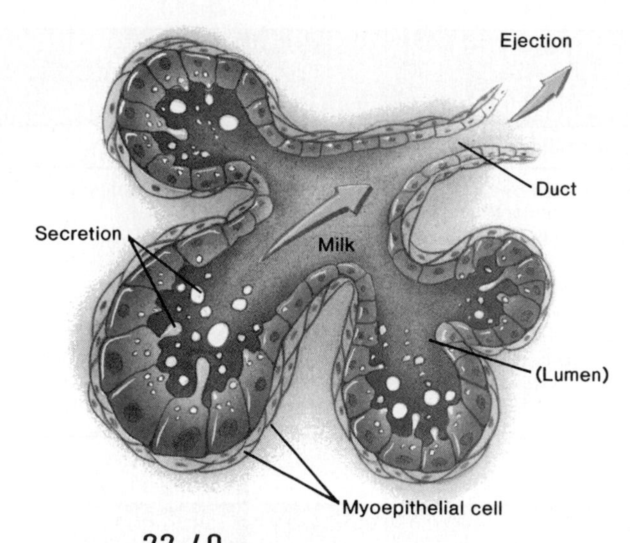


**Figure 22.40**
Myoepithelial cells eject milk from an alveolar gland.

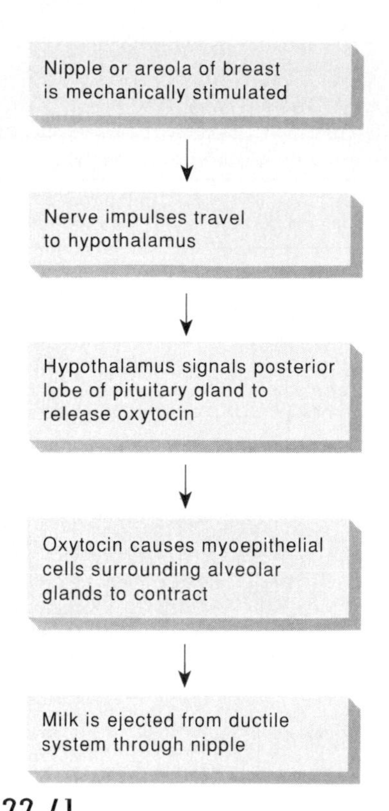


**Figure 22.41**
Mechanism that ejects milk from the breasts.

However, if milk is not removed regularly, the hypothalamus inhibits the secretion of prolactin, and within about one week, the mammary glands lose their capacity to produce milk.

> A woman who is breast-feeding feels her milk "let down," or flood her breasts, when her infant suckles. If the baby nurses on a very regular schedule, she may feel the letdown shortly before the baby is due to nurse. The connection between mind and hormonal control of lactation is so strong that if a nursing mother simply hears a baby cry, her milk may flow. If this occurs in public, she can keep from wetting her shirt by pressing her arms strongly against her chest.

To wean a nursing child, it is best to stop breastfeeding gradually, by eliminating one feeding per day each week, for example. If a woman stops nursing abruptly, her breasts will become painfully engorged for several days.

A woman who is breast-feeding usually does not ovulate for several months. This may be because prolactin suppresses release of gonadotropins from the anterior pituitary gland. Then, FSH is released and the menstrual cycle is activated. However, if a new mother does not wish to repeat her recent childbirth experience soon, she or her partner should begin practicing contraception, because she will be fertile during the two weeks prior to the return of her menstrual period.

Chart 22.7 summarizes the hormonal control of milk production, and chart 22.8 lists some agents that adversely affect lactation or harm the child. Clinical Application 22.5 explains the benefits of breast-feeding.

1. Describe the structure of a mammary gland.
2. How does pregnancy affect the mammary glands?
3. What stimulates the mammary glands to produce milk?
4. What causes milk to flow into the ductile system of a mammary gland?
5. What happens to milk production if the milk is not regularly removed from the breast?

## Birth Control

*Birth control* is the voluntary regulation of the number of offspring produced and the time they will be conceived. This control requires a method of **contraception** designed to avoid fertilization of an egg cell following sexual intercourse or to prevent implantation of a blastocyst. Chart 22.9 describes several contraceptive approaches and devices and indicates their effectiveness.

### Coitus Interruptus

*Coitus interruptus* is withdrawing the penis from the vagina before ejaculation, preventing entry of sperm cells into the female reproductive tract. This method of contraception often proves unsatisfactory and may

## CHART 22.7 Hormonal Control of the Mammary Glands

| Before Pregnancy (beginning of puberty) | Following Childbirth |
|---|---|
| Ovarian hormones secreted during menstrual cycles stimulate alveolar glands and ducts of mammary glands to develop. | 1. Placental hormonal concentrations decline, so that the action of prolactin is no longer inhibited.<br>2. The breasts begin producing milk.<br>3. Mechanical stimulation of the breasts releases oxytocin from the posterior pituitary gland.<br>4. Oxytocin stimulates ejection of milk from ducts.<br>5. As long as milk is removed, more prolactin is released; if milk is not removed, milk production ceases. |
| **During Pregnancy**<br>1. Estrogen causes the ductile system to grow and branch.<br>2. Progesterone stimulates development of alveolar glands.<br>3. Placental lactogen promotes development of the breasts.<br>4. Prolactin is secreted throughout pregnancy, but milk production is inhibited by placental progesterone. | |

## CHART 22.8 Agents Contraindicated During Breast-Feeding

| Agent | Use | Effect on Lactation or Baby |
|---|---|---|
| Doxorubicin, methotrexate | Cancer chemotherapy | Immune suppression |
| Cyclosporine | Immune suppression in transplant patients | Immune suppression |
| Radioactive isotopes | Cancer diagnosis and therapy | Radioactivity in milk |
| Phenobarbitol | Anticonvulsant | Sedation, spasms on weaning |
| Oral contraceptives | Birth control | Decreased milk production |
| Caffeine (large amounts) | Additive | Irritability, poor sleeping |
| Cocaine | Drug of abuse | Infant becomes intoxicated |
| Ethanol (alcohol) (large amounts) | Drug of abuse | Weak, drowsy; infant decreases in length but gains weight; decreased milk ejection reflex |
| Heroin | Drug of abuse | Tremors, restlessness, vomiting, poor feeding |
| Nicotine | Drug of abuse | Diarrhea, shock, increased heart rate; lowered milk production |
| Phencyclidine | Drug of abuse | Hallucinations |

result in pregnancy, since some males find it difficult to withdraw just prior to ejaculation. Also, small quantities of semen containing sperm cells may reach the vagina before ejaculation occurs.

### Rhythm Method

The *rhythm method* (also called timed coitus or natural family planning) requires abstinence from sexual intercourse a few days before and a few days after ovulation. The rhythm method results in a relatively high rate of pregnancy because pinpointing "safe" times to have intercourse is difficult. Another disadvantage of the rhythm method is that it requires adherence to a particular pattern of behavior and restricts spontaneity in sexual activity.

**1** Why is coitus interruptus unreliable?

**2** Describe the idea behind the rhythm method of contraception.

**3** What factors make the rhythm method less reliable than some other methods of contraception?

The effectiveness of the rhythm method can sometimes be increased by measuring and recording the woman's body temperature when she awakes each morning for several months. Body temperature typically rises about 0.6° Fahrenheit immediately following ovulation. However, this technique does not work for all women. More helpful may be an "ovulation predictor kit" that detects the surge in LH preceding ovulation.

## Clinical Application 22.5

### Human Milk—The Perfect Food for Human Babies

The female human body manufactures milk that is a perfect food for a human newborn in several ways. Human milk is rich in the lipid needed for rapid brain growth, and it is low in protein. Cow's milk is the reverse, with three times as much protein as human milk. Much of this protein is casein, which is fine to spur a calf's rapid muscle growth but forms hard-to-digest curds in a human baby's stomach. The protein in human milk has a better balance of essential amino acids than does the protein in cow's milk.

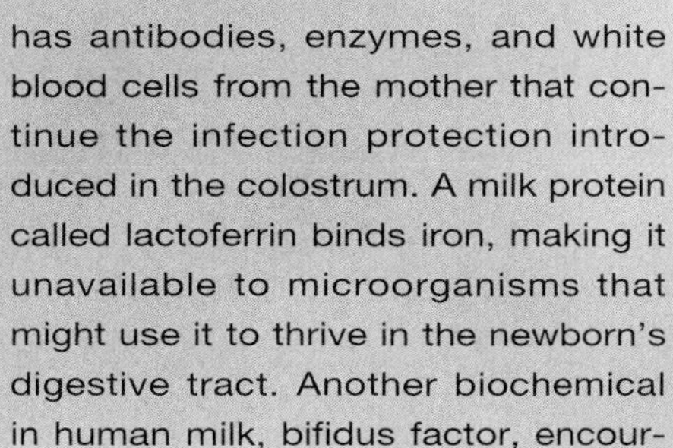

Human milk protects a newborn from many infections. For the first few days after giving birth, a new mother's breasts produce colostrum, which has less sugar and fat than mature milk but more protein, and is rich in antibodies. The antibodies protect the baby from such infections as Salmonella poisoning and polio. When the milk matures by a week to 10 days, it has antibodies, enzymes, and white blood cells from the mother that continue the infection protection introduced in the colostrum. A milk protein called lactoferrin binds iron, making it unavailable to microorganisms that might use it to thrive in the newborn's digestive tract. Another biochemical in human milk, bifidus factor, encourages the growth of the bacteria *Lactobacillus bifidus,* which manufactures acids in the baby's digestive system that kill harmful bacteria.

A breast-fed baby typically nurses until he or she is full, not until a certain number of ounces have been drunk, which may explain why breast-fed babies are less likely to be obese than bottle-fed infants. Babies nurtured on human milk are also less likely to develop allergies to cow's milk.

But breast-feeding is not the choice for all women. It may be impossible to be present for each feeding or to provide milk. Also, many drugs taken by the mother enter breast milk and can affect the baby. A nursing mother must eat about 500 calories per day more than usual to meet the energy requirements of milk production—but she also loses weight faster than a mother who bottle-feeds, because the fat reserves set aside during pregnancy are used to manufacture the milk. Another disadvantage of breast-feeding is that the father cannot do it.

An alternative to breast-feeding is infant formula, which is usually cow's milk plus fats, proteins, carbohydrates, vitamins, and minerals added to make it as much like breast milk as possible. Although infant formula is nutritionally sound, the foul-smelling and bulkier bowel movements of the bottle-fed baby compared to the odorless, loose, and less abundant feces of a breast-fed baby indicate that breast milk is a more digestible first food than infant formula.

### Mechanical Barriers

*Mechanical barriers* prevent sperm cells from entering the female reproductive tract during sexual intercourse. One such device used by males is a *condom.* It consists of a thin latex or natural membrane sheath that is placed over the erect penis before intercourse to prevent semen from entering the vagina upon ejaculation (fig. 22.42*a*).

A recently introduced *female condom* resembles a small plastic bag. A woman inserts it into her vagina prior to intercourse. The device blocks sperm from reaching the cervix.

A condom is relatively inexpensive, and it may also help protect the user against contracting sexually transmitted diseases. However, men often feel that a condom decreases the sensitivity of the penis during intercourse. Also, its use interrupts the sex act.

Another mechanical barrier is the *diaphragm.* It is a cup-shaped structure with a flexible ring forming the rim. The diaphragm is inserted into the vagina so that it covers the cervix, preventing entry of sperm cells into the uterus (fig. 22.42*b*).

To be effective, a diaphragm must be fitted for size by a physician, be inserted properly, and be used

## CHART 22.9 BIRTH CONTROL METHODS

| | Method | Mechanism | Advantages | Disadvantages | Pregnancies per year per 100 women* |
|---|---|---|---|---|---|
| | None | | | | 85 |
| Barrier and Spermicidal | Condom | Worn over penis, keeps sperm out of vagina | Protection against sexually transmitted diseases | Disrupts spontaneity, can break, reduces sensation in male | 2–12 |
| Barrier and Spermicidal | Condom and spermicide | Worn over penis, keeps sperm out of vagina, and kills sperm that escape | Protection against sexually transmitted diseases | Disrupts spontaneity, reduces sensation in male | 2–5 |
| Barrier and Spermicidal | Diaphragm and spermicide | Kills sperm and blocks uterus | Inexpensive | Disrupts spontaneity, messy, needs to be fitted by doctor | 6–18 |
| Barrier and Spermicidal | Cervical cap and spermicide | Kills sperm and blocks uterus | Inexpensive, can be left in 24 hours | May slip out of place, messy, needs to be fitted by doctor | 6–18 |
| Barrier and Spermicidal | Spermicidal foam or jelly | Kills sperm and blocks vagina | Inexpensive | Messy | 3–21 |
| Barrier and Spermicidal | Spermicidal suppository | Kills sperm and blocks vagina | Easy to use and carry | Irritates 25% of users, male and female | 3–15 |
| Barrier and Spermicidal | Contraceptive sponge | Kills, blocks, and absorbs sperm in vagina | Can be left in for 24 hours | Expensive | 6–28 |
| Hormonal | Combination birth control pill | Prevents ovulation and implantation, thickens cervical mucus | Does not interrupt spontaneity, lowers risk of some cancers, decreases menstrual flow | Raises risk of cardiovascular disease in some women, causes weight gain and breast tenderness | 3 |
| Hormonal | Minipill | Blocks implantation, deactivates sperm, thickens cervical mucus | Fewer side effects | Weight gain | 1–10 |
| Hormonal | Medroxy-progesterone acetate (Depo-Provera) | Prevents ovulation, alters uterine lining | Easy to use | Menstrual changes, weight gain | 0.3 |
| Hormonal | Progesterone implant | Prevents ovulation, thickens cervical mucus | Easy to use | Menstrual changes | 0.3 |
| Behavioral | Rhythm method | No intercourse during fertile times | No cost | Difficult to do, hard to predict timing | 20 |
| Behavioral | Withdrawal (coitus interruptus) | Removal of penis from vagina before ejaculation | No cost | Difficult to do | 4–18 |
| Surgical | Vasectomy | Sperm cells never reach penis | Permanent, does not interrupt spontaneity | Requires minor surgery | 0.15 |
| Surgical | Tubal ligation | Egg cells never reach uterus | Permanent, does not interrupt spontaneity | Requires surgery, entails some risk of infection | 0.4 |
| Other | Intrauterine device | Prevents implantation | Does not interrupt spontaneity | Severe menstrual cramps, increases risk of infection | 3 |

*The lower figures apply when the contraceptive device or technique is used correctly. The higher figures take into account human error.

in conjunction with a chemical spermicide that is applied to the surface adjacent to the cervix and to the rim of the diaphragm. The device must be left in position for several hours following sexual intercourse. A diaphragm can be inserted into the vagina up to 6 hours before sexual contact is to occur.

Similar to but smaller than the diaphragm is the *cervical cap,* which adheres to the cervix by suction. A woman inserts it with her fingers before intercourse. Cervical caps have been used for centuries in different cultures, and have been made of such varied substances as beeswax, lemon halves, paper, and opium poppy fibers.

A *contraceptive sponge* is another contraceptive device combining physical and chemical barriers. A woman moistens the sponge just prior to insertion to activate spermicide in it. A fabric loop attached to the sponge enables the woman to remove it, at least 6

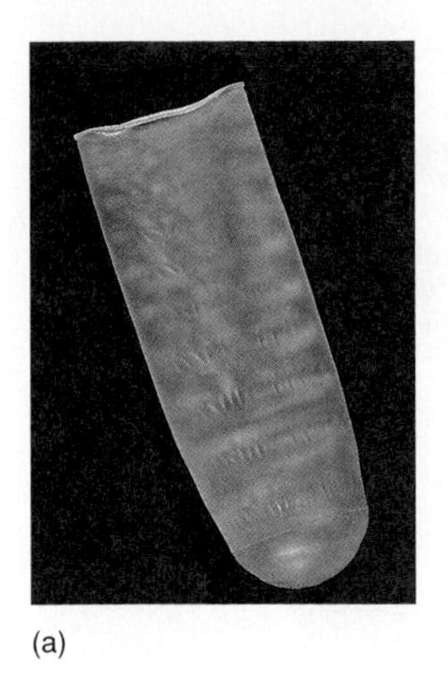

(a)

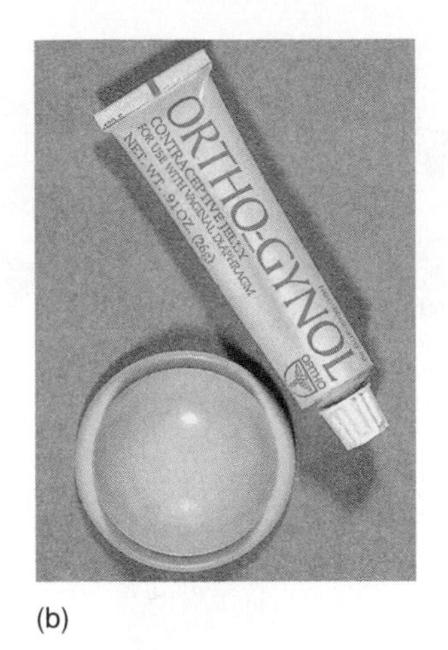


(b)

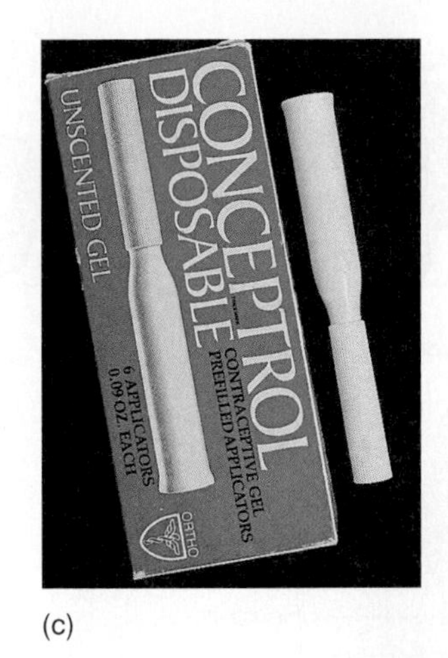


(c)

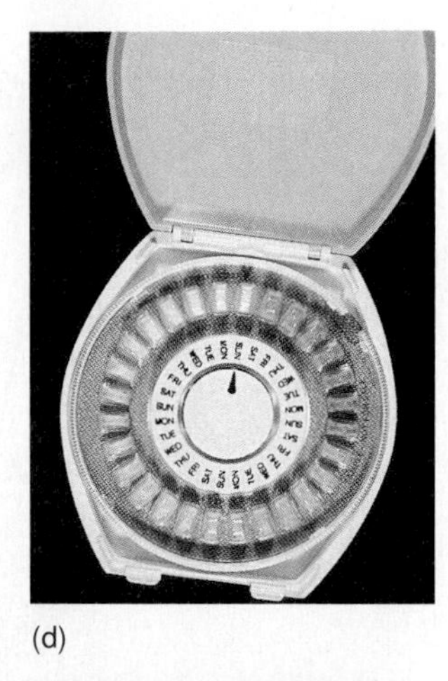

(d)

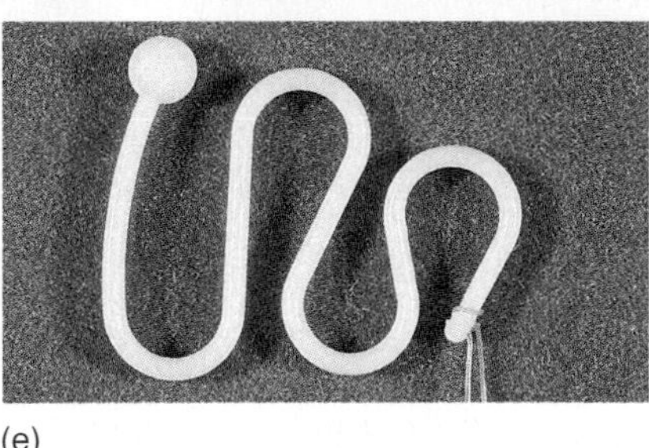

(e)

**FIGURE 22.42**
Devices and substances used for birth control. (*a*) Condom; (*b*) diaphragm; (*c*) spermicidal gel; (*d*) oral contraceptive; (*e*) IUD.

hours after intercourse. Drugstores sell contraceptive sponges without a doctor's prescription. It too has a long history. Centuries ago, Europeans used sponges soaked in lemon juice, which killed sperm.

### Chemical Barriers

*Chemical barrier* contraceptives include creams, foams, and jellies with spermicidal properties. Within the vagina, such chemicals create an environment that is unfavorable for sperm cells (fig. 22.42*c*).

Chemical barriers are fairly easy to use, but have a relatively high failure rate when used alone. They are most effective when used with a diaphragm.

### Oral Contraceptives

An *oral contraceptive,* or birth control pill, contains synthetic estrogen-like and progesterone-like substances. When taken daily by a woman, these drugs disrupt the normal pattern of gonadotropin secretion and prevent the surge in LH release that triggers ovulation. They also interfere with buildup of the uterine lining that is necessary for implantation (fig. 22.42*d*).

Oral contraceptives, if used correctly, prevent pregnancy nearly 100% of the time. However, they may cause nausea, retention of body fluids, increased pigmentation of the skin, and breast tenderness. Also, some women, particularly those over 35 years of age who smoke, may develop intravascular blood clots, liver disorders, or high blood pressure when using certain types of oral contraceptives.

### Injectable Contraception

An intramuscular injection of Depo-Provera (medroxyprogesterone acetate) protects against pregnancy for three months by preventing maturation and release of a secondary oocyte. It also alters the uterine lining, making it less hospitable for a blastocyst. Because Depo-Provera is long-acting, it takes 10 to 18 months after the last injection for the effects to wear off.

Use of Depo-Provera requires a doctor's care, because it has side effects and risks. The most common side effect is weight gain. Women with a history of breast cancer, depression, kidney disease, high blood pressure, migraine headaches, asthma, epilepsy, or diabetes, or strong family histories of these conditions, should probably not use this form of birth control.

### Contraceptive Implants

A *contraceptive implant* is a set of small progesterone-containing capsules or rods, which are inserted surgically under the skin of a woman's arm or scapular region. The progesterone, which is released slowly from the implant, prevents ovulation in much the same way as do oral contraceptives. A contraceptive

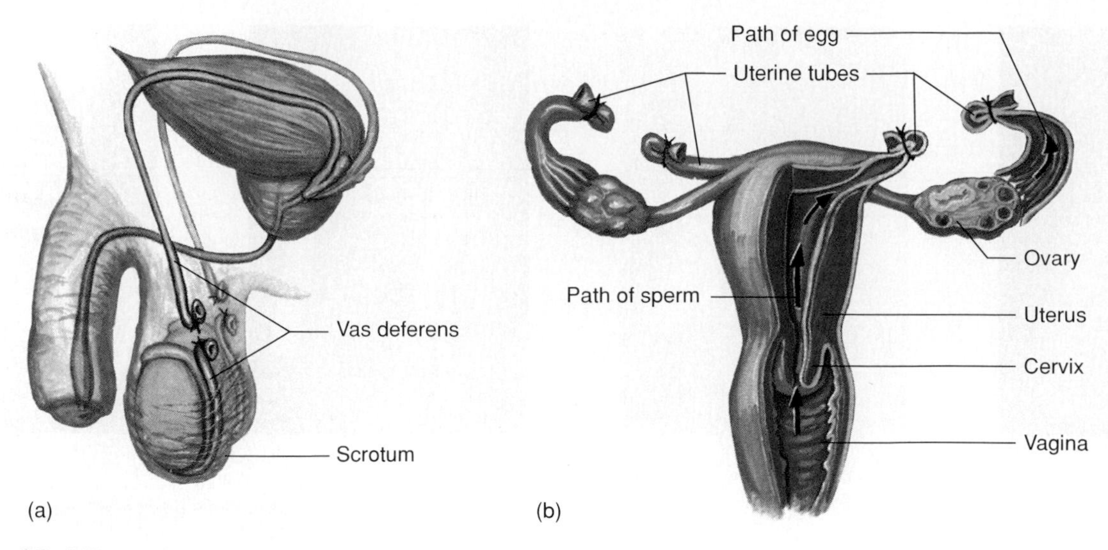


**FIGURE 22.43**
(*a*) Vasectomy removes a portion of each vas deferens. (*b*) Tubal ligation removes a portion of each uterine tube.

implant is effective for a period of up to 5 years, and its contraceptive action can be reversed by removing the device.

> A large dose of high-potency estrogens can prevent implantation of a blastocyst in the uterus. Such a "morning-after pill," taken shortly after unprotected intercourse, promotes powerful contractions of smooth muscle in a woman's reproductive tract. This may dislodge and expel a fertilized egg or blastocyst. However, if a blastocyst has already implanted, this treatment may injure it.

1. Describe two methods of contraception that use mechanical barriers.
2. How can the effectiveness of chemical contraceptives be increased?
3. What substances are contained in oral contraceptives?
4. How is pregnancy prevented by an oral contraceptive? An injectable contraceptive? A contraceptive implant?

## Intrauterine Devices

An *intrauterine device,* or *IUD,* is a small solid object that a physician places within the uterine cavity. An IUD interferes with implantation, perhaps by inflaming the uterine tissues (fig. 22.42*e*).

Unfortunately, an IUD may be expelled from the uterus spontaneously or produce abdominal pain or excessive menstrual bleeding. It may also injure the uterus or produce other serious health problems, and should be checked at regular intervals by a physician. A few babies have been born with IUDs attached to them.

## Surgical Methods

*Surgical methods* of contraception sterilize the male or female.

In the male, a physician removes a small section of each vas deferens near the epididymis, and ties the cut ends of the ducts. This is a *vasectomy,* and it is a relatively simple operation that produces few side effects, although it may cause some pain for a week or two.

After a vasectomy, sperm cells cannot leave the epididymis, thus they are not included in the semen. However, sperm cells may already be present in portions of the ducts distal to the cuts. Consequently, the sperm count may not reach zero for several weeks.

The corresponding procedure in the female is called *tubal ligation.* The uterine tubes are cut and tied so that sperm cells cannot reach an egg cell.

Neither a vasectomy nor a tubal ligation produces any changes in hormonal concentrations or sexual drives. These sterilization procedures, shown in figure 22.43, provide the most reliable forms of contraception. Reversing them requires microsurgery.

1. How does an IUD prevent pregnancy?
2. Describe the surgical methods of contraception for a male and for a female.

## CHART 22.10 Some Sexually Transmitted Diseases

| Disease | Cause | Symptoms | Number of Reported Cases (U.S.) | Effects on Fetus | Treatment | Complications |
|---|---|---|---|---|---|---|
| Acquired immune deficiency syndrome | Human immuno-deficiency virus | Fever, weakness, infections, cancer | > 14 million (infected) | Exposure to AIDS virus and other infections | Drugs to treat or delay symptoms; no cure | Dementia |
| Chlamydia infection | Bacteria of genus *Chlamydia* | Painful urination and intercourse, mucus discharge from penis or vagina | 3–10 million | Prematurity, blindness, pneumonia | Antibiotics | Pelvic inflammatory disease, infertility, arthritis, ectopic pregnancy |
| Genital herpes | Herpes simplex virus type II | Genital sores, fever | 20 million | Brain damage, stillbirth | Antiviral drug (acyclovir) | Increased risk of cervical cancer |
| Genital warts | Human papilloma virus | Warts on genitals | 1 million | None known | Chemical or surgical removal | Increased risk of cervical cancer |
| Gonorrhea | *Neisseria gonorrhoeae* bacteria | In women, usually none; in men, painful urination | 2 million | Blindness, stillbirth | Antibiotics | Arthritis, rash, infertility, pelvic inflammatory disease |
| Syphilis | *Treponema pallidum* bacteria | Initial chancre sore usually on genitals or mouth; rash 6 months later; several years with no symptoms as infection spreads; finally damage to heart, liver, nerves, brain | 90,000 | Miscarriage, prematurity, birth defects, stillbirth | Antibiotics | Dementia |

## Sexually Transmitted Diseases

The twenty recognized **sexually transmitted diseases** (STDs) are often called "silent infections" because the early stages may not produce symptoms, especially in women (chart 22.10). By the time symptoms appear, it is often too late to prevent complications or the spread of the infection to sexual partners. Because many STDs have similar symptoms, and some of the symptoms are also seen in diseases or allergies that are not sexually related, it is wise to consult a physician if one or a combination of these symptoms appears:

1. Burning sensation during urination
2. Pain in the lower abdomen
3. Fever or swollen glands in the neck
4. Discharge from the vagina or penis
5. Pain, itch, or inflammation in the genital or anal area
6. Pain during intercourse
7. Sores, blisters, bumps, or a rash anywhere on the body, particularly the mouth or genitals
8. Itchy, runny eyes

One possible complication of the STDs gonorrhea and chlamydia is **pelvic inflammatory disease,** in which bacteria enter the vagina and spread throughout the reproductive organs. The disease begins with intermittent cramps, followed by sudden fever, chills, weakness, and severe cramps. Hospitalization and intravenous antibiotics can stop the infection. The uterus and uterine tubes are often scarred, resulting in infertility and increased risk of ectopic pregnancy.

The most frightening sexually transmitted disease is *acquired immune deficiency syndrome* (AIDS), which is a steady deterioration of the body's immune defenses caused by a virus. The body becomes overrun by infection and often cancer, diseases that are usually conquered by the immune system. The AIDS virus (human immunodeficiency virus, or HIV) is passed from one person to another in body fluids such as semen, blood, and milk. It is most frequently transmitted during unprotected intercourse or by using a needle containing contaminated blood.

1. Why are sexually transmitted diseases often called "silent infections"?
2. Why are sexually transmitted diseases sometimes difficult to diagnose?
3. What are some common symptoms of sexually transmitted diseases?

## Clinical Terms Related to the Reproductive Systems

**abortion** (ah-bor′shun) The spontaneous or deliberate termination of pregnancy; a spontaneous abortion is commonly termed a miscarriage.

**amenorrhea** (a-men″o-re′ah) The absence of menstrual flow, usually due to a disturbance in hormonal concentrations.

**cesarean section** (sĕ-sa′re-an sek′shun) The delivery of a fetus through an abdominal incision.

**conization** (ko″nĭ-za′shun) The surgical removal of a cone of tissue from the cervix for examination.

**curettage** (ku″rĕ-tahzh′) A surgical procedure in which the cervix is dilated and the endometrium of the uterus is scraped (commonly called D and C for dilation and curettage).

**dysmenorrhea** (dis″men-o-re′ah) Painful menstruation.

**eclampsia** (ē-klamp′se-ah) A condition characterized by convulsions and coma that sometimes accompanies toxemia of pregnancy.

**endometritis** (en″do-mē-tri′tis) An inflammation of the uterine lining.

**epididymitis** (ep″ĭ-did″i-mi′tis) An inflammation of the epididymis.

**gestation** (jes-ta′shun) The entire period of pregnancy.

**hematometra** (hem″ah-to-me′trah) An accumulation of menstrual blood within the uterine cavity.

**hydrocele** (hi′dro-seal) An enlarged scrotum caused by accumulation of fluid along the spermatic cord.

**hyperemesis gravidarum** (hi″per-em′ĕ-sis grav′i-dar-um) Vomiting associated with pregnancy; morning sickness.

**hypospadias** (hi″po-spay′dee-us) A male developmental anomaly in which the urethra opens on the underside of the penis.

**hysterectomy** (his″te-rek′to-me) The surgical removal of the uterus.

**mastitis** (mas″ti′tis) An inflammation of a mammary gland.

**oophorectomy** (o″of-o-rek′to-me) The surgical removal of an ovary.

**oophoritis** (o″of-o-ri′tis) An inflammation of an ovary.

**orchiectomy** (or″ke-ek′to-me) The surgical removal of a testis.

**orchitis** (or-ki′tis) An inflammation of a testis.

**prostatectomy** (pros″tah-tek′to-me) The surgical removal of a portion or all of the prostate gland.

**prostatitis** (pros″tah-ti′tis) An inflammation of the prostate gland.

**salpingectomy** (sal″pin-jek′to-me) The surgical removal of a uterine tube.

**toxemia of pregnancy** (tok-se′me-ah) A group of metabolic disturbances that may occur during pregnancy.

**vaginitis** (vaj″ĭ-ni′tis) An inflammation of the vaginal lining.

**varicocele** (var′ĭ-ko-sēl″) Distension of the veins within the spermatic cord.

# InnerConnections
## Reproductive Systems

Production of gametes, fertilization, development of the fetus, and childbirth are essential for survival of the species

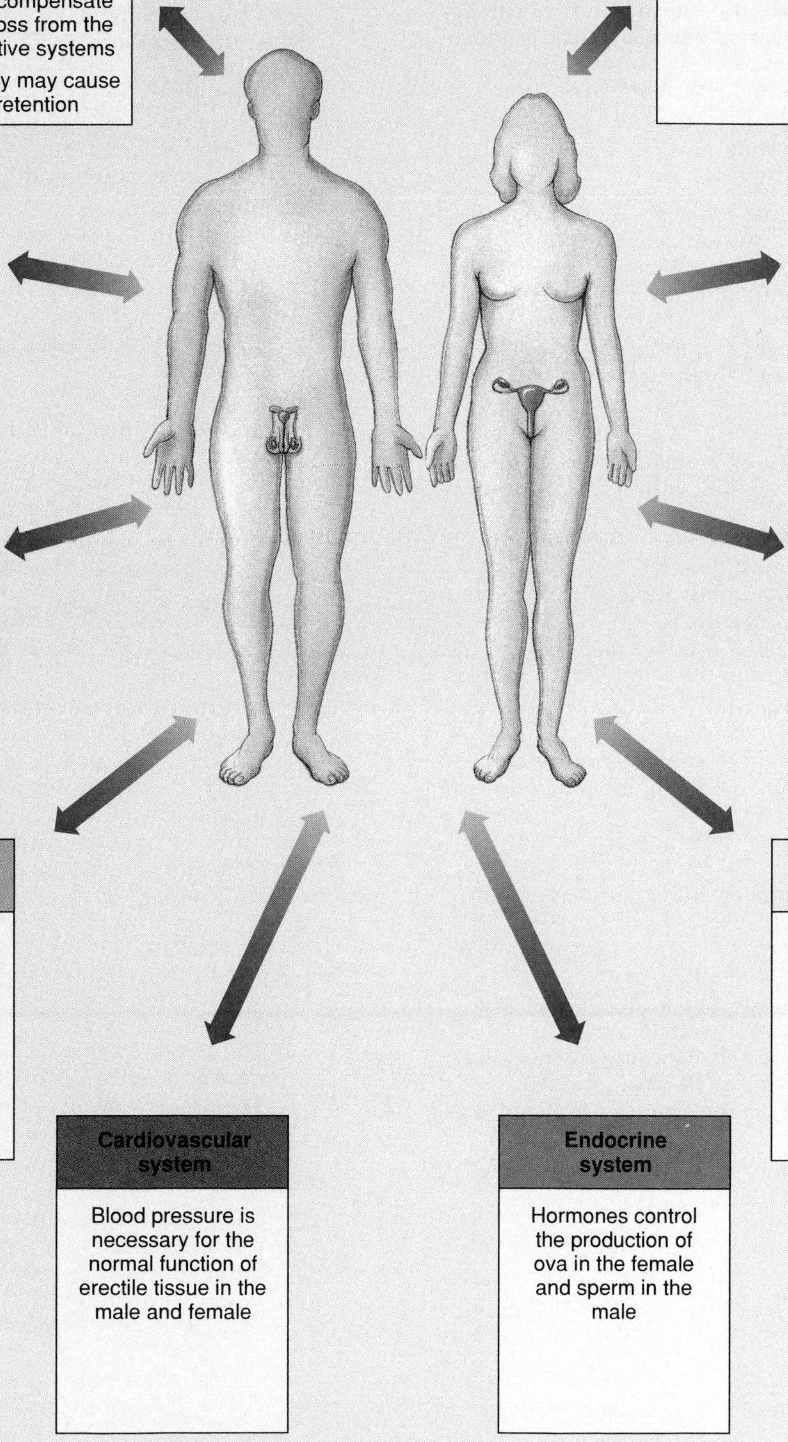

**Urinary system**

Male urinary and reproductive systems share common structures

Kidneys compensate for fluid loss from the reproductive systems

Pregnancy may cause fluid retention

**Integumentary system**

Skin sensory receptors play a role in sexual pleasure

**Respiratory system**

During pregnancy, oxygen is provided to the fetus and carbon dioxide is removed from the fetus through the placenta

**Skeletal system**

Bones can be a temporary source of calcium during lactation

**Digestive system**

Proper nutrition is essential for the formation of normal gametes and normal development of fetus during pregnancy

**Muscular system**

Skeletal, cardiac, and smooth muscles all play a role in reproductive processes and sexual activity

**Lymphatic system**

Special mechanisms inhibit the female immune system in its attack of sperm from the male as foreign invaders

**Nervous system**

The nervous system plays a major role in sexual activity and sexual pleasure

**Cardiovascular system**

Blood pressure is necessary for the normal function of erectile tissue in the male and female

**Endocrine system**

Hormones control the production of ova in the female and sperm in the male

## Introduction (page 853)

Various reproductive organs produce sex cells and sex hormones, sustain these cells, or transport them from place to place.

## Organs of the Male Reproductive System (page 853)

The male reproductive organs produce and maintain sperm cells, transport these cells, and produce male sex hormones. The primary male sex organs are the testes, which produce sperm cells and male sex hormones. Accessory organs include the internal and external reproductive organs.

## Testes (page 853)

1. Descent of the testes
   a. Testes originate behind the parietal peritoneum near the level of the developing kidneys.
   b. The gubernaculum guides the descent of the testes into the lower abdominal cavity and through the inguinal canal.
   c. Undescended testes fail to produce sperm cells because of the relatively high abdominal temperature.
2. Structure of the testes
   a. The testes are composed of lobules separated by connective tissue and filled with seminiferous tubules.
   b. The seminiferous tubules unite to form the rete testis that joins the epididymis.
   c. The seminiferous tubules are lined with epithelium, which produces sperm cells.
   d. The interstitial cells that produce male sex hormones occur between the seminiferous tubules.
3. Formation of sperm cells
   a. The epithelium lining the seminiferous tubules includes sustentacular cells and spermatogenic cells.
      (1) The sustentacular cells support and nourish the spermatogenic cells.
      (2) The spermatogenic cells give rise to spermatogonia.
   b. The process of spermatogenesis produces sperm cells from spermatogonia.
      (1) Meiosis reduces the number of chromosomes in sperm cells by one-half (46 to 23).
      (2) Spermatogenesis produces four sperm cells from each primary spermatocyte.
   c. Membranous processes of adjacent sustentacular cells form a barrier within the epithelium.
      (1) The barrier separates early and advanced stages of spermatogenesis.
      (2) It helps provide a favorable environment for differentiating cells.
4. Structure of a sperm cell
   a. Sperm head contains a nucleus with 23 chromosomes.
   b. Sperm body contains many mitochondria.
   c. Sperm tail propels the cell.

## Male Internal Accessory Organs (page 859)

1. Epididymis
   a. The epididymis is a tightly coiled tube on the outside of the testis that leads into the vas deferens.
   b. It stores and nourishes immature sperm cells and promotes their maturation.
2. Vas deferens
   a. The vas deferens is a muscular tube that forms part of the spermatic cord.
   b. It passes through the inguinal canal, enters the abdominal cavity, courses medially into the pelvic cavity, and ends behind the urinary bladder.
   c. It fuses with the duct from the seminal vesicle to form the ejaculatory duct.
3. Seminal vesicle
   a. The seminal vesicle is a saclike structure attached to the vas deferens.
   b. It secretes an alkaline fluid that contains nutrients and prostaglandins.
   c. This secretion is added to sperm cells during emission.
4. Prostate gland
   a. This gland surrounds the urethra just below the urinary bladder.
   b. It secretes a thin, milky fluid, which neutralizes and enhances the motility of sperm cells.
5. Bulbourethral glands
   a. These glands are two small structures beneath the prostate gland.
   b. They secrete a fluid that lubricates the penis in preparation for sexual intercourse.
6. Semen
   a. Semen is composed of sperm cells and secretions of the seminal vesicles, prostate gland, and bulbourethral glands.
   b. This fluid is slightly alkaline and contains nutrients and prostaglandins.
   c. Semen activates sperm cells, but these sperm cells are unable to fertilize egg cells until they enter the female reproductive tract.

## Male External Reproductive Organs (page 863)

1. Scrotum
   a. The scrotum is a pouch of skin and subcutaneous tissue that encloses the testes.
   b. The dartos muscle in the scrotal wall causes the skin of the scrotum to be held close to the testes or to hang loosely.
2. Penis
   a. The penis conveys urine and semen.
   b. It is specialized to become erect for insertion into the vagina during sexual intercourse.
   c. Its body is composed of three columns of erectile tissue surrounded by connective tissue.

d. The root of the penis is attached to the pelvic arch and membranes of the perineum.

3. Erection, orgasm, and ejaculation
   a. During erection, the vascular spaces within the erectile tissue become engorged with blood as arteries dilate and veins are compressed.
   b. Orgasm is the culmination of sexual stimulation and is accompanied by emission and ejaculation.
   c. Semen moves along the reproductive tract as smooth muscle in the walls of the tubular structures contract, stimulated by a reflex.
   d. Following ejaculation, the penis becomes flaccid.

## Hormonal Control of Male Reproductive Functions (page 866)

1. Hypothalamic and pituitary hormones
   The male body remains reproductively immature until the hypothalamus releases GnRH, which stimulates the anterior pituitary gland to release gonadotropins.
   a. FSH stimulates spermatogenesis.
   b. LH (ICSH) stimulates the interstitial cells to produce male sex hormones.
   c. Inhibin prevents oversecretion of FSH.
2. Male sex hormones
   a. Male sex hormones are called androgens.
   b. Testosterone is the most important androgen.
   c. Testosterone is converted into dihydrotestosterone in some organs.
   d. Androgens that fail to become fixed in tissues are metabolized in the liver and excreted.
   e. Androgen production increases rapidly at puberty.
3. Actions of testosterone
   a. Testosterone stimulates the development of the male reproductive organs and causes the testes to descend.
   b. It is responsible for the development and maintenance of male secondary sexual characteristics.
4. Regulation of male sex hormones
   a. A negative feedback mechanism regulates testosterone concentration.
      (1) As the concentration of testosterone rises, the hypothalamus is inhibited and the anterior pituitary secretion of gonadotropins is reduced.
      (2) As the concentration of testosterone falls, the hypothalamus signals the anterior pituitary to secrete gonadotropins.
   b. The concentration of testosterone remains relatively stable from day to day.

## Organs of the Female Reproductive System (page 868)

The primary female sex organs are the ovaries, which produce female sex cells and sex hormones. Accessory organs are internal and external.

## Ovaries (page 868)

1. Ovary attachments
   a. Several ligaments hold the ovaries in position.
   b. These ligaments include broad, suspensory, and ovarian ligaments.
2. Ovary descent
   a. The ovaries descend from behind the parietal peritoneum near the developing kidneys.
   b. They are attached to the pelvic wall just below the pelvic brim.
3. Ovary structure
   a. The ovaries are subdivided into a medulla and a cortex.
   b. The medulla is composed of connective tissue, blood vessels, lymphatic vessels, and nerves.
   c. The cortex contains ovarian follicles and is covered by cuboidal epithelium.
4. Primordial follicles
   a. During development, groups of cells in the ovarian cortex form millions of primordial follicles.
   b. Each primordial follicle contains a primary oocyte and a layer of flattened epithelial cells.
   c. The primary oocyte begins to undergo meiosis, but the process is soon halted and is not continued until puberty.
   d. The number of oocytes steadily declines throughout the life of a female.
5. Oogenesis
   a. Beginning at puberty, some oocytes are stimulated to continue meiosis.
   b. When a primary oocyte undergoes oogenesis, it gives rise to a secondary oocyte in which the original chromosome number is reduced by one-half (from 46 to 23).
   c. A secondary oocyte can be fertilized to produce a zygote.
6. Follicle maturation
   a. At puberty, FSH initiates the maturation of follicles.
   b. During maturation, the oocyte enlarges, the follicular cells proliferate, and a fluid-filled cavity appears and produces a secondary follicle.
   c. Ovarian cells surrounding the follicle form two layers.
   d. Usually only one follicle reaches full development.
7. Ovulation
   a. Ovulation is the release of an oocyte from an ovary.
   b. The oocyte is released when its follicle ruptures.
   c. After ovulation, the oocyte is drawn into the opening of the uterine tube.

## Female Internal Accessory Organs (page 873)

1. Uterine tubes
   a. These tubes convey egg cells toward the uterus.
   b. The end of each uterine tube is expanded, and its margin bears irregular extensions.
   c. Ciliated cells that line the tube and peristaltic contractions in the wall of the tube move an egg cell into the tube's opening.
2. Uterus
   a. The uterus receives the embryo and sustains it during development.
   b. The vagina partially encloses the cervix.
   c. The uterine wall includes the endometrium, myometrium, and perimetrium.

3. Vagina
   a. The vagina connects the uterus to the vestibule.
   b. It receives the erect penis, conveys uterine secretions to the outside, and provides an open channel for the fetus during birth.
   c. The vaginal orifice is partially closed by a thin membrane, the hymen.
   d. Its wall consists of a mucosa, muscularis, and outer fibrous coat.

## Female External Reproductive Organs (page 876)

1. Labia majora
   a. The labia majora are rounded folds of adipose tissue and skin that enclose and protect the other external reproductive parts.
   b. The upper ends form a rounded elevation over the symphysis pubis.
2. Labia minora
   a. The labia minora are flattened, longitudinal folds between the labia majora.
   b. They are well supplied with blood vessels.
3. Clitoris
   a. The clitoris is a small projection at the anterior end of the vulva; it corresponds to the male penis.
   b. It is composed of two columns of erectile tissue.
   c. Its root is attached to the sides of the pubic arch.
4. Vestibule
   a. The vestibule is the space between the labia majora that encloses the vaginal and urethral openings.
   b. The vestibular glands secrete mucus into the vestibule during sexual stimulation.
5. Erection, lubrication, and orgasm
   a. During periods of sexual stimulation, the erectile tissues of the clitoris and vestibular bulbs become engorged with blood and swollen.
   b. The vestibular glands secrete mucus into the vestibule and vagina.
   c. During orgasm, the muscles of the perineum, uterine wall, and uterine tubes contract rhythmically.

## Hormonal Control of Female Reproductive Functions (page 878)

Hormones from the hypothalamus, anterior pituitary gland, and ovaries play important roles in the control of sex cell maturation, and the development and maintenance of female secondary sexual characteristics.

1. Female sex hormones
   a. A female body remains reproductively immature until about 10 years of age when gonadotropin secretion increases.
   b. The most important female sex hormones are estrogen and progesterone.
      (1) Estrogen is responsible for the development and maintenance of most female secondary sexual characteristics.
      (2) Progesterone causes changes in the uterus.
2. Female reproductive cycle
   a. The menstrual cycle is characterized by regularly recurring changes in the uterine lining culminating in menstrual flow.
   b. A menstrual cycle is initiated by FSH, which stimulates maturation of a follicle.
   c. Granulosa cells of a maturing follicle secrete estrogen, which is responsible for maintaining the secondary sexual traits and thickening the uterine lining.
   d. Ovulation is triggered when the anterior pituitary gland releases a relatively large amount of LH.
   e. Following ovulation, the follicular cells and thecal cells give rise to the corpus luteum.
      (1) The corpus luteum secretes progesterone, which causes the uterine lining to become more vascular and glandular.
      (2) If an oocyte is not fertilized, the corpus luteum begins to degenerate.
      (3) As the concentrations of estrogen and progesterone decline, the uterine lining disintegrates, causing menstrual flow.
   f. During this cycle, estrogen and progesterone inhibit the release of LH and FSH; as the concentrations of these hormones decline, the anterior pituitary secretes FSH and LH again, stimulating a new menstrual cycle.
3. Menopause
   a. Eventually the ovaries cease responding to FSH, and cycling ceases.
   b. Menopause is characterized by a low estrogen concentration and a continuous secretion of FSH and LH.
   c. The female reproductive organs undergo varying degrees of regressive changes.

## Pregnancy (page 882)

1. Transport of sex cells
   a. Ciliary action aids movement of the egg cell to the uterine tube.
   b. A sperm cell moves by its tail lashing and muscular contraction in the female reproductive tract.
2. Fertilization
   a. With the aid of an enzyme, a sperm cell penetrates the zona pellucida.
   b. When a sperm cell penetrates an egg cell membrane, changes in the egg cell membrane and the zona pellucida prevent entry of additional sperm.
   c. Fusion of the nuclei of a sperm and an egg cell complete fertilization.
   d. The product of fertilization is a zygote with 46 chromosomes.
3. Early embryonic development
   a. Cells undergo mitosis, giving rise to smaller and smaller cells.
   b. The developing offspring (preembryo) moves down the uterine tube to the uterus, where it implants in the endometrium.
   c. The offspring is called an embryo from the second through the eighth week of development; thereafter it is called a fetus.
   d. Eventually some of the embryonic and maternal cells form a placenta.
4. Hormonal changes during pregnancy
   a. Embryonic cells produce hCG that maintains the corpus luteum.
   b. Placental tissue produces high concentrations of estrogen and progesterone.

(1) Estrogen and progesterone maintain the uterine wall and inhibit secretion of FSH and LH.
(2) Progesterone and relaxin inhibit contractions of uterine muscles.
(3) Estrogen causes enlargement of the vagina.
(4) Relaxin helps relax the ligaments of the pelvic joints.

c. The placenta secretes placental lactogen that stimulates the development of the breasts and mammary glands.
d. During pregnancy, increasing secretion of aldosterone promotes retention of sodium and body fluid, and increasing secretion of parathyroid hormone helps maintain a high concentration of maternal blood calcium.

5. Other changes during pregnancy
   a. The uterus enlarges greatly.
   b. The woman's blood volume, cardiac output, breathing rate, and urine production increase.
   c. The woman's dietary needs increase, but if intake is inadequate, fetal tissues have priority for use of available nutrients.
6. Birth
   a. Pregnancy usually lasts 40 weeks.
   b. During pregnancy, placental progesterone inhibits uterine contractions.
   c. A variety of factors are involved with the birth process.
      (1) A decreasing concentration of progesterone and the release of prostaglandins may initiate the birth process.
      (2) The posterior pituitary gland releases oxytocin.
      (3) Uterine muscles are stimulated to contract, and labor begins.
      (4) A positive feedback mechanism causes stronger contractions and greater release of oxytocin.
   d. Following the birth of the infant, placental tissues are expelled.

## Mammary Glands (page 891)

1. Location of the glands
   a. The mammary glands are located in the subcutaneous tissue of the anterior thorax within the breasts.
   b. The breasts extend between the second and sixth ribs, and from sternum to axillae.
2. Structure of the glands
   a. The mammary glands are composed of lobes that contain tubular glands.
   b. The lobes are separated by dense connective and adipose tissues.
   c. The mammary glands are connected to the nipple by ducts.
3. Development of the breasts
   a. Male breasts remain nonfunctional.
   b. Estrogen stimulates female breast development.
      (1) Alveolar glands and ducts enlarge.
      (2) Fat is deposited around and within the breasts.
   c. During pregnancy, the breasts change.
      (1) Estrogen causes the ductile system to grow.
      (2) Progesterone causes development of alveolar glands.
      (3) Prolactin is released during pregnancy, but progesterone inhibits milk production.
4. Milk production and secretion
   a. Following childbirth, the concentrations of placental hormones decline.
      (1) The action of prolactin is no longer blocked.
      (2) The mammary glands begin to secrete milk.
   b. Reflex response to mechanical stimulation of the nipple causes the posterior pituitary to release oxytocin, which causes milk to be ejected from the alveolar ducts.
   c. As long as milk is removed from glands, more milk is produced; if milk is not removed, production ceases.
   d. During the period of milk production, the menstrual cycle is partially inhibited.

## Birth Control (page 895)

Voluntary regulation of the number of children produced and the time they are conceived is called birth control. This usually involves some method of contraception.

1. Coitus interruptus
   a. Coitus interruptus is withdrawal of the penis from the vagina before ejaculation.
   b. Some semen may be expelled from the penis before ejaculation.
2. Rhythm method
   a. Abstinence from sexual intercourse a few days before and after ovulation is the rhythm method.
   b. It is almost impossible to predict the time of ovulation accurately.
3. Mechanical barriers
   a. Males and females can use condoms.
   b. Females use diaphragms.
4. Chemical barriers
   a. Spermicidal creams, foams, and jellies are chemical barriers to conception.
   b. These provide an unfavorable environment in the vagina for sperm survival.
5. Oral contraceptives
   a. Tablets that contain synthetic estrogen-like and progesterone-like substances are taken by the woman.
   b. They disrupt the normal pattern of gonadotropin secretion, and prevent ovulation and the normal build-up of the uterine lining.
   c. When used correctly, this method is almost 100% effective.
   d. Some women develop undesirable side effects.
6. Injectable contraceptives
   a. Intramuscular injection with medroxyprogesterone acetate every three months.
   b. High levels of hormone act similarly to oral contraceptives to prevent pregnancy.
   c. Very effective if administered promptly at the end of the three months.
   d. Women may experience side effects; in some women use is contraindicated.
7. Contraceptive implants
   a. A contraceptive implant consists of a set of progesterone-containing capsules or rods that are inserted under the skin.
   b. Progesterone released from the implant prevents ovulation.
   c. The implant is effective for years and its action can be reversed by having it removed.

8. Intrauterine devices
   a. An IUD is a solid object inserted in the uterine cavity.
   b. It is thought to prevent pregnancy by interfering with implantation.
   c. It may be expelled spontaneously or produce undesirable side effects.
9. Surgical methods
   a. These are sterilization procedures.
      (1) Vasectomy is performed in males.
      (2) Tubal ligation is performed in females.
   b. Surgical methods are the most reliable forms of contraception.

## Sexually Transmitted Diseases (page 901)

1. Sexually transmitted diseases are passed during sexual contact, and may go undetected for years.
2. The twenty such disorders share certain symptoms.

# Critical Thinking Questions

1. What changes, if any, might occur in the secondary sexual characteristics of an adult male following removal of one testis? Following removal of both testes? Following removal of the prostate gland?
2. How would you explain the fact that new mothers sometimes experience cramps in their lower abdomens when they begin to nurse their babies?
3. If a woman who is considering having a tubal ligation asks, "Will the operation cause me to go through my change of life early?" how would you answer?
4. What effect would it have on a woman's menstrual cycles if a single ovary were removed surgically? What effect would it have if both ovaries were removed?
5. Which methods of contraception are theoretically the most effective in preventing unwanted pregnancies? Which methods are theoretically the least effective?
6. As a male reaches adulthood, what will be the consequences if his testes have remained undescended since birth? Why?
7. Why does injecting a sperm cell into an egg cell not result in fertilization?
8. What types of contraceptives provide the greatest protection against sexually transmitted diseases?

# Review Exercises

1. List the general functions of the male reproductive system.
2. Distinguish between the primary and accessory male reproductive organs.
3. Describe the descent of the testes.
4. Define *cryptorchidism.*
5. Describe the structure of a testis.
6. Explain the function of the sustentacular cells in the testis.
7. List the major steps in spermatogenesis.
8. Describe a sperm cell.
9. Describe the epididymis, and explain its function.
10. Trace the path of the vas deferens from the epididymis to the ejaculatory duct.
11. On a diagram, locate the seminal vesicles, and describe the composition of their secretion.
12. On a diagram, locate the prostate gland, and describe the composition of its secretion.
13. On a diagram, locate the bulbourethral glands, and explain the function of their secretion.
14. Describe the composition of semen.
15. Define *capacitation.*
16. Describe the structure of the scrotum.
17. Describe the structure of the penis.
18. Explain the mechanism that produces an erection of the penis.
19. Distinguish between emission and ejaculation.
20. Explain the mechanism of ejaculation.
21. Explain the role of GnRH in the control of male reproductive functions.
22. Distinguish between androgen and testosterone.
23. Define *puberty.*
24. Describe the actions of testosterone.
25. List several male secondary sexual characteristics.
26. Explain the regulation of testosterone concentration.
27. List the general functions of the female reproductive system.
28. Distinguish between the primary and accessory female reproductive organs.
29. Describe how the ovaries are held in position.
30. Describe the descent of the ovaries.
31. Describe the structure of an ovary.
32. Define *primordial follicle.*
33. List the major steps in oogenesis.
34. Distinguish between a primary and a secondary follicle.
35. Describe a mature follicle.
36. Define *ovulation.*

37. On a diagram, locate the uterine tubes, and explain their function.
38. Describe the structure of the uterus.
39. Describe the structure of the vagina.
40. Distinguish between the labia majora and the labia minora.
41. On a diagram, locate the clitoris, and describe its structure.
42. Define *vestibule.*
43. Describe the process of erection in the female reproductive organs.
44. Define *orgasm.*
45. Explain the role of GnRH in regulating female reproductive functions.
46. List several female secondary sexual characteristics.
47. Define *menstrual cycle.*
48. Explain how a menstrual cycle is initiated.
49. Summarize the major events in a menstrual cycle.
50. Define *menopause.*
51. Describe how male and female sex cells are transported within the female reproductive tract.
52. Describe the process of fertilization.
53. List the major functions of the placenta.
54. Explain the major hormonal changes that occur in the maternal body during pregnancy.
55. Describe the major nonhormonal changes that occur in the maternal body during pregnancy.
56. Describe the role of progesterone in initiating the birth process.
57. Discuss the events that occur during the birth process.
58. Describe the structure of a mammary gland.
59. Explain the roles of prolactin and oxytocin in milk production and secretion.
60. Define *contraception.*
61. List several methods of contraception, and explain how each prevents pregnancy.
62. List several sexually transmitted diseases.

23

CHAPTER

# HUMAN GROWTH AND DEVELOPMENT

GROWTH AND DEVELOPMENT OCCUR THROUGHOUT LIFE, BUT NO PERIOD IS AS FUN TO WATCH AS INFANCY.

## CHAPTER OBJECTIVES

AFTER YOU HAVE STUDIED THIS CHAPTER, YOU SHOULD BE ABLE TO:

1. Distinguish between growth and development.
2. Describe the major events that occur during the period of cleavage.
3. Explain how the primary germ layers originate and list the structures produced by each layer.
4. Describe the formation and function of the placenta.
5. Define *fetus* and describe the major events that occur during the fetal stage of development.
6. Trace the general path of blood through the fetal circulatory system.
7. Describe the major circulatory and physiological adjustments that occur in the newborn.
8. Name the stages of development that occur between the neonatal period and death, and list the general characteristics of each stage.

## KEY TERMS

**allantois** (ah-lan′to-is)
**amnion** (am′ne-on)
**chorion** (ko′re-on)
**cleavage** (klēv′ij)
**embryo** (em′bre-o)
**fetus** (fe′tus)
**germ layer** (jerm la′er)
**neonatal** (ne″o-na′tal)
**placenta** (plah-sen′tah)
**postnatal** (pōst-na′tal)
**prenatal** (pre-na′tal)
**senescence** (se-nes′ens)
**umbilical cord** (um-bil′ĭ-kal kord)
**zygote** (zi′gōt)

One of the greatest joys of being a woman is feeling the first stirrings of a new life. It usually happens sometime during the fifth month of pregnancy. At first, fetal movements feel like a tremulous fluttering in one's middle, happening so fast that by the time it registers, the feeling has passed. But the next time it happens, a woman usually notices—and that is truly a magical moment. A missed menstrual period, a colored circle appearing on a pregnancy test kit, even nausea in the morning all indicate pregnancy. But nothing makes impending parenthood seem as real as the first detectable movements of a fetus within a woman's body. Fathers-to-be must wait another month or so for the unforgettable experience, when a hand pressed to the woman's abdomen feels the kicks and jabs that were once barely perceptible flutterings.

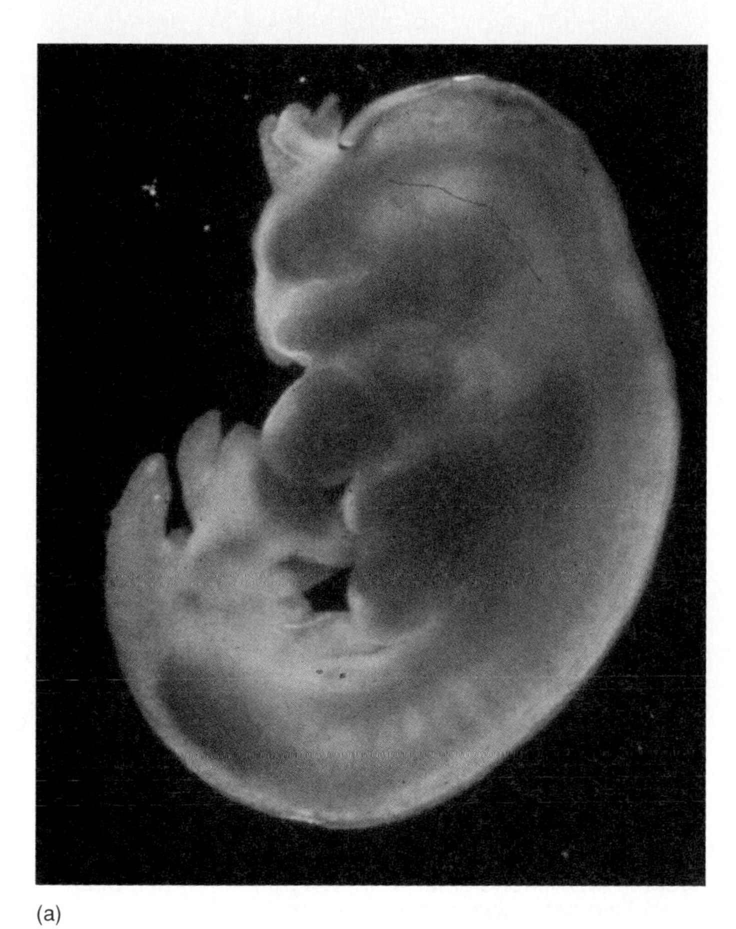

(a)

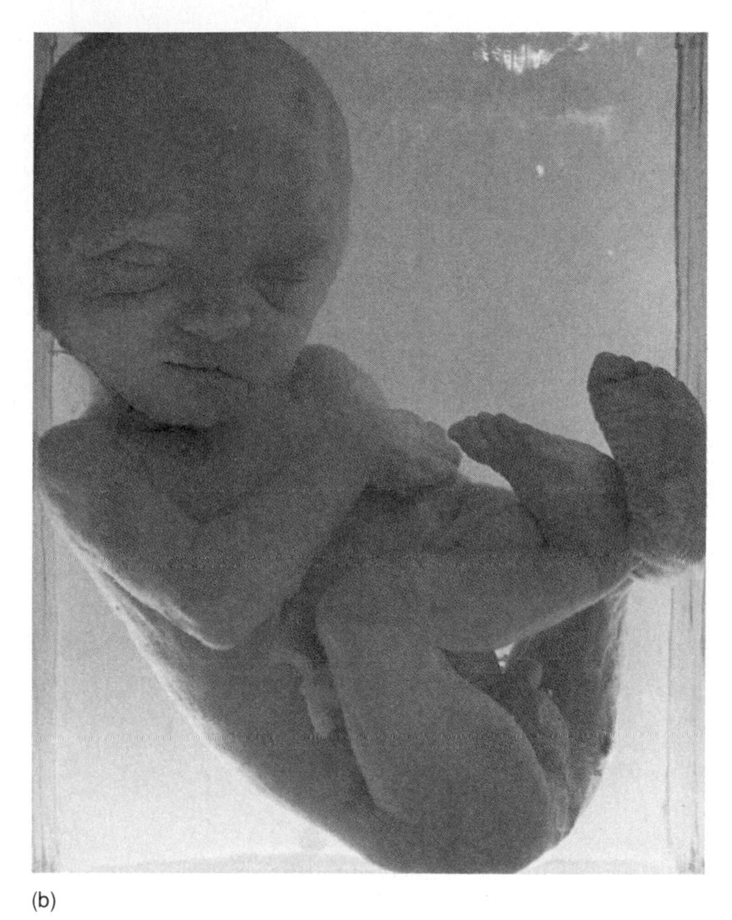

(b)

**FIGURE 23.1**

The contrast between a human embryo at 28 days (*a*) and a six-month-old fetus (*b*) shows evidence of both tremendous growth and profound changes in development.

Humans grow and develop. Growth is an increase in size. It entails increase in cell numbers as a result of mitosis, followed by enlargement of the newly formed cells and of the body.

Development, which includes growth, is the continuous process by which an individual changes from one life phase to another (fig. 23.1). These life phases include a **prenatal period,** which begins with the fertilization of an egg cell and ends at birth, and a **postnatal period,** which begins at birth and ends with death.

## PRENATAL PERIOD

The prenatal period of development usually lasts for thirty-eight weeks from conception and can be divided into a period of cleavage, an embryonic stage, and a fetal stage. The prenatal organism during the period of cleavage (the first two weeks) is sometimes called a preembryo.

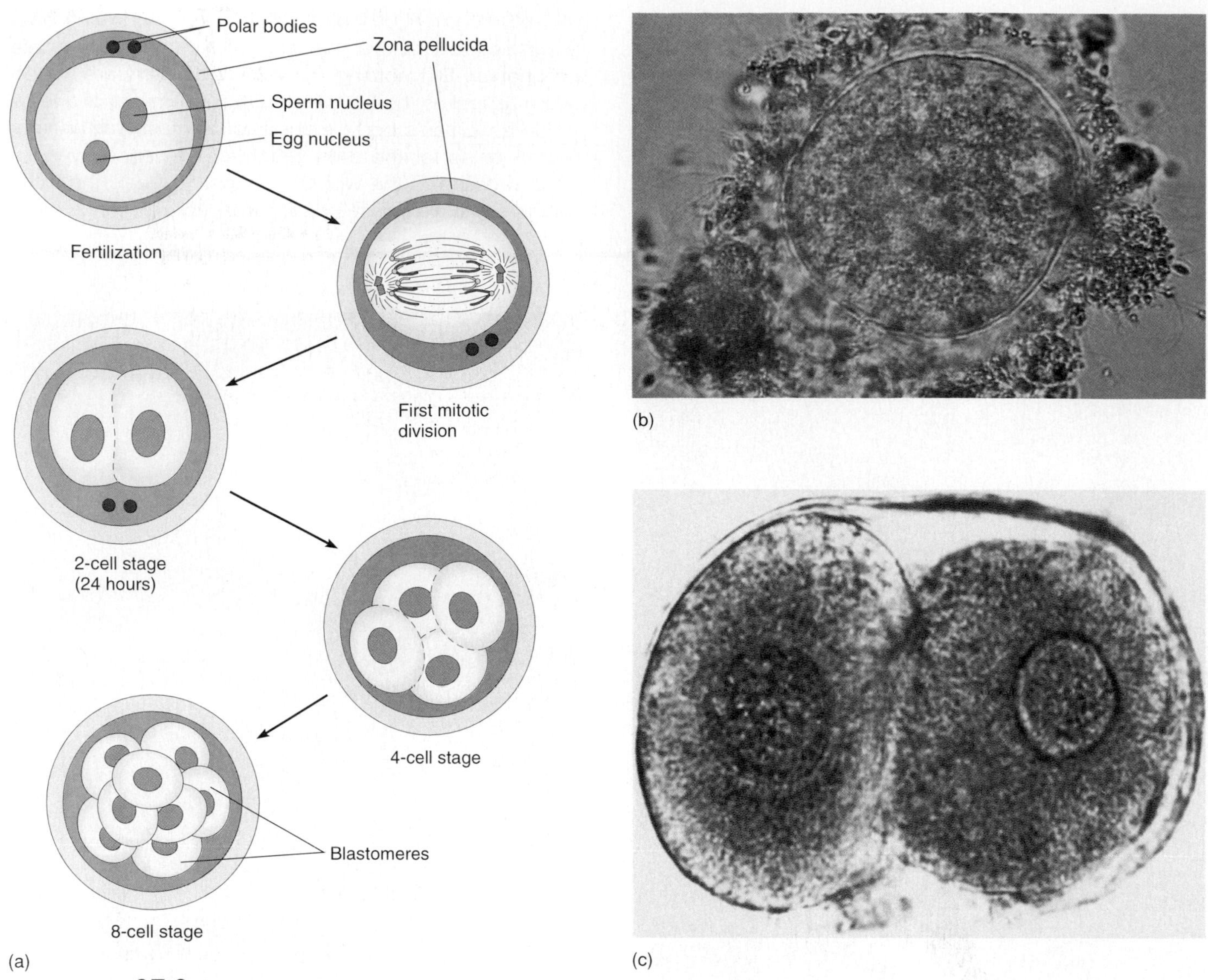


**FIGURE 23.2**
*(a)* During the period of cleavage, the cells divide by mitosis and become smaller and smaller. *(b)* A light micrograph of a human egg cell surrounded by follicular cells and sperm cells (25×). *(c)* Two-cell stage of development (1,000x).

## Period of Cleavage

Conception occurs when the genetic packages of sperm and egg merge, forming a *zygote.* Thirty hours later, the zygote undergoes mitosis, giving rise to two new cells. These cells, in turn, divide to form four cells, which then divide into eight cells, and so forth. With each subsequent division, the resulting cells are smaller and smaller. This distribution of the zygote's cytoplasmic contents into smaller and smaller cells is called **cleavage,** and the cells produced in this way are called *blastomeres* (fig. 23.2). Researchers can manipulate eight-celled preembryos (Clinical Application 23.1).

The approximate time of conception (fertilization) can be calculated by adding 14 days to the date of the onset of the last menstrual period. The expected time of birth can be estimated by adding 266 days to the fertilization date. Most fetuses are born within 10 to 15 days of this calculated time.

Obstetricians estimate the date of conception by scanning the embryo with ultrasound and comparing the crown-to-rump length to known values that are the average for each day of gestation. This approach is inaccurate if an embryo is smaller or larger than usual due to a medical problem.

Clinical Application 23.1

## Preimplantation Genetic Diagnosis

Chloe O'Brien is a very special, healthy little girl. When she was just a ball of eight cells, growing in a laboratory dish, she was tested to see if she would develop cystic fibrosis, the disorder that her parents carried and had passed to her brother. He had a severe case and had to be hospitalized frequently for respiratory difficulty.

One of Chloe-to-be's eight cells was removed and tested to see if it matched a DNA probe for the cystic fibrosis gene (fig. 23A). When that one cell was found to be free of the defective gene, doctors implanted the remaining seven-celled ball into Chloe's mother's uterus, where prenatal development completed. Chloe was born in March 1992. The procedure is called preimplantation genetic diagnosis because it is performed before the eight-celled prenatal human implants in the uterus. Several other healthy children have since been born following this prenatal procedure.

People fear that preimplantation genetic diagnosis might someday be abused, enabling parents to pick and choose trivial traits in their offspring, such as those affecting appearance rather than health. The technique is so technically difficult and costly, however, that such use is unlikely to occur.

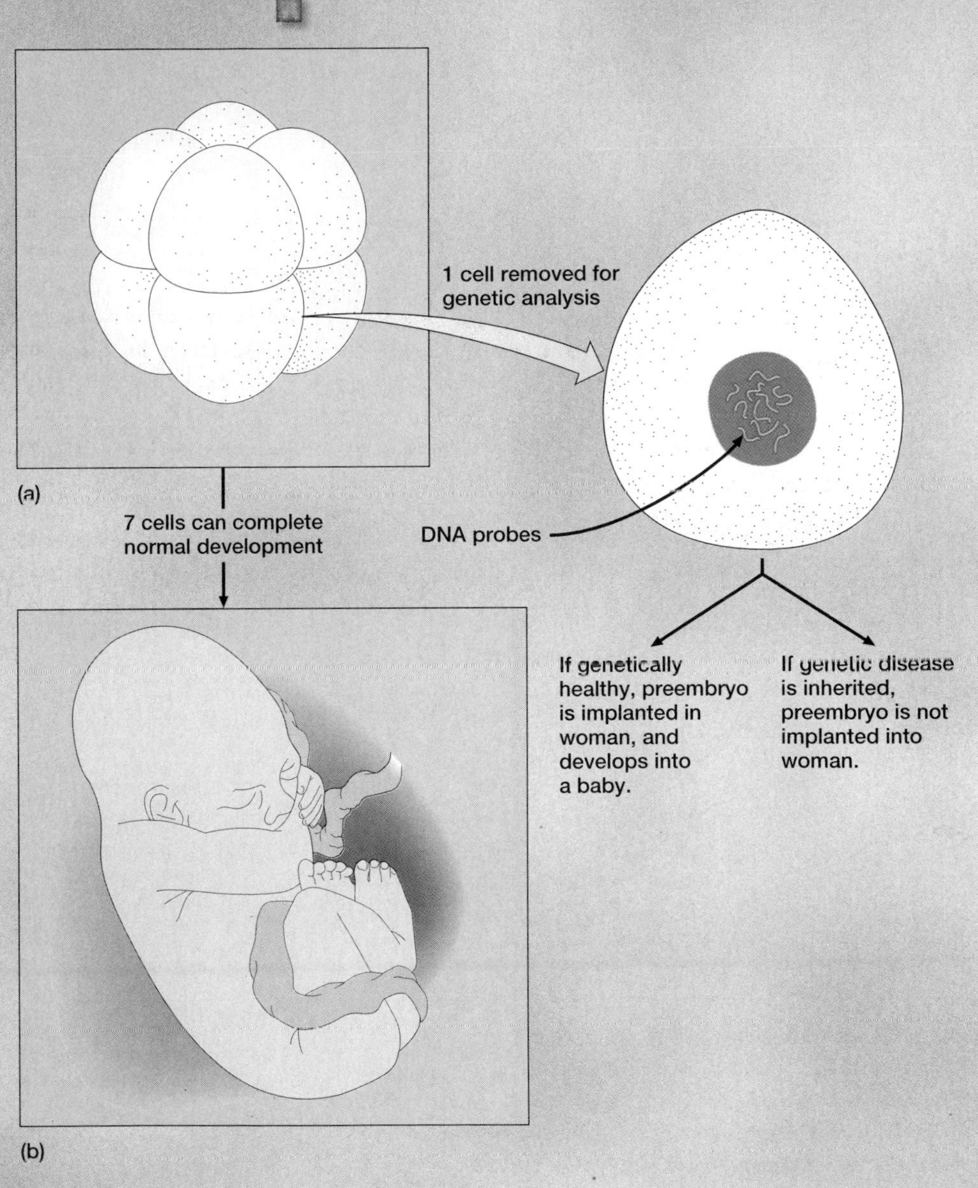


**Figure 23A** Preimplantation genetic diagnosis probes disease-causing genes in an eight-celled preembryo.

(a) Morula (3 days)

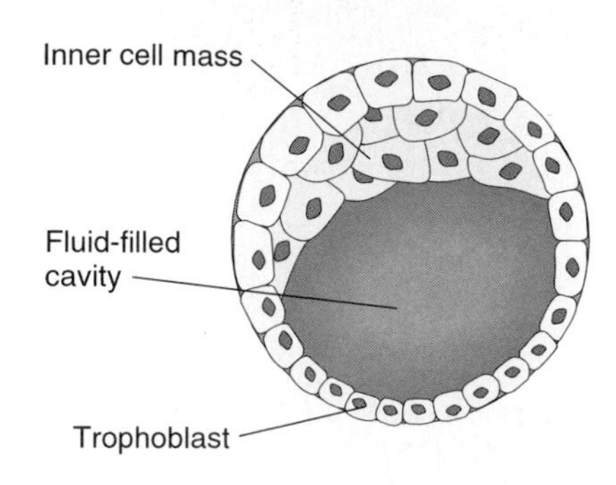


(b) Section of blastocyst (6 days)

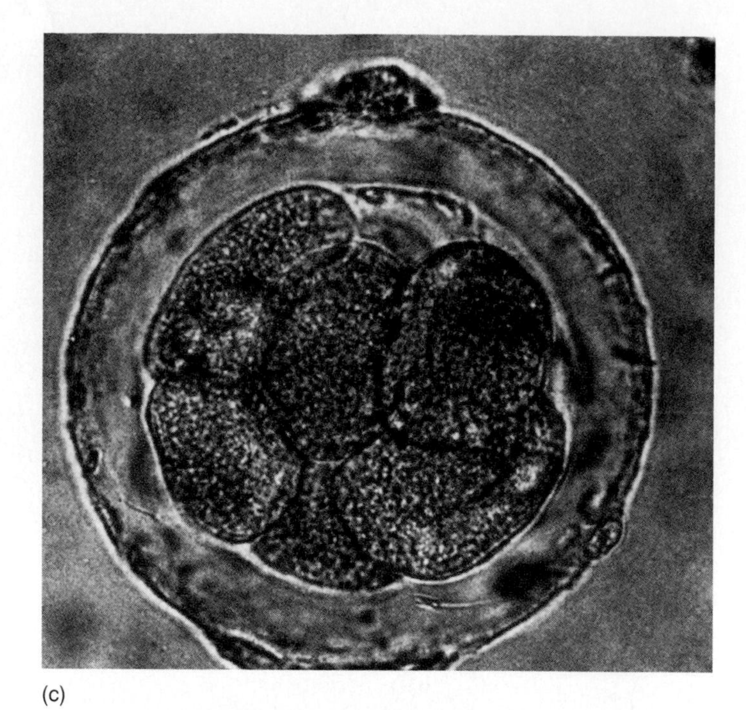

(c)

**FIGURE 23.3**

(*a*) A morula is a solid ball of cells. (*b*) A blastocyst has a fluid-filled cavity. (*c*) Light micrograph of a human morula.

The mass of cells formed by cleavage is still enclosed in the zona pellucida of the original egg cell. In about three days, it consists of a solid ball of about sixteen cells called a **morula.**

The morula enters the uterine cavity and remains unattached for about three days. During this time, the zona pellucida degenerates, and the morula develops a fluid-filled central cavity. Once this cavity appears, the morula becomes a hollow ball of cells called a **blastocyst** (fig. 23.3).

Within the blastocyst, cells in one region group together to form an *inner cell mass* that eventually gives rise to the **embryo proper**—the body of the developing offspring. The cells forming the wall of the blastocyst make up the *trophoblast,* which forms structures that assist the embryo in its development.

Sometimes two ovarian follicles release egg cells simultaneously, and if both are fertilized, the resulting zygotes can develop into fraternal (dizygotic) twins. Such twins are no more alike genetically than any brothers or sisters. Twins may also develop from a single fertilized egg (monozygotic twins). This may happen if two inner cell masses form within a blastocyst and each produces an embryo. Twins of this type usually share a single placenta, and they are identical genetically. Thus, they are always the same sex and are very similar in appearance.

About the sixth day, the blastocyst begins to attach to the uterine lining (fig. 23.4), aided by its secretion of proteolytic enzymes that digest a portion of the endometrium. The blastocyst sinks slowly into the resulting depression, becoming completely buried in the uterine lining. At the same time, the uterine lining is stimulated to thicken below the implanting blastocyst, and cells of the trophoblast begin to produce tiny, fingerlike processes (microvilli) that grow into the endometrium. This process of **implantation** begins near the end of the first week and is completed during the second week of development (fig. 23.5).

The trophoblast secretes the hormone hCG, which maintains the corpus luteum during the early stages of pregnancy and helps to protect the blastocyst against being rejected as foreign (nonself) by the maternal immune system. This hormone also stimulates synthesis of other hormones from the developing placenta. Clinical Application 23.2 discusses an alternate way for a pregnancy to begin—in the laboratory.

1. Distinguish between growth and development.
2. What changes characterize the period of cleavage?
3. How does a blastocyst attach to the endometrium?
4. In what ways does the endometrium respond to the activities of the blastocyst?

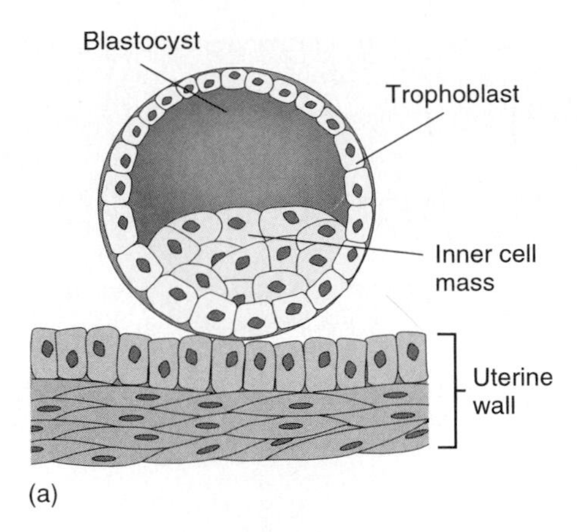


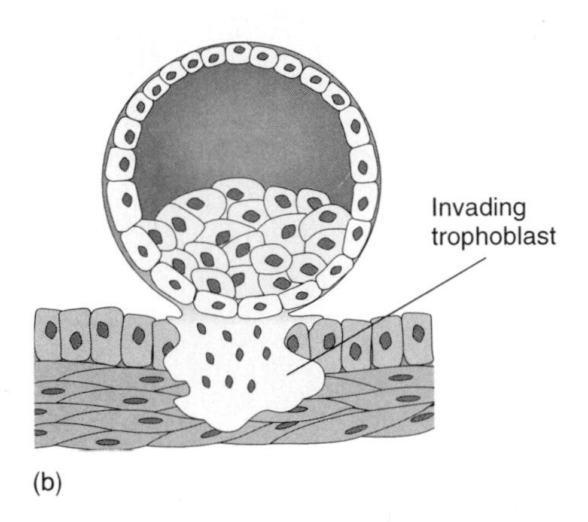


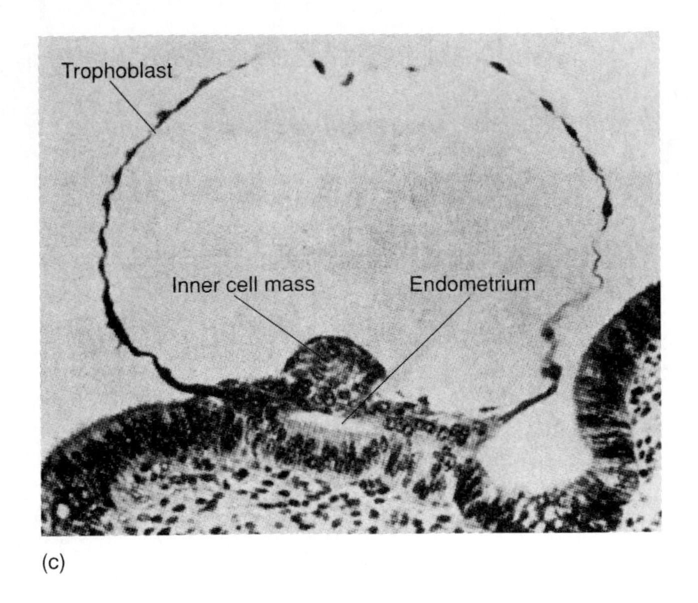


**FIGURE 23.4**

(*a*) About the sixth day of development, the blastocyst contacts the uterine wall and (*b*) begins to implant. (*c*) Light micrograph of a blastocyst from a monkey in contact with the endothelium of the uterine wall (300x).

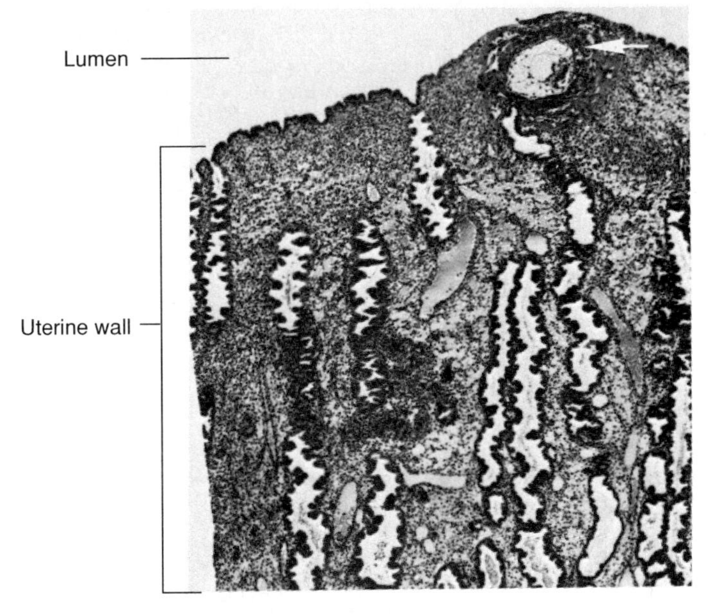


**FIGURE 23.5**

Light micrograph of a human preembryo (arrow) implanting in the uterine wall (18x).

## Embryonic Stage

The **embryonic stage** extends from the end of the second week through the eighth week of development. During this time, the placenta forms, the main internal organs develop, and the major external body structures appear.

During the second week of prenatal development, the blastocyst completes implantation, and the inner cell mass changes. A space, called the *amniotic cavity,* forms between the inner cell mass and the portion of the trophoblast that "invades" the endometrium. The inner cell mass then flattens and is called the **embryonic disk.** By the start of the third week, the structure is considered an embryo because it has layers. A connecting stalk begins to form, attaching the embryo to the developing placenta.

The embryonic disk initially consists of two distinct layers: an outer *ectoderm* and an inner *endoderm.* A short time later, a third layer of cells, the *mesoderm,* forms between the ectoderm and endoderm. These three layers of cells are called the **primary germ layers** of the primordial embryo, or **gastrula.** All organs form from these germ layers (fig. 23.6).

Gastrulation is an important process in prenatal development because a cell's fate is determined by which layer it is in. *Ectodermal cells* give rise to the nervous system, portions of special sensory organs, the epidermis, hair, nails, glands of the skin, and linings of the mouth and anal canal. *Mesodermal cells* form all types of muscle tissue, bone tissue, bone marrow, blood, blood vessels, lymphatic vessels,

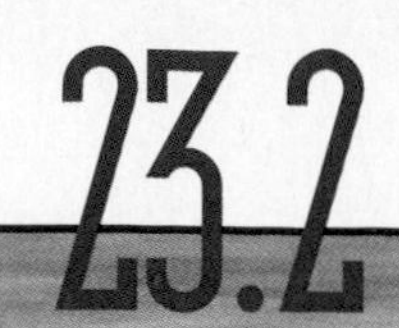

CLINICAL APPLICATION 23.2

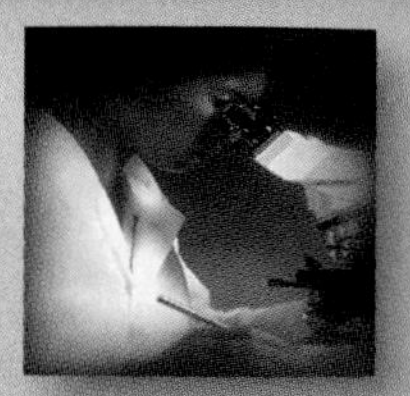

## Assisted Reproductive Technologies

Michele and Ray L'Esperance wanted children badly, but her uterine tubes had been removed due to scarring. Happily, a procedure called in vitro fertilization (IVF) solved the couple's problem.

First, Michele received human menopausal gonadotropin to stimulate development of ovarian follicles. When an ultrasound scan showed the follicles had grown to a certain diameter, she received human chorionic gonadotropin to induce ovulation. Then her physician used an optical instrument called a laparoscope to examine the interior of her abdomen and harvest the largest oocytes from an ovary. The oocytes were incubated at 37° C in a medium buffered at pH 7.4. When the oocytes had matured, they were mixed in a laboratory dish with Ray's sperm, which had been washed to remove various inhibitory factors. Secretions from Michele's reproductive tract were added to capacitate the sperm.

FIGURE 23B
IVF worked too well for Michele and Ray L'Esperance. Five fertilized ova implanted in Michele's uterus are now Erica, Alexandria, Veronica, Danielle, and Raymond.

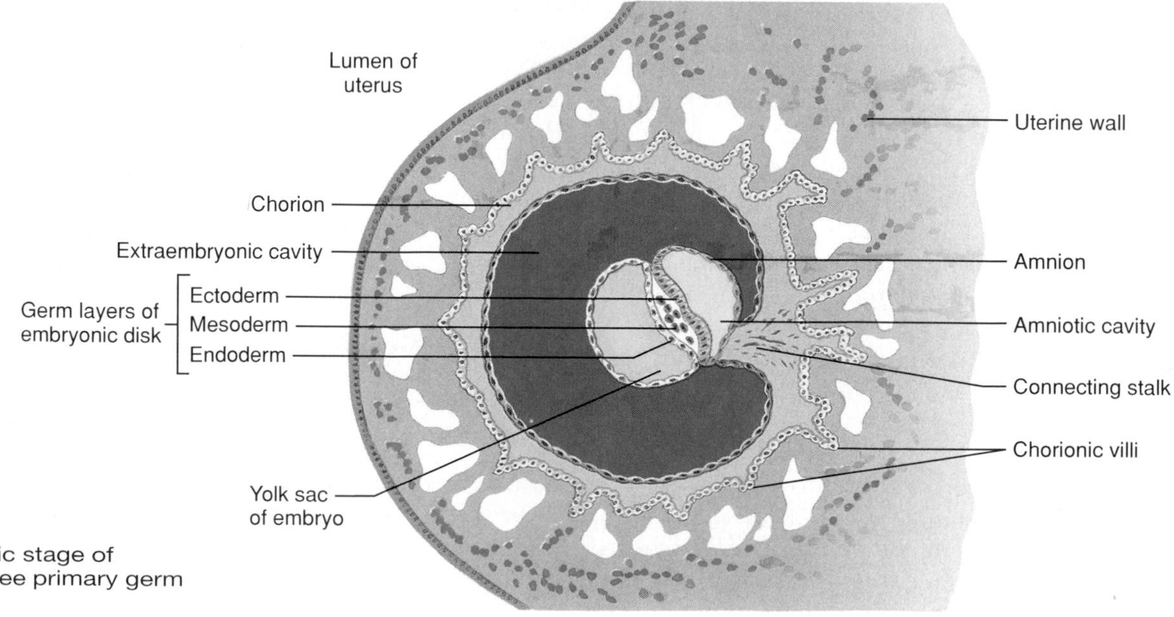


FIGURE 23.6
Early in the embryonic stage of development, the three primary germ layers form.

Next, fertilized eggs were selected and incubated in a special medium for about 60 hours. At this stage, five eight- to sixteen-cell preembryos were transferred through Michele's cervix and into her uterus with the aid of a specially designed catheter. So many preembryos were transferred to increase the odds that one or two would complete development. Michele received progesterone to prepare her uterus to receive the preembryos.

IVF worked almost too well for the L'Esperances—they had quintuplets (fig. 23B)! Success rates for IVF vary from clinic to clinic, ranging from 0% to 40%, with the average about 14%. It costs thousands of dollars to become pregnant via IVF. Chart 23A describes other assisted reproductive technologies.

### CHART 23A Variations on the Reproductive Theme

| Technique | Procedure | Condition It Treats |
|---|---|---|
| Artificial insemination | Donated sperm are placed near a woman's cervix. | Male infertility—lack of sperm or pooled specimens with low sperm count |
| Surrogate mother | An oocyte fertilized in vitro is implanted in a woman other than the one who donated the oocyte. The surrogate, or "gestational mother," gives the newborn to the "genetic mother" and her partner, the sperm donor. | Female infertility—lack of a uterus |
| Gamete intrafallopian transfer (GIFT) | Oocytes are removed from a woman's ovary, then placed along with donated sperm into a uterine tube. | Female infertility—bypasses blocked uterine tube |
| Zygote intrafallopian transfer (ZIFT) | An oocyte fertilized in vitro is placed in a uterine tube. It travels to the uterus on its own. | Female infertility—bypasses blocked uterine tube |
| Embryo adoption | A woman is artificially inseminated with sperm from a man whose partner cannot ovulate healthy oocytes. If the woman conceives, the morula is flushed from her uterus and implanted in the uterus of the sperm donor's partner. | Female infertility—a woman has nonfunctional ovaries, but a healthy uterus |

connective tissues, internal reproductive organs, kidneys, and the epithelial linings of the body cavities. *Endodermal cells* produce the epithelial linings of the digestive tract, respiratory tract, urinary bladder, and urethra (fig. 23.7).

As the embryo continues to implant in the uterus, slender projections grow out from the trophoblast into the surrounding endometrium. These extensions, called **chorionic villi,** branch. By the end of the fourth week, they are well formed.

Also during the fourth week of development, the flat embryonic disk transforms into a cylindrical structure (fig. 23.8). By this time, the head and jaws appear, the heart is beating and forcing blood through blood vessels, and tiny buds, which will give rise to the upper and lower limbs, are forming (fig. 23.9).

During the fifth through the seventh weeks, as shown in figure 23.10, the head grows rapidly and becomes rounded and erect. The face, which is developing the eyes, nose, and mouth, becomes more humanlike. The upper and lower limbs elongate, and fingers and toes appear (fig. 23.11). By the end of the seventh week, all the main internal organs are established, and as these structures enlarge, the body takes on a humanlike appearance.

During the development of the chorionic villi, the embryonic blood vessels appear within them, and these vessels are continuous with those passing through the connecting stalk to the body of the embryo. At the same time, irregular spaces called **lacunae** form around and between the villi. These spaces fill with maternal blood that escapes from eroded endometrial blood vessels.

A thin membrane separates embryonic blood within the capillary of a chorionic villus from maternal blood in a lacuna. This membrane, called

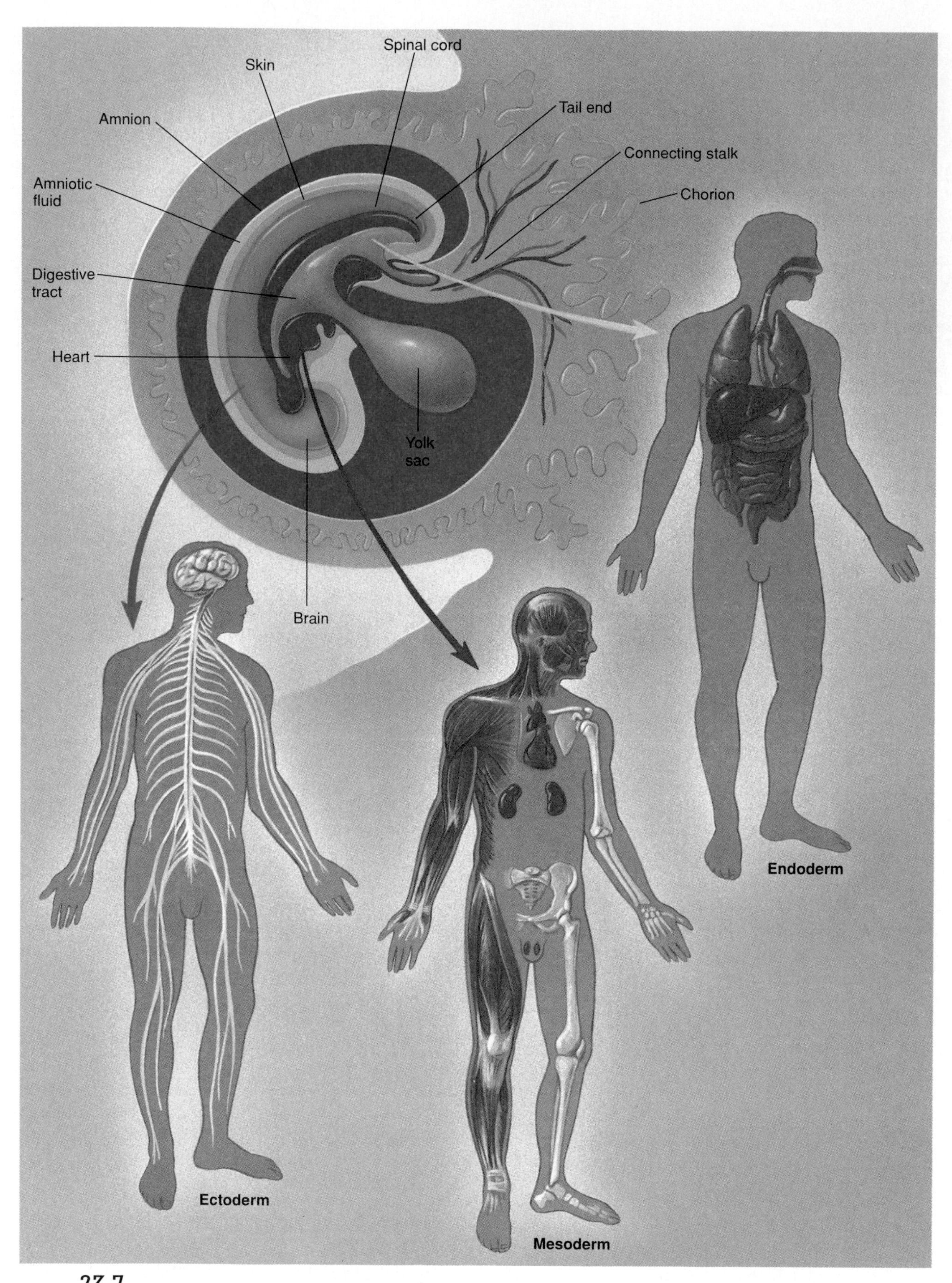


**FIGURE 23.7**
Each of the primary germ layers forms a particular set of organs.

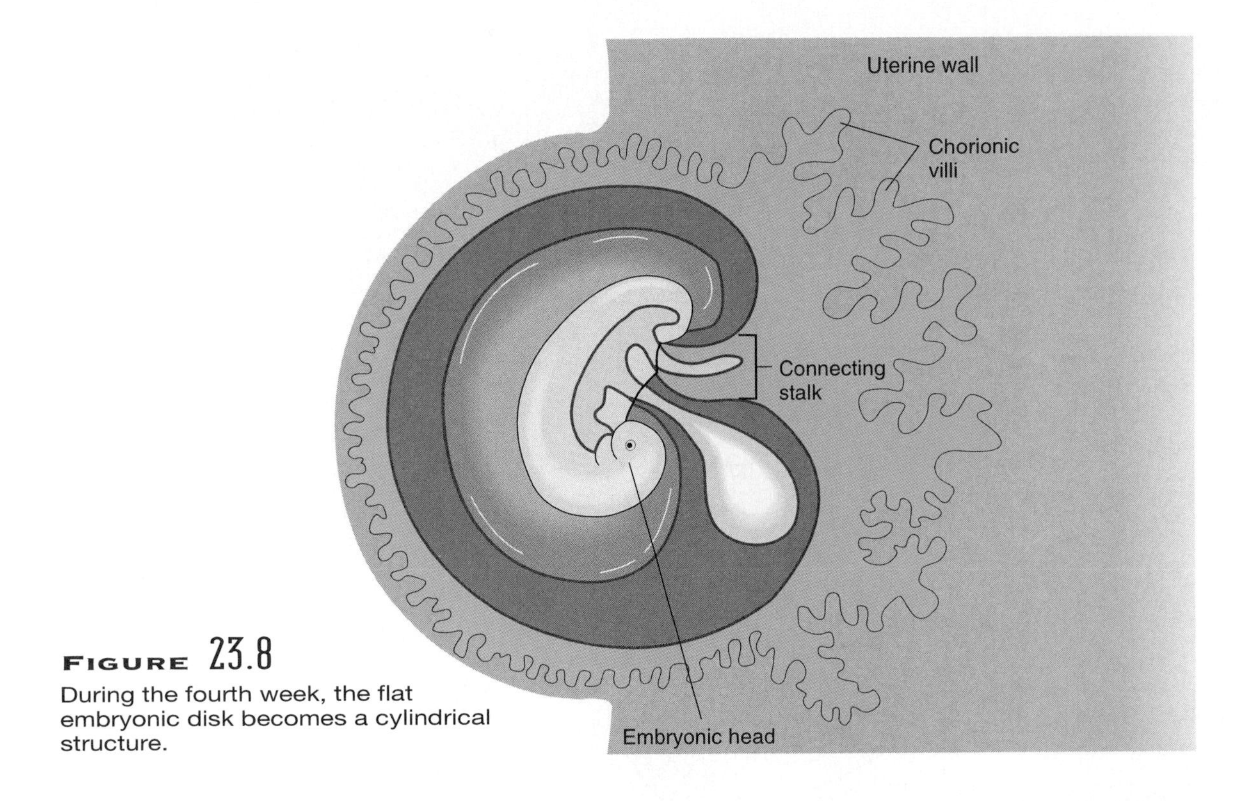


**FIGURE 23.8**
During the fourth week, the flat embryonic disk becomes a cylindrical structure.

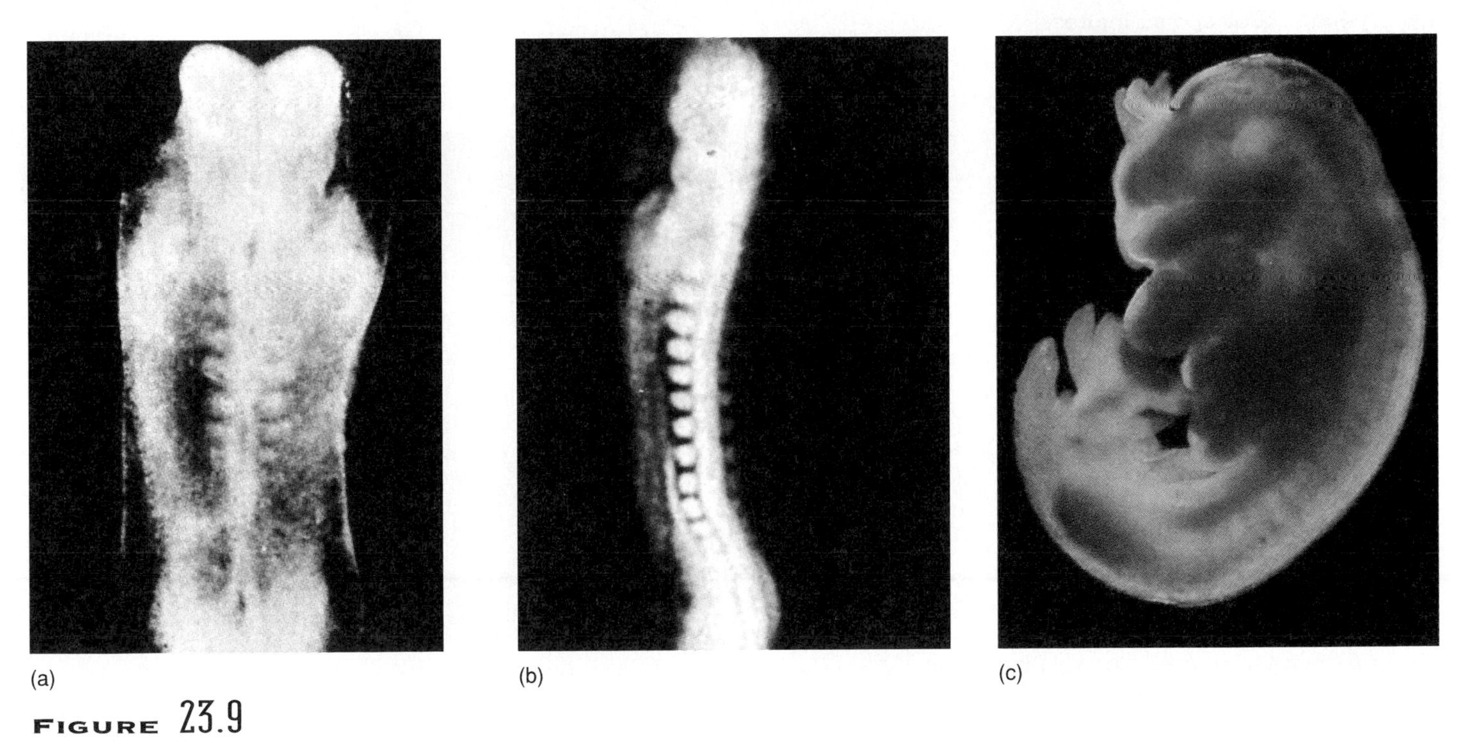


**FIGURE 23.9**
(*a*) A human embryo at three weeks; (*b*) at 3.5 weeks; (*c*) at about four weeks.

the **placental membrane,** is composed of the epithelium of the villus and the epithelium of the capillary (fig. 23.12). Through this membrane, exchanges take place between the maternal blood and the embryonic blood. Oxygen and nutrients diffuse from the maternal blood into the embryonic blood, and carbon dioxide and other wastes diffuse from the embryonic blood into the maternal blood. Various substances also move through the placental membrane by active transport and pinocytosis.

A prenatal test called chorionic villus sampling (CVS) examines the chromosomes in chorionic villus cells to obtain information on fetal health. The idea behind the test is that villus cells are genetically identical to fetal cells because they derive from the same fertilized egg. Because the test carries a risk of causing miscarriage, it is usually offered to women who have previously had a child with a detectable chromosome abnormality. CVS is performed at the tenth week of gestation.

1. What major events occur during the embryonic stage of development?
2. What tissues and structures develop from ectoderm? From mesoderm? From endoderm?
3. Describe the structure of a chorionic villus.
4. What is the function of the placental membrane?
5. How are substances exchanged between the embryonic blood and the maternal blood?

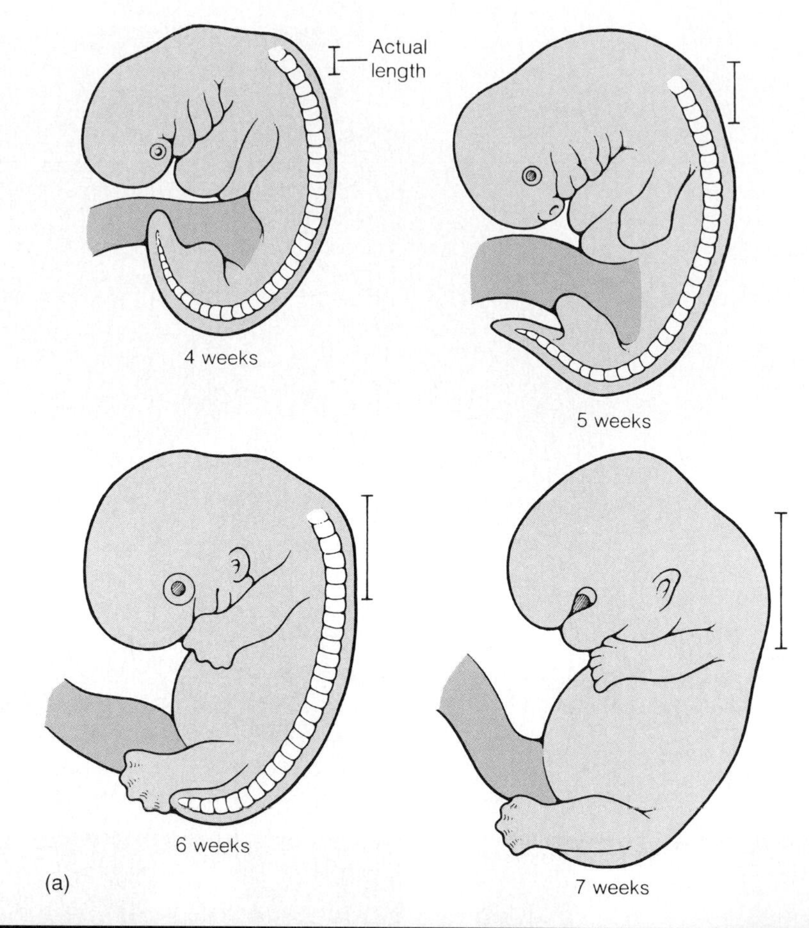


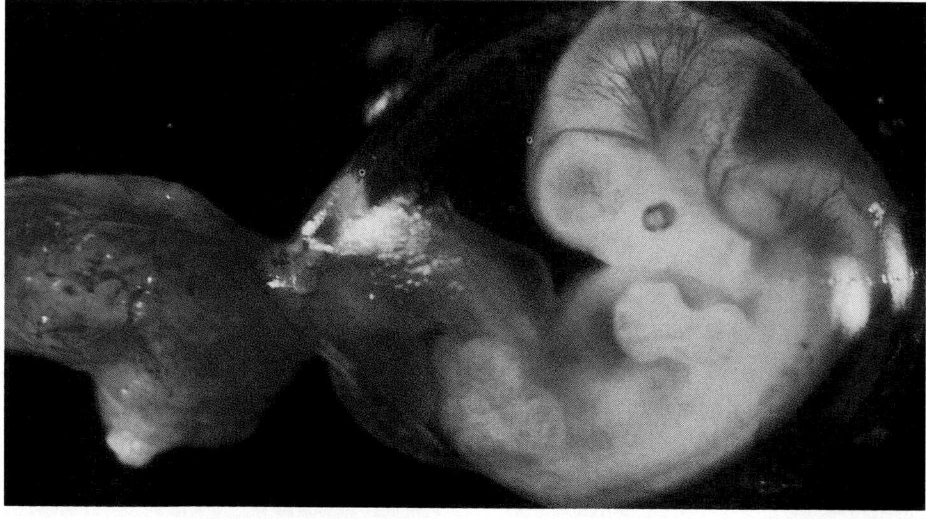


**FIGURE 23.10**

(*a*) In the fifth through the seventh weeks of gestation, the embryonic body and face develop a more humanlike appearance. (*b*) A human embryo after about six weeks of development.

Until about the end of the eighth week, the chorionic villi cover the entire surface of the former trophoblast, which is now called the **chorion.** However, as the embryo and the chorion surrounding it enlarge, only those villi that remain in contact with the endometrium endure. The others degenerate, and the portions of the chorion to which they were attached become smooth. Thus, the region of the chorion still in contact with the uterine wall is restricted to a disk-shaped area that becomes the **placenta.**

The embryonic portion of the placenta is composed of the chorion and its villi; the maternal portion is composed of the area of the uterine wall (decidua basalis) to which the villi are attached. When it is fully formed, the placenta appears as a reddish-brown disk, about 20 centimeters long and 2.5 centimeters thick. It usually weighs about 0.5 kilogram. Figure 23.13 shows the structure of the placenta.

**FIGURE 23.11**

Changes occurring during the fifth (*a–c*), sixth (*d*), and seventh (*e–g*) weeks of development.

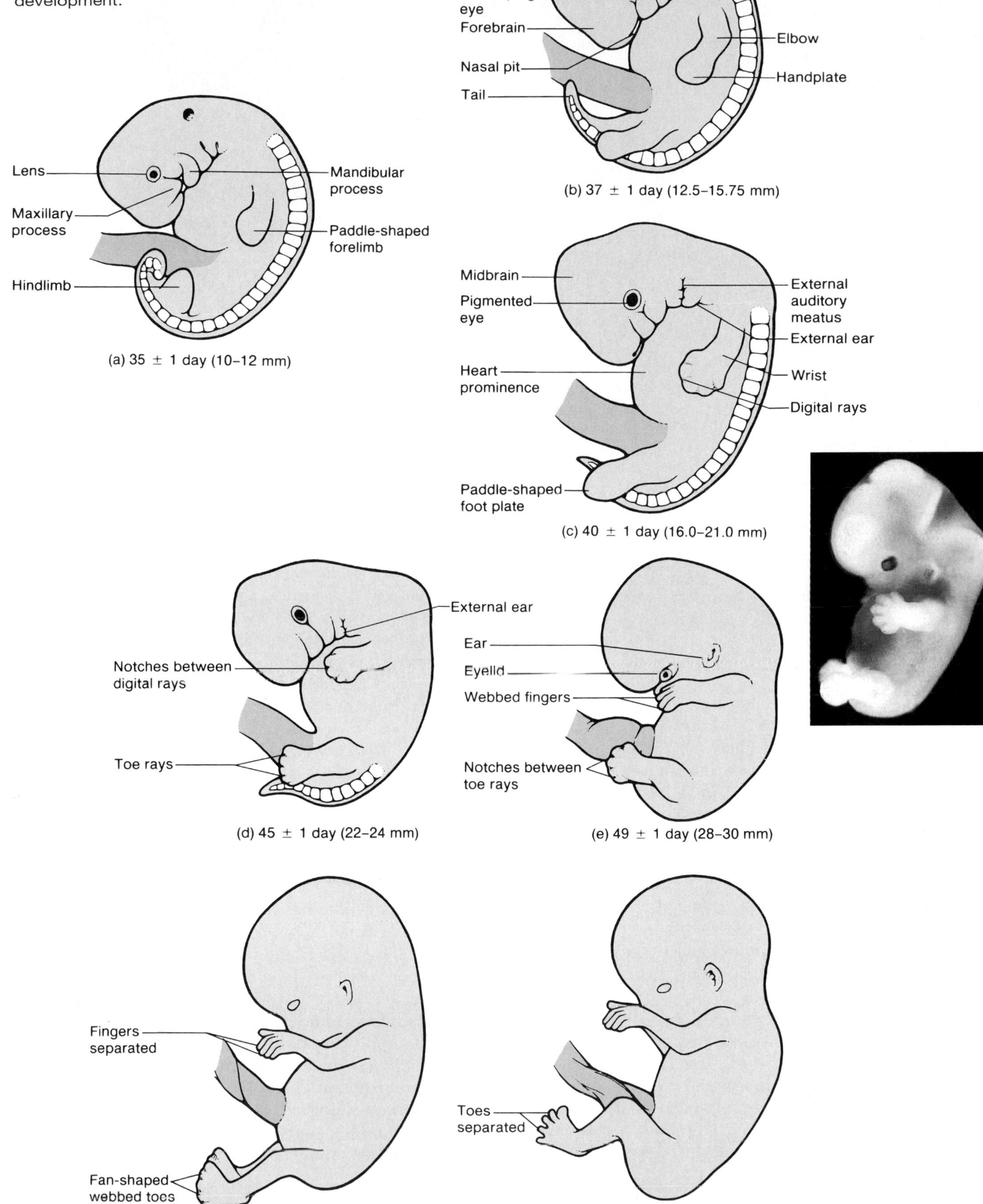

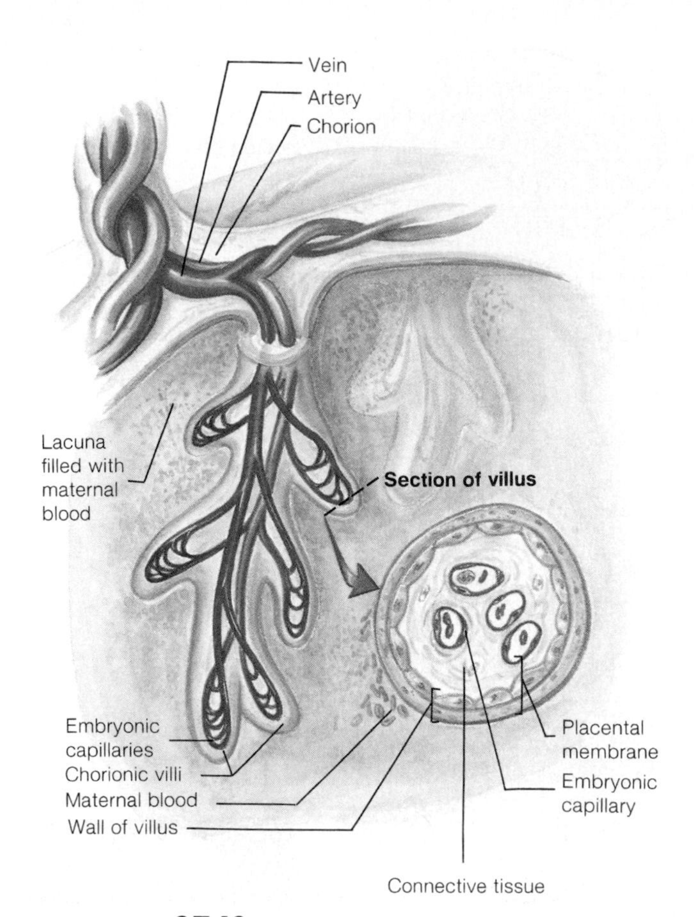


**FIGURE 23.12**

As is illustrated in the section of villus (lower part of figure), the placental membrane consists of the epithelial wall of an embryonic capillary and the epithelial wall of a chorionic villus.

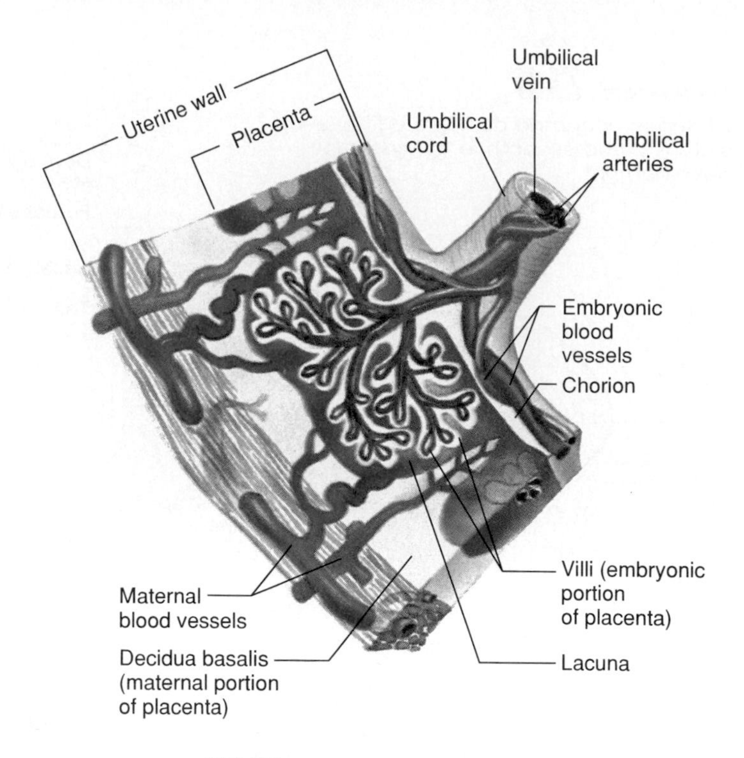


**FIGURE 23.13**

The placenta consists of an embryonic portion and a maternal portion.

While the placenta is forming from the chorion, another membrane, called the **amnion,** develops around the embryo. This second membrane begins to appear during the second week. Its margin is attached around the edge of the embryonic disk, and fluid called **amniotic fluid** fills the space between the amnion and the embryonic disk.

The developing placenta synthesizes progesterone from cholesterol in the maternal blood. Cells associated with the developing fetal adrenal glands use the placental progesterone to synthesize estrogens. The estrogens, in turn, promote changes in the maternal uterus and breasts, and influence the metabolism and development of various fetal organs.

As the embryo becomes more cylindrical, the margins of the amnion fold, enclosing the embryo in the amnion and amniotic fluid. The amnion envelops the tissues on the underside of the embryo, particularly the connecting stalk, by which it is attached to the chorion and the developing placenta. In this manner, the **umbilical cord** forms (fig. 23.14).

> A generation ago, biologists thought that the placenta screened all harmful substances from reaching a fetus. Today we know that if a pregnant woman takes an addictive substance, her newborn may suffer from withdrawal symptoms when amounts of the chemical it is accustomed to receiving suddenly plummet. Newborn addiction occurs with certain addictive drugs of abuse, such as heroin; with certain prescription drugs used to treat anxiety; and even with very large doses of vitamin C. Although vitamin C is not addictive, if a fetus is accustomed to megadoses, after birth the sudden drop in vitamin C level may bring on symptoms of vitamin C deficiency.

The fully developed umbilical cord is about 1 centimeter in diameter and about 55 centimeters in length. It begins at the umbilicus of the embryo and inserts into the center of the placenta. The cord contains three blood vessels—two *umbilical arteries* and one *umbilical vein*—that transport blood between the embryo and the placenta (fig. 23.15).

The umbilical cord also suspends the embryo in the *amniotic cavity,* and the amniotic fluid provides a watery environment in which the embryo can grow freely without being compressed by surrounding tissues. The amniotic fluid also protects the embryo from being jarred by the movements of the woman's body.

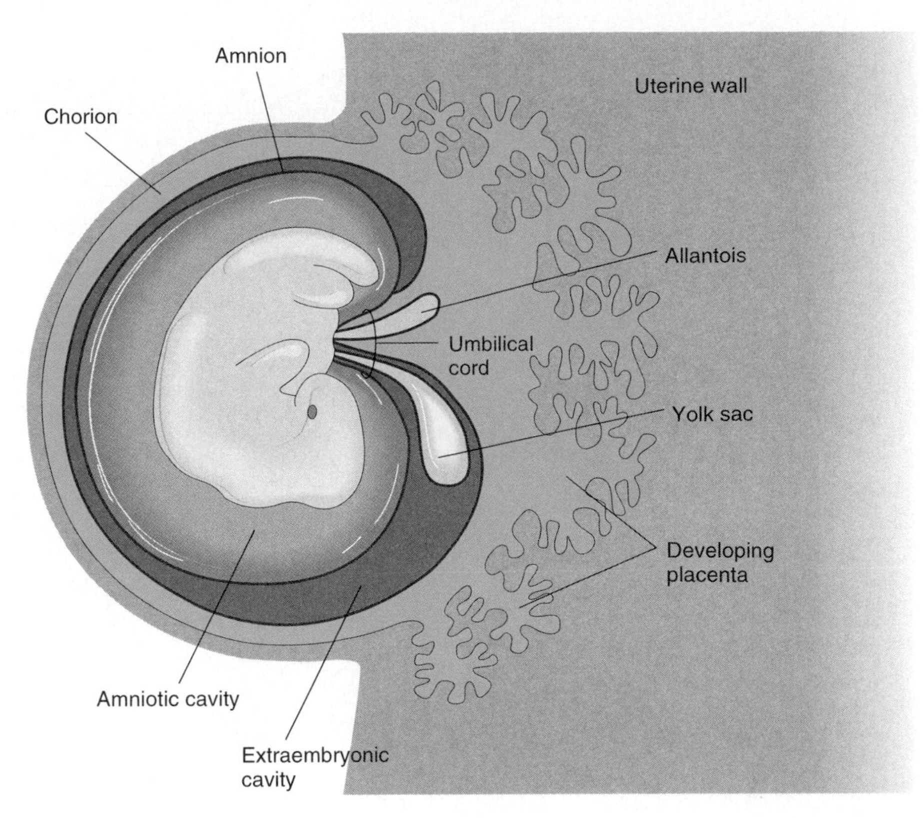


**Figure 23.14**
As the amnion develops, it surrounds the embryo, and the umbilical cord begins to form from structures in the connecting stalk.

In addition to the amnion and chorion, two other embryonic membranes appear during development. They are the yolk sac and the allantois.

The **yolk sac** appears during the second week, and it is attached to the underside of the embryonic disk (see fig. 23.14). It forms blood cells in the early stages of development and gives rise to the cells that later become sex cells. The yolk sac also gives rise to the stem cells of the immune system. Portions of the yolk sac form the embryonic digestive tube as well. Part of the membrane derived from the yolk sac becomes incorporated into the umbilical cord, while the remainder lies in the cavity between the chorion and the amnion near the placenta.

The **allantois** forms during the third week as a tube extending from the early yolk sac into the connecting stalk of the embryo. It, too, forms blood cells and gives rise to the umbilical arteries and vein (see figs. 23.14 and 23.15).

Eventually, the amniotic cavity becomes so enlarged that the membrane of the amnion contacts the thicker chorion around it, and the two membranes fuse into an *amniochorionic membrane* (see fig. 23.15).

The *embryonic stage* ends at the end of the eighth week. It is the most critical period of development, for during this time the embryo implants within the uterine wall and all the essential external and internal body parts form. Any disturbances in the developmental processes occurring during the embryonic stage are likely to result in major malformations or malfunctions. This is why early prenatal care is very important.

By the beginning of the eighth week, the embryo is usually 30 millimeters long and weighs less than 5 grams. Although its body is quite unfinished, it is recognizable as a human being (fig. 23.16).

Factors that cause congenital malformations by affecting an embryo during its period of rapid growth and development are called **teratogens.** Such agents include drugs, viruses, radiation, and even large amounts of otherwise healthful substances, such as fat-soluble vitamins. Each prenatal structure has a time in development, called its *critical period,* when it is sensitive to teratogens (fig. 23.17).

A critical period may extend over many months or be just a day or two. Neural tube defects, for example, are traced to day 28 in development, when a sheet of ectoderm called the neural tube normally folds into a tube, which then develops into the central nervous system. Occasionally something intervenes to disrupt this process, and an opening remains in the spine (spina bifida) or in the brain (anencephaly). If a woman has had a child with a neural tube defect, she

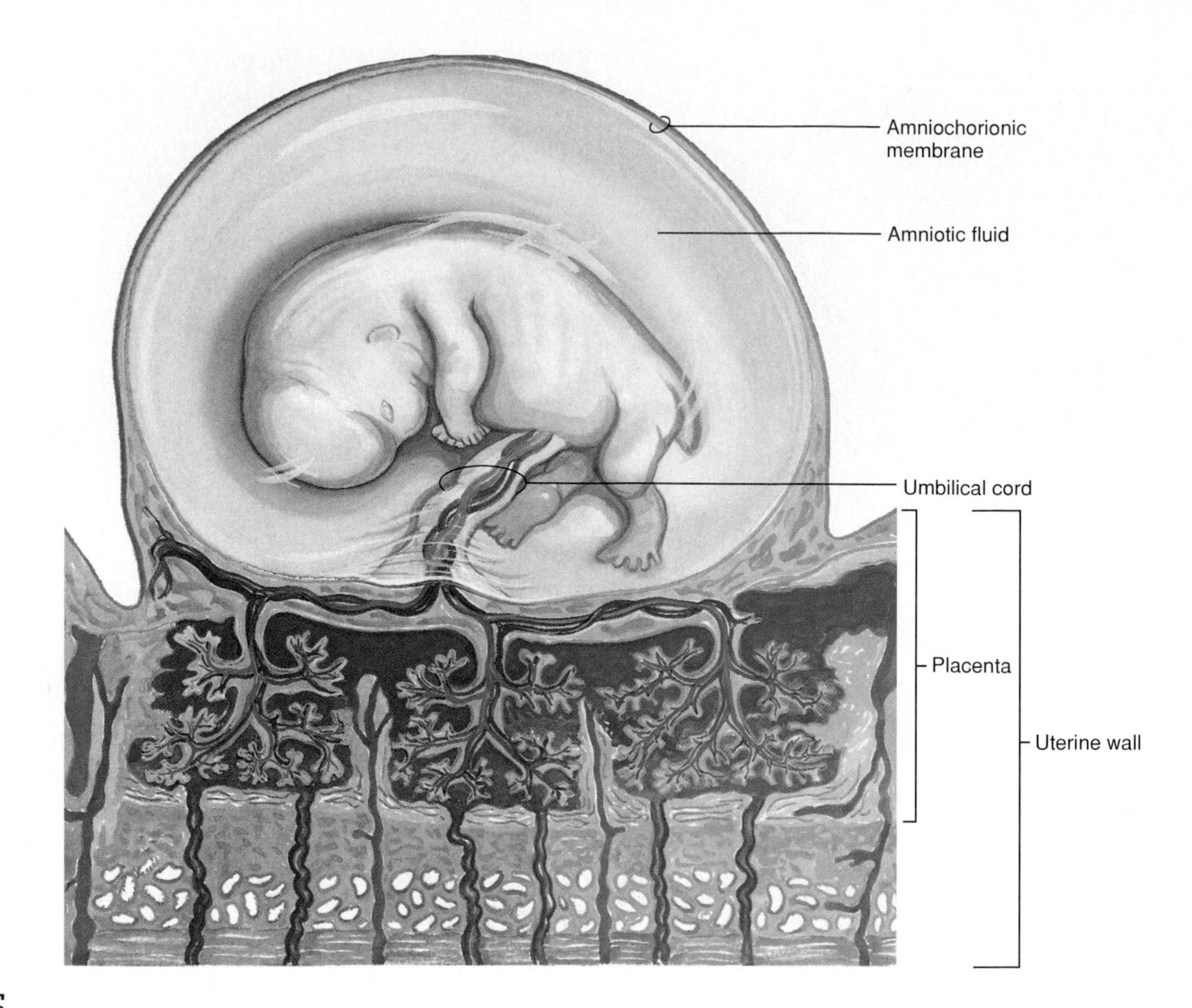


**FIGURE 23.15**
The developing placenta as it appears during the seventh week of development.

can reduce the 5% recurrence risk by two-thirds if she takes supplements of the vitamin folic acid just before and during the first trimester of pregnancy. This apparent protective effect suggests that folic acid may be crucial for normal development of the central nervous system.

The critical period for the brain begins when the anterior neural tube begins to swell into a brain, and continues throughout gestation. This is why so many teratogens affect the brain. Clinical Application 23.3 discusses some teratogens and their effects.

1. Describe the development of the amnion.
2. What blood vessels are in the umbilical cord?
3. What is the function of amniotic fluid?
4. What types of cells and other structures derive from the yolk sac?
5. How do teratogens cause birth defects?

## Fetal Stage

The **fetal stage** begins at the end of the eighth week of development and lasts until birth. During this period, growth is rapid, and body proportions change considerably. At the beginning of the fetal stage, the head is disproportionately large, and the lower limbs are relatively short (fig. 23.18). Gradually, proportions more like those of a child appear.

During the third month, the growth in body length accelerates, but growth of the head slows. The upper limbs achieve the relative length they will maintain throughout development, and ossification centers appear in most of the bones. By the twelfth week, the external reproductive organs are distinguishable as male or female. Figure 23.19 illustrates how these external reproductive organs of the male and female differentiate from precursor structures.

In the fourth month, the body grows very rapidly and reaches a length of up to 20 centimeters and weighs about 170 grams. The lower limbs lengthen considerably, and the skeleton continues to ossify. The fetus has hair, eyebrows, lashes, nipples, and nails, and may even scratch itself.

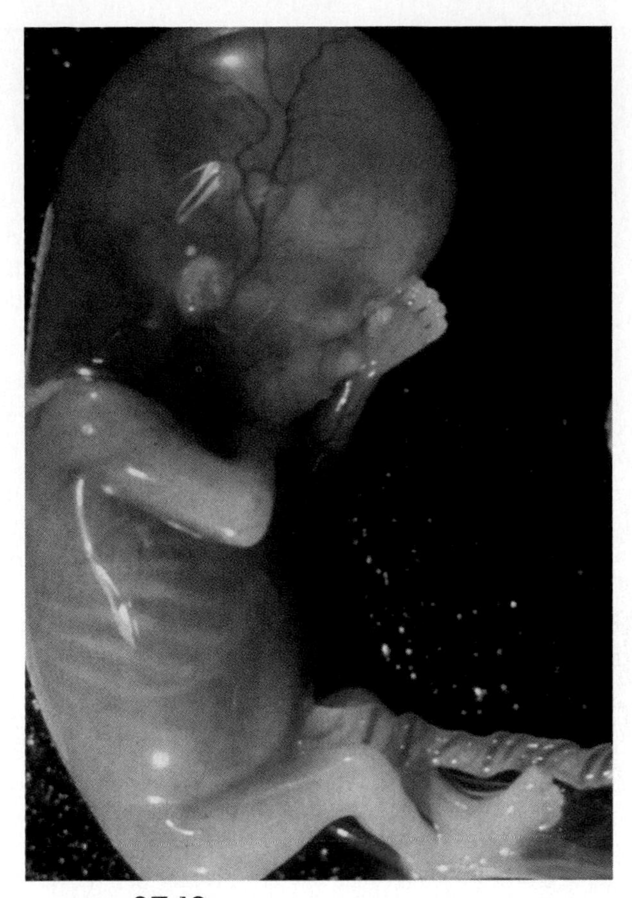

**FIGURE 23.16**

By the beginning of the eighth week of development, the embryonic body is recognizable as a human.

Amniocentesis is a prenatal test performed after the fourteenth week of gestation. A physician inserts a needle into the amniotic sac and withdraws about 5 milliliters of fluid. Fetal fibroblasts in the fluid are cultured and their chromosomes checked. It takes about a week to grow the chromosomes. A faster technique uses DNA probes to highlight specific chromosomes.

Amniocentesis carries about a 0.5% chance of being followed by miscarriage, and so it is usually performed on women over the age of 35, when the risk of the fetus having a chromosomal anomaly equals the risk of the procedure. Women of any age who have already had a child with a detectable chromosomal abnormality are also candidates for amniocentesis.

An experimental prenatal test, fetal cell sorting, is safer than either chorionic villus sampling or amniocentesis because it detects the rare fetal cells that make their way into the maternal bloodstream.

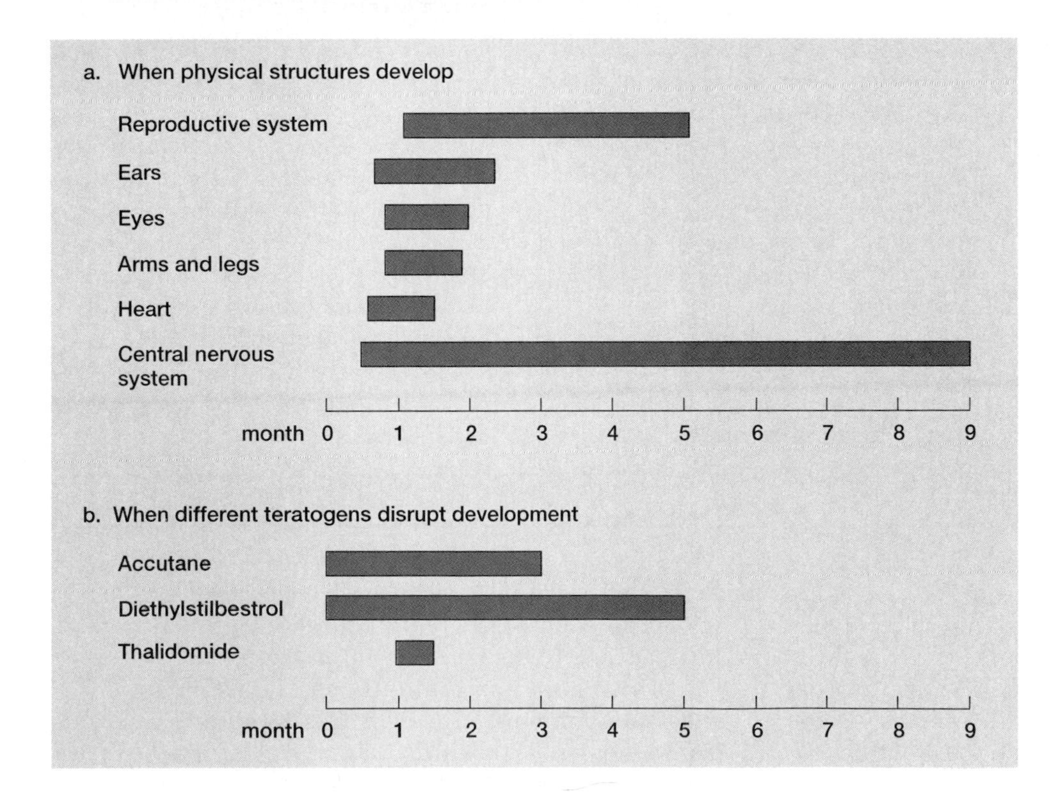


**FIGURE 23.17**

Structures in the developing embryo and fetus are sensitive to specific teratogens at different times in gestation.

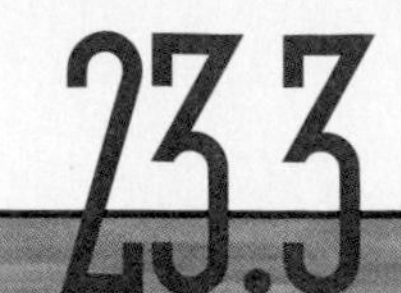

Clinical Application

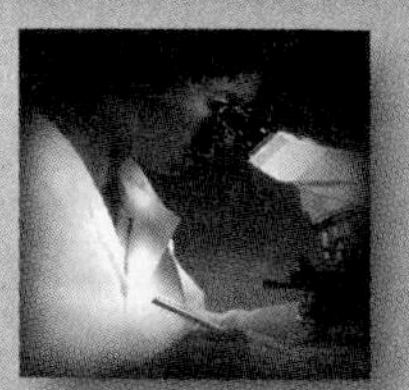

## Some Causes of Birth Defects

### Thalidomide

The idea that the placenta protects the embryo and fetus from harmful substances was tragically disproven between 1957 and 1961, when 10,000 children in Europe were born with flippers in place of limbs. Doctors soon identified a mild tranquilizer, thalidomide, which all of the mothers of deformed infants had taken early in pregnancy, during the time of limb formation. The United States was spared a thalidomide disaster because an astute government physician noted adverse effects of the drug on monkeys in experiments, and she halted testing.

### Rubella

At about the same time that the medical community was noting the severity of the thalidomide crisis, another teratogen, a virus, was sweeping through the United States. In the early 1960s, a rubella epidemic here caused 20,000 birth defects and 30,000 stillbirths.

### Alcohol

A pregnant woman who has just one or two alcoholic drinks a day, or perhaps a large amount at a crucial time in prenatal development, risks *fetal alcohol syndrome* in her unborn child. Because the effects of small amounts of alcohol at different stages of pregnancy are not yet well understood, and because each woman metabolizes alcohol slightly differently, it is best to avoid drinking alcohol entirely when pregnant, or when trying to become pregnant.

A child with fetal alcohol syndrome has a characteristic small head, misshapen eyes, and a flat face and nose (fig. 23C). He or she grows slowly before and after birth. Intellect is impaired, ranging from minor learning disabilities to mental retardation.

Teens and young adults with fetal alcohol syndrome are short and have small heads. Many individuals remain at early grade school level. They often lack social and communication skills, such as understanding the consequences of actions, forming friendships, taking initiative, and interpreting social cues.

Problems in children of alcoholic mothers were noted by Aristotle more than 23 centuries ago. In the United States today, fetal alcohol syndrome is the third most common cause of mental retardation in newborns, and 1 to 3 in every 1,000 infants has the syndrome—more than 40,000 born each year.

### Cigarettes

Chemicals in cigarette smoke stress a fetus. Carbon monoxide crosses the placenta and plugs up the sites on the fetus's hemoglobin molecules that would normally bind oxygen. Other chemicals in smoke prevent nutrients from reaching the fetus. Studies comparing placentas of smokers and nonsmokers show that smoke-exposed placentas lack important growth factors. The result of all of these assaults is poor growth before and after birth. Cigarette smoking during pregnancy is linked to spontaneous abortion, stillbirth, prematurity, and low birth weight.

### Nutrients

Certain nutrients in large amounts, particularly vitamins, act in the body as drugs. The acne medication *isotretinoin* (Accutane) is a derivative of vitamin A that causes spontaneous abortions and defects of the heart, nervous system, and face. The tragic effects of this drug were noted exactly 9 months after dermatologists began prescribing it to young women in the early 1980s. Today, the drug package bears prominent warnings, and it is never prescribed to pregnant women. A vitamin A-based drug used

In the fifth month, growth rate decreases somewhat. The lower limbs achieve their final relative proportions. The skeletal muscles become active, and the pregnant woman may experience the first exciting detectable fetal movements. Some hair appears on the fetal head, and the skin becomes covered with fine, downy hair called lanugo. The skin is also coated with a cheesy mixture of sebum from the sebaceous glands and dead epidermal cells (vernix caseosa). The fetus, weighing about 450 grams and about 30 centimeters long, curls into the curved fetal position.

During the sixth month, the fetal body gains a substantial amount of weight. The eyebrows and eyelashes appear. The skin is quite wrinkled and translucent. Blood vessels in the skin give the fetus a reddish appearance.

to treat psoriasis, as well as excesses of vitamin A itself, also cause birth defects. This is because some forms of vitamin A are stored in body fat for up to 3 years after ingestion.

Malnutrition in a pregnant woman threatens the fetus. Obstetrical records of pregnant women before, during, and after World War II link inadequate nutrition early in pregnancy to an increase in the incidence of spontaneous abortion. The aborted fetuses had very little brain tissue. Poor nutrition later in pregnancy affects development of the placenta. The infant has a low birth weight and is at high risk for short stature, tooth decay, delayed sexual development, learning disabilities, and possibly mental retardation.

### Occupational Hazards

Some teratogens are encountered in the workplace. Increased rates of spontaneous abortion and birth defects have been noted among women who work with textile dyes, lead, certain photographic chemicals, semiconductor materials, mercury, and cadmium. We do not know much about the role of the male in environmentally caused birth defects. Men whose jobs expose them to sustained heat, such as smelter workers, glass manufacturers, and bakers, may produce sperm that can fertilize an oocyte and possibly lead to spontaneous abortion or a birth defect. A virus or a toxic chemical carried in semen may also cause a birth defect.

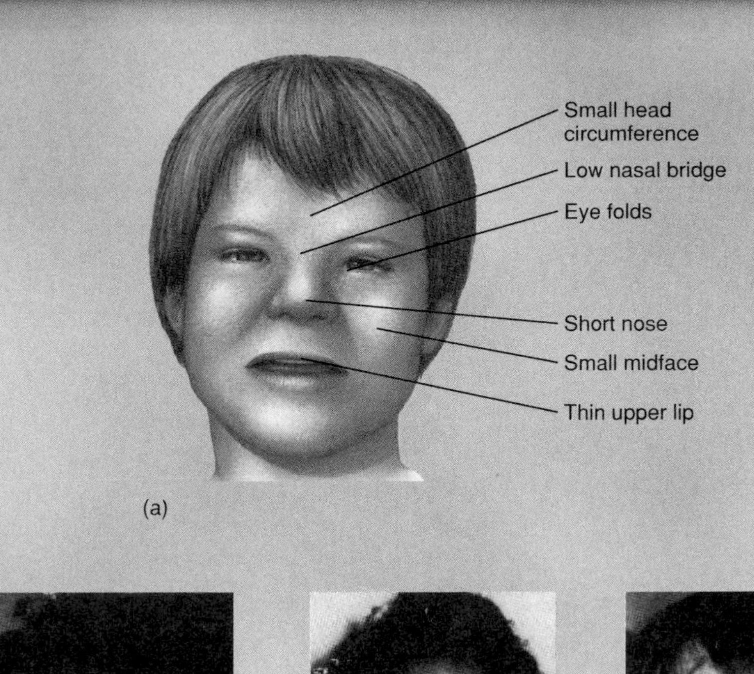


(a)

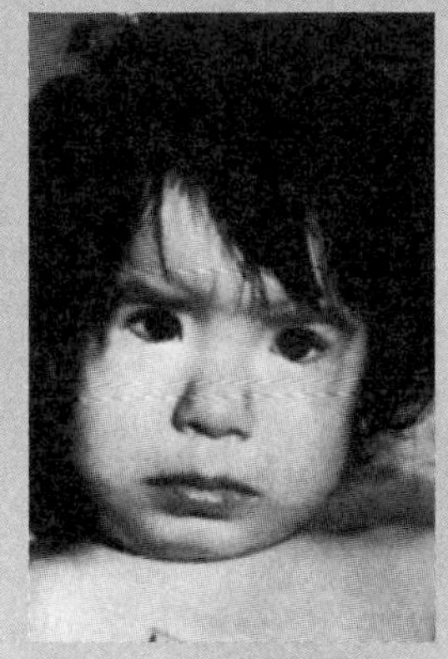

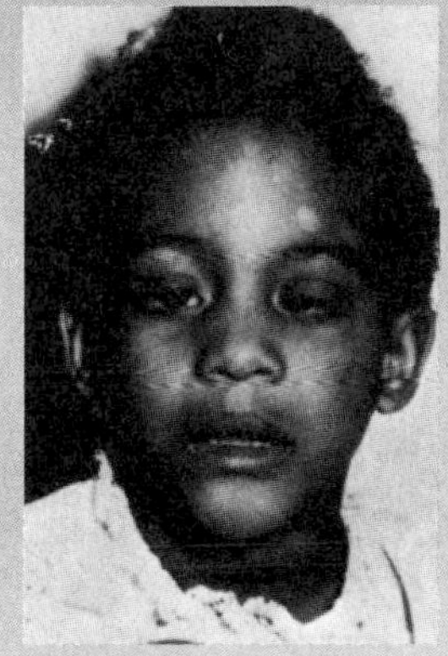

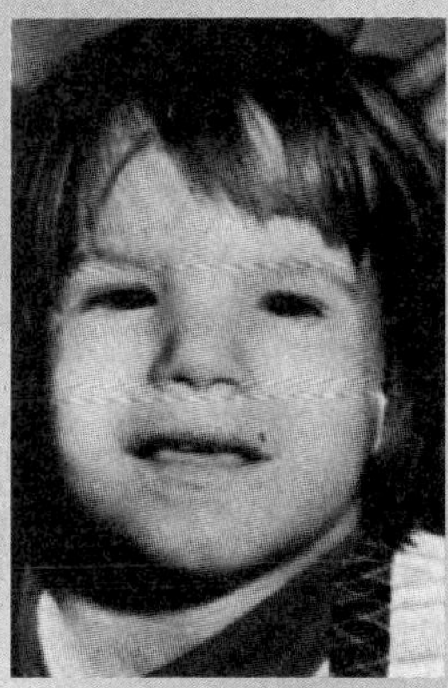

(b)

**Figure 23C**

Fetal alcohol syndrome. Some children whose mothers drank alcohol during pregnancy have characteristic flat faces (*a*) that are strikingly similar in children of different races (*b*). Women who drink excessively while pregnant have a 30% to 45% chance of having a child who is affected to some degree by prenatal exposure to alcohol. Two mixed drinks per day seems to be the level at which damage occurs, but researchers are not yet certain of this.

In the seventh month, the skin becomes smoother as fat is deposited in the subcutaneous tissues. The eyelids, which fused together during the third month, reopen. At the end of this month, a fetus is about 40 centimeters in length.

In the final trimester, fetal brain cells rapidly form networks, as organs elaborate and grow. A layer of fat is laid down beneath the skin. The testes of males descend from regions near the developing kidneys, through the inguinal canal, and into the scrotum (see chapter 22). The digestive and respiratory systems mature last, which is why infants born prematurely often have difficulty digesting milk and breathing.

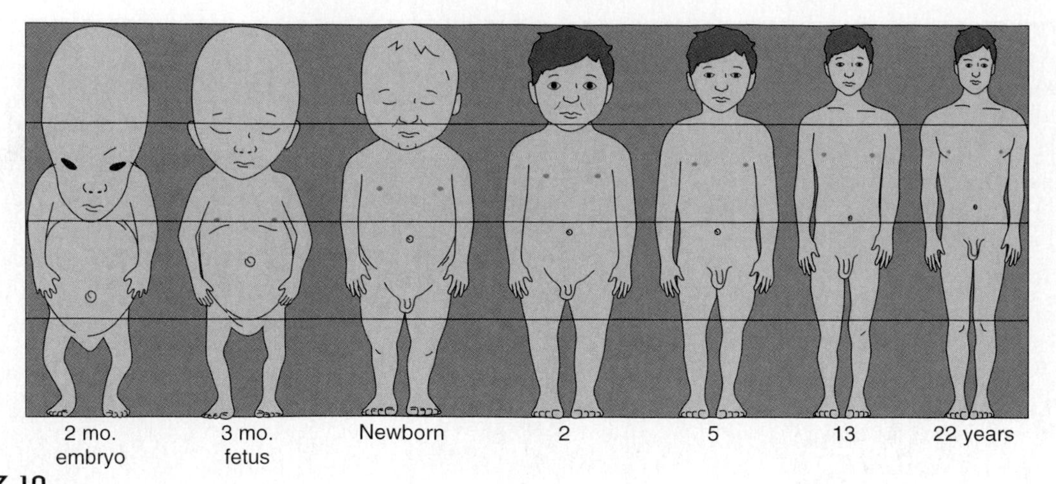


**Figure 23.18**
During development, body proportions change considerably.

If a fetus is born prematurely, its chance of surviving increases directly with its age and weight. One factor that affects the chance of survival is the development of the lungs. Survival chance increases if the lungs are sufficiently developed so that they have the thin respiratory membranes necessary for rapid exchange of oxygen and carbon dioxide, and if they produce enough surfactant to reduce the alveolar surface tension (see chapter 19). A fetus of less than 24 weeks or weighing less than 600 grams at birth seldom survives, even when given intensive medical care. Neonatology is the field of medicine that deals with premature and ill newborns.

Approximately 266 days after a single sperm burrowed its way into an oocyte, a baby is ready to be born. It is about 50 centimeters long and weighs 2.7 to 3.6 kilograms. The skin has lost its downy hair, but is still coated with sebum and dead epidermal cells. The scalp is usually covered with hair; the fingers and toes have well-developed nails; and the skull bones are largely ossified. As figure 23.20 shows, the fetus is usually positioned upside down with its head toward the cervix (*vertex position*).

The birth of a live, healthy baby is against the odds, considering human development from the beginning. Of every 100 secondary oocytes that are exposed to sperm, 84 are fertilized. Of these, 69 implant in the uterus, 42 survive 1 week or longer, 37 survive 6 weeks or longer, and only 31 are born alive. Of those that do not survive to birth, about half have chromosomal abnormalities that are too severe to maintain life.

Chart 23.1 summarizes the stages of prenatal development.

**The Fetus as a Patient**
Little Blake Schultz made medical history when he underwent major surgery 7 weeks before his birth. Ultrasound had revealed that his stomach, spleen, and intestines protruded through a hole in his diaphragm, the muscle sheet separating the abdomen from the chest. This defect would have suffocated him shortly after birth, were it not for pioneering surgery by Michael Harrison at the University of California at San Francisco. Through an incision in the pregnant woman, the surgical team exposed Blake's left side, gently tucked his organs in place, and patched the hole with Gore-Tex, a synthetic material used in clothing.

Some prenatal medical problems can be treated by administering drugs to the pregnant woman or by altering her diet. An undersized fetus can receive a nutritional boost by putting the pregnant woman on a high-protein diet. It is also possible to treat prenatal medical problems directly: A tube inserted into the uterus can drain the dangerously swollen bladder of a fetus with a blocked urinary tract, providing relief until the problem can be surgically corrected at birth. A similar procedure can remove excess fluid from the brain of a hydrocephalic fetus.

1. What major changes occur during the fetal stage of development?
2. When can the sex of a fetus be determined?
3. How is a fetus usually positioned within the uterus at the end of the pregnancy?

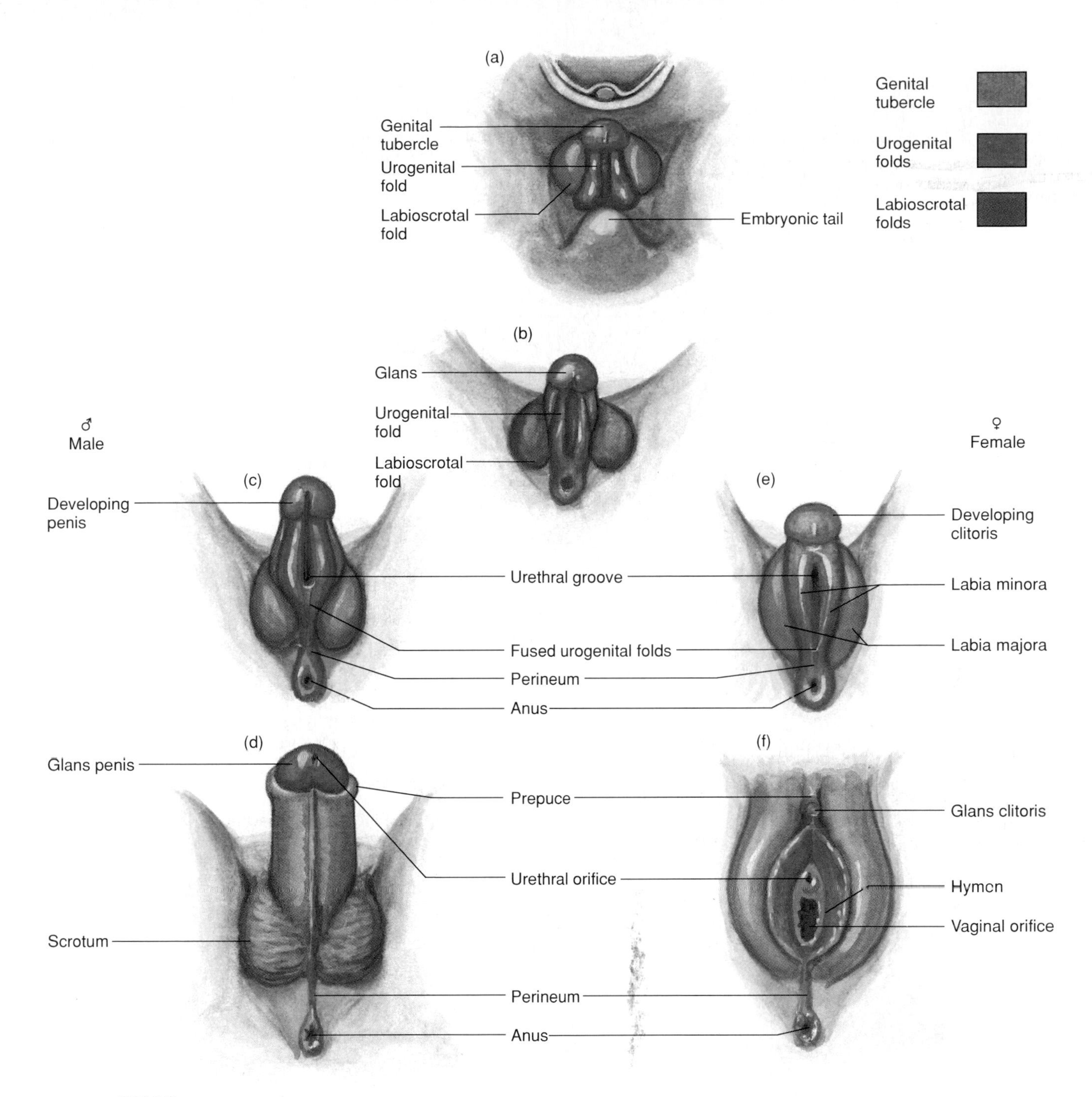


**FIGURE 23.19**

(*a* and *b*) The genital tubercle, urogenital fold, and labioscrotal folds that appear during the fourth week of development may differentiate into (*c* and *d*) male external reproductive organs or (*e* and *f*) female external reproductive organs.

## Fetal Blood and Circulation

Throughout fetal development, the maternal blood supplies oxygen and nutrients, and carries away wastes. These substances diffuse between the maternal and fetal blood through the placental membrane, and they are carried to and from the fetal body by the umbilical blood vessels (fig. 23.21). Consequently, the fetal blood and vascular system must adapt to intrauterine existence. For example, the concentration of oxygen-carrying hemoglobin in the fetal blood is about 50% greater than in the maternal blood, and fetal hemoglobin has a greater attraction for oxygen than an adult's hemoglobin. Thus, at a particular oxygen pressure, fetal hemoglobin can carry 20%–30% more oxygen than adult hemoglobin.

## CHART 23.1 Stages of Prenatal Development

| Stage | Time Period | Major Events |
|---|---|---|
| Period of cleavage | First week | Cells undergo mitosis, blastocyst forms; inner cell mass appears; blastocyst becomes implanted in uterine wall<br>Size: 1/4 inch (0.63 centimeters), weight: 1/120 ounce (0.21 grams) |
| Embryonic stage | Second through eighth week | Inner cell mass becomes embryonic disk; primary germ layers form, embryo proper becomes cylindrical; main internal organs and external body structures appear; placenta and umbilical cord form, embryo proper is suspended in amniotic fluid<br>Size: 1 inch (2.5 centimeters), weight: 1/30 ounce (0.8 grams) |
| Fetal stage | Ninth through twelfth week | Ossification centers appear in bones, sex organs differentiate, nerves and muscles become coordinated so that the fetus can move its upper and lower limbs<br>Size: 4 inches (10 centimeters), weight: 1 ounce (28 grams) |
| | Thirteenth through sixteenth week | Body grows rapidly; ossification continues<br>Size: 8 inches (20 centimeters), weight: 6 ounces (170 grams) |
| | Seventeenth through twentieth week | Muscle movements are stronger and woman may be aware of slight flutterings; skin is covered with fine downy hair (lanugo) and coated with sebum mixed with dead epidermal cells (vernix caseosa)<br>Size: 12 inches (30.5 centimeters), weight: 1 pound (454 grams) |
| | Twenty-first through thirty-eighth week | Body gains weight, subcutaneous fat deposited; eyebrows and lashes appear; eyelids reopen; testes descend.<br>Size: 21 inches (53 centimeters), weight: 6 to 10 pounds (2.7 to 4.5 kilograms) |

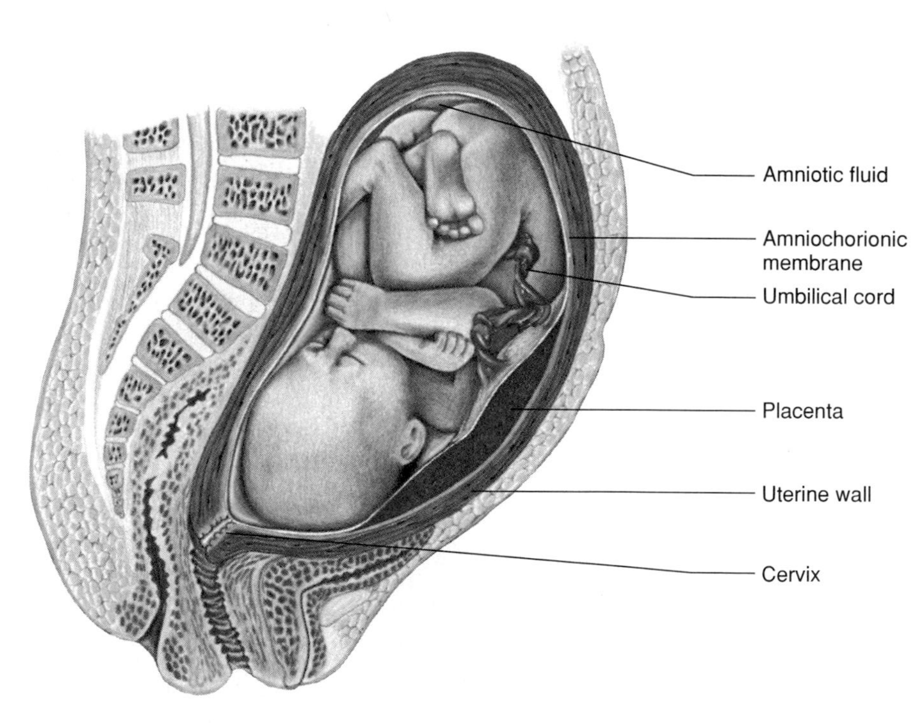


**FIGURE 23.20**
A full-term fetus usually becomes positioned with its head near the cervix.

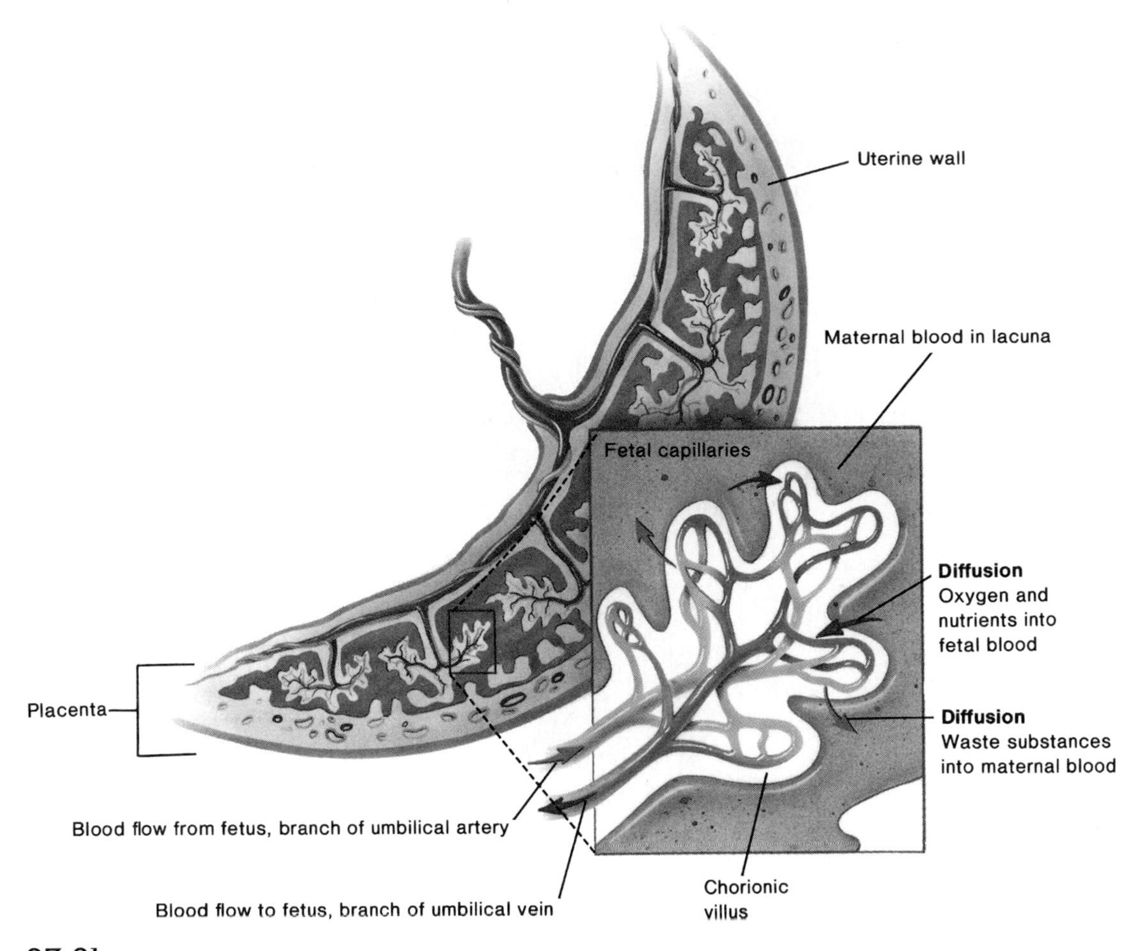


**FIGURE 23.21**
Oxygen and nutrients diffuse into the fetal blood from the maternal blood. Wastes diffuse into the maternal blood from the fetal blood.

In the fetal circulatory system, the *umbilical vein* transports blood rich in oxygen and nutrients from the placenta to the fetal body. This vein enters the body through the umbilical ring and travels along the anterior abdominal wall to the liver. About half the blood it carries passes into the liver, and the rest enters a vessel called the **ductus venosus,** which bypasses the liver.

The ductus venosus travels a short distance and joins the inferior vena cava. There, the oxygenated blood from the placenta mixes with deoxygenated blood from the lower parts of the fetal body. This mixture continues through the vena cava to the right atrium.

In an adult heart, the blood from the right atrium enters the right ventricle and is pumped through the pulmonary trunk and pulmonary arteries to the lungs. In the fetus, however, the lungs are nonfunctional, and the blood largely bypasses them. As blood from the inferior vena cava enters the fetal right atrium, much of it is shunted directly into the left atrium through an opening in the atrial septum. This opening is called the **foramen ovale,** and the blood passes through it because the blood pressure in the right atrium is somewhat greater than that in the left atrium. Furthermore, a small valve (septum primum) located on the left side of the atrial septum overlies the foramen ovale and helps prevent blood from moving in the reverse direction.

The rest of the fetal blood entering the right atrium, including a large proportion of the deoxygenated blood entering from the superior vena cava, passes into the right ventricle and out through the pulmonary trunk. Only a small volume of blood enters the pulmonary circuit, because the lungs are collapsed, and their blood vessels have a high resistance to blood flow. However, enough blood reaches the lung tissues to sustain them.

Most of the blood in the pulmonary trunk bypasses the lungs by entering a fetal vessel called the **ductus arteriosus,** which connects the pulmonary trunk to the descending portion of the aortic arch. As a result of this connection, the blood with a relatively low oxygen concentration, which is returning to the heart

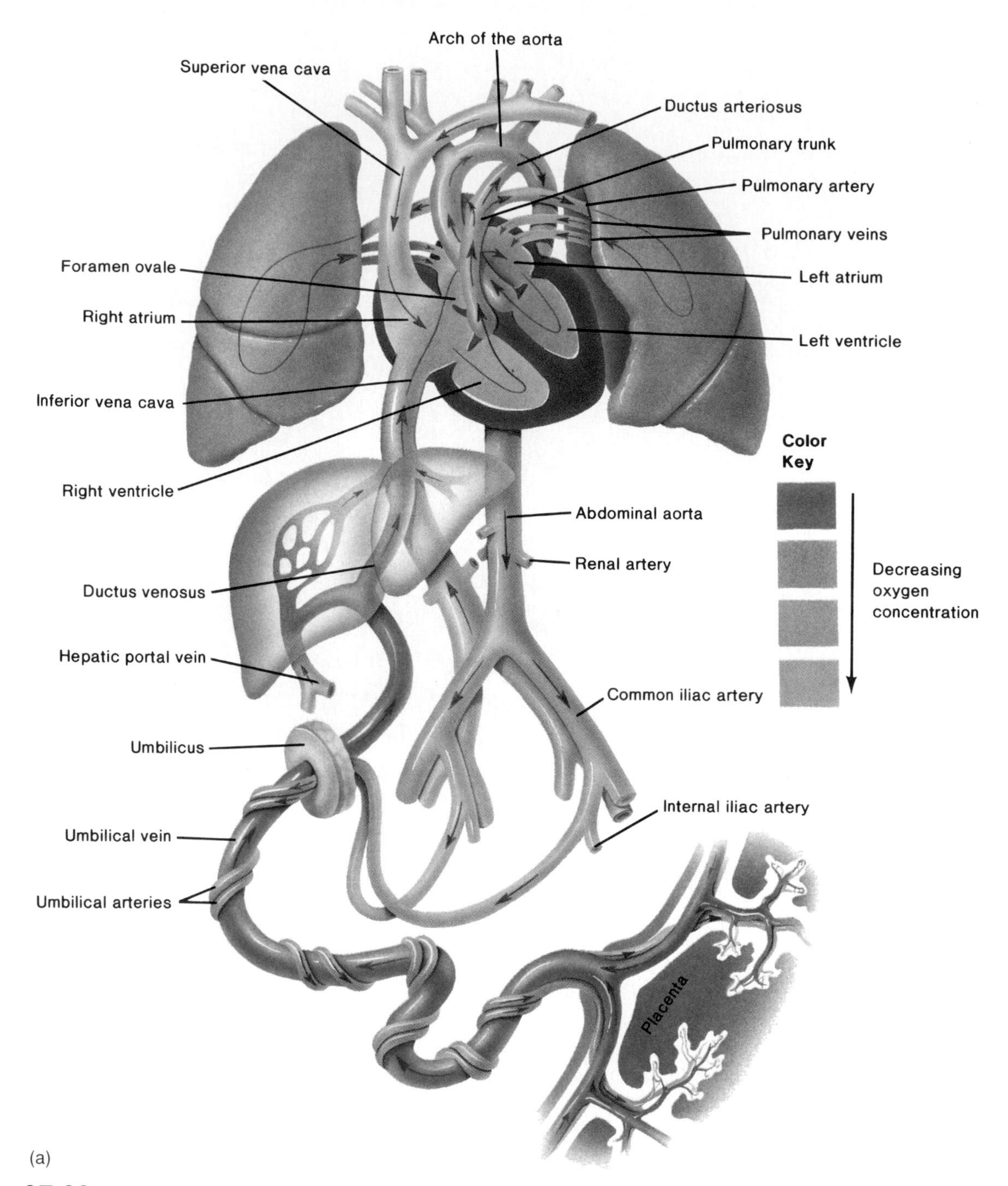


**FIGURE 23.22**
The general pattern of fetal circulation is shown anatomically (*a*) and schematically (*b*).

through the superior vena cava, bypasses the lungs. At the same time, it is prevented from entering the portion of the aorta that branches to the heart and brain.

The more highly oxygenated blood that enters the left atrium through the foramen ovale mixes with a small amount of deoxygenated blood returning from the pulmonary veins. This mixture moves into the left ventricle and is pumped into the aorta. Some of it reaches the myocardium through the coronary arteries, and some reaches the brain tissues through the carotid arteries.

The blood carried by the descending aorta is partially oxygenated and partially deoxygenated. Some of it is carried into the branches of the aorta that lead to the lower regions of the body. The rest passes into the *umbilical arteries,* which branch from the internal iliac arteries and lead to the placenta. There the blood is reoxygenated (fig. 23.22).

The umbilical cord usually contains two arteries and one vein. A small percentage of newborns have only one umbilical artery. This condition is

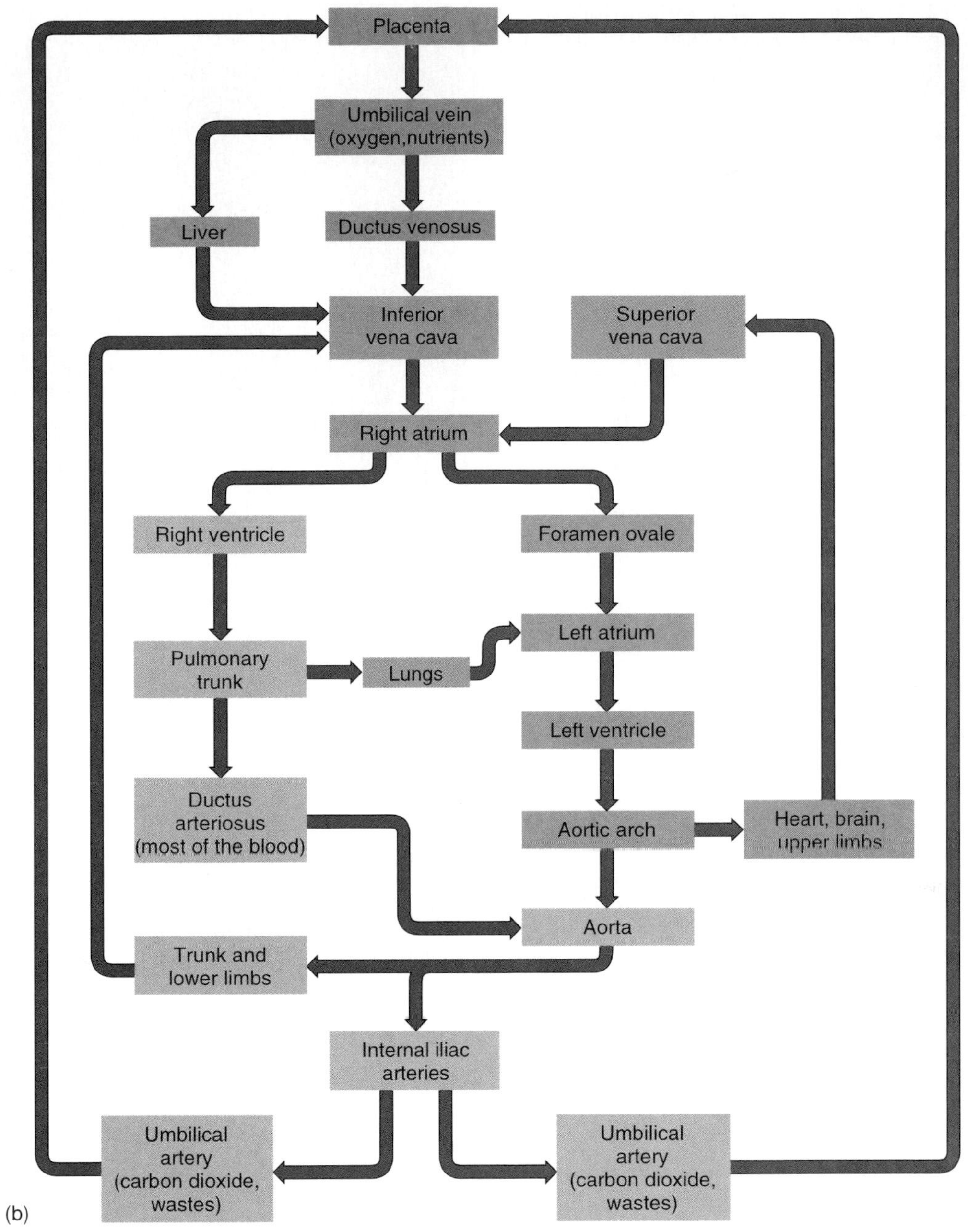


**FIGURE 23.22** *Continued*

often associated with other cardiovascular or gastrointestinal disorders. Because of the possibility of these conditions, the vessels within the severed cord are routinely counted following a birth.

Chart 23.2 summarizes the major features of fetal circulation. At the time of birth, important adjustments must occur in the circulatory system when the placenta ceases to function and the newborn begins to breathe.

1. How does the pattern of fetal circulation differ from that of an adult?
2. Which umbilical vessel carries oxygen-rich blood to the fetus?
3. What is the function of the ductus venosus?
4. How does fetal circulation allow blood to bypass the lungs?
5. What characteristic of the fetal lungs tends to shunt blood away from them?

> Blood in the umbilical cord at birth is a rich source of hematopoietic stem cells. Umbilical blood cells can be "banked," and used for an autologous bone marrow transplant should the individual later in life develop cancer of the blood cells or severe anemia.

## Chart 23.2 Fetal Circulatory Adaptations

| Adaptation | Function |
|---|---|
| Fetal blood | Has greater oxygen-carrying capacity than adult blood |
| Umbilical vein | Carries oxygenated blood from the placenta to the fetus |
| Ductus venosus | Conducts about half the blood from the umbilical vein directly to the inferior vena cava, thus bypassing the liver |
| Foramen ovale | Conveys a large proportion of the blood entering the right atrium from the inferior vena cava, through the atrial septum, and into the left atrium, thus bypassing the lungs |
| Ductus arteriosus | Conducts some blood from the pulmonary trunk to the aorta, thus bypassing the lungs |
| Umbilical arteries | Carry the blood from the internal iliac arteries to the placenta |

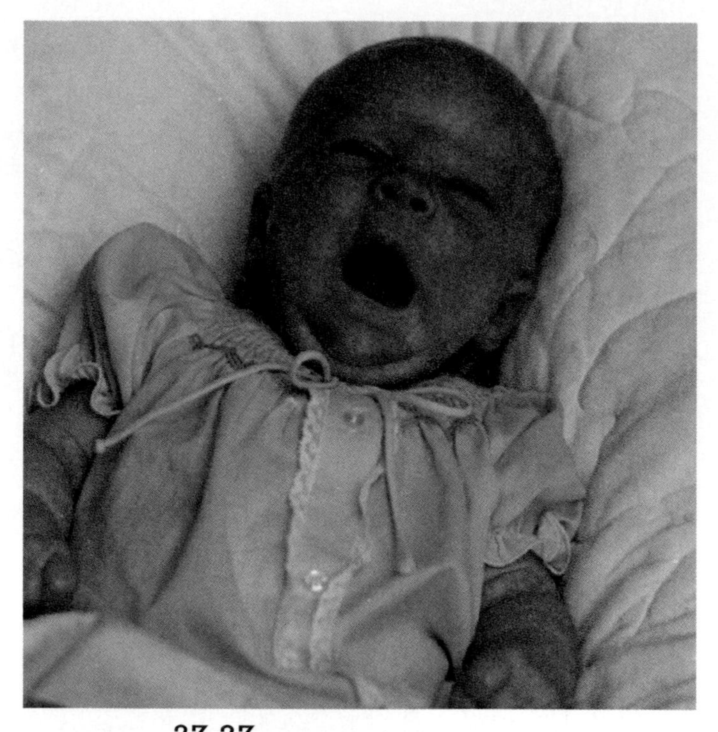

**Figure 23.23**
The neonatal period extends from birth to the end of the fourth week after birth.

## Postnatal Period

The postnatal period of development lasts from birth until death. It can be divided into the neonatal period, infancy, childhood, adolescence, adulthood, and senescence.

### Neonatal Period

The **neonatal period,** which extends from birth to the end of the first four weeks, begins very abruptly at birth (fig. 23.23). At that moment, physiological adjustments must be made quickly, because the newborn must suddenly do for itself what the pregnant woman's body has been doing for it. Thus, the newborn (neonate) must respire, obtain and digest nutrients, excrete wastes, and regulate body temperature. However, a newborn's most immediate need is to obtain oxygen and excrete carbon dioxide, so its first breath is critical.

The first breath must be particularly forceful, because the newborn's lungs are collapsed, and the airways are small, offering considerable resistance to air movement. Also, surface tension tends to hold the moist membranes of the lungs together. However, the lungs of a full-term fetus continuously secrete surfactant (see chapter 19), which reduces surface tension. Thus, after the first powerful breath begins to expand the lungs, breathing becomes easier.

> Mechanical stimulation of a newborn's lungs, caused by the pregnant woman's muscular uterine contractions during natural birth, seems to increase the quantity of surfactant secreted. This may explain why respiratory distress syndrome occurs more commonly in infants whose births do not include labor.

A newborn's first breath is stimulated by increasing concentration of carbon dioxide, decreasing pH, low oxygen concentration, drop in body temperature, and mechanical stimulation that occurs during and after birth. Also, in response to the stress the fetus experiences during the birth process, the fetal blood concentrations of epinephrine and norepinephrine rise significantly (see chapter 13). These hormones promote normal breathing by increasing the secretion of surfactant and dilating the airways.

For energy, the fetus depends primarily on glucose and fatty acids in the pregnant woman's blood. The newborn, on the other hand, is suddenly without an external source of nutrients. The woman will not produce milk for two to three days, by which time the infant's gastrointestinal tract will be able to digest it. However, the mother's breasts secrete *colostrum,* a fluid rich in nutrients and antibodies, until the mature milk comes in—an adaptation to the state of the newborn's digestive physiology. The newborn has a relatively high rate of metabolism, and its liver, which is not fully mature, may be unable to supply enough glucose to support its metabolic needs. Consequently, the newborn typically utilizes stored fat as an energy source.

1. Define the neonatal period of development.
2. Why must the first breath of an infant be particularly forceful?

3 What factors stimulate the first breath?

4 What does a newborn use for energy during the first few days after birth?

A newborn's kidneys are usually unable to produce concentrated urine, so they excrete a relatively dilute fluid. For this reason, the newborn may become dehydrated and develop a water and electrolyte imbalance. Also, some of the newborn's homeostatic control mechanisms may not function adequately. For example, the temperature regulating system may be unable to maintain a constant body temperature. As a result, body temperature may respond to slight stimuli by fluctuating above or below the normal level during the first few days of life.

When the placenta ceases to function and breathing begins, changes also occur in the newborn's circulatory system. For example, following birth, the umbilical vessels constrict. The arteries close first, and if the umbilical cord is not clamped or severed for a minute or so, blood continues to flow from the placenta to the newborn through the umbilical vein, adding to the newborn's blood volume.

The proximal portions of the umbilical arteries persist in the adult as the *superior vesical arteries* that supply blood to the urinary bladder. The more distal portions become solid cords (lateral umbilical ligaments). The umbilical vein becomes the cordlike *ligamentum teres* that extends from the umbilicus to the liver in an adult. The ductus venosus constricts shortly after birth and appears in the adult as a fibrous cord (ligamentum venosum) superficially embedded in the wall of the liver.

The foramen ovale closes as a result of blood pressure changes in the right and left atria as the fetal vessels constrict. As blood ceases to flow from the umbilical vein into the inferior vena cava, the blood pressure in the right atrium falls. Also, as the lungs expand with the first breathing movements, resistance to blood flow through the pulmonary circuit decreases, more blood enters the left atrium through the pulmonary veins, and blood pressure in the left atrium increases.

As the pressure in the left atrium rises and that in the right atrium falls, the valve (septum primum) on the left side of the atrial septum closes the foramen ovale. In most individuals, this valve gradually fuses with the tissues along the margin of the foramen. In an adult, the site of the previous opening is marked by a depression called the *fossa ovalis.*

The ductus arteriosus, like other fetal vessels, constricts after birth. After the ductus arteriosus closes, blood can no longer bypass the lungs by moving from the pulmonary trunk directly into the aorta. In an adult, the ductus arteriosus is represented by a cord called the *ligamentum arteriosum.*

The ductus arteriosus sometimes fails to close completely. This condition is called patent ductus arteriosus (PDA). Although the cause of this condition is usually unknown, it often occurs in newborns whose mothers were infected with rubella virus (German measles) during the first three months of pregnancy.

After birth, the metabolic rate and oxygen consumption of the neonatal tissues increase, in large part because of the need to maintain body temperature. If the ductus arteriosus remains open, the neonate's blood oxygen concentration may stay too low to adequately supply body tissues, including the myocardium. If not corrected surgically the heart may fail, even though the myocardium is normal.

Changes in the newborn's circulatory system are gradual. Although constriction of the ductus arteriosus may be functionally complete within fifteen minutes, the permanent closure of the foramen ovale may take up to a year. These circulatory changes are illustrated in figure 23.24 and summarized in chart 23.3.

1 How do the kidneys of a newborn differ from those of an adult?

2 What portions of the umbilical arteries continue to function into adulthood?

3 What is the fate of the foramen ovale? Of the ductus arteriosus?

4 When is closure of the ductus arteriosus functionally complete?

## Infancy

The period of continual development extending from the end of the first four weeks to one year is called **infancy.** During this time, the infant grows rapidly and may triple its birth weight. Its teeth begin to erupt through the gums, and its muscular and nervous systems mature so that coordinated muscular activities become possible. The infant is soon able to follow objects visually; reach for and grasp objects; and sit, creep, and stand.

Infancy also brings the beginning of the ability to communicate. The infant learns to smile, laugh, and respond to some sounds. By the end of the first year, the infant may be able to say two or three words. Which parent's name is first spoken is sometimes a matter of great interest; often it is the name of a beloved pet.

## CHART 23.3 Circulatory Adjustments in the Newborn

| Structure | Adjustment | In the Adult |
|---|---|---|
| Umbilical vein | Constricts | Becomes ligamentum teres that extends from the umbilicus to the liver |
| Ductus venosus | Constricts | Becomes ligamentum venosum that is superficially embedded in the wall of the liver |
| Foramen ovale | Closes by valvelike septum primum as blood pressure in right atrium decreases and pressure in left atrium increases | Valve fuses along margin of foramen ovale and is marked by a depression called the fossa ovalis |
| Ductus arteriosus | Constricts | Becomes ligamentum arteriosum that extends from the pulmonary trunk to the aorta |
| Umbilical arteries | Distal portions constrict | Distal portions become lateral umbilical ligaments; proximal portions function as superior vesical arteries |

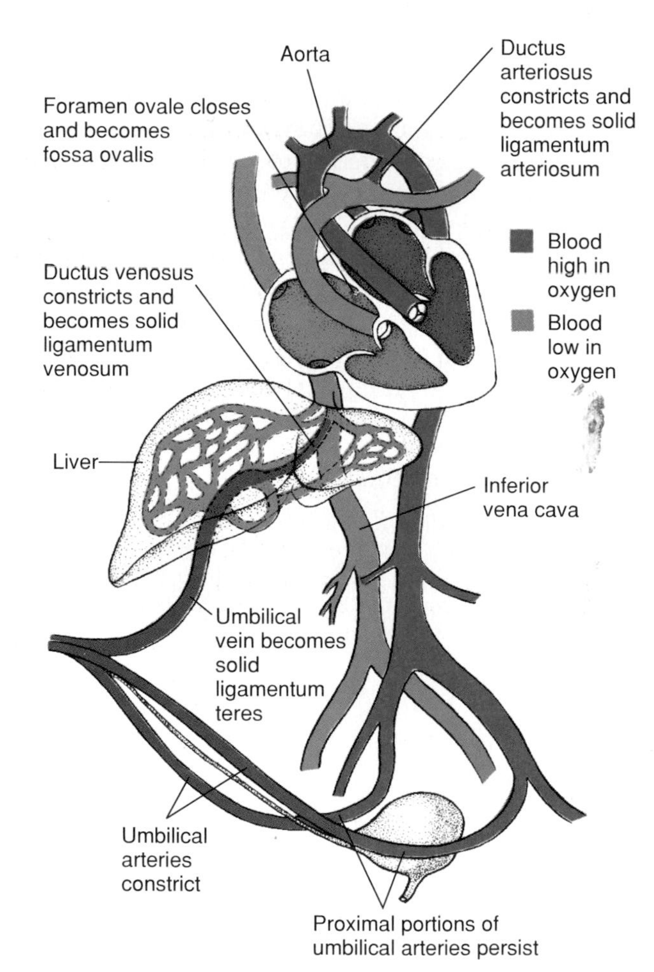


**FIGURE 23.24**
Major changes that occur in the newborn's circulatory system.

Because infancy (as well as childhood) is a period of rapid growth, the infant has particular nutritional needs. In addition to an energy source, the body requires proteins to provide the amino acids necessary to form new tissues; calcium and vitamin D to promote the development and ossification of skeletal structures; iron to support blood cell formation; and vitamin C for production of structural tissues such as cartilage and bone. By the time an infant is four months old, most of the circulating hemoglobin is the adult type.

## Childhood

**Childhood** begins at the end of the first year and ends at puberty. During this period, growth continues at a high rate. The primary teeth appear, and then secondary teeth replace them. The child develops a high degree of voluntary muscular control and learns to walk, run, and climb. Bladder and bowel controls are established. The child learns to communicate effectively by speaking, and later, usually learns to read, write, and reason objectively. At the same time, the child is maturing emotionally.

1. Define infancy.
2. What developmental changes characterize infancy?
3. Define childhood.
4. What developmental changes characterize childhood?

## Adolescence

**Adolescence** is the period of development between puberty and adulthood. It is a time of anatomical and physiological changes that result in reproductively functional individuals. Most of these changes are hormonally controlled, and they include the appearance of secondary sexual characteristics as well as growth spurts in the muscular and skeletal systems.

Females usually experience these changes somewhat earlier than males, so that early in adolescence, females may be taller and stronger than their male peers. On the other hand, females attain full growth at earlier ages, and in late adolescence, the average male is taller and stronger than the average female.

The periods of rapid growth in adolescence, which usually begin between the ages of 11 and 13 in females and between 13 and 15 in males, cause increased demands for certain nutrients. It is not uncommon for a teenager to consume a huge plate of food, go back for more—and still remain thin. In addition to energy sources, foods must provide ample amounts of proteins, vitamins, and minerals to support growth of new tissues. Adolescence also brings increasing levels of motor skills, intellectual ability, and emotional maturity.

Aging at the cellular, tissue, and organ level is a lifelong process. Human cells begin to die even before an individual is born. Fingers and toes, for example, are gradually carved from weblike extremities by cells in the webbing that die.

By the age of 2 years, a child's brain cells are as numerous as they will ever be. The brain has been growing over the past couple of years at the same rate as it did in the last 6 months in the uterus. At the age of 10, a person's hearing is the best that it will ever be. Shortly before sexual maturity is reached, the *thymus gland* reaches its greatest size (about the size of a walnut) and then begins to shrink gradually until it is almost microscopic by about the seventh decade of life. The organ's declining activity may contribute to the increasing susceptibility to certain illnesses seen with advancing age.

## Adulthood

**Adulthood** (maturity) extends from adolescence to old age. By age 18, the human male is producing the highest level of the sex hormone testosterone that he will ever have in his lifetime, and sex drive is strong. In the 20s, muscle strength peaks in both sexes. Hair is its fullest, with each hair as thick as it will ever be. By the end of the third decade of life, obvious signs of aging may first appear as a loss in the elasticity of facial skin, producing small wrinkles around the mouth and eyes. Height is already starting to decrease, but not yet at a detectable level.

The age of 30 seems to be a developmental turning point. Hearing often becomes less acute. Heart muscle begins to thicken. The elasticity of the ligaments between the small bones in the back lessens, setting the stage for the slumping posture that becomes apparent in later years. Some researchers estimate that beginning roughly at age 30, the human body becomes functionally less efficient by about 0.8% every year.

During their 40s, many people weigh 10 to 20 pounds (4.5 to 9 kilograms) more than they did at the age of 20, thanks to a slowing of metabolism and decrease in activity level. They may be 1/8 inch (0.3 centimeter) shorter, too. Hair may be graying as melanin-producing biochemical pathways lose efficiency, and some hair may fall out. Vision may become farsighted or nearsighted. The immune system is less efficient, making the body more prone to infection and cancer. Skeletal muscles tend to lose strength as more and more connective tissue appears within the muscles; the circulatory system is strained as the lumens of arterioles and arteries narrow due to accumulations of fatty deposits; skin loosens and wrinkles as elastic fibers in the dermis break down.

The early 50s bring further declines in the functioning of the human body. Nail growth slows, taste buds die, and the skin continues to lose its elasticity. For most people, the ability to see close objects becomes impaired, but for the nearsighted, vision improves. Women stop menstruating, although this does not necessarily mean an end to or loss of interest in sex. Decreased activity of the pancreas may lead to diabetes. By the decade's end, muscle mass and weight begin to decrease. A male produces less semen but is still sexually active. His voice may become higher as his vocal cords degenerate. A man has half the strength in his upper limb muscles and half the lung function as he did at age 25. He is about 3/4 inch (2 centimeters) shorter.

The 60-year-old may experience minor memory losses. A few million of the person's billions of brain cells have been lost over his or her lifetime, but for the most part, intellect remains quite sharp. By age 70, height decreases a full inch (2.5 centimeters). Sagging skin and loss of connective tissue, combined with continued growth of cartilage, make the nose, ears, and eyes more prominent. Figure 23.25 outlines some of the anatomical and physiological changes that accompany aging.

## Senescence

**Senescence** is the process of growing old. It is a continuation of the degenerative changes that begin during adulthood. As a result, the body becomes less and less able to cope with the demands placed on it by the individual and by the environment.

Senescence is a result of the normal wear-and-tear of body parts over many years. For example, the cartilages covering the ends of bones at joints may wear away, leaving the joints stiff and painful.

Other degenerative changes are caused by disease processes that interfere with vital functions, such as gas exchanges or blood circulation. Metabolic rate and distribution of body fluids may change. The rate of cell

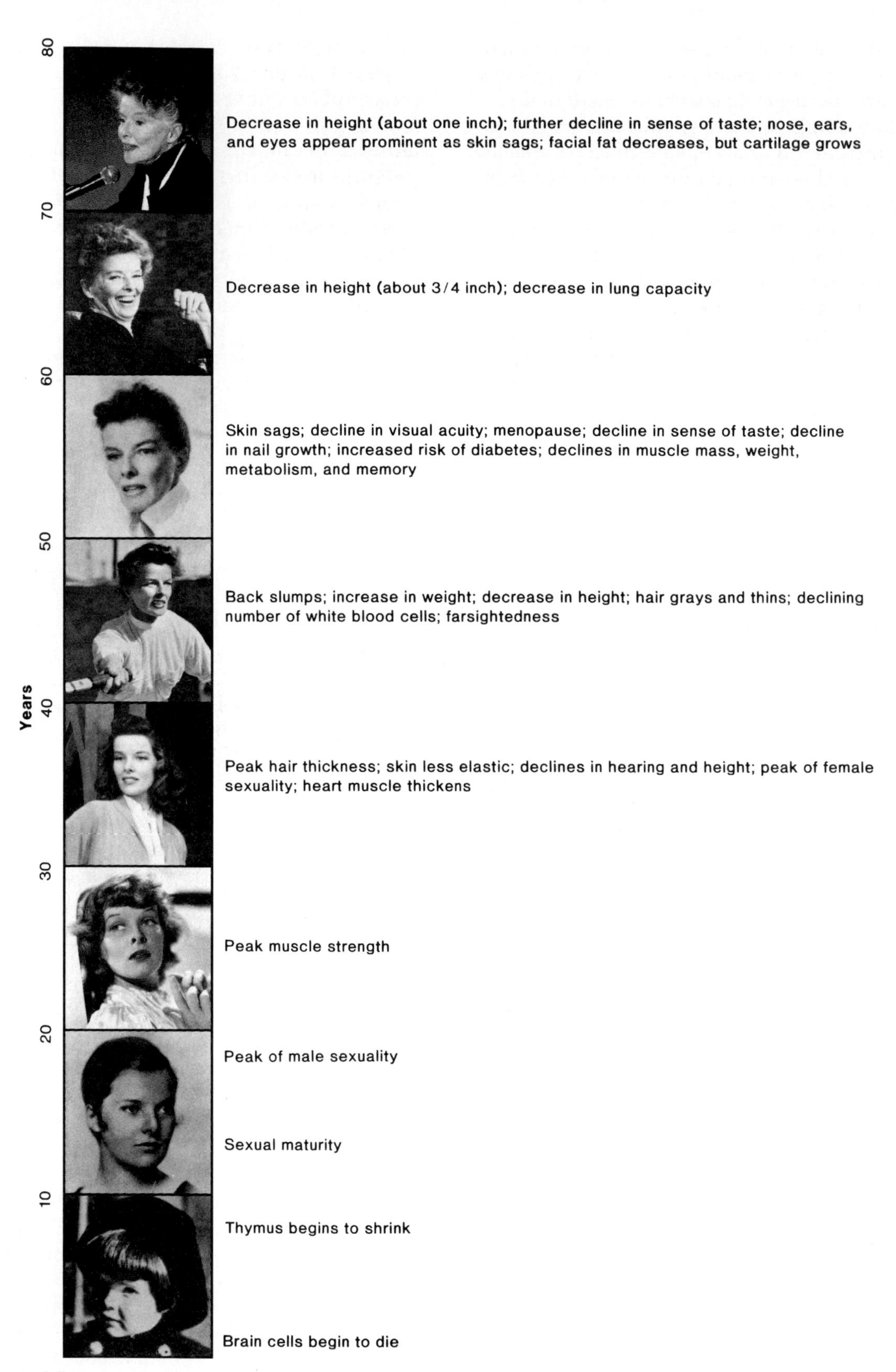


**FIGURE 23.25**
Although many biological changes ensue as we grow older, photographs of actress Katharine Hepburn at various stages of her life indicate that we can age with great grace and beauty.

division may decline, and the immune responses may weaken. As a result, the person becomes less able to repair damaged tissue and more susceptible to disease.

Decreasing efficiency of the central nervous system accompanies senescence. As a result, the person may lose some intellectual functions. Also, the physiological coordinating capacity of the nervous system may decrease, and homeostatic mechanisms may fail to operate effectively.

Sensory functions usually decrease with age also. Thus, perhaps one-third of older persons have significant hearing losses; difficulty focusing on objects at varying distances; and problems with depth perception. Typically, the senses of taste and smell diminish with age, and many older persons have decreased abilities to perceive environmental temperature accurately. They may also have impaired perception of touch and pressure as well as decreased sensitivity to painful stimuli.

Death usually results, not from these degenerative changes, but from mechanical disturbances in the cardiovascular system, failure of the immune system, or disease processes that affect vital organs.

Chart 23.4 summarizes the major phases of postnatal life and their characteristics, while chart 23.5 lists some of the aging-related changes.

1. What changes characterize adolescence?
2. Define adulthood.
3. What changes occur during adulthood?
4. What changes accompany senescence?

## Causes of Aging

The aging process is difficult to analyze because of the intricate interactions of the body's organ systems. Breakdown of one structure ultimately affects the functioning of others. The medical field of gerontology examines the biological changes of aging at the molecular, cellular, organismal, and population levels. Aging is passive and active.

### Passive Aging

Aging as a passive process entails breakdown of structures and slowing of functions. At the molecular level, passive aging is seen in the degeneration of the elastin and collagen proteins of connective tissue, causing skin to sag and muscle tissue to lose its firmness.

During a long lifetime, biochemical abnormalities accumulate. Mistakes occur throughout life when DNA replicates in dividing cells. Usually repair enzymes correct this damage immediately. But over many years, exposure to chemicals, viruses, and radiation disrupt DNA repair mechanisms, so that the error burden becomes too great to be fixed. The cell may die as a result of faulty genetic instructions.

Another sign of passive aging at the biochemical level is the breakdown of lipids. As aging membranes leak during lipid degeneration, a fatty, brown pigment called lipofuscin accumulates. Mitochondria also begin to break down in older cells, decreasing the supply of chemical energy to power the cell's functions.

The cellular degradation associated with aging may be set into action by highly reactive chemicals called **free radicals.** A molecule that is a free radical has an unpaired electron in its outermost valence shell. This causes the molecule to grab electrons from other molecules, destabilizing them, and a chain reaction of chemical instability begins that could kill the cell. Free radicals are a by-product of normal metabolism, and also form by exposure to radiation or toxic chemicals. Enzymes that usually inactivate free radicals diminish in number and activity in the later years. One such enzyme is *superoxide dismutase* (SOD).

> Some health-food stores promote superoxide dismutase (SOD) as an antiaging remedy. Even though the enzyme is a natural free radical fighter, there is no evidence that it stalls aging on a whole-body level.

### Active Aging

Aging also entails new activities or the appearance of new substances. Lipofucsin granules, for example, may be considered an active sign of aging, but they result from the passive breakdown of lipids. Another example of active aging is autoimmunity, in which the immune system turns against the body, attacking its cells as if they were invading organisms.

Active aging actually begins before birth, as certain cells die as part of the developmental program encoded in the genes. This process of programmed cell death, called **apoptosis,** occurs regularly in the embryo, degrading certain structures to pave the way for new ones. The number of neurons in the fetal brain, for example, is halved as those that make certain synaptic connections are spared from death. Apoptosis also occurs in the fetal thymus, where T cells that do not recognize "self" cell surfaces die, thereby building the immune system. In the adult, apoptosis removes brain, blood, and skin cells. Throughout life, apoptosis makes it possible for organs to maintain their characteristic shapes.

Mitosis and apoptosis are opposite, but complementary, processes. That is, as organs grow, the

## CHART 23.4 Stages in Postnatal Development

| Stage | Time Period | Major Events |
|---|---|---|
| Neonatal period | Birth to end of fourth week | Newborn begins to carry on respiration, obtain nutrients, digest nutrients, excrete wastes, regulate body temperature, and make circulatory adjustments |
| Infancy | End of fourth week to one year | Growth rate is high; teeth begin to erupt; muscular and nervous systems mature so that coordinated activities are possible; communication begins |
| Childhood | One year to puberty | Growth rate is high; deciduous teeth erupt and are replaced by permanent teeth; high degree of muscular control is achieved; bladder and bowel controls are established; intellectual abilities mature |
| Adolescence | Puberty to adulthood | Person becomes reproductively functional and emotionally more mature; growth spurts occur in skeletal and muscular systems; high levels of motor skills are developed; intellectual abilities increase |
| Adulthood | Adolescence to old age | Person remains relatively unchanged anatomically and physiologically; degenerative changes begin to occur |
| Senescence | Old age to death | Degenerative changes continue; body becomes less and less able to cope with the demands placed upon it; death usually results from mechanical disturbances in the cardiovascular system or from disease processes that affect vital organs |

## CHART 23.5 Aging-related Changes

| Organ System | Aging-related Changes |
|---|---|
| Integumentary system | Degenerative loss of collagenous and elastic fibers in dermis; decreased production of pigment in hair follicles; reduced activity of sweat and sebaceous glands |
| | Skin tends to become thinner, wrinkled, and dry; hair turns gray and then white |
| Skeletal system | Degenerative loss of bone matrix |
| | Bones become thinner, less dense, and more likely to fracture; stature may shorten due to compression of intervertebral disks and vertebrae |
| Muscular system | Loss of skeletal muscle fibers; degenerative changes in neuromuscular junctions |
| | Loss of muscular strength |
| Nervous system | Degenerative changes in neurons; loss of dendrites and synaptic connections; accumulation of lipofuscin in neurons; decreases in sensory sensitivities |
| | Decreasing efficiency in processing and recalling information; decreasing ability to communicate; diminished senses of smell and taste; loss of elasticity of lenses and consequent loss of ability to accommodate for close vision |
| Endocrine system | Reduced hormonal secretions |
| | Decreased metabolic rate; reduced ability to cope with stress; reduced ability to maintain homeostasis |
| Digestive system | Decreased motility in gastrointestinal tract; reduced secretion of digestive juices |
| | Reduced efficiency of digestion |
| Respiratory system | Degenerative loss of elastic fibers in lungs; reduced number of alveoli |
| | Reduced vital capacity; increase in dead air space; reduced ability to clear airways by coughing |
| Cardiovascular system | Degenerative changes in cardiac muscle; decrease in lumen diameters of arteries and arterioles |
| | Decreased cardiac output; increased resistance to blood flow; increased blood pressure |
| Lymphatic system | Decrease in efficiency of immune system |
| | Increased incidence of infections and neoplastic diseases; increased incidence of autoimmune diseases |
| Excretory system | Degenerative changes in kidneys; reduction in number of functional nephrons |
| | Reductions in filtration rate, tubular secretion, and reabsorption |
| Reproductive systems | |
| Female | Degenerative changes in ovaries; decrease in secretion of sex hormones |
| | Menopause; regression of secondary sexual characteristics |
| Male | Reduced secretion of sex hormones; enlargement of prostate gland; decrease in sexual energy |

## Clinical Application 23.4

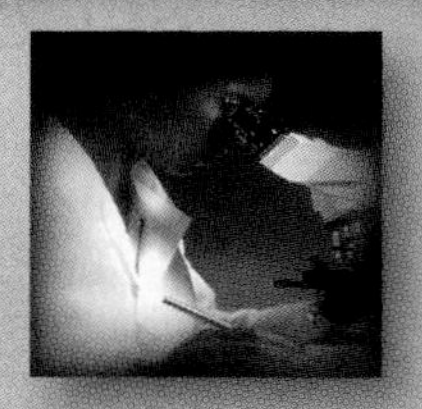

# Old Before Their Time

Eight-year-old Fransie Geringer of South Africa and nine-year-old Mickey Hays of Texas thought they were the only "elderly children" in the world, until they met at Disneyland in 1981. Each had Hutchinson-Gilford syndrome, the most severe form of *progeria,* or inherited rapid aging (fig. 23D).

In Hutchinson-Gilford syndrome, a child appears normal at birth, but by the first birthday, growth is obviously retarded. Within just a few years, the child acquires a shocking appearance of wrinkles, baldness, and the prominent facial features characteristic of advanced age. The body is aging on the inside as well, as arteries clog with fatty deposits. The child usually dies of a heart attack or a stroke by the age of 12, although some patients live into their 20s. Only a few dozen cases of this syndrome have ever been reported. In an "adult" form of progeria called *Werner's syndrome,* which becomes apparent before the twentieth birthday, death from old age usually occurs in the 40s.

The cells of progeria patients show profound aging-related changes. Normal cells in culture divide only fifty times before dying. Cells from progeria patients die in culture after only ten to thirty divisions, as if they were programmed to die prematurely. Certain structures seen in normal cultured cells as they near the fifty-division limit (glycogen particles, lipofuscin granules, many lysosomes and vacuoles, and a few ribosomes) appear in the cells of progeria patients early on. Understanding the mechanisms that cause these diseased cells to move through the aging process at an abnormally rapid pace may help us to understand the biology of normal aging.

**Figure 23D**
Progeria causes rapid aging. Fransie Geringer and Mickey Hays met a third child with progeria at Disneyland in 1981. People from the children's hometowns had raised money to grant them their wishes of visiting the park.

number of cells in some regions increases, but in others it decreases. Cell death, then, is not a phenomenon that is restricted to the aged. It is a normal part of life.

Clinical Application 23.4 discusses genetic disorders that greatly accelerate aging.

## The Human Life Span

In the age-old quest for longer life, people have sampled everything from turtle soup to owl meat to human blood. A Russian-French microbiologist, Ilya Mechnikov, believed that a life span of 150 years could be achieved with the help of a steady diet of milk cultured with bacteria. He thought that the bacteria would live in the large intestine and somehow increase the human life span. (He died at 71.) Ironically, many people have died in pursuit of a literal "fountain of youth."

The human *life span*—the length of time that a human can theoretically live—is 120 years. Of course, most people succumb to disease or injury long before that point. *Life expectancy* is a more realistic projection of how long an individual will live, based on epidemiological information. In the United States, life expectancy for males is 79 years and for females 83 years.

Life expectancy approaches life span as technology conquers diseases. Technology also alters the most prevalent killers. Development of antibiotic drugs removed some infectious diseases such as pneumonia and tuberculosis from the top of the "leading causes of death" list, a position that was filled by heart disease. Cancer is currently overtaking heart disease as the most common cause of death. By the year 2020, with a third of the U.S. population over 65 years of age, Alzheimer's disease will likely earn the top spot.

1. Why is it difficult to sort out the causes of aging?
2. How is aging a passive process?
3. How is aging an active process?
4. What are common causes of death?

The many successes of medical science over the past century have greatly altered the nature of the illnesses most likely to cause our deaths. However, the state of current infectious diseases is perhaps a warning that we should never grow complacent about our ability to conquer illness. Who could have predicted that a mere virus could decimate the powerful human immune system, in the form of AIDS? Or that another virus, hantavirus passed from a rodent, could kill a human in just a few days? The resurgence of infectious diseases that we thought we had conquered, such as measles and tuberculosis, further underscores the fact that humans are just one form of life on earth. Although we can alter our environment more than other species can, we are still subject to forces of nature beyond our control.

## Clinical Terms Related to Growth and Development

**abruptio placentae** (ab-rup′shē-o plah-cen′ta) A premature separation of the placenta from the uterine wall.

**amniocentesis** (am″ne-o-sen-te′sis) A technique in which a sample of amniotic fluid is withdrawn from the amniotic cavity by inserting a hollow needle through the pregnant woman's abdominal wall.

**dizygotic twins** (di″zi-got′ik twinz) Twins resulting from the fertilization of two ova by two sperm cells.

**hydatid mole** (hi′dah-tid mōl) A type of uterine tumor that originates from placental tissue.

**hydramnios** (hi-dram′ne-os) Excess amniotic fluid.

**intrauterine transfusion** (in″trah-u′ter-in trans-fu′zhun) Transfusion administered by injecting blood into the fetal peritoneal cavity before birth.

**lochia** (lo′ke-ah) Vaginal discharge following childbirth.

**meconium** (mĕ-ko′ne-um) The first fecal discharge of a newborn.

**monozygotic twins** (mon″o-zi-got′ik twinz) Twins resulting from the fertilization of one ovum by one sperm cell.

**perinatology** (per″ĭ-na-tol′o-je) Branch of medicine concerned with the fetus after twenty-five weeks of development and with the newborn for the first four weeks after birth.

**postpartum** (pōst-par′tum) Occurring after birth.

**teratology** (ter″ah-tol′o-je) The study of abnormal development and congenital malformations.

**trimester** (tri-mes′ter) Each third of the total period of pregnancy.

**ultrasonography** (ul″trah-son-og′rah-fe) Technique used to visualize the size and position of fetal structures by means of ultrasonic waves.

## Chapter Summary

### Introduction (page 911)

Growth refers to an increase in size; development is the process of changing from one phase of life to another.

### Prenatal Period (page 911)

1. Period of cleavage
   a. Fertilization occurs in a uterine tube and results in a zygote.
   b. The zygote undergoes mitosis, and the newly formed cells divide.
   c. Each subsequent division produces smaller and smaller cells.
   d. A solid ball of cells (morula) is formed, and it becomes a hollow ball called a blastocyst.
   e. The inner cell mass that gives rise to the embryo proper forms within the blastocyst.

f. The blastocyst becomes implanted in the uterine wall.
   (1) Enzymes digest the endometrium around the blastocyst.
   (2) Fingerlike processes from the blastocyst penetrate into the endometrium.
g. The period of cleavage lasts through the first week of development.
h. The trophoblast secretes hCG, which helps maintain the corpus luteum, helps protect the blastocyst against being rejected, and stimulates the developing placenta to secrete hormones.

2. Embryonic stage
   a. The embryonic stage extends from the second through the eighth weeks.
   b. It is characterized by the development of the placenta and the main internal and external body structures.
   c. The cells of the inner cell mass fold inward, forming a gastrula which has two and then three primary germ layers.
      (1) Ectoderm gives rise to the nervous system, portions of the skin, the lining of the mouth, and the lining of the anal canal.
      (2) Mesoderm gives rise to muscles, bones, blood vessels, lymphatic vessels, reproductive organs, kidneys, and linings of body cavities.
      (3) Endoderm gives rise to linings of the digestive tract, respiratory tract, urinary bladder, and urethra.
   d. The embryonic disk becomes cylindrical and is attached to the developing placenta by the connecting stalk.
   e. The embryo develops head, face, upper limbs, lower limbs, and mouth, and appears more humanlike.
   f. Chorionic villi develop and are surrounded by spaces filled with maternal blood.
   g. The placental membrane consists of the epithelium of the villi and the epithelium of the capillaries inside the villi.
      (1) Oxygen and nutrients diffuse from the maternal blood through the placental membrane and into the fetal blood.
      (2) Carbon dioxide and other wastes diffuse from the fetal blood through the placental membrane and into the maternal blood.
   h. The placenta develops in the disk-shaped area where the chorion contacts the uterine wall.
      (1) The embryonic portion consists of the chorion and its villi.
      (2) The maternal portion consists of the uterine wall and attached villi.
   i. A fluid-filled amnion develops around the embryo.
   j. The umbilical cord is formed as the amnion envelops the tissues attached to the underside of the embryo.
      (1) The umbilical cord includes two arteries and a vein.
      (2) It suspends the embryo in the amniotic cavity.
   k. The chorion and amnion fuse.
   l. The yolk sac forms on the underside of the embryonic disk.
      (1) It gives rise to blood cells and cells that later form sex cells.
      (2) It helps form the digestive tube.
   m. The allantois extends from the yolk sac into the connecting stalk.
      (1) It forms blood cells.
      (2) It gives rise to the umbilical vessels.
   n. By the beginning of the eighth week, the embryo is recognizable as a human.

3. Fetal stage
   a. This stage extends from the end of the eighth week and continues until birth.
   b. Existing structures grow and mature; only a few new parts appear.
   c. The body enlarges, upper and lower limbs reach final relative proportions, the skin is covered with sebum and dead epidermal cells, the skeleton continues to ossify, muscles contract, and fat is deposited in subcutaneous tissue.
   d. The fetus is full term at the end of the ninth month, which equals approximately 266 days.
      (1) It is about 50 centimeters long and weighs 6–8 pounds.
      (2) It is positioned with its head toward the cervix.

4. Fetal blood and circulation
   a. Umbilical vessels carry blood between the placenta and the fetus.
   b. Fetal blood carries a greater concentration of oxygen than does maternal blood.
   c. Blood enters the fetus through the umbilical vein and partially bypasses the liver by means of the ductus venosus.
   d. Blood enters the right atrium and partially bypasses the lungs by means of the foramen ovale.
   e. Blood entering the pulmonary trunk partially bypasses the lungs by means of the ductus arteriosus.
   f. Blood enters the umbilical arteries from the internal iliac arteries.

## Postnatal Period (page 934)

1. Neonatal period
   a. This period extends from birth to the end of the fourth week.
   b. The newborn must begin to respire, obtain nutrients, excrete wastes, and regulate its body temperature.
   c. The first breath must be powerful in order to expand the lungs.
      (1) Surfactant reduces surface tension.
      (2) A variety of factors stimulate the first breath.
   d. The liver is immature and unable to supply sufficient glucose, so the newborn depends primarily on stored fat as an energy source.
   e. Immature kidneys cannot concentrate urine very well.
      (1) The newborn may become dehydrated.
      (2) Water and electrolyte imbalances may develop.
   f. Homeostatic mechanisms may function imperfectly and body temperature may be unstable.
   g. The circulatory system changes when placental circulation ceases.
      (1) Umbilical vessels constrict.
      (2) The ductus venosus constricts.
      (3) The foramen ovale is closed by a valve as blood pressure in the right atrium falls and pressure in the left atrium rises.
      (4) The ductus arteriosus constricts.

2. Infancy
   a. Infancy extends from the end of the fourth week to one year of age.
   b. Infancy is a period of rapid growth.
      (1) The muscular and nervous systems mature, and coordinated activities become possible.
      (2) Communication begins.
   c. Rapid growth depends on an adequate intake of proteins, vitamins, and minerals in addition to energy sources.
3. Childhood
   a. Childhood extends from the end of the first year to puberty.
   b. It is characterized by rapid growth, development of muscular control, and establishment of bladder and bowel control.
4. Adolescence
   a. Adolescence extends from puberty to adulthood.
   b. It is characterized by physiological and anatomical changes that result in a reproductively functional individual.
   c. Females may be taller and stronger than males in early adolescence, but the situation reverses in late adolescence.
   d. Adolescents develop high levels of motor skills, their intellectual abilities increase, and they continue to mature emotionally.
5. Adulthood
   a. Adulthood extends from adolescence to old age.
   b. The adult remains relatively unchanged physiologically and anatomically for many years.
   c. After age thirty, degenerative changes usually begin to occur.
      (1) Skeletal muscles lose strength.
      (2) The circulatory system becomes less efficient.
      (3) The skin loses its elasticity.
      (4) The capacity to produce sex cells declines.
6. Senescence
   a. Senescence is the process of growing old.
   b. Degenerative changes continue, and the body becomes less able to cope with demands placed upon it.
   c. Changes occur because of prolonged use, effects of disease, and cellular alterations.
   d. An aging person usually experiences losses in intellectual functions, sensory functions, and physiological coordinating capacities.
   e. Death usually results from mechanical disturbances in the cardiovascular system or from disease processes that affect vital organs.

## Causes of Aging (page 939)

1. Passive aging
   a. Passive aging entails breakdown of structures and slowing or failure of functions.
   b. Connective tissue breaks down.
   c. DNA errors accumulate.
   d. Lipid breakdown in aging membranes releases lipofuscin.
   e. Free radical damage escalates.
2. Active aging
   a. In autoimmunity, the immune system attacks the body.
   b. Apoptosis is a form of programmed cell death. It occurs throughout life, shaping organs.
3. The human life span
   a. The theoretical maximum life span is 120 years.
   b. Life expectancy, based on real populations, is 79 years for males and 83 years for females.
   c. Medical technology makes life expectancy more closely approach life span.

**Explorations in Human Anatomy and Physiology CD-ROM**
The module accompanying Chapter 23 is #3 Life Span and Lifestyle.

## Critical Thinking Questions

1. How would you explain the observation that twins resulting from a single fertilized egg cell can exchange body parts by tissue or organ transplant procedures without experiencing rejection reactions?
2. One of the more common congenital cardiac disorders is a ventricular septum defect in which an opening remains between the right and left ventricles. What problem would such a defect create as blood moves through the heart?
3. What symptoms may appear in a newborn if its ductus arteriosus fails to close?
4. Why is it sometimes difficult to determine the racial origin of a newborn by its skin color during the first few days following birth?
5. Why is it important for a middle-aged adult who has neglected physical activity for many years to have a physical examination before beginning an exercise program?
6. What precautions do you think should be taken to avoid conflicts between "gestational" (surrogate) mothers and genetic mothers?
7. If an aged relative came to live with you, what special provisions could you make in your household environment and routines that would demonstrate your understanding of the changes brought on by aging processes?

## Review Exercises

1. Define *growth* and *development.*
2. Describe the process of cleavage.
3. Distinguish between a blastomere and a blastocyst.
4. Describe the formation of the inner cell mass, and explain its significance.
5. Describe the process of implantation.
6. List three functions of hCG.
7. Explain how the primary germ layers form.
8. List the major body parts derived from ectoderm.
9. List the major body parts derived from mesoderm.
10. List the major body parts derived from endoderm.
11. Describe the formation of the placenta, and explain its functions.
12. Define *placental membrane.*
13. Distinguish between the chorion and the amnion.
14. Explain the function of amniotic fluid.
15. Describe the formation of the umbilical cord.
16. Explain how the yolk sac and the allantois are related, and list the functions of each.
17. Explain why the embryonic period of development is so critical.
18. Define *fetus.*
19. List the major changes that occur during the fetal stage of development.
20. Distinguish between preimplantation genetic diagnosis, chorionic villus sampling, and amniocentesis.
21. Describe a full-term fetus.
22. Compare the properties of fetal hemoglobin with those of adult hemoglobin.
23. Explain how the fetal circulatory system is adapted for intrauterine life.
24. Trace the pathway of blood from the placenta to the fetus and back to the placenta.
25. Distinguish between a newborn and an infant.
26. Explain why a newborn's first breath must be particularly forceful.
27. List some of the factors that stimulate the first breath.
28. Explain why newborns tend to develop water and electrolyte imbalances.
29. Describe the circulatory changes that occur in the newborn.
30. Describe the characteristics of an infant.
31. Distinguish between a child and an adolescent.
32. Define *adulthood.*
33. List some of the degenerative changes that begin during adulthood.
34. Define *senescence.*
35. List some of the factors that seem to promote senescence.
36. Cite evidence of passive aging and active aging.

24

CHAPTER

# GENETICS

HEREDITY IS OBVIOUS WHEN TWO FAMILY MEMBERS STRONGLY RESEMBLE EACH OTHER, AS THESE SIBLINGS DO. GENETICS ALSO AFFECTS US IN MANY NOT-SO-OBVIOUS WAYS.

## CHAPTER OBJECTIVES

AFTER YOU HAVE STUDIED THIS CHAPTER, YOU SHOULD BE ABLE TO:

1. Explain how gene discoveries are relevant to the study of anatomy and physiology and to health care.
2. Distinguish between genes and chromosomes.
3. Outline the process of meiosis and explain why meiotic products have different combinations of parental genes.
4. Define the two types of chromosomes.
5. Explain how genes can have many alleles (variants), but a person can have only two alleles of a particular gene.
6. Describe how alleles interact.
7. Distinguish between modes of inheritance by which traits pass from parents to offspring.
8. Explain how gene expression varies among individuals.
9. Describe how multiple genes and the environment interact to produce complex traits.
10. Describe how genes and chromosomes control the inheritance of gender and how traits are transmitted on the X chromosome.
11. Describe how gender affects gene expression.
12. Explain how deviations in chromosome number or arrangement can harm health.
13. Explain how conditions caused by extra or missing chromosomes reflect a meiotic error.
14. Describe the challenges that the ability to diagnose inherited disease presents.
15. Explain how gene therapy works.

## KEY TERMS

**allele** (ah-lēl′)
**autosome** (aw′to-sŏm)
**dominant** (dom′i-nant)
**gene** (jēn)
**genetics** (jĕ-net′iks)
**genotype** (je′no-tīp)
**heredity** (he-red′i-te)
**heterozygous** (het″er-o-zi′gus)
**homozygous** (ho″mo-zi′gus)
**meiosis** (mi-o′sis)
**multiple allele** (mul′ti-pl ah-lēl′)
**mutation** (mu-ta′shun)
**nondisjunction** (non″dis-jungk′shun)
**phenotype** (fe′no-tīp)
**recessive** (re-ses′iv)
**sex chromosome** (seks kro′mo-sōm)

The year is 2010, and the young pregnant woman has blood drawn for a routine prenatal exam. Among her cells are a few immature red blood cells (erythroblasts) from the fetus. These will be separated out and then scrutinized for clues to the health of the person-to-be. An initial peek at the chromosomes, using a fluorescent dye that homes in on the Y chromosome denoting maleness, shows that the fetus is male. At this news, the woman and her mate decide to name him Orwell. A closer look at the specific genes comprising his chromosomes reveals much more.

Happily, Orwell has not inherited any of the more common inherited illnesses, nor has he inherited the gene that caused the early senility of Alzheimer's disease in a grandparent. Despite Orwell's apparently healthy background, some of his blood cells in the umbilical cord will be set aside and stored at birth, should he need a bone marrow transplant later in life. The information from the prenatal test also indicates that Orwell has inherited certain potentially unhealthy characteristics, but he can minimize their effects, and perhaps extend his life, by modifying his lifestyle in particular ways.

A panel of tests to type the genes whose protein products could predispose Orwell to heart disease—clotting factors, apoproteins, angiotension-converting enzyme—make it clear that a lifetime of low-fat foods combined with regular exercise can extend his life. The same measures may help prevent, or delay, colon cancer, for which he has inherited a pair of susceptibility genes. Finally, Orwell's parents know that their son should avoid drinking alcohol, smoking cigarettes, and abusing drugs, because a neurotransmitter receptor variant that strongly correlates to addictive behavior is also a part of his genetic makeup.

In 1990, when Cynthia Cutshall was eight years old, she became the first person to receive gene therapy. At the National Institutes of Health in Bethesda, Maryland, she received an infusion of her own white blood cells, bolstered with the gene specifying the enzyme adenosine deaminase, which her cells lacked due to an inborn error of metabolism. Correcting the genetic defect enabled her body to develop an immune system, and as this photo taken 3 years later shows, she has done quite well.

## The New Role of Genetics in Medicine

Orwell's scenario takes place in the future, but every one of the tests described is possible today, although all are experimental. The avalanche of genetic discoveries from the human genome project (see chapter 4) is providing a wealth of new diagnostic tests and therapies, for rare as well as common disorders. **Genetics,** the study of inheritance, will play an ever-increasing role in future health care.

New genetic information has explained many physiological processes at the cellular and molecular levels, as we have seen throughout this book. Figure 24.1 recaps some of these examples along a timeline, with the numbers referring to chapters where the disorder is discussed. For further descriptions of the disorders mentioned, see chart 24.6 at the chapter's end.

Genes encode traits other than illness, including eye and hair color, singing ability, and hard-to-define personality traits. Clinical Application 24.1 highlights a few interesting human inherited traits.

**Heredity** is the transmission of genetic information from parents to offspring. This transfer occurs through DNA molecules in the nuclei of eggs and sperm that provide chemical instructions for human development. The environment, however, often influences how genes are expressed. The environment includes the chemical, physical, social, and biological factors surrounding an individual that influence his or

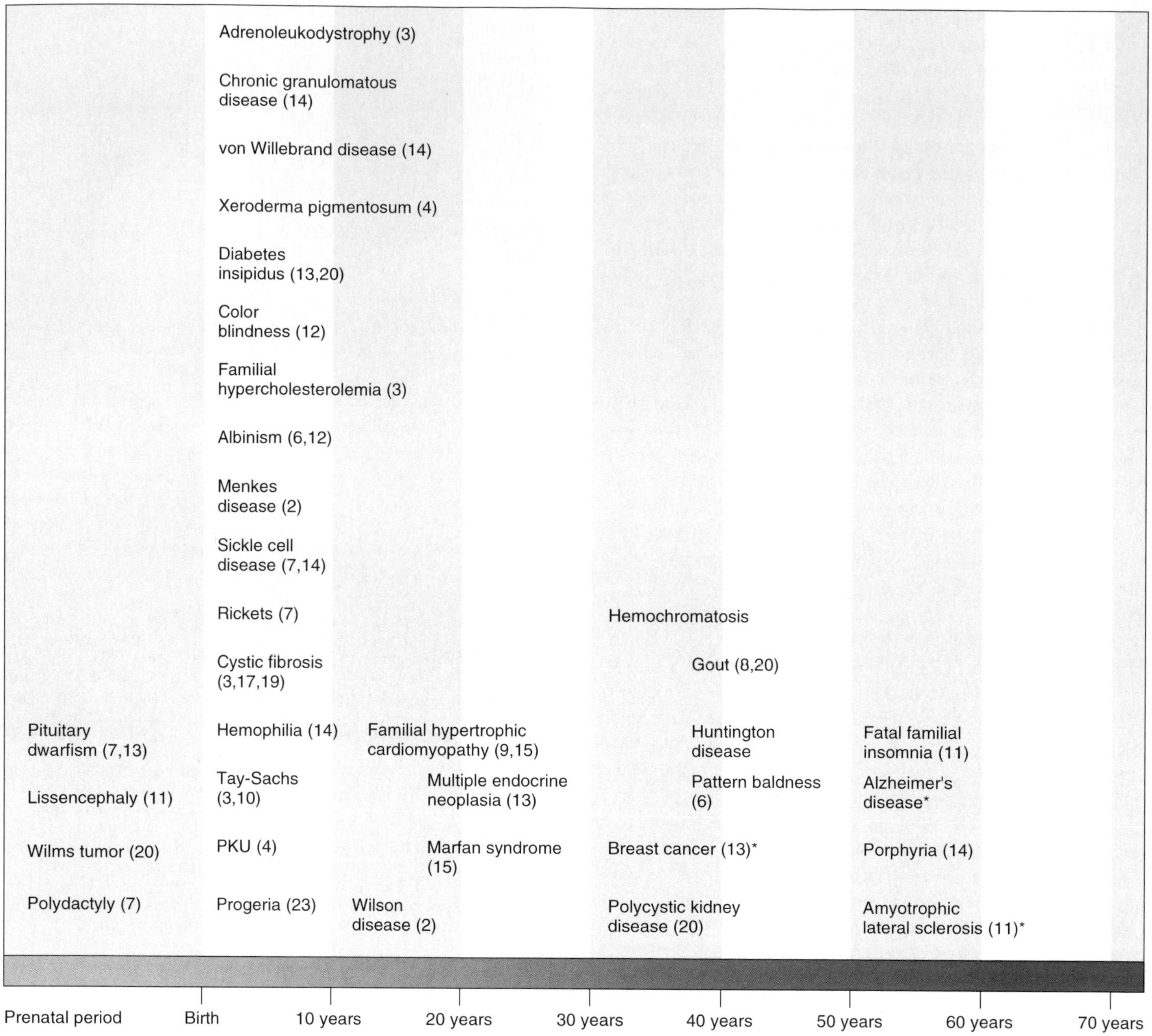

Prenatal period:

Pituitary dwarfism (7,13)

Lissencephaly (11)

Wilms tumor (20)

Polydactyly (7)

Birth:

Adrenoleukodystrophy (3)

Chronic granulomatous disease (14)

von Willebrand disease (14)

Xeroderma pigmentosum (4)

Diabetes insipidus (13,20)

Color blindness (12)

Familial hypercholesterolemia (3)

Albinism (6,12)

Menkes disease (2)

Sickle cell disease (7,14)

Rickets (7)

Cystic fibrosis (3,17,19)

Hemophilia (14)

Tay-Sachs (3,10)

PKU (4)

Progeria (23)

10 years:

Familial hypertrophic cardiomyopathy (9,15)

Wilson disease (2)

20 years:

Multiple endocrine neoplasia (13)

Marfan syndrome (15)

30 years:

Hemochromatosis

Breast cancer (13)*

Polycystic kidney disease (20)

40 years:

Gout (8,20)

Huntington disease

Pattern baldness (6)

50 years:

Fatal familial insomnia (11)

Alzheimer's disease*

Porphyria (14)

Amyotrophic lateral sclerosis (11)*

Prenatal period | Birth | 10 years | 20 years | 30 years | 40 years | 50 years | 60 years | 70 years

* These conditions may also result from nongenetic causes.

**FIGURE 24.1**
Genetic disorders described throughout this book are arranged along a timeline, illustrating that many conditions affect children, but some may strike before birth or well into adulthood. Numbers in parentheses indicate chapters where each disorder is discussed.

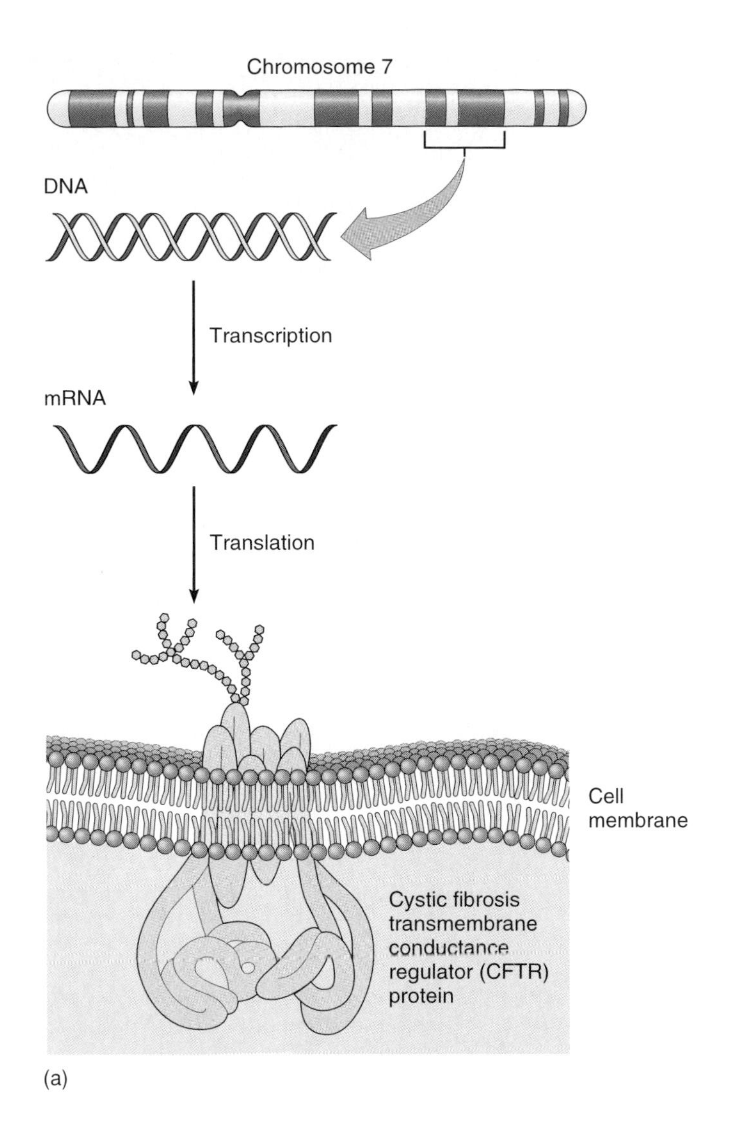


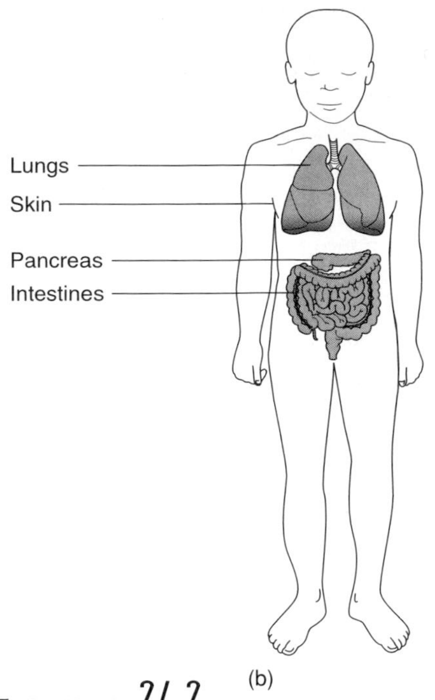


**FIGURE 24.2**

From gene to protein to person. (*a*) The gene encoding the CFTR protein, and causing cystic fibrosis when mutant, resides on the seventh largest chromosome. CFTR protein folds into a channel that regulates the flow of chloride ions into and out of cells lining the respiratory tract, pancreas, intestines, and possibly elsewhere. (*b*) In cystic fibrosis, the CFTR protein is abnormal, usually missing an amino acid. Its shape is altered, which entraps the chloride ions inside cells. Water entering these cells leaves behind very thick mucus and other secretions in the places highlighted in the illustration. It is the sticky secretions that cause the symptoms of the illness.

Source: Data from M. C. Iannuzi and F. S. Collins, "Reverse Genetics and Cystic Fibrosis" in *American Journal of Respiratory Cellular and Molecular Biology,* 2:309–316, 1990.

her characteristics. For example, a person inheriting a gene conferring susceptibility to lung cancer may never develop the illness if he or she lives away from air pollution and never smokes.

## Genes and Chromosomes

A **gene** is the nucleotide sequence of a DNA molecule that tells a cell how to construct a specific protein molecule. A gene may also function by controlling (turning on or off) another gene. Gene products—proteins—ultimately produce the trait associated with the gene, as figure 24.2 illustrates for the common inherited illness *cystic fibrosis* (CF). We will return to this condition throughout the chapter to illustrate various concepts.

All of the genes in a human cell—estimated to number about 70,000—constitute the human **genome.** These genes are distributed among 23 pairs of chromosomes. Recall from chapter 3 that *chromosomes* consist of proteins (histones) as well as a continuous, highly wound thread of DNA. Chromosomes appear as rod-shaped bodies in the nucleus when a cell divides (fig. 24.3). The number of chromosomes within cells varies with species. A human fertilized egg contains 46 chromosomes, 23 from the sperm and 23 from the egg.

## Chromosomes and Sex Cell Production

Siblings, except for identical twins, are never exactly alike physically. The reason for this lies in **meiosis,** the special form of cell division that produces sperm and eggs, which are also called *gametes* or *germ cells.* An individual's gametes (sperm or eggs) differ from

CLINICAL APPLICATION 24.1

# It's All in the Genes

Do you have uncombable hair, misshapen toes or teeth, a pigmented tongue tip, or an inability to smell skunk? Do you lack teeth, eyebrows, eyelashes, nasal bones, thumbnails, or fingerprints? Can you roll your tongue or wiggle your ears? If so, you may find your unusual trait described in a book called *Mendelian Inheritance in Man,* compiled by Johns Hopkins University geneticist Victor McKusick, which catalogs more than 5,000 known human genetic variants. Most of the entries consist of family histories, clinical descriptions, and hypotheses of how the trait is transmitted. Amidst the medical terminology can be found some fascinating inherited traits in humans, from top to toes.

Genes control whether hair is blond, brown, or black, whether or not it has red highlights, and whether it is straight, curly, or kinky. Widow's peaks, cowlicks, a whorl in the eyebrow, and white forelocks run in families, as do hairs with triangular cross sections. Some people have multicolored hairs like cats, and others have hair in odd places, such as on the elbows, nosetip, knuckles, palms of the hands, or soles of the feet. Teeth can be missing or extra, protuberant or fused, present at birth, or "shovel-shaped" or "snow-capped." One can have a grooved tongue, duckbill lips, flared ears, egg-shaped pupils, three rows of eyelashes, spotted nails, or "broad thumbs and great toes." Extra breasts have been observed in humans and guinea pigs, and one family's claim to fame is a double nail on the littlest toes.

Unusual genetic variants can affect metabolism, sometimes resulting in disease or other times producing harmless yet quite noticeable effects. Members of some families experience "urinary excretion of odoriferous component of asparagus" or "urinary excretion of beet pigment" after eating the vegetables in question. In "blue diaper syndrome," an infant's inherited inability to break down an amino acid results in urine that turns blue on contact with air.

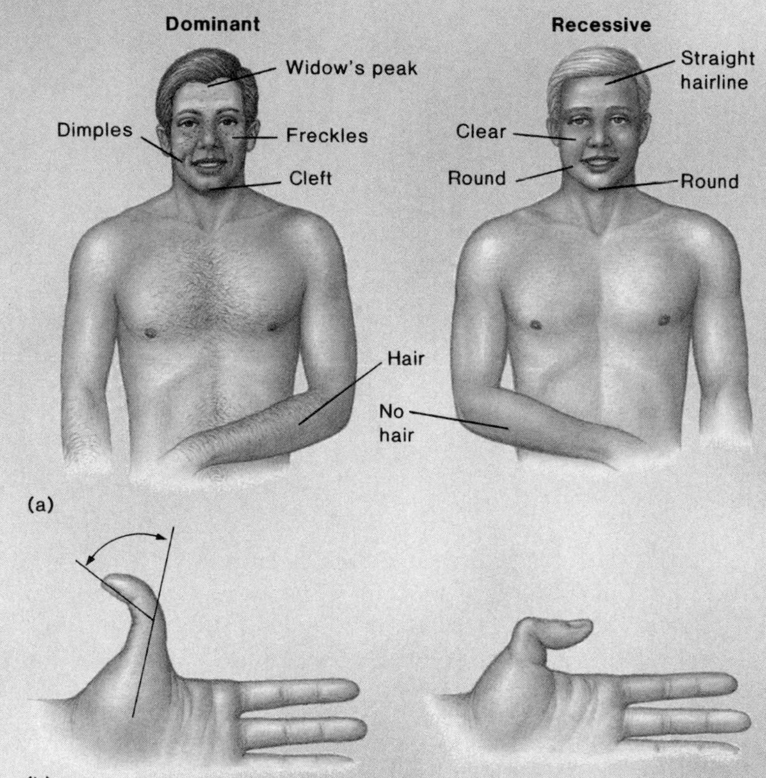


FIGURE 24A

Inheritance of some common traits: (*a*) Freckles, dimples, widow's peak, hairy elbows, and a cleft chin; (*b*) the ability to bend the thumb backward or forward; (*c*) widow's peak; and (*d*) attached or unattached earlobes.

One father and son were plagued with an inability to open their mouths completely. Some families suffer from "dysmelodia," the inability to carry a tune. Those who have inherited "odor blindness" are unable to smell either musk, skunk, cyanide, or freesia flowers. Motion sickness, migraine headaches, and stuttering may be inherited. Uncontrollable sneezing may be due to inherited hayfever or to Achoo syndrome (an acronym for "autosomal dominant compelling helioophthalmic outburst syndrome").

Perhaps the most bizarre inherited illness is the "jumping Frenchmen of Maine syndrome." This is an exaggerated startle reflex that was first noted among French-Canadian lumbermen from the Moosehead Lake area of Maine, whose ancestors were from the Beauce region of Quebec. The disorder was first described at a meeting of the American Neurological Association in 1878 and was confirmed with the aid of videotape in 1980. McKusick describes the disorder as follows:

"If given a short, sudden, quick command, the affected person would respond with the appropriate action, often echoing the words of command. . . . For example, if one of them was abruptly asked to strike another, he would do so without hesitation, even if it was his mother and he had an ax in his hand."

Some more common genetic quirks are illustrated in figure 24A.

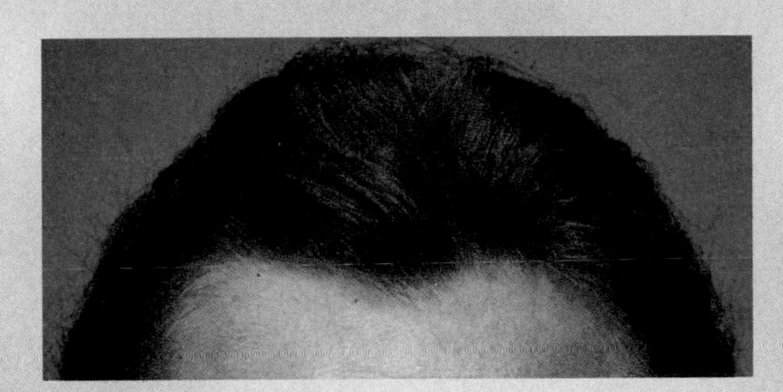

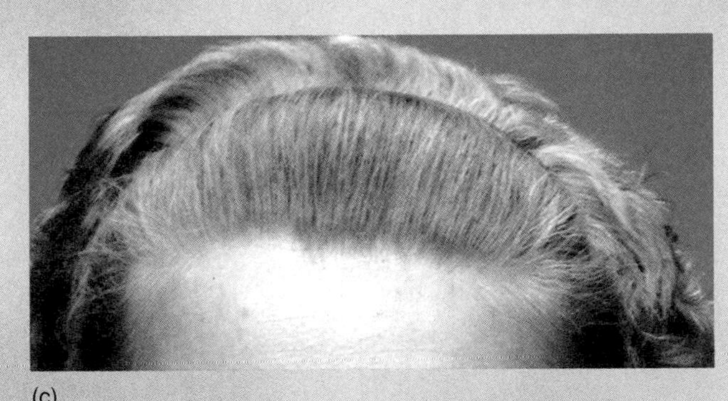

(c)

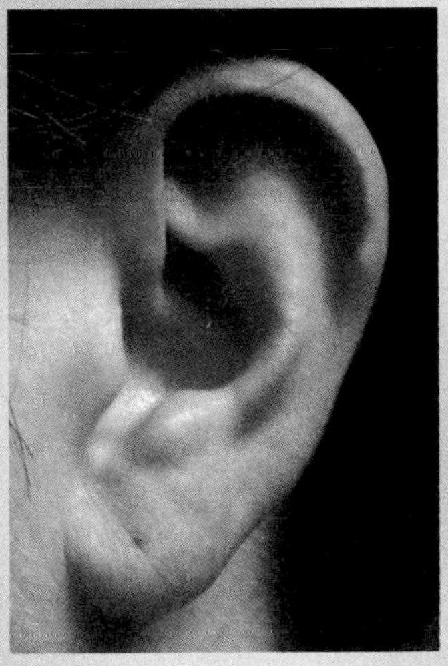

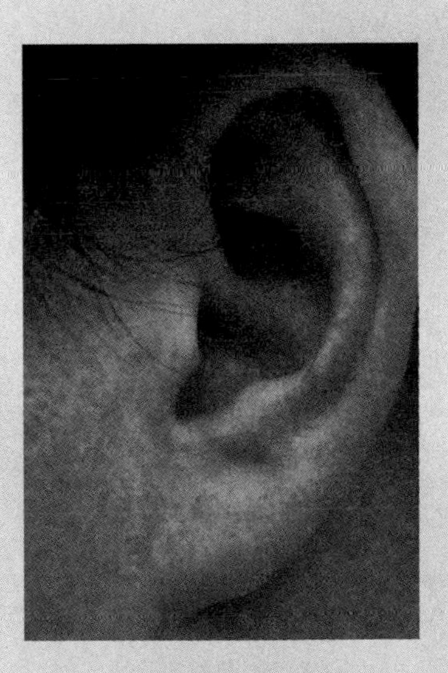

(d)

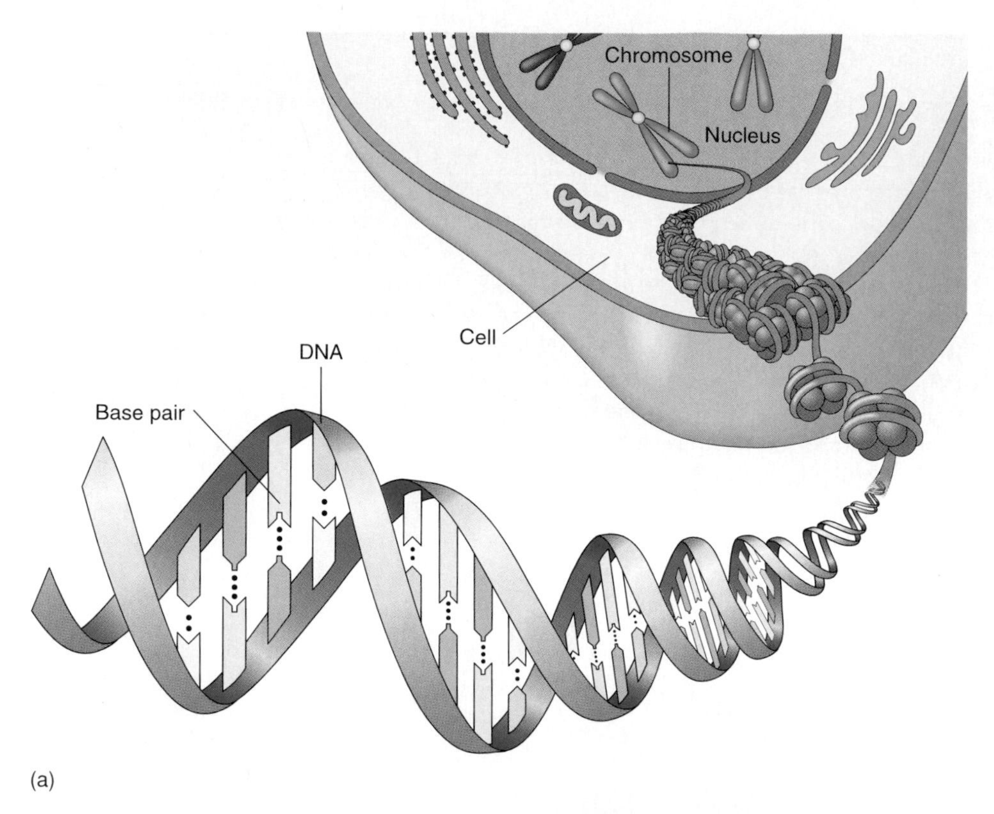


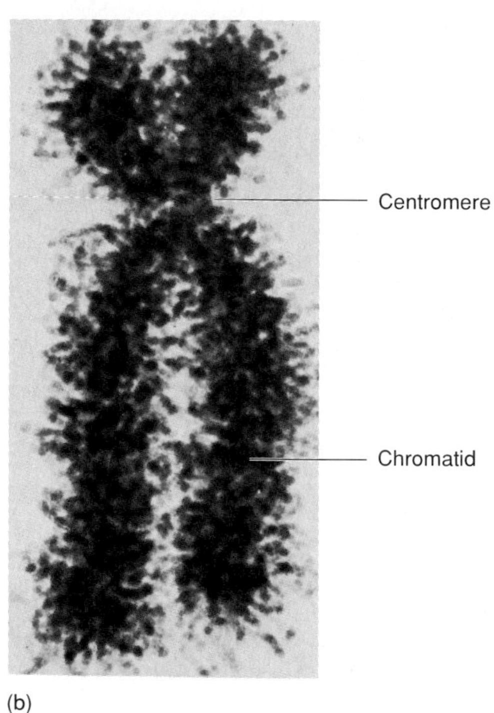


**FIGURE 24.3**

(*a*) Chromosomes consist of a continuous DNA double helix and associated proteins. They are located in the cell's nucleus. (*b*) A transmission electron micrograph of a chromosome in its replicated form, where it is dividing. Each half of the chromosome is a chromatid. Note the constriction, called the centromere (38,824×).

each other in that each contains a unique combination of the person's genes. That is, a man's sperm have variable combinations of genes, as do a woman's eggs.

## Meiosis

Meiosis is the part of *spermatogenesis* and *oogenesis* that halves the chromosome number (see chapter 22). This process includes two successive divisions, called the *first* and *second meiotic divisions.* The first meiotic division (*meiosis I*) separates homologous chromosome pairs. Homologous pairs are the same, gene for gene. At this stage, each homologous chromosome is replicated, so that it consists of two complete DNA strands called *chromatids.* The chromatids of a replicated chromosome are attached at special regions called the *centromeres.* Each of the cells that undergo the second meiotic division (*meiosis II*)

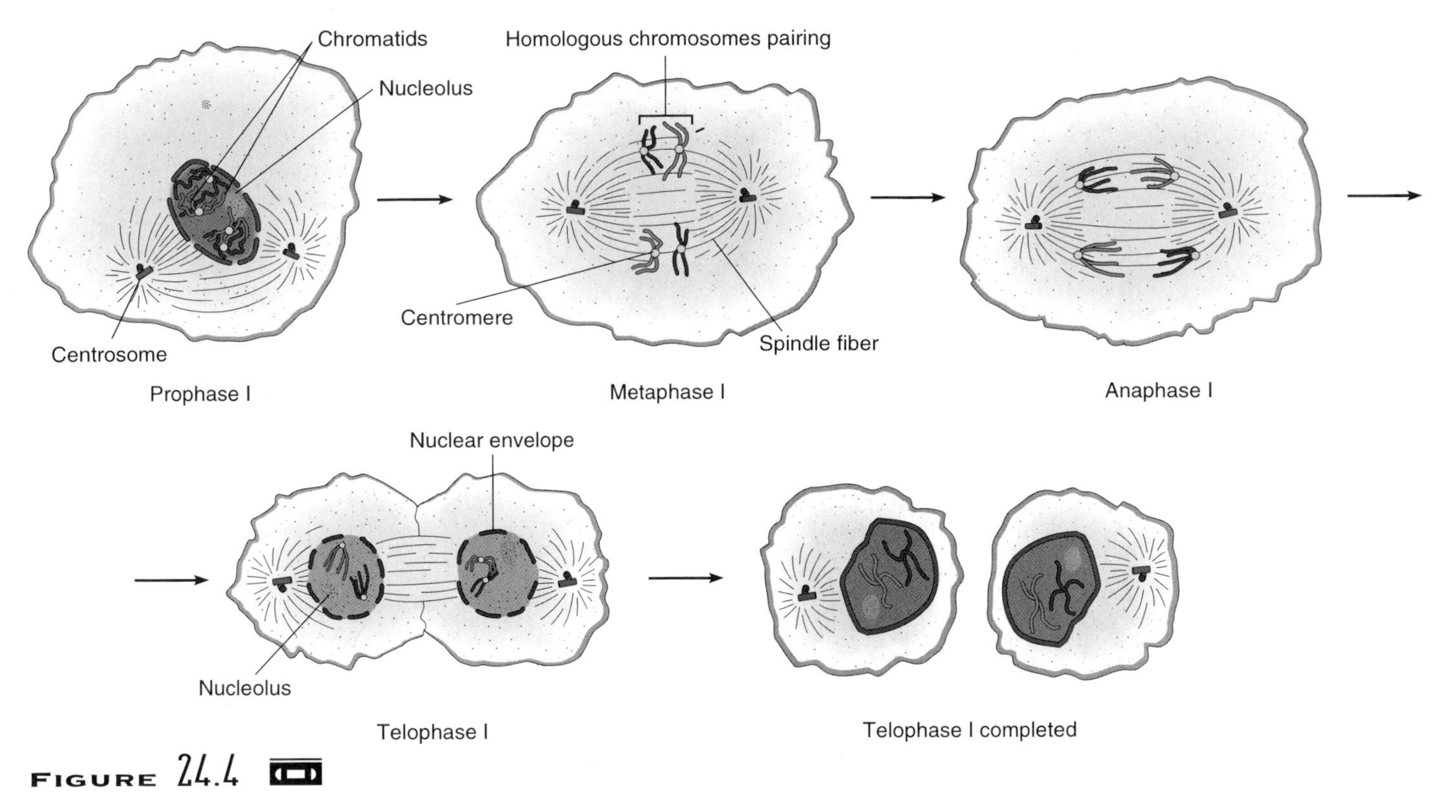


**FIGURE 24.4**
Stages in the first meiotic division.

have one representative of each homologous pair, and are thus haploid (having one copy of each chromosome). This second division separates chromatids, producing cells that are still haploid, but whose chromosomes are no longer replicated. After meiosis II, each of the chromatids is considered a full-fledged chromosome.

The steps of this complex process are clearer when considered in order. However, keep in mind that, like mitosis, meiosis is a continuous process. Dividing it into steps simply makes it easier to follow. Refer to figure 24.4 throughout the following discussion.

1. **Prophase I.** Individual chromosomes appear as thin threads within the nucleus, then shorten and thicken. Nucleoli disappear, the nuclear membrane temporarily disassembles, and microtubules begin to build the spindle that will separate the chromosomes. The chromosomes are replicated.

   As prophase continues, homologous chromosomes attract, lie side by side, and tightly intertwine. During this pairing, called *synapsis,* the chromatids of the homologous chromosomes contact one another at various points. Often, the chromatids break in one or more places and exchange parts, forming chromatids with new combinations of genetic information (fig. 24.5). For example, since one chromosome of a homologous pair is from a person's mother and the other is from the father, an exchange, or **crossover,** between homologous chromosomes produces chromatids that contain genetic information from both maternal and paternal sources.

2. **Metaphase I.** During the first metaphase, chromosome pairs line up about midway between the poles of the developing spindle, held under great tension, like two groups of people playing tug-of-war. Each pair consists of two chromosomes, which equals four chromatids. Each chromosome attaches to spindle fibers from one pole. The chromosome alignment is random with respect to maternal and paternal origin of the chromosomes. That is, each of the 23 chromosomes contributed from the mother may be on the left or the right, and the same is true for the paternal chromosomes—it is similar to the number of ways that 23 pairs of children could line up, while maintaining the pairs. Chromosomes can line up with respect to each other in many, many ways.

3. **Anaphase I.** Homologous chromosome pairs separate, each replicated member moving to one end of the spindle. Thus, each new cell receives only one replicated member of a homologous pair of chromosomes, reducing the chromosome number by one-half.

4. **Telophase I.** The original cell divides in two. Nuclear membranes form around the chromosomes, nucleoli reappear, and the spindle fibers disassemble into their constituent microtubules. After a short *interphase,* the second meiotic division begins.

   Meiosis II (fig. 24.6) is very similar to a mitotic division. During *prophase II,* chromosomes condense so that they reappear, still replicated. They move into positions midway between the poles of the developing spindle. In *metaphase II,* the replicated chromosomes attach to spindle fibers. In *anaphase II,* centromeres separate, freeing the chromatids to move to opposite poles of the spindles. The former chromatids are now considered to be chromosomes. In *telophase II,* each of the two cells resulting from meiosis I divides to form two cells. Therefore, each cell undergoing meiosis has the potential to produce four germ cells. In males, these mature into four sperm cells. In females, three of the products of meiosis are "cast aside" as polar bodies and one cell becomes the egg.

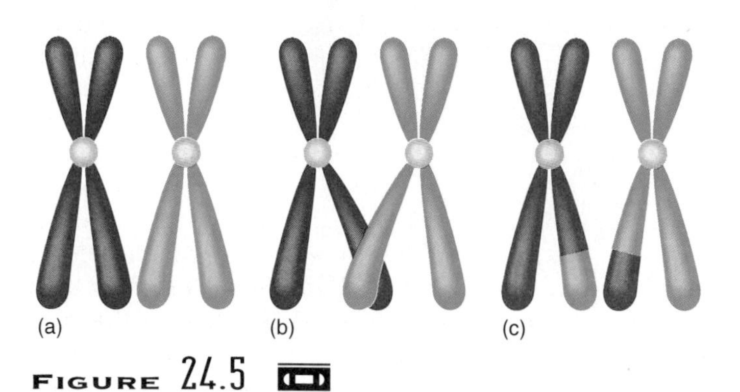


FIGURE 24.5

(*a*) Pairing of homologous chromosomes; (*b*) chromatids crossing over; (*c*) results of crossing over.

## Meiosis Leads to Genetic Variability

Meiosis generates astounding genetic variety. Any one of a person's more than 8 million possible combinations of 23 chromosomes can combine with any one of the more than 8 million combinations of his or her mate, raising the potential variability to more than 70 trillion genetically unique individuals! Crossing over contributes even more genetic variability. Figure 24.7 illustrates in a simplified manner how maternal and paternal traits become reassorted during meiosis.

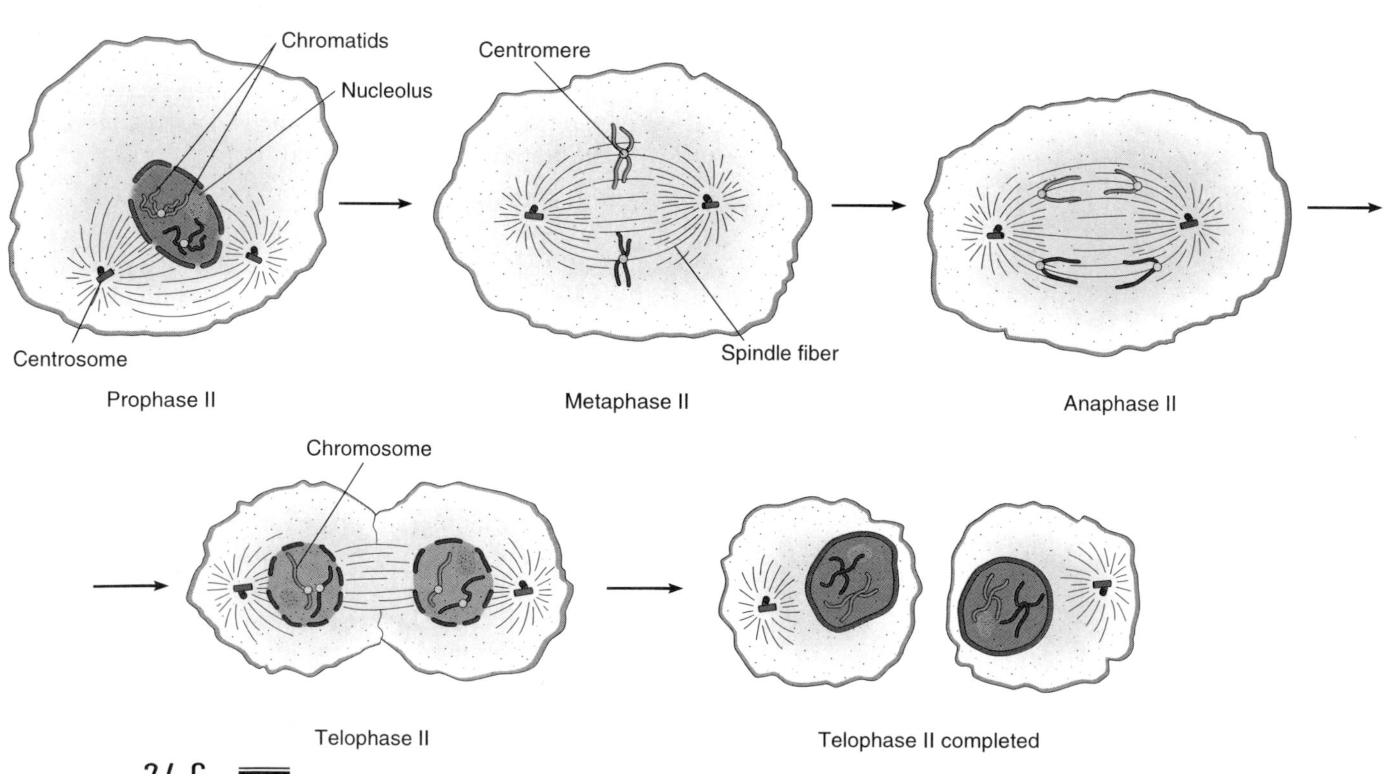


FIGURE 24.6

Stages in the second meiotic division.

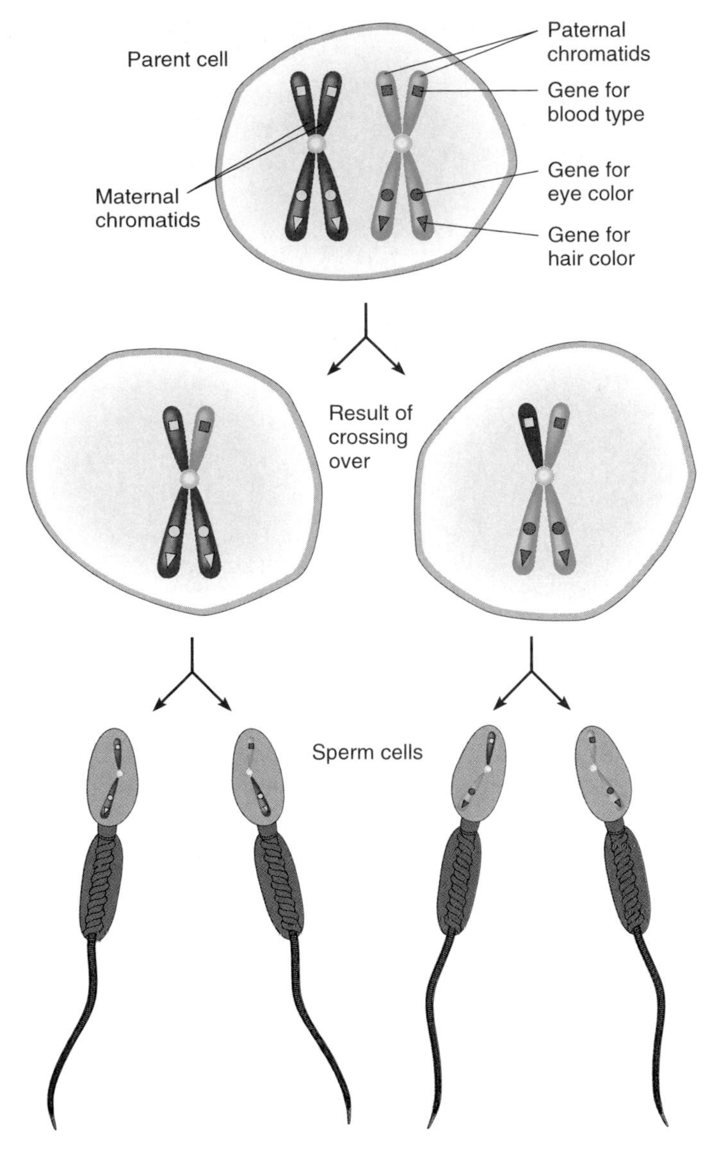


**FIGURE 24.7**
As a result of crossing over, the genetic information held within the sperm cells and the egg cells varies from cell to cell.

1. What are genes and chromosomes?
2. What is the role of meiosis in transmission of traits?
3. Describe the major events that occur during meiosis.
4. How does meiosis provide genetic variability?

## Chromosomes and Genes Come in Pairs

From the moment of conception, a human cell is *diploid,* containing two copies of each of the 23 different chromosomes. Geneticists construct chromosome charts called **karyotypes** that display the 23 chromosome pairs in size order (fig. 24.8). Pairs 1 through 22 are **autosomes,** chromosomes not involved in determining sex. The other two chromosomes, the X and the Y, determine sex and are called **sex chromosomes.** They will be discussed later in the chapter.

Until recently, geneticists constructed karyotypes by separating white blood cells from a drop of blood, allowing them to splash down onto a microscope slide, and finding and photographing a view of a cell where the chromosomes are spread apart. The photo would be developed and enlarged, and images of individual chromosomes would be cut out and pasted in size order. Today, most laboratories use a computerized karyotype system that automatically selects a well-spread cell and aligns the chromosomes into a chart (fig. 24.9).

Since we have two copies of each chromosome, we also have two copies of each gene, each located at the same position on the chromosome. Sometimes the members of a gene pair are alike, their DNA sequences specifying the same amino acid sequence of the protein product. However, because a gene consists of hundreds of nucleotide building blocks, it may exist in several variant forms, called **alleles.** Thus, while homologous pairs of chromosomes are the same, gene for gene, they may have different alleles of those genes. An individual who has two identical alleles of a gene is said to be **homozygous** for that gene. A person with two different alleles for a gene is said to be **heterozygous** for it. A gene may have a huge number of alleles, but an individual person can only have two alleles for a particular gene.

Clinical Application 24.2 discusses how localizing genes to particular chromosomes is providing new types of medical tests.

> The allele that causes most cases of cystic fibrosis was discovered in 1989, and researchers immediately began developing a test to detect it. However, other alleles were soon discovered. Today, hundreds of mutations (changes) in the cystic fibrosis gene are known. Different allele combinations produce different combinations and severities of symptoms, although most cases are quite serious.

1. What is the relationship between genes and chromosomes?
2. What is an allele?
3. What does it mean to be homozygous or heterozygous for a particular gene?

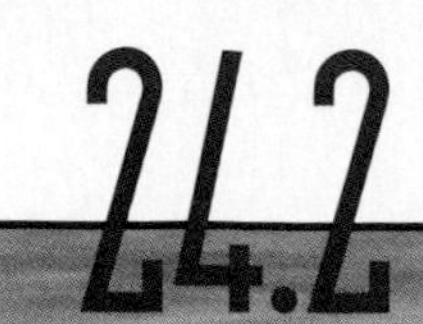

CLINICAL APPLICATION 24.2

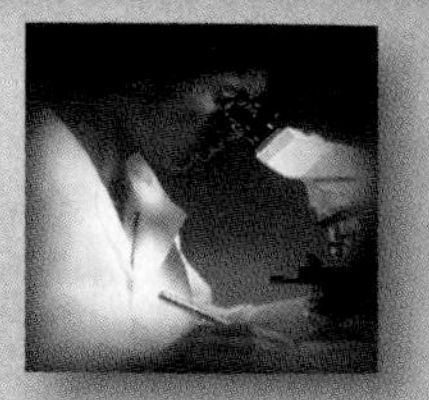

## Maps, Markers, and Medicine

Geneticists have "mapped" genes to particular chromosomes since genetic studies began, early in the twentieth century. Until the 1980s, most maps were constructed indirectly. Working with fruit flies, researchers observed that frequently two traits occurred together, or became separated by crossing over. The closer two genes were located to each other along a chromosome, the less often crossing over occurred between them, simply because there was less space. Researchers took such crossover frequencies and converted them into map distances, constructing intricate "linkage" maps. To do this in humans required studying very large families, and so was not usually possible.

Today, researchers use a variety of biochemical techniques to first localize a gene to a particular part of a specific chromosome, and then to actually obtain the DNA sequence of the gene. Often such studies begin with people who have a particular inherited illness, and a chromosome that is broken at a particular point, leading researchers to the probable site of the disease-causing gene.

As genes associated with a particular chromosome accumulate, researchers build maps, or ideograms, to depict gene locations on the chromosome. Figure 24B shows an ideogram for the entire human genome. These "physical" maps tend to agree with the gene orders established with older linkage maps. Each October, *Science* magazine publishes ideograms for all human chromosomes, tracking progress of the human genome project. It is interesting to see how much more crowded the depictions become each year.

Often discovery of a disease-causing gene in humans is preceded by detecting a genetic marker, which is a DNA sequence located very close to the gene of interest on its chromosome. Within a family, the presence of the marker indicates a high probability that the disease-causing gene has also been inherited, even if symptoms have not yet appeared. The first marker discovered was in 1983, for *Huntington disease,* which causes neurological degeneration in middle age. It took 10 years to "walk" down chromosome 4 to finally find the disease-causing gene. However, a marker for an inherited form of breast cancer was found in 1993, and the causative gene was discovered just a year later.

Knowing the sequences of genetic markers and disease-causing genes makes possible presymptomatic diagnoses, and sometimes even lifesaving treatments. This is the case for a rare form of thyroid cancer. In 1994, researchers tested children who had lost a parent or sibling to the cancer and removed the thyroids of those who had inherited the gene and were therefore destined, with 100% certainty, to develop the cancer. Some healthy women with strong family histories of breast cancer who inherit a causative gene elect to have their breasts removed, even though their chance of developing the cancer is 85%, not 100%.

The health practitioners of tomorrow will increasingly face situations where genetic tests indicate a future illness in a presently healthy person. It will be interesting to see how insurance companies handle these "healthy ill" people.

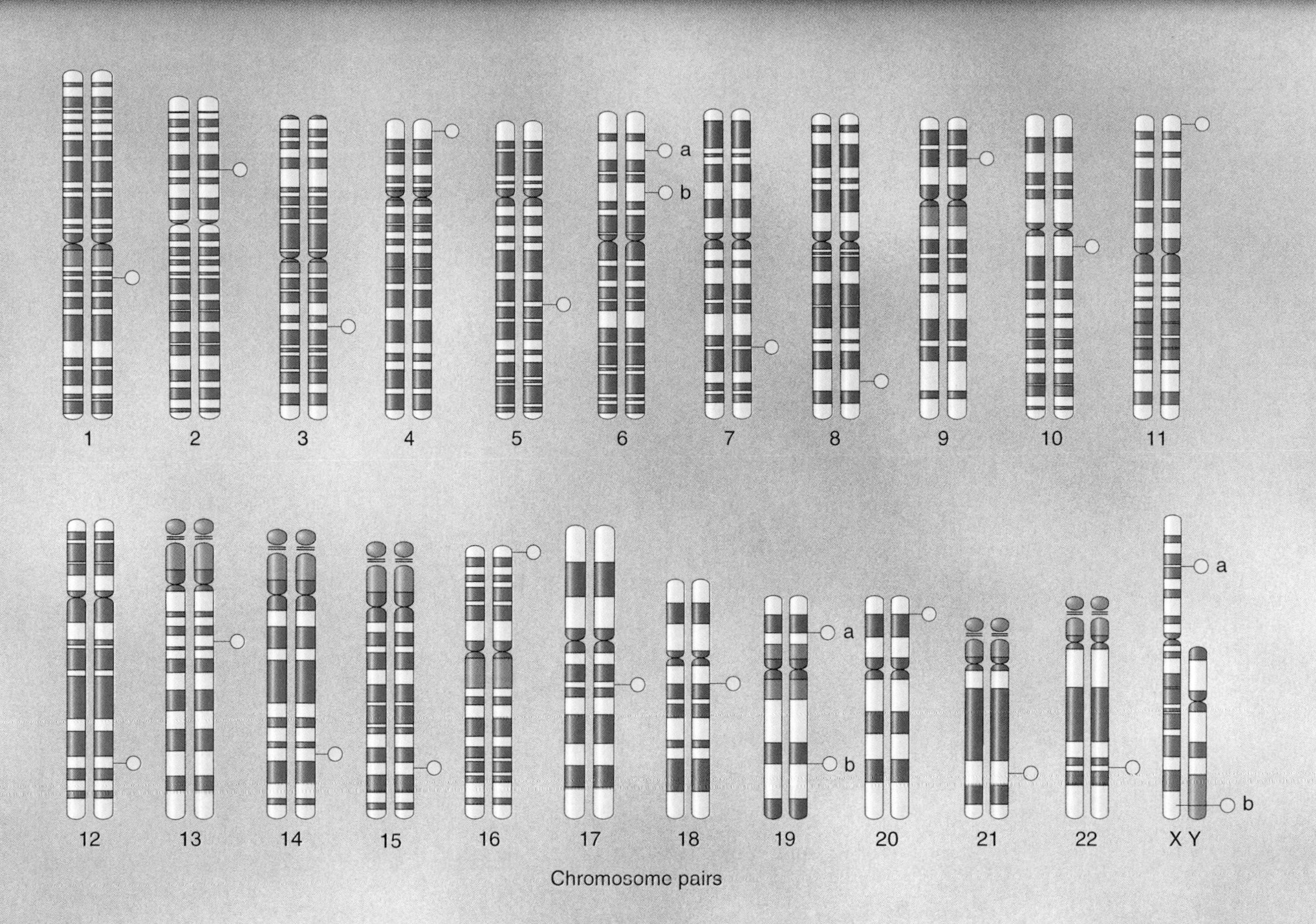


1. Gaucher Disease
2. Familial Colon Cancer*
3. Retinitis Pigmentosa*
4. Huntington Disease
5. Familial Polyposis of the Colon

6a. Spinocerebellar Ataxia
6b. Hemochromatosis

7. Cystic Fibrosis
8. Multiple Exostoses*
9. Malignant Melanoma
10. Multiple Endocrine Neoplasia, Type 2
11. Sickle Cell Disease
12. PKU (phenylketonuria)
13. Retinoblastoma
14. Alzheimer's Disease*
15. Tay-Sachs Disease
16. Polycystic Kidney Disease
17. Breast Cancer*
18. Amyloidosis

19a. Familial Hypercholesterolemia
19b. Myotonic Dystrophy

20. ADA Immune Deficiency
21. Amyotrophic Lateral Sclerosis*
22. Neurofibromatosis, Type 2

Xa. Muscular Dystrophy
Xb. Hemophilia

*Gene responsible for only some cases.

**FIGURE 24B**
Each year the human genome map becomes more crowded, as researchers map genes to chromosomes, then decipher the genes' roles in human anatomy and physiology, in sickness and in health. In this depiction of the human genome, the circles indicate the sites of specific genes on chromosomes.

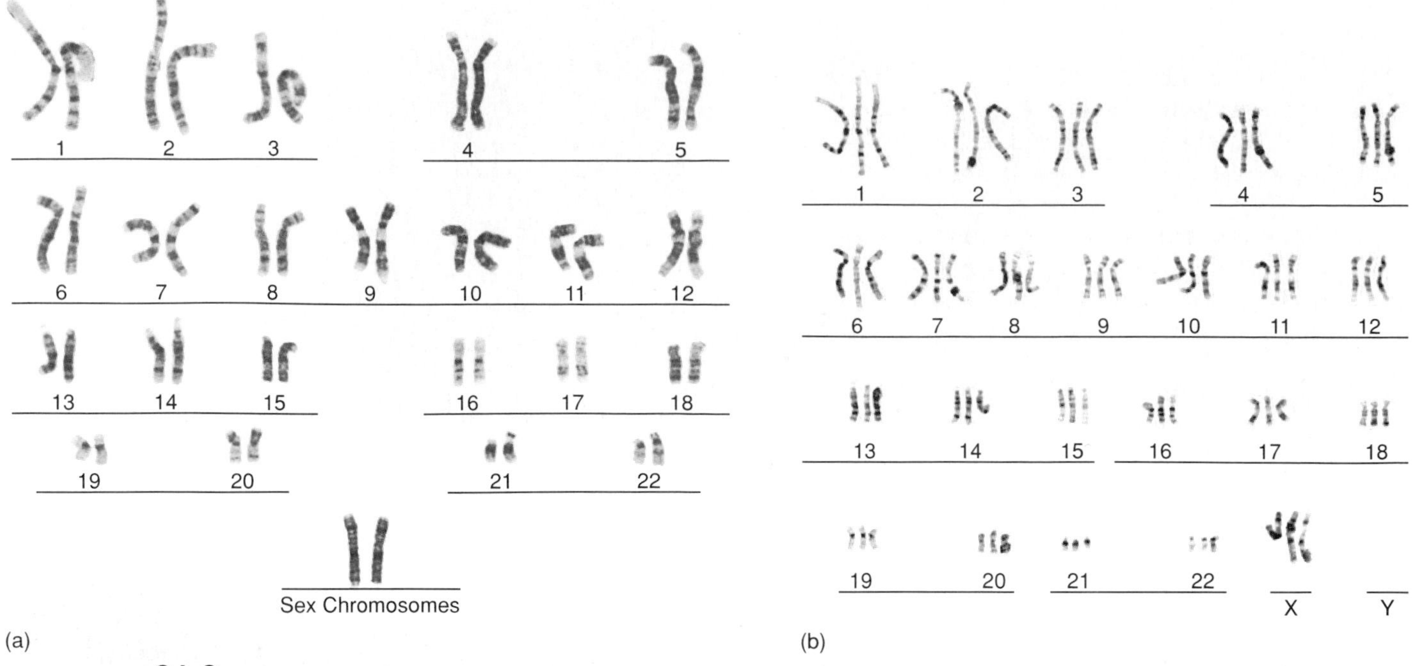


**FIGURE 24.8**

(*a*) A normal human karyotype, with the 22 pairs of autosomes aligned in size order, plus the sex chromosomes. (*b*) The problem indicated by this unusual karyotype is drastic. Individuals with three copies of each chromosome (triploids) account for 17% of all spontaneous abortions and 3% of stillbirths and newborn deaths.

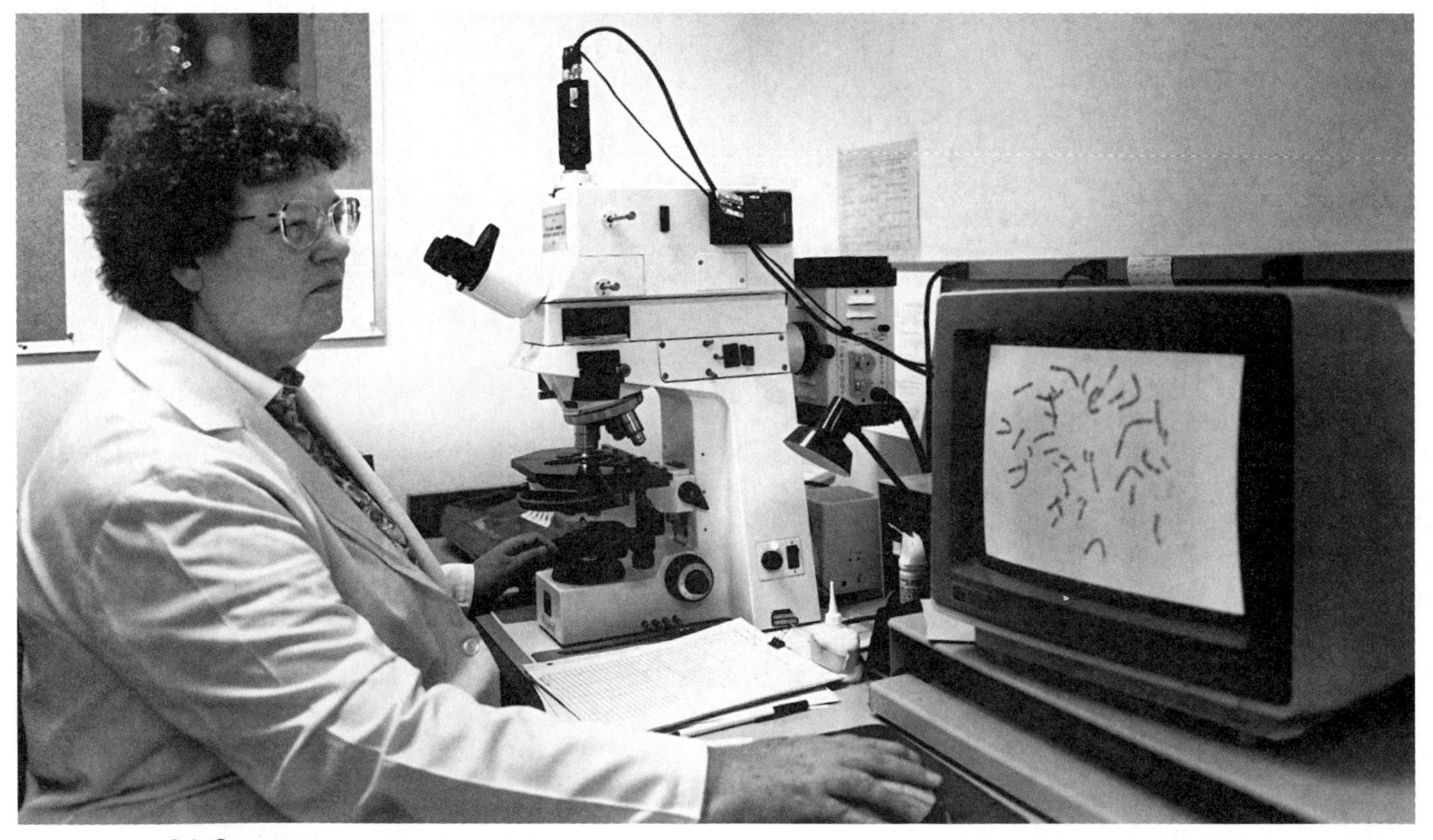

**FIGURE 24.9**

To study human chromosomes, stains are applied to white blood cells, which create banded patterns on the chromosomes. Chromosomes are distinguished by their size and banding patterns. A computerized karyotype device, shown here, arranges the chromosomes in a cell into the standard size order chart.

## Modes of Inheritance

The particular combination of genes present in a person's cells constitutes the **genotype.** The appearance of the individual that develops as a result of the ways the genes are expressed is termed the **phenotype.** An allele is **wild-type** if its associated phenotype is either normal function or the most common expression in a particular population. An allele that is a change from wild-type, perhaps producing an uncommon phenotype, is **mutant.** Disease-causing alleles are mutant.

### Dominant and Recessive Inheritance

Often in heterozygotes, one allele determines the phenotype. Such an allele whose action masks that of another allele is termed **dominant.** The allele whose expression is masked is **recessive.** A gene causing a disease can be recessive or dominant. It may also be *autosomal* (carried on a non-sex chromosome) or *sex-linked* (carried on the X chromosome).

Whether a trait is dominant or recessive, autosomal or sex-linked is called its *mode of inheritance.* Mode of inheritance has important consequences in predicting the chance that offspring will inherit an illness or trait. These rules emerge:

1. An autosomal condition is equally likely to affect both sexes. Sex-linked characteristics affect males much more often than females, a point we will return to later in the chapter.
2. A person most likely inherits a recessive condition from two parents who are each heterozygotes (carriers). The parents are usually healthy. For this reason, recessive conditions "skip" generations.
3. A person who inherits a dominant condition has at least one affected parent. Therefore, dominant conditions do not skip generations.

Cystic fibrosis (CF) is an example of an autosomal recessive disorder. The wild-type allele for the CFTR gene, which is dominant over the disease-causing allele, specifies chloride channels on cells of the pancreas, respiratory tract, intestine, and elsewhere (see fig. 24.2). Disease-causing recessive mutant alleles disrupt the structure and possibly the function of the chloride channels. An individual inheriting two such alleles has cystic fibrosis. He or she is homozygous recessive. A person inheriting only one recessive mutant allele plus a dominant wild-type allele is a carrier. He or she will pass on the disease-causing allele in half of his or her germ cells. A person who has two wild-type alleles is homozygous dominant for the gene. Note that although three genotypes are possible, there are two phenotypes, because carriers as well as homozygous dominant individuals are healthy.

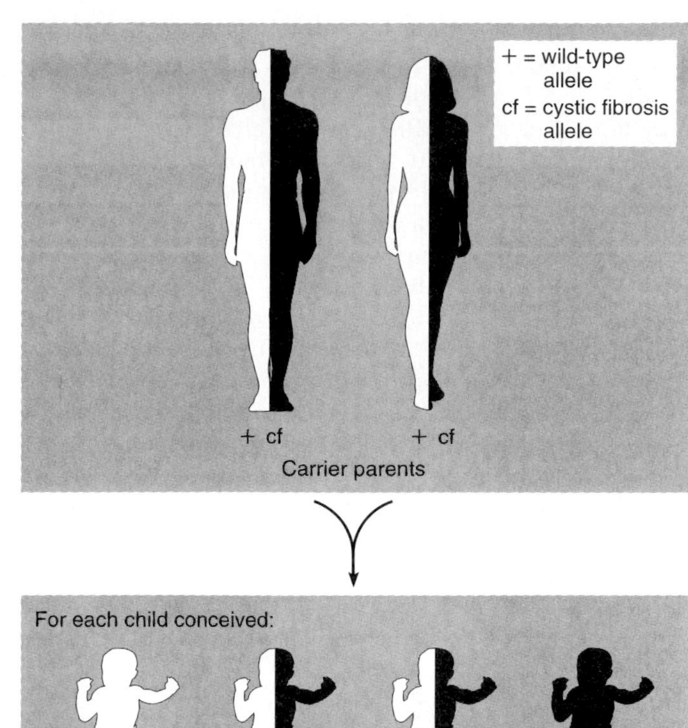


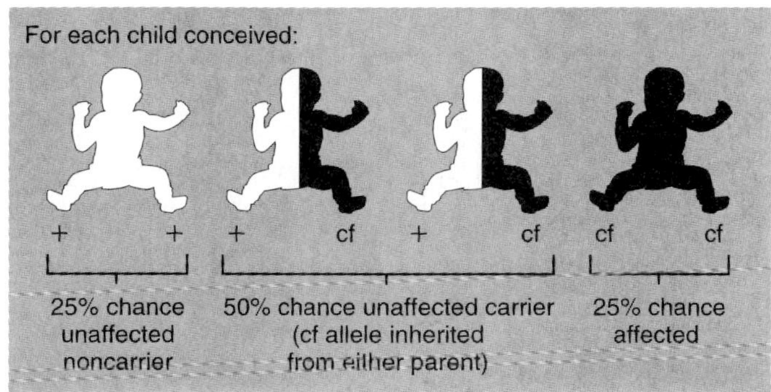


**Figure 24.10**
Inheritance of cystic fibrosis from carrier parents illustrates the autosomal recessive mode of inheritance. Sexes are affected with equal frequency, and symptoms usually begin in early childhood. Autosomal recessive conditions can skip generations, as carriers pass on the mutant allele.

Using logic, understanding how chromosomes and genes are apportioned into germ cells in meiosis, and knowing that the allele that causes cystic fibrosis is autosomal recessive, we can predict genotypes and phenotypes of the next generation. Figure 24.10 illustrates the situation of two people who are each heterozygous for a CF-causing allele. Half of the man's sperm contain the mutant allele, as do half of the woman's eggs. Since sperm and eggs combine at random, each child that these people conceive has a

- 1/4 chance of inheriting two wild-type alleles (homozygous dominant, healthy, and not a carrier)
- 1/2 chance of inheriting a mutant allele from either parent (heterozygous and a carrier, but healthy)
- 1/4 chance of inheriting a mutant allele from each parent (homozygous recessive, has CF)

Medical consequences are more dire when a disease-causing allele is dominant. In an autosomal recessive illness, an affected person's parents are carriers—they do not have the illness. In an autosomal dominant condition, however, an affected person has an affected parent. He or she need inherit only one copy of the mutant allele to have the associated phenotype; in contrast, expression of an autosomal recessive condition requires two copies of the mutant allele.

An example of an autosomal dominant condition is Huntington disease. Symptoms usually begin in the late 30s or early 40s, and include loss of coordination, uncontrollable dancelike movements, and personality changes. Figure 24.11 shows the inheritance pattern. Here, if one parent has the mutant allele, half of his or her germ cells will have it. Assuming the other parent does not have a mutant allele, each child conceived has a 1 in 2 chance of inheriting the gene and, eventually, the condition.

Most of the 5,000 or so known human inherited disorders are autosomal recessive. These tend to produce symptoms very early in life, sometimes before birth. Autosomal dominant conditions often have an adult onset.

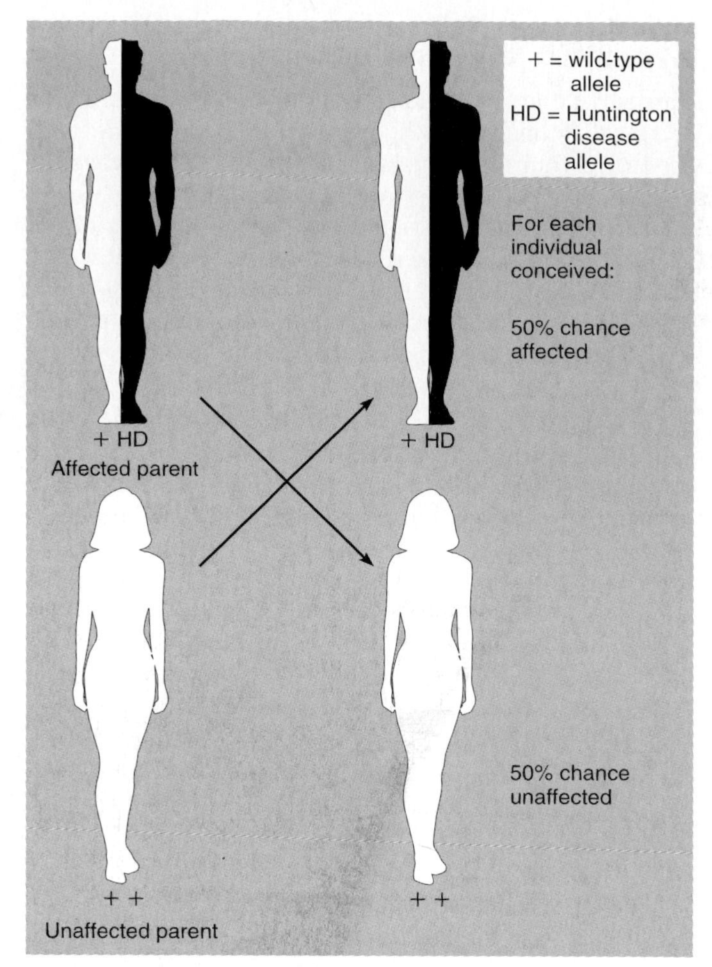


**FIGURE 24.11**

Inheritance of Huntington disease from a parent who will be affected in middle age illustrates the autosomal dominant mode of inheritance. Sexes are affected with equal frequency. Autosomal dominant conditions affect every generation until, by chance, none of the children in a generation inherits the mutant allele.

Why would recessive alleles that cause illness remain in a population, if they so endanger health? One reason may be that carrying a genetic disease can protect against an infectious disease. The basis of this phenomenon lies in anatomy and physiology. For example, carriers of *sickle cell disease,* an inherited anemia, do not contract malaria. In sickle cell disease, a tiny genetic alteration causes the gene's product—the beta globin chain of hemoglobin—to aggregate, bending the red blood cell containing it into a sickle shape that can block blood flow when oxygen level is low. Carriers have only a few sickled cells, but these apparently are enough to somehow make red blood cells inhospitable to malaria parasites.

Similarly, carriers of *cystic fibrosis* resist cholera, in which a bacterial toxin opens chloride channels in the small intestine, causing rapidly fatal diarrhea. Carriers of CF have some abnormal chloride channels, which renders cholera toxin ineffective.

Carriers of *Tay-Sachs disease* may resist tuberculosis. This association was first noted in the Jewish ghettos of World War II, where some healthy relatives of children who died of Tay-Sachs disease did not contract tuberculosis, even after repeated exposure. Tay-Sachs disease causes excess myelin to accumulate on nerve cells. How this protects against tuberculosis is not known.

1. Distinguish between genotype and phenotype.
2. Distinguish between wild-type and mutant.
3. What are the consequences of autosomal recessive and autosomal dominant inheritance?

## Different Dominance Relationships

Some genes show **incomplete dominance,** in which the heterozygous phenotype is intermediate between that of either homozygote. In **familial hypercholesterolemia (FH),** a person with two disease-causing alleles completely lacks receptors on liver cells that take up cholesterol from the bloodstream. A person with one disease-causing allele has half the normal number of cholesterol receptors. Someone with two wild-type

(a)

(b)

**FIGURE 24.12**

(*a*) Falsely colored scanning electron micrograph of normal red blood cells (5,555×). (*b*) Red blood cells from a person with sickle cell disease (5,555×).

alleles has the normal number of receptors. The phenotypes parallel the number of receptors—those with two mutant alleles die as children of heart attacks, those with one mutant allele die in young or middle adulthood, and those with two wild-type alleles have healthy hearts.

Incomplete dominance is seen in carriers of *sickle cell disease* at the molecular and cellular levels. Some of their hemoglobin molecules are abnormal, and under low oxygen conditions, some of their red blood cells may sickle (fig. 24.12).

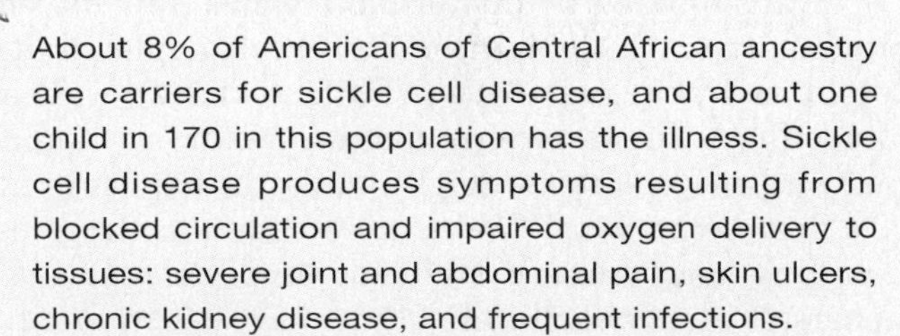

About 8% of Americans of Central African ancestry are carriers for sickle cell disease, and about one child in 170 in this population has the illness. Sickle cell disease produces symptoms resulting from blocked circulation and impaired oxygen delivery to tissues: severe joint and abdominal pain, skin ulcers, chronic kidney disease, and frequent infections.

Being a carrier of sickle cell disease very rarely, if ever, produces symptoms. However, in the early 1970s, when many people were tested for the disorder, people who were carriers were said to have "sickle cell trait." Because this implied that they were ill, the name led to discrimination. Today, most states test all newborns for sickle cell disease, because early treatment with penicillin can prevent severe infections and sometimes forestall symptoms.

Different alleles that are both expressed in a heterozygote are **codominant.** Two of the three alleles of the I gene, which determines ABO blood type, are codominant. People of blood type A have a molecule called antigen *A* on the surfaces of their red blood cells. Blood type B corresponds to red blood cells with antigen *B*. A person with type AB has red blood cells with both the *A* and *B* antigens, and the red cells of a person with type O blood have neither antigen.

The I gene encodes the enzymes that place the *A* and *B* antigens on red blood cell surfaces. The three alleles are $I^A$, $I^B$, and $i$. People with type A blood may be either genotype $I^AI^A$ or $I^Ai$; type B corresponds to $I^BI^B$ or $I^Bi$; type AB to $I^AI^B$; and type O to $ii$.

## GENE EXPRESSION

### Penetrance and Expressivity

The same allele combination can produce different degrees of the phenotype in different individuals, even siblings, because of outside influences such as nutrition, toxic exposures, other illnesses, and other genes. Most disease-causing allele combinations are **completely penetrant,** which means that everyone who inherits it has some symptoms. A genotype is **incompletely penetrant** if some individuals do not express the phenotype. Polydactyly, having extra fingers or toes, is incompletely penetrant (see fig. 7.48). Some people who inherit the dominant allele have more than five digits on a hand or foot, yet others who are known to have the allele (because they have an affected parent and child) have the normal number of fingers and toes. The penetrance of a gene is described numerically. If 80 of 100 people who have inherited the dominant polydactyly allele have extra digits, the allele is 80% penetrant.

A phenotype is **variably expressive** if the symptoms vary in intensity in different people. One person with polydactyly might have an extra digit on both hands and a foot; another might have two extra digits on both hands and both feet; a third person might have just one extra fingertip on a finger. *Penetrance* refers to the all-or-none expression of a genotype in an individual; *expressivity* refers to the severity of a phenotype.

### Pleiotropy

Some genetic disorders produce several symptoms. When family members have different symptoms, they can appear to have different illnesses. This phenomenon is called **pleiotropy.**

Pleiotropy is seen in genetic diseases affecting a single protein that is found in different parts of the body, or in the case of an enzyme that participates in more than one reaction. This is the case for *Marfan syndrome,* an autosomal dominant defect in an elastic connective tissue protein called fibrillin. The protein is abundant in the lens of the eye, in the aorta, and in the bones of the limbs, fingers, and ribs. Knowing this, the Marfan syndrome symptoms of lens dislocation, long limbs, spindly fingers, and a caved-in chest make sense. The most serious symptom is a life-threatening weakening in the aorta wall, sometimes causing the vessel to suddenly burst. However, if the weakening is found early, that part of the vessel wall can be replaced with a synthetic graft, saving the person's life.

The American Revolution has been indirectly blamed on a pleiotropic illness, *porphyria,* that affected British King George III's ability to rule. At age 50 he first experienced recurring cycles of symptoms, including abdominal pain, red urine, fever, and weakness, then insomnia, headache, visual disturbances, restlessness, delirium, convulsions, and stupor. The symptoms always appeared in this order. He was eventually dethroned because of his illness.

The king's crimson urine led twentieth century physicians to diagnose porphyria. In this inborn error of metabolism, the organic portions of the hemoglobin molecule that surround iron atoms, called porphyrin rings, are excreted in the urine instead of being broken down in cells. The result is red urine. Looking back at the royal family tree, researchers identified several relatives who experienced only some of the symptoms that plagued George III. One had red urine, another abdominal pain. The many guises of this pleiotropic illness had baffled physicians for decades.

### Genetic Heterogeneity

The same phenotype may result from the actions of different genes, a phenomenon called **genetic heterogeneity.** For example, nearly two hundred forms of hereditary deafness are known. There are eleven types of clotting disorders, because there are many clotting factors and enzymes that control the biochemical pathways that lead to blood clotting. Any of several genes may also cause cleft palate, albinism, and diabetes insipidus.

1. What is incomplete dominance?
2. What is codominance?
3. How do penetrance, expressivity, and pleiotropy affect gene expression?

---

## Complex Traits

Most of the inherited disorders mentioned so far are *monogenic*—that is, they are determined by a single gene, and their expression is usually not influenced greatly by the environment. However, most characteristics and disorders reflect the environment as well as genes.

Traits determined by more than one gene are termed *polygenic.* Usually, several genes each contribute an equal, small degree toward molding the overall phenotype, which may vary greatly among individuals. Height and skin color are polygenic traits (fig. 24.13 and chart 24.1).

Traits molded by one or more genes plus the environment are termed *multifactorial* or complex traits. Height and skin color are multifactorial as well as polygenic, because environmental factors influence them: nutrition affects height, and sun exposure affects skin color. Most of the more common illnesses, including heart disease, diabetes, hypertension, and cancers, are multifactorial. Clinical Application 24.3 discusses how twins help geneticists understand complex traits.

1. How does polygenic inheritance make possible many variations of a trait?
2. How may the environment influence gene expression?

**FIGURE 24.13**

Height is a polygenic, multifactorial trait. When 175 soldiers were asked to line up according to their heights, they formed a bell-shaped, continuous distribution, characteristic of a polygenic trait.

**CHART 24.1 POLYGENIC MODEL OF SKIN COLOR INHERITANCE**

| Genotype | Number of Pigment Genes |
|---|---|
| *AABBCC* | 6 |
| *AaBBCC, AABbCC, AABBCc* | 5 |
| *aaBBCC, AAbbCC, AABBcc, AaBbCC, AaBBCc, AABbCc* | 4 |
| *AaBbCc, aaBbCC, AAbbCc, AabbCC, AABbcc, aaBBCc, AaBBcc* | 3 |
| *AaBbcc, AabbCc, aaBbCc, AAbbcc, aaBBcc, aabbCC* | 2 |
| *Aabbcc, aaBbcc, aabbCc* | 1 |
| *aabbcc* | 0 |

Studies that classify skin color by measuring the amount of light reflected from the skin surface suggest that three or four or more different genes, probably with several alleles each, are involved in producing pigment in the skin. The greater the number of pigment-specifying genes, the darker the skin. The inheritance of skin color is most likely even more complicated than this model of three genes.

## MATTERS OF SEX

Human somatic (nonsex) cells include an X and a Y chromosome in males, and two X chromosomes in females. All eggs carry a single X chromosome, and sperm carry either an X or a Y chromosome. Sex is determined at conception: a Y-bearing sperm fertilizing an egg conceives a male, and an X-bearing sperm conceives a female (fig. 24.14).

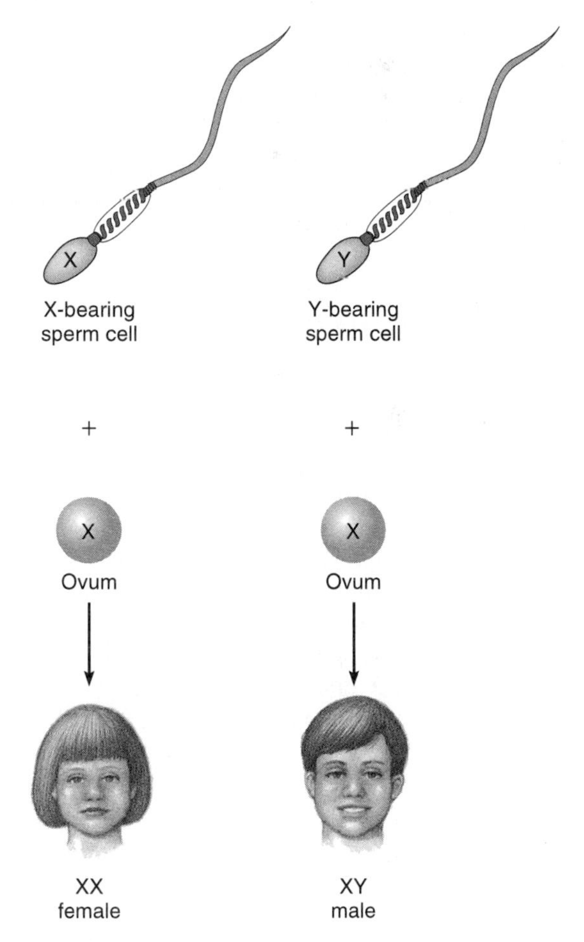


**FIGURE 24.14**

Sex determination. An egg contributes an X chromosome, and a sperm either an X or a Y. If a Y-bearing sperm cell fertilizes an egg, the zygote is male (XY). If an X-bearing sperm cell fertilizes an egg, the zygote is female (XX).

Clinical Application 24.3

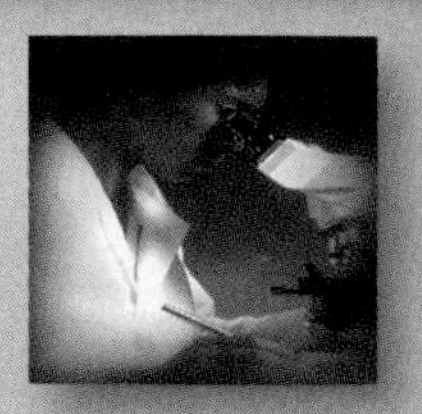

## Adoption and Twin Studies

Geneticists enlist two types of people to tease apart the inherited and environmental components of multifactorial traits—adopted individuals and twins.

A person who is adopted usually shares environmental influences with the adoptive parents, and genes with the biological parents. By comparing an adopted person's characteristics with those of both sets of parents, researchers can attempt to sort out inherited from environmental influences. Studies examining causes of death, for example, find that a person's risk of dying of infection before age 50 is more similar to the risk of the biological parents doing so than to that of adoptive parents. Susceptibility to infection, then, is largely determined by genetics, most likely inherited variations in immune system function.

Identical twins arise from the same fertilized egg and are therefore genetically identical. Fraternal twins share half of their genes, and are no more alike genetically than are any two siblings. A trait has a large genetic component if it affects both members of identical twin pairs with greater frequency than it affects both members of fraternal twin pairs.

Even more informative are identical twins who were separated at birth and reared in different environments. Any distinctive traits they share are likely to be rooted in their genes. For example, a pair of identical male twins raised in different countries in different religions were astounded when they were reunited as adults to find that they both laugh when someone sneezes and flush the toilet before using it. A pair of girls reunited in late adolescence found they were both afraid of swimming. Twins who met for the first time in their 30s both paused for 30 seconds after being asked a question, each rotated a gold necklace she was wearing three times, and then answered. Coincidence, or genetics?

### Sex Determination

Association of the Y chromosome with maleness has been known since 1959, but researchers did not identify the precise genes responsible for maleness until 1990. The **SRY gene** (sex-determining region of the Y) encodes a type of protein called a *transcription factor,* which switches on other genes that direct development of male structures in the embryo, while suppressing formation of female structures. Figure 24.15 shows the X and Y chromosomes.

Some unusual individuals helped researchers identify the SRY gene—men who have two X chromosomes, and women who have one X and one Y chromosome, which is the reverse of normal. A close look at the composition of these unusual peoples' sex chromosomes revealed that the XX males actually had a small piece of a Y chromosome, and the XY females lacked a small part of the Y chromosome. The part of the Y chromosome that was present in the XX males was the same part that was missing in the XY females.

### Sex Chromosomes and Sex-linked Traits

The X and Y chromosomes carry genes, but they are inherited in different patterns than are autosomal genes because of the differing sex chromosome constitutions in males and females. The X chromosome has more than one thousand genes; the Y chromosome has very few genes. Traits transmitted on the X chromosome are said to be **sex-linked.**

Any gene on the X chromosome of a male is expressed in his phenotype, because there is no second allele to mask its expression. An allele on an X chromosome in a female may or may not be expressed depending upon whether it is dominant or recessive, and on the nature of the allele on the second X chromosome. The human male is said to be **hemizygous** for sex-linked traits because he has half the number of genes that the female has. *Red-green color blindness* and the most common form of the clotting disorder *hemophilia* are examples of recessive sex-linked traits.

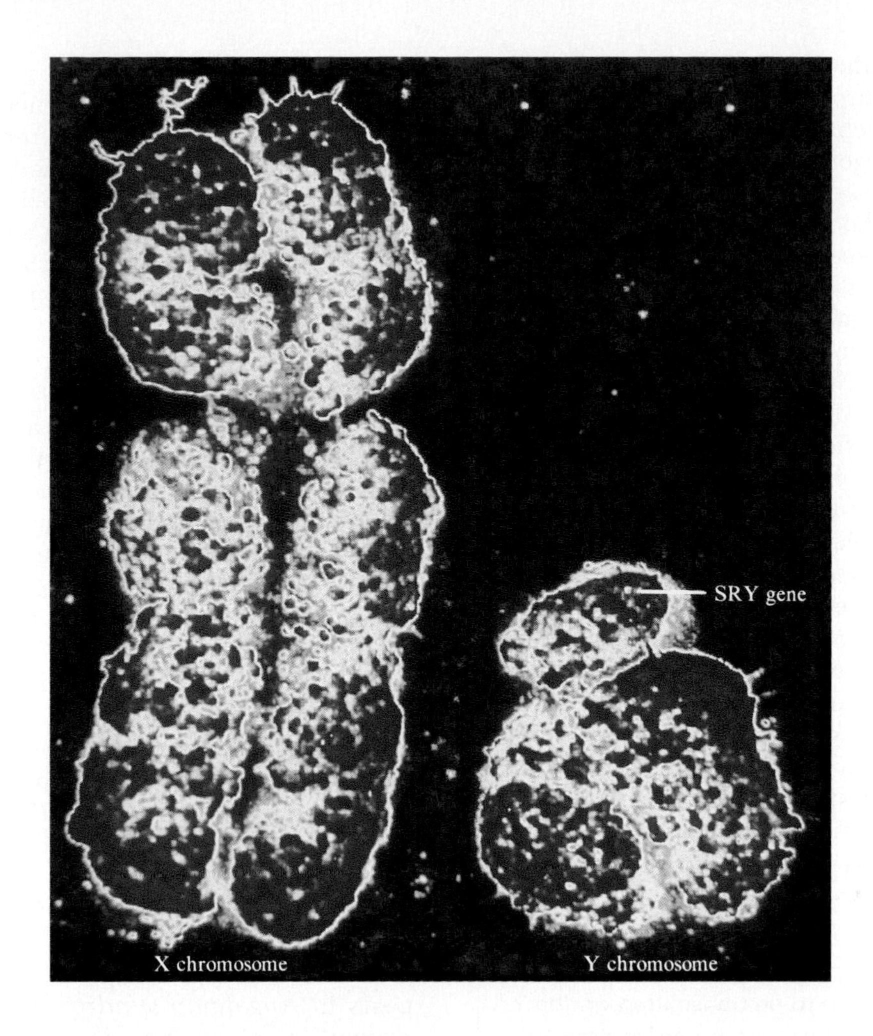


**Figure 24.15**

The X and Y chromosome. The SRY gene, at the top of the short arm of the Y chromosome, sets into motion the cascade of gene activity that directs development of a male (31,000×).

It would appear that the sexes are unequal, as far as sex-linked traits go. To remedy the inequity, the cells of female mammalian embryos undergo a process called *X inactivation,* which shuts off one X chromosome in each somatic cell. A gene called XIST on the X chromosome controls the shutdown. Which of a female's X chromosomes is silenced—the one she inherited from her mother or the one from her father—occurs randomly. Therefore, a female is a mosaic, with her father's X chromosome expressed in some cells, and her mother's in others. This can be vividly illustrated for heterozygous, sex-linked genes.

Consider a woman who is a carrier (a heterozygote) for *Duchenne muscular dystrophy.* This means that she has a wild-type allele for the dystrophin gene on one X chromosome, and a disease-causing allele on the other. Cells in which the X chromosome bearing the wild-type allele is inactivated do not produce the gene's protein product, dystrophin. However, cells in which the mutant allele is inactivated do produce dystrophin. When a stain for dystrophin is applied to a sample of her muscle tissue, some cells take up the stain, turning blue, and others do not. This is the basis of a carrier detection test for this illness. If by chance many of such a woman's wild-type dystrophin alleles are turned off in her muscle cells, she may experience mild muscle weakness.

A male always inherits his Y chromosome from his father and his X chromosome from his mother. A female inherits one X chromosome from each parent. If a mother is heterozygous for a particular sex-linked gene, her son has a 50% chance of inheriting either allele from her. Sex-linked genes are therefore passed from mother to son. Because a male does not receive an X chromosome from his father (he inherits the Y chromosome from his father), a sex-linked trait is not passed from father to son.

Consider the sex-linked inheritance pattern of hemophilia A, a blood clotting disorder. It is passed from carrier mother to affected son with a risk of 50%, because he can inherit either her normal allele or the mutant one. A daughter has a 50% chance of inheriting the hemophilia allele and being a carrier like her mother, and a 50% chance of not carrying the allele.

A daughter can inherit a sex-linked recessive disorder or trait if her father is affected and her mother is a carrier. She inherits one affected X chromosome from each parent. Without a biochemical test, though, a woman would not know that she is a carrier of a sex-linked recessive trait unless she has an affected son.

A sex-linked recessive trait is more likely to be seen in females if it isn't serious enough to prevent an affected man from fathering children. This is the case for color blindness.

Very few traits are known to be transmitted on the Y chromosome, probably because it carries very few genes. One such Y-linked trait is hairy ears.

Dominant disease-causing alleles on the X chromosome are extremely rare. Generally in such cases, males are much more severely affected than females, who have a second X to offer a protective effect. In a condition called *incontinentia pigmenti,* for example, an affected girl has swirls of pigment on her skin where melanin in the epidermis leaks into the dermis. She may have abnormal teeth, sparse hair, visual problems, and seizures. However, males inheriting the dominant gene on their X chromosomes are so severely affected that they do not survive to be born.

## Gender Effects on Phenotype

Sex-linked traits generally are more prevalent in males than females. Traits that are not sex-linked, however, can be expressed differently in males and females, due to differences between the sexes.

### Sex-limited Traits

A **sex-limited** trait affects a structure or function of the body that is present in only males or only females. Such a gene may be sex-linked or autosomal. Beard growth and breast size are sex-limited traits. A woman cannot grow a beard because she does not manufacture the hormones needed for facial hair growth. She can, however, pass to her sons the genes specifying heavy beard growth.

### Sex-influenced Traits

In **sex-influenced inheritance,** an allele is dominant in one sex but recessive in the other. Again, such a gene may be sex-linked or autosomal. This difference in expression can be caused by hormonal differences between the sexes. For example, a gene for hair growth pattern has two alleles, one that produces hair all over the head and another that causes pattern baldness (fig. 24.16). The baldness allele is dominant in males but recessive in females, which is why more men than women are bald. A heterozygous male is bald, but a heterozygous female is not. What is the genotype of a bald woman?

In *genomic imprinting,* for unknown reasons, the expression of a disorder differs depending upon which parent transmits the disease-causing gene or chromosome. The phenotype may differ in degree of severity, in age of onset, or even in the nature of the symptoms.

A fascinating example of genomic imprinting appears in *Angelman syndrome* and *Prader-Willi syndrome.* In these two disorders affecting the same region of chromosome 15, different sets of symptoms result depending upon the sex of the parent transmitting the gene. Prader-Willi is apparently inherited from the mother. A child with Prader-Willi syndrome is obese, has small hands and feet, is mentally retarded, and does not develop signs of puberty. Angelman syndrome, inherited from the father, is sometimes called "happy puppet syndrome," describing the phenotype. The child has an extended tongue, poor muscle coordination that gives a "floppy" appearance, and a large jaw, laughs uncontrollably, and has peculiar convulsions in which the upper limbs flap. Geneticists do not yet understand how the same genes can have such different effects.

Clinical Application 24.4 discusses a disorder that is more severe in males—fragile X syndrome.

1. Which chromosomes and genes determine sex?
2. What is sex-linked inheritance?
3. Why do sex-linked recessive conditions appear most commonly in males?
4. How can gender affect gene expression?

(a)

(b)

(c)

(d)

**Figure 24.16**

Pattern baldness is a sex-influenced trait, and was a genetic trademark of the illustrious Adams family. (*a*) John Adams (1735–1826) was the second president of the United States. He was the father of (*b*) John Quincy Adams (1767–1848), who was the sixth president. John Quincy was the father of (*c*) Charles Francis Adams (1807–1886), who was a diplomat and the father of (*d*) Henry Adams (1838–1918), who was a historian.

## Chromosome Disorders

Deviations from the normal human chromosome number of 46 produce syndromes, because of the excess or deficit of genes. Rearrangement of chromosomes, such as an inversion of a section of a chromosome, or two nonhomologous chromosomes exchanging parts, may also cause symptoms. This may happen if the rearrangement disrupts a vital gene, or if it results in "unbalanced" gametes containing too little or too much genetic material. Following is a closer look at specific types of chromosome aberrations.

### Polyploidy

The most drastic upset in chromosome number is the presence of an entire extra set, a condition called **polyploidy.** This results from formation of a diploid (rather than a normal haploid) gamete. For example, if a haploid sperm fertilizes a diploid egg, the fertilized egg is *triploid,* with three copies of each chromosome (see fig. 24.8*b*). Most human polyploids die as embryos or fetuses, but occasionally an infant survives for a few days, with defects in nearly all organs. However, many agriculturally important plants are polyploids.

### Aneuploidy

Cells missing a chromosome or having an extra one are called **aneuploid.** A normal chromosome number is termed **euploid.** Symptoms resulting from aneuploidy depend upon which chromosome is missing or extra. Autosomal aneuploidy often results in mental retardation, possibly because so many genes deal with brain function. Sex chromosome aneuploidy is less severe. Chart 24.2 describes those polyploids that can survive to be born.

Extra genetic material is apparently less dangerous than missing material, and this is why most children born with the wrong number of chromosomes have an extra one (called a **trisomy**) rather than a missing one (called a **monosomy**). Trisomies and

CLINICAL APPLICATION 24.4

## Fragile X Syndrome and Expanding Genes

An unusually breakable X chromosome is associated with a form of mental retardation that affects 1 in 1,500 individuals, most of them male. *Fragile X syndrome* has an interesting history.

In the 1940s, geneticists noticed that more males than females were mentally retarded. Might sex-linked inheritance explain this observation? In 1969, a clue emerged to the genetic basis of one form of sex-linked mental retardation, in a family with two affected brothers. Chromosomes from the brothers and their mother revealed a most unusual X chromosome. One tip of it dangled, separated from the rest by a thin thread (fig. 24C). When grown under certain culture conditions, this part of the X chromosome was very prone to breaking—hence, the name fragile X.

Youngsters with fragile X syndrome look normal, but by adolescence, characteristic features appear. These include a very long, narrow face (fig. 24D), a long jaw, and protruding ears. The testicles are very large. Mental impairment varies greatly, and may include mental retardation, learning disabilities, poor speech, hyperactivity, shyness, social anxiety, and a short attention span.

In 1991 researchers discovered the molecular basis of fragile X, and in so doing stumbled upon a completely unknown way that genes can change: They can expand.

In unaffected individuals, the fragile X area contains a stretch of DNA consisting of five to fifty repeats of the sequence CGG. (See chapter 4 and chart 4.1 for an explanation of the genetic code.) In people with full symptoms and fragile X chromosomes, this region is greatly expanded, harboring 230 to more than two thousand CGG repeats. This expansion attracts methyl groups ($CH_3$) to the fragile X gene, shutting it down and causing the symptoms. Most interesting is that mothers of affected boys often have a "premutation"—an expansion of sixty to two hundred repeats. They usually do not have symptoms, but may be mentally slow. Occasionally a male has the premutation, which means that he can pass the expanding gene to a daughter, who will then have a 50% chance of passing it to a son, who will be affected.

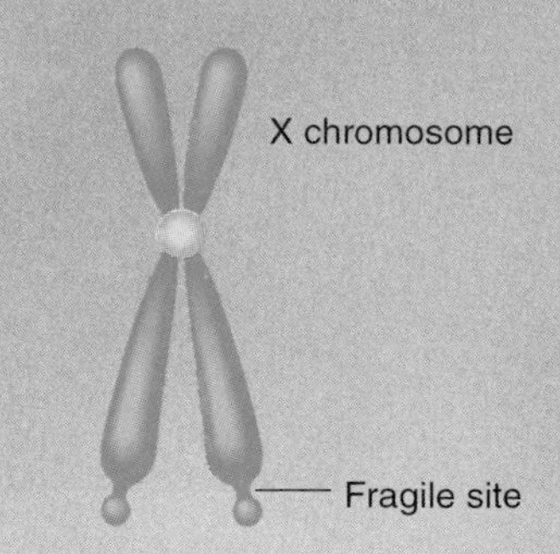


FIGURE 24C
A constriction appears in the X chromosome of individuals who have symptoms of fragile X syndrome when the cells containing them are grown in a medium that lacks thymidine and folic acid.

Fragile X syndrome, as the first of the "triplet repeat" disorders, opened up a whole new area of genetic investigation. Expanding genes have since been implicated in Huntington disease, the neuromuscular disorder myasthenia gravis, and more than a dozen other conditions, all of which affect brain function. On a whole-body level, expanding genes explain a common phenomenon called anticipation, in which symptoms of an inherited illness worsen with each generation.

monosomies are named according to the involved chromosome, and the associated syndrome is often named for the investigator who first described it. Down syndrome, for example, is also known as trisomy 21 (Clinical Application 24.5).

The meiotic error that results in aneuploidy is called **nondisjunction** (fig. 24.17). In normal meiosis, pairs of homologous chromosomes separate and each of the resulting gametes therefore contains only one member of each pair. In nondisjunction, a chromosome pair fails to separate, either at the first or at the second meiotic division, producing a sperm or egg that has two copies of a particular chromosome or none, rather than the normal one copy. When such a gamete fuses with its mate at fertilization, the zygote has either 47 or 45 chromosomes, instead of the normal 46.

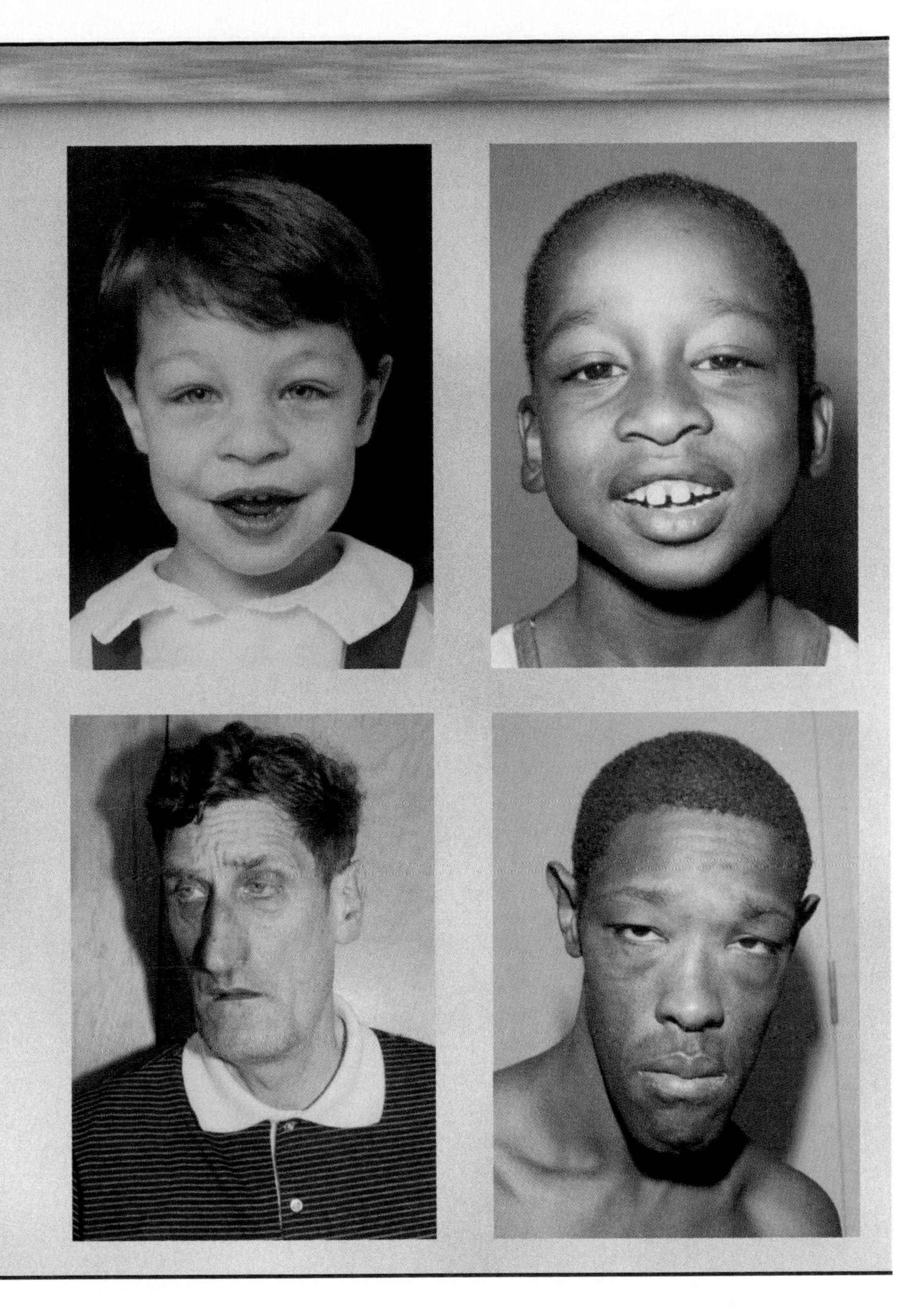

**FIGURE 24D**
The characteristic facial structure and features of individuals who have fragile X syndrome become more pronounced with age.

Today geneticists know that a man with an extra Y chromosome may be very tall, have acne, and also perhaps have speech and reading problems. They did not always understand this. A study published in 1965 identified twelve men of 197 at a Scottish high-security mental hospital who had chromosome abnormalities. Of the twelve, seven had the same anomaly, an extra Y chromosome. This led to the widely publicized idea that an extra Y chromosome causes aggressiveness and criminal behavior. For a short time, hospital nurseries in England, Canada, Denmark, and Boston screened newborn boys for the telltale extra chromosome, then sent psychologists to help parents cope with their "congenital criminals." It is more likely that such boys become aggressive because adults expect aggressive behavior from such large children.

## CHART 24.2 ANEUPLOIDS SURVIVING BIRTH

| Chromosome Constitution | Syndrome | Incidence | Phenotype |
|---|---|---|---|
| Trisomy 21 | Down | 1/770 | Mental retardation<br>Abnormal pattern of palm creases<br>Flat face<br>Sparse, straight hair<br>Short stature<br>High risk of:<br>cardiac anomalies<br>leukemia<br>cataracts<br>digestive blockages |
| Trisomy 13 | Patau | 1/15,000 | Mental and physical retardation<br>Skull and facial abnormalities<br>Defects in all organ systems<br>Cleft lip<br>Large, triangular nose<br>Extra digits |
| Trisomy 18 | Edward | 1/4,000–8,000 | Mental and physical retardation<br>Skull and facial abnormalities<br>Defects in all organ systems<br>Extreme muscle tone |
| Trisomy 22 | | Very few reported | Multiple defects<br>Large ears and nose<br>Narrow face<br>Small jaw<br>Excess muscle tone |
| Monosomy 21 | | Very few reported | Mental and physical retardation<br>Skull, jaw, and facial abnormalities |
| XO | Turner | 1/2,500–10,000 females | No sexual maturity<br>Short stature<br>Webbed neck<br>Wide-spaced nipples<br>Narrow aorta<br>Pigmented moles |
| XXY | Klinefelter | 1/500–2,000 males | Lack of secondary sexual characteristics<br>Breast swelling<br>No sperm |
| XXX | Superfemale | 1/1,000–2,000 | Tall and thin<br>Menstrual irregularity |
| XYY | Jacobs | 1/1,000 males | Tall |

1. Why do deviations from the normal chromosome number of 46 affect health?
2. Define polyploidy.
3. Define aneuploidy.
4. How do extra sets of chromosomes, or extra individual chromosomes, arise?

## MEDICAL GENETICS

This is an exciting time in the field of human genetics, with new discoveries reaching medical journals and headlines weekly. This avalanche of new genetic information has implications for the diagnosis, treatment, and better understanding of hundreds, if not thousands, of disorders.

### Diagnosis

Once a gene is localized to a specific part of a particular chromosome, researchers can develop tests to diagnose the associated condition. These tests are unlike other diagnostic tests in three ways. First, they can be performed on any cell type, because all of the cells of an individual contain all of the genes. Thus, a test to detect an allele that could cause cystic fibrosis, which predominantly affects the lungs, can be performed on cells scraped from the inside of the cheek, on white blood cells, or on cells shed from a fetus into amniotic fluid.

Second, the fact that all cells have all the genes means that a disease whose symptoms do not usually appear until middle age can be diagnosed much earlier—even in an eight-celled preembryo, as chapter 23 showed (see page 913). A presymptomatic diagnosis is more a prediction than it is a demonstration of symptoms.

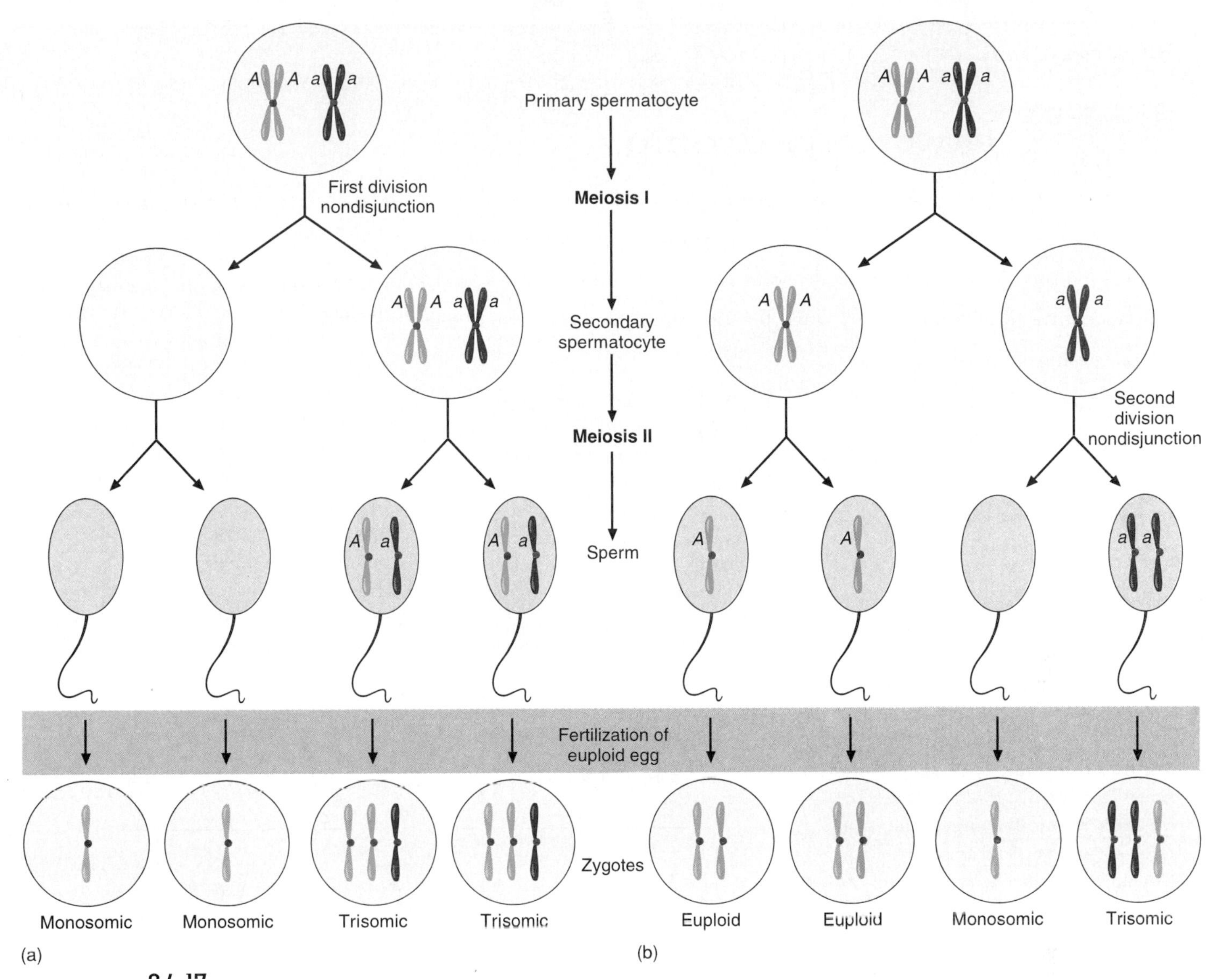


**FIGURE 24.17**

Extra or missing chromosomes constitute aneuploidy. Unequal division of chromosome pairs into sperm and egg cells can occur at either the first or the second meiotic division. (*a*) A single pair of chromosomes is unevenly partitioned into the two cells arising from the first division of meiosis in a male. The result: two sperm cells that have two copies of the chromosome, and two sperm cells that have no copies of that chromosome. When a sperm cell with two copies of the chromosome fertilizes a normal egg cell, the zygote produced is trisomic for that chromosome; when a sperm cell lacking the chromosome fertilizes a normal egg cell, the zygote is monosomic for that chromosome. Symptoms depend upon which chromosome is involved. (*b*) This nondisjunction occurs at the second meiotic division. Because the two products of the first division are unaffected, two of the mature sperm are normal and two aneuploids. Egg cells can undergo nondisjunction as well, leading to zygotes with extra or missing chromosomes when they are fertilized by normal sperm cells.

Third, a genetic disease diagnosis—presymptomatically or after a person experiences symptoms—carries with it a burden of guilt not seen in acquired illnesses, because an affected person may have passed the gene causing his or her illness to children.

## Genetic Counseling

Because of the unique ethical questions and dilemmas that can result from genetic testing, most medical geneticists concur that a person undergoing genetic testing should meet with a **genetic counselor,** who can explain the facts, available tests, and options. A genetic counselor is trained in psychology, statistics, counseling techniques, genetics, molecular biology, and related areas. Working with an accurate diagnosis and as complete a family history as can be obtained (Clinical Application 24.6), the counselor determines recurrence risks for certain conditions in specific relatives, and provides information on the illness so that families can make informed medical decisions. Chart 24.3 describes some actual situations that a genetic counselor faces.

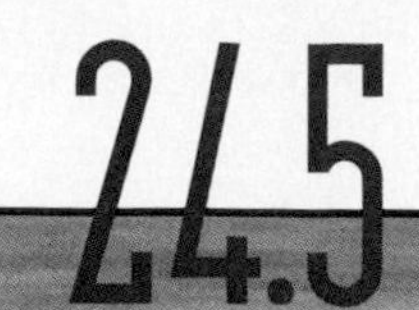

CLINICAL APPLICATION 24.5

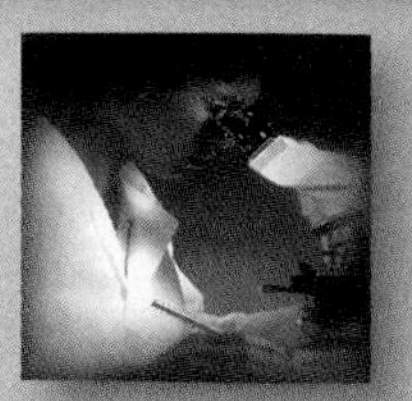

## Down Syndrome

The most common autosomal aneuploid is *trisomy 21* (fig. 24E). The characteristic slanted eyes and flat face of the Down syndrome patient prompted Sir John Langdon Haydon Down to coin the inaccurate term "mongolism" when he described the syndrome in the 1880s. As the medical superintendent of a facility for the profoundly mentally retarded, Down noted that about 10% of his patients resembled people of the Mongolian race. The resemblance is coincidental. Males and females of all races can have Down syndrome (fig. 24F).

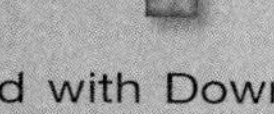

A person with Down syndrome is short and has straight, sparse hair and a tongue protruding through thick lips. The hands have an abnormal pattern of creases, the joints are loose, and reflexes and muscle tone are poor. Developmental milestones (such as sitting, standing, and walking), are slow, and toilet training may take several years. These people tend to have warm, loving personalities and enjoy art and music. Intelligence varies greatly, from profound mental retardation to following simple directions, reading and using a computer. One young man with Down syndrome graduated from a junior college in California, and another starred in a television series.

Down syndrome is associated with many physical problems. Nearly 50% of affected people die before their first birthdays, often of heart or kidney defects, or a suppressed immune system that enables infections to rage out of control. Blockages in the digestive system are common and must be corrected surgically shortly after birth. A child with Down syndrome is 15 times more likely to develop leukemia (a cancer of the white blood cells) than a healthy child.

Down syndrome individuals who live past age 40 develop the black fibers and tangles of amyloid protein in their brains that are also seen in the brains of people who have died of Alzheimer's disease. Alzheimer's disease strikes 2 million American adults, causing severe memory loss and impaired reasoning. Both disorders seem to involve accelerated aging of part of the brain and accumulation of the sticky amyloid protein.

By looking at people who have a third copy of only part of chromosome 21, researchers are zeroing in on the genes whose malfunction causes the symptoms of Down syndrome. This tiny chromosome contains only 1.5% of the genome, or about 1,050 genes. The distal third of the long arm is consistently present in people with Down syndrome who have only part of the extra chromosome, a region that may house a few hundred genes. It appears that

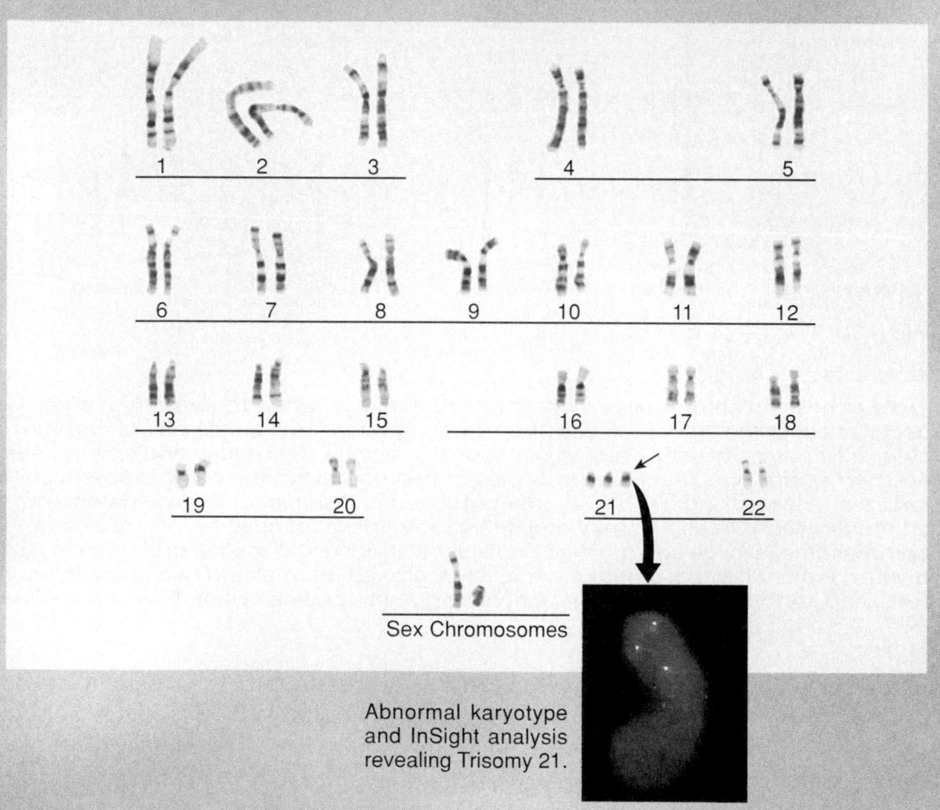


**FIGURE 24E**
The karyotype for an individual with Down syndrome has an extra chromosome 21. Another way to detect the condition is to use a DNA probe, which is a piece of DNA normally found on chromosome 21, with a fluorescent molecule attached to it. Three flashes, rather than the normal two, indicate an extra chromosome 21.

**Figure 24F**

Down syndrome is usually caused by an extra chromosome 21. Symptoms vary greatly. About half of all patients die early of malformed hearts or kidneys. Some undergo heart surgery or suffer frequently from common illnesses. However, many people with Down syndrome lead happy lives. Unfortunately, prenatal diagnosis of Down syndrome cannot predict how severely affected the child will be.

certain parts of the chromosome are more crucial than others to the Down syndrome phenotype, because some people with large parts of chromosome 21 missing have normal intelligence.

The likelihood of giving birth to a child with Down syndrome increases dramatically with the age of the mother. For women under 30, the chances of conceiving such a child are 1 in 3,000. By age 48, the number jumps to 1 in 9 (fig. 24G). The age factor in Down syndrome may be due to the fact that meiosis in the female is completed after conception. The older a woman is, the longer her oocytes have been arrested on the brink of completing meiosis. During this time, the oocytes may have been exposed to chromosome-damaging chemicals or radiation. Other trisomies are more likely to occur among the offspring of older women, too.

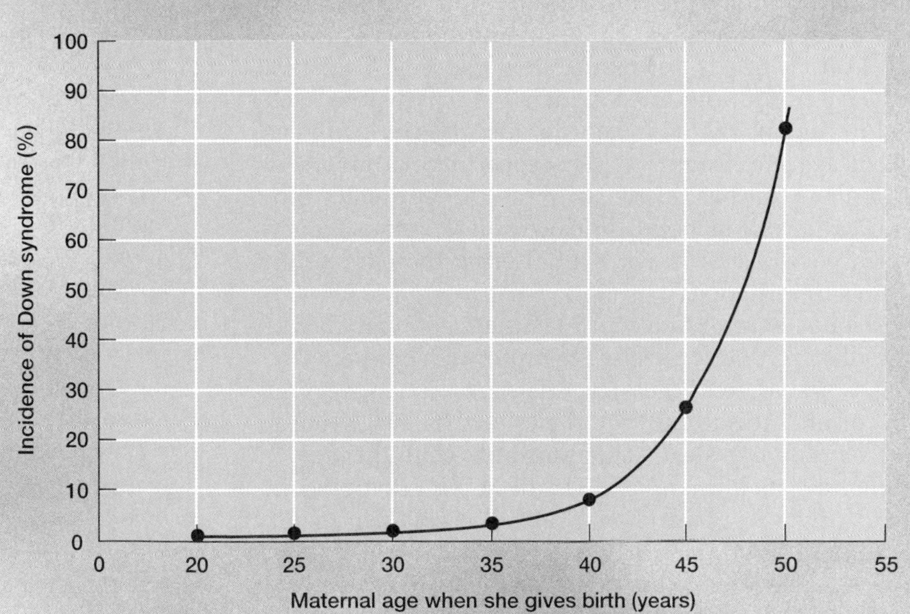


**Figure 24G**

Down syndrome risk increases with the age of the mother. The incidence of Down syndrome in the general population is about 1 in every 770 births. Among women over the age of 35 years, however, the risk of delivering a child with Down syndrome increases.

## CHART 24.3 GENETIC COUNSELING CASES

1. A mother of a 6-year-old has just found out that her son, whom she thought was merely clumsy, has Duchenne muscular dystrophy. She also has a 2-year-old son but is not certain she wants to know whether he has also inherited the disease, because she feels that knowing so would affect her ability to be a good parent. She also feels guilty because she knows she carries the gene that made her son ill.
2. A young man is distressed because his father has just been diagnosed with Huntington disease. His father's mother probably had the condition, although she was institutionalized and her diagnosis was uncertain. The man needs help in deciding whether to take a test that will tell him if he has inherited the dominant gene that causes the condition. He is only 23, and symptoms do not usually appear until the late thirties. He thinks that he wants the information to help make a decision concerning having children.
3. A young couple meet with a genetic counselor because the woman's brother has a son with hemophilia. They are relieved to hear that their children's risk is not elevated, because the affected nephew received the hemophilia gene from his mother, who is not a blood relative.
4. A woman has a very mild case of an autosomal dominant disorder, neurofibromatosis I. Her only symptoms are a few brownish patches on her skin and two benign lumps under the skin on her upper left thigh. She is surprised to learn from the genetic counselor that if the condition appears in a child, it might not be so mild.
5. Amniocentesis indicates that a fetus has three X chromosomes. The parents-to-be are distraught, and can hardly believe the genetic counselor when she tells them that the child will probably be healthy.
6. A woman undergoes a genetic test that indicates that she has not inherited the gene that caused breast cancer in her mother and sisters. She becomes depressed because she feels guilty.

## Gene Therapy

Cynthia Cutshall and Ashanthi DeSilva made headlines in 1990 when they became the first two people to officially receive **gene therapy.** Each girl received white blood cells bearing the gene for adenosine deaminase, which their bodies could not produce and which robbed them of an immune system.

Cutshall and DeSilva's treatment replaced genes, thus operating at the level of the genotype. However, geneticists have been able to treat some genetic disorders at the phenotype level for several years. A boy with hemophilia receives a clotting factor. A child with cystic fibrosis sprinkles powdered cow digestive enzymes onto applesauce, which she eats before each meal to replace the enzymes that her clogged pancreas cannot secrete. People with certain inborn errors of metabolism follow restrictive diets to counteract biochemical buildups (see Clinical Application 4.4). Even wearing eyeglasses is a way of altering the expression of one's inheritance.

Gene therapy is a more lasting treatment for genetic disease than correcting symptoms. There are two basic types of gene therapy. Effecting a change in all cells of an individual would require intervention at the fertilized egg stage. This **heritable gene therapy** is impractical in humans and is not being attempted in them. However, researchers use it to create *transgenic* experimental animals that serve as models of human disease, such as mice that have muscular dystrophy.

Researchers are making tremendous strides in **nonheritable gene therapy,** which targets genes in affected cells, such as young Cynthia and Ashanthi's genetically engineered white blood cells. However, nonheritable gene therapy is not passed to subsequent generations, as is heritable gene therapy. An early form of nonheritable gene therapy, bone marrow transplant, has been performed for many years, but with a high mortality rate. This procedure replaces blood stem cell populations in bone marrow, but leaves the patient vulnerable to infection for several months because the recipient's bone marrow must be destroyed first.

Laura Cay Boren has witnessed firsthand the evolution of therapies for her inherited adenosine deaminase (ADA) deficiency, which left her without an immune system since her birth in July 1982. In 1983 and again in 1984, she received bone marrow transplants from her father, which did not cure her underlying problem. Blood transfusions helped for awhile. But by the end of 1985, Laura was near death. Fortunately, she was chosen to receive a new treatment—a form of the missing enzyme, altered so that it would remain in her bloodstream long enough to help her. It did. Within hours her ADA level increased 20-fold. By 3 months her blood no longer contained toxins, although immunity had not yet appeared. But by 6 months, she had a functioning immune system! Laura must receive frequent injections of ADA to maintain immunity, and physicians are not sure how long the treatments will continue to work. So Laura now awaits gene therapy—the same that cured Cynthia Cutshall and Ashanthi DeSilva.

Both the successes and failures of bone marrow transplants have prompted researchers to alter genes in other types of cells (fig. 24.18). A look at some of the ongoing work in gene therapy provides a nice ending to our survey of human genetics, as well as

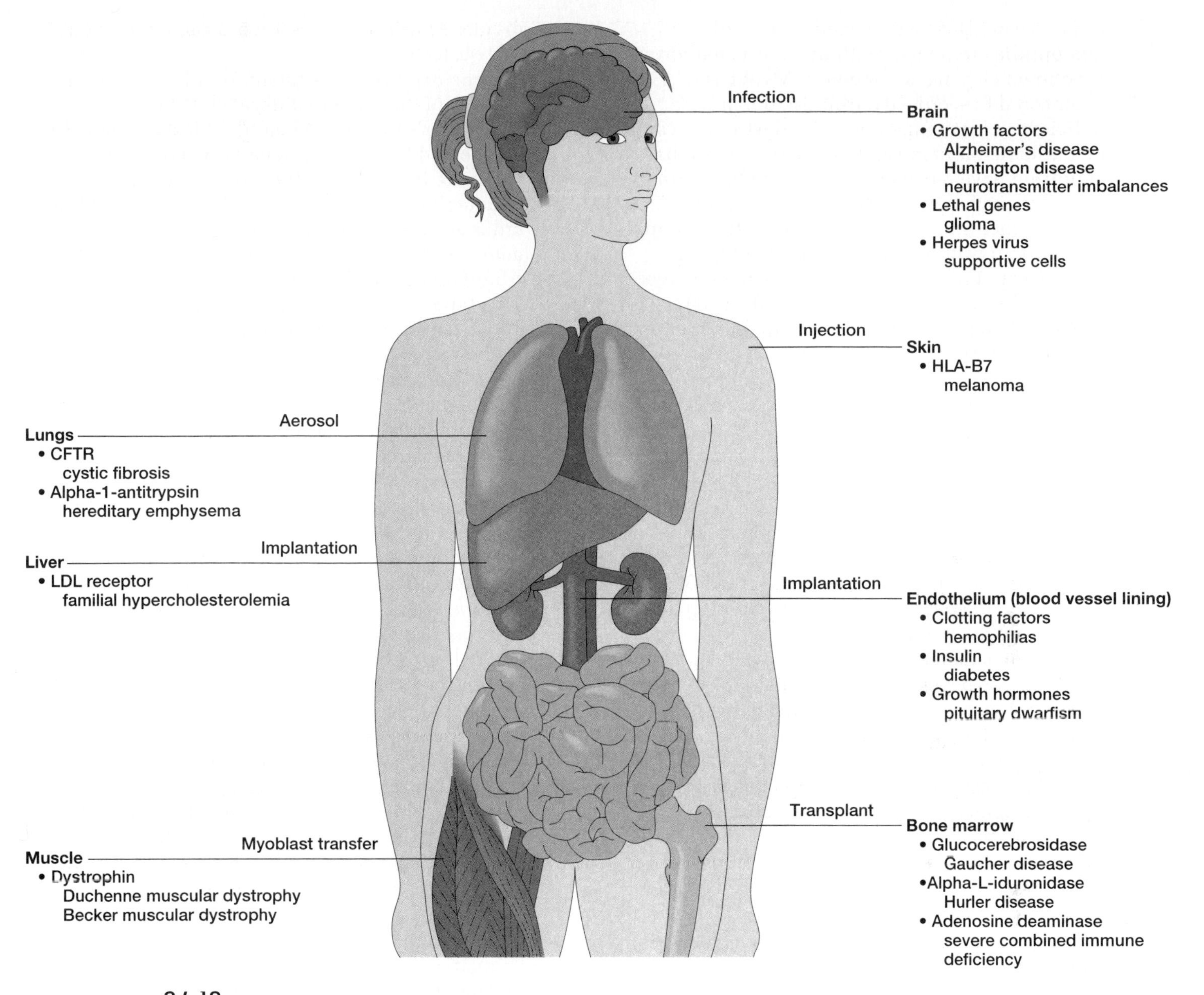


**FIGURE 24.18**
Sites of gene therapy and the methods used to introduce normal DNA.

reviews organ systems discussed throughout the book. Most of these efforts are still experimental. A decade from now, gene therapy may be a routine part of medical practice.

**Skin** Skin grafts genetically engineered to secrete needed proteins, such as clotting factors or enzymes, may be a new route to drug delivery. So far, a human gene encoding human growth hormone has been inserted into human skin cells growing in culture. When grafted onto mice from a strain that does not reject foreign tissue, the grafts grow into normal appearing mouse epidermis—but they secrete human growth hormone. This is a first step to human gene therapy on skin.

**Endothelium** Recall that endothelium forms the walls of capillaries and lines other blood vessels. This tissue can be engineered to secrete a needed substance right into the bloodstream. In a person with diabetes, for example, altered endothelium might secrete insulin.

**Muscle** Immature skeletal muscle cells, called myoblasts, bearing therapeutic genes inserted into their DNA, can be incorporated into the natural muscle in blood vessel walls, where they could secrete a needed protein into the bloodstream. Dogs with a canine form of hemophilia are treated with myoblasts harboring functional clotting factor genes. Myoblasts engineered to secrete dystrophin, the protein deficient in Duchenne muscular dystrophy, can be implanted into muscle. Some boys with the condition have received healthy myoblasts from their fathers, but researchers are not certain whether their improvement is due to the therapy or to immunosuppressant drugs used to help them accept the implants. The trick in effectively treating this condition is to restore dystrophin function in enough muscles to help the child regain movement.

**Liver** The liver is a very important focus of gene therapy because it controls many bodily functions. The first liver gene therapy targets heart disease. Normal liver cells have low density lipoprotein (LDL) receptors on their surfaces, which bind cholesterol in the bloodstream and bring it into the cell. When liver cells lack LDL receptors, cholesterol accumulates on artery interiors. Liver cells genetically engineered to have more LDL receptors can relieve the cholesterol buildup that causes cardiovascular disease.

Such gene therapy could be lifesaving for children who have inherited familial hypercholesterolemia. Their liver cells are completely devoid of LDL receptors. At present the only approved treatment is a liver transplant or continually cleansing the blood of cholesterol. They die of heart failure before age 20. Gene therapy for these individuals removes about 15% of the liver, then inserts the LDL receptor gene. The corrected liver cells are then injected into a blood vessel in the patient's liver. Here, the cells attach to the vessel walls and produce LDL receptors that bind cholesterol.

**Lungs** The respiratory tract is a prime candidate for gene therapy because an aerosol can directly reach its lining cells, making it unnecessary to remove cells, alter them, and reimplant them. To treat cystic fibrosis, a modified virus that causes the common cold is genetically engineered to not cause cold symptoms and to contain a functional CFTR gene. Once inhaled, lung lining cells take up the gene and produce the protein missing or mutant in the inherited illness. Gene therapy is also being used to provide an enzyme whose absence causes a hereditary form of emphysema. The gene is introduced into fibroblasts or directly in an aerosol of tiny bubbles (liposomes).

**Nerve Tissue** Gene therapy on neurons presents a problem, because these cells do not normally divide. Altering other cell types can circumvent this obstacle.

Consider a gene treatment for mice injured in the part of the brain that degenerates in Alzheimer's disease. When fibroblasts genetically engineered to contain a gene for nerve growth factor are implanted at the wound site, they secrete the growth factor, which neurons need in order to enlarge. The neighboring traumatized neurons grow and secrete their neurotransmitter, stimulated by the bath of nerve growth factor. This type of treatment may one day boost neurotransmitter levels in Huntington disease, Alzheimer's disease, and clinical depression, which are caused by deficits of these important biochemicals.

Another route to nerve cell gene therapy is to send a missing gene in attached to the herpes simplex virus, which remains in nerve cells after infection. Such a herpes gene carrier could alter a neuron's ability to secrete neurotransmitters.

**Gene Therapy against Cancer** Viruses may provide a treatment for a deadly type of brain tumor called a *glioma,* which affects the glial cells of the nervous system. Unlike neurons, glia can divide, and cancerous glia do so very rapidly, usually causing death within a year even with aggressive treatment. A new approach to treatment is to infect fibroblasts with a virus bearing a gene from a herpes virus that makes the cells sensitive to a drug called ganciclovir. The engineered fibroblasts are implanted near the tumor. Then, the virus infects nearby rapidly dividing glia—the cancer cells. Give the patient ganciclovir, and only the cells harboring the virus die—not healthy brain cells.

Another genetic approach to battling cancer is to enable tumor cells to produce immune system biochemicals, or to mark them so that they are more easily recognized by the immune system. An experimental treatment for the skin cancer *melanoma,* for example, engineers tumor cells to display an antigen called HLA-B7, which stimulates the immune system to attack the cell (fig. 24.19). Other cancers may in the future be treated by altering cells in blood vessels that feed a tumor to secrete immune system biochemicals. Charts 24.4 and 24.5 highlight some concerns researchers have about delivering gene therapy.

Not long ago, genetic diseases were thought to be extremely rare. The gene discoveries of the past decade, however, have revealed the contributions of genes to virtually every physiological process.

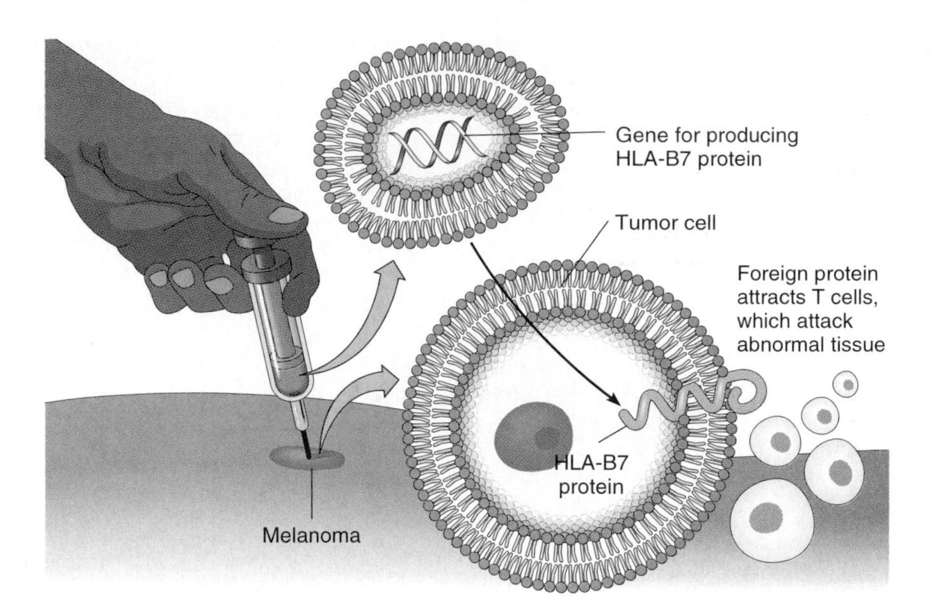


**FIGURE 24.19**
A gene encoding a cell surface protein that attracts the immune system's tumor-killing T cells is injected directly into a melanoma tumor.

### CHART 24.4 REQUIREMENTS FOR APPROVAL OF CLINICAL TRIALS FOR GENE THERAPY

1. Knowledge of defect and how it causes symptoms.
2. An animal model.
3. Success in human cells growing in vitro.
4. Either no alternate therapies, or a group of patients for whom existing therapies are not possible or have not worked.
5. Experiments must be safe.

Therefore, genes play a role in many, if not all, illnesses. Susceptibility to infection depends to an extent upon one's immune system, which is inherited. Cancers often reflect genetic alterations in somatic cells. Many genes encode proteins that affect the cardiovascular system. The enzymes that control virtually all biochemical pathways are the products of genes.

As the human genome project continues to offer genetic explanations for normal anatomy and physiology, as well as pathology, health practitioners will embrace this new information, this new molecular and cellular way of looking at the functioning of the human body. For by the year 2010, genetic tests are likely to be a part of everyone's life, from conception to old age.

### CHART 24.5 GENE THERAPY CONCERNS

1. Which cells should be treated?
2. What proportion of the targeted cell population must be corrected to alleviate or halt progression of symptoms?
3. Is overexpression of the therapeutic gene dangerous?
4. If the engineered gene "escapes" and infiltrates other tissues, is there danger?
5. How long will the affected cells function?
6. Will the immune system attack the introduced cells?

1. What are the unique challenges associated with diagnosing genetic disease?
2. What is genetic counseling?
3. Why is heritable gene therapy in humans not practical?
4. What are some of the ways that nonheritable gene therapy is being conducted?

## CLINICAL APPLICATION

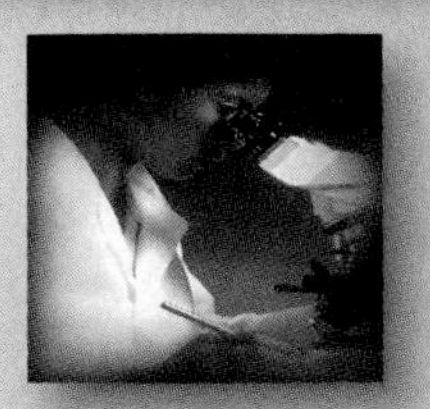

# Pedigree Analysis

Families are the primary tool of the human geneticist, and the bigger the better. The more children in a generation, the easier it is to discern patterns, or modes, of inheritance. Family relationships and phenotypes are simultaneously displayed in a standard chart called a pedigree. A pedigree in humans serves the same purpose as one in purebred dogs or cats or thoroughbred horses—keeping track of relationships and traits.

**Symbols:**

= normal female, male

= female, male who express trait

= female, male who carry an allele for the trait but do not express it (carriers)

= dead female, male

= sex unspecified

= aborted or stillborn individuals

**Lines:**

= generation

= parents

= adoption

= siblings

= identical twins

= fraternal twins

= parents closely related

= person who prompted pedigree analysis

**Lines:**

Roman numerals = generations

Arabic numerals = individuals

**FIGURE 24H**

Symbols used in pedigree construction are connected to form a pedigree chart, which displays the inheritance patterns of particular traits.

**FIGURE 24I**

A pedigree for an autosomal recessive trait. Albinism (absence of pigment in the skin and hair) does not show up in this family until the third generation. Individuals II-2 and II-3 must each be carriers.

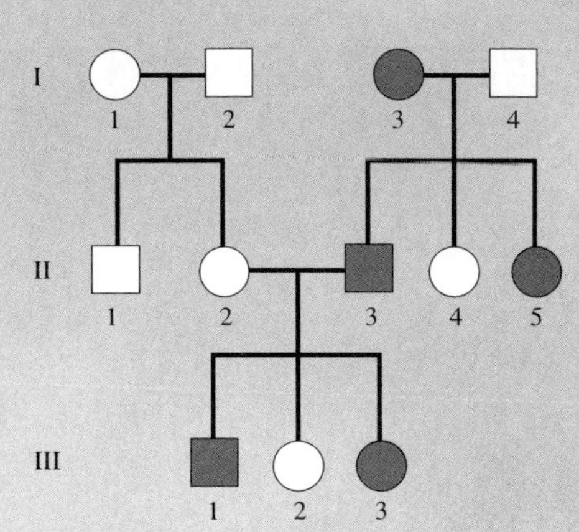


**FIGURE 24J**

A pedigree for an autosomal dominant trait, such as Huntington disease. Males and females are affected, and generations are not skipped.

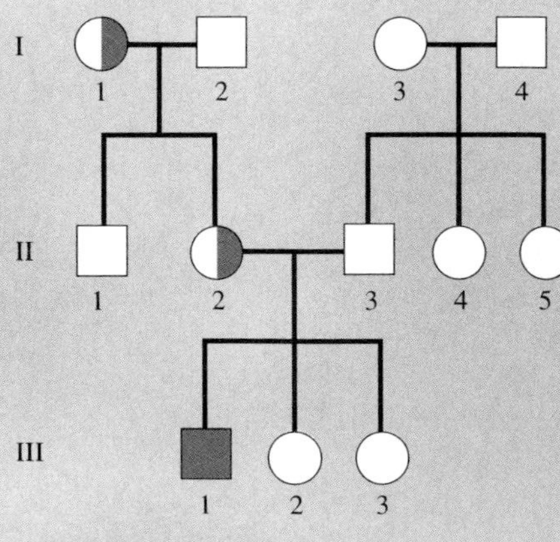


**FIGURE 24K**

A pedigree for a sex-linked trait, such as hemophilia. The affected male inherits the causative mutant allele on the X chromosome, which he can only inherit from his mother, who is a carrier. She inherited the allele from her mother, also a carrier.

A pedigree is built of shapes connected by lines. Vertical lines represent generations; horizontal lines connecting shapes at their centers depict parents; shapes connected by vertical lines joined horizontally above them represent siblings. Squares indicate males, circles females, and diamonds individuals of unknown sex. Figure 24.H shows these and other commonly used pedigree symbols. Colored shapes indicate individuals who express the trait being studied and half-filled shapes are known carriers.

Pedigrees can reveal modes of inheritance. An *autosomal recessive* trait by definition can affect both sexes and can skip generations, because carriers show no symptoms (fig. 24I). If a condition is known to be autosomal recessive, carrier status can be inferred for individuals who have affected (homozygous recessive) children. An *autosomal dominant* trait affects every generation and can affect both sexes, but transmission stops whenever an individual does not inherit the causative gene (fig. 24J). A *sex-linked trait* is passed from carrier mothers to affected sons. A female can inherit a disease-causing gene on the X chromosome from her mother (who is also a carrier), or from her father if he was well enough to have children (fig. 24K).

Following are some problems utilizing pedigrees.

(1)

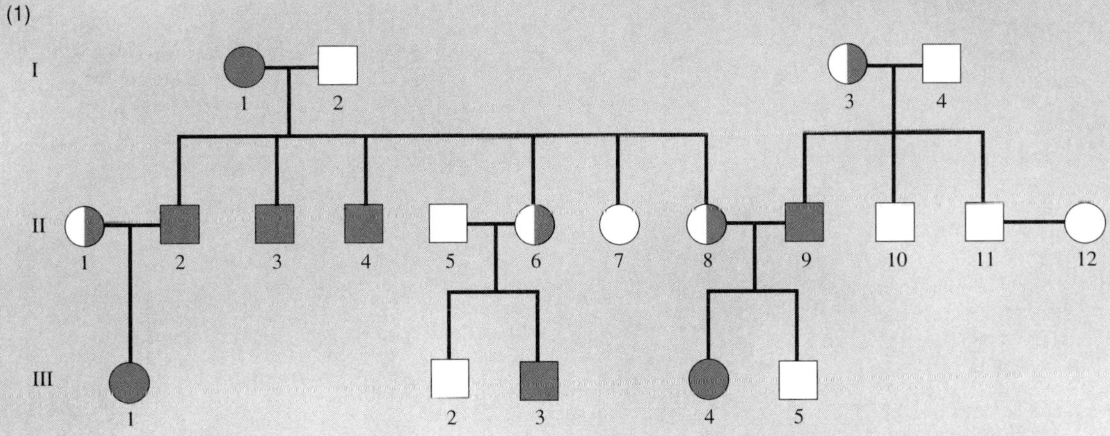

(2)

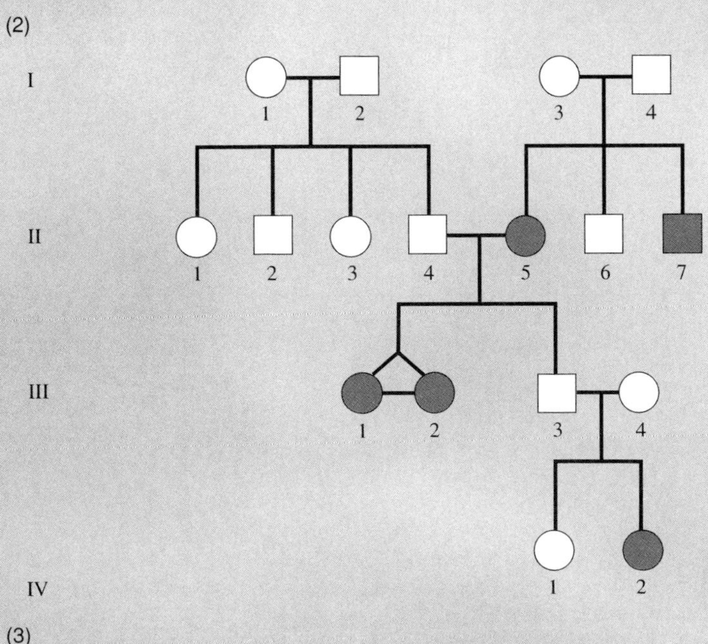

(3)

**Pedigree Problems**

1. Retinitis pigmentosa causes night blindness, a constricted visual field, and pigment lumps in the eye. In 84% of affected families, the condition is inherited as an autosomal recessive; in 10% of families it is an autosomal dominant; in 6% of families, it is sex-linked recessive. Can you tell what the mode of inheritance is for retinitis pigmentosa in the family depicted in this pedigree? If so, what is it? If not, why not?
2. Metacarpal 4–5 fusion is a sex-linked recessive condition in which certain finger bones are fused. It occurs in many members of the Flabudgett family, depicted in this pedigree:
   - **a.** Why are so many females affected, considering that this is a sex-linked condition?
   - **b.** What is the risk that individual III1 has an affected son?
   - **c.** What is the risk that individual III5 has an affected son?
3. Chands syndrome is an autosomal recessive condition characterized by very curly hair, underdeveloped nails, and abnormally shaped eyelids. In the following pedigree, which individuals must be carriers?

## CHART 24.6 SELECTED INHERITED DISORDERS

| Disorder | Cause and Symptoms | Mode of Inheritance* |
|---|---|---|
| Adrenoleukodystrophy | Accumulation of certain fatty acids causes seizures, nervous system degeneration, and death in childhood | xlr |
| Albinism | Lack of melanin pigment in skin and hair | ar |
| Alzheimer's disease | Progressive loss of memory and reasoning ability, beginning in late middle age | AD |
| Amyotrophic lateral sclerosis | Progressive muscle weakening, beginning in adulthood | AD |
| Breast cancer | Tumors in breast | AD |
| Chronic granulomatous disease | Phagocytes cannot destroy pathogenic bacteria, causing frequent infection | xlr |
| Color blindness | Inability to see red or green | xlr |
| Cystic fibrosis | Abnormal chloride channels in cell membranes clogs lungs, impairs pancreatic secretion, causes respiratory infection and male infertility | ar |
| Diabetes insipidus | Deficiency of antidiuretic hormone leads to copious urination | xlr |
| Duchenne muscular dystrophy | Lack of dystrophin protein collapses skeletal muscle cells, destroying muscles | xlr |
| Familial hypercholesterolemia | Lack of LDL receptors on liver cells leads to extreme cholesterol buildup | ar |
| Familial hypertrophic cardiomyopathy | Sudden death in young adulthood from heart failure | AD |
| Fatal familial insomnia | Degeneration of brain's sleep center causes neurological degeneration | AD |
| Gout | Enzyme deficiency causes buildup of uric acid in joints | AD |
| Hemochromatosis | Iron deposits in liver, heart, endocrine glands; may not produce symptoms | ar |
| Hemophilia | Absence of a clotting factor causes slow wound healing, internal bleeding | xlr |
| Huntington disease | Adult onset loss of coordination, uncontrollable movements, personality changes | AD |
| Lissencephaly | Lack of brain convolutions | ar |
| Marfan syndrome | Lack of fibrillin protein causes long limbs, lens dislocation, weakened aorta wall | AD |
| Menkes disease | Cells cannot utilize copper, causing twisted hair, weak arteries, scurvy, poor growth, brain degeneration | xlr |
| Multiple endocrine neoplasia | Cancers of endocrine glands | AD |
| Osteogenesis imperfecta | Collagen defect causes easily broken bones, possibly before birth | ar |
| Pattern baldness | Hair loss | AD (♂) ar (♀) |
| Phenylketonuria | Inborn error that causes phenylalanine to accumulate, causing mental retardation | ar |
| Pituitary dwarfism | Short stature due to lack of growth hormone | ar |
| Polycystic kidney disease | Adult-onset cysts in kidneys | AD |
| Polydactyly | Extra digits | AD |
| Porphyria | Abnormal porphyrin ring in hemoglobin leads to accumulation of hemoglobin breakdown products, producing red urine, abdominal pain, neurological impairment | ar |
| Progeria | Extreme premature aging | ar |
| Retinitis pigmentosa | Visual field constriction, night blindness, pigment clumps in eye | xlr |
| Rickets | Inability of cells to respond to vitamin D, weakening bones | ar |
| Tay-Sachs disease | Enzyme deficiency causes lipid to accumulate on nerve cells, impairing neurological development | ar |
| von Willebrand disease | Absent clotting factor causes prolonged bleeding | ar |
| Wilms tumor | Prenatal or childhood kidney cancer | AD |
| Wilson disease | Excess copper damages brain and liver, causing abdominal pain, headache, neurological symptoms | ar |
| Xeroderma pigmentosum | Absent DNA repair leads to freckling and high risk of skin cancer | ar |

*xlr = sex-linked recessive
ar = autosomal recessive
AD = autosomal dominant

## CHAPTER SUMMARY

### Introduction (page 947)

Individual characteristics result from the interaction of heredity (transmission of genetic information from parents to offspring) and the environment (chemical, physical, social, and biological influences in an individual's surroundings). Genetics will play an increasingly important role in health care, as new discoveries explain physiological and pathological processes.

## Genes and Chromosomes (page 949)

1. A gene is a portion of a DNA molecule that contains information for producing one kind of protein molecule, or controls the activities of other genes.
2. All genes in a human cell constitute the genome.
3. The 70,000 or so human genes are distributed among 23 pairs of chromosomes, half of which are inherited from the mother and half from the father.

## Chromosomes and Sex Cell Production (page 949)

1. Meiosis
   a. Meiosis consists of two successive divisions, each progressing through prophase, metaphase, anaphase, and telophase.
   b. In the first meiotic division, homologous, replicated chromosomes separate and their number is halved. A replicated chromosome consists of two chromatids held together by a centromere.
   c. In the second meiotic division, the chromatids part, producing four haploid cells from each diploid cell undergoing meiosis.
   d. Meiosis yields sex cells. Each male cell undergoing meiosis produces four mature sperm cells; each female cell produces one oocyte and three polar bodies.
2. Meiosis leads to genetic variability
   a. Variation results from random alignment of maternally and paternally-derived chromosomes in metaphase I.
   b. Crossing over mixes up contributions from the previous generation.
3. Chromosomes and genes come in pairs
   a. Chromosome charts are called karyotypes.
   b. Chromosomes 1 through 22, numbered in decreasing size order, are autosomes. They do not determine sex.
   c. The X and Y chromosomes are sex chromosomes. They determine sex.
   d. Because chromosomes are paired, the genes they carry are also paired.
   e. An alternate form of a gene is called an allele. An individual can have two different alleles for a particular gene. The gene itself can have many alleles, because a gene consists of many building blocks, any of which may be altered.
   f. An individual with a pair of identical alleles for a particular gene is homozygous; if the alleles are different, the individual is heterozygous.

## Modes of Inheritance (page 959)

The combination of genes present in an individual's cells constitutes a genotype; the appearance of the individual is its phenotype. A wild-type allele provides normal or the most common function. A mutant allele causes disease or an unusual trait; it represents a change from the wild-type condition.

1. Dominant and recessive inheritance
   a. In the heterozygous condition, an allele that is expressed when the other is not is dominant. The masked allele is recessive.
   b. Recessive and dominant genes may be autosomal or sex-linked.
   c. An autosomal recessive condition affects both sexes, and may skip generations. The homozygous dominant and heterozygous individuals have normal phenotypes. The homozygous recessive individual has the condition. The heterozygote is a carrier. An affected individual inherits one mutant allele from each parent.
   d. An autosomal dominant condition affects both sexes and does not skip generations. A person inherits it from one parent, who is affected.
2. Different dominance relationships
   a. In incomplete dominance, a heterozygote has a phenotype intermediate between those of both homozygotes.
   b. In codominance, each of the alleles in the heterozygote is expressed.

## Gene Expression (page 961)

1. Penetrance and expressivity
   a. A genotype is incompletely penetrant if not all individuals inheriting it express the phenotype.
   b. A genotype is variably expressive if it is expressed to different degrees in different individuals.
2. Pleiotropy
   a. A pleiotropic disorder has several symptoms, different subsets of which are expressed among individuals.
   b. Pleiotropy reflects a gene product that is part of more than one biochemical reaction, or is found in several organs or structures.
3. Genetic heterogeneity
   a. Genetic heterogeneity refers to a phenotype that can be caused by alterations in more than one gene.
   b. The same symptoms may result from alterations in genes whose products are enzymes in the same biochemical pathway.

## Complex Traits (page 962)

1. A trait caused by the action of a single gene is monogenic.
2. A trait caused by the action of more than one gene is polygenic.
3. A trait caused by the action of one or more genes and the environment is multifactorial or complex.
4. Height, skin color, and many common illnesses are complex traits.

## Matters of Sex (page 963)

A female has two X chromosomes; a male has one X and one Y chromosome. The X chromosome has many genes; the Y chromosome has very few genes.

1. Sex determination
   a. A male zygote forms when a Y-bearing sperm fertilizes an egg. A female zygote forms when an X-bearing sperm fertilizes an egg.
   b. A gene on the Y chromosome, called SRY, switches on genes in the embryo that promote development of male characteristics and suppresses genes that promote development of female characteristics.
   c. Researchers discovered the SRY gene by studying XY females lacking SRY, and XX males with a small bit of a Y chromosome.

2. Sex chromosomes and sex-linked traits
   a. Genes on the sex chromosomes are inherited differently than those on autosomes because the sexes differ in sex chromosome constitution.
   b. Traits transmitted on the X chromosome are termed sex-linked.
   c. Males are hemizygous for sex-linked traits; that is, they can have only one copy of a sex-linked gene, because they have only one X chromosome.
   d. Females can be heterozygous or homozygous for genes on the X chromosome, because they have two copies of it.
   e. A male inherits a sex-linked trait from a carrier mother. These traits are more common in males than in females.
   f. A female inherits a sex-linked mutant gene from her carrier mother, and/or from her father if the associated trait does not impair his ability to have children.
   g. Dominant sex-linked traits are rare. Affected males die before birth.

## Gender Effects on Phenotype (page 966)

Gender can affect gene expression.

1. Sex-limited traits
   a. Such a trait affects a structure or function seen in only one sex.
   b. Such a trait may be inherited according to any mode of inheritance.
2. Sex-influenced traits
   a. A sex-influenced trait is dominant in one sex and recessive in the other.
   b. In genomic imprinting, the severity, age of onset, or nature of symptoms varies according to which parent transmits the causative gene.

## Chromosome Disorders (page 967)

Extra, missing, or rearranged chromosomes or parts of them can cause syndromes, because they either cause an imbalance of genetic material or disrupt a vital gene.

1. Polyploidy
   a. Polyploidy is an extra chromosome set.
   b. Polyploidy results from fertilization of a diploid gamete.
   c. Human polyploids do not survive beyond a few days of birth.
2. Aneuploidy
   a. Cells with an extra or missing chromosome are aneuploid. Cells with the normal chromosome number are euploid.
   b. A cell with an extra chromosome is trisomic. A cell with a missing chromosome is monosomic. Individuals with trisomies are more likely to survive to be born than those with monosomies.
   c. Aneuploidy results from a meiotic error called nondisjunction, in which a chromosome pair does not separate, either in meiosis I or meiosis II, producing a gamete with a missing or extra chromosome. At fertilization, a monosomic or trisomic zygote results.

## Medical Genetics (page 970)

Frequent discoveries of new genes are providing new diagnostic tests and new treatments.

1. Diagnosis
   a. Diagnostic tests for inherited conditions present special challenges.
   b. Such tests can be performed on any cell type.
   c. Because diagnostic tests for genetic disease can be performed at any point in life, they can presymptomatically detect illness.
   d. Genetic disease often brings guilt feelings of responsibility for relatives' illness.
2. Genetic counseling
   a. A genetic counselor is trained in science and psychology to explain inheritance patterns, risks, and options to families with inherited disease.
   b. A genetic counselor must have an accurate diagnosis and a complete family history to provide useful information to patients.
3. Gene therapy
   a. Gene therapy does not just treat symptoms but corrects the genetic defect causing symptoms.
   b. Heritable gene therapy alters all genes in an individual, and therefore must be done on a fertilized egg. It is not being pursued in humans but is useful in research on other animals.
   c. Nonheritable gene therapy replaces or corrects defective genes in somatic cells, often those where symptoms occur.
   d. Gene therapy is being tested in skin, endothelium, muscle, liver, lungs, and to treat cancer.

**Explorations in Human Anatomy and Physiology CD-ROM**
The modules accompanying Chapter 24 are #1 Cystic Fibrosis and #15 Heredity in Families.

## CRITICAL THINKING QUESTIONS

1. Should all fetuses or newborns be karyotyped? What would some advantages and disadvantages be of mandatory karyotyping at either of these stages?
2. A young couple is devastated when their second child is born and has PKU. Their older child is healthy and no one else in the family has PKU. How is this possible?
3. A balding man undergoes a treatment which transfers some of the hair from the sides of his head, where it is still plentiful, to the top. Is he altering his phenotype or his genotype?
4. Bob and Joan know from a blood test that they are each heterozygous for the autosomal recessive gene that causes sickle cell disease. If their first three children are healthy, what is the probability that their fourth child will have the disease?

5. A man with type AB blood has children with a woman who has type O blood. What are the chances that a child that they conceive has blood that is type A? B? AB? O?
6. A person has one X chromosome, and a Y chromosome missing the SRY gene. Describe this person's phenotype.
7. Traits that appear more frequently in one sex compared to the other may be caused by genes that are inherited in a sex-linked, sex-limited, or sex-influenced fashion. How might these possibilities be distinguished?
8. Cirrhosis of the liver, emphysema, and heart disease are all conditions that can be caused by a faulty gene, or by a dangerous lifestyle habit (drinking alcohol, smoking, following a poor diet). When gene therapies become available for these conditions, should people with gene-caused disease be given priority in receiving the treatments? If not, what other criteria should be used for deciding who should receive a limited medical resource?

## Review Exercises

1. Identify two major factors that influence the development of individual characteristics.
2. Define *gene; chromosome; genome.*
3. Discuss the origin of the 46 chromosomes in a human zygote.
4. Define *homologous chromosomes.*
5. Outline the process of meiosis.
6. Explain the significance of synapsis during meiosis.
7. List two ways that meiosis provides genetic variability.
8. Distinguish between:
   - homozygote and heterozygote
   - autosome and sex chromosome
   - mutant and wild-type
   - homozygous dominant, homozygous recessive, and heterozygous
   - phenotype and genotype
   - incomplete dominance and codominance
9. Explain how a gene can have many alleles.
10. Describe the components of a complex trait.
11. Describe how the environment can influence gene expression.
12. Explain how genes and chromosomes determine gender.
13. Explain how the presence of the SRY gene causes embryonic reproductive organs to develop as male structures.
14. Define sex-linked genes.
15. Explain why recessive sex-linked genes are expressed more frequently in males than females.
16. Explain why a male cannot inherit a sex-linked trait from his father.
17. Explain why X-linked dominant traits are not seen in males.
18. Explain how an individual with an extra set of chromosomes arises.
19. Explain how nondisjunction leads to aneuploidy.
20. List three ways that diagnosis of genetic disease differs from diagnosis of other disorders.
21. Describe the services that a genetic counselor provides.
22. Describe why heritable gene therapy is impractical in humans.
23. Explain how nonheritable gene therapy is being attempted in various human tissues.

# HUMAN CADAVERS

*The following set of illustrations includes medial sections, horizontal sections, and regional dissections of human cadavers. These photographs will help you visualize the spatial and proportional relationships between the major anatomic structures of actual specimens. The photographs can also serve as the basis for a review of the information you have gained from your study of the human organism.*

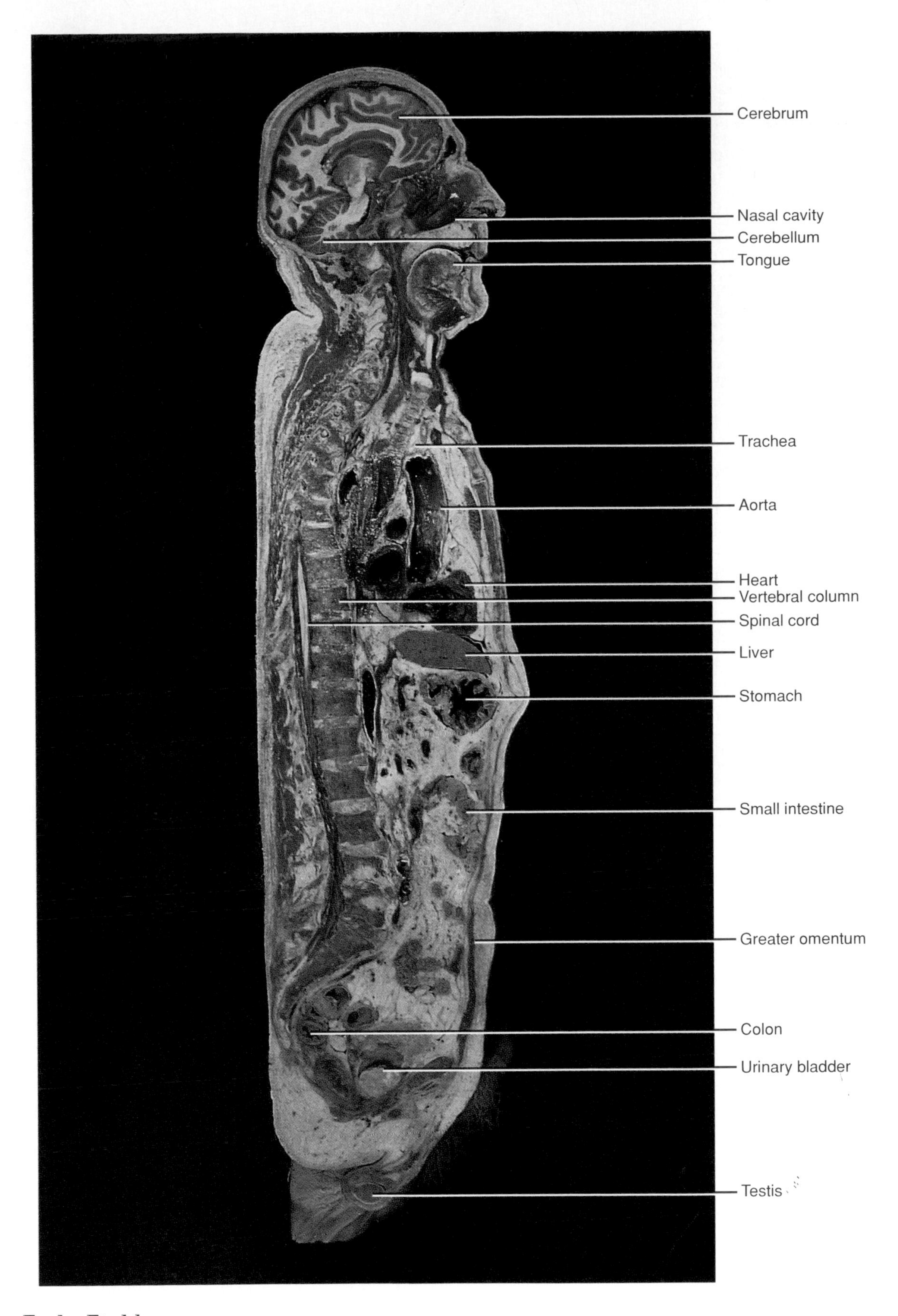


**PLATE Forty-Eight**
Sagittal section of the head and trunk.

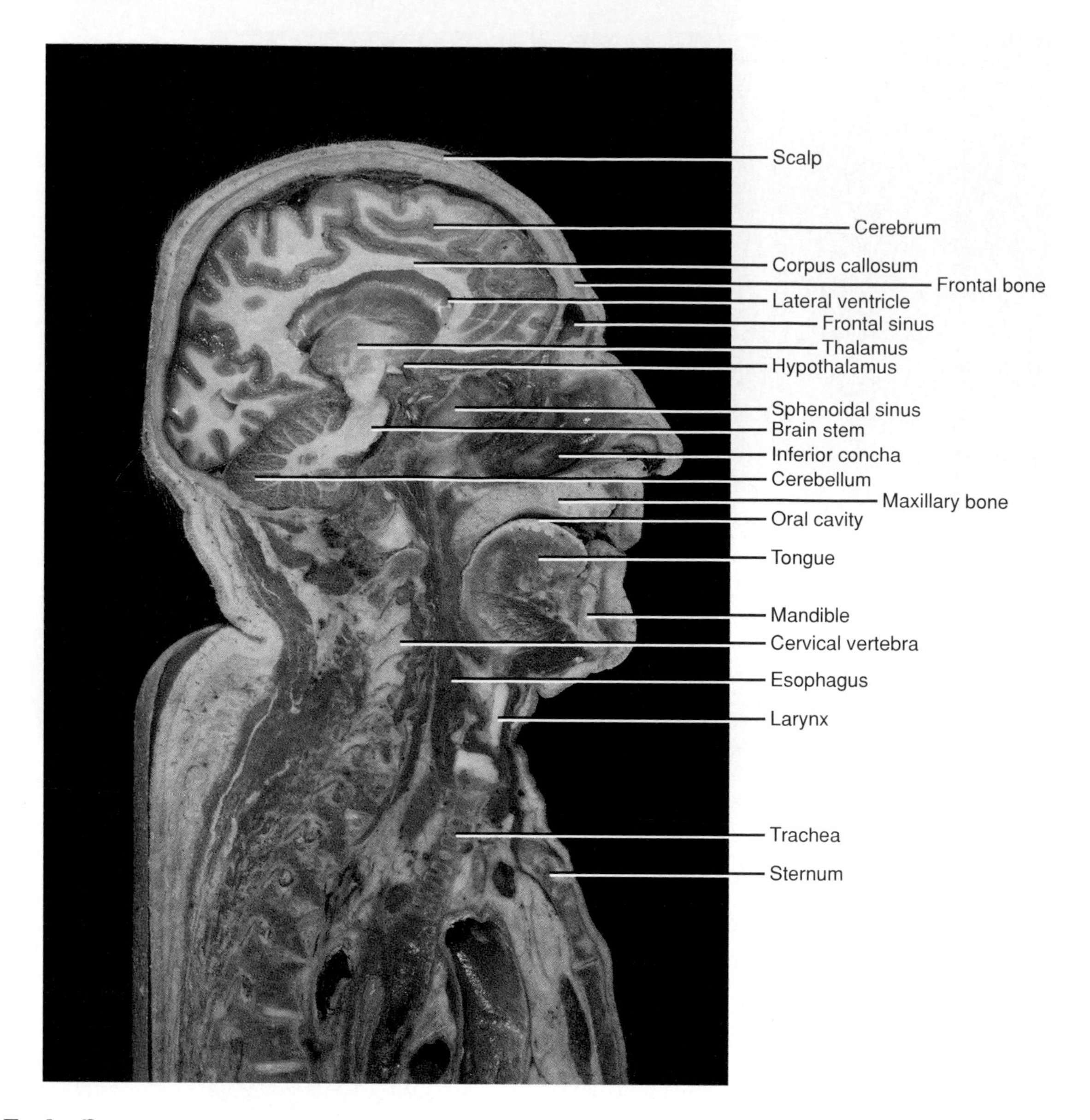


**PLATE Forty-Nine**

Sagittal section of the head and neck.

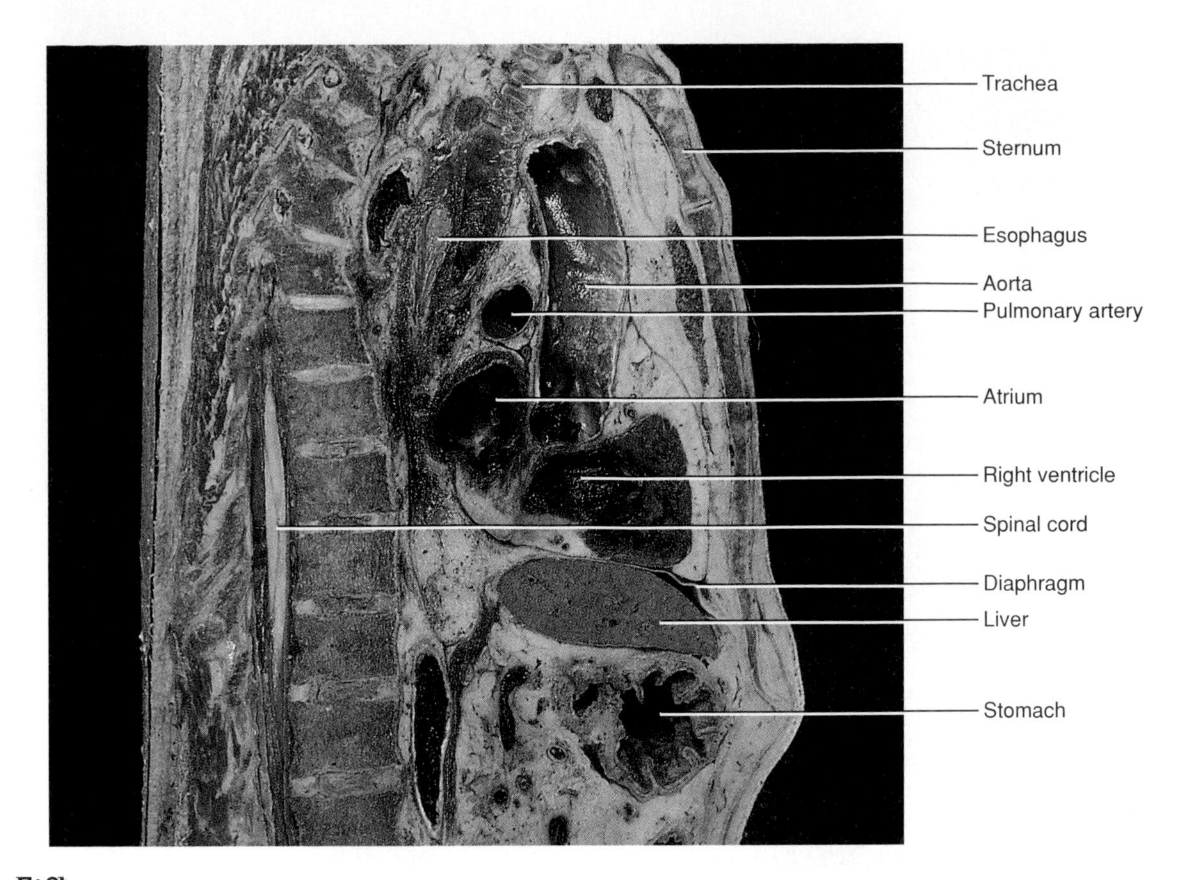


**PLATE Fifty**

Viscera of the thoracic cavity, sagittal section.

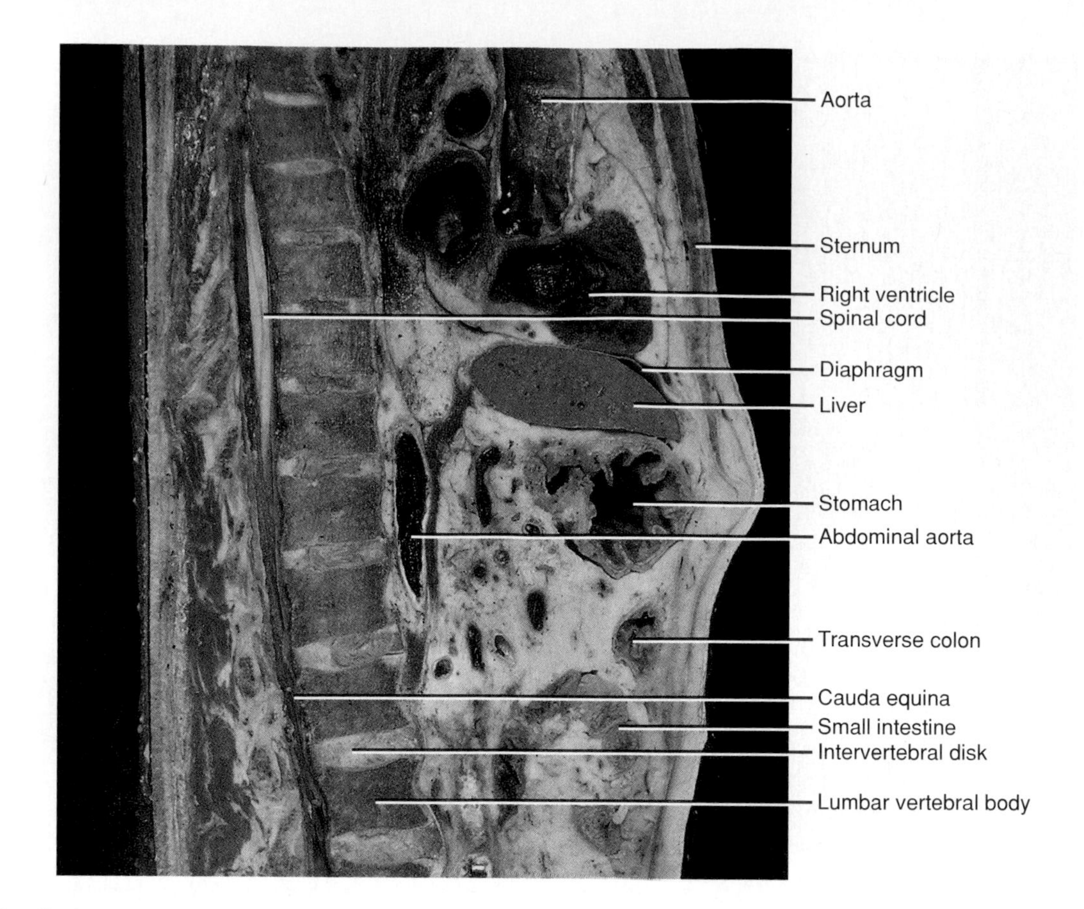


**PLATE Fifty-One**
Viscera of the abdominal cavity, sagittal section.

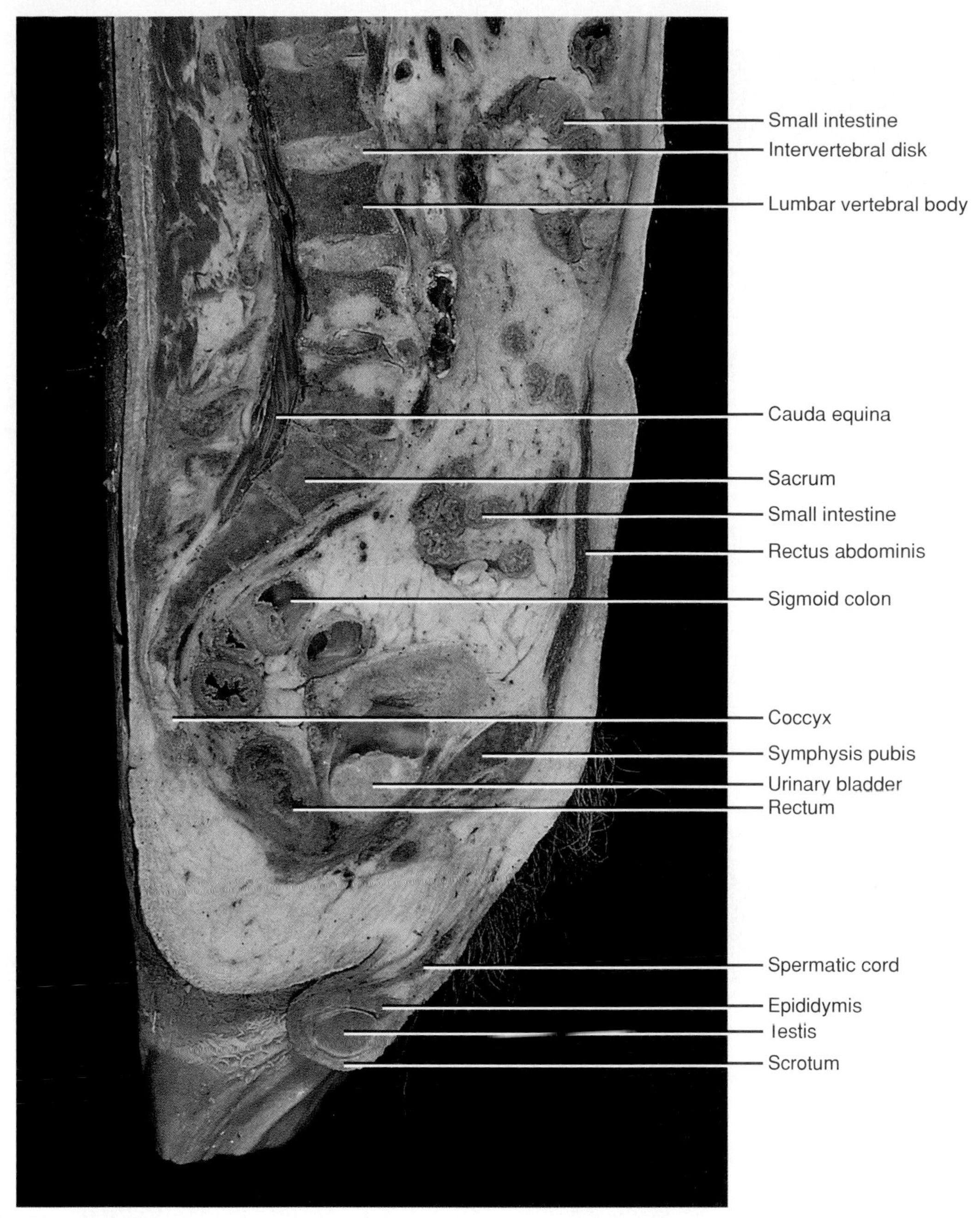


**PLATE Fifty-Two**

Viscera of the pelvic cavity, sagittal section.

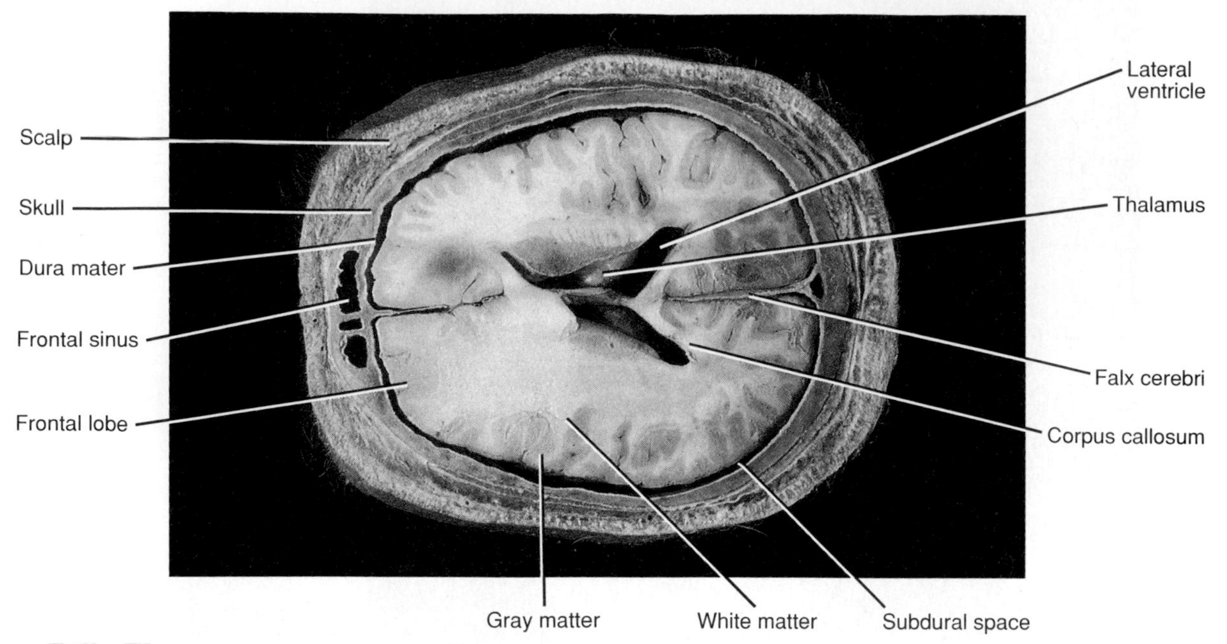


**PLATE Fifty-Three**
Horizontal section of the head above the eyes, superior view.

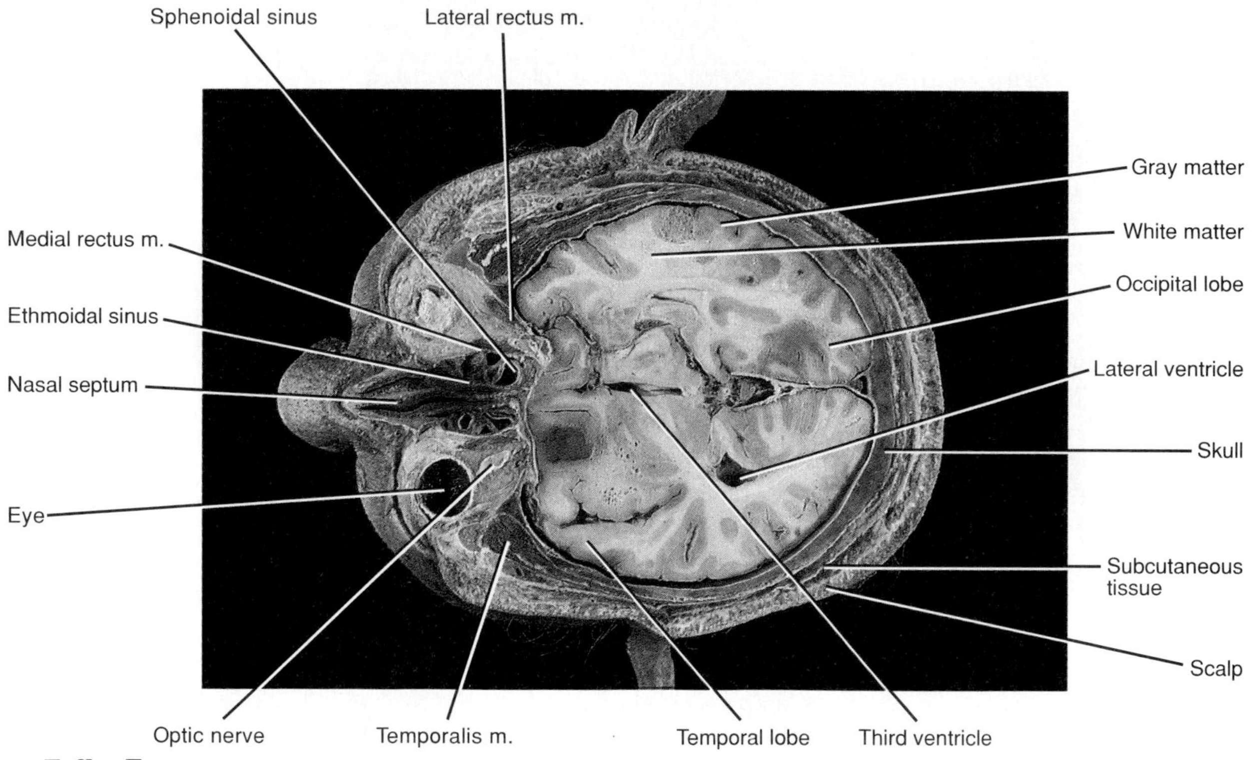


**PLATE Fifty-Four**
Horizontal section of the head at the level of the eyes, superior view.

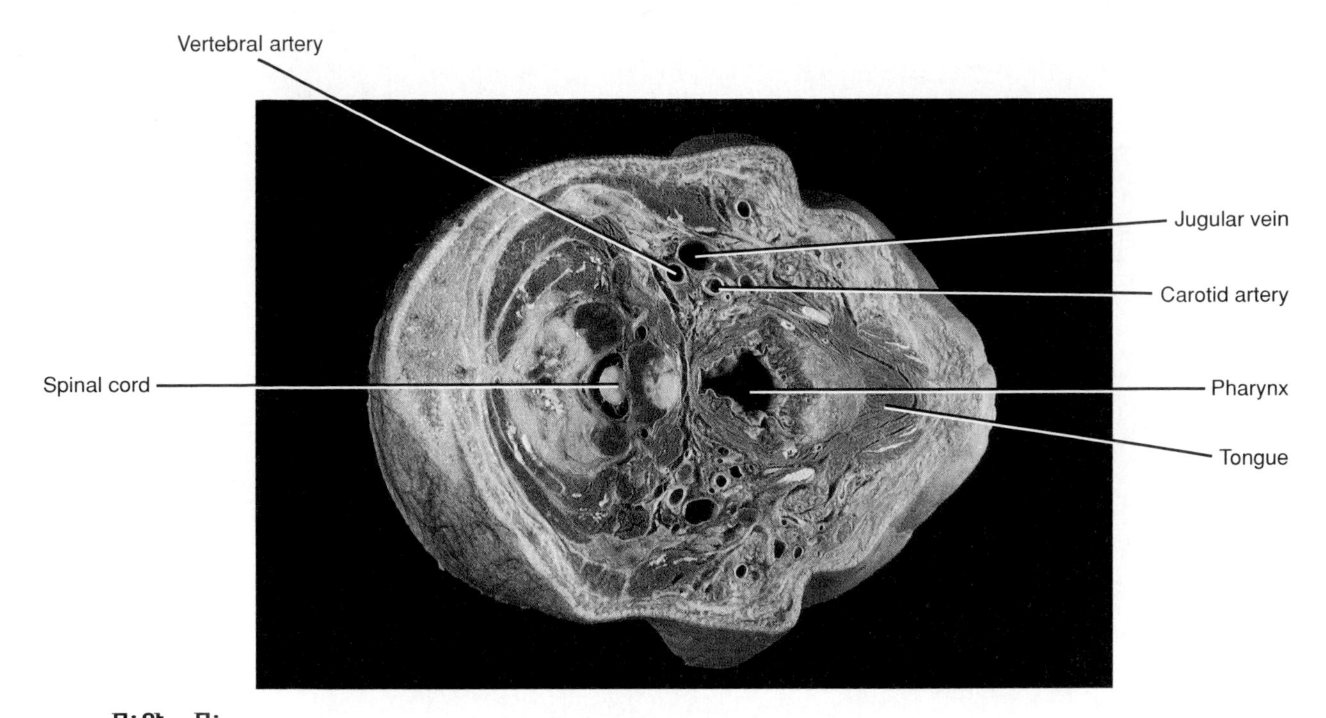


**PLATE** Fifty-Five

Horizontal section of the neck, inferior view.

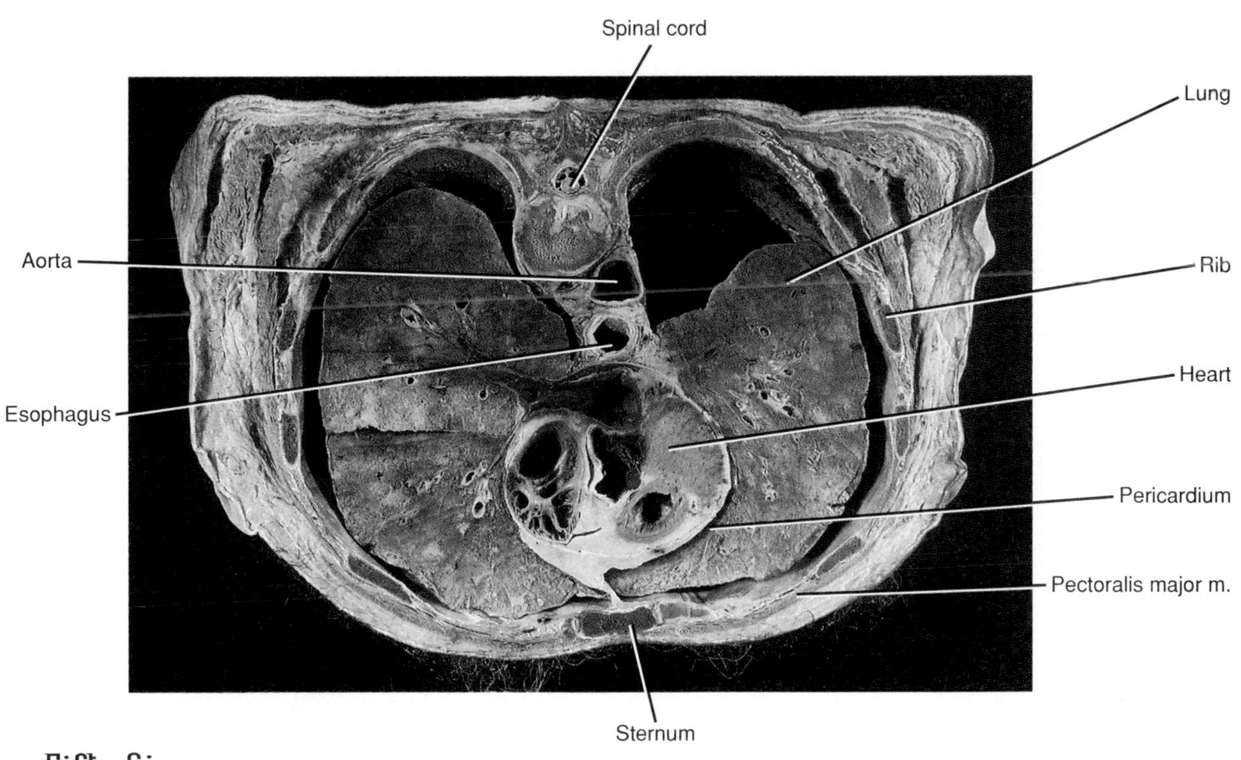


**PLATE** Fifty-Six

Horizontal section of the thorax through the base of the heart, inferior view.

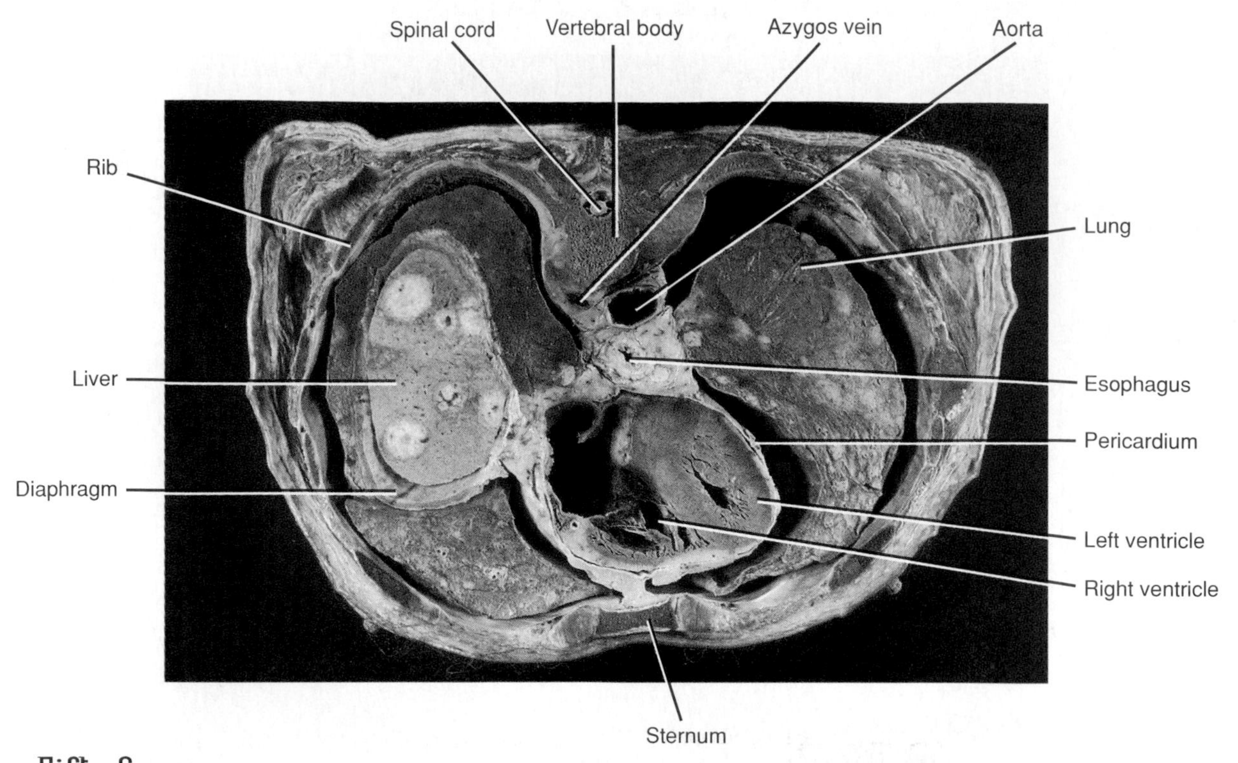


**PLATE Fifty-Seven**

Horizontal section of the thorax through the heart, inferior view.

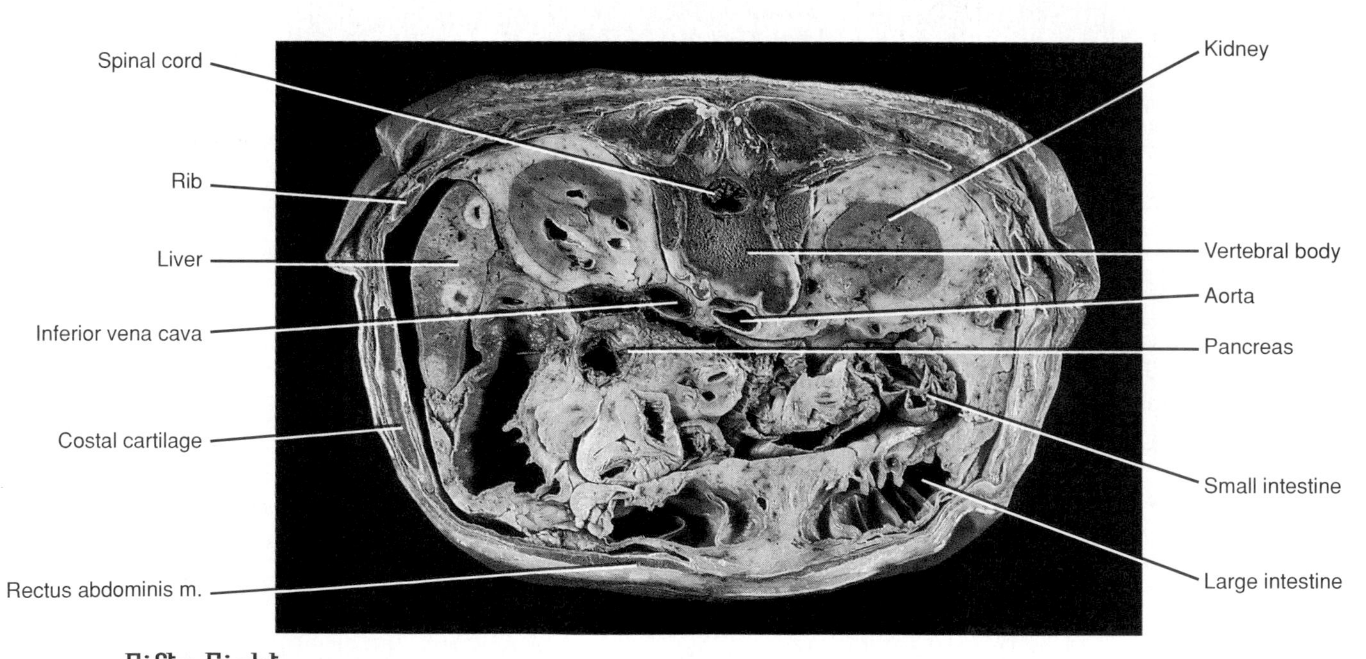


**PLATE Fifty-Eight**

Horizontal section of the abdomen through the kidneys, inferior view.

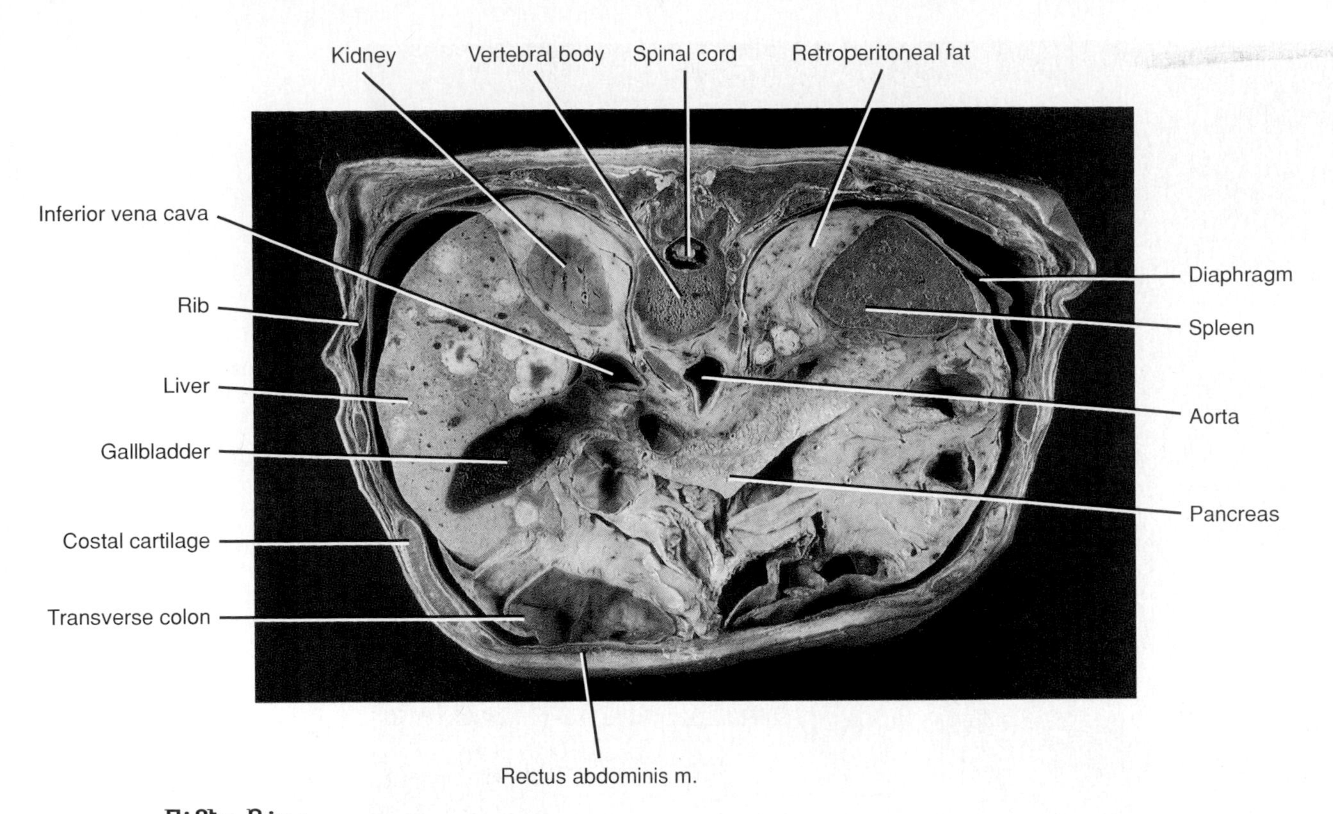


**PLATE Fifty-Nine**
Horizontal section of the abdomen through the pancreas, inferior view.

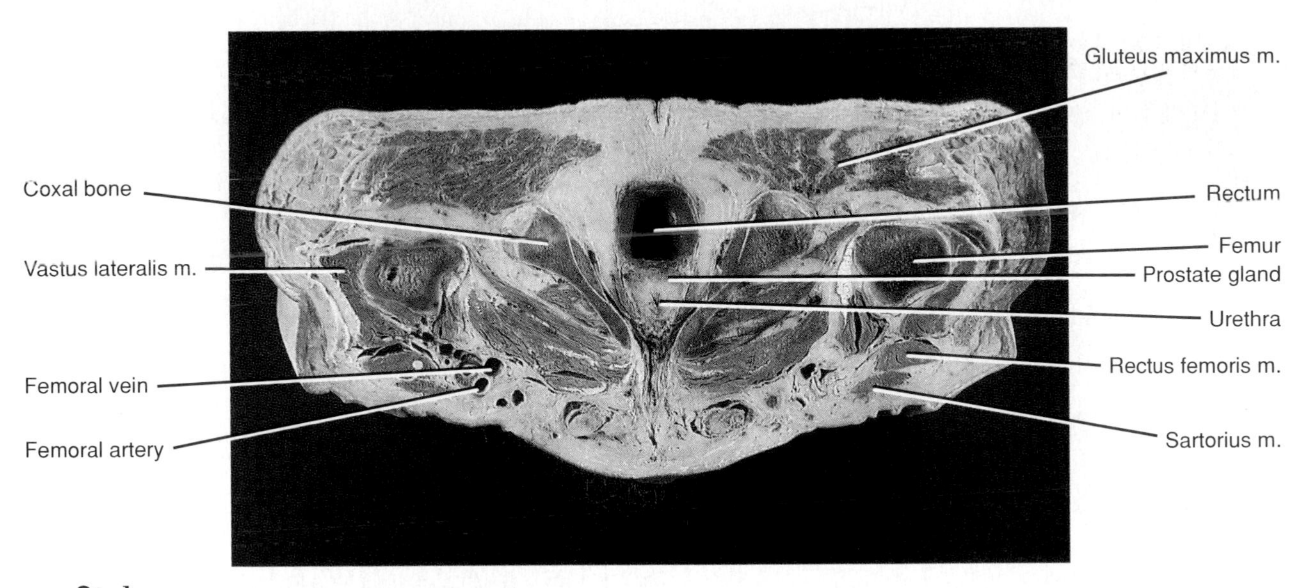


**PLATE Sixty**
Horizontal section of the male pelvic cavity, superior view.

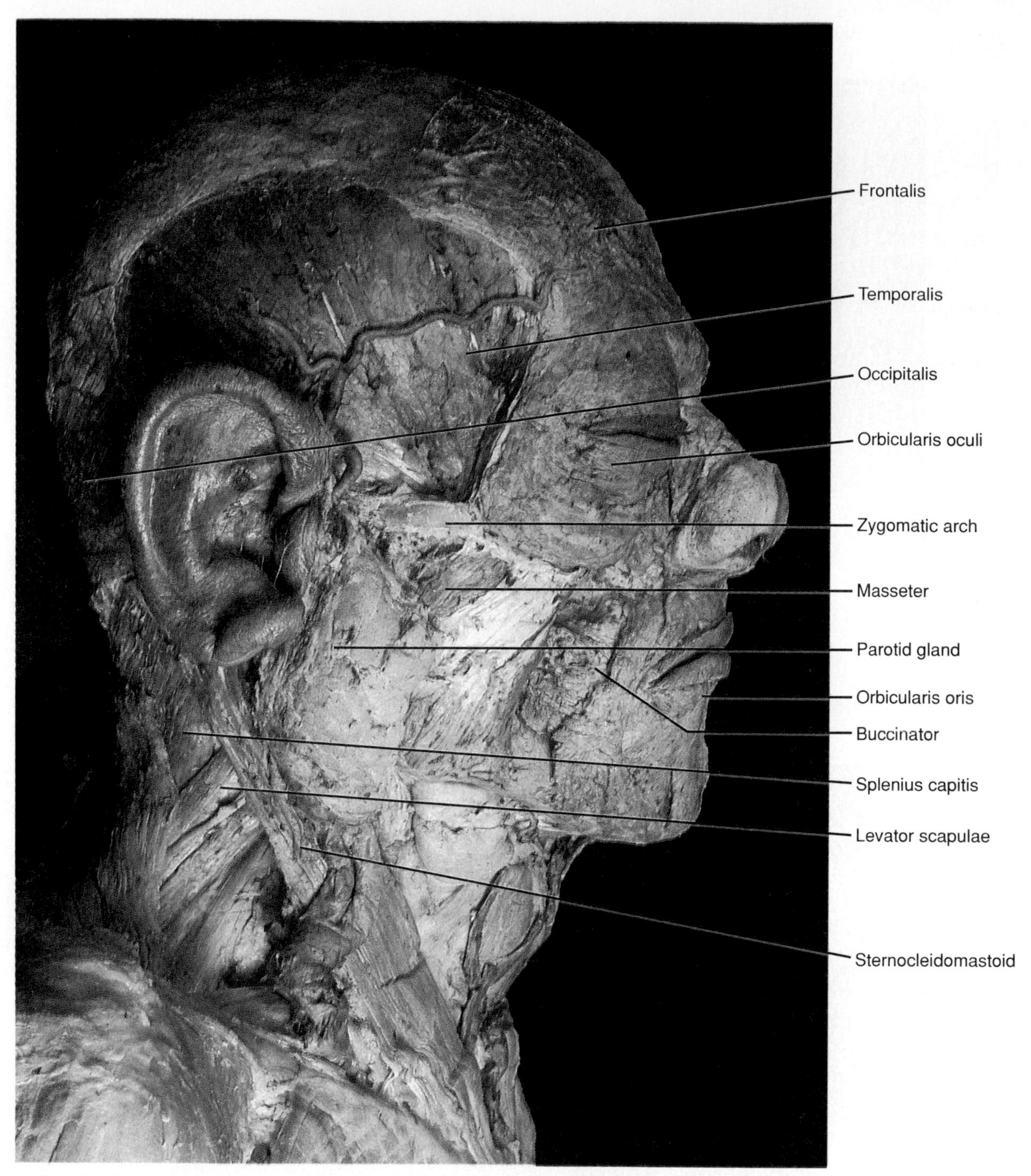


**PLATE Sixty-One**
Lateral view of the head.

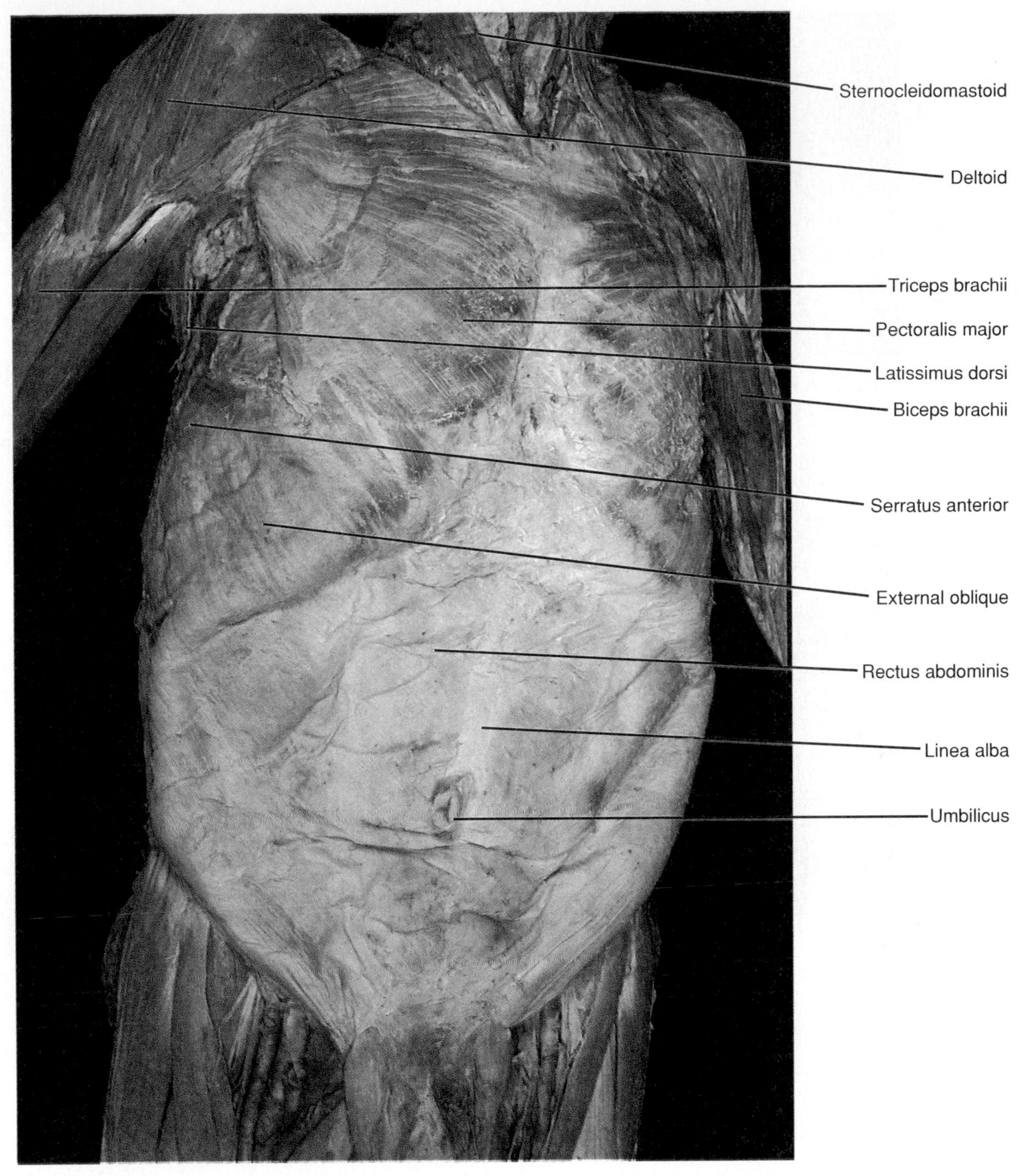


**PLATE Sixty-Two**
Anterior view of the trunk.

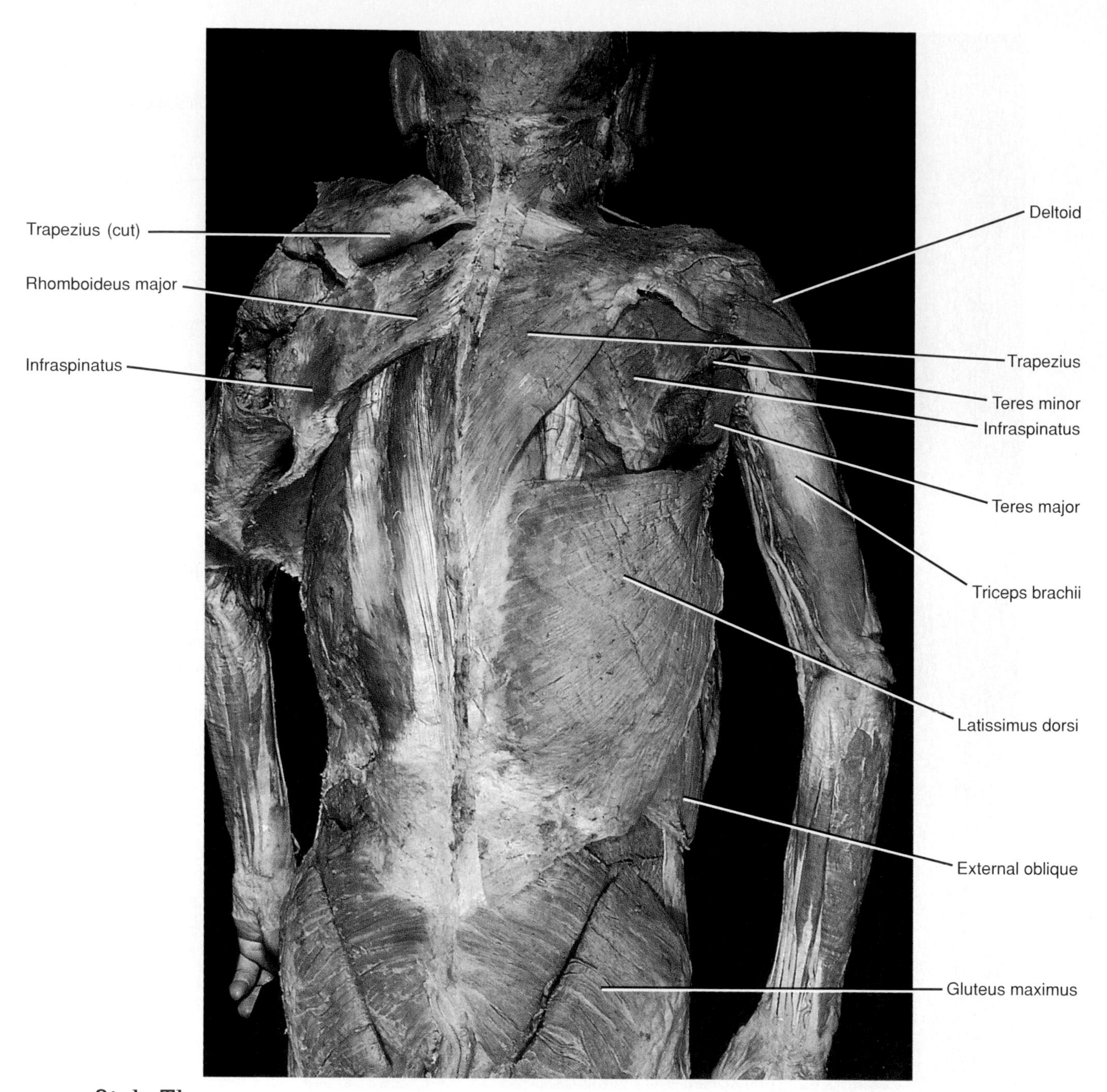


PLATE Sixty-Three

Posterior view of the trunk, with deep thoracic muscles exposed on the left.

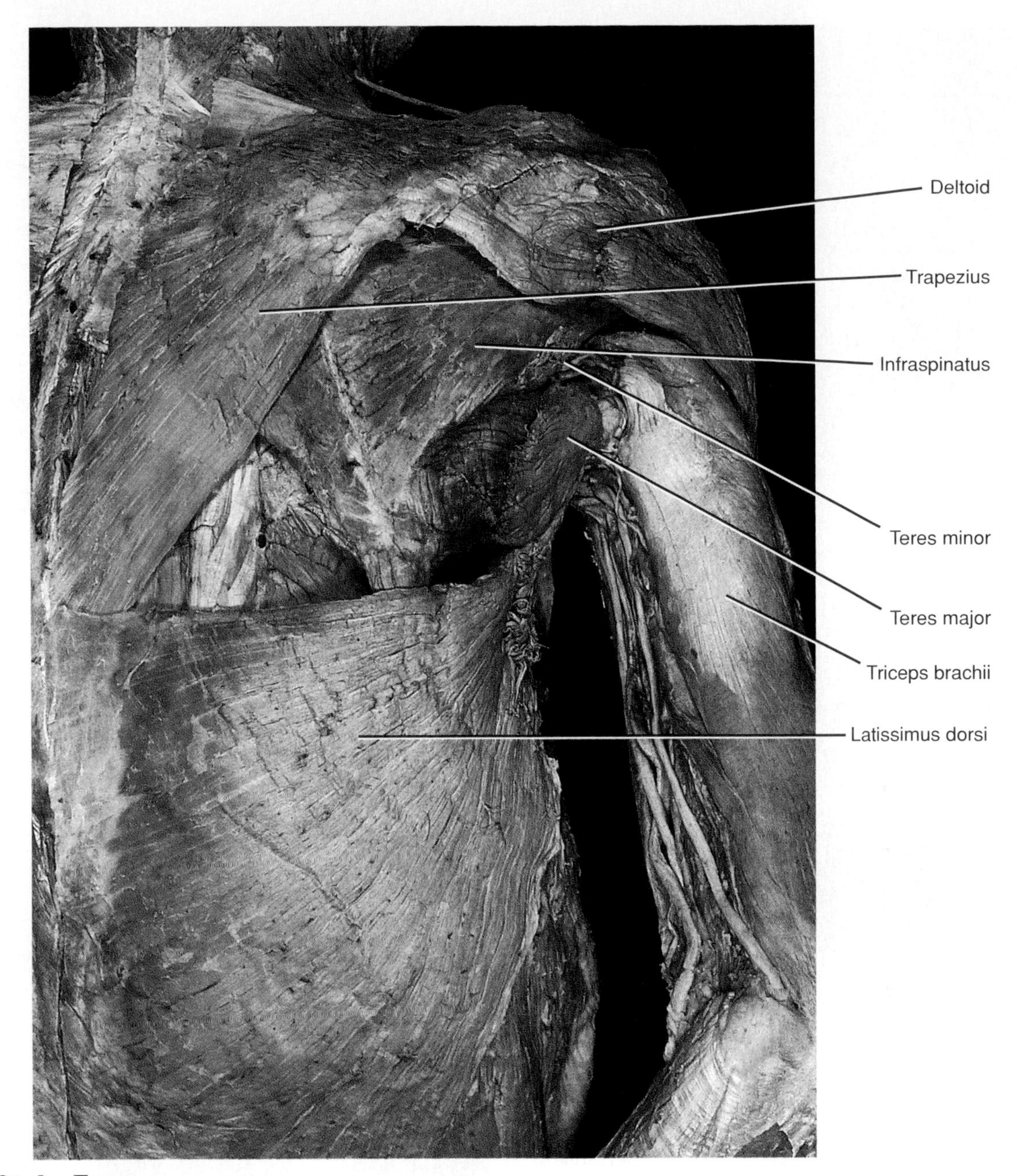


**PLATE Sixty-Four**
Posterior view of the right thorax and arm.

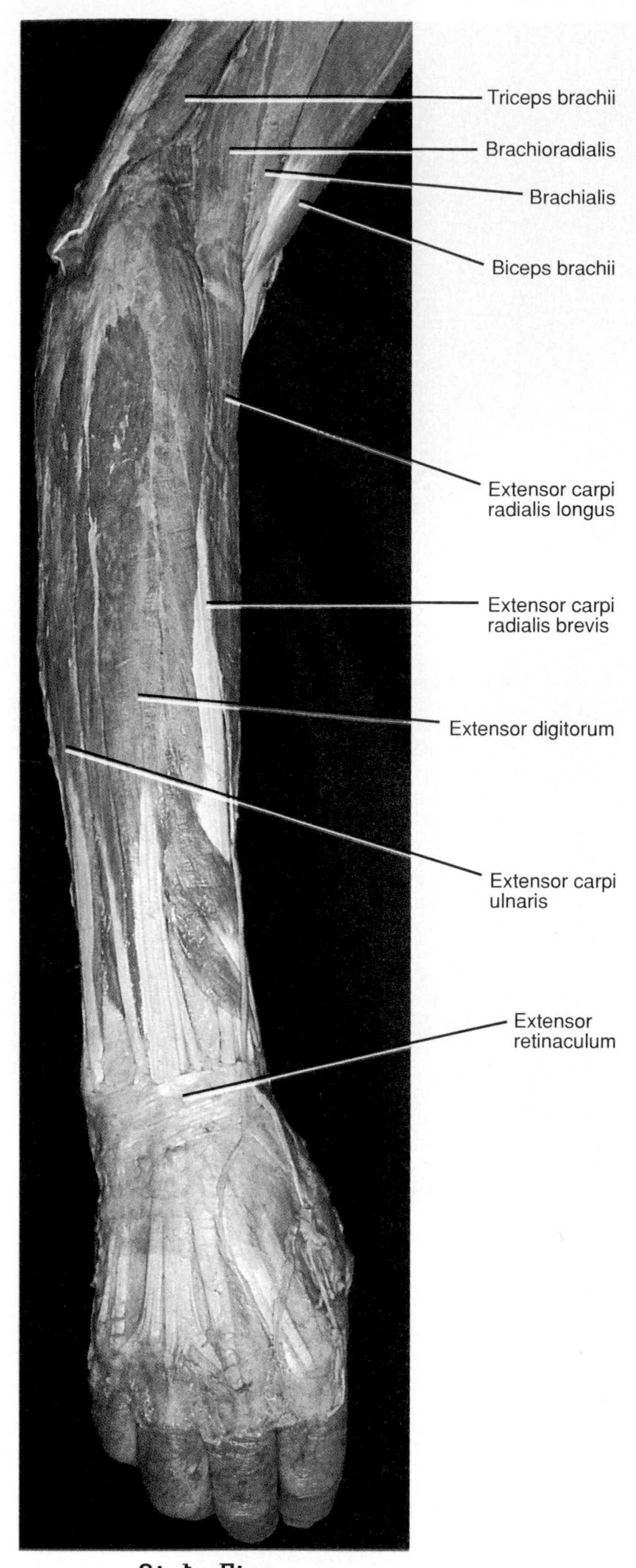


**Plate Sixty-Five**
Posterior view of the right forearm and hand.

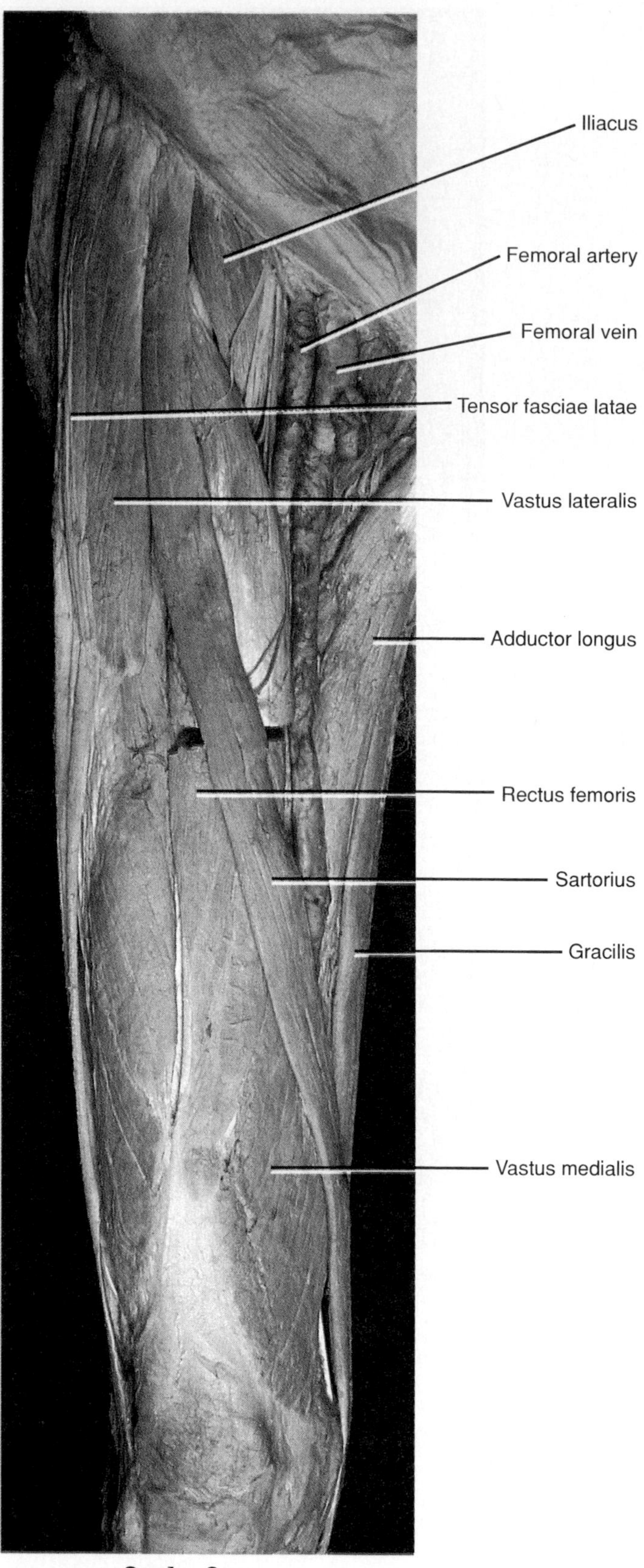


**Plate Sixty-Six**
Anterior view of the right thigh.

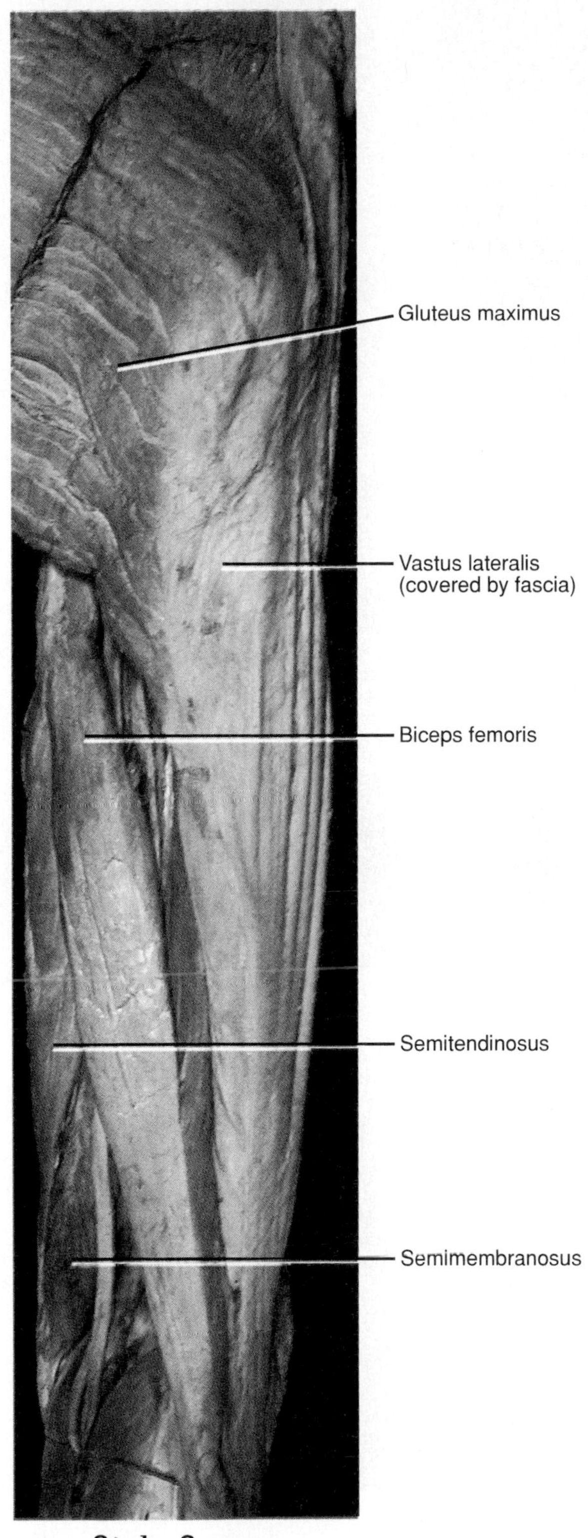


**PLATE** Sixty-Seven
Posterior view of the right thigh.

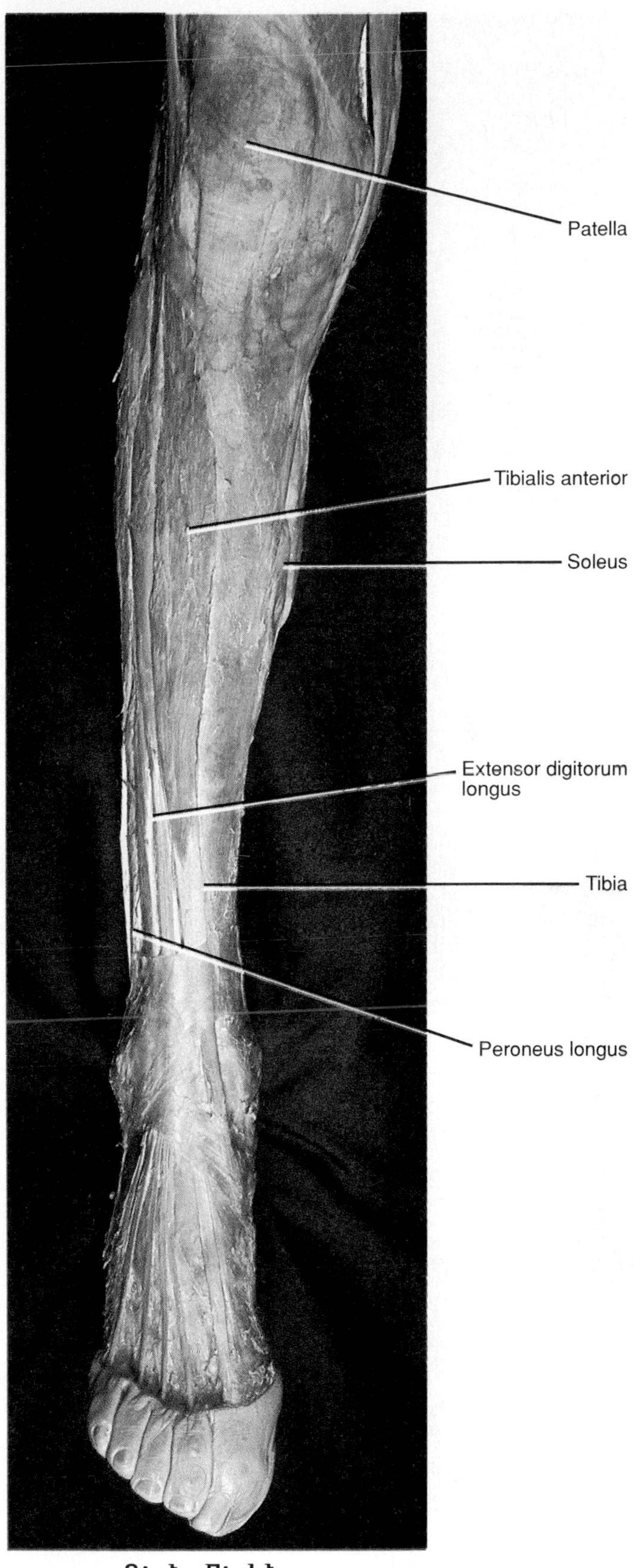


**PLATE** Sixty-Eight
Anterior view of the right leg.

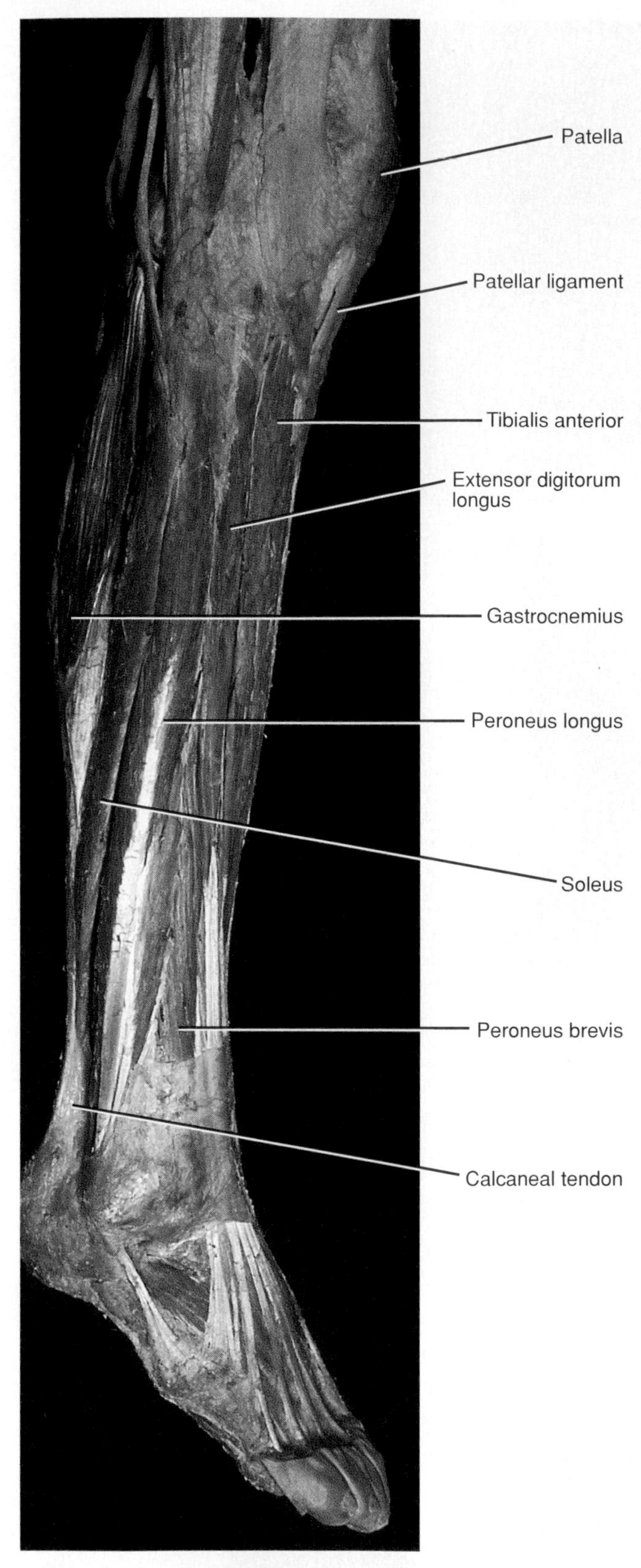


**PLATE Sixty-Nine**
Lateral view of the right leg.

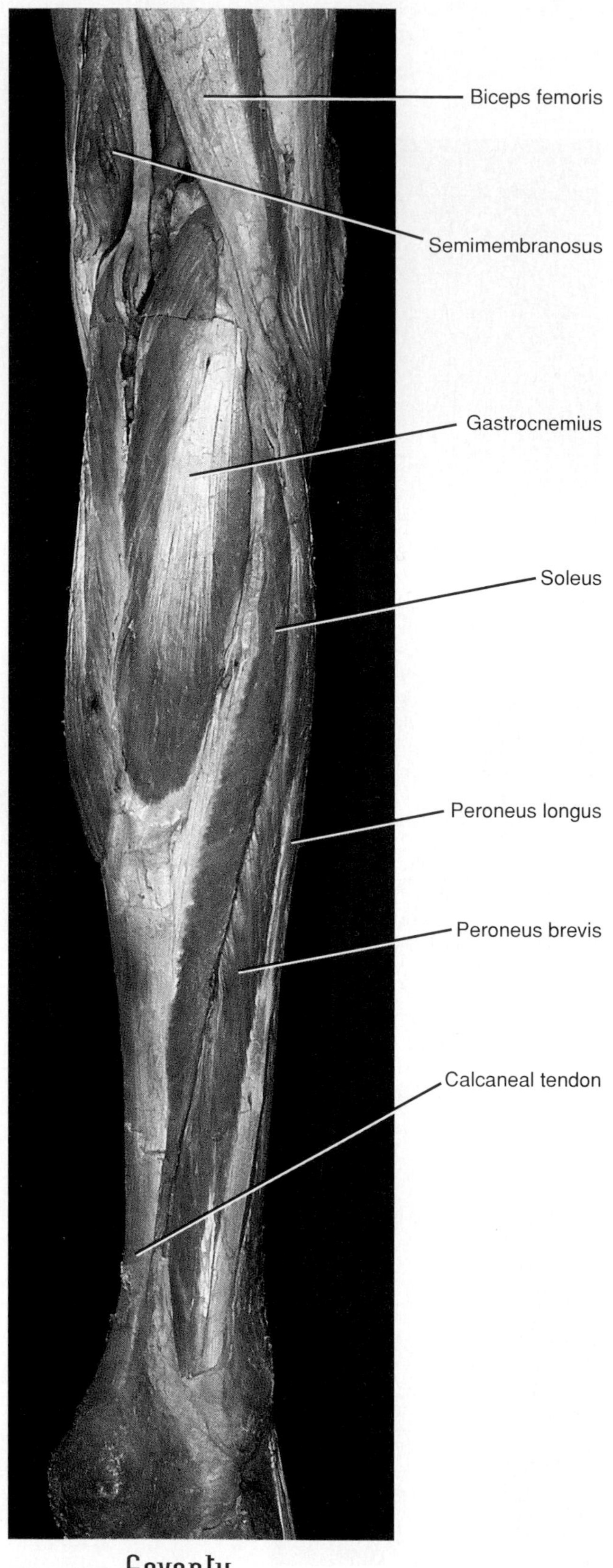


**PLATE Seventy**
Posterior view of the right leg.

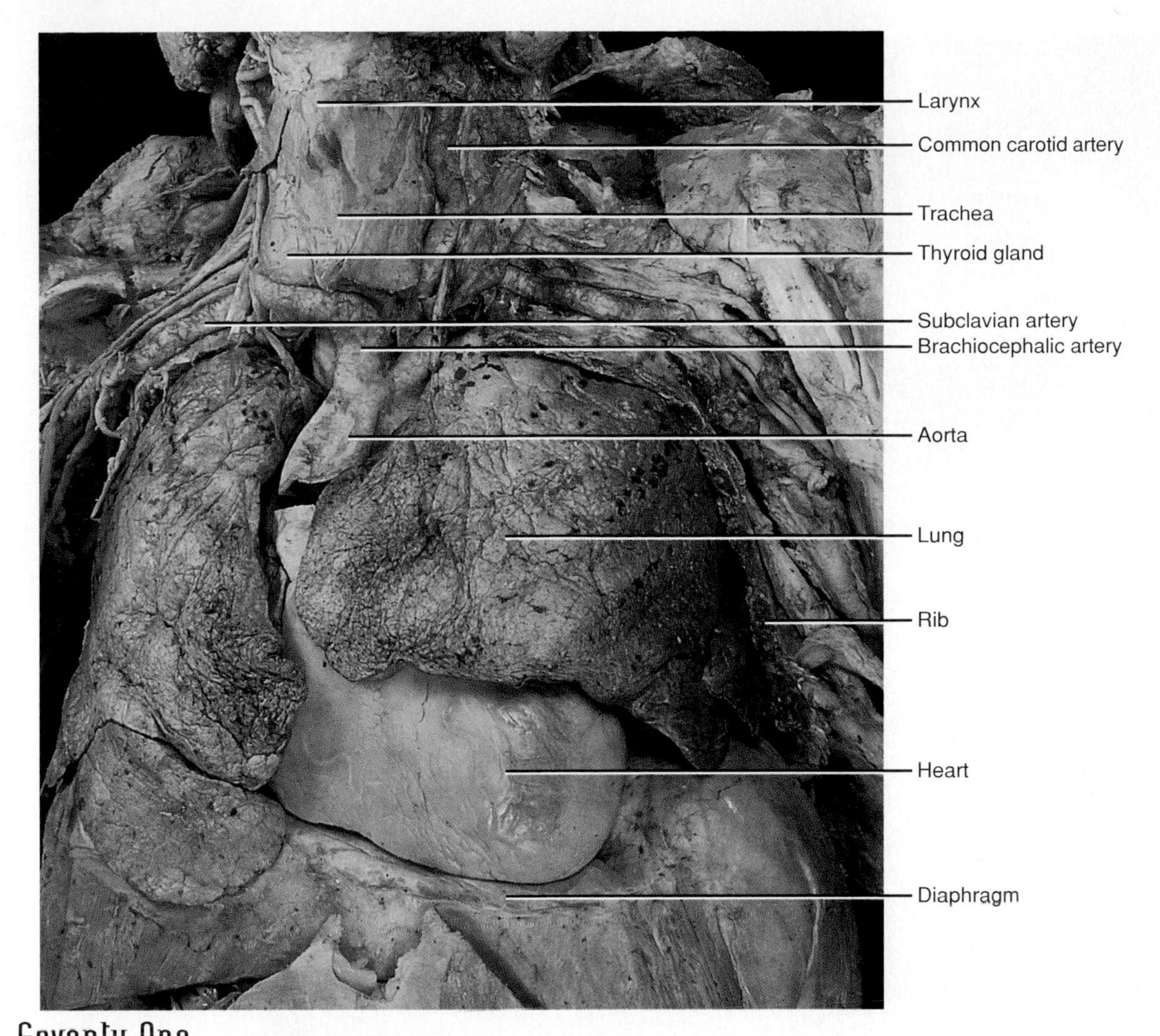


**PLATE** Seventy-One

Thoracic viscera, ventral view. (Brachiocephalic vein has been removed to expose the aorta.)

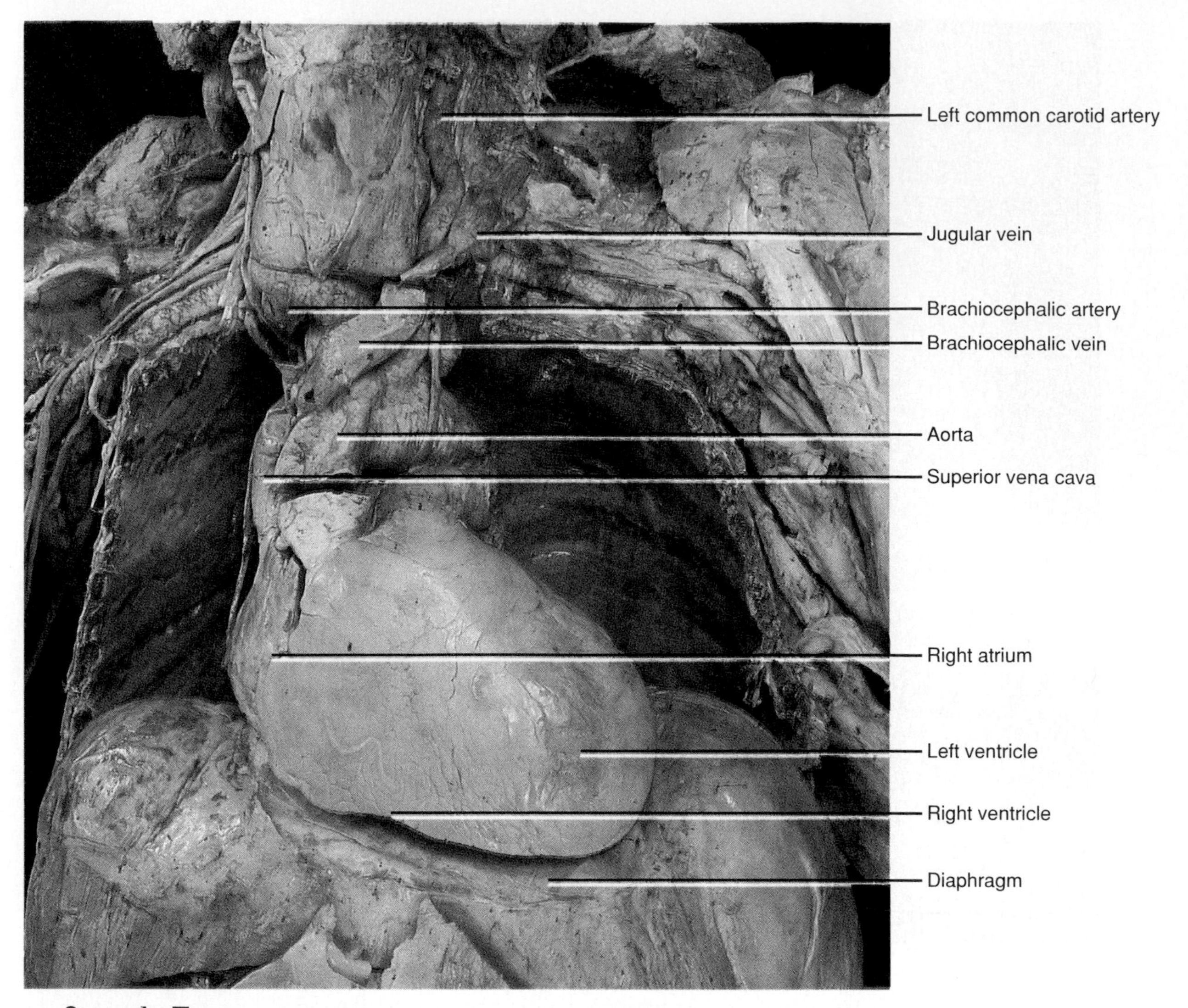


**PLATE Seventy-Two**

Thorax with the lungs removed, ventral view.

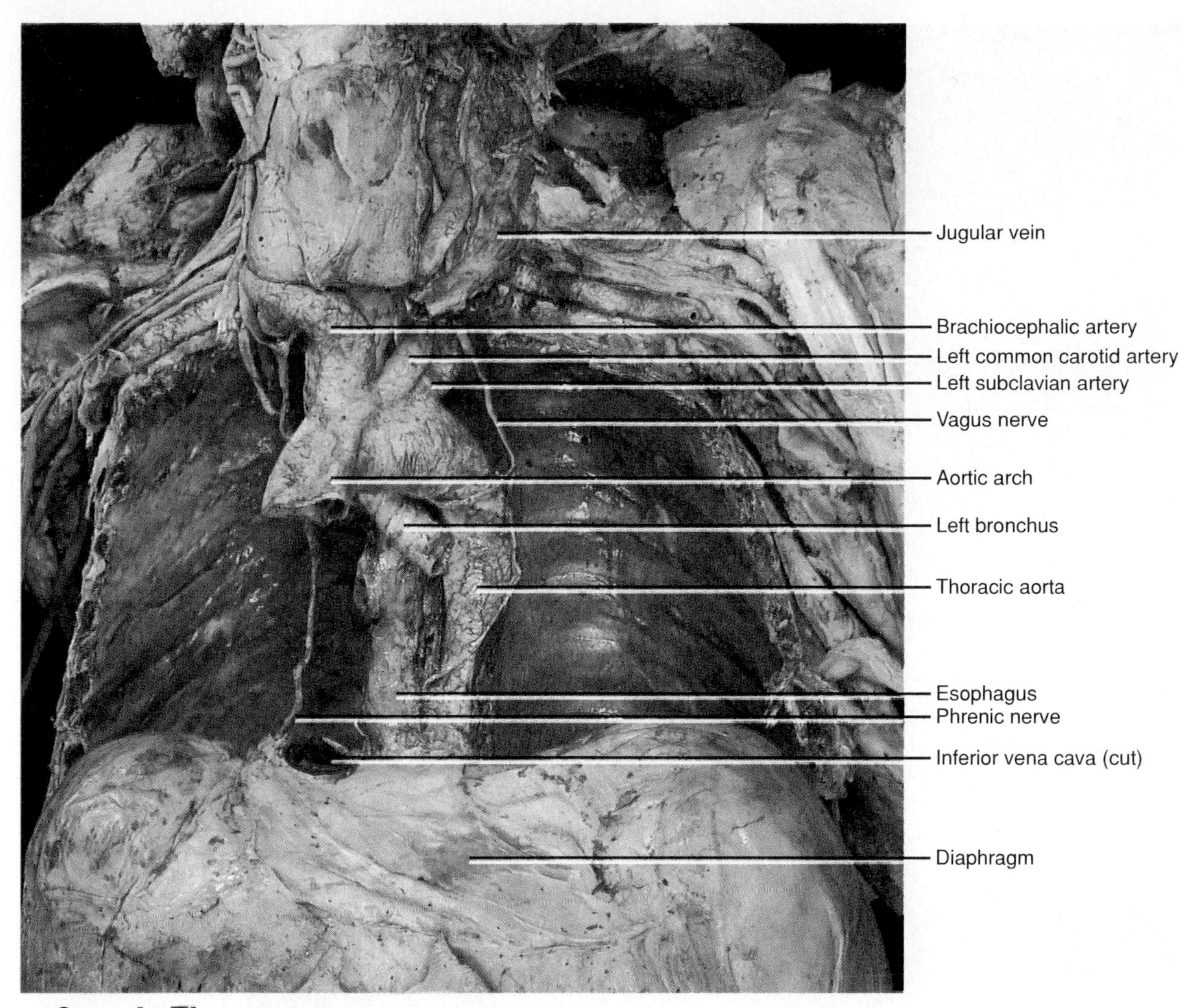


**PLATE Seventy-Three**
Thorax with the heart and lungs removed, ventral view.

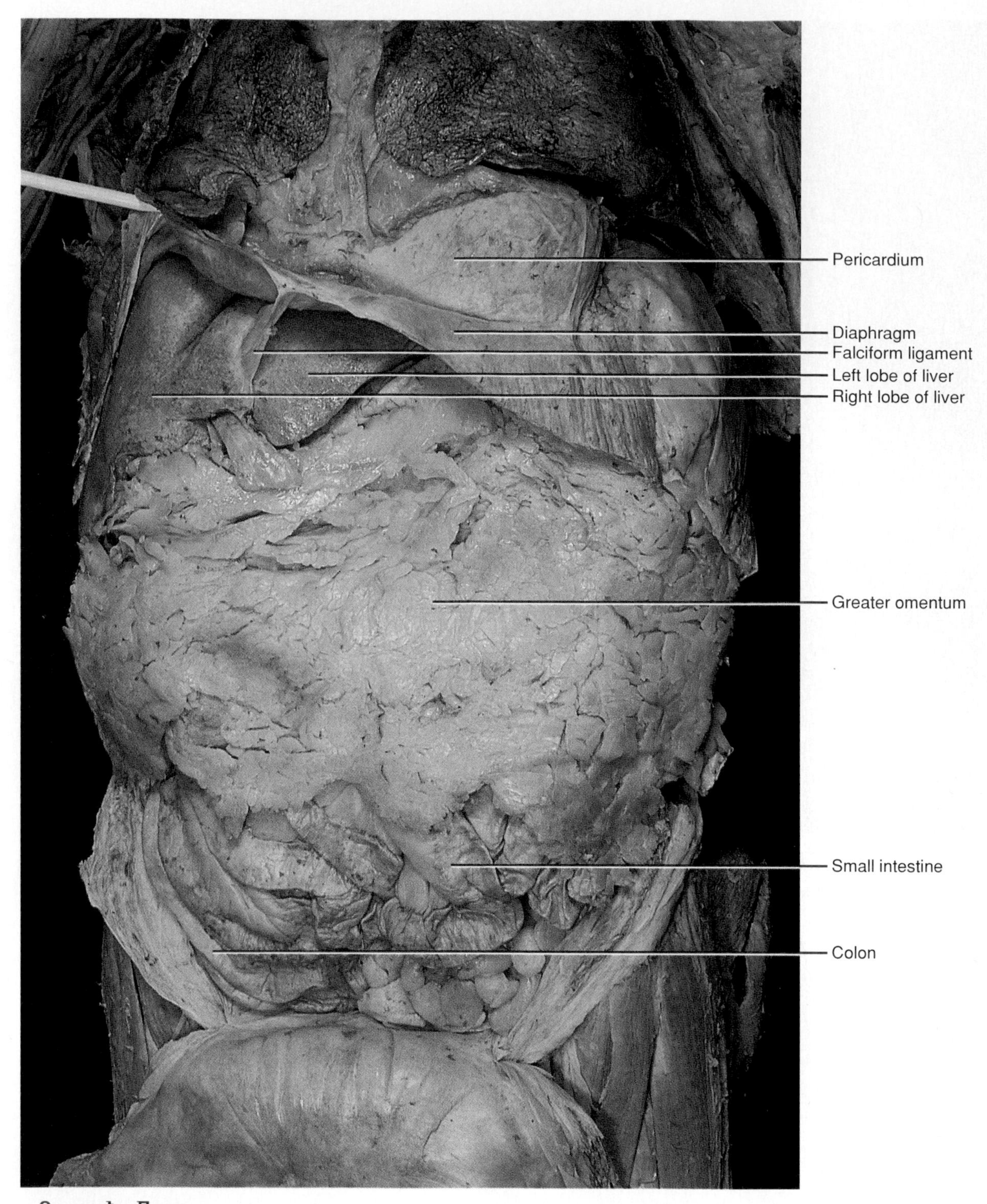


**PLATE Seventy-Four**
Abdominal viscera, ventral view.

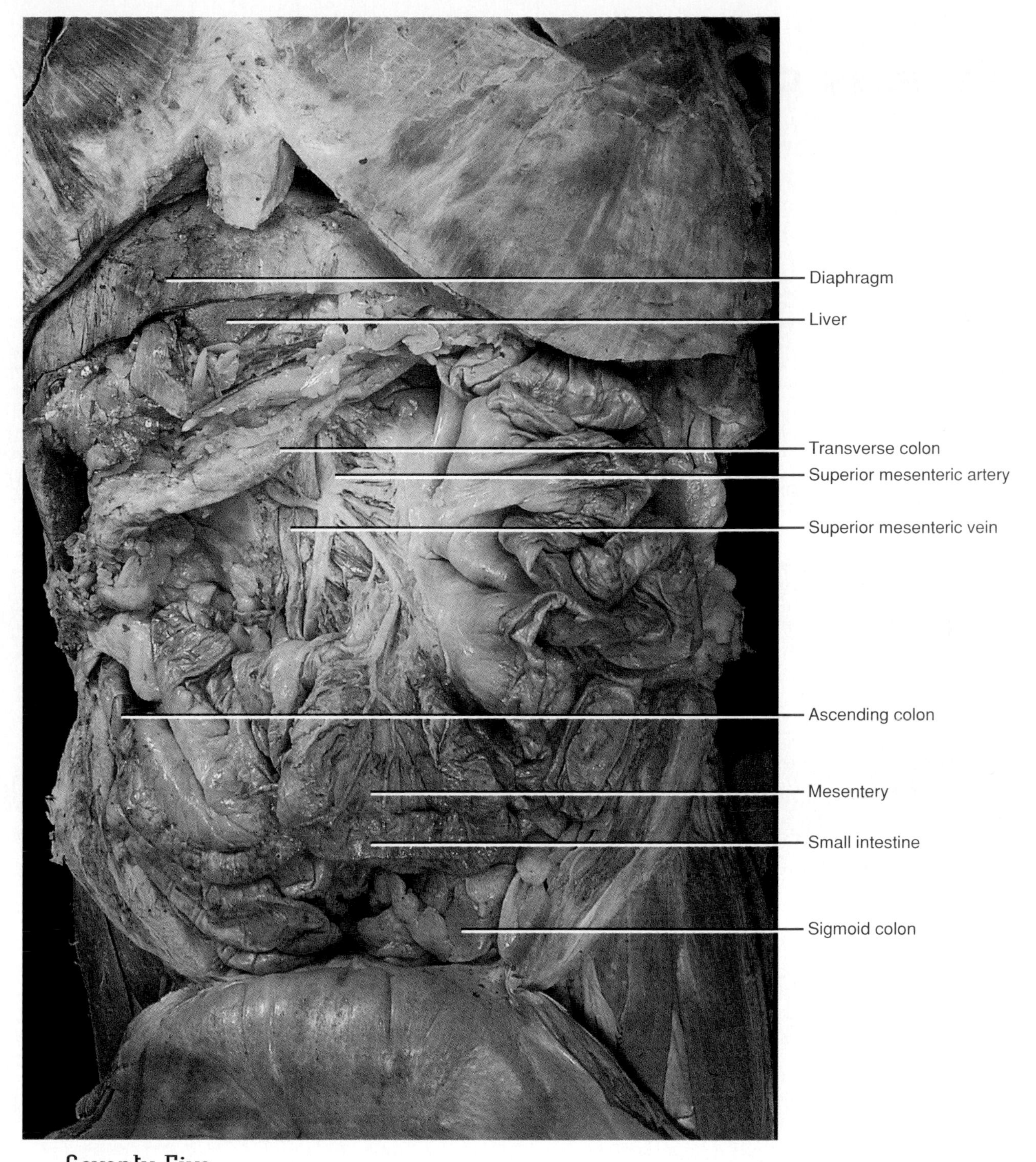


**PLATE Seventy-Five**
Abdominal viscera with the greater omentum removed. (Small intestine has been displaced to the left.)

# APPENDIX A

## Periodic Table of Elements

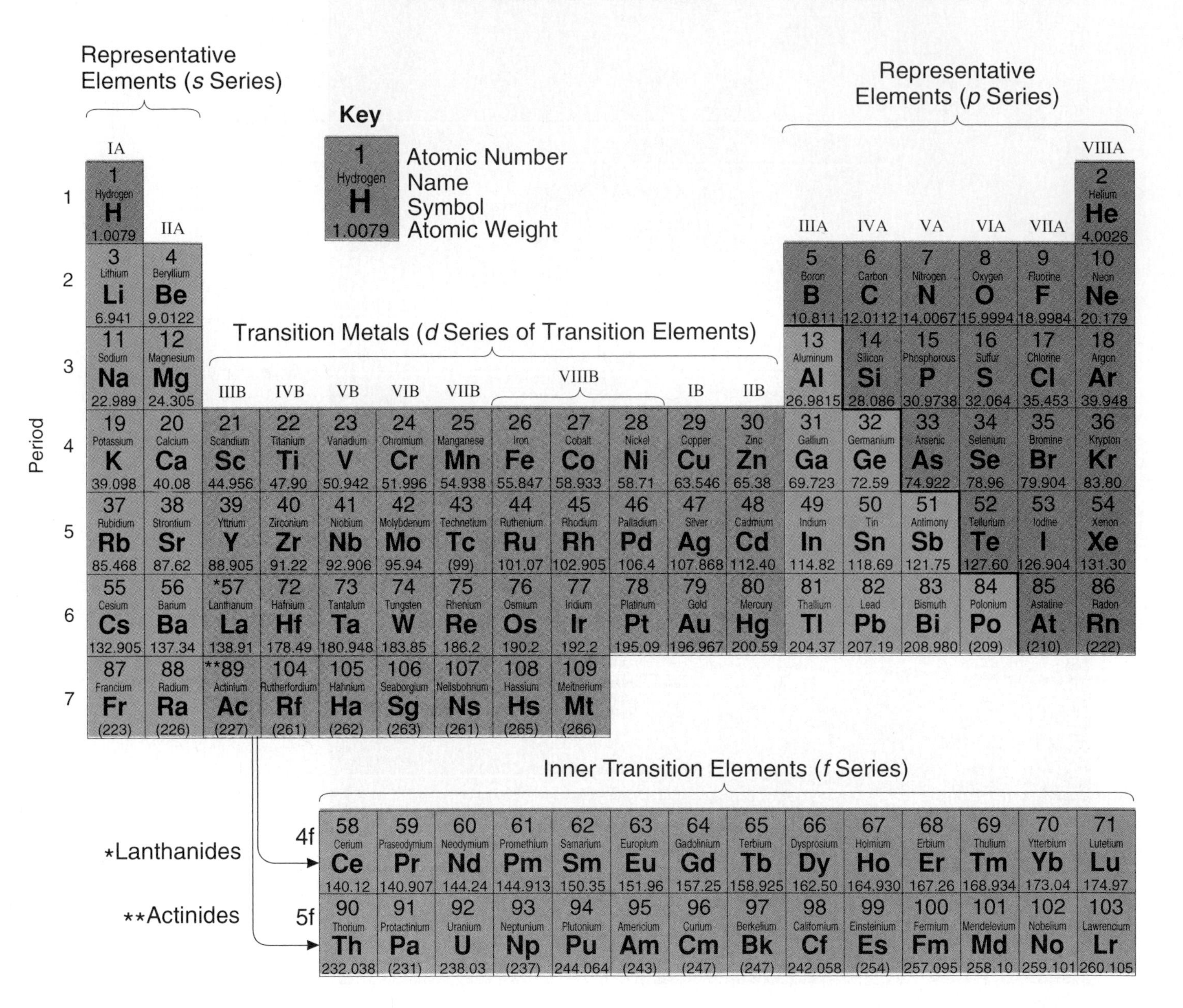


Key: 1 / Hydrogen / H / 1.0079 = Atomic Number / Name / Symbol / Atomic Weight

| Period | IA | IIA | IIIB | IVB | VB | VIB | VIIB | | VIIIB | | IB | IIB | IIIA | IVA | VA | VIA | VIIA | VIIIA |
|---|---|---|---|---|---|---|---|---|---|---|---|---|---|---|---|---|---|---|
| 1 | 1 Hydrogen H 1.0079 | | | | | | | | | | | | | | | | | 2 Helium He 4.0026 |
| 2 | 3 Lithium Li 6.941 | 4 Beryllium Be 9.0122 | | | | | | | | | | | 5 Boron B 10.811 | 6 Carbon C 12.0112 | 7 Nitrogen N 14.0067 | 8 Oxygen O 15.9994 | 9 Fluorine F 18.9984 | 10 Neon Ne 20.179 |
| 3 | 11 Sodium Na 22.989 | 12 Magnesium Mg 24.305 | | | | | | | | | | | 13 Aluminum Al 26.9815 | 14 Silicon Si 28.086 | 15 Phosphorous P 30.9738 | 16 Sulfur S 32.064 | 17 Chlorine Cl 35.453 | 18 Argon Ar 39.948 |
| 4 | 19 Potassium K 39.098 | 20 Calcium Ca 40.08 | 21 Scandium Sc 44.956 | 22 Titanium Ti 47.90 | 23 Vanadium V 50.942 | 24 Chromium Cr 51.996 | 25 Manganese Mn 54.938 | 26 Iron Fe 55.847 | 27 Cobalt Co 58.933 | 28 Nickel Ni 58.71 | 29 Copper Cu 63.546 | 30 Zinc Zn 65.38 | 31 Gallium Ga 69.723 | 32 Germanium Ge 72.59 | 33 Arsenic As 74.922 | 34 Selenium Se 78.96 | 35 Bromine Br 79.904 | 36 Krypton Kr 83.80 |
| 5 | 37 Rubidium Rb 85.468 | 38 Strontium Sr 87.62 | 39 Yttrium Y 88.905 | 40 Zirconium Zr 91.22 | 41 Niobium Nb 92.906 | 42 Molybdenum Mo 95.94 | 43 Technetium Tc (99) | 44 Ruthenium Ru 101.07 | 45 Rhodium Rh 102.905 | 46 Palladium Pd 106.4 | 47 Silver Ag 107.868 | 48 Cadmium Cd 112.40 | 49 Indium In 114.82 | 50 Tin Sn 118.69 | 51 Antimony Sb 121.75 | 52 Tellurium Te 127.60 | 53 Iodine I 126.904 | 54 Xenon Xe 131.30 |
| 6 | 55 Cesium Cs 132.905 | 56 Barium Ba 137.34 | *57 Lanthanum La 138.91 | 72 Hafnium Hf 178.49 | 73 Tantalum Ta 180.948 | 74 Tungsten W 183.85 | 75 Rhenium Re 186.2 | 76 Osmium Os 190.2 | 77 Iridium Ir 192.2 | 78 Platinum Pt 195.09 | 79 Gold Au 196.967 | 80 Mercury Hg 200.59 | 81 Thallium Tl 204.37 | 82 Lead Pb 207.19 | 83 Bismuth Bi 208.980 | 84 Polonium Po (209) | 85 Astatine At (210) | 86 Radon Rn (222) |
| 7 | 87 Francium Fr (223) | 88 Radium Ra (226) | **89 Actinium Ac (227) | 104 Rutherfordium Rf (261) | 105 Hahnium Ha (262) | 106 Seaborgium Sg (263) | 107 Neilsbohrium Ns (261) | 108 Hassium Hs (265) | 109 Meitnerium Mt (266) | | | | | | | | | |

Inner Transition Elements (f Series)

| | | | | | | | | | | | | | | |
|---|---|---|---|---|---|---|---|---|---|---|---|---|---|---|
| *Lanthanides 4f | 58 Cerium Ce 140.12 | 59 Praseodymium Pr 140.907 | 60 Neodymium Nd 144.24 | 61 Promethium Pm 144.913 | 62 Samarium Sm 150.35 | 63 Europium Eu 151.96 | 64 Gadolinium Gd 157.25 | 65 Terbium Tb 158.925 | 66 Dysprosium Dy 162.50 | 67 Holmium Ho 164.930 | 68 Erbium Er 167.26 | 69 Thulium Tm 168.934 | 70 Ytterbium Yb 173.04 | 71 Lutetium Lu 174.97 |
| **Actinides 5f | 90 Thorium Th 232.038 | 91 Protactinium Pa (231) | 92 Uranium U 238.03 | 93 Neptunium Np (237) | 94 Plutonium Pu 244.064 | 95 Americium Am (243) | 96 Curium Cm (247) | 97 Berkelium Bk (247) | 98 Californium Cf 242.058 | 99 Einsteinium Es (254) | 100 Fermium Fm 257.095 | 101 Mendelevium Md 258.10 | 102 Nobelium No 259.101 | 103 Lawrencium Lr 260.105 |

# APPENDIX B

## Units of Measurement and Their Equivalents

### Apothecaries' Weights and Their Metric Equivalents

**1 grain (gr) =**
*0.05 scruple (s)*
*0.017 dram (dr)*
*0.002 ounce (oz)*
*0.0002 pound (lb)*
*0.065 gram (g)*
*65. milligrams (mg)*

**1 scruple (s) =**
*20. grains (gr)*
*0.33 dram (dr)*
*0.042 ounce (oz)*
*0.004 pound (lb)*
*1.3 grams (g)*
*1300. milligrams (mg)*

**1 dram (dr) =**
*60. grains (gr)*
*3. scruples (s)*
*0.13 ounce (oz)*
*0.010 pound (lb)*
*3.9 grams (g)*
*3,900. milligrams (mg)*

**1 ounce (oz) =**
*480. grains (gr)*
*24. scruples (s)*
*0.08 pound (lb)*
*31.1 grams (g)*
*31,100. milligrams (mg)*

**1 pound (lb) =**
*5,760. grains (gr)*
*288. scruples (s)*
*96. drams (dr)*
*12. ounces (oz)*
*373. grams (g)*
*373,000. milligrams (mg)*

### Apothecaries' Volumes and Their Metric Equivalents

**1 minim (min) =**
*0.017 fluid dram (fl dr)*
*0.002 fluid ounce (fl oz)*
*0.0001 pint (pt)*
*0.06 milliliter (ml)*
*0.06 cubic centimeter (cc)*

**1 fluid dram (fl dr) =**
*60. minims (min)*
*0.13 fluid ounce (fl oz)*
*0.008 pint (pt)*
*3.70 milliliters (ml)*
*3.70 cubic centimeters (cc)*

**1 fluid ounce (fl oz) =**
*480. minims (min)*
*8. fluid drams (fl dr)*
*0.06 pint (pt)*
*29.6 milliliters (ml)*
*29.6 cubic centimeters (cc)*

**1 pint (pt) =**
*7,680. minims (min)*
*128. fluid drams (fl dr)*
*16. fluid ounces (fl oz)*
*473. milliliters (ml)*
*473. cubic centimeters (cc)*

### Metric Weights and Their Apothecaries' Equivalents

**1 gram (g) =**
*0.001 kilogram (kg)*
*1,000. milligrams (mg)*
*1,000,000. micrograms (μg)*
*1,000,000,000. nanograms (ng)*
*1,000,000,000,000. picograms (pg)*
*15.4 grains (gr)*
*0.032 ounce (oz)*

**1 kilogram (kg) =**
*1,000. grams (g)*
*1,000,000. milligrams (mg)*
*1,000,000,000. micrograms (μg)*
*32. ounces (oz)*
*2.7 pounds (lb)*

**1 milligram (mg) =**
*0.000001 kilogram (kg)*
*0.001 gram (g)*
*1,000. micrograms (μg)*
*0.0154 grains (gr)*
*0.000032 ounce (oz)*

### Metric Volumes and Their Apothecaries' Equivalents

**1 liter (l) =**
*1,000. milliliters (ml)*
*1,000. cubic centimeters (cc)*
*2.1 pints (pt)*
*270. fluid drams (fl dr)*
*34. fluid ounces (fl oz)*

**1 milliliter (ml) =**
*0.001 liter (l)*
*1. cubic centimeter (cc)*
*16.2 minims (min)*
*0.27 fluid dram (fl dr)*
*0.034 fluid ounce (fl oz)*

### Approximate Equivalents of Household Measures

**1 teaspoon (tsp) =**
*4. milliliters (ml)*
*4. cubic centimeters (cc)*
*1. fluid dram (fl dr)*

**1 tablespoon (tbsp) =**
*15. milliliters (ml)*
*15. cubic centimeters (cc)*
*0.5 fluid ounce (fl oz)*
*3.7 teaspoons (tsp)*

**1 cup (c) =**
*240. milliliters (ml)*
*240. cubic centimeters (cc)*
*8. fluid ounces (fl oz)*
*0.5 pint (pt)*
*16. tablespoons (tbsp)*

**1 quart (qt) =**
*960. milliliters (ml)*
*960. cubic centimeters (cc)*
*2. pints (pt)*
*4. cups (c)*
*32. fluid ounces (fl oz)*

## Conversion of Units from One Form to Another

Refer to the preceding equivalency lists when converting one unit to another equivalent unit.

To convert a unit shown in bold type to one of the equivalent units listed immediately below it, multiply the first number (bold type unit) by the appropriate equivalent unit listed below it.

## Sample Problems:

1. Convert 320 grains into scruples (1 gr = 0.05 s).

$$\mathbf{320\ gr \times \frac{0.05\ s}{1\ gr} = 16.0\ s}$$

2. Convert 320 grains into drams (1 gr = 0.017 dr).

$$\mathbf{320\ gr \times \frac{0.017\ dr}{1\ gr} = 5.44\ dr}$$

3. Convert 320 grains into grams (1 gr = 0.065 g).

$$\mathbf{320\ gr \times \frac{0.065\ g}{1\ gr} = 20.8\ g}$$

## Body Temperatures in °Fahrenheit and °Celsius

| °F | °C | °F | °C |
|---|---|---|---|
| 95.0 | 35.0 | 100.0 | 37.8 |
| 95.2 | 35.1 | 100.2 | 37.9 |
| 95.4 | 35.2 | 100.4 | 38.0 |
| 95.6 | 35.3 | 100.6 | 38.1 |
| 95.8 | 35.4 | 100.8 | 38.2 |
| 96.0 | 35.5 | 101.0 | 38.3 |
| 96.2 | 35.7 | 101.2 | 38.4 |
| 96.4 | 35.8 | 101.4 | 38.6 |
| 96.6 | 35.9 | 101.6 | 38.7 |
| 96.8 | 36.0 | 101.8 | 38.8 |
| 97.0 | 36.1 | 102.0 | 38.9 |
| 97.2 | 36.2 | 102.2 | 39.0 |
| 97.4 | 36.3 | 102.4 | 39.1 |
| 97.6 | 36.4 | 102.6 | 39.2 |
| 97.8 | 36.6 | 102.8 | 39.3 |
| 98.0 | 36.7 | 103.0 | 39.4 |
| 98.2 | 36.8 | 103.2 | 39.6 |
| 98.4 | 36.9 | 103.4 | 39.7 |
| 98.6 | 37.0 | 103.6 | 39.8 |
| 98.8 | 37.1 | 103.8 | 39.9 |
| 99.0 | 37.2 | 104.0 | 40.0 |
| 99.2 | 37.3 | 104.2 | 40.1 |
| 99.4 | 37.4 | 104.4 | 40.2 |
| 99.6 | 37.6 | 104.6 | 40.3 |
| 99.8 | 37.7 | 104.8 | 40.4 |
| | | 105.0 | 40.6 |

**To convert °F to °C**

Subtract 32 from °F and multiply by 5/9.

___ °F – 32 × 5/9 = ___ °C

**To convert °C to °F**

Multiply °C by 9/5 and add 32.

___ °C × 9/5 + 32 = ___ °F

# APPENDIX C

## Laboratory Tests of Clinical Importance

**Common Tests Performed on Blood**

| *Test* | *Normal Range* (adult)* | *Clinical Significance* |
|---|---|---|
| Albumin (serum) | 3.2–5.5 gm/100 ml | Values increase in multiple myeloma and decrease with proteinuria and as a result of severe burns. |
| Albumin-globulin ratio, or A/G ratio (serum) | 1.5:1 to 2.5:1 | Ratio of albumin to globulin is lowered in kidney diseases and malnutrition. |
| Ammonia | 12–55 μ mol/l | Values increase in severe liver disease, pneumonia, shock, and congestive heart failure. |
| Amylase (serum) | 4–25 units/ml | Values increase in acute pancreatitis, intestinal obstructions, and mumps. They decrease in chronic pancreatitis, cirrhosis of the liver, and toxemia of pregnancy. |
| Bilirubin, total (serum) | 0–1.0 mg/100 ml | Values increase in conditions causing red blood cell destruction or biliary obstruction. |
| Blood urea nitrogen, or BUN (plasma or serum) | 8–25 mg/100 ml | Values increase in various kidney disorders and decrease in liver failure and during pregnancy. |
| Calcium (serum) | 8.5–10.5 mg/100 ml | Values increase in hyperparathyroidism, hypervitaminosis D, and respiratory conditions that cause a rise in $CO_2$ concentration. They decrease in hypoparathyroidism, malnutrition, and severe diarrhea. |
| Carbon dioxide (serum) | 24–30 mEq/l | Values increase in respiratory diseases, intestinal obstruction, and vomiting. They decrease in acidosis, nephritis, and diarrhea. |
| Chloride (serum) | 100–106 mEq/l | Values increase in nephritis, Cushing's syndrome, dehydration, and hyperventilation. They decrease in metabolic acidosis, Addison's disease, diarrhea, and following severe burns. |
| Cholesterol, total (serum) | 120–220 mg/100 ml (below 200 mg/100 ml recommended by the American Heart Association) | Values increase in diabetes mellitus and hypothyroidism. They decrease in pernicious anemia, hyperthyroidism, and acute infections. |
| Cholesterol, high-density lipoprotein (HDL) | Women: 30–80 mg/100 ml<br>Men: 30–70 mg/100 ml | Values increase in liver disease. Decreased values are associated with an increased risk of atherosclerosis. |
| Cholesterol, low-density lipoprotein (LDL) | 62–185 mg/100 ml | Increased values are associated with an increased risk of atherosclerosis. |
| Creatine (serum) | 0.2–0.8 mg/100 ml | Values increase in muscular dystrophy, nephritis, severe damage to muscle tissue, and during pregnancy. |
| Creatinine (serum) | 0.6–1.5 mg/100 ml | Values increase in various kidney diseases. |
| Erythrocyte count, or red cell count (whole blood) | Men: 4,600,000–6,200,000/cu mm<br>Women: 4,200,000–5,400,000/cu mm<br>Children: 4,500,000–5,100,000/cu mm (varies with age) | Values increase as a result of severe dehydration or diarrhea, and decrease in anemia, leukemia, and following severe hemorrhage. |

*These values may vary with hospital, physician, and type of equipment used to make measurements.

| Common Tests Performed on Blood—*continued* | | |
|---|---|---|
| ***Test*** | ***Normal Range* (adult)*** | ***Clinical Significance*** |
| Ferritin (serum) | Men: 10–270 μg/100 ml<br>Women: 5–280 μg/100 ml | Values correlate with total body iron store. They decrease with iron deficiency. |
| Globulin (serum) | 2.3–3.5 gm/100 ml | Values increase as a result of chronic infections. |
| Glucose (plasma) | 70–110 mg/100 ml | Values increase in diabetes mellitus, liver diseases, nephritis, hyperthyroidism, and pregnancy. They decrease in hyperinsulinism, hypothyroidism, and Addison's disease. |
| Hematocrit (whole blood) | Men: 40–54 ml/100 ml<br>Women: 37–47 ml/100 ml<br>Children: 35–49 ml/100 ml<br>(varies with age) | Values increase in polycythemia due to dehydration or shock. They decrease in anemia and following severe hemorrhage. |
| Hemoglobin (whole blood) | Men: 14–18 gm/100 ml<br>Women: 12–16 gm/100 ml<br>Children: 11.2–16.5 gm/100 ml<br>(varies with age) | Values increase in polycythemia, obstructive pulmonary diseases, congestive heart failure, and at high altitudes. They decrease in anemia, pregnancy, and as a result of severe hemorrhage or excessive fluid intake. |
| Iron (serum) | 50–150 μg/100 ml | Values increase in various anemias and liver disease. They decrease in iron deficiency anemia. |
| Iron-binding capacity (serum) | 250–410 μg/100 ml | Values increase in iron deficiency anemia and pregnancy. They decrease in pernicious anemia, liver disease, and chronic infections. |
| Lactic acid (whole blood) | 0.6–1.8 mEq/l | Values increase with muscular activity and in congestive heart failure, severe hemorrhage, and shock. |
| Lactic dehydrogenase, or LDH (serum) | 45–90 U/l | Values increase in pernicious anemia, myocardial infarction, liver disease, acute leukemia, and widespread carcinoma. |
| Lipids, total (serum) | 450–850 mg/100 ml | Values increase in hypothyroidism, diabetes mellitus, and nephritis. They decrease in hyperthyroidism. |
| Magnesium | 1.3–2.1 mEq/l | Values increase in renal failure, hypothyroidism, and Addison's disease. They decrease in renal disease, liver disease, and pancreatitis. |
| Mean corpuscular hemoglobin (MCH) | 26–32 pg/RBC | Values increase in macrocytic anemia. They decrease in microcytic anemia. |
| Mean corpuscular volume (MCV) | 86–98 μ mm³/RBC | Values increase in liver disease and pernicious anemia. They decrease in iron deficiency anemia. |
| Osmolality | 275–295 osmol/kg | Values increase in dehydration, hypercalcemia, and diabetes mellitus. They decrease in hyponatremia, Addison's disease, and water intoxication. |
| Oxygen saturation (whole blood) | Arterial: 96–100%<br>Venous: 60–85% | Values increase in polycythemia and decrease in anemia and obstructive pulmonary diseases. |
| pH (whole blood) | 7.35–7.45 | Values increase due to vomiting, Cushing's syndrome, and hyperventilation. They decrease as a result of hypoventilation, severe diarrhea, Addison's disease, and diabetic acidosis. |
| Phosphatase acid (serum) | Women: 0.01–0.56 Sigma U/ml<br>Men: 0.13–0.63 Sigma U/ml | Values increase in cancer of the prostate gland, hyperparathyroidism, certain liver diseases, myocardial infarction, and pulmonary embolism. |
| Phosphatase, alkaline (serum) | 13–39 U/l | Values increase in hyperparathyroidism (and in other conditions that promote resorption of bone), liver diseases, and pregnancy. |

*These values may vary with hospital, physician, and type of equipment used to make measurements.

**Common Tests Performed on Blood—*continued***

| *Test* | *Normal Range* (adult)* | *Clinical Significance* |
|---|---|---|
| Phosphorus (serum) | 3.0–4.5 mg/100 ml | Values increase in kidney diseases, hypoparathyroidism, acromegaly, and hypervitaminosis D. They decrease in hyperparathyroidism. |
| Platelet count (whole blood) | 130,000–360,000/cu mm | Values increase in polycythemia and certain anemias. They decrease in acute leukemia and aplastic anemia. |
| Potassium (serum) | 3.5–5.0 mEq/l | Values increase in Addison's disease, hypoventilation, and conditions that cause severe cellular destruction. They decrease in diarrhea, vomiting, diabetic acidosis, and chronic kidney disease. |
| Protein, total (serum) | 6.0–8.4 gm/100 ml | Values increase in severe dehydration and shock. They decrease in severe malnutrition and hemorrhage. |
| Prothrombin time (serum) | 12–14 sec (one stage) | Values increase in certain hemorrhagic diseases, liver disease, vitamin K deficiency, and following the use of various drugs. |
| Red cell distribution width (RDW) | 8.5–11.5 microns | Variation in cell width changes with pernicious anemia. |
| Sedimentation rate, erythrocyte (whole blood) | Men: 1–13 mm/hr<br>Women: 1–20 mm/hr | Values increase in infectious diseases, menstruation, pregnancy, and as a result of severe tissue damage. |
| Serum glutamic pyruvic transaminase (SGPT) | Women: 4–17 U/l<br>Men: 6–24 U/l | Values increase in liver disease, pancreatitis, and acute myocardial infarction. |
| Sodium (serum) | 135–145 mEq/l | Values increase in nephritis and severe dehydration. They decrease in Addison's disease, myxedema, kidney disease, and diarrhea. |
| Thromboplastin time, partial (plasma) | 35–45 sec | Values increase in deficiencies of blood factors VIII, IX, and X. |
| Thyroid-stimulating hormone (TSH) | 0.5–5.0 μU/ml | Values increase in hypothyroidism and decrease in hyperthyroidism. |
| Thyroxine, or $T_4$ (serum) | 4–12 μg/100 ml | Values increase in hyperthyroidism and pregnancy. They decrease in hypothyroidism. |
| Transaminases, or SGOT (serum) | 7–27 units/ml | Values increase in myocardial infarction, liver disease, and diseases of skeletal muscles. |
| Triglycerides | 40–150 mg/100 ml | Values increase in liver disease, nephrotic syndrome, hypothyroidism, and pancreatitis. They decrease in malnutrition and hyperthyroidism. |
| Triiodothyronine, or T3 (serum) | 75–195 ng/100 ml | Values increase in hyperthyroidism and decrease in hypothyroidism. |
| Uric acid (serum) | Men: 2.5–8.0 mg/100 ml<br>Women: 1.5–6.0 mg/100 ml | Values increase in gout, leukemia, pneumonia, toxemia of pregnancy, and as a result of severe tissue damage. |
| White blood cell count, differential (whole blood) | Neutrophils 54–62%<br>Eosinophils 1–3%<br>Basophils <1%<br>Lymphocytes 25–33%<br>Monocytes 3–9% | Neutrophils increase in bacterial diseases; lymphocytes and monocytes increase in viral diseases; eosinophils increase in collagen diseases, allergies, and in the presence of intestinal parasites. |
| White blood cell count, total (whole blood) | 5,000–10,000/cu mm | Values increase in acute infections, acute leukemia, and following menstruation. They decrease in aplastic anemia and as a result of drug toxicity. |

*These values may vary with hospital, physician, and type of equipment used to make measurements.

| Common Tests Performed on Urine | | |
|---|---|---|
| *Test* | *Normal Range** | *Clinical Significance* |
| Acetone and acetoacetate | 0 | Values increase in diabetic acidosis. |
| Albumin, qualitative | 0 to trace | Values increase in kidney disease, hypertension, and heart failure. |
| Ammonia | 20–70 mEq/l | Values increase in diabetes mellitus and liver diseases. |
| Bacterial count | Under 10,000/ml | Values increase in urinary tract infection. |
| Bile and bilirubin | 0 | Values increase in melanoma and biliary tract obstruction. |
| Calcium | Under 300 mg/24 hr | Values increase in hyperparathyroidism and decrease in hypoparathyroidism. |
| Creatinine (24 hours) | 15–25 mg/kg body weight/day | Values increase in infections, and decrease in muscular atrophy, anemia, leukemia, and kidney diseases. |
| Creatinine clearance (24 hours) | 100–140 ml/min | Values increase in renal diseases. |
| Glucose | 0 | Values increase in diabetes mellitus and various pituitary gland disorders. |
| Hemoglobin | 0 | Blood may occur in urine as a result of extensive burns, crushing injuries, hemolytic anemia, or blood transfusion reactions. |
| 17-hydroxycorticosteroids | 3–8 mg/24 hr | Values increase in Cushing's syndrome and decrease in Addison's disease. |
| Osmolality | 850 osmol/kg | Values increase in hepatic cirrhosis, congestive heart failure, and Addison's disease. They decrease in hypokalemia, hypercalcemia, and diabetes insipidus. |
| pH | 4.6–8.0 | Values increase in urinary tract infections and chronic renal failure. They decrease in diabetes mellitus, emphysema, and starvation. |
| Phenylpyruvic acid | 0 | Values increase in phenylketonuria. |
| Specific gravity (SG) | 1.003–1.035 | Values increase in diabetes mellitus, nephrosis, and dehydration. They decrease in diabetes insipidus, glomerulonephritis, and severe renal injury. |
| Urea | 25–35 gm/24 hr | Values increase as a result of excessive protein breakdown. They decrease as a result of impaired renal function. |
| Urea clearance | Over 40 ml blood cleared of urea/min | Values increase in renal diseases. |
| Uric acid | 0.6–1.0 gm/24 hr as urate | Values increase in gout and decrease in various kidney diseases. |
| Urobilinogen | 0–4 mg/24 hr | Values increase in liver diseases and hemolytic anemia. They decrease in complete biliary obstruction and severe diarrhea. |

*These values may vary with hospital, physician, and type of equipment used to make measurements.

# Appendix D

## Aids to Understanding Words

***a-*** negative prefix: amorphous
***ab-*** away from: abduction
***acet-*** vinegar: acetabulum
***ad-*** to: adduction
***adip-*** fat: adipose tissue
***aer-*** air: aerobic respiration
***af-*** to: afferent arteriole
***agglutin-*** to glue together: agglutination
***alb-*** white: albino
***aliment-*** food: alimentary canal
***allant-*** sausage: allantois
***alve-*** trough, channel, cavity: alveolus
***an-***[1] *see* a-
***an-***[2] *see* ana-
***ana-*** up, positive: anabolic
***andr-*** man: androgens
***angi-*** vessel: angiogram
***annul-*** ring: annular ligament
***ante-*** before: antebrachium
***append-*** to attach, hang, or fix to: appendicular
***-ar*** pertaining to: lobar bronchus
***arbor-*** tree: arbor vitae
***arc-*** bow: arciform artery
***areol-*** open space: areola
***arthr-*** joint: arthrology
***astr-*** star: astrocyte
***athero-*** porridge: atherosclerosis
***atri-*** entrance: atrium
***aud-*** hearing: auditory canal
***aur-*** ear: auricle
***auto-*** self: autoimmune
***ax-*** axis: axial skeleton
***axill-*** armpit: axillary artery
***bas-*** base: basilar artery
***bi-*** two: bicuspid valve
***bicep-*** two heads: biceps brachii
***bil-*** bile: bilirubin
***bio-*** life: biology
***-blast*** bud, child, a growing thing in early stages: osteoblast
***blephar-*** eyelid: blepharitis
***brachi-*** arm: brachial region
***brady-*** slow: bradycardia
***bronch-*** windpipe: bronchus
***burs-*** bag, purse: bursa
***calat-*** something inserted: intercalated disk
***calc-*** heel: calcaneus
***calor-*** heat: calorie
***calyc-*** cup: major calyx
***canal-*** channel: canaliculus
***carcin-*** crab, cancer: carcinoma
***cardi-*** heart: pericardium
***cari-*** decay: dental caries
***carin-*** hull, keel: carina
***carot-*** carrot: carotene
***carpus-*** wrist: carpals
***cata-*** down: catabolic
***caud-*** tail: cauda equina
***cav-*** hollow: corpora cavernosa
***cec-*** blind: cecum
***centr-*** point, center: centromere
***cephal-*** head: cephalic region
***cerebr-*** brain: cerebrum
***cervic-*** neck: cervical vertebrae
***chiasm-*** cross-shaped: optic chiasma
***chondr-*** cartilage: chondrocyte
***chori-*** protective fetal membrane: chorion
***chrom-*** color: chromatin
***chym-*** juice: chyme
***-clas*** break: osteoclast
***clav-*** bar: clavicle
***cleav-*** to divide: cleavage
***co-*** with, together: coenzyme
***cochlea*** snail: cochlea
***col-*** lower intestine: colon
***condyl-*** knuckle: condyle
***contra-*** against, counter: contraception
***corac-*** raven: coracoid process
***corn-*** horn: cornea
***corpor-*** body: corpus callosum
***cortic-*** bark, rind: cortex
***crani-*** skull: cranial
***cribrum-*** sieve: cribriform plate
***cric-*** ring-shaped: cricoid cartilage
***-crin*** distinguish, separate off: endocrine
***crista*** crest: crista galli
***crur-*** shin, leg: crura
***cusp-*** point, apex: tricuspid valve
***cut-*** skin: subcutaneous
***cyst-*** bladder: cystitis
***cyt-*** cell: cytoplasm
***dart-*** like skin: dartos muscle
***de-*** to remove: deamination
***decidu-*** falling off: deciduous
***deglut-*** to get rid of: deglutination
***dendr-*** tree: dendrite
***derm-*** skin: dermis
***detrus-*** to force away: detrusor muscle
***di-*** two: diencephalon
***diastol-*** dilation: diastole
***digit-*** finger: digital artery
***diuret-*** to pass urine: diuretic
***dors-*** back: dorsal
***ect-*** outside: ectoderm
***ectop-*** out of place: ectopic beat
***ede-*** swelling: edema
***ejacul-*** to shoot forth: ejaculation
***embol-*** stopper: embolus
***-emia*** in the blood: hypoproteinemia
***encephal-*** the brain: encephalitis
***end-*** inside: endoplasmic reticulum
***ependym-*** tunic: ependyma
***epi-*** upon, after, in addition: epithelial tissue
***erg-*** work, deed: synergist
***erythr-*** red: erythrocyte
***ethmo-*** sieve: ethmoid bone
***exo-*** outside: exocrine gland
***extra-*** outside of, beyond: extracellular
***falx-*** a sickle: falx cerebelli
***fasci-*** band: fasciculus
***femur-*** thigh: femur
***fimb-*** fringe: fimbriae
***flacc-*** flabby: flaccid
***flagell-*** whip: flagellum
***follic-*** small bag: hair follicle
***fore-*** before, the front part of: forebrain
***foss-*** ditch, trench: fossa ovalis
***fovea-*** pit: fovea capitis
***frenul-*** bridle, restraint: frenulum
***funi-*** small cord or fiber: funiculus
***gangli-*** swelling: ganglion
***gastr-*** stomach: gastric gland
***-gen*** become, be produced: allergen
***genesis*** origin: spermatogenesis
***gladi-*** sword: gladiolus
***glen-*** joint socket: glenoid cavity
***-glia*** glue: neuroglia
***glom-*** little ball: glomerulus
***glut-*** buttocks: gluteus maximus
***glyc-*** sweet: glycogen
***gracil-*** slender: gracilis
***-gram*** letter, drawing: myogram
***gubern-*** to steer, to guide: gubernaculum
***gust-*** to taste: gustatory receptor
***gyn-*** woman: gynecology
***hem-*** blood: hemoglobin
***hemi-*** half: hemiplegia
***hepar-*** liver: heparin
***hepat-*** liver: hepatic duct
***hetero-*** other, different: heterozygous
***hiat-*** opening: esophageal hiatus
***hist-*** web, tissue: histology
***hol-*** entire, whole: holocrine gland
***hom-*** same, common: homeostasis
***horm-*** impetus, impulse: hormone

***humor-*** moisture, fluid: humoral
***hy-*** U-shaped: hyoid bone
***hyal-*** glass: hyaline cartilage
***hyper-*** above, beyond: hypertonic
***hypo-*** under, below: hypotonic
***im-*** negative prefix: imbalance
***immun-*** free, exempt: immunity
***in-*** *see* im-
***inflamm-*** to set on fire: inflammation
***infra-*** beneath: infraorbital nerve
***inhal-*** to breathe in: inhalation
***insul-*** island: insula
***inter-*** among, between: interphase
***intra-*** inside: intracellular
***irid-*** rainbow, colored circle: iris
***is-*** equal: isotope
***jugul-*** neck, throat: jugular vein
***junct-*** yoke, join: junctional fiber
***juxta-*** near, beside: juxtaglomerular nephron
***kerat-*** horn: keratin
***kyph-*** hump, humpbacked: kyphosis
***labi-*** lip: labia
***labr-*** lip: glenoidal labrum
***labyrinth-*** maze: bony labyrinth
***lacrima-*** tears: lacrimal gland
***lact-*** milk: prolactin
***lacun-*** pool: lacuna
***lamell-*** thin plate: lamella
***lanug-*** down: lanugo
***laten-*** hidden: latent
***-lemm*** rind, peel: neurolemma
***leuk-*** white: leukocyte
***lingu-*** tongue: lingual tonsil
***lip-*** fat: lipids
***-logy*** study of: physiology
***lord-*** bent back: lordosis
***lun-*** moon: semilunar valve
***lute-*** yellow: macula lutea
***ly-*** loose, dissolve: lysosome
***-lyt*** dissolvable: electrolyte
***macr-*** large: macrophage
***macula*** spot, stain: macula lutea
***mal-*** bad, abnormal: malnutrition
***malle-*** hammer: malleus
***mamm-*** breast: mammary gland
***man-*** hand: manubrium
***mandib-*** jaw: mandible
***mast-*** breast: mastitis
***maxill-*** jawbone: maxillary bone
***meat-*** a passage: auditory meatus
***medi-*** middle: adrenal medulla
***melan-*** black: melanin
***men-*** month: menstrual
***mening-*** membrane: meninges
***ment-*** mind: mental foramen
***mes-*** middle: mesoderm
***meta-*** after, beyond, accompanying: metabolism
***meter*** measure: calorimeter
***micr-*** small: microfilament
***mict-*** to pass urine: micturition
***mid-*** middle: midbrain
***milli-*** one-thousandth: millimeter
***mit-*** thread: mitosis
***mon-*** one: monozygotic
***mons-*** an eminence: mons pubis
***morul-*** mulberry: morula
***mot-*** move: motor neuron
***multi-*** many: multipolar neuron
***mut-*** change: mutation
***my-*** muscle: myofibril
***nas-*** nose: nasal
***nat-*** to be born: prenatal
***ne-*** new, young: neonatal period
***nephr-*** kidney: nephron
***neur-*** nerve: neuron
***neuter*** neither one nor the other: neutral
***nod-*** knot: nodule
***nucle-*** kernel: nucleus
***nutri-*** nourish: nutrient
***obes-*** fat: obesity
***occipit-*** back of the head: occipital lobe
***oculi-*** eye: orbicularis oculi
***odont-*** tooth: odontoid process
***-oid*** form: odontoid process
***olfact-*** to smell: olfactory
***olig-*** few, small: oligodendrocyte
***oo-*** egg: oogenesis
***or-*** mouth: oral cavity
***orb-*** circle: orbital
***-osis*** abnormal condition: leukocytosis
***oss-*** bone: osseous tissue
***-ous*** having qualities of: cancerous
***ov-*** egg: synovial fluid
***palpebra*** eyelid: levator palpebrae superioris
***papill-*** nipple: papillary muscle
***para-*** beside: parathyroid glands
***pariet-*** wall: parietal membrane
***path-*** disease, sickness: pathogen
***pell-*** skin: pellagra
***pelvi-*** basin: pelvic cavity
***peri-*** around: pericardial membrane
***pept-*** digest: peptic ulcer
***phag-*** eat: phagocytosis
***phen-*** show, be seen: phenotype
***phleb-*** vein: phlebitis
***phot-*** light: photoreceptor
***phren-*** mind, midriff: phrenic nerve
***pia-*** pious, gentle: pia mater
***pino-*** drink: pinocytosis
***pleur-*** rib, side: pleural membrane
***plex-*** strike: choroid plexus
***plic-*** fold: plicae circularis
***poie-*** make, produce: hematopoiesis
***poly-*** many: polyunsaturated
***popl-*** ham (knee): popliteal artery
***por-*** passage, channel: pore
***post-*** after: postnatal period
***pre-*** before: prepatellar bursa
***prime*** first: primordial follicle
***pro-*** before: prophase
***prox-*** nearest: proximal
***pseud-*** false: pseudostratified epithelium
***pter-*** wing: pterygoid process
***puber-*** adult: puberty
***pulmo-*** lung: pulmonary artery
***pyl-*** door, orifice: pyloric sphincter
***quadr-*** four: quadriceps femoris
***ramus*** branch: gray ramus
***rect-*** straight: rectum
***ren-*** kidney: renal cortex
***ret-*** net: rete testis
***reticul-*** network: sarcoplasmic reticulum
***rhin-*** nose: rhinitis
***sacchar-*** sugar: monosaccharide
***saltator*** dancer: saltatory conduction
***sarc-*** flesh: sarcoplasm
***scler-*** hard: sclera
***seb-*** grease: sebaceous gland
***sect-*** cut: section
***sella-*** saddle-shaped: sella turcica
***semi-*** half: semitendinosus
***sen-*** old: senescence
***sens-*** feeling: sensory neuron
***sin-*** hollow: sinus
***-some*** body: ribosome
***sorpt-*** to soak up: absorption
***squam-*** scale: squamous epithelium
***sta-*** halt, make stand: hemostasis
***strat-*** spread out: substrate
***stria-*** groove: striated muscle
***sub-*** under: substrate
***sulc-*** furrow: sulcus
***super-*** above: superior
***supra-*** above: supraspinatus muscle
***sutur-*** sewing: suture
***syn-*** together: synthesis
***syndesm-*** binding together: syndesmosis
***systol-*** contraction: systole
***tachy-*** rapid: tachycardia
***tal-*** ankle: talus
***tetan-*** stiff: tetanic
***thalam-*** chamber: thalamus
***theo-*** sheath: theca externa
***therm-*** heat: thermoreceptor
***thromb-*** clot: thrombocyte
***thyr-*** shield: thyroid gland
***toc-*** childbirth: oxytocin
***tom-*** cut: anatomy
***ton-*** stretch: isotonic
***trans-*** across: transverse
***tri-*** three: tricuspid valve

***trigon-*** triangle: trigone
***trop-*** turn, react: adrenocorticotropic
***troph-*** nurture: hypertrophy
***tuber-*** swelling: tuberculosis
***tympan-*** drum: tympanic membrane
***umbil-*** navel: umbilical cord
***un-*** one: unipolar neuron
***ur-*** urine: ketonuria
***uter-*** womb: uterine tube
***vag-*** to wander: vagus nerve
***valent*** having power: electrovalent
***vas-*** vessel: vasa recta
***vesic-*** bladder: vesicle
***vill-*** hair: villus
***visc-*** internal organs: viscera
***vitre-*** glass: vitreous humor
***voluntar-*** of free will: voluntary muscle
***xiph-*** sword: xiphoid process
***zon-*** belt: zona pellucida
***zym-*** ferment: enzyme

# GLOSSARY

Each word in this glossary is followed by a phonetic guide to pronunciation. In this guide, any unmarked vowel that ends a syllable or stands alone as a syllable has the long sound. Thus, the word *play* would be spelled *pla*. Any unmarked vowel that is followed by a consonant has the short sound. The word *tough*, for instance, would be spelled *tuf*. If a long vowel appears in the middle of a syllable (followed by a consonant), it is marked with the macron (¯), the sign for a long vowel. Thus, the word *plate* would be phonetically spelled *plāt*. Similarly, if a vowel stands alone or ends a syllable, but has the short sound, it is marked with a breve (˘).

## A

**abdomen** (ab-do′men) Portion of the body between the diaphragm and the pelvis.

**abdominal** (ab-dom′ĭ-nal) Pertaining to the abdomen.

**abdominal cavity** (ab-dom′ ĭ-nal kav′ ĭ-te) The space between the diaphragm and the pelvic inlet that contains the abdominal viscera.

**abdominopelvic cavity** (ab-dom″ ĭ-no-pel′vik kav′ĭ-te) The space between the diaphragm and the lower portion of the trunk of the body.

**abduction** (ab-duk′shun) Movement of a body part away from the midline.

**absorption** (ab-sorp′shun) The taking in of substances by cells or membranes.

**accessory organs** (ak-ses′o-re or′ganz) Organs that supplement the functions of other organs; accessory organs of the digestive and reproductive systems.

**accommodation** (ah-kom″o-da′shun) Adjustment of the lens for close vision.

**acetone** (as′e-tōn) One of the ketone bodies produced as a result of the oxidation of fats.

**acetylcholine** (as″ĕ-til-ko′lēn) A type of neurotransmitter, which is a biochemical secreted at the axon ends of many neurons. It transmits nerve impulses across synapses.

**acetyl coenzyme A** (as′ĕ-til ko-en′zīm) An intermediate compound produced during the oxidation of carbohydrates and fats.

**acid** (as′id) A substance that ionizes in water to release hydrogen ions.

**acid-base buffer system** (as′id-bās buf′er sis′tem) A pair of chemicals, one a weak acid, the other a weak base, that resists pH changes.

**acidosis** (as″ ĭ-do′sis) A relative increase in the acidity of body fluids.

**acromial** (ah-kro′me-al) Pertaining to the shoulder.

**ACTH** Adrenocorticotropic hormone.

**actin** (ak′tin) A protein in a muscle fiber that, together with myosin, is responsible for contraction and relaxation.

**action potential** (ak′shun po-ten′shal) The sequence of electrical changes occurring when a nerve cell membrane is exposed to a stimulus that exceeds its threshold.

**activation energy** (ak″tĭ-va′shun en′er-je) Energy required to initiate a chemical reaction.

**active site** (ak′tiv sīt) Region of an enzyme molecule that combines temporarily with a substrate.

**active transport** (ak′tiv trans′port) Process that requires an expenditure of energy to move a substance across a cell membrane; usually moved against the concentration gradient.

**acupuncture** (ak′u-pungk″chūr) Procedure in which needles are inserted into various tissues to control pain sensations.

**adaptation** (ad″ap-ta′shun) Adjustment to environmental conditions.

**adduction** (ah-duk′shun) Movement of a body part toward the midline.

**adenoids** (ad ĕ-noids) The pharyngeal tonsils located in the nasopharynx.

**adenosine diphosphate** (ah-den′o-sēn di-fos′fāt) ADP; molecule produced when the terminal phosphate is lost from a molecule of adenosine triphosphate.

**adenosine triphosphate** (ah-den′o-sēn tri-fos′fāt) An organic molecule that stores energy and releases energy for use in cellular processes.

**adenylate cyclase** (ah-den′ĭ-lāt si′klās) An enzyme that is activated when certain hormones combine with receptors on cell membranes, causing ATP to release two phosphates and circularize, forming cyclic AMP.

**ADH** Antidiuretic hormone.

**adipose tissue** (ad′ĭ-pōs tish′u) Fat-storing tissue.

**adolescence** (ad″o-les′ens) Period of life between puberty and adulthood.

**ADP** Adenosine diphosphate.

**adrenal cortex** (ah-dre′nal kor′teks) The outer portion of the adrenal gland.

**adrenal glands** (ah-dre′nal glandz) Endocrine glands located on the superior portions of the kidneys.

**adrenaline** (ah-dren′ah-lin) Epinephrine. A hormone produced by the adrenal glands.

**adrenal medulla** (ah-dre′nal me-dul′ah) The inner portion of the adrenal gland.

**adrenergic fiber** (ad″ren-er′jik fi′ber) A nerve fiber that secretes norepinephrine at the axon terminal.

**adrenocorticotropic hormone** (ah-dre″no-kor″te-ko-trōp′ik hor′mōn) ACTH; hormone secreted by the anterior lobe of the pituitary gland that stimulates activity in the adrenal cortex.

**aerobic respiration** (a″er-ōb′ik res″pĭ-ra′shun) Phase of cellular respiration that requires oxygen.

**afferent** (af′er-ent) Conducting toward a center. For example, an afferent arteriole conveys blood to the glomerulus of a nephron within the kidney.

**agglutination** (ah-gloo″ti-na′shun) Clumping of blood cells in response to a reaction between an antibody and an antigen.

**agranulocyte** (a-gran′u-lo-sīt) A nongranular leukocyte.

**albumin** (al-bu′min) A plasma protein that helps regulate the osmotic concentration of blood.

**aldosterone** (al-dos′ter-ōn) A hormone, secreted by the adrenal cortex, that regulates sodium and potassium ion concentrations and water balance.

**alimentary canal** (al″i-men′tar-e kah-nal′) The tubular portion of the digestive tract that leads from the mouth to the anus.

**alkaline** (al′kah-līn) Pertaining to or having the properties of a base or alkali; basic.

**alkaline tide** (al′kah-lin tīd) An increase in the blood concentration of bicarbonate ions following a meal.

**alkaloid** (al′kah-loid) A group of organic substances that are usually bitter in taste and have toxic effects.

**alkalosis** (al″kah-lo′sis) A relative increase in the alkalinity of body fluids.

**allantois** (ah-lan′to-is) A structure that appears during embryonic development and functions in the formation of umbilical blood vessels.

**allele** (ah-lēl) Different forms of a gene.

**allergen** (al′er-jen) A foreign substance capable of stimulating an allergic reaction.

**all-or-none response** (al′or-nun′ re-spons′) Phenomenon in which a muscle fiber contracts completely when it is exposed to a stimulus of threshold strength.

**alpha receptor** (al′fah re-sep′tor) Receptor on effector cell membrane that combines with epinephrine or norepinephrine.

**alpha-tocopherol** (al″fah-to-kof′er-ol) Vitamin E.

**alveolar ducts** (al-ve′o-lar dukts′) Fine tubes that carry air to the air sacs of the lungs.

**alveolar pores** (al-ve′o-lar pōrz) Minute openings in the walls of air sacs, which permit air to pass from one alveolus to another.

**alveolar process** (al-ve′o-lar pros′es) Projection on the border of the jaw in which the bony sockets of the teeth are located.

**alveolus** (al-ve′o-lus) An air sac of a lung; a saclike structure (pl. *alveoli*).

**amacrine cell** (am′ah-krin sel) A retinal neuron whose fibers pass laterally between other retinal cells.

**amine** (am′in) A type of nitrogen-containing organic compound, including the hormones secreted by the adrenal medulla.

**amino acid** (ah-me′no as′id) An organic compound of relatively small molecular size that contains an amino group ($-NH_2$) and a carboxyl group ($-COOH$); the structural unit of a protein molecule.

**amniocentesis** (am″ne-o-sen-te′sis) A procedure in which a sample of amniotic fluid is removed through the abdominal wall of a pregnant woman. Fetal cells in it are cultured and examined to check the chromosome complement.

**amnion** (am′ne-on) An extra embryonic membrane that encircles a developing fetus and contains amniotic fluid.

**amniotic cavity** (am″ne-ot′ik kav′ĭ-te) Fluid-filled space enclosed by the amnion.

**amniotic fluid** (am″ne-ot′ik floo′id) Fluid within the amniotic cavity that surrounds the developing fetus.

**ampulla** (am-pul′ah) An expansion at the end of each semicircular canal that contains a crista ampullaris.

**amylase** (am′ĭ-lās) An enzyme that hydrolyzes starch.

**anabolism** (ah-nab′o-liz″em) Metabolic process by which larger molecules are synthesized from smaller ones; anabolic metabolism.

**anaerobic respiration** (an-a″er-ōb′ik res″pĭ-ra′shun) Phase of cellular respiration that occurs in the absence of oxygen.

**anal canal** (a′nal kah-nal′) The most distal two or three inches of the large intestine that open to the outside as the anus.

**anaphase** (an′ah-fāz) Stage in mitosis when duplicate chromosomes move to opposite poles of the cell.

**anastomosis** (ah-nas″to-mo′sis) A union of nerve fibers or blood vessels to form a network.

**anatomical position** (an″ah-tom′ĭ-kal po-zish′un) A body posture with the body erect, the face forward, the arms at the sides with the palms facing forward, and the toes pointing straight ahead.

**anatomy** (ah-nat′o-me) Branch of science dealing with the form and structure of body parts.

**androgen** (an′dro-jen) A male sex hormone such as testosterone.

**anemia** (ah-ne′me-ah) A condition of red blood cell or hemoglobin deficiency.

**aneuploid** (an′u-ploid) A cell with one or more extra or missing chromosomes.

**aneurysm** (an′u-rizm) A saclike expansion of a blood vessel wall.

**angiotensin** (an″je-o-ten′sin) A vasoconstricting biochemical that is produced when blood flow to the kidneys is reduced, elevating blood pressure.

**angiotensinogen** (an″je-o-ten-sin′o-jen) A serum globulin the liver secretes that renin converts to angiotensin.

**anion** (an′i-on) An atom carrying a negative charge due to an extra electron.

**anorexia nervosa** (an″o-rek′se-ah ner-vo′sa) Disorder caused by the fear of becoming obese; includes loss of appetite and inability to maintain a normal minimum body weight.

**anoxia** (an-ok′se-ah) Abnormally low oxygen concentration of the tissues.

**antagonist** (an-tag′o-nist) A muscle that acts in opposition to a prime mover.

**antebrachium** (an″te-bra′ke-um) The forearm.

**antecubital** (an″te-ku′bi-tal) The region in front of the elbow joint.

**anterior** (an-te′re-or) Pertaining to the front.

**anterior pituitary** (an-te′re-or pi-tu′i-tār″e) The front lobe of the pituitary gland.

**antibody** (an′ti-bod″e) A protein that B cells of the immune system produce in response to the presence of a nonself antigen; it reacts with the antigen.

**anticoagulant** (an″tĭ-ko-ag′u-lant) A biochemical that inhibits blood clotting.

**anticodon** (an″ti-ko′don) Three contiguous nucleotides of a transfer RNA molecule that are complementary to a specific mRNA codon.

**antidiuretic hormone** (an″tĭ-di″u-ret′ik hor′mōn) Hormone released from the posterior lobe of the pituitary gland that enhances the conservation of water by the kidneys; ADH.

**antigen** (an′tĭ-jen) A chemical that stimulates cells to produce antibodies.

**antigen binding site** (an′tĭ-jen bīn′ding sīt) Specialized ends of antibodies that bind specific antigens.

**antigen-presenting cell** (an′tĭ-jen-pre-sen′ting cel) The cell that displays the antigen to the cells of the immune system so they can defend the body against that particular antigen.

**antioxidant** (an′tĭ-ok′sĭ-dant) A substance that inhibits oxidation of another substance.

**antithrombin** (an″tĭ-throm′bin) A substance that inhibits the action of thrombin and thus inhibits blood clotting.

**anus** (a′nus) Inferior outlet of the digestive tube.

**aorta** (a-or′tah) Major systemic artery that receives blood from the left ventricle.

**aortic body** (a-or′tik bod′e) A structure associated with the wall of the aorta that contains a group of chemoreceptors.

**aortic sinus** (a-or′tik si′nus) Swelling in the wall of the aorta that contains baroreceptors.

**aortic valve** (a-or′tik valv) Flaplike structures in the wall of the aorta near its origin that prevent blood from returning to the left ventricle of the heart.

**apneustic area** (ap-nu′stik a′re-ah) A portion of the respiratory control center located in the pons.

**apocrine gland** (ap′o-krin gland) A type of sweat gland that responds during periods of emotional stress.

**aponeurosis** (ap″o-nu-ro′sis) A sheetlike tendon by which certain muscles are attached to other parts.

**appendicular** (ap″en-dik′u-lar) Pertaining to the arms or legs.

**appendix** (ah-pen′diks) A small, tubular appendage that extends outward from the cecum of the large intestine; vermiform appendix.

**aqueous humor** (a′kwe-us hu′mor) Watery fluid that fills the anterior and posterior chambers of the eye.

**arachnoid granulation** (ah-rak′noid gran″u-la′shun) Fingerlike structures that project from the subarachnoid space of the meninges into blood-filled dural sinuses and reabsorb cerebrospinal fluid.

**arachnoid mater** (ah-rak′noid ma′ter) Delicate, weblike middle layer of the meninges.

**arbor vitae** (ar′bor vi′ta) Treelike pattern of white matter in a section of cerebellum.

**areola** (ah-re′o-lah) Pigmented region surrounding the nipple of the mammary gland or breast.

**arrector pili muscle** (ah-rek′tor pil′i mus′l) Smooth muscle in the skin associated with a hair follicle.

**arrhythmia** (ah-rith′me-ah) Abnormal heart action characterized by a loss of rhythm.

**arterial pathway** (ar-te′re-al path′wa) Course blood follows as it flows from the heart to body cells.

**arteriole** (ar-te′re-ōl) A small branch of an artery that communicates with a capillary network.

**arteriosclerosis** (ar-te″re-o-sklĕ-ro′sis) Condition in which the walls of arteries thicken and lose their elasticity; hardening of the arteries.

**artery** (ar′ter-e) A vessel that transports blood away from the heart.

**arthritis** (ar-thri′tis) Joint inflammation.

**articular cartilage** (ar-tik′u-lar kar′tĭ-lij) Hyaline cartilage that covers the ends of bones in synovial joints.

**articulation** (ar-tik″u-la′shun) The joining of structures at a joint.

**ascending colon** (ah-send′ing ko′lon) Portion of the large intestine that passes upward on the right side of the abdomen from the cecum to the lower edge of the liver.

**ascending tracts** (ah-send′ing trakts) Groups of nerve fibers in the spinal cord that transmit sensory impulses upward to the brain.

**ascites** (ah-si′tez) Serous fluid accumulation in the abdominal cavity.

**ascorbic acid** (as-kor′bik as′id) One of the water-soluble vitamins; vitamin C.

**assimilation** (ah-sim″ĭ-la′shun) The action of chemically changing absorbed substances.

**association area** (ah-so″se-a′shun a′re-ah) Region of the cerebral cortex related to memory, reasoning, judgment, and emotional feelings.

**astigmatism** (ah-stig′mah-tizm) Visual defect due to errors in refraction caused by abnormal curvatures in the surface of the cornea or lens.

**astrocyte** (as′tro-sīt) A type of neuroglial cell that connects neurons to blood vessels.

**atherosclerosis** (ath″er-o-sklĕ-ro′sis) Condition in which fatty substances accumulate on the inner linings of arteries.

**atmospheric pressure** (at″mos-fer′ik presh′ur) Pressure exerted by the weight of the air; about 760 mm of mercury at sea level.

**atom** (at′om) Smallest particle of an element that has the properties of that element.

**atomic number** (ah-tom′ik num′ber) Number equal to the number of protons in an atom of an element.

**atomic weight** (ah-tom′ik wāt) Number approximately equal to the number of protons plus the number of neutrons in an atom of an element.

**ATP** Adenosine triphosphate, the biological energy molecule.

**ATPase** Enzyme that causes ATP molecules to release the energy stored in the terminal phosphate bonds.

**atrial natriuretic peptide** (a′trē-al-na″trē-u-ret′ik pep′tīd) A family of polypeptide hormones that increase sodium excretion.

**atrioventricular bundle** (a″tre-o-ven-trik′u-lar bun′dl) Group of specialized fibers that conduct impulses from the atrioventricular node to the ventricular muscle of the heart; A-V bundle.

**atrioventricular node** (a″tre-o-ven-trik′u-lar nōd) Specialized mass of cardiac muscle fibers located in the interatrial septum of the heart; functions in transmitting cardiac impulses from the sinoatrial node to the ventricular walls; A-V node.

**atrioventricular orifice** (a″tre-o-ven-trik′u-lar or′i-fis) Opening between the atrium and the ventricle on one side of the heart.

**atrioventricular sulcus** (a″tre-o-ven-trik′u-lar sul′kus) A groove on the surface of the heart that marks the division between an atrium and a ventricle.

**atrioventricular valve** (a″tre-o-ven-trik′u-lar valv) Cardiac valve located between an atrium and a ventricle.

**atrium** (a′tre-um) A chamber of the heart that receives blood from veins (pl. *atria*).

**atrophy** (at′ro-fe) A wasting away or decrease in size of an organ or tissue.

**audiometer** (aw″de-om′ĕ-ter) An instrument used to measure the acuity of hearing.

**auditory** (aw′di-to″re) Pertaining to the ear or to the sense of hearing.

**auditory ossicle** (aw′di-to″re os′i-kl) A bone of the middle ear.

**auditory tube** (aw′di-to″re toob) The tube that connects the middle ear cavity to the pharynx; eustachian tube.

**auricle** (aw′ri-kl) An earlike structure; the portion of the heart that forms the wall of an atrium.

**autoantibody** (aw″to-an′tĭ-bod″e) An antibody produced against oneself.

**autocrine** (aw′to-krin) A hormone that acts on the same cell that secreted it.

**autoimmunity** (aw″to-ĭ-mū′ni-tē) An immune response against a person's own tissues; autoallergy.

**autonomic nervous system** (aw″to-nom′ik ner′vus sis′tem) Portion of the nervous system that controls the actions of the visceral organs and skin.

**autoregulation** (aw″to-reg″u-la′shun) Ability of an organ or tissue to maintain a constant blood flow in spite of changing arterial blood pressure.

**autosome** (aw′to-sōm) A chromosome other than a sex chromosome.

**A-V bundle** (bun′dl) A group of fibers that conduct cardiac impulses from the A-V node to the Purkinje fibers; bundle of His.

**A-V node** (nōd) Atrioventricular node.

**axial skeleton** (ak′se-al skel′ĕ-ton) Portion of the skeleton that supports and protects the organs of the head, neck, and trunk.

**axillary** (ak′sĭ-ler″e) Pertaining to the armpit.

**axon** (ak′son) A nerve fiber that conducts a nerve impulse away from a neuron cell body.

**axonal transport** (ak′so-nal trans′port) Transport of substances from the neuron cell body to an axon terminal.

## B

**baroreceptor** (bar″o-re-sep′tor) Sensory receptor in the blood vessel wall stimulated by changes in pressure (pressoreceptor).

**basal ganglion** (bas′al gang′gle-on) Mass of gray matter located deep within a cerebral hemisphere of the brain.

**basal metabolic rate** (ba′sal met″ah-bol′ic rāt) Rate at which metabolic reactions occur when the body is at rest.

**base** (bās) A substance that ionizes in water to release hydroxyl ions ($OH^-$) or other ions that combine with hydrogen ions.

**basement membrane** (bās′ment mem′brān) A layer of nonliving material that anchors epithelial tissue to underlying connective tissue.

→ **basophil** (ba′so-fil) White blood cell containing cytoplasmic granules that stain with basophilic dye.

**beta oxidation** (ba′tah ok″sĭ-da′shun) Chemical process by which fatty acids are converted to molecules of acetyl coenzyme A.

**beta receptor** (ba′tah re-sep′tor) Receptor on effector cell membrane that combines mainly with epinephrine and only slightly with norepinephrine.

**bicarbonate buffer system** (bi-kar′bo-nāt buf′er sis′tem) A mixture of carbonic acid and sodium bicarbonate to weaken a strong base and a strong acid, respectively; to resist a change in pH.

**bicarbonate ion** (bi-kar′bon-āt i′on) $HCO_3^-$.

**bicuspid tooth** (bi-kus′pid tooth) A premolar that is specialized for grinding hard particles of food.

**bicuspid valve** (bi-kus′pid valv) Heart valve located between the left atrium and the left ventricle; mitral valve.

**bile** (bīl) Fluid secreted by the liver and stored in the gallbladder.

**bilirubin** (bil″ĭ-roo′bin) A bile pigment produced from hemoglobin breakdown.

**biliverdin** (bil″ĭ-ver′din) A bile pigment produced from hemoglobin breakdown.

**biochemistry** (bi″o-kem′is-tre) Branch of science dealing with the chemistry of living organisms.

**biofeedback** (bi″o-fēd′bak) Procedure in which electronic equipment is used to help a person learn to consciously control certain visceral responses.

**biotin** (bi′o-tin) A water-soluble vitamin; a member of the vitamin B complex.

**bipolar neuron** (bi-po′lar nu′ron) A nerve cell whose cell body has only two processes, one serving as an axon and the other as a dendrite.

**blastocyst** (blas′to-sist) An early stage of preembryonic development that consists of a hollow ball of cells.

**B lymphocyte** (lim′fo-sīt) Lymphocyte that reacts against foreign substances in the body by producing and secreting antibodies; B cell.

**BMR** Basal metabolic rate.

**bond** (bond) Connection between atoms in a compound.

**Bowman's capsule** (bo′manz kap′sūl) Proximal portion of a renal tubule that encloses the glomerulus of a nephron; glomerular capsule.

**brachial** (bra′ke-al) Pertaining to the arm.

**bradycardia** (brad″e-kar′de-ah) An abnormally slow heart rate or pulse rate.

**brain stem** (brān stem) Portion of the brain that includes the midbrain, pons, and medulla oblongata.

**brain wave** (brān wāv) Recording of fluctuating electrical activity in the brain.

**Broca's area** (bro′kahz a′re-ah) Region of the frontal lobe that coordinates complex muscular actions of the mouth, tongue, and larynx, making speech possible.

**bronchial tree** (brong′ke-al trē) The bronchi and their branches that carry air from the trachea to the alveoli of the lungs.

**bronchiole** (brong′ke-ōl) A small branch of a bronchus within the lung.

**bronchus** (brong′kus) A branch of the trachea that leads to a lung (pl. *bronchi*).

**buccal** (buk′al) Pertaining to the mouth and the inner lining of the cheeks.

**buffer** (buf′er) A substance that can react with a strong acid or base to form a weaker acid or base, and thus resist a change in pH.

**bulbourethral glands** (bul″bo-u-re′thral glandz) Glands that secrete a viscous fluid into the male urethra at times of sexual excitement; Cowper's glands.

**bulimia** (bu-lim′e-ah) A disorder of binge eating followed by purging.

**bursa** (bur′sah) A saclike, fluid-filled structure, lined with synovial membrane, near a joint.

**bursitis** (bur-si′tis) Inflammation of a bursa.

## C

**calcification** (kal″sĭ-fĭ-ka′shun) Deposition of calcium salts in a tissue.

**calcitonin** (kal″sĭ-to′nin) Hormone secreted by the thyroid gland that helps regulate the level of blood calcium.

**calorie** (kal′o-re) A unit used to measure heat energy and the energy contents of foods.

**calorimeter** (kal″o-rim′ĕ-ter) A device used to measure the heat energy content of foods; bomb calorimeter.

**canaliculus** (kan″ah-lik′u-lus) Microscopic canals that connect the lacunae of bone tissue.

**cancellous bone** (kan′sĕ-lus bōn) Bone tissue with a latticework structure; spongy bone.

**capacitation** (kah-pas″i-ta′shun) Activation of a sperm cell to fertilize an egg cell.

**capillary** (kap′ĭ-ler″e) A small blood vessel that connects an arteriole and a venule.

**carbohydrate** (kar″bo-hi′drāt) An organic compound that contains carbon, hydrogen, and oxygen, with a 2:1 ratio of hydrogen to oxygen atoms.

**carbonic anhydrase** (kar-bon′ik an-hi′drās) Enzyme that catalyzes the reaction between carbon dioxide and water to form carbonic acid.

**carbon monoxide** (kar′bon mon-ok′sīd) A toxic gas that combines readily with hemoglobin to form a relatively stable compound; CO.

**carboxyhemoglobin** (kar″bok″se-he″mo-glo′bin) Compound formed by the union of carbon dioxide and hemoglobin.

**carboxypeptidase** (kar-bok″se-pep′ti-dās) A protein-splitting enzyme in pancreatic juice.

**cardiac center** (kar′de-ak sen′ter) A group of neurons in the medulla oblongata that controls heart rate.

**cardiac conduction system** (kar′de-ak kon-duk′shun sis′tem) System of specialized cardiac muscle fibers that conducts cardiac impulses from the S-A node into the myocardium.

**cardiac cycle** (kar′de-ak si′kl) A series of myocardial contractions that constitute a complete heartbeat.

**cardiac muscle** (kar′de-ak mus′el) Specialized type of muscle tissue found only in the heart.

**cardiac output** (kar′de-ak owt′poot) A measurement calculated by multiplying the stroke volume in milliliters by the heart rate in beats per minute.

**cardiac sphincter** (kar′de-ak sfingk′ter) A ring of muscle located at the junction of the esophagus and stomach that prevents food from re-entering the esophagus when the stomach contracts; gastroesophageal sphincter.

**cardiac veins** (kar′de-ak vāns) Blood vessels that return blood from the venules of the myocardium to the coronary sinus.

**cardiovascular** (kar″de-o-vas′ku-lar) Pertaining to the heart and blood vessels.

**carina** (kah-ri′nah) A cartilaginous ridge located between the openings of the right and left bronchi.

**carotene** (kar′o-tēn) A yellow, orange, or reddish pigment in plants and a precursor of vitamin A.

**carotid bodies** (kah-rot′id bod′ēz) Masses of chemoreceptors located in the wall of the internal carotid artery near the carotid sinus.

**carpals** (kar′pals) Bones of the wrist.

**carpus** (kar′pus) The wrist; the wrist bones as a group.

**cartilage** (kar′tĭ-lij) Type of connective tissue in which cells are located within lacunae and are separated by a semi-solid matrix.

**catabolism** (ka-tab′o-lizm) Metabolic process that breaks down large molecules into smaller ones; catabolic metabolism.

**catalase** (kat′ah-lās) An enzyme that catalyzes the decomposition of hydrogen peroxide.

**catalyst** (kat′ah-list) A chemical that increases the rate of a chemical reaction, but is not permanently altered by the reaction.

**cataract** (kat′ah-rakt) Loss of transparency of the lens of the eye.

**catecholamine** (kat″ĕ-kol′am-in) A type of organic compound that includes epinephrine and norepinephrine.

**cauda equina** (kaw′da ek-wīn′a) A group of spinal nerves that extends below the distal end of the spinal cord.

**cecum** (se′kum) A pouchlike portion of the large intestine attached to the small intestine.

**celiac** (se′le-ak) Pertaining to the abdomen.

**cell** (sel) The structural and functional unit of an organism.

**cell adhesion molecules** (sel ad-hee′zhon mol′ĕ-kūlz) Proteins that guide cellular movement within the body.

**cell body** (sel bod′e) Portion of a nerve cell that includes a cytoplasmic mass and a nucleus, and from which the nerve fibers extend.

**cell cycle** (sel sī-kl) The life cycle of a cell consisting of G1 (growth), S (DNA synthesis), G2 (growth), and mitosis (division).

**cell-mediated immunity** (sel me′de-ā-tĭd ĭ-mu′nĭ-te) The body's attack on foreign antigens carried out by T-lymphocytes and their secreted products.

**cell membrane** (sel mem′brān) The selectively permeable outer boundary of a cell consisting of a phospholipid bilayer embedded with proteins.

**cellular respiration** (sel′u-lar res″pĭ-ra′shun) Process that releases energy from organic compounds in cells.

**cellulose** (sel′u-lōs) A polysaccharide very abundant in plant tissues that human enzymes cannot break down.

**cementum** (se-men′tum) Bonelike material that fastens the root of a tooth into its bony socket.

**central canal** (sen′tral kah-nal′) Tube within the spinal cord that is continuous with the ventricles of the brain and contains cerebrospinal fluid.

**central nervous system** (sen′tral ner′vus sis′tem) Portion of the nervous system that consists of the brain and spinal cord; CNS.
**centriole** (sen′tre-ōl) A cellular structure built of microtubules that organizes the mitotic spindle.
**centromere** (sen′tro-mēr) Portion of a chromosome to which spindle fibers attach during mitosis.
**centrosome** (sen′tro-sōm) Cellular organelle consisting of two centrioles.
**cephalic** (sĕ-fal′ik) Pertaining to the head.
**cerebellar cortex** (ser″ĕ-bel′ar kor′teks) The outer layer of the cerebellum.
**cerebellar peduncles** (ser″ĕ-bel′ar pe-dung′kl) A bundle of nerve fibers connecting the cerebellum and the brain stem.
**cerebellum** (ser″ĕ-bel′um) Portion of the brain that coordinates skeletal muscle movement.
**cerebral aqueduct** (ser′ĕ-bral ak′wĕ-dukt″) Tube that connects the third and fourth ventricles of the brain.
**cerebral cortex** (ser′ĕ-bral kor′teks) Outer layer of the cerebrum.
**cerebral hemisphere** (ser′ĕ-bral hem′ĭ-sfēr) One of the large, paired structures that together constitute the cerebrum of the brain.
**cerebrospinal fluid** (ser″ĕ-bro-spi′nal floo′id) Fluid occupying the ventricles of the brain, the subarachnoid space of the meninges, and the central canal of the spinal cord.
**cerebrovascular accident** (ser″ĕ-bro-vas′ku-lar ak′si-dent) Sudden interruption of blood flow to the brain; a stroke.
**cerebrum** (ser′ĕ-brum) Portion of the brain that occupies the upper part of the cranial cavity and provides higher mental functions.
**cerumen** (sĕ-roo′men) Waxlike substance produced by cells that line the external ear canal.
**cervical** (ser′vĭ-kal) Pertaining to the neck or to the cervix of the uterus.
**cervix** (ser′viks) Narrow, inferior end of the uterus that leads into the vagina.
**chemoreceptor** (ke″mo-re-sep′tor) A receptor that is stimulated by the binding of certain chemicals.
**chemotaxis** (ke″mo-tak′sis) Attraction of leukocytes to chemicals released from damaged cells.
**chief cell** (chēf sel) Cell of gastric gland that secretes various digestive enzymes, including pepsinogen.
**chloride shift** (klo′rīd shift) Movement of chloride ions from the blood plasma into red blood cells as bicarbonate ions diffuse out of the red blood cells into the plasma.
**cholecystokinin** (ko″le-sis″to-ki′nin) Hormone the small intestine secretes that stimulates release of pancreatic juice from the pancreas and bile from the gallbladder.
**cholesterol** (ko-les′ter-ol) A lipid produced by body cells used to synthesize steroid hormones and excreted into the bile.
**cholinergic fiber** (ko″lin-er′jik fi′ber) A nerve fiber that secretes acetylcholine at the axon terminal.
**cholinesterase** (ko″lin-es′ter-ās) An enzyme that catalyzes breakdown of acetylcholine.
**chondrin** (kon′drin) A protein in the intercellular substance of cartilage.
**chondrocyte** (kon′dro-sīt) A cartilage cell.
**chorion** (ko′re-on) Extra-embryonic membrane that forms the outermost covering around a fetus and contributes to formation of the placenta.
**chorionic villi** (ko″re-on′ik vil′i) Projections that extend from the outer surface of the chorion and help attach an embryo to the uterine wall.
**choroid coat** (ko′roid kōt) The vascular, pigmented middle layer of the wall of the eye.
**choroid plexus** (ko′roid plek′sus) Mass of specialized capillaries from which cerebrospinal fluid is secreted into a ventricle of the brain.
**chromaffin cell** (kro-maf′in sel) Cell of the adrenal medulla that produces, stores, and secretes epinephrine and norepinephrine.
**chromatid** (kro′mah-tid) One half of a replicated chromosome or a single unreplicated chromosome.
**chromatin** (kro′mah-tin) DNA and complexed protein that condenses to form chromosomes during mitosis.
**chromosome** (kro′mo-sōm) Rodlike structure that condenses from chromatin in a cell's nucleus during mitosis.
**chylomicron** (kil″o-mi′kron) A microscopic droplet of fat in the blood following fat digestion.
**chyme** (kīm) Semifluid mass of partially digested food that passes from the stomach to the small intestine.
**chymotrypsin** (ki″mo-trip′sin) A protein-splitting enzyme in pancreatic juice.
**cilia** (sil′e-ah) Microscopic, hairlike processes on the exposed surfaces of certain epithelial cells.
**ciliary body** (sil′e-er″e bod′e) Structure associated with the choroid layer of the eye that secretes aqueous humor and contains the ciliary muscle.
**circadian rhythm** (ser″kah-de′an rithm) A pattern of repeated behavior associated with the cycles of night and day.
**circle of Willis** (sir′kl uv wil′is) An arterial ring located on the ventral surface of the brain.
**circular muscles** (ser′ku-lar mus′lz) Muscles whose fibers are arranged in circular patterns, usually around an opening or in the wall of a tube; sphincter muscles.
**circumduction** (ser″kum-duk′shun) Movement of a body part, such as a limb, so that the end follows a circular path.
**cisternae** (sis-ter′ne) Enlarged portions of the sarcoplasmic reticulum near the actin and myosin filaments of a muscle fiber.
**citric acid cycle** (sit′rik as′id si′kl) A series of chemical reactions that oxidizes certain molecules, releasing energy; Krebs cycle.
**cleavage** (klēv′ij) The early successive divisions of the blastocyst cells into smaller and smaller cells.
**clitoris** (kli′to-ris) Small erectile organ located in the anterior portion of the vulva; corresponds to the penis.
**clone** (klōn) A group of cells that originate from a single cell and are therefore genetically identical.
**CNS** Central nervous system.
**coagulation** (ko-ag″u-la′shun) Bloodclotting.
**cocarboxylase** (ko″kar-bok′sĭ-lās) A coenzyme synthesized from thiamine that oxidizes carbohydrates.
**cochlea** (kok′le-ah) Portion of the inner ear that contains the receptors of hearing.

**codon** (ko′don) A set of three nucleotides of a messenger RNA molecule corresponding to a particular amino acid.
**coenzyme** (ko-en′zīm) A nonprotein substance that is necessary for the activity of a particular enzyme.
**coenzyme A** (ko-en′zīm) Acetyl coenzyme A.
**cofactor** (ko′fak-tor) A nonprotein substance that must be combined with an enzyme for activity.
**collagen** (kol′ah-jen) Protein in the white fibers of connective tissues and in bone matrix.
**collateral** (ko-lat′er-al) A branch of a nerve fiber or blood vessel.
**collecting duct** (ko-lek′ting dukt) In the kidneys, a straight tubule that receives fluid from several nephrons.
**colon** (kolon) The large intestine.
**colony stimulating factor** (ko′le-ne stim′yu-lay″ting fak′tor) A protein that stimulates differentiation and maturation of white blood cells.
**color blindness** (kul′er blīnd′nes) An inherited inability to distinguish certain colors.
**colostrum** (ko-los′trum) The first secretion of a woman's mammary glands after she gives birth.
**common bile duct** (kom′mon bīl dukt) Tube that transports bile from the cystic duct to the duodenum.
**compact bone** (kom′pakt bōn) Dense tissue in which cells are arranged in Haversian systems with no apparent spaces.
**compartment** (com-part′ment) A space occupied by a group of muscles, blood vessels, and nerves that is enclosed by fasciae.
**complement** (kom′plĕ-ment) A group of enzymes that are activated by the combination of antibody with antigen and enhance the reaction against foreign substances within the body.
**completely penetrant** (kom-plēt′le pen′e-trent) In genetics, indicates that the frequency of expression of a genotype is 100%.
**complete protein** (kom-plēt′ pro′tēn) A protein that contains adequate amounts of the essential amino acids to maintain body tissues and to promote normal growth and development.
**compound** (kom′pownd) A substance composed of two or more chemically bonded elements.
**concussion** (kon-kush′un) Loss of consciousness due to a violent blow to the head.
**condom** (kon′dum) A latex sheath used to cover the penis during sexual intercourse; used as a contraceptive and to minimize the risk of transmitting infection.
**conduction** (kon-duk′shun) Movement of body heat into the molecules of cooler objects in contact with the body surface.
**condyle** (kon′dīl) A rounded process of a bone, usually at the articular end.
**cones** (kōns) Color receptors in the retina of the eye.
**conformation** (kon-for-ma′shun) The three-dimensional form of a protein, determined by its amino acid sequence and attractions and repulsions between amino acids.
**congenital** (kon-jen′ĭ-tal) A medical condition present at the time of birth not caused by a genetic defect.
**conjunctiva** (kon″junk-ti′vah) Membranous covering on the anterior surface of the eye.
**connective tissue** (kō-nek′tiv tish′u) One of the basic types of tissue that includes bone, cartilage, blood, loose and fibrous connective tissue.
**contraception** (kon″trah-sep′shun) A behavior or device that prevents fertilization.
**contralateral** (kon″trah-lat′er-al) Positioned on the opposite side of something else.
**convection** (kon-vek′shun) The transmission of heat from one substance to another through the circulation of heated air particles.
**convergence** (kon-ver′jens) Nerve impulses arriving at the same neuron.
**convolution** (kon″vo-lu′shun) An elevation on a structure's surface caused by infolding of it upon itself.
**cornea** (kor′ne-ah) Transparent anterior portion of the outer layer of the eye wall.
**coronary artery** (kor′o-na″re ar′ter-e) An artery that supplies blood to the wall of the heart.
**coronary sinus** (kor′o-na″re si′nus) A large vessel on the posterior surface of the heart into which the cardiac veins drain.
**corpus callosum** (kor′pus kah-lo′sum) A mass of white matter within the brain, composed of nerve fibers connecting the right and left cerebral hemispheres.
**corpus luteum** (kor′pus lu′te-um) Structure that forms from the tissues of a ruptured ovarian follicle and secretes female hormones.
**corpus striatum** (kor′pus stri-a′tum) Portion of the cerebrum that includes certain basal ganglia.
**cortex** (kor′teks) Outer layer of an organ such as the adrenal gland, cerebrum, or kidney.
**cortical nephron** (kor′tĭ-kl nef′ron) A nephron with its corpuscle located in the renal cortex.
**cortisol** (kor′ti-sol) A glucocorticoid secreted by the adrenal cortex.
**costal** (kos′tal) Pertaining to the ribs.
**countercurrent mechanism** (kown′ter-ker″ent me′kĕ-nĭ″zm) Process by which the kidneys concentrate urine.
**covalent bond** (ko′va-lent bond) Chemical bond formed by electron sharing between atoms.
**coxal** (kok′sel) Pertaining to the hip.
**cranial** (kra′ne-al) Pertaining to the cranium.
**cranial cavity** (kran′e-al kav′i-te) A hollow space in the cranium containing the brain.
**cranial nerve** (kra′ne-al nerv) Nerve that arises from the brain.
**creatine phosphate** (kre′ah-tin fos′fāt) A muscle biochemical that stores energy.
**crenation** (kre-na′shun) Shrinkage of a cell caused by contact with a hypertonic solution.
**crest** (krest) A ridgelike projection of a bone.
**cretinism** (kre′tĭ-nizm) A condition resulting from lack of thyroid secretion in an infant.
**cricoid cartilage** (kri′koid kar′tĭ-lij) A ringlike cartilage that forms the lower end of the larynx.
**crista ampullaris** (kris′tah am-pul-lah′ris) Sensory organ located within a semicircular canal that functions in the sense of dynamic equilibrium.
**crossing over** (kros′ing o′ver) The exchange of genetic material between homologous chromosomes during meiosis.
**crossmatching** (kros′mach″ing) A procedure used to determine whether donor and recipient blood samples will agglutinate.
**crural** (krur′al) Pertaining to the leg.
**cubital** (ku′bi-tal) Pertaining to the forearm.
**cuspid** (kus′pid) A canine tooth.
**cutaneous** (ku-ta′ne-us) Pertaining to the skin.
**cyanocobalamin** (si″ah-no-ko-bal′ah-min) Vitamin $B_{12}$.

**cyanosis** (si″ah-no′sis) Bluish skin coloration due to decreased blood oxygen concentration.
**cyclic AMP** (sik′lik AMP) A circularized derivative of ATP that responds to messages entering a cell and triggers the cell's response.
**cyclosporine** (si″klo-spor′in) A drug that suppresses the action of helper T cells, preventing rejection of transplanted tissue.
**cystic duct** (sis′tik dukt) Tube that connects the gallbladder to the common bile duct.
**cytochrome** (si′to-krōm) Protein within the inner mitochondrial membrane that is an electron carrier in aerobic respiration (electron transport system).
**cytocrine secretion** (si′to-krin se-kre′shun) Transfer of granules of melanin from melanocytes into adjacent epithelial cells.
**cytokine** (si′to-kīn) A type of protein secreted by a T lymphocyte that attacks viruses, virally infected cells, and cancer cells.
**cytoplasm** (si′to-plazm) The contents of a cell excluding the nucleus and cell membrane.
**cytoskeleton** (si′to-skel″e-ton) A system of protein tubules and filaments that reinforces a cell's three-dimensional form and provides scaffolding and transport tracts for organelles.

## D

**deamination** (de-am″ĭ-na′shun) Removing amino groups ($NH_2$) from amino acids.
**deciduous teeth** (de-sid′u-us tēth) Teeth that are shed and replaced by permanent teeth; primary teeth.
**decomposition** (de-kom″po-zish′un) The breakdown of molecules into simpler compounds.
**defecation** (def″ĕ-ka′shun) The discharge of feces from the rectum through the anus.
**dehydration** (de″hi-dra′shun) Excessive water loss.
**dehydration synthesis** (de″hi-dra′shun sin′thĕ-sis) Anabolic process that joins small molecules; synthesis.
**dendrite** (den′drīt) Nerve fiber that transmits impulses toward a neuron cell body.
**densitometer** (den″si-tom′e-ter) An instrument used to measure the density of bone tissue.
**dental caries** (den′tal kar′ēz) Decalcification and decay of teeth.
**dentin** (den′tin) Bonelike substance that forms the bulk of a tooth.
**deoxyhemoglobin** (de-ok″sĭ-he″mo-glo′bin) Hemoglobin to which oxygen is not bound.
**deoxyribonucleic acid** (dē-ok′si-rī″bō-nu-klē″ik as′id) The genetic material; a double-stranded polymer of nucleotides, each containing a phosphate group, a nitrogenous base (adenine, thymine, guanine, or cytosine), and the sugar deoxyribose.
**depolarization** (de-po″lar-ĭ-za′shun) The loss of an electrical charge on the surface of a membrane.
**dermatome** (der′mah-tōm) An area of the body supplied by sensory nerve fibers associated with a particular dorsal root of a spinal nerve.
**dermis** (der′mis) The thick layer of the skin beneath the epidermis.
**descending colon** (de-send′ing ko′lon) Portion of the large intestine that passes downward along the left side of the abdominal cavity to the brim of the pelvis.
**descending tracts** (de-send′ing trakts) Groups of nerve fibers that carry nerve impulses downward from the brain through the spinal cord.
**desmosome** (des′mo-sōm) A specialized junction between cells, which serves as a "spot weld."
**detrusor muscle** (de-trūz′or mus′l) Muscular wall of the urinary bladder.
**dextrose** (dek′strōs) Glucose.
**diabetes insipidus** (di″ah-be′tēz in-sip′ĭ-dus) Extremely copious urine produced due to a deficiency of antidiuretic hormone.
**diabetes mellitus** (di″ah-be′tēz mel-li′tus) High blood glucose level and glucose in the urine due to a deficiency of insulin.
**dialysis** (di-al′ĭ-sis) Separation of smaller molecules from larger ones in a liquid.
**diapedesis** (di″ah-pĕ-de′sis) Squeezing of leukocytes between the cells of blood vessel walls.
**diaphragm** (di′ah-fram) A sheetlike structure composed largely of skeletal muscle and connective tissue that separates the thoracic and abdominal cavities; also a caplike contraceptive device inserted in the vagina.
**diaphysis** (di-af′ĭ-sis) The shaft of a long bone.
**diastole** (di-as′to-le) Phase of the cardiac cycle when a heart chamber wall relaxes.
**diastolic pressure** (di-a-stol′ik presh′ur) Arterial blood pressure during the diastolic phase of the cardiac cycle.
**diencephalon** (di″en-sef′ah-lon) Portion of the brain in the region of the third ventricle that includes the thalamus and hypothalamus.
**differentiation** (dif″er-en″she-a′shun) Cell specialization.
**diffusion** (dĭ-fu′zhun) Random movement of molecules from a region of higher concentration toward one of lower concentration.
**digestion** (di-jest′yun) Breaking down of large nutrient molecules into smaller molecules that can be absorbed; hydrolysis.
**digital** (di′ji-tal) Pertaining to the finger.
**dihydrotestosterone** (di-hi″dro-tes-tos′ter-ōn) Hormone produced from testosterone that stimulates certain cells of the male reproductive system.
**dipeptide** (di-pep′tīd) A molecule composed of two joined amino acids.
**disaccharide** (di-sak′ah-rīd) A sugar produced by the union of two monosaccharides.
**distal** (dis′tal) Further from the midline or origin; opposite of proximal.
**diuretic** (di″u-ret′ik) A substance that increases urine production.
**divergence** (di-ver′jens) A spreading apart.
**DNA** Deoxyribonucleic acid.
**dominant gene** (dom′ĭ-nant jēn) The member of a gene pair that is expressed while its allele is not expressed.
**dorsal cavity** (dor′sal kav′i-te) A hollow space in the posterior portion of the body containing the cranial cavity and vertebral canal.
**dorsal root** (dor′sal rōot) The sensory branch of a spinal nerve by which it joins the spinal cord.
**dorsal root ganglion** (dor′sal rōot gang′gle-on) Mass of sensory neuron cell bodies located in the dorsal root of a spinal nerve.
**dorsiflexion** (dor″si-flek′shun) Bending the foot upward.

**dorsum** (dor′sum) Pertaining to the back surface of a body part.

**ductus arteriosus** (duk′tus ar-te″re-o′sus) Blood vessel that connects the pulmonary artery and the aorta in a fetus.

**ductus venosus** (duk′tus ven-o′sus) Blood vessel that connects the umbilical vein and the inferior vena cava in a fetus.

**duodenum** (du″o-de′num) The first portion of the small intestine that leads from the stomach to the jejunum.

**dural sinus** (du′ral si′nus) Blood-filled channel formed by the splitting of the dura mater into two layers.

**dura mater** (du′rah ma′ter) Tough outer layer of the meninges.

**dynamic equilibrium** (di-nam′ik e″kwĭ-lib′re-um) Maintenance of balance when the head and body are suddenly moved or rotated.

**dystrophin** (dis′tre-fin) A protein comprising only 0.002% of the total protein in skeletal muscle that supports the cell membrane. Its absence causes muscular dystrophy.

## E

**eccrine gland** (ek′rin gland) Sweat gland that maintains body temperature.

**ECG** Electrocardiogram; EKG.

**ectoderm** (ek′to-derm) The outermost primary germ layer.

**edema** (ĕ-de′mah) Accumulation of fluid within the tissue spaces.

**effector** (ĕ-fek′tor) A muscle or gland that responds to stimulation.

**efferent** (ef′er-ent) Conducting away from the center. For example, an efferent arteriole conducts blood away from the glomerulus of a nephron.

**ejaculation** (e-jak″u-la′shun) Discharge of sperm-containing semen from the male urethra.

**ejaculatory duct** (e-jak′u-lah-to″re dukt) Tube, formed by the joining of the vas deferens and the tube from the seminal vesicle, that transports sperm to the urethra.

**elastin** (e-las′tin) Protein that comprises the yellow, elastic fibers of connective tissue.

**electrocardiogram** (e-lek″tro-kar′de-o-gram″) A recording of the electrical activity associated with the heartbeat; ECG or EKG.

**electrolyte** (e-lek′tro-līt) A substance that ionizes in water solution.

**electrolyte balance** (e-lek′tro-līt bal′ans) Condition when the quantities of electrolytes entering the body equal those leaving it.

**electron** (e-lek′tron) A small, negatively charged particle that revolves around the nucleus of an atom.

**electron transport system** (e-lek′tron trans′pohrt sis′tem) A series of oxidation-reduction reactions that takes high energy electrons from the citric acid cycle and forms water and energy.

**electrovalent bond** (e-lek″tro-va′lent bond) Chemical bond formed between two ions as a result of the transfer of electrons.

**element** (el′ĕ-ment) A basic chemical substance.

**elevation** (el-e-vā-shun) Upward movement of a part of the body.

**embolus** (em′bo-lus) A substance, such as a blood clot or bubble of gas, that is carried by the blood and obstructs a blood vessel.

**embryo** (em′bre-o) A prenatal stage of development after germ layers form but before the rudiments of all organs are present.

**embryonic disk** (em″brē-on′ik disk) A flattened area in the preembryo from which the embryo arises.

**emission** (e-mish′un) The movement of sperm cells from the vas deferens into the ejaculatory duct and urethra.

**emphysema** (em″fĭ-se′mah) Abnormal enlargement of the air sacs of the lungs.

**emulsification** (e-mul″sĭ-fĭ′ka′shun) Breaking up of fat globules into smaller droplets by the action of bile salts.

**enamel** (e-nam′el) Hard covering on the exposed surface of a tooth.

**endocardium** (en″do-kar′de-um) Inner lining of the heart chambers.

**endochondral bone** (en′do-kon′dral bōn) Bone that begins as hyaline cartilage that is subsequently replaced by bone tissue.

**endocrine gland** (en′do-krin gland) A gland that secretes hormones directly into the blood or body fluids.

**endocytosis** (en″do-si-to′sis) Physiological process by which substances may move through a cell membrane and involves the formation of tiny, cytoplasmic vacuoles.

**endoderm** (en′do-derm) The innermost primary germ layer.

**endolymph** (en′do-limf) Fluid contained within the membranous labyrinth of the inner ear.

**endometrium** (en″do-me′tre-um) The inner lining of the uterus.

**endomysium** (en″do-mis′e-um) The sheath of connective tissue surrounding each skeletal muscle fiber.

**endoneurium** (en″do-nu′re-um) Layer of loose connective tissue that surrounds individual nerve fibers.

**endoplasmic reticulum** (en-do-plaz′mic rĕ-tik′u-lum) Organelle composed of a system of connected membranous tubules and vesicles along which protein synthesis occurs.

**endorphin** (en-dor′fin) A neuropeptide synthesized in the pituitary gland that suppresses pain.

**endosteum** (en-dos′tē-um) Tissue lining the medullary cavity within the bone.

**endothelium** (en″do-the′le-um) The layer of epithelial cells that forms the inner lining of blood vessels and heart chambers.

**energy** (en′er-je) An ability to cause something to move and thus to do work.

**energy balance** (en′er-je bal′ans) When the caloric intake of the body equals its caloric output.

**enkephalin** (en-kef′ah-lin) A neuropeptide that occurs in the brain and spinal cord; it inhibits pain impulses.

**enterogastric reflex** (en-ter-o-gas′trik re′fleks) Inhibition of gastric (stomach) peristalsis and secretions when food enters the small intestine.

**enzyme** (en′zīm) A protein that catalyzes a specific biochemical reaction.

**eosinophil** (e″o-sin′o-fil) White blood cell containing cytoplasmic granules that stain with acidic dye.

**ependyma** (ĕ-pen′dĭ-mah) A membrane composed of neuroglial cells that lines the ventricles of the brain.

**epicardium** (ep″ĭ-kar′de-um) The visceral portion of the pericardium on the surface of the heart.

**epicondyle** (ep″ĭ-kon′dīl) A projection of a bone located above a condyle.

**epidermis** (ep″ĭ-der′mis) Outer epithelial layer of the skin.

**epididymis** (ep″ĭ-did′ĭ-mis) Highly coiled tubule that leads from the seminiferous tubules of the testis to the vas deferens.
**epidural space** (ep″ĭ-du′ral spās) The space between the dural sheath of the spinal cord and the bone of the vertebral canal.
**epigastric region** (ep″ĭ-gas′trik re′jun) The upper middle portion of the abdomen.
**epiglottis** (ep″ĭ-glot′is) Flaplike cartilaginous structure located at the back of the tongue near the entrance to the trachea.
**epimysium** (ep″ĭ-mis′e-um) The outer sheath of connective tissue surrounding a skeletal muscle.
**epinephrine** (ep″ĭ-nef′rin) A hormone the adrenal medulla secretes during times of stress.
**epineurium** (ep″ĭ-nu′re-um) Outermost layer of connective tissue surrounding a nerve.
**epiphyseal disk** (ep″ĭ-fiz′e-al disk) Cartilaginous layer within the epiphysis of a long bone that grows.
**epiphysis** (e-pif′ĭ-sis) The end of a long bone.
**epiploic appendage** (ep″i-plo′ik ah-pen′dij) Small collections of fat in the serous layer of the large intestinal wall.
**epithelium** (ep″ĭ-the′le-um) Tissue type that covers all free body surfaces.
**equilibrium** (e″kwĭ-lib′re um) A state of balance between two opposing forces.
**erythroblast** (ĕ-rith′ro-blast) An immature red blood cell.
**erythroblastosis fetalis** (ĕ-rith″ro-blas-to′sis fe-tal′is) A life-threatening condition of massive agglutination of the blood in the fetus or neonate due to the mother's anti-Rh antibodies reacting with the baby's Rh-positive red blood cells.
**erythrocyte** (ĕ-rith′ro-sīt) A red blood cell.
**erythropoiesis** (ĕ-rith″ro-poi-e′sis) Red blood cell formation.
**erythropoietin** (ĕ-rith″ro-poi′ĕ-tin) (EPo) A kidney hormone that promotes red blood cell formation.
**esophageal hiatus** (ĕ-sof″ah-je′al hi-a′tus) Opening in the diaphragm through which the esophagus passes.
**esophagus** (ĕ-sof′ah-gus) Tubular portion of the digestive tract that leads from the pharynx to the stomach.
**essential amino acid** (ĕ-sen′shal ah-me′no as′id) Amino acid required for health that body cells cannot synthesize in adequate amounts.
**essential fatty acid** (ĕ-sen′shal fat′e as′id) Fatty acid required for health that body cells cannot synthesize in adequate amounts.
**essential nutrient** (ĕ-sen′shal nu′trē-ent) A nutrient necessary for growth, normal functioning, and maintaining life that the diet must supply because the body cannot synthesize it.
**estrogen** (es′tro-jen) Hormone that stimulates the development of female secondary sexual characteristics.
**euploid** (u′ploid) Having a balanced set of chromosomes.
**eustachian tube** (yoo-sta′she-an tūb) Tube that connects the middle ear to the pharynx; auditory tube.
**evaporation** (e″vap′o-ra-shun) Changing a liquid into a gas.
**eversion** (e-ver′zhun) Outward turning movement of the sole of the foot.
**exchange reaction** (eks-chānj re-ak′shun) A chemical reaction in which parts of two kinds of molecules trade positions.
**excretion** (ek-skre′shun) Elimination of metabolic wastes.
**exocrine gland** (ek′so-krin gland) A gland that secretes its products into a duct or onto a body surface.
**expiration** (ek″spĭ-ra′shun) Expulsion of air from the lungs.
**extension** (ek-sten′shun) Movement increasing the angle between parts at a joint.
**extracellular** (ek″strah-sel′u-lar) Outside of cells; refers to the organism's internal environment, body fluids outside the individual cells.
**extrapyramidal tract** (ek″strah-pi-ram′i-dal trakt) Nerve tracts, other than the corticospinal tracts, that transmit impulses from the cerebral cortex into the spinal cord.
**extremity** (ek-strem′ĭ-te) A limb; an arm or leg.

## F

**facet** (fas′et) A small, flattened surface of a bone.
**facilitated diffusion** (fah-sil′ĭ-tāt′id dĭ-fu′zhun) Diffusion in which substances move through membranes from a region of higher concentration to a region of lower concentration by carrier molecules.
**facilitation** (fah-sil″ĭ-tā′shun) The hastening of any natural process, increasing efficiency of the natural process.
**fallopian tube** (fah-lo′pe-an tūb) Tube that transports an egg cell from the region of the ovary to the uterus; oviduct or uterine tube.
**fascia** (fash′e-ah) A sheet of fibrous connective tissue that encloses a muscle.
**fasciculus** (fah-sik′u-lus) A small bundle of muscle fibers.
**fat** (fat) Adipose tissue; or an organic substance whose molecules contain glycerol and fatty acids.
**fatty acid** (fat′e as′id) A building block of a fat molecule.
**fatty acid oxidase** (fat′e as′id ok′si-days″) An enzyme that catalyzes the removal of hydrogen or electrons from a fatty acid molecule.
**feces** (fe′sēz) Material expelled from the digestive tract during defecation.
**femoral** (fem′or-al) Pertaining to the thigh.
**ferritin** (fer′ĭ-tin) An iron-protein complex that stores iron in liver cells.
**fertilization** (fer″tĭ-lĭ-za′shun) The union of an egg cell and a sperm cell.
**fetoscopy** (fe-tos′ko-pe) The direct observation of a fetus using a fiber optic device.
**fetus** (fe′tus) A human embryo after eight weeks of development.
**fever** (fe′ver) Elevation of body temperature above normal.
**fibril** (fi′bril) A tiny fiber or filament.
**fibrillation** (fi″brĭ-la′shun) Uncoordinated contraction of muscle fibers.
**fibrin** (fi′brin) Insoluble, fibrous protein formed from fibrinogen during blood coagulation.
**fibrinogen** (fi-brin′o-jen) Plasma protein converted into fibrin during blood coagulation.
**fibrinolysin** (fi″brĭ-nol′ĭ-sin) A protein-splitting enzyme that can digest fibrin in a blood clot.
**fibroblast** (fi′bro-blast) Cell that produces fibers and other intercellular materials in connective tissues.
**fibrocartilage** (fi″bro-kar′ti-lij) Strongest and most durable cartilage; made up of cartilage cells and many collagenous fibers.

**filtration** (fil-tra′shun) Movement of material through a membrane as a result of hydrostatic pressure.

**filtration pressure** (fil-tra′shun presh′ur) Equal to the hydrostatic pressure of the blood entering the glomerulus minus the pressure of the opposing forces (the hydrostatic pressure within the glomerular capsule and the plasma osmotic pressure of the blood in the glomerulus).

**fissure** (fish′ūr) A narrow cleft separating parts, such as the lobes of the cerebrum.

**flaccid paralysis** (flak′sid pah-ral′ĭ-sis) Total loss of muscle tone when nerve fibers are damaged.

**flagella** (flah-jel′ah) Relatively long, motile processes that extend from the surface of a cell.

**flexion** (flek′shun) Bending at a joint to decrease the angle between bones.

**follicle** (fol′ĭ-kl) A pouchlike depression or cavity.

**follicle-stimulating hormone** (fol′ĭ-kl stim′u-la″ting hor′mōn) A substance secreted by the anterior pituitary gland that stimulates development of an ovarian follicle in a female or production of sperm cells in a male; FSH.

**follicular cells** (fŏ-lik′u-lar selz) Ovarian cells that surround a developing egg cell and secrete female sex hormones.

**fontanel** (fon″tah-nel′) Membranous region between certain cranial bones in the skull of a fetus or infant.

**foramen** (fo-ra′men) An opening, usually in a bone or membrane (pl. *foramina*).

**foramen magnum** (fo-ra′men mag′num) Opening in the occipital bone of the skull through which the spinal cord passes.

**foramen ovale** (fo-ra′men o-val′e) Opening in the interatrial septum of the fetal heart.

**forebrain** (fōr′brān) The anteriormost portion of the developing brain that gives rise to the cerebrum and basal ganglia.

**formula** (fōr′mu-lah) A group of symbols and numbers used to express the composition of a compound.

**fossa** (fos′ah) A depression in a bone or other part.

**fovea** (fo′ve-ah) A tiny pit or depression.

**fovea centralis** (fo′ve-ah sen-tral′is) Region of the retina, consisting of densely packed cones, that provides the greatest visual acuity.

**fracture** (frak′chur) A break in a bone.

**free radical** (frē rad′eh-kel) Highly reactive by-product of metabolism that can damage tissue.

**frenulum** (fren′u-lum) A fold of tissue that anchors and limits movement of a body part.

**frontal** (frun′tal) Pertaining to the forehead.

**FSH** Follicle-stimulating hormone.

**functional syncytium** (funk′shun-al sin-sish′e-um) Merging cells performing together as a unit; those of the heart are joined electrically. A syncytium lacks cell boundaries, appearing as a multinucleated structure.

## G

**galactose** (gah-lak′tōs) A monosaccharide component of the disaccharide lactose.

**gallbladder** (gawl′blad-er) Saclike organ associated with the liver that stores and concentrates bile.

**gamete** (gam′ēt) A sex cell; either an egg cell or a sperm cell.

**ganglion** (gang′gle-on) A mass of neuron cell bodies, usually outside the central nervous system (pl. *ganglia*).

**gastric gland** (gas′trik gland) Gland within the stomach wall that secretes gastric juice.

**gastric juice** (gas′trik jōōs) Secretion of the gastric glands within the stomach.

**gastric lipase** (gas′trik lī′pās) Fat-splitting enzyme in gastric juice.

**gastrin** (gas′trin) Hormone secreted by the stomach lining that stimulates secretion of gastric juice.

**gastrula** (gas′troo-lah) Embryonic stage following the blastula; cells differentiate into three layers: the endoderm, the mesoderm, and the ectoderm.

**gene** (jēn) Portion of a DNA molecule that encodes the information to synthesize a protein, a control sequence, or tRNA or rRNA. The unit of inheritance.

**genetic code** (jĕ-net′ik kōd) Information for synthesizing proteins that is encoded in the nucleotide sequence of DNA molecules.

**genetic heterogeneity** (jĕ-net′ik het″er-o-je-ne′ĭ-te) Different genotypes that have identical phenotypes.

**genetics** (jĕ-net′iks) The study of heredity.

**genital** (jen′i-tal) Pertaining to the genitalia (internal and external organs of reproduction).

**genome** (jeh′nōm) All of the DNA in a cell of an organism.

**genotype** (je′no-tīp) The combination of genes present within a zygote or within the cells of an individual.

**germinal epithelium** (jer′mĭ-nal ep″ĭ-the′le-um) Tissue within an ovary that gives rise to sex cells.

**germ layers** (jerm la′ers) Layers of cells within an embryo that form the body organs during development: ectoderm, mesoderm, and endoderm.

**glans penis** (glanz pe′nis) Enlarged mass of corpus spongiosum at the end of the penis; may be covered by the foreskin.

**globin** (glo′bin) The protein portion of a hemoglobin molecule.

**globulin** (glob′u-lin) A type of protein in blood plasma.

**glomerulus** (glo-mer′u-lus) A capillary tuft located within the glomerular capsule of a nephron.

**glottis** (glot′is) Slitlike opening between the true vocal folds or vocal cords.

**glucagon** (gloo′kah-gon) Hormone secreted by the pancreatic islets of Langerhans that releases glucose from glycogen.

**glucocorticoid** (gloo″ko-kor′tĭ-koid) Any one of a group of hormones secreted by the adrenal cortex that influences carbohydrate, fat, and protein metabolism.

**gluconeogenesis** (gloo″ko-ne″o-jen′ĕ-sis) Synthesis of glucose from noncarbohydrates such as amino acids.

**glucose** (gloo′kōs) A monosaccharide in the blood that is the primary source of cellular energy.

**glucosuria** (gloo′ko-sur′e-ah) Presence of glucose in urine.

**gluteal** (gloo′te-al) Pertaining to the buttocks.

**glycerol** (glis′er-ol) An organic compound that is a building block for fat molecules.

**glycogen** (gli′ko-jen) A polysaccharide that stores glucose in the liver and muscles.

**glycolysis** (gli-kol′ĭ-sis) The conversion of glucose to pyruvic acid during cellular respiration.

**glycoprotein** (gli″ko-pro′te-in) A compound composed of a carbohydrate combined with a protein.

**goblet cell** (gob′let sel) An epithelial cell that is specialized to secrete mucus.

**goiter** (goi′ter) An enlarged thyroid gland.
**Golgi apparatus** (gol′jē ap″ah-ra′tus) An organelle that prepares cellular products for secretion.
**Golgi tendon organ** (gol″jē ten′dun or′gan) Sensory receptors in tendons close to muscle attachments that are involved in reflexes that help maintain posture.
**gomphosis** (gom-fo′sis) Type of joint in which a cone-shaped process is fastened in a bony socket.
**gonad** (go′nad) A sex-cell-producing organ; an ovary or testis.
**gonadotropin** (go-nad″o-trōp′in) A hormone that stimulates activity in the gonads.
**granulocyte** (gran′u-lo-sīt) A leukocyte that contains granules in its cytoplasm.
**gray matter** (grā mat′er) Region of the central nervous system that generally lacks myelin and thus appears gray.
**gray ramus** (grā ra′mus) A short nerve containing postganglionic axons returning to a spinal nerve.
**groin** (groin) Region of the body between the abdomen and thighs.
**growth** (grōth) Process by which a structure enlarges.
**growth hormone** (grōth hōr′mōn) A hormone released by the anterior lobe of the pituitary gland that promotes the growth of the organism; GH or somatotropin.
**gubernaculum** (goo″ber′nak′u-lum) A structure that guides another structure.

## H

**hair cell** (hār sel) Mechanoreceptor in the inner ear that lies between the basilar membrane and the tectorial membrane and triggers action potentials in fibers of the auditory nerve.
**hair follicle** (hār fol′ĭ-kl) Tubelike depression in the skin in which a hair develops.
**half-life** (haf′līf) The time it takes for one-half of the radioactivity of an isotope to be released.
**hapten** (hap′ten) A small molecule that, in combination with a larger one, forms an antigenic substance and can later stimulate an immune reaction by itself.
**haustra** (haws′trah) Pouches in the wall of the large intestine.
**hematocrit** (he-mat′o-krit) The volume percentage of red blood cells within a sample of whole blood.
**hematoma** (he″mah-to′mah) A mass of coagulated blood within tissues or a body cavity.
**hematopoiesis** (hem″ah-to-poi-e′sis) The production of blood and blood cells; hemopoiesis.
**heme** (hēm) The iron-containing portion of a hemoglobin molecule.
**hemizygous** (hem″ ĭ-zi′gus) A gene carried on the Y chromosome in humans.
**hemocytoblast** (he″mo-si′to-blast) A cell that gives rise to blood cells.
**hemoglobin** (he″mo-glo′bin) Pigment of red blood cells responsible for the transport of oxygen.
**hemolysis** (he-mol′ ĭ-sis) The rupture of red blood cells accompanied by the release of hemoglobin.
**hemopoiesis** (he″mo-poi-e′sis) The production of blood and blood cells; hematopoiesis.
**hemorrhage** (hem′ō-rij) Loss of blood from the circulatory system; bleeding.
**hemostasis** (he″mo-sta′sis) The stoppage of bleeding.
**heparin** (hep′ah-rin) A substance that interferes with the formation of a blood clot; an anticoagulant.
**hepatic** (hĕ-pat′ik) Pertaining to the liver.
**hepatic lobule** (hĕ-pat′ik lob′ūl) A functional unit of the liver.
**hepatic sinusoid** (hĕ-pat′ik si′nŭ-soid) Vascular channel within the liver.
**heredity** (hĕ-red′ ĭ-te) The transmission of genetic information from parent to offspring.
**heritable gene therapy** (her′ ĭ-tah-bl jēn ther′ah-pe) Manipulation of genes to cure a medical condition.
**heterozygote** (het″er-o-zi′gōt) An individual who possesses different alleles in a gene pair.
**hilum** (hi′lum) A depression where vessels, nerves, and other structures (bronchus, ureter, etc.) enter an organ.
**hilus** (hi′lus) Hilum.
**hindbrain** (hīnd′brān) Posterior-most portion of the developing brain that gives rise to the cerebellum, pons, and medulla oblongata.
**hippocampus** (hip″o-kam′pus) A part of the cerebral cortex where memories form.
**histamine** (his′tah-min) A substance released from stressed cells.
**histology** (his-tol′o-je) The study of the structure and function of tissues.
**homeostasis** (ho″me-o-sta′sis) A state of equilibrium in which the internal environment of the body remains in the normal range.
**homozygote** (ho″mo-zi′gōt) An individual possessing identical genes in a gene pair.
**hormone** (hor′mōn) A substance secreted by an endocrine gland that is transmitted in the blood or body fluids.
**humoral immunity** (hu′mor-al ĭ-mu′nĭ-te) Destruction of cells bearing foreign (nonself) antigens by circulating antibodies.
**hydrolysis** (hi-drol′ĭ-sis) Splitting of a molecule into smaller portions by addition of a water molecule.
**hydrostatic pressure** (hi″dro-stat′ik presh′ur) Pressure exerted by fluids, such as blood pressure.
**hydroxyapatite** (hi-drok″se-ap′ah-tīt) A type of crystalline calcium phosphate found in bone matrix.
**hydroxyl ion** (hi-drok′sil i′on) $OH^-$.
**hymen** (hi′men) A membranous fold of tissue that partially covers the vaginal opening.
**hyperextension** (hi″per-ek-sten′shun) Extreme extension; continuing extension beyond the anatomical position.
**hyperglycemia** (hi″per-gli-se′me-ah) Excess blood glucose.
**hyperkalemia** (hi″per-kah-le′me-ah) Elevated concentration of blood potassium.
**hypernatremia** (hi″per-nah-tre′me-ah) Elevated concentration of blood sodium.
**hyperparathyroidism** (hi″per-par″ah-thi′roi-dizm) Excess secretion of parathyroid hormone.
**hyperplasia** (hi″per-pla′ze-ah) Increased production and growth of new cells.
**hyperpolarization** (hi″per-po″lar-i-za′shun) An increase in the negativity of the resting potential of a cell membrane.
**hypertension** (hi″per-ten′shun) Elevated blood pressure.
**hyperthyroidism** (hi″per-thi′roi-dizm) Oversecretion of thyroid hormones.
**hypertonic** (hi″per-ton′ik) Condition in which a solution contains a greater concentration of dissolved particles than the solution with which it is compared.
**hypertrophy** (hi-per′tro-fe) Enlargement of an organ or tissue.

**hyperventilation** (hi″per-ven″tĭ-la′shun) Breathing that is abnormally deep and prolonged.
**hypervitaminosis** (hi″per-vi″tah-mĭ-no′sis) Excessive intake of vitamins.
**hypochondriac region** (hi″po-kon′dre-ak re′jun) The portion of the abdomen on either side of the middle or epigastric region.
**hypodermis** (hi″po-der′mis) Mainly composed of fat, this loose layer is directly beneath the dermis; subcutaneous.
**hypogastric region** (hi″po-gas′trik re′jun) The lower middle portion of the abdomen.
**hypoglycemia** (hi″po-gli-se′me-ah) Abnormally low concentration of blood glucose.
**hypokalemia** (hi″po-kah-le′me-ah) A low concentration of blood potassium.
**hyponatremia** (hi″po-nah-tre′me-ah) A low concentration of blood sodium.
**hypoparathyroidism** (hi″po-par″ah-thi′roi-dizm) An undersecretion of parathyroid hormone.
**hypophysis** (hi-pof′i-sis) The pituitary gland.
**hypoproteinemia** (hi″po-pro″te-ĭ-ne′me-ah) A low concentration of blood proteins.
**hypothalamus** (hi″po-thal′ah-mus) A portion of the brain located below the thalamus and forming the floor of the third ventricle.
**hypothyroidism** (hi″po-thi′roi-dizm) A low secretion of thyroid hormones.
**hypotonic** (hi″po-ton′ik) Condition in which a solution contains a lesser concentration of dissolved particles than the solution to which it is compared.
**hypoxia** (hi-pok′se-ah) A deficiency of oxygen in the tissues.

## I

**idiotype** (id′e-o-tīp′) The parts of an antibody's antigen binding site that are complementary in conformation to a particular antigen.
**ileocecal valve** (il″e-o-se′kal valv) Sphincter valve located at the distal end of the ileum where it joins the cecum.
**ileum** (il′e-um) Portion of the small intestine between the jejunum and the cecum.
**iliac region** (il′e-ak re′jun) Portion of the abdomen on either side of the lower middle, or hypogastric, region.
**ilium** (il′e-um) One of the bones of a coxal bone or hipbone.
**immunity** (ĭ-mu′nĭ-te) Resistance to the effects of specific disease-causing agents.
**immunoglobulin** (im″u-no-glob′u-lin) Globular plasma proteins that function as antibodies.
**immunosuppressive drugs** (im″u-no-sŭ-pres′iv drugz) Substances that suppress the immune response against transplanted tissue.
**implantation** (im″plan-ta′shun) The embedding of a preembryo in the lining of the uterus.
**impulse** (im′puls) A wave of depolarization conducted along a nerve fiber or muscle fiber.
**incisor** (in-si′zor) One of the front teeth that is adapted for cutting food.
**inclusion** (in-kloo′zhun) A mass of lifeless chemical substance within the cytoplasm of a cell.
**incomplete dominance** (in″kom-plēt′ do′meh-nents) A heterozygote whose phenotype is intermediate between the phenotypes of the two homozygotes.
**incompletely penetrant** (in″kom-plēt′le pen′e-trent) When the frequency of genotype expression is less than 100%.
**incomplete protein** (in″kom-plēt′ pro′tēn) A protein that lacks adequate amounts of essential amino acids.
**infection** (in-fek′shun) The invasion and multiplication of microorganisms in body tissues.
**inferior** (in-fēr′e-or) Situated below something else; pertaining to the lower surface of a part.
**inflammation** (in″flah-ma′shun) A tissue response to stress that is characterized by dilation of blood vessels and an accumulation of fluid in the affected region.
**infrared ray** (in″frah-red′ ra) A form of radiation energy, with wavelengths longer than visible light, by which heat moves from warmer surfaces to cooler surroundings.
**infundibulum** (in″fun-dib′u-lum) The stalk attaching the pituitary gland to the base of the brain.
**ingestion** (in-jes′chun) The taking of food or liquid into the body by way of the mouth.
**inguinal** (ing′gwĭ-nal) Pertaining to the groin region.
**inguinal canal** (ing′gwĭ-nal kah-nal′) Passage in the lower abdominal wall through which a testis descends into the scrotum.
**inhibin** (in′hib′in) A hormone secreted by cells of the testes and ovaries that inhibits the secretion of FSH from the anterior pituitary gland.
**inorganic** (in″or-gan′ik) Chemical substances that lack carbon and hydrogen.
**insertion** (in-ser′shun) The end of a muscle attached to a movable part.
**inspiration** (in″spĭ-ra′shun) Breathing in; inhalation.
**insula** (in′su-lah) A cerebral lobe located deep within the lateral sulcus.
**insulin** (in′su-lin) A hormone secreted by the pancreatic islets of Langerhans that controls carbohydrate metabolism.
**integumentary** (in-teg-u-men′tar-e) Pertaining to the skin and its accessory organs.
**intercalated disk** (in-ter″kah-lāt′ed disk) Membranous boundary between adjacent cardiac muscle cells.
**intercellular** (in″ter-sel′u-lar) Between cells.
**intercellular fluid** (in″ter-sel′u-lar floo′id) Tissue fluid located between cells other than blood cells.
**interferon** (in″ter-fēr′on) A class of immune system chemicals (cytokines) that inhibit multiplication of viruses and growth of tumors.
**interleukin** (in″ter-lu′kin) A class of immune system chemicals (cytokines) with varied effects on the body.
**interneuron** (in″ter-nu′ron) A neuron located between a sensory neuron and a motor neuron; intercalated; internuncial, or association neuron.
**interphase** (in′ter-fāz) Period between two cell divisions when a cell is carrying on its normal functions and prepares for division.
**interstitial cell** (in″ter-stish′al sel) A hormone-secreting cell located between the seminiferous tubules of the testis.
**interstitial fluid** (in″ter-stish′al floo′id) Same as intercellular fluid.

**intervertebral disk** (in″ter-ver′tĕ-bral disk) A layer of fibrocartilage located between the bodies of adjacent vertebrae.
**intestinal gland** (in-tes′tĭ-nal gland) Tubular gland located at the base of a villus within the intestinal wall.
**intestinal juice** (in-tes′tĭ-nal jōōs) The secretion of the intestinal glands.
**intracellular** (in″trah-sel′u-lar) Within cells.
**intracellular fluid** (in″trah-sel′u-lar floo′id) Fluid within cells.
**intracellular junction** (in″trah-sel′u-lar jungk′shun) A connection between the membranes of adjacent cells.
**intramembranous bone** (in″trah-mem′brah-nus bōn) Bone that forms from membranelike layers of primitive connective tissue.
**intrathecal injection** (in″trah-the′kal in-jek′shun) The infusion of a substance directly into the cerebrospinal fluid.
**intrauterine device** (in″trah-u′ter-in de-vīs) A solid object placed in the uterine cavity for purposes of contraception; IUD.
**intrinsic factor** (in-trin′sik fak′tor) A substance produced by the gastric glands that promotes absorption of vitamin $B_{12}$.
**inversion** (in-ver′zhun) Movement in which the sole of the foot is turned inward.
**involuntary** (in-vol′un-tār″e) Not consciously controlled; functions automatically.
**iodopsin** (i″o-dop′sin) A light-sensitive pigment within the cones of the retina.
**ion** (i′on) An atom or molecule with an electrical charge.
**ionic bond** (i-on′ik bond) A chemical bond formed by the transfer of electrons from one atom to another.
**ionization** (i′on-i-za′shun) Dissociation into ions.
**ipsilateral** (ip″sĭ-lat′er-al) On the same side.
**iris** (i′ris) Colored muscular portion of the eye that surrounds the pupil and regulates its size.
**irritability** (ir″ĭ-tah-bil′ĭ-te) The ability of an organism to react to changes in its environment.
**ischemia** (is-ke′me-ah) A deficiency of blood in a body part.
**isometric contraction** (i″so-met′rik kon-trak′shun) Muscular contraction in which the muscle fails to shorten.
**isotonic contraction** (i″so-ton′ik kon-trak′shun) Muscular contraction in which the muscle shortens.
**isotonic solution** (i″so-ton′ik so-lu′shun) A solution that has the same concentration of dissolved particles as the solution with which it is compared.
**isotope** (i′so-tōp) An atom that has the same number of protons as other atoms of an element but has a different number of neutrons in its nucleus.
**IUD** An intrauterine device.

## J

**jejunum** (jĕ-joo′num) Portion of the small intestine located between the duodenum and the ileum.
**joint** (joint) The union of two or more bones; an articulation.
**joint capsule** (joint kap′sul) An envelope, attached to the end of each bone at the joint, enclosing the cavity of a synovial joint.
**juxtaglomerular apparatus** (juks″tah-glo-mer′u-lār ap″ah-ra′tus) Structure located in the walls of arterioles near the glomerulus that regulates renal blood flow.
**juxtamedullary nephron** (juks″tah-med′u-lār-e nef′ron) A nephron with its corpuscle located near the renal medulla.

## K

**karyotype** (kar′ē-o-tīp) A set of chromosomes characteristic of a species arranged in homologous pairs. The human karyotype has 23 chromosome pairs.
**keratin** (ker′ah-tin) Protein present in the epidermis, hair, and nails.
**keratinization** (ker″ah-tin″ĭ-za′shun) The process by which cells form fibrils of keratin and harden.
**ketogenesis** (ke″to-jen′ĕ-sis) Formation of ketone bodies.
**ketone body** (ke′tōn bod′e) Type of compound produced during fat catabolism, including acetone, acetoacetic acid, and betahydroxybutyric acid.
**ketosis** (ke″to′sis) An abnormal elevation of ketone bodies in body fluids.
**kilocalorie** (kil′o-kal″o-re) One thousand calories.
**kilogram** (kil′o-gram) A unit of weight equivalent to 1,000 grams.
**kinase** (ki′nas) An enzyme that converts an inactive or precursor form of another enzyme to an active form by adding a phosphate group.
**Krebs cycle** (krebz sī′kl) The citric acid cycle.
**Kupffer cell** (koop′fer sel) Large, fixed phagocyte in the liver that removes bacterial cells from the blood.
**kwashiorkor** (kwash″e-or′kor) Starvation resulting from a switch from breast milk to food deficient in nutrients.
**kyphosis** (ki-fo′sis) An abnormally increased convex curvature in the thoracic portion of the vertebral column.

## L

**labor** (la′bor) The process of childbirth.
**labyrinth** (lab′ĭ-rinth) The system of connecting tubes within the inner ear, including the cochlea, vestibule, and semicircular canals.
**lacrimal gland** (lak′rĭ-mal gland) Tear-secreting gland.
**lactase** (lak′tās) Enzyme that catalyzes breakdown of lactose into glucose and galactose.
**lactate** (lak′tāt) Lactic acid.
**lactation** (lak-ta′shun) Production of milk by the mammary glands.
**lacteal** (lak′te-al) A lymphatic capillary associated with a villus of the small intestine.
**lactic acid** (lak′tik as′id) An organic compound formed from pyruvic acid during anaerobic respiration.
**lactose** (lak′tōs) A disaccharide that occurs in milk; milk sugar.
**lacuna** (lah-ku′nah) A hollow cavity.
**lamella** (lah-mel′ah) A layer of matrix in bone tissue.
**large intestine** (lahrj in-tes′tin) The part of the gastrointestinal tract extending from the ileum to the anus; divided into the cecum, colon, rectum, and anal canal.
**laryngopharynx** (lah-ring″go-far′ingks) The lower portion of the pharynx near the opening to the larynx.
**larynx** (lar′ingks) Structure located between the pharynx and trachea that houses the vocal cords.
**latent period** (la′tent pe′re-od) Time between the application of a stimulus and the beginning of a response in a muscle fiber.
**lateral** (lat′er-al) Pertaining to the side.
**leukocyte** (lu′ko-sīt) A white blood cell.
**leukocytosis** (lu″ko-si-to′sis) Too many white blood cells.

**leukopenia** (lu″ko-pe′ne-ah) Too few leukocytes in the blood.
**lever** (lev′er) A simple mechanical device consisting of a rod, fulcrum, weight, and a source of energy that is applied to some point on the rod.
**ligament** (lig′ah-ment) A cord or sheet of connective tissue binding two or more bones at a joint.
**limbic system** (lim′bik sis′tem) A group of connected structures within the brain that produces emotional feelings.
**linea alba** (lin′e-ah al′bah) A narrow band of tendinous connective tissue in the midline of the anterior abdominal wall.
**lingual** (ling′gwal) Pertaining to the tongue.
**lipase** (lī′pās) A fat-digesting enzyme.
**lipid** (lip′id) A fat, oil, or fatlike compound that usually has fatty acids in its molecular structure.
**lipoprotein** (lip″o-pro′te-in) A complex of lipid and protein.
**liver** (liv′er) A large, dark red organ in the upper part of the abdomen on the right side that detoxifies blood, stores glycogen and fat-soluble vitamins, and synthesizes proteins.
**lobule** (lob′ul) A small, well-defined portion of an organ.
**lordosis** (lor-do′sis) An abnormally increased concave curvature in the lumbar portion of the vertebral column.
**lower limb** (loh′er lim) Inferior appendage consisting of the thigh, leg, ankle, and foot.
**lumbar** (lum′bar) Pertaining to the region of the loins.
**lumen** (lu′men) Space within a tubular structure such as a blood vessel or intestine.
**luteinizing hormone** (lu′te-in-īz″ing hor′mōn) A hormone secreted by the anterior pituitary gland that controls the formation of the corpus luteum in females and the secretion of testosterone in males; LH (ICSH in males).
**lymph** (limf) Fluid transported by the lymphatic vessels.
**lymph node** (limf nōd) A mass of lymphoid tissue located along the course of a lymphatic vessel.
**lymphocyte** (lim′fo-sīt) A type of white blood cell that provides immunity.
**lysosome** (li′so-sōm) Organelle that contains enzymes that degrade worn cell parts.

## M

**macrocyte** (mak′ro-sīt) A large red blood cell.
**macromineral** (mak′ro-min″er-al) An inorganic substance that is necessary for metabolism and is one of a group that accounts for 75% of the mineral elements within the body; major mineral.
**macrophage** (mak′ro-fāj) A large phagocytic cell.
**macroscopic** (mak″ro-skop′ik) Large enough to be seen with the unaided eye.
**macula** (mak′u-lah) A group of hair cells and supporting cells associated with an organ of static equilibrium.
**macula lutea** (mak′u-lah lu′te-ah) A yellowish depression in the retina of the eye that is associated with acute vision.
**malabsorption** (mal″ab-sorp′shun) Failure to absorb nutrients following digestion.
**malignant** (mah-lig′nant) The power to threaten life; cancerous.
**malnutrition** (mal″nu-trish′un) Physical symptoms resulting from lack of specific nutrients.
**maltase** (mawl′tās) An enzyme that catalyzes conversion of maltose into glucose.
**maltose** (mawl′tōs) A disaccharide composed of two glucose molecules.
**mammary** (mam′ar-e) Pertaining to the breast.
**mammillary body** (mam′ĭ-lar″e bod′e) Two small, rounded bodies posterior to the hypothalamus involved with reflexes associated with the sense of smell.
**marasmus** (mah-raz′mus) Starvation due to profound nutrient deficiency.
**marrow** (mar′o) Connective tissue that occupies the spaces within bones and includes stem cells.
**mast cell** (mast sel) A cell to which antibodies, formed in response to allergens, attach, bursting the cell and releasing allergy mediators, which cause symptoms.
**mastication** (mast″ĭ-ka′shun) Chewing movements.
**matrix** (ma′triks) The intercellular substance of connective tissue.
**matter** (mat′er) Anything that has weight and occupies space.
**meatus** (me-a′tus) A passageway or channel, or the external opening of such a passageway.
**mechanoreceptor** (mek″ah-no-re-sep′tor) A sensory receptor that is sensitive to mechanical stimulation such as changes in pressure or tension.
**medial** (me′de-al) Toward or near the midline.
**mediastinum** (me″de-ah-sti′num) Tissues and organs of the thoracic cavity that form a septum between the lungs.
**medulla** (mĕ-dul′ah) The inner portion of an organ.
**medulla oblongata** (mĕ-dul′ah ob″long-gah′tah) Portion of the brain stem located between the pons and the spinal cord.
**medullary cavity** (med′u-lār″e kav′ĭ-te) Cavity within the diaphysis of a long bone containing marrow.
**medullary rhythmicity area** (med′u-lār″e rith-mi′si-te air′e-a) Area of the brain stem that controls the basic rhythm of inspiration and expiration.
**megakaryocyte** (meg″ah-kar′e-o-sīt) A large bone marrow cell that gives rise to blood platelets.
**meiosis** (mi-o′sis) Cell division halves the genetic material, resulting in egg and sperm cells (gametes).
**Meissner′s corpuscle** (mīs′nerz kor′pus-l) Sensory receptor close to the surface of the skin that is sensitive to light touch.
**melanin** (mel′ah-nin) Dark pigment found in skin and hair.
**melanocyte** (mel′ah-no-sīt) Melanin-producing cell.
**melatonin** (mel″ah-to′nin) A hormone secreted by the pineal gland.
**memory cell** (mem′o-re sel) B-lymphocyte or T-lymphocyte produced in a primary immune response that remains dormant and can be activated rapidly if the same antigen is encountered in the future.
**menarche** (mĕ-nar′ke) The first menstrual period.
**meninges** (mĕ-nin′jēz) Three membranes that cover the brain and spinal cord (sing. *meninx*).
**meniscus** (men-is′kus) Fibrocartilage that separates the articulating surfaces of bones in the knee (pl. *menisci*).
**menopause** (men′o-pawz) Termination of the menstrual cycle.
**menstrual cycle** (men′stroo-al si′kl) Recurring changes in the uterine lining of a woman of reproductive age.
**menstruation** (men″stroo-a′shun) Loss of blood and tissue from the uterine lining at the end of a female reproductive cycle.

**mental** (men′tal) Pertaining to the mind; pertaining to the chin body region.
**mesentery** (mes′en-ter″e) A fold of peritoneal membrane that attaches an abdominal organ to the abdominal wall.
**mesoderm** (mez′o-derm) The middle primary germ layer.
**messenger RNA** (mes′in-jer RNA) RNA that transmits information for a protein's amino acid sequence from the nucleus of a cell to the cytoplasm.
**metabolic rate** (met″ah-bol′ic rāt) The rate at which chemical changes occur within the body.
**metabolism** (mĕ-tab′o-lizm) All of the chemical changes that occur within cells considered together.
**metacarpals** (met″ah-kar′pals) Bones of the hand between the wrist and finger bones.
**metaphase** (met′ah-fāz) Stage in mitosis when chromosomes align in the middle of the spindle.
**metastasis** (mĕ-tas′tah-sis) The spread of disease from one body region to another; a characteristic of cancer.
**metatarsals** (met″ah-tar′sals) Bones of the foot between the ankle and toe bones.
**microfilament** (mi″kro-fil′ah-ment) Tiny rod of protein in the cytoplasm that provides structural support or movement.
**microglia** (mi-krog′le-ah) Neuroglial cells that support neurons and phagocytize.
**micromineral** (mi′kro min″er al) An essential mineral present in a minute amount within the body; trace element.
**microtubule** (mi″kro-tu′būl) A minute, hollow rod of the protein tubulin.
**microvilli** (mi″kro-vil′i) Tiny, cylindrical processes that extend from some epithelial cell membranes and increase the membrane surface area.
**micturition** (mik″tu-rish′un) Urination.
**midbrain** (mid′brān) A small region of the brain stem located between the diencephalon and the pons.
**mineralocorticoid** (min″er-al-o-kor′tĭ-koid) Hormones the adrenal cortex secretes that influence the concentrations of electrolytes in body fluids.
**mitochondrion** (mi″to-kon′dre-on) Organelle housing enzymes that catalyze reactions of aerobic respiration (pl. *mitochondria*).
**mitosis** (mi-to′sis) Division of a somatic cell to form two genetically identical cells.
**mitral valve** (mi′tral valv) Heart valve located between the left atrium and the left ventricle; bicuspid valve.
**mixed nerve** (mikst nerv) Nerve that includes both sensory and motor nerve fibers.
**molar** (mo′lar) A rear tooth with a flattened surface adapted for grinding food.
**molding** (mold′ing) Changing in shape of the fetal skull during birth.
**molecular formula** (mo-lek′u-lar for′mu-lah) An abbreviation for the number of atoms of each element in a compound.
**molecule** (mol′ĕ-kūl) A particle composed of two or more atoms joined together.
**monoamine inhibitor** (mon″o-am′in in-hib′i-tor) A substance that inhibits the action of the enzyme monoamine oxidase.
**monocyte** (mon′o-sīit) A type of white blood cell that functions as a phagocyte.
**monosaccharide** (mon″o-sak′ah-rīd) A single sugar, such as glucose or fructose.
**monosomy** (mon′o-so″me) A cell missing one chromosome.
**morula** (mor′u-lah) An early stage in preembryonic development; a solid ball of cells.
**motor area** (mo′tor a′re-ah) A region of the brain from which impulses to muscles or glands originate.
**motor end plate** (mo′tor end plāt) Specialized portion of a muscle fiber membrane at a neuromuscular junction.
**motor nerve** (mo′tor nerv) A nerve that consists of motor nerve fibers.
**motor neuron** (mo′tor nu′ron) A neuron that transmits impulses from the central nervous system to an effector.
**motor unit** (mo′tor unit) A motor neuron and the muscle fibers associated with it.
**mucosa** (mu-ko′sah) The membrane that lines tubes and body cavities that open to the outside of the body; mucous membrane.
**mucous cell** (mu′kus sel) Glandular cell that secretes mucus.
**mucous membrane** (mu′kus mem′brān) Mucosa.
**mucus** (mu′kus) Fluid secretion of the mucous cells.
**multiple alleles** (mul′tĭ-pl ah-lēls′) Different forms of a gene.
**multiple motor unit summation** (mul′tĭ-pl mo′tor u′nit sum-mā′shun) A sustained muscle contraction of increasing strength in response to input from many motor units.
**multipolar neuron** (mul″tĭ-po′lar nu′ron) Nerve cell that has many processes arising from its cell body.
**muscle impulse** (mus′el im′puls) Impulse that travels along the sarcolemma to the transverse tubules.
**muscle spindle** (mus′el spin′dul) Modified skeletal muscle fiber that can respond to changes in muscle length.
**muscle tone** (mus′el tōn) The contraction of some fibers in skeletal muscle at any given time.
**mutagen** (mu′tah-jen) An agent that can cause mutations.
**mutation** (mu-ta′shun) A change in a gene.
**myelin** (mi′ĕ-lin) Fatty material that forms a sheathlike covering around some nerve fibers.
**myocardium** (mi″o-kar′de-um) Muscle tissue of the heart.
**myofibril** (mi″o-fi′bril) Contractile fibers within muscle cells.
**myoglobin** (mi″o-glo′bin) A pigmented compound in muscle tissue that stores oxygen.
**myogram** (mi′o-gram) A recording of a muscular contraction.
**myometrium** (mi″o-me′tre-um) The layer of smooth muscle tissue within the uterine wall.
**myoneural junction** (mi″o-nu′ral jungk′shun) Site of union between a motor neuron axon and a muscle fiber.
**myopia** (mi-o′pe-ah) Nearsightedness.
**myosin** (mi′o-sin) A protein that, together with actin, produces muscular contraction and relaxation.
**myxedema** (mik″sĕ-de′mah) A deficiency of thyroid hormones in an adult.

## N

**nasal cavity** (na′zal kav′ ĭ-te) Space within the nose.
**nasal concha** (na′zal kong′kah) Shell-like bone extending outward from the wall of the nasal cavity; a turbinate bone.

**nasal septum** (na′zal sep′tum) A wall of bone and cartilage that separates the nasal cavity into two portions.

**nasopharynx** (na″zo-far′ingks) Portion of the pharynx associated with the nasal cavity.

**natural killer cell** (nat′u-ral kil′er sel) Lymphocyte that causes an infected or cancerous cell to burst.

**negative feedback** (neg′ah-tiv fēd′bak) A mechanism activated by an imbalance that corrects it.

**neonatal** (ne″o-na′tal) The first four weeks of life.

**nephron** (nef′ron) The functional unit of a kidney, consisting of a renal corpuscle and a renal tubule.

**nerve** (nerv) A bundle of nerve fibers.

**nerve fiber** (nerv fi′ber) An axon.

**nerve impulse** (nerv im′puls) The electrochemical process of depolarization and repolarization along a nerve fiber.

**nerve tract** (nerv trakt) A long bundle of nerve fibers having the same origin, function and termination.

**neurilemma** (nur″ĭ-lem′ah) Sheath on the outside of some nerve fibers formed from Schwann cells.

**neurofibrils** (nu″ro-fi′brils) Fine cytoplasmic threads that extend from the cell body into the processes of neurons.

**neuroglial cell** (nu-rog′le-ahl sel) Specialized cell of the nervous system that produces myelin, communicates between cells, and maintains the ionic environment, as well as provides other functions.

**neuromodulator** (nu″ro-mod′u-lā-tor) A substance that alters a neuron's response to a neurotransmitter.

**neuromuscular junction** (nu″ro-mus′ku-lar jungk′shun) Point of contact between a nerve and muscle cell.

**neuron** (nu′ron) A nerve cell that consists of a cell body and its processes.

**neuropeptide** (nu″ro-pep′tīd) A peptide in the brain that functions as a neurotransmitter or neuromodulator.

**neurosecretory cell** (nu″ro-se-kre′to-re sel) Cell in the hypothalamus that functions as a neuron at one end but like an endocrine cell at the other, by receiving messages and secreting the hormones ADH and oxytocin.

**neurotransmitter** (nu″ro-trans-mit′er) Chemical secreted by the end of an axon that stimulates a muscle fiber to contract or a neuron to fire an impulse.

**neutral** (nu′tral) Neither acid nor alkaline; pH 7.

**neutron** (nu′tron) An electrically neutral particle found in an atomic nucleus.

**neutrophil** (nu′tro-fil) A type of phagocytic leukocyte.

**niacin** (ni′ah-sin) A vitamin of the B-complex group; nicotinic acid.

**niacinamide** (ni″ah-sin′ah-mīd) The physiologically active form of niacin.

**nicotinic acid** (nik″o-tin′ik as′id) Niacin.

**Nissl bodies** (nis′l bod′ēz) Membranous sacs within the cytoplasm of nerve cells that have ribosomes attached to their surfaces.

**nitrogen balance** (ni′tro-jen bal′ans) Condition in which the quantity of nitrogen ingested equals the quantity excreted.

**node of Ranvier** (nōd of Ron′vee-ay) A short region of exposed (unmyelinated) axon between adjacent Schwann cells on neurons of the peripheral nervous system.

**nodule** (nod′ul) A small mass of tissue detected by touch.

**nondisjunction** (non″dis-jungk′shun) Failure of a pair of chromosomes to separate during meiosis.

**nonelectrolyte** (non″e-lek′tro-līt) A substance that does not dissociate into ions when it is dissolved.

**nonheritable gene therapy** (non-her′i-tah-bl jēn ther′ah-pe) Manipulation of genes in somatic cells to correct the effects of a mutation.

**nonprotein nitrogenous substance** (non-pro′te-in ni-troj′ĕ-nus sub′stans) A substance, such as urea or uric acid, that contains nitrogen but is not a protein.

**norepinephrine** (nor″ep-ĭ-nef′rin) A neurotransmitter released from the axons of some nerve fibers.

**nuclease** (nu′kle-ās) An enzyme that catalyzes decomposition of nucleic acids.

**nucleic acid** (nu-kle′ik as′id) A substance composed of nucleotides bonded together; RNA or DNA.

**nucleolus** (nu-kle′o-lus) A small structure within the nucleus of a cell that contains RNA and proteins (pl. *nucleoli*).

**nucleoplasm** (nu′kle-o-plazm″) The contents of the nucleus of a cell.

**nucleosome** (nu′kle-o-sōm) A beadlike part of a chromatin fiber consisting of DNA wrapped around protein octets.

**nucleotide** (nu′kle-o-tīd″) A building block of a nucleic acid molecule, consisting of a sugar, a nitrogenous base, and a phosphate group.

**nucleus** (nu′kle-us) A cellular organelle that is enclosed by a double-layered, porous membrane and contains DNA; the dense core of an atom that is composed of protons and neutrons (pl. *nuclei*).

**nutrient** (nu′tre-ent) A chemical substance that must be supplied to the body from the environment.

**nutrition** (nu-trish′un) The study of the sources, actions, and interactions of nutrients.

## O

**obesity** (o-bēs′ĭ-te) Excess adipose tissue; exceeding desirable weight by more than 20%.

**occipital** (ok-sip′ĭ-tal) Pertaining to the lower, back portion of the head.

**olfactory** (ol-fak′to-re) Pertaining to the sense of smell.

**olfactory nerves** (ol-fak′to-re nervz) The first pair of cranial nerves, which conduct impulses associated with the sense of smell.

**oligodendrocyte** (ol″ĭ-go-den′dro-sīt) A type of neuroglial cell that connects neurons to blood vessels and forms myelin.

**oncogene** (ong′ko-jēn) A gene that normally controls cell division but when overexpressed leads to cancer.

**oocyte** (o′o-sīt) An immature egg cell.

**oogenesis** (o″o-jen′ĕ-sis) Differentiation of an egg cell.

**ophthalmic** (of-thal′mik) Pertaining to the eye.

**optic** (op′tik) Pertaining to the eye.

**optic chiasma** (op′tik ki-az′mah) X-shaped structure on the underside of the brain formed by a partial crossing over of fibers in the optic nerves.

**optic disk** (op′tik disk) Region in the retina of the eye where nerve fibers leave to become part of the optic nerve.

**oral** (o′ral) Pertaining to the mouth.

**orbital** (or′bĭ-tal) Pertaining to the body region of the eyeball; a region, in the atom, containing the electrons.

**organ** (or′gan) A structure consisting of a group of tissues that performs a specialized function.
**organ of Corti** (or′gan uv kor′te) An organ in the cochlear duct that contains the receptors for hearing. It consists of hair cells and supporting cells.
**organelle** (or″gah-nel′) A part of a cell that performs a specialized function.
**organic** (or-gan′ik) Carbon-containing molecules.
**organism** (or′gah-nizm) An individual living thing.
**orgasm** (or′gaz-em) The culmination of sexual excitement.
**orifice** (or′ĭ-fis) An opening.
**origin** (or′ ĭ-jin) End of a muscle that is attached to a relatively immovable part.
**oropharynx** (o″ro-far′ingks) Portion of the pharynx in the posterior part of the oral cavity.
**osmoreceptor** (oz″mo-re-sep′tor) Receptor that is sensitive to changes in the osmotic pressure of body fluids.
**osmosis** (oz-mo′sis) Diffusion of water through a selectively permeable membrane in response to a concentration gradient.
**osmotic** (oz-mot′ik) Pertaining to osmosis.
**osmotic pressure** (oz-mot′ik presh′ur) The amount of pressure needed to stop osmosis; the potential pressure of a solution caused by nondiffusible solute particles in the solution.
**osseous tissue** (os′e-us tish′u) Bone tissue.
**ossification** (os″ ĭ-f ĭ-ka′shun) The formation of bone tissue.
**osteoblast** (os′te-o-blast″) A bone-forming cell.
**osteoclast** (os′te-o-klast″) A cell that erodes bone.
**osteocyte** (os′te-o-sīt) A mature bone cell.
**osteon** (os′te-on) A cylinder-shaped unit containing bone cells that surround an osteonic canal; Haversian system.
**osteonic canal** (os′te-o-nik ka-nal′) A tiny channel in bone tissue that contains a blood vessel; Haversian canal.
**osteoporosis** (os″te-o-po-ro′sis) A condition in which bones break easily because calcium is removed from them faster than it is replaced.
**otic** (o′tik) Pertaining to the ear.
**otolith** (o′to-lith) A small particle of calcium carbonate associated with the receptors of equilibrium.
**otosclerosis** (o″to-sklĕ-ro′sis) Abnormal formation of spongy bone within the ear that may interfere with the transmission of sound vibrations to hearing receptors.
**oval window** (o′val win′do) Opening between the stapes and the inner ear.
**ovarian** (o-va′re-an) Pertaining to the ovary.
**ovary** (o′var-e) The primary reproductive organ of a female; an egg-cell-producing organ.
**oviduct** (o′vĭ-dukt) A tube that leads from the ovary to the uterus; uterine tube or fallopian tube.
**ovulation** (o″vu-la′shun) The release of an egg cell from a mature ovarian follicle.
**ovum** (o′vum) A mature egg cell.
**oxidase** (ok′sĭ-dās) An enzyme that promotes oxidation.
**oxidation** (ok″sĭ-da′shun) Process by which oxygen is combined with another chemical; the removal of hydrogen or the loss of electrons; the opposite of reduction.
**oxidative phosphorylation** (ok′sĭ-da-tiv fos″fo-ri-la′shun) The process of transferring electrons to form a high energy phosphate bond by introducing a phosphate group to ADP and forming ATP.
**oxygen debt** (ok′sĭ-jen det) The amount of oxygen that must be supplied following physical exercise to convert accumulated lactic acid to glucose.
**oxyhemoglobin** (ok″sĭ-he″mo-glo′bin) Compound formed when oxygen combines with hemoglobin.
**oxytocin** (ok″sĭ-to′sin) Hormone released by the posterior lobe of the pituitary gland that contracts smooth muscles in the uterus and mammary glands.

## P

**pacemaker** (pās′māk-er) Mass of specialized cardiac muscle tissue that controls the rhythm of the heartbeat; the sinoatrial node.
**Pacinian corpuscle** (pah-sin′e-an kor′pusl) Nerve endings deep in the dermis providing perception of pressure.
**pain receptor** (pān re″sep′tor) Sensory nerve ending associated with the feeling of pain.
**palate** (pal′at) The roof of the mouth.
**palatine** (pal′ah-tīn) Pertaining to the palate.
**palmar** (pahl′mar) Pertaining to the palm of the hand.
**pancreas** (pan′kre-as) Glandular organ in the abdominal cavity that secretes hormones and digestive enzymes.
**pancreatic** (pan″kre-at′ik) Pertaining to the pancreas.
**pantothenic acid** (pan″to-the′nik as′id) A vitamin of the B-complex group.
**papilla** (pah-pil′ah) Tiny, nipplelike projection.
**papillary muscle** (pap′ ĭ-ler″e mus′el) Muscle that extends inward from the ventricular walls of the heart and to which the chordae tendineae attach.
**paracrine** (par′ah-krin) A type of endocrine secretion in which the hormone affects nearby cells.
**paradoxical sleep** (par″ah-dok′se-kal slēp) Sleep in which some areas of the brain are active, producing dreams and rapid eye movements.
**paralysis** (pah-ral′i-sis) Loss of ability to control voluntary muscular movements, usually due to a disorder of the nervous system.
**parasympathetic division** (par″ah-sim″pah-thet′ik dĭ-vizh′un) Portion of the autonomic nervous system that arises from the brain and sacral region of the spinal cord.
**parathyroid glands** (par″ah-thi′roid glandz) Small endocrine glands embedded in the posterior portion of the thyroid gland.
**parathyroid hormone** (par″ah-thi′roid hor′mōn) Hormone secreted by the parathyroid glands that helps regulate the level of blood calcium and phosphate ions; PTH.
**paravertebral ganglia** (par″ah-ver′tĕ-bral gang′gle-ah) Sympathetic ganglia that form chains along the sides of the vertebral column.
**parietal** (pah-ri′ĕ-tal) Pertaining to the wall of an organ or cavity.
**parietal cell** (pah-ri′ĕ-tal sel) Cell of a gastric gland that secretes hydrochloric acid and intrinsic factor.
**parietal pleura** (pah-ri′ĕ-tal ploo′rah) Membrane that lines the inner wall of the thoracic cavity.
**parotid glands** (pah-rot′id glandz) Large salivary glands located on the sides of the face just in front and below the ears.
**partial pressure** (par′shal presh′ur) The pressure one gas produces in a mixture of gases.
**parturition** (par″tu-rish′un) Childbirth.

**pathogen** (path′o-jen) A disease-causing agent.
**pathology** (pah-thol′o-je) The study of disease.
**pectoral** (pek′tor-al) Pertaining to the chest.
**pectoral girdle** (pek′tor-al ger′dl) Portion of the skeleton that provides support and attachment for the arms.
**pedal** (ped′al) Pertaining to the foot.
**pedigree** (ped′ĭ-gre) A chart showing the relationships of relatives and which ones have a particular trait.
**pelvic** (pel′vik) Pertaining to the pelvis.
**pelvic girdle** (pel′vik ger′dl) Portion of the skeleton to which the legs are attached.
**pelvic inflammatory disease** (pel′vik in-flam′ah-tor′e dĭ-zēz′) An ascending infection of the upper female genital tract.
**pelvis** (pel′vis) Bony ring formed by the sacrum and coxal bones.
**penis** (pe′nis) External reproductive organ of the male through which the urethra passes.
**pepsin** (pep′sin) Protein-splitting enzyme secreted by the gastric glands of the stomach.
**pepsinogen** (pep-sin′o-jen) Inactive form of pepsin.
**peptidase** (pep′tĭ-dās) An enzyme that catalyzes the breakdown of polypeptides.
**peptide** (pep′tīd) Compound composed of two or more amino acid molecules bonded together.
**peptide bond** (pep′tīd bond) Bond that forms between the carboxyl group of one amino acid and the amino group of another.
**perception** (per-sep′shun) Mental awareness of sensory stimulation.
**pericardial** (per″ĭ-kar′de-al) Pertaining to the pericardium.
**pericardium** (per″ĭ-kar′de-um) Serous membrane that surrounds the heart.
**perichondrium** (per″ĭ-kon′dre-um) Layer of fibrous connective tissue that encloses cartilaginous structures.
**perilymph** (per′ĭ-limf) Fluid contained in the space between the membranous and osseous labyrinths of the inner ear.
**perimetrium** (per-ĭ-me′tre-um) The outer serosal layer of the uterine wall.
**perimysium** (per″ĭi-mis′e-um) Sheath of connective tissue that encloses a bundle of striated muscle fibers.
**perineal** (per″ĭ-ne′al) Pertaining to the perineum.
**perineum** (per″ĭ-ne′um) Body region between the scrotum or urethral opening and the anus.
**perineurium** (per″ĭ-nu′re-um) Layer of connective tissue that encloses a bundle of nerve fibers within a nerve.
**periodontal ligament** (per″e-o-don′tal lig′ah-ment) Fibrous membrane that surrounds a tooth and attaches it to the bone of the jaw.
**periosteum** (per″e-os′te-um) Covering of fibrous connective tissue on the surface of a bone.
**peripheral** (pĕ-rif′er-al) Pertaining to parts located near the surface or toward the outside.
**peripheral nervous system** (pĕ-rif′er-al ner′vus sis′tem) The portions of the nervous system outside the central nervous system.
**peripheral protein** (pe-rif′er-al pro′te-in) A globular protein associated with the inner surface of the cell membrane.
**peripheral resistance** (pĕ-rif′er-al re-zis′tans) Resistance to blood flow due to friction between the blood and the walls of the blood vessels.
**peristalsis** (per″ĭ-stal′sis) Rhythmic waves of muscular contraction that occur in the walls of certain tubular organs.
**peritoneal** (per″ĭ-to-ne′al) Pertaining to the peritoneum.
**peritoneal cavity** (per″ĭ-to-ne′al kav′ĭ-te) The potential space between the parietal and visceral peritoneal membranes.
**peritoneum** (per″ĭ-to-ne′um) A serous membrane that lines the abdominal cavity and encloses the abdominal viscera.
**peritubular capillary** (per″ĭ-tu′bu-lar kap′ĭ-ler″e) Capillary that surrounds a renal tubule and functions in reabsorption and secretion during urine formation.
**permeable** (per′me-ah-bl) Open to passage or penetration.
**peroxisome** (pĕ-roks′ĭ-sōm) Membranous cytoplasmic vesicle that contains enzymes that catalyze reactions that produce and decompose hydrogen peroxide.
**pH** The negative logarithm of the hydrogen ion concentration used to indicate the acid or alkaline condition of a solution; values range from 0 to 14.
**phagocytosis** (fag″o-si-to′sis) Process by which a cell engulfs and digests solid substances.
**phalanx** (fa′langks) A bone of a finger or toe (pl. *phalanges*).
**pharynx** (far′ingks) Portion of the digestive tube between the mouth and the esophagus.
**phenotype** (fe′no-tīp) The appearance of an individual due to the action of a particular set of genes.
**phosphate buffer system** (fos′fāt buf′er sis′tem) Consists of a mix of sodium monohydrogen phosphate and sodium dihydrogen phosphate to weaken a strong base and a strong acid, respectively; to resist changes in pH.
**phospholipid** (fos″fo-lip′id) A lipid that contains two fatty acid molecules and a phosphate group combined with a glycerol molecule.
**photoreceptor** (fo″to-re-sep′tor) A nerve ending that is sensitive to light energy.
**physiology** (fiz″e-ol′o-je) The branch of science that studies body functions.
**pia mater** (pi′ah ma′ter) Inner layer of meninges that encloses the brain and spinal cord.
**pineal gland** (pin′e-al gland) A small structure located in the central part of the brain.
**pinocytosis** (pin″o-si-to′sis) Process by which a cell engulfs droplets of fluid from its surroundings.
**pituitary gland** (pĭ-tu′ĭ-tār″e gland) Endocrine gland attached to the base of the brain that consists of anterior and posterior lobes; the hypophysis.
**placenta** (plah-sen′tah) Structure attaching a fetus to the uterine wall, providing a conduit to receive nutrients and excrete wastes.
**placental lactogen** (plah-sen′tahl lak′to-jen) Hormone secreted by the placenta to inhibit maternal insulin activity during pregnancy.
**plantar** (plan′tar) Pertaining to the sole of the foot.
**plantar flexion** (plan′tar flek′shun) Bending the foot downward.
**plasma** (plaz′mah) Fluid portion of circulating blood.
**plasma cell** (plaz′mah sel) Antibody-producing cell forms as a result of the proliferation of sensitized B cells.
**plasma protein** (plaz′mah pro′te-in) Proteins dissolved in blood plasma.

**platelet** (plāt′let) Cytoplasmic fragment formed in the bone marrow that functions in blood coagulation.

**pleiotropy** (plī-ah′tre-pē) A gene that has several expressions (phenotypes).

**pleural** (ploo′ral) Pertaining to the pleura or membranes investing the lungs.

**pleural cavity** (ploo′ral kav′ ĭ-te) Potential space between the pleural membranes.

**pleural membranes** (ploo′ral mem′brānz) Serous membranes that enclose the lungs.

**plexus** (plek′sus) A network of interlaced nerves or blood vessels.

**pneumotaxic area** (nu″mo-tax′ik a′re-ah) A portion of the respiratory control center in the pons of the brain.

**polar body** (po′lar bod′e) Small, nonfunctional cell that is a product of meiosis in the female.

**polarization** (po″lar-ĭ-za′shun) Development of an electrical charge on the surface of a cell membrane due to an unequal distribution of ions on either side of the membrane.

**polycythemia** (pol″e-si-the′me-ah) Excess red blood cells.

**polydactyly** (pol″e-dak′ti-le) Extra fingers or toes.

**polymorphonuclear leukocyte** (pol″e-mor″fo-nu′kle-ar lu′ko-sīt) A white blood cell (leukocyte) with an irregularly lobed nucleus.

**polynucleotide** (pol″e-noo′-kle-o-tīd) A compound formed by the union of many nucleotides; a nucleic acid.

**polypeptide** (pol″e-pep′tīd) A compound formed by the union of many amino acid molecules.

**polyploidy** (pol′e-ploi″de) A condition in which a cell has one or more extra sets of chromosomes.

**polysaccharide** (pol″e-sak′ah-rīd) A carbohydrate composed of many monosaccharides joined together.

**pons** (ponz) A portion of the brain stem above the medulla oblongata and below the midbrain.

**popliteal** (pop″lĭ-te′al) Pertaining to the region behind the knee.

**positive chemotaxis** (poz′ ĭ-tiv ke″mo-tak′sis) Movement of a cell toward the greater concentration of a substance.

**positive feedback** (poz′ ĭ-tiv fēd′bak) Process by which changes cause more changes of a similar type, producing unstable conditions.

**posterior** (pos-tēr′e-or) Toward the back; opposite of anterior.

**postganglionic fiber** (pōst″gang-gle-on′ik fi′ber) Autonomic nerve fiber located on the distal side of a ganglion.

**postnatal** (pōst-na′tal) After birth.

**postsynaptic neuron** (pōst″sĭ-nap′tik nu′ron) One of two adjacent neurons transmitting an impulse; cell situated after the synapse is crossed.

**postsynaptic potential** (pōst″sĭ-nap′tik po-ten′shal) Membrane polarization is increased (excitatory) or decreased (inhibitory) in the postsynaptic neuron with repeated stimulation over an excitatory or inhibitory pathway so that the neuron will either fire or have diminished responsiveness.

**precursor** (pre-ker′sor) Substance from which another substance forms.

**preganglionic fiber** (pre″gang-gle-on′ik fi′ber) Autonomic nerve fiber located on the proximal side of a ganglion.

**pregnancy** (preg′nan-se) The condition in which a female has a developing offspring in her uterus.

**prenatal** (pre-na′tal) Before birth.

**presbycusis** (pres″bĭ-ku′sis) Loss of hearing that accompanies old age.

**presbyopia** (pres″be-o′pe ah) Loss of the eye's ability to accommodate due to declining elasticity in the lens; farsightedness of age.

**presynaptic neuron** (pre″sĭ-nap′tik nu′ron) One of two adjacent neurons transmitting an impulse; cell situated before the synapse is crossed.

**primary germ layers** (pri′ma-re jerm lā′erz) Three layers (endoderm, mesoderm, and ectoderm) of embryonic cells that develop into specific tissues and organs.

**primary immune response** (pri′ma-re ĭ-mūn′ re-spons′) The immune system's response to its first encounter with a foreign antigen.

**primary reproductive organs** (pri′ma-re re″pro-duk′tiv or′ganz) Sex-cell-producing parts; testes in males and ovaries in females.

**prime mover** (prīm mōōv′er) Muscle that is mainly responsible for a particular body movement.

**profibrinolysin** (pro″fi-brĭ-no-li′sin) The inactive form of fibrinolysin.

**progeny** (proj′e-ne) Offspring; new cells resulting from cell division.

**progesterone** (pro-jes′tĕ-rōn) A female hormone secreted by the corpus luteum of the ovary and by the placenta.

**projection** (pro-jek′shun) Process by which the brain causes a sensation to seem to come from the region of the body being stimulated.

**prolactin** (pro-lak′tin) Hormone secreted by the anterior pituitary gland that stimulates the production of milk from the mammary glands; PRL.

**pronation** (pro-na′shun) Movement of the palm of the hand downward or backward.

**prophase** (pro′fāz) Stage of mitosis when chromosomes become visible.

**proprioceptor** (pro″pre-o-sep′tor) A sensory nerve ending that is sensitive to changes in tension of a muscle or tendon.

**prostacyclin** (pros″tah-si′klin) A substance released from endothelial cells that inhibits platelet adherence.

**prostaglandins** (pros″tah-glan′dins) A group of compounds that have powerful, hormonelike effects.

**prostate gland** (pros′tāt gland) Gland surrounding the male urethra below the urinary bladder that adds its secretion to semen during ejaculation.

**protein** (pro′tēn) Nitrogen-containing organic compound composed of amino acid molecules joined together.

**protein buffer system** (pro′tēn buf′er sis′tem) The amino acids of a protein accept or donate hydrogen ions to keep the concentration of hydrogen ions in solution constant; resist changes in pH.

**prothrombin** (pro-throm′bin) Plasma protein that functions in the formation of blood clots.

**proton** (pro′ton) A positively charged particle in an atomic nucleus.

**protraction** (pro-trak′shun) A forward movement of a body part.

**proximal** (prok′sĭ-mal) Closer to the midline or origin; opposite of distal.

**pseudostratified** (soo″do-strat′ĭ-fīd) A single layer of epithelium appearing as more than one layer because the nuclei occupy different positions in the cells.
**puberty** (pu′ber-te) Stage of development in which the reproductive organs become functional.
**pulmonary** (pul′mo-ner″e) Pertaining to the lungs.
**pulmonary circuit** (pul′mo-ner″e ser′kit) System of blood vessels that carries blood between the heart and the lungs.
**pulmonary valve** (pul′mo-ner″e valv) Valve leading from the right ventricle to the pulmonary trunk (artery); pulmonary semilunar valve.
**pulse** (puls) The surge of blood felt through the walls of arteries due to the contraction of the ventricles of the heart.
**pupil** (pu′pil) Opening in the iris through which light enters the eye.
**Purkinje fibers** (pur-kin′je fi′berz) Specialized muscle fibers that conduct the cardiac impulse from the A-V bundle into the ventricular walls.
**pyloric sphincter muscle** (pi-lor′ik sfingk′ter mus′l) A ring of muscle located between the stomach and the duodenum; controls food entry into the duodenum.
**pyramidal cell** (pĭ-ram′ĭ-dal sel) A large, pyramid-shaped neuron found within the cerebral cortex.
**pyridoxine** (pir′ĭ-dok′sēn) A vitamin of the B-complex group; vitamin $B_6$.
**pyrogen** (pi′ro-jen) A biochemical that causes an increase in body temperature.
**pyruvic acid** (pi-roo′vik as′id) An intermediate product of carbohydrate oxidation.

## R

**radiation** (ra″de-a′shun) A form of energy that includes visible light, ultraviolet light, and X rays; the means by which body heat is lost in the form of infrared rays.
**radioactive** (ra″de-o-ak′tiv) An atom that releases energy at a constant rate.
**rate-limiting enzyme** (rāt lim′ĭ-ting en′zīm) An enzyme, usually present in limited amount, that controls the rate of a metabolic pathway by regulating one of its steps.
**reabsorption** (re″ab-sorp′shun) Selective reuptake of substances into or across tissues.
**receptor** (re″sep′tor) A structure, usually built of protein, at the distal end of a sensory dendrite that is sensitive to stimulation.
**receptor-mediated endocytosis** (re″sep′tor-me′de-ā-tid en″do-si-to′sis) A selective uptake of molecules into a cell by binding to a specific receptor.
**recessive gene** (re-ses′iv jēn) An allele of a gene pair that is not expressed while the other member of the pair is expressed.
**recruitment** (re-kro͞ot′ment) Increase in number of motor units activated as intensity of stimulation increases.
**rectum** (rek′tum) The terminal end of the digestive tube between the sigmoid colon and the anus.
**red blood cell** (red blud sel) A disc-shaped cell, lacking a nucleus, that is packed with the oxygen-carrying molecule hemoglobin.
**red marrow** (red mar′o) Blood-cell-forming tissue located in spaces within bones.
**red muscle** (red mus′el) Slow-contracting postural muscles that contain an abundance of myoglobin.
**reduction** (re-duk′shun) A chemical reaction in which electrons are gained; to properly realign a fractured bone.
**referred pain** (re-ferd′ pān) Pain that feels as if it is originating from a part other than the site being stimulated.
**reflex** (re′fleks) A rapid, automatic response to a stimulus.
**reflex arc** (re′fleks ark) A nerve pathway, consisting of a sensory neuron, interneuron, and motor neuron, that forms the structural and functional bases for a reflex.
**refraction** (re-frak′shun) A bending of light as it passes from one medium into another medium with a different density.
**refractory period** (re-frak′to-re pe′re-od) Time period following stimulation during which a neuron or muscle fiber will not respond to a stimulus.
**relaxin** (re-lak′sin) Hormone from the corpus luteum that inhibits uterine contractions during pregnancy.
**releasing hormone** (re-le′sing hor′mōn) A substance secreted by the hypothalamus whose target cells are in the anterior pituitary gland.
**renal** (re′nal) Pertaining to the kidney.
**renal corpuscle** (re′nal kor′pusl) Part of a nephron that consists of a glomerulus and a glomerular capsule; Malpighian corpuscle.
**renal cortex** (re′nal kor′teks) The outer portion of a kidney.
**renal medulla** (re′nal mĕ-dul′ah) The inner portion of a kidney.
**renal pelvis** (re′nal pel′vis) The cavity in a kidney.
**renal tubule** (re′nal tu′būl) Portion of a nephron that extends from the renal corpuscle to the collecting duct.
**renin** (re′nin) Enzyme released from the kidneys that triggers a rise in blood pressure.
**replication** (rep″lĭ-ka′shun) Reproduction of an exact copy of DNA.
**repolarization** (re-po″lar-ĭ-za′shun) Returning the cell membrane potential to resting potential.
**reproduction** (re″pro-duk′shun) Offspring formation.
**residual volume** (re-zid′u-al vol′ūm) The amount of air remaining in the lungs after the most forceful expiration.
**resorption** (re-sorp′shun) Decomposition of a structure as a result of physiological activity.
**respiration** (res″pĭ-ra′shun) Cellular process that releases energy from nutrients; breathing.
**respiratory center** (re-spi′rah-to″re sen′ter) Portion of the brain stem that controls the depth and rate of breathing.
**respiratory membrane** (re-spi′rah-to″re mem′brān) Membrane composed of a capillary wall and an alveolar wall through which gases are exchanged between the blood and the air.
**response** (re-spons′) The action resulting from a stimulus.
**resting potential** (res′ting po-ten′shal) The difference in electrical charge between the inside and outside of an undisturbed nerve cell membrane.
**reticular formation** (rĕ-tik′u-lar fōr-ma′shun) A complex network of nerve fibers within the brain stem that arouses the cerebrum.

**reticulocyte** (rĕ-tik′u-lo-sīt) An immature red blood cell that has a network of fibrils in its cytoplasm.

**reticuloendothelial tissue** (rĕ-tik″u-lo-en″do-the′le-al tish′u) Tissue composed of widely scattered phagocytic cells.

**retina** (ret′ĭ-nah) Inner layer of the eye wall that contains the visual receptors.

**retinal** (ret′ĭ-nal) A form of vitamin A; retinene.

**retinene** (ret′ĭ-nēn) Chemical precursor of rhodopsin.

**retraction** (rĕ-trak′shun) Movement of a part toward the back.

**retroperitoneal** (ret″ro-per″ĭ-to-ne′al) Located behind the peritoneum.

**rhodopsin** (ro-dop′sin) Light-sensitive biochemical in the rods of the retina; visual purple.

**rhythmicity area** (rith-mis′ ĭ-te a′re-ah) A portion of the respiratory control center located in the medulla.

**riboflavin** (ri″bo-fla′vin) A vitamin of the B-complex group; vitamin $B_2$.

**ribonucleic acid** (ri″bo-nu-kle′ik as′id) A nucleic acid that contains ribose sugar; RNA.

**ribose** (ri′bōs) A five-carbon sugar in RNA.

**ribosomal RNA** (ri-bo-sōm′al) A type of RNA that forms part of the ribosome, which is a structural support for protein synthesis.

**ribosome** (ri′bo-sōm) An organelle composed of RNA and protein that is a structural support for protein synthesis.

**RNA** Ribonucleic acid.

**rod** (rod) A type of light receptor that provides colorless vision.

**rotation** (ro-ta′shun) Movement turning a body part on its longitudinal axis.

**rouleaux** (roo-lo′) Groups of red blood cells stuck together like stacks of coins.

**round window** (rownd win′do) A membrane-covered opening between the inner ear and the middle ear.

**rugae** (roo′je) Thick folds in the inner wall of the stomach that disappear when the stomach is distended.

## S

**saccule** (sak′ūl) A saclike cavity that makes up part of the membranous labyrinth of the inner ear.

**sacral** (sa′kral) Pertaining to the five fused (pelvic) vertebrae at the distal end of the spinal column.

**sagittal** (saj′i-tal) A plane or section that divides a structure into right and left portions.

**salivary gland** (sal′ĭ-ver-e gland) A gland, associated with the mouth, that secretes saliva.

**salt** (sawlt) A compound produced by a reaction between an acid and a base.

**saltatory conduction** (sal′tah-tor-e kon-duk′shun) Nerve impulse conduction that seems to jump from one node to the next.

**S-A node** (nōd) Sinoatrial node.

**sarcolemma** (sar″ko-lem′ah) The cell membrane of a muscle fiber.

**sarcomere** (sar′ko-mēr) The structural and functional unit of a myofibril.

**sarcoplasm** (sar′ko-plazm) The cytoplasm within a muscle fiber.

**sarcoplasmic reticulum** (sar″ko-plaz′mik rĕ-tik′u-lum) Membranous network of channels and tubules within a muscle fiber, corresponding to the endoplasmic reticulum of other cells.

**saturated fatty acid** (sat′u-rāt″ed fat′e as′id) Fatty acid molecule that lacks double bonds between the atoms of its carbon chain.

**Schwann cell** (shwahn sel) A type of neuroglial cell that surrounds a fiber of a peripheral nerve, forming the neurilemmal sheath and myelin.

**sclera** (skle′rah) White fibrous outer layer of the eyeball.

**scoliosis** (sko″le-o′sis) Abnormal lateral curvature of the vertebral column.

**scrotum** (skro′tum) A pouch of skin that encloses the testes.

**sebaceous gland** (sĕ-ba′shus gland) Gland of the skin that secretes sebum.

**sebum** (se′bum) Oily secretion of the sebaceous glands.

**secondary immune response** (sek′un-der″e i-mun′ re-spons′) The immune system's response to subsequent encounters with a foreign antigen.

**secretin** (se-kre′tin) Hormone secreted from the small intestine that stimulates the pancreas to release pancreatic juice.

**secretion** (se-kre′shun) Substances produced in and released from a gland cell.

**semen** (se′men) Fluid discharged from the male reproductive tract at ejaculation that contains sperm cells and secretions.

**semicircular canal** (sem″ĭ-ser′ku-lar kah-nal′) Tubular structure within the inner ear that contains the receptors providing the sense of dynamic equilibrium.

**seminal vesicle** (sem′ ĭ-nal ves′ĭ-kel) One of a pair of pouches that adds fructose and prostaglandins to sperm as semen forms.

**seminiferous tubule** (sem″ĭ-nif′er-us tu′būl) Tubule within the testes where sperm cells form.

**semipermeable** (sem″ĭ-per′me-ah-bl) A membrane that allows some molecules through but not others.

**senescence** (sĕ-nes′ens) Aging.

**sensation** (sen-sa′shun) A feeling resulting from the brain's interpretation of sensory nerve impulses.

**sensory adaptation** (sen′so-re ad″ap-ta′shun) The phenomenon of a sensation becoming less noticeable once it has been recognized by constant repeated stimulation.

**sensory area** (sen′so-re a′re-ah) A portion of the cerebral cortex that receives and interprets sensory nerve impulses.

**sensory nerve** (sen′so-re nerv) A nerve composed of sensory nerve fibers.

**sensory neuron** (sen′so-re nu′ron) A neuron that transmits an impulse from a receptor to the central nervous system.

**sensory receptor** (sen′so-re re″sep′tor) A specialized dendrite that is specific to detecting a particular sensation and firing an action potential in response, which is transmitted to the central nervous system.

**serosa** (sēr-o′sah) Serous membrane composing the outer layer in the walls of the organs of digestion.

**serotonin** (se″ro-to′nin) A vasoconstrictor that blood platelets release when blood vessels break, controlling bleeding.

**serous cell** (se′rus sel) A glandular cell that secretes a watery fluid with a high enzyme content.

**serous fluid** (se′rus floo′id) The secretion of a serous membrane.

**serous membrane** (se′rus mem′brān) Membrane that lines a cavity without an opening to the outside of the body.

**serum** (se′rum) The fluid portion of coagulated blood.

**sesamoid bone** (ses′ah-moid bōn) A round bone that may occur in tendons adjacent to joints.

**sex chromosome** (seks kro′mo-sōm) A chromosome that carries genes responsible for the development of characteristics associated with maleness or femaleness; an X or Y chromosome.

**sex-influenced inheritance** (seks-in′floo-enst in-her′ ĭ-tens) Transmission of a trait that is dominant in one sex but recessive in the other.

**sex-limited inheritance** (seks-lim′it-ed in-her′ ĭ-tens) Transmission of a trait expressed in one sex only.

**sex-linked inheritance** (seks′-linkt in-her′ĭ-tens) Transmission of a trait controlled by a gene on a sex chromosome (X or Y).

**sexually transmitted disease** (sek′shoo-ah-le trans-mi′ted dĭ-zēz′) Infection transmitted from one individual to another by direct contact during sexual activity; STD.

**sigmoid colon** (sig′moid ko′lon) S-shaped portion of the large intestine between the descending colon and the rectum.

**simple sugar** (sim′pl shoog′ar) A monosaccharide.

**signal transduction** (sig′nahl trans-duk′shun) A series of biochemical reactions that allows cells to receive and respond to messages coming in through the cell membrane.

**sinoatrial node** (si″no-a′tre-al nōd) Specialized tissue in the wall of the right atrium that initiates cardiac cycles; the pacemaker; S-A node.

**sinus** (si′nus) A cavity or space in a bone or other body part.

**skeletal muscle** (skel′ĕ-tal mus′l) Type of muscle tissue found in muscles attached to bones.

**sliding filament theory** (slī′ding fil′eh-ment the′o-re) Muscles contract when the thin (actin) and thick (myosin) filaments move past each other, shortening the skeletal muscle cells.

**small intestine** (smawl in-tes′tin) Part of the digestive tract extending from the stomach to the cecum; consisting of the duodenum, jejunum, and ileum.

**smooth muscle** (smo͞oth mus′el) Type of muscle tissue found in the walls of hollow visceral organs; visceral muscle.

**sodium pump** (so′de-um pump) The active transport mechanism that concentrates sodium ions on the outside of a cell membrane.

**solute** (sol′ūt) The substance dissolved in a solution.

**solvent** (sol′vent) The liquid portion of a solution in which a solute is dissolved.

**somatic** (so-mat′ik) Pertaining to the body.

**somatic cell** (so-mat′ik sel) Any cell of the body other than the sex cells.

**somatomedin** (so″mah-to-me′din) A substance released from the liver in response to growth hormone that promotes the growth of cartilage.

**somatostatin** (so-mat′o-sta′tin) A hormone secreted by the islet of Langerhans cells that inhibits the release of growth hormone.

**somatotropin** (so″mah-to-tro′pin) Growth hormone.

**spastic paralysis** (spas′tik pah-ral′ĭ-sis) A form of paralysis characterized by an increase in muscular tone without atrophy of the muscles involved.

**special sense** (spesh′al sens) Sense that involves receptors associated with specialized sensory organs, such as the eyes and ears.

**species resistance** (spe′sēz re-zis′tans) A natural ability of one type of organism to resist infection by microorganisms that might cause disease in another type of organism.

**spermatic cord** (sper-mat′ik kord) A structure consisting of blood vessels, nerves, the vas deferens, and other vessels extending from the abdominal inguinal ring to the testis.

**spermatid** (sper′mah-tid) An intermediate stage in the formation of sperm cells.

**spermatocyte** (sper-mat′o-sīt) An early stage in the formation of sperm cells.

**spermatogenesis** (sper″mah-to-jen′ĕ-sis) The production of sperm cells.

**spermatogonium** (sper″mah-to-go′ne-um) Undifferentiated spermatogenic cell found in the germinal epithelium of a seminiferous tubule.

**spermatozoa** (sper″mah-to-zo′ah) Male reproductive cells; sperm cells.

**sphincter** (sfingk′ter) A circular muscle that closes an opening or the lumen of a tubular structure.

**sphygmomanometer** (sfig′mo-mah-nom′ĕ-ter) Instrument used for measuring blood pressure.

**spinal** (spi′nal) Pertaining to the spinal cord or to the vertebral canal.

**spinal cord** (spi′nal kord) Portion of the central nervous system extending downward from the brain stem through the vertebral canal.

**spinal nerve** (spi′nal nerv) Nerve that arises from the spinal cord.

**spleen** (splēn) A large, glandular organ located in the upper left region of the abdomen.

**spongy bone** (spunj′e bōn) Bone that consists of bars and plates separated by irregular spaces; cancellous bone.

**squamous** (skwa′mus) Flat or platelike.

**staircase effect** (stair′kays i-fekt′) A gradual increase in contractile strength of a muscle in response to repeated stimuli of the same intensity.

**starch** (starch) A polysaccharide that is common in foods of plant origin.

**static equilibrium** (stat′ik e′kwĭ-lib′re-um) The maintenance of balance when the head and body are motionless.

**stereocilia** (ste″re-o-sil′e-ah) The hairlike processes of the hair cells within the organ of Corti.

**stereoscopic vision** (ster″e-o-skop′ik vizh′un) Objects perceived as three-dimensional; depth perception.

**sternal** (ster′nal) Pertaining to the sternum.

**steroid** (ste′roid) A type of organic molecule including complex rings of carbon and hydrogen atoms.

**stimulus** (stim′u-lus) A change in the environmental conditions that is followed by a response by an organism or cell (pl. *stimuli*).

**stomach** (stum′ak) Digestive organ located between the esophagus and the small intestine.

**strabismus** (strah-biz′mus) Lack of visual coordination; crossed eyes.

**stratified** (strat′ĭ-fīd) Arranged in layers.

**stratum corneum** (stra′tum kor′ne-um) Outer horny layer of the epidermis.

**stratum germinativum** (stra′tum jer′mĭ-na″ti-vum) The deepest layer of the epidermis in which the cells divide; stratum basale.

**stress** (stres) Response to factors perceived as capable of threatening life.

**stressor** (stres′or) A factor capable of stimulating a stress response.

**stretch receptor** (strech re-sep′ter) Sensory nerve ending that responds to tension.

**stretch reflex** (strech re′fleks) Muscle contraction in response to stretching the muscle.

**stroke volume** (strōk vol′ūm) The amount of blood discharged from the ventricle with each heartbeat.

**structural formula** (struk′cher-al fōr′mu-lah) A representation of the way atoms bond to form a molecule, using the symbols for each element and lines to indicate chemical bonds.

**subarachnoid space** (sub″ah-rak′noid spās) The space within the meninges between the arachnoid mater and the pia mater.

**subcutaneous** (sub″ku-ta′ne-us) Beneath the skin.

**sublingual** (sub-ling′gwal) Beneath the tongue.

**submaxillary** (sub-mak′sĭ-ler″e) Below the maxilla.

**submucosa** (sub″mu-ko′sah) Layer of connective tissue underneath a mucous membrane.

**substrate** (sub′strāt) The substance upon which an enzyme acts.

**sucrase** (su′krās) Digestive enzyme that catalyzes the breakdown of sucrose.

**sucrose** (soo′krōs) A disaccharide; table sugar.

**sulcus** (sul′kus) A shallow groove, such as that between convolutions on the surface of the brain (pl. *sulci*).

**summation** (sum-ma′shun) Phenomena in which the degree of change in membrane potential is directly related to the intensity of stimulation.

**superficial** (soo′per-fish′al) Near the surface.

**superior** (su-pe′re-or) Pertaining to a structure that is higher than another structure.

**supination** (soo″pĭ-na′shun) Rotation of the forearm so that the palm faces upward when the arm is outstretched.

**surface tension** (ser′fas ten′shun) Force that holds moist membranes together due to an attraction that water molecules have for one another.

**surfactant** (ser-fak′tant) Substance produced by the lungs that reduces the surface tension within the alveoli.

**suture** (soo′cher) An immovable joint, such as that between flat bones of the skull.

**sympathetic nervous system** (sim″pah-thet′ik ner′vus sis′tem) Portion of the autonomic nervous system that arises from the thoracic and lumbar regions of the spinal cord.

**symphysis** (sim′fĭ-sis) A slightly movable joint between bones separated by a pad of fibrocartilage.

**synapse** (sin′aps) The junction between the axon of one neuron and the dendrite or cell body of another neuron.

**synaptic cleft** (sĭ-nap′tik kleft) A narrow extracellular space between the presynaptic and postsynaptic neurons.

**synaptic knob** (sĭ-nap′tik nob) Tiny enlargement at the end of an axon that secretes a neurotransmitter.

**synaptic transmission** (sĭ-nap′tik trans-mish′un) Communication of an impulse from one neuron to the next.

**synchondrosis** (sin″kon-dro′sis) Type of joint in which bones are united by bands of hyaline cartilage.

**syncytium** (sin-sish′e-um) A mass of merging cells.

**syndesmosis** (sin″des-mo′sis) Type of joint in which the bones are united by relatively long fibers of connective tissue.

**syndrome** (sin′drōm) A group of symptoms that characterize a disease condition.

**synergist** (sin′er-jist) A muscle that assists the action of a prime mover.

**synovial fluid** (sĭ-no′ve-al floo′id) Fluid secreted by the synovial membrane.

**synovial joint** (sĭ-no′ve-al joint) A freely movable joint.

**synovial membrane** (sĭ-no′ve-al mem′brān) Membrane that forms the inner lining of the capsule of a freely movable joint.

**synthesis** (sin′thĕ-sis) The process by which substances unite to form more complex substances.

**system** (sis′tem) A group of organs that act together to carry on a specialized function.

**systemic circuit** (sis-tem′ik ser′kit) The vessels that conduct blood between the heart and all body tissues except the lungs.

**systole** (sis′to-le) Phase of the cardiac cycle during which a heart chamber wall contracts.

**systolic pressure** (sis-tol′ik presh′ur) Arterial blood pressure during the systolic phase of the cardiac cycle.

## T

**tachycardia** (tak″e-kar′de-ah) An abnormally rapid heartbeat.

**target tissue** (tar′get tish′u) Specific tissue on which a hormone exerts its effect.

**tarsus** (tar′sus) The ankle bones.

**taste bud** (tāst bud) Organ containing receptors associated with the sense of taste.

**telophase** (tel′o-fāz) Stage in mitosis when newly formed cells separate.

**tendon** (ten′don) A cordlike or bandlike mass of white fibrous connective tissue that connects a muscle to a bone.

**teratogen** (ter′ah-to-jen) A chemical or other environmental agent that causes a birth defect.

**testis** (tes′tis) Primary reproductive organ of a male; a sperm-cell-producing organ (pl. *testes*).

**testosterone** (tes-tos′tĕ-rōn) Male sex hormone secreted by the interstitial cells of the testes.

**tetanus** (tet′ah-nus) A continuous, forceful muscular contraction (tetanic contraction) without relaxation.

**thalamus** (thal′ah-mus) A mass of gray matter located at the base of the cerebrum in the wall of the third ventricle.

**thermoreceptor** (ther″mo-re-sep′tor) A sensory receptor that is sensitive to changes in temperature; a heat receptor.

**thiamine** (thi′ah-min) Vitamin $B_1$.

**thoracic** (tho-ras′ik) Pertaining to the chest.

**threshold potential** (thresh′old po-ten′shal) Level of potential at which an action potential or nerve impulse is produced.

**threshold stimulus** (thresh′old stim′u-lus) The level of stimulation that must be exceeded to elicit a nerve impulse or a muscle contraction.

**thrombin** (throm′bin) Blood clotting enzyme that catalyzes formation of fibrin from fibrinogen.

**thrombocyte** (throm′bo-sīt) A blood platelet.

**thrombocytopenia** (throm″bo-si″to-pe′ne-ah) A low number of platelets in the circulating blood.

**thrombus** (throm′bus) A blood clot in a blood vessel that remains at its site of formation.

**thymosin** (thi′mo-sin) A group of peptides secreted by the thymus gland that increases production of certain types of white blood cells.

**thymus** (thi′mus) A glandular organ located in the mediastinum behind the sternum and between the lungs.

**thyroglobulin** (thi″ro-glob′u-lin) Biochemical secreted by cells of the thyroid gland that stores thyroid hormones.

**thyroid gland** (thi′roid gland) Endocrine gland located just below the larynx and in front of the trachea that secretes thyroid hormones.

**thyrotropin** (thi″ro-trōp′in) Hormone secreted by the anterior pituitary gland that stimulates the thyroid gland to secrete hormones; TSH.

**thyroxine** (thi-rok′sin) A hormone secreted by the thyroid gland.

**tidal volume** (tīd′al vol′ūm) Amount of air that enters the lungs during a normal, quiet inspiration.

**tissue** (tish′u) A group of similar cells that performs a specialized function.

**T cell** (T sel) Lymphocytes that interact directly with antigen-bearing cells and particles and secrete cytokines, producing cell-mediated immunity.

**tonsils** (ton′silz) Collections of lymphatic tissue in the throat.

**trabecula** (trah-bek′u-lah) Branching bony plate that separates irregular spaces within spongy bone.

**trachea** (tra′ke-ah) Tubular organ that leads from the larynx to the bronchi.

**transcellular fluid** (trans″sel′u-lar floo′id) A portion of the extracellular fluid, including the fluid within special body cavities.

**transcription** (trans-krip′shun) Manufacturing RNA from DNA.

**transferrin** (trans-fer′rin) Blood plasma protein that transports iron.

**transfer RNA** (trans′fer RNA) Molecule of RNA that carries an amino acid to a ribosome in the process of protein synthesis.

**translation** (trans-la′shun) Assembly of an amino acid chain according to the sequence of base triplets in a molecule of mRNA.

**transverse** (trans-vers′) At right angles to the long axis of a part; crosswise.

**transverse colon** (trans-vers′ ko′lon) Portion of the large intestine that extends across the abdomen from right to left below the stomach.

**transverse tubule** (trans-vers′ tu′būl) Membranous channel that extends inward from a muscle fiber membrane and passes through the fiber.

**tricuspid valve** (tri-kus′pid valv) Heart valve located between the right atrium and the right ventricle.

**trigger zone** (trig′ger zōn) Site on the neuron where an action potential is generated.

**triglyceride** (tri-glis′er-īd) A lipid composed of three fatty acids combined with a glycerol molecule.

**triiodothyronine** (tri″i-o″do-thi′ro-nēn) A type of thyroid hormone.

**trisomy** (tri′so-me) Condition in which a cell contains three chromosomes of a particular type instead of two.

**trochanter** (tro-kan′ter) A broad process on a bone.

**trochlea** (trok′le-ah) A pulley-shaped structure.

**trophoblast** (trof′o-blast) The outer cells of a blastocyst that help form the placenta and other extra embryonic membranes.

**tropic hormone** (trōp′ik hor′mōn) A hormone that has an endocrine gland as its target tissue.

**tropomyosin** (tro″po-mi′o-sin) Protein that blocks muscle contraction until calcium ions are present.

**troponin** (tro′po-nin) Protein that functions with tropomyosin to block muscle contraction until calcium ions are present.

**trypsin** (trip′sin) An enzyme in pancreatic juice that breaks down protein molecules.

**trypsinogen** (trip-sin′o-jen) Substance pancreatic cells secrete that is enzymatically cleaved to yield trypsin.

**tubercle** (tu′ber-kl) A small, rounded process on a bone.

**tuberosity** (tu″bĕ-ros′ ĭ-te) An elevation or protuberance on a bone.

**tumor** (too′mor) A tissue mass formed when cells lose division control.

**twitch** (twich) A brief muscular contraction followed by relaxation.

**tympanic membrane** (tim-pan′ik mem′brān) A thin membrane that covers the auditory canal and separates the external ear from the middle ear; the eardrum.

## U

**umbilical cord** (um-bil′ĭ-kal kord) Cordlike structure that connects the fetus to the placenta.

**umbilical region** (um-bil′ĭ-kal re′jun) The central portion of the abdomen.

**umbilicus** (um-bil′ ĭ-kus) Region to which the umbilical cord was attached; the navel.

**unipolar neuron** (un″ĭ-po′lar nu′ron) A neuron that has a single nerve fiber extending from its cell body.

**unsaturated fatty acid** (un-sat′u-rāt″ed fat′e as′id) Fatty acid molecule that has one or more double bonds between the atoms of its carbon chain.

**upper limb** (uh′per lim) The superior appendage consisting of the arm, forearm, wrist, and hand.

**urea** (u-re′ah) A nonprotein nitrogenous substance produced as a result of protein metabolism.

**ureter** (u-re′ter) A muscular tube that carries urine from the kidney to the urinary bladder.

**urethra** (u-re′thrah) Tube leading from the urinary bladder to the outside of the body.

**uric acid** (u′rik as′id) End product of nucleic acid metabolism in the body.

**urine** (u′rin) Wastes and excess water removed from the blood and excreted by the kidneys into the ureters to the urinary bladder and out of the body through the urethra.

**uterine** (u′ter-in) Pertaining to the uterus.

**uterine tube** (u′ter-in tūb) Tube that extends from the uterus on each side toward an ovary and transports sex cells; fallopian tube or oviduct.

**uterus** (u′ter-us) Hollow muscular organ within the female pelvis in which a fetus develops.

**utricle** (u′trĭ-kl) An enlarged portion of the membranous labyrinth of the inner ear.

**uvula** (u′vu-lah) A fleshy portion of the soft palate that hangs down above the root of the tongue.

## V

**vaccine** (vak′sēn) A substance that contains antigens used to stimulate an immune response.

**vacuole** (vak′u-ōl) A space or cavity within the cytoplasm of a cell.

**vagina** (vah-ji′nah) Tubular organ that leads from the uterus to the vestibule of the female reproductive tract.

**Valsalva's maneuver** (val-sal′vahz mah-noo′ver) Increasing the intrathoracic pressure by forcing air from the lungs against a closed glottis.

**varicose veins** (var′ĭ-kos vānz) Abnormally swollen and enlarged veins, especially in the legs.

**vasa recta** (va′sah rek′tah) A branch of the peritubular capillary that receives blood from the efferent arterioles of juxtamedullary nephrons.

**vascular** (vas′ku-lar) Pertaining to blood vessels.

**vas deferens** (vas def′er-ens) Tube that leads from the epididymis to the urethra of the male reproductive tract (pl. *vasa deferentia*).

**vasoconstriction** (vas″o-kon-strik′shun) A decrease in the diameter of a blood vessel.

**vasodilation** (vas″o-di-la′shun) An increase in the diameter of a blood vessel.

**vasomotor center** (vas″o-mo′tor sen′ter) Neurons in the brain stem that control the diameter of the arteries.

**vasopressin** (vas″o-pres′in) Antidiuretic hormone.

**vein** (vān) A vessel that carries blood toward the heart.

**vena cava** (vēn′ah kāv′ah) One of two large veins that convey deoxygenated blood to the right atrium of the heart.

**ventral** (ven′tral) Pertaining to the front or anterior.

**ventral root** (ven′tral rōōt) Motor branch of a spinal nerve by which it attaches to the spinal cord.

**ventricle** (ven′trĭ-kl) A cavity, such as those of the brain that are filled with cerebrospinal fluid, or those of the heart that contain blood.

**venule** (ven′ūl) A vessel that carries blood from capillaries to a vein.

**vermiform appendix** (ver′mĭ-form ah-pen′diks) Appendix.

**vertebral** (ver′te-bral) Pertaining to the bones of the spinal column.

**vesicle** (ves′ĭ-kal) Membranous, cytoplasmic sac formed by an infolding of the cell membrane.

**vestibule** (ves′tĭ-būl) A space at the opening to a canal.

**villus** (vil′us) Tiny, fingerlike projection that extends outward from the inner lining of the small intestine (pl. *villi*).

**visceral** (vis′er-al) Pertaining to the contents of a body cavity.

**visceral peritoneum** (vis′er-al per″ĭ-to-ne′um) Membrane that covers the surfaces of organs within the abdominal cavity.

**visceral pleura** (vis′er-al ploo′rah) Membrane that covers the surfaces of the lungs.

**viscosity** (vis-kos′ĭi-te) The tendency for a fluid to resist flowing due to the internal friction of its molecules.

**vital capacity** (vi′tal kah-pas′i-te) The maximum amount of air a person can exhale after taking the deepest breath possible.

**vitamin** (vi′tah-min) An organic compound other than a carbohydrate, lipid, or protein that is needed for normal metabolism but that the body cannot synthesize in adequate amounts.

**vitreous body** (vit′re-us bod′e) The collagenous fibers and fluid that occupy the posterior cavity of the eye.

**vitreous humor** (vit′re-us hu′mor) The substance that occupies the space between the lens and the retina of the eye.

**vocal cords** (vo′kal kordz) Folds of tissue within the larynx that produce sounds when they vibrate.

**Volkmann's canal** (fōlk′mahnz kah-nal′) A transverse channel that connects Haversian canals within compact bone.

**volt** (vōlt) Unit used to measure differences in electrical potential.

**voluntary** (vol′un-tār″e) Capable of being consciously controlled.

**vulva** (vul′vah) The external reproductive parts of the female that surround the opening of the vagina.

## W

**water balance** (wot′er bal′ans) Condition in which the volume of water entering the body is equal to the volume leaving it.

**water of metabolism** (wot′er uv mĕ-tab′o-lizm) Water produced as a by-product of metabolic processes.

**wave summation** (wāv sum-ma′shun) A sustained muscle contraction that occurs when a series of twitches fuse together; summation of twitches.

**white blood cell** (whīt blud sel) A cell that helps fight infection.

**white muscle** (whīt mus′el) Fast-contracting skeletal muscle.

**wild type** (wīld tīp) A phenotype or allele that is the most common for a certain gene in a population.

## X

**X-linked trait** (eks-linkt′ trāt) Trait determined by a gene located on an X chromosome.

**x-ray** (eks′ ray) Used as a verb, to photograph using radiation.

**X ray** (eks ray) Used as a noun, a photograph produced by radiation. May also be used as an adjective, X-ray.

## Y

**yellow marrow** (yel′o mar′o) Fat storage tissue found in the cavities within certain bones.

**yolk sac** (yōk sak) An extraembryonic membrane connected to the embryo by a long, narrow tube.

## Z

**zonular fiber** (zon′u-lar fi′ber) A delicate fiber within the eye that extends from the ciliary process to the lens capsule.

**zygote** (zi′gōt) Cell produced by the fusion of an egg and sperm; a fertilized egg cell.

**zymogen granule** (zi-mo′jen gran′ūl) A cellular structure that stores inactive forms of protein-splitting enzymes in a pancreatic cell.

# CREDITS

## PHOTOGRAPHS

### Chapter 1

**Opener:** © S. Gazin/The Image Works, Inc.; **p. 2:** © L. Kolvoord/The Image Works, Inc.; **1.1:** From *De Humani Corporis Fabrica,* second book, 1543, by Andreas Vesalius; **1.3:** © Times Mirror Higher Education Group, Inc./Carol D. Jacobson, Ph.D., Department of Veterinary Anatomy, Iowa State University; **1.18*a:*** © Patrick J. Lynch/Photo Researchers, Inc.; **1.18*b:*** © Biophoto Associates/Photo Researchers, Inc.; **1.18c:** © A. Glauberman/Photo Researchers, Inc.; **1.A:** © Alexander Tsiaras/Photo Researchers Inc.; **1.B:** © Lutheran Hospital/Peter Arnold, Inc.; **1.C:** © SPL/Photo Researchers, Inc.

### Chapter 2

**Opener:** © Leonard Lessin/Peter Arnold, Inc.; **2.A:** © Mark Antman/The Image Works, Inc.; **2.B*a:*** © SIU/Peter Arnold, Inc.; **2.22*a–d:*** Courtesy of John W. Hole, Jr.; Courtesy of Ealing Corporation; **2.22*e:*** Courtesy of John W. Hole, Jr.; **2.22*f:*** Courtesy of Ealing Corporation; **2.D*a:*** © SPL/Photo Researchers, Inc.; **2.D*b:*** © CNRI/SPL/Photo Researchers, Inc.; **2.E:** From M. I. Posner, "Seeing the mind," *Science,* 262: 29 Oct. 1993, page 673, fig. 1 © AAAS. Photograph by Marcus Raichle.

### Chapter 3

**Opener:** © SPL/Photo Researchers, Inc.; **3.4:** © W. Omerod/Visuals Unlimited; **3.5*a:*** © Ed Reschke; **3.5*b:*** © Keith Porter; **3.5c:** © A. Liepens/SPL/Photo Researchers, Inc.; **3.6*a:*** © Gordon Leedale/Biophoto Associates; **3.10*a:*** © D. W. Fawcett/Photo Researchers, Inc.; **3.11*a:*** © Gordon Leedale/Biophoto Associates; **3.13*a:*** © Keith Porter/Photo Researchers, Inc.; **3.14:** © K. G. Murti/Visuals Unlimited; **3.15*a:*** © Gordon Leedale/Biophoto Associates; **3.16*a:*** Courtesy of American Lung Association; **3.17:** © Manfred Kage/Peter Arnold, Inc.; **3.18:** © Keith Porter; **3.19*b:*** From Joseph G. Gall, Microtubule fine structure. *Journal of Cell Biology* 31: (1966) pp. 639–643; **3.20*a:*** © Stephen L. Wolfe; **3.25 *a–c*:** © David M. Phillips/Visuals Unlimited; **3.34*b*, 3.36*b*, 3.37*b*; 3.38*b*, 3.39*b*, 3.40*b:*** © Ed Reschke; **3.41 *a–c*:** © R. G. Kessel and C. Y. Shih, *Scanning Electron Microscopy in Biology.* Springer-Verlag, 1976.; **3.43:** © Dr. Tony Brain/SPL/Photo Researchers, Inc.

### Chapter 4

**Opener:** © Bruce Curtis/Peter Arnold, Inc.; **4.22*b:*** Courtesy of Ealing Corporation; **4.*a:*** March of Dimes Birth Defects Foundation

### Chapter 5

**Opener:** © Robert Becker/Custom Medical Stock; **5.1*b*, 5.2*b*, 5.3*b:*** © Ed Reschke; © Manfred Kage/Peter Arnold, Inc.; **5.4:** © D.W. Fawcett/Photo Researchers, Inc.; **5.5*b:*** © Ed Reschke; **5.6*b:*** © Fred Hossler/Visuals Unlimited; **5.7*b*, 5.8*b:*** © Times Mirror Higher Education Group, Inc./Carol D. Jacobson, Ph.D., Department of Veterinary Anatomy, Iowa State University; **5.9*a:*** © Ed Reschke; **5.12:** © Ed Reschke/Peter Arnold, Inc.; **5.13:** © David M. Phillips/Visuals Unlimited; **5.14:** Jean Paul Revel; **5.15:** © Veronica Burmeister/Visuals Unlimited; **5.16:** © P. Motta/SPL/Photo Researchers, Inc.; **5.17:** © St. Mary's Hospital Medical School/SPL/Photo Researchers, Inc.; **5.18*b*, 5.19*b*, 5.20*b:*** © Ed Reschke; **5.21*b*, 5.22*b:*** © Alfred Owczarzak; **5.23*b*, 5.24*b:*** © Ed Reschke; **5.25*b:*** © Times Mirror Higher Education Group, Inc./Carol D. Jacobson, Ph.D., Department of Veterinary Anatomy, Iowa State University; **5.26*b:*** © Victor B. Eichler, Ph.D.; **5.27*b:*** © Ed Reschke/Peter Arnold, Inc.; **5.28*b*, 5.29*b:*** © Ed Reschke; **5.30*b:*** © Manfred Kage/Peter Arnold, Inc.; **5.31*b:*** © Ed Reschke; **5.A:** Courtesy of Organogenesis Inc.; **5.B:** Courtesy of CytoTherapeutics, Inc.

### Chapter 6

**Opener:** © Alan Carey/The Image Works, Inc.; **6.1:** © Times Mirror Higher Education Group, Inc./Carol D. Jacobson, Ph.D., Department of Veterinary Anatomy, Iowa State University; **6.3*b:*** © Ed Reschke; **6.4:** © Victor B. Eichler, Ph.D.; **6.5*a:*** © M. Schliwa/Visuals Unlimited; **6.A:** © Kenneth Greer/Visuals Unlimited; **6.B:** © SPL/Photo Researchers, Inc.; **6.6*b:*** © Victor B. Eichler, Ph.D.; **6.7:** © CNRI/SPL/Photo Researchers, Inc.; **6.C:** © Photo Edit; **6.9:** © Per H. Kjeldsen, University of Michigan; **6.11:** © Times Mirror Higher Education Group, Inc./Carol D. Jacobson, Ph.D., Department of Veterinary Anatomy, Iowa State University; **6.13:** Field Museum of Natural History, #A118, Chicago

### Chapter 7

**Opener:** AP/Wide World Photos; **7.3*a:*** © Lester Bergman & Associates; **7.3*b*, 7.3*c:*** Courtesy of John W. Hole, Jr.; **7.4:** © R. G. Kessel and R. H. Kardon, *A Text-Atlas of Scanning Electron Microscopy,* 1979.; **7.6*a:*** © Biophoto Associates/Photo Researchers, Inc.; **7.6*b:*** Courtesy of Dr. Francis Collins; **7.7:** © Biophoto Associates/Photo Researchers, Inc.; **7.9*b:*** © Victor B. Eichler, Ph.D.; **7.10:** © Biophoto Associates/Photo Researchers Inc.; **7.11:** © James Shaffer; **7.12:** Courtesy of

John W. Hole, Jr.; **7.32*a*:** © Jan Halaska/Photo Researchers, Inc.; **7.32*b*:** © Science Source/Photo Researchers, Inc.; **7.37:** © Kent Van De Graaff, Ph.D./Utah Valley Hospital, Department of Radiology; **7.40*b*:** © Victor B. Eichler, Ph.D.; **7.42*b*:** Courtesy of Eastman Kodak; **7.44*d*, 7.47*b*:** © Martin Rotker; **7.48:** © Lester V. Bergman and Associates; **7.49*c*, 7.52*a*, 7.55*a*:** © Martin Rotker; **7.56*b*:** © Ted Conde

## Chapter 8

**Opener:** © Joseph Nettis/Photo Researchers, Inc.; **8.3*a*,*b*:** Courtesy of John W. Hole, Jr.; **8.13*b*, 8.15*b*:** © Paul Reiman; **8.17*a*,*b*:** Courtesy of John W. Hole, Jr.; **8.18*b*, 8.20*b*:** © Paul Reiman; **8.B:** The Bettmann Archives

## Chapter 9

**Opener:** © Bob Daemmrich/The Image Works, Inc.; **9.3:** © Times Mirror Higher Education Group, Inc./Carol D. Jacobson, Ph.D., Department of Veterinary Anatomy, Iowa State University; **9.5*a*:** © H. E. Huxley; **9.8*b*:** © Times Mirror Higher Education Group, Inc./Carol D. Jacobson, Ph.D., Department of Veterinary Anatomy, Iowa State University; **9.11*a*:** © H. E. Huxley; **9.18:** © Dr. Paul Heidger, University of Iowa College of Medicine; **p. 349 all:** John C. Mese

## Chapter 10

**Opener:** © Dr. Mark Mattson, University of Kentucky College of Medicine; **10.1:** © Ed Reschke; **10.5*a*:** © Dr. Dennis Emery, Department of Zoology and Genetics, Iowa State University; **10.5*b*:** © Biophoto Associates/Photo Researchers, Inc.; **10.8:** © R. G. Kessel and R.H. Kardon, *A Text-Atlas of Scanning Electron Microscopy,* 1979.; **10.17*c*:** © Don Fawcett/Photo Researchers, Inc.

## Chapter 11

**Opener:** © John Allison/Peter Arnold, Inc.; **p. 392:** From H. Damasio, T. Grabowski, R. Frank, A. M. Galaburda, and A. R. Damasio, "The return of Phineas Gage: clues about the brain from the skull of a famous patient," *Science* 264: 20 May 1994, cover. © AAAS; **11.4*b*:** © Per H. Kjeldsen, University of Michigan; **11.C both:** © Kent Van De Graaff, Ph.D./Utah Valley Hospital, Department of Radiology; **11.19:** © Martin Rotker/Photo Researchers, Inc.; **11.21:** © R. G. Kessel and R. H. Kardon, *Tissues and Organs: A Text-Atlas of Scanning Electron Microscopy,* 1979.

## Chapter 12

**Opener:** © Bob Coyle; **12.2*a*,*b*:** © Ed Reschke; **12.7:** © Dwight Kuhn; **12.9:** © Victor B. Eichler, Ph.D.; **12.16*a*:** © John D. Cunningham/Visuals Unlimited; **12.16*b*:** © Fred Hossler/Visuals Unlimited; **12.21:** © Dean E. Hillman; **12.29:** © SPL/Photo Researchers, Inc.; **12.35:** © Per H. Kjeldsen, University of Michigan; **12.36*a*:** © Carroll Weiss/Camera M. D. Studios; **12.40*c*:** © Frank S. Werblin

## Chapter 13

**Opener:** Courtesy of ELi Lilly and Company Archives; **13.A:** © Gianni Giansanti/Sygma; **13.12:** © John D. Cunningham/Visuals Unlimited; **13.B*a–d*:** Albert Mendeloff. Acromegaly, diabetes, hypermetabolism, proteinuria, and heart failure. *American Journal of Medicine* 20:1 (Jan. 1956) pg. 135. Reprinted with permission from *American Journal of Medicine*.; **13.17:** © Fred Hossler/Visuals Unlimited; **13.19:** From R. H. Kampmeier and T. M. Blake, *Physical Examination in Health and Disease* © 1952 F.A. Davis Company, Philadelphia.; **13.20:** From R. H. Kampmeier and T. M. Blake, *Physical Examination in Health and Disease* © 1952 F. A. Davis Company, Philadelphia.; **13.21:** © Lester V. Bergman and Associates; **13.23:** © BioPhoto Associates/Photo Researchers, Inc.; **13.27*a*:** © John D. Cunningham/Visuals Unlimited; **13.27*b*:** © Times Mirror Higher Education Group, Inc./Carol D. Jacobson, Ph.D., Department of Veterinary Anatomy, Iowa State University; **13.33:** © Ed Reschke

## Chapter 14

**Opener:** © Ken Eward/Science Source/Photo Researchers, Inc.; **14.3*b*:** © J. & L. Weber/Peter Arnold, Inc.; **14.4*b*:** © Bill Longcore/Photo Researchers, Inc.; **14.8*a*:** © Ed Reschke; **14.8*b*:** © Ed Reschke; **14.9:** © Times Mirror Higher Education Group, Inc./Carol D. Jacobson, Ph.D., Department of Veterinary Anatomy, Iowa State University; **14.10, 14.11:** © Ed Reschke; **14.12:** © Times Mirror Higher Education Group, Inc./Carol D. Jacobson, Ph.D., Department of Veterinary Anatomy, Iowa State University; **14.13, 14.A*a*:** © Ed Reschke/Peter Arnold, Inc.; **14.A*b*:** © Ed Reschke; **14.19:** © SPL/Photo Researchers, Inc.; **14.21*a*,*b*:** © Carrol Weiss/Camera M. D. Studios; **14.B:** National Library of Medicine; **14.C:** © Biopharm (USA) Limited, 1994; **14.D:** Courtesy Dr. Leland Clark

## Chapter 15

**Opener:** © D. W. Fawcett/Photo Researchers, Inc.; **p. 566 top:** From Leonard V. Crowley, *Introduction to Human Disease,* 3rd ed. © 1992 Jones and Bartlett Publishers, Boston. Reprinted by permission.; **p. 566 bottom:** From Leonard V. Crowley, *Introduction to Human Disease,* 3rd ed., © 1992 Jones and Bartlett Publishers, Boston. Reprinted by permission.; **15.2:** © A. & F. Michler/Peter Arnold, Inc.; **15.7:** © Times Mirror Higher Education Group, Inc./University of Michigan Biomedical Communications; **15.8:** © Times Mirror Higher Education Group, Inc./Karl Ruben, photographer; **15.14:** Courtesy of Eastman Kodak; **15.A:** © Bob Coyle; **15.F:** © Richard Menard; **15.25:** © R. G. Kessel and R. H. Kardon, *Tissues and Organs: A Text-Atlas of Scanning Electron Microscopy,* 1979.; **15.27*b*,*c*:** © D. W. Fawcett; **15.28:** © T. Kuwabara, from Bloom, W. and Fawcett, D. W. *Textbook of Histology,* 10th ed., W. B. Saunders Co., 1975.; **15.30*a*:** © Ed Reschke; **15.30*b*:** © Victor B. Eichler, Ph.D.; **15.J*a–d*:** American Heart Association; **15.35:** From Patricia Phelps and J. Luft, "Electron study of relaxation and constriction in frog arterioles." *American Journal of Anatomy* 125:409, fig. 2–3. © 1969, Wiley-Liss, a division of John Wiley and Sons, Inc. Reprinted by permission.; **15.44B:** © Kent Van De Graaff, Ph.D./Utah Valley Hospital, Department of Radiology; **15.46:** From J. Dankmeijer, H. G. Lambers, and J. M. F. Landsmeer, *Practische Ontleedkunde.* Bohn, Scheltema & Holkema, Publishers, Netherlands.; **15.N*a*:** AP/Wide World Photos

## Chapter 16

**Opener:** © The Photo Works/Photo Researchers, Inc.; **16.3:** © Ed Reschke; **16.5:** Courtesy of Eastman Kodak; **16.9*b:*** © John Watney/Photo Researchers, Inc.; **16.10:** © Yocochi, C. and Rohen, J. W. *Photographic Anatomy of the Human Body,* 2nd ed., 1978. © Igaku-Shoin, Ltd.; **16.13:** © John D. Cunningham/Visuals Unlimited; **16.15:** © Biophoto Associates/Photo Researchers, Inc.; **16.17:** Courtesy of Sloan-Kettering Institute for Cancer Research/ Etienne de Harven, M.D. and Nina Lampen; **16.A:** Courtesy of Schering-Plough Corporation/Phillip A. Harrington, photographer; **16.21*b:*** © Manfred Kage/Peter Arnold, Inc.; **16.24*b:*** © Lennart Nilsson/Boehringer Ingelheim International Gmbh; **16.B:** © Zeva Oelbaum/Peter Arnold, Inc.

## Chapter 17

**Opener:** © P. Motta/SPL/Photo Researchers, Inc.; **17.8:** Courtesy of John W. Hole, Jr.; **17.9*b:*** © Biophoto Associates/Photo Researchers, Inc.; **17.*a:*** © SPL/Photo Researchers, Inc.; **17.12*a:*** © Bruce Iverson; **17.12*b:*** © Bio-Photo Associates/Photo Researchers, Inc.; **17.12*c:*** © Bruce Iverson; **17.16:** © Ed Reschke; **17.18:** © Kent Van De Graaff, Ph.D./Utah Valley Hospital, Department of Radiology; **17.20:** © Ed Reschke; **17.29*c:*** © Victor B. Eichler, Ph.D.; **17.31:** © Carroll Weiss/Camera M. D. Studios; **17.34:** Armed Forces Institute of Pathology; **17.38:** © Manfred Kage/Peter Arnold, Inc.; **17.39*b:*** © K. R. Porter; **17.46:** © James Shaffer; **17.48:** © Ed Reschke/Peter Arnold, Inc.; **17.49:** © Ed Reschke; **17.B:** © Susan Leavines/Photo Researchers, Inc.; **17.C:** From Leonard V. Crowley, *Introduction to Human Disease,* 3rd ed., fig. 23.22. © 1992 Jones and Bartlett Publishers, Boston. Reprinted by permission.

## Chapter 18

**Opener:** © Tony Freeman/Photo Edit; **p. 721:** AP/Wide World Photos; **18.6*a:*** © Bob Daemmrich/The Image Works; **18.6*b:*** © Bruce Curtis/Peter Arnold, Inc.; **18.8:** World Health Organization; **18.15*b,c:*** © G. W. Willis/Biological Photo Service; **18.18:** Reuters/Bettmann

## Chapter 19

**Opener:** © Manfred Kage/Peter Arnold, Inc.; **p. 757:** © Mulvehill/The Image Works; **19.4*a,b:*** Courtesy of Eastman Kodak; **19.A both:** Used with permission of American Cancer Society, Inc.; **19.B:** © Moredun Animal Health/SPL/Photo Researchers, Inc.; **19.7*c:*** © CNRI/Phototake, NYC; **19.10:** © Ed Reschke; **19.13:** © John Watney Photo Library; **19.15:** © Dwight Kuhn; **19.17:** © R. G. Kessel and R.H. Kardon, *Tissues and Organs: A Text-Atlas of Scanning Electron Microscopy,* 1979.; **19.18:** Courtesy of American Lung Association; **19.26:** © Edward Lettau/Photo Researchers, Inc.; **19.C both:** © Victor B. Eichler, Ph.D.; **19.31:** Courtesy of American Lung Association; **19.33:** From John F. Murray, *The Normal Lung,* 2nd ed., fig. 12.7. © 1986, W. B. Saunders Company; **19.D both:** © CNRI/Phototake, NYC

## Chapter 20

**Opener:** © Dennis Kunkel/Phototake, NYC; **20.2:** © CNRI/SPL/Photo Researchers, Inc.; **20.6*b:*** © Lester V. Bergman & Associates; **20.7*a:*** © R. G. Kessel and R. H. Kardon, *Tissues and Organs: A Text-Atlas of Scanning Electron Microscopy,* 1979.; **20.7*b:*** Courtesy of R. B. Wilson, M. D./Eppley Institute for Research in Cancer, University of Nebraska Medical Center; **20.8:** © David M. Phillips/Visuals Unlimited; **20.10*a:*** © Biophoto Associates/Photo Researchers, Inc.; **20.10*b:*** © Manfred Kage/Peter Arnold, Inc.; **20.27:** © Per H. Kjeldsen, University of Michigan; **20.30:** © John D. Cunningham/Visuals Unlimited; **20.31:** © Ed Reschke

## Chapter 21

**Opener:** © Bruce Curtis/Peter Arnold, Inc.

## Chapter 22

**Opener:** © Bob Schuchman/Phototake, NYC; **22.4:** © R. G. Kessel and R. H. Kardon, *Tissues and Organs: A Text-Atlas of Scanning Electron Microscopy,* 1979.; **22.5*a:*** © Biophoto Associates/Photo Researchers, Inc.; **22.7:** © David M. Phillips/The Population Council/Photo Researchers, Inc.; **22.8:** © Ed Reschke; **22.9*a:*** © R. G. Kessel and R. H. Kardon, *Tissues and Organs: A Text-Atlas of Scanning Electron Microscopy,* 1979; **22.9*b:*** © Ed Reschke; **22.10:** © Manfred Kage/Peter Arnold, Inc.; **22.17*b:*** © R. J. Blandau; **22.18:** © Ed Reschke; **22.19*b:*** © Ed Reschke/Peter Arnold, Inc.; **22.20:** From W. Bloom and D. W. Fawcett, *A Textbook of Histology,* 10th ed., W. B. Saunders Co., 1975 © D. W. Fawcett; **22.21:** © Dr. Landrum B. Shettles; **22.24*a:*** © Ed Reschke; **22.24*b:*** © Don Fawcett/Photo Researchers, Inc.; **22.25:** © Times Mirror Higher Education Group, Inc./Carol D. Jacobson, Ph.D., Department of Veterinary Anatomy, Iowa State University; **22.31:** From M. Tegner and D. Epel, "Sea urchin sperm," *Science,* 179:685–688, 16 Feb. 1973 © AAAS; **22.C:** Courtesy of Southern Illinois University School of Medicine; **22.39*a,b:*** © Biophoto Associates/Photo Researchers, Inc.; **22.42*a–e:*** © Bob Coyle

## Chapter 23

**Opener:** © Erika Stone/Peter Arnold, Inc.; **23.1*a:*** © Petit Format/Nestle/Photo Researchers, Inc.; **23.1*b:*** © Richard Nowitz/Phototake, NYC; **23.2*b:*** © Alexander Tsiaras/Photo Researchers, Inc.; **23.2*c:*** © Omikron/Photo Researchers, Inc.; **23.3*c:*** © Petit Format/Nestle/Photo Researchers, Inc.; **23.4*c*, 23.5:** Courtesy of Ronan O'Rahilly, M.D., Carnegie Institute of Washington/Department of Embryology, Davis Division; **23.B:** *People Weekly* © Taro Yamasaki, 1992; **23.9*a,b:*** © Dr. Landrum B. Shettles; **23.9*c:*** © Petite Format/Nestle/Photo Researchers, Inc.; **23.10*b:*** © Donald Yaeger/Camera M. D. Studios; **23.11 E2:** © Dr. Landrum B. Shettles; **23.16:** © Donald Yaeger/Camera M. D. Studios; **23.C*b:*** From Anne Pytkowicz Streissguth. *Science* 209:353–61 (18 July 1980) © AAAS.; **23.23:** © Donna Jernigan; **23.25 (top to bottom):** UPI/Bettmann; UPI/Bettmann; UPI/Bettmann; UPI/Bettmann; UPI/Bettmann; AP/Wide World Photos; The Bettmann Archives; **23.D:** Eddie Adams/AP/Wide World Photos

## Chapter 24

**Opener:** © Jeff Greenberg/Photo Researchers, Inc.; **p. 947:** AP/Wide

World Photos; **24.A*c* both:** © Bob Coyle; **24.A*d* both:** © Tom Ballard/EKM-Nepenthe; **24.3*b*:** From E. J. Dupraw. *DNA and Chromosomes,* © 1970, Holt, Rinehart & Winston, Inc.; **24.8*a*:** Courtesy of Integrated Genetics; **24.8*b*:** Courtesy Dr. Frederick Elder, Department of Pediatrics, Medical Genetics, University of Texas Medical School, Houston, Texas; **24.9:** Courtesy of Cytogenetics Laboratory at Eleanor Roosevelt Institute, Denver, Co.; **24.12*a,b*:** © Bill Longcore/Photo Researchers, Inc.; **24.13:** Library of Congress; **24.15:** © BioPhoto Associates/Photo Researchers, Inc.; **24.16*a–d*** The Bettmann Archive; **24.D all:** From R. Simensen and R. Curtis Rogers, "Fragile X Syndrome," *American Family Physician,* 39(5):186, May 89.; **24.E both:** Courtesy of Integrated Genetics; **24.F:** © Laura Dwight/Peter Arnold, Inc.

### Color Plates

**10–36** Courtesy of John W. Hole, Jr.; **37–47:** © Dr. Sheril D. Burton

## LINE ART

### Chapter 2

**2B(*b*)** From Kent M. Van De Graaff and Stuart Ira Fox, *Concepts of Human Anatomy and Physiology,* 2d ed. Copyright © 1989 Wm. C. Brown Communications, Inc., Dubuque, Iowa. Reprinted by permission of Times Mirror Higher Education Group, Inc., Dubuque, Iowa. All Rights Reserved.

### Chapter 3

**3.9, 3.12:** From Ricki Lewis, *Life,* 2d ed. Copyright © 1995 Wm. C. Brown Communications, Inc., Dubuque, Iowa. Reprinted by permission of Times Mirror Higher Education Group, Inc., Dubuque, Iowa. All Rights Reserved. **3.28, 3.33:** From Stuart Ira Fox, *Human Physiology,* 4th ed. Copyright © 1993 Wm. C. Brown Communications, Inc., Dubuque, Iowa. Reprinted by permission of Times Mirror Higher Education Group, Inc., Dubuque, Iowa. All Rights Reserved.

### Chapter 4

**4.16** From Ricki Lewis, *Life,* 2d ed. Copyright © 1995 Wm. C. Brown Communications, Inc., Dubuque, Iowa. Reprinted by permission of Times Mirror Higher Education Group, Inc., Dubuque, Iowa. All Rights Reserved. **Charts 4A, 4.4:** From Ricki Lewis, *Human Genetics.* Copyright © 1994 Wm. C. Brown Communications, Inc., Dubuque, Iowa. Reprinted by permission of Times Mirror Higher Education Group, Inc., Dubuque, Iowa. All Rights Reserved. **4.30:** From Ricki Lewis, *Human Genetics.* Copyright © 1994 Wm. C. Brown Communications, Inc., Dubuque, Iowa. Reprinted by permission of Times Mirror Higher Education Group, Inc., Dubuque, Iowa. All Rights Reserved.

### Chapter 5

**5.10:** From Kent M. Van De Graaff, *Human Anatomy,* 4th ed. Copyright © 1995 Wm. C. Brown Communications, Inc., Dubuque, Iowa. Reprinted by permission of Times Mirror Higher Education Group, Inc., Dubuque, Iowa. All Rights Reserved.

### Chapter 6

**6.2, 6.15:** From Kent M. Van De Graaff and Stuart Ira Fox, *Concepts of Human Anatomy and Physiology,* 4th ed. Copyright © 1995 Wm. C. Brown Communications, Inc., Dubuque, Iowa. Reprinted by permission of Times Mirror Higher Education Group, Inc., Dubuque, Iowa. All Rights Reserved.

### Chapter 7

**7.17, 7.19, 7.22, 7.26, 7.29:** From Kent M. Van De Graaff, *Human Anatomy,* 3d ed. Copyright © 1992 Wm. C. Brown Communications, Inc., Dubuque, Iowa. Reprinted by permission of Times Mirror Higher Education Group, Inc., Dubuque, Iowa. All Rights Reserved.

### Chapter 9

**9.1, 9.20, 9.21:** From Kent M. Van De Graaff and Stuart Ira Fox, *Concepts of Human Anatomy and Physiology,* 4th ed. Copyright © 1995 Wm. C. Brown Communications, Inc., Dubuque, Iowa. Reprinted by permission of Times Mirror Higher Education Group, Inc., Dubuque, Iowa. All Rights Reserved. **9.5*b*** From Sylvia S. Mader, *Human Biology,* 4th ed. Copyright © 1995 Wm. C. Brown Communications, Inc., Dubuque, Iowa. Reprinted by permission of Times Mirror Higher Education Group, Inc., Dubuque, Iowa. All Rights Reserved. **9.32*e*** From Kent M. Van De Graaff, *Human Anatomy,* 2d ed. Copyright © 1988 Wm. C. Brown Communications, Inc., Dubuque, Iowa. Reprinted by permission of Times Mirror Higher Education Group, Inc., Dubuque, Iowa. All Rights Reserved.

### Chapter 10

**10.2** From Kent M. Van De Graaff and Stuart Ira Fox, *Concepts of Human Anatomy and Physiology,* 2d ed. Copyright © 1989 Wm. C. Brown Communications, Inc., Dubuque, Iowa. Reprinted by permission of Times Mirror Higher Education Group, Inc., Dubuque, Iowa. All Rights Reserved. **10.4*b*** From Stuart Ira Fox, *Human Physiology,* 3d ed. Copyright © 1990 Wm. C. Brown Communications, Inc., Dubuque, Iowa. Reprinted by permission of Times Mirror Higher Education Group, Inc., Dubuque, Iowa. All Rights Reserved. **10.6** From Kent M. Van De Graaff, *Human Anatomy,* 4th ed. Copyright © 1995 Wm. C. Brown Communications, Inc., Dubuque, Iowa. Reprinted by permission of Times Mirror Higher Education Group, Inc., Dubuque, Iowa. All Rights Reserved.

### Chapter 11

**11.4*a*, 11.23, 11.24, 11.27, 11.28, 11.34:** From Kent M. Van De Graaff, *Human Anatomy,* 3d ed. Copyright © 1992 Wm. C. Brown Communications, Inc., Dubuque, Iowa. Reprinted by permission of Times Mirror Higher Education Group, Inc., Dubuque, Iowa. All Rights Reserved. **11.10*a* & *b*:** From Kent M. Van De Graaff, *Human Anatomy,* 2d ed. Copyright © 1988 Wm. C. Brown Communications, Inc., Dubuque, Iowa. Reprinted by permission of Times Mirror Higher Education Group, Inc., Dubuque, Iowa. All Rights Reserved. **11.11:** From Kent M. Van De Graaff, *Human Anatomy,* 4th ed. Copyright © 1995 Wm. C. Brown Communications, Inc., Dubuque, Iowa. Reprinted by permission of Times Mirror Higher Education Group, Inc., Dubuque, Iowa. All Rights Reserved. **11.13:** From Kent M. Van De Graaff and Stuart Ira Fox, *Concepts of Human Anatomy and Physiology,* 2d ed. Copyright © 1989 Wm. C. Brown Communications, Inc., Dubuque, Iowa.

Reprinted by permission of Times Mirror Higher Education Group, Inc., Dubuque, Iowa. All Rights Reserved.

## Chapter 12

**Excerpts,** pp. 441, 448, 449, 451, 483: From "When Smell and Taste Go Awry" in *FDA Consumer,* November 1991, pages 29–32. **12.4:** From Kent M. Van De Graaff and Stuart Ira Fox, *Concepts of Human Anatomy and Physiology,* 2d ed. Copyright © 1989 Wm. C. Brown Communications, Inc., Dubuque, Iowa. Reprinted by permission of Times Mirror Higher Education Group, Inc., Dubuque, Iowa. All Rights Reserved. **12.15*a*:** From Kent M. Van De Graaff and Stuart Ira Fox, *Concepts of Human Anatomy and Physiology,* 3d ed. Copyright © 1992 Wm. C. Brown Communications, Inc., Dubuque, Iowa. Reprinted by permission of Times Mirror Higher Education Group, Inc., Dubuque, Iowa. All Rights Reserved. **12.15*b*, 12.18:** From Stuart Ira Fox, *Human Physiology,* 4th ed. Copyright © 1993 Wm. C. Brown Communications, Inc., Dubuque, Iowa. Reprinted by permission of Times Mirror Higher Education Group, Inc., Dubuque, Iowa. All Rights Reserved. **12.17** From Stuart Ira Fox, *Human Physiology,* 3d ed. Copyright © 1990 Wm. C. Brown Communications, Inc., Dubuque, Iowa. Reprinted by permission of Times Mirror Higher Education Group, Inc., Dubuque, Iowa. All Rights Reserved. **12.32, 12.33:** From Kent M. Van De Graaff and Stuart Ira Fox, *Concepts of Human Anatomy and Physiology,* 4th ed. Copyright © 1995 Wm. C. Brown Communications, Inc., Dubuque, Iowa. Reprinted by permission of Times Mirror Higher Education Group, Inc., Dubuque, Iowa. All Rights Reserved.

## Chapter 13

**13.1*b*, 13.13:** From Kent M. Van De Graaff and Stuart Ira Fox, *Concepts of Human Anatomy and Physiology,* 4th ed. Copyright © 1995 Wm. C. Brown Communications, Inc., Dubuque, Iowa. Reprinted by permission of Times Mirror Higher Education Group, Inc., Dubuque, Iowa. All Rights Reserved.

## Chapter 14

**Chart 14.4:** From Ricki Lewis, *Life,* 2d ed. Copyright © 1995 Wm. C. Brown Communications, Inc., Dubuque, Iowa. Reprinted by permission of Times Mirror Higher Education Group, Inc., Dubuque, Iowa. All Rights Reserved. **14.16:** From Ricki Lewis, *Life,* 2d ed. Copyright © 1995 Wm. C. Brown Communications, Inc., Dubuque, Iowa. Reprinted by permission of Times Mirror Higher Education Group, Inc., Dubuque, Iowa. All Rights Reserved. **14.24:** From Ricki Lewis, *Human Genetics.* Copyright © 1994 Wm. C. Brown Communications, Inc., Dubuque, Iowa. Reprinted by permission of Times Mirror Higher Education Group, Inc., Dubuque, Iowa. All Rights Reserved.

## Chapter 15

**15.13:** From Kent M. Van De Graaff, *Human Anatomy,* 4th ed. Copyright © 1995 Wm. C. Brown Communications, Inc., Dubuque, Iowa. Reprinted by permission of Times Mirror Higher Education Group, Inc., Dubuque, Iowa. All Rights Reserved. **15.22, 15D, 15.26, 15.47:** From Kent M. Van De Graaff and Stuart Ira Fox, *Concepts of Human Anatomy and Physiology,* 4th ed. Copyright © 1995 Wm. C. Brown Communications, Inc., Dubuque, Iowa. Reprinted by permission of Times Mirror Higher Education Group, Inc., Dubuque, Iowa. All Rights Reserved. **15.32:** From Stuart Ira Fox, *Human Physiology,* 5th ed. Copyright © 1996 Times Mirror Higher Education Group, Inc., Dubuque, Iowa. All Rights Reserved. Reprinted by permission. **15K:** From Stuart Ira Fox, *Human Physiology,* 3d ed. Copyright © 1990 Wm. C. Brown Communications, Inc., Dubuque, Iowa. Reprinted by permission of Times Mirror Higher Education Group, Inc., Dubuque, Iowa. All Rights Reserved. **15.43:** From Kent M. Van De Graaff, *Human Anatomy,* 2d ed. Copyright © 1988 Wm. C. Brown Communications, Inc., Dubuque, Iowa. Reprinted by permission of Times Mirror Higher Education Group, Inc., Dubuque, Iowa. All Rights Reserved. **15.50:** From Kent M. Van De Graaff and Stuart Ira Fox, *Concepts of Human Anatomy and Physiology.* Copyright © 1986 Wm. C. Brown Communications, Inc., Dubuque, Iowa. Reprinted by permission of Times Mirror Higher Education Group, Inc., Dubuque, Iowa. All Rights Reserved.

## Chapter 16

**16.2:** From Kent M. Van De Graaff, *Human Anatomy,* 3d ed. Copyright © 1992 Wm. C. Brown Communications, Inc., Dubuque, Iowa. Reprinted by permission of Times Mirror Higher Education Group, Inc., Dubuque, Iowa. All Rights Reserved. **Chart 16.9:** From Ricki Lewis, *Human Genetics.* Copyright © 1994 Wm. C. Brown Communications, Inc., Dubuque, Iowa. Reprinted by permission of Times Mirror Higher Education Group, Inc., Dubuque, Iowa. All Rights Reserved.

## Chapter 17

**17.1:** From Kent M. Van De Graaff, *Human Anatomy,* 3d ed. Copyright © 1992 Wm. C. Brown Communications, Inc., Dubuque, Iowa. Reprinted by permission of Times Mirror Higher Education Group, Inc., Dubuque, Iowa. All Rights Reserved. **17.3, 17.45, 17.47:** From Kent M. Van De Graaff and Stuart Ira Fox, *Concepts of Human Anatomy and Physiology,* 4th edition. Copyright © 1995 Wm. C. Brown Communications, Inc., Dubuque, Iowa. Reprinted by permission of Times Mirror Higher Education Group, Inc., Dubuque, Iowa. All Rights Reserved. **17.10, 17.24, 17.29*a* & *b*:** From Kent M. Van De Graaff, *Human Anatomy,* 4th ed. Copyright © 1995 Wm. C. Brown Communications, Inc., Dubuque, Iowa. Reprinted by permission of Times Mirror Higher Education Group, Inc., Dubuque, Iowa. All Rights Reserved.

## Chapter 18

**Chart 18.5:** From Ricki Lewis, *Life,* 2d ed. Copyright © 1995 Wm. C. Brown Communications, Inc., Dubuque, Iowa. Reprinted by permission of Times Mirror Higher Education Group, Inc., Dubuque, Iowa. All Rights Reserved. **18A:** From B. T. Burton and W. R. Foster: "Health Implications of Obesity, an NIH Consensus Conference." Copyright © The American Dietetic Association. Reprinted by permission from *Journal of the American Dietetic Association,* Vol. 85: 1985, 1117–21. **18.19:** From Ricki Lewis, *Life.* Copyright © 1992 Wm. C. Brown Communications, Inc., Dubuque, Iowa. Reprinted by permission of Times Mirror Higher Education Group, Inc., Dubuque, Iowa. All Rights Reserved.

## Chapter 19

**19.1:** From Kent M. Van De Graaff, *Human Anatomy,* 3d ed. Copyright © 1992 Wm. C. Brown Communications, Inc., Dubuque, Iowa. Reprinted by permission of Times Mirror Higher

Education Group, Inc., Dubuque, Iowa. All Rights Reserved. **19.11:** From Kent M. Van De Graaff and Stuart Ira Fox, *Concepts of Human Anatomy and Physiology,* 2d edition. Copyright © 1989 Wm. C. Brown Communications, Inc., Dubuque, Iowa. Reprinted by permission of Times Mirror Higher Education Group, Inc., Dubuque, Iowa. All Rights Reserved. **19.29:** From Kent M. Van De Graaff, *Human Anatomy,* 4th ed. Copyright © 1995 Wm. C. Brown Communications, Inc., Dubuque, Iowa. Reprinted by permission of Times Mirror Higher Education Group, Inc., Dubuque, Iowa. All Rights Reserved.

### Chapter 20

**20.5:** From Kent M. Van De Graaff, *Human Anatomy,* 2d ed. Copyright © 1988 Wm. C. Brown Communications, Inc., Dubuque, Iowa. Reprinted by permission of Times Mirror Higher Education Group, Inc., Dubuque, Iowa. All Rights Reserved. **20.11:** From Kent M. Van De Graaff and Stuart Ira Fox, *Concepts of Human Anatomy and Physiology,* 4th edition. Copyright © 1995 Wm. C. Brown Communications, Inc., Dubuque, Iowa. Reprinted by permission of Times Mirror Higher Education Group, Inc., Dubuque, Iowa. All Rights Reserved. **20.32*b*:** From Kent M. Van De Graaff, *Human Anatomy,* 3d ed. Copyright © 1992 Wm. C. Brown Communications, Inc., Dubuque, Iowa. Reprinted by permission of Times Mirror Higher Education Group, Inc., Dubuque, Iowa. All Rights Reserved.

### Chapter 22

**22.2:** From Kent M. Van De Graaff and Stuart Ira Fox, *Concepts of Human Anatomy and Physiology,* 3d ed. Copyright © 1992 Wm. C. Brown Communications, Inc., Dubuque, Iowa. Reprinted by permission of Times Mirror Higher Education Group, Inc., Dubuque, Iowa. All Rights Reserved. **22.3** From Kent M. Van De Graaff, *Human Anatomy,* 2d ed. Copyright © 1988 Wm. C. Brown Communications, Inc., Dubuque, Iowa. Reprinted by permission of Times Mirror Higher Education Group, Inc., Dubuque, Iowa. All Rights Reserved. **22A, 22B:** From Ricki Lewis, *Human Genetics.* Copyright © 1994 Wm. C. Brown Communications, Inc., Dubuque, Iowa. Reprinted by permission of Times Mirror Higher Education Group, Inc., Dubuque, Iowa. All Rights Reserved. **22.11, 22.23, 22.30, 22.37, 22.38:** From Kent M. Van De Graaff and Stuart Ira Fox, *Concepts of Human Anatomy and Physiology,* 4th edition. Copyright © 1995 Wm. C. Brown Communications, Inc., Dubuque, Iowa. Reprinted by permission of Times Mirror Higher Education Group, Inc., Dubuque, Iowa. All Rights Reserved. **Charts 22B, 22C:** From Ricki Lewis, *Human Genetics.* Copyright © 1994 Wm. C. Brown Communications, Inc., Dubuque, Iowa. Reprinted by permission of Times Mirror Higher Education Group, Inc., Dubuque, Iowa. All Rights Reserved. **Charts 22.9, 22.10:** From Ricki Lewis, *Life,* 2d ed. Copyright © 1995 Wm. C. Brown Communications, Inc., Dubuque, Iowa. Reprinted by permission of Times Mirror Higher Education Group, Inc., Dubuque, Iowa. All Rights Reserved.

### Chapter 23

**23A:** From Ricki Lewis, *Human Genetics.* Copyright © 1994 Wm. C. Brown Communications, Inc., Dubuque, Iowa. Reprinted by permission of Times Mirror Higher Education Group, Inc., Dubuque, Iowa. All Rights Reserved. **23.11, 23.19:** From Kent M. Van De Graaff, *Human Anatomy,* 4th ed. Copyright © 1995 Wm. C. Brown Communications, Inc., Dubuque, Iowa. Reprinted by permission of Times Mirror Higher Education Group, Inc., Dubuque, Iowa. All Rights Reserved. **23.15:** From Kent M. Van De Graaff and Stuart Ira Fox, *Concepts of Human Anatomy and Physiology,* 4th edition. Copyright © 1995 Wm. C. Brown Communications, Inc., Dubuque, Iowa. Reprinted by permission of Times Mirror Higher Education Group, Inc., Dubuque, Iowa. All Rights Reserved. **23.17:** From Ricki Lewis, *Life,* 2d ed. Copyright © 1995 Wm. C. Brown Communications, Inc., Dubuque, Iowa. Reprinted by permission of Times Mirror Higher Education Group, Inc., Dubuque, Iowa. All Rights Reserved. **23C(*a*):** From Ricki Lewis, *Life.* Copyright © 1992 Wm. C. Brown Communications, Inc., Dubuque, Iowa. Reprinted by permission of Times Mirror Higher Education Group, Inc., Dubuque, Iowa. All Rights Reserved.

### Chapter 24

**24A, 24.10, 24.14, 24C, 24.17, 24.18, 24.19, 24H:** From Ricki Lewis, *Human Genetics.* Copyright © 1994 Wm. C. Brown Communications, Inc., Dubuque, Iowa. Reprinted by permission of Times Mirror Higher Education Group, Inc., Dubuque, Iowa. All Rights Reserved. **Charts 24.1, 24.2, 24.4, 24.5:** From Ricki Lewis, *Human Genetics.* Copyright © 1994 Wm. C. Brown Communications, Inc., Dubuque, Iowa. Reprinted by permission of Times Mirror Higher Education Group, Inc., Dubuque, Iowa. All Rights Reserved. **24.G:** From *Human Genetics* 2/E by Singer. Copyright © 1985 by W. H. Freeman and Company. Used with permission.

## ILLUSTRATORS

New illustrations to this edition:

**Todd Buck:** 7.25 icon; 7.40*a*

**Chris Creek:** 1.6; 1.20; 1.21

**Peg Gerrity:** 1.12; 1.13; 1.14; 9.2; 9.4; 10.3; 10.20; 11.2*a,b;* 15.6; 16.12; 17.30; 22.33

**Illustrious, Inc.:** 1.4; 1.5; 1.7; 1.8; 1.16; 1.17; 1.18*a–c;* 1.19; 2.1; 2.2; 2.3; 2.4; Clinical Application 2.2, 2C; 2.5; 2.6; 2.7; 2.8; 2.9; 2.10; 2.11; 2.12; 2.13; 2.14; 2.15; 2.16; 2.17; 2.18; 2.19; 2.20; 2.21; 3.1; 3.2; 3.3; 3.6; 3.7; 3.8; 3.9; 3.10 icon, *b,c;* 3.11 icon, *b;* 3.13 icon, *b;* 3.15 icon, *b;* 3.16 icon, *b;* 3.20; 3.21; 3.22; 3.23; 3.24; 3.26; 3.27; 3.28; 3.29; 3.30; 3.31; 3.32; 3.34*a;* 3.35; 3.36*a;* 3.37*a;* 3.38*a;* 3.39*a;* 3.40*a;* 3.44; 4.1; 4.2; 4.3; 4.4; 4.6; 4.7; 4.8; 4.9; 4.10; 4.11; 4.12; 4.13; 4.14; 4.15; 4.17; 4.18; 4.19; 4.20; 4.21; 4.22*a;* 4.23; 4.24; 4.25; 4.27; 4.28; 4.29; 4.30; 5.1 icon, *a;* 5.2 icon; 5.3 icon; 5.5 icon; 5.6 icon, *a;* 5.7 icon, *a;* 5.8 icon, *a;* 5.11; 5.18 icon; 5.20 icon; 15.21 icon, *a;* 5.22 icon, *a;* 5.23 icon; 5.24 icon; 5.25 icon, *a;* 5.27 icon, *a;* 5.28 icon; 5.29 icon; 5.30 icon, *a;* 5.31 icon; 6.5 *b;* 6.12, InnerConnections page 182; 7.1 icon; 7.15, 7.20 icon; 7.38 icon; 7.39 icon; 7.42 icon; InnerConnections page 241; 8.1; 8.2; Clinical Application 8.2, 8A; 9.12; 9.13; 9.14; 9.15; 9.16; InnerConnections page 340; 10.10; 10.11; 10.12; 10.13; 10.14; 10.18; InnerConnections page 383; 11.8; 11.18; 12.15 icon; 12.22 icon; 12.23 icon and line art; 12.24 icon; 13.3; 13.4; 13.5; 13.6; 13.7; 13.9 icon; 13.10; 13.11; 13.13; 13.14; 13.15; 13.18; 13.25; 13.28; 13.29; 13.30; 13.31; 13.34; 13.35; 13.36; InnerConnection page 524; 14.3*a;* 14.5; 14.6; 15.1; 15.7 icon; 15.8 icon; 15.9 icon; 15.16; 15.37; 15.38; 15.40; 15,42; 15.57; Clinical Application 15.8, 15N*b*, InnerConnections page 629, 16.20; 16.21*a;* 16.22; 16.24*a;* InnerConnections page 663; 17.19 icon; 17.21; 17.25; 17.41; 17.42; 17.43;

InnerConnections page 715; 18.1; 18.2; 18.3; 18.4; 18.7; 18.9; 18.10; 18.11; 18.12; 18.13; 18.14; 18.15*a;* 18.17; 19.12; 19.14; 19.16; 19.21; 19.22; 19.25; 19.30; 19.34; 19.35; 19.36; 19.37; 19.38; 19.39; 19.40; 19.41; 19.42; InnerConnections page 791; 20.4; 20.9; 20.13; 20.14; 20.18; 20.19; 20.20; 20.22; 20.23; 20.24; 20.25; 20.26; InnerConnections page 826; 21.10; 22.5*b;* 22.14; 22.27; 22.28; 22.32; InnerConnections page 903; 23.2*a;* 23.3; 23.4; 23.8; 23.14; 23.22 *b;* 23.18; 24.1; 24.2; 24.3*a;* 24.4; 24.5; 24.6; 24.7; 24.10; Clinical Application 24.2, 24B; 24.11; 24.14 (top portion of art); Clinical Application 24.4, 24C; 24.19

**Elizabeth Morales:** 3.19*a;* 6.6*a;* 6.10; 12.1

**Thomas Waldrop:** 9.23; 9.37; 12.12

# INDEX

Note: Page numbers in *italics* designate figures; page numbers followed by t designate tables (charts).

## A

## C

## D



## E

## F

## G

## H

## I

## J



## K



## L

## M

## N

## O



## P

## Q



## R

## S

## T

## U



## V

# AIDS TO UNDERSTANDING WORDS

***a-*** negative prefix: amorphous
***ab-*** away from: abduction
***acet-*** vinegar: acetabulum
***ad-*** to: adduction
***adip-*** fat: adipose tissue
***aer-*** air: aerobic respiration
***af-*** to: afferent arteriole
***agglutin-*** to glue together: agglutination
***alb-*** white: albino
***aliment-*** food: alimentary canal
***allant-*** sausage: allantois
***alve-*** trough, channel, cavity: alveolus
***an-***[1] *see* a-
***an-***[2] *see* ana-
***ana-*** up, positive: anabolic
***andr-*** man: androgens
***angi-*** vessel: angiogram
***annul-*** ring: annular ligament
***ante-*** before: antebrachium
***append-*** to attach, hang, or fix to: appendicular
***-ar*** pertaining to: lobar bronchus
***arbor-*** tree: arbor vitae
***arc-*** bow: arciform artery
***areol-*** open space: areola
***arthr-*** joint: arthrology
***astr-*** star: astrocyte
***athero-*** porridge: atherosclerosis
***atri-*** entrance: atrium
***aud-*** hearing: auditory canal
***aur-*** ear: auricle
***auto-*** self: autoimmune
***ax-*** axis: axial skeleton
***axill-*** armpit: axillary artery
***bas-*** base: basilar artery
***bi-*** two: bicuspid valve
***bicep-*** two heads: biceps brachii
***bil-*** bile: bilirubin
***bio-*** life: biology
***-blast*** bud, child, a growing thing in early stages: osteoblast
***blephar-*** eyelid: blepharitis
***brachi-*** arm: brachial region
***brady-*** slow: bradycardia
***bronch-*** windpipe: bronchus
***burs-*** bag, purse: bursa
***calat-*** something inserted: intercalated disk
***calc-*** heel: calcaneus
***calor-*** heat: calorie
***calyc-*** cup: major calyx
***canal-*** channel: canaliculus
***carcin-*** crab, cancer: carcinoma
***cardi-*** heart: pericardium
***cari-*** decay: dental caries
***carin-*** hull, keel: carina
***carot-*** carrot: carotene
***carpus-*** wrist: carpals
***cata-*** down: catabolic
***caud-*** tail: cauda equina
***cav-*** hollow: corpora cavernosa
***cec-*** blind: cecum
***centr-*** point, center: centromere
***cephal-*** head: cephalic region
***cerebr-*** brain: cerebrum
***cervic-*** neck: cervical vertebrae
***chiasm-*** cross-shaped: optic chiasma
***chondr-*** cartilage: chondrocyte
***chori-*** protective fetal membrane: chorion
***chrom-*** color: chromatin
***chym-*** juice: chyme
***-clas*** break: osteoclast
***clav-*** bar: clavicle
***cleav-*** to divide: cleavage
***co-*** with, together: coenzyme
***cochlea*** snail: cochlea
***col-*** lower intestine: colon
***condyl-*** knuckle: condyle
***contra-*** against, counter: contraception
***corac-*** raven: coracoid process
***corn-*** horn: cornea
***corpor-*** body: corpus callosum
***cortic-*** bark, rind: cortex
***crani-*** skull: cranial
***cribrum-*** sieve: cribriform plate
***cric-*** ring-shaped: cricoid cartilage
***-crin*** distinguish, separate off: endocrine
***crista*** crest: crista galli
***crur-*** shin, leg: crura
***cusp-*** point, apex: tricuspid valve
***cut-*** skin: subcutaneous
***cyst-*** bladder: cystitis
***cyt-*** cell: cytoplasm
***dart-*** like skin: dartos muscle
***de-*** to remove: deamination
***decidu-*** falling off: deciduous
***deglut-*** to get rid of: deglutination
***dendr-*** tree: dendrite
***derm-*** skin: dermis
***detrus-*** to force away: detrusor muscle
***di-*** two: diencephalon
***diastol-*** dilation: diastole
***digit-*** finger: digital artery
***diuret-*** to pass urine: diuretic
***dors-*** back: dorsal
***ect-*** outside: ectoderm
***ectop-*** out of place: ectopic heat
***ede-*** swelling: edema
***ejacul-*** to shoot forth: ejaculation
***embol-*** stopper: embolus
***-emia*** in the blood: hypoproteinemia
***encephal-*** the brain: encephalitis
***end-*** inside: endoplasmic reticulum
***ependym-*** tunic: ependyma
***epi-*** upon, after, in addition: epithelial tissue
***erg-*** work, deed: synergist
***erythr-*** red: erythrocyte
***ethmo-*** sieve: ethmoid bone
***exo-*** outside: exocrine gland
***extra-*** outside of, beyond: extracellular
***falx-*** a sickle: falx cerebelli
***fasci-*** band: fasciculus
***femur-*** thigh: femur
***fimb-*** fringe: fimbriae
***flacc-*** flabby: flaccid
***flagell-*** whip: flagellum
***follic-*** small bag: hair follicle
***fore-*** before, the front part of: forebrain
***foss-*** ditch, trench: fossa ovalis
***fovea-*** pit: fovea capitis
***frenul-*** bridle, restraint: frenulum
***funi-*** small cord or fiber: funiculus
***gangli-*** swelling: ganglion
***gastr-*** stomach: gastric gland
***-gen*** become, be produced: allergen
***genesis*** origin: spermatogenesis
***gladi-*** sword: gladiolus
***glen-*** joint socket: glenoid cavity
***-glia*** glue: neuroglia
***glom-*** little ball: glomerulus
***glut-*** buttocks: gluteus maximus
***glyc-*** sweet: glycogen
***gracil-*** slender: gracilis
***-gram*** letter, drawing: myogram
***gubern-*** to steer, to guide: gubernaculum
***gust-*** to taste: gustatory receptor
***gyn-*** woman: gynecology
***hem-*** blood: hemoglobin
***hemi-*** half: hemiplegia
***hepar-*** liver: heparin
***hepat-*** liver: hepatic duct
***hetero-*** other, different: heterozygous
***hiat-*** opening: esophageal hiatus
***hist-*** web, tissue: histology
***hol-*** entire, whole: holocrine gland
***hom-*** same, common: homeostasis
***horm-*** impetus, impulse: hormone
***humor-*** moisture, fluid: humoral